# VOLUME

A milliliter is about
one fifth of a teaspoon

A liter is becoming the standard
size for soft drink, wine, and
liquor bottles in the U.S. If
its price is the same as for the
old quart bottle, you're in luck
because a liter is equal to
slightly more than a quart

The gas tank of a large car
holds 64 to 75 liters of gasoline

## Volume Conversion

1 tsp   = 5 ml
1 tbsp  = 15 ml
1 fl oz = 30 ml
1 cup   = 0.24 liter
1 pt    = 0.47 liter
1 qt    = 0.95 liter
1 gal   = 3.8 liter
1 ml    = 0.03 fl oz
1 liter = 2.1 pt
1 liter = 1.06 qt
1 liter = 0.26 gal

## Temperature Conversion

$$°C = \frac{(°F - 32) \times 5}{9}$$

$$°F = \frac{°C \times 9}{5} + 32$$

### Interval Equivalents

| °C | °F |
|---|---|
| 1° | 1.8° |
| 5° | 9° |
| 10° | 18° |

# TEMPERATURE

Fahrenheit Scale — Centigrade (Celsius) Scale

Boiling point of water
100 °C (212 °F)

Highest temperature recorded
in the U.S.
57 °C (134 °F)

Water scalds
54 °C (130 °F)

Human body temperature
37 °C (98.6 °F)

Room temperature
20–25 °C (68–77 °F)
It's easy to remember that
16 °C = 61 °F

Freezing point of water
0 °C (32 °F)

Coldest area of a freezer
−23 °C (−10 °F)

# BIOLOGY

*Fourth Edition*

# BIOLOGY

*Fourth Edition*

KAREN ARMS

PAMELA S. CAMP

SAUNDERS COLLEGE PUBLISHING

*Harcourt Brace College Publishers*

Fort Worth   Philadelphia   San Diego   New York   Orlando

San Antonio   Toronto   Montreal   London   Sydney   Tokyo

Text Typeface: Garamond
Compositor: York Graphic Services
Acquisitions Editor: Julie Levin Alexander
Developmental Editor: Lee Marcott
Managing Editor: Carol Field
Project Editor: Bonnie Boehme
Copy Editor: Janice Moore
Manager of Art and Design: Carol Bleistine
Art Director: Christine Schueler
Art Assistant: Sue Kinney
Text Designer: Rebecca Lemna
Cover Designer: Rebecca Lemna
Text Artwork: Rolin Graphics
Layout Artwork: Tracy Baldwin and Merce Wilczek
Director of EDP: Tim Frelick
Production Manager: Joanne M. Cassetti
Marketing Manager: Susan Westmoreland

Cover Credit: New Zealand forest fungus (*Mycena*). (© Brian Enting/Westlight)
Frontispiece: Trinidad day geckos on a bromeliad. (© Zig Leszczynski/Animals Animals)

Credits for Part Openers:
Part One: Desmids (© M. I. Walker/Science Source/Photo Researchers)
Part Two: Ear of corn
Part Three: Diatoms (© Robert Brons/BPS)
Part Four: Chantarelles (*Cantharellus cibarius*) and blewits (*Lepista nuda*)
(© D. Cavagnaro/DRK Photo)
Part Five: Golden toads in breeding aggregation (© Michael Fogden/DRK Photo)
Part Six: Oriental day lilies (*Hemerocallis*) (© S. Nielsen/DRK Photo)
Part Seven: Coral reef (© Norbert Wu)

Printed in the United States of America

Biology, Fourth Edition
ISBN 0-03-050003-6

Library of Congress Catalog Card Number: 94-065949

5678901234   032   10 987654321

# PREFACE

The first edition of *Biology* grew from our attempts to introduce our students to the essentials of biology without relying on fact-filled lectures. These attempts taught us a lot about how students learn and what excites or bores them. Each new edition has required many changes to keep the book current, but our goals have not changed: to help students grasp biological principles in a way that permits them to understand the living world and their own role in it, to solve problems, to prepare for future studies, and to become aware of the rewards of life as a scientist.

In this edition, we have of course updated the entire book to reflect new discoveries and theories in biology. Equally important, we have reconsidered the overall organization and rethought the presentation of every topic, looking for ways to change wording, emphasis, or order that would make the book more effective to learn and teach from. We hope that this new edition will prove to be an improved tool for students in a general biology course and a useful reference for teacher and student in years to come.

## OUR GOALS FOR THE FOURTH EDITION

Beyond the obvious updating, we identified three main goals for this revision: to make the book more conceptual; to stress the process and techniques of scientific investigation; and to redesign the entire art program to support these goals.

### A More Conceptual Approach

Our first objective was to strengthen the treatment of the conceptual threads that pervade biology:

♦ evolution

♦ the cellular basis of life

♦ matter and energy flow in organisms and their environment

♦ structure/function relationships

♦ the genetic basis of life, from protein structure and metabolism at the molecular level, to embryonic development and the structure, function, and behavior of individual organisms

♦ homeostasis and negative feedback controls

♦ the roles of physical factors, competition, and predation in shaping populations in ecological and evolutionary time

In addition to these time-honored major concepts of biology, recent literature provides mounting evidence for the pervasive role of symbiosis in evolution and ecology. Symbiosis figures most prominently in the evolution of eukaryotic cells but is now cropping up everywhere in the literature, in associations between organisms of all descriptions. Accordingly, we have woven this emerging theme throughout the book, with special prominence in Parts Three and Four, on evolution and the diversity of life.

We took several steps to increase the emphasis on concepts. First, we redesigned chapter openings to enhance their role in establishing a frame of reference for the chapter to come. The learning objectives are now at the front, calling students' attention to the important points. Next comes an introduction that puts the chapter's subject matter into perspective, ties it to material in other chapters, and defines a few basic terms. The introduction is followed and summarized by bullet statements in the chapter's key concepts, a feature new to this edition.

Thus prepared, the student takes on the chapter itself. Here we have scrutinized details and tried to keep only those that support and illustrate the main concepts (and their major exceptions). For example, we omit names and pictures of some intermediate stages in plant life cycles, opting instead to trace evolutionary trends and help students recognize structures they may actually find in nature, such as moss sporophytes. We also take the opportunity to mention areas of controversy, alternative hypotheses, and what we don't yet know about particular topics.

Great horned owl.

v

We are sometimes asked why we include numerical facts such as the width of a DNA molecule or the generation time of cells. We believe such items do serve an important purpose: to give budding biologists some feel for the scale of things. It is not important to memorize that the DNA double helix is 2 nm across (rather than 4 nm), but it IS important to have a sense that it's not on the order of, say, 200 nm. To aid this sort of learning, we have included size bars on micrographs.

We have introduced Latin and Greek roots of many more terms. Biology has an extensive vocabulary of its own, and learning to recognize these roots can help students not only to remember what various terms mean but also perhaps to figure out the meanings of new terms that combine now-familiar roots in new ways. These word roots are especially useful in the chapters on diversity, as aids to remembering definitive features of various phyla.

Chapter summaries are now more extensive and follow a new format. Each begins with a short introductory paragraph or two, reinforcing the overall chapter concept. The chapter's main points follow in a list of numbered paragraphs.

## Science as Process

An emphasis on procedures in science becomes more important as schools find themselves strapped for funds to provide lab experience, relying more on demonstrations by the lecturer, videos, and computer simulations.

This edition introduces a new learning feature, boxes entitled "How Do We Know?" These should not be considered "optional" but, rather, important experiments or techniques that educated biologists should know and will perhaps use in their future careers. These topics are set off in the box format for three reasons: to highlight them as topics, to provide an adequate explanation without interrupting the flow of the main text, and to ensure that explanations and supporting pictures appear together on the same pages.

Some "How Do We Know?" boxes cover modern techniques such as DNA sequencing. Others draw from historical material. These let students glimpse the "roots" of biology. More importantly, they demonstrate how some basic tenets of biology were established using simple, everyday equipment guided by that most important of scientific tools: the human brain.

It is essential for students to recognize that the logic of an experiment is well within their understanding. These boxes are designed to carry this message:

◆ the uses of isotopes in biology (Chapter 2)

◆ freeze-fracture method of preparing specimens (Chapter 4)

◆ the methods used to unravel the citric acid cycle (Chapter 7)

◆ experimental milestones in our understanding of photosynthesis (Chapter 8)

◆ how the Calvin-Benson cycle was discovered (Chapter 8)

◆ Meselson and Stahl's and Taylor's evidence for semiconservative DNA replication (Chapter 9)

◆ experiments that shed new light on the evolution of introns (Chapter 10)

◆ the interacting effects of histones and transcription factors (Chapter 11)

◆ the importance of selecting an appropriate experimental organism (Chapter 12)

◆ how the nucleotide sequence of DNA is determined (Chapter 13)

◆ how molecular analysis elucidated the evolution of globin genes (Chapter 22)

◆ the production and uses of monoclonal antibodies (Chapter 34)

◆ patch clamping technique (Chapter 37)

◆ transpiration pull in plants (Chapter 43)

◆ discovery of auxin and the technique of bioassay (Chapter 45)

◆ how productivity is measured (Chapter 48)

## The New Art Program

As video technology proliferates, many students have become much more visual learners. This makes the design of the art program even more important: it must be colorful and attractive, but must also enhance understanding of the text. As we prepared the manuscript, we reviewed each piece of supporting art, working with the artists not just to redraw every piece more attractively but to reconcept most of them. In many cases, two or more small pieces are combined into a larger layout to show their relationships more clearly. We also continue to pair interpretive art with photos. When the photo alone is perfectly comprehensible, we label the photo itself. For clarity, some figures are designed with a table-like format, and many tables are now illustrated with comparative drawings.

One goal of the art program was to color-code structures for uniform appearance in a group of related pictures: all sugars brown, proteins blue, and ribosomes purple.

A second goal was to place the labels to fall in with natural patterns of reading (generally, top left to bottom right). Figures are laid out so that the viewer first assimilates the location of major landmarks such as the plasma membrane and cytoplasm, then moves on to details such as specific membrane molecules or cytoplasmic organelles. Or, the labels may lead the eye in order from the outer to inner layers of a structure, or from start to finish of a process. In a diagram of an earthworm, the labels from all digestive system structures are grouped together, not interspersed with those for the nervous and circulatory systems, which have their own groups.

Our researchers found hundreds of superb new photographs. Our final choice was not always the most beautiful or colorful, but the one illustrating structures, actions, or context most clearly. We also discovered that some old favorites are still best at making particular points.

## WHAT'S NEW IN THIS EDITION?

### Content and Organization

We have retained the general organization of the previous edition with its early coverage of evolution. Also as in the past, this book can be used with many different course outlines. We ourselves have taught the subject in many different orders, starting with ecology if it was more convenient to conduct field trips in the fall, and teaching genetics and biochemistry in the spring if most students were taking organic chemistry in the fall. Chapters are designed to allow flexibility in creating the course syllabus, and sections within chapters are numbered for easy assignment or deletion of these parts.

In this edition we adopt the increasingly popular system wherein multicellular algae, slime molds, and oomycotes are classified in kingdom Protista. Hence, protists are no longer defined simply as unicellular eukaryotes; they also include these simple multicellular forms. Kingdom Fungi is now restricted to the lineage of organisms with absorptive nutrition and chitinous walls, and kingdom Plantae to organisms with cellulose walls and embryonic development (formerly called the land plants). Recent research also indicates a splitting of some phyla is in order, and we have done this.

This fourth edition contains two new chapters: Chapter 13, "Genetic Engineering," and Chapter 50, "Human Ecology and Natural Resources." Former Chapter 52, "Evolutionary Ecology," has been reorganized and moved forward as Chapter 18, "Adaptation and Coevolution," with the addition of a strong new section on symbiotic relationships.

## SURVEY OF FOURTH EDITION CHAPTERS

Chapter 1 introduces the main themes of the book: the fundamental concepts of biology, especially evolution, and the scientific method, now including a new section on correlation studies in addition to the experimental method.

### Part One: Cells

An extensive knowledge of chemistry is not necessary to study biology, but many teachers have discovered that students flounder when studying cell and molecular biology unless they are familiar with some of the structures and properties of molecules. Chapter 2 does not need to be taught in class. It can be read by students who need a refresher in chemistry to prepare them to understand the biologically important molecules discussed in Chapter 3. Our understanding of membranes has blossomed in recent years with the happy result that the chapters on membranes and cell structure are now a logical story of how cells work rather than a dry listing of the cell's organelles. Chapters 6 through 8 explore the energy transactions in cells. While cellular respiration (Chapter 7) and photosynthesis (Chapter 8) have their own chapters, some teachers may prefer to omit the details and concentrate on the principles of energy flow in organisms, the focus of Chapter 6.

### Part Two: Genetic Information

Part Two develops the concept of genetic information and its expression from the molecular level up. This makes the inheritance and expression of genes at the level of individual organisms less of a "black box" to students and allows us to illustrate and tie together all levels of genetics in the final chapter.

Chapters 9 through 13 take an historical approach to molecular biology. The high point of eukaryotic gene expression is embryonic development, covered in Chapters 11 and 12. This process has made possible the evolution of higher eukaryotes. Developments in our understanding of gene expression are finally throwing new light on this fascinating subject. New Chapter 13 gives an overview of biotechnology and introduces students to such topics as DNA fingerprinting and the plant genetics behind the development of Flav'r Sav'r tomatoes.

Chapter 15, "Mendelian Genetics," still builds on meiosis in Chapter 14 ("Reproduction of Eukaryotic Cells") so that students see inheritance patterns as the manifestation of the underlying processes of meiosis and fertilization. The presentation of genetic crosses is broken into smaller, easier-to-follow steps. New essays present an expanded picture of Mendel's life and work and a discussion of genome imprinting. Chapter 16, "Inheritance Patterns and Gene Expression," of course includes recent advances in understanding several genetic diseases as well as an update on sex determination.

### Part Three: Evolution

This part is devoted to evolution, its mechanisms, and results. Chapter 17 lays out the evidence for evolution and examples that help the student grasp how natural selection works. New Chapter 18 discusses examples of adaptations, especially those involving symbiotic relationships and coevolution with other species. Population genetics and speciation have been merged into one chapter that emphasizes the rapidly changing viewpoints of experts in this area. Chapter 20 examines the concept that evolutionary success is defined in terms of reproduction of genomes and shows how this has selected for various reproductive adaptations, including social life, altruism, and behavior toward kin. A thoroughly revised Chapter 21 on the origin of life ends this part and includes the controversy on the RNA world and a new section on the history of life.

## Part Four: Diversity of Life

Chapter 22 details the revolution in the way we classify organisms, embracing cladistic methods, and discussing the uses and pitfalls of molecular phylogeny. Chapters 23 through 30 cover viruses and the kingdoms of organisms, with an emphasis on the evolutionary breakthroughs that mark each stage of the story. The redefinition of the kingdom Protista and recent elucidation of the origin and evolutionary relationships among different groups called for major rewriting and reordering of large portions of Chapter 25. Multicellular algae, slime molds, and oomycotes have been moved into this kingdom, and several phyla have been split. Choanoflagellates are introduced as the probable ancestors of animals.

Chapter 27, "The Plant Kingdom," reflects many changes in both knowledge and theories about the origin of plants and their features. These include a strong case for charophyte green algae as the ancestral group, independent origin of alternation of generations, a new section on the origin of seeds, and rewritten sections on bryophytes, vascular plants, and seedless vascular plants (formerly referred to as lower vascular plants). All classes in the third edition have been elevated to divisions, and divisions Hepatophyta, Psilotophyta, and Ginkgophyta are now separated out.

## Part Five: Animal Biology

Major changes in animal biology include an updating of Chapter 31 on nutrition, a complete rewriting of Chapter 34 on immunology, and the merging of the chapters on neurons and the nervous systems. Chapter 37, "Nervous Systems," contains a new overview section for those who prefer to skip the sections on electrical properties of neurons. Understanding of animal behavior and the selective pressures that shape its development has grown rapidly in recent years. We include a discussion of foraging behavior and the light it sheds on cost-benefit analysis, as well as the controversial topic of whether and how animals use magnetoreceptors for navigation.

## Part Six: Plant Biology

Chapter 42's introduction emphasizes the concept of "plantness." The chapter's new material includes a section on tissue systems and cell types, placing each in the context of familiar plant parts. Integration of monocots throughout the chapter provides a more balanced monocot/dicot coverage.

Chapter 45, "Regulation and Response in Plants," points out the emerging evidence for similarity of chemical control mechanisms in animals and plants and introduces recently discovered chemical regulators in plants: polyamines, polypeptides, and salicylic acid as well as oligosaccharins, which were new in the third edition. Other new sections cover nastic and stress responses.

## Part Seven: Ecology

The importance of ecology increases as the human population approaches Earth's carrying capacity. We start with a chapter on the distribution of organisms in the biosphere and continue with ecosystems, populations, and communities, ending with a new Chapter 50 on human ecology and natural resources, which shows that we are subject to the same pressures as any other organism. We introduce some of the mathematical techniques of ecology and consider the theory of island biogeography and species diversity, topics of increasing importance as the biodiversity of our planet declines.

## OTHER FEATURES NEW TO THIS EDITION

As before, **Essay** boxes highlight material of general interest that goes beyond the material in the chapter or ties together material from more than one chapter. Topics that are the subject of new essays in this edition are:

- everyday results of protein denaturation (Chapter 3)
- speculations on symbioses in the evolution of eukaryotes (Chapter 5)
- DNA fingerprinting (Chapter 13)
- genome imprinting (Chapter 15)
- Mendel's life and research methods (Chapter 15)
- the controversy over certifying the sex of athletes (Chapter 16)
- drugs obtained from plants (Chapter 18)
- the short-term selective advantage of sexual reproduction (Chapter 20)
- the Gaia hypothesis (Chapter 21)
- the mystery of scrapie (Chapter 23)
- a possible symbiotic origin for flagella (Chapter 24)
- phytoplankton (Chapter 25)
- poisonous fungi (Chapter 26)
- the closest living algal relative of plants (Chapter 27)
- *Cinchona* and quinine (Chapter 27)
- the K-T extinctions (Chapter 30)
- test tube babies and surrogate mothers (Chapter 36)
- chimpanzee societies (Chapter 41)
- grains, the staff of life (Chapter 42)
- soil destruction (Chapter 44)

- the Everglades (Chapter 47)
- invasions of exotic plants (Chapter 49)
- the Tragedy of the Commons (Chapter 50)

**In More Detail** boxes, as their name suggests, set aside the details of glycolysis, the Hardy-Weinberg Law, and mathematical models of population growth. Students can address these separately, meanwhile concentrating on the essential thread of the main text, and teachers have the option whether to require mastery of this material.

## PEDAGOGY

Many features of the text have always proved popular with users. The learning aids are, of course, back again in this edition. Changes to the learning objectives, chapter introductions, key concepts, and chapter summaries were all discussed earlier. In addition, each chapter again contains a self-quiz (with answers in the back of the book), questions for discussion, and suggested readings. Many users particularly like the tables, and in this edition many of the tables are illustrated with simple art.

## SUPPLEMENTS

A textbook is only one of the resources that instructors and students use. A number of other items enhance the usefulness of the fourth edition of *Biology*.

**Study Guide** by Janann Jenner, Talladega College, includes chapter outlines, key terms, definitions, and self-tests that correspond to text sections.

**Instructor's Manual** by Bernard Frye, University of Texas, Arlington, contains key words, a chapter overview, lecture outlines, additional suggested readings, and teaching strategies for each chapter.

A **Laboratory Manual** written by Russell V. Skavaril, Mary M. Finnen, and Steven M. Lawton, all of Ohio State University, and an accompanying *Laboratory Instructor's Manual* are available.

**General Biology Laboratory Manual,** Second Edition, by Carolyn Eberhard, Cornell University, is a lab manual for nonmajors or mixed majors/nonmajors in a general biology course. An **Instructor's Manual** to accompany the Laboratory Manual is also available.

**Test Bank** provides 2000 printed test questions in several formats and levels of difficulty. Instructors can add to, alter, or edit to customize their own tests.

**ExaMaster™ Computerized Test Bank** enables instructors to edit, revise, add to, or delete the printed **Test Bank.** Available in IBM and Macintosh forms.

**Overhead Transparency Acetates** feature 250 pieces of art from the text, using labels with large type for easy classroom viewing.

**Electron Micrograph Overhead Transparencies** display 150 electron micrographs from the text.

**Saunders General Biology Sequence Overhead Transparencies** are a set of 50 sequential overhead transparencies containing topics displayed in a series of stages via layers.

**Bio-Art** reproduces 150 selected pieces of art from the text as black-and-white unlabeled line drawings, encouraging students to learn the labeling process and to take notes. **Bio-Art** also provides a handy study tool or can be used as a test item.

**Infinite Voyage Videos/Videodiscs** from PBS bring the subject of biology to life in the classroom by offering great adventures of scientific exploration and discovery.

**Saunders General Biology Videodisc III + Directory** offers more than 1500 live-action clips and still images from four Saunders biology textbooks, including *Biology,* Fourth Edition. A **Directory** with descriptions, barcode labels, and reference numbers for each image accompanies the **Videodisc.**

**LectureActive™ Software Package Version III** contains all video clip and still images from the **Saunders Videodisc,** allowing instructors to create custom lectures quickly and easily. Lectures can be read from the computer screen or printed with barcodes for all **Videodisc** instructions. Available for Macintosh and IBM formats and accompanied by a **User's Guide.**

**Lecture Outline on Disk,** an ASCII version of the lecture outline, enables instructors to edit, expand the outline, and create study guideline handouts for students. Available for IBM and Macintosh computers.

**BIO-XL,** a unique, computer-assisted tutorial software package, is available in two modes. Test Mode quizzes students about chapter material and assesses student's knowledge; Tutor Mode adds pedagogical support through immediate feedback to responses. Corresponds to specific page references in the text. Available in IBM 5.25 and 3.5, Macintosh, and Windows.

Possibly the greatest strength of the book is that it has been read and criticized by hundreds of thousands of biology students, teaching assistants, and experienced teachers of general biology. With the benefit of their experience to add to our own, we have identified difficult areas, devoted more space to them than is usual, and, we hope, have helped students over the conceptual hurdles they have encountered.

Now the fourth edition of *Biology* is out of our hands and into yours. We hope our labors make yours easier and more enjoyable.

**Karen Arms**
*Savannah, Georgia*

**Pamela S. Camp**
*Chapel Hill, North Carolina*

# ACKNOWLEDGMENTS

The authors' names may appear on the cover of this book, but dozens of people are really responsible for it. A textbook's success depends on teachers and students finding it easy and pleasant to use, as well as effective. The first edition drew on our own teaching experience and the comments of expert reviewers in particular fields. Now the book has been much more widely reviewed and has faced its truest test in classrooms all over the world. All those teachers and students who have taken the trouble to write to us, pointing out mistakes and suggesting changes, have contributed to this edition.

We are particularly grateful to our official reviewers, nearly all of whom are biology teachers. Their enthusiastic reception of the new material we wrote for this edition cheered us enormously while we wrestled with the final revisions, and their suggestions for additional improvements have been invaluable. For your contributions to this fourth edition, our thanks to: L. Rao Ayyagari, *Lindenwood College;* Penelope H. Bauer, *Colorado State University;* Stephen Benson, *California State University, Hayward;* P. K. Bhattacharya, *Indiana University Northwest;* Michael Breed, *University of Colorado, Boulder;* J. H. U. Brown, *University of Houston;* Theodore Burk, *Creighton University;* Vicki Cameron, *Ithaca College;* Dorothy Chappell, *Wheaton College;* Keith Clay, *Indiana University;* James Coggins, *University of Wisconsin, Milwaukee;* William Cordes, *Loyola University, Chicago;* Donald Cronkite, *Hope College;* Kenneth J. Curry, *University of Southern Mississippi;* Darleen DeMason, *University of California, Riverside;* Stephen J. Dina, *St. Louis University;* James Eckblad, *Luther College;* Robert Egan, *Golden West College;* Paul Elliott, *The Florida State University;* Sharon Eversman, *Montana State University;* Stuart Feinstein, *University of California, Santa Barbara;* Donald P. French, *University of Maryland, Eastern Shore;* Chris George, *California State Polytechnic University, Pomona;* Patricia A. Grove, *College of Mt. St. Vincent;* W. Holt Harner, *Broward Community College, Central Campus;* James Howell, *Allegany Community College;* Terry L. Hufford, *The George Washington University;* Linda-Margaret Hunt, *Notre Dame University;* Austin Hughes, *The Pennsylvania State University;* Dan Johnson, *East Tennessee State University;* Charles Leavell, *Fullerton College;* Charles H. Mallery, *University of Miami;* William Marks, *Villanova University;* John R. Menninger, *Iowa State University;* Dian O. Petty, *Kingwood College;* David M. Polcyn, *California State University, San Bernardino;* Mimi Sayed, *Michigan State University;* Brian R. Schmaefsky, *Kingwood College;* Eric Scully, *Towson State University;* Karen Steudel, *University of Wisconsin, Madison;* Marshall Sundberg, *Louisiana State University;* Salvatore Tavormina, *Austin Community College;* Kathy Thompson, *Louisiana State University;* Richard M. Trelease, *Arizona State University;* John F. Utley, *University of New Orleans;* Thomas Wentworth, *North Carolina State University;* H. Patrick Woolley, *East Central College.*

In addition to these reviewers are other people who have helped us in a variety of ways. They read chapters, fed us and educated us, argued with us about biology, teaching, and grammar, contributed photographs, drawings, reprints, and moral support: Valerie Antoine of the U.S. Metric Association, Margaret Arion, William Arms, Peter Brussard, Rita Calvo, Wilfred A. Côté, Carolyn Eberhard, Paul and Richard Feeny, Virginia Fry, Joe Gall, Jerome Gross, Andrew Knoll, Melvin Kreithen, Kenneth Lohman, Robert J. Lynch, James Mauseth, Marian and John McGrath, Tibby and Fred McLafferty, Alan J. Neumann, Thomas Perrin, Robert J. Raikow, John Saidla, Judy and David Sharp, David Shotton, Gail Stetten, and J. Thorsen.

There is much more to a large textbook than writing. In this edition in particular, we have leaned heavily on the expertise of artists S. Jane Whiteley, John Norton, and Carlyn Iverson, who translated our ideas into art sketches. Photo researchers Nisa Geller and Amy Ellis Dunleavy unflaggingly pursued the right photograph for each slot. These hard workers have remained cheerful and helpful even while deadlines were abolishing vacations and social life.

A yellow-headed gecko encounter.   (© Paul Freed/Animals Animals)

It has again been a pleasure to work with the people at Saunders College Publishing. Our heartfelt thanks to Julie Levin Alexander for having the faith to sponsor this new edition. She has cheered us on our way, providing invaluable input on the areas to which we should pay special attention. She also brought on board our wonderful Developmental Editor, Lee Marcott, whose tireless work, clear-sighted editorial instinct, and constant cheer have made our

work easier and more enjoyable. Chris Schueler, Art Director, Bonnie Boehme, Project Editor, and Joanne Cassetti, Production Manager, warrant special thanks for their dedication and cooperation, which have made even the hectic final stages of completing the book a remarkably pleasant experience.

We would also like to acknowledge once again the considerable contributions of all those who reviewed and otherwise helped to prepare the previous editions of *Biology*.

## REVIEWERS OF ARMS AND CAMP, EDITIONS 1, 2, AND 3

KRAIG ADLER
*Cornell University*
RICHARD ADLER
*University of Michigan, Dearborn*
JOHN ALCOCK
*Arizona State University*
BETTY ALLAMONG
*Ball State University*
ROBERT L. AMY
*Southwestern University*
VERNON L. AVILA
*San Diego State University*
CAROL M. BAILEY
*Dean Junior College*
DENNIS BARRETT
*University of Denver*
WILLIAM E. BARSTOW
*University of Georgia*
SALLY BAUER
*Hudson Valley Community College*
JOHN C. BELTON
*California State University, Hayward*
WILLIAM M. BETHAL
*St. Louis University*
CHARLES K. BIERNBAUM
*The College of Charleston*
JOHN P. BIHN
*LaGuardia Community College*
ANTONIE W. BLACKLER
*Cornell University*
PATRICIA BONAMO
*State University of New York at Binghamton*
LOIS M. BORGMAN
*Mount Ida Junior College*
ROGER R. BOWERS
*California State University, Los Angeles*
EDMUND D. BRODIE, JR.
*Adelphi University*
PETER F. BRUSSARD
*Cornell University*
NEAL D. BUFFALOE
*University of Central Arkansas*

MAC A. CALLAHAM
*North Georgia College*
BRIAN CAPON
*California State University, Los Angeles*
C. BLAINE CARPENTER
*Clayton Junior College*
ARTHUR G. CARROLL
*Oklahoma State University*
JOHN L. CARUSO
*University of Cincinnati*
ROBERT H. CATLETT
*University of Colorado, Colorado Springs*
DOROTHY F. CHAPPELL
*Wheaton College*
SHEPLEY S. CHEN
*University of Illinois, Chicago Circle*
FRANCES S. CHEW
*Tufts University*
SIMON L. CHUNG
*Trinity College*
MILDRED A. COLLINS
*Stillman College*
RICHARD L. COLLINS
*Louisiana State University*
DEBORAH F. COOPERSTEIN
*Adelphi University*
ROLAND REECE COREY
*U.S. Naval Academy*
JAMES R. DARWOOD
*Westchester Community College*
PETER DAVIES
*Cornell University*
D. G. DAVIS
*University of Alabama*
JOHN D. DAVIS
*Mississippi University for Women*
NORMAN T. DAVIS
*University of Connecticut*
JEAN DESAIX
*University of North Carolina at Chapel Hill*

ALFRED G. DIBOLL
*Macon Junior College*
STEPHEN J. DINA
*St. Louis University*
F. PAUL DOERDER
*University of Pittsburgh*
JAMES K. DOOLEY
*Adelphi University*
WALTER RUDD DOUGLASS
*Fordham University*
MARVIN DRUGER
*Syracuse University*
ROBERT EGAN
*University of Nebraska at Omaha*
LLOYD E. EIGHME
*Pacific Union College*
GEORGE F. ESTABROOK
*University of Michigan*
WILLIAM L. EVANS
*University of Arkansas*
JOHN FARMER
*University of Oklahoma*
GERALD FASSELL
*Cayuga Community College*
PAUL FEENY
*Cornell University*
ALBERT E. FELDMAN
*Dutchess Community College*
ALFRED F. FINOCCHIO
*St. Bonaventure University*
ALLEN D. FORSYTHE
*Holyoke Community College*
CHARLES E. FOSTER
*State University College at Potsdam*
CARL S. FRANKEL
*Pennsylvania State University at Hazleton*
EDWIN C. FRANKS
*Western Illinois University*
DAVID FROMSON
*California State University, Fullerton*

C. W. GADDIS
*University of Arizona*
HERSCHEL W. GARNER
*Tarleton State University*
TIM GASKIN
*Cuyahoga Community College*
LAWRENCE GILBERT
*University of Texas at Austin*
WILLIAM H. GILBERT
*Simpson College*
GORDON L. GODSHALK
*University of Southern Mississippi*
JULIUS HAND GOODEN
*Bowie State College*
JUDITH GOODENOUGH
*University of Massachusetts at Amherst*
JACK L. GOTTSCHANG
*University of Cincinnati*
JAMES L. GRECO
*State University of New York at Albany*
ARNOLD J. GREER
*St. Louis Community College at Meramec*
JOHN R. GREGG
*Duke University*
PERRY B. HACKETT
*University of Minnesota, Twin Cities*
STEVEN N. HANDEL
*Yale University*
DAVID A. HASKELL
*Smith College*
STEPHEN C. HEDMAN
*University of Minnesota at Duluth*
RICHARD F. HELLER
*Bronx Community College*
FRANK HEPPNER
*University of Rhode Island*
TERRY M. HILL
*Southwestern University at Memphis*
PETER HINKLE
*Cornell University*
KENNETH M. HOFF
*Cleveland State University*
RICHARD J. HOFFMAN
*University of Pittsburgh*
RUSSELL C. HOLLINGSWORTH
*Tarrant County Junior College*
JOHN G. HOLT
*Iowa State University*
LINDA-MARGARET HUNT
*University of Notre Dame*
DANIEL R. HYSTROM
*Bakersfield College*
IRWIN R. ISQUITH
*Fairleigh Dickinson University*

ANDRÉ JAGENDORF
*Cornell University*
ALAN J. JAWORSKI
*University of Georgia*
E. O. JONES
*Oakwood College*
EDMUND D. KEISER
*University of Mississippi*
GEORGE H. KEIFFER
*University of Illinois, Urbana-Champaign*
SAMUEL KIRKWOOD
*University of Minnesota, Twin Cities*
JOHN A. W. KIRSCH
*Yale University*
ROBERT M. KITCHIN
*University of Wyoming*
DAVID KLINGENER
*University of Massachusetts*
HENRY M. KNIZESKI, JR.
*Mercy College*
SHIRLEY J. KURTZBERG
*Westchester Community College*
RICHARD J. LACEY
*Adelphi University*
CLIFFORD E. LAMOTTE
*Iowa State University*
LAWRENCE A. LARSON
*Ohio University*
ARTHUR LAVIGNE
*Middlesex Community College*
GEORGIA E. LESH-LAURIE
*Cleveland State University*
LISA LEVINSON
*Nassau Community College*
THOMAS A. LONERGAN
*University of New Orleans*
WILLIAM F. LOOMIS, JR.
*University of California at San Diego*
CARMITA E. LOVE
*Community College of Philadelphia*
JOHN D. LYON
*Adirondack Community College*
CHARLES H. MALLERY
*University of Miami*
ARTHUR MANGE
*University of Massachusetts at Amherst*
JAMES MARINACCIO
*Somerset County College*
W. WALLACE MARTIN
*Randolph-Macon College*
WILLIAM H. MASON
*Auburn University*
LEATHEM MEHAFFEY III
*Vassar College*

HENRY MERCHANT
*George Washington University*
MICHAEL C. MIX
*Oregon State University*
KEITH MOFFAT
*Cornell University*
RUSSELL K. MONSON
*University of Colorado, Boulder*
LOUISE M. MORGAN
*Mercer University*
J. THOMAS MULLINS
*University of Florida*
JAMES T. MURRELL
*Mississippi University for Women*
HENRY J. MUSCHIO
*Dutchess Community College*
JOSEPH J. NAPOLITANO
*Adelphi University*
MILTON NATHANSON
*Queens College*
FRANK G. NORDLIE
*University of Florida*
ALMA MOON NOVOTNY
*University of Houston*
DANIEL R. NYSTROM
*Bakersfield College*
JOHN D. O'CONNOR
*University of California, Los Angeles*
LOWELL P. ORR
*Kent State University*
LYNNE J. OSBORN
*Middlesex Community College*
MARLENE K. PALMER
*Vassar College*
MARTHA CONSTANTINE PATON
*Princeton University*
DONALD I. PATT
*Boston University*
LLOYD M. PEDERSON
*San Joaquin Delta College*
PETER S. PETRAITIS
*University of Pennsylvania*
RICHARD P. PETRIELLO
*Saint Peter's College*
L. JACK PIERCE
*Mountain View College*
PATRICIA PIETROPAOLO
*Community College of the Finger Lakes*
RUTHANNE B. PITKIN
*Allegheny College*
MICHAEL V. PLUMMER
*Harding University*
CAROLINE M. POND
*The Open University*

GENE A. PRATT
*University of Wyoming*
C. J. PROBST, JR.
*University of New Orleans*
JEAN E. PUGH
*Christopher Newport College*
RALPH S. QUATRANO
*Oregon State University*
DENNIS C. RADABAUGH
*Ohio Wesleyan University*
RUDOLF A. RAFF
*Indiana University*
ROBERT J. RAIKOW
*University of Pittsburgh*
FRANCIS V. RANZONI
*Vassar College*
BASIL ROBINSON
*Holyoke Community College*
CHARLES F. RODELL
*St. John's University*
KAREL L. ROGERS
*Adams State College*
MARTIN SACKS
*City University of New York, City College*
ROGER H. SAWYER
*University of South Carolina*
MITCHEL SAYARE
*University of Connecticut*
HOWARD A. SCHNEIDERMAN
*University of California, Irvine*
ALLAN A. SCHOENHERR
*Fullerton College*
JOHN W. SECHRIST
*Wheaton College*
JOHN L. SHANE
*Middlesex Community College*
GERTRUDE D. SHAY
*University of Dayton*
HARRY L. SHERMAN
*Mississippi University for Women*

CHARLES G. SMITH
*Lake Erie College*
GARY A. SMITH
*Tarrant County Junior College*
MARGA H. SMITH
*Wittenberg University*
NORMAN SMITH
*San Joaquin Delta College*
WAYNE L. SMITH
*University of Tampa*
ARLIEN STEINER
*Tidewater Community College*
DAVID STETLER
*Virginia Polytechnic Institute and State University*
KAREN L. STEUDEL
*University of Wisconsin, Madison*
DARRELL R. STOKES
*Emory University*
BOYD R. STRAIN
*Duke University*
CARL A. STRANG
*Dickinson College*
DARYL SWEENEY
*University of Illinois, Urbana-Champaign*
STAN R. SZAREK
*Arizona State University*
JANE B. TAYLOR
*Northern Virginia Community College*
THOMAS M. TERRY
*University of Connecticut*
C. DALE THERRIEN
*Pennsylvania State University*
JOHN THORNTON
*Oklahoma State University*
ELIZABETH K. TOMASZEWSKI
*Texas A&M University*
BIK-KWOON TYE
*Cornell University*

JOSEPH W. VANABLE, JR.
*Purdue University*
GERALD L. VANAMBURG
*Concordia College*
MARVALEE H. WAKE
*University of California, Berkeley*
JOHN M. WALKER
*Vincennes University*
EILEEN WALSH
*Westchester Community College*
BARBARA ANN WALTON
*University of Tennessee at Chattanooga*
ROSAMUND WENDT
*Community College of Philadelphia*
NORMAN K. WESSELLS
*Stanford University*
TED WEINHEIMER
*California State College, Bakersfield*
JAMES R. WILLMAN
*St. Louis Community College at Meramec*
NORMAN WILLIAMS
*University of Iowa*
JOHN T. WINDELL
*University of Colorado*
RICHARD WINN
*Mira Costa College*
CLARENCE C. WOLFE
*Northern Virginia Community College*
JOSEPH M. WOOD
*University of Missouri, Columbia*
NEWELL A. YOUNGGREN
*University of Arizona*
THEO ZEMEK
*College of Du Page*
JOHN L. ZIMMERMAN
*Kansas State University*

# CONTENTS OVERVIEW

A cheetah.

# CONTENTS

Tomato clownfish and sea anemone. (Carl Roessler/Animals Animals)

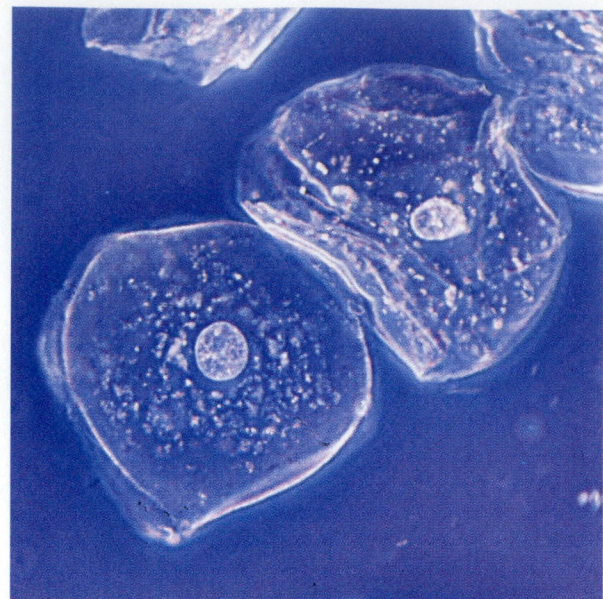

Living human cheek cells. (Biophoto Associates)

Ice plant.

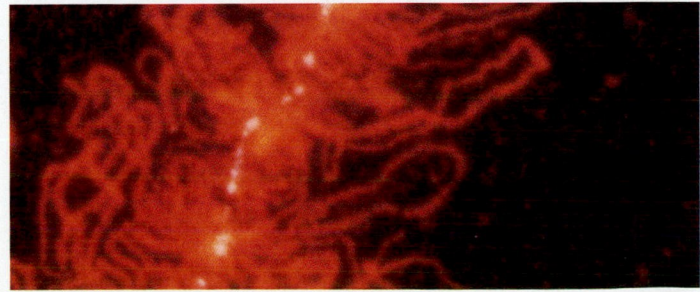

RNA synthesis in a lampbrush chromosome. (M. B. Roth and J. G. Gall)

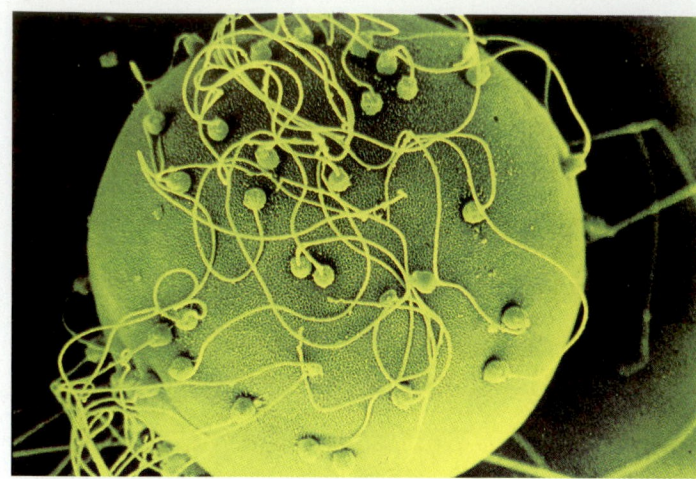

Clam egg surrounded by sperm.   (David M. Phillips/Visuals Unlimited)

Incomplete dominance in snap-dragon plants.

Giraffe.

Stromatolites. (William E. Ferguson)

Bighorn rams. (Pat and Tom Leeson/Photo Researchers)

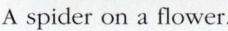

A spider on a flower.

Mushroom.

Water lilies.

Great blue heron. (Florida Audubon Society)

Nudibranch.  (Jeff Rotman)

Texas horned lizard.  (Florida Audubon Society)

Neurons. (Biophoto Associates)

Black-maned lions.

Cicada.

Leaf arrangements.

# PART SIX  Plant Biology  905

Pitcher plant.  (Runk/Schoenberger from Grant Heilman)

Golden barrel and saguaro cactus in Arizona.

Bougainvillea.

Sally Light-foot crab on the Galápagos Islands.

# Introduction

## O B J E C T I V E S

*When you have studied this chapter, you should be able to:*

1. Define the following terms and use them in context: organism, evolution, natural selection, selective pressure, adaptation.

2. List three steps in the scientific method, and apply them to investigating a sample scientific problem.
3. List eight characteristics of living things.

P eople have always been interested in the living things around them. They had to be: all food, clothing, and fuel came from plants and animals and so did dangers such as poisons, predation, and disease. Eventually, the study of living things moved from the realm of survival skills to scholarship and became the discipline of natural history (Figure 1–1).

Then in 1858 two British naturalists put forward an idea that shook the human view of the living world. The two were Charles Darwin and Alfred Russel Wallace, and the idea was how humans and other living things evolve. No longer could humans regard other forms of life as merely neighbors on Earth. They are, in fact, our kin. All **organisms**—the individual animals, plants, fungi, algae, protozoa, and bacteria on Earth—are related. Slice them thin and examine them with a microscope, put them in a test tube and analyze their chemistry, feed them sugar and track how they use it for energy, and you find that all organisms are remarkably similar.

The task of biology is to study this unity of living organisms, describe their immense diversity, outline the history of life on Earth from ancient to modern forms, and learn more about organisms, including ourselves. How are all organisms alike? How do they differ? How do they work, what do they do, how are they made? How do they affect their physical surroundings, the other organisms around them, and human life?

Now is an exciting time for biologists. New techniques, developed in the last 20 years, are shedding fascinating light on questions that have puzzled biologists for decades, and

many decisions affecting our future depend upon biological discoveries. Most of the biologists who have ever lived are still alive and reporting the results of their research in thousands of scientific articles every year. No one person can possibly become expert in any but a small segment of this vast body of knowledge. But luckily for students and professional biologists alike, the world of life is unified by a number of concepts and generalizations. It is these general principles that students must master and that professionals must constantly bear in mind.

*individual living thing made up of one or more cells*

**FIGURE 1–1** Natural history: students examine plants and small animals collected on a southern beach.

KEY CONCEPTS

- ◆ Organisms evolve by means of natural selection.
- ◆ Scientific method attempts to determine the causes of natural phenomena.
- ◆ There is unity in life: all organisms have a common ancestry and hence share many features of their chemistry, cell structure and function, genetic information systems, and processes of reproduction.
- ◆ Life is diverse: genetic variation and changing environments result in the evolution of different species of organisms.

## 1–A    FUNDAMENTAL CONCEPTS OF BIOLOGY

Biology is the branch of science that studies living things: their structure, function, reproduction, and interactions with one another and with their nonliving environment.

It is difficult to define "life" or "living," although we have no trouble seeing that cats and trees are alive, whereas stones and water are not. The usual way of overcoming this problem is to say that living things share a group of characteristics not found in nonliving things. These characteristics of living things are the basic concepts in biology that we shall explore in this book:

1. **Living things are highly ordered.** The chemicals that make up a living organism are much more complex and highly organized than those in nonliving things. An organism maintains a chemical makeup very different from that of its nonliving environment. All organisms contain very similar kinds of chemical building blocks, and they share the same pathways of **metabolism,** the processes by which these building blocks are made and destroyed. This chemical organization is reflected in the overall structure and function of the organism's body.

2. **Living things are organized into units called cells.** A cell is a packet of highly organized living material enclosed by a membrane. Cells are the units of structure, function, and reproduction in organisms. Many small organisms, such as bacteria and amoebas, consist of one cell each. Larger organisms, such as grasses and humans, contain hundreds of millions of cells, each so small that we must use a microscope to see them (Figure 1–2). Each cell is a biochemical factory that shows all the features of life. The reproduction of all organisms involves divisions of cells, forming new cells.

3. **Living things use energy from their environments.** Most organisms depend, directly or indirectly, on energy from the sun. Green plants use the sun's energy to make food, which supports the plants themselves. It is also used by all organisms that eat plants and eventually by those that eat the plant-eaters. Organisms use the energy they take in to maintain and increase the high degree of orderliness of their bodies, to grow, and to reproduce.

4. **Living organisms respond to stimuli.** Most animals respond rapidly to environmental changes by making some sort of movement—exploring, running away, or even rolling into a ball. Plants respond more slowly but still actively: stems and leaves bend toward light, and roots grow downward. The capacity to respond to environmental stimuli is universal among living things.

5. **Living things develop.** Everything changes with time, but living organisms change in particularly complex ways called development. A nonliving crystal grows by addition of identical or similar units, but a plant or animal develops new structures, such as leaves or teeth, that often differ chemically and structurally from the cells that produced them.

6. **Living things reproduce themselves.** New cells arise only from the division of other cells. New organisms arise only from the reproduction of other, similar, organisms.

7. **Living things contain genetic information.** An organism's genetic material—its chromosomes and genes—contains information specifying the possible range of the organism's development, structure, function, and response to its environment. An organism passes on genetic information to its offspring, and this is why offspring are similar to their parents. Genetic information does vary somewhat, though, so parents and offspring are usually similar but not identical.

8. **Organisms change their environments.** Since life evolved, living things have altered Earth from a planet where modern organisms could not have survived to the planet teeming with life that we know today. For instance, green plants produce the oxygen in the air we breathe, which we need to survive. Because of human numbers and technology, our own species now affects the environment more dramatically and permanently than do other organisms. Some believe that we are actually reversing billions of years of evolution and turning Earth back into a planet unable to support most forms of life.

9. **Living things evolve and are adapted to their environments.** Today's organisms have arisen by evolution, the descent and modification of organisms from more ancient forms of life. Evolution proceeds in such a way that living things and their components are well suited to their ways of life. Fish, earthworms, and frogs are all so constructed that we can predict roughly how they live merely by examining them. The adaptation of organisms to their environments is one result of evolution.

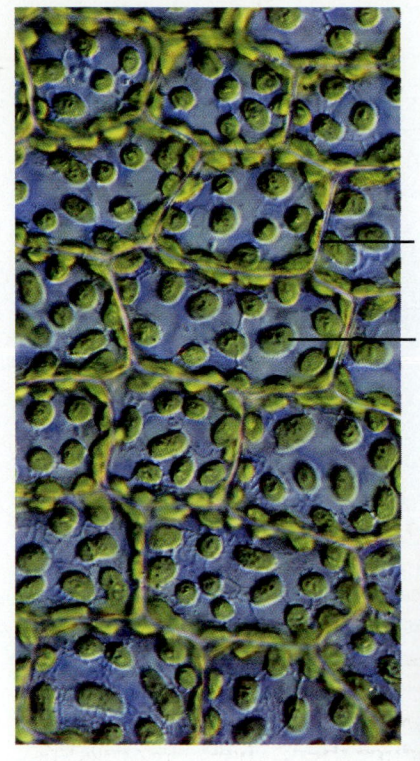

— Cell wall

— Chloroplast

(a)

**FIGURE 1–2** Characteristics of living things. (a) Organisms are composed of cells, usually too small to be seen with the naked eye. The green chloroplasts in these moss cells make food. (b,c,d) Living things develop. This cicada is undergoing its final molt from a nymph with tiny wingpads to an adult with two pairs of large, transparent wings. The crumpled wings expand and straighten as air is pumped into them. The wings and the rest of the external skeleton must then dry and harden before the cicada can fly.   (a, Biophoto Associates)

(b)

(c)

(d)

An organism's environment invariably includes other organisms. Living things adapt to life with other organisms in many ways, evolving ways to escape some, exploit others, and avoid or overcome competition for resources. Many organisms live in intimate long-term **symbioses** (singular: **symbiosis**) with members of other species (Figure 1–3). The partners in a symbiotic relationship supply each other with resources such as food, shelter, or transport.

*Close association between members of two or more species*

## 1–B   EVOLUTION AND NATURAL SELECTION

A living organism is the product of interactions between its genetic information and its environment. This interaction is the basis for the most important generalization in biology, that organisms evolve by means of natural selection.

**Evolution** is the origin of organisms by descent and modification from more ancient forms of life. For instance, human beings evolved from now-extinct animals that looked something like apes, and this happened through accumulation of changes from generation to generation. In more

**FIGURE 1–3** Symbiosis. This tomato clownfish lives symbiotically with a sea anemone in a Solomon Islands coral reef. The anemone's stinging tentacles protect both animals from predators. However, butterfly fish are immune to the stings and will eat the anemone if a clownfish is not present to drive them away. (Carl Roessler/Animals Animals)

modern terms, we can say that **evolution** is the process by which the members of a population of organisms come to differ genetically from their ancestors.

Like many other great ideas in science, evolution makes sense of many observations of the natural world. Soon after Charles Darwin proposed how organisms evolved, his friend and champion, Thomas Huxley, remarked, "How extremely stupid not to have thought of that!"

The mechanism of evolution was deduced from three familiar observations:

1. Organisms are variable. Even the most closely related individuals differ in some respects.

2. Some of the differences among organisms are inherited. Inherited variations are caused by differences in genetic material. Because organisms inherit their genes from their parents, parents and offspring tend to resemble each other more closely than do more distant relatives, whose genetic material is less similar.

3. More organisms are produced than live to grow up and reproduce. Fish and insects may produce hundreds of eggs, oak trees thousands of acorns, but only a few of these survive to reproduce in their turn.

Some inherited variations are bound to affect the chances that an individual will live to reproduce. Individuals with some genetic variations will therefore produce more offspring (which inherit this genetic material). This is **natural selection,** the production of more offspring by individuals with some genes than with others. Natural selection produces **evolution,** which we can define in yet another, more rigorous way: a change in the proportions of different genes from one generation of a population to the next.

To take an example of natural selection producing evolution, the length and thickness of an animal's hair is largely determined by its genes. A very cold winter may kill many individuals with short, sparse hair. Individuals with longer, thicker fur are more likely to survive the winter and reproduce in the following spring. Because more animals with thicker fur breed and pass on the genetic material that dictates the growth of thick hair, a larger proportion of individuals in the next generation of the population will have genes for thick fur (Figure 1–4). The genetic makeup of the population has changed somewhat from one generation to the next, and that is evolution. The agent of natural selection in this case is low temperature, which acts as a **selective pressure** against those individuals with short, sparse hair.

The result of natural selection is that populations undergo **adaptation,** a process of accumulating changes appropriate to their environments, over the course of many generations. The selective pressures acting on a population "select" those genetic characteristics that are adapted, or

**FIGURE 1–4**   Natural selection. Muskoxen in Alaska have evolved long, thick fur that protects them from the arctic cold. (Johnny Johnson/Animals, Animals)

well suited, to the environment. For instance, through selection, populations living in cold areas evolve so as to become better adapted to withstand the cold.

When we say that selection causes organisms to become adapted to their environments, we should note that "environment" in this context is a catchall word meaning much more than merely whether an organism lives in a forest rather than a desert and whether it can obtain enough food. Environment includes all the external factors that act throughout the organism's life and affect the number of offspring it produces.

Let us, for example, consider a frog. Whether it successfully meets the pressure of its environment depends on the speed and normality of its embryonic development, whether bacteria penetrate the jelly coat of the egg and destroy it during development, whether as a tadpole it can find enough food and avoid being eaten by a predator, whether the pond in which it lives as a tadpole dries up before it becomes a frog, and whether as a small frog it avoids death by disease or predators (Figure 1–5).

To make things more complicated, environmental pressures are frequently contradictory. For instance, a hot summer may benefit our frog because frog embryos develop faster at higher temperatures, but it also increases the chance that the tadpole's pond will dry up before it is ready to live on land. And environmental pressures often change. The frog must have characteristics that allow it to withstand both the heat of summer and the cold of winter. It should sit still to be safe from some predators and move quickly to escape from others; and so forth. So the frog's genetic makeup is a compromise brought about by selection for a number of opposing characteristics.

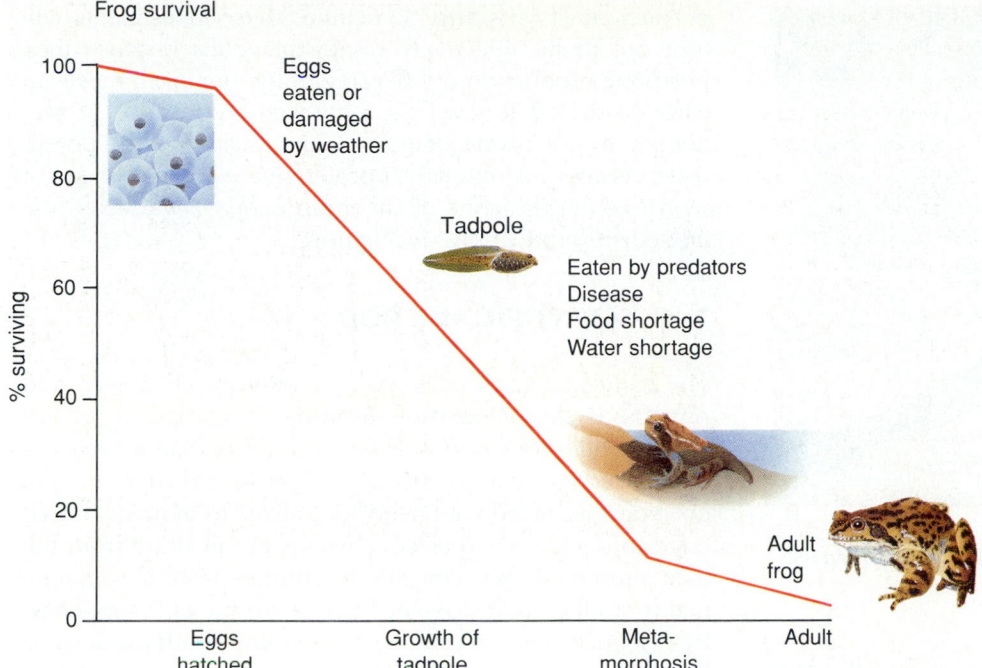

Frog survival

**FIGURE 1–5** Frog survival. This graph shows the percentage of frogs surviving to each stage of the life cycle. Few eggs develop into tadpoles that live to reproduce.

## Adaptations

We have just defined adaptation as the process by which populations become better suited to their environments as a result of natural selection. Biologists also use the word in a second way. An **adaptation** is any genetically determined characteristic that has been selected for and that occurs in a large part of the population because it increases an individual's chance of reproducing successfully. Note that this may include the capacity to change depending on environmental conditions, as with the ability to learn. Much of biology involves the study of the adaptations of organisms.

Adaptations may be broadly classed as anatomical, physiological, or behavioral. **Anatomical** adaptations involve the organism's physical structure. For instance, a penguin's flippers and webbed feet are anatomical adaptations that permit it to swim (Figure 1–6). An organism's **physiology** is all of the internal workings of its body: the biochemistry of its cells and the processes that allow it to digest food, exchange gases, excrete wastes, reproduce, move, and sense and respond to the outside world. An example of physiological adaptation is the ability of a camel to walk for days through the desert, where most other organisms would die of dehydration. An example of an impressive be-

**FIGURE 1–6** Penguin adaptations to catching fish. The bird swims using wings modified as flippers, and strong webbed feet, to propel itself through the water. The body's covering of feathers, and a layer of fat under the skin, streamline the body and insulate it from the cold water. This is a Humboldt penguin, from the coast of Peru.

**FIGURE 1–7**    An Australian bee-eater eating a bee.    (Hans and Judy Beste/Animals Animals)

havioral adaptation is the ability of some birds called bee eaters to catch bees and manipulate them with their beaks and feet so as to eat them without being stung (Figure 1–7).

## Energy and Natural Selection

Living things must take in and use energy to maintain their bodies, to grow, to obtain more energy, and to reproduce. The evolutionarily successful individual is one that leaves descendants, bearing its genetic information, in future generations of the population. Therefore, natural selection favors those individuals that can channel the most energy into producing offspring. The use of energy in other activities such as feeding, fighting, or growing is selectively advantageous only insofar as these activities result in the organism's producing more offspring.

Each individual has an "energy income," all of the energy that it acquires during its lifetime. It also has an "energy budget," its allotment of different amounts of energy to various activities. The most evolutionarily successful organisms are those most effective at converting energy to offspring.

This does not mean that organisms use all their energy directly to produce offspring. For example, suppose that a tree converts some of its energy into growing a large root system. The energy thus spent cannot be used to produce offspring. Its large root system may enable the tree to obtain a lot of water and minerals from the soil and so to produce more leaves, another diversion of energy away from the production of offspring. However, its many leaves may enable the tree to make more food than it would have otherwise and so allow it to recoup some of its previous energy expenditure by producing more offspring in the end.

Thus organisms make energy investments that may ultimately yield energy gains that can be reinvested in the production of offspring. Sometimes these investments will turn out to be selectively disadvantageous because they postpone production of offspring. If the organism meets an early death, it will never get a chance to reproduce. So any item in an organism's energy budget must have the potential to confer an ultimate reproductive gain that makes it worth taking the risks of diverting energy away from the immediate production of offspring.

## 1–C  SCIENTIFIC METHOD

The hallmark of sciences, such as biology, chemistry, and physics, is the **scientific method,** a logical way of answering questions about cause and effect. Similar reasoning is used by business people, athletes, and each of us in everyday life. We do not need specialized training or knowledge to decide whether conclusions are justified from the data presented. We can ask for further tests if someone makes a claim that does not appear to be well supported by evidence, and we can agree or disagree with predictions based on such claims. We can improve the way we do these things ourselves if we first understand how scientists arrive at conclusions about the natural world through the same kind of process.

In principle, the scientific method has three main steps (although in practice scientists work in many different ways). The first and most important step is to collect **observations,** not only by sight, but perhaps using other senses too (hearing, smell, taste, and touch). Scientists often use instruments to extend human senses or to detect things our senses cannot. Examples are microscopes, radar, voltmeters, and oscilloscopes (Figure 1–8). Second, the sci-

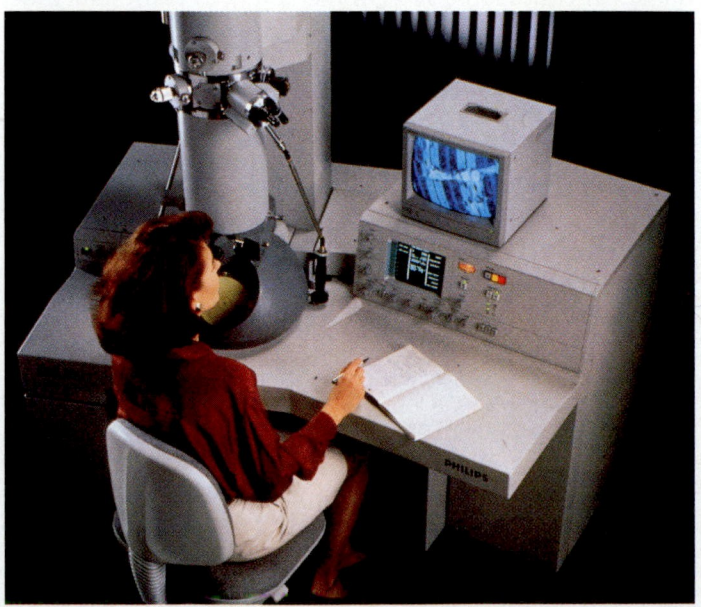

**FIGURE 1–8**    A scientist using an electron microscope to study muscle fibers.    (Philips Electronic Instruments Company)

entist thinks of several alternative **hypotheses** (singular: **hypothesis**), proposed answers to questions about what has been observed. The third step is **experimentation,** performing tests designed to show that one or more of the hypotheses is more or less likely to be incorrect. As a result of these experiments, the scientist should be able to draw some conclusions about *why* the original observed events occur. Let us see how this works in practice.

## Observations and Hypotheses

Scientists usually start with observations that stimulate questions. Some years ago, one of your authors was part of a group of biologists discussing the clusters of butterflies that seemed to be everywhere that June (Figure 1–9).

"Today," said one, "I saw about 20 yellow sulfur butterflies by a stream and some black swallowtail butterflies on a manure heap. What are they doing?"

"It's called 'puddling behavior,' replied another. "You find puddling butterflies in places such as the edges of drying puddles, or sand bars. I don't think anyone knows what they are doing. Another odd thing is that in many species only the males puddle."

These observations led us to ask what the butterflies were doing and why. To answer these questions, we had to think of some hypotheses that would account for the observations. That evening, the hypotheses came thick and fast from our armchair scientists.

"An article I read suggested it was a method of population control. Coming together permits the males to count each other. A newcomer can see if there is likely to be enough land for him to set up a territory in the area. Puddling saves them having to fight over territories."

"That sounds wrong to me," replied one of the company. "How can a butterfly figure out the density of males in the area from a group like that? Besides, swallowtails do fight for territories—I've seen them."

"I think it is more likely they're feeding," another contributed. "It was called 'puddling' in the first place because the butterflies often seem to be sucking something up from the ground with their proboscies [tongues]."

"I wonder if they are after nitrogen? In our lab we've shown that butterfly caterpillars grow faster if you feed them extra nitrogen, and there is lots of nitrogen in a manure pile."

"But not in sand," came the objection. "And if they are after nitrogen, you'd expect females to puddle, not males. Extra nitrogen in the female's eggs might be useful to the caterpillars when they hatch."

"It sounds to me," chipped in another, "as if they're after salts—perhaps salts containing sodium. All the puddling places contain quite a lot of salts: manure piles have salts from urine, and puddles have salts at the edges, left behind by evaporation of water. Lots of animals that feed on plants are short of sodium because most plants contain so little of it. We put out salt blocks for cows and horses and end up attracting deer and rabbits as well. Perhaps male butterflies need more sodium than females do."

We could test these alternative hypotheses only by doing experiments. Some hypotheses are no use because they cannot be tested. For instance, the hypothesis "puddling butterflies count each other" is probably untestable because it is hard to imagine an experiment that could show us whether or not another animal has counted its neighbors. Even a testable hypothesis usually cannot be tested directly. We must first develop a testable **prediction** from it. From

**FIGURE 1–9** Sulfur butterflies puddling on a sandbank. (Keith Brown)

the hypothesis that butterflies suck up sodium when they puddle, we predicted that if we put out trays containing sodium, butterflies would puddle on them. The hypothesis that butterflies suck up nitrogen generated the prediction that butterflies would puddle on trays of amino acids, molecules that contain nitrogen. These predictions can be tested, and, in this case, both can be tested at the same time, in the same experiment.

## Experiments

We must design experiments so that their results are as clear-cut as possible. For this reason, experiments have to include **control treatments** as well as **experimental treatments.** The two differ only by the factor(s) being investigated. For instance, to test our hypotheses, we had to show that butterflies would puddle on an experimental tray containing amino acids or one containing sodium but would not puddle on control trays that were identical except that they did not contain the amino acids or sodium.

Suppose we put out three trays—one containing sodium, another containing amino acids, and a control tray containing something butterflies are most unlikely to eat, such as plain sand or sand and water. We would predict that if butterflies are attracted to puddle on sodium, they would come to puddle on the sodium tray but not on the other two. If they are attracted to amino acids, they would puddle only on that tray. If they are attracted to both, they would puddle on both of these but not on the control tray, and if they are attracted to neither, they would not puddle on the trays at all. Note that there are many other possible reasons for the last result. Butterflies might not come because they never see the trays, or because they avoid trays or the human watchers nearby, or for any number of other reasons. If no butterflies came to puddle on our trays, we would learn nothing.

So that our experiment would not fail for lack of butterflies, we put our trays on a sandbank by a lake where tiger swallowtail butterflies often puddled in large numbers. We filled the trays with clean sand for the butterflies to stand on, and in each tray we pinned a dead male tiger swallowtail as a decoy, because we thought butterflies might be attracted to puddling places by seeing other butterflies there. We put out ten trays of sand and poured the same volume of solution (substances dissolved in water) into each one (Figure 1–10). Then we sat nearby, with binoculars, notebooks, and watches, to see what would happen (Figure 1–11).

Soon dozens of tiger swallowtails were hovering over the trays. Whenever a butterfly landed on a tray, it stuck its proboscis into the sand. At times, as many as 30 butterflies were on a tray together. Most spent a few seconds on several trays, but they puddled (which we defined as staying for more than 15 seconds) on only a few trays: all those

(a)

(b)

**FIGURE 1–10**   The puddling experiment. (a) Observers watching puddling trays. (b) Butterflies visiting the trays. In the middle of each tray is pinned a dead butterfly to serve as a decoy.   (Paul Feeny)

containing sodium in any form, and those containing amino acids (see Figure 1–10).

We were satisfied that these results were accurate because we had taken another precaution: the people recording the butterflies' visits did not know which tray contained which solution. Making an experiment "blind" in this way is important. Psychologists have shown that, even in a carefully controlled experiment, experimenters tend to find the results they want to find. This is also why scientists think up as many hypotheses as possible to explain their observations. It is easy to bend the facts to fit your only hypothesis without even realizing it.

Those who favored the hypothesis that butterflies puddle on sodium were disappointed that they also came to amino acids. But prejudice can sometimes be useful, even in science! Not only were we disappointed by the results, we were inclined to think they were wrong. Back we went

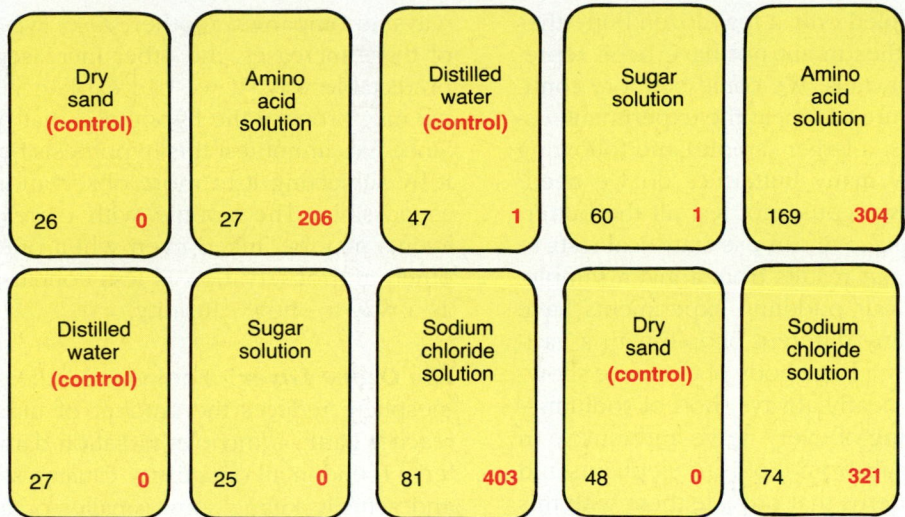

**FIGURE 1–11** Arrangement of trays on one day of the puddling experiment. Each tray contained the same volume of sand. Each of eight trays also contained 1.5 liters of water or solution. Different solutions were placed in different trays on subsequent days. (Sugar was tested because swallowtail butterflies eat sugar-filled nectar from flowers, and therefore we wondered if they might be attracted to puddle on sugar.) The black number on each tray shows the number of "sampling" visits (lasting less than 15 seconds) by butterflies. Colored numbers show the number of butterfly-minutes spent puddling on the tray in visits lasting more than 15 seconds. The numbers make it obvious that butterflies puddled on the trays containing sodium and those containing amino acids but not on any of the other trays.

to our bottle of amino acids. We now made an observation that we should have made before doing the experiment: the label said, "Prepared in sodium citrate." According to popular myth, scientists are calm and objective, but we were very excited as a technician analyzed our amino acids: they were chock full of sodium! There followed frantic phone calls and special deliveries to obtain amino acids free from sodium. At last came a suspenseful experiment, which showed that butterflies did not puddle on our new, sodium-free amino acids.

We had now conducted a well-controlled scientific experiment. What conclusions could we draw? Had we proved the hypothesis that butterflies puddle because it permits them to obtain sodium? No. We had not even shown that the butterflies actually drank the sodium solution. All we had shown was that male tiger swallowtail butterflies would puddle on sand containing sodium salts but not on sand containing various other solutions. Many more hypotheses and experiments were needed if we were to learn more.

## Limitations of Experiments

One peculiarity of the scientific method is that a hypothesis can never formally be proved but can only be disproved. A correct hypothesis leads to predictions that are borne out

by experiments, but an incorrect hypothesis may also produce correct predictions (predictions that are right, but for the wrong reason). Therefore, if the results of an experiment agree with the prediction, we are still not sure that the hypothesis is correct. For instance, the hypothesis that butterflies puddle because they need sodium is not proven by the experimental finding that butterflies puddle on sodium. They might puddle because wherever there is sodium in nature there is also nitrogen or something else they need. We have not even disproved the hypothesis that puddling is a means for the butterflies to "count" each other. They might puddle on sodium merely as a convenient rendezvous (although the fact that the butterflies appear to feed when they puddle makes this hypothesis unlikely). The more alternative hypotheses we disprove or cast doubt on, however, the greater the likelihood that the hypothesis that remains is correct.

Scientists also hesitate to accept the results of an experiment until it has been repeated. Repeating an experiment guards against two kinds of errors. The first is **human error** (a polite term for mistakes). We might have inadvertently switched the solutions, written results in the wrong column of our data notebook, or alarmed the butterflies. Even in this simple experiment, the possibilities are endless. Second, any experiment is subject to **sampling error,** error due to using a relatively small number of subjects. Our experiment sampled only a few dozen butterflies on six days. These butterflies might not have been representative of all tiger swallowtails. We could be more confident of our results if we were to repeat the experiment, using more butterflies (that is, a larger sample) and following the same procedure. How many butterflies do we need? The more the better, but we could not test all the butterflies in the world. In practice, we can use statistical tests to tell how "sure" we are of our results from a given sample.

In fact, variations on our puddling experiments have now been repeated by many different people with a variety of animals. There is now a large body of evidence showing that many animals are nearly always short of sodium—which is vital to the working of every nerve and muscle in the body. As a result, animals from moths to elephants and humans have behavior patterns that provide them with this mineral.

Historically, salt has been a valuable commodity in the trade of human populations unable to obtain salt from the sea. Soldiers in the army of ancient Rome received an allowance of salt called a *salarium*. Later the payment was changed to money to buy salt—hence our word "salary" and the expression "not worth his salt." In New Guinea to this day mountain villages with mineral springs evaporate the water to produce "salt," which can be traded for large quantities of food or clay pots.

A hypothesis supported by many different lines of evidence from repeated experiments is generally regarded as a **theory,** and after even further testing it comes to be generally accepted. An interesting example is the "theory of evolution." A hundred years ago, it would have been accurate to describe as a theory, supported by several lines of evidence, the idea that the organisms on Earth have arisen by evolution. Today, the evidence for evolution is overwhelming. We shall see in this book that we can actually observe evolution happening around us. Evolution is no longer merely a theory.

## Correlation Studies

Many scientific questions cannot be studied by the type of experiments we used to study puddling behavior, for practical or ethical reasons. For example, it would be unethical to test a vaccine against AIDS by injecting intact HIV (human immunodeficiency virus) into vaccinated and unvaccinated people. Geologists who study Earth's crust, or paleontologists who study dinosaur fossils, work with events that occurred millions of years ago: it is too late for experiments. We also cannot study long-term changes in Earth's climate by doing experiments because we have no control planet and because the experiment is too big to do.

To test their hypotheses, scientists in such fields depend on **correlations,** reliable associations between two events. We may observe many occasions on which one event always accompanies another. Also, we may find that if one of these increases, the other increases (or decreases) in a predictable way. If we can explain why this might be so, we may propose the hypothesis that one causes the other. Since we cannot test this hypothesis by experiment, we test it by subjecting it to more observations of as many kinds as possible. The trouble with correlation studies is that events may be linked even when one does not cause the other, and observation is less conclusive than experiment as a way to show causality.

***The Ozone Layer***    For instance, the ozone layer in the atmosphere reduces the amount of ultraviolet radiation that reaches Earth. Ultraviolet radiation damages the genetic material found in all organisms, causing skin cancer in humans and actually killing some smaller organisms. Ozone forms in the upper atmosphere when intense ultraviolet light from the sun acts on oxygen, converting some of it into ozone. Once formed, the ozone itself is a good absorber of ultraviolet radiation and keeps much of it from reaching Earth's surface.

In the early 1980s researchers discovered a hole in the ozone layer in the atmosphere over Antarctica. Nowadays, the ozone layer is monitored by weather balloons and NASA's Upper Atmosphere Research Satellite. Samples of air in various parts of the atmosphere are also collected by the high-flying ER-2, a converted U-2 spy plane.

Scientists studying atmospheric gases had already found that ozone can be destroyed by chemicals containing chlorine, in the presence of sunlight. They hypothesized that holes in the ozone layer form wherever the right kinds of particles are present in large numbers together with chlorine-containing compounds such as **chlorofluorocarbons (CFCs),** pollutants found everywhere in the atmosphere.

CFCs are artificial chemicals that were widely used for producing foam rubber, for cleaning electrical components, and for refrigeration. They have long been used in the production of aerosol sprays. CFCs are ideal aerosol propellants because they do not react chemically with whatever is being sprayed. But this same lack of chemical reactivity means that they are not broken down in the air but slowly make their way to the upper atmosphere. Scientists predicted that as industry and consumers released more and more stable chlorine compounds into the air, destruction of the ozone layer would accelerate.

This prediction was borne out by observation. The concentration of chlorine compounds in the atmosphere increased, and holes in the ozone layer grew larger and more numerous (Figure 1–12). Governments were convinced by the correlation evidence and negotiated international treaties to reduce the output of ozone-destroying chemicals. Industry has responded by cutting the production of CFCs and allied chemicals even more rapidly than mandated by law. The rate of chlorine accumulation in the atmosphere has slowed.

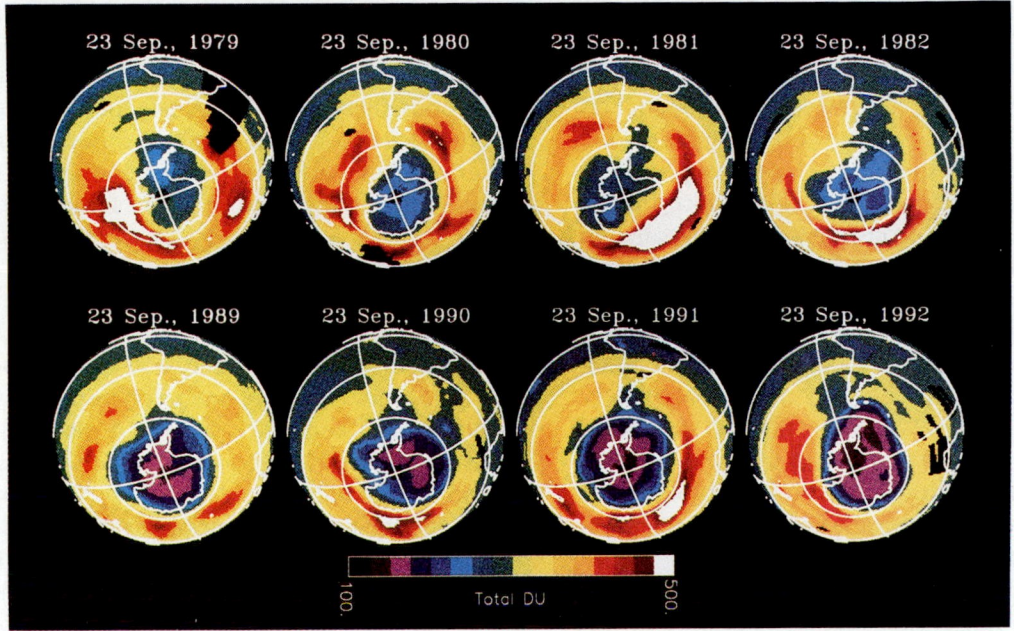

23 Sep., 1979  23 Sep., 1980  23 Sep., 1981  23 Sep., 1982

23 Sep., 1989  23 Sep., 1990  23 Sep., 1991  23 Sep., 1992

Total DU

**FIGURE 1–12** Ozone depletion: computer views of ozone concentration near the South Pole (where the white lines cross toward the bottom of each view). By 1992, the blue area of low ozone levels increased to more than twice the area of the United States and was 15% greater than in 1991. In 1989, the first area of extreme depletion (purple) appeared. By 1992, it had approximately doubled in size. Ozone concentration is expressed in Dobson Units (DU). The U.S. *Nimbus* satellite has measured ozone with a spectrometer since 1978. In 1991, the task was taken over by a similar instrument on the Russian *Meteor-3* satellite. (NASA)

The test of the hypothesis about ozone destruction that most scientists would consider conclusive is yet to come in the form of yet another correlation. As production of volatile chlorine compounds declines, we should reach the point where their level in the atmosphere will decline. Since ozone will still be formed, we would predict that the ozone level in the atmosphere will increase and holes in the ozone layer become smaller and less numerous.

## 1–D  IT'S A FACT?

"It's a scientific fact" is often presented as the clincher to an argument. Most scientists, however, would argue that any scientific finding is open to question (Figure 1–13). The

**FIGURE 1–13** When is a fact not a fact? Nineteenth-century doctors were taught that men and women breathed differently: men used their diaphragms (the sheet of muscle below the rib cage) to expand their chests, whereas women raised the ribs near the top of the chest. Finally, a woman doctor, Clelia Duel Mosher, found that women breathed in this way because their clothes were so tight that the diaphragm could not move far enough to pull air into the lungs. Some women even had their lower ribs removed surgically so that they could lace their waists more tightly. (Bettman Archives)

doubts and uncertainties inherent in scientific method make it impossible to be 100% sure that a scientific discovery is "right."

"Frogs breed in the spring" and "spiders have eight legs" look like facts at first glance, but they are really predictions about what will happen in the future, based on past experience. "This is a table" may also seem like a fact, but it is really a statement resulting from an agreement: all have agreed to call that sort of object a table.

"Facts" are also less sure than they seem because they depend on our faith in our senses. Suppose several people look at two photographs, one of a table and one of an object floating in a lake. Everyone may agree that the first photo clearly shows a table. When they look at the second one, however, some may say, "That is a Loch Ness monster," but the others may disagree. When technology, in the form of a camera, microscope, or oscilloscope, intervenes between our senses and an object, as it often must in scientific research, the problem of interpreting what we see or hear or smell becomes even more difficult. Thus, a "fact" is really a piece of information that we believe in strongly or that seems highly likely to be repeated without change.

The history of science abounds with dogmas that turned out to be wrong, although for a time they were widely accepted. Indeed, many statements in this book will undoubtedly prove untrue in the future. This is one reason why the cautious person or society will not place too much faith or invest too heavily in a new scientific discovery until it has been well tested.

Although scientific findings are less reliable than most people realize, scientists do believe that their methods discover useful information, and that careful study increases the probability that science's generalizations about nature come close to reality. Public support for science rests on the belief that a better understanding of the natural world increases our ability to promote human well-being.

## 1–E  HUMANS AND ENVIRONMENT

Nowadays, a practical interest in biology is often prompted by environmental concerns. For instance, we seek to feed everyone on Earth. We learn that planting trees can make our houses and cities cooler in summer and save on air-conditioning bills.

Our environmental problems stem from the human population explosion (Figure 1–14). We live in extraordinary times, witnessing an event that has never happened before and will never happen again: the tripling of Earth's human population in less than a century, from fewer than 2 billion people in 1900 to more than 6 billion by 2000. This can never happen again because Earth cannot produce the resources to support triple the present human population.

Like all other living things, we depend on other organisms: to produce the food we eat and the oxygen we breathe, to destroy our wastes and purify the water we drink. As we drive other organisms from more and more of the globe, we imperil our own survival. As we enter the

**FIGURE 1–14**  Population explosion. The more than 220,000 human babies that are born every day are part of the problem and many of them will grow up to become part of the solution.  (Girl Scouts of U.S.A.)

twenty-first century, our species faces one of our most important evolutionary milestones. If humans are to survive, we must change our behavior from a species that destroys its environment to one that does not.

## SUMMARY

Biology is the science that studies living things, using scientific method. Biological research has produced some fundamental concepts of biology.

1. All living things consist of one or more cells. Cells take energy from their environments and use it to maintain a high degree of chemical and structural order and for activities such as maintenance, response to stimuli, growth, development, and reproduction to produce more cells and individuals.
2. Living things contain information in the form of their genetic material. This information dictates how organisms develop, survive, and reproduce, and determines the characteristics they can pass on to their offspring.
3. Organisms change their environments both in the short term and throughout the history of life on Earth.
4. Living things evolve, adapting to their environments and giving rise to new types of organisms.
5. The chief agent of evolution is natural selection, the phenomenon by which individuals with certain genetic traits are more likely to survive and reproduce, thereby increasing the proportion of their own genetic information in future generations of the population. Natural selection ensures that those individuals most effective in converting energy to offspring will be evolutionarily successful. This ensures that a population of organisms becomes well adapted to its environment.
6. Scientific knowledge is developed by subjecting problems to the scientific method. First, scientists make observations. Then they formulate alternative hypotheses that might explain the observations, and they test the hypotheses by experiments designed to disprove one or more of the hypotheses and therefore to strengthen the evidence for those that remain. When biological questions cannot be studied by direct experimentation, they are studied by searching for correlations that may reveal cause and effect.
7. Scientific discoveries and theories are useful, but they are always open to question. Time and again in the history of science, accepted dogmas have turned out to be wrong, and even today scientists are busy discarding or remodeling some of the "facts" presented in this book.
8. Human life depends on the activities of millions of organisms that share Earth with us. Rapid growth of the human population in the twentieth century threatens many of these life support systems and the future of the human species.

## Questions for Discussion

1. After every hard rain you find dead earthworms lying on the sidewalk. What experiments would you perform to show the cause of death?
2. Many characteristics of life can be found in some nonliving things. Can you think of examples of these?
3. What might you expect was the selective pressure that resulted in each of the following adaptations?
   an elephant's trunk
   the scent of honeysuckle
   a leopard's spots
   the bark of a tree
   human language
4. Some people believe that the world will never run out of resources because research and technology will find replacements for depleted natural resources. Do you agree? Why?
5. If we imagine the evolution of the human species toward a sustainable society, we encounter a theoretical problem. In general, the people who reproduce most rapidly today are those who are poor, with the least education, and therefore often with the most environmentally destructive agriculture and little concern for environmental problems. Are these, therefore, the most evolutionarily successful people? If so, how can we imagine that humanity will evolve toward an environmentally responsible society with a low rate of reproduction?
6. Scientists often say that science is neither good nor bad; only the use of science, by scientists or by society, has moral consequences. For example, the discovery that the atom could be split was merely a scientific discovery. It was the decision to use this knowledge to build an atom bomb that produced the moral dilemma of whether it was ever right to use such a weapon. Do you think that in practice scientists must take, and society should force them to take, more moral responsibility for the consequences of their research? Is it always possible to foresee that a particular area of research will eventually have consequences?

## Suggested Readings

Arms, K. *Environmental Science,* 2d ed. Philadelphia: Saunders College Publishing, 1994. An introductory textbook on the environment.

Arms, K., P. Feeny and R. C. Lederhouse. "Sodium: stimulus for puddling behavior by tiger swallowtail butterflies, *Papilio glaucus." Science* 185:372, 1974. The puddling experiments described in this chapter.

Ehrlich, P. R., and A. H. Ehrlich. *The Population Explosion.* New York: Simon and Schuster, 1990. The environmental and human disasters caused by explosive growth of the human population.

Feshbach, M., and A. Friendly. *Ecocide in the USSR: Health and Nature under Siege.* New York: Basic Books, 1992. The amazing story of land rendered uninhabitable and health destroyed by decades of agricultural and industrial pollution in the former Soviet Union.

Mayr, E. *The Growth of Biological Thought.* Boston: Belknap Press of Harvard University Press, 1982. A historical perspective from an eminent evolutionary geneticist.

Roszak, T. *Where the Wasteland Ends.* Garden City, N.Y.: Doubleday, 1973. Critique of modern science by a man who believes science dominates western society and causes much of its malaise.

PART
ONE

*Cells*

C H A P T E R

2

# Some Basic Chemistry

## O B J E C T I V E S

*When you have studied this chapter, you should be able to:*

1. Define or recognize the characteristics of the following, and use this knowledge to answer questions about the relationships among them:

   atom, proton, neutron, electron, isotope
   ion, molecule, single bond, double bond, polar, nonpolar
   mole, molecular mass
   solution, solvent, solute, concentration
   reactant, product
   dissociation, acid, base, pH scale, buffer

2. Recognize examples of ionic, covalent, and hydrogen bonds, and explain the differences between them.
3. Write the correct molecular formulas for water, carbon dioxide, oxygen gas, and table salt.
4. Given the atomic masses of the atoms involved, calculate the molecular mass of any compound listed in Objective #3; given the chemical formula and atomic masses for any other compound, calculate its molecular mass.
5. List and discuss six reasons why water plays an important role in living systems.

---

A chemist once defined a human being as "20 gallons of water and $5 worth of assorted chemicals." This definition emphasizes the fact that the material in living bodies, including our own, ultimately comes from some of the very common substances that make up the nonliving world. However, living organisms differ from nonliving things in the structure and organization of their chemicals. The $5 price tag applies to chemicals in a fairly simple form. In a living body, these materials would be assembled into the large, highly organized molecules we shall meet in the next chapter: enzymes, hormones, muscle proteins, fats, the DNA of the genetic material, and so on. If you were to buy these in the same amounts found in your body, a chemical supply company would charge you an estimated $6 million!

The chemistry of life has two notable features: first, living things are composed mainly of water; and second, the molecules characteristic of living things have skeletons of carbon. These carbon-containing substances have complex structures that give them many interesting properties.

In this chapter we shall look at some basic properties and behavior of matter, and especially of water, as background for our study of biology. Chapter 3 covers the carbon-containing substances made by living organisms.

## K E Y   C O N C E P T S

- The chemistry of living and nonliving things follows the same rules.
- The chemistry of life has two notable features:
  1. Living things are composed mainly of water.

2. The large chemical components characteristic of living things have structures based on carbon skeletons.
- The unusual properties of water make it essential to life.

## 2–A    CHEMICAL ELEMENTS OF LIFE

Many substances in the natural world can be broken down into other substances. For example, under certain conditions water breaks down into two gases, hydrogen and oxygen, and rust separates into iron and oxygen. Hydrogen, oxygen, and iron are examples of **chemical elements,** substances that cannot be broken down into other kinds of substances by ordinary chemical reactions. (However, some elements change into others by radioactive decay.)

Chemists have discovered more than 100 elements, each with a unique set of chemical properties. Living organisms use only about 20 elements (Table 2–1). These are not the most common elements, but those with properties that have

**FIGURE 2–1**   Thirsty cape buffalo. These animals trek long distances across the dry savanna to scarce water holes, where they replenish their water supply. They also eat soil at the edges, which is rich in minerals carried downhill by rainwater runoff and left behind when the water evaporated.

### TABLE 2–1

### Chemical Elements Found in Animals, Their Approximate Abundance by Mass, and Their Atomic Masses

| Element | Symbol* | Mass (Percent) | Atomic Mass (Daltons)† |
|---------|---------|----------------|------------------------|
| Oxygen | O | 62 | 16.0 |
| Carbon | C | 20 | 12.0 |
| Hydrogen | H | 10 | 1.0 |
| Nitrogen | N | 3.3 | 14.0 |
| Calcium | Ca | 2.5 | 40.0 |
| Phosphorus | P | 1.0 | 31.0 |
| Sulfur | S | 0.25 | 32.0 |
| Potassium | K | 0.25 | 39.0 |
| Chlorine | Cl | 0.2 | 35.5 |
| Sodium | Na | 0.10 | 23.0 |
| Magnesium | Mg | 0.07 | 24.5 |
| Iodine | I | 0.01 | 127.0 |
| Iron | Fe | 0.01 | 56.0 |
|  |  | 99.69 |  |

**Trace Elements (*needed in very small amounts*)**

| Element | Symbol | | Atomic Mass |
|---------|--------|---|-------------|
| Copper | Cu | | 63.5 |
| Manganese | Mn | | 55.0 |
| Molybdenum | Mo | | 96.0 |
| Cobalt | Co | | 59.0 |
| Boron | B | | 11.0 |
| Zinc | Zn | | 65.5 |
| Fluorine | F | | 19.0 |
| Selenium | Se | | 79.0 |
| Chromium | Cr | | 52.0 |

*Each element is assigned a one- or two-letter symbol that is used as a chemical "shorthand" in writing chemical formulas and equations.

†Where atomic mass is not an integer, the mass given is the average of commonly occurring isotopes (atoms of the element with different masses). The unit of atomic mass is the dalton, named for English chemist and physicist John Dalton, who developed the first table of atomic masses.

made them useful to organisms. For example, silicon is the most common element in Earth's crust, about 300 times more abundant (by mass) than carbon. Yet carbon plays an indispensable role in every living thing, whereas silicon is a major component of very few. Evolution has selected those organisms whose chemical compositions are best at carrying on the activities of life. Carbon became a vital part of life because its properties make it peculiarly suitable to combine with other elements and form thousands of different biological substances.

Organisms selectively take in the chemicals they need. For example, a plant may contain certain minerals at concentrations much higher than those found in the soil because the roots absorb these minerals much more than others. Animals may travel many miles to salt deposits where they can eat sodium, or to water holes (Figure 2–1).

## 2–B    STRUCTURE OF ATOMS

The element carbon occurs in two familiar pure forms: graphite (part of the "lead" in some pencils) and diamond. In theory, you could divide a diamond into smaller and smaller pieces until you had separated it into atoms of carbon. An **atom** is the smallest unit of an element that retains all of the element's properties. Atoms are the units of matter, and they are extremely small. A carbon atom's diameter is about 0.14 nanometer (nm). (1 nm = $10^{-9}$ m, that is, one billionth of a meter, or one millionth [$10^{-6}$] of a millimeter.)

Atoms contain three main kinds of particles:

| Particle | Electric Charge | Atomic Mass (Daltons) | Location |
|---|---|---|---|
| **Proton** | +1 | 1 | Nucleus |
| **Neutron** | 0 | 1 | Nucleus |
| **Electron** | −1 | 0 | Electron shells |

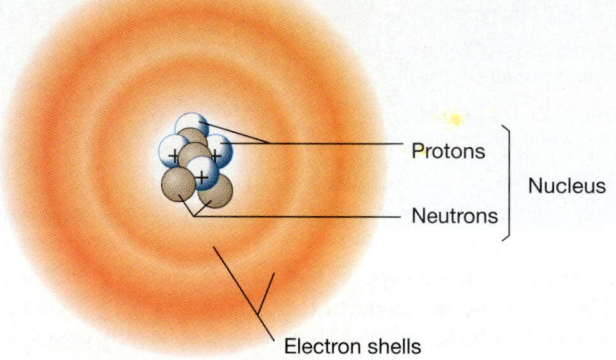

Protons
Neutrons
} Nucleus

Electron shells

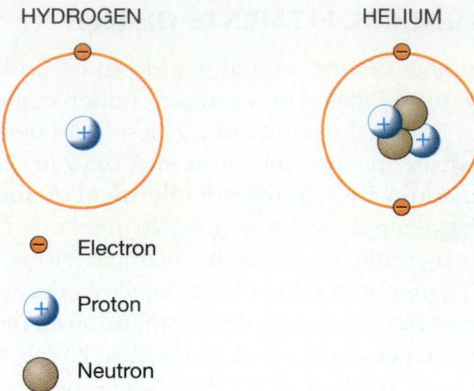

Electron

Proton

Neutron

**FIGURE 2−2** Bohr models of hydrogen and helium atoms (not drawn to scale). Hydrogen atoms are the smallest and simplest atoms, only 0.1 nm in diameter.

Picture an atom enlarged to the size of a football field, with an orange in the center. The orange represents the atom's **nucleus,** composed of its protons and neutrons clustered together. The nucleus contains all of the atom's positive charge(s) and virtually all of its mass.

The rest of the football field is space where gnat-size electrons whiz about rapidly. An electron's mass is so small that it is usually considered to be zero. Because electrons each carry a negative charge, they are attracted to the positively charged protons in the nucleus, and this attraction helps hold the atom together. An atom contains equal numbers of electrons and protons, and so its net electric charge is zero.

Atoms of the element hydrogen are the smallest and simplest atoms. A hydrogen atom contains one positively charged proton, with one negatively charged electron zipping around it. Most hydrogen atoms do not have neutrons, but the atomic nuclei of all other elements do contain neutrons. For example, a helium atom's nucleus contains two protons and two neutrons (Figure 2−2).

The number of protons in the nucleus determines what element the atom is. This is the element's **atomic number;** each of its atoms has this many protons. Each element has a different atomic number: hydrogen 1, helium 2, and so on. This number determines how many electrons the atom has, which in turn determines the atom's overall size and its unique properties—that is, how atoms of the element interact with other atoms.

To find an atom's **atomic mass,** we add up the numbers of its protons and neutrons, each having 1 atomic mass unit, known as a **dalton. Mass** is a measure of the quantity of matter. It differs from **weight,** which is a measure of how strongly a mass is attracted by the force of gravity. Your weight on the moon would be only one-sixth that on Earth, but your mass would be the same.

Some elements occur in two or more different forms, or **isotopes:** atoms of the same element with different numbers of neutrons, and hence different atomic mass. For example, all carbon atoms have six protons (this defines them as carbon atoms) and most have six neutrons, for an atomic mass of 12. However, one in every trillion carbon atoms has eight neutrons, for a mass of 14. These isotopes are named carbon 12 and carbon 14 (the numbers refer to atomic mass).

Like many other rare isotopes, carbon 14 is **radioactive;** that is, its atomic nuclei are more or less unstable and will eventually decompose to form atoms of other elements, emitting radioactive energy in the process (*How Do We Know? Isotopes: Chemical Tools for Biology*).

In 1913 physicist Niels Bohr proposed a useful model of atomic structure, in which electrons move in **electron shells** at specific distances from the nucleus (see Figures 2–2, 2–3). Each shell is associated with a specific level of energy, and an electron's energy is determined by the shell it occupies. The first shell, nearest the nucleus, is the one where electrons have the least energy. Electrons move to the lowest possible energy level, and so the first shell must be filled before electrons begin to occupy the second shell. This innermost shell can hold only two electrons; the second can hold eight, and the third can also hold eight when it is the outermost shell.

The most stable atoms are those with filled outermost electron shells. A helium atom has two electrons (see Figure 2–2). Because its electron shell is filled, it is stable and "inert," existing as a single atom that does not interact with other atoms; that is, it does not readily undergo chemical reactions. This is why helium gas is used in blimps and balloons instead of lighter hydrogen gas, which tends to react explosively with oxygen.

On the other hand, carbon has six protons, six neutrons, and six electrons. It has two electrons in its first shell,

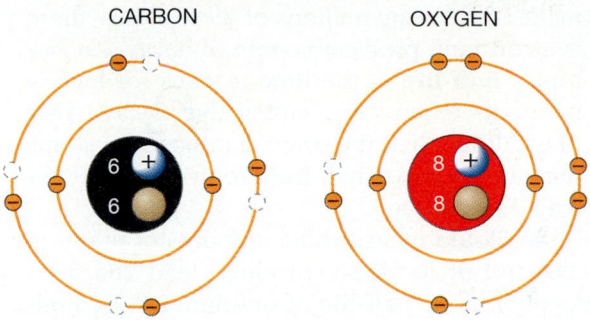

CARBON    OXYGEN

**FIGURE 2–3** Bohr models for carbon and oxygen. Carbon has room for four more electrons in its outer shell (empty dotted circles). Oxygen can accept two more electrons before its outer shell has a full complement of eight.

but only four in its second, outer shell. This shell has room for four more electrons (Figure 2–3). Hence it may react with other atoms.

The Bohr model of atoms is vastly oversimplified, because electrons do not move in a set path at a set distance from the nucleus. However, it is good for showing that electrons occupy distinct energy levels and fill the lowest first, and we use it here to keep our electron-shell bookkeeping straight.

In fact, electrons are really in constant, random motion. However, the various electrons in an atom do spend most of their time in predictable regions within the electron shells. The electron cloud model shows an atom's electron shells as fuzzy shading, as in some figures in the next section. The density of shading indicates the proportion of time electrons spend in various regions and therefore the probability of finding electrons there at any particular instant. The electrons in a filled shell occur in pairs.

## 2–C BONDS BETWEEN ATOMS

The atoms of most elements do not have exactly enough electrons to fill their outermost shells. These atoms take part in chemical reactions in which they gain, lose, or share electrons with one or more other atoms, thereby forming **bonds.** Both atoms involved in a bond end up with stable, filled outer electron shells.

The average position occupied by electrons involved in a bond depends on the **electronegativity** of each atom, that is, how strongly the atom attracts electrons. In general, two features tend to make atoms more electronegative: small size (the outermost electron shell lies closer to the nucleus and so the electrons are attracted more strongly), and nearly filled outermost electron shells (giving them a strong tendency to acquire enough electrons to complete the shell).

Hence the most electronegative elements are those in the upper right corner of the periodic table (see Appendix A).

Three types of bonds between atoms are important in living things:

1. **Ionic bonds.** An **ionic bond** is a strong electrical attraction between a positively charged ion and a negatively charged ion. **Ions** are electrically charged particles formed when an atom gives up one or more of its outermost electrons to another atom, which has a much higher electronegativity. Because electrons are negatively charged, atoms that give up electrons end up as ions with net positive charges, and atoms that receive electrons become negatively charged ions.

As a result of this atomic give and take, the newly formed ions end up with stable outermost energy shells. For instance, a sodium atom has one electron in its outermost shell. If this electron leaves, the resulting sodium ion will have a new outer shell, with the stable number of eight electrons. On the other hand, a chlorine atom has seven electrons in its outermost shell. If it accepts an electron from a sodium atom, it will have a stable outer shell of eight (Figure 2–4).

A sodium ion has 11 protons but only 10 electrons, for a net charge of +1. The ionic form of chlorine, called a chloride ion, has 18 electrons but only 17 protons, for a net charge of −1. The oppositely charged sodium ($Na^+$) and chloride ($Cl^-$) ions are attracted to each other, and this attraction results in an ionic bond. When many sodium and chloride ions bond in this way, they form crystals of sodium chloride ($NaCl$), also called table salt (see Figure 2–16). (Note that an ion is represented by its chemical symbol [see Table 2–1] followed by a superscript showing its charge.)

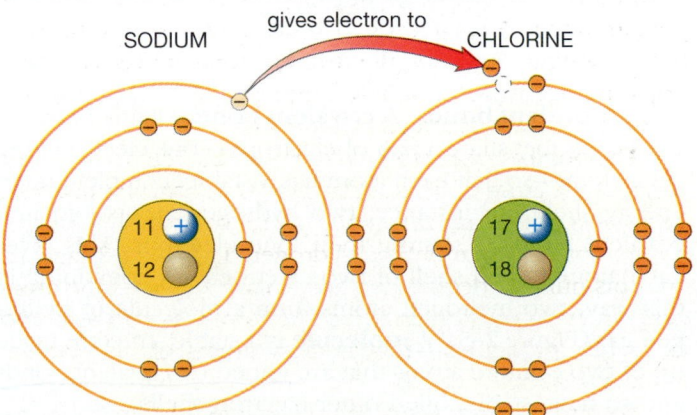

gives electron to
SODIUM                CHLORINE

**FIGURE 2–4** Formation of an ionic bond. Sodium becomes stable by giving up the lone electron in its outer shell. Chlorine adds one electron to its outer shell of seven and achieves a stable, filled shell of eight. After losing a negatively charged electron, sodium is a positively charged ion. By accepting the electron, chlorine becomes negatively charged.

## ISOTOPES: CHEMICAL TOOLS FOR BIOLOGY

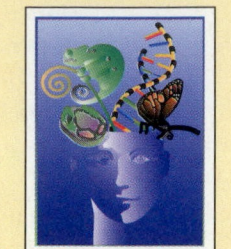

How long ago did dinosaurs live? How do plants make food from carbon dioxide? Is the Shroud of Turin genuine or one of the many fake religious relics made by charlatans in the Middle Ages? These are some of the questions we can seek to answer using isotopes.

Isotopes are handy tools for tracing the fate of chemicals in living organisms. For instance, to find out how plants make food from carbon dioxide, we could put a plant in an airtight box with carbon dioxide containing a higher-than-normal percentage of the rare radioactive isotope carbon 14. The carbon 14 "labels" this carbon, distinguishing it from the many carbon atoms already in the plant, which are mostly of the common isotope, carbon 12. We then follow this carbon 14 through a series of chemical reactions by taking frequent samples from the plant and analyzing which compound carbon 14 appears in first, second, and so on (see Figure 8–B).

The fate of radioactive isotopes can be followed by using a Geiger counter, photographic film, or medical imaging equipment to detect the energy emitted during radioactive decay (see Figure 43–16). Both radioactive and nonradioactive isotopes used as labels can be traced by the difference in mass, using sensitive analytical instruments.

Radioactive isotopes can also be used for dating fossils or archeological finds. Each atom of a particular radioactive isotope has the same probability of undergoing radioactive decay per unit of time. So, in a chunk of matter containing millions of these atoms, there is a constant, predictable rate of decay. An isotope's **half-life** is the time it takes for half of its atoms to undergo radioactive decay. After one half-life, half the original radioactive atoms remain; after two half-lives, only a quarter remain, and so on.

We could measure the amounts of uranium 238 and of its decay product, lead 206, in a rock. Using the known half-life of uranium 238 (4.5 billion years), we could calculate approximately when the rock was formed. (This analysis requires care: other isotopes of lead may be present, both "normal" lead 204 and additional isotopes derived by decay of other radioactive elements.) Rocks containing the common mineral mica can be dated using radioactive potassium 40, with a half-life of 1.25 billion years, and its decay product argon 40.

Since all living things contain carbon, carbon dating is often used to determine the age of biological remains. A living organism's body cycles carbon continuously. The proportion of carbon 14 to carbon 12 remains constant, with negligible loss by radioactive decay. At death this cycling stops, and as carbon 14 decays, its ratio to unchanging carbon 12 decreases. To find out how long an organism has been dead, we first measure the proportion of carbon 14 to carbon 12 in its remains. Using these data, the half-life of carbon 14, and the typical ratio of carbon 14 to carbon 12 in similar living organisms, we can calculate the age of our sample.

However, carbon 14 has such a short half-life (5730 years) that dating by radioactive carbon is reliable only

---

2. **Covalent bonds. A covalent bond** is a link between two atoms that share a pair of electrons—one electron from each atom—so that each atom has a stable, complete outer energy shell. For instance, two hydrogen atoms join in a covalent bond by sharing their lone electrons. This gives each atom a filled shell of two electrons. By combining in this way, two hydrogen atoms form a molecule of hydrogen gas (Figure 2–5). **A molecule** is a stable structure made up of two or more atoms that are joined by covalent bonds and so have stable, filled outer electron shells.

In a **double covalent bond,** each atom contributes two electrons, for a total of two pairs of shared electrons. For example, in Figure 2–3 we saw that a carbon atom requires four more electrons to fill its outer shell of eight, whereas an oxygen atom requires two more electrons. A molecule of carbon dioxide contains two oxygen atoms, each double-bonded to a carbon atom, which lies between them.

**HYDROGEN GAS (H₂)**

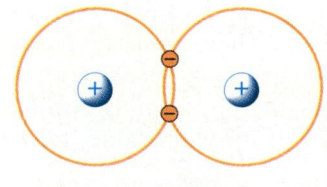

Bohr model

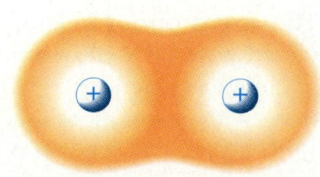

Electron cloud model

**FIGURE 2–5**  Two ways to show a covalent bond between two hydrogen atoms. Each hydrogen atom completes its shell of two electrons by sharing its electron with another hydrogen atom. In the electron cloud model, density-of shading shows the proportion of time spent by the rapidly moving electrons in various regions of the atoms' electron shells. The electrons in a covalent bond spend most of their time between the nuclei that share them.

for objects less than 30,000 years old. In terms of the long history of life on Earth (more than 3.5 billion years), such organisms disappeared but a moment ago.*

This last 30,000 years, however, does include the emergence of human societies and cultures. Carbon dating is often used to study the remains of humans, the animals they slew, the timbers of buildings, or charcoal from fires (Figure 2–A).

In 1988 the Archbishop of Turin requested carbon dating to settle a 600-year debate within the Roman Catholic Church on the authenticity of the Shroud of Turin, the supposed burial sheet of Jesus Christ. The shroud's linen fabric came from fibers of flax plants.

A 1 × 7-cm strip of the shroud was divided among three laboratories for carbon analysis by extremely sensitive techniques. The result: the shroud's fibers dated from closer to 1356, the year the shroud was "discovered," than to the first century AD, when Jesus Christ died.

In 1990 investigators found that elephants from different parts of Africa show different ratios of isotopes in their bones and tusks, reflecting characteristic differences in food plants in different habitats. Grasses and trees differ in the percentage of the nonradioactive isotope carbon 13 they capture from the air. Plants from different regions also differ in their isotope ratios of strontium and lead, according to the age of the underlying rocks that form the area's soil. All these differences in isotope ratios are passed on to elephants that eat the plants. Analysis of isotope ratios can be used to determine where ivory tusks came from.

*If radioactive carbon 14 constantly decays, and has such a short half-life compared with Earth's age, why isn't it all gone by now? A constant barrage of cosmic radiation converts some of the nitrogen 14 in the atmosphere into new atoms of carbon 14. This offsets radioactive decay.

**FIGURE 2–A** Iceman. In 1991 mountain climbers in the Austrian Alps discovered the freeze-dried, mummified remains of a human body in a melting glacier. Carbon dating of hay used as insulation in the man's boots showed that he lived about 5300 years ago, in the late Stone Age.  (Sygma)

The carbon atom shares two electrons with each oxygen atom, and so all three atoms fill their outer shells of eight (Figure 2–6). A double bond is stronger than a single bond and holds the nuclei of the bonded atoms closer to each other.

Two atoms of the same element have the same electronegativity. Therefore, when they bond covalently, they attract the shared pair of electrons equally, and so the average position of the shared electrons is midway between them (see Figure 2–5). If the two atoms are of different elements, the more electronegative atom attracts electrons more strongly, and so the shared electrons spend more time near this atom, creating a partial negative charge ($\delta^-$) in its vicinity. Meanwhile, the atom at the other end of the bond bears a partial positive charge ($\delta^+$). Such an electrically asymmetrical covalent bond is called a **polar covalent bond,** usually shortened to **polar bond** (Figure 2–7).

CARBON DIOXIDE

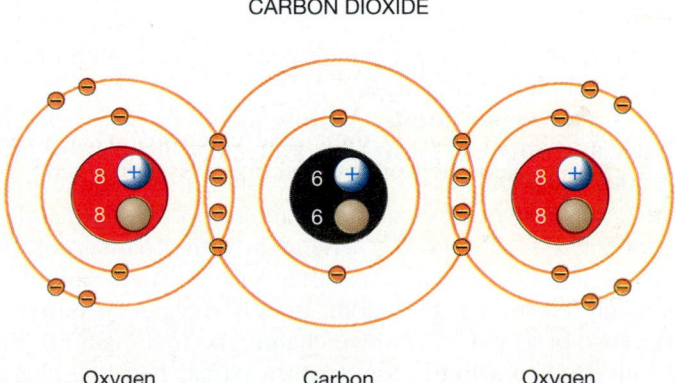

Oxygen        Carbon        Oxygen

**FIGURE 2–6** Double covalent bonds. In a carbon dioxide molecule, two oxygen atoms each share two pairs of electrons with a carbon atom, sandwiched between them. This forms two double bonds, with each atom filling its outer shell of eight electrons.

HYDROGEN CHLORIDE (HCl)

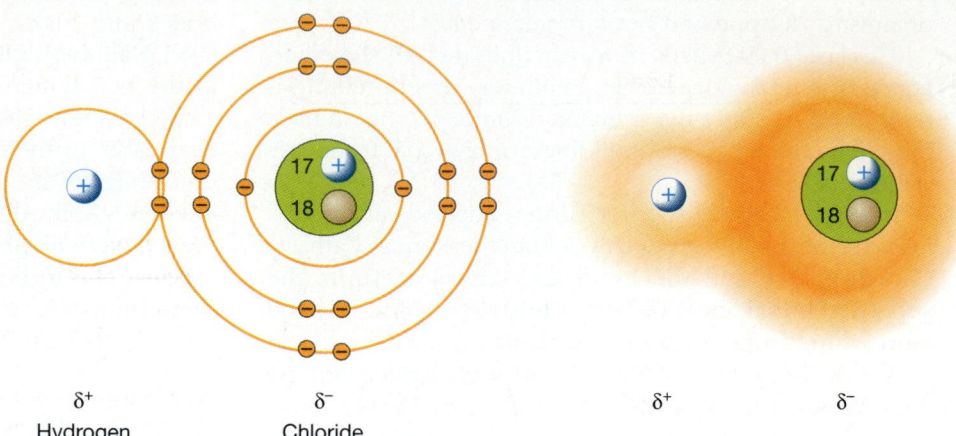

**FIGURE 2–7** Two ways to show the polar molecule of hydrogen chloride (HCl). The electron pair shared in the polar covalent bond linking the atoms is attracted more strongly by the chlorine nucleus than by the hydrogen nucleus. As a result, the chlorine has a partial negative charge ($\delta^-$), and the hydrogen atom has partly lost its electron, leaving it with a partial positive charge ($\delta^+$). ($\delta$ is the lower case form of the Greek letter delta.)

Oxygen and nitrogen are much more electronegative than hydrogen. So, when either bonds with hydrogen, the bond is polar. The oxygen (or nitrogen) bears a partial negative charge, the hydrogen a partial positive charge. On the other hand, carbon and hydrogen are about equally electronegative. Therefore, a carbon-to-hydrogen bond is **nonpolar,** with the average position of the shared electrons about midway between the two atomic nuclei, and with no difference in electrical charge between the two atoms. Both polar and nonpolar covalent bonds play vital roles in the chemistry of life (Chapters 3 and 4).

We can arrange bonds in a series according to the distribution of electrons. In nonpolar covalent bonds, the position of the shared electrons, averaged over time, is symmetrical between the two atomic nuclei. In polar covalent bonds, the electrons spend more time near one end of the bond. The bond is lopsided, with opposite partial electrical charges at its ends. Ionic bonding is the extreme case. Here, one atom gives up one or more electrons to another atom, resulting in two separate particles, each with one or more full electrical charges.

3. **Hydrogen bonds.** A consequence of polar bonding is that the partial electrical charges interact with others nearby. A **hydrogen bond** is a weak and often transient electrical attraction between two atoms bearing opposite partial electrical charges ($\delta^+$ and $\delta^-$). The atom with a partial positive charge is a hydrogen atom with a polar covalent bond to a strongly electronegative atom, usually oxygen or nitrogen. Because of its partial positive charge, the hydrogen nucleus is attracted to a third atom, with a partial negative charge (again usually oxygen or nitrogen), and this forms the hydrogen bond (Figure 2–8). Hydrogen bonds form between atoms that belong to different molecules or between atoms that are on different parts of a large molecule.

Although we use the term "bond" for all three of these atomic interactions, they differ greatly in strength. Compared with an ionic or covalent bond, a single hydrogen bond is very weak and easily broken (Table 2–2). However, the combined force of numerous hydrogen bonds is what holds many biological structures together.

The atoms of each element gain, lose, or share a particular number of electrons when they form stable electron shells. Hence the numbers and types of bonds any atom can form are predictable. The four most common elements in living organisms are the smallest ones capable of forming one, two, three, and four stable covalent bonds (Table 2–3).

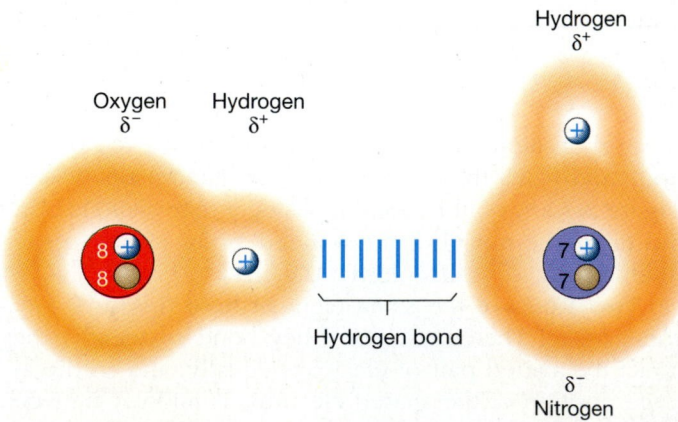

**FIGURE 2–8** Hydrogen bonding. A hydrogen bond (blue dashes) is a weak attraction between a polar-bonded hydrogen with a partial positive charge ($\delta^+$) and a polar-bonded atom of nitrogen or oxygen with a partial negative charge ($\delta^-$). Hydrogen bonds form between atoms on different molecules or between different parts of a large molecule. Note that the hydrogen bond is in a straight line with the hydrogen atom's polar covalent bond to oxygen.

## TABLE 2-2

### Strength of Some Chemical Bonds

| Type | Bond | Strength (kJ*/Mole)† |
|------|------|--------------------|
| Ionic | Na—Cl | 410 |
| | H—Cl | 431 |
| | Mg—O | 381 |
| | H—F | 569 |
| Covalent | C—H | 414 |
| | C—O | 352 |
| | C=O | 745 |
| | C—C | 347 |
| | C=C | 611 |
| | C≡C | 837 |
| | C—N | 293 |
| | N≡N | 946 |
| Hydrogen | | 20 to 40, depending on conditions |

*kJ = kilojoule. The joule, named for Dalton's student James P. Joule, is a unit of energy equivalent to that of a 2-kilogram object moving with a velocity of 1 meter per second.

†A mole is a numerical measure (analogous to a dozen). One mole = $6.022 \times 10^{23}$ units (e.g., $6.022 \times 10^{23}$ hydrogen chloride molecules). The strength of a particular kind of bond is given in terms of the amount of heat energy needed to break that type of bond in a mole of a substance.

## TABLE 2-3

### Numbers of Bonds Formed by the Most Common Elements in Living Organisms

| Element | Number of Bonds |
|---------|-----------------|
| Carbon   (C) | 4 |
| Hydrogen   (H) | 1 |
| Oxygen   (O) | 2 |
| Nitrogen   (N) | 3 |

## 2–D  COMPOUNDS AND MOLECULES

Some molecules contain atoms of only one element, as in hydrogen gas. However, many molecules contain atoms of different elements, as in hydrogen chloride and carbon dioxide.

A **compound** is a substance made up of atoms of two or more different elements, in specific proportions, and with a specific pattern of bonds. A compound's properties differ from those of its component elements. A molecule of a compound is the smallest unit that retains all the compound's properties, just as an atom is the smallest unit that retains all the element's properties. However, molecules that contain atoms of only one element are *not* compounds. Ionically bonded compounds, such as sodium chloride, are said to consist of ions instead of molecules.

A **molecular formula** is a shorthand way to show the kinds and numbers of atoms in a molecule, using the symbols for elements (see Table 2–1). A molecule of hydrogen gas, also called molecular hydrogen, contains two hydrogen atoms, $H_2$. Similarly, molecular oxygen, in the air we breathe, is $O_2$. The formula for carbon dioxide, $CO_2$, says that each molecule has one carbon atom and two oxygen atoms. Likewise, in the formula for water, $H_2O$, the subscript 2 shows that a water molecule has two hydrogen atoms (H) as well as an oxygen atom (O).

The formula for sodium chloride, NaCl, says that table salt contains sodium and chloride ions in a 1:1 ratio.

**Structural formulas** take more space than molecular formulas but show the arrangement of atoms and bonds as well as the numbers and kinds of atoms. For instance, the structural formula for water, H—O—H, shows that each hydrogen atom is separately attached to the oxygen atom; the lines between atoms represent covalent bonds. In carbon dioxide, each oxygen is double-bonded to the carbon atom: O=C=O. When two different compounds have the same molecular formula, only a structural formula will distinguish between them (Figure 2–9).

The **molecular mass** of a molecule is the sum of the atomic masses of all its atoms. Using the atomic masses from Table 2–1, we can determine that the molecular mass of water ($H_2O$) is 18 daltons: $2 \times 1$ for the two hydrogens, +16 for the oxygen.

Scientists have a unit of quantity for counting out molecules in the proportions needed for a chemical experiment (just as we count in dozens when we buy donuts for a party). The counting unit for molecules is the **gram molecular weight** (1 **mole**)—the amount found by computing the molecular mass of a substance and weighing out that number of grams. For example, a mole of water has a

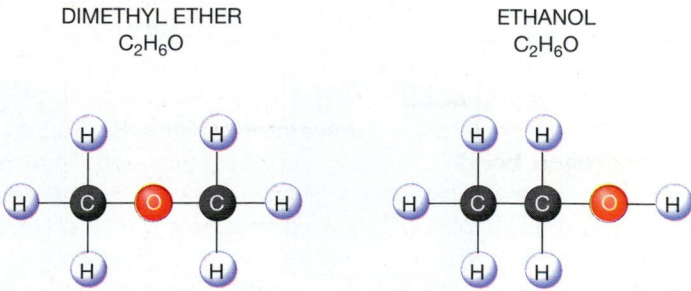

DIMETHYL ETHER
$C_2H_6O$

ETHANOL
$C_2H_6O$

**FIGURE 2–9** Molecular versus structural formulas. Dimethyl ether and ethanol have the same molecular formula, $C_2H_6O$. However, their structural formulas differ, since the same atoms can be arranged in different ways. Ethanol, alias ethyl alcohol, is the type of alcohol in alcoholic beverages and in some blends of motor fuel.

mass of 18 grams. A mole of any substance contains 6.022 × 10²³ molecules. The mole is a useful quantity because it is based on the number of molecules; a mole of table sugar (342 grams) and a mole of ethanol (46 grams) contain the same number of molecules. By contrast, because the molecular mass of sugar is 342 daltons and that of ethanol only 46 daltons, 1 gram of ethanol contains more than seven times as many molecules as 1 gram of sugar.

In biology we usually measure substances in moles because this gives us an easy way to compare the ratios of molecules of different substances. Because the chemistry of life takes place in a watery environment, biological experiments often call for dissolving the desired quantity of a substance in water. A **solution** consists of a **solvent,** usually water, plus the substances dissolved in it, called **solutes.** The **concentration** of a solution is a measure of the proportion of solutes it contains. For example, a **one molar (1M)** solution contains 1 mole of solute in a total of 1 liter of solution. To make a one molar solution of table sugar, you would put 342 grams of sugar in a beaker and add enough water to bring the final solution volume to 1 liter. Concentration is symbolized by square brackets: [sugar] means "concentration of sugar."

## 2–E  MOVEMENT OF MOLECULES

All atoms and molecules have energy and so they are in constant, random motion. In solids, the molecules occupy fixed places with respect to each other, and each vibrates in its own space, much like the passengers in a crowded bus bumping into each other. In a liquid the molecules are still quite close together, and they constantly jostle one another, but they can slide past each other and so change places and even travel slowly and erratically into another

area. A gas consists mostly of space, and the scattered molecules move quickly and freely, occasionally colliding with one another.

This spontaneous movement of molecules accounts for **diffusion,** the process whereby molecules of two or more substances move about and become evenly mixed. Consider a sugar cube in a beaker of water (Figure 2–10). A sugar molecule moves in one direction until it bumps into some other molecule, either another sugar molecule or a water molecule. Both molecules bounce off in new directions. The sugar molecule may bounce back toward the sugar cube, but most of the possible new directions will take it still farther from the original cube. Hence its overall, random path will probably take it away from the cube. In the sugar molecule population as a whole, the net movement is also outward because more molecules move farther away from the cube than bounce back toward it.

As sugar molecules move away from the cube, they set up a **concentration gradient,** a gradual decrease in the concentration of sugar with distance from the cube (Figure 2–10b). On the whole, sugar molecules tend to move toward the area where they are less concentrated, that is, down this concentration gradient, until they are evenly dispersed throughout the solution. At the same time, the water molecules also move down their own gradient and spread evenly throughout the beaker, including the part originally occupied by the sugar cube.

In any substance, some molecules move faster than others. The faster a particle moves, the more **kinetic energy,** or energy of motion, it has. **Temperature** is a measure of the average kinetic energy of molecules; the faster the average speed, the higher the reading of the thermometer. Heating a substance increases the energy of its molecules and so increases their average speed (and their tempera-

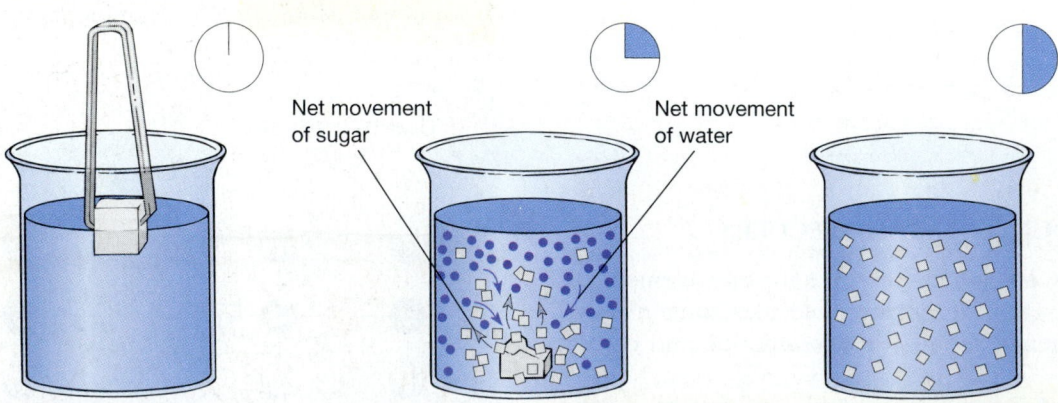

Net movement of sugar

Net movement of water

**FIGURE 2–10**  Diffusion. (a) When a sugar cube is placed in a beaker of water, sugar molecules gradually diffuse away from the cube, and water molecules diffuse into the region where the cube was. (b) Both sugar and water show a concentration gradient. Overall, the *net* movement of each substance is down its concentration gradient, into the area originally occupied by the other. (c) Eventually, the molecules of sugar and water are spread evenly throughout the solution. They continue to move randomly, maintaining this even distribution.

ture). If we add enough heat to a solid such as a block of chocolate, the molecules will begin to move so fast that the solid melts into a liquid. The very fastest molecules will even reach escape velocity and vaporize into the gaseous state, entering the air and diffusing throughout the house. It is when they are in this gaseous state that we smell them.

## 2–F    CHEMICAL REACTIONS

When molecules or ions bump into each other they usually remain intact and bounce off in new directions. However, if molecules with high internal energy collide forcefully at a specific angle, they may undergo a change. The energy of the impact distorts the arrangement of electrons, raising the molecules to an unstable, high-energy transition state. Next, one of two things can happen. Either the molecules settle back to their original state, or the electrons rearrange themselves further, forming a new set of bonds and therefore making new substances. This is called a **chemical reaction.** The energy needed to raise the molecules to the transition state is the **activation energy.** At normal temperatures on Earth's surface, few molecules have enough energy, and so few collisions produce reactions, but heating can substantially increase the reaction rate by increasing the kinetic energy of molecules.

Reactions can be written as chemical equations, like this one for the burning of marsh gas (methane):

$$CH_4 + 2\,O_2 \longrightarrow CO_2 + 2\,H_2O$$

$$\underbrace{\text{methane} \quad \text{oxygen}}_{\text{reactants}} \qquad \underbrace{\overset{\text{carbon}}{\text{dioxide}} \quad \text{water}}_{\text{products}}$$

In place of an "equals" sign, a chemical equation has an arrow, which we read to mean "yields." The **reactants** (starting materials) are written on the left and the **products** on the right, after the arrow. This particular equation says that two molecules of oxygen combine with each molecule of methane, and that for each carbon dioxide molecule produced, two water molecules are also formed. Notice that the equation is *balanced*: the products contain all the atoms of each element that went into the reaction, although they are now combined into different molecules. Thus, a chemical reaction follows the basic physical Law of Conservation of Matter. The number of molecules indicates the proportions of reactants and products: two moles of oxygen are required to burn one mole of methane completely, and two moles of water are produced for each mole of carbon dioxide.

The arrows in a chemical equation may point in both directions:

$$CO_2 + H_2O \rightleftharpoons H_2CO_3$$

$$\underset{\text{dioxide}}{\text{carbon}} \quad \text{water} \qquad \underset{\text{acid}}{\text{carbonic}}$$

This means that the reaction can go from left to right (forward) or from right to left (backward), depending on the conditions. Such a reaction is said to be reversible.

## 2–G    WATER

A continuous and highly organized set of chemical reactions is a hallmark of a living cell, and so is a high content of water—usually 75% or more. This is no coincidence. Most of the chemical reactions of life must take place in aqueous (watery) solution. A very small percentage of water molecules actually participate as raw materials in certain cellular reactions. Most of the water is needed to provide a suitable environment.

Although water is one of the most common compounds on Earth, it has several uncommon properties. These properties have proved so important to living organisms that biologists cannot imagine the existence of life as we know it on any planet without an abundant supply of water (Figure 2–11).

The unique properties of water result from its molecular structure and distribution of electrical charge. In a water molecule, an oxygen atom is covalently bonded to two

**FIGURE 2–11** Water and life. Earth's organisms depend on the planet's abundant supply of water, seen as swirling clouds and blue oceans in this distant view. By contrast, our own moon is bare rock, waterless and airless, inhospitable to life as we know it. This portrait of Earth and the moon was taken by the *Galileo* spacecraft 6.2 million km from Earth's surface.  (NASA)

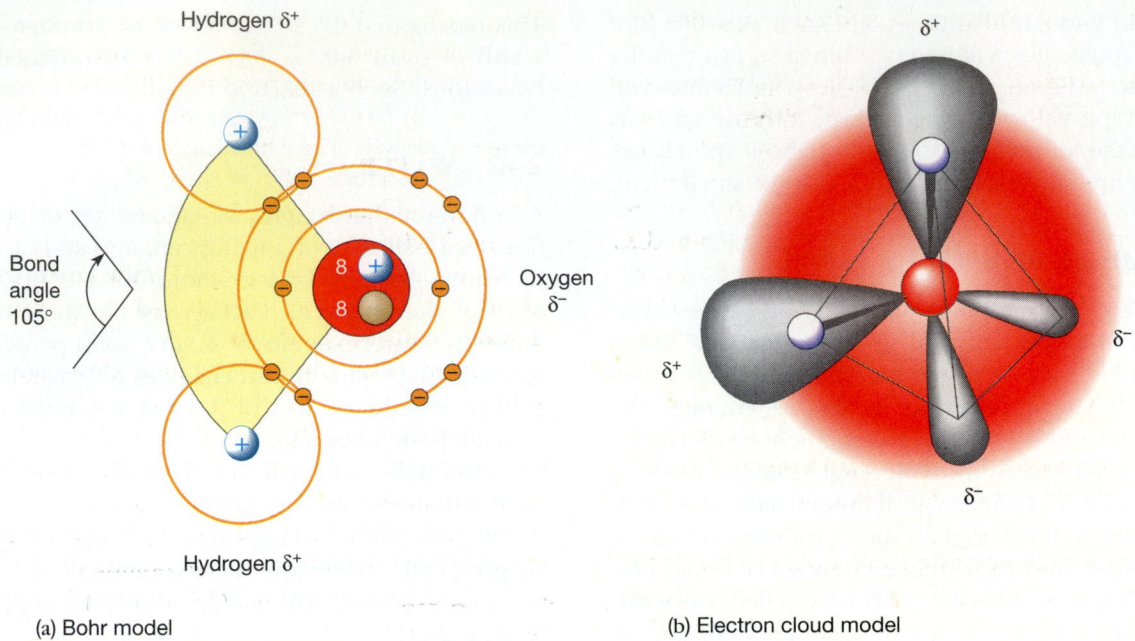

Hydrogen δ⁺

Bond
angle
105°

8 +
8

Oxygen
δ⁻

Hydrogen δ⁺

(a) Bohr model

δ⁺

δ⁻

δ⁺

δ⁻

(b) Electron cloud model

**FIGURE 2–12**  Two ways of showing a water molecule. (a) One atom of oxygen is covalently bonded to two atoms of hydrogen. The molecule is polar because the oxygen nucleus attracts the shared electrons to the point of the angle between the bonds. This gives the oxygen a partial negative charge and leaves each hydrogen atom with a partial positive charge. (b) In the three-dimensional electron cloud model, the four pairs of electrons in oxygen's outer shell point to the corners of a tetrahedron, with hydrogen nuclei at two corners, unbonded electron pairs hovering near the other two, and the oxygen nucleus in the center. Each corner bears a partial electric

atoms of hydrogen (Figure 2–12). The molecule is polar be-cause it has an angle of 105° between the two bonds, and the electronegative oxygen attracts the electrons to the point of the angle. This gives a partial negative charge ($\delta^-$) on the oxygen atom and a partial positive charge ($\delta^+$) on each hydrogen atom. These partial charges are distributed roughly in the shape of a tetrahedron, with the oxygen nu-cleus at the center, the two partially positive hydrogens at two corners, and oxygen's two partially negative unshared outer electron pairs at the other two corners. This arrange-ment puts the four pairs of electrons in oxygen's outer shell as far as possible from each other.

A water molecule's partially positive hydrogens are at-tracted to the partially negative oxygens of other molecules, including other water molecules, and so water molecules attach to one another by hydrogen bonds. At any given time, each water molecule can form as many as four hy-drogen bonds (Figure 2–13). In liquid water, these weak hydrogen bonds form and break rapidly as the molecules shift around. Although each hydrogen bond has a very short lifetime (about $10^{-11}$ second), large numbers of them act-ing together can contribute considerable stability to a group of molecules. Hydrogen bonding is why most of the water at Earth's surface is liquid, instead of a gas like molecular

oxygen, which has a greater molecular mass but is nonpo-lar, and therefore does not form hydrogen bonds.

The structure of water molecules, with their polar elec-trical nature and the resulting formation of hydrogen bonds

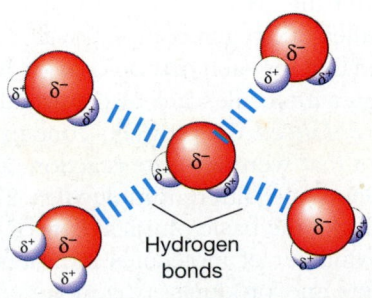

δ⁻

δ⁺

δ⁻

δ⁺

δ⁻

δ⁺

δ⁻

δ⁺

δ⁻

δ⁺

δ⁻

δ⁺

Hydrogen
bonds

**FIGURE 2–13**  Hydrogen-bonded water molecules. A water mol-ecule's hydrogen atoms (white) bear partial positive charges and so are attracted to the partial negative charges on other water mol-ecules' oxygen atoms (red). In liquid water the resulting hydro-gen bonds (blue dashes) last only about a hundred billionth of a second before they break, and new ones form to other molecules. At any time, one water molecule may be hydrogen-bonded to as many as four others.

between them, endows water with several properties important to life:

1. **Water is cohesive and adhesive. Cohesion** is the holding together of like substances; **adhesion** is the holding of different substances to each other. You can fill a glass of water slightly above its brim; a water strider can run across the surface of a pond (Figure 2–14). Such feats are possible because of water's **surface tension,** which makes the surface appear to be covered by a "skin." Surface tension results from the cohesion of water molecules to one another by their hydrogen bonds. Water is more cohesive than any other liquid except mercury. The polar molecules of water also adhere strongly to any electrically charged surface. The adhesive and cohesive nature of water accounts for its **capillary action,** the movement up a piece of porous paper, through the fine pores in leaves or soil, or into a germinating seed.

2. **Water has a high specific heat.** Specific heat is the amount of heat energy needed to raise the temperature of a gram of a substance by 1 °C; for water, it is 4.186 joules (J).

The specific heat of a substance depends mainly on how many particles (molecules or ions) it contains per gram. Recall that, to raise the temperature of a substance, we must make all these molecules move faster. Also recall that polar bonds, such as those in water molecules, have some degree of ionic character. Hence, the three atoms in a water molecule act a lot like independent ions instead of moving together as a single rigid unit. When these atoms absorb heat energy they can vibrate as though they were free. As a result, it takes about twice as much heat to raise the temperature of water as we would predict on the basis of the actual number of molecules present.

Water's high specific heat gives it a large **thermal capacity:** it takes a lot of heat to raise the temperature of water, and much heat must be lost to lower its temperature. Compared with the air above it, a body of water warms more slowly in the spring and cools more slowly in the fall. For aquatic organisms, this means more gradual changes in the environmental temperature. Because living organisms are made up largely of water, they also gain and lose heat relatively slowly. The chemical reactions in an organism's body may produce a lot of heat. Were it not for the high capacity of water to absorb heat, an organism's temperature might rise so far that life would cease.

Water also has a high **thermal conductivity:** heat added at one point rapidly spreads throughout the body of water. This keeps the heat produced in an organism's water-filled body from generating destructive local hot spots.

3. **Water has a high boiling point.** It takes a lot of heat energy to overcome all the hydrogen bonds between water molecules and so change water from a liquid to a gas, in which each molecule is separate. Temperatures on Earth's surface seldom reach the boiling point of water, and so living organisms need not face the prospect of boiling away.

4. **Water is a good evaporative coolant.** Water molecules change from the liquid state (water) to the gaseous state (water vapor) by absorbing enough heat to escape from the main body of the liquid into the air. The molecules that reach escape velocity and leave the body carry away this considerable heat energy. Many land-dwelling organisms dispose of excess body heat by this means. Sweating in humans and panting in dogs are examples of evaporative cooling.

5. **Water has a high freezing point and has a lower density as a solid than as a liquid.** These properties have both advantages and disadvantages for living things. As warm water cools, its molecules lose energy and move more slowly, and hydrogen bonding holds them closer together. So, the mass of water contracts and becomes more dense. However, water has the peculiar property that its maximum density occurs at 4 °C. As water cools from 4 °C to its freezing point, 0 °C, it expands again, becoming less dense as the molecules begin to move apart and form the crystal lattice of ice (Figure 2–15).

Ice is less dense than liquid water because its molecules are packed less closely; therefore an ice crystal is larger than the volume of water it replaces, and ice floats in water. This has an advantage for aquatic organisms: in winter, floating ice forms an insulating blanket between the water and the cold air above. This slows the formation of more ice from the remaining water, and so protects organisms living below the ice from freezing. When spring comes, the sun shines directly on the ice and melts it.

Because water expands when it freezes, ice crystals formed within an organism may destroy its delicate internal structure and cause death. This is the fate of the limp, black tomato and dahlia plants in our gardens after a hard frost. Only their seeds, with too little water to form ice, survive winter to produce the next generation.

**FIGURE 2–14** Surface tension. The cohesion of water molecules holds them together so strongly that a water strider can walk on the surface without sinking.  (Dennis Drenner)

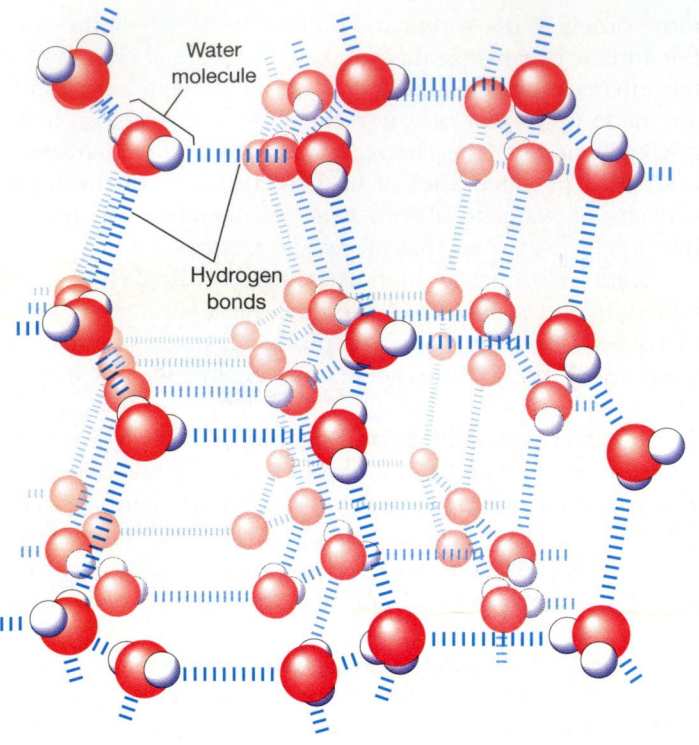

Water
molecule

Hydrogen
bonds

(a) Ice crystal

(b)

**FIGURE 2–15**   Ice. (a) The arrangement of water molecules in an ice crystal. Each molecule becomes hydrogen-bonded to four others in the crystal structure. (Red spheres = oxygen atoms; white spheres = hydrogen atoms.) Because the molecules in ice crystals are farther apart than those in liquid water, ice forms on top of water. (b) These six-sided ice crystals clearly reflect the angles of hydrogen bonds on the molecular level.   (b, Bruce Matheson)

Some organisms have adaptations that allow them to survive freezing temperatures. Winter wheat and some insects, fish, and frogs are among organisms with natural antifreezes—often glycerol, which is also used in automobile antifreeze (see Figure 3–6). Many organisms produce proteins that act as antifreeze. These large molecules attach to forming ice crystals and prevent them from growing to a dangerous size. Other organisms, like some plants, some frogs, and even painted turtle hatchlings, tolerate freezing—that is, small ice crystals form between cells, but glycerol keeps the cell interiors liquid.

6. **Water is a solvent.** More substances dissolve in water than in any other known liquid. When a substance dissolves, its individual molecules or ions separate from one another and mingle with molecules of the solvent (in this case, water). Because of their partial electrical charges, the polar molecules of water are attracted to charged ions and to partially charged polar molecules. So water readily surrounds and dissolves these solutes (Figure 2–16). This illustrates the useful rule of thumb, "like dissolves like": similarly, nonpolar liquids such as benzene and carbon tetrachloride dissolve nonpolar substances.

Nonpolar molecules, such as those composed mainly of carbon and hydrogen, do not dissolve in water because they lack electrical charge to interact with water molecules. While water and its solutes form one big, "friendly" crowd of molecules, all connected by electrical attractions, these "shy" nonpolar molecules are shoved aside. Here the nonpolar molecules form groups, not so much by mutual attraction as by default—they end up grouped together because they are all excluded from the mass of water molecules. This is what happens in salad dressing when polar water and nonpolar oil separate from each other.

Instead of dissolving in water, then, nonpolar molecules form interfaces with it (Figure 2–17). Similar interfaces are the basis for the existence of membranes in cells. Hence the inability of water to dissolve nonpolar substances is also necessary to life.

## 2–H   DISSOCIATION AND THE pH SCALE

In order to understand some major biochemical processes, you must realize that many substances come apart, or **dissociate,** into ions when they dissolve in water (or other

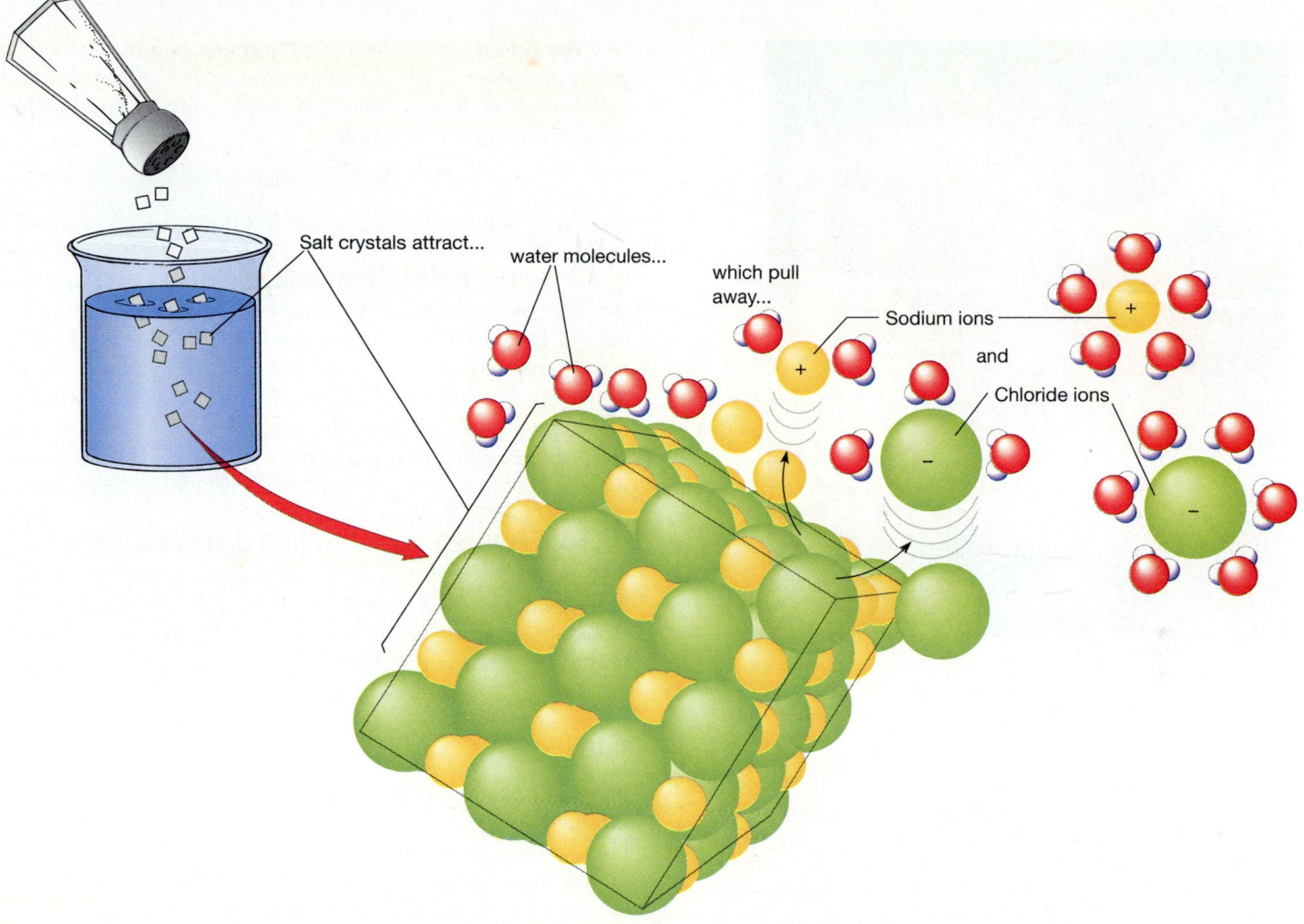

**FIGURE 2–16**  Sodium chloride (NaCl) dissolving in water. Read from the top left of the picture toward the bottom right. In salt crystals, sodium ($Na^+$) and chloride ($Cl^-$) ions are attracted to one another because of their opposite electrical charges. In water, positively charged sodium ions (+) attract the partial negative charges of oxygen atoms in water molecules. Likewise, chloride ions (−) attract the partially positive hydrogens of water molecules. All the tiny electrical tugs from water molecules pull $Na^+$ and $Cl^-$ away from each other. Each ion becomes surrounded by a "shell" of water molecules, which shields it from the other's attraction.

polar solvents). This occurs because the resulting ions have stable electron shells. Some compounds dissociate completely, others only partially—some of their molecules remain intact, and some are ionized. Water itself dissociates partially, most commonly into hydrogen ions ($H^+$) and hydroxide ions ($OH^-$):

$$H_2O \longrightarrow H^+ + OH^-$$

When this happens, the $H^+$ generally remains attached to a neighboring water molecule to which it was already hydrogen-bonded. This forms $H_3O^+$ (hydronium ion). However, biologists usually speak only of $H^+$, as if it were on its own.

Because water molecules carry both partial negative and partial positive charges, they can assist dissociation by form-

ing "shells" around ions. The watery shells shield the ions from the attraction of oppositely charged ions in the solution and allow them to move independently of one another. With these shells of water molecules, hydrated ions behave as if they were larger and move more slowly than we would expect from the size of the ion itself (see Figure 2–16).

Chemists classify substances by the particles they yield when they dissociate in water. An **acid** is a substance that releases $H^+$ when it dissociates in water. For example, when hydrogen chloride gas (HCl) is dissolved in water to form hydrochloric acid, it yields hydrogen and chloride ions ($H^+$ and $Cl^-$) in solution.

A **base** (also called an **alkali**) is a substance that releases hydroxide ions ($OH^-$) in water, or one that accepts $H^+$. The base sodium hydroxide (NaOH), the active ingre-

The acidity or alkalinity of a solution is indicated by its **pH,** a measure of the concentration of H$^+$. The pH value is the negative logarithm of the molar concentration of hydrogen ions in the solution:

$$pH = -\log[H^+]$$

That is, a solution with a concentration of $10^{-5}$ moles of H$^+$ per liter of solution has a pH of 5. The pH scale goes from 0 to 14. A pH of 7 is neutral, neither acidic nor basic. Pure water is neutral because it gives off equal numbers of H$^+$ and OH$^-$ when it dissociates. Values of pH below 7 are acidic, and those above 7 are alkaline (Figure 2–18). Because the pH scale is logarithmic, a solution of pH 5 is ten times more acidic than a solution of pH 6, and so forth. (The fact that *lower* pH values mean *greater* acidity often confuses people. It is worth repeating this to yourself until it comes naturally.)

Most chemical reactions of life occur most rapidly at a near-neutral pH. In our own bodies, the pH of blood and most other fluids is about 7.4. A notable exception is the stomach contents during digestion of a meal: the stomach lining secretes hydrochloric acid, with a pH of 1 or less.

## Buffers

If we add drops of acid or base to pure water, its pH changes rapidly. If we add acid or base to blood instead of water, however, we find that the pH remains steady until we have added many times more acid or base than would change the pH of an equal volume of water. This is because blood and other body fluids contain **buffers,** substances that tend to keep the pH constant by taking up or releasing H$^+$ ions (or OH$^-$ ions). One of the most important buffers in living organisms is bicarbonate ion, HCO$_3^-$, which takes up excesses of either H$^+$ or OH$^-$:

Water drop

Flower surface

**FIGURE 2–17**    An interface. These petals produce a waxy surface coating. The wax is nonpolar, and so it will not mix with water—that is, it is waterproof. When it rains, wax and water "keep to themselves": the water stands up in round droplets on the petals. Each drop is held in shape by the surface tension of water, which results from the attraction among water molecules.

$$H^+ + HCO_3^- \rightleftharpoons H_2CO_3 \rightleftharpoons H_2O + CO_2$$

bicarbonate ion        carbonic acid        water        carbon dioxide

$$OH^- + HCO_3^- \rightleftharpoons CO_3^{2-} + H_2O$$

bicarbonate ion        carbonate ion        water

dient in some drain cleaners, dissociates into sodium and hydroxide ions (Na$^+$ and OH$^-$) in solution. A base that accepts H$^+$ is ammonia (NH$_3$), which thereby becomes ammonium ion (NH$_4^+$). (Note that by these definitions water is both an acid and a base!)

A **salt** is a substance in which the H$^+$ of an acid has been replaced by another positively charged ion. A salt dissociates into oppositely charged ions; for example, the salt sodium chloride (NaCl) dissociates into sodium and chloride ions (Na$^+$ and Cl$^-$).

In each case, notice that the bicarbonate ion removes the added H$^+$ or OH$^-$ ions from the solution. Hence these added ions do not alter the solution's pH. If some other reaction removes H$^+$ or OH$^-$ from the solution, the reactions can reverse and provide H$^+$ or OH$^-$ to make up the deficit. Thus the pH remains constant as long as the solution contains enough bicarbonate ion to absorb any added H$^+$ or OH$^-$, and enough carbonic acid or carbonate ion to replace any H$^+$ or OH$^-$ lost from the system.

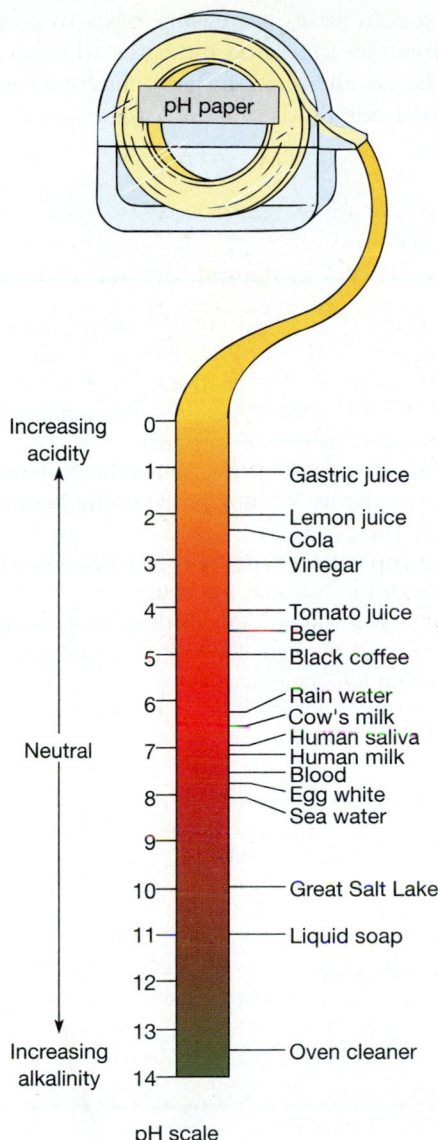

Increasing
acidity

| pH | Substance |
|----|-----------|
| 0 | |
| 1 | Gastric juice |
| 2 | Lemon juice |
| | Cola |
| 3 | Vinegar |
| 4 | Tomato juice |
| | Beer |
| 5 | Black coffee |
| 6 | Rain water |
| | Cow's milk |
| | Human saliva |
| 7 | Human milk |
| | Blood |
| 8 | Egg white |
| | Sea water |
| 9 | |
| 10 | Great Salt Lake |
| 11 | Liquid soap |
| 12 | |
| 13 | Oven cleaner |
| 14 | |

Neutral

Increasing
alkalinity

pH scale

**FIGURE 2–18** The pH scale, with the pH readings of some familiar substances.

The first equation shows that the carbonic acid formed from bicarbonate and hydrogen ions can break down to water and carbon dioxide. Carbon dioxide is a major product of cellular respiration in living organisms, and we usually think of it as such, forgetting its role as an important buffer.

Many of the body's chemical reactions produce acids or bases, but buffers in the surrounding fluids prevent wild swings in pH. Buffers are vital because the chemical reactions of living organisms work best when they occur at a particular pH. This is partly because enzymes, the mole-cules that mediate chemical reactions in living organisms, usually require a particular narrow range of pH in order to function properly (Section 3–F).

## SUMMARY

Living organisms are subject to the same physical and chemical laws that govern nonliving systems. Like nonliving matter, organisms are made up of atoms, which bond in various ways to form ions and molecules. The most abundant substance in living organisms is water. Organisms also contain many kinds of molecules with carbon skeletons, as well as various salts.

1. Of more than 100 different chemical elements known, about 20 are essential to living things. The unique chemical properties of these elements are necessary to the activities of life.
2. The number of protons in an atom's nucleus defines what element it is and how many electrons it has. The number of electrons, in turn, determines how many electron shells the atom has and how nearly filled they are. Electrons move to the lowest available energy level and tend to form stable, filled shells. This behavior of electrons determines the bonding properties of each element: the numbers and kinds of bonds its atoms form.
3. Ionic bonds form when one atom takes one or more electrons from another atom, and the resulting ions are attracted to each other by their opposite electric charges. Covalent bonds form when atoms share electron pairs. Covalent bonds may be polar or nonpolar, depending on the average position of the shared electrons between the ends of the bond. Hydrogen bonds are much weaker electrical attractions between partial positive and partial negative charges on polarly bonded atoms of different molecules or on different parts of a large molecule.
4. Molecules are in constant, random motion. As a result, they may diffuse into new areas and become mixed with other molecules.
5. A forceful collision between molecules may result in a chemical reaction, in which electrons rearrange into new patterns of bonds between atoms, forming different compounds. Living organisms constantly carry out a variety of chemical reactions, forming different compounds as required.
6. Water is the most abundant substance in living things and is absolutely necessary for life as we know it. The water molecule's structure and hydrogen-bonding ability give water a unique set of properties that make it essential to life: water creeps into small spaces by capillary action; it absorbs heat and disperses it throughout the body; it carries away body heat when it va-

porizes from the body surface; it is denser as a liquid than as a solid; it dissolves polar and ionic substances; and it forms interfaces with nonpolar substances.
7. Many substances dissociate into ions when they dissolve in water.

8. The pH of a solution is a measure of its hydrogen ion concentration; the pH value indicates whether a solution is acidic or alkaline. Buffers, chiefly bicarbonate ion, keep the body fluids of living organisms at a nearly constant pH.

## Self-Quiz

In questions 3, 4, and 7, select the correct word from the choices in parentheses. For multiple choice questions, choose the one *best* answer.

1. Atoms of the same element that contain different numbers of neutrons are known as _____.
2. A positive ion has:
   a. more protons than electrons
   b. more electrons than protons
   c. equal numbers of neutrons and electrons
   d. equal numbers of protons and electrons
   e. more neutrons than electrons
3. What kind of bond involves the sharing of electrons between atoms such that each atom completes its outer electron shell? (covalent, ionic)
4. In a water molecule, the hydrogen atoms are joined to the oxygen atom by (ionic, covalent, hydrogen) bonds.
5. Compute the mass of a mole of hydrogen chloride (HCl) and of carbon dioxide ($CO_2$) (see Table 2–1).
6. In the dissociation of NaCl in water:
   a. water exerts forces that induce dissociation
   b. water is a passive solvent, accepting particles that dissociate because of their own internal forces
   c. water molecules lose hydrogen ions

   d. twice as many $H^+$ ions are formed as $Na^+$ ions
   e. equal numbers of $H^+$ and $Na^+$ ions are formed
7. If the pH of a solution changes from 2 to 5, it has become more (acidic, basic); its hydrogen ion concentration has (increased, decreased, remained constant).

Matching: for each event listed below, select the property of water responsible from the list (a through h) at the bottom of the column.

_____  8. Heat applied to the bottom of a kettle spreads evenly through the water in the kettle
_____  9. Ionic and polar substances dissolve in water
_____ 10. A water strider can stand on the surface of water
_____ 11. Freezing kills begonia plants
_____ 12. Lake water remains warm in the autumn after the air above it cools

   a. adhesion
   b. high boiling point
   c. cohesion
   d. denser as liquid than as solid
   e. evaporative coolant
   f. polar molecules
   g. specific heat
   h. thermal conductivity

## Questions for Discussion

1. Why are atoms considered the units of matter, when they actually consist of even smaller particles (protons, neutrons, and electrons)?
2. Do oxygen molecules diffuse more quickly through air or through water? Explain.
3. If plants' roots absorb substances from the soil selectively, why do plants growing on soil contaminated with copper or lead take in so much of these toxic metals that they die?
4. Why does ethanol have a higher boiling point than dimethyl ether, which has the same molecular mass? (See Figure 2–9.)
5. What effect does wind have on the movement of molecules?

Using your knowledge of molecular movement and evaporation, explain the phenomenon of the "wind chill factor."
6. Imagine that water, like most other substances, was denser as a solid than as a liquid. How would the freezing of water in winter and the melting of ice in spring be different? How would these differences affect organisms living in lakes?
7. Fabrics are sometimes made "water repellent" by coating them with substances that cause water to form beads instead of spreading out on the fabric surface. What do you suppose is happening on a molecular level when a surface repels water in this way?

## Suggested Readings

Hill, J. W. *Chemistry for Changing Times*, 5th ed. New York: Macmillan, 1988. An excellent and entertaining "chemistry for poets."

Storey, K. B., and J. M. Storey. "Frozen and alive." *Scientific American*, December 1990. How animals survive cold winters.

# Biological Chemistry:
# Variations on Four Themes

## O B J E C T I V E S

*When you have studied this chapter, you should be able to:*

1. Give or recognize the characteristics of the following, and use this knowledge to answer questions about the relationships among them:

   organic compound, monomer, polymer, macromolecule, structural molecule, hormone

   carboxyl group, amino group, hydrocarbon chain

   lipid, fatty acid, wax, triacylglycerol, phospholipid, steroid

   carbohydrate, monosaccharide, pentose sugar, hexose sugar, disaccharide, polysaccharide

   nucleotide, nucleic acid

   amino acid, dipeptide, peptide bond, polypeptide, protein

2. List four main classes of biological macromolecules, state the roles of each in living organisms, and name the type(s) of subunits from which each is synthesized and the chemical elements typical of each class of molecules.

3. Classify the following in the appropriate classes of molecules listed in Objective 1, and briefly state their function in living

organisms: glycerol, cholesterol, glucose, glycogen, starch, cellulose, ATP.

4. Given a diagram of the structures of glycerol and a fatty acid, or two sugars, or two amino acids, draw and explain a condensation reaction between them, and name the classes of compounds to which the reactants and products belong; or, given the products of such a condensation reaction, draw a diagram of the hydrolysis reaction that they would undergo.

5. Define enzyme, catalyst, and activation energy. Explain what an enzyme's active site is and its significance. Explain the effect of enzyme or substrate concentration, temperature, pH, and competitive or noncompetitive inhibitors on the rate of an enzyme-mediated reaction.

6. Define metabolism and briefly describe how metabolic pathways are regulated by negative feedback acting on allosteric enzymes.

---

**A**ll organisms, from the smallest bacteria to the largest trees and the most complex animals, use the same basic set of molecules to build and run their bodies (Figure 3–1). This universal biological chemistry reflects the common ancestry of all living things. Each kind of molecule was selected early in the evolution of life because its structure gives it distinctive properties that make it good for performing certain functions. Ancient organisms passed on the ability to make and use these molecules to their descendants—all the life on Earth today. These molecular building blocks of life have several features in common.

First, the chemistry of life is based on the element carbon. What made carbon the natural selection for this central role? Carbon is lightweight, and each carbon atom forms four covalent bonds with other atoms. This permits carbon atoms to bond together into chains or rings, forming "skele-

**FIGURE 3–1** Two heads, one chemistry. All organisms, from cabbages to kids, are made from the same set of chemical building blocks.

tons" for the large molecules of life. And there are still bonding sites left for atoms of other elements, notably hydrogen and oxygen. Nitrogen, sulfur, and phosphorus are also often part of biological molecules.

Molecules with carbon skeletons and covalent carbon-hydrogen bonds are called **organic molecules** because they are made chiefly by living organisms. Only the simplest carbon compounds, such as carbon dioxide ($CO_2$) and carbonate ions ($CO_3^{2-}$), are considered inorganic. All substances that do not contain carbon are also classed as inorganic, although many of them, such as water and many kinds of ions, are vital components of living things.

Organisms make a variety of small organic molecules. Many of these serve as **monomers** (mono=one), building units that can be joined together to make larger molecules. Some of these larger molecules are made from only a few monomers, but others contain so many monomers that they are called **polymers** (poly=many). The biggest polymers are known as **macromolecules** (macro=big), arbitrarily defined as having a molecular mass of 50,000 daltons or more.

Living organisms can assemble an enormous array of different polymers from a relatively small number of common biological monomers. Familiar natural polymers include wool, silk, rubber, starch, and cotton. Since about 1900, chemists have been making the artificial organic polymers known as "plastics" by joining small organic monomers, such as dimethyl ether (see Figure 2–9) or ethyl acetate, in various ways—a case of art, or at least industry, imitating nature.

Living things make four main classes of organic compounds:

1. **Lipids:** nonpolar substances, which do not dissolve in water—including waxes, fats, oils, and steroids.
2. **Carbohydrates:** sugars, starches, cellulose, and related compounds.
3. **Nucleic acids:** the genetic material (containing instructions for making proteins) and molecules that help assemble proteins.
4. **Proteins:** molecules that make up silk, hair, tendons, and cartilage; carry out cell movements and muscle contraction; act as hormones; transport substances in the blood; fight infections; and perform many other crucial functions. One important group of proteins is the enzymes, which carry out the cell's thousands of biochemical reactions.

This chapter introduces all four groups and considers enzymes in some detail. In each group we shall see that the molecular structures of the monomers determine their chemical features, which in turn make them suited to performing particular functions in living organisms.

---

**K E Y   C O N C E P T S**

♦ All living organisms share a common chemical heritage. The distinctive chemistry of life is based on molecules with skeletons made up of strings and rings of carbon atoms. These organic molecules also contain hydrogen and oxygen, and often nitrogen, sulfur, or phosphorus.

♦ The structure of a molecule determines its functional properties.

♦ Organisms make and use thousands of kinds of organic molecules, which fall into four main classes: lipids, carbohydrates, nucleic acids, and proteins.

♦ Biological macromolecules are composed of many similar or identical monomer subunits joined together.

♦ A cell's enzymes carry out specific reactions at much greater rates than they would normally occur at the temperatures in living organisms. Enzymes convert organic molecules into different forms, step by step, in complex, controlled pathways.

---

## 3–A    STRUCTURE OF ORGANIC MOLECULES

Some figures in this chapter show several molecules with similar structures and functions. They are included, not for you to memorize, but so you can see a sample of biological molecules.

Even the small monomers in this chapter contain many atoms, often of several different elements. This seemingly complicated structure is made up of only a few functional parts. When you meet a new molecule, you must first get an idea of its size and shape by examining its carbon skeleton. Then look at the other atoms or groups of atoms—the molecule's functional groups.

### Carbon Skeletons

Carbon is the only element that can form enough different, complex compounds to make up the variety of molecules found in living organisms. These compounds are stable at the temperatures found in living organisms, yet can readily be remodeled by enzyme-mediated reactions.

A carbon atom can form four covalent bonds (Section 2–C). When there are four different atoms bonded to carbon, they occupy the corners of a tetrahedron with the carbon atom at its center; when there are three (one double-bonded to carbon), they lie in a plane with the carbon atom; and when there are two, they form a straight line (Figure 3–2).

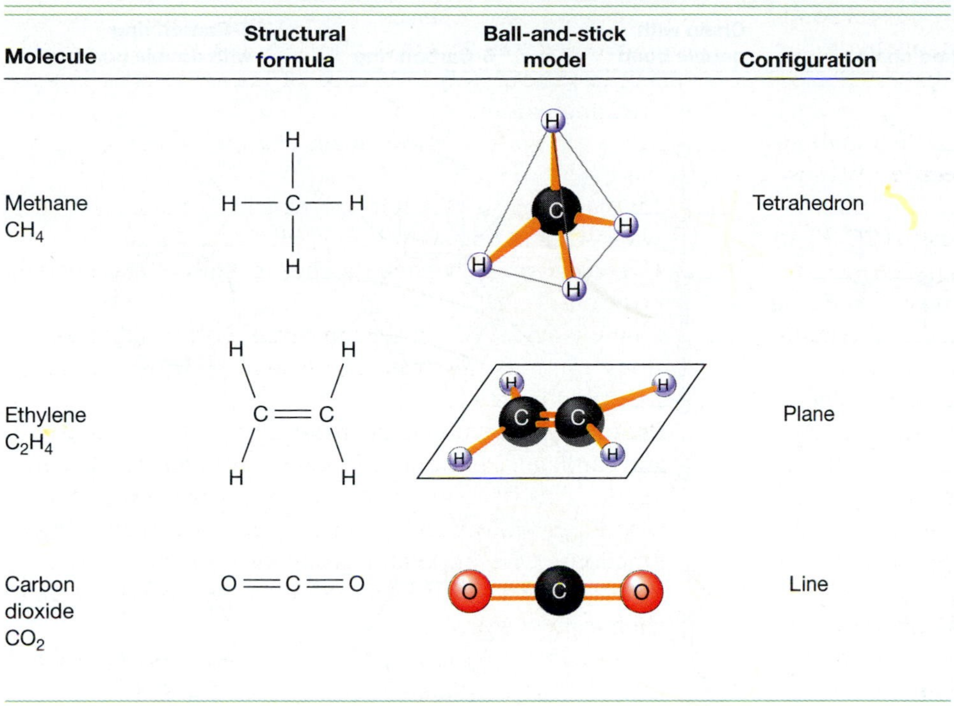

| Molecule | Structural formula | Ball-and-stick model | Configuration |
|---|---|---|---|
| Methane $CH_4$ | H—C—H (with H above and H below) | | Tetrahedron |
| Ethylene $C_2H_4$ | C=C (with H's attached) | | Plane |
| Carbon dioxide $CO_2$ | O=C=O | | Line |

**FIGURE 3–2** Carbon and its bonds. When a carbon atom (black) forms four single covalent bonds, they point to the four corners of a tetrahedron. This maximizes the distance between the atoms bonded to the carbon atom. Methane ($CH_4$) is a gas produced by bacteria as they break down dead, waterlogged vegetation.

When carbon is bonded to three other atoms (one of them by a double bond), it lies in a plane with them, as in the gaseous plant hormone ethylene.

In carbon dioxide ($CO_2$), each oxygen atom is joined to the carbon atom by a double bond, and the molecule is linear.

Carbon atoms can be joined together to form chains. Because four single bonds around carbon are tetrahedral, a so-called straight chain of single-bonded carbons actually forms a zigzag pattern. Sometimes carbon chains have other carbon chains as side branches, and sometimes the ends of carbon chains are joined, forming rings. Such chains and rings of carbon are called the "skeletons" of organic molecules, with atoms, or groups of atoms, of other elements attached to them.

Since all organic molecules have carbon skeletons, it is often easier to simplify drawings of structural formulas by omitting the C's that stand for carbon atoms. By convention, carbon atoms are assumed to be at all corners of a zigzag chain or a ring unless another atom is shown. Also, any carbon shown with fewer than four bonds to other atoms is assumed to be bonded to enough hydrogen atoms to complete a total of four bonds (Figure 3–3).

## Functional Groups

To simplify the great variety of biological molecules, we can sort them into families having the same **functional groups:** clusters of atoms that give a molecule specific chemical properties. For instance, compounds containing **carboxyl groups** (—COOH), such as acetic acid and amino acids, are called **organic acids** because they tend to release hydrogen ions in solution:

$$-COOH \rightarrow -COO^- + H^+ \qquad \text{(Section 2–H)}$$

As we saw in Section 2–C, covalent carbon-hydrogen bonds are nonpolar (electrically symmetrical). Therefore, a functional group containing only C and H, such as the methyl group (—$CH_3$), is also nonpolar and tends to make molecules less soluble in water.

Other functional groups contain polar bonds, particularly O—H, N—H, and C=O. In each case, the O or N bears a partial negative charge, and the H or C carries a partial positive charge. Polar functional groups interact with other polar groups or with charged ions. In particular, they attract water molecules, giving the molecule in which they occur a tendency to dissolve in water (Section 2–G).

Each kind of functional group takes part in specific kinds of reactions. For example, a molecule with a carbon bonded to a hydroxyl group (C—OH) is called an alcohol. An alcohol will react with a molecule containing a carboxyl group to form an ester linkage (Section 3–B). Hence, functional groups contribute to a molecule's solubility properties and determine what kinds of chemical reactions it will undergo. Table 3–1 shows the main functional groups. Rather than trying to memorize it, use it as a reference as you study this chapter.

## Looking at Molecular Structures

We shall inspect two sample molecules, each with six carbon atoms, to see what to look for when meeting a new molecule.

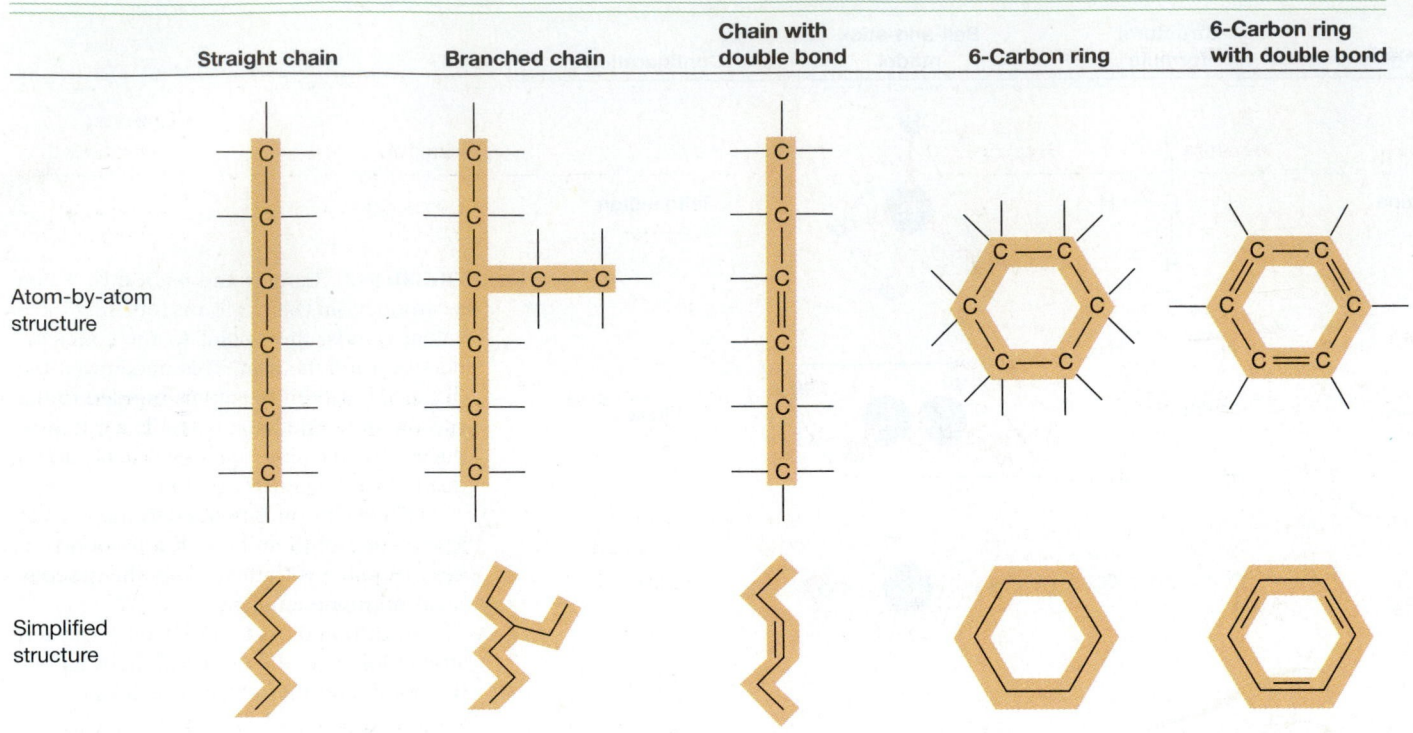

| Straight chain | Branched chain | Chain with double bond | 6-Carbon ring | 6-Carbon ring with double bond |

Atom-by-atom structure

Simplified structure

**FIGURE 3–3**   Carbon skeletons. Carbon atoms bond together in many ways, forming the "carbon skeletons" of organic molecules (color). (top row) The unconnected lines sticking out from each carbon atom (C) can bond to atoms of other elements, commonly hydrogen, oxygen, nitrogen, or sulfur. Because a carbon atom forms tetrahedral bonds, the "straight" chain actually forms a zigzag in space, as shown by the simplified structure.

First, let's look at a representative of the lipids, in this case a small fatty acid with six carbons:

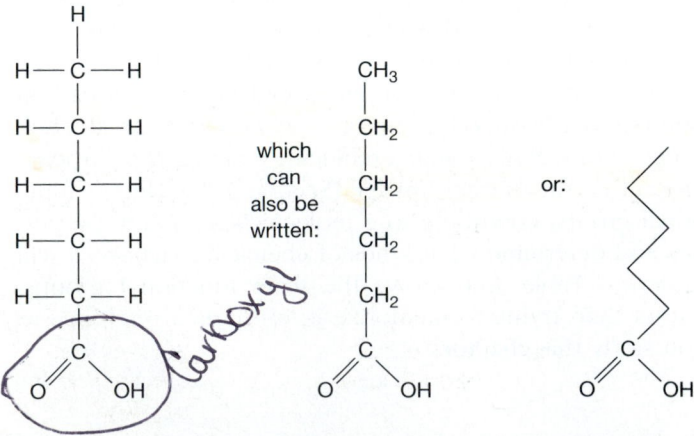

We can simplify this assembly of 20 atoms by looking at the molecule's features one at a time:

1. Each carbon atom (C) forms four covalent bonds to other atoms.
2. The fatty acid's six carbon atoms are linked to one another, forming a chain, the molecule's carbon skeleton.
3. Five of these carbons are attached only to hydrogen (H) atoms. This **hydrocarbon chain** is nonpolar and will not easily dissolve in water.
4. The sixth carbon is part of a carboxyl group (—COOH). This carbon is double-bonded to an oxygen and also bonded to —OH. This carboxyl group is the acidic part of the fatty acid: it can ionize to give —COO⁻ and H⁺. Because the carboxyl group contains the polar C=O and —OH groups, we can tell that it tends to dissolve in water even though the rest of the molecule tends not to. This results in the unique behavior of lipid molecules (Section 3–B).

Since carbon forms tetrahedral bonds, the "straight" chain of carbon atoms actually forms a zigzag in space, as shown by the version of the fatty acid at the far right.

**T A B L E    3 – 1**

**Some Common Functional Groups**

*PRACTICE*

| Group | Formula | Example | Found in | Chemical Characteristics |
|---|---|---|---|---|
| Methyl | —$CH_3$ | 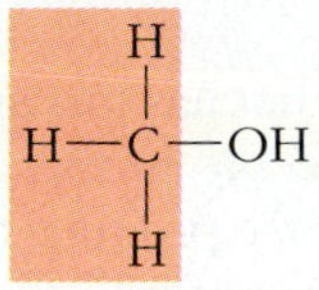 [C = 2.5, H = 2.1] H–C–OH structure | Methanol (wood alcohol) | Various organic molecules | Nonpolar, lowers water-solubility |
| Hydrocarbon chain | —$(CH_2)_nCH_3$ | 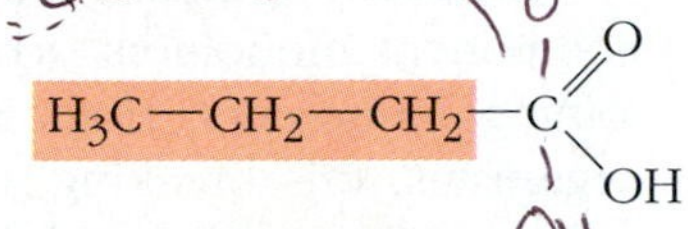 $H_3C$—$CH_2$—$CH_2$—C(=O)(OH) | Butyric acid (in milk) | Fatty acids, some amino acids, and so on | Nonpolar, insoluble in water |
| Carboxyl | 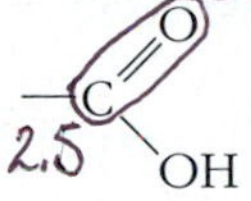 C=O [3.5] / 2.5 OH | 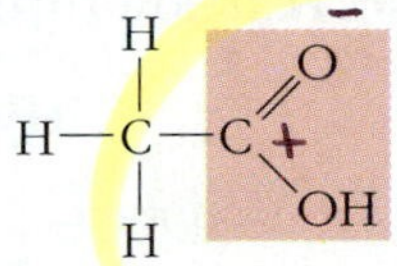 H–C–C(O⁻)(OH) acetic acid | Acetic acid (in vinegar) | Fatty acids, amino acids, organic acids | Polar, acidic, releasing $H^+$ in solution |
| Hydroxyl | —OH | 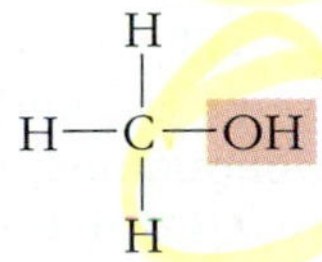 H–C–OH [O = 3.5, H = 2.1] | Methanol | Alcohols, sugars | Polar, forms hydrogen bonds |
| Ester linkage | 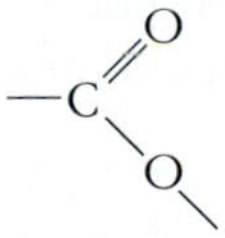 —C(=O)—O— | 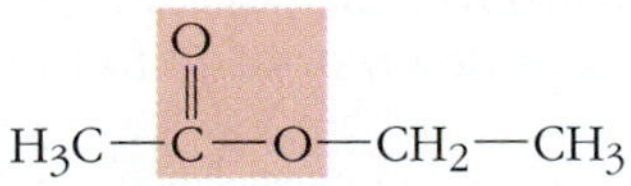 $H_3C$—C(=O)—O—$CH_2$—$CH_3$ | Ethyl acetate (lacquer thinner) | Linkage of alcohol to organic acid | Polar |
| Carbonyl | —C=O | 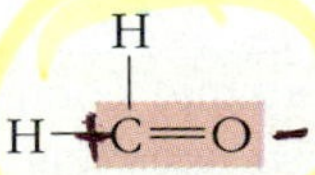 H—C=O; H–C–C(=O)–C–H acetone | Formaldehyde; Acetone | Aldehydes and ketones, such as sugars | Polar |
| Amino | —$NH_2$ | 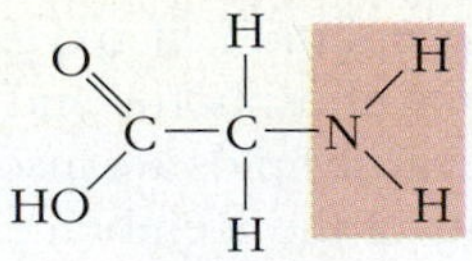 glycine structure C(O)(HO)–C–N(H)(H) | Glycine (an amino acid) | Amino acids, amino sugars, nitrogenous bases | Polar, forms hydrogen bonds |
| Phosphate | 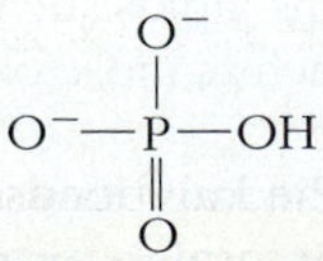 $O^-$—P(=O)($O^-$)—OH | 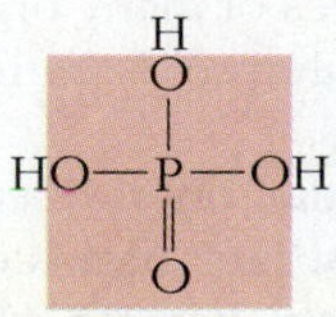 HO—P(=O)(OH)(H) | Phosphoric acid | Nucleotides, phospholipids | Polar, acidic, releasing $H^+$ in solution |

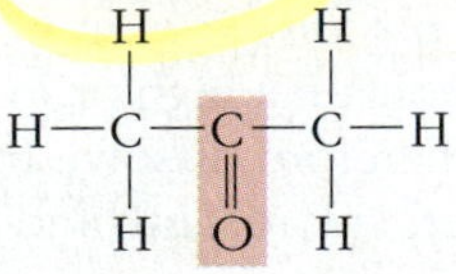

Handwritten annotations: "(H—C=O carboxyl)", "OH", "C=2.5 H=2.1", "3.5", "2.5", "O=3.5 H=2.1"

Now consider another six-carbon molecule, glucose, a simple sugar, which is a carbohydrate:

GLUCOSE
($C_6H_{12}O_6$)

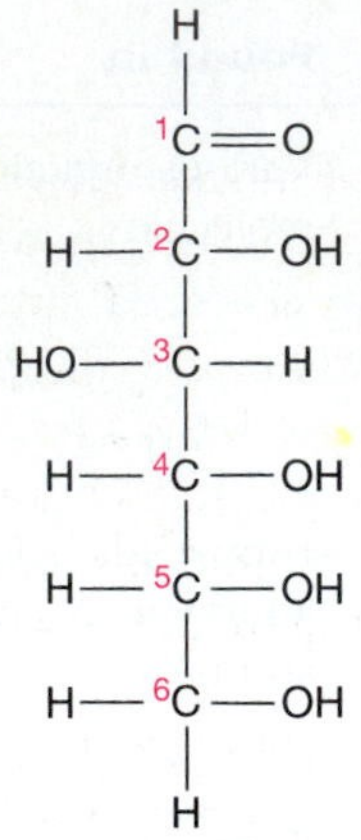

*[handwritten margin note: 6 carbon chain]*
*[handwritten margin note: H—C=O → carboxyl group]*

1. Again, each carbon forms four bonds, and a six-carbon chain forms the molecule's skeleton.
2. The first carbon atom, numbered 1, is double-bonded to an oxygen atom, forming a **carbonyl** group. Because the carbonyl group occurs on an end carbon which is also attached to a hydrogen, glucose is an **aldehyde.**
3. All of the other carbon atoms have hydroxyl groups (—OH), on the right in carbons 2, 4, 5, and 6, and on the left in carbon 3. Hydroxyl groups are the hallmark of alcohols, and so the sugar glucose is both an aldehyde and a multiple alcohol. This molecule, bristling with polar functional groups (the five —OH groups and the C=O), dissolves readily in water.
4. The remaining bonds attach to hydrogen atoms.

So a glucose molecule is really fairly simple: it contains a chain of six C atoms, the first forming a C=O bond and the rest attached to —H on one side and —OH on the other. The remaining positions are filled by —H. When you look at a structural formula, start by finding the carbon chain and see what groups are attached to it. If there is one C=O and a lot of —H and —OH groups, you are probably looking at a sugar.

## Building Biological Polymers

In living organisms, monomers such as fatty acids and glucose are joined into larger molecules. This occurs by somewhat roundabout mechanisms. The same overall reaction occurs more directly under laboratory conditions. This lab reaction is called a **condensation** or **dehydration synthesis reaction:** two molecules join as one loses an —H and the other an —OH, which themselves join together and form a water molecule. The terms "condensation" and "dehydration" refer to this loss of water. In this chapter we

show the simpler laboratory reactions, which occur by the general equation:

**Dehydration synthesis:**

$$\text{Monomer}-\text{H} + \text{HO}-\text{Monomer} \longrightarrow$$
$$\text{Monomer}-\text{Monomer} + H_2O$$

A monomer that has lost an atom or two by combining with another molecule in this way is called a **residue.** For example, the first product of this reaction is a dimer (di = two) consisting of two monomer residues.

Reactions breaking larger molecules back into their component monomers occur more readily than reactions joining monomers. Both in the laboratory and in living organisms, this breaking apart occurs by adding a water molecule into the bond linking the residues. This is a **hydrolysis reaction** (hydro = water; lysis = loosening):

**Hydrolysis:**

$$\text{Monomer}-\text{Monomer} + H_2O \longrightarrow$$
$$\text{Monomer}-\text{H} + \text{HO}-\text{Monomer}$$

When we eat, our digestive systems hydrolyze the polymers in our food into monomers, which are then distributed throughout the body. Our cells may assemble these monomers into our own macromolecules. Sooner or later, these too are hydrolyzed, as part of the body's continuous turnover of molecules, and the monomers are either broken down or used to build other macromolecules.

## 3–B   LIPIDS

**Lipids** are organic compounds that vary in structure but share one distinguishing property: they are nonpolar and so do not dissolve appreciably in water. This is because lipids contain a high proportion of carbon-hydrogen bonds. They therefore dissolve in nonpolar organic solvents, such as ether, chloroform, and benzene, but not in water, which is polar.

Lipids contain mostly carbon and hydrogen, with a very small proportion of oxygen. Some lipids also contain the elements phosphorus and nitrogen.

Because lipids are insoluble in water, they are vital components of the membranes that separate aqueous (watery) compartments from one another in living organisms. On the outer surfaces of many organisms, lipids such as waxes and oils form waterproof coatings such as those on leaves, wool, and feathers.

Lipids also offer an excellent way to store energy, for two reasons. First, lipids contain a high proportion of carbon-hydrogen bonds, which are rich in stored energy. Weight for weight, lipids yield more than twice the energy of carbohydrates, the other group generally used to supply energy. In Section 3–A, compare the proportions of carbon-hydrogen bonds in the fatty acid, a lipid, and glucose, a carbohydrate. Compared with lipids, carbohydrates contain a lot of oxygen, which adds to the molecule's weight but not to its energy content.

Second, because lipids do not attract water, they can be stored in concentrated form. In contrast, polar carbohydrates must inevitably be surrounded by water. When carbohydrate plus its storage water is compared with an equal weight of lipid, the lipid contains six times more energy.

This is undoubtedly why lipids became increasingly important food reserves in the bodies of animals during the course of evolution, especially in birds and mammals, which use a great deal of energy. Without lipid reserves, for instance, the annual nonstop migrations of many birds that winter in warmer climates would be impossible. A small bird may almost double its body weight in the fall as it stores fat reserves before the long flight. To carry the same energy reserve in carbohydrate, the bird's weight would have to increase so much that it could not fly. Likewise, small seeds, especially those dispersed by wind, frequently carry lipids rather than carbohydrates as their food reserves.

Small lipid molecules can combine to form larger ones, but these do not contain enough subunits to qualify as polymers or macromolecules.

## Fatty Acids

Fatty acids are the simplest lipids. A **fatty acid** consists of a long hydrocarbon chain with a carboxyl group at one end (Figure 3–4). In a **saturated** fatty acid, all carbon atoms of the hydrocarbon chain are "filled" with as many hydrogens as they can hold. A fatty acid with one or more double bonds in its hydrocarbon chain is **unsaturated:** it could hold more hydrogens if one of the two bonds were broken and two hydrogen atoms were attached to the carbons instead (Figure 3–4b).

The bonds in a carboxyl group are polar, and so this end of a fatty acid is **hydrophilic** (hydro = water; philic = loving); that is, it attracts water molecules and forms hydrogen bonds with them. However, the carbon-hydrogen bonds in the hydrocarbon chain are nonpolar, and so the chain is **hydrophobic** ("water-fearing"). A fatty acid's hydrophilic end will dissolve in aqueous solutions, and the hydrophobic end will dissolve in nonpolar organic com-

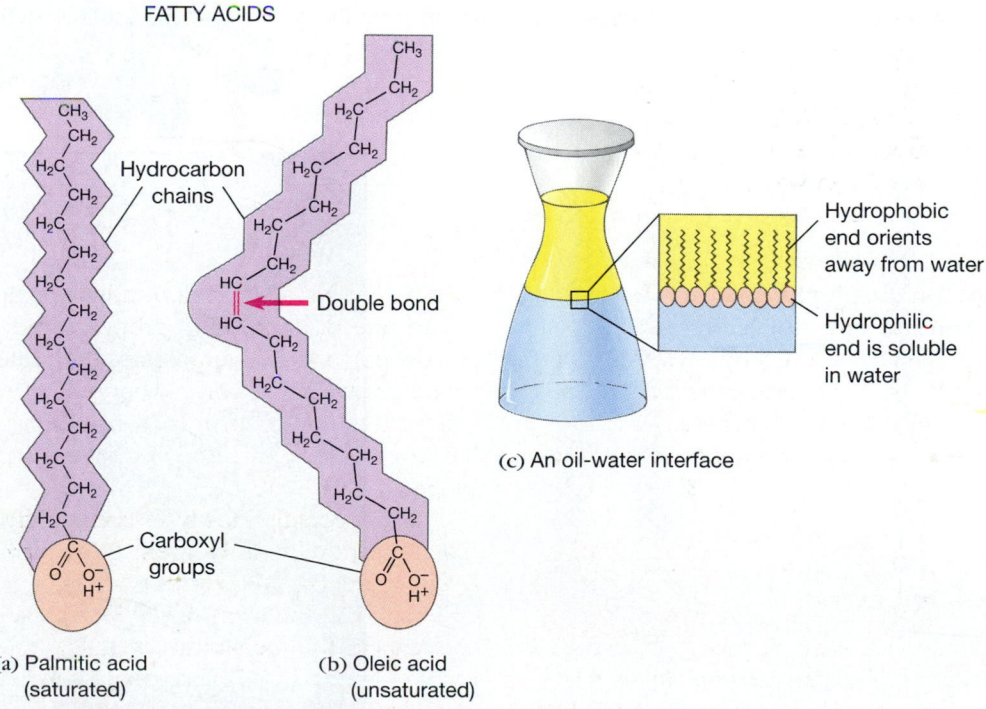

FATTY ACIDS

Hydrocarbon chains

Double bond

Carboxyl groups

(a) Palmitic acid (saturated)

(b) Oleic acid (unsaturated)

Hydrophobic end orients away from water

Hydrophilic end is soluble in water

(c) An oil-water interface

**FIGURE 3–4** Fatty acids. (a, b) The hydrophilic carboxyl groups, which make these compounds acids, are shown in orange. The rest of each molecule is a nonpolar hydrocarbon chain (purple). The most common fatty acids have even numbers of carbon atoms in chains 14 to 22 carbon atoms long. The red arrow points to the double bond that makes oleic acid unsaturated. (c) Because the hydrophilic carboxyl group is soluble in water, whereas the long hydrophobic hydrocarbon chain is not, the molecules orient themselves at interfaces between water and nonpolar organic substances such as oil, or between water and air, with their hydrophilic ionic ends in the water.

pounds (Figure 3–4c). This behavior makes some lipids containing fatty acids important parts of the membranes that divide living systems into compartments.

Fatty acids seldom occur free, but are usually combined with other molecules to form substances such as **glycolipids** (carbohydrate + lipid) or **lipoproteins** (lipid + protein). They are also parts of many larger lipids.

## Waxes

A **wax** is a lipid formed by joining two monomers: one very long fatty acid and one very long fatty alcohol, composed of a hydrocarbon chain with C—OH at one end. The fatty acid and alcohol residues are held together by an **ester linkage** (Figure 3–5 and Table 3–1). The long hydrocarbon chains (24 to 36 carbons each) make waxes solid, with high melting points.

Waxes provide a waterproof coating on the outer surfaces of many organisms, for example, on leaves, fruit, skin, hair, feathers, or the armor of insects. Economically important waxes include carnauba wax (from a palm tree) and beeswax, both used in polishes and candles, and lanolin (from wool), used in cosmetics. Although these external waxes of land organisms are most obvious to us, waxes are also important as long-term energy stores in the bodies of marine animals, such as copepods, diminutive kin of shrimp.

## Triacylglycerols (Triglycerides)

A **triacylglycerol** (old name: **triglyceride**) is a molecule made by joining three fatty acids to the three alcohol groups of glycerol, forming ester linkages (Figure 3–6).

Triacylglycerols that are solid at room temperature are commonly called **fats** (such as butter, lard, or suet), whereas

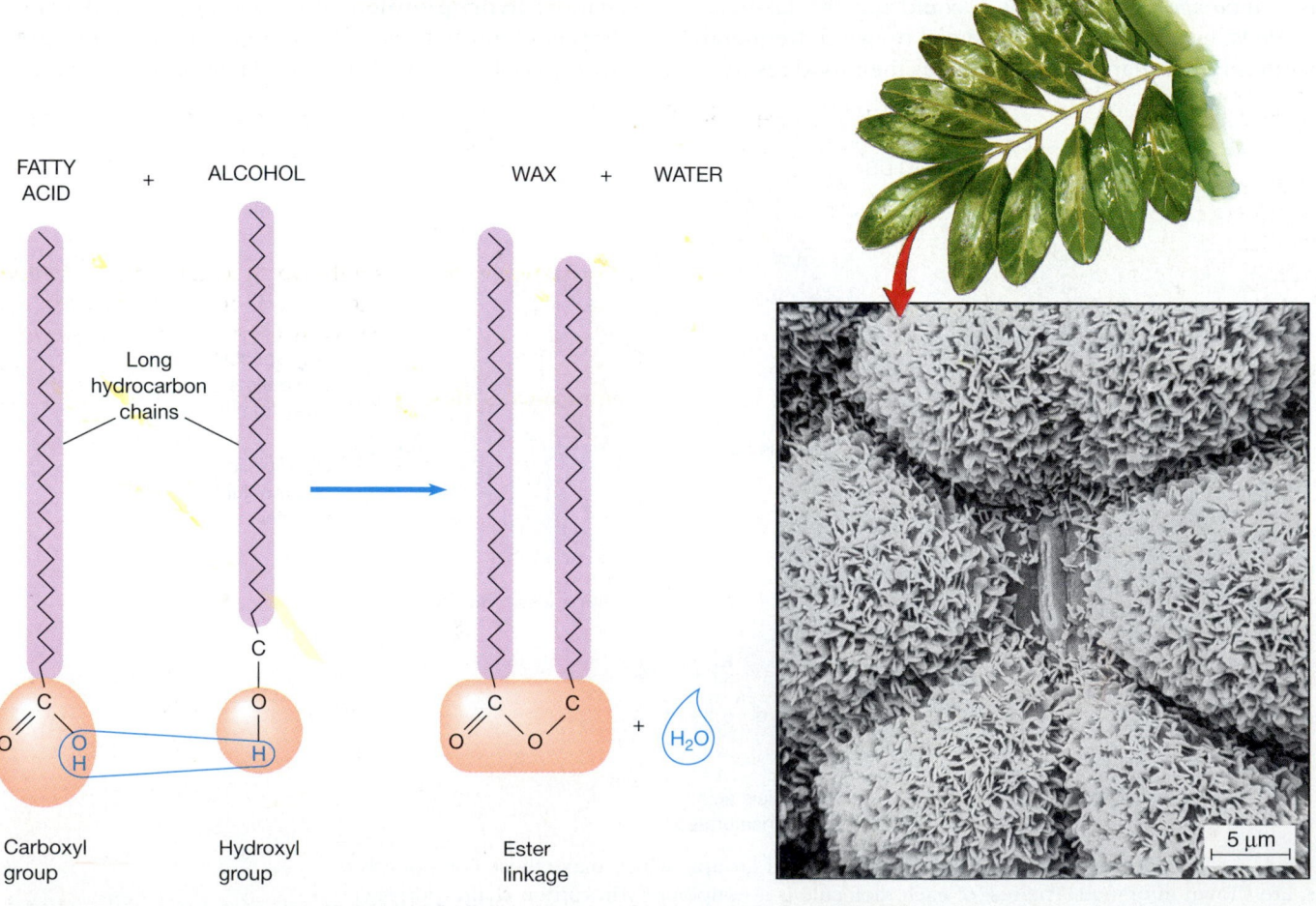

(a)

(b) Wax flakes on leaf surface

**FIGURE 3–5** Wax. (a) A wax is formed by joining a long fatty acid and a long alcohol, resulting in a covalent attachment called an ester linkage between the two residues. (b) Flakes of wax form a waterproof covering on the surface of a leaf and prevent evaporation of the plant's water into the surrounding dry air. Water conservation is further enhanced by the closing of the slit-like pore in the center of this photo, which leads to the moist leaf interior.   (b, Biophoto Associates)

GLYCEROL + 3 STEARIC ACIDS                    GLYCERYL TRISTEARATE, A TRIACYLGLYCEROL + WATER

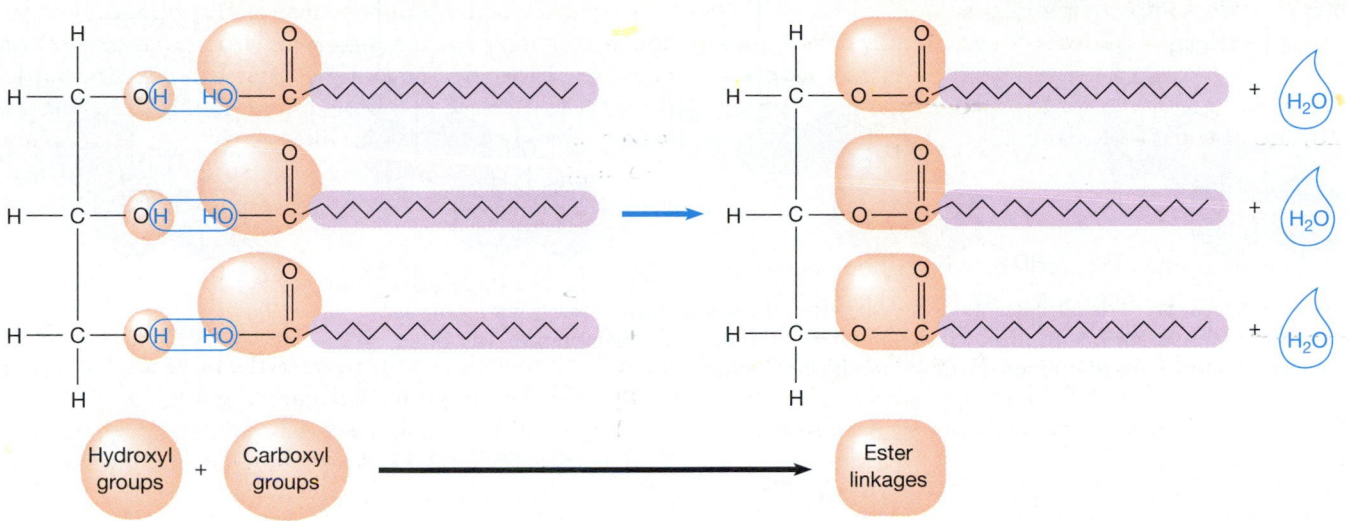

**FIGURE 3–6** Formation of a triacylglycerol. Three fatty acids (alike or different: here, three molecules of stearic acid) are attached to a molecule of glycerol, with the removal of water (blue). Three ester linkages form as each of glycerol's alcohol groups (—OH) reacts with the carboxyl group (—COOH) of a fatty acid.

those that are liquid are called **oils** (such as olive, corn, and peanut oils). Fats are solid because they contain mostly saturated fatty acids, which tend to lie straight and so can pack close together with other fat molecules. Oils generally contain more unsaturated fatty acids than fats do. Unsaturated fatty acids have rigid elbows (see Figure 3–4b), which will not fit snugly against other molecules. The loose packing of the molecules gives oils their fluidity. Oils occur more commonly in plants than in animals. However, animals living in cold areas, such as Arctic and Antarctic fish, usually

produce a relatively high proportion of unsaturated fatty acids, which keep their bodies flexible.

## Phospholipids

**Phospholipids** are similar to triacylglycerols, except that one (or two) of the fatty acids is replaced by a phosphate group, which in turn is usually linked to a nitrogen-containing group (Figure 3–7). Phospholipids are **structural molecules,** that is, building materials that contribute

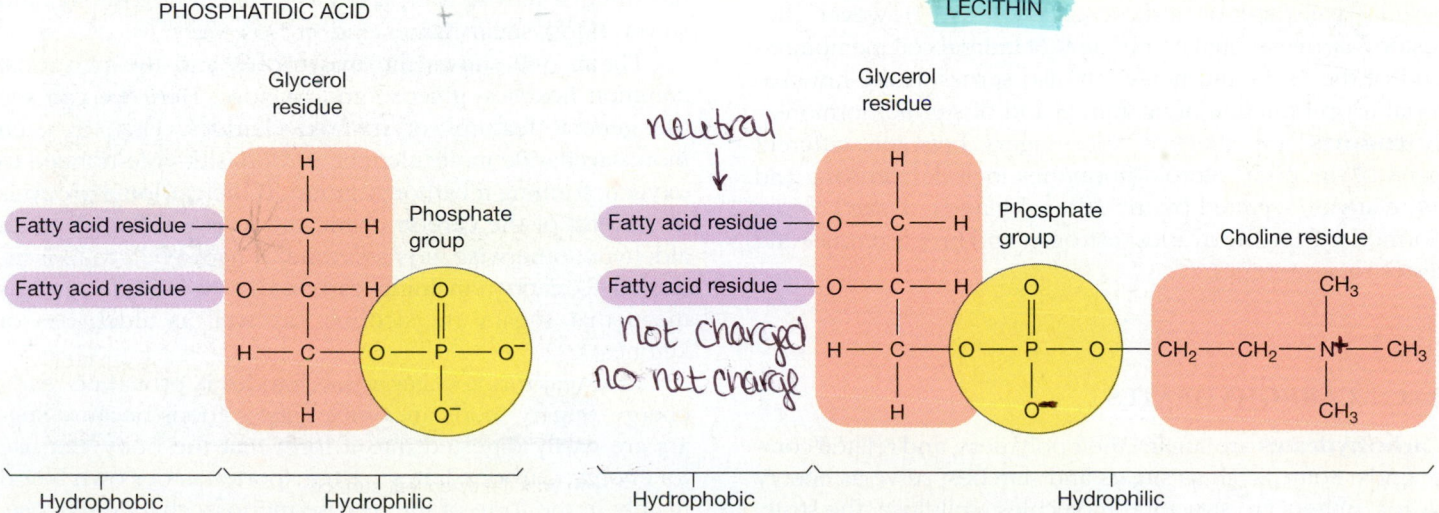

**FIGURE 3–7** Phospholipids. Many phospholipids are derivatives of phosphatidic acid, in which glycerol (orange) has joined to two fatty acids (purple) and a phosphate group (yellow). Lecithin, a phospholipid found in the membranes of all cells, is formed by joining choline to phosphatidic acid.

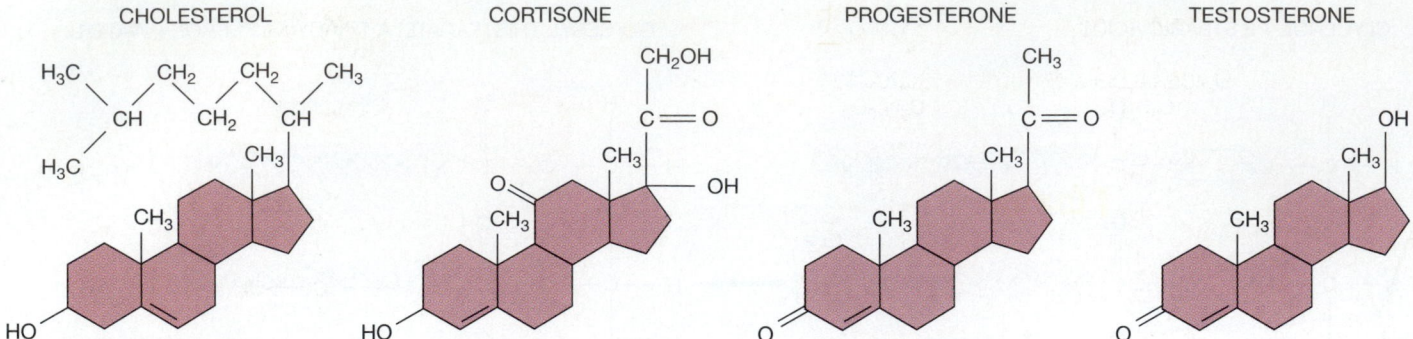

CHOLESTEROL    CORTISONE    PROGESTERONE    TESTOSTERONE

**FIGURE 3–8**   Steroids. Steroids all have the same skeleton of carbon rings (color), with various functional groups. Cholesterol is an important component of animal cell membranes. The other steroid molecules shown are hormones, made by modifying cholesterol.

to the shape of a cell or of a multicellular body. They are the chief lipid components of biological membranes, with their polar phosphate and nitrogenous groups facing aqueous areas and their fatty acid tails buried in the membrane's nonpolar interior.

## Steroids

**Steroids** differ from other lipids in structure, but qualify as lipids because they are insoluble in water. The basic steroid skeleton consists of four contiguous carbon rings, with different functional groups attached in different steroids (Figure 3–8).

The most abundant steroid, cholesterol, gets a lot of bad press because deposits of cholesterol in the body may result in gallstones or cardiovascular disease. However, cholesterol is an essential component of animal cell membranes and of the brain and nerves. It also serves as the raw material for production of vitamin D and of steroid hormones. **Hormones** are chemical messengers between different parts of the body. Steroid hormones include cortisone and its relatives, secreted by the adrenal glands, as well as sex hormones from the ovaries (estrogen and progesterone) and testes (testosterone).

## 3–C   CARBOHYDRATES

**Carbohydrates** are sugars, their polymers, and related compounds. Some, such as sugars and starches, serve as energy stores. Others are structural molecules; cellulose, the structural carbohydrate in plant cell walls, is the most abundant organic compound on Earth.

The name carbohydrate, meaning "water of carbon," comes from the general formula for this group, $CH_2O$. As in lipids, the principal chemical elements are C, H, and O, but carbohydrates have a higher proportion of oxygen; in lipids, the typical unit in a molecule is $CH_2$.

The simplest carbohydrates, the monosaccharides, often occur singly, but some of them are also the monomer subunits used to build larger carbohydrates. They often occur joined together in pairs, forming disaccharides, or polymerized into macromolecules called polysaccharides.

## Monosaccharides

**Monosaccharides** are the simple sugars. Most have molecular formulas that are multiples of $CH_2O$. They contain three to nine carbons—usually three, five, or six. Simple sugars can be classified by how many carbons they contain: a triose has three, a tetrose four, a pentose five, a hexose six, and so on. (Most sugar names end in "-ose.")

Figure 3–9 shows the two trioses and the two most common hexoses, glucose and fructose. Here we can see the general features of monosaccharides. First, in each monosaccharide molecule, one carbon is double-bonded to oxygen, forming a carbonyl group. If the carbonyl group is at the end of the carbon chain, the monosaccharide is an aldehyde; otherwise, it is a ketone. A hydroxyl group is attached to each remaining carbon. These C—OH groups mean that sugars are alcohols (as well as aldehydes or ketones).

One major role of monosaccharides is providing ready energy. Sugary foods are quick-energy foods because sugars are easily digested into a form that the body can use for energy. Carbohydrates are transported in our own blood mainly in the form of the hexose monosaccharide glucose, which cells take up and use for energy or as a raw material to make other molecules. Certain pentoses are also common in our bodies because they are used to make nucleic acids (Section 3–D).

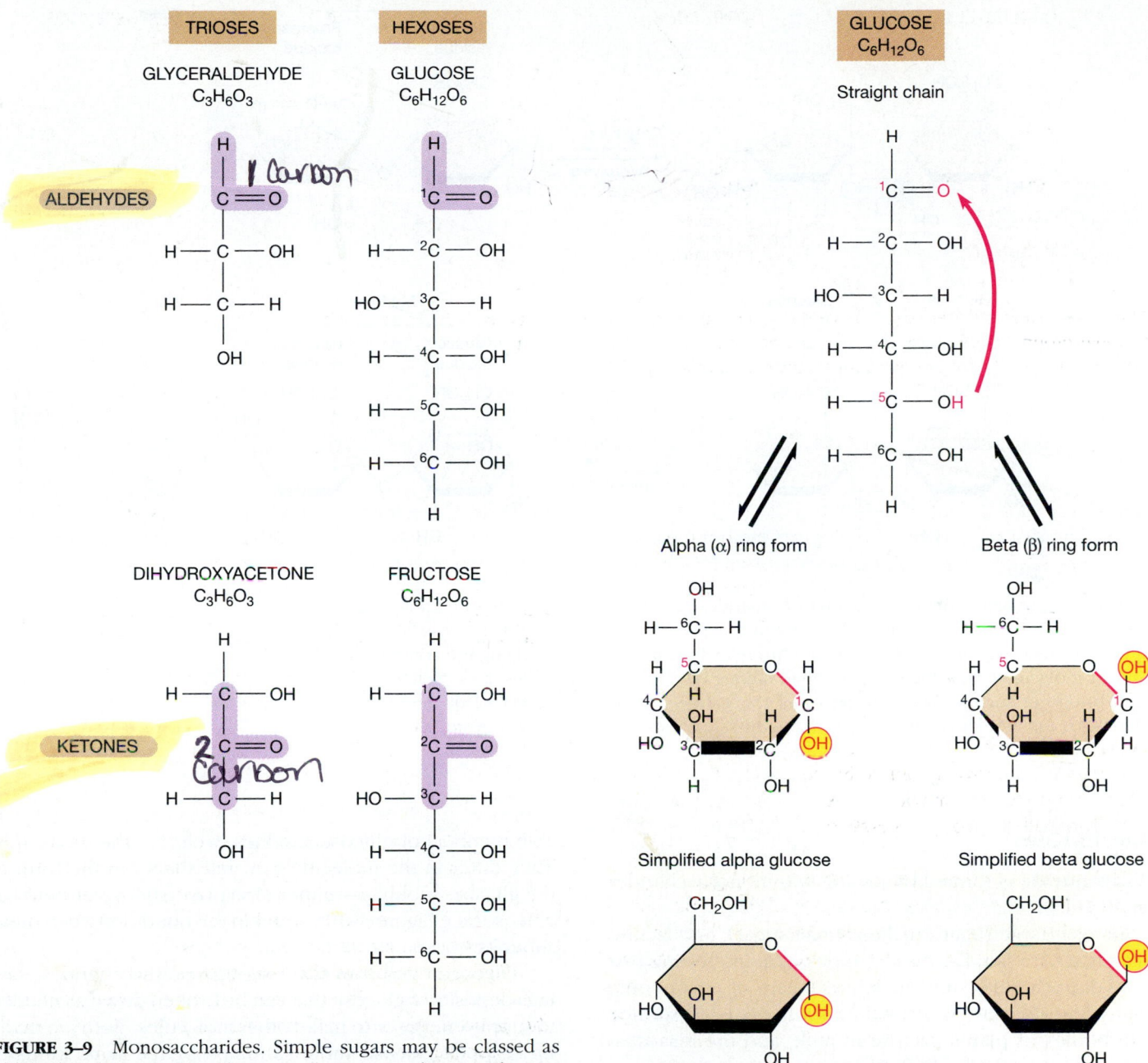

FIGURE 3–9 Monosaccharides. Simple sugars may be classed as aldehydes (top row) or ketones (bottom row), according to whether their carbonyl group (C=O) involves an end carbon or not. (Colored screens show characteristic aldehyde and ketone configurations.) Monosaccharides may also be classed according to the number of carbons. There are only two trioses (left), an aldehyde and a ketone. The aldehyde glucose and the ketone fructose are the two most common hexose monosaccharides. By convention, the carbon atoms are numbered starting from the end of the molecule which is more oxidized—that is, the one containing the higher ratio of oxygen to (carbon + hydrogen).

When a monosaccharide with five or more carbons is in a watery environment such as a living body, most of the molecules undergo a rearrangement of bonds and take on the shape of a ring (Figure 3–10).

FIGURE 3–10 Forms of glucose. When straight-chain glucose (top) dissolves in water, the molecule bends back on itself, bringing the —OH group on carbon #5 next to the =O on carbon #1 (red arrow). The hydrogen atom "hops" from one oxygen to the other, and the leftover bonds of C #1 and O #5 join, so that the molecule forms a six-sided ring with O at one corner. (center) In alpha glucose, the —OH of carbon #1 is below the ring (red); in beta glucose, it is above the ring. In solution, glucose changes from one form to the other by way of the straight-chain intermediate. The heavier bond lines show that this edge of the ring projects out of the page toward you, whereas the opposite edge lies behind the page. (bottom) Simplified structures take less space. Again, carbon atoms lie at all corners except the upper right one, occupied by oxygen. Hydrogen atoms filling out the four bonds of carbon atoms are omitted.

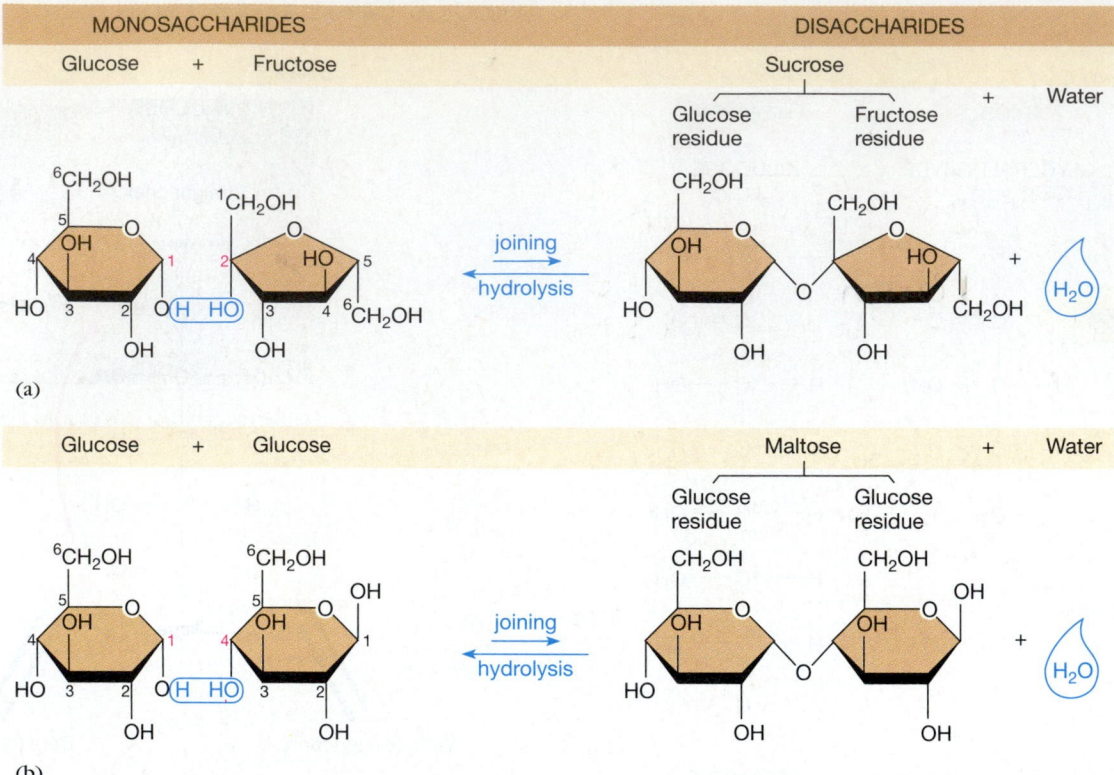

**FIGURE 3–11** Formation of two common disaccharides. (a) The disaccharide sucrose (table sugar) is formed from the monosaccharides glucose and fructose. (b) The disaccharide maltose is formed from two glucose monomers. Both reactions are reversible by hydrolysis: adding the components of a water molecule across the oxygen bridge linking the monosaccharide residues. Note that the oxygen bridge in sucrose joins carbon #1 of the glucose residue to carbon #2 of the fructose residue, whereas in maltose the oxygen bridge links carbon #1 of one glucose residue to carbon #4 of the other.

## Disaccharides

A **disaccharide** is formed by joining two monosaccharides (Figure 3–11).

Sucrose (table sugar), maltose (malt sugar), and lactose (milk sugar) are familiar disaccharides. It is interesting that several disaccharides serve mainly as means of transporting carbohydrate. Sucrose is the main carbohydrate transported in the bodies of plants, lactose in milk, and the disaccharide trehalose in the body fluids of insects.

Trehalose and other disaccharides also help protect membranes and proteins from disruption in small invertebrate animals that can survive severe dehydration. The sugars' many—OH groups take the place of water molecules in stabilizing protein structures and sheets of membrane phospholipids.

## Polysaccharides

**Polysaccharides** are polymers composed of hundreds to thousands of monosaccharide residues. Three important polysaccharides made of glucose monomers occur in living things: glycogen, starch, and cellulose. All three contain vari-

able numbers of glucose residues, well into the thousands. They differ in the molecule's overall shape, in the form of the glucose subunits—either alpha ($\alpha$) glucose or beta ($\beta$) glucose (see Figure 3–10)—and in the bonds between these subunits.

Glycogen and starch are storage polysaccharides, savings deposits of glucose that can be broken down as needed to supply energy or to make other molecules. Both are made up of alpha glucose subunits joined by the same kinds of linkages.

Animals store energy in the form of the polysaccharide **glycogen.** The liver and muscles remove glucose from the blood and assemble it into glycogen, which is stored until its glucose units are needed again. Glycogen is made up of alpha glucose subunits joined by oxygen bridges linking carbon #1 of one glucose residue to carbon #4 of its neighbor; these are called alpha-1,4 linkages. Glycogen molecules branch about every ten glucose residues (Figure 3–12).

The energy-storage polysaccharide in plants is **starch.** Starch consists of two kinds of polymers of alpha glucose. Amylose is a long, unbranched chain of glucose residues joined by alpha-1,4 linkages. Amylopectin is a branched polymer similar to glycogen but with less frequent branching.

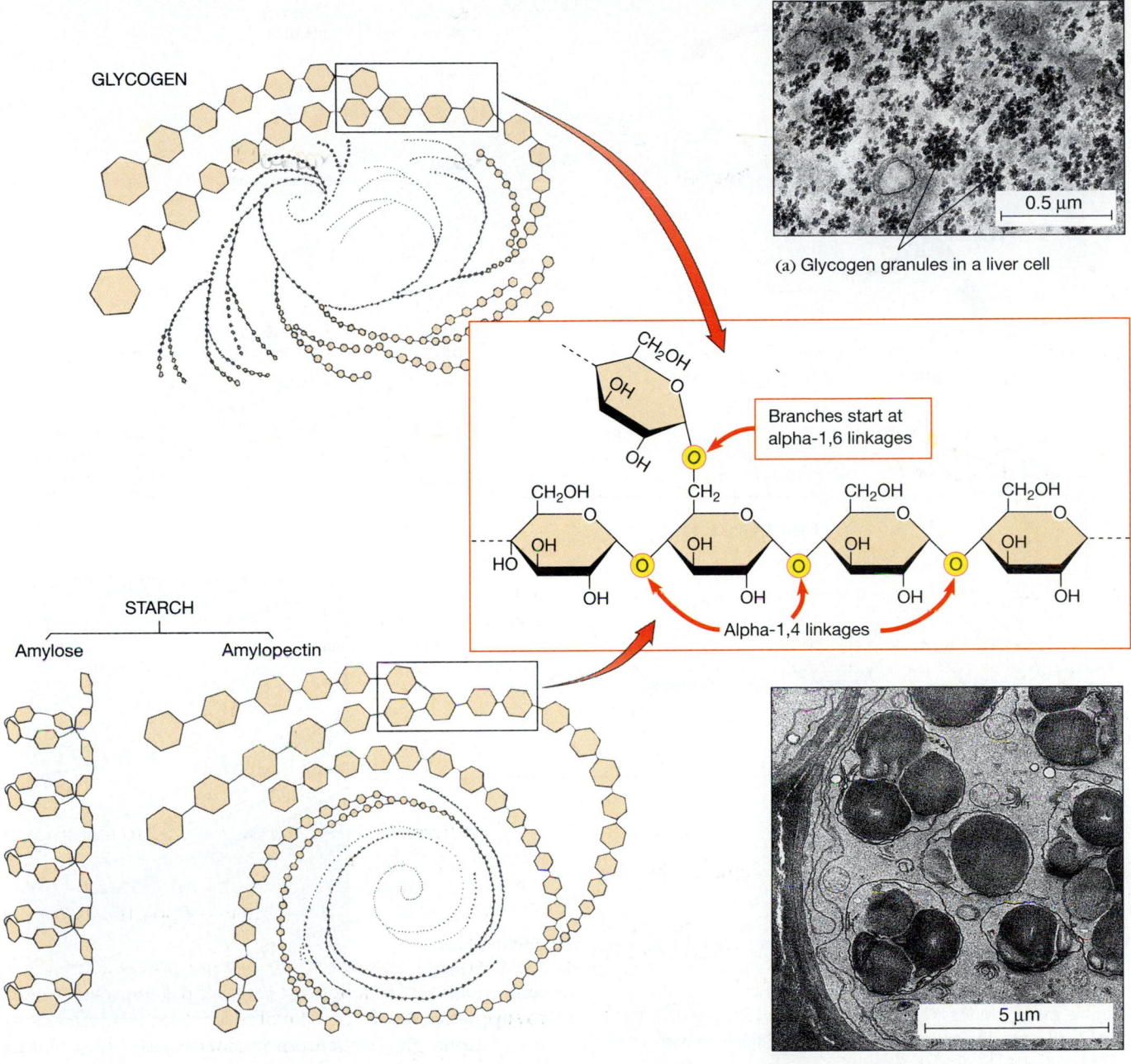

GLYCOGEN

0.5 μm

(a) Glycogen granules in a liver cell

CH₂OH

OH

OH

Branches start at
alpha-1,6 linkages

CH₂OH            CH₂            CH₂OH            CH₂OH

HO    OH            OH            OH            OH

OH            OH            OH            OH

Alpha-1,4 linkages

STARCH

Amylose            Amylopectin

5 μm

(b) Starch grains in a root cell

**FIGURE 3–12**   Storage polysaccharides. Animals store glucose as granules of the polysaccharide glycogen, and plants store it as grains of starch, which consists of amylopectin plus amylose. All these molecules contain chains of glucose residues joined by alpha-1,4 linkages. In glycogen and amylopectin, branches arise by alpha-1,6 linkages between glucose residues. Glycogen is more highly branched than amylopectin. The unbranched amylose chain assumes a spiral (helix) shape. (1 μm = 0.001 mm)   (a, Biophoto Associates; b, Barrie Juniper)

When you boil potatoes, the water becomes cloudy as amylose dissolves in it, whereas amylopectin stays in the potatoes and is later digested (hydrolyzed) to glucose subunits in your intestine. Left to itself, a living potato (or any other starch-storing plant) would eventually break down its starch into glucose. It would use some of this glucose to supply energy and some as a building material for its growth and reproduction.

The major structural material made from glucose in plants is the polysaccharide **cellulose.** Cellulose is made up of long, straight chains of beta glucose monomers. The beta-1,4 linkages joining these monomers make the glucose

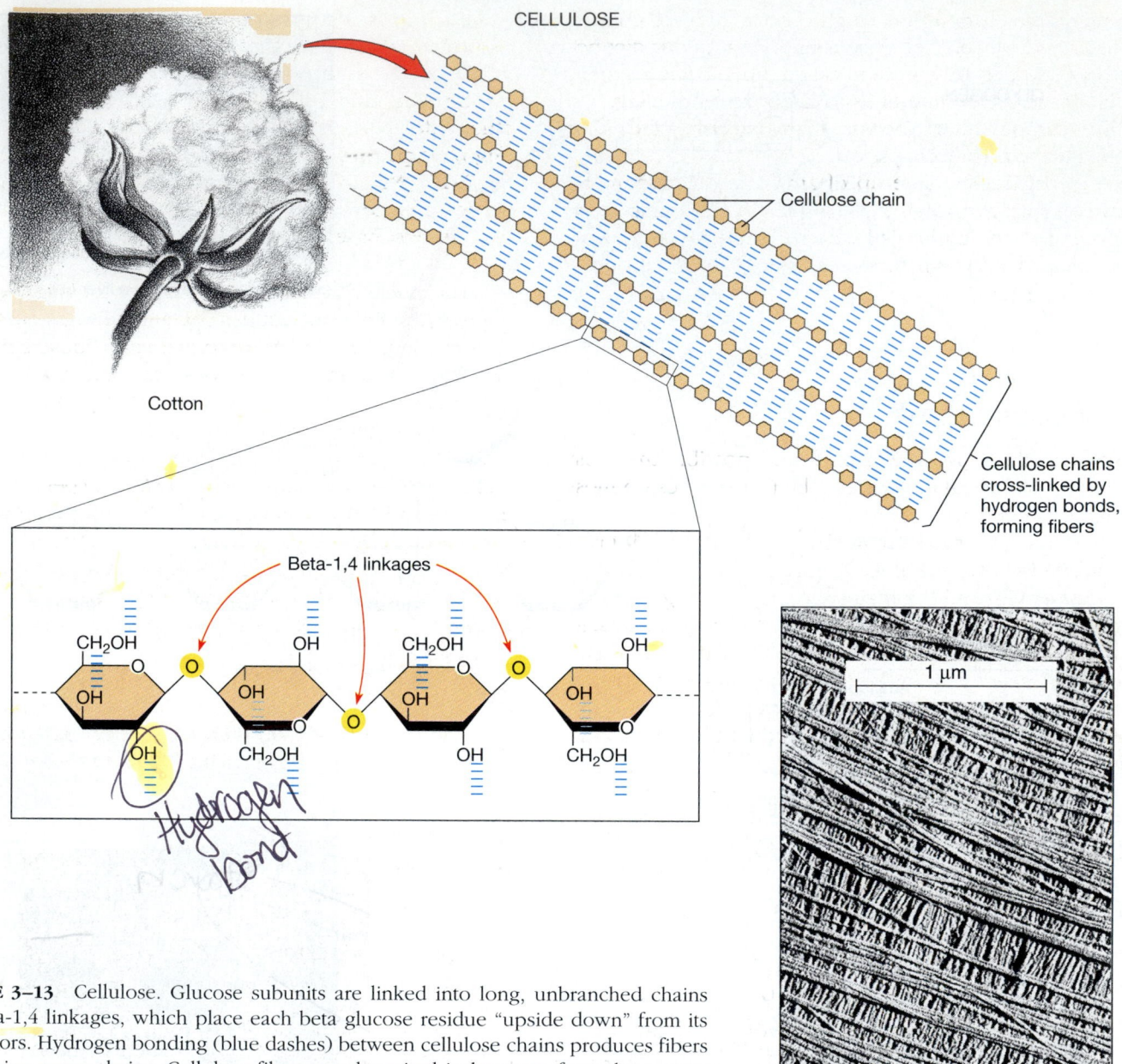

**FIGURE 3–13** Cellulose. Glucose subunits are linked into long, unbranched chains by beta-1,4 linkages, which place each beta glucose residue "upside down" from its neighbors. Hydrogen bonding (blue dashes) between cellulose chains produces fibers containing many chains. Cellulose fibers are deposited in layers to form the protective cell wall outside a plant cell (photo).   (Biophoto Associates)

Cellulose fibers in cell wall

residues alternately right-side-up and upside-down (Figure 3–13). Whereas alpha linkages twist the polymers of glycogen and starch into spirals, beta linkages make the strands of cellulose linear.

Free —OH groups on the beta glucose residues in cellulose form hydrogen bonds to neighboring cellulose chains. In this way many chains are linked together, forming cellulose fibers. The fibers of cotton consist almost entirely of cellulose.

Each plant cell surrounds itself with a tough external cell wall made up of several layers of cellulose fibers, of-

ten glued together and reinforced with other substances. The cell walls help to stiffen and support the plant body (consider celery, which is rich in cellulose). The cell wall also contains small amounts of several other structural polysaccharides.

The human digestive tract makes enzymes that hydrolyze the alpha linkages of starch, but because beta linkages have different shapes, these enzymes cannot hydrolyze cellulose. In fact, few organisms make enzymes that can digest cellulose. Because we cannot digest the cellulose in our food, it passes through the digestive tract unchanged, and our bodies cannot use its glucose to provide energy. However, indigestible cellulose is important in our diet because it provides bulk (often called fiber or roughage), which stimulates the intestines to keep things moving.

The carbohydrates include many sugars with other groups attached. For example, the important structural polysaccharide **chitin** is made up of amino sugar monomers, containing nitrogen (Figure 3–14). Chitin is an important component of the armor-like external skeletons of arthropods (crabs, insects, spiders, and so on) and of the cell walls of most fungi.

## 3–D  NUCLEIC ACIDS

**Nucleic acids** include the largest, and possibly the most fascinating, biological molecules. Their story takes up most of Chapters 9 and 10.

There are two kinds of nucleic acids. **Deoxyribonucleic acid (DNA)** is the genetic material of all organisms and of many viruses. It contains the organism's genetic information, including the order in which different amino acid monomers are joined to form proteins. DNA also carries the information needed to make the other nucleic acid, **ribonucleic acid (RNA).** RNA then participates in making proteins (Chapter 10). The genetic material of some viruses is RNA rather than DNA.

Nucleic acids are built from subunits called **nucleotides,** which have three distinct parts: a pentose sugar in the center, with one to three phosphate groups joined to one side, and a ring-shaped nitrogenous base on the other side. The bases are derivatives of either the double-ring base **purine** or the single-ring base **pyrimidine** (Figure 3–15). Nucleic acids are acidic because their phosphate groups are more strongly acidic (releasing $H^+$) than their nitrogenous bases are alkaline (accepting $H^+$) (Section 2–H).

The nucleic acids are named after the sugars in their nucleotides: RNA nucleotides contain the sugar **ribose,** and DNA nucleotides contain **deoxyribose** (ribose stripped of one oxygen atom). Nucleotides are named according to which sugar and which base they contain and their number of phosphate groups (Table 3–2).

Besides serving as the monomers of nucleic acids, several nucleotides have other roles. Some are coenzymes, molecules needed for enzymes to work properly. In addition, **adenosine triphosphate (ATP)** and guanosine triphosphate (GTP) supply the energy for many chemical reactions. The nucleotide cyclic adenosine monophosphate (cAMP) is an important intermediary in the function of some hormones (Section 40-C).

In a nucleic acid polymer, the single phosphate group of one nucleotide is linked to the sugar of the next, forming a long backbone of alternating sugar and phosphate residues, with the bases sticking out at one side.

CHITIN

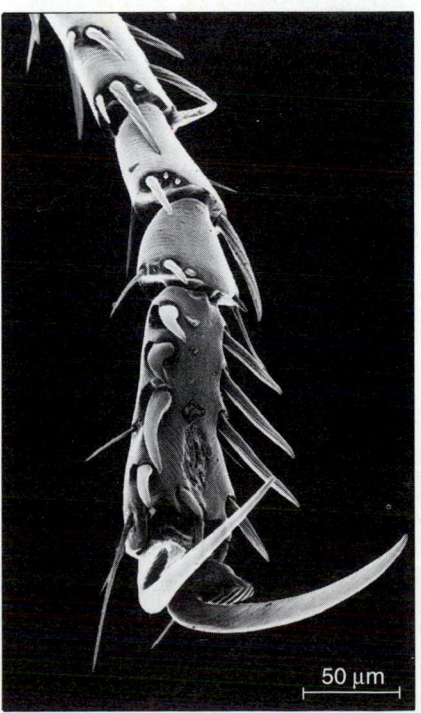

**FIGURE 3–14**  Chitin. This structural polysaccharide is made up of amino sugars derived from glucose. Chitin makes up the external skeletons of insects, spiders, crabs, and their relatives. (b) The foot of a flea, encased in its jointed chitinous armor.   (b, Biophoto Associates)

(b) Flea foot

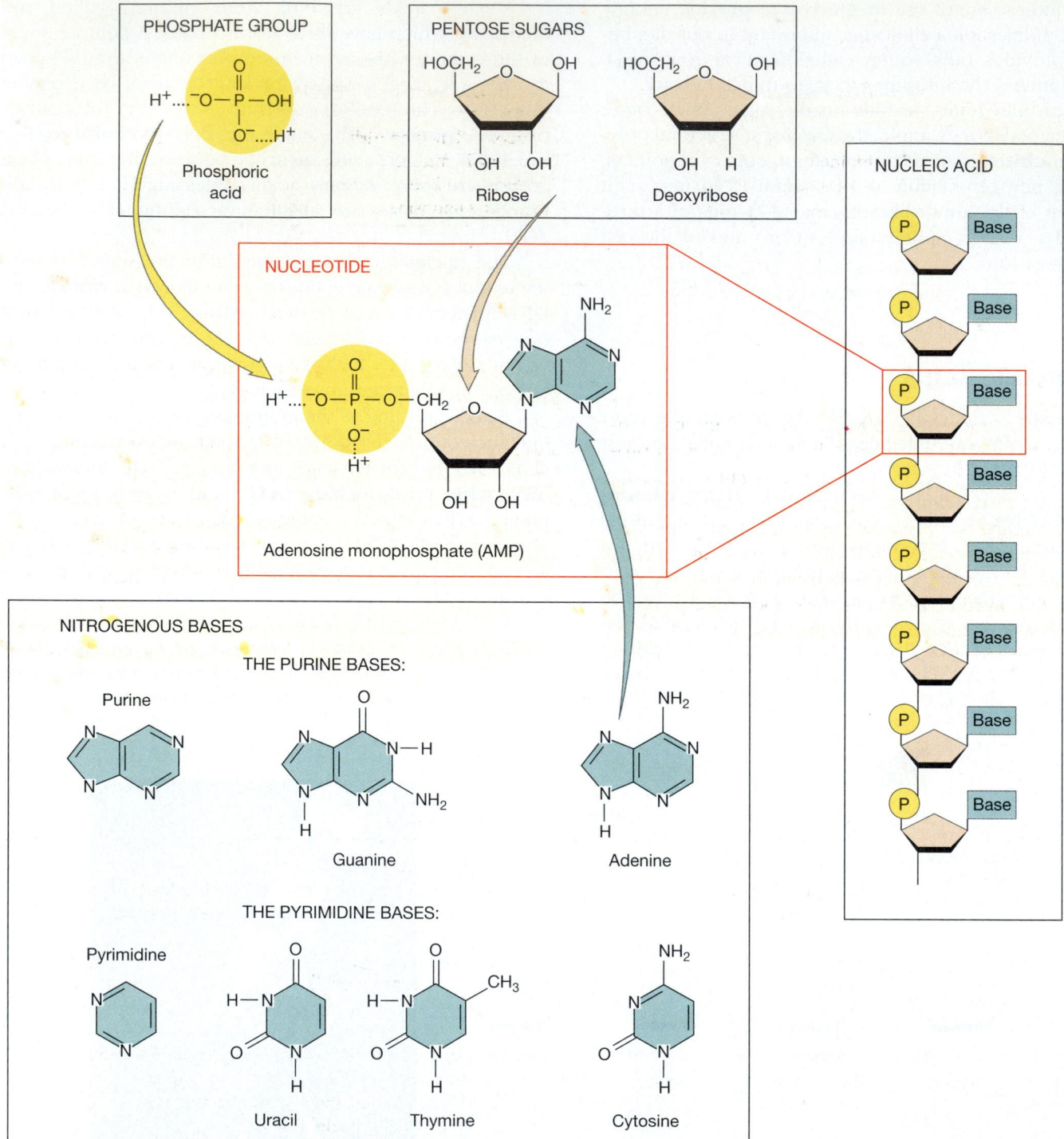

**FIGURE 3–15** Nucleotides and nucleic acids. A nucleotide (center box) has three parts: one to three phosphate groups (yellow), a sugar (brown), and a base (blue-green). The ring-shaped nitrogenous bases (lower box) are derived from one of two parent compounds, purine and pyrimidine. A nucleic acid polymer (far right box) consists of nucleotide monomers, linked with the phosphate group of one attached to the sugar of the next.

**TABLE 3-2**

**The Common Nucleotides**

| Needed in the Synthesis of RNA | Needed in the Synthesis of DNA |
| --- | --- |
| ATP (adenosine triphosphate)* | dATP (deoxyadenosine triphosphate) |
| UTP (uridine triphosphate) | dTTP (deoxythymidine triphosphate) |
| CTP (cytidine triphosphate) | dCTP (deoxycytidine triphosphate) |
| GTP (guanosine triphosphate) | dGTP (deoxyguanosine triphosphate) |

*Each triphosphate can give up one or two of its phosphate groups to become a diphosphate or monophosphate. For example, uridine triphosphate (UTP) can become uridine monophosphate (UMP).

## 3–E  PROTEINS

Proteins make up more than 50% of the dry weight of animals and bacteria. This abundance is in keeping with proteins' many important functions in living organisms (Table 3–3). Structural proteins make up hair, fingernails, silk, and many other things. Soluble proteins in the body fluids of animals include antibodies, which help combat disease, as well as transport proteins such as hemoglobin, the oxygen-carrying protein in red blood cells. Other kinds of proteins are responsible for muscle contraction, and yet others for the detection of chemical substances that we can smell or taste. Insulin is one of a number of small protein hormones. Proteins are also components of biological membranes, where they serve as receivers of chemicals outside the cell and regulate the passage of many substances through membranes. Inside cells, DNA-binding proteins called transcription factors help determine which of the DNA's genetic information is used to make proteins. But the most numerous class of proteins is the enzymes, which speed up specific chemical reactions in living organisms and hence control their chemistry (Section 3–F).

The monomers of proteins are **amino acids.** In a diagram of a generalized amino acid, the chief landmark is the

**TABLE 3-3**

**Some Functions of Proteins**

| Type | Example | Function |
| --- | --- | --- |
| Enzymes | Amylase | Converts starch to glucose |
|  | RNA polymerase | Joins nucleotides into RNA molecules |
|  | Transaminase | Transfers amino group from an amino acid to another molecule |
| Structural proteins | Viral coat proteins | Forms outer covering of virus |
|  | Keratin | Forms hair, nails, horns, hoofs |
|  | Collagen | Forms tendons, cartilage |
| Hormones | Insulin, glucagon | Regulate glucose metabolism |
|  | Oxytocin | Regulates milk production in female mammals |
|  | Vasopressin | Increases retention of water by kidney |
| Contractile proteins | Actin, myosin | Muscle contraction; cell movement |
| Storage proteins | Casein | A nutrient protein in milk |
|  | Ferritin | Stores iron in spleen and egg yolk |
| Transport proteins | Hemoglobin | Carries $O_2$ in blood |
|  | Serum albumin | Carries fatty acids in blood |
|  | Cytochrome c | Transfers electrons |
| Immunological proteins | Gamma globulins (circulating antibodies) | Form complexes with foreign proteins |
| Toxins | Neurotoxins | Block nerve function |
| Membrane receptor | Insulin receptor protein | Binds insulin and stimulates glucose uptake by cell |
| Membrane transport proteins | Glucose transporter | Admits glucose into cell |
|  | Sodium channel | Provides path for sodium ions into cell |
| Transcription factors | Nuclear hormone receptors | Regulate which cells make which proteins |

*Amino Acids*

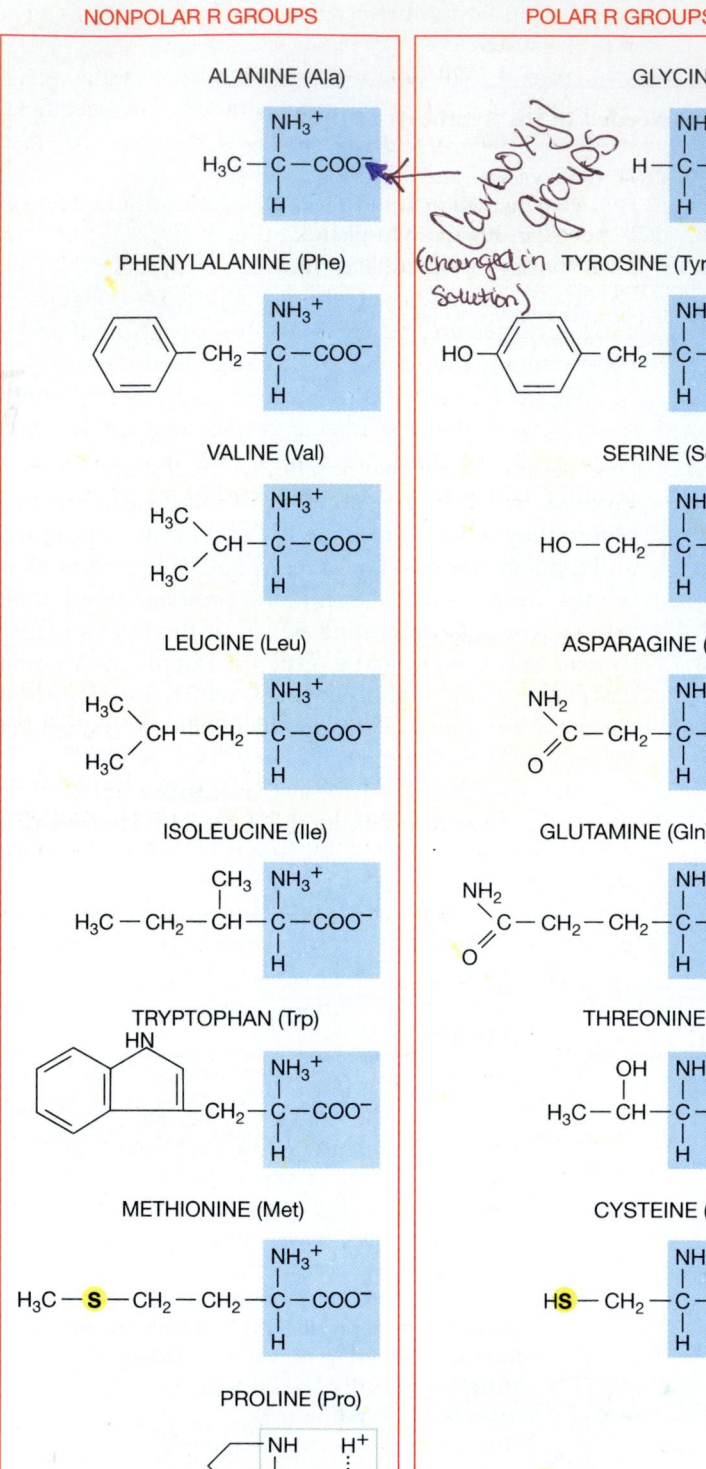

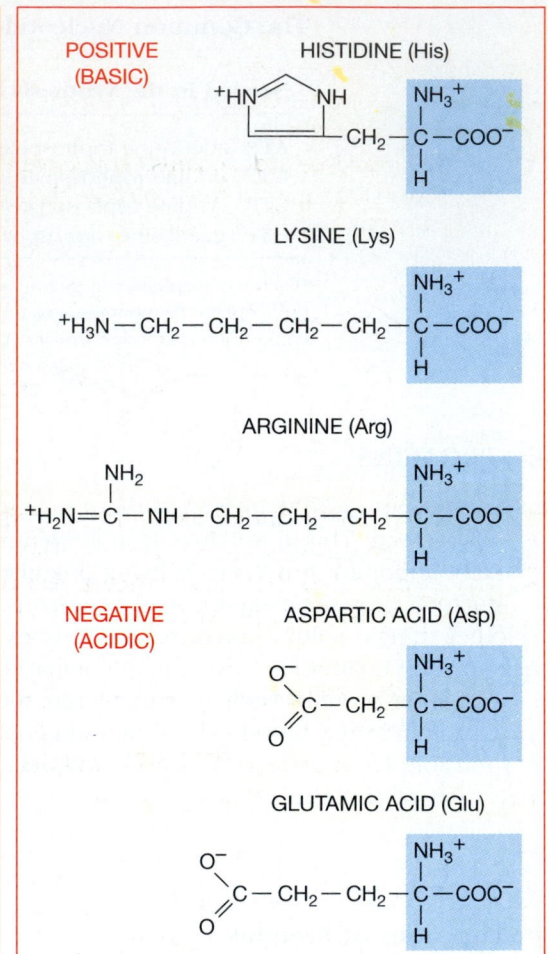

**NONPOLAR R GROUPS**

ALANINE (Ala)

PHENYLALANINE (Phe)

VALINE (Val)

LEUCINE (Leu)

ISOLEUCINE (Ile)

TRYPTOPHAN (Trp)

METHIONINE (Met)

PROLINE (Pro)

**POLAR R GROUPS**

GLYCINE (Gly)

*Carboxyl Groups (changes in solution)*

TYROSINE (Tyr)

SERINE (Ser)

ASPARAGINE (Asn)

GLUTAMINE (Gln)

THREONINE (Thr)

CYSTEINE (Cys)

**IONIC R GROUPS**

**POSITIVE (BASIC)**

HISTIDINE (His)

LYSINE (Lys)

ARGININE (Arg)

**NEGATIVE (ACIDIC)**

ASPARTIC ACID (Asp)

GLUTAMIC ACID (Glu)

**FIGURE 3–16** The 20 amino acids found in proteins. The parts shown in blue boxes are the same for all amino acids. The parts on white backgrounds are the R groups, which make each kind of amino acid different. The R groups may be nonpolar, polar, or ionic (carrying an electric charge, as shown, at the pH inside living organisms). Some R groups contain sulfur (S, highlighted in yellow), nitrogen (N), or ring structures.

$\alpha$ (alpha) carbon atom (shown in red below). Attached to this carbon's four bonds are:

1. An amino group (—NH$_2$).
2. A carboxyl group (—COOH).
3. A hydrogen atom (—H)
4. One of 20 possible side chains, collectively called R groups. This is the part of the molecule that makes one amino acid different from another.

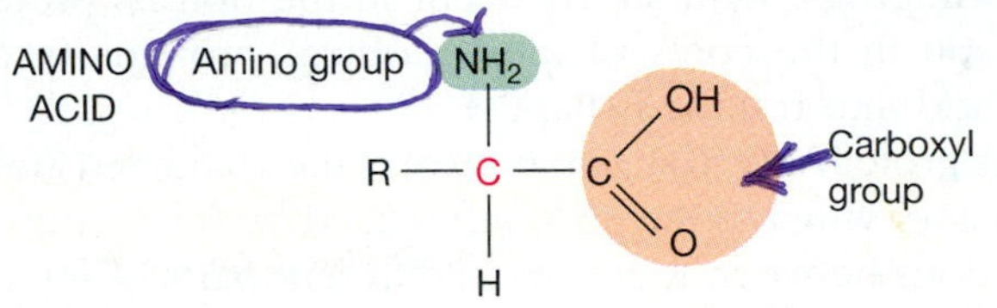

In aqueous solution (including cell and body fluids), the acidic carboxyl group releases H$^+$, and the alkaline amino group takes up H$^+$, so that the molecule has two oppositely charged areas: —COO$^-$ and —NH$_3$$^+$:

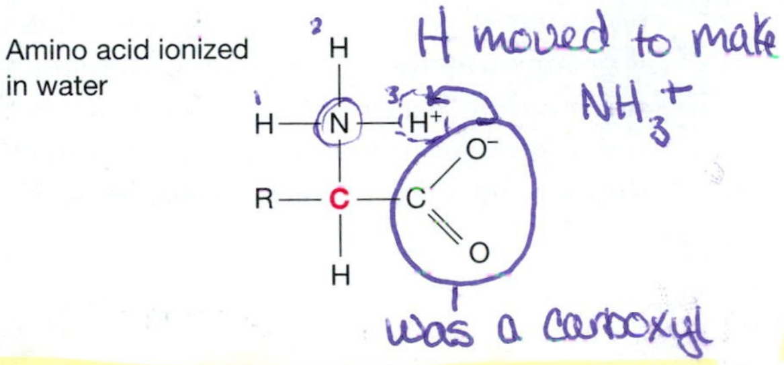

*H moved to make NH$_3^+$*

*was a carboxyl*

Some amino acids carry additional charges because they also have ionic R groups.

The proteins of all organisms are assembled from the same 20 common amino acids. All contain atoms of carbon, oxygen, hydrogen, and nitrogen, and some also contain sulfur (Figure 3–16).

Various organisms also produce several unusual amino acids with special functions. These include some chemical messengers in the brain and some building units in bacterial cell walls.

From the 20 common amino acids, living cells build thousands of kinds of different proteins. The specific amino acids present, and their order, determine the protein's properties.

Two amino acids are joined by a covalent bond called a **peptide bond,** which links the carboxyl carbon of one amino acid and the amino nitrogen of another. The linking of two amino acids forms a **dipeptide** (Figure 3–17). **Polypeptides** are long unbranched strings of amino acids; most contain 150 to 750 amino acid residues.

The amino acid residue at one end of a peptide has an amino group that has not taken part in peptide bond formation; this is called the amino, or N-terminal, end of the peptide. Likewise, the opposite end of the peptide is called the C-terminal.

## Protein Structure

A **protein** is a functional unit composed of one or more polypeptides folded into a characteristic shape that permits them to perform some job. For example, the protein insulin is a hormone that stimulates removal of glucose from the bloodstream; it is made up of two linked polypeptides.

Biochemists analyze the structure of proteins at several levels:

A protein's **primary structure** is its unique sequence of amino acids, dictated by inherited genetic information. For instance, each of insulin's two polypeptides contains particular amino acid residues lined up in a specific order (Figure 3–18).

Most polypeptides have some regions that form regular twists or pleats. These shapes, known as **secondary structure,** result from regular patterns of hydrogen bonds between the amino acid residues in these regions. Each hydrogen bond joins the partially positive hydrogen of the

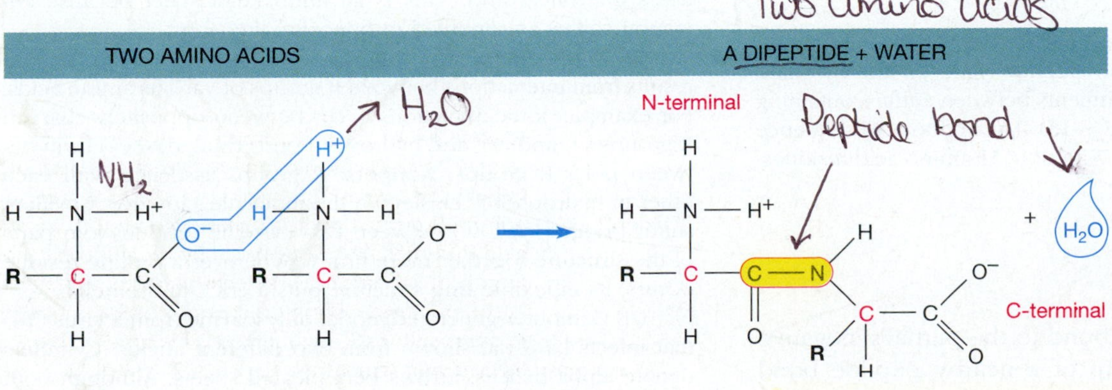

*Two amino acids*

**FIGURE 3–17** Formation of a peptide bond. In these two amino acids, the alpha carbon atoms are shown in red. The atoms that will form a molecule of water are blue. A peptide bond (highlighted in yellow) is a covalent bond between the carboxyl carbon of one amino acid residue and the amino nitrogen of another.

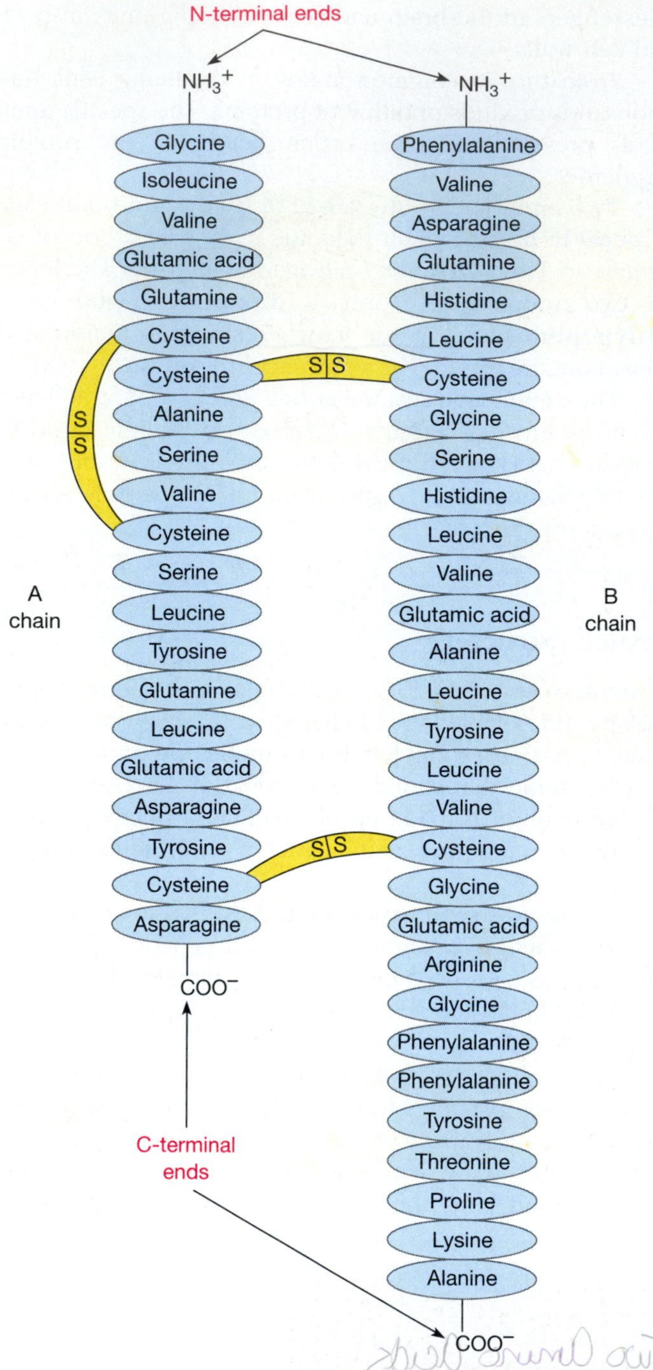

**FIGURE 3–18** Primary structure: insulin. This small protein hormone is made up of two short polypeptide chains, A and B, joined by (yellow) sulfur bridges, attachments between sulfur-containing cysteine amino acids (see Figure 3–16). The amino acid sequence shown is for insulin from cattle. With only 51 amino acid residues, insulin is small as proteins go.

—NH group of one peptide bond to the partially negative oxygen of the —C≡O group of a nearby peptide bond (Figure 3–19). A common secondary structure is a coil called an alpha helix (α-helix), formed by hydrogen bonds between amino acid residues in the same part of a polypeptide chain.

A second important type of secondary structure is the beta (β) pleated sheet. Here, the amino acid residues involved are on different polypeptide chains or in segments of the same chain that bend back and so lie close to one another (Figure 3–19b). Because alpha helices are coiled like springs, they can be stretched, but beta pleated sheets are rigid structures, with the polypeptides lying side by side in flat zigzags. Beta sheets occur in the fibrous protein of silk and in the cores of most globular proteins (proteins that fold into rounded shapes).

A protein's **tertiary structure** is the characteristic overall shape, which is strongly influenced by four types of interactions between R groups in different parts of the chain:

1. Ionic bonds (Section 2–C) between R groups with positive charges and those with negative charges (see Figure 3–16).
2. Hydrogen bonds (Section 2–C) between polar R groups bearing partial positive and partial negative charges.
3. **Hydrophobic interactions,** the clustering together of nonpolar R groups. The main force keeping nonpolar R groups together is the surface tension exerted by the surrounding water (Section 2–G). Water squeezes nonpolar groups together, just as it forces nonpolar oil droplets into the smallest possible space.

**FIGURE 3–19** Secondary and tertiary structure of proteins. This ribbon model shows only the peptide backbone of the enzyme lysozyme from chicken egg white. It hydrolyzes cell walls of invading bacteria.

Secondary structure. (a) An alpha helix and (b) a beta pleated sheet. These are the two main types of regular structure in regions of a protein where hydrogen bonds (blue dashes) form as shown between the hydrogen and oxygen atoms that lie next to peptide bonds (see Figure 3–17). By convention, strands in beta pleated sheets are marked by arrows, which point toward the polypeptide's C-terminal end. This is an antiparallel sheet because adjoining chains are oriented in opposite directions.

(c) Tertiary structure. The three-dimensional folding pattern results from interactions between R groups of various amino acids. For example, ionic attractions occur between oppositely charged R groups (+ and −), and hydrogen bonds (blue dashes) form between polar R groups. Nonpolar R groups associate with each other in hydrophobic clusters in the molecule's interior. Covalent sulfur bridges (yellow) between two cysteine residues join parts of the structure together more firmly. Wherever a proline residue occurs, its inflexible ring structure puts a crick in the helix.

(d) Computer-generated model of lysozyme from a virus (T4) that infects bacteria, shown from two different angles. Cylinders denote alpha helices, arrows beta pleated sheets. Although both chicken and T4 lysozyme perform the same function, their structures are somewhat different.   (Protein Data Bank)

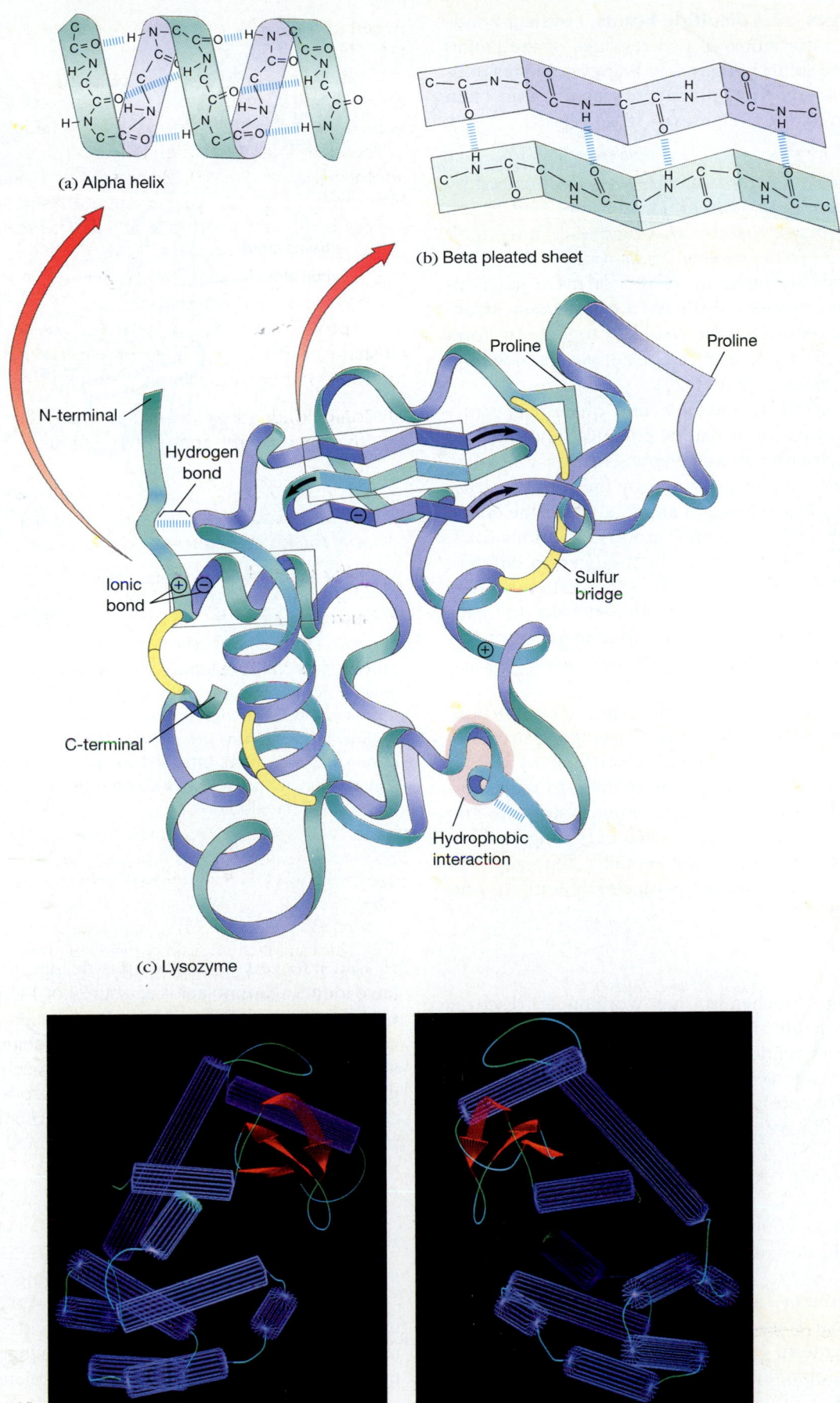

(a) Alpha helix

(b) Beta pleated sheet

Proline

Proline

N-terminal

Hydrogen bond

Ionic bond

C-terminal

Sulfur bridge

Hydrophobic interaction

(c) Lysozyme

(d)

4. **Sulfur bridges,** alias **disulfide bonds,** covalent bonds linking the sulfur atoms of two residues of the amino acid cysteine. Sulfur bridges may link cysteine residues on two parts of the same polypeptide chain or may join one chain to another (Figures 3–18 and 3–19).

A polypeptide's tertiary structure is also influenced by proline (see Figure 3–16, bottom left). Proline contains an inflexible ring structure that causes a rigid kink in the molecule wherever a proline residue occurs (Figure 3–19).

Many proteins are made up of two or more polypeptide chains. These proteins also have a **quaternary structure:** the structure in which the chains fit together to form a complete, functional protein. For example, hemoglobin has four polypeptides (Figure 3–20).

A protein's secondary, tertiary, and quaternary structures are not random. In its natural chemical and physical environment, each protein has a characteristic, but flexible, shape. This is at least partly dictated by the primary structure—that is, by the sequence of amino acids in the chain. So we see that amino acids with R groups of various sizes, shapes, and electrical charges can be arranged in different sequences to "sculpt" chemical spaces in a variety of (tertiary) shapes, which perform many different tasks. In Chapter 10 we shall see how instructions carried by nucleic acids are used to join amino acids in the correct order for each protein.

A protein's tertiary structure determines its function. Polypeptide molecules can be made to lose their structure, and hence their function, by gentle heating or by certain chemical treatments. When they are returned to more normal surroundings, many regain their original structure and function. However, with harsher treatment (stronger chemicals or higher temperature), polypeptides generally lose their shape (and their function) permanently and are said to be **denatured.**

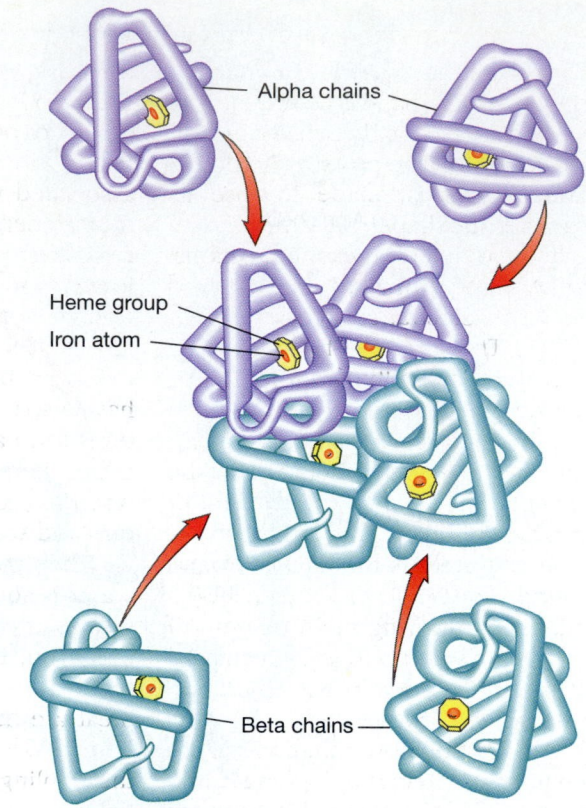

**FIGURE 3–20** Hemoglobin. The complete protein (center) consists of four intricately folded polypeptide chains—two alpha chains (shown from different sides) and two beta chains. Only the outlines of the chains are shown. The heme groups (yellow) are nonprotein structures attached to the polypeptide chains. They contain iron atoms (red), which bind the oxygen molecules transported by hemoglobin.

***Protein Folding*** Biochemists are working to discover how the final structure of a protein relates to the primary structure of its polypeptides. This is important for two reasons. First, it is easier to determine the sequence of amino acids (primary structure) than to analyze a protein's tertiary structure. It would save a lot of work if we could predict the tertiary structure of a protein from its primary structure. Second, recombinant DNA techniques now make it possible to produce artificial proteins with any desired amino acid sequence. If we could reliably predict how they would fold up, we could design proteins to perform many useful functions.

Many of a protein's amino acids can be substituted by others with similar R groups, without having much effect on its folding pattern. For example, the alpha and beta chains of hemoglobin, and the related protein myoglobin, all fold into very similar shapes despite the fact that they have identical amino acids at only 24 of 141 positions. However, having specific amino acids in certain positions is crucial to the protein's overall shape. A change of only one amino acid makes the difference between normal hemoglobin and the kind of hemoglobin responsible for sickle cell anemia (Section 16–B). Protein structures are masterpieces of molecular engineering, fashioned by hundreds of millions of years of natural selection.

## Repeating Protein Subunits

Many structures are made up of repeating protein subunits—identical or very similar pieces (Figure 3–21). These structural protein subunits have regions with complementary shapes and electrical charges that nestle together as the subunits assemble themselves into the completed structure. For

A protein's overall three-dimensional shape is determined by interactions between its amino acid residues. The protein can be denatured—that is, made to lose its characteristic shape—by various treatments that disrupt these interactions. For example, a human hair contains hundreds of strands of the protein keratin. The strength of a hair comes from abundant sulfur bridges linking residues of the amino acid cysteine in neighboring keratin strands. When hair is given a permanent wave, the first solution used breaks the sulfur bridges. This allows the hair structure to be distorted as the hair is wound around curlers. Then another solution is applied, allowing the formation of new sets of sulfur bridges, which hold the keratin strands in the configuration imparted by the curlers.

Perhaps the most common way to denature a protein is by heating it, and the most familiar place this happens is the kitchen. Proteins denatured by cooking are more digestible, perhaps because these less tightly folded molecules give protein-hydrolyzing digestive enzymes better access to the peptide bonds they cleave.

Food contains many globular proteins, with water molecules surrounding the outside of the protein and also lying among some of its internal loops and folds. At high temperatures the protein's atoms, and the associated water molecules, have so much energy that their motion disrupts the hydrophobic, hydrogen, and ionic bonds that give the protein its normal shape. The protein unfolds and the loose ends form new bonds to other protein molecules, which have also been denatured. As the proteins form a meshwork with one another, there is less room available for water molecules, so water is squeezed out, and some of it evaporates away.

When you roast meat, much of the water is squeezed out into the tissue spaces. If you carve a roast fresh from the oven, this juice runs out as the knife slices through, and the slices of meat are quite dry. However, if you let the roast sit for 15 to 20 minutes, the cooling proteins undergo a partial reversal of their denaturation, allowing water to move back among them. The result: moister meat.

The same thing happens when you cook eggs. The unfolding of globular proteins allows them to interact with each other and eventually form one big, tangled, solid network, moist with infiltrating water molecules. But if cooking is not stopped at this crucial moment, the mesh tightens further and squeezes the water out. Overcooking eggs has two possible outcomes: the proteins coagulate in lumps, floating in the squeezed-out liquid, or they form a single, rubbery mass, with the water either separated and floating on top or simply evaporated off altogether.

Another interesting form of denaturation occurs with some of the globular proteins of egg whites. When you whip egg whites, you force air in next to the proteins, exposing them to air on one side and liquid on the other. The proteins' hydrophilic regions are attracted to the liquid, while the hydrophobic regions tend to associate with the air pockets instead. So the proteins unfold, and the hydrophilic stretches of different proteins bond to each other. They form a lacy network, reinforcing the liquid wall of the bubble around the trapped air. This structure is not strong enough to withstand baking, however. In the heat of the oven, the air bubbles expand, and they can burst the protein bonds in their walls, collapsing the whole structure. This does not happen in a properly cooked meringue, soufflé, or angel cake, thanks to still other proteins, which did not participate in forming the foam itself. These proteins are denatured by heat, and they form a stronger meshwork that stabilizes the bubble walls before they can be ruptured by escaping hot air.

Gelatin also consists of a loose protein meshwork—in this case, denatured collagen—with water held in the spaces between proteins. The directions on gelatin packages warn cooks against adding fresh or frozen pineapple. Pineapple contains a protein-digesting enzyme, which chops the gelatin proteins into pieces too short to form a gelled meshwork. Gelatin made with fresh pineapple never sets but remains a soupy liquid. Canned pineapple is fine to use because heat applied in the canning process denatures the enzyme, so it cannot attack the gelatin molecules (Figure 3–A).

**FIGURE 3–A** It's digested! A culinary disaster: enzymes in fresh pineapple have digested the proteins in the gelatin in the center, so that they are too short to form the meshwork that makes gelatin set. Heat used during the canning process denatures these enzymes, and so gelatin containing canned pineapple (right) sets as well as the fruit-free control (left).

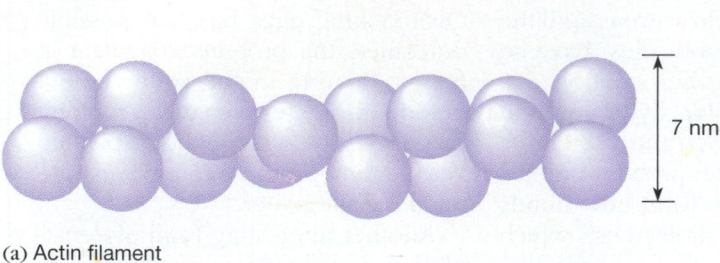

(a) Actin filament

(b) Collagen triple helix

(c) Collagen fibers

**FIGURE 3–21**  Structures made from repeating protein subunits. (a) An actin filament consists of many globular proteins in a twisting pattern. (b) A collagen helix consists of three intertwined polypeptide molecules. (c) Collagen fibers. Each fiber is made up of thousands of triple helices like the one in (b). Actin is the most abundant protein inside cells. Collagen is the most abundant protein in the bodies of vertebrates, lying outside cells.   (c, Jerome Gross)

example, globular molecules of the protein actin fit together in helical filaments that play vital roles in cellular motions such as muscle contraction.

The most abundant protein in animals is collagen, which makes up fibers in skin, bones, tendons, ligaments, and cartilage. Collagen consists of three polypeptides wound around each other into a triple helix. Covalent cross-linkages tie collagen molecules together, forming tremendously strong fibers. Repeating protein subunits also make up hair, the protein coats of viruses (Section 23–A), and the microtubules of cilia and flagella (Section 5–Q).

## 3–F   ENZYMES

Left to themselves, most of the molecules in a living organism would react at negligible rates because most collisions at body temperature do not attain the transition state. In the laboratory, chemists often increase the reaction rate by raising the temperature, which makes the molecules go faster and therefore increases the force of their collisions. Yet organisms carry out their reactions at high rates at the moderate temperatures of living bodies,

This remarkable ability is due to **enzymes,** proteins that act as catalysts. A **catalyst** is a substance that increases the rate of a chemical reaction, taking part in the reaction with-

out being permanently changed. But this is only half the importance of enzymes. The other half is control of *which* reactions occur. Each enzyme catalyzes only a particular reaction, and therefore enzyme action determines which reactions occur in a living organism and which do not. The whole of biological chemistry exists because of these specific enzyme catalysts and is controlled by their activity.

Biochemists have identified almost 2000 different enzymes. Each enzyme catalyzes a particular reaction involving particular reactant molecules, called the enzyme's **substrates.** What are some of these reactions? Some enzymes join monomers into larger molecules or release monomers by hydrolyzing large molecules. Other enzymes transfer functional groups, for example, moving amino or methyl groups from one molecule to another. In phosphorylation, a phosphate group is moved from ATP or another nucleotide to a second molecule, "activating" this molecule so that it can participate in further biochemical reactions. Oxidation-reduction reactions transfer electrons (or hydrogen atoms) from one molecule to another. These are especially important as means of storing or releasing energy during the manufacture and breakdown of food (photosynthesis and respiration), and we shall study them in some detail in Chapters 6, 7, and 8.

Enzymes are given common names according to their substrates and the kinds of reactions they catalyze. For example, RNA polymerase links nucleotides to form RNA polymers, and glucose 6-phosphate dehydrogenase removes hydrogen atoms from glucose molecules that have phosphate groups attached to carbon atom #6. All enzyme names end with the suffix "-ase," although some enzymes still usually go by the "old" names used before modern rules for naming enzymes were invented. Examples are digestive enzymes such as trypsin and pepsin.

As an example of enzyme activity, let us consider a reaction familiar to cat owners, the hydrolysis of urea from a cat's urine into carbon dioxide and ammonia, which gives its characteristic odor to a litterbox in need of cleaning:

$$H_2N-\underset{\underset{O}{\|}}{C}-NH_2 + H_2O \longrightarrow CO_2 + 2 NH_3$$

| urea | water | carbon dioxide | ammonia |

$\underbrace{\hspace{3cm}}_{\text{substrates}}$    $\underbrace{\hspace{3cm}}_{\text{products}}$

This reaction is catalyzed by the enzyme **urease,** produced by bacteria that settle out of the air and grow and reproduce in the litterbox. At room temperature and pH 8 (slightly alkaline), a molecule of urease can catalyze the hydrolysis of about 30,000 molecules of urea per second. Without a catalyst, this reaction would take about 3 million years. So urease makes the reaction go at more than a trillion times its natural rate. Some enzymes work faster than urease, some more slowly.

How does an enzyme speed up a reaction? First let us review how a reaction occurs without a catalyst. For urea and water to react by themselves, they must collide at the proper angle, and both molecules must have enough internal energy to reach a reactive transition state. At room temperature, few molecules have enough energy, and so few reactions occur. The energy needed to reach the transition state is called the **activation energy.**

An enzyme speeds up a reaction by lowering its activation energy (Figure 3–22). It is thought that the enzyme does this by stabilizing the transition state. How? The enzyme actually combines with its substrates (urea and water) and holds them in the correct position for the reaction. Furthermore, the enzyme binds more strongly to the transition state than to the reactants themselves. That is, in the process of binding, parts of the enzyme interact with the substrates' bonds and "loosen" them to the transition state. At the same time the enzyme molds itself more closely to the changing substrate(s). As a result, the enzyme-substrate complex reaches the reactive transition state even with relatively low energy. Thus, enzymes permit organisms to carry out reactions rapidly, at the relatively low temperatures found in their bodies.

An enzyme lowers the energy barriers to both the forward and reverse reactions to the same extent. Hence it does not alter a reaction's equilibrium point (the point at which the forward and reverse reaction rates are equal). It

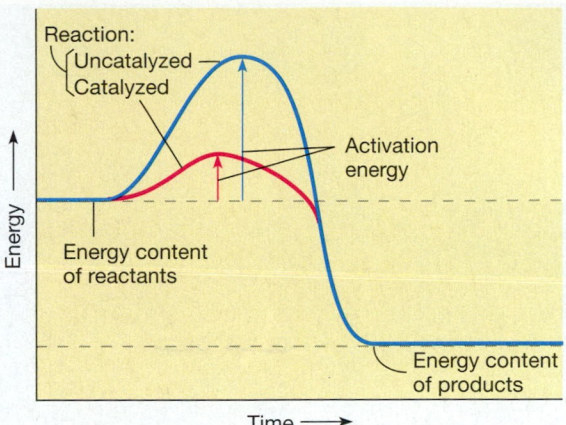

**FIGURE 3–22** Lowering of activation energy by a catalyst. The activation energy is an energy barrier that the reactants must overcome before they can form the product. The reaction proceeds because the products have lower energy than the reactants; the natural tendency is to "slide down the energy hill."

merely increases the speed with which the reaction approaches equilibrium (Section 6–B). Furthermore, since a catalyst—such as an enzyme—is not permanently changed by participating in a reaction, it emerges exactly as it started, ready to catalyze another reaction.

## Enzyme-Substrate Complexes

Enzymes are said to be specific, meaning that each enzyme catalyzes only a particular reaction of one or a few kinds of substrates. This is because the substrate binds specifically to an area called the enzyme's **active site,** a small groove formed as the protein folds up. The size, shape, and electrical charge of amino acid R groups clustered around the active site form a space complementary to the substrate's size, shape, and electrical charge, and this determines which substrates can bind. For example, an active site containing a hydrophobic and a negatively charged group would be expected to bind a substrate molecule having a hydrophobic and a positively charged group, respectively, in places where they can interact with the enzyme's corresponding groups. Thus, complementary enzymes and substrates bind by specific point-to-point interactions. This means that each enzyme can bind only one or a few very similar kinds of substrates and that it always orients them in the same way.

The close correlation of enzyme and substrate shape has been compared to a "lock and key" fit, but this analogy conjures up a misleading image of unyielding hardware. Actually, both enzyme and substrate(s) change shape slightly when they combine, a phenomenon known as "induced fit" (Figure 3–23). The most flexible part of the enzyme is the active site. Biochemists speculate that this flex-

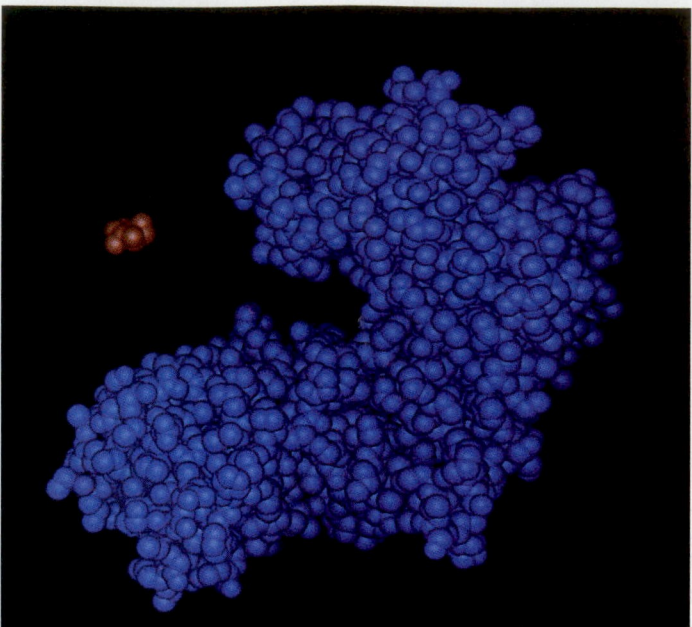

(a) Enzyme and substrate molecules (space-filling models)

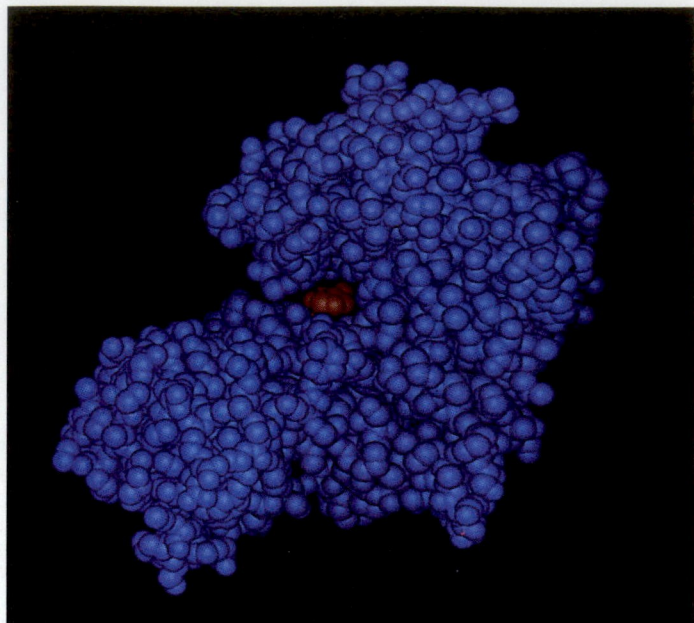

(b) Enzyme with substrate bound at active site.

**FIGURE 3–23** Induced fit between enzyme and substrate. (a) A glucose molecule (orange) approaching the enzyme hexokinase (blue), which catalyzes the addition of phosphate groups to hexose sugars. (b) Glucose has bound at the active site, which closes snugly around this substrate. (Thomas A. Steitz, Yale University)

ibility permits the active site to mold itself around the substrate, squeezing out the water molecules that lie between them so that the two bind over a large contact surface. When the reaction is over, the enzyme releases the products and resumes its original shape, ready to catalyze another reaction.

Some enzymes do not bind their substrates unless their active sites contain additional ions or nonprotein molecules. **Prosthetic groups** are attached to the protein by covalent bonds and are therefore effectively part of the enzyme. We have already seen an example of a prosthetic group: the iron-containing heme group of the protein hemoglobin, which happens to be a transport protein, not an enzyme (see Figure 3–20). Substances held to the protein by other kinds of bonds are called enzyme **cofactors.** Some cofactors are ions, such as nickel ($Ni^{2+}$) in urease or zinc ($Zn^{2+}$) in RNA polymerase; other cofactors are nonprotein organic molecules, known as **coenzymes.** Many vitamins are coenzymes or parts of coenzymes. We need very little of them because coenzymes are not destroyed in a chemical reaction but, like enzymes themselves, can be used over and over again.

## Factors That Affect Enzyme Activity

*Enzyme Characteristics*   A single enzyme molecule can process only a limited amount of substrate in a given time. If the concentration of substrate is low, enzyme molecules spend some time empty, waiting to encounter and bind a substrate. We can increase the rate of the reaction by adding more substrate. However, eventually we will reach a point when all the enzyme molecules are bound to substrate. Now the reaction rate depends, not on how fast substrate molecules bind to enzymes, but on how fast the enzymes can process bound substrate. When this happens the reaction rate approaches a maximum limit, called **$V_{max}$** for "maximum velocity" (Figure 3–24).

This rate can be expressed as the turnover number, that is, the number of reactions catalyzed by one enzyme molecule per unit time. We have seen that the turnover number for urease at room temperature is 30,000 per second. The enzyme carbonic anhydrase, which combines carbon dioxide and water into carbonic acid ($H_2CO_3$), has a very high turnover number: 600,000 reactions per second. A very slow enzyme is lysozyme, which catalyzes only one reaction every two seconds (see Figure 3–19).

Another characteristic often used to describe an enzyme is its **$K_M$**, defined as the substrate concentration at which the reaction rate is one-half the maximum rate (Figure 3–24). $K_M$ is a measure of the enzyme's affinity for the substrate—that is, how readily enzyme-substrate binding occurs. A low $K_M$ value means that the reaction can proceed rapidly even at low substrate concentration. This generally happens in the case of enzymes that bind their substrates very tightly. An enzyme that has more than one substrate can have very different $K_M$ values for each substrate.

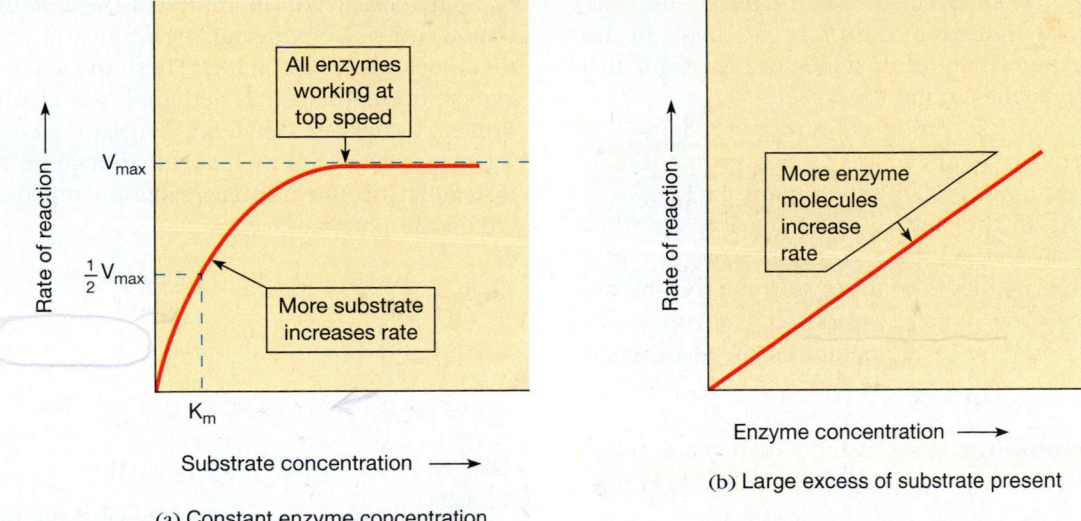

(a) Constant enzyme concentration

(b) Large excess of substrate present

**FIGURE 3–24**  Effect of substrate and enzyme concentrations on the rate of an enzyme-mediated reaction. Raising the concentration of either increases the rate at which substrate is converted to product, but at higher substrate concentrations all enzyme molecules are working at top speed, and the reaction curve levels off (a). $V_{max}$ is the maximum velocity of the reaction at a given enzyme concentration. $K_M$ is the substrate concentration at which the reaction reaches half its maximum rate ($\frac{1}{2}V_{max}$); the greater the affinity of the enzyme for the substrate, the lower the $K_M$. (b) Raising the enzyme concentration raises the reaction rate indefinitely, since it increases the likelihood of a collision between enzyme and substrate molecules.

**pH**  The pH of the surrounding solution affects the activity of enzymes. Enzymes are electrically charged because many of their amino acid residues bear R groups that ionize in aqueous solution. The pH of the solution may change these charges. For example, the $H^+$ ions in an acid solution tend to combine with an enzyme's negatively charged R groups, neutralizing them. This may disrupt ionic bonds in the enzyme's folding pattern (see Figure 3–19) or change the active site so that it does not bind substrate so well. This is why pH affects enzyme activity (Figure 3–25).

Most proteins work best when the pH is approximately neutral, but some are adapted to extreme pH. For instance, organisms living in the acidic or alkaline mineral springs of Yellowstone National Park have enzymes that work at the pH of their particular environment. Digestive enzymes of the human stomach work best at an extremely acidic pH, around 1.5 to 2.0. When these enzymes pass into the small intestine with the food, they become inactive because sodium bicarbonate is secreted into the small intestine, raising the pH to about 8.

**Temperature**  The rate of enzymatic reactions is also affected by temperature. At higher temperatures, molecules move faster, collide harder and more often, and so are more likely to react. (This is true whether the reaction is catalyzed by an enzyme or not.) On the other hand, temperatures above about 60 °C permanently destroy enzymes' three-dimensional structure, denaturing them so that they can no

longer function. Heating preserves food by destroying the enzyme activity of organisms that cause decay.

At the other extreme, chemical reactions proceed slowly at low temperatures because molecules move so slowly that

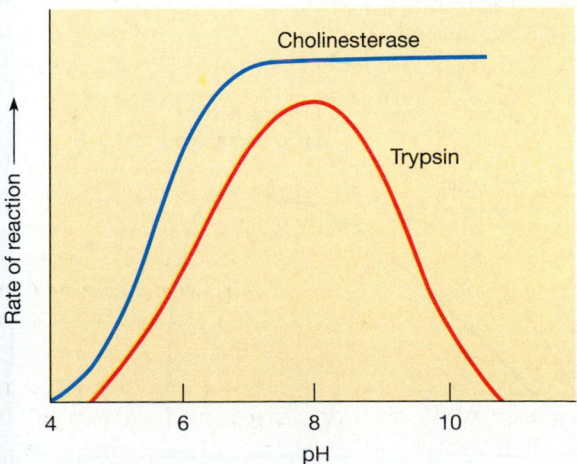

**FIGURE 3–25**  Effect of pH on the activity of two enzymes of mice. Trypsin is a digestive enzyme that hydrolyzes proteins in the small intestine. It works best (that is, its activity is highest) at the pH of 8 found in the intestine. Cholinesterase hydrolyzes substances that are important in the nervous system. The graph line shows that it works well over a range of basic pHs but quickly loses activity when the pH is acidic.

few productive collisions occur between enzyme and substrate molecules. Refrigeration also preserves food, in this case by slowing the activity of enzymes in organisms that cause decay, or enzymes in the food itself.

Natural selection has produced enzymes adapted to function in the range of temperatures an organism normally encounters. For example, some of the organisms known as cyanobacteria live on the surface of glacial ice, and they are adapted to temperatures close to the freezing point of water (0 °C). Other members of this group inhabit the hot springs of Yellowstone, which, besides having unusually high or low pH, may be at temperatures of 80 to 85 °C (Figure 3–26).

▶ ***Inhibitors*** An **inhibitor** is a substance that binds selectively to a specific enzyme and decreases its reaction rate. Many drugs and toxins act as enzyme inhibitors.

**Competitive inhibitors** are compounds so similar in structure to an enzyme's substrate that they compete with substrate molecules for the enzyme's active site. The reaction rate decreases because enzyme molecules spend some time holding inhibitor instead of converting substrate to product. Often, an enzyme's product acts as a competitive inhibitor, since it is quite similar to the substrate. Competitive inhibition can be overcome by supplying a vast excess of substrate, so that the enzyme is virtually certain to bind substrate rather than inhibitor.

Ethylene glycol (in antifreeze) and methanol (wood alcohol) are poisons because enzymes in the body convert them into toxic substances. These molecules mimic the enzymes' normal substrate, ethanol, the alcohol in alcoholic drinks. Victims of ethylene glycol or methanol poisoning are given ethanol. With luck, ethanol keeps the enzyme tied up catalyzing the normal reaction until the body has excreted the potential toxins.

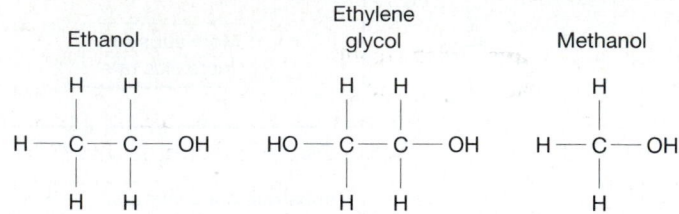

Penicillin is an example of a substrate mimic that binds irreversibly to an enzyme's active site, permanently blocking the enzyme. Normally, an amino acid residue in the enzyme's active site forms a temporary covalent bond with the substrate during the reaction. Penicillin enters the reaction and makes the covalent bond but then fails to break free. It kills bacteria because they need the now-inactive enzyme to make their protective outer walls.

**Noncompetitive inhibitors** bind to the enzyme at sites other than the active site. They distort the enzyme's shape so that it no longer functions.

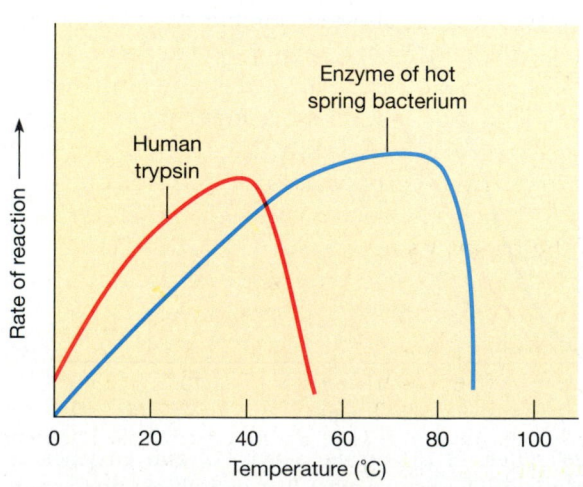

**FIGURE 3–26** Effect of temperature on two enzymes. The enzymes normally function at different temperatures: 37.6 °C in the human body and about 80 °C in hot spring bacteria, such as the ones forming the colorful bands in this photo of a hot spring. Each enzyme is most active in the temperature range it normally encounters. The curves fall off steeply at high temperatures as the enzymes are denatured and their function completely destroyed.

## 3–G  METABOLISM

A living organism is a busy biochemical factory. Here organic monomers such as sugars, amino acids, and nucleotides are converted from one form to another, new macromolecules are built and old ones dismantled, and the energy required to run these processes is extracted from food molecules. Each reaction is catalyzed by an enzyme.

An organism's **metabolism** is the total of all these biochemical reactions. The reactions are organized into **metabolic pathways,** each having several enzyme-mediated steps such that the products released by one enzyme are the substrates of the next enzyme. Each metabolic pathway carries out a particular overall task. For example, in animals the pentose phosphate pathway converts a six-carbon sugar (glucose 6-phosphate) into a five-carbon sugar (ribose 5-phosphate), which is needed to make nucleotides. This pathway uses four different enzymes. Other metabolic pathways have more or fewer enzymes.

The various metabolic pathways interconnect in a staggeringly complex pattern reminiscent of the street map of a sizeable city. This metabolic map has busy "main highways" and lightly used "side streets"; pathways that supply substrates to various branches and pathways that merge as several molecules are assembled into one; even "traffic circles" with multiple entry and exit routes where substrates and products are shunted into various pathways as needed.

A living organism contains an enormous number of substrates, products, and enzymes, not to mention bystanders such as structural molecules. Furthermore, the concentration of each substrate is very low. These tiny amounts of so many different substrates must all be channeled into the appropriate metabolic pathways, filling the organism's needs with no shortages nor surpluses. How can these intricate biochemical traffic patterns be controlled?

One way of bringing order into this chaos is to confine different metabolic pathways to specific locations. This may keep a metabolic pathway separate from others that might divert its substrates or undo its accomplishments, and also helps it work faster. For instance, the enzymes that catalyze a series of reactions may be arranged in large **multienzyme complexes** (Figure 3–27). This increases efficiency because the product released by one enzyme is quickly bound as substrate by its neighbor, instead of diffusing away. Many multienzyme complexes are attached to or embedded in biological membranes or other structures. Membranes may also enclose compartments where particular kinds of molecules are confined and concentrated.

### Control of Metabolic Pathways

Most metabolic pathways are controlled by turning on or off the first enzyme in the pathway. This speeds or slows the pathway's output of product.

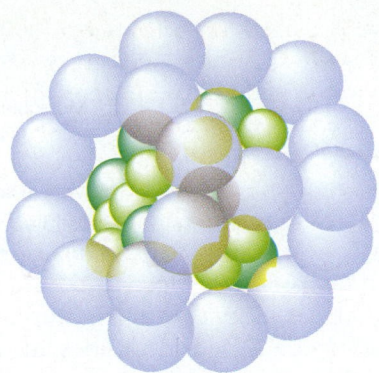

**FIGURE 3–27**  A multienzyme complex. This cluster contains three different kinds of enzymes, shown in different colors. There are many copies of each enzyme in the complex. This is the pyruvate dehydrogenase complex, which catalyzes the reactions summarized in Figure 7–6.

The first enzyme in a pathway is usually **allosteric** (allos = other; stereos = space). This means that it can exist in both active and inactive forms, with somewhat different shapes, and it changes from one form to another.

Most allosteric enzymes have at least one **regulatory site,** which binds molecules that regulate the enzyme's activity. The binding of regulatory molecules changes the enzyme's shape and hence its activity.

The active form of the enzyme binds substrate at its active site and converts the substrate into the form needed by the pathway's next enzyme. The active enzyme is converted to an inactive form when another kind of molecule, known as an allosteric inhibitor, binds to a regulatory site elsewhere in the enzyme. Often the end product of the metabolic pathway is the allosteric inhibitor that fits the regulatory site. If the pathway is making too many final product molecules, some of them bind to the regulatory sites of some of the allosteric enzyme molecules. These enzymes are turned off by being converted to the inactive form, and so the entire pathway's production line slows down. Thus, the final product of a metabolic pathway can act as an inhibitor of the pathway's first reaction.

Later, as the pathway's product is used up by other pathways, the opposite allosteric transition occurs. More and more end product molecules unbind from enzyme regulatory sites, and these enzyme molecules revert to the active form, sending more molecules along the pathway. This example illustrates the common metabolic principle of **negative feedback control:** the regulation of the rate of a process as the level of its product feeds back on an earlier step (Figure 3–28).

The opposite kind of control also occurs: inactive allosteric enzymes are turned on when stimulatory molecules bind to their regulatory sites. In this case, the stimulatory

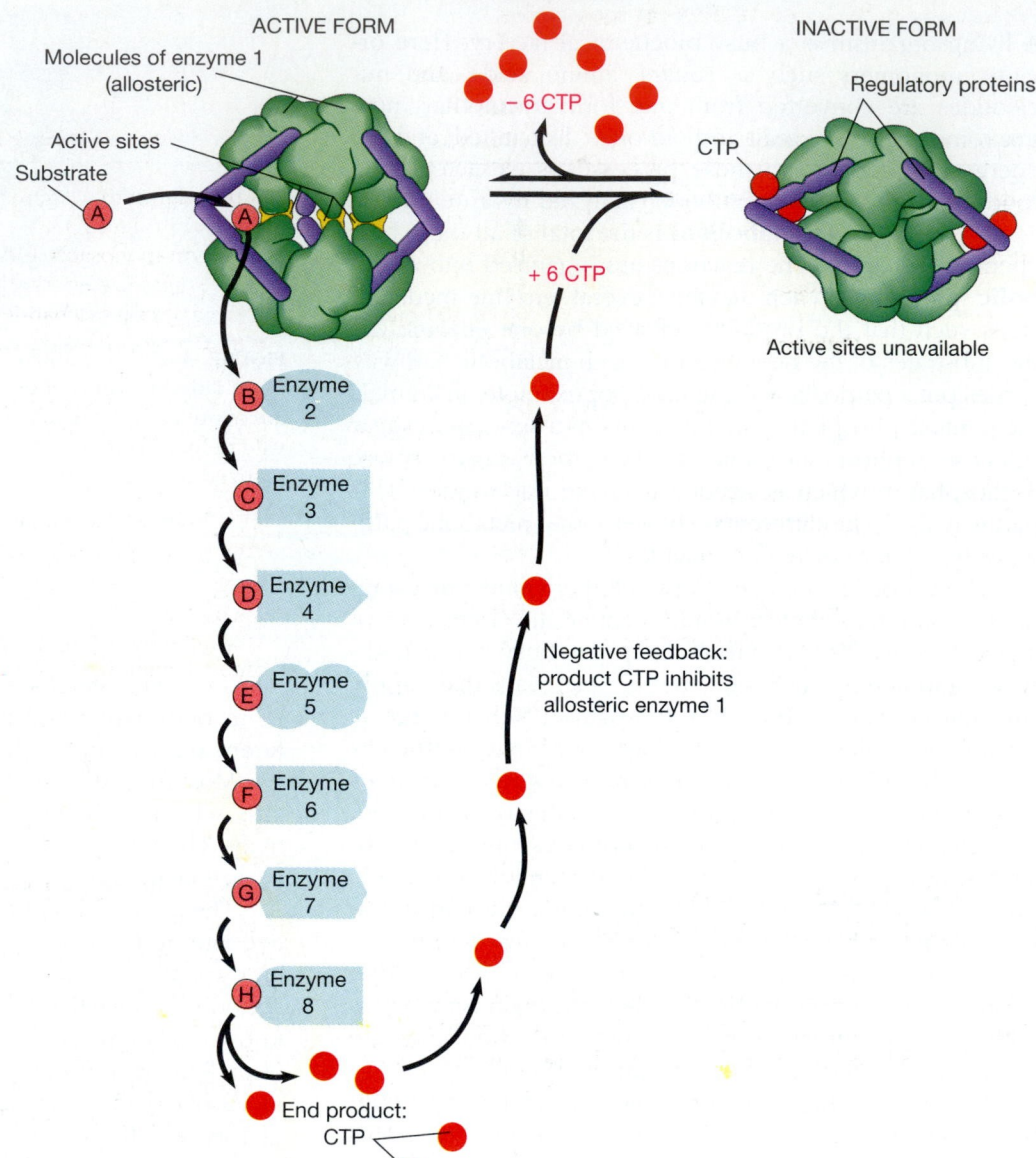

**FIGURE 3–28** Regulation of a metabolic pathway. The pathway that makes the nucleotide cytidine triphosphate (CTP) is regulated by turning on and off protein complexes containing six copies of enzyme 1 (aspartate transcarbamoylase), which is allosteric. The raw material for the pathway is this enzyme's substrate, A, which is converted to B, C, D, and so on, by a series of enzymatic reactions. The end product CTP acts as an inhibitor of allosteric enzyme 1. Hence, when excess CTP accumulates, some of it feeds back and combines with regulatory proteins in the complex, which shut down enzyme 1, thereby slowing the operation of the entire pathway. ATP competes with CTP for the regulatory sites.

molecule is often the substrate for the pathway's first reaction. Other substrate molecules then bind to the active site and start through the pathway. Hence, a buildup of the pathway's starting material turns the pathway on.

The molecules that turn allosteric enzymes on and off are not always the substrates or products of the metabolic pathway. During the course of evolution, many allosteric enzymes have come to be controlled by quite different substances, which may be molecules from other metabolic pathways, hormones, inorganic ions, or regulatory proteins.

In keeping with their exquisite ability to respond to changes in the chemical environment, allosteric proteins have complex structures. All known allosteric proteins have more than one polypeptide subunit, often several copies of the allosteric enzyme as well as other protein subunits that act as regulators. These regulatory proteins may themselves be allosteric, playing their roles only when they have bound (or released) appropriate chemicals.

Some enzymes become activated after another enzyme operates on them, making a covalent change. For example, many enzymes switch to the active form when another enzyme attaches a phosphate group to them. Many enzymes involved in digestion and in blood clotting are produced in an inactive form. They become active when another enzyme hydrolyzes off part of the peptide chain.

The end result of all this is an efficient system for coordinating the overall flow of traffic on the highways and byways of metabolism.

**T A B L E    3 – 4**

**Relative Sizes of Some Biological Molecules**

| Group | Substance | Molecular Mass (Daltons) | References or Comments |
|---|---|---|---|
| Lipids | Glyceryl tristearate | 890 | A triacylglycerol (Figure 3–6) |
| | Cholesterol | 373 | A steroid (Figure 3–8) |
| Carbohydrates | Glucose | 180 | Common monosaccharide (Figure 3–10) |
| | Sucrose | 342 | Table sugar (a disaccharide) (Figure 3–11) |
| | Starch | up to 100 million | A storage polysaccharide (Figure 3–12) |
| | Cellulose | 50,000 to 2,500,000 | A structural polysaccharide, 300 to 15,000 glucose residues (Figure 3–13) |
| Nucleic acids | Adenosine monophosphate | 347 | A nucleotide (Figure 3–15) |
| | DNA | 1.7 million to hundreds of millions | Contains instructions for primary structure of proteins |
| Proteins | Amino acids | 75 to 204 | (Figure 3–16) |
| | Insulin (cow) | 5700 | Two polypeptide chains (Figure 3–18) |
| | Hemoglobin (human) | 4,500 | Four polypeptide chains + four prosthetic groups; carries $O_2$ in blood (Figure 3–20) |
| | DNA polymerase I | 109,000 | Single polypeptide chain; enzyme that repairs DNA molecules |
| | Ribulose bisphosphate carboxylase | 550,000 | Sixteen subunits; enzyme of photosynthesis (Section 8–F); most common enzyme on Earth |

## 3–H  COMPARING CLASSES OF BIOLOGICAL MOLECULES

The molecules discussed in this chapter vary enormously in size. The monomers—fatty acids, monosaccharides, nucleotides, and amino acids—all have molecular masses of [Small]

**T A B L E    3 – 5**

**Chemical Composition (Excluding Water) of the Bacterium *Escherichia coli***

| Type of Molecule | Percent of Total Dry Mass | Comments |
|---|---|---|
| Small molecules | 10 | Inorganic ions, monomers, coenzymes |
| Polysaccharides and lipids | 16 | Protective outer wall and membrane; some glycogen stored inside cell |
| DNA | 4 | One or two molecules per bacterium; each molecule is about 1 mm long and highly folded; the bacterium itself is only about 0.002 mm long |
| RNA | 20 | About 3000 different kinds |
| Proteins | 50 | About 2500 different kinds: about $\frac{1}{3}$ structural proteins, $\frac{2}{3}$ enzymes |

less than 400 (found by adding the atomic masses of all their atoms; see Table 2–1). Most lipids are also small; in fact, the common lipids are not even considered polymers because they have so few subunits. Polysaccharides are much larger, and each polysaccharide may vary in size a great deal. In contrast, every living organism contains many different nucleic acid and protein molecules, each with its own characteristic size (Table 3–4).

The four classes of biological molecules are also found in different amounts in living organisms. In keeping with their many and crucial functions, it should be no surprise that proteins are the most abundant organic compounds inside most cells. As we become more familiar with the roles of the many substances in living organisms in succeeding chapters, we shall gain a better understanding of the relative abundances shown in Table 3–5.

## SUMMARY

Aside from water, the main chemical components of living organisms are organic molecules. Organic molecules are based on carbon, a versatile element that can form molecular skeletons of myriad sizes and shapes. The functional groups attached to the carbon skeleton determine the molecule's solubility and the kinds of reactions it undergoes.

Most of the organic molecules in a living body are macro-molecules. These polymers are made up of residues of many monomers. A polymer contains monomers that are either identical or variations on a basic molecular plan.

Enzymes catalyze the linking of monomers into larger molecules and the hydrolysis of these large molecules back into their component monomers by the addition of a water molecule between the subunits, which thus become separated.

Biological molecules fall into four main groups: lipids, carbohydrates, nucleic acids, and proteins. Unlike members of the other three groups, lipids are nonpolar and so do not dissolve in water. Lipids do not form molecules large enough to be called polymers, but the other three groups contain polymers formed from monomers:

|  | | **Joining together** | |
| Group | Monomers | **Hydrolysis** | Polymers |
| Carbohydrates | Monosaccharides | ⇌ | Polysaccharides |
| Proteins | Amino acids | ⇌ | Polypeptides |
| Nucleic acids | Nucleotides | ⇌ | DNA, RNA |

1. Lipids and carbohydrates are composed mainly of carbon, hydrogen, and oxygen, but the nonpolar lipids contain much less oxygen than the polar carbohydrates. Some lipids and carbohydrates are important energy-storage compounds that may be metabolized to release energy. Some lipids are important hormones. Structural lipids are components of all biological membranes.

Structural polysaccharides include cellulose in plants and chitin in arthropods.

2. Nucleic acids and proteins direct an organism's growth, activity, and reproduction. Nucleic acids contain the elements carbon, hydrogen, oxygen, nitrogen, and phosphorus. They contain instructions for building proteins, help assemble amino acid monomers into proteins, and supply energy to many biochemical reactions.

3. Proteins contain carbon, hydrogen, oxygen, nitrogen, and some sulfur. Important proteins include enzymes, structural and transport proteins, membrane proteins, DNA-binding proteins, and hormones.

4. Enzymes are catalytic proteins. By lowering the energy of activation, enzymes enable organisms to carry out specific chemical reactions quickly at the relatively low temperatures of their bodies. Each of the 2000 known enzymes mediates particular reactions between specific substrates. The activity of enzymes is affected by enzyme and substrate concentration, cofactors, pH, temperature, and inhibitors.

5. Enzymatic reactions are organized into the various metabolic pathways that convert one kind of molecule to another, build up or break down polymers, and break down food to release energy. Production in metabolic pathways is often under negative feedback control: allosteric enzymes at the beginnings of metabolic pathways are activated or inactivated according to the organism's metabolic needs. Still other enzymes are activated by covalent changes.

## Self-Quiz

Make a summary table of this chapter for yourself by filling in the blanks numbered 1 to 13. (For example, in #1, fill in the class that contains the elements C, H, O, N, P; in #2, name that class's monomer subunits, and so on.)

**T A B L E   3 – A**

**Summary of the Major Classes of Biological Compounds**

| Class | Chemical Elements | Monomer Subunits | Main Roles |
|---|---|---|---|
| 1. _____ | C, H, O, N, P | 2. _____ | 3. a. _____ <br> b. _____ |
| 4. _____ | 5. _____ | Fatty acids, glycerol, and so on | 6. a. _____ <br> b. _____ <br> c. _____ |
| Proteins | 7. _____ | 8. _____ | 9. a. _____ <br> b. _____ <br> c. _____ |
| 10. _____ | 11. _____ | 12. _____ | 13. a. _____ <br> b. _____ |

14. Which of the following is an amino acid?

a.

b.

c.

d.

e.

15. What is the R group of the amino acid in Question 14?
16. Draw a hydrolysis reaction using the molecule below:

17. The molecule shown in Question 16 is a:
    a. fatty acid  c. disaccharide  e. pentose
    b. dipeptide  d. hexose sugar
18. Which of the following is *not* made up of hexose sugar subunits?
    a. sucrose  d. insulin
    b. starch  e. cellulose
    c. glycogen
19. Increasing temperature from 0 to 25 °C (increases/decreases/stops) an enzyme-mediated reaction because _____. Cooking food (increases/decreases/stops) reactions mediated by enzymes in the food because the enzymes become _____.
20. Which of the following statements about enzymes is *true?* Enzymes:
    a. are altered permanently in the reaction they catalyze
    b. make the equilibrium of the reaction more favorable for the organism
    c. increase the energy of the reactant molecules
    d. lower the energy of activation of a reaction
    e. lower the energy of the products

## Questions for Discussion

1. Science fiction tales sometimes feature life forms based on silicon rather than carbon. Silicon is much more abundant on Earth than carbon, and like carbon its atoms can bond to four other atoms. Bonds between two silicon atoms are unstable in the presence of $O_2$, but bonds between silicon and oxygen atoms are extremely stable and difficult to break. What implications would these properties have for silicon-based life forms?
2. You go on a journey, taking your cat along but leaving its (used) litter box at home. Draw a graph to show the rate of hydrolysis of urea to ammonia and carbon dioxide during your absence (see Section 3–F).
3. During the winter, you are too lazy to take your cat's litter box outside to empty it, until finally the stench is so overpowering you take the box outside and leave it in the snow. What happens to the rate of hydrolysis of urea by the enzyme urease, and why?
4. Which molecule in Table 3–1 would be most likely to act as a competitive inhibitor of the enzyme urease?
5. When you cut an apple or banana, phenol oxidase enzymes in the injured areas quickly begin a "wound reaction," which results in the cut surfaces turning brown. Why does sprinkling lemon juice on sliced fruit prevent it from discoloring in this way?
6. Explain why cellulose does not dissolve well in water despite its many polar —OH groups.
7. If most organisms do not produce cellulose-digesting enzymes, and cellulose is the most abundant organic compound on Earth, why aren't we up to our ears in cellulose?

## Suggested Readings

Alberts, B., et al. *Molecular Biology of the Cell,* 2d ed. New York: Garland Publishing, 1989. An advanced cell biology book, excellently illustrated, useful as a reference for Chapters 3 to 11 of this book.

Doolittle, R. F. "Proteins." *Scientific American,* October 1985. One of the many articles in an issue entitled *The Molecules of Life.*

Richards, F. M. "The protein folding problem." *Scientific American,* January 1991.

Sharon, N. "Carbohydrates." *Scientific American,* November 1980. Covers historic and modern discoveries about carbohydrates and their roles in organisms.

Stryer, L. *Biochemistry,* 3d ed. San Francisco: W. H. Freeman, 1988.

# Cells and Their Membranes

## OBJECTIVES

*When you have studied this chapter, you should be able to:*

1. Define and use the following terms correctly:
   homeostasis
   unicellular, multicellular
   selectively permeable
   concentration gradient, electrochemical gradient
   membrane potential

2. Explain what this statement means: cells are the structural, functional, and reproductive units of life.

3. Describe the structure of the plasma membrane, explain why it is called a fluid mosaic, and relate its structure to the ability to carry on the processes mentioned in Objective 4.

4. Define and explain the following processes, and state or identify the characteristics that distinguish them from one another: movement of substances through lipid bilayers; osmosis; facilitated diffusion via (1) protein channels and (2) carrier proteins; active transport; endocytosis; exocytosis.

5. Name or recognize the types of substances that enter the cell by each of the processes listed in Objective 4.

6. Given (1) a description or picture of two solutions separated by a membrane, and (2) the permeability properties of the membrane, predict (a) which solution will have the lower osmotic potential, and (b) the total and net movement of water, solutes, or both in the system.

7. Describe the effect of placing a plant or animal cell in distilled water or in a concentrated salt solution, using these terms as appropriate: osmotic potential, osmotic pressure, isotonic, hypertonic, hypotonic, lysis, plasmolysis, turgor.

8. Describe the functions of the central vacuole and the cell wall in the water relations of plant cells.

9. List the important features of the sodium-potassium pump and give and explain three reasons why it is important to a living cell.

10. Describe and compare the structures of the following specialized membrane areas, and state their functions: tight junctions, adherens junctions, desmosomes, gap junctions, plasmodesmata.

---

I took a good clear piece of Cork and with a Pen-knife sharpen'd as keen as a razor . . . cut off . . . an exceeding thin piece of it, and placing it on a black object Plate . . . and casting the light on it with a deep planoconvex Glass, I could exceedingly plainly perceive it to be all perforated and porous . . . these pores, or cells, were not very deep, but consisted of a great many little Boxes, separated out of one continued long pore by certain Diaphragms . . . Nor is this kind of texture peculiar to Cork only; for upon examination with my Microscope, I have found that the pith of an Elder, or almost any other Tree, the inner pulp or pith of the Cany hollow stalks of several other Vegetables: as of Fennel, Carrets, Daucus, Bur-docks, Teasels, Fearn . . . & c. have much such a kind of Schematisme, as I have lately shewn that of Cork.

With these words, written in 1665, Robert Hooke first reported the existence of cells. What Hooke really saw in the bark of the cork oak tree were nonliving **cell walls** surrounding empty pores that once housed living cells (Figure 4–1). Other early microscopists soon observed cells in all kinds of plants. It turned out that animals too had similar units, but animal cells were harder to see because they do not have cell walls. Observers also reported the existence of many tiny organisms consisting of only one cell each.

More than 170 years later, in 1838 and 1839, botanist Matthias Schleiden and zoologist Theodor Schwann originated what is often called the cell theory:

1. All organisms are made up of one or more cells; and
2. Cells are the fundamental units of life—the smallest entities that can be called "living."

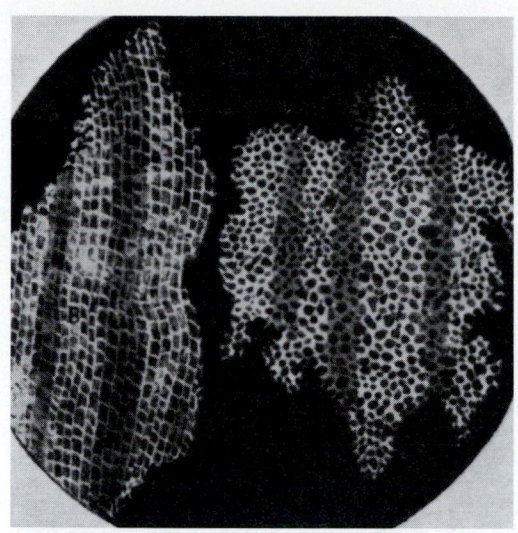

**FIGURE 4–1** Cell portrait. Robert Hooke drew this illustration, published in 1665 in his book *Micrographia*. It shows thin slices of cork as seen through his microscope. The tiny dark pores reminded him of the cells where monks slept in their monastery dormitories.

In 1855 Rudolf Virchow added a third statement:

3. Cells arise only by division of other cells.

In other words, the cell theory states that cells are the fundamental structural, functional, and reproductive units of life.

Why do organisms need cells? Metabolic reactions require a chemical environment different from any found in the nonliving world, with higher concentrations of some substances and lower concentrations of others. Cells are miniature life-support chambers that maintain this special environment. A living cell keeps its chemical composition within narrow limits, a condition called **homeostasis** (homeo = same; stasis = standing). In the controlled environment of a cell, the activities of life occur: acquiring energy; using this energy to maintain the cell's internal chemical environment, to build organic molecules, and to grow; and reproducing by division into two new cells.

Many organisms are **unicellular** (consisting of only one cell). However, a cell's size is limited. This is because the cell must keep up a lively trade in chemicals with its environment, taking in raw materials for its metabolic reactions and expelling its waste products. This trade occurs through the cell's outer surface. As a cell grows, its surface area, which imports and exports materials, does not increase as rapidly as its volume, which uses the materials. Also, the surface comes to lie farther from the cell's innermost areas, and so it takes longer for substances to diffuse between the two. (This time increases as the *square* of the distance from the surface to the center.) At some point, the cell divides into two new cells, each with a lower surface area to volume ratio and shorter surface-to-center distance (Figure 4–2).

Because of these size constraints, large organisms such as animals and plants are **multicellular,** composed of many cells formed by repeated division from one original cell.

All cells must perform certain basic housekeeping tasks. In addition, each cell of a multicellular organism makes a specialized contribution to the body as a whole. For example, a muscle cell in the heart is specialized to contract and help pump blood. Since it lives deep inside the body,

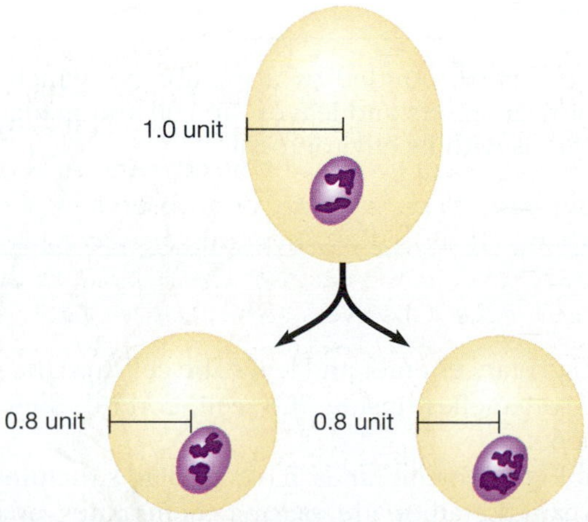

| | Radius | Surface area | Volume |
|---|---|---|---|
| 1.0 unit | 1.0 unit | 12.56 unit$^2$ | 4.17 unit$^3$ |
| 0.8 unit    0.8 unit | 0.8 unit | 16.08 unit$^2$ | 4.17 unit$^3$ |
| | 20% shorter distance to center of each cell | 28% more surface area in two cells | Same volume divided into two cells |

**FIGURE 4–2** Downsizing. When a more-or-less spherical cell reproduces, the same total volume is divided between two new cells, each having a radius 20% shorter. The total surface area of the two cells is 28% larger.

it cannot capture its own food nor obtain oxygen from the air, but must rely on other specialized cells, such as those of the intestines, lungs, and blood, to provide it with food and oxygen. Thus there is division of labor among the cells of a multicellular organism.

Unicellular organisms are also highly specialized for their own ways of life, and many are much more complex than the cells of most plants and animals. A specialized cell is usually distinguished by exaggeration or modification of features common to most types of cells, rather than by possessing structures or chemicals that other cells lack. So we can think of specialized cells as variations on basic cell structure and function.

Most cells have three main parts (Figure 4–3):

1. The **plasma membrane** (also called the **cell membrane**) covers the outside of the cell and controls what enters and leaves. (In plants, this lies just inside the nonliving cell wall.)
2. The **cytoplasm** (cyto = cell; plasma = form, mold) contains water, various salts, and organic molecules, including many metabolic enzymes. It also contains various **organelles,** larger structures that perform particular tasks. Many of these "little organs" are enclosed by membranes very similar to the plasma membrane.
3. The **cell nucleus** (in bacteria, the **nuclear area**) contains the cell's genetic material (DNA and associated RNA and proteins). The genetic material contains directions for making the cell's proteins.

In this chapter we examine the structure and function of the plasma membrane and other biological membranes. In the next chapter we discuss the remaining cell components and look at the differences among the cells of bacteria, animals, and plants.

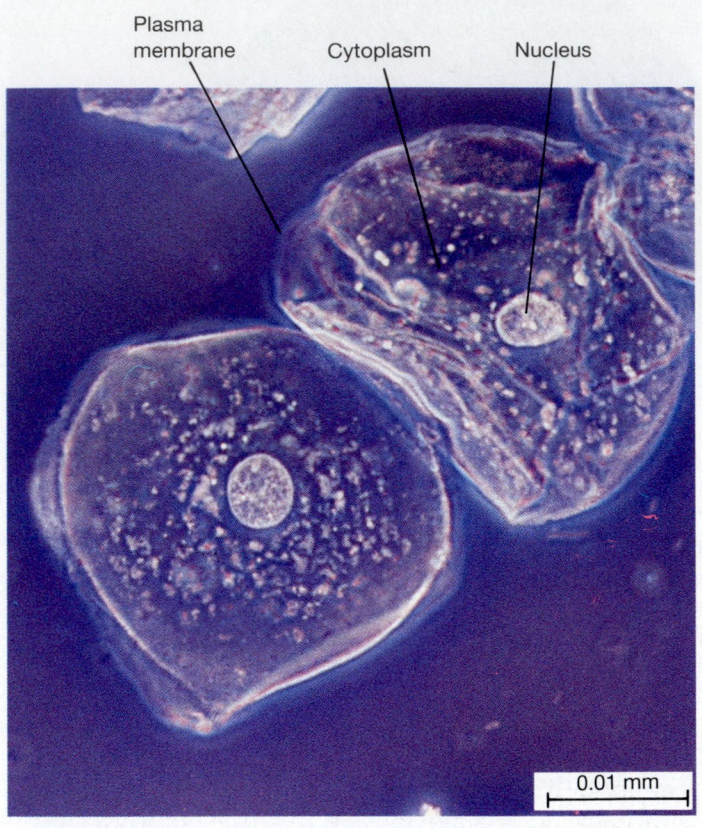

**FIGURE 4–3** Living human cheek cells. Layers of these flattened epithelial cells line the mouth. These cells were removed by scraping the inside of the cheek with a toothpick and then were placed in salt solution on a slide and viewed through a light microscope. Each cell is surrounded by a plasma membrane and contains a large oval nucleus. Cytoplasm occupies the rest of the cell. (Biophoto Associates)

## KEY CONCEPTS

- Cells are the basic structural, functional, and reproductive units of life.
- To remain alive, a cell must maintain homeostasis, keeping its internal chemical composition within the narrow limits suitable for life.
- Every cell is surrounded by a membrane, which helps control what enters and leaves the cell and mediates its interactions with its environment.

## 4–A BIOLOGICAL MEMBRANES

Molecules and ions are in constant, random motion, and so they tend to diffuse down their concentration gradients and spread out uniformly (Section 2–E). A cell cannot permit substances to diffuse in and out freely. But neither can it seal itself off from the world and stop gaining and losing substances by diffusion; it must continuously exchange substances with its environment. The plasma membrane controls what substances enter and leave the cell, and the membranes of organelles inside the cell control their own compartments.

All biological membranes have similar structures and functions, whether they are plasma membranes or membranes of organelles inside the cell. Biological membranes consist mainly of lipids and proteins, with the kinds of these molecules varying from one type of membrane to another. Typically the ratio of lipids to proteins is about half and

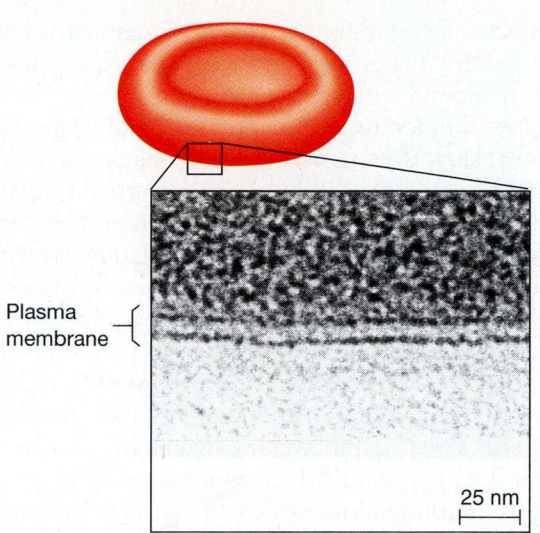

**FIGURE 4–4**   The two-layered structure of the plasma membrane. This photograph was taken with an electron microscope (Section 5–A). Hydrophilic groups at the membrane surfaces show as two dark lines, with the membrane's hydrophobic interior a light area between them.   (Biophoto Associates)

half by mass, but since protein molecules are much larger than lipids, there are many more lipid than protein molecules. Biological membranes are very thin, only about 7.5 to 10 nanometers (nm) thick (1 nm = $10^{-9}$ m = $10^{-6}$ mm).

In electron micrographs, a membrane appears as a continuous double line (Figure 4–4).

## 4–B  MEMBRANE STRUCTURE

The basis of biological membranes is a thin sheet of lipid just two molecules thick, called a **lipid bilayer.** This bilayer forms as a result of the properties of the lipid molecules. In fact, lipids can form bilayers spontaneously in the absence of cells.

The most abundant lipids in biological membranes are phospholipids. Plasma membranes also contain cholesterol (except in bacteria) and **glycolipids** (lipid + carbohydrate). These are all long, asymmetrical molecules with one hydrophilic ("water-loving," or polar) end and one hydrophobic ("water-fearing," or nonpolar) end (Figure 4–5). When surrounded by water, these molecules tend to form groups with their hydrophilic heads exposed to the water and their hydrophobic tails huddled together, minimizing their contact with water. They can do this in one of two ways: either by forming spheres or by forming bilayers, with the hydrophobic tails sandwiched between the hydrophilic heads. Not only do lipid bilayers form spontaneously, but also they tend to form closed sacs, leaving no free edges where hydrophobic tails would have to touch water (Figure 4–6).

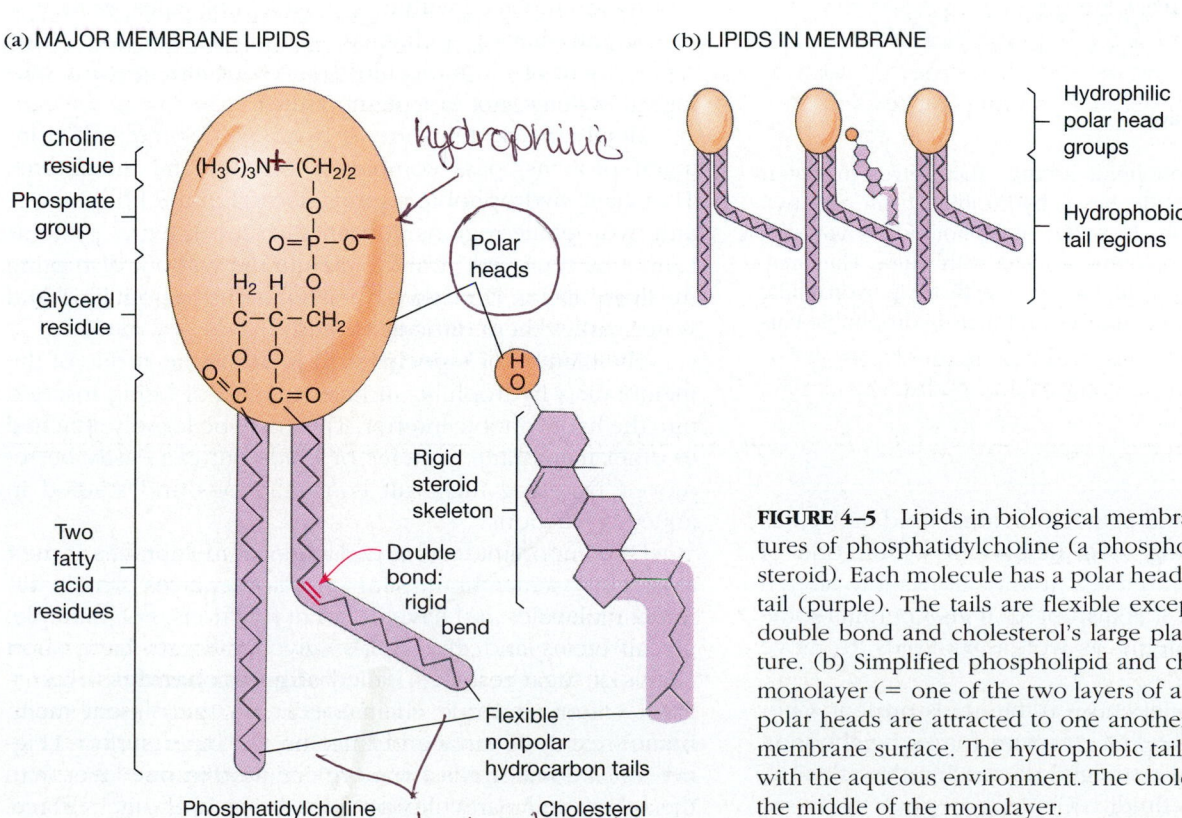

**FIGURE 4–5**   Lipids in biological membranes. (a) Molecular structures of phosphatidylcholine (a phospholipid) and cholesterol (a steroid). Each molecule has a polar head (orange) and a nonpolar tail (purple). The tails are flexible except for the phospholipid's double bond and cholesterol's large plate-like steroid ring structure. (b) Simplified phospholipid and cholesterol molecules in a monolayer (= one of the two layers of a plasma membrane). The polar heads are attracted to one another and to the water at the membrane surface. The hydrophobic tails make minimum contact with the aqueous environment. The cholesterol skeleton stabilizes the middle of the monolayer.

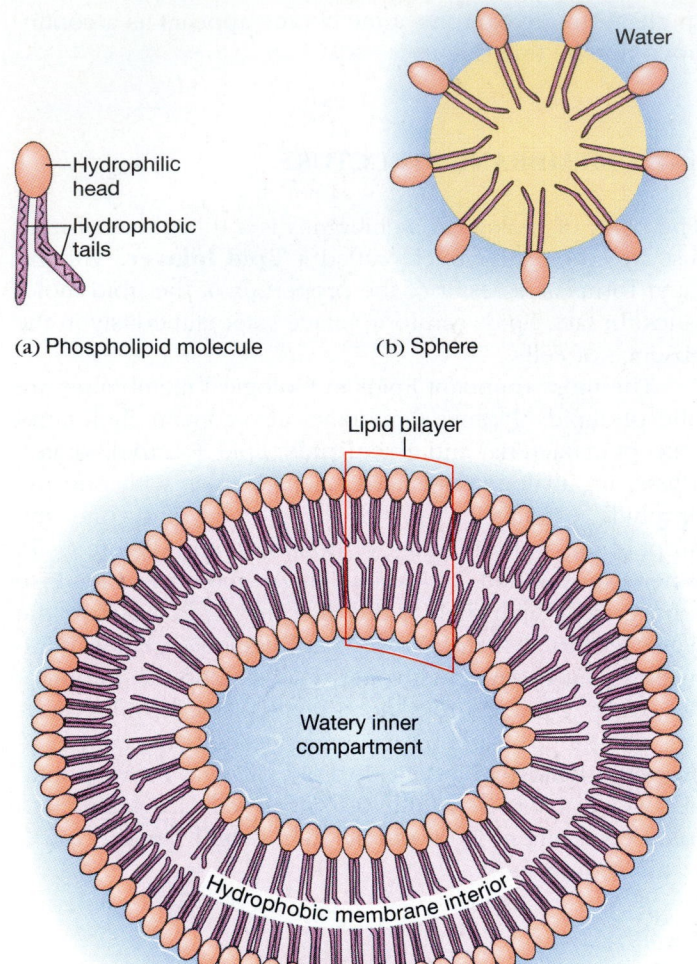

Water

Hydrophilic head

Hydrophobic tails

(a) Phospholipid molecule

(b) Sphere

Lipid bilayer

Watery inner compartment

Hydrophobic membrane interior

Watery environment

(c) Sac enclosed by lipid bilayer

**FIGURE 4–6** How phospholipids arrange themselves in water. (a) A phospholipid molecule has a hydrophilic head and two hydrophobic fatty acid tails. In water, phospholipids orient with their hydrophobic tails in minimum contact with water. They may form (b) a sphere, or (c) a double layer with their hydrophilic heads at the inner and outer surfaces and their hydrophobic tails inside.

Lipid bilayers are fluid in that individual lipid molecules can move about among their neighbors within one of the two layers. The membrane's fluidity allows it to stretch and to reseal itself after it is disrupted. It also permits some of the membrane's proteins, as well as its lipids, to move within it.

Membrane fluidity depends on the lipids present. Cholesterol's rigid, plate-like ring structure forces neighboring hydrocarbon chains to lie straight, decreasing their flexibility and the membrane's fluidity. Cholesterol gives the mem-

brane mechanical stability. Animal cells unable to make cholesterol cannot be grown in tissue culture because their plasma membranes break, and the cells die.

At low temperatures, membrane lipids (like all other molecules) slow down and eventually pack together into a solid. Organisms in cold environments remain flexible because their membrane lipids contain different fatty acids. These may have shorter hydrocarbon chains, which move faster because they are smaller, or more double bonds, which bend the molecules and make them harder to pack together. Although cholesterol decreases membrane fluidity at the normal body temperature of mammals, it increases fluidity in the cold by interrupting other lipids' regular packing pattern. Cells in hibernating mammals produce extra cholesterol, which keeps their membranes fluid.

Lipids are difficult to work with, and so we know very little about the specific lipids in membranes and their roles. Why are certain lipids more common in one lipid layer than in the other? How do specific lipids affect the function of membrane proteins? These are some questions about membranes that biologists are studying.

Figure 4–7 summarizes our current understanding of membrane structure. This is called the **fluid mosaic model** because the lipid layers form a two-dimensional fluid containing a mosaic of various proteins and **glycoproteins** (protein + carbohydrate). Some proteins move around in the membrane; others are relatively immobile because they are attached to structures in the cytoplasm. The membrane is fluid in another sense: its molecules are continually removed and replaced with newly made molecules, each type of molecule having a characteristic rate of turnover. (This replacement of old molecules by new occurs in many biological systems, not just membranes.)

Membrane proteins are of two types. Some, called integral proteins, pass completely through the membrane. They have hydrophobic regions nestled in the lipid bilayer and hydrophilic regions at both surfaces. Integral proteins cannot be removed from the membrane without disrupting the lipid bilayer. Even then, their solubility in both lipid and water makes them difficult to study.

The membrane's **peripheral proteins** lie at one of the membrane's hydrophilic surfaces instead of being inserted into the hydrophobic interior. They may be loosely attached to other membrane proteins or lipids but can easily be removed by using high salt concentrations and studied in aqueous solutions.

One important feature of biological membranes is their asymmetry: a membrane's inner and outer faces contain different molecules and have different functions. For example, glycoproteins and glycolipids (glyco = sugar) have short chains of sugar residues, called **oligosaccharides** (oligo = few). Oligosaccharide chains occur on the plasma membrane's external surface but not on the inner surface (Figure 4–7). Some lipids are more common in one layer than the other, and particular proteins lie at only one surface.

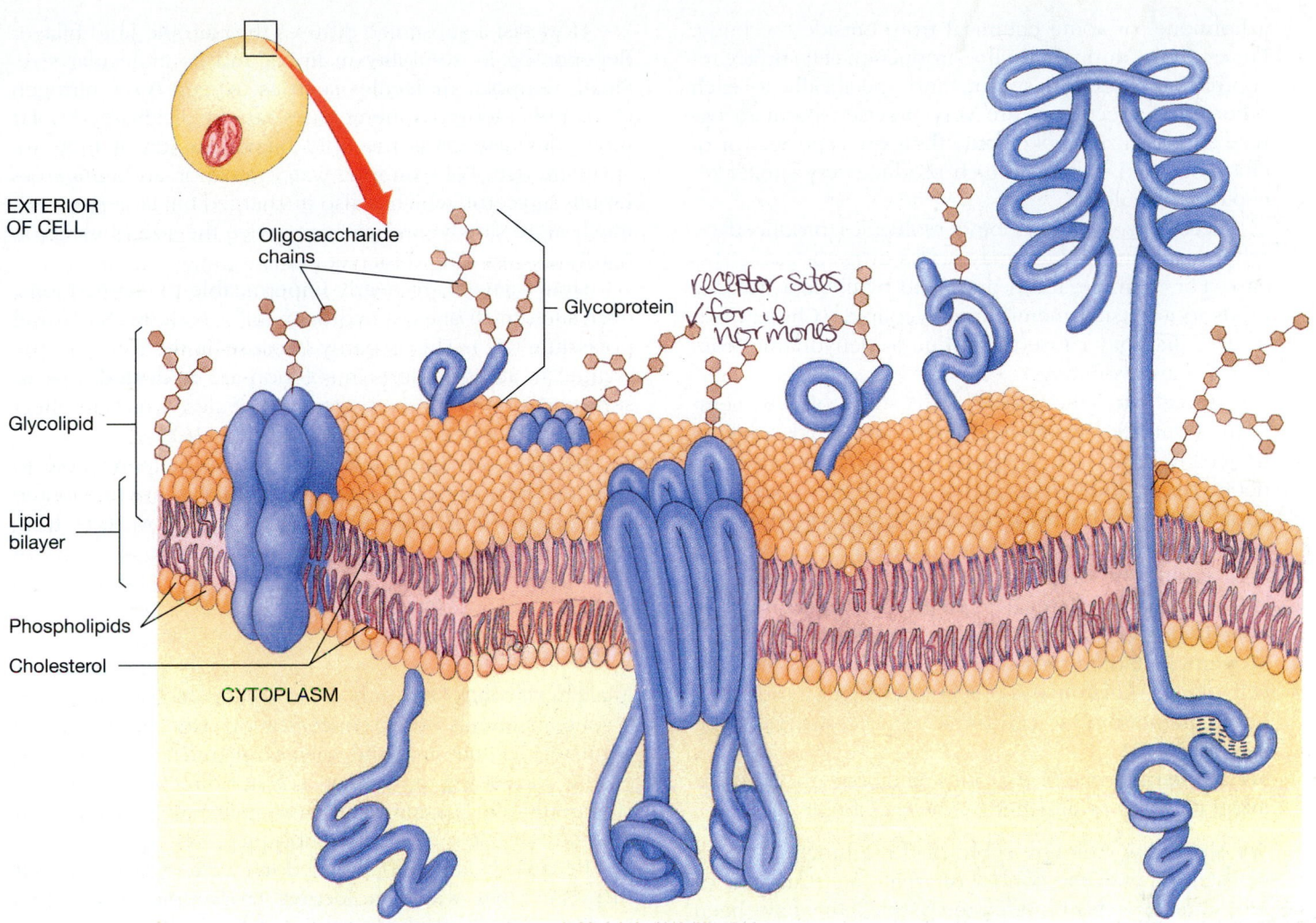

EXTERIOR
OF CELL

Oligosaccharide
chains

Glycoprotein

*receptor sites for il hormones*

Glycolipid

Lipid
bilayer

Phospholipids

Cholesterol

CYTOPLASM

**FIGURE 4–7**  Plasma membrane structure. Lipid molecules form a bilayer, with their hydrophilic, polar heads (orange) at the membrane's inner and outer surfaces and their hydrophobic, nonpolar ends (purple) in the membrane interior. Some proteins and glycoproteins extend completely through the bilayer, whereas others lie at one surface or the other. Some phospholipids and some proteins have oligosaccharides attached on the outer surface of the membrane.

Integral proteins have distinct inner and outer ends and must be aligned right to work properly.

## 4–C   ROLES OF THE PLASMA MEMBRANE

The plasma membrane lies at the frontier between the cell and its environment. This strategic position is the key to many of its roles. The membrane's major functions are:

1. **Containment** and **separation.** The membrane's lipid bilayer forms a continuous, closed physical barrier holding the cell's contents in. The bilayer's hydrophobic interior largely prevents the aqueous solutions of the cytoplasm and the external environment from mingling.

2. **Material exchange.** One of the membrane's most vital roles is maintaining homeostasis by controlling what enters and leaves the cell. The membrane is **selectively permeable,** permitting some substances to pass through it more freely than others, and completely preventing the passage of other substances, to which it is **impermeable.** Both the lipid bilayer and some membrane proteins affect this selective permeability. The membrane also actively moves some substances into or out of the cell.

3. **Information detection.** Many glycoproteins and glycolipids on the membrane surface act as the cell's antennae. These **receptors** detect information, in the form of molecules outside the cell, which they bind selectively (much as an enzyme binds substrates). The molecules bound may be part of another cell's membrane,

a hormone, or some chemical from outside the body. For example, during sexual reproduction cell-surface receptors on eggs and sperm bind specifically to each other. These receptors are very precise: sperm recognize and fertilize eggs of only their own species (or of closely related species, which produce very similar receptor molecules).

Receptors may also bind molecules produced by other organisms, which may act as drugs or cause disease. For example, toxin produced by cholera bacteria binds to a plasma membrane glycolipid. (Cholera is a disease that may cause death due to dehydration from diarrhea and vomiting.)

4. **Identification.** Like an ID card, some of the membrane's surface glycoproteins and glycolipids identify the cell to other cells' receptors as coming from a particular tissue, individual, and sometimes even species.

5. **Attachment** and **reinforcement.** Proteins in the membrane attach to structural proteins both in the cytoplasm and outside the cell. These proteins contribute to the cell's shape, mechanical strength, and attachment to neighboring cells or other structures outside the cell.

6. **Movement.** Some of the structural proteins inside the cell act as a skeleton and push or pull against the membrane, permitting the cell to move or to change shape (Section 5–Q).

7. **Metabolism.** Some of the cell's metabolic enzymes are integrated into or attached to the membrane.

We shall now examine some of these functions more closely, starting with the control of what enters and leaves the cell. Some aspects of membrane permeability have been understood for a long time. However, not until the 1970s did it become clear that substances cross biological membranes in only three ways: some dissolve through the lipid layers; some pass through the lipid layers aided by membrane proteins; and some move within a sac formed from part of the membrane.

## 4–D  DIFFUSION THROUGH THE LIPID BILAYER: SMALL UNCHARGED MOLECULES

Molecules and ions diffuse down their concentration gradients. A membrane's lipid bilayer is relatively permeable to small uncharged molecules, such as those of oxygen and nitrogen. These cross biological membranes by diffusing between the lipid molecules, each moving down its own concentration gradient. In essence, these substances dissolve in the lipid on one side of the membrane and emerge at the opposite face. To separate this process from the effects of membrane proteins, researchers work with artificial lipid bilayers, which behave essentially like lipids in intact biological membranes.

How fast a substance diffuses through the lipid bilayer depends on its solubility in lipids and its molecular size. Small, nonpolar molecules such as oxygen ($O_2$), nitrogen ($N_2$), and ether cross membranes rapidly. Uncharged polar molecules also cross the lipid bilayer rapidly if they are small enough. For example, water, ethanol, and urea cross rapidly; glycerol, which is also uncharged but larger, crosses much more slowly; and glucose, twice the size of glycerol, hardly crosses a lipid bilayer at all.

Lipid bilayers are nearly impermeable to charged ions, even such small ones as hydrogen ($H^+$), sodium ($Na^+$), and potassium ($K^+$). This is partly because of the ions' electric charge. In addition, ions in solution are **hydrated,** that is, surrounded by a layer of water molecules, which in effect makes them much larger (see Figure 2–16).

Because the lipid bilayer is relatively impermeable to ions and to many polar molecules, the plasma membrane prevents most of the cell's water-soluble contents from escaping.

## Osmosis

A cell must control not only the absolute amounts of solutes it contains but also their concentration. Hence it must also regulate its content of the solvent, water. However, cells cannot control the movement of water directly. Water travels through the plasma membrane quite freely—in fact, at a higher rate than any other substance.

Because water and lipids don't mix, it is somewhat surprising that water molecules cross lipid bilayers so readily. This is partly because of the water molecule's small size, but it may also be that the molecule's unique bipolar structure (see Figure 2–12) somehow permits it to pass the membrane's hydrophilic outer layers especially easily.

**Osmosis** is the process by which water moves through a selectively permeable membrane. Osmosis is a special case of diffusion because it involves the movement of a *solvent* (water) rather than a *solute* and because the water is moving *through a membrane.* As usual, the net diffusion of water is down its concentration gradient, from a higher to a lower concentration of *water*—that is, from a dilute to a concentrated solution.

A simple way to demonstrate osmosis is to separate distilled (pure) water from an aqueous solution by a membrane permeable to water but not to the solute (Figure 4–8). As time passes, the volume of the solution increases and that of the distilled water decreases. This happens because water is moving by osmosis from the water, across the membrane, and into the solution.

Why does this happen? Water molecules can cross the membrane in either direction. However, the water molecules in the solution are slowed down in two ways: by collisions with solutes and by electrical attractions to solutes.

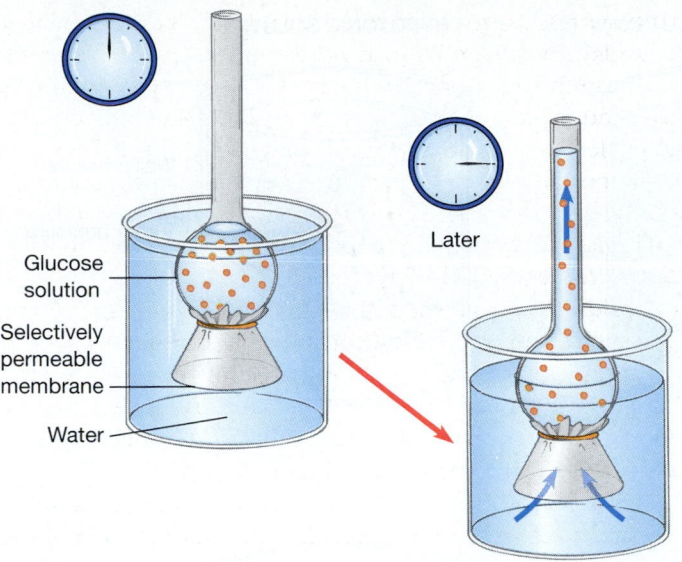

(a) Water moving by osmosis

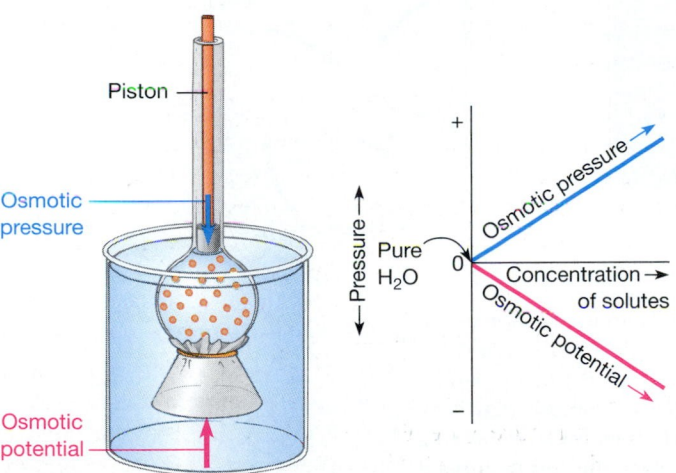

(b) Osmotic pressure and osmotic potential

**FIGURE 4–8** An osmotic system. (a) The bottom of the thistle tube is covered by a membrane permeable to water but not to glucose. The wide end of the tube is filled with glucose solution and immersed in pure water to the same level. Later, the solution has risen in the tube because water has moved in through the membrane. This continues until the weight of solution in the tube pushes water back out at the same rate as it enters. (b) The osmotic pressure of a solution is the pressure that must be applied to the piston to prevent net entry of water. The opposite force, osmotic potential, is the solution's tendency to gain water through the membrane from pure water on the other side. As a solution becomes more concentrated, its osmotic potential decreases.

As a result, more water molecules move into the solution than move out. This raises the height of the solution in the tube. The weight of the solution in the column exerts **hydrostatic pressure,** which builds up until it pushes wa-

ter out at the same rate that it enters. The solution will remain at this level.

We can predict which way water will move through a membrane if we know the osmotic potentials of the two solutions on either side. A solution's **osmotic potential** is its tendency to gain water when separated from pure water by an ideal selectively permeable membrane. A stricter definition of osmotic potential is that it is the negative of the **osmotic pressure,** the pressure we must apply to a solution to prevent it from gaining water when it is separated from pure water by an ideal selectively permeable membrane (Figure 4–8). The osmotic potential of pure water is zero; relative to water, any solution has an osmotic potential expressed in negative terms. The more concentrated the solution, the lower (*more negative*) its osmotic potential and the greater its tendency to gain water from a solution with a higher (*less negative*) osmotic potential. Osmotic potential is the driving force of osmosis, because water moves down its concentration (and energy) gradient, that is, in the direction of the lower osmotic potential.

The osmotic potential in a system depends on the total concentration of particles of all solutes present and on how much the particles attract and slow down the water molecules. Ionic substances dissociate into more than one particle in aqueous solution: NaCl dissociates into two particles, $MgCl_2$ into three, and so on. The higher the concentration of particles in a solution, the lower the concentration of water and the lower the osmotic potential. If the solute particles can pass through the membrane, then the solution's osmotic potential gradually changes as solute particles enter or leave.

Because water moves through the plasma membrane by osmosis, a cell can control its water content only indirectly. The cell can create a difference in osmotic potential across its membrane by moving solutes from one side of the membrane to the other (Section 4–E). Water then enters or leaves the cell, moving by osmosis toward the side of the membrane with the lower osmotic potential.

***Cells as Osmotic Systems***    Cells behave as osmotic systems. A living cell has a selectively permeable plasma membrane, enclosing the aqueous solution inside the cell. To remain alive, the cell must be covered by at least a thin layer of water, which also contains solutes. If the internal and external solutions are in osmotic balance, no net exchange of water occurs between them, and the cell is said to be living in an **isotonic** (iso = same; tonus = tension) solution. Solutions that are isotonic with the body fluids are used for such purposes as washing contact lenses and injecting drugs into the bloodstream.

Figure 4–9 shows why this is important. If the solution outside the cell is made more concentrated, so that the cell loses water to its environment, this external solution is said to be **hypertonic** (hyper = exceeding) to the cell contents.

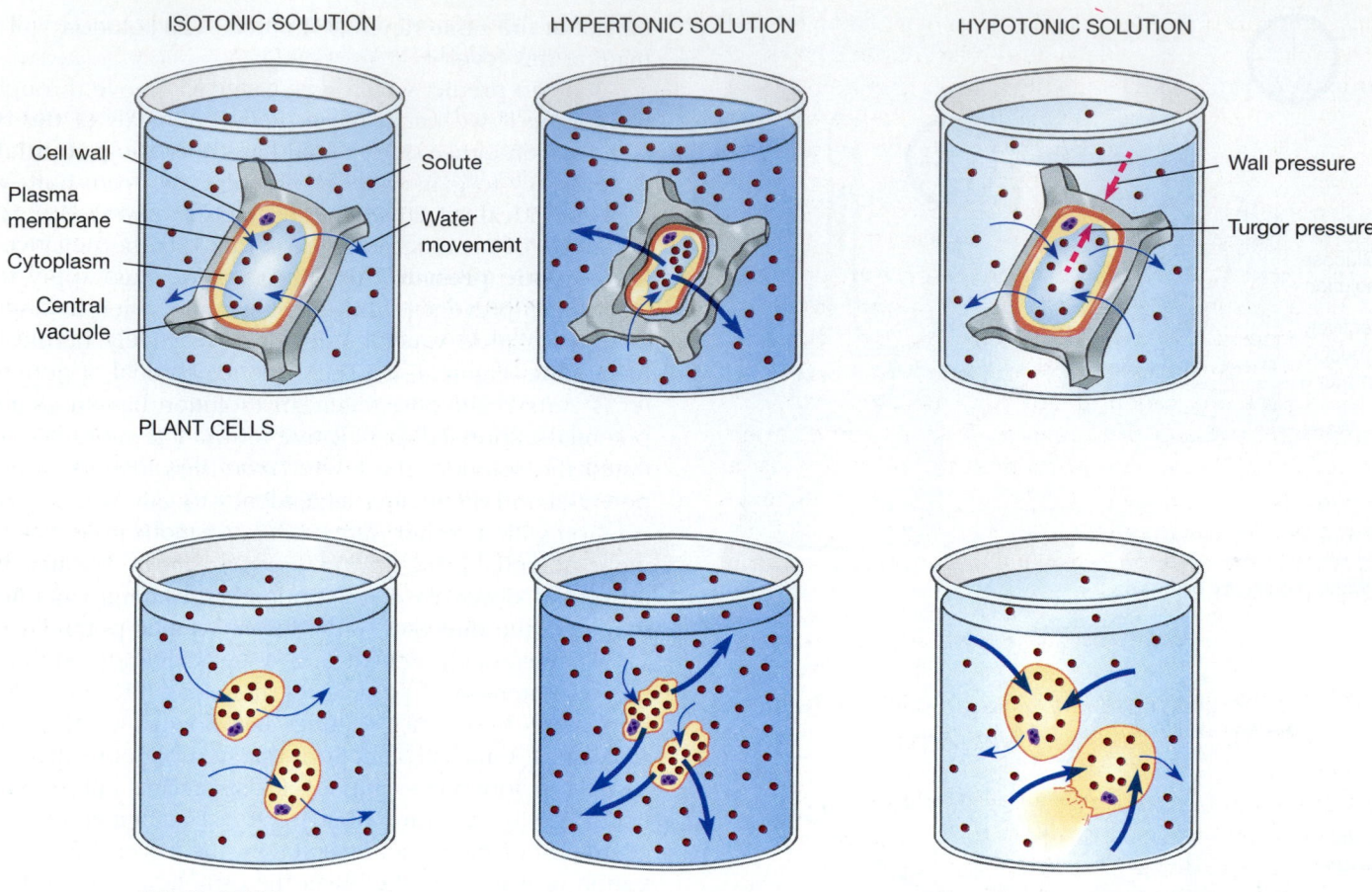

ISOTONIC SOLUTION          HYPERTONIC SOLUTION          HYPOTONIC SOLUTION

Cell wall
Plasma membrane
Cytoplasm
Central vacuole

Solute
Water movement

Wall pressure
Turgor pressure

PLANT CELLS

ANIMAL CELLS

**FIGURE 4–9**  Osmotic behavior of plant and animal cells. (left) Cells in an isotonic solution are in osmotic balance with their environment and lose as much water as they gain by osmosis (blue arrows). Most of the plant cell's water is stored in a large central vacuole, which is surrounded by its own membrane.  (center) In a hypertonic solution, cells lose more water by osmosis than they gain. In the plant cell, the vacuole loses fluid and the cell shrinks away from its wall. (right) In a hypotonic solution, there is a greater tendency for water to enter than to leave the cell. In a plant cell, the pressure of water entering the cell is opposed by the cell wall, so the cell gains only a limited amount of water before the movement of water equalizes. However, an animal cell may swell so much that it bursts.

In a hypertonic solution, a plant cell shrinks away from its wall and the external solution seeps into the space between the plasma membrane and cell wall. This shrinkage is called **plasmolysis,** and the plant cell is said to be **plasmolyzed.**

If a cell is placed in a solution dilute enough for the cell to gain water from outside, this environment is said to be **hypotonic** (hypo = lower) to the cell. Plant cells in a hypotonic solution can gain only a limited amount of water before their rigid cell walls prevent them from expanding further. After this, the **wall pressure** of the cell wall squeezes water back out at the same rate that it enters. In the same hypotonic solution, an animal cell may swell so far that it ruptures the plasma membrane, allowing the cell

contents to escape. This process is known as **lysis** (bursting) of the cell.

Many animals live in fresh water, which is hypotonic to their cells' cytoplasm. Why don't these animals take up so much water by osmosis that they swell up and burst? Most of a freshwater animal's body surface is covered by a layer of rather impermeable material, which retards water uptake. Such layers include the mucus of fish and worms or the chitinous armor of aquatic insects and spiders. In addition, these organisms have well-developed excretory systems that form large volumes of very dilute urine, thereby ridding their bodies of excess water while conserving precious salts (Chapter 35).

*Water Relations in Plant Cells*    Most of a plant cell's volume is occupied by a large, fluid-filled **central vacuole,** surrounded by a selectively permeable membrane. Most plant cells swell or shrink by gaining or losing water from this central vacuole. The **cell sap** inside the vacuole is a solution of salts, sugar, and various proteins. The low osmotic potential of the sap allows the vacuole to take up water by osmosis.

In nonwoody plants, the vacuoles play an important part in supporting the body. The cell walls are strong but not rigid enough to hold the plant upright by themselves. As each cell takes up water by osmosis, it builds an internal fluid pressure, called **turgor pressure,** and a cell exerting turgor pressure against its wall is said to be **turgid.** The cells' turgor pressure helps hold the plant up, just as air pressure in each compartment supports an air mattress. Wilting occurs when a plant loses so much water by evaporation that the cells do not maintain their turgor pressure but hang limply together.

## 4–E    TRANSPORT BY MEMBRANE PROTEINS: POLAR MOLECULES AND IONS

Most molecules that cross biological membranes do so with the help of proteins. Examples include various ions, glucose, and amino acids. These substances move by way of a variety of **membrane transport proteins.** A membrane transport protein contains several alpha helices that form the walls of an aqueous pore through the lipid bilayer. Each protein is more or less specific in the types of substances it ushers through the membrane.

There are two categories of membrane transport proteins. **Channel proteins** provide a pore through which solutes can diffuse across the membrane. The pore's structure permits only solutes of a certain size and electrical charge to squeeze through. **Carrier proteins** actually bind the solutes they transport, much as an enzyme binds its substrate. The carrier is then thought to change shape, opening a path to the opposite face of the membrane, and the solute unbinds, having crossed the membrane.

Channel proteins and some carriers permit substances to move through the membrane down their diffusion gradients. Other carriers engage in active transport, using energy to move substances against their gradients.

### Facilitated Diffusion

In the process of **facilitated diffusion,** a channel or a carrier protein provides a way for a particular solute to move through the membrane down its diffusion gradient. Since the substance crosses the membrane faster than it otherwise would, facilitated diffusion in effect increases the membrane's permeability to the substance.

*Channel Proteins*    Channel proteins form water-filled pores that give solutes a route to diffuse through the bilayer, avoiding the membrane's hydrophobic interior. In laboratory tests, these channels are often not particularly specific but permit the passage of anything of the right size and charge—generally inorganic ions. However, in a living organism most of the ions that fit into the channel are of one particular type. Hence the channels are often named for the predominant ion passing through them.

One well-studied channel admits mostly sodium ions ($Na^+$) into the cell (although potassium ions can also pass through it). Around each entrance of this sodium channel lies a ring of negatively charged amino acids, which attract positively charged ions to pass through the channel and repel negatively charged ions such as chloride (Figure 4–10). This particular type of channel is called a **gated channel** because it behaves as if it has a gate that opens and closes.

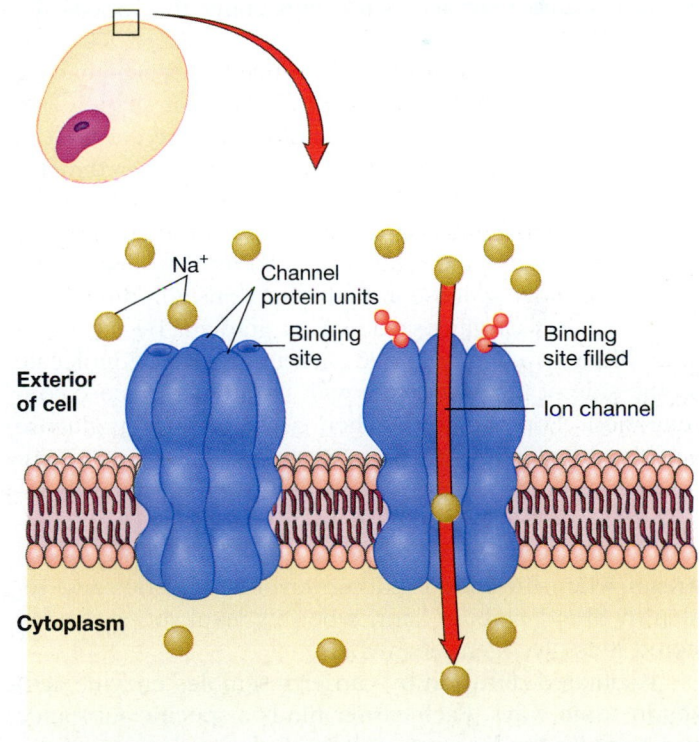

**FIGURE 4–10** Channel proteins. (left) The channel passes through the center of a group of five protein subunits spanning the membrane's lipid bilayer. (right) Two subunits removed to reveal the channel. This is a gated channel from muscle. It opens when it binds chemical messengers (red) released by a nearby nerve cell. This briefly changes the shapes of the channel proteins such that sodium ions rush into the cell through the channel, signaling the muscle to contract.

Part of the channel protein is a receptor that binds a chemical messenger. When this happens, the channel opens briefly, permitting $Na^+$ to diffuse through the membrane. Hence, the membrane's permeability to $Na^+$ changes depending on whether its sodium channels are open or closed. Gated channels are vital to the working of nerves and muscles, among other things (Chapters 37 and 39).

When a defective gene causes a defect in a channel protein, the result may be devastating. Cystic fibrosis is a genetic disorder in which thick mucus clogs the lungs and other organs. It is caused by defective chloride channel proteins, which do not permit chloride ions to leave cells as they should. Normally, water follows chloride by osmosis and gives the mucus its proper, thin consistency.

***Carrier Proteins***    Facilitated diffusion by carrier is more specific than by channel proteins because carriers bind their passengers, usually small organic molecules such as sugars and amino acids. Some biochemists think that facilitated diffusion by carrier is essentially the same as movement by channel proteins, with the distinction that the substance to be transported may act as the messenger that opens the channel.

Studies of glucose carriers show that the carrier changes rapidly back and forth between two different shapes: one is thought to have the glucose binding site open to the cell exterior, the other to the cytoplasm. According to the current model, the carrier binds glucose on one side of the membrane, changes shape, releases the glucose on the other side of the membrane, and reverts to its first shape. The carrier can move glucose in both directions. On the whole, though, it moves glucose down its gradient because it is more likely to encounter and pick up a glucose molecule on the side of the membrane with the higher glucose concentration. For example, when a cell is using glucose quickly, the glucose concentration inside the cell falls. Glucose is then more plentiful outside the cell, and more glucose is moved into the cell than out. On the other hand, cells in the liver not only remove glucose from the bloodstream when the blood glucose level is high but also replenish blood glucose later, when its level drops, having stored it as glycogen meanwhile.

Facilitated diffusion by carrier resembles enzyme activity in some ways. Each carrier binds a specific substance at a specific binding site. When all the molecules of the carrier have bound passenger molecules, that substance is moving through the membrane as fast as it can ($V_{max}$ in Figure 4–11). This contrasts with movement via channel proteins, in which the rate of transfer keeps increasing with the solute's diffusion gradient and may be much faster than carrier transport. In another similarity to enzymes, carriers can be blocked by competitive inhibitors, which compete for the carriers' binding sites. However, the enzyme-substrate analogy is limited, in that carriers usually do not

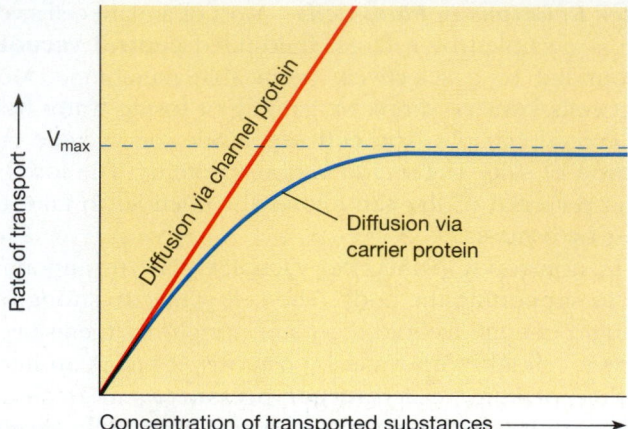

**FIGURE 4–11**    Rates of facilitated diffusion through a membrane. The rate of diffusion through channel proteins depends only on the diffusion gradient of the substance(s) involved on the two sides of the membrane. In contrast, the rate of transport by carrier protein reaches a maximum ($V_{max}$) when the carrier is saturated—that is, when all molecules of carrier are working as fast as they can.

change the structure of the solutes they carry—only their location.

## Active Transport

**Active transport** also uses carrier proteins, but it differs from facilitated diffusion by carriers in that it can move substances against their concentration gradients by the expenditure of energy. The energy comes either from ATP or from a gradient of ions ($Na^+$ or $H^+$) across the membrane. Because there is a strong tendency for ions to move down such a gradient, the gradient serves as a source of energy, much like the energy of a waterfall.

For example, the plasma membranes of many cells contain "calcium pumps," which use ATP energy in active transport of calcium ions ($Ca^{2+}$) out of the cell and so keep the $Ca^{2+}$ concentration much lower inside than outside the cell. Another active transport pump permits some cells in the stomach wall to secrete stomach acid. This pump uses energy from ATP to export hydrogen ions ($H^+$) from the cell into the stomach fluid against a concentration gradient of about a million to one! Still another active transport pump occurs in some cancer cells that are not harmed by chemotherapy (Figure 4–12). Drug-resistant malaria parasites have a similar pump that rids them of the antimalaria medicine chloroquine.

By far the most important and best studied active transport mechanism, though, is the sodium-potassium pump.

***The Sodium-Potassium Pump***    The **sodium-potassium pump,** often called the **sodium pump,** uses energy from ATP to transfer sodium ($Na^+$) out of the cell and potassium

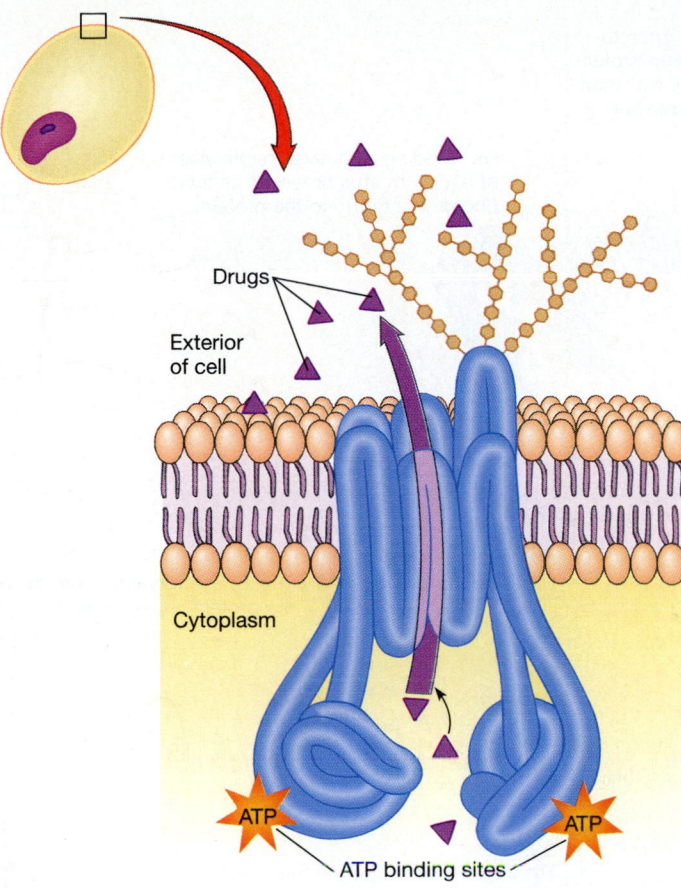

Drugs

Exterior of cell

Cytoplasm

ATP

ATP

ATP binding sites

**FIGURE 4–12** The active transport protein that ejects drugs from cancer cells resistant to chemotherapy. The protein chain folds in such a way that parts of it weave through the membrane 12 times, forming a ring around a central tunnel. ATP attaches to sites where the protein juts into the cytoplasm and provides energy needed to move drugs out through the central tunnel. Unlike most other pumps, this one is nonspecific. It exports various drugs that are not chemically related to each other, making these cells resistant to many drugs.

($K^+$) in, moving both ions against their concentration gradients.

There is more sodium outside than inside a cell. How do we know that sodium is actively transported out of the cell, rather than simply kept out by a plasma membrane that is very impermeable to sodium ions? First, when cells are poisoned so that their enzymes cannot make ATP, sodium enters the cell until its concentration is the same inside as out. Second, the sodium in cells can be labeled by incubating the cells in a solution with radioactive sodium, so that their internal sodium becomes radioactive. If they are then transferred to a medium with nonradioactive sodium, they gradually lose the radioactive sodium to the outside and take in nonradioactive sodium. These experi-

ments show that sodium ions normally move into and out of the cell.

Another experiment shows that the outward movement of sodium requires energy. If red blood cells are stored for a while, sodium enters the cells until its concentration is the same inside and out. If glucose is now added to the cells, they expel the sodium that has entered. Since cells use glucose to supply them with energy, mainly in the form of ATP, the conclusion is that the cells normally use energy to pump sodium out. Based on such evidence, researchers deduced the existence of a mechanism to pump sodium.

Experiments on red blood cells also showed that if the concentration of potassium ions outside the cell is reduced, so is the movement of sodium ions out of the cell. This led to the discovery that the pumping of sodium is linked to the pumping of potassium: for every three sodium ions pumped out of the cell, two potassium ions enter. Furthermore, each exchange of three sodium for two potassium uses energy from one molecule of ATP (Figure 4–13). Because this transport protein also acts as an enzyme that hydrolyzes ATP, the sodium pump is also called $Na^+$-$K^+$ ATPase.

The sodium-potassium pump is enormously important. An estimated one third of all our energy goes to power the sodium pumps in our cells! The ability of nerves to conduct electrical impulses, of muscles to contract, of the digestive tract to absorb food, and of the kidneys to form urine, all depend on the working of this ionic pump.

For example, the sodium-potassium pump plays a major role in controlling the cell's water content. Cells cannot control their water content directly. Instead, they move solutes through the plasma membrane by active transport, and then water follows by osmosis (Section 4–D). Since sodium and potassium are among the most common solutes in living organisms, the pump is often used to move water in this way.

This vital pump has two more important roles, discussed in the next two sections.

***Cotransport*** **Cotransport** is the transport of one substance coupled to that of another. Many active transport carriers are cotransporters powered by the concentration gradient of an ion that is one of the substances transported.

For instance, the $Na^+$ gradient across the plasma membrane powers the active transport of glucose and amino acids into many animal cells. The glucose transporter has binding sites for both glucose and $Na^+$, which it can pick up at either side of the membrane. However, because the concentration of $Na^+$ is usually much higher outside than inside the cell, the carrier is more likely to bind $Na^+$ when it faces outward. But it cannot transport $Na^+$ through the membrane until a glucose molecule also binds. Then the carrier rapidly changes shape and releases both $Na^+$ and glucose into the cell. The driving force for this transport is

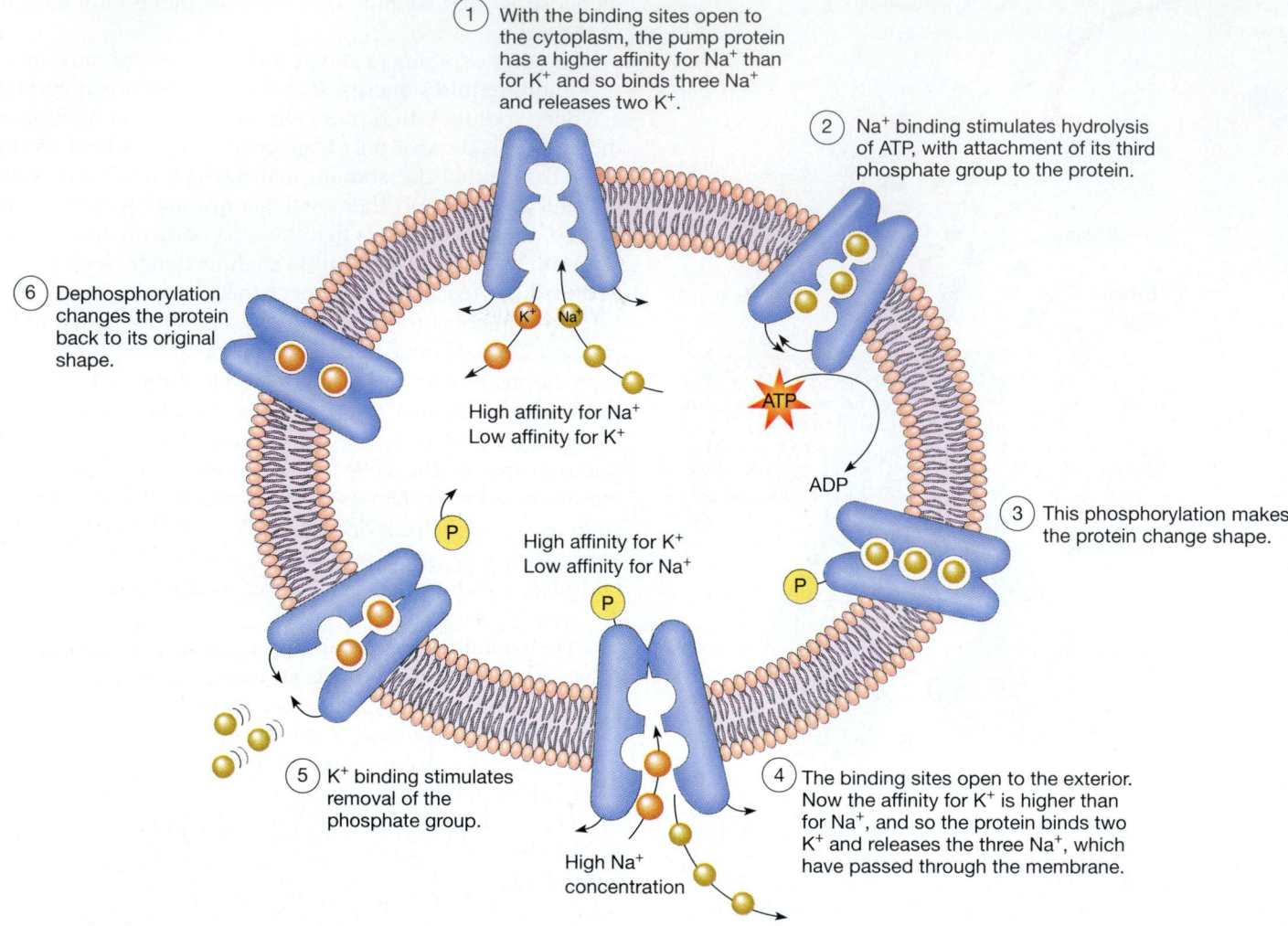

① With the binding sites open to the cytoplasm, the pump protein has a higher affinity for Na⁺ than for K⁺ and so binds three Na⁺ and releases two K⁺.

② Na⁺ binding stimulates hydrolysis of ATP, with attachment of its third phosphate group to the protein.

③ This phosphorylation makes the protein change shape.

④ The binding sites open to the exterior. Now the affinity for K⁺ is higher than for Na⁺, and so the protein binds two K⁺ and releases the three Na⁺, which have passed through the membrane.

⑤ K⁺ binding stimulates removal of the phosphate group.

⑥ Dephosphorylation changes the protein back to its original shape.

High affinity for Na⁺
Low affinity for K⁺

High affinity for K⁺
Low affinity for Na⁺

High Na⁺ concentration

**FIGURE 4–13**  The sodium-potassium pump. This allosteric protein (blue) switches between different shapes, with different affinity for sodium (Na⁺) and potassium (K⁺) ions.

the Na⁺ gradient: the steeper this gradient, the higher the rate of glucose transport into the cell.

This cotransport mechanism is one of the reasons the sodium-potassium pump needs to keep pumping Na⁺ out of a cell: Na⁺ re-enters the cell during the cotransport of sugars or amino acids into the cell. By maintaining the sodium concentration gradient across the plasma membrane, the sodium-potassium pump indirectly supplies the energy for this cotransport mechanism (Figure 4–14). Sodium also re-enters the cell by routes such as the gated channels described earlier, again providing work for the sodium pump.

Most ion-driven transport systems across plasma membranes in animals are powered by Na⁺ gradients; in bacteria and plants, most are driven by H⁺ gradients.

***Membrane Potentials***   We have seen that membranes maintain different concentrations of substances on either

side. Because these substances include electrically charged ions, the electrical charges on the two sides of the membrane may also differ. The plasma membrane's transport proteins usually move ions in such a way that the inside of the cell is electrically negative compared with the outside. As a result, there is an **electrical potential difference,** or **membrane potential,** across the membrane. Membrane potentials exist in most, if not all, cells, but they play an especially important role in the working of nerves and muscles (Chapters 37 and 39).

Most of the negative charge inside a cell comes from proteins and other organic molecules too large to escape through the membrane. Most of the positive charge outside the membrane comes from Na⁺ pushed out by the sodium-potassium pump (Table 4–1). Furthermore, another membrane protein, the **K⁺ leak channel,** permits K⁺ to diffuse out of the cell down its concentration gradient, making the

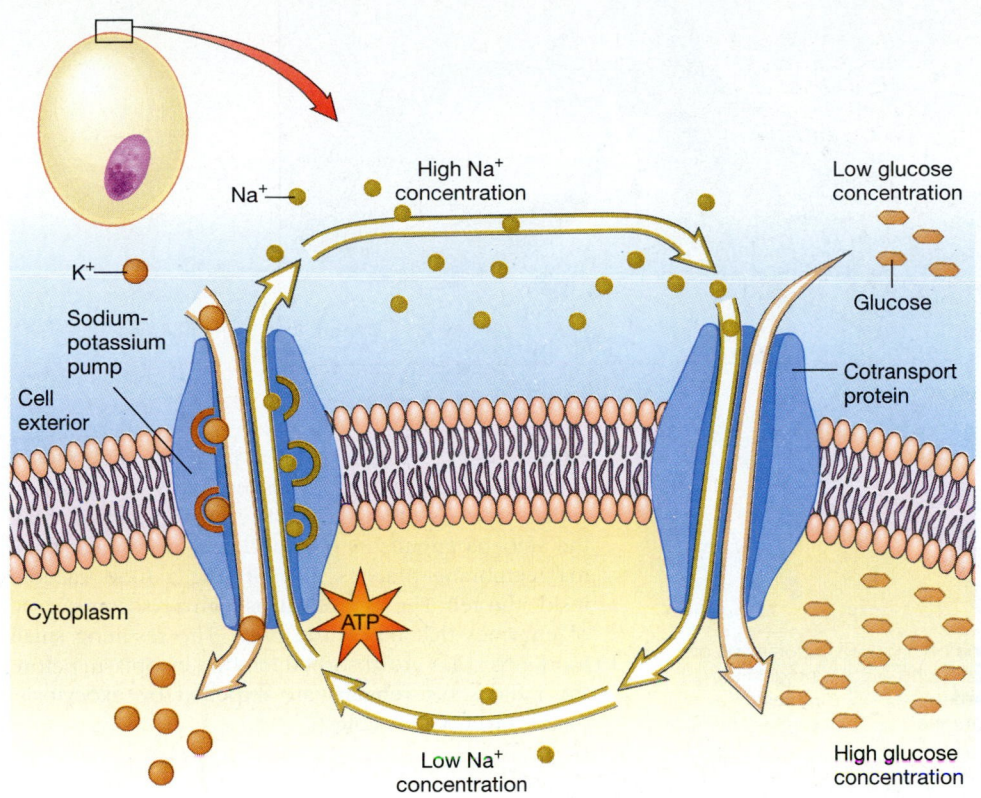

**FIGURE 4–14** Active transport of glucose indirectly powered by the sodium-potassium pump. Using energy from ATP, the pump maintains a high concentration of sodium outside the cell by active transport. In some types of cells, the flow of sodium ions down this steep gradient into the cell then supplies the energy for the active cotransport of glucose into the cell, against the glucose gradient.

cell's interior even more negative. At some point, the interior becomes so negative (attracting $K^+$), and the exterior so positive (repelling $K^+$), that net movement of $K^+$ ceases. This point is the cell's **resting membrane potential,** from −20 to −200 millivolts (negative inside) depending on the species and cell type.

Now we can see that the diffusion of a given ion through a membrane depends on two factors: (1) the ion's own concentration gradient and (2) the membrane potential, which is related to the sum of all the electrical charges

present. Both these gradients together constitute the ion's **electrochemical gradient,** and this helps to determine the ion's rate of diffusion through the membrane. For example, a positively charged ion has a greater tendency to move into a cell (or remain in a cell), attracted by the excess negative charge there, than would be predicted from its concentration gradient alone. Similarly, a negatively charged ion has a greater tendency to remain outside the cell than we would expect from its concentration gradient because it is attracted by the external positive charge and repelled by the negative charge inside.

**TABLE 4–1**

**Concentrations of Various Substances Inside and Outside a Mammalian Muscle Cell**

| Electrical Charge | Substance | Concentration (mM) | |
| --- | --- | --- | --- |
| | | Inside Cell | Outside Cell |
| Positive (cations) | $Na^+$ | 12 | 145 |
| | $K^+$ | 150 | 5 |
| | $Mg^{2+}$ | 30 | 1 |
| | $Ca^{2+}$ | 1 | 4 |
| Negative (anions) | $Cl^-$ | 4 | 120 |
| | $HCO_3^-$ | 27 | 8 |
| | Organic anions (proteins, etc.) | 155 | 7 |

## 4–F  BULK TRANSPORT: LARGE PARTICLES

Sometimes cells must import or export very large molecules, or even bigger particles of matter. Items of this size cannot pass through membranes either by penetrating the lipid bilayer or by way of transport proteins. Rather, this transport of large items, often called bulk transport, depends on the fluid membrane's ability to change shape and fuse with, or pinch off, small membrane-enclosed sacs, called **vesicles,** and to reseal itself automatically.

In the process of **endocytosis** (endon = within), cells take in material from their surroundings. The unicellular organism *Amoeba* engulfs its food by extending projections called pseudopods from the cell body and surrounding the food (Figure 4–15). The pseudopods then fuse so that

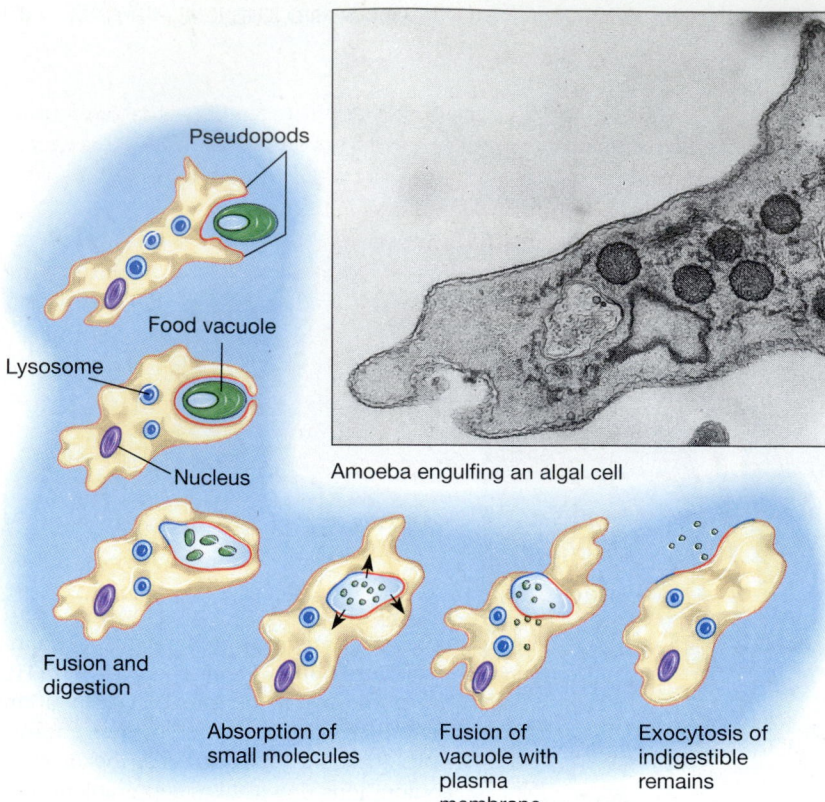

Pseudopods

Lysosome

Food vacuole

Nucleus

Fusion and digestion

Amoeba engulfing an algal cell

1 µm

Absorption of small molecules

Fusion of vacuole with plasma membrane

Exocytosis of indigestible remains

**FIGURE 4–15** Phagocytosis, a form of endocytosis. (photo) An *Amoeba* extends pseudopods around a unicellular photosynthetic alga. The drawings show how the amoeba engulfs its prey and how part of the plasma membrane pinches off, forming a food vacuole inside the cell. The vacuole fuses with a lysosome, a sac of enzymes that digest the prey. The resulting small food molecules are absorbed into the cytoplasm before the indigestible remains are expelled by exocytosis. (photo, Biophoto Associates)

the food becomes enclosed in a membrane-bounded sac in the amoeba's interior. This kind of endocytosis is known as **phagocytosis** ("cell eating")—the engulfing of a large particle, such as an entire bacterium or a fragment of a disintegrating cell, into a large, sac-like **vacuole** in the cell. Phagocytosis is a major feeding method of many unicellular organisms and simple multicellular animals.

In most animals, phagocytosis plays a part in defense against disease. For example, some of a vertebrate's white blood cells are phagocytes, which engulf and digest invading bacteria. Phagocytes also clean away the body's dead cells. In your body, phagocytes remove 100 billion worn-out red blood cells per day.

Phagocytosis of, say, a bacterium involves binding of receptors on the phagocyte's plasma membrane to complementary molecules on the bacterium's surface. Some strains of *Streptococcus* ("strep") bacteria surround themselves with a capsule of carbohydrates that inhibits binding and ingestion by phagocytes. These strains are apt to cause illness, whereas phagocytes easily dispose of strains lacking the capsule, before they can cause disease. In contrast, the bacteria that cause leprosy and tuberculosis are easily engulfed but have adaptations making them resistant to being digested by phagocytes.

**Receptor-mediated endocytosis** takes up smaller items such as proteins or low-density lipoprotein (LDL) particles selectively. These particles must first bind to specific protein receptors on the plasma membrane. The membrane then invaginates, forming a depression around a number of these receptors and the molecules they have bound. Finally the membrane pinches off a vesicle, containing the loaded receptors, into the cytoplasm. Low-density lipoprotein particles taken into the cell in this fashion contain cholesterol and fatty acids for membranes. Growing egg cells take in yolk proteins by a similar mechanism (Figure 4–16). Hence, some of the membrane's receptors serve to identify and grapple needed substances, permitting the membrane to engulf them in bulk.

**Pinocytosis** ("cell drinking") is a kind of endocytosis in which a cell engulfs fluid, plus whatever solutes it contains. This occurs in amoebas and in some kinds of animal cells.

Materials can be discharged from a cell as well as engulfed. In **exocytosis,** the membrane of an internal vesicle or vacuole fuses with the plasma membrane, which then opens and allows the vesicle's contents to escape from the cell. Substances released in this way may be indigestible food particles (see Figure 4–15) or secretions such as hormones. This process has been studied using **mast cells,** part of the body's defense system, which release histamine from large vesicles. When a mast cell is stimulated, exocytosis is so rapid that the membranes of these vesicles fusing with the plasma membrane may triple its area within 20 seconds.

Figure 4–15 shows how a patch of plasma membrane becomes a vacuolar membrane in the cytoplasm by endocytosis and eventually returns to the plasma membrane during exocytosis. Likewise, membrane material that merges

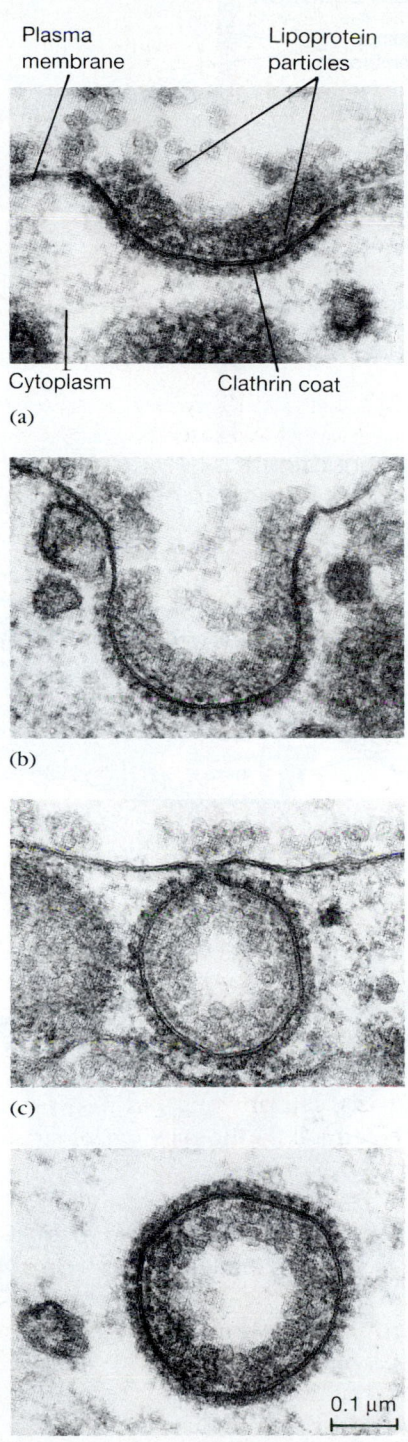

Plasma membrane

Lipoprotein particles

Cytoplasm

Clathrin coat

(a)

(b)

(c)

0.1 μm

(d)

**FIGURE 4–16** Receptor-mediated endocytosis. These photographs show the stages by which the yolk of a forming chicken egg takes in lipoproteins. (a) Lipoprotein particles bind to receptors on the plasma membrane, while molecules of the protein clathrin assemble into a basket-like structure just under the membrane, folding it inward (b). (c) Inside the cell, the membrane pinches off, (d) forming a sac of lipoprotein called a yolk platelet. (M. M. Perry and A. B. Gilbert, *J. Cell Sci.* 39:257, 1979)

with the plasma membrane during exocytosis is reclaimed by endocytosis and recycled into more vacuoles. This membrane-recycling process, called the endocytic cycle, can be completed within about a half hour.

We still do not know exactly how membranous vesicles merge with the plasma membrane during exocytosis. The negatively charged hydrophilic surfaces of the two membranes should repel each other. Hence, it should take energy to force them together and to separate the lipid molecules in each bilayer enough to produce free edges so that the two can fuse into one. Similar problems exist with pinching off vesicles from plasma membranes during endocytosis.

Artificial lipid-walled vesicles called liposomes are used to package water-soluble substances such as chemotherapy drugs and deliver them directly into the cytoplasm of cells by endocytosis. Researchers also use liposomes to transplant DNA with anticancer potential into tumor cells of cancer patients. Early results show some promise that this will be a way to kill cancer cells or stimulate cancer-fighting immune cells selectively.

## 4–G    ATTACHMENTS BETWEEN CELLS

The many cells of a multicellular organism's body must be held together somehow. Since the plasma membrane is at the cell's outer surface, it participates in this important interaction between the cell and its environment.

In plants, each cell builds and lives inside its own cell wall, which is attached to its neighbors' walls by glue-like polysaccharides. Some cell wall components are made inside the cell and dispatched to the exterior by exocytosis, but the wall's cellulose fibers are made outside the cell by enzymes that are part of the plasma membrane.

An animal's body also contains a lot of protein fibers and polysaccharide chains made inside cells and moved out by exocytosis. Cells attach to these nonliving extracellular (outside cells) materials by some of their membrane proteins.

Animal cells may also form three main kinds of close attachments to their neighbors: tight junctions, which seal the outer edges of the cells' membranes together; and adherens junctions and desmosomes, which provide mechanical strength. All three kinds of attachments occur in many cell types, but here we will illustrate them using the cells lining the small intestine. These cells form a sheet, one cell thick, called an epithelium.

The contents of the intestine must not be permitted to seep into the body without passing "inspection" by the membranes of the intestinal epithelial cells. **Tight junctions** are areas where the cells' plasma membranes are sealed together, forming such a tight barrier that even small molecules cannot move through the spaces between the cells; whatever enters the body must pass through the cells. In

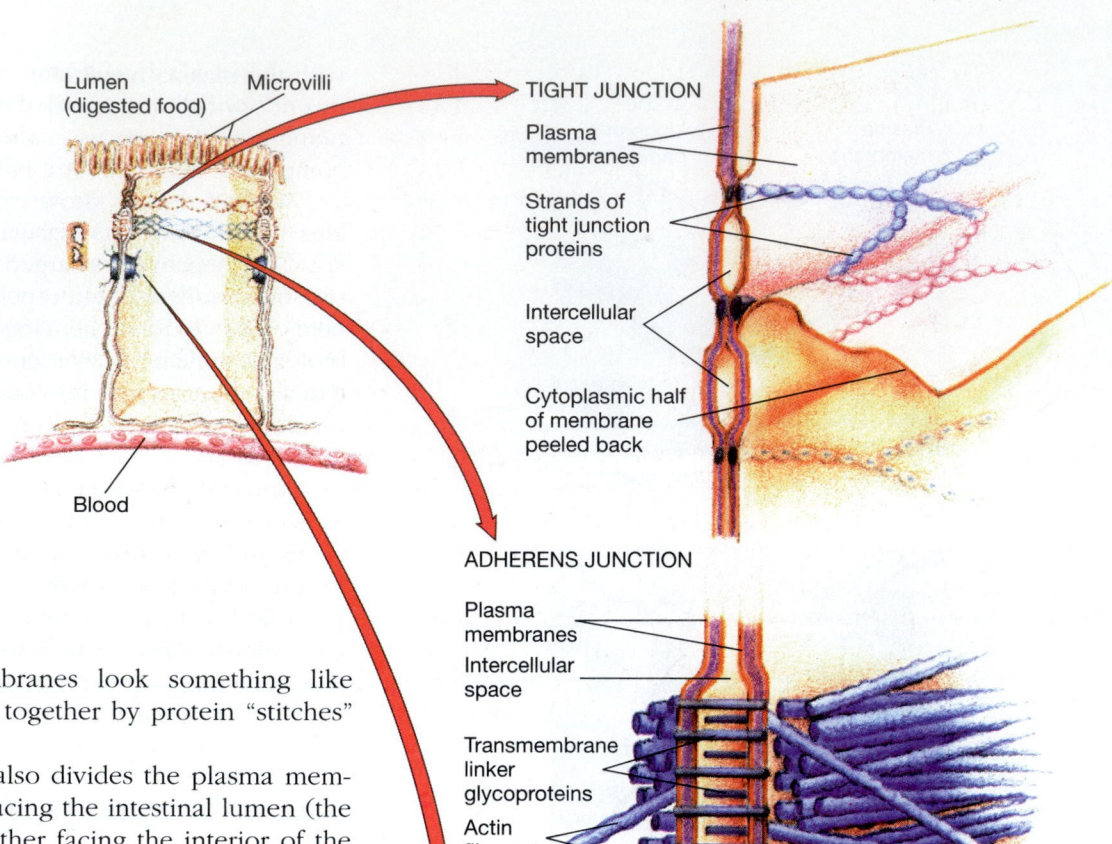

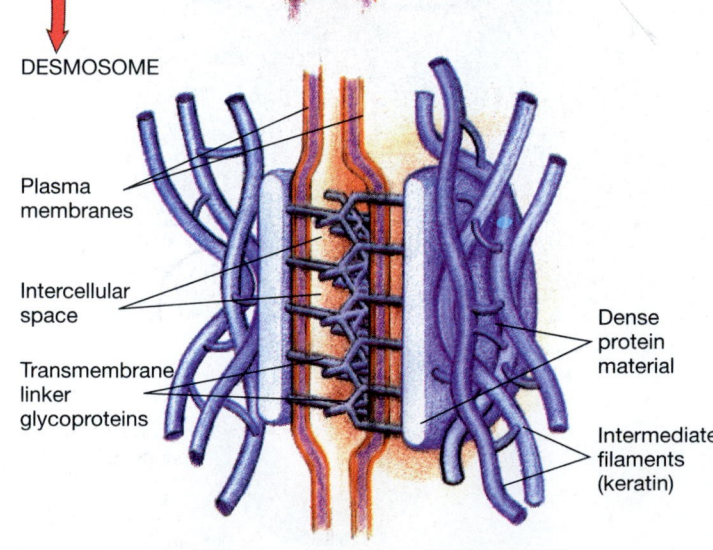

a tight junction, the membranes look something like two pieces of fabric quilted together by protein "stitches" (Figure 4–17).

The tight junction area also divides the plasma membrane into two zones: one facing the intestinal lumen (the hollow inside a tube), the other facing the interior of the body. Each zone contains a different set of membrane proteins, which must stay in that zone if the cell is to perform its job properly: the proteins next to the lumen take food from the intestine into the cell, and those on the body side release food from the cell into the bloodstream. Letting these membrane proteins diffuse into the wrong zone could create real problems! However, the tight junctions prevent this.

Beyond the tight junctions, the membranes of epithelial cells are attached together by **adherens junctions,** and a bit further on by **desmosomes.** Both serve the mechanical function of holding cells in the epithelial layer firmly together despite the churning of the intestinal wall and the sloshing of partly digested food. In both these attachments, the membranes of neighboring cells are held together by proteins that extend through each membrane and link up in the intercellular (= between cells) space. On the cytoplasmic side of each membrane is a dense plate of proteins, which link to structural protein filaments in the cytoplasm (Figure 4–17). Adherens junctions attach to bundles of actin filaments (see Figure 3–21a), desmosomes to intermediate filaments (Section 5–Q).

With its cells joined snugly into a continuous sheet, the intestinal epithelium acts as a sort of super-membrane separating the living body from the nonliving food in the intestinal lumen. The membrane surface facing the lumen absorbs food molecules selectively. The surface area available for this absorption is increased about 20-fold by the expansion of these membrane surfaces into hundreds of tiny finger-like projections called microvilli. The opposite surface of the epithelium passes the absorbed food into the body.

**FIGURE 4–17** Close encounters of three kinds. (top left) Attachments between cells in the intestinal epithelium keep digested food from seeping between cells into the blood and from there to the rest of the body. (top right) In tight junctions, proteins extend through the lipid bilayer and attach to their counterparts in the membrane of a neighboring cell. (below) In adherens junctions and desmosomes, the attachment between cells is strengthened by structural proteins that link together between the plasma membranes and also connect to filaments of reinforcing proteins (actin or keratin) inside each cell. Desmosomes have been compared to rivets, whereas adherens junctions are more like Velcro.

This discussion illustrates one case of cell specialization. Intestinal epithelial cells are not specialized by having components that other cells lack. Rather, they have an expansion of a common structure, the plasma membrane, and a special organization of common types of transport molecules and membrane attachments in a way that permits them to carry out a specialized task.

## 4–H  COMMUNICATION BETWEEN CELLS

The cells of a multicellular organism communicate with each other by chemical signals. Cells send messages over a distance by a sort of chemical mail: one cell releases messenger molecules such as hormones, which are addressed to specific receptors in other cells. When the messenger molecules arrive and bind to their receptors, they cause some change inside the receiving cell. Many kinds of cells communicate with their next-door neighbors more quickly by a sort of "back-fence" exchange of substances, using direct cytoplasmic connections, with no membranes to cross.

Many animal cells have direct cytoplasm-to-cytoplasm connections with their neighbors at **gap junctions.** Here, each cell has a patch of membrane containing an array of pipe-like channel proteins. Each pipe consists of a ring of six roughly cylindrical proteins that sticks through the plasma membrane and butts against a similar pipe in the adjacent cell's membrane (Figure 4–18).

Gap junctions permit ions and small molecules to pass directly from cell to cell without leaking into the intercellular space. This cell-to-cell connection can be shown using microelectrodes, very small probes that conduct electric current. When these electrodes are placed inside two adjacent cells linked by gap junctions, the electrical resistance between the electrodes is very low. This shows that electric current, in the form of moving ions, can move unimpeded between the cells, and the cells are said to be electrically coupled. In contrast, if the electrodes are placed so that one is inside and the other outside a cell, the electrical resistance measured is high. This shows that the flow of ions through the membrane is highly controlled.

In some tissues, the function of electrical coupling between cells via gap junctions is clear. For instance, electric

(a) A gap junction

**FIGURE 4–18**  A gap junction. (a) Electron micrograph of a gap junction revealed by freeze fracture of a plasma membrane (see *How Do We Know? Freeze-Fracture Technique*). Each bump is one end of a channel protein pipe connecting to the next cell. (b) Section through two plasma membranes joined by a gap junction. Unlike tight junctions, gap junction pipes leave a gap between the two plasma membranes; hence their name. The center pipe in the front row is seen from outside; the others have some proteins removed to show the tunnel.  (a, E. Anderson, *J. Morphol.* 156:339, 1978)

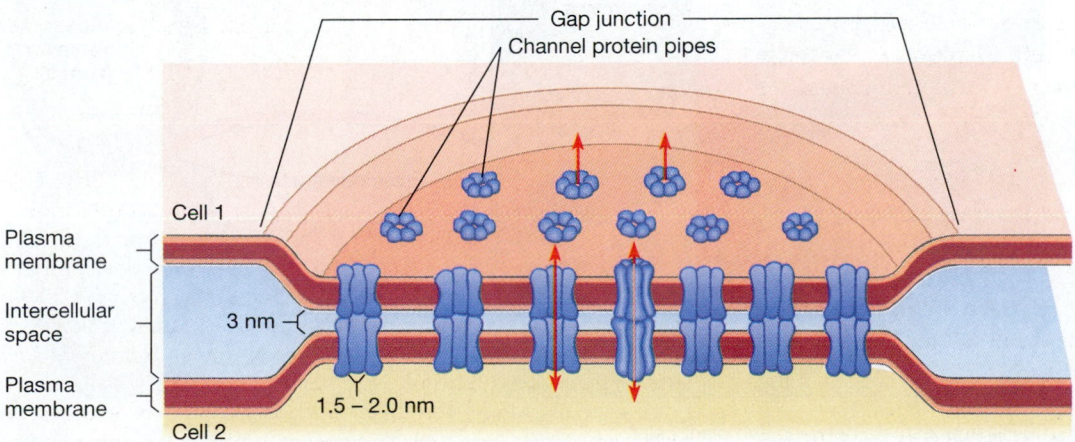

(b) Cross section of side view

### FREEZE-FRACTURE TECHNIQUE

The plasma membrane is so thin that it could not be seen or photographed until electron microscopes became widely available in the 1950s. Even then, a typical thin cross section of a cell revealed little about the plasma membrane beyond its presence (see Figure 4–4).

Researchers learned more about membrane structure using a technique called freeze fracture. First, a tissue sample is frozen at −150 °C. It is then put into a chamber under high vacuum and fractured with a knife. The cells split along a zone of weakness, often right between the two lipid layers of a membrane because there is practically no attraction holding the lipid molecules' nonpolar tails together (Figure 4–A). For example, when a plasma membrane splits in this way, one lipid layer remains covering the cytoplasm and the other lifts off with the cell's external medium. The membrane's integral proteins remain attached to one of the two lipid layers, jutting out from it as little bumps and leaving matching dimples in the opposite surface.

Next, a platinum–carbon replica is made. Atoms of platinum are evaporated onto the exposed surface to form a thin coating, which is then reinforced by evaporating a carbon layer onto it. The thawed biological material is then dissolved away with a strong acid or base. The platinum-carbon replica is transparent to electrons and can be examined

**FIGURE 4–A** Freeze fracture. The knife cleaves a frozen cell along lines of weakness, often between lipid layers of membranes. Some proteins remain embedded in each lipid layer.

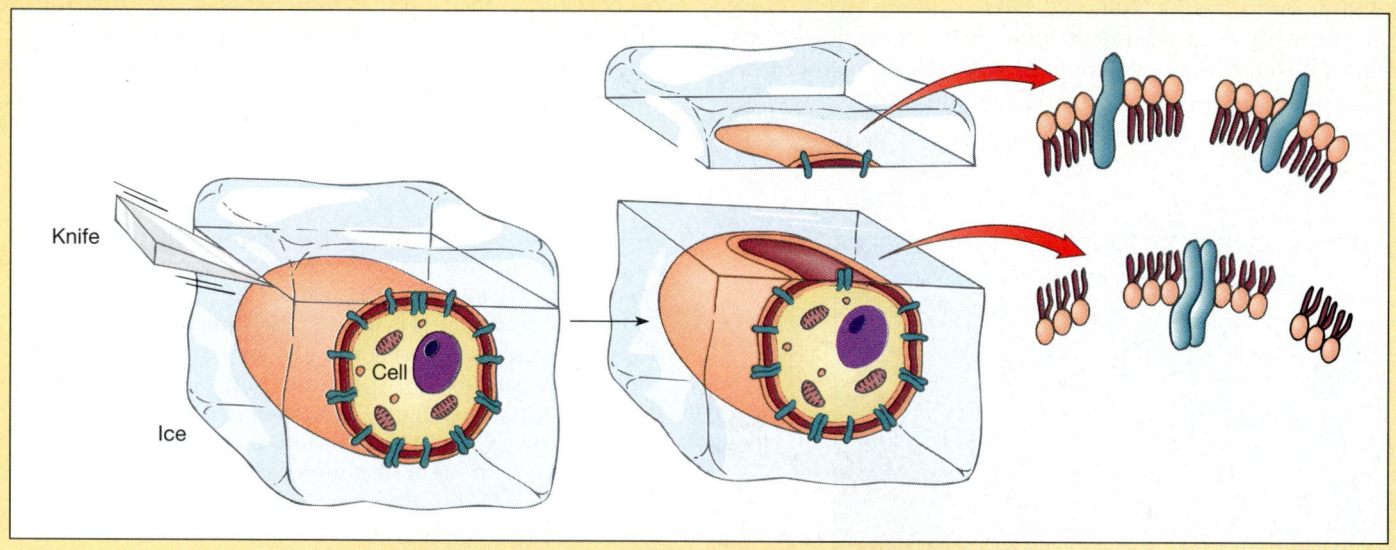

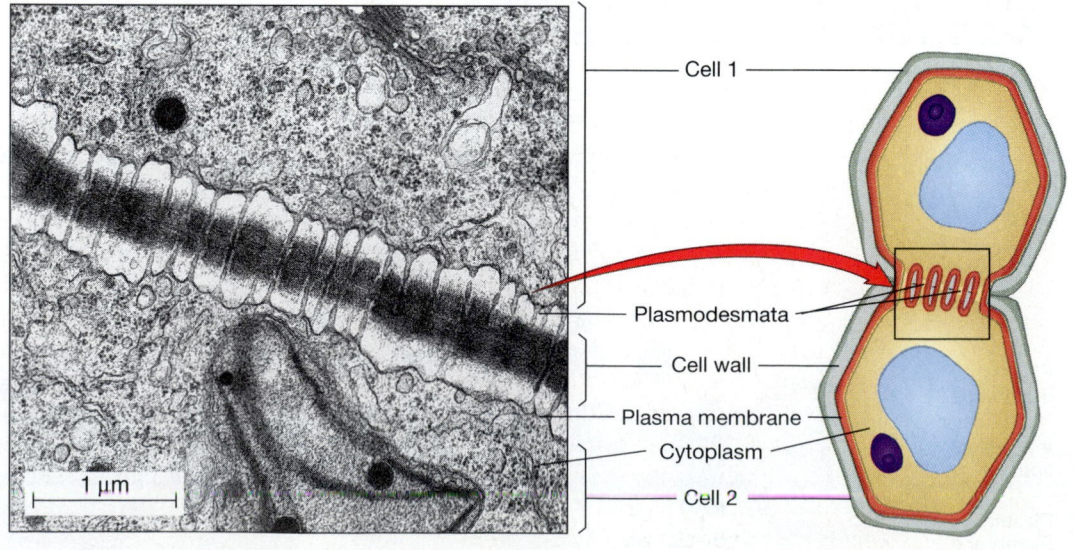

**FIGURE 4–19** Plasmodesmata. The drawing shows how the plasma membranes and cytoplasm of the two neighboring cells continue through the pores in the cell walls. (a, Biophoto Associates)

1 µm

(a) Plasmodesmata between two plant cells

(b)

Cell 1

Plasmodesmata

Cell wall

Plasma membrane

Cytoplasm

Cell 2

with an electron microscope. In this way researchers can see the contours of the membrane surface and the distribution of some important proteins. This technique was immensely useful in studying the structure of gap junctions (see Figure 4–18a) and of junctions between intestinal epithelial cells (Figure 4–B).

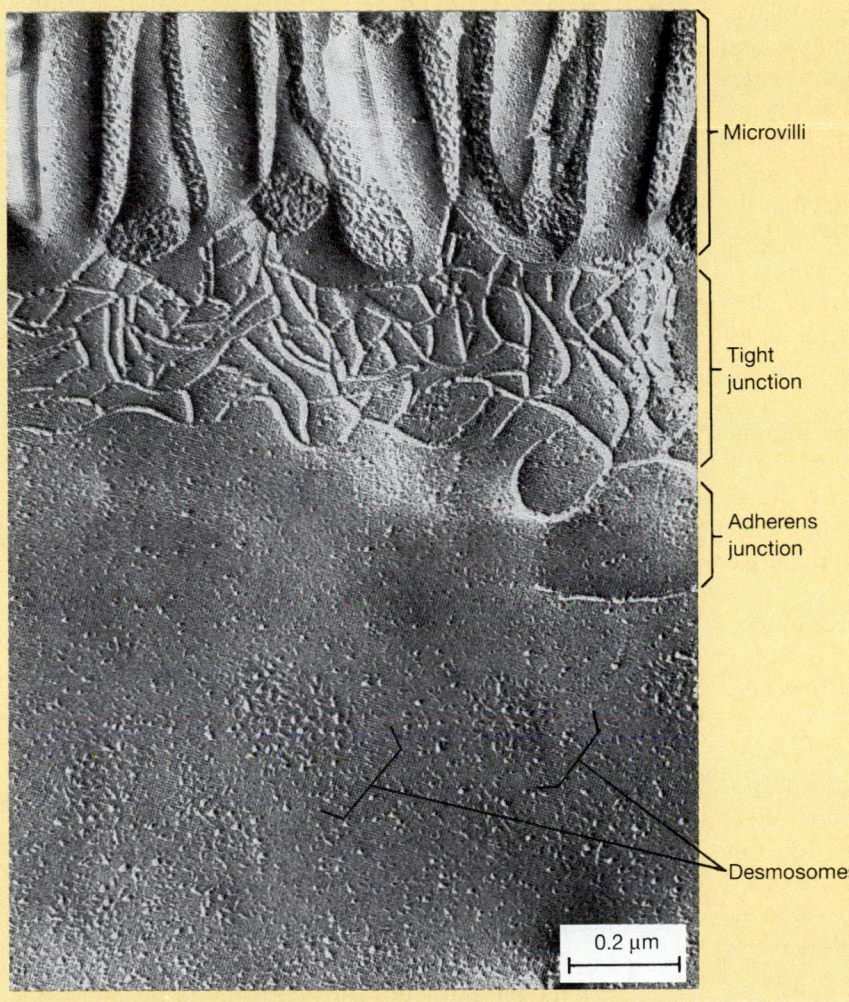

— Microvilli

— Tight junction

— Adherens junction

— Desmosomes

0.2 μm

**FIGURE 4–B** Freeze-fracture preparation of an intestinal epithelial cell. Proteins of tight junctions form a pattern of intersecting ridges; those of adherens junctions form thin, faintly visible filaments; and those of desmosomes form isolated patches of large particles. Photographs such as this helped researchers to deduce the structures of these junctions shown in Figure 4–17. (L. Andrew Staehelin, University of Colorado)

currents of moving ions play an important role in muscle contraction. In the heart, electrical coupling by gap junctions between adjacent muscle cells helps to coordinate their contractions during each heartbeat. However, the role of coupling is not yet clear in other cases. For example, early animal embryos contain many gap junctions. Embryos injected with substances that inactivate the junctions develop abnormally, but we do not yet know what signals go awry when this happens.

In plants, the cytoplasm of neighboring cells is often connected by strands of cytoplasm called **plasmodesmata** (singular: **plasmodesma**). These cytoplasmic bridges pass through openings in the cell walls between the two cells, and both the cytoplasm and the plasma membranes of these cells are essentially continuous with each other (Figure 4–19).

## SUMMARY

Organisms are composed of one or more cells, the units of life. Under the general direction of its genetic material, a cell carries on its metabolism, grows, and may eventually reproduce by dividing in two. All the time a cell is conducting these internal affairs, it must also interact with its environment. For example, it must maintain internal chemical homeostasis while trading molecules with the environment.

The plasma membrane is responsible for these interactions between the cell and its environment. The membrane serves as a barrier between the cell and the outside world and it regulates what enters and leaves the cell. It is selectively permeable to many types of small molecules and ions, yet prevents the loss of the cell's macromolecules such as nucleic acids and proteins. The plasma membrane also detects and responds to changes in the environment, identifies the cell to other cells, and attaches the cell to its neighbors or to supporting structures outside the cell.

1. A biological membrane consists of a fluid lipid bilayer with various proteins floating in it, some mobile and some anchored. This basic structure has two properties crucial to membrane function. First, lipid bilayers spontaneously form closed compartments, thereby keeping the solutions inside and outside the membrane separate. Second, membranes are asymmetrical, having distinct cytoplasmic and exterior sides, with different lipid and protein components and with distinct functions.

2. Biological membranes are selectively permeable. Most are freely permeable to water and to small, lipid-soluble molecules, which diffuse through the lipid layers down their concentration gradients.

3. Cells gain and lose water by osmosis. The membrane does not control the movement of water molecules directly; rather, some of its proteins carry on active transport of solutes, creating an osmotic potential difference that induces osmosis. The cell wall of a plant cell exerts pressure that limits the entry of water.

4. Most ions and small polar molecules can cross a membrane only with the aid of membrane transport proteins. Each protein admits a particular solute or a few very similar solutes.

5. In facilitated diffusion, channel proteins and carrier proteins provide for the diffusion of specific ions and polar molecules through the membrane down their electrochemical gradients. Channel proteins form aqueous channels through the membrane, and some are gated so that they open in response to a specific stimulus. Carrier proteins are more specific than channels because they bind their passenger molecules.

6. Some carrier proteins move small organic molecules by facilitated diffusion; others mediate active transport, which can move a solute against its electrochemical gra-

dient. Active transport requires energy, provided either by ATP or by an electrochemical gradient of ions such as $Na^+$ or $H^+$.

7. The sodium-potassium pump, powered by ATP, pumps $Na^+$ out of a cell and $K^+$ in. This pump is ultimately responsible for the membrane potential of animal cells, and the electrochemical gradient of sodium that it creates also provides energy for the active transport of solutes such as glucose and amino acids. By actively transporting sodium and potassium ions through the plasma membrane, the sodium pump also regulates the cell's water content.

8. A cell takes in macromolecules or larger particles by endocytosis. The membrane surrounds the particles and pinches off to become a vesicle or vacuole inside the cell. Substances can be discharged from many cells by the opposite process, exocytosis. Membrane material cycles between the plasma membrane and some membranous compartments in the cytoplasm by endocytosis and exocytosis.

9. The plasma membranes of adjacent animal cells are sometimes attached to each other. Tight junctions between some animal cells seal their membranes together and prevent seepage of substances between the cells. Adherens junctions and desmosomes provide mechanical strength by attaching the membranes of adjacent cells to each other and to structural proteins in the cytoplasm.

10. Cells communicate with each other over distances by means of chemical messengers released by one cell and bound by membrane receptors of others. Neighboring cells often communicate directly by cytoplasmic connections. Gap junctions contain an array of pipes joining the cytoplasm of adjacent animal cells and providing for direct transfer of ions and small molecules from cell to cell. In plants, direct transfer between cytoplasm of adjacent cells occurs by way of plasmodesmata.

## Self-Quiz

1. Louis Pasteur placed sterile broth in two sterile containers, A and B. Container A had a straight mouth open to the atmosphere, whereas container B, which was also open, had a bent neck to prevent any particles entering the container from the air (see Figure 21–2). Which of these results would support the cell theory?
   a. After several weeks both A and B were teeming with organisms.
   b. After several weeks A was teeming with organisms, but B contained no life.
   c. After several weeks B was teeming with organisms, but A contained no life.

Match the substances from the list (a through d) to the way(s) they cross membranes:

_____ 2. Active transport              a. Ions
_____ 3. Diffusion through bilayer     b. Macromolecules
_____ 4. Diffusion through channels    c. Small uncharged
_____ 5. Endocytosis                      molecules
_____ 6. Osmosis                       d. Water

Which process(es) in Questions 2 through 6 depend(s) on:

_____ 7. The membrane's fluidity
_____ 8. Membrane proteins

9. A nerve cell sends messages to other cells by means of a special transmitter substance. Vesicles containing transmitter

molecules fuse with the nerve cell's plasma membrane and then open, releasing the transmitter outside the cell. This is an example of:
a. exocytosis
b. endocytosis
c. active transport
d. facilitated diffusion

10. The amino acid glutamine can cross the plasma membrane of cells either with or against its electrochemical gradient. The rate at which glutamine enters a cell depends on the concentration of sodium outside the cell and does not increase as the concentration of glutamine outside the cell increases. This is probably an example of:
a. facilitated diffusion
b. entry through a gated channel controlled by sodium ions
c. endocytosis
d. active transport powered by an electrochemical gradient
e. active transport powered by ATP

11. An area where the plasma membranes of neighboring cells are pressed directly to each other is a(n):
a. tight junction
b. gap junction
c. adherens junction
d. desmosome

In Questions 12, 15, and 16, choose the correct word from each set in parentheses.

12. The higher the concentration of a solution, the (higher, lower) its osmotic potential and the (more, less) water will enter it through a membrane permeable to water but not to solutes.

13. In the tube in the figure below, the membrane is permeable to both water and glucose. The tube is filled with a solution of glucose and placed in a beaker of water. Describe the sequence of events that will take place in this system, and draw the final water level in the tube and in the beaker.
(*Hint*: the water molecules are smaller than glucose molecules and diffuse faster.)

14. The **U**-tube in the figure below is divided in the middle by a membrane that is impermeable to starch but permeable to water. A 10% starch solution is put into the right-hand half of the tube and an equal amount of 6% starch solution is put into the left-hand half of the tube.

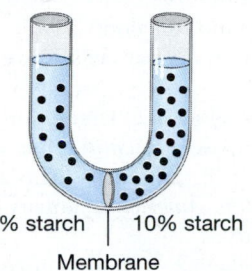

6% starch | 10% starch

Membrane

In this situation:
a. water will move from the right to the left
b. water will move from the left to the right
c. starch will move from the right to the left
d. water will move in both directions, but more from left to right than right to left
e. water will move in both directions, but more from right to left than left to right

15. *Cambarus* is an animal that excretes a very dilute urine. Therefore you would expect that *Cambarus* lives in an environment that is (hypertonic, isotonic, hypotonic) to its body fluids. Its habitat is most likely which of the following:
a. a freshwater pond
b. the ocean
c. Great Salt Lake
d. the intestine of a horse

16. At the produce counter in the supermarket, you pick up a head of lettuce whose outer leaves are wilted. You take it home and place it in cold water; the outer leaves become crisper. The vacuoles in the cells of these leaves are now (larger, smaller) than they were when you put the lettuce into your shopping cart.

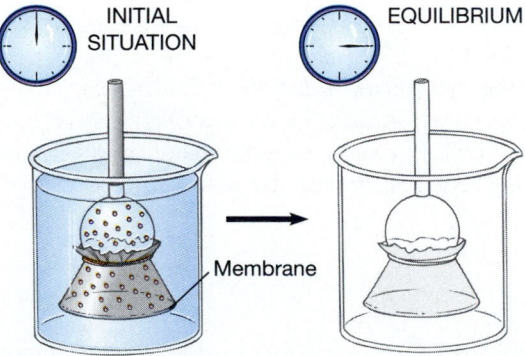

INITIAL SITUATION    EQUILIBRIUM

Membrane

## Questions for Discussion

1. Why is it important for cells to maintain chemical homeostasis?

2. Hydrogen cyanide (HCN) and carbon monoxide (CO) are poisons that penetrate cell membranes readily. Can you think of an explanation for the fact that cells have not evolved adaptations to keep these molecules out?

3. Phospholipids and integral membrane proteins seldom "flip-flop," that is, somersault in the membrane so that the end originally exposed at one surface of the membrane becomes exposed at the other surface. Can you explain why this is so?

4. In many developing countries, severe diarrhea often results in death from loss of body water: the intestinal cells leak ions

into the lumen, and water follows in such volume that the victim feels an intense urge to void. In oral rehydration therapy, the patient replaces lost body water and salts by drinking a solution containing sodium chloride, potassium chloride, and glucose. Explain why the treatment does not work so well if the glucose is omitted.

5. If cells act as osmotic systems, why don't we swell up and burst when we go swimming in fresh water, which is hypotonic to our blood and to the solutions inside and outside our cells?

## Suggested Readings

Bretscher, M. S. "The molecules of the cell membrane." *Scientific American,* October 1985. Good general overview of membrane structure and function.

Bretscher, M. S. "How animal cells move." *Scientific American,* December 1987.

Lienhard, G. E., J. W. Slot, D. E. James, and M. M. Mueckler. "How cells absorb glucose." *Scientific American,* January 1992.

Lodish, H. F., and J. E. Rothman. "The assembly of cell membranes." *Scientific American,* January 1979. Examines membrane asymmetry.

Loewy, A. G., P. Siekevitz, J. R. Menninger, and J. A. N. Gallant. *Cell Structure and Function,* 3d ed. Philadelphia: Saunders College Publishing, 1991. Chapter 7 provides an interesting history of membrane study.

Sharon, N., and H. Lis. "Carbohydrates in cell recognition." *Scientific American,* January 1993.

Singer, S. J., and G. Nicolson. "The fluid mosaic model of the structure of cell membranes." *Science* 175:720, 1972. The original description of the plasma membrane as a lipid bilayer containing proteins.

Staehelin, L. A., and B. E. Hull. "Junctions between living cells." *Scientific American,* May 1978. How electron microscopy reveals the way cells attach to their neighbors.

5

# Cell Structure and Function

**O B J E C T I V E S**

*When you have studied this chapter, you should be able to:*

1. Outline how a light microscope, a transmission electron microscope, and a scanning electron microscope work.
2. List at least four components of cells that can be seen with a light microscope and four that can be seen with an electron microscope but not with a light microscope.
3. Describe the main differences between a eukaryotic and a prokaryotic cell.
4. Give at least one function of each of the following structures and state whether each would be found in prokaryotic, plant, or animal cells:

*nuclear area
*nucleus, *nuclear envelope, *chromosome, *nucleolus
*ribosome, *endoplasmic reticulum, *Golgi complex, lysosome
peroxisome, *mitochondrion
plastid, *chloroplast, *cell wall, *vacuole
cytoskeleton, microtubules, *cilium or flagellum, centriole,
intermediate filaments, microfilament

5. Given a photograph taken using a light microscope or transmission electron microscope, identify the plasma membrane, cytoplasm, and structures marked with * in Objective 4.

---

In Chapter 4 we saw that cells are the structural, functional, and reproductive units of life. We also examined the plasma membrane—the thin, flexible covering of the cell—and saw how it mediates the exchange of substances between the cell and its environment. Now let's look further inside the cell.

The cell's viscous, fluid cytoplasm contains metabolic enzymes, many other molecules, and ions. In the cytoplasm of virtually all cells we find a nucleus or nuclear area, containing the cell's genetic material (DNA); and many ribosomes, structures that make the cell's proteins according to instructions from the DNA.

What else we find inside the cell depends on which of two basic cell types we look at: prokaryotic or eukaryotic. Bacterial cells are **prokaryotic** (pro = before; karyon = nucleus): they have no nuclear membranes separating the genetic material from the cytoplasm and therefore have no nucleus. Prokaryotic cells are believed to have evolved earlier than eukaryotic cells because prokaryotic cells are generally smaller and simpler in structure. Usually prokaryotic cells have a rigid external cell wall. Most bacteria live as independent cells, even though they may remain attached

after cell division and form clumps or strings of many cells. Bacteria (including the photosynthetic cyanobacteria) are often simply called **prokaryotes.**

A **eukaryotic cell** (eu = good, true) contains a nucleus, which is bounded by membranes, as well as other membrane-bounded organelles, parts of the cell with specialized functions. Eukaryotic cells make up the bodies of all organisms other than bacteria. These **eukaryotes** include many unicellular organisms and the multicellular algae, fungi, plants, and animals.

As an example of a eukaryotic cell, Figure 5–1 shows a human white blood cell called a lymphocyte. The lymphocyte's plasma membrane takes in food molecules and expels wastes. Enzymes in the cytoplasm and in mitochondria break down sugars and use their energy to make ATP (adenosine triphosphate), the energy source for many biochemical reactions. The lymphocyte's ribosomes use amino acids taken in through the plasma membrane, ATP from mitochondria, and RNA molecules from the nucleus to make proteins. Some of these proteins perform the "housekeeping" activities needed in every cell. Others are made especially by this cell as its contribution to the whole

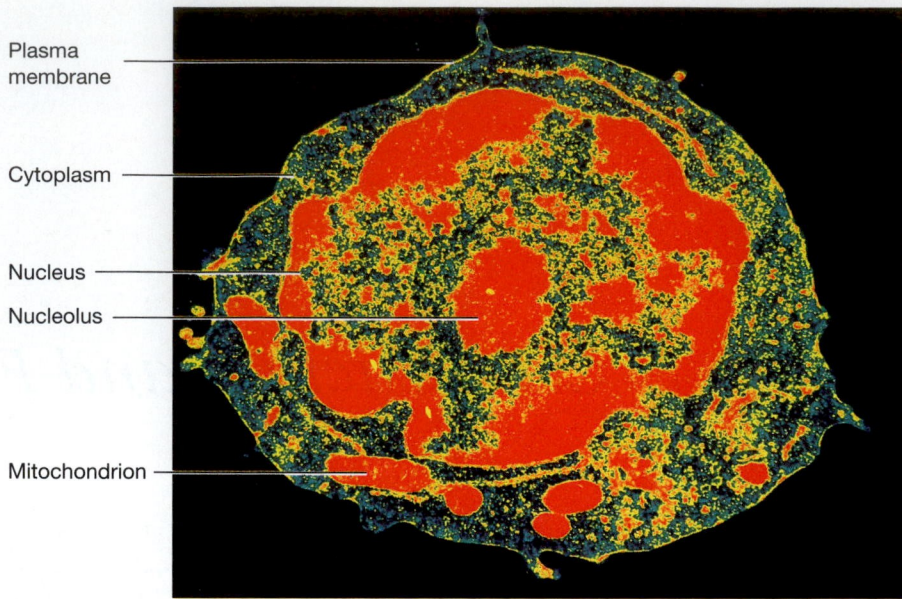

Plasma membrane

Cytoplasm

Nucleus

Nucleolus

Mitochondrion

Human lymphocyte (TEM, false color)

**FIGURE 5–1**  A eukaryotic cell. This human lymphocyte, a white blood cell, contains a large nucleus and smaller mitochondria. Such membrane-bounded organelles are characteristic of eukaryotic cells. The cell's diameter is about 8 micrometers (μm) (see Figure 5–2).  (Biophoto Associates/Science Source/Photo Researchers)

body. For example, special receptor proteins in the lymphocyte's plasma membrane act as sentinels, ready to detect viruses or other foreign material that might damage the body.

Because our lymphocyte's job is patrolling the body, it has no fixed address. Most other kinds of cells are attached to their neighbors or to a framework of material outside the cell, forming an organized structure of many similar cells, called a tissue. Various kinds of tissues arranged in particular ways form organs, such as the intestine. Blood is also a tissue, consisting of a fluid containing cells.

In this chapter we first look at some techniques used to study cell structure and function. Then we consider prokaryotic cells briefly, before turning to eukaryotic cells.

## KEY CONCEPTS

- ◆ Cells can be divided into two groups—prokaryotic and eukaryotic—based on differences in their organization and size.
- ◆ All cells contain components essential for life: a plasma membrane, genetic material (DNA), ribosomes, and cytoplasm.
- ◆ In addition to the basic cell components, eukaryotic cells have a membrane-bounded nucleus, containing the genetic material. They also have many other membrane-bounded organelles.

- ◆ Eukaryotic organisms may be unicellular or multicellular. Each cell of a multicellular organism must carry on its own life processes and in addition perform some specialized task that contributes to the body as a whole.
- ◆ In most multicellular organisms, cells are organized into tissues, tissues into organs, and organs into the various organ systems of the body.

## 5–A  MICROSCOPY

Although cells vary considerably in size, most plant and animal cells are very small, between 5 and 50 micrometers (μm) in diameter (Figure 5–2). Most cells are too small to be seen without magnification, and much of our increasing knowledge of cells has depended upon the gradual improvement of microscopes over the centuries.

## Light Microscopes (LM)

Italian monks developed the art of grinding lenses in the fourteenth century, to make spectacles for monks who copied and illustrated exquisite manuscripts by hand. By 1590 Dutch lens grinders Hans and Zacharias Janssen had mounted two lenses in a tube to produce the first **compound microscope** (one with two main lenses).

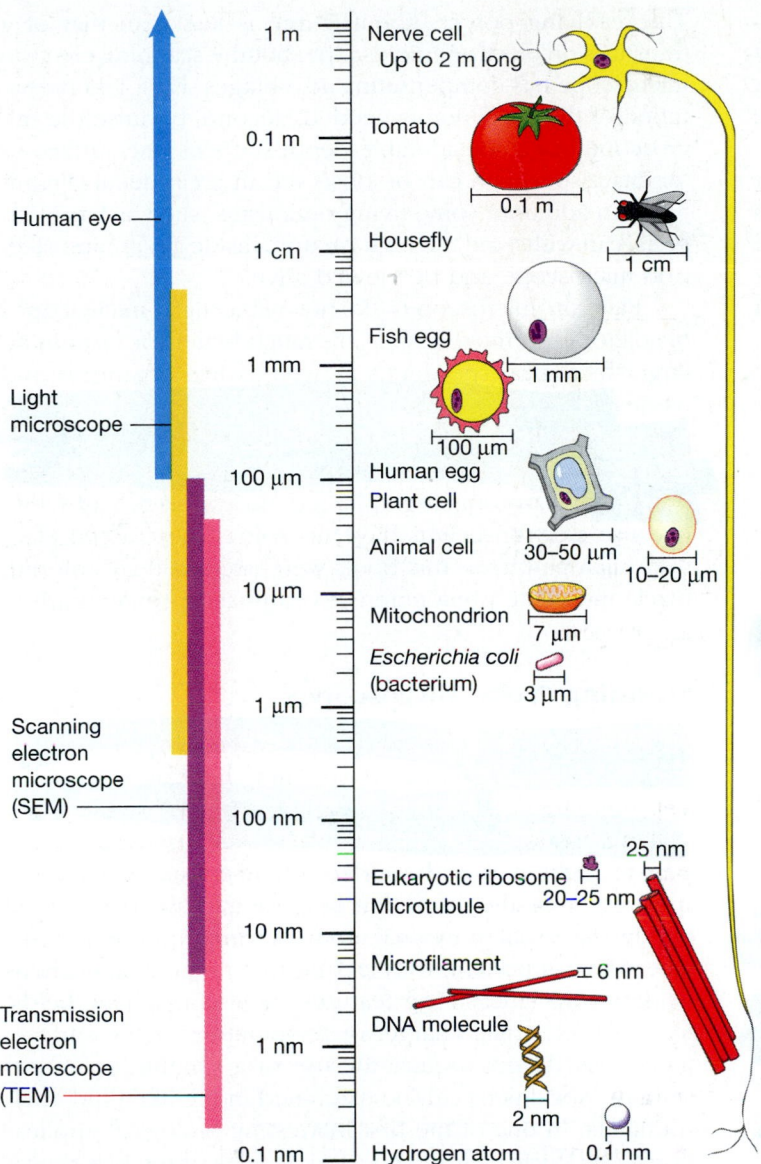

Human eye

Light
microscope

Scanning
electron
microscope
(SEM)

Transmission
electron
microscope
(TEM)

| | |
|---|---|
| 1 m | Nerve cell |
| | Up to 2 m long |
| 0.1 m | Tomato |
| | 0.1 m |
| 1 cm | Housefly |
| | 1 cm |
| | Fish egg |
| 1 mm | 1 mm |
| | 100 μm |
| 100 μm | Human egg |
| | Plant cell |
| | Animal cell     30–50 μm |
| | 10–20 μm |
| 10 μm | Mitochondrion |
| | 7 μm |
| | Escherichia coli |
| | (bacterium)     3 μm |
| 1 μm | |
| 100 nm | 25 nm |
| | Eukaryotic ribosome |
| | Microtubule     20–25 nm |
| 10 nm | Microfilament     6 nm |
| | DNA molecule |
| 1 nm | |
| | 2 nm |
| 0.1 nm | Hydrogen atom     0.1 nm |

**FIGURE 5–2**   Units of length used in measuring cellular structures:

centimeter (cm) = one-hundredth of a meter ($10^{-2}$ m)
millimeter (mm) = one-thousandth of a meter ($10^{-3}$ m)
micrometer (μm) = one-millionth of a meter ($10^{-6}$ m; $10^{-3}$ mm)
nanometer (nm) = one-billionth of a meter ($10^{-9}$ m)

The colored bars (left) indicate the range of object sizes that can be viewed with various optical systems. The resolving power of each is the lower limit of the range. For example, the resolving power of the human eye is 100 μm (0.1 mm) at a distance of 10 cm.

The pictures show the sizes of some important biological structures.

In the compound light microscopes used today, the **objective lens,** close to the object or specimen being viewed, forms a magnified image of the specimen. The image is further magnified by the **ocular lens,** which lies close to the viewer's eye (see Figure 5–3). Other lenses focus the light beam without magnification.

Surprisingly, the most important factor determining how small an object can be viewed with a microscope is not the microscope's magnifying power but its **resolving power:** its ability to distinguish the separateness of two objects that are close together. The resolving power of the human eye, for instance, is about 0.1 mm, meaning that we can distinguish two dots as separate when they are at least 0.1 mm apart. If a microscope's lenses cannot resolve detail in an object, they produce a fuzzy image, and magnification produces only a larger fuzzy image. However, the better the

resolving power, the greater the magnification that can usefully enlarge the detail distinguished by the lenses. A good microscope using white light can resolve two objects separated by 0.4 μm.

The resolving power of a lens system is limited by the **diffraction,** or scattering, of light as it passes through the lens. Diffraction at the objective lens may make the edges of the image so fuzzy that two small specimens close together appear to overlap and cannot be resolved as separate.

Resolving power can be increased by reducing diffraction as the light passes through the objective lens. This can be done by increasing the diameter of the lens, or by using light of shorter wavelength. (Light behaves as though it travels in waves. The distance between peaks of adjacent waves is the wavelength.) Shorter wavelengths of light are diffracted less than long wavelengths. For example, the ultraviolet microscope, using light with a wavelength about one quarter the average wavelength of white light, has four times better resolving power than a microscope using white light.

A microscope cannot be used to view any object smaller than its resolving power. In other words, if the "objects" to be resolved are the two sides of one particle, the particle must be at least as large as the resolving power to be visible using the microscope. Thus, an object smaller than 0.4 $\mu$m will not be visible through an ordinary light microscope. We can use a light microscope to view entire cells and large cell components such as nuclei, but to see many smaller structures we must use microscopes that use shorter wavelengths, such as electron microscopes.

## Electron Microscopes

***Transmission Electron Microscopes (TEM)***    Transmission electron microscopes pass electrons, instead of light, through the specimen. We often think of electrons as particles, but they can also be considered as electromagnetic waves. The best transmission electron microscopes today have a resolution of about 0.2 nm. However, in theory a typical electron beam, with a wavelength of 0.005 nm, could be used to achieve a resolving power of 0.0025 nm.

The features of a transmission electron microscope are similar to those of a light microscope, but assembled upside down (Figure 5–3). The lenses are not glass but electromagnets, which can deflect the negatively charged electrons. A beam of electrons is produced by heating a tungsten filament in the electron gun. The beam is accelerated through the **condenser lens,** which focuses it. The electrons then pass through the specimen and the objective lens. The **projector lens** in an electron microscope is the equivalent of the ocular lens in a light microscope. Because the human eye cannot detect electrons, the projector lens focuses the electron beam on a photographic plate or fluorescent screen, where it produces a visible image.

One disadvantage of transmission electron microscopes is that electrons are easily deflected or absorbed by molecules in the air or in the specimen itself. For this reason specimens must be observed in an almost complete vacuum and must also be sliced very thin (50 to 100 nm). In contrast, specimens for light microscopy are seldom sliced thinner than 4000 nm.

***Scanning Electron Microscopes (SEM)***    Scanning electron microscopes were first manufactured in the 1960s. Here electrons do not pass through the specimen. Instead, electrons bombarding the specimen cause atoms at its surface to emit lower-energy **secondary electrons,** which are collected and used to vary the intensity of a spot on a television screen that scans in synchrony with the electron beam.

The resolving power, about 5 nm, is less than that of a transmission electron microscope, but the scanning electron microscope has compensating advantages. First, less preparation of the specimen is needed. Second, because the microscope has a considerable depth of focus, the surface of an intact specimen can be observed in great detail (Figure 5–3). In addition, some living organisms, such as hardy insects, can withstand the high vacuum inside a scanning electron microscope and be viewed alive.

Electron microscopes do not make light microscopes obsolete. Light microscopes are much better for examining larger biological specimens as well as live organisms and tissues. Perhaps the most important advantage of a light microscope, though, is that it produces a colored image. Photographs taken through microscopes are called **micrographs.** Electron micrographs are always black and white because electron beams have no color. The colored electron micrographs in this book were produced by coloring black and white photographs to emphasize certain parts.

## Scanning Probe Microscopes

Several types of **scanning probe microscopes** were invented in the 1980s. Instead of aiming light or electrons at a sample, these microscopes feel their way across an object's surface with an extremely fine needle, much as a blind person might explore the surface of the ground with a cane. In some cases, the resolution of these microscopes is good enough to reveal individual atoms sticking up from the surface. More important, they permit us to look at specimens in new ways, measuring features such as magnetic fields, ion flow, electrical charge, or temperature of the surface. Since they do not require the use of a vacuum or special staining, specimens can be examined in life-like fluid environments. In one of the first interesting biological applications, investigators watched as molecules of the blood protein fibrin reacted to form a sheet, a key step in blood clotting.

## Specimen Preparation and Staining

A microscope produces an image with light and dark areas because light or electrons pass through low-density parts of the specimen but are absorbed or deflected by parts with higher density. Light microscopes also show the specimen's colors. Because most of the contents of living cells are transparent and colorless, it is often hard to distinguish details in these cells with an ordinary compound light microscope such as you might use in your laboratory class. However, various special (and expensive) microscopes enhance differences in density, and therefore in refraction (deflection of light waves) (Figure 5–4).

To increase the contrast in the final image, most specimens are specially prepared and stained for microscopy.

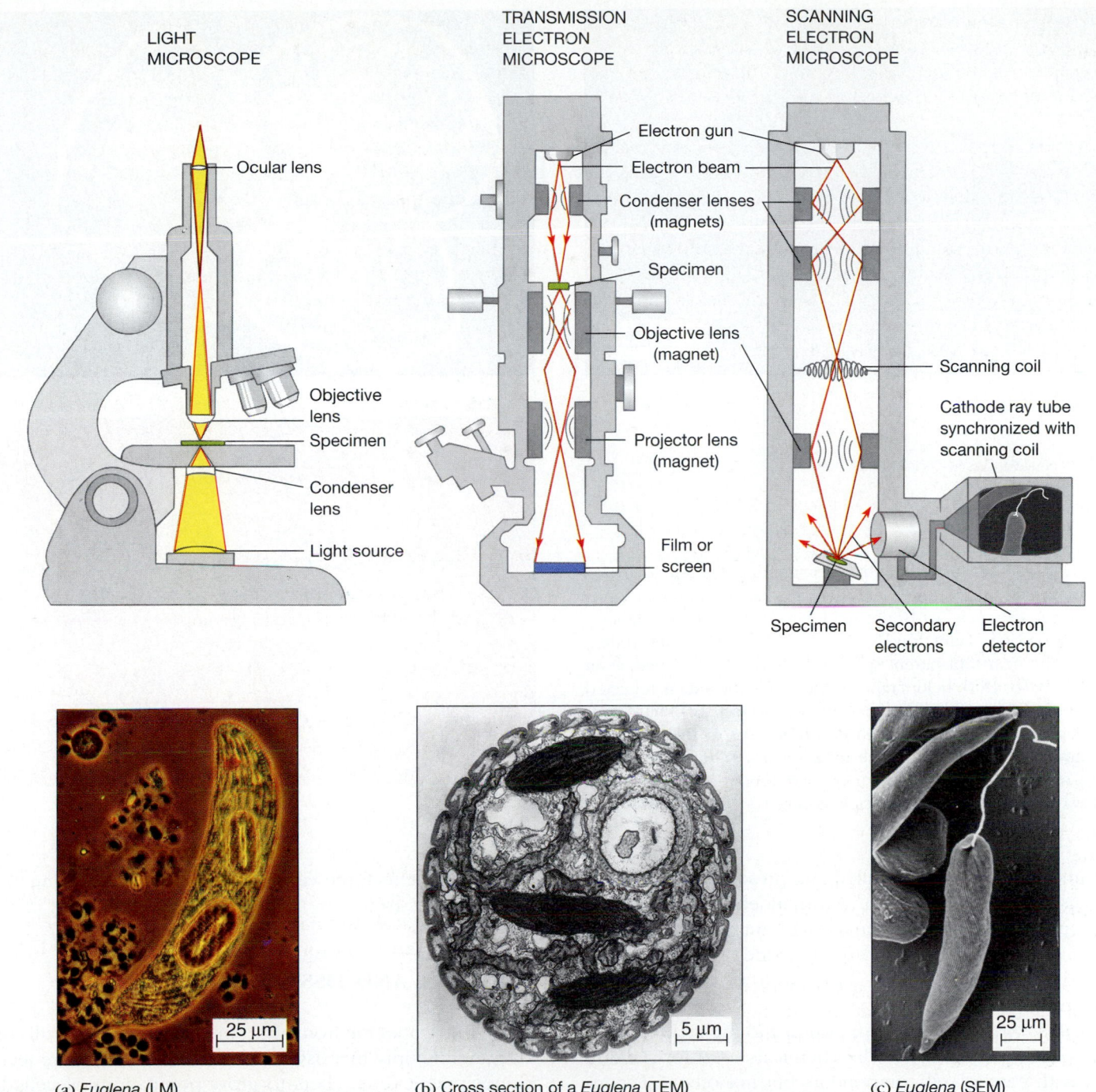

LIGHT MICROSCOPE

- Ocular lens
- Objective lens
- Specimen
- Condenser lens
- Light source

TRANSMISSION ELECTRON MICROSCOPE

- Electron gun
- Electron beam
- Condenser lenses (magnets)
- Specimen
- Objective lens (magnet)
- Projector lens (magnet)
- Film or screen

SCANNING ELECTRON MICROSCOPE

- Scanning coil
- Cathode ray tube synchronized with scanning coil
- Specimen
- Secondary electrons
- Electron detector

(a) *Euglena* (LM) — 25 μm

(b) Cross section of a *Euglena* (TEM) — 5 μm

(c) *Euglena* (SEM) — 25 μm

**FIGURE 5–3** Microscopes. Comparison of a light microscope with a transmission electron microscope and a scanning electron microscope. In the scanning electron microscope (right), the scanning coils deflect a fine beam of electrons so that it travels rapidly across the specimen, in synchrony with the spot on a cathode ray tube. The user views the specimen as a picture on the television screen.

The photographs show views of the protist *Euglena* taken with each kind of microscope. The light micrograph shows a whole cell, the TEM a thin cross section (slice) through the middle of a *Euglena,* and the SEM the cell's outer surface, with its pattern of fine ridges and its flagellum (the white line at the top).   (Biophoto Associates)

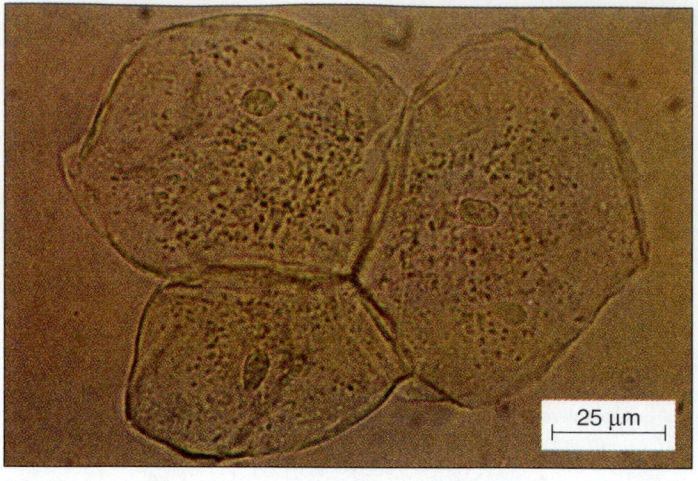

(a)

(b)

(c)

**FIGURE 5–4** Comparative light micrographs. The same living, unstained human cheek cells are viewed through different kinds of light microscopes. (a) Bright field, similar to a student laboratory microscope. Light travels in a line from the light source through the near-transparent specimen to the viewer's eye, as in Figure 5–3. (b) Dark field. Light shining from one side is reflected to the viewer by some of the specimen's surfaces. (c) Phase contrast. Light rays passing through different parts of the specimen are slowed in proportion to the local density. The microscope's optics recombine light so that these differences in phase of the light waves appear as differences in contrast.   (Biophoto Associates)

Specimens are first **fixed** (killed and preserved) in fixatives such as an aqueous solution of formaldehyde for light microscopy or glutaraldehyde and osmium tetroxide for electron microscopy. Then they are embedded in wax or plastic and **sectioned** into thin slices with a glass, metal, or diamond knife.

Stains for microscopy give contrast to the image by absorbing light or electrons. The chemicals used for staining react with certain cell components but not others. For instance, an acidic blue stain called hematoxylin reacts mainly with alkaline proteins attached to DNA in eukaryotic cells and therefore is used to stain genetic material for light microscopy.

For transmission electron microscopy, structures are stained with heavy metal ions, which absorb electrons and so produce dark areas in the final image. A specimen might be stained with lead solution, which reacts with acidic structures to leave a deposit of lead.

For scanning electron microscopy, the surface of the dried specimen is usually coated with a very thin layer of gold, platinum, or some other good emitter of secondary electrons. In effect, the viewer "sees" the metal coating, not the specimen itself.

## 5–B   CELL AND TISSUE CULTURE

Given the proper environment, some kinds of cells will survive and multiply in a tissue culture dish. Cell culture techniques permit researchers to grow populations of cells to study (Figure 5–5). Cell biologists can also fuse cells of different types together, to form hybrid cells. This technique has provided much useful information on how cell membranes break and form and on how a cell's genes operate to produce its proteins.

Cell culture techniques were developed for use in research, but they have found many practical applications. In some cases, researchers use tissue cultures instead of entire plants or animals to determine the effects of drugs or other chemicals. Tissue culture is also used to grow plants, particularly those that grow very slowly from seed. For example, orchid plants are produced by growing groups of

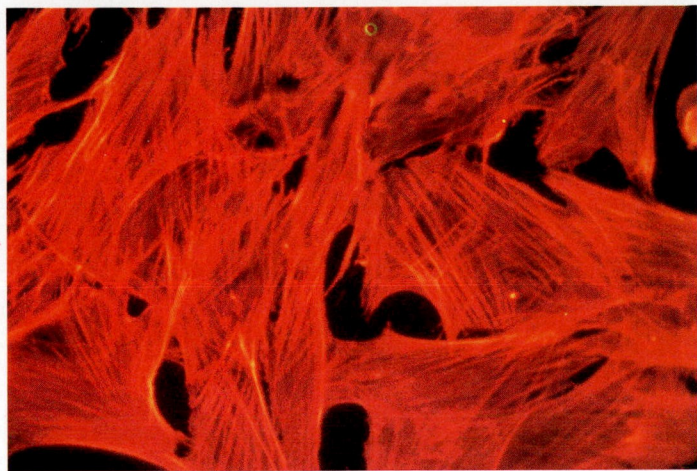

**FIGURE 5–5**  Cells in culture. These are human fibroblasts, connective tissue cells that produce a matrix of collagen fibers around themselves and can develop into a number of more specialized cell types. Filaments of the protein actin inside the cells have been stained with red fluorescent dye.   (Biophoto Associates)

cells until they form small plants that can be transplanted to pots.

Cultured cells from the outer layer of the skin are used to treat burn victims. Because the body rejects cells from other people, the victim's undamaged skin is used to grow new skin cells in culture to replace burned skin.

## 5–C   CELL FRACTIONATION

Researchers can use microscopes to study the structure and distribution of organelles, and even of some macromolecules, in a cell. However, to study the biochemical function of specific organelles, such as ribosomes, without interference from other cellular processes, they must isolate these organelles from the rest of the cell. This is done by taking cells from an organism or cell culture, mixing them with a sugar or salt solution, and then breaking them open with a tissue homogenizer or an electric blender.

The resulting suspension is subjected to **differential centrifugation,** a procedure that separates particles according to their size and density (Figure 5–6). A centrifuge

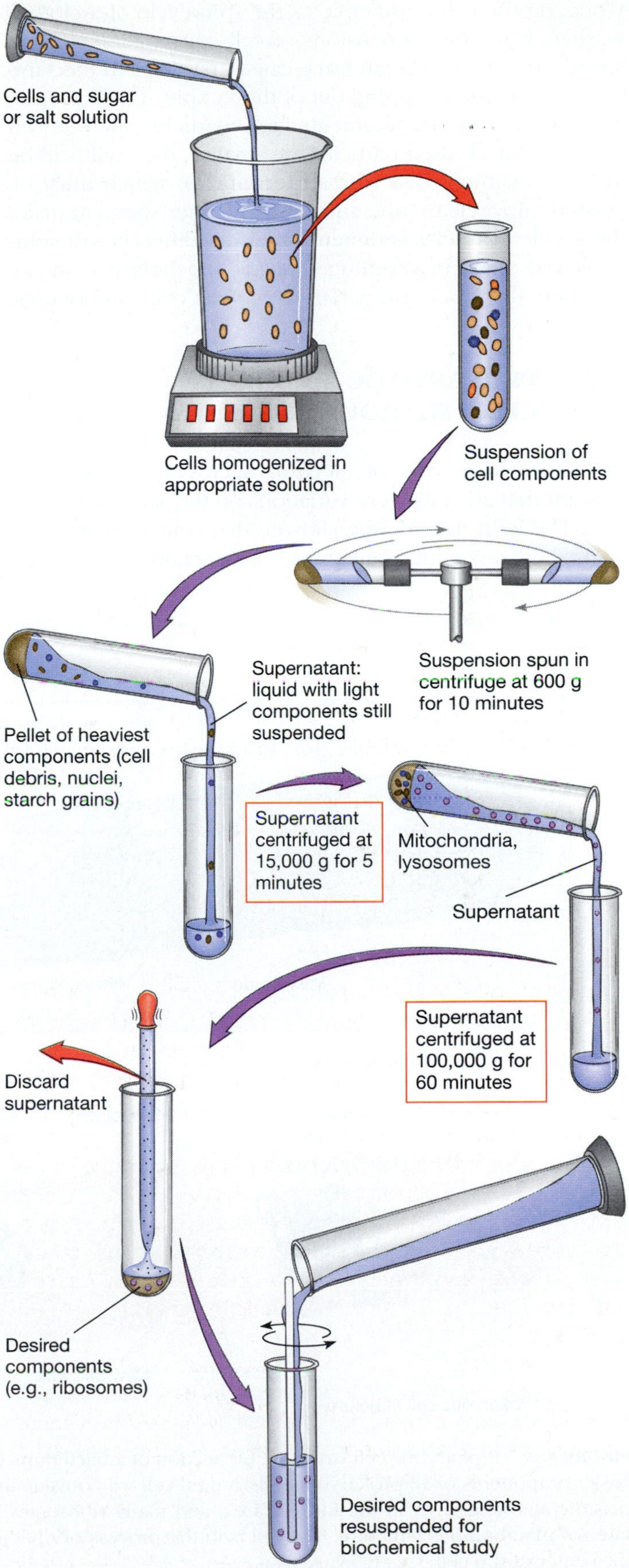

**FIGURE 5–6**  Separating cell components by differential centrifugation. The cells are broken apart in a buffered salt or sugar solution. (The solutes prevent osmotic swelling or shrinking of the membrane-bounded organelles.) The mixture is then poured into centrifuge tubes and spun in a centrifuge. Large particles settle into a pellet at the bottom of the tube; smaller, lighter ones remain suspended in the supernatant, which can be poured into another tube and spun at higher speeds to make them settle out. The pellet containing the desired components is then resuspended in fresh solution for biochemical experiments.

Cells and sugar or salt solution

Cells homogenized in appropriate solution

Suspension of cell components

Suspension spun in centrifuge at 600 g for 10 minutes

Supernatant: liquid with light components still suspended

Pellet of heaviest components (cell debris, nuclei, starch grains)

Supernatant centrifuged at 15,000 g for 5 minutes

Mitochondria, lysosomes

Supernatant

Supernatant centrifuged at 100,000 g for 60 minutes

Discard supernatant

Desired components (e.g., ribosomes)

Desired components resuspended for biochemical study

works on the same principle as the spin cycle of a clothes washer. It rotates suspensions of cell components at high speed, so that centrifugal force causes particles in the sample to sediment, dropping out of the sample. The first spin, done at low speed, sediments large particles such as cell nuclei. If the desired particles are smaller, they will still be in the remaining fluid (the **supernatant**), which must be poured into a clean tube and spun at higher speed to make the smaller particles sediment. By using different salt solutions and different centrifuge speeds, biochemists can obtain essentially pure preparations of most cell components.

## 5–D    PROKARYOTIC CELLS: CELLS WITHOUT NUCLEI

Before the invention of electron microscopes, biologists thought that all cells were variations on the same basic pattern. The light microscope showed that cells of plants, animals, and unicellular eukaryotes all contain many of the

same kinds of organelles. However, most bacteria are so small that light microscopes do not show their internal structure.

The electron microscope revealed that bacterial cells differ from those of higher organisms. In particular, the prokaryotic cells of bacteria lack the membrane-bounded organelles found in eukaryotic cells. Most importantly, they do not have a nucleus. Prokaryotic DNA occurs as a single, circular molecule of double-stranded DNA folded up in a **nuclear area** in the cytoplasm (Figure 5–7). Unlike the DNA of eukaryotes, the DNA of prokaryotes has very little protein attached to it.

The cytoplasm of prokaryotes contains many ribosomes, which carry on protein synthesis. These are smaller than the ribosomes of eukaryotic cells. (Section 5–G discusses ribosomes in more detail.)

The plasma membrane, present in all cells, is the only membrane in many prokaryotes. However, in some prokaryotes an internal membrane called the **mesosome** appears continuous with the plasma membrane (Figure 5–7). Its

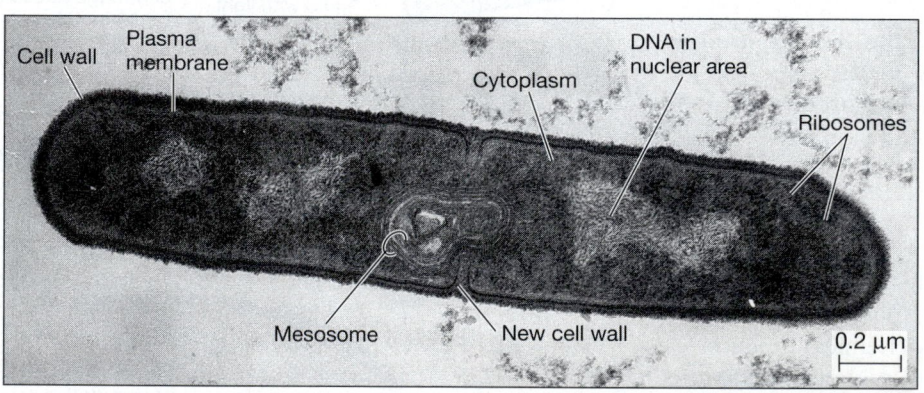

(a) A dividing bacterial cell (TEM)

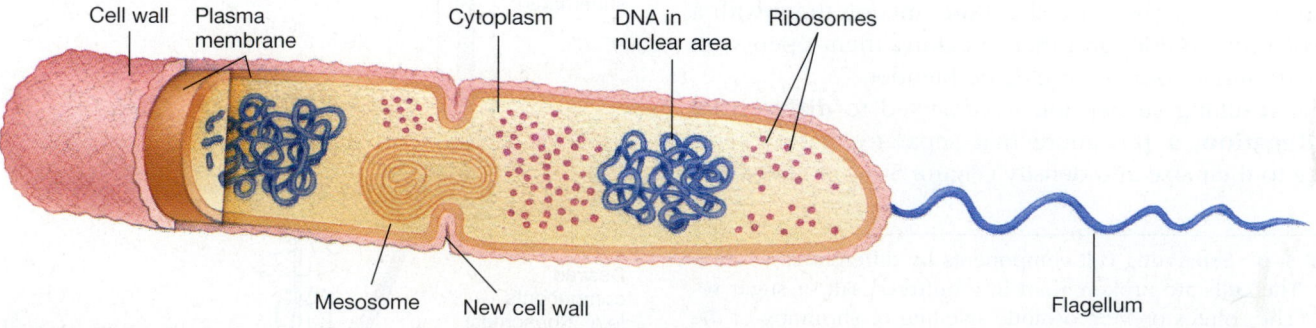

(b) Prokaryotic cell structure

**FIGURE 5–7**    Prokaryotic cell structure. (a) Section of a bacterium. (b) Interpretive art showing the basic components of all prokaryotic cells: a rigid cell wall outside the plasma membrane, and cytoplasm containing DNA in the nuclear area, and many ribosomes. Note the mesosome, the only internal membranous structure. The cell is in the process of dividing, as its wall grows inward in the center of the cell.    (a, Biophoto Associates)

function is unknown. In fact, some biologists doubt that it exists in living bacteria. Instead it may be an artifact, a pattern that looks like a structure but is actually produced by the process of fixing the cell for microscopy.

Photosynthetic bacteria contain many membranous vesicles (sacs) equipped with molecules needed to carry out photosynthesis (the manufacture of food, using light energy). These photosynthetic membranes lie in the cytoplasm, whereas in eukaryotic plant cells they occur in organelles called chloroplasts, separated from the cytoplasm by additional membranes.

The cytoplasm of a prospering bacterial cell contains granules of storage polymers, such as glycogen or other food reserves, and polyphosphate (a phosphorus reserve).

Most bacteria are surrounded by a thick cell wall, with essentially the same functions as a plant cell wall: it protects the cell, gives it shape, and keeps it from bursting in hypotonic media (Section 4–D). A bacterial cell wall is thought to consist of one huge, bag-like polymer of amino sugars (sugars containing amino groups) and several unusual amino acids, different from those in proteins. Penicillin and related drugs interfere with the building of this wall and therefore inhibit the growth of bacteria, but not of eukaryotes, which neither need nor make these kinds of polymers. In many bacteria, the cell wall is surrounded by an outer membrane, which prevents loss of enzymes that build the cell wall and entry of large molecules from the environment. Toxic substances occur in the outer membrane and cell wall of some bacteria. Many bacterial diseases, such as cholera, are caused by these toxic chemicals.

Some bacteria swim by means of one or more simple **flagella** (singular: **flagellum**). A bacterial flagellum is composed of three strands of the globular protein flagellin, wound together (see Figure 24–2). The flagellum rotates like a corkscrew and propels the cell through its liquid environment, powered by the energy of a steep gradient of $H^+$ ions across the plasma membrane (Section 6–E).

Chapter 24 examines the structure of prokaryotic cells in more detail and discusses how eukaryotes may have originated from prokaryotes.

## 5–E    EUKARYOTIC CELLS: CELLS WITH NUCLEI

Eukaryotic cells differ from prokaryotic cells in several ways. By looking at the comparison of cell types in Table 5–1 we can list four main differences:

1. **Size.** Most eukaryotic cells are much larger than prokaryotic cells.
2. **Compartments.** Unlike prokaryotic cells, eukaryotic cells typically contain many membrane-bounded organelles, notably the nucleus, which keeps the cell's genetic material separate from the cytoplasm. The membranes around organelles divide a eukaryotic cell into compartments, each housing molecules that perform a specific task. When a task is complete, the products move to other areas where they are used or processed further. Many of the cell's membranous compartments are linked, either physically or by membranous vesicles that bud off from one compartment and carry some of its contents to another.
3. **Cytoskeleton.** A eukaryotic cell contains a framework of protein fibers running through the cytoplasm, giving shape and support to the cell, and serving as tracks for the movement of organelles within the cell.
4. **Cohabitants.** Two kinds of membrane-bounded organelles, mitochondria and plastids, are thought to have evolved from prokaryotic cells that took up residence inside larger eukaryotic cells. Both organelles play roles in energy metabolism. Mitochondria provide energy in the form of ATP. Chloroplasts, found only in algae and plants, are plastids that make food by photosynthesis. Other kinds of plastids store food in plants. In return, eukaryotic cells provide mitochondria and plastids with raw materials and a sheltered environment. So in some ways it is useful to view a eukaryotic cell not as a single unit of life but as a community of interdependent cells.

All of these features are related: eukaryotic cells can be larger than prokaryotic cells because of their membranous compartments, their cytoskeletons, and their energy-processing organelles. In Chapter 4 we saw that cell size is limited by the surface area available to exchange substances with the environment and by the surface-to-center diffusion distance. In eukaryotic cells, some substances are carried between the plasma membrane and cell interior by membranous vesicles. This bulk transport, traveling along fibers of the cytoskeleton, moves molecules around the cell faster and more directly than diffusion would. Hence, the cell can support a larger volume of cytoplasm without expanding the plasma membrane's surface to the same extent. However, this larger volume still requires more total membrane surface for some vital metabolic activities. This is provided by the large surface area of the cell's many internal membranes.

Besides forming a rapid transport network, the cytoskeleton also provides support and structural reinforcement for the larger eukaryotic cell, especially in the case of animal cells, which do not have the support of cell walls.

In addition to the ribosomes, membrane-bounded organelles, and cytoskeleton, eukaryotic cells may contain food stored in lipid droplets and glycogen granules (Figures 5–8 and 5–9 on pages 99 and 100, respectively).

Cells also contain other components too small to distinguish even with the electron microscope. These make up the **cytosol,** the soluble part of the cytoplasm, a watery suspension containing ions and molecules, and surround-

**T A B L E   5 – 1**

## Comparison of Eukaryotic and Prokaryotic Cells

| Feature | Function | Eukaryotic Cells | | Prokaryotic Cells |
| | | Animal | Plant | |
|---|---|---|---|---|
| Average size | | 10–20 $\mu$m | 30–50 $\mu$m | 1–10 $\mu$m |
| Cell wall | Surrounding, supporting, and protecting cell | − | + (mainly cellulose) | + (amino sugar, amino acid polymers) |
| Plasma membrane | Enclosing cytoplasm; exchanging substances with environment | + | + | + |
| Nucleus (nuclear area) | Housing genetic material | + | + | Nuclear area |
| Nuclear envelope | Separating chromosomes from cytoplasm | + | + | No membranes around DNA |
| Chromosomes | Storing genetic information | Many, linear | Many, linear | One, circular |
| Nucleoli | Producing ribosomes | + | + | − |
| Ribosomes | Making proteins | + | + | + (smaller, different) |
| Endoplasmic reticulum | Segregating proteins to be secreted; acting as site of new membrane synthesis | + | + | − |
| Golgi complex | Modifying, sorting, and packaging cell products | + | + | − |
| Lysosomes | Digesting food and worn-out cell components | + (in many cells) | + (some cells) | − |
| Peroxisomes | Oxidizing toxins | + | + | − |
| Mitochondria | Breaking down food to produce ATP for energy (respiration) | + | + | − |
| Plastids | Making or storing food | − | + | − |
| Vacuoles | Storing fluid, food, pigments | + (some) | + (most) | − |
| Cytoskeleton | Providing cell shape; movement | + | + | − |
| Microtubules | Providing cell shape, forming spindle for chromosome separation during cell division | + | + | − |
| Centrioles | Organizing microtubules and basal bodies of cilia | + | Only in lower plants | − |
| Cilia, flagella | Moving cell or moving fluid past cell | + (tubulin) | Only in lower plants | + (in some; flagellin) |
| Intermediate filaments | Strengthening cytoskeleton | + | + | − |
| Microfilaments | Moving cell, changing cell shape | + | + | − |

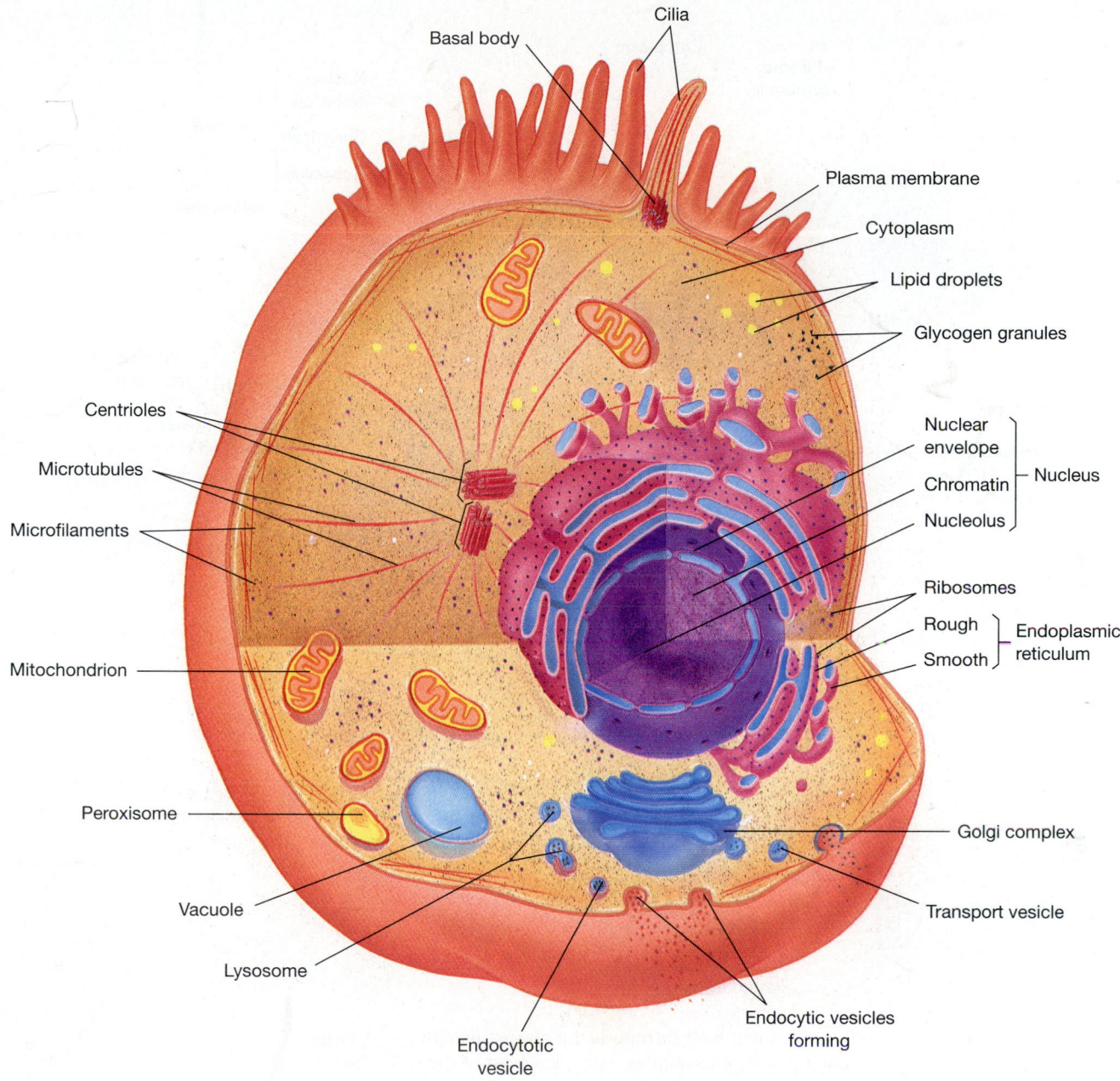

**FIGURE 5–8**   A generalized animal cell.

ing the visible structures in the cytoplasm. The cytosol takes up about half the volume of an animal cell. Most of a cell's general metabolism and protein synthesis occur here, and so it is not surprising that the cytosol is about 20 percent protein—mostly metabolic enzymes.

In the rest of this chapter we shall examine the most common structures in eukaryotic cells and the elements of the cytoskeleton, and then consider the organization of cells into tissues.

## 5–F   THE NUCLEUS: GENETIC MESSAGE CENTER

A eukaryotic cell's **nucleus** houses its genetic material, which contains instructions for making RNA and proteins. The nucleus is usually the most obvious structure in an animal cell when we look at it with a light microscope (see Figure 5–25). A plant cell nucleus is often less obvious but is visible with proper staining.

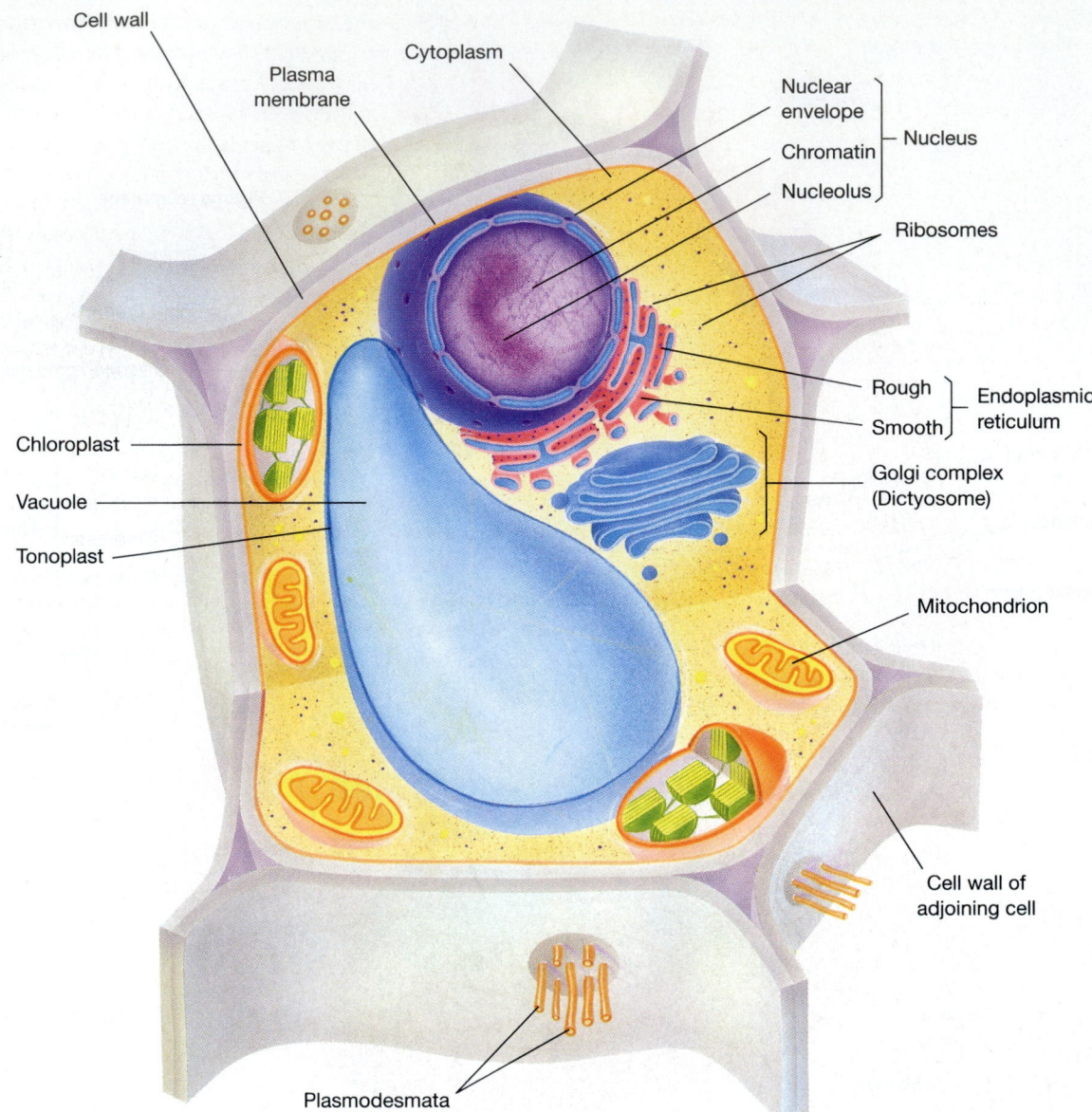

**FIGURE 5–9**  A generalized plant cell. A thick cell wall surrounds the plasma membrane. A large vacuole, a membrane-bounded sac containing various solutes, occupies most of the cell. Chloroplasts are prominent in photosynthetic plant cells like this one. Other components are much as in animal cells.

The genetic material in the nucleus of a eukaryotic cell is organized into chromosomes. Each **chromosome** consists of a long double strand of DNA with associated RNA and proteins. When the nucleus divides, the chromosomes coil up into short, thread-like structures. While chromosomes are coiled up like this, their information cannot be used. Most of the time, though, the nucleus is not dividing, and the chromosomes uncoil in a loose, indistinct tangle called **chromatin,** in which some of their information is accessible (Figure 5–10).

DNA in the chromatin determines what RNA molecules are made in the nucleus. Some of the RNA travels to the cytoplasm, where it directs protein synthesis. Therefore, DNA also determines what enzymes and other proteins the cell makes. This, in turn, affects most of the cell's structures and activities. Hence, the nucleus has been called the cell's "control center." However, control is a two-way street. Substances in the cytoplasm enter the nucleus and influence the DNA, thus changing which RNAs, and hence which proteins, are produced, depending on the cell's needs.

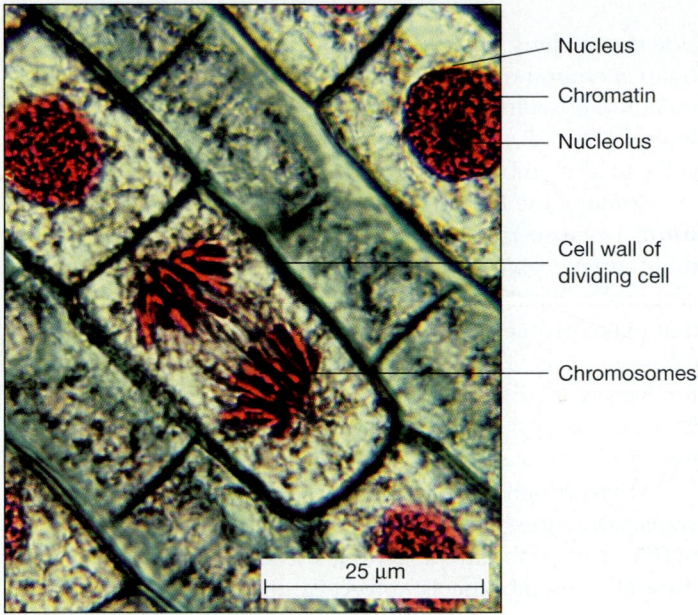

Onion root tip cells (LM, X400)

**FIGURE 5–10**   Nuclei and chromosomes. Chromatin is stained red in these cells from an onion root tip. Circular areas are cell nuclei, with their red chromatin spread in an indistinct mass containing darker blobs, the nucleoli. Individual chromosomes appear as red threads in the dividing cell (center). (Chuck Brown/ Photo Researchers)

With the light microscope, the only obvious features in the nuclei are usually the **nucleoli** (singular: **nucleolus**), areas where ribosomes are synthesized. The information needed for making ribosomes lies in parts of several chromosomes. After the cell divides, each of these becomes active and is soon surrounded by a dense mass of RNA and protein. These areas grow and may merge into one ribosome-producing region. This dense material disperses when the cell divides, and so the nucleolus disappears (Figure 5–10).

The nucleus is surrounded by a double membrane, the **nuclear envelope** or **nuclear membrane,** which is perforated by hundreds of pores (Figure 5–11). Through these pores, RNA and ribosomes made in the nucleus go out into the cytoplasm, and nuclear proteins made in the cytoplasm come in. (All of these are very large as well as hydrophilic, so they could not pass through the lipid layers of the nuclear envelope.) Each pore has an open channel about 9 nm across, surrounded by proteins in a doughnut-like **nuclear pore complex,** which apparently controls what passes through the pore. For instance, it is thought that RNA molecules must contain specific nucleotide sequences if they are to leave the nucleus, and only proteins with specific signal sequences are taken in. Some of the things that go

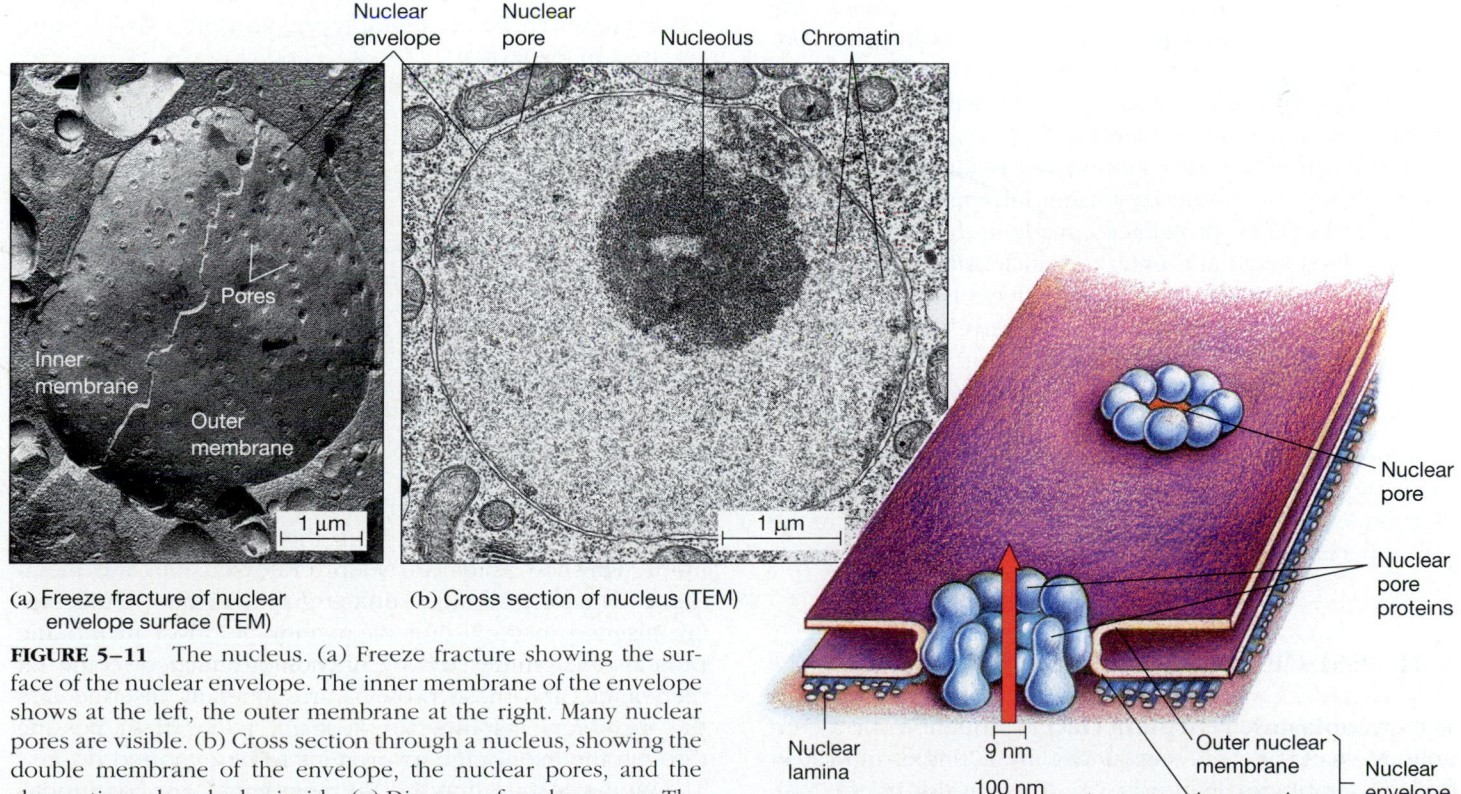

(a) Freeze fracture of nuclear envelope surface (TEM)

(b) Cross section of nucleus (TEM)

(c) Closeup of nuclear pores

**FIGURE 5–11**   The nucleus. (a) Freeze fracture showing the surface of the nuclear envelope. The inner membrane of the envelope shows at the left, the outer membrane at the right. Many nuclear pores are visible. (b) Cross section through a nucleus, showing the double membrane of the envelope, the nuclear pores, and the chromatin and nucleolus inside. (c) Diagram of nuclear pores. The pores are partly blocked by proteins of the nuclear pore complex. (Biophoto Associates)

through the pores are much larger than 9 nm, so apparently the pores can open wider at times. The active transport of many of these items through nuclear pores uses energy from ATP, but we do not yet know how it works.

The inner and outer membranes of the nuclear envelope connect with each other through the nuclear pores, and the outer membrane, in turn, is continuous with a system of membranes in the cytoplasm, the endoplasmic reticulum (Section 5–H). These interconnections allow the nuclear envelope to shrink or grow rapidly by losing material to, or gaining it from, the endoplasmic reticulum. In most eukaryotes, the nuclear envelope breaks down into small vesicles that disperse during cell division and re-form the nuclear envelope afterwards.

The inner layer of the nuclear envelope is lined by a meshwork of protein fibers, the **nuclear lamina,** which maintains the shape of the nucleus.

## 5–G  RIBOSOMES: PROTEIN ASSEMBLERS

**Ribosomes** are the sites of protein synthesis and so are necessary components of virtually all cells. Ribosomes are only about 23 nm in diameter, so that even in electron micrographs they look like little more than small granules (see Figure 5–12). Most of our knowledge of ribosomes comes from laboratory studies using ribosomes prepared by differential centrifugation (Section 5–C). Ribosomes are roughly half RNA and half protein, and each ribosome consists of one large and one small subunit. The ribosomes of prokaryotes and eukaryotes are essentially similar, but prokaryotic ribosomes are smaller.

Eukaryotic ribosomal subunits are made in the nucleolus. Their RNA is synthesized using information carried by nuclear DNA. Their proteins are made in the cytoplasm (at existing ribosomes) and enter the nucleus through the nuclear pores. After RNA and proteins have been assembled into ribosomal subunits, the subunits leave the nucleus via the pores and start work in the cytoplasm. Chapter 10 discusses the role of ribosomes in protein synthesis.

A cell may contain up to half a million ribosomes, depending on how much protein the cell makes. Some ribosomes appear to be attached to part of the cytoskeleton (Section 5–Q), and some are attached to membranes in the cell, especially those of the endoplasmic reticulum.

## 5–H  ENDOPLASMIC RETICULUM

The **endoplasmic reticulum (ER)** is thought to be a continuous, maze-like sac surrounded by a single, intricately folded membrane (Figure 5–12). The interior of this sac, the ER **lumen,** provides the cell with a storage compartment separate from the cytosol. Part of the ER lies just out-

side the nucleus, since its membrane is continuous with the outer membrane of the nuclear envelope and its lumen is continuous with the space between the envelope's membranes. The ER usually accounts for more than half of the cell's total membrane.

Some of the ER is known as **rough endoplasmic reticulum** because ribosomes attached to the outer (cytoplasmic) surface give it a bumpy appearance in electron micrographs (Figure 5–12). There is lots of rough ER in cells that make proteins which will be secreted from the cell (for example, cells of the pancreas that make digestive enzymes for export to the small intestine). In fact, this association between ER and ribosomes makes it possible for these proteins to leave a cell.

When proteins are folded into their final three-dimensional structures, they cannot pass through the plasma membrane. The problem of getting these large molecules out through a membrane is solved by the ER. As the proteins are synthesized, the end formed first contains an export signal that causes the ribosome to attach to the ER membrane. Here the protein chains are pushed through the ER membrane into the lumen before they begin to fold up.

Moving through the ER lumen, the proteins arrive in an area of **smooth endoplasmic reticulum,** with no ribosomes attached to it. Here, patches of the ER membrane bud off and form transport vesicles, sacs carrying proteins from the ER lumen (Figure 5–13). These vesicles move to a Golgi complex (Section 5–I, next) where the proteins are modified further before being repackaged into vesicles and released by exocytosis.

Most cells contain only a little smooth ER, the area where transport vesicles form. However, some cells have more, especially cells that synthesize a lot of lipid (such as steroids) or lipoprotein, and cells that detoxify harmful substances. Faced with large quantities of certain drugs, the smooth ER of some liver cells expands enormously, decreasing again when the drug has been rendered harmless. This suggests that the enzymes involved in detoxification and in lipid synthesis can function only when they are attached to membranes. ER provides such a membrane surface, and new membrane material can be added as necessary to accommodate more enzymes.

Most of the cell's new membrane is produced in the endoplasmic reticulum. The ER membrane contains enzymes that use substrates from the cytosol to make new lipids. Finished lipids are simply released into the membrane. Thus, membrane surface grows as new molecules are inserted into existing ER membrane. Most membrane proteins are synthesized by ribosomes attached to the ER membrane, and these proteins are generally inserted into the membrane as they are formed, rather than passing through the membrane as proteins to be exported do.

We have seen that the ER membranes are continuous with the nuclear envelope and that parts of the ER can pinch off as vesicles, which fuse with the membranes of Golgi

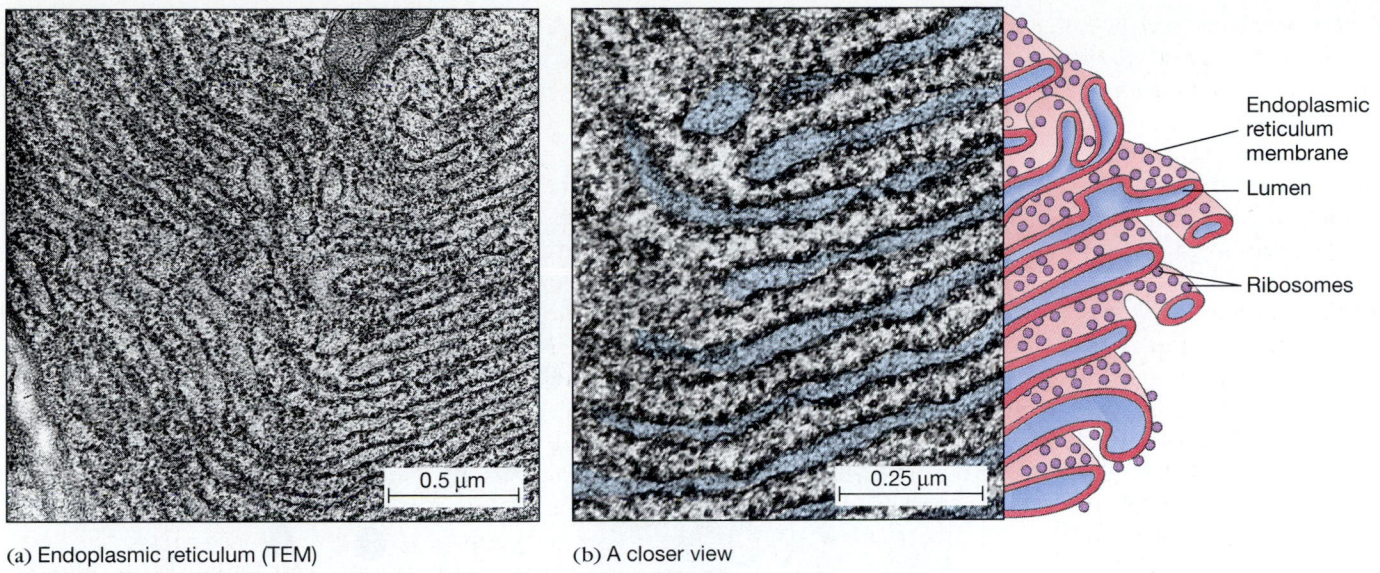

(a) Endoplasmic reticulum (TEM)    (b) A closer view

Endoplasmic
reticulum
membrane

Lumen

Ribosomes

**FIGURE 5–12** Rough endoplasmic reticulum. (a) The many ribosomes on the membrane's outer surface give the bumpy appearance. This cell is from the pancreas of a mouse. (b) A higher-power view with an artist's interpretation of three-dimensional structure.

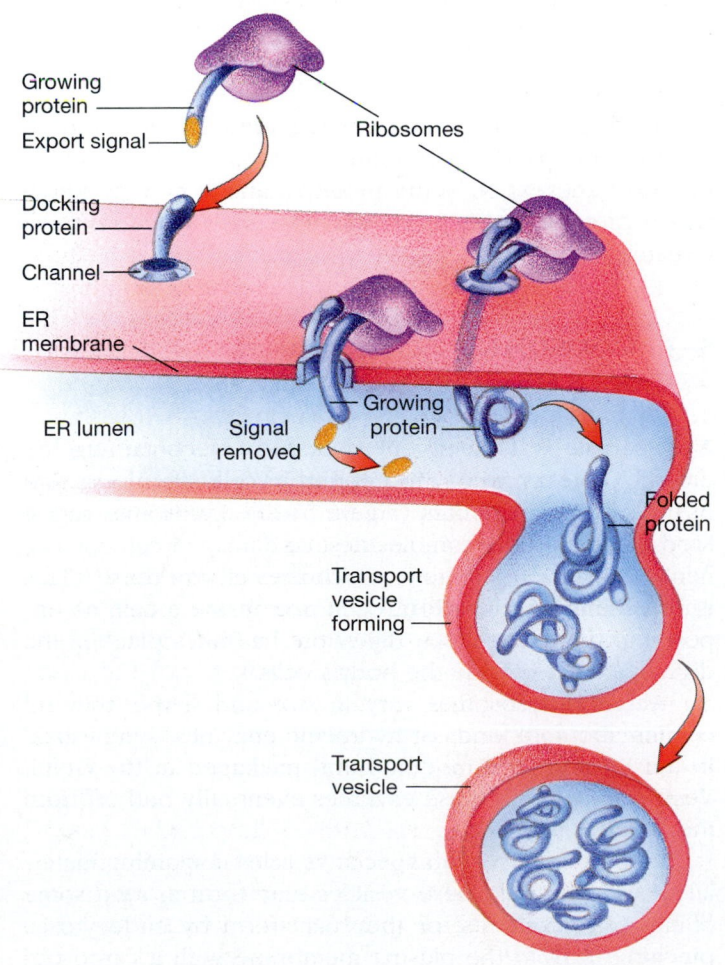

complexes and go from there to still other membranes. In this way new membrane material can be added to existing membranes either by direct transfer from the ER or by fusion of vesicles to existing membranes (see Figure 5–16). However, the different membranes in a cell contain different kinds and proportions of lipids and proteins, and we do not yet know how the molecules end up in the right place.

## 5–I    GOLGI COMPLEX: MOLECULAR MODIFICATION AND SORTING AREA

The transport vesicles pinched off from the ER soon fuse with larger sacs that are part of a **Golgi complex** (nickname: Golgi): a stack of flat, membranous sacs like a pile of pita bread. It is surrounded by many smaller, round transport vesicles carrying molecules to or from these large sacs

**FIGURE 5–13** How protein synthesis on rough endoplasmic reticulum permits a hydrophilic protein to cross a membrane. As protein synthesis begins, a sequence of amino acids near the front end signals that the protein is to be exported. Ribosomes making proteins with this signal attach to docking proteins associated with channels in the ER membrane. The growing proteins pass through these channels into the ER lumen. Later, transport vesicles pinched off from the ER membrane carry the proteins to Golgi complexes.

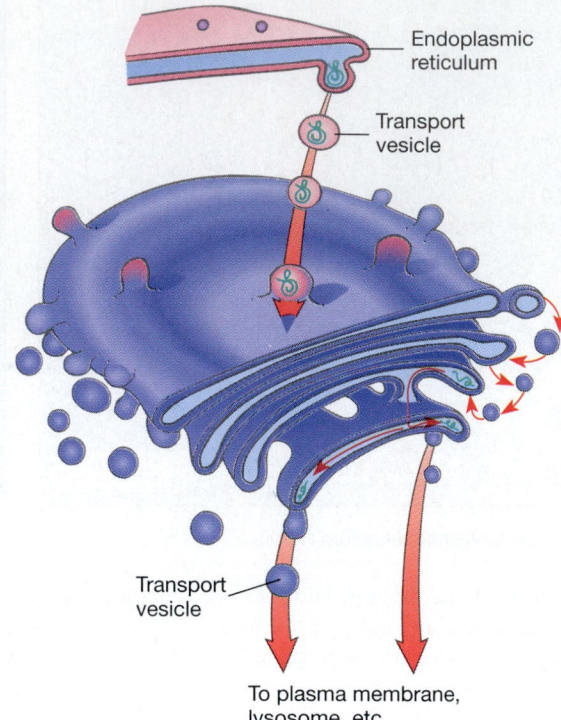

**FIGURE 5–14** Golgi complexes. (a) Two Golgi complexes at right angles to each other. The membranes that look like a pile of sliced pita bread (top) are a stack of Golgi vesicles cut longitudinally (from top to bottom of the stack). The circular structure at the bottom of the photo is one vesicle seen from above. (b) Drawing of a Golgi complex, sliced at the right to show the characteristic structure seen in electron micrographs such as (a). (a, Biophoto Associates)

(a) Two Golgi complexes (TEM)

(b) Proteins moving through a Golgi complex

Endoplasmic reticulum

Transport vesicle

Transport vesicle

To plasma membrane, lysosome, etc.

(Figure 5–14). Like the ER, the Golgi often lies near the nucleus. A cell may have one large Golgi complex or up to hundreds of smaller ones. In plant cells, the Golgi is often called the **dictyosome.**

The Golgi's role is to modify, sort, and package molecules such as proteins and glycoproteins made in other parts of the cell. These molecules then travel on to their final destinations in other organelles or outside the cell. Molecules being processed in the Golgi move from one Golgi sac to the next in sequence, carried by transport vesicles. In each sac, the molecules are modified by enzymes and then enclosed in another transport vesicle, which fuses with a Golgi sac containing enzymes for the next biochemical steps. The last sac finishes, sorts, and packages the molecules into transport vesicles according to "address labels" that have been attached in the Golgi. For example, proteins for lysosomes are labeled with sugar (mannose) phosphate groups at their first stop in the Golgi. In the last sac these proteins bind to membrane receptors specific for mannose phosphate and become organized to bud off in a lysosomal vesicle.

Molecules to be exported from the cell are enclosed in **secretory vesicles,** which pinch off from the Golgi sac membranes, move to the plasma membrane, and discharge their contents by exocytosis. These molecules include things like mucus and digestive enzymes, secreted by cells lining the digestive tract, or hormones from gland cells. Other vesi-

cles leaving the Golgi carry new proteins and lipids to be added to the plasma membrane itself. How cells produce, package, and export some proteins, and how they know which proteins to keep, are exciting areas of current cell research.

## 5–J   LYSOSOMES: SACS OF HYDROLYTIC ENZYMES

A **lysosome** is a membrane-bounded sac containing hydrolytic (digestive) enzymes, found in cells of animals and in unicellular eukaryotes (Figure 5–15). Lysosomes digest food, unneeded macromolecules, or damaged cell components into their monomers, which the cell can reuse. (This is **intracellular digestion,** occurring inside a cell, as opposed to the extracellular digestion that takes place in the digestive tract, outside the body's cells.)

Although lysosomes vary in size and shape, they all contain about 40 kinds of hydrolytic enzymes, synthesized in the ER and then modified and packaged in the Golgi. Vesicles filled with these enzymes eventually bud off from the Golgi as lysosomes.

Lysosomes fuse with special vesicles containing material to be digested. These vesicles may form around some of the cell's contents, or they may form by endocytosis, pinching in from the plasma membrane with a cargo ob-

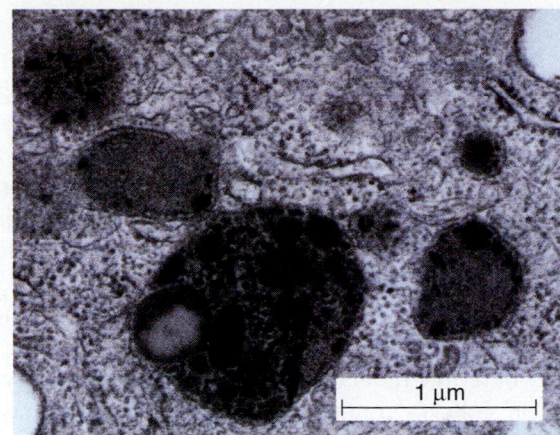

**FIGURE 5-15** Lysosomes. The five dark membrane-bounded organelles are lysosomes in a human cell.   (Biophoto Associates)

off from Golgi complexes expel materials by exocytosis, and the vesicle membrane becomes part of the plasma membrane. Membrane material is returned to structures inside the cell by endocytosis. Inside the cell, lysosomes budded from the Golgi and endocytotic vesicles from the plasma membrane fuse, and their membrane material too may later travel elsewhere. These interrelated internal membranes are sometimes collectively called the **endomembrane system** (Figure 5-16).

At all times the contents of membrane-bounded organelles are separate from the cytosol. In some ways, the

tained from outside the cell. Lysosomes digest a variety of things: food molecules, disease-causing viruses brought into the cell by endocytosis, damaged organelles, or macromolecules in the cell. The entire cell may even die and chunks of it be engulfed by surviving cells and digested by their lysosomes. This is what happens when a tadpole changes into a frog: the cells of the tail are digested, and the molecules released are reused by other cells.

The appropriate transport vesicles and lysosomes must be able to recognize each other. It is vital that the vesicle fuse with a lysosome and not with, say, part of the ER. We do not yet know how this is controlled. It is only one of many cases requiring specific recognition between membrane surfaces.

If the lysosomal enzymes were to escape into the cytoplasm, they could digest just about everything in the cell. This does not happen because the enzymes work best at a pH of about 5, found inside lysosomes, whereas the pH in the cytoplasm is about 7.2. Such a difference in pH would slow down any escaped lysosomal enzymes and protect the cell in case of lysosomal accidents.

## 5-K   THE ENDOMEMBRANE SYSTEM

We have seen that many membranes in the cell are connected, either physically or by way of transport vesicles that bud off from one membrane and fuse with another. The outer membrane of the nuclear envelope is continuous with the endoplasmic reticulum. The far side of the ER buds off transport vesicles, containing substances produced in the ER, which merge with Golgi sacs. Some vesicles that bud

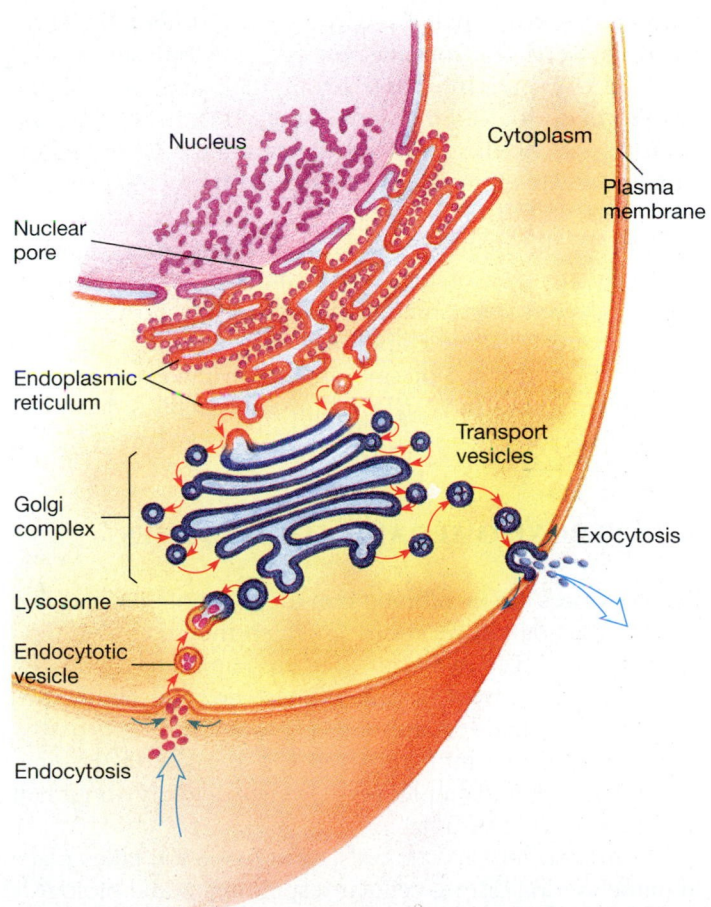

**FIGURE 5-16** Endomembrane system. The endoplasmic reticulum is continuous with the outer membrane of the nuclear envelope, which surrounds the nucleus. Transport vesicles bud off from the endoplasmic reticulum and carry substances to the membranous sacs of Golgi complexes. Vesicles from Golgi complexes travel to lysosomes or expel materials by exocytosis. The arrows show how membrane material is exchanged among the nuclear envelope, plasma membrane, and many membrane-bounded compartments in the cytoplasm. Areas inside compartments, shown in blue, are equivalent to the exterior of the cell.

interiors of these organelles are equivalent to the exterior of the cell. This is most obvious in the case of endocytotic vesicles, which carry captured bits of the outside world into the cell, but it also applies to the other organelles mentioned. For example, the oligosaccharides of glycoproteins always lie outside the cell or in the interiors of membrane-bounded organelles such as the ER and Golgi, where they are synthesized, but never in the cytosol.

From this discussion, we can see why the nuclear envelope is a double membrane. The part of the nucleus equivalent to the cell's exterior is not the compartment housing the chromosomes, but the one between the envelope's inner and outer membranes, which is continuous with the ER lumen. During cell division, the envelope breaks up into vesicles, each enclosing some of this "external" compartment, and the chromosomes mingle for a time with the cytoplasm. Later we shall see why a cell must keep the chromosomes separate from the rest of the cytoplasm most of the time (Chapter 10), while letting them mingle with the cytoplasm during cell division (Chapter 14). In contrast, the contents of all the cell's other membrane-bounded organelles always remain separated from the cytoplasm and never mix with it.

Not all the membrane-bounded organelles of eukaryotic cells are part of the endomembrane system. Peroxisomes, glyoxysomes, mitochondria, and plastids (Sections 5–L, M, and N) obtain lipid and protein molecules individually from the cytoplasm, rather than incorporating existing membranes in vesicles as other organelles do.

## 5–L  PEROXISOMES AND GLYOXYSOMES

**Peroxisomes** are membrane-bounded sacs of enzymes that carry out oxidation reactions in which they combine oxygen with various substrates. They are named for hydrogen peroxide ($H_2O_2$), which some of these enzymes make. Another peroxisome enzyme, catalase, uses hydrogen peroxide to detoxify harmful substances, especially in the liver and kidneys. For example, peroxisomes detoxify about half the ethanol we consume.

In modern eukaryotic cells, most oxidation takes place in mitochondria during cellular respiration. Some biologists think peroxisomes are the remnants of ancient oxidizing organelles that have largely been supplanted by mitochondria. The advantage of mitochondria over peroxisomes is that the major product of their oxidation reactions is ATP, which the cell can use for energy.

**Glyoxysomes** are similar to peroxisomes but contain additional enzymes, of the "glyoxylate pathway," which helps convert fatty acids to carbohydrates. Glyoxysomes are most common in the seedlings of plants. Here they convert lipids stored in the seed into carbohydrates that can be used to build structures such as cell walls.

## 5–M  MITOCHONDRIA: MILLS MAKING ATP

**Mitochondria** (singular: **mitochondrion**) are large organelles sometimes called the powerhouses of eukaryotic cells because they produce most of the cell's energy supply of ATP. They do this by **cellular respiration,** the process that uses oxygen in the breakdown of small organic molecules to carbon dioxide and water. ATP molecules leave the mitochondrion and supply energy for most of the cell's activities, and mitochondria often lie near structures that need this energy, such as flagella, which use a lot of energy when they move. Cells that use a lot of energy, such as muscle cells in animals, or cells in the growing root tips of plants, have many mitochondria. A busy biochemical factory such as a liver cell may contain about 2500 mitochondria, which occupy about one fifth of the cell's volume.

A mitochondrion has two membranes, an outer membrane separating the mitochondrion from the cytoplasm and a highly folded inner membrane (Figure 5–17). These are not part of the endomembrane system (Section 5–K).

Mitochondria contain their own genetic machinery (DNA, RNA, and ribosomes) and make some of their own proteins and lipids. They also reproduce: new mitochondria are formed by division of existing ones, and cells cannot make mitochondria from raw materials. In all these features, they are more like bacteria than like eukaryotic cells. Hence most biologists think that mitochondria evolved from prokaryotic cells living symbiotically inside larger cells, a theory we explore in Section 24–H.

Generally, a eukaryotic organism inherits half the DNA in its nucleus from each parent. However, the same is not true of mitochondrial DNA, nor of chloroplast DNA in photosynthetic eukaryotes. Male sperm and pollen contribute nuclear chromosomes to their offspring, but in most species all the mitochondria or chloroplasts in a fertilized egg come from the egg cytoplasm of the female parent. Therefore the genetic information in these organelles comes only from the mother.

## 5–N  PLASTIDS: FOOD FACTORIES AND STOREHOUSES

Most organelles described so far occur in both plant and animal cells, and their functions are essential to nearly all cells. (The exceptions are lysosomes and glyoxysomes.) However, if we were to examine cells with a microscope, three features would tell us that they came from a plant rather than an animal: plastids, large vacuoles, and cell walls.

**Plastids** are organelles that bear striking similarities to some prokaryotic cells. Like mitochondria, they contain DNA, RNA, and ribosomes, and they reproduce. Plastids are bounded by two outer membranes separating them from

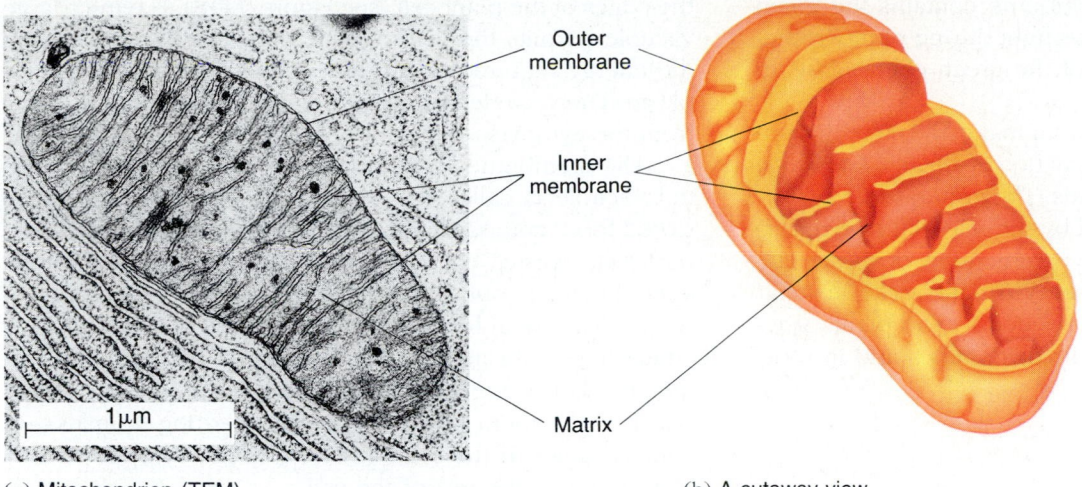

(a) Mitochondrion (TEM)

(b) A cutaway view

**FIGURE 5–17**   Mitochondria. (a) Section of a mitochondrion from the pancreas of a bat. The inner membrane forms narrow folds extending deep into the matrix inside the organelle; the outer membrane is much smoother. ATP produced by the mitochondrion powers protein synthesis by the many ribosomes in the surrounding cytoplasm and attached to the ER. (b) Interpretive drawing of mitochondrial structure.   (a, Keith Porter)

the cytoplasm, and they also contain a separate system of internal membranes. None of these membranes is part of the endomembrane system (Section 5–K). There is good evidence that plastids are descended from free-living photosynthetic bacteria that set up housekeeping inside larger cells. Whereas virtually all eukaryotic cells contain mitochondria, plastids are found only in photosynthetic eu-

karyotes—plants and algae. A plastid may develop into a chloroplast, a chromoplast, or a leucoplast, depending on the type of cell in which it occurs.

**Chloroplasts** are the green plastids that carry out photosynthesis—the manufacture of food using carbon dioxide, water, and light energy. The internal membrane of a chloroplast is highly folded (Figure 5–18). Chloroplasts are

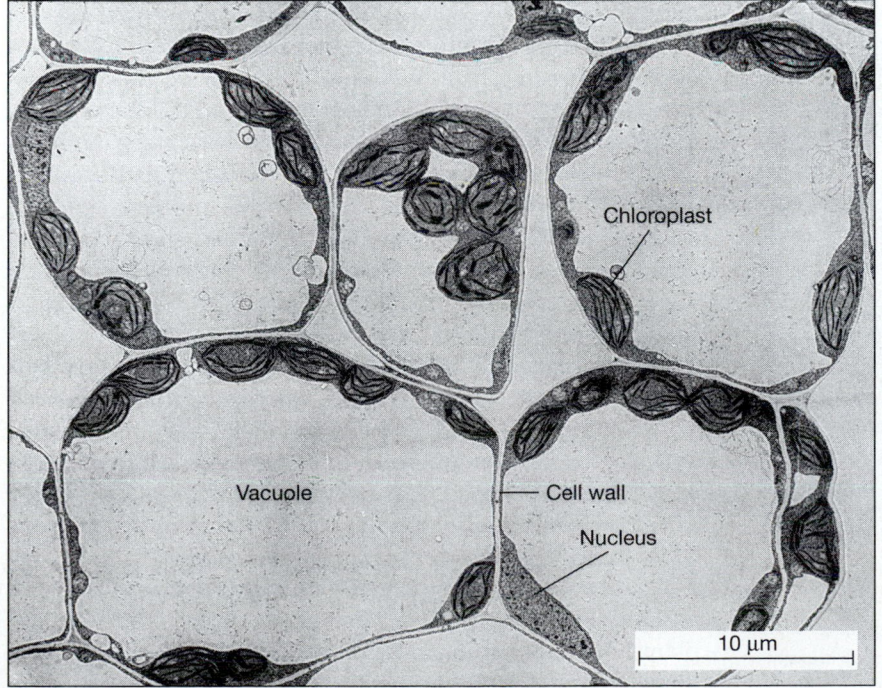

(a) Spinach leaf cells, low-power TEM

**FIGURE 5–18**   Plant cells. (a) These spinach leaf cells show the characteristic features of plant cells: cell walls, large vacuoles, and plastids (in this case, many chloroplasts). (b) Drawing of a chloroplast, showing the two outer membranes and the highly folded green photosynthetic membrane inside.   (a, Biophoto Associates)

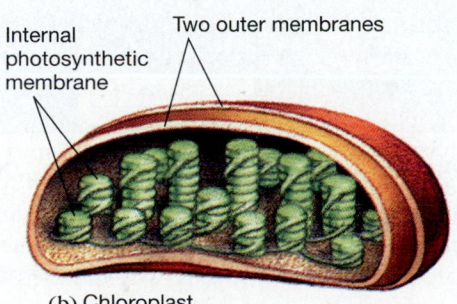

(b) Chloroplast

green because the internal membrane contains the green pigment chlorophyll, which traps light during photosynthesis. Leaves are green because of the green chloroplasts in some of their cells.

**Chromoplasts** are plastids that make and store the yellow and orange pigments of many flowers, fruits, and roots. **Leucoplasts** are colorless plastids. They often serve as storage areas and may be classified by the material stored: carbohydrates, fats, proteins, or a combination of these. The most common leucoplasts, the **amyloplasts,** take in sugar and store it as starch until it is needed by the plant. Amyloplasts are abundant in the cells of potatoes and in roots (Figure 5–19).

## 5–O  VACUOLES: SACS OF FLUID

A **vacuole** is a membrane-bounded sac of fluid. Vacuoles occur in many cells but are particularly large and common in plant cells. Here they are thought to begin as Golgi vesicles, which fuse and expand until they fill most of the plant cell's volume. As a result the nucleus, plastids, mitochondria, and other cytoplasmic organelles are crowded around

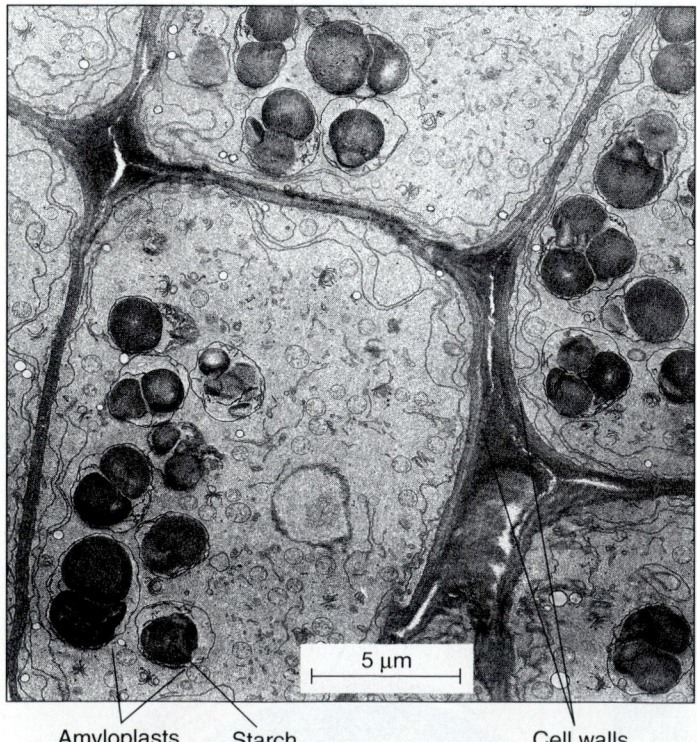

Amyloplasts    Starch                     Cell walls

**FIGURE 5–19**  Amyloplasts. These cells from a root tip of a corn plant contain many starch-storing amyloplasts. The cell walls are made up of layers of cellulose fibers and other substances.  (Barrie Juniper)

the edges of the plant cell (see Figure 5–18). Having a large vacuole permits the cell to grow bigger without spending the energy required to produce a lot of protein-rich cytoplasm. The vacuole also stores water that can replace losses from the cytoplasm during dry periods.

The membrane surrounding the plant cell's large central vacuole is called the **tonoplast.** The tonoplast holds stored food, anthocyanin pigments (deep red, blue, or purple), salts, wastes, and other substances inside the vacuole. Vacuoles may contain digestive enzymes and act as lysosomes. The vacuole is also convenient as a storage locker where toxic substances can be segregated from the rest of the cell. For instance, some acacia trees store cyanides—which make them poisonous to plant-eating animals—in their vacuoles. If the cyanides were in the cytoplasm, they would poison the rest of the cell.

## 5–P  CELL WALLS: PROTECTION AND SUPPORT

A plant cell's plasma membrane is surrounded by a thick but porous **cell wall.** The cell wall holds the cell in shape, and the cell walls of adjacent plant cells are cemented firmly to one another, defining the plant's general structure. Composed of cellulose and other fibers, the cell wall is porous enough to allow water and small dissolved substances to pass through it, tough enough to give the plant structure and support, and flexible enough to permit the plant to bend in the wind without breaking.

When a plant cell divides, the two new cells build a common partition, the **middle lamella,** between them. This middle lamella is composed largely of polysaccharides called **pectins,** the substances that make fruit set into a jam or jelly. Ripening fruit becomes soft as its pectins are broken down by the enzyme pectinase. This explains why extra pectin must be added to jam made with overripe fruit.

Each plant cell builds an elastic **primary cell wall** on its own side of the middle lamella. The primary cell wall contains thin fibers of cellulose which can slide past each other, allowing the wall to stretch as the cell grows (see Figure 3–13). These fibers are embedded in a matrix of branched polysaccharides and glycoproteins.

Many plant cells lay down a more rigid **secondary cell wall** when they are fully grown. Wood is composed mainly of secondary cell walls in which the cellulose has been reinforced by a strengthening material called **lignin.** Cells whose main function is support, such as those of a tree trunk, contain more reinforcing lignin in their secondary cell walls than do other types of cells.

Cell walls contribute largely to the structural support of a plant's cells and of the entire plant body. In soft-bodied plants, such as lettuce, the cell walls are thinner and more flexible than in woody plants, and the cell walls alone cannot support the plant. The cells' large central vacuoles must also be filled with water, so that the cells exert an outward (turgor) pressure on their walls. Losing too much water decreases the turgor pressure that keeps the cells firm, and the plant wilts. Turgor pressure is also the force that expands the primary cell wall during growth.

Because every plant cell is cemented to its neighbors, the cells and organs of plants cannot move much with respect to one another. This, along with the relative rigidity of cell walls, accounts for many of the special characteristics of plants and plant cells.

Cell walls are found not only in plants but also in algae and in most fungi and bacteria. However, these differ from plant cell walls in their chemical composition.

## 5–Q THE CYTOSKELETON: CELL SHAPE AND MOVEMENT

Living eukaryotic cells bustle with constant motion. Membrane material is exchanged among the plasma membrane, vesicles, Golgi complexes, endoplasmic reticulum, and nuclear envelope. In plant cells, such as those of a pondweed, the green chloroplasts circulate around the central vacuole, moved by **cytoplasmic streaming.** Heart muscle cells, whether in an intact heart or isolated in culture dishes, contract several times a minute. Cells growing in a culture dish may change shape and move around. And nerve cells in an embryo grow long, thin extensions to other nerve cells and to the fingers, toes, and other distant parts of the body.

Such movements and changes of shape are not yet well understood. However, this is an area of active study now that researchers have found ways to use electron microscopes and video cameras to study the delicate structures involved.

These modern methods have revealed that the cytoplasm of all eukaryotic cells contains a network of assorted protein filaments attached to the plasma membrane and to some organelles. This is called the **cytoskeleton** because it provides a framework for cell shape and movement. However, unlike our own skeletons, much of it is not permanent but seems more like scaffolding: its components can be disassembled, moved to new locations, and used in new structures. The cytoskeleton is highly developed in muscle cells, where the arrangement of protein filaments permits these cells to shorten and produce muscle contraction. As with other specialized cell types, muscle cells merely contain exaggerated versions of the protein framework found in other eukaryotic cells.

The cytoskeleton is made up of at least three types of fibers: microtubules, intermediate filaments, and microfilaments (Table 5–2). Microtubules and microfilaments are made up of globular protein subunits that can be dismantled and reassembled as required, but the intermediate filaments are composed of fibrous proteins built into stable structures.

Some workers have suggested that a three-dimensional lattice of very thin, tapering **microtrabeculae** extends throughout the cytoplasm and forms a major network in which the rest of the cytoskeleton is suspended. However,

TABLE 5–2

### Elements of the Cytoskeleton

| Properties | Microtubules | Intermediate Filaments | Microfilaments |
|---|---|---|---|
| Diameter (nm) | 20 to 25 | 8 to 10 | 4 to 6 |
| Protein composition | Tubulin | Keratin (epithelial cells) Vimentin (many cell types) Desmin (muscle) Many others | Actin |
| Permanence | Labile | Stable | Labile |
| Roles | Defining cell shape | Strengthening cell | Contraction, notably in muscle |
| | Forming mitotic spindle | | Moving membrane |
| | Forming framework of centrioles, cilia, flagella, basal bodies | | Giving cell shape |
| | Serving as tracks for moving organelles | | Attaching cell to external matrix by way of membrane proteins |
| | | | Resisting mechanical stress |
| | | | Serving as tracks for moving organelles |

many other people think that the network-like fibers are artifacts formed during fixing and staining for electron microscopy.

## Microtubules

The thickest filaments in the cytoskeleton, the **microtubules,** form the cell's general framework. They are hollow tubes, 20 to 25 nm in diameter (Figure 5–20). The tubes' walls are made up of thousands of molecules of the globular protein **tubulin,** which is always present in the cytoplasm and can be assembled and dismantled rapidly as needed. In most cells, the cytoskeleton's microtubules radiate out from a microtubule organizing center known as the **cell center,** or **centrosome,** near the cell nucleus. The tubules lengthen as tubulin subunits are added at their free ends.

This array of microtubules organized by the cell center normally breaks down just before cell division. Then, early in cell division, a new array of microtubules, the **spindle,** appears. The spindle helps distribute chromosomes equally to the new cells being formed. Then it in turn breaks down, and new arrays of microtubules are assembled, starting from the cell centers of the two newly formed cells.

Besides providing a skeletal framework, microtubules serve as tracks for the movement of organelles. Proteins as-sociated with the microtubules use energy from ATP to move organelles along these tracks.

In addition to this framework of microtubules, many eukaryotic cells have structures composed of more complex arrangements of microtubules: cilia or flagella, including their basal bodies, and centrioles.

***Cilia and Flagella***    **Cilia** (singular: **cilium**) and flagella are slender, thread-like organelles on the surfaces of many eukaryotic cells. Cilia are generally shorter and more numerous than flagella, but both have the same basic structure, built on a scaffolding of microtubules.

Cilia and flagella are organelles of locomotion, serving either to propel the cell through its environment or to move something past the cell. For example, many protists move by the beating of their cilia or flagella, and a human sperm moves by thrashing the single flagellum that forms its tail. Cilia also move mucus and debris up and out of the human air passages, thereby helping to keep the lungs clear.

Electron micrographs of eukaryotic cilia and flagella show that each contains a characteristic so-called **9 + 2** arrangement of microtubules: nine paired microtubules in a circle, with two single microtubules in the center (Figure 5–21). The plasma membrane covers the outside of the entire structure. More than 100 kinds of proteins link the microtubules and coordinate the wave-like motions that pass along the length of the whole structure.

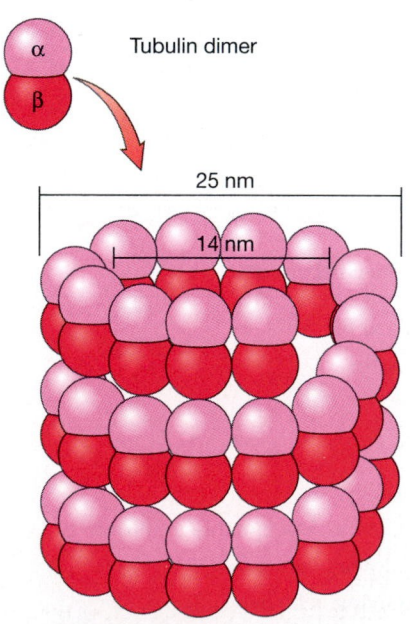

(a) Microtubule structure

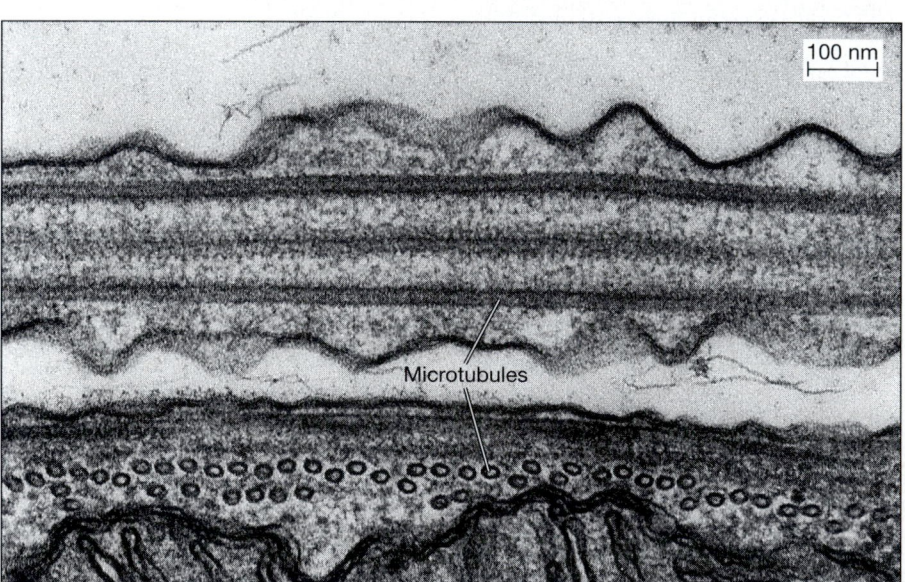

(b) Microtubules in longitudinal and cross section (TEM)

**FIGURE 5–20**   Microtubules. (a) A tubule's walls are made up of dimers (two-molecule units) containing one molecule each of $\alpha$- and $\beta$-tubulin, arranged in a spiral pattern. (b) Lengthwise section (top) and cross section (bottom) of microtubules in a single-celled eukaryote.   (b, Biophoto Associates)

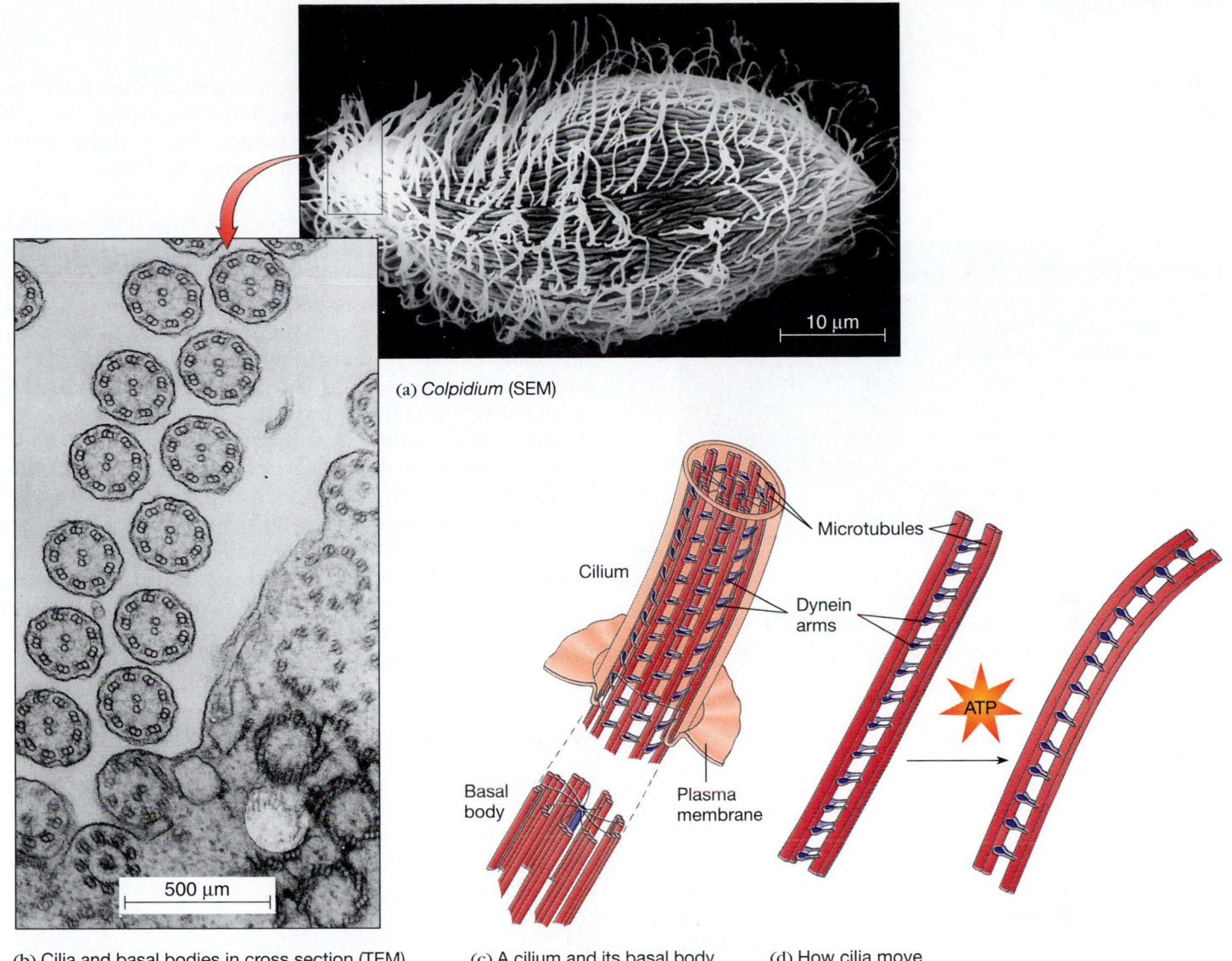

(a) *Colpidium* (SEM)

10 µm

500 µm

(b) Cilia and basal bodies in cross section (TEM)

(c) A cilium and its basal body

(d) How cilia move

Cilium

Microtubules

Dynein arms

Basal body

Plasma membrane

ATP

**FIGURE 5–21**   Cilia. (a) The single-celled eukaryote *Colpidium* moves by the beating of its many cilia. (b) Cross section of a field of cilia such as those shown in the box in part (a). The section has cut straight across several cilia outside the cell (top left) and across basal bodies inside the cell (bottom right). (c) Drawing of a cilium and its basal body. The cilium has a (9 + 2) microtubule arrangement: nine double microtubules around two single tubules. The basal body pattern is (9 + 0): nine triple microtubules arranged in a circle. The structure in the middle of the basal body is not a microtubule; its makeup is unknown. (d) The dynein arms on each microtubule doublet attach to the next doublet and use energy from ATP to produce a sliding force. This bends the cilium, causing it to move through the surrounding fluid.   (a, b, Biophoto Associates)

A cilium (or flagellum) moves by the action of "arms" of the protein dynein that extend from one microtubule of each pair. These dynein arms attach briefly to a microtubule in the next pair and use energy from ATP to produce a sliding force. This bends the whole 9 + 2 bundle of microtubules, and the cilium pushes against the fluid outside the cell (Figure 5–21).

A cilium or flagellum grows from a **basal body** just inside the cell. The basal body consists of a circle of nine microtubule triplets (instead of pairs), with no microtubules in the center. This is called a **9 + 0** arrangement. The basal body's nine triplets serve as a template (pattern) for the growth of the nine doublets in the cilium itself. It is not known how the two central tubules of the cilium originate.

***Centrioles***   The cell center of eukaryotic cells other than higher plant cells contains two **centrioles**, oriented at right angles to each other. Each centriole has the same 9 + 0 arrangement of microtubule triplets as the basal body of a cilium (Figure 5–22).

111

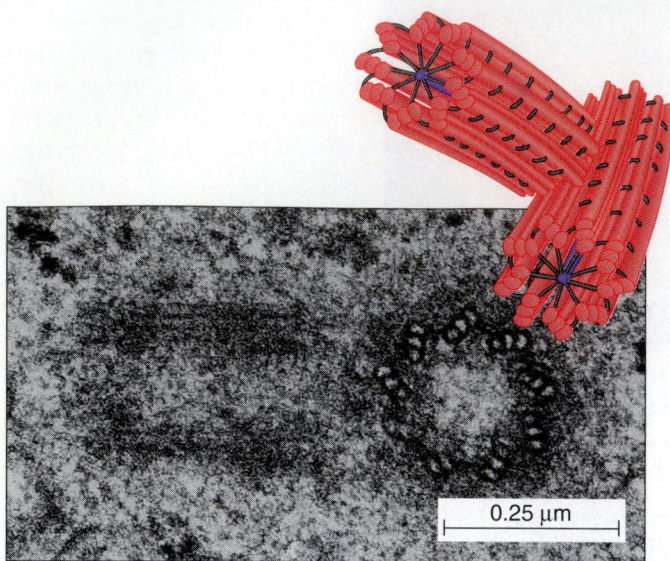

**FIGURE 5–22** Centrioles. Centrioles lie at right angles to one another. In this TEM, the section has cut lengthwise through one centriole (left) and crosswise through the other (right). (Keith Porter/Photo Researchers)

Before cell division, the centrioles move apart, and each serves as a template for the assembly of a new centriole perpendicular to itself. The two pairs of centrioles then separate as the microtubules of the spindle are assembled between them. Each pair of centrioles ends up in one of the two new cells. It is tempting to conclude that the centrioles are necessary for cell division. However, the cells of higher plants divide perfectly well even though they apparently contain no centrioles or similar structures. Hence, the role of centrioles in cell division is not yet clear. Interestingly, plants that lack centrioles, such as conifers and flowering plants, also do not have cilia or flagella at any stage of their life cycle. Could these two facts be related?

In some cells, the centrioles direct the formation of the basal bodies of cilia. The basal body then serves as the template for the assembly of the microtubules of the cilium itself. In the unicellular alga *Chlamydomonas,* the same structures do double duty: before cell division, the two flagella are resorbed (absorbed back) into the cell, and their basal bodies act as centrioles in cell division. However, in many cells with a lot of cilia, the basal bodies do not appear to be assembled using centrioles as templates.

## Intermediate Filaments

**Intermediate filaments** are rope-like protein fibers about 8 to 10 nm in diameter, intermediate between microtubules and microfilaments (Figure 5–23). They are made up of

long, fibrous proteins, which are stable and insoluble. Intermediate filaments are thought to strengthen the cell, thereby helping to maintain its shape. This is important because cells are often stretched or squeezed when parts of the body move.

In animals, different types of cells have different kinds of proteins in their intermediate filaments, whereas microtubules and microfilaments always have the same proteins. Long, thin nerve cells are held in shape by filaments made of three different polypeptides. **Keratins** are a diverse family of intermediate filament proteins found in epithelial cells covering the surfaces of various organs, in epidermal cells in the skin, and in hair, fingernails, and feathers. Because keratin filaments differ according to cell type, they can be used to determine the origin of cancerous epithelial cells that have migrated to other parts of the body.

The various intermediate filament proteins are thought to be adapted to withstand the sorts of mechanical stress experienced by each kind of tissue. However, these different proteins all apparently contain **homologous regions,** with similar or identical amino acid sequences, which are probably the regions permitting them to form filaments.

## Microfilaments

**Microfilaments** are even thinner than intermediate filaments, only about 4 to 6 nm in diameter. They are also called **actin filaments** because they are made up of the globular protein actin (see Figure 3–21). Actin was famil-

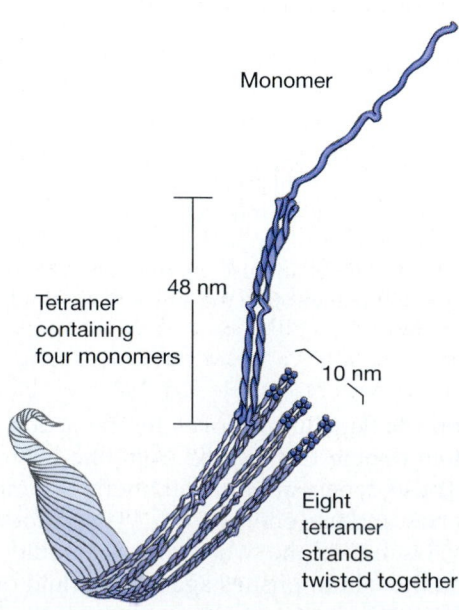

Monomer

Tetramer containing four monomers

48 nm

10 nm

Eight tetramer strands twisted together

**FIGURE 5–23** Intermediate filaments. The basic units are long, fibrous protein molecules, which twist together as shown, forming a strong rope.

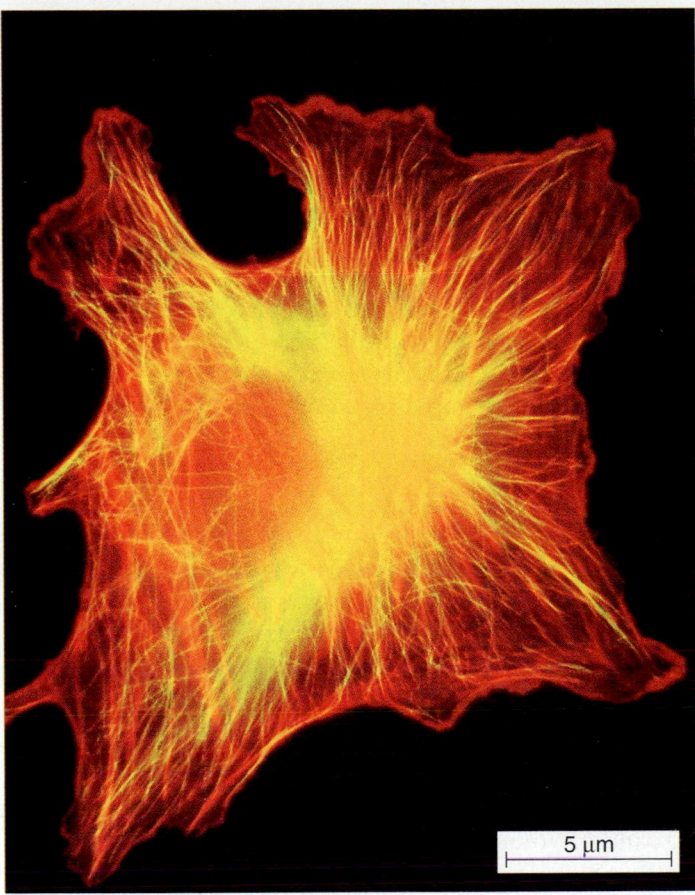

**FIGURE 5–24** The cytoskeleton. This cell has been stained with fluorescent dye so that actin filaments glow red, microtubules yellow-green. Note the differences in their distribution. The oval area outlines the nucleus, which does not contain either element. (Dr. M. Schliwa/Visuals Unlimited)

iar as one of the contractile proteins in the elaborate cytoskeletons of muscle cells long before its more general role in the protein framework of eukaryotic cells was recognized. We now know that it is the most abundant cytoskeletal protein and also the most abundant protein inside many eukaryotic cells. As with the tubulin protein of microtubules, free actin molecules in the cytoplasm can be assembled into filaments as needed and later disassembled into actin subunits, which can be reused.

Microfilaments are especially plentiful just under the plasma membrane, at contacts with neighboring cells. Where the membrane is under stress, the microfilaments are arranged in bundles (Figure 5–24). Microfilaments are attached to the membrane by way of other proteins and can strengthen, pull, or push the membrane during changes in cell shape. Associated with the microfilaments in many cells are small amounts of **myosin,** the protein that interacts with actin in muscle cells to produce contraction.

Actin filaments and myosin are responsible for many cellular events that involve motion or contraction, such as endocytosis and exocytosis and the division of animal cells into two new cells. Actin filaments can also serve as tracks for the movement of organelles, which are propelled by myosin, using energy from ATP. Microfilaments also play an active role in cytoplasmic streaming in plant cells and in ameboid movement and other changes in cell shape.

Drugs called cytochalasins, derived from fungi, prevent actin molecules from joining into microfilaments. This disrupts many cell movements, including division of the cytoplasm (but not separation of the chromosomes, which requires microtubules, not microfilaments). These drugs also do not affect muscle contraction, which requires existing actin filaments but does not involve assembling new ones.

In order to move parts of the cell, microfilaments must be attached to the plasma membrane by proteins. For instance, the protein **vinculin** attaches microfilaments to some integral membrane proteins. In muscle cells, with their abundant, highly ordered actin filaments, vinculin is organized in a striped lattice, lying just beneath the plasma membrane and enveloping the whole cell. **Spectrin** is a protein involved in attaching microfilaments to the plasma membrane in red blood cells, and various other spectrin-like proteins perform the same function in other cells.

## 5–R TISSUES AND ORGANS: CELL WORK PARTIES

All but the simplest multicellular organisms contain a range of different types of cells, many of them specialized for different functions, and hence with different shapes, sizes, and cell chemistry. Cells of one or a few types form groups called **tissues,** held together in characteristic patterns and performing a particular function.

**Organs** are functional units of the body made up of more than one type of tissue; examples are eyes, kidneys, muscles, leaves, or roots.

### Animal Tissues

An animal cell's plasma membrane is covered by a layer of carbohydrate, the **glycocalyx,** made up of the oligosaccharides attached to the membrane's glycoproteins and glycolipids (see Figure 4–7). This sugar coating apparently plays a role in cell recognition and binding and so helps cells communicate and organize themselves in the body.

Every animal cell is surrounded by a space containing **extracellular fluid** as well as an irregular network of fibers, the **extracellular matrix.** The extracellular fluid is the cell's immediate environment. It is the source of nearly all the substances the cell takes in, as well as being the immediate sink for the cell's wastes.

The extracellular matrix helps to hold cells together in tissues, and it provides an organized network within which cells can migrate and interact with one another. Cells attach to the matrix by way of proteins in the plasma membrane that connect actin microfilaments in the cell's cytoskeleton to fibronectin fibers in the matrix. Fibronectin in turn attaches to other matrix proteins: fibrin and the vertebrate body's most abundant protein, collagen (see Figure 3–21).

Animal tissues are generally divided into four main types:

1. **Epithelial tissue** forms coverings and linings. These cover the outside of the body and of each internal organ. They also line the cavities of tubes such as the digestive tract, lungs, mouth, and vagina (Figure 5–25). In keeping with this function, epithelial cells form sheets one to several cells thick, with the cells tightly packed together. The cells seal themselves to their neighbors by tight junctions and strengthen these attachments with desmosome "rivets" (Section 4–G). Substances may enter or leave the body through epithelia—gases through the lung epithelium and food through the digestive tract epithelium, for example. Many epithelial tissues also secrete substances. The lining of the human digestive tract secretes mucus, and the epithelium of a snail secretes its shell.

2. **Connective tissue** is the most abundant type of animal tissue. Cartilage, bone, and adipose (fat) tissue are examples of connective tissues. Connective tissue characteristically has a great deal of extracellular matrix secreted by the cells and made up of fluid, fibers, and gelatinous substances.

3. **Nervous tissue** contains nerve cells, which have the special property of irritability: the ability to conduct electrical impulses in response to stimuli (Chapter 37). Nervous tissue also contains various other types of cells closely associated with the nerve cells.

4. **Muscle tissue** is made up of cells that can both conduct electrical impulses and contract (Chapter 39).

## Plant Tissues

Plant cells surround themselves with thick but porous cell walls, which are firmly cemented to the walls of neighboring cells. Many cells also have direct cytoplasmic connections to each other via plasmodesmata.

There are four main types of plant tissues:

1. **Dermal** ("skin") **tissue** covers the outside of the plant. For example, the **epidermis,** one or two cells thick, forms a tight-fitting sheet on the outer surfaces of leaves and of young stems and roots (Figure 5–26).

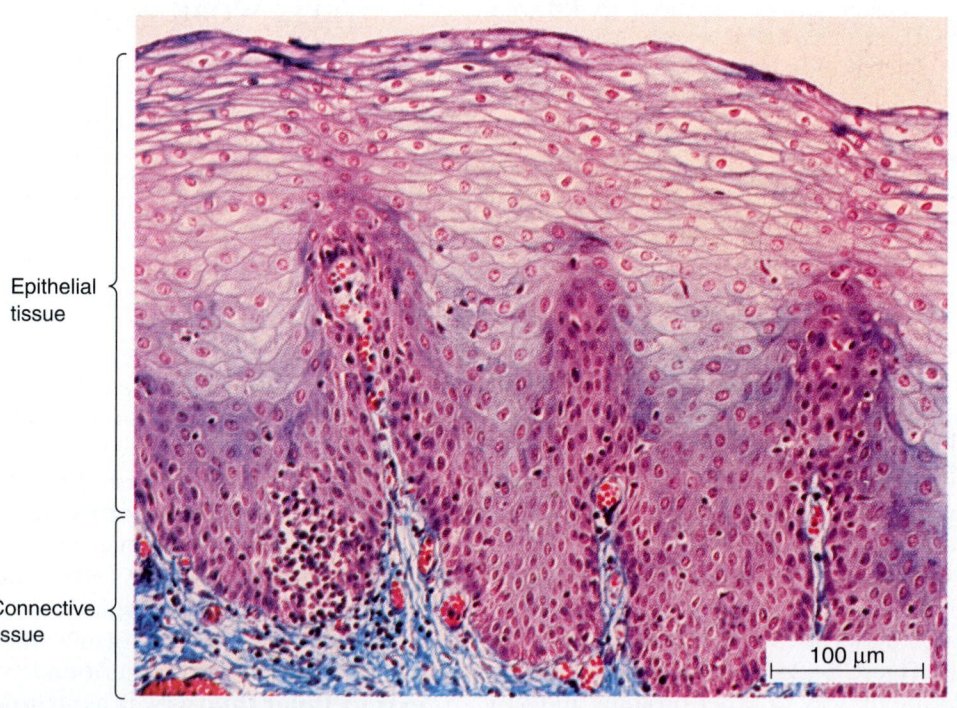

Epithelial tissue

Connective tissue

100 µm

**FIGURE 5–25**  Epithelial and connective tissue. In this cross section of the human vagina, cell nuclei and plasma membranes are stained purple. The close-packed cells deep in the epithelial tissue divide constantly, replacing cells that die and wear off from the surface of the vaginal lining (top). Beneath the epithelium lies connective tissue containing many fibers, stained blue.   (Biophoto Associates)

Epidermis

Bundle of vascular tissue

Ground tissue

Hollow center of stem

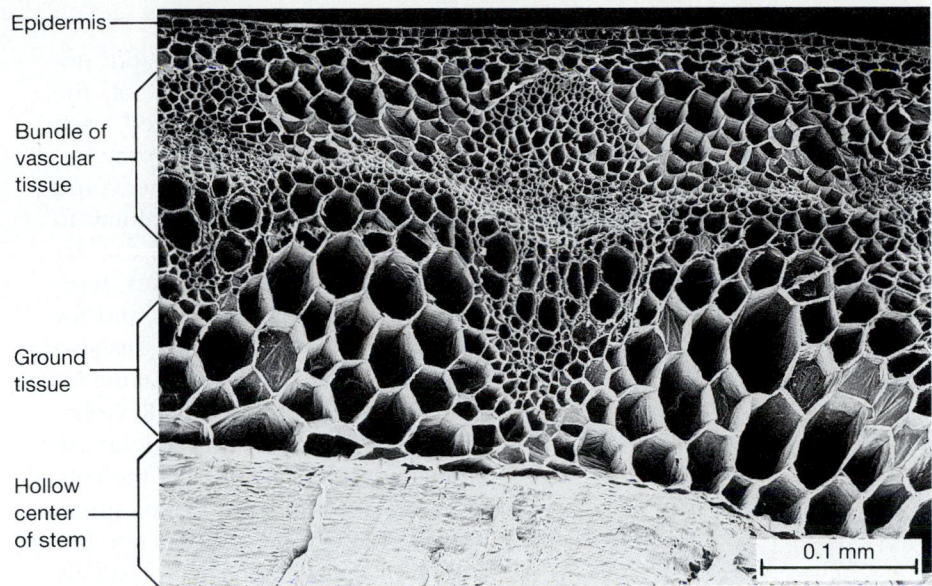

0.1 mm

**FIGURE 5–26** Plant tissues. This low-power scanning electron micrograph shows part of a hollow stem of a pea plant in cross section. The brick-like cells of the epidermis protect the outer surface. The bundles of vascular tissue contain pipe-like cells that transport food and water from one part of the plant to another. Ground tissue, made up of thin-walled parenchyma cells, occupies the rest of the stem. (Biophoto Associates)

2. **Vascular tissue** transports water, food, hormones, and other substances between different parts of the plant. It makes up the veins in leaves and the wood of trees. A plant is "woody" because cells in the vascular tissue have deposited a great deal of strengthening lignin in their secondary cell walls.

3. **Ground tissue** fills the spaces between the epidermis and vascular tissue inside leaves and in nonwoody stems and roots. It consists mostly of thin-walled **parenchyma** cells, which are often loosely packed, with many spaces between the cells. Some parenchyma cells contain chloroplasts, a specialization for photosynthesis, or plastids for food storage.

4. **Meristems** are tissues made up of cells that are ready to divide and develop into the other three types of tissue whenever the plant grows new parts.

## SUMMARY

A cell must carry out all the processes of life, and each cell of a multicellular organism also performs some specialized function(s) for the body as a whole.

1. Microscopes are used to study the structure and position of cell components.

2. Cell chemistry and physiology can be studied using cells in tissue culture or purified fractions of cell parts isolated by centrifugation.

3. The prokaryotic cell of a bacterium contains the basic necessities of life:
   - The plasma membrane, a double layer of lipid and associated protein, which separates the cell from its environment and controls the passage of substances into and out of the cell.
   - The cytoplasm, containing small molecules and ions, food storage deposits, and metabolic enzymes.
   - Ribosomes, the structures where proteins are made, which also lie in the cytoplasm.
   - The bacterium's DNA, which lies in a nuclear area in the cytoplasm.
   - A cell wall, which gives the cell shape and provides protection, usually surrounds the cell, outside the plasma membrane.

4. Eukaryotic cells have a plasma membrane, cytoplasm, and ribosomes. In addition, they contain many membrane-bounded organelles, separate interior compartments specialized to carry out particular jobs. They also have a cytoskeleton made up of various proteins. Important structures found in eukaryotic cells are:
   - The nucleus, containing the genetic material in the form of the DNA of the chromosomes. DNA and RNA are made in the nucleus. The nucleolus is the area in the nucleus where ribosomes are made. The nucleus is surrounded by a nuclear envelope consisting of two membranes and pierced by large pores. The outer membrane is continuous with the endoplasmic reticulum.
   - Endoplasmic reticulum, a system of membranes that provides surfaces where many chemical reactions occur, including synthesis of most of the cell's new membrane material and synthesis of proteins for export. These proteins enter the ER lumen as they are made.
   - Golgi complexes, stacks of membranous sacs in which proteins and other materials are modified and packaged, some for lysosomes, some for secretion to the exterior of the cell.
   - Lysosomes, membrane-bounded sacs of hydrolytic enzymes. Lysosomes fuse with vesicles containing

According to the fossil evidence we have now, prokaryotes existed for more than 1.5 billion years before eukaryotic cells evolved about 1.8 billion years ago. We cannot hope to discover definitely how eukaryotes evolved, but we can make some educated guesses.

It seems that eukaryotic cells must have evolved from prokaryotic ancestors. Various modern prokaryotes have three sorts of membrane adaptations suggesting possible steps in the evolution of eukaryotes: first, expansion of the plasma membrane into the cytoplasm; second, specialization of patches of membrane for particular tasks. In many cases, these two adaptations go hand in hand. The third adaptation carries both to an extreme: the cytoplasm contains specialized membranous vesicles apparently disconnected from the plasma membrane and from each other (see Figure 24–5).

Given these possibilities, it is not difficult to imagine an inwardly expanding membrane engulfing the DNA, thereby forming a nuclear envelope (Figure 5–A). This would explain why the nuclear envelope is a double membrane. It also suggests a possible origin for the endoplasmic reticulum, which is continuous with the nuclear envelope's outer membrane and has a lumen continuous

with the space between the envelope's membranes. The ER may have evolved from an area of the membrane surface left over after the rest had surrounded the DNA. At some point this internal membrane pinched off from the plasma membrane. Also, the nuclear and endoplasmic reticulum areas of the membrane became specialized, each containing a different set of proteins.

By extension of this argument (and of the internal membrane's ability to pinch off compartments) we can account for the Golgi complexes, lysosomes, and transport vesicles, and for endo- and exocytosis. All of this also would account for the fact that the interior compartments of the ER, Golgi, lysosomes, vesicles, and vacuoles are equivalent to the exterior of the cell, whereas the nuclear compartment housing the chromosomes is not, but communicates with the cytoplasm via the nuclear pores.

None of this discussion applies to mitochondria and plastids. These organelles do not take part in the exchange of membrane material within the endomembrane system, and their

**Prokaryotic cell**

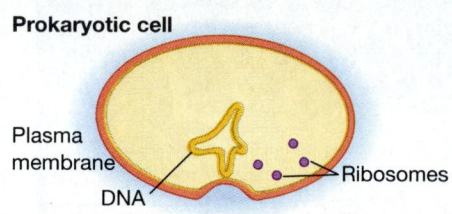

Plasma membrane
DNA
Ribosomes
Plasma membrane expands and invaginates...

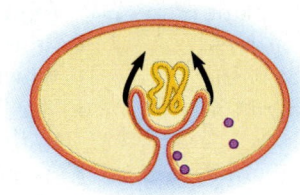

...engulfs DNA and pinches off

**Eukaryotic cell**

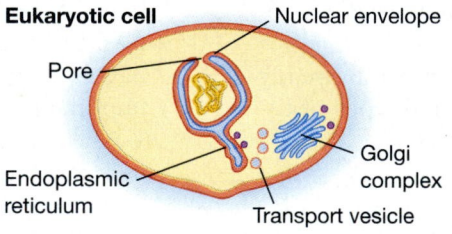

Nuclear envelope
Pore
Endoplasmic reticulum
Golgi complex
Transport vesicle

**FIGURE 5–A** Possible evolution of the nucleus and endomembrane system. (1, 2) The plasma membrane of a prokaryote, with sites for DNA and ribosome attachment, expands into the cytoplasm. (3) Eventually it surrounds the DNA, forming the double membrane layers of the nuclear envelope. The outer layer is continuous with the endoplasmic reticulum. Golgi complexes and transport vesicles form from sacs pinched off from parts of this membrane. The interiors of these organelles are in some ways part of the "extracellular" medium (blue).

food, foreign matter, or worn-out cellular structures, which are broken down by the lysosomal enzymes.

- The endomembrane system, consisting of the plasma membrane, nuclear envelope, endoplasmic reticulum, Golgi complexes, and lysosomes, which all exchange membranous material directly or by way of transport vesicles.

- Peroxisomes and glyoxysomes, membrane-bounded sacs of oxidative enzymes.

- Mitochondria, organelles that produce most of the cell's energy supply of ATP. They are thought to be descended from bacteria living inside larger cells.

5. Plant cells typically contain all these structures except lysosomes. They also have:
   - Plastids, which are also thought to be descendants of photosynthetic bacteria living in cells. The most important plastids are the chloroplasts, which carry out photosynthesis.
   - Vacuoles, membrane-bounded sacs of fluid, especially large and prominent in plant cells.
   - A cell wall, made largely of cellulose fibers, a porous but fairly tough protective and supportive structure outside the plasma membrane.

6. A eukaryotic cell's cytoskeleton is responsible for shape

interiors are not equivalent to the exterior of the cell. Instead, they are apparently descendants of bacteria that took up residence in primitive eukaryotic cells (Figure 5–B).

In order to take in something as large as another cell, early eukaryotes must have lacked cell walls, and they must have had flexible plasma membranes able to carry out endocytosis. This would have permitted them to live by engulfing their neighbors whole, a new mode of nutrition that enabled them to acquire food in bulk and therefore to reach a larger size. In contrast, prokaryotes with cell walls could take in only molecules small enough to diffuse through the cell wall spaces.

During the history of life on Earth, many major evolutionary advances were made by predators. Acquiring mitochondria, and later chloroplasts, may have been the first and second of these predatory innovations. Both events were certainly major coups: mitochondria are very efficient at extracting energy from food and converting it into ATP, and chloroplasts can make food by trapping an abundant energy source—sunlight. Having these organelles would have given early eukaryotic cells a major advantage. Section 24-H explores the origin of these energy-metabolizing organelles in more detail.

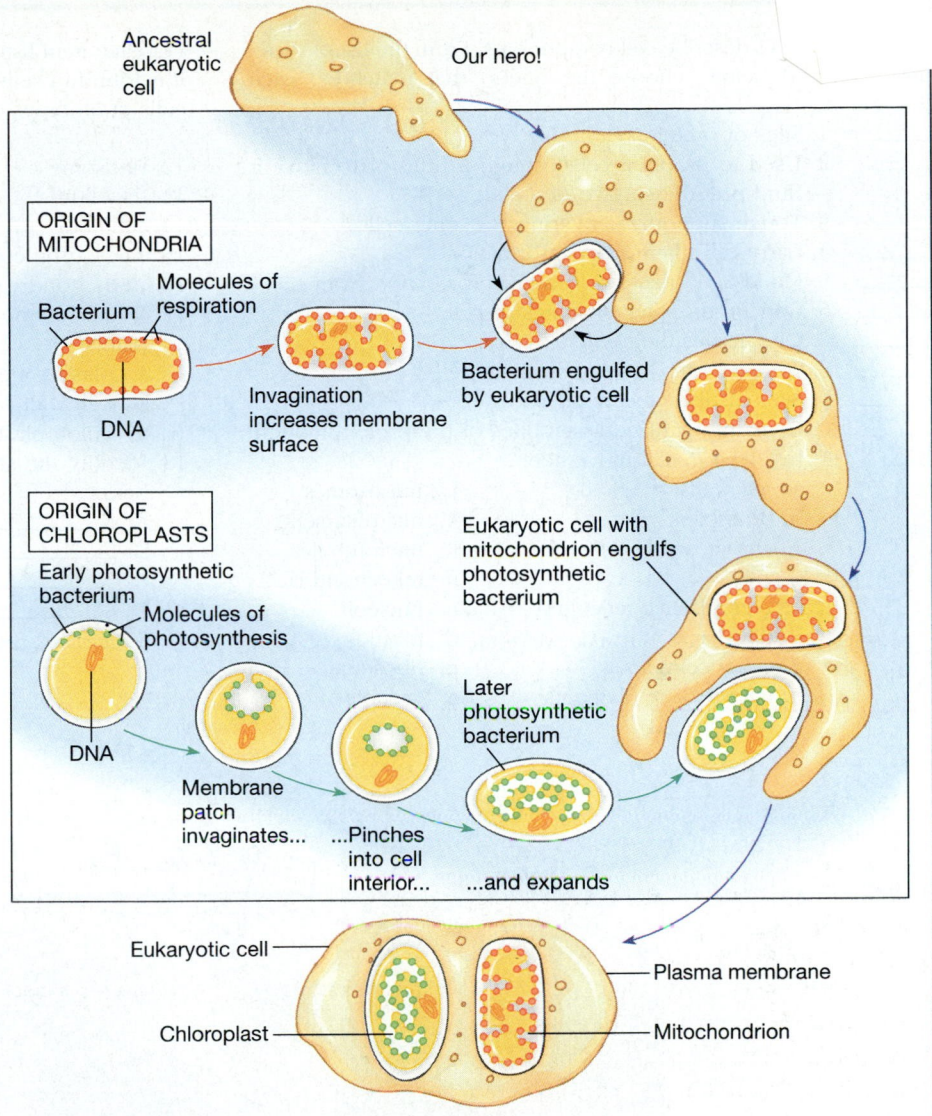

**FIGURE 5–B** A theory of the bacterial origin of mitochondria and chloroplasts.

and movement. It is attached to the plasma membrane and probably to some internal organelles. It is made up of a variety of protein fibers, listed in order of decreasing thickness:

- Microtubules, composed of tubulin subunits that can be assembled and disassembled at need. Microtubules radiate from the cell center near the nucleus and form the cell's overall framework. They also form the framework of centrioles in the cell center (except in higher plants) and of cilia or flagella and their basal bodies. Cilia and flagella are thread-like projections at the cell surface that move the cell itself or move substances past the cell surface.

- Intermediate filaments, composed of stable fibrous proteins, which provide mechanical strength.
- Microfilaments, composed of actin subunits associated with small amounts of myosin. These contractile proteins provide the forces necessary for shape changes and movement. Actin filaments also serve as tracks for transport of organelles and anchor the cell to the surrounding matrix by way of some membrane proteins.

7. In multicellular organisms, cells are organized into tissues, such as epithelium and connective tissue. Several types of tissue may be organized to form organs, such as kidneys or roots, each with its own function.

## Self-Quiz

Questions 1 to 11 describe cell components. From the list of structures (a to q) below, choose the one(s) that match(es) each description.

_____ 1. Sites of protein synthesis.
_____ 2. Used to propel a cell through a fluid or to move a fluid past the surface of a cell.
_____ 3. Tough supportive coverings of some cells.
_____ 4. Carry cell's hereditary information.
_____ 5. Modify and package products for export from cell.
_____ 6. Contain digestive enzymes of cell.
_____ 7. Give leaves their color.
_____ 8. Storage compartments in plant cells.
_____ 9. Sites of ribosome synthesis.
_____ 10. Contain organized assemblies of the protein tubulin
_____ 11. Involved in cell movement.

a. cell walls
b. centrioles
c. chromosomes
d. cilia
e. endoplasmic reticulum
f. flagella
g. Golgi complexes
h. intermediate filaments
i. lysosomes
j. mesosomes
k. microfilaments
l. microtubules
m. mitochondria
n. nucleoli
o. plastids
p. ribosomes
q. vacuoles

For each item listed below, place check marks indicating whether it is found in cells of animals, cells of plants, prokaryotic cells, or cells of all types of organisms.

|  | Animal | Plant | Prokaryote | All |
|---|---|---|---|---|
| 12. ribosome | _____ | _____ | _____ | ___ |
| 13. flagellum | _____ | _____ | _____ | ___ |
| 14. cell wall | _____ | _____ | _____ | ___ |
| 15. microtubule | _____ | _____ | _____ | ___ |
| 16. mitochondrion | _____ | _____ | _____ | ___ |

17. Which of the following is _not_ found in the cells of higher plants?
a. plasma membrane     d. ribosome
b. cell wall     e. centriole
c. chloroplast

18. Identify the structures indicated by letters in the photograph below.

A. _____     F. _____
B. _____     G. _____
C. _____     H. _____
D. _____     I. _____
E. _____

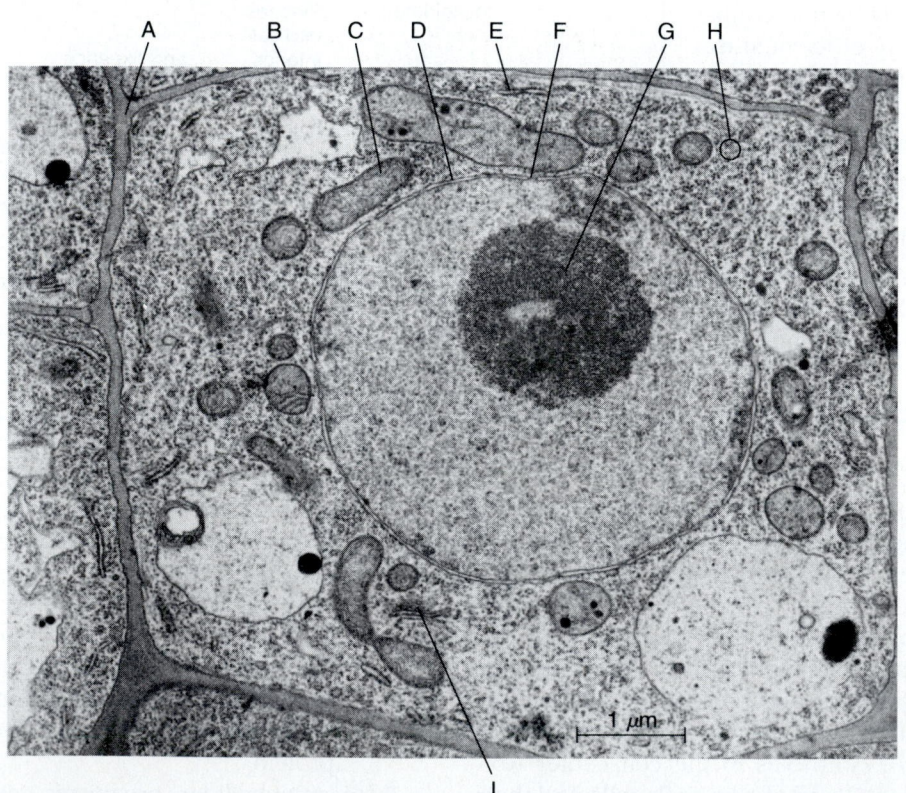

1 μm

## Questions for Discussion

1. Would you expect cells that produce hair to contain more ribosomes than cells that store fat? Why?
2. It has been said that animals, as we know them, could not exist if they had cell walls. Why not?
3. What is the advantage to cells of keeping microtubule protein subunits on hand in the cytoplasm, rather than making them anew from amino acids each time they are needed for mitotic spindles or other structures?
4. Tobacco smoke reduces the activity of cilia in the air passages between the throat and lungs. How does this contribute to "smoker's cough" and lung disease?

## Suggested Readings

(Some references from Chapters 3 and 4 will also be useful here.)

Alberts, B., D. Bray, J. Lewis, M. Raff, K. Roberts, and J. D. Watson. *Molecular Biology of the Cell,* 2d ed. New York: Garland Publishing, 1989. A comprehensive text.

Allen, R. D. "The microtubule as an intracellular engine." *Scientific American,* February 1987.

DeDuve, C. *A Guided Tour of the Living Cell,* student edition. New York: Scientific American Books, 1984.

Green, H. "Cultured cells for the treatment of disease." *Scientific American,* November 1991.

Kessel, R. G., and R. H. Kardon. *Tissues and Organs: A Text-Atlas of Scanning Electron Microscopy.* San Francisco: W. H. Freeman Co., 1979.

Orci, L., J. Vassalli, and A. Perrelet. "The insulin factory." *Scientific American,* September 1988. Traces the path taken by this important protein hormone from the production in the nucleus of RNA instructions for making it to its export from the cell by exocytosis.

Rothman, J. E. "The compartmental organization of the Golgi apparatus." *Scientific American,* September 1985.

Weber, K., and M. Osborn. "The molecules of the cell matrix." *Scientific American,* October 1985. The cytoskeleton and methods for studying it.

Wickramasinghe, H. K. "Scanned-probe microscopes." *Scientific American,* October 1989. How scanning probe microscopes produce images.

# Energy and Living Cells

**O B J E C T I V E S**

*When you have studied this chapter, you should be able to:*

1. In addition to the terms mentioned in the other objectives, use or interpret the following terms correctly:
   potential energy, kinetic energy
   free energy, entropy
   autotroph, heterotroph
   ADP, $P_i$
2. State the first and second laws of thermodynamics.
3. List the basic starting materials and end products of photosynthesis and respiration, and describe the roles of photosynthesis and respiration in the energy economy of the living world.
4. Recognize examples of oxidation and reduction reactions.

Given such reactions, identify which substances are oxidized and which reduced as a result of the reaction.
5. Explain how living organisms carry out apparently endergonic chemical reactions.
6. Discuss the role of ATP in the energy economy of living organisms.
7. Distinguish between substrate-level phosphorylation and chemiosmotic phosphorylation.
8. Describe the chemiosmotic theory of ATP synthesis, including the role of the electron transport system, the membrane housing this system, the electrochemical gradient, and ATP synthetase.

---

**B**ombs, earthquakes, fires, and hurricanes make front-page headlines. Yet, from a scientific point of view, the notable thing about the collapsed walls, eroded beaches, burning timbers, and even lost lives is not that they occurred, but that so much happened so fast, and so soon. Inevitably, all these things would have disintegrated sooner or later anyway. This principle is called the **second law of thermodynamics;** it states that the amount of disorder, or **entropy,** in the universe must increase, and simultaneously the energy that can do useful work decreases.

In this context, the apparently unexpected items are ones usually relegated to inside pages: birth announcements, a bumper crop of turnips, a cute photo of a seven-year-old growing new front teeth. At first glance, these— and all other—living organisms appear to run counter to the law of increasing disorder in that their bodies maintain and even increase molecular orderliness. A living body has a chemical composition very different from that of its en-

vironment, and it maintains this internal chemistry within narrow limits. An organism also apparently increases orderliness by joining simple molecules together to form more complex ones, and by growing and reproducing, thereby increasing the amount of matter that is organized into living cytoplasm.

In fact, the existence of life *is* compatible with the second law of thermodynamics. The law applies, not to an isolated object like a bomb or an organism, but rather to an *object plus its surroundings.* An organism maintains its own orderliness at the expense of its environment: the total entropy of the organism plus its environment increases even though the organism's own entropy does not.

The organism acquires energy from the environment and uses it to combat the universal tendency toward increasing disorder (Figure 6–1). As this energy proceeds on its inevitable downhill path, it is directed so as to extract useful work, for example, to synthesize molecules needed for growth, repair, and reproduction (chemical work); to

**FIGURE 6–1** Maintaining order. These seaweeds, resembling tiny palm trees, maintain the orderliness of their bodies despite the constant pounding of the surf. They use the sun's energy in photosynthesis and make their own food, which they can then use to supply energy needed to maintain their bodies.

move substances into, within, and out of the body (transport work); and often to move structures in the body, or the body itself (mechanical work).

The energy to perform these feats comes mainly from food. The living world depends on two energy-processing pathways to meet most of its energy needs: photosynthesis and respiration.

The sun is the ultimate source of energy for virtually all life. During photosynthesis, green plants capture solar energy and store it in the form of chemical bonds in food molecules, usually carbohydrates. The energy stored in food can later be used by the plant or by an animal that has eaten the plant. During cellular respiration, food molecules are broken down, releasing energy to drive energy-requiring processes. Because this useful energy inevitably decreases, while living organisms continue to require energy, they must constantly acquire fresh supplies of energy.

This chapter introduces the rules of energy transactions and shows how they apply to the chemical reactions—including the steps of respiration and photosynthesis—carried out by cells. We shall see that all cells have a few versatile tricks that provide them with energy in usable doses. The next two chapters examine respiration and photosynthesis in more detail.

## KEY CONCEPTS

- Energy flows continuously through living things.
- Organisms use energy to combat the inevitable tendency toward increasing disorder.
- The sun is the ultimate energy source for virtually all life on Earth.

- Energy enters the living world during photosynthesis, when green plants capture solar energy and use it to build food molecules from carbon dioxide and water.
- Energy is released and made available to do useful things during respiration as food is broken back down to carbon dioxide and water.

## 6–A  ENERGY TRANSFORMATIONS

We can define energy as the capacity to do work, that is, to cause a change. Energy occurs in many familiar forms, such as heat, light, chemical, and electrical energy. Energy can be **transformed,** or changed from one form into another. For example, the filament of a light bulb converts the energy of an electric current into light energy and heat energy.

Energy transformations are governed by physical laws called the **laws of thermodynamics** (therme = heat; dunamis = power). This book does not attempt to teach thermodynamics, but only to link common biological processes to some of the fundamental concepts of thermodynamics.

The first law of thermodynamics states that energy can be neither created nor destroyed; it can only be transformed from one form to another. This is sometimes known as the law of conservation of energy.

The second law of thermodynamics states that in any energy transaction, some useful energy is converted to a useless form, and there is an increase in entropy. Therefore the energy available to do useful work decreases at each transaction. In other words, useful energy always proceeds one way—"downhill." In most cases, some useful energy is degraded from the form of an orderly flow that can do work into the useless form of random molecular motion—that is, heat. So, although the quantity of energy remains constant, the quality is always decreasing.

The laws of thermodynamics apply to everything, including living organisms. As an example of the second law, some molecules diffuse into and out of cells at random—an increase in entropy and decrease in orderliness. To combat this tendency, cells continuously require more energy, which they obtain from the environment. They can then transform this energy to forms they can use to maintain orderliness.

Plants obtain energy from the sun, animals from eating other organisms (or their products, such as nectar or feces). In good times, organisms capture enough energy to store some for times of energy shortage. So energy flows continuously through living things. Organisms must constantly process this energy—obtaining it, storing it, releasing it from food, and using the released energy to drive energy-requiring activities.

The type of useful energy released during energy transactions in living organisms is called **free energy:** energy available to do useful work at a constant temperature and pressure. Heat energy is also released, but organisms cannot harness heat energy to drive their chemical reactions.

Energy can exist in kinetic or potential form. **Kinetic energy** is the energy of things in *motion,* such as a rolling stone, running water, or a vibrating molecule. Only kinetic energy can actually perform work. **Potential energy** is stored energy, and it is associated with *position:* it is the possible energy that can be released to do work if the position of something is permitted to change. A rock perched on a cliff has potential energy, which can be changed into kinetic energy if the rock is pushed over the edge. Electrical potential energy is present when oppositely charged particles are held apart by some barrier, which prevents them from moving together in response to the force of electrical attraction between them. Chemical potential energy is stored in bonds between atoms in a molecule. Each of these forms of potential energy can do work if its components are permitted to move into new positions—that is, if the stored potential energy is transformed into kinetic energy (Figure 6–2).

The driving force of an energy transaction is the decrease in free energy during the energy transformation. Conditions permitting, the object and its surroundings will interact so as to proceed to a lower level of free energy. This occurs spontaneously, meaning "without external input," although it is often misunderstood to mean "quickly." For example, when the gates of a dam are opened, the potential energy of still water is converted to the kinetic energy of moving water. The water runs downhill, both literally and in terms of energy. As this energy decrease occurs, free energy is released. We can tap and use some of this released energy if we put a paddle wheel or turbine in the water's path.

The same idea applies to other systems, including living cells: a cell can use energy only as it is released and "runs downhill" to a lower energy level. However, the cell can never trap and use all of the energy released during an energy transaction: some is always lost as heat.

Living cells use energy in all the forms shown in Figure 6–2. For example, nerve cells perform electrical work as they transmit an electrical nerve impulse, perhaps from

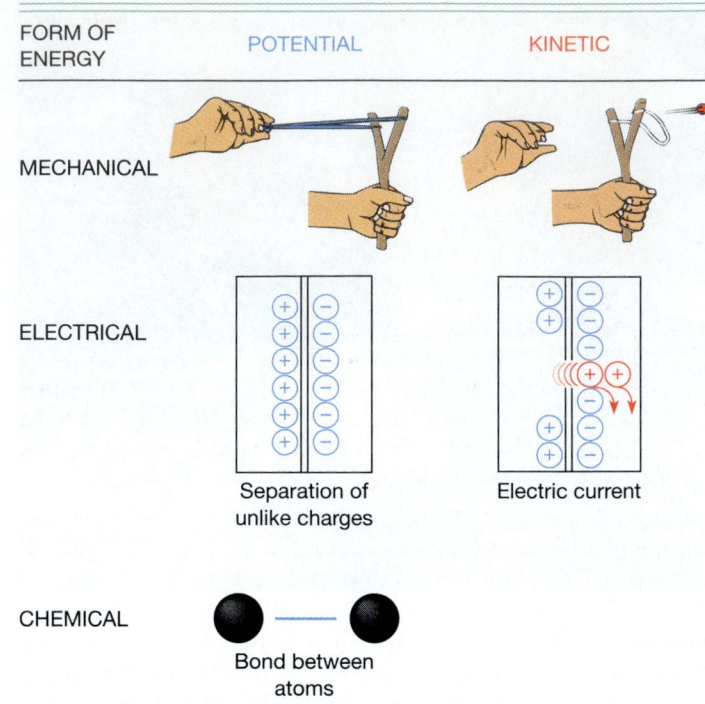

| FORM OF ENERGY | POTENTIAL | KINETIC |
|---|---|---|

MECHANICAL

ELECTRICAL

Separation of unlike charges    Electric current

CHEMICAL

Bond between atoms

**FIGURE 6–2** Some important forms of energy in biological systems. Both potential and kinetic examples of mechanical and electrical energy are shown. Potential energy is transformed to kinetic energy if the marble is released from the slingshot or if electric charge is permitted to move down the electrochemical gradient. Chemical energy exists in potential form, stored in chemical bonds. It can be released when the bond is broken, but chemical potential energy does not have a kinetic counterpart.

the brain to a muscle, telling the muscle to contract. A nerve cell sets up an electrical potential, similar to the one in Figure 6–2, by separating positively and negatively charged ions across the barrier of its plasma membrane. This membrane potential has a degree of order: there is a higher concentration of positively charged ions outside the membrane, and the inside of the cell is more negative. The orderly state of the membrane potential contains stored electrical energy:

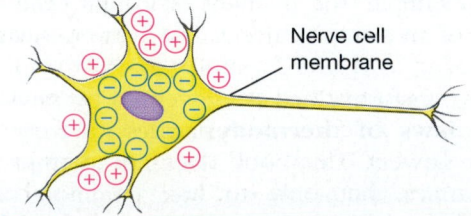

Nerve cell membrane

Membrane potential: stored electrochemical energy

This membrane potential is transformed into the kinetic electrical energy of a nerve impulse when some of the ions are permitted to move through the membrane, down their electrochemical gradient:

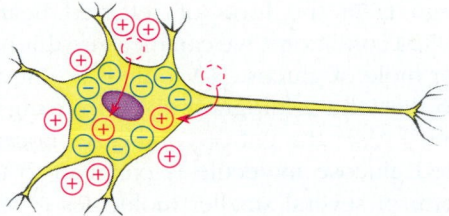

Kinetic electrical energy: flow of charged particles

The force moving the ions is the attraction between oppositely charged particles. This is free energy, and it can do work, as we saw in the case of a Na$^+$ gradient that drives the active transport of glucose into intestinal cells (Section 4–E). However, as the oppositely charged ions come together, this free energy of the moving ions is dissipated, leaving only the intrinsic, random kinetic energy (heat) of the ions themselves. Also, the entropy has increased, because the ions are now mixed more randomly. To maintain its membrane potential, the cell must expend energy on the active transport of ions against their electrochemical gradients.

Muscles perform mechanical work, using chemical potential energy stored in molecules of ATP (adenosine triphosphate). Breaking a chemical bond in ATP releases energy, which is briefly stored as mechanical potential energy by changing the shape of certain muscle proteins. Soon this mechanical potential energy is converted to kinetic energy as the proteins spring back to their original form, causing the muscle to contract. Again, the cell must continually expend energy to keep muscle contraction going. Fresh supplies of ATP are needed constantly, and making new ATP is one of the many kinds of chemical work the cell must perform.

## 6–B CHEMICAL REACTIONS AND ENERGY

Cells perform a lot of chemical work in the many reactions of their metabolism. As in other energy transactions, the driving force of a chemical reaction is the decrease in free energy that occurs as reactants are converted to products.

The free energy of a chemical compound is a function of the bonds between its atoms. We cannot measure the total free energy content of a molecule. However, we can measure the *change in free energy* as the molecule undergoes a chemical reaction, releasing some of the energy stored in its broken bonds and storing energy as it forms new bonds. This change in free energy is symbolized as $\Delta G$ (G stands for physicist Josiah Willard Gibbs, who studied free energy).

The change in free energy during a reaction is the amount of free energy added or lost as reactants are con-

verted to products. A reaction that occurs spontaneously, with the release (that is, a decrease) of free energy, is said to be **exergonic** (ex = out of; ergon = work) (Figure 6–3). In **endergonic** reactions (endon = within), the free energy increases as the reactants are converted to products. Hence,

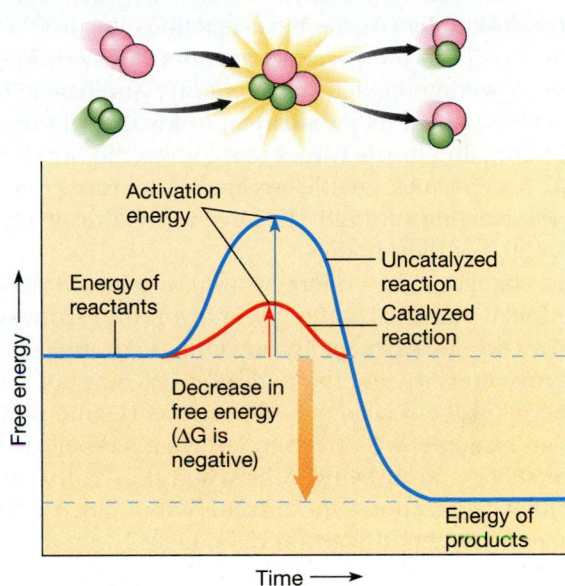

(a) Exergonic reaction

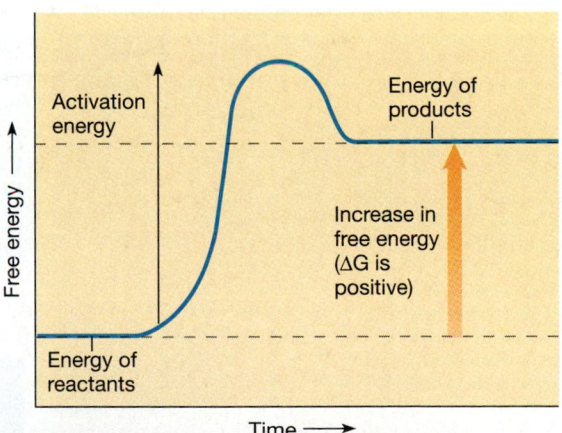

(b) Endergonic reaction

**FIGURE 6–3** Changes in free energy during exergonic and endergonic reactions. (a) A chemical reaction occurs spontaneously only if the free energy of the products is less than that of the reactants—that is, the change in free energy ($\Delta G$) must be negative. Such a reaction is exergonic, releasing energy. However, it may not actually happen unless additional activation energy is supplied to bring the reactants to a transition state from which they can form products. An enzyme lowers the amount of activation energy needed to make the reaction occur (red curve). (b) An endergonic reaction involves an increase in free energy ($\Delta G$ is positive). Such a reaction will not proceed spontaneously, but must be driven by input of energy.

endergonic reactions will not occur without an input of additional energy from some other source, usually an exergonic reaction.

Even though a reaction is exergonic and will occur spontaneously, it still needs **activation energy** to start. At the temperatures found in living organisms, few molecular collisions have enough energy to reach this activation energy threshold. Hence, chemical reactions occur at a very low rate except in the presence of enzymes, which act as catalysts, lowering the activation energy. An enzyme binds the reactants, called its **substrates,** holds them in the correct position, and exerts forces that loosen the bonds to be broken. So enzymes enable organisms to carry out their metabolic reactions at high rates even at the relatively low temperatures of their bodies.

The change in free energy during a reaction depends on the initial and final states and not on the pathway the reactants take in between. By lowering activation energy, an enzyme alters the energy path of a reaction but not the free energy of the reactants and products (Figure 6–3).

As an example, we can compare what happens in non-living systems and in living cells when they carry out the same chemical reaction—the breakdown of glucose, in the presence of oxygen, to carbon dioxide and water:

$$C_6H_{12}O_6 + 6\,O_2 \rightarrow 6\,CO_2 + 6\,H_2O + heat$$

To perform this reaction in the laboratory, we must apply a healthy dose of activation energy, perhaps by lighting the glucose with a match. Most of the ensuing free-energy change is in the form of released heat energy. Under controlled conditions, we can measure this heat: 2817 kilojoules per mole of glucose. (A joule is a unit of energy equivalent to that of a 2-kilogram object moving at 1 meter per second.) Also, the entropy increases because each highly ordered glucose molecule is changed to the more disorderly form of several smaller molecules (7 molecules enter the reaction; 12 come out). Furthermore, these changes are one-way: the chances that the random movements of the carbon dioxide and water molecules will reassemble even one glucose molecule are vanishingly small. This is true even if these products retain more than enough kinetic energy for the reverse reaction to occur. Hence the reaction is said to be irreversible.

When living cells perform the same overall reaction, the energy follows a different path but eventually reaches the same final state: the same increase in entropy occurs (the same products are formed) and ultimately so does the same release of heat (Figure 6–4). The main difference is that the cell traps a good deal of the energy along the way and uses it to perform work. This is possible because it breaks this one highly exergonic reaction into a number of smaller, enzyme-catalyzed steps.

**FIGURE 6–4** Oxygen combines with organic molecules to release energy. (a) A fire releases energy from the organic molecules of dry grass rapidly and with little control, giving off a burst of heat. (b) Living organisms, such as these Grevy's zebras, do a slow burn: the same overall reaction occurs, but in controlled steps. At some steps the released energy is captured in a form that can be used to do the mechanical, chemical, or electrical work of the organism's cells. The gradual release of heat does not disrupt the structure of enzymes or other cell components.    (Peter Arnold, Inc.: a, S. J. Krasemann; b, Gerard Lacz)

(a)

(b)

Some of the free energy released by breaking down glucose is stored as chemical potential energy in ATP molecules. This energy can later be released and do work such as active transport, assembly of monomers into polymers, or muscle contraction. However, cells cannot convert all the free energy from glucose into potential energy stored in ATP. In order to obtain this energy, some free energy must be converted to heat, which cannot do useful work in a cell.

When energy is later released from ATP itself, some of it may be stored temporarily in the form of new chemical bonds or a membrane potential, but again the conversion is much less than 100%, and some heat is released. Eventually even the energy of the newly made bond or membrane potential will be used, and so all the free energy originally released from glucose will have reached the same final state as in the laboratory reaction. For energy-requiring processes to continue, the organism must constantly acquire new supplies of energy.

Strongly exergonic reactions, such as the burning of glucose to carbon dioxide and water, proceed until almost all the reactants are used up. These reactions are practically irreversible. However, where the free-energy change in a reaction is not great, the reaction has a tendency to go in both directions and is termed reversible:

glucose-phosphate $\rightleftarrows$ fructose-phosphate

Under standard biochemical reaction conditions (25 °C, one atmosphere pressure, pH 7.0, 1 molar concentrations of reactants and products), this reaction is weakly exergonic in the direction of the longer arrow; the opposite direction is endergonic.

The free-energy change during a reaction depends partly on the nature of the reactants and products. For example, glucose and similar organic molecules contain a lot of stored energy, whereas carbon dioxide and water contain very little. Therefore, the oxidation of organic matter to $CO_2$ and $H_2O$ releases a lot of energy. In contrast, glucose-phosphate and fructose-phosphate have similar energy content, and so the conversion of one to the other involves very little change in free energy. We can get an idea of the nature of reactants and products by performing a reaction under standard conditions to measure the standard free-energy change ($\Delta G^{\circ\prime}$).

The actual free-energy change ($\Delta G$) of a reaction depends not only on the nature of the reactants and products, but also on their relative concentrations and on reaction conditions such as temperature and pH. As a reaction proceeds, it releases less energy as the reactants are used, and products accumulate, until the reactant and product concentrations reach a point where the change in free energy becomes zero. At this **equilibrium point,** the rates of the forward and backward reactions are equal, and so there is no further net change in the concentrations of reactants and products.

The role of enzymes is to make all of this happen faster. An enzyme lowers the energy barriers to both the forward and reverse reactions, and so it does not alter a reaction's equilibrium point.

Reactions seldom reach equilibrium in living organisms because other reactions constantly add new reactants or remove the products, and this keeps the reaction going. Furthermore, under certain conditions reactions can be reversed from their "standard" direction. For example, our sugar-phosphate reaction can be pushed to the right if other reactions constantly produce glucose-phosphate. It can also be pulled to the right if other reactions are using fructose-phosphate. Indeed, exactly this happens in living cells: this reaction is sandwiched between two exergonic ones, one making glucose-phosphate and thus exerting a push, and the other using fructose-phosphate and thus exerting a pull (see reactions 1, 2, and 3 on page 142). Therefore the concentrations of both sugar phosphates are kept such that the reaction is in fact weakly exergonic going to the *right*.

The farther a chemical reaction is from equilibrium, the more free energy it releases. Some of the chemical reactions in cells are held far from equilibrium. These reactions are controlled by allosteric enzymes, which can be turned on and off (Section 3–G). For example, an allosteric enzyme may be turned off until its substrate accumulates and its product has largely been removed by other metabolic reactions. When the enzyme is finally turned on, the concentrations of these molecules are so far from equilibrium that the reaction will proceed with a large decrease in free energy.

This regulation permits some reactions to be used as a source of energy. For example, the concentration of ATP is held far above equilibrium with its products, ADP and phosphate groups. Other examples are exergonic reactions 1 and 3 on page 142, which help drive the metabolic pathway that breaks down glucose in the cell.

The fact that regulated reactions release a lot of free energy means that they are highly exergonic, and therefore irreversible, under cellular conditions (Figure 6–5). These reactions occur at strategic points in metabolic pathways. In this way, cells maintain some metabolic steps steep enough to direct the overall flow of energy and material through the cell's metabolic pathways. Some of these steps also provide energy for the cell's chemical work. To continue to extract energy from these reactions, a cell must keep substrates and products in disequilibrium by obtaining new energy sources and expelling end products.

## 6–C PHOTOSYNTHESIS AND RESPIRATION

The energy to run most cellular processes comes from organic food molecules, which contain chemical potential energy in the form of the bonds between their atoms. This

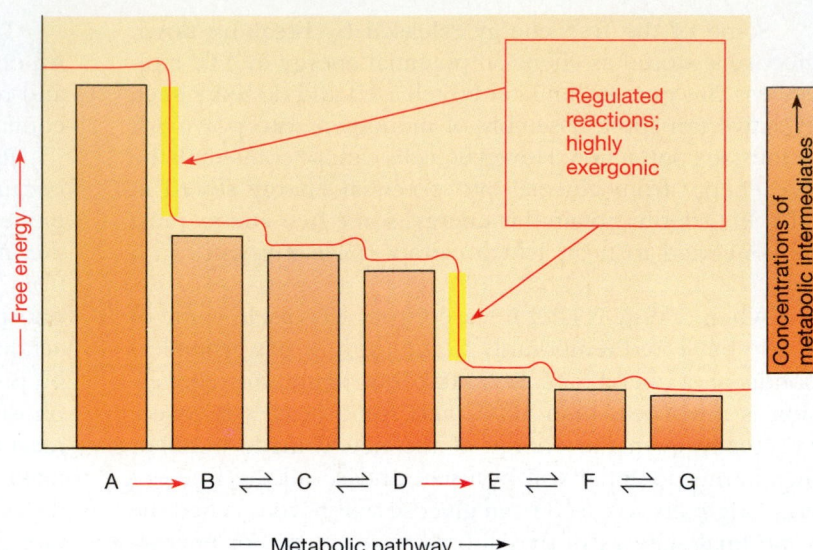

**FIGURE 6–5** Flow of energy and material in a metabolic pathway. The pathway's raw material A is converted to final product G by enzymes that catalyze the reactions shown below the horizontal axis. Reactions A → B and D → E are catalyzed by allosteric enzymes. Regulation of these enzymes keeps the concentrations of their substrates and products out of equilibrium. As a result, these reactions are highly exergonic, as shown by their large decrease in free energy (yellow highlights and one-way reaction arrows). These essentially one-way reactions drive the whole pathway to the right.

energy can be released by breaking the bonds, and the free energy so obtained (which is less than the free energy it took to make the bonds in the first place) can be used by the cell.

Almost all life on Earth today depends on the sun because photosynthesis is powered by solar energy. **Photosynthesis** is the process by which solar energy is captured and stored in the form of new bonds in organic molecules, usually carbohydrates. The raw materials for most photosynthesis are the simple, low-energy inorganic molecules of carbon dioxide and water, which are assembled to form more complex, higher-energy food molecules. Oxygen is given off as a by-product, and virtually all the $O_2$ in the air today came from photosynthesis.

All photosynthetic organisms—plants, algae, and some bacteria—are called **autotrophs** (auto = self; trophe = food) because they do not require food molecules from other organisms to meet their energy needs. The food made by these photosynthetic organisms is used by almost all the other life on Earth. (There are also some autotrophic bacteria that make food using sources other than the sun for energy [Section 24–C], but their contribution to the world's food supply is negligible.)

Most organisms, including photosynthetic ones, break down food molecules by **cellular respiration,** releasing energy that their cells can use. Respiration typically uses oxygen in the complete breakdown of food molecules, usually carbohydrates, and gives off carbon dioxide and water (plus the released energy) as its end products. Thus, respiration has the overall effect of undoing photosynthesis. The carbon dioxide and water that went into photosynthesis are returned to the environment, and the energy released is used to drive the cell's energy-requiring processes. Autotrophic organisms use respiration to break down the food they have made as they need its stored energy. Animals and

other organisms that cannot make their own food molecules are called **heterotrophs** (hetero = other) because they must obtain their food from other organisms (Figure 6–6).

Photosynthesis produces carbohydrate molecules, and respiration breaks them down. Simplified overall equations are:

$$\text{photosynthesis: } CO_2 + H_2O + \text{energy} \longrightarrow CH_2O + O_2$$

$$\text{respiration: } CH_2O + O_2 \longrightarrow CO_2 + H_2O + \text{energy}$$

These equations appear to be direct opposites: the raw materials of each are the end products of the other, and photosynthesis uses energy, whereas respiration releases it. However, the two processes are not simply the opposite of one another. They are similar in some ways and different in others.

## 6–D   OXIDATION-REDUCTION REACTIONS

One major similarity between the metabolic pathways of photosynthesis and respiration is the large number of oxidation-reduction reactions. Some of these reactions release a lot of energy, enough to drive endergonic reactions that require considerable energy input.

**Oxidation-reduction** (nicknamed **redox**) reactions involve the transfer of one or more electrons ($e^-$) from an electron donor molecule (or ion) to an electron acceptor. The molecule that loses the electron is **oxidized,** and the one that gains the electron is **reduced.**

Oxidation and reduction are complementary: for every oxidation, there is a corresponding reduction, because in cells electrons don't float around on their own. Redox re-

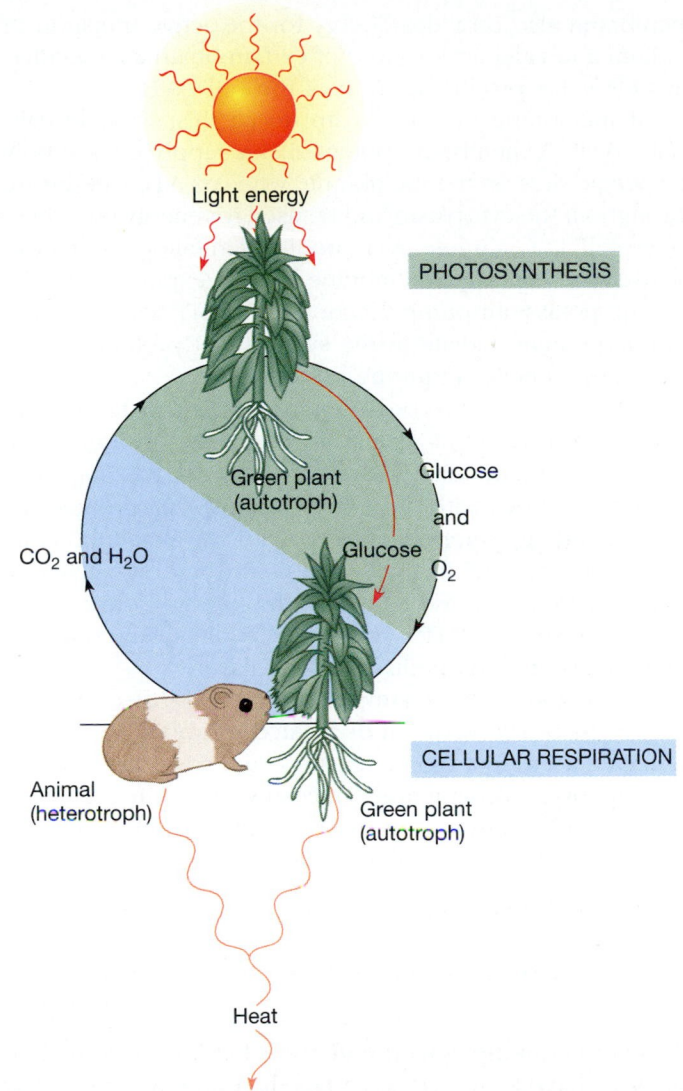

Light energy

PHOTOSYNTHESIS

Green plant
(autotroph)

Glucose
and
Glucose

$CO_2$ and $H_2O$

$O_2$

CELLULAR RESPIRATION

Animal
(heterotroph)

Green plant
(autotroph)

Heat

**FIGURE 6–6** Flow of matter (black arrows) and energy (red arrows) in living things. Although matter cycles indefinitely between photosynthesis and respiration, energy does not. The sun must constantly supply new energy.

gen atom, which contains an electron. Adding oxygen also oxidizes a substance. Oxidation takes its name from oxygen, which is often the electron acceptor. Oxygen is highly **electronegative;** that is, it attracts electrons strongly. Therefore, oxygen is a good electron acceptor, able to remove electrons from a wide variety of substances. Conversely, reduction is the gain of one or more electrons or entire hydrogen atoms.

Looking again at the overall equations of the energy-processing pathways of living cells, respiration and photosynthesis, we see that they are redox equations. Respiration is the oxidation of small organic molecules, such as carbohydrates:

$$CH_2O + O_2 \longrightarrow CO_2 + H_2O + energy$$

$O_2$ is a strong oxidant. Each of its atoms can accept two hydrogen atoms (which include two electrons) to form water. The transfer of hydrogen atoms from organic molecules to oxygen releases a great deal of free energy. In cellular respiration, this energy is released in small, usable quantities by breaking up this big redox reaction into a series of smaller ones.

Oxygen atoms hold electrons so strongly that water does not give up electrons readily. Therefore, photosynthesis requires a considerable energy input to make water give up its hydrogen atoms (containing electrons), which in effect are ultimately used to reduce carbon dioxide:

$$CO_2 + H_2O + energy \longrightarrow CH_2O + O_2$$

(It is important not to confuse the *dissociation* of water into $H^+$ and $OH^-$ with the *splitting* of water into 2 H and O.

actions are generally regarded as the sum of two half-reactions:

| Half-reaction | $Ae^- \longrightarrow A + e^-$ (oxidation) |
|---|---|
| Half-reaction | $B + e^- \longrightarrow Be^-$ (reduction) |

Net reaction $Ae^- + B \longrightarrow A + Be^-$

electron    electron    has been    has been
donor      acceptor   oxidized    reduced
(reductant)  (oxidant)

The simplest case of oxidation is loss of an electron, as from an ion of iron: $Fe^{2+} \longrightarrow Fe^{3+} + e^-$ (Figure 6–7). Substances are also oxidized if they lose an entire hydro-

**FIGURE 6–7** Oxidation-reduction reactions. Oxygen from the air has accepted electrons from the metal of this car and combined with it, forming rust. Meanwhile, plants have grown by reducing carbon from $CO_2$ as they form organic molecules during photosynthesis.

Because oxygen is so electronegative, it easily accepts the electron left behind when only a hydrogen ion leaves a water molecule during dissociation. But oxygen's avid hold on electrons is also what makes it so difficult to remove a hydrogen *atom*—$H^+ + e^-$—when water is split.)

In general, energy-rich molecules are highly reduced (hydrogen-rich). Energy-poor molecules are oxidized. Food molecules contain a lot of hydrogen. Energy is removed from food molecules by systematically stripping off their hydrogens in a series of redox reactions.

Because both respiration and photosynthesis consist of series of redox reactions, we can trace the flow of energy in these processes by following the pathway of hydrogen atoms and electrons as they are transferred from one molecule to the next in the series. Indeed, this will be one of our chief activities in Chapters 7 and 8. At some steps in these reaction series, energy to do the cell's work is stored in the form of energy intermediates.

## 6–E   ENERGY INTERMEDIATES

As cells maintain and increase their own organization, they need to perform many endergonic chemical reactions, which go "uphill" in terms of energy. Enzymes can catalyze reactions but they cannot alter the energy changes that occur during a reaction nor make an endergonic reaction proceed without the necessary added energy. How then do organisms carry out endergonic reactions?

The answer is that energy-requiring endergonic reactions are coupled to energy-releasing exergonic ones by way of common intermediates. That is, the reactions are combined in such a way that an overall exergonic reaction occurs. As long as the total of free-energy changes for the individual reactions gives an overall decrease in free energy, the reaction series can occur spontaneously—*if* a pathway exists between them. Enzymes provide such a pathway by binding the reactants and lowering the activation energy barriers in each reaction.

Reactions that release a lot of energy, such as the oxidation of organic molecules during respiration, are carried out in small steps that release energy gradually. At certain steps, some of the energy released is stored in the form of **energy intermediates:** "middlemen" that couple exergonic to endergonic reactions by transferring energy between them.

The usual energy intermediates in living cells are an electrochemical gradient of ions across a membrane and ATP. Both supply energy to a variety of energy transactions. For example, in bacteria the rotation of flagella during locomotion is powered by the energy of $H^+$ moving from the external environment through the plasma membrane and into the cell down its electrochemical gradient. In both bacteria and mitochondria, movement of $H^+$ in through the

membrane also provides energy for the active transport of sodium and calcium ions and of certain organic molecules, as well as for producing most of the cell's ATP.

If membrane potentials can do all this, why do cells make ATP? A membrane potential can supply energy only at specific sites on the membrane, whereas ATP can diffuse throughout the cytoplasm and be used for energy anywhere in the cell. For example, ATP produced in mitochondria can be used at the plasma membrane for active transport by the sodium-potassium pump (Section 4–E). ATP also serves as an energy intermediate in the synthesis of proteins and in many other energy-requiring processes.

ATP is believed to be an energy intermediate in every living organism, and it is certainly the most common chemical energy intermediate. However, it is not the only one: other nucleotides with three or two phosphate groups also serve as energy intermediates for various metabolic reactions. For example, in the synthesis of the polysaccharide glycogen from glucose subunits, the energy donor is uridine triphosphate (UTP). Guanosine triphosphate (GTP) is another energy intermediate in animals.

Let us look at an example of ATP's role as an energy intermediate. The addition of a glucose subunit to a starch molecule is an endergonic reaction, and so it cannot occur spontaneously. A plant adds glucose to starch in a two-step sequence of exergonic reactions, using energy from ATP (Ⓟ represents a phosphate group):

1. ATP + glucose–Ⓟ ⇌ ADP-glucose + Ⓟ–Ⓟ

2. ADP-glucose + starch molecule ⇌ larger starch molecule + ADP

These reactions are both energetically feasible because their products have lower free energies than their reactants. The sequence is energetically "downhill" and can occur, catalyzed by enzymes, in a cell. This pair of exergonic reactions gives the overall *appearance* of having used an exergonic reaction (hydrolysis of the energy intermediate ATP) to drive an endergonic one (adding a monomer subunit to a polymer). Indeed, for convenience, biochemists often write equations coupling endergonic and exergonic reactions:

$$\text{ATP} \xrightarrow{\text{Exergonic}} \text{ADP} + \text{P}_i$$
$$\text{glucose + starch} \xrightarrow{\text{Endergonic}} \text{larger starch molecule}$$

## 6–F   ATP

The role of ATP in the cell's energy economy has been compared to that of cash in the human economy. ATP energy is spent to provide energy for activities such as muscle contraction, active transport, waste excretion, and synthesis of new macromolecules. ATP is also used during photosynthesis as an energy intermediate in the building of carbo-hydrate molecules, which are the cell's savings account of stored energy, broken down at need to make the ready cash of ATP. The central position of ATP in the economy of life can be shown like this:

$$\text{Solar energy} \longrightarrow \text{ATP} \longleftrightarrow \text{Food molecules}$$

Growth, Reproduction, Movement, Transport, Heat, etc.

Both photosynthesis and respiration, then, can be regarded as means to the same end: providing a steady supply of ATP.

An ATP molecule consists of an adenine residue, a ribose sugar residue, and three phosphate groups. Breaking the bonds shown as squiggles in Figure 6–8, which attach the last two phosphate groups to the molecule, releases a great deal of energy. Energy is usually released from ATP by removing the end phosphate group, yielding ADP (adenosine diphosphate), an inorganic phosphate group (abbreviated as $\mathbf{P_i}$), and energy:

$$\text{ATP} + \text{H}_2\text{O} \longrightarrow \text{ADP} + \text{P}_i + \text{energy}$$

The free energy released by this reaction under standard laboratory conditions (symbolized $\Delta G°'$) is about 30.6 kilo-joules per mole of ATP hydrolyzed ($\Delta G°' = -30.6$ kJ/mole). However, in the cell the concentration of ATP is held so far above the equilibrium level that the reaction actually releases 46 to 54 kJ per mole ($\Delta G = -46$ to $-54$ kJ/mole) (Section 6-B). This makes the reaction very useful to drive the cell's energy-requiring processes.

This reaction releases a lot of energy because ADP and $P_i$ are much more stable than ATP is. ATP is unstable because it contains three phosphate groups, with a total of four negative charges, jammed together. Removing the end phosphate relieves some of the force of repulsion between these charges by permitting the products to spring apart. This phosphate ($P_i$) is further stabilized by forming more hydrogen bonds with water molecules.

ADP may be further converted to AMP (adenosine monophosphate) plus $P_i$, releasing another 30.6 kJ per mole (under standard conditions). At times, the two terminal phosphates of ATP are broken off together.

The enzymes that use ATP energy are loosely called **ATPases.** These enzymes do not simply hydrolyze ATP to free ADP and $P_i$, but instead transfer part of the ATP molecule to another molecule. In the example in the last section, ADP was attached to glucose; other reactions attach AMP to a substrate molecule. More commonly, ATPase enzymes carry out **phosphorylation** reactions, which transfer the end phosphate of ATP to another molecule. Whether a molecule receives ADP, AMP, or phosphate, it contains more energy and is more reactive. Hence, it can undergo reactions that were energetically impossible before the reaction. These reactions eventually release the products formed from ATP—for example, ADP and phosphate.

The number of ATP molecules in a cell is relatively small, and all the energy-requiring reactions of metabolism are continually breaking down ATP. Meanwhile, the energy-yielding reactions (mainly those of cellular respiration) must constantly renew the supply of ATP. A typical mammalian cell turns over its ATP supply completely every minute or two. Each second it uses about $10^7$ molecules of ATP and regenerates the same number. ATP is regenerated by phosphorylating ADP. This may occur by two mechanisms.

## ATP Synthesis by Substrate-Level Phosphorylation

The removal of a phosphate group from some so-called high-energy organic molecules yields enough free energy to phosphorylate ADP to ATP. **Substrate-level phosphorylation** is an enzyme-mediated transfer of a phosphate group from such a high-energy substrate molecule to ADP, forming ATP (Figure 6–9). The energy stored in the high-energy substrate came from a previous energy-releasing oxidation reaction (usually removal of hydrogen).

ADENOSINE TRIPHOSPHATE (ATP)

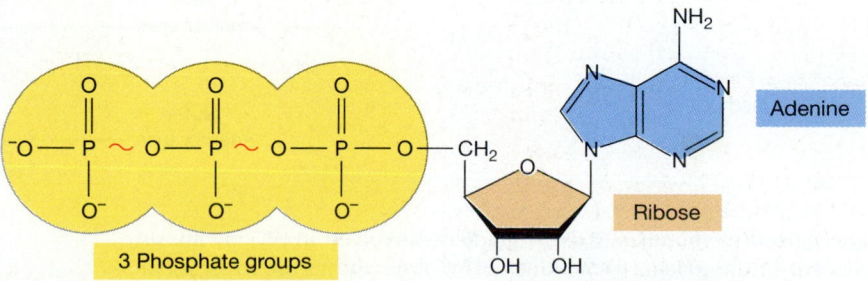

**FIGURE 6–8** Adenosine triphosphate (ATP). Removal of the last one or two phosphate groups, by breaking the bonds shown as red squiggles, releases a great deal of energy. At the pH inside living cells, ATP's phosphate groups give up $H^+$ to the surrounding solution and carry negatively charged oxygen atoms, as shown.

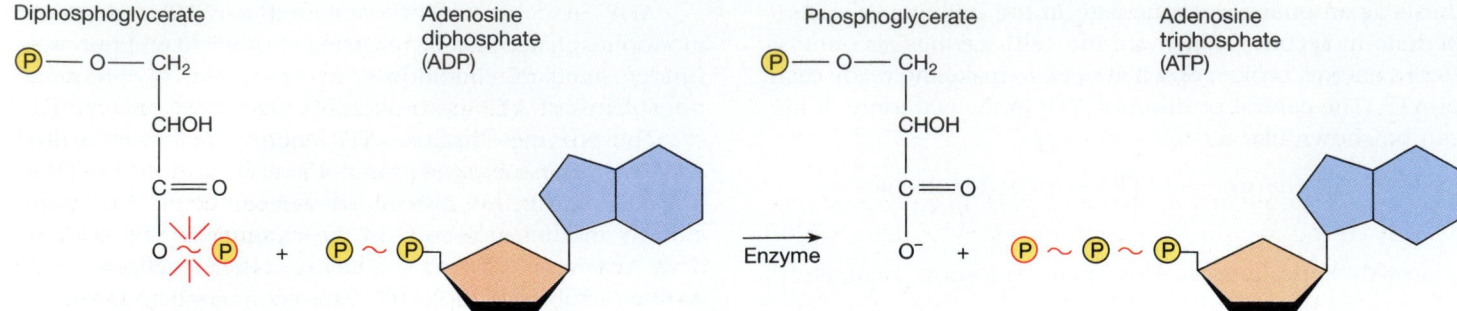

**FIGURE 6–9**    Substrate-level phosphorylation of ADP to ATP, using diphosphoglycerate as a phosphate-group donor.

## ATP Synthesis by Chemiosmosis

In most cells, substrate-level phosphorylation produces only a minor fraction of the total ATP. Most ATP is made using the other energy intermediate, an electrochemical gradient of hydrogen ions ($H^+$) across specialized membranes (Figure 6–10):

1. The plasma membranes of bacteria (in respiration).
2. The inner membranes of mitochondria (in respiration).
3. The inner membranes of chloroplasts (in photosynthesis).

(We will refer to all of these membranes simply as "the membranes" in the rest of the chapter.)

Associated with these membranes are sets of proteins, called **electron transport systems.** These electron transport proteins carry out a series of redox reactions, releasing free energy that drives ATP synthesis. In the 1950s and 1960s biochemists thought that this energy went to form some high-energy compound, which in turn donated energy to make ATP by substrate-level phosphorylation. However, after three decades of fruitless search for the intermediate, this "chemical coupling" hypothesis was

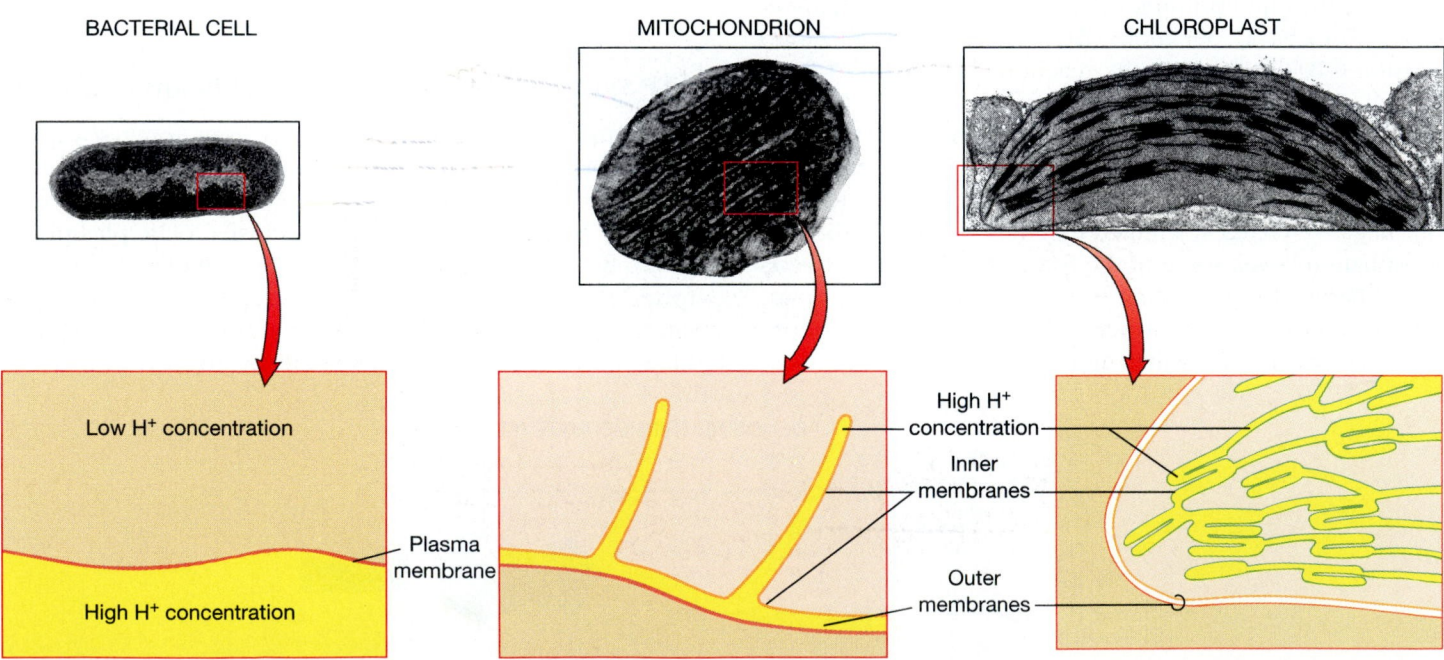

**FIGURE 6–10**    Positions of specialized membranes and $H^+$ gradients involved in ATP synthesis. Electron transport molecules occur in the plasma membranes of bacteria and in the inner membranes of mitochondria and chloroplasts. The electron transport systems accumulate $H^+$ on one side of these membranes (yellow areas): outside the membrane in bacteria and mitochondria, but inside in chloroplasts. Low $H^+$ concentrations are found on the other side of these membranes. (a, L. Santo, from Santo, L., H. Hohl, and H. Frank, *J. Bacteriol.* 99:824, 1969; c, Herbert W. Israel, Cornell University)

abandoned in favor of a very different mechanism, proposed by Peter Mitchell in 1961. His idea rapidly gained supporting evidence, and he received a Nobel Prize for his work in 1978.

Mitchell's scheme has two steps:

1. The electron transport system builds up an energy supply in the form of an electrochemical gradient of $H^+$ ions across a membrane.
2. This energy is used to make ATP as $H^+$ ions move through the membrane down their concentration gradient.

The membrane's electron transport system creates the store of electrochemical energy. It takes electrons from hydrogen atoms and passes them through a series of redox reactions. This supplies energy to move $H^+$ ions from one side of the membrane to the other. Since the membrane forms a continuous barrier impermeable to $H^+$, this results in an $H^+$ gradient across the membrane.

The energy stored in this gradient can then do work as $H^+$ flows down its electrochemical gradient through special channel proteins in the membrane. At the far end of some of these channels, an **ATP synthetase** enzyme harnesses the energy to make ATP (Figure 6–11). As $H^+$ passes through the membrane during this ATP synthesis, the electrochemical gradient is reduced. Hence, the electron transport system must keep working to maintain the gradient. This way of making ATP is called **chemiosmosis** (chemi = chemical, in this case $H^+$; osmosis = passage through a membrane).

In respiration, organic molecules in food are oxidized by removal of their hydrogen atoms. The hydrogens then proceed to the electron transport system. Here their electrons are passed through a series of redox reactions, and the $H^+$ gradient is formed and used to make ATP. This is why the energy-releasing redox reactions of respiration are so important: they provide the cell with most of its ATP, which in turn drives most of its energy-requiring metabolic processes.

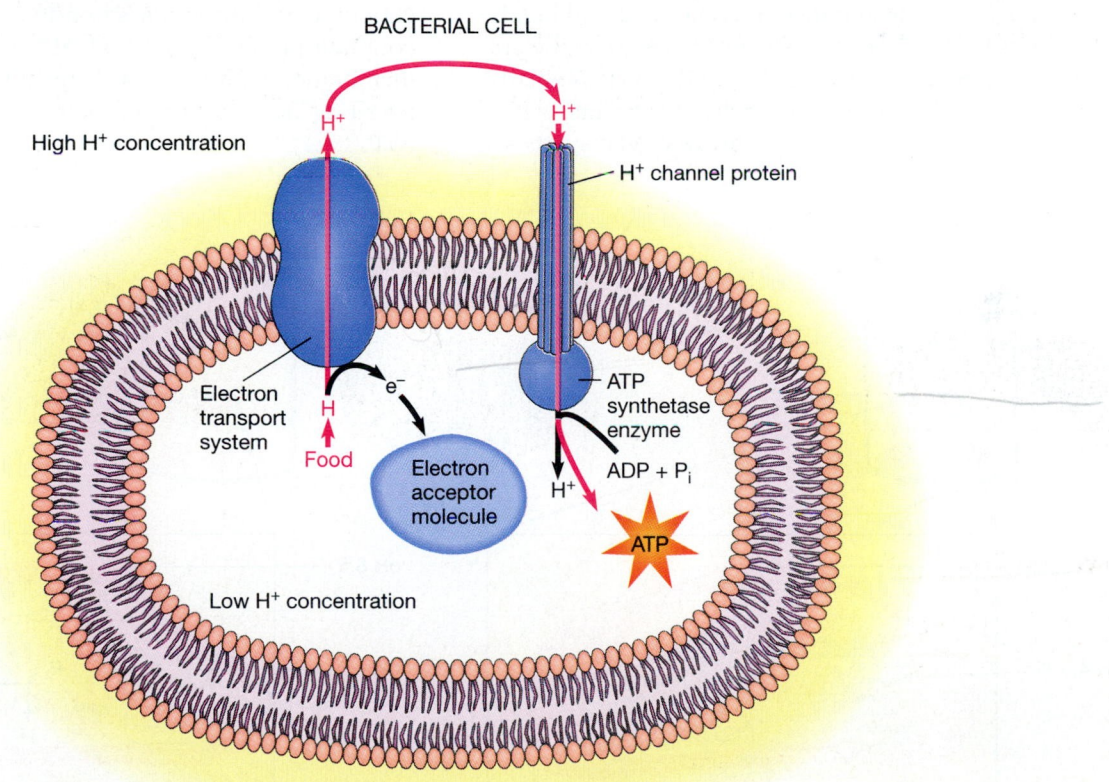

**FIGURE 6–11**   Chemiosmotic ATP synthesis. This simplified diagram shows the basic process using a bacterial cell as an example. Red arrows show the path of energy from food molecules to ATP. The electron transport system in the plasma membrane accepts hydrogen atoms, passes their electrons ($e^-$) to electron acceptor molecules, and expels $H^+$, creating a high $H^+$ concentration outside the cell. The energy of the resulting electrochemical gradient is released and used when $H^+$ moves back through the membrane, down the gradient. $H^+$ passes through channel proteins associated with ATP synthetase enzymes, where ADP and $P_i$ join and form ATP.

***Evidence for the Chemiosmotic Theory*** What is the evidence supporting the chemiosmotic theory? First, we can predict from the theory that electron transport in a respiring mitochondrion or bacterium, or in a photosynthesizing chloroplast or bacterium, should produce a difference in the $H^+$ concentration (that is, pH) on the two sides of the membrane. Researchers have found that such a change in pH does in fact occur when electron transport is proceeding. In chloroplasts a difference of as much as 3.5 pH units (more than a 1000-fold difference in $H^+$ concentration) has been measured across the membrane.

Because the theory holds that the $H^+$ gradient across the membrane is the source of energy for ATP synthesis, we can predict that an artificial pH gradient will also stimulate ATP synthesis. This has been confirmed in chloroplasts. Normally, electron transport in chloroplasts is driven by light energy, and so they produce pH gradients and make ATP only in the light. However, chloroplasts given an artificial pH gradient also make ATP (Figure 6–12).

The chemiosmotic theory also explains why ATP synthesis stops if the membrane is broken, even though the redox reactions of electron transport continue: breaking the membrane lets $H^+$ ions move freely and so destroys the gradient needed to provide the energy to make ATP. Similarly, the chemiosmotic theory explains why detergents, which make membranes more permeable, prevent ATP synthesis.

In respiration, the breakdown of food molecules provides the energy needed to drive electron transport and form the $H^+$ gradient. However, if substances called **uncouplers** are present in the lipid layers of membranes, they combine with $H^+$ and carry it through the membrane to the other side. This dissipates the $H^+$ gradient without using its energy to make ATP. If too much $H^+$ escapes from the gradient without doing work, a well-nourished cell can literally starve to death. Dinitrophenols, yellow substances once used as food additives to make baked goods look as if they contained more eggs than they really did, are such uncouplers. They were prescribed for a time as a cure for obesity, but were abandoned after several deaths were attributed to them in 1929.

Uncoupling also occurs in the cells of brown fat, a special heat-generating tissue in many newborn mammals (including humans) and in hibernating animals. During hibernation, animals are inactive and require much less ATP than usual. However, they do need heat to stay alive, even though the body temperature is permitted to drop much lower than normal. Brown fat cells are packed with mitochondria, and these mitochondrial membranes contain uncoupling proteins that allow $H^+$ to flow back through the membranes without producing ATP. The energy is dissipated as heat instead of being used to make unneeded ATP.

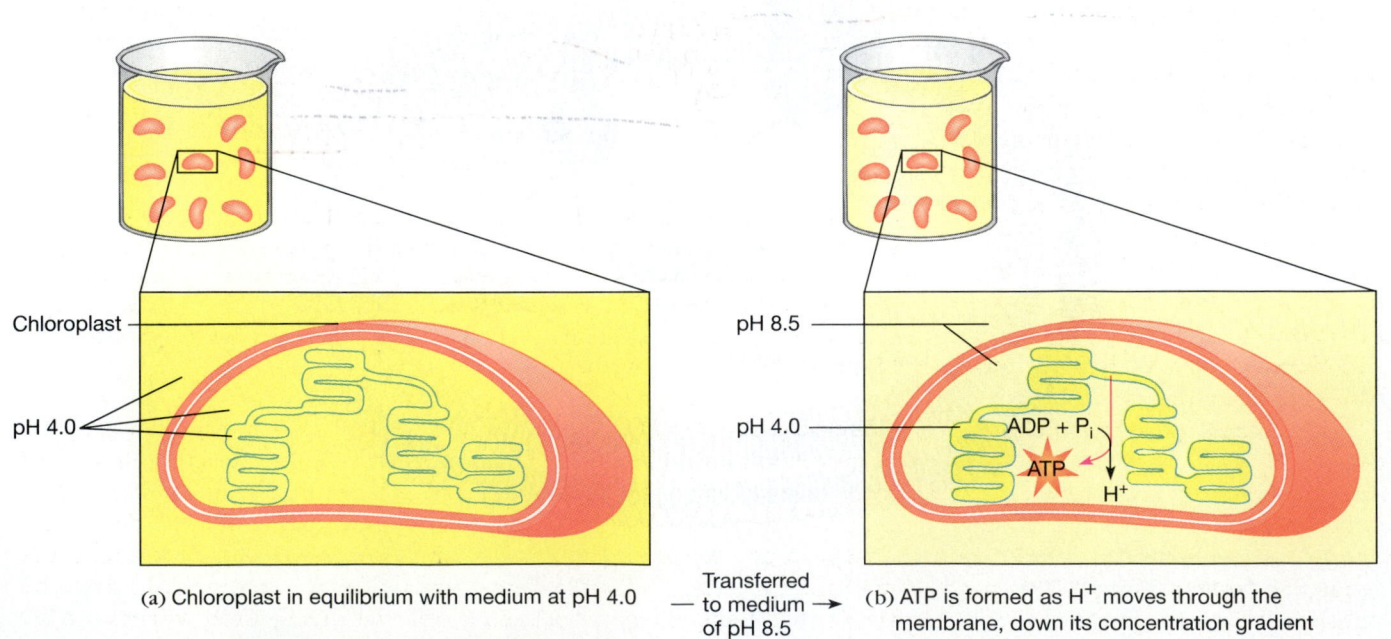

(a) Chloroplast in equilibrium with medium at pH 4.0

— Transferred to medium → of pH 8.5

(b) ATP is formed as $H^+$ moves through the membrane, down its concentration gradient

**FIGURE 6–12** Demonstration of chemiosmotic ATP synthesis in chloroplasts. Isolated chloroplasts first reach equilibrium with a solution of pH 4.0 in the dark. Still in the dark, they are moved to a solution of pH 8.5. For a short time, $H^+$ moves out through the membrane, down its concentration gradient, and drives ATP synthesis, until the gradient is dissipated. ATP synthesis in chloroplasts usually occurs only in the light, because light energy drives the accumulation of $H^+$ inside the chloroplast membranes.

## ESSAY:  *The Solar-Powered Purple Proton Pump*

**W**hat's purple, grows best in a 4.3 molar salt solution, and dies if the salt concentration falls below 3.0 molar (which is about five times saltier than sea water)? It's the purple salt bacterium, member of an ancient group, the extreme halophiles (halo = salt; philos = loving). These bacteria live in super-salty environments such as industrial salt evaporation ponds and Utah's Great Salt Lake.

The warmer and saltier water is, the less dissolved $O_2$ it holds, and the little $O_2$ present is quickly used in bacterial respiration. The respiratory electron transport system requires $O_2$, and so a scarcity of $O_2$ reduces ATP production. At low $O_2$ levels, a purple salt bacterium harnesses light energy as well as respiration to expel $H^+$ from the cell and so maintains the gradient needed for ATP synthesis.

Light energy is captured by patches of purple material that may occupy up to half the plasma membrane's surface. These patches contain bacterial rhodopsin, a purple pigment similar to the visual pigment rhodopsin in the retina of the human eye. When bacterial rhodopsin is struck by light, it gains and then loses a proton ($H^+$), but in a peculiar way: it gains a proton from inside the cell but gives it up outside, thereby contributing to the external $H^+$ gradient (Figure 6–A).

The bacteria produce rhodopsin only when they require supplementary energy: when low oxygen levels limit respiration and a lot of light energy is available to work the rhodopsin "proton pump." This puts purple salt bacteria into the peculiar category of photoheterotrophs: organisms that rely on other organisms for food but can use light to meet some of their energy needs. (These bacteria are *not* photosynthetic.)

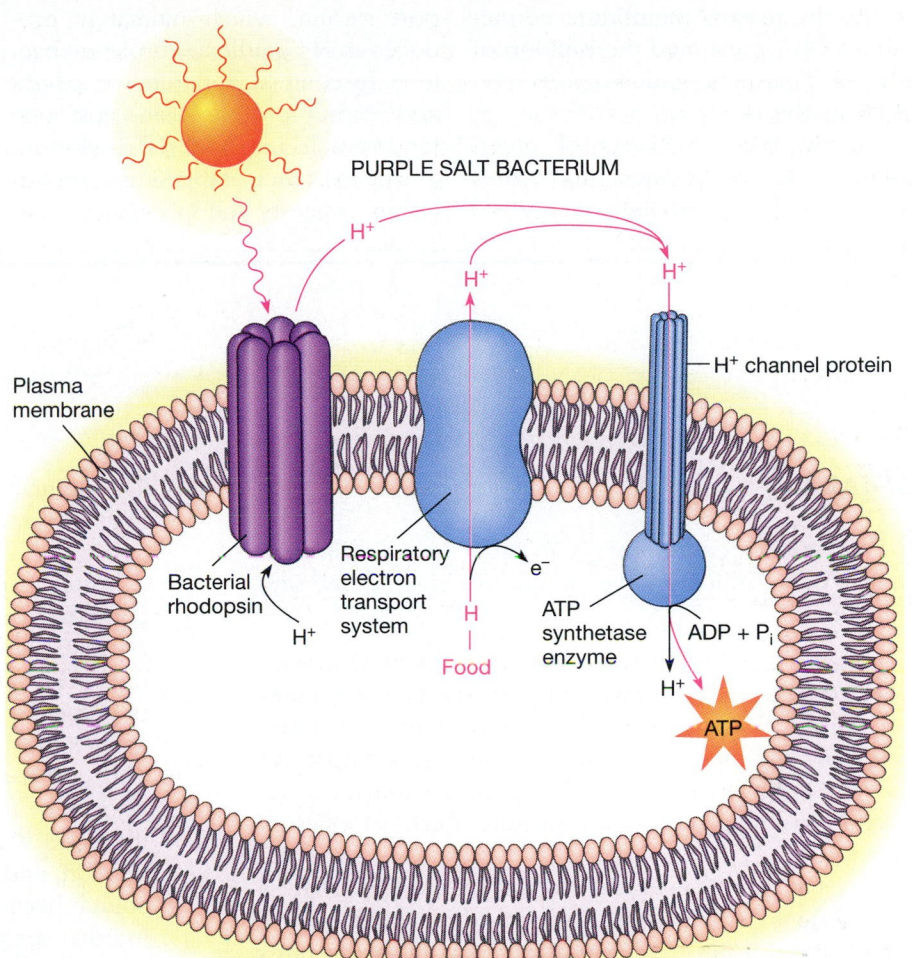

**FIGURE 6–A**  ATP synthesis in purple salt bacteria. Red arrows show the path of energy. These bacteria have two means of exporting $H^+$: by way of bacterial rhodopsin, using light energy, or by way of the respiratory electron transport system, using energy from oxidation of food molecules. In either case, the $H^+$ stockpiled outside the cell is used to make ATP as $H^+$ moves down its concentration gradient through channels connected with ATP synthetase enzymes.

How did investigators discover that purple salt bacteria can use light energy to supplement their chemiosmotic ATP synthesis? If the bacteria are kept in an oxygen-free environment in the laboratory, their respiratory electron transport systems can't work. In darkness, the ATP content of these bacteria falls to very low levels, but in the light they produce normal levels of ATP. When cells kept in darkness were illuminated, $H^+$ appeared in the surrounding medium and then disappeared (that is, the pH of the medium decreased and then increased again). As $H^+$ decreased, the level of ATP in the cells rose: the energy of $H^+$ moving back into the cell was being used to make ATP. Adding substances that block respiratory electron transport did not affect this light-driven ATP production.

*(Continued)*

**E S S A Y** *(Continued)*

However, adding uncouplers that make the plasma membrane permeable to $H^+$ prevented the buildup of the $H^+$ gradient and so prevented ATP synthesis.

In the photosynthesis of green plants, various pigment molecules work together to trap light energy and pass it on to a complex electron transport system, which ultimately produces an $H^+$ gradient. Purple salt bacteria, by contrast, apparently trap light and export protons using just one kind of molecule, bacterial rhodopsin.

The existence of this simple proton pump suggests the possibility that primitive cells may have used similar mechanisms to harness the sun's energy for ATP synthesis long before modern electron transport systems evolved. Being able to use the virtually unlimited supply of solar energy would have given a tremendous impetus to the evolution of early life.

## SUMMARY

Living organisms require energy in order to maintain their chemical composition, move, repair damage, grow, and reproduce.

1. Energy cannot be created nor destroyed. However, each time energy is converted from one form to another, some free energy is degraded into an unusable form, and entropy increases, making it necessary to use yet more energy to restore orderliness. To remain alive, organisms must constantly acquire fresh supplies of energy.

2. A cell's metabolic enzymes direct the downhill progress of energy in usable steps. Some metabolic reactions are held away from equilibrium, and so these reactions release enough free energy to perform useful work. An important example is removal of phosphate groups from ATP (see #6 below).

3. The central energy-processing pathways of life are photosynthesis and respiration. In photosynthesis, the sun's energy is captured and stored in the chemical bonds of food molecules, which can later be broken down during respiration, releasing the trapped energy:

   photosynthesis: $CO_2 + H_2O + energy \rightarrow CH_2O + O_2$

   respiration: $CH_2O + O_2 \rightarrow CO_2 + H_2O + energy$

4. Both photosynthesis and respiration are essentially series of oxidation-reduction reactions, transferring electrons (or hydrogen atoms) from one molecule to the next. Many of these redox reaction steps release a great deal of free energy.

5. The ultimate task of both photosynthesis and respiration is to release energy by means of redox reactions and trap it in the form of energy intermediates, which can then supply energy for other reactions. The most common energy intermediates are ATP and electrochemical gradients of ions across membranes. In fact, most ATP is formed by chemiosmosis, using the energy of an electrochemical gradient of $H^+$ to join phosphate groups to ADP.

6. The cell uses the free energy released by removing phosphate from ATP as a source of energy for much of the cell's work, such as active transport, muscle contraction, and building polymers. ATP is a useful energy source because of the nature of the molecule, with its internal stresses, and because the ATP concentration is kept well above its equilibrium with ADP and $P_i$.

7. Some ATP is made by substrate-level phosphorylation, but most comes from chemiosmosis, which occurs at respiratory and photosynthetic membranes. In chemiosmotic ATP synthesis, electron transport systems use energy released by their redox reactions to build an electrochemical gradient of $H^+$ ions across the membrane. This gradient is used to make ATP by allowing $H^+$ to pass through the membrane, down its concentration gradient, via protein channels associated with ATP synthetase enzymes. The chemiosmotic theory is supported by experimental findings that ATP synthesis occurs in the presence of pH gradients across intact membranes and that destroying the gradient or the membrane virtually halts ATP formation.

## Self-Quiz

1. True or False? The cells of plants contain mitochondria and carry on cellular respiration.
2. Does the following equation describe the overall process of photosynthesis or of respiration?

$$CO_2 + H_2O \longrightarrow C_6H_{12}O_6 + O_2$$

3. In the above equation, is carbon oxidized or reduced as a result of the reaction?
4. For each reaction shown below, tell which substance has been oxidized and which reduced as a result of the reaction.
   a. $Fe + O_2 \longrightarrow Fe_2^{3+}O_3^{2-}$
   b. $FMN + NADH + H^+ \longrightarrow FMNH_2 + NAD^+$
   c. $FMNH_2 + FeS \longrightarrow FMN + FeS^{2-} + 2\,H^+$
5. ATP is made by joining _____ and _____ . This reaction is (endergonic, exergonic). Hence it requires _____ in addition to the reactants.
6. Cells are lysed (broken up) and centrifuged, and the fraction containing their membranes is discarded. When both $P_i$ containing radioactive phosphorus and ADP are added to the suspension of nonmembranous cell contents, some radioactive ATP is formed. This probably occurs by (substrate-level, chemiosmotic) phosphorylation.
7. The electron transport system:
   a. makes ATP
   b. contains ATP synthetase
   c. builds an $H^+$ gradient across the membrane
   d. can work against the second law of thermodynamics
   e. cannot perform redox transfers if the membrane is broken
8. The inner membranes of mitochondria and chloroplasts:
   a. are relatively impermeable to $H^+$
   b. have ATP synthetase enzymes attached to only one face
   c. contain molecules of the electron transport system
   d. form closed compartments
   e. all of the above

## Questions for Discussion

1. List as many energy-requiring activities carried out by living organisms as you can.
2. Explain why water puts out a fire (which is a highly exergonic oxidation of carbon), yet the oxidation of carbon can occur in a living organism, which is mostly water.
3. Why must plants carry on respiration as well as photosynthesis?
4. Plants are believed to convert less than 1% of the light energy reaching Earth from the sun into the form of potential energy stored in the chemical bonds of food molecules. What happens to the rest of the energy?
5. In this chapter we saw that organisms cannot use heat energy to drive their energy-requiring processes. Does this mean that the heat released by metabolism is of no use to them? Why or why not?
6. The ATP synthetase enzymes of chemiosmosis must bind a new molecule of ADP before the previously made ATP is released. How might this property help regulate the cell's production of ATP?
7. As shown in Figure 6–10, $H^+$ accumulates inside the innermost membrane of chloroplasts and so is trapped within a very restricted space. However, the electron transport systems of bacteria are located in the plasma membrane, and the $H^+$ gradient builds up outside the cell itself. Why doesn't this $H^+$ drift away into the wide world, thereby dissipating the $H^+$ gradient and preventing the cell from using it as a source of energy?

## Suggested Readings

Atkins, P. W. *The Second Law.* New York: Scientific American Books, 1984.

Hinkle, P. C., and R. E. McCarty. "How cells make ATP." *Scientific American,* March 1978. Compares chemiosmotic ATP synthesis in bacteria, mitochondria, and chloroplasts. [*Note:* To understand this article, you must know that biochemists ignore the outer mitochondrial membrane. When they say, "outside the mitochondrion," they mean "outside the inner membrane."]

Stoeckenius, W. "The purple membrane of salt-loving bacteria." *Scientific American,* June 1976.

C H A P T E R

7

# Food as Fuel:
# Cellular Respiration and Fermentation

**O B J E C T I V E S**

*When you have studied this chapter, you should be able to:*

1. Recognize examples of coenzymes and explain their functions.
2. Explain why most organisms need oxygen and state how carbon dioxide and water are produced by cellular respiration.
3. Name the starting materials and the important end products of:
    a. glycolysis
    b. preparation of acetyl CoA
    c. the citric acid cycle
    d. oxidative phosphorylation
4. Explain the functions of the four processes listed in Objective 3 in the scheme of cellular respiration.
5. State where in eukaryotic and in prokaryotic cells each process listed in Objective 3 takes place.

6. Explain why the electron transport system is important to cells.
7. Describe how mitochondria carry out ATP synthesis.
8. Compare and contrast the processes of alcoholic fermentation by a wine yeast and lactate fermentation by a muscle.
9. Explain how the ability to carry out lactate fermentation, and to acquire an oxygen debt, is a useful adaptation for a muscle.
10. Compare respiration and fermentation with respect to the net yield of ATP.
11. Explain why we get fat when we eat more food than we use.

L iving cells constantly use energy for activities such as movement, protein synthesis, active transport, and cell division. Most of the energy to power these processes comes from ATP and other energy intermediates. A cell must constantly replace ATP. It does this by breaking down organic food molecules, releasing energy that is used to join ADP and phosphate to form ATP. The breakdown of food to release energy occurs by two kinds of processes: respiration and fermentation.

Most organisms can carry on **cellular respiration,** the stepwise oxidation of food molecules using an inorganic substance as the final acceptor of electrons. In most species, respiration is **aerobic:** using molecular oxygen ($O_2$) as the final electron acceptor. The end products are the low-energy molecules carbon dioxide and water. The overall equation for aerobic respiration is:

$$\text{organic molecules} + O_2 \rightarrow CO_2 + H_2O + \text{energy}$$

The same equation describes the combustion of wood or paper: a fire also uses $O_2$ in the breakdown of organic molecules and produces carbon dioxide and water. But whereas a fire oxidizes organic molecules in one big, uncontrolled reaction, respiration oxidizes food in a series of smaller, controlled steps, each releasing a little of the food molecule's energy. This permits cells to capture and store more energy as ATP than they could if the energy were released in one big burst. ATP produced during respiration is then available to do work in the organism (Figure 7–1).

Some cells live in **anaerobic** conditions, with little or no $O_2$ available. Here, some kinds of bacteria carry out **anaerobic respiration,** which uses other inorganic substances, such as nitrate ($NO_3^-$) or sulfate ($SO_4^{2-}$), instead of $O_2$ as a final electron acceptor.

Other cells living in anaerobic conditions carry out one of many pathways of **fermentation**—the breakdown of food molecules in which the final electron acceptors are organic rather than inorganic molecules.

(a)

(b)

**FIGURE 7–1**  Respiration supplies energy for many activities. (a) These skimmers and terns use energy in muscle contraction. (b) Fruits, such as these cherry tomatoes, respire rapidly as they ripen. This provides energy for the chemical processes that make the fruit softer, sweeter, and more colorful.   (a, Steve Bisson)

In this chapter we first meet several coenzymes that help capture the energy of food molecules during respiration. Then we shall see how cells break down the most commonly used food molecule, glucose, to carbon dioxide and water. After this we examine two kinds of fermentation. Finally, we shall see how organisms obtain energy from molecules other than glucose—other sugars, and fats and proteins—and how the pathways of respiration connect with other metabolic pathways.

## KEY CONCEPTS

◆ Cellular respiration is a series of reactions that oxidizes food molecules in small steps. In the process, the food molecules are broken down and their stored energy is released. Some of this energy is stored as chemical potential energy in ATP.

◆ Fermentation makes less ATP than respiration and breaks down food molecules only partially.

◆ Because respiration provides most of the energy for metabolism, measuring oxygen consumption is a convenient way to gauge energy use and hence the level of activity.

◆ Other metabolic pathways, such as those for proteins and lipids, feed molecules into the respiratory pathway, and so any organic molecule can be used to supply energy.

## 7–A  COENZYMES

Some enzymes involved in fermentation and respiration catalyze reactions with the help of smaller, nonprotein organic molecules called **coenzymes.** Coenzymes act as shuttle molecules, carrying substances from one enzyme-catalyzed reaction to another. They also carry energy: passing the substance they carry to the second reaction releases a lot of free energy, which drives the reaction.

Many coenzymes are made from vitamins. Coenzymes last for a long time and can be used again and again, so they are required only in very small quantities. This is why we need only small amounts of vitamins in our diets.

Some coenzymes carry electrons between redox reactions, usually as part of hydrogen atoms. These hydrogen-carrying coenzymes include **nicotinamide adenine dinucleotide,** commonly called **NAD$^+$; flavin adenine dinucleotide (FAD)** and **flavin mononucleotide (FMN);** and **coenzyme Q** (Table 7–1).

ATP is also a coenzyme, one that carries phosphate groups. Adenine-containing nucleotides are also part of coenzyme A, NAD$^+$, and FAD.

## TABLE 7−1

### Coenzymes Used in Respiration

| Coenzyme | Acronym | Function | Made from This Vitamin*: |
|---|---|---|---|
| Nicotinamide adenine dinucleotide | $NAD^+$ | Carry hydrogen atoms to electron transport system | Niacin |
| Flavin adenine dinucleotide | FAD | | Riboflavin |
| Flavin mononucleotide | FMN | Coenzyme of NADH dehydrogenase, first carrier of electron transport system; accepts hydrogens from $NADH + H^+$ | Riboflavin |
| Coenzyme Q (Ubiquinone) | Q | Coenzyme in electron transport system; accepts hydrogens coming from $FADH_2$ | |
| Coenzyme A | CoA | Carries acetyl group to citric acid cycle | Pantothenic acid |

*See Table 31–2 for more information on vitamins and vitamin deficiency.

Many steps in the oxidation of food molecules remove hydrogen atoms and attach them to hydrogen-carrying coenzymes. These coenzymes in turn pass the hydrogens' electrons to an electron transport system, where their energy is released during respiration.

Most of the hydrogens removed from food molecules are carried by the coenzyme nicotinamide adenine dinucleotide ($NAD^+$). Each $NAD^+$ picks up one hydrogen atom and the electron of another, leaving the remaining $H^+$ in solution (Figure 7–2). The reduced form of $NAD^+$ may be written NADH or, more completely, $NADH + H^+$.

For convenience, we say NADH carries hydrogens, although strictly speaking it carries only one hydrogen atom plus the electron of another. The other hydrogen-carrying coenzymes (FAD, FMN, and Q) can carry two whole hydrogen atoms.

## 7−B OVERVIEW: BREAKDOWN OF GLUCOSE UNDER AEROBIC CONDITIONS

To see how cells release energy from food, we begin with a molecule of glucose, a six-carbon monosaccharide. Under aerobic conditions, glucose is broken down to carbon dioxide and water in a series of reactions. Some of the energy once stored in the glucose molecule's chemical bonds is eventually used to make ATP. Most of the energy transfer occurs by removing the glucose's hydrogen atoms and passing their electrons through a series of redox reactions in a pathway called the electron transport system, ending at $O_2$. So to understand this energy flow we must follow the movements of hydrogen atoms and their components: electrons ($e^-$) and hydrogen ions ($H^+$).

NICOTINAMIDE ADENINE DINUCLEOTIDE ($NAD^+$)

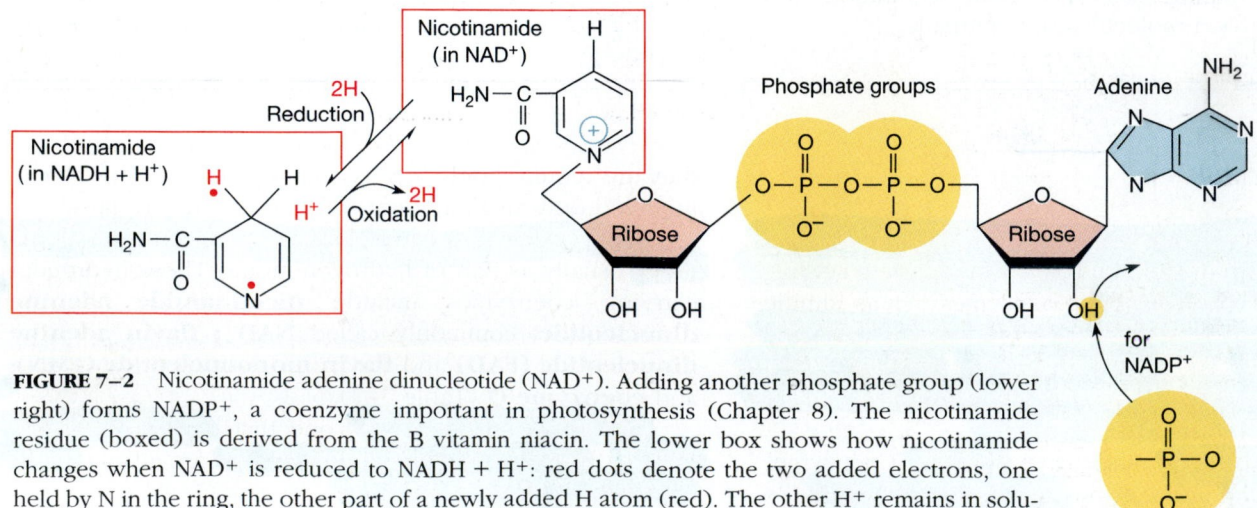

**FIGURE 7–2** Nicotinamide adenine dinucleotide ($NAD^+$). Adding another phosphate group (lower right) forms $NADP^+$, a coenzyme important in photosynthesis (Chapter 8). The nicotinamide residue (boxed) is derived from the B vitamin niacin. The lower box shows how nicotinamide changes when $NAD^+$ is reduced to $NADH + H^+$: red dots denote the two added electrons, one held by N in the ring, the other part of a newly added H atom (red). The other $H^+$ remains in solution.

The overall reaction for the complete oxidation of glucose is:

$$C_6H_{12}O_6 + 6\ O_2 \rightarrow 6\ CO_2 + 6\ H_2O + \text{energy (ATP)}$$

This summarizes a long, complex process, which we can divide into four shorter metabolic pathways:

1. **Glycolysis** (glyco = sugar; lysis = unbinding). Glycolysis breaks down the six-carbon sugar glucose ($C_6$) to two three-carbon **pyruvate** molecules ($C_3$). Two important products are made along the way: a small amount of ATP, and hydrogen atoms. The hydrogens are picked up by the coenzyme nicotinamide adenine dinucleotide ($NAD^+$), forming NADH (Section 7–A).

Glycolysis is an anaerobic process that is not really part of respiration, but rather prepares sugars to enter respiration. Respiration itself consists of three aerobic pathways:

2. **Preparation for the citric acid cycle.** Each pyruvate ($C_3$) gives up one carbon atom as $CO_2$, leaving a two-carbon **acetyl** group ($C_2$). Hydrogens are also released and picked up by coenzyme $NAD^+$, forming more NADH.

3. **Citric acid cycle.** The acetyl group ($C_2$) is attached to another molecule ($C_4$), forming citric acid ($C_6$). This molecule then goes through a series of reactions in which it loses two carbon atoms (equivalent to the acetyl group) as $CO_2$. The remaining $C_4$ molecule eventually accepts another acetyl group, and so the whole pathway is a cycle. The cycle forms some ATP. More importantly, many hydrogen atoms are plucked off during the cycle and picked up by coenzymes $NAD^+$ and FAD (flavin adenine dinucleotide), forming NADH and $FADH_2$.

4. **Electron transport and chemiosmotic ATP synthesis.** NADH and $FADH_2$ from the first three steps pass the hydrogens they carry to an electron transport system, housed in a membrane. The system passes the hydrogens' electrons through a series of redox reactions. This releases energy, which is used for the active transport of $H^+$ to one side of the membrane, forming an $H^+$ gradient across the membrane. This gradient in turn supplies energy for making ATP from ADP and $P_i$. Electrons leaving the system join with oxygen and with still other $H^+$, forming water. The overall process that couples oxidation of hydrogen-laden coenzymes with ATP synthesis is called **oxidative phosphorylation.**

Figure 7–3 and Table 7–2 outline the events in the aerobic breakdown of glucose to carbon dioxide and water.

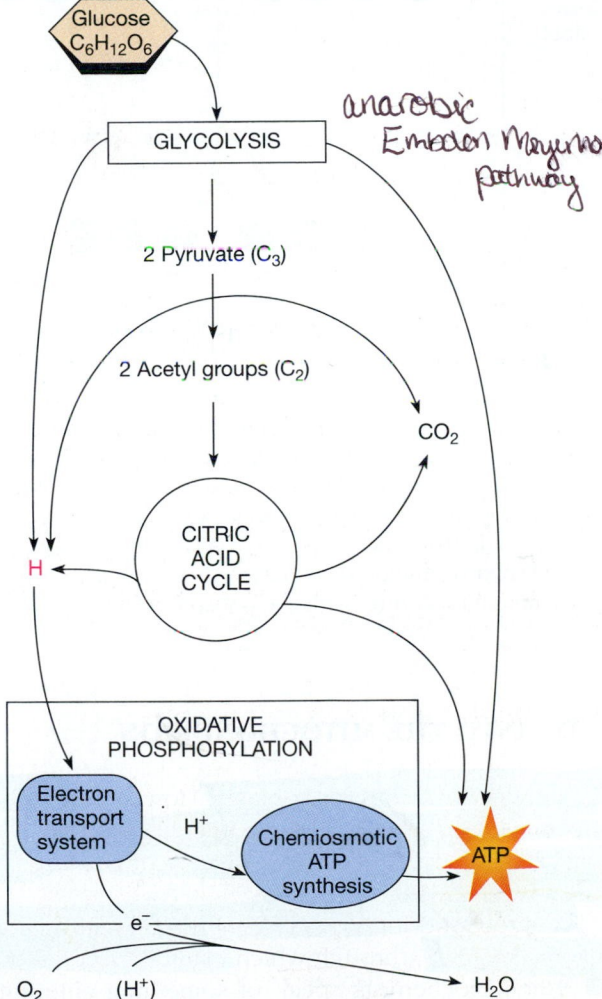

**FIGURE 7–3** Overview of the breakdown of glucose under aerobic conditions. Inputs (glucose and $O_2$) appear at the far left, top and bottom; products ($CO_2$, ATP, and $H_2O$) at the far right. ATP stores much of the energy released from glucose. Glycolysis and the citric acid cycle make some ATP directly. They also send hydrogen atoms to the electron transport system. Electron transport sets up a hydrogen ion gradient, which drives the chemiosmotic ATP synthesis that makes most of the ATP derived from glucose.

**T A B L E   7 – 2**

**Overview of Glucose Breakdown**

| Process | Inputs | Outputs |
|---|---|---|
| Glycolysis | Glucose | Pyruvate<br>NADH<br>ATP |
| Preparation for the citric acid cycle | Pyruvate | $CO_2$<br>Acetyl CoA<br>NADH |
| Citric acid cycle | Acetyl CoA | $CO_2$<br>ATP<br>NADH<br>$FADH_2$ |
| Electron transport and chemiosmotic ATP synthesis | NADH<br>$FADH_2$ → $H^+$ gradient → <br>$O_2$ | ATP<br><br>$H_2O$ |

## 7–C  GLYCOLYSIS

Glycolysis is a series of reactions that break down sugars in virtually all living cells. It serves as the sole source of ATP in many kinds of (anaerobic) fermentation (Section 7–F). However, all aerobic organisms also possess the citric acid cycle and electron transport pathways and use them to make most of their ATP when oxygen is available. In organisms with these respiratory pathways, the main function of glycolysis is to produce three-carbon pyruvate molecules, which can be prepared for the citric acid cycle.

The enzymes of glycolysis are found in the cytoplasm. Interestingly, the first half of glycolysis actually *uses* energy—two molecules of ATP. Each ATP donates a phosphate group to the six-carbon sugar, which is then split into two three-carbon molecules, each with a phosphate group (Figure 7–4). The cell must make this early "investment" of ATP to start a project that will eventually bring in an ATP "profit."

In the second half of glycolysis, each three-carbon molecule receives another phosphate group, not from ATP but from inorganic phosphate ($P_i$) in the cytosol. At the same time, each three-carbon molecule also gives up hydrogen to $NAD^+$. These two NADH molecules from glycolysis carry the first hydrogens we must follow to their energy destination.

We now have two three-carbon molecules, each with two phosphate groups. Eventually all four phosphate groups are transferred to ADP molecules, forming a total of four ATP. Two three-carbon molecules of pyruvate are left at the end of glycolysis.

To summarize, glycolysis involves the conversion of a six-carbon molecule (glucose) into two three-carbon molecules (pyruvate):

$$C_6H_{12}O_6 + 2\ ATP + 4\ ADP + 2\ P_i + 2\ NAD^+ \longrightarrow$$
$$2\ C_3H_4O_3 + 2\ ADP + 4\ ATP + 2\ NADH + 2\ H^+$$

The important products of glycolysis are:

1. **Energy in the form of ATP.** For each glucose molecule broken down, glycolysis produces a net energy gain of two molecules of ATP (four ATP have been formed but two were previously used up).

2. **Energy in the form of NADH.** Under aerobic conditions, the two $NADH + H^+$ from glycolysis can pass their hydrogens to the electron transport system, giving energy to produce more ATP by oxidative phosphorylation.

3. **Pyruvate.** Glycolysis forms two pyruvates from each glucose. Each pyruvate can be modified to form acetyl groups for the citric acid cycle, where further energy is extracted from these fragments of the original glucose molecule.

The individual reactions of glycolysis appear in *In More Detail: The Steps of Glycolysis,* pages 142 and 143.

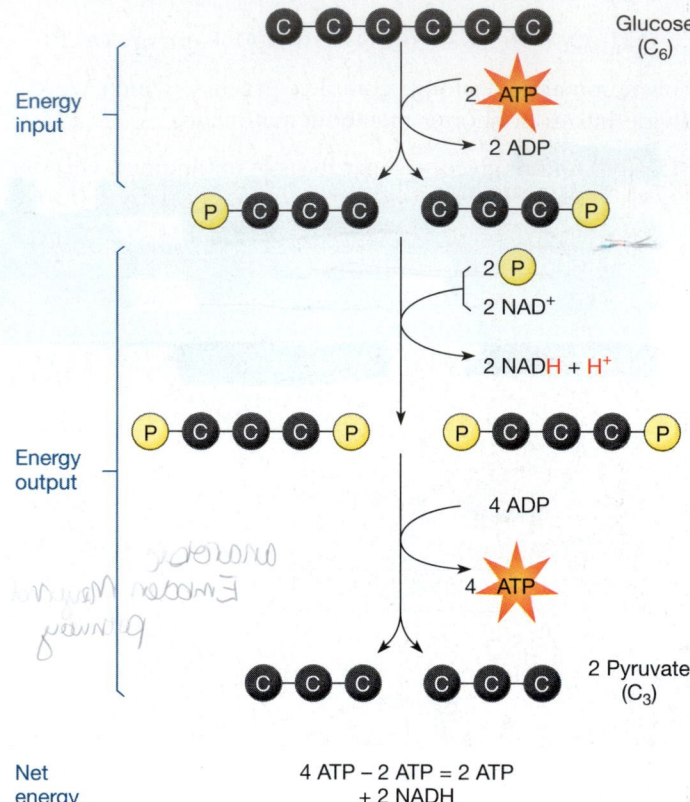

OVERVIEW OF GLYCOLYSIS

**FIGURE 7–4** Overview of glycolysis. In the first series of reactions, glucose receives two phosphate groups from ATP and is broken into two three-carbon molecules. The rest of glycolysis releases more than enough energy to replace the two ATP. Four ATP are made (for a net gain of two ATP in glycolysis), plus two reduced (hydrogen-laden) coenzymes NADH, which supply energy for more ATP during oxidative phosphorylation.

## 7–D  INTO THE MITOCHONDRION

Glycolysis occurs in the cytoplasm. However, in eukaryotic cells respiration occurs in the mitochondria, which take in pyruvate and hydrogens from glycolysis and use them to make most of the cell's ATP.

A mitochondrion's outer membrane contains large-pore channel proteins, through which many molecules pass easily. When biochemists speak of something entering a mitochondrion, they mean crossing the selective inner membrane, not the permeable outer membrane.

The highly folded inner mitochondrial membrane encloses a compartment containing a protein-rich solution called the **mitochondrial matrix.** Many enzymes of the citric acid cycle are dissolved in the matrix, and the rest are attached to the inner face of the inner membrane. This membrane also contains electron transport molecules, which cre-

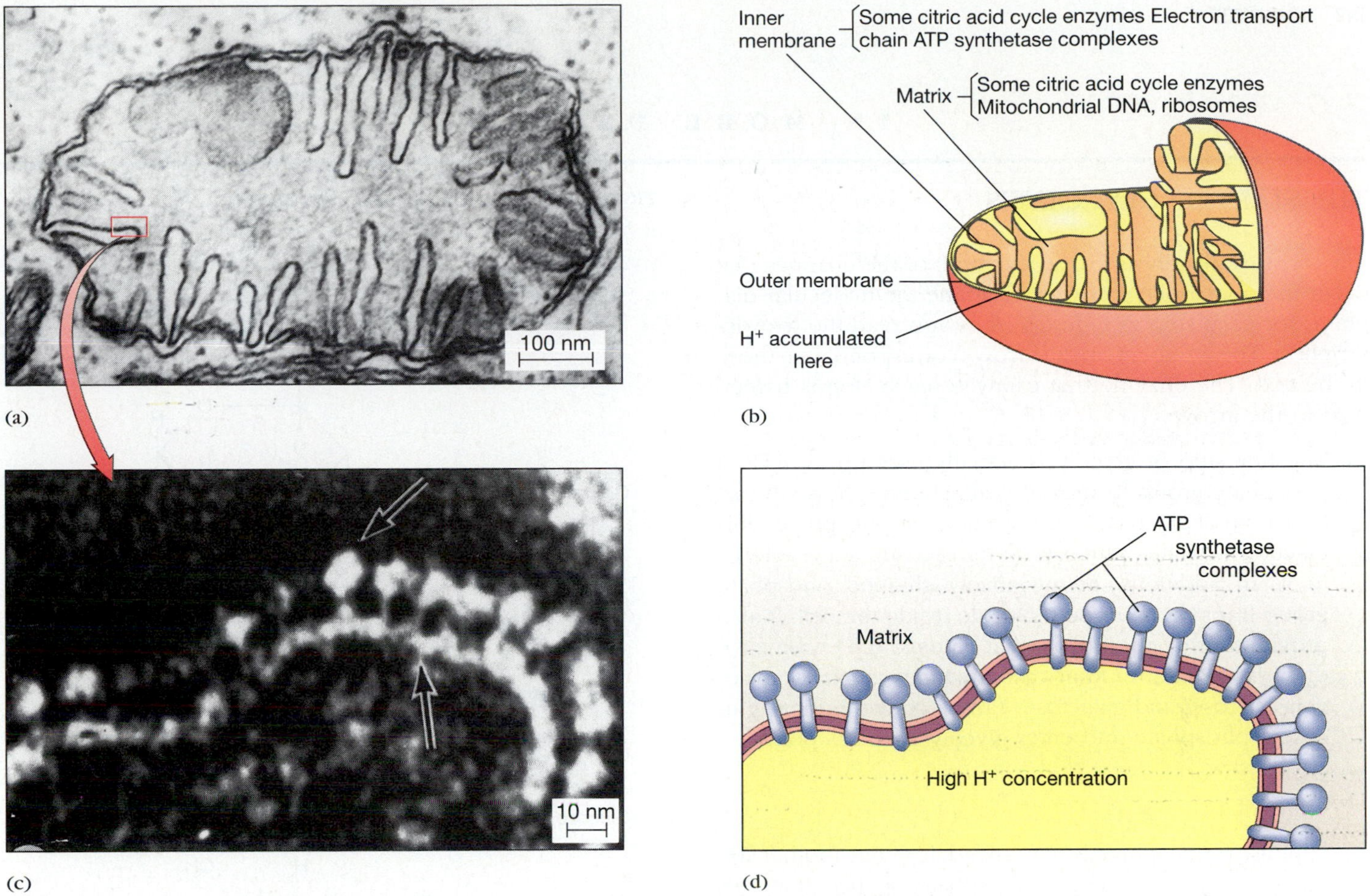

**FIGURE 7–5**  Mitochondrial structure. (a) Section of a mitochondrion from a plant cell and (b) a three-dimensional cutaway drawing of the same mitochondrion. Mitochondria from different kinds of organisms have somewhat different structures; compare this one with the animal mitochondrion in Figure 5–17. (c) Closeup of the inner mitochondrial membrane. The upper arrow points to one of the ATP synthetase complexes, which lie on the inner surface and protrude into the matrix. The lower arrow points to the membrane's outer face, where H⁺ builds up. (d) Interpretive drawing of (c). (a, Biophoto Associates; c, H. Fernández-Morán/Photo Researchers)

In diagram (b):
- Inner membrane: {Some citric acid cycle enzymes Electron transport chain ATP synthetase complexes
- Matrix: {Some citric acid cycle enzymes Mitochondrial DNA, ribosomes
- Outer membrane
- H⁺ accumulated here

In diagram (a): 100 nm

In diagram (c): 10 nm

In diagram (d): ATP synthetase complexes, Matrix, High H⁺ concentration

ate an electrochemical gradient by pushing $H^+$ ions outside the membrane. The inner membrane is also the site of large protein complexes containing ATP synthetase enzymes, which make ATP (Figure 7–5).

In aerobic bacteria, which lack mitochondria, these membrane molecules occur in the plasma membrane. Indeed, the inner mitochondrial membrane is believed to correspond to the plasma membrane of live-in bacteria that evolved into mitochondria.

## 7–E   RESPIRATION

Respiration is a series of redox reactions, typically using oxygen as a final electron acceptor, that breaks down organic molecules and releases their energy. Respiration occurs inside the mitochondria of eukaryotes (both plants and animals) and in the cytoplasm and plasma membranes of bacteria.

### Preparation of Acetyl Coenzyme A

Our two pyruvate molecules, formed by glycolysis in the cytoplasm, cross the inner mitochondrial membrane into the matrix. Here they bind to a complex composed of many enzymes, which carry out a series of reactions on each pyruvate molecule (Figure 7–6):

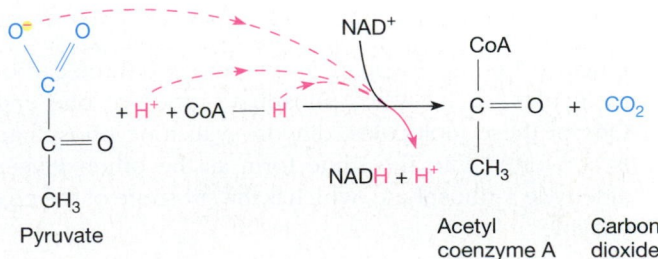

**FIGURE 7–6**  Preparation of acetyl CoA. Pyruvate is oxidized and decarboxylated. $NAD^+$ picks up a hydrogen atom from coenzyme A and an electron from pyruvate, while an $H^+$ follows in solution, forming $NADH + H^+$. Pyruvate's carboxyl carbon and oxygens leave as a molecule of carbon dioxide ($CO_2$). The remaining two-carbon acetyl group binds to coenzyme A, forming acetyl CoA, which proceeds to the citric acid cycle.

*Text continues on page 144*

**IN MORE DETAIL**

## THE STEPS OF GLYCOLYSIS

The steps of glycolysis are an example of the changes that occur in a metabolic pathway. Examine the molecular diagrams for each step in Figure 7–A as you read the text description. The numbers on the arrows correspond with those in the text. The enzyme that catalyzes each step is named next to the arrow.

1. The first step of glycolysis actually uses up an ATP: a phosphate group from ATP is attached to the sixth carbon atom of glucose, forming glucose 6-phosphate. This reaction activates glucose by transferring some energy to it. In addition, the negatively charged phosphate group traps the glucose molecule inside the cell. It also provides a recognition site that binds to the enzyme for the next reaction. Other carbohydrates, such as glycogen, sucrose, and galactose, can be converted into glucose 6-phosphate and enter glycolysis at this point.

2. Glucose 6-phosphate is rearranged, forming another six-carbon sugar, fructose 6-phosphate. Notice that this leaves the molecule's first carbon poking out from the sugar's ring structure.

3. A second ATP is invested, donating another phosphate group, which is attached to the newly exposed carbon atom. This produces fructose 1,6-bisphosphate.

4. Fructose 1,6-bisphosphate is split into two three-carbon molecules, each with a phosphate group at one end. One of these molecules, dihydroxyacetone phosphate, is converted into the same form as the other, glyceraldehyde 3-phosphate, which is the substrate of the next enzyme.

Up to this point, the reaction sequence has invested energy, in the form of two ATP. Beginning with the next step, the cell begins to extract its energy profit.

**FIGURE 7–A** The reactions of glycolysis.

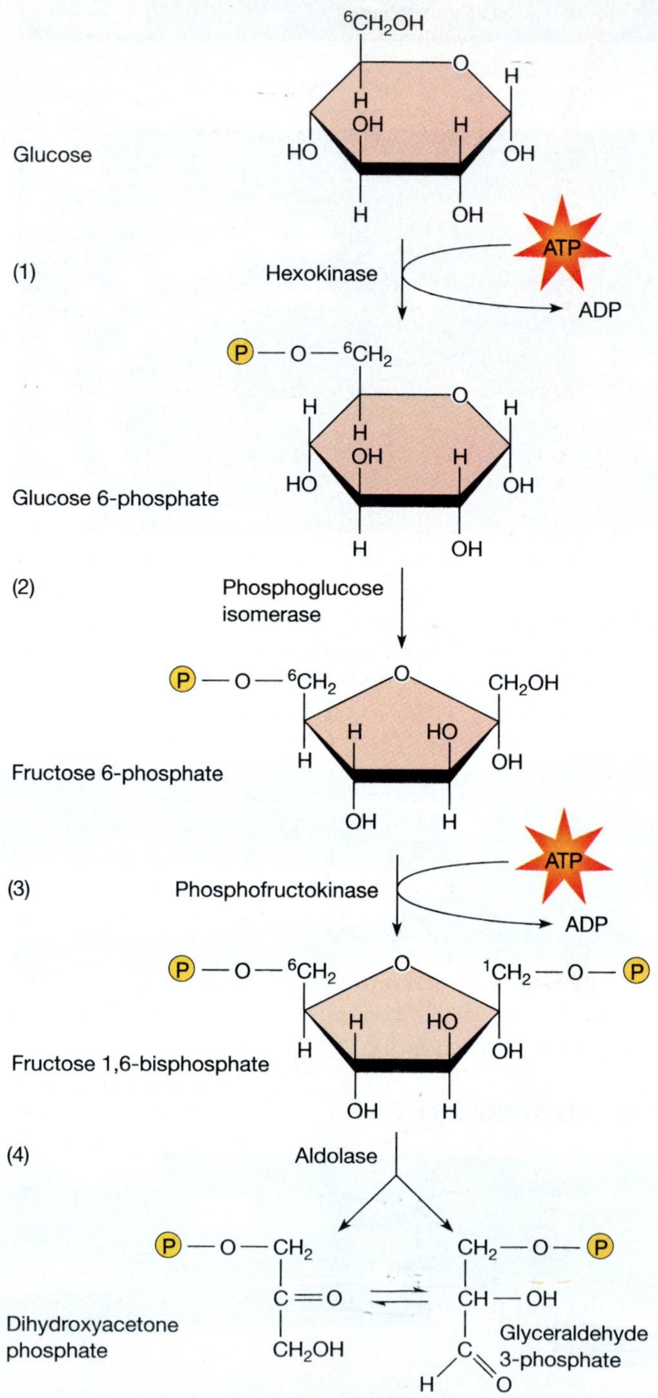

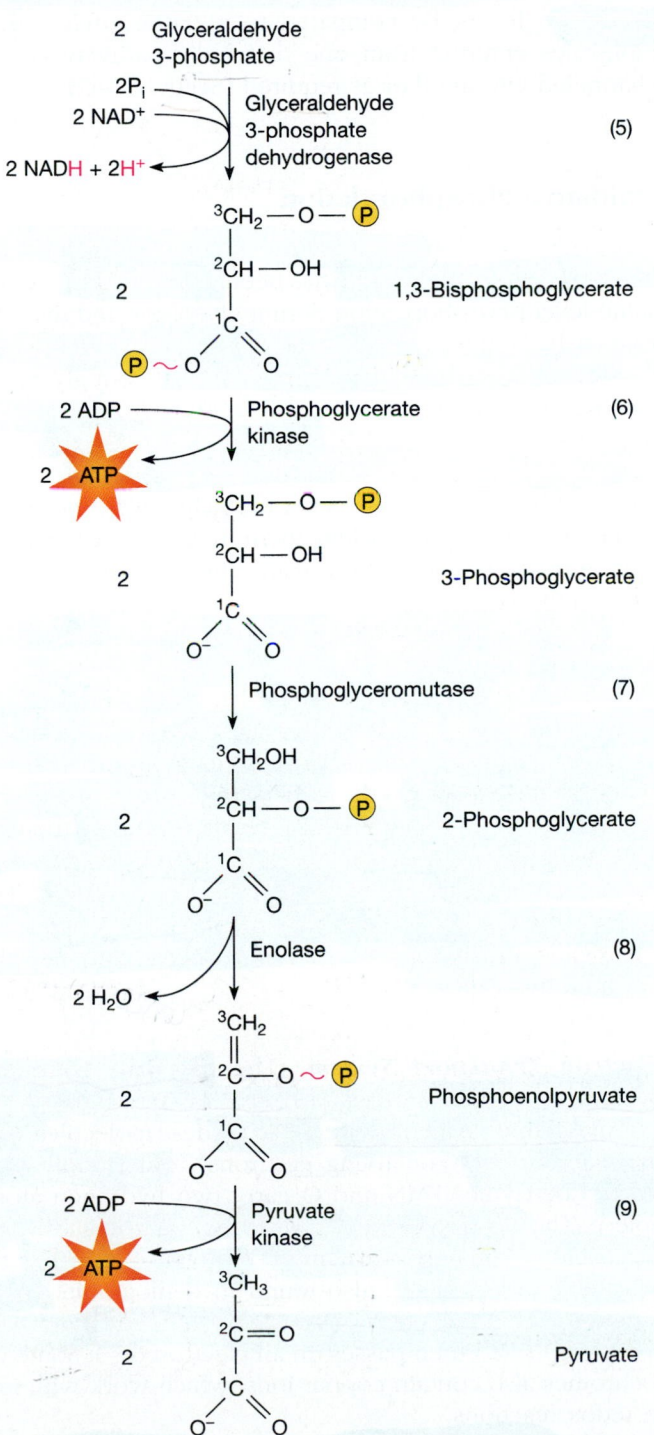

5. Two glyceraldehyde 3-phosphate molecules are oxidized and phosphorylated. The oxidation provides energy for the phosphorylation. This reaction requires two inorganic phosphate groups ($P_i$) from the cytosol and two molecules of coenzyme $NAD^+$. A phosphate group is added to each molecule of glyceraldehyde 3-phosphate, and two hydrogen atoms are removed, one from glyceraldehyde 3-phosphate and one from $P_i$. These two hydrogens reduce $NAD^+$ to $NADH + H^+$. The product of this reaction is 1,3-bisphosphoglycerate. This is called a high-energy phosphate compound because breaking the bond written as a squiggle ($\sim$) is a highly exergonic reaction that yields enough energy to transfer the phosphate group to ADP, forming ATP.

6. In the next step, these newly added phosphate groups are transferred from the high-energy phosphate compound to ADP, a substrate-level phosphorylation (Section 6–F). Note that steps #5 and #6 of glycolysis are responsible for the net gain of ATP during glycolysis. These steps provide a pathway for adding an unattached phosphate group to ADP.

7. The remaining phosphate group of each phosphoglycerate molecule is transferred to the molecule's center carbon.

8. A molecule of water is removed from each 2-phosphoglycerate molecule, forming phosphoenolpyruvate. This is another high-energy phosphate compound: removing its phosphate group (at the $\sim$ bond) is another highly exergonic reaction.

9. The remaining phosphate group is transferred from phosphoenolpyruvate to ADP (another substrate-level phosphorylation), leaving the three-carbon compound pyruvate. This reaction repays the ATP energy used in steps #1 and #3.

1. The pyruvate loses one carbon and two oxygens as a carbon dioxide molecule.
2. The remaining two-carbon acetyl group is attached to a coenzyme A (CoA) molecule, forming **acetyl CoA.**
3. Meanwhile, a molecule of NAD$^+$ is reduced to NADH.

The acetyl group, now attached to CoA, is ready to enter the citric acid cycle.

## Citric Acid Cycle

The citric acid cycle is named for citric acid, formed by joining an acetyl group to a four-carbon molecule. Actually, this is a slight misnomer because at the pH in the mitochondrial matrix, most molecules of citric acid dissociate to H$^+$ and citrate, a negatively charged ion. (The other acids made during the cycle do likewise.) This cycle is also called the **Krebs cycle,** after Sir Hans Krebs, who first worked out the cycle in 1937 and received a Nobel Prize for this work in 1953.

Figure 7–7 shows the reactions of the citric acid cycle. Coenzyme A transfers its two-carbon acetyl group to a four-carbon molecule, oxaloacetate, forming a six-carbon compound, citrate. (Coenzyme A goes back and picks up another acetyl group.) The six-carbon molecule gives up two of its carbons, in the form of carbon dioxide. The remaining four-carbon compound is eventually converted into a new molecule of oxaloacetate, ready to accept another two-carbon acetyl group from acetyl coenzyme A.

We will not describe each reaction in the citric acid cycle in words, but you should take a few minutes to compare the reactants and products for each reaction in Figure 7–7. Here are the important features:

1. Hydrogen atoms are removed at various stages of the citric acid cycle and picked up by coenzymes NAD$^+$ and FAD, forming NADH and FADH$_2$. In terms of energy, this is the most important outcome of the cycle, because these hydrogens are used in oxidative phosphorylation to power the formation of most of the ATP derived from the original glucose molecule.
2. One molecule of ATP is formed by substrate-level phosphorylation during each turn of the cycle.
3. Because each glucose molecule gives rise to two pyruvates, each yielding one acetyl group, it takes two turns of the citric acid cycle to break down the remains of one molecule of glucose.
4. After these two cycles, the equivalents of all six carbons from the original glucose have been released as carbon dioxide. For each pyruvate processed, one CO$_2$ is formed when pyruvate is converted to an acetyl group, and two are formed in the citric acid cycle. This carbon dioxide leaves the cell. In our bodies, it enters the blood, which carries it to the lungs, where it is breathed out as a waste product.

The citric acid cycle may be summarized (the products of the equation are arranged directly under the corresponding reactants so you can compare them easily):

oxaloacetate + acetyl    CoA + ADP + P$_i$ + 3 NAD$^+$ + FAD →

oxaloacetate + 2 CO$_2$ + CoA + ATP +    3 NADH + FADH$_2$

The citric acid cycle is also an important metabolic exchange. It can be compared to a traffic circle, where molecules entering from one metabolic pathway can be channeled into another as required (Section 7–G).

## Oxidative Phosphorylation

Our molecule of glucose has now been completely dismantled. Some of its energy has been stored in ATP by substrate-level phosphorylation during glycolysis and the citric acid cycle. However, most of the energy is now in the form of hydrogen carried by coenzymes NADH (from glycolysis, making acetyl CoA, and the citric acid cycle) and FADH$_2$ (from the citric acid cycle). Hydrogen has little mass, and it yields a lot of energy when it is oxidized by O$_2$. This is why hydrogen is used as rocket fuel. However, more than a billion years before rockets were invented, aerobic bacteria and mitochondria were using this reaction to power ATP synthesis.

Oxidative phosphorylation couples the oxidation of hydrogen to ATP synthesis by using an electrochemical gradient as an energy intermediate. The gradient is formed when coenzymes pass their hydrogen atoms to the molecules of the electron transport system in the inner mitochondrial membrane. As the hydrogens' electrons are passed through the system in a series of redox reactions, their energy is gradually released and used to pump H$^+$ out through the membrane, creating an electrochemical gradient. This gradient then provides energy to phosphorylate ADP to ATP. Meanwhile, coenzymes NAD$^+$ and FAD go back and pick up more hydrogens.

*Electron Transport System*    The **electron transport system** consists of a series of carrier molecules in the inner mitochondrial membrane. Some of these molecules carry hydrogen atoms (containing electrons), others only electrons. Coenzymes FMN and Q carry two hydrogen atoms apiece. The electron carriers include several iron-sulfur proteins and several **cytochromes,** proteins containing the prosthetic group heme, also found in hemoglobin. Heme contains iron, which changes from Fe$^{3+}$ to Fe$^{2+}$ and back as it accepts and then passes on an electron (e$^-$). Some cytochromes also contain copper ions, which work with iron in redox reactions.

The electron transport molecules are grouped into three types of protein complexes, with increasing affinity for elec-

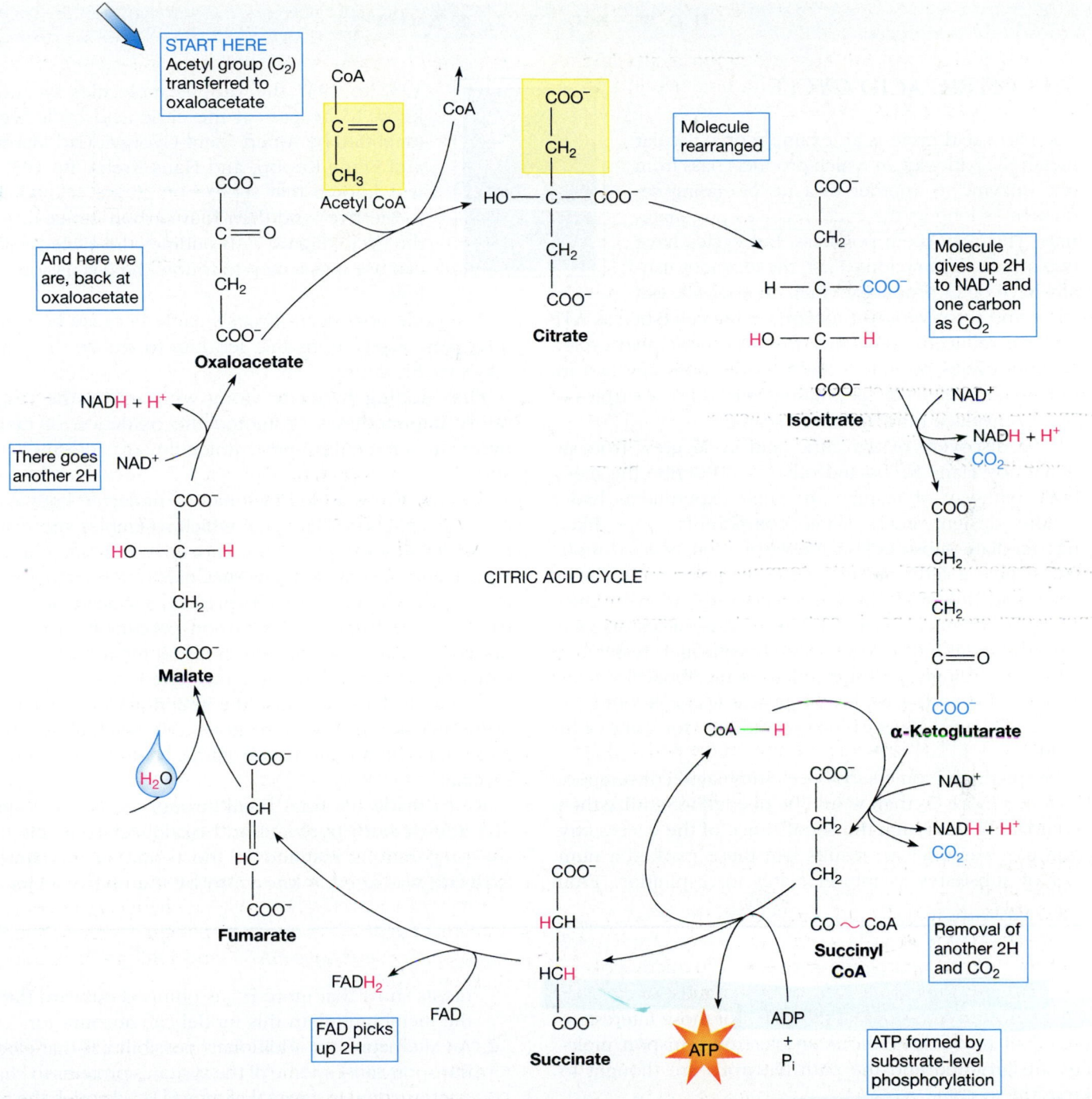

**FIGURE 7–7**  The citric acid cycle. Begin at "START HERE," in the top left-hand corner, and follow the cycle clockwise, reading each event in turn (blue boxes).

# THE CITRIC ACID CYCLE

The citric acid cycle is an example of a circular metabolic pathway, in which products pass from one enzyme to another in a never-ending sequence as long as the cycle receives raw materials. This may seem pointless, but cycles have two important functions. First, the enzymes usually have other products, which do not cycle but leave and serve various functions in the cell (such as ATP and the reduced coenzymes from the citric acid cycle). Second, cycles provide a pathway between any two intermediate molecules, permitting cells to use a surplus of one to produce others that are scarce.

The discovery of the citric acid cycle grew from attempts by Hans Krebs and others to describe the metabolic pathway of respiration. Their experiments had a simple design: guess what compounds were likely intermediate substrates in the respiration of food molecules; give a little of one of these substances to living cells; and determine how it affects respiration by measuring how much $O_2$ the cells use. It is easiest to measure the uptake of oxygen in cells with high respiration rates. Accordingly, investigators used metabolically active tissues. Minced pigeon breast muscle was a favorite, but Krebs also used slices of liver and kidney for some of his work.

Surprisingly, some substances stimulated consumption of much more $O_2$ than would be needed to oxidize them completely. Although the significance of the excess oxygen use was unclear, results like these marked a number of substrates as intermediates in respiration. From 1932 to 1937 the various molecules we now know to be part of the citric acid cycle were identified by Albert Szent-Gyorgyi, Carl Martius and Franz Knoop, and Hans Krebs. By 1937 it was known that six-carbon citrate is oxidized to the five- and then four-carbon molecules of the cycle. Figure 7–B outlines the cycle so you can use it as a map to follow the discussion below.

The cycle now bears Krebs's name because he made three key observations that led him to realize the pathway's cyclic nature.

First, adding pyruvate along with any of the respiratory intermediates promoted the oxidation of many pyruvate molecules per molecule of intermediate added.

Second, Krebs added malonate, a molecule known to poison respiration. Malonate, which resembles succinate, is the classic example of a competitive inhibitor, blocking the enzyme that converts succinate to fumarate. Krebs found that when malonate is present, succinate accumulates, and its five- and six-carbon precursors also back up. The oxidation of pyruvate is also inhibited. However, when fumarate is added, it is converted to succinate despite the blockage! Because the inhibitor blocks catalysis of the reaction in both directions, this result shows that there must be an alternative pathway from fumarate to succinate.

Krebs made his final breakthrough when he found that adding both pyruvate and oxaloacetate to his tissue preparations resulted in the formation of citrate. Pyruvate was already known to be the end product of

---

trons. Electrons pass along in sequence, ending at oxygen, with the highest electron affinity of all. For those interested, Figure 7–8 shows how the main electron transport molecules are grouped and the path electrons are thought to take in the system.

We do not yet know exactly how $H^+$ ions are moved through the membrane, but biochemists see two possibilities:

1. In the electron transport system, a hydrogen carrier transports hydrogen atoms from the matrix to the inner membrane's outer surface. Here an electron carrier accepts only the electrons, stranding $H^+$ outside the membrane without a ride back in. Meanwhile, back at the matrix side, the electrons are joined to more $H^+$, formed by dissociation of $H_2O$ in the mitochondrial matrix. The process repeats, expelling more $H^+$. However, experiments show that more $H^+$ is pumped outward through the membrane than this model can account for.

2. An alternative (or additional) possibility is that electron transport causes some of the system's proteins to change conformation in a way that moves $H^+$ through the membrane.

In any case, electron transport builds a high concentration of $H^+$ outside the membrane. When electron transport is inhibited, no $H^+$ gradient is formed. For example, some poisons and antibiotics act by inhibiting certain electron transport molecules. The first segment of the system is inhibited by the insecticide rotenone, the second by the antibiotic antimycin A, and the third by cyanide and by carbon monoxide.

We finally come to the role of oxygen in cellular respiration. In our bodies, oxygen is breathed into the lungs

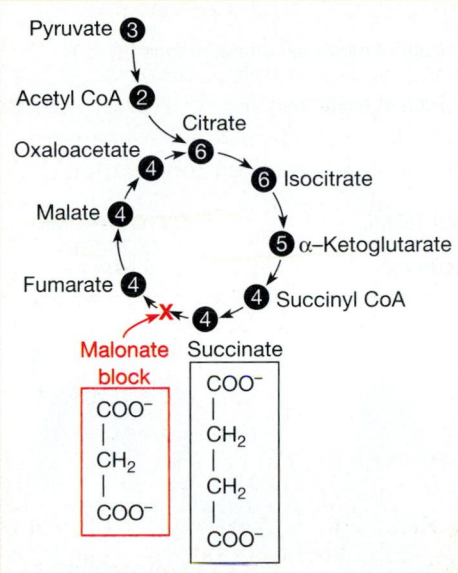

**FIGURE 7–B** Outline of the citric acid cycle. Numbers of carbons in each molecule are shown in black circles. Malonate blocks the conversion of succinate to fumarate because it is so similar to succinate that it competes for succinate's binding sites on the enzyme succinate dehydrogenase.

glycolysis. Hence, this finding linked two vital pathways: the one that breaks down carbohydrates (glycolysis) and the one that oxidizes the molecules with two or three carboxyl groups (respiration). It also showed that citrate, which can be oxidized to oxaloacetate, can be regenerated from this oxaloacetate by adding pyruvate. Therefore, Krebs saw, the whole pathway is a cycle, which in effect oxidizes carbohydrates (pyruvate) fed into it.

This explained how fumarate could be converted to succinate when the direct enzymic step was poisoned by malonate: it was processed all the way around the circle. It also explained the excess oxidation seen when any of the cycle's intermediates was added to respiring tissue. The material was not simply oxidized but was converted to additional oxaloacetate, which escorted more pyruvate into the citric acid cycle. By adding more oxaloacetate, the cell's capacity to oxidize pyruvate was expanded, and this accounted for much of the extra oxidation. And since the cycle's intermediates come back for more pyruvate, it explained why each one of these molecules added promotes the oxidation of many additional molecules of pyruvate.

Krebs was not surprised to discover the cyclic nature of this respiratory pathway. In 1932 he had discovered the urea cycle (Section 35–A). But providing the key pieces for understanding this vital metabolic pathway earned him the honor of having his name attached to the cycle. He also received a Nobel Prize for this work in 1953. This prize was shared by Fritz Lipmann, who had introduced the idea of ATP as a universal energy medium in cells, regenerated by the oxidation of foodstuffs.

and taken up by the blood, which carries it to the cells. Oxygen diffuses into a cell and on into the inner membrane of a mitochondrion. Here it becomes sandwiched between a heme-group iron and a copper ion in the last cytochrome of the electron transport system. These ions hold it in place for the last step of respiration.

Electrons arriving at the end of the electron transport system have given up most of their energy in the system's redox reactions. Oxygen takes these electrons, plus some $H^+$ (again, from dissociation of water), forming new water molecules:

$$4 \, e^- + 4 \, H^+ + O_2 \rightarrow 2 \, H_2O$$

This reaction completes the oxidation part of oxidative phosphorylation. The phosphorylation part produces ATP from ADP and $P_i$, using energy stored by oxidation: the hydrogen ion gradient across the inner mitochondrial membrane.

**The $H^+$ Gradient and Phosphorylation**    The $H^+$ gradient is a store of electrochemical potential energy. That is, it has both an electrical and a chemical component. The electrical part comes from the difference in positive electrical charges on the two sides of the membrane (the membrane potential, Section 4–E). The chemical part comes from the concentration difference of $H^+$ (that is, a difference in pH). The combined energy of these two components, expressed electrically, is about 220 millivolts. Hydrogen ions are protons, and so this combined force tending to move them back through the inner mitochondrial membrane, down their concentration gradient, is called a **proton-motive force.**

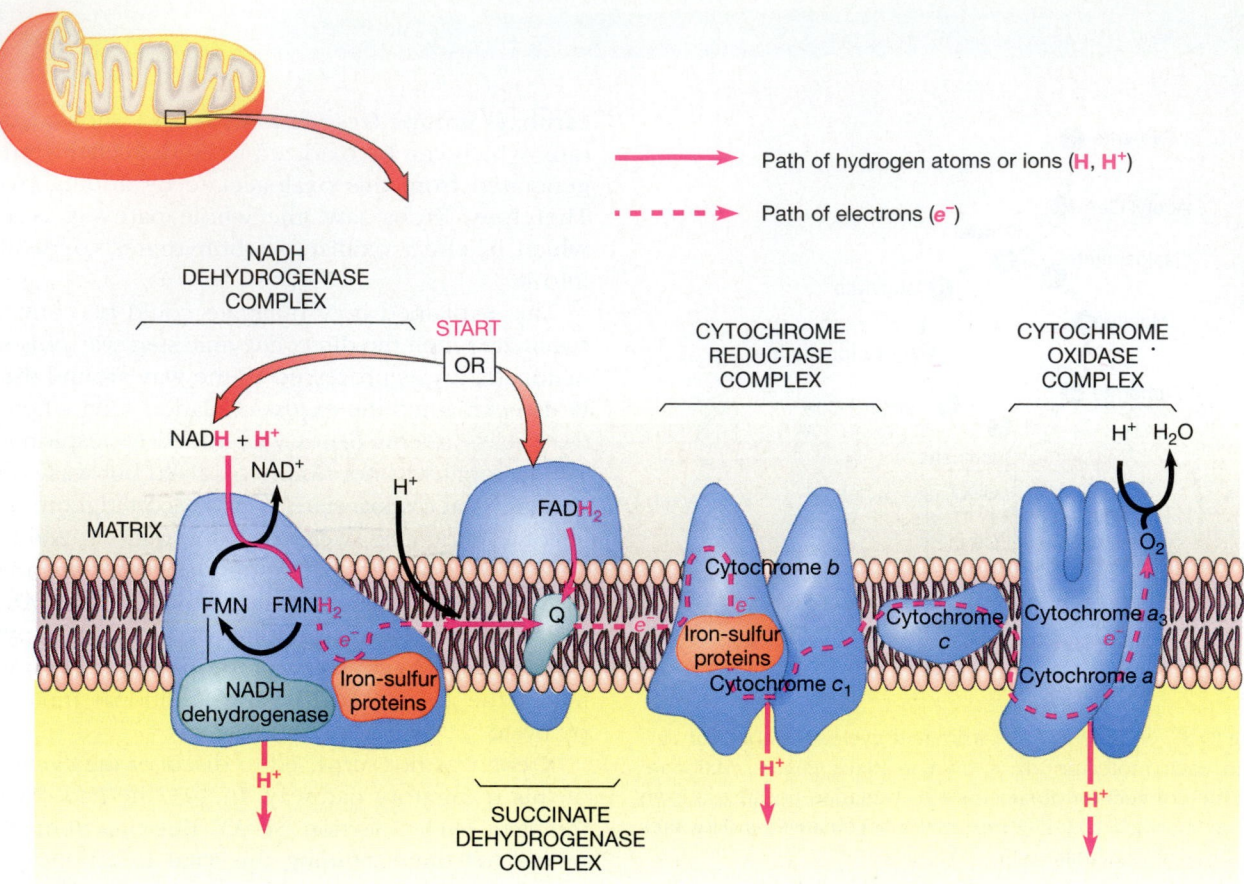

Path of hydrogen atoms or ions (**H**, **H⁺**)

Path of electrons (**e⁻**)

**FIGURE 7–8**    Electron transport in the inner mitochondrial membrane. Electron transport molecules form three kinds of large complexes, each able to export $H^+$: the NADH dehydrogenase, cytochrome reductase, and cytochrome oxidase complexes. Coenzyme Q and cytochrome $c$ are smaller, mobile molecules carrying electrons between the large complexes in the order shown. At the end of the chain, $O_2$ held in a cytochrome oxidase complex accepts the spent electrons, forming water.

Hydrogen enters the electron transport system from either NADH + $H^+$ or $FADH_2$. NADH + $H^+$ passes two hydrogen atoms ($2H^+ + 2e^-$) to coenzyme FMN in an NADH dehydrogenase complex.

The succinate dehydrogenase complex is part of the citric acid cycle. Hydrogens coming from coenzyme FAD in this complex do not pass through an NADH dehydrogenase complex but enter the electron transport system at coenzyme Q. Hence they bypass one of the system's three $H^+$-pumping sites.

The membrane's lipid bilayer is virtually impermeable to $H^+$ ions. However, the membrane contains channel proteins that permit $H^+$ to pass back through the membrane to the inside. At the channel's inner end sits a complex of proteins including ATP synthetase, an enzyme that joins ADP and $P_i$, forming ATP (Figure 7–9).

The coupling between the flow of $H^+$ down its gradient and phosphorylation of ADP is called chemiosmotic ATP synthesis. ATP synthetase actually makes ATP unaided by the flow of $H^+$. However, it cannot *release* this ATP until $H^+$ ions flow through the channel toward the matrix, down their electrochemical gradient. Hence, the $H^+$ gradient powers ATP synthesis indirectly by freeing the enzyme's active site to make more ATP.

With each ATP released from the enzyme, the $H^+$ gradient loses some of its stored energy. The gradient must be constantly renewed by the continuous flow of electrons arriving via NADH and $FADH_2$ and the electron transport system.

The exact number of $H^+$ transported to the $H^+$ gradient per electron pair transported by the system is still unclear. So is the number of $H^+$ needed to release each ATP from ATP synthetase. Indeed, these numbers may be variable.

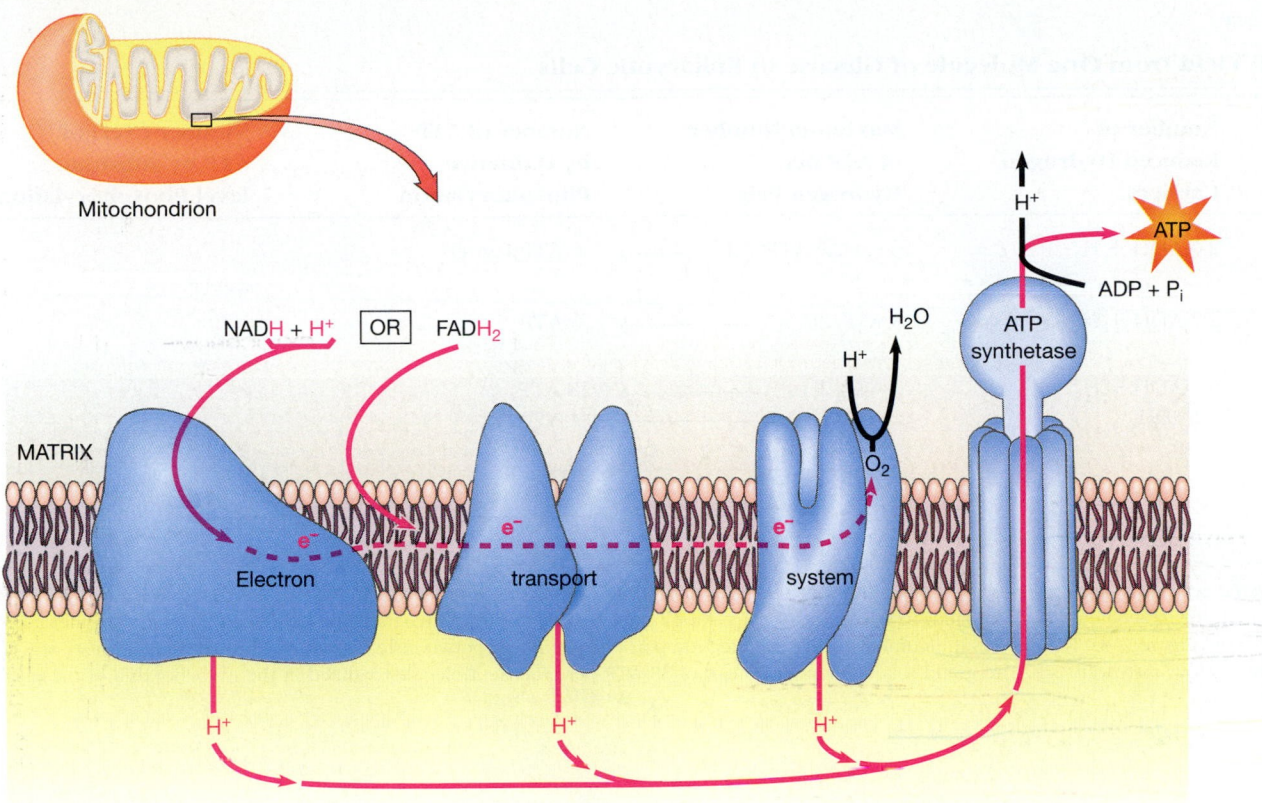

**FIGURE 7–9**  Overview of oxidative phosphorylation. Red arrows show the path of energy. The electron transport system forms an electrochemical gradient of $H^+$ by exporting $H^+$ into the space between mitochondrial membranes, as shown in Figure 7–8. $H^+$ then moves back through the membrane, down its gradient, through a channel protein associated with an ATP synthetase complex, which joins ADP and $P_i$, forming ATP.

The events of oxidative phosphorylation can be summarized:

1. **Oxidation.** Energy is stockpiled in the form of an $H^+$ electrochemical gradient across the inner mitochondrial membrane, via a series of oxidation-reduction reactions. The electron transport system accepts hydrogens from NADH and $FADH_2$ (formed during glycolysis, preparation of acetyl CoA, and the citric acid cycle). The system passes the hydrogens' electrons through a series of redox reactions and uses the energy so released to pump $H^+$ through the membrane against its electrochemical gradient. At the end of the system, the electrons are accepted by oxygen, which also picks up $H^+$, to form water.

2. **Phosphorylation.** The energy of the $H^+$ gradient is used in the phosphorylation of ADP to ATP. This happens as $H^+$ recrosses the membrane via channel proteins connected to ATP synthetase enzymes.

## The Energy Yield of Glucose

How much ATP does a cell obtain from one glucose molecule? We have seen that glycolysis produces four ATP but uses two ATP to start the process. So the net gain from glycolysis is two ATP per glucose broken down. The citric acid cycle produces another two ATP, for a total of four ATP per glucose by substrate-level phosphorylation.

Experiments show that a pair of electrons passed from NADH through the electron transport system to oxygen provides the energy for synthesis of up to three ATP. When $FADH_2$ passes its hydrogens to the system, the maximum is two ATP made per electron pair donated by $FADH_2$. So we can calculate that the maximum possible number of ATPs derived from one glucose molecule in glycolysis, the citric acid cycle, and oxidative phosphorylation combined is 38, as shown in Table 7–3.

The grand total of 38 ATP per glucose molecule is a maximum estimate. Substrate-level phosphorylation (in gly-

## Maximum ATP Yield from One Molecule of Glucose in Eukaryotic Cells

| Process | Number of Reduced Hydrogen Carriers | Maximum Number of ATP per Hydrogen Pair | Number of ATP by Oxidative Phosphorylation | Number (Net) of ATP by Substrate-level Phosphorylation |
|---|---|---|---|---|
| Glycolysis | 2 NADH + H$^+$ | 3 (or 2)* ATP/2H $\longrightarrow$ | 6 ATP (or 4) | |
| | | | | 2 ATP |
| Pyruvate to acetyl CoA | 2 NADH + H$^+$ | 3 ATP/2H $\longrightarrow$ | 6 ATP | |
| Citric acid cycle | 6 NADH + H$^+$ | 3 ATP/2H $\longrightarrow$ | 18 ATP | |
| | 2 FADH$_2$ | 2 ATP/2H $\longrightarrow$ | 4 ATP | |
| | | | | 2 ATP |
| | | | 34 ATP (or 32) | 4 ATP |
| | GRAND TOTAL: | | 38 ATP (or 36) | |

*NADH produced in the cytoplasm during glycolysis cannot enter the mitochondrion, but it passes hydrogen to shuttle molecules that can. In the brain and skeletal muscles, this shuttle molecule hands hydrogens to coenzyme Q, thus skipping the first site for the active transport of H$^+$ out through the inner mitochondrial membrane (see Figure 7–8). This lowers the number of ATP per electron pair to a maximum of two instead of three, reducing the number of ATP in the first row above to four instead of six. The grand total is then 36 instead of 38 ATP per glucose molecule oxidized in the mitochondria of these tissues.

colysis and the citric acid cycle) does give a net yield of 4 ATP per glucose. However, the actual number of ATP from oxidative phosphorylation may (like your car's estimated gas consumption) be lower than 34, for several reasons. First, the membrane is not completely impermeable to H$^+$; slow H$^+$ leakage back through the membrane without ATP synthesis is inevitable. Second, some of the H$^+$ gradient's electrochemical energy is used, not by ATP synthetase, but by proteins that transport substances such as sodium and calcium through the membrane. Some energy is used to bring phosphate and pyruvate into the matrix, where they are used to make ATP, and to export newly made ATP to the cytoplasm, where most of the cell's ATP is used.

Bacteria apparently have only two, rather than three, electron-transporting and H$^+$-pumping protein complexes, which are housed in their plasma membranes. Hence, they make fewer ATP per glucose used.

## 7–F   FERMENTATION

Cellular respiration (including the citric acid cycle) demands a constant supply of oxygen to accept electrons from the electron transport system. If a cell runs out of oxygen, all the electron transport molecules are soon stuck in the reduced form, holding electrons, and the system stops working. The H$^+$ gradient is quickly used up and then is no longer available to drive ATP synthesis.

Some cells, such as brain cells, die rapidly if deprived of oxygen because they cannot survive for long without a constant, large supply of ATP. Other cells tolerate oxygen deprivation and can make do with the small amounts of ATP made by the anaerobic process of **fermentation,** which uses organic molecules as the final electron acceptors. Indeed, some anaerobic bacteria meet all their energy needs in this way. Other organisms can survive indefinitely in the absence of oxygen; these include many bacteria, fungi, and invertebrates (animals without backbones, such as worms).

We shall look at two kinds of fermentation. Both use glycolysis, which makes a net of two ATP in the breakdown of each glucose molecule to two pyruvates. In glycolysis, NAD$^+$ must be converted to NADH before any ATP can be made (see Figure 7–A). But when electron transport stops for lack of oxygen, NADH cannot pass electrons on to the electron transport system. Cells contain very little NAD$^+$, and all of it will soon be converted to NADH. Without NAD$^+$ to go back and accept more hydrogen, glycolysis too will grind to a halt.

In fermentation, pyruvate produced by glycolysis serves as an alternative acceptor of hydrogen from NADH. This frees NAD$^+$ to go back and accept more hydrogen from glycolysis, so glycolysis keeps making its small amount of ATP.

Different organisms have somewhat different fermentation pathways. Some carry out **alcoholic fermentation,** discovered by Louis Pasteur during his study of the chemistry of wines. In winemaking, juice pressed from grapes is inoculated with yeasts, which are unicellular fungi. Yeasts break down the sugars in the juice to pyruvate by glycol-

## E S S A Y :  *Pasteur and Yeasts*

The word enzyme means "in yeast." Most early studies on enzymes and their actions were attempts to understand the alcoholic fermentation by which wine is made. As early as 1785 the Academy of Florence offered a prize for a theory of fermentation that could be applied to keeping wine in better condition while it was transported. However, no real light was shed on the subject until the French wine industry asked Louis Pasteur to investigate the condition called *"l'amer"* that destroyed large quantities of the best Burgundy every year.

From his experiments, Pasteur concluded that fermentation occurred only when living yeast was present. Justus von Liebig, an influential chemist, thought otherwise and performed many experiments in which he killed yeast cells by boiling them and then tested them to see if they would ferment sugar. They would not, and enzymes, also called "fer-

**FIGURE 7–C**  Louis Pasteur (1822–1895). (Smithsonian Institution)

ments," came to be considered as catalysts that would not function outside a living cell. We now know that enzymes can function outside cells. Liebig, in boiling the yeast cells, had not only killed them but also denatured their enzymes so that they no longer functioned.

Pasteur discovered that *l'amer*, which turned wine sour, was caused by bacteria. Microscopic examination showed that the wine turned sour when it contained more bacteria than yeast cells. *L'amer* could be prevented by greater cleanliness, including sulfur sterilization, which is now standard practice in many winemaking steps. Pasteur also showed why it is important to exclude air during fermentation: wine yeasts produce alcohol under anaerobic conditions, but if oxygen is present, other yeasts and bacteria that convert alcohol into acetic acid ($CH_3CH_2OH \rightarrow CH_3COOH$) will turn the wine to vinegar.

Pasteur loved good wines and devoted many years to studies of their fermentation and aging. His book *Études sur le Vin*, published in 1866, revolutionized winemaking, giving it a scientific basis for the first time.

---

ysis. Then each pyruvate molecule is dismantled into a molecule of carbon dioxide and a molecule of the two-carbon compound acetaldehyde (Figure 7–10). Acetaldehyde is next reduced by accepting two hydrogens from NADH + H$^+$, forming the two-carbon alcohol ethanol (ethyl alcohol), the active ingredient in alcoholic beverages. This transfer of hydrogen frees NAD$^+$ to go back to glycolysis and be reduced again, allowing the yeast cell to keep making ATP.

If fermentation continues until the yeast cells have used up all the sugar around them, a dry wine results. Sweet wines contain sugar because the yeast cells produced enough alcohol to inhibit fermentation before they had used all the sugar. Stoppering a wine bottle before fermentation has finished yields a bubbly liquid because carbon dioxide is still being given off. Such was the young wine that stretched and split the wine skins in the New Testament story. To make a fizzy wine like champagne you use a very strong bottle and cork it before fermentation has finished. Carbon dioxide dissolves in the wine under pressure and is released as bubbles when the bottle is opened.

Yeasts produce alcohol only when little or no oxygen is present. Given plenty of oxygen, they use respiration to break sugar down completely to carbon dioxide and water. When a vat of wine is fermenting rapidly, it produces carbon dioxide fast enough to drive off the air over the wine, and this prevents oxygen from dissolving in the wine. But when fermentation slows down, wine must be sealed up immediately, before oxygen can enter. Otherwise, bacteria that fall into the wine from the air will use the oxygen to convert the alcohol into acetic acid ($CH_3COOH$). Acetic acid is the acid in vinegar, and this is how wine vinegar is made.

Alcoholic fermentation by yeasts is also used to produce ethanol for fuels. Currently, many researchers are trying to develop yeasts that convert more of the sugars in corn or other materials to ethanol.

Variations on the alcoholic fermentation pathway occur in some yeasts, in other microorganisms, and in many invertebrates. Strangely enough, the common goldfish, a vertebrate, can survive for several days without O$_2$ by car-

(a) Budding yeast cells (LM)

Pyruvate          Acetaldehyde                    Ethanol

**FIGURE 7–10** Alcoholic fermentation. (a) Yeasts reproducing rapidly by budding. Some buds grow so fast that they bud again before they separate from the parent. (b) Alcoholic fermentation as performed in yeast and some other organisms. Pyruvate produced in glycolysis is decarboxylated. The resulting two-carbon compound, acetaldehyde, accepts two hydrogens from NADH + H$^+$. This releases NAD$^+$, which is needed for glycolysis to continue and produce ATP at a low rate while the cell is deprived of oxygen.   (a, Biophoto Associates/Photo Researchers)

rying on alcoholic fermentation, provided it is kept in cold water. At low temperatures, the fish's metabolism slows down so that it requires little energy.

Fermentation in higher animals, including ourselves, is most common in muscle tissue during strenuous exercise. The muscles make ATP by respiration, using oxygen as fast as the bloodstream can deliver it, but this does not provide enough ATP. So the muscles make a small additional supply of ATP by fermentation. Again, pyruvate from glycolysis is reduced by accepting hydrogens from NADH + H$^+$, but in muscles pyruvate is converted into another three-carbon compound, lactate (Figure 7–11). The NAD$^+$ so freed picks up more hydrogen, allowing glycolysis to proceed.

As lactate accumulates, it causes fatigue and muscle pain. Interestingly, some strains of influenza ("flu") and other diseases somehow trigger a switch to fermentation in the muscles at levels of exertion lower than usual. The resulting lactate accounts for the tiredness and aches that may persist after recovery from an illness.

Lactate produced in the muscles is picked up by the blood. However, the body cannot tolerate a great buildup of lactate, but must eventually oxidize it back to pyruvate. After strenuous exercise, we still keep breathing heavily for a time, to repay the body's **oxygen debt.** The liver takes up lactate and converts it back to pyruvate. ATP is restored to its normal level by processing pyruvate through the citric acid cycle and electron transport system.

The remaining pyruvate is converted back to glucose. (This process reverses many steps of glycolysis but has alternative pathways to get back up energetically steep reactions.) The glucose re-enters the blood and is carried back to the muscles, where it is converted to glycogen and stored.

Fermentation producing lactate as an end product is also carried out by various bacteria. Some of these are cultured and used to produce fermented foods such as cultured buttermilk and sour cream, yogurt, sauerkraut, pickles, green olives, and some sausages.

Most cells can carry on glycolysis, but not all organisms can carry on respiration. For this reason, and because Earth's atmosphere probably contained little O$_2$ when life first evolved, it is generally believed that glycolysis and fermentation evolved before aerobic pathways.

## 7–G   ALTERNATIVE FOOD MOLECULES

Presented with a smorgasbord of common food molecules, most cells use glucose to make ATP. However, since all organic molecules contain stored energy, any of them may be broken down to release the energy needed to make ATP.

Carbohydrates are processed by way of glycolysis. Polysaccharides, from food or the body's glycogen stores, can be broken down to glucose. Monosaccharides other than glucose can be converted into glucose or fructose and fed into glycolysis. Thus, claims that other "natural" sugars are less fattening than sucrose or glucose are untrue; a cell treats them all alike. However, it is true that sucrose (table sugar) is worse than other sugars for your teeth: bacteria that cause tooth decay can use sucrose, but not other sugars, to make a "glue" by which they stick to tooth surfaces.

Fats and proteins are metabolized in various ways, but they too ultimately reach either glycolysis or the citric acid cycle (Figure 7–12). Fats are broken down to glycerol and fatty acids. Glycerol is converted to glyceraldehyde 3-phosphate and enters glycolysis at that point. Fatty acids are broken down into two-carbon acetyl groups, which combine with coenzyme A and enter the citric acid cycle. Here their

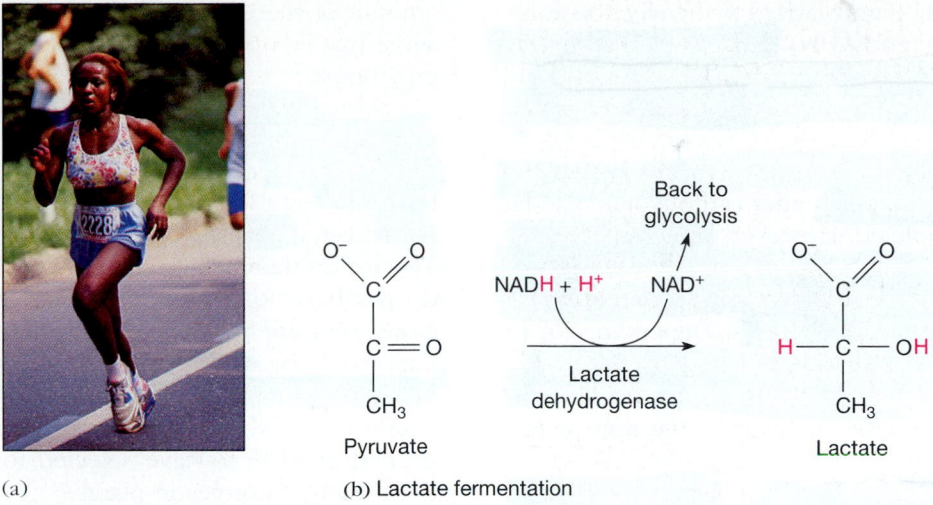

**FIGURE 7–11**  Lactate fermentation in muscle. (a) Strenuous exercise is anaerobic. (b) When the cell lacks $O_2$, NADH accumulates. Pyruvate produced by glycolysis accepts hydrogens from NADH + H⁺ and becomes lactate, freeing NAD⁺ to return to glycolysis and accept more hydrogens.    (a, courtesy New York Road Runners Club)

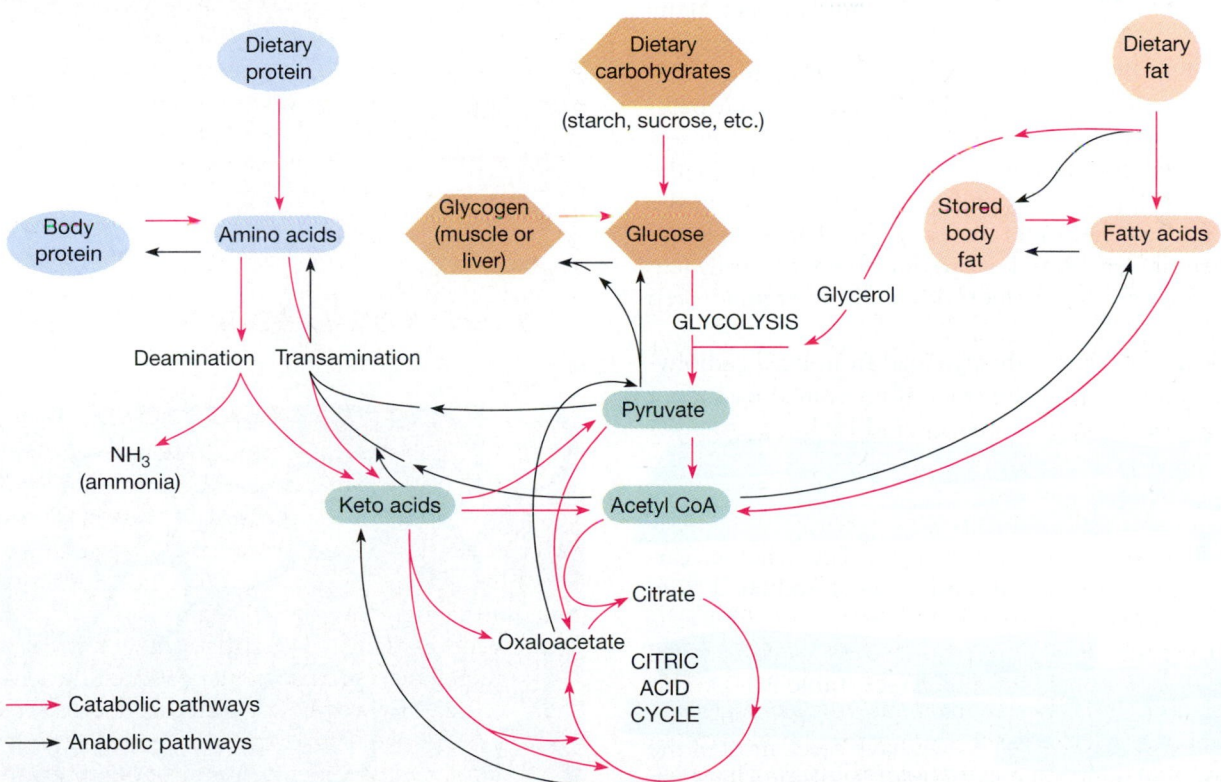

**FIGURE 7–12**  Some major metabolic pathways. Proteins, carbohydrates, and fats in the diet may become part of the body or may be processed to release energy. The most usual way for a cell to obtain energy is via glycolysis (from glucose to pyruvate, center), which breaks down carbohydrates to release energy. Amino acids may be converted to keto acids by deamination (removal of the amino group) or transamination (transfer of the amino group to another molecule). Fatty acids and some keto acids enter cellular respiration at the level of acetyl CoA. Other keto acids enter as pyruvate or as molecules in the citric acid cycle. Both proteins and carbohydrates can also follow pathways that result in accumulation of body fat.

hydrogens are stripped off and carried to the electron transport system by NADH and FADH$_2$. Fatty acids are built by the reverse process, linking two-carbon acetyl units, and this is why most fatty acids contain an even number of carbon atoms.

An organism's own protein is not used for energy except during advanced starvation, after carbohydrate and fat reserves have been depleted. However, proteins in an animal's food are hydrolyzed into amino acids. Any amino acids not needed to build new proteins are **deaminated** by the removal of their amino groups. Depending on its structure, the rest of the molecule is converted into pyruvate, acetyl CoA, or one of the intermediates of the citric acid cycle. It then enters the pathway at the appropriate place.

In summary, the citric acid cycle and electron transport system are a final common pathway for the breakdown of just about any organic molecule to yield energy. All the energy-releasing molecular breakdown pathways of metabolism are collectively called **catabolism.**

In contrast, **anabolism** consists of all the energy-using, molecule-building pathways. The citric acid cycle is a grand junction where catabolic and anabolic pathways meet. Many of the cycle's molecules are diverted into body-building processes when they can be spared from respiration. Figure 7–12 shows how protein, carbohydrate, and fat pathways are linked and shows that many anabolic pathways resemble the reverse of catabolic ones and use many of the same enzymes. However, usually one or two steps in the anabolic pathway use different enzymes. These enzymes catalyze "detour" reactions that provide an energetically feasible route back around a steep downhill (exergonic) step in the catabolic pathway.

Note especially that pathways lead from both carbohydrates and proteins to fat deposits. If an animal eats more food than it uses to release energy and build macromolecules, the excess is converted to fat, by way of acetyl CoA. Acetyl CoA is the basic building block for making fatty acids, which are then incorporated into triacylglycerols and stored as body fat. So any excess food in the diet, whether carbohydrate, protein, or fat, can end up as stored fat, a "savings deposit" of energy.

Gram for gram, fats contain more than twice as much energy as carbohydrates or proteins (see Table 31–1). This is because fats contain a higher hydrogen:oxygen ratio than do carbohydrates or proteins. As we have seen, most of the energy from food molecules is obtained by passing the electrons from hydrogen atoms along the electron transport system. So the substances richest in energy are those with the highest proportions of hydrogen, by mass.

Fats are nonpolar, and so they tend to repel water and form concentrated fat droplets. The nonpolar nature and high energy content of fats make it advantageous for animals to store most of their energy reserves as fat (Figure 7–13a). However, the liver and muscles store a small amount of the carbohydrate glycogen as a short-term reserve that can be used when the body needs a quick energy boost.

Some plants store a lot of fat in seeds. However, plants generally store most of their energy reserves in the form of carbohydrates, especially starch (Figure 7–13b). The many hydroxyl (—OH) groups of carbohydrates make them polar, and so they attract and hold much water around them. This makes them heavy and bulky to store, but since plants do not have to move around and drag this food supply along, they are not inconvenienced by the weight as an animal would be. Carbohydrates are also much easier to mobilize into the energy-release pathways than are fats. (This is one reason why it is so hard to lose fat weight.) These properties seem to have selected for storage of carbohydrate energy reserves in plants.

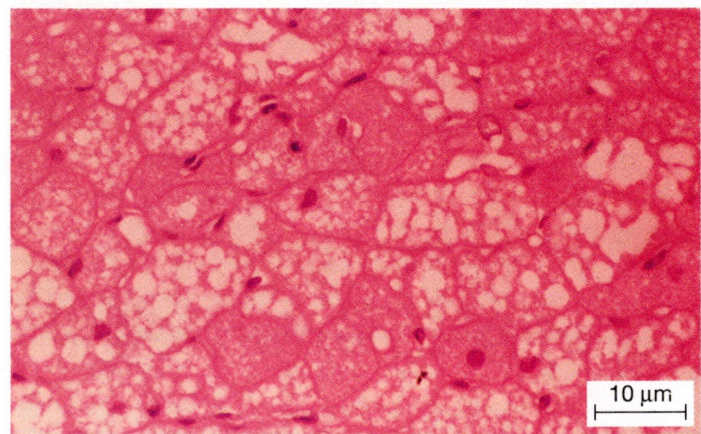

(a)

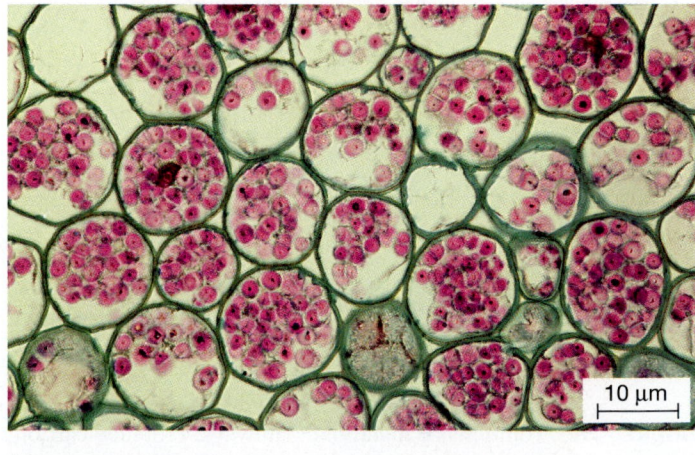

(b)

**FIGURE 7–13** Energy storage. (a) Fat globules appear as pale areas in these cells from brown fat, a heat-generating tissue found in many newborn mammals and in hibernating mammals. (b) Starch, stained purple, is stored in numerous amyloplasts in these cells from a buttercup root.   (a, Biophoto Associates; b, Ed Reschke)

## SUMMARY

Cellular respiration is the process by which cells extract free energy from the energy stored in the chemical bonds of food molecules (usually glucose). This is done in a series of catabolic pathways featuring redox reactions, and using oxygen as the final electron acceptor. The energy so released is used to regenerate the cell's supply of ATP. ATP, in turn, donates the energy to various energy-requiring processes, such as metabolic reactions, active transport, muscle contraction, or production of new polymers by anabolic pathways.

Under aerobic conditions, the carbon skeleton of glucose is dismantled to $CO_2$, and some energy from glucose is stored in ATP by substrate-level phosphorylation. However, most is stored in the form of hydrogen-laden coenzymes NADH and $FADH_2$, which pass hydrogen atoms to the electron transport system. Here their electrons' energy is used to pump $H^+$ through the inner mitochondrial membrane, where an $H^+$-based electrochemical gradient is used in the production of ATP. $O_2$ at the end of the electron transport system accepts the electrons and adds $H^+$, forming water.

1. During glycolysis, glucose is broken down anaerobically to two molecules of pyruvate, which is processed further in respiration. Also, $NAD^+$ is reduced to $NADH + H^+$, which passes two hydrogen atoms to the electron transport system, where their electrons' energy can be extracted. Glycolysis also yields ATP by substrate-level phosphorylation.

2. Glycolysis occurs in the cytoplasm. In prokaryotes, the molecules of respiration are found in the cytoplasm and plasma membrane. In eukaryotes, these molecules lie in the mitochondrial matrix and inner mitochondrial membrane.

3. Each pyruvate formed during glycolysis loses a carbon dioxide and becomes an acetyl group, which combines with coenzyme A.

4. Coenzyme A transfers the acetyl group to the citric acid cycle, where the acetyl group combines with the four-carbon compound oxaloacetate, forming citrate. During one turn of the cycle, the equivalents of the acetyl group's two carbons are removed as carbon dioxide, one of the end products of respiration. Some ATP is also produced by substrate-level phosphorylation. $NAD^+$ and FAD are reduced to $NADH + H^+$ and $FADH_2$, which carry hydrogens to the electron transport system. Oxaloacetate and coenzyme A are regenerated and can go through the cycle again.

5. Most of the ATP derived from respiration is produced by oxidative phosphorylation: electron transport coupled to chemiosmotic ATP synthesis. $NADH + H^+$ and $FADH_2$ (from glycolysis, formation of acetyl CoA, and the citric acid cycle) pass pairs of hydrogen atoms to the electron transport system in the inner mitochondrial membrane. The electron transport system uses the energy of these atoms' electrons to form an electrochemical gradient of $H^+$ that can be used in the phosphorylation of ADP to ATP. At the end of the electron transport system, the electrons combine with oxygen and $H^+$ to form water, the other end product of respiration.

6. Neither glycolysis nor the citric acid cycle requires oxygen directly. However, if a cell is short of the electron transport system's final electron acceptor, oxygen, most of its $NAD^+$ will be tied up as NADH, unable to release its electrons to the electron transport system. Under such anaerobic conditions, some cells continue to produce ATP from glycolysis by carrying out fermentation. Pyruvate accepts electrons from $NADH + H^+$, forming ethanol or lactate. This releases $NAD^+$ and permits glycolysis to continue. No such mechanism exists for the citric acid cycle, which therefore cannot function under anaerobic conditions.

7. Many other metabolic pathways feed into glycolysis, the citric acid cycle, and the electron transport system, enabling cells to use many organic compounds other than glucose as food sources to generate usable energy in the form of ATP.

### TABLE 7-4

**Comparison of Glycolytic Fermentation and Aerobic Breakdown of Glucose**

| Feature | Glycolytic Fermentation | Aerobic Breakdown |
|---|---|---|
| Raw materials | Glucose | Glucose, $O_2$ |
| Final electron acceptor | Pyruvate (from glucose) | $O_2$ |
| End products | $CO_2$, ethanol (yeast) Lactate (muscle) | $CO_2$, $H_2O$ |
| Net energy yield | 2 ATP | Up to 38 ATP |

## Self-Quiz

1. $NAD^+$ functions in cellular respiration as a(n):
   - a. energy intermediate
   - b. enzyme
   - c. coenzyme
   - d. oxidizable substrate
   - e. hydrogen donor
2. Which of the following statements is *not* true?
   - a. During cellular respiration, more ATP is formed by oxidative phosphorylation than by substrate-level phosphorylation.
   - b. In eukaryotes, the formation of ATP by oxidative phosphorylation requires that the inner mitochondrial membrane remain intact.
   - c. $NAD^+$ is a carrier molecule that travels down the electron transport system to release ATP during oxidative phosphorylation.
   - d. Cellular respiration produces more ATP than does fermentation.
   - e. The yield of ATP per glucose is the same for alcoholic and lactate fermentation.
3. In eukaryotes, the electron transport system and the enzymes of the citric acid cycle are located in _____, whereas the enzymes of glycolysis are located in _____ .
4. The role of oxygen in cellular respiration is to act as _____ .
5. The respiratory electron transport system is found in the _____ membrane in eukaryotes and in the _____ membrane in prokaryotes.
6. Give the end products of the following reaction sequences:
   - a. glycolysis
   - b. the citric acid cycle
   - c. yeast fermentation
   - d. electron transport system
   - e. lactate fermentation in muscle
7. While a muscle is in the process of reducing an oxygen debt:
   - a. lactate is converted into pyruvate
   - b. all the $NAD^+$ is in the reduced form
   - c. pyruvate is converted into lactate
   - d. NADH acts as an oxygen acceptor
8. True or False? Both yeast cells and muscle make two ATP (net) per glucose molecule fermented anaerobically.
9. True or False? Carbohydrate is unnecessary in the human diet, since the products of fat and protein breakdown can enter the citric acid cycle to generate energy.

## Questions for Discussion

1. Vitamins are substances that organisms need in small amounts in the diet because they cannot synthesize them. You have encountered several vitamins in this chapter (e.g., niacin and pantothenic acid). What do they have in common? Why are they vital, and why are they required only in very small amounts?
2. Why is it wrong to call the two ATP used during the first half of glycolysis activation energy?
3. Cyanide inactivates cytochromes. Why is it poisonous?
4. How many hydrogen atoms are there in a glucose molecule? How many hydrogen atoms are transported to the electron transport system by coenzymes during the aerobic breakdown of one molecule of glucose? Where do the additional hydrogen atoms come from? (See Figures 7–A, 7–6, and 7–7.)
5. Explain what the term "aerobic exercise" means with respect to metabolic processes.
6. Why do foods that are rich in fat tend to be expensive (in dollars and cents) compared with carbohydrates?

## Suggested Readings

Krebs, H. A. "The history of the tricarboxylic acid cycle." *Perspectives in Biology and Medicine,* 14:154, 1970. An engaging account of how Sir Hans Krebs worked out the citric acid cycle.

Stryer, L. *Biochemistry,* 3d ed. San Francisco: W. H. Freeman, 1988. A standard biochemistry textbook with excellent descriptions of respiration and photosynthesis.

C H A P T E R

8

# Photosynthesis

**O B J E C T I V E S**

*When you have studied this chapter, you should be able to:*

1. Name or recognize the necessary raw materials of photosynthesis and the main end products.
2. Name the three main groups of photosynthetic pigments and state their functions.
3. State which colors of light are most effective in promoting photosynthesis, and explain why.
4. Describe or sketch the structure of a leaf. Show where the chloroplasts are, and explain how the raw materials for photosynthesis arrive there and how the end products leave.
5. Describe or sketch the structure of a chloroplast. Explain where the following are found and the importance of their location to their roles in photosynthesis: chlorophyll, electron transport system, ATP synthetase enzymes, $H^+$ reservoir, $C_3$ cycle enzymes.

6. Name the raw materials and end products of the electron transport reactions, chemiosmotic ATP synthesis, and the $C_3$ cycle. Predict how altering light intensity or temperature will affect each.
7. State what drives carbon fixation and describe what happens in the $C_3$ cycle.
8. Summarize the important steps in energy transfer during photosynthesis.
9. List three ecological variants of photosynthesis, and explain why each is advantageous to plants growing in certain habitats.

---

A story about wondrous plants that turn sunlight, air, and water into sugar might seem like a fairy tale, but in fact it's a chapter in a biology textbook. This is the story of **photosynthesis,** the process whereby most autotrophs capture the sun's energy and store it in the form of chemical bonds in carbohydrate molecules. The carbohydrates are made by joining carbon dioxide to an organic molecule and then reducing it, using hydrogen extracted from water. The green pigment chlorophyll captures light energy to power the process. Photosynthesis consists of a complex series of reactions. The overall reaction can be summarized:

$$CO_2 + H_2O \xrightarrow{\text{light, chlorophyll}} CH_2O + O_2$$

Organisms that carry out photosynthesis according to this equation are loosely called **green plants:** eukaryotic algae and land plants, and prokaryotic cyanobacteria and prochlorophytes (Figure 8–1). These are not the only pho-

tosynthetic organisms. Some bacteria photosynthesize using different pigments and hydrogen sources, and they do not produce $O_2$ as a by-product. This chapter covers only the oxygen-producing photosynthesis performed by green plants. We consider other kinds of photosynthesis in prokaryotes in Chapter 24.

Photosynthesis occurs in the chloroplasts of eukaryotic plant cells. In prokaryotes, it occurs partly in specialized photosynthetic membranes in the cytoplasm and partly in the cytoplasm itself.

In our study of respiration, we traced the downhill path of energy by following electrons from hydrogen atoms. Hydrogens were stripped from organic molecules, and their electrons were then passed through the electron transport system in a series of redox reactions until they combined with oxygen to form water, a low-energy molecule.

The basis of photosynthesis is the opposite reaction: splitting water into hydrogen and oxygen and forcing electrons up an energy hill. This highly endergonic process requires an appreciable energy input. The splitting of water

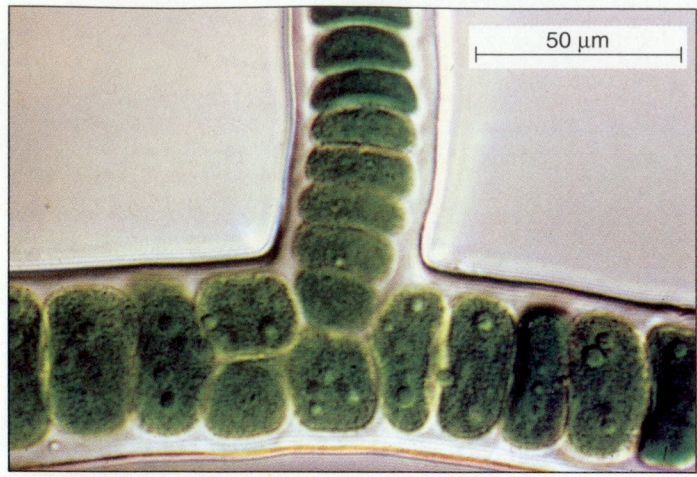

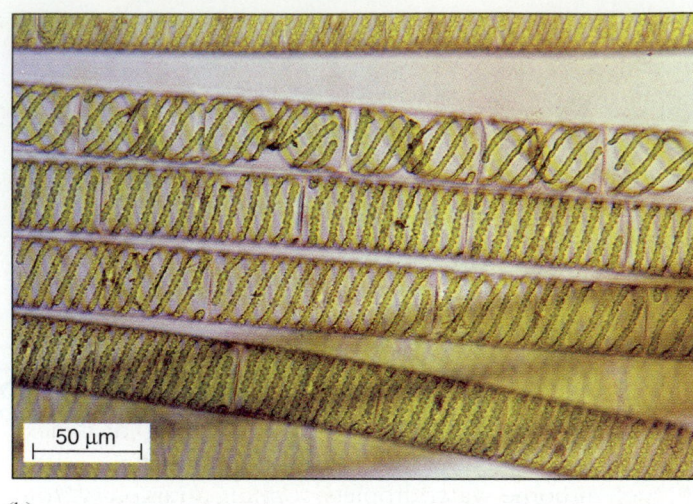

(a)

(b)

**FIGURE 8–1** Green plants. Organisms that carry out oxygen-producing photosynthesis include (a) prokaryotic cyanobacteria, such as this filament of *Stigonema;* (b) eukaryotic algae, such as *Spirogyra;* and (c) land plants, such as this grape vine.   (a, Biophoto Associates; b, Dwight Kuhn, DRK)

(c)

is, ultimately, the reaction that light energy drives in photosynthesis. The hydrogen so released supplies the energy to make ATP and to reduce carbon dioxide in the formation of carbohydrates.

Solar energy is the most abundant source of energy on Earth, but it has one big drawback: it is not always available when a plant needs energy. During photosynthesis, green plants trap this fleeting source of energy and transform it into the chemical potential energy of food molecules, which can be stored indefinitely. (The organic molecules of the coal and oil we burn for energy today contain "fossil sunshine," stored by photosynthesis hundreds of millions of years ago.) Once created, a plant's food stores can later be broken down by respiration to provide energy when and where the plant needs it.

Photosynthesis is vital to life as we know it for two reasons. First, all of our food and the food of almost every other organism comes, directly or indirectly, from photosynthesis. (The exceptions are chemoautotrophic bacteria, Section 24–C.) Second, virtually all the oxygen in the air today (about 20% of the atmosphere) came from photosynthesis, and plants continue to replenish the oxygen supply. This oxygen is essential for every organism (including plants) that relies on aerobic respiration to break down food and produce ATP.

Our dependence on plants for food has stimulated research on how photosynthesis works and how we can manipulate plants and their environments to produce more food. Also, learning the secrets of electron flow in chloroplasts may help us to build more efficient solar collectors for electric power. Although we have learned a great deal about photosynthesis, plants still hold many secrets that our most sophisticated equipment has yet to unravel, much less duplicate.

In this chapter, we first examine the properties of light, the energy source for photosynthesis, and the molecules plants use to trap light energy. Then we shall look at the structure of chloroplasts and the reactions of photosynthesis. Finally, we shall study factors that affect the rate of photosynthesis and see adaptations that enable plants in different habitats to flourish even though they must carry out photosynthesis under less than ideal conditions.

## KEY CONCEPTS

♦ Photosynthesis converts light energy into the more permanent form of energy stored in chemical bonds of carbohydrates. The light energy is trapped by chlorophyll, and carbon dioxide and water are used to make carbohydrates.

♦ All our food, and the food of nearly all other organisms, comes directly or indirectly from photosynthesis.

♦ The oxygen in the air comes from photosynthesis. The vast majority of organisms need this oxygen for respiration.

## 8–A  LIGHT ENERGY

Visible light, the light we can see, is made up of light of many colors. We can observe this when sunlight passes through airborne water droplets, forming a rainbow. However, visible light is only part of the vast range of electromagnetic radiation emitted by the sun.

Electromagnetic radiation behaves as if it travels in waves, and it comes in a range of wavelengths. A wavelength is the distance between comparable points on adjacent waves:

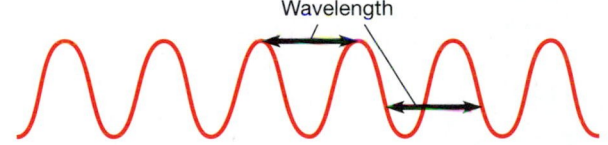

Electromagnetic radiation ranges over a spectrum from gamma rays and x-rays, with very short wavelengths, through visible light and microwaves, to radio waves with very long wavelengths (Figure 8–2). Visible light extends from violet light, with a wavelength of about 400 nanometers (nm), to red light of about 750 nm. This is also the range that plants use for photosynthesis.

A **photon** is the basic unit of electromagnetic radiation energy. The amount of energy carried by a photon depends on its wavelength: the shorter the wavelength, the higher the energy. Thus, a photon of violet or blue light has more energy than a photon of red light.

It is not by chance that photosynthesis uses visible light rather than other parts of the electromagnetic spectrum. First, most of the solar energy reaching Earth's surface is visible light. Second, visible light contains the right amount of energy. Radiation with wavelengths shorter than violet light contains so much energy that it breaks bonds and destroys many biological molecules. On the other hand, radiation with wavelengths longer than about 750 nm contains little energy, and most of it is rapidly absorbed by water before it even gets to a plant's photosynthetic machinery. Low-energy radiation has little effect on the molecules in an organism except to heat them up. Only visible light, with intermediate wavelengths, has enough energy to cause chemical changes without destroying biological molecules.

## 8–B  TRAPPING LIGHT ENERGY: PHOTOSYNTHETIC PIGMENTS

**Photochemical reactions** are chemical reactions activated by light energy. They are familiar in photography: light striking the film causes reactions that produce an image. The photochemical reactions of photosynthesis are redox reactions that occur when light energy is absorbed by the plant's photosynthetic pigments.

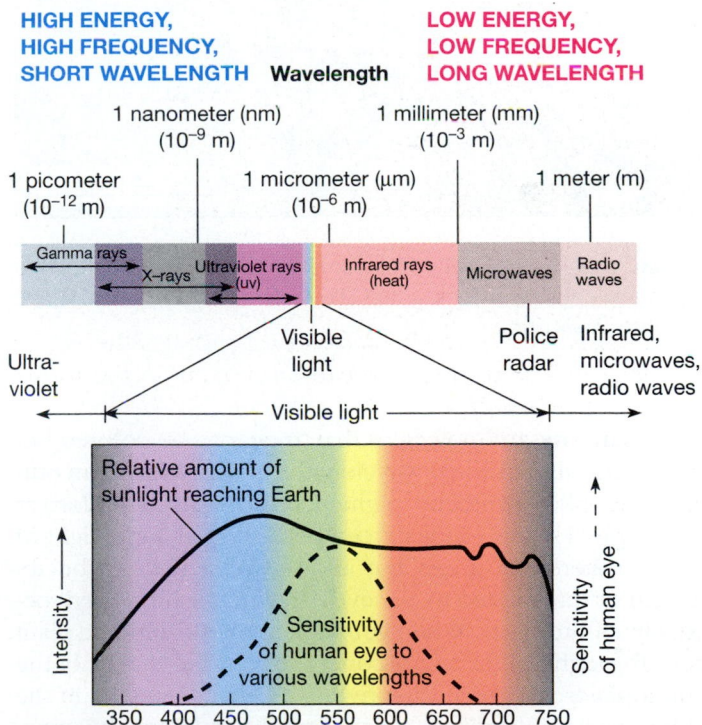

**FIGURE 8–2**  The electromagnetic spectrum. Our eyes can detect light with wavelengths of about 400 to 750 nanometers, and so this range is called visible light. Wavelengths just shorter than visible light are called ultraviolet. Even shorter are x-rays (which overlap with ultraviolet and gamma rays), then gamma rays. Infrared waves, microwaves, and radio waves have longer wavelengths than the visible spectrum.

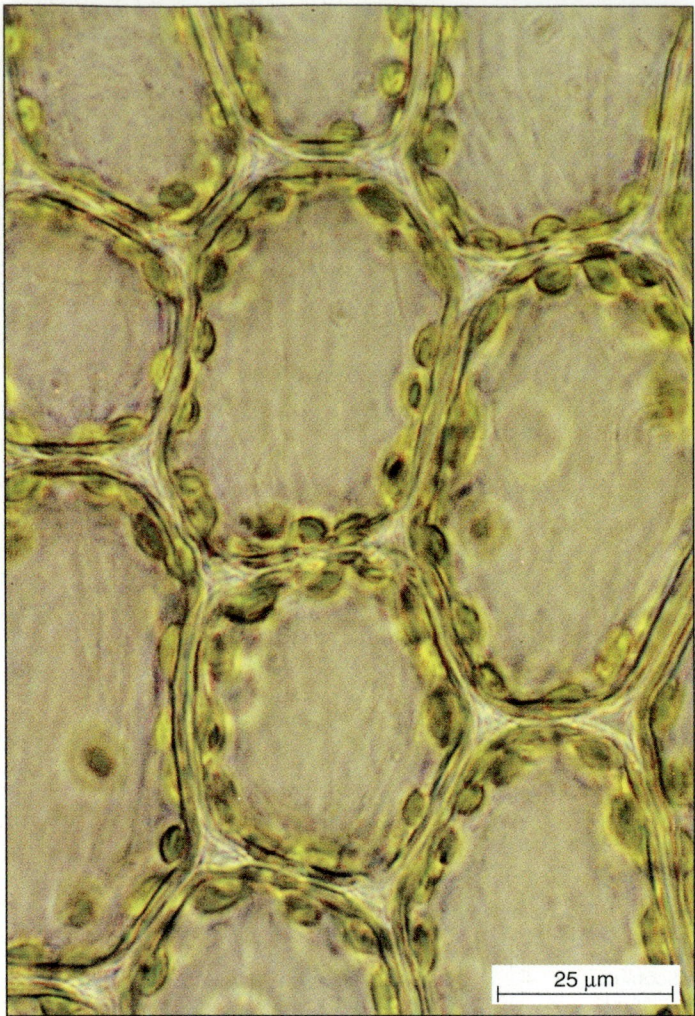

**FIGURE 8–3** Living cells from a leaf. The green chloroplasts make the entire leaf look green. The rest of the cell is colorless. (Biophoto Associates)

**Pigments** are molecules that appear to be colored because they absorb some wavelengths of light more than others. The photosynthetic pigment **chlorophyll** (chloro = green; phyll = leaf) looks green because it absorbs light of colors other than green but allows green light to be reflected or transmitted to our eyes. In plants, chlorophyll occurs in chloroplasts. When you look at photosynthetic plant cells through a light microscope, you can see that only the chloroplasts are green (Figure 8–3). Other pigments in the chloroplast reflect different wavelengths and tinge the green with various hues.

We can tell which wavelengths of light a pigment captures most effectively by finding the peaks of its **absorption spectrum.** This is a graph prepared by shining different wavelengths of light through a solution of purified pigment and measuring how much light of each wavelength the pigment absorbs (Figure 8–4).

To find out which pigment plays the major role in photosynthesis, we can measure the rate of photosynthesis in an intact plant illuminated with different wavelengths of light. We then compare the photosynthetic **action spectrum** plotted from these data with the absorption spectra of various purified pigments to see which absorption spectrum resembles the photosynthetic activity curve most closely. In practice, we find that the absorption spectrum of chlorophyll bears the strongest resemblance to the action spectrum.

In 1883 T. W. Engelmann found evidence that chlorophyll plays a role in photosynthesis. He used *Spirogyra,* an alga with long, spiral chloroplasts (see Figure 8–1b). He put the alga in water on a microscope slide with some aerobic (oxygen-requiring) bacteria. By using a prism to disperse the beam of light into a spectrum, he was able to expose different parts of the chloroplast to different wave-

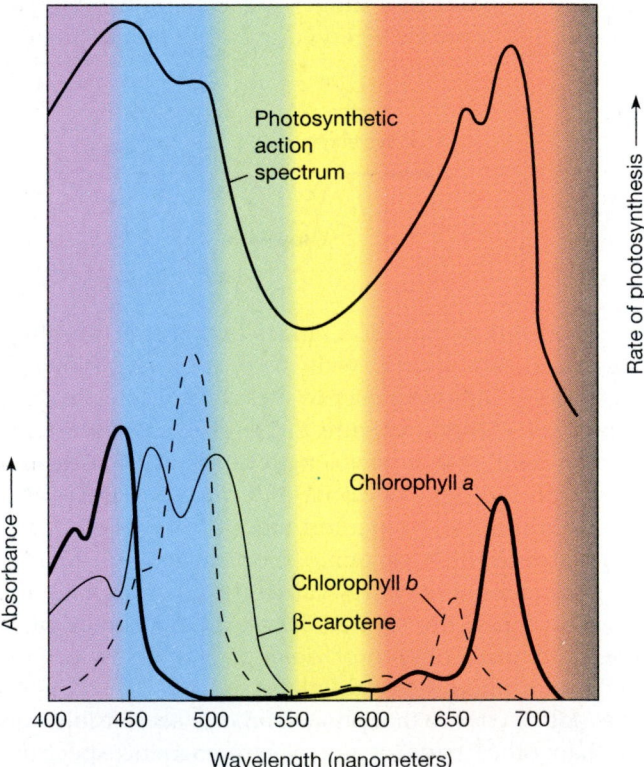

**FIGURE 8–4** Absorption and action spectra. The lower curves show absorption spectra for three pigments. Chlorophyll *a* occurs in all green plants, chlorophyll *b* in all land plants and some algae. β-carotene is common in land plants.

The action spectrum (top curve) shows the rate of photosynthesis when a plant is exposed to light of various wavelengths. It closely follows the absorption spectra of the chlorophylls.

Absorbance, shown on the vertical axis for the absorption spectra, is found by shining light through a solution of purified pigment and measuring the percentage of light transmitted to a meter on the other side. From this is calculated:

absorbance = 2 − log (% transmittance)

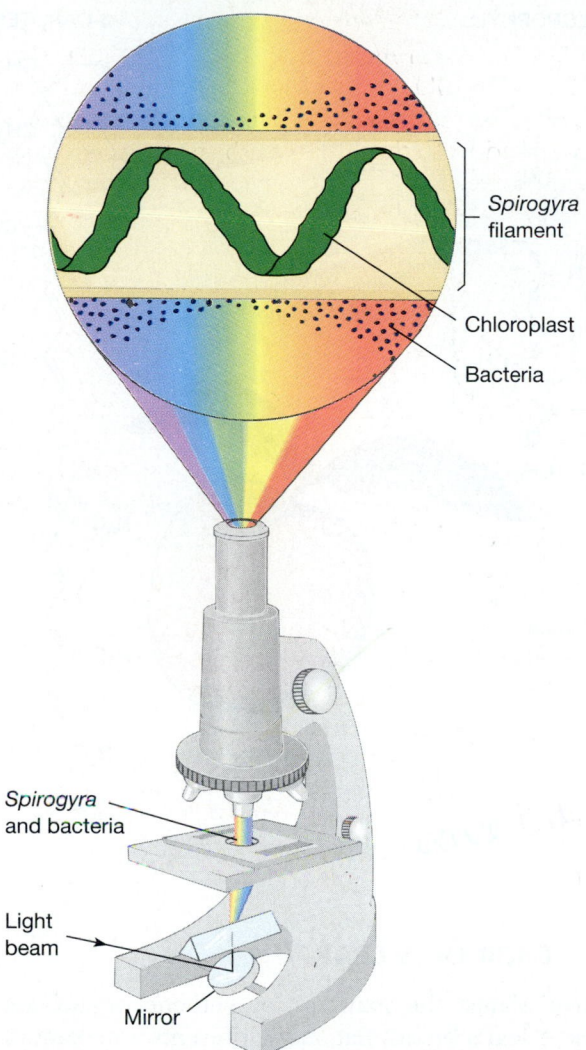

**FIGURE 8–5** Engelmann's experiment. To determine which wavelengths of light best support photosynthesis, Engelmann inserted a prism into the beam of light reflected from the microscope's mirror. The prism dispersed the light into a spectrum, directed at strands of the alga *Spirogyra* on the microscope stage. Oxygen released by photosynthesis in the *Spirogyra* chloroplasts attracted aerobic bacteria, which congregated where red and blue light fell on the alga. Engelmann concluded that red and blue light support higher rates of photosynthesis than do other wavelengths.

lengths of light (Figure 8–5). Engelmann expected that some wavelengths of light would support more photosynthesis than others. The parts of the chloroplast exposed to these wavelengths would give off oxygen most rapidly, and this oxygen would attract the most oxygen-requiring bacteria. He observed that the bacteria clustered around the alga where it was exposed to red and blue light, indicating that photosynthesis was occurring fastest at these wavelengths. Since chlorophyll absorbs red and blue light (and so appears green), this result strongly suggested a major role in photosynthesis for the green chlorophylls.

There are several types of chlorophyll, with similar molecular structures. A chlorophyll molecule has two main parts. At one end is a multi-ring structure with a magnesium ion ($Mg^{2+}$) in the center. This is where light energy is trapped. The molecule's long nonpolar "tail" anchors it in the lipid of a photosynthetic membrane (Figure 8–6a). Chlorophyll *a* is the main photosynthetic pigment in all green plants.

Plants do not waste the wavelengths not absorbed by chlorophyll *a*. These wavelengths are absorbed by various **accessory pigments,** which absorb photons and transfer their energy to chlorophyll *a*. For example, many green plants contain chlorophyll *b* or *c,* which absorb different wavelengths in the blue and red parts of the spectrum (see Figure 8–4).

**Carotenoids** are another important group of accessory pigments in all green plants. They absorb blue and green wavelengths and so impart yellow or orange hues to the plant. Carotenoids usually consist of long hydrocarbon chains with a ring structure at each end (Figure 8–6b). They are thought to absorb energy and pass it to chlorophyll. More important, carotenoids protect chlorophyll. In very bright light, the energy absorbed by chlorophyll cannot all be used in photosynthesis. Instead, some $O_2$ molecules are partially reduced, forming highly reactive "free radicals" that can destroy other molecules. Carotenoids bind these radicals, thereby sparing chlorophyll from damage. The sacrifice of carotenoids is worthwhile because chlorophyll takes more energy to make and so is a more expensive molecule for the plant to replace.

In many plants, chlorophyll is broken down in the autumn, and its magnesium and nitrogen are moved to storage areas elsewhere in the plant before the leaves drop. This conserves hard-to-obtain elements, which can be used again the following spring. The breakdown of chlorophyll reveals the yellow, orange, or brown colors of carotenoids in autumn leaves. The red color of some autumn leaves comes from anthocyanin pigments. These are not photosynthetic pigments. Rather, they are produced when sugar is trapped in the leaf as the tree walls off the leaf before it is shed.

In addition to chlorophyll *a* and carotenoids, red algae and cyanobacteria contain accessory pigments of another group, the **phycobilins:** red phycoerythrin predominates in red algae, blue phycocyanin in cyanobacteria. These pigments absorb light of wavelengths that chlorophyll does not absorb strongly, and they transfer the energy to chlorophyll *a* for use in photosynthesis.

## Photosystems

Photosynthetic pigments are the stars of photosynthesis because they take part in photochemical reactions, transforming light energy into a form that can be used to make carbohydrates.

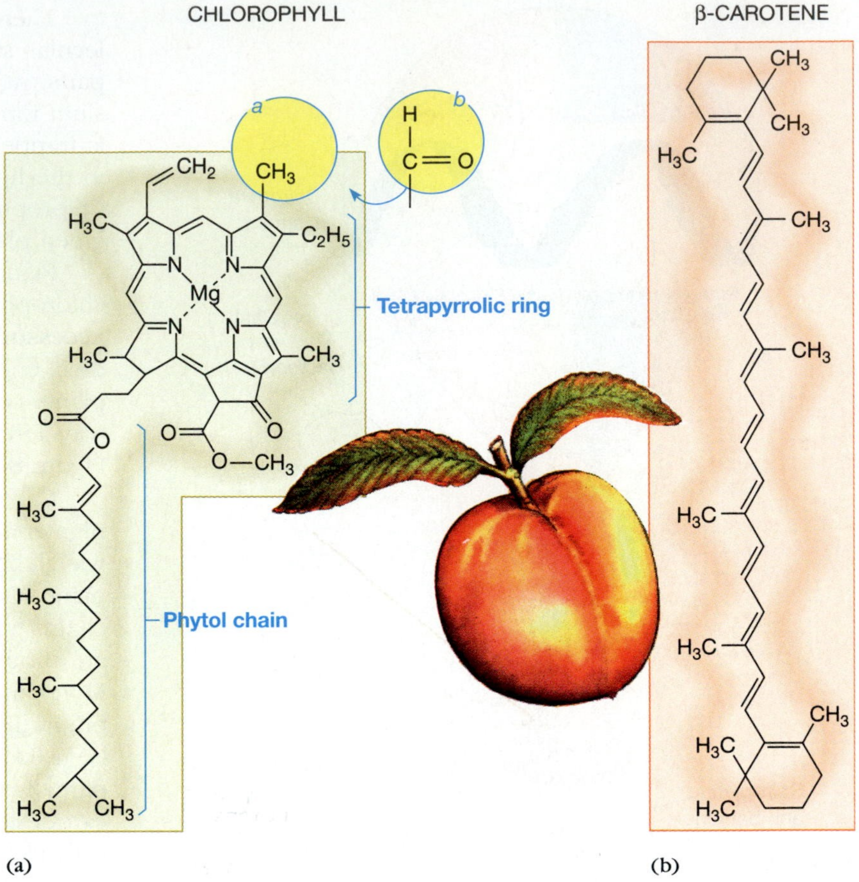

CHLOROPHYLL

β-CAROTENE

**FIGURE 8–6** Photosynthetic pigments. (a) Molecular structure of chlorophyll *a*. Swapping the —CH₃ group circled for —CHO would give chlorophyll *b*. The tetrapyrrolic ring is the molecule's "business end," which absorbs light energy. The long, nonpolar phytol chain "tail" lies among the lipids of the photosynthetic membrane. (b) Structure of *β*-carotene. Carotenoids are present but are masked by chlorophyll in leaves and other green organs. The carotenoids become apparent when chlorophyll is destroyed in autumn leaves and in ripening yellow or orange fruits.

(a)

(b)

In a chloroplast, the photosynthetic pigments are found in particles called **photosystems,** each containing hundreds of pigment molecules bound to several proteins. Most of the pigments serve as **antenna pigments,** which absorb light energy and pass it to a so-called special pair of chlorophyll *a* molecules in a protein complex, the photosystem's **reaction center.** In most green plants, the antenna pigments are chlorophylls *a* and *b*, with some carotenoids. Having lots of antenna pigments optimizes the photosystems' ability to intercept light energy, which is then funneled to the reaction center.

Two kinds of photosystems take part in photosynthesis. The chlorophyll *a* molecules in their reaction centers have different absorption peaks, thought to be due to differences in the surrounding proteins. In Photosystem I, this chlorophyll *a* is called **P700** (short for pigment 700) because it has an absorption peak in the long red wavelengths at about 700 nm. The chlorophyll *a* in the Photosystem II reaction center is called **P680.**

The absorption peaks at 680 and 700 nm are at the long-wavelength, lower-energy end of the visible spectrum (see Figure 8–2). Hence, light of shorter wavelengths and higher energy can be absorbed by the antenna pigments and passed downhill, in terms of energy, to chlorophylls in the reaction centers, which channel the energy into photochemical reactions.

## 8–C  TOUR OF A LEAF

In most plants, the main photosynthetic organs are the leaves. A leaf's broad, flat shape presents a maximum surface area for light absorption at a minimum weight. The tightly packed cells of the **epidermis** form the leaf's outer surfaces and secrete a waxy, waterproof covering, the **cuticle.** The cuticle prevents loss of water vapor from the leaf into the air but also impedes the passage of carbon dioxide and oxygen. Gas exchange between the air and the leaf's interior occurs through pores in the epidermis called **stomata** ("mouths") (Figure 8–7). In most plants, the sto-

**FIGURE 8–7** Anatomy of photosynthesis. (a) In higher plants, photosynthesis occurs in leaves. Water taken in by the roots passes up the stem and through the leaf veins to the cells. (b) Leaf tissues. Photosynthetic cells, which contain chloroplasts (small green dots), are sandwiched between the upper and lower epidermis. Stomata in the lower epidermis admit CO₂ needed for photosynthesis and permit the by-product O₂ to leave. Water vapor also escapes by this route. (c) A chloroplast from tobacco (TEM) and an interpretive drawing of its three-dimensional structure. The photosynthetic thylakoid membranes are green because they contain chlorophyll. (d) Closeup of the thylakoid membranes.   (c, d, Herbert Israel, Cornell University)

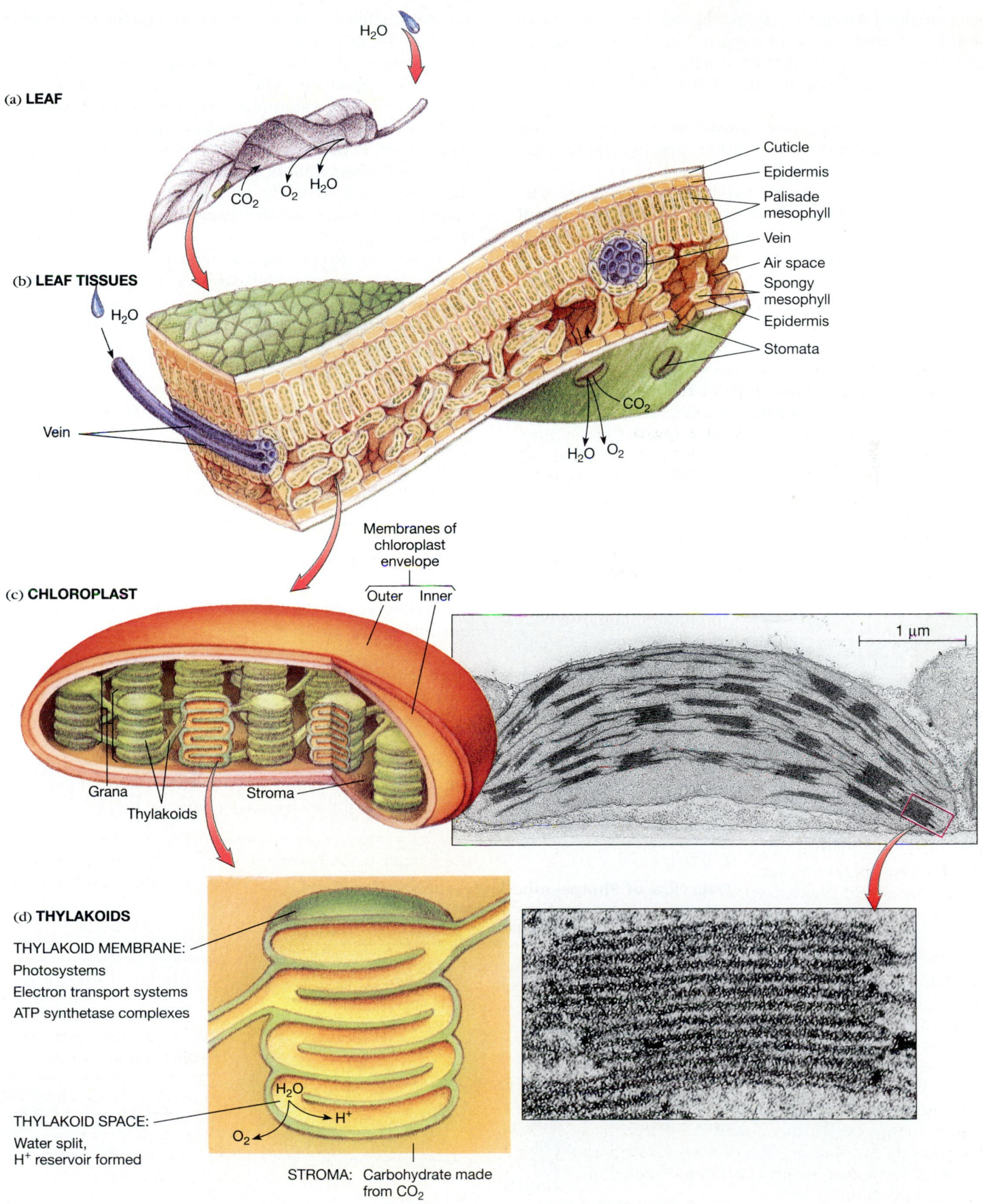

$H_2O$

(a) **LEAF**

$CO_2$    $O_2$    $H_2O$

(b) **LEAF TISSUES**

$H_2O$

Vein

Cuticle
Epidermis
Palisade mesophyll
Vein
Air space
Spongy mesophyll
Epidermis
Stomata

$CO_2$

$H_2O$    $O_2$

Membranes of chloroplast envelope

Outer    Inner

(c) **CHLOROPLAST**

1 μm

Grana

Thylakoids    Stroma

(d) **THYLAKOIDS**

THYLAKOID MEMBRANE:

Photosystems

Electron transport systems

ATP synthetase complexes

THYLAKOID SPACE:

Water split,
$H^+$ reservoir formed

$H_2O$

$H^+$

$O_2$

STROMA:   Carbohydrate made
from $CO_2$

mata are open during the day, permitting the uptake of carbon dioxide and release of oxygen, but also allowing water vapor to escape. Stomata are usually closed at night, and they also close during the day if the plant is losing too much water.

The **veins,** made up of vascular tissue, conduct water into the leaf and carry newly made carbohydrate to other parts of the plant.

Photosynthetic cells form two layers inside the leaf. The closely packed columnar cells of the **palisade mesophyll** stand above the loosely arranged, irregular cells of the **spongy mesophyll** (meso = middle). Air spaces in the spongy mesophyll permit rapid diffusion of gases between the stomata and the photosynthetic cells. Each photosynthetic cell contains many chloroplasts.

A chloroplast's structure is intimately related to its function. The outer envelope consists of two membranes. Inside lies a series of **photosynthetic membranes** arranged in flattened sacs called **thylakoids** (Figure 8–7). Some of the thylakoids occur in stacks, called **grana** ("grains") because a light microscope shows them as little specks. All the thylakoids in one chloroplast are thought to be continuous and to enclose a continuous interior compartment. This thylakoid space serves as a reservoir where hydrogen ions ($H^+$) are accumulated to supply energy for the chemiosmotic synthesis of ATP, a process we met in Chapters 6 and 7. Embedded in the thylakoid membranes are molecules of three important groups:

1. The photosystems, containing the photosynthetic pigments and reaction centers.
2. The chloroplast's electron transport systems.
3. The protein complexes involved in chemiosmotic ATP synthesis.

The arrangement of the thylakoid membranes provides a large surface area to capture light energy but encloses a relatively small volume. As a result, the membrane's many transport molecules can rapidly build a high concentration of $H^+$ in the thylakoid space.

A protein-rich solution, the **stroma,** surrounds the thylakoids. The stroma contains enzymes that make carbohydrate, as well as the chloroplast's DNA, RNA, and ribosomes.

A eukaryotic green cell may have from 1 to 40 or more chloroplasts, depending on the species, age, and health of the cell. *Chlamydomonas,* a single-celled, flagellated green alga often used to study photosynthesis, has only one chloroplast. Spinach, another laboratory favorite, has many chloroplasts per cell.

In the prokaryotic cyanobacteria and prochlorophytes, thylakoids lie in the cytoplasm (see Figure 24–5). (It is thought that some ancient members of these groups became residents inside eukaryotic cells and gave rise to chloroplasts.)

## 8–D OVERVIEW OF PHOTOSYNTHESIS

The reactions of photosynthesis can be divided into two main stages: those that capture light energy and those that use this energy to make carbohydrate.

### Energy Capture

Energy capture in photosynthesis involves three steps, which all occur in the thylakoid membranes (Figure 8–8 and Table 8–1):

***Light Absorption***  Light energy boosts electrons in chlorophyll *a* to higher energy levels—so high that the electrons can leave the molecule.

***Electron Transport***  The chloroplast's electron transport systems accept these high-energy electrons from chlorophyll and use some of the electrons' energy to build a $H^+$ reser-

---

**TABLE 8–1**

**Overview of Photosynthetic Reaction Series**

| Process | Inputs | Outputs |
|---|---|---|
| Light absorption | Light energy | |
| | Pigments | Electrons |
| Electron transport | Electrons ⎫ NADP$^+$ ⎬ | NADPH |
| | H$_2$O | ⎧ O$_2$ ⎨ ⎩ H$^+$ reservoir |
| Chemiosmotic ATP synthesis | H$^+$ reservoir | |
| | ADP + P$_i$ | ATP |
| Carbon fixation | Ribulose bisphosphate ⎫ CO$_2$ ⎬ | Sugars |
| | ATP | ADP + P$_i$ |
| | NADPH | NADP$^+$ |

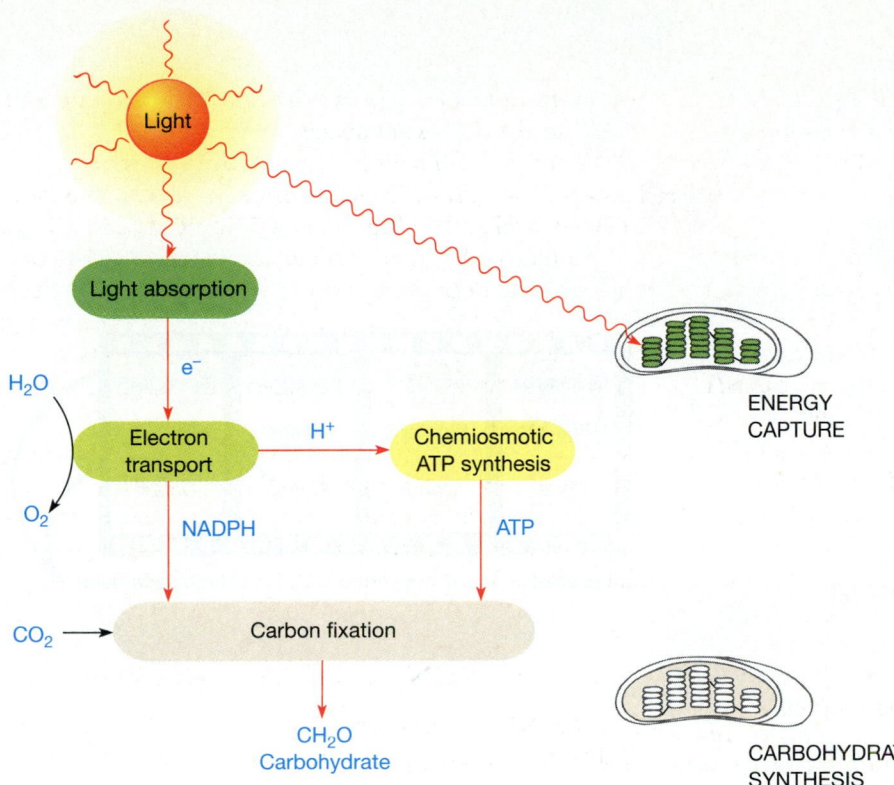

**FIGURE 8–8**  Overview of photosynthesis. Red arrows trace the path of energy from light to carbohydrate molecules. Energy-capturing steps occur in the thylakoid membranes, carbohydrate synthesis in the stroma of chloroplasts.

voir in the thylakoid space. The electrons finally pass to the coenzyme $NADP^+$, and some of their energy is used to reduce it to NADPH. Chlorophyll's lost electrons are replaced by splitting water molecules ($H_2O$) and taking electrons from the hydrogen atoms. $H^+$ is left behind in the reservoir. Oxygen atoms from split water molecules join and form $O_2$.

***Chemiosmotic ATP Synthesis***   $H^+$ from the thylakoid space moves through the membrane down its concentration gradient via channels associated with ATP synthetase enzymes. This releases energy to power the synthesis of ATP from ADP and $P_i$.

Energy capture is now complete. The energy in light has been captured as chemical energy in ATP and NADPH.

## Carbon Fixation

Carbon fixation occurs in the chloroplast's stroma. Carbon becomes "fixed" when a gas (carbon dioxide) is incorporated into a solid (carbohydrate). ATP and NADPH, produced during the energy-capturing reactions, provide the energy and hydrogen needed to fix the carbon.

If a plant has already captured light energy as chemical energy in NADPH and ATP, why does it need carbon fixation too? First, ATP and NADPH are short-lived. A food-store of carbohydrate to be used whenever energy is needed

is much more useful. Second, carbon fixation builds up carbon skeletons from which other organic molecules can be made.

## 8–E   THE ENERGY-CAPTURING REACTIONS

The energy-capturing reactions take place in the thylakoid membranes. Energy is captured in two light absorption events and is then used to power two sets of electron transport reactions. One set of electron transport reactions produces NADPH, and the other indirectly produces ATP.

### Photochemical Reactions: Light Absorption

The "kickoff" for photosynthesis occurs when light energy reaches chlorophyll *a* in a photosystem reaction center, exciting one of its electrons to a higher energy level. This initial photochemical reaction powers the rest of photosynthesis: as the electron proceeds downhill, in energetic terms, its energy drives a series of reactions.

Before this happens, a photon of the proper wavelength must be absorbed by an antenna pigment, boosting one of the molecule's electrons to a higher energy level. The excited electron quickly falls back to its normal lower energy level, re-emitting energy to a neighboring antenna pigment, and so on, until the energy reaches a reaction center chlorophyll *a*. Here, instead of the excited electron falling back to its normal level, the added energy permits the elec-

**FIGURE 8–9**   Absorption of light energy. Red arrows show the path of energy. (a) Antenna pigments absorb light energy and transform it to the energy of excited electrons. This energy is passed from one molecule to the next as electrons return to their normal energy level. When the energy reaches a reaction-center chlorophyll, the excited electron has enough energy to leave and pass to the photosystem's electron acceptor molecule.

(b) A photochemical reaction transforms light energy to electrical potential energy. Light energy absorbed by a reaction-center pigment boosts an electron to a higher energy level. The excited pigment (*) donates this electron to an acceptor and replaces the lost electron by taking one from a nearby donor. In the last frame, the pigment has returned to its original state, but some of the light energy has been conserved by separating positive and negative electrical charges.

tron to be passed "downhill" (in terms of energy) to a neighboring molecule in the reaction center complex (Figure 8–9).

This electron transfer is a redox reaction, one that is too endergonic to occur without an input of energy. Hence, light energy has been used to power an endergonic redox reaction and has been transformed into electrical energy in the form of separated electrical charges: having lost an electron, chlorophyll bears a positive charge, and the molecule now holding the electron bears a negative charge.

## Electron Transport

The high-energy electron passes to several molecules in the reaction center complex and then to the electron transport system. Here a series of exergonic redox reactions puts the energy to use.

Electrons may follow one of two electron transport pathways: one is noncyclic, the other cyclic.

*Noncyclic Electron Transport*   The overall effect of noncyclic electron flow is to split water inside the thylakoid space and transfer its electrons through the membrane to the stroma, where they reduce $NADP^+$. This occurs by way of the two photosystems and two sets of electron transport reactions, all working in series:

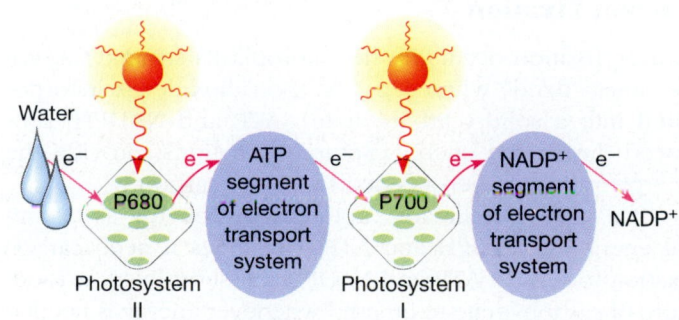

Because water does not give up electrons easily, nor NADP$^+$ accept them readily, the transfer is highly endergonic. It requires an energy input of more than 210 kJ per mole. (In comparison, the synthesis of ATP from ADP and P$_i$ takes about 42 to 54 kJ per mole.) Light supplies the required energy, but this energetic hill is so high that the electrons cannot be pushed all the way up in one jump. The electrons are boosted by light energy at two points along the way —one push from each photosystem (red arrows on page 166 and in Figure 8–10). After each push they pass through a segment of the electron transport system and lose some energy, which is used to power photosynthesis (pink arrows).

A diagram of the energy level at each step has a zigzag pattern, commonly called the Z scheme (Figure 8–10). Electrons make the trip in stages, and some are going through the final stage, passing to NADP$^+$, while others are just leaving water molecules. Let us follow these reactions, starting with the *final* leg of the electrons' journey.

1. **Photosystem I to NADP$^+$.** A photon of light striking Photosystem I boosts an electron from chlorophyll P700 to a higher energy level. The electron is ultimately accepted by the NADPH segment of the electron transport system, which uses some of its energy to reduce NADP$^+$ in the stroma to NADPH. NADP$^+$ accepts two electrons leaving the system and takes a hydrogen ion (H$^+$) from the stroma. (NADP$^+$ is NAD$^+$ with an extra phosphate group; see Figure 7–2.)

2. **Photosystem II to Photosystem I.** This step replaces P700's lost electron and transports H$^+$ from the stroma into the thylakoid space.

A photon striking Photosystem II boosts an electron from P680 to a higher energy level. This electron passes to the other part of the electron transport system, which delivers it to P700 in Photosystem I, replacing its lost electron.

As the electron passes down this part of the electron transport system, the energy it gained from light is, in

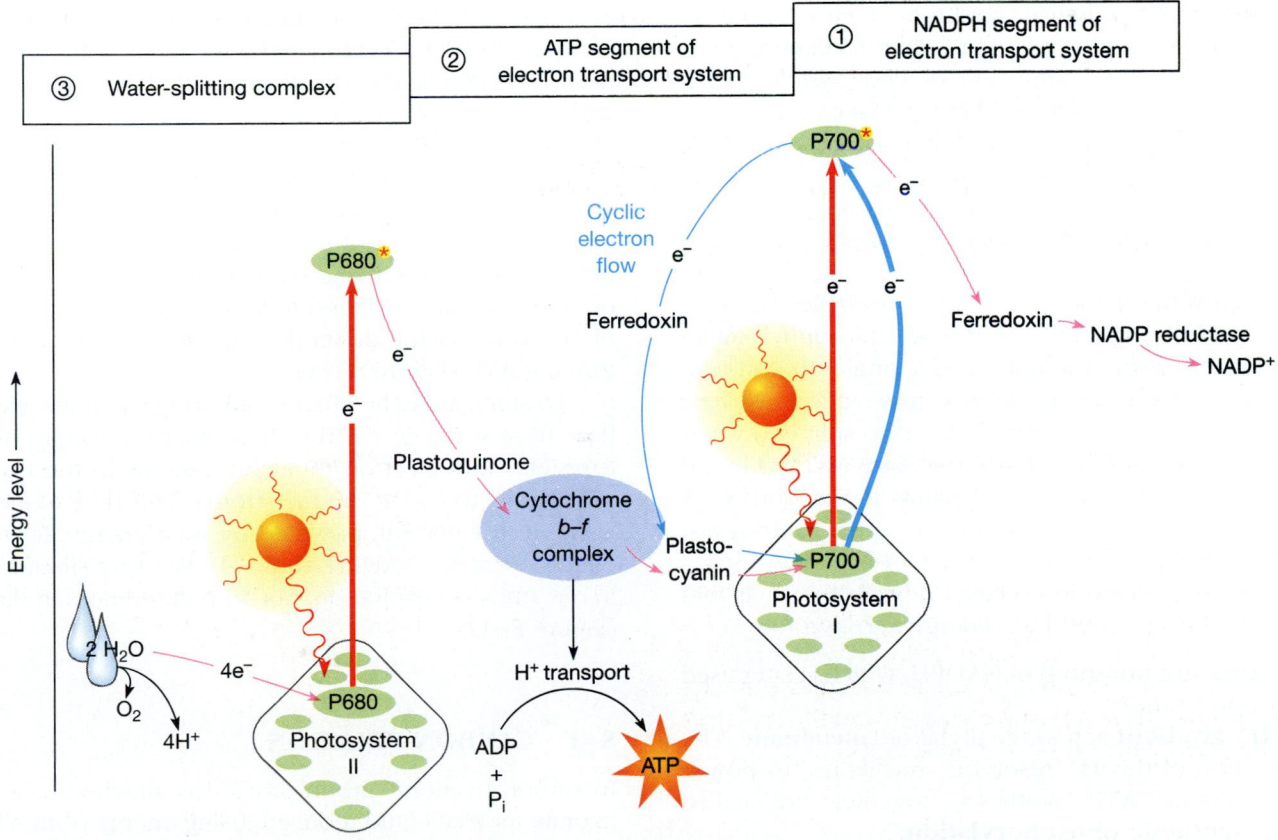

**FIGURE 8–10** The path of energy in the photosynthetic electron transport system. Red and pink arrows show the path of electrons in noncyclic flow. Overall, electrons move from water (far left), through the photosystems and electron transport systems, to NADP$^+$ (far right). This transfer is steeply "uphill" in terms of energy and requires two boosts from light along the way. This results in a zigzag pattern, commonly called the Z scheme (red and pink arrows). The steps at the top of the graph are numbered to correspond to the text, so the graph must be read backwards. Blue arrows show the cyclic electron flow path.

effect, used for active transport of $H^+$ from the stroma into the thylakoid space. Since this $H^+$ reservoir is used to make ATP, the electron transport molecules between Photosystems II and I may be thought of as the ATP segment of the electron transport system.

3. **Water to Photosystem II.** Now P680 in Photosystem II is missing an electron. The splitting of water molecules from the thylakoid space replaces electrons lost by P680 and also adds more $H^+$ to the $H^+$ reservoir:

$$2\ H_2O \rightarrow 4\ e^- + 4\ H^+ + O_2$$

The electrons ($e^-$) replace those lost by P680 in Photosystem II. $H^+$ remains in the thylakoid space, as part of the $H^+$ reservoir. $O_2$ is a by-product of photosynthesis and diffuses out of the cell.

Recent evidence suggests that the water-splitting reaction is catalyzed by part of the Photosystem II reaction center complex, with the aid of the manganese ions ($Mn^{2+}$) it contains. The complex binds two water molecules from the thylakoid space and releases four electrons, one by one, as P680 needs replacements. Meanwhile, four $H^+$ are released to the reservoir. When four $e^-$ and four $H^+$ have been removed, the remaining oxygen atoms join, forming $O_2$.

The water-splitting component of Photosystem II is a triumph of molecular evolution. Photosynthetic bacteria cannot extract hydrogen from water but must obtain it from molecules such as $H_2$, $H_2S$, or organic acids. They are therefore confined to habitats that supply these scarce raw materials, such as the anaerobic bottoms of marshes, bogs, and swamps.

Although water is the most readily available source of hydrogen on Earth, it is difficult to use. Chlorophyll molecules trap one photon at a time, and a single photon cannot split a water molecule. In essence, green plants use four photons (one per excited P680 electron) to split two water molecules. This ability to use an abundant and easily obtained resource, water, has given plants tremendous evolutionary success and permitted them to spread across vast areas of the globe.

In summary, noncyclic electron flow in the thylakoid membranes has converted light energy into two forms:

1. The **reducing potential** of NADPH, which is later used in carbon fixation.
2. The **$H^+$ gradient** across the thylakoid membrane. The generation of this $H^+$ reservoir, and its use to power chemiosmotic ATP synthesis, together are called **photosynthetic phosphorylation.**

Some herbicides, such as paraquat and the triazines, kill plants by interfering with photosynthetic electron transport and preventing the buildup of the $H^+$ gradient needed to make ATP.

***Cyclic Electron Flow: More $H^+$ to the Reservoir***    Electrons sometimes follow a different electron transport route.

This pathway is called **cyclic electron flow** because chlorophyll P700, in Photosystem I, serves as both electron donor and final electron acceptor (Figure 8–10, blue arrows). It occurs when most of the stroma's $NADP^+$ has been reduced, leaving little available to accept electrons. Instead, the electrons are passed to the ATP segment of the electron transport system, where some of their energy powers transport of more $H^+$ to the $H^+$ reservoir. The electrons are then returned to P700. This cyclic pathway does not create NADPH nor generate oxygen.

## Chemiosmotic ATP Synthesis

Electron transport in the thylakoid membranes increases the concentration of $H^+$ in the thylakoid space, producing a pH about 3.5 units lower than that in the stroma. (Since the pH scale is logarithmic, this means that the $H^+$ concentration is more than a thousand times higher inside the thylakoids than outside.) The energy of this gradient is almost entirely in the form of the difference in chemical concentration. There is virtually no difference in electrical charge because the thylakoid membrane permits negatively charged $Cl^-$ ions to flow freely into the reservoir, attracted by the positively charged $H^+$. (In effect, the thylakoid space is filled with hydrochloric acid.)

This steep chemical gradient supplies about 20 kJ of energy per mole of $H^+$, which powers the chemiosmotic synthesis of ATP. It takes more than one $H^+$ to join ADP and $P_i$. $H^+$ flows out of the reservoir through special channel proteins in the thylakoid membrane. At the outer ends of these channels, ATP synthetase enzymes use the energy of $H^+$ ions moving down their gradient to join $P_i$ to ADP, forming ATP (Figure 8–11).

To summarize the energy-capturing reactions: electrons flow from water in the thylakoid space, through the photosystems and electron transport systems in the thylakoid membrane, to $NADP^+$ in the stroma. This flow of electrons is an electric current, powered by solar energy. It provides energy to make NADPH and ATP. We saw similar events in our study of respiration, but with differences in direction (Figure 8–12).

## 8–F    CARBON FIXATION

In carbon fixation, carbon dioxide is attached to a carbohydrate molecule and reduced, using energy from ATP and hydrogen from NADPH, both produced in the energy-capturing reactions. The enzymes that catalyze these reactions occur in the stroma of chloroplasts (or in the cytoplasm of photosynthetic prokaryotes).

Carbon fixation begins with the attachment of carbon dioxide to **ribulose bisphosphate (RuBP),** a five-carbon sugar with two phosphate groups. The resulting six-carbon structure is unstable and immediately hydrolyzes to two

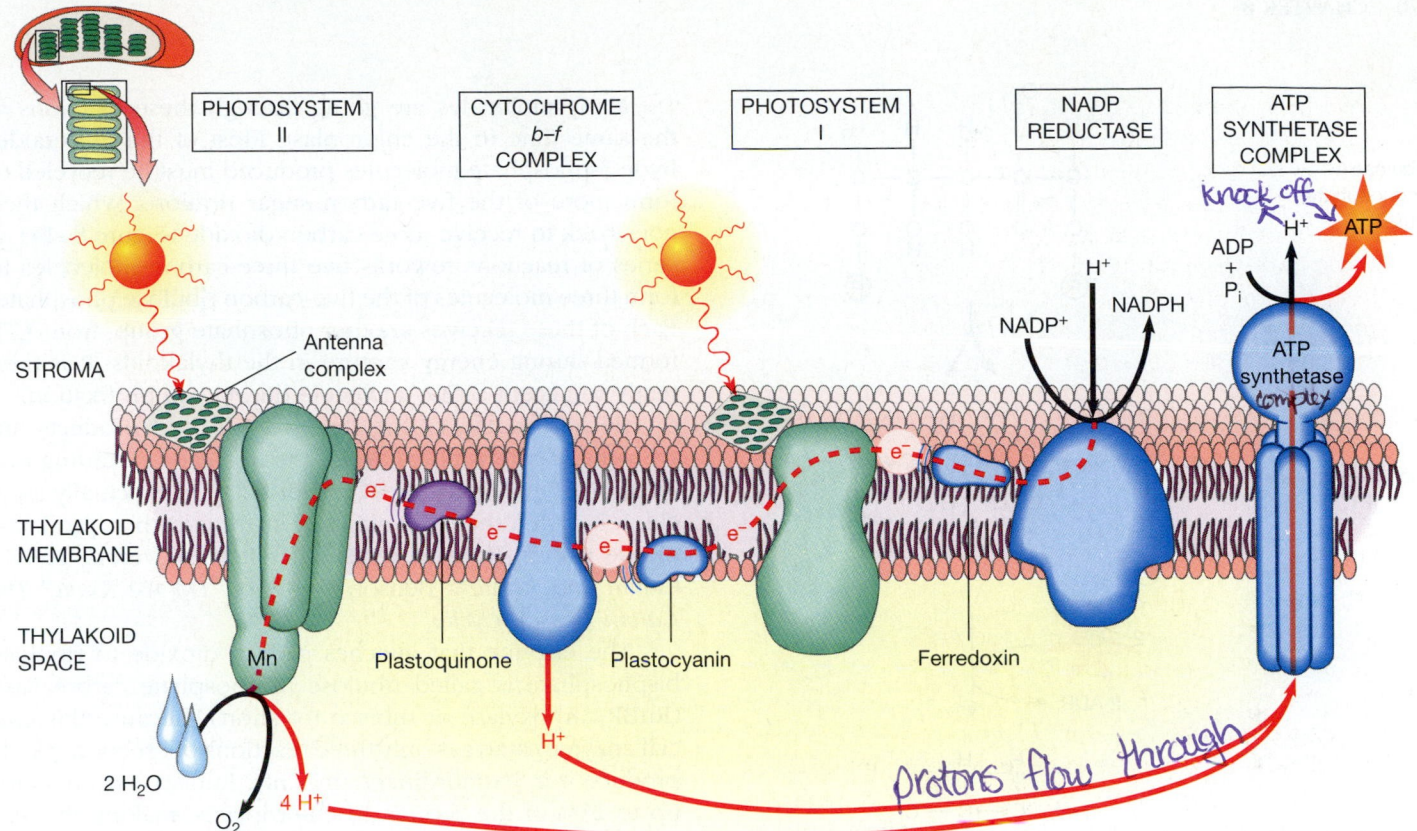

**FIGURE 8–11** Summary of energy capture in photosynthesis. The boxes across the top name the large protein complexes directly below (green and blue). The first four from the left take part in electron flow. Plastoquinone, plastocyanin, and ferredoxin are smaller, mobile carriers transporting electrons between these large complexes, as shown. Light initiates electron flow in the photosystems. As a result of electron flow, NADPH and $O_2$ are formed, and a high concentration of $H^+$ accumulates in the thylakoid space. This $H^+$ powers the chemiosmotic synthesis of ATP by ATP synthetase complexes (far right).

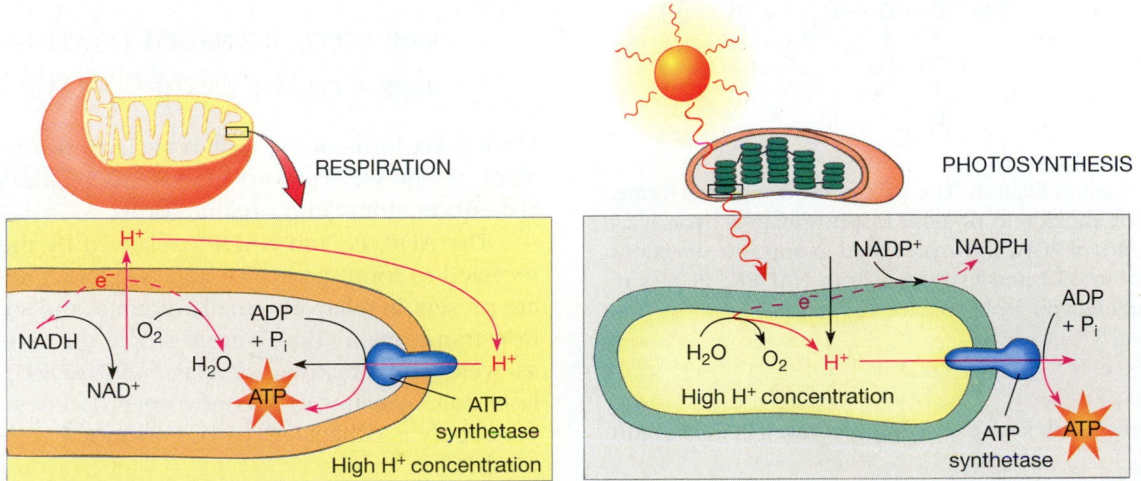

**FIGURE 8–12** Comparison of electron flow and chemiosmotic ATP synthesis in respiration (in mitochondria) and photosynthesis (in chloroplasts). Whereas in respiration electrons are passed *away from* hydrogen carriers (such as $NAD^+$) and *toward* oxygen, in photosynthesis electrons pass to $NADP^+$ and *away from* oxygen. Furthermore, in respiration $H^+$ accumulates outside the inner mitochondrial membrane and ATP synthetase lies on the *inner* surface of the inner membrane, whereas in photosynthesis $H^+$ accumulates inside the thylakoid membranes and ATP synthetase lies on these membranes' *outer* surfaces, where ATP is released directly into the stroma.

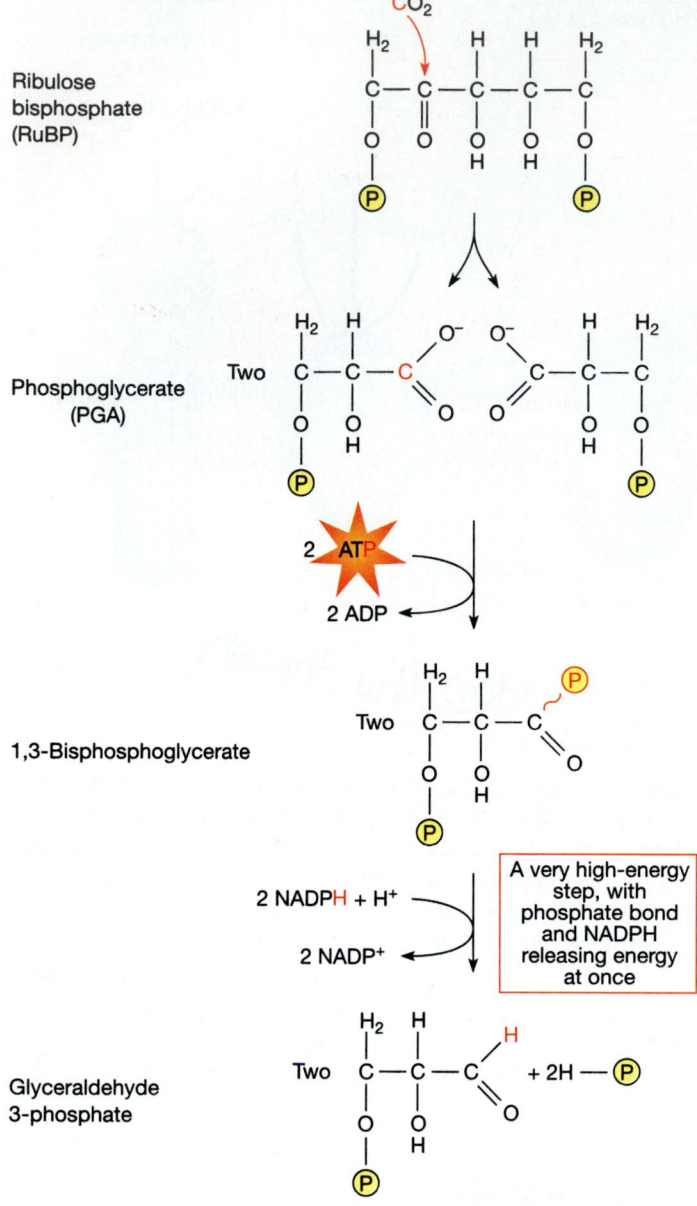

**FIGURE 8–13**   Carbon fixation. The six-carbon compound formed by joining carbon dioxide to ribulose bisphosphate is hydrolyzed into two molecules of PGA, shown oriented in opposite directions. Using energy from ATP and hydrogen from NADPH, PGA is reduced to glyceraldehyde 3-phosphate.

three-carbon molecules of phosphoglycerate (PGA) (Figure 8–13).

These two PGA molecules are next reduced to glyceraldehyde 3-phosphate, a triose sugar phosphate. This takes two steps. First, each PGA molecule receives more energy as a second phosphate group is donated by ATP, forming a high-energy molecule. In the next step, the new phosphate group is hydrolyzed, releasing this energy, as the molecule is reduced to glyceraldehyde 3-phosphate by hydrogen from NADPH.

Many molecules are going through these reactions at the same time in the chloroplast. Most of the glyceraldehyde 3-phosphate molecules produced must be recycled to form more of the five-carbon sugar ribulose, which then goes back to receive more carbon dioxide (Figure 8–14). A series of reactions reworks five three-carbon molecules to form three molecules of the five-carbon ribulose phosphate. Each of these receives another phosphate group, from ATP formed during energy capture in the thylakoids. It is then ready to accept a $CO_2$ molecule during carbon fixation.

Because some of the three-carbon end products are converted to new molecules of the five-carbon starting material, the whole process of carbon fixation is actually a cycle. It is called the **$C_3$ cycle** after its three-carbon products, or the **Calvin-Benson cycle** after its discoverers, Melvin Calvin and Andrew Benson (see *How Do We Know? The Calvin-Benson Cycle*).

The enzyme that attaches carbon dioxide to ribulose bisphosphate is called ribulose bisphosphate carboxylase (RuBP carboxylase or rubisco for short). Because this crucial enzyme catalyzes only three reactions per second, plants produce it in extraordinary amounts. Rubisco accounts for up to 25% of the protein in chloroplasts, making this enzyme far and away the most abundant protein in green tissue and therefore in the world! It is estimated that there are 20 pounds of this protein for every person on Earth.

In summary, the $C_3$ cycle begins with carbon dioxide and a five-carbon molecule, RuBP, which react to form two three-carbon molecules (PGA). Each PGA molecule uses an ATP and an NADPH as it is converted to glyceraldehyde 3-phosphate. A third ATP is required to phosphorylate ribulose phosphate to regenerate the starting molecule, ribulose bisphosphate. The overall equation is:

$$RuBP + CO_2 + 2\,NADPH + 3\,ATP \rightarrow$$

$$RuBP + CH_2O + 2\,NADP^+ + 3\,ADP + 3\,P_i$$

It takes six turns of the Calvin cycle to produce the equivalent of one (six-carbon) glucose molecule, that is, to fix six carbon atoms into organic form.

The ADP, $P_i$, and $NADP^+$ released by the $C_3$ cycle are recycled to form more ATP and NADPH. These substances are present in relatively small amounts, and so if either electron transport or the $C_3$ cycle stops, the other soon stops as well. The stockpile of ATP and NADPH, for instance, lasts only a matter of seconds once darkness begins; after the supply is exhausted, carbon fixation can no longer proceed.

Although most of the three-carbon products of carbon fixation are recycled to regenerate more ribulose bisphosphate, one sixth of them represent new carbon skeletons and can be used to make other new molecules. For instance, two three-carbon molecules can be joined into six-carbon sugars. These hexose sugars can then be polymerized into starch, an energy-storage compound, or cellulose, which makes up the cell wall. Or, they can be processed

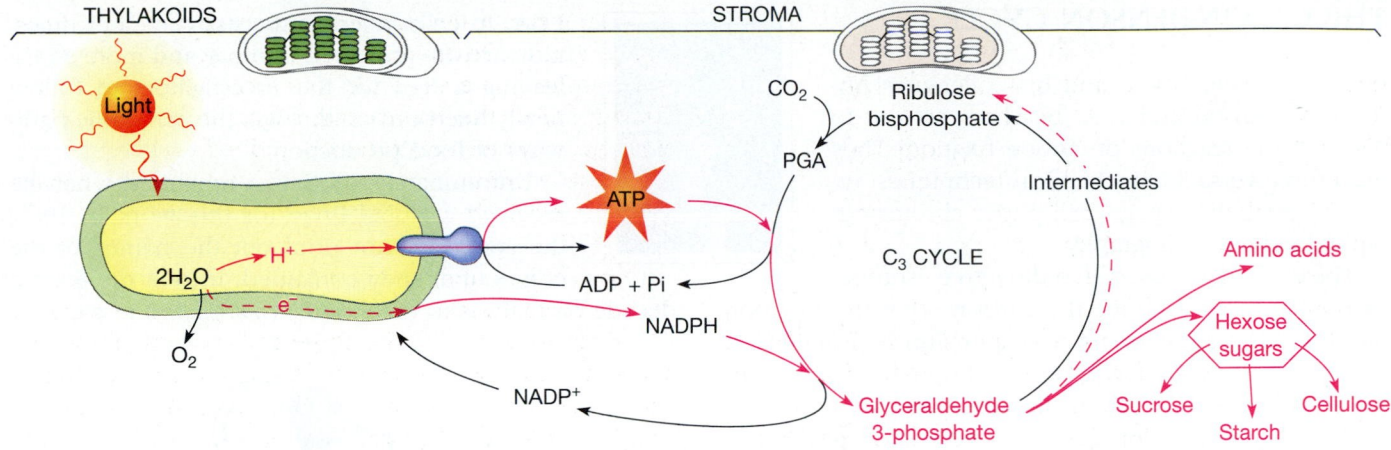

**FIGURE 8–14**   The $C_3$ cycle and its connections with the energy-capturing reactions in the thylakoid membranes. ATP and NADPH, and ADP, $P_i$, and $NADP^+$, shuttle between the two interdependent sets of reactions (energy-capturing in the thylakoid membranes; and carbon fixation in the stroma). Red arrows trace the course of energy from sunlight through temporary energy intermediates to the more permanent energy-storage form of six-carbon sugars and their polymers.

to make the disaccharide sucrose, which is transported to other parts of the plant. Other three-carbon molecules go into the synthesis of amino acids by the addition of nitrogen-containing groups. Newly fixed carbon may also be used to make fatty acids or nitrogenous bases for nucleotides. Plastids are the sites where all these molecules are made in plant cells.

## 8–G   THE RATE OF PHOTOSYNTHESIS

Photosynthesis comprises many different chemical reactions. Several factors affect the rate of some of its individual reactions and so determine the rate of the whole process.

Photosynthesis begins with photochemical reactions, which trap light energy and go faster with increasing light intensity (Figure 8–15). However, the other reactions of photosynthesis are like those in previous chapters, called **thermochemical reactions** because their rates are increased by heat. Because photosynthesis is made up of both photochemical and thermochemical reactions, both light and temperature influence its rate. In fact, by measuring the rate of photosynthesis at different temperatures and light intensities, F. F. Blackman deduced the existence of both photochemical and thermochemical reactions in 1905, long before any of the reactions were actually identified.

In dim light, the rate of photosynthesis is limited by shortage of light. The photochemical reactions that initiate energy capture go slowly, and NADPH and ATP are not made as fast as carbon fixation could use them. As the light becomes brighter, the rate of photosynthesis increases until ATP and NADPH become so plentiful that carbon

fixation cannot keep up, and the rate of photosynthesis levels off. This is what happens on bright, cool days. On warmer days, the rate of photosynthesis increases still further because the rate of the thermochemical reactions of carbon fixation increases.

The rate of a reaction can also be limited by scarcity of raw materials or excess of products. On bright, warm days, light and temperature are optimum for photosynthesis, but its rate is often limited by a low concentration of the raw material carbon dioxide. Adding more carbon diox-

**FIGURE 8–15**   Catching sunbeams. In bright light, the energy-capturing reactions of photosynthesis proceed rapidly, as evidenced by the vigorous bubbling of the by-product oxygen from these aquatic ferns and duckweeds.   (W. Ormerod/Visuals Unlimited)

## THE CALVIN-BENSON CYCLE

In 1945, Melvin Calvin and his colleagues Andrew A. Benson and J. A. Bassham set out to discover the reactions of carbon fixation. They used three versatile biochemical techniques: radioactive labeling (page 20), paper chromatography, and autoradiography.

These workers used a radioactive isotope, carbon 14 ($^{14}C$), to trace the carbon atom from carbon dioxide. First, they set up a suspension of *Chlorella*, a unicellular alga, under good conditions for photosynthesis, but without carbon dioxide. Then they injected $^{14}CO_2$ into the reaction vessel. At very short time intervals thereafter, they dropped some of the photosynthesizing algae into boiling methanol, which immediately denatured their enzymes and prevented further reactions (Figure 8–A).

All the small organic molecules were then extracted from the cells and separated by two-dimensional paper chromatography (Figure 8–B). In this procedure, the various substances in a mixture move up the paper at different rates, depending on their solubility in the solvents used. Therefore they end up in different positions on the paper. In the finished chromatogram, each spot contains a different compound from the original mixture.

After the paper had dried, it was placed on a sheet of unexposed photographic film and left in a dark place. As radioactive carbon decays, it emits particles that expose the film, so that the spots containing carbon 14 take their own pictures, a process known as autoradiography.

By determining the sequence in which the radioactively labeled compounds appeared, the researchers could follow the metabolic pathway leading from $CO_2$ to various organic compounds in the cells. When the algae had no $CO_2$ available, five-carbon ribulose bisphosphate accumulated. When $^{14}CO_2$ was added, ribulose bisphosphate disappeared and radioactive ($^{14}C$-containing) molecules of the three-carbon phosphoglycerate (PGA) appeared. In chromatograms from cells killed after just 5 seconds with $^{14}CO_2$, PGA was the only radioactively labeled compound they found. From this they deduced that $^{14}CO_2$ and ribulose bisphosphate had reacted, producing PGA. In cells killed at successively later times, radioactivity appeared in more and more spots, leaving a trail for the investigators to follow newly fixed carbon through the metabolic pathways of food production.

Unfortunately, as Calvin remarked when he received a Nobel Prize for this work in 1961, the spots did not print out the names of the compounds they contained. It took the team a decade of analysis to determine what each one was.

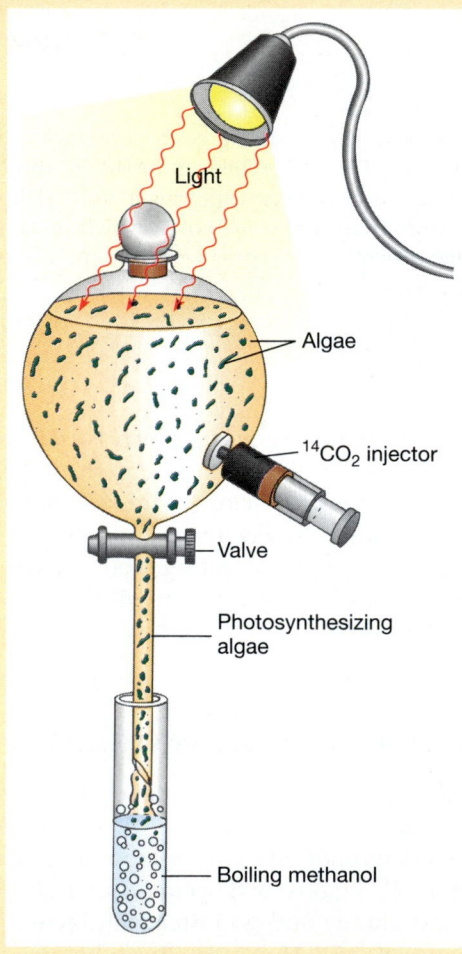

**FIGURE 8–A** Apparatus for following the path of carbon in photosynthesis.

ide to the air in a greenhouse increases the rate of photosynthesis (and so of plant growth) of some greenhouse crops. Above a certain concentration, however, carbon dioxide inhibits photosynthesis.

Because modest increases in $CO_2$ enhance photosynthesis, the rising level of $CO_2$ in Earth's atmosphere might be expected to make plants grow faster. Plants might even be expected to curb further increases in $CO_2$ by using more of it in photosynthesis. However, several studies have shown that the initial spurt seen in greenhouse plants given more $CO_2$ is not necessarily sustained over longer periods in natural plant communities.

The other raw material of photosynthesis, water, is so abundant in living tissue that the amount used in photosynthesis is negligible. However, a shortage of water does limit photosynthesis indirectly by causing the stomata to

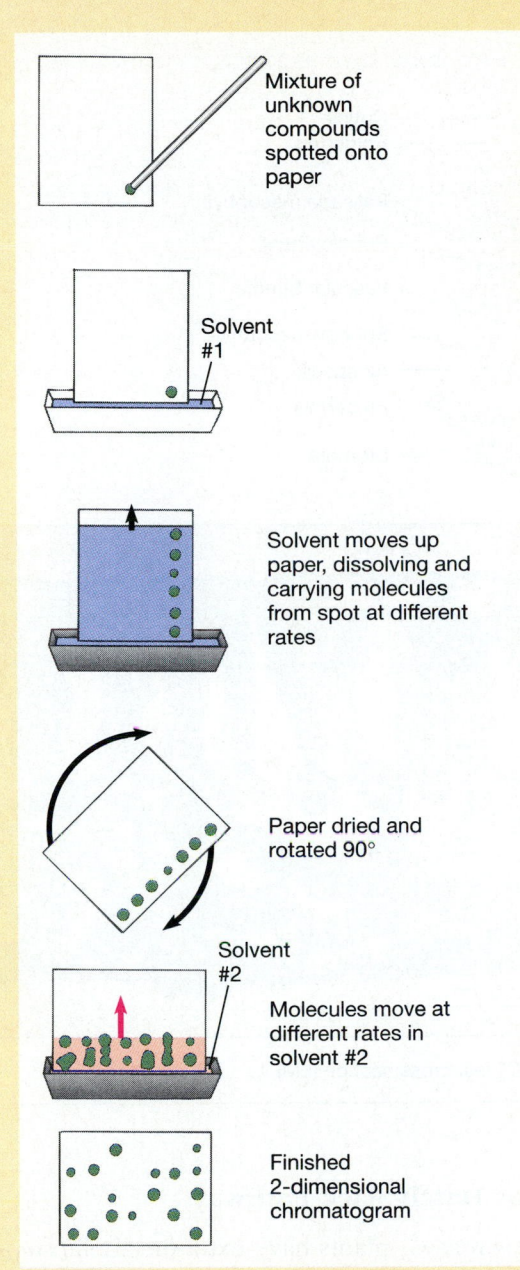

Mixture of unknown compounds spotted onto paper

Solvent #1

Solvent moves up paper, dissolving and carrying molecules from spot at different rates

Paper dried and rotated 90°

Solvent #2

Molecules move at different rates in solvent #2

Finished 2-dimensional chromatogram

**FIGURE 8–B**  Two-dimensional paper chromatography.

close, conserving water but also reducing the uptake of carbon dioxide and therefore the rate of carbon fixation.

The carbohydrate end products of photosynthesis are quickly changed into other chemical forms or removed from the chloroplast, and so they do not accumulate and inhibit carbon fixation.

The other end product, oxygen, diffuses out of the cell and leaves the plant. However, when oxygen is being pro-

duced rapidly, enough of it may accumulate to slow photosynthesis in several ways. First, oxygen combines with electron transport molecules, inhibiting electron transport. At very high levels, it also oxidizes and destroys photosynthetic pigments (Section 8–B). However, oxygen's most important effect is its role in photorespiration.

## 8–H  PHOTORESPIRATION

Photorespiration is an odd series of reactions in photosynthetic cells. It appears to serve no useful function. As its name implies, it occurs in the presence of light, and like ordinary respiration, it oxidizes organic compounds, using oxygen, and releases carbon dioxide. However, unlike ordinary respiration, photorespiration does not involve electron transport and does not produce ATP. In fact, since it uses ATP and releases carbon dioxide from organic molecules without making any new energy source, photorespiration appears a waste of energy.

Photorespiration apparently occurs because $CO_2$ and $O_2$ have such similar molecular structures that they compete for the binding site on the enzyme RuBP carboxylase. When the enzyme binds oxygen, it oxidizes RuBP instead of adding a carboxyl group to it. This releases one molecule of PGA, which remains in the $C_3$ cycle, and a two-carbon molecule, which leaves the chloroplast and undergoes further reactions in peroxisomes (Section 5–L) and mitochondria. Here, some carbon is released as carbon dioxide, and the rest is salvaged, eventually returning to a chloroplast to take part in photosynthesis. The membranes of chloroplasts, peroxisomes, and mitochondria almost touch each other, an intimate arrangement that aids this cooperation.

On a warm, bright day, photosynthesis uses carbon dioxide and releases oxygen rapidly. The high temperature, abundance of oxygen, and scarcity of carbon dioxide promote photorespiration, and up to half of the carbon dioxide fixed in photosynthesis may be lost again in photorespiration instead of remaining as stored food energy. Because this reduces the amount of plant material produced by agricultural crops, there is a great deal of research into photorespiration and the possibility of preventing it.

Does photorespiration have any effects favorable to the plant? It may be that evolution cannot produce a rubisco enzyme better able to distinguish carbon dioxide from oxygen. If this is so, photorespiration may be a pathway that minimizes the damage resulting from RuBP oxidation. It does this by (1) removing the two-carbon product, which inhibits the $C_3$ carbon fixation reactions, and (2) salvaging at least some of the organic carbon from this molecule, which would otherwise be lost.

On the other hand, some biochemists propose that photorespiration has advantages. For example, using up $O_2$ in photorespiration may prevent oxygen from reaching levels high enough to destroy photosynthetic pigments, which

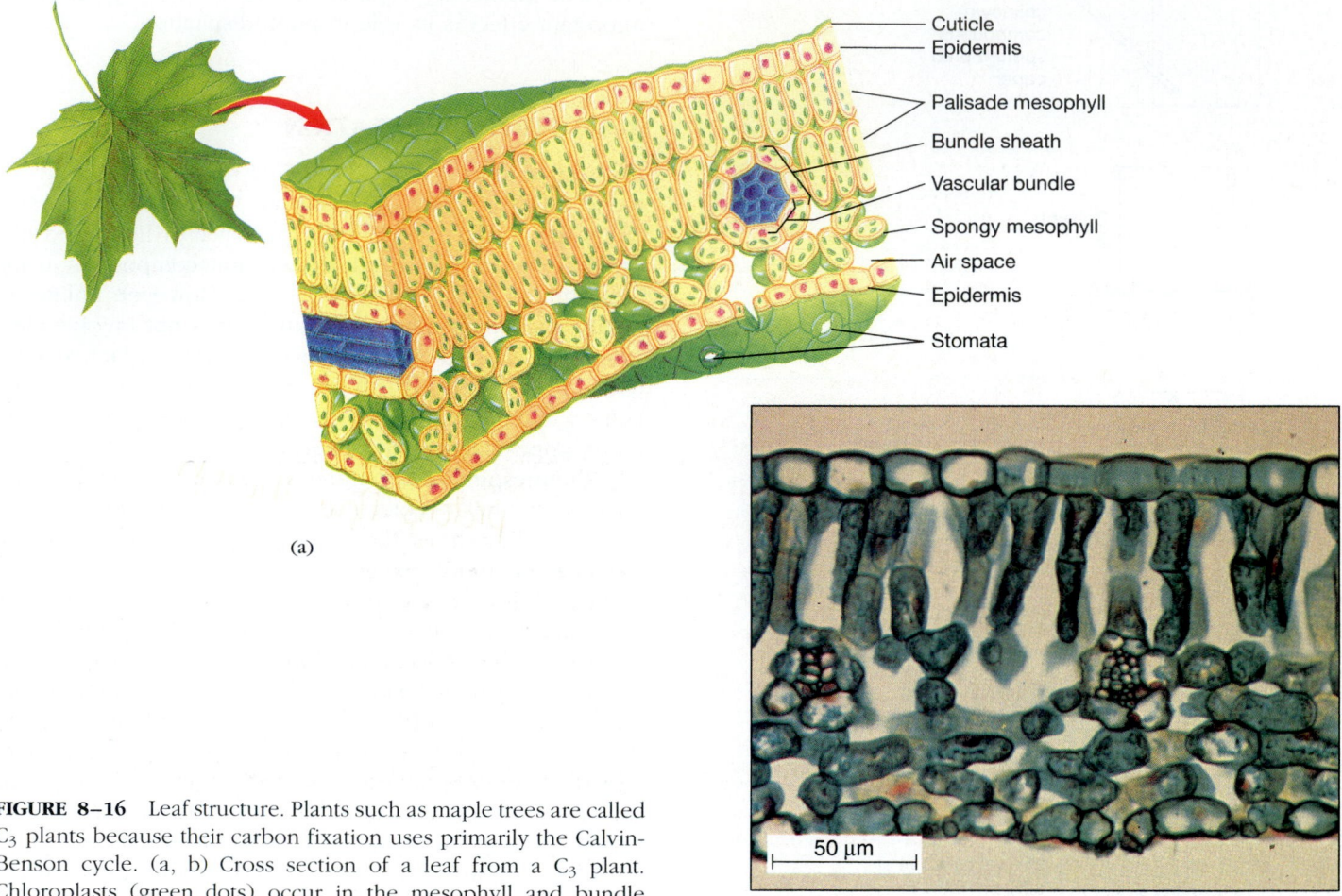

(a)

**FIGURE 8–16**  Leaf structure. Plants such as maple trees are called $C_3$ plants because their carbon fixation uses primarily the Calvin-Benson cycle. (a, b) Cross section of a leaf from a $C_3$ plant. Chloroplasts (green dots) occur in the mesophyll and bundle sheath cells and in the cells on either side of stomata.

Cuticle
Epidermis
Palisade mesophyll
Bundle sheath
Vascular bundle
Spongy mesophyll
Air space
Epidermis
Stomata

50 µm

(b) $C_3$ leaf cross section (LM)

would cripple photosynthesis. The temporary setback to photosynthesis due to photorespiration might cost the plant less than having to produce replacement pigment molecules. It has also been suggested that photorespiration is important because it releases $CO_2$, thereby keeping at least small amounts of this gas available. This keeps the allosteric rubisco enzyme in the active form and keeps the $C_3$ cycle from shutting down completely. However, many biologists are not convinced by these ideas.

## 8–I    ECOLOGICAL ASPECTS OF PHOTOSYNTHESIS

Green plants in different habitats have evolved variations on photosynthesis adapted to particular environments.

## The $C_4$ or Hatch-Slack Pathway

Plants known as $C_4$ plants have extra metabolic steps preceding the $C_3$ cycle. These plants have especially high rates of photosynthesis in sunny, hot, somewhat dry conditions, which is surprising because these conditions promote photorespiration. However, $C_4$ plants have adaptations that permit them to concentrate carbon dioxide around their rubisco enzymes while keeping oxygen away. This enhances carbon fixation and minimizes photorespiration.

In the early 1960s, Australian botanists M. D. Hatch and C. R. Slack worked out the metabolic reactions that permit $C_4$ plants to thrive. When the plants are given radioactive $CO_2$ for photosynthesis, the first organic compound to be labeled is oxaloacetate, which has four carbons (hence the name $C_4$ plants). Oxaloacetate is formed by adding carbon dioxide to a three-carbon molecule, phosphoenolpyruvate.

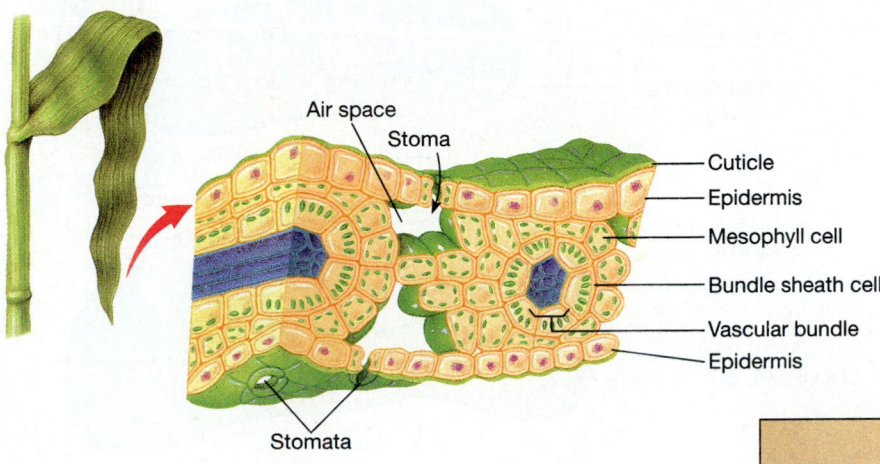

Air space
Stoma
Cuticle
Epidermis
Mesophyll cell
Bundle sheath cell
Vascular bundle
Epidermis

Stomata

(c)

**FIGURE 8–16** *(Continued)* (c, d) Corn (maize) is a $C_4$ plant, and its photosynthetic cells are arranged differently from those in a $C_3$ plant. Larger bundle sheath cells with prominent chloroplasts surround each vascular bundle. These cells carry out $C_3$ carbon fixation. The surrounding mesophyll cells use a $C_4$ pathway to fix carbon.   (b, d, J. Robert Waaland, University of Washington/BPS)

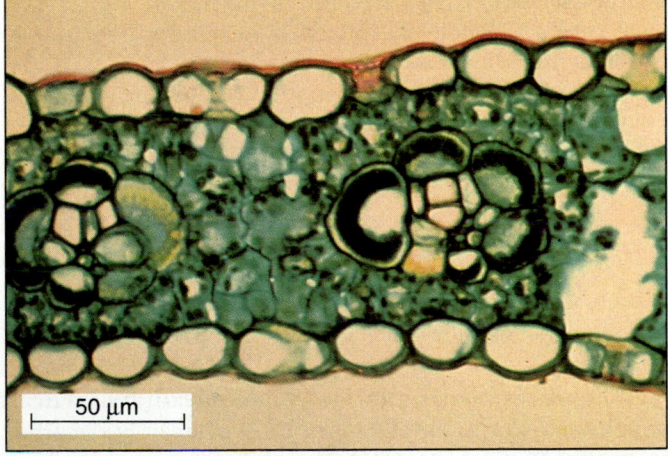

50 µm

(d) $C_4$ leaf cross section (corn) (LM)

The enzyme catalyzing this reaction is found in most plants, not just $C_4$ species. $C_4$ photosynthesis occurs in plants of tropical origin, although many $C_4$ plants now live in temperate areas. $C_4$ plants include crabgrass, some other important weeds, and agricultural plants such as sugar cane, corn (maize), millet, and sorghum.

There is evidence that $C_4$ photosynthesis evolved independently several times in plants exposed to similar environmental conditions. First, $C_4$ plants are scattered among many different plant families, which also contain $C_3$ species. Second, $C_4$ pathways vary from one group to another.

The evolution of $C_4$ photosynthesis involved coordination of existing metabolic reactions in new ways and usually rearrangement of leaf tissues as well.

The leaf tissues of most $C_4$ plants show a special arrangement, called **Kranz** ("wreath") anatomy because of the prominent photosynthetic **bundle sheath cells** packed in a circle around the vascular bundles. The chloroplasts of these cells contain all the leaf's rubisco. Mesophyll cells packed around the bundle sheath cells shield the rubisco from $O_2$ in the leaf's small air spaces (Figure 8–16).

Carbon dioxide is fixed in the mesophyll cells by joining it to phosphoenolpyruvate, a three-carbon compound. The resulting four-carbon oxaloacetate is converted into other compounds, which are moved to the bundle sheath cells by way of plasmodesmata (cytoplasmic connections). Here the carbon dioxide is removed and re-fixed by the $C_3$ cycle (Figure 8–17).

Overall, the $C_4$ pathway increases photosynthesis and decreases photorespiration. Several factors contribute to this outcome. First, the enzyme that fixes carbon in the mesophyll cells does not use $CO_2$ but bicarbonate ion ($HCO_3^-$), which differs enough from $O_2$ that the two do not compete for the enzyme's active site. Second, the $C_4$ pathway builds

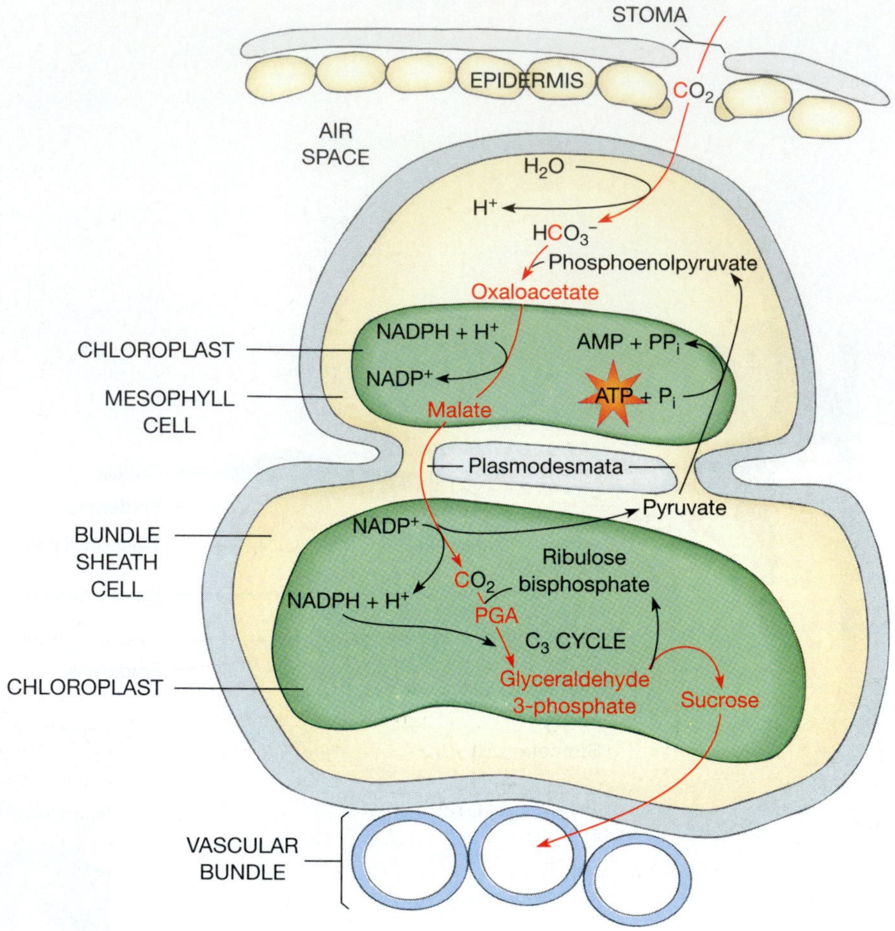

**FIGURE 8–17** Carbon fixation in a $C_4$ leaf. Red arrows and lettering trace the path of new carbon from $CO_2$ to sucrose, which is transported to other parts of the plant through the vascular bundle. $CO_2$ enters a mesophyll cell from one of the leaf's air spaces and combines with water, forming bicarbonate ion ($HCO_3^-$). Carbon fixation occurs as part of this ion becomes a carboxyl group joined to phosphoenolpyruvate, a three-carbon compound. The resulting four-carbon oxaloacetate is reduced to malate by hydrogens from $NADH + H^+$. Malate travels from the mesophyll cell to a bundle sheath cell by way of a plasmodesma. Here it is oxidized, forming a new molecule of NADPH and releasing the carboxyl group in the form of $CO_2$. The $CO_2$ and NADPH are used to re-fix carbon by way of the $C_3$ cycle (Figures 8–13 and 8–14).

a high $CO_2$ concentration in the bundle sheath cells, increasing the chances that rubisco enzymes will bind $CO_2$ rather than $O_2$. Third, in many $C_4$ plants the $C_4$ compound that shuttles from mesophyll to bundle sheath cells is malate, which yields both NADPH and $CO_2$ when it is decarboxylated. In effect, this NADPH is formed in mesophyll cells, reduces oxaloacetate to malate, and is carried to bundle sheath cells where it is regenerated (see Figure 8–17). This NADPH can be used when the $CO_2$ is re-fixed by the $C_3$ cycle, and so bundle sheath cells need not produce so much NADPH as mesophyll cells do. Bundle sheath chloroplasts contain fewer Photosystem II complexes than normal, and so they carry on less noncyclic electron flow and generate less oxygen. However, these chloroplasts do produce a lot of ATP by vigorous cyclic electron flow.

All of this has a cost: it takes more energy. The $C_4$ pathway uses five ATP per molecule of carbon dioxide fixed instead of the three ATP used in $C_3$ photosynthesis alone. However, this is one of many cases in evolution where speed beat efficiency. At high temperatures $C_4$ plants can photosynthesize faster than $C_3$ plants, and the sacrifice of efficiency for speed allows these plants to grow and reproduce faster.

This profligate use of energy pays off only when energy is abundant compared with other resources. $C_4$ photosynthesis has evolved in plants of warm, sunny, some-what dry tropical and subtropical climates, such as grasslands, where there is plenty of light energy to drive ATP synthesis but water stress occurs frequently. In such a situation, the $C_4$ plant's stomata may close partway, reducing loss of water vapor from the leaves. This also reduces uptake of $CO_2$ from the air, but the $C_4$ pathway allows the plant to capture enough carbon dioxide for a high rate of photosynthesis anyway.

The rapid growth of $C_4$ plants has attracted the attention of researchers interested in increasing the productivity of crops. Because this rapid growth depends on a high rate of photosynthesis in bright light, $C_4$ plants are at a relative disadvantage in cool or shady situations, and $C_3$ and $C_4$ plants coexist, with neither having a noticeable edge, in many habitats (Figures 8–18 and 8–19).

## Crassulacean Acid Metabolism

Many plants primarily adapted to desert living have evolved variations of $C_4$ photosynthesis collectively called **crassulacean acid metabolism (CAM).** The name comes from the plant family Crassulaceae (e.g., hens-and-chicks), but CAM has also evolved in several other families and is found in cacti, euphorbs, and other succulents—plants with fleshy, water-storing stems or leaves (Figure 8–20).

*Text continues on page 180*

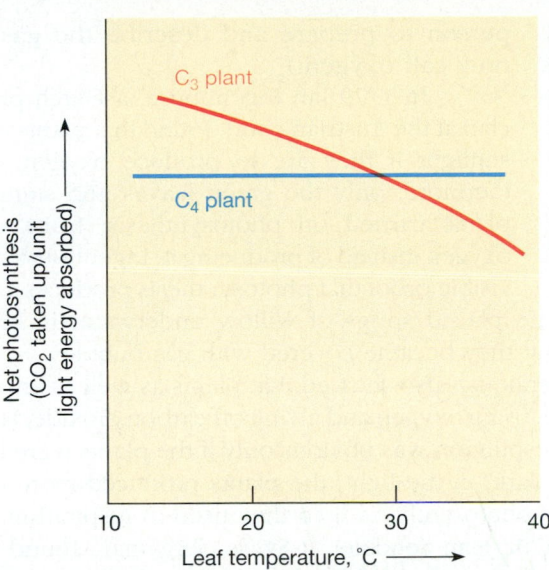

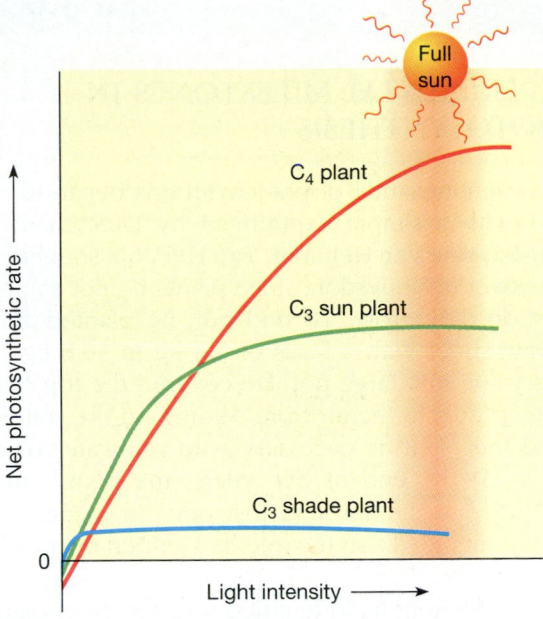

**FIGURE 8–18** Comparison of net photosynthesis in a $C_3$ and a $C_4$ plant at increasing temperatures. $C_4$ plants have higher temperature optima for photosynthesis than do $C_3$ plants from similar habitats. This is because $C_3$ plants have high rates of photorespiration at high temperatures, thereby losing much of the carbon fixed by photosynthesis. (Redrawn after J. R. Ehleringer, *Oecologia* 31(3):255, 1978.)

**FIGURE 8–19** Photosynthetic rates of $C_3$ shade plants, $C_3$ sun plants, and $C_4$ (sun) plants at different light intensities. Shade plants use light more efficiently than sun plants at low light intensities, but they reach their maximum photosynthetic rate (saturation) at relatively low light intensity. Saturation of $C_3$ sun plants occurs at much higher light intensities. Even here, $C_4$ plants still show increased rates of photosynthesis.

(b)

**FIGURE 8–20** Water-saving plants. (a) Sugar cane, an important $C_4$ crop plant. (b) Ice plant, a CAM plant indigenous to South Africa, was introduced to California and now covers countless acres along the California coastline. Note the thick, succulent leaves.

177

# EXPERIMENTAL MILESTONES IN PHOTOSYNTHESIS

The scientific study of photosynthesis began in 1648 with a simple experiment by Dutchman Jean-Baptiste Van Helmont. Van Helmont sought to answer this question: since plants do not eat, how do they grow? To find out, he planted a willow shoot with a mass of 2.3 kg in 90.8 kg of dry soil in a large pot. He covered the top of the soil to keep dust in the air from settling on the soil, and watered the plant as necessary with rainwater or distilled water. At the end of five years, the plant's mass was 76.8 kg, but the soil had lost only 56 grams. Van Helmont concluded that the tree had gained over 74 kg from water alone.

Van Helmont had identified one source of plant material, but he was wrong in thinking that the tree's increased mass came only from water. He had ignored one other possible source of nourishment: the air.

In 1771, English scientist and clergyman Joseph Priestley published evidence that animals and plants alter the composition of the air around them in complementary ways. He burned a candle in a covered jar until it went out. If a plant then grew in the jar for several days, a candle would once more burn in the jar. Priestley concluded that "there was something attending vegetation, which restored the air . . . that had been injured by the burning of candles." He found that exactly the same effect occurred if he used a mouse instead of a candle. The mouse and the candle both altered the air inside the jar in some fashion that could be reversed by a plant (Figure 8–C). Today, we know that a photosynthesizing plant gives off oxygen and that a mouse or a burning candle uses oxygen from the air. (In later experiments, Priestley was the first person to prepare and describe the gas we now call oxygen.)

In 1779 Jan Ingenhousz, a Dutch physician at the Austrian court, found that plants need sunlight if they are to produce oxygen. Furthermore, only the green leaves and stems of plants carried on photosynthesis; fruits used oxygen instead of producing it. Ingenhousz gave visible proof that photosynthesis produces a gas: when he placed sprigs of willow underwater in bright sunlight, they became covered with gas bubbles.

Ingenhousz also showed that plants as well as animals respire, using oxygen and giving off carbon dioxide. However, respiration was obvious only if the plants were kept in the dark; in the light, the plants produced more oxygen by photosynthesis than they used in respiration.

In 1782 Jean Senebier, a Swiss clergyman, found that plants use carbon dioxide when they produce oxygen and suggested that carbon dioxide contributed to the nourishment of the plant: he recognized that air was a source of plant nutrition. In 1804 Nicholas Theodore de Saussure, also a Swiss, found that the amount of carbon dioxide taken up did not nearly account for the increase in dry mass of a growing plant. He decided that the remainder of the increased mass must come from water.

By 1796 Ingenhousz was able to write a general equation for photosynthesis:

$$\text{carbon dioxide} \xrightarrow[\text{light}]{\text{green plant}} \text{organic matter} + \text{oxygen}$$

In 1864 the German plant physiologist Julius Sachs placed a photosynthesizing leaf on a microscope stage and watched starch grains growing inside the chloroplasts. He suggested that at least part of the organic matter produced by photosynthesis was carbohydrate.

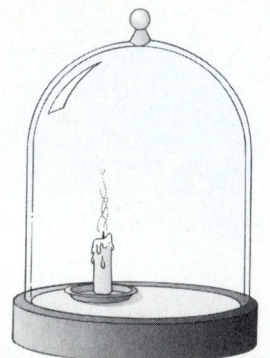

A candle in a closed jar goes out.

A candle with a plant stays alight.

A mouse in a closed jar dies.

A mouse with a plant lives.

**FIGURE 8–C**   The results of Priestley's experiments.

By the end of the nineteenth century, the plant could be regarded as a "black box" that took in water and carbon dioxide and, in the presence of light, converted them to carbohydrate and oxygen.

The carbon atoms in the carbohydrate produced during photosynthesis must come from carbon dioxide, and the hydrogen must come from water. However, both carbon dioxide and oxygen gas contain oxygen atoms, and because both are gases it was long assumed that the oxygen given off during photosynthesis came from carbon dioxide. However, we now know that the oxygen gas released during photosynthesis comes from water. Evidence for this conclusion came from two sources.

In the early 1930s, C. B. van Niel found that purple sulfur bacteria use light to make carbohydrates. They require hydrogen sulfide as a raw material and give off sulfur, which accumulates in the cells as yellow globules. Van Niel, while still a graduate student, speculated that the reactions in sulfur bacteria were analogous to photosynthesis in green plants.

a) General reaction for sulfur bacteria:

$$CO_2 + 2\,H_2S \xrightarrow{\text{light}} (CH_2O) + H_2O + 2\,S$$

b) General reaction for green plants:

$$CO_2 + 2\,H_2O \xrightarrow{\text{light}} (CH_2O) + H_2O + O_2$$

In 1941 Samuel Ruben and Martin Kamen produced evidence that supported van Niel's hypothesis. They used the heavy isotope $^{18}O$ to distinguish the oxygen of water from the oxygen of carbon dioxide. When they gave a plant water labeled with $^{18}O$ and unlabeled carbon dioxide (containing the common isotope $^{16}O$), labeled oxygen was given off (shown in heavy type below):

$$6\,CO_2 + 12\,H_2\mathbf{O} \longrightarrow C_6H_{12}O_6 + 6\,\mathbf{O}_2 + 6\,H_2O$$

The experiment can also be done the other way around:

$$6\,CO_2 + 12\,H_2O \longrightarrow C_6H_{12}\mathbf{O}_6 + 6\,O_2 + 6\,H_2\mathbf{O}$$

## TABLE 8–A

## A Chronology of the Study of Photosynthesis

| Year | Investigator | Major Findings |
|---|---|---|
| 1648 | Van Helmont | Plants gain weight from water |
| 1771 | Priestley | Plants restore air injured by animals |
| 1779 | Ingenhousz | Plants need light to restore air |
|  |  | Plants also respire |
| 1782 | Senebier | Carbon dioxide nourishes plants |
| 1796 | Ingenhousz | Incomplete equation for photosynthesis |
| 1804 | de Saussure | Water contributes to plant nutrition |
| 1864 | Sachs | Photosynthesis produces carbohydrate |
| 1883 | Engelmann | Action spectra of algae follow absorption spectra of their pigments |
| 1898 |  | Term "photosynthesis" coined |
| 1905 | Blackman | Photosynthesis has both photochemical and thermochemical reactions |
| 1913 | Willstätter | Structure of chlorophyll |
| 1930 | van Niel | Oxygen may come from water |
| 1932 | Emerson, Arnold | Pigments occur in photosynthetic units (photosystems) |
| 1938 | Hill | Illuminated chloroplasts reduce dyes and produce oxygen |
| 1941 | Ruben and Kamen | Oxygen comes from water |
| 1948+ | Calvin | $C_3$ carbon fixation and following steps |
| 1951 | Vishniac and Ochoa | Light + chloroplasts + $NADP^+ \rightarrow NADPH + O_2$ |
| 1954 | Arnon | Chloroplasts produce ATP |
| 1957 | Emerson | Two photosystems exist |
| 1960+ | Jagendorf | $H^+$ gradient drives ATP synthesis in chloroplasts (see Figure 6–12) |
| 1960s | Kortschak, Hatch, Slack | $C_4$ photosynthesis |
| 1969–70 | Joliot, Kok | Two water molecules release four electrons in series, forming $O_2$ |

...ata of CAM plants are closed during the hot ...g water loss. During the cooler, more humid ...omata are opened, and carbon dioxide is taken u... ...ed into organic acids. During the day, when the stomata are closed, carbon dioxide is removed from the acids, and light energy is used to re-fix it by way of the $C_3$ pathway. Both $C_4$ photosynthesis and CAM store carbon dioxide for later use, but in CAM, activity is divided between day and night instead of between different types of cells.

Although CAM plants receive plenty of sun, they must conserve water very strictly, and so they have very low photosynthetic rates. However, their ability to conserve water enables them to survive in habitats too dry for the "fast food" plants that would otherwise crowd out these slow growers.

## Sun Plants and Shade Plants

$C_3$ plants can be separated into sun plants and shade plants, adapted to different levels of sunlight. While **sun plants,** such as soybeans, cotton, and tomatoes, have increasing rates of photosynthesis as light intensity increases, **shade plants** do not. Shade plants, including many ferns, orchids, African violets, and philodendrons, simply do not photosynthesize rapidly even in bright light. On the other hand, shade plants are much more efficient at using dim light (see Figure 8–19).

Some forest trees, such as oaks and maples, have both shade leaves and sun leaves: sun leaves near the top and shade leaves near the bottom. Sun leaves tend to be smaller and thicker, with more pronounced lobes than shade leaves. Sun leaves also have more palisade mesophyll cells, rubisco enzymes, and stomata per unit of surface area—all features contributing to their high rates of photosynthesis. Shade leaves are thinner, with fewer photosynthetic cells in a unit of leaf area, but their ratio of chlorophyll to rubisco is higher than in sun leaves.

In many plant species, the cells of the leaves' upper epidermis become more convex in the shade, acting as miniature lenses that gather dim light and focus it on the photosynthetic tissue.

## SUMMARY

Photosynthesis is the process in which green plants store the energy of sunlight by converting carbon dioxide and water into organic compounds:

$$CO_2 + H_2O \xrightarrow{\text{light, chlorophyll}} CH_2O + O_2$$

These organic compounds are used by plants, and by the animals that eat plants, to build cells and to power other energy-requiring processes. Organic molecules are eventually broken back down to carbon dioxide and water via respiration, and these materials can then be recycled in photosynthesis. Because energy cannot be recycled, however, the sunlight that drives photosynthesis is the ultimate source of energy for nearly all life on Earth.

1. Photosynthesis may be considered in two parts:
   - **Energy capture.** Solar energy is trapped by chlorophyll and other photosynthetic pigments in the thylakoid membranes, initiating a flow of electrons through the membrane's electron transport system. The overall flow of electrons is from water molecules split inside the thylakoid space, through photosystems and electron transport systems in the thylakoid membrane, to $NADP^+$ in the stroma. This flow of electrons reduces $NADP^+$ to NADPH and creates the $H^+$ reservoir used to make ATP. Oxygen from water is released as a by-product. The NADPH and ATP are released into the stroma of chloroplasts, where they are used to fix carbon dioxide.
   - **Carbon fixation.** During the $C_3$ cycle, carbon dioxide becomes attached to a five-carbon sugar, ribulose bisphosphate, which then breaks to yield two three-carbon phosphoglycerate molecules. ATP and NADPH are used to reduce these to glyceraldehyde 3-phosphate. This three-carbon product may be made into structural or energy-storing molecules or may be processed, using more ATP, to make more ribulose for carbon fixation.

   In $C_4$ and CAM plants, carbon dioxide is temporarily added to a three-carbon molecule, forming a four-carbon compound, from which it is later removed and re-fixed by the $C_3$ pathway.

2. The rate of photosynthesis is often limited by shortage of light energy, low temperatures, or scarcity of the raw material carbon dioxide. Buildup of the end product oxygen interferes with photosynthesis in several ways. One of its major effects is its role in photorespiration, which in effect undoes some of the plant's photosynthetic carbon fixation.

3. A photosynthesizing plant captures light energy in two parts of the reaction sequence. In each case, the light energy boosts an electron to a higher energy level, and the energy so transferred is then channeled into increasingly more permanent forms of energy storage:

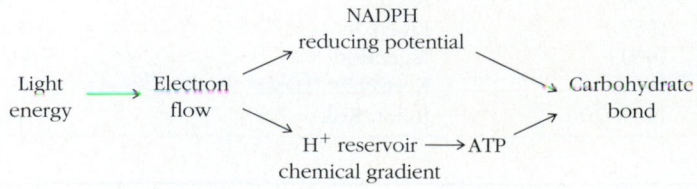

## Self-Quiz

1. Red and blue light support the highest rates of photosynthesis because:
   a. these are the only wavelengths reaching Earth from the sun
   b. these are the only wavelengths that carotenoids cannot absorb
   c. chlorophyll absorbs these wavelengths more than other wavelengths
   d. these wavelengths have the highest energy in the visible spectrum
   e. these wavelengths activate the ATP synthetase enzyme

2. The role of phycobilins in photosynthesis is to:
   a. absorb and pass energy to chlorophyll $a$
   b. donate electrons to the electron transport system
   c. fix carbon dioxide
   d. carry hydrogen or electrons
   e. all of the above

3. The oxygen from $H_2O$ is incorporated into:
   a. oxygen gas
   b. water
   c. carbohydrates
   d. NADPH
   e. ATP

4. Oxygen produced during photosynthesis leaves a leaf by _____ , carbohydrates by _____ .

Match each item with its location in the chloroplast:
_____ 5. Chlorophyll
_____ 6. Enzymes for carbon fixation
_____ 7. ATP synthetase
_____ 8. $H^+$ reservoir for ATP synthesis

a. stroma
b. chloroplast envelope
c. thylakoid membranes
d. thylakoid space

Match (give all correct answers):
_____ 9. ribulose bisphosphate
_____ 10. $NADP^+$
_____ 11. glyceraldehyde 3-phosphate
_____ 12. $O_2$
_____ 13. $CO_2$
_____ 14. $ADP + P_i$

a. raw material of electron transport reactions
b. end product of electron transport reactions
c. raw material for chemiosmotic ATP synthesis
d. end product of chemiosmotic ATP synthesis
e. raw material of $C_3$ cycle
f. end product of $C_3$ cycle

## Questions for Discussion

1. Why do trees absorb their chlorophylls, but not their carotenoids, in the autumn?
2. If oxygen is given off as a result of electron transport in photosynthetic membranes, why does the rate of oxygen emission level off when the rate of the $C_3$ cycle limits the rate of photosynthesis?
3. A. J. Kluyver stated that a continuous, orderly flow of electrons is the most fundamental characteristic of living things. Explain this statement in terms of your study of Chapters 6 through 8.

## Suggested Readings

Bazzaz, F. A., and E. D. Fajer. "Plant life in a $CO_2$-rich world." *Scientific American,* January 1992. How rising carbon dioxide levels affect photosynthesis and growth of plants.

Galston, A. W., P. J. Davies, and R. L. Satter. *The Life of the Green Plant,* 3d ed. Englewood Cliffs, NJ: Prentice-Hall, 1980. A text with many fascinating tidbits.

Govindjee, and W. J. Coleman. "How plants make oxygen." *Scientific American,* February 1990. What researchers have found out about Photosystem II, and what they still don't know.

Youvan, D. C., and B. L. Marrs. "Molecular mechanisms of photosynthesis." *Scientific American,* June 1984.

PART
TWO

*Information
Coding and
Transfer*

# DNA and Genetic Information

## OBJECTIVES

*When you have studied this chapter, you should be able to:*

1. Use these terms correctly:
   genetic information
   gene, chromosome, genome
   3' end and 5' end of a DNA molecule, antiparallel strands
   DNA replication, template, replication origin, DNA helicase,
   DNA polymerase, DNA ligase
   chromatin, histones, centromere, telomere
   transposable elements, noncoding DNA
2. Describe and explain the evidence that DNA is the genetic material, using these studies as evidence:
   (a) bacterial transformation
   (b) infection of bacteria by bacteriophages
   (c) the quantity of DNA in body cells and in reproductive cells of a species
   (d) comparison of the base composition of DNA in cells from members of the same and different species
3. Describe the structure of a nucleotide.

4. Describe the structure of a molecule of DNA, and explain why the number of adenine bases in the molecule equals the number of thymine bases, and the number of guanine bases equals the number of cytosine bases.
5. Briefly describe the replication of DNA, mentioning the importance of these structural features of the molecule:
   (a) existence of two DNA strands in the helix
   (b) covalent bonding between sugar and phosphate groups of each strand's backbone
   (c) hydrogen bonding of nitrogenous bases on opposite strands
6. Define mutation, list some kinds of mutations, and explain the importance of mutations.
7. Compare and contrast the structures of a eukaryotic chromosome and prokaryotic DNA.
8. Compare and contrast the genomes of prokaryotes and higher eukaryotes with respect to their content of coding versus noncoding DNA, transposable elements, and repetitive DNA.

---

**E**ach of us originated from a single cell—an egg, fertilized by a sperm. What directed this fertilized egg to divide into more and more cells and the resulting mass of cells to move around, grow, absorb nourishment, and take shape as a unique individual? What makes each of us distinct from other people but gives us all considerable similarity as members of the human species? And what allows friends hovering over a new baby to proclaim that it has its father's nose and its mother's grin?

The answer to all these questions is **genetic information:** the inherited instructions for building proteins and other molecules, including *when* in the life of a cell or organism each should be made, and *which kind* of cell should make it. Proteins, in turn, build and operate cells and hence the body as a whole. The units of genetic information are

called **genes.** Genes govern all inherited characteristics such as hair color, blood type, and embryonic development.

During the nineteenth century, biologists learned that an individual inherits half its chromosomes from each parent. When they studied cells dividing to form eggs and sperm, they noticed that each egg or sperm ends up with half the number of chromosomes found in the nucleus of the original cell. When egg and sperm unite at fertilization, only the nucleus of the sperm, containing chromosomes, enters the egg. The new individual receives a full set of chromosomes: half from its mother's egg and half from its father's sperm.

This behavior of chromosomes was exactly what biologists expected of structures bearing genetic information. They became convinced that chromosomes do carry genes and genetic information. However, one question remained: chromosomes are composed of two substances, DNA (short for deoxyribonucleic acid, Section 3–D) and proteins. Which carries the genetic information?

Biologists knew that proteins are a diverse group of polymers, made up of 20 kinds of amino acid monomers. On the other hand, DNA polymers contain only four kinds of nucleotide monomers. It seemed as though only proteins were sufficiently complex and specific to carry the variety of genetic information a cell must contain. Not until the 1940s did scientists realize that this was wrong: DNA contains the genetic information.

A chromosome contains a single DNA molecule bearing many genes. Each gene is made up of hundreds or thousands of nucleotide monomers that act as a functional unit. A gene's function may be regulating the activity of other genes or carrying instructions for making RNA molecules required for protein synthesis. The most familiar category is **structural genes,** which contain instructions for making proteins: the order of their nucleotides serves as a "genetic code" of directions for the order of nucleotides in RNA molecules, which in turn dictate the order in which amino acids should be joined to make the corresponding protein.

So DNA codes for protein structure, and a protein's structure, in turn, determines its function. A protein's job may be something like catalyzing the synthesis of eye pigments or becoming part of a hair. So the gene's information is said to be **expressed** in the form of protein, which in turn contributes to the structure or function of the whole organism.

In this chapter we shall examine the evidence that the genetic material is in fact DNA and see how the molecular structure of DNA was worked out. This structure is very simple, yet elegant, and it has the remarkable property that it dictates the production of exact copies of itself—copies that are then passed on to future generations of cells. We shall see how DNA is copied and how it is packaged and organized into chromosomes.

---

### KEY CONCEPTS

♦ In sexual reproduction, a new individual inherits half of its genetic information from each parent in the form of DNA in the chromosomes.

♦ The units of genetic information are genes, which are sections of DNA molecules.

♦ A cell's DNA controls what proteins are made and in this way controls most of the cell's activity.

♦ A DNA molecule contains the information needed to produce exact copies of itself.

---

## 9–A    EVIDENCE THAT DNA IS THE GENETIC MATERIAL

### The Riddle of Bacterial Transformation

The first evidence that genes were not made of proteins came in 1928. Fred Griffith, a British Public Health Service scientist, was studying bacterial pneumonia, a serious disease in the days before antibiotics. He worked with two strains of bacteria, containing different genetic information. The genetic information of one strain caused these bacteria to produce external capsules that protected them from attack by an animal's immune system. This strain was **virulent:** when injected into mice, it caused fatal disease. The other strain, which could not produce capsules, was **nonvirulent** and did not kill the mice.

Griffith discovered that nonvirulent bacteria could be changed to the virulent form. If virulent bacteria were killed by heating they did not cause disease, but if they were killed and then injected into mice along with living, nonvirulent bacteria, some of the mice died (Figure 9–1). The dead mice contained living virulent bacteria even though none had been injected. Griffith concluded that some of the genetic material from the dead virulent bacteria had entered the living nonvirulent ones, making them virulent. This phenomenon was named **bacterial transformation,** the transfer of genetic information from one bacterium into another.

As is usual in science, the results of Griffith's experiment raised a new question: what had transformed the bacteria? Griffith had killed his bacteria by heating them. Heat denatures proteins, and so it seemed that whatever had transformed the bacteria could not have been protein.

In the United States, Oswald Avery and his colleagues worked for a decade to discover what caused this transformation. During this time they grew tons of virulent bacteria and separated them into carbohydrates, lipids, proteins, RNA, and DNA. Then they added samples of each substance to suspensions of living, nonvirulent bacteria. In 1944 they reported that the transformation studied by Griffith could be produced by DNA from virulent bacteria but not by their proteins nor any other substance. (By this time, Griffith was dead, killed when he refused to leave his laboratory during the bombing of London in World War II. Meanwhile, Avery had nearly lost his research funding when it became obvious that pneumonia could be controlled by antibiotics.)

Treatment with DNA had given the living bacteria a genetic characteristic they did not have before. We now know that the transforming DNA becomes incorporated into the

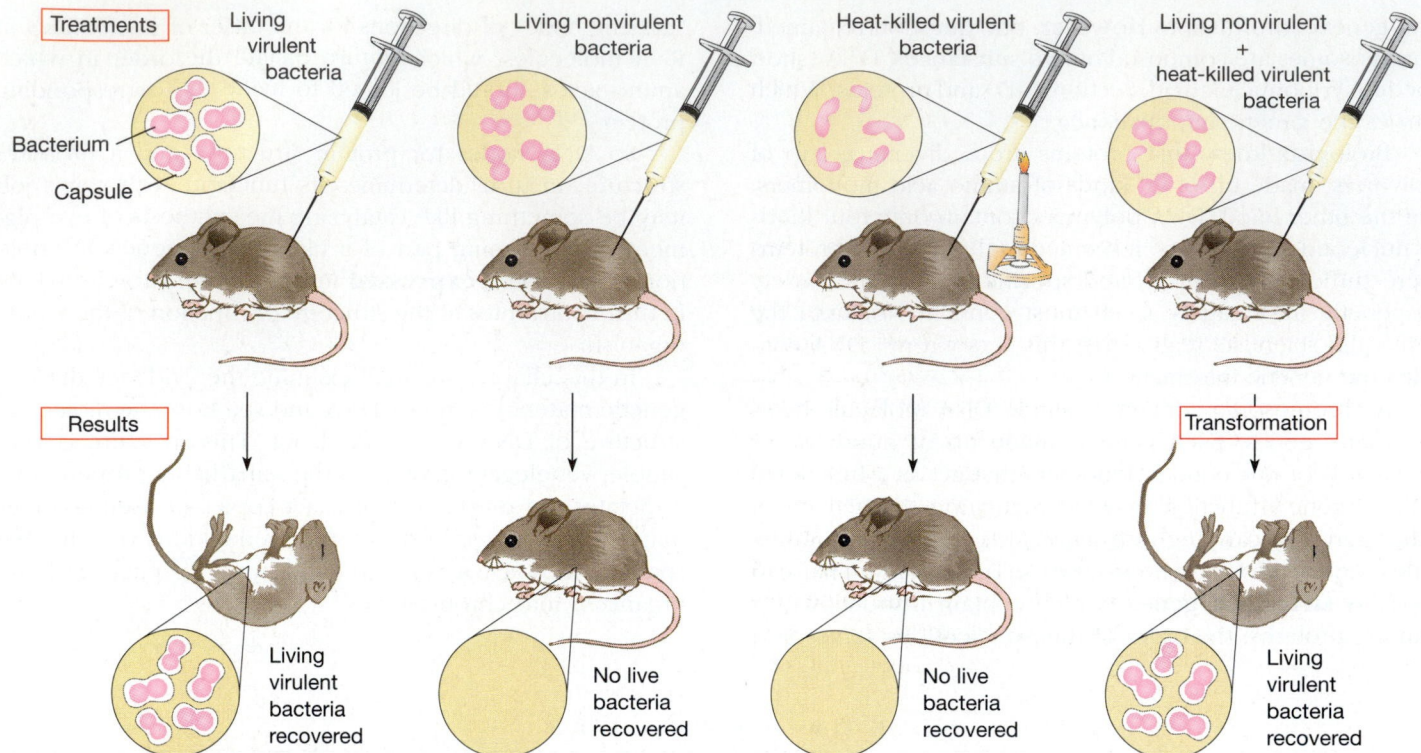

**FIGURE 9–1**   Griffith's experiments on bacterial transformation. Living virulent bacteria injected into mice cause fatal disease. Living nonvirulent or heat-killed virulent bacteria do not cause disease, but mice injected with both do develop disease. Something from the heat-killed bacteria transforms the nonvirulent bacteria into virulent cells.

recipient cell's DNA, thereby endowing the cell with the genetic traits carried by the transforming DNA.

The conclusion that DNA is the genetic material was not immediately accepted. At that time, DNA was widely assumed to consist of a sequence of its four nucleotide subunits repeated over and over, an arrangement too simple to contain much information. Biochemists pointed out that Avery's DNA contained a little protein; perhaps this protein was really the transforming material. In addition, most biologists at that time did not regard bacteria as "real" organisms. If DNA was the genetic material of bacteria, so what? Many felt this information was irrelevant to understanding eukaryote genetics. We now know that DNA is in fact the genetic material of all eukaryotes, as well as of bacteria and some viruses. (In other viruses, the genetic material is RNA.)

Indeed, most of the major discoveries in molecular biology were made using bacteria, which can be grown quickly and easily for experiments. Chromosomal proteins and sheer size make eukaryotic DNA harder to work with, and good techniques were not developed until the 1970s.

## Bacteriophages

More evidence that genetic material is DNA was obtained from studies of **bacteriophages** (**phages** for short), viruses that parasitize bacteria. A phage takes over a bacterium's metabolic machinery, causing it to produce many new phages and then to burst, releasing these phages to infect more bacteria.

A phage consists of a DNA molecule inside a protein coat. In 1952 Alfred Hershey and Martha Chase tested the hypothesis that phages do not enter bacteria intact; rather, the protein coat attaches to the cell wall and injects its DNA into the bacterium. If this is so, and the bacterium then produces new phages, the DNA must be the phage's genetic material, and its protein coat is merely a combination packing crate and micro-syringe.

Hershey and Chase distinguished between the phages' protein and DNA using the radioactive isotopes sulfur 35 and phosphorus 32 as labels (*How Do We Know? Isotopes: Chemical Tools for Biology*, Chapter 2). Proteins contain sulfur but not phosphorus, whereas DNA contains phosphorus but not sulfur. Hershey and Chase added either sulfur 35 or phosphorus 32 to the liquid nutrient media in which they grew cultures of bacteria and phages. This gave them a batch of new phages having proteins labeled with radioactive sulfur and a batch having DNA labeled with radioactive phosphorus.

Now ready to do the experiment, Hershey and Chase added these radioactive phages to fresh cultures of bacteria and gave them a short time to infect the bacteria. Then they spun these cultures in a Waring blender to shear away the phages attached to the bacterial cell walls. Next, they used centrifugation to settle the bacteria out of the liquid. After this treatment, the radioactive sulfur was found to be in the liquid culture medium: the phage proteins had not entered the bacteria. On the other hand, the radioactive phosphorus was in the pellet of bacteria: the phage DNA had entered the bacteria (Figure 9–2). Furthermore, both groups of bacteria produced new phages. This was strong evidence that the phage genetic material consisted of DNA but not protein. Even stronger evidence was that some new phages in the phosphorus group were radioactive, showing that DNA from the original phages had been inherited by the new generation.

It is interesting to note that these experiments would not have worked with higher organisms because considerable quantities of protein are tightly bound to the chromosomal DNA of eukaryotes.

## The Quantity of DNA in Cells

Circumstantial evidence that DNA is the genetic material in eukaryotes came from measuring the amount of DNA in different cells. For example, the body (nonreproductive) cells from the liver, kidneys, and other organs of a chicken contain roughly the same amount of DNA as each other and twice as much as sperm, which are reproductive cells. Since two reproductive cells—sperm and egg—combine to form the new individual, each reproductive cell must have half as much genetic information as a body cell, or else the amount of genetic information would double in each generation. This distribution of DNA is what one would expect of the genetic material. The amount of protein in cells, on the other hand, varies a lot from one tissue to another and is not necessarily lower in reproductive cells. This makes it less likely that protein is the genetic material.

In addition, DNA shows very little turnover compared with other cell components. Proteins and RNA are constantly being made and destroyed, but DNA, once made, is remarkably stable.

## Proportions of Bases in DNA

All DNA is made up of the same four nucleotides—those containing the nitrogenous bases adenine (A), thymine (T), guanine (G), and cytosine (C). However, these nucleotides

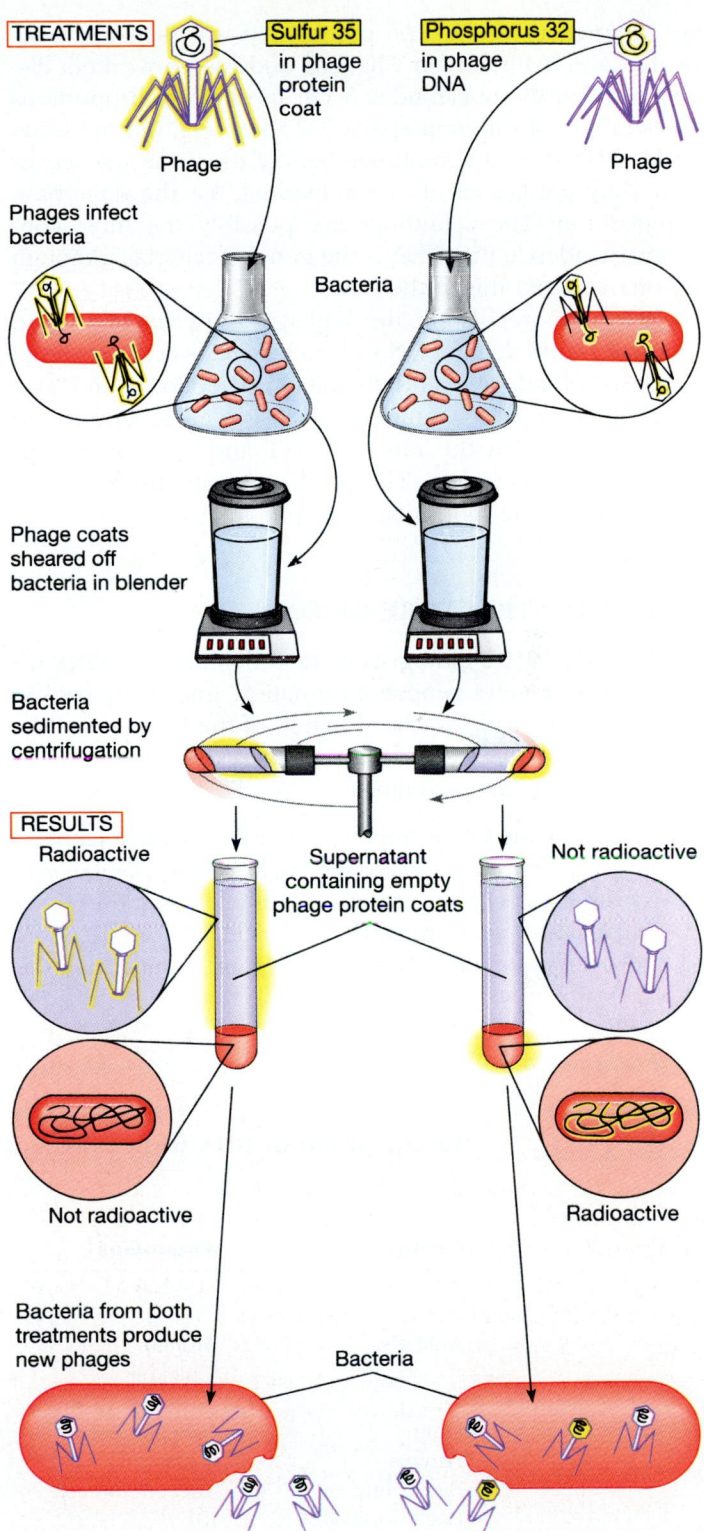

**FIGURE 9–2** Evidence that DNA is the genetic material of bacteriophages. Hershey and Chase labeled phage protein coats with radioactive sulfur or phage DNA with radioactive phosphorus. Radioactivity is shown as a yellow glow in each treatment. The labeled viral protein coat remains outside the bacteria, whereas the labeled DNA enters the cells. In both treatments many new phages are produced inside the bacteria, indicating that DNA, not protein, is the phage hereditary material.

are not present in equal proportions, as was first supposed. In the late 1940s, Erwin Chargaff and his co-workers discovered that the nucleotides occur in different proportions in members of different species (Table 9–1). On the other hand, DNA from different members of the same species, or from different tissues of one individual, has the same base composition. These findings are possibly the most convincing evidence that DNA is the genetic material, although no one realized this at the time.

Furthermore, within the limits of experimental error, Chargaff found that the DNA of each species contains equal numbers of adenine and thymine nucleotides, and also equal numbers of guanine and cytosine nucleotides: note in Table 9–1 that the ratios of A/T and of G/C are approximately equal to 1. This finding eventually became a major clue to the molecular structure of DNA.

## 9–B    THE STRUCTURE OF DNA

By the early 1950s, biologists were convinced that DNA indeed carries a cell's genetic information, and many people were trying to work out the structure of the DNA molecule. Any model of DNA structure had to take several experimental findings into account:

1. DNA is made up of nucleotides. Each DNA nucleotide has three parts: a five-carbon sugar (deoxyribose); one to three phosphate groups, covalently bonded to the sugar's fifth carbon atom; and one of four possible nitrogen-containing bases, covalently bonded to the

sugar's first carbon atom. The base may be one of the single-ring pyrimidine derivatives thymine (T) or cytosine (C), or one of the double-ring purine derivatives adenine (A) or guanine (G) (see Figure 3–15).

2. The nucleotides are linked together in a strand of DNA: the phosphate group attached to the 5' (pronounced "five prime") carbon of the deoxyribose sugar of one nucleotide joins to the 3' carbon on the sugar of an adjacent nucleotide (Figure 9–3). This forms a string of alternating phosphate and sugar groups, called the sugar-phosphate backbone, held together by covalent bonds. One end of the backbone, called the 3' end, has a free hydroxyl (—OH) group, whereas the other, 5' end has a free phosphate group. The purine and pyrimidine bases stick out to one side of the sugar-phosphate backbone.

3. Chargaff had shown that the number of nucleotides containing adenine (A) equals the number containing thymine (T), and the numbers containing guanine (G) and cytosine (C) are also equal to each other. In the shorthand popular with biologists, A = T, and G = C (Table 9–1).

4. The most direct evidence for the structure of DNA came from x-ray diffraction pictures, made by passing x-rays through crystals of DNA. This produces a pattern of dots that gives information about the molecule's shape. In 1952 Rosalind Franklin produced photographs of crystals of highly purified DNA. Her results showed that DNA is twisted into a spiral, or **helix,** with the bases perpendicular to the length of the fiber, like a twisted

T A B L E   9 – 1

**Composition of DNA from Different Organisms**

| Group | Organism | Percent of Nucleotide Molecules | | | | Base Ratios* | |
|---|---|---|---|---|---|---|---|
| | | *A* | *T* | *G* | *C* | *A/T* | *G/C* |
| Animals: | Human | 30.9 | 29.4 | 19.9 | 19.8 | 1.05 | 1.00 |
| | Chicken | 28.8 | 29.2 | 20.5 | 21.5 | 1.02 | 0.95 |
| | Locust | 29.3 | 29.3 | 20.5 | 20.7 | 1.00 | 1.00 |
| Plant: | Wheat | 27.3 | 27.1 | 22.7 | 22.8 | 1.01 | 1.00 |
| Fungus: | Yeast | 31.3 | 32.9 | 18.7 | 17.1 | 0.95 | 1.09 |
| Bacterium: | *Escherichia coli* | 24.7 | 23.6 | 26.0 | 25.7 | 1.04 | 1.01 |
| Bacteriophages: | T4 | 26.0 | 26.0 | 24.0 | 24.0 | 1.00 | 1.00 |
| | Lambda | 21.3 | 22.9 | 28.6 | 27.2 | 0.92 | 1.05 |
| | φX174 | 24.6 | 32.7 | 24.1 | 18.5 | 0.75 | 1.30 |

*Except for bacteriophage φX174, these ratios are all approximately 1. Within the limits of experimental error, the number of nucleotides containing A and T are equal, and likewise for G and C. However, note that the proportion of A and T differs from the proportion of G and C from one species to another. φX174 does not show the same ratio because its DNA is single-stranded, not double-stranded like the others.

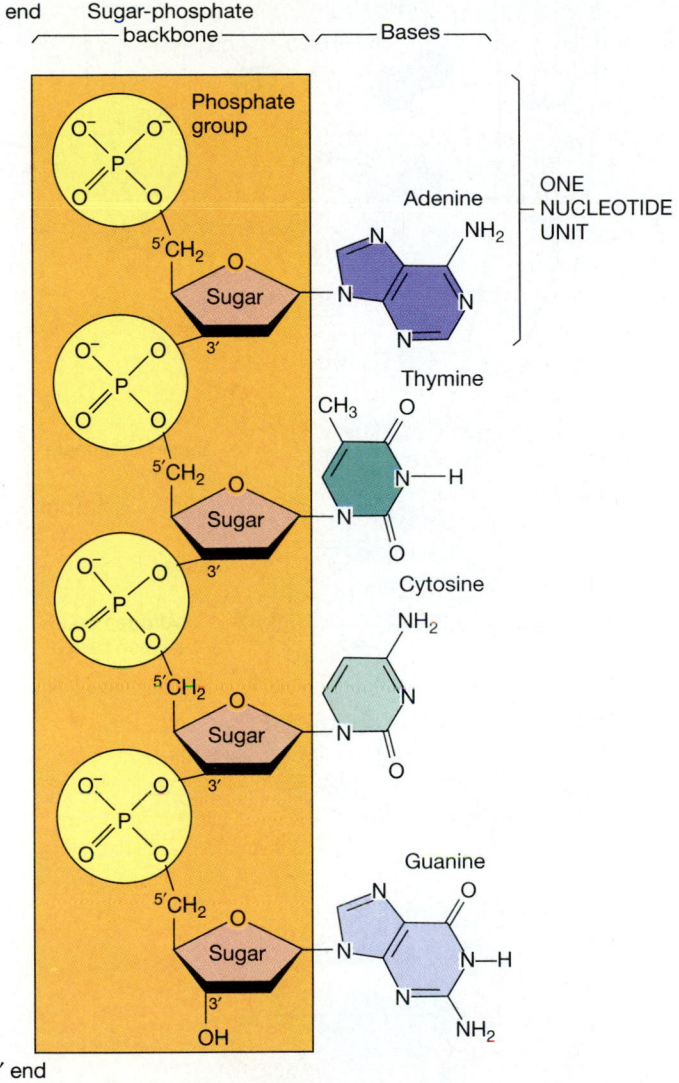

## A DNA STRAND

5′ end — Sugar-phosphate backbone — Bases

Phosphate group

Adenine

Thymine

Cytosine

Guanine

ONE NUCLEOTIDE UNIT

3′ end

**FIGURE 9–3** A short stretch of DNA. Each nucleotide subunit contains a phosphate group, a sugar, and one of four nitrogenous bases: adenine, thymine, cytosine, and guanine. The carbons in the sugar are numbered 1′ to 5′: the prime (′) sign distinguishes the atoms in the sugar ring from those in the nitrogenous base, which are numbered without this sign. The phosphate group at the 5′ end of the molecule is not attached to another nucleotide, nor is the sugar at the 3′ end.

ladder. These pictures also gave evidence that the sugar-phosphate backbone is on the outside of the helix, with the bases inside. One turn of the helix contains ten nucleotides. Furthermore, the diameter of the helix showed that it must be composed of more than one strand of DNA.

Two main questions remained: how many strands are there in the DNA molecule, and how are they joined to-

gether? The answers to these questions turned out to be the most interesting aspects of the structure of DNA.

At the time Franklin made her pictures, there was a race going on to put all the available data together into a consistent model of DNA structure. Chemist Linus Pauling published a model with three strands. However, this model contained an error, and he was working on a new version. Maurice Wilkins, head of the laboratory where Franklin worked in London, was also concentrating on the problem. However, the first people to fit all the evidence together in a workable form were James Watson, then a young postdoctoral researcher, and Francis Crick, a physicist-turned-biochemist. They used a set of scale models of nucleotides to build possible structures until they found one that fitted all the data.

The Watson and Crick model of DNA structure consists of two strands of DNA. (To a biologist, the number two is satisfying because both cells and chromosomes reproduce by the formation of two new entities from the original one.) The two strands are arranged like a ladder, with the ladder's sides being the sugar-phosphate backbones of the two strands and the rungs being the bases.

A rung consists of either adenine paired to thymine, or guanine paired to cytosine. The atoms in each pair match up in such a way that hydrogen bonds form between the bases: two hydrogen bonds between adenine and thymine, three between guanine and cytosine (Figure 9–4). These hydrogen bonds hold the two bases in the rung together. A and T, and G and C, form the most stable combinations of hydrogen bonds. Therefore, the bases in these pairs are said to be **complementary.** This explains Chargaff's finding that A = T and G = C in the DNA of any species.

Since each pair consists of one single-ring and one double-ring base, all the rungs of the ladder are the same width, and the backbones of the two DNA strands are always the same distance from one another. In each rung, either base may be on either backbone strand.

Watson and Crick also saw that for hydrogen bonds to form properly between the base pairs in DNA, the two nucleotide strands of the DNA molecule had to run in opposite directions. They are said to be **antiparallel,** with the free 5′-phosphate groups of the two strands at opposite ends of the molecule, facing each other's free 3′ hydroxyl groups. In Figure 9–4, if you move your finger in the 5′-to-3′ direction along one strand, you are going in the other strand's 3′-to-5′ direction.

Finally, the whole ladder of DNA is twisted, with ten nucleotide pairs per turn, to form the spiral detected by Franklin's x-ray photographs. Because the spiral is composed of two strands wound around each other, the DNA molecule is a **double helix** (Figure 9–5).

Tremendously excited by this simple yet elegant structure, Watson and Crick published their model. Their paper was only two pages long, but it became a cornerstone of

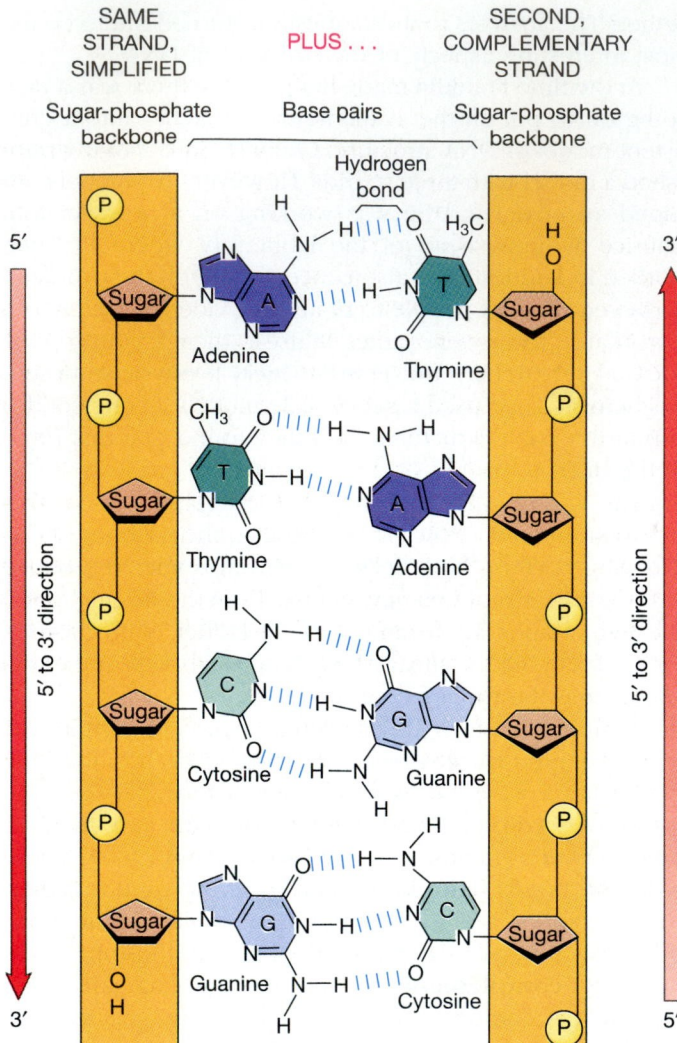

**FIGURE 9–4** Two-stranded structure of DNA. Here the strand from Figure 9–3 is shown with its partner. The two strands form a ladder-like structure with their sugar-phosphate backbones on the outside of the molecule and the nitrogenous bases forming rungs as they meet in the middle. In each rung, the bases on opposite strands are held together by hydrogen bonds. Adenine always pairs with thymine, cytosine with guanine.

modern molecular genetics. As Watson remarked, "It was too pretty not to be true." In 1962 Watson, Crick, and Wilkins received a Nobel Prize for their work on the structure of DNA.

Both prokaryotes and eukaryotes contain DNA with this double helix structure. A prokaryote's genetic material is a circular double helix of DNA. The circular bacterial DNA is folded many times and occupies a nuclear area about one tenth of the cell's volume.

In contrast, eukaryotic DNA is organized into a number of chromosomes, each containing one long linear DNA double helix bound to proteins (Section 9–G). Eukaryotic cells contain more DNA than prokaryotic cells do.

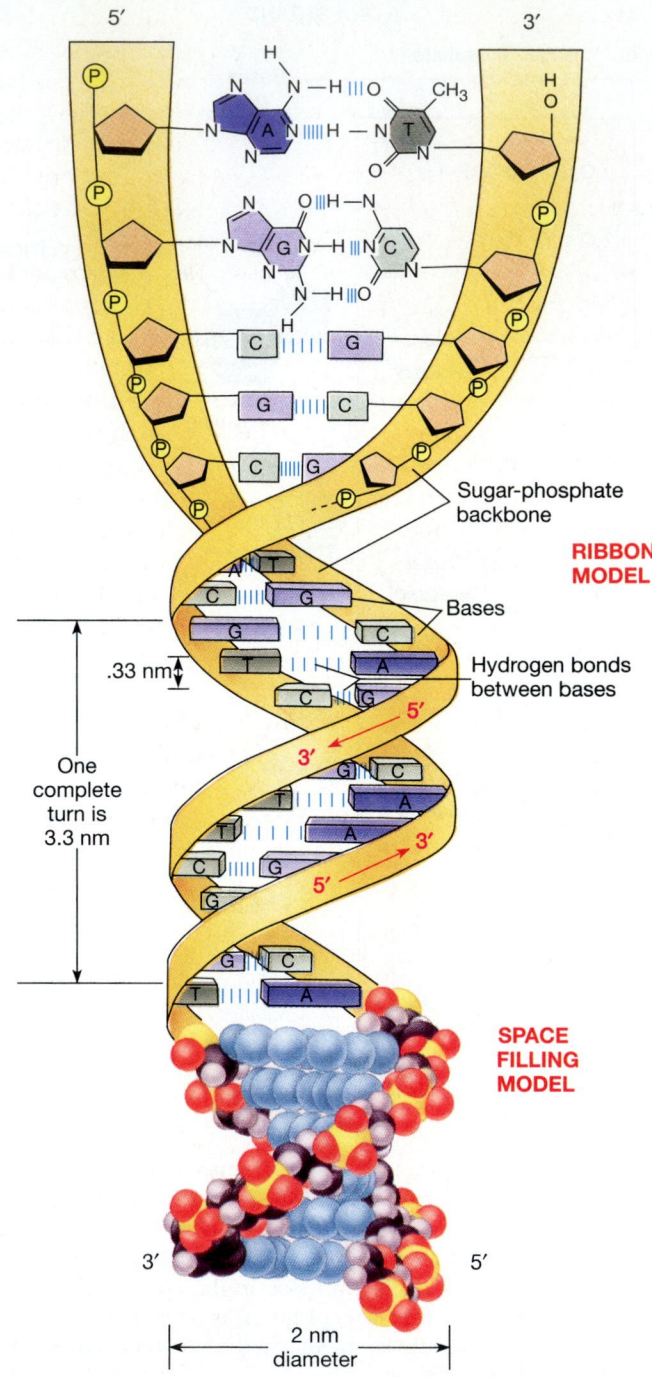

**FIGURE 9–5** The DNA double helix. This diagram depicts four different models of DNA structure. The two strands are antiparallel, meaning that one has its 3' end, and the other its 5' end, at the top of the page. Blue dashes indicate the hydrogen bonds that hold complementary pairs of bases on the two strands together. A full turn of the helix takes ten base pairs.

## 9–C  DNA REPLICATION

Before a cell divides and forms two new cells, its DNA is **replicated,** or duplicated. Each new cell will then receive a complete copy of the original cell's genetic information.

Watson and Crick pointed out that the double-stranded, base-paired structure of the DNA molecule provides a built-in way to copy the genetic information accurately. Because the two strands have complementary base pairs, the nucleotide sequence of each strand automatically supplies the information needed to produce its partner. For example, if part of one strand runs 5′ A—A—T—G—C—C 3′, its partner *must* run 3′ T—T—A—C—G—G 5′.

If the two strands of a DNA molecule are separated, each can be used as a mold, or **template,** to produce a complementary strand. The template and its complement together then form a new DNA double helix, identical with the original molecule. This process is called **semiconservative** ("half-conserved") replication because each new double helix contains one old (conserved) and one newly formed strand. Watson and Crick suggested that this was in fact how DNA replicated. Their paper stimulated a flurry of experiments to test this hypothesis, which was confirmed in a classic experiment by Matthew Meselson and Franklin Stahl in 1958 (*How Do We Know? Semiconservative DNA Replication*).

Meselson and Stahl worked with bacteria, but it is now clear that the DNA of all organisms is replicated semiconservatively. The two strands are separated along the weak hydrogen bonds linking the paired bases, much like opening a zipper. Each strand then serves as the template to make a new strand by linking together nucleotides with complementary bases.

## Proteins of DNA Replication

DNA is replicated by more than 20 enzymes and other proteins acting in concert.

For replication to occur, the double helix must be unwound. The two strands must also be separated by breaking the hydrogen bonds between the paired bases. (For an idea of the complications involved, try unwinding the strands of half a meter of two-ply yarn.) The strands must be held apart, to expose the bases so that each old strand can serve as a template for a new strand. New nucleotide partners can then hydrogen-bond to the bases on each template strand and be linked into a new strand complementary to the template.

Before replication begins, proteins pry the double helix open at specific points called **replication origins. DNA helicase** enzymes then bind and move along the double helix in both directions from the origin, separating the two strands (see Figure 9–7). The two points where helicase enzymes are working at any particular time, and therefore where unwinding and replication are occurring, are called **replication forks.**

The enzyme **DNA polymerase** grasps a template strand exposed by helicase and moves along making a complementary new DNA strand. It adds nucleotides, one by one, to the 3′ end of this new DNA strand (Figure 9–6). To be added to the new strand, a nucleotide must be paired by hydrogen bonding to a base exposed on the opposite, tem-

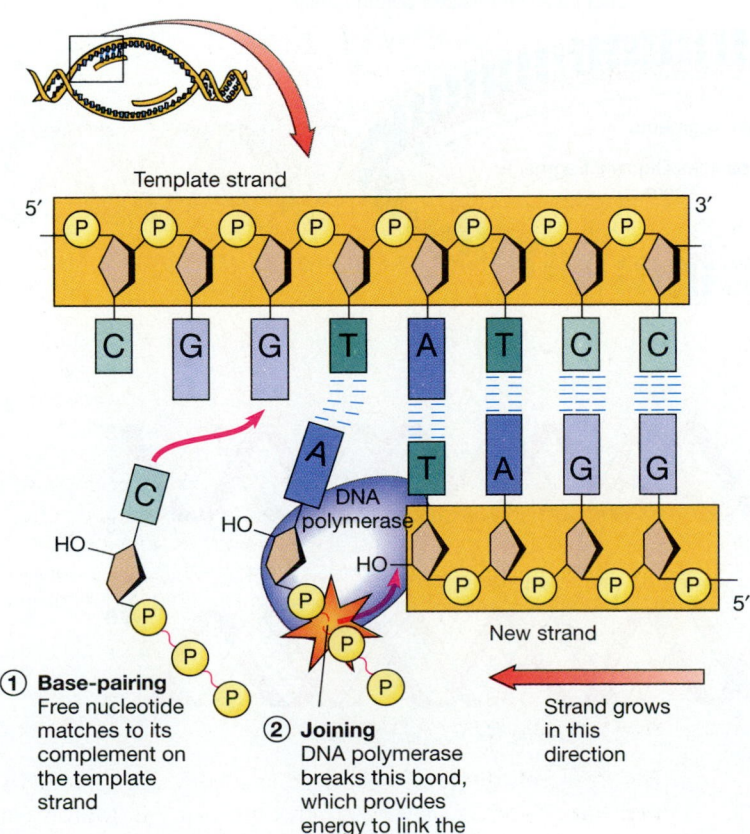

Template strand

**FIGURE 9–6** DNA replication. New nucleotides first base-pair with their complements in the template strand. Then DNA polymerase joins the 3′ end of the new strand to each paired nucleotide in turn. The energy for this reaction comes from hydrolyzing the last two phosphate groups from the nucleotide being added.

① **Base-pairing**
Free nucleotide matches to its complement on the template strand

② **Joining**
DNA polymerase breaks this bond, which provides energy to link the nucleotides

Strand grows in this direction

New strand

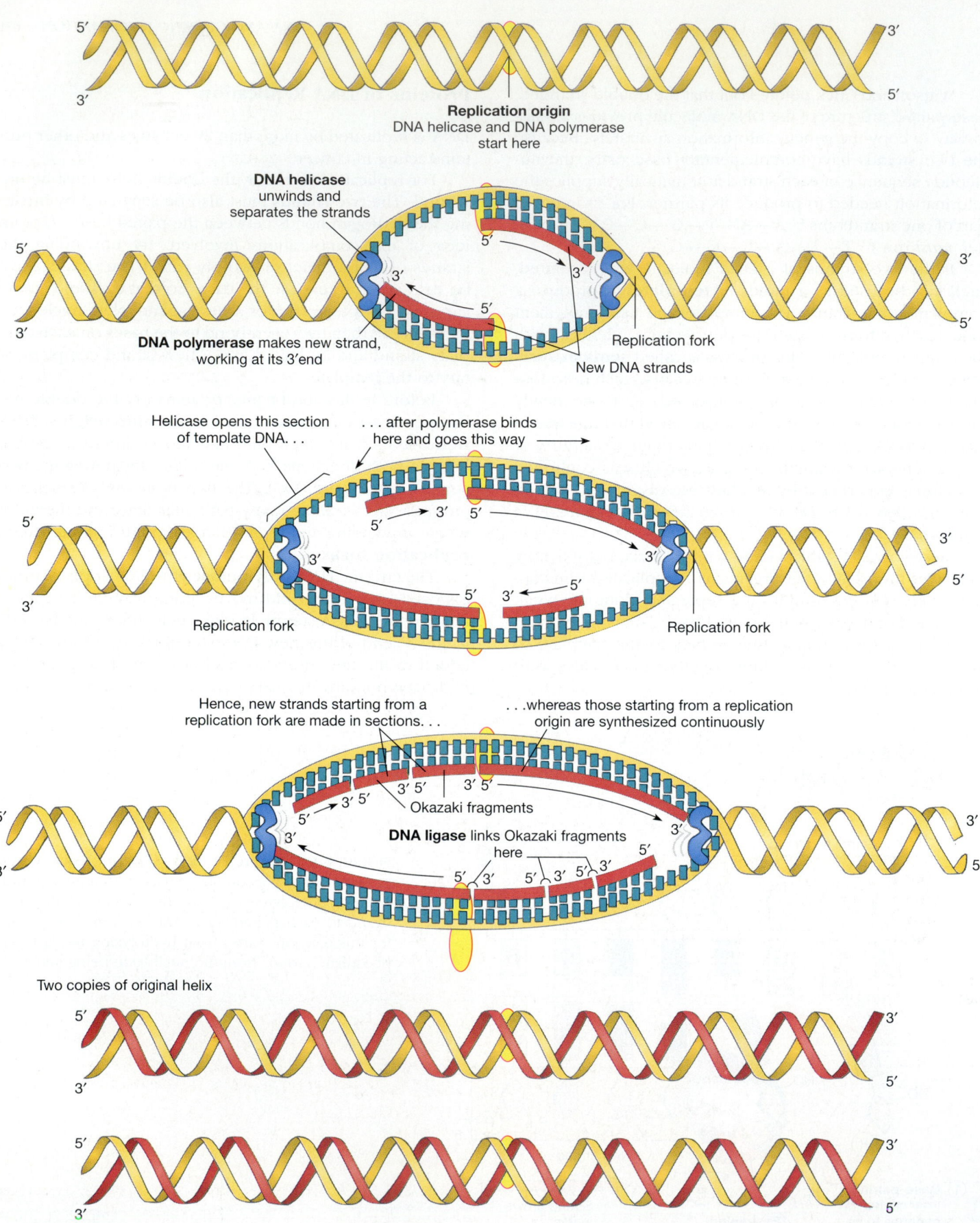

**Replication origin**
DNA helicase and DNA polymerase
start here

**DNA helicase**
unwinds and
separates the strands

**DNA polymerase** makes new strand,
working at its 3′end

Replication fork

New DNA strands

Helicase opens this section
of template DNA. . .

. . . after polymerase binds
here and goes this way

Replication fork

Replication fork

Hence, new strands starting from a
replication fork are made in sections. . .

. . .whereas those starting from a replication
origin are synthesized continuously

Okazaki fragments

**DNA ligase** links Okazaki fragments
here

Two copies of original helix

**FIGURE 9–7**  Replication of DNA.

plate, DNA strand. It must also have three phosphate groups, because the energy required to link the monomer to the growing DNA strand is provided by hydrolyzing the last two of the monomer's three phosphate groups. All four kinds of DNA nucleotides release about the same amount of energy in this reaction.

New nucleotides are added at a rate of about 500 to 1000 per second in bacteria. In eukaryotes, this rate is only about 50 nucleotides per second, perhaps because chromosomal proteins bound to DNA slow the process.

DNA polymerase can work in only one direction, from the 5′ end toward the 3′ end of the new strand it is synthesizing. This poses an interesting problem. The two strands of the DNA double helix are antiparallel (Section 9–B). Therefore, as a helicase moves along, separating the molecule's two strands, it works toward the 3′ end of one strand and the 5′ end of the other (Figure 9–7).

As polymerase moves along the 3′-to-5′ template strand, it makes a continuous new strand, adding nucleotides to the 3′ end of the new strand as it moves away from the replication origin. However, this won't work on the opposite strand because DNA polymerase can't work from the 3′ toward the 5′ end of the strand it is synthesizing. This strand is made in shorter pieces, called **Okazaki fragments** after their discoverer, Reiji Okazaki. These fragments have perhaps 1000 to 2000 nucleotides in bacteria, and 100 to 200 nucleotides in eukaryotes. As helicase opens stretches of template DNA, DNA polymerase assembles these pieces working in its usual (5′-to-3′) direction on the new strand. Then the pieces are joined together by another enzyme, **DNA ligase** (Figure 9–7).

In a prokaryote, the circular DNA double helix is attached to the plasma membrane at one point, which is the replication origin. From this single origin, two replication forks travel in opposite directions around the molecule until they meet. Later, the plasma membrane at the replication origin expands, separating the two new DNA molecules attached to it.

By contrast, in the single, linear DNA molecule of a eukaryotic chromosome, replication may start at as many as 1000 replication origins. From each origin, two replication forks travel away from each other until they meet another replication fork. Having many replication origins enables eukaryotes to compensate somewhat for their low speed of replication.

DNA replication is extraordinarily accurate. DNA polymerase makes very few errors, and most of these are quickly corrected. In bacteria, DNA polymerase "proofreads" the nucleotide it has just added to the new DNA strand. If this nucleotide contains the wrong base, the enzyme backs up, removes the nucleotide, and attaches one with the correct base in its place. A separate protein complex also performs a similar proofreading function. It seems likely that eukaryotes have proofreading mechanisms like both of these prokaryotic ones. Thanks to the accuracy of replication and proofreading, a cell's DNA is copied with less than one mistake per billion nucleotides added to the growing chain. In the bacterium *Escherichia coli,* with a DNA molecule of 4 million base pairs, this translates to less than one copying error in every 250 cells replicating their DNA.

## 9–D  DNA REPAIR

All biological polymers, including DNA, are subject to damage. Any damage to DNA might alter its information content and produce disastrous changes in the cell's proteins. News headlines make us aware that DNA can be damaged by various chemicals and by radiation (Section 9–E). But you might be surprised to learn that DNA is also in danger from the body's own heat and the aqueous environment inside cells. The DNA in each of your cells loses an estimated 5000 purine bases (A and G) each day because body heat breaks the bonds linking them to the sugar deoxyribose.

Although thousands of changes occur in a cell's DNA every day, the cell accumulates only about two or three *stable* changes (mutations) each year. The vast majority of changes are eliminated with remarkable efficiency by a squad of 20 or more kinds of DNA repair enzymes. These enzymes, working together, recognize and remove a damaged area of a DNA strand, replacing it with nucleotides complementary to those on the opposite, undamaged strand (Figure 9–8).

DNA repair depends on the existence of two copies of the genetic information, one in each strand of the double helix. As long as one strand remains undamaged, the repair enzymes can use it as a template to replace a damaged segment in its partner. Most damage is repaired unless both strands are altered beyond recognition at the same time. The genetic material of some very small viruses is single-stranded DNA, which cannot be repaired. These viruses show high rates of genetic change because of damage to their DNA. The double helix is, therefore, vital to genetic stability.

In humans, the rare inherited disease xeroderma pigmentosum results from having defective genes for DNA repair enzymes. As a result, skin cells cannot repair DNA damaged by the sun's ultraviolet rays. People with this condition often develop fatal skin cancer and die by age 30.

Mutations in genes for DNA repair enzymes also contribute to some hereditary forms of colon cancer.

## 9–E  MUTATIONS

**Mutations** are inheritable changes in DNA molecules. They may result from uncorrected errors in replication, from failure to repair damage correctly, or from spontaneous rearrangements as part of the normal function of some segments of DNA (Table 9–2). Whatever the change, the new,

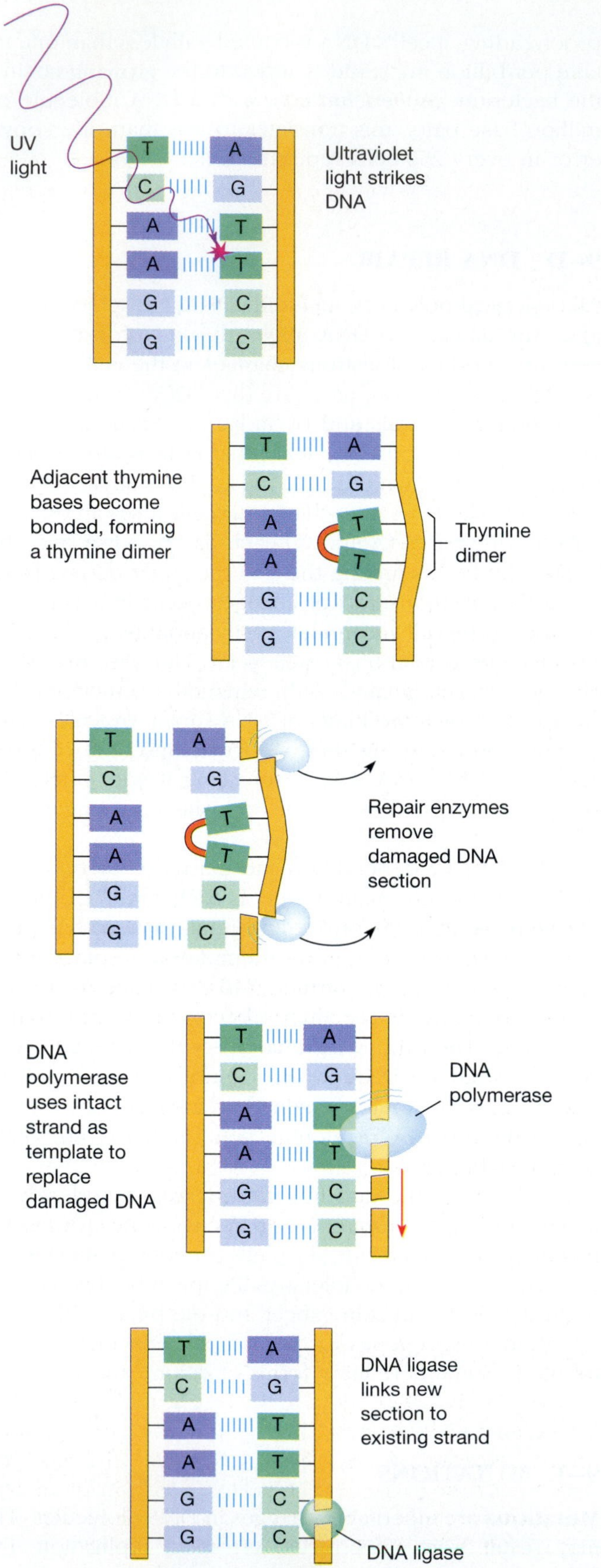

UV light

Ultraviolet light strikes DNA

Adjacent thymine bases become bonded, forming a thymine dimer

Thymine dimer

Repair enzymes remove damaged DNA section

DNA polymerase uses intact strand as template to replace damaged DNA

DNA polymerase

DNA ligase links new section to existing strand

DNA ligase

**FIGURE 9–8**   Repair of a DNA strand damaged by ultraviolet light.

**TABLE 9–2**

**Some Types of Mutations**

**Substitution** of one nucleotide for another (most common)
**Insertion** of one or more nucleotides into DNA sequence
**Deletion** of one or more nucleotides from DNA sequence
**Inversion** of part of nucleotide sequence (so that part of the DNA is "backwards")
**Breakage** of chromosome and loss of fragment
**Attachment** of (part of) one chromosome to another
**Loss** of chromosome(s)
**Extra copy** of chromosome(s)
**Polyploidy** (duplication of entire set of chromosomes)

"wrong" DNA sequence is copied just as faithfully as "right" sequences.

Because a sequence of nucleotides in DNA carries the code for making a protein, a mutation can result in production of a protein with a different amino acid sequence and hence perhaps altered effectiveness. The extent of change in the mutated DNA is not necessarily correlated with the effect it has on the organism. For example, a change of one nucleotide pair for another may have effects so slight as to be undetectable, or so severe as to cause death (Section 10–C).

Although some mutations occur in the course of "business as usual," many others are brought about by **mutagenic agents,** often called **mutagens.** Various kinds of radiation cause mutations (*Essay: Radiation and Cell Division,* Chapter 14). X-rays and radioactive particles may cause breaks in the DNA molecule and have been implicated as causes of some kinds of cancer. Certain chemicals also alter DNA (see Table 11–1). Some are similar to nucleotides and may be accidentally incorporated into DNA. Others react with bases in the DNA strand, changing the chemical groups involved in hydrogen bonding and so changing the nucleotide's base-pairing properties. In either case, when the strand is replicated, the wrong nucleotide is paired to the altered member of the old strand (Figure 9–9).

Mutations are inherited when they are copied during replication and then passed on to a cell's descendants. Mutation in a body cell, called **somatic mutation,** may change the hereditary characteristics of the cell and of body parts made up of that cell's descendants. For example, some cancers are known to be the result of a change of a single base pair, which activates a normal gene to an abnormal degree and so makes the cell and its descendants malignant. Mutations in cells destined to form eggs or sperm **(germ cell mutations)** can be passed on to an organism's offspring. In humans, typical mutation rates for a gene range from 1 to 250 per million eggs or sperm, depending on the gene involved. Some genes mutate more frequently than others.

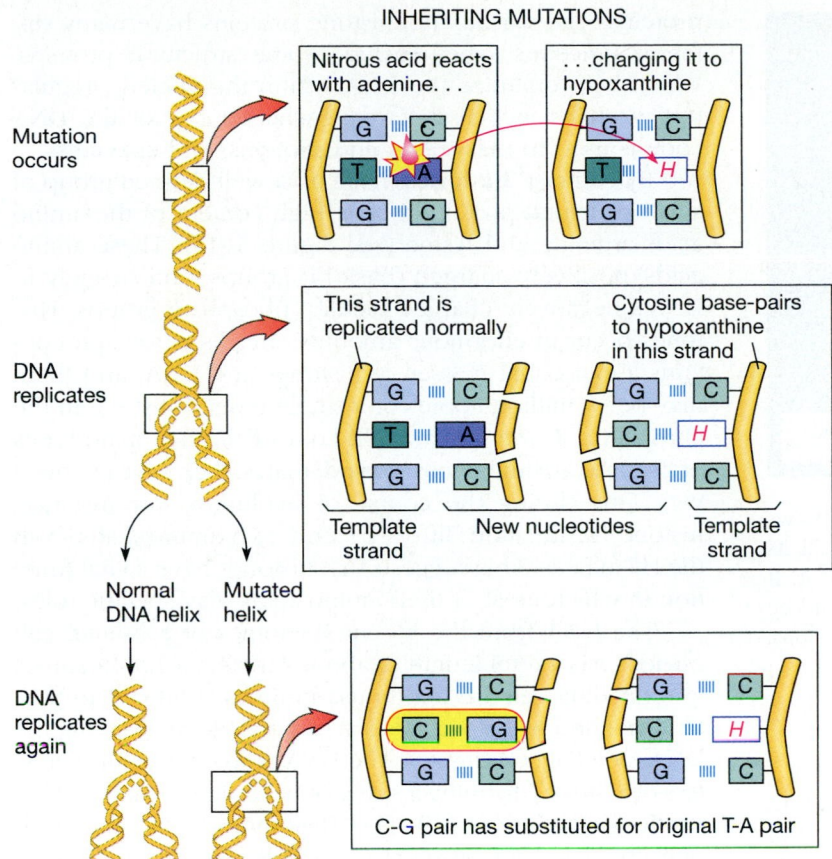

INHERITING MUTATIONS

**Mutation occurs**

Nitrous acid reacts with adenine. . .

. . .changing it to hypoxanthine

**DNA replicates**

This strand is replicated normally

Cytosine base-pairs to hypoxanthine in this strand

Template strand    New nucleotides    Template strand

Normal DNA helix    Mutated helix

**DNA replicates again**

C-G pair has substituted for original T-A pair

**FIGURE 9–9** Mutation: inheritance of an unrepaired change in DNA. A chemically changed base pairs with a different partner during replication. When this DNA is replicated again, the new partner (C) pairs with its usual mate, completing the change to a new pair of bases at that position. This mutation is passed to all future generations of DNA.

Mutations in the genes for DNA repair enzymes may affect how well damage is repaired and hence mutation rate. So, like any other inherited characteristic, mutation rate itself is subject to natural selection. For example, different strains of the bacterium *Escherichia coli* have different mutation rates, due to variations in the structure of one of the proofreading proteins.

A small amount of mutation is an advantage because it produces variation in the genetic material, and genetic variations are the raw material of evolution (Section 1–B). However, the genetic material of modern organisms has resulted from hundreds of millions of years of evolution. Most of the changes that occur now are more apt to harm than to help the delicate balance of living cells and so to be selected against.

Hence, both fidelity of replication and the rare occurrence of mutations are essential features of DNA. Accurate replication reproduces a successful organism's instructions for leading a particular kind of life in a specific environment, whereas mutation gives the potential for innovation.

During the course of evolution, as organisms pass genetic material from generation to generation, mutations arise and are expressed as features of the organism, which in turn are subject to natural selection. If the organism survives, it may pass on its mutated gene to its offspring, and in this way genetic changes may accumulate over many generations. Thus, organisms come to have somewhat different genetic material from that of their ancestors.

Furthermore, two or more different lines of organisms descended from the same ancestors will have similar, but not identical, genetic material, and the closeness of the relationship will be reflected in the degree of genetic similarity. We can now determine the sequence of nucleotides in various genes. By comparing the structure of genes from different organisms, we can sometimes tell how closely they are related and construct a family tree showing their evolutionary relationships.

## 9–F  DNA IN PROKARYOTES

If the circular DNA double helix of the bacterium *Escherichia coli* were cut and stretched into a straight line, it would be about 1.4 mm long, whereas the cell itself is only about 2 $\mu$m long! The circular bacterial DNA is folded many times and occupies a nuclear area about one tenth of the cell's volume (Figure 9–10). This DNA molecule is attached to the plasma membrane at one or more points.

Some bacteria also contain one or several additional, much smaller, circular DNA molecules called **plasmids** (see

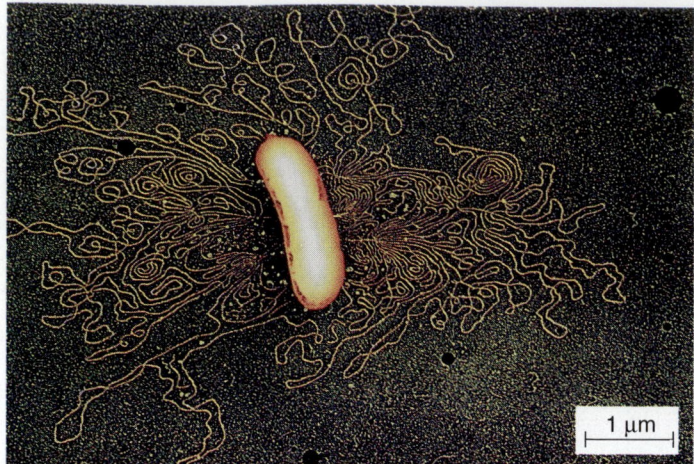

**FIGURE 9–10** Prokaryotic DNA. This electron micrograph shows a broken cell of the bacterium *Escherichia coli,* with part of its single, circular DNA molecule spilling out in loops around it. If the entire DNA were stretched out in a straight line, it would be about a thousand times as long as the bacterium itself.   (Dr. Gopal Murti/Photo Researchers/Science Photo Library)

Figure 13–3). (Plasmids are also occasionally found in yeasts, which are eukaryotes.) Plasmids are widely used in genetic engineering. A bacterium replicates any plasmids it contains at the same time as its main DNA. Scientists take advantage of this activity by introducing artificial plasmids into bacteria in order to obtain many copies of particular genes they contain.

The DNA of mitochondria and chloroplasts, organelles of eukaryotic cells, is very similar to that of prokaryotic cells. For instance, the DNA of these organelles is circular and is not complexed with the proteins that are always found with eukaryotic DNA. (This is part of the evidence that mitochondria and chloroplasts originated as prokaryotes-in-residence within eukaryotic cells [Section 24–H].)

## 9–G  EUKARYOTIC CHROMOSOMES

Eukaryotic cells contain many linear chromosomes. For example, human body cells normally contain 46 chromosomes, with a total of about 6 billion nucleotide pairs divided among them. Each chromosome contains a single DNA molecule extending from one end of the chromosome to the other, but coiled and folded many times. The DNA is associated with various proteins, forming a substance called **chromatin,** which contains roughly equal amounts of DNA and proteins and also much RNA.

The chromosomal proteins may be divided into histones (discussed below) and assorted nonhistone chromosomal proteins. The various nonhistone proteins have many different functions. They include some structural proteins, which help organize the DNA within the nucleus; regulatory proteins, which determine whether or not the DNA code is used to make RNA and proteins; and enzymes.

By contrast, **histones** make up a well-defined group of small structural proteins with a high content of the amino acids arginine and lysine (see Figure 3–16). These amino acids' positively charged (basic) R groups bind strongly to DNA's negatively charged (acidic) phosphate groups. Histones occur in enormous amounts: a cell's chromatin contains about equal masses of histones and DNA, and there may be 60 million copies of a single type of histone in the chromatin of one human cell! Four of the five main types of histones are highly **conserved;** that is, they have changed very little during the course of evolution. For instance, histone H4 in cattle differs by only two amino acids from the H4 in peas. This suggests that histones have a vital function in which most of their amino acids play specific roles.

The DNA from the 46 chromosomes of a human cell nucleus has a total length of about 2 meters. Chromosomes in a dividing cell are ten thousand times shorter than this. Histones and other proteins are responsible for packing the DNA molecules so tightly. The DNA is wound around clusters of histones, forming a string of bead-like particles called **nucleosomes** (Figure 9–11). This shortens the molecule only about tenfold. Most of the nucleosome "beads," in turn, are wound up still more tightly into a supercoiled helical fiber about 30 nm in diameter. This 30-nanometer fiber is still about 200 times longer than the diameter of a cell nucleus. It is further folded into large loops that shorten the chromosome even more. Chromatin packed as tightly as possible is said to be **condensed.** When the DNA is in this state, enzymes apparently cannot get at it. The package must be unwound, at least into a string of nucleosomes, before the DNA can serve as a template for DNA or RNA synthesis.

Stretches of DNA between nucleosomes may bind regulatory molecules that switch genes "on" and "off." Genes are said to be "on" when their code is being used for RNA (and protein) synthesis. Some genes are turned on all the time (except during cell division), but others are not.

### Special Chromosome Areas

Each chromosome contains hundreds or thousands of genes. Losing so much genetic information could spell disaster, and in fact cells with the wrong number of chromosomes often die or develop abnormally, sometimes causing cancer. However, chromosomes have special structures that normally keep them from being lost or destroyed.

The **centromere** ("center part") is a specialized area that provides a handle by which the chromosome can be moved around. This makes it possible for chromosomes to

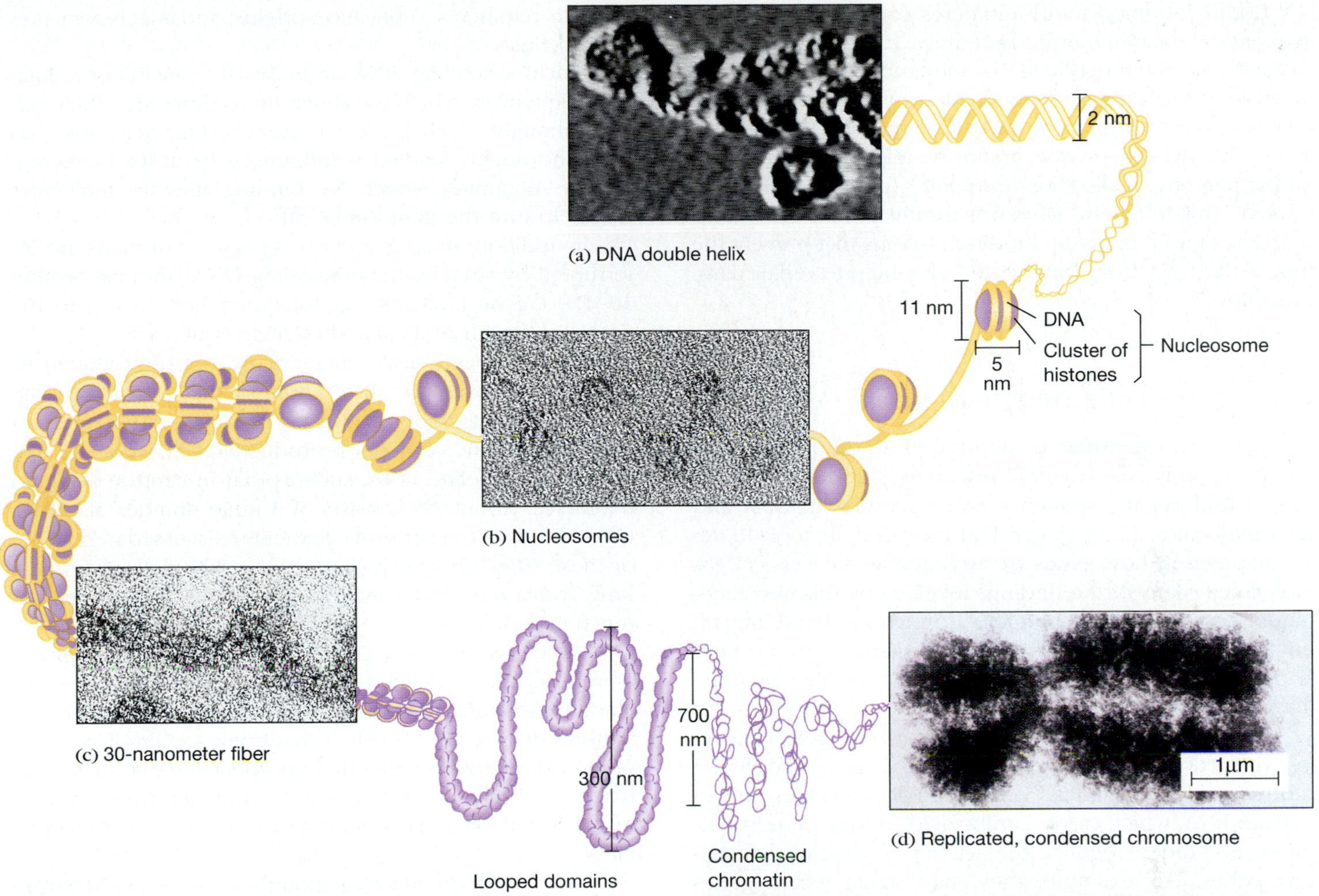

(a) DNA double helix

2 nm

11 nm

DNA

5 nm

Cluster of histones

Nucleosome

(b) Nucleosomes

(c) 30-nanometer fiber

700 nm

300 nm

Looped domains

Condensed chromatin

1 μm

(d) Replicated, condensed chromosome

**FIGURE 9–11** Packing of DNA in eukaryotic chromatin. (a) Scanning tunneling electron micrograph of the DNA double helix. (b) In eukaryotes DNA is wound around clusters of histone proteins, forming nucleosomes. (c) Nucleosomes wind up tightly into fibers 30 nm wide, which in turn form large loops packed together in a highly condensed chromosome in a dividing cell (d).   (a, Wigbert Siekhaus, Lawrence Livermore National Laboratory, University of California, Berkeley; b, courtesy of D. E. Olins and A. L. Olins; c, Barbara Hamkalo; d, D. W. Fawcett/Photo Researchers)

be separated into two identical groups, one for each new nucleus, before the cell divides. In a cell about to divide, the centromere appears as the narrow "waist" between two chromosomes replicas, called **chromatids.** As its name suggests, it often occurs near the center of the chromosome, but it can lie anywhere along the chromosome's length (Figure 9–12).

   **Telomeres** ("end parts"), in effect, tie off the loose ends of the eukaryotic chromosome's linear DNA molecule and so stabilize the chromosome against fraying and attack by DNA-hydrolyzing enzymes. If this DNA is removed, the chromosomes disintegrate. Hence, telomeres are vital to the chromosome's survival.

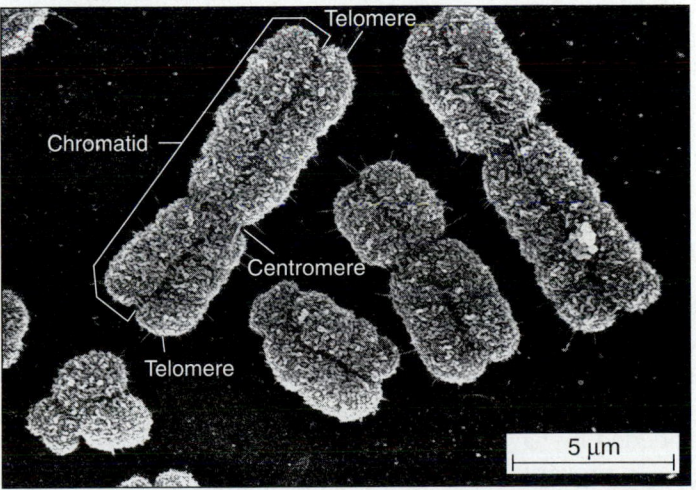

Telomere

Chromatid

Centromere

Telomere

5 μm

**FIGURE 9–12** Replicated chromosomes. In a cell about to divide, each chromosome consists of two chromatids, each containing a complete replica of the chromosome's DNA molecule, highly condensed as shown in Figure 9–11d. The chromatids are still joined at the centromere, and each has telomeres at both ends.   (Biophoto Associates)

Both centromeres and telomeres consist of short DNA sequences repeated hundreds of times, called satellite DNA. In each case, the repetitive DNA sequences produce a characteristic three-dimensional structure that binds specific proteins. For example, the repetitive nucleotide sequence T—T—A—G—G—G occurs in the telomeres of humans and apparently in all other vertebrates (animals with backbones). This telomeric DNA apparently protects the chromosome at least partly by binding proteins that prevent the tips of the DNA from fraying and also help to repair damaged tips.

## 9–H    STRUCTURE OF THE GENOME

An individual's **genome** is the total of all the DNA in any one of its body cells. In 1977 researchers developed methods to find out the sequence of nucleotides in DNA and RNA molecules (Chapter 13). This provided the tools to describe precisely how genes are arranged within a cell's DNA molecules. Many of the findings revealed by this new technique, however, were completely unexpected and are still unexplained. It is oddly true that in many ways we now know less about how genes are organized within a genome than we thought we did in 1974!

Some parts of the genome are structural genes, carrying instructions for making proteins. Other genes dictate the sequence of nucleotides in the RNA of ribosomes and in transfer RNA, which carries amino acids during protein synthesis. Still others regulate the activity of some of their fellow genes. For example, they may signal where genes begin and end, or tell which kinds of cells should make the protein encoded by a gene. (For instance, red blood cells make hemoglobin but other cells generally do not.) These areas are important in controlling protein synthesis. The human genome is now estimated to contain 100,000 or more genes.

Most genes occur in only one copy per reproductive cell (egg or sperm). Hence an individual's genome contains two copies of the gene (or variations derived from mutation), one from each parent. However, every eukaryotic cell carries many copies of the genes for ribosomal RNA and histones. For example, the human genome contains 400 copies of ribosomal RNA genes and 30 to 40 copies of histone genes. This is adaptive because the cell needs large numbers of these molecules, and their production is speeded by having multiple copies of these genes.

In prokaryotes, genes coding for RNA and protein synthesis make up most of the genome. However, in many plants and animals, including humans, such coding DNA plus the regulatory signal sites for these genes makes up less than 20% of the genome. Biologists are only beginning to learn the functions of the rest of the genome.

At least a small fraction of this **noncoding DNA** forms important parts of chromosome structure, such as the cen-

tromere, telomeres, replication origins, and attachment sites for proteins.

Much noncoding DNA undoubtedly consists of regulatory sequences, which are much more numerous than was once thought. Each human protein-coding gene that has been thoroughly studied is influenced by at least five regulatory sequences, which are binding sites for molecules that help turn the gene on or off.

In addition, most eukaryotic genes for proteins are interrupted by stretches of noncoding DNA (introns, Section 10–D). A gene's introns may total anywhere from 2 to 100 times the length of its actual coding sequences!

Eukaryotic cells also contain many repeating sequences of DNA. For example, about 30% of the human genome consists of nucleotide sequences in DNA that are present in more than one copy per reproductive cell. About 10% of our DNA is satellite DNA, such as that in centromeres and telomeres. About 9% consists of a huge number of copies of two kinds of transposable elements (discussed next). One of these (the *Alu* sequence) is about 300 nucleotide pairs long and is repeated a million times in each body cell. Its function is unknown.

Mutations in which DNA contains abnormal numbers of repeats are associated with some genetic diseases, including some breast, bladder, and colorectal cancers; Huntington's disease; and fragile-X syndrome, a hereditary form of mental retardation. Researchers speculate that these mutations in numbers of repeats occur when defective enzymes lose their place while operating on sections of repetitive DNA.

Much noncoding DNA is thought to be descended from ancient viruses and essentially parasitic, with no function except to be replicated along with the rest of the genome.

### Mobile Genetic Elements

Geneticists once envisioned the genome as a fixed number of genes arranged in specific sequences on the chromosomes, but we now know that in many cases genes move around. This has important consequences. First, in the process of moving, some DNA may be duplicated or lost, changing the genome's size as well as the position of some of its genes. Second, a change in a gene's position is important because genes often affect their neighbors' activity.

In the 1940s, Barbara McClintock first cast doubt on the picture of a static genome. She discovered that corn (maize) plants have **transposable elements** (alias "jumping genes"), segments of DNA that can move from one part of the genome to another.

McClintock studied Indian corn, the kind with different-colored kernels often used for autumn decorations. She found that many of the color variations are caused by unstable mutations that occur when transposable elements insert into a gene and later leave. For example, the gene *C*

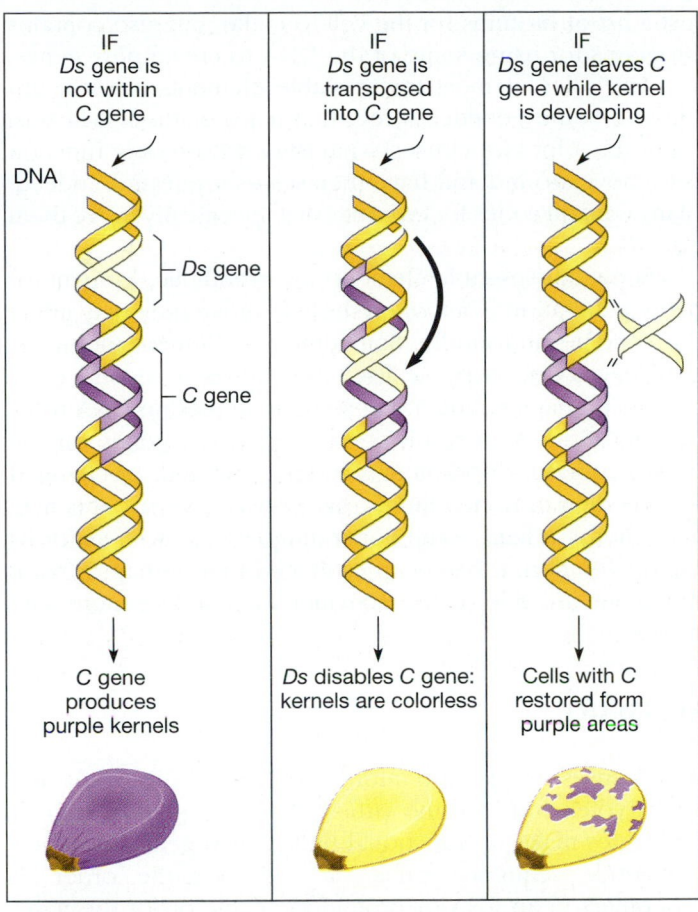

(a)

(b)

**FIGURE 9–13** Transposable elements in corn. (a) If the transposable element *Ds* (disabler gene) is separate from *C, C* is expressed, and a purple kernel is produced. If *Ds* is inserted into *C, C* is disabled, and therefore no pigment is produced. Hence, the kernel formed is colorless. The purple spots on some of the kernels in (b) result when another transposable element, *Ac* (activator gene), causes *Ds* to move out of *C* during development of the kernel. (b) The mixture of solid-color, colorless, and spotted kernels shows that *C, Ds,* and *Ac* behave differently during the development of individual kernels.

normally makes colored (purple) kernels, but sometimes the transposable element *Ds* (disabler gene) drops into the middle of *C,* and a colorless (white) kernel develops. After staying for a variable time, *Ds* may leave and *C* resumes functioning. If *C* is restored while the kernel is developing, the cell and its descendants form a purple spot. The earlier this occurs, the more descendants the cell has by the time the kernel matures, and the larger the purple area (Figure 9–13). Although *Ds* is a transposable element, it cannot move by itself. Another transposable element, *Ac* (activator), must be present in the genome for *Ds* to move around.

For many years geneticists thought transposable elements were peculiar to corn, with no general significance. In the 1970s, however, it became clear that jumping genes are both widespread and important, and in 1983 McClintock received a Nobel Prize for her work.

Another major group of mobile genetic elements is some viruses. A virus consists of genetic material packed into a protein shipping crate that carries the viral genome from one cell to another. When a virus invades a cell, its genes may be spliced into the cell's genome. The splicing enzyme action was first described in bacteriophage lambda (λ), which invades *Escherichia coli* (Figure 9–14).

Some bacteria have small mobile elements called **transposons,** which can move from the cell's main DNA molecule into a smaller plasmid and back again (Section 9–F). One of the best-studied bacterial transposons is about 5000 nucleotides long and carries three genes. Two of these code for the enzymes that move the transposon, and the third is a passenger gene which makes any bacterium containing it resistant to the antibiotic ampicillin. This transposon may be replicated and passed from one bacterium to another, rapidly making a large population of bacteria resistant to ampicillin.

It is now thought that both viral genes and mobile genetic elements are scattered through the genomes of all cells. Furthermore, a cell often contains many copies of a single element. In fact, at least 10% of a higher eukaryote's genome (including our own) is thought to consist of mobile genetic elements. Each is occasionally activated to move by unknown signals.

Viruses and mobile genetic elements can move between individuals not only of the same species, but also of different species. Because they may carry genes from one individual and incorporate them into the genome of another, they provide a means of transferring genes between organisms that cannot breed together. Geneticists are finding evidence that genes are moved between species much more often than anyone had imagined.

McClintock's transposable elements play a role in controlling gene activity during the embryonic development of corn plants. It is possible that transposable elements play similar roles in other organisms.

Higher animals have a different kind of DNA rearrangement, which occurs in specific DNA sequences that

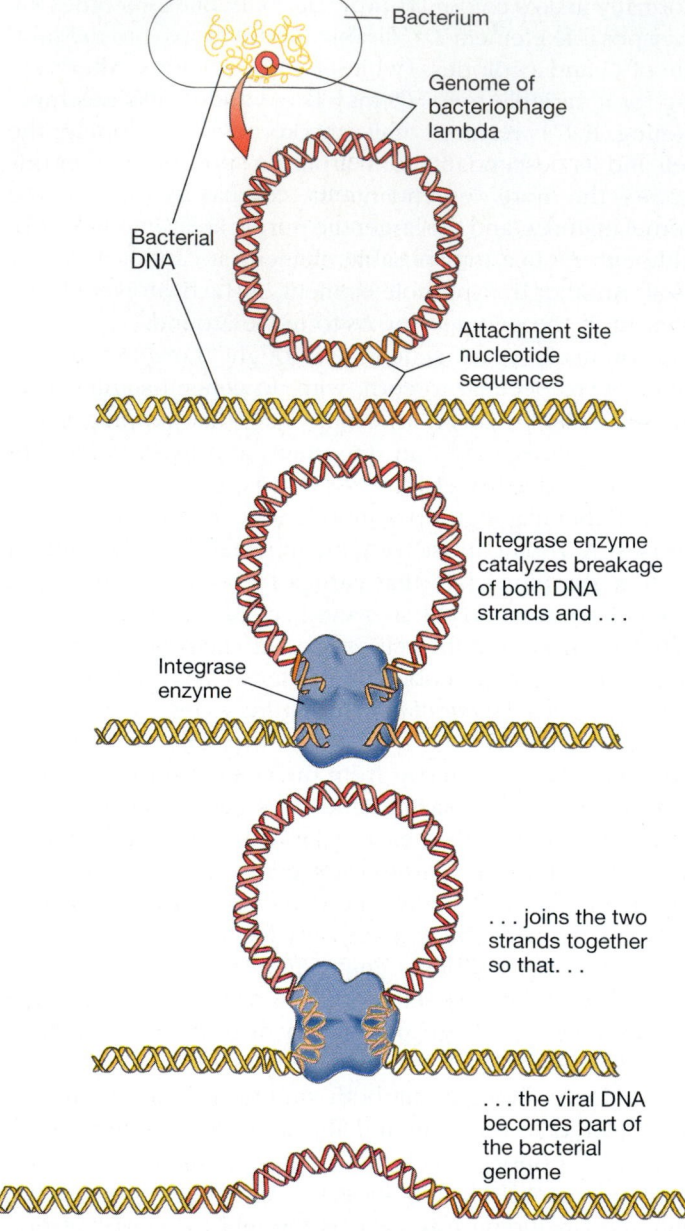

Bacterium

Genome of
bacteriophage
lambda

Bacterial
DNA

Attachment site
nucleotide
sequences

Integrase enzyme
catalyzes breakage
of both DNA
strands and . . .

Integrase
enzyme

. . . joins the two
strands together
so that. . .

. . . the viral DNA
becomes part of
the bacterial
genome

**FIGURE 9–14**    The DNA of bacteriophage lambda being inserted into the DNA of a bacterial cell. Occasionally, after the phages have multiplied in a bacterial cell, one of the usually linear phage DNA molecules forms a circle that can become part of the bacterial genome. The viral genome codes for the integrase enzyme that integrates the phage genome into the bacterial DNA. Integrase recognizes specific recombination sequences of nucleotides in the DNA of both genomes and breaks and rejoins both double helices at these sequences.

help create the thousands of different antibody genes of the immune system (Section 34–E).

The existence of transposable elements and DNA rearrangement mechanisms that play a vital role in the organism's normal development shows that the genome is not just a list of proteins for the cell to make, but also contains programs for using some of the DNA to create new genes.

The roles of most transposable elements are still unknown. Some geneticists feel that a lot of them will turn out to be a lot like viruses—parasites whose only function is to move around and have themselves replicated. Indeed, many contain codes for enzymes that specifically move them elsewhere.

Some transposable elements apparently lie dormant for many generations, but when they do move they may affect the genome profoundly. Sometimes an element multiplies and leaves one copy of itself behind as a second copy moves, perhaps taking along some neighboring DNA to its new location. As a result of such activities, genes can be moved around, duplicated, lost, split, merged, or changed from coding to noncoding forms, and vice versa. This may account for at least some noncoding DNA as well as for its rapid evolution. It has been estimated that perhaps 50% of mutations are due to the activities of mobile genetic elements.

## SUMMARY

DNA carries the genetic information in all eukaryotes and prokaryotes and in some viruses. Some segments of DNA molecules make up functional units called genes. A gene's nucleotide sequence carries a code for the order of monomers in an RNA or protein molecule or for the regulation of other genes.

Since an organism's genetic information governs the possible range of its structure, function, and behavior, it is vital for each cell and each individual to pass on copies of its DNA to its offspring. The structure of DNA makes the molecule easy to copy faithfully. At the same time, the rare occurrence of mutations in DNA makes evolution possible.

1. The evidence that DNA is the genetic material in all organisms, from bacteria to oak trees, came from several lines of inquiry:

   ♦ DNA is the substance that transfers genetic information from one cell to another during bacterial transformation.

   ♦ When a phage takes over the genetic machinery of a bacterium, only its DNA enters the cell; its protein coat remains outside.

   ♦ In most plants and animals, all of the body cells of individuals of the same species contain the same amount of DNA; reproductive cells contain half this amount of DNA. The cells of members of different species contain different amounts of DNA.

   ♦ The DNA of all members of any one species has the same proportions of A and T, and of C and G, which differ from the proportions in other species.

*Text continues on page 204*

## SEMICONSERVATIVE REPLICATION OF DNA

The DNA double helix structure described by Watson and Crick has features that immediately suggested how it might be replicated: the two strands could easily be "unzipped" from each other along the weak hydrogen bonds linking the paired bases. Each row of the zipper's exposed base-pair "teeth" then serves as the template to make a new strand by linking together nucleotides with complementary bases. Biologists soon tested this idea by experiment.

In 1958 Matthew Meselson and Franklin Stahl published the first convincing evidence of how DNA replicates. Their work is a classic model of a scientific experiment.

First, Meselson and Stahl considered three alternative hypotheses, any one of which would account for the observations of DNA replication then available (Figure 9–A):

1. **Conservative replication.** A DNA double helix acts as a template for a completely new double-stranded molecule. The original DNA molecule is conserved

*(Continued)*

| HYPOTHESIS: | PREDICTED DNA REPLICATIONS | | |
|---|---|---|---|
| | Parent cell's DNA | DNA of first new generation | DNA of second new generation |
| **CONSERVATIVE** Parental strands remain intact and act as template for new double helix | | — Both strands — $^{15}$N — Both strands — $^{14}$N | |
| **SEMICONSERVATIVE** Parental strands separate and each acts as template for its new complement | | $^{15}$N strand and $^{14}$N strand | |
| **DISPERSIVE** Parental strands break up into short segments, which act as templates for new strands. Daughter helices have varying amounts of parental nucleotides | | $^{15}$N - $^{14}$N mixed strands | |

**FIGURE 9–A**   The alternative hypotheses of DNA replication tested by Meselson and Stahl. DNA of parent cells is labeled with $^{15}$N, the heavy isotope of nitrogen, whereas all new DNA made during replication of these parental molecules contains the normal isotope $^{14}$N.

*(continued)*

intact and goes into one new cell, the new molecule into the other.

2. **Semiconservative replication.** The two strands of the original molecule are separated by disrupting the hydrogen bonds between the base pairs. Each strand then serves as a template for formation of one new strand. New hydrogen bonds between the nitrogenous bases link one old and one new strand into a complete double helix. At cell division, each new cell inherits a DNA molecule that is a **hybrid,** consisting of one new and one old (template) strand.

3. **Dispersive replication.** The parental DNA is broken up into short segments used as templates for the formation of new segments, which are then somehow joined together. At cell division, each new cell inherits a DNA molecule with some old and some new nucleotides in each strand, in variable proportions.

In a beautifully simple series of experiments, Meselson and Stahl disproved the first and third hypotheses and provided strong support for the theory of semiconservative replication. (Remember that a hypothesis can never be proved, because we can never be sure we have taken all possible factors into account; however, it is possible to *disprove* a hypothesis if the experimental results do not agree with predictions of what should happen if the hypothesis is true.)

To distinguish between "old" and "new" DNA, Meselson and Stahl used two different isotopes of nitrogen: $^{14}N$, the more common isotope, and $^{15}N$, a heavier isotope. First, Meselson and Stahl grew bacteria in a nutrient medium containing $^{15}N$ for several generations, so that virtually all the bacteria had DNA containing $^{15}N$. Next, they transferred the bacteria to a nutrient medium containing $^{14}N$. They then removed samples of cells after there had been enough time to produce one, two, and then three new generations. The cells of each generation were broken open, and their DNA was purified. The DNA from each generation could then be analyzed to find out the distribution of old $^{15}N$ and new $^{14}N$ DNA.

To do this, Meselson and Stahl used a technique called **equilibrium density gradient centrifugation,** in which DNA of different masses, and hence different densities, ends up at different levels in the centrifuge tubes. They placed each batch of isolated DNA in a solution of the salt cesium chloride in a centrifuge tube and then spun all the tubes in a centrifuge (see Figure 5–6). As the tubes spun, the salt formed a density gradient, with the highest density at the bottom of the tube. At the same time, DNA floated to the level in the tube where its density was in equilibrium with the salt solution. Hence, the highest-density DNA, containing $^{15}N$, migrated closer to the bottom than did $^{14}N$ DNA. After spinning the tubes, the investigators could actually see bands of DNA of different densities in the tubes.

Figure 9–B shows the patterns of bands Meselson and Stahl found. In the parental generation, grown in $^{15}N$, all the DNA contained $^{15}N$. However, when these bacteria were then switched to $^{14}N$ medium, the DNA of the first new generation settled in a position between that of pure $^{14}N$ DNA and pure $^{15}N$ DNA, showing that it contained both $^{14}N$ and $^{15}N$.

This result disproved the possibility that replication is conservative: if it were, we would expect two separate bands, one of new $^{14}N$ DNA and one of parental $^{15}N$ DNA. Thus, the first hypothesis was eliminated by the experimental evidence.

The DNA from the second new generation of bacteria permitted Meselson and Stahl to distinguish between the two remaining hypotheses. This DNA formed two distinct bands, each with about half of the DNA. One band contained both $^{14}N$ and $^{15}N$, as in the first new generation, and the other contained only $^{14}N$. This result disproved the hypothesis of dispersive replication, which would produce one diffuse band lying between the two actually found. However, this pattern was consistent with semiconservative replication.

Therefore, Meselson and Stahl concluded that of their three hypotheses, only semiconservative replication was supported by the experimental evidence.

A year earlier, in 1957, J. Herbert Taylor had found evidence for semiconservative replication in eukaryotes. He used dividing plant root cells, which replicate DNA rapidly, and gave them radioactive thymine (T) nucleotides to label their chromosomal DNA. Taylor also treated the cells with colchicine, a chemical that prevents cell division but not DNA replication. Thus, he could tell which generation of chromosomes he was looking at by counting the chromosomes in each cell and dividing by the usual number per cell. He could also tell where the radioactive T nucleotides were by autoradiography: after squashing the cells on glass slides, he covered them with photographic film, and radioactive emissions darkened the film against the parts of chromosomes containing radioactive T.

In cells that incorporated radioactive T during one round of replication, each chromatid contains one copy of the original chromosome's DNA. Both chromatids contained radioactive label, which appeared as dark dots on the autoradiograph:

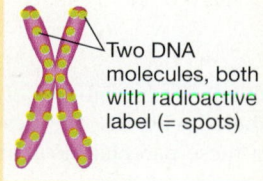

Two DNA molecules, both with radioactive label (= spots)

*(Continued)*

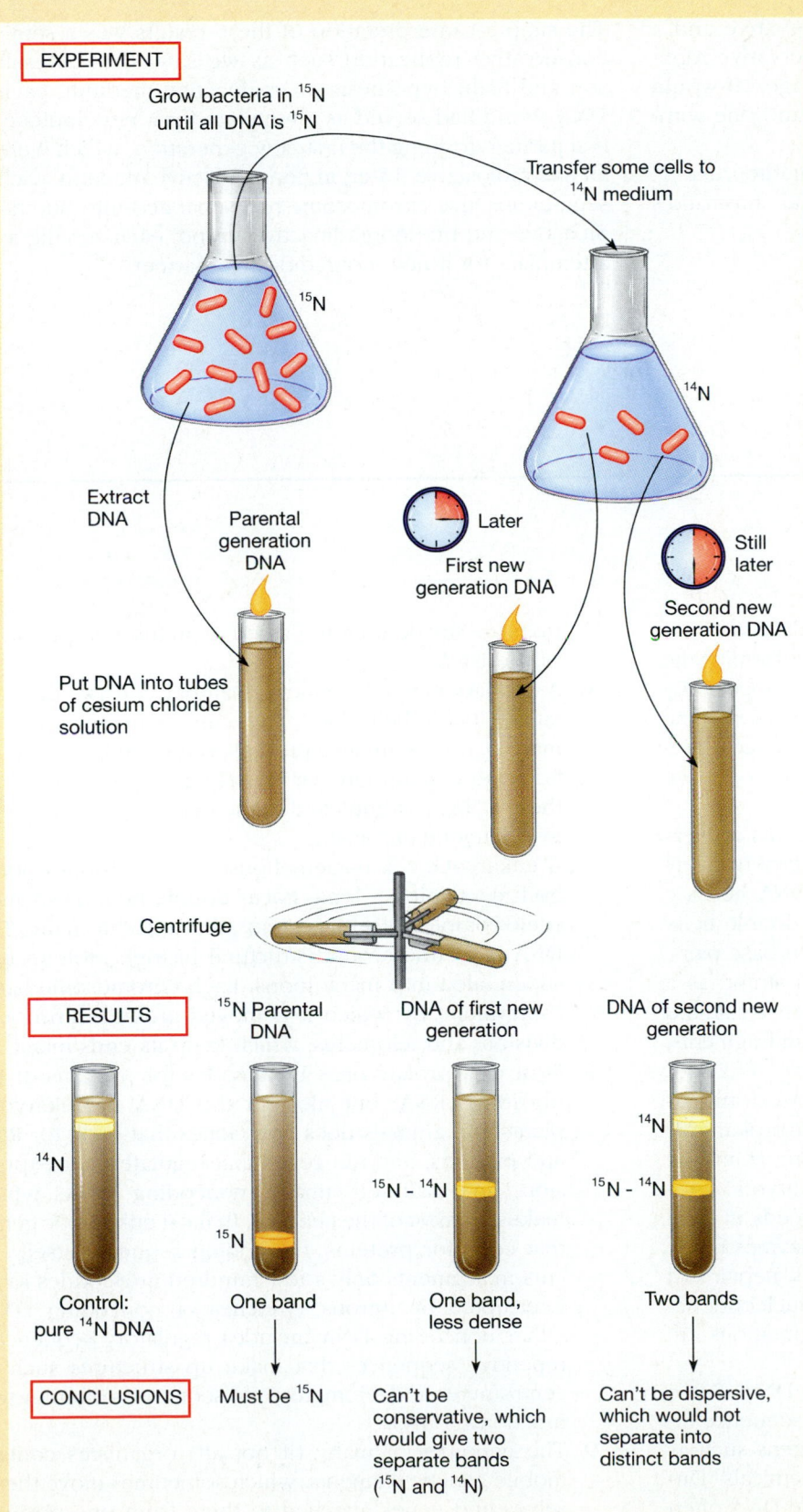

**EXPERIMENT**

Grow bacteria in ¹⁵N until all DNA is ¹⁵N

Transfer some cells to ¹⁴N medium

¹⁵N

¹⁴N

Extract DNA

Parental generation DNA

Later

First new generation DNA

Still later

Second new generation DNA

Put DNA into tubes of cesium chloride solution

Centrifuge

**RESULTS**

¹⁵N Parental DNA

DNA of first new generation

DNA of second new generation

¹⁴N

¹⁵N

¹⁵N - ¹⁴N

¹⁴N

¹⁵N - ¹⁴N

Control: pure ¹⁴N DNA

One band

One band, less dense

Two bands

**CONCLUSIONS**

Must be ¹⁵N

Can't be conservative, which would give two separate bands (¹⁵N and ¹⁴N)

Can't be dispersive, which would not separate into distinct bands

**FIGURE 9–B** Meselson and Stahl's experiment. The pattern of DNA bands in centrifuge tubes for the first new generation ruled out conservative replication, while that in the second new generation ruled out a dispersive mechanism (see Figure 9–A).

*(continued)*

This is consistent with both a semiconservative and a dispersive mechanism, but not with conservative replication, which could be ruled out at this stage. (It would have produced one chromatid with label and one without.)

A second round of replication, in nonradioactive T, produced replicated chromosomes with one chromatid's DNA containing radioactive T and one not:

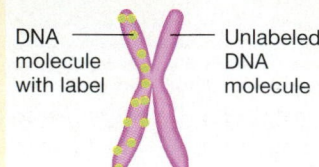

DNA molecule with label — Unlabeled DNA molecule

The simplest interpretation of these results was a semiconservative replication such as we saw in the Meselson and Stahl hypothesis: in radioactive medium, each DNA strand had served as a template for a new radioactive partner, forming the first new generation, which were all half-radioactive. Later, in nonradioactive medium, each semi-radioactive chromosome had separated into one radioactive and one nonradioactive strand, each serving as a template for a new nonradioactive partner.

## SUMMARY *(Continued from page 200)*

2. The DNA molecule is a double helix, with two antiparallel sugar-phosphate backbone strands forming the sides of a twisted ladder. The strands are connected by crosswise "rungs" consisting of the base pairs adenine and thymine, or guanine and cytosine, with each base hydrogen-bonded to its complement on the opposite strand.

3. DNA contains the information that dictates its replication. The DNA molecule is replicated semiconservatively by the cooperation of several enzymes. DNA helicase enzymes separate the two strands of the double helix by disrupting the hydrogen bonds between base pairs. DNA polymerase enzymes then use each strand as a template for the formation of a complementary strand of DNA. Some new strands are assembled in fragments, which are joined together by DNA ligase. Enzymes proofread the newly formed DNA and correct any errors of replication. The circular DNA of a prokaryotic cell is replicated in both directions, starting from one replication origin. Replication of linear eukaryotic chromosomes starts at many origins and proceeds in both directions until neighboring replication enzymes meet.

4. Damaged DNA is also repaired by enzymes. Repair and proofreading mechanisms ensure that the nucleotide sequence of DNA is very stable and that mutations are rare.

5. Mutations are inheritable changes in the DNA. Some mutations are local changes in nucleotide sequence resulting from unrepaired damage by mutagens such as x-rays, ultraviolet radiation, or various chemicals. Duplication, loss, or movement of parts of the DNA often result from the activities of the various mobile genetic elements found in all genomes. In either case, muta-

tions are (by definition) passed on in future replications of the DNA.

6. A prokaryotic cell's genetic material consists of a circular, double-helix DNA molecule. Sometimes one or more plasmids are also present. Mitochondria and plastids also contain circular molecules of DNA. None of these DNAs is complexed to the histone proteins found in eukaryotic chromatin.

7. A eukaryotic cell nucleus houses many chromosomes, each containing a long, linear double helix DNA molecule complexed with proteins, forming chromatin. The DNA is wound around structural histone proteins and supercoiled into many loops. Each chromosome has a centromere, by which it is moved around during cell division, and telomeres, which keep its ends intact.

8. Most of a prokaryote's DNA codes for proteins or for ribosomal RNA, but most of the DNA in eukaryotic plants and animals does not. Genes that code for RNA and proteins, and the genes that regulate their expression, are scattered among noncoding DNA, which makes up most of the genome. In most eukaryotic genes that code for proteins, the coding sequence itself occurs in segments only a few hundred nucleotides long, interrupted by introns, stretches of noncoding DNA. Other noncoding DNA includes regulatory sequences, repetitive sequences that make up structures such as centromeres and telomeres, and sequences of unknown function (if any).

9. The genomes of many (if not all) organisms contain mobile genetic elements, which sometimes move themselves and genes attached to them from one position in the genome to another. Their movement may multiply, divide, or delete other genes.

## Self-Quiz

1. DNA is believed to be the genetic material because:
   a. all the body cells of an individual seem to have identical amounts and compositions of DNA, whereas reproductive cells have half the amount of DNA found in body cells
   b. the proteins are the same from cell to cell in an individual, but the DNA differs; thus, the DNA must be the material that makes different tissues different
   c. DNA is the largest type of macromolecule found in living organisms
   d. DNA is found in the cell nucleus
2. A nucleotide consists of:
   a. A, G, T, and C
   b. nitrogenous bases
   c. a sugar, a phosphate group, and a nitrogen-containing ring compound
   d. a sugar-phosphate backbone
3. Draw the complementary strand for this DNA template:

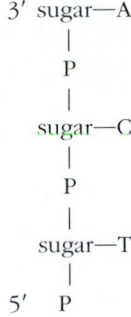

```
3′ sugar—A
       |
       P
       |
     sugar—C
       |
       P
       |
     sugar—T
       |
5′     P
```

4. In a DNA molecule, the sugars:
   a. bond covalently to phosphate groups
   b. bond covalently to nitrogenous bases
   c. bond to nitrogenous bases by hydrogen bonds
   d. bond to both phosphate groups and nitrogenous bases by covalent bonds
   e. bond to phosphate groups by ionic bonds and to nitrogenous bases by hydrogen bonds

5. The number of adenine bases in a DNA molecule equals the number of thymine bases because:
   a. whenever DNA polymerase places a thymine base into a new DNA strand, it always puts an adenine directly after it
   b. a DNA strand consists of alternating adenine and thymine bases
   c. DNA polymerase places adenine in the new strand opposite thymine in the template strand, and thymine in the new strand opposite adenine in the template strand
   d. DNA contains equal numbers of each of the four nitrogenous bases
6. A mutation may result from:
   a. addition or loss of one or more nucleotides in a DNA strand
   b. change of one nucleotide to another
   c. loss of an entire chromosome
   d. part of a DNA strand getting turned around "backwards"
   e. all of the above
7. What is the advantage of the fact that nitrogenous bases in DNA form hydrogen bonds to each other, rather than covalent bonds?

Tell whether each of the following would be found in the nuclei of your own cells, in the cells of the intestinal bacterium *Escherichia coli,* or both:

_____ 8. Circular DNA molecules
_____ 9. DNA in 46 linear molecules
_____ 10. DNA with only one replication origin
_____ 11. Genes carrying instructions for proteins and ribosomal RNA
_____ 12. Histones closely bound to DNA
_____ 13. DNA containing transposable elements
_____ 14. More coding than noncoding DNA

15. The human genome contains repetitive DNA in the form of:
    a. genes coding for ribosomal components
    b. genes coding for histones
    c. satellite DNA near the centromeres
    d. telomeres
    e. all of the above

## Questions for Discussion

1. Why is it necessary for eggs, sperm, and pollen to contain only half the amount of genetic material found in the other cells of the body?
2. Not all organisms have identical DNA in every cell. In some roundworms, some body cells lack entire chromosomes. How is this advantageous? How can these worms pass on a complete set of genetic information to their offspring?
3. Given a fresh chicken liver, how would you determine the amount of DNA per chicken liver cell (as opposed to per gram of tissue)?
4. Why is it necessary to limit the amount of x-rays a person is exposed to over a given period of time? Which organs must be especially well shielded from x-ray exposure?

## Suggested Readings

Avery, O. T., C. M. MacLeod, and M. McCarty. "Studies on the chemical nature of the substance inducing transformation of pneumococcal types." *Journal of Experimental Medicine* 79:137, 1944. The paper that convinced many people that DNA was the genetic material.

Dickerson, R. E. "The DNA helix and how it is read." *Scientific American,* December 1983. Fairly advanced level; excellent illustrations.

Fedoroff, N. V. "Transposable genetic elements in maize." *Scientific American,* June 1984.

Lewin, R. "A naturalist of the genome." *Science* 222:402, 1983. Describes Barbara McClintock's work on transposable elements.

Loewy, A., et al. *Cell Structure and Function: An Integrated Approach,* 3d ed. Philadelphia: Saunders College Publishing, 1991. Chapters 10 and 11 provide good background material about DNA experiments mentioned in this chapter.

McCarty, M. *The Transforming Principle: Discovering that Genes Are Made of DNA.* New York: W. W. Norton, 1986. An important book in the history of biology. McCarty's autobiographical account of the discovery that Griffith's pneumococcal transforming principle was DNA.

Meselson, M., and F. W. Stahl. "The replication of DNA in *E. coli.*" *Proceedings of the National Academy of Sciences (U.S.)* 44:671, 1958. The classic paper demonstrating semiconservative replication in DNA.

Moyzis, R. K. "The human telomere." *Scientific American,* August 1991.

Murray, A. W., and J. W. Szostak. "Artificial chromosomes." *Scientific American,* November 1987.

Radman, M., and R. Wagner. "The high fidelity of DNA duplication." *Scientific American,* August 1988.

Watson, J. D. *The Double Helix.* New York: Atheneum, 1968. A personal story of the discovery of DNA structure.

Watson, J. D., and F. H. C. Crick. "Molecular structure of nucleic acids. A structure of deoxyribose nucleic acid." *Nature* 171:737, 1953. Probably the shortest Nobel Prize–winning paper in biology. It is less than three pages.

# CHAPTER

## 10

# RNA and Protein Synthesis

## OBJECTIVES

*When you have studied this chapter, you should be able to:*

1. List three differences between DNA and RNA.
2. Describe the genetic code and explain the evidence that it is a triplet code.
3. Given a DNA coding strand and a table of codons, determine the complementary mRNA strand, the codons that would be involved in peptide formation from that mRNA sequence, and the amino acid sequence that would be translated.
4. Describe the roles of DNA, mRNA, tRNA, spliceosomes, ribosomes, and amino acids in protein synthesis.
5. Sketch a transfer RNA molecule and indicate where the anticodon and aminoacyl attachment sites are.
6. Describe the initiation of protein synthesis and the three steps in elongation of a polypeptide.

---

The genetic information in DNA determines the inherited characteristics of an organism. But how is this genetic information expressed as fur or feathers, chlorophyll, metabolic enzymes, or any other inherited character?

In 1909 physician Archibald Garrod suggested that genes control the synthesis of enzymes. He studied what he called "inborn errors of metabolism," inherited diseases or conditions, such as albinism and cystinuria, caused by the absence of particular enzymes (Section 16–C). If each gene causes production of one enzyme, then damage to a gene could result in an individual born without the enzyme controlled by that gene. This would explain these inherited diseases.

Experiments by George Beadle and Edward Tatum in the 1940s supported this idea. They studied strains of the pink bread mold *Neurospora* with various **mutations,** inheritable changes in their genetic material. Beadle and Tatum found that each mutant had lost the ability to produce one enzyme. We now know that the production of all proteins, not just enzymes, is controlled by genetic information in DNA.

By this time biologists were coming to realize that DNA is the genetic material. DNA and proteins are polymers with specific sequences of monomers. People started to suggest that the order of nucleotide monomers in DNA determines the order of amino acid monomers in polypeptides and therefore in proteins.

Biochemists were also finding that the other kind of nucleic acid, ribonucleic acid (RNA), is synthesized rapidly in cells that produce a lot of protein, such as oocytes—cells that develop into eggs (Figure 10–1). This suggested that

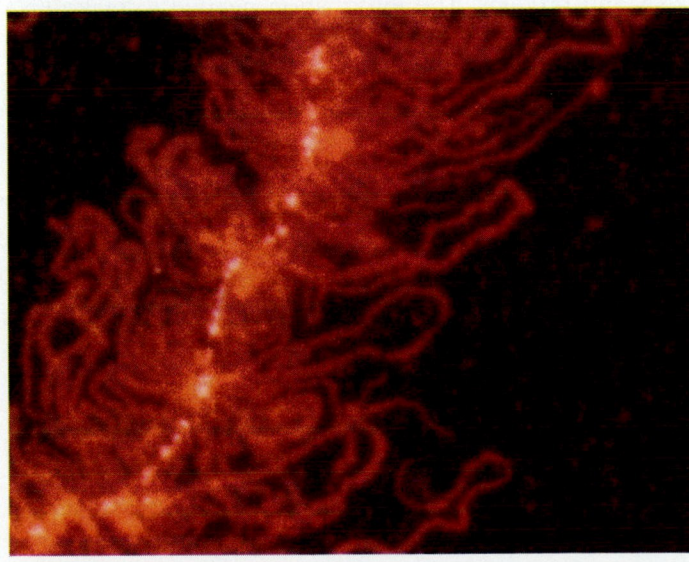

**FIGURE 10–1** RNA synthesis. Part of a lampbrush chromosome in an oocyte from a newt, showing some of the many loops of RNA and DNA where RNA is being synthesized. The white dots are inactive DNA; the red loops are areas of DNA where RNA synthesis is taking place. (M. B. Roth and J. G. Gall)

RNA also plays a role in protein synthesis. In a eukaryotic cell, DNA remains in the nucleus, whereas RNA is found both in the nucleus and in the cytoplasm, where proteins are synthesized. It seemed likely that at least some of the RNA acts as a messenger, carrying genetic information from nuclear DNA into the cytoplasm. We now know that this is in fact the case.

DNA serves as a template for the synthesis of a complementary strand of RNA, by a process much like the synthesis of a new strand of DNA during DNA replication (Section 9–C). The order of nucleotides in DNA dictates the order of nucleotides in RNA. RNA synthesis is called **transcription** ("written across") because it rewrites the genetic message coded in DNA, in the form of a complementary RNA molecule.

The nucleotide sequences of some RNA molecules, in turn, determine the order in which amino acids are joined to form proteins by ribosomes in the cytoplasm. The conversion of the genetic information carried by these RNA molecules into the amino acid sequence of a protein is known as **translation** because the information is converted from the nucleotide "language" of nucleic acids into the amino acid "language" of proteins. The transfer of information from DNA to RNA, or from DNA to RNA to proteins, is **gene expression.**

Important as genes are, it has always been rather difficult to define a gene, partly because our understanding of the term keeps changing. A **gene** is an inherited length of DNA with a particular function. For most of the genes discussed in this chapter, that function is carrying the information needed to produce an RNA molecule or a protein. Genes that code for proteins are called **structural genes.** But there are also other kinds of genes. For instance, some regulatory genes bind specific proteins that hold the chromosome together or control the activity of other genes.

In this chapter, we consider how genetic information is transcribed and translated into protein structure during protein synthesis.

## KEY CONCEPTS

- Proteins are made on ribosomes using genetic information from DNA. The sequence of nucleotides in DNA determines the sequence of amino acids in proteins.

- The genetic information for protein synthesis is carried from DNA to ribosomes by RNA molecules.
- DNA contains instructions for making other kinds of RNA that cooperate in making proteins.

## 10–A OVERVIEW OF PROTEIN SYNTHESIS

Proteins are made by cooperation between many kinds of RNA and proteins. The information needed to make RNA and protein molecules is carried in the DNA. Francis Crick described what is known as the "central dogma" of molecular biology: the "genetic code" for a protein is transcribed into RNA, which is then translated into the sequence of amino acids in a protein:

$$\text{DNA} \xrightarrow{\text{Transcription}} \text{RNA} \xrightarrow{\text{Translation}} \text{Protein}$$

**Messenger RNA (mRNA)** carries the code for the sequence of amino acids in a polypeptide to **ribosomes,** structures composed of protein and RNA, where protein synthesis takes place. **Transfer RNA (tRNA)** molecules bring the necessary amino acids to the ribosome, where they are joined to the growing polypeptide one by one (Figure 10–2). When the polypeptide is complete, it is released. One or more polypeptides are folded up and joined into a functional protein.

## 10–B RNA

Like DNA, RNA molecules are long polymers made up of nucleotides. Each nucleotide subunit consists of a sugar molecule, a nitrogenous base, and a phosphate group (see Figure 9–3). A nucleotide's phosphate group is bonded to the 3' carbon of the sugar in the next nucleotide, forming the sugar-phosphate backbone of a nucleic acid.

RNA differs from DNA in several respects (Figure 10–3):

1. The sugar in RNA is ribose, whereas that in DNA is deoxyribose, which contains one less oxygen atom, hence the names ribonucleic acid and deoxyribonucleic acid.
2. RNA and DNA contain different sets of nitrogenous bases. Both contain adenine (A), guanine (G), and cytosine (C), but RNA contains uracil (U) instead of thymine (T). Like thymine, uracil base-pairs with adenine.
3. RNA usually consists of a single strand of nucleotides, whereas DNA is usually double-stranded, with two complementary chains of nucleotides wound into a double helix.

In some RNA molecules, the single strand develops secondary structure: it folds back on itself, and complementary sequences of nucleotides base-pair, forming double-stranded sections known as **hairpins** (see Figure 10–3). The loops at the ends of hairpins contain unpaired nucleotides. Hairpins give these RNA molecules a definite structure necessary for their function in protein synthesis. This structure also makes them stable and resistant to RNA-

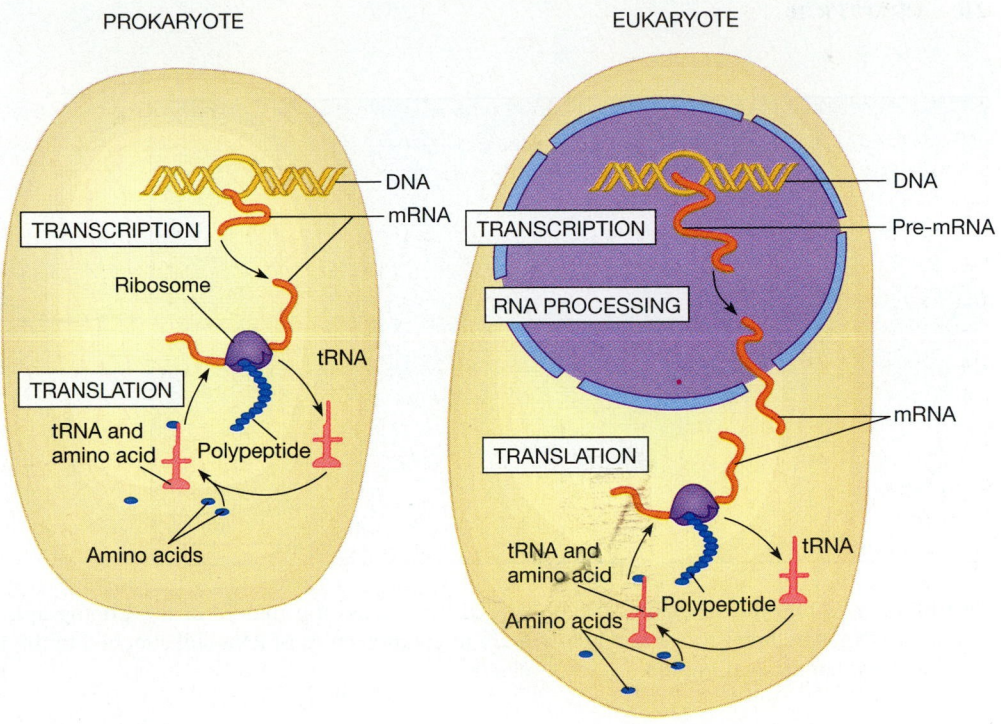

**FIGURE 10–2** The flow of genetic information. These diagrams show the main difference between protein synthesis in prokaryotes and eukaryotes: in eukaryotes, RNA is processed in the nucleus after it is transcribed.

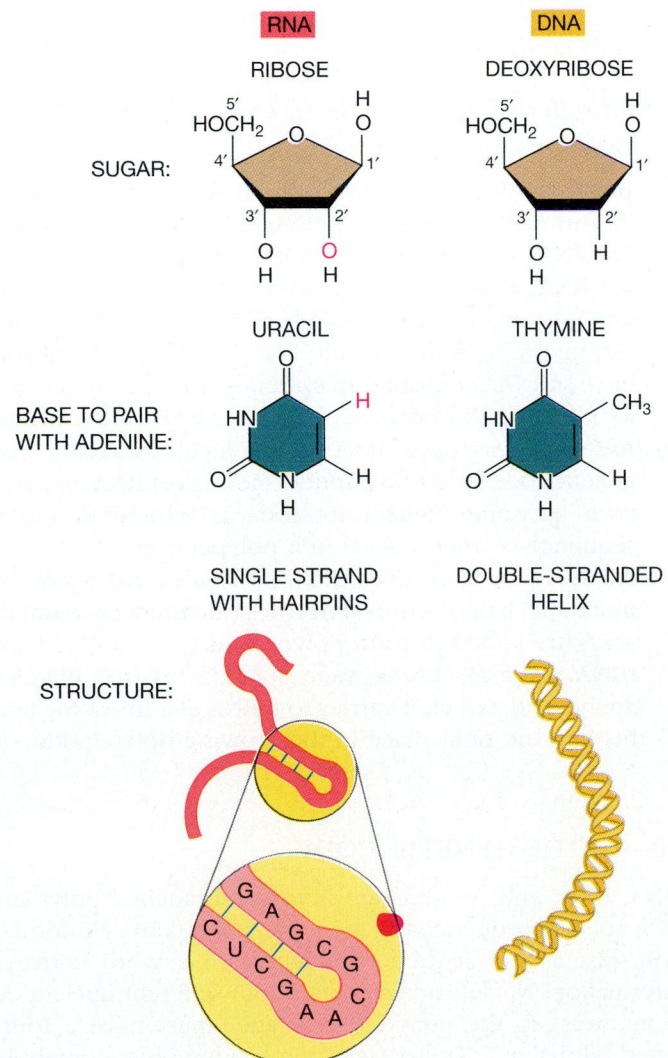

**FIGURE 10–3** Differences between RNA and DNA.

hydrolyzing enzymes. In contrast, messenger RNA, which does not usually have a stabilizing secondary structure, is broken down after being used briefly for protein synthesis.

## Transcription of DNA into RNA

The transcription of DNA into RNA begins when the enzyme **RNA polymerase** binds to a sequence of nucleotides on DNA called the **promoter.** The two strands of DNA are separated. One strand serves as the template for the formation of a complementary strand of RNA. RNA polymerase moves along this DNA strand, making the RNA strand by linking the complementary ribonucleotides (containing A, C, G, or U). RNA polymerase works in the 3′-to-5′ direction along DNA, assembling RNA in the 5′-to-3′ direction (Figure 10–4).

This process is much like the replication of DNA (see Figure 9–6). However, there is one important difference: whereas DNA replication, once begun, usually copies all the DNA in the cell, RNA synthesis transcribes only selected portions of the DNA. When RNA polymerase reaches a termination signal on the DNA, it leaves the DNA, and the RNA also detaches.

In prokaryotes, there is only one type of RNA polymerase, but in eukaryotes there are three. RNA polymerase II transcribes genes that will be translated into proteins. RNA polymerase I and RNA polymerase III transcribe genes for other kinds of RNA. Each polymerase recognizes a different promoter on the DNA.

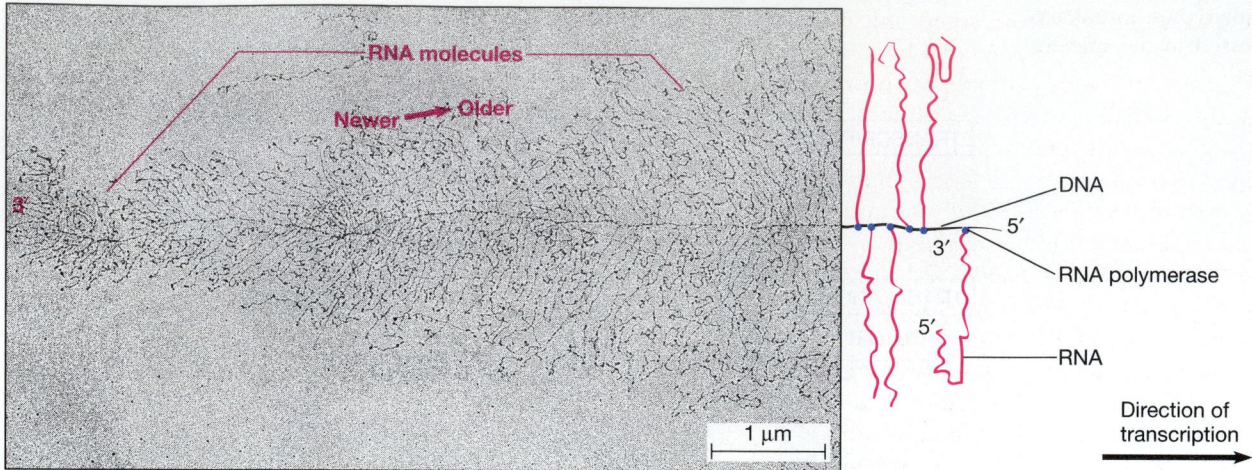

**FIGURE 10-4** Gene transcription. The horizontal line across the middle of the photograph is a strand of DNA. The lines above and below it are many molecules of RNA still attached to the RNA polymerases that are making them. (O. L. Miller)

## Types of RNA

Research since the 1970s makes it increasingly obvious that RNA participates in many aspects of protein synthesis once believed to be the province of DNA and proteins. These

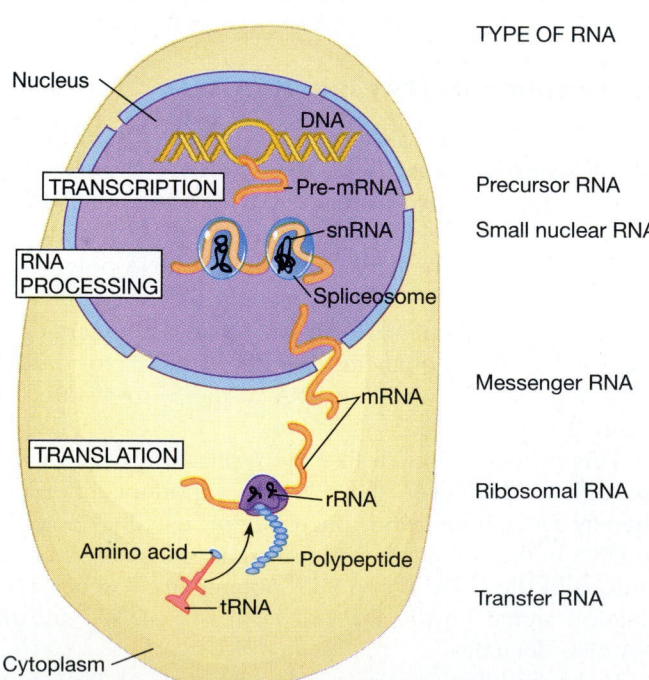

**FIGURE 10-5** The functions of different RNAs in protein synthesis in eukaryotes. All the RNAs are made in the nucleus, many of them as pre-RNAs, which are processed by spliceosomes. Mature mRNA, tRNA, and ribosome subunits then move into the cytoplasm and take part in protein synthesis.

are the different types of RNA involved in protein synthesis and illustrated in Figure 10-5:

1. **pre-RNA.** Precursor-RNA, any newly transcribed RNA that must be processed before it becomes functional. Pre-RNAs include pre-mRNA, the precursor to messenger RNA, as well as precursors of other types of RNA.
2. **snRNA.** Small nuclear RNAs, which take part in converting pre-RNA into mature RNA molecules. In eukaryotes, snRNAs are found in **spliceosomes,** granules that look like small ribosomes but never leave the nucleus.
3. **mRNA.** Messenger RNA, a molecule carrying the genetic code for a polypeptide. Messenger RNA attaches to a ribosome, where its code is translated into a sequence of amino acids in a polypeptide.
4. **rRNA.** Ribosomal RNAs, RNA molecules that make up more than half of a ribosome, the structure where amino acids are joined to form polypeptides.
5. **tRNA.** Transfer RNAs, each specific for one kind of amino acid, which it carries to mRNA at a ribosome and fits into the right place in the growing polypeptide.

## 10-C   THE GENETIC CODE

DNA, RNA, and proteins are linear, unbranched polymers with specific sequences of monomers that convey information, just as the sequence of letters in a word conveys information. Nucleic acids contain four different nucleotide monomers, so the genetic "language" must have a four-letter "alphabet." These four nucleotides must somehow make up at least 20 different "words," representing the 20 amino acids used in protein synthesis.

The words of the genetic code cannot be only one letter long, because in that case there would be only four possible code words, and proteins could contain only four different amino acids. Similarly, the words cannot be two nucleotides long, because four letters arranged in all possible combinations of two gives only $4^2 = 16$ different code words, still not enough to specify 20 different amino acids (Figure 10–6). The four nucleotides arranged in triplets, however, produce $4^3 = 64$ different code words, more than enough to produce a unique code word for each amino acid. The smallest theoretical size for a code word in DNA is, therefore, three nucleotides.

Francis Crick and others tested this triplet code theory by adding different numbers of nucleotides into the DNA of a bacteriophage virus. They reasoned as follows: if a code word consists of three nucleotides, and assuming it is read from one end with no "punctuation marks," then introducing one or two nucleotides into the middle of a gene will change its coded message after that point. For example, if one or two nucleotides containing guanine (G) are added, the DNA message

CAT—CAT—CAT

becomes

CA**G**—TCA—TCA—T

or

CA**G**—**G**TC—ATC—AT

However, introducing three nucleotides should merely create a short disruption, after which the message will read like the original version:

CAT—**GGG**—CAT—CAT

or

CA**G**—**GG**T—CAT—CAT

or

C**GG**—**G**AT—CAT—CAT

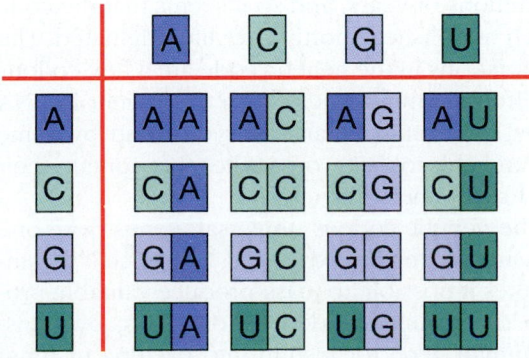

|  | A | C | G | U |
|---|---|---|---|---|
| A | AA | AC | AG | AU |
| C | CA | CC | CG | CU |
| G | GA | GC | GG | GU |
| U | UA | UC | UG | UU |

**FIGURE 10–6** Two-letter code words: the four different kinds of nucleotides in RNA can be arranged in pairs to form these 16 combinations.

Experiments bore out this prediction. When a string of three extra nucleotides was added into the DNA, a slightly altered but functional protein was produced. Additions of one, two, or four nucleotides changed all subsequent code words, and so the protein ended in an abnormal string of amino acids. This usually made the protein, and hence the bacteriophage, unable to function.

In the 1960s, biochemists worked out the genetic code by making artificial mRNAs with known sequences of nucleotides. When these RNAs were mixed with transfer RNAs, amino acids, ribosomes, enzymes, and other substances needed for protein synthesis, they were translated into peptides. Marshall Nirenberg and Heinrich Matthaei reported that RNA containing only uracil nucleotides coded for the formation of a polypeptide containing only the amino acid phenylalanine. They reasoned that the mRNA code word for phenylalanine must be UUU. The DNA code for phenylalanine must be the complement of this: AAA. By 1965 all the amino acid code words had been worked out. In 1968 Nirenberg and H. Gobind Khorana received a Nobel Prize for this work.

The genetic code words, or **codons,** appear in Table 10–1. From this table, note several features of the genetic code:

1. The codons shown are the code words found in messenger RNA, not in DNA. The DNA code triplets are the complements of those shown. For example, the codon GGG (bottom right corner), specifying the amino acid glycine, is transcribed from the complementary DNA code CCC.

2. Three of the 64 triplets do not code for amino acids: UAA, UAG, and UGA are *Stop* codons, signaling the end of a polypeptide.

3. The code is **degenerate,** meaning that most amino acids are specified by more than one codon. This degeneracy is advantageous. For one thing, it minimizes the effects of a mutation, which is a change in the sequence of nucleotides. If the code were not degenerate, 20 codons would code for amino acids, and 44 would code for nothing. Probably, they would act as *Stop* codons. Therefore, most mutations would lead to *Stop* codons, which would end the polypeptide prematurely. Shortening a polypeptide usually leads to inactive proteins, whereas substituting one amino acid for another may be harmless. A mutation in which one nucleotide is changed to another is a **point mutation** (see Figure 9–9). With the degenerate code, most point mutations will exchange one codon for another codon that also specifies an amino acid.

4. A codon's third base is often less specific than the first two. For instance, all of the codons for the amino acid proline have CC as the first two bases. From Table 10–1, you can see that a point mutation that changes the third nucleotide in a codon often will not change the amino acid placed into the polypeptide at that spot.

**T A B L E  1 0 – 1**

## Codons Found in Messenger RNA

### Instructions

1. Find the letter of the first base of the codon in the column at the left.
2. Go across this row until you are in the column headed by the letter of the second base.
3. Then find the third base, marked at the far right of the table.
4. In each box, the words in black are the amino acids specified by the codons in blue.

| First Base (5') | Second Base | | | | Third Base (3') |
|---|---|---|---|---|---|
| | **U** | **C** | **A** | **G** | |
| **U** | UUU UUC ] Phenylalanine<br>UUA UUG ] Leucine | UCU UCC UCA UCG ] Serine | UAU UAC ] Tyrosine<br>UAA  STOP†<br>UAG  STOP† | UGU UGC ] Cysteine<br>UGA  STOP†<br>UGG  Tryptophan | U C A G |
| **C** | CUU CUC CUA CUG ] Leucine | CCU CCC CCA CCG ] Proline | CAU CAC ] Histidine<br>CAA CAG ] Glutamine | CGU CGC CGA CGG ] Arginine | U C A G |
| **A** | AUU AUC AUA ] Isoleucine<br>AUG  Methionine (START)* | ACU ACC ACA ACG ] Threonine | AAU AAC ] Asparagine<br>AAA AAG ] Lysine | AGU AGC ] Serine<br>AGA AGG ] Arginine | U C A G |
| **G** | GUU GUC GUA GUG ] Valine | GCU GCC GCA GCG ] Alanine | GAU GAC ] Aspartic acid<br>GAA GAG ] Glutamic acid | GGU GGC GGA GGG ] Glycine | U C A G |

*The codon AUG initiates synthesis of a polypeptide and calls for methionine as the first amino acid.

†The three STOP codons signal positions where the ribosome stops reading and terminates the polypeptide chain.

5. The genetic code contains no punctuation or spacer that signals the beginning or end of a codon. This means the code must be read from a particular starting point, the beginning of the "reading frame" for the message. Otherwise the whole sequence will be read incorrectly. For instance, the RNA sequence UCUAGAGCUA will produce the amino acid sequence Serine—Arginine —Alanine if read from left to right. But if the reading of the RNA sequence starts at the second nucleotide (C) instead of at the beginning, it will produce the completely different amino acid sequence Leucine—Glutamic acid—Leucine. Mutations that force the code to be read from the wrong place are called **frameshift mutations** because they shift the reading frame of the message.

6. The starting point for a reading frame is the initiation codon AUG. This also codes for the amino acid methionine, so every newly formed polypeptide starts with methionine. This does not mean that all proteins start with methionine. The methionine (and sometimes other amino acids) may be removed after the polypeptide is synthesized.

The genetic code is essentially universal, a legacy of the relatedness of all forms of life. The major groups of organisms have had separate evolutionary histories for hundreds of millions of years, and so it seems that the code must have been established shortly after life originated. The only known variations in the genetic code are a few codons that specify different amino acids when they occur in RNA synthesized by some unicellular eukaryotes and by some mitochondria and chloroplasts, organelles that contain their own DNA and ribosomes.

Because the genetic code is universal, genes from one species of organism can be expressed in the cells of another. This makes it possible to mass-produce valuable proteins, such as the insulin needed by diabetics, by transplanting the human gene for insulin into bacteria or fungi (Section 13–F). The microorganisms are grown in large quantities and synthesize the human protein, which can later be purified and used.

## 10–D   RNA PROCESSING

In prokaryotes, newly transcribed mRNA can be translated as soon as it is synthesized. Ribosomes may attach to one end of the mRNA and start translating it into protein even before the other end detaches from the DNA (Figure 10–7). In eukaryotes, ribosomes function only in the cytoplasm, so mRNA must pass out of the nucleus into the cytoplasm before it can be translated.

In the 1960s, researchers discovered that much RNA is broken down in the nucleus soon after it is synthesized and never reaches the cytoplasm. Some of this nuclear RNA turned out to be pieces removed from mRNA and rRNA molecules. This led to the discovery that many RNA molecules, of all types, are transcribed as pre-RNAs that must be processed before they become functional. Processing is particularly common in eukaryotes, probably because eukaryotes have more complex controls over protein synthesis than do prokaryotes (Chapter 11). But this is not a clearcut distinction between prokaryotes and eukaryotes: some RNA is processed in prokaryotes, and some eukaryotic RNAs are not processed.

In eukaryotes, as soon as the first (5′) end of a pre-mRNA has been transcribed, a "cap" is added to it. This is a special nucleotide (7-methylguanosine) that will eventually attach the mature mRNA to a ribosome. When the other (3′) end of the pre-mRNA is released, a length of it is removed by an enzyme and discarded. The remaining 3′ end is then bound by an enzyme that attaches a "tail" consisting of up to 200 adenine nucleotides. This poly-A tail appears to protect mRNA from breakdown by enzymes in the cytoplasm: mRNAs with tails may last for hours or days, whereas those without tails, such as mRNAs for histone proteins, are broken down in minutes. The tail is also handy for molecular biologists. Since A base-pairs with T or U, synthetic poly-T or poly-U strands attached to a solid surface can be used to "catch the RNA by the tail," thereby separating pre-mRNA and mRNA from other types of RNA in the cell.

### Splicing: RNA Catalysts (Ribozymes)

The second step in RNA processing consists of splicing pieces out of the middle of the pre-RNA. In 1977 researchers discovered that pre-mRNA molecules transcribed from a structural gene are often much longer than the mRNA molecule needed to code for its protein. Molecular biologists were astonished to find that most eukaryotic genes contain strings of nucleotides that do not code for amino acids interspersed with coding sections. The noncoding sections, **introns,** separate coding sections called **exons.** Introns have been found in the genes of viruses and all groups of organisms, although they are most common in plants and vertebrates. They occur in genes coding for transfer and other types of RNA as well as in those coding for proteins.

We tend to imagine a gene as a string of nucleotides along one strand of the DNA double helix, but the presence of introns means that many genes are actually much more complicated. For instance, in yeast mitochondria, the gene for the electron transport system enzyme NADH dehydrogenase is scattered in five segments on both strands of the circular DNA molecule. Genes are known that contain more than 60 introns and exons, lying in various places on both strands of the double-stranded DNA. Think of the complications involved in transcribing such a gene!

The key to understanding how introns are spliced out of RNA molecules came in 1981. Thomas Cech and Sidney Altman discovered introns that splice themselves out of pre-mRNA molecules. This remarkable discovery showed that some RNAs (ribozymes) can act as catalysts, contrary to the long-held assumption that proteins (enzymes) were the only catalysts in biological systems. In 1989 Cech and Altman received a Nobel Prize for this work.

In a eukaryote, spliceosomes splice pre-RNA in the nucleus. A spliceosome forms from four subunits made up of RNA and protein and known as snRNPs or **snurps** (small nuclear ribonucleoprotein particles). Even as an RNA molecule is being transcribed, four snurps attach to it, assembling into a spliceosome large enough to be seen with the

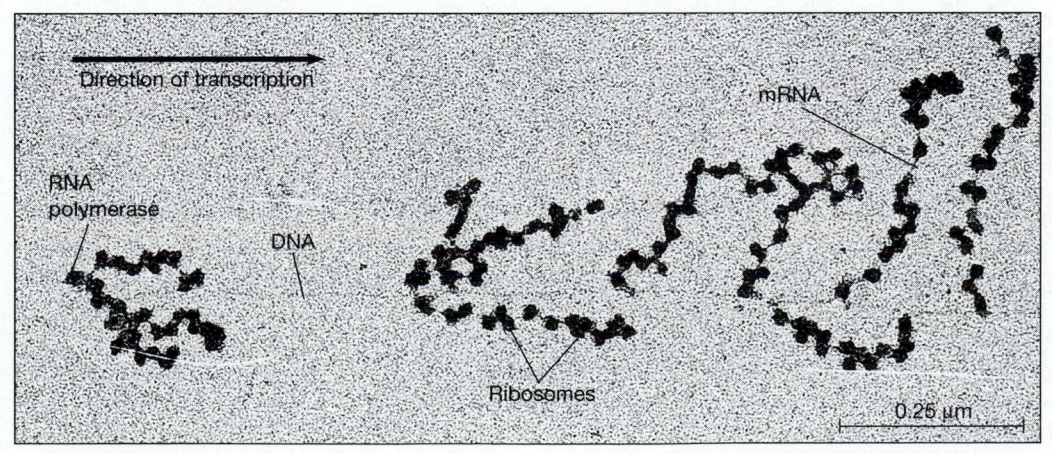

**FIGURE 10–7** Simultaneous transcription and translation. RNA polymerase molecules are transcribing DNA from the bacterium *Escherichia coli* into mRNA. As the 5′ end of each mRNA is formed, a ribosome binds to it and starts to translate it. (The polypeptide molecules cannot be seen.) As the ribosome translates, it moves along the mRNA. As soon as there is room behind it, another ribosome attaches to the initiation site. (Miller, O. L., Jr., B. A. Hamkalo, and C. A. Thomas, Jr., *Science* 169:392, 1970)

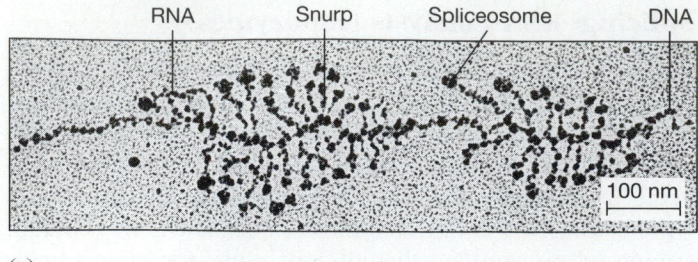

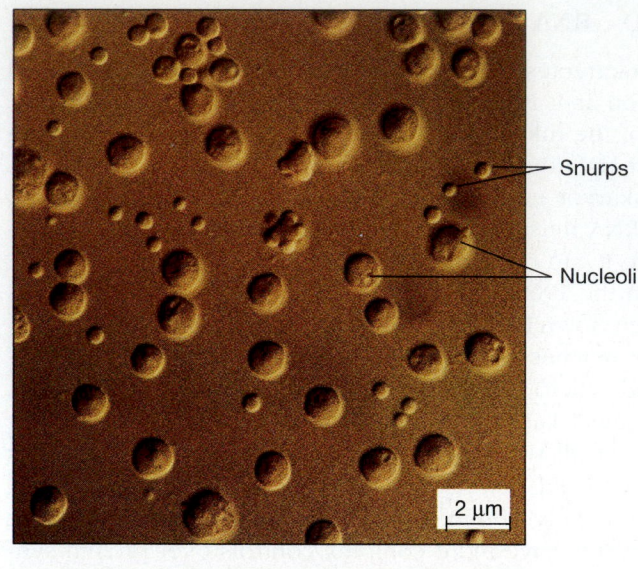

**FIGURE 10–8** Spliceosomes and snurps. (a) Spliceosomes in action show up as large dots on RNA molecules that extend above and below the horizontal DNA molecule. Snurps, the subunits of spliceosomes, are visible as smaller dots. (b) Snurps. Part of a lampbrush chromosome from a newt. (a, Yvonne Osheim; b, Z. Wu and J. G. Gall)

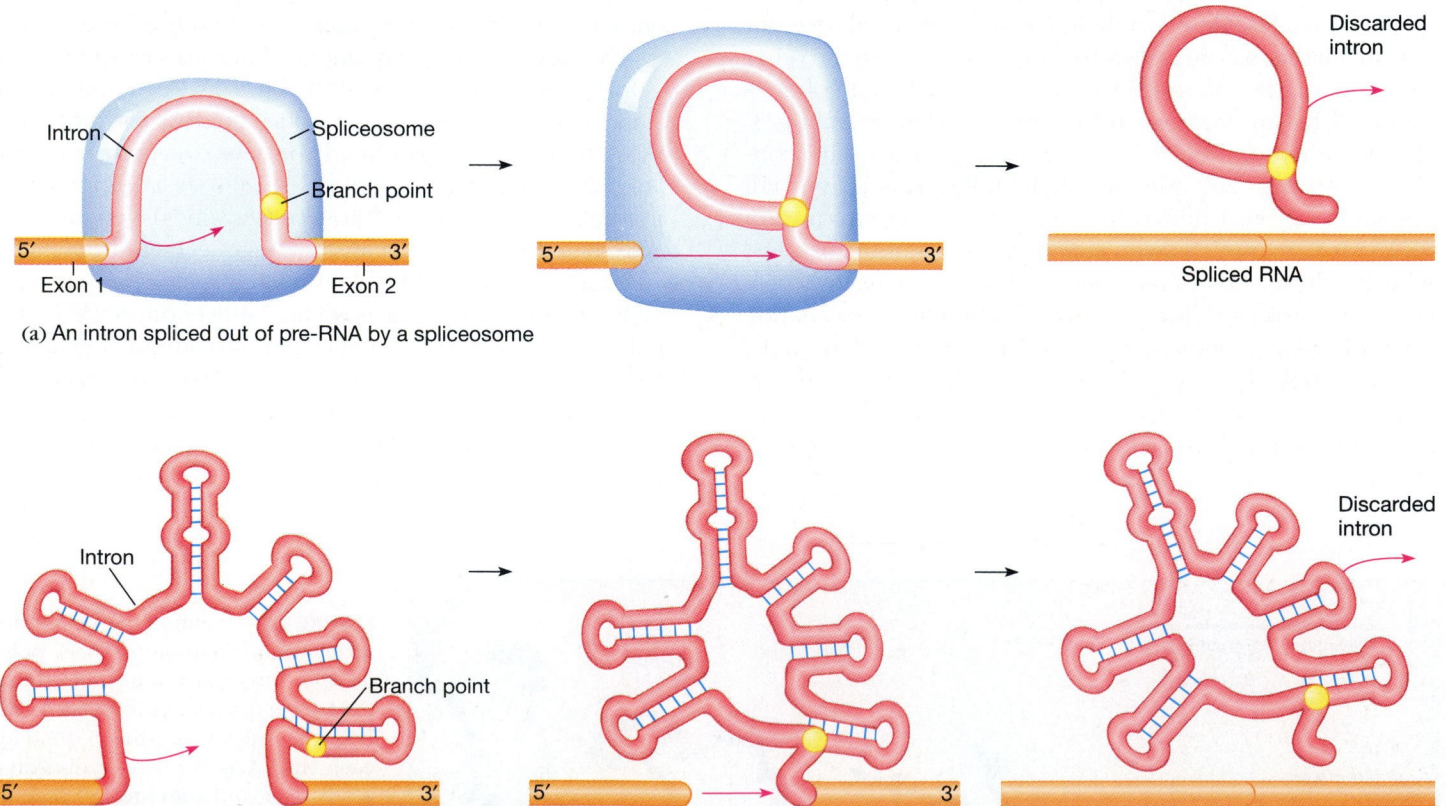

(a) An intron spliced out of pre-RNA by a spliceosome

(b) A self-splicing intron spliced out of pre-RNA

**FIGURE 10–9** Splicing introns out of RNA. (a) A spliceosome attaches one end of the intron to a branch point near the other end. This causes the remaining RNA to join together. (b) The hairpin structures of a self-splicing intron bring the end of the intron near to the branch point, and the intron itself catalyzes formation of the loop and joining of the two exons.

electron microscope (Figure 10–8). Both RNA and protein in spliceosomes take part in splicing reactions. Spliceosomes must finish splicing and disassemble into snurps before the mature RNA is free to leave the nucleus.

Both self-splicing introns and introns removed by spliceosomes are converted into loops of RNA (Figure 10–9). This suggests that splicing is essentially the same process, with or without spliceosomes.

Why then do eukaryotes spend energy and material making spliceosomes? The most obvious difference between a self-splicing system and one with spliceosomes is that a spliceosome can splice an intron of almost any size and nucleotide sequence. Organisms with spliceosomes can survive even if mutations change an intron so that it no longer forms the hairpin loops necessary for self-splicing. Once freed from these constraints, introns can undergo mutations, resulting in new structures that perform novel functions (*How Do We Know? Evolution of Introns*).

A spliceosome does for an intron what a self-splicing intron does for itself: folds it up in such a way that the ends can splice. For splicing to occur, introns in eukaryotic nuclear genes need only contain a sequence of three to six signal nucleotides at each end marking the splice sites and

similar sequences marking the branch point. Genetic diseases may result from mutations that change these splicing signals, making them unrecognizable by spliceosomes. An example is some kinds of thalassemia, a condition characterized by a severe deficit in the production of oxygen-carrying hemoglobin in red blood cells (Section 16–B).

## 10–E    TRANSFER RNA

The codons in mRNA do not recognize the corresponding amino acid directly and put it in place. Transfer RNA (tRNA) is needed to bring the correct amino acid and fit it in place at its codon. For each kind of amino acid, there is one (or more) specific tRNA molecule.

All tRNA molecules have the same general shape, with three main hairpins (Figure 10–10). Each tRNA bears an **anticodon,** a triplet of nucleotides that can base-pair with the appropriate mRNA codon. Thus, the methionine-tRNA has the anticodon UAC, which pairs with the methionine codon AUG. Amino acids for which there are several different codons may have more than one kind of tRNA, each bearing the appropriate anticodon, although in many cases one

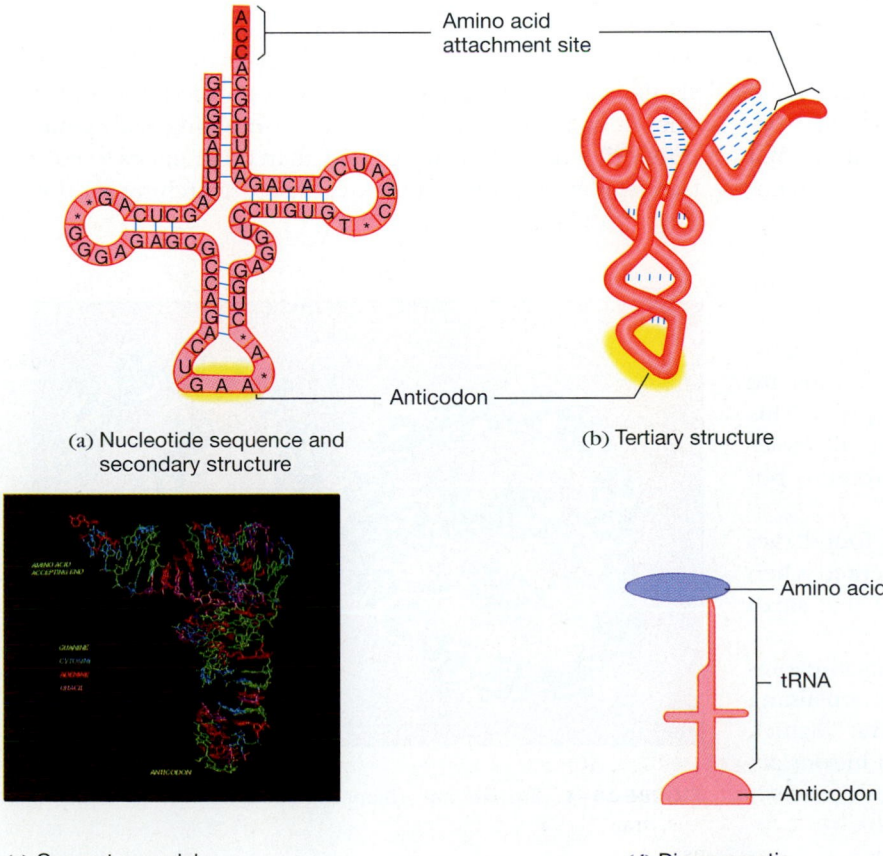

(a) Nucleotide sequence and secondary structure

Amino acid attachment site

Anticodon

(b) Tertiary structure

(c) Computer model

Amino acid

tRNA

Anticodon

(d) Diagrammatic

**FIGURE 10–10** Transfer RNA. (a) The structure of a tRNA molecule. Unusual nucleotides that have been modified during tRNA synthesis are indicated by asterisks. (b) An outline of the molecule's three-dimensional shape. Hydrogen bonds between base-paired parts of the molecule are shown as blue dashes. All tRNA molecules have roughly the same shape, but their nucleotide sequence differs somewhat. (c) A computer model of a tRNA. (d) The diagrammatic representation of a tRNA that we use in this chapter. (c, Judith A. Callaway, Protein Data Bank, Brookhaven National Laboratory)

# EVOLUTION OF INTRONS

Since 1977, when introns were discovered, there has been much speculation about the origin and functions of these useless-looking pieces of RNA. Now some answers are beginning to emerge.

Because the genetic code is universal, prokaryotes and eukaryotes are thought to share a common ancestry. But although hundreds of introns were soon identified in plants and animals, introns had never been found in bacteria, with the exception of the peculiar group known as archaeobacteria.

Then, in 1990, researchers found introns in other prokaryotes. These were in leucine-tRNA genes of several species of cyanobacteria, including *Anabaena*, a species often found as a blue-green scum on polluted ponds (Figure 10–A). Cyanobacteria are photosynthetic, and chloroplasts are descended from cyanobacteria that moved into eukaryotic cells. Indeed, the newly discovered introns resemble those in the leucine-tRNA gene in the chloroplasts of liverworts (members of an ancient plant group). These two introns occur in the same position in both organisms and differ by only a few nucleotides. This supports the idea that introns did indeed originate in prokaryotes.

Some biologists assumed that introns were mere genetic "junk," like that found in all genomes. For instance, all of us contain genetic material from viruses that have infected us over the years. Introns seemed to vary in a random fashion even between closely related species, supporting the theory that introns have no useful function but are the remains of primitive life forms that became incorporated into cells and evolved with them. But why did they become so common and widespread in eukaryotes?

In 1993 Ben Koop and Leroy Hood found that some genes have introns that do not vary between species. The introns in a gene for a vital immune system protein (the T-cell receptor) are almost identical in mice and humans. John Mattick suggested that this meant these introns are not genetic junk but serve to regulate transcription. This was a revolutionary idea because at the time all molecules known to control transcription were proteins, but introns are nucleic acids.

Then Jeanne Lawrence and Rosalind Lee found two examples of genes whose transcription is prevented when RNA molecules bind to DNA, making it clear that small RNA molecules can control transcription.

These findings suggest a new theory of intron evolution. Eukaryotes presumably inherited genes containing introns from their prokaryotic ancestors. But Mattick points out that there is strong selection against introns accumulating in prokaryotes. Prokaryotes have no nuclear envelope preventing contact between unprocessed, new RNA molecules and ribosomes. Ready or not, prokaryotic mRNA is translated as soon as it is transcribed. RNA containing a new intron would be translated into useless protein, a waste of energy that would probably doom the cell. Only self-splicing introns, which remove themselves from RNA as soon as it is formed, have survived in prokaryotes. In eukaryotes, on the other hand, introns are removed in the nucleus and so do not interfere with protein synthesis in the cytoplasm.

Eukaryotes have spliceosomes that remove introns from a pre-RNA molecule in the nucleus no matter what the intron's internal structure (Section 10–D). As a result, mutations in introns could accumulate, and the introns could evolve in various directions. Some of them took on roles as gene regulators, able to bind to DNA in the nucleus and control transcription. This gave eukaryotes a whole new method of regulating transcription that was not available to prokaryotes: the RNA of introns. It may be useful for a regulatory intron to lie within the gene it regulates. This permits negative feedback control of transcription: when a lot of mRNA has been produced from a particular gene, a lot of the regulator will also be present to switch off further transcription.

Not everyone accepts Mattick's hypothesis that introns gave eukaryotes a new system for regulating gene activity. But we can be sure that this idea will be tested in the years to come. We would predict from the hypothesis that genes that must be finely controlled and occur only in eukaryotes, such as genes controlling embryonic development, will be more likely than other genes to contain regulatory introns. Researchers are searching for this kind of evidence.

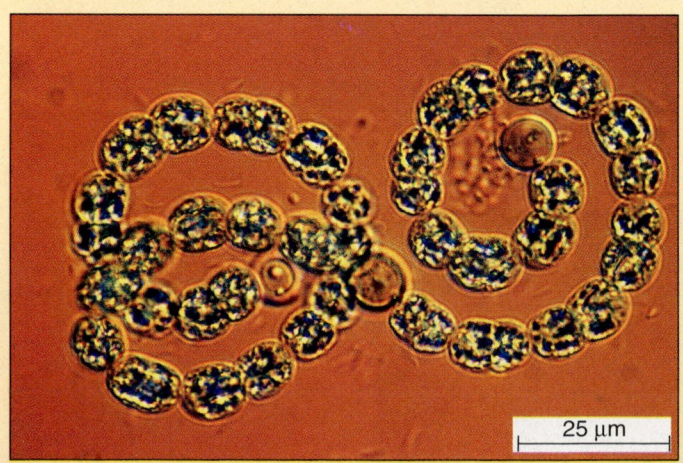

**FIGURE 10–A** *Anabaena.*   (Biophoto Associates)

tRNA pairs with more than one codon by binding less tightly to the third base than the first two.

Another important part of a tRNA is the 3′ end, where the amino acid is attached. This end always consists of the three bases CCA.

Amino acids are attached to tRNAs by **aminoacyl synthetase** enzymes, each of which recognizes molecules of a single amino acid and the corresponding tRNA. The energy to join the amino acid and tRNA comes from the hydrolysis of ATP to AMP (rather than the more usual ADP). Some of this energy is stored in the ester linkage binding the amino acid to its tRNA. Breaking this linkage eventually provides the energy for joining the amino acid to the growing protein chain. A tRNA with an amino acid attached to it is an **aminoacyl tRNA.**

An ingenious experiment showed that it is the tRNA, not the amino acid attached to it, that determines where the amino acid is added during protein synthesis. The amino acid cysteine, attached to cysteine-tRNA, was chemically converted into another amino acid, alanine. When the tRNA carrying the altered amino acid was used in protein synthesis, the alanine was inserted into the protein chain wherever cysteine would normally have appeared.

## 10–F   RIBOSOMAL RNA AND RIBOSOMES

Ribosomes, the sites of protein synthesis, contain several types of ribosomal RNA (rRNA) and about 70 kinds of proteins. In eukaryotes, the production of ribosomes takes place in an area of the nucleus called the nucleolus. The nucleolus bustles with activity, sometimes churning out several hundred thousand ribosomes per hour. Such rapid production is possible because a eukaryotic cell's DNA contains up to 600 copies of the rRNA genes.

A eukaryotic ribosome contains one each of four RNA molecules of different sizes. The genes for the three largest rRNAs are all in a line on the DNA, separated by noncoding sequences. RNA polymerase I transcribes the entire sequence into one long rRNA precursor. Then the spacers between pre-rRNA sequences, and introns within the sequences, are cut out, leaving three separate rRNA molecules (Figure 10–11). This arrangement is an efficient way to ensure that these rRNAs are made in the equal amounts needed for the assembly of ribosomes. The gene for the smallest rRNA is located among the tRNA genes and is transcribed by RNA polymerase III.

Ribosomal RNA was long assumed to be merely a scaffolding for the enzymes that did the real work in a ribosome. However, in 1991 Harry Noller and his co-workers announced that ribosomal RNA, without proteins, could catalyze formation of peptide bonds between amino acids. They discovered this by using enzymes to digest away the ribosomal proteins. For good measure, they then washed the ribosomes with phenol, an organic solvent that does not affect nucleic acids but denatures proteins. Now any protein remnants should be inactivated.

The protein-free ribosomes were then placed in salt solutions in test tubes with an energy supply, messenger RNA, and transfer RNAs with attached amino acids, but not a single protein. When analyzed later, the test tubes were found to contain polypeptides.

Noller pointed out that this was not cast-iron evidence that rRNA had catalyzed peptide bond formation, because he could not rule out the possibility that traces of protein remained in the ribosomes. Noller's group is now trying to synthesize rRNA molecules that have never been in contact with ribosomal protein to determine whether such RNA can still assemble polypeptides.

In any case, protein synthesis occurs more rapidly when ribosomal proteins are present along with rRNA. Proteins are generally more efficient than RNA at catalyzing reactions. Many biologists think that RNA molecules were the original catalysts in protein synthesis and that in the course of evolution enzymes have come to boost or replace the action of RNA catalysts (Section 21–H). In modern ribo-

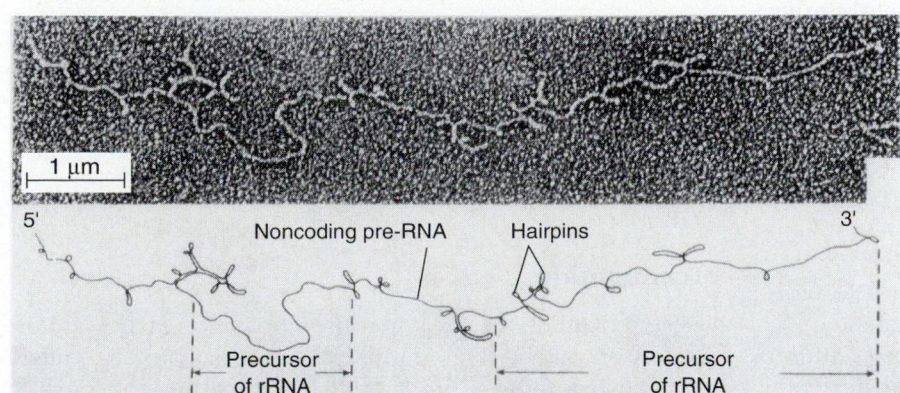

**FIGURE 10–11**  Pre-ribosomal RNA in a eukaryote. This electron micrograph and interpretive drawing show part of the pre-RNA, including two segments that will be cut out during formation of the two largest rRNA molecules. The secondary structure of hairpins in rRNA appears to be the same in all eukaryotes. Note that although there are a few hairpins in the longer section (right), the shorter section (left) is featureless.   (P. Wellauer and I. David)

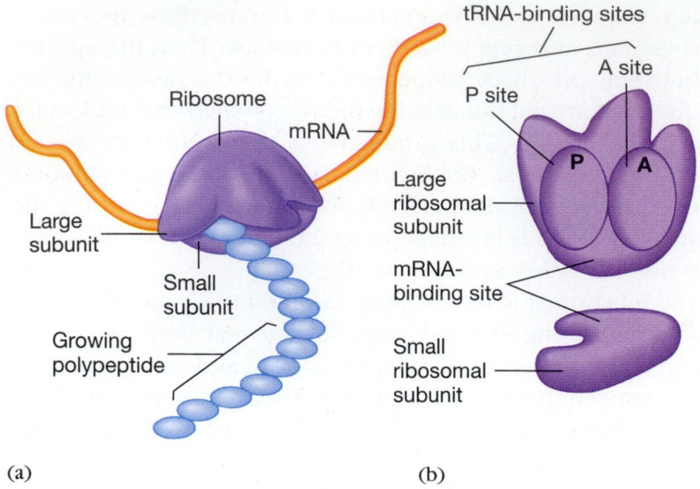

FIGURE 10–12  Ribosome structure. (a) The two subunits joined for protein synthesis. (b) The two subunits opened up like a clam to show the binding sites inside that take part in protein synthesis.

somes, it appears that proteins and RNA collaborate to organize the molecules needed for protein synthesis and to form peptide bonds.

A ribosome consists of two subunits, one large and one smaller (Figure 10–12). When not engaged in protein synthesis, the subunits separate and move around independently. The small subunit contains one rRNA molecule, and the large one contains the other three rRNAs.

In eukaryotic cells, ribosomal proteins are made in the cytoplasm and then pass through the nuclear envelope and reach the nucleolus, where rRNA and proteins are assembled into ribosomal subunits. The subunits are incapable of protein synthesis while they are in the nucleus. They undergo some unknown activation as they pass through the nuclear envelope into the cytoplasm.

Ribosome synthesis is similar in prokaryotes, plastids, and mitochondria except that their rRNAs and ribosomal subunits are somewhat smaller than those of eukaryotes. Also, because there is no nuclear envelope, the subunits are active as soon as they are formed.

Ribosomes are so small that their shape is hard to distinguish, even with the most powerful electron microscopes. It is becoming clear, however, that the ribosomes of different groups of organisms, and possibly even of different species, differ from each other in minor ways.

## 10–G   PROTEIN SYNTHESIS

A ribosome makes a polypeptide as a messenger RNA molecule moves past its tRNA-binding sites. The genetic code in the mRNA codons is translated into a sequence of amino acids, which are brought in one by one by transfer RNA

molecules. The basic reaction of protein synthesis is the formation of a peptide bond between two amino acids (see Figure 3–17). This reaction is repeated as each amino acid in turn is added to the growing polypeptide. Energy for formation of the peptide bond is released by breaking the ester linkage between the amino acid and its tRNA.

### Initiation

Protein synthesis starts when mRNA binds to a small ribosomal subunit and the first codon (AUG) is positioned for initiation of protein synthesis. The AUG codon base-pairs with the anticodon of a tRNA carrying methionine. This methionine becomes the first amino acid in the polypeptide. Now a large ribosomal subunit binds to the complex, and the reactions of protein synthesis itself can begin.

A ribosome has two sites where tRNA can bind. The initiation codon, AUG, on the mRNA is positioned at the first of these sites on the ribosome, the **peptidyl (P) site** (Figure 10–13). The mRNA codon for the second amino acid is at the second site, the **aminoacyl (A) site.**

### Elongation

From this point, the polypeptide elongates by a cycle of three steps (Figure 10–14):

1. **Binding of the next tRNA.** The mRNA codon at the A site binds a tRNA with a complementary anticodon. The amino acid carried by this tRNA will become the next amino acid in the polypeptide.
2. **Peptide bond formation.** The amino acid at the P site is attached to the one at the A site, apparently by a collaboration between rRNA and proteins (Section 10–F). Peptide bond formation leaves the tRNA at the P site empty and the one at the A site holding the growing peptide. The empty tRNA leaves the ribosome and eventually picks up another molecule of its amino acid.
3. **Translocation.** A molecule of the nucleotide GTP provides the energy for the mRNA (and the tRNA still attached to it) to move from the A to the P site. This translocation brings the next mRNA codon to the ribosome's A site. The cycle of elongation repeats as the anticodon of the appropriate tRNA binds to this new codon, bringing the next amino acid into position at the A site. This is a rapid process: in the bacterium *Escherichia coli*, a polypeptide is synthesized at the rate of about 20 amino acids per second.

### Termination

Protein synthesis stops when the ribosome reaches a *Stop* codon on mRNA (Figure 10–15). A special protein, called a **releasing factor,** binds to the *Stop* codon and pushes

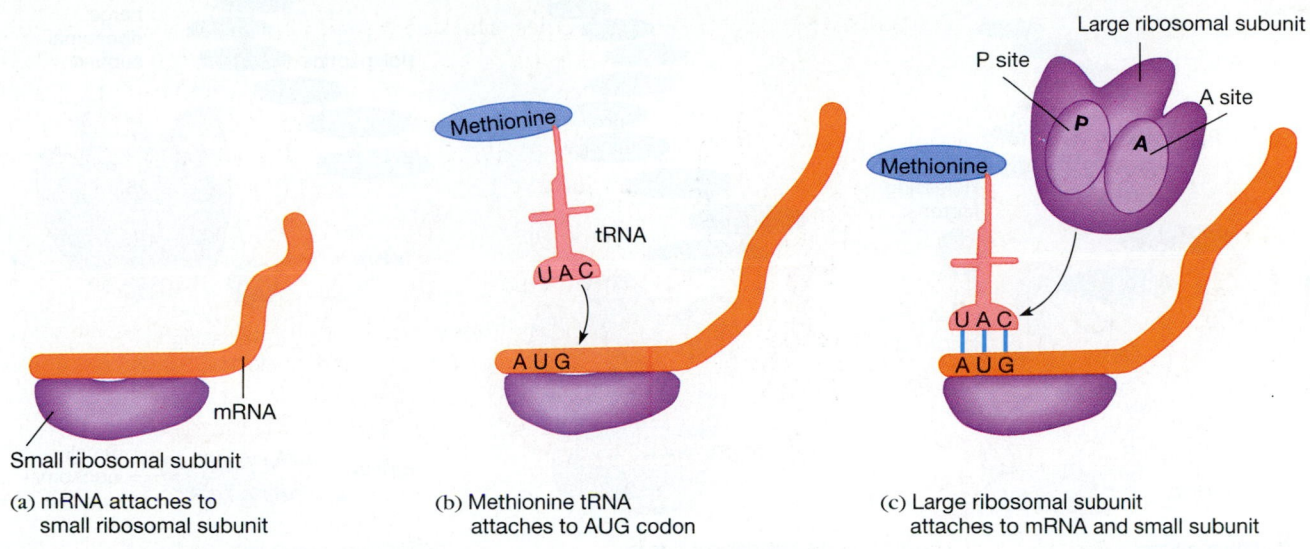

**FIGURE 10–13**   The initiation of protein synthesis.

the mRNA off the ribosome, and the ribosomal subunits separate. In prokaryotes, the mRNA may contain the codes for more than one protein. Different ribosomes read and translate each message.

As the 5′ end of the mRNA emerges from the ribosome, it may bind to another small ribosomal subunit, which initiates protein synthesis again. Each mRNA molecule typically has several to more than 100 ribosomes attached to it and transcribing its message as they move along. One mRNA

with many ribosomes attached to it forms a cluster called a **polyribosome** or **polysome** (see Figure 10–7).

Protein synthesis uses a lot of energy—in many cells, more than any other synthetic pathway. Four phosphate bonds are split to make each peptide bond: two when ATP is hydrolyzed to AMP during the attachment of an amino acid to its tRNA, a third as the aminoacyl-tRNA binds to the ribosome, and a fourth when GTP is hydrolyzed during translocation.

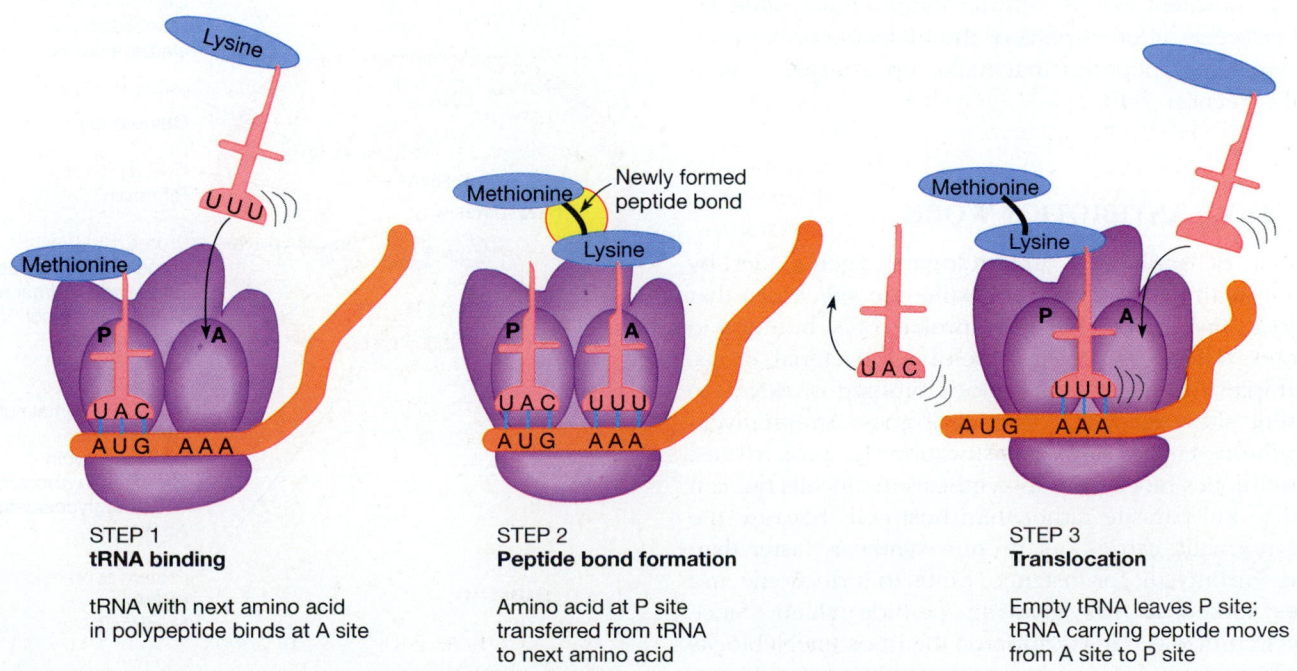

**FIGURE 10–14**   The three steps in elongation of a peptide chain.

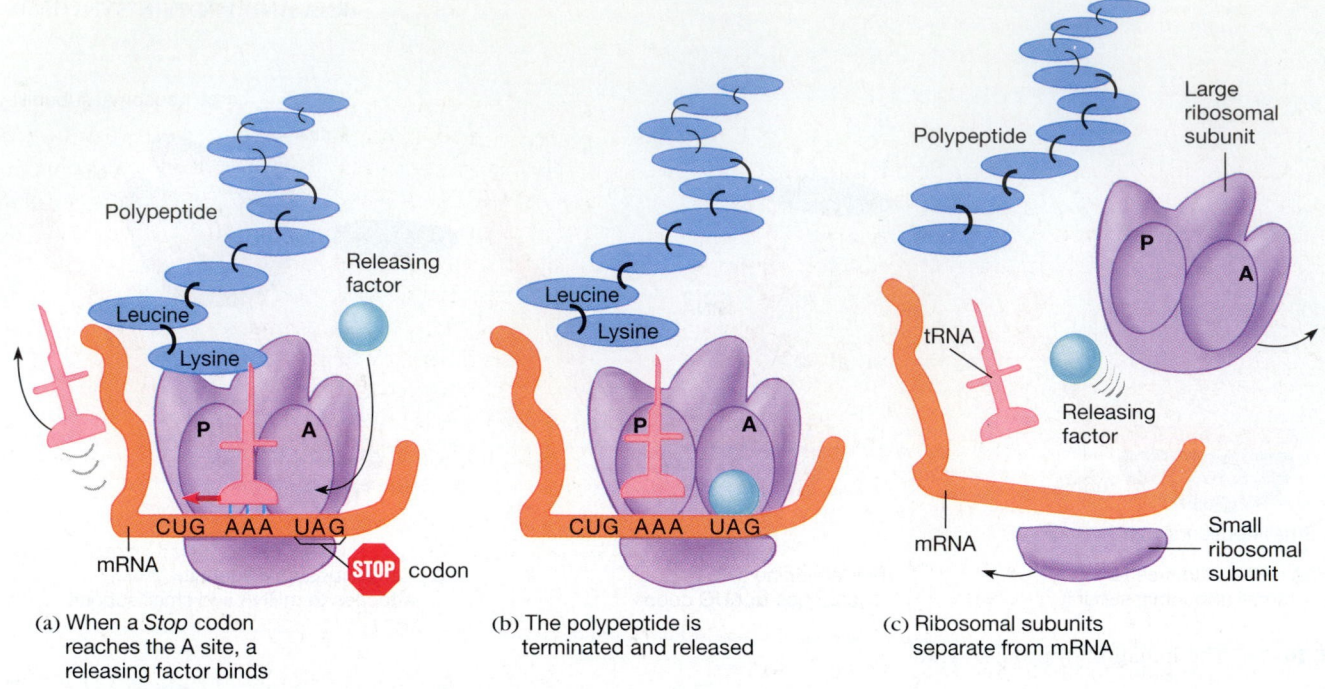

(a) When a *Stop* codon reaches the A site, a releasing factor binds

(b) The polypeptide is terminated and released

(c) Ribosomal subunits separate from mRNA

**FIGURE 10–15** Termination of protein synthesis.

## Beyond the Ribosome

Newly synthesized proteins must usually be processed further after they are released from the ribosome. Generally the first methionine, and often some of its neighbors, must be removed. Sometimes additional trimming is also needed. Various chemical groups may be added or changed, and this is how some proteins come to contain amino acids other than the 20 used in protein synthesis. Prosthetic groups such as the heme group in hemoglobin may be attached by covalent bonds. Sulfur bridges must often be formed between different parts of the molecule or between two separate polypeptides that make up a larger protein molecule (Section 3–E).

## 10–H HOW ANTIBIOTICS WORK

A number of antibiotics (antibiological agents) act by interfering with gene expression. Some are specific in that they block protein synthesis in prokaryotes but not in eukaryotes. These agents are useful antibacterial drugs. For example, the tetracyclines block binding of tRNAs to attachment sites on prokaryotic ribosomes. Streptomycin and erythromycin prevent translocation in prokaryotes. Other antibiotics block protein synthesis in all cells but can be used to kill parasite rather than host cells because the parasite normally carries out protein synthesis faster than the host. Puromycin, for instance, binds to a ribosome and becomes linked to the growing peptide chain. Since puromycin prevents translocation on the ribosome, it blocks further formation of the polypeptide at this point. Figure 10–16 shows the sites of action of a number of antibiotics.

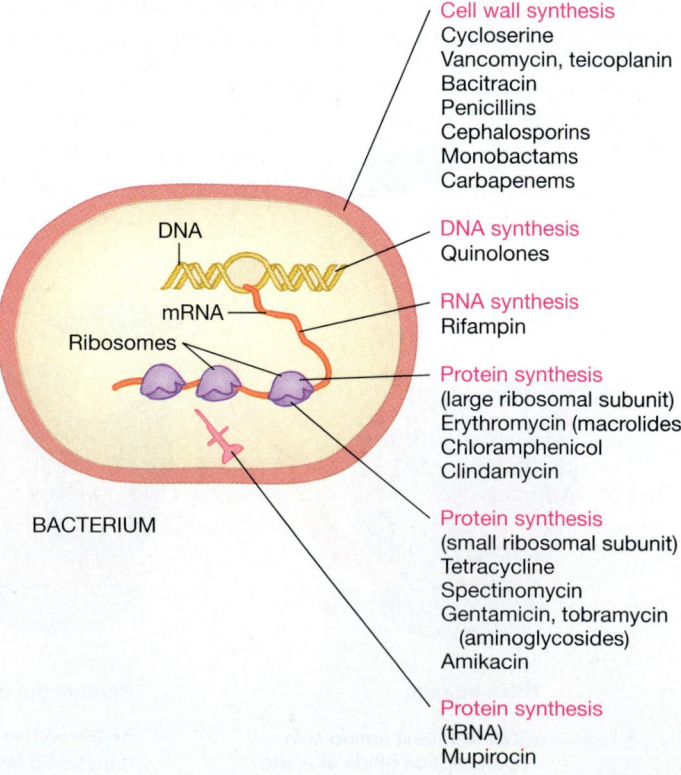

Cell wall synthesis
Cycloserine
Vancomycin, teicoplanin
Bacitracin
Penicillins
Cephalosporins
Monobactams
Carbapenems

DNA synthesis
Quinolones

RNA synthesis
Rifampin

Protein synthesis
(large ribosomal subunit)
Erythromycin (macrolides)
Chloramphenicol
Clindamycin

Protein synthesis
(small ribosomal subunit)
Tetracycline
Spectinomycin
Gentamicin, tobramycin
(aminoglycosides)
Amikacin

Protein synthesis
(tRNA)
Mupirocin

**FIGURE 10–16** Sites of action of some antibiotics on a prokaryotic cell.

## SUMMARY

Proteins are essential molecules of both cell structure and metabolism. A cell's structural genes specify the proteins it can make. The order of nucleotides in their DNA is transcribed into a complementary nucleotide sequence in messenger RNA. This mRNA binds to a ribosome, where its nucleotide sequence is translated into the order of amino acids in a polypeptide. Transfer RNA, ribosomal RNA, and various proteins also take part in translation.

1. All RNA is transcribed by RNA polymerase enzymes, starting from a promoter sequence on a DNA template. Thus the nucleotide sequence of RNA is complementary to that in DNA. Unlike DNA, RNA is usually single-stranded, but sections of the molecule may base-pair to give parts of the molecule a hairpin secondary structure.

2. The genetic code consists of three-nucleotide codons in mRNA, each specifying a particular amino acid or coding for polypeptide termination. The code is degenerate, in that most amino acids are coded for by more than one codon. The code's only "punctuation marks" are codons that signal the beginning and end of the polypeptide.

3. Mutations are passed on when the DNA replicates and are also transcribed into RNA; they may result in changes in the protein produced. Frameshift mutations generally affect final protein structure more than point mutations do.

4. Most mRNA in prokaryotes can be translated by ribosomes as soon as it is synthesized. Particularly in eukaryotes, many other RNAs are transcribed as pre-RNAs, which must be processed. In eukaryotes, pre-mRNA is processed by adding a cap and tail and splicing out introns, joining the exons into a functional mRNA molecule.

5. Some introns are self-splicing, with the RNA itself catalyzing removal of the intron. However, in most eukaryotic genes introns are spliced out by spliceosomes, which contain RNA and proteins and work by a mechanism similar to that in self-splicing introns.

6. Transfer RNA carries amino acids to the ribosome engaged in protein synthesis and fits them into their proper position in the polypeptide.

7. A functional ribosome consists of one small and one large subunit, each containing rRNA and proteins. The ribosome holds mRNA and tRNAs in place for protein synthesis and catalyzes formation of peptide bonds between amino acids. Ribosomes in eukaryotic cytoplasm are larger than those in prokaryotes, mitochondria, and plastids.

8. Protein synthesis involves four stages:
   - Initiation. An mRNA attaches to a small ribosomal subunit, a large subunit also binds, and a tRNA carrying the first amino acid (methionine) binds to the first codon.
   - Elongation. One by one, tRNAs carrying amino acids attach to the mRNA-ribosome complex by base-pairing between the mRNA codon and the tRNA anticodon. Each amino acid is joined to the growing polypeptide by formation of a peptide bond. Then, during translocation, the ribosome moves along the mRNA, bringing the next codon onto the ribosome, where it can bind the anticodon of the tRNA carrying the next amino acid.
   - Termination. When a *Stop* codon reaches the ribosome, the completed polypeptide is released.
   - Post-translation processing. Most polypeptides must be trimmed, finished, and folded after the ribosome releases them. They may also join with other polypeptides, forming a complete protein.

9. Many antibiotics work by interfering with aspects of protein synthesis, slowing down the growth and reproduction of disease-causing parasites and permitting host defenses to prevail.

## Self-Quiz

1. Using the base-pairing rules, fill in the bases to be found on the RNA strands transcribed from the following DNA strands:
   a. DNA: A–G–G–C–C–T–G–C–T–T–A
      RNA:_____
   b. DNA: T–G–G–C–A–G–C–T–A–C
      RNA:_____
   c. DNA: T–T–T–A–C–G–C–A–C–C
      RNA:_____

2. Write out the amino acid sequences that would be translated when the following mRNA molecules combine with a ribosome:
   a. A–U–G–C–A–U–A–G–A–A–G–G–C–C–U–A–U–U–G–U–A
   b. C–A–U–G–U–U–U–C–U–U–A–A–A–G–G–U–C–G–U–U

3. Write out the mRNA sequence that would be transcribed from the following strand of DNA and the amino acid sequence that would be translated when the mRNA combines with a ribosome:

   T–A–C–A–A–G–T–A–C–T–T–G–T–T–T–C–T–T

4. Suppose the two guanine (G) nucleotides in Question 3 were changed to cytosine (C) nucleotides. How would these mutations affect the amino acid sequence translated from the mRNA?

5. Suppose the two G nucleotides were removed from the DNA in Question 3. How would this affect the amino acid sequence translated from the mRNA?

6. According to current ideas concerning protein synthesis:
   a. transfer RNA molecules specific for particular amino acids are synthesized along a messenger RNA template in the cytoplasm
   b. amino acids line up with their mRNA codons on the ribosome and are then linked together by transfer RNA
   c. enzymes that catalyze protein-synthesizing reactions in the cytoplasm are transcribed from promoters
   d. transfer RNA molecules transport mRNA from the nucleus to the ribosomes
   e. messenger RNA provides information that determines the sequence in which amino acids are linked during translation

7. List three differences between the structures of DNA and RNA:

   | DNA | RNA |
   | --- | --- |
   | _____ | _____ |
   | _____ | _____ |
   | _____ | _____ |

8. Transfer RNA is synthesized:
   a. on a DNA template
   b. from a messenger RNA template on a ribosome
   c. on ribosomes without a template
   d. in the nucleolus by the interaction of messenger RNA and chromosomal DNA

## Questions for Discussion

1. Why is it important for each type of tRNA to have its own aminoacyl synthetase enzyme to bind it to an amino acid?
2. Suppose a bacterial cell contained a mutation that changed one of the nucleotides in an anticodon of tRNA. How might this mutation affect protein synthesis?
3. We have seen why the genetic code could not consist of codons with fewer than three nucleotides each. What might have selected against codons of more than three nucleotides?
4. Why are antibiotics ineffective in treating diseases caused by viruses?

## Suggested Readings

Darnell, J. E., Jr. "RNA." *Scientific American,* October 1985. The production and function of RNA and what they tell us about the place of RNA in molecular evolution.

Guthrie, C. "Messenger RNA splicing in yeast: clues to why the spliceosome is a ribonucleoprotein." *Science* 253:157, 1991.

Loewy, A. G., et al. *Cell Structure and Function*. Philadelphia: Saunders College Publishing, 1991. A useful cell biology text.

Nirenberg, M. W., and J. H. Matthaei. "The dependence of cell-free protein synthesis in *E. coli* upon naturally occurring or synthetic polyribonucleotides." *Proceedings of the National Academy of Sciences* (U.S.) 47:1588, 1961. The classic paper on poly-U and mRNA.

Noller, H. F. "Ribosomal RNA and translation." *Annual Review of Biochemistry* 60:191, 1991. A review of RNA catalysis during translation.

Nomura, M. "The control of ribosome synthesis." *Scientific American,* January 1984.

Steitz, J. A. "Snurps." *Scientific American*, June 1988. The story of the discovery of snurps and their role in splicing eukaryotic RNA.

# Gene Expression and Cell Differentiation

## OBJECTIVES

*When you have studied this chapter, you should be able to:*

1. Use the following terms in context: differentiation, gene transcription, operon, promoter, transcription factor, homeobox gene, oncogene.
2. Describe evidence that in most species of plants and animals all the cells contain the same genetic material.
3. Describe the structure of a polytene chromosome; state where such chromosomes are found and why they are useful in studies of genetic activity.
4. Describe how food molecules induce the synthesis of specific enzymes in prokaryotes.
5. Describe what is known of the regulation of transcription in eukaryotes.
6. Describe how steroid hormones activate transcription.
7. Describe X–chromosome inactivation and its effect on gene activity in the cells of a female mammal.
8. Describe what is known about how cancers are caused.

A cell's DNA contains hundreds of genes, but at any one time only some of these are expressed by being used to transcribe an RNA molecule, synthesize a protein, or regulate the transcription of another gene. The "housekeeping" genes for enzymes involved in metabolic pathways such as glycolysis and ribosome synthesis are active in most cells most of the time, but other genes are expressed only in particular types of cells at particular times. One of the most fascinating areas of biology is our fast-growing understanding of how gene expression is controlled.

During the 1950s, experiments revealed that protein synthesis in prokaryotes is controlled by proteins encoded by regulatory genes (Section 11–C). The control of gene expression in eukaryotes has proved more difficult to understand because eukaryotes have additional means of regulating gene activity. These appear to stem from one of the most important differences between eukaryotes and prokaryotes: most eukaryotes undergo embryonic development, whereas prokaryotes do not.

During embryonic development, genetically identical cells descended from one original cell **differentiate,** that is, mature and become different from one another as different genes are switched on and off in different cells. Differentiation results in cell specialization, leading to division

of labor among cells, which is the hallmark of a multicellular organism (Figure 11–1). Without differentiation, mul-

**FIGURE 11–1** Differentiation in a test tube. A microscopic clump of plant cells has been placed on nutrient agar medium. The cells divide and differentiate into all the cell types and organs of the adult plant. (Cornell Biotechnology Program)

ticellular organisms would be nothing more than colonies of identical cells. We can view differentiation as cascades of events: each gene, or suite of genes, is switched on in its turn, and then something resulting from that gene expression, or some environmental influence, switches on the next gene in the program. In terms of gene expression, development is an exceedingly complicated process.

The turning on and off of genes to produce particular differentiated cells is one of the most important things that happens in an embryo and it now appears that many of the same processes are involved in the development of cancers. These are the processes covered in this chapter. In Chapter 12 we shall discuss embryonic development in more detail.

## KEY CONCEPTS

♦ All the body cells in most multicellular organisms contain the same genetic material.

♦ Differentiation of one cell type from another involves the turning on and off of different genes in different types of cells at particular times in development.

♦ Gene expression is controlled by regulatory genes, many of which code for protein transcription factors, which bind to regulatory sites on DNA and slow down or speed up transcription.

♦ Control of transcription usually involves the interaction of transcription factors with cofactors from the cell's environment, which work together, turning transcription on or off.

♦ Cancer is abnormal invasive cell proliferation that may occur when internal or environmental stimuli cause mutations or alter the expression of particular genes.

## 11–A    GENETIC EQUIVALENCE OF BODY CELLS

Differentiated eukaryotic cells, such as a liver and a muscle cell, synthesize different sets of proteins because different genes are expressed in the two. Is this because the two cells contain different genes? Or do both cells contain the same genes but express different ones? Several lines of evidence have convinced biologists that each cell of an organism contains the same genetic information as the **zygote,** the fertilized egg from which it developed.

There are minor exceptions to this rule. For instance, in the parasitic worm *Ascaris,* some of the chromosomes are destroyed in the **somatic cells,** those cells that make up most of the body, but not in the **germ cells,** those that give rise to reproductive cells (eggs and sperm). In many species, genes for ribosomal RNA are replicated many times in the **oocytes** (the germ cells that will give rise to eggs), which have to make large numbers of ribosomes. In addition, mutations and viruses accumulate in body cells, causing the DNA of some cells to differ slightly from that of others.

However, the general rule is that all the cells in the body of an organism, whether embryo or adult, contain identical genomes and that differences between adult cells depend on gene expression, not on presence or absence of genes. On a molecular level, when the amount of DNA in body cells is measured, it is found to be the same in every cell. Further evidence comes from the fact that genes occur even in differentiated cells where they are not expressed. For instance, hybridization experiments (Section 13–A) have detected the human insulin gene in nerve, muscle, and all other kinds of human cells examined, as well as in the pancreas where it is normally expressed. The genomes of all an organism's cells are not merely physi-

cally alike, they are also functionally equivalent in many cases, in that the genome of one cell can be induced to take on functions normally performed by the genome of another type of cell.

### The Evidence in Plants

In the 1950s, Frederick C. Steward and his colleagues succeeded in growing whole carrot plants from single root cells in tissue culture (Figure 11–2). This showed that these differentiated cells could function like embryonic cells and contained the genetic information for development of all the adult cell types.

Many other plants can be grown from single cells in this way, and breeders take advantage of this to reproduce genetically identical plants. Chrysanthemums are grown for market by taking a particularly desirable plant, chopping it into small pieces, and culturing each until it is a plant large enough to be potted up and sold. Unfortunately for plant breeders, some crop plants contain genes that prevent them from regenerating from single cells. Not until the 1980s did researchers discover how to grow rice from small pieces of tissue, and biologists are now attempting to breed corn without these inhibitory genes so that it too can be grown in tissue culture.

### Nuclear Transplantation

No one has yet grown an entire animal from a single differentiated cell. However, experiments implying that the nuclei in some differentiated cells have not lost the genetic information needed for differentiation into other types of cells have been done with frogs. These **nuclear trans-**

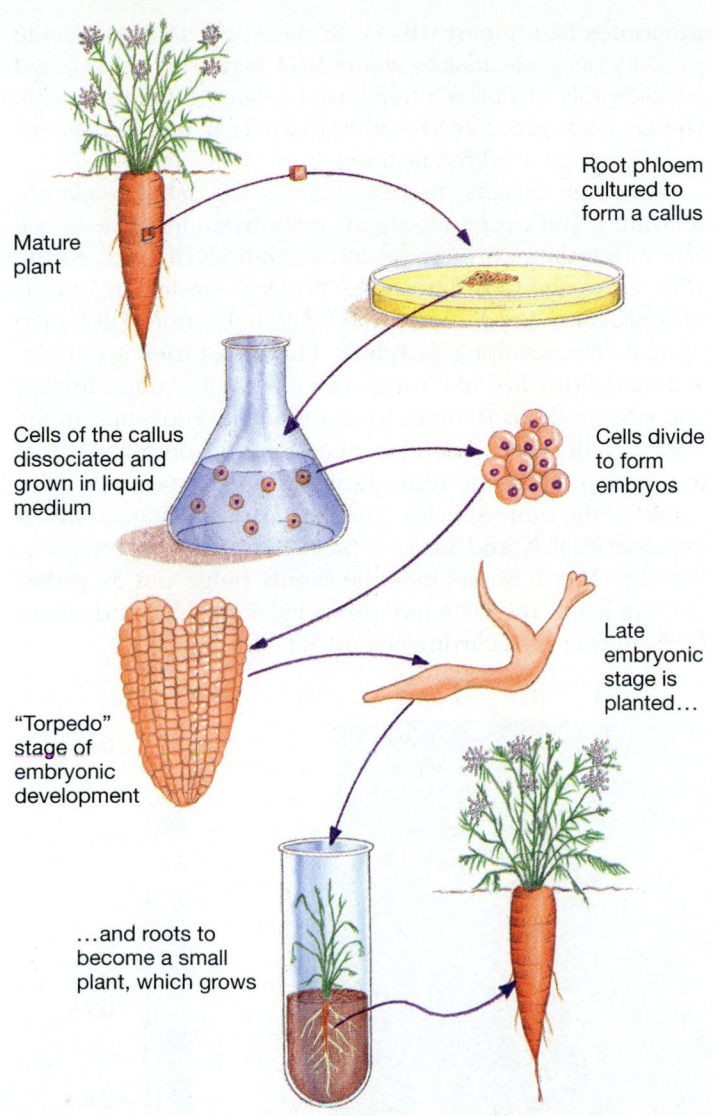

**FIGURE 11–2**  Steward's experiment on carrots. A mature carrot plant develops from cells isolated from the root.

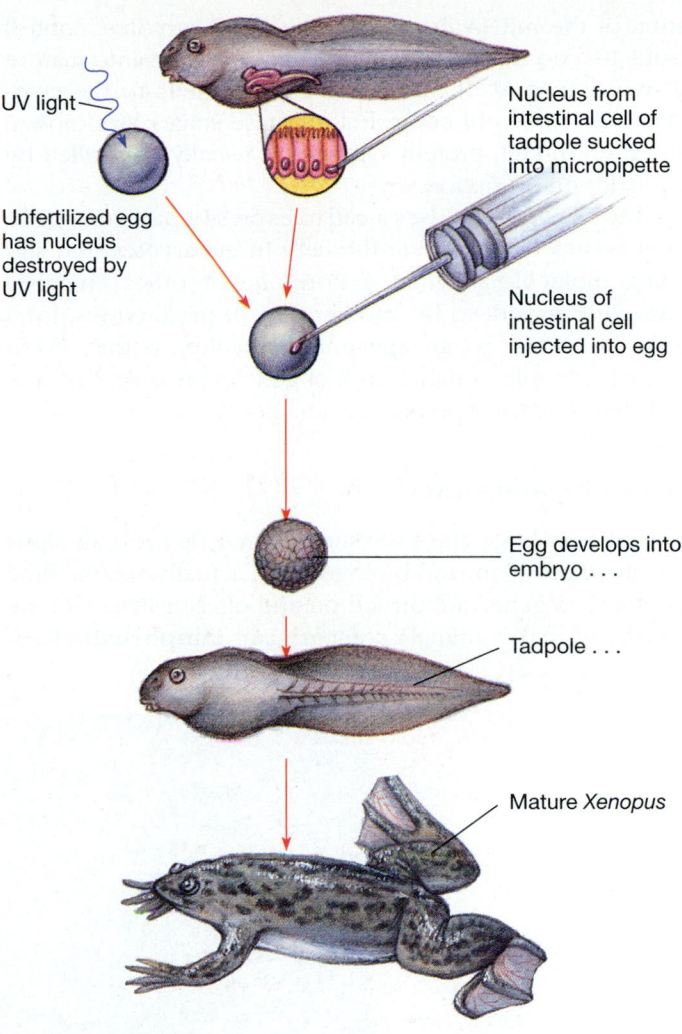

**FIGURE 11–3**  Nuclear transplantation in the South African clawed frog, *Xenopus*. A nucleus from a cell in a tadpole's intestine (chosen because it is large enough to manipulate) is injected into an egg without a nucleus and supports development of a mature frog.

**plantation** experiments were designed to determine whether any genetic information present in the zygote is missing from the nucleus of a differentiated cell.

The technique is to remove the nucleus from an egg, replace it with the nucleus of a differentiated cell from another individual of the same species, and see if this nucleus will take over the role of the zygote nucleus and support development of another complete animal. In the 1950s, sexually mature frogs were grown from artificial zygotes, made by implanting nuclei from the intestinal cells of a frog tadpole into eggs of the same species (Figure 11–3). The nuclei transplanted from such differentiated cells are "totipotent," meaning that they can express all the genetic information needed to support development.

## 11–B  DIFFERENTIAL GENE EXPRESSION

There is, then, much evidence that all the cells in a multicellular organism contain the same genetic information. During embryonic development and during periods of change in later life, such as wound-healing, molting, puberty, and pregnancy, differential gene expression causes cells to become specialized for their various functions. Unicellular eukaryotes and prokaryotes also show differential gene expression, producing some proteins all the time and others only as needed.

In principle, protein synthesis may be controlled at any one of its stages: modification of the DNA, transcription of DNA into RNA, attachment of a ribosome to mRNA, or translation of the mRNA by a ribosome. In eukaryotes, control could also occur during processing of pre-RNA into mature RNA or transport of mRNA from the nucleus to the cytoplasm. Examples of control at all these stages are known. However, in fact, protein synthesis is usually controlled by regulating transcription.

The signal that causes a cell to express a particular gene often comes from outside the cell. In eukaryotes, the signaling molecule may be a hormone or other chemical messenger produced by another cell. In prokaryotes, food molecules often act as signaling molecules, as they do in the first example of the control of gene expression that was analyzed, the *Lac* operon (Section 11–C).

## Giant Chromosomes

In some very large chromosomes, microscopy reveals signs of gene expression, and biologists can actually see the time sequence as genes are turned on and off. For instance, the oocytes of many animals contain giant **lampbrush chro-**

**mosomes** (see Figure 10–1). In these, the DNA molecule partially unravels at sites where RNA synthesis occurs, and considerable detail is often visible. Sites of transcription change with time, as we would expect if different genes are expressed at different times.

In some insects, tissues such as the salivary glands, intestines, and excretory organs grow by an increase in cell size rather than an increase in the number of cells. As the cells grow, their chromosomes undergo up to ten rounds of replication, producing up to 1000 times more DNA than normal. The resulting **polytene chromosomes** are multi-stranded structures like ropes and may be 100 times thicker and stretched out 10 times longer than chromosomes in the insect's other cells. Each polytene chromosome has a pattern of bands that is reproduced in all polytene chromosomes of the same species. The dark bands contain a higher density of DNA and histone proteins than do the regions between them. Sometimes the bands bulge out as **puffs,** areas in which the DNA has unraveled somewhat and where transcription is occurring (Figure 11–4).

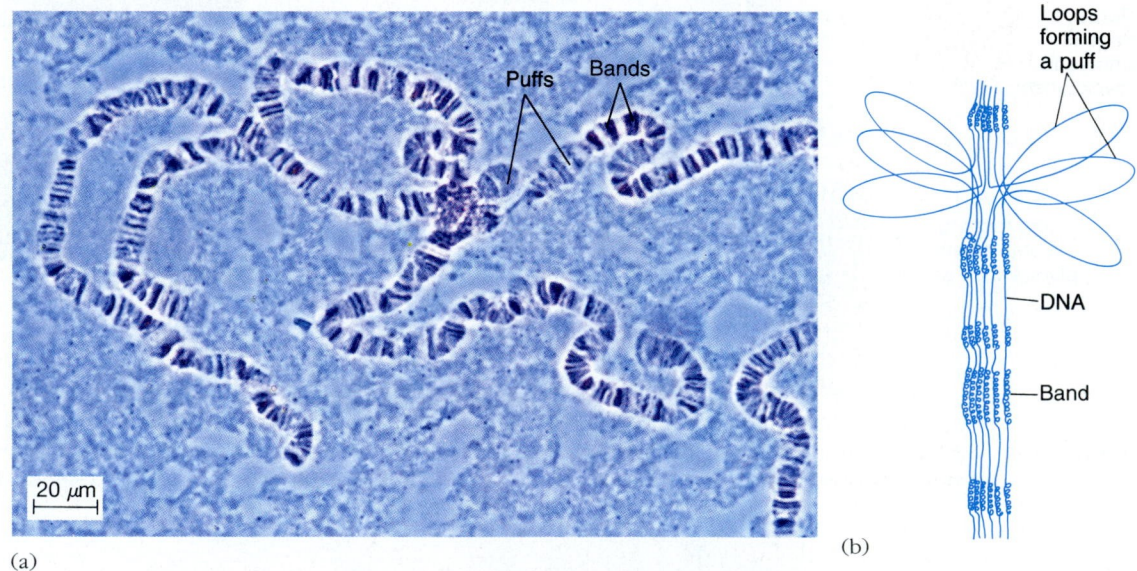

(a)

(b)

**FIGURE 11–4**   Giant polytene chromosomes. (a) Polytene chromosomes from the salivary gland of a fruit fly larva *(Drosophila)*. Some bands are expanded into puffs. (b) Drawing of part of a polytene chromosome. The chromosome contains hundreds of strands of DNA, only a few of them shown here. The chromosome appears banded, and one band is expanded into a puff, an area where transcription is occurring. (c) One chromosome from a salivary gland of a midge. The long, thin single lines are lengths of DNA containing inactive genes that are not being transcribed. The short loops and wiggly lines sticking out from the DNA line are a mixture of DNA with RNA molecules that are being transcribed.   (a, P. J. Bryant, University of California, Irvine/BPS; c, from M. M. Lamb and B. Daneholt. *Cell* 17:838, 1979. Copyright Massachusetts Institute of Technology, published by the MIT Press)

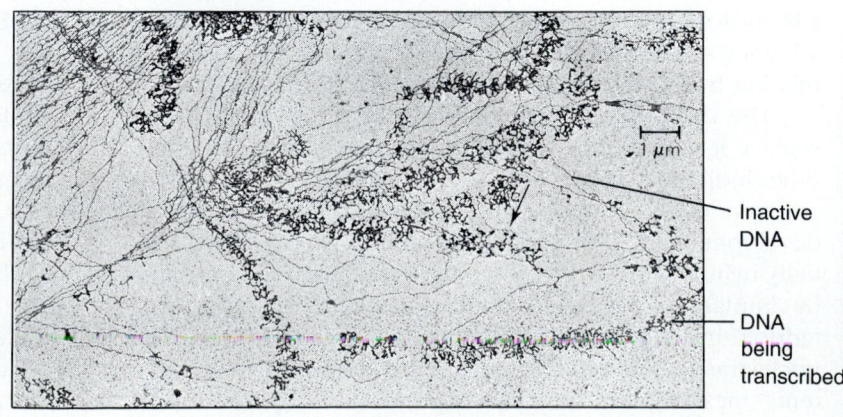

(c)

If the insect hormone ecdysone is injected into a larva at a time when it is not due to molt its external skeleton, puffs appear and disappear on its polytene chromosomes in a sequence identical with that found during normal molting. In other words, the effect of the hormone is to activate transcription of specific genes in a specific sequence, as one would expect from our discussion of how differentiation occurs.

## 11–C   CONTROL OF TRANSCRIPTION IN PROKARYOTES

Work by François Jacob and Jacques Monod on bacteria, published in 1961, led to the realization that protein synthesis is usually controlled by genes that affect transcription and thereby control the expression of other genes. Many regulatory genes code for **transcription factors,** small proteins that bind to particular **regulatory sites** on DNA and slow or speed up transcription of another gene or set of genes. (Transcription factors are also sometimes known as gene regulatory proteins.)

### The *Lac* Operon

Jacob and Monod studied the ***Lac* operon,** a group of genes controlling lactose metabolism in *Escherichia coli,* a bacterium found in the human intestine. When we drink milk, these bacteria produce enzymes that break down the milk sugar lactose into glucose and galactose. These enzymes are not produced all the time, which would use up valuable energy and materials. The bacterium contains a **repressor protein,** a transcription factor that usually prevents transcription of the *Lac* operon.

RNA polymerase initiates transcription by binding to a **promoter** sequence on the DNA strand to be transcribed. In the *Lac* operon, the repressor transcription factor is usually bound to an **operator,** a sequence of 21 DNA base pairs that overlaps the promoter on the downstream side. ("Downstream" refers to the direction RNA polymerase travels, toward the 5′ end of the DNA it is copying, and a structural prokaryotic gene always lies downstream of its promoter. "Upstream" refers to the opposite, 3′, direction on the DNA.) When the repressor is bound to the operator, RNA polymerase cannot move from the promoter, and the operon's structural genes cannot be transcribed (Figure 11–5).

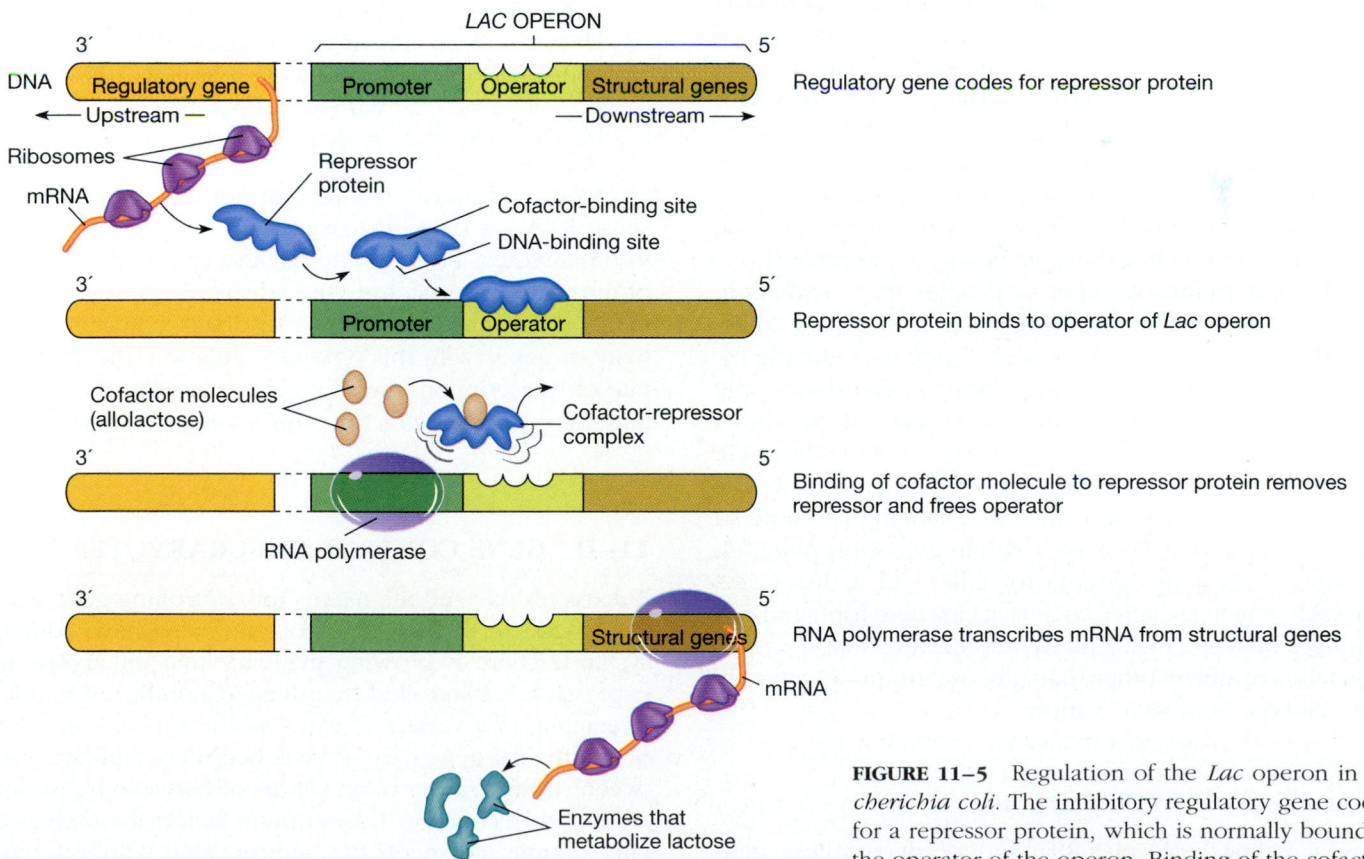

**FIGURE 11–5** Regulation of the *Lac* operon in *Escherichia coli.* The inhibitory regulatory gene codes for a repressor protein, which is normally bound to the operator of the operon. Binding of the cofactor, allolactose, removes the repressor protein from the operator site. This permits RNA polymerase to bind to the promoter and transcribe the structural genes coding for the enzymes that metabolize lactose.

This transcription factor can be removed from the *Lac* operator by a **cofactor,** a signaling molecule. In this case, some of the lactose in milk is converted to the cofactor, which is another sugar, allolactose. Allolactose binds to the repressor, inducing an allosteric change that loosens its hold on the operator. Now RNA polymerase can bind to the promoter properly and move on to transcribe the *Lac* operon's structural genes. The messenger RNA transcribed is translated into proteins that transport lactose into the cell and metabolize it. When the available lactose has been hydrolyzed, the concentration of the cofactor allolactose falls again, freeing the repressor to bind to the operator again, blocking transcription.

In prokaryotes, many genes are arranged in operons containing the instructions for producing proteins for a particular metabolic path. Interaction of a single transcription factor with its cofactor can turn the whole set of genes on or off and ensure that the correct quantity of each protein is produced.

## Catabolite Activator Protein

Whereas the *Lac* operon is controlled by a repressor transcription factor, which inhibits transcription, **activator proteins,** transcription factors that increase the rate of transcription, are also known. A well-studied example in *Escherichia coli* is the **catabolite activator protein (CAP),** a transcription factor that helps control *Lac* and several other operons. In the *E. coli* genome, DNA sites where CAP binds lie upstream of the *Lac* operon and of genes that code for several other sugar-metabolizing enzymes. CAP must be bound to these sites for these genes to be transcribed.

What determines whether CAP binds to its regulatory sites on the DNA? Not surprisingly, the answer is a cofactor. CAP's function is to ensure that *E. coli* uses glucose for food whenever it is available, switching to other sugars only when there is not enough glucose. As glucose runs low, the concentration of the nucleotide cyclic adenosine monophosphate (cyclic AMP) in the cell increases. Cyclic AMP is a cofactor that binds to CAP, changing its shape so that it can bind to its DNA sites. When glucose is plentiful, the concentration of the cofactor falls, and it dissociates from CAP, which switches back to its inactive form and can no longer bind to DNA. This system ensures that enzymes to metabolize sugars other than glucose are produced only when glucose is in short supply.

## Other Controls in Prokaryotes

Many signaling molecules that induce or suppress transcription in bacteria are known. For instance, toxic chemicals may induce the production of enzymes that break them down, and cofactors within the cell stimulate the production of proteins needed for cell division.

We can summarize the ways transcription may be controlled in prokaryotes:

1. **Repressor transcription factors.** In the *Lac* operon, a cofactor turns on transcription by loosening a repressor protein's hold on the operator, permitting RNA polymerase to bind. Other repressors work in the opposite way: they bind to DNA and prevent transcription only when they themselves have bound a cofactor.

2. **Activator transcription factors.** Activators increase the chance of RNA polymerase binding a particular promoter. Like repressors, activators often bind specific cofactors that either increase or decrease their affinity for DNA and thereby cause them to turn genes on or off respectively. In fact, some transcription factors function as both repressors and activators: they bind at several DNA sites, repressing transcription of some genes and stimulating transcription of others.

3. **Antisense RNA.** Repressors and activators are protein transcription factors. However, in some cases, the control molecule is RNA, not a protein. Usually, RNA polymerase transcribes only the coding strand of DNA and then releases the mRNA. In some genes, however, RNA polymerase may also transcribe the other, noncoding, DNA strand, producing an **antisense RNA** molecule that codes for nothing. When this happens, the antisense RNA can base-pair with the mRNA, turning it into a double-stranded RNA, which cannot be translated.

More and more examples of gene regulation by antisense RNA are being discovered, in eukaryotes as well as in prokaryotes, although little is known about the function of most of these systems. One advantage of antisense RNA is that it can provide negative feedback control of the activity of genes with this type of regulation: the greater the rate of transcription, the more antisense RNA is produced, and the greater its effect in suppressing gene expression.

## 11–D   GENE CONTROL IN EUKARYOTES

Eukaryotes have all the means for controlling gene expression found in prokaryotes, and they also have additional systems. There is growing evidence that eukaryotic gene expression is controlled by a limited number of regulators interacting in a variety of ways. For instance, transcription of a particular gene may be switched on by interactions between several transcription factors or between transcription factors and a cofactor. Transcription factors themselves contain subunits, segments that are repeated within the molecule and occur in many different transcription factors. As well as these repeating subunits, each transcription factor contains unique segments that determine its particular activity.

The result of this "building block" system is that a relatively small number of transcription factors can control the expression of the thousands of genes a eukaryote may contain. Not only do these genes have to be switched on and off at particular times, but they must also be transcribed and translated at different rates depending on the circumstances.

Protein synthesis in both prokaryotes and eukaryotes is controlled by transcription factors that bind to DNA and switch RNA synthesis on or off. In addition, we shall see that eukaryotes have two ways of regulating gene expression that are not found in prokaryotes. First, eukaryotic introns appear to give these organisms new types of RNA that regulate gene expression. Second, eukaryotic chromosomes carry large amounts of histones not found in prokaryotes, and these play roles in regulating gene expression.

## Regulatory Sites

Transcription in a eukaryote requires that RNA polymerase and several other proteins assemble into an **RNA polymerase complex** bound to the promoter. When all are assembled, the other proteins move the polymerase, positioning it so that it can launch itself along the DNA. Binding of these proteins to DNA under the right conditions in a test tube is, by itself, enough to produce a low level of RNA synthesis.

A gene is not transcribed rapidly, however, unless transcription factors are also bound to various other regulatory sites in the genome. For instance, in the fruit fly *Drosophila* a gene called *hunchback* has six identical regulatory sites downstream of the structural gene. The transcription factor "bicoid" binds to these sites. In cells where bicoid is concentrated, all six regulatory sites are occupied, and *hunchback* is transcribed strongly. Where bicoid is less concentrated, fewer of the binding sites are occupied, resulting in weaker expression of *hunchback*. Eventually, in cells with bicoid concentrations below a certain threshold, *hunchback* is not transcribed at all.

Most eukaryotic genes are thought to be affected by at least five regulatory sites. These may be all the same, like the bicoid-binding sites, or different, and they may be in any position: upstream or downstream of the promoter, in the middle of the gene itself, on the opposite, noncoding DNA strand, or far from the gene they control. As in prokaryotes, there are transcription factors that inhibit transcription when they bind to regulatory sites as well as those that enhance transcription.

## 11–E    TRANSCRIPTION FACTOR STRUCTURE

A transcription factor usually has at least three parts. If the transcription factor interacts with the RNA polymerase complex, it has an amino acid sequence that recognizes this complex. Second, most transcription factors have sites that bind one or more cofactors or other transcription factors. Third, it has a DNA-binding site, an amino acid sequence that recognizes the sequence of nucleotides at its regulatory site on the chromosome.

Some transcription factors also have areas other than DNA-binding sites that permit them to cling to DNA, by poking into or curling around the double-stranded DNA molecule. Many transcription factors do this by sticking together in pairs, forming dimers, a structure that grips DNA. One example is found in both eukaryotic and prokaryotic transcription factors, including CAP in *E. coli* (Section 11–C). This is the **helix-turn-helix,** a structure in which a bent section links two alpha helices, which are common secondary structures in proteins. Two helix-turn-helix transcription factors can bind tightly to each other, and the pair binds to one complete turn of a DNA double helix (Figure 11–6).

## Leucine Zippers and Zinc Fingers

The most common amino acid sequence that allows transcription factors to form dimers is the "leucine zipper," discovered in 1987 by Stephen McKnight and his colleagues. Trying to determine the three-dimensional structure of a transcription factor, they coiled a model of part of the protein into an alpha helix. They found that residues of the amino acid leucine in every seventh position fell in a line

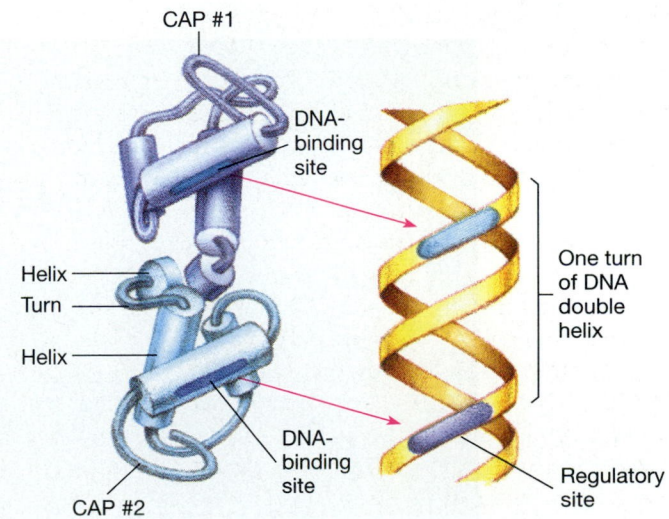

**FIGURE 11–6** Helix-turn-helix structure permits CAP transcription factor to bind DNA. Each CAP molecule consists of four alpha helices linked by short turns. In the presence of the cofactor cyclic AMP, two CAPs form a structure in which DNA-binding sites on the CAPs are oriented so they can attach to regulatory sites on DNA that are separated by one turn of the DNA double helix. This structure falls apart when cyclic AMP is not present.

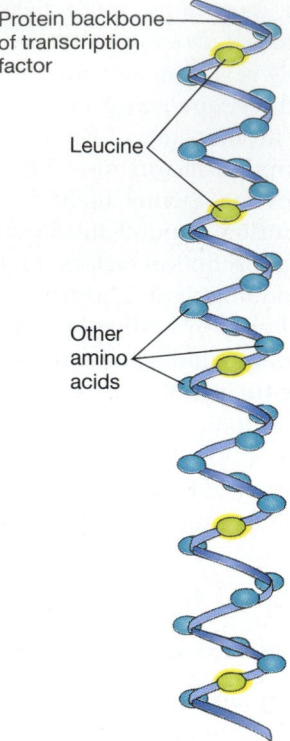

Protein backbone of transcription factor

Leucine

Other amino acids

**FIGURE 11–7** One side of a leucine zipper. In this part of the transcription factor, every seventh amino acid is a leucine. When a model of the protein is coiled into an alpha helix, the leucines line up in a row, as shown here.

down the helix (Figure 11–7). Leucine's R group is a nonpolar knob (see Figure 3–16). These hydrophobic knobs on one helix poke into holes formed by spaces between the knobs on another helix, making the two into a dimer.

Mutations that disrupt the leucine zipper prevent transcription factors from binding to DNA even though the zipper is not the DNA-binding site. To discover why the zipper is needed for binding, researchers compared the amino acid sequences of transcription factors from plants, fungi, and mammals, which have had separate evolutionary histories for a very long time. They reasoned that the amino acids necessary to attach transcription factors to DNA might have remained the same during evolution because the molecules would not work without them.

All the transcription factors they examined contained zippers with five leucines in the same positions on the molecule (Figure 11–8). In the DNA-binding region, the proteins also all contained one asparagine, an amino acid that causes a bend in an alpha helix. On either side of the bend are amino acids with basic R groups, which bind to acidic DNA. The positions of the basic amino acids are such that they nestle into the major groove on the DNA. The dimer holds onto DNA by a grip similar to the scissor grip of a wrestler with legs locked around the opponent's torso (Figure 11–8b). You can see why the leucine zipper is necessary to binding by imagining a pair of scissors or pliers with only one side.

Another protein configuration that permits a transcription factor to attach to DNA is the **zinc finger,** discovered

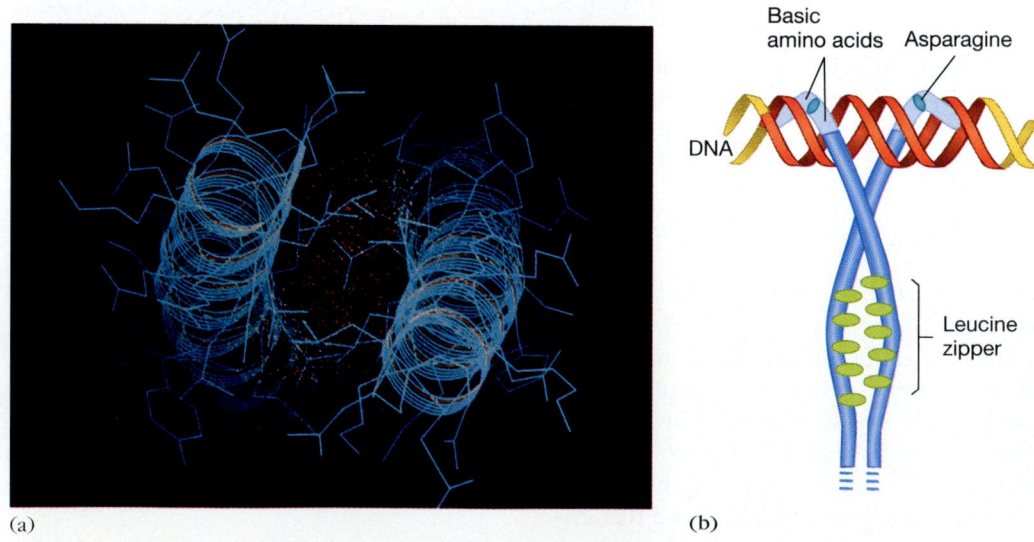

(a)

(b)

Basic amino acids   Asparagine

DNA

Leucine zipper

**FIGURE 11–8** Transcription factor dimers containing leucine zippers. (a) Computer model of an end view of the two parts of a leucine zipper (with the rest of the molecule removed). The alpha helix of each molecule in the zipper region is shown in pale blue. The stippling between the two helices illustrates the forces holding the hydrophobic leucines together. (b) How a dimer with a leucine zipper binds to regulatory sites on DNA. The asparagine near the end of each transcription factor causes a bend in the protein helix, which permits it to hook onto the DNA double helix.   (a, Tom Alber)

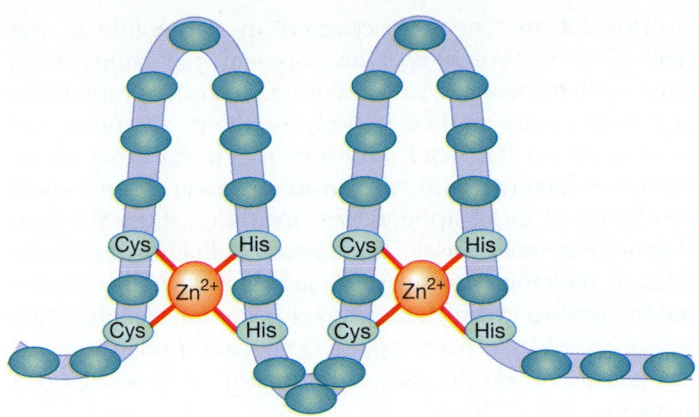

**FIGURE 11-9** Zinc fingers in a transcription factor. The amino acids cysteine and histidine bond to zinc ions, creating loops in part of the protein.

in 1985 by Daniela Rhodes and Aaron Klug. The zinc finger is a loop of about 30 amino acids held together by bonds between polar amino acids such as cysteine and a zinc ion (Figure 11-9). Proteins with as many as 37 zinc fingers sticking out from them are known.

## Cofactors

Cofactors influence the binding of transcription factors to DNA in many ways. One well-studied example is steroid hormones, which act as cofactors that permit transcription factors to form dimers. (Hormones may also affect cells in ways that do not involve transcription, discussed in Chapter 40.)

Steroid hormones (such as the sex hormones estrogen and progesterone) are lipids, synthesized from cholesterol. Hence, they are lipid-soluble and can cross the plasma membrane and enter a cell. Here they bind to **nuclear hormone receptors,** transcription factors that can bind to DNA and initiate transcription only when they are also bound to a specific steroid or thyroid hormone or vitamin. The cell may respond to the hormone in several stages. In the case of the insect molting hormone ecdysone, the ecdysone-receptor complex activates transcription of a few genes, some of which themselves encode transcription factors. These cause a secondary response by initiating RNA synthesis at as many as 100 more sites on the DNA.

All known nuclear hormone receptors contain two zinc fingers. One of the fingers carries the DNA-binding site, and the other hooks onto a second hormone receptor molecule to form a dimer (Figure 11-10).

The discovery that nuclear hormone receptors contain zinc fingers explains why sexual development is delayed in children whose diet contains too little zinc. Without enough

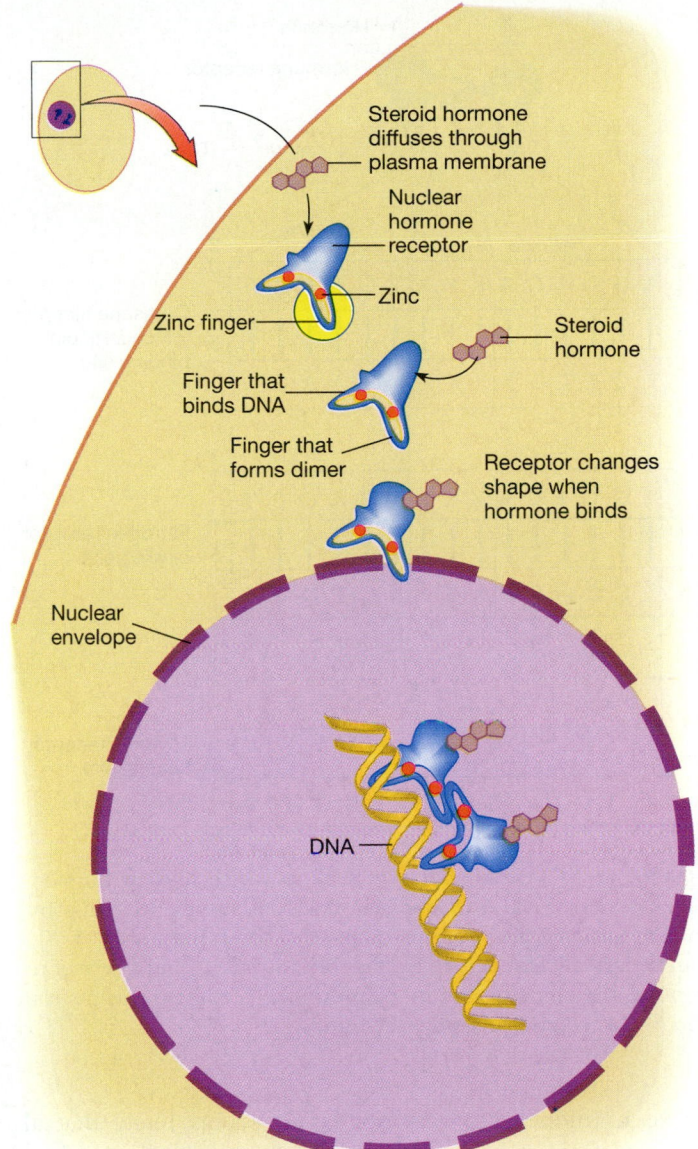

**FIGURE 11-10** How a nuclear hormone receptor works. Probably, the receptor must bind its hormone before it can bind to another receptor or to DNA. We do not know whether the dimer forms before or after attachment to DNA.

zinc, the body cannot produce a normal level of receptors for the sex hormones that cause puberty.

## DNA-Binding Sites

A transcription factor can recognize a remarkably short regulatory sequence in DNA if it is stabilized by forming a dimer. Figure 11-11 shows the DNA sequences bound by three nuclear hormone receptors. Each is only six nucleotide pairs long (12 pairs per dimer). The binding sites for the

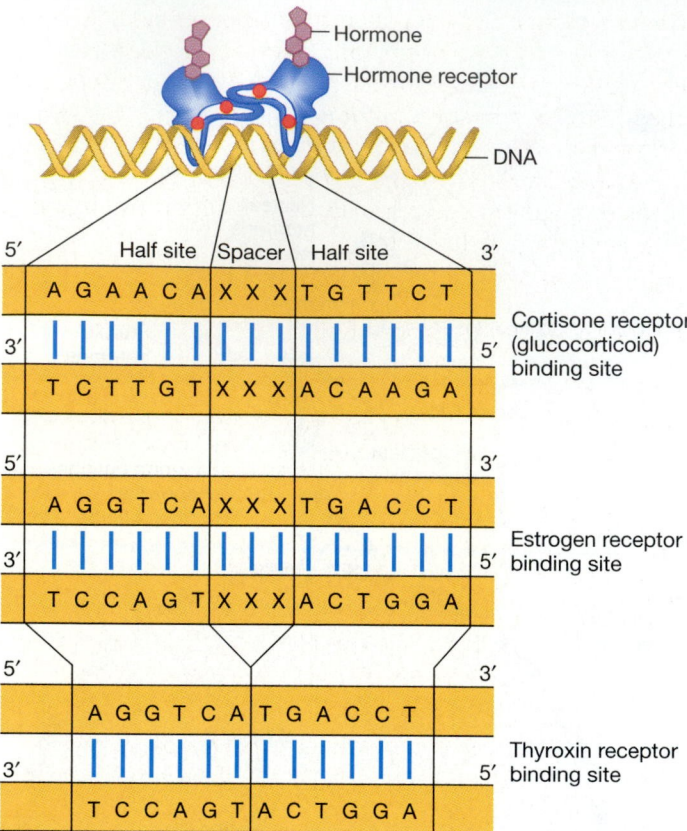

**FIGURE 11–11** DNA regulatory sites to which nuclear hormone receptors bind. These are the DNA nucleotide sequences (in mammals) that bind the receptor dimers for three hormones. X X X can be any nucleotides. (Note that because they bind a pair of receptors, the binding sites are palindromes, reading the same from the 5′ (or 3′) end on opposite strands.)

three hormones are very similar. It seems likely that all evolved from the same gene, which was duplicated long ago and then mutated into slightly different forms performing different functions.

## Homeoboxes

**Homeoboxes** are DNA sequences of about 180 nucleotide pairs, found within the genes for many transcription factors. Part of the homeobox codes for the amino acid sequence that forms the transcription factor's DNA-binding site.

Genes containing homeoboxes, **homeobox genes,** appear to be the master genes of development, coding for transcription factors that determine the overall pattern of the multicellular body. It seems likely that the relatively few transcription factors coded by homeobox genes, working together in different combinations, will prove to be the "letters" of the "command words" that control embryonic development (Section 12–F).

Homeoboxes were discovered in *Drosophila* in the 1980s. During *Drosophila's* development, regulatory genes cause each appendage to develop in a form appropriate to its position on a particular body segment: antennae and mouthparts on the head, and legs in the right places on thoracic segments. Many of the same structural genes must be expressed in all appendages, including those for production of nerves, muscle, and the chitin that makes up the external skeleton. However, the amount of various tissues and the pattern in which they develop are different for each type of appendage. Transcription factors encoded by homeobox genes turn on the right combination of genes in each segment.

Geneticists had found mutations that interfere with this development, turning off some of the genes needed to produce one appendage and turning on those needed for another. The genes responsible for these wholesale switches in transcription are **homeotic genes.** The nucleotide sequences of several homeotic genes were analyzed by Amy Weiner and Matthew Scott, and by Walter Gehring, using methods described in Chapter 13. They discovered that all homeotic genes contain very similar homeoboxes. The homeotic genes that control appendage development in *Drosophila* are examples of the many homeobox genes that apparently control development in all eukaryotes.

Homeoboxes in different eukaryotes differ only slightly in nucleotide sequence. They are evolutionarily conserved, preserved almost unaltered through millions of years of evolution. Similar homeoboxes have been found in genes of humans, mice, insects, and recently in jellyfish, thought to be descended from some of the earliest animals to evolve. Homeoboxes occur not just in animals, but also in plants and fungi.

Not only are homeobox genes similar in all eukaryotes that have been examined, but the number and order of these genes is highly conserved, at least in animals. For instance, *Drosophila* contains 10 homeobox genes on one chromosome. Mice and humans contain essentially identical strings of homeobox genes, arranged in the same order as in *Drosophila*. The main difference is that during evolution the string of 10 genes (or the chromosomes they are on) has been duplicated twice, so that mammals contain 40 of these homeobox genes, 10 on each of 4 chromosomes.

Proof that some homeobox genes are essentially the same in insects and mammals came from injecting the transcription factor coded by a mouse homeobox gene into a *Drosophila* embryo. Injected into the middle of the embryo, the protein had no effect. Injected into the head end, it caused the front end of the embryo to develop as if it were the middle, producing the spiky hairs usually found on the underside of the thorax instead of the normal head skeleton. This means that at least one transcription factor that controls differentiation in the embryo is essentially the same in insects and mammals.

Most transcription factors are very small proteins, but most of them have another string of amino acids beyond the few dozen needed to bind DNA, cofactors, and each other. The probable function of these remaining amino acids brings us to the next section of the story.

## 11–F  CHROMOSOMAL PROTEINS

Most of the early work on gene control in eukaryotes was performed *in vitro* ("in glass"), in flasks and test tubes, isolated from the organism. However, by the late 1980s it was clear that gene expression *in vitro*, using DNA without the chromosomal proteins that normally surround it, differed in some ways from that in a living cell. To take two examples:

1. **Long-range activation.** Some regulatory sites lie as far as 1500 nucleotides from the promoters of the genes they control. No one had ever managed to activate a regulatory site *in vitro* if it lay more than about 200 nucleotides from the promoter, even though this process obviously happens in the intact cell.
2. **Threshold activation.** In an embryo, genes are often switched from off to on when the concentration of a cofactor builds up to some threshold concentration (as in the bicoid/*hunchback* example in Section 11–D). *In vitro*, this did not happen. Instead, as the concentration of a cofactor built up, transcription of the gene increased steadily, never switching suddenly from off to on.

Could these differences be due to the absence of chromosomal proteins from DNA *in vitro?* Researchers suggested that chromosomal histones played an active role in gene expression, as well as coiling DNA into nucleosomes (Section 9–G). Recently, several groups of researchers have demonstrated that adding histones to DNA *in vitro* causes gene activation that is much more life-like than that using naked DNA.

***The Action of Histones***  In eukaryotes, it may take five or more transcription factors to switch a gene on or off, and many of these proteins bind far from the gene's promoter. How do the proteins interact with each other and with the promoter? At least in some cases, it appears that several transcription factors have to attach to each other to affect gene expression. A transcription factor may attach to another transcription factor, or to a histone, and physically link two parts of the chromosome. Sometimes this produces a visible loop in the DNA.

If you add histones to naked DNA, the DNA wraps around them, forming nucleosomes. Nucleosomes do not form at random positions on a chromosome; they are particularly likely to include a promoter. This blocks transcription because RNA polymerase and its associated proteins cannot bind to a promoter wound in a nucleosome.

However, once RNA polymerase has bound to DNA, it can transcribe DNA downstream, whether nucleosomes are present or not. Experiments on how histones affect transcription are described in *How Do We Know? The Role of Histones in Transcription.*

So far, research suggests that transcription of a gene whose promoter lies in a nucleosome is often controlled in two steps (Figure 11–12):

I  HISTONE-DEPENDENT ACTIVATION

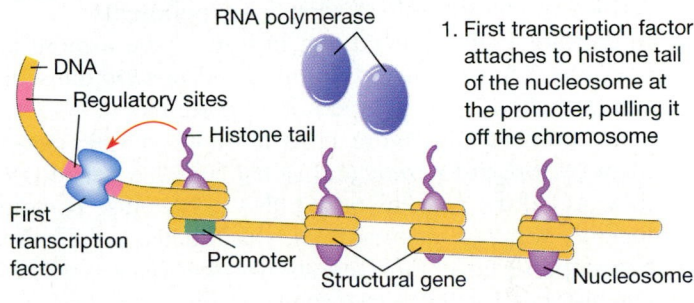

1. First transcription factor attaches to histone tail of the nucleosome at the promoter, pulling it off the chromosome

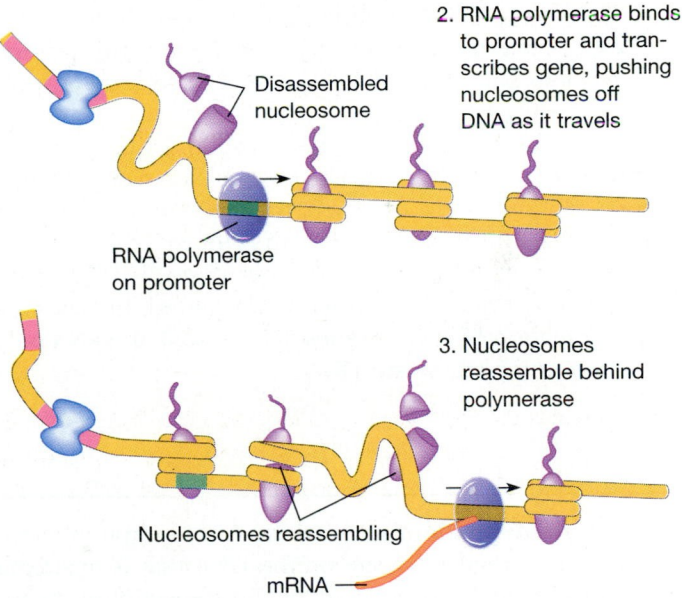

2. RNA polymerase binds to promoter and transcribes gene, pushing nucleosomes off DNA as it travels

3. Nucleosomes reassemble behind polymerase

II  HISTONE-INDEPENDENT ACTIVATION

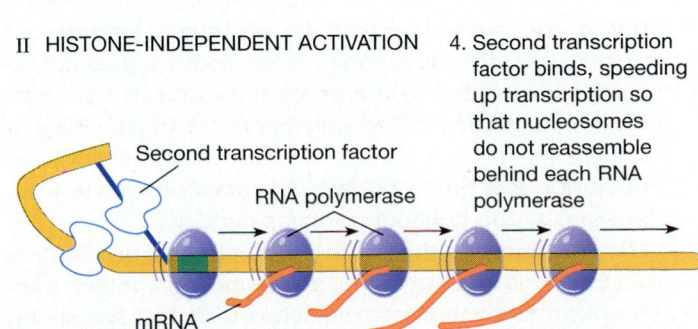

4. Second transcription factor binds, speeding up transcription so that nucleosomes do not reassemble behind each RNA polymerase

**FIGURE 11–12**  Summary of what we know about the role of histones in gene expression.

## THE ROLE OF HISTONES IN TRANSCRIPTION

Experiments in the 1960s suggested that the histones in nucleosomes help to control gene expression in eukaryotes. Biologists thought that the role of histones was to turn off, more or less permanently, the many genes that are not transcribed in mature cells. A series of experiments in the late 1980s showed that histones have a more active role. The experiments all involved *in vitro* transcription of naked DNA, stripped of its proteins.

Experiment 1 in Figure 11–A illustrates a 1986 experiment by Joseph Knezetic and Donal Luse. To naked DNA they added the proteins of the RNA polymerase complex together with histones and found that the histones blocked transcription. Roger Kornberg showed that this was because the histones formed nucleosomes at promoter sites, preventing the polymerase complex from binding to the DNA.

These experiments showed that when histones and RNA polymerase compete for a promoter, the histones win. Does this mean that histones can knock the polymerase off DNA?

Experiment 2 answered this question. Donald Brown's group added RNA polymerase to naked DNA, starting transcription (Step 1). Then they added histones and found that transcription continued even when the histones assembled into nucleosomes (Step 2). The results showed that once transcription had started, histones and the formation of nucleosomes did not push the RNA polymerase complex off the DNA.

Experiments 1 and 2 showed that histones and RNA polymerase compete for promoters. What then is the role of a transcription factor? In 1988 Michael Grunstein's students Min Hang and Ung-Jin Kim removed most of the nucleosomes from yeast DNA and measured transcription of individual genes. Transcription of those genes normally switched on by transcription factors speeded up. In contrast, transcription of housekeeping genes, those the cell needs all the time, continued unaffected. Apparently the promoters of housekeeping genes are not normally bound by nucleosomes, so removing the nucleosomes has no effect. This suggested that transcription factors might act by removing nucleosomes from promoters.

Experiment 3 tested this theory. It is the same as Experiment 1 except that transcription factor is added to the naked DNA at the same time as histones and RNA polymerase. This time transcription occurred. Binding of transcription factor to the DNA prevented histones from binding (or removed them when they did bind). As a result, RNA polymerase could bind to the DNA and transcribe it.

The conclusion from these three experiments is that transcription of many genes is prevented by the binding of histones to their promoters. Transcription factors switch on these genes by removing nucleosomes from the promoters so that RNA polymerase can bind and transcribe the gene. In practice, nucleosomes, RNA polymerase, and transcription factors compete for promoters, and the balance between them determines whether the gene is transcribed or not.

1. **Histone-dependent activation.** A transcription factor binds to a regulatory site on the DNA that is upstream from the nucleosome. Part of the transcription factor then binds to the "tail" of a histone molecule sticking out of the nucleosome at the promoter, causing the nucleosome to disassemble. Now an RNA polymerase complex can bind to the promoter and start transcribing the gene. Once the polymerase is transcribing, it knocks any nucleosomes it encounters off the DNA, an operation that slows it down. The nucleosomes re-form as soon as the polymerase has passed.

2. **Histone-independent control.** Once the promoter is accessible to transcription, the same or another transcription factor binds a regulatory site and speeds up transcription by increasing the rate at which additional RNA polymerase molecules attach to the promoter. When polymerase molecules are traveling down the DNA close together, the nucleosomes do not have time

to re-form between them. This explains the absence of nucleosomes on lengths of DNA where transcription is occurring rapidly.

## 11–G   REGULATORY RNA

Nearly all the hundreds of molecules known to control transcription are proteins. Only recently have a number of regulatory RNAs been discovered, and some of these turn out to be introns (Section 10–D). John Mattick suggested that the ability to evolve a variety of different introns gave eukaryotes a new method of regulating gene expression that prokaryotes do not possess and that permitted the evolution of embryonic development. The theory is supported by the fact that the first few RNA regulators to be analyzed control major and long-lasting switches in development. It will be tested as more regulatory RNAs are discovered.

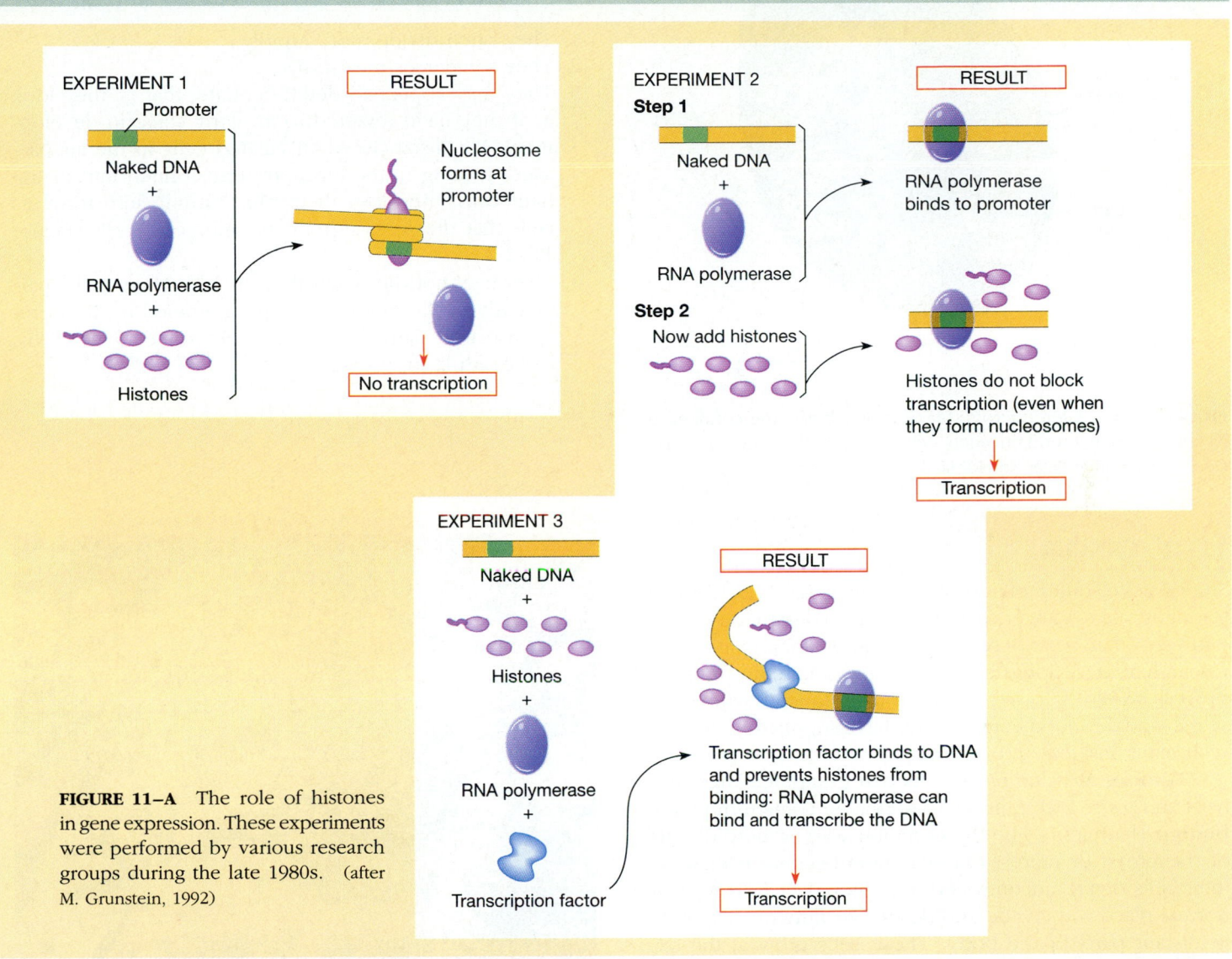

**FIGURE 11–A**   The role of histones in gene expression. These experiments were performed by various research groups during the late 1980s.   (after M. Grunstein, 1992)

## X–Chromosome Inactivation

Women and other female mammals have two chromosomes called **X chromosomes** in each cell, whereas males have only one. In females, one of the two X chromosomes in each cell is condensed, tightly coiled, with most of its genes turned off. This turned-off X chromosome is called a **Barr body,** and it can be seen with a light microscope (Figure 11–13).

At the beginning of embryonic development, genes from both X chromosomes are transcribed. In the germ cells, both X chromosomes remain active. But in the body cells, most of one X chromosome is inactivated. Which chromosome is inactivated appears to depend on chance and varies from one cell to another. Once X–inactivation has occurred, all of the cell's descendants will have the same X chromosome condensed. In other words, once the genes are switched off in a Barr body, they remain that way through many cell divisions.

In 1992 researchers found that the *XIST* gene, which converts one X chromosome into a Barr body, achieves its effect without ever making a protein. In 1994 a team led by Jeanne Lawrence found evidence that *XIST* transcribes an RNA molecule that binds to the X chromosome, converting it into an inactive Barr body.

Whether the X–chromosome inactivator is transcribed from an intron is not yet known. But a regulatory gene that is an intron has been discovered in the roundworm *Caenorhabditis elegans*. It acts much like the antisense RNA of prokaryotes (Section 11–C). A gene that controls embryonic development encodes a small RNA that binds to the messenger RNA of another gene, preventing it from being translated into a protein.

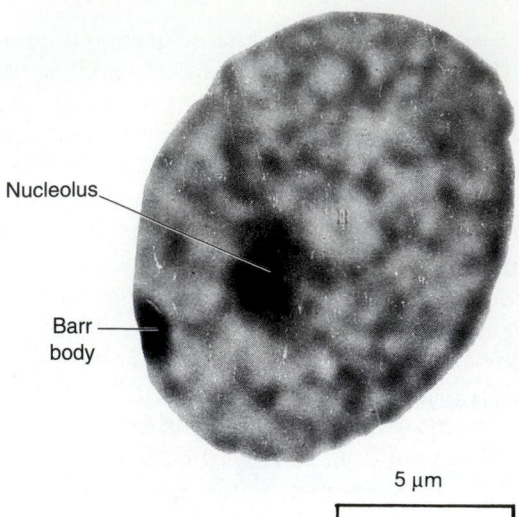

Nucleolus

Barr
body

5 μm

**FIGURE 11–13**  Light micrograph of a Barr body, the condensed X chromosome found in each cell of a female mammal.  (Dr. Dorothy Warburton/Peter Arnold, Inc.)

## 11–H   CANCER

Studies of cell differentiation and of cancer have been linked for many years. Cancer often results from disruptions in the pathways that control cell division and differentiation, so researchers have long felt that an understanding of either process would help explain the other. Indeed, cancer research and developmental biology have supplied each other with many insights and techniques.

**Tumors** are clumps of cells that escape cellular controls and grow and multiply abnormally. Some tumors are **benign** (harmless), like the common wart or fibroid cysts of the breast or uterus. Other tumors become malignant: their cells divide uncontrolled, and they detach from their normal place and invade and destroy healthy tissues, often in distant parts of the body. These cells grow at the expense of their neighbors, sometimes ending in death of the organism. At other times, malignant cells are killed or suppressed by the defense mechanisms of other cells in the body. A **cancer** is a malignant tumor. Cancer is not a single disease but any one of a number of conditions in which cells form such tumors.

Cancers develop in steps, which are not necessarily the same from one case of cancer to another, even for the same kind of cancer. The first step occurs when one cell undergoes a mutation. This mutation will be passed on to the cell's offspring. Later, additional changes may occur in cells of this lineage and also be passed on. Eventually, a cell bearing accumulated changes sustains one more that transforms it into a cancer cell, which divides abnormally and forms a tumor. Cancer cells may **metastasize,** detaching from their neighbors and traveling to other parts of the body, where they invade healthy tissue and start new tumors. This whole process usually takes years.

As cancerous cells develop, they come to differ from their normal neighbors in one or more of four main ways:

1. They often divide more rapidly.
2. They sometimes metastasize.
3. They may appear to dedifferentiate; that is, they look as if they have reverted to an early stage in development. For instance, when ciliated cells in the air passages leading to the lungs are transformed into malignant cells, they lose their cilia, turning into formless cells that divide as rapidly as embryonic cells (Figure 11–14).
4. They may not die when they should. Many cells, particularly in the embryo, are programmed to die in response to a particular stimulus. In some cancer cells, cell death is suppressed.

**FIGURE 11–14**  False color scanning electron micrograph of the most common form of human lung cancer, in the bronchi. The normal cells are covered with cilia, which sweep dust particles up into the throat. The cancer cells are orange and do not have cilia. (Photo Researchers)

## What Causes Cancer?

Just as cancer is not a single disease, it also has no one cause. Many factors usually contribute to a single cancer. Because cancers take years to develop, the traditional way to discover the cause of a cancer is by correlation studies: comparing a group of people with cancer with a matched group of healthy people to see what factors are more common in the backgrounds of people with cancer. As long ago as 1775, Sir Percival Potts noted that chimney sweeps were more likely than most men to develop cancer of the scrotum, suggesting that the soot in chimneys caused this kind of cancer. In the 1960s cigarette smoking was linked to most cases of lung cancer, and it is now known to increase the risk of several other types of cancer as well.

Some cancers occur in the presence of **oncogenic** (cancer-causing) viruses. Some cancers also run in families, and genes that predispose individuals to particular cancers have been identified. However, people who inherit these genes do not always develop the cancer, showing that additional factors are necessary for the cancer to develop.

In theory, the changes that turn a normal cell into a cancer cell might result from a mutation in the DNA or from a change in gene expression. The vast majority of cancers that have been studied actually involve mutations, although the end result may be a change in gene expression. For instance, Wilms' tumor, a childhood cancer of the kidney, is caused by a mutation in the gene for the transcription factor that normally shuts down growth of the kidney when it has reached full size. In people with the mutation, just one amino acid in a zinc finger is altered: the transcription factor cannot bind to its regulatory site, and so the kidney continues to grow, eventually killing the child.

A **carcinogen** is a factor that increases the risk of cancer (Table 11–1). Not surprisingly, most known carcinogens are also **mutagens,** agents that cause genetic mutations.

## TABLE 11–1

### Some Known Carcinogens

| Carcinogen | Comments |
| --- | --- |
| Asbestos dust Chromium compounds Some petroleum products | Workers exposed to these have a high risk of lung cancer, especially if they smoke |
| Tobacco | 12–15% of cigarette smokers die of lung cancer; especially risky in combination with asbestos or radon |
| Radon | Radioactive gas produced by breakdown of uranium; present in many porous soils containing uranium. May seep up through soil and become trapped in a house; underground miners may also be exposed; risk factor for lung cancer and leukemia |
| Arsenic | Element that is a common water pollutant. Found in soil in low concentrations; concentrated in surface and ground water by irrigation and industrial processes; risk factor for many types of cancer |
| Vinyl chloride | Used in manufacture of plastics; risk factor for one type of liver cancer |
| DDT | Organo-halogen pesticide, now banned in the United States. Exposure increases risk of breast cancer |
| Estrogen | Mammalian hormone. In large amounts, increases risk of uterine cancer unless the hormone progesterone is also present |
| X-rays | Many people have unnecessary medical x-rays; ionizing radiation is a risk factor for leukemia |
| Benzene | Once a common solvent in laboratories; main sources now are smoking, auto exhausts, household solvents |
| Nitrates and nitrites | Converted into carcinogenic nitrosamines in digestive tract. Common as a food preservative and in most green vegetables. Also common water pollutants in agricultural areas where nitrogen fertilizer or manure runs off farmland. |
| Aflatoxins | Produced by fungus *Aspergillus flavus* when growing on food such as peanuts and corn. Risk factor for liver cancer, especially combined with hepatitis B infection |
| Ethylene dibromide | Pesticide used to kill insects and roundworms; common pollutant in ground water |
| *Helicobacter pylori* | Infection with this bacterium (common in Asia and Africa) increases the risk of stomach cancer |

For instance, carcinogens include ionizing radiation such as x-rays, which often break chromosomes; chemical carcinogens, which tend to cause point mutations; and viruses, which introduce foreign DNA into a cell. To avoid exposing the public to unnecessary risk of mutations, including those that are carcinogenic, the U.S. Food and Drug Administration requires that manufacturers test drugs and food additives for mutagenic properties before they can be approved for public use. The commonly used **Ames test** detects small, local mutations in the DNA of bacteria. However, it does not detect some other types of mutations that can cause cancer.

Most mutations do not cause cancer. Every gene in a human being changes many times in a lifetime. Usually, the body's DNA repair systems prevent these changes from becoming mutations that are inherited at cell division. There are also cells in the immune system that recognize and destroy cells damaged by mutations, including cancerous and pre-cancerous cells.

Furthermore, a single heritable mutation is not enough to cause cancer. If it were, cancers would be equally likely to occur at any age. In fact, the incidence of nearly all types of cancer rises steeply with age. From such statistics, it has been calculated that between three and seven independent mutations are required to turn a normal cell into a cancerous cell. The control of cell division has multiple backups: if one part fails, another prevents improper cell division, and so it takes several mutations for a cell to escape all these controls and become cancerous. Carcinogens in the environment merely increase the rate of mutation, increasing the likelihood that a sufficient set of mutations will accumulate in a single cell.

## Genetic Basis of Cancer

Mutations that can convert a normal cell to a pre-cancerous state occur in two kinds of genes. One kind consists of **tumor suppressor genes.** Few of these genes are known. Most of them code for small proteins that inhibit cell division. An example is the kidney transcription factor that is abnormal in cases of Wilms' tumor.

The other, and more common, genes found in cancerous cells are **oncogenes,** genes that cause cancer, discovered by Michael Bishop and Harold Varmus. Oncogenes form by mutation of 60 or so known **proto-oncogenes,** normal genes that code for growth factors, transcription factors, or proteins involved in cell death or cell adhesion. If these genes mutate so that they do not function normally, the cell may become cancerous.

A proto-oncogene is often converted to an oncogene by a mutation that causes the gene to be overexpressed. Sometimes such a mutation does not change the proto-oncogene but produces extra copies of it or brings the gene under the control of a transcription factor that increases its expression. Overexpression means that too much of the gene's protein is made, which may make the cell divide more rapidly than usual and form a tumor.

An example of a proto-oncogene is the *ras* gene, which codes for Ras protein. Ras is a messenger in the intracellular pathways by which cells divide in response to growth factors. When *ras* mutates into an oncogene, it produces a Ras protein that no longer responds to growth factors but constantly promotes cell division, whether the growth factor is present or not.

## Development of Cancer

Even after the mutations that cause a cancer have occurred, the cancer may take years to develop. The incidence of lung cancer does not increase until after about 20 years of heavy smoking. Those exposed to radiation when atomic bombs were dropped on Hiroshima and Nagasaki in 1945 did not have an increased incidence of leukemia caused by the radiation until more than five years later.

**Tumor promoters** are stimuli that are carcinogenic, although they do not cause mutations. They act during the slow process of cancer development that may follow genetic transformation of a normal to a pre-cancerous cell. Common examples of tumor promoters are sex hormones: many tumors of the reproductive organs develop only in the presence of appropriate levels and combinations of particular sex hormones. This is probably why the older a woman is before her first full-term pregnancy, and the less time she breast-feeds, the higher her chances are of developing breast cancer later in life. Pregnancy and breast-feeding apparently cause permanent changes in gene expression in breast cells, which alter responses of the cells to hormones later. These changes are believed to be part of tumor development, not of the genetic mutations that initiate cancer.

A specific combination of factors is usually necessary for a cancer to develop. This is illustrated by the fact that rates for different cancers vary among countries. When groups of people move from one country to another, their risk of cancer comes to resemble that in the host country, not in the country of origin. From such data, it has been calculated that 80% of cancers could be prevented by avoiding combinations of circumstances that are known to increase the risk of particular types of cancer.

A potential cancer may be stopped by surgical removal or irradiation of the tumor while it is still a localized group of cells. Once it metastasizes throughout the body, it is very difficult to locate and destroy all the cells that may be able to cause tumors, and so a metastasized cancer is almost impossible to cure. Hence, the easiest way to decrease the death rate from cancer is not treatment but teaching people to avoid risk factors. For instance, the number of new cases of lung cancer and liver cancer can be reduced by avoiding prolonged exposure to cigarette smoke and vinyl chloride. In many cases, however, avoiding risk factors is not practical. For instance, most of us will continue to live with our own sex hormones, whatever problems they may cause.

Cancers are the second most common cause of death in the United States, accounting for 20% of all deaths. This figure scares many people but, to put it into perspective, note that an American has almost as great a chance of dying of homicide as of cancer. Deaths from cancer have increased in the twentieth century. This is partly because cancers tend to develop in later life, and people are living longer instead of dying of infectious bacterial diseases in early life as they used to. (Life expectancy in the United States is now almost 80 years, compared with about 45 years in 1900.) However, there is now a convincing body of evidence that cancer rates are increasing because carcinogens are becoming more common in our environment.

## Oncogenic Viruses

One type of genetic change that can lead to cancer may occur when oncogenic viruses enter a cell, bringing with them genes that affect the control of cell division or cell death. For example, Rous sarcoma virus of chickens introduces a gene coding for a protein very similar to a normal growth factor that stimulates cell division.

The other kind of change brought about by oncogenic viruses is placing normal cellular genes under control of viral promoters, so that the gene is transcribed and translated more often than it should be, and the cell contains an excess of the protein product. This idea was confirmed by attaching viral promoters to a proto-oncogene isolated from a normal cell, copying the resulting molecule, and introducing the copies into other normal cells: they became cancerous.

Cancers caused solely by viruses are not common in humans, probably because the immune system is efficient at killing body cells infected with viruses before they cause damage. In fact the only virus shown to act alone to cause a human cancer is HIV, the AIDS virus. The virus inserts its genetic material into a cell's genome and switches on a nearby oncogene that causes non-B-cell lymphoma, a cancer of the immune system.

Nevertheless, viruses are known to be risk factors for several human cancers (Table 11–2). For instance, children with Burkitt's lymphoma, a lymph cancer common in parts of Africa, invariably contain Epstein-Barr (EB) virus, which causes a rearrangement of DNA within their chromosomes. However, Burkitt's lymphoma is not common in the United States or northern Europe, even though the EB virus is widespread. (The virus is a member of the herpes group and also causes infectious mononucleosis and depression.) This suggests that whatever other factors besides the EB virus contribute to the development of Burkitt's lymphoma are more common in Africa than in the United States.

**TABLE   11 – 2**

## Viruses Associated with Human Cancers

| Virus | Increases Risk of These Tumors | Main Areas of Incidence | Comments |
|---|---|---|---|
| Papilloma virus (many strains) | Warts (benign) Carcinoma* of cervix and penis | Worldwide Worldwide | Smoking increases risk of cancer; virus spread by sexual activity |
| Hepatitis B virus | Liver cancer | Tropical Africa, SE Asia; increasing rapidly worldwide | Virus spread by sexual activity and by needle-sharing; risk of cancer increased by aflatoxin from fungus on food and by alcoholism, smoking, other viruses |
| HTLV-I (Human T-cell leukemia virus) | Adult T-cell leukemia and lymphoma | Japan, Caribbean | Poorly understood |
| HIV-I (AIDS virus: Human immunodeficiency virus) | Causes non–B-cell lymphoma; increases risk of Kaposi's sarcoma† | Central Africa | Virus spread by sexual activity, needle-sharing; risk of infection increases with sexually transmitted diseases |
| Epstein-Barr virus | Kaposi's sarcoma, Burkitt's lymphoma, carcinoma of nasal passages | Africa | Unknown factors contribute to cancer |

*A carcinoma is a cancer arising from epithelial tissue.

†A sarcoma is a cancer arising from muscle or connective tissue.

## SUMMARY

Most protein synthesis in eukaryotes and prokaryotes is regulated by controlling gene transcription. Research is rapidly adding to our understanding of how gene expression is controlled at this and other points in the process.

1. With rare exceptions, all the cells of an adult plant or animal contain the same genome as the zygote. Evidence includes the fact that single differentiated cells of some plants will develop into mature plants and that nuclei from some differentiated cells can take the place of the zygote nucleus and support the development of some animals.

2. During embryonic development, cells with identical genes come to differ from one another as different genes are expressed or switched off in different cells. Visible evidence of this comes from studies of giant polytene chromosomes in some flies and lampbrush chromosomes in amphibian oocytes.

3. Transcription starts when RNA polymerase binds to the promoter of a gene and is positioned to move along the gene. Protein synthesis is switched on and off when the binding of RNA polymerase is prevented or facilitated. In prokaryotes, transcription of particular genes is switched on or off when transcription factors bind or release the operator region of an operon. Cofactors, which are often food molecules, control the binding of transcription factors to operators.

4. In addition to repressor and activator transcription factors, protein synthesis in prokaryotes is sometimes controlled by synthesis of antisense RNA, which binds complementary messenger RNA and prevents its translation.

5. Eukaryotes have homeobox genes that control the overall pattern of differentiation. These code for transcription factors that have massive effects, causing dozens of genes to be switched on and off in various tissues and in specific sequences.

6. Eukaryotes also have methods of regulating gene expression not found in prokaryotes: new types of regulatory RNA molecules that bind DNA or messenger RNA, preventing transcription or translation, respectively, and histone proteins that block binding of RNA polymerase to promoters.

7. In eukaryotes, transcription factors that speed or slow transcription bind to regulatory sites on DNA. Many transcription factors, such as nuclear hormone receptors, bind DNA by forming dimers, whose two halves clasp the DNA. Helix-turn-helix, leucine zippers, and zinc fingers are among the protein secondary structures that permit transcription factors to form dimers.

8. Differences between transcription *in vitro* and in a cell led to the realization that histones not only wind DNA in nucleosomes but are also normally involved in the control of transcription of at least some genes. Nucleosomes are more likely to form at promoters than at other sites on the DNA, and binding between a transcription factor and a histone is sometimes necessary to release the promoter from a nucleosome, after which transcription can occur.

9. X–chromosome inactivation during development of female mammals is an example of gene regulation controlled by an RNA.

10. Cancer may develop when a cell has accumulated several mutations of tumor suppressor genes or proto-oncogenes that cause the controls of cell division, differentiation, death, or adhesion to function abnormally. Mutations that may lead to cancer are caused by various chemical mutagens, ionizing radiation, and new DNA introduced into the cell by viruses. The actions of carcinogens, including tumor promoters such as sex hormones, often contribute to the development of cancer.

## Self-Quiz

1. During differentiation, cells with the same DNA:
   a. must develop similarly
   b. divide at equal rates
   c. contain different genes
   d. may transcribe different genes

2. Protein synthesis is most often controlled by:
   a. alterations in ribosome action
   b. binding of transcription factors to regulatory sites
   c. increasing the rate of tRNA synthesis
   d. controlling access of mRNA to the cytoplasm

3. Steroid hormones are:
   a. transcription factors
   b. cofactors that permit transcription factors to bind to DNA
   c. mutagens
   d. proto-oncogenes

4. How do food molecules induce prokaryotic cells to make enzymes that metabolize the food? Food molecules:
   a. cause a cofactor to bind to DNA and attract RNA polymerase, which transcribes mRNA coding for the needed enzymes
   b. cause repressor proteins to leave or activators to bind the DNA, which can then be transcribed to mRNA
   c. bind to tRNA, which carries them to the ribosomes for protein synthesis
   d. bind to an RNA polymerase complex and activate it
   e. All of the above

5. Homeoboxes are:
   a. transcription factors that cause histones to release promoters from nucleosomes
   b. the genes coding for nuclear hormone receptors

c. small segments of DNA, found in all eukaryotes, which code for parts of transcription factors that control major switches in developmental pathways

d. mutations that cause major changes in development, such as formation of an abdomen where a thorax should be

6. Which of the following is *not* true as far as we know from recent advances in the understanding of gene expression in eukaryotes?

a. An RNA polymerase complex must bind the promoter of a structural gene before RNA polymerase can transcribe the gene.

b. Binding of transcription factors to DNA speeds up or slows down transcription.

c. Some transcription factors act by binding histones.

d. Many transcription factors bind to DNA in the form of dimers, consisting of two molecules bound to each other.

e. Nucleosomes must be completely removed from DNA before RNA polymerase can transcribe it.

7. Which of the following is *not* true of nuclear hormone receptors?

a. The receptors for several steroid hormones and thyroid hormones all contain two zinc fingers.

b. Each receptor molecule contains at least three binding sites for other molecules.

c. The receptors are embedded in the plasma membrane, where they can interact with hormones circulating in the blood.

d. The receptor-hormone complex binds to DNA as a dimer.

8. Which of the following is true?

a. Cancer can develop as the result of one large mutation in a cell.

b. Chemical carcinogens that are not mutagens sometimes act alone to cause cancer.

c. Oncogenes are tumor promoters that have evolved from normal genes.

d. Proto-oncogenes are normal genes that control functions such as cell division, cell death, and formation of growth factors.

## Questions for Discussion

1. Ecdysone stimulates the same puffing sequence in DNA in some tissues not directly involved in molting as in those that are. Does this disprove the theory that ecdysone triggers molting by causing transcription of specific genes?

2. During development of the parasitic roundworm *Ascaris* and in some gall midges, some somatic cells destroy entire chromosomes and therefore genetic information that was present in the zygote. What might be the selective advantage to these organisms of losing these chromosomes?

3. Even though some 20% of Americans die of cancer, life expectancy in developed countries is about 80 years. Would the enormous sums spent on cancer research, therefore, be better spent on diseases such as AIDS that kill most of their victims when they are much younger, or on diseases such as Alzheimer's that damage the quality of the patients' (and their relatives') lives for much longer than cancer does?

## Suggested Readings

Alberts, B., et al. *Molecular Biology of the Cell,* 2d ed. New York: Garland Publishing, 1989. Chapter 21 is an excellent discussion of cancer.

Browder, L. W., C. A. Erickson, and W. R. Jeffery. *Developmental Biology,* 3d ed. Philadelphia: Saunders College Publishing, 1991. An embryology textbook with a good section on differentiation.

Grunstein, M. "Histones as regulators of genes." *Scientific American,* October 1992. The experiments that elucidated the role of histones in gene regulation.

Liotta, L. A. "Cancer cell invasion and metastasis." *Scientific American,* February 1992.

McGinnis, W., and M. Kuziora. "The molecular architects of body design." *Scientific American,* February 1994. How homeobox genes control differentiation.

McKnight, S. L. "Molecular zippers in gene regulation." *Scientific American,* April 1991. The discovery of leucine zippers and how they permit transcription factors to bind DNA.

Rhodes, D., and A. Klug. "Zinc fingers." *Scientific American,* February 1993. The discovery of zinc fingers and of the remarkable similarity in different organisms of nuclear hormone receptors and the DNA regulatory sequences to which they bind.

# Embryonic Development in Animals

## OBJECTIVES

*When you have studied this chapter, you should be able to:*

1. Use the following correctly: zygote, morula, blastocoel, blastula, gastrula, neurula, differentiation, determination, embryonic induction, homeobox, apoptosis
2. Describe the process of fertilization and state its function and importance.
3. Name the four main stages of embryonic development and list definitive features of each.

4. Describe evidence that cleavage is controlled by factors in the egg cytoplasm.
5. List or recognize body parts formed from each of the germ layers (ectoderm, mesoderm, and endoderm).
6. Describe an experiment that shows the difference between determination and differentiation.
7. Outline what is known about determination.

E mbryonic development is the remarkable process by which each of us originated from a fertilized egg only just large enough to see without a microscope. Development is the means by which the genes we inherit from our parents are expressed, and it is also a reflection of our evolutionary history. Despite more than a hundred years of studying this complex process, biologists are only beginning to understand it.

Although developmental changes occur in all organisms, they are most obvious in multicellular plants and animals. These originate as single cells, such as spores and fertilized eggs, or as buds that later separate from the parent. In every case, the organism must undergo developmental changes before it matures. Embryonic development starts after sexual reproduction in both plants and animals. The single-celled **zygote** (fertilized egg) becomes an **embryo,** a multicellular developmental stage that cannot survive outside the seed, egg, or mother's body. Development has been studied more intensively in animals than in plants. The development of plant embryos is discussed in Chapter 46.

It is becoming increasingly clear that development is fundamentally similar in all eukaryotes and depends on genes and mechanisms that evolved in early organisms. Development involves several types of processes, which unfold and interact as genetic information expresses itself. The first of these is **differentiation** into cell types. From one fertilized zygote, which is one type of cell, cells as dif-

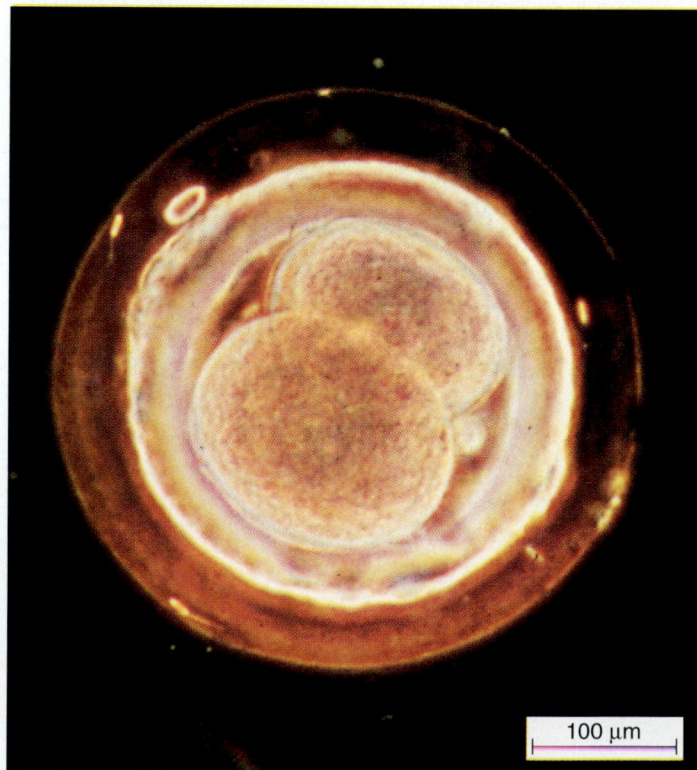

**FIGURE 12–1** The first cell division of a mammal (mouse) zygote, surrounded by its protective membranes. (Biophoto Associates)

ferent as liver, muscle, nerve, and skin cells are produced. These cells differ from one another in that they have synthesized different enzymes and structural proteins. This must be because different genes have become active in different cells during development.

The second aspect of embryonic development is **growth.** At various stages in development, the embryo receives food from the outside and so can increase in overall size. Before food reaches the embryo, no overall growth occurs, but by this time the number of cells in the embryo has increased enormously by cell division (Figure 12–1).

A third aspect of embryonic development is the formation of shape. This consists, on a molecular level, of the rearrangement of proteins and other molecules into structures within cells and, on a larger scale, of the movement of cells into specific patterns that form organs. Formation of shape involves **induction,** the process whereby one set of cells influences the development of others. It also involves programmed cell death and cell movement. Cell death is particularly obvious in organs such as hands and feet that contain indentations, some of which are formed by the death of specific cells. Its cytoskeleton permits a cell to move, and many cells move around in an embryo. The interesting questions are what determines the path a cell follows when it moves and how it "knows" when it has reached its destination, where it synthesizes adhesive molecules that anchor it firmly to neighboring cells.

In this chapter, we first describe the main stages in the development of an animal embryo and then examine what is known of the cellular and molecular processes responsible.

K E Y   C O N C E P T S

♦ Development of an adult from a zygote involves processes including cell division, gene expression, embryonic induction, cell movement, and cell death.

♦ Early development of an animal embryo is controlled mainly by genetic information from the mother, stored as mRNA in the egg cytoplasm and translated during cleavage.

♦ The fate of an embryonic cell is normally determined in a progressive series of steps, each occurring before the cell shows any outward sign of differentiation.

♦ Development can be thought of as a series of events that trigger later events.

# THE STAGES OF ANIMAL DEVELOPMENT

## 12–A   FERTILIZATION

Embryonic development of most animals is preceded by **fertilization,** the union of two reproductive cells called **gametes** to form a zygote. In animals, the two gametes are of different size and appearance. During evolution, the female gamete, the **egg,** became the main repository of supplies of food, ribosomes, messenger RNA (mRNA), and other cytoplasmic components for the embryo's early development. The egg also provides all the zygote's mitochondria. As a result, all the mitochondrial genes an embryo inherits come from its mother, not its father. Because the egg contains all this luggage, it is too large to move itself around. An actively swimming **sperm,** the male gamete, is a necessary complement to the nonmotile egg, and the two have evolved together. (We consider how gametes form in Section 14–G.)

For fertilization to occur, egg and sperm must recognize one another. When a sperm of the same species touches the thick coat that surrounds an egg, it sticks there, immobilized by receptor proteins on the egg surface that bind receptors on the sperm (Figure 12–2). This system ensures that sperm fertilize only eggs of their own species (or, occasionally, of closely related species with similar receptors).

Thousands of sperm usually stick to the egg coat before fertilization can occur. Enzymes secreted by the sperm digest the outer coat of the egg until one sperm reaches and fuses with the egg's plasma membrane. The membrane engulfs the head of a sperm and pulls the sperm nucleus into

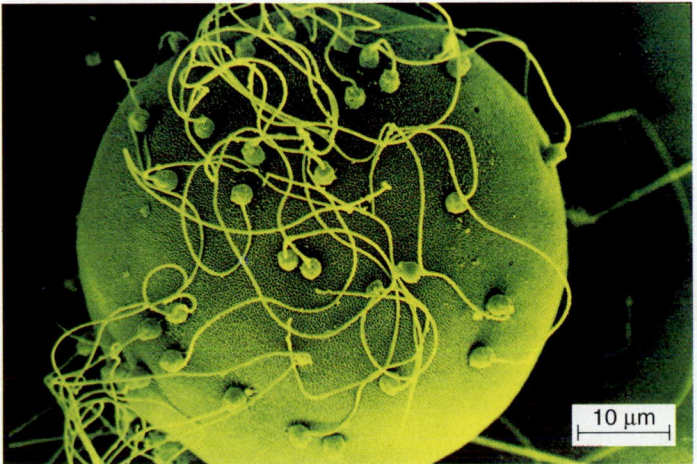

10 μm

**FIGURE 12–2**  Fertilization: a clam egg surrounded by sperm. Only one sperm will eventually fertilize the egg, although it cannot do so without the assistance of many others.  (David M. Phillips/Visuals Unlimited)

the egg (Figure 12–3). Within seconds, electrical and chemical reactions run through the egg membranes, preventing the entry of further sperm.

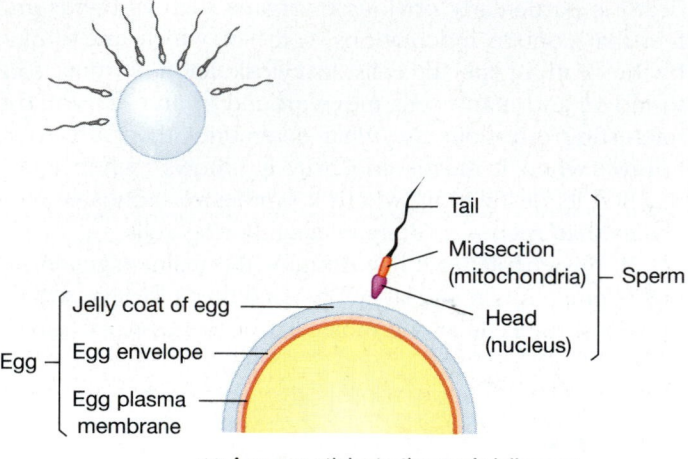

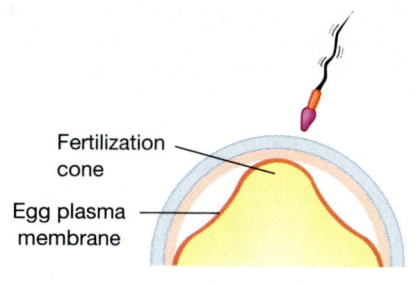

(a) A sperm sticks to the egg's jelly coat

(b) At the point of sperm contact, a fertilization cone forms in the egg

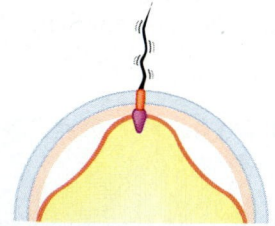

(c) The fertilization cone engulfs the sperm nucleus

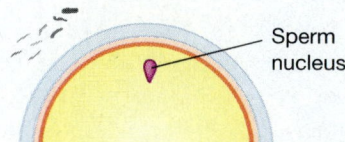

(d) Microfilaments in the egg pull the sperm nucleus in as the rest of the sperm disintegrates

**FIGURE 12–3**   Fertilization in a frog. (a) A sperm sticks to the coat that surrounds the egg. (b) The egg forms a fertilization cone, which (c) engulfs the sperm nucleus and then (d) retracts, as microfilaments in the egg cytoplasm shorten.

Fertilization contributes chromosomes to the zygote and prevents the entry of other sperm, and it has two other effects as well. First, it activates the egg to synthesize proteins rapidly, using its previously stockpiled ribosomes and mRNA. Second, it induces the egg to start dividing. In many species, activation and cell division can occur without fertilization. Frogs' eggs can be activated and will divide after being pricked with a needle or given an electric shock, and sea urchin eggs can be activated by being placed in a hypertonic solution. Some eggs can be activated even if they lack the egg nucleus. It appears that everything needed for protein synthesis in early development is present in the egg cytoplasm before fertilization and that the chromosomes contributed to the zygote by the sperm are not necessary for these early embryonic events.

## 12–B   CLEAVAGE

**Cleavage** is the process by which a zygote divides rapidly after fertilization, forming 2, 4, 8, and then 16 cells, and so on. The embryo becomes a solid clump of cells called a **morula** ("mulberry") (Figure 12–4). Eventually, a fluid-filled

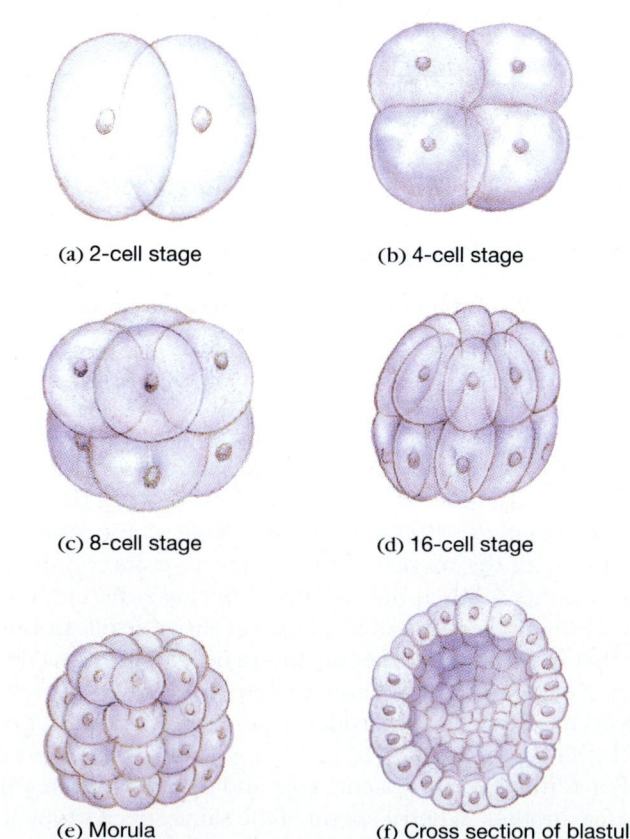

(a) 2-cell stage

(b) 4-cell stage

(c) 8-cell stage

(d) 16-cell stage

(e) Morula

(f) Cross section of blastula

**FIGURE 12–4**   Cleavage in a sea urchin. (a) Cleavage starts when the zygote divides into two cells. (b) to (e) The cells continue to divide symmetrically, forming a morula. Eventually a hollow space forms in the middle of the embryo, defining the blastula stage (f).

cavity, the **blastocoel** (blast = bud; coel = hollow), appears in the center of the embryo in most animals. After the blastocoel has formed, the embryo is known as a blastula.

Cleavage is unusual in that the cells do not increase in size between cell divisions: the zygote divides into ever-smaller cells. Cleavage not only produces a large number of cells, it also segregates parts of the egg cytoplasm into different cells. We shall see that this distribution of cytoplasmic molecules determines how cells develop later.

Cleavage follows different patterns in different animals, and the pattern can generally be predicted from the amount of yolk in the egg. **Yolk** is a store of energy and nutrients for the embryo, containing lipids, carbohydrates, and proteins. In most birds and reptiles it is yellow because it contains a yellow iron-storage protein. In eggs with little yolk, such as those of mammals and many invertebrates, cleavage produces many cells of roughly the same size. Frog eggs contain more yolk than mammalian eggs, and the **cleavage furrow,** formed as the cells divide, passes more slowly through the yolk at the bottom of the egg than through the clearer area above (Figure 12–5). As a result, the top (animal pole) of the morula gets ahead of the bot-

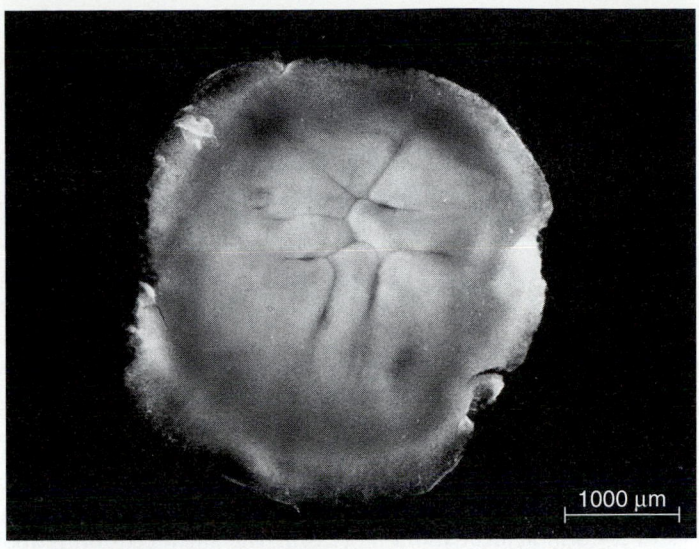

(a) Egg surface after two divisions

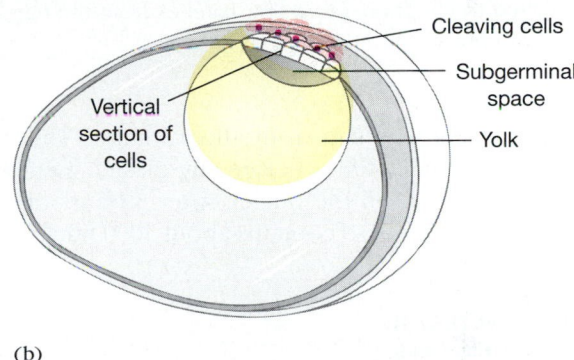

(b)

**FIGURE 12–6** Cleavage in the very yolky embryo of a bird. (a) Light micrograph of a chicken egg surface after two cleavage divisions. The cleavage furrows move so slowly through the yolk that no complete cells have yet formed. (b) Side view of the top of the egg during later cleavage. (The subgerminal space is a fluid-filled cavity that separates the developing embryo from the yolk.) (a, Murray Bakst and S.K. Gupta/USDA)

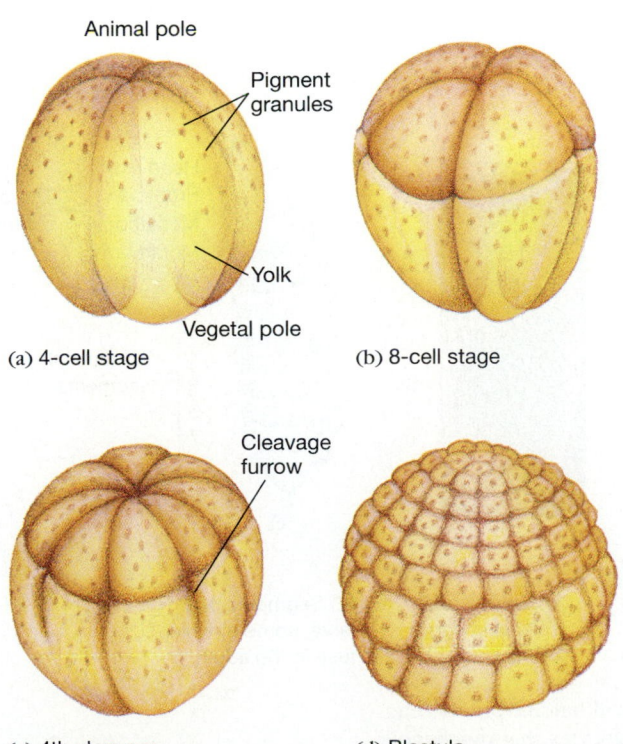

(a) 4-cell stage

(b) 8-cell stage

(c) 4th cleavage

(d) Blastula

**FIGURE 12–5** Cleavage in the fairly yolky egg of a frog. Yolk slows down cleavage in the vegetal half of the morula. As a result, there are more cells in the animal than in the vegetal half of the embryo by the time the blastula forms (d).

tom (vegetal pole), and many small cells have been formed in the top half of the embryo while there are still only a few larger ones at the bottom. Bird and reptile eggs contain so much yolk (which is why people and other animals eat them) that the cleavage furrow cannot pass through the yolk at all, and the egg cleaves only in a small area on top of the yolk (Figure 12–6).

All these animals have **radial cleavage,** in which the cleavage furrow divides the egg into radial segments like those of an orange. Many invertebrates have **spiral cleavage,** in which newly formed cells lie in a spiral pattern (Figure 12–7).

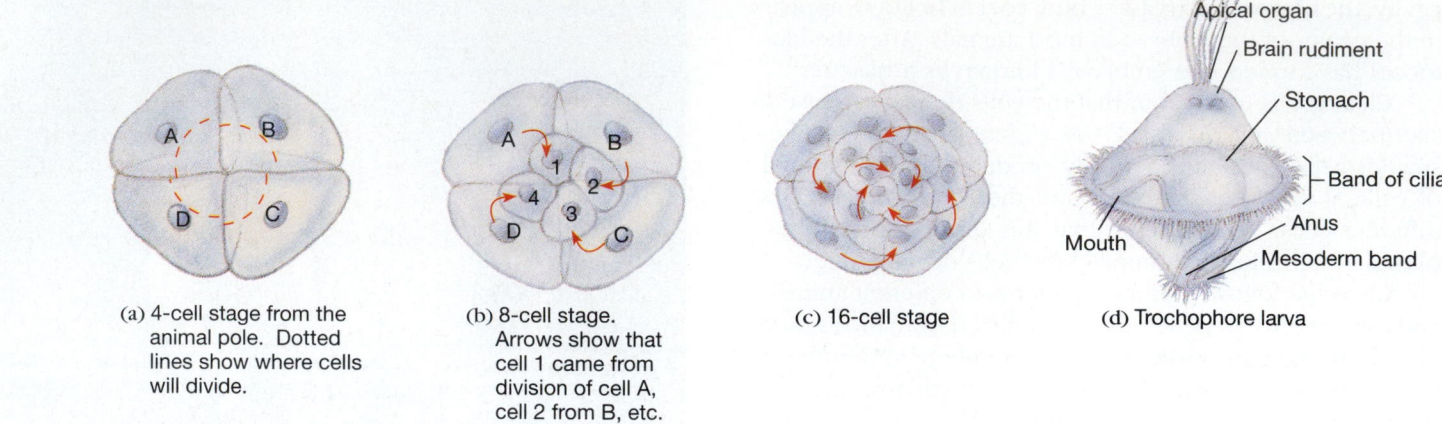

(a) 4-cell stage from the animal pole. Dotted lines show where cells will divide.

(b) 8-cell stage. Arrows show that cell 1 came from division of cell A, cell 2 from B, etc.

(c) 16-cell stage

(d) Trochophore larva

**FIGURE 12–7** Spiral cleavage in an annelid (segmented worm) or mollusc (clam, snail, and others). The spiral pattern first appears at the third cleavage division, when each of the four cells divides to form a small and a large cell. The small cell forms above and to one side of the large cell. In the case shown here, the small cell is rotated clockwise with respect to the large cell, creating a right-handed spiral. Left-handed spirals also occur, often in the same species. On the right is the ciliated trochophore larva that hatches from the egg of marine annelids and molluscs and later metamorphoses into the adult.

Cleavage in insects shows another variation. The zygote nucleus divides many times before any plasma membranes are laid down between the nuclei. After 3 hours of development a fruit fly embryo contains about 5000 nuclei. Then, membranes grow in from the plasma membrane and separate all these nuclei and the cytoplasm around them into cells lying at the surface of the egg and surrounding the yolk in the center of the egg (Figure 12–8).

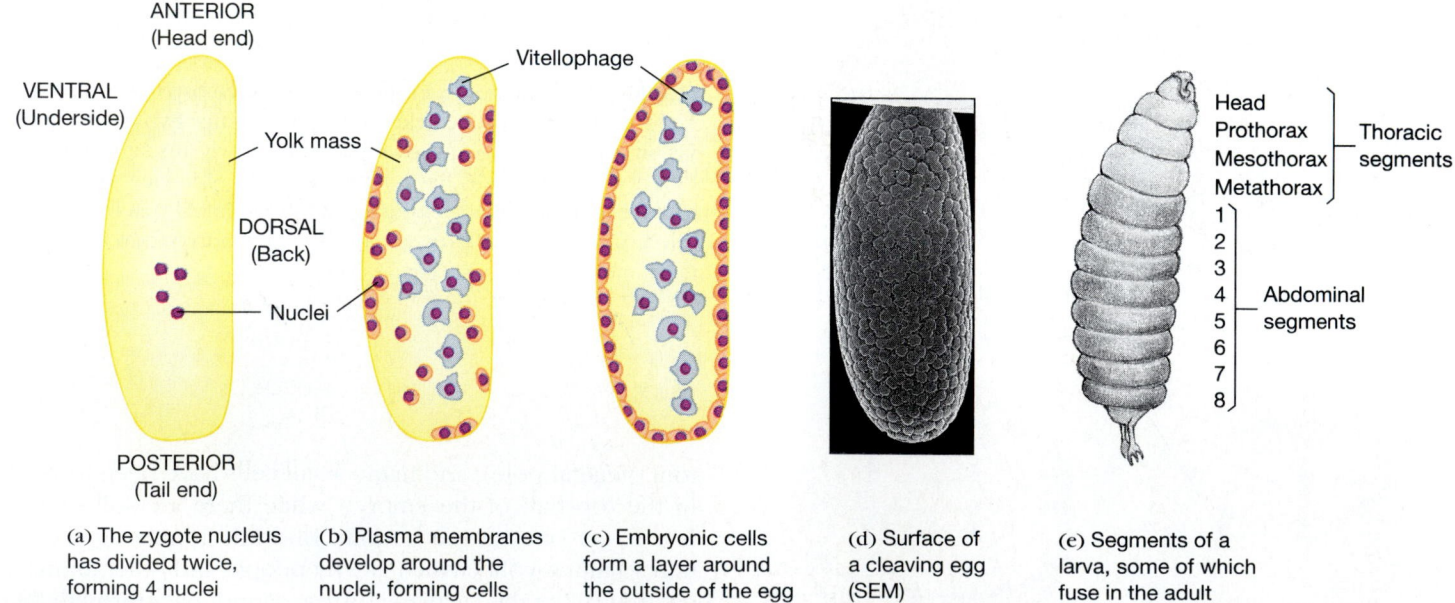

(a) The zygote nucleus has divided twice, forming 4 nuclei

(b) Plasma membranes develop around the nuclei, forming cells

(c) Embryonic cells form a layer around the outside of the egg

(d) Surface of a cleaving egg (SEM)

(e) Segments of a larva, some of which fuse in the adult

**FIGURE 12–8** Cleavage in an insect embryo together with the larva that will hatch from the egg. (a) to (c) Vertical sections through cleavage stages. Vitellophages are phagocytes that digest yolk and deliver the nutrients to the cells of the developing embryo, which lie at the outer edge of the egg. (d) Scanning electron micrograph of the surface of a cleaving *Drosophila* egg. (e) Segments of a *Drosophila* larva. In the adult, some of the segments fuse, and wings, legs, and other appendages obscure some segments. (d, F. R. Turner and A. P. Mahowald, *Developmental Biology* 50:99, 1976)

## 12–C   GASTRULATION

After a blastula has formed during cleavage, the next stage in embryonic development is **gastrulation,** during which the cells rearrange themselves into distinct layers. Cell division continues during gastrulation, but the most obvious process is cell movement. A cavity called the **blastopore** forms at the side of the blastula. Cells from the surface of the blastula move through the blastopore into the hollow interior (Figure 12–9).

During gastrulation, the single-layered, hollow blastula becomes a **gastrula.** The cells arrange themselves into three layers called the **germ layers,** containing a new central cavity, the **gastrocoel** (or **archenteron**), which will eventually form the lumen (cavity) of the digestive tract (Figure 12–10).

The outermost germ layer is the **ectoderm.** This will form the **neurectoderm,** which develops into the nervous system, and the **epidermis,** which gives rise to skin, hair, nails, sweat glands, and other structures on the outside of the body. Ectoderm develops from the **animal hemisphere** of the egg, the top half, which in yolky eggs contains less yolk.

The yolk-filled **vegetal hemisphere** at the bottom of the egg forms the innermost germ layer, the **endoderm.** From the endoderm will come cells filled with yolk lining the gut, the digestive glands, and similar internal structures. The third germ layer, the **mesoderm,** forms between the endoderm and the ectoderm, at the equator of the egg. The mesoderm will form the skeleton, muscles, reproductive organs, and allied structures in the adult.

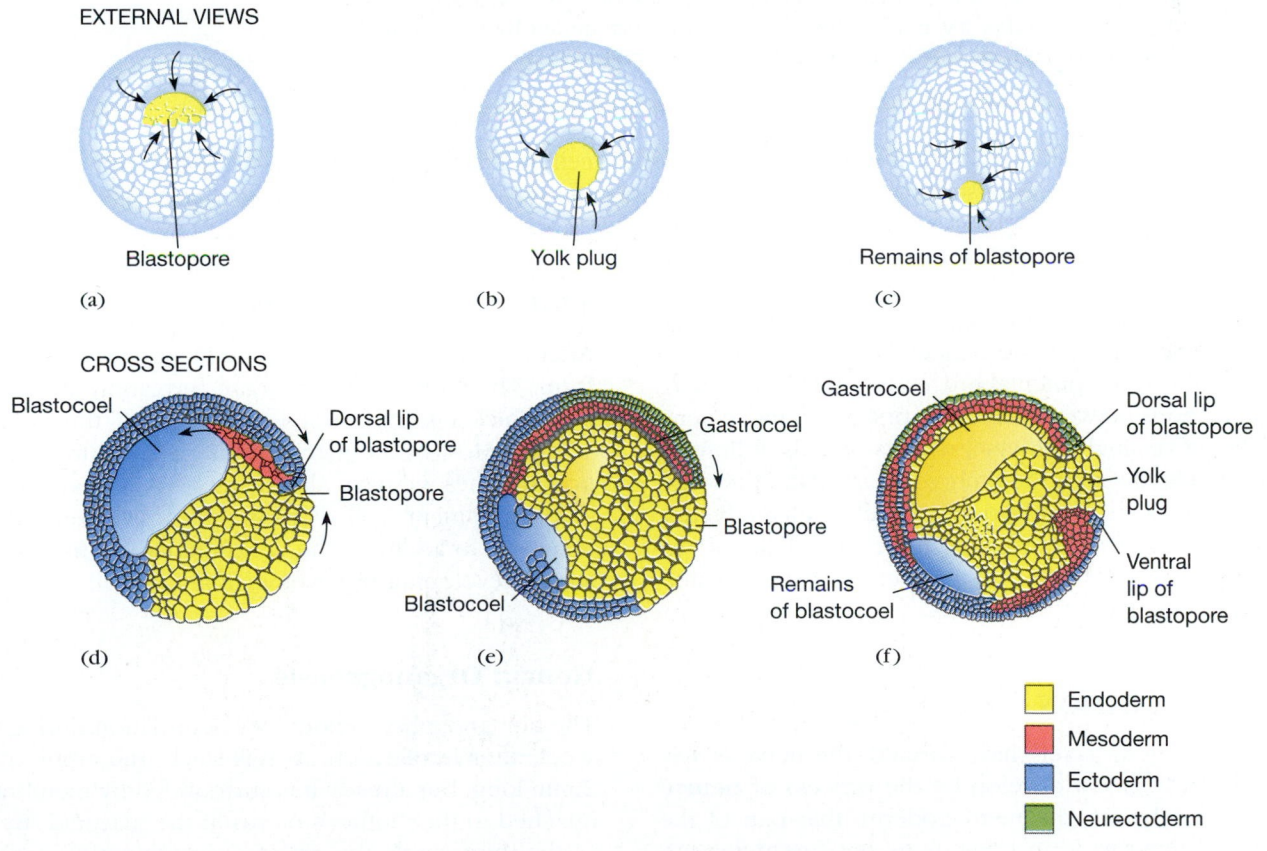

**FIGURE 12–9**   Gastrulation in a frog. (a) to (c) External views with arrows showing movement of cells from the surface in through the blastopore. The blastopore contains a plug of large cells filled with yolk which bulge out from inside the embryo. (d) to (f) Longitudinal sections. (d) Yolky cells at the bottom of the embryo (yellow) move through the blastopore into the embryo, where they will become endodermal cells, lining the digestive tract. Small cells at the top of the embryo move over the dorsal lip of the blastopore into the embryo. The first cells to enter will become the mesoderm (red), which fills the blastocoel. (e) The small cells that remain at the top (blue and green) spread out until they cover the surface of the embryo. These become ectoderm cells (blue), which will form the skin and other external structures, and neurectoderm (green), which will form the nervous system.

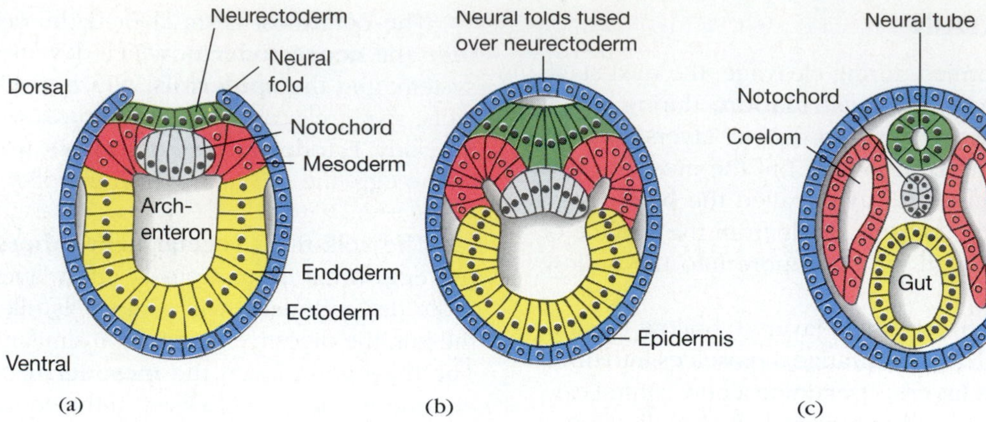

**FIGURE 12–10**   Fate of the germ layers in a vertebrate. (a) At the end of gastrulation, ectoderm surrounds the embryo, with neurectoderm beginning to sink below the ectoderm on the embryo's dorsal surface. The notochord is a mesodermal rod running the length of the embryo. In most vertebrates it is later replaced by vertebrae. The endoderm is growing up to surround the archenteron, the cavity of the future digestive tract. (b) Ectoderm has grown up to cover the neurectoderm. Pouches that will become the coelom are forming in the mesoderm. (c) Neurectoderm has rolled up to form the neural tube, the gut has formed as a tube surrounded by endoderm, and the coelom has formed within the mesoderm on each side of the body.

At a later stage, yet another cavity forms in the embryo, this time in the middle of the mesoderm. This is the **coelom,** the body cavity in which internal organs are suspended. In humans, the thoracic and abdominal cavities are parts of the coelom.

Cell division continues throughout gastrulation, with the cells again getting smaller and smaller. But gastrulation also introduces new developmental processes, most of which involve interactions between cells. The first is cell movement. Some cells move singly, others as sheets of cells, following particular tracks through the embryo. Gastrulation also introduces the process of induction, whereby cells influence one another. For instance, cells in the dorsal lip of the blastopore cause cells lying above them to form the nervous system (Section 12–H).

## 12–D   NEURULATION

After the three germ layers have formed, the nervous system and head begin to develop by the process of **neurulation.** In vertebrates, the neurectoderm, that part of the ectoderm destined to form nervous tissue, first differentiates as two parallel folds that rise up and then join, forming the neural tube, which develops into the nervous system (Figure 12–11). Somewhat different processes produce the nervous system in most invertebrate embryos.

In neurulation, we see cells changing shape, something that occurs frequently during development. For instance, the cells that form the wall of the neural tube change from part of a flat sheet of ectoderm into pie-shaped cells lining a tube. This change occurs because microfilaments inside each cell contract, shortening the part of the plasma membrane that will become the inside of the tube.

## 12–E   ORGANOGENESIS

After neurulation, all of the body's organ systems start to form. The early stages of organ formation are easy to see in a chick embryo as it lies on top of the yolk (Figure 12–12). Most of the mechanisms essential to organ formation are ones we have already encountered at earlier stages of development: cell division, cell movement, and differentiation. In addition, the programmed deaths of cells help shape developing organs (Section 12–G).

### Human Organogenesis

The human embryo undergoes neurulation during the third week after fertilization. At this stage, the embryo is about 2 mm long, but already it is surrounded by membranes and attached to the mother's uterus at the placenta. By the end of the third week, the embryo is undergoing organogenesis, and the major organ systems begin to form: the nervous, digestive, and circulatory systems. The heart, shaped like a lumpy tube, starts to pulsate. Drugs and diseases are most apt to damage the embryo at this stage, when the organ systems are forming. After about three months of development, the embryo, although it is still small (about 30 mm long), is more or less fully formed and less susceptible to malformations. The drug thalidomide, prescribed as a tranquilizer for pregnant women during the 1960s,

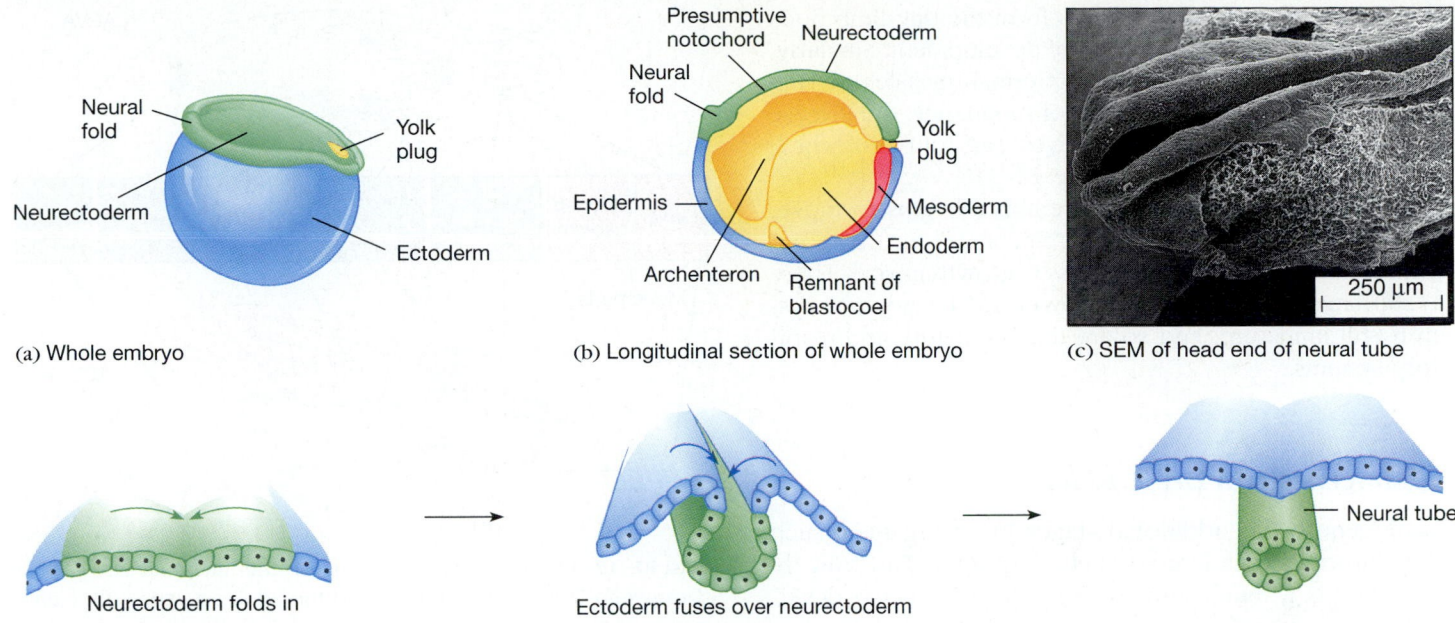

(a) Whole embryo

(b) Longitudinal section of whole embryo

(c) SEM of head end of neural tube

Neural fold

Neurectoderm

Yolk plug

Ectoderm

Presumptive notochord

Neurectoderm

Neural fold

Yolk plug

Epidermis

Mesoderm

Endoderm

Archenteron

Remnant of blastocoel

250 µm

Neurectoderm folds in

Ectoderm fuses over neurectoderm

Neural tube

(d) Section through neural fold

**FIGURE 12–11** Neurulation. (a) External view of a frog neurula. (b) Longitudinal section of the same stage. The embryo elongates, and neural folds fuse in the dorsal midline to form the neural tube. (c) Scanning electron micrograph of part of the back of a chick neurula. The lip-like ridges of the neural tube are moving toward each other at the head end of the embryo. (d) Sections through an embryo to show how the neurectoderm (green) folds in to form the neural tube covered by ectoderm (blue). The lip-like ridges in (c) can be seen in the middle diagram.   (c, Kathryn Tosney)

**FIGURE 12–12** External view of organogenesis in a chick embryo after 30 hours of development. (a) Light micrograph of the embryo lying on top of the yolk. (b) Interpretive drawing. Near the tail, the neural tube is still closing, and the notochord can be seen between its folds. The brain develops by growth of the neural tube's walls around a central cavity. The heart, which also forms from a tube, develops early compared with other organs because it is needed to bring nourishment from the yolk below it, through the large veins seen here, to the developing embryo. It is already beating at this stage. Ten somites, muscle blocks of the body wall, lie on either side of the spinal cord. (a, Photo Researchers)

(a)

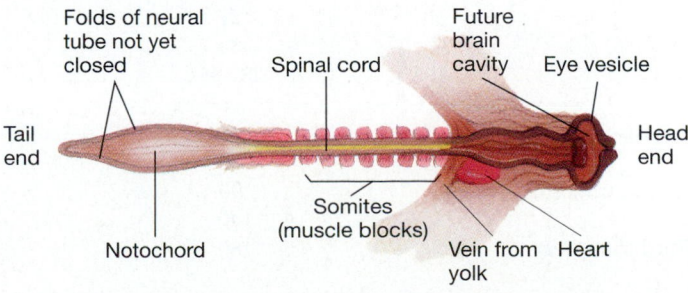

(b)

Folds of neural tube not yet closed

Spinal cord

Future brain cavity

Eye vesicle

Tail end

Head end

Somites (muscle blocks)

Notochord

Vein from yolk

Heart

stunted development of the limbs from the tiny limb buds during the fourth and fifth weeks of development. Similarly, if a pregnant woman has rubella (German measles) during the fourth through twelfth weeks of pregnancy, the disease may damage the embryo's heart, eyes, ears, or brain, which are developing at that time (Table 12–1).

During the third month, the embryo begins to move, and the mother may feel its movements. From this point onward, the most obvious progress is growth in size. There are still changes taking place, however. The nervous system is still immature, and so are the circulatory and respiratory systems.

## Organogenesis in *Drosophila*

Organogenesis has additional stages in an organism such as an insect that undergoes metamorphosis. The fruit fly *Drosophila* is a much-studied example. *Drosophila* develops from a zygote through organogenesis and then hatches from the egg as a larva (maggot). There are three larval stages, separated by molts. The final stage forms a pupa, an inactive stage that undergoes metamorphosis and molts for the last time, emerging as the adult, the **imago,** with a body plan completely different from that of the larva.

The imago develops largely from groups of undifferentiated cells called **imaginal discs,** which are fully formed in the final larval stage. There are 19 imaginal discs, arranged in nine pairs with an extra one in the middle. All the imaginal disc cells look alike but each develops into a different adult structure. Left and right eyes and antennae develop from one pair of discs, the first pair of legs from another,

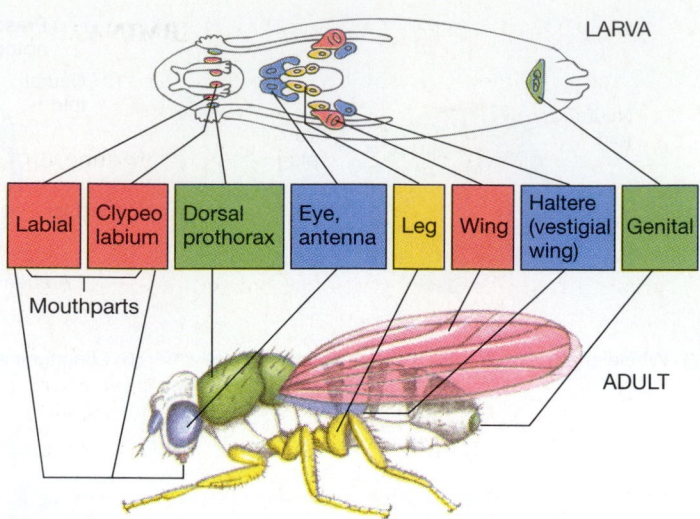

**FIGURE 12–13** Simplified diagram of the imaginal discs of a *Drosophila* larva (colored, top) and the adult structures they give rise to (bottom).

and so on (Figure 12–13). This differentiation is caused by hormones that circulate briefly during metamorphosis. The rest of the larval cells die.

## MECHANISMS OF DEVELOPMENT

In the next sections we explore some of the cellular and molecular mechanisms involved in development.

**TABLE 12–1**

**Outline of Human Development**

| Medical Name | Days After Fertilization* | What's Happening |
|---|---|---|
| First trimester | 0–8 | Cleavage |
| Embryo | 6 | Implantation in uterus |
| | 21 | Neurulation |
| | 24 | Nervous system, gut, and blood vessels start to develop |
| | 28–35 | Embryo most susceptible to damage by rubella (German measles), drugs, etc. |
| Fetus | 45 | Testes differentiating in male |
| | 70 | Eggs maturing in female |
| Second trimester | 90 | All major organ systems formed but tiny and immature |
| | 140 | Heart can be heard with a stethoscope |
| Third trimester | 180 | Temperature regulation, central nervous system, and lungs maturing; fetus most susceptible to damage by alcohol |
| | 270 | Birth (parturition) |

*Approximate; varies a lot.

## 12–F  DIFFERENTIATION AND DETERMINATION

When we consider how similar cells come to differ from one another in the embryo, two main questions arise:

1. How does division of the single zygote produce embryonic cells that are not identical but can differentiate into all the kinds of cells found in later embryos and adults? After all, if a cell in an adult liver divides, it gives rise to two new liver cells, not to two different kinds of cells.

2. When is the fate of a cell determined? One cell in a blastula looks much like the others. Is its fate, as a liver or limb cell, already determined, or might it be induced to turn into something else by events later in development?

Early answers to these questions came from transplanting part of an embryo to an abnormal position and from studies of **hybrid** embryos formed from the gametes of two species. Modern work is refining the answers by following the effects of different genes.

### Control of Cleavage

Biologists have long suspected that molecules located in different parts of the egg cytoplasm control differentiation in the early embryo. An egg is a large cell, often with visible differences in the cytoplasm in different areas. When the zygote divides into two, four, and then eight cells, these cells will contain different amounts of any substance that is distributed unevenly in the egg. For instance, in *Drosophila*, cells that come to contain cytoplasm from one pole (end) of the egg will become **germ cells,** cells that give rise to gametes. Shining ultraviolet light on that pole of the egg destroys something needed to produce germ cells, and the resulting adult is sterile because it produces no gametes. The damage can be reversed by injecting polar cytoplasm from a normal egg into an ultraviolet-treated egg. This makes it clear that the determining factor is part of the cytoplasm, not of a nucleus.

The egg of the roundworm *Caenorhabditis elegans* has granules scattered through the cytoplasm. After fertilization, these granules move to one end of the egg, and all end up in one cell after the first cell division (Figure 12–14). At the second division, they also end up in a single cell, and in the adult they are found only in germ cells. It seems that these granules are a visible sign of chemical signals in the egg cytoplasm that determine the fates of cells. Later experiments have confirmed this theory.

This evidence of the importance of substances in the egg cytoplasm means that early development is controlled largely by genes of the mother, who produces the egg. This had long been suspected from studies of hybrid embryos produced by artificial fertilization between two species of frogs. Most hybrid embryos do not grow into normal animals because the genes of the two species cannot work to-

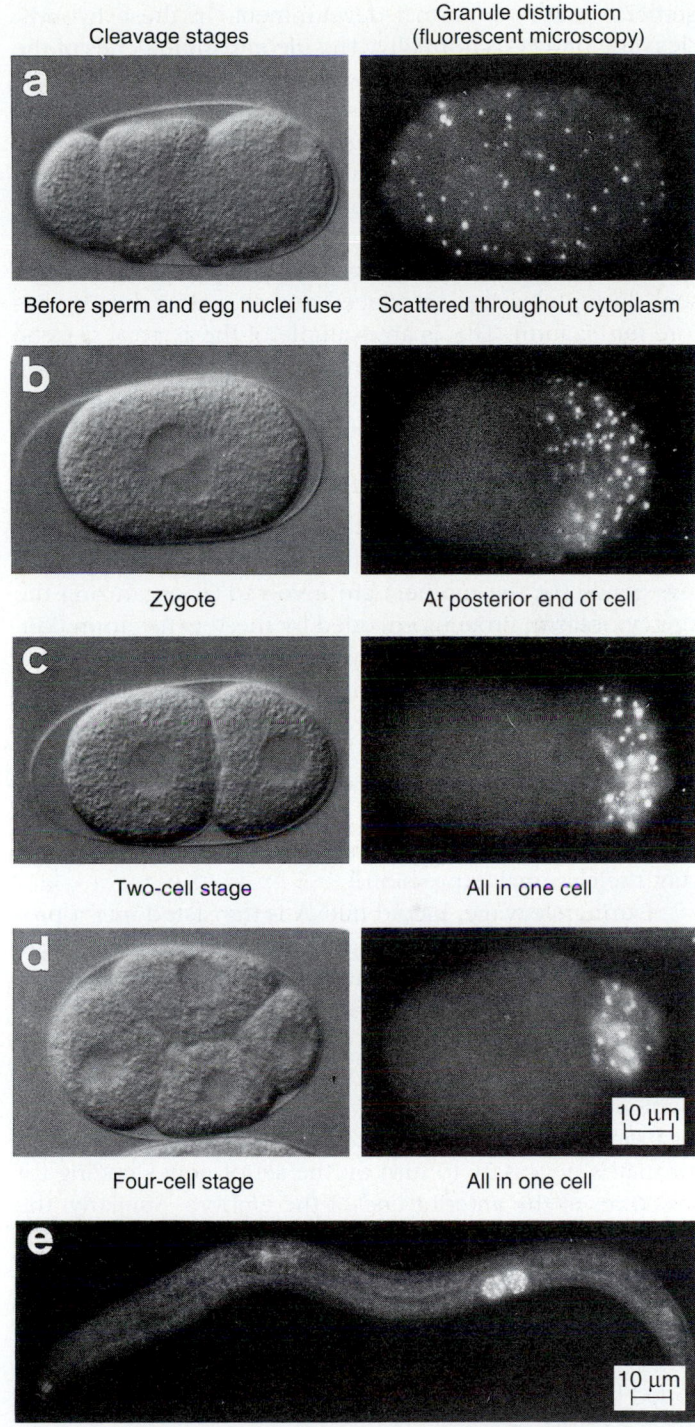

**Cleavage stages** | **Granule distribution (fluorescent microscopy)**

**a** Before sperm and egg nuclei fuse | Scattered throughout cytoplasm

**b** Zygote | At posterior end of cell

**c** Two-cell stage | All in one cell

**d** Four-cell stage | All in one cell

**e** Adult worm | All in germ cells

**FIGURE 12–14**  Segregation of cytoplasmic granules into one cell during cleavage of *Caenorhabditis elegans.* On the left are photographs of the cleavage stages; on the right, the granules (fluorescent dots) at the same stages. In (a) the egg is fertilized but sperm and egg nuclei have not yet fused. Granules are scattered throughout the cytoplasm. In the zygote (b), the granules have moved to the posterior end of the cell. As a result, they all end up in one cell of the 2-cell stage (c) and in one cell of the 4-cell stage (d). (e) The granules in the adult, localized in the germ cells, which will give rise to the gametes.  (Susan Strome)

gether to support normal development. In these hybrids, cleavage proceeds normally. The incompatible genes of the embryo do not interfere with development until later.

The hypothesis that genes of the mother control cleavage can be tested by injecting a normal fertilized egg with actinomycin D, a drug that inhibits transcription. Cleavage proceeds normally, but the embryo dies later. On the other hand, inhibitors of protein synthesis do interfere with cleavage. This means that messenger RNA must already be present during cleavage, produced and stored in the egg before fertilization. This is an example of the control of gene expression by controlling translation rather than transcription: mRNA is transcribed during egg formation but not translated until fertilization activates the events of cleavage.

## Protein Gradients

Studies of *Drosophila* mutants have shown that relatively few genes (of the mother) are involved in organizing the egg cytoplasm. Proteins encoded by these genes form concentration gradients in the egg cytoplasm. As a result, each part of the cytoplasm is distinguished by containing different concentrations of each of these proteins. The first step is for nurse cells in the mother's ovary to transcribe mRNA from a gene called *bicoid* and secrete it into the anterior (front) end of the developing egg. (Like most genes, these developmental genes are named after mutations that prevent their normal expression.)

During cleavage, bicoid mRNA is translated into a protein that diffuses down the egg. Mutant females without the bicoid gene produce eggs that develop into embryos without a head or thorax. Removing cytoplasm from the anterior end of a normal egg also results in an embryo without head or thorax. Furthermore, when this cytoplasm is transplanted elsewhere in the egg, a head develops in the new location. It appears that bicoid protein is a transcription factor that is necessary to turn on the set of genes coding for structures at the anterior end of the embryo. Similarly, the action of seven maternal genes is necessary for normal development of the posterior segments of the larva.

Thus, gradients of proteins diffusing from each end of the egg determine the head and tail ends of the embryo. In 1991, researchers showed that the protein dorsal, also coded by a maternal gene, defines the future dorsal (back) surface of the embryo.

Toward the end of cleavage, these proteins that organize the early embryo switch on the first embryonic genes. In this way, control of development passes from the mother's genes to those of the embryo.

## Embryonic Genes

Relatively few genes determine the main features of the embryo. This has been apparent since the discovery of **homeotic mutations** in *Drosophila*. These are mutations in genes that contain homeoboxes, DNA sequences encoding parts of transcription factors that control development in multicellular organisms (Section 11–E). Homeotic mutations therefore cause major changes in development. For instance, the homeotic mutation *antennapedia* causes cells from an imaginal disc that would normally form an antenna to develop into a leg instead (Figure 12–15).

The interactions of early developmental genes have been worked out in some detail in *Drosophila*. The overall pattern is that each gene produces a protein transcription factor that stimulates or inhibits transcription of the next gene in the series. As the series progresses, the embryo is divided into areas and finally into segments. At the end of the series, the final transcription factors switch on homeotic selector genes, which activate all the genes needed to produce a particular structure, such as a leg or wing.

The first transcription factors in the *Drosophila* series are proteins translated in the egg from mRNA from *bicoid* and at least two other maternal genes. Staining for the bicoid protein shows the protein transcription factor diffusing down the egg (Figure 12–16a and b). The bicoid protein, by a combination of activation and repression, stimulates transcription of the **gap genes,** the first embryonic genes involved in determining the embryo's overall pattern. Four broad stripes of protein are established across the cleaving embryo, two of the protein product of the *hunchback* gene, separated by stripes of protein from *Krüppel* and *knirps* genes (Figure 12–16c). These proteins are also transcription factors, and they activate other genes that divide the embryo into narrower stripes until eventually 14 narrow bands of protein from the *engrailed* gene appear (Figure 12–16d).

While all this is taking place, a separate series of gene interactions stimulated by transcription factor from the maternal *dorsal* gene is dividing the embryo up in the dorsal-

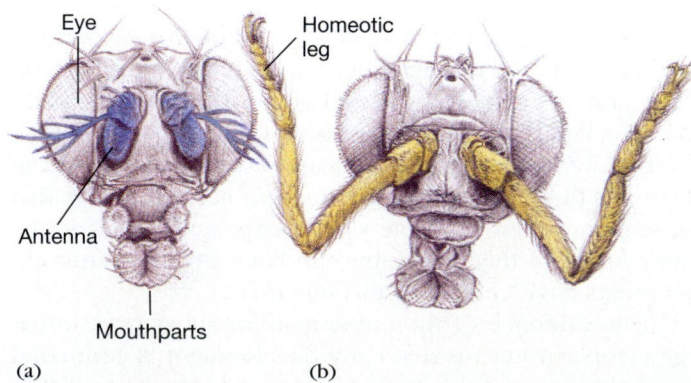

Eye

Homeotic leg

Antenna

Mouthparts

(a)                     (b)

**FIGURE 12–15** Effect of a homeotic mutation. (a) Front view of the head of a normal *Drosophila*. (b) Head of a *Drosophila* bearing the homeotic mutation *antennapedia*, which converts the imaginal discs that usually form antennae into discs that form legs.

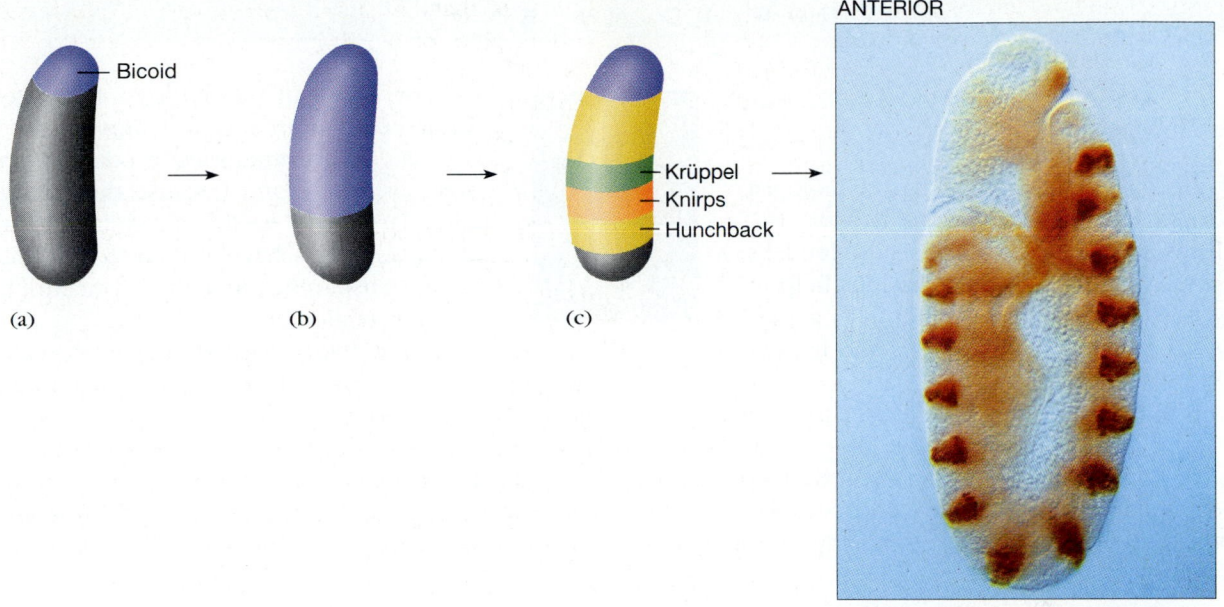

ANTERIOR

— Bicoid

— Krüppel
— Knirps
— Hunchback

(a)          (b)          (c)

(d) Embryo stained with antibodies
to engrailed protein

**FIGURE 12–16** Genes coding for transcription factors express themselves progressively in *Drosophila*. These pictures are obtained by staining embryos with antibodies that bind specific proteins. The transcription factors are called by the names of mutations that prevent their formation. (a) and (b) The distribution of bicoid protein, translated from mRNA secreted into the anterior end of the egg by cells in the ovary. Bicoid diffuses down the egg, determining the anterior end of the embryo and switching on transcription of all the other proteins shown. (c) The positions of three proteins whose transcription is induced by bicoid protein. (d) An embryo stained with antibody to the protein encoded by the *engrailed* gene.   (d, Thomas Kornberg)

ventral direction (dividing upper from lower surfaces). In this, as in the bicoid series, the relative concentrations of the various transcription factors in different parts of the embryo apparently determine which gap genes are transcribed and which are suppressed.

## Determination

The fate of a cell in an early embryo, how the cell differentiates, is **determined** by which genes are expressed in which cells. Determination is a progressive process, which gradually restricts the fate of a cell as ever more specific genes are activated. As a result, a cell's fate may be determined long before the cell shows any outward sign of differentiation.

Different species are said to have determinate or indeterminate development, depending on how early in cleavage determination begins. An **indeterminate** embryo, such as that of a sea urchin or amphibian, develops normally even if a cell is removed during early development. If the cells of a 4-cell embryo are separated, each grows into a normal embryo. Determination begins at the next division, when cells at the animal pole come to differ from those at the vegetal pole. This can be shown by dividing the 8-cell

embryo in half either vertically or horizontally with a glass needle. Vertical division leaves the embryo with both vegetal and animal cells, and both halves develop normally. Horizontal division deprives each embryo of substances in the other half and results in abnormal development (Figure 12–17, page 255).

In contrast, **determinate** embryos, such as those of roundworms and molluscs, show some determination from the first cell division, and the embryo is abnormal if any cell is destroyed during cleavage.

Determination in later development can be demonstrated by transplantation experiments. For instance, in an early amphibian embryo a block of cells that would normally become brain cells can be transplanted to another area from which cells that would normally form skin have been removed. The transplanted cells form skin. They differentiate in accordance with their new, rather than their original, position in the embryo. We can conclude that these cells had not been determined as brain cells before their transplantation. However, if the same experiment is made three days later in development, the transplanted cells develop as brain. We conclude that they had become determined as brain cells during the three-day period, although they had not yet differentiated into brain cells.

## CHOOSING EXPERIMENTAL ORGANISMS

One of the most important things a biologist does is to choose which organism to work on. We shall see in Chapter 15 that Gregor Mendel's choice of pea plants was a tremendous help in working out the inheritance patterns of genes, whereas his work with hawkweeds and honeybees resulted in confusion.

Organisms for experiments are often chosen by chance. You happen to end up in a lab where everyone else is working on frogs or fruit flies, and you do too, because they seem suitable for your work and because your coworkers' experience can help solve many problems you will encounter.

Early work in embryology was often performed on sea urchins and amphibians, largely because these organisms produce many embryos at a time and because the embryos are fairly large and easily raised in the lab. Sea urchin embryos have the added advantage of being transparent, since they contain little yolk or pigmentation.

By the 1940s, workers began to realize that development was largely the expression of an organism's genes. They noted that little was known about the genetics of sea urchins or frogs and predicted that work on the fruit fly *Drosophila* would prove rewarding because more was known about the genetics of *Drosophila* than of any other eukaryote. Some people disagreed. They pointed out that insect embryology is so unlike that of

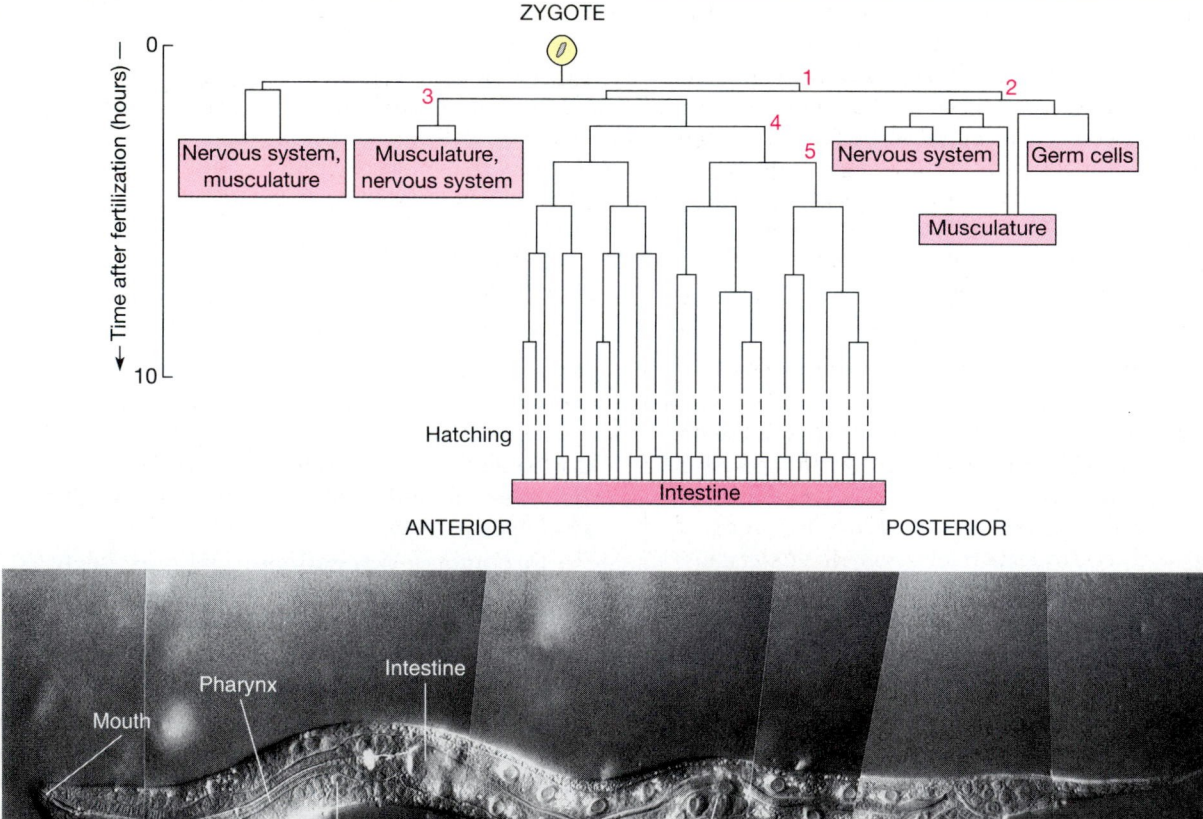

*Caenorhabditis elegans* (LM)

vertebrates that it was usually not even covered in embryology textbooks; work on *Drosophila* embryos was unlikely to apply to mammals and would not answer questions about human embryology. A few workers, however, started to work with *Drosophila* embryos. As we can see from this chapter, their work has proved very fruitful.

Then, in the 1970s, a number of workers decided that all the animal embryos being studied had too many cells and were too complicated to be easily understood. So they embarked on a search for a new, smaller, and simpler experimental animal—an amazing undertaking because they would have to learn about its genetics and embryology from scratch.

This search lighted on *Caenorhabditis elegans,* a roundworm (nematode) common in soil. *Caenorhabditis* is easily raised in the lab, and it is transparent. But its most valuable characteristic to embryologists is that it contains only a few hundred cells, many of them visible in the live animal under a low-powered microscope.

Despite early doubts, *Caenorhabditis* has proven to be all that its champions hoped. By 1983, workers had mapped the origin in the embryo of all 1090 cells formed during development. Figure 12–A shows part of that map. Having such a cell lineage map means that every change in every cell during development can be traced. The nervous system, for instance, contains only 302 nerve cells (compared with billions in a mammal), and we now know how the connections of every nerve cell to every other nerve cell are formed, something that is impossible in more complex organisms.

Fears that research done on roundworms might not apply to vertebrates have proven to be unfounded. In fact, the more we know about development, the clearer it becomes that the basic mechanisms evolved hundreds of millions of years ago and are found in all eukaryotes. *Caenorhabditis* is even proving itself as a guinea pig for the Human Genome Project, the attempt to analyze human genomes. Researchers expect to know the complete nucleotide sequence of a *Caenorhabditis* genome by 2000.

**FIGURE 12–A** *Caenorhabditis elegans.* (Top) Part of the cell lineage map, showing the origin of the intestine. Each cell division is shown by a horizontal line; the first few divisions are numbered in red. Thus, the cell that gives rise to all the cells of the intestine separates at the third cell division from a cell that will give rise to muscles and part of the nervous system. (Bottom) The adult.   (John Sulston)

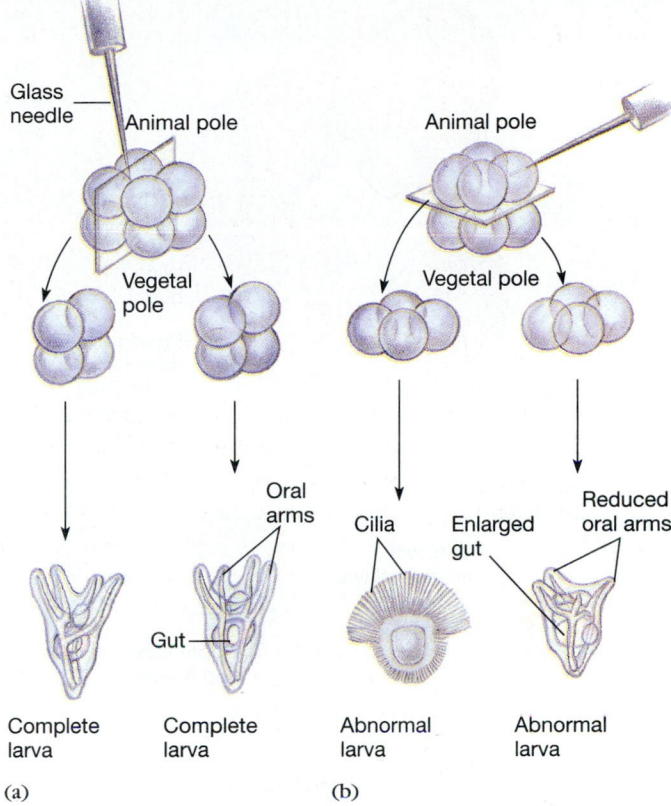

**FIGURE 12–17** The beginning of determination in a sea urchin (which has indeterminate development). If the embryo is cut in half at the 8-cell stage, it develops into two normal larvae only if each half receives cells from both the animal and vegetal halves of the embryo (a). Cells from either pole alone (b) develop into abnormal larvae.

In chick embryos, the wings and legs begin as pairs of limb buds growing out from the trunk. At first, the cells of the buds are similar to each other. If we took tissue from the base of a limb bud which would normally form a thigh and grafted it into the tip of a wing bud, we might expect that the graft would develop as thigh if its fate was determined before it was transplanted, or as wing tip if it was undetermined at that time. In fact, it does neither: the transplanted tissue develops into a toe, appropriate to the tip of a leg, not to the base of a leg or the tip of a wing (Figure 12–18). This shows that the cells had been determined as leg rather than wing before they were transplanted, but which part of the leg they would form had not yet been determined. In other words, determination builds up, step by step, as a tissue develops.

Tracing the fate of marked cells in *Drosophila* embryos has revealed that embryos are divided into compartments (*How Do We Know? Choosing Experimental Organisms*). For instance, long before the wing buds form, the cells that will form the wing are determined to belong either to the posterior or to the anterior compartment of the wing. The cells

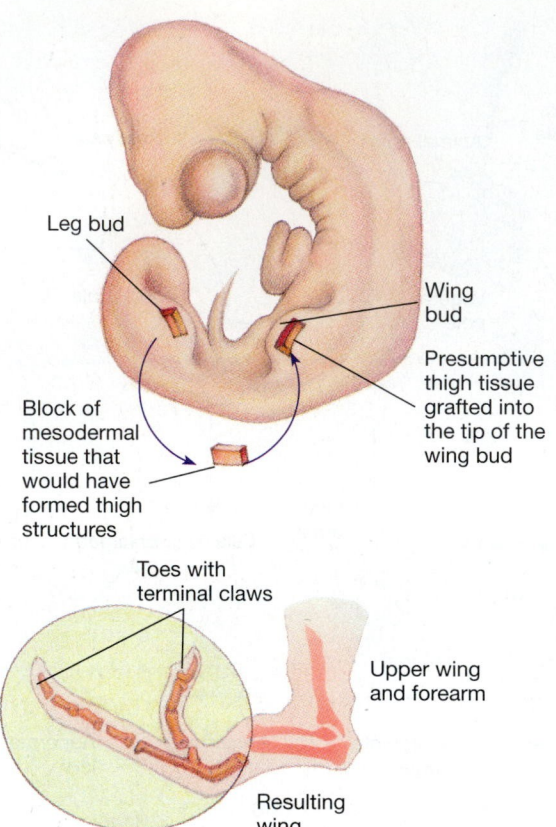

**FIGURE 12–18** *Determination in chick development. A block of mesoderm that would have formed thigh structures is transplanted from the hind limb bud to the tip of a wing bud. The tissue develops into toes, characteristic of the tip of a hind limb, not of the thigh from which it originated nor of the wing tip to which it was transplanted.*

in a particular compartment of an embryo acquire a compartment "address" as control genes, such as homeobox genes, are switched on (or off). Relatively few control genes are needed to determine how a cell will differentiate, because more than one gene affects each cell. One gene, for

example, may say "anterior half of segment." It can coexist with other control genes determining which segment and whether the cell develops into muscle or skeleton. Control genes, then, seem to act like code letters that can be combined in various ways to determine the future of a cell.

In summary, the general plan for the body is specified very early in development. New levels of detail are added progressively. Determination appears to be the activation of control genes that cannot be switched off again except in exceptional circumstances. These genes determine later development of the cell and its descendants by controlling the activity of many structural genes.

## 12–G   CELL DIVISION AND CELL DEATH

Cell division is a vital part of differentiation and determination. A cleavage cell determined as belonging to the head of the embryo cannot give rise to all the different structures of the head until it has divided into many cells that differentiate in various ways.

We are beginning to understand that cell death is an essential counterpoint to cell division. An embryo often produces too many cells, and then those that are not needed die. Consider Figure 12–19, showing development of a chick's foot. The toes become separate digits after the death of cells that originally lay between them. In a small proportion of embryos these cells do not die, and the embryo ends up with cells between the toes—appearing web-footed. Programmed cell death is an active process requiring protein synthesis. It has been given the name **apoptosis** (apo = away from; ptosis = fall), to distinguish it from cell death that results from injury or similar destruction.

The death of cells is also vital to development of the vertebrate nervous system, which has millions of precise connections between **neurons** (nerve cells). An embryo's genes do not always dictate how many of a particular cell type should be made. In other words, the number of cell

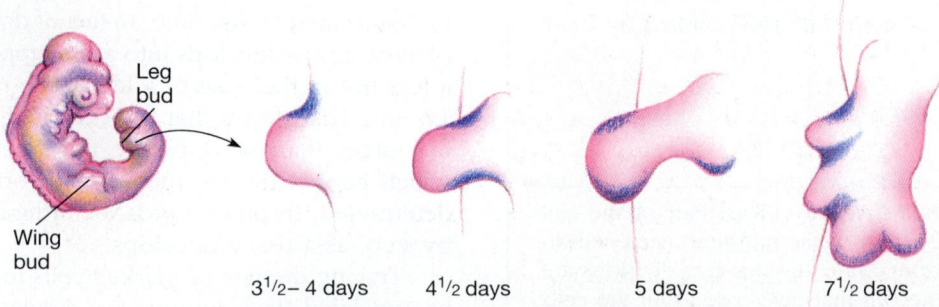

3½– 4 days          4½ days          5 days          7½ days

**FIGURE 12–19**  Cell death shapes tissues during development of a chick leg. The developing foot is stained with a blue stain for macrophages, the cells that phagocytose dying cells. The stain therefore marks areas where massive cell die-off occurs, refining the shape of the limb formed by the cells left alive.

divisions in the nervous system (and many other tissues) is not pre-programmed. Instead, specific stimuli eventually switch off cell division. It would be fatal to switch off cell division too soon because differentiated neurons do not usually divide, so there is no way to make more of them. Instead, more than enough cells are made in the first place and excess cells killed later.

We know how this occurs in the sympathetic nervous system, which controls functions such as movements of the gut. Sympathetic neurons near the spinal cord reach the organs they will control by growing long extensions toward **nerve growth factor (NGF),** a protein produced by the internal organs. Some neurons absorb enough of this protein to live; others do not, and they die.

Then in the 1990s came an experiment that produced a conceptual breakthrough: drugs that inhibit protein synthesis prevent neurons from dying when deprived of NGF. This showed that the neurons are all programmed to die. Getting enough NGF prevents them from committing suicide because NGF switches off the genes for death.

Programmed cell death occurs in the development of many organisms and appears to be controlled by similar genes. In the roundworm *Caenorhabditis*, a total of 1090 cells are formed during development, and 131 of these are programmed to die. The protein products of two genes are known to trigger cell death.

*Caenorhabditis* also has another gene with the opposite effect: cells that express it are protected from death. Researchers have also found one vertebrate gene that prevents apoptosis. This gene saves lymphocytes (a class of white blood cells) from apoptosis and thereby causes a cancer called B-cell leukemia. This gene also prevents apoptosis of neurons in the absence of NGF. If this gene is inserted into *Caenorhabditis* embryos it prevents programmed cell death in that organism as well. The vertebrate and *Caenorhabditis* genes that prevent cell death have similar nucleotide sequences. By the time you read this book, you can be fairly sure that someone will have discovered the gene that causes apoptosis in vertebrate neurons. And it seems safe to bet that the gene will be similar in nucleotide sequence to the apoptosis genes in *Caenorhabditis*.

## 12–H  EMBRYONIC INDUCTION

Induction is a complex interaction between cells, in which one cell in an embryo alters the fate of another. For instance, the lens of the eye is formed from ectoderm of the head as a result of contact with part of the brain, which grows out and forms a bulge called the optic vesicle. If some barrier such as a piece of cellophane is placed between the growing optic vesicle and the ectoderm, the lens never develops. If optic vesicles are transplanted from their normal site in the head to elsewhere in the body, they induce lens tissue to form in any ectoderm that they touch.

They cannot, however, induce tissue other than ectoderm to form a lens. Hence, the fate of a tissue is determined both by the activity of its own genes and by its environment: genetic activity determines that a cell has become ectoderm rather than mesoderm or endoderm, but unless it touches the optic vesicle, ectoderm will not develop into a lens.

More than 70 years ago, German embryologist Hans Spemann discovered the earliest example of induction in vertebrates, the process by which ectoderm, mesoderm, and endoderm are induced to form in the gastrula. Using early amphibian embryos, he found that if he transplanted tissue from the dorsal lip of the blastopore to a second embryo, the recipient developed a second head-tail axis. The embryo might even develop two heads. The dorsal lip of the blastopore became known as the "primary organizer" because its cells appeared to induce other cells to form the main axis between head and tail. Embryologists have spent much of the twentieth century trying to figure out how the organizer works.

Recently, several growth factors (proteins) that mimic the effect of the organizer have been discovered in amphibians. When injected into the ventral surface of the gastrula, they cause formation of a double-headed embryo (Figure 12–20). Because induction in the gastrula is such a major event, workers suspected that homeobox genes might be active in the organizer but not in surrounding cells. When they searched, they found that cells in the organizer contained a protein transcription factor, part of which is essentially identical to the bicoid protein involved in formation of the head-tail axis in *Drosophila*. With this discovery, it appears that embryologists are finally on the track of the cellular events that determine formation of the germ layers in the vertebrate embryo.

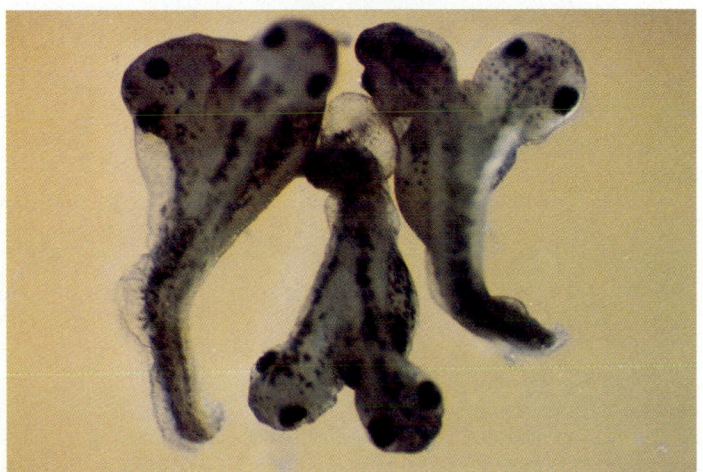

**FIGURE 12–20**  Two-headed embryos of the African clawed toad, *Xenopus laevis*, formed after a growth factor from the primary organizer was injected into the gastrula.  (Jan Christian and Randall Moon)

## 12–1   PATTERN AND POSITION

Determination has a lot to do with a cell's position because cells' fates are usually determined by the positions they occupy in the embryo. We have seen that a cell during cleavage "knows" where it is in the embryo by the presence in its cytoplasm of substances it inherited from a certain part of the egg cytoplasm. However, this does not explain how a cell moving around in a later embryo "knows" where it is. For years, embryologists assumed that position in the later embryo is identified in the same way as position in the egg: by a system that distinguishes top from bottom, right from left, front from back. However, this did not seem sufficient to explain what was known about position from studies on regeneration.

Many organisms, particularly plants and lower animals, have the ability to replace or regenerate tissues and organs that have been cut off. **Regeneration** is the process by which an amputated organ regrows. Most animals, includ-

ing humans, can regenerate parts of some organs. For instance, the human liver is capable of regenerating parts that have been removed. However, other animals have more impressive powers of regeneration. There have been many studies of regeneration of lost legs in amphibians such as newts and salamanders. A planarian worm can even regenerate its head and most other parts of the body (Figure 12–21).

Regeneration starts with formation, at the site of injury, of a **blastema**—a bud containing a mass of cells that look very like one another. Blastemal cells may arise by dedifferentiation of existing differentiated cells. The regenerating organ forms as the blastemal cells divide and differentiate into the various cell and tissue types (e.g., bone, muscle, blood vessel) that make up the new limb.

One vital stimulus to regeneration is the presence of nervous tissue. Neither a planarian nor an amphibian can regenerate an organ unless many nerve fibers reach the blastema. For instance, frogs cannot normally regenerate a

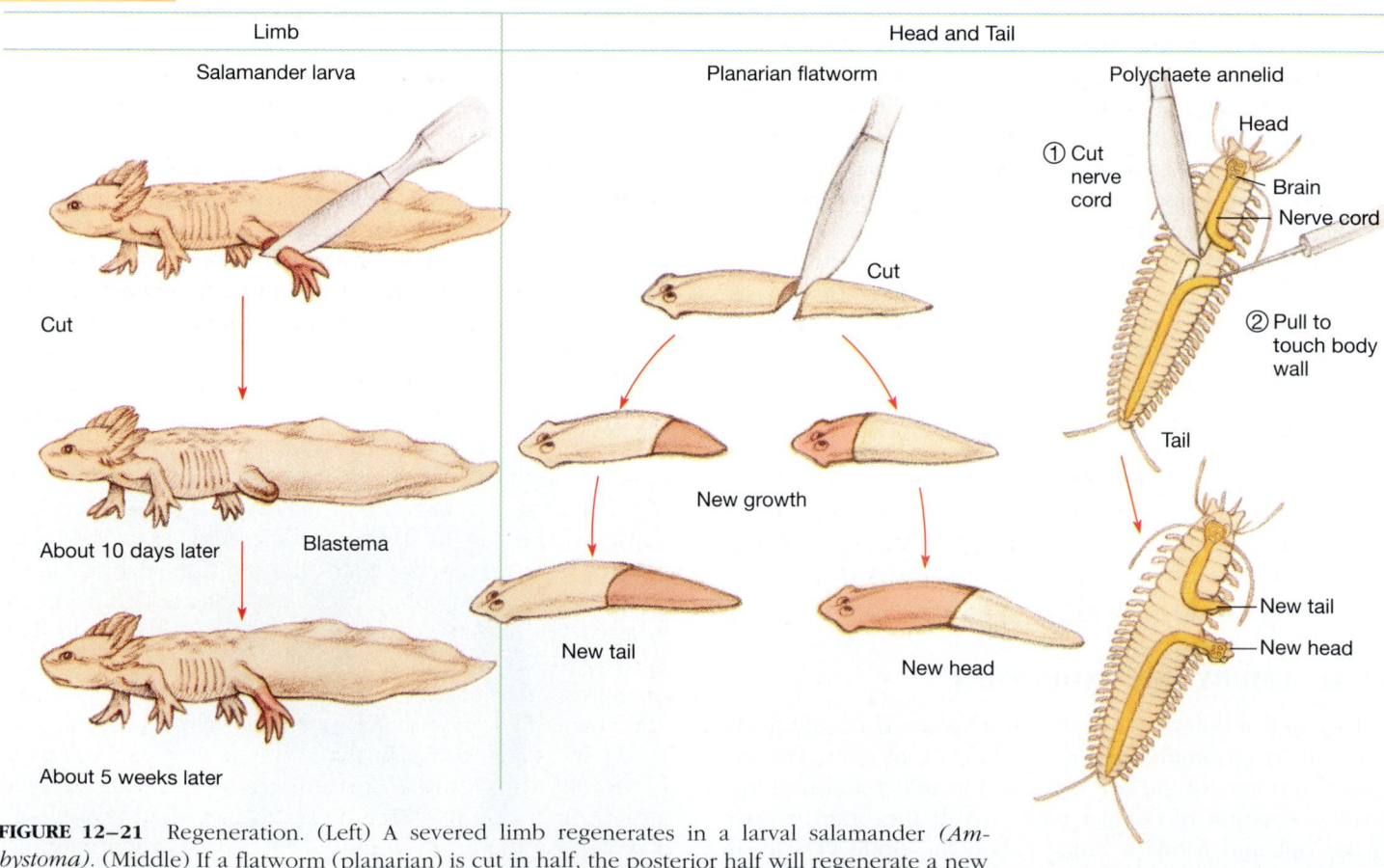

REGENERATION

**FIGURE 12–21**   Regeneration. (Left) A severed limb regenerates in a larval salamander *(Ambystoma)*. (Middle) If a flatworm (planarian) is cut in half, the posterior half will regenerate a new front end, and the anterior half will regenerate a new rear end. (Right) The effect of nerves on regeneration. The nerve cord of this polychaete worm has been severed and the cut ends moved so they touch the body wall. The part of the cord on the posterior side of the cut induces formation of a new head from the body wall, whereas the part of the cord anterior to the cut induces formation of a new tail.

limb. But if additional nerves are grafted into the limb of a frog, that limb will regenerate if it is later severed. Nerves secrete substances that are necessary if regeneration is to occur. It is still not known what influences in the blastema cause some of its cells to differentiate into muscle cells, for example, and others into the bone of the regenerating limb.

The studies on regeneration suggested that position in a differentiating limb is determined by two coordinates. One describes position as either **proximal** (close to the body) or **distal** (far from the body). The other describes position as anterior or posterior. In the 1990s, studies of differentiation of *Drosophila* imaginal discs are lending support to this theory. Mutants that develop with part of a limb missing show that cells in the disc that will develop into a limb are arranged in concentric circles. The disc develops by extending into a cone whose apex will be the end of the leg (Figure 12–22). Nearer the base of the cone are circles that will develop into the tibia and femur and, finally, into the part of the body (thorax) to which the leg is attached.

The order of cell determination is, first, which segment of the body the cell belongs to and which appendage it will form. This is determined very early in development by bicoid and related proteins during cleavage. A later stage of determination is which concentric circle the cell belongs to (determining which part of the appendage it will form). It seems clear that a similar system operates in vertebrates, because this order of determination explains why cells transplanted from the base of a bird's leg to a wing tip develop into a toe (Section 12–F). Before transplantation, the cells are determined as belonging to a leg rather than a wing.

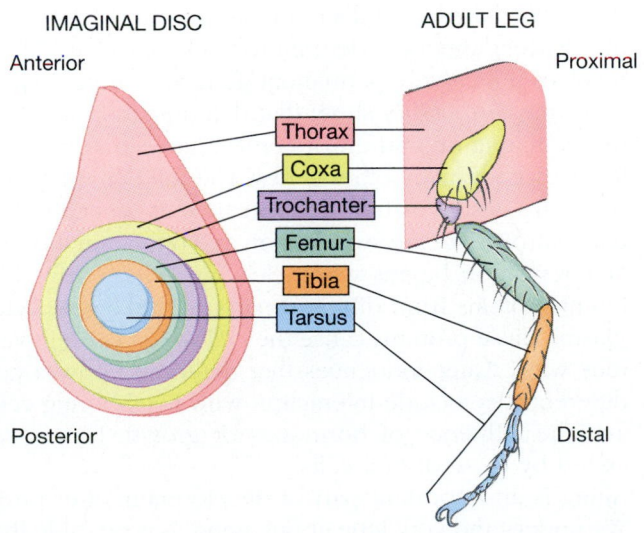

**FIGURE 12–22** Comparison of the arrangement of cells in an imaginal disc of *Drosophila* with the leg that develops from the disc. Cells that will develop into the distal tip of the leg lie in the center of the disc, whereas those that will develop into the (proximal) base of the leg and the thorax lie at the outer edge of the disc.

When transplanted from the outside (proximal, thigh) of the limb bud to the inside (tip) they are moved from an outside circle to an inner one and consequently develop as toe rather than thigh.

## 12–J   MATURATION AND AGING

The term "embryonic development" usually refers only to that period in an animal's life before it leaves the protection of the egg or of the mother's body at hatching or birth. Animals continue to change in various ways after this time. For instance, different mammals are born at different stages of development. A mouse is born hairless, blind, and incapable of regulating its body temperature. The foal of a horse, on the other hand, can run around and use all of its senses within an hour of birth. Further development also occurs after birth, in that most animals do not become sexually mature until later in their lives. In addition, some animals undergo metamorphosis, a change in body form, before they are sexually mature. Both metamorphosis and sexual maturation involve the same processes of genetic differentiation, growth, and change of form as those that occurred in the embryo.

Aging is another change that occurs after embryonic development is complete. "As soon as we are born," the Apocrypha points out, "we begin to draw to our end," for development and aging merge imperceptibly into each other. Why and how do we age and eventually die?

**Aging** is the sum of changes that accumulate with time and make an organism more likely to die. It begins even before an individual officially completes development and sexual maturity. Slower healing, for instance, is one of the cellular signs of aging, and even a human teenager's broken bones heal less rapidly than those of a young child.

Aging and death are genetically programmed. Species have characteristic lifespans that can be altered by selection. An elephant dies of old age when it is about 100 years old, a human at about 80 years, and some insects at a few days or weeks. Aging assures death at this characteristic age if disease or predators do not kill the animal first. Some biologists think the genes that cause aging may have been selected because they confer advantages early in life that outweigh their later disadvantages. For instance, the female sex hormone estrogen is necessary for female reproduction but increases the risk of uterine cancer in later life. Still others view aging as a selective tradeoff between different genes, with those that increase reproduction favored over those involved in repairing tissue damage. Because organisms usually die of predation, disease, or accident within a certain length of time, there is no selective pressure for mechanisms that repair wear and tear on the body that accumulates after the likely lifespan.

We do not know how aging occurs; there are dozens of different theories. Although we can describe many processes that are characteristic of aging, it has proved impossible to decide which of these is cause and which effect. For instance, the immune system, which defends the body against disease, becomes less efficient with age. Is this why disease becomes more common with age, or does disease cause the immune system's deterioration?

As people grow older, their bodies cope less effectively with stress or disease. Because the ability to survive changing conditions depends largely on the immune, nervous, and hormonal systems, researchers have concentrated on these systems, using the hypothesis that degeneration here results in the loss of adaptability seen in the rest of the body: slower healing, hardening connective tissue, brittle bones, and so forth.

Two factors with rather generalized effects are known to influence aging. One is oxidation damage to biological molecules which inevitably occurs in the course of normal metabolism. Enzymes that protect the body against oxidation are especially active in long-lived strains of *Drosophila* and *Caenorhabditis,* and there is a correlation between a species' lifespan and the overall amount of these enzymes that its members produce. The second factor is glucose, which undergoes uncatalyzed reactions with some of the body's proteins and is thought to cause the stiffening of connective tissue and heart muscle that comes with age.

Another theory holds that aging is a cellular process resulting from the accumulation of mutations in all the body's cells. This does not explain aging in all animals, however. During their lifetimes, mammals accumulate proportionately more mutations in the DNA of their cells than insects do (allowing for insects' much shorter lives). Similarly, cells die throughout our lives, but this alone cannot explain aging. Human beings who lose cells through disease or accident have survived with many more cells missing from vital organs (such as the liver, kidney, brain, and lungs) than are normally destroyed during aging.

Searches for a single cause of aging have all ended in failure. It seems possible that the "aging genes" do not control one single system but instead control a multiplicity of minor degenerations. Because the body's systems all interact with one another, minor deficiencies anywhere in the body can accumulate to produce aging of the body as a whole and of its individual systems.

## SUMMARY

An animal starts life as a zygote, which must undergo embryonic development before it becomes self-sufficient. Development involves several different processes: cell division; differential gene activity in different cells so that they dif-

ferentiate into various types of cells in the adult; differential growth and movement of cells; and programmed cell death.

1. Fertilization contributes the sperm's chromosomes to the zygote, activates protein synthesis from maternal mRNA stored in the egg, and induces cell division.
2. During cleavage, the embryo divides rapidly into ever-smaller cells and forms a hollow blastula. Cleavage is controlled by the genes of the mother, not of the embryo, by way of mRNA and proteins stored in the cytoplasm as the egg was formed. Cleavage looks very different in different animals. The amount of yolk in the egg is one factor that determines how cleavage occurs.
3. During gastrulation, the embryo's genes become active. The cells rearrange themselves into three germ layers surrounding the archenteron cavity:
   • Ectoderm, which will develop into the skin and other external structures as well as the nervous system.
   • Mesoderm, which forms internal organs such as the kidneys, heart, and blood vessels. A split in the mesoderm will eventually form the coelom.
   • Endoderm, which forms the digestive tract and organs associated with it.
4. During neurulation in vertebrates and sea urchins, the neurectoderm rolls up and forms the neural tube, which develops into the brain and the rest of the nervous system.
5. Organogenesis produces all the organs of the body by a complicated series of cell interactions.
6. Studies in which parts of an embryo are transplanted into new positions reveal that the way a cell will develop may be determined before the cell differentiates into a specialized cell. Determination is inherited by all the descendants of a determined cell, showing that it involves a relatively permanent genetic change. Experiments in *Drosophila* show that determination involves the switching on and off of control genes that regulate the activity of many other genes. The full determination of a cell's destiny is built up over time as several different control genes are switched on and off within the cell.
7. Differentiation begins when cleaving cells receive different proteins from different parts of the zygote cytoplasm. These proteins cause the cells to develop in various ways. Later influences that determine how a cell differentiates include interaction with neighboring cells and the influence of hormones or growth factors secreted by more distant cells.
8. Aging is an important part of development after birth. We understand very little about aging. It is possible that aging has many causes whose interactions are complex and difficult to disentangle or, less likely, that there is some major key to the process of aging that has yet to be discovered.

## Self-Quiz

1. Fertilization normally does *not*:
   a. prevent more than one sperm from entering the egg
   b. stimulate the egg to synthesize messenger RNA
   c. stimulate the egg to synthesize proteins
   d. contribute chromosomes to the zygote
   e. stimulate cell division

Match the correct stage(s) of development with each of the following characteristics:

C=Cleavage
N=Neurulation
G=Gastrulation
O=Organogenesis

_____ 2. Pattern depends on amount and distribution of yolk
_____ 3. Not prevented by inhibition of RNA synthesis
_____ 4. Embryonic nerve tube first forms
_____ 5. Results in formation of skeleton and muscles from mesoderm
_____ 6. Rapid cell division with no increase in size
_____ 7. Influenced by embryo's genome
_____ 8. Produces gastrocoel

9. Which of the following is *not* evidence that the pattern of cleavage is determined by factors in the egg cytoplasm rather than by the genes of the zygote?
   a. Cleavage is prevented by injection of inhibitors of protein synthesis
   b. Cleavage is not affected if the zygote is injected with inhibitors of RNA synthesis
   c. Hybrid zygotes go through normal cleavage but die at the beginning of gastrulation
   d. The direction of the third cleavage division in the snail *Cepaea* (with spiral cleavage) is determined by the mother's genes

10. The blastopore:
    a. is obliterated during gastrulation
    b. induces formation of the lens of the eye
    c. is the hollow space inside the blastula
    d. is the hollow where the embryo implants in the uterus
    e. is the hole through which some external cells pass to the inside of the embryo

11. In a chick embryo, if tissue from the base of a limb bud, which would normally form the thigh, is grafted into the tip of a wing bud, it develops into a toe. This experiment shows that:
    a. the fate of a cell is determined at the time it differentiates
    b. determination is not a one-step process, but occurs in more than one step
    c. it is impossible to tell by merely looking at it whether or not the fate of an embryonic cell is determined
    d. an imaginal disc cell may differentiate in any one of several different ways

12. When the optic vesicle of a frog embryo is transplanted from its normal site in the head to a position where it touches ectoderm elsewhere on the body, a lens forms from the ectoderm. This is an illustration of the fact that:
    a. the path by which a cell differentiates may be altered by its contact with other cells
    b. embryonic cells may be determined before they differentiate
    c. primary embryonic induction occurs during gastrulation
    d. the determination of a tissue can be altered by transplanting the tissue to a new site in the embryo

## Questions for Discussion

1. What is the adaptive advantage of an embryo's being provided with preformed RNA transcribed from its mother's genes?
2. The debate over abortion sometimes centers on the question of when life begins. Now that you have learned a little embryology, when would you say human life begins? Can further study of embryonic development give a more precise answer to the question?
3. Why is it necessary for scientists to study development in more than one species of organism?

## Suggested Readings

Beardsley, T. "Smart genes." *Scientific American,* August 1991. A summary of recent work on transcription, including how cells in the embryo know where they are.

Browder, L. W., C. E. Erickson, and W. R. Jeffer. *Developmental Biology,* 3d ed. Philadelphia: Saunders College Publishing, 1991. A straightforward, well-illustrated text.

Bryant, P. J. "The polar coordinate model goes molecular." *Science* 259: 471, 1993. Determination of position in the imaginal disc of a *Drosophila* leg.

Caldwell, M. "How does a single cell become a whole body?" *Discover,* November 1992. An overview of differentiation.

Cerami, A., H. Vlassara, and M. Brownlee. "Glucose and aging." *Scientific American,* May 1987.

Conklin, E. G. "The orientation and cell lineage of the ascidian egg." *Journal of the Academy of Natural Sciences (Philadelphia)* 2:13, 1905. A delightfully written early account, showing the ingenuity of a scientist asking a question difficult to answer with the limited techniques available at the time.

DeRobertis, E. M., G. Oliver, and C. Wright. "Homeobox genes and the vertebrate body plan." *Scientific American,* July 1990.

Gilbert, S. F. *Developmental Biology,* 3d ed. Sunderland, MA: Sinauer Associates, 1991. A good general text.

Rusting, R. L. "Why do we age?" *Scientific American,* December 1992.

Wassarman, P. M. "Fertilization in mammals." *Scientific American,* December 1988.

13

# Genetic Engineering

## OBJECTIVES

*When you have studied this chapter, you should be able to:*

1. Use the following terms in context:
   plasmid, vector, recombinant DNA, polymerase chain reaction, transgenic, reverse transcriptase, Human Genome Project
2. Describe nucleic acid hybridization and what it can be used for.

3. Explain why the discovery of restriction enzymes was so important to genetic research.
4. Describe the role of plasmids and viruses as vectors in gene transplantation experiments.
5. Describe how a transgenic plant is created.

In 1993 Promega, a biotechnology company in Wisconsin, asked a court to declare that the patent held by Hoffman-La Roche on *Taq* DNA polymerase was invalid. The *Taq* patent was one of several that Hoffman-La Roche had purchased for $300 million. *Taq* is DNA polymerase from the bacterium *Thermus aquaticus,* and is an essential part of a DNA-duplication machine (Section 13–D). If Promega wins this case, any company that can isolate *Taq* from the bacterium will be able to market it, and the price of gene machines will tumble. As you can tell from this case, biotechnology has become big business, with the bitter battles that potential profits produce. Biotechnology is now used to solve criminal cases, analyze ancient DNA from human bodies and fossilized plants, transplant useful genes into agricultural crops, and correct human genetic defects.

For centuries, people have bred and selected crops and livestock to have features useful to humans. Animals and plants with desired characteristics are used as the parents of the next generation. But selective breeding is a slow process because most plants and animals take months or years to grow large enough to reproduce. Biotechnology can speed up production of plants and animals with desirable genetic characteristics.

Fred Griffith's discovery of bacterial transformation showed that bacteria could take up loose pieces of DNA, express them, and pass them on to future generations (Section 9–A). What if people could select desirable genes and

insert them into bacteria? Suppose bacteria could be given a gene for an enzyme that produces a molecule needed by industry. It might then be possible to use bacteria to produce this compound. Bacteria reproduce rapidly. Within weeks, enough bacteria could be grown to produce large amounts of the molecule. This might be much less expensive than producing the compound by industrial processes such as extracting it from a plant or animal.

**FIGURE 13–1** Transgenic plants, containing genes transplanted from other organisms, growing in a greenhouse. (Cornell University Biotechnology Program)

In the 1950s, such possibilities were the dreams of bio-chemists and science fiction writers. Today they are reality. We can isolate a gene and produce millions of copies of it. We can analyze these copies to find the nucleotide sequence of the gene. In some cases, we can transfer functioning genes into the cells of plants and animals to produce **transgenic** organisms, containing genes from more than one source (Figure 13–1). The hybrid genome produced is called **recombinant DNA** because it combines DNA from more than one individual.

Biotechnology derives from the revolution in molecular biology that started in the 1970s with the discovery of ways to analyze the sequence of nucleotides in DNA. The technology is sometimes called **genetic engineering**, the manipulation of genes. In this chapter, we look at a variety of techniques used to manipulate genetic material. Then we shall see how these techniques are applied to the practical problems of genetic engineering and are used to answer many previously unanswerable questions.

## KEY CONCEPTS

- Hybridization occurs between any two single-stranded nucleic acids with some complementary nucleotide sequences and may be used to compare nucleic acids.
- The discovery of restriction enzymes, which break DNA into pieces with known nucleotide sequences at the ends, was the most important single step in the development of genetic engineering.
- Recombinant DNA is produced by treating DNA from two sources with the same restriction enzyme and letting the sticky ends combine.

- The nucleotide sequence of DNA can be analyzed by fragmenting DNA with restriction enzymes and analyzing the fragments by electrophoresis.
- Lengths of DNA can be duplicated rapidly by the polymerase chain reaction.
- The polymerase chain reaction, together with hybridization, has led to an explosion of research into evolutionary relationships deduced from similarities between DNA nucleotide sequences.
- Products of genetic engineering include disease-resistant plants as well as plants and bacteria that produce useful molecules.

## 13–A    NUCLEIC ACID HYBRIDIZATION

Single strands of nucleic acids will **hybridize,** binding to each other by base-pairing between segments with complementary sequences of bases. If DNA is heated to 100°C, the hydrogen bonds linking base pairs are disrupted, and the two strands of the double helix separate into single-stranded DNA. On cooling, complementary base pairs (A-T and C-G) find each other, and the DNA sticks, or **anneals,** to another single strand to form a double helix once again. Any two single strands of DNA or RNA (or one of each) can anneal, provided they have at least short complementary base sequences. The extent to which the strands stick depends on how well their base sequences match. This can be measured by how much heat is required to separate them again.

Hybridization has two particularly important uses. First, hybridization is used to determine evolutionary relationships between species. The more closely two species are related, the more similar their DNA sequences, and therefore the greater the extent of annealing between their DNAs. Second, if you have DNA from a known gene, hybridization can be used to determine whether that gene is present in a DNA sample, and if so in how many copies. A **probe** is prepared, consisting of a single-stranded sequence of nucleotides complementary to one strand of the gene in question, using radioactive nucleotides to label the gene. The probe is then mixed with the DNA sample. The concentra-

tion of the gene in the DNA is indicated by the rate of formation of radioactive double helixes, from which the number of copies of the gene can be calculated.

A technique called *in situ* ("in place") hybridization may be used to find out which cells in a tissue are making a particular protein. Researchers first make a radioactive probe complementary to mRNA encoding the protein. They add this probe to a tissue sample. Here it hybridizes with the mRNA, which is present only in cells that make the protein. These cells can be detected by exposing slices of the tissue to photographic film for autoradiography (*How Do We Know? Isotopes,* Chapter 2).

Hybridization studies require quite large quantities of DNA. The invention of methods of duplicating genomes and individual genes has made hybridization a much more useful technique.

## 13–B    RESTRICTION ENZYMES

The first step in the revolution that led to genetic engineering was the discovery of **restriction enzymes (restriction endonucleases),** produced by bacteria as a defense against viruses. Restriction enzymes bind specific nucleotide sequences in the viral DNA and break the DNA into useless fragments. Why don't the enzymes destroy the bacterium's own DNA? The DNA is protected by **modification enzymes,** which bind the same DNA sequences at-

## DETERMINING THE NUCLEOTIDE SEQUENCE OF DNA

Once a DNA fragment has been amplified, the nucleotide sequence can be determined easily and quickly. One commonly used method was devised by Allan Maxam and Walter Gilbert. In this method, the 5' end of each fragment is first labeled by treating the sample of the fragments with phosphate containing radioactive $^{32}$P. Each fragment will now look, for example, like this:

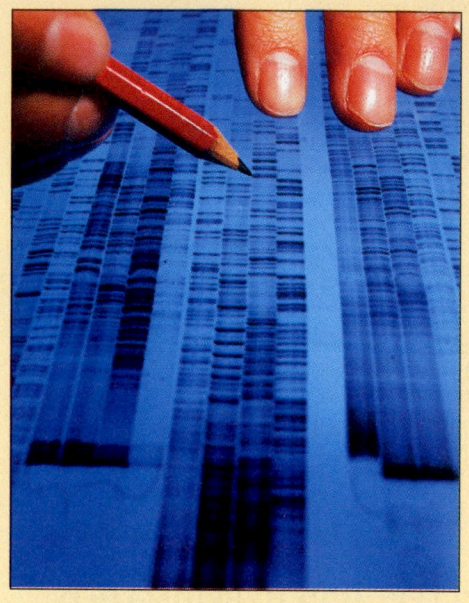

5' end $^{32}$P—TAGCCGATCATCGTACC     3' end

The fragments are now treated briefly with a chemical that breaks DNA fragments on the 5' side of an A nucleotide. Because the treatment is brief, not all the A sites in the DNA will be broken. The treatment will leave some of the original labeled fragments intact and produce DNA pieces with the following sequences:

$^{32}$P—T **BREAK** AGCCGATCATCGTACC
$^{32}$P—TAGCCG **BREAK** ATCATCGTACC
$^{32}$P—TAGCCGATC **BREAK** ATCGTACC
$^{32}$P—TAGCCGATCATCGT **BREAK** ACC

These pieces are now separated by electrophoresis through a gel on a glass plate (Figure 13–A). Electrophoresis is similar to chromatography (Figure 8–B, page 173) but uses an electric current rather than a moving solvent to move the molecules in the sample. This technique separates the DNA fragments by length (that is, by number of nucleotides in each fragment). The shorter the fragment, the faster it moves and the farther it travels from the point at which the mixture of fragments was originally placed

on the gel. The result is examined by a method that detects only the radioactively labeled pieces, so that the unlabeled pieces on the right-hand side of the list above do not interfere with the result. Electrophoresis will produce a result like that shown in Figure 13–B, with the original labeled fragment and the four labeled broken pieces separated from one another.

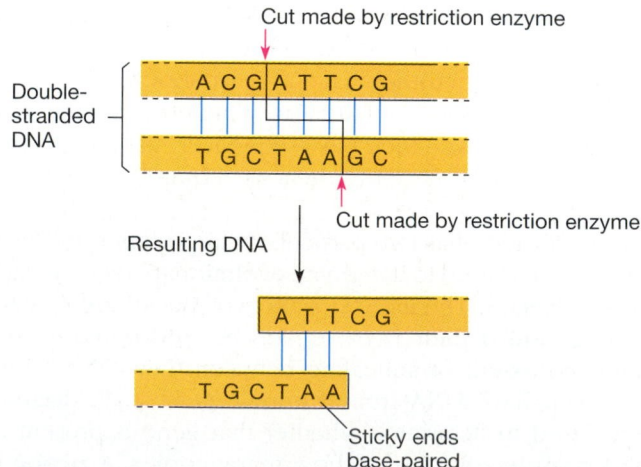

**FIGURE 13–A** Sequencing DNA. A scientist observes three electrophoresis gels, each with four rows (A, C, T, and G), under ultraviolet light, which reveals the position of the nucleotides in a DNA molecule.  (Sinclair Stammers, Science Source/Photo Researchers)

---

tacked by restriction enzymes and add methyl groups (—CH$_3$) to some of the nucleotides. The methyl groups protect the bacterial DNA by blocking the binding of restriction enzymes.

Hundreds of restriction enzymes are now known. Each recognizes a specific sequence of nucleotides in DNA and snips the DNA at or near this recognition site. Restriction enzymes provide a way to break DNA into pieces with known nucleotide sequences at their ends. Many of them produce a staggered cut, in which the double helix is left with two single-stranded ends. These are called "sticky ends" because they will base-pair with any other single-stranded end produced by the same restriction enzyme, and the cut ends can then be joined into a continuous strand by the enzyme DNA ligase (Figure 13–2). In this way researchers can splice a section of DNA into another DNA molecule. In 1978 Werner Arber, Hamilton Smith, and Daniel Nathans received a Nobel Prize for their experiments on bacterial restriction enzymes.

**FIGURE 13–2** The action of one restriction enzyme. The enzyme cleaves double-stranded DNA between A and G nucleotides in a C—G—A sequence. It leaves sticky ends—single-stranded ATT and TAA that will stick to each other.

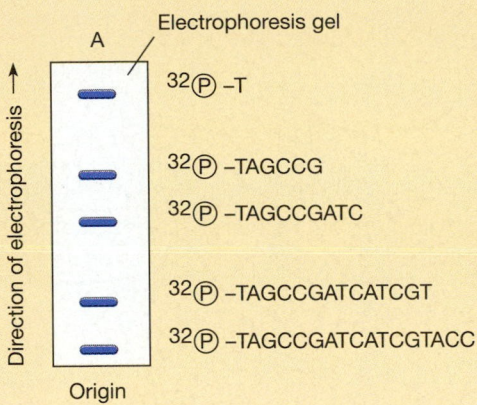

FIGURE 13–B The pattern from electrophoresis of the nucleotide fragments produced as described in the text.

only one nucleotide, is on the gel treated with the A chemical. This tells us that the second nucleotide is A. Since the other radioactive fragments on this A gel are at positions showing that they contain 6, 9, and 14 nucleotides, respectively, we know that A is also at the seventh, tenth, and fifteenth positions in the original DNA sample. We can read off the positions of the other nucleotides on the other three gels in a similar way, to get the rest of the chain.

The ability to "read" a DNA sequence by these simple techniques appears almost miraculous when we remember that as recently as 1975 most people thought the enormous size and chemical monotony of DNA made determining its sequence seem like a hopeless dream.

Three more samples of the DNA fragment are now treated by the same procedure, but using different chemicals to break the DNA in each batch at one of the other three nucleotides (T, C, and G). When the finished electrophoresis gels for all four samples of the DNA fragment are put side by side, they will look like Figure 13–C.

The position of each radioactive fragment in the gel tells us how many nucleotides are in the fragment. The first nucleotide, at the 5′ end of the original sample, cannot be detected by this method because there is no nucleotide on its 5′ side. The T chemical will not distinguish it from the intact fragment. With this exception, the nucleotide sequence of the original DNA sample can be read directly from the gels. The smallest fragment, containing

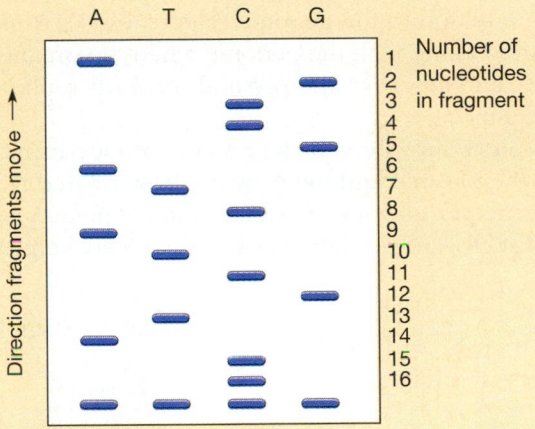

FIGURE 13–C Electrophoresis patterns for four batches of nucleotide fragments. From these we can read the sequence of nucleotides in the DNA molecule used as an example in the text, except for the first nucleotide.

The specificity of restriction enzymes provided Frederick Sanger and Walter Gilbert with the key tools for determining the sequence of nucleotides in DNA (*How Do We Know? Determining the Nucleotide Sequence of DNA*). From being the most difficult macromolecule in the cell to analyze, DNA became, almost overnight, the easiest. Whereas once the structure of DNA could be determined only indirectly from the structure of proteins, now it is often simpler to determine the amino acid sequence of a protein indirectly from the nucleotide sequence of the DNA that codes for it. In 1980, a scant three years after they published their techniques, these men shared a Nobel Prize for their DNA sequencing methods.

## 13–C  FINDING GENES

The main barrier to studies of a single gene is separating the gene from the rest of the cell's DNA and then getting enough copies of the gene to work with.

Genes may be obtained in various ways. In the "shotgun" approach, cells are broken up and the DNA of the entire genome is isolated and cleaved with a restriction enzyme. A mammalian genome treated in this way may produce millions of DNA fragments. The entire batch is duplicated as described in Section 13–D. Researchers must then apply various selection techniques to isolate the DNA sequences they want to work on.

A more elegant way is to start with cells in which the gene of interest is active. For example, cells in the pancreas produce the protein hormone insulin, and the precursors of red blood cells produce the oxygen-transporting protein hemoglobin. These cells contain a great deal of messenger RNA transcribed from the desired gene. Investigators can isolate this messenger RNA and produce the complementary DNA by using the enzyme **reverse transcriptase.** This enzyme, from RNA viruses, synthesizes DNA on an RNA template, unlike the transcription enzymes of most organisms, which synthesize RNA on a DNA template.

## 13–D GENE AMPLIFICATION

DNA has to be **amplified,** duplicated in some way, to produce enough copies of it to study or to use for protein synthesis. There are now several ways to do this.

### Cloning

The first method to be developed was cloning. A **clone** is a group of individuals—DNA molecules, cells, or organisms—derived from the reproduction of a single set of genetic information and therefore genetically identical. DNA to be cloned is first inserted into a **vector** (carrier) that will take it into a cell. Usually the vector is a virus or a plasmid, a small circle of DNA found in many bacteria and yeasts in addition to the genome (Figure 13–3). A plasmid is replicated along with the genome when the organism reproduces, and any DNA the plasmid carries is replicated at the same time.

The procedure for splicing a gene into a plasmid is as follows: the plasmid and the gene are both treated with the same restriction enzyme to create complementary single-stranded sticky ends. When the two are mixed, some of the

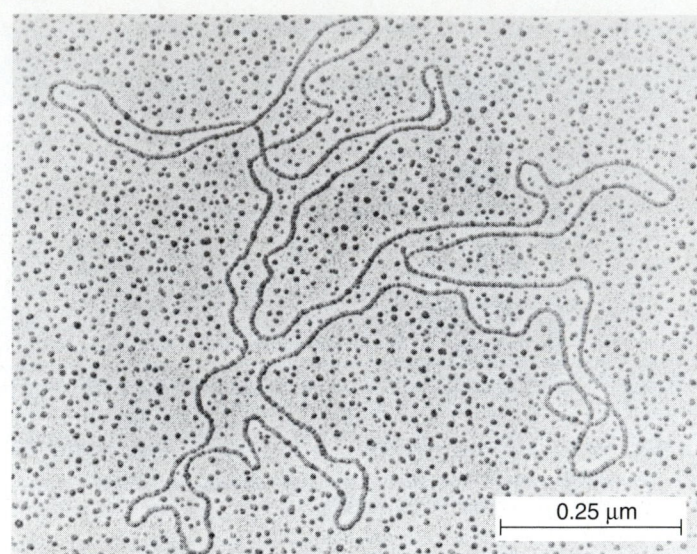

**FIGURE 13–3** A plasmid from a bacterium. (Biophoto Associates)

plasmids' sticky ends will base-pair with those of the desired gene, and the two can be joined by DNA ligase (Figure 13–4).

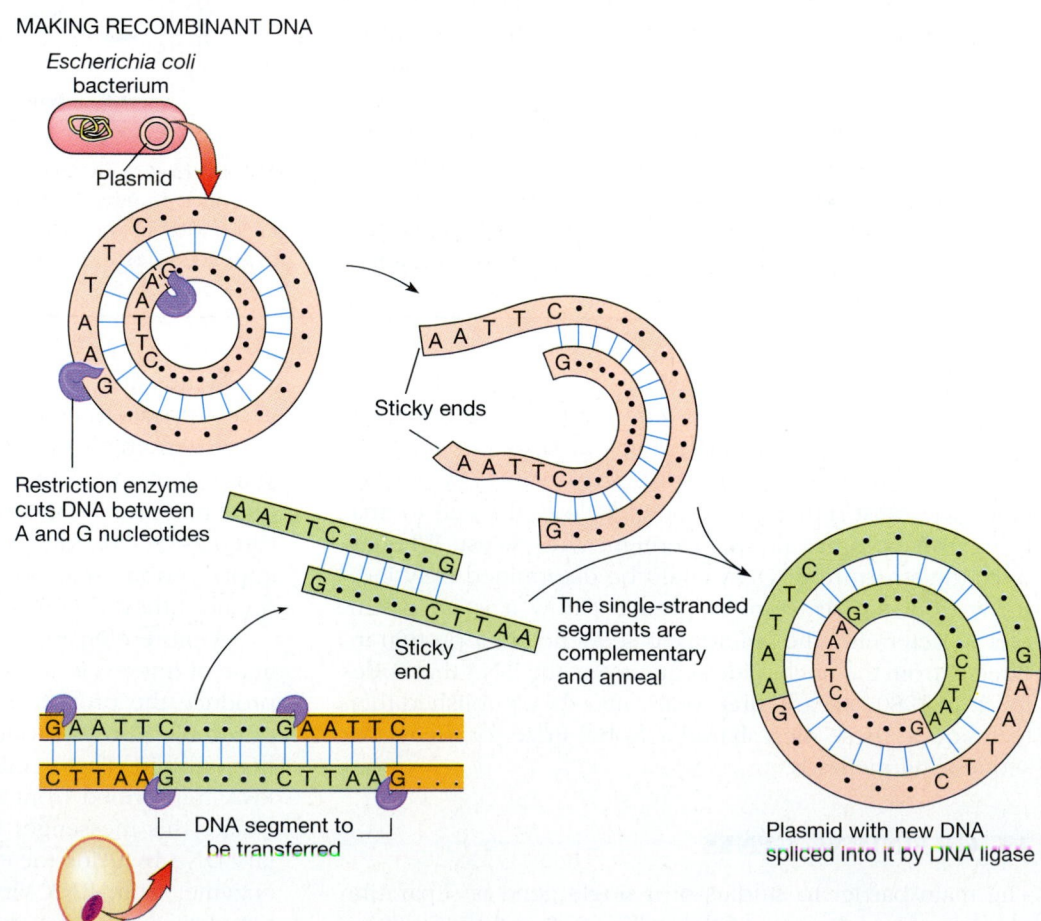

MAKING RECOMBINANT DNA

*Escherichia coli* bacterium

Plasmid

Restriction enzyme cuts DNA between A and G nucleotides

Sticky ends

Sticky end

The single-stranded segments are complementary and anneal

DNA segment to be transferred

Human cells

Plasmid with new DNA spliced into it by DNA ligase

**FIGURE 13–4** Introducing a DNA sequence into a plasmid. The plasmid and the target DNA are treated with the same restriction enzyme, which cuts both strands, leaving sticky ends. DNA ligase is used to splice the two DNA molecules after the single-stranded sticky ends have annealed.

Next, the recombinant plasmids are added to a culture of bacterial (or yeast) cells that have been treated to make them more permeable to DNA. Some of the plasmids are taken up by cells, and as the cells grow and divide they replicate the plasmid along with their own genomes.

The few bacteria containing recombinant plasmids must then be isolated from the other bacteria. This is done by working with plasmids bearing markers that can be used to select them. For instance, the plasmid used in Figure 13–5 contains a gene conferring resistance to the antibiotic ampicillin. Growing the bacteria in medium containing ampicillin will kill all bacteria that do not contain the plasmid.

Since bacteria and yeasts produce many new generations each day, culturing bacteria containing the recombinant plasmid soon produces huge numbers of copies of the gene. Now the cells can be broken open and the plasmids recovered. Treatment with the same restriction enzyme used previously releases the cloned gene from the plasmid, and genes and plasmids can then be separated by differential centrifugation.

The investigator now has a large number of identical DNA molecules that can be analyzed to determine their nucleotide sequence or prepared for transplantation into another cell. This preparation might include adding a promoter or regulatory sequence so that the gene can be used to synthesize RNA when it is introduced into a cell.

## Polymerase Chain Reaction

If the nucleotide sequences on either side of a gene (or any DNA sequence) are known or partly known, the gene can be amplified using the **polymerase chain reaction (PCR),** invented by Kary Mullis in 1983. The reaction permits a worker to make multiple copies of a DNA sequence, even from a single-stranded piece of DNA that also contains many unwanted genes.

Double-stranded DNA is first heated to separate it into single strands. Sequences on either side of the target gene are then hybridized with **primer,** artificially made single-stranded nucleotide sequences. A primer is usually made 18 nucleotides long, short enough to be easy to make and long enough so that it will usually bind only to the desired DNA sequence. (A much shorter primer of, say, four nu-

cleotides would find its complementary bases in dozens of places in any reasonably long piece of DNA and would bind in many places.)

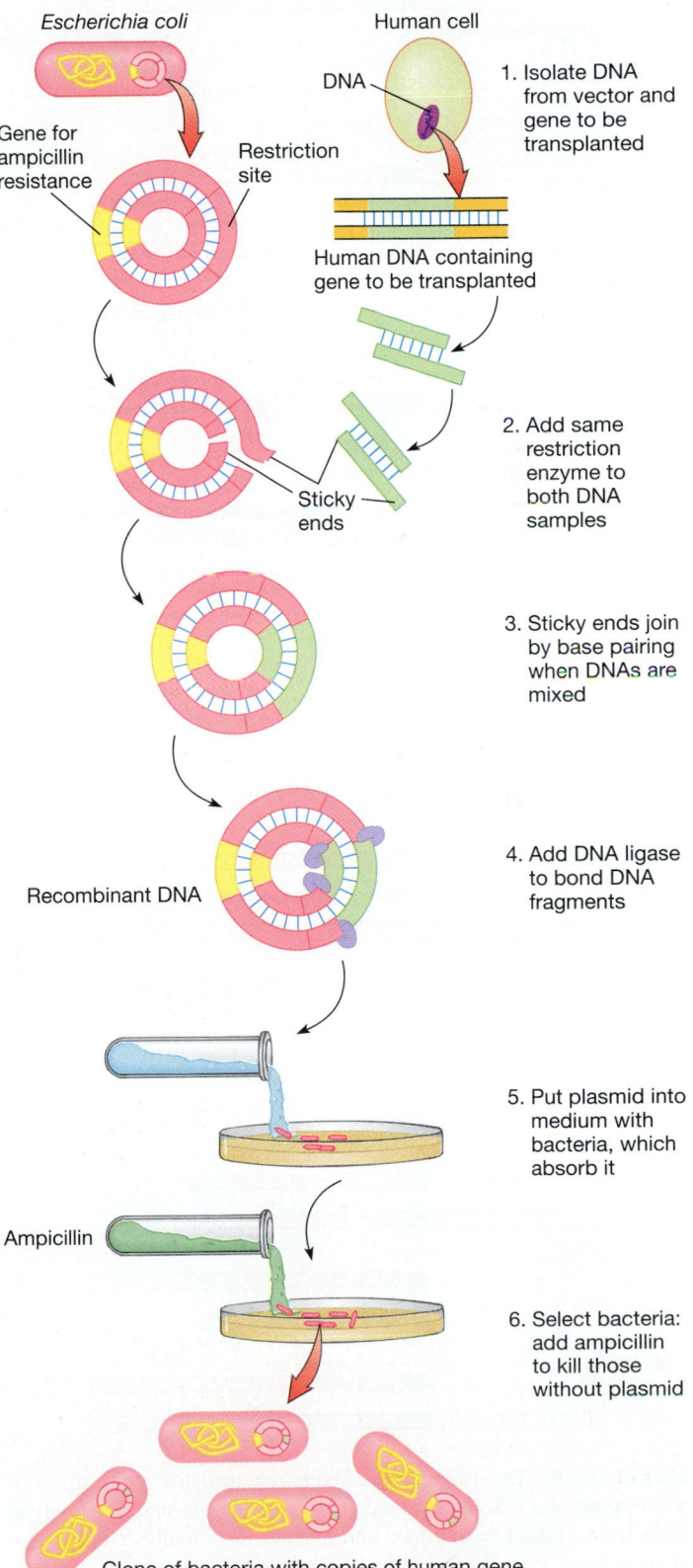

**FIGURE 13–5** Making and cloning recombinant DNA. The procedure shown here produces multiple copies of a human gene. The same restriction enzyme is used to fragment the chromosome containing the gene and the plasmid DNA. This leaves sticky single-stranded ends that base-pair, and the gene is spliced into the plasmid. The plasmid is then cloned by introducing it into a bacterium, which will replicate the plasmid with its own genome each time it reproduces. The gene can then be isolated, in bulk, from the plasmids in the bacteria, using the same restriction enzyme that was used to produce it in the first place.

*Escherichia coli*    Human cell

DNA

1. Isolate DNA from vector and gene to be transplanted

Gene for ampicillin resistance    Restriction site

Human DNA containing gene to be transplanted

2. Add same restriction enzyme to both DNA samples

Sticky ends

3. Sticky ends join by base pairing when DNAs are mixed

Recombinant DNA

4. Add DNA ligase to bond DNA fragments

5. Put plasmid into medium with bacteria, which absorb it

Ampicillin

6. Select bacteria: add ampicillin to kill those without plasmid

Clone of bacteria with copies of human gene

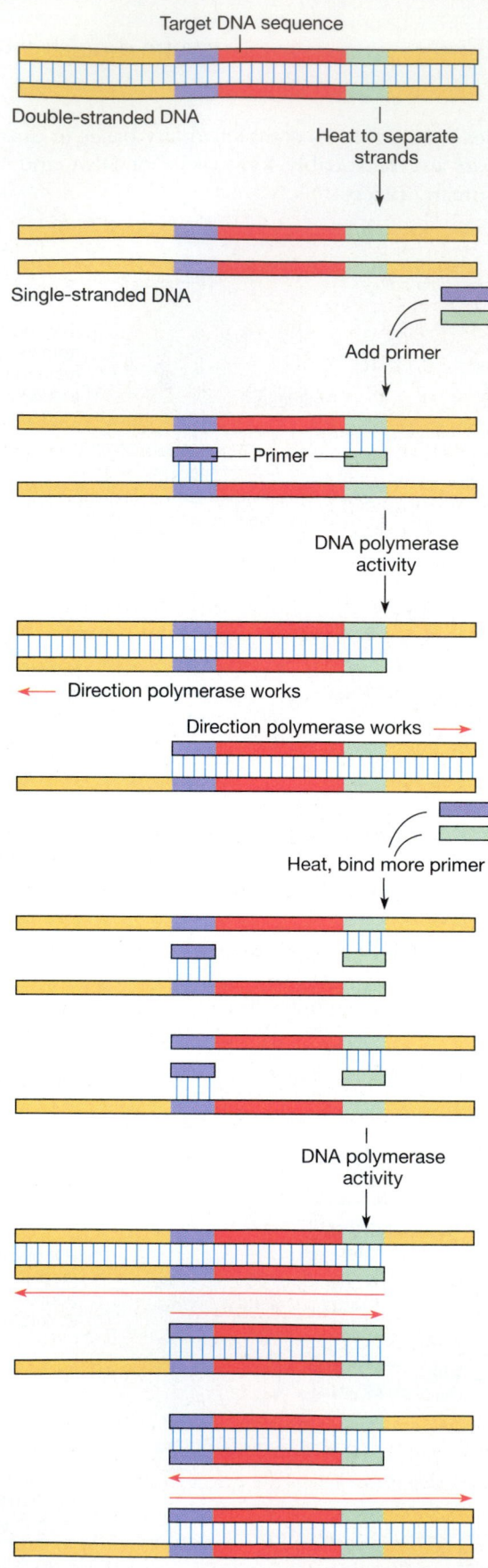

Target DNA sequence

Double-stranded DNA

Heat to separate strands

Single-stranded DNA

Add primer

Primer

DNA polymerase activity

← Direction polymerase works

Direction polymerase works →

Heat, bind more primer

DNA polymerase activity

**FIGURE 13–6** The polymerase chain reaction for making multiple copies of a DNA sequence. The target sequence is shown in red. DNA, target sequence, and primer are really longer than shown here. Primer is usually 18 nucleotides long.

Primer gets its name from the fact that it primes the action of DNA polymerase, which acts only from a double-stranded starting point. When DNA polymerase is added to the primed DNA, it starts from the primer toward the 3′ end and produces complementary strands to all the single-stranded DNA it encounters, making it double-stranded again.

The cycle of heating to produce single strands, and then adding primer and DNA polymerase to make the second strands, is repeated over and over again. Figure 13–6 shows two cycles. You can see that the desired gene multiplies much faster than the rest of the DNA. After a few hours, the reaction can be stopped and the copies of the gene separated from the rest of the DNA by size: the desired gene copies are the short pieces in the collection of DNA molecules.

The polymerase chain reaction has been patented and machines built that perform the steps automatically. If you read of "machine-made" DNA, it probably comes from such a machine (Figure 13–7). Several refinements have been added over the years that increase the speed and efficiency of the technique.

Originally, the process was slow and expensive because the heat used in the reaction denatured the DNA polymerase, destroying its structure and activity. Fresh polymerase had to be added during each replication cycle. Now researchers use heat-tolerant DNA polymerase from bacteria such as those found in hot springs. Enzymes from these organisms remain active at the temperatures used in the reaction and do not have to be added each time. (*Taq* polymerase, the subject of the lawsuit described at the beginning of this chapter, is one of these heat-tolerant polymerases.)

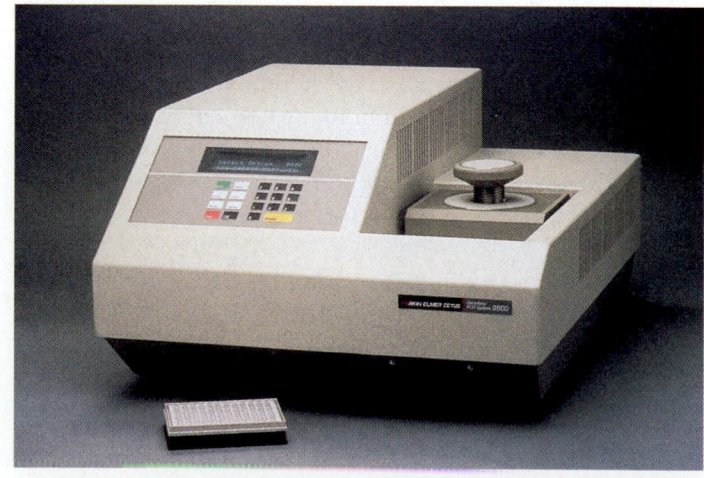

**FIGURE 13–7** A "gene machine" that performs the polymerase chain reaction automatically. (Perkin-Elmer Corp.)

## The Uses of Polymerase Chain Reaction

Invention of the polymerase chain reaction has transformed the study of DNA because it permits researchers to make large quantities of the DNA sequence they want to study in just hours instead of weeks. Among the uses of polymerase chain reaction:

1. **Solving criminal cases.** Tiny samples of blood, tissue, or semen found at a crime scene can be used to identify suspects (*Essay: DNA Fingerprinting*).

2. **Analyzing ancient DNA.** The polymerase chain reaction has turned museums into hunting grounds for molecular clues to evolution. Even a tiny sample of DNA, from a source such as a mummified human body, a 20-million-year-old magnolia leaf, or a fly preserved in amber, is now enough for researchers to analyze its structure.

3. **Tracing human lineages.** Allan Wilson and his colleagues analyzed the genomes of human mitochondria for a study that suggested that all modern humans inherited their mitochondrial genes from one person who lived in Africa between 40,000 and 200,000 years ago (Section 30–J).

   For his original study, published in 1986, Wilson analyzed mitochondrial genomes from blood samples taken from 200 individuals of different human populations around the world, a mammoth task before the days of automated polymerase chain reactions. In 1990 the group used the automated reaction to repeat the study on twice as many samples—in a few months instead of several years.

4. **Detecting viruses.** Primers that contain sequences known to occur in particular genes can be used as probes to go on "fishing expeditions" in the genome, searching for related genes. For instance, one group constructed primers complementary to part of the viral gene for reverse transcriptase. Then they searched the human genome for genes containing this sequence, and found dozens of them. This tends to confirm the theory that viruses may become permanently incorporated into the genomes of their hosts. The genes that were found would be the relics of ancient RNA virus infections that are now passed on as parts of the normal human genome. We usually think of genetic variation as arising from mutations within the genome, but this discovery means that new DNA is sometimes introduced from outside the organism.

## Ligase Chain Reaction

The polymerase chain reaction replicates the DNA between two markers, but it tells you little about the DNA. The **ligase chain reaction,** a variation using DNA ligase from hot springs bacteria, is used to detect mutations in the DNA.

The starting point for the ligase chain reaction is a copy of a gene whose nucleotide sequence is known, together with the same gene isolated from an individual of unknown genotype. The reaction amplifies those parts of the two molecules that are identical. If one of the molecules contains mutations, the reaction makes DNA with breaks at the site of each mutation.

In 1991 Francis Barany suffered third-degree burns when he slipped into a hot spring in Yellowstone National Park while collecting bacteria with which to improve the ligase chain reaction. In the same year he used the reaction to confirm that a single nucleotide change in DNA distinguishes a carrier of sickle cell anemia from someone without the mutation. Because it is so sensitive that it can detect a change of one nucleotide, the ligase chain reaction is being used to investigate:

1. **Genetic disorders.** The ligase chain reaction is used to screen human DNA for mutations, including those that cause beta-thalassemia, hemophilia, Tay-Sachs disease, phenylketonuria, Duchenne muscular dystrophy, and the most common mutations that cause cystic fibrosis (all discussed in Chapter 16). The reaction can also be used to detect genetic defects cheaply and quickly in early embryos.

2. **Safer vaccines.** A harmless form of the polio virus is used to make polio vaccine, but occasionally this mutates into a virulent form. Such mutations used to cause polio in a few vaccine recipients every year. By using the ligase chain reaction to screen for mutations, manufacturers can now make the vaccine completely safe.

3. **Identifying cancers.** The ligase chain reaction is used to screen human DNA for point mutations found in some types of cancer.

4. **Diagnosing HIV.** The reaction is used to distinguish the virus that causes AIDS from related viruses.

## 13–E THE HUMAN GENOME PROJECT

Techniques for sequencing DNA made it possible to contemplate mapping the 3-billion nucleotide pair sequence of an entire human genome. Scientists all over the world are working on this task, known as the Human Genome Project.

The method originally used to map the human genome involves amplifying and then sequencing a random selection of overlapping DNA fragments from a chromosome and then piecing them together like a linear jigsaw puzzle. This procedure contains two time-consuming steps: amplifying the DNA fragments by cloning in plasmids to produce enough DNA for sequencing, and figuring out where any segment you have analyzed fits into a chromosome. In 1993 researchers agreed that there was no hope of completing the genome project within budget unless they turned to more efficient ways of sequencing. Luckily, technical advances may have made this possible.

Some researchers now advocate switching to **primer walking,** a method in which sequencing moves steadily along a piece of DNA, from one known sequence labeled by a primer to the next. Although the DNA must still be amplified for sequencing, the second bottleneck is eliminated because the starting point is known. As a result, you do not have to figure out where the sequence lies on the chromosome.

The bottleneck in primer walking is the slow process of producing the 18-base primers—DNA fragments with known sequences, which in the past have been tailor-made to bind the right sections of the DNA. In 1994 several groups were experimenting with producing each primer from three hexamers—strings of six nucleotides each. The four nucleotides in DNA can be arranged in sixes in only 4096 possible ways. It is perfectly possible to make all of these and store them until needed.

Suppose you wanted to start a primer walk from a sequence AAAAAATTTTTTCCCCCC. You would need the complementary primer TTTTTTAAAAAAGGGGGG. Could it be made just by adding the hexamers TTTTTT, AAAAAA, and GGGGGG to the DNA? Preliminary experiments showed that it could. Researchers added the appropriate hexamers to single-stranded DNA and found that they did indeed bind side by side to the desired starting point. Of course, the hexamers may also bind to other parts of the DNA, but this does not matter as long as they do not bind in the middle of the section being amplified. If they do, the method will not work.

Several groups are now working to automate primer walking using primers made from hexamers. If a machine will make the primers, drop the right ones into the reaction when you press the appropriate buttons, and then clone the DNA by the polymerase chain reaction, workers estimate that 10 people might be able to sequence a bacterial genome in a few weeks—a task that would take several years with 1994 technology. Skeptics doubt that the technique will work with the human genome because mammalian DNA contains so many repeated sequences. This may mean that each hexamer will bind in several positions in many parts of the genome. Enthusiasts acknowledge the problems but are sure they can be solved. If they are right, the Human Genome Project may be much further along by the time you read this book than seemed likely a few years ago.

## 13–F PRODUCTION OF PROTEINS

The first practical application of recombinant DNA technology was the production of proteins. The gene for the protein is introduced into a yeast or bacterium, which is then grown in large quantities. The microorganism reproduces the gene and synthesizes the protein, secreting it into the medium. The protein can then be purified and marketed. In 1982 the U.S. Food and Drug Administration licensed the first recombinant-DNA protein—human insulin, a hormone needed daily by millions of people with diabetes. The human insulin gene was introduced into a plasmid of the bacterium *Escherichia coli,* which was then grown in large vats.

Researchers have also developed microorganisms containing the genes for human growth hormone, which children need to grow to normal height, and for interferons, proteins that interfere with replication of viruses and are sometimes used to treat viral diseases, including warts and some cancers. In the 1980s drug company researchers introduced 15 human interferon genes into bacteria, and relatively abundant and inexpensive interferons are now available.

If you own a cat, you may already have purchased genetically engineered vaccine against feline leukemia, another of the first commercially available proteins produced by genetic engineering.

## Identifying Defective Proteins by Reverse Genetics

Some individuals are born with mutant genes that cause inherited diseases such as Duchenne muscular dystrophy, Huntington's disease, or cystic fibrosis. The disease symptoms result from the actions of abnormal amounts or types of protein produced by the mutant genes.

In the past, it has been difficult to distinguish an abnormal protein from the thousands of normal proteins in a cell (especially if we do not know the protein's function, as is often the case). But DNA sequencing is changing that. Workers can now determine the amino acid sequence of a protein from the nucleotide sequence of the DNA that codes for it. If the gene for the disease can be located, it can be isolated and cloned to produce multiple copies of the gene. These copies can then be transcribed and translated to produce enough of the abnormal protein to analyze. The structure and properties of the protein provide clues to the most effective treatment of the disease.

Researchers used this method to detect the protein responsible for chronic granulomatous disease (CGD), whose victims suffer from chronic bacterial infections. The disease was thought to be caused by a defective cytochrome in the immune system. However, the genetic detectives found this was wrong. Reverse genetics showed that the CGD gene produces a protein too large to be a cytochrome. Researchers are now attempting to work out the protein's structure and function. Eventually, they hope to cure CGD by transplanting the gene for the normal protein into CGD patients.

## 13–G  GENE TRANSPLANTS

One of the goals of genetic engineering has always been transplanting genes from one plant or animal to another. This could be used to cure genetic disorders or to add desirable genes to domesticated plants and animals more rapidly than can be done by classic breeding techniques.

In 1982 the first gene transplants in multicellular organisms introduced a gene from one strain of fruit flies into the embryos of another. The gene was inserted into cells destined to become eggs or sperm. When the embryos grew up, they passed on the transplanted gene, which endowed their offspring with ruby-red eyes instead of the normal brown. Other researchers grew plant cells containing introduced genes into entire plants, showing that the new genes were passed on to the plants' descendants through sexual reproduction (Figure 13–8). The genes transplanted in these experiments had no practical use; they were cho-sen because it was easy to detect their presence and so to determine that the transplant had succeeded.

Many transgenic organisms have now been produced. In 1987 the first field trials of a plant containing a pesticide gene took place using tobacco plants. The pesticide gene comes from the bacterium *Bacillus thuringiensis* (BT), the active ingredient of sprays for gypsy moths and caterpillars that eat vegetables. The bacterium produces a toxin that kills caterpillars but is not toxic to most other insects, nor to mammals or birds. The problem with spraying BT on plants is that it rapidly washes off, and it also degrades in sunlight. These problems are overcome in transgenic plants because the toxin is produced continuously within the leaf.

### Agriculture

In the early 1980s researchers found that a plasmid of *Agrobacterium tumefaciens,* a nitrogen-fixing bacterium that infects plant cells, was an efficient vector for carrying new genes into plants. This tumor-inducing (Ti) plasmid normally causes crown gall tumors (Figure 13–9). A modified form of the plasmid, which does not cause tumors, is now the standard vector for gene transfer in plants. *A. tumefaciens* containing the plasmid with the transplanted gene is permitted to infect plant tissue, and whole plants are grown

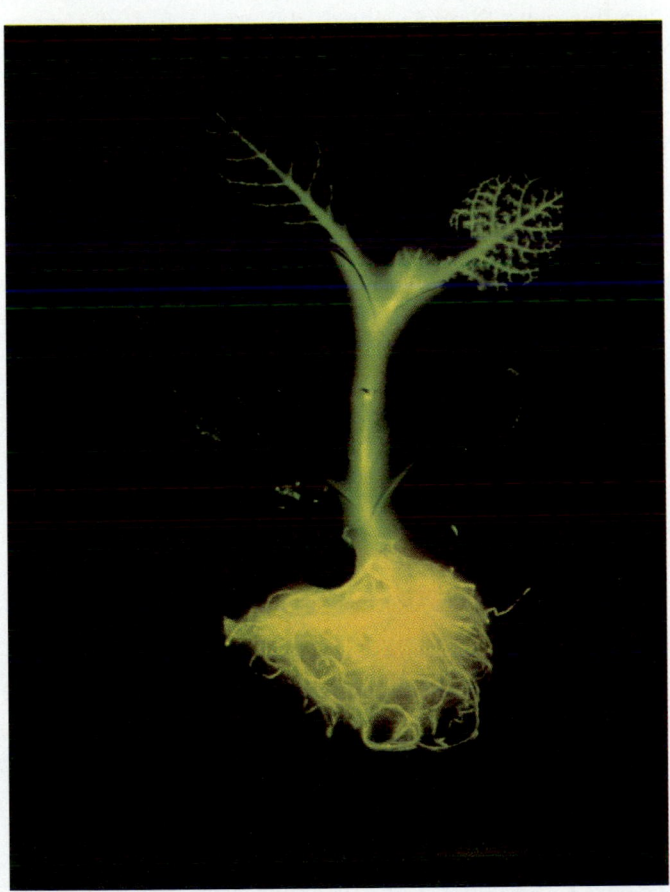

**FIGURE 13–8**  A luminous tobacco plant grown to demonstrate gene transplantation in plants. Fireflies produce light when the enzyme luciferase acts on luciferin. The gene for luciferase has been transplanted from a firefly into this tobacco plant. When the plant is "watered" with luciferin it produces light. This shows that the transplanted gene is active, causing the tobacco plant to produce luciferase.   (Marlene DeLuca, University of California, San Diego. From *Science* 234:856–859, 14 November 1986)

**FIGURE 13–9**  Crown gall tumors on an oak tree, produced by infection with *Agrobacterium tumefaciens* containing a Ti plasmid.

AGROBACTERIUM METHOD          DNA GUN METHOD

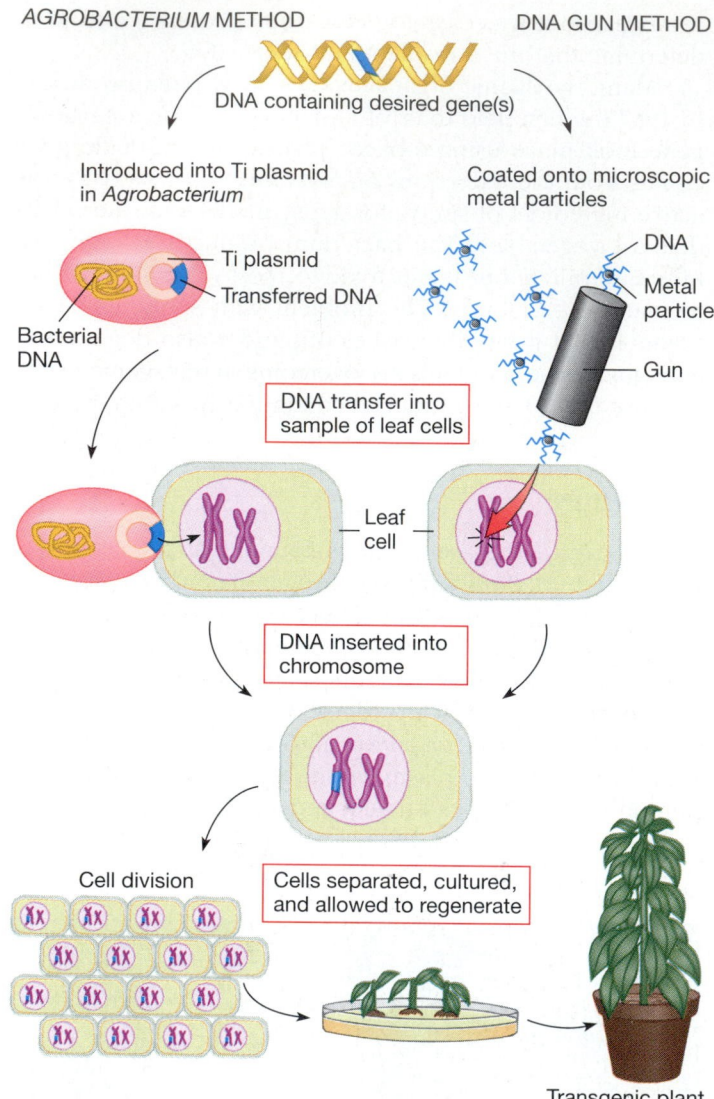

DNA containing desired gene(s)

Introduced into Ti plasmid
in *Agrobacterium*

Coated onto microscopic
metal particles

Ti plasmid
Transferred DNA
Bacterial
DNA

DNA
Metal
particle

Gun

DNA transfer into
sample of leaf cells

Leaf
cell

DNA inserted into
chromosome

Cell division

Cells separated, cultured,
and allowed to regenerate

Transgenic plant

**FIGURE 13–10**  The two main methods of transplanting genes into plants. On the left, the reliable *Agrobacterium* method; on the right, the gun method, used for plants that *Agrobacterium* does not infect.

from the tissue. Plants with genetic resistance to many diseases have been produced in this way (Figure 13–10).

A major disadvantage of the Ti plasmid as a gene transplant vector is that *Agrobacterium tumefaciens* infects mainly dicotyledonous "broad-leaved" plants. Of the four most important human food plants—wheat, rice, maize (corn), and potatoes—only the potato is dicotyledonous. The rest are members of the monocotyledonous grass family. Hence, this method cannot be used to produce transgenic plants of the three species that would make the most difference to attempts to feed the human population. A much less efficient "gene gun" approach has to be used instead: microscopic fragments of metal coated with DNA are

shot at plant cells, penetrating the cell wall and plasma membrane. Some of the cells will be transformed by the DNA and can then be cultured (Figure 13–11).

An added complication is that many of the genes responsible for photosynthesis lie in the genome of the chloroplast, where photosynthesis occurs, and not in the nucleus. Many of these genes work properly only when they are inside the chloroplast. Gene guns can introduce DNA into chloroplasts, but the success rate is very low, probably because chloroplasts are so small.

Calgene Fresh, Inc. produced a genetically engineered "Flavr Savr" tomato, designed to be the first tasty supermarket tomato. Tomatoes are the vegetable most widely grown by home gardeners because tomatoes have little taste unless they are allowed to ripen on the plant. But vine-ripened tomatoes are so soft that they squash easily during shipping. Consequently, most tomatoes are picked while still hard and then ripened in the store. Calgene's tomato contains an introduced antisense gene for a fruit-softening enzyme. Messenger RNA from this gene hybridizes with the mRNA from the normal gene for the enzyme and prevents it being translated. Because the fruit contains little of the enzyme that normally softens ripe tomatoes, the fruit can ripen on the plant and develop flavor while remaining hard enough to be shipped. One might

**FIGURE 13–11**  Young transgenic rice plants produced by the gun method.  (Cornell University Biotechnology Program)

hesitate to eat a tomato containing a toxin that makes it pest-resistant, but a gene for antisense RNA that is never translated into protein should be harmless. Nevertheless, more than a thousand chefs in the United States have joined a group that promises never to offer its customers genetically engineered fruit or vegetables.

Agricultural researchers are also attempting to isolate and transplant the genes that enable members of the legume (pea and bean) family to house nitrogen-fixing *Rhizobium* bacteria in their roots (see Figure 24–9). These bacteria convert nitrogen into a form that plants can absorb. Because they have access to this extra nitrogen, legumes can be grown without nitrogen fertilizer. They also contain more protein than do other plants, making them high-protein food for humans. Agriculture would benefit if plants other than legumes could be grown without nitrogen fertilizer, which takes a lot of fossil fuel to produce and causes water pollution when it runs off fields into rivers.

In addition, breeders aim to produce a human food plant containing "perfect" protein. Legumes contain a lot of protein, but they contain little of several essential amino acids. These amino acids are produced in large amounts by plants such as potatoes, maize, and wheat. If breeders could produce potato, wheat, or maize varieties with *Rhizobium* associations, these plants might be a perfect human food and would be enormously helpful in feeding the human population. Despite the great potential value of this research, the companies that might bring such crops to market show little interest in this work because they do not see any profit in it for many years to come.

Some of the first genetically engineered products to reach the market are varieties of soybeans and canola (also called rape) that can withstand high doses of herbicide (weed-killer) (Figure 13–12). Farmers want herbicide-resistant crops so that they can weed their fields by simply spraying them with herbicides. Plant breeders have been trying to produce herbicide-resistant crops for decades, and genetic engineering has now achieved that goal much more rapidly. Useful as they may be in North America and Europe, however, herbicide-resistant crops pose dangers to agriculture in developing countries. Major crops such as wheat, rice, potatoes, and soybeans originated in South America and Asia, and wild relatives of these plants often grow in fields as weeds alongside cultivated crops. Scientists worry that herbicide-resistant crop plants might breed with their wild relatives, producing hybrid weeds resistant to herbicides. Then herbicides would be essentially useless to farmers in these countries.

## Plants That Produce Polyester

A new line in transgenic crops is plants that produce almost any organic product you can think of. A vat of bacteria or yeasts can produce a few grams of human growth

**FIGURE 13–12** Herbicide resistance in cotton. On the right is a transgenic herbicide-resistant plant. On the left is a normal cotton plant. Both have been sprayed with the herbicide Bromoxynil. As a result, the normal plant is dying; the transgenic plant remains healthy. (Calgene)

hormone, but a field of wheat could produce tons of it. In addition, the protein produced by a plant is often more like the human protein than that produced by a bacterium. This is because many mammalian proteins normally have oligosaccharide chains added to them after they are synthesized. Plants have enzymes that add the saccharides, but bacteria do not.

Because of these advantages, transgenic versions of easily grown plants such as the mustard *Arabidopsis*, tobacco, and potato have now been made with genes that cause them to synthesize many molecules. These include human albumin, interferon, and disease-fighting antibodies (which their inventors call "plantibodies"); bacterial alpha-amylase (used in food processing); and polymers, including a type of polyester (Figure 13–13). Polymers from plants have advantages for the American economy: the United States has to import nearly half its oil, much of which goes to manufacture plastics such as polyester. If agriculture could grow the polymers, it would reduce the bill for imported oil.

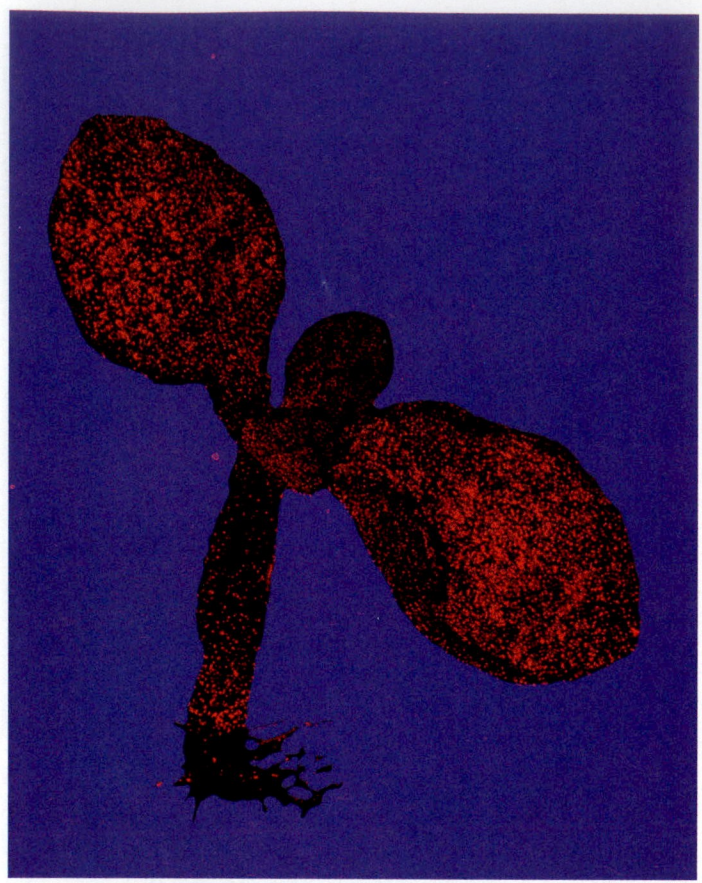

**FIGURE 13–13** A polyester-producing transgenic plant. This seedling of *Arabidopsis thaliana* has been stained with lipophilic Nile Blue A to show granules of the polyester polyhydrobutyrate as spots of red fluorescence scattered throughout the plant. (Yves Poinier, Carnegie Institution)

The latest twist in transgenic plants is to introduce genes that redirect metabolic pathways already present. For instance, Calgene scientists introduced regulatory genes that induced canola plants to make only one of the two oils they normally produce, instead of making both at once. High-starch potatoes, which are used to produce high-fructose syrup and ethanol, and a canola plant that synthesizes valuable jojoba oil are also in the pipeline.

## 13–H   GENE TRANSPLANTS IN ANIMALS

Gene transplants into animals work much less often than those into plants or bacteria, for two reasons. First, we do not yet have equally good vectors for carrying genes into animal cells. Second, most animal cells cannot be grown into new organisms as the cells of many plant species can. This makes it more difficult to grow large numbers of animals than of plants.

The usual method of gene transplants in animals is illustrated by an experiment involving the gene for the beta chains of hemoglobin (see Figure 3–20). The human gene for beta chains was injected into zygotes of mice with beta-thalassemia, a fatal hereditary disease in which the red blood cells produce defective beta chains (or none). Victims suffer from acute anemia (lack of hemoglobin) and usually die young. Apparently, the transplanted gene was incorporated into the zygotes' genomes, because the zygotes grew into adult mice with normal hemoglobin and passed the transplanted gene on to their offspring when they reproduced.

Despite this remarkable result, we still cannot replace defective genes in animal zygotes at will. Not all the transgenic mice were cured of their thalassemia. Some produced no normal beta chains; others produced too many. Expression of the gene varied enormously. This is because different host cells spliced the gene into different places in their own chromosomes. Genes are not expressed normally unless surrounded by the appropriate regulatory sequences.

Transplants of genes from other species have not so far produced very useful results in animals. Early examples included injecting growth hormone genes from cattle or humans into pigs in the hope of producing animals that fattened faster (Figure 13–14). Although the pigs did grow faster, they had a variety of abnormalities that made them useless as breeding stock. A more promising experiment by Scottish researchers involved transplanting genes for medically valuable human genes, such as blood clotting factors, into sheep in such a way that the sheep secreted the protein in their milk, so that it could be harvested without killing the animal.

Some progress has been made. Researchers have now discovered how to insert a mutant gene into the correct position on a mouse chromosome. This technique has been used to produce strains of mice that act as models for human diseases, for instance, mice containing the mutations responsible for cystic fibrosis and many types of cancer (Figure 13–15). These mice are proving very useful in the study of disease, but they are not transgenic—the introduced genes are mutant mouse genes.

## Transplanting Many Genes

Plasmids and viruses cannot carry genes longer than about 50,000 nucleotides. Many genes that people would like to transplant are much longer, including genes lacking in people with hemophilia A and muscular dystrophy. Even some of the short genes that have already been transplanted do not express themselves normally without regulatory sequences that lie too far away to fit into the vector along with the gene they control. Also, some desirable characteristics are governed by complexes of many genes. Examples are genes that enable plants to resist drought, tolerate salty soil, or host nitrogen-fixing bacteria. Researchers have dis-

(a)

**FIGURE 13–14** Transgenic pig. (a) The first transgenic pig born at the U.S. Department of Agriculture laboratory in Beltsville, Maryland. The pig developed from a fertilized egg injected with DNA containing growth hormone genes from cattle as shown in (b). (Robert Wall, U.S. Department of Agriculture)

(b)

covered how to assemble vectors that can carry these large quantities of genetic material: yeast artificial chromosomes (YACs).

Chromosomes carry hundreds of genes. As we saw in Chapter 9, they also contain three kinds of DNA sequences that enable them to be replicated and passed on to new cells at cell division: replication origins where DNA polymerase attaches, centromeres, and telomeres. In unicellular yeasts, these sequences are shorter than in other eukary-

otes. They are the basic pieces in the kit researchers use to build artificial chromosomes, splicing in genes to be transplanted into new hosts. The artificial chromosomes are cloned in yeast cells because they are too long to be amplified by the polymerase chain reaction. Finally, the yeasts are broken open and the artificial chromosomes recovered.

What are the uses of such artificial chromosomes? Researchers have injected artificial chromosomes into naked plant cells with their walls removed and into zygotes or early embryonic cells of mice. Some of the treated cells have been reared to adulthood and passed on their new genes to their offspring. One group even decided to bypass the laborious task of recovering artificial chromosomes from yeasts. They simply stripped yeasts of their cell walls and fused each naked yeast cell with a cell from an early mouse embryo. When injected into mouse embryos at a similar stage of development, these cells developed into tissues with joint mouse/yeast genomes.

Researchers still have technical problems to overcome. Only a small percentage of cells treated with artificial chromosomes actually survive to adulthood. Also, very long DNA sequences tend to break and then to contain many mistakes in nucleotide sequences.

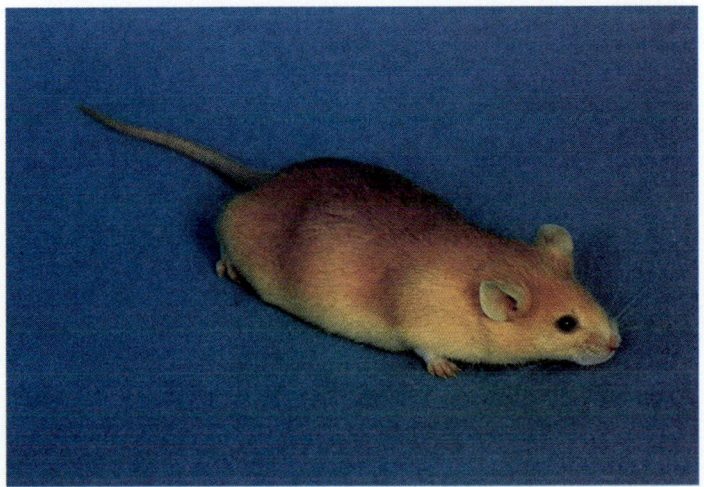

**FIGURE 13–15** A genetically engineered mouse used for research on genetic diseases. This yellow obese mouse contains a transplanted mutant gene that causes its orange coat color and makes it more susceptible to tumors. (Jackson Lab)

## 13–I   TOWARD HUMAN GENE THERAPY

Gene therapy means treating disease by gene transplants. For instance, diabetics do not produce enough insulin. Why not treat them by transplanting insulin genes into their pancreatic cells instead of by frequent insulin injections? And why not transplant a normal gene into a baby or embryo known to have inherited a genetic defect? Although production of transgenic mice and pigs might suggest that the techniques needed for gene therapy are known, this is not the case. When researchers produce transgenic animals, they can kill the many animals in which the gene transplant does not work or produces deformity, but doctors could not do this with human patients. In addition, the transplanted genes are injected into the egg, which is a far cry from transplanting genes into specific cells in an organism.

Many technical problems remain before gene therapy will be permitted in humans except in cases of life or death. The list of gene therapy trials under way as of 1993 (Table 13–1) shows that the therapy was authorized only for people who were already terminally ill or had no chance for normal life without the therapy. The first four examples in Table 13–1 introduce genes that are specific for the condition. All the other examples are nonspecific treatments and would seem to have little chance of succeeding.

## Finding Appropriate Vectors

Human gene therapy will remain unreliable until scientists discover better vectors than those used today to transfer genes into animals, which are very inefficient: only about 10% of cells exposed to the vector actually acquire the gene.

The usual vectors are viruses. Alternatively, the DNA to be inserted can be adsorbed onto the surface of a liposome, an artificial vesicle capable of crossing the plasma membrane. Cells are removed from the body and exposed to viruses or liposomes carrying the DNA to be transplanted. Then the transformed cells are injected back into the body (Figure 13–16). Occasionally this method works. For instance, in 1990 two sisters born without the gene for adenosine deaminase, an immune system enzyme, received cultures of their own white blood cells containing the gene for the enzyme. It worked. Today the girls live normal lives instead of having to live in isolation because they had no resistance to disease.

However, this successful transplant is an exception. Even in the few cells that do acquire a gene from a vector, the gene does not usually become integrated into the genome but lies around loose in the cytoplasm and is eventually broken down. Viruses are favored as vectors because they often join the genes they contain to the host genome.

In theory, viruses could also provide a way of inserting a gene into a specific type of cell without removing the cells from the body, because viruses enter cells by binding to cell-surface receptors, which differ from one cell type to another. This is why HIV binds to immune cells but not to muscle cells and why the common cold virus binds to cells in the respiratory system. However, we cannot yet transfer genes to particular cells in the body because the cell-surface receptors for most viruses are not known.

TABLE 13–1

**Gene Therapy Trials Authorized in the United States by January 1993**

| Disease | Gene Inserted |
|---|---|
| Adenosine deaminase deficiency | Adenosine deaminase into white blood cells |
| Hemophilia B | Blood clotting factor |
| Lung cancer | Antisense *ras/p53* (*ras* is an oncogene; see Section 11–H; *p53* suppresses cell division and is inactive in lung cancer cells) |
| AIDS | 1. HIV *env* (a gene that inactivates HIV) <br> 2. Thymidine kinase (an enzyme that converts the drug ganciclovir into a poison that kills brain tumor cells in mice) |
| Ovarian cancer | Thymidine kinase |
| Brain tumor | Thymidine kinase |
| Malignant melanoma (skin cancer) | Human leukocyte antigen-B7 (cell-surface receptor that induces rejection by immune system: designed to induce the body to reject the tumor; Section 34–D) |
| Advanced cancers (2 cases) | Tumor necrosis factor (protein that kills tumor cells) |
| Fatty liver degeneration | High-density lipoprotein receptor |
| Advanced cancers (2 cases) | Interleukin-2 (chemical messenger that intensifies immune response; Section 40–C) |
| Malignant melanoma | Interleukin-2 |
| Neuroblastoma (nerve cell cancer) | Interleukin-2 |
| Kidney cancer | Interleukin-2 |

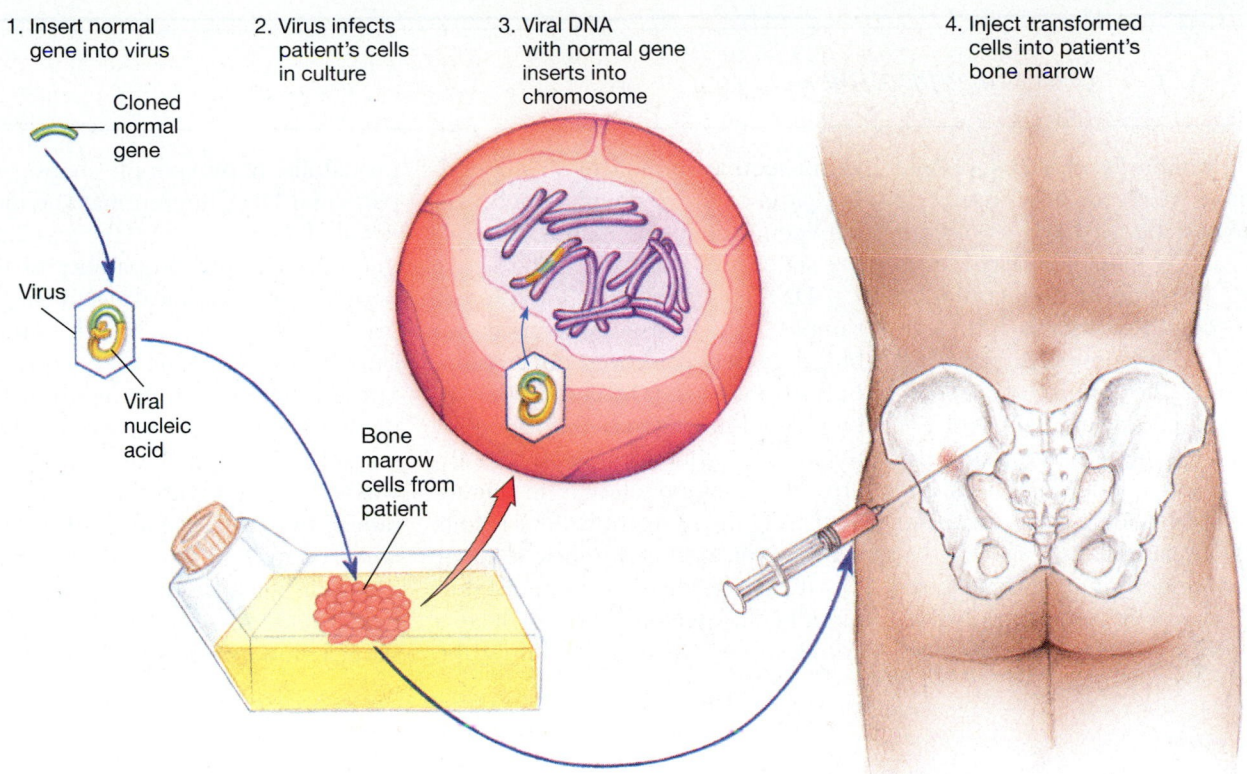

1. Insert normal gene into virus

Cloned normal gene

Virus

Viral nucleic acid

2. Virus infects patient's cells in culture

Bone marrow cells from patient

3. Viral DNA with normal gene inserts into chromosome

4. Inject transformed cells into patient's bone marrow

**FIGURE 13–16** Human gene therapy. In this particular procedure, cells are removed from the patient's bone marrow and cultured. An RNA transcript of the gene the patient lacks is introduced into an RNA virus, which is injected into the patient's cell. RNA viruses transcribe their RNA genomes with reverse transcriptase and insert them into the host genome, in this case with the missing gene as well. After further culture to ensure that the gene is present and is being replicated, the cells are injected back into the patient—in this case into the bone marrow, where many blood cells, which circulate throughout the body, normally divide.

## 13–J  HOW SAFE IS GENETIC ENGINEERING?

Many early fears about the dangers of genetic engineering have proved to be unfounded, thanks to efforts that prevent the release of potentially dangerous transgenic organisms into the wild. For instance, the bacteria used in many recombinant DNA experiments are *Escherichia coli,* a species universally found in the human intestine. However, the strains used in the laboratory have lived outside human bodies for thousands of generations. They have evolved in such a way that they can no longer survive outside their test-tube homes. The danger is further reduced by regulation of laboratories doing recombinant DNA research. Government, scientific organizations, and citizens' groups all participate in drawing up the rules that researchers must follow.

Genetically engineered organisms have undergone hundreds of field tests, using isolated areas to minimize the danger of escapes. Before a treatment can be used on human patients, it must be tested on cultured cells and on animals and be approved by government agencies. Questions remain about who should receive treatment. Currently, children whose bodies do not make growth hormone are given injections of genetically engineered hormone so that they will grow as tall as other people instead of becoming dwarfs. However, some parents are having their normal children who are shorter than average treated with growth hormone as well, even though researchers do not know whether this will increase the children's height or whether the injections will have adverse effects. Is this an appropriate use of expensive and limited medical resources? What will happen when we discover genes that contribute to intelligence, docility, or creativity? Will parents be permitted to have their children treated with the products of these genes in an attempt to mold normal children into geniuses?

Careful regulation will continue to be needed if new dangers are not to arise in the future. The possibility that

*Text continues on page 280*

## E S S A Y : *DNA Fingerprinting*

For hundreds of years people have realized how useful it would be to have a surefire way of telling one individual from another. The information could be used to identify criminals from evidence left at the scene of a crime and to pinpoint the origin of illegally imported plants and animals. With the advent of techniques permitting the analysis of nucleotide sequences, it is now possible to identify an organism from a DNA sample.

"DNA fingerprinting" detects restriction fragment length polymorphisms (RFLPs). These are differences in nucleotide sequences in noncoding parts of the genome, which vary between individuals more than coding sequences do.

DNA is isolated from the forensic sample, which usually consists of blood, semen, hair roots, dental pulp, or bone marrow. It is then broken up by restriction enzymes and tested for nucleotide sequences by the Southern blot method shown in Figure 13–D. Figure 13–E shows the results from a typical RFLP forensic test.

If DNA evidence is to be used to identify the person who left it at a crime scene, we need to know the probability of two people sharing any particular DNA fingerprint. The more detailed the fingerprint, the less likely this is. Thus, the probability of two people with identical DNA fingerprints is one in a few hundred thousand for one RFLP marker sequence, but less than one in a million if five marker sequences are used. For each marker, the membrane must be washed, treated with the appropriate radioactive nucleotide probe, then autoradiographed, and each autoradiograph must be exposed for up to two weeks. As a result, RFLP analysis is a slow process.

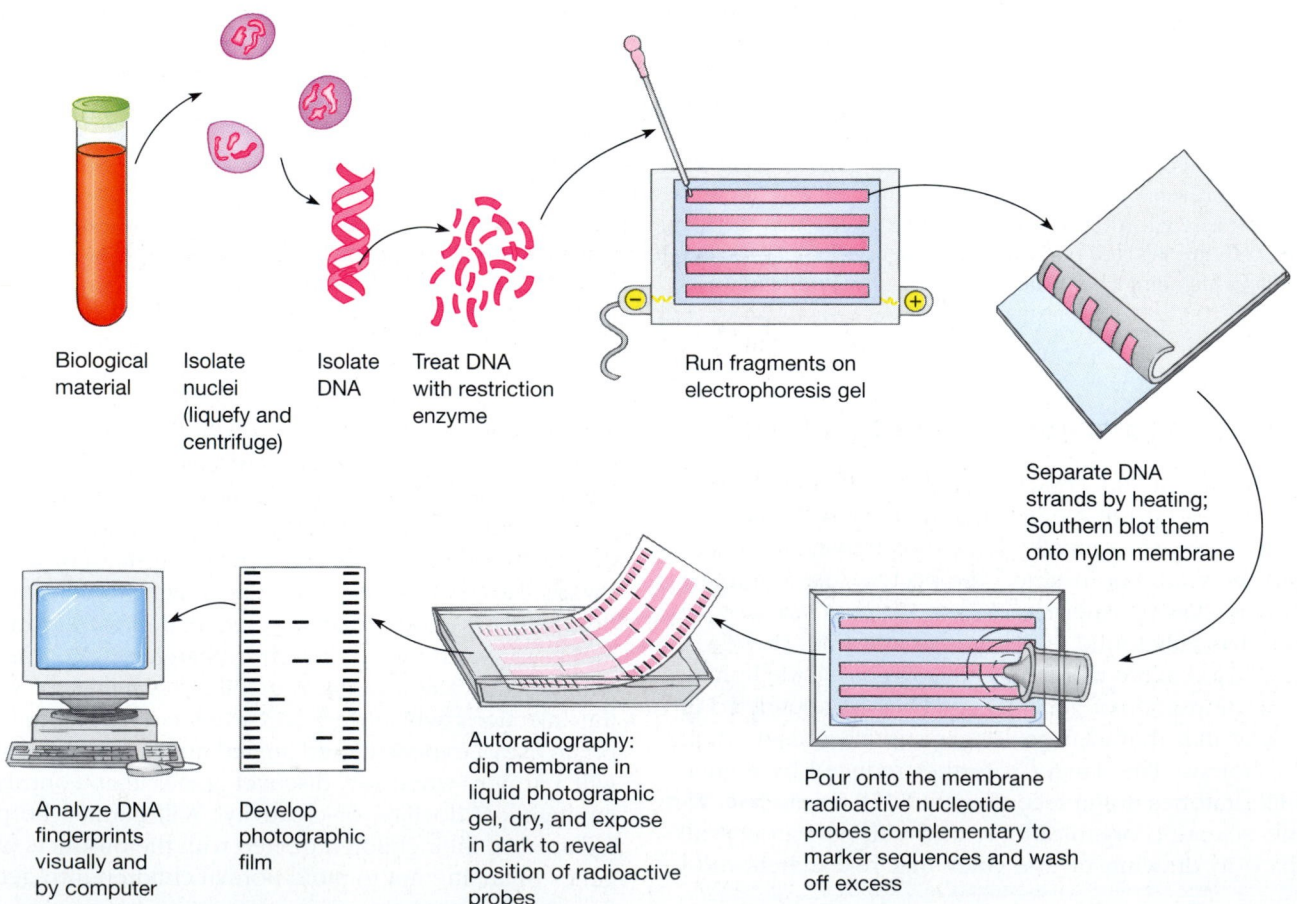

Biological material

Isolate nuclei (liquefy and centrifuge)

Isolate DNA

Treat DNA with restriction enzyme

Run fragments on electrophoresis gel

Separate DNA strands by heating; Southern blot them onto nylon membrane

Pour onto the membrane radioactive nucleotide probes complementary to marker sequences and wash off excess

Autoradiography: dip membrane in liquid photographic gel, dry, and expose in dark to reveal position of radioactive probes

Develop photographic film

Analyze DNA fingerprints visually and by computer

**FIGURE 13–D**    The technique of DNA "fingerprinting" by the Southern blot method (named after its inventor).

Some DNA samples are so small or so degraded that the RFLP markers are missing. In this case, the polymerase chain reaction (PCR) can be used to amplify the DNA and produce a large enough sample for electrophoresis (Section 13-D). PCR analysis produces a less detailed "fingerprint" of the genome than RFLP analysis but it can analyze a tiny amount of DNA in a few days.

The marker sequences used in DNA fingerprinting are chosen because they are highly variable even between members of the same family (unless the two people are identical twins, who have the same genetic makeup). For instance, in the case of *Caldwell v. State of Georgia* (1990), a brother and sister were attacked. The sister was raped and murdered; the brother was injured but recovered. Their father was charged and de-fended himself by claiming that the brother had raped and murdered his sister and had been injured in the process. Conventional blood testing might or might not have distinguished blood and semen samples from these members of the same family, but DNA fingerprinting did. DNA from semen in the sister's vagina matched that of the father, not the brother, and the father was convicted of rape, murder, and attempted murder.

In the United States, DNA fingerprinting has also been used to release from prison more than 20 people who had been wrongly convicted of rape or murder. These people were convicted before DNA fingerprinting was first accepted in court in 1987, usually on the basis of eyewitness testimony, which is notoriously unreliable. Later analysis showed that blood or semen from the crime scene came from someone other than the convicted person or the victim.

The more closely related two individuals are, the more similar their genomes. As a result, DNA fingerprinting is a very accurate test for paternity in cases where a woman claims that a man owes child support because he fathered her child. For the same reason, DNA fingerprinting is also being used to help suppress illegal international trade in endangered species. The members of a population of elephants or tigers in an area are genetically distinct from those in other areas. Customs officials can sometimes use DNA analysis to pinpoint the origin of a particular shipment of elephant tusks, whale meat, or items from other endangered species.

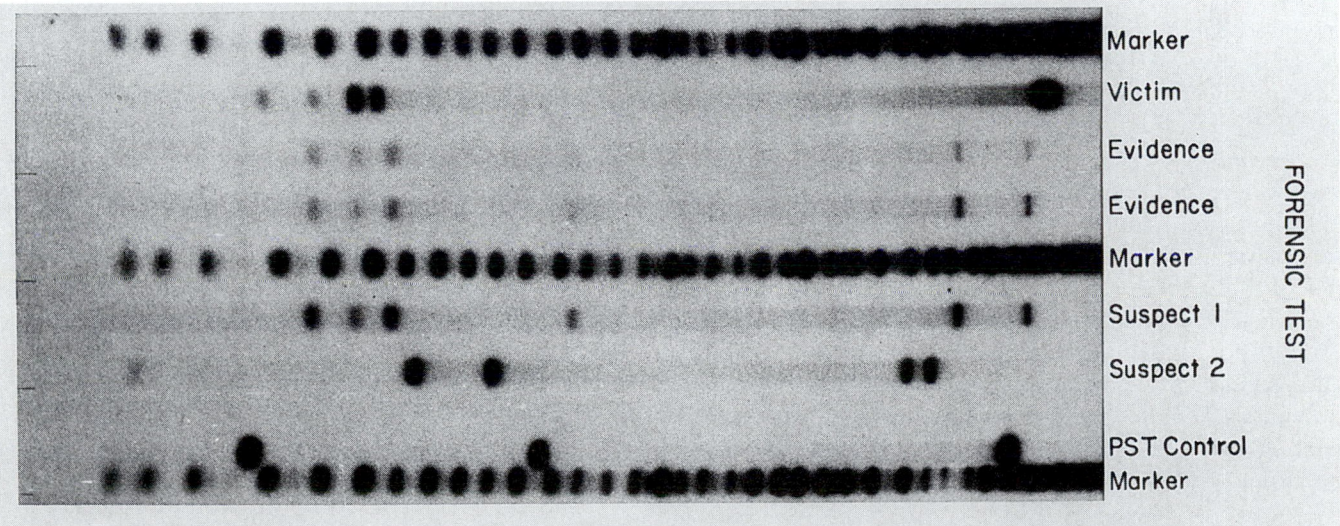

**FIGURE 13–E**  Results of a forensic test in a rape case. This is an autoradiograph of the electrophoresis of several samples of DNA fragments. The gel contains DNA from blood samples of the victim and two suspects and evidence: two samples of DNA from semen taken from vaginal swabs of the victim. The marker is a mixture of viral DNAs used to determine the size of DNA samples. The PST control is DNA from cultured human cells. The fingerprint from the evidence DNA matches that from suspect 1 but is unlike DNA from the victim or suspect 2.  (Ivan Balazs, Lifecodes Corporation)

we might one day be able to control the genetic makeup of a human being poses ethical problems that are new to our experience and must be faced. The human race is not, and probably never will be, ready to cope with genetic control over the population. For instance, one study showed that if people could choose how many children to have and the sex of each, three quarters of the babies born would be boys and only one quarter girls.

The likelihood that we shall one day be able to transplant genes from one human to another opens up many possibilities—both good and bad. Organizations such as churches and medical societies are studying the moral and ethical implications of genetic engineering techniques such as this one now, before the techniques are ready to be used.

## SUMMARY

Since about 1970, techniques permitting us to analyze DNA sequences and to alter an organism's genome have been developed rapidly.

1. Nucleic acid hybridization between complementary sequences in single-stranded nucleotide molecules can be used to detect genes and to determine evolutionary relationships.
2. Bacterial restriction enzymes break DNA at specific nucleotide sequences, leaving single-stranded ends. They are used to isolate DNA sequences for cloning, transplantation, or nucleotide sequencing. DNA sequencing methods determine the nucleotide sequence of DNA by breaking the molecule into fragments, labeling the fragments, and separating them by electrophoresis.
3. Genes can be isolated by a shotgun approach, which involves cleaving the entire genome with restriction enzymes. Alternatively, a structural gene can be identified using mRNA from cells that produce large quantities of the protein it encodes as a template for reverse transcriptase to synthesize complementary DNA.
4. Amplification techniques produce many copies of a DNA sequence for experiments or transplantation. The polymerase chain reaction copies genes in a cycle of heating and DNA polymerase action. A gene can also be cloned by using a virus or plasmid to carry it into the cells of microorganisms, which then reproduce in culture.

5. Uses of the polymerase chain reaction have included analyzing DNA from living and museum specimens for evolutionary studies and identifying particular genes in the human genome.
6. The ligase chain reaction is a variant of the polymerase chain reaction that permits detailed comparison of two DNA sequences. It is used to screen for mutations that cause human genetic disorders, to eliminate mutant viruses from polio vaccine, and to detect mutations characteristic of some types of cancer.
7. The Human Genome Project is an ambitious scheme to sequence all the DNA in a human genome. The method originally used involved cloning and sequencing of shotgun fragments of DNA. The project will not be completed according to schedule or without vast cost overruns unless primer walking or some similar technique can be automated to speed up the work.
8. The first commercial application of genetic engineering was the production of useful proteins from recombinant DNA in microorganisms.
9. Transgenic plants are produced by introducing genes into plasmids in bacteria that will infect cells in culture. The procedure has been used to introduce disease resistance and other desirable qualities into crop plants. Transgenic plants can be used to produce proteins and other molecules in much larger quantities than microorganisms can produce. Unfortunately, the bacterial plasmid cannot be used to introduce genes into the most important human food plants. Instead, the gun method must be used. It is not yet possible to transplant genes reliably into chloroplasts, which contain many of the genes that control photosynthesis.
10. Gene transplants in animals are not as advanced as in plants because the viruses usually used are inefficient vectors, and the introduced genes are seldom expressed normally. Human gene therapy still faces many barriers, although there have been some minor successes. Gene transplants have been used to introduce genes for genetic disorders into animals, which are then used as models for studying the disorder.
11. There are many potential dangers in genetic engineering, but the most obvious problems have been avoided to date by careful regulation of procedures.

## Self-Quiz

1. Which of the following would be used in preparing recombinant DNA?
   a. plasmids
   b. DNA ligase
   c. restriction enzymes
   d. DNA from two different sources
   e. all of the above

2. Two fragments of single-stranded DNA will stick together if:
   a. both contain A—T—T—C sequences of nucleotides
   b. each contains a regulatory gene
   c. the two contain complementary nucleotide sequences
   d. part of each is double-stranded
   e. both are heated before being mixed

3. Plasmids and viruses are useful in genetic engineering because:
   a. they contain nucleic acids
   b. they are small
   c. foreign genes can be spliced into them
   d. they readily invade new host cells, bringing in the genes they carry
   e. all of the above

4. If two different chromosomes are treated with the same restriction enzyme that leaves sticky ends:
   a. the enzyme will break the chromosomes into fragments of equal length
   b. heating will restore the original double-stranded structure of the two chromosomes
   c. the genes isolated by the treatment will not support the synthesis of normal amounts of protein when they are transplanted into a virus
   d. fragments of the two different chromosomes will subsequently stick together to form hybrid DNA molecules

5. Which of the following *cannot* be used to produce multiple copies of a human gene?
   a. cycles of heating and DNA polymerase activity, provided the gene has double-stranded markers on either side
   b. transplanting the gene into the genome of a microorganism that reproduces rapidly
   c. gene amplification
   d. retrovirus translocation

6. Which of the following statements is untrue?
   a. It should be possible to cure some genetic diseases by transplanting normal replacement genes into some of the patient's cells.
   b. Gene transplantation can be used to produce clones of genetically identical animals.
   c. Adult potato plants can be grown from single potato cells.
   d. Human genes have been transplanted into mice and the proteins for which the genes code synthesized in the recipient mice.

## Questions for Discussion

1. Is transplanting genes into people for medical reasons ethically equivalent to transplanting a replacement kidney into someone whose own kidneys have failed?
2. Why do workers on the Human Genome Project using the shotgun approach amplify their DNA fragments by cloning in plasmids instead of by the polymerase chain reaction?
3. Is genetic engineering likely to be used to produce organisms for biological warfare? What kind of organisms? Is there any action we should take on this matter?
4. What limits would you place on gene transplant research involving human eggs and embryos? Why?
5. Manipulation of the human genome raises many horrifying possibilities, including populations with many more of one sex, or clones of identical individuals. How far does religious teaching, the source of many of our moral values, give us guidelines about what we should and should not permit?
6. Can you think of any dangers and benefits, other than those listed in this chapter, that might result from research on recombinant DNA?

## Suggested Readings

Gasser, C. S., and R. T. Fraley. "Transgenic crops." *Scientific American,* June 1992. Explains how transgenic plants are engineered and some of the kinds of plants that may become commercially available in the 1990s.

Gilbert, W., and L. Villa-Komaroff. "Useful proteins from recombinant bacteria." *Scientific American,* April 1980. Gives further details of recombinant DNA techniques.

Maxam, A. M., and W. Gilbert. "A new method for sequencing DNA." *Proceedings of the National Academy of Sciences (U.S.)* 74:560, 1977. The Nobel Prize–winning description of a frequently used method for determining the sequence of nucleotides in DNA. This paper joins those by Avery et al. and Watson and Crick as one of the few hundred papers that are definitely "classics" in biology.

McElfresh, K. C., D. Vining-Forde, and I. Balazs. "DNA-based identity testing in forensic science." *BioScience* 43:149, 1993. Describes techniques of DNA fingerprinting and the controversy about whether the techniques are sufficiently accurate and well performed to be used as conclusive evidence of identity.

Mulligan, R. C. "The basic science of gene therapy." *Science* 260: 926, 1993. Describes the technical problems that remain to be solved before human gene transplants will be much use.

Mullis, K. B. "The unusual origin of the polymerase chain reaction." *Scientific American,* April 1990. A personal account.

Verma, I. M. "Gene therapy." *Scientific American,* November 1990. Discusses prospects for human gene therapy.

Weinberg, R. A. "The molecules of life." *Scientific American,* October 1985. Summarizes recombinant DNA technology and its history.

# Reproduction of Eukaryotic Cells

**O B J E C T I V E S**

*When you have studied this chapter, you should be able to:*

1. Define the following terms, use them in context, and answer questions about the relationships among them:
   centromere, sister chromatids, homologous chromosomes
   haploid (N), diploid (2N), tetraploid
   somatic cell, germ cell
   interphase
   kinetochore, mitotic spindle
   cytokinesis
   gamete, fertilization, zygote
   synapsis, tetrad, chiasma
2. Define and describe the four stages in the cell cycle.
3. Briefly state the functions of nutrition, growth factors, and cyclin-dependent kinase in the control of cell division.
4. Recognize a cell in prophase, prometaphase, metaphase, anaphase, or telophase of mitosis.

5. State why substances that interfere with microtubule function or formation interfere with cell division; name one such substance and state how it is useful to plant breeders.
6. Explain how cytokinesis occurs in animal cells and in plant cells.
7. Define syncytium and coenocyte and give one example of how such a structure can be formed.
8. Describe the roles of mitosis and meiosis in an organism's life cycle.
9. Compare and contrast what happens to chromosomes during mitosis with what happens to them in meiosis.
10. Describe synapsis, crossing over, and the reassortment of chromosomes during meiosis, and explain their biological importance.
11. Describe nondisjunction and translocation of chromosomes.
12. Compare and contrast the processes of spermatogenesis and oogenesis in animals.

---

**L**ife is handed down from one generation of organisms to the next in the form of new cells. A unicellular organism produces more of its kind by dividing in two. A multicellular organism begins life as a single cell, and repeated cell divisions produce the many cells of the body. Eventually, some cells in the reproductive organs divide and form reproductive cells, which give rise to the next generation.

Before a cell divides, it must contain enough components for two new cells. This includes two sets of the cell's blueprint for life: the genetic information coding for all the kinds of proteins the cell can produce. In a eukaryotic cell, this genetic information is carried in the DNA of the many chromosomes in the nucleus. Before dividing, the cell must copy this DNA, collate two complete sets of chromosome blueprints, and package each set in its own nuclear envelope, one for each of the two new cells. In addition, both new cells must receive cytoplasm with all the equipment to express this genetic information: ribosomes to make

proteins, mitochondria to supply the necessary energy, and so on.

The most complicated part of cell division is the separation of the nucleus into two new nuclei. A time-lapse movie of this process shows the replicated chromosomes being divided accurately into two complete sets. This feat, achieved not by intelligence but solely by interactions among countless macromolecules, attests to the awesome results of evolutionary happenstance.

There are two types of nuclear division: mitosis and meiosis. **Mitosis** produces two new nuclei, each with a set of chromosomes identical to those in the original nucleus. This ensures that a complete set of the parent cell's genetic information goes to each daughter cell.[1] In unicellular

---

[1]Some traditional terms used in cell division are feminine words: mother, daughter, sister. This doesn't mean that the structures involved are female rather than male. The biologists who invented the terminology used feminine terms because the female's role in reproduction is usually greater than the male's.

organisms, mitosis produces two new individuals that are genetically identical. A multicellular organism starts as one cell. During embryonic development, repeated mitotic cell divisions produce many new cells, which are genetically identical to each other and to the original cell (barring mutations). As the organism develops, its genetically identical cells differentiate and form specialized tissues by expressing different combinations of the genes they all contain (Chapter 12). Mitosis continues throughout life as the body grows, repairs itself, and replaces old cells with new.

**Meiosis** produces daughter nuclei with half the original genetic information, arranged in new combinations. Meiosis is necessary for sexual reproduction. In animals it gives rise to the **gametes,** the sexual reproductive cells: sperm and eggs (Figure 14–1). In plants meiosis produces **spores,** reproductive cells that divide mitotically and produce structures that give rise to the gametes (Chapter 27). At fertilization, two gametes combine and form a **zygote,** a fertilized egg, with a full complement of chromosomes.

Meiosis is an integral part of sexual reproductive cycles. By halving the number of chromosomes, meiosis ensures that the chromosome complement remains the same instead of doubling from generation to generation.

Mitosis and meiosis technically refer only to division of the nucleus. These nuclear divisions are usually followed by **cytokinesis,** the division of the entire cell. Mitosis and meiosis occur only in eukaryotes. We shall consider how prokaryotic cells divide in Section 24–B.

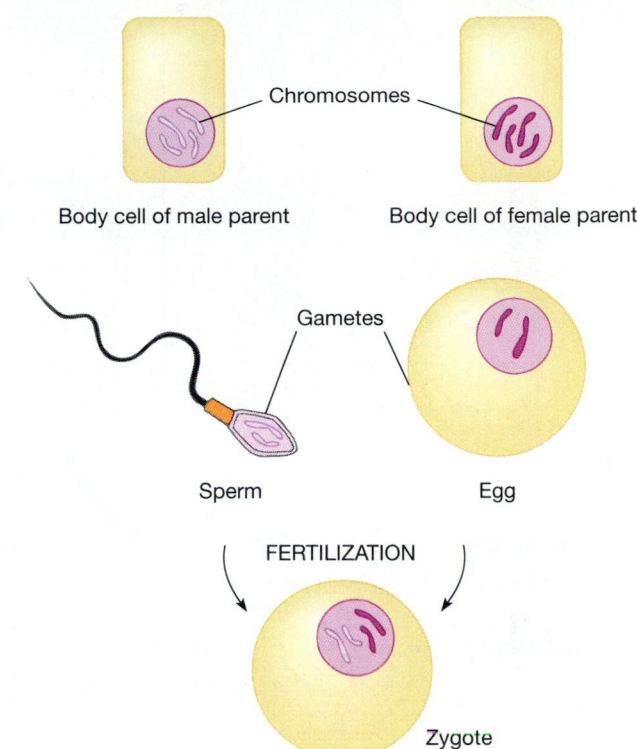

**FIGURE 14–1** Body cells and reproductive cells. As a result of meiosis, sperm and eggs (gametes) contain only half as many chromosomes as the other cells in an animal's body. After a sperm fuses with an egg, the zygote (fertilized egg) contains the characteristic number of chromosomes found in normal body cells of the species.

## KEY CONCEPTS

- A cell reproduces by dividing into two new cells.
- Before a eukaryotic cell divides, its duplicated chromosomes are distributed precisely into two new nuclei.
- Mitosis produces two new nuclei that both contain the same genetic information as the original nucleus.

- Meiosis produces new nuclei with new genetic combinations and with only half the number of chromosomes found in the original nucleus.

## 14–A  EUKARYOTIC CHROMOSOMES

Viewed through a microscope, a cell's chromosomes usually appear as a single, diffuse mass of chromatin. The individual chromosomes can be distinguished only before and during cell division, when they are **condensed**—that is, coiled up tightly into short, thread-like structures. By this time, the chromosomes have already been replicated. Each consists of two copies, attached to each other at a region called the **centromere.** As long as the two copies remain attached, they are called **sister chromatids** (Figure 14–2).

### Haploid and Diploid Chromosome Numbers

In most plants and animals, chromosomes from the body cells can be matched up in pairs. One member of each

pair came from the mother's egg, the other from the father's sperm. The two chromosomes of a pair are called **homologous chromosomes,** or simply **homologues.** Most homologous chromosomes look alike: they are the same length, their centromeres are in the same position, they show the same pattern of light and dark bands when stained, and they carry genes for the same inherited characteristics, lined up on the chromosome in the same order. For example, human chromosome #1 contains the genes for the Rh blood factor and for salivary amylase, a starch-digesting enzyme in the saliva. Each of these genes occupies a particular location, known as its **locus** (plural: **loci**), on chromosome #1 (Figure 14–2).

A person's two homologous chromosomes #1 may contain identical genes at these loci or slightly different versions of the same gene. For instance, at the Rh locus some

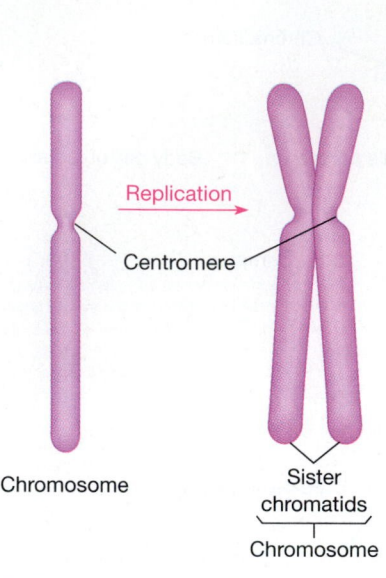

(a) Chromosome replication

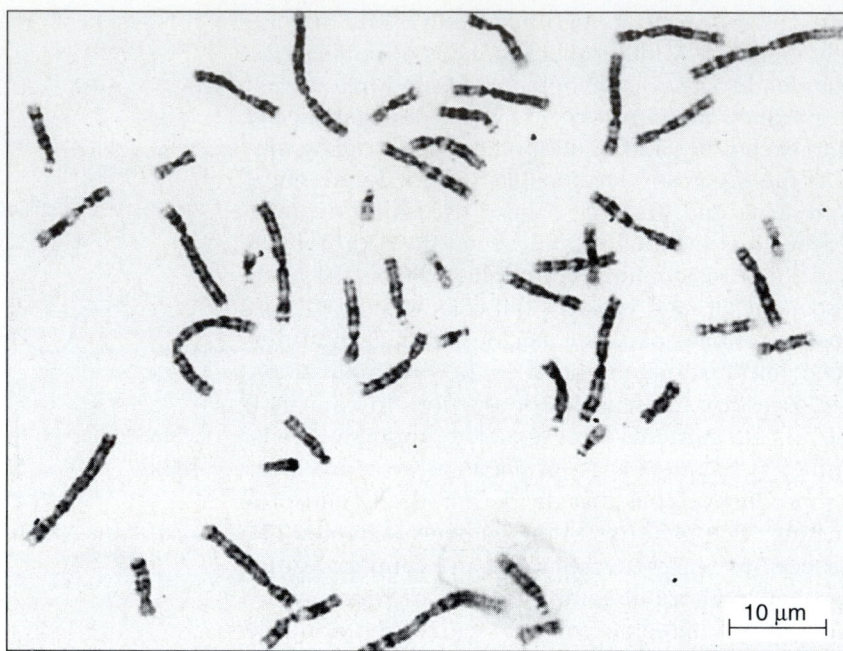

(b) Replicated chromosomes from one cell of a human male (LM)

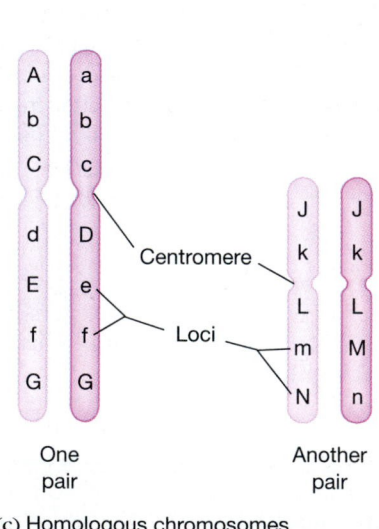

(c) Homologous chromosomes

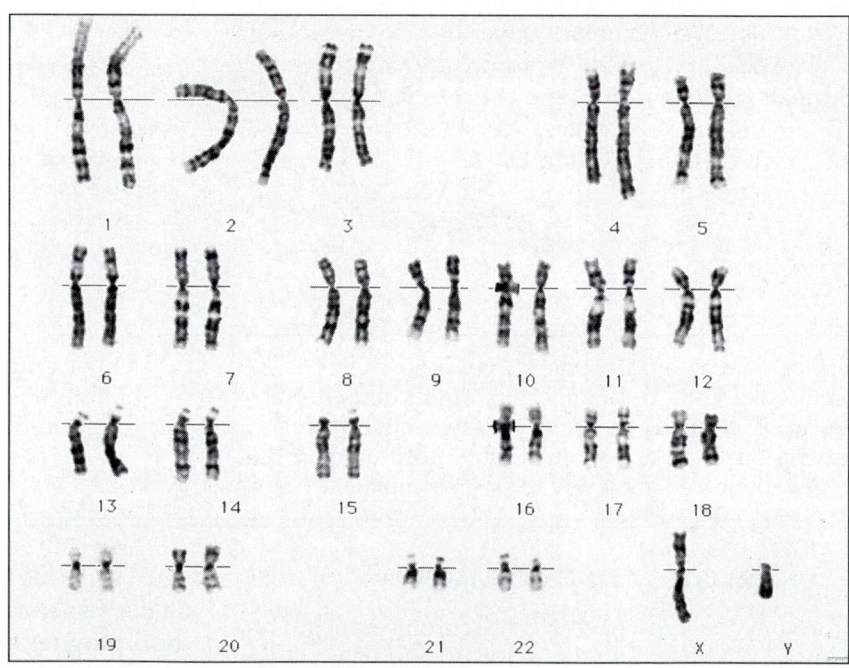

(d) Karyotype: 23 homologous chromosome pairs from a human male

**FIGURE 14–2**   Counting chromosomes. (a) A replicated chromosome consists of two chromatids, each containing a complete copy of the chromosome's DNA. The two remain joined at the centromere. (b) The 46 replicated chromosomes from a single cell of a human male. (c) Two pairs of homologous chromosomes. The members of each pair have the same genetic loci but may have the same or different versions of a particular gene at any one locus. For example, at the A locus on the longer pair, *a* might code for a different form of the protein dictated by *A*, whereas both chromosomes have the same form of the gene (*b*) at the B locus, and so on. (d) A karyotype: the chromosomes from (b) arranged in 23 homologous pairs. In the lower right corner, notice the different lengths of the X and Y chromosomes. The pair indicates that the individual is male. Both (b) and (d) were computer generated.   (b, d, Prenatal Diagnostic Center Cytogenetics Laboratory, The Johns Hopkins School of Medicine)

chromosomes have a gene for the protein that makes a person Rh-positive, and some have a gene coding for a different version of this protein (Rh-negative).

The number of chromosome pairs varies from one species to another. In humans each body cell contains 46 chromosomes, which can be arranged in 23 homologous pairs according to their length and the position of the centromere. In human males, 22 of these pairs contain look-alike chromosomes, but the twenty-third pair is odd, with two unlike chromosomes, called X and Y (Figure 14–2d). In the cells of a human female, both chromosomes in the twenty-third pair are X chromosomes. The X and Y chromosomes are called **sex chromosomes** because this pair of chromosomes differs between the sexes and because they play a role in determining their owner's sex (Section 16–F). The other 22 pairs of chromosomes are called **autosomal chromosomes,** or **autosomes** (auto = self; soma = body).

Having pairs of chromosomes is important in the life cycle of eukaryotic organisms. At meiosis, the two members of each homologous chromosome pair are separated into different nuclei. As a result, a gamete contains one member of each pair of chromosomes, for a complete set containing half the number of chromosomes. For example, each human egg or sperm contains 23 chromosomes, one from each of the 23 pairs. When an egg and sperm join at fertilization, the new individual receives one member of each pair of chromosomes from its mother, and one member of each pair from its father, for a total of 23 complete pairs of chromosomes.

A cell that contains pairs of homologous chromosomes is said to be **diploid,** that is, having two sets of chromosomes. The diploid number in humans is 46, or 23 pairs. A cell that contains one set of unpaired chromosomes is said to be **haploid,** containing half the diploid chromosome number. An egg or sperm is haploid, and in humans the haploid number of chromosomes is 23. The haploid number of chromosomes in a species is generally designated as **N:** N = 23 in humans, and **2N** (diploid) = 46. In corn plants, N = 10, and in the fruit fly *Drosophila,* N = 4 (Table 14–1).

In most animals, the zygote and the cells that arise from it by mitotic division contain the diploid number of chromosomes. Some cells in the ovaries or testes eventually undergo meiosis, which produces haploid nuclei. Only the egg and sperm cells have haploid nuclei. Cells that can undergo meiosis are **germ cells;** the rest of the body's cells are called body or **somatic** (soma = body) **cells.**

Not all organisms are diploid. Those that are normally haploid include moss plants, many algae and fungi, and some invertebrates (animals without backbones), such as male honeybees. Some organisms, especially many plants, are **tetraploid** (4N), with four homologous chromosomes of each type. Triploid (3N), hexaploid (6N), and other ploidy numbers are also found, particularly among plants.

**TABLE 14–1**

### Haploid (N) and Diploid (2N) Numbers of Chromosomes in Some Organisms

| Organism | N | 2N |
|---|---|---|
| Pea plant | 7 | 14 |
| Corn | 10 | 20 |
| Potato | 24 | 48 |
| Fruit fly | 4 | 8 |
| Chicken | 39 | 78 |
| Cat | 19 | 38 |
| Dog | 39 | 78 |
| Chimpanzee | 24 | 48 |
| Human | 23 | 46 |

## 14–B THE CELL CYCLE

A cell's lifespan begins when the cell is formed by division and ends when the cell itself divides or dies. In unicellular eukaryotes the time between divisions varies from about 2 hours for yeast (a single-celled fungus) to a few days for *Amoeba* in laboratory culture. Cells of early animal embryos may also divide rapidly, as often as every 10 minutes in *Drosophila*. Most dividing cells of the adult have lifespans of about 8 hours to more than 100 days, or even much longer, depending on the cell type.

Some types of cells cannot divide once they have reached their final differentiated state and must eventually die. Examples are nerve, skeletal muscle, and red blood cells in mammals, and some cells in the transport systems of plants. Most other cells are capable of dividing. In tissues such as the blood, skin, and lining of the digestive tract, some cells die from normal wear and tear and are replaced by division of other cells. In other tissues, cell division may be infrequent unless the body has been damaged and new cells are needed to replace those lost.

Cells that can and do divide have a typical life cycle, called the **cell cycle,** lasting from the time the cell is formed by division until it divides in its turn (Figure 14–3). The cell cycle has four distinct periods. During the period of mitosis **(M),** the nucleus and cytoplasm divide and form two new cells. The rest of the cell cycle, known as **interphase,** is divided into the remaining three periods. The period from a new cell's "birth" until it begins to replicate its DNA is called the first gap period, or **G$_1$**. During this time, the cell is usually carrying on the business of life and growing to about double its original size. The middle period, the **S** (synthesis) period, is the time of DNA synthesis, when the chromosomes are replicated in preparation for the next cell division. The second gap period, **G$_2$**, lasts from the end of DNA synthesis until the cell divides.

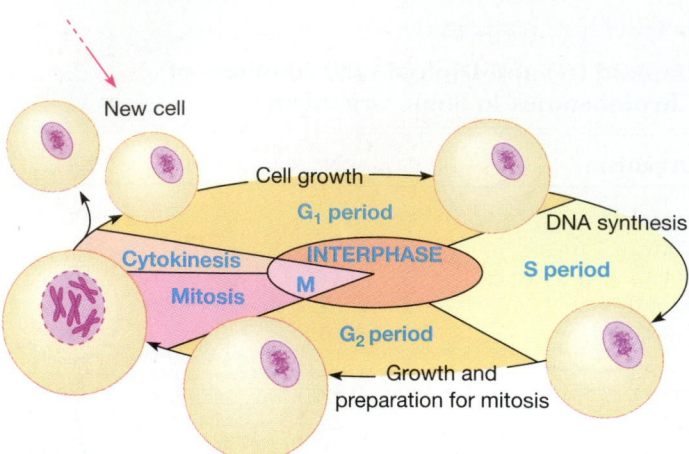

**FIGURE 14–3** A typical cell cycle. Start at the top left with the new cell at the red arrow. The two gap periods (G₁ and G₂) are separated by the synthesis (S) period, during which the chromosomes are replicated. Mitosis, or nuclear division, is followed by cell division (cytokinesis).

Since DNA is replicated during the S period, a cell in G₂ contains twice the normal amount of DNA. However, it has the same number of chromosomes, even though each now consists of two chromatids, and so a cell in G₂ is still considered diploid. The total length of the cell cycle, as well as the percentage of time spent in each period, varies widely.

The existence of the three distinct periods of interphase (G₁, S, and G₂) was detected by experiments using radioactive hydrogen to label thymidine nucleotides, that is, those containing the base thymine (T). Cells incorporate these nucleotides into DNA but not into other macromolecules. Radioactive thymidine was put into a laboratory culture of cells for about 30 minutes and then washed away and replaced with nonradioactive thymidine.

Next, samples of the culture were taken at intervals, and the DNA of dividing cells was checked for radioactivity. The cells entering mitosis in the first few hours after exposure to labeled thymidine did not contain radioactive DNA. They must already have finished replicating their DNA when the labeled thymidine was introduced: this shows that a gap period (G₂) occurs between the end of the S period and the beginning of mitosis (Figure 14–4). The cells entering mitosis somewhat later did have radioactive DNA. These cells were in the S period while radioactive thymidine was present. The cells dividing still later again have no radioactive DNA: they had not yet started to replicate their DNA while radioactive thymidine was present. These cells show that a G₁ period exists before the S period.

Synthesis of RNA and proteins shows typical patterns during the cell cycle. A cell makes some proteins more or less continuously, but others only in certain parts of the cycle. For example, the S period is the only time when the

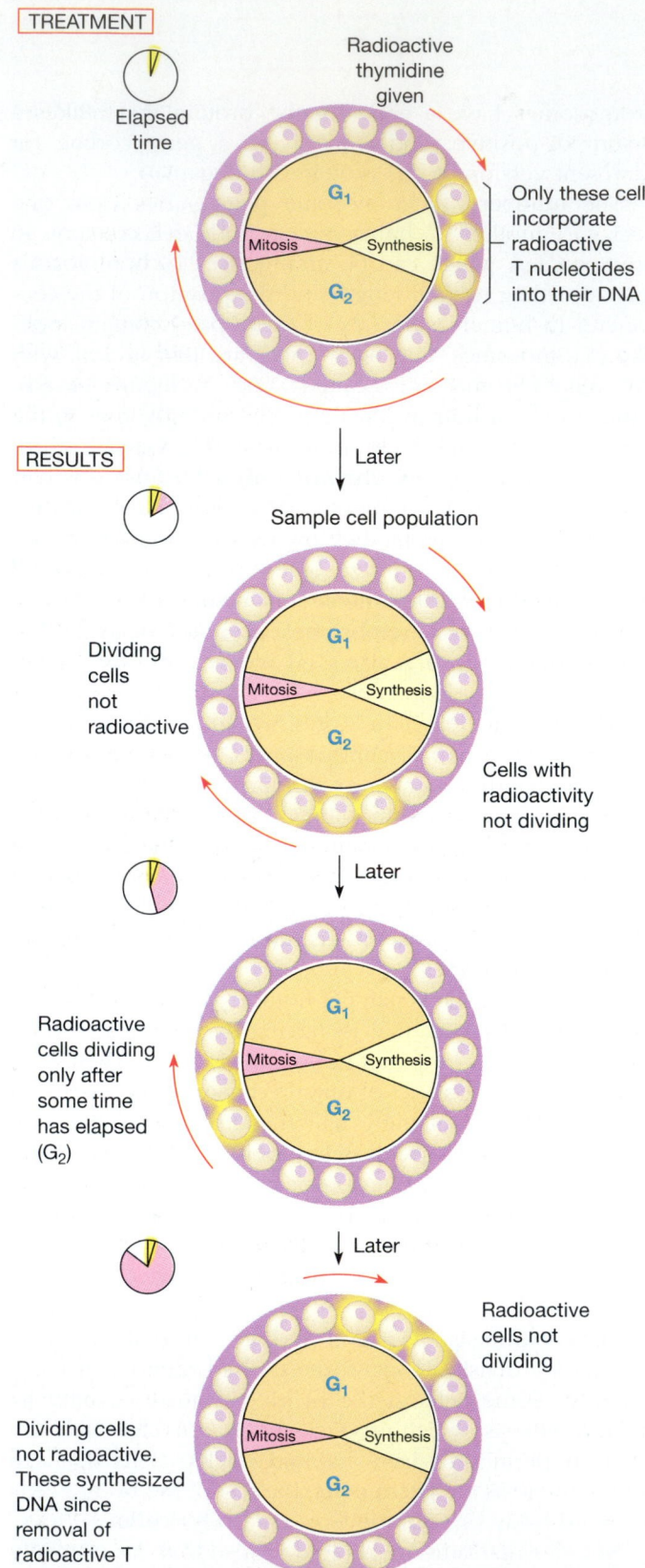

**FIGURE 14–4** Determining the periods of the cell cycle. A cell culture is given radioactive thymidine (T) nucleotides for a brief period (yellow section of clock) and then sampled at intervals to determine when radioactive cells divide. This method demonstrates the existence of G₁, S, G₂, and M periods and the length of each, which varies among different cell types.

cell makes the enzymes that replicate DNA and the histone proteins for the new chromatids.

The length of $G_1$ is the most variable part of the cell cycle. The key point in the cycle occurs when the cell switches from $G_1$ into the S period. After this point of no return, the cell is committed to proceed through the next mitosis.

During the S period, DNA replication begins at replication origins in the DNA and proceeds until all the DNA is copied. Histone proteins for the new chromatids are also made, and the cell centers and centrioles double (except in those plants that lack centrioles) (Section 5–Q).

During $G_2$, the cell is thought to make final preparations for division, including synthesis of some of the proteins used in mitosis itself. Adding chemicals that inhibit protein synthesis to cells in $G_2$ prevents them from making the second critical transition, from $G_2$ to mitosis.

A newly formed cell will usually not divide until it has approximately doubled in size. To do this, it must absorb nutrients and use them to produce more ribosomes, enzymes, cytoskeleton molecules, and membrane material. The cell's mitochondria and plastids reproduce themselves by dividing in two.

Such growth does not occur in the early embryonic cells of animals. Before fertilization, the egg has already grown much larger than most other animal cells, amassing enough food and building materials to carry the embryo through several divisions. During this period, cells need not spend time making new ribosomes, messenger RNA, histones, and DNA replication enzymes, but can divide rapidly as DNA synthesis (S) and mitosis (M) alternate without intervening gap periods. For example, a frog embryo developing from the fertilized egg proceeds rapidly through 12 division cycles, forming progressively smaller cells. When the embryo has used up its "head start" molecules from the egg, cell division slows down because the cells must start producing their own RNA and proteins as well as replicating their DNA between successive rounds of division.

## Control of Cell Division

Many unicellular organisms and fungi grow and divide as fast as they can obtain enough nutrients. By definition, this produces evolutionary success because the sooner a cell divides, the more descendants it will leave in a given time.

In contrast, multicellular organisms must have controls on the number of each type of cell in the body, and the cells control one another's division. One of the distinctions between cancerous and normal cells is that some cancer cells escape these controls. Understanding what controls cell division, and how to repress the division of cancer cells, is therefore a pressing medical problem. Control of cell division is a complex process that we are only beginning to understand.

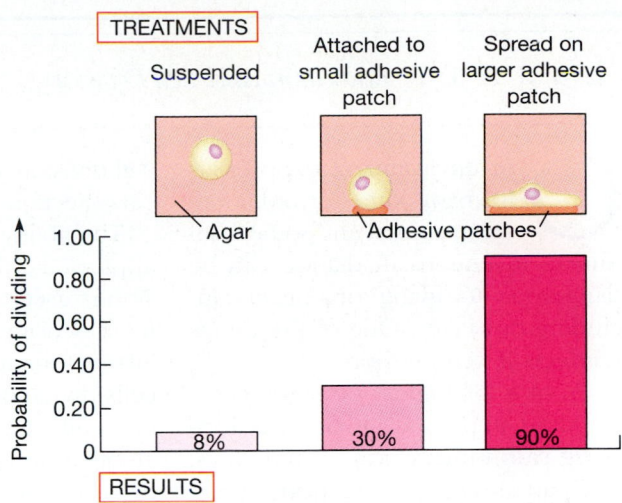

**FIGURE 14–5** Effect of attachment on cell division in animal cells. Cells suspended in culture medium with no attachment have a lower probability of dividing than those permitted to attach to patches of adhesive material. Larger patches permit cells to spread and attach over more of their surface, increasing their probability of dividing.

In animals, cell division seems to require at least two things in addition to nutrients. First, the cell must be attached to something, such as neighboring cells or the membranes or fibers between cells (Figure 14–5). Once an attached cell does begin to divide, however, it loses much of its attachment and assumes a rounded shape until division is complete. The two daughter cells then settle into the parent cell's position, reattach to their surroundings, and resume their typical shape.

The second requirement for animal cell division is that a cell's membrane receptors must bind a critical number of one or more kinds of **growth factors**—specific proteins found in the body fluids in minute amounts. Cells compete for molecules of the right kinds of growth factors for their cell type. So, at any one time, only a limited number of cells in any tissue have enough growth factors to begin dividing. For example, immune system cells called T cells produce plasma-membrane receptors for the growth factor interleukin-2. T cells can switch from $G_1$ to S only when the plasma membrane contains enough receptors with interleukin-2 bound to them.

A cell's ability to switch from $G_1$ to S depends on its obtaining enough nutrients and the right combination of growth factors. These conditions control the cell's buildup of a crucial protein, **cyclin.** The concentration of cyclin changes during the cell cycle, rising during interphase and dropping at mitosis. Cyclin combines with another protein that is always present, the enzyme **cyclin-dependent kinase.** This binding activates the kinase to phosphorylate several other proteins, switching them in turn to their active form and starting the S period.

## ESSAY: *Radiation and Cell Division*

Our environment exposes us to many kinds of radiation, some natural and some from human inventions. Radiation can be dangerous to living organisms, including ourselves. One of the things it affects is cell division.

From a biological point of view, the most important radiation is **ionizing radiation:** x-rays, gamma rays, and particles such as neutrons, alpha particles (helium nuclei), and the like. These are given off by the decay of radioactive elements in Earth's crust and in space. Ionizing radiation has enough energy to ionize substances it strikes, by knocking off electrons. For instance, ionized water, $H_2O^+$, is very reactive and can cause unusual reactions in the cell.

In large enough doses, ionizing radiation causes so much disruption that it can kill a cell. In much smaller doses, the most important effect of radiation is to cause breaks in DNA molecules. If a DNA molecule is damaged too badly for repair enzymes to restore it, it cannot be replicated, and this prevents cell division. On the other hand, the damage may show up as mutations that can be replicated and passed on. Some mutations of this type in somatic cells are believed to be responsible for the unrestrained

cell division and invasiveness of cancers (Section 11–H).

The ability of ionizing radiation in appropriate doses to block cell division is used to treat cancer. Cells are most sensitive to radiation damage just before mitosis. Because cancer cells divide more often than the normal cells around them, bombarding an organ with radiation kills or blocks division in many more cancerous than normal cells.

**Ultraviolet radiation (UV),** part of the electromagnetic spectrum (Figure 14–A), is emitted by the sun and by "black light" bulbs. Nucleic acids absorb UV and can be permanently damaged by it. Ultraviolet radiation can kill cells, and in fact it is used to kill bacteria on laboratory equipment that cannot be sterilized by heat or solvents.

Less drastically, UV can cause mutations, including mutations that make cells divide more rapidly, as happens in skin cancer, or that make them stop dividing. Slowing of cell division is the reason light-colored skin ages so rapidly when exposed to much sunlight (or to sun lamps or tanning booths): damaged cells are not replaced as fast as they otherwise would be. Darker-skinned people are less prone to premature skin aging

and skin cancer because the skin's dark pigment, melanin, absorbs ultraviolet rays and prevents them from penetrating to the DNA of living cells. UV also causes cataracts in the lens of the eye. Ultraviolet radiation does not cause damage to deeper organs of the body because, unlike x-rays and gamma rays, it is rapidly absorbed by water in living tissue and so does not penetrate beyond the skin.

Both ionizing and ultraviolet radiation are part of the natural environment. It has been estimated that more than 80% of the "average" American's exposure to ionizing radiation is natural, or background, radiation from rocks, cosmic radiation, decay of radioactive elements in our own bodies, and the like. About two thirds of this comes from breathing radon gas, and its decay products, in indoor air. This natural radiation averages 260 to 300 millirems per year. For comparison, a person receiving a dose of 350,000 millirems (350 rems) in a month (rather than a year) is thought to have a 50-50 chance of dying from this exposure. (A rem is a unit of ionizing radiation defined by its biological effect.)

"Civilization" exposes us to additional radiation. Most of this comes from medical and dental x-rays.

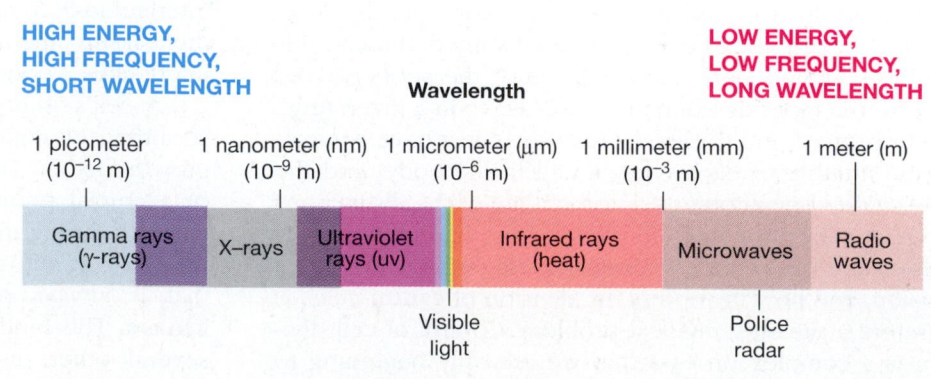

**FIGURE 14–A** The electromagnetic spectrum.

About 1% comes from the fallout from nuclear weapons testing and from nuclear power stations and their waste. About 3% comes from such domestic sources as water, building materials, natural gas, color televisions, and smoke detectors. Most experts think we would be well advised to reduce our exposure to medical x-rays.

It is, however, exceedingly difficult to say how much radiation beyond the inevitable background level should be considered a health hazard. First, it is hard to estimate the background level of radiation. Both the kinds and the amounts of radiation a person receives vary with geography, occupation, personal habits, and so forth. Second, the degree of damage resulting from exposure to a particular dose of radiation varies depending on which particular tissues and molecules are struck by the radiation, something we cannot predict.

A third variable, very important but poorly understood, is the ability of affected tissue to correct the damage. Cells contain enzymes that can repair either breaks in DNA or mis-pairing caused by radiation damage to particular nucleotides in the molecule. Interestingly, some enzymes that can repair damage caused by ultraviolet radiation are activated by light, the very agent that causes the damage! This repair is not perfect, however, and this is why radiation can sometimes cause cancers and other disorders.

Studies in the early 1990s found that workers who had long-term exposure to radiation on the job were more likely than previously thought to develop cancers many years later. Accordingly, exposure limits permitted in the workplace were lowered by 60%. Fortunately, most workers in situations with good radiation protection were already well under the new limit.

Later, the cell produces a different form of cyclin. Active cyclin-dependent kinase containing this molecule phosphorylates proteins that switch the cell from $G_2$ to M: proteins responsible for chromosome condensation, mitotic spindle formation, and breakdown of the nuclear envelope. It also activates enzymes that destroy cyclin and therefore deactivate the kinase complex itself, turning mitosis off. Adding this activated kinase causes cells in any part of the cell cycle to enter mitosis.

Cell division is affected not only by factors normally found within the body but also by substances used as drugs or present in the environment. Some of these substances slow or stop cell division. Some drugs used in chemotherapy, the chemical treatment of cancer, block cell division or kill only dividing cells. These drugs tend to do more damage to cancer cells, which proliferate without the restraints that control the division of normal cells. For instance, the nitrogen mustards, such as cyclophosphamide, alter DNA chemically and so prevent normal replication. Vinblastine and vincristine, extracted from Madagascar periwinkle plants, prevent cell division directly. Methotrexate interferes with general cell metabolism, damaging cancer cells more than normal cells.

## 14–C    MITOSIS

During interphase, the chromatin spreads out in a loose mass, and the individual chromosomes cannot be distinguished. The chromosomes are replicated during the S period.

Mitosis is the series of events in which these replicated chromosomes are separated into two equal groups. Two daughter nuclei are formed, each containing a complete set of the genetic information present in the original nucleus. Cells of any ploidy (haploid, diploid, tetraploid, and so on) can undergo mitosis.

Microscopists described the movements of chromosomes during mitosis over a century ago. Although we can now observe mitosis in more detail, we are still far from explaining exactly how it occurs.

Mitosis is a continuous process, but for convenience it is divided into five stages according to the appearance of the chromosomes as viewed through a light microscope: prophase, prometaphase, metaphase, anaphase, and telophase.

**Prophase**    Looking through a light microscope, you can first tell that a cell is about to divide when it enters prophase. The loose mass of interphase chromatin condenses into distinct chromosomes, visible as sets of sister chromatids (Figure 14–6). Condensation is an impressive process. It is comparable to taking a thin strand some 200 m long and coiling it into a cylinder about 1 mm wide by 8 mm long. A complex of proteins, called the **kinetochore,** assembles

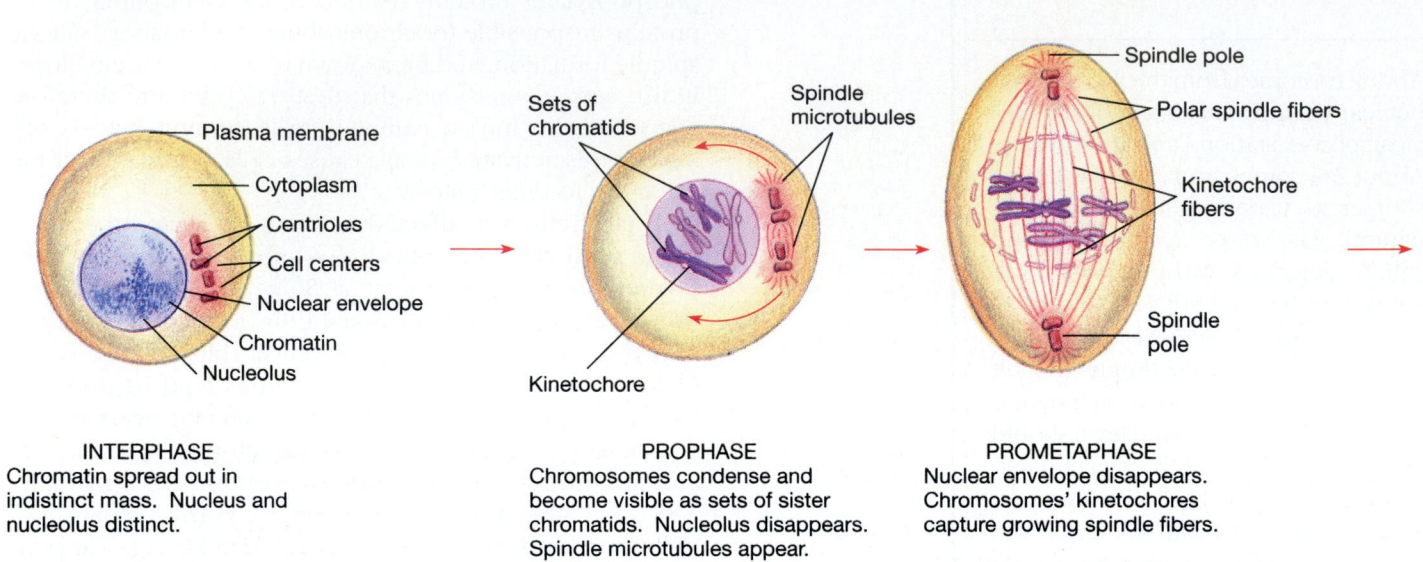

**INTERPHASE**
Chromatin spread out in
indistinct mass.  Nucleus and
nucleolus distinct.

**PROPHASE**
Chromosomes condense and
become visible as sets of sister
chromatids.  Nucleolus disappears.
Spindle microtubules appear.

**PROMETAPHASE**
Nuclear envelope disappears.
Chromosomes' kinetochores
capture growing spindle fibers.

**FIGURE 14–6** Mitosis in an animal cell. The nucleus divides into two daughter nuclei with identical genetic information. The first cell shown is in interphase, before mitosis begins. The next five cells show the stages of mitosis: prophase, prometaphase, metaphase, anaphase, and telophase. The last cell has two nuclei re-entering interphase as the cytoplasm undergoes cytokinesis.

on each chromatid, at the outer face of the centromere region (see Figure 14–7).

During prophase, the nucleolus, which is the site of ribosome synthesis, usually disappears because the material in the nucleolus becomes scattered through the nucleus.

Another notable change during prophase is the beginning of a framework of microtubules, the **mitotic spindle,** which will eventually take part in the movement of the chromosomes. As mitosis begins, microtubules of the cytoskeleton break down into their tubulin protein subunits. (This loss of the cytoskeleton is why dividing cells become rounded.) They then reassemble into the microtubules of the spindle, called **spindle fibers.**

A cell about to undergo mitosis contains two cell centers, which duplicated during the S period. These become the ends, or **poles,** of the spindle. As the spindle microtubules assemble from tubulin subunits, they push the poles of the spindle apart. In most eukaryotes, except higher plants and some unicells, each pole is occupied by a pair of centrioles, also composed of microtubules. The centrioles do not seem to be necessary to mitosis; destroying them with a laser beam has no effect on the subsequent mitosis, whereas destroying the surrounding cloud of cell center material prevents mitosis.

*Prometaphase*    During the next stage, prometaphase, the nuclear membrane disappears (except in some unicellular eukaryotes and fungi). It breaks down into small vesicles and is re-formed from them after mitosis.

As this barrier breaks down, the spindle fibers in the cytoplasm and the chromosomes, formerly in the nucleus, can interact. (Organisms whose nuclear envelope does not break down have various other arrangements permitting this interaction.) A growing spindle fiber may be captured by a kinetochore on one of the sets of chromatids, thereby becoming a kinetochore fiber. Each kinetochore typically captures many fibers, and the two kinetochores of sister chromatids capture fibers growing from opposite poles (Figure 14–7). Fibers extending toward the opposite pole but not captured by kinetochores are called polar fibers.

*Metaphase*    During the metaphase stage of mitosis, each kinetochore is pulled toward its fibers' pole: the farther this is, the stronger the pull. Hence, the pair of sister chromatids is pulled harder toward the pole that is farther away, and so it moves in that direction. The forces in this tug-of-war balance out midway between the poles. As a result, all of the pairs of sister chromatids become lined up in an imaginary plane at the equator of the spindle, held by the ten-

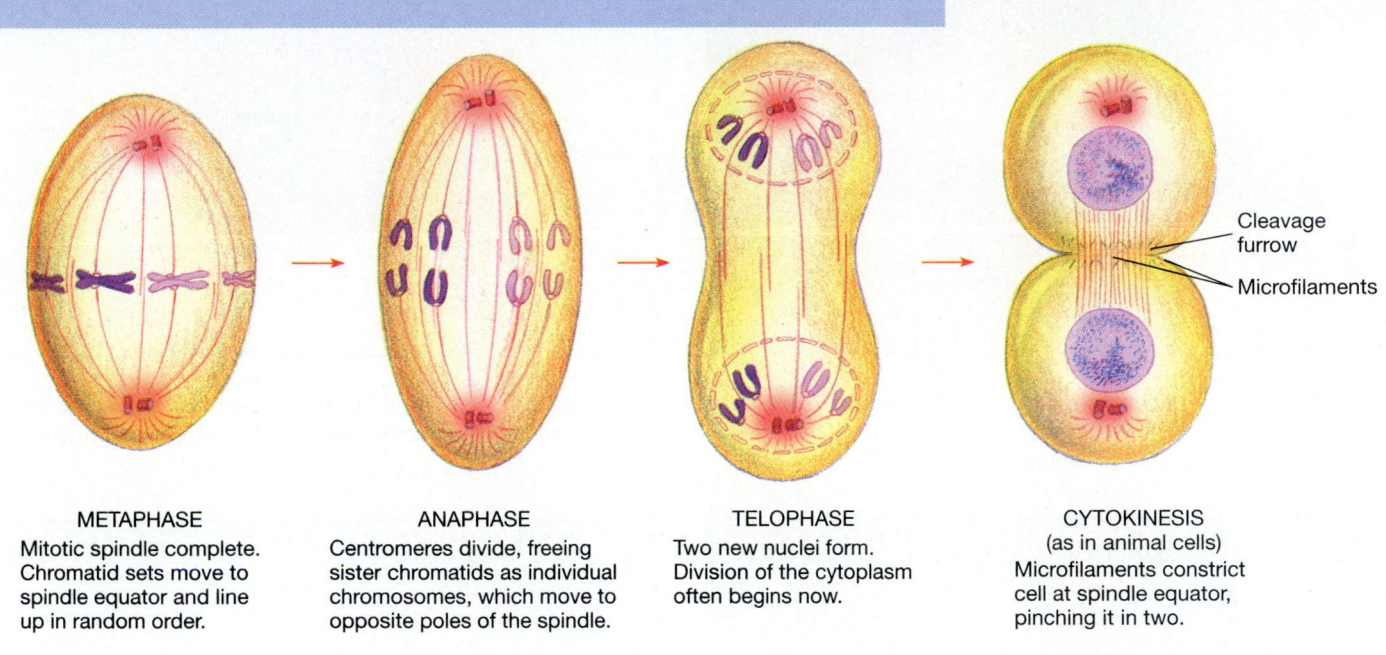

**METAPHASE**
Mitotic spindle complete. Chromatid sets move to spindle equator and line up in random order.

**ANAPHASE**
Centromeres divide, freeing sister chromatids as individual chromosomes, which move to opposite poles of the spindle.

**TELOPHASE**
Two new nuclei form. Division of the cytoplasm often begins now.

**CYTOKINESIS**
(as in animal cells)
Microfilaments constrict cell at spindle equator, pinching it in two.

sion of the equal but opposite pulls. This chromosome lineup, called a **metaphase array,** is the sure sign of a cell in metaphase (Figure 14–6).

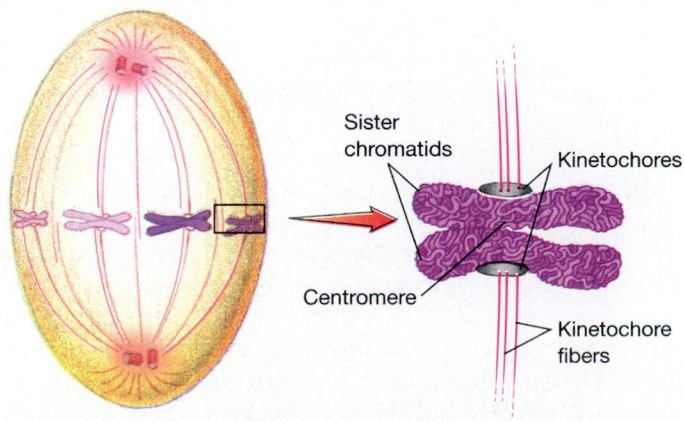

**FIGURE 14–7**  Attachment of chromosomes to the mitotic spindle. A kinetochore forms on each sister chromatid's centromere area. Each kinetochore typically attaches to many spindle fibers radiating from one pole, although the drawing shows only a few.

***Anaphase***   Anaphase is the stage when the chromosomes are separated into two groups. It begins abruptly: vesicles release calcium ions into the cytoplasm, and all at once, each set of sister chromatids separates into two independent chromosomes, which are pulled to opposite poles of the spindle. Each chromosome ends up near one pole of the mitotic spindle, with its sister near the opposite pole. As a result, there is a complete set of chromosomes at each pole, the basis for a new nucleus.

Two kinds of movement can usually be observed during anaphase: the chromosomes are pulled toward the poles, and the poles are pushed farther apart, making the cell longer.

The chromosomes move jerkily toward the poles as the kinetochore fibers shorten by losing the subunits nearest the kinetochore. (Exactly how this works without the chromosome's falling off its fibers is not understood.) Adding substances that stabilize microtubules prevents the movement of chromosomes during anaphase, whereas adding substances that promote disassembly of the tubules speeds it up.

Separation of the poles seems to be partly due to a sliding mechanism between the microtubules. Microtubules extending from opposite poles have opposite orientations,

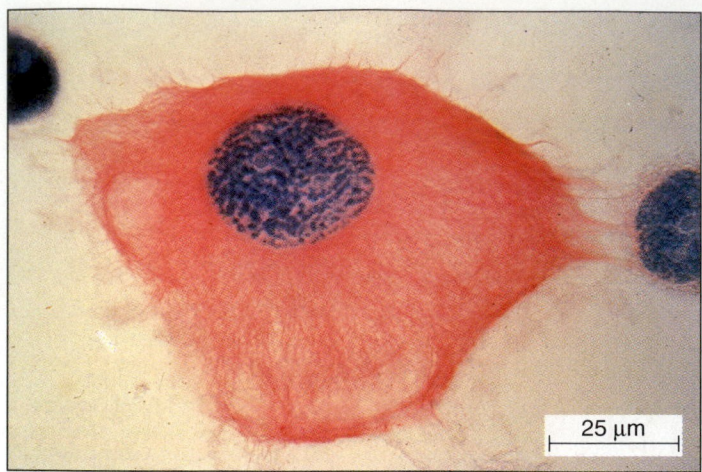

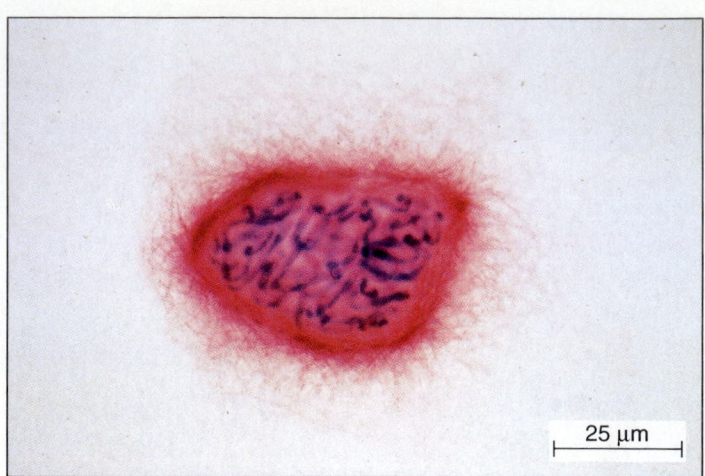

(a) In a cell entering early prophase, the chromatin is dispersed throughout the nucleus and just beginning to condense, and the microtubules still show the interphase arrangement.

(b) A cell in prophase, with the chromatin condensing into distinctly visible chromosomes and the microtubules reorganizing.

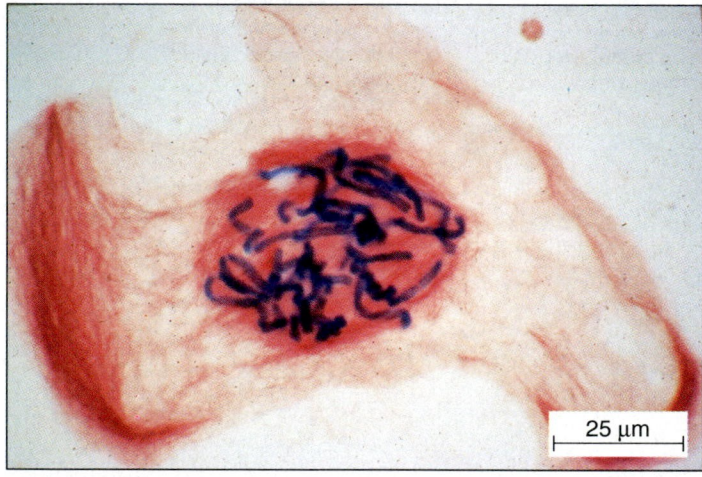

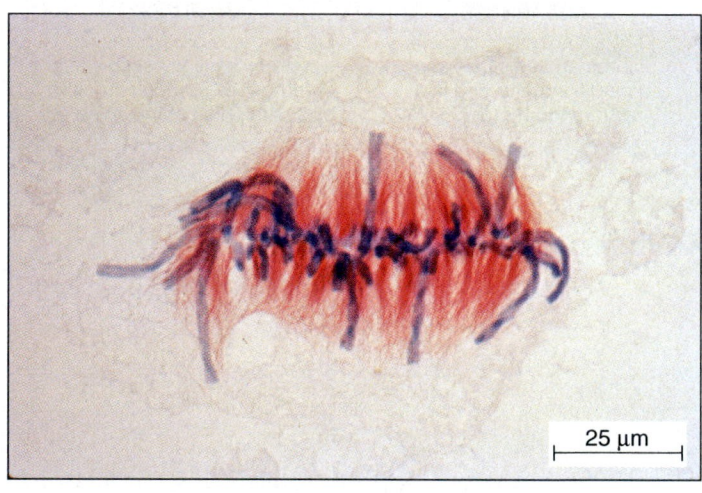

(c) Prometaphase. The nuclear envelope has disappeared, and the chromosomes' kinetochores have attached to microtubules of the newly formed spindle.

(d) Metaphase. The sets of sister chromatids have lined up with their kinetochores all at the equator.

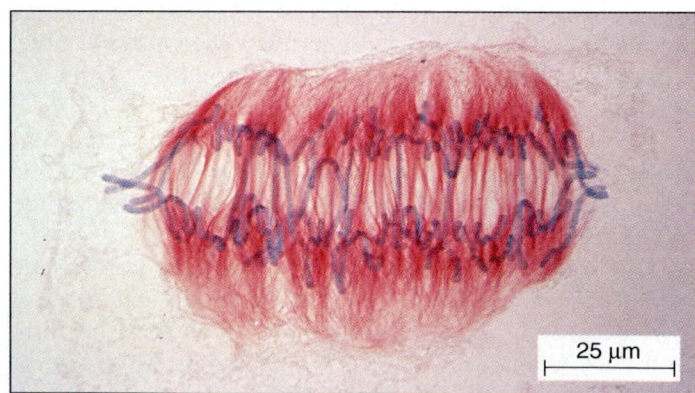

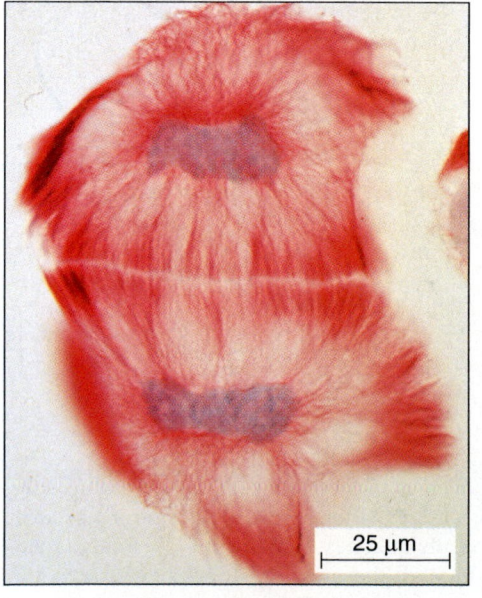

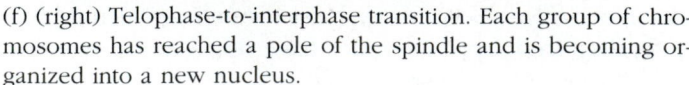

(e) Anaphase. The sister chromatids have separated into individual chromosomes, which are separating into two groups along the spindle.

(f) (right) Telophase-to-interphase transition. Each group of chromosomes has reached a pole of the spindle and is becoming organized into a new nucleus.

and they overlap in the middle of the cell. It is thought that some proteins form cross-bridges between two oppositely oriented fibers and use energy from ATP to push each microtubule toward its own pole, thereby lengthening the distance between poles. This effect is enhanced by the lengthening of the polar fibers as additional tubulin subunits are added to the end of the microtubule farthest from the pole.

***Telophase***  In the last stage of mitosis, telophase, two daughter nuclei are organized (Figure 14–8). The chromosomes, now in two groups at the poles of the mitotic spindle, uncoil into masses of chromatin. Vesicles from the old nuclear envelope associate with the chromosomes and fuse together, forming a new nuclear envelope around each group of chromosomes. Transcription of ribosomal RNA resumes, and nucleoli form in each daughter nucleus.

The events of mitosis ensure that all the sister chromatids synthesized during DNA replication are separated precisely into two sets, so that each daughter cell receives one copy of each chromosome.

**Colchicine,** a chemical derived from the autumn crocus plant, binds to tubulin molecules and prevents their assembly into microtubules. Treating cells with colchicine prevents the formation of the mitotic spindle and blocks cell division. However, sister chromatids can still separate from each other, thus doubling the number of chromosomes in the cell. In this way, diploid cells can become tetraploid. If part of a diploid plant is treated with colchicine and then cultured or permitted to form seeds, a tetraploid plant can be grown. Tetraploid plants are often larger and more vigorous than their diploid ancestors. Many cultivated vegetables and flowers are tetraploids that have arisen either naturally or by treatment with colchicine. Water-processed decaffeinated coffee comes from a diploid species with strongly flavored beans. The milder species used for regular coffee is tetraploid.

Colchicine is also used to stop cell division at metaphase so that condensed chromosomes can be collected for analysis, for instance, to make a karyotype like the one in Figure 14–2d.

**FIGURE 14–8**  Mitosis. These plant cells, from the food-storing endosperm tissue in developing seeds of the blood lily, have the largest spindles known, ten times longer than those in human cells. No walls are present at this early stage in seed formation. Microtubules are stained red with immunonogold stain, and chromosomes are counterstained with toluidine blue. (All LM, ×250) (Andrew S. Bajer, University of Oregon)

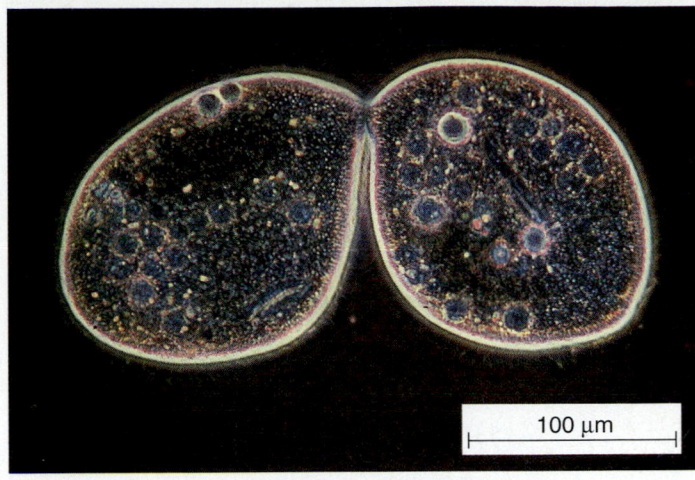

100 µm

**FIGURE 14–9**  Cytokinesis in the protist *Paramecium*. This unicellular organism undergoes cytokinesis as microfilaments constrict the cytoplasm, similar to division of an animal cell.  (Biophoto Associates)

## 14–D  CYTOKINESIS

In the process of **cytokinesis,** the cytoplasm of the original cell is divided to form two new cells, each housing one of the newly formed nuclei (Figure 14–9).

In animal cells, cytokinesis begins during early anaphase. A ring of microfilaments, made up of the contractile proteins actin and myosin, forms just beneath the plasma membrane. These filaments constrict the cell to form a **cleavage furrow** and eventually pinch the cytoplasm in two (see Figure 14–6). This ring of filaments forms around the equator of the mitotic spindle. If the spindle is experimentally nudged into a different position early in mitosis, before the filaments assemble, the cleavage furrow will then appear around the new position of the spindle equator. The spindle usually forms in the middle of the cell, and so the resulting daughter cells have roughly equal sizes. However, some embryonic cells normally divide unequally, producing new cells of quite different sizes.

Plant cells are surrounded by a rigid cell wall, and cytokinesis occurs by a completely different method. The Golgi complexes release vesicles containing material to make a new partition between the cells-to-be. The vesicles move along the spindle microtubules to the middle of the cell. Here they fuse, and their contents form a flat disc, the **cell plate,** surrounded by a membrane made up of the fused vesicles. Both the cell plate and the membrane grow as more vesicles add on (Figure 14–10). Soon the cell plate extends completely across the cell, cutting it in two. The material in the cell plate goes to form the **middle lamella,** the common partition between the two daughter cells, and the membrane on either side of the cell plate becomes part of the daughter cells' plasma membranes. Each daughter cell builds a new cell wall on its side of the middle lamella.

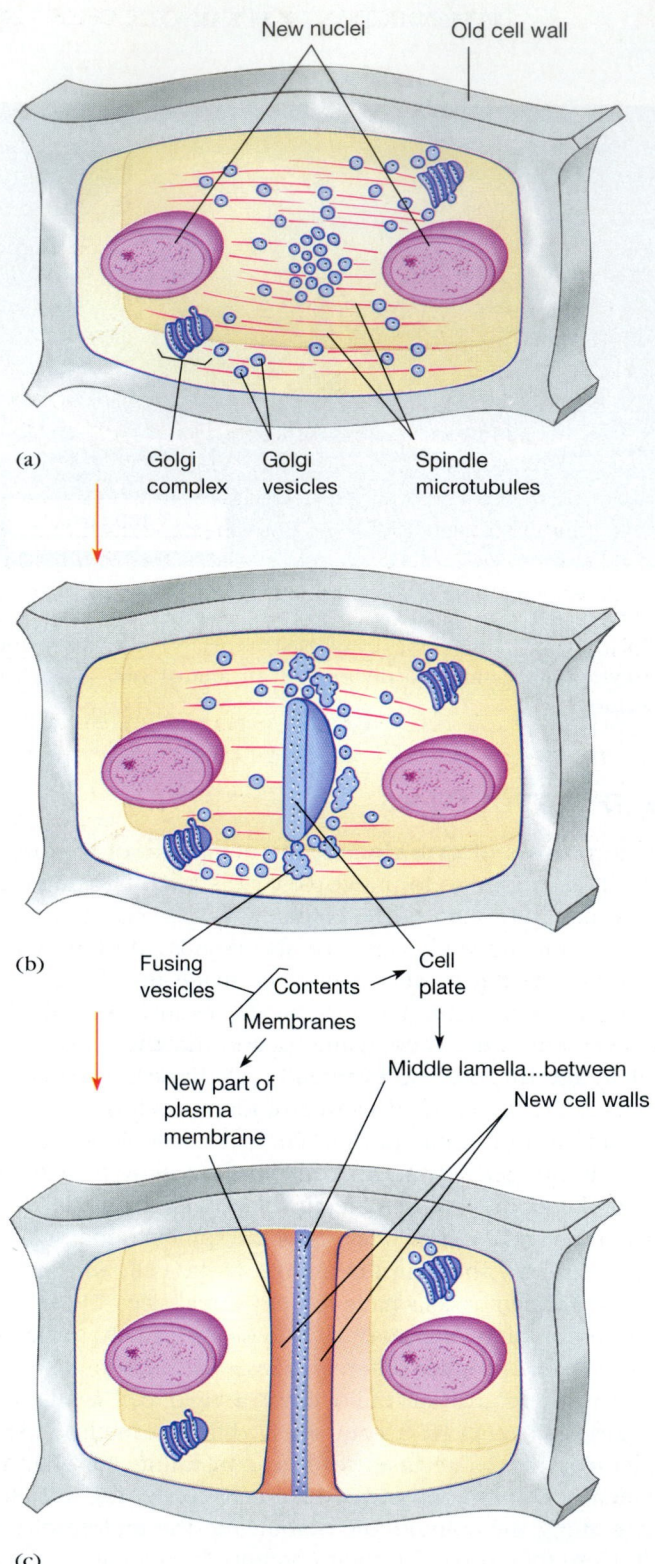

(a)

New nuclei    Old cell wall

Golgi complex    Golgi vesicles    Spindle microtubules

(b)

Fusing vesicles    Contents    Cell plate

Membranes

New part of plasma membrane

Middle lamella...between

New cell walls

(c)

**FIGURE 14–10** Cytokinesis in a plant cell. (a) Vesicles containing material for the cell plate bud off from Golgi complexes and travel along the lingering spindle microtubules to the equator, where they fuse. (b) Material from the vesicle contents is built into a disc-like cell plate, which forms the middle lamella between the two daughter cells. The fused vesicle membranes on either side of the plate will become the plasma membranes at the ends of the new cells. (c) Each new cell lays down a wall between the middle lamella and the new portion of its plasma membrane.

Cytokinesis divides up not only the cytoplasm itself but also the various structures within it: ribosomes, Golgi complexes, mitochondria, plastids, cytoskeleton molecules, and so forth. Most cells contain many of each kind of structure, distributed throughout the cytoplasm, so that each new cell is bound to receive some of every component it needs. In many cases, various cytoplasmic components associate with the chromosomes or spindle microtubules, and so are distributed to both new cells. Some unicellular algae normally have only one chloroplast, which divides when the cell does, and each daughter cell receives one of the two new chloroplasts.

### Syncytia and Coenocytes

In certain tissues, nuclear division is not followed by cytokinesis. The result—called a **syncytium** in animals and a **coenocyte** in plants, algae, and fungi—is a cell-like unit containing more than one nucleus within a single plasma membrane. This condition may be permanent or only temporary, until cell formation catches up with mitosis.

One example of such a syncytium occurs in the embryos of insects (see Figure 12–8). After an insect egg has been fertilized, the diploid zygote nucleus undergoes several mitotic divisions that are not followed by cytokinesis. Only after the developing egg contains a number of nuclei (sometimes several thousand) are the first plasma membranes between them formed.

A similar situation sometimes occurs in the endosperm tissue that surrounds and nourishes the embryo of a flowering plant. An extreme case is the coconut; its "milk" is a nutrient fluid filled with nuclei. Eventually the nuclei become surrounded by membranes and cell walls, forming the coconut "meat."

### 14–E MEIOSIS

Meiosis is the process of nuclear division in which new combinations of genes are produced. This occurs in two ways. First, meiosis swaps different versions of genes between the maternal and paternal chromosomes of a homologous pair in a diploid cell. Second, it sorts the resulting chromosomes into haploid sets containing new chromosome combinations, which are usually different from those the organism inherited from its parents. After meiosis, fertilization mixes genes up in a third way: two gametes join, uniting their haploid sets of chromosomes into a new diploid combination. Eventually this diploid combination will undergo meiosis in its turn, reshuffling the genes and chromosomes for yet another generation of offspring.

The simplest way to form haploid nuclei from a diploid nucleus would be by skipping DNA synthesis and dividing the diploid cell's chromosomes between two nuclei, forming two haploid nuclei. This does not happen. Instead, the S period occurs during interphase before meiosis as well as before mitosis, presumably because meiosis evolved from mitosis. Therefore, a nucleus enters meiosis with enough DNA to make four haploid nuclei. However, a nucleus cannot divide into more than two daughter nuclei at any one division. So it takes two divisions during meiosis to reduce each nucleus to the haploid number of chromosomes.

The two divisions in meiosis are called meiosis I and meiosis II. Like mitosis, both meiotic divisions involve formation of a spindle and movement of chromosomes to the spindle's equator before being separated into two groups and pulled to the poles. So meiosis looks very similar to mitosis at first glance (see Figure 14–11). The names of the stages are also similar. However, meiosis has some additional features not found in mitosis.

Because meiosis produces haploid nuclei from diploid nuclei, it must provide a way for the cell's homologous chromosome pairs to be parceled out precisely into two groups, each group containing exactly one member of each homologous chromosome pair. We can summarize the movements of one pair of homologous chromosomes through meiosis:

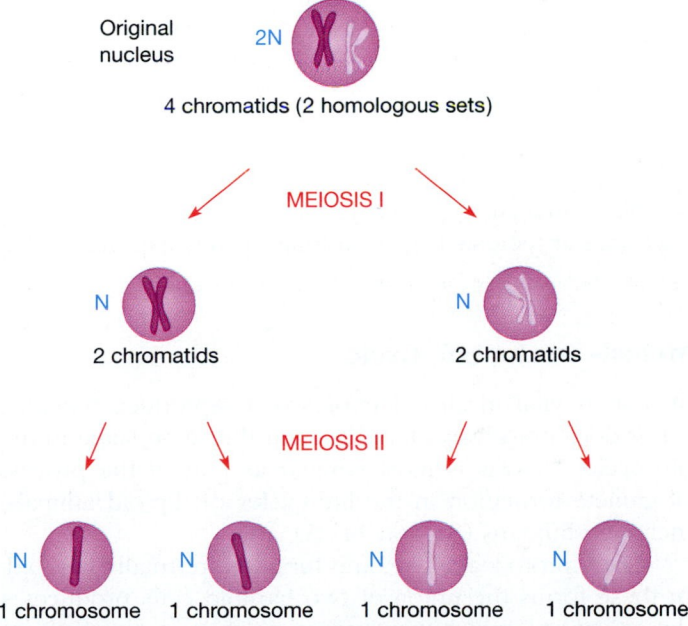

Original nucleus

2N

4 chromatids (2 homologous sets)

MEIOSIS I

N — 2 chromatids

N — 2 chromatids

MEIOSIS II

N — 1 chromosome

N — 1 chromosome

N — 1 chromosome

N — 1 chromosome

The special events that allow this precise sorting of homologues into different nuclei occur during prophase of meiosis I.

## Meiosis I

***Prophase I***    Each chromosome somehow finds its homologue among all the other chromosomes in the nucleus, and the two line up next to each other with point-by-point precision, in a poorly understood process called **synapsis.** Proteins that associate with the chromosomes at this time seem to be responsible for aligning the chromosomes beside, but not touching, each other. Because the chromosomes have already been replicated, the resulting group consists of four chromatids (two pairs of sister chromatids) and is called a **tetrad.**

Within each tetrad, corresponding lengths of chromatids are exchanged between the maternal and paternal homologous chromosomes, a phenomenon called **crossing over** (Figure 14–11). This snipping and cross-splicing of the DNA is one source of genetic variation that occurs as a result of meiosis and sexual reproduction. Each chromatid may cross over with either chromatid of the homologous chromosome, and crossovers usually occur at precisely the same point on the two chromatids involved. For a while, the chromatids remain joined at the crossover exchange point, called a **chiasma** ("cross"; plural, **chiasmata**). Each pair of homologous chromosomes forms at least one chiasma and generally more. Chiasmata hold the homologous chromosomes of a pair together, and the centromeres hold the two sister chromatids of each chromosome together; hence, the entire tetrad moves as a unit when the chromosomes attach to the spindle fibers (Figure 14–11).

The complex events of prophase I take a long time, making this the longest stage of meiosis. Meiosis frequently takes days to complete instead of the hours or minutes required for mitosis.

***Metaphase I***    In metaphase I, all the tetrads line up at the spindle equator. The centromere of each set of sister chromatids has a kinetochore attached to spindle fibers from one pole. The homologous set, just across the equator, is attached to fibers from the opposite pole. This arrangement, much like couples lined up opposite their partners for a barn dance, allows the homologous partners to be separated from one another at anaphase I.

***Anaphase I***    As anaphase I begins, the chiasmata come apart, whereas the centromeres remain intact. Hence, each set of sister chromatids moves as a unit toward one spindle pole, while the homologous set of chromatids moves to the opposite pole. In each set of chromatids, part of the original DNA has been replaced by the corresponding section(s) from the homologous set during crossing over.

***Telophase I***    The group of chromosomes at each pole is organized into a new nucleus.

The precise choreography of meiosis I ensures that each daughter nucleus receives one member of each homologous chromosome pair. Meiosis I is often called "reduction division" because it reduces a diploid nucleus to two haploid ones, each with half the original number of chromosomes.

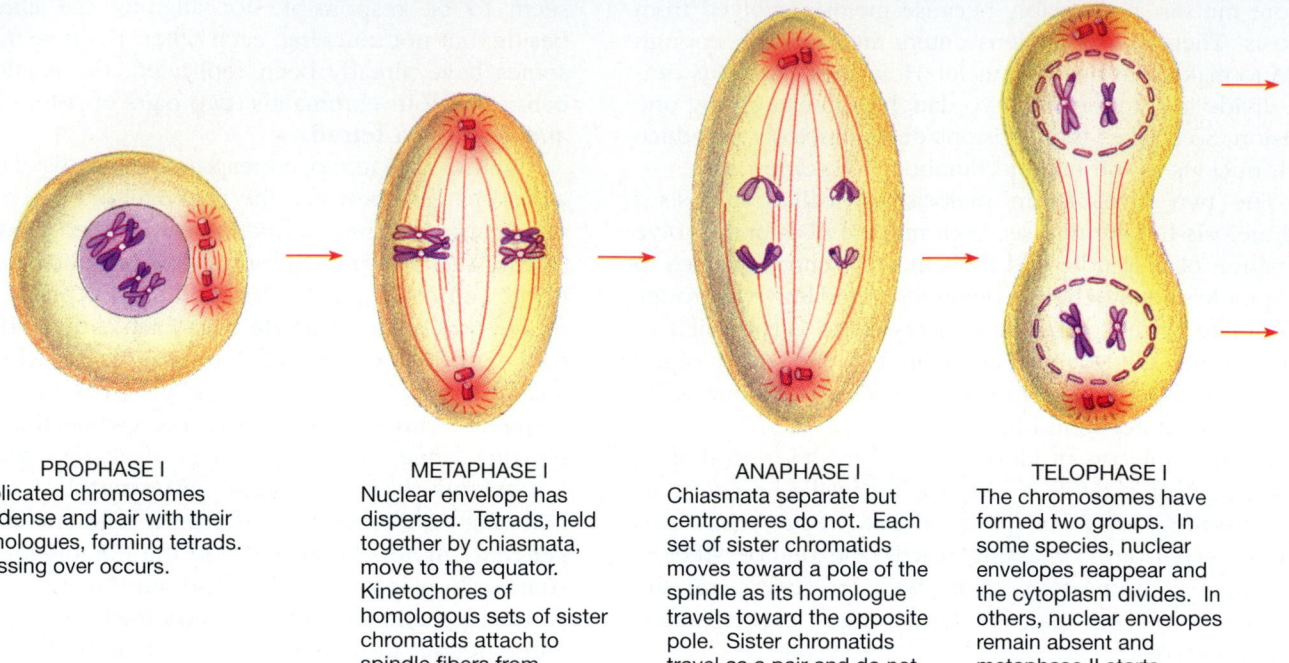

**MEIOSIS I** | Homologous chromosomes separate from their partners:

**PROPHASE I**
Replicated chromosomes condense and pair with their homologues, forming tetrads. Crossing over occurs.

**METAPHASE I**
Nuclear envelope has dispersed. Tetrads, held together by chiasmata, move to the equator. Kinetochores of homologous sets of sister chromatids attach to spindle fibers from opposite poles.

**ANAPHASE I**
Chiasmata separate but centromeres do not. Each set of sister chromatids moves toward a pole of the spindle as its homologue travels toward the opposite pole. Sister chromatids travel as a pair and do not separate until anaphase II.

**TELOPHASE I**
The chromosomes have formed two groups. In some species, nuclear envelopes reappear and the cytoplasm divides. In others, nuclear envelopes remain absent and metaphase II starts immediately.

**FIGURE 14–11**  Meiosis. The events of meiosis as it would occur in the formation of male gametes in an animal.

The interphase between meiosis I and II is so brief that in some species the chromosomes appear to go directly from telophase I to prophase II. No DNA is synthesized between divisions because the chromosomes have already been replicated.

## Meiosis II

Meiosis II is essentially a mitotic division of a haploid nucleus into two new haploid nuclei, and it occurs much more quickly than meiosis I (Figure 14–11).

***Prophase II***  A new spindle forms in each of the two new cells. The kinetochores of sister chromatids attach to spindle fibers from opposite poles.

***Metaphase II***  The kinetochore fibers tug all the sets of sister chromatids into a metaphase array at the spindle equator.

***Anaphase II***  The centromeres finally divide, releasing the sister chromatids as individual chromosomes. The chromosomes then separate into two groups.

***Telophase II***  The chromosomes become organized into two haploid nuclei.

Since meiosis I produced two nuclei, the division of each one at meiosis II gives a total of four haploid nuclei.

## Meiosis in the Life Cycle

Meiosis is vital in all eukaryotes that reproduce sexually, but it does not always take place at the same stage in the life cycle. Meiosis is most familiar as part of the process of gamete formation in the life cycles of diploid animals, including humans (Section 14–G).

Many protozoa, algae, and fungi are normally haploid. In these forms the union of two haploid cells produces a diploid zygote, which then undergoes meiosis, forming four new haploid individuals.

In diploid plants, meiosis produces haploid reproductive cells called spores. The spores grow into multicellular haploid structures, which produce haploid eggs and sperm by mitosis. The diploid zygote formed at fertilization then grows into a diploid plant. We shall study these rather unfamiliar kinds of life cycles in Chapter 27.

By reducing the chromosome complement from diploid to haploid, meiosis ensures that sexually reproducing species maintain the same chromosome number from generation to generation.

**MEIOSIS II** Each nucleus divides again, producing four haploid nuclei. Centromeres divide, and sister chromatids become separate chromosomes.

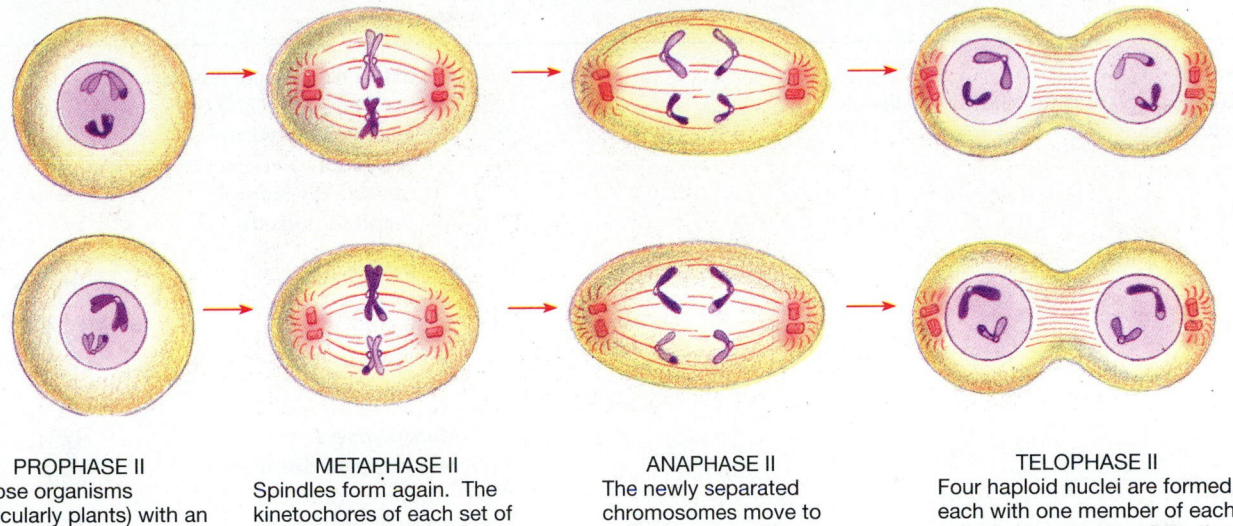

**PROPHASE II**
In those organisms (particularly plants) with an interphase between the two meiotic divisions, the chromosomes must condense again before the second division.

**METAPHASE II**
Spindles form again. The kinetochores of each set of chromatids attach to spindle fibers from opposite poles. The centromeres then divide, just as they do in metaphase of mitosis, and each set of sister chromatids becomes two separate chromosomes.

**ANAPHASE II**
The newly separated chromosomes move to opposite poles of the spindle.

**TELOPHASE II**
Four haploid nuclei are formed, each with one member of each pair of chromosomes from the original nucleus that entered meiosis. Nuclear envelopes form and cytokinesis occurs.

## 14–F   GENETIC REASSORTMENT

The most important consequence of meiosis is that it reshuffles the genetic material. This forms new combinations of genes and chromosomes that become the genetic information of the next generation. Meiosis produces this **genetic reassortment** in two ways. First, it produces new combinations of genes on chromosomes. This is an instance of **genetic recombination,** the splicing of DNA molecules in new ways. Second, meiosis produces new assortments of chromosomes.

### Crossing Over: New Combinations of Genes on Chromosomes

Chromosomes with new genetic combinations are produced during meiosis by crossing over, the type of genetic recombination that exchanges equivalent segments of DNA between homologous chromosomes. The location of a crossover along the length of the DNA varies, and so crossing over rearranges a variable number of genes between two homologous chromosomes.

The swap is mediated by large proteinaceous particles called recombination nodules, which attach between the two synapsed chromosomes during prophase I. Two chromatids, one from each homologue, are snipped at the same location in the DNA sequence, and each segment is rejoined to the opposite chromatid. Normally, each chromatid emerges from this amazingly precise but poorly understood process with the same amount of DNA (Figure 14–12).

For example, suppose a person had one chromosome #1 with a gene for normal salivary amylase and a gene for Rh-positive blood. On the homologous chromosome #1, there might be a mutated gene coding for a defective salivary amylase and a gene for Rh-negative blood. If a crossover occurred between these two loci, the result would be one chromatid with genes for normal amylase and Rh-negative blood, and one chromatid with genes for defective amylase and Rh-positive blood. Since one of the original chromosomes was inherited from each parent, crossing over combines genes from both parents in each of the recombinant chromatids. (The other two chromatids retain the original combinations of genes because they did not participate in the crossover.) In humans, crossing over occurs an average of two or three times in each pair of homologous chromosomes during gamete formation. When more than one crossover occurs, these may involve the same two chromatids or different combinations of chromatids (Figure 14–12c).

T A B L E   1 4 – 2

**Differences Between Mitosis and Meiosis**

| Mitosis | Meiosis |
|---|---|
| Occurs in somatic cells | Occurs in reproductive cells |
| Occurs in haploid (N) and diploid (2N) cells | Occurs in diploid (2N) cells |
| Nucleus divides once | Nucleus divides twice: |
| |   I  reduction division (2N ⟶ N) |
| |   II  mitotic division of haploid nucleus |

*Prophase I*
Synapsis, tetrad formation, and crossing over

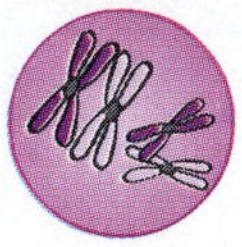

*Metaphase I*
Sets of sister chromatids line up in tetrads, held to their homologues by chiasmata

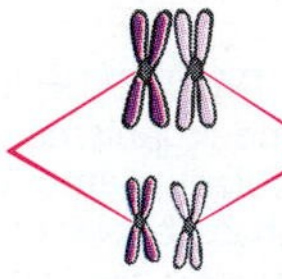

*Anaphase I*
Homologues separate, but centromeres do not divide

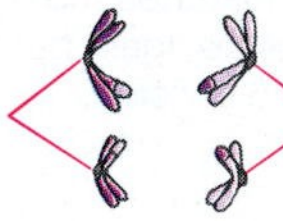

*Prophase*
No synapsis

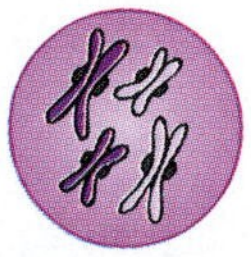

*Prophase II*

*Metaphase*

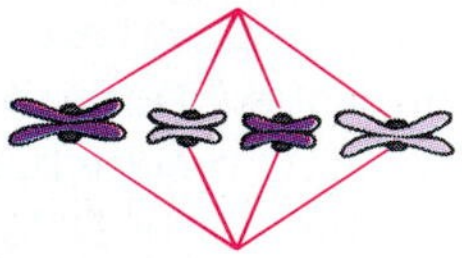

Sets of sister chromatids line up singly

*Metaphase II*

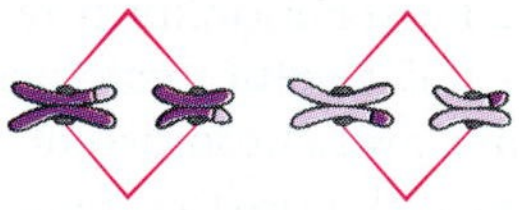

*Anaphase*

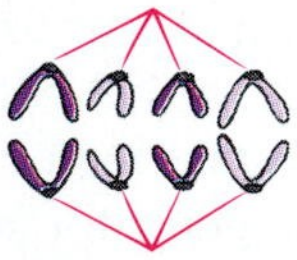

Sister chromatids separate as centromeres divide

*Anaphase II*

*Outcome*

*Outcome*

Produces two daughter nuclei with same ploidy (N, 2N, and so on) as original nucleus
Daughter nuclei genetically identical to each other and to original cell, with same chromosome content
Produces new individuals in unicells; new cells in growth and repair of multicellular organisms

Produces four haploid daughter nuclei

Daughter nuclei genetically different from each other and from original cell, with half its chromosome content
Produces gametes in animals, spores in plants and fungi

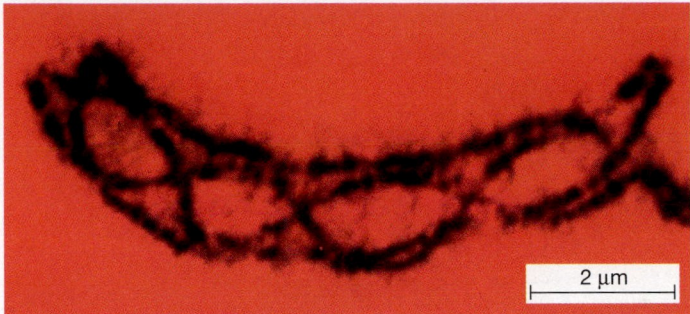

(a) Crossing over in a tetrad (LM)

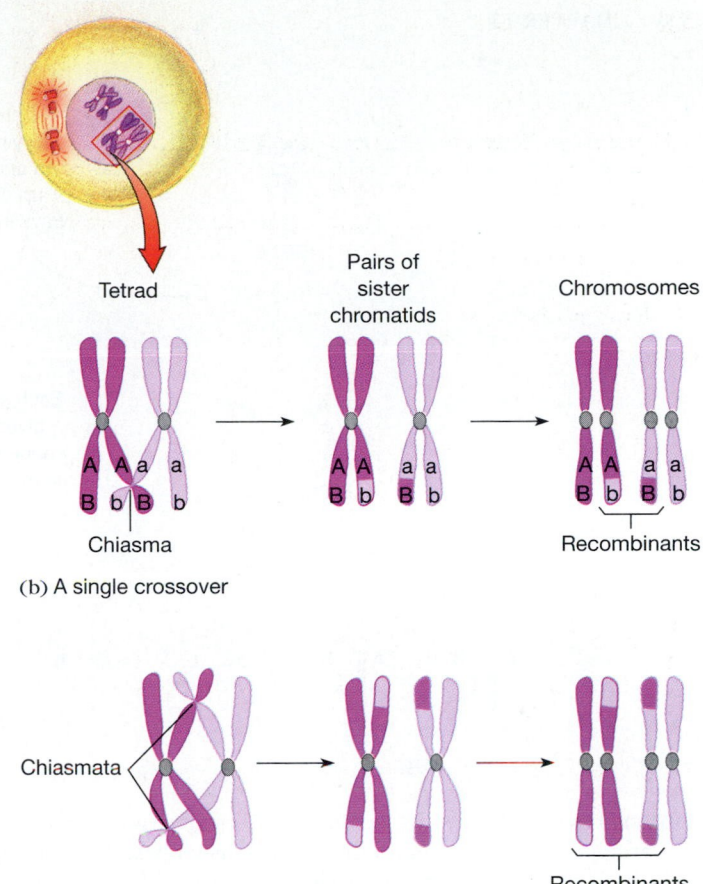

(b) A single crossover

(c) Crossovers of different chromatids

**FIGURE 14–12** Crossing over. (a) Several crossovers between homologous chromatids. (b) Chromatids may form chiasmata, where part of each chromatid is broken off and rejoined onto the other chromatid. Eventually this produces four chromosomes with different gene combinations (AB, Ab, aB, and ab). (c) If more than one crossover occurs in the same tetrad, different chromatids may be involved in each crossover event. Here, each dark chromatid is involved in one crossover. The pale chromatid on the left takes part in two crossovers, its sister in none. Three chromatids end up as recombinants. Crossing over occurs during meiotic prophase I and rarely during mitosis, but not during prophase II of meiosis. (a, B. John, Cabisco/VU)

Despite the remarkable accuracy of crossing over, mistakes do occur. For example, the human eye contains three light receptor pigments: blue, red, and green. The genes for the red and green pigments are next-door neighbors on the X chromosome, and their DNA sequences are 98% identical. Therefore, it's easy for mistakes to occur in the homologous pairing of two X chromosomes during meiosis. If crossing over occurs in the mismatched region, one chromosome may end up without a green-pigment gene or with a jumbled gene that expresses itself improperly

(Figure 14–13). The 8% of Caucasian (white) males who are red-green color-blind received just such a defective X chromosome at fertilization.

## New Assortments of Chromosomes

Meiosis produces new assortments of chromosomes because chromatid tetrads can line up at metaphase I with either set of chromatids on either side of the spindle equator. Then,

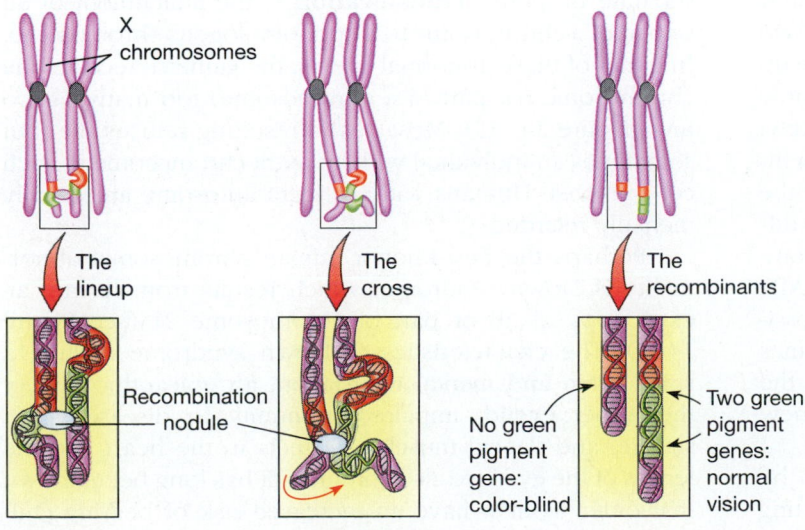

**FIGURE 14–13** Crossover error. The high degree of similarity between genes for red and green visual pigments permits them to match up at synapsis, and the leftover genes form unpaired loops. The recombination nodule then swaps the DNA between the two chromatids at the point where it attaches. One chromatid ends up with no green pigment gene, the other with two.

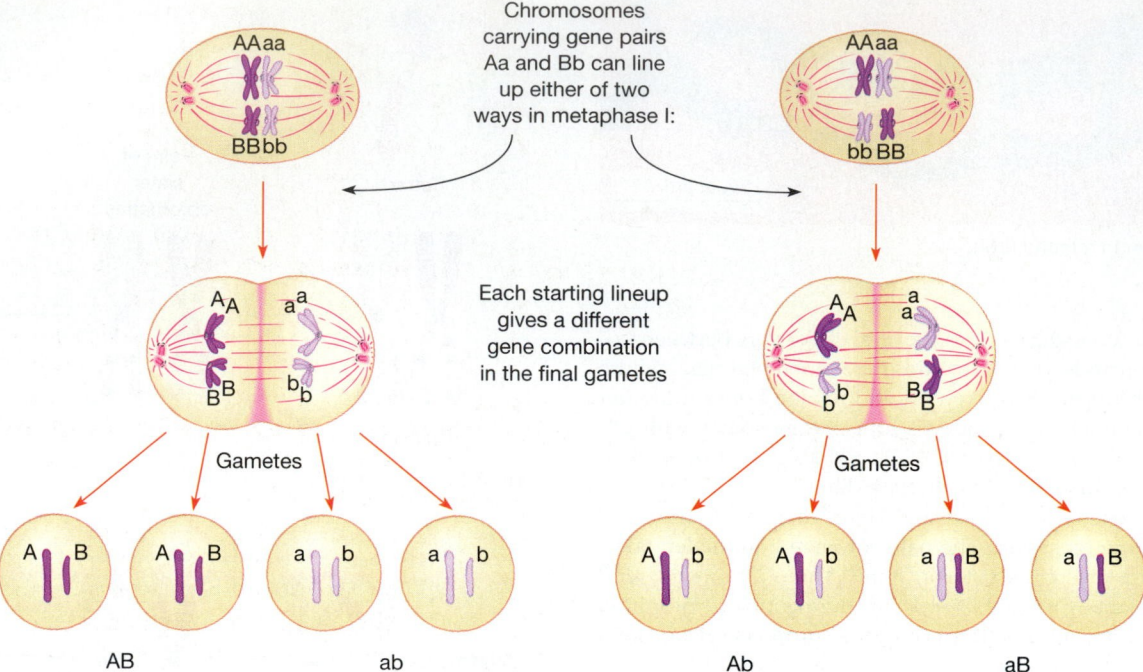

**FIGURE 14–14** The luck of the lineup. Genetic reassortment is partly due to the various possible arrangements of chromosomes during metaphase I of meiosis. Two chromosome pairs can line up in either of two ways, producing four possible combinations in the resulting gametes.

in anaphase I, each set of chromatids is separated from the homologous set and goes into a new nucleus with the members of other homologous pairs that lined up on the same side of the equator as itself (Figure 14–14). Because the lining up at metaphase I is random, any one chromosome has an equal chance of ending up in a new cell with either member of any other pair of homologous chromosomes.

For instance, consider an organism that inherited chromosomes A and B from its mother and the respective homologues a and b from its father. During metaphase I, the tetrad AAaa can line up with either chromosome nearer either pole, and the same is true of BBbb. There are two possible ways that these two chromosome pairs can line up in metaphase I, giving a total of four different possible chromosome combinations in the gametes, all equally likely: AB, ab, Ab, and aB (Figure 14–14). Likewise, if an organism has three chromosome pairs (such as ABC from the mother and abc from the father), they can line up four different ways, giving gametes with eight chromosome combinations: ABC and abc; AbC and aBc; Abc and aBC; ABc and abC. In fact, the number of new combinations possible is $2^N$, where N is the haploid number of chromosomes for the species. Thus, the chromosomes received from the organism's parents are mixed up in a wide variety of new combinations in its gametes.

Meiosis, then, produces genetic reassortment both by swapping genes between chromosomes and by sorting the chromosomes into new combinations. Additional reassortment occurs during the random fusion of gametes at fertilization. In this way a steady supply of new genetic combinations arises in sexually reproducing species, furnishing raw material for evolution.

## Nondisjunction and Translocation

Occasionally, chromosomes behave abnormally during meiosis. **Nondisjunction** is the failure of chromosomes to separate properly. **Translocation** is the attachment of all or part of a chromosome to a nonhomologous chromosome. In either of these abnormal events, the gametes receive one chromosome (or part of a chromosome) too many or too few (Figure 14–15). Most of the resulting fetuses die, but sometimes an individual with an extra chromosome in each cell survives. Humans with an extra autosome are usually mentally retarded.

Perhaps the best-known human chromosomal abnormality is Down syndrome, which results from having an extra copy of all or part of chromosome 21 (see Figure 14–2). The characteristics of Down syndrome include a small brain and mental retardation, an epicanthic fold of the upper eyelid, impaired immunity to disease, short stature, and flaccid muscles. Defects in the heart and the lenses of the eyes are also common. It has long been known that older women have an increased risk of bearing chil-

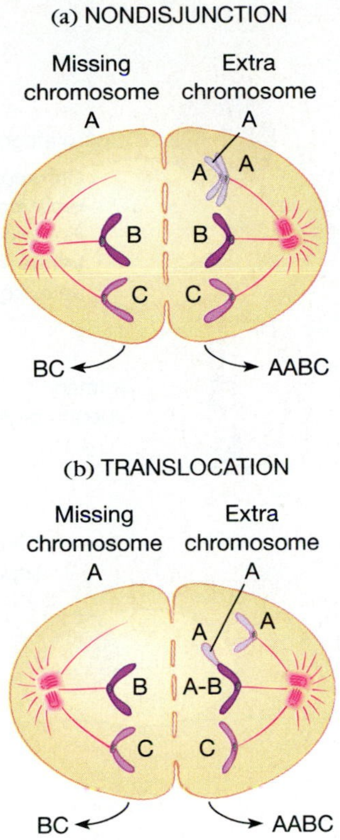

FIGURE 14–15 Nondisjunction and translocation. The cells are in the second division of meiosis (see Figure 14–11). (a) In chromosome A, the sister chromatids fail to separate, instead remaining attached and moving to the same pole. Eventually they end up in the same gamete (arrows). (b) One copy of chromosome A becomes attached to chromosome B, which moves into a gamete along with a separate copy of chromosome A. In both cases, the gametes formed on the left will lack chromosome A, whereas those on the right will contain two copies of chromosome A.

dren with Down syndrome, and recent studies show that the risk also increases slightly for older fathers. Translocations of chromosome 21, or variants of chromosome 21 with a tendency to nondisjunction, run in some families and produce Down syndrome children regardless of the parents' ages. About 1 baby in 700 live births has Down syndrome.

Sometimes the sex chromosomes fail to segregate properly during meiosis. In humans, this results in such abnormal sets of sex chromosomes as XXY (Klinefelter syndrome), XYY, XXX, or a single X chromosome (Turner syndrome) (see Table 16–2).

## 14–G  GAMETE FORMATION IN ANIMALS

The formation of gametes is similar in most animals, although details vary among species. However, the two processes—formation of sperm and of eggs—differ somewhat from each other. Let us begin with the production of sperm, which is in some ways simpler.

Sperm or **spermatozoa** (singular: **spermatozoon**) are the male gametes. A sperm is very small, with little cytoplasm, and in nearly all species it is motile, swimming with the flagellum that forms its tail.

Sperm are produced in the testes by the process of **spermatogenesis.** Each testis is made up of masses of tiny tubes, the **seminiferous tubules.** The stages of spermatogenesis occur in an orderly sequence in the cells lining these tubules. The male's diploid germ cells, **spermatogonia,** lie near the periphery of the tubules. They divide continuously by mitosis, producing more spermatogonia. This repeated division pushes some of the cells inward, and as these cells progress toward the center, they undergo meiosis and differentiate into sperm (Figure 14–16).

The cells that actually undergo meiosis are **spermatocytes.** Primary spermatocytes go through meiosis I, which produces two secondary spermatocytes. Each secondary spermatocyte contains one member of each pair of homologous chromosomes, in the form of a set of sister chromatids. During meiosis II, the two secondary spermatocytes divide again and produce a total of four haploid spermatids.

Although meiosis is now complete, the spermatids must undergo further differentiation into spermatozoa. In the process, each loses most of its cytoplasmic mass. A mature sperm has a head, which contains the nucleus with its haploid set of chromosomes; a long tail, or flagellum, which propels the sperm through its fluid surroundings; and between these a midpiece containing many mitochondria, which supply the ATP necessary for flagellar motion.

**Oogenesis** is the formation of female gametes, the eggs or **ova** (singular: **ovum**). Whereas sperm are often the smallest cells in a male animal's body, eggs are the largest cells in a female. This is because the egg cell is the main source of stored food, ribosomes, messenger RNA, and other cytoplasmic components that support the embryo's early development. Oogenesis ensures that the mature egg contains as much as possible of these components: meiotic nuclear division is accompanied by unequal cytokinesis, so that the original diploid cell produces only one large ovum and a number of tiny cells called **polar bodies** (Figure 14–17).

Oogenesis is more complicated than spermatogenesis. In the ovary, cells called **oogonia** divide by mitosis for a time. Eventually, the oogonia stop dividing and differentiate into primary oocytes. The DNA is replicated, and the primary oocytes enter prophase I, proceeding through the formation of tetrads and crossing over. In humans and other mammals, all this occurs while the female is still an embryo, and she will never produce any more potential eggs. At this point in prophase I, meiosis is arrested for days or years, depending on the species, and it does not resume until the female reaches sexual maturity. During this time the cell absorbs nutrients from neighboring somatic cells

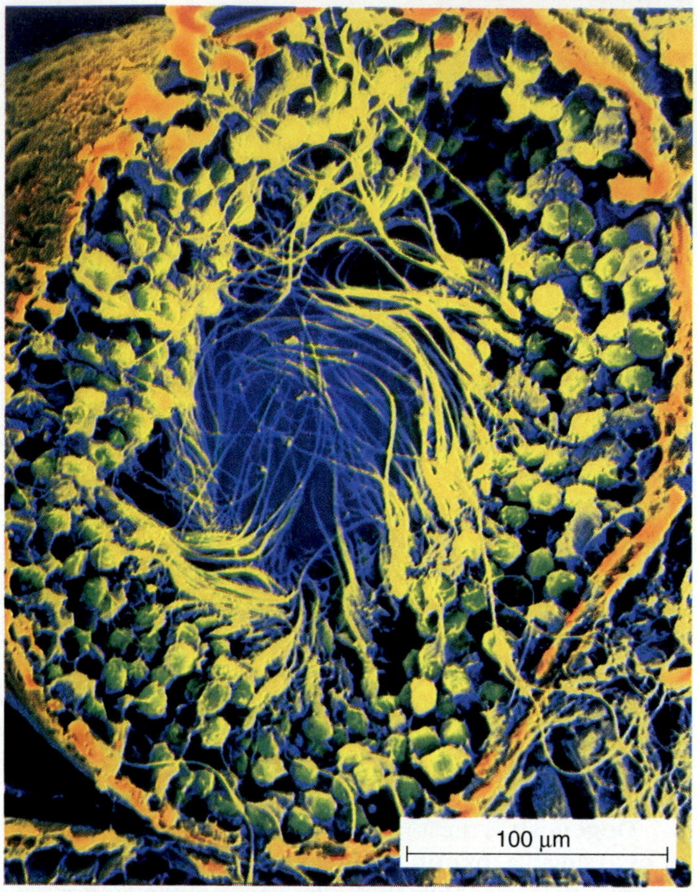

(a) Human seminiferous tubule, cross section (SEM).

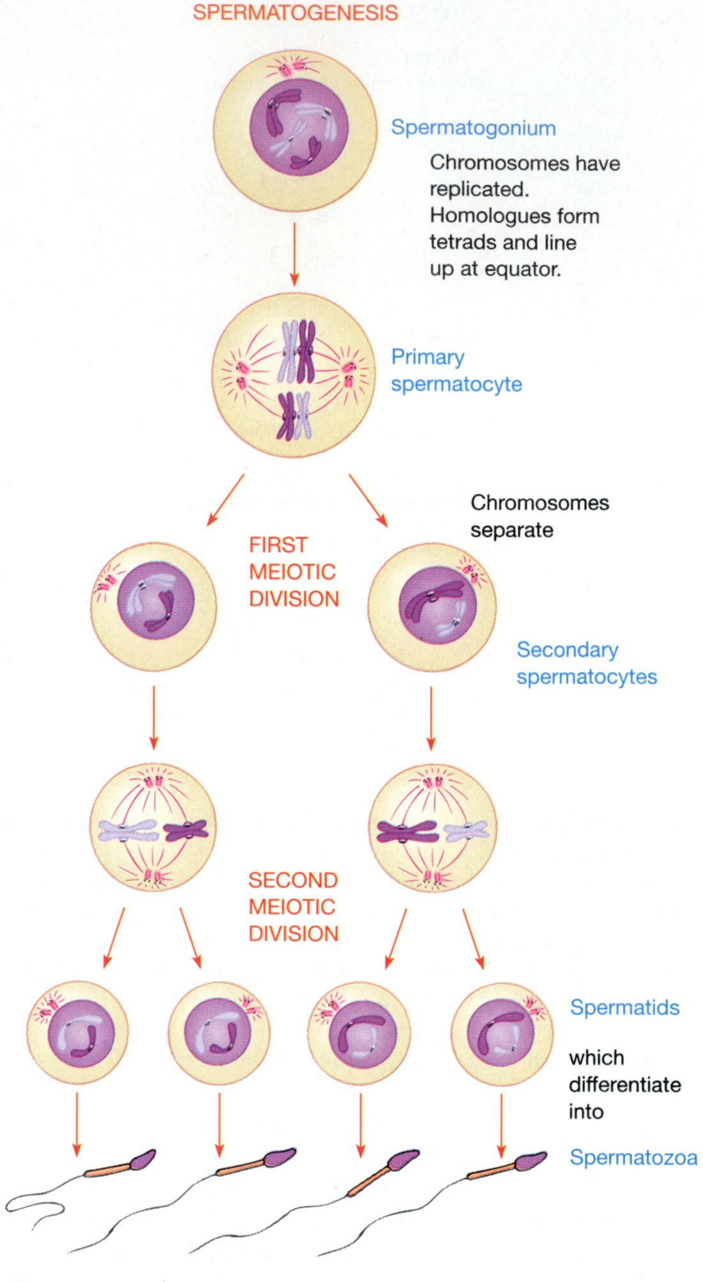

**FIGURE 14–16**  Spermatogenesis. (a) Cross section of a seminiferous tubule in a human testis. Cells move from the periphery toward the center of the tubule as they undergo the development outlined in (b). (b) Development of a spermatogonium into four spermatozoa (sperm) through meiosis and differentiation. (a, Secchi, Lecaque, Roussel, Uclaf, CNRI/Science Photo Library/Photo Researchers)

SPERMATOGENESIS

Spermatogonium

Chromosomes have replicated. Homologues form tetrads and line up at equator.

Primary spermatocyte

FIRST MEIOTIC DIVISION

Chromosomes separate

Secondary spermatocytes

SECOND MEIOTIC DIVISION

Spermatids

which differentiate into

Spermatozoa

(b)

and stockpiles the materials needed for early embryonic development. The chromosomes, still in their replicated form, are used to transcribe the RNA needed for these activities.

Eventually, the female becomes sexually mature and one or more eggs undergo maturation at regular intervals, resuming progress through meiosis. In meiosis I, the chromosomes separate in the usual manner, but during cytokinesis the cytoplasm divides unequally. One nucleus is pinched off with a minimum of cytoplasm, forming the first polar body, while the other nucleus is left with most of the

cytoplasm, in a cell called the secondary oocyte. This goes through the second meiotic division. Again, two haploid nuclei are formed according to the normal events of meiosis, but cytokinesis is extremely unequal, forming a tiny second polar body and an enormous ovum. Depending on the species, the first polar body may go through meiosis II and produce two haploid polar bodies. However, all the polar bodies are really just a means of shedding excess chromosomes from the developing egg, and they soon disintegrate.

Spermatogenesis proceeds to completion whether or not there is any chance of the resulting sperms' fertilizing

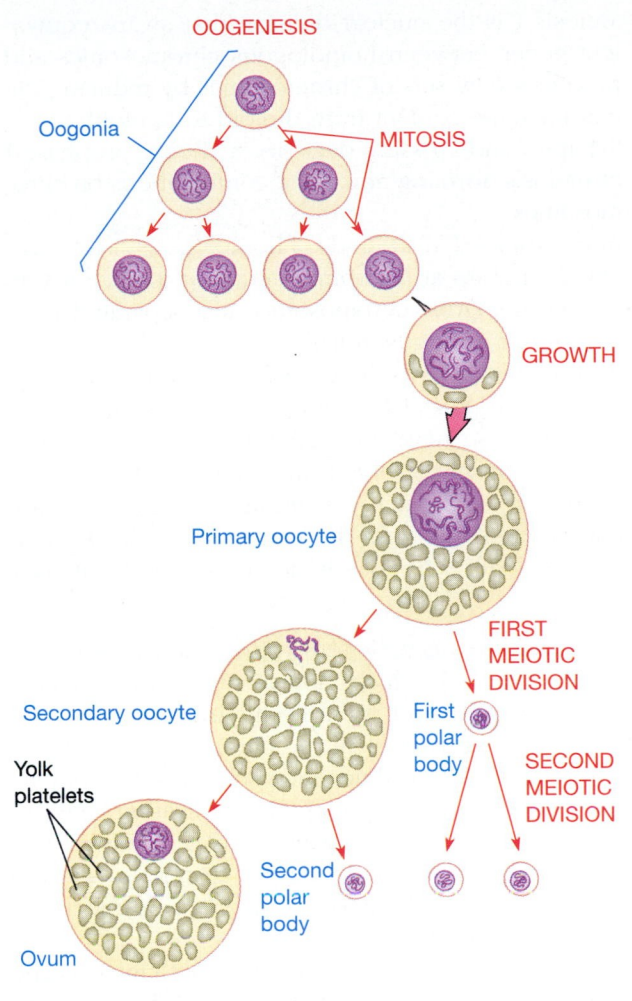

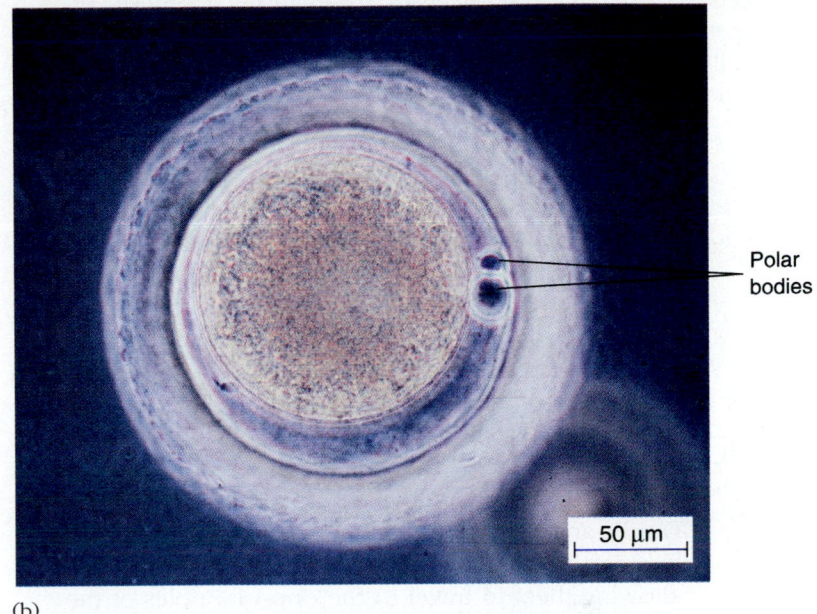

(b)

**FIGURE 14–17** Oogenesis in a vertebrate. (a) The oogonia divide and form more oogonia by mitosis. Eventually, oogonia grow and differentiate into primary oocytes. The nuclear divisions of meiosis are like those in spermatogenesis (Figure 14–16), but cytokinesis is unequal, forming a large cell and a tiny polar body at each division. (b) The fertilized egg of a rabbit. Two polar bodies can be seen.    (b, Biophoto Associates)

an egg. However, oogenesis usually becomes arrested, at various stages in different animals. In humans, when a woman ovulates, a secondary oocyte (not an ovum) is released from the ovary. Only if the secondary oocyte is penetrated by a sperm will the second meiotic division occur. This produces a polar body and an ovum. The chromosomes of the ovum can then combine with those from the sperm, producing the zygote nucleus.

## SUMMARY

Cells are the reproductive units of life: new cells are produced when existing cells divide in two. Eukaryotic cell divisions are of two kinds. In mitotic division, a cell of any ploidy gives rise to two daughter cells, each with a complete set of genetic information identical to that of the original cell. In meiotic division, which is a vital element of sexual reproduction, new combinations of chromosomes and genes are produced, packaged in nuclei containing half

the original number of chromosomes. This occurs when a diploid nucleus divides twice, forming four haploid daughter nuclei.

1. Most plants and animals are diploid, with paired chromosomes. During sexual reproduction, each parent passes a haploid nucleus containing one member of each chromosome pair to each offspring.

2. The cell cycle is the time from one mitotic division to the next. It can be divided into interphase—$G_1$, the S period (when DNA synthesis occurs), and $G_2$—and mitosis. The length of the cell cycle varies from as little as 15 minutes in some early embryos to several months or longer. The initiation of DNA synthesis (S) is the key event committing the cell to undergo division. This occurs when the level of cyclin builds up and activates cyclin-dependent kinase, which initiates both the S period and, later, mitosis.

3. Mitosis is a nuclear division in which precise events ensure that the two daughter nuclei inherit chromosomes identical to those of the parent nucleus. Thus, a uni-

cellular organism produces two genetically identical offspring by mitosis, and all of a multicellular organism's body cells also contain identical genetic information (unless mutations occur).

- During prophase of mitosis, the replicated chromosomes, each consisting of two sister chromatids, condense and become visible under the light microscope. The nucleolus disperses, and microtubules are assembled into the mitotic spindle.
- During prometaphase, the nuclear envelope breaks down, permitting the chromosomes to interact with the spindle microtubules. In each set of sister chromatids, the two kinetochores attach to spindle fibers from opposite poles.
- In metaphase, all of the sets of sister chromatids are lined up at the equator of the spindle.
- During anaphase, each centromere splits into two, releasing the sister chromatids from one another and allowing them to travel to the opposite poles of the spindle.
- During telophase, the chromosomes at each pole form a nucleus as the nuclear membrane re-forms, the chromosomes unravel from their condensed form, and a new nucleolus develops.

4. Mitosis is usually accompanied by cytokinesis, the division of the cytoplasm and its components into two separate cells. In animal cells, a band of microfilaments pinches the cell in two. In plants, the two daughter cells build a common partition, and then each produces a new panel of cell wall on its own side. Mitosis without cytokinesis forms a syncytium or a coenocyte, containing many nuclei within one membrane.

5. Meiosis is the type of nuclear division that produces new gene and chromosome combinations as it forms four haploid nuclei from a diploid nucleus. Meiosis halves the number of chromosomes in a cell in such a way that each daughter nucleus receives one member of each pair of homologous chromosomes. Thus, meiosis shuffles genetic information and prevents the chromosome number from doubling in each generation of organisms that reproduce sexually.

- Meiosis I is the nuclear division that swaps equivalent genes between homologous chromosomes and produces new sets of chromosomes by reducing the chromosome content from diploid to haploid.
- Synapsis and crossing over occur during prophase I of meiosis, forming new gene combinations on chromosomes.
- At metaphase I, the tetrads of homologous sister chromatids line up at the spindle equator in such a way that homologous chromosomes are separated from each other during anaphase I.
- Each of the two resulting nuclei contains one member of each pair of homologous chromosomes.
- Usually these new sets contain a mixture of chromosomes from the organism's two parents.
- Meiosis II is essentially a mitotic division of a haploid cell. At this time the centromeres divide, permitting sister chromatids to move into different nuclei.

6. Meiosis produces genetic reassortment in two ways: by crossing over, in which homologous chromosomes exchange genes; and by forming new chromosome combinations as a result of the chromosome lineup at metaphase I. Additional genetic variety results from the random combination of gametes at fertilization. Translocation or nondisjunction during meiosis results in cells with too much or too little genetic material, usually a fatal condition. Survivors often have severe mental or physical abnormalities.

7. Gamete formation in animals involves both meiosis and differentiation to form specialized reproductive cells. Each spermatocyte gives rise to four sperm, the male gametes, which are stripped down to the bare necessities: a haploid set of genes and the locomotory apparatus to deliver it to the egg. Oogenesis involves unequal cytokinesis, producing only one large egg swollen with material destined to support the early embryo, and tiny polar bodies, containing little more than the excess chromosomes being shed from the forming egg.

## Self-Quiz

1. A cell cycle is:
   a. the time from the formation of a cell until its death
   b. the series of events that takes place from the formation of a cell until it divides again
   c. the sequence of events that assures each daughter cell of a set of chromosomes identical to that of its parent cell (mitosis)
   d. the growth of a cell until it is large enough to divide again

2. For the species depicted in Figure 14–1, what is the value of N? of 2N?

3. A diploid somatic cell:
   a. cannot undergo division again
   b. can undergo mitosis but not meiosis
   c. can undergo mitosis or meiosis
   d. can undergo meiosis but not mitosis

4. The photograph below shows dividing cells in an onion root tip. Identify the stage of mitosis in the cells marked A through E.

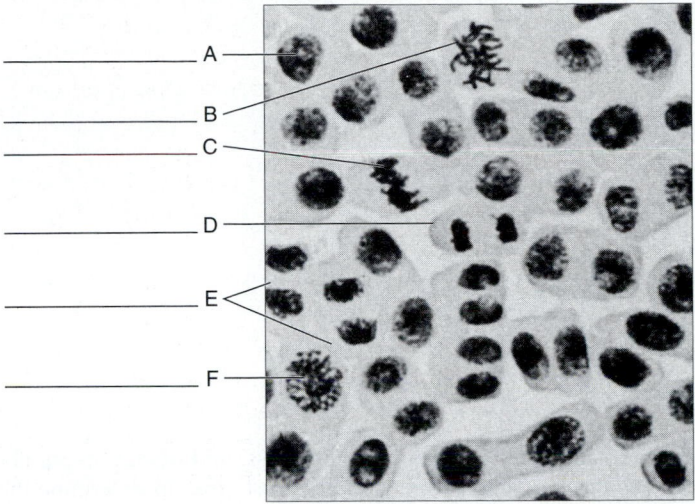

5. A cell is in metaphase if:
   a. its chromosomes are visible as distinct thread-like structures
   b. the nuclear membrane is not visible
   c. the chromosomes are lined up at the equator of the cell
   d. the chromosomes are separated into two distinct groups attached to the spindle
   e. the chromosomes are found in two compact groups in two small patches of cytoplasm that are in the process of separating into two distinct cells
6. A cell in prophase of mitosis can be distinguished from a cell in prophase I of meiosis by:
   a. the presence of only half as many chromosomes in the meiotic cell

b. the formation of tetrads in the meiotic cell
   c. the presence of twice as many chromosomes in the meiotic cell
7. The function of mitotic cell division in the life cycle of an organism is:
   a. reproduction of identical individuals if the organism is unicellular
   b. growth of an individual if the organism is multicellular
   c. repair of injured tissue
   d. all of the above
8. Substances that interfere with microtubule function interfere with cell division because:
   a. microtubules must be distributed equally to the new cells
   b. microtubules are involved in the precise separation of the chromosomes, which ensures that a complete set of chromosomes gets into each daughter cell
   c. without microtubules, cytokinesis cannot take place, and a syncytium is formed
   d. microtubules are essential to the disappearance of the nuclear membrane, and without them the chromosomes cannot separate into two new nuclei
9. Both oogenesis and spermatogenesis involve equal division of the _____ . However, unequal division of the _____ occurs during production of _____ , whereas in production of _____ this division is equal.
10. The importance of crossing over during meiosis is:
   a. it assures that one member of each homologous pair ends up in each new nucleus
   b. it results in chromosomes containing new combinations of genes
   c. it results in nuclei with too much or too little genetic material
   d. it ensures that the developing egg receives most of the cytoplasm from the oocyte

## Questions for Discussion

1. Banana plants are triploid (3N). Can you explain why they are unable to reproduce sexually? How do you suppose they do reproduce?
2. Why is it necessary for cytokinesis to occur in such a way that each daughter cell receives some ribosomes, mitochondria, and, in plants, plastids?
3. Since the genetic information is carried equally by egg and sperm, what do you suppose to be the selective advantage of the inequality of size that has evolved between the tiny mobile sperm and the large immobile egg?
4. How would leakage from nuclear waste repositories affect the organisms that come into contact with it? Why is it difficult to design safe nuclear waste disposal facilities?

## Suggested Readings

Alberts, B., et al. *Molecular Biology of the Cell,* 2d ed. New York: Garland Publishing, 1989. Chapters 11, 13, and 15.

Glover, D. M., C. Gonzalez, and J. W. Raff. "The centrosome." *Scientific American,* June 1993. The role of the cell center material in cell division.

McIntosh, J. R., and K. L. McDonald. "The mitotic spindle." *Scientific American,* October 1989.

Murray, A. W., and M. W. Kirschner. "What controls the cell cycle." *Scientific American,* March 1991.

Nathans, J. "The genes for color vision." *Scientific American,* February 1989. How we see color, how this ability evolved, and how mistakes in crossing over during meiosis result in color blindness.

Patterson, D. "The causes of Down syndrome." *Scientific American,* August 1987.

Upton, A. C. "The biological effects of low-level ionizing radiation." *Scientific American,* February 1982.

# Mendelian Genetics

## O B J E C T I V E S

*When you have studied this chapter, you should be able to:*

1. Define and use these terms:
   parental (P), first filial ($F_1$) and second filial ($F_2$) generations
   alleles, homozygous, heterozygous
   segregation, independent assortment
   codominance, incomplete dominance
   homologous chromosomes, locus, linkage groups
2. Define and compare the terms phenotype and genotype and their relationship to the terms dominant and recessive.
3. Use a Punnett square to illustrate a monohybrid or independently assorting dihybrid cross; use a "branching" diagram to illustrate a dihybrid cross, and work out the genotypic and phenotypic ratios expected from such crosses.

4. Explain what is meant by a test cross, and discuss its significance as a genetic tool. Design a test cross to determine the genotype of an organism with a dominant phenotype.
5. Correlate the pattern of inheritance of genetic characteristics in breeding experiments with the behavior of chromosomes during meiosis and fertilization.
6. Given data from problem situations, identify linkage and calculate map distances between linked loci.
7. Explain the biological significance of tetrad formation and crossing over during meiosis.
8. Use the rules of probability to solve genetics problems such as those at the end of this chapter.
9. In your own words, state the rules of inheritance that form the basis of Mendelian genetics.

---

Humans have long understood that plants and animals inherit some characteristics from their parents. Prehistoric people doubtless recognized a child's resemblance to its parents, bred calves from the cows that gave the most milk, and saved the most productive grain for seed. Eventually people tried breeding together two plants or animals with different desirable features, hoping that the offspring would inherit all the desired features. Sometimes this worked, sometimes not. When it did, later generations often did not retain all the desired features. The consistent breeding of desirable strains of plants and animals was not put on a scientific basis until the beginning of the twentieth century.

**Genetics** is the study of inherited variations among organisms and the transmission of these variations to offspring. The hereditary characteristics involved are also called **characters** or **traits**.

Genetics is based on the work of Gregor Mendel, a monk (and later the abbot) at the monastery of Brünn (now called Brno, in the Czech Republic). Mendel's work was reported at a meeting of the Brünn Society for the Study of Natural Science in 1865 and was published the following year.

Mendel's paper presented a new and carefully documented model of inheritance, but his work received little attention until after his death (*Essay: Gregor Mendel*). It was rediscovered in 1900 almost simultaneously by Hugo De Vries in the Netherlands, Carl Correns in Germany, and Erich Tschermak in Austria.

In the meantime, the movements of chromosomes during mitosis and meiosis were observed and described. Scientists realized that chromosomes behave precisely like the hereditary traits Mendel had studied. They concluded that these traits are dictated by segments of chromosomes, which we now call genes. Genes are replicated and passed on to new cells as parts of chromosomal DNA molecules, and this is how offspring inherit genetic traits from their parents. In sexual reproduction, a new individual receives half of its chromosomes (and genes) from its mother's egg and half from its father's sperm.

In this chapter we shall study Mendel's breeding experiments and see how they revealed the patterns of inheritance as genetic traits were passed from one generation to the next. We then go beyond his work to other patterns that also result from the behavior of chromosomes during meiosis.

## KEY CONCEPTS

♦ In most familiar organisms, each individual has a pair of genes for each hereditary trait, one from each parent.

♦ Each of the individual's gametes (eggs or sperm), and hence each offspring, receives one of the two genes for each trait.

♦ This pattern of inheritance of genes from generation to generation reflects the behavior of chromosomes in meiosis and fertilization.

## 15–A   A SIMPLE BREEDING EXPERIMENT: THE MONOHYBRID CROSS

Mendel worked with garden peas. Pea flowers contain both male and female parts, and normally each flower is self-pollinated: pollen from the male parts fertilizes eggs in the female part of the same flower. Because a plant is both male and female parent to the seeds it produces, the off-spring often bear a greater resemblance to the parent than usual. Mendel identified a number of strains of peas that were pure-breeding for particular traits, meaning that offspring always inherited the same traits seen in their parent.

To study inheritance, Mendel made crosses between two pure-breeding strains of pea plants with different versions of a trait, such as different flower colors. He crossbred these two strains by transferring pollen between their flowers. Then he followed the inheritance of these traits in the offspring.

In one set of experiments, Mendel used two pure-breeding strains of pea plants: one having red flowers, the other having white flowers. These pure-bred plants are called the **parental,** or **P, generation.**

A cross between different parental strains, such as these, produces genetically mixed offspring known as **hybrids.** These hybrid offspring belong to the **first filial (F₁) generation** (filius, filia = son, daughter). All these F₁ plants produced red flowers, regardless of whether they grew from seeds produced by red-flowered or white-flowered parental plants.

Mendel allowed these F₁ red flowers to self-pollinate, and from them he collected more than 900 seeds of the **second filial (F₂) generation.** About three fourths of these F₂ seeds grew into red-flowered plants, and about one fourth grew into white-flowered plants (Figure 15–1).

### Gene Pairs

These results could be explained by assuming that a plant's flower color is governed by a pair of genes. The plant passes one of these two genes to each gamete. When two gametes unite at fertilization, the offspring receives two flower-color genes, one from each parent.

The genes dictating a genetic trait, such as flower color, are not all alike but can occur in two or more alternative forms, called **alleles**—in this case, the red-flower allele and the white-flower allele. Each diploid pea plant may have two alleles for red flowers, or two alleles for white, or one of each. Each haploid gamete contains only one of the plant's two alleles.

In Mendel's crosses, the original P generation were pure-breeding, meaning that each red-flowered plant had two red-flower alleles, and each white-flowered plant had two white-flower alleles. When they reproduced, each gamete received only one of these alleles:

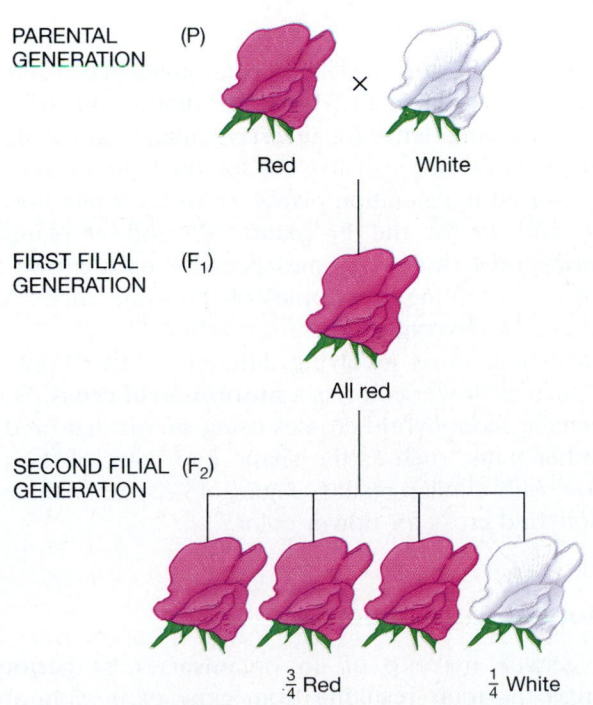

**PARENTAL (P) GENERATION**

Red × White

**FIRST FILIAL (F₁) GENERATION**

All red

**SECOND FILIAL (F₂) GENERATION**

$\frac{3}{4}$ Red      $\frac{1}{4}$ White

**FIGURE 15–1** Diagram of a cross involving flower color. The P generation were pure-breeding red-flowered and pure-breeding white-flowered pea plants. All the F₁ offspring were red-flowered. Self-pollination of these F₁ offspring yielded an F₂ generation of about $\frac{3}{4}$ red-flowered and $\frac{1}{4}$ white-flowered plants. (The multiplication sign between the two flowers at the top is the symbol for a mating.)

**P generation:**

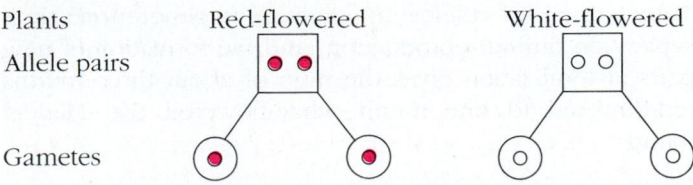

| Plants | Red-flowered | White-flowered |
|---|---|---|
| Allele pairs | ●● | ○○ |
| Gametes | ● ● | ○ ○ |

When these plants self-fertilize, two gametes carrying similar alleles unite (red + red; or white + white). The offspring always receive pairs of alleles just like the parent's and therefore inherit the parent's flower color.

When the two pure-breeding strains are crossed, each hybrid $F_1$ offspring inherits one red-flower allele and one white-flower allele:

**P generation:**

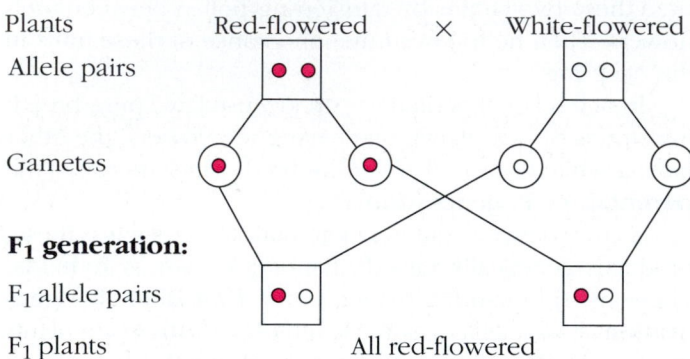

**F₁ generation:**

F₁ allele pairs

F₁ plants    All red-flowered

All these $F_1$ plants had red flowers, but self-crossing them produced some white-flowered plants in the $F_2$ generation. Mendel saw that the white-flower allele must have been present in the $F_1$ plants but masked by the red-flower allele. The allele that expresses itself when two different alleles occur together is **dominant,** and the other, masked allele is **recessive.**

White flowers reappear in the $F_2$ generation because the two alleles in the $F_1$ generation separate during reproduction and then form new pairs in the $F_2$ generation. An $F_1$ gamete is equally likely to contain a red-flower or a white-flower allele and to unite with another gamete containing either allele at fertilization:

**F₁ generation:**

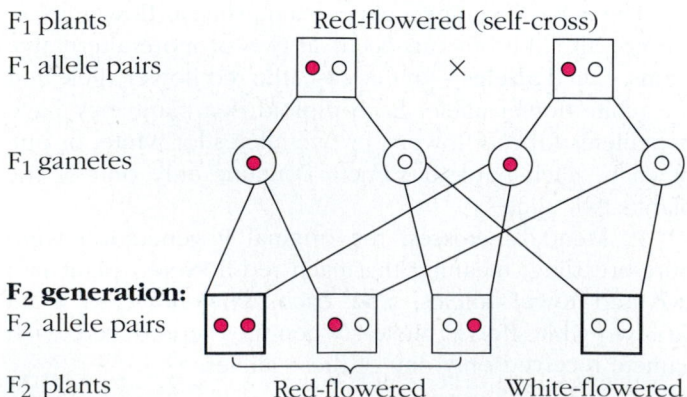

This pairing of alleles in each new generation, their separation during reproduction, and the formation of new pairs at fertilization gives the ratio of about three fourths red-flowered to one fourth white-flowered that Mendel found.

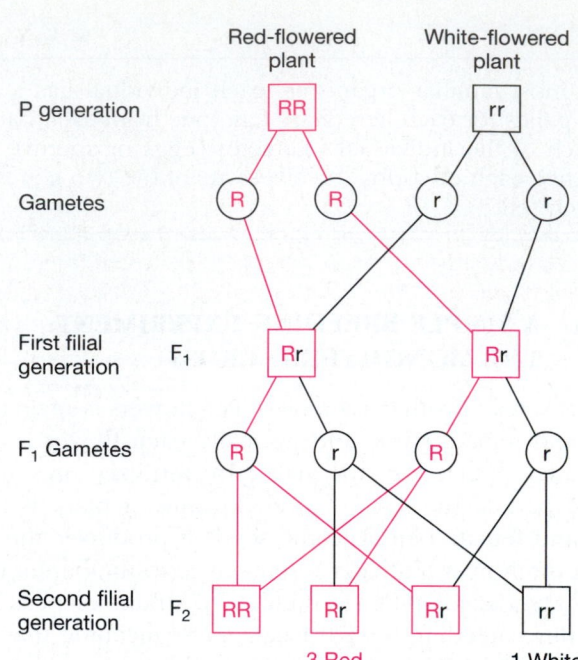

**FIGURE 15–2**  Diagram of a monohybrid cross. This is the same cross shown in Figure 15–1 but now showing the alleles found in the plants and in their gametes. The P generation plants are homozygous for red and white flower color alleles, respectively.

Geneticists often use letters to designate genes and their alleles: capital letters for dominant alleles and the lower case of the same letter for recessive alleles. In the flower-color example, we can use *RR* for the pair of alleles in red-flowered P generation plants, *rr* for the white-flowered ones, and *Rr* for the $F_1$ plants. *RR* and *rr* plants are **homozygotes** (homo = same) because each grows from a zygote containing two copies of the same allele. An *Rr* plant is a **heterozygote** (hetero = other).

A genetic cross involving different forms of only one trait, such as flower color, is a **monohybrid cross.** Mendel did similar monohybrid crosses using strains that bred true for other traits, such as the shape and color of the peas, always with similar results. Figure 15–2 summarizes the monohybrid cross for flower color.

## Genotype and Phenotype

The genetic makeup of an organism is its **genotype,** whereas the traits resulting from gene expression are its **phenotype.** Because of dominance, two plants with different genotypes may have the same phenotype: pea plants with genotypes *RR* and *Rr* both have the red-flower phenotype. The term phenotype is often used to refer to an organism's appearance, but it also applies to any other observable gene expression, such as development, behavior, or biochemistry. For example, blood type is a phenotype that can be observed by making biochemical tests on blood samples that appear identical.

An individual's genotype is fixed at the time of fertilization. However, the phenotype may vary because it results from the interaction of genes with one another and with factors in the environment. For example, a plant may be short because it has "dwarf" genes or because it is so poorly nourished that it cannot grow to the height dictated by its "tall" genes.

An individual with a dominant phenotype may have a genotype that is either **homozygous dominant** (homozygous for the dominant allele) or heterozygous. An individual with a recessive phenotype, however, must have a genotype that is **homozygous recessive** (homozygous for the recessive allele) (Figure 15–3). Therefore, in a monohybrid cross involving a dominant-recessive allele pair, the ratio of genotypes among the F$_2$ offspring differs from the ratio of phenotypes.

For example, in the flower-color cross, we expect F$_2$ genotypes in a ratio of about 1$RR$:2$Rr$:1$rr$, because there are two ways to obtain the $Rr$ combination and only one way to obtain each of the others (see Figure 15–2). In contrast, we would expect phenotypes of about 3 red-flowered:1 white-flowered, because plants with genotypes $RR$ and $Rr$ all have red flowers.

For convenience, we can write the genotype of a red-flowered plant as $R\_\_$. The underline means either that the second allele is unknown or that we are lumping together both $RR$ and $Rr$ plants, which have the same phenotype.

## Law of Segregation and Meiosis

The paired alleles for a trait separate from each other when an organism makes reproductive cells, and each allele occurs in half the gametes. Geneticists call this the **law of**

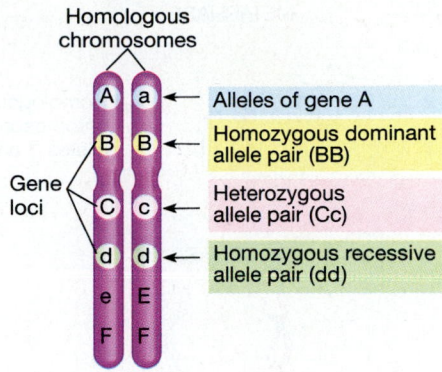

**FIGURE 15–4**  The vocabulary of genetics. This diagram shows the relationships among some terms introduced in the text.

**segregation.** Gametes containing single alleles then combine at random and form new allele pairs at fertilization.

This happens because genes are actually segments of chromosomes. Every diploid body cell contains pairs of homologous chromosomes. Since chromosomes come in pairs, so do the genes they carry.

A gene's **locus** is the location where it occurs in a chromosome's DNA: we find one of the alleles of that particular gene in this place. For example, in pea plants, either the $R$ or the $r$ allele occurs at the flower-color locus on one chromosome, and either allele may also occur at the same locus on the homologous chromosome. If both homologues have the same allele, the plant is homozygous for flower color ($RR$ or $rr$). If the two chromosomes have different alleles, the individual is heterozygous at that locus ($Rr$) (Figure 15–4).

The events of meiosis account for the law of segregation: during meiosis, the members of a pair of homologous chromosomes become separated into different nuclei, and so do the genes carried by these chromosomes. For example, in a pea plant heterozygous for flower color, homologous chromosomes bearing alleles $R$ and $r$ are separated into different nuclei during meiosis, and so they end up in different cells (Figure 15–5).

## 15–B  PREDICTING THE OUTCOME OF A GENETIC CROSS

We can predict the probable outcome of a genetic cross if we know the genotypes of the parents. One way to do this is by drawing a **Punnett square** (named after geneticist Reginald Crundall Punnett). To construct a Punnett square, we determine the alleles and their frequencies among the gametes of each parent. We write those of one parent above the boxes across the top of the square and those of the

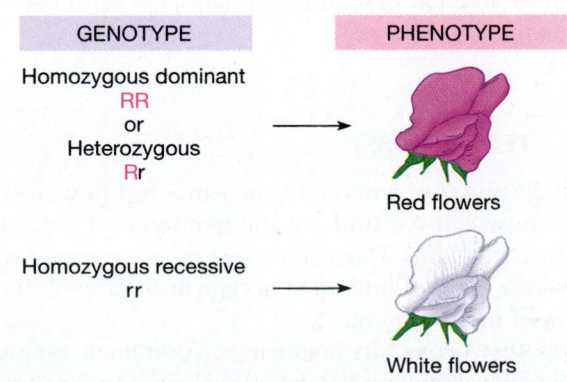

**FIGURE 15–3**  Genotype and phenotype. Individuals of dominant phenotype result from more than one possible genotype, but the recessive phenotype indicates a homozygous recessive genotype.

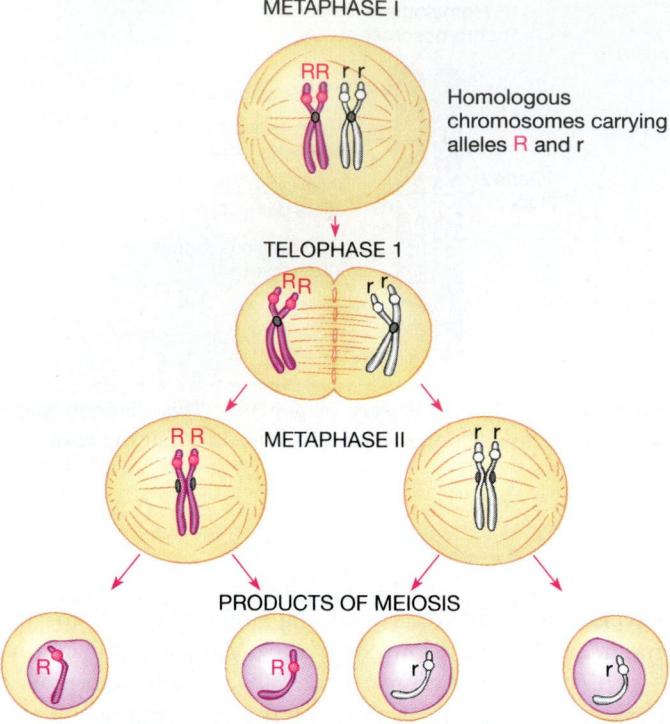

**FIGURE 15–5** The law of segregation reflects the events of meiosis. Homologous chromosomes are replicated before entering meiosis. In the first meiotic division, homologues are separated so that each member of the pair ends up in a separate nucleus, and hence so do the paired alleles (*R* and *r*) they carry. At the second meiotic division, the sister chromatids separate from one another, forming a total of four nuclei. Figure 14–11 shows meiosis in more detail.

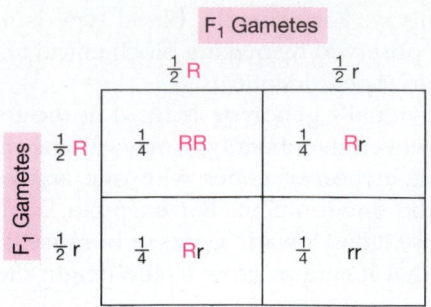

**FIGURE 15–6** A Punnett square. This one diagrams the self-crossing of the heterozygous $F_1$ generation shown in Figure 15–2.

other parent down the side (Figure 15–6). We then fill in all the boxes by combining the alleles shown at the top of the column with those shown at the left of the row, multiplying the frequencies to find the probability of each combination. These combinations in the boxes show the possible genotypes of the offspring and the ratio in which they are expected to occur.

In the flower-color cross, the $F_2$ generation includes genotypes *RR, Rr,* and *rr* in a ratio of 1:2:1. If we know that one allele is dominant to the other, we can also predict the $F_2$ phenotypes. In this case, *R* is dominant to *r,* and so the expected ratio of red-flowered to white-flowered plants is 3:1.

We could also calculate this directly from the probabilities of each type of gamete. The gametes produced by each $F_1$ heterozygous parent can be written ($\frac{1}{2}R + \frac{1}{2}r$). To find the distribution of genotypes in the next ($F_2$) generation, we multiply the frequencies of gametes of each parent together:

$$\underbrace{(\tfrac{1}{2}R + \tfrac{1}{2}r)}_{\substack{\text{male}\\\text{gametes}}} \underbrace{(\tfrac{1}{2}R + \tfrac{1}{2}r)}_{\substack{\text{female}\\\text{gametes}}} = \underbrace{\tfrac{1}{4}RR + \tfrac{1}{4}Rr + \tfrac{1}{4}rR + \tfrac{1}{4}rr}_{\substack{\text{offspring}\\\text{genotypes}}}$$

Adding the two heterozygous genotypes ($\frac{1}{4}Rr + \frac{1}{4}rR = \frac{1}{2}Rr$) gives $\frac{1}{4}RR + \frac{1}{2}Rr + \frac{1}{4}rr$. This is the same 1:2:1 ratio as the genotypes found using the Punnett square. Again, if *R* is dominant to *r*, then *RR* and *Rr* are both red-flowered, and the offspring phenotypes are $\frac{3}{4}$ red-flowered (*R\_\_*) : $\frac{1}{4}$ white-flowered (*rr*), or 3 red:1 white.

So, in a monohybrid cross where one allele is dominant and the other recessive, we expect the $F_2$ generation to show 3 dominant phenotypes:1 recessive phenotype (or $\frac{3}{4}:\frac{1}{4}$). The genotype ratios are 1 homozygous dominant: 2 heterozygous:1 homozygous recessive.

The ratios of offspring expected from a genetic cross are statistical probabilities, not firm guarantees. Because chance plays a large role in the events of meiosis, fertilization, development, and offspring survival, genetic crosses rarely produce exactly the ratio of offspring expected in each category. However, the more offspring there are, the closer the results are likely to come to expectation. This is why geneticists obtain hundreds of offspring from crosses whenever possible.

Mendel performed monohybrid crosses for seven different traits of pea plants, each appearing in two distinct forms in different pure-breeding strains. Table 15–1 shows how many individuals of each phenotype he obtained in each cross. In each case, the $F_2$ phenotype ratio is close to 3 dominant:1 recessive.

## 15–C   TEST CROSS

If both *RR* and *Rr* plants have the same red-flowered phenotype, how can we find out the genotype of a particular red-flowered plant? The usual method is to cross such a plant with a plant of known genotype and observe the phenotypes of the offspring.

In a **test cross,** an organism of dominant phenotype but unknown genotype is crossed with one that is homozygous recessive for the trait in question. In peas, a white-flowered plant has the homozygous recessive genotype, *rr,* and can pass only an *r* allele to each offspring. Therefore,

**TABLE 15–1**

## Mendel's Monohybrid Crosses Involving Seven Pairs of Traits

| Parental Characters* | | | Results of Crosses | | |
|---|---|---|---|---|---|
| *Dominant Trait* | × | *Recessive Trait* | $F_1$ | $F_2$ | $F_2$ Ratio† |
| Red flowers and gray-brown seed coats‡ | | White flowers and white seed coats | All red | 705 red:224 white | 3.15:1 |
| Smooth seeds | | Wrinkled seeds | All smooth | 5474 smooth:1850 wrinkled | 2.96:1 |
| Yellow seeds | | Green seeds | All yellow | 6022 yellow:2001 green | 3.01:1 |
| Inflated pods | | Constricted pods | All inflated | 882 inflated:299 constricted | 2.95:1 |
| Green pods | | Yellow pods | All green | 428 green:152 yellow | 2.82:1 |
| Axial flowers | | Terminal flowers | All axial | 651 axial:207 terminal | 3.14:1 |
| Long stems | | Short stems | All long | 787 long:277 short | 2.84:1 |

*All plants of the parental generation were pure-breeding.

†For all pairs of traits, the ratios of dominant to recessive phenotypes in the $F_2$ generation approximate the expected ratio of 3:1.

‡The same gene pair governs color in both the flower and the seed coat (the "skin" of the pea).

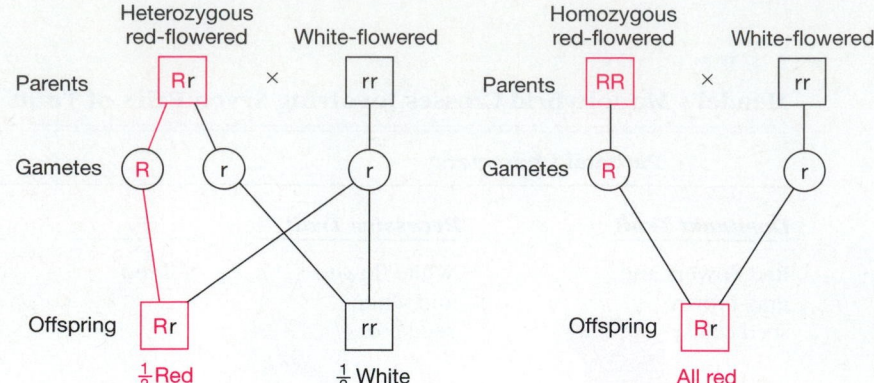

**FIGURE 15–7** Test cross. A red-flowered plant of dominant phenotype but unknown genotype is crossed with a white-flowered (homozygous recessive) plant. By examining the offspring, it should be possible to deduce the genotype of the red-flowered parent plant.

the phenotypes of the offspring depend on which allele they receive from the red-flowered parent. If the red-flowered plant were actually heterozygous ($Rr$), we would expect half of the offspring of the test cross to be red-flowered and half white-flowered. However, if the red-flowered parent were homozygous ($RR$), it could pass only the $R$ allele to its offspring and they would all be heterozygous, with red flowers (Figure 15–7).

So, if a red-flowered × white-flowered test cross produces some white-flowered offspring, the red-flowered parent must be heterozygous, because white-flowered offspring must have received an $r$ allele from each parent. However, if all the offspring have red flowers, we cannot be absolutely certain that the red-flowered parent was homozygous $RR$. It is possible, though unlikely, for a plant with genotype $Rr$ to have all red-flowered offspring. But the more offspring—all red-flowered—we get from such a cross, the more nearly certain we can be that the parent with the dominant phenotype is not heterozygous.

## 15–D THE DIHYBRID CROSS: INDEPENDENT ASSORTMENT

In addition to his monohybrid crosses, Mendel performed **dihybrid** crosses, in which he followed two pairs of alleles. In one experiment, Mendel crossed plants homozygous for seeds that were both smooth and yellow with plants homozygous for wrinkled green seeds. All the $F_1$ offspring were smooth and yellow, showing that smooth was dominant to wrinkled and yellow was dominant to green.

Self-fertilization of the $F_1$ plants produced an $F_2$ generation of seeds with the following phenotypes:

  315 smooth yellow   101 wrinkled yellow
  108 smooth green    32 wrinkled green

To find the ratio among these $F_2$ phenotypes, we take the number of offspring in the smallest category (32) and divide it into the number of offspring in each category. Then we round the result to the nearest whole number. We find that the $F_2$ phenotypic ratio is about 9:3:3:1. This ratio is typical of a dihybrid cross in which both pairs of alleles show a dominant-recessive relationship.

Based on these data, Mendel asserted that the seed texture and seed color traits are independent of each other. The hybrid plants must be forming gametes with all possible combinations of alleles for seed texture and color. In this process of **independent assortment,** each pair of alleles behaves as it would in a monohybrid cross, with no effect on the other pair. For example, when we consider only smooth versus wrinkled seeds, we find: $315 + 108 = 423$ smooth, and $101 + 32 = 133$ wrinkled. This gives a ratio of $423:133 = 3.18:1$, quite close to the 3:1 ratio of a monohybrid cross, which is what it is. Similarly, the inheritance of yellow versus green seeds behaves like a monohybrid cross. In other words, this dihybrid cross is the product of two separate monohybrid crosses:

(3 smooth + 1 wrinkled) (3 yellow + 1 green) =
9 smooth yellow + 3 smooth green
+ 3 wrinkled yellow + 1 wrinkled green

In this example, let $S$ = smooth, $s$ = wrinkled, $Y$ = yellow and $y$ = green. The P generation plants must have had genotypes $SSYY$ (smooth, yellow) and $ssyy$ (wrinkled, green). Because each gamete receives just one member of each gene pair, these parents must have produced gametes $SY$ and $sy$, respectively. All members of the $F_1$ generation have the genotype $SsYy$, giving a smooth yellow phenotype (Figure 15–8a).

Self-fertilization of the $F_1$ plants produces a more complex situation. According to the "law of independent assortment," the members of each gene pair are sorted into gametes independently of the members of the other gene pair. This occurs because these genes are carried on different pairs of chromosomes. We can see why they assort independently from our study of meiosis: when the tetrads of homologous chromatids line up for the first division of meiosis (at metaphase I), the four alleles can be arranged in either of two ways (Figure 15–8b). If $S$ and $Y$ line up opposite $s$ and $y$, the gametes formed are $SY$ and $sy$. If $S$ and $y$ line up opposite $s$ and $Y$, the gametes $Sy$ and $sY$ form. Either arrangement is equally likely, and so each $F_1$ plant is expected to produce four kinds of gametes—$SY$, $sy$, $Sy$, and $sY$—in equal proportions. Each gamete always receives *one member of each pair of genes.*

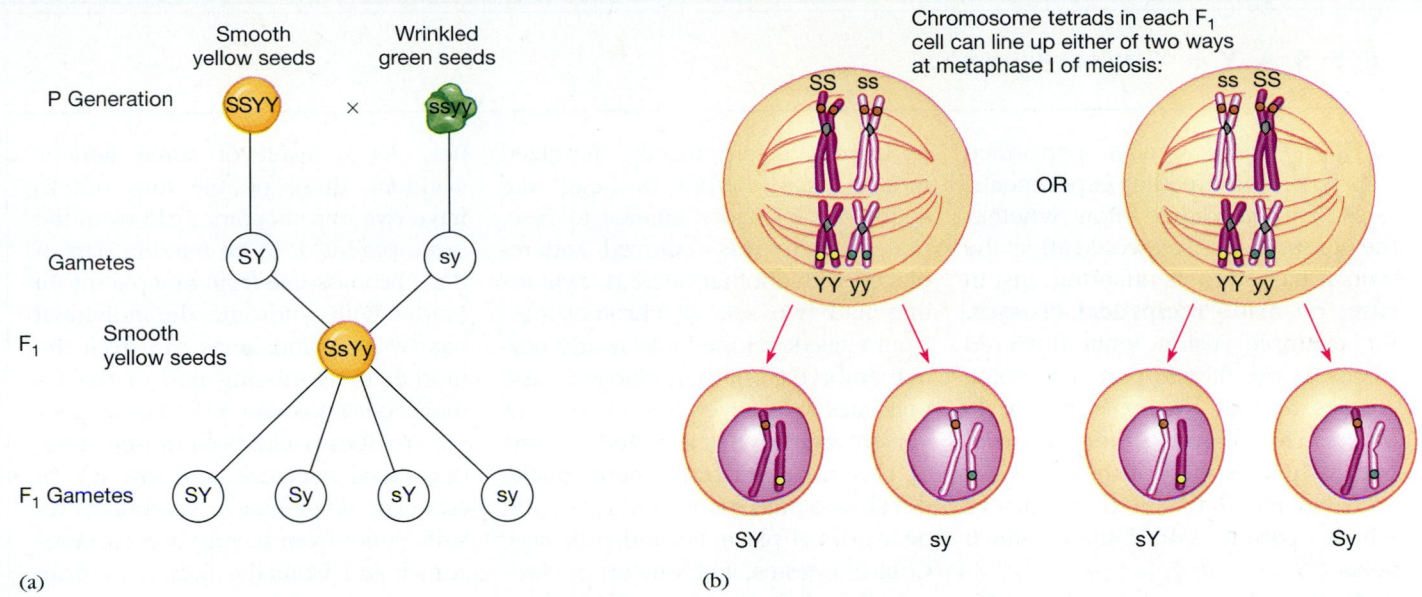

(a)

(b)

**FIGURE 15–8**   Independent assortment. In dihybrid individuals, members of the two gene pairs move independently during gamete formation. (a) The dihybrid ($F_1$) individuals produce four different types of gametes in equal proportions. Some of the $F_1$ gametes have the same gene combinations as those received from the P generation gametes (*SY* and *sy*), but two new combinations are also produced (*Sy* and *sY*). (b) Two different possible tetrad alignments at metaphase I of meiosis result in the production of a total of four types of gametes. Figure 15–9 shows the $F_2$ generation.

To find all possible genetic combinations in the $F_2$ offspring, we can write the $F_1$ gametes formed by independent assortment along the sides of a Punnett square. Since any female gamete can be fertilized by any male gamete, there is a total of nine possible genotypes, falling into four phenotypes, in the $F_2$ generation (Figure 15–9 and Table 15–2).

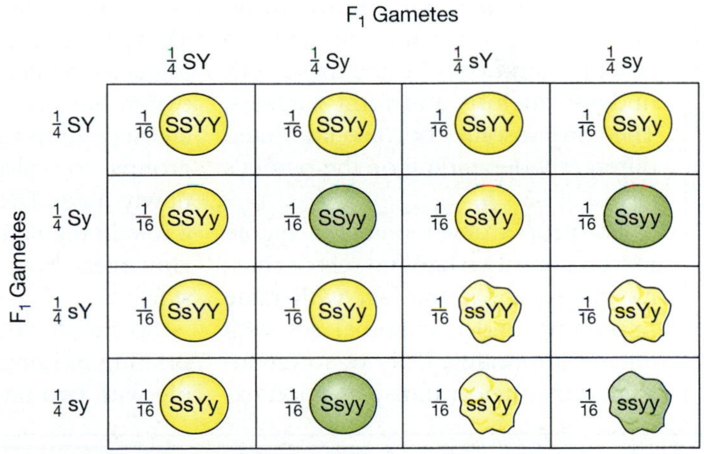

**FIGURE 15–9**   Punnett square for the dihybrid cross in Figure 15-8. Gametes produced by the $F_1$ generation are arranged along the top and side of the square. The letters in the boxes represent genotypes of the $F_2$ generation, and the drawings of peas show the associated phenotypes: smooth yellow, wrinkled yellow, smooth green, and wrinkled green.

$F_1$ Gametes

| | $\frac{1}{4}$ SY | $\frac{1}{4}$ Sy | $\frac{1}{4}$ sY | $\frac{1}{4}$ sy |
|---|---|---|---|---|
| $\frac{1}{4}$ SY | $\frac{1}{16}$ SSYY | $\frac{1}{16}$ SSYy | $\frac{1}{16}$ SsYY | $\frac{1}{16}$ SsYy |
| $\frac{1}{4}$ Sy | $\frac{1}{16}$ SSYy | $\frac{1}{16}$ SSyy | $\frac{1}{16}$ SsYy | $\frac{1}{16}$ Ssyy |
| $\frac{1}{4}$ sY | $\frac{1}{16}$ SsYY | $\frac{1}{16}$ SsYy | $\frac{1}{16}$ ssYY | $\frac{1}{16}$ ssYy |
| $\frac{1}{4}$ sy | $\frac{1}{16}$ SsYy | $\frac{1}{16}$ Ssyy | $\frac{1}{16}$ ssYy | $\frac{1}{16}$ ssyy |

**TABLE 15–2**

**Phenotypes and Genotypes, and Their Ratios, Resulting from a Self-Cross of *SsYy* Individuals**

| Phenotype | Ratio | Genotype |
|---|---|---|
| Smooth, yellow | $\frac{9}{16}$ | $\frac{1}{16}SSYY : \frac{2}{16}SSYy : \frac{2}{16}SsYY : \frac{4}{16}SsYy$ or $S\_\_Y\_\_$ |
| Smooth, green | $\frac{3}{16}$ | $\frac{1}{16}SSyy : \frac{2}{16}Ssyy$   or $S\_\_yy$ |
| Wrinkled, yellow | $\frac{3}{16}$ | $\frac{2}{16}ssYy : \frac{1}{16}ssYY$   or $ssY\_\_$ |
| Wrinkled, green | $\frac{1}{16}$ | $\frac{1}{16}ssyy$   **must be** *ssyy* |

## E S S A Y :  *Genome Imprinting*

When Mendel performed his breeding experiments, he didn't know whether the parent plant's sex would affect the traits it passed to its offspring. Just in case, he made **reciprocal crosses,** for example, using white-flowered plants as the female parent in some crosses and as the male parent in others. He found no difference between the offspring of these crosses and concluded that it didn't matter which parent contributed which gene.

Most crosses performed in this century gave similar results. For years geneticists assumed that the copies of a particular allele were equivalent, regardless of which parent it came from. However, we now know that this is not always true: some genes are expressed differently depending on whether they come from the male or the female parent. This is because of a phenomenon called **imprinting:** inactivation of certain genes in the parent's germ cells before these cells undergo meiosis and form gametes. The chromosomes become imprinted differently depending on the parent's sex.

For proper development, a zygote must receive two haploid sets of chromosomes: a paternal set with male imprinting and a maternal set with female imprinting. This was demonstrated by nuclear transplantation ex-

periments using freshly fertilized mouse eggs. Before the egg and sperm nuclei had a chance to fuse, one of them was removed and replaced with another nucleus. Zygotes that had two sets of chromosomes from eggs developed into nearly normal embryos, but their placentas and yolk sacs were severely stunted. Conversely, zygotes that received two sets of chromosomes from sperm nuclei developed into puny embryos with near-normal placentas and yolk sacs. Control zygotes, with one set of chromosomes from an egg and one from a sperm, developed normally.

An individual begins life with one set of male- and one set of female-imprinted chromosomes. How then does it pass on chromosomes that all have the correct imprint for its sex? In the germ cells, the parental imprinting is erased from all the chromosomes, which are then re-imprinted according to the sex of the individual. There is good evidence that imprinting is associated with differences between the two sexes in DNA methylation patterns (the addition of methyl groups to cytosine nucleotides in newly made DNA; methylation plays a role in regulating gene expression). However, this may not be the whole story.

So far the best evidence for imprinting in humans comes from people with two rare genetic abnormali-

ties. As a result of some genetic accident, these people turn out to have two chromosomes #15 from the same parent or to be missing part of that chromosome from one parent. In Prader-Willi syndrome, the individual has two chromosomes #15 from the mother or is missing part of the father's chromosome #15. These people are obese, with poor muscle tone, odd facial structure, and low IQ. In contrast, Angelman's syndrome results either from having two chromosomes #15 from the father, or from missing part of the mother's chromosome #15. These people are severely retarded, with puppet-like movements and bouts of uncontrollable laughter.

Some inherited diseases also show patterns suggesting the influence of imprinted genes. For example, some cancer suppressor genes are thought to be imprinted in paternal chromosomes. This leaves the body's cells with only the maternal cancer suppressor genes active. Any mutation that inactivates this maternal gene moves the affected cell a step closer to becoming cancerous. Juvenile diabetes seems to set in earlier and more severely when it's inherited from the father. A paternal chromosome is also responsible for 90% of cases of juvenile onset of Huntington's disease, which involves deteriorated nerve function (Section 16–B).

---

The Punnett square for a dihybrid cross is somewhat cumbersome. It is sometimes more convenient to use the "branching" method for working out the results of such a cross. This technique takes advantage of the fact that each pair of genes can be considered as a separate cross. So we first set up separate columns for the two crosses: $Ss \times Ss$ and $Yy \times Yy$. If we want to determine the phenotypic ratios, we write 3 smooth and 1 wrinkled under $Ss \times Ss$. Then, under $Yy \times Yy$ we write the phenotypic ratio 3 yellow : 1 green *twice,* once beside "3 smooth" and once beside "1 wrinkled":

Phenotypic Ratios for the Cross $SsYy \times SsYy$:

$$\underbrace{Ss \times Ss}_{\downarrow} \qquad \underbrace{Yy \times Yy}_{\downarrow}$$

3 smooth ⟨ ×—3 yellow = 9 smooth yellow
           ×—1 green  = 3 smooth green

1 wrinkled ⟨ ×—3 yellow = 3 wrinkled yellow
            ×—1 green  = 1 wrinkled green

The procedure for obtaining genotypes for this cross is similar. However, since each of the component monohybrid crosses will produce three separate genotypes, the outcome of the second cross must be written down three times:

Genotypic Ratios from the Cross $SsYy \times SsYy$:

$$\underbrace{Ss \times Ss}_{\downarrow} \qquad \underbrace{Yy \times Yy}_{\downarrow}$$

$$1SS \begin{cases} \times-1YY = 1SSYY \\ \times-2Yy = 2SSYy \\ \times-1yy = 1SSyy \end{cases}$$

$$2Ss \begin{cases} \times-1YY = 2SsYY \\ \times-2Yy = 4SsYy \\ \times-1yy = 2Ssyy \end{cases}$$

$$1ss \begin{cases} \times-1YY = 1ssYY \\ \times-2Yy = 2ssYy \\ \times-1yy = 1ssyy \end{cases}$$

We now know that independent assortment is not the invariable rule. It applies only to genetic loci on different chromosomes. Section 15–F shows what happens if the loci are on the same chromosome.

## 15–E  INCOMPLETE DOMINANCE AND CODOMINANCE

If instead of peas we cross pure-breeding red-flowered and white-flowered snapdragons, all the $F_1$ offspring are pink. If these self-fertilize, the red and white flower colors

reappear in the $F_2$ generation, although half the $F_2$ offspring are again pink. In this case, the red-flower and white-flower alleles show **incomplete dominance**: the heterozygote has a phenotype intermediate between those of the two homozygotes. This happens because pink flowers produce less pigment than red flowers, and white flowers produce none.

As in the case of peas, pure-breeding red-flowered or white-flowered snapdragons are homozygous. In a cross between the two, the $F_1$ plants are all heterozygous for red and white flower color alleles, and they all have pink flowers. Half their gametes will contain the allele for red flowers, and the other half will contain the allele for white. The $F_2$ phenotype and genotype ratios will both be 1:2:1, just the same as the $F_2$ *genotype* ratios for any other monohybrid cross (Figure 15–10). Since the heterozygote produces pink flowers, the expected phenotype ratio for flower color in the $F_2$ plants is:

1 red:2 pink:1 white

The term **codominance** refers to a situation in which both alleles are expressed in the heterozygote's phenotype. A case in point is sickle cell anemia: heterozygous peoples' red blood cells contain both normal and sickle hemoglobin (an oxygen-carrying protein, Section 16–B).

In both incomplete dominance and codominance, heterozygotes have a different phenotype from homozygotes for either allele. These sorts of relationships between alleles are much more common than the dominant/recessive rela-

**FIGURE 15–10**  Incomplete dominance. In snapdragon plants, the red- and white-flower alleles show incomplete dominance: heterozygotes have pink flowers and can be produced by the crosses shown on the right.

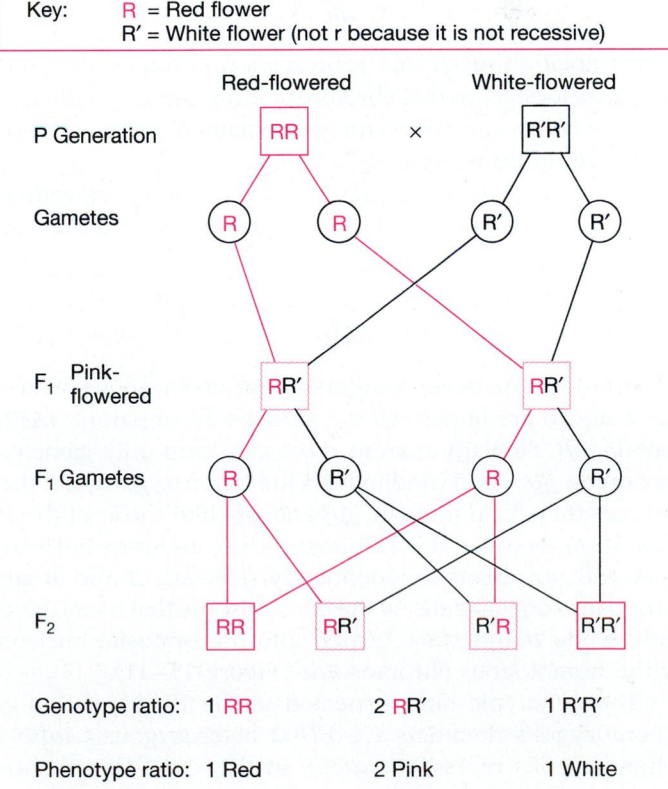

Key:    R = Red flower
        R' = White flower (not r because it is not recessive)

| | Red-flowered | | | White-flowered | |
|---|---|---|---|---|---|
| P Generation | | RR | × | R'R' | |
| Gametes | R | R | R' | R' | |
| $F_1$  Pink-flowered | | RR' | | RR' | |
| $F_1$ Gametes | R | R' | R | R' | |
| $F_2$ | RR | RR' | R'R | R'R' | |
| Genotype ratio: | 1 RR | | 2 RR' | | 1 R'R' |
| Phenotype ratio: | 1 Red | | 2 Pink | | 1 White |

## 15-F  LINKAGE GROUPS

In individuals heterozygous for two pairs of genes (for example, *SsYy*), Mendel found equal frequencies of four gametes: *SY, Sy, sY,* and *sy*. This led to the law of independent assortment. However, later researchers found many pairs of genes that did not assort independently: many more $F_2$ offspring than expected had the two combinations found in the P generation, and many fewer than expected had the other two combinations.

How can we explain this? Looking back at Figure 15-8(b), we can see that the alleles *S* and *s* assort independently from *Y* and *y* because the S and Y loci are on different pairs of homologous chromosomes. But what happens if a cross involves two of the many loci on the same pair of chromosomes?

A chromosome moves as a unit during meiosis. Hence, we would expect loci on the same chromosome to stay together and end up in the same haploid nucleus, rather than assorting independently, as genes on different chromosome pairs do. In other words, they will act as though they are linked, and so they are called **linked loci.** Linkage is common because every chromosome carries many genes.

We can draw two homologous chromosomes from an individual heterozygous at two loci, A and B, like this:

$$\dfrac{A \qquad\qquad B}{a \qquad\qquad b}$$

Geneticists use a shorter way to show these linked loci:

$$\dfrac{AB}{ab}$$

In this notation, each line represents a chromosome, and the alleles located on that chromosome are the letters directly above or below the line. Gametes containing these chromosomes would be written: *AB* and *ab*.

It is also possible to have a heterozygous individual with chromosomes in which *A* and *b* are linked, and likewise *a* and *B*:

$$\dfrac{Ab}{aB}$$

In a cross between parents *AABB* and *aabb* in which loci A and B are linked, all the gametes from parent *AABB* contain *AB*. Similarly, parent *aabb* can form only gametes containing *ab*. The $F_1$ individuals are heterozygous, but the gametes they form must be *AB* and *ab*, like those of the P generation, because *A* is still linked to *B*, and *a* to *b*. These pairs will not assort independently. Instead, *A* and *B* are carried into one nucleus by the chromosome that bears them both, while *a* and *b* are carried into the opposite nucleus by the homologous chromosome (Figure 15-11).

The genotypic ratio expected in the $F_2$ generation is 1 homozygous dominant (*AABB*):2 heterozygous (*AaBb*): 1 homozygous recessive (*aabb*). In this case, the pheno-

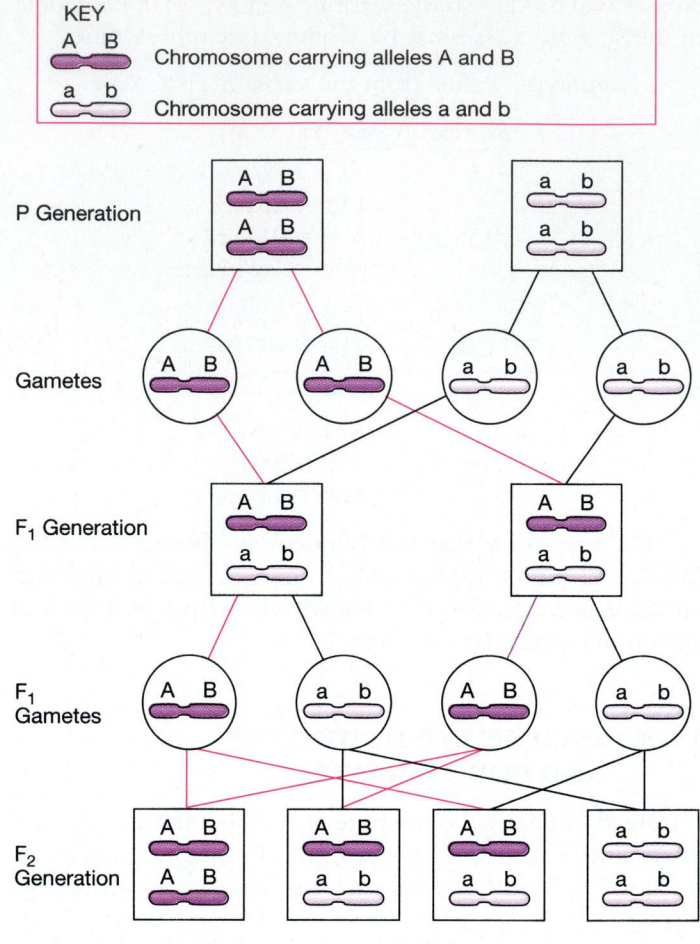

**FIGURE 15-11** Dihybrid cross involving linked loci. Because the A and B loci are on the same chromosome, the alleles on each chromosome move as a unit rather than assorting independently (compare with the S and Y pairs in Figure 15-8b). Hence the genotype ratio in the $F_2$ generation is the same as for a monohybrid cross rather than for an unlinked dihybrid cross (assuming that no crossing over occurs between the A and B loci).

typic ratio is 3 dominant:1 recessive, the same ratio expected of a monohybrid cross. In a sense, this *is* a monohybrid cross, although it involves two gene pairs; the fact that the genes are linked makes them assort like one gene pair.

The genotypic and phenotypic ratios from a cross involving linked loci differ from those expected if the loci were not linked. An excess of parental genotypes is the clue that two gene pairs are linked. Hence, we can identify linked genes by looking at the ratios of offspring obtained in the $F_2$ generation. If the ratio is 9:3:3:1, for a two-character cross, the genes are assorting independently, and they are probably located on different chromosomes. However, if we find a different ratio, we are looking at linked loci.

The term **linkage group** refers to all the loci with inheritance patterns that show they are linked to each other, so a linkage group is really all the loci on one chromosome. This may be hundreds of genes.

## 15-G  CROSSING OVER

As you may expect from studying meiosis in Chapter 14, the genes in a linkage group do not remain together forever. Consider this cross involving linked loci in the fruit fly *Drosophila*. The male was homozygous recessive for two alleles, purple eye (*p*) and black body (*b*), at linked loci. For both traits, the female's phenotype was **wild-type** (the prevailing form in natural populations), but she was heterozygous for *p* and *b*. Hence, her wild-type alleles (*P* and *B*) were dominant to *p* and *b*, as most wild-type alleles are to their less common alleles. The male fly has the genotype $\frac{pb}{pb}$, and so all his gametes carry a *pb* chromosome. The female might be either $\frac{PB}{pb}$ or $\frac{Pb}{pB}$. This mating yielded the following offspring:

| Phenotypes | Genotypes |
|---|---|
| 151 wild-type eye and body colors | $\frac{PB}{pb}$ |
| 8 purple eyes, wild-type body | $\frac{pB}{pb}$ |
| 10 wild-type eyes, black body | $\frac{Pb}{pb}$ |
| 131 purple eyes, black body | $\frac{pb}{pb}$ |

recombinants { 8 purple eyes, wild-type body; 10 wild-type eyes, black body }

Total:  300 offspring

Most of the offspring (the wild-types and those with purple eyes and black bodies) inherited either a *PB* or a *pb* chromosome, respectively, from the female parent. Therefore, these must be the alleles linked on her chromosomes. However, 18 of the offspring inherited chromosomes with new allele combinations (*Pb* and *pB*) not found in either parent. These offspring are known as **recombinants.** Somehow, the alleles for purple eye and for normal body color have become combined on the same chromosome, as have the alleles for normal eye and for black body. This happens when crossing over takes place.

**Crossing over** occurs when homologous chromatids exchange pieces of DNA during meiosis, rearranging alleles that were previously on the same chromosome so that they end up on opposite chromosomes, and vice versa (see Figure 14–12). The resulting recombinant chromosomes bear new combinations of alleles. Research with the fungus *Neurospora* has shown that all four chromatids in a tetrad may cross over and exchange genetic material.

Crossing over involves splicing chromatids at corresponding points, so that they usually do not gain nor lose genetic material. Each crossover produces two complementary

recombinant homologous chromosomes. For example, when crossing over occurs between *AB* and *ab*, both the chromosomes *Ab* and *aB* are produced. These chromosomes eventually end up in about equal numbers of offspring, as in the *Drosophila* example.

If two genes are close together, they will stay on the same chromosome and be inherited together more often than not. However, the farther apart two genes on a chromosome are, the more likely it is that a crossover will occur between them. We can estimate the relative distance between two linked loci by performing a large number of test crosses, counting how many offspring have each phenotype, and calculating the percentage of recombinant offspring. This percentage is taken to be an estimate of the frequency of crossing over:

$$\text{Frequency of recombination} = \frac{\text{number of recombinants}}{\text{total number of offspring}} \times 100\%$$

For our *Drosophila* example:

$$\text{Frequency of recombination} = \frac{18}{300} \times 100\% = 6\%$$

A small frequency of recombination, indicating a short distance, tends to be more accurate than a larger one. This is because some double crossovers may occur between the alleles being studied, and they may end up on the same chromosome even though a segment of the chromosome between them has been exchanged. Consider Figure 15–12:

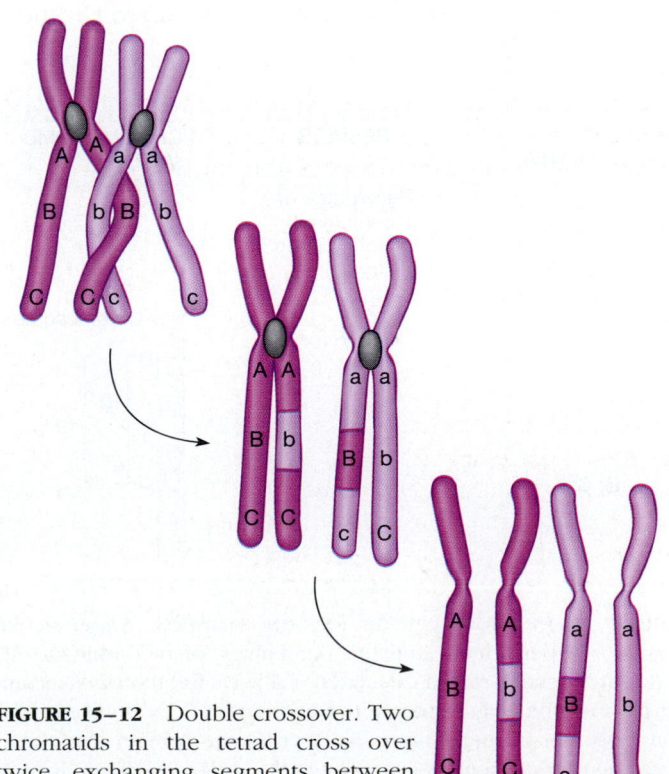

**FIGURE 15–12**  Double crossover. Two chromatids in the tetrad cross over twice, exchanging segments between the A and C loci.

if the *B* and *b* alleles were not present, the double crossover between the A and C loci could not be detected among the offspring of a test cross. These offspring would contain <u>AC</u> and <u>ac</u>, the parental combinations. They would be counted as parental rather than recombinant, giving a lower frequency of recombination than actually occurred. However, following all three loci at once makes the double crossovers detectable because some offspring have the combinations *AbC* and *aBc*, which differ from the parental combinations. Crossovers tend not to occur very close together, so if the loci are very close there is little chance of a double crossover's occurring between them.

## Chromosome Mapping

By compiling crossover data from a large number of crosses, each involving just a few linked loci that are close together, geneticists have been able to construct **chromosome maps** for several organisms. That is, they have determined which loci are together on which chromosomes, in what linear order, and approximately how far apart they are (Figure 15–13).

A 1% frequency of recombination between two loci is said to equal a distance of one **map unit** between them. Thus, two linked loci are ten map units apart if experiments show a 10% frequency of recombination between them. The most accurate maps are those using crosses involving as many different loci on the chromosome as possible.

The first genetic map, for five loci on a chromosome of *Drosophila*, was constructed in 1911 by Alfred H. Sturte-

vant while he was an undergraduate in the laboratory of geneticist Thomas Hunt Morgan.

Thanks to crossover mapping, we now know a great deal about the chromosome maps of such organisms as *Drosophila*, laboratory mice, corn, and the pink bread mold *Neurospora*, all popular subjects for genetic experiments (Figure 15–14). Somewhat different techniques have also allowed geneticists to prepare genetic maps for some viruses and for the circular DNA of some bacteria.

Ethics forbids setting up controlled crosses of humans. Therefore, few features of the human chromosome map were known until recently. However, it is now being filled in rapidly. A large number of researchers are working on the formidable undertaking of mapping the entire human genome, using genetic engineering techniques to copy and sequence segments of DNA from human chromosomes, as well as powerful computers to handle the resulting data (Section 13–E).

By now, thousands of genes have been mapped to their chromosomes in the human genome, and many have been localized to specific parts of their chromosome. One finding that surprised some researchers was that genes with related functions are often located near each other on a chromosome. With the broad outlines of a human chromosome map established, research is filling in more detail between major landmarks already known.

## 15–H  PROBABILITY

The Punnett square is cumbersome to use for a dihybrid cross, and even the "branching" diagram becomes extremely complex for crosses involving three or more gene pairs. It is often more convenient to calculate the probability of particular offspring types directly, instead of drawing all of the possibilities.

The probability (*P*) of a particular outcome of an event is the fraction of times that outcome is expected to occur in the total of all events. For example, in a cross $Rr \times Rr$, the probability of any one offspring's having red flowers is three fourths and the probability of any one offspring's having white flowers is one fourth. In other words, these outcomes are expected to happen three times in every four and one time in every four, respectively. The ratio actually obtained will be close to the predicted probability only when a large number of offspring is obtained.

The probability of one kind of outcome, such as red flowers, is defined as:

$$P = \frac{\text{number of one kind of event}}{\text{total of all events}} = \frac{\text{red flowers}}{\text{red + white flowers}}$$

Rules of probability:

1. All probabilities fall on a scale from zero to one. If an event is certain to occur, its probability is one ($P = 1$).

| BREEDING EXPERIMENTS | RESULTS | CHROMOSOME MAP |
|---|---|---|
| Cross involving genetic loci: | Percentage of recombinant offspring (map distance) | |
| A, B | 10 | |
| A, C | 5 | |
| B, C | 15 | |

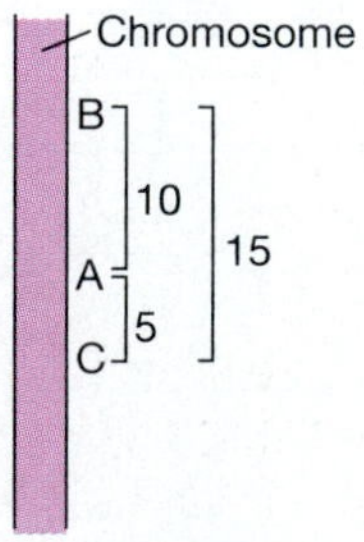

**FIGURE 15–13** How genetic loci are mapped. Experimental crosses are performed, and the percentage of recombinant offspring from each cross is calculated. Each 1% frequency of recombination is arbitrarily set equal to one map unit of distance on the chromosome. By performing many crosses to obtain many map distances, experimenters can deduce the order of genetic loci on the chromosome and their relative distances from each other.

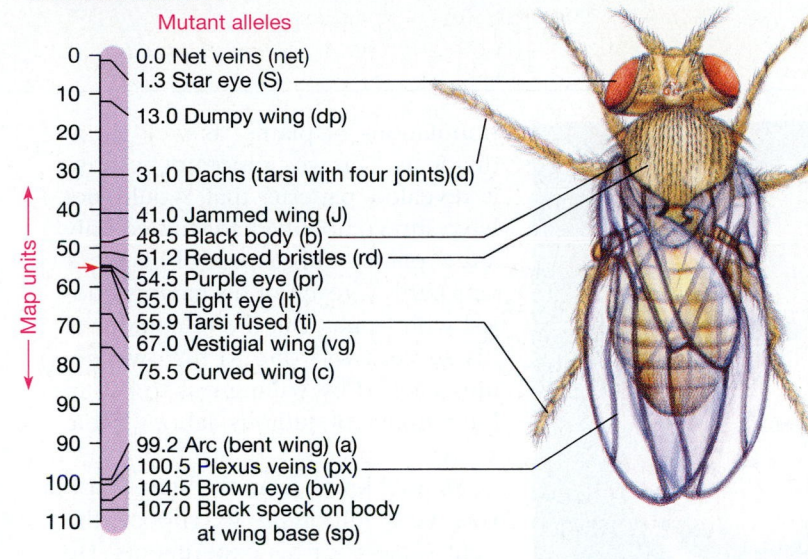

**CHROMOSOME 2**

Mutant alleles

| Map units | |
|---|---|
| 0 | 0.0 Net veins (net) |
| | 1.3 Star eye (S) |
| 10 | |
| | 13.0 Dumpy wing (dp) |
| 20 | |
| 30 | 31.0 Dachs (tarsi with four joints)(d) |
| 40 | 41.0 Jammed wing (J) |
| | 48.5 Black body (b) |
| 50 | 51.2 Reduced bristles (rd) |
| | 54.5 Purple eye (pr) |
| 60 | 55.0 Light eye (lt) |
| | 55.9 Tarsi fused (ti) |
| 70 | 67.0 Vestigial wing (vg) |
| 80 | 75.5 Curved wing (c) |
| 90 | |
| 90 | 99.2 Arc (bent wing) (a) |
| | 100.5 Plexus veins (px) |
| 100 | 104.5 Brown eye (bw) |
| | 107.0 Black speck on body |
| 110 | at wing base (sp) |

**FIGURE 15–14** A map of some of the genetic loci on chromosome #2 of the fruit fly *Drosophila*. The map shows mutant alleles that have been identified at the loci shown, along with their distance in map units from one end of the chromosome. Note that two of the mutations, star eye and jammed wing, are dominant, but that most are recessive. Also note that loci are not arranged in any particular anatomical order, although for the sake of neatness we have drawn lines to the affected body parts so that they don't crisscross. The arrow indicates the position of the centromere.

If it will not occur, its probability is zero. If an event might occur, its probability must be somewhere between zero and one.

2. The probabilities of all possible outcomes of an event must add up to 1, assuming the outcomes are mutually exclusive. For example, an event either will or will not occur, so the probabilities of these two outcomes add to one: (will + will not) = 1.

3. The probability of independent events' occurring together is found by multiplying their separate probabilities.

The easiest way to grasp how these rules work is to apply them to some problems.

**Q.** What is the probability that a tossed coin will turn up heads?
**A.** $P = \frac{1}{2}$, because heads is one of two equally probable results (definition of probability).
**Q.** What is the probability that it will not turn up heads?
**A.** $P = \frac{1}{2}$. Because (heads + not heads) = 1, we start with 1 and subtract the probability of heads: $1 - \frac{1}{2} = \frac{1}{2}$ (rule 2).
**Q.** When a coin is tossed, what is the probability that it will land with either heads or tails up?
**A.** $P = \frac{1}{2} + \frac{1}{2} = 1$, because the only possible result is heads or tails (rule 2).
**Q.** What is the probability that a coin will come up heads two tosses in a row?
**A.** $P = \frac{1}{2} \times \frac{1}{2} = \frac{1}{4}$, because the outcome of each toss is independent of the other.
**Q.** If the probability that you will be on time for class is $\frac{7}{10}$, the probability that you read the assignment is $\frac{8}{10}$, and the probability that the teacher will call on you is $\frac{1}{4}$, what is the probability that you will make a good impression by being prompt and prepared?

**A.** (Rule 3) Because these events are independent of each other, the three probabilities must be multiplied: $P = \frac{7}{10} \times \frac{8}{10} \times \frac{1}{4} = \frac{56}{400} = 0.14$.
**Q.** In a self-cross of a pea plant heterozygous for both seed color and seed texture $SsYy$, what is the probability of an offspring's being smooth and green ($S\_\_yy$)?
**A.** The loci are unlinked, so the chance of obtaining a smooth seed ($S\_\_$) is $\frac{3}{4}$, and the chance of obtaining a green seed ($yy$) is $\frac{1}{4}$. Since these outcomes are independent of one another, they must be multiplied to yield the probability that the seed will be both smooth and green: $\frac{3}{4} \times \frac{1}{4} = \frac{3}{16}$. (Compare with the Punnett square, Figure 15–9.)

To determine the probabilities of a number of events, as is often called for in genetics problems, it is more convenient to use algebra. For example, in determining the possible outcomes of three coins tossed into the air at once, the long way would be to write all the possible combinations of heads and tails:

| | | |
|---|---|---|
| HHH all heads $= \frac{1}{8}$ | | TTT all tails $= \frac{1}{8}$ |

| HHT | | | TTH | |
|---|---|---|---|---|
| HTH | } 2 heads, 1 tail $= \frac{3}{8}$ | | THT | } 2 tails, 1 head $= \frac{3}{8}$ |
| THH | | | HTT | |

The probabilities of these four combinations can be obtained algebraically from the binomial expansion: let $p$ = probability of heads and $q$ = probability of tails; $p = q = \frac{1}{2}$. For two coins, the probability of each combination can be found by squaring the binomial ($p + q$):

$$(p + q)^2 = p^2 + 2pq + q^2.$$

For three coins, the binomial is cubed, and so on:

$$(p + q)^3 = p^3 + 3p^2q + 3pq^2 + q^3$$

*Text continues on page 322*

## E S S A Y :  *Gregor Mendel*

It is one of the ironies of science history that between 1860 and 1900 many biologists (including Charles Darwin) struggled in vain with the question of how characters are inherited, which Gregor Mendel had already solved. Mendel's 1866 paper put forth a simple mathematical model containing the foundations of the rules of heredity. His work has withstood the test of time, although others added to it after it was rediscovered in the first decade of the twentieth century.

Why did it take biologists 34 years to realize that Mendel had solved the problem?

First, Mendel himself may not have realized that he had discovered the general rules of heredity because this was not exactly what he had set out to do. In Mendel's day, plant breeders had begun to study the inheritance of variation in plants such as melons and peas. Artificial hybridization to obtain novel or useful varieties of organisms was practiced by many agriculturists. However, most professional biologists were more interested in evolution and the differences among species. Franz Unger, Mendel's botany professor at the University of Vienna, believed that new species evolved from variants within existing species. Others were investigating the idea that new plant species evolve by hybridization, but their experiments produced confusing results.

Mendel set out to find out how many different types of descendants hybrids could produce, and their proportions, in each generation. In the process he discovered the essential principles of heredity. However, his paper did not consider his results from this perspective. Rather, he focused on the quantitative relationships he deduced and on evolutionary questions of speciation.

**FIGURE 15–A**  Gregor Mendel, 1822–1884. (V. Orel, Mendelianum of the Moravian Museum)

Previous investigators of this problem had no grasp of scientific method. In contrast, Mendel applied the methods he had learned as a student and teacher of experimental physics, especially the need to obtain numerical data from a large sample. He was fascinated with numbers and kept records of weather, sunspots, and other phenomena throughout his life.

Mendel considered the methods he used to be his chief contribution:

- He kept track of which generation each plant belonged to.
- He determined how many different forms (genotypes) of offspring were produced by hybrids and their descendants in each generation.
- He kept track of the ratios among these different forms and used them as clues when genotypes were obscured by dominant alleles.

Because Mendel was interested in species and evolution, he studied populations of plants, as well as individuals. This was important because it revealed patterns that would not have shown up if he had studied only a few plants. This aspect of his work was partly a result of his having studied with Unger, but the remainder of his success was due to his own genius backed by willingness to put in long hours of tedious labor (Table 15–A).

Mendel knew from the start that it was very important to choose the right subject for his experiments. He decided that he needed a plant in which (a) reproduction could be readily controlled, (b) there were pure-breeding varieties with distinctly different traits, and (c) the offspring of crosses between different varieties were just as fertile as their parents.

The reproduction of peas is easy to control because of their flower structure. Most familiar flowers have male parts **(stamens)** and female parts **(pistils)** exposed to the air, and the pistils can receive pollen blown or rubbed off neighboring plants or carried by insects. However, in peas and their relatives a modified petal, the **keel,** completely surrounds the reproductive parts. Pollen from other flowers cannot enter, and each flower normally pollinates itself (Figure 15–B).

Mendel could permit self-pollination or he could cross-pollinate by hand, taking pollen from flowers of one variety of peas and placing it on the pistils of flowers of another variety. In order to do this, Mendel had to open one flower and pluck a stamen. Then he had to open another flower and dust some of the first flower's pollen onto the second flower's pistil. (To be certain that the pistil was not fertilized by pollen from its own flower, Mendel had opened these flowers earlier and amputated the stamens before they produced pollen.)

In order to obtain large numbers of offspring, Mendel hand-pollinated hundreds of flowers. This painstaking labor ensured that Mendel knew the parentage of every seed he collected. Equally tedious but important, he kept meticulous records of these crosses and their outcomes.

Mendel had studied mathematics and probability. Hence, he realized the importance of obtaining a large number of offspring in his experiments to minimize the effects of "sampling error," which may result from looking at too few cases. He also began by studying just one genetic trait at a time, and he followed each trait through many generations. In this way he was able to discern the patterns of dominance and recessiveness, segregation, and the inheritance of one gene for each character from each parent. When he came to study two traits at a time, Mendel's mathematical ability quickly helped him to grasp the essentials of independent assortment.

Perhaps Mendel's most important contribution, however, was his recognition of the discrete nature of inherited characters. This is far from obvious. Many hereditary traits, such as human height, intelligence, and skin color, are continuous over a broad range. Today we realize that this is because each of these traits is controlled by many pairs of alleles, which interact with each other and make each offspring's phenotype somewhat different from those of its parents. However, virtually all biologists before 1900 had the mistaken notion that the inheritable characters of parents blended in their offspring. They also thought that each character was determined by an indefinite number of particles (genes) in each cell. If this

were so, no consistent ratios would ever be found in crosses, and it is hard to see how any theory of genetics could ever be developed.

This blending theory presented a problem for Charles Darwin and other evolutionists. The theory of evolution by natural selection required that inherited variations be maintained from generation to generation, giving natural selection different traits to "choose" among. If variations blended

*(Continued)*

*(Continued)*

## TABLE 15–A

### Summary of Reasons for Mendel's Success

1. Chose pea plants, which normally self-fertilize (simple to use)
2. Used pure-breeding strains
3. Chose traits with distinctly different forms
4. Followed one trait at a time
5. Followed trait for many generations
6. Obtained large numbers of offspring
7. Counted the number of offspring in each category, and analyzed these results mathematically
8. Luck: although some of the pure-breeding strains he used are governed by loci on the same chromosomes, he did not happen to use any of these in the same dihybrid cross. If he had, the linked loci would not have assorted independently, and he would not have obtained typical dihybrid cross ratios in the $F_2$ generation.

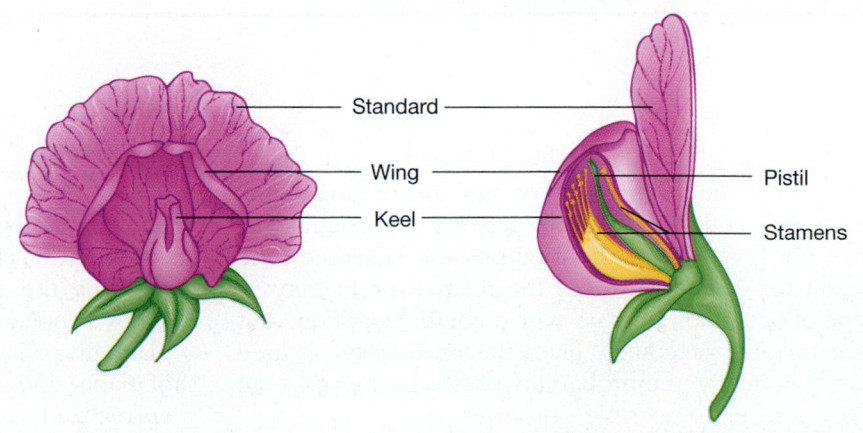

(a) Front view                    (b) Side view, some petals removed

**FIGURE 15–B**  Flower of the garden pea. (a) External view; the petals are modified into shapes which suggest their names: standards, wings, and keel. (b) Cutaway view of flower from one side, showing the position of the reproductive parts (stamens and pistil). Since these parts are enclosed by the keel, the pea flower normally self-pollinates.

*(continued)*

with each other in each generation, they would eventually merge into some great average, and differences among individuals would disappear.

Why did these mistakes persist even after Mendel's work was published? Mendel's modesty did not help. He made little effort to publicize his work and once referred to his seven years' labor, involving more than 30,000 plants, as "one isolated experiment"!

It also seems that even nineteenth-century biologists did not keep up with their reading. Although Mendel published very little, his most important paper went to 115 libraries. Ironically, a copy of Mendel's treatise was found among Darwin's papers, but the pages were still uncut: Darwin had never read it! Darwin never knew that Mendel's work would have removed one of the chief objections to the theory of evolution by natural selection.

A major reason Mendel was ignored was the arrogance of a professional biologist toward an amateur. Mendel sent his paper to an influential botanist, Carl Nägeli, with a cheerful and enthusiastic letter. Nägeli either did not understand Mendel's theory or, more likely, rejected it because it conflicted with his own theory of blending inheritance. Instead of supporting Mendel, Nägeli suggested that Mendel should repeat his pea plant experiments using hawkweeds. We now know that hawkweeds do not always follow the rules of sexual reproduction: they sometimes produce seeds without benefit of pollen, in which case the offspring have no male parent. It would obviously be extremely confusing to try to sort out patterns of heredity if you think you know which plants are parents but really don't! Unfortunately, Mendel took Nägeli's advice, to his great confusion. Nägeli's influential book on evolution and inheritance, published in 1884, did not mention Mendel's work.

In addition to the hawkweed debacle, Mendel attempted to study inheritance in honeybees, which also have an aberrant sexual system (Section 20–F). After two such disasters, his earlier success with peas may well have seemed interesting but a minor fluke, irrelevant to the general problem of inheritance. It would be small wonder if Mendel became discouraged about the true significance of his earlier work on peas.

Mendel became abbot of the monastery where he spent most of his life, a position that left little time for research. For years much of his energy went to resisting a tax on monasteries which he regarded as unjust. It was not until after his death that his work was recognized and became the foundation of modern genetics.

Although we usually think of the basis of Mendelian genetics as being all Mendel's own work, later authorities deserve credit for some of the basic concepts and for almost all the terminology we use today. The terms "dominant" and "recessive" were coined by Mendel, but there is evidence that he didn't have the concept of gene pairs or alleles (although he worked with both and could predict and explain the results of crosses properly). He also phrased his results in terms of combinations of abstract mathematical symbols: he did not indicate any notion of a physical entity responsible for carrying inherited characters from parent to offspring.

---

Here, $p^3$ represents the probability of three heads in a row, $p^2q$ represents the probability of two heads and a tail (in any order), and so on. Adding together the coefficients $(1 + 3 + 3 + 1 = 8)$ gives the total possible outcomes; the coefficient for each term gives the relative probability of that type of outcome. Thus $p^3$, with a coefficient of one out of eight possible outcomes, gives the probability of three heads $= \frac{1}{8}$. Similarly, the probability of two heads and a tail is $3p^2q$, out of eight possible outcomes, or $\frac{3}{8}$.

**Q.** In a cross $Rr \times rr$, what is the probability that the three peas in a pod will grow into two red-flowered plants and one white-flowered plant?

**A.** The cross will produce $Rr$ and $rr$ plants in an expected 50:50 ratio. From the binomial expansion, $\frac{3}{8}$ of three-seeded pods would be expected to produce 2 red-flowered plants: 1 white-flowered plant.

## SUMMARY

The experiments of Gregor Mendel were the foundation of the modern science of genetics. Mendel succeeded in discovering the rules of inheritance largely because of his shrewd choice of an experimental organism, the garden pea plant; his painstaking breeding of large numbers (hundreds) of plants; and his quantitative analysis of results. Later workers realized that the behavior of genetically determined traits in Mendel's breeding experiments paralleled the behavior of chromosomes during meiosis. This parallelism provides part of the evidence that genes are carried on chromosomes.

1. Hereditary traits are determined by discrete units, called genes, which are passed from parent to offspring during reproduction.

2. A diploid plant or animal contains pairs of genes that determine its genetic characteristics.

3. Genes for a trait may occur in different allelic forms. In a heterozygous individual, the two different alleles in a gene pair may interact in different genotype/phenotype relationships:

- ◆ Dominant/recessive. One allele is dominant and masks the presence of the recessive allele.
- ◆ Incomplete dominance. One allele expresses itself, but not as much as two copies of the allele would. The phenotype is intermediate between the homozygous conditions for each allele.
- ◆ Codominance. Both alleles express themselves in the phenotype.

4. During meiosis, the two members of each allele pair separate from one another and pass into different cells (law of segregation).

5. At fertilization, each offspring receives two alleles for each characteristic, one from each parent. Gametes combine randomly with respect to the alleles they contain.

6. The genes from each parent remain distinct in the offspring, and recessive alleles may reappear and express themselves in the phenotype of later generations even if they are masked by dominant alleles in some individuals in intervening generations.

7. During meiosis, the genes of one pair assort independently of genes of other pairs, so long as they are located on different chromosomes (law of independent assortment).

8. Genes located on the same chromosome are linked and are inherited together except when they are separated by crossing over during meiosis.

9. The probability that the offspring of given parents will inherit a particular set of alleles can be determined using a Punnett square, a branching diagram, or the rules of probability.

## Self-Quiz

The following problems will test your understanding of the ideas in this chapter.

1. In humans, the ability to taste phenylthiourea (PTU) is dominant. "Tasters" ($TT$) or ($Tt$) perceive an extremely bitter taste from very dilute solutions of PTU, while "nontasters" ($tt$) experience no sensation even at much higher concentrations.
   a. What are the genotypes of Mr. and Mrs. Gagglebud, who can taste PTU and who have three children, one of whom is a nontaster?
   What offspring phenotypes would be expected from the following crosses, and in what ratios?
   b. heterozygote × heterozygote
   c. homozygous taster × heterozygote
   d. heterozygote × nontaster

2. The allele for axial flowers in peas is dominant to the allele for flowers borne terminally. What phenotypic ratios would you expect among the offspring of a cross between a known heterozygous axial-flowered plant and one whose flowers were terminal?

3. Two *Drosophila* (fruit flies) with normal wings are crossed. Among 123 offspring, 91 have normal wings and 32 have "dumpy" wings.
   a. What inheritance pattern is shown by the normal and dumpy alleles?
   b. What were the genotypes of the two parents?

4. If a dumpy-winged female (from Problem 3) is crossed with her father, how many normal-winged flies will be expected among 80 offspring?

5. A number of plant species have a recessive allele for albinism; homozygous albino (white) individuals are unable to synthesize chlorophyll. If a tobacco plant heterozygous for albinism is allowed to self-pollinate and 500 of its seeds germinate:
   a. how many of these offspring will be expected to have the same genotype as the parent plant?
   b. how many seedlings will be expected to be white?

6. Sniffles, a male mouse with a colored coat, was mated with Esmeralda, an alluring albino. The resulting litter of six young all had colored fur. The next time around, Esmeralda was mated with Whiskers, who was the same color as Sniffles. Some of Esmeralda's next litter were white. (One gene pair determines whether a mouse is colored or albino.)
   a. What are the probable genotypes of Sniffles, Whiskers, and Esmeralda?
   b. If a male of the first litter were mated with a colored female of the second litter, what phenotypic ratio might be expected among the offspring?
   c. What would the expected results be if a male from the first litter mated with an albino female from the second litter?

7. A kennel owner has a magnificent Irish setter, which he wants to hire out for stud. He knows that one of its ancestors was Erin-go-braugh, who carried a recessive allele for atrophy of the retina. In its homozygous state, this gene produces blindness. Before he can charge a stud fee, he must check to make sure his dog does not carry this allele. How can he go about this?

8. In *Drosophila*, the allele for dachs (short-legged, $d$) is recessive to its allele for normal leg length ($D$), and the allele for hairy body ($h$) is recessive to its allele for normal body ($H$). Make a Punnett square for each of the following crosses:
   a. $DdHh × Ddhh$
   b. $DDHh × Ddhh$
   Make a branching diagram for the genotypes of offspring resulting from each of the following crosses:
   c. $DdHh × ddhh$
   d. $DdHh × DDHh$
   e. What proportion of the offspring from cross (b) would be expected to show the normal wild-type phenotype for both traits?

9. A peony plant with straight stamens and red petals was crossed with another plant having straight stamens and streaky petals. The seeds were collected and germinated, and the following offspring were obtained:

    62 straight stamens, red petals
    59 straight stamens, streaky petals
    18 incurved stamens, red petals
    22 incurved stamens, streaky petals

    a. Which allele in each pair (straight vs. incurved stamens, red vs. streaky petals) is dominant?
    b. What were the genotypes of the parental plants?
    c. What further crosses would you make in order to get a definite answer for Part a?

10. In tomato plants, the gene for purple stems ($A$) is dominant to its allele for green stems ($a$), and the gene for red fruit ($R$) is dominant to its allele for yellow fruit ($r$). If two tomato plants heterozygous for both traits are crossed, state what proportion of the offspring are expected to have:

    a. purple stems and yellow fruits
    b. green stems and red fruits
    c. purple stems and red fruits

11. If 640 seeds resulting from the cross in Problem 10 are collected and planted, determine how many are expected to grow into plants with:

    a. red fruit
    b. green stems
    c. both green stems and yellow fruits

12. If one of the parents from Problem 10 is crossed with a green-stemmed plant heterozygous for red fruits, what proportion of the offspring would you expect to have:

    a. purple stems and yellow fruits?
    b. green stems and yellow fruits?
    c. green stems and red fruits?

13. Pooh had a colony of tiggers whose stripes went across the body. His American pen-pal, Yogi, sent him a tigger whose stripes ran lengthwise. When Pooh crossed it with one of his own animals, he obtained plaid tiggers. Interbreeding among the plaid tiggers produced litters of a majority of plaid members, but some crosswise- and lengthwise-striped animals were also produced. Diagram the crosses made by Pooh, showing the genotypes of the tiggers that account for the coat patterns observed.

14. In cattle, the gene for straight coat ($S$) is dominant to its allele for curly coat ($s$). The gene pairs for red ($RR$) or white ($R'R'$) coat color show incomplete dominance; heterozygotes have a roan coat ($RR'$) (red lightened by intermixed white hairs).

    a. If a curly red cow is mated to a homozygous straight white bull, what will the genotype and phenotype of the calf be?
    b. If the calf is mated to a roan animal with curly hair, what are the possible offspring phenotypes?

15. A farmer has three groups of cows: white ones in the clover patch, red ones in the alfalfa field, and roan in the cornfield. He has a roan bull, Ferdinand, who services the cows in all three fields. (Refer to Problem 14 for more information.)

    a. What color calves should he expect in each field, and in what ratios?
    b. Ferdinand dies from a bee sting, and the farmer decides to make his herd of cows exclusively roan coat in memory of his beloved bull. He sells all the red and white cows, and vows to sell any red or white calves born later. What color bull should he buy to replace Ferdinand, if he wants to sell as many calves as possible?

16. The allele for pea comb ($P$) in chickens is dominant to the allele for single comb ($p$), but the alleles for black ($B$) and white ($B'$) feather color show incomplete dominance, $BB'$ individuals having "blue" feathers. If birds heterozygous for both pairs of genes are mated, determine what proportion of the offspring are expected to be:

    a. single-combed
    b. blue-feathered
    c. white-feathered
    d. white-feathered and pea-combed
    e. blue-feathered and single-combed

17. In a plant heterozygous for two pairs of genes ($AaBb$), state the chance that a pollen grain it produces will carry:

    a. an $A$ allele
    b. an $a$ allele and a $b$ allele
    c. an $a$ allele and a $B$ allele
    d. a $B$ allele or a $b$ allele

18. If the plant in Problem 17 self-pollinates, figure the probability that a seed will contain:

    a. two $a$ alleles
    b. an $A$ allele and an $a$ allele
    c. two $a$ alleles and two $B$ alleles
    d. all four alleles ($AaBb$)

19. Mr. and Ms. Miller have two sons. What is the probability that their third child will be a boy?

20. The man in Problem 7 has mated his Irish setter to two bitches known to be heterozygous for the recessive allele for retinal atrophy. Between the two litters, nine pups are obtained, none of which shows retinal atrophy. How certain is the owner that his dog lacks the retinal atrophy allele?

21. In *Drosophila,* the allele for miniature wing ($m$) is recessive to the allele for normal wing ($M$), and the gene for vermilion eye ($v$) is recessive to the allele for normal eye ($V$). A female heterozygous for vermilion eye and miniature wing was mated to a vermilion-eyed, miniature-winged male. The following offspring were collected:

    140 normal wing, normal eyes
      3 normal wing, vermilion eyes
      6 miniature wing, normal eyes
    151 miniature wing, vermilion eyes

    a. What were the linkage groups of the female parent?
    b. What is the frequency of recombination between $v$ and $m$?

22. A female *Drosophila* heterozygous for the recessive alleles sable body ($s$) and miniature wing ($m$, Problem 21) was mated with a sable-bodied, miniature-winged male, and the following offspring were obtained:

    250 normal body, normal wings
     15 normal body, miniature wings
     20 sable body, normal wings
    215 sable body, miniature wings

    a. Diagram the linkage groups of the female parent.
    b. Draw the relative positions of the loci of the $v$, $s$, and $m$ alleles (using also your answer to Problem 21).
    c. What further cross must be made in order to answer Part (b) conclusively?
    d. How could the three loci have been mapped in one experiment?

23. In *Drosophila,* the gene for red eye is dominant to its allele for purple eye, and the gene for long wings is dominant to its allele for dumpy wings. A fly heterozygous for both traits is crossed with a fly having purple eyes and dumpy wings. The $F_1$ offspring are:

    109 red eyes, normal wings
    114 red eyes, dumpy wings
    122 purple eyes, normal wings
    116 purple eyes, dumpy wings

    Would you expect that the two loci involved are on the same chromosome or different chromosomes?

24. A young woman had a brother who died in infancy of a rare genetic disease. The disease is caused by a recessive allele found in the heterozygous condition in about 0.0001 of the population. It is lethal only in the homozygous condition.
    a. What are her chances of having a child with this disease?
    b. This woman is thinking of marrying her first cousin. What are her chances of having a child with the disease if she goes through with this marriage?

25. Lois Lane Kent's husband has passed on to their daughter three strange traits: x-ray vision, sensitivity to the mineral kryptonite, and muscles like steel. A book that Mr. Kent was given by his mother states that the genes for these three traits are dominant and that the vision and muscle loci are 16 map units apart on the same chromosome. The kryptonite locus is on another chromosome. Assuming that an Earthman will someday marry her daughter, Lois is worried about the chances that her grandchildren will inherit the genes that have made their mother such a difficult child to raise. What is the probability that Ms. Kent's grandchildren will have:
    a. all three traits?
    b. x-ray vision and sensitivity to kryptonite but normal muscles?

26. In sweet peas, flower color is determined by two pairs of genes. Plants with at least one dominant allele in both pairs (*P__ C__*) have purple flowers; lack of either *P* or *C* produces white flowers. If a plant heterozygous for both loci self-pollinates, what will be the proportions of purple flowers and of white flowers in the offspring?

## Questions for Discussion

1. In performing a cross to determine the genotype of an organism having a dominant phenotype (*A __* ), why is it preferable to mate it with a homozygous recessive individual rather than a known heterozygote?

2. Figure 15–8 shows that two pairs of chromosomes can line up two different ways at metaphase I. Therefore, an individual heterozygous for two gene pairs on different pairs of chromosomes (*AaBb*) can form four different kinds of gametes. Consider an individual heterozygous for three gene pairs (*AaBbCc*) on three different chromosome pairs. How many ways can the chromosomes line up at metaphase I, and how many different kinds of gametes will be formed? How many different kinds of gametes are possible from a human being, with 23 chromosome pairs?

3. In Figure 15–11, if the chromosome linkages were *Ab* and *aB* instead of *AB* and *ab*, how would the genotype and phenotype ratios of the $F_2$ generation differ?

4. Explain why genes linked at a map distance of 50 units or more behave as if they are on different chromosomes. (Hint: Determine the kinds and proportions of gametes formed by an organism of genotype *AaBb* assuming [1] that the A and B loci are unlinked and [2] that the loci are linked at a map distance of 50 units.) How could you tell that they are on the same chromosome?

5. Evaluate the saying, "alike as two peas in a pod," in light of your study of this chapter.

## Suggested Readings

Harrison, D. *Problems in Genetics with Notes and Examples.* Reading, MA: Addison-Wesley, 1970. A useful review and practice book.

Mendel, G. J. *Experiments in Plant Hybridisation.* Edinburgh, Scotland: Oliver and Boyd, 1965. An English translation of Mendel's original paper, together with comments and a biography of Mendel by others.

Miller, J. A. "Mendel's peas: a matter of genius or of guile?" *Science News* 125:108, February 18, 1984. Discusses the accusation by modern statisticians that Mendel fudged his data.

Sapienza, C. "Parental imprinting of genes." *Scientific American,* October 1990.

Strickberger, M. W. *Genetics,* 3d edition. New York: Macmillan, 1984.

White, R., and J.-M. Lalouel. "Chromosome mapping with DNA markers." *Scientific American,* February 1988.

# Inheritance Patterns and Gene Expression

## O B J E C T I V E S

*When you have studied this chapter, you should be able to:*

1. Explain how mutations may affect the protein encoded by a gene and how this is related to phenotypic expression of mutant alleles.

2. Given data from an appropriate breeding experiment, recognize lethal alleles producing 1:2:1 and 2:1 ratios in crosses between two heterozygotes, and demonstrate knowledge of the inheritance patterns expected from parents carrying lethal alleles by working out crosses correctly.

3. State the possible genotypes of people with blood types A, B, AB, and O, and use your knowledge of these genotypes to solve problems.

4. Explain what is meant by the term multiple alleles and how this differs from polygenic characters; give or recognize examples of each.

5. State the pattern of sex determination (sex chromosome complement of each sex) and inheritance of sex-linked alleles for mammals, birds, and *Drosophila;* use this information in working out sex-linkage problems; and recognize the phenomenon of sex linkage when presented with data showing these patterns.

6. Explain the difference between sex-linked and sex-influenced characteristics, and give examples of each.

7. List at least five factors that may affect the expression of a particular gene in an organism.

8. Describe the inheritance pattern found in the human genetic disorders hemophilia, red-green color blindness, sickle cell anemia, Tay-Sachs disease, and phenylketonuria.

Early in this century, geneticists realized that many abnormalities of humans and other organisms are the phenotypic expression of mutant alleles of genes. Much research has been devoted to finding out how these mutations have changed the DNA, how they are inherited and expressed, how their expression is affected by environmental factors, and how medical treatment can help the individuals who inherited them.

Genes are lengths of DNA that act as units of hereditary information. Many genes code for the sequences of amino acids in polypeptides and proteins. Many mutations change the DNA of a gene in such a way that it ends up coding for a different version of the protein or perhaps no protein

at all. This in turn may alter the organism's metabolism, structure, or physiology, and this is how a mutated DNA sequence makes a difference in an organism's phenotype.

In Chapter 15 we met the basic inheritance pattern for pairs of alleles and saw how the resulting genotype ratios reflect the behavior of chromosomes and genes in meiosis and fertilization. This chapter introduces several variations on this basic pattern, with new genotype and phenotype ratios that provide clues to the nature of particular genes. It also considers how some different alleles produce different phenotypes and looks at some of the factors that control phenotypic expression of genes.

## K E Y   C O N C E P T S

♦ A mutation may result in a change in the protein encoded by the mutated gene.

♦ The effect of a mutation depends on how much it changes the protein encoded by the gene and on how important the protein is to life.

♦ A gene's expression depends on the other genes present in the genome and on factors in the organism's external environment.

## 16–A  PHENOTYPIC EXPRESSION OF MUTATIONS

A **mutation** is a rare, random, and heritable change in a cell's genetic material (Section 9–E). The mutation becomes part of the genotype of the cell and of all its descendants, and this may result in an abnormal phenotype. Mutations in somatic (body) cells may cause damage, including cancer, to the parts of the body that arise from the mutated cells. Mutations in germ (reproductive) cells may cause no noticeable abnormality in the individual in which they occur, but they may be passed on to its offspring and be expressed in the offspring's phenotype.

Probably most mutations have so little effect on a protein and its function that they go undetected without molecular analysis. Most mutations detected by casual inspection have deleterious effects on a protein's structure and function. Individuals with dominant, deleterious alleles are usually removed from the population by selection. Most deleterious mutations that persist in a population, therefore, are those producing recessive alleles (see Figure 15–14).

Often it is difficult to explain on a molecular level why a mutation is recessive, but in some cases the reason is clear. For example, one study analyzed seven different mutations from men with **hemophilia,** a genetic condition in which proteins that cause blood to clot do not work properly. As a result, the person bleeds profusely when injured (Section 16–G). In three of the mutant alleles, amino acid codons had changed to *Stop* codons, and so only part of a protein needed for clotting was produced. Three other mutations had deleted thousands of nucleotides from the gene. In these six cases, it is likely that no functional protein was produced. The seventh mutation had substituted one amino acid for another, resulting in a milder form of hemophilia.

An allele that codes for no protein or for an inactive protein will not be expressed in the phenotype and so will be at least partly recessive to an allele that codes for a functional protein molecule. In a heterozygote, the normal allele directs production of the protein, whereas the recessive allele contributes no functional protein. If enough functional protein is produced, the individual's phenotype is normal, and the normal allele is dominant. In an individual homozygous for the inactive allele, no functional protein is made. Therefore, the trait is not expressed, and the recessive phenotype is the absence of that trait (Figure 16–1). For example, an albino plant results from absence of chlorophyll, and a dwarf plant results from absence of growth hormones.

If an individual with only one copy of the normal allele does not make enough functional protein to produce the normal phenotype, the heterozygote differs noticeably from either homozygote. Hence the normal allele shows incomplete dominance.

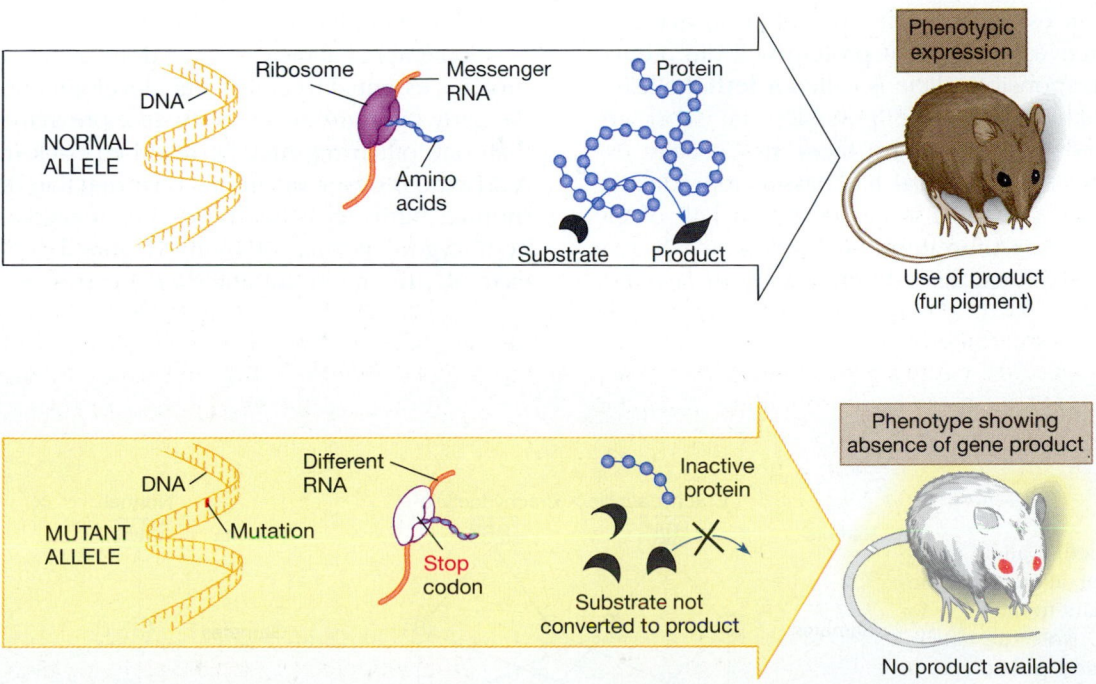

**FIGURE 16–1**  How some mutations may result in recessive alleles. The DNA in this example codes for an enzyme needed to make a pigment in the fur. The mutant makes an incomplete enzyme, and so the animal does not produce the pigment. An albino animal results.

In codominance, both alleles code for functional proteins, but these may have different properties. For example, two alleles may code for enzymes with different sensitivity to temperature or affinity for substrate. As a result, the heterozygote's phenotype may again differ from that of either homozygote.

In the past, it has been very difficult to distinguish an abnormal protein from the thousands of normal proteins in the cells of individuals with a genetic disease (especially if we do not know the protein's function, as is often the case). But genetic engineering is changing that. Once the gene responsible for a disease has been identified, it can be isolated and amplified, then transcribed and translated to produce enough of the abnormal protein to analyze (Section 13–F). The protein's structure and properties may provide clues to the most effective treatment of the disease.

In 1986 researchers used this method to detect the protein responsible for chronic granulomatous disease (CGD), which makes people susceptible to chronic bacterial infections. Eventually, researchers hope to cure CGD by transplanting the allele for the normal protein into people with CGD. Many other genes have now been detected by similar techniques and analyzed to deduce the nature of the proteins they encode. These include one gene responsible for inherited Alzheimer's disease and genes for Duchenne's muscular dystrophy and cystic fibrosis.

## 16–B   LETHAL ALLELES

Suppose a mutation destroys a crucial part of the genetic code for a protein essential to life. An organism that fails to produce an active form of that protein will die prematurely, and the responsible allele is called a **lethal** allele.

Dominant lethal alleles are possible, but most are rapidly eliminated. Exceptions are those not usually expressed until after the individual has passed reproductive age, in which case the allele is passed on to half of the offspring, on average. (An example is Huntington's disease in humans, not usually expressed until age 35 or later.)

Recessive lethal alleles, on the other hand, are eliminated by selection only when they occur in homozygotes. These alleles usually occur heterozygously, masked by a dominant allele that permits the individual to survive and pass on the recessive lethal allele to future generations. A lethal allele may even become quite common if it is closely linked to an advantageous allele of another gene or if the heterozygous condition has some advantage, as in the case of sickle hemoglobin, discussed shortly. It has been calculated that the average human is heterozygous for perhaps three to five lethal recessive alleles. This is part of the reason that marriages between close relatives produce a disproportionate frequency of offspring with lethal inherited traits.

Sometimes just one copy of a normal allele does not make enough of its protein to produce the normal phenotype. In this case the normal allele shows incomplete dominance to the lethal allele, and the heterozygote has a different phenotype from either homozygote. An example in humans is the lethal allele that causes the middle bone in the fingers of heterozygotes to be unusually short, a condition called brachydactyly (brachy = short; dactyl = finger or toe). This makes the fingers appear to have only two bones instead of three. In homozygotes, this allele results in abnormal development of the skeleton. Homozygous babies lack fingers and have other skeletal defects that cause death in infancy.

In a marriage between two brachydactylic people, each child has a one-fourth chance of being homozygous for the lethal allele and dying as an infant; a one-half chance of being a brachydactylic heterozygote; and a one-fourth chance of not inheriting an allele for brachydactyly (Figure 16–2). This 1:2:1 offspring ratio is typical of a monohybrid cross involving incomplete dominance.

Some lethal alleles are mutations of genes that code for proteins essential to embryonic development. Embryos that die early miscarry or, in the case of pregnancies with more than one offspring, may be resorbed back into the uterus. A 2:1 ratio is observed among offspring that develop to term (normal birth age): two-thirds heterozygotes to one-third homozygous normal offspring (Figure 16–3). In mice, for example, the short-tail allele ($T'$) causes early embryonic

**FIGURE 16–2** A lethal allele in humans. The normal allele, *B*, is incompletely dominant to the allele for brachydactyly, *B'*. Note the characteristic genotype and phenotype ratios for crosses of (heterozygous × heterozygous) and (normal × heterozygous).

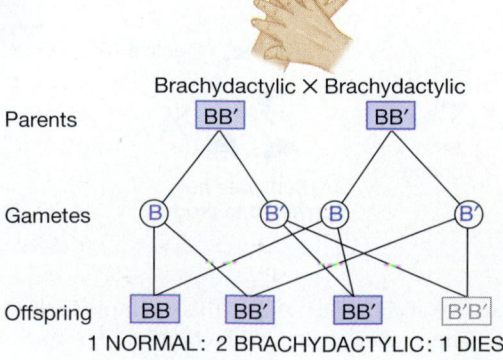

Brachydactylic × Brachydactylic

Parents    BB'    BB'

Gametes    B    B'    B    B'

Offspring    BB    BB'    BB'    B'B'

1 NORMAL: 2 BRACHYDACTYLIC: 1 DIES

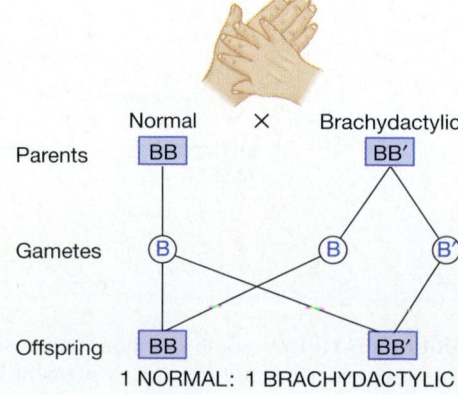

Normal    ×    Brachydactylic

Parents    BB    BB'

Gametes    B    B    B'

Offspring    BB    BB'

1 NORMAL: 1 BRACHYDACTYLIC

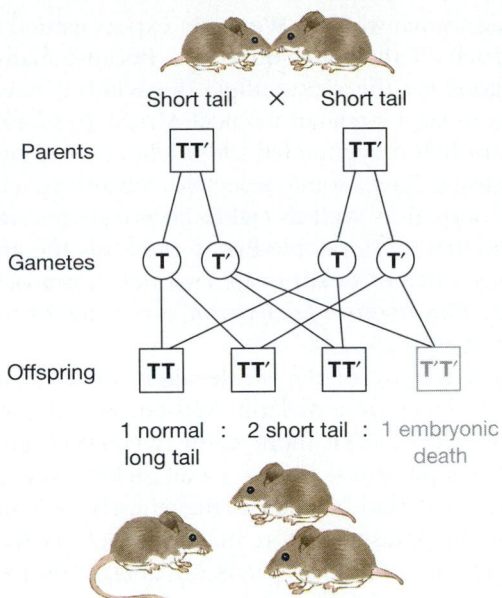

**FIGURE 16–4**  A Manx cat. The shortened spinal column curves upward more than in normal cats, and there is no tail. (Larry Johnson/Johnson Photography)

**FIGURE 16–3**  A cross involving a lethal allele that shows a 2:1 ratio among the offspring. The short-tail allele ( $T'$ ) in mice causes death and resorption of homozygous embryos early in development (gray). These never appear among offspring born to short-tailed parents. One third of the offspring are normal long-tailed, and two thirds are short-tailed.

death in the homozygote. The embryo is then resorbed. If such embryos are taken from the uterus early in pregnancy, before they can be resorbed, they are seen to have no backbone and none of the mesoderm tissue normally destined to form the muscles, kidneys, and many other important organs. Heterozygotes ( $TT'$ ) have shorter tails than wild-type mice ( $TT$ ).

Manx cats are heterozygous for a similar lethal allele. The backbone is so short that the cat has no tail. The last vertebrae of the back and the last part of the digestive tract may be abnormal, and in this case the cat may have problems that prevent it from living out a full nine lives (Figure 16–4).

***Sickle Cell Anemia***    In humans, the allele responsible for sickle cell anemia is often lethal in the homozygous condition. The gene involves codes for the beta ( $\beta$ ) polypeptide chain of hemoglobin, the oxygen-carrying protein found in red blood cells and responsible for their red color (see Figure 3–20). The sickle allele results from a point mutation: a change in just one nucleotide pair, which in this case substitutes valine for glutamic acid as the sixth amino acid in the hemoglobin beta chain (see Table 10–1 and Figure 3–16).

This seemingly small change has drastic consequences. When red blood cells containing sickle hemoglobin are exposed to low oxygen levels, the hemoglobin molecules aggregate and form rigid fibers. These fibers distort the cells into odd shapes, such as sickles (Figure 16–5). The sickled cells become stuck in the capillaries, the narrowest blood vessels, rather than bending and squeezing through in single file as normal red cells do. The stuck cells impede circulation to the areas supplied by the blocked capillaries. The sickled cells also break down easily, leaving the victim with fewer red blood cells than normal, a condition known as anemia. Poor circulation and anemia deprive the tissues of needed oxygen, producing symptoms such as tiredness, headaches, muscle cramps, poor growth, and eventually perhaps failure of organs such as the heart and kidneys.

An individual homozygous for a deleterious recessive allele is an **affected individual,** whereas a heterozygote is a **carrier.** People heterozygous for the sickle allele are sometimes referred to as "having sickle cell trait." This phrase is unfortunate, since it suggests that the carrier is less fit than the normal homozygote, which is not usually the case. The sickle allele occurs most commonly (but not exclusively) in black people. In the United States, about 1 in 400 black newborns is homozygous for the sickle allele.

The sickle and normal alleles are codominant: heterozygotes produce both normal and sickle beta chains. Their red blood cells sickle only when the oxygen level is extremely low. For instance, a study showed that black military recruits who were carriers of sickle cell trait were 28 times more apt to die from the strenuous exercise of basic training than were homozygous normal black recruits. Without special blood tests, heterozygotes such as these may be unaware that they are among the 8% of American black people who carry the sickle allele.

People homozygous for the sickle allele are more severely affected because all of their beta chains are abnormal. About half of them die by the age of 20. Furthermore, women in this group have fewer babies than do heterozygous or ho-

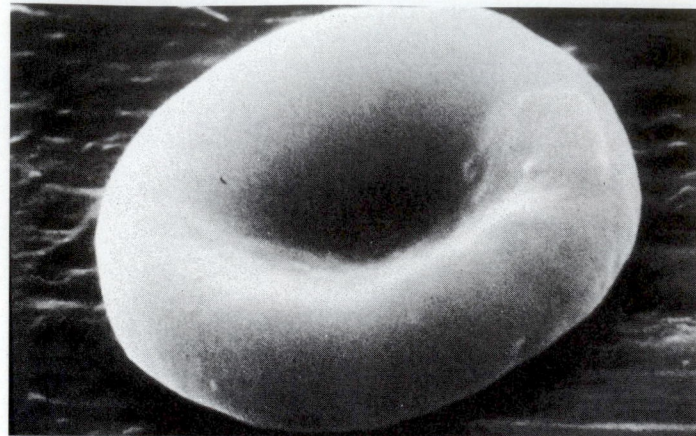

(a)    Normal human red blood cell (SEM)

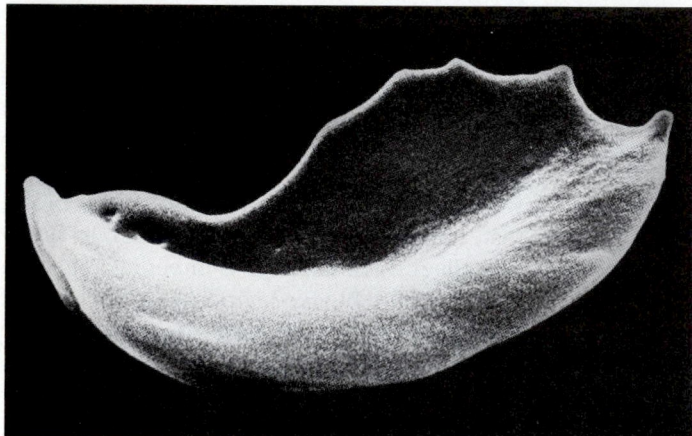

(b)    Mildly sickled red blood cell (SEM)

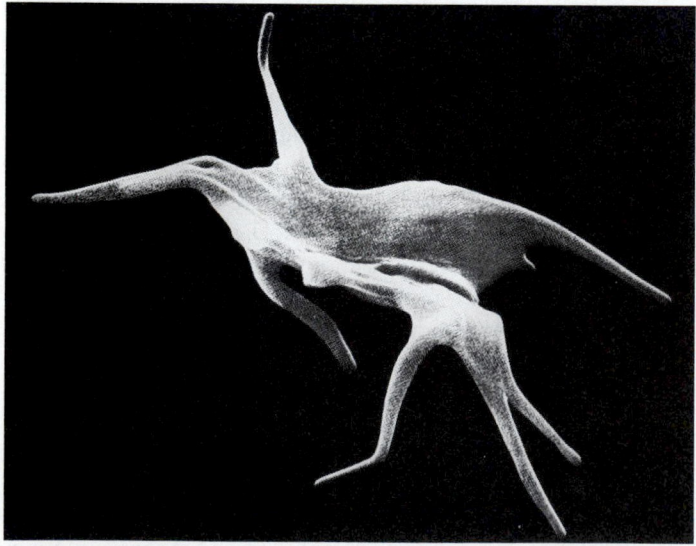

(c)    Severely sickled red blood cell (SEM)

**FIGURE 16–5** Sickle cell anemia. When the blood is low in oxygen, the red blood cells become contorted into unusual shapes. (a, Stanley Flegler / Visuals Unlimited; b, c, courtesy *Johns Hopkins Magazine*)

mozygous normal women. We might expect natural selection to keep such a lethal allele quite rare, because many people homozygous for the sickle allele die without having children. Yet in large areas of tropical Africa, 20 to 40% of the people are heterozygous for the allele. This suggests that heterozygotes have some selective advantage compared with the normal as well as sickle homozygotes. In 1953 it was noted that these people live in precisely the areas with the highest rates of death from a virulent form of malaria, caused by *Plasmodium falciparum,* a parasite of red blood cells (see Figure 25–13).

Having a copy of the sickle allele lowers a person's chances of developing malaria. Red blood cells containing sickle hemoglobin sickle more readily when they are infected with malaria parasites. When a cell sickles, the parasites inside it die. The body's defenses may then be able to destroy the remaining parasites before malaria develops. In malaria-infested regions, therefore, it is advantageous to be heterozygous for the sickle allele, which protects against a common deadly disease, even though the sickle allele is usually lethal in the homozygous state.

The same explanation may account for the high frequency of **thalassemias,** a group of genetic conditions in which too little hemoglobin is produced, in districts of Italy, Greece, and other areas where malaria was once common.

So far there are no effective drugs to prevent sickling in homozygous patients. Genetic engineering provides approaches that could help patients with sickle cell anemia or thalassemia. One way is to try to turn on the genes for gamma ($\gamma$) chains of hemoglobin, normally expressed only in the fetus. If these genes could stay turned on after birth, the gamma chains produced would combine with alpha ($\alpha$) chains and form near-normal hemoglobin. Researchers recently isolated stem cells, which produce all the blood cells, from bone marrow. If a patient's stem cells could be isolated, given transplants of normal beta chain alleles, cultured, and returned to the patient's bone marrow, they would provide a lifelong cure. However, these homozygous patients would still pass on a copy of the sickle or thalassemia allele to each of their children.

***Tay-Sachs Disease***    Tay-Sachs disease, a metabolic disorder resulting in deterioration of the brain and death by about the age of four, is also the result of a lethal recessive allele. A homozygous recessive child lacks the enzyme hexosaminidase, which metabolizes a lipid in the brain's nerve cells. Without this enzyme, the lipid accumulates and destroys the cells' ability to function. So far, this condition is untreatable, but genetic tests that detect it very early in embryonic development are now widely used. The highest frequency of this allele occurs among people of East European Jewish extraction: one in 30 members of this group is a carrier (heterozygous) for this disorder. However, about one third of the Tay-Sachs cases in the United States are among non-Jewish people.

***Cystic Fibrosis***   The most common recessive lethal alleles in the Caucasian population of the United States are those responsible for cystic fibrosis. About 1 in 20 Caucasians is a carrier, and 1 in 2000 babies is affected by this disorder. This disease is caused by a defect in a protein that controls a membrane channel for the movement of chloride ions. Normally, chloride ions leave cells in the lung surface via these channels, and water follows by osmosis.

In cystic fibrosis victims, the movement of chloride (and hence of water) is impeded. As a result, mucus on the lung surface is unusually thick, because it contains little water. Victims suffer from poor lung function, and many die of respiratory infections by age 21, with few surviving past age 30. In addition, thick mucus in the digestive tract may interfere with digestion and absorption of food, and victims may become poorly nourished despite an adequate diet. The defective channel activity also produces abnormally salty sweat, which is often the first clue in diagnosing cystic fibrosis.

In 1989 cystic fibrosis alleles and their normal counterparts were isolated and analyzed. About 70% of cystic fibrosis patients have an allele missing the code for one amino acid (phenylalanine) in the protein. About 40 other defects occur among the remaining cystic fibrosis alleles. Genetic engineers are experimenting with gene therapy techniques to transplant normal alleles into lung cells using vectors such as viruses or liposomes (Sections 13–I and 4–F).

***Huntington's Disease***   Huntington's disease is unusual because it is caused by a dominant allele. Hence, a person with only one allele for this condition will develop the disease. Its symptoms include involuntary twitching, degeneration of part of the brain, depression, and irritability. The disease progresses for 10 to 20 years, finally causing death. The first symptoms usually do not appear until the person is 35 to 45 years of age. By that time, most victims have children, who in turn have a 50:50 chance of having inherited the allele.

In 1993 researchers pinpointed the Huntington locus, which had long been known to lie near one tip of chromosome #4. Normal alleles at this locus have 11 to 24 repeats of the base sequence CAG, whereas in patients with Huntington's disease the allele contains 42 to 86 of these repeats. Furthermore, the more repeats, the earlier in life disease symptoms appear.

With the locus identified, researchers must now determine the gene's function, what goes wrong in Huntington patients, and what treatments might help them.

Extra repeats of other base sequences have also been found in some other genetic conditions, including fragile X syndrome, which causes a form of mental retardation, and some breast, bladder, and colon cancers. It will be interesting to see whether the extra repeats have the same kind of effect in all these cases.

***Marfan's Syndrome***   Marfan's syndrome is another condition that is sometimes lethal. It too is caused by a dominant allele, in this case coding for a defective structural protein: fibrillin, a component of fibers in the extracellular matrix that knits connective tissue together. People with Marfan's syndrome are often tall, long-limbed, and loose-jointed, with the lenses of the eyes dislocated and with weak walls in the aorta, the main blood vessel carrying blood from the heart to the body's tissues. The cause of death is often rupture of the aorta during strenuous physical exertion. Marfan's syndrome affects 1 in every 10,000 people, many of whom do not realize they have it. It is conjectured that Abraham Lincoln had this condition.

These examples show that defects in genes encoding many kinds of proteins can be life-threatening.

## 16–C   INBORN ERRORS OF METABOLISM

Many genes code for proteins that are enzymes for a step in one of the body's metabolic pathways. When such a gene mutates, the new code may produce a defective enzyme unable to carry out its metabolic reaction at a normal rate. The resulting genetic abnormality is an **inborn error of metabolism.** The metabolic disorder of Tay-Sachs disease is lethal, but others are less severe, and some do little or no apparent harm to affected individuals.

**Phenylketonuria (PKU)** and **albinism** are two human hereditary disorders resulting from defective alleles for enzymes that happen to be on the same metabolic pathway (Figure 16–6).

PKU-affected individuals are homozygous recessives who lack the enzyme that normally converts the amino acid phenylalanine to another amino acid, tyrosine. Without this enzyme, phenylalanine builds up, perhaps to 50 times its normal level. Minor metabolic pathways convert some of this phenylalanine to various other products, such as phenylpyruvic acid, which is excreted in the urine, giving it a characteristic odor.

High concentrations of phenylalanine and its products inhibit the activity of many metabolic enzymes. This damages various organs, especially the brain, and without treatment children with PKU become mentally retarded.

PKU can now be controlled by a special diet low in phenylalanine during childhood. This prevents most brain damage, but some patients may still have learning disabilities. Since this treatment must begin within a few weeks of birth, many states now require that newborns receive a blood test for PKU (and for several other metabolic disorders). When brain development is complete, PKU patients can adopt a normal diet.

If a woman homozygous for PKU becomes pregnant, the high phenylalanine level in her blood is transferred to the fetus through the placenta. This puts the fetus at risk of mental retardation or microcephaly (small head). Some

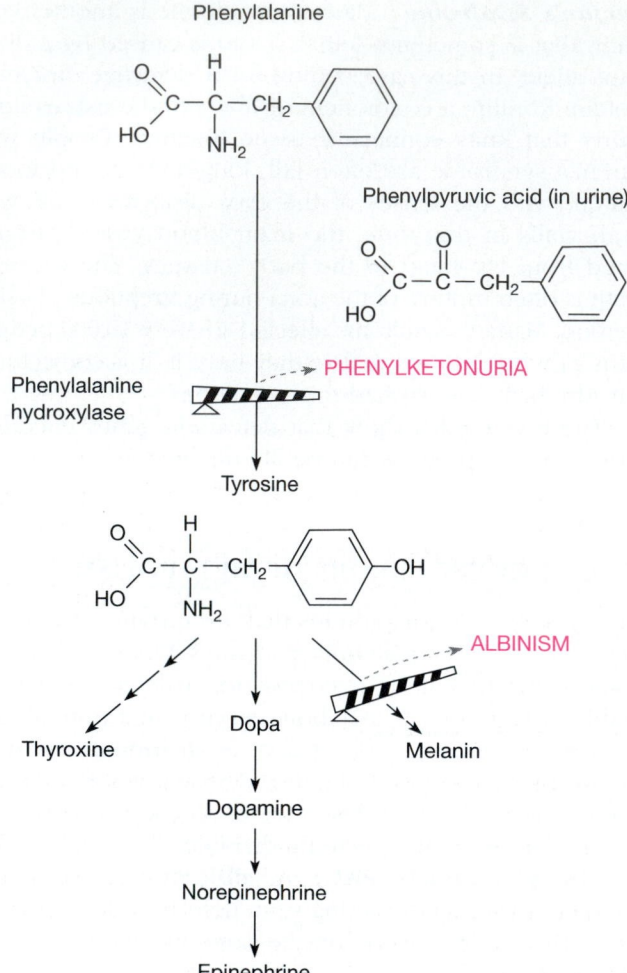

Phenylalanine

Phenylpyruvic acid (in urine)

PHENYLKETONURIA

Phenylalanine
hydroxylase

Tyrosine

Thyroxine

Dopa

ALBINISM

Melanin

Dopamine

Norepinephrine

Epinephrine

**FIGURE 16–6** Inborn errors of metabolism. This diagram shows the metabolic pathway that converts the amino acid phenylalanine to tyrosine, which in turn can be converted to several other substances. The solid arrows are enzyme-catalyzed steps in the pathway. "Metabolic blocks" (shown as striped barriers) result from the absence of the corresponding enzymes. This interrupts the pathway at that point and leads to abnormal genetic conditions, such as phenylketonuria and albinism (dashed arrows).

**FIGURE 16–7** Albinism, a homozygous recessive genetic condition. Here an albino peacock spreads his tail in a mating display while a normally colored male strolls past.   (D. J. Cross/BPS)

that normally converts tyrosine to melanin. People with the other common kind of albinism are homozygous recessive for an abnormal allele of a different gene; these people do make the tyrosine-to-melanin enzyme, but for unknown reasons this enzyme produces almost no melanin pigment in their bodies. Some marriages between two albino people have produced normally pigmented children, indicating that one spouse was homozygous recessive for the first allele, and the other spouse was homozygous recessive for the second. If both spouses are homozygous recessive for the same allele, their children are all albino.

You may wonder whether victims of PKU are also albino, since they cannot make the tyrosine that is eventually converted to melanin. The answer is no, because tyrosine can be obtained in the diet as well as from conversion of phenylalanine. However, people homozygous for PKU usually have light coloring because phenylalanine products inhibit the pigment-forming enzymes. Of course, a person could be homozygous recessive for both PKU and albinism.

## 16–D   MULTIPLE ALLELES

Up to this point, we have talked as if either of two alleles can occupy a particular locus on a chromosome. This is an oversimplification. A gene contains hundreds of nucleotides, and changes in different parts of the gene in different individuals mean that the population contains **multiple alleles** at that locus. Each allele may code for a different version of a protein (or for no functional protein) and so produce a somewhat different phenotype. Any one individual can contain no more than two of the many alleles of a gene, one on each chromosome of the homologous pair carrying that genetic locus. Various pairings of all these

such women have returned to a low-phenylalanine diet during pregnancy, but it is not yet clear whether this eliminates the risks to the fetus. Since the mother is homozygous for PKU, her children must inherit one copy of the recessive allele from her. Hence, they will all be PKU carriers (or homozygotes if they also receive a PKU allele from their father).

Albinism is a condition characterized by absence of melanin, the dark pigment that makes eyes, hair, and skin brown or black. True albinos have white hair (or feathers, as in Figure 16–7) and very light skin and eyes. There are two common types of albinism in humans. In one form, people homozygous for a recessive allele lack an enzyme

alleles produce an array of genotypes and phenotypes in the population as a whole.

## Human ABO Blood Groups

A familiar case of different phenotypes resulting from multiple alleles is that of the human ABO blood groups, with three common alleles: $I^A$, $I^B$, and $i$. $I^A$ and $I^B$ code for two different enzymes, each of which attaches a different sugar to a protein on the surface of red blood cells. The enzyme coded by $I^A$ attaches acetylgalactosamine (a derivative of the sugar galactose), forming antigen A, whereas the $I^B$ enzyme attaches galactose instead, forming antigen B. (An antigen is a substance that provokes an immune response when it is introduced into an individual normally lacking that antigen, as during a blood transfusion.) Both enzymes are produced in a person having both the $I^A$ and $I^B$ alleles. That is, $I^A$ and $I^B$ are codominant, and people with both alleles have red cells bearing both antigens. The $i$ allele does not code for an enzyme; it is recessive to both $I^A$ and $I^B$. Table 16–1 shows possible ABO genotypes and phenotypes.

Table 16–1 also shows the serum antibody proteins found in people of each blood group. For example, people without antigen A on their red cells produce anti-A antibody in their blood. If they are given a transfusion of blood with cells bearing antigen A, their anti-A antibody binds these foreign red cells into clumps, which block parts of their circulation. This is why blood types must be matched when a person receives a blood transfusion.

The ABO and rhesus (Rh positive or negative) blood groups are the best known and the most important medically because incompatible ABO and Rh blood groups should not be mixed when giving blood transfusions. However, more than 20 different human chromosome loci are known to carry genes coding for various blood proteins.

In the past, blood groups were sometimes used to decide questions of parentage, such as in paternity lawsuits or suspected mixups of babies in a hospital. Only a few drops of blood are needed to determine the blood types of the child and its supposed parents. This genetic evidence reveals whether a particular person or couple could have had a child of a particular blood type. Such evidence can never be used to determine that a particular person definitely is the parent of a particular child, but it will often rule him or her out. For instance, a man with blood type AB has the genotype $I^AI^B$, and so he could not have fathered a baby with blood type O. A baby with blood type O must have the genotype $ii$, and its father must therefore have at least one $i$ allele to pass on (see Table 16–1). If the baby's blood type is A or B, this man could have been its father. However, many other men in the world have blood types such that they could have fathered the baby, and so the baby's parentage can never be established conclusively based on ABO blood group evidence alone.

Modern paternity testing uses DNA fingerprinting or proteins coded by other series of multiple alleles, which are more variable than the ABO series, in fact, so much so that there is almost no chance of two people having identical genotypes (except identical twins).

## 16–E POLYGENIC CHARACTERS

A **polygenic character** is a phenotypic trait influenced by more than one pair of genes, which occur at two or more different loci, on the same or different chromosome pairs (Figure 16–8). Familiar examples in humans include height, intelligence, body build, and hair and skin color, all determined by the interactions of many genes.

Human skin color is a polygenic trait. There is some debate whether three or four different gene pairs are involved. Very dark-skinned people have alleles coding for production of melanin at all their skin-color loci, whereas light-skinned people have alleles not contributing to melanin production at these loci. For example, if three gene pairs are involved, a very dark-skinned person would have six alleles for melanin (*AABBCC*), whereas a very light-skinned person would have none (*aabbcc*). The alleles are thought to be additive: the more alleles for melanin production a person has, the more melanin is produced, and the darker the skin.

A similar example is known in wheat. Gene pairs at three loci determine the color of the kernels. Figure 16–9 shows a cross of two strains of wheat that differ in these three gene pairs: *AABBCC* × *aabbcc*. Again, the alleles have additive effects: each capital-letter allele (*A*, *B*, or *C*) contributes a small increment to kernel pigmentation, and the kernel-color phenotype is determined by the total number of these similar alleles at the three different loci.

Most polygenic characters result from expression of many gene pairs whose products interact in some way. For example, the genetic basis of intelligence is not "smart" versus "stupid" alleles but rather various alleles at loci encoding proteins that take part in the growth and metabolism of the brain's nerve cells. Each protein may make

**TABLE 16–1**

**Genotypes and Serum Antibodies in Human ABO Blood Groups**

| Blood Group | Genotype | Serum Antibody |
|---|---|---|
| A | $I^AI^A$ or $I^Ai$ | anti-B |
| B | $I^BI^B$ or $I^Bi$ | anti-A |
| AB | $I^AI^B$ | none |
| O | $ii$ | anti-A and anti-B |

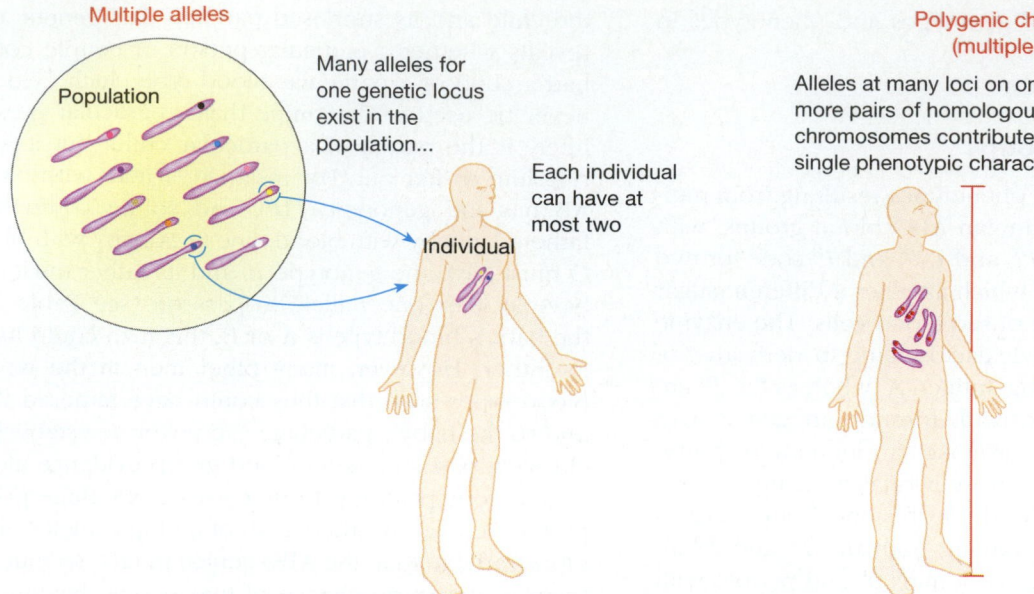

**FIGURE 16–8** Multiple alleles and polygenic characters. In the case of multiple alleles, many alleles exist at the same chromosomal locus in different members of the population, with a maximum of two of these alleles in any one individual. Polygenic characters involve many loci that all affect the same phenotypic character. In this example, one individual has alleles that all produce more growth in height, and so this person is taller than the one whose alleles produce less growth.

only a small contribution to the phenotype we call intelligence, and it may be difficult to determine which loci play a role and their relative importance.

As is the case for many genetic traits, environment may further muddy the waters by affecting the expression of polygenic characters. For instance, whether a person has light or dark skin, the color can become lighter or darker depending on exposure to the sun. Traits such as human height and the size of an ear of corn are strongly influenced by environmental factors, such as nutrition, as well as by the many genes that determine the possible range of variation in the phenotype.

Because polygenic traits vary over a wide range, and offspring phenotypes are often intermediate between those of parents from opposite extremes, these traits frequently appear to blend. For example, human height varies continuously over a wide range, rather than by, say, 2-centimeter steps, and a tall and a short parent often have children of medium height. Polygenic characters such as this are so common that early observers were misled into believing that characters of parents blend in their offspring.

## 16–F   SEX DETERMINATION

In many species, the most obvious difference in phenotype between individuals is their sex. It is also one of the most influential differences because sex hormones affect the ex-

pression of many other genes, including some not directly involved in sexual reproduction. In most familiar animals, sex is genetically determined, and about half the individuals are male and half female.

The simplest cross to produce half males and half females (a 1:1 ratio in the offspring) is one between a homozygote and a heterozygote. In many species, this is what happens: one sex is heterozygous and one homozygous for a pair of chromosomes, called the **sex chromosomes.** In humans and most other mammals, males have the pair of different sex chromosomes: one X and one Y chromosome. Females have two X chromosomes. Many insects, including the fruit fly *Drosophila,* also have XX females and XY males. Higher plants often have reproductive organs of both sexes in the same individual, but those with separate sexes tend to have XX females and XY males.

During the formation of gametes in a man (or in XY males of other species), the X and Y chromosomes segregate and end up in different sperm. All eggs produced by an XX female contain one of her two X chromosomes. At fertilization, an X-bearing egg may combine with an X-bearing sperm (forming a female) or with a Y-bearing sperm (forming a male). This is why we say that the father determines the sex of the baby. Females are **homogametic,** formed from an egg and sperm bearing the same kind of sex chromosome (X); males are **heterogametic,** resulting from union of an X-bearing egg and a Y-bearing sperm (Figure 16–10).

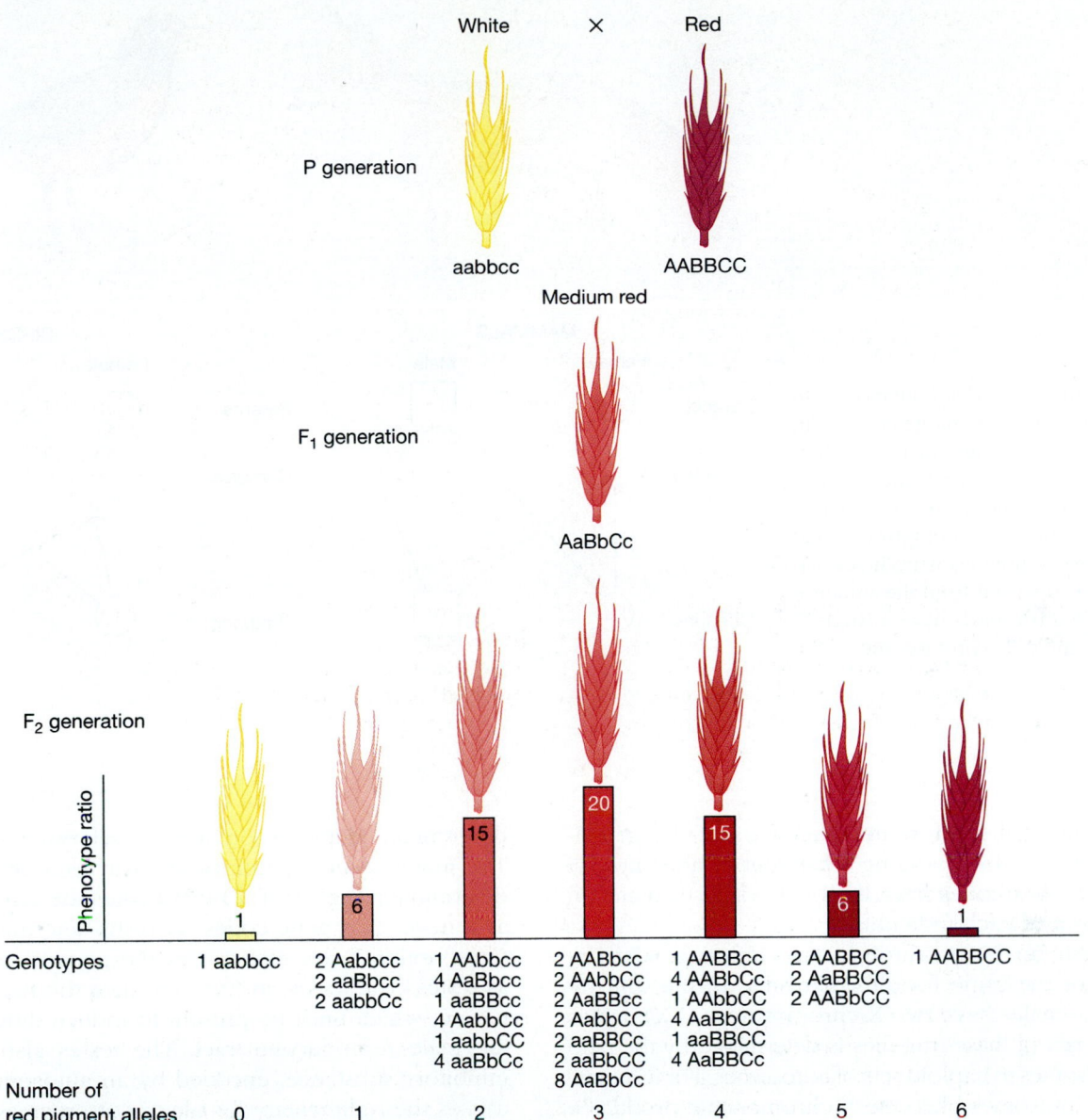

**FIGURE 16–9** A cross involving a polygenic trait. The two parental strains of wheat differ in three gene pairs for kernel color. One strain has six alleles for uncolored kernels *(aabbcc),* and the other strain has six alleles for red color in the kernels *(AABBCC).* F₁ plants have three red-color alleles *(AaBbCc),* and so their color is midway between those of the two parental strains. F₂ plants have kernels of seven different colors (white and six shades of red), depending on whether they have zero, one, two, three, four, five, or six red-color alleles.

Birds and some reptiles are the other way around: females are the heterogametic sex, with one Z and one W chromosome, whereas males are homogametic ZZ. So in these species the sex chromosome carried by the egg determines the offspring's sex. Moths, butterflies, some fish, and amphibians are like birds in having ZZ males and ZW females.

Having a particular sex chromosome complement causes an embryo to develop into a male or female by the process of **sex differentiation.** Although one might assume that the sex chromosomes carry the genes involved in sex differentiation, the story is not so simple. Sex is a polygenic character, with at least 19 participating loci. Most of the genes involved, such as those needed to produce the sex hormones, lie on the **autosomes,** those chromosomes that are not sex chromosomes. This means that all individuals have most of the genes needed to develop into a member of either sex. For example, embryonic birds and mammals

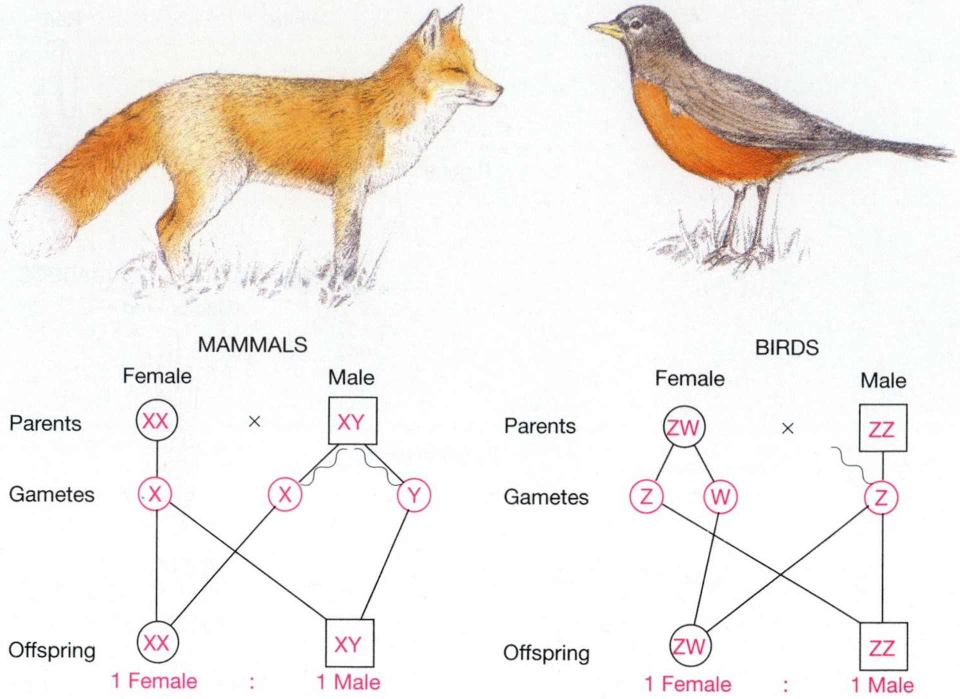

FIGURE 16–10 Sex determination. In mammals, females are homogametic, with a pair of X chromosomes; males are heterogametic, with one X and one Y chromosome. In birds, males have a pair of like chromosomes, called Z to emphasize that the male/female homogametic/heterogametic system is reversed from the situation in mammals. Female birds have a single Z chromosome and a W chromosome.

have been induced to grow into members of the genetically "wrong" sex by hormone treatment. Under normal circumstances, hormones from the ovary or testis maintain the correct sex of each individual.

How, then, do the sex chromosomes influence sex? The answer is not the same for all organisms. In the fruit fly *Drosophila*, females have two X chromosomes (XX); males are XY. The sex of these fruit flies is determined by the ratio of X chromosomes to haploid sets of autosomes. For instance, two sets of autosomes plus one X chromosome produce a normal male; two sets of autosomes plus two X chromosomes produce a normal female, whether or not a Y chromosome is also present:

Male      2 sets autosomes + XY (or X; or XYY)

Female    2 sets autosomes + XX (or XXY)

X chromosomes carry mainly genes that cause femaleness, whereas autosomes tend to cause maleness. A *Drosophila* female's second X chromosome ensures that she will be female. A fly with two sets of autosomes + XXX is an abnormal, often sterile, superfemale. The Y chromosome is needed for production of fertile sperm but not for sex determination.

A different mechanism of sex differentiation operates in humans and most other mammals. The early embryo develops rudimentary "indifferent gonads," which can become either testes or ovaries (Figure 16–11). Which way they

differentiate depends on the sex chromosomes present. The mammalian Y chromosome carries a gene, *SRY* (*Sex-determining Region of Y*) that makes the embryo develop as a male. This gene codes for testis-determining factor, a transcription factor that causes the gonads to differentiate as testes. The testes, in turn, produce the hormone testosterone, which must be present to induce differentiation of the male reproductive tract. The testes also produce an inhibitory substance, encoded by an autosomal gene, that causes the rudimentary female tract to regress.

In human embryos, the testes begin differentiating in the sixth week of development. If this does not occur, the gonads differentiate into ovaries in the following week. In this case, the rest of the female reproductive tract develops automatically, without hormonal signals from the ovaries. If a mammalian embryo's ovaries or testes are removed before the reproductive tract differentiates, the embryo develops a female tract.

So, in mammals, the Y chromosome carries at least one gene that gives the embryo its first "push" toward becoming a male. Having a Y chromosome leads to maleness, whereas embryos with only X chromosomes become phenotypically female. This is true even in the rare cases of people born with more or fewer than two sex chromosomes (Table 16–2). These individuals result from failure of the sex chromosomes to segregate normally during meiosis, forming gametes with one too many or one too few sex chromosomes.

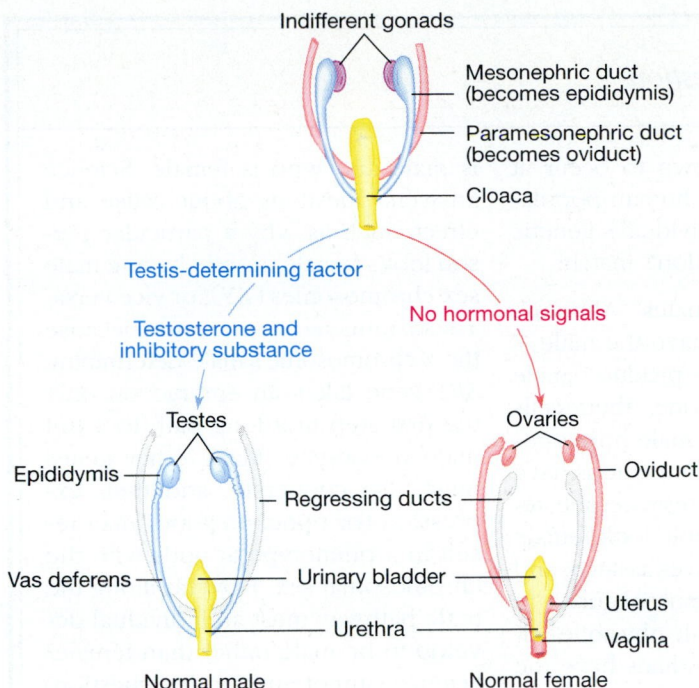

**FIGURE 16–11** Sex differentiation in human embryos. Early embryos have indifferent gonads. Near them lie two sets of undifferentiated ducts (mesonephric and paramesonephric ducts). The protein encoded by the Y chromosome's *SRY* gene causes the gonads to develop into testes. The testes produce the male hormone testosterone, which stimulates the mesonephric ducts to differentiate into the male reproductive tract (the epididymis and vas deferens). Testes also produce an inhibitory substance that causes regression of the paramesonephric ducts. Without these chemical stimuli, the gonads develop into ovaries and the paramesonephric ducts into the oviducts of the female tract. The mesonephric ducts also regress.

**TABLE 16–2**

**Phenotypes for Various Sex Chromosome Complements in Humans**

| Sex Chromosomes | Phenotype* |
| --- | --- |
| XX | Normal female |
| XY | Normal male |
| XXX | Female; fertile or sterile; usually normal |
| X (Turner syndrome) | Female; short; sterile, no ovaries |
| XXY (Klinefelter syndrome) | Male; sterile; possible mental retardation |
| XXXY | Male |
| XYY | Male; tall, acne-prone; fertile |

*Defects in various genes involved in hormone production can alter the phenotype normally exhibited by a particular sex chromosome combination.

Various unusual methods of sex determination occur in some organisms. In Hymenoptera (the group of insects including wasps, bees, and ants), females are diploid but males are haploid, developing from unfertilized eggs.

In some animals, sex is determined by environmental factors. In the American alligator, snapping turtle, and several other reptiles, sex is determined by the environmental temperature during a particular stage of embryonic development. This discovery led to speculation about the mysterious extinction of the dinosaurs and other ancient reptile groups. If these reptiles had a similar mode of sex determination, a prolonged change in climate could have resulted in production of offspring of only one sex. This would have doomed any species whose members could reproduce only sexually.

Environmental determination of sex may prove useful to animals that cannot move far to find a mate. For instance, in the marine worm *Bonellia viridis,* the developing larva swimming in the sea belongs to neither sex. Eventually it drifts to the bottom and becomes an adult. If it settles down alone, it develops into a relatively large female, but if it lands near an existing female, it is attracted to her, and she produces a chemical that causes the larva to develop into a microscopic male. The male migrates into the female's excretory organ and lives there as a parasite.

The snail-like slipper shell, *Crepidula,* lives in stacks of individuals. Young individuals are males, which turn into females as they grow larger and older. The male reproductive tract degenerates, and a female tract develops in its place. Chemicals appear to influence sex determination in this situation as well: if a stack consists entirely of males, some of them will turn into females. In contrast, some fish, such as the saddleback wrasse and bluestreak cleaner, may start out as females, but when there are no larger males around, the largest females become males. Some female frogs have also changed into fertile males when they were held in all-female tanks in the laboratory. All of these systems guarantee that when two or three are gathered together, some will be male and some female.

## 16–G   SEX LINKAGE

Like other chromosomes, the sex chromosomes carry genes. A number of genes have been found on the human Y chromosome, including *SRY,* coding for testis-determining factor; one for H-Y antigen (or its regulatory protein), a cell-surface protein expressed in males; and one that affects size of the teeth. H-Y antigen or a closely linked gene is thought to control spermatogenesis.

These genes have no homologues on the X chromosome, but at least one quarter of the Y chromosome is homologous to part of the X chromosome. The homologous regions permit the sex chromosomes to pair at the beginning of meiosis and then segregate into different sperm

## E S S A Y :  *Male or Female? An Olympian Question*

Should an athlete's sex be ascertained by glancing in the pants or by screening the genes? This question has international athletic competition in a quandary. It started with rumors that some nations had sent men disguised as women to the Olympic games. (Only one male Olympian is actually known to have posed as a female: a member of the 1936 German team claimed the Nazis forced him to masquerade as a female high jumper, and he placed fourth.)

Because men have higher levels of male sex hormones, they tend to be taller than women and have larger, stronger muscles. In the population as a whole, men's and women's height, strength, and other features have overlapping distribution curves, as we expect from such polygenic characters. Nevertheless, inequalities of male and female physique are real, and that is why the sexes compete separately in most athletic events.

The International Olympic Committee decided that women (but not men) must prove their sex. Before 1968, female Olympians had to parade nude past doctors, who checked for female genitalia. In 1968 a more scientific test was adopted: since women should have two X chromosomes, cells scraped from inside the mouth were checked to be sure they contained Barr bodies (the inactivated second X chromosome, Section 11–G).

This criterion, however, fails to classify individuals with aberrant sex chromosomes such as X and XXY properly (see Table 16–2). Therefore, the Olympic committee explored using a DNA probe to search for a Y-linked gene, which would disqualify its bearers from female competitions.

These tests use increasingly modern and sophisticated technology, and yet all leave questions. Consider the following cases, known to occur at low frequency in the human population, in which an individual's genetic and phenotypic sex don't match:

- XY and XXY females. Although these individuals have the male Y chromosome and produce male levels of testosterone, their cells cannot respond to male hormones because autosomal mutations have left them without testosterone receptors. These people look female and regard themselves as such, and they have female musculature.

- XX females with male physique. Although these individuals have the female sex chromosome complement, abnormalities in hormone production have produced a male-like physique, including an enlarged clitoris resembling a penis. Some individuals of this type have won Olympic gold medals in female contests.

- XX males. As a result of a meiotic error, some individuals have part of a Y chromosome containing the male-determining *SRY* gene attached to an X chromosome.

- XY males with a uterus.

The difficulty in trying to find a more "scientific" test is that science cannot answer the question of who is male and who is female. Science answers questions about cause and effect, such as why a particular person looks female despite having male sex chromosomes (XY), or vice versa. These unusual cases result because the Y chromosome's male-determining *SRY* gene takes an embryo on only the first step in a long path to a full male phenotype. Many other genes must also cooperate, and their expression (or nonexpression) may result in a phenotype at odds with the chromosomal sex. How far along the male pathway must an individual develop to be male rather than female? Science cannot answer this question.

Imagine someone telling you it's a scientific fact that you're not the sex you've always thought you were—or that your best friend or sweetheart isn't. Add to the shattering of your self-image the loss of a career you've trained hard for, of your athletic scholarship, and of your friends who may feel confused and betrayed by this information, and you'll see how devastating a chromosomal criterion of sex can be. For this reason, chromosomal testing is no longer favored by most sports organizers, and the problem of distinguishing male from female athletes remains.

cells. Because genes in the homologous areas of the sex chromosomes are paired, they behave like autosomal genes. A locus for another cell-surface protein is known to lie in these homologous sections, and many other homologous loci will be discovered now that both chromosomes are being mapped by DNA sequencing techniques.

The much longer mammalian X chromosome has large **nonhomologous portions,** where the Y chromosome has no matching loci. Nonhomologous loci on the sex chromosomes are said to be **sex-linked:** mostly X-linked, on the long nonhomologous portion of the X chromosome, although *SRY* is an example of a Y-linked locus. In contrast to autosomal genes, sex-linked genes show a pattern of inheritance that depends on the sex of the parents and offspring.

The hallmark of a sex-linked recessive trait is that it appears more often in the heterogametic sex. For example, in male mammals, any recessive allele on a nonhomologous part of the X chromosome will be expressed in the phenotype, since there is no homologous locus on the Y chromosome that could carry a dominant allele to mask it. Therefore, it is possible for a single recessive allele to express itself in the male's phenotype. However, a female must have two copies of such a recessive allele before it shows in her phenotype. Hence, recessive X-linked phenotypes are more common in male mammals than in females. Because many recessive alleles are deleterious, this is one reason why more male than female mammals of any age die.

Recessive X-linked traits in humans include red-green color blindness, hemophilia, and the Duchenne type of muscular dystrophy.

Suppose a woman homozygous for normal color vision marries a color-blind man: all their children will have normal color vision because the woman passes on an X chromosome with a normal allele to each child (Figure 16–12). Imagine that the daughters in this family marry men with normal color vision (like their brothers), and the sons marry women who are carriers of the allele for color blindness (like their sisters). In the next (F₂) generation, we would find the ratio of three children with normal vision to one color-blind, as expected from a monohybrid cross. However, our results have this added twist: all the color-blind children are male! Girls can also be color-blind, but only if they inherit an X chromosome with an allele for color blindness from their fathers as well as from their mothers. Therefore, color-blind girls are much rarer than color-blind boys.

Certain forms of hemophilia are also caused by a recessive allele on the X chromosome. Persons with hemophilia (almost always males) lack a protein needed to make blood clot, and so may bleed to death from minor internal injuries or even a slight cut. Unlike female carriers, a male with a hemophilia allele has no second X chromosome bearing a normal allele to code for this clotting protein. Hence, his hemophilia allele will appear in his phenotype. Hemo-

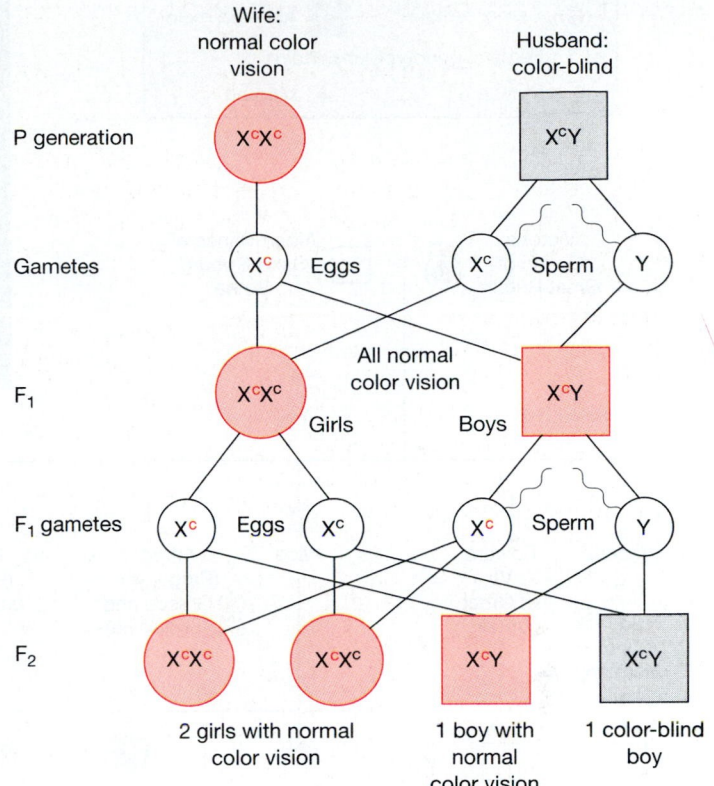

**FIGURE 16–12** Sex linkage. Inheritance of red-green color blindness is controlled by genes on the X chromosome. In a marriage of a woman homozygous for normal color vision and a color-blind man, all the children have normal color vision. If the children marry people with the same genotypes as their siblings, half of the male offspring are expected to be color-blind, inheriting the maternal X chromosomes bearing the color blindness allele. Girls who inherit this maternal X chromosome receive an X chromosome with a normal allele from their fathers, and so they are not color-blind.

philia can now be controlled (but not cured) by regular injections of clotting factor, obtained by extraction from normal blood or by genetic engineering. As a result, more males with hemophilia now live to grow up and reproduce. More hemophiliac females are appearing in the population, as some men with hemophilia marry women carriers and both pass these alleles to their children.

Queen Victoria of England was the world's most famous carrier of hemophilia. Her hemophiliac son, Leopold, Duke of Albany, and her two carrier daughters, Princesses Alice and Beatrice, spread the allele through the royal houses of Europe, including those of Russia, Prussia, and Spain. For a time hemophilia was called the "royal disease," but none of Queen Victoria's modern descendants appears to have inherited the allele (Figure 16–13).

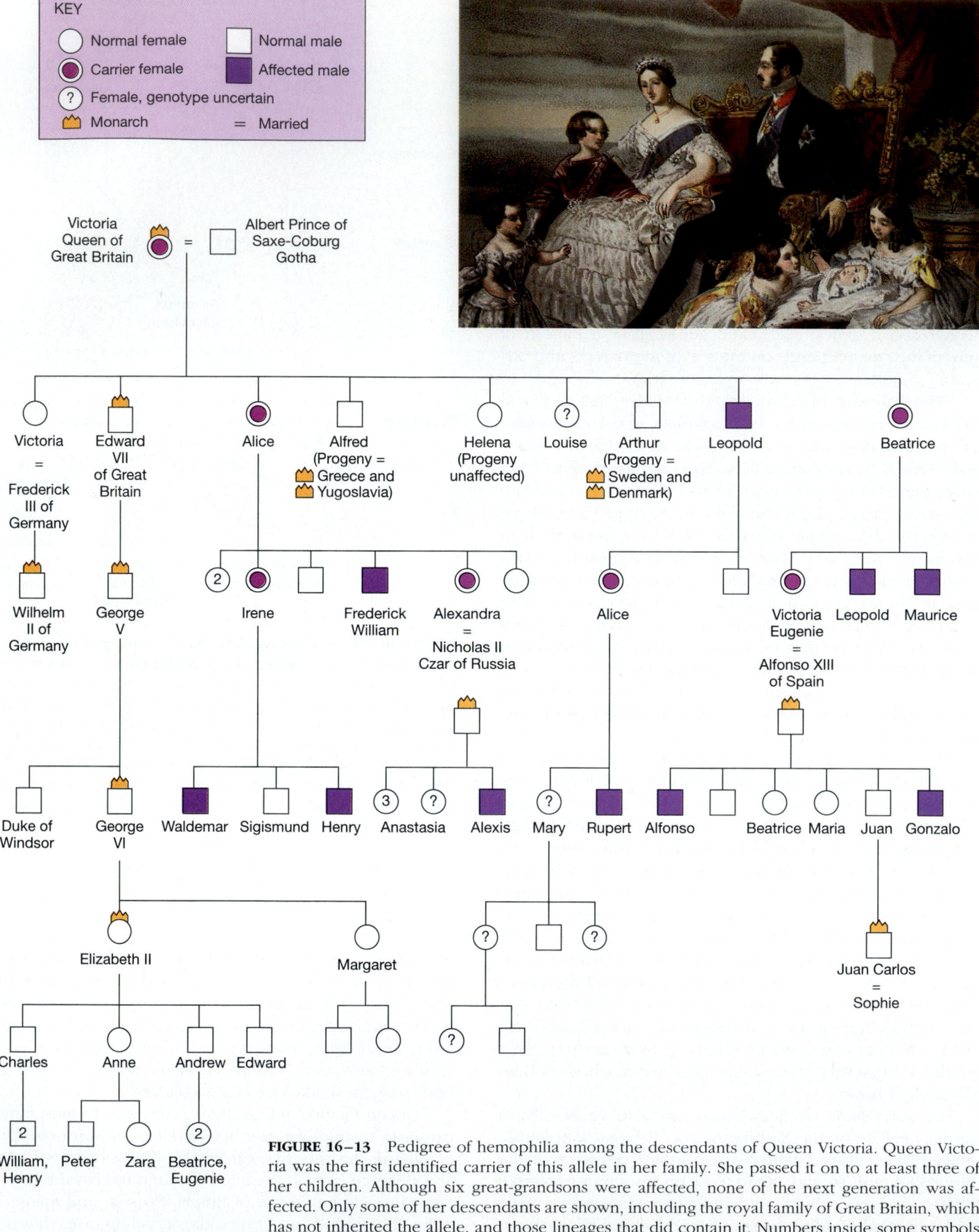

Victoria Queen of Great Britain = Albert Prince of Saxe-Coburg Gotha

Victoria = Frederick III of Germany

Edward VII of Great Britain

Alice

Alfred (Progeny = Greece and Yugoslavia)

Helena (Progeny unaffected)

Louise

Arthur (Progeny = Sweden and Denmark)

Leopold

Beatrice

Wilhelm II of Germany

George V

Irene

Frederick William

Alexandra = Nicholas II Czar of Russia

Alice

Victoria Eugenie = Alfonso XIII of Spain

Leopold

Maurice

Duke of Windsor

George VI

Waldemar

Sigismund

Henry

Anastasia

Alexis

Mary

Rupert

Alfonso

Beatrice

Maria

Juan

Gonzalo

Elizabeth II

Margaret

Juan Carlos = Sophie

Charles

Anne

Andrew

Edward

William, Henry

Peter

Zara

Beatrice, Eugenie

**FIGURE 16–13** Pedigree of hemophilia among the descendants of Queen Victoria. Queen Victoria was the first identified carrier of this allele in her family. She passed it on to at least three of her children. Although six great-grandsons were affected, none of the next generation was affected. Only some of her descendants are shown, including the royal family of Great Britain, which has not inherited the allele, and those lineages that did contain it. Numbers inside some symbols denote more than one offspring of that phenotype. The photograph shows Queen Victoria, Prince Albert, and their first five children.

Another common example of sex linkage is one of the many loci that affect coat color in cats. This locus on the X chromosome can be occupied by an orange or a non-orange allele. The sex-linked orange allele diverts molecules to a metabolic pathway that makes them into orange pigment instead of black. A male cat with the orange allele on his X chromosome, or a female with orange on both X chromosomes, is orange or yellow. The non-orange allele allows black pigment to form, and the non-orange male or homozygous non-orange female is black (or brown or gray, depending on its other gene pairs). A cat with an orange allele on one X chromosome and a non-orange allele on the other has a mottled pattern of orange and black spots called tortoiseshell (Figure 16–14). Such a cat is almost always female, since only females normally have two X chromosomes.

The mottled pattern of a tortoiseshell cat results from X–chromosome inactivation: early in the embryonic development of female placental mammals, one of the two X chromosomes in each somatic cell becomes inactive, and that X chromosome is inactive in all the cell's descendants. If the chromosome with the orange allele remains active, the cell's descendants form a patch of orange fur, and if the chromosome with the non-orange allele is active, the patch is black. Since the X chromosomes are inactivated randomly, tortoiseshell cats vary in the amount and pattern of the colors in their coats. Coat color in cats is affected by a number of different loci (*Essay: Genetics of Cats and Dogs,* page 345).

## 16–H    SEX-INFLUENCED GENES

The main role of sex hormones is influencing the reproductive system and related organs, but these hormones also affect many other characters. **Sex-influenced genes** are those whose expression depends on the level of sex hormones. These genes are usually located on the autosomes, and so there is no difference in genotype between the two sexes. However, males and females with the same genotype may differ greatly in phenotype because the expression of the genes depends on the levels of sex hormones and hence on the individual's sex. For example, a bull may have alleles for high milk production, but he will not produce milk because he has only low levels of female hormones. However, these genes would make him a useful sire for a dairy herd.

In humans, the allele for male pattern baldness (Figure 16–15) is autosomal, and its expression is influenced by the presence of male hormones. A man will become bald if he has only one allele for baldness. In other words, the allele acts as a dominant in men because male sex hormones stimulate expression of the baldness allele. However, in women, the allele acts as a recessive, so that a female must have two alleles for baldness before she loses her hair.

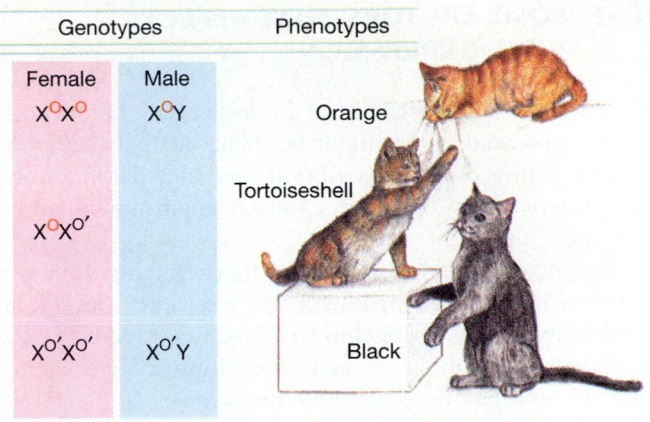

**FIGURE 16–14**   Effect of the sex-linked orange allele on coat color in cats. In the presence of the orange allele (colored O), the fur is orange. The non-orange allele (*O′*) permits expression of some shade of black, coded by various autosomal genes. Patches of orange and non-orange fur in heterozygous (tortoiseshell) females result from the random inactivation of one X chromosome in each cell during embryonic development.

Gout is another trait expressed much more in the presence of male than of female sex hormones: 0.18% of the U.S. population suffers from gout, and 95% of these are men. People with gout have painful deposits of uric acid salts in the tissues, especially in the joints of the big toes. In Victorian literature, gout figured largely as a reason for the temper tantrums of irascible old men. Avoiding red wine and rich and spicy foods was supposed to alleviate the condition, but this treatment tried the tempers of its victims still further. Gout can now be treated, but it is still a painful nuisance to many sufferers.

**FIGURE 16–15**   Some factors that affect gene expression. Male sex hormones promote the expression of the allele for baldness in men. The child's blond hair is turning darker as she grows older, a common age-related inherited trait.

## 16–I SOME FACTORS THAT AFFECT GENE EXPRESSION

All the genes possessed by an individual determine its genetic potential: what might be. What actually happens is another matter. The transcription and translation of genes are influenced by both the external environment and the other genes present.

Some genes code for proteins that regulate other genes or interact with their products. For example, damage to a regulatory gene is suspected to cause nearly half the cases of severe hemophilia because the clotting factor gene in these people has no detectable mutations.

Other genes code for enzymes whose products influence gene expression. An example is enzyme-produced sex hormones in the previous section.

Hormone production varies with age, and so age may play a part in gene expression. Consider the many changes at puberty, such as voice change and growth of the testes in males, breast enlargement and the deposition of body fat in a characteristic pattern in females, and growth of hair in the armpits and pubic area in both sexes.

Many traits are controlled mainly by one gene pair but are also influenced by the products of other, **modifier genes.** It was long believed that eye color in humans was controlled by a single pair of genes, with brown eyes dominant to blue. It is now known that there are also at least two pairs of modifier genes involved, and it is possible, though uncommon, for blue-eyed parents to have brown-eyed children. The piebald gene in cats also seems to modify the expression of other genes (Figure 16–16).

The external environment also plays an important role, both in embryonic development and in later life. In the last few decades several drugs taken by pregnant women have turned out to cause improper development of the fetus or cancers later in the child's life.

Many other external factors also affect gene expression. For instance, a good diet is necessary if a person is to reach the height or mental capacity made possible by his or her genes. In many countries, young adults tower above their parents or grandparents as a result of improved nutrition. Farmers and gardeners know that proper nourishment is just as important for plants. One species of caterpillar even develops a phenotype that mimics the appearance of the food it eats, thereby blending in with it (Figure 16–17).

Light also influences gene expression. During the development of plants, many genes responsible for producing photosynthetic pigments and enzymes are turned on by exposure to light. A human being produces more melanin and becomes darker when exposed to bright sunlight for a time.

Temperature affects the expression of some genes. Himalayan rabbits and Siamese cats are normally light-colored, with dark feet, ears, nose, and tail. The darker fur color on these parts of the body is due to the activity of an

**FIGURE 16–16** The effects of a modifier. In a calico (three-color) cat, the piebald gene, coding for patches of white fur, somehow affects the size of colored patches next to white fur: these orange and black patches are larger than those on the cat's back, where no white patches occur.

enzyme that is unstable at higher temperatures. The extremities of these animals are cool enough for the enzyme to function and produce dark fur, but the body itself is warm enough to inactivate the enzyme (Figure 16–18). As we saw before, temperature can even determine an individual's sex in several species of reptiles!

## 16–J GENETIC COUNSELING AND FETAL TESTING

Genetic abnormalities can bring pain and suffering to the victims and to their families. Parents of victims may feel guilty that they passed a defective allele to their child, and in some cases this may even lead to alcoholism, drug addiction, denial of paternity and rejection, or divorce. The time, energy, and money needed to care for afflicted children may also deprive the family's other children of a normal home life.

Genetic counseling can help couples to determine their chances of having children afflicted by a particular genetic defect, an event that is more likely if the couple or their relatives have already had such a child. Blood tests or genetic analysis of chromosomes from blood cells can now determine whether prospective parents are carriers for traits such as Tay-Sachs disease, cystic fibrosis, sickle cell anemia, or thalassemia or are destined to develop Huntington's disease.

Some couples, knowing that each child born to them has a 25% chance of being homozygous recessive for a condition that will bring years of suffering or incapacity followed

(a)    Caterpillar fed oak flowers

(b)    Sibling fed oak leaves

**FIGURE 16–17**    Phenotypes dependent on diet. (a) This oak-feeding caterpillar, fed a diet of oak flowers, has developed a resemblance to one of the male flower clusters, complete with dots that look like the flowers' pollen sacs. (b) A sibling of the first caterpillar, fed oak leaves, resembles the oak tree's twigs. In nature, caterpillars that hatch in the spring look like the flowers they eat; those hatched after the flowers wither eat leaves and become twig mimics.    (Erick Greene)

by an early death, choose adoption instead of taking that risk. Others begin pregnancy and have the fetus tested to determine whether it is affected; if it is, they may choose abortion and hope that a later fetus will be normal or heterozygous. The availability of genetic testing and abortion has increased the birth rate among at-risk couples.

Some genetic abnormalities are due not to mutant alleles but to abnormal separation of chromosomes during meiosis, resulting in one chromosome too many or too few. One example is Down syndrome, which results from the presence of an extra chromosome #21 (Section 14–F). Extra copies of other chromosomes can also cause extreme abnormalities, although they are rarer.

Genetic defects can be detected by examining a sample of fetal cells for chromosomal abnormalities, such as those resulting from translocation or nondisjunction of chromosomes during meiosis. The cells can also be cultured in the laboratory to produce enough cells for geneticists to test the DNA for disorders such as phenylketonuria, sickle cell anemia, thalassemia, cystic fibrosis, Huntington's disease, and Marfan's syndrome. Samples of fluid from the amniotic sac surrounding the fetus can also be analyzed for chemicals indicating metabolic defects. Figure 16–19 shows two methods used to test a fetus for genetic abnormalities.

In 1992 researchers took single cells from each of five eight-cell "test tube" embryos and screened them for cystic fibrosis before implantation. This procedure is expensive and limited to similar test tube embryos.

Recently, some human fetuses have been treated before birth for deficiency of the B vitamin biotin, an inborn error of metabolism, and for blockage of the urethra (the tube that empties the urinary bladder). We can expect that more genetic disorders and other conditions will be detected and treated before birth in the future.

Shaved area

Later

**FIGURE 16–18**    Effect of skin temperature on expression of coat color genes in the Himalayan rabbit. Black fur grows on parts of the body with skin temperature below 33 °C. If fur is shaved from a warmer part of the body and an ice pack is applied while the fur grows back, the new fur is also black.

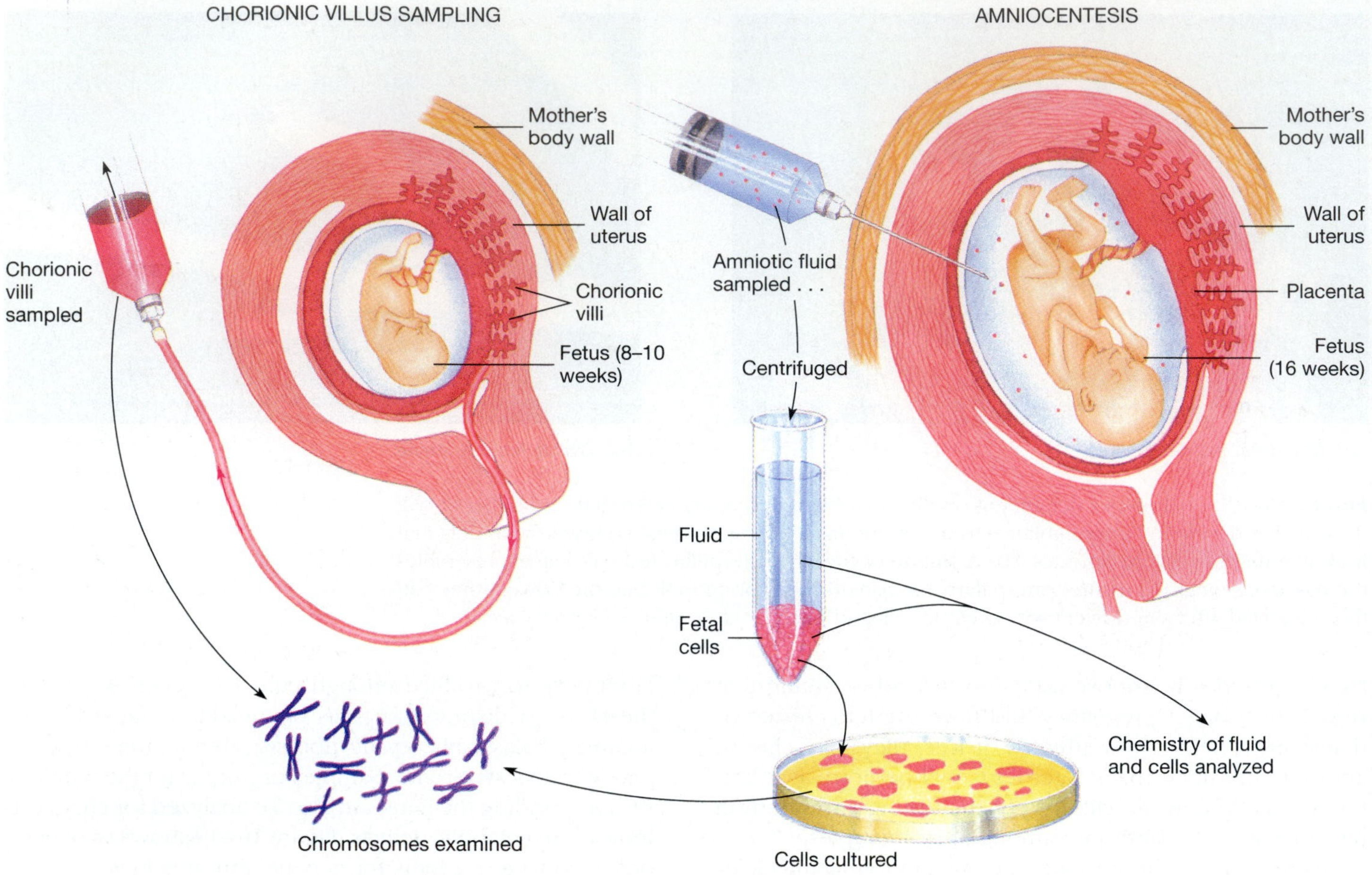

CHORIONIC VILLUS SAMPLING

AMNIOCENTESIS

Chorionic villi sampled

Mother's body wall

Wall of uterus

Amniotic fluid sampled . . .

Chorionic villi

Fetus (8–10 weeks)

Centrifuged

Mother's body wall

Wall of uterus

Placenta

Fetus (16 weeks)

Fluid

Fetal cells

Chemistry of fluid and cells analyzed

Chromosomes examined

Cells cultured

**FIGURE 16–19** Diagnosing genetic defects in a fetus. (a) Chorionic villus sampling. By three weeks after fertilization, the fetus's chorionic membrane has developed branched villi, part of the life-support system attached to the wall of the uterus. At 8 to 10 weeks there are enough fetal cells here that some can be sucked up for genetic analysis. (b) Amniocentesis. A needle is inserted through the mother's abdominal wall and uterus into the amniotic sac, and fluid containing sloughed fetal cells is withdrawn into a syringe. Fetal cells are cultured and examined for chromosomal abnormalities. The fluid and cells can also be tested chemically for metabolic defects.

## SUMMARY

Genes express themselves by coding for the sequences of amino acids in polypeptides or proteins: enzymes, structural proteins, blood proteins, and so on. The effect of an allele depends on how it changes the structure and function of the encoded protein and on how important that protein is in maintaining life.

1. Some mutations result in lethal alleles, which cause premature death. Most familiar lethal alleles are recessive and cause death only in homozygotes.
2. Changes in less vital proteins may cause metabolic disorders (inborn errors of metabolism), such as albinism and phenylketonuria.
3. Multiple alleles of a gene usually exist in a population, for example, in the human ABO blood group.

4. Polygenic characters are determined by the interaction of the alleles at several different genetic loci. Polygenic characters occur in a wide range of phenotypes.
5. In most familiar organisms, sex is determined by a pair of unlike sex chromosomes, with one sex being homogametic and the other heterogametic. In humans and most other mammals and in *Drosophila*, females have the sex chromosome combination XX and males are XY, whereas in birds females are ZW and males ZZ. In many other organisms, sex is determined by environmental factors, including presence or absence of members of one sex, and individuals may even change and become members of the scarcer sex.
6. Traits carried on nonhomologous portions of the sex chromosomes are said to be sex-linked. Among mammals, X-linked traits appear more often in the phenotype of males because the sexes have different genotypes

## ESSAY: *Genetics of Cats and Dogs*

Human beings probably domesticated dogs about 14,000 years ago, and the Egyptians of 1600 BC revered cats as sacred. Selective breeding through thousands of years has produced dozens of breeds of cats and dogs descended from the (presumably) more uniform ancestral wild forms. In this century we have begun to understand some of the genes that give rise to such variety.

The most obvious differences among cats are in coat color and pattern. As in many other mammals, color is based on two pigments, black and orange (often called yellow). The enormous variety we see in cats comes from the interplay of several different pairs of genes that affect the production and distribution of these pigments. We have room to discuss only a few (Table 16–A).

The ancestral "wild-type" cat probably resembled the present-day mackerel tabby: black stripes on an agouti background, like the Manx cat in Figure 16–4. The term agouti means that the hair shaft is dark (black) with a yellow band or tip, a common pattern giving an inconspicuous drab color to the fur of many wild mammals, such as rabbits, ground hogs, and mice. One pair of genes governs the cat's tabby stripes and another pair governs the agouti background color. Agouti is dominant to non-agouti, in which the hair shaft is uniformly colored. If we changed a mackerel tabby's agouti allele(s) to homozygous non-agouti, the cat would be solid black because the black stripes dictated by its tabby alleles would not show up against the black background. (Sometimes faint stripes can be seen in black kittens.)

Another pair of genes influencing coat color is the sex-linked orange and non-orange alleles, discussed in Section 16–G.

(a)

(b)

White fur is also very common in cats. The allele for piebald causes spots of white fur, with the homozygote having more extensive white areas than the heterozygote. There are about ten different degrees of white spotting, ranging from a small area on the chest to a mostly white coat with a few colored patches on the head, back, and tail. One pair of alleles—piebald versus solid-color—probably cannot account for so much

*(Continued)*

**FIGURE 16–A** Try your hand at determining the genotypes of these cats. It is not possible to tell the alleles present for every genetic locus listed in Table 16–A, but work out as many as you can. (a) This long-haired black cat is the mother of (b), a long-haired gray cat with white markings. The gray cat's genotype provides some clues about the genes present in her mother. Answers appear with the self-quiz answers for Chapter 16.

### TABLE 16–A

**Alleles Coding for Coat Color Different from the Wild-Type in Cats***

| Allele* | Phenotype |
|---|---|
| $a$ | Non-agouti (hair shaft uniform color) (recessive to $A$, agouti) |
| $b$ | Brown (recessive to $B$, black) |
| $c^b$ | Burmese (recessive to $C$, full color) |
| $c^s$ | Siamese (recessive to $C$ and $c^b$) |
| $c^a$ | Blue-eyed albino (true pink-eyed albinos almost unknown) |
| $d$ | Dilution of color (black diluted to gray, orange to cream) (recessive to $D$, full color) |
| $O$ and $O'$ | Orange and non-orange, respectively (X-linked, codominant) |
| $S$ | Piebald (white spotting) |
| $T^a$ | Abyssinian (faint markings on face, feet, tail) |
| $T^\dagger$ | Mackerel tabby (wild-type, striped) |
| $t^b$ | Blotched tabby |
| $W$ | Dominant white (sometimes associated with blue eyes, deafness) |

*Capital letters indicate dominants, lower case recessive. If there are more than two forms, the alternatives are listed in order from most to least dominant.

$^\dagger T$ is wild-type but is included in the table because it is recessive to $T^a$ but dominant to $t^b$.

*(Continued)*

variation. It is unclear how the extent of spotting is controlled.

A completely white cat can occur in either of two ways, which are governed by two different gene pairs. First, the cat may have a "dominant white" allele, which prevents the pigment encoded by other genes from being deposited in the hairs and hence prevents the expression of any other genes for coat color or pattern. (About two thirds of cats with this allele have blue eyes and one third are deaf, but it is unclear why this is so.)

The second way for a cat to be white is for it not to make any pigment in the first place. The blue-eyed albino is homozygous for a recessive allele ($c^a c^a$), but it looks very much like a blue-eyed cat with the dominant white allele ($W$). The $c^a$ (blue-eyed albino) allele is part of a series of multiple alleles ranging from albino to full-colored. The most familiar allele between these extremes is $c^s$ (Siamese), which codes for a temperature-sensitive enzyme (Section 16–I). The Siamese allele is dominant to albino but recessive to full color.

Hair length in cats is governed mainly by one gene pair, with short hair dominant to long. A mutation that gives rise to extra toes on the front feet, so that the feet are abnormally large, seems to be dominant.

Many characteristics of dogs (and indeed of other mammals) are determined by gene pairs similar to those described for cats. Short hair is dominant to long, and black is dominant to brown. Piebald spotting occurs in many degrees, and here the variations seem to be due to modifier genes that affect the expression of the main gene pair coding for piebald. Deafness in dogs accompanies an allele for the coat color pattern called merle: dark color dappled on a dilute background. Dogs can also have X-linked hemophilia similar to human hemophilia (Section 16–G).

The most striking differences among dogs, however, are their varied body forms. In contrast to most other kinds of mammals, the dog species seems to harbor great genetic potential for variation in size, shape, and behavior. This has allowed the selective development of breeds suit-

able for a wide range of tasks: dachshunds were bred to dig badgers out of their lairs, pointers to help hunters locate gamebirds, bulldogs to bait bulls and bears (a once-popular sport) by biting their noses, and sheepdogs to help shepherds get the flock into the fold.

The differences in size, shape, behavior, and temperament among the various breeds of dogs tend to be under polygenic control, and we know little about the actual genes involved. Selective breeding has modified many traits to such extremes that they endanger health (Figure 16–B). For example, the relatively long spinal columns of dachshunds are prone to slipped discs, and the "pushed-in" faces of bulldogs are accompanied by drooling, crowding of the teeth in the short jaws, and breathing difficulties due to malformed air passages. Modern dog breeders are realizing that they must relax their standards for certain traits in order to maintain the health of the breed.

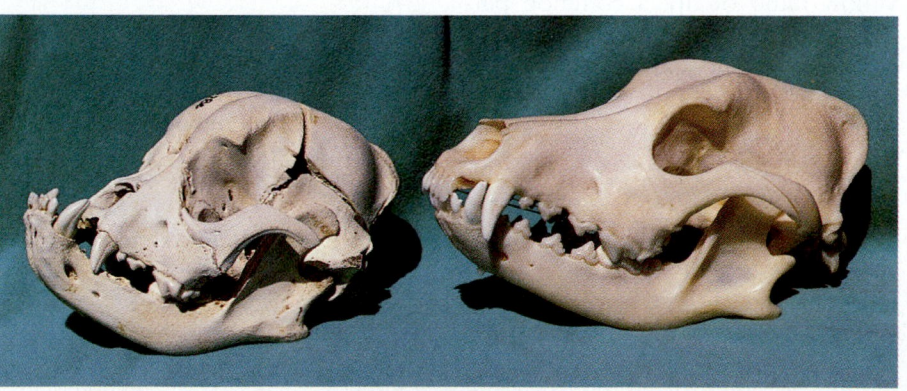

**FIGURE 16–B** Differences between dog breeds. (left) The skull of a boxer; (right) the skull of a shepherd mix. The boxer's upper jaw is shorter because of a genetic abnormality in which one bone (maxilla) grows less than usual. As a result, the lower jaw, which is normal length, protrudes beyond the upper, and the teeth do not meet properly.   (Courtesy of Dr. John Saidla, Cornell School of Veterinary Medicine)

at these loci: males are haploid at X-linked loci, whereas females are diploid.

7. In contrast to sex-linked traits, sex-influenced characters are carried on the autosomes (usually). Although members of both sexes may have the same genotype, the trait occurs more commonly in the phenotype of one sex because its expression depends on the balance of sex hormones.

8. An individual's phenotype depends on what mix of genes it has, how these genes are influenced by the products of other genes (enzymes, enzyme products such as hormones or pigments, or nonenzyme proteins), and what factors it encounters in its external environment. External factors influencing gene expression include nutrition, light, and temperature.

## Self-Quiz

1. In the homozygous condition, a recessive lethal allele in cattle produces "amputated" calves with malformations of the limbs, skull, and internal organs. These calves die soon after birth.
   a. What proportion of the normal offspring from a cross of two heterozygotes would be expected to be carriers for this trait?
   b. How could a farmer eliminate this trait from his herd if some "amputated" calves have been born to his cows?
2. Review the information on brachydactyly, Section 16-B.
   a. If two brachydactylic people marry, what are their chances of having a child with normal fingers?
   b. If a brachydactylic person marries a person with normal fingers, what phenotypic ratios can be expected among their offspring?
3. A geneticist studying the various loci that govern coat color in mice is trying to develop true-breeding strains of each possible coat color. He carries out several generations of matings among mice with yellow coats and always obtains some offspring with other colors of coats.
   a. What does this indicate about the genotype of yellow mice?
   b. The geneticist tallies up his results over several generations and finds that he has obtained a total of 184 yellow mice and 95 of other colors.
   What does this suggest about the nature of the yellow allele?
   c. Why did the geneticist never obtain a homozygous yellow mouse?
   d. How could he prove what became of the homozygous yellow offspring?
4. A farmer orders two dozen baby chicks. When they arrive, he notices that some of them have legs noticeably shorter than normal. The farmer crosses some of these short-legged birds with normal-legged ones, and also makes crosses between the short-legged birds. The following offspring are obtained:

   normal × short: 47 short legs,
   43 normal leg length
   short × short: 39 short legs,
   19 normal leg length

   a. What should the farmer conclude about the short-legged genotype?
   b. What happens to the offspring homozygous for the short-leg allele?
5. Shown opposite is a pedigree of ABO blood groups for several generations of humans. Circles represent females, and squares represent males. Marriages are shown by horizontal lines directly connecting two people, and children are connected to their parents by a vertical line down from the marriage line. For example, (b) and (c) are married to each other, and (d) is one of their two sons. Give the possible genotype(s) for each individual marked with a letter. (*Hint:* start at the bottom of the diagram and use what you know about children's genotypes to deduce those of their parents.)
6. Ms. Smith and Ms. Jones gave birth to baby boys (named John and Tom, respectively) on the same day in a large city hospital. After Ms. Smith took her baby home, she began to suspect that it was Ms. Jones's baby and that the hospital had somehow mixed the infants up. Blood tests revealed that Mr. Smith

has blood type O, MN, and Rh$^+$, Ms. Smith has blood type B, N, Rh$^+$, and John Smith has blood type B, M, Rh$^-$. Mr. Jones has blood type A, M, Rh$^+$, Ms. Jones has blood type AB, MN, Rh$^+$, and Tom Jones has blood type O, MN, Rh$^+$. The Rh$^+$ allele is dominant to the Rh$^-$ allele; the M and N alleles are codominant. Had a mixup occurred?

7. In rabbits, coat color is determined by multiple alleles, with normal coat color ($C$) dominant to chinchilla ($c^{cb}$), which is dominant to Himalayan ($c^b$), which is dominant to albino ($c$). What offspring phenotypes are expected from the following crosses, and in what ratios?
   a. $Cc^b \times c^{cb}c^b$     b. $c^{cb}c \times c^bc$     c. $c^{cb}c \times c^{cb}c$
8. In chickens a sex-linked dominant allele causes a feather pattern known as "barred." If a barred hen is mated with a non-barred rooster, what will be the feather pattern and sex of the offspring?
9. What are the expected genotypic ratios among the children of a woman whose father was a hemophiliac and whose husband is normal?
10. Under what circumstances is it possible for both a father and his son to be hemophiliacs?
11. Red-green color blindness in humans is an X-linked recessive trait.
    a. If about one in every 12 human males shows this trait, how common is it among human females?
    b. In a large family in which all the daughters have normal vision and all the sons are color-blind, what are the probable genotypes of the parents?
    c. If a normal-sighted woman whose father was color-blind marries a color-blind man, what is the probability that their son will be color-blind?
    d. What is the probability that a daughter born to the couple in (c) will be color-blind?
12. If a species of mammal has some members which carry a sex-linked lethal trait that causes early death and resorption of the embryo, what sex ratio would be expected among the offspring of a female carrier and a normal male?
13. It is often said that men inherit baldness from their maternal grandfathers via their mothers. In light of what you have learned about this trait, is this a valid statement? Explain.

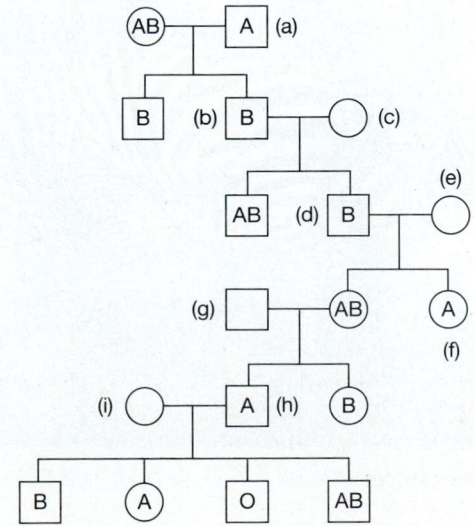

## Questions for Discussion

1. One problem in genetic counseling is that people who learn they are carriers for genetic diseases such as hemophilia, Tay-Sachs disease, sickle cell anemia, or phenylketonuria may consider this a terrible stigma. Men have been known to deny paternity of their children and divorce their wives for infidelity when told that the child had inherited a deleterious recessive allele from each parent. What kinds of arguments and counseling would you use, if you were a genetic counselor, in an attempt to induce a healthier, more productive response to such a discovery?

2. People who are carriers for sickle cell anemia face the risk that their red blood cells may sickle when they are in environments with low oxygen levels. Hence, these people are probably exposed to greater than usual risks if they become divers, jet pilots, or mountaineers. Otherwise they have no physical handicaps. Nevertheless, many have been denied access to various professions as a result of ignorance and prejudice against genetic disorders. Since this is the case, the start of nationwide screening for sickle cell carriers resulted in loss of jobs and in other forms of discrimination. Is it better for carriers to remain in ignorance of genetic conditions for which there is no cure at the moment? If not, why not?

3. Until the advent of modern technology, hemophiliac males usually died before they reached reproductive age. Now, they can be provided with "clotting factor," a blood extract that permits them to lead normal lives and live to have children. The treatment costs about $5000 to $10,000 a year per person, and there are about 20,000 hemophiliacs living in the United States. Can society or should society insist that such men be sterilized, so that they cannot perpetuate their disease, if taxpayers have to pay the bill for their medication?

4. Table 16–2 shows that individuals with a single X chromosome are known to occur, but not individuals with only a Y chromosome. Why do you think this is?

5. Every so often the Ann Landers newspaper column has a letter from a mother whose husband or in-laws have been chiding her for having daughters instead of sons. Is this censure justified? Why?

6. Name some factors besides those mentioned in the chapter that may influence gene expression.

7. Propose an explanation of the genetic basis for the phenotypes shown in each of these photographs:

(b) Variations of color and pattern in the leaves of one plant.

(a) The lion's mane, and the black tips of its hairs.

(c) The zebra's stripes.

(d) Three different leaf shapes in sassafras.

## Suggested Readings

Crews, D. "Animal sexuality." *Scientific American,* January 1994. A look at what we might learn about the evolution of sexuality from the effect of hormones, temperature, and genes on the sexual differentiation of vertebrates.

Friedman, M. J., and W. Trager. "The biochemistry of resistance to malaria." *Scientific American,* March 1981.

Friedmann, T. "Prenatal diagnosis of genetic disease." *Scientific American,* November 1971. Thoughtful article explaining the technology of amniocentesis and pointing out the ethical problems it presents.

Greene, E. "A diet-induced developmental polymorphism in a caterpillar." *Science* 243:643, 1989. Easily understood experiments show that what this caterpillar looks like depends on what it eats.

Hartl, D. L. *Human Genetics.* New York: Harper and Row, 1983.

Hutt, F. B. *Genetics for Dog Breeders.* San Francisco: W. H. Freeman, 1979.

Lawn, R. M., and G. A. Vehar. "The molecular genetics of hemophilia." *Scientific American,* March 1986.

Marx, J. L. "The cystic fibrosis gene is found." *Science* 245:923, 1989. The saga of an exciting genetic milestone told in nontechnical terms.

Murray, J. D. "How the leopard gets its spots." *Scientific American,* March 1988. A speculative model of how coat patterns might develop in spotted and striped animals.

Robinson, R. *Genetics for Cat Breeders,* 2d ed. New York: Pergamon Press, 1977. A nontechnical book of interest to cat lovers.

Sayers, Dorothy L. *Have His Carcase.* London: Harcourt, Brace, Jovanovich, 1932. A mystery novel about a human genetic trait.

# Evolution and Natural Selection

## OBJECTIVES

*When you have studied this chapter, you should be able to:*

1. Define evolution in your own words.
2. Describe Lamarck's theory of how evolution occurs, and explain why Lysenko believed in this theory.
3. Describe the roles of Darwin and Wallace in formulating the theory of evolution, and state why their theory was more convincing than the descriptions of evolution put forward by their predecessors.
4. Define artificial selection; homologous, analogous, and vestigial structures; adaptive radiation; and endemic species; and give or recognize examples of each.
5. Explain how the following provide evidence for the occurrence of evolution, and give or recognize examples in each category: artificial selection, the fossil record, comparative anatomy, embryology, and biogeography.
6. List or recognize conditions that favor fossilization and conditions that make it unlikely that an organism will be fossilized.
7. State the four observations leading to the conclusion that evolution occurs under the influence of natural selection.
8. Define natural selection, and state what it means to be evolutionarily successful. Outline how an adaptation may result from selection, mentioning the role played by each of the following: gene, genetic variation, natural selection, and population.
9. Describe Kettlewell's experiments with the peppered moth, and explain how they provided evidence that industrial melanism in this moth evolved under the influence of natural selection.
10. Describe Lack's experiments with Swiss starlings, and state what these experiments tell us about the effect of natural selection.
11. Explain why selection does not result in a population of identical, perfectly adapted organisms.

---

**W**e discussed genetics (Chapters 15 and 16) before evolution because knowing genetics makes evolution easier to understand. Nineteenth-century biologists did not have this advantage. Charles Darwin and Alfred Russel Wallace first proposed the theory of evolution by means of natural selection in 1858, seven years before Mendel reported his work on genetics. And as we saw in Chapter 15, Mendel's findings did not become generally known until 1901. By then, most biologists had accepted the idea that evolution was the correct explanation of how organisms came to be as they are, even though many people (including Darwin himself) were troubled that natural selection did not appear compatible with what was then known about genetics.

Our study of genetics showed how genes are passed from grandparents to parents to offspring. Evolution views such events on a much larger scale. Instead of looking at

pairs of genes passed through a few generations of one family, it considers all the genes passed through many generations within an entire population. During this time, the proportions of alleles and genotypes in the population may change. Sometimes this causes changes in the population so slight as to pass unnoticed, but at other times genetic changes result in organisms that differ markedly from their ancestors.

Change in the genetic makeup of a population over time is **evolution.** In more general terms, **evolution** is the origin of organisms by descent and modification from previously existing forms of life. When members of a population have evolved enough differences from their ancestors, we may consider them a new species.

Notice that the unit that evolves—that changes over time—is the population, not its individual members. An individual cannot evolve. Each individual's genetic makeup is fixed at the time of fertilization. The individual is a temporary vessel for some of the population's genes, and the evolution of the population results from some individuals but not others reproducing and so contributing genes to the next generation.

The main mechanism of evolution is **natural selection,** the differential reproduction of genotypes. By "differential reproduction," we mean that individuals with some genotypes produce more offspring than those with other genetic combinations. Hence, the alleles they carry are likely to become more common in future generations. This change in the proportions of different genetic alleles in the population over time is evolution. Darwin and Wallace suggested the occurrence of natural selection, noting that some of the inherited variations among organisms must affect an individual's chances of living to reproduce.

## KEY CONCEPTS

- ◆ The theory of evolution states that organisms arise by descent and modification from previously existing organisms.
- ◆ Evolution is a change in proportions of one or more alleles in a population from one generation to another.

- ◆ Natural selection, the differential reproduction of genotypes, is the most important cause of evolution.
- ◆ Populations of organisms evolve adaptations to their environments.

## 17–A  HISTORY OF THE THEORY OF EVOLUTION

For thousands of years, most people believed that each separate species of organism had been specially created. From time to time philosophers proposed that the living world changed through the ages, but this idea gained little acceptance in the Western world.

From about 1750 on, however, evidence supporting evolution accumulated, and many people became increasingly convinced that evolution does occur. Georges-Louis Le Clerc de Buffon and Jean Baptiste de Lamarck were two influential French biologists who believed that species develop progressively and change in a changing environment. In England, geologists James Hutton, William Smith, and Charles Lyell studied fossils of extinct animals embedded in different layers of rock and became convinced that different organisms had lived at different times.

### Lamarckism

French biologist Jean Baptiste de Lamarck studied invertebrate fossils and classification. In 1809, the year of Darwin's birth, he elaborated the beliefs of his fellow scientists into his own proposal of evolution. This did not gain wide acceptance because the mechanisms he proposed were not convincing, and he did not present much supporting evidence. According to Lamarck, evolution is destined to produce "higher" organisms, with human beings as its ultimate goal. Because organisms appear so well adapted to their ways of life, Lamarck's mechanism of evolution included inheritance of acquired characters: an organism's lifestyle could bring about changes that it passed on to its offspring.

Lamarck's most famous example was the long neck of the giraffe (Figure 17–1). He suggested that giraffes had evolved long necks because they strained to reach leaves

**FIGURE 17–1** Why does the giraffe have a long neck? Lamarck suggested that its ancestors stretched their necks to browse on the leaves of trees and that this increase in length was passed on to succeeding generations.

growing above their heads as they ate, thereby stretching their necks; this added length was passed on to their offspring. This idea dovetailed nicely with pre-1900 beliefs that different parts of the body contributed to inheritance by sending tiny particles through the bloodstream to the eggs and sperm in the reproductive organs. Particles from a giraffe's lengthened neck would enter the gametes and endow the offspring with longer necks.

Because no one yet knew how genes were inherited, Darwin did not refute this idea when he put forward natural selection as the mechanism of evolution. He thought that the inheritance of acquired characters might play a minor role in evolution, although we now know that this is not how inheritance works.

The political importance of Lamarckism is an interesting story. The idea that acquired characteristics could be inherited was taken up by T. D. Lysenko, a Soviet agriculturist. He said that he could breed better varieties of grains by giving the parent plants better conditions. This implied that improving social conditions such as health care and education would eventually produce an improved race of people with better genetic potential for health and intelligence. In the 1930s this appealed to a revolutionary society much more than the writings of Western biologists who said the only way to "improve" the human race was eugenic programs: encouraging healthier, more intelligent people to reproduce and discouraging the reproduction of others. So, for ideological reasons, Lysenko rose to the top in Soviet science, and many biologists who opposed his ideas lost their jobs and even their lives. By the 1950s, however, other Soviet scientists had disproved many of Lysenko's ideas, and he was finally discredited. But the idea that acquired characters may be inherited had an important impact on society and is undoubtedly quite widely believed today.

## Darwin and Wallace

The theory of evolution by natural selection was put forward in a joint presentation of the views of Charles Darwin and Alfred Russel Wallace before the Linnaean Society of London in 1858. As we have seen, Darwin and Wallace were not the first to suggest that evolution occurs. Their names are linked with the concept because their proposal of natural selection as the mechanism for evolution was so convincing.

We know little about Wallace's early life, but it seems likely that Darwin as a young man believed in special creation. This was part of the faith of most Christian denominations of his day, and Darwin at one time began training for the clergy. However, years of observation and reading presented Darwin with evidence that led him to another explanation for the origin of species.

Although Darwin and Wallace never met, both came to the same conclusion about the mechanism of evolution as

a result of remarkably similar experiences. First, both men were influenced by reading the work of geologist Charles Lyell and economist and clergyman Thomas Robert Malthus. In his book *Principles of Geology,* Lyell wrote that the world was an ancient arena in which rock formations were slowly produced, changed, and destroyed. He recognized that organisms compete and saw that the spread of one species may diminish or eliminate others, and he even discussed the extinction of species caused by human activities. All the information needed to formulate the theory of evolution was present in Lyell's work.

Malthus similarly argued that there is competition among organisms. His *Essay on the Principle of Population,* published in 1789, argued that every human population outgrows its food supply and is eventually reduced by starvation, disease, and war. Populations tend to increase geometrically, whereas food supplies at best increase only arithmetically. Darwin later made famous Malthus's phrase, "the struggle for existence" among individuals, which they both saw as an inevitable result of the discrepancy between rapid population growth and limited food supply.

In addition, both Wallace and Darwin observed plant and animal life in several parts of the world. Wallace traveled in South America, and later in the Malay Archipelago, the most biologically diverse area in the world. It was here, in 1854, that the idea of natural selection came to him as he lay in bed with a fever. In the 1830s Darwin obtained a position on *H.M.S. Beagle,* a British naval ship embarking on a five-year mapping and collecting expedition. This trip took Darwin to South America and the nearby Galápagos Archipelago (Figure 17–2), where he collected much of the evidence he later used to support the theory of evolution by natural selection.

In 1845 Darwin published *The Voyage of the Beagle,* an engaging and readable account of his travels. In it he commented on his observations of the many species of finches unique to the Galápagos Islands: "Seeing this gradation and diversity of structure in one small, intimately related group of birds, one might really fancy that from an original paucity of birds in this archipelago, one species had been taken and modified for different ends."

This book shows that Darwin already held the clue to how evolutionary change was brought about, a mechanism he was not to put forward for public discussion for more than ten years. He said, "Some check is constantly preventing the too rapid increase of every organized being left in a state of nature. The supply of food, on an average, remains constant; yet the tendency in every animal to increase by propagation is geometrical." Darwin realized that the potential population explosions of organisms are constantly checked by some force, which he came to call "natural selection": "Nature"—environmental factors such as predators, disease, starvation, bad weather, and so on—kills many individuals before they can reproduce and so "selects" which shall be ancestors of future generations.

(a)

(b)

(c)

**FIGURE 17–2** The Galápagos Islands. (a) In this hot, dry, isolated archipelago, Darwin discovered a wealth of endemic species (found nowhere else on Earth), such as the tree-size *Opuntia* cactus, whose flowers are shown in (b). The pads of this cactus are a favorite food of the land iguana, such as the male shown in (c). Darwin wrote, "We could not for some time find a spot free from their burrows on which to pitch our single tent . . . they are ugly animals, of a yellowish orange beneath, and of a brownish red colour above . . . These lizards, when cooked, yield a white meat, which is liked by those whose stomachs soar above all prejudices."

Upon his return to England in 1837, Darwin settled down to a lifetime of writing and thought. By the next year, he had formulated the theory of evolution by means of natural selection, but he decided to ponder it and gather supporting evidence before presenting it publicly. Twenty years later, in 1858, he received a manuscript from Wallace, in which he read his own painstakingly documented theory. Wallace, meditating on his own experiences and on Malthus's essay, had conceived the idea of natural selection and written his paper in three days. Darwin passed Wallace's paper to Lyell and to botanist Joseph Dalton Hooker, who persuaded Darwin to let them present a version of his theory with Wallace's paper at a meeting of the Linnaean Society.

Darwin then worked feverishly to finish his definitive work, *The Origin of Species by Means of Natural Selection,* which was published in 1859. In it, he marshaled an impressive array of evidence to support his theory, collected over a quarter of a century of observation and inquiry. The book sparked immense controversy, as befits the most original and important biology book ever written. Although evolution was accepted in Darwin's day, not until the twentieth century did most biologists fully accept natural selection.

Many people thought that evolution occurs by sudden transitions, and the mechanism of natural selection seemed too gradual.

Another obstacle to the acceptance of natural selection was the widespread belief in blending inheritance. Individuals who inherited different forms of a character from their parents were thought to pass on an intermediate blend of the two to their offspring. This would tend to reduce the amount of inherited variation in a population in each generation, eliminating the raw material that natural selection acts on. Blending inheritance and natural selection appeared incompatible.

Mendel's discovery that genes are inherited as discrete units was not generally known before 1901. This work showed that different parental alleles may produce an intermediate phenotype in the offspring yet be passed on to the third and future generations in their original form, thus preserving genetic variation in each generation. For example, pink snapdragons inherit one red-flower and one white-flower allele from their parents, and they pass on one or the other of these alleles to each offspring, which may have white, pink, or red flowers. In a mechanism of blending inheritance, the offspring of pink snapdragons would all be expected to inherit pink-flower alleles and have pink flowers. The inheritance of discrete genetic alleles explained why the reproduction of some individuals but not others might cause major changes in the population.

## 17–B   THE EVIDENCE FOR EVOLUTION BY NATURAL SELECTION

Several different lines of evidence convinced Darwin and Wallace, and many of their contemporaries, that modern organisms arose by evolution from more ancient forms of life. Today, every branch of biology supplies evidence for evolution.

### The Evidence from Artificial Selection

Darwin illustrated how selection causes evolution with examples drawn from the selective breeding of domestic plants and animals. Human breeders and farmers practice **artificial selection,** determining which members of a population shall reproduce and which shall not. Seed is saved from only the largest, prettiest flowers and the tastiest melons. Dairy farmers mate the cows that produce the most milk with bulls whose mothers were good milk producers (Figure 17–3). Modern hybrid corn is very different from its inbred parents. Selective breeding for particular phenotypes soon produces a population that differs markedly from the rest of the species. The striking changes produced over relatively few generations are powerful proof that organisms can evolve.

However, this evolution results from deliberate manipulation by breeders with definite ends in view. It is not easy to show that a similar process causes changes in natural populations. A weakness of Darwin's evidence for evolution was that he never provided a convincing demonstration that selection occurs in nature. His detractors pointed out that unlike human breeders, nature has no mind, no goal nor purpose. How could a haphazard series of accidents result in organisms that appeared as though they were designed specifically for the place they hold in nature? Examples of selection in wild populations were not worked out until a century later (Section 17–C).

### The Evidence from the Fossil Record

Fossil hunting was a popular recreation in nineteenth-century England. Drawings in Victorian magazines portray ladies in long skirts and gentlemen in jackets and ties scrambling over rocks, with geological hammers in their hands and fanatical gleams in their eyes. Most important fossils of the time were found by amateurs such as these.

These finds captured the popular imagination, and newspapers printed articles and letters arguing about the theological and scientific implications of fossils. Some people suggested that God had fashioned the fossils and scattered them in the rocks to delight fossil hunters. However, geologists were beginning to discern a very different explanation.

Usually, when an organism dies, scavengers and decomposer organisms rapidly destroy it. However, on occasion a body falls into an acid bog or is buried under a layer of mud that shuts out oxygen, conditions that retard decay and may permit the body to be preserved. A **fossil** is any preserved evidence of life long past: a body or body part; an impression of the surface of the body such as a footprint; substances such as oil, the chemical remains of organisms; or even coprolites (preserved excrement).

Very few fossils retain their original organic matter. For example, suppose a river carried mud to the sea, where it buried slow-moving animals, shutting out oxygen so that they decayed very slowly. Successive layers of mud piled up, their pressure converting the fossil bed to rock. Meanwhile, the animals had decayed slowly, leaving behind their imprints in the rock, or their soft parts had rotted, leaving the hard parts. The original minerals of hard parts such as shell or bone would have been replaced, ion by ion, by more stable chemicals dissolved in the water seeping through the mud and rock.

Rocks laid down by the settling and packing of small particles are called **sedimentary rocks;** examples are shale, sandstone, and limestone. These rocks may be buried still further, or they may be caught between other masses of rock and crushed and crumpled as Earth's crust shifts. Under intense heat and pressure, sedimentary rocks un-

(a)                                                    (b)

**FIGURE 17–3**  Products of artificial selection. (a) Some of the many spectacular varieties bred by dahlia enthusiasts. (b) Swiss cheese on the hoof in an alpine pasture.

dergo internal rearrangements and become extremely hard **metamorphic rocks,** so called because they have been greatly changed. Examples are slate (from shale) and marble (from limestone). The metamorphic process usually destroys fossils. Because many of the oldest sedimentary deposits have been converted to metamorphic rocks, much of the most ancient fossil record has been destroyed.

Only a minute proportion of the animals and plants alive at any one time were preserved in the fossil record. Animals with hard skeletons have a much better chance of being fossilized than those without. Thus, animals with bones or shells are better represented in the fossil record than soft-bodied worms are. Organisms living in areas subject to sedimentation are more likely to be preserved than those in other areas. Sedimentation is most widespread on the sea floor, and it is also fairly continuous in slow-moving rivers and in lakes and swamps. A terrestrial (land) animal that dies by drowning or whose body is swept down a river after death is more likely to be fossilized than one that never ventures near a large body of water. Forests produce few fossils, largely because the acidity of the leaf litter on a forest floor is likely to dissolve the skeleton away rather quickly. All of these factors make any collection of fossils subject to a very large sampling error. Hence, the fossil record gives a biased history of life.

We sometimes see fairly complete descriptions not only of the structure but also of the life and habits of an extinct organism. How do we obtain this information just from the animal's fossilized parts? First, certain features of an animal's skeleton, such as the placement and size of the limbs, pro-

vide information about its habitat and locomotion. The size and shape of the teeth may give evidence of the animal's diet. Such things as a heap of bones in the position of the animal's stomach may tell us what it ate. Finding a large animal skeleton with a group of small but similar skeletons in its pelvic region strongly suggests that the young developed inside their mothers' bodies (Figure 17–4). The type of deposit containing the fossil, its relationship to nearby rock formations, and the other fossils present can also reveal much about the habitat of the fossilized organism: whether it was from the mountains or the plains, from a freshwater or marine environment.

Many people recognized that the growing collections of fossils in North America and Europe provided strong evidence that organisms had changed over time. First, nearly all fossils come from species that are now extinct. Second, some species are older than others. Many fossils occur in rock formations made up of several layers. Geologists realized that the bottom **stratum** (layer) had usually been formed first and contained older fossils than the overlying strata (Figure 17–5). Determining relative ages of fossils in this way made it clear that some groups of species were older than others. In this century, the discovery of radioactivity made it possible to calculate the actual age of rock formations by using isotopes (*How Do We Know? Isotopes,* Chapter 2).

This book's endpaper shows some of the evidence collected from rocks of various ages. For instance, Paleozoic rocks contain no birds, mammals, nor flowering plants. Various types of fish, on the other hand, are found in Or-

**FIGURE 17–4**   Fossil mother-to-have-been. This skeleton of a whale-like reptile, *Ichthyosaurus quadricissus*, was fossilized along with her unborn young. The arrows point to the skull and backbone of one of the young. The skull of another lies below the bend in the adult's tail. Several other young are still in the mother's body cavity.   (Photo 35041 by Anderson, courtesy Department of Library Services, American Museum of Natural History)

dovician and all more recent times, but not in Precambrian rocks. The abundance of members of different groups also changes with the age of the rock. Fossil reptiles are common in Jurassic and Cretaceous rocks, but become rare in Quaternary rocks, whereas the opposite is true for birds and mammals. From all this, it appears that some groups of organisms originated earlier than others and that some more recent groups have largely replaced more ancient ones over the course of time.

In a few cases, fossils allow us to reconstruct the genealogy, or family tree, of a particular group. For instance, we can trace the origin of mammals from reptiles in great detail in the fossil record. We can also follow the change of a dinosaur-like reptile of the Permian to an ancient bird with feathers and teeth in the Jurassic (Figure 17–6a), to modern birds with feathers and no teeth.

The classic case of fossil genealogy is the story of horse evolution, published by Othniel C. Marsh in 1879. Marsh described older and older fossils linking modern horses with their dog-sized ancestors found in Eocene rocks. Through the fossil record he traced the major changes in the teeth, legs, and feet of ancestral horses. Today we know that the evolutionary tree of horses also contains many other branches representing extinct species with varied size and form (Figure 17–7). Many similar examples have now been described in which the ancestry of modern species can be traced through successive rock strata, with the youngest rocks containing those fossils most like the modern forms.

## The Evidence from Comparative Anatomy

Even without fossil evidence that different organisms lived at different times in the past, we might suspect that organisms had evolved by comparing the structures of species alive today. Not surprisingly, similar kinds of organisms have very similar structures. For example, different members of the cat family have highly similar skeletons, teeth, muscles, and so on. The same is true of different species of bats or of whales.

Moreover, comparing the bones of cats, bats, and whales reveals that these three groups of animals all have skeletons composed of similar groups of bones despite their adaptations to very different ways of life. For example, the

**FIGURE 17–5**   Fossil-bearing rock strata in the Grand Canyon.

(a)

(b)

**FIGURE 17–6**  Fossils. (a) *Archaeopteryx*. This rare fossil from the Jurassic period is the remains of one of the first known birds. Note the impressions left by the feathers, characteristic of birds, in the area of the forelimbs and the tail. The skeleton is lizard-like, with a long tail containing many vertebrae. (b) Petrified logs. During the late Triassic period, this tree was buried in mud on a floodplain in what is now Arizona. Water seeping through the wood deposited mineral ions in place of the original organic matter, preserving the structure of the tissues. Today erosion is uncovering these long-buried logs in the Petrified Forest.

forelimb bones of all these animals are arranged in the same pattern: a bat's wing, a cat's foreleg, and a whale's flipper all contain bones identifiable as humerus, radius, ulna, and so on (Figure 17–8). Indeed, all mammals, birds, reptiles, and adult amphibians (frogs, newts, and their relatives) have forelimbs containing the same bones, although the limbs perform different functions in animals as different as a pigeon, a penguin, a turtle, and a human. Furthermore, all of these forelimb bones originate from the same structures in the embryo. Such structures, with the same evolutionary and embryonic origin but occurring in different species, are said to be **homologous** to each other. Homologous organs may perform the same or different functions.

The converse of homologous organs are **analogous** organs, which have similar functions but different development and structure and appear to be unrelated. The wings of birds and of insects, for instance, may both be used for flying, but they have completely different structure and origin (Figure 17–9).

Darwin saw that homologous and analogous organs posed problems to the creationist view. It made no sense to create several different types of wings. Even the homologous wings of birds and bats differ somewhat in structure.

Surely one design must be superior to the other. Why create several different sorts of wings?

Similarly, why do so many animals contain apparently inefficient homologous structures? Why do whales have flippers containing heavy bones, like those of terrestrial mammals, instead of the lighter, folding fins of a fish, apparently so much better suited to aquatic propulsion?

Darwin came to realize that evolution made sense of all these paradoxes. Organisms were not created from a clean slate; they arose from ancestors with characteristics already determined by their own evolutionary histories. Whales have bony flippers because they evolved from terrestrial mammals with bony forelimbs. Insects have no bones to support their wings because they evolved from animals with external skeletons but no bones. To Darwin, the very imperfection of adaptations, the feeling that so many of them could have been designed better, became the most convincing evidence that evolution has occurred.

Anatomy provides a further argument for evolution in the form of **vestigial** structures: organs useless to their present owners but homologous to structures that serve important functions in other species. The most familiar example is the human appendix, a small sac near the junc-

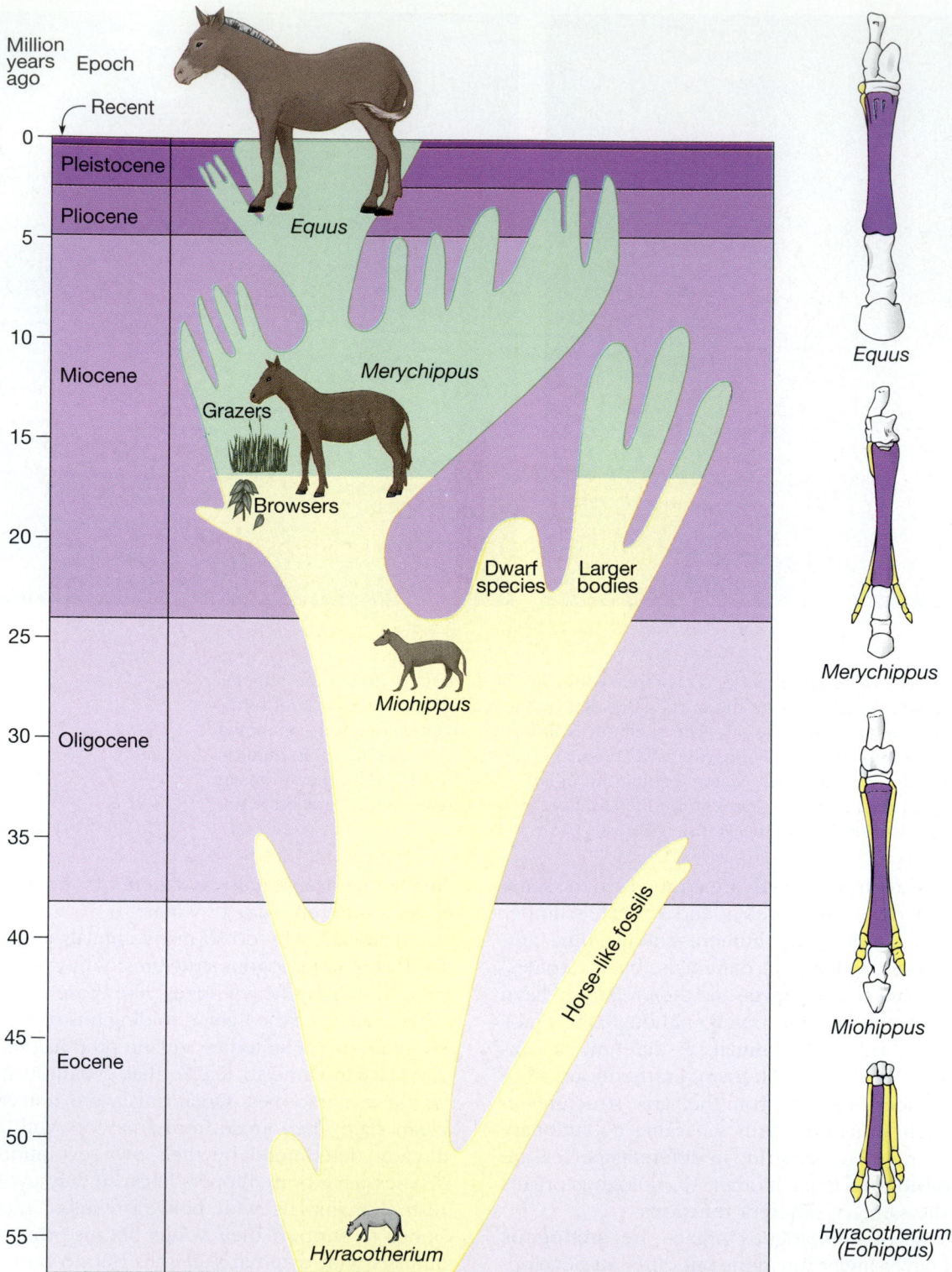

**FIGURE 17—7**  Evolution of horses. The ancestral *Hyracotherium* gave rise to many species. Most became extinct, but some have survived to the present. The legs of the four species shown here exhibit progressive loss of the side toes and development of the central toe, as the animals' stance changes to more "standing on tiptoe" (a horse's hoofs are its toenails). The animals are also progressively larger, although some evolutionary side branches shown produced species smaller than their ancestors.

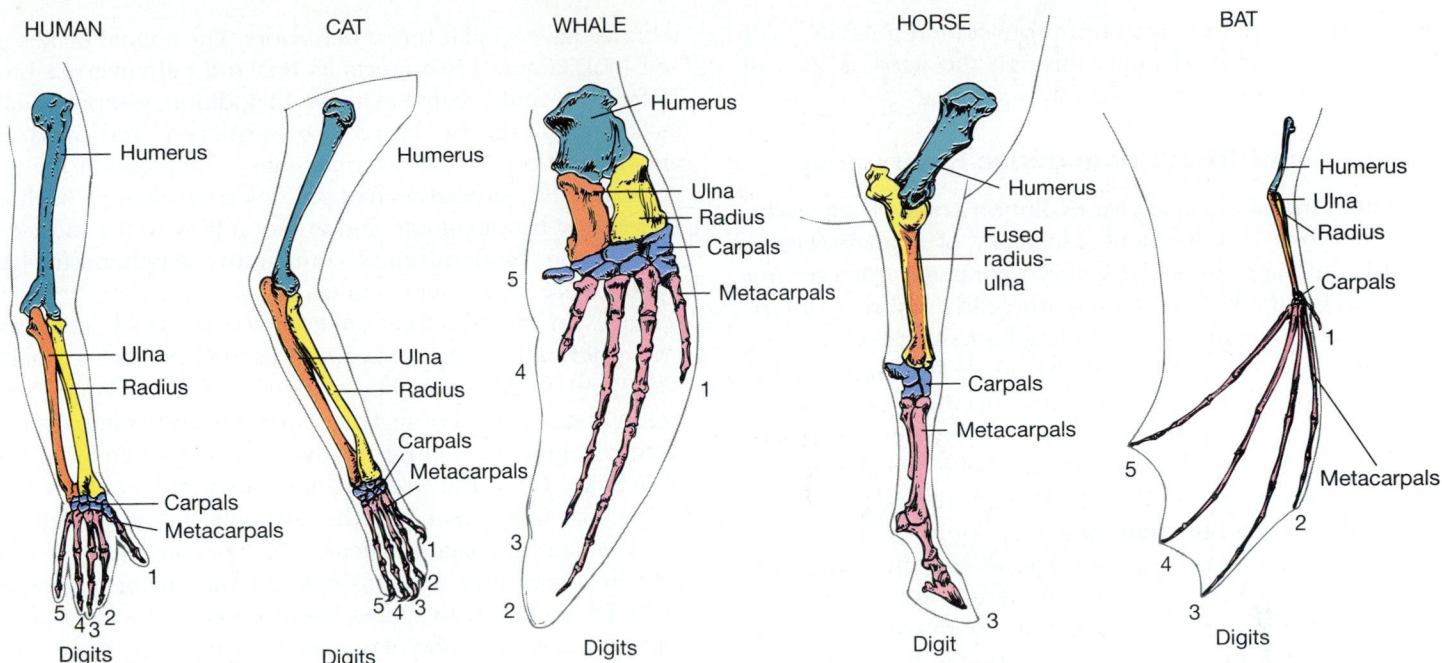

**FIGURE 17–8**   The bones of the forelimbs of five mammals. All have the same basic pattern, but in different species the various bones evolved different proportions as the limbs became adapted to different functions. The digits (fingers) are numbered 1 through 5, but note that the horse has lost all but the greatly enlarged third digit.

tion of the small and large intestines (see Figure 31–7). The appendix is homologous to the caecum, a large blind chamber in which leaves and grasses are digested in many other mammals. Another example is the existence of minute pelvic and hind limb bones in the skeletons of whales and of boa constrictors, even though these animals have no external hind limbs. A whale fossil from 40 million years ago shows that the 16-meter-long giant did have external hind limbs less than a meter long, too short to be used in locomotion. A still older whale, from 57 million years ago, had larger but weak hind limbs, which probably were used much like a sea lion's both on land and in water. This provides evi-

(a)                                     (b)

**FIGURE 17–9**   Analogous structures. The wings of an elephant hawk moth (a) and a blue tit (b) are both broad, flat, lightweight structures used in flight. However, they have very different developmental and evolutionary origins.   (Stephen Dalton/Photo Researchers)

dence that whales evolved from four-footed animals, with the reduction of hind limbs through the ages as the tail became the main organ used for swimming.

## The Evidence from Comparative Embryology

A similar line of evidence for evolution comes from studying embryonic development, especially of animals (Figure 17–10). In many cases, the embryo contains structures not found in the adult. For example, the early embryos of reptiles, birds, and mammals, including humans, develop a row of vestigial gill slits just behind the head. This suggests that these groups of animals descended from fishes, in which the embryonic gill slits develop into those of the adult. Similarly, the embryos of baleen (whalebone) whales and of birds develop tooth buds, even though the adult animals are toothless. All vertebrate embryos sport a tail that projects beyond the anus, a feature that persists into adulthood in most species but is lost by adult frogs, great apes, and humans. Sometimes human babies are born with short tails, or with several nipples in two rows down the front of the body, features common in other mammals.

Embryology also reveals that certain apparently "new" features of vertebrates developed, not from scratch, but from the remodeling of ancestral structures. For instance, some of the bony arches that support the gills in fishes become, in vertebrates that evolved later, the lower jaw and the ear bones. Furthermore, all early vertebrate embryos contain several pairs of blood vessels in the gill area. In fishes, these develop into the arteries that carry blood to the gills, but in other vertebrates some of them degenerate, while others become modified to carry blood through new routes, such as to the evolutionarily "new" respiratory organs, the lungs.

The development of many embryonic structures makes sense only in light of the organism's evolutionary history.

## Other Comparative Evidence

The arguments for evolution based on comparative anatomy have parallels in other fields of biology, such as comparative animal behavior and comparative physiology (the physical principles on which organisms work). For example, vertebrates have similar threat behaviors. The animal makes itself look larger: a fish erects its fins and gill covers, a bird its feathers, and a mammal its fur. In addition, weapons such as teeth, beak, or horns are displayed and held in attack position. These external signs of aggression are accompanied by similar internal physiological changes such as increase in heartbeat rate and in blood flow to the muscles.

In more modern times, comparative biochemistry has joined these ranks, with studies of the similarities and differences in the structures of homologous DNA, proteins, and other biological chemicals. Molecular evidence usually bears out fossil and other evidence of the relationships among organisms. For instance, humans and chimpanzees, long recognized as close relatives, have proteins that are 99% alike. DNA and proteins are much less similar when they come from organisms that are only distantly related.

Comparing macromolecules has become a powerful tool in elucidating relationships among organisms, especially for groups with sparse fossil records, few anatomical features, and no embryology, such as bacteria, fungi, and unicellular eukaryotes. We shall examine this sort of evidence in Section 22–E and later chapters.

## The Evidence from Biogeography

Biogeography is the study of the geographical distribution of organisms. In their travels, both Darwin and Wallace noticed that the present-day distribution of organisms made no sense seen from a creationist point of view but could be explained by evolution. Why did the Galápagos Archipelago, a group of small islands 960 km off the west coast of South America, contain more different species of finches than the entire South American continent? Another puzzle was the distribution of mammals. Why were marsupial (pouched) mammals found only in Australia and South America? (Opossums are marsupials, but have colonized North America from South America.) Why did Australia contain so few of the placental mammals found throughout the rest of the world?

In Australia, mammals are represented mainly by marsupials. There are also three species of monotremes (egg-

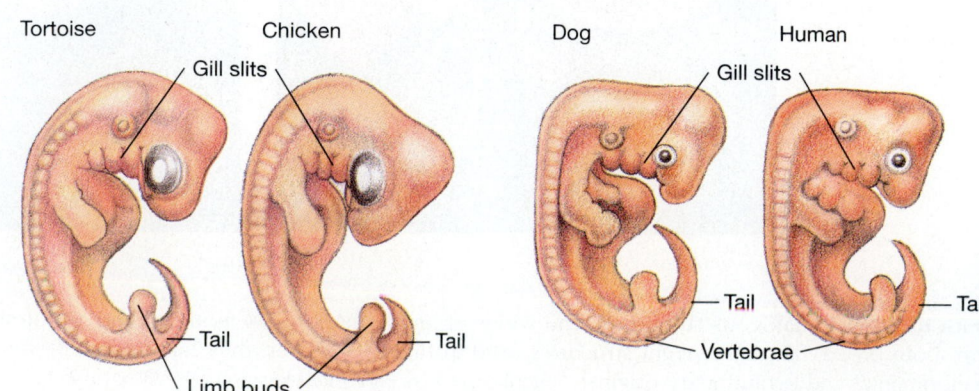

**FIGURE 17–10** Similarities among vertebrate embryos. Structures that are homologous in all vertebrates are obvious in these early embryos. Each has a column of vertebrae, limb buds, and a tail that projects behind the anus, as well as traces of gill slits in the neck region.

layers: the duck-billed platypus and two species of echidnas). The only placentals were bats and rodents until human immigrants brought other forms in relatively recent times. The marsupials underwent their own adaptive radiation and gave rise to a variety of species resembling their placental equivalents elsewhere. Australian marsupials include the rabbit-like bandicoot, the woodchuck-like wombat, the extinct Tasmanian "wolf," and the squirrel-like flying phalanger, as well as unique forms such as kangaroos. Rabbits and bandicoots, wombats and woodchucks have undergone **convergent evolution,** the evolution of the same adaptations in unrelated organisms, presumably as a result of natural selection in similar environments (Figure 17–11).

PLACENTALS

MARSUPIALS

Mole

Marsupial mole

Jerboa

Jerboa marsupial mouse

Flying squirrel

Flying phalanger

Anteater

Numbat

Woodchuck

Wombat

Wolf

Tasmanian wolf

**FIGURE 17–11** Convergent evolution. As a result of the similar selective pressures of their habitats and ways of life, many species of marsupials in Australia and of placentals on other continents evolved remarkably similar structure and behavior. A few striking examples are shown here. Marsupials are pouched mammals. Placentals are named for the placenta, the organ in which food, oxygen, and wastes are exchanged between a pregnant female and the fetus.

Both Darwin and Wallace became convinced that this and dozens of other puzzles could be explained as the result of the evolutionary histories of these modern organisms, including where their ancestors lived. From an ancestral group living in a particular place, descendant populations could spread, or radiate, into other areas. In doing so, they would encounter new environmental conditions and evolve new adaptations. Such an evolutionary process, giving rise to new species adapted to new habitats and ways of life, is called **adaptive radiation.**

The evolution of major groups such as flowering plants, reptiles, birds, and mammals was adaptive radiation on a grand scale. Some species resulting from the adaptive radiation of one group became the ancestors in the adaptive radiation of a succeeding group. Most of the species have since become extinct. Thus most reptiles have disappeared, but not before they gave rise to the great adaptive radiations of the mammals and birds, as well as to modern representatives of the reptiles themselves, such as snakes, lizards, and crocodiles.

***Islands: Laboratories of Evolution***   Like Wallace, Darwin was profoundly struck by the flora and fauna of islands, particularly in the Galápagos Archipelago. What caught Darwin's attention was the remarkable numbers of **endemic** species, species found nowhere else, even on other apparently similar islands nearby. That the tiny, relatively barren Galápagos Islands (and other islands visited by the *Beagle,* shown in Figure 17–12) housed large num-

**FIGURE 17–12**   Voyage of the *Beagle.* The route followed by *H.M.S. Beagle* on her voyage around the world, 1831 to 1836. Most of the time was spent surveying the coast of South America, but the ship also called at many islands. Darwin was able to observe, collect, and compare mainland and island organisms. The photograph shows a replica of the *Beagle.* (Christopher Ralling)

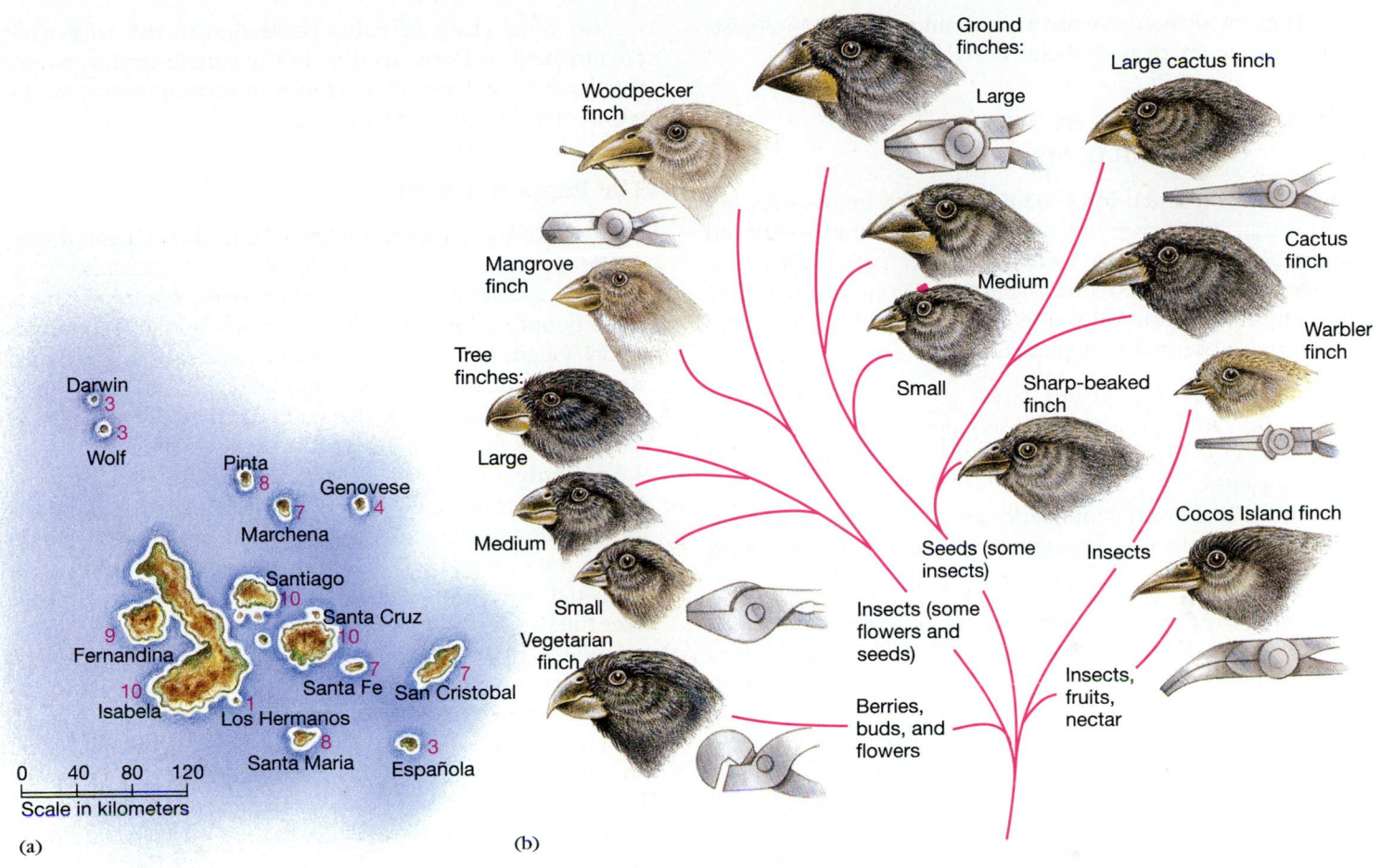

**FIGURE 17–13** Darwin's finches. (a) Map of the Galápagos Islands. Red numbers show how many species of Darwin's finches inhabit each island. (b) Heads of Darwin's finches showing different bill sizes and shapes and the human tool that each type of bill resembles. A bird's bill is a built-in tool for obtaining food. Darwin's finches have undergone adaptive radiation, and each species has evolved a bill structure adapted to a certain type of diet.

bers of endemic species seemed a profligate waste of effort by the Creator.

Darwin grasped that the existence of so many endemic organisms could be explained by assuming that members of mainland species had emigrated to the islands, and there evolved into new forms of life. Local conditions on islands differ from those on the mainland, so natural selection on an island inevitably produces different adaptations. In addition, because members of competing species often have not emigrated to the island, the island pioneers may give rise to several new species through adaptive radiation: different populations of their descendants become adapted to different habitats or ways of life that were not available to their ancestors on the mainland.

Such was the case with the 13 species of Darwin's Galápagos finches: the ancestral finches must have come from the South American mainland, and their descendants underwent adaptive radiation in the Galápagos Islands, be-

coming adapted to living in different habitats and to eating different sorts of food (Figure 17–13). A fourteenth species evolved on Cocos Island, about 500 km away.

In the Hawaiian Islands, about 3000 insect species have been found. It is thought that these islands were invaded by some 250 insect species and that the current tally results from adaptive radiation of the invaders. Only a few dozen species of the fruit fly genus *Drosophila* live in the western United States, whereas several hundred species in this group inhabit Hawaii. Similarly, of the 1700 species of Hawaiian plants, some 270 are thought to be colonizers; the 1430 others are probably endemic, formed by adaptive radiation from the few pioneer species. In the Galápagos Islands, closer to the mainland source of immigrant species, more than 30% of the plants are endemic species. This is an extraordinary amount of speciation for such relatively small areas.

Darwin and Wallace became convinced that only an evolutionary origin for species could reasonably explain the distribution of modern plants and animals, and biologists since have agreed with them.

## 17–C   EVOLUTION BY MEANS OF NATURAL SELECTION

All of this evidence for evolution is quite impressive, but the feature of Darwin and Wallace's theory that convinced most people they were right is the mechanism they proposed for evolution: natural selection. Evolution by means of natural selection follows inescapably from a few simple and often observed biological facts:

1. Individuals of a species vary.
2. Some variations are genetically determined.
3. More individuals are produced than live to grow up and reproduce.
4. Individuals with some genes are more likely to survive and reproduce in a particular environment than those with others.

Conclusion: From these four premises it follows that those genes that make their owners more likely to grow up and reproduce will become more common in a population from one generation to the next.

To take an example, a minute percentage of the people now living have genetic alleles that make them resistant to HIV (human immunodeficiency virus). These people will survive and reproduce even if they are exposed to the virus, whereas others without the relevant genes will succumb to AIDS as children and young adults, without having reproduced. And only by reproducing does an individual pass on his or her inherited characteristics. If an organism does not reproduce, it plays no direct role in the evolution of future generations: its genes have been selected against.

Inherited traits that improve an organism's chances of living and reproducing will be more common in the next generation than traits that decrease its chances of reproducing. Various combinations of genes will be naturally selected for or against, from one generation to the next, depending on how they affect the chance of reproducing. For natural selection to cause a change in a population from one generation to the next (that is, to cause evolution), it is not necessary that all genes affect survival or reproduction; the same result occurs if just some genes make an individual more likely to grow up and reproduce.

We now understand that other factors also cause evolution (Chapter 19), but natural selection is far and away the most important.

Natural selection is a simple idea: some genotypes are reproduced more frequently than others. However, it is not so simple to grasp how natural selection affects populations and brings about evolution. This is one area of biology where thinking about a subject will teach you more than reading about it. (This chapter's Questions for Discussion list some examples to think about.)

No good cases of natural selection in the wild were documented in Darwin's day, but twentieth-century scientists have provided such evidence in several instances. We shall consider some of them here.

### The Peppered Moth

The case of the peppered moth in England is a classic example of natural selection in the wild, documented by observation and experiment. In nineteenth-century England, many people collected moths and butterflies, and collectors avidly sought rare specimens of the peppered moth, which were a dark charcoal color rather than the usual pale, mottled gray. We now know that each moth's genes determine whether it is the normal pale form or the dark form, called the **melanic** form after the black pigment melanin. By looking at collections made from about 1850 to 1950, biologists found that melanic moths became more and more common during the century, and pale ones scarcer, particularly near industrial cities. This change in a population of organisms over time is, in itself, evolution.

By the 1930s it was realized that other species of moths were also becoming darker near industrial cities, although the pale forms remained more common in the countryside. Geneticist E. B. Ford proposed that natural selection had brought about the change to a darker form in industrial areas, a phenomenon known as **industrial melanism.**

Moths are generally nocturnal: they fly, feed, and mate at night. During the day they rest on tree trunks or other surfaces, protected from predators by camouflage. Ford proposed that before industrial pollution, the typical pale gray form of the peppered moth had been well camouflaged against pale, lichen-covered tree trunks. In polluted areas, however, where industrial smoke had killed the lichens and blackened the tree trunks, the typical pale form stood out in contrast to this dark background (Figure 17–14). Here, many more pale than melanic moths would be found and eaten by predators. The most likely predators were birds, which hunt by sight, and against whom camouflage, or lack of it, would be important.

In the 1950s Bernard Kettlewell decided to test Ford's hypothesis experimentally, using the **mark-release-recapture** technique. He bred large numbers of the melanic and pale forms of the moth, marked them, and released them in two different environments: one an unpolluted rural area in Dorset, where the melanic form was more visible, the other near the polluted industrial city of Birmingham, where the pale form was easier to see against the blackened tree trunks. Kettlewell then recaptured as many of the marked moths as possible on subsequent days.

If equal numbers of both forms were released and equal numbers fell victim to predators after their release, equal numbers of the two sorts should be recaptured (as long as both behaved in such a way that they were equally likely to be recaptured). Any variation from a 50:50 ratio of the

(a)  (b)

**FIGURE 17–14** The two forms of the peppered moth. (a) A melanic moth is highly visible against the pale lichens of a tree trunk in an unpolluted area, whereas the normal pale form is well camouflaged. (b) Moths on a blackened tree trunk covered with soot, which has killed all the lichens. Against this background, the pale form of the moth is much more visible than the melanic one. (Michael Tweedie/Photo Researchers)

two forms recaptured would indicate that one form had suffered a higher mortality than the other after their release.

The percentage of melanic moths recovered was twice that of pale moths in the industrial area, but only half that of pale moths in the unpolluted countryside. This agreed with the prediction that the pale moths were more likely to survive (and so to be recaptured) in the country, and melanic moths were more likely to survive near the city.

Are these results due to differential camouflage protecting the moths against their predators? To find out, Kettlewell hid in a blind and watched moths he had placed on tree trunks. On one occasion, equal numbers of gray and black moths were released in unpolluted Dorset, and watchers in the blind recorded 164 melanic and only 26 gray moths captured by birds.

Table 17–1 gives the numerical results of Kettlewell's experiments. In a polluted area, a larger proportion of melanic than of gray moths is likely to live long enough to reproduce. Since the color of the moths is inherited, the next generation will contain proportionally more melanic moths. In other words, the frequency of the alleles for black color increases in the population with time—and that is evolution.

The selective pressure that brings about this evolution is clear: in polluted areas, pale moths are selected against because birds eat a higher percentage of pale than of melanic moths. Natural selection over many generations has produced populations of the peppered moth that are well adapted to survive in their environments, populations whose characteristics change as the environment changes.

On the basis of this evidence, we would predict that if pollution were reduced, melanic moths would become rarer and pale ones more common in industrial areas. In fact, the Clean Air Act of 1952 reduced air pollution in England, and collections of peppered moths from industrial Manchester in the next 20 years revealed a dramatic decrease in the proportion of melanic individuals in the moth population. The ability to predict events in this way is the most impressive evidence that can be produced for a scientific theory.

## Tolerance of Toxic Metals by Plants

Another well-documented example of natural selection is found in plants that grow on soil containing wastes from copper, zinc, or lead mines. These metals are toxic (except in tiny amounts), and most plants cannot grow on the contaminated soil. The plants that do grow around mines belong to species of grasses and weeds also found in nearby pastures, but not all members of these species can tolerate toxic metals.

In one experiment, seeds from a nontolerant population of grasses were scattered on soil contaminated with

**TABLE 17–1**

**Numbers of Pale and Melanic Peppered Moths Recaptured After the Release of Marked Individuals in Two Areas**

| Location | Year | | Pale | Melanic |
|---|---|---|---|---|
| Dorset (unpolluted) | 1953 | Released | 469 | 473 |
| | | Recaptured | 62 | 30 |
| | | **% recaptured** | **13.2** | **6.3** |
| Birmingham (polluted) | 1953 | Released | 137 | 447 |
| | | Recaptured | 18 | 123 |
| | | **% recaptured** | **13.1** | **27.5** |
| Birmingham (polluted) | 1955 | Released | 64 | 154 |
| | | Recaptured | 16 | 82 |
| | | **% recaptured** | **25.0** | **53.3** |

copper. Only 1 in 7000 survived and grew; the copper was clearly a strong selective force!

Janis Antonovics planted seeds from metal-tolerant populations of grass on contaminated soil and found that tolerance to metals varied more among the seeds than among adult plants. In other words, each new generation (the seeds) showed considerable genetic variation, and the presence of the metal in the soil selected against individuals that lacked the genes conferring tolerance. The metals inhibited the germination of seeds only slightly; most deaths occurred among the seedlings because they could not establish roots. Only highly tolerant plants survived to adulthood and contributed their genes for metal tolerance to the next generation.

## 17–D  GENETIC CONTRIBUTION TO FUTURE GENERATIONS

The phrase "survival of the fittest," often used in discussions of evolution, suggests that the important outcome of natural selection is survival; it is not. Rather, it is contribution of genes to future generations. Survival is important, because an individual cannot contribute genes to the next generation unless it survives long enough to reproduce, but even reproduction does not guarantee evolutionary success.

Table 17–2 shows how many young starlings survived for three months after hatching. The female starlings laying

the most eggs did not have the greatest evolutionary success: large broods were selected against, in that hardly any of the young survived. Females laying four or five eggs per brood had a higher number of offspring surviving for at least three months after hatching.

David Lack showed that young birds from larger broods weighed less than those from smaller broods, presumably because the parents could not provide enough food for so many nestlings. Food shortage was probably a major cause of death among young from larger broods. Table 17–2 also shows that nestlings from the most common brood sizes have the lowest mortality, as we would predict from the action of natural selection. It seems reasonable to suppose that in years when there is more (or less) food available to the birds than in the year Lack studied, selection would favor birds with broods larger (or smaller) than average.

Of course, a starling's reproductive success often depends on more than one brood. Selection optimizes reproductive success over a lifetime, and the adaptations that produce this success are myriad.

Selection has led to reproductive strategies that are very different from species to species. For example, a herring may lay 3 million eggs at a time; most of them die, but a few survive to carry on the parents' genes. At the other extreme, humans lavish about 20 years of parental care on each offspring.

## 17–E  ADAPTATIONS

When we say that the environment exerts selective pressure that results in populations of organisms adapted to their environment, we use "environment" to mean all factors, other than the organism's genes, that can affect whether it lives to reproduce or not. This includes its environment as an embryo, a juvenile, and an adult.

For example, consider an acorn growing into an oak tree. Whether it survives the selective pressures of its environment depends on whether temperature and rainfall permit normal embryonic development and germination, whether bacterial or fungal infections destroy it as a seed or seedling, whether as a seedling it has enough stored food for rapid growth, whether it escapes being eaten, whether the soil in which it grows can support a large plant, and whether, as a small tree, it avoids death by disease, trampling, or browsing. To make things more complicated, selective pressures may frequently be contradictory. For instance, a hot summer is in one way advantageous because the acorn can grow faster at higher temperatures, but hot weather also increases the chance that the soil will dry out before the root system grows deep enough to obtain a reliable supply of water.

In this section we consider several sets of adaptations that are especially illuminating because biologists have worked out the selective forces involved in their evolution.

**TABLE 17–2**

### Survival in Swiss Starlings in Relation to Number of Eggs Laid*

| Brood Size (Number of Eggs in Nest) | Number of Young Marked | Recoveries per 100 Young Marked[†] |
|---|---|---|
| 1 | 65 | 0 |
| 2 | 328 | 1.8 |
| 3 | 1278 | 2.0 |
| 4 | 3956 | 2.1 |
| 5 | 6175 | 2.1 |
| 6 | 3156 | 1.7 |
| 7 | 651 | 1.5 |
| 8 | 120 | 0.8 |
| 9, 10 | 28 | 0 |

*The number of eggs laid during one nesting period is genetically regulated and, like other genetic variations, is acted upon by natural selection. David Lack marked all the nestlings in all the nests he could find and then recaptured them months later when they had left the nest.

[†]The only recoveries scored are those for birds more than three months old when they were recaptured. Source: Lack, D. Ecology 2, 1948.

## Resistance to Pesticides and Antibiotics

Several dramatic examples of natural selection in action, and the resulting adaptations, are provided by the evolution of resistance to pesticides and antibiotics.

A scale insect feeds on citrus trees in California. In the early 1900s, growers killed the scale by spraying the trees with cyanide. But in 1914 some scale insects survived the spraying because they had a single gene, newly apparent in the population, that permitted them to break down cyanide into harmless compounds. As spraying continued, more insects with the new gene than without it survived to reproduce, and they passed on the gene to their offspring. Before long the whole population was resistant to the spray.

Because scale insects, like many other insects, have more than one generation a year, they evolve quickly. To combat the evolution of resistance, growers are encouraged to spray only when necessary and to use different chemicals in different months or years.

Similarly, resistance to the rat poison warfarin appeared in rats about a decade after its introduction in the 1950s. Resistant rats now live in many parts of the United States and Europe.

Bacteria also evolve resistance to antibiotics meant to kill them. When a bacterial population meets a particular drug, individuals susceptible to that drug are killed (Figure 17–15). Sometimes a population contains one or a few individuals with mutations conferring resistance to the drug; they survive, and they multiply particularly rapidly because the death of susceptible bacteria reduces competition for food. In addition, many genes conferring resistance to antibiotics are carried in plasmids, which can be duplicated and passed to other members of the population, giving them genetic resistance too (Section 9–F). Soon genes for resistance become widespread. Some strains of the bacteria that cause tuberculosis and the sexually transmitted disease gonorrhea can no longer be killed by any known drug.

Because antibiotics and disease-causing bacteria frequently meet in hospitals, it is not surprising that hospitals often harbor drug-resistant bacteria. In many countries, women are now encouraged to give birth at home if possible, because mother and infant are safer from bacterial infection at home than in the hospital.

Most countries have outlawed the use of antibiotics in cattle feed. Cattle fatten faster if fed antibiotics, but they also become breeding grounds for antibiotic-resistant bacteria. Antibiotics are still added to cattle feed in the United States, and drug-resistant bacteria in cattle are becoming increasingly common. Such bacteria have caused many cases of *Salmonella* diarrhea, some of them fatal, in people who eat infected hamburger.

Disease-causing organisms don't always prevail. In 1915 a disease killed nearly all the oysters in Malpeque Bay of Prince Edward Island, Canada. The few survivors reestablished the population. Fifteen years later, the disease-causing organism was still present, but most of the oyster population now had genetic resistance to it; only 1 oyster in 1000 was susceptible. By 1938 the oyster harvest was higher than before the disease struck, and when the disease appeared elsewhere, oysters from Malpeque Bay were sent to contribute their genetic resistance to the newly afflicted populations.

In all these cases, note that individuals did not evolve: they either did or did not happen to have genes that permitted them to survive the selective pressure of cyanide, antibiotics, or disease. The population evolved a change in its overall genetic makeup by means of natural selection: the reproduction of resistant genotypes more than susceptible ones.

These examples of the evolution of adaptations illustrate merely a few of the selective pressures that are always acting on all organisms and the adaptations that have evolved as a result of natural selection among randomly produced variations of genes.

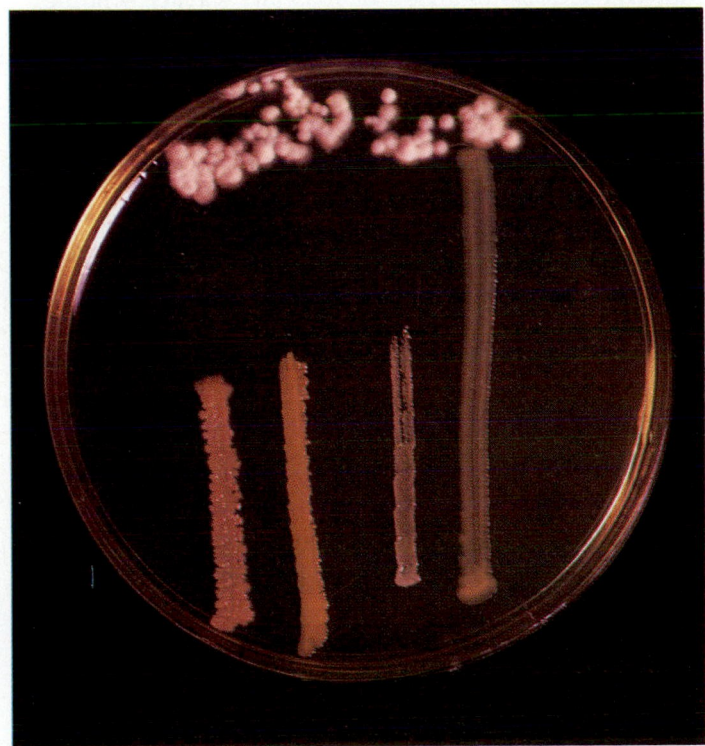

**FIGURE 17–15** Bacterial resistance to antibiotics. The fuzzy dots across the top of this dish are clumps of the fungus *Penicillium,* which produces the antibiotic penicillin. The penicillin spreads out through the dish. The four lines are rows of different varieties of bacteria. Three of the bacterial varieties have been killed as the penicillin reached them; the fourth (far right) is penicillin-resistant, and it continues to grow. (Biophoto Associates)

## E S S A Y : *Charles Darwin*[1]

*C*harles Darwin was born in 1809 into the sort of upper-middle-class English society that Jane Austen brought to life in her novels.

My father sent me to Edinburgh University where I stayed for two years. I became convinced that my father would leave me property enough to subsist on with some comfort; my belief was sufficient to check any strenuous effort to learn medicine. The instruction at Edinburgh was altogether by lectures, and these were intolerably dull. Dr. Duncan's lectures on Materia Medica at 8 o'clock on a winter's morning are something fearful to remember.

It has proved one of the greatest evils of my life that I was not urged to practice dissection, for I should soon have got over my disgust, and the practice would have been invaluable for all my future work.

*Darwin dropped out of medical school and went to Cambridge to study theology.*

From my passion for shooting and hunting, I got into a sporting set. We used often to dine together in the evening and we sometimes drank too much, with jolly singing and playing at cards afterwards. I know that I ought to feel ashamed of days and evenings thus spent, but as some of my friends were very pleasant, and we were all in the highest spirits, I cannot help looking back on these times with much pleasure.

But no pursuit at Cambridge gave me so much pleasure as collecting beetles. No poet ever felt more delight at seeing his first poem published than I did at seeing the magic words "Captured by C. Darwin, Esq." on the label for an insect.

In 1831 on returning home I found a letter informing me that Captain Fitz-Roy was willing to give up part of his own cabin to any young man who would go with him, without pay, as naturalist to the voyage of the *Beagle*. Afterwards, I heard that I had run a very narrow risk of being rejected because of the shape of my nose! Fitz-Roy was an ardent disciple of Lavater, and was convinced that he could judge of a man's character by the outline of his features.

The voyage of the *Beagle* has been by far the most important event of my life. As far as I can judge, I worked to the utmost during the voyage from the mere pleasure of investigation. But I was also ambitious to take a fair place among scientific men.

During the voyage, I had been deeply impressed by discovering great fossil animals like existing armadillos. It was evident that such facts as these could only be explained on the supposition that species gradually became modified. It was equally evident that neither the surrounding conditions, nor the will of the organisms could account for the innumerable cases in which organisms of every kind are beautifully adapted to their habitats. I soon perceived that selection was the keystone to man's success in making useful races of animals and plants. But how selection could be applied to organisms living in a state of nature remained for some time a mystery to me. In October 1838, I happened to read for amusement Malthus on population and, being well prepared to appreciate the struggle for existence which everywhere goes on, it at once struck me that under these circumstances favourable variations would tend to be preserved and unfavourable ones destroyed. The result of this would be formation of a new species.

*This was in 1838. It was nearly 20 years before Darwin published his theory, although in the meantime he wrote his ideas down in unpublished essays and published books and articles on a myriad of other biological subjects. Darwin offers no explanation for his long delay except to say:* I had at last got a theory by which to work; but I was so anxious to avoid prejudice, that I determined not for some time to write even the briefest sketch of it.

Early in 1856, Lyell advised me to write out my views pretty fully and I began at once to do so. But my plans were overthrown, for early in the summer of 1858 Mr. Wallace sent me an essay, and this essay contained exactly the same theory as mine. [*Wallace had written his essay in three days!*]

*Lyell urged that Wallace's essay and the abstract of Darwin's manuscript be published together.* I was at first very unwilling to consent, as I thought Mr. Wallace might consider my doing so unjustifiable, for I did not then know how generous and noble was his disposition. Nevertheless, our joint production excited very little attention and the only notice I can remember was by Professor Haughton, whose verdict was that all that was new in them was false, and all that was old was true. This shows how necessary it is that any new view should be explained at considerable length in order to arouse public attention.

*Finally Darwin described evolution and natural selection in* The Origin of Species, *published in 1859. This version of the theory, full of details and examples, immediately attracted attention, and the book became a bestseller. It sparked controversy and public debate, which has continued*

[1]Most of this essay is in Darwin's own words. Our additions are in italics. Excerpts are from *The Autobiography of Charles Darwin* (Nora Barlow, ed.). London: Collins, 1958.

*ever since. In particular it bothered, and continues to offend, some Christians. This is for two reasons. First, the theory of evolution contradicts a literal interpretation of the biblical story that Earth and its organisms were created in seven days. This offends people who believe in the literal truth of the Bible. Second, the theory superficially leads to a rather deterministic view of life, since neither human nor divine agency is needed for evolution to occur. Darwin himself became trapped in this second point of view. His mind rejected any reality that could not be tested by observation or experiment.*

*Toward the end of his life he wrote: "I gradually came to disbelieve in Christianity as a divine revelation." His wife, Emma Wedgewood, fought him relentlessly on this point. She wrote to him, "May not the habit in scientific pursuits of believing nothing till it is proved, influence your mind too much in other things which cannot be proved in the same way, and which if true are likely to be above our comprehension?" She pointed out to Darwin that bringing scientific method to bear on religious beliefs is a pointless exercise because the two are philosophically distinct realms.*

*By and large, biologists accepted Darwin and Wallace's theory of evolution with open arms, since it explained so much. There have always been those who resisted the appeal of evolution and every now and then declare "Darwin was wrong," in the hope of some profitable publicity, usually revealing that they do not understand Darwinism. Of course, modern evolutionists have refined many aspects of Darwin's work, but there can be no question that the theory of evolution by natural selection has been the single most fruitful piece of biological thinking ever produced.*

(a)   Charles Darwin

(b)   Alfred Russel Wallace

**FIGURE 17–A**   Fathers of the theory of evolution by means of natural selection   (Biophoto Associates, Linnaean Society, London)

## SUMMARY

Organisms evolve by descent and modification from pre-existing forms. Members of any species differ from one another, and some of their differences are inherited. Natural selection is the differential reproduction of individuals (and their genotypes); it is the main cause of evolution, a change in the genetic makeup of a population from one generation to the next.

1. The theory of evolution by means of natural selection was put forward in 1858 by Charles Darwin and Alfred Russel Wallace. Their thinking was stimulated by the writings of Charles Lyell and Thomas Robert Malthus and by observations they made during their own travels. Darwin and Wallace did not invent the concept of evolution, but they were the first to propose that evolution was shaped by natural selection.

2. Many lines of evidence support the theory of evolution:
   - Artificial selection by human farmers and breeders produces rapid evolution in domesticated plants and animals.
   - The fossil record and biogeography show that the present-day structure and distribution of organisms are most simply explained in terms of evolutionary history.

- Comparative anatomy, biochemistry, and embryology provide evidence that various structures in ancestral organisms became modified in their descendants and adapted to different functions. Some have even been lost when a new way of life rendered them unnecessary.

3. The evolution of wild populations by means of natural selection was not convincingly shown until the twentieth century, when Bernard Kettlewell showed that predation by birds was the selective pressure that led to the evolution of melanic populations of the peppered moth in polluted areas of England. Many other examples have now been established. Various environmental factors act as selective pressures, including nonliving factors such as climate and toxic chemicals and living organisms such as predators, food sources, and disease-causing organisms.

4. The anatomical, physiological, and behavioral traits that survive natural selection may be thought of as adaptations that fit an organism to live and reproduce in its particular environment. Adaptations are many and various. The only consistent effect of selection is that it maximizes genetic contribution to future generations.

## Self-Quiz

1. In light of the definition of evolution, which of the following is *not* capable of evolving?
   a. a population of deer
   b. the color of a population of moths
   c. your biology teacher
   d. a population of cattle
   e. the millions of bacteria in your large intestine

2. Which of the following did Kettlewell conclude from his studies on industrial melanism in moths?
   a. a melanic moth lays more eggs than a pale moth in industrial areas
   b. melanic moths are more resistant to pollution than are pale moths
   c. pollution caused some moths to become darker than others
   d. melanic moths are more likely to survive in polluted areas than are pale moths
   e. birds prefer the taste of melanic moths over pale moths

3. Which bird is most evolutionarily successful?
   a. lays 9 eggs, 8 hatch and 2 reproduce
   b. lays 2 eggs, 2 hatch and 2 reproduce
   c. lays 5 eggs, 5 hatch and 3 reproduce
   d. lays 9 eggs, 9 hatch and 2 reproduce
   e. lays 7 eggs, 5 hatch and 4 reproduce

4. Suppose that you have a pack of 50 assorted dogs. You select the largest male and the largest female from the group, mate them, and sterilize the other dogs. Assuming that food supplies remain adequate, you can expect that, in the next generation of dogs:
   a. the young dogs will be, on the average, larger than their two parents

   b. the young dogs will be, on the average, larger than the older members of the pack
   c. the young dogs will be the same average size as the older dogs
   d. all of the young dogs will be larger than the older dogs

5. Explain how Darwin would have accounted for the evolution of the long necks of giraffes.

6. Fossilization is favored by:
   a. lack of hard, abrasive projections
   b. absence of oxygen
   c. an abundance of nutrients
   d. warm temperatures
   e. strong sunlight

7. Penicillin and other antibiotics were introduced in the 1940s and were effective in combating infections caused by *Staphylococcus* bacteria. In 1958, however, there were several outbreaks of *Staphylococcus* infection. People with the infections did not respond to treatment with any antibiotic, and many people died. The most likely explanation for this situation is:
   a. the bacteria reproduced in hosts that were not contaminated by antibiotics
   b. bacteria from other animals (such as deer, birds, and cats) migrated into human hosts
   c. the bacteria exposed to nonlethal doses of antibiotics quickly learned to avoid them
   d. each generation of bacteria acquired the ability to use the antibiotics as nutrients
   e. bacteria containing a gene for antibiotic resistance survived and multiplied, and these were the forms causing the lethal infections

## Questions for Discussion

For Questions 1 to 5, consider Table 17–2.

1. From what brood size do the greatest number of young survive?
2. Is this also the most frequent family size (assume that Lack marked every bird he could find)?
3. What do you suppose is the disadvantage to a starling of laying a very small clutch of eggs?
4. Suppose the environment changed so that only half as much food was available to the starlings. Would you expect a gradual change in the most frequent brood size? How would this change be brought about?
5. Which female starlings will leave more young per head in the population and hence make the greatest contribution to the genes of the next generation?
6. Are all causes of death natural selection? When people die in an earthquake, have they been selected against?
7. Embryologist Charles H. Waddington treated fly larvae with heat shock. As a result of this treatment, some of the adult flies showed the abnormal condition "crossveinless" (some of their wing veins were missing). After many generations of this treatment, he let a generation of flies develop without heat treatment and many of them were also crossveinless. Does this experiment provide convincing proof of Lamarckism? If not, what other explanation can you suggest, and what experiments would you perform to test your suggestion?
8. Is human evolution subject to selective pressures? Why or why not?
9. Is there any time in its life cycle when an organism is not subject to selective pressure? Are gametes subject to selective pressure? Are eggs? Embryos? Postreproductive individuals? Is there selective pressure on young animals that are fed and protected by their parents?
10. Some insects lay eggs on more than one species of larval food plant. There is some evidence that a female is more likely to lay her eggs on the plant species on which she grew as a larva than on any other kind of plant. Is this an example of Lamarckian inheritance? Why?
11. Scientists are now breeding crop plants to have built-in chemical defenses against insect pests. In your opinion, how well will this work?
12. What is the adaptive advantage to a plant of a contact irritant (such as the oil on poison ivy leaves that makes a rash on the skin of passing animals)?

## Suggested Readings

Antonovics, J., A. D. Bradshaw, and R. G. Turner. "Heavy metal tolerance in plants." *Advances in Ecological Research* 7:1, 1971.

Berry, R. J., and A. Hallam. *The Encyclopedia of Animal Evolution.* New York: Facts on File, 1987. Very readable history of animal life and of evolutionary thought from ancient Greece to the present controversy about punctuated equilibrium.

Cook, L. M., G. S. Mani, and M. E. Varley. "Postindustrial melanism in the peppered moth." *Science* 231:611, 1986. Followup studies show how a reduction in air pollution in England since 1952 has affected the ratio of melanic to typical individuals in peppered moth populations.

Darwin, C. *The Origin of Species by Means of Natural Selection.* New York: The Modern Library, Random House, 1982. A reprint of the 1859 first edition in one volume together with the sequel, *The Descent of Man.*

Eiseley, L. C. *Darwin's Century; Evolution and the Men who Discovered It.* Garden City, NY: Doubleday, 1958. An interesting account of the history of the theory of evolution.

Eiseley, L. C. "Charles Darwin." *Scientific American,* February 1956. A short biography of Darwin and discussion of his work.

Eiseley, L. C. "Alfred Russel Wallace." *Scientific American,* February 1959. A fascinating account of the life of Wallace, how he came to believe in the evolution of species by natural selection, and his views on human evolution.

Mayr, E. "Evolution." *Scientific American,* September 1978. A prominent evolutionist recounts the history of evolutionary theory and explains current ideas about how it works. This article introduces an entire issue devoted to evolution.

Moorehead, A. *Darwin and the Beagle.* New York: Harper and Row, 1969. A short, beautifully illustrated account of Darwin's travels based on Darwin's diaries.

Morris, S. C., and H. B. Whittington. "The animals of the Burgess shale." *Scientific American,* July 1979. An account of an unusually rich fossil find.

Nelkin, D. *The Creation Controversy: Science or Scripture in the Schools.* New York: W. W. Norton, 1982. The controversy between creationism and evolution. Nelkin examines the tactics by which creationists attempt to impose their views on educational systems. She contends that the battle is political (rather than scientific or religious). Judge Overton's decision in the Arkansas case is discussed and reprinted.

Stebbins, G. L., and F. J. Ayala. "The evolution of Darwinism." *Scientific American,* July 1985.

Weiner, J. *The Beak of the Finch.* New York: A. A. Knopf, 1994. An account of research documenting current evolution of Galápagos finches.

Young, D. *The Discovery of Evolution.* New York: Cambridge University Press, 1992.

C H A P T E R

◆ 18

# Adaptation and Coevolution

## OBJECTIVES

*When you have studied this chapter, you should be able to:*

1. Use the following terms in context: symbiosis, search image, camouflage.
2. Explain the advantages and disadvantages to a plant of producing secondary compounds.
3. Describe experiments showing that aposematic coloration and Müllerian and Batesian mimicry confer protection on their owners.
4. Describe the main differences between Müllerian and Batesian mimicry.
5. Explain why a Batesian mimic is considered "parasitic" on its model's reputation.
6. List the four main principles of camouflage, and give an example of each.

7. Explain why appropriate behavior is essential to the adaptive advantage of protective coloration.
8. Describe how you would determine whether an animal's coloration means that the animal is cryptic, aposematic, a Batesian model, a Batesian mimic, or a member of a Müllerian mimicry complex.
9. Tell how you would design and carry out an experiment to show that a particular physical or chemical characteristic of a plant may function as a feeding attractant, egg-laying stimulus, or repellent for herbivores; describe such an experiment that has already been done.
10. List the benefits of mutualism to acacias and to the ants that live on them.

---

Organisms are equipped in various ways to cope with their environments. Flying animals have wings; climbing plants produce tendrils and suckers that enable them to cling to a support. We would not expect to find wings on an animal that lives underground or tendrils on a plant that lives floating in a pond. Wings and tendrils are **adaptations,** features that enable organisms to live within their environments.

When we examine particular adaptations, we can often deduce the selective pressures that have been at work. For example, when we find a plant using a lot of energy making toxic compounds, we may reasonably guess that herbivores have been important agents of selection in the evolution of that plant species.

When the environment changes, only organisms with adaptations suited to the new situation survive. Those with adaptations that are no longer suitable become extinct. Furthermore, the changing environment often selects for modification of an existing adaptation. For instance, in Chapter 17 we saw that the existing camouflage of peppered moths,

which blends with lichens on tree bark, evolved to a darker color when increased pollution darkened the tree trunks where the moths hide.

When we examine adaptations, we find that every organism has adaptations resulting from four main categories of selection (Figure 18–1):

1. Survival in the physical and chemical environment
2. Reproduction: contributing genes to the next generation
3. Predation and parasitism
4. Acquisition of energy and nutrients

The selective pressures in many of these cases are exerted by other organisms. For instance, birds eating poorly camouflaged peppered moths exert selection for dark coloration among moths in polluted areas. We can see that an evolutionary "arms race" tends to develop between predators and prey. Better camouflage in peppered moths selects for birds with sharper eyesight, which in turn selects for better camouflaged moths, and so on. This can be thought

374

of as **coevolution,** the reciprocal evolution of two (or more) populations as a result of their selective pressures on one another.

We start by examining an extreme case of coevolution and go on to consider some of the main categories of adaptations found in organisms.

## KEY CONCEPTS

♦ Among the most powerful selective forces shaping the evolution of organisms are other living things.

♦ Many species have adaptations that permit them to live symbiotically with members of other species. Indeed, some early symbiotic relationships became major contributors to the evolutionary success of eukaryotes.

♦ Predators and prey (including parasites and their hosts, and herbivores and their food plants) often coevolve, exerting selective pressures on each other.

♦ Predation selects for methods of defending against and escaping predators. Defenses may be physical or chemical. Escape may involve hiding from a predator or producing huge numbers of offspring at unpredictable times.

## 18–A  SYMBIOSIS

The most profound cases of coevolution involve **symbiosis** ("together living"), an intimate, long-term relationship between members of two (or more) species which may affect each other's evolution for millions of years. For instance, large populations of *Escherichia coli* bacteria live symbiotically in the large intestine of every human being.

Symbiosis is often subdivided into three categories, according to the degree of benefit derived by the partners:

1. **Parasitism.** One species, the parasite, benefits at the expense of the other, the host species, as in the case of viruses or tapeworms and their hosts.
2. **Commensalism** (Com = together; mensa = table). One species benefits, and the other is not harmed. This category takes its name from cases such as some small fish that live attached to or near the bodies of large fish, or among the tentacles of sea anemones, and feed on scraps that "fall from the table" of these predators (Figure 18–2).
3. **Mutualism.** Both species benefit from the association, as in the case of oxpecker birds that eat ticks and other parasites from the skin of grazing mammals. The two members of the relationship are **symbionts.**

These categories give us useful terms for communicating about symbiotic relationships. However, the relationships

**FIGURE 18–1**  Adaptations. This cactus has a water-storing stem, an adaptation to life in a dry climate; brightly colored flowers that attract insect pollinators for reproduction; spines that reduce the depredations of predators; and green color due to chlorophyll, the photosynthetic pigment used to make its food.

**FIGURE 18–2**  Commensalism. These striped pilot fish off the coast of California await scraps of food from a 45-kilogram sunfish. (Richard Herrmann, International Wildlife)

that exist in nature are not so simple. Without detailed experiments, we may not be able to tell whether one organism harms or benefits another, that is, increases or decreases its evolutionary success. Furthermore, harm and benefit vary in degree, and both may even occur together. Although a parasite's net effect on its host may be harmful, it may also benefit the host by protecting it from invasion by a second parasite. Or a relationship may change depending on circumstances: for example, bacteria living on our skin protect us from disease-causing organisms, but they can cause dangerous infections if they get into cuts.

Symbiosis is a major, recurring theme in evolution, one we shall meet many times in later chapters. We shall see not only disease-causing parasites but also groups of organisms whose very survival depends on mutualistic partnerships: corals, lichens, virtually all plants, and a wide range of animals that rely on symbionts living in the gut to digest their food and produce nutrients.

## Selective Pressures for Symbiosis

The most common benefits symbionts gain are protection from predation, shelter, and food. A constantly renewed food supply is the most common benefit enjoyed by at least one symbiotic partner (and sometimes by both). For example, bacteria living in our intestines obtain some of our digested food and supply us with nutrients, including some vitamins. They also gain shelter from the external environment: they could not survive on their own in sunshine and dry air.

Because of the intimacy of symbiosis, each partner usually exerts great selective pressure on the other's evolution. However, the very intimacy of symbiosis may make it hard to perceive. We seldom see one partner without the other in nature and do not realize their interdependence until we remove one partner in an experiment or try—and fail—to grow only one of the partners in isolation. For instance, foresters discovered that nearly all trees need particular species of fungi growing around their roots. When tree seedlings are grown in soil without fungi, they die or grow poorly.

## Parasitism

Parasites extract their food from living hosts. External parasites (**ectoparasites**) attach to the outside of the host's body. These include leeches, ticks, lice, and fleas. **Endoparasites** live inside the host's body: viruses, many bacteria, and parasitic worms are examples.

One common adaptation of parasites is prolific reproduction. Appropriate hosts may be few and far between, and most of a parasite's offspring die without finding a suitable new host. Many parasitic worms compensate for these losses by producing hundreds of thousands to millions of eggs, the large numbers improving the chances that at least a few find a host and survive.

Female rabbit fleas have another adaptation that gives their offspring a better chance of success: they are most attracted to pregnant rabbits. The flea feeds on the rabbit's blood and lays eggs, which hatch into larvae that live on scraps of skin or dung in the burrow of the mother-to-be. When the young rabbits leave their mother, the next generation of fleas hops aboard, provided with new hosts and a means of dispersal.

Intestinal parasites of animals are surrounded by digested food, and other parasites feed on nutritious blood. There is not much for the parasite's digestion to do, and most parasites have reduced digestive systems. The energy freed by this savings is devoted to expansion of the reproductive system. In fact, tapeworms have no digestive system at all: they absorb all their food through the body wall, and their bodies are little more than egg factories (see Figure 28–13).

Parasite life cycles are often complex and involve more than one host. Some immature parasites change their host's behavior, making it more apt to be eaten by a host in which the adult parasite can develop. For example, immature worms that form cysts in muscles may make the host slower and less likely to escape a predator in which the worm can develop to adulthood. Larvae of the canine tapeworm invade the nervous system of hosts such as sheep, causing them to totter around in circles and become separated from the herd. They are easily picked off by wolves, which then provide the tapeworm with its adult home. Some immature parasites cause their ant hosts to clamp their jaws onto blades of grass, where they are eaten by grazing animals, the host of the adult parasite.

***Evolution of Virulence***    The worst outbreaks of diseases occur when parasites—be they viruses, bacteria, protists, fungi, or animals—first encounter a new host population. Parasites such as the plague and syphilis bacteria killed large proportions of several human populations when they were first introduced from other countries. Such a first encounter selects for those hosts with defense mechanisms against the parasite and often for those parasites that are less virulent.

There are four species of parasites that cause human malaria. The most deadly, *Plasmodium falciparum,* is more closely related to malaria parasites of birds than to the other three human malaria organisms. *P. falciparum* probably began to infect humans only in the last 10,000 years, so recently that we haven't yet evolved defenses comparable to those protecting us against the other three types, which have infected humans for much longer.

The genetic changes that enable host and parasite to coexist may be very small. In Chapter 16 we saw that sickle hemoglobin, which protects against *P. falciparum* malaria, results from a mutation of just one DNA base pair. On the parasite's side, *Yersinia pestis,* which causes deadly plague, differs by only two small mutations from *Y. pseudotuber-culosis,* which causes only mild disease symptoms.

Evolving decreased virulence is advantageous to a parasite that is transmitted directly from one host to another and that cannot survive long outside a host. For example, the viruses that cause the common cold produce more descendants when we feel well enough to continue our daily routine, sneezing new generations of viruses onto other people, than when we crawl into bed with a box of tissues until the immune system destroys all the viruses.

However, when virulence enhances the parasite's reproductive success, selection will preserve it even though it may eventually kill enough hosts to cause the parasite's extinction. For example, malaria parasites reproduce extensively in the host's body. This makes the host too sick to swat mosquitoes and also ensures that a mosquito will suck up many parasites with the host's blood and pass on a substantial dose to the next host she bites.

Parasites that can survive long periods outside a host's body also tend to be virulent. For example, the often deadly smallpox virus could last for ten years in the environment, waiting for new hosts to move into an area after the original population was decimated by a smallpox epidemic.

***Bird Brood Parasites***   Surprisingly, about 80 species of birds qualify as parasites because as eggs and nestlings they are cared for and fed by birds of another (host) species, to the detriment of the host's reproductive success. These so-called **brood parasites** include many cuckoos, honeyguides in Africa, and cowbirds in the Americas.

The European cuckoo is a highly adapted brood parasite of many other species, with a distinct genetic race of cuckoos specializing in each host species. This specialization evolved because in most cases selection favors cuckoos that lay eggs as much like the host's as possible, making it hard for the host to detect the parasitism of its nest.

A cuckoo finds nesting birds of her host species, which she can identify because she learned the appearance of the pair that raised her. After a host female begins laying her own clutch, the cuckoo watches until both host birds are away from the nest and then removes one or more host eggs, which she eats, and lays one of her own. If she finds another cuckoo's egg already there, she removes it before laying hers. This is important because the naked, blind cuckoo hatchling pushes any other eggs or young birds it touches over the side of the nest (Figure 18–3). This guarantees that it receives all the food its foster parents bring to feed young. If a nest contains two cuckoo eggs, the first laid will probably hatch first and kill the other.

A cuckoo cannot just eat all the host eggs and lay her own because the host birds will abandon a nest which they perceive as looted of all but one of their own eggs. By leaving some host eggs, the cuckoo ensures that her own egg will be incubated by the hosts, and it is up to the cuckoo hatchling itself to slay the nestmates its mother

**FIGURE 18–3**   Brood parasitism. A cuckoo hatchling ejects the egg of its foster parents, a pair of dummocks.   (J. A. L. Cooke/Oxford Scientific Films/Animals Animals)

spared. It can safely do this because foster parents will not abandon a lone chick even though they will desert a lone egg. A cuckoo may lay a dozen or more eggs a year in different host nests.

The nest's owners sometimes recognize and eject the cuckoo's egg. There is strong selection for this behavior, as it prevents their wasting resources raising a nestling that does not bear their own genes. Cuckoos are larger than the birds they parasitize, and the foster parents may literally work themselves to death raising one huge, hungry young cuckoo.

By placing artificial bird eggs in nests, behaviorists found that hosts reject eggs not resembling those of their own species. Thus, hosts exert selective pressure for cuckoo eggs resembling their own as closely as possible. Species of birds too small to feed a cuckoo hatchling show little rejection of strange eggs, whereas some suitable but unparasitized species show strong rejection. Possibly these species evolved such good cuckoo-egg detection that they eliminated the race of cuckoos that parasitized them.

Brood parasitism boosts the reproductive success of the parasite female because she does not invest energy incubating eggs and bringing food to the young but can use this energy to produce more eggs. In addition, placing eggs in different nests boosts the chances that at least some will escape predators such as snakes, which usually eat all the eggs in any nest they plunder. Birds of many species do occasionally lay an egg in a nest of neighbors of their own species if the chance arises. If this tendency is inherited and selected by survival of the fostered offspring, and the tendency to build a nest and incubate eggs is also lost, brood parasitism eventually becomes the species' only means of reproduction.

## Commensalism

Commensalism is a catchall category used for any relationship in which a biologist sees a clear benefit for one species but no obvious effect on the other. An example is the relationship of sea anemones and clownfish. This was long classified as "commensalism" because the fish is protected from predators among the anemone's stinging tentacles and eats food scraps that the anemone drops, whereas the anemone does not appear to benefit. However, when their clownfish were removed, the anemones were eaten by butterfly fish. When a clownfish is present, it drives butterfly fish away from the anemone so the relationship is actually one of mutual defense.

Examples of commensalism include cattle egrets, which follow grazing mammals (Figure 18–4). The egrets eat insects that fly away from plants disturbed by the mammals, and insects attracted to dung. The relationship benefits the egrets, but it has no apparent effect on the mammals.

Some cases of commensalism may have evolved from parasite-host relationships where the parasite has become less and less virulent. Some may result from casual encounters so beneficial to one party that that species learned or evolved behavior that permits it to seek out a member of the other species. The scavenger fish that accompany some sharks and other large fish and feed on dropped food may fit this pattern. So too may cattle egrets.

## Mutualism

Mutualism originated among ancient prokaryotes. It permitted organisms to pool their genetic resources. For exam-

**FIGURE 18–4** Cattle egrets with a wild pony in a field on Assateague Island, Virginia.

ple, each partner might have had genes encoding enzymes not found in the other. Living close together permitted each to pick up valuable molecules diffusing out of the other.

Some ancient eukaryotic cells formed mutualistic symbioses with bacteria that had oxidative or photosynthetic metabolic pathways not found in the eukaryotes themselves. These bacteria eventually evolved into mitochondria and plastids, respectively (*Essay: Evolution of Eukaryotic Cells,* Chapter 5, and Sections 24–G and H). Eukaryotes owe much of their evolutionary success to these organelles, which we now regard as standard eukaryotic equipment.

Mutualism may involve the sharing of biochemical resources, structures, and even behavior. For instance, an animal's gut bacteria may make cellulose-digesting enzymes that the host cannot produce. The bacteria are sheltered and fed, and the host gains the use of enzymes that permit it to digest food it could not otherwise use. Sea anemones do not have structures or behaviors that would permit them to drive away butterfly fish, which are unharmed by the anemone's stings. Mutualism with a clownfish gives the anemone the use of the clownfish's aggressive behavior.

## 18–B   DEFENSES AGAINST PREDATION

In contrast to symbioses, most adaptations involve species that interact often but not continuously. For instance, organisms have adaptations reducing their risk from predators—which we can define very broadly to include parasites and plant-eating herbivores as well as meat-eating carnivores. Adaptations against predators are of two general kinds, defense and escape.

Thorns and spines are plant defenses against vertebrate predators. Thorns are not much defense against small insects, as any rose gardener knows, but many plants have spiny or hairy leaves that discourage such small predators. For instance, varieties of potatoes and beans with hairy leaves suffer less damage from leaf hoppers, a major insect pest, than do varieties with smooth leaves (Figure 18–5). A coating of silica, the material in sand and glass, makes an effective defense for many grasses. Leaf toughness or stringiness and cells containing sharp, needle-shaped crystals known as raphides may also induce herbivores to look elsewhere for a meal.

Physical defenses are sometimes inducible, that is, produced only when they are needed. For instance, *Brachionus* is a small aquatic animal with short spines that protect it against some predators but not against one called *Asplanchna.* However, *Brachionus* can detect a chemical that *Asplanchna* releases into the water, and this stimulus causes the next generation of *Brachionus* to develop with longer spines, making them too big to be eaten by *Asplanchna.* Such induction has the advantage that energy is spent on defense only when it is actually needed.

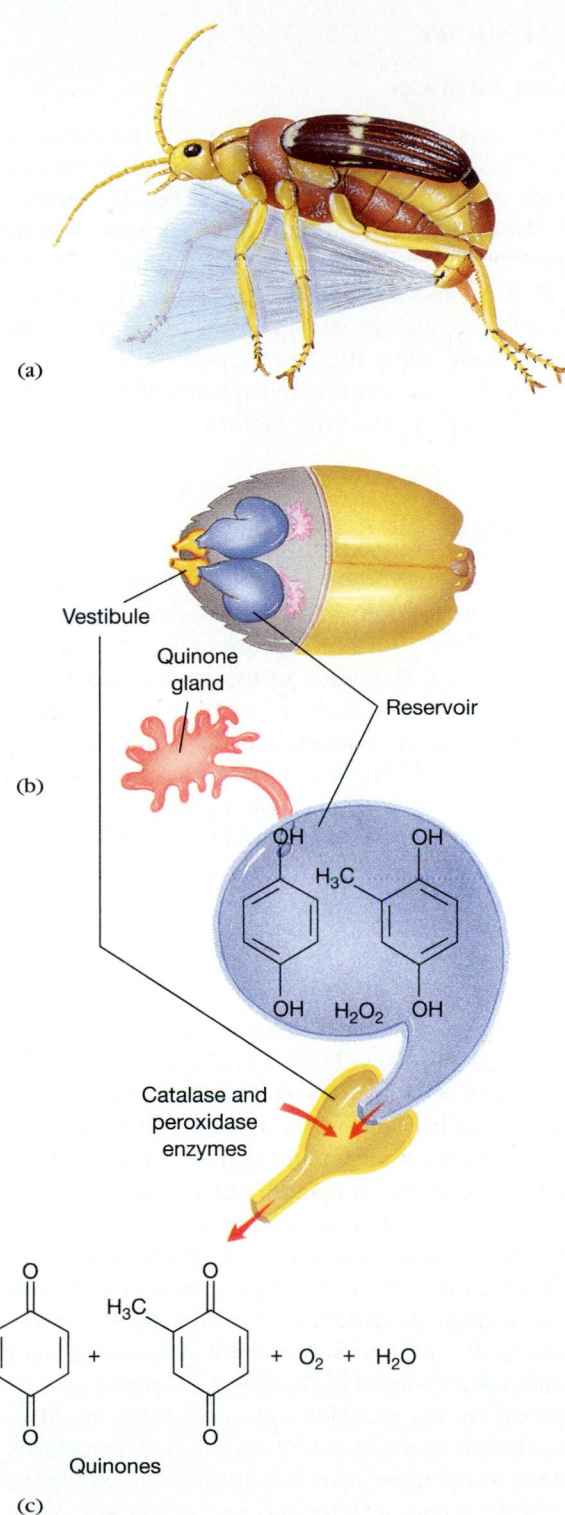

(a)

(b)

Vestibule

Quinone gland

Reservoir

Catalase and peroxidase enzymes

Quinones

(c)

**FIGURE 18–7** Defensive apparatus of the bombardier beetle (*Brachinus*). (a) The tip of the abdomen can be rotated to aim the spray at a would-be predator. (b) Dorsal view, with part of the back removed to reveal a pair of structures that join at the tip of the abdomen. (c) Diagram of one structure showing its quinone gland; the reservoir containing an aqueous solution of hydroquinone, methylhydroquinone, and hydrogen peroxide; and the vestibule in which the heat-releasing reaction occurs.

**FIGURE 18–8** A monarch butterfly caterpillar feeding on milkweed.

in the mucus that covers their skin. Some of these toxins are incredibly poisonous. A few micrograms of batrachotoxin, from the frog used by Colombian Indians to make poison darts, will kill a human being. Tetrodotoxin is equally poisonous. Puffer fish is considered a delicacy in Japan, and chefs who prepare it have to be specially licensed to ensure that they know how to remove the parts of the fish that contain the toxin. Nevertheless, accidents will happen, and several people die every year from tetrodotoxin poisoning.

Animals sometimes obtain their chemical defenses at second hand. Many insects have adaptations that permit them not only to tolerate toxins in their food plants, but also to reuse these toxins for their own defense. For example, monarch butterflies are toxic because they contain cardiac glycosides (sugar-steroid compounds that affect the heart) from the milkweed plants they ate as larvae (Figure 18–8).

## 18–D  APOSEMATIC COLORATION

Many animals with effective chemical defenses, such as wasps, skunks, monarch butterflies, and poisonous fish and frogs, are strikingly colored. Because the coloration alerts potential predators to stay away and not attack, it is called **aposematic coloration** (apo = away; sematic = warning).

**FIGURE 18–9** Aposematic coloration. This tropical frog is protected by toxic compounds in the mucus secreted by its skin. (Fundación Neotrópica de Costa Rica)

Aposematic coloration protects best against predators that can learn, especially vertebrates. Jane and Lincoln Brower showed that a toad stung once by a bumblebee will not touch anything looking like a bumblebee for a long time afterward. If you have ever seen a cat try to attack a toad or frog, you have seen this learning process at work (Figure 18-9). Similarly, dogs will seldom tangle with a skunk more than once.

The only defense of some aposematically colored animals is that they are distasteful to most animals. Ladybug beetles and cinnabar caterpillars are examples. Here the predator's learning ability plays a vital role in the success of the aposematic coloration, since some individuals will be killed by inexperienced predators. If the predator did not learn, after one or two trials, to leave cinnabar caterpillars alone, many caterpillars would be killed even if they tasted so nasty that the predator spat them out.

### Aposematic Sound

As bats hunt moths at night, they produce ultrasonic squeaks which they use as radar (*Essay: Moths and Bats,* Chapter 38). Dorothy Dunning and Kenneth Roeder found, to their surprise, that some moths respond to bat sounds by flexing their legs, producing ultrasonic clicks. This seemed peculiar because the bats can hear the clicks and so find the moths more easily. However, Dunning found that bats refuse to eat moths that click, either alive or dead. The moths are protected by a distasteful chemical. Clicking serves as a signal of this, allowing bats to learn to avoid these moths without approaching and perhaps injuring them. This is an example of aposematic sound—the acoustic equivalent of aposematic coloration.

## 18–E  MIMICRY

### Müllerian Mimicry

During the course of evolution, members of several different well-protected species have come to resemble one another. This phenomenon was pointed out by nineteenth-century naturalist Fritz Müller and is now known as **Müllerian mimicry.** For instance, a number of different species of bees and wasps have black and yellow stripes. The selective advantage of Müllerian mimicry is that the more individuals carry the same aposematic pattern, the lower the chance that any one individual will be killed while a predator learns to avoid the pattern.

### Batesian Mimicry

If a bird may learn to avoid eating all insects with black and yellow stripes, then there might be an enormous advantage to a tasty, unprotected insect if it too had black and yellow stripes. **Batesian mimicry** (named for zoologist Henry Walter Bates) is the resemblance of a **mimic,** an unprotected species, to a **model,** an aposematically colored, protected species. There is a family of flies whose members mimic bees and wasps. You yourself have probably reacted to Batesian mimicry by moving carefully away from such a harmless fly wearing black and yellow stripes (Figure 18–10).

Mimicry is often very precise, down to details of anatomy and behavior. Most flies are actually quite different from bees and wasps. Flies have only one pair of wings, whereas other winged insects have two pairs. Most flies also lack the "waist" of the body between thorax and abdomen that is found in bees and wasps, but flies that mimic bees and wasps usually look as though they do have a waist because the body either has an actual constriction or has coloring that looks like a constriction. A mimic fly's wings may also be so large that they look like two pairs. These flies are often found hovering over flowers, with rapidly beating wings, just where one would expect to find bees.

No protection in nature is ever absolute. Any animal will always have predators that can learn how to avoid its sting or that have evolved the biochemical pathways to cope with its toxic chemicals. Although most birds avoid wasps and bees, shrikes and bee-eaters have behavior patterns that permit them to eat these insects without being stung. Evolving the ability to eat a defended species is advantageous because it opens up a food source for which there may be little competition.

***Batesian Mimicry in Nature***   How can we decide if two similar-looking animals are examples of Batesian mimicry, and which species is the model and which the mimic? One way would be to perform a series of experiments, but other clues may often be used.

(a) Bee

(b) Fly

(c)

(d)

**FIGURE 18–10** Two bees or not two bees? Batesian mimicry. Resemblance of a mimic fly (b and d) to a bee (a and c) is good enough to fool some of the people some of the time. Careful examination, however, shows that there are differences between the bee and the fly. The fly has large eyes and stubby, club-shaped antennae. The bee's eyes are smaller and her antennae more slender. A bee also has two pairs of wings to a fly's one pair, but this may be hard to see. (c, John Dudan/Phototake)

First, a model usually looks and behaves like closely related species, whereas a mimic may be very different from its relatives. Thus, if an insect looks and acts like a wasp, but on close examination turns out to be a fly, it is a fair guess that it is a mimic.

Second, unusual behavior is a good clue to mimicry. There are black-and-yellow-striped flies that, if you touch them, rapidly curve their abdomens around and stab your hand. In fact they have no sting, so the gesture is meaningless, but the fly looks so much like a wasp about to sting that you are likely to shake it off your hand without harming it. Most moths fly by night and rest by day, but moths that mimic butterflies fly by day with the butterflies they mimic.

Third, we must consider a much-argued point about the theory of mimicry: how many mimics can a given population of models support? We might expect always to find fewer mimics than models in an area because the mimic's existence detracts from the protection the model gets from its aposematic coloration. If many of the prey attacked by a predator are palatable Batesian mimics, the predator is likely to attack more members of the model species before it finally learns to avoid prey with that color pattern. Thus, as we might predict, models are usually more common than mimics.

Caution is the watchword in mimicry studies. The close mimicry of orange and black monarch butterflies by sup-posedly edible viceroy butterflies was used as a classic example of Batesian mimicry in textbooks, including previous editions of this one. This conclusion was based on experiments as well as on the observation that viceroys are closely related to tasty admiral butterflies and do not feed on toxic plants as monarch larvae do (Figure 18–11). Then, in 1991, David Ritland showed that red-winged blackbirds tasted, but would not eat, either species after he removed the wings. Apparently, both species are unpalatable. This means that viceroys and monarchs are Müllerian mimics, the viceroy having evolved a closer resemblance to monarchs than to its admiral relatives under selection for the mutual protection the resemblance gives both butterflies.

**FIGURE 18–11** Monarch and viceroy butterflies. The monarch butterfly (larger) is mimicked by the viceroy (smaller). Both are somewhat protected by toxic chemicals. (Thomas C. Emmel)

## 18–F    ESCAPING PREDATION

So far we have considered various ways in which organisms defend themselves against predation. Escape is the other major way of reducing losses to predation. Escape may include hiding, fleeing, or finding safety in numbers. Each has its drawbacks. In hiding, an individual loses time that could have been spent feeding; an animal that flees must expend energy in developing long legs or wings and the muscles to work them.

## Camouflage

Many organisms hide under rocks or leaves or in burrows, but others have coloration that hides them even in plain sight. Camouflage is a means of disguising things so that they are difficult to perceive. The camouflage of many animals (and plants) protects them from discovery by their predators or prey. For example, it is difficult to see a green frog in the grass on the bank of a pond or a speckled trout against the gravel of a streambed. Experiments have shown that camouflage is in fact of selective advantage to its owner. We have studied one such case, the evolution of melanism in moths: moths blending into their background are less likely to be eaten by birds than moths that contrast with their background.

Because vision is the dominant sense in humans, visual camouflage has been the most studied. Other ways of hiding are advantageous in interactions with animals to whom other senses are more important. Some animals camouflage their smell or the sounds they make.

We recognize an animal by three main visible features: its silhouette, its eyes, and its bulk, or appearance of being rounded. The silhouette in Figure 18–12 might be almost anything. Adding an eye and ears makes the silhouette recognizable as an animal. The importance of bulk is clear from the fact that we would not confuse even a perfectly colored cardboard cutout of a cow with a real cow because it would not have any of the shadows that the bulk of a real cow would make. An animal's camouflage must disguise its bulk, its silhouette, and its eyes.

Bulk is nearly always disguised by **countershading** (Figure 18–13). If an object is the same color all over, its underside appears darker when light falls on it from above. Hence it appears rounded. The vast majority of animals have light-colored bellies and dark backs. Light falling from above makes them look uniformly colored and therefore flat. Animals that habitually live upside down, such as the three-toed sloth, backswimmers (aquatic bugs), and some caterpillars, have inverted countershading, with light backs and dark bellies. The military camouflage first used in World War I was designed by zoologists such as Alistair Hardy who had studied animal camouflage: a camouflaged gun or plane is also painted with a pattern that countershades it.

Silhouette is disguised by **disruptive coloration.** Some parts of the body appear the same color and intensity as

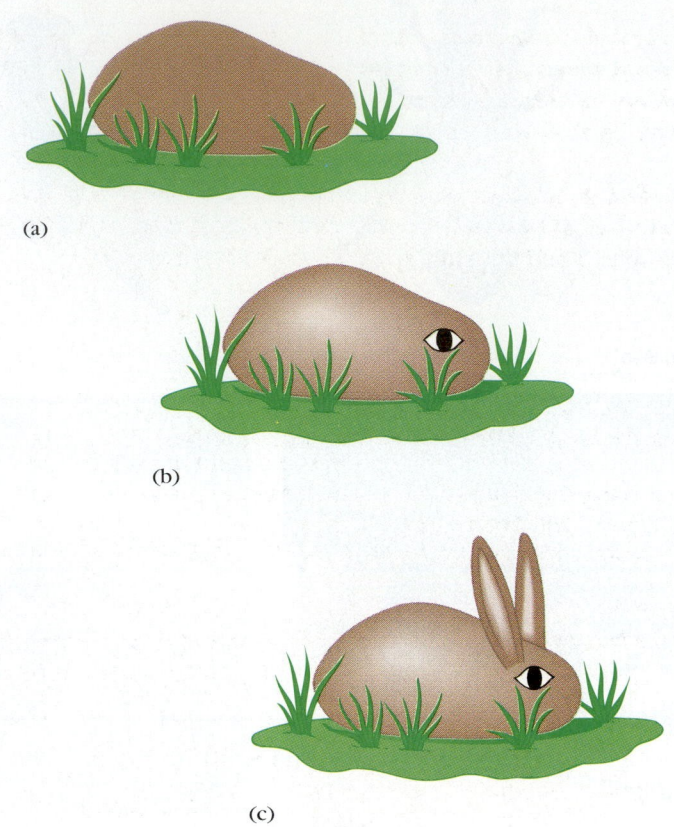

**FIGURE 18–12**    Silhouette, eye, and bulk are the main features that permit us to recognize an animal.

the normal background, while others contrast with it (Figure 18–14). Under these conditions, some parts of the object stand out, whereas others seem to disappear. The result is a pattern of splotches rather than a recognizable animal.

Camouflaging the eye is important for two reasons. First, where there is an eye, there is an animal. Second, an eye is always near the brain, one of an animal's most vital organs. Eyes may be disguised in various ways (Figure 18–15).

Many animals have false eyes on some other part of the body. Eyespots on the wings of moths are quite common (Figure 18–16). Experiments have shown that a bird is more likely to attack a wing with spots, which can be damaged without killing the moth, than the moth's real head: predators are apparently fooled into thinking eyespots are on the head. Some insects startle predators by rapidly showing and then hiding their eyespots. A predator glimpsing the eyespots is apparently fooled into thinking this is not a moth but a much larger and potentially dangerous animal, such as might possess two large eyes so far apart.

Eyespots are common in tropical brush-footed butterflies whose members feed on the ground on rotting fruit. Members of the same family that feed on flowers above the ground do not have eyespots. Entomologists suggest that the difference protects the fruit-feeding butterflies from their own drunkenness! Fungi in rotting fruit produce alcohol,

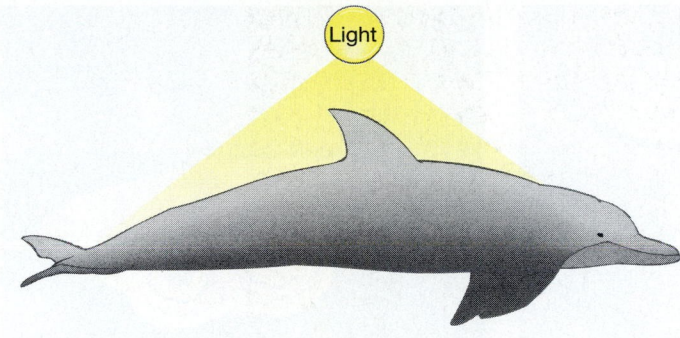

(a) Uniformly colored animal lit from above

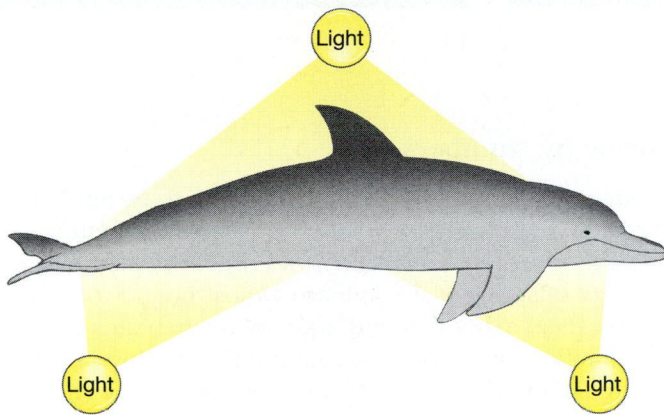

(b) Countershaded animal lit uniformly from all sides

(c) Countershaded animal lit from above

**FIGURE 18–13** Countershading. (a) Appearance of a uniformly colored object lit from above. (b) Appearance of a countershaded object lit uniformly from all sides. (c) Appearance of a countershaded object lit from above.

which the butterflies ingest until they are incapable of the jerky flight by which they normally escape predators. Perhaps these butterflies depend on their large eyespots to deter predators, allowing them time to sober up in safety.

Camouflage is useless unless appropriate behavior patterns evolve along with coloration. For instance, a "leaf" wandering up and down a twig is apt to be noticed by a predator. Most butterflies are not camouflaged; they fly by day and are bound to be visible whatever their coloration.

**FIGURE 18–14** Disruptive coloration. When it rests against a black background, the black border of this butterfly becomes invisible, leaving only a serrated green area which does not remind a predator of a butterfly. (Biophoto Associates)

(a)

(b)

**FIGURE 18–15** Disguising the eye. (a) The eye is the most recognizable part of any animal and is made less obvious in many animals by such devices as a black stripe across the eye. (b) This moth carries eye disguise to extremes by having a false head (yellow) on the wing tips. Predatory birds peck at the false head, frequently saving the real head at the other end of the body from damage. (Biophoto Associates)

**FIGURE 18-16** Eyespots. When this moth (a) is disturbed, it moves its upper wings sideways to reveal remarkably realistic eyespots on its lower wings (b).

(a)

(b)

Camouflage is much more common in moths, most of which are nocturnal. During the day they rest motionless and camouflaged on some appropriate surface.

***Functions of Camouflage*** What is the advantage of visual camouflage? First, **cryptic coloration**—coloration that hides an animal against its background—might have been selected for in animals whose main predators hunt by sight. If this is true, we would expect that animals with few predators would be less likely to be cryptically colored. This is in fact the case. Large birds such as swans and gulls, which have few predators, are conspicuous, but their small, vulnerable young are cryptically colored. The vulnerable eider duck, incubating her eggs on an open nest, is also cryptically colored.

Plants, too, can be camouflaged. Stone plants are almost invisible against the desert gravel in which they live (Figure 18-17).

## Escape by Numbers

Another adaptation that allows organisms to escape predation is the presence of so many prey that the predators cannot find or eat them all. That is, an organism may produce so many offspring that some are almost bound to survive. For instance, many fish and aquatic invertebrates produce vast numbers of offspring, compensating for losses caused by predation. The nymphs (immature forms) of the periodic cicada take 17 years to grow (13 years in the deep South), living in the soil and feeding mainly on fluids from the roots of trees. At the end of the 17 years, all of the nymphs in the population emerge within a few days, climb up on plants, telephone poles, and other supports, and molt into winged adults (Figure 18-18). Then they mate, the fe-

**FIGURE 18-17** Camouflaged plant. This stone plant, *Lithops,* is well disguised against the gravel in which it lives. (James Mauseth)

**FIGURE 18-18** Seventeen-year cicada nymphs swarm up a tree trunk.

males lay their eggs, and the next generation develops over another 17 years. The adaptive value of this life cycle seems to be escape from predation. During cicada outbreaks, predators such as blue jays and rodents stuff themselves but have little effect on the number of cicadas. In the intervening years there is no supply of cicadas to permit predators to build up populations that could inflict significant losses. Many plants reduce predation by a similar method, producing seeds only occasionally and in vast numbers (Section 46–F).

## 18–G  ADAPTATIONS FOR OBTAINING FOOD

Animals have three main ways of feeding: staying in a good spot and letting the food come, moving around and hunting for food, or parasitizing another organism's ability to obtain food.

Roughly half of all animals are hunters ranging from mountain lions and giraffes that roam over many square kilometers to nematodes that hunt bacteria in a few cubic centimeters of soil. Active visual hunters, such as ladybugs and blue jays, increase their feeding success by forming **search images,** mental "pictures" of the object they are looking for. We ourselves form search images. When we scan a bookshelf for a biology book that we think is green, we may overlook the book if it is actually red, even though it is sitting there with "Biology" written in large letters on its spine. A search image increases the predator's overall success by giving it fewer decisions to make: the predator need not examine and identify every object to see if it is good to eat, but merely scan for items of a certain type.

Camouflage and mimicry are common among "sit and wait" predators. Some praying mantids, for example, look like twigs, leaves, or even flowers. Unsuspecting insects come close enough for the mantid to strike and capture them, and predators often overlook the mantids (Figure 18–19).

The light flashes of fireflies are part of their courtship behavior, and males are attracted to a female by her pattern of flashes. But females of the genus *Photuris* have turned this fact to predatory advantage. They produce flashes that mimic those produced by females of another genus, *Photinus. Photinus* males are attracted to the *Photuris* female, who promptly eats them (Figure 18–20). When she is in a more romantic mood, the *Photuris* female flashes in the courtship pattern that attracts *Photuris* males.

## 18–H  COEVOLUTION BETWEEN PREDATORS AND PREY

The tactics of defense and deception that we have considered are just part of larger patterns of coevolution. Parasites evolve adaptations that make them better at finding and colonizing their hosts. This, in turn, selects for hosts that

**FIGURE 18–19**  A praying mantis blends in with the vegetation as it waits for other insects to come within range.

are harder for parasites to find or to invade successfully. Mimicry is an adaptation that gives an edible prey species a better chance of being avoided by a potential predator. In turn, this selects for predators that are increasingly able to distinguish edible from inedible species.

The most thoroughly studied examples of coevolution between plants and animals are interactions between flow-

**FIGURE 18–20**  Aggressive mimicry. A female *Photuris* firefly eating a *Photinus* male, which she has attracted by mimicking the pattern of light flashes produced by female *Photinus.*   (James E. Lloyd)

ering plants and insects. Insects are the most important animal pollinators of flowers (*Essay: Coevolution of Plants and Their Pollinators,* Chapter 46). Also, many insects have evolved as specialized feeders on the leaves, seeds, or other parts of only one or a few species of plants. Thus, insects and flowering plants exhibit relatively simple, easily studied patterns of coevolution. Some of these have been investigated experimentally.

## Mutualism with Ants

Acacias are small tropical trees of the legume (pea and bean) family. Many acacias have spines that protect them against herbivorous mammals, and some Central American "ant acacias" are protected by a mutualistic relationship with the ants that live on them. Why ants? First, there are thousands of species of ants in tropical areas. Second, many ants have stings that can disable predators, and third, ants are social insects: an ant can recruit other members of its colony for assistance.

Ant acacia trees produce specialized structures that benefit the ants. These include swollen, hollow thorns in which the ants live, nectar glands on their leaves, and Beltian bodies (Figure 18–21). Beltian bodies are swollen, nutrient-rich leaf tips, which the ants cut off and feed to their larvae. The ants also collect nectar secreted by glands on the stems of the acacias.

It is advantageous for the acacia to have tenant ant colonies because the ants reduce the damage that herbivores do to the tree. Daniel Janzen found that when he removed the ants from an ant acacia, the tree usually died after a few months. Fungi grew on it, and its shoot tips were eaten. As a result, it grew more slowly and became overgrown with vines. When ants are present, they react to anything that touches the tree. They remove dust, fungal spores, pollen grains, and spider webs. They destroy the seedlings of other plants that sprout under their tree and sting other insects or mammals that try to eat the tree.

A few species of insects have evolved defenses that permit them to survive on an acacia tree defended by ants. Some of these seem immune to ant stings and ignore the ants; others can pick up the ants and throw them off the tree; still others have hard cuticles that an ant's sting cannot penetrate.

The coevolution of ant acacias and their insect populations probably went something like this: ants invaded an acacia and fed on the leaf parts and nectar of the tree. The ants also killed other, competing insects, allowing the acacia

(a)

(b)

**FIGURE 18–21** Ants and acacias in Central America. (a) An acacia ant feeding on the yellowish-brown "Beltian bodies" on an acacia. A nectar gland is also visible on the stem below the ant. (b) This armyworm caterpillar, placed on an ant acacia, was stung to death in minutes by the resident ants. Also seen here are the hollow swollen thorns in which acacia ants make their nests. (Paul Feeny)

to grow faster and to produce more offspring. Acacias that were more attractive to ants reproduced more rapidly. The availability of food and shelter, in turn, exerted selection on the ants to protect the tree with increasing efficiency. Of the insects that fed on the tree before the ants arrived, most species were expelled, but a few evolved defenses against ant attacks.

Acacia species that are not defended by ants have cyanides and other defensive chemicals in their leaves. These chemicals have been lost in the ant acacias, presumably because they are no longer needed for protection in the presence of ants.

## Cabbages and Caterpillars

Members of the cabbage family and the insects that eat them are a well-studied case of coevolution between predator and prey species. When you cook cabbage, broccoli, mustard greens, or most other members of the crucifer family, they give off the odor of mustard oils—the group of secondary chemicals characteristic of crucifers. In the intact plant, the mustard oils are usually bonded to sugar molecules, forming glucosinolates, which are much less toxic and can be stored without damaging the tissues they defend. When a cell is damaged—as it is when an insect bites into it—an enzyme cleaves off the sugar molecule, which is analogous to pulling the pin on a hand grenade, and the mustard oil is released (Figure 18–22).

How toxic are mustard oils? They are plainly not very poisonous to humans or to cabbage white butterfly caterpillars, which sometimes wipe out entire plantings of crucifers in the home garden. The main difficulty in answering this question is that insects usually do not eat anything except their normal food plants. You cannot take a caterpillar from an oak tree and plunk it on a cabbage to see if it will be poisoned, because the caterpillar will not eat cabbage. In an experiment to get around this problem, black swallowtail butterfly larvae, which normally eat plants of the carrot family, were raised on carrot leaves cultured in solutions with various concentrations of glucosinolates. The larvae were therefore eating their usual carrot diet, plus secondary chemicals from plants that they do not normally eat.

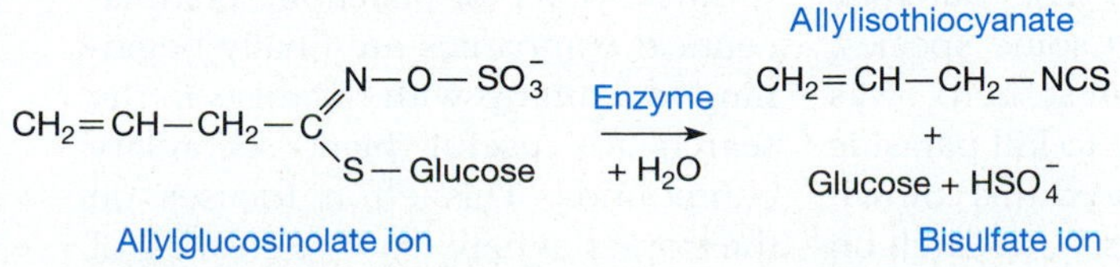

As the concentration of glucosinolates in the carrot leaves increased, the larvae ate more slowly, but lost weight much more rapidly than would be expected from their feeding rate. At glucosinolate concentrations that occur naturally in crucifer plants, the larvae lost so much fluid in their feces that they soon died. Clearly, then, glucosinolates can be an effective defense against insects that do not normally attack the plants that contain them.

## Counteradaptations to Secondary Chemicals

Since every plant is eaten by some herbivore, it is clear that herbivores have evolved counteradaptations permitting them to eat some species of plants that are toxic to most other animals. For instance, experiments show that cabbage white butterfly caterpillars grow normally on crucifer leaves in which glucosinolate levels have been artificially increased to more than ten times the concentrations found in nature.

Some of the counteradaptations that insects have evolved against plants' chemical defenses are remarkably sophisticated. For instance, the seeds of some tropical legumes contain canavanine, an amino acid similar to arginine (Figure 18–23). In most animals, canavanine can replace arginine during protein synthesis. This produces proteins that do not function properly, and so canavanine is extremely toxic. The larvae of some Costa Rican beetles, however, eat legume seeds that contain canavanine. The arginine-tRNA synthetase enzyme of these larvae is more specific than that of other animals and will join only arginine, but not canavanine, to the arginine-tRNA. As a result, the beetle excludes canavanine from its proteins and instead uses it as food.

## Secondary Compounds That Attract Herbivores

Many insects use their food plants' secondary compounds as cues by which they locate the plants. For example, some flea beetles home in on the odor of their crucifer food plants

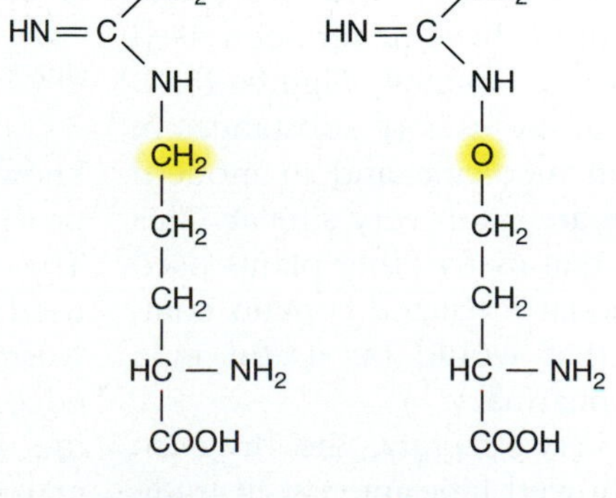

FIGURE 18–22  Mustering mustard. When tissues of crucifer plants are damaged, glucosinolates come into contact with enzymes that hydrolyze them to volatile and toxic mustard oils. Allylisothiocyanate, formed from allylglucosinolate, is responsible for the odor of cooking cabbage and the pungent taste of horseradish.

FIGURE 18–23  The rare amino acid canavanine is similar to arginine, a common amino acid in the proteins of all organisms.

*Text continued on page 392*

## E S S A Y :  *Drugs from Plants*

Most drugs and medicines are plant secondary compounds. For instance, digitalin, extracted from foxglove plants, is an important heart stimulant. An overdose can be fatal. In fact, plant secondary compounds are the active ingredients of 25% of the prescription drugs sold in the United States, and the World Health Organization estimates that 80% of the world's people depend on plant compounds for nearly all their medicines (Figure 18–A).

Other animals too depend on plants for medicine. Dogs that have eaten rotten food or have tapeworms in their stomachs sometimes search out and eat grass, which makes them vomit the worms or rotten food. Researchers noticed that chimpanzees in Tanzania occasionally eat the leaves of a local shrub, swallowing the leaves whole without chewing as if they tasted bad—which they do to humans. The shrub has long been used in the folk medicine of the area's human inhabitants. Now researchers have discovered that it contains a sulfur compound, thiarurbrine, which does indeed kill parasitic worms as Tanzanian folk medicine has long claimed.

Norman Farnsworth of the University of Illinois collected a list of 120 chemicals that are extracted directly from plants and used in medicines in various parts of the world. At least 46 of these drugs have never been used in the United States. Figure 18–B shows that the uses of substances in traditional medicine and in modern medicine are often very similar. This suggests that many other plants used in traditional medicine contain compounds that would be useful in a modern pharmacy.

Until very recently, the drug industry showed little interest in traditional plant medicines. Native North Americans were seldom consulted by early settlers, and much of their

**FIGURE 18–A**   A herbal medicine shop in a Costa Rican market. Traditional herbal medicine is quite effective. Costa Ricans live several years longer, on average, than U.S. citizens.

knowledge has been lost, but they provided the keys to some species. The mayapple, a forest herb, was used by the Cherokee to kill parasitic worms, and they passed this knowledge on to the colonists. Research on the secondary compounds of the mayapple has led to antiviral agents, a drug used to treat testicular cancer, and a means of protecting crops from potato beetles.

After years of inaction, pharmaceutical companies are finally beginning to team up with botanists in the search for useful plant secondary compounds. This effort focuses on the tropics, where massive clearing of forest endangers millions of plant species, most of which have not even been described by botanists (Section 50–G).

| Compound | Plant Source | Use in Medicine | |
|---|---|---|---|
| | | *Modern* | *Herbal* |
| Agrimophol | Common agrimony (*Agrimonia eupatoria*) | Parasitic worm infections | Same |
| Anabasine | Tumbleweed (*Anabasis aphylla*) | Skeletal muscle relaxant | Not used |
| Codeine | Opium poppy (*Papaver somniferum*) | Pain-killer: prevents coughing | Pain-killer; sedative |
| Danthron | Senna (*Cassia*) | Laxative | Same |
| Digitalin | Foxglove (*Digitalis purpurea*) | Heart stimulant | Same |
| Etoposide | Mayapple (*Podophyllum peltatum*) | Antitumor agent | Cancer |
| Gossypol | Cotton (*Gossypium*) | Male contraceptive | Decreased fertility noted |
| Lobeline | Indian tobacco (*Lobelia inflata*) | Respiratory stimulant | Expectorant |
| Papain | Papaya (*Carica papaya*) | Digesting protein, mucus | Digestive disorders |
| Pseudoephedrine | Ma-Huang (*Ephedra sinica*) | Bronchodilator | Bronchitis |
| Quinidine | Cinchona (*Cinchona*) | Controls heart arrhythmia | Malaria |
| Quinine | Cinchona (*Cinchona*) | Antimalaria, reduce fever | Malaria |
| Salicin | White willow (*Salix alba*) | Pain-killer (in asprin) | Pain-killer |
| Sanguinarine | Bloodroot (*Sanguinaria canadensis*) | Plaque-inhibitor (red) | No medical use; red dye |
| Scopolamine | Thornapple (*Datura metel*) | Sedative | Same |
| Theobromine | Cocoa, cacao (*Theobroma cacao*) | Diuretic, dilate blood vessels | Diuretic (increase urine production) |
| Trichosanthin | Chinese snake gourd (*Tricosanthes*) | Abortifacient (induces abortion) | Same |
| Vincristine | Madagascar periwinkle (*Catharanthus roseus*) | Antitumor | Not used |

Data from Farnsworth, N. R. "Screening plants for new medicines." In Wilson, E. O., ed., *Biodiversity*. Washington, DC: National Academy Press, 1988.

Senna

Opium poppy

Thornapple

**FIGURE 18–B** Some plant secondary compounds and their uses in modern medicine and traditional herbal medicine.

**FIGURE 18–24** Trapping cabbage flea beetles with mustard oil traps. The glass vial on the left contains a 1% aqueous solution of allylisothiocyanate. Many beetles have been attracted by this odor and remain stuck to the glue lining the inside of the carton. The vial on the right is a control, containing only solvent and glue, and has caught almost no beetles.   (Paul Feeny)

in a field containing many other kinds of plants. In order to show that these beetles are indeed finding their food by scent, the odor must be separated from the plant. When this is done, traps containing mustard oils and a sticky glue to catch the beetles will attract just as many flea beetles as will a real plot of crucifers. Few beetles are caught in control traps containing glue alone (Figure 18–24).

Another insect that moves toward the scent of mustard oil in experiments is a wasp that does not eat crucifers but parasitizes an aphid that often does. The wasp lays each egg in an aphid, and its growing larva eats the aphid from inside. Moving toward mustard oils permits the wasp to find aphid hosts for its offspring. Experiments have shown that the aphid is almost free from parasitism by this wasp when it is on a plant other than a crucifer.

These and many other experiments have shown that mustard oils not only protect crucifers from being eaten by many herbivores but also organize the lives of a variety of animals living on or near crucifers. Thus, mustard oils are feeding stimulants to flea beetles and egg-laying stimulants to the cabbage white butterfly. Both are attracted by the mustard oils that crucifers produce. On the other hand, mustard oils repel those herbivores that cannot detoxify mustard oils and eat crucifers.

## Interference with Hormone Control

Some plants produce secondary compounds with structures similar to insect hormones. An overdose of the insect molting hormone ecdysone, produced by the polypody fern and

by some conifers and flowering plants, speeds up some phases of insect development so much that other phases, which are not much influenced by the hormone, cannot keep up. The result is a misshapen individual that soon dies.

Juvenile hormone of insects normally prevents metamorphosis before the insect is mature. The presence of substances similar to juvenile hormone in plants was discovered in an interesting way. A Czech scientist visited Harvard, bringing with him insects raised for several generations in his laboratory in Europe. When he attempted to raise them at Harvard, the insects grew into bigger nymphs than usual, but never molted into adults. Investigation revealed that the paper towels placed in the bottom of the insect cultures were the culprits and that all American newspapers and journals had the same effect. Manufacturers traced the paper back to balsam fir trees, which, when tested, also prevented maturation. The active substance was extracted from a few hundred paper towels. It had a juvenile hormone–like effect only on members of the red bug family, a group containing several destructive pests of cotton and trees in the South. We can surmise that somewhere along the evolutionary line these insects may have exerted selective pressure against the balsam fir and some other conifers, such as the eastern hemlock, Pacific yew, and larch, which also contain this compound.

Several plants are now known to produce secondary compounds that defend against herbivores by interfering with the hormones that control insect growth and metamorphosis. One such compound, found in the bedding plant *Ageratum*, was named precocene because it causes premature metamorphosis in larval milkweed bugs by suppressing the secretion of juvenile hormone (Figure 18–25). Female milkweed bugs that metamorphose prematurely are all sterile. Hence precocene reduces the size of the next generation of bugs.

**FIGURE 18–25** *Ageratum*, a plant defended against insects by precocene, which suppresses the secretion of juvenile hormone.

# SUMMARY

Every organism is subject to selective pressures exerted by other organisms. Hence, every organism coevolves with a number of other species, including its predators, prey, and any symbiotic partners. The long-term intimacy typical of symbiosis grades into specialized predator-prey relationships, such as insects that eat only one plant species and in turn are one of the plant's main predators, to more general cases of species in the same habitat that interact regularly but not continuously.

1. Symbiosis is an extreme example of coevolution in which two species live in such an intimate association that at least one of them exerts overwhelming selective pressure on the other. At least one partner is usually completely dependent on the other for shelter, defense, or food.

2. Predation (including parasitism and herbivory) has selected for adaptations that defend the prey or permit it to escape. Plants defend themselves from being consumed by herbivores by physical defenses, such as spines, tough cuticles, or silica; by toxic secondary chemicals; or by a combination of these. Plants expend considerable energy in producing these defenses. Many animals also have protective toxic chemicals, in some cases extracted from their food plants.

3. Animals with chemical defenses are often aposematically colored, thereby gaining defenses against predators that hunt by sight and can learn to avoid aposematically colored animals.

4. Different species of chemically defended animals often have similar aposematic color patterns, reducing the chance of an individual's being killed while a predator learns not to attack organisms with that appearance.

5. Some poorly defended species, particularly insects, have evolved as mimics of the appearance and behavior of chemically defended species, thereby gaining protection from predation.

6. Other animals escape predators by hiding, camouflage, fleeing, or reproducing in pulses of such large numbers that predators could not possibly consume all of the individuals. Plants have counterparts of all these strategies. Hiding and camouflage are also frequent adaptations of sit-and-wait predators.

7. The best-studied examples of coevolution come from the evolutionary interactions between flowering plants and the insects that feed on them and pollinate them, such as acacia trees and ants and crucifer plants and the various insects that depend upon them directly or indirectly.

## Self-Quiz

1. The spines of a cactus are an example of:
   a. avoiding predation by defense
   b. avoiding predation by escape
   c. a chemical deterrent to predation
   d. Batesian mimicry
   e. cryptic coloration

2. True or False? A Batesian mimic is considered parasitic on its model's reputation because the existence of a mimic is detrimental to members of the model species.

3. Which of the following is true of a Batesian mimicry complex but not of a Müllerian mimicry complex?
   a. The species in the complex are found in the same area at the same time.
   b. Resemblance between species in the complex is as detailed as possible.
   c. Protection of the species involved depends on the predator's ability to learn.
   d. The species are aposematically colored.
   e. Appropriate behavior patterns enhance the effects of similar coloration

4. Pretend you are a birdwatcher seeking the shy and elusive Connecticut warbler, a small woodland bird. You wish to be as inconspicuous as possible. Which of the following will probably *not* help you achieve your aim?
   a. wearing clothing of a mottled green and brown pattern
   b. painting a strip of brown mascara across your face from ear to ear, taking care not to get it in your brown eyes

   c. wearing a hat with leaves or twigs attached to it
   d. wearing a dark shirt or jacket and light-colored slacks
   e. sitting or standing perfectly still

5. A black bug with bright red-orange triangles on its wings lives on plants that have gray-green leaves and dull pink flowers. From this information, you conclude that the bug's coloration is an example of:
   a. cryptic coloration
   b. aposematic coloration
   c. disruptive coloration
   d. countershading

6. Insect species A and B look very much alike. Uninitiated toads quickly learn to leave insects of species A alone. The same toads are then offered species B. The toads don't even try to catch it. Species B is:
   a. a Batesian model
   b. a Batesian mimic
   c. a Müllerian mimic
   d. not enough information given to answer the question

7. Design experiments that would help you find the correct answer to Question 6.

8. For each adaptation below, tell whether it would be more advantageous to an animal that was an active hunter or a sit-and-wait predator.
   _____ a. camouflage
   _____ b. mimicry
   _____ c. search image

9. Which of the following is a *disadvantage* of producing secondary chemicals?
   a. Herbivores that can detoxify them use them to "home in" on the plant for feeding or egg laying
   b. The chemical damages the plant
   c. The chemical may attract pollinators
   d. The chemical may prevent the growth of other plants nearby
10. Many conifers combat insect herbivores by producing juvenile hormone. This is an effective defense because:
    a. juvenile hormone is toxic to insects
    b. juvenile hormone prevents larvae from maturing into adults and producing another generation of herbivorous larvae
    c. juvenile hormone keeps the insects from digesting their food efficiently
    d. insects are repelled by the odor of juvenile hormone, thinking that there are already too many young insects present
11. An advantage of being a herbivore that feeds only on one or a few related plant species is:
    a. only a few kinds of detoxifying enzymes are needed
    b. there is no competition for food
    c. it is easy to find enough food
    d. there will be other herbivores competing for the food source

## Questions for Discussion

1. Why is Batesian mimicry so rare? What would happen if every undefended species mimicked a defended one?
2. Richard Southwood has found a positive correlation between the abundance of members of various plant groups over geological history and the number of specialist herbivores associated with each group. Can you explain this?
3. Vincent Dethier has stated that a host plant probably exerts a greater selective pressure on a herbivorous insect specialized to feed on it than the insect exerts on the plant. What are the selective pressures that are likely to be acting (1) on the insect, (2) on the plant? Do you agree with Dethier?
4. *Dentaria* is a crucifer that lives in shady woods instead of in the sunny open fields where most crucifers are found. Few herbivores are observed feeding on *Dentaria* in its woody habitat. What would you expect to happen if *Dentaria* were planted in a field? Why?
5. Why might some insects not be very susceptible to manufactured insecticides?
6. Why do insect populations that have become resistant to insecticides lose their resistance if insecticide spraying is discontinued for a long time?

## Suggested Readings

Cox, P. A., and M. J. Balick. "The ethnobotanical approach to drug discovery." *Scientific American,* June 1994. Consulting native herbalists in the search for new drugs.

Davies, N. B., and M. Brooke. "Coevolution of the cuckoo and its hosts." *Scientific American,* January 1991. Fascinating account of the arms race between mimicry of host eggs by the cuckoo and the host's ability to discriminate its own from other eggs and so avoid brood parasitism.

DeVries, P. J. "Singing caterpillars, ants, and symbiosis." *Scientific American,* October 1992.

Ewald, P. W. "The evolution of virulence." *Scientific American,* April 1993. Explains what factors select for increasing and decreasing virulence of disease-causing parasites and what this means in our battle against diseases, particularly AIDS.

Janzen, D. H. "Coevolution of mutualism between ants and acacias in Central America." *Evolution* 20:249, 1966.

Rennie, J. "Living together." *Scientific American,* January 1992. Describes some bizarre adaptations resulting from parasite-host coevolution, including the argument that parasites were the selective pressure behind the evolution of sex.

Rosenthal, G. A. "A seed-eating beetle's adaptations to a poisonous seed." *Scientific American,* November 1983. The canavanine story.

Rosenthal, G. A. "The chemical defenses of higher plants." *Scientific American,* January 1986. A survey of the various chemical means by which plants defend themselves against herbivores.

Whittaker, R. H., and P. P. Feeny. "Allelochemics: chemical interactions between species." *Science* 171:757, 1971. Includes a discussion of the coevolution of crucifers and their herbivores.

Young, A. M. "The evolution of eyespots in tropical butterflies in response to feeding on rotting fruit: an hypothesis." *Journal of the New York Entomological Society* 87:66, 1979.

CHAPTER

19

# Population Genetics
# and Speciation

## OBJECTIVES

*When you have studied this chapter, you should be able to:*

1. Define and use in context: gene pool, evolution, cline, gene flow, founder effect, coadapted gene complex, species.
2. State and explain the significance of the Hardy-Weinberg law and the five situations that do not meet Hardy-Weinberg conditions.
3. Given the necessary data, use the Hardy-Weinberg equation to determine the frequencies of two alleles, and of the genotypes they produce, in a population.
4. Explain why genetic drift is more likely to cause evolution in a small than in a large population.
5. Differentiate among stabilizing, directional, and disruptive selection, and give or recognize examples of selective pressures that may cause each.

6. Give one reason for heterozygote advantage.
7. Define polymorphism, and discuss three ways it can be maintained.
8. List three factors that tend to increase and three that tend to decrease genetic variation in a population, and state how each acts.
9. Explain why rapid evolutionary change tends to occur in small populations, whereas widespread, broadly adapted species may remain much the same for millions of years.
10. Distinguish between allopatric and sympatric speciation, and describe how they may occur. List or recognize examples of each type.

I n Chapter 17 we saw that individual organisms do not evolve; populations evolve. The difference between ancient dinosaurs and their descendants, modern birds, lies in the different genes of members of the two populations.

The importance of populations in evolution was emphasized by R. A. Fisher, Sewall Wright, and J. B. S. Haldane. During the 1920s, their work on genetic variation laid the foundation of **population genetics,** the study of the frequencies of alleles in populations (that is, the relative proportions of alternative alleles). The emphasis on populations rather than individuals transformed evolutionary thinking into a more modern form, often known as Neo-Darwinism or *The Modern Synthesis,* the title of a book on the subject by J. S. Huxley.

A **population** consists of all the members of a species that occupy a particular area at the same time—for example, the perch population of a lake or the fir tree population of a valley (Figure 19–1). The members of a population are much more likely to breed with one another than with members of other populations of the same species. There-

**FIGURE 19–1** A population of Douglas fir trees in the Cascade Mountains. (Richard Feeny)

395

fore, populations form breeding groups, and a group of genes tends to stay within the same population generation after generation.

Population geneticists' mathematical models show how alleles and genotypes are preserved within populations and the effects of forces that change allele frequencies. In 1908 G. H. Hardy and W. Weinberg pointed out that the gametes contributed by all the members of a population to the next generation can be considered as containing a giant **gene pool** from which offspring receive their various genomes at random. In the absence of natural selection or other factors that change frequencies of alleles in the gene pool,

these frequencies tend to be preserved from one generation to the next, and the population does not change genetically; that is, it does not evolve. If we can discover how a population's gene pool changes with time when it does change, we shall understand how evolution occurs.

Different populations of the same species sometimes become isolated from one another. When this happens, one of the most important processes in evolution may occur: formation of a new species, a group whose members, by definition, seldom exchange genes with members of other species.

### K E Y    C O N C E P T S

- ◆ Populations, not individuals, evolve.
- ◆ Evolution will not occur in a (hypothetical) large, genetically isolated population with no net mutation, no selection pressures, and no mating preference.

- ◆ Most new species form by evolution of an isolated population, which may diverge rapidly from the parent population.
- ◆ Polyploidy may lead to the formation of a new species (genetic isolation) without geographic isolation.

## 19–A    THE HARDY-WEINBERG LAW

One way to see how a population evolves is to construct a model of a population that does *not* change genetically from one generation to the next, and then see how a real population differs from this model. The **Hardy-Weinberg law** provides such a model (*In More Detail: The Hardy-Weinberg Law*). It shows that evolution does not occur in a large population of diploid organisms where mating is random and there is no mutation nor natural selection.

Consider a population whose gene pool contains two alternative alleles, *A* and *a*, either of which can occupy one locus on a chromosome. Every member of the population has one of three possible genotypes: *AA*, *Aa*, or *aa*. The Hardy-Weinberg law shows that the next generation will contain the two alleles at the same frequencies. Indeed the frequencies of the alleles and the genotypes will remain the same through all successive generations, if the population meets *all of the following conditions:*

1. **No net mutation.** The alleles in question must not mutate (or, if they do, the rate of mutation of *A* to *a* must equal the mutation rate from *a* to *A*).
2. **No mating preferences.** The population must reproduce sexually, and mating must be random with respect to genotype (so that, for instance, an *AA* female does not prefer *aa* to *AA* or *Aa* males when she mates).
3. **Large size.** The population must be large, because the law is based on statistical probabilities. Small popula-

tions are more subject to sampling errors, in which the individuals chosen as a sample are not representative of the population as a whole.

4. **No gene flow.** The population must be isolated so that there is no exchange of genes (gene flow) by migration of individuals or gametes between the population and any other population.
5. **No selection.** There must be no natural selection with respect to the alleles in question (i.e., no genotype has a reproductive advantage over the others).

If all these conditions are fulfilled, *A* and *a* will remain in the population indefinitely at the same frequencies, and the three genotypes (*AA*, *Aa*, and *aa*) will also remain the same: there will be no evolution.

The Hardy-Weinberg law expresses the fact that sexual reproduction, with its reshuffling of genes, is not by itself enough to cause evolution. Evolution is a change in allele frequencies from one generation to the next, and under Hardy-Weinberg conditions there is no change.

The most useful application of the law is that, since it states the conditions under which evolution will not occur, it also states that if these conditions are not met, evolution is likely to occur. In other words, evolution usually occurs if mating is not random, if the population is small, if there is gene flow between populations, or if natural selection favors some genotypes over others.

# THE HARDY-WEINBERG LAW

The Hardy-Weinberg law states that the frequencies of alleles $A$ and $a$ will remain the same from generation to generation in a population if all the conditions listed in Section 19–A are met. To satisfy ourselves that this is true:

Let $p$ = the frequency of allele $A$ in the population (that is, the proportion of all alleles at this locus that are $A$)

Let $q$ = the frequency of allele $a$

Since all chromosome loci for this gene in all members of the population must be occupied by either the $A$ or the $a$ allele, $p + q = 1$. Thus, if $a$ occurs at 20% of the loci ($q$ = frequency of $a = 0.2$), the other 80% of the loci must be occupied by $A$ ($p$ = frequency of $A = 0.8$), and the two frequencies together equal one ($0.2 + 0.8 = 1$).

Now let us see what happens during reproduction. The frequencies of the two alleles $A$ and $a$ in the gametes produced by the population are the same as the frequencies of the alleles. $AA$ homozygotes, of either sex, produce only $A$ gametes; $aa$ homozygotes produce only $a$ gametes; $Aa$ heterozygotes produce equal numbers of $A$ and $a$ gametes.

What are the chances that an $A$ sperm will fertilize an $A$ egg? Since the frequency of $A$ gametes is $p$, the frequency with which $A$ sperm fertilize $A$ eggs is $p \times p = p^2$. Similarly, the frequency of $a \times a$ fertilizations is $q^2$. The frequency of fertilizations between $A$ and $a$ gametes will be $(p \times q) + (q \times p)$ (since $A$ sperm, of frequency $p$, will encounter $a$ eggs, of frequency $q$, with a frequency of $p \times q$, and vice versa). Therefore, the expected frequencies of the three genotypes in the next generation is:

$$AA \quad p^2$$
$$Aa \quad 2pq$$
$$aa \quad q^2$$

Because the frequencies of the three genotypes must add up to 1, $p^2 + 2pq + q^2 = 1$.

We could do the same calculations using algebra:

Allele frequencies: $\qquad\qquad p + q = 1$
Squaring both sides of the equation gives: $(p + q)^2 = (1)^2$
Genotype frequencies: $\qquad\quad p^2 + 2pq + q^2 = 1$

Now that we know the expected frequencies of the three genotypes in the new generation, we can check the Hardy-Weinberg prediction that the frequencies of $A$ and $a$ are still $p$ and $q$, respectively. The frequency of $A$ is $p^2$ (from the $AA$ homozygotes) $+ \frac{1}{2}(2pq)$ (from the heterozygotes; half of their alleles are $A$, and half are $a$). So the frequency of $A$ in our new generation is

$$p^2 + \tfrac{1}{2}(2pq) = p^2 + pq$$

Since $q = 1 - p$,

$$p^2 + p(1 - p) = p^2 + p - p^2 = p$$

So the frequency of $A$ is still $p$.

A similar calculation shows that the frequency of the $a$ allele in this new generation must be $q$. In other words, as predicted by the Hardy-Weinberg law, the frequencies of $A$ and $a$ have not changed.

To see how this might work in practice, we will work through an example. Suppose we have a population made up of 600 $AA$ males, 400 $aa$ males, 600 $AA$ females, and 400 $aa$ females. If the population meets the conditions of the Hardy-Weinberg equation, what will be the proportions of the three genotypes in the next generation, and what will be the new allele frequencies of $A$ and $a$? Since 60% of the members of each sex are $AA$, 60% of the gametes, both eggs and sperm, carry allele $A$. Similarly, since the other 40% of the population are $aa$, the other 40% of the gametes of both sexes carry allele $a$. So the frequency of allele $A = 0.6 = p$, and the frequency of allele $a = 0.4 = q$. Note that $p + q = 0.6 + 0.4 = 1$. The frequencies of the three genotypes in the next generation can therefore be calculated as follows:

| | | |
|---|---|---|
| Frequency of | $AA = p^2 = 0.6 \times 0.6$ | $= 0.36$ |
| Frequency of | $Aa = 2pq = 2 \times 0.6 \times 0.4$ | $= 0.48$ |
| Frequency of | $aa = q^2 = 0.4 \times 0.4$ | $= 0.16$ |
| Total: | $p^2 + 2pq + q^2$ | $= 1.00$ |

If we assume that this new generation contains 1000 individuals, then the most probable numbers of the three genotypes are 360 $AA$ individuals, 480 $Aa$ heterozygotes, and 160 $aa$ individuals. Note that the genotype frequencies have changed from the starting population. Has this also happened to the allele frequencies? Noting that 1000 individuals contain 2000 alleles, we can calculate the allele frequency for the new generation as follows:

$$\text{Frequency of allele } A = \frac{(2 \times 360) + 480}{2000} = \frac{1200}{2000} = 0.6$$

$$\text{Frequency of allele } a = \frac{(2 \times 160) + 480}{2000} = \frac{800}{2000} = 0.4$$

The allele frequencies have therefore not changed. Nor would they change if this process were repeated for an infinite number of successive generations.

The Hardy-Weinberg law applies equally when the population contains three or more alleles of the same gene instead of the two discussed here. For instance, the genotype frequencies of three alleles are expressed by $(p + q + r)^2$, where $p$, $q$, and $r$ represent the frequencies of the three alleles in the gene pool.

*(Continued)*

**IN  MORE  DETAIL** *(Continued)*

### Recessive Alleles

A practical application of the Hardy-Weinberg equation is calculating how many people in any generation of a human population are carriers for a particular recessive allele. If we know the number of babies born annually with diseases such as sickle cell anemia or phenylketonuria (PKU), which are expressed only in the homozygous condition, we can estimate how many people in the parental population are heterozygotes for the allele and therefore carriers of the disease. This value should be recalculated in each generation, since there is selective pressure against these particular homozygous recessive genotypes, and hence the alleles do not meet Hardy-Weinberg conditions.

The equation also shows that a deleterious recessive allele, even one so unfavorable as to be lethal, will hardly ever be completely eliminated from a large population. The lethal phenotype will be expressed only when the allele occurs in the homozygous recessive condition; these individuals will always die. But many more heterozygous than homozygous recessive individuals will be formed in each generation. The lower the frequency of the allele in the population, the lower the frequency of recessive homozygotes relative to heterozygotes (Figure 19–A). Thus, however rare the homozygous recessive genotype becomes, there are still many heterozygous carriers in the population.

In contrast, a lethal dominant allele will be removed from the population immediately every time it arises by mutation because it will be expressed in the phenotype of every individual that carries it and will cause its carrier to die. The allele will arise only by mutation and will therefore become extremely rare.

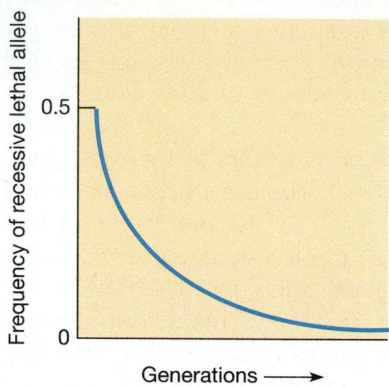

**FIGURE 19–A**   The frequency of a recessive lethal allele in a population of diploid organisms. Even if we start with a frequency of 0.5 for the lethal recessive allele (that is, with a population in which every individual is a carrier), the frequency will drop rapidly in succeeding generations. This is because many copies of the allele will be eliminated by selection as all homozygous recessive individuals die without having a chance to reproduce and pass the allele to future generations. However, heterozygous carriers of the allele will persist in the population indefinitely. They become so rare that matings between two carriers are extremely infrequent, and therefore few homozygous recessive individuals are produced.

## 19–B   CAUSES OF EVOLUTION

Let us now discuss cases that do not meet the Hardy-Weinberg conditions and that may therefore produce evolution.

### Mutation

Mutation is the ultimate source of genetic variation. However, it is such a rare event that mutation acting alone probably never causes a stable change in gene frequencies in a population.

Suppose allele $A$ is **fixed** in a population, meaning that all individuals have the genotype $AA$. Then a new mutation changes $A$ to $a$ in one individual. Ronald A. Fisher calculated that there is a 95% chance that $a$ will be lost from the population within 30 generations for the following reasons:

1. The $Aa$ individual may not survive to reproduce.

2. Even if $Aa$ reproduces, studies show that 40% of matings in most stable populations (including human populations) produce no surviving offspring ($a$ is lost) or only one surviving offspring. Even if there is one surviving offspring, there is still a 50% chance that $a$ will be lost. (Half the $Aa$ parent's gametes contain $a$, so there is a 50% chance that the offspring will not receive $a$, in which case $a$ will be lost from the population when the $Aa$ parent dies.)

3. Even if $Aa$ has two surviving offspring, there is a 25% chance that neither will inherit $a$.

Fisher calculated that the chance of a new mutant allele disappearing from the population in one generation is more than 33%. But even if it is not lost by random events, the new mutation will take a very long time to have a stable effect on the gene pool. For one thing, the reverse mutation ($a$ to $A$) will occur. If mutation is the only factor changing allele frequency, the eventual frequency of $a$ in the gene pool depends on the relative rates at which $A$ mutates to $a$ and $a$ mutates back to $A$. At the usual mutation rates of one in every $10^4$ to $10^9$ individuals, it takes hundreds of generations for $A$ and $a$ to reach equilibrium in the population under the influence of mutation alone. Because mutation rates change with time, the two alleles will probably never reach equilibrium. Hence, mutation alone is never likely to account for observed allele frequencies in a population.

## Mating Preference

Mating that is not random with respect to genotype can bring about evolutionary change by selecting for one genotype over another. If females consistently choose to mate with $AA$ males, for example, they exert selection in favor of the $AA$ genotype. Examples are known from studies on the fruit fly *Drosophila* where females prefer to mate with males heterozygous for certain alleles to mating with either homozygote.

## Genetic Drift

The Hardy-Weinberg law holds true only for large populations because the law depends on probability, which applies more accurately to a large sample than to a small one. Sewall Wright pointed out that in small populations evolution can occur simply by chance: random events may bring death or parenthood to some individuals regardless of their genetic makeup. The resulting change in the gene pool is called **genetic drift** (Figure 19–2).

Consider a population of five individuals, in which only two breed. The chances are good that any particular allele is represented in only one member of the population. If this individual does not breed, the allele will not be present in the next generation. If the individual does breed, the frequency of the allele may increase in the next generation. In either case, since a change in allele frequency from one generation to the next has occurred, evolution has taken place. In contrast, if the population is large—say we multiply each of these numbers by 1000—then the 2000 who breed are likely to include many of the 1000 who contain the allele. The next generation will contain roughly the same proportion of the allele, and so evolution by genetic drift is much less likely in a large population than in a small one.

There has been much argument about the extent of genetic drift in natural populations because it is hard to measure. Drift has significant effects only if the population is small and if one allele has little or no selective advantage over another. Since breeding populations of many species consist of fewer than 50 individuals, one might get the impression that drift is often important. On the other hand,

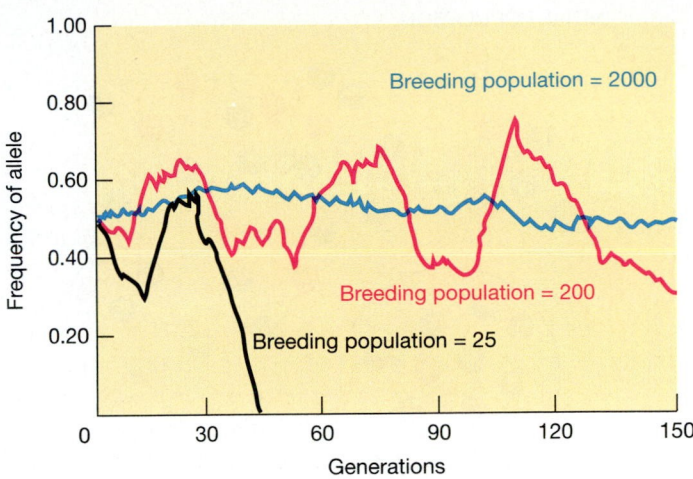

**FIGURE 19–2** Genetic drift due to sampling error with different population sizes. The graphs show the results of computer experiments. These showed the effects of size of the breeding population on the frequency of one allele. The computer modeled three populations of different sizes, each starting with the allele at a frequency of 0.5.    (Data courtesy of S. C. Smith)

there are hundreds of species, including loggerhead sea turtles, black swallowtail butterflies, and many fish, in which selection is known to eliminate more than 95% of the population each year, many of them deformed juveniles that die at a young age. In these cases, nonrandom changes in the gene pool caused by natural selection tend to overshadow the effects of drift.

One example of genetic drift is the **founder effect,** the change in allele frequencies when a new population arises from one or a few individuals. When a few individuals leave a population, the chances are good that they will not contain all the alleles found in the population. If these few individuals become the founders of a new population elsewhere, the gene pools of the old and new populations will differ.

For example, human populations in southern Africa show great diversity at some genetic loci compared with European populations, and Native Americans have even lower diversity. This is almost certainly a result of the founder effect. Humans evolved in Africa. Then groups containing many, but not all, of the genotypes found in Africa migrated north and east to become the ancestors of European and Asian populations. Finally, much smaller numbers crossed from Asia to America and became the ancestors of Native Americans, a group strongly affected by the founder effect and with little genetic diversity. This lack of genetic diversity is a major reason that so many Native Americans died from diseases introduced to the Americas by European immigrants.

The founder effect may also come into play during a **population bottleneck,** when the number of individuals

INITIAL POPULATION

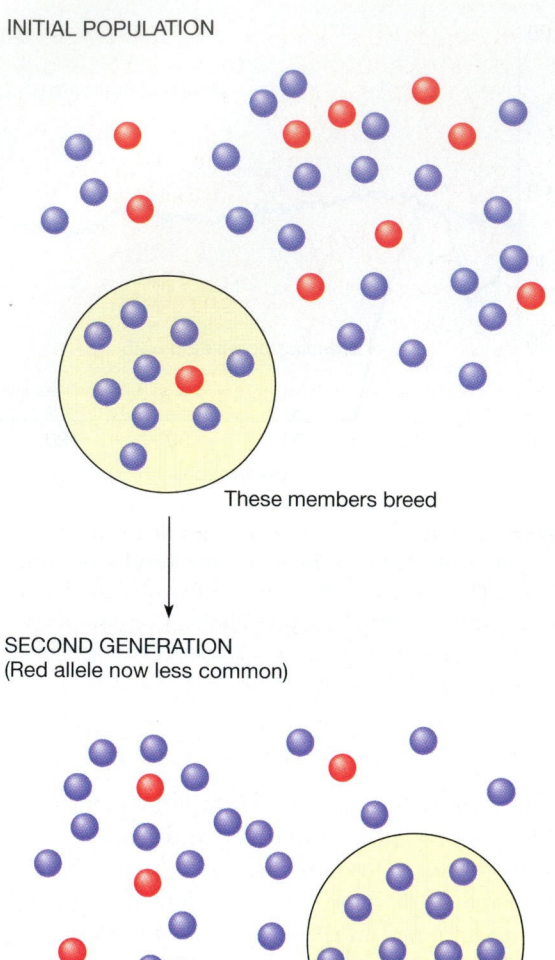

These members breed

SECOND GENERATION
(Red allele now less common)

These members breed

**FIGURE 19–3**   A model illustrating genetic drift and the founder effect. The marbles represent individuals. Red individuals carry a rare allele. Suppose that in any one year only 25% of individuals breed—those in the circle. By chance, only one of them carries the red allele, and so the allele is rare in the next generation. The breeding population in the second generation is quite likely to contain no individuals with the rare allele: gene fixation of the common (wild-type) allele will have occurred.

falls. This often happens to endangered species and is fairly common in isolated populations of most species (Figure 19–3). Reduced to a few individuals, the population may show genetic abnormalities resulting from the random loss of alleles. For instance, investigating why cheetahs breed poorly in North American zoos, biologists discovered that both captive and wild cheetah populations contain little genetic variation: less than even such inbred populations as laboratory mice. They theorize that cheetahs went through a population bottleneck about 10,000 years ago. Very few individuals, with a limited gene pool, survived to become

**FIGURE 19–4**   A cheetah.

the ancestors of modern populations (Figure 19–4). Most wild cats (including the endangered Florida panther) suffer from similar lack of genetic diversity for the same reason.

It is often impossible to tell how much of the genetic difference between two populations results from the founder effect and how much from different selective pressures in two environments. The founder effect surely had a great influence on the 28 closely related populations of silversword plants, which probably all evolved from one original seed carried to the Hawaiian Islands by a bird, and on domestic hamsters, most of which are descended from one pregnant female captured in Syria.

## Gene Flow

A lost allele may reappear in a population by mutation, but it is more likely to be reintroduced by immigration of individuals (or pollen or sperm) carrying the allele from a neighboring population, resulting in **gene flow** between the two populations (Figure 19–5).

Animals may migrate from one area to another, contributing their genes to another gene pool; a tornado may disperse seeds beyond the boundaries of the parent population; or humans may carry organisms from one population to another. As an example of the evolutionary change caused by gene flow, genetic tests show that an average of 20 to 30% of the genes of American blacks are of European

**FIGURE 19–5** Gene flow: a yellow cloud of pollen blows away from a fir tree. Most will remain within the population's area, but some may be blown farther, into another population.

origin and 70 to 80% of African origin. This is the result of interbreeding between populations of European and African origin during the 15 or fewer generations the two have existed together in the United States.

The greater the distance between populations of the same species, the greater the genetic differences between them usually are. Maple trees in Vermont differ from maple trees in Michigan more than they differ from those in New Hampshire. This phenomenon, in which a character shows a gradient of variation across an area, is called a **cline** (Figure 19–6). For instance, song sparrows on the west coast of North America show a cline of increasing size and decreasing contrast in markings from California to Alaska.

There are two main reasons why characters within a species may show clinal variation. First, gene flow between adjacent populations means that the gene pools of populations that are close together share more alleles than those that are farther apart. Second, environmental features, such as climate, often vary along gradients. In Figure 19–6, height up the mountain changes steadily. Because these environmental factors act as selective pressures, the phenotypic characters that are best adapted to such pressures will also vary in a gradient.

Gene flow between populations tends to increase the similarity between all the populations of a species. Natural selection has the opposite effect. It tends to tailor every population to its particular habitat. Clines are one possible outcome of these two conflicting forces.

## Natural Selection

Natural selection is the nonrandom differential survival and reproduction of genotypes from one generation to the next. When a selective force, such as predation or competition, is at work, some of the different phenotypes in a population are more likely to survive and reproduce than others. If the phenotypic characteristics selected for are at least partly under genetic control—as they usually are—then the genes responsible for them will be represented at higher frequencies in the next generation.

In the simplest case of one pair of alleles, if having one allele confers even a slight, but consistent, reproductive advantage, the allele's frequency in the population will increase from one generation to the next. It is said to have a greater **fitness** than the less favored allele. In nearly all cases that have been analyzed, natural selection is by far the most potent evolutionary force changing gene frequencies in a population.

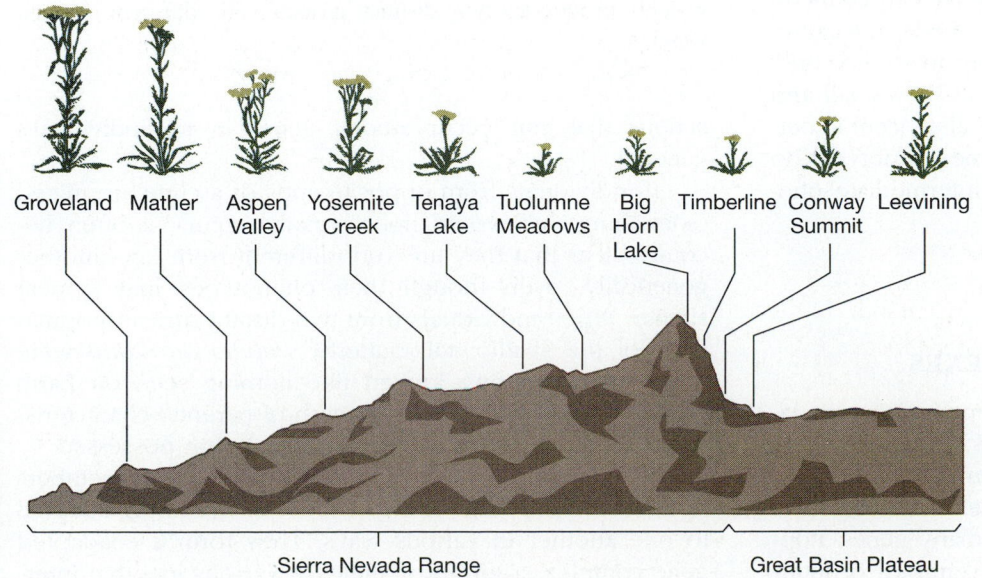

Groveland  Mather  Aspen Valley  Yosemite Creek  Tenaya Lake  Tuolumne Meadows  Big Horn Lake  Timberline  Conway Summit  Leevining

Sierra Nevada Range          Great Basin Plateau

**FIGURE 19–6** A cline. *Achillea millefolium* (yarrow) plants collected from populations in the Sierra Nevada of California and Nevada and then grown under uniform conditions reveal differences that must be genetic.

Probably the most common form of natural selection is **stabilizing selection,** when average phenotypes have a selective advantage over extremes in either direction. In a classic 1899 study, H. C. Bumpus measured nine characters of sparrows that died in a winter storm and those that survived it. He found that the sparrows that survived had intermediate weight, length, wingspan, and other phenotypic characters, whereas the sparrows killed by the storm had a much wider range of characters, both smaller and larger than those of the survivors. Similarly, babies of average birth weight (3.5 kg) have a much higher chance of surviving to age five than babies with weights above or below this average. In one study of more than 6500 babies, 70% of deaths before one month of age could be ascribed to selection against babies weighing more or less than the "optimal" 3.5 kg.

**Directional selection** occurs when the phenotypes at one extreme have a selective advantage. It is usually the temporary result of environmental change, when only extreme phenotypes happen to be adapted to the new conditions. Figure 19–7 shows the actions of these kinds of selection, using a population with a normal (bell-shaped) distribution of phenotypes from a polygenic trait—a trait controlled by many genes, such as human height or the weight of seeds.

Let us consider an example. In a population of seeds, if seeds of average size have a better chance of growing than seeds that are unusually large or small, and if seed size is inherited, then stabilizing selection is acting, and the next generation will contain a lower proportion of unusually large or small seeds. This might happen if large seeds have such thick coats that water penetrates them only slowly, and small ones with very thin coats germinate so soon that they are killed by frost. On the other hand, if birds tend to eat large seeds and ignore smaller ones, they will exert directional selection in favor of small seeds.

**Disruptive selection** takes place when the extremes of a range of phenotypes are favored relative to intermediate phenotypes. It might happen to our seeds, for example, if a particular kind of beetle specialized in feeding only on seeds of intermediate size, ignoring the very small and very large seeds. Disruptive selection may also occur in heterogeneous environments in which extreme phenotypes do well under certain conditions, whereas intermediate phenotypes are never well adapted.

## 19–C    COADAPTED GENE COMPLEXES

For simplicity we often speak of the genetics of populations in terms of a pair of alleles at a locus, but genes rarely exert their effects in such a simple manner. Each gene may influence more than one phenotypic character, and most phenotypic characters are influenced by many genes. Population geneticists have found dramatic examples of inter-

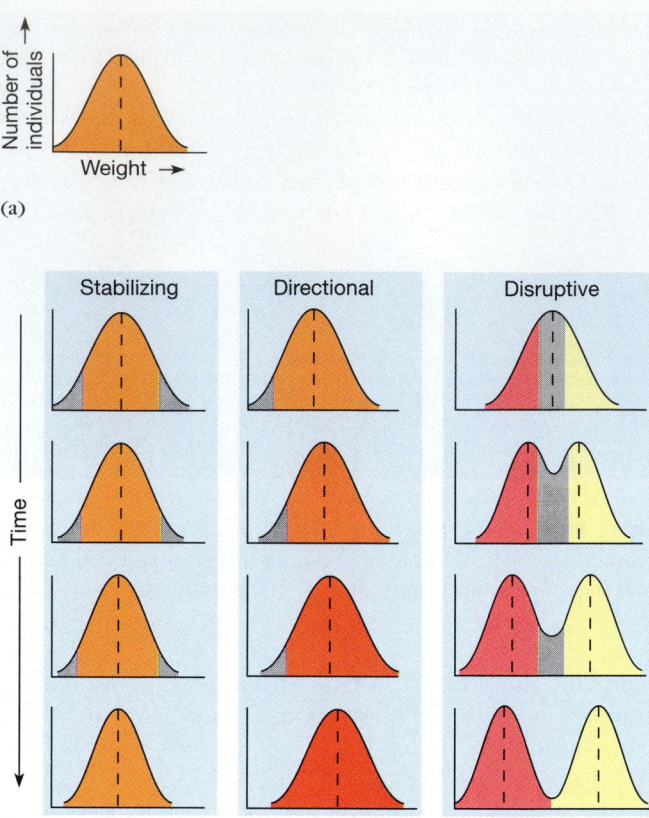

(a)

(b)

**FIGURE 19–7**   Types of natural selection. The effects of selection on a polygenic character, such as weight, in a population. (a) Before selection, the population has a normal distribution of individuals of different weights. In (b) the individuals eliminated by selection in each generation are shown in gray. Stabilizing selection eliminates very heavy or very light individuals, leading to a population with less weight variation than in (a). Directional selection in this case eliminates light individuals, leading to a population in which the median weight (dotted lines) is higher than in (a). Disruptive selection (rarer) eliminates individuals of median weight, producing two distinct groups with different median weights.

actions that can occur among genes in an individual's genome.

If individuals from opposite ends of a cline are mated (something that would not normally occur), it often becomes clear that they are very different from one another genetically, even though their phenotypes may appear similar. When individuals from two distant African populations of the swallowtail butterfly *Papilio dardanus* were mated, the offspring looked like nothing seen on Earth before. They displayed a mishmash of parental characteristics and of color patterns that neither parent possessed.

When genes occur together for a long time in members of the same population, they coevolve, becoming adapted to one another in various ways. They form a coadapted gene complex, a group of alleles at various loci that interact and produce individuals well adapted to their environment. Genes from distant areas are not adapted to each other in this way. When they come together in the offspring of a mating, they may produce individuals with bizarre phenotypes. What genetic mechanisms permit genes to interact more harmoniously with one another when they have evolved within the same gene pool than when they have not?

## Evolution of Dominance and Recessiveness

Alleles may become adapted to one another by changes in their relative degree of dominance. For example, the allele that produces melanism in the peppered moth (Section 17–C) was once less dominant than it is today. In moth collections from the nineteenth century, moths that were presumably heterozygous for this allele are paler than presumed homozygotes. Today, heterozygotes look just like homozygotes: the dominance of the melanic over the non-melanic allele has increased. Similarly, artificial selection in laboratory populations of *Drosophila* has shown that the degree of dominance or recessiveness of most alleles can be modified by artificial selection.

If the degree of dominance can be altered by selection, what causes dominance? Experiments on the lesser yellow-underwing moth (*Triphaena comes*) provide one answer. A melanic allele is dominant in populations of this moth on islands in Scotland. However, E. B. Ford crossed members of populations from different islands and found that the melanic allele was not dominant in their offspring. In other words, the allele is dominant only in the complex of genes in which it normally occurs. Apparently genes elsewhere in the genome alter its expression, causing dominance.

## 19–D    WHAT PROMOTES AND MAINTAINS VARIABILITY IN POPULATIONS?

Through evolution, organisms become better adapted to their environment. Why, then, do all the members of a population not become more and more alike genetically, as selection eliminates the less fit alleles? The answer must be that factors promoting genetic variation are constantly at work. Mutation is an obvious source of variation, but it is such a rare event that there must be others.

## Heterozygote Advantage and Hybrids

Often individuals heterozygous at a particular locus are more common in a population than the Hardy-Weinberg equilibrium would predict. This suggests the heterozygotes enjoy a selective advantage, called **heterozygote advantage,** over homozygotes. The best documented example in humans is the case of sickle cell anemia, where the heterozygote is at a selective advantage over either homozygote in areas with a high incidence of malaria. In these areas, normal homozygotes survive to adulthood at only 88% the rate of heterozygotes, and sickle homozygotes survive only 13% as well, most dying from the lethal effects of sickle cell anemia before adulthood. Heterozygote advantage maintains genetic variability in these populations: where the heterozygote is at an advantage over either homozygote, selection ensures continued variation at the locus.

Heterozygote advantage can arise when each allele contributes its beneficial effects to the heterozygous phenotype. This is the case with heterozygotes for sickle cell anemia in areas where malaria is common: these individuals have the advantages of both near-normal hemoglobin function and resistance to malaria. Sometimes the two alleles in a heterozygote code for enzymes with slightly different optimum temperatures, an advantage to organisms exposed to varying environmental temperatures.

Heterozygotes are rare in populations with a high degree of **inbreeding,** that is, mating between close relatives or even self-fertilization. This is because close relatives inherit many of the same alleles from their common ancestors. Individuals from inbred populations are often homozygous for nearly all their alleles. These often include disadvantageous homozygous recessive pairs which reduce health and vigor.

Matings between members of two different strains produce **hybrid** offspring, which may be superior to their parents in many ways. Mongrels tend to be healthier than "pure-bred" dogs, which suffer from a high frequency of genetic ailments. Hybrid corn is valued for its reliable uniformity as well as for the specific qualities of the parental lines (Figure 19–8).

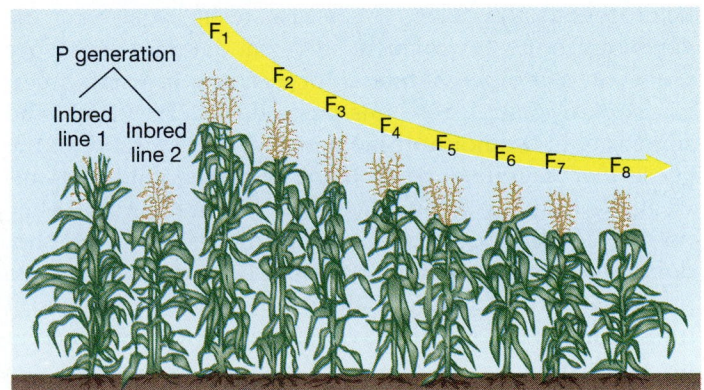

**FIGURE 19–8**  Hybrid vigor. Hybrid corn (the $F_1$ generation) is heterozygous at many loci and enjoys many genetic advantages over either parental line. When the $F_1$ and succeeding generations reproduce, many individuals will become homozygous for various traits. This results in a gradual decrease in the desirable genetic qualities of the average plant.

What is the genetic basis of this **hybrid vigor?** Since hybrids are heterozygous at many loci where their parents were homozygous, their superiority might result from heterozygote advantage in many gene pairs. Perhaps more often, hybrid superiority results simply from the fact that hybrids have a high percentage of loci with at least one dominant, advantageous allele, which masks the disadvantageous effects of its recessive partner.

Because of its pitfalls, inbreeding is often selected against, and adaptations that promote outbreeding are common. For example, many plants have adaptations promoting cross-pollination rather than self-pollination. Most human societies have taboos against incest. In many animals, such as monkeys and lions, young males leave the social group of their birth and, with luck, eventually join another group and breed with the resident females. By the time their daughters grow up, they will often have moved on. In chimpanzees and some rodents, it is the females that emigrate to join other groups. These mechanisms reduce the chances that disadvantageous recessive alleles will become fixed in the population.

## Polymorphism

One of the most common types of variation in populations is **genetic polymorphism** (poly = many; morph = form), the occurrence in the same place at the same time of two or more genetic variants in such proportions that the rarest of them cannot be maintained by recurrent mutation alone. Polymorphism may occur for a time while a favorable allele steadily replaces alternative alleles, eventually reaching fixation. More commonly, a balanced polymorphism persists indefinitely.

Balanced polymorphism can result from disruptive selection, the opposite effects of selection and gene flow, and heterozygote advantage. It can also result from recurring variations in the environment such that different genotypes are favored at different times. For instance, in North America, some Canada geese migrate south every winter, and some stay in the north. If the winter is severe, more of the geese that migrate survive. If the winter is mild, the survival rate is higher among those that stay in the north. They do not face the perilous journey or the risk of being shot during the southern hunting season. Because the selective pressures for migrating or staying at home are not constant from year to year, both behavior patterns are maintained in the population.

Sex is probably the most widespread polymorphism. Most individuals are either male or female, and both morphs are almost always present in a population at the same time.

The human ABO blood group is a polymorphism involving three main alleles (see Table 16–1). Surveys throughout the world reveal variations in the proportions of each blood type in different human populations. For

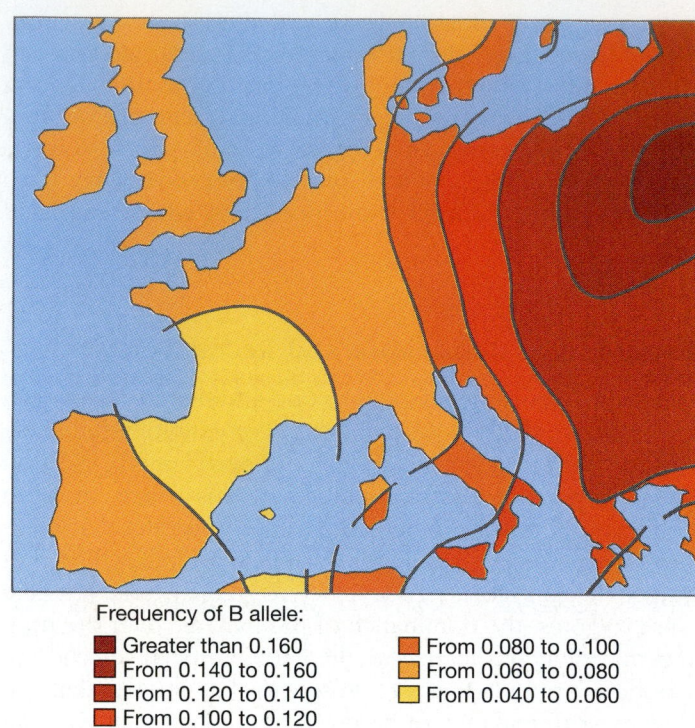

Frequency of B allele:

■ Greater than 0.160
■ From 0.140 to 0.160
■ From 0.120 to 0.140
■ From 0.100 to 0.120
■ From 0.080 to 0.100
■ From 0.060 to 0.080
■ From 0.040 to 0.060

**FIGURE 19–9** Frequency of the B allele of the human ABO blood group system in various European populations. In general, the frequency of the allele decreases progressively from east to west. Populations in central Asia and India, to the east of the area shown, have especially high frequencies of this allele.

instance, the frequency of the B allele ($I^B$) is highest in the central part of Eurasia—India, Mongolia, and western Siberia—and generally decreases with increasing distance from these areas (Figure 19–9). This allele is not found in native Australians or Native Americans, except among Alaskan Eskimos.

To explain the distribution of ABO blood groups, we might suggest gene flow as populations mix, a gradient of selective pressures, or perhaps a combination of these factors. People with blood group A tend to develop more stomach cancer, pernicious anemia, and diabetes; people of group O are more liable to ulcers and pituitary tumors. There is also evidence that people of blood type A are more susceptible to smallpox, because the A antigen is immunologically similar to the smallpox virus, making it harder for their bodies to produce effective defenses against smallpox. This fits in with the finding that in India, where smallpox was long endemic, only about 27% of the people contain the A allele, compared with 46% in England and 48% in Germany. Smallpox was officially declared eradicated in the late 1970s, so it will be interesting for future scientists to observe any changes in the frequency of blood groups over the next few generations in India, as might be expected if smallpox was one of several conflicting selective pressures determining blood group frequency.

Stable polymorphisms can also result from **frequency-dependent selection,** in which phenotypes have different selective advantages depending on their frequency in a population. For example, many generalized predators form "search images" for particular phenotypes in their prey. While hunting with a particular search image, they tend to kill more individuals with that phenotype than would be expected from its frequency in the population. Birds that feed on the peppered moth show this behavior. When both melanic and normal morphs of the moth were observed in places where both forms were more or less equally visible, individual birds were found to feed only on one morph until it became quite rare relative to the other. The birds would then switch their search image and start to feed only on the other morph. Humans often select pet cats or fish for their novelty, choosing an uncommon breed or color. Such frequency-dependent selection tends to favor the phenotypes that are rare at the time, and thus to produce a balanced polymorphism.

## A Butterfly Population Explosion

When a population grows rapidly in size, it invariably becomes more genetically diverse. E. B. Ford studied the marsh fritillary butterfly (*Euphydryas aurinia*) for 15 years, including a four-year period when the butterfly increased dramatically in numbers and genetic variability. During this period, scarcely two of the butterflies looked alike, and some were so deformed that they could not fly. Both before this population explosion, when the butterflies were very rare, and afterward, when they were quite common, essentially all the butterflies caught looked identical, although the phenotype before the population explosion looked different from the phenotype afterward.

During population explosions, fewer members die and more reproduce than normal, usually because selective pressures are relaxed. In the case of the human population explosion of the twentieth century, deaths from infectious diseases have declined as a result of improved hygiene and antibiotics. With selection reduced, many more individuals live to maturity, including those, like butterflies that cannot fly, with deleterious genetic combinations whose phenotypes were previously selected against.

Whether the population is large or small to start with, genetic variation is greater at times when selective pressures are relaxed. At times of stringent selective pressure, a greater proportion of individuals with deviations from the best-adapted genotypes will die or will produce few offspring.

Forces that tend to increase and to decrease genetic variation act on all populations (Table 19–1), and the amount of variation found at any one time is presumably a result of all these factors.

## 19–E  SELECTIVELY NEUTRAL MUTATIONS

Most biologists would probably agree that the mechanisms discussed so far suffice to account for the polymorphism found in natural populations. Others disagree, and think that natural levels of polymorphism can be explained only by the additional presence of selectively neutral alleles.

A **selectively neutral allele** is one with no selective advantage or disadvantage relative to alternative alleles. Once such an allele had arisen by mutation, it could progress through the population's gene pool either by being linked to an advantageous allele or by genetic drift. Calculations show that, on the average, neutral alleles would drift in the gene pool for long periods and contribute strongly to the polymorphism of the population.

The most obvious way for a selectively neutral allele to arise is by a mutation that does not affect the proteins an organism produces. For instance, a mutation within a noncoding intron could alter the gene without altering its effect. Similarly, recall that the genetic code is redundant, and many amino acids are encoded by several codons with different third bases (see Table 10–1). In these codons, a point mutation of the third base in a structural gene changes the gene but not the resulting protein and might well be selectively neutral.

It is less obvious that a mutation can be selectively neutral if it changes protein structure. The best evidence for such mutations comes from the many differences between the cytochrome *c* of different species. A comparison of evolutionary history with the amino acid sequences of cytochrome *c* shows that some amino acids have been substituted for others at a steady rate in the course of evolution. This constant rate of substitution is unlikely (so it is argued) to result from selection because selection would be more likely to produce uneven rates of evolution.

A regular rate of amino acid substitution *could* be produced by selective neutrality, but there are difficulties with this argument. Some have pointed out that constant rates of amino acid substitution over the hundreds of millions of years of evolution could obscure enormous variation in that rate over shorter time periods. Others consider it a logical fallacy to call a change selectively neutral merely because we do not know of a selective pressure that could favor the change. The contribution of selectively neutral or selectively trivial genes to natural polymorphism remains an open question.

## 19–F  WHAT IS A SPECIES?

One or more populations of organisms capable of breeding with one another but not with members of other such groups constitute a species according to the traditional definition. Thus, species live in **reproductive isolation** from one another, even though they may live side by side. The

**TABLE 19-1**

**Factors That Increase and Decrease Genetic Variation in a Natural Population**

| Change | Factor | Effect |
|--------|--------|--------|
| Increasing variation | Mutation | Introduces variation |
| | Sexual reproduction | Genetic reassortment occurs at gamete formation and at fertilization |
| | Polymorphism, disruptive natural selection, and heterozygote advantage | Retain more than one allele at a locus in the population (prevent gene fixation) |
| | Immigration and outbreeding (gene flow) | May introduce new alleles or allele combinations |
| | Increased population size | Occurs when selective pressures are relaxed; hence, more genetic variants survive in the breeding population |
| | Geographic variation | Adaptation to several different habitats increases variation |
| Decreasing variation | Stabilizing and directional natural selection | Limits number of alleles passed on to the gene pool of the next generation |
| | Inbreeding | Reduces number of heterozygotes (promotes gene fixation) |
| | Emigration | May remove genotypes from gene pool |
| | Decreased population size | Usually due to increased selection, leaving less variation in breeding population (also, loss of variation by genetic drift is more likely in smaller populations) |

consequence of reproductive isolation is genetic isolation: populations that do not interbreed cannot exchange genes. This means that the gene pools of different species are isolated from each other and evolve separately.

A more modern definition states that a **species** consists of one or more populations that share a common gene pool. This definition emphasizes the genetic continuity of related individuals and populations rather than the discontinuity of unrelated ones. The members of a species that reproduces sexually usually share a common gene pool because there is gene flow among them as a result of interbreeding. Members of a species that reproduces only asexually share a common gene pool because they are descended from a common ancestor.

As long as gene flow can be shown between two populations, they belong to the same species even if their members cannot breed together. For instance, stretching across Europe is a string of populations of the European cherry fruit fly, with interbreeding and gene flow between adjacent populations. However, when flies from eastern and western Europe are brought together to breed, the resulting eggs fail to develop. Nevertheless, the two populations

are members of one species because there is, at least in theory, some exchange of genes between them, and so they share a common gene pool.

In practice, it is seldom possible to test directly whether two populations share the same gene pool. However, the range of phenotypes a gene pool can produce is limited, and therefore members of the same species usually look alike. The eighteenth-century naturalist Carl Linnaeus developed the first working definition of a species, based on **morphological** (structural, anatomical) differences. If two organisms were sufficiently different, they were considered to belong to different species. In practice, this is how most organisms are classified today.

Morphological characters are convenient to work with, and they allow for clear communication. People who discover a new species can write a precise definition of it. They can also select **type specimens** of the species, individuals that are preserved in a museum for future reference. The morphological characters that distinguish species are also the natural criteria to use in constructing a dichotomous key, one of the most important tools of a field biologist (Figure 19–10). However, organisms can be very dif-

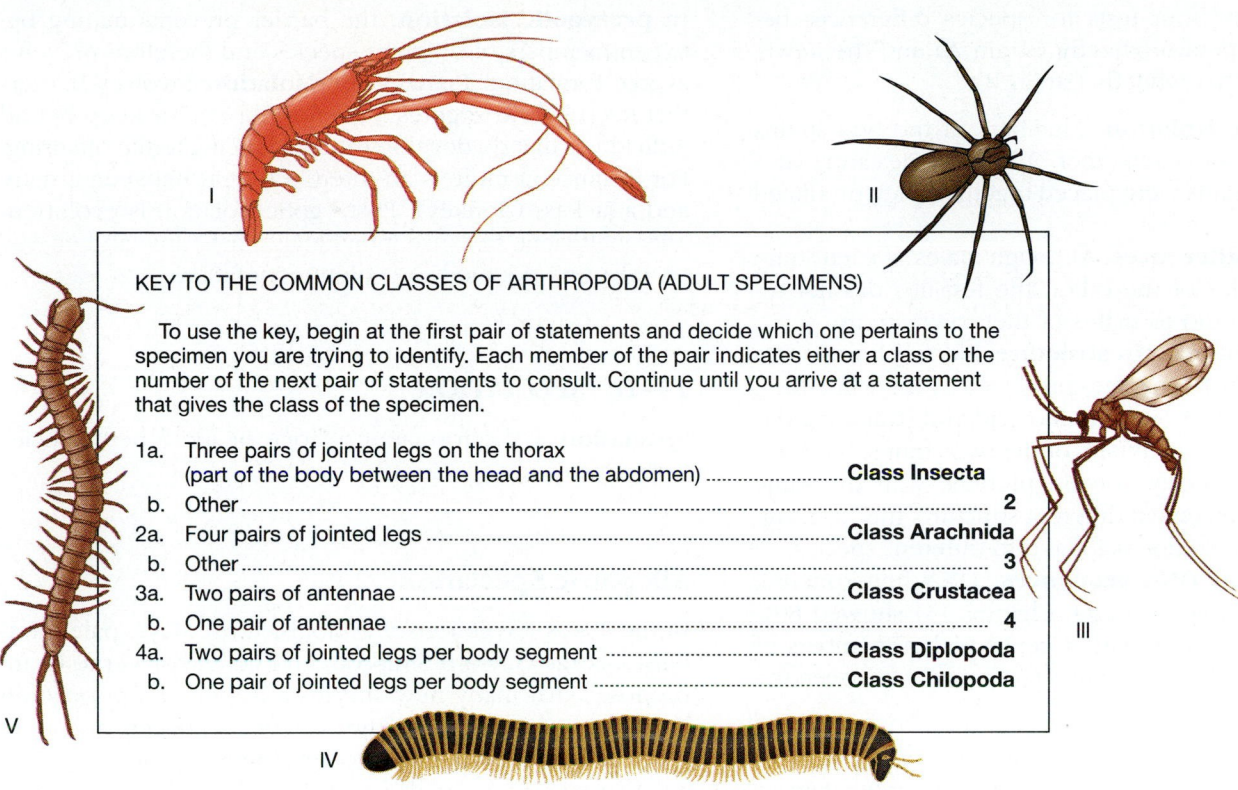

KEY TO THE COMMON CLASSES OF ARTHROPODA (ADULT SPECIMENS)

To use the key, begin at the first pair of statements and decide which one pertains to the specimen you are trying to identify. Each member of the pair indicates either a class or the number of the next pair of statements to consult. Continue until you arrive at a statement that gives the class of the specimen.

1a. Three pairs of jointed legs on the thorax
(part of the body between the head and the abdomen) ......................... **Class Insecta**
 b. Other ............................................................................................................ 2
2a. Four pairs of jointed legs ............................................................ **Class Arachnida**
 b. Other ............................................................................................................ 3
3a. Two pairs of antennae .................................................................. **Class Crustacea**
 b. One pair of antennae ...................................................................................... 4
4a. Two pairs of jointed legs per body segment ------------------------ **Class Diplopoda**
 b. One pair of jointed legs per body segment -------------------------- **Class Chilopoda**

**FIGURE 19–10** A dichotomous key, so called because each step gives two alternative choices. To see how it works, follow the instructions given at the top. The pictures are all of animals from the large group known as arthropods, which have chitinous external skeletons and jointed appendages. Arthropods can be subdivided into a number of classes. Use this key to identify the classes of the specimens drawn around the edges of the key. For the answers, see the Chapter 19 Self-Quiz answers at the back of the book.

ferent and yet belong to the same species, as in the case of eclectus parrots, where the male is green and the female red: the two were originally described as members of different species. And organisms can be outwardly identical and yet belong to different species, as the next example shows. Such apparently identical species are called **sibling species.** They are usually closely related and have a recent common ancestor.

## Tests for Species

In 1991, whiteflies that had escaped from poinsettia greenhouses in Florida were destroying crops in Arizona and California (Figure 19–11). They looked just like the sweet potato whitefly, *Bemisia tabaci,* but they devastated crops that *B. tabaci* does not attack. Entomologists (scientists who study insects) began to suspect that they might belong to a new

(a)

(b)

**FIGURE 19–11** A new species of whitefly (a). The tiny pest causes many conditions that disfigure plants, including irregular ripening of tomatoes (b). (T. M. Perring)

species. They used four tests for species differences between the sweet potato whitefly (strain A) and the newly discovered poinsettia whitefly (strain B):

1. **Reproductive isolation.** Members of the two strains do not breed with each other. When females and males of different strains were placed together, they produced no offspring.
2. **Behavioral differences.** Although males of each strain courted females of the other, the females did not respond as they did to males of their own strain.
3. **Differences in protein structure.** Electrophoresis was used to identify 18 enzymes produced by the white-flies. The enzymes all showed amino acid and struc-tural differences between members of the two strains. Since related species usually contain proteins with the same function but somewhat different structure, this also suggested that the strains belonged to different species.
4. **Differences in DNA sequences.** DNA fingerprinting (*Essay: DNA Fingerprinting*, Chapter 13) showed 80% similarity of selected marker genes within members of the same strain but less than 10% similarity between the two strains.

The entomologists concluded that the two strains of whiteflies belonged to different species. Research suggests that the poinsettia whitefly was accidentally introduced into the United States with imported poinsettias in the 1980s. The practical importance of this conclusion lies in the search for parasites or predators that might control these white-flies. Among the most useful biological controls for insect pests are tiny parasitic wasps that lay their eggs in insect larvae. Such a wasp was used successfully to control an outbreak of the ash whitefly that invaded California in 1991. Many of these wasps are species-specific, so when searching for a wasp it is important to know which species of whitefly you are dealing with.

## 19–G  REPRODUCTIVE ISOLATION

The whiteflies just discussed have already formed separate species. However, some populations are only partially isolated from their neighbors, genetically or sexually. These borderline cases illustrate that speciation is a process: new species may arise by the evolution of barriers to gene exchange between populations that eventually become reproductively isolated from each other.

When two populations live in the same place, they are said to be **sympatric.** If they live in different places, they are **allopatric** (allos = other; patria = homeland). In nature, sympatric populations of related species do not interbreed because their reproductive mechanisms are not compatible.

Table 19–2 lists the types of reproductive barriers that usually prevent members of different species from breeding with one another. These barriers fall into two categories.

In **prezygotic isolation,** the barrier prevents mating between members of different species and therefore prevents zygote formation. **Postzygotic isolation** involves barriers that keep hybrid zygotes from developing, or keep hybrid individuals that do develop from producing fertile offspring. For instance, a mule is an infertile hybrid between a mare and a jackass (donkey). Postzygotic isolation is evolutionarily inefficient, because individuals waste time and energy on offspring doomed to evolutionary failure.

## 19–H  SPECIATION

**Speciation** is the formation of one or more new species from an existing species.

### Allopatric Speciation

In the 1940s, evolutionary biologist Ernst Mayr, paleontologist George Gaylord Simpson, and geneticist Verne Grant proposed that many new species arise when a population becomes separated from the rest of its species and then changes so much that it becomes a new species.

Suppose a few seeds or a few birds are blown onto an island, where they grow or breed. The most likely fate of this new population is extinction because its members are not well adapted to their new island home. But sometimes a few individuals survive and found a new, isolated population. Hawaii's endangered state bird, the nene, is believed to have evolved in this way from a few Canada geese, native to North America. Although the nene looks like a goose, it is adapted to life in rugged uplands, far from water (Figure 19–12).

An isolated population will tend to become genetically distinct from the rest of the species for two main reasons. First, because of the founder effect the new population will have a unique gene pool from the start (Section 19–B). Second, because the new population is subject to a new set of selection pressures, it will evolve adaptations to its new home. Its reproduction will be one of the many features that may change in response to this selection.

Eventually, the differences between the original and new populations may become so great that the two can no longer interbreed. The test of speciation comes if the two populations ever again become sympatric. If they do not then interbreed and produce fertile offspring, the two populations are considered separate species. This is probably how Darwin's finches, and indeed most species on small islands, originated.

The 13 species of Darwin's finches in the Galápagos and a fourteenth on Cocos Island, 500 km away, differ mainly in the kinds of foods they eat and the sizes and shapes of their beaks (see Figure 17–14). The kinds of selective pressures probably involved in their evolution

**TABLE 19-2**

**Barriers That Prevent Interbreeding Between Members of Different Species**

| Barrier | Effect | Example |
|---|---|---|
| **Prezygotic isolation** | **Prevents mating (and therefore zygote formation) between species** | |
| Habitat differences | Individuals of two species never meet | Darwin's finches living on different islands |
| Different breeding times | Members of different species not in breeding condition at the same time | Species of frogs that mate and lay eggs in the same pond at different times |
| Mechanical barriers | Shape of genitalia (copulatory structures) prevents fertilization by members of other species | Many insect and butterfly species |
| Behavioral specificity | Mating cannot occur without species-specific behavior | Whitefly species |
| **Postzygotic isolation** | **Prevents successful reproduction after zygote formation** | |
| Hybrid inviability | Hybrid offspring dies before reproductive age | Many species of frogs |
| Hybrid sterility | Hybrid offspring survives but is sterile | Mule (mare × jackass), lion × tiger |
| Hybrid breakdown | Hybrid offspring fertile, but many of its offspring are not | Species of sulfur butterfly |

were shown by studies started in the 1970s. Peter Grant and Peter Boag had counted the population of medium ground finches on one island and measured their beaks. Only the largest individuals survived a 1977 drought that killed 85% of the population. Apparently only the birds with the largest beaks could crack open the large, hard seeds that were the only available food during the drought.

(a)

(b)

**FIGURE 19-12** Parent and derived species. (a) A Canada goose. (b) A nene. (a, Florida Audubon Society; b, M. J. Rauzon/VIREO)

## ESSAY: *Continental Drift*

Geological changes produce new barriers between populations and change climates, both of which promote speciation. The gradual drifting apart of the continents was the most dramatic geological change of them all, affecting the evolution of thousands of species.

While many geologists still scoffed at the theory of continental drift, some biologists believed in it—because of what they knew about the distribution of species. For instance, lungfishes are restricted to freshwater habitats and do not move far, yet three living species, while clearly related, occur in South America, South Africa, and Australia, respectively. Biologists were never convinced that ancestral lungfishes crossed oceans. They thought it was more likely that South America, Africa, and Australia were once joined. Similarly, flightless birds—the ostrich in Africa, the rhea in South America, and the emu in Australia—share closely related species of flightless insect parasites in their feathers. How did this strange situation arise?

The distribution of some more ancient groups of organisms is even more puzzling. Fossils of *Glossopteris*, a genus of plants, occur in South America, Africa, India, and Australia (Figure 19–B). To account for these and many other distributions, theories involving land bridges between the continents developed. These were replaced by the belief that the continents themselves had moved.

During the 1960s the Earth sciences underwent a revolution with the general acceptance of the theory of **plate tectonics,** which includes the idea that Earth's outermost layer is made up of rigid plates, which are pushed past, over, and under one another. The plates move because they are floating on the mantle of molten rock beneath them. Because the continents are parts of these plates, they too move around Earth's surface, a process called **continental drift.**

Geological evidence, from patterns of rock formations and patterns of magnetism in rocks, supports the fossil evidence suggesting that in Permian and Triassic times, the southern continents (South America, Africa, Antarctica, and Australia), along with India, were united as one large mass, Gondwana (Figure 19–C). North America, Europe, and Asia were meanwhile united as a northern land mass, Laurasia. Even earlier, these two supercontinents were themselves united for a time as a single world land mass, named Pangaea by Alfred Wegener in the 1920s. The rest of Earth was covered by ocean—the ancestor of today's Pacific. The Atlantic, Indian, and Antarctic oceans began as rifts in Pangaea when the plates beneath it separated about 200 million years ago.

Gondwana and Laurasia separated during the Jurassic period. Then Gondwana began to split up into the

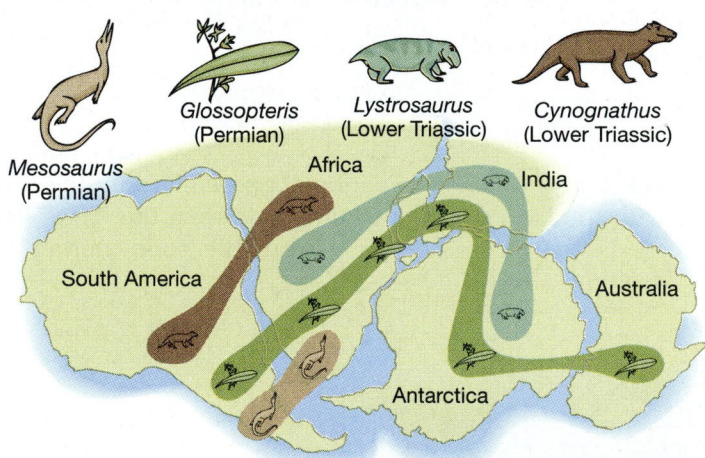

*Glossopteris* (Permian)

*Lystrosaurus* (Lower Triassic)

*Cynognathus* (Lower Triassic)

*Mesosaurus* (Permian)

Africa

India

South America

Australia

Antarctica

**FIGURE 19–B** The distributions of four fossil species found on more than one of the southern continents. This distribution suggested that the southern continents were once joined.   (Redrawn from Colbert, E. H. *Wandering Lands and Animals.* New York: E. P. Dutton, 1973. Copyright 1973 by Edwin H. Colbert. Reprinted by permission of E. P. Dutton)

Each species of Darwin's finch probably evolved in isolation on an island where locally abundant foods exerted similar, long-term selective pressure that determined how the beak evolved. Later, members of most of these populations emigrated to other islands, and so today many of the islands house several species of finches.

It may occur to you that new species might also form when one large, widespread population becomes split into two, for instance by an earthquake or when continents drift apart (*Essay: Continental Drift*). Apparently, however, this does not usually occur. For instance, North American and European populations of sycamore trees have been sepa-

present-day southern continents and India. By this time the dinosaurs and conifers were widespread, early mammals and birds were well established, and flowering plants had evolved. After the continents separated, populations of these groups evolved in different ways on the separate, new continents. Marsupial (pouched) mammals underwent wide adaptive radiation in Australia and South America, while placental mammals came to dominate the other land masses. Different species of conifers, birds, and flowering plants evolved on different continents.

India drifted north and collided with Asia during the Oligocene, the collision pushing up the Himalaya Mountains. At about the same time, Laurasia was splitting apart. By this time bats, carnivores, primates, rodents, and many other groups of mammals had evolved in Laurasia. Thus it is not surprising that North American, Asian, and European mammals are related to each other more closely than any of them are related to the mammals of South America or Australia.

Central America was pushed up from the ocean 3 to 5 million years ago, linking North and South America. Only a few South American animals, such as the opossum (a marsupial) and the armadillo, crossed Central America and succeeded in colonizing North America. By contrast, many placental mammals invaded South America, causing the extinction of several South American marsupial families.

These ancient wanderings of continents and their living passengers explain the biogeographical distribution of many organisms today.

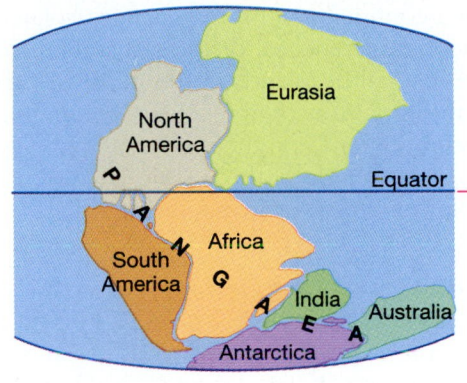

(a) Pangaea:
    200 million years ago

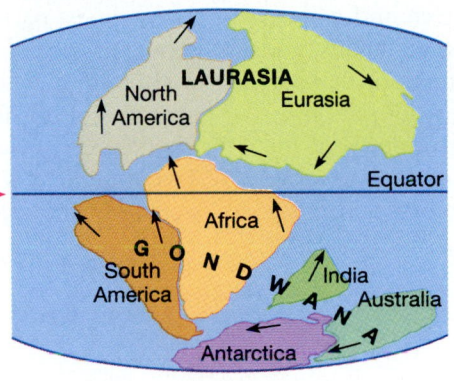

(b) Laurasia and Gondwana:
    180 million years ago

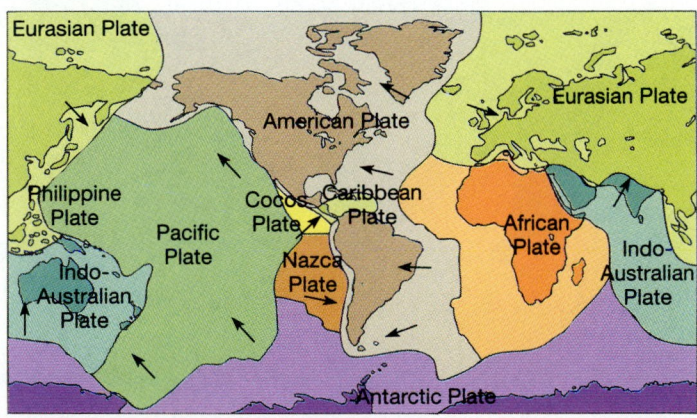

(c) Earth's tectonic plates and the direction in which they are moving

**FIGURE 19–C** Continental drift. (a) The continents during the late Paleozoic. (b) The continents during the Mesozoic, when Pangaea had broken up. (c) Tectonic plates and continents today, showing the directions in which the plates are moving. Where plates push into each other, force and friction create geological upheavals such as volcanic eruptions and earthquakes. (The San Andreas fault is part of the line where the Pacific and American plates meet.)

rated for at least 30 million years, but they have not formed separate species. The two are very similar morphologically, and they can interbreed and produce normal, fertile offspring. It seems that species containing large numbers of individuals and covering wide areas of a continent may remain essentially unaltered for millions of years. This is because they are subject to gene flow and stabilizing selection. The genetic changes that lead to the formation of new species are much more likely to occur under the influence of directional selection as a small population adapts to a new habitat.

***Pleistocene Glaciations***   A dramatic example of speciation in isolated populations comes from the effects of the great glaciations of the Pleistocene epoch in North America. Over the past million years, and ending only a few thousand years ago, four major glaciations covered much of the Northern Hemisphere with ice, often thousands of feet thick.

Each advance of the glaciers southward pinched off many isolated populations. Each retreat of the glaciers northward permitted some of these isolated, and now different, populations to come back into contact again. In some cases reproductive isolation had proceeded far enough to produce two or more species from one species present before the glaciation. In many other cases, reproductive isolation was not complete, and indeed we can still see various levels of hybridization between adjacent western and eastern populations of many North American birds, mammals, and insects, each presumably representing populations that survived in different areas during the last major glaciation.

The wood warblers of North America underwent impressive allopatric speciation. Two dozen or more species can be found hunting insects in the foliage of a single forest in parts of North America. The ancestral black-throated green warbler was probably distributed across the continent before the Pleistocene. At each glacial advance, its range was pushed down to the southeast forest refuge, pinching off a population in the west that subsequently formed a new species, reproductively isolated by the time the glaciers retreated (Figure 19–13). By the time of the next glacial advance, the black-throated green warbler had again extended across most of the continent and was again pinched into two populations by the next glacial advance, and so on. Similar speciation from other ancestral warblers gave rise to a total of 46 species of forest-adapted wood warblers in North America today and probably to many species that have since become extinct.

This example shows that allopatric "islands" may be of many types: part of a continent cut off by glaciers, isolated mountain peaks, deep sea trenches, or the last undrained pothole in a prairie may all be islands in the midst of vast stretches of hostile territory, from their residents' viewpoint. The degree of isolation depends on the residents' mobility. Twenty miles to the next prairie pothole is no barrier to a duck, but impossible for a snail or frog to cover.

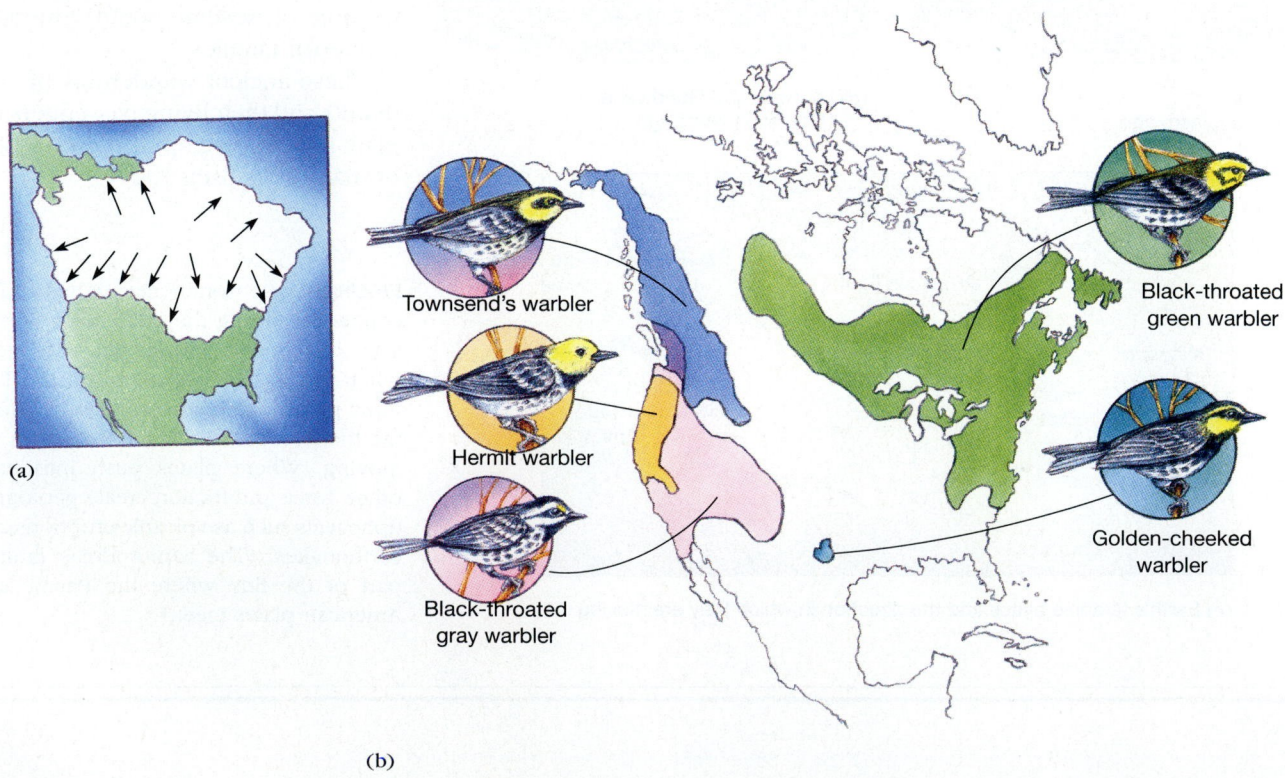

**FIGURE 19–13**   Speciation during the four Pleistocene glaciations. (a) A map showing the extent of the Wisconsin ice sheet, last of the ice sheets that pushed south into the United States. (b) Breeding areas of members of the black-throated green warbler group. Note that there is very little overlap between breeding ranges. These five species probably all evolved from a single ancestral species.   (Redrawn from Lancaster, D. *The Living Bird* vol. 3. The Cornell Laboratory of Ornithology, 1964.)

## Sympatric Speciation

**Sympatric speciation** is the production of new species within a single population. It is not common in species that reproduce sexually because it requires that barriers to reproduction evolve despite the gene flow normally found between members of a population. Consequently, sympatric speciation usually involves a change in the number of chromosomes such that an individual is prevented from breeding with other members of the population.

A new species can form instantaneously by **polyploidy,** multiplication of the normal chromosome number. This is common in plants. If the plant can reproduce asexually, it can form a new population despite its inability to breed with individuals of the parent species.

An **autopolyploid** arises within a species when the chromosome number doubles. This can happen when chromosomes fail to segregate at meiosis, producing diploid (2N) instead of haploid (N) gametes. Self-fertilization produces a zygote, which grows into a tetraploid (4N) plant that cannot breed with its parent.

Also common is **allopolyploidy,** which starts with two related species interbreeding and producing hybrid offspring. Hybrids are usually sterile because the incompatible chromosomes of the two species cannot undergo meiosis. However, the sterile hybrid may reproduce asexually. Then the chromosome number doubles in one of the resulting plants, producing an individual having pairs of homologous chromosomes, which can undergo meiosis. If the plant can fertilize itself, it can reproduce sexually, giving rise to a new species that cannot breed with either parent species.

Polyploid plants are often larger than their parents and better able to tolerate adverse environmental conditions, such as cold or drought. Indeed, the number of polyploid species increases with increasing latitude (that is, with colder climate) (Figure 19–14). Probably more than a third of all plant species have arisen by polyploidy, although the phenomenon is rare in conifers (cone-bearing plants such as pines, firs, and spruces). This may be one reason there are so few species of conifers.

Farmers and plant breeders often select polyploids as crop plants. Modern varieties of cotton, sugar cane, wheat, potatoes, coffee, barley, and most other major crops are all natural polyploids (Figure 19–15).

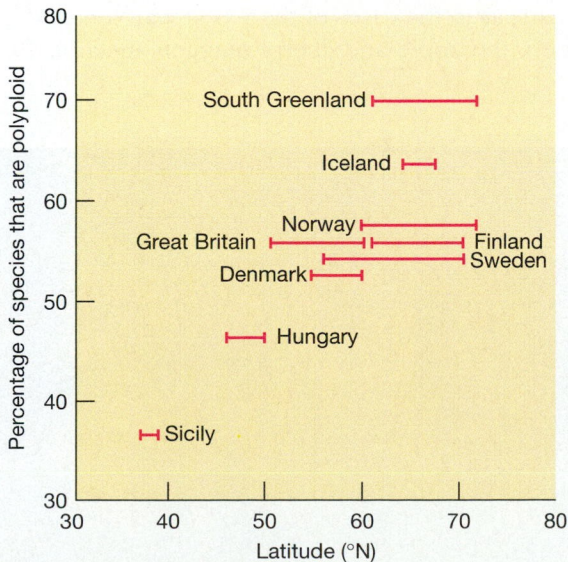

**FIGURE 19–14**  A survey of the plants in different areas of Europe shows a general increase in the percentage of plant species that are polyploid from Sicily in the south to southern Greenland in the north. (There are hardly any plants in northern Greenland, which is covered with ice.)

(a)

(b)

**FIGURE 19–15**  Polyploid plant species. (a) Sugar cane. (b) Coffee.

***New Fruits***   Researchers have discovered how to produce polyploids artificially, by treating plants with chemicals such as colchicine, extracted from the autumn crocus, and vincristine, extracted from the Madagascar periwinkle. These produce nondisjunction of the chromosomes at meiosis, leading to polyploidy.

Geneticists Robert Knight and Ann Amis have used this method to create a new fruit whose juice may well be on the market by the time you read this book. They started in the 1970s to breed a cold-hardy passion fruit by crossing the tropical purple-fruited passion vine (*Passiflora edulis*) with the orange-fruited maypop (*P. incarnata*), which is native to most of the eastern United States. Early hybrids reproduced poorly and had small fruit. By treating the hybrids with colchicine, the workers doubled the chromosome number and produced vines that are not only hardy as far north as North Carolina but also produce fruit the size of oranges. The new plant will appeal to juice growers in Florida and California who now lose millions of dollars' worth of tender citrus fruit to freezes every winter.

***Sympatric Speciation in Animals***   Speciation by polyploidy may occur occasionally in sexually reproducing animals. Some species of earthworms, fish, and frogs differ from one another in ploidy and may have arisen in this way. Many animal species that are permanently parthenogenetic (reproducing by means of unfertilized eggs) are certainly polyploid: various beetles, moths, shrimp, goldfish, and salamanders fall into this category.

Experiments with *Drosophila* have shown that disruptive selection can produce sympatric speciation without polyploidy in laboratory animals. In each generation, Aubrey Manning removed individuals with intermediate numbers of bristles (a polygenic character), and only those with high and low bristle numbers were allowed to breed. In just 13 generations, there was reproductive isolation between the two groups. Females would mate only with males who had a similar bristle number. All of the offspring resembled their parents, and no more hybrids between the two groups were formed. (Bristle number is not a character normally involved in mate choice.)

Can sympatric speciation occur within natural populations of animals? In many groups of animals it undoubtedly cannot; in others it probably can. For instance, adults of the apple maggot, *Rhagoletis pomonella,* are attracted to apple trees, which they find by visual and chemical cues. Around 1960, apple maggots colonized cherry trees in Wisconsin, where apple and cherry orchards grow side by side. The cherry race now emerges in early summer, at the peak ripening time for cherries but before apples ripen. This produces some reproductive isolation from the parent apple race, which continues to emerge in late summer.

These two races—populations with distinct, genetically based host preferences—are not yet "good" species, since they continue to hybridize somewhat. Gene flow between them has been reduced so greatly, however, that it is probably only a matter of time before complete reproductive isolation evolves. The apple race itself first appeared on apples in the Hudson Valley in 1864, presumably originating from another race that attacks the fruits of wild hawthorn trees.

## 19–I   HOW QUICKLY DO NEW SPECIES FORM?

Biologists sometimes divide evolution into microevolution and macroevolution. **Microevolution** is changes in the frequencies of alleles in gene pools, the kind of evolution discussed earlier in this chapter. **Macroevolution** is the evolution of new groups of species, such as flowering plants or dinosaurs. The main way of investigating macroevolution is to study the fossil record of life on Earth. This reveals patterns that we might not predict from what we know of microevolution. For example, traditional evolutionary theory holds that new species evolve by gradual changes resulting from different selective pressures on populations in different environments. However, when we measure the rate at which new species appear in the fossil record, we find that evolutionary changes may be fast or slow or anywhere in between.

Many species have undergone little evolution in millions of years (Figure 19–16). For example, half of the fossil seashells from 7 million years ago apparently belonged to species still living today, and nearly all species of mosses fossilized 4 to 5 million years ago are still with us. The sycamores discussed in Section 19–H are another example. In other cases, many new species have evolved in less than 5000 years, as in the cases of the wood warblers and, probably soon, the apple and cherry maggots (Section 19–H).

**FIGURE 19–16**   The fossil of a fern that died about 300 million years ago shown beside one frond of a living fern. The two are strikingly similar.   (Robert J. Lynch)

Evolutionary biologists Stephen Jay Gould and Niles Eldredge named this variation in rates of evolution **punctuated equilibrium.** According to their view, a species may exist for a long time more or less unchanged under the influence of stabilizing selection. Then one of its small isolated populations evolves into a new species with a particularly successful coadapted gene complex. This new species spreads rapidly, giving rise to many new populations and either driving the parent species to extinction or existing with it.

On the macroevolutionary level, the appearance of new major groups of organisms usually follows the evolution of novel adaptations. These often include reproductive adaptations, such as flowers in flowering plants and waterproof eggs in reptiles. These evolutionary breakthroughs proved so successful that their owners underwent wide adaptive radiation, spreading across the globe and spawning hundreds of populations, many of which evolved into new species. This swift burst of speciation appears as rapid change in the fossil record. When organisms with the new adaptations had spread to most of the habitats they were equipped to exploit, speciation became less frequent, reflected by slower change in the group's fossil record.

Since evolutionary novelties usually arise in small populations, the chances of the intermediate forms being preserved as fossils, and of our finding them if they were preserved, are small. This often gives the appearance of sudden change from one species to another in the group's fossil record.

## SUMMARY

Natural selection acts on individual phenotypes, but only populations can evolve. This is because the population is the smallest unit with a gene pool in which the frequencies of genes can change.

1. The Hardy-Weinberg law shows that the proportions of different alleles and genotypes in a large population will remain the same as long as there is no net mutation, mating is random with respect to genotype, and there is no gene flow nor selection.

2. Evolution can be caused, in theory, by a number of events:
   a. Net mutation, which replaces one allele by another
   b. Mating preference with respect to genotypes
   c. Genetic drift, the random occurrence of genetic changes due to sampling errors in small populations. Examples are the founder effect and population bottlenecks, both of which tend to reduce genetic variation.
   d. Gene flow, which moves genes from one population to another
   e. Natural (or artificial) selection

3. Natural selection is the main cause of evolution. It takes two common forms: stabilizing selection, in which deviations from the average phenotype are selected against; and directional selection, in which phenotypes at one extreme of the range are selected for, usually temporarily as a result of environmental change. Less common is disruptive selection, in which extremes of a phenotype are advantageous compared with the intermediate form.

4. A population's gene pool evolves as a coadapted gene complex in which the normal expression of a gene depends on other genes that have evolved with it in the population. Dominance and recessiveness of alleles is one possible result of genes coevolving in a coadapted gene complex.

5. Genetic variation in populations is increased by factors such as mutation, gene flow, heterozygote advantage, and population growth. Some biologists believe that genetic variation within populations is increased by the presence of selectively neutral genes that drift randomly in the gene pool. Variation is decreased by factors including most forms of selection, inbreeding, genetic drift to fixation, and decreasing population size.

6. Heterozygous individuals often have a selective advantage over either homozygote. This heterozygote advantage contributes to genetic variation in a population because it increases the chance that none of the alleles at a locus will become fixed. Inbreeding is often selected against because it tends to reduce fitness by decreasing heterozygosity.

7. Balanced polymorphism is a common form of genetic variation. It may result from heterozygote advantage, disruptive selection, frequently changing selective pressures, frequency-dependent selection, or the opposing effects of selection and gene flow.

8. Members of a species share a common gene pool and can interbreed. Their common gene pool gives members of a species similar morphology, and this is the usual basis for classification and identification of species. Species may also be distinguished by their different behavior, proteins, and DNA, all of which reflect genetic differences.

9. Reproductive isolation between species may be due to prezygotic mechanisms, which prevent breeding between members of different species, or postzygotic mechanisms, which prevent successful reproduction of hybrid offspring.

10. Allopatric speciation occurs when two populations of a species become separated and gene flow no longer occurs between them. This is especially common when a few individuals colonize an island. The isolated population evolves under the influence of local selective pressures and may become so different that it is considered a new species. Among the events known to have isolated species and resulted in speciation are the Pleistocene

glaciations, which were responsible for widespread speciation in North America.

11. Sympatric speciation usually occurs as a result of polyploidy, which is particularly common in plants. Polyploid individuals often have selective advantages over their diploid ancestors in harsh conditions such as cold weather. Most cultivated plants are natural polyploids. Speciation by polyploidy may also occur in animals. Sympatric speciation in animals also appears likely in some cases where individuals are very specialized, for instance in their food preferences.

12. The traditional concept of speciation as a result of gradual and continuous change in the gene pool is questioned by those who believe in punctuated equilibrium. It now seems likely that species often form in small isolated populations. Such a population may diverge rapidly from the parent population and, in the unlikely case that it does not become extinct, form a new species that undergoes rapid adaptive radiation.

## Self-Quiz

1. The Hardy-Weinberg law states that when two alleles are considered:
   a. dominant alleles will always be more frequent in the population than recessive alleles
   b. heterozygotes will be twice as common as homozygotes in a population
   c. members of adjacent populations are not able to interbreed with one another
   d. large populations are well adapted to their environment
   e. allelic frequencies cannot change in a large population unless one is at a selective advantage over the other or mating is nonrandom

2. The Hardy-Weinberg law allows us to predict that:
   a. sexual reproduction is necessary for evolution
   b. sexual reproduction may be a cause of evolution
   c. sexual reproduction plays no role in evolution
   d. sexual reproduction will cause evolution if individuals prefer mates with one genotype over those with other genotypes

3. a. In certain parts of Africa, people with one normal and one sickle hemoglobin allele enjoy heterozygote advantage over either homozygote. Thus the population is experiencing (directional, disruptive, stabilizing) selection.
   b. Suppose that mosquito control measures completely eliminate the threat of malaria from an area where it was once prevalent. If you followed the allele frequencies for the next 30 generations, what changes would you expect to see in the frequencies of the normal and sickle alleles once selection against the homozygous normal individuals is removed?
   c. If, at the time malaria is eliminated, the adult population consists of 10% homozygous normal individuals and 90% heterozygous carriers, and if the people marry without regard to this trait, what will be the percentages of the three genotypes—homozygous normal, heterozygous, and homozygous sickle—in the children born in the next generation?

4. Selection will not eliminate a lethal recessive allele from a large population of diploid organisms because:

a. there will always be some heterozygous carriers for the allele
b. gene fixation will occur in the population
c. heterozygotes are at a selective advantage
d. the allele will have some good effects, and these will become dominant
e. the rate of mutation to the lethal allele is higher in a larger population

5. Genetic drift is more likely to cause evolution in a smaller population because:
   a. mating is nonrandom in small populations
   b. random events are more apt to happen to small populations
   c. there is no natural selection in small populations
   d. deviations from statistical averages are more likely to be seen in small populations than in large ones

6. Which of the following is *not* a partial explanation for heterozygote advantage?
   a. The Hardy-Weinberg equilibrium states that there will always be twice as many heterozygotes as homozygotes
   b. Two alleles may express themselves in slightly different ways that are adapted to different parts of the life cycle
   c. Two different alleles may each operate best under different environmental conditions to which individuals are apt to be exposed
   d. The presence of one allele may alleviate the detrimental consequences caused by the presence of the other

7. In each of the following situations, tell whether genetic variability in a population would increase, decrease, or remain the same.
   a. increased mutation rate
   b. decreased natural selection
   c. increased variability in the environment
   d. sexual reproduction

8. A rapidly growing population experiences increasing genetic variability because:
   a. there is a greater chance for polymorphism in large populations
   b. the mutation rate is higher for large populations

c.  there is less selective pressure

d.  a large population can support more genetic experimentation

e.  heterozygote superiority is selectively favored

9.  Tell whether you would use morphological characters or the ability to interbreed to decide how many species exist in each situation described below.

_____ a.  Fossils that appear identical are found in rocks on the two coasts of North America

_____ b.  Grub-like white larvae are found in a nest of ants

_____ c.  A colony of ants contains individuals of many different body structures

_____ d.  You find two parthenogenetic female aphids (parthenogenetic organisms reproduce via unfertilized eggs)

_____ e.  The Rongovian government will not permit the export of the rare red-faced blooper for breeding experiments with a similar population in the Yukon; however, CIA agents have managed to obtain (from undisclosed sources) photographs of several specimens, which they have turned over to scientists.

10.  New species may originate by:

a.  doubling of the genetic material of a flowering plant

b.  gradual accumulation of changes selected for by local conditions

c.  a mutation that prevents reproduction with most other members of the species

d.  all of the above

e.  a and b only

## Questions for Discussion

1.  Phenylketonuria (PKU), a genetic disorder due to homozygosity for a certain recessive allele, occurs in about 1 person in 15,000. Roughly what proportion of the population is heterozygous for the disease?

2.  Huntington's disease is a genetic condition caused by a dominant allele. The brain deteriorates, and the victim loses control over both mind and muscles. A period of insanity accompanied by jerky movements of the face and limbs is finally followed by death. (Interestingly, at least seven women accused of witchcraft in New England during the 1600s were related to families now known to contain Huntington's disease.) However, the condition usually does not set in until the victim's 30s or 40s, and by then he or she has produced children, half of whom will also receive the allele and are therefore doomed to suffering the disease. Can selection operate against such a late-acting genetic trait?

3.  What, in biological terms, is a "race" of people or any other animal?

4.  Why are the apple and cherry races of the apple maggot becoming separate species within a decade or two, whereas the gray and black forms of the peppered moth have not become separate species in more than a century?

5.  Evolutionist Theodosius Dobzhansky wrote that a totally uniform physical environment could support only one species of organism. Do you agree? Why or why not?

6.  The Galápagos Archipelago consists of about a dozen main islands and many more smaller ones, all within about 64 km of the nearest neighbor. Thirteen species of Darwin's finches occur in the archipelago, with three to eleven of these species found on each of the main islands. Many evolutionists feel that if there had been only one island, there would only be one species of finch. This is supported by the fact that Cocos Island, isolated from both the Galápagos and Central America by several hundred kilometers of open ocean, has only one, unique species of finch. How would the existence of an archipelago, instead of a single island, promote speciation?

7.  What characteristics would you expect to see in the genetic makeup of a species that has the ability to colonize a variety of habitats that may become available?

## Suggested Readings

Colbert, E. H. *Wandering Lands and Animals*. New York: E. P. Dutton, 1973. A very readable introduction to continental drift.

Dawkins, R. *The Selfish Gene*. New York: Oxford University Press, 1976. The racy story of natural selection from the gene's point of view; written as science fiction, with some provocative thoughts on such topics as genesmanship and the battle of the sexes; excellent discussion of the evolution of altruism.

Grant, P. "Natural selection and Darwin's finches." *Scientific American*, October 1991. A study that detected evolution in a population of finches.

Perring, T. M., A. D. Cooper, R. J. Rodriguez, C. A. Farrar, and T. S. Bellows. "Identification of a whitefly species by genomic and behavioral studies." *Science* 259: 74, 1993. The story of the poinsettia whitefly.

Stebbins, G. L., and F. J. Ayala. "The evolution of Darwinism." *Scientific American*, July 1985. Thumbnail sketch of how our view of evolution has changed since Darwin's time and how modern molecular techniques have helped.

Ward, P. D. *On Methuselah's Trail: Living Fossils and the Great Extinctions*. San Francisco: W. H. Freeman, 1992. A young evolutionist's entertaining personal account of his worldwide travels studying living fossils: species that have survived from remote ages apparently unchanged.

C H A P T E R

20

# Evolution and Reproduction

## OBJECTIVES

*When you have studied this chapter, you should be able to:*

1. State the advantages and disadvantages of sexual reproduction contrasted with asexual reproduction.
2. Discuss the advantages enjoyed by organisms that combine sexual and asexual reproduction in the life cycle.
3. Explain why sexual reproduction promotes more rapid evolution than asexual reproduction does.
4. Outline the life cycle of a haploid eukaryote such as *Ulothrix*, mentioning the roles of mitosis, meiosis, and fertilization. Compare and contrast this with the human life cycle.
5. Define isogamy, oogamy, oviparity, ovoviviparity, viviparity, hermaphrodism; give examples of organisms showing each.
6. Give at least one advantage each of oogamy, diploidy, and hermaphrodism.

7. Give some reasons why selective pressures acting on a female may be different from those acting on a male, and describe some resulting differences between the two sexes in a species.
8. Give evidence for the theory that female choice almost invariably exists in mating systems.
9. Distinguish among polygamy, polyandry, polygyny, and monogamy, and discuss conditions under which polygyny and monogamy evolve.
10. Describe what is meant by altruistic behavior and kin selection, and explain how alleles for altruistic behavior may spread through a population.

An organism's genes may enable it to obtain food, avoid being eaten, and cope with the climate, but ultimately there is only one measure of its evolutionary success: the proportion of its genes present in future generations. Without genes that make it reproduce, all the adaptations resulting from its other genes are useless, in evolutionary terms. In fact, it is sometimes useful, though inaccurate, to view the organism itself as a machine built by its genes as a means of multiplying and perpetuating themselves.

At its core, the reproduction of organisms is the reproduction of cells, which is very similar in all organisms (Chapter 14). The genes responsible are highly conserved during evolution because they form the very basis of evolutionary success: any deviation may spell doom. Many organisms have additional structures or behaviors that may be considered reproductive adaptations because their main role is improving the chances of completing reproduction. These vary more because they are shaped by each species' ecology.

This chapter looks at some patterns of reproduction found in the living world and the kinds of selective pressures that permit the genes responsible for them to increase. Most of the chapter is devoted to sexual reproduction, a complex, energetically wasteful process. What advantages of sexual reproduction more than compensate for these disadvantages and make it so widespread among organisms? How did sexual reproduction evolve in the first place? Beyond these problems lies the question of why a species has one type of sexual system rather than another.

Biologists have tried to draw generalizations relating reproductive patterns to the type of organism and its ecology. But be warned at the outset that many exceptions exist, and the watchword in this field is controversy. In fact, this chapter contains the basis for some lively arguments—indulge!

In this discussion, keep in mind the selective pressures acting on genes that govern reproduction. Successful reproduction automatically selects for perpetuation of the genes that brought it about—and the various modes of reproduction in the living world are indeed wonderful.

♦ The bottom line of evolution is the proportion of an organism's genes present in future generations.

♦ The individual organism can be viewed as a vehicle by which genes propagate themselves from one generation to the next.

♦ Sexual reproduction produces more genetic variation than asexual reproduction does. This is probably what permitted the adaptive radiation of sexual organisms into

such a great diversity of forms, and therefore why sex is so widespread.

♦ The ecology of a species plays a major role in determining the type of sexual system it develops.

♦ A species' sexual system determines how members of the species are related to each other and thus the types of behaviors between individuals that are likely to evolve.

## 20–A IS SEX NECESSARY?

At first glance it seems that sex *is* necessary. Most animals, and many plants, rely exclusively on sexual reproduction to perpetuate their genes. On the other hand, asexual reproduction is common among eukaryotes such as protozoa, algae, and fungi, although most of them also engage in sex at times. In fact, some of these organisms, such as *Paramecium,* require an occasional bout of sex or the lineage will become extinct. However, many organisms apparently reproduce only by asexual means.

**Asexual reproduction** encompasses any means of multiplying that does not involve both meiosis and fertilization. Many unicellular organisms reproduce asexually simply by dividing into two identical, smaller cells. Many plants can reproduce vegetatively, forming new individuals from roots, stems, or leaves (Section 46–J). Asexual reproduction in some plants and simple invertebrate animals occurs by forming new individuals attached to the parent at first and later becoming independent (Figure 20–1).

In some organisms, new individuals are produced asexually from unfertilized eggs. In animals, this is called **parthenogenesis** (parthenos = virgin), and in plants **agamospermy** (a = without; gamos = marriage; sperma = seed). The most familiar example is the common dandelion, *Taraxacum officinale,* which sets seeds by agamospermy; its pollen is sterile. Other agamospermous plants include hawkweeds and bluegrasses. Parthenogenesis is found among aphids (see Figure 20–3), rotifers (small aquatic animals, Section 28–H), and even a dozen species of lizards.

What are the selective advantages and disadvantages of sexual and asexual reproduction? And why hasn't one or the other proven superior and become universal during the course of evolution? We can examine these questions by considering two main aspects of each type of reproduction: its energy cost and its genetic consequences for evolutionary success.

Asexual reproduction often converts energy to offspring more efficiently. It usually produces only one or a few offspring at a time, and they are often relatively large. For example, a unicellular organism divides into two, each half the parent's size, and a multicellular organism's asexual offspring often remain attached to the parent until they reach

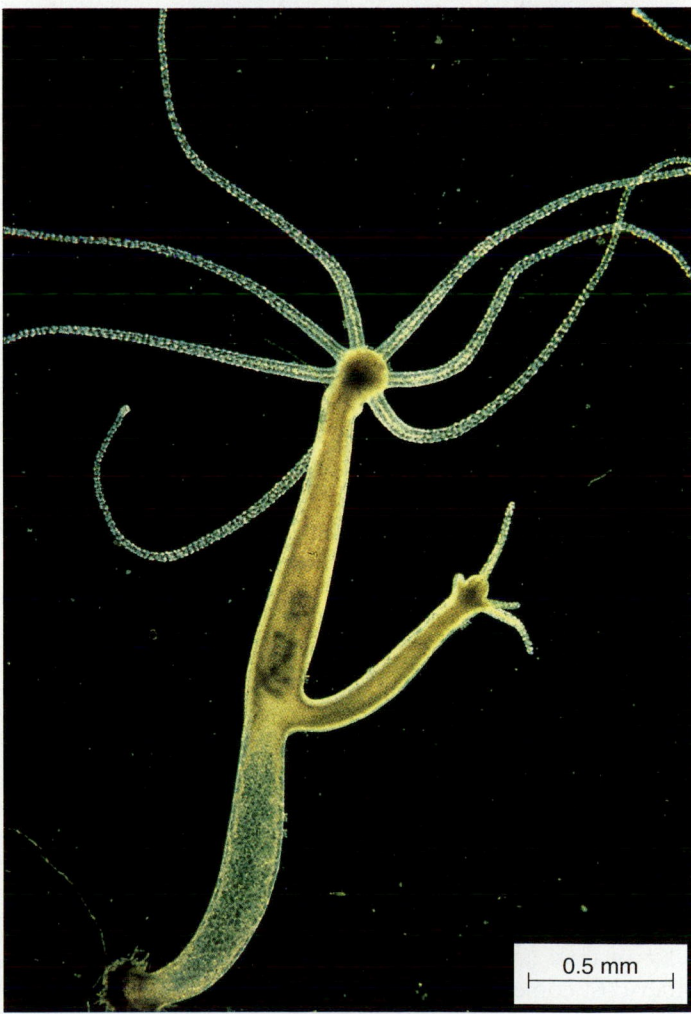

**FIGURE 20–1** Chip off the old block. *Hydra,* a commonly studied invertebrate animal, produces offspring asexually by budding. When the bud is large enough, it breaks off as a new individual. (Biophoto Associates)

Asexual
offspring

Sexual
offspring

**FIGURE 20–2** Sexual versus asexual reproduction. A strawberry plant produces many small offspring sexually. In addition to the energy used to produce the offspring themselves, the plant uses energy making flower petals, nectar, and the flesh of the strawberry, which attract animals that pollinate the flowers and disperse the new offspring. All the energy used in asexual reproduction goes into a few large offspring almost certain to survive. (G. I. Bernard/Earth Scenes)

a fair size (Figures 20–1 and 20–2). Hence, each offspring represents a considerable investment of energy: the organic material of the offspring's body contains stored food energy acquired by the parent, and more energy goes to form this material into a new individual. However, the energy is efficiently used, in that nearly all of it goes directly into the offspring's growth. Furthermore, the relatively large size of these offspring gives them a good chance of surviving to reproductive age.

By contrast, sexual reproduction wastes a lot of energy. If sperm, pollen, or eggs are released into the water or air to find each other by chance, millions of them fail to find mates or fall prey to other organisms, and the energy spent to make them is wasted. With internal fertilization, sperm are introduced directly into the female's body near the eggs. Fewer gametes are lost, but the organism must invest energy in other ways. For example, plants may produce flowers and nectar, thereby attracting animals that carry pollen to the female parts of other flowers, and animals must spend a lot of time and energy finding and courting mates. Another cost comes from the genetics of sexual reproduction: some offspring are genetically unfit, and the energy spent to produce them is also wasted.

Asexual and sexual reproduction also differ in their genetic consequences because of the difference between mitosis and meiosis. In most organisms that reproduce asex-

ually, the offspring result from mitosis. Hence, the parent passes copies of all its genes to each offspring. The very fact that the parent grew large enough to reproduce virtually guarantees that its descendants will also be well adapted to the local environment, assuming that it remains the same. The offspring are therefore likely to be successful too.

By contrast, in sexual reproduction, based on meiosis, each offspring an organism produces carries only half the parent's genes. The other half of the offspring's genes come from another individual at fertilization. Therefore, half a sexual organism's investment in reproduction goes to subsidize the perpetuation of another individual's genes. In addition, as a result of meiosis and fertilization, some offspring end up homozygous for deleterious recessive alleles that doom them to an early death, and others have genomes less suited to the environment than those of either parent. By passing genes among lineages, sex also permits the faster spread of "selfish genes," those that contribute no advantage to the organism but merely require time and materials to be replicated during reproduction. (This may explain why eukaryotes have so much apparently nonfunctional genetic material.)

To summarize, asexual reproduction is apt to waste less energy than sexual reproduction does. Also, offspring for offspring, an asexual parent's evolutionary success is twice as great, because it passes all its genes to each offspring, compared with only half its genes for a sexual parent. So sexual reproduction has two strikes against it.

Why, then, do so many organisms use so much energy and waste so many cells reproducing sexually? Sexual reproduction must have tremendous adaptive value, or it could not have become so widespread.

## 20–B  SELECTION FOR SEXUAL REPRODUCTION

The main advantage of sexual over asexual reproduction is that sexual reproduction produces more genetic variation in a population. An asexual organism and its descendants are all alike genetically and are said to form a **clone** (Figure 20–3). The only genetic variation in such a population comes from mutations.

Sexual organisms are also subject to mutation, but the main sources of their often enormous variability are meiosis and fertilization. Each of us inherited only half our mother's genes and half our father's genes. These genes were reshuffled by genetic recombination of genes on chromosomes and reassortment of chromosome sets during meiosis, and by formation of a new assortment at fertilization. This is why we show many genetic differences from our parents and siblings (brothers and sisters) and why we are even more different from people who are not close relatives.

Some organisms can reproduce both with and without sex. Many protozoa, algae, and small invertebrates such as aphids reproduce asexually during the summer but sexually when the growing season ends. Experiments have shown that many of these organisms switch from asexual to sexual reproduction because environmental conditions become less favorable for growth. Depriving them of nutrients, warmth, light, or oxygen, or allowing the population to become overcrowded, will often switch them from asexual to sexual reproduction.

It makes sense that environmental conditions should cause the switch from no sex to sex in this way. Under this system, an organism that is well adapted to its environment can quickly produce many, equally well-adapted copies of itself while conditions remain constant during the summer. When conditions then change in the fall, the organism reproduces sexually. This creates many genetically different offspring, some of which will probably survive. In these species, sexual reproduction usually gives rise to a dormant egg or cyst in a thick, weatherproof covering, which protects the enclosed individual until good growing conditions return. (Why shouldn't the *asexual* individual just encyst until spring? We will see a possible advantage of sexual reproduction associated with dormancy in the next section.)

Many kinds of bacteria, protozoa, and algae have apparently existed virtually unchanged for more than 500 million years. Such organisms have genes enabling them to tolerate a wide range of conditions, often by extended periods of dormancy. It is hard to imagine an environmental change so catastrophic and widespread as to cause their extinction. On the other hand, most plants and animals are more specialized and can live only in the limited number of places that supply their particular needs. Individual species usually survive for only a few million years (which is not long in evolutionary terms).

A small minority of such plant and animal species have only asexual reproduction. These evolved relatively recently from sexual ancestors, and their isolated gene pools derive from a single (female) ancestor. Their only source of ge-

**FIGURE 20–3**  A clone. The aphids on this branch are sap-sucking insects that produce many generations in a summer by parthenogenesis: unfertilized eggs develop in the mother's body and are born as miniature versions of her. These offspring soon produce their own daughters, genetically identical to their sisters, cousins, and aunts, who are all descended from the same ancestress. Eventually, overcrowding or other cues may cause females to lay eggs, which develop into winged males and females that disperse to new locations.

netic variation is mutation. Unlike their sexual relatives, these species will change little genetically, and isolated populations exposed to new environmental conditions undergo little adaptive radiation. Such an asexual species (the common dandelion is a good example) may do very well for a while but is ultimately doomed to extinction much more surely than is a species that reproduces sexually. A change in the environment that kills dandelions will kill *all* dandelions in the area because they are all very similar genetically (Figure 20–4). Hence, these asexual species are short-lived in evolutionary terms.

By contrast, sexual reproduction breaks up and restructures the genome. Sometimes this produces combinations better adapted to new environmental conditions, which may even become founders of new species. If a group of organisms does not give rise to new species, it may become extinct. We think of the dinosaurs as a large group of animals that became extinct. But today, thanks in part to adaptive radiation made possible by sexual reproduction, some of these reptiles' descendants, the birds, are alive and well and living all over the world.

However, the possible long-term evolutionary benefits of sexual reproduction cannot explain how sex evolved in the first place, nor explain why sex is retained in the short term despite its disadvantages. Selection acts on individuals and their genes here and now, not on the possible long-term survival of the species. There is a lot of controversy among biologists as to what selective advantage favors sexual reproduction in the near term; many ideas have been suggested (*Essay: Sex: What's the Use?*, page 436).

Sexual reproduction, then, may be inefficient for widespread organisms that tolerate a broad range of environmental conditions. The members of more specialized species are in constant danger of dying, leaving no descendants, if they do not have the genetic variability that sexual reproduction provides.

## 20–C  ORIGIN OF SEXUAL REPRODUCTION

Sexual reproduction, involving meiosis and fertilization at some stage in the life cycle, occurs only in eukaryotes (but not in all of them). Genetic recombination does occur in prokaryotes: a bacterium may replace some of its DNA with DNA from another individual. However, the amount of DNA transferred between individuals varies, and we do not know how commonly this happens in nature.

Sexual reproduction is so complex that it cannot have sprung suddenly from nothing. Rather, it must have arisen from previously existing processes, each with its own selective advantages. Sex probably evolved by elaborating these simpler processes and linking them into a new program.

We humans view sex as meiosis of diploid cells (having pairs of homologous chromosomes), forming haploid gametes (having unpaired chromosomes). Two haploid gametes then unite at fertilization, restoring the diploid state. The diploid zygote (fertilized egg) grows and differentiates into a new individual.

However, the first sexual organisms most likely evolved from asexual haploid cells. Hence, the haploid stage was the norm in the earliest sexual life cycles, as in many unicellular eukaryotes today. On rare occasions these haploid cells united. This fertilization produced a transient diploid zygote stage, which soon underwent meiosis, once more forming the haploid cells that carried on the main business of living, growing, and reproducing asexually by mitosis. So in these organisms sexual reproduction began with fertilization and ended with meiosis, not the other way around, as in humans and other animals.

One speculation about the origin of fertilization came from chance observations of unicellular eukaryotic symbionts in the intestine of a dying termite. The host termite had stopped eating, and so its symbionts had no food. In this

**FIGURE 20–4** Dandelions. The pollen of these widespread and successful plants is sterile. The eggs develop without fertilization. Therefore, each seed carries the same genes as the parent plant.

extremity, one of the cells partially engulfed another of the same type but didn't digest it fully. Instead, the two cells fused, as did their nuclei: an event that at first looked like feeding turned out to be fertilization. Lack of food induced these cells to pool their resources.

The hypothesis that fertilization began as the union of two starving cells is supported by parallels in the life cycles of other protozoa and algae. In many of these organisms, sexual reproduction occurs in response to scarcity of an essential nutrient, water, or warmth. So sex may have originated as the merger of two cells into one, with just enough food reserves to form a cyst and outlast hard times in dormancy. When good growing conditions return, the zygote breaks dormancy and undergoes meiosis, producing four cells that resume growth and reproduce asexually, by mitosis.

In its temporary united state, the zygote's assets included two sets of chromosomes. Each set could be used as a source of intact genetic information to repair mutation damage that the other had accumulated during the preceding asexual generations. While this was occurring the chromosomes could not be used to direct RNA and protein synthesis, but this did not matter because the dormant zygote had very low metabolic activity. Hence the zygote used its off-season down time as a chance to make DNA repairs.

The genetic recombination at the beginning of meiosis probably originated in such DNA repair. During meiosis, homologous chromosomes from the two parents line up together and exchange segments of DNA. The machinery that does this probably evolved from the mechanisms of DNA repair and prokaryotic DNA recombination, which use many enzymes that do the same kinds of operations.

The two nuclear divisions of meiosis resemble mitosis —in fact, the second division *is* a mitotic division of the two haploid nuclei formed by the first division. The first division surely evolved as a mitotic division governed by some novel control mechanism.

## Evolution of Gametes

In the first sexual organisms, entire haploid individuals served as gametes. We saw this in the termite-gut symbionts, and it also occurs in many other unicellular organisms.

A major trend in the early evolution of sexual reproduction was the development of more specialized gametes. For example, some of the (haploid) cells in a filament of the alga *Ulothrix* form gametes by dividing into many tiny cells with flagella. These gametes resemble the alga's asexual spores, but they are smaller and have fewer flagella (Figure 20–5). All the *Ulothrix* filaments in an area form and release gametes at the same time, in response to the same environmental cues, and there may be thousands swarming in the water at once. When two gametes collide, they stick together and fuse, forming a zygote.

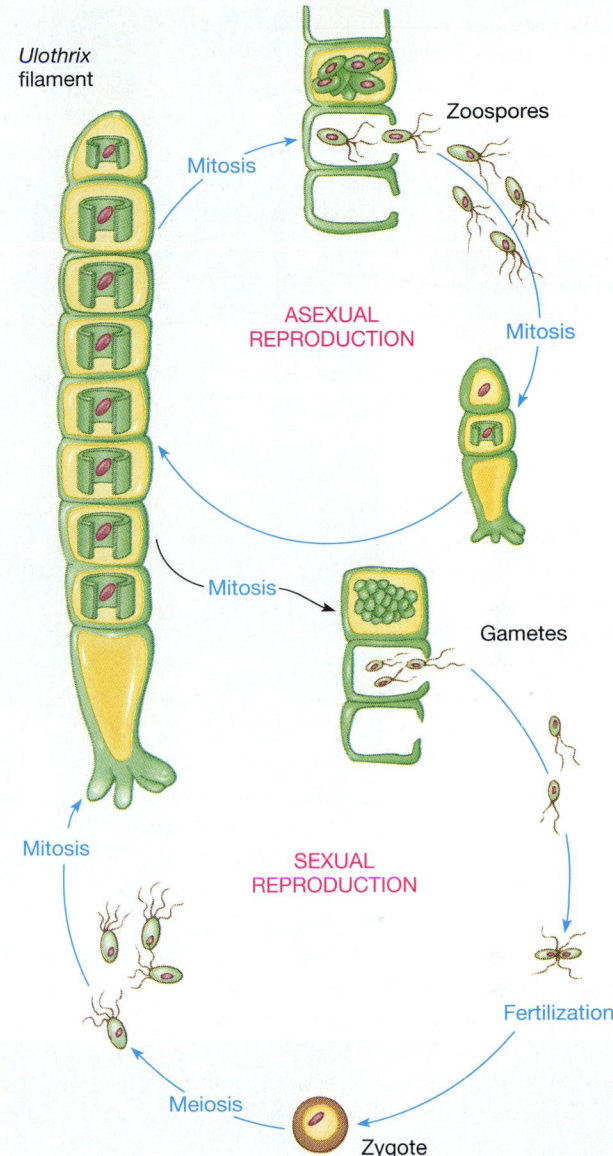

**FIGURE 20–5**  Reproduction in *Ulothrix*. In asexual reproduction (top), a filament produces zoospores (zoo = animal, denoting the spores' active swimming by means of their flagella; spore = reproductive cell). Because zoospores are produced by mitosis, each germinates into a filament with the same genetic makeup as the parent filament. In sexual reproduction (below), gametes fuse and form a diploid zygote, which spends the winter in a dormant state. It then undergoes meiosis and forms four zoospores of new genetic types. Each can grow into a new filament.

Sexual reproduction involving apparently identical gametes, as in *Ulothrix* and the termite-gut symbionts, is **isogamy** (iso = same). It occurs in various relatively simple eukaryotes.

The most familiar type of sexual reproduction, **oogamy** (oo = egg), involves a large, nonmotile gamete provided

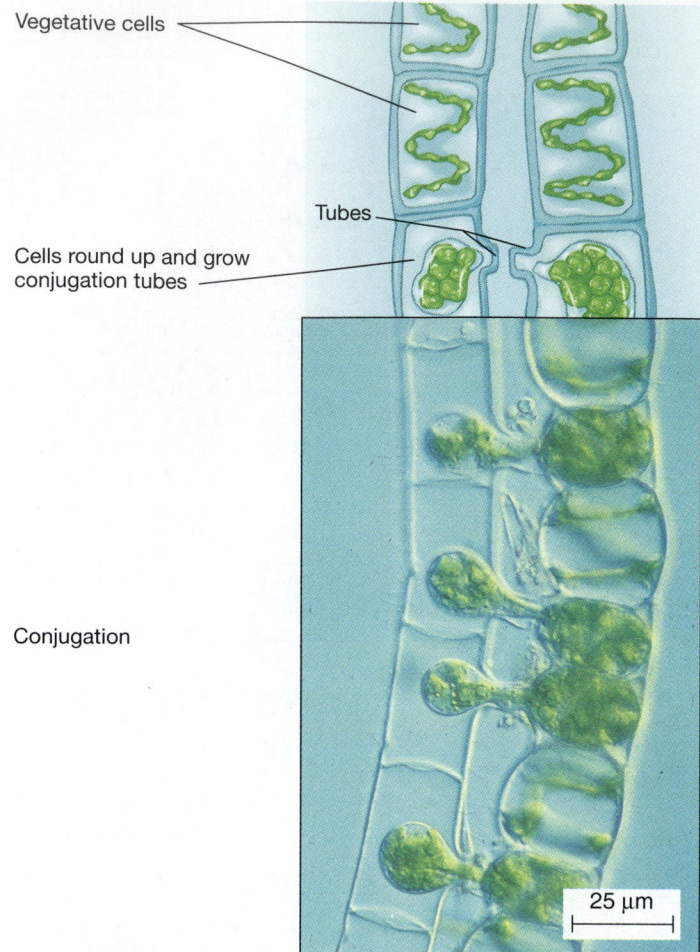

Vegetative cells

Tubes

Cells round up and grow
conjugation tubes

Conjugation

25 µm

(a) Isogamy in *Spirogyra*

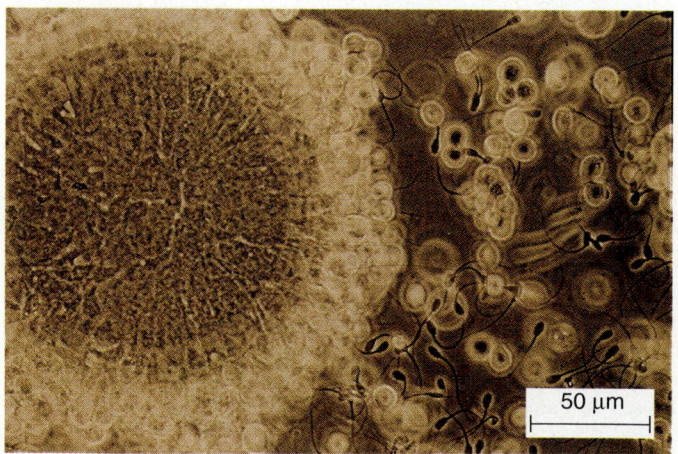

50 µm

(b) Oogamy: rabbit egg and sperm

**FIGURE 20–6** Isogamy and oogamy. (a) Filaments of the alga *Spirogyra* in the act of conjugation. Vegetative cells (top) round up, and the cells from one filament move into cells of a neighboring filament by way of tubes that grow between the two. Each newly arrived cell then fuses with the resident cell, forming a zygote. (b) An egg surrounded by many sperm. (a, Biophoto Associates; b, SIU Biomedical Communications/Photo Researchers)

with much food and a smaller gamete that must move, or be moved, to the larger gamete. By convention, we call the larger gamete **female** and the smaller one **male** (Figure 20–6). In the course of evolution, oogamy became the most widespread form of sexual reproduction. All sexually reproducing plants and animals are oogamous.

It seems likely that oogamy began to evolve when some early organisms started to produce larger gametes, containing more stored food than usual. These fused with the usual small gametes from other individuals. Even today there are algae that reproduce in this way: the zygote is formed when a large flagellated gamete fuses with a smaller one. This fusion of unlike gametes is **anisogamy** (an = not); anisogamy includes oogamy:

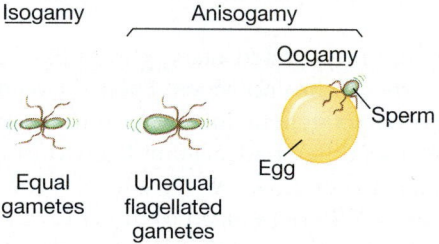

Isogamy     Anisogamy

Oogamy

Sperm

Egg

Equal
gametes

Unequal
flagellated
gametes

The selective advantage of such a situation was undoubtedly that spores formed from the resulting zygotes contained more stored food and therefore had a head start over smaller spores. Eventually this trend produced full oogamy: the larger gamete became so laden with nutrients that flagella could not move it, and it sat where it was until a motile gamete found it.

In plants and animals, the egg contains not only maternal chromosomes and stored food but also messenger RNA carrying the information necessary to direct the early stages of embryonic development. The major advantage of this system is that the embryo can be larger and better developed before it must provide its own food. Most animals could probably never have evolved without an egg containing stored food and genetic information for embryonic development. It is hard to imagine even a worm developing if the tiny embryo had to form a mouth and feed itself when it contained only two or four cells.

Thus, an egg that contains food and information, and is therefore too large to be motile, is of enormous selective advantage. But a nonmotile egg is no use in sexual reproduction without a motile sperm that can reach it. So sperm and egg vitally complement each other and have evolved together.

***Beyond Oogamy*** In many organisms, the fertilized egg contains all the nutrients the offspring will have until it finishes embryonic development and starts life on its own. **Oviparity** is the mode of reproduction in which a female lays her eggs and they develop and hatch outside her body; it occurs, for example, in most insects and in frogs and birds.

Less common is **ovoviviparity,** in which the female retains the fertilized eggs within her body during their embryonic development and releases the young after they hatch. Many sharks and snakes reproduce in this way. This gives the eggs more protection from climate and predators, but no more food than was originally stored in the egg (ovi = egg; vivi = alive; parity = giving birth).

By contrast, in **viviparity** the embryos have a physiological connection to the mother's body during development, and the mother provides them with additional food as they grow. This enables her to put energy into her offspring gradually instead of having to save it all up ahead of time. It may also permit the embryos to reach a larger final size before they become independent. Viviparity occurs in mammals, and reproduction in plants also fits this description.

Viviparous mammalian females continue to provide food to their offspring after giving birth. In some oviparous and ovoviviparous species, the young also receive care after hatching.

## Diploidy

Another trend in the evolution of sexually reproducing organisms was the tendency toward diploidy. Instead of undergoing meiosis as it broke dormancy, the diploid zygote stage persisted, grew, and divided by mitosis, producing more diploid cells. In many algae and plants the life cycle alternates between multicellular haploid and diploid stages.

Many other organisms now spend all their time diploid except during the short-lived gamete stage. Although most unicellular organisms are haploid, some are diploid, and most animals are also diploid. In the time between sexual reproduction, diploid unicellular organisms divide by mitosis and produce more diploid individuals asexually. By contrast, in animals the cells produced by division of the diploid zygote remain together and keep dividing, forming the many cells of the body.

Diploidy has the advantage of providing both more, and more diverse, genetic information. With two copies of a gene for a useful RNA or protein, a cell can transcribe both and churn out the product faster. In addition, if one copy mutates to a lethal form, the other is available to mask it. Or perhaps a mutation of one copy produces a code for a second, favorable but somewhat different protein. This gives the individual a heterozygote advantage and permits the species to accumulate two (or more) useful alleles in the gene pool.

The main disadvantage of diploidy is that it permits lethal alleles to accumulate in the species' gene pool. However, a lethal allele with no redeeming features is kept at low levels and generally does not impose an intolerable burden (see Figure 19–A, page 398).

## 20–D  MECHANISMS ENSURING FERTILIZATION

When an organism reproduces asexually, it is "doing its own thing," independent of other individuals. (Many individuals in a population may reproduce asexually at the same time, but this is because they are all obtaining enough energy to put into reproduction.) In sexual reproduction, however, each new individual comes from the union of two cells, usually contributed by two different individuals. The production and distribution of these cells must be coordinated. One common adaptation is having a breeding season, when members of a population all come into breeding condition in response to some environmental cue such as temperature or daylength (duration of daylight). Another common adaptation is reproductive behavior, including chemical secretions by some members of the population that stimulate others to come into breeding condition. For example, the locusts of Africa respond to chemicals secreted by the first male to reach sexual maturity.

Organisms that live fixed in place, or that cannot move far to find mates, may have various adaptations increasing their chance of sexual reproduction. In *Ulothrix*, environmental factors cause the entire population to produce gametes at the same time. Release of gametes into the water, by one or both sexes, also occurs in many other algae, in some marine worms, and in sea urchins. Motile gametes find mates either in the water or still associated with the (female) parent.

A second kind of adaptation that increases the chances of finding a mate is the ability of some animals to develop as either sex or to change sex. Still another is having reproductive organs of both sexes in the same individual, as most plants do. In animals this is called **hermaphrodism,** and it is common in earthworms, leeches, and snails. For organisms containing reproductive organs of both sexes, any other member of the species living nearby is a potential mate. They may be able to self-fertilize, but there are often mechanisms that ensure crossing with other individuals, so that inbreeding does not eliminate the genetic variation produced by sexual reproduction (Figure 20–7).

## 20–E  ROLES OF MALE AND FEMALE

In the rest of this chapter we shall discuss sexual systems and their effects upon social structure in animals.

The female is, in a sense, a limiting resource in sexual reproduction. At the simplest level, because eggs are bigger than sperm, eggs take more energy to produce. A male's evolutionary success is usually limited, not by how many sperm he can produce, but by how many eggs he can deliver his sperm to. The female's evolutionary success is limited by how many of her eggs survive to become part of the breeding population.

For some females this means laying as many eggs as her resources permit and leaving the offspring to raise themselves. For example, a female herring (a fish) may lay a million tiny eggs a year, and she leaves them to hatch and fend for themselves. At the other extreme, a female mammal gestates her developing young inside her own body and nurtures them for weeks or even years after giving birth. Which strategy is favored by evolution depends on the ecology of the species.

The fact that a female's parental investment in each offspring is usually greater than the male's has a fascinating consequence: the selective pressures acting on a female may conflict with those acting on a male of the same species. Whereas a male may raise his chances of fathering surviving offspring by copulating with as many females as pos-

sible, it may pay a female to be much more choosy. She produces fewer eggs and so has fewer second chances to reproduce successfully if her first mate is genetically unfit.

Not surprisingly, therefore, females of all animal species studied do discriminate in choosing mates. For example, in an experiment with fruit flies (*Drosophila*), only 4% of the females failed to reproduce, whereas 24% of the males failed to copulate even once. These celibate males courted females just as vigorously as did successful males, but no female ever accepted them.

Selection favors the female who discriminates and copulates only with genetically fit males. On the other hand, it may be to a male's advantage to appear genetically fit even when he is not, because females may then be deceived into mating with him. This has been envisioned as an evolutionary battle of the sexes, with skilled salesmanship among the males and an equally well-developed sales resistance and discrimination among the females.

## Sexual Differences

A female's reproductive success is not usually limited by her ability to find mates. It is more likely to be limited by how many eggs she can produce and her ability to rear her young. A male who demonstrates that he can contribute to raising offspring will be attractive to females. For instance, among birds, the male of choice is often the holder and defender of a territory that provides food and shelter needed by the female and her young. Defending a territory is also of direct selective advantage to the male, for the territory he holds promotes the welfare of his offspring. Once a female finds a good territory, she need not bother about extensive preliminaries, but may mate with the resident male without further courtship. In such a situation, a male's ability to hold and defend a territory determines his breeding success. Males may also compete for control over valued resources other than territories, because possession makes them irresistible to females. For instance, a man with wealth and status attracts some women no matter how unattractive he is physically (and the same is true for wealthy women, who have few female counterparts in nature).

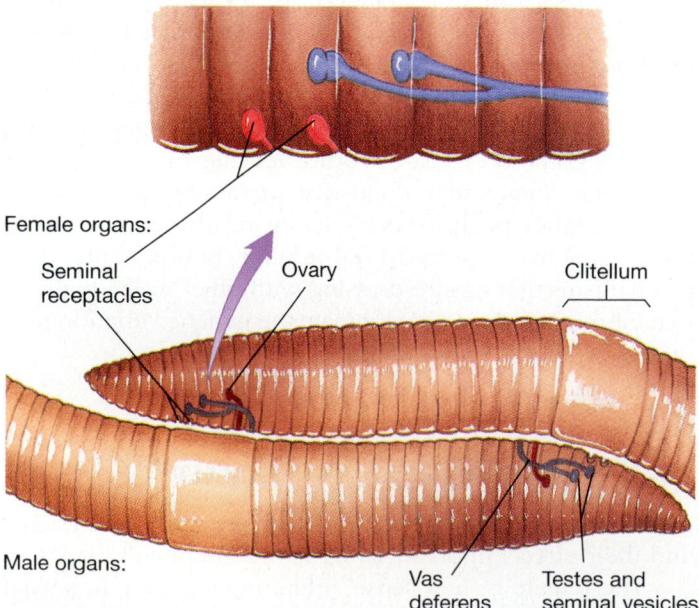

Female organs:

Seminal
receptacles          Ovary          Clitellum

Male organs:

Vas          Testes and
deferens     seminal vesicles

(b)

**FIGURE 20–7**    Mutual copulation between two earthworms. Earthworms are hermaphroditic, and they exchange sperm. Hence, each uses its partner's sperm to fertilize its own eggs. Sperm do not move directly from the male to the female openings but pass along a groove on the underside of the body.    (a, Roger K. Burnard/Biological Photo Service)

The different sexual roles of male and female may lead to their having different appearances, a phenomenon known as **sexual dimorphism** (di = two; morph = form). One type of sexual dimorphism is the possession of weapons by the male but not by the female. Features such as large antlers or horns in many hoofed mammals, long tusks in boars, and the enormous size of male seals give a male an edge in combat against other males for mates or breeding territories (Figure 20–8). Hence there is selective pressure for the males, but not the females, to possess these traits. These male weapons are more common in species in which successful males mate with more than one female, since winning in combat is generally the only way a male can get a chance to mate.

**FIGURE 20–8** Rocky Mountain bighorn rams. The males' massive horns are used to fight other males for dominance and the right to mate and to identify him to females. Horns take a lot of energy to produce and are heavy to carry.    (Pat and Tom Leeson/Photo Researchers)

(a)

(b)

In **monogamous** species—those whose members have just one mate at a time—a male is almost bound to find a mate without having to compete with other males. So he is not under selective pressure to evolve the weapons that are the hallmarks of a male destined either for glory or, much more likely, for early death.

However, monogamous as well as other species often show another kind of sexual dimorphism, in which males are more brightly colored than females or have cumbersome decorations, such as the peacock's tail (Figure 20–9). Because females are vulnerable as they sit on their eggs, it is advantageous for them to have drab, inconspicuous coloring. This camouflage works; when the male is the more conspicuous sex, predators kill more males than females. When defending a territory, a male may flaunt vivid colors or unusual, exaggerated postures, making him more visible not only to other males who might think of invading but also to females, who may notice what a nice territory he has.

A male's evolutionary success depends not only on his survival but also on females' choosing him as a mate. Therefore, females who choose showy males exert selective pressure for such traits, despite the added danger to the males, and whether or not the males also have better survival genes and resources. A showy male's sons are likely to inherit the feature that attracted their mother and so to be attractive to females of the next generation. This is how such features evolve and persist.

Charles Darwin coined the term **sexual selection,** which he applied to the evolution of characters that enhance mating success (as opposed to survival) and are found only in one sex and passed on to same-sex offspring. In

**FIGURE 20–9** Male decorations. (a) A peacock displays his tail as part of his courtship of the drably colored female. (b) This satin bowerbird in the rain forest of eastern Australia may have the best of both worlds. His coloring is inconspicuous (though not as drab as the female's), but he collects colorful objects to decorate his bower of twigs. Here he displays a snail shell to a prospective mate, who watches from the bower. The many blue plastic objects this bird has collected seem to be especially prized, probably because blue is a rare color in nature. The bird's ability to build and decorate a colorful bower for courtship, and defend it against rivals who try to wreck it, makes him attractive to females. (a, D. Cavagnaro/DRK Photo; b, Michael Fogden/DRK Photo)

contrast, he viewed natural selection as increasing the "fitness" of organisms to survival and reproduction in their environment. Modern biologists tend to define natural selection more generally, as the differential reproduction of genotypes, in which case sexual selection is merely a form of natural selection.

Some biologists contend that female choice of showy males may not be as counterproductive as it first seems. A male's bright feathers or long tail handicap him but also advertise his age and competence: surviving despite this extra cost demonstrates his genetic superiority, which the female would do well to secure for her offspring.

An interesting variation on males' use of color to advertise their valuable property and genes to females is found among some bowerbirds and weaverbirds. For example, the male African village weaverbird is dull colored, but he builds a colorful nest and jumps up and down beside it saying, in effect, not "look at me" but "look at the gorgeous nest I have built for you." If no females are attracted, and if the color of the nest starts to fade, he will tear it to pieces and build another one. Meanwhile, when he is not engaged in reproductive behavior, his dull colors make him inconspicuous to predators.

**FIGURE 20–10** Sexually monomorphic, monogamous animals: Fischer's lovebirds. (William Dilger)

## Mating Systems

Each animal species has a characteristic courtship behavior. One of its functions is to ensure that prospective mates recognize each other as members of the same species, so that they do not copulate with a member of the wrong species. In **polygamous** species—those in which each animal may mate with more than one other—males frequently have a vast sex drive and little discrimination. They will court almost anything vaguely appropriate, and females must recognize and pick out the right male. The male's appearance, physique, and courtship behavior assist in this.

As a corollary, in monogamous animals the sex drive in both sexes is about equal, the sexes are often indistinguishable in appearance and behavior, and courtship is mutual (Figure 20–10). The male also discriminates as much as the female in his choice of mate. Like the female, he has only one chance to choose a mate, and it will be just as advantageous for him as for the female to choose a mate who will produce healthy, successful offspring.

Polygamy may be divided into polyandry and the much more common polygyny.

*Polygyny* **Polygyny** is a system in which one male mates with more than one female. It may evolve where a female gets a better share of some limited resource for her offspring by joining a mated pair, or a male and his harem, than by mating with an unmated male. The resource she gains may be more food or better protection from predators in a group than in a pair, or peace and quiet and a helping hand to raise the young.

In a polygynous social system, one or a few males live in a group with a number of females. The dominant male is usually the only one that mates at the females' most fertile time, and he must defend his harem not only against predators but also against other males who try to depose him as head of the harem. Defending his dominant position requires a lot of energy and constant vigilance. This takes such a toll that in many species the dominant male is displaced by a rival several times in a year.

Dominance is worth fighting for because dominant males have an enormous reproductive advantage. For instance, male elephant seals guard harems of females as they haul out of the ocean onto rocky coasts to bear their pups and then to mate. In one study, 4% of the males were responsible for 88% of copulations observed in such a polygynous group (Figure 20–11).

Another selective force for harems with one or a few males may be that females do not have to share food and other resources with males whom they do not need for copulation. In such situations, the females help the resident male to defend their territory. On the other hand, where it's "all hands to the barricades" to fight off predators or help feed the young, nonreproductive males may be wel-

**FIGURE 20–11**   Elephant seals. Space is at a premium on this crowded breeding beach; a male's enormous size and his teeth are the weapons used to push rivals out. The dominant male sires most of the year's offspring.   (Frans Lanting/Minden Pictures)

come members of the group. Both kinds of social organization occur among various monkeys and apes.

A second kind of polygyny may evolve if resources are widely distributed and hard to defend, and the male's help is not needed for raising the offspring. Females of these species can find their own resources. In many such species, males defend special mating territories that do not perform any of the other usual functions of territories. Members of both sexes attend a traditional display site or rendezvous called a **lek,** where males compete for high-status mating territories and females mate with the high-status males. This system permits females to mate with the available males who are fittest (perhaps more in terms of mating than of survival).

For instance, male swallowtail butterflies fight for and defend territories at the tops of hills. This is purely a mating area, and any male that can defend it will mate with many females. Females fly up the hill, fighting off males who attempt to mate with them on the way. If a female makes it to the top, she mates with the dominant male there. As a result, her sons get good fighting genes (even if these genes have no other advantages) and so have a good chance of fathering the next generation of butterflies. This system also boosts the (top) male's success: in one study, the male defending the most sought-after territory at the top of the hill mated with 11 of the 16 females that flew into the area.

All other factors being equal, polygyny is a more favorable system than monogamy for the few males who succeed in reproducing. However, polygyny is possible only where the female does not need the male's full-time help in bringing up the young. Its presence or absence, therefore, usually depends on the advantages or disadvantages to the female.

*Polyandry*   **Polyandry,** the mating of one female with more than one male, is much less common than polygyny, but it has evolved at least five times in different groups of birds. Consider the jacana, a long-legged bird that runs around on the waterlily pads covering some lakes in Central America. The females defend small territories where they lay eggs and then abandon them to their mates to incubate and raise. A female may mate with, and lay eggs for, several males in one breeding season.

There has been much debate about how polyandry evolves. The most widely accepted idea is that it evolves from a monogamous situation, common in birds, in which both parents share equally in nest-building, incubation, and parental care of the young. If it became advantageous for only one parent to be at the nest at any one time, pure chance might determine which leaves and which stays.

Another idea notes that polyandry may evolve in environments where eggs and young are frequently destroyed. This is the case with spotted sandpipers, shorebirds that breed along streams, where flooding often destroys all the eggs or young in a nest. Males join a territorial female, mate with her, and incubate the resulting clutch of eggs. The female's energy is more usefully devoted to laying replacement eggs than to incubating existing clutches. She can devote most of her energy to producing more eggs, which will be fertilized and incubated by the original male if the original clutch has been destroyed, or by another male if it has not. The female gains a sort of insurance by not having all her eggs in one basket. Females compete for males, which are generally scarce because at any one time many of them are busy incubating eggs.

Polyandry also occurs among red and red-necked phalaropes, which breed in the Arctic. Here they have only six weeks to recover from the arduous migration, nest, and raise the young to readiness for the fall migration. Females are larger, more aggressive, and more colorful than males, and they produce very little of the hormone responsible for the urge to incubate eggs (Figure 20–12). The ancestral female who first lost her nesting instinct was at no disad-

**FIGURE 20–12**   Red phalaropes. The female is larger and more strongly marked.   (M. A. Chappell/Animals Animals)

vantage because her mate still incubated the eggs, and the young hatch able to run, swim, and feed themselves with minor guidance from an adult. Turning over these duties to the male left the female free to court and lay eggs for another mate, if one were available, and to restock her energy supply. A female phalarope must eat around the clock if she is to produce her clutch of four eggs and store fat reserves for the fall migration.

***Monogamy***    **Monogamy** is the system in which each individual takes just one mate, for a whole breeding season or for life. In species where it takes both parents' efforts to raise offspring successfully, monogamy is invariably the most advantageous mating system for the female. Selection for the male to be monogamous occurs when he gains greater evolutionary success by helping one female raise their mutual young than by pursuing additional females.

Monogamy (more or less) is the most common form of human sexual system, although both types of polygamy arise under various circumstances. We can infer that monogamy predominates because the human infant matures more slowly than most other animals. It seems likely that, during much of human history, both parents had to do their share if they were to raise many offspring to sexual maturity. It is in the man's interest as much as the woman's that his offspring reach maturity, and so he, too, would be better off monogamous unless he could provide for the children of more than one wife.

The slow maturation of humans is related to the large size of the human brain and the long time we take to educate. Every human society has a large body of culture, which is a major factor in the survival and reproduction of its members. However, this culture is passed on, not in the genes but by example and by language.

An aspect of culture unique to humans probably reinforced the tendency toward monogamy in early humans: the possession of material goods, which are of vast importance to human evolutionary success. It would be strongly advantageous to bequeath things like clothes, a cave or house, or land to your genetic children, and the majority of men throughout history have not amassed enough to split up among the children of more than one wife.

Monogamous animals that live with their mates may form an attachment known as a pair bond. They keep in close contact, cooperate in many behaviors, including parental, and seldom quarrel. In exchange for this commitment to his mate and her young, the male relies on her fidelity to him. If other males sire her offspring, he is wasting his energy, in evolutionary terms, helping raise the young (Figure 20–13).

Close study of monogamous animals shows that many do not conform to this idyllic picture because it is often selectively advantageous to "cheat" on a mate. Although 92% of bird species are "officially" monogamous, copulation with nonmates has been observed in more than 150 bird species,

**FIGURE 20–13**   Monogamous family life. Like most other birds, swans mate for life, and both parents care for their young.

often forced on a female by a neighbor more dominant than her own mate. This type of behavior is advantageous to the interloping male because it gives him a chance to father extra offspring at little extra cost to himself. This offspring will often be raised by others with no further effort on his part. The female gains nothing from such an encounter: whether her offspring are fathered by one male or more, she bears the brunt of producing and rearing offspring.

In some cases, the male's faithful attendance seems to be not devotion to his mate but insurance of her fidelity to him. If another male forces her to copulate, her mate may peck her cloacal opening, forcing her to eject some of his rival's sperm. He then immediately copulates with her, and his own sperm compete with the rival's. Once she completes her clutch of eggs, he may be off to snatch a copulation with a still-fertile female neighbor.

During this little drama, the pair has left their nest and their already-laid eggs unguarded. Female neighbors sometimes sneak in and lay their own eggs to be reared by the residents. This permits females to increase their own evolutionary success with no cost beyond that of making the egg. In a study of starlings, researchers analyzed DNA from the adults and young in each nest and found that most of the young were the offspring of both resident adults. However, 2 to 8% were offspring of the resident female by other males, with fewer being offspring of nonresident females. Hence, it is more advantageous for the male to guard his mate from other males than his nest from other females.

***Ecology and Mating Systems***    An important factor in determining the mating system of a species is its ecology, the relationship of the organism to its environment. The distribution of food, water, nesting sites, and shelter affects the distribution and social behavior of individuals. Here we give some generalizations, but good biologists can always find exceptions.

(a)

(b)

**FIGURE 20–14** Ecology and mating systems. (a) A monogamous pair of black-backed jackals hunts and feeds together, defending their territory against other pairs. (b) Members of a baboon troop, a social group that travels together, feeding mostly on plentiful plant food and defending themselves against jackals and other predators.   (b, Erwin and Peggy Bauer/Natural Selection)

Where food is scarce and found in small, isolated pockets, individuals tend to be solitary and come together only for a short time, in the breeding season. (Bears, badgers, and moose behave like this.) Couples may come together only to mate, or they may form short-lived pairs or colonies while they raise the young.

Alternatively, some animals may live as monogamous pairs defending a territory with widely scattered resources. In this way they can help each other hunt or watch for predators, and they need not spend time and energy searching for a mate when the time comes. Most wild members of the dog family fall into this group, as do various other mammals. Territorial animals may be monogamous by default: both mates chase others of their own sex out of the territory, leaving each other no choice but fidelity. When several members of these species are caged together, individuals are often willing to mate with more than one animal of the opposite sex.

On the other hand, most monkeys and apes can live in troops because their diet is mainly plentiful plant food. The all-important (in evolutionary terms) females and young can eat well and still enjoy the protection of living in groups (Figure 20–14). As we would expect from this, in cases where food is sometimes inadequate, groups containing only one male are the rule. Males then compete to enter a troop, since this is the only way they can breed.

An interesting example of this came from a study of baboon troops in Africa. Most troops live on the open savanna and travel from one feeding ground to another. They are not limited by food and water, but they are exposed to many predators. Here, females must stay close to their young to protect them, and males protect the troop as a whole, a role that makes extra males welcome. The troop can afford to let them eat some of the food, and the extra males also mate on occasion.

In contrast, one troop lived in a woodland area that contained food and water all year. This troop treated their woodlot as a territory, and females as well as males defended its boundaries. Individual members came and went from year to year, but the troop's size stayed the same, with fewer males than in any other troop in the study. This is almost certainly because the territory provided food for a limited number of individuals, and so nonreproductive young males were regularly driven away.

## 20–F    SELFISHNESS AND ALTRUISM

Mating systems not only reflect the species' ecology and evolutionary history but in turn have evolutionary consequences of their own. For one thing, the mating system has a profound effect upon how closely members of a species are related to one another. For instance, in a polygynous herd of horses where only one male mates, all of one year's offspring are more closely related than they would be in a monogamous group where the year's offspring have different fathers. Having a closer degree of relatedness can have some unusual evolutionary results. One of these, suggested by biologists such as William Hamilton and Robert Trivers, is selection for **altruistic behavior,** which is behavior that favors the reproductive success not of the altruistic individual but of another member of the species. (As far as we know, other animals do not consciously think of their behavior as altruistic, but they may act in ways that we would describe by this term.)

To understand altruism, we must think of selection as acting not on individuals but on genes or gene complexes. Genes last much longer than individuals, which are their temporary mobile homes. An altruistic allele may spread through a population at the expense of a specific individual that carries it. In particular, an allele that favors altruistic behavior toward close relatives may improve the chances that it will survive in the population because there is a good chance that closely related individuals also contain copies of the allele. To take an extreme example, if an individual dies to save ten close relatives, one copy of the "kin-altruism" allele is lost but many other copies are saved.

This means that the rate at which a particular allele increases in a population depends on the rate at which it is reproduced by its bearer *plus* the rate at which copies of the allele are reproduced by relatives whom the bearer may help. This kind of indirect selection, for genes present in related individuals, is known as **kin selection.**

The more closely individuals are related, the more likely altruism is to evolve between them because the more likely they are to share some of the same genes. William Hamilton pointed out that the degree of kinship necessary for kin selection to operate could be calculated by considering the **index of relatedness:** the probability that altruist and beneficiary share a particular allele. An individual in a sexually reproducing species inherited any particular allele from either its mother or its father. Because sperm and eggs each contain 50% of the parent's genes, the chance that the allele is also present in a sibling is 50%, and therefore the degree of relatedness of siblings is $\frac{1}{2}$. Note that this is the *average* degree of relatedness. By the luck of the meiotic draw, an allele may be present in more or fewer siblings. However, the relatedness between a parent and its offspring is exactly $\frac{1}{2}$, because half the parent's genetic material is actually inherited by the offspring.

Indices of relatedness can be calculated for other relatives. For identical twins, it is 1. In a monogamous species, the chance that a grandparent, uncle, aunt, nephew, or niece also has a particular allele of yours is $\frac{1}{4}$. The index of relatedness with a great grandchild or first cousin is $\frac{1}{8}$. Between second cousins (children of first cousins) it is $\frac{1}{32}$.

Third cousins (children of second cousins) are much less special; their index of relatedness is only $\frac{1}{128}$, probably not much greater than the chances of sharing the allele with any other individual in the whole population.

We would therefore expect to observe less altruistic behavior toward relatives who are less close. An allele that prompted an individual to lay down his or her life for a relative would have to save more than two offspring or siblings, more than four uncles and aunts, or more than eight first cousins, and so on, to be selectively advantageous. Otherwise it would not survive, on the average, in enough other bodies to compensate for its loss in the altruist.

Of course, organisms do not calculate their degree of relatedness to others before acting. Selection favors alleles that operate as if such calculations had been done. And in many cases, altruism is futile because the beneficiary does not contain the altruistic allele. However, it is the long-term statistical pattern that counts: alleles promoting the appropriate level of altruism toward other individuals are likely to survive. Altruistic suicide is not common (indeed, it is heavily selected against). What counts more often is the statistical risk of death or loss. Even a third cousin may be worth saving if the risk to the altruist is small. The most common acts of altruism, such as taking a turn at sentry duty, cost the altruist little but contribute markedly to its relatives' welfare (Figure 20–15).

Another common altruistic behavior is helping rear a relative's offspring. Young male birds sometimes forego the chance to mate, instead remaining with their parents and helping to feed the next brood. This significantly raises the nestlings' chances of reaching maturity. Because of such kin selection, the helping son's evolutionary success is not entirely assessed by counting how many of his own offspring he rears. Rather, it is measured as his **inclusive fitness,** the

**FIGURE 20–15**  Altruistic behavior. California ground squirrels live in small colonies. These rodents recognize and greet other members of the colony, and individuals take turns watching for danger.

number he rears weighted by their relationship to him (the closer the relationship, the more points he gets).

How do organisms know who their relatives are? Normally, animals living in a small group are related. However, recognition may be more specific. Research shows that organisms as diverse as honeybees, tadpoles, rodents, and monkeys may recognize their kin, even if they have never met them before. Furthermore, they can distinguish different degrees of kinship. For example, female ground squirrels fight less with their full sisters than with littermates fathered by a different male. In many cases, the clue to kinship is odor.

In most animals, altruism between siblings is not so marked as between parents and their offspring. The reason may be partly that the parent/child relationship is highly predictable. There is little probability that the relationship is not really as it seems. In most species, a mother can be more sure that her offspring are her own than can a father. Fathers are vulnerable to deception (unbeknownst to him, his mate may have copulated with another male). Therefore, fathers may be expected to put less effort into caring for young than mothers do. Similarly, maternal grandmothers are more sure of their grandchildren than are paternal grandmothers; a grandmother can be sure of her daughter's children, but her son might have been deceived into rearing young that do not contain his (and therefore her) genes.

Reproduction is costly. Females in particular invest considerable energy, nutrients, and time in offspring. What determines exactly how much, on the average, the evolutionarily successful parent spends on any particular offspring? Does (s)he devote equal time and resources to each?

Robert L. Trivers suggested studying such questions by comparing **parental investment,** defined as any investment in an individual offspring that increases its chance of surviving (and future reproductive success) at the cost of the parent's ability to invest in other offspring. For example, if an eagle feeds a fish to one of two nestlings, the major part of the parental investment that this fish represents is the increased probability that the other nestling will die because it did not receive the fish. Another part of the parental investment is the decreased ability of the female to lay other eggs and rear other young because of the time and energy she invested in catching the fish (Figure 20–16).

Clearly, parental investment must not be spread too thinly over many offspring, because then few or none of them will receive enough food to survive. Equally clearly, devoting too much investment to too few offspring will be selected against because more prolific parents will leave more offspring in the next generation.

How is parental investment distributed among existing offspring, all of whom are equally related to the mother? Because the parental investment required to raise the smallest offspring is greater than that required to finish raising a larger elder brother or sister, it would clearly pay a parent

**FIGURE 20–16** Parental investment. A female bald eagle feeding her young. (Lon E. Lauber/Natural Selection)

to concentrate on saving the elder offspring first if food is scarce. On the other hand, if food is ample, it would pay to feed the smallest first, because the others are better able to survive a short period without attention. Offspring will presumably be pushed out to care for themselves when the parental investment involved in further attention to them would be more advantageously spent on raising new offspring.

Altruism between siblings differs somewhat from that between parents and offspring because the age differences are less and because the certainty of the relationship is less secure. Nevertheless, siblings share, on the average, 50% of their genes, and altruism would be expected. Should an offspring always try to grab food for itself, or should it share? Since an offspring is 100% related to itself, and only about 50% related to a sibling, the offspring "should" try to grab food for itself unless this action costs its siblings at least twice as much as grabbing the food would benefit itself. This might occur, for example, when the sibling is much younger and would be placed in great jeopardy by the elder sibling's grabbing all the food.

Field studies show that siblings coexist peacefully in many species. However, in some species the older and larger sibling attacks and kills the other(s). This appears to be the norm in about two dozen bird species studied, including the common egret and masked boobies. In some cases, killing of a younger sibling is triggered by hunger. As a result, parents raise as many young as they can provide with adequate food: more in a good year, fewer when food is scarce. Producing an extra egg or two provides the parents with insurance in case the first one fails to hatch, is defective, or is killed by predators.

Among mammals, sibling murder occurs among spotted hyenas and even among unborn embryos of pronghorns. When the tadpoles of spadefoot toads turn cannibal, they spare their siblings (which they can identify by a taste test) if unrelated tadpoles are readily available.

## Does Altruistic Behavior Exist?

Parental behavior is altruistic. For example, suckling her young costs a female mammal energy and benefits her nothing directly, but she is enhancing the survival of her own genes, which are carried by her offspring. Parental behavior was a prerequisite to the evolution of animals such as birds and mammals, which produce young that do not survive unless their parents invest a lot of energy in caring for, feeding, and protecting them. There is also strong selection for such behavior, because individuals who do not help raise young do not contribute genes to future generations. Alleles for parental behavior pass directly from parents to offspring; there is no need to invoke the indirect mechanism of kin selection to account for this kind of altruism.

However, some biologists question whether other altruistic behaviors truly exist. Many other cases that look like altruism, when examined more closely, have turned out in fact to benefit the actor at the same or a later time. For example, our young male helper bird may get more food for himself because his experienced father holds a better territory than he could get on his own, and he may inherit the territory someday.

In one study, monkeys who discovered food and called others to share it actually got more to eat. If they did not call but were found out, they were attacked and forced to give up the food.

Researchers found that vampire bats sometimes regurgitate blood and feed another individual who had no luck finding a meal. Because a bat starves to death after two nights in a row without eating, sharing blood may save a life. Closer study showed that blood is most often shared with a relative, usually the donor's offspring. However, sometimes the recipient is an unrelated individual who has been a recent roost mate, and who is likely to return the favor if the donor fails to find a meal in the near future.

This sort of tit for tat situation cannot be considered pure altruism, but is sometimes called reciprocal altruism. This category also includes cooperation among individuals in group behavior such as hunting or defense, which benefits all (Figure 20–17). Here individuals promote the success of each other's genes. In the long run this also benefits their own genes because everyone produces offspring who will cooperate in the next generation.

Many biologists feel that humans operate primarily through selfish motives: we perform helpful acts either because we count on future gain for ourselves when the recipient returns the favor, or because of a selfish desire to

**FIGURE 20–17**    Reciprocal altruism. White pelicans feed in groups, herding fish toward shallow water where the birds can easily scoop them up.

feel good about helping or to avoid feeling guilty for not helping. These views can be reconciled. We would expect altruistic genes to operate by making the actor feel virtuous for helping or guilty for not helping.

Our brief survey of altruism and selfishness shows that we can account for many common patterns of human behavior in terms of their likely effect on the evolutionary success of their owners' genes, either directly or by kin selection. However, in most cases we have little idea how much of our behavior is determined by hypothetical "altruist" genes and how much stems from our ability to think, learn, and assess costs and benefits deliberately before choosing a course of action.

## The Social Insects

Biologists have long been intrigued by the degree of cooperation between individuals in colonies of social insects and by the specialized division of labor. Most of the colony's members are sterile workers, who will never leave any offspring. Instead, they devote their lives to raising the offspring of other individuals. This is altruism on a grand scale.

Colonies of wasps, ants, termites, and social bees, especially honeybees, are impressive societies. Food is shared communally, and information is exchanged in elaborate ways, including a great range of different chemical signals and the elaborate "dances" by which worker honeybees indicate the direction and distance of a food source to their nestmates. Workers are fearless in the defense of a colony, and many even sacrifice their lives on its behalf.

A colony of social insects is a huge family, sometimes numbering several million, all descended from the same mother—the reproductive female, or queen (Figure 20–18). Among the ants, bees, and wasps (all in the insect order Hymenoptera), workers are infertile females. In the termites (order Isoptera), workers also include infertile males. How do the workers gain an advantage great enough for such altruistic, even Kamikaze, individuals to have evolved?

In ants, bees, and wasps, the answer lies in a curious aspect of sex determination. The single mature queen in

**FIGURE 20–18** A honeybee society. The queen can be distinguished from the workers surrounding her by her longer abdomen. She is laying eggs in empty cells of the hive. The workers will feed and tend these new sisters as the larvae grow. (Biophoto Associates)

the colony receives enough sperm during her mating flight to last her for the rest of her life of perhaps ten years or more. She uses these sperm to fertilize eggs as she lays them, producing female offspring, most of which develop as workers (a few become new queens).

However, not all of the queen's eggs are fertilized. The unfertilized ones develop into males, which thus have no father. A male has only a single (haploid) set of chromosomes, derived from his mother, and his degree of relatedness to her is 100%. On the other hand, his diploid mother gave him only half of her genes, and so, strange as it may seem, *her* relatedness to *him* is only 50%.

A colony contains workers who are full sisters, sharing the same (haploid) father, whose sperm all contain identical haploid sets of chromosomes. Therefore, any allele a worker obtained from her father is also present in each of her sisters. However, there is only a 50% chance that any

allele a worker obtained from her mother will also be present in any one sister. In other words, the relatedness between sisters averages $\frac{3}{4}$ (not $\frac{1}{2}$ as in normal sexual animals).

So, remarkably, a hymenopteran worker is related more closely to her sisters than to her mother! She can therefore further the interests of her genes more by farming her mother as a sister-making machine than she can by reproducing. More accurately, an allele that promotes making sisters will replicate more rapidly than will an allele for making offspring. This selects for sterility of workers, which has evolved independently at least 11 times in the Hymenoptera and once elsewhere (the termites), an extraordinary example of the complex effects that can be produced by a particular sexual system.

Some biologists draw a parallel between an insect society and the body of a multicellular animal. An animal's body can be viewed as a society of cells all contributing specialized tasks that eventually enable some cells in the ovaries or testes to reproduce. This perpetuates the genes shared by all the body's cells. Likewise, the analogy goes, insect societies are like a multicellular organism that happens to exist in many independently mobile pieces. Thus, the queen is an ovary on legs, and a worker a roving set of hands or a traveling gut.

### Naked Mole Rats

A fascinating parallel to social insects occurs in the naked mole rat of dry eastern Africa (Figure 20–19). Colonies of these mammals consist of 75 or more individuals. Only one female—the queen—reproduces, using sperm from one to three male consorts. The workers build the colony's tunnels, locate food (plant tubers: swollen underground plant stems storing starch and water), and fight predatory snakes.

Curious about how such a situation arose, researchers analyzed the DNA of various individuals and found that all members of a colony are virtually identical genetically. Hence, an individual's evolutionary success does not depend on producing offspring of its own.

**FIGURE 20–19** Naked mole rat workers cooperating to dig a tunnel. (Raymond A. Mendez/Animals Animals)

## E S S A Y : *Sex: What's the Use?*

Sex is one of the greatest puzzles in evolution. Compared with asexual reproduction, it is slow, complex, costly, and wasteful (Section 20–A). Yet nearly all multicellular organisms can or must reproduce sexually. Even most unicellular organisms, which may reproduce asexually for hundreds of generations, still keep sex as an option. What powerful evolutionary advantage retains sex despite its drawbacks?

The standard textbook answer is based on ideas set out by August Weismann in 1889: sex acts as a powerful engine of genetic variability, the raw material from which natural selection shapes adaptations to the changing environment and eventually evolution of new species.

However, a species may remain apparently unchanged for thousands of generations before giving rise to new species in response to some radical change in the environment. Meanwhile, any individual that mutated to asexual reproduction could outreproduce the sexual members of its species and probably replace them.

Natural selection cannot maintain sexual reproduction for millions of years until the population needs novel genomes to survive an environmental cataclysm. Why doesn't sex get lost more often between these evolutionary spurts? Surely sexual variation has some short-term advantage that selects for its survival from one generation to the next, as well as this standard long-term advantage.

Biologists have proposed several explanations, in which sex is assigned various roles:

1. **Mutation clearance.** By chance, sex inevitably produces some new gene combinations that are heavily loaded with unfavorable alleles. Natural selection sends these individuals to the evolutionary trash can, thereby removing copies of deleterious genes from the species' gene pool.

2. **Genome renovator.** On the other side of the coin from the clearance function, some new gene combinations produced by sex reconstitute well-adapted gene combinations that have been damaged by unfavorable mutation(s).

3. **Marriage broker for mutations.** Sex provides a quick way for two new, favorable mutations that arose in separate individuals to be united into one genome. (Even if these two individuals do not mate, their descendants may, especially if selection increases the frequency of each allele in the meantime.) Genomes containing this favorable combination of alleles are likely to arise sooner by sexual recombination than by a second mutation in an individual that already has one of them.

4. **Double dipper.** Sex provides a quick way to produce individuals with two copies of a new, favorable allele, which may double the production of the gene's product. In the asexual alternative, an individual with one copy of such an allele must await a highly unlikely mutation to produce a second copy. This idea is a variant of the previous one.

5. **Fine tuner.** By providing genetic variations, sex permits a population to take advantage of environmental variations. Individuals with different genomes use parts of the habitat with different sets of environ-

## SUMMARY

Although the variety of reproductive adaptations seems bewildering, they all exist because they achieve the same final outcome: the passing of genes to future generations. This permits them to persist no matter how bizarre or self-destructive they may be.

1. The members of all species have a small amount of genetic variation as a result of mutation. This is the only variation among members of species that reproduce solely by asexual means. Members of sexual species are much more variable because genes are reshuffled during meiosis and form new combinations at fertilization.

2. Asexual reproduction uses energy more efficiently than sexual reproduction and is common in many wide-spread species that can survive changing conditions. For more localized and specialized species, the energy wasted in sexual reproduction is worthwhile because at least some of the genetically different individuals in a sexually reproducing species can usually survive and evolve in changed conditions. However, this is generally a long-term advantage of sex. Its selective value in the short term is controversial.

3. Sexual reproduction may have originated as a way in which two starving cells could survive hard times by pooling nutrients, meanwhile repairing damaged DNA. When good times returned, they formed joint offspring by meiosis, a modification of mitosis.

4. The advantage of having one gamete stuffed with a food supply and messenger RNA for the development of the

mental conditions. In contrast, asexual clone-mates would compete among themselves for a more localized area to which they are all best adapted.

6. **Moving target.** Genetic variation helps some members of a population evade parasites (including those that cause diseases). Parasites and their hosts engage in a continuous co-evolutionary genetic chase. The most successful parasite genomes tend to be those adapted to infecting the most common host genomes. But heavy parasitism lowers a host's evolutionary success, permitting other, less susceptible host genomes to increase more—until new parasite genomes catch up with them in turn! In this view, the role of sex is to churn out variation for its own sake, with no particular direction, rather than to generate the "onward and upward" evolutionary trend of the standard view of sex. This idea's champions are Leigh Van Valen, Robert Trivers, and William D. Hamilton. Van Valen called it the Red Queen's race, after the character in *Through the Looking Glass* who ran to stay in the same place.

This view is supported by a study that found higher levels of parasitism in asexual than in sexual populations of minnows (a kind of small fish). The sexual populations also had greater variance in their parasite loads. This also explains why Native American populations, which have less genetic variation than most other races of humans, suffered heavy death tolls from diseases such as malaria, smallpox, and tuberculosis introduced by European settlers.

Further support comes from the case of the fungus *Epichloe typhina,* which infects tufts of wild grasses and sedges. The fungus destroys the plants' flowers, the sexual organs, while permitting asexual production of new shoots. This ensures that the host plant will produce more of the same genome the fungus has already successfully invaded.

These ideas are not mutually exclusive; indeed, some are similar or overlapping. Several or perhaps all of them may be correct, depending on which genetic loci and which evolutionary situations are studied. For example, the moving target role of sex may apply best to alleles for resistance to disease and parasites, which exert heavy selective pressure on some populations. On the other hand, the clearance model applies best to lethal recessive mutations. So, like many (if not most) other adaptations, sex may turn out on close scrutiny to have more than one advantage, and it's often hard to guess which of many effects is most important.

Lynn Margulis holds a dissenting viewpoint. She thinks sex has no selective advantage but is simply a legacy handed down from early eukaryotes that lived more than a billion years ago. She contends we should be looking, not at the survival of sexual reproduction, but of *sexually reproducing organisms.* Because they pass on genes for sex to their offspring, sexual reproduction continues to thrive. Perhaps mutations from sexual to asexual reproduction are simply very rare.

new individual selected for the evolution of oogamy, the most common form of sexual reproduction today. Oviparous and ovoviviparous females save up all the food they will provide the offspring and then produce eggs containing this food supply. Viviparous females produce small eggs and gradually furnish more food as their embryos develop.

5. In most species, two individuals must act in concert to achieve sexual reproduction. Adaptations such as specific mating seasons, environmentally determined sex, and hermaphrodism help to ensure that mating (fertilization) can occur.

6. The selective pressures on the two sexes often conflict. Because it takes more energy to produce eggs than sperm, and the female often provides more parental care, females are usually more selective than males in their choice of mate. Males tend to seek many mates (even in many nominally monogamous species).

7. Sexual dimorphism may arise from conflicting selective pressures on the two sexes or from males' and females' different roles in reproduction. The possession of weapons by the male tends to be greater in polygamous species.

8. A species' sexual system is determined by the amount of energy each sex puts into producing and rearing offspring and by ecological factors such as the distribution of food, prevalence of predators, and so on. Polygamy can evolve when one sex (usually the female) can raise the young without assistance from the other. Monogamy prevails when both parents must contribute

care. In monogamous species, the members of both sexes must choose their mates with discrimination, and the appearance, parental roles, and behavior of the two sexes are more similar.

9. A species' sexual system determines how closely members of a population are likely to be related to each other. According to the theory of kin selection, selection can favor an allele that makes an individual perform altruistic behaviors costly to itself. If this behavior enhances the reproduction of other copies of the allele in the individual's relatives, the allele can spread through the population. Altruistic behavior is most likely to arise in species in which closely related individuals spend much time together. Parental behavior is the best example. Reciprocal altruism is also common.

10. Insect societies, in which many sterile individuals labor to raise the young of a single fertile female, apparently evolved because the genetic relationship among the society's members is closer than usual. The same is true of naked mole rats.

## Self-Quiz

1. An advantage of sexual reproduction over asexual reproduction is that it:
   a. increases the mutation rate
   b. increases genetic variability in a population
   c. produces larger offspring
   d. reduces the offsprings' risk of death during development
   e. gives organisms something to do on Saturday night

2. Some organisms reproduce asexually when environmental conditions are (favorable, unfavorable) and sexually when conditions are (favorable, unfavorable) for growth.

3. The main advantage of oogamy is that it:
   a. has separate male and female sexes
   b. provides for the nourishment of the developing zygote
   c. assures cross-fertilization
   d. involves two nonmotile gametes

4. It is generally true that males are an abundant resource and that they may be most evolutionarily successful by mating with as many females as possible. One possible exception to this generalization might occur when:
   a. there are many more females than males
   b. there are about equal numbers of males and females
   c. the father's care is required to raise the offspring
   d. there are many predators
   e. the male holds a territory against other males

5. Studies on female choice of mates in *Drosophila* showed that:
   a. more males than females reproduce
   b. females who reproduce leave more offspring per individual than do males who reproduce
   c. males who do not reproduce fail because they never court females

   d. males who do not reproduce fail because females will not accept them
   e. females who do not reproduce fail because males do not court them

6. When food is distributed in such a way that an animal must spend a large part of its day wandering from one place to another to find enough to eat, what type of mating system would you expect it to have?
   a. monogamy
   b. polyandry
   c. polygamy
   d. polygyny

7. Monogamy is most likely to be found among birds and mammals whose young are:
   a. born or hatched helpless and in need of much parental care
   b. born or hatched precocial (able to care for themselves, like a horse)
   c. nourished on milk
   d. part of a litter or clutch of more than five
   e. gestated inside the female's body from conception to birth

8. A bachelor human male is most likely to enhance his evolutionary success by altruistic behavior toward the children of:
   a. his sister
   b. his brother
   c. his mother
   d. his grandmother
   e. his niece

## Questions for Discussion

1. Does the desirability of genetic variations imply that monogamy in humans is a counter-evolutionary tendency? Why? (*Hint:* compare the average human family size with the number of possible genetically different eggs or sperm that a person could produce; for humans, with 23 pairs of gene-bearing chromosomes, each person could produce $2^{23}$ different kinds of reproductive cells.)

2. What are the selective advantages of courtship rituals? Do humans have courtship rituals? Is courtship in humans mutual or is it carried out predominantly by one sex? Why?

3. Female walruses must bear their young on land, but suitable stretches of beach are scarce. Similarly, nest sites for gulls are scarce. Both have crowded breeding grounds. Walruses eat molluscs (mussels, clams, and so on), whereas gulls will eat

almost anything—including the egg or chick next door. Explain what mating system you would expect each of these animals to show.

4. What is the advantage of hermaphrodism over separation of the sexes? What selective pressures might have led to evolution of species with separate sexes?

5. Why are there only two sexes in most sexually reproducing species?

6. Some individuals don't reciprocate a favor when it is their turn. What might be some possible outcomes if such a nonaltruistic allele were present in a population together with an altruistic allele?

7. Explain sibling rivalry for parental favors and parental impartiality toward offspring in terms of degree of relatedness.

8. Trivers and Hare weighed the fertile males and females produced in colonies of 20 species of ants and found that the investment in females was three times the investment in males (by mass). Can you explain how this situation may have been selected for?

9. How might the invention of birth-control methods and labor-saving household appliances affect the monogamous mating system of humans?

10. Martin Daly and Margo Wilson analyzed statistics on homicides within families in the United States, Canada, and elsewhere. Can you explain the genetic or evolutionary theory that might account for the following examples of their findings? How could we determine whether these behaviors in fact have a genetic basis?

   a. In nonindustrial societies, infants are most often killed by parents for the following reasons: (1) doubt that the child is the parent's own, (2) conviction that the child is weak and unlikely to produce offspring as an adult, and (3) external pressures such as food scarcity and burdensome demands from older siblings that reduce the baby's chance of surviving.

   b. In industrial societies, parents are more likely to kill infants than to kill older children.

   c. Mothers are more likely than fathers to kill infants.

   d. Disproportionate numbers of child killings are by stepparents.

## Suggested Readings

Beckoff, M., and M. C. Wells. "The social ecology of coyotes." *Scientific American,* April 1980.

Beehler, B. M. "The birds of paradise." *Scientific American,* December 1989.

Bell, G. *Sex and Death in Protozoa: The History of an Obsession.* New York: Cambridge University Press, 1988. The author applies modern models of the evolution of gene pools and ideas about the function of sex to answer a century-old question: does sexual reproduction rejuvenate protozoan cultures enfeebled by many generations of asexual propagation?

Campbell, B., ed. *Sexual Selection and the Descent of Man, 1871–1971.* Chicago: Aldine, 1972. A collection of essays by some prominent students of the topic.

Clutton-Brock, T. H. "Reproductive success in red deer." *Scientific American,* February 1985. A case study in a polygynous species illustrating many principles discussed in this chapter.

Cole, C. J. "Unisexual lizards." *Scientific American,* January 1984. An interesting study of parthenogenetic vertebrates, relating to many topics covered in the last several chapters.

Ford, B. "A family affair." *National Wildlife,* August/September 1984. How animals recognize their relatives.

Irion, R. "Killer siblings." *International Wildlife,* March/April 1994. A summary of field studies on many animal species.

Sherman, P. W., J. U. M. Jarvis, and S. H. Braude. "Naked mole rats." *Scientific American,* August 1992.

Thornhill, R. "Sexual selection in the black-tipped hanging-fly." *Scientific American,* June 1980. Observations of bizarre mating habits in this insect support Darwin's theory of sexual selection.

Westneat, D. F., and P. W. Sherman. "When monogamy isn't." *The Living Bird Quarterly,* Summer 1990.

Wilson, E. O. *Sociobiology: The New Synthesis.* Cambridge, MA: Belknap Press, 1975. The classic work on the subject of sociobiology (the biology and evolution of social behavior). It has stirred controversy because its author—famous for his studies of insect societies—treats humans just like other animals when considering the origin of behavior patterns. Good discussion of altruism.

# Origin of Life

## OBJECTIVES

*When you have studied this chapter, you should be able to:*

1. Define spontaneous generation and give two reasons why it is unlikely to occur under the conditions existing on Earth today.
2. Discuss the importance of time in the theory of the origin of life presented in this chapter.
3. Compare the environment in which life is believed to have arisen with the present environment on Earth.
4. Trace the steps by which life may have originated on Earth, from the formation of organic monomers through the rise of eukaryotic organisms.
5. State what evidence we have that these steps were possible, describing the experimental work of Miller and Fox.
6. List (a) kinds of places and (b) sources of energy on the prebiotic Earth that could have been important in the formation of organic molecules and prebiotic systems.
7. Explain why proteins and nucleic acids would have had to evolve together.
8. Describe the order and steps by which fermentation (glycolysis), respiration, and oxygen-producing photosynthesis may have evolved.
9. Compare fermentation and respiration, and describe their significance for the origin and evolution of life.
10. State the significance of autotrophy for the evolution of early life.
11. Describe changes in the environment that resulted from the presence of living organisms, and explain how the evolution of modern organisms depended on the change to modern environmental conditions.
12. State the role played by mass extinctions in the evolution of life.

---

I f we could trace the ancestry of living organisms, we should find a long line of cells stretching back billions of years. Each cell came from division of a previously existing cell . . . but where did the first cell come from?

Some people suggest that the first organisms came to Earth in spaceships or falling meteorites, but this only moves the question of how life began to a more distant arena, beyond our reach to study. Most scientists think that life on Earth started here.

We can't go back to the early Earth to see exactly what it was like and how life began, but we can make some educated guesses. We can also search present-day Earth for evidence of how and when life arose.

Geologists who study ancient rocks have found evidence that the early Earth differed greatly from what we see now—even if we were to subtract all the living things. In fact, conditions then would have killed the vast majority of modern organisms almost instantly. The atmosphere was toxic, the heat intense (Figure 21–1).

Yet this poisonous inferno gave rise to life: most scientists today believe that chance chemical events, spanning hundreds of millions of years, built the simple compounds of primordial Earth into organic monomers, then polymers, then aggregates of many polymers. By a few billion years ago, some of these aggregates had become cells. This scenario was first proposed in 1924 by a Russian, Alexander Oparin. In 1928 Oparin's ideas, not yet translated from the Russian, were echoed by J. B. S. Haldane in England.

Research supports this hypothesis. Scientists have simulated the **prebiotic** ("before life") world, using educated guesses about its temperature, chemical components, and energy sources. Amazingly, the nonliving systems formed in the laboratory show many features typical of life.

What properties would we expect of a chemical aggregate on its way to becoming a cell? In earlier chapters, we saw the basic features shared by all present-day cells. The forerunners of true cells should have shown at least hints of these:

1. A large number and variety of organic molecules, many of them joined into long polymer chains.

2. A lipid-protein membrane that separates the cell from its environment, selectively controls the exchange of materials between the two, and maintains a difference in the concentration of molecules and ions between them.

3. Proteins that help to exchange substances with the environment, form cell structures, and most important, catalyze the cell's metabolic reactions.

4. Nucleic acids that contain precise instructions for making proteins.

5. Ribosomes, the apparatus where proteins are made according to the nucleic acids' instructions.

In this chapter we consider what prebiotic Earth was like and outline how nonliving chemicals may have become organized, step by step, into living cells. This subject is highly speculative, perhaps more so than any other in the book. Research in this area is lively with ideas and arguments about how the features of life began. Future work will fill gaps in the picture, but we can never know whether it tells us what really happened.

**FIGURE 21–1** A challenging environment. This landscape of Venus shows what early Earth might have looked like. Venus is Earth's nearest neighbor and has about the same size, density, and physical composition. However, its position closer to the sun, and its dense "greenhouse" atmosphere (96.5% $CO_2$), keep the surface at an average 482 °C. Sulfurous clouds cover the surface. The photograph was prepared from imaging radar data obtained by the Magellan spacecraft mission, using a computer to simulate perspective and exaggerate topographic relief to about 20 times greater than the actual terrain. The color is based on color images recorded in the early 1980s by the Soviet *Venera 13* and *14* landers. (NASA)

## KEY CONCEPTS

♦ Chemical and physical conditions on early Earth differed from those we see today.

♦ Chemical reactions in the nonliving world sometimes produce organic chemicals. When living organisms are absent, these organic molecules can assemble into larger aggregates. On early Earth, some aggregates eventually arrived at the organization, function, and reproduction characteristic of life.

♦ Living organisms themselves are responsible for many of the differences between early and present-day environmental conditions on Earth.

## 21–A SPONTANEOUS GENERATION

For centuries people believed that living organisms are often produced from nonliving matter, a process called **spontaneous generation.** The Greek philosopher Aristotle wrote that frogs and insects were generated from moist soil, and people frequently observed that stagnant watering troughs produced worms and algae and that spoiled meat produced maggots.

Francesco Redi, an Italian, dealt the first important blow to this notion of spontaneous generation in 1668. Redi placed dead snakes and eels in glass jars and then covered some of the jars with fine muslin, leaving the others open to the air. Flies soon arrived and laid eggs on the dead animals in the open jars; the eggs hatched into maggots. No maggots appeared in the covered jars, but flies did settle on the muslin and lay eggs there. Similar experiments soon convinced most people that plants and animals come only from their parents' seeds or eggs and do not arise from nonliving things.

Meanwhile, in the late 1600s, the Dutch lens grinder Anton van Leeuwenhoek built a microscope and discovered tiny organisms, invisible to the unaided eye, living in ponds, ditches, soil, and many other places—even in his own body. He proposed that these, too, came from the reproduction of others of their kind.

The Abbé Lazzaro Spallanzani agreed with van Leeuwenhoek. In the eighteenth century, he set out to show that microorganisms such as bacteria and fungi do not arise spontaneously. He prepared flasks of nutrient gravy, sealed and heated them, and found that most of them remained free from living microorganisms. When he later broke the seals and allowed air to enter, microorganisms soon appeared. From these experiments, Spallanzani contended that microorganisms could not arise spontaneously. His opponents claimed that heating the flasks had destroyed "vital mole-

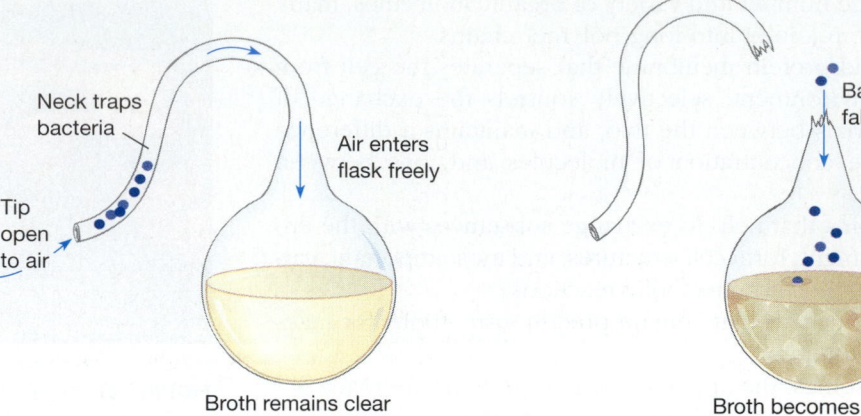

**FIGURE 21–2** Pasteur's swan-necked flasks. Air entered the flask's open tip freely, but not fast enough to carry bacteria through the curved neck along with it. Bacteria were trapped at the bottom of the curve, while air continued on into the flask. Only if the neck was broken off could bacteria enter the flask and putrify the nutrient broth.

Neck traps bacteria

Tip open to air

Air enters flask freely

Broth remains clear

Bacteria fall in

Broth becomes cloudy

cules" that float around in the air until they enter matter, giving it life. Since Spallanzani's sealed, heated flasks sometimes came out teeming with microorganisms, the issue remained in doubt. (We now know that some bacteria make resting spores that survive heating and grow afterward, accounting for Spallanzani's erratic results.)

The theory of spontaneous generation of microorganisms was finally laid to rest in the mid-nineteenth century by the experiments of Louis Pasteur in France and John Tyndall in England. They showed that bacteria travel through the air and that if air is purified before it enters a flask of sterilized broth, no bacteria will appear in the broth. Pasteur drew the necks of his glass flasks out into S-shaped curves. Air could enter a flask, but it did not travel quickly enough to carry bacteria along with it. Any cells in the air were trapped at the bottom of the curve (Figure 21–2).

Tyndall sterilized the air entering his flasks by passing it through a flame or through absorbent cotton. These treatments, too, removed bacteria from the air and kept the broth clear. By the late 1870s, all but a few diehards agreed that living organisms, of whatever size, come from reproduction of previously existing organisms. This brings us back to the question: where did the first living organisms come from?

## 21–B  CONDITIONS FOR THE ORIGIN OF LIFE

Louis Pasteur is often credited with disproving the idea of spontaneous generation. But Pasteur himself once remarked that his fruitless 20-year search for spontaneous generation did not convince him that it was impossible. What Pasteur showed was that life did not arise in his flasks under the conditions he used (sterilized broth, clean air) in the time he waited. He did not show that life could *never* arise from nonliving matter under *any* set of conditions.

Indeed, we now believe that life did arise from nonliving matter, but under conditions very different from those we see around us today, and over a span of hundreds of millions of years. Many scientists see the origin of life as

an inevitable stage in the evolution of matter and believe that it has probably happened time and again in many parts of the universe where conditions were suitable.

Under what conditions can life arise? There are four basic requirements: appropriate chemicals, including liquid water, various inorganic ions, and organic compounds; an energy source; little or no oxygen gas ($O_2$); and long stretches of time.

Of the necessary chemicals, liquid water is abundant on Earth—in fact, this is the only place in the universe we're sure has liquid water. The inorganic substances occur in rocks, volcanic gases, and the atmosphere. The four most common elements in organic molecules (hydrogen, carbon, oxygen, and nitrogen) are among the six most abundant in the cosmos, and they are very abundant in Earth's air and oceans. Phosphorus and sulfur rank among the ten most common elements in Earth's crust.

How were organic molecules produced from these simple chemicals without the enzymes of living organisms, which catalyze most of these reactions today? Before we answer this question, let us look at the last two conditions: time and scarcity of $O_2$.

*Time*  It may take millions of years for a given quantity of some chemical to undergo a reaction that an enzyme could catalyze in a second or two. In the prebiotic era, chemicals must have reacted extremely slowly, sometimes speeded a bit by natural energy sources and by catalysts such as metal ions. Once simple organic chemicals had formed, they must have come together in ever larger, more complex structures. Scientists suppose that life evolved in a series of steps, none of which was wildly unlikely. However, the chances of *all* of them happening in the correct sequence are minuscule. (Remember, all their individual probabilities must be multiplied together [Section 15–H].)

Given enough time, however, even very improbable events are bound to occur. For example, if the probability that an event will occur in a year is one in a thousand, the probability that it will *not* occur is 0.999; the probability that

**TABLE 21-1**

**Probability That an Event Will Not Happen**

| Given this probability: | in 1 year | 0.999 |
|---|---|---|
| then: | in 2 years | 0.998 |
| | in 3 years | 0.997 |
| | in 4 years | 0.996 |
| | in 1024 years | 0.359 |
| | in 2048 years | 0.129 |
| | in 4096 years | 0.017 |
| | in 8128 years | 0.000276 |

it will not happen in two years is $(0.999)^2$; in three years, $(0.999)^3$, and so on. Table 21-1 shows that there is a very small probability that the event will not happen at least once in 8128 years. Conversely, there is a very high probability (0.9997) that it *will* happen *at least once*—and once may have been enough for the origin of life on Earth.

However improbable the events required for the origin of life, they had plenty of time to happen. Geological evidence dates Earth's formation at more than 4.5 billion years ago, and signs of life appear in rocks laid down a billion years later (that is, about 3.5 billion years ago). So, unlikely as living systems are, they had so much time to evolve that their origin was probably inevitable!

***Absence of Molecular Oxygen*** The origin of life almost certainly required an atmosphere with little or no oxygen gas because $O_2$ is a powerful oxidizing agent that breaks down organic molecules. So organic molecules exposed to $O_2$ on early Earth would not have lasted long enough to form more complex structures. This is one reason why spontaneous generation from organic matter does not occur on Earth today. (Another is that free organic molecules are usually absorbed and used as food by bacteria and fungi even before oxygen can damage them.)

There is geological evidence that Earth's atmosphere originally contained much less $O_2$ than it does now. This is supported by evolutionary trees based on comparative biochemistry of living bacteria. Anaerobic bacteria, which can live without $O_2$, evolved before oxygen-requiring aerobic forms. However, small amounts of $O_2$ may have been present on early Earth.

## 21-C PREBIOTIC EARTH

Astronomers think Earth formed by solidification and accretion of matter from space a little over 4.5 billion years ago. Chaos prevailed as meteorites bombarded the forming planet, adding more material. About 4.3 billion years ago, conditions began to stabilize.

Scientists once thought early Earth had a **reducing atmosphere**, made up of gases containing the reducing agent hydrogen: $H_2$, $H_2O$ (water vapor), $NH_3$ (ammonia), and $CH_4$ (methane). Hydrogen is enormously abundant in the solar system, especially in the gases that formed the sun and planets; even today the atmospheres of Jupiter and Saturn are mostly molecular hydrogen, water vapor, and ammonia. The composition of gases now emerging from volcanoes also supports the idea of a primitive reducing atmosphere.

However, many scientists now contend that $H_2$, being so light, escaped into space before or shortly after Earth formed. Sunlight, much brighter on Earth than on the outer planets, decomposed ammonia into $H_2$ (which also escaped) and nitrogen gas ($N_2$). Similarly, methane was replaced by carbon dioxide ($CO_2$).

By the time life was evolving, the atmosphere was only mildly reducing. The most abundant gas was probably water vapor, which later condensed to form the oceans. Next were carbon dioxide and carbon monoxide (CO), which later became locked in rocks in the form of carbonates, for example, in limestone, where they are today. Also present was $N_2$, which makes up about 80% of the air today. Sunlight striking water or carbon dioxide released small amounts of $O_2$, but this soon reacted with iron in the oceans or rocks. Only when all the iron was oxidized did $O_2$ start to accumulate in the air. The oxidizing atmosphere we breathe today is about one fifth $O_2$, produced almost exclusively by the photosynthesis of green plants.

We don't know enough about early Earth to give a specific recipe of ingredients for life. However, the early atmosphere, oceans, and rocks must have provided a variety of chemical environments where different mixtures of molecules could react. Researchers have tested many possibilities in attempts to produce organic monomers **abiotically** ("without life": neither within living cells nor under their influence).

## 21-D PRODUCTION OF ORGANIC MONOMERS

In 1953 Stanley Miller, then a graduate student, built a small-scale model of conditions on early Earth, including an "ocean" and a primitive reducing "atmosphere" (Figure 21-3). Electrodes in the "atmosphere" chamber produced electrical discharges, representing lightning, a possible source of energy to drive chemical reactions on early Earth. After a week, the "ocean" contained many small organic compounds, including amino acids, formed in the "atmosphere" and carried down by the condensation of "rain."

Miller's amino acids surprised and excited other scientists. Amino acids are the building blocks of proteins, and biochemists had recently recognized the enormous variety of proteins and their importance in the activities of living

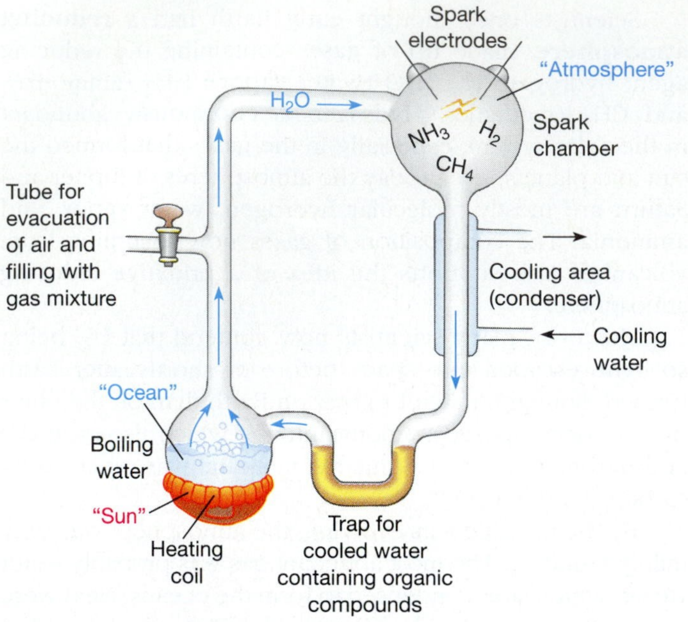

**FIGURE 21–3** Miller's apparatus for simulating prebiotic conditions. The "atmosphere" was a mixture of hydrogen gas ($H_2$), methane ($CH_4$), and ammonia ($NH_3$) in a glass chamber the size of a soccer ball. Sparks from electrodes represented lightning, a source of energy. The "ocean" was boiled to produce water vapor, which could provide oxygen to the reaction and also carry organic molecules back to the sea in "rainfall."

cells. Indeed, proteins were regarded as *the* class of substances necessary for life.

Miller performed this simulation when the prebiotic atmosphere was thought to have been highly reducing. Many people have done variations on his procedure, using different guesses about "atmosphere" gas mixtures, and sometimes including hydrogen sulfide ($H_2S$), carbon dioxide, or inorganic salts. Energy sources used, besides electric sparks, include heat, bright sunlight, ultraviolet light, and radioactivity, all possible sources of energy on prebiotic Earth. As long as $O_2$ is excluded, these simulations yield a variety of organic molecules. Gas mixtures based on recent ideas about the prebiotic atmosphere yield less of some substances, but sometimes more amino acids.

Astronomy and geology also provide evidence that organic monomers can form without the agency of living organisms. Several molecules that are raw materials or intermediates in the synthesis of organic monomers occur in stars, dust clouds, space, and the atmospheres of other planets. The European Space Agency's *Giotto* spacecraft, which flew past Comet Halley in 1986, even detected polymers of formaldehyde, the precursor of sugars, aldehydes, alcohols, and acids. Some meteorites, chunks of matter that fall from space, contain a wide variety of more complex organic compounds (Table 21–2). Small amounts of six common amino acids were also found in material brought back

from the moon. Even now, small amounts of organic compounds are formed on Earth abiotically when hot metallic carbides in volcanic gases and lava react with water, forming hydrocarbons.

All this evidence supports the claim that organic compounds could have formed on prebiotic Earth by the action of available forms of energy. Without oxygen to destroy them or organisms to absorb them, these compounds would have accumulated. Haldane suggested that eventually the sea had the composition of a "hot, dilute soup."

Today, many scientists favor another view: organic molecules formed not a soup, but a sludge, concentrated because of their attraction to the complex, charged surfaces of clay or other minerals in the pores of soil or underwater rocks (Figure 21–4). These minerals could serve both as templates holding organic molecules in place and as catalysts of chemical reactions. Indeed, the chemical industry today uses mineral catalysts to produce many organic compounds. Many enzymes also use mineral ions as cofactors, thought to be a holdover from times when the ions themselves were the catalysts of the reactions.

## 21–E    FORMATION OF POLYMERS

In the next step of chemical evolution, some organic monomers polymerized into larger molecules. These condensation reactions, too, required energy. In addition, there must be little water present because water is a raw material for the much faster reverse reaction:

$$\text{monomers} \underset{+H_2O}{\overset{-H_2O}{\rightleftarrows}} \text{polymers}$$

Sidney Fox found that heating a dry mixture of amino acids to about 60 °C produced **proteinoids,** protein-like molecules of about 100 amino acid residues each. The heat quickly evaporated water released as the amino acids linked together, preventing the proteinoids from hydrolyzing back into amino acids. Such events might have happened on early Earth as soil or tide pools containing amino acid solutions were dried up by the sun. Clay and other minerals would have enhanced this polymerization. Many kinds of monomers adhere to the surfaces of mineral particles, increasing their local concentration. Polymers form readily when the minerals are dried and then warmed. The alkaline pH of the early ocean would have boosted polymerization of sugars.

Perhaps organic polymers formed without any of these special conditions. The "primitive atmosphere" gases used to make monomers also form aldehydes and hydrogen cyanide ($HC \equiv N$), which can combine and form a variety of organic compounds. Hydrogen cyanide can also polymerize and then react with water, forming short peptides

## TABLE 21–2
### Chemicals Important to the Origin of Life

| Stars, Dust Clouds, Interstellar Space, Other Planets' Atmospheres | Carbonaceous Chondrites (Meteorites) | Laboratory Simulations of Prebiotic Earth |
|---|---|---|
| Ammonia ($NH_3$) | Hydrocarbons | 18 Amino acids used in proteins (no histidine or arginine) |
| Hydrogen cyanide ($HC \equiv N$) | Amino acids | Formaldehyde, other aldehydes |
| Cyanogen ($N \equiv C—C \equiv N$) | Alcohols | Alcohols |
| Methane ($CH_4$) | Sugars | Organic acids |
| Ethane ($H_3C—CH_3$) | Nitrogenous bases | Nucleotide bases |
| Ethylene ($H_2C = CH_2$) | Chains and rings containing sulfur | Phospholipids (able to form vesicles) |
| Acetylene ($HC \equiv CH$) | Porphyrin derivatives (related to chlorophyll ring) | |
| Carbon monoxide ($CO$) | Pristane and phytane (related to chlorophyll tail) | |
| Formaldehyde ($H_2CO$) | Lipid-like molecules that self-assemble into fluid-enclosing films | |
| Ethanol ($CH_3CH_2OH$) | | |

directly, rather than by making individual amino acids and then linking them.

In solution, short peptide chains tend to hydrolyze into amino acids. However, longer chains remain intact, stabilized by interactions between different parts of the molecule in a manner similar to secondary and tertiary structure in proteins.

Some proteinoids exhibit enzyme-like properties. For example, they can catalyze some chemical reactions, and their catalytic properties can be destroyed by overheating and by chemicals that inhibit enzymes.

We should not be too surprised that random strings of amino acids act as catalysts. A protein's catalytic function depends largely on its folding pattern, which in turn depends

(a) Shallow, mineral-rich hot spring     (b) A concentrated sludge

**FIGURE 21–4** Concentration of organic molecules. (a) Shallow depressions, such as tide pools or these hot springs terraces, can quickly accumulate minerals and organic solutes brought in by rain or seepage. The heated water evaporates, leaving a more concentrated solution, or even a dry crust at the edges. (b) Clay and other minerals have irregular, electrically charged surfaces. Dissolved ions or polar molecules become concentrated at the interface between the minerals and the surrounding solution. Minerals not only hold other molecules, but also act as catalysts for chemical reactions between them.

on its amino acid sequence. However, many sequences fold into the same general shape. For example, the "same" proteins (such as cytochrome *c*) from different organisms vary in amino acid sequence yet fold into similar shapes all performing the same function. A large population of random proteinoids is bound to contain some with at least weakly catalytic shapes.

The conclusion must be that molecules similar to enzymes, which are so vital to life, could have been produced on Earth before living organisms existed.

## 21–F  FORMATION OF AGGREGATES

When proteinoids or natural polymers are mixed in water, they aggregate and form larger structures (Figure 21–5). Any lipids present assemble themselves into membrane-like coatings around the outside. This helps keep the whole collection of molecules together as a unit, which is set apart from its surroundings by distinct boundaries.

This structure sounds something like a cell, and aggregates also perform some cell-like functions. Aggregates are selectively permeable and accumulate some kinds of monomers, but not others, from the surrounding medium. They also catalyze certain reactions in their interiors, often enhancing polymerization of monomers, and they become larger, eventually breaking up into smaller aggregates.

Not surprisingly, aggregates have been proposed as precursors of the first living cells. Units such as these are prerequisites for the evolution of life because selection operates by "choosing" among discrete units that can replicate

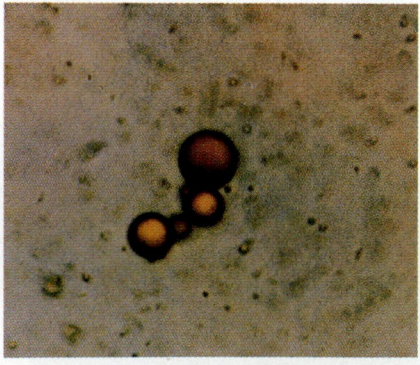

**FIGURE 21–5** Proteinoid microspheres. These microspheres formed when water was added to proteinoids. Each is 1 to 2 micrometers in diameter. The microspheres have a double layer of proteins as a boundary between the interior and the external environment. This boundary is selectively permeable and admits polynucleotides (in theory, the precursors of nucleic acids) very readily. Like living cells, microspheres have internal structure: watery areas, lipid-like areas, and boundary-layer areas provide sites for distinct chemical activities. (SEM by Steven Brooke; color copy arranged by Richard Le Duc)

or at least persist, and it seems likely that the units must have been combinations of interacting molecules with life-like properties, rather than individual polymers.

However, a vast gulf separates such simple molecular aggregates from living cells. The aggregates have extremely limited function in their "membranes" and "metabolism," and they cannot reliably reproduce any useful molecules they contain.

Oparin wrote: "The path followed by nature from the original systems of aggregates to the most primitive bacteria . . . was not in the least shorter or simpler than the path from the amoeba to man." (Poetic license: this does not necessarily mean that Oparin viewed amoebas as direct ancestors of humans.)

What were these aggregates like? And how did at least one attain living status by evolving interdependent metabolic and genetic systems that perpetuated each other?

We have little good evidence to fill this blank in our picture of the origin of life. No laboratory-made aggregate seems capable of evolving further. Even very simple pre-living aggregates must have been too complex to be studied in a controlled way, even if we knew the ingredients to make them. Instead, researchers break the problem into pieces they *can* investigate in the laboratory, such as replication of nucleic acids or the basis of the triplet codes for amino acids.

## 21–G  BEGINNINGS OF METABOLISM

Aggregates that have been studied in the laboratory have a rudimentary metabolism of just a few reactions, a far cry from the thousands in a living cell, with their speed, specificity, and feedback regulation. Of course, these aggregates also lack the mechanisms for protein synthesis and replication of genetic material that permit cells to make many copies of their metabolic enzymes and other proteins.

Aggregates arrived at the threshold of life by **chemical selection,** a process similar to natural selection but acting on nonliving systems. At first, chemical selection probably favored mere longevity: aggregates with the most stable combinations of chemicals outlasted their less stable neighbors. When they broke up, their components might recombine with debris from other aggregates, forming new units that would undergo the test of time.

Selection would have favored aggregates with catalysts for chemical reactions that made them more stable. At first, the organic soup probably contained abundant substrates for these reactions, compared with the few aggregates using them. However, as these successful, stable aggregates grew and fragmented, and as new ones assembled, the raw materials became scarcer. For example, suppose crucial molecule A became rare as more and more aggregates took it up faster than abiotic forces produced it. Any aggregate with a catalyst that converted abundant molecule B to now-scarce A would have an advantage—until B too became scarce. Now an aggregate that evolved a catalyst for the reaction C——→B as well as B——→A would survive, while B-users would disappear, and so on. It is supposed that some metabolic pathways became ever longer in this way, evolving "backwards," from useful end products to less directly useful raw materials.

Aggregates (or cells) that could catalyze the most and fastest useful reactions would have accumulated molecules from the broth, grown, and fragmented (or reproduced) faster than their competitors.

## Origin of Energy Metabolism

One of a metabolic system's first requirements is a source of energy to drive various reactions. Today's organisms have two universal energy sources: ATP and other nucleotides, and membrane potentials based on concentration gradients of ions (Section 6–E). Both probably date from prebiotic times.

All organisms today have hydrogen ion ($H^+$) pumps housed in membranes. Perhaps similar pumps provided energy that aggregates or primitive cells used to transport materials through their membranes. Such pumps might have been driven by light, an abundant energy source used by simple $H^+$ pumps in some bacteria today. The electron transport systems of respiration and photosynthesis include $H^+$ pumps as part of more elaborate energy-trapping pathways.

Early cells probably obtained ATP or its components from the primordial soup. Adenine is the simplest of the nucleotide bases and the easiest to form abiotically, from hydrogen cyanide and water. Adding a ribose molecule and three phosphate groups at the proper positions would produce ATP. The nucleotides GTP, CTP, and UTP might also have been useful energy donors (as they are in modern organisms' metabolism), but abiotic processes probably made smaller amounts of them.

Thus, ATP was probably a relatively common and available molecule that hydrolyzed easily to yield energy. ATP has been found to promote some kinds of reactions in laboratory aggregates. As ATP in the external soup ran low, selection favored cells that could use energy from their $H^+$ membrane potentials, or from metabolic reactions, to make ATP.

The need for an ATP-regenerating system is thought to have selected for anaerobic pathways of fermentation. Glycolysis is extremely ancient, judging by the fact that nearly all modern organisms use this pathway to ferment glucose to pyruvate. The oxidative phosphorylation found in aerobic organisms provides a highly efficient way to extract further energy from the products of fermentation. This aerobic respiration is thought to have arisen relatively late in the evolution of metabolism.

## 21–H   ORIGIN OF GENETIC INFORMATION

Self-organized aggregates of polymers have some similarities to modern cells, but they cannot be called "living" because they cannot reproduce. True, they may split into smaller aggregates, but these may end up lacking some molecule(s) crucial to the success of the parent's rudimentary metabolism.

In contrast, modern cells have genetic information, in the form of DNA, containing the instructions for making all their RNA and proteins. When they reproduce, they pass a set of this information to each offspring. We can diagram this information flow in modern organisms as:

$$DNA \xrightarrow{\text{transcription}} RNA \xrightarrow{\text{translation}} protein$$
replication

In this remarkable system, DNA, RNA, and proteins show division of labor. DNA, a very stable molecule, serves as a file copy of the genetic information required to make many copies of RNAs. It also spends some time being replicated before cell division. RNA plays many roles in protein synthesis. Proteins, in turn, perform the actual work of the cell's metabolism, exchanging substances with the environment, catalyzing metabolic reactions, and even assembling nucleic acids.

When we ask how such a complex system could have evolved, we find a problem. In the synthesis of nucleic acids, the nucleotide monomers are joined together by enzymes. However, these enzymes must first be made according to instructions provided by already-existing nucleic acids! Which came first, the enzymes or the nucleic acids?

Researchers have explored several possible answers:
1. **Proteins first.** Amino acids form and polymerize much more easily than nucleotides do under prebiotic conditions. Furthermore, some of the resulting proteinoids would catalyze reactions leading closer to life (perhaps including making and polymerizing nucleotides). This was the first avenue explored, but most workers now regard it as a dead end. Its main problem is that proteinoids show no sign of self-replication and so cannot multiply and evolve into more lifelike forms.
2. **RNA first.** In the 1980s, another possible answer was suggested by the discovery that RNA can not only carry genetic information but also act as a catalyst. (Thomas R. Cech and Sidney Altman received a 1989 Nobel Prize for this work.) Experiments show that RNA polymers can form without enzymes. In addition, short RNA strands complementary to existing RNA polymers form in the presence of zinc ions (which are part of all modern RNA polymerase enzymes). RNA can also catalyze RNA splicing and the joining of short RNA chains into longer ones. This and other evidence suggests that the

first genetic information was RNA and that it could replicate without aid from proteins.

The "RNA world" model of early life proposes that RNA was the genetic material of the first cells. It replicated itself and catalyzed the many reactions of a complex metabolism. Gradually, segments of catalytic RNA molecules were replaced by more efficient protein strands. Today, RNA catalyzes only a few kinds of reactions; the vast majority are catalyzed by proteins.

This scenario has three problems. First, RNA nucleotide monomers do not form readily under abiotic conditions. Second, both polymerization and replication of RNA require highly purified solutions and precise conditions most unlikely to have existed on prebiotic Earth. Third, RNA apparently catalyzes a much narrower range of reactions than proteins and proteinoids do. Although some workers are searching for ways around these problems, with modest progress, some prefer a third alternative.

3. **Coevolution of RNA and proteins.** Perhaps some early aggregates contained not only RNA but also proteinoid catalysts that helped the aggregate survive and helped the RNA reproduce. The RNA and proteinoids would inevitably interact to some extent. As time went on, chemical selection would preserve, not favorable proteinoids nor favorable RNAs alone, but rather those aggregates in which the two cooperated and increased the aggregate's longevity.

Whether or not it began this way, the coevolution of proteins and nucleic acids has been going on for billions of years. An early and crucial part of this process was forging the intimate link between the two: the genetic code.

## Origin of the Genetic Code

The genetic code may have begun with the tendency of certain amino acids and nucleotides to associate with each other, as experiments have shown. If this helped amino acids or nucleotides join together, then their order in the new polymers was not random but influenced by their companions. This implies that the genetic code is not random but based on slight chemical affinities.

That said, the origin of the genetic code and its translation to proteins is still a weak area in our understanding of the origin of life. Probably we shall learn more with further research on transfer RNA and ribosomes, the sites where nucleic acids and amino acids interact in modern protein synthesis.

Even a rudimentary system of protein synthesis would have helped a prebiotic aggregate greatly because it could produce at least rough copies of proteins that catalyzed favorable reactions. At first, replication of nucleic acids was probably quite inaccurate. Frequent changes in the order of nucleotides would have produced new amino acid sequences in proteins, some of them improved catalysts. Indeed, there would be little selection for accuracy in replication and translation until unfavorable mistakes began to greatly outnumber favorable ones.

The coevolution of nucleic acids and proteins must have involved a complex feedback cycle of many cooperating primitive "genes" and "enzymes."

It now seems likely that the first genetic information was RNA, as it is in some viruses, not the DNA used by all modern cells. The 2'-hydroxyl group of the sugar ribose in RNA's ribonucleotides can serve as a "handle" holding amino acids in place until they polymerize, but the sugar deoxyribose in DNA lacks this group:

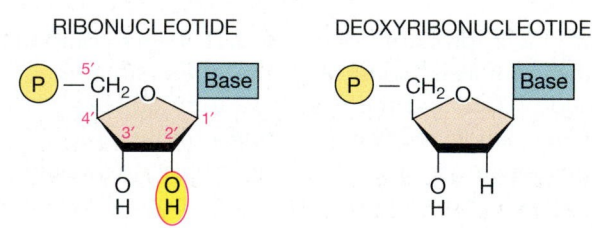

RIBONUCLEOTIDE          DEOXYRIBONUCLEOTIDE

Modern cells make deoxyribonucleotides by removing this hydroxyl group from ribonucleotides, and ribonucleotides

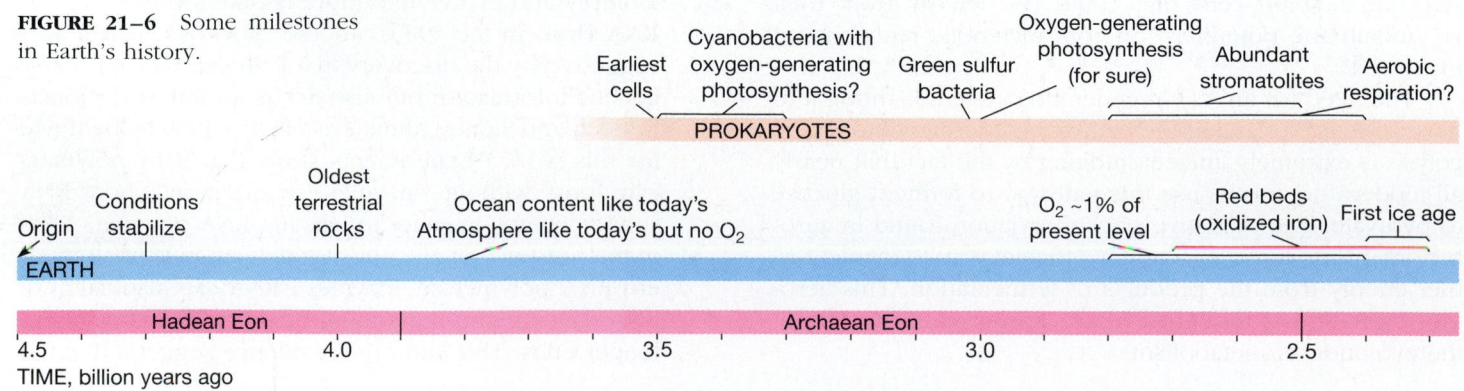

**FIGURE 21–6** Some milestones in Earth's history.

may also have come first in evolution. Furthermore, some viruses have an enzyme that uses RNA as a template for DNA synthesis, rather than vice versa. All this evidence suggests that DNA was added onto the front end of an evolving RNA-protein synthesis system.

DNA's double-stranded structure gave it two advantages over RNA as genetic material: DNA is more stable than RNA, and it is easier to copy faithfully. RNA molecules remained as the "go-betweens" that carry information and instructions from DNA to protein synthesis.

Reproduction and metabolism must have evolved together, although we have discussed them separately. Reproduction relies on metabolism to provide the energy and raw materials required to replicate the genetic information and to produce many copies of proteins. Metabolism, in turn, relies on genetic information to direct the production of all the enzyme catalysts needed to make these raw materials.

Once an aggregate evolved reproducible protein synthesis, gene replication, and division into two parts with equivalent sets of genes, it qualified as a living cell, however crude. The story of the origin of life was complete. Note, however, that our survey of these events has been mostly hypothesis, based on modern molecular biology, and so far has little experimental support. Is this because researchers haven't yet found the right set of conditions or because they are starting with raw materials that are already too complex? Some feel research needs to begin with simpler self-replicating catalytic substances, perhaps small or-

ganic molecules or even inorganic crystals. These could have attracted RNA and protein helpers, which eventually became so proficient that the original replicators are now lost to us.

## 21-I HETEROTROPHS AND AUTOTROPHS

The advent of the first true organisms marked the end of the era of chemical selection and the beginning of the era of natural selection. Survival was no longer enough; competition grew more intense, and primitive cells evolved better ways to get energy and convert it into offspring.

Living organisms have two main modes of nutrition. **Heterotrophs** feed on organic molecules made outside their own bodies, whereas **autotrophs** make their own food from inorganic molecules. The most ancient fossils known are anaerobic bacteria, including what we think are both heterotrophs and autotrophs (Figure 21-6). Assuming the primordial soup existed, we might expect that the first organisms were heterotrophs, feeding on the soup, on also-ran aggregates, and on dead neighbors.

Autotrophy freed the living world from dependence on the slow production of food by random abiotic reactions. An autotroph's organized metabolic assembly line made food much faster, and in greater abundance, supporting the cell's growth and reproduction. Inevitably, some autotrophs also provided a new food supply to their heterotrophic neighbors, whose populations also grew. Hence, the evo-

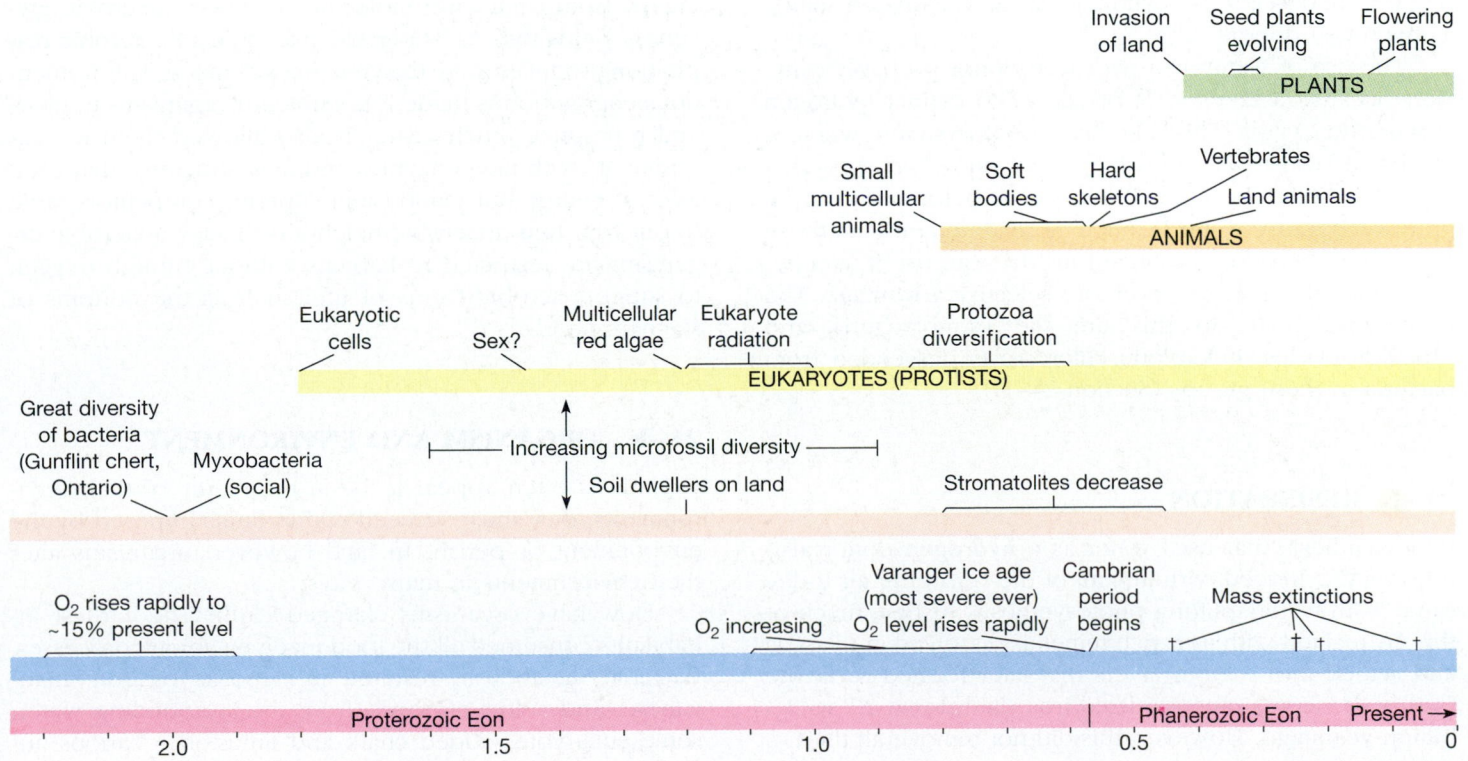

lution of autotrophy vastly increased the amount of life Earth could support.

The most familiar autotrophs today are green plants, but their photosynthesis is a very advanced form of autotrophy. Other, less successful evolutionary "experiments" in do-it-yourself food production are still found among some modern bacteria. **Chemosynthetic (chemoautotrophic)** bacteria use energy released by various inorganic chemical oxidation reactions to fix $CO_2$ into organic compounds, even in the absence of light.

Various kinds of bacteria also have a range of ability to trap light energy. By studying these bacteria, we can outline the possible steps in the evolution of photosynthesis. At first, light energy was probably used to create an $H^+$ membrane potential, which supplied energy to transport substances through the membrane and to make ATP. Today, photoheterotrophs use such gradients to meet some of their energy needs, but they must also take in food molecules from their surroundings (*Essay: The Solar-Powered Purple Proton Pump,* Chapter 6). Some of these bacteria can also fix $CO_2$, and the next step would have been using light energy to obtain hydrogen atoms, which could then reduce $CO_2$ as it was fixed—the hallmark of true autotrophic photosynthesis.

Like some photosynthetic bacteria today, early photosynthesizers probably extracted hydrogen from easy donors, such as their own organic waste products, hydrogen gas ($H_2$), or hydrogen sulfide ($H_2S$). The purple nonsulfur bacteria (Athiorhodaceae) transfer hydrogen from the end products of glycolysis to carbon dioxide, thus using waste products as energy sources. Purple and green sulfur bacteria (Thiorhodaceae) use hydrogen gas and hydrogen sulfide as hydrogen donors.

The use of water as a hydrogen donor probably came later because it takes a lot of energy to extract hydrogen atoms from water. This takes two photosystems, whereas the bacteria just discussed have only one (Photosystem I for green sulfur bacteria, Photosystem II for both purple groups). Because water is much more abundant than those other hydrogen donors, organisms that can use it as a hydrogen source have an enormous selective advantage. The prokaryotes that do this are the cyanobacteria and prochlorophytes. Eukaryotic chloroplasts descended from bacteria in these groups (Section 24–H).

## 21–J   RESPIRATION

Photosynthesis that uses water as a hydrogen donor also releases $O_2$. Indeed, virtually all of the $O_2$ in the air today came from water-splitting photosynthesis. At first, much of this $O_2$ reacted with iron-rich minerals dissolved in the seas and settled into bottom layers that later formed rock, the "red beds," containing oxidized iron, laid down 2.7 to 2.3 billion years ago. However, this did not remove all the $O_2$.

Oxygen accumulating in the atmosphere produced an environmental crisis because $O_2$ destroys flavins, important coenzymes in biochemical pathways. Without their flavins organisms could not survive, and many probably perished. However, some retreated to anaerobic habitats, the kinds of places where their descendants live today. For still others, oxygen poisoning was probably the selective pressure that brought about the next important evolutionary advance, aerobic respiration.

Respiration is the oxidation of organic molecules using inorganic substances as final electron acceptors. Before there was any $O_2$ to speak of, some bacteria had evolved the metabolic pathways of anaerobic respiration. They used nitrate ($NO_3$) or the sulfate released by photosynthetic green sulfur bacteria as the final electron acceptor. Aerobic respiration, using $O_2$ as the final electron acceptor, probably evolved before $O_2$ reached 1% of its present atmospheric concentration.

Studies of living bacteria show that all the molecules used in respiration probably already existed in ancient anaerobic bacteria but played different roles. For example, some modern autotrophic bacteria use the citric acid cycle to assimilate and reduce carbon dioxide. The electron transport molecules of respiration probably evolved from the homologous ones of photosynthesis. In respiration, both these pathways run backward from their original direction. In addition, they are hooked up in new ways, and some molecules have been modified or added. Emerging aerobic bacteria had more than a billion years to perfect these changes as $O_2$ slowly accumulated to modern levels.

Respiration permitted organisms to release much more energy from each food molecule. Whereas anaerobic glycolysis yields two ATPs per glucose molecule, aerobic respiration produces more than ten times as much. This tremendous energy bonus made it possible for organisms to grow and reproduce much faster. It also allowed them to "experiment" with new enzymes and new structures that used a lot of energy but made them superior competitors, able to outstrip their anaerobic neighbors. Today anaerobic organisms are restricted to habitats without enough oxygen to support aerobic forms of life, such as the bottoms of stagnant ponds.

## 21–K   ORGANISM AND ENVIRONMENT

Organisms often appear to be at the mercy of their environments: they must succeed within limits imposed by the environment or perish. In fact, however, organisms alter their environments in many ways.

How have organisms changed Earth? The earliest inhabitants consumed all the food made by abiotic processes. Bacterial metabolism resulted in deposits of iron, manganese, and sulfur removed from the surrounding water; some eukaryotes added chalk and limestone. Various au-

# ESSAY: *The Gaia Hypothesis*

Earth's atmosphere is very different from those of neighboring planets Venus and Mars, where $CO_2$ is the main gas and there is hardly any $O_2$ or water (Figure 21–A). The reason for the difference is that Earth's living inhabitants have altered their surroundings in many ways.

Earth's peculiarities convert Earth into a suitable home for living organisms, quite unlike the hot, waterless, and suffocating environment of Venus and the cold, nearly waterless, and equally suffocating environment of Mars. This fact led James Lovelock to suggest that Earth's organisms actively adjust physical conditions in the planet's surface and lower atmosphere in such a way as to keep them suitable for life. This suggestion is known as the Gaia hypothesis, after the Greek goddess of the earth.

For instance, temperatures on Earth appear to have remained relatively constant for the past 3.5 billion years, although the sun has warmed up during that time. Why has the temperature on Earth not increased correspondingly? Probably because the level of $CO_2$ in the atmosphere has fallen. $CO_2$ is a "greenhouse gas" that traps heat near Earth's surface, preventing it from escaping into space. The Gaia hypothesis proposes that as the solar system warms, marine organisms convert more $CO_2$ into cal-

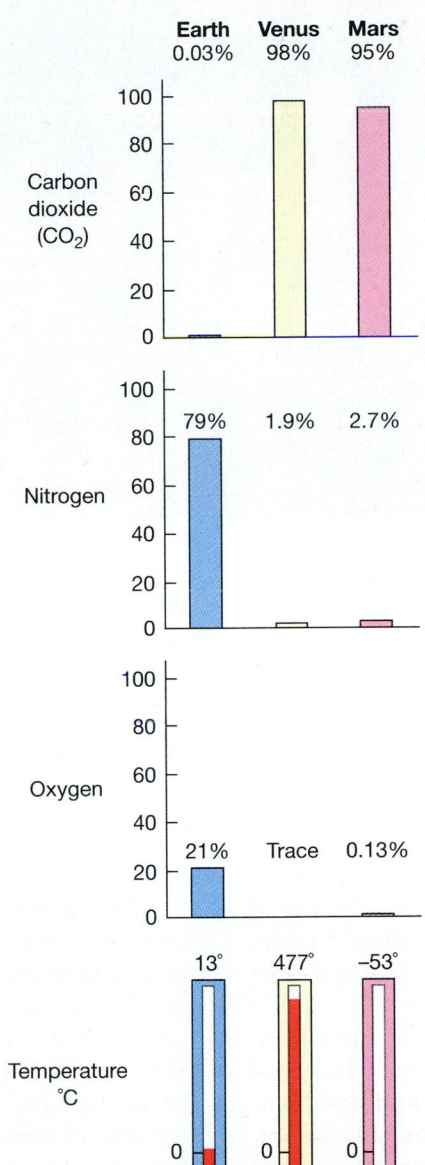

cium carbonate shells, which sink to the sea floor, removing $CO_2$ from the atmosphere. This is a feedback loop, in which life responds to changing conditions by actions that bring them back toward their original values.

The Gaia hypothesis is not generally accepted by scientists, partly because it is so difficult to test. The theory about $CO_2$, for instance, predicts that if the temperature on Earth falls, organisms will fix less $CO_2$ in the ocean, thus raising the temperature, but there is no way to test this.

It is also difficult to imagine possible mechanisms for Gaia. Lynn Margulis and Lovelock have suggested several possibilities. In the case of $CO_2$, suppose that rising levels of this gas, a vital resource for photosynthesis, permitted increases in populations of autotrophs compared with those of heterotrophs, which add $CO_2$ to the atmosphere. The $CO_2$ content of the atmosphere and oceans would fall once more.

**FIGURE 21–A** Comparison of atmosphere composition and temperature on Earth and its nearest planetary neighbors, Venus and Mars.

---

totrophs deposited materials that later became oil, coal, and peat. They also removed much of the atmosphere's carbon dioxide, and water-splitting autotrophs added $O_2$ to the air.

The $O_2$ in turn formed the ozone layer of the atmosphere. Ozone ($O_3$) forms from $O_2$ in the presence of ultraviolet light from the sun. The ozone layer acts as a filter, preventing much of the ultraviolet light from reaching Earth's surface. Ultraviolet light (UV for short) damages proteins and nucleic acids. The first organisms lived in water,

which absorbs the energy of UV and so shields its inhabitants; they probably also had mechanisms for repairing damage caused by UV, just as all organisms do today. Nevertheless, before the ozone layer formed, UV was probably one of several factors that kept many early organisms from moving out of water onto land.

Small organisms such as cyanobacteria probably spread to land first, moving outward from the tidal zone or other areas subject to periodic drying. Here they promoted soil

(a)                    (b)                    (c)

**FIGURE 21–7** Stromatolites. (a) Stromatolites consist of layers of cyanobacteria, here seen growing in Shark Bay, Australia. This hypersaline (super-salty) habitat is too harsh to support eukaryotes, which would break up the mats by burrowing in them or by eating the cells. (b) Similar cell layers appear in these Late Proterozoic fossil stromatolites from East Greenland. This view shows a slice down through the layers of cells. (c) Microscopic view of some cells from this fossil stromatolite. The alternating vertical and horizontal alignment of filaments reflects microbial behavior. (a, William E. Ferguson; b, c, Andrew H. Knoll)

formation as their acidic wastes speeded the weathering of rocks. Their sticky polysaccharide coatings glued the resulting particles together, preventing erosion, and also helped hold water.

On land, vast resources of sunlight and minerals were available to plants that could adapt to drier areas, and both plant and animal pioneers of terrestrial life found little competition. The trees and grasses that now clothe much of the land continue to add $O_2$ to the atmosphere. They also change the patterns of water flow from the land to the seas and speed the formation of soil from rock. So organism and environment have molded each other during the history of life on our planet.

## 21–L   HISTORY OF LIFE ON EARTH

Until 1954, the earliest known fossils were from the Cambrian Period of geological time, which began about 570 million years ago. Most major groups of organisms had evolved by this time, and their origins were a mystery because most Precambrian rocks have been either eroded or altered by heat and pressure great enough to destroy fossils.

By using microscopes to examine thin sections of rock, scientists have found many fossils of tiny organisms even in Precambrian deposits. Fossils believed to be photosynthetic cyanobacteria date from almost 3.5 billion years ago. Stromatolites (mats of photosynthetic bacteria) became abundant about 2.8 billion years ago (Figure 21–7). A 1-billion-year-old rock formation contains prokaryotes (and possibly eukaryotes) of about 30 different species. Some of the presumed cyanobacteria in this assemblage look exactly like forms alive today!

For about the first half of the history of life on Earth, "life" meant bacteria. During this time, bacteria evolved the most diverse metabolic abilities of any group of organisms.

The earliest bacteria were anaerobic, and many were autotrophic. Nucleotide sequencing of ribosomal RNA from living bacteria shows that several lines of photosynthetic bacteria gave rise to heterotrophic forms. Similarly, as $O_2$ accumulated in the atmosphere, several aerobic lineages evolved independently from anaerobic ancestors.

As bacteria with different metabolic abilities evolved, they probably formed coalitions in which one cell's waste became its neighbor's food. The ability to live in close symbiotic association contributed not only to the continuing

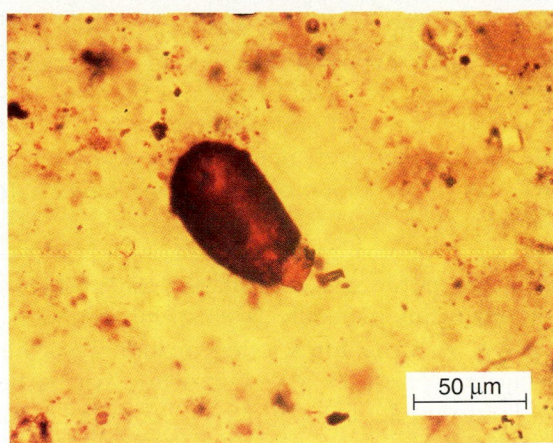

**FIGURE 21–8**  An early heterotrophic protist. This organism, which lived 700 million years ago, was found in rock from the Back-lundtoppen Formation, Spitsbergen.  (Andrew H. Knoll. Knoll and Calder, *Palaeontology* 26:467, 1983)

success of bacteria but also to the success of eukaryotes, which adopted some bacteria as symbionts inside their own cells.

## Origin of Eukaryotes

Large cells interpreted as eukaryotes appear in the fossil record about 1.8 billion years ago. These fossils do not reveal internal cell structure or lifestyle (Figure 21–8). However, from studying the features shared by primitive eukaryotes living today, we believe that early eukaryotes evolved from anaerobic, heterotrophic bacteria without cell walls. Inside the cell, they probably had a nuclear envelope and some other membrane-bounded organelles, with division of labor among several specialized areas. These cells could grow larger thanks to a new mode of nutrition, phagocytosis: with no cell wall, the flexible plasma membrane could engulf food in bulk rather than absorbing individual molecules.

Most eukaryotes living today also have more advanced features, notably sexual reproduction, mitochondria, and often chloroplasts. Originally the advantage of sexual reproduction was probably that it regenerated well-adapted genomes damaged by mutation (Section 20–C). However, by reshuffling genes into a variety of new combinations, it also drove the rapid evolution of eukaryotes. The fossil record shows that eukaryotes evolved much faster than prokaryotes, diversifying into organisms with a variety of sizes, shapes, and lifestyles.

Trying out new structures in this spate of evolution used energy. Some eukaryotes acquired energy-handling metabolic pathways of respiration and photosynthesis by keeping bacteria as symbiotic tenants instead of digesting them

as meals. Mitochondria originated as aerobic bacteria that became live-in power plants paying their rent in ATP. This provided so much extra energy that eukaryotes with mitochondria quickly outstripped those without, increasing in numbers and evolving into a wide variety of forms. Later, some of them adopted photosynthetic bacteria, which evolved into chloroplasts. This happened several times independently, founding different lineages of photosynthetic eukaryotes.

## Animals

From about 800 to 700 million years ago the outer protective layers of microfossils grew thicker, and the diversity of stromatolites declined. These changes suggest the presence of predators, either large unicellular eukaryotes or microscopic early animals and fungi. At about this time, tectonic events were breaking up continents, opening ocean basins, and burying more organic matter than usual. Because this matter was not consumed and respired, more oxygen accumulated in the atmosphere. This permitted the evolution of larger, more active organisms, which could now meet their metabolic demands for oxygen despite longer diffusion paths to their innermost cells.

Macroscopic animals made their fossil debut about 700 million years ago in the form of tracks, burrows, and impressions of their flat, soft bodies (Figure 21–9). Most of them probably gained extra energy by housing symbiotic algae or chemosynthetic bacteria. The shallow, nutrient-poor seas at that time would have exerted selective pressures favoring such symbiotic associations.

Then about 570 million years ago, as the Cambrian period began, a burst of adaptive radiation produced all the major body plans found in modern animals, plus several

**FIGURE 21–9**  An early animal. This fossil, thought to be similar to modern jellyfish, was found in Precambrian rocks in the Edi-acara Hills of Australia.  (William E. Ferguson)

that are now extinct. Predation must have become fierce, judging by the sudden increase in kinds of animals with hard shells, sharp spines, or deep burrows.

This radiation occurred so rapidly because pioneer animals were inventing new ways of making a living and so faced no direct competition from other organisms. In addition, the animals themselves became a new resource, opening opportunities for the evolution of new predators, parasites, and so on: the radiation of animals fed itself in a spiraling explosion of new forms of life. Another factor promoting adaptive radiation at these early stages was that animals probably had more genetic flexibility than they do now. They had not yet evolved all the interlocking programs of embryonic development that make wholesale genetic changes less possible today.

## The Move to Land

The next big event in evolution was the move to land, a habitat offering abundant light energy and minerals but very short of life's most crucial ingredient, water. Some bacteria may have moved into damp, sheltered pockets of rock or soil billions of years ago. However, the first stubby plants spread inland from the edge of the water only about 420 million years ago. Animals followed shortly: the oldest fossil land animals are centipedes and spider-like creatures dated at 414 million years old.

We have already seen that the formation of the ozone layer was probably one factor permitting organisms to emerge onto land. For both plants and animals, terrestrial life also demanded adaptations that conserve precious body water. A waterproof body covering minimizes loss of water vapor to the dry air. This is especially important for reproductive units, such as spores or eggs, because their small volume compared with surface area makes them especially susceptible to loss of water. Terrestrial organisms also lose water by evaporation as they exchange gases with the air. They usually have internal gas exchange surfaces that communicate to the exterior by small openings, such as stomata in plants or nasal openings in terrestrial vertebrates. Another requirement for life on land was extra structural support, since the body would no longer be buoyed up by water. Chapters 27 and 30 discuss adaptations to life on land in more detail.

## Symbiosis in the Evolution of Life

The formation of symbiotic relationships is a recurrent theme in the history of life. Today many bacteria live together symbiotically, sharing the advantages of their complementary metabolic abilities, and it is likely that many of these relationships date from time immemorial. Many bacteria also live in symbioses with various eukaryotes. Earlier we saw that most eukaryotic cells themselves can be regarded as symbiotic mergers between a bacterial host cell and its mitochondria, and often chloroplasts as well. Some biologists think eukaryotes' peroxisomes and cilia (or flagella) also had a symbiotic origin (*Essay: Eukaryotic Flagella,* Chapter 24).

Symbiotic relationships have also contributed to other evolutionary breakthroughs. The adaptive radiation of a new group of organisms generally begins with founders who evolve a new set of features, often including a symbiotic relationship that produces a new structure with a helpful function. For example, many groups of unicellular eukaryotes have distinctive organelles, not found in other organisms, with symbiotic origins. Vascular plants (ferns, trees, shrubs, and herbs) owe much of their ability to live on land to symbiotic fungi intimately associated with their roots. Ruminant mammals (cattle, sheep, goats, and so on) evolved from ancestors that developed a special part of the stomach housing symbiotic fermenting bacteria able to digest cellulose, the major source of energy in the ruminant diet.

## Extinction

The successive appearance of new and improved kinds of organisms is only half the unfolding saga of life on Earth. The fossil record also tells grim tales of extinction, often widespread and occurring within a short time, in geological terms.

One study traced all the animals inhabiting an area of western North America through geological time. This revealed a pattern in which many different species tended to appear or disappear at the same time. Most species apparently vanished because some of their members evolved into one or more new species. This suggested that a long-term change in the environment had selected for somewhat different sets of adaptations in the area's residents. However, a few lineages disappeared completely, apparently unable to evolve adaptations to the new environment.

At several times in Earth's history more dramatic environmental cataclysms have caused mass extinctions, wiping out the majority of lineages in a large region or even worldwide. We don't know what caused these episodes, but suggestions include large meteorite impacts, huge wildfires, massive volcanic eruptions, and movements of tectonic plates, all capable of changing climate and other features of the environment abruptly, giving organisms too little time to adapt. Very likely, each episode of extinction had its own cause (or combination of them).

The greatest extinction of all time occurred 250 million years ago, at the end of the Permian period, when the land masses merged and formed the huge supercontinent Pangaea (Figure 19–C, page 411). The resulting drop in sea

**FIGURE 21–10**   Trilobites. Only fossils bear witness to the existence of this large group of marine animals. These are *Phacops rana,* from the mid-Devonian, about 365 million years ago.   (William E. Ferguson)

level left continental shelves high and dry, killing 95% of marine invertebrates (ocean-dwelling animals without backbones). The victims included the last of the trilobites, once a major group of invertebrates, which had already been decimated by three previous mass extinctions (Figure 21–10).

Another famous extinction occurred during several million years between the Cretaceous and Tertiary periods, 66 million years ago. It claimed half of all life on Earth, including the dinosaurs (*Essay: Mesozoic Murder,* Chapter 30). In one marine rock formation studied, a sea urchin was the only species of large animal to survive from the Cretaceous into the Tertiary layer.

Today we appear to be in the midst of another period of widespread extinction, this one caused mostly by human activity. Some of this may have been due to hunting in prehistoric times, but most is occurring right now as we destroy vast areas of some habitats, permit toxic materials to accumulate in others, overfish marine populations, and transport species into new areas where they devour, outcompete, or fatally infect local residents.

By wiping out so many established forms of life, a mass extinction opens new opportunities. As conditions settle down, the survivors (and new immigrants) undergo a wave of adaptive radiation and repopulate the area with a new set of organisms. Although these new species may differ from the old in many features, they are often functionally equivalent, occupying ways of life left vacant by the demise of their predecessors. For example, only tiny mouse-sized mammals existed before the Tertiary period, but they quickly

radiated and produced large grazers, predators, and seagoing forms, replacing extinct dinosaurs that formerly filled these niches.

This pattern of evolutionary turnover—adaptive radiation, extinction, and new adaptive radiations of survivors—is characteristic of eukaryotes. In contrast, many prokaryotic lineages have persisted, with no apparent change, for half a billion years or more.

## SUMMARY

We shall never know exactly how life on Earth began. However, by combining observations, experiments, and guesswork, scientists have pieced together a description of how it may have occurred.

All life now on Earth probably descended from one or more cells that arose spontaneously under the very different conditions that existed here billions of years ago. Gases in the mildly reducing atmosphere of that eon reacted and formed small organic molecules. These gradually polymerized, formed macromolecular aggregates, and evolved systems of metabolism, information transfer, and reproduction, eventually becoming living organisms. Later, photosynthetic cyanobacteria added oxygen to the air. The present oxidizing environment and the abundant existing life preclude the spontaneous generation of new organisms on Earth today.

1. Evidence from geology and astronomy suggests that the primitive Earth had a mildly reducing atmosphere, composed of the gases in today's atmosphere except for oxygen. Such an atmosphere would have been conducive to the formation and stabilization of organic compounds.

2. Natural energy sources would cause these atmospheric gases to react with each other. This formed organic molecules, which dissolved in bodies of water or formed a sludge on mineral surfaces.

3. Mild heat or minerals may have catalyzed the joining of organic monomers into polymers. Many of the resulting proteinoids and RNAs themselves had catalytic properties.

4. Randomly formed polymers would have formed aggregates, the proposed precursors of cells. These aggregates would have had a self-assembled, selectively permeable lipid coating and a primitive metabolism carried out by proteinoid or RNA catalysts.

5. Metabolic pathways probably evolved "backwards" as aggregates added catalysts able to convert more and more different raw materials into useful molecules. Both ATP and hydrogen ion pumps probably provided energy to this early metabolism, as they do to all modern cells.

6. Genetic information and metabolism are interdependent and must have evolved together in an increasingly precise relationship. RNA was probably the first genetic material but was eventually replaced by DNA, which is more stable and easier to copy. This left RNA with the role of making proteins.

7. The evolution of autotrophy, apparently near the dawn of life, provided the living world with an abundant and self-renewing supply of food, which in turn supported large populations of heterotrophs. The food supply expanded even more when some photosynthetic organisms evolved a way to split a widely available hydrogen donor, water.

8. Molecular oxygen, the byproduct of this water-splitting photosynthesis, exerted selective pressure for the evolution of aerobic respiration. By extracting more energy from food molecules, respiration permitted organisms to grow and reproduce faster and evolve more complex structures.

9. Organisms have changed their environment from an Earth of barren water and rock under a mildly reducing atmosphere to one of teeming oceans and verdant landscapes in an oxidizing atmosphere. Each environmental change caused by organisms exerted selective pressures to adapt to the new environment, which in turn changed the environment even more. Thus, living organisms and their environment have shaped each other during the evolution of life on Earth.

10. Bacteria evolved about 3.5 billion years ago and remained the planet's sole inhabitants for nearly 2 billion years. During this time, the most important events were the evolution of water-splitting photosynthesis and of aerobic respiration.

11. Eukaryotic cells appeared about 1.8 billion years ago. Phagocytosis and the symbiotic acquisition of energy-processing organelles provided them with a lot of energy, and sexual reproduction provided genetic variability. Thanks to this combination, eukaryotes evolved rapidly into a variety of forms. A dramatic burst of adaptive radiation produced the founders of all the main animal groups between about 700 and 600 million years ago. Land plants first appeared about 420 million years ago, followed by land animals.

12. Throughout the history of life, many evolutionary breakthroughs have resulted from the formation of new symbiotic relationships.

13. Life on Earth has experienced many widespread cataclysms which caused mass extinction. Each of these episodes was followed by adaptive radiation of the survivors, filling the vacancies with new forms of life.

## Self-Quiz

1. For each gas listed below, tell whether it was *more* or *less* abundant in the mildly reducing atmosphere of early Earth than it is today:

    _____ oxygen
    _____ carbon dioxide
    _____ water vapor

2. In Stanley Miller's classic experiment:
    a. nucleic acids were formed
    b. ultraviolet radiation was used
    c. oxygen was one of the starting ingredients
    d. water was strictly excluded from the system
    e. amino acids were formed

3. Number the following structures and processes in the order in which they are believed to have evolved:

    _____ aerobic respiration
    _____ polymers
    _____ water-splitting photosynthesis
    _____ organic monomers
    _____ acquisition of intracellular organelles
    _____ fermentation

4. RNA rather than DNA is believed to have been the first form of genetic information because:
    a. RNA is more stable than DNA
    b. RNA is single-stranded, whereas DNA has two strands
    c. RNA catalyzes some steps in its own synthesis and replication

    d. RNA mutates less than DNA and is easier to copy without mistakes
    e. DNA is complementary to RNA

5. Early autotrophs were important to the evolution of life on Earth because:
    a. they blocked harmful ultraviolet rays from the sun
    b. they provided a self-renewing food supply
    c. they rid the environment of toxic substances
    d. they provided oxygen for respiration
    e. all of the above

6. List two changes in the environment that resulted from evolution of water-splitting photosynthesis. What effects did these have on the later evolution of living organisms?

7. Respiration was important to early life on Earth because:
    a. it rid the environment of toxic ozone
    b. it provided much more energy than photosynthesis
    c. it provided much more energy than fermentation
    d. it permitted experimentation with the genetic code
    e. all of the above

8. Mass extinctions:
    a. completely eliminate all life in an area
    b. completely eliminate many species
    c. eliminate most individual organisms but leave a few survivors of most species
    d. occur within the space of a few weeks

## Questions for Discussion

1. Are the laboratory simulations that were presented in this chapter (e.g., Miller's reducing atmosphere simulation and Fox's proteinoids) experiments or observations? Are there any controls?

2. Suppose that a scientist claimed to have produced life from nonliving materials under laboratory conditions. What criteria must such an "organism" meet before *you* agreed that it was truly living?

3. Some organic compounds formed in simulations such as Miller's are not found in living organisms. Can you explain this?

4. Simulations of RNA replication in the absence of enzymes show mutation rates several orders of magnitude higher than the rates for DNA in modern cells. How might such high mutation rates have affected prebiotic evolution? Why are mutation rates so much lower today?

5. If the Gaia hypothesis is correct, why didn't organisms compensate for environmental disturbances and thereby prevent the major ice ages of Earth's past?

## Suggested Readings

Cairns-Smith, A. G. "The first organisms." *Scientific American,* June 1985. Argues for an alternative view of the origin of life, very different from the one presented in this chapter.

Cavalier-Smith, T. "The evolution of cells." In S. Osawa and T. Honjo, eds. *Evolution of Life: Fossils, Molecules, and Culture.* New York: Springer-Verlag, 1991. Explores what some obscure modern cell types tell us about the history of cells.

DeDuve, C. *Blueprint for a Cell: The Nature and Origin of Life.* Burlington, NC: Neil Patterson Publishers, 1991. Written for those with a grasp of elementary biology, this is the product of an eminent cell biologist's survey of research literature across many fields.

Dickerson, R. E. "Cytochrome *c* and the evolution of energy metabolism." *Scientific American,* March 1980. Builds an evolutionary tree for kinds of photosynthesis and respiration in prokaryotes.

Eigen, M., W. Gardiner, P. Schuster, and R. Winkler-Oswatitsch. "The origin of genetic information." *Scientific American,* April 1981. Interesting experiments on formation of RNA without templates and some more advanced ideas about the origin of the genetic code and genomes.

Farley, J. *The Spontaneous Generation Controversy from Descartes to Oparin.* Baltimore: Johns Hopkins University Press, 1977. A historical account of philosophical and scientific controversy over the origin of life; full of fascinating tidbits of information and interesting ideas about the development of scientific thought.

Gesteland, R. F., and J. F. Atkins, eds. *The RNA World: The Nature of Modern RNA Suggests a Prebiotic RNA World.* Cold Spring Harbor, NY: Cold Spring Harbor Laboratory Press, 1993.

Hartman, H., J. G. Lawless, and P. Morrison. *Search for the Universal Ancestors.* Boston: Blackwell Scientific Publications, 1987. What we have learned about life's beginnings from rocks, fossils, space, and laboratory experiments.

Horgan, J. "In the beginning..." *Scientific American,* February 1991. Thumbnail sketch of the history and hangups of the modern view of the origin of life on Earth.

Knoll, A. H. "End of the Proterozoic era." *Scientific American,* October 1991.

McMenamin, M. A. S. "The emergence of animals." *Scientific American,* April 1987.

Rebek, J., Jr. "Synthetic self-replicating molecules." *Scientific American,* July 1994. Studies of possible precursors of living systems.

Scott, A. *The Creation of Life.* New York: Basil Blackwell Ltd., 1986. Summary and criticism of modern ideas about the origin of life.

Waldrop, M. "Did life really start out in an RNA world?" *Science* 246:1248, 1989. Arguments against the model of early life based on RNA without protein-like molecules.

PART
FOUR

*Diversity*

# Classification of Organisms

## OBJECTIVES

*From your study of this chapter, you should be able to:*

1. Give the meaning of the following words and use them correctly: taxonomy, taxon, morphology, type specimen, species, phylogeny, monophyletic, polyphyletic.
2. Describe Linnaeus's contribution to taxonomy and the basis for his classification.
3. State the theoretical basis for modern biological classification, and discuss some of the problems of constructing satisfactory classification schemes.

4. Write the Latin binomial of an organism correctly.
5. List, in order, the seven main hierarchical taxonomic levels into which organisms are placed.
6. Explain what is meant by saying that characters are ancestral, derived, homologous, analogous, conservative, or convergent.
7. List the five kingdoms of organisms used in this book, and state the criteria used to assign species to each kingdom.

H undreds of years ago, Europeans began to explore the far corners of the world, bringing back specimens of thousands of plants and animals unknown at home. Europe is inhabited by fewer types of organisms than any other continent except Antarctica, so the wealth of organisms found elsewhere astonished explorers. Later, early microscopists discovered a myriad of tiny organisms living around them almost everywhere they looked. Eventually the number of different kinds of living things known became almost overwhelming.

Today, we classify organisms using a system invented by Carl Linnaeus, a Swedish botanist (Figure 22–1). Linnaeus based his classification on morphology (structure,

**FIGURE 22–1** Carl Linnaeus (1707–1778), who described thousands of species of plants and animals, described himself thus: "Brown-eyed, nimble, hasty, did everything promptly." Linnaeus classified plants by their sexual parts with group names such as Polyandria, meaning "twenty or more males with the same female." This emphasis on sex shocked some of his contemporaries. The Bishop of Carlisle wrote: "To tell you that nothing could exceed the gross prurience of Linnaeus's mind is perfectly needless," and Goethe worried about the embarrassment chaste young people might suffer when reading botany textbooks. (Linnaean Society of London)

anatomy). For example, two species of plants with similar flower structure fell close together in his scheme. After Darwin's theory of evolution gained acceptance, biologists decided that classifying organisms according to their evolutionary relationships would be more natural. Fortunately, the morphological features used by Linnaeus usually reflect evolutionary relationships. Hence, biologists have been able to retain many of the names used in Linnaeus's thorough and meticulous system.

The task of finding, naming, and classifying organisms is far from complete. Biologists estimate that there are at least 10 million different species of organisms in the world, but so far only about 1.5 million have been described by scientists. Today we are destroying natural habitats so rapidly—by clearing, draining, filling, damming, and polluting—that many of the remaining species will probably be extinct before they can be described.

In this chapter we consider how living things are classified and how biologists approach the problem of deciding where in the classification scheme particular organisms belong. We then describe the five-kingdom system of classification that we use in this book.

### KEY CONCEPTS

♦ Organisms are classified in a hierarchical system that aims to indicate evolutionary relationships.

♦ The modern cladistic system aims to classify organisms in such a way that each group contains the common ancestor of all its members.

## 22–A  BINOMIAL NOMENCLATURE

The basic unit for classifying organisms is the species. A **species** (plural: species) is a group of organisms that share a common gene pool. Biologists following Linnaeus's method of describing organisms adopted the habit of giving each species a Latin **binomial,** a two-word name. The first word in this binomial designates the genus to which the species belongs. A **genus** (plural: genera; adjective: generic) contains one species or a group of related species. The second word in the binomial is the **specific epithet,** an adjective denoting the species itself. For example, the binomial for the snowy egret is *Egretta thula* (Figure 22–2). This name indicates that the egret has been placed in the genus *Egretta* and, within that genus, has been given the specific epithet *thula*. The complete binomial, *Egretta thula,* is needed to distinguish the snowy egret from other species in the same genus, such as *Egretta tricolor,* the Louisiana or tricolor heron.

The name of a genus can be used only for one or a number of related species. Because specific epithets are adjectives, however, the same one may be combined with different generic names and used for a number of unrelated organisms. *Erythronium americanum,* the trout lily; *Euarctos americanus,* the American black bear; *Hepatica americana,* a relative of the buttercup; and *Coccyzus americanus,* the

(a) Snowy egret, *Egretta thula*

(b) Tricolor heron, *Egretta tricolor*

(c) Great blue heron, *Ardea herodias*

**FIGURE 22–2**  Members of the bird family Ardeidae. (b, c, Florida Audubon Society)

T A B L E   2 2 – 1

## Classification of Human Beings

| Taxon | Name | Characteristics |
|-------|------|-----------------|
| Kingdom | Animalia | Multicellular heterotrophs with embryonic development via a blastula |
| Phylum | Chordata | Animals with a dorsal hollow nerve tube, a notochord, and gill slits at some stage of the life cycle |
| Class | Mammalia | Chordates with only one bone in each side of lower jaw, hair, mammary glands, warm-blooded |
| Order | Primates | Mammals adapted to arboreal living, with flattened nails and mobile fingers, excellent vision |
| Family | Hominidae | Primates with upright posture, small flat faces, binocular color vision |
| Genus | *Homo*\* | Hominids with short arms, large brain, light jaw and small teeth, long childhood |
| Species | *Homo sapiens*† | With high forehead, body hair reduced, prominent chin, well-developed language |

\**Homo* Latin: man; †*sapiens* Latin: knowing, wise

yellow-billed cuckoo, all have adjectives meaning "American" as the second word in their binomials. The name of a newly described species must be officially approved by one of the International Commissions on Nomenclature. Different commissions oversee the naming of plants, animals, and microorganisms. Their task is to prevent problems such as two genera with the same name.

## 22–B  TAXONOMY

**Taxonomy** is the branch of biology concerned with the classification of organisms. Linnaeus classified organisms into a hierarchy of ever more inclusive categories, a system borrowed from the Swedish military. The most inclusive categories are the **kingdoms.** The other main categories, in descending order, are **phylum, class, order, family, genus,** and **species.** (You can remember this order by memorizing the sentence "**K**ing **P**hil **c**ame **o**ver **f**or **G**ina's **sp**ecial.") Intermediate categories, such as superfamilies and suborders, are also used occasionally.

A **taxon** (plural: **taxa**) is a group of organisms defined by biological classification, such as a species or phylum. For example, Ardeidae (a family) is a taxon including the species in the genus *Egretta* as well as those in other genera of herons, such as the little blue heron, *Florida caerulea,* and the great blue heron, *Ardea herodias.* Some taxa contain only one group at the next lower level. For example, there are many families that contain only one genus and one species.

Table 22–1 gives the classification for human beings, and here you can see the seven main levels of the hierarchical classification. Table 22–2 gives guidelines for the use of binomials and names of taxa.

Linnaeus named and classified the more than 11,000 plants and animals known to him in his massive books *Systema Naturae* and *Species Plantarum.* He believed that a species could be described by listing the morphological characteristics of a "perfect" member of the species. This led to the practice of selecting and preserving a typical individual of each newly described species to become the official example, or **type specimen,** of the species. Today,

T A B L E   2 2 – 2

## Conventions for Using Binomials and Names of Higher Taxa

**Capitalize:**
1. Genus, but not species (unless it is a proper noun)
2. Latin names of taxa above genus level, but not their English counterparts

**Examples**
*Homo sapiens*
Hominidae, hominids

**Italicize or underline:**
Genus and species (binomial), but not Latin or English taxon name above genus level

<u>Homo sapiens</u> or *Homo sapiens,* Hominidae

**Spell out:**
1. The generic name the first time you use it in each paragraph
2. The specific epithet every time you use it

*Homo sapiens*
*H. sapiens*

**Abbreviate:**
1. The generic name to its first letter at the second and subsequent mentions in the same paragraph
2. When you know the genus but not the species of the organism(s) you are discussing.
(In this case, spell out the genus name, even if you have already mentioned it previously in the same paragraph)

*H. sapiens*
*Egretta* sp. (one unknown species)
*Egretta* spp. (more than one species of *Egretta*)

people who describe new species preserve several specimens showing a typical range of the species' phenotypes. If later workers need to know whether they are really working on the same species that the original author described, they compare their experimental organisms with the type specimen(s).

Nowadays, some microorganisms are preserved alive by freezing them in liquid nitrogen ($-196\,°C$). Researchers can then obtain portions of a culture of a particular species, and even of particular genetic strains within a species, for experimentation and comparison.

Thousands of the species and taxon names introduced by Linnaeus are still in use. However, an understanding of evolution has led biologists to classify organisms in a way that is, at least in theory, different from that used by Linnaeus. Linnaeus's classification was an artificial system, designed to be helpful in organizing and retrieving information about organisms. We shall see that one school of thought today advocates precisely this goal for modern classification (Section 22–D).

The alternative approach, preferred by most biologists, is to classify organisms by **phylogeny,** their evolutionary history. The goal of such a **phylogenetic system** is to produce a classification that is easy to use and that also provides information on patterns of evolution within taxa. The method

amounts to drawing an evolutionary family tree for an organism (Figure 22–3). In many cases, this leads to the same result as an artificial classification, because organisms that have evolved from a common ancestor are more likely to be similar than those that have not.

## 22–C   SYSTEMATICS AND ITS TOOLS

**Systematics** is the study of the relationships among organisms, which means the reconstruction of phylogenies. However, even an accurate phylogenetic history of several organisms usually does not permit a taxonomist to classify the organisms in a way that everyone will accept. Phylogeny depicts the continuous process of evolution, but the taxonomist's business is to carve this up into discrete sets of organisms, for human convenience. There are always disagreements about how, where, and why parts of the evolutionary tree should be separated. For instance, how closely must organisms be related to fall within the same order? Are all species of the cat family phylogenetically close enough to be lumped into the same genus, and if not, how many genera should we use? How do we define insect families so that they are comparable to families of plants? There are no simple answers to such questions. As a result, there

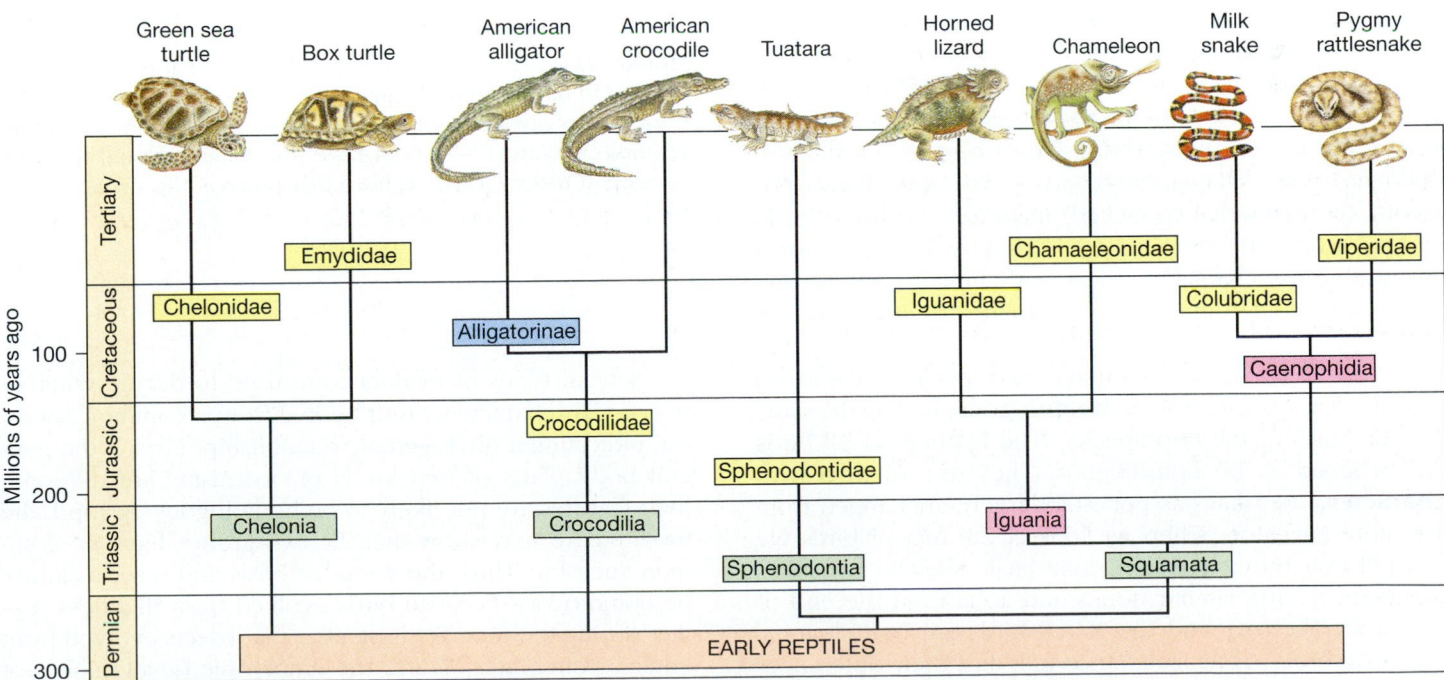

**FIGURE 22–3**   A phylogenetic tree of the species of reptiles shown at the top. Several levels of taxon are shown: orders in green, suborders in red, families in yellow, and one subfamily in blue. A phylogenetic tree is a genealogy of the likely evolutionary relationships among taxa. It also shows how long ago each evolutionary split occurred. Evidence for the tree comes from the fossil record, from comparative anatomy of living species, and from analysis of their proteins and DNAs.

are many different systems of classification in use today, with different taxon names, and different cutoff points between taxa.

Let us first consider the information systematists use to construct phylogenies.

## Interpreting The Characteristics of Organisms

Every organism has a variety of features, often giving conflicting information about its phylogenetic position. Biologists need ways to interpret the significance of different features, giving some more weight than others.

When we say two species are closely related phylogenetically, we mean that they have a common ancestor. For example, it seems that modern birds are all descended from one ancestral species, *Archaeopteryx lithographica,* from the Jurassic Period. *Archaeopteryx* is the first known organism with feathers, which apparently evolved only once. All organisms with feathers are therefore believed to be related and are classified as birds.

Evolution, however, seldom hands us a simple character such as feathers. More often, the characters of an organism give conflicting information about its phylogenetic position, and systematists have to give some characters more weight than others. For instance, if your biology teacher asked you to classify a flea, a frog, and a kangaroo in a phylogenetic scheme, we hope you would ignore the fact that all three have hind legs strongly developed for jumping, and concentrate more on such features as the general plan of the skeleton, the type of body covering, and the mode of reproduction. By this line of reasoning you might arrive at the conclusions of most biologists: fleas belong among the insects, close to the flies; frogs among the amphibians, with salamanders and newts; and kangaroos among the marsupial (pouched) mammals, with the koala and opossum. The jumping legs shared by these organisms are **analogous** characters, those with similar function and appearance but different evolutionary origins (Section 17–B).

In contrast, characters found in two species and derived from a common ancestor are **homologous,** having the same genetic basis in the two species. The feathers of all birds are believed to be homologous. They are an **ancestral character,** one that has persisted largely unchanged from a remote ancestor. When we look at the feet of birds we can tell that they share the same basic skeletal structure, with toes having similar bones and a claw on the end of each toe. The legs and feet of all birds are homologous, the basic bone structure is ancestral, and their differences are **derived characters,** modifications that evolved more recently as adaptations to different ways of life. The derived characters of various birds range from the webbed feet of ducks and cormorants, adapted to swimming and diving, to the heavily clawed feet of ospreys and hawks, adapted to grasping prey (Figure 22–4).

OSPREY: catching fish

CORMORANT: diving and swimming

JACANA: walking on floating waterlily leaves

PTARMIGAN: walking on snow

OSTRICH: fast running on ground

**FIGURE 22–4** Derived characters. Feet of various birds are adapted to different environmental activities. (Note that feet of an ostrich show the same sort of adaptations to fast running as those of horses (Figure 17–7): both have lost some ancestral toes and developed a thick claw or hoof that protects the main running toe.)

Several types of evidence are used to decide whether two similar characters are inherited from a common ancestor, indicating a phylogenetic relationship. First, if the general body plans of two kinds of organisms are different, their features are not likely to be homologous. Second, the fossil record may show that the two groups have no common ancestor. Thus, the wings of birds and insects cannot be homologous because birds evolved from flightless reptiles during the Jurassic, long after the insects evolved from other invertebrates. Third, the embryonic development of homologous organs is invariably similar, and this can often be used to tell homology from analogy.

Another difficulty in determining whether organisms are closely related is sorting out cases of **convergent evolution,** the development of similar adaptations as a response to similar environmental pressures. Are all snakes descended

(a)

(b)

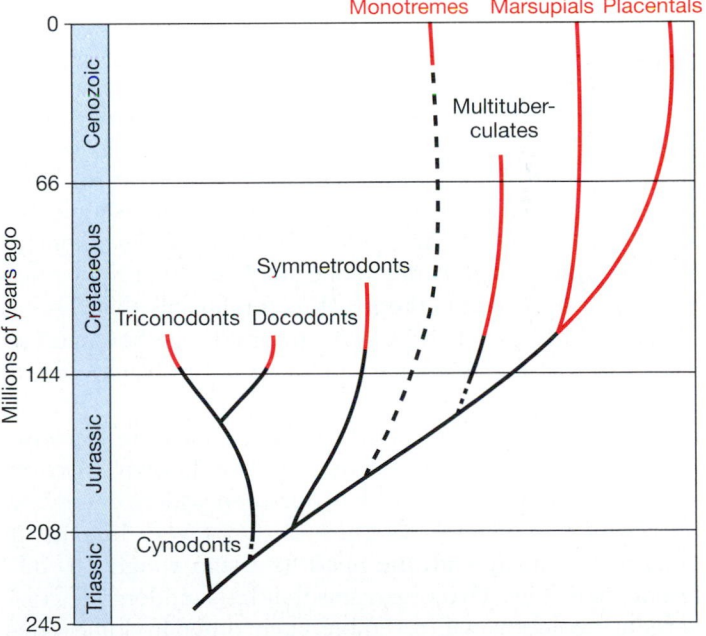

**FIGURE 22–5** Convergent evolution. (a) Euphorb and (b) cactus evolved in different parts of the world, but both show similar adaptations to their desert habitats: fleshy stems that store water and carry on photosynthesis, and leaves specialized as spines that deter animals from eating the plants to obtain water.

from a legless common ancestor, or has leglessness evolved more than once in different groups of reptiles (as now seems likely)? Striking examples of convergent evolution occur in the plant world. Many desert-dwelling members of the New World family Cactaceae resemble desert-adapted members of the Old World family Euphorbiaceae. In both groups, the advantage of being able to store water led to the evolution of thick, water-storing stems and of spiny leaves that deter animals from using the stems as the source of their own water (Figure 22–5). Among animals, the New World and Old World vultures are also examples of convergent evolution of strong hooked beaks and small naked heads adapted to probing into animal carcasses for food. Old World vultures are most closely related to hawks and eagles, New World vultures to storks.

## Monophyletic or Polyphyletic?

One of the most useful taxa is a **clade** of organisms, that is, a common ancestor and all the species descended from it. A clade is **monophyletic,** meaning that it represents one evolutionary line. In practice, many taxa in use today are **polyphyletic,** made up of several evolutionary lines, not including their common ancestor. For instance, the class Mammalia contains all vertebrates with only one bone in each side of the lower jaw. By this definition, mammals probably arose at least three, and possibly five, times from different groups of reptiles. Mammals also differ from reptiles in the joint between jaw and skull and in the way the limbs are attached to the rest of the skeleton. The differences arose as fast-moving, dog-like mammals evolved from reptiles with less efficient locomotion and with jaws prone to being broken by struggling prey. Apparently, several groups of reptiles underwent **parallel evolution:** the groups all evolved in the same direction and gave rise to descendants classified as mammals. Mammals, then, are not

members of a monophyletic clade but of a polyphyletic **grade** of organization attained more than once by related, but different, evolutionary lines (Figure 22–6).

**FIGURE 22–6** Grades and clades in the origin of mammals. Animals that reached the mammal grade of organization are represented by red lines. Names of the three groups of modern mammals are in red. The dotted line indicates that the origin of monotremes is much debated. Monotremes are those mammals that lay eggs and suckle their young (e.g., duck-billed platypus). Marsupials are those mammals whose young undergo part of their development attached to a teat in the mother's pouch (e.g., opossum, kangaroo, koala). Placentals are the familiar mammals in which the young begin independent existence when they leave the uterus (e.g., cattle, humans).

## 22–D  TAXONOMIC METHODS

The traditional method of classifying a newly discovered organism requires several steps. First the organism is tentatively assigned to a taxon with apparently similar organisms. It can be moved later in accordance with any new information. For instance, a study of the life cycle of a new species may show that its larvae resemble those of another taxon. Among invertebrates, similarities between larvae often give better clues to relationships than do resemblances between adults. Or perhaps the fossil record may reveal that organisms are more (or less) closely related than had been thought. Hares and rabbits (mammalian order Lagomorpha) were once classified with the similar-looking rodents (rats, mice, and their relatives), but the fossil record shows that the two groups had separate origins and have undergone convergent evolution. Similarly, molecular analysis, behavior, geographical distribution, and many other features may be added to the list of features used to classify organisms.

Even with all this information, taxonomists still face the old problem of how much weight to give to the various characters that may point to different conclusions about how the organism should be classified. Today there are three main schools of thought on this question.

### Phenetics

Pheneticists advocate an artificial system of classification. They argue that we can never be certain that a phylogeny is correct, and therefore we should not even try to base classification on phylogeny. Instead, organisms should be classified for convenience, as we classify books in a library, according to their similarity to one another—that is, by phenetic (= phenotypic) criteria. Thus, organisms that resemble each other closely should be grouped together, even in clear cases of convergent evolution from unrelated ancestors.

Phenetic criteria are used to classify bacteria, because the phylogeny of most bacteria is not known (Section 24–D). Otherwise phenetics has not been widely embraced by taxonomists, largely because it is not very helpful. It does not do away with the need to assign weights to different characters. Two organisms that have undergone convergent evolution will resemble each other in some phenotypic characters but not in others. Why emphasize the similarities rather than the differences? Phenetics provides no criteria for such decisions.

### Phylogenetic Systematics

Most of the classification in use today has been produced by phylogenetic systematists, but there are two rather different approaches to the same goal of a classification that reflects phylogeny.

***Evolutionary Systematics***  Evolutionary systematists hold that classification should simultaneously reflect genealogy (common ancestry) and genetic relationships (shared characters). Thus, species with shared characters (for instance, cacti and euphorbs) will be classified together only if these characters have a common genetic basis, which is the case if there is a common ancestor.

Sometimes genealogy and genetics lead to conflicting conclusions. For instance, crocodilians and birds have a more recent common ancestor (in the Triassic) than do crocodilians and other reptiles. Therefore, on genealogical grounds, crocodilians should be classified with birds instead of with other reptiles. However, crocodilians are in fact classified with reptiles on the genetic grounds that crocodilians show many ancestral reptilian characteristics, whereas birds have few ancestral (reptile) and dozens of derived (bird) characters. Where genealogy and genetics conflict, then, evolutionary systematists go by genetics.

***Cladistics***  The most influential contribution to taxonomy in the twentieth century has been the discipline of cladistics. Cladists eliminate the conflict between genealogy and genetics by coming down firmly on the side of genealogy: taxa should be monophyletic, containing a common ancestor and its descendants, and polyphyletic taxa should be broken up. A cladistic classification consists of a nested series of monophyletic groups. For instance, birds and reptiles have a common ancestor and should belong to the same taxon, which might be called a class. Within that class birds, dinosaurs, and crocodilians, which have a common ancestor, might form one order; lizards and snakes, which have a different common ancestor, a second order; and turtles a third order.

The heart of the cladistic method is listing ancestral and derived states of a number of independent homologous characters. For land vertebrates, one character might be number of toes on the forelimbs. The ancestral state in most fossil amphibians, five digits, is found in humans, whereas derived states include the single digit of horses and the four digits of frogs. If there have been no reversals of evolution, any derived state appears in all the descendants of the species in which it first arose, and the group is monophyletic. Thus, a monophyletic group of frogs can be identified by finding the first fossil amphibian with four digits on the forelimbs and classifying all four-fingered frogs in the same group.

The general rule is that species sharing the highest number of derived characters make up monophyletic groups. A cladistic classification is usually written in the form of a diagram called a Hennigian comb, after its inventor William Hennig (Figure 22–7). The comb shape provides room at the side to write in the new derived characters that distinguish the taxa and room at the top for taxon names.

It is important, in cladistic as well as evolutionary taxonomies, to use **conservative** characters, characters that

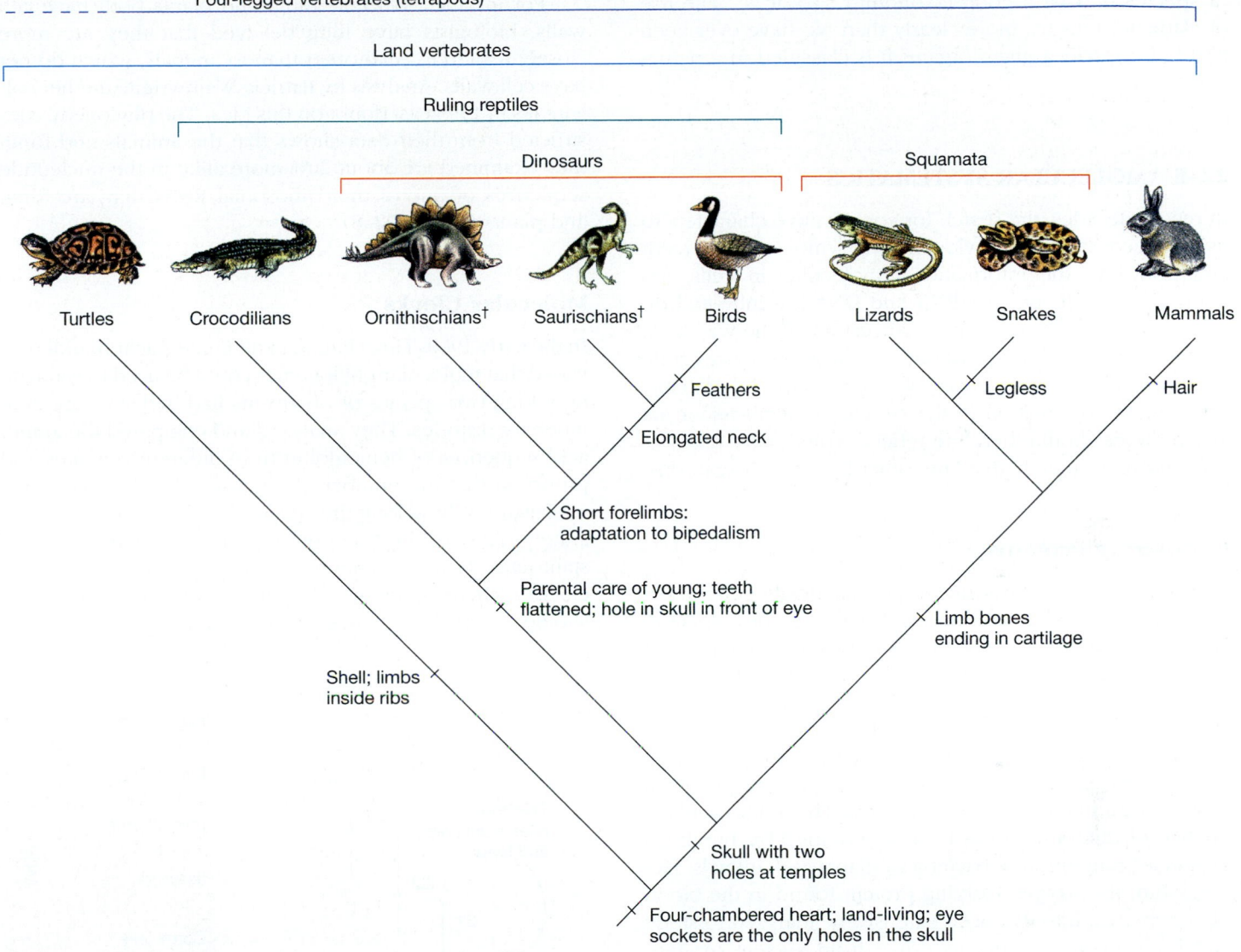

**FIGURE 22–7** Cladistic classification. This Hennigian comb diagrams the phylogenetic relationships of reptiles, birds, and mammals. The comb branches wherever a new derived character appears. The character is written after the branchpoint. Nested boxes at the top of the diagram give the taxon names. Since cladistics emphasizes common ancestry, birds are classified with ornithischian and saurischian dinosaurs despite the many derived characters of birds. By convention, daggers indicate extinct taxa.

evolve slowly. For instance, the shape, size, and number of front teeth is the conservative character used to define the mammalian order Rodentia (the rodents). Beavers, kangaroo rats, and many ancient fossil species all have similar teeth and are therefore all classified as rodents. Characters suspected to be adaptations to specific environments are usually not conservative: color patterns, leaf shape, and type and thickness of hair are the kinds of characters that will probably vary a lot in the course of evolution and should therefore be ignored for purposes of classification.

The most useful guide to conservative, ancestral characters is to compare members of a group with related non-

members. For instance, six legs are believed to be ancestral within the order Lepidoptera (butterflies and moths) because members of the other insect orders also have six legs, and cladists cleave to the conservative assumption that a character has not evolved more than once unless there is clear evidence to the contrary.

The most important contribution of cladistics to modern taxonomy is that cladists state explicitly how they construct phylogenies, and they list criteria for judging whether characters are ancestral or derived. They are attempting to add precision to classification, so that we can eventually escape from evolutionist Theodosius Dobzhansky's definition:

"a species is what a good taxonomist says it is." The use of cladistics tells us, more clearly than we have ever been told before, *why* a given organism is classified in a particular taxon.

## 22–E   MOLECULAR SYSTEMATICS

In recent decades, the search for conservative characters to use in reconstructing phylogenies has moved from teeth and bones to the informational molecules in cells. Sequences of nucleotides in RNA and DNA are inherited directly from an unbroken line of ancestors all the way back to the beginning of life. Furthermore, these sequences code for corresponding sequences of amino acids in proteins. The genetic information in the structures of nucleic acids and proteins should therefore reflect evolutionary relationships more accurately than any other feature of organisms.

### Comparing Proteins

A protein's amino acid sequence is genetically determined and tends to be the same in most members of a species. However, the structure of the same protein often varies slightly among different species. To use proteins to construct phylogenies, biologists choose proteins that occur in all the organisms to be studied and that perform some vital function and so are likely to be conservative.

A common choice is cytochrome c, an early-evolved protein found in all aerobic organisms. The amino acid sequence of this protein has been determined for hundreds of species ranging from bacteria to plants and animals. Hemoglobin, the oxygen-carrying protein found in the blood of vertebrates, has also been extensively analyzed to study vertebrate evolution, and its origin from ancient oxygen-carrying molecules has been traced (*How Do We Know? Molecular Phylogeny of Globin Genes*).

### RNA and DNA

The most direct reflection of the relatedness of two species lies in their genomes. Methods described in Chapter 13 permit biologists to compare the nucleotide sequences of individual genes or other sequences of DNA.

In recent years, a consensus has developed that the best molecules to use for genetic analysis of distantly related taxa are ribosomal RNAs because they are extremely conservative. (This only works for distantly related taxa because closely related species and even orders generally have identical ribosomal RNA.) Every organism contains ribosomes, and ribosome structure can evolve only slowly because ribosomes are so crucial to life: a mutation that interfered with protein synthesis by ribosomes would be eliminated rapidly by natural selection.

For instance, because fungi and plants both have cell walls, biologists have long believed that they are more closely related to each other than to animals, which do not have cell walls. Analysis by Patricia Wainwright and her colleagues in 1993 cast doubt on this idea. The phylogeny constructed from their data shows that the animals and fungi they examined are about 20% more alike in the nucleotide sequences of one of their ribosomal RNAs than are fungi and plants (Figure 22–8).

## Molecular Clocks

In the early 1960s Linus Pauling and Emile Zuckerkandl proposed that molecular phylogenies could be used to estimate how long two species of organisms had had separate evolutionary histories. They analyzed and compared the amino acid sequences of hemoglobin from different primates and proposed that the number of amino acid differences gave a measure of how long the species had been evolving separately. This will be true only if molecules change at constant rates during evolution, so that they act as "molecular clocks." There seems no obvious reason why a molecule should accumulate changes in amino acid or nucleotide se-

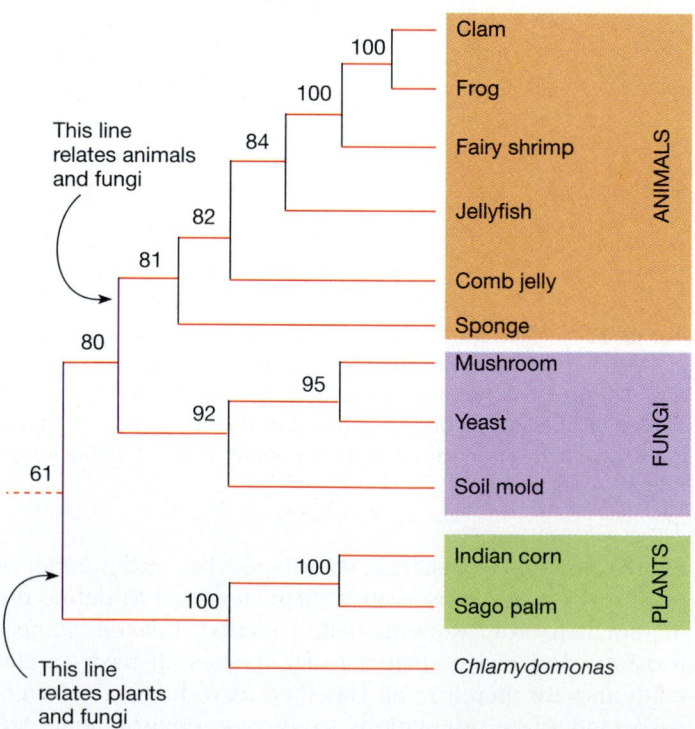

**FIGURE 22–8** Relationships among animals, fungi, and plants inferred from nucleotide sequences of ribosomal RNA molecules. The numbers show the degree of similarity between the molecules at the next fork to the right. For instance, the rRNA molecules analyzed were 100% similar (identical) in all the plants. Note that animals and fungi are more similar than plants and fungi.

quence at a constant rate but, there is evidence that some do—at least for limited periods.

For example, primates are basically arboreal (tree-living). The closest living relatives of humans are the apes: gibbon, orangutan, chimpanzee, and gorilla, which show more signs of arboreal ancestry than do humans. Humans have shorter arms, not as well adapted to swinging from branch to branch. In the 1960s, human ancestors were believed to have formed a separate evolutionary line before modern species of apes evolved (Figure 22–9a).

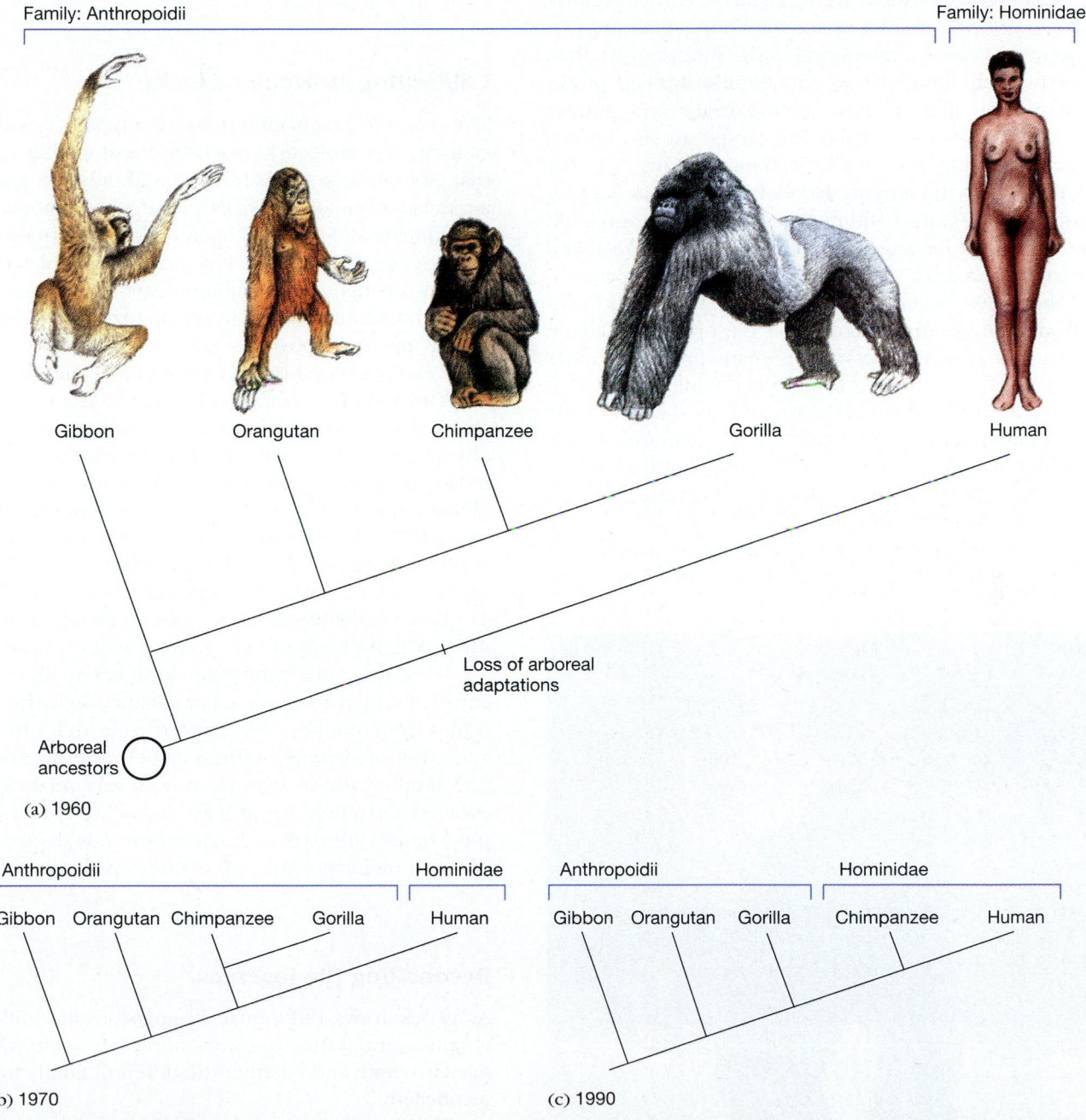

**FIGURE 22–9** Phylogenetic relationships of apes and humans accepted by biologists at different dates. (a) In 1960, based on anatomical evidence, ape and human lines were believed to have diverged relatively early. (b) By 1970, based on evidence from protein structure, chimpanzees, gorillas, and humans had been shown to be the most closely related of the five. (c) By 1990, analysis of DNA and proteins had shown chimpanzees and humans to be so closely related that many biologists would classify them in the same family.

Then, in the 1960s, Morris Goodman analyzed blood proteins and concluded that chimpanzees, humans, and gorillas are more closely related than any of them is to gibbons or orangutans (Figure 22–9b). This became generally accepted, and the anatomical evidence was reinterpreted: it was not surprising that humans had lost the long arms of their ancestors when they became bipedal (walking on two feet). The great apes had retained the long arms of arboreal ancestors because these were still adaptive, not because arm length is a conservative character. Orangutans and gibbons generally live in trees and have longer arms than ground-dwelling chimpanzees and gorillas. Instead of becoming bipedal like humans, chimpanzees and gorillas evolved a faster method of moving on the ground known as knuckle-walking: they walk on their hind feet and the knuckles of their still-lengthy forelimbs, with suitable specializations of the bones and muscles of the wrist and hand (Figure 22–10). It became accepted that chimpanzees and gorillas were the most closely related of the five species.

In the 1980s, several groups analyzed DNA and proteins from many primates and upset the phylogeny again, declaring that chimpanzees and humans are much more closely related than either is to gibbons, gorillas, or orangutans. Human and chimpanzee cytochrome *c* differs by only 1 in 141 amino acids, the proteins of the two species are 99% similar overall, and enormous numbers of their genes are identical. This phylogeny, shown in Figure 22–9c, has now become generally accepted.

Does this mean that when anatomical and molecular evidence conflict, the molecular evidence wins? Not necessarily. Fossil evidence cannot be ignored. In the 1960s, protein analysis suggested that gibbons diverged from the ancestors of humans and other apes about 5 million years ago. However, anthropologists pointed out that this could not be true because fossils nearly identical to modern gibbons dated from 20 million years ago. This means that the common ancestor of gibbons and other apes must have lived more than 20 million years ago, not 5 million years ago.

## Calibrating Molecular Clocks

The secret to reconciling these two types of evidence is to calibrate the molecular clock by fossil evidence. Knowing that gibbons appeared about 20 million years ago, and that a particular homologous DNA sequence in modern gibbons and humans differs by 40 nucleotides, we can estimate that, on average, each DNA sequence accumulated two nucleotide changes every million years. If the same sequence differs by 12 nucleotides in chimpanzees and humans, it is reasonable to suppose that these two species have had separate evolutionary histories for about 6 million years.

This kind of calibration has led to the conclusion that molecular clocks are not well named because they do not always run at the same speed, and their speed varies from molecule to molecule. We have already seen that hemoglobin molecules have changed much more rapidly than ribosomal RNAs. Similarly, the rate of change in DNA has slowed during primate evolution: chimpanzees and humans, separate for the last 6 or 7 million years, have about twice as many similarities in their DNA as chimpanzees and gorillas, which diverged only about 2 million years earlier.

Even the same molecule does not really evolve at a constant rate in all species. For instance, cytochrome *c* molecules from monkeys and chimpanzees differ by six amino acids, but cytochrome *c* from camels and whales is identical, although these two are not closely related, but have evolved separately for at least 70 million years. Molecular phylogenies are most accurate when they depend on many different molecules that are evolutionarily conservative.

## Reconciling Phylogenies

As all this shows, different techniques produce different phylogenies, and it then becomes necessary to decide which is most accurate and whether the different family trees can be reconciled.

The problem of weighing different characters does not disappear merely because we can now compare genomes directly. Genealogy can be obscured unless taxonomists take into account many peculiarities of the ways genomes evolve. For instance, the DNA of most organisms contains "hot spots" where mutation occurs much more frequently

**FIGURE 22–10**  Knuckle-walking by a gorilla.

than it does in other parts of the genome. In addition, genes sometimes move from one species to another. These features of genomes may produce results that make organisms appear more (or less) closely related than they really are.

An example in the early 1990s involved comparison of nucleotide sequences in several genes of crocodiles, birds, and mammals. This showed birds to be much more similar to mammals than to crocodiles although all other evidence, from anatomy and the fossil record, shows that the common ancestor of birds and crocodiles, lived millions of years later than that of birds and mammals. Charles Marshall studied this strange result and found that mutations caused by copying errors in DNA do not all occur with the same frequency. Substitutions that replace a thymine nucleotide with one containing cytosine occur much more often than any other possible substitution. When Marshall programmed a computer to weight nucleotide substitutions according to their likelihood, the genetic analysis agreed with the generally accepted theory that crocodiles and birds are more closely related than birds and mammals.

An additional problem with molecular phylogenies is that molecules for analysis can be extracted only from organisms that are alive or fairly recently dead. Occasionally fragments of DNA are preserved in organisms that died millions of years ago, but this is exceedingly rare. Molecular systematics can tell us nothing about the phylogenies of long-extinct organisms—which make up most of all the species that have ever lived.

## 22–F   THE FIVE KINGDOMS

The most inclusive taxa are the kingdoms. Linnaeus's system of classification had two kingdoms, the plants (Plantae) and animals (Animalia). This seemed reasonable in his day, since land plants and animals were the most familiar organisms, and they were clearly very different. Plants did not move around; they did not eat, but seemed to need only water in order to grow. Animals were **motile;** that is, they could move from place to place. Animals had to eat plants or each other in order to stay alive. On a microscopic level, plants could be seen to have cell walls, which animal cells lacked. Fungi seemed to be aberrant plants, since they had cell walls and root-like structures that absorbed food from living or dead organisms, but lacked the green pigments of the other "plants."

However, as more organisms were discovered and studied, the division into plant and animal kingdoms seemed to become more and more artificial, and the zone between the two became more and more confusing. Some organisms, such as the one-celled *Euglena*, seemed to fit both descriptions. *Euglena* (see Figure 25–17) has a rather stiff covering, not as thick as a plant cell wall but certainly affording more protection than an animal's plasma membrane. *Euglena* also has chloroplasts and can carry on photosyn-

thesis when exposed to light. However, it has a flagellum with which it can swim—an animal-like characteristic—and it can engulf other organisms and digest them as food, just like an animal. Bacteria present another problem since they have cell walls but often also possess flagella that make them motile; most cannot make their own food, but some can carry on photosynthesis.

Recent attempts to revise biological classification at the kingdom level have been many and varied, and you will find several conflicting schemes in use. In this book, we use a modified version of the five-kingdom system popularized by ecologist Robert Whittaker.

The first organisms on Earth probably belonged to kingdom Prokaryotae, which contains all species of organisms with prokaryotic cells. The other four kingdoms, made up of organisms with eukaryotic cells, are separated according to two main criteria: complexity of tissues and mode of nutrition. Kingdom Protista is composed of unicellular eukaryotes and of comparatively simple multicellular forms that do not qualify for one of the other three kingdoms. Kingdoms Plantae, Animalia, and Fungi are distinguished largely by their nutrition. Members of the kingdom Plantae are photosynthetic. Members of the kingdom Animalia are **ingestive;** that is, they engulf or swallow their food and digest it internally. The Fungi are **absorptive:** they absorb organic molecules from outside their bodies directly through their exterior plasma membranes.

### Kingdom Prokaryotae (Monera)

All prokaryotic organisms—that is, the bacteria and cyanobacteria—are placed in the kingdom Prokaryotae (formerly called Monera). Prokaryotes lack nuclei, mitochondria, chloroplasts, and other membrane-bounded organelles found in eukaryotic cells. Their genome consists of one circular DNA double helix with very little protein, and they divide and reproduce without the nuclear divisions of meiosis or mitosis. Prokaryotes also differ from eukaryotes biochemically in such things as their cell wall material, the size and makeup of their ribosomes, and some of their metabolic pathways (Figure 22–11). Although members of this kingdom generally have the smallest and simplest cell structure, they are masters of metabolism: as a group, they have more diverse biochemical abilities than eukaryotes.

### Kingdom Protista

Kingdom Protista contains the unicellular eukaryotic organisms and all the multicellular forms that are not members of the other three kingdoms. As eukaryotes, they have nuclei containing linear chromosomes that can go through mitosis and, in most forms, meiosis. The first eukaryotes were undoubtedly protists, some of which gave rise to the

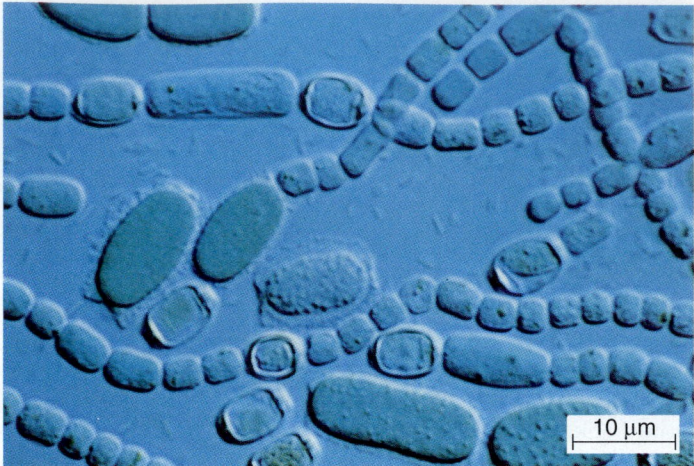

**FIGURE 22–11**    Prokaryotae. Filaments of *Cylindrospermum,* a cyanobacterium, show more cellular differentiation than most other cyanobacteria.    (Biophoto Associates)

fungi, animals, and plants that dominate life on Earth today. Protists show modes of life that parallel those of higher eukaryotes, some being photosynthetic, some ingestive, and some absorptive, some motile and some nonmotile, some with cell walls and some without. All combinations of these characters are represented in various members of this group. It is probable that the eukaryotic condition evolved more than once, in which case protists are polyphyletic. Although protists have neither the metabolic diversity of bacteria nor the embryonic development of plants and animals, as a group they have the greatest diversity of cell anatomy and of specialized organelles, and some of them have the most complex cells of any organisms (Figure 22–12). In contrast, the comparatively uniform chemistry and cell structure of

fungi, plants, and animals has lent itself to a wide variety of larger multicellular structures.

## Kingdom Fungi

Fungi cannot make their own food like plants nor ingest it like animals. Instead, they absorb food through their body surfaces (Figure 22–13). In many cases, they secrete digestive enzymes that digest food outside their bodies before they can absorb it. Fungi have often been classified as plants because they are nonmotile and because they have an external wall that resembles the cell wall of a plant cell. Whittaker felt that fungi should be separated from plants, partly because they are completely absorptive, whereas most plants are only secondarily absorptive. (Plants absorb water, minerals, and carbon dioxide, but make their own organic molecules.) In addition, fungi and plants differ in cell wall composition, body plan, and reproduction. The recent evidence from ribosomal RNA suggests that Whittaker was right in placing plants and fungi in different kingdoms (Section 22–E).

## Kingdom Plantae

In Linnaeus's classification, Plantae could have been called "the walled kingdom" because it contained all the organisms with cell walls. However, twentieth-century biologists have whittled it down, removing the bacteria, fungi, and unicellular photosynthetic protists. Many biologists now feel that multicellular algae should also be removed and classified in the kingdom Protista with their one-celled relatives, which is where we place them in this book.

This new definition means that the plant kingdom now contains only the organisms we generally think of as plants:

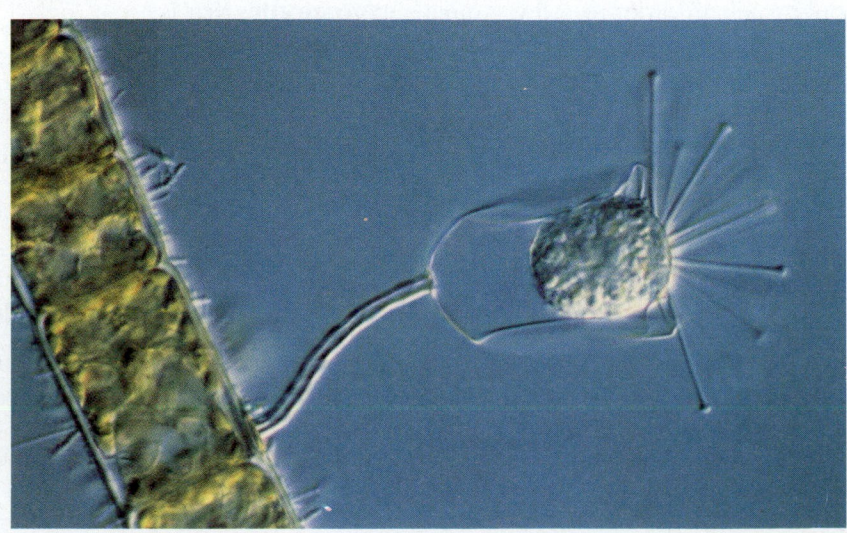

**FIGURE 22–12**    Protista. This single-celled suctorian has a complex structure. It captures prey with its tentacles. Its stalk is attached to another protist: an alga consisting of a string of photosynthetic cells.    (P. W. Johnson and J. Sieburth/BPS).

**FIGURE 22–13** Fungi. A mushroom is the reproductive structure. The main body consists of underground threads, some of which are visible at the base of the stalk.

mosses, ferns, grasses, trees, and shrubs. All are multicellular eukaryotes with cell walls containing cellulose. Most contain chlorophyll and carry on photosynthesis inside chloroplasts, although a few species have lost their chlorophyll and obtain all of their nutrients by absorption. Plants can be thought of as partly absorptive because they obtain water and minerals by absorbing them from their surroundings through the cell walls of their roots and leaves. Most plants live on land, although some have readapted to life in water. They have complex life cycles with alternation of sexual and asexual reproduction, and the sexually produced multicellular embryo develops inside the multicellular female reproductive organ.

## Kingdom Animalia

Animals are multicellular, eukaryotic organisms that obtain food mainly by ingestion. Like plants, they undergo complex cellular differentiation during their embryonic development. Most animals can move, and this permits them to acquire food from their environment (Figure 22–14). All but

the simplest animals produce gametes (eggs and sperm) in multicellular organs, and the fertilized eggs develop into multicellular embryos.

## The Three-Kingdom System

The main problem with the five-kingdom system is that the protists (and possibly plants and prokaryotes) are polyphyletic. Breaking up the world of life into truly monophyletic groups would result in more different kingdoms than we wish to bother with for most purposes. Despite its inadequacies, the five-kingdom system is widely used today, not because it is natural but because it is convenient.

A classification containing only three kingdoms has gained popularity in recent years. The kingdoms are Archaea, Bacteria, and Eucarya. Eucarya contains all eukaryotes (protists, fungi, animals, and plants). Archaea contains the Archaeobacteria, a group that includes bacteria adapted to life in hot springs or to extremely acidic or salty environments (Section 24–D). Bacteria contains all other bacteria.

This new classification resulted from the influence of cladistics and from studies on ribosomal RNA and mechanisms of protein synthesis during the 1980s. Some cladists argue that the five kingdoms contain too many polyphyletic groups. Other biologists concluded that archaeobacteria are so different from other prokaryotes and from eukaryotes that they belong in a separate kingdom.

Some who believe that the five kingdoms are too useful to abolish choose to treat Archaea, Bacteria, and Eucarya as superkingdoms, or **domains.** Others argue that if the goal is to produce taxa that are clades, there is evidence that both Bacteria and Eucarya are polyphyletic, and there is no point in replacing five polyphyletic kingdoms with three.

**FIGURE 22–14** Representatives of kingdoms Plantae and Animalia.

## MOLECULAR PHYLOGENY OF GLOBIN GENES

As an example of what we can learn about phylogeny by analyzing molecular structure, let us consider what has been learned about the globins, polypeptides made up of 140 to 150 amino acids that are found in all vertebrates and some invertebrates.

Globins are best known as the polypeptides that make up oxygen-carrying hemoglobin in the blood. In adult humans and great apes, most hemoglobin molecules consist of two α– and two β–globin chains (see Figure 3–20). In addition, a small amount of adult hemoglobin contains δ–globin in place of β–globin. The fetus produces γ–globin rather than β–globin, and its hemoglobin has a higher affinity for oxygen than does adult hemoglobin. This facilitates gas exchange between mother and fetus. Another globin found in vertebrates is myoglobin, an oxygen-storing protein in muscle.

Because globins are found in virtually all vertebrates, analyzing their structure (or the DNA sequences that code for them) sheds light on the evolution of vertebrates and of the globin molecule itself. Table 22–A shows the number of amino acid differences between human and horse globins and the myoglobin of whales.

By looking at the table we can see that the greatest differences in amino acid sequences lie between whale myoglobin and all the horse and human globins. The greatest similarities are not, as we might expect, between the various human globins. Although human β–globin and δ–glo-

**T A B L E   2 2 – A**

### Number of Differences in Amino Acid Sequences Between Various Globin Chains*

|  | Horse α–globin | Human α–globin | Horse β–globin | Human β–globin | Human δ–globin | Human γ–globin | Whale myoglobin |
|---|---|---|---|---|---|---|---|
| **Horse α–globin** | 0 | 13 | 84 | 86 | 87 | 87 | 118 |
| **Human α–globin** | 13 | 0 | 87 | 84 | 85 | 89 | 115 |
| **Horse β–globin** | 84 | 87 | 0 | 25 | 26 | 39 | 119 |
| **Human β–globin** | 86 | 84 | 25 | 0 | 10 | 39 | 117 |
| **Human δ–globin** | 87 | 85 | 26 | 10 | 0 | 41 | 118 |
| **Human γ–globin** | 87 | 89 | 39 | 39 | 41 | 0 | 121 |
| **Whale myoglobin** | 118 | 115 | 119 | 117 | 118 | 121 | 0 |

*To find the number of amino acid differences between, say, human β–globin and whale myoglobin, look in the square where the column for human β–globin intersects the row for whale myoglobin: 117 amino acid differences.

## SUMMARY

Organisms are named and classified to facilitate communication about them. Modern methods of classification aim to reflect evolutionary relationships.

1. Taxonomy is the branch of biology concerned with relationships among organisms and with their classification. The basic unit of classification is the species. Each species is given a unique Latin binomial, consisting of the genus name and specific epithet. Newly described species are defined both verbally and by preserved type specimens.

2. Species are grouped into progressively more inclusive taxa. The main levels in the taxonomic hierarchy, from most to least inclusive, are: kingdom, phylum, class, order, family, genus, and species. A taxon in each higher level contains one or more taxa of the next lower level. Taxa are artificial units that cut across the natural con-

bin are similar (pink squares), horse α–globin and human α–globin (yellow squares) are much more similar than human δ–globin and human α–globin (green squares).

The most logical explanation for these data is shown in the phylogeny in Figure 22–A. The evolutionary ancestor of globins is an oxygen-carrying protein found in many marine worms, insects, and some fish. The gene for this myoglobin-like molecule must have been present in the common ancestor of horses, whales, and humans. Sometime during vertebrate evolution, this gene was duplicated. This had two advantages. First, a major mutation in one of the genes would no longer prove lethal. Second, the two proteins could evolve independently. As the genes underwent mutations, these produced amino acid changes in the proteins they uncoded. One protein became better suited to oxygen storage in muscle (myoglobin) and the other to

oxygen transport in blood (hemoglobin). Myoglobin and hemoglobin have had separate evolutionary histories ever since.

The greatest phylogenetic distance shown in Table 22–A is the 100+ amino acid difference between myoglobin and hemoglobin. Next comes the difference of 80+ amino acids between α–globin (horse or human) and all the other globins. It appears that the first hemoglobin molecule was duplicated, giving rise to both α– and β–globin, in a common ancestor of horses and humans. Two further duplications of globin genes occurred, permitting the evolution of human δ– and γ–globins. From the amino acid differences shown in the table, we can deduce that two duplications of the β–globin gene permitted first the evolution of γ–globin and later of δ–globin.

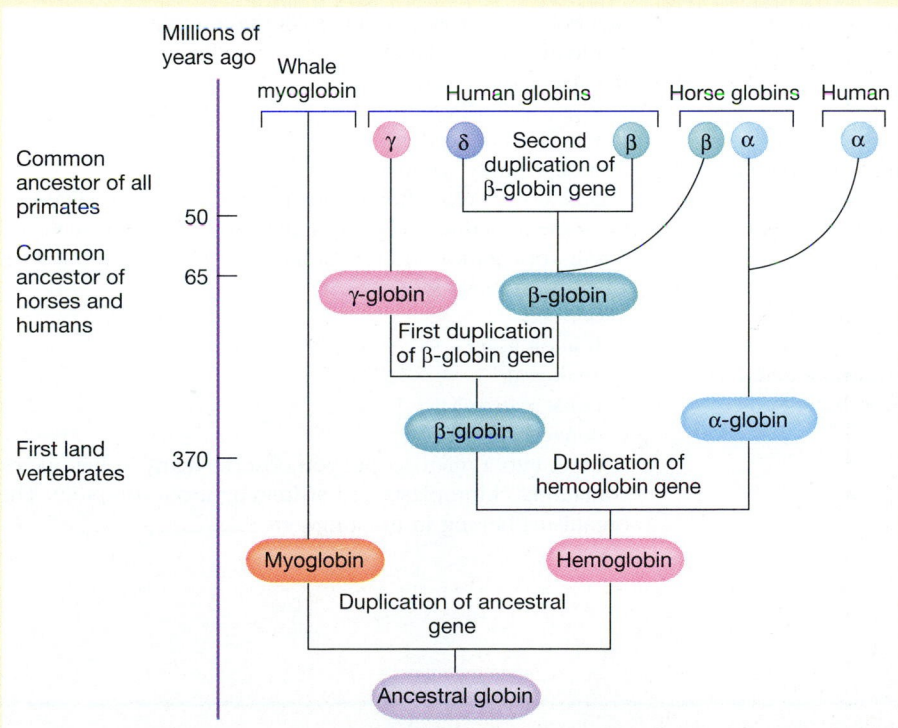

**FIGURE 22–A** Evolution of globin genes deduced from amino acid differences between globins of horses, humans, and whales. Dates and events from fossil evidence are shown on the left.

tinuity of evolution. Biologists have many conflicting views about how the rules of taxonomy should be applied and where the lines should be drawn to define taxa.

3. In theory, living things are classified by phylogenetic relationships, but these are often difficult to disentangle, and the sheer number of existing species precludes drawing up a phylogenetic tree encompassing all

known organisms. In practice, therefore, living things are usually classified by morphology. Other features, such as physiology, biochemistry, behavior, and geographic distribution, are also used.

4. Evolutionary relationships may be obscured by similarities that do not reflect phylogeny, such as those produced by convergent evolution. To determine evolutionary relationships, systematists must identify

conservative characters, which do not change rapidly, permitting them to distinguish between homologous and analogous, ancestral and derived characters.

5. Pheneticists advocate using phenotypic characters as the criteria for setting up a classification scheme, whereas cladists insist on phylogenetic criteria. Evolutionary systematists use both phylogeny and the genetic similarities that give rise to similar morphology, and it is this system, imperfect as it is, that has given us the many different and somewhat conflicting classification schemes in use today.

6. The modern trend in systematics is to base phylogenies on analysis of differences between proteins and nucleic acids in different organisms. As with traditional systematics, there are many potential sources of error, and accurate results depend on choosing conservative molecules and understanding how they change during the course of evolution. Sometimes, molecules evolve at fairly constant rates for periods of time. This permits changes in the molecule to be treated as a molecular clock that can be used to indicate the evolutionary distance between organisms.

7. This book uses a taxonomic system that divides organisms into five kingdoms: Prokaryotae, Protista, Fungi, Plantae, and Animalia. This classification is based largely on the mode of nutrition and cellular organization of organisms.

## Self-Quiz

1. Modern classification of most groups is based on:
   a. taxonomy
   b. phylogeny
   c. phenetics
   d. fossils
   e. autotrophy
2. All of the following make it difficult to construct acceptable classification schemes *except:*
   a. deciding where to impose artificial cutoffs in the midst of a naturally continuous series of organisms
   b. convergent evolution
   c. differences in rates of evolution for different characters
   d. persistence of conservative characters
   e. the large numbers of species that must be accommodated
3. The Latin binomial for the common dog is properly written:
   a. canis familiaris
   b. Canis Familiaris
   c. Canis familiaris
   d. *Canis familiaris*
   e. *canis familiaris*

4. In which of the following lists are the levels of the taxonomic hierarchy *not* arranged in correct descending order?
   a. phylum, order, family
   b. class, family, genus
   c. class, order, family
   d. family, class, order
   e. order, family, genus
5. Characters of two different organisms that have evolved from the same structure in an ancestral form but now have very different appearance and function can properly be termed (give all correct answers):
   a. derived
   b. homologous
   c. analogous
   d. conservative
   e. convergent
6. Peering into a microscope, you observe many individual cells containing chloroplasts and swimming around rapidly. These organisms belong in the kingdom _____.

## Questions for Discussion

1. Although modern taxonomy tries to classify organisms on the basis of their phylogenetic relationships, in practice most organisms are classified according to their morphology. Why is this so?
2. Why is the classification that Linnaeus produced before the theory of evolution so similar to that of modern evolutionary biologists?
3. Why is it so difficult for biologists to agree on a classificatory scheme that a dozen different classifications may be in use at one time?
4. If you set up a new scheme for classifying organisms, what criteria would you use to divide them into kingdoms (nutrition? cell structure? size? habitat?) and why?

## Suggested Readings

Margulis, L., and K. V. Schwartz. *Five Kingdoms: An Illustrated Guide to the Phyla of Life on Earth,* 2d ed. San Francisco: W. H. Freeman and Company, 1988.

May, R. M. "How many species inhabit the Earth?" *Scientific American,* October 1992. Discusses various approaches to estimating this unknown quantity.

O'Brien, S. J. "The ancestry of the giant panda." *Scientific American,* November 1987. How molecular data helped resolve conflicting anatomical and chromosomal evidence about pandas' relationships to bears and raccoons.

Pääbo, S. "Ancient DNA." *Scientific American,* November 1993. The difficulties and rewards of finding DNA fragments for studies of molecular phylogenies in long-dead organisms.

Sibley, C. G., and J. E. Ahlquist. "Reconstructing bird phylogeny by comparing DNAs." *Scientific American,* February 1986. Bird DNA held some surprises for researchers and ornithologists!

Strickberger, M. W. *Evolution.* Boston, MA: Jones and Bartlett Publishers, 1990. Contains excellent chapters on classification and molecular phylogenies.

Whittaker, R. H. "New concepts of kingdoms of organisms." *Science* 163:150, 1969. The paper that popularized the five-kingdom system of classification, with discussion of the rationale for such a system and the problem of drawing boundaries.

23

# Viruses

### O B J E C T I V E S

*When you have studied this chapter, you should be able to:*

1. Discuss similarities and differences between viruses and cellular organisms.
2. Describe the general structure of a virus.
3. Describe lytic and lysogenic cycles and the continuous production of new virus particles by budding from host cells.
4. Explain how the genome of an RNA virus is replicated.
5. Explain why a virus can infect only certain kinds of cells in certain host species.
6. Explain one mechanism by which an oncogenic virus causes a tumor.

V iruses are tiny particles composed largely of nucleic acid and protein but lacking many features of cells. They were discovered in the late nineteenth century, when it was observed that the infectious agents causing smallpox and tobacco mosaic disease were too small to be seen with the light microscope and would even pass through filters that stopped all known bacteria.

Viruses occupy a strange limbo somewhere between the living and nonliving worlds. Like living organisms, they possess genetic material, composed of nucleic acids and capable of mutation and recombination. Viruses can therefore evolve and adapt to their changing environments. On the other hand, viruses do not have a cellular structure, and they have no ribosomes nor the metabolic machinery for protein synthesis and energy generation.

Lacking these components, viruses are necessarily parasitic. They can reproduce only inside living host cells, and even here their reproduction is unique. Cells reproduce by growing and eventually dividing into two new cells, each containing a complete set of the components needed for life. By contrast, viruses are disassembled and separated into their components: nucleic acid genomes and protein coats. The virus takes over the host cell and causes its metabolic machinery to produce a few dozen to hundreds of new viral genomes and thousands of protein subunits to make new viral coats. Then these components are assembled into

new virus particles, the same size as the original one: unlike cells, viruses do not grow (Table 23–1).

Another bizarre feature of viruses is that many of them can be crystallized, a common property of minerals and even of some organic molecules, but certainly not of living cells. Furthermore, crystallized viruses, when wetted and exposed to living host cells, soon establish infections and get back to the business of causing the cell to produce more virus particles.

Because of all these odd characteristics, viruses are not classified in the five kingdoms of organisms. Nevertheless, since viruses are active only inside living cells, and indeed may have devastating effects on their hosts, the study of viruses is the province of biology.

Much of our knowledge of viruses comes from work on **bacteriophages** (often shortened to **phages**), viruses that infect only bacteria. Phages have long been produced by the millions in cultures of bacteria. With modern methods of culturing animal cells, we can now study many of the viruses that infect humans and other animals, a field of obvious economic importance as well as scientific interest.

Many viral diseases are age-old scourges of humans and domestic stock and crops. However, scientists have found a few positive uses for viruses. Because some viruses have very small genomes and are reproduced rapidly, they have been good subjects for studying the nature of genetic ma-

**TABLE 23–1**

**Differences Between Viruses and Cells**

| Character | Viruses | Cells |
|---|---|---|
| Structure | Virus particle: nucleic acid core inside protein capsid | Cell containing nucleic acids, lipid-protein membrane, ribosomes, cytoplasm, etc. |
| Nucleic acids | DNA or RNA, but not both | Both DNA and RNA |
| Enzymes | One or a few; e.g., lysozyme (digests bacterial cell wall), polymerase (replicates viral genome) | Many enzymes; diverse functions |
| Metabolism | None; relies on host cell metabolism for monomers, protein synthesis machinery, and some enzymes of nucleic acid synthesis | Makes own ribosomes and enzymes needed for synthesis of proteins, nucleic acids, etc. |
| Reproduction | Nucleic acid genome and capsid proteins produced separately, then assembled into virus particles | Division into two similar cells following growth |

terial and the control of gene expression. In addition, viruses are sometimes used as vectors for gene transplants. Medical researchers hope to find ways to treat some human disorders by modifying viruses to carry therapeutic genes into specific cells. For example, it may someday be possible to alleviate Alzheimer's, Parkinson's, or Huntington's disease by using herpes simplex viruses to carry certain genes into the nerve cells of the brain.

## KEY CONCEPTS

◆ A virus consists of a nucleic acid genome enclosed in a protein coat, which in some forms is surrounded by a membrane.

◆ Viruses rely on the living cells they invade and parasitize to provide the energy and materials needed for their reproduction.

◆ Because viruses contain genetic material, they can evolve and become adapted to their environment.

◆ Viruses cause many devastating diseases of humans, other animals, plants, and even bacteria.

## 23–A    VIRUS STRUCTURE

Viruses are extremely small, ranging from about 15 to a few hundred nanometers in diameter. Only the largest surpass the smallest bacteria in size. The "basic" virus particle consists of a nucleic acid molecule, containing the viral genome, and a surrounding protein coat, the **capsid.** In addition, some viruses have a membranous outer envelope of glycoprotein and lipid covering the capsid; most of these enveloped viruses infect animals. Virus particles may also contain one or a few enzymes—those needed for invasion of a cell and for replication of the viral nucleic acid.

The capsid protects the genetic material in its passage from one host cell to another. Proteins in the outermost capsid or envelope layer also bind specifically to receptor molecules on the host cell's surface, the first step in invading a cell, which is why each virus can invade only one or a few cell types.

The capsid is made up of a number of protein subunits, and their organization determines the virus particle's shape: most viruses are either helical or polyhedral, or a combination of the two. A more complex structure occurs in the T-even bacteriophages—T2, T4, and T6—which infect the bacterium *Escherichia coli* (Figure 23–1). Some viruses can self-assemble when their nucleic acids are mixed with capsid subunits under appropriate conditions. Others cannot self-assemble, but must be assembled by components that are not found in the finished product.

The viral genome consists of a molecule of DNA or RNA, but not both, which may be either single- or double-stranded. Some RNA viral genomes consist of more than one molecule. For example, influenza viruses have eight. RNA genomes are linear, whereas DNA genomes may be either linear or circular. The number of genes varies, from four in the small bacteriophage Qβ to about 250 in the large poxviruses.

In something as small as a virus particle, space is extremely limited. There is room only for genes that code for necessities, such as viral coat proteins and nucleic acid polymerases, enzymes needed to take over the host cell,

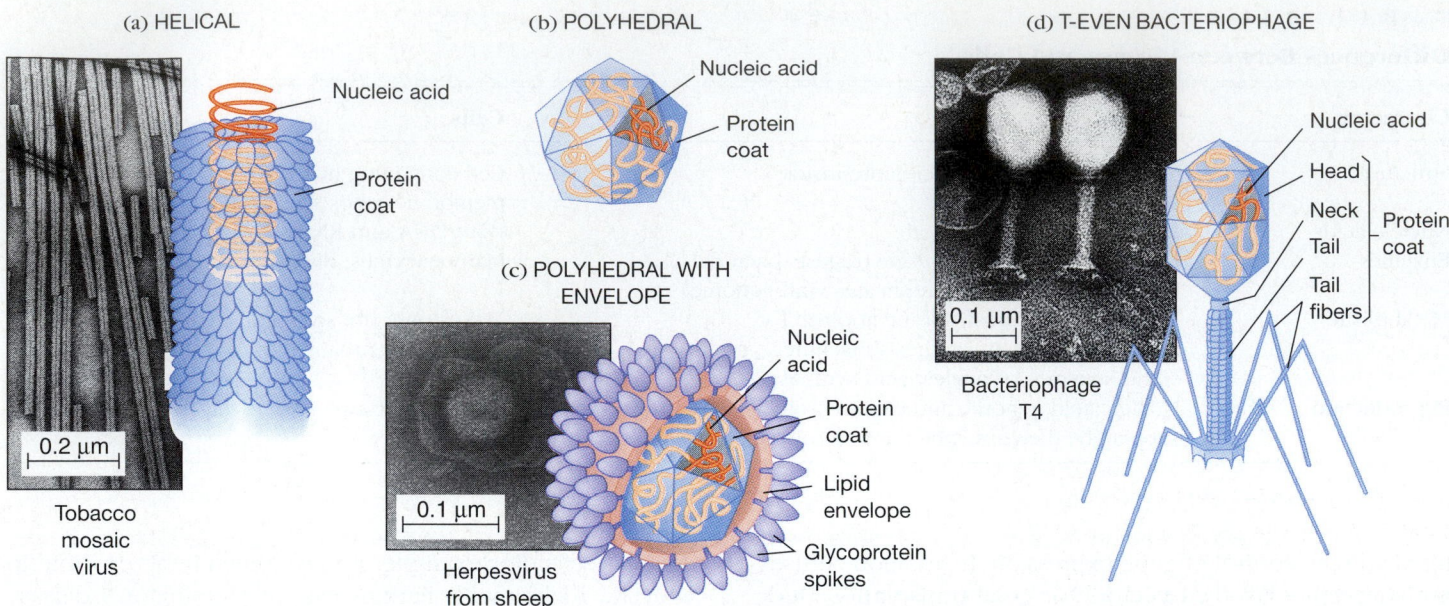

(a) HELICAL

Nucleic acid

Protein coat

0.2 μm

Tobacco mosaic virus

(b) POLYHEDRAL

Nucleic acid

Protein coat

(c) POLYHEDRAL WITH ENVELOPE

0.1 μm

Herpesvirus from sheep

Nucleic acid

Protein coat

Lipid envelope

Glycoprotein spikes

(d) T-EVEN BACTERIOPHAGE

0.1 μm

Bacteriophage T4

Nucleic acid

Head

Neck

Tail

Tail fibers

Protein coat

**FIGURE 23–1** Virus structure. (a) In helical viruses, the nucleic acid is wound inside a coat of repeating protein subunits arranged in a helical pattern. (b) A polyhedral virus has a capsid in the shape of an icosahedron, a geometric solid having 20 faces. (c) Enveloped viruses have a membranous outer envelope around a helical or (as here) polyhedral capsid. (d) T-even bacteriophages, which attack *Escherichia coli,* have more complex capsids. The head encloses the DNA genome. The tail fibers attach to the bacterial cell wall and contract to inject the DNA into the host cell. (a, Biophoto Associates; c, J. Thorsen, University of Guelph, Ontario Veterinary College; d, Carolina Biological Supply Company)

and regulatory genes for production of new viral components. In enveloped viruses, the envelope proteins are encoded by the viral genome, whereas the lipids are appropriated from the host cell's plasma membrane.

Why can't the virus simply use the host's polymerase enzymes to replicate the viral genome? The smallest DNA viruses of animals do. However, a cell's DNA polymerase works in such a way that DNA is copied only once per cell cycle. A DNA virus must ensure that its genome is copied many times in a much shorter period. Therefore, among the larger DNA viruses, a viral polymerase is necessary, specific to the viral DNA and able to produce multiple copies. For RNA viruses, there are more complications, as we shall see (Section 23–C). RNA viruses are generally replicated in the host's cytoplasm, DNA viruses in the nucleus, with a few exceptions.

Viruses are classified by their nucleic acid (DNA or RNA), by capsid shape and size, and by presence or absence of an envelope. Table 23–2 lists major groups of viruses that infect animals.

## 23–B  VIRAL REPRODUCTION

Three main types of reproductive cycles have been found among viruses.

### The Lytic Cycle

A **lytic cycle** occurs when a virus invades a cell, is reproduced, and is dispersed when the cell breaks open, or **lyses.** The virus destroys the cell's DNA, takes over its metabolic machinery, and causes the cell to make new virus particles. Lysis of the host cell releases these particles, which disperse and infect new host cells.

For example, when a T-even phage collides with an *Escherichia coli* cell, it attaches to the bacterium by specific binding between its tail fibers and receptors on the outer surface of the *E. coli* cell (Figure 23–2). An enzyme (phage lysozyme) in the phage tail then breaks down part of the bacterial cell wall. The phage's tail sheath contracts, injecting the viral DNA into the cell.

Soon the bacterium's DNA is broken down, and only viral DNA is transcribed into messenger RNA. The host cell's metabolic machinery still functions, but it now produces phage DNA and proteins. The phage DNA encodes proteins such as enzymes that break down the host DNA and enzymes that assemble the completed phage protein into a coat. New phage DNA and coat proteins are assembled into whole virus particles (Figure 23–3). Finally, a phage-encoded lysozyme digests the bacterial cell wall from within. About 20 minutes after infection, the cell lyses, releasing perhaps 200 new phage particles.

**T A B L E   2 3 – 2**

## Major Groups of Viruses That Infect Animals

| Group | Size (nm) | Capsid | Comments | Some Diseases Caused |
|---|---|---|---|---|
| **Viruses with DNA Genomes (Most Double-Stranded)** | | | | |
| Poxviruses | 100 × 200 × 300 | Helical, brick-shaped | Very large; complex structure; DNA replicated in host cytoplasm | Smallpox, cowpox, myxomatosis in rabbits, diseases in fowl |
| Herpesviruses | 150–200 | Polyhedral, enveloped | Often cause latent infections | Human oral and genital herpes infections, varicella zoster, Epstein-Barr infections, tumors, etc. |
| Adenoviruses | 70–80 | Polyhedral | Very variable base content | Human respiratory and intestinal infections, conjunctivitis, sore throat, tumors, etc. |
| Papovaviruses | 45–55 | Polyhedral | Small circular DNA | Human warts; cancers in other animals |
| Parvoviruses | 20 | Polyhedral | Very small; some have single-stranded DNA | Infections in rodents, pigs, arthropods; in humans with other viruses |
| **Viruses with RNA Genomes (Most Single-Stranded)** | | | | |
| Paramyxoviruses | 100–300 | Helical, enveloped | | Human rubeola, mumps; canine distemper; Newcastle disease of chickens |
| Myxoviruses | 80–120 | Helical, enveloped | Genome has more than one molecule | Influenza of humans, other animals |
| Retroviruses | 100 | Polyhedral, enveloped | Contain reverse transcriptase | Rous sarcoma of chickens; mouse mammary tumor; feline leukemia; AIDS (acquired immune deficiency syndrome); adult T-cell leukemia |
| Rhabdoviruses | 70 × 180 | Helical, bullet-shaped, enveloped | | Rabies, various infections of mammals, fish, insects |
| Reoviruses | 70–80 | Polyhedral | Double-stranded RNA; much like some plant viruses | Vomiting and diarrhea, fatal to children in developing nations; Colorado tick fever; infections in other animals |
| Togaviruses | 50–60 | Polyhedral, enveloped | Large, diverse group | Human rubella, yellow fever, dengue, equine encephalitis, etc. Semliki Forest virus in this group |
| Picornaviruses | 20–30 | Polyhedral | Small | Intestinal infections (enteroviruses), poliomyelitis, common cold (rhinoviruses), hepatitis A, foot-and-mouth disease |

A cell invaded by a lytic virus is almost invariably killed by it within a very short time. Hence, lytic viruses are virulent (extremely damaging). Lytic cycles are typical of many phages and also of viruses that cause colds, poliomyelitis ("polio"), and viral meningitis.

## The Lysogenic Cycle

In contrast to the virulent lytic cycle is the **lysogenic cycle** typical of other phages. These **temperate** phages may either go through a lytic cycle and destroy their host cell, or may instead enter a dormant phase in which their DNA is joined to the host's and replicated with it in each cell generation (Figure 23–3). A host cell containing such a temperate phage is called a lysogenic cell. Various external stimuli can cause a lysogenic cell's phage DNA to enter the lytic cycle, and the cell soon dies, releasing many intact phages. Herpes viruses in animals, such as those that cause cold sores in humans, show similar cycles.

Some lysogenic bacterial cells are medically important. *Corynebacterium diphtheriae,* the bacterium that causes diphtheria, synthesizes the toxin responsible for the disease only when it contains a particular phage, which carries the

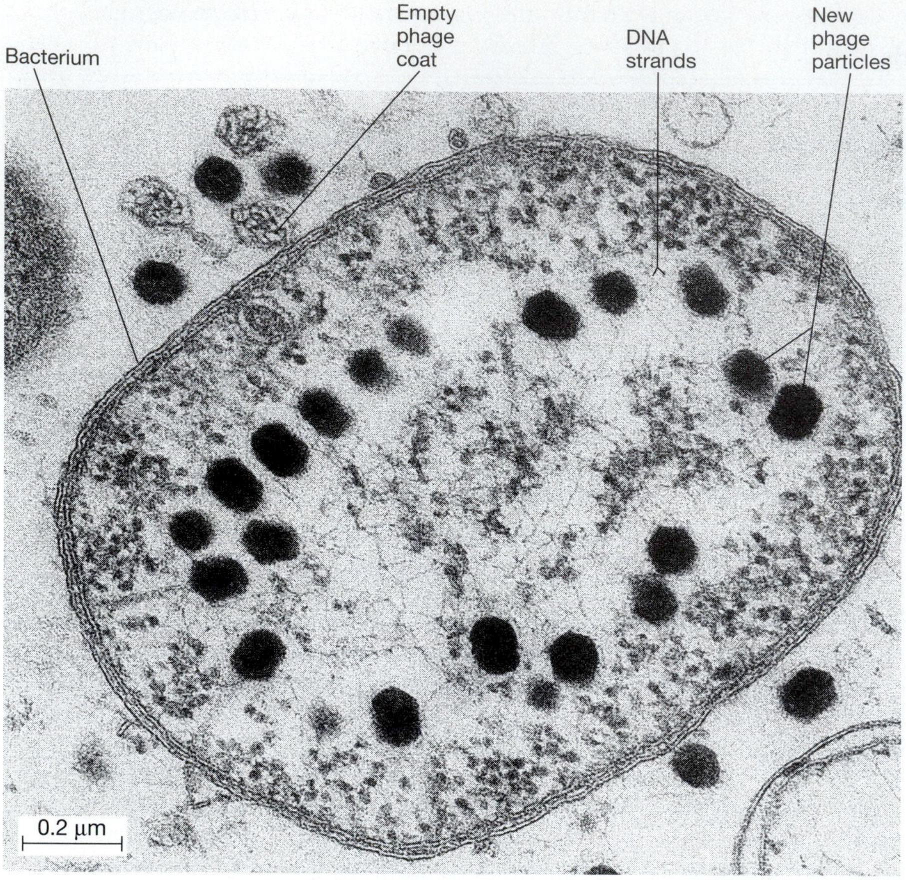

Bacterium

Empty
phage
coat

DNA
strands

New
phage
particles

**FIGURE 23–2**  A cell of the bacterium *Escherichia coli* infected with bacteriophage T2. (TEM)   (L. D.) Simon, *Virology* 38:287, 1969)

0.2 μm

**FIGURE 23–3** Reproduction of bacteriophages. In a lytic cycle, the phage takes over the host cell and destroys it. In a lysogenic cycle, the host may survive for many generations with the phage genetic material incorporated into the host genome, until some condition triggers the phage to become lytic.

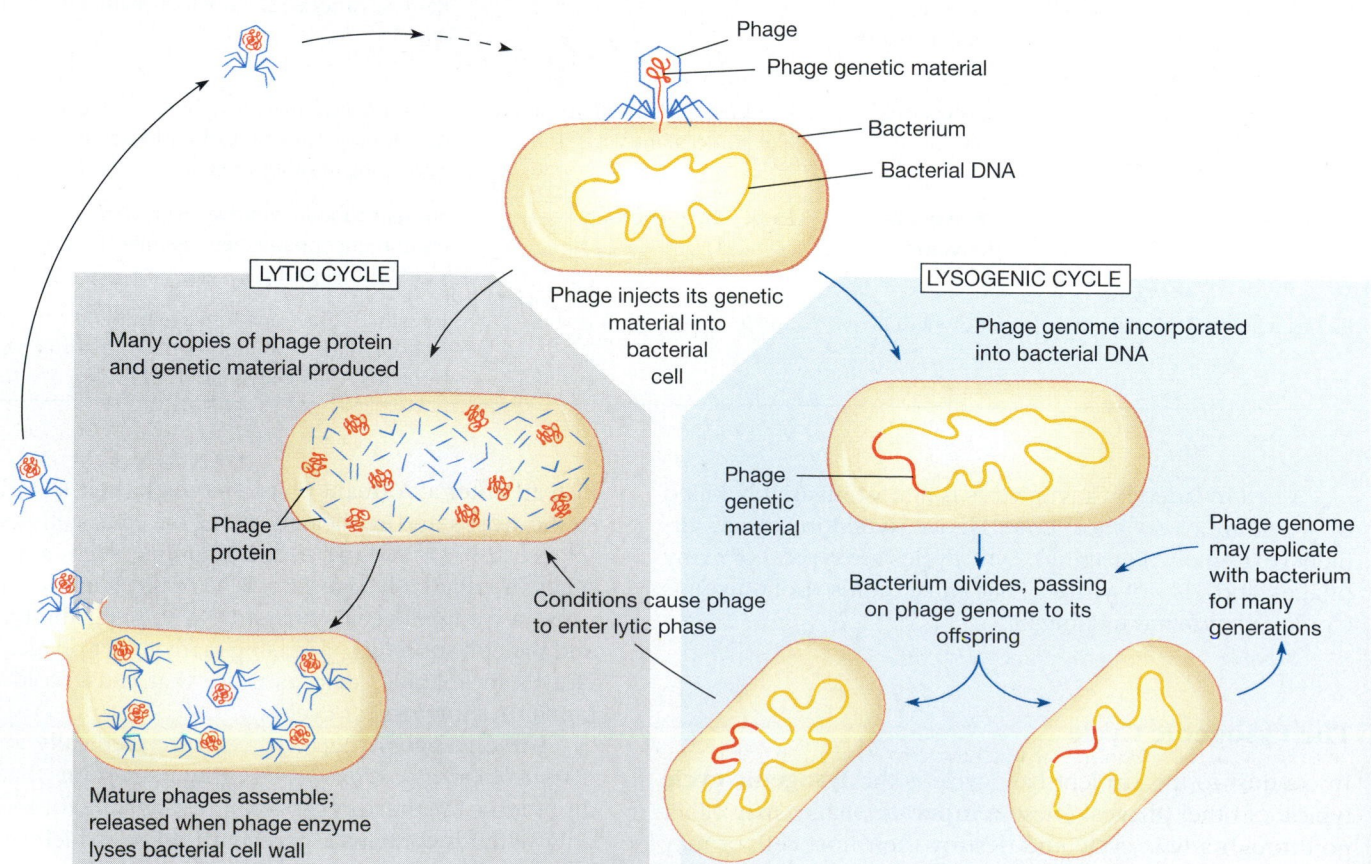

Phage

Phage genetic material

Bacterium

Bacterial DNA

Phage injects its genetic material into bacterial cell

LYTIC CYCLE

LYSOGENIC CYCLE

Many copies of phage protein and genetic material produced

Phage genome incorporated into bacterial DNA

Phage protein

Phage genetic material

Conditions cause phage to enter lytic phase

Bacterium divides, passing on phage genome to its offspring

Phage genome may replicate with bacterium for many generations

Mature phages assemble; released when phage enzyme lyses bacterial cell wall

gene encoding the toxin. The same is true of *Clostridium botulinum* (which causes botulism) and *Streptococcus pyogenes* (which causes scarlet fever).

Phages released when a lysogenic cell finally lyses may carry along a piece of the bacterial DNA, which is later inserted into the DNA of their next hosts. This process, called **transduction** ("leading across"), produces genetic recombination in the second bacterium. Viruses that can carry DNA from one cell to another are used in some gene transplant procedures (see Figure 13–16) and in studies of bacterial genetics.

## Budding from Living Host Cells

In lytic and lysogenic cycles, virus particles are assembled inside the host cell and released when the cell lyses. A few bacteriophages, and many animal viruses, are produced and released continuously by budding from intact host cells.

A well-studied example is the Semliki Forest virus, an RNA virus with a membranous envelope surrounding its capsid. This virus enters its animal host cell by endocytosis: glycoproteins in the viral envelope bind to receptor pro-

teins in the cell's plasma membrane, which engulfs the virus into a vesicle in the cytoplasm. Here the vesicle fuses with a lysosome, a membranous sac of digestive enzymes, which ordinarily ought to digest the vesicle's contents (Section 5–J). However, the viral envelope quickly fuses with the lysosomal membrane and opens a passageway into the cytoplasm for the **nucleocapsid:** the capsid with its enclosed nucleic acid.

Next the capsid disassembles, releasing the RNA into the cytoplasm. Here the RNA directs the host cell to replicate the RNA genome and to produce proteins for the viral capsid and envelope. Capsid proteins are made in the cytoplasm. Envelope proteins are produced in rough endoplasmic reticulum, finished in Golgi complexes, and incorporated into the plasma membrane.

New copies of the viral genome and capsid proteins combine in the cytoplasm. Then these new nucleocapsids move to the host's plasma membrane, where the viral capsid and envelope proteins bind together and the plasma membrane bulges out, forming the envelope as the virus buds off from the host cell (Figure 23–4). The cell may or may not be killed.

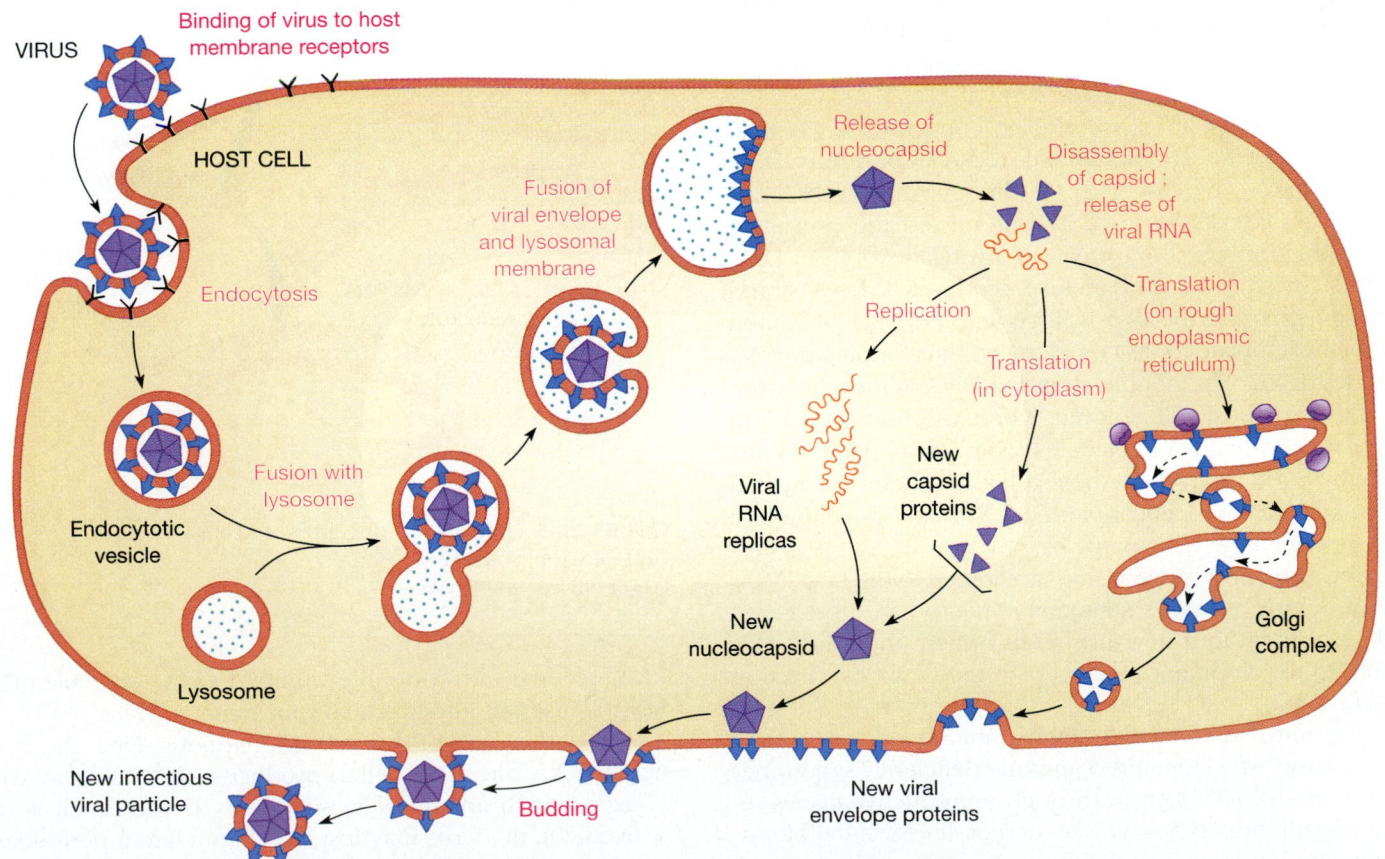

**FIGURE 23–4** Continuous virus production. At the top left, a virus enters an animal cell and causes it to make many new viral genomes and protein coats, which combine in the cytoplasm and form nucleocapsids. These move to the plasma membrane, where they associate with new viral envelope proteins, also produced by the host under the direction of viral RNA. The finished virus particle buds off from the plasma membrane (bottom).

Other enveloped animal viruses that bud from their host cells in this way include influenza, measles, mumps, and rabies viruses.

## 23–C RNA VIRUSES

Viruses with DNA genomes use many host cell enzymes to replicate and transcribe their DNA. However, uninfected cells do not contain RNA-copying enzymes. Hence, RNA viruses must have their own enzymes for replication.

In some RNA viruses, the genome binds to host ribosomes and serves as messenger RNA for the synthesis of RNA-copying enzymes. For example, polio virus RNA first acts as messenger RNA coding for an RNA replicase enzyme and viral coat proteins. It then serves as a template as the newly made RNA replicase makes more RNA genomes.

Other viral RNA genomes cannot act as messenger RNA but must first be copied into complementary RNA that can. Because a new host will not have the necessary enzymes, these are made on the previous host's ribosomes and packaged in the virus particle.

In the reproduction of single-stranded viral RNA, this genome strand usually serves as the template for a complementary RNA strand. This second strand then acts as a template for new copies of the viral genome, identical to the original RNA strand (Figure 23–5).

**Retroviruses** are RNA viruses containing a peculiar enzyme, **reverse transcriptase (RNA-dependent DNA polymerase)**. The discovery of this enzyme in 1970 astonished biologists because it contradicted the prevailing dogma that biological information always flows from DNA to RNA. In fact, this polymerase produces a DNA strand complementary to the virus's RNA genome! This DNA then acts as the template for its complement, thus forming double-stranded DNA. The double-stranded DNA is then incorporated into the host cell genome, where it is transcribed into many new viral RNA molecules. New virus particles disperse by budding from the host cell's membrane. If the cell is not killed, its descendants inherit the viral infection and also produce new virus particles.

Because retrovirus genomes are joined to the host DNA, new retrovirus particles sometimes incorporate host genes and carry them into new host cells, where they may contribute to development of cancers in the new host (Section 23–D).

The retrovirus called human immunodeficiency virus (HIV) causes AIDS (acquired immune deficiency syndrome) (Figure 23–6). HIV can go through reproductive cycles resembling all three types we saw earlier. It enters the bloodstream and binds specifically to receptors on a T helper cell (part of the immune system; Chapter 34). Inside the cell, the virus loses its protein coat, releasing its RNA genome and reverse transcriptase enzymes. The enzyme copies the

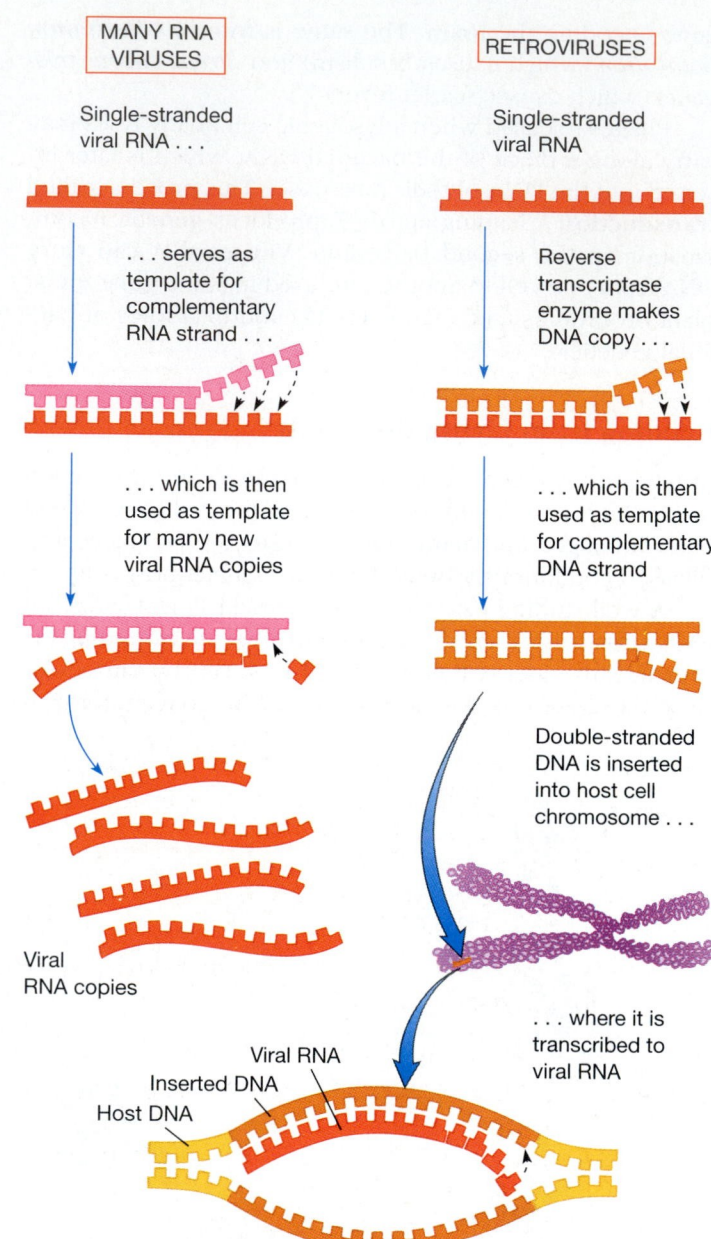

**FIGURE 23–5** Replication of genomes of RNA viruses.
(left) RNA ⟶ RNA ⟶ RNA.
(right) RNA ⟶ DNA ⟶ DNA ⟶ RNA.

RNA genome into a double-stranded DNA molecule. This DNA is inserted into a cell chromosome.

After this, one of three things happens. First, the virus may quickly direct the cell to produce a flood of new AIDS viruses, which are released so rapidly that the cell is destroyed. Or, the virus may instead enter a latent period lasting up to ten years before it is activated and causes the cell to churn out a destructive horde of viruses. Third, the virus may cause a persistent infection, with production of new viruses slow enough so that few host cells are killed. AIDS

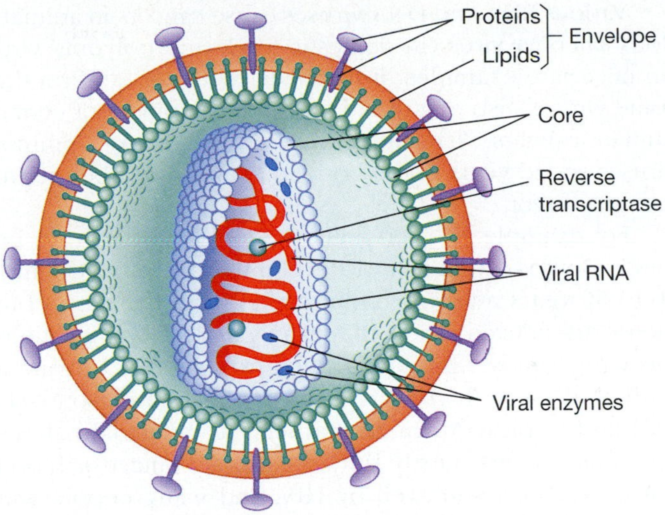

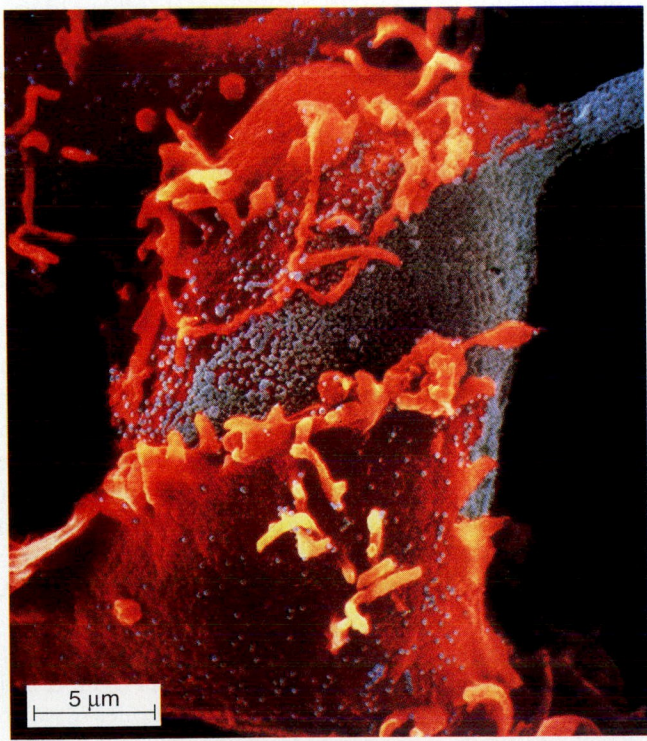

Proteins ⎤
          ⎦— Envelope
Lipids ⎤

Core

Reverse
transcriptase

Viral RNA

Viral enzymes

**FIGURE 23–6**   Human immunodeficiency virus (HIV).

occurs when so many T helper cells have been destroyed that the body's immune system can no longer fight off diseases.

Although reverse transcriptase was discovered in retroviruses, it is not unique to them. DNA viruses such as hepatitis B virus produce more of their DNA genomes by reverse transcription of an RNA intermediate. Some transposons (Section 9–H) also encode reverse transcriptase.

## 23–D   VIRAL DISEASES OF PLANTS AND ANIMALS

Before a virus can invade a cell, its outer proteins must bind to protein receptors on the host cell surface. Because of this specificity, each kind of virus can infect only host cells of particular species and of particular kinds of tissues. Polio viruses can infect only humans and a few other primates (monkeys and apes); the common cold virus attacks the cells lining the human respiratory tract but not our other tissues.

Virus diseases of plants include tobacco mosaic, tobacco necrosis, alfalfa mosaic, rice dwarf, and wound tumor. Plant viruses may be spread by wind or insects. It is usually impossible to cure diseased plants. Instead, farmers try to prevent viruses from spreading, by practicing soil hygiene and by burning infected plants, and breeders develop virus-resistant strains of important crop plants.

Viral diseases of animals include rabies, chickenpox, polio, colds, influenza, warts, AIDS, and some forms of cancer. Various herpes viruses that cause eukaryotic cells to become lysogenic may contribute to some human cancers. Different herpes viruses cause cold sores, sexually transmitted infections, and mononucleosis. Even after the symp-

toms of the infection have disappeared, the virus apparently remains in a lysogenic form, and can enter the lytic cycle when the person is ill or stressed, causing a fresh outbreak of cold sores or genital sores.

Most viruses cause disease by disturbing the infected cell's metabolism and eventually destroying it. The virus is reproduced in the dying cell (Figure 23–7). The new virus particles then invade neighboring cells. The symptoms of viral infections, such as fever and swollen lymph nodes, are caused not by the viruses but by the activity of the body's immune system, which destroys most viruses before they can cause serious damage (Chapter 34).

In the 1980s, researchers found that viruses sometimes have subtler effects. On occasion, they may cause persistent infections without leaving the usual evidence, namely anti-viral antibodies in the blood and structural changes to the infected cells. Instead, the virus may diminish the expression of some nonvital host cell genes, such as those encoding hormones or other chemical messengers. The resulting chemical deficiency in turn causes conditions such as stunted growth, reduced function of the brain or immune system, or diabetes. For example, about one fourth of human fetuses infected with rubella (German measles) later develop diabetes.

Because viruses rely so heavily on host cell machinery for their reproduction, it is extremely difficult to find drugs

5 μm

**FIGURE 23–7**   A human cell rupturing as it releases newly formed viruses (blue) into the surrounding fluid.   (Photo by Lennart Nilsson, © Boehringer Ingelheim International, GmbH)

that will destroy them without also damaging the host. The antiviral drug acyclovir, used externally to treat herpes simplex infections and warts, works by interfering more with the virus's DNA polymerase than with the host's. However, some strains of herpes have evolved resistance to this drug since its introduction in 1982.

Our best defense against virus diseases is still prevention: good hygiene, vaccination, and quarantine of infectious cases. Thanks to such practices, the age-old scourge of smallpox became extinct in the late 1970s, and new vaccines put the viruses causing polio and rubella on the endangered list in the United States.

However, viruses still pose a real threat to human welfare. The measles virus bounced back from near-extinction in the United States after many people decided that it was not necessary to have their children vaccinated. Today, 2 million people become infected with polio virus each year in parts of the world unable to afford vaccine. At least 10 million people worldwide are believed to be infected with HIV, which is fatal once it develops into AIDS. About 20% of those who develop AIDS also have cytomegalovirus infections in the retina of the eye. This causes them to go blind before they die. Many HIV-infected people also have hepatitis B virus infections.

In 1980 no one could have predicted this devastating epidemic of AIDS and its companion viral and bacterial infections. Many other dangerous viruses lurk in remote areas of the world. They are likely to cause deadly epidemics in the future if they come into contact with new populations of humans as a result of activities such as forest clearing, highway building, travel, or movement of local rural populations to big cities. Another danger is that, at any time, a common virus could mutate into a deadlier form, such as the strain of influenza that killed an estimated 20 million people in 1918.

### Viruses and Cancer

"Cancer" is a lay term for a variety of abnormal growths, more properly called **tumors.** Some tumors are caused by viruses, which are therefore called **oncogenic** ("tumor-causing") viruses. As with the temperate viruses discussed before, the viral genome becomes integrated into the host's and is inherited with it. This integrated oncogenic virus may **transform** a host cell into a cancerous state. (This process is not related to bacterial transformation, Section 9–A.) Transformed cells often undergo drastic changes in structure and metabolism such that they and their descendants become unresponsive to the normal controls over cell division. The viral genome remains and multiplies with the host cell's as the tumor grows. More than 20 different vertebrate oncogenes (cancer-causing genes) have been found in various retroviruses, which introduce them into new host cells (Section 11–H).

Various RNA and DNA viruses cause cancers in animals; Rous sarcoma virus (in domestic fowl) and polyoma virus (in mice) are examples. It has been harder to confirm that some viruses also contribute to the development of certain human cancers. This is partly because many common viruses spend years in the body in a latent state before initiating a tumor.

For example, the 300 million people infected with hepatitis B virus are at risk of developing cancer of the liver 30 to 50 years after infection, making this the most common virus-induced cancer in humans. In developing nations, this virus is most often transmitted from mother to child at birth, and more than 10% of the population of southeast Asia and tropical Africa are carriers. In industrial nations, the virus spreads mostly through sexual contact or blood, the same routes followed by HIV, and many people with AIDS harbor both viruses. Health care workers are also at risk because their work brings them into contact with infected blood. The first genetically engineered vaccine available for use in humans is directed against this virus, and all health care workers are being urged to become immunized.

Tables 23–2 and 11–2 list viruses associated with cancers.

### 23–E    VIROIDS

**Viroids** are short, single strands of RNA, even smaller than viral genomes, which cause infectious diseases in various kinds of plants. Like viruses, viroids can reproduce only in living cells. Viroids have no capsids—not surprisingly, since their total length of only 250 to 400 nucleotides is too short to encode the proteins to make such a coat. Indeed, infected host cells do not appear to contain any proteins coded by the viroid genome. Furthermore, viroid RNA is not translated into protein by ribosomes in test-tube systems, nor does it interfere with the translation of host messenger RNA molecules in the system. Because the viroid seems not to code for any proteins, the host's own enzymes must be used to replicate the viroid's RNA. This is done by RNA-to-RNA replication. Therefore, it is presumed that the host cell must normally contain an as yet unknown RNA-copying enzyme.

If viroids do not introduce any foreign proteins into their host cells, how do they cause disease? We don't yet know, but since they are generally found in the cell nucleus, there are two main hypotheses: that they interfere with intron splicing or that they act as RNA regulators of host genes (Section 11–G). In fact, it has been proposed that viroids are escaped introns, because the two contain some nucleotide sequences in common.

Viroid infections are spread by pollen, by humans and their tools, and at low rates by aphids (see Figure 20–3).

Many wild and cultivated plants harbor viroids without showing symptoms of disease. When these viroids invade susceptible crops growing nearby, they can wreak havoc. For instance, a large proportion of grapevines contains viroids, which do not harm them but cause a devastating stunt disease if they infect fields of hops. Viroids have killed more than 10 million coconut trees in the Philippines, and they decimated the United States chrysanthemum trade in the 1950s. Viroids also cause spindle tuber of potatoes and exocortis of citrus trees (Figure 23–8).

## 23–F    VIRUSES AND EVOLUTION

The peculiarities of viruses raise the question of their evolutionary origin. Several answers have been proposed. The first was that viruses are evolutionary relics, descended from ancestors that never evolved into true cells. When biologists realized that viruses depend totally on the complex metabolic machinery of living cells, this idea was discarded.

Second, viruses may be reduced cells, which became parasites inside other cells and eventually jettisoned most of their own cell components and genes. These were, after all, readily available in their host cells. Some fairly large viruses, containing dozens of genes and surrounded by a lipid-protein membrane, might be viewed as stripped-down cells.

A third view is that viruses are neither retarded pre-cells nor regressed cells, but renegade genes that must return "home" to be replicated. This view is supported by the fact that the genetic similarity between virus and host is much greater than between one virus and another. According to this scenario, viral genes originated as host genes. They increase their own evolutionary success at the expense of their former peers by escaping the usual "one for all, and all for one" routine in which the entire genome is replicated together. This view also fits with the fact that viruses often pick up bits of the host genome and transfer it to another host cell, where it may become part of the genome.

A fourth possibility explains the genetic similarity between virus and host by assuming that viruses have captured and incorporated useful host cell genes during their evolution. In this view, viruses may have started as small independently replicating plasmids such as those found today in some bacteria, yeast, and mammalian cells (see Figure 13–3).

Whatever the case, viruses clearly play a role in the evolution of cellular organisms. First, they exert selective pressure by killing or weakening large numbers of hosts. Second, by moving genes from one host to another and by sometimes becoming part of the host genome, viruses cause what are essentially large mutations. Viruses even transfer bits of genetic material between species. Studies show that viral genes have become permanent parts of most species'

(a)

(b)

**FIGURE 23–8**    Viroid disease. (a) Left: healthy citron leaves. Right: leaf curl produced by exocortis viroid. (b) The viroid causes scaling of the bark of this orange tree.    (L. W. Timmer, University of Florida Citrus Research and Education Center)

genomes; perhaps viral genes have contributed useful proteins to modern cells.

Viruses themselves often evolve very rapidly. RNA viruses, notably those that cause colds, influenza, and AIDS, mutate especially often. This is because RNA polymerases do not proofread and correct their handiwork as DNA polymerases do (Section 9–C). Viruses may also undergo genetic recombination, as happened in 1977 when two strains of influenza struck one person at the same time and gave rise to new virus particles containing genes from each strain. Frequent genetic change gives viruses a great selective advantage: some new mutant viruses can always find some

A mysterious serial killer is on the loose. Its victims: thousands of domestic sheep, goats, cattle, and mink; captive mule deer and elk; even humans. They do not die at once, but appear normal for 2.5 to 8 years. Then they begin to behave strangely and succumb within months. At autopsy, their brains show spongy degeneration, with deposits of fibrous proteins.

The suspect: a prion, a small infectious particle consisting of an abnormal form of a cell protein and apparently not containing any nucleic acid. "Prion" stands for proteinaceous infectious particle.

Sheep and goats infected with these particles develop a disease that is called scrapie because the animals scrape themselves against fixed objects (Figure 23–A). Similar diseases occur in other species of mammals. "Mad cow disease" claimed more than 28,500 cattle in Britain between 1986 and 1991. This outbreak was traced to feed containing undercooked sheep by-products. In the United States, a similar disease occurred in mink fed raw cattle carcasses.

Two rare diseases of the human central nervous system, Creutzfeldt-Jakob disease and kuru, are attributed to similar particles. Creutzfeldt-Jakob disease usually begins with dementia (mental deterioration), kuru with loss of coordination. Kuru, found among some natives of New Guinea, was probably transmitted by handling the brains of deceased, diseased relatives as part of the funeral rites. This is no longer done, and cases of kuru are decreasing.

Other people have contracted the disease from medical procedures, such as injection of hormones extracted from the human pituitary (a gland attached to the brain), grafts of meninges (tissues covering the brain), or unsterilized instruments used in brain surgery.

Researchers seeking the causes of these diseases found that infected animals contained glycoprotein particles with one major polypeptide. They analyzed part of its amino acid sequence, made the corresponding DNA, and used it as a probe in nucleic acid hybridization experiments (Section 13–A). They found that the host cell genome itself contains a gene coding for this protein, which normally occurs on the outer surfaces of neurons.

Apparently, a prion acts as a template directing normal protein molecules that become partially unfolded to refold like itself rather than resuming their original shape. In the prion conformation much more of the molecule is arranged in beta pleated sheets (see Figure 3–19). The insolubility of this form makes the change in shape irreversible. Each time a normal protein is converted into a prion, it not only adds to the number of prions but also makes it more likely that other unfolded molecules will encounter prions and be converted too.

This is how prion particles are reproduced.

Prion diseases sometimes run in families. Individuals who inherit certain alleles are more susceptible to the disease because they produce a version of the protein that is more likely to fold abnormally, setting off the fatal chain of events. These particles, too, cause disease if they are injected into new hosts, and the closer the match of amino acid sequences between prions and normal proteins, the faster disease develops.

The mystery of prions is twofold. First, prions are apparently a case of nongenetic biological information, a subject we know almost nothing about. How do they cause the reshaping of normal proteins into replicas of themselves? The second mystery about these particles is how they cause disease. It is uncertain whether the protein fibrils found in the brains of victims are composed of prions themselves or result from prions disrupting normal functions.

**FIGURE 23–A** A sheep infected with scrapie. (Stanley Prusiner, University of California at San Francisco Neurology Research)

hosts that have not yet built up immunity to the proteins encoded by their particular genes. This makes it difficult to develop preventive vaccines or cures for these viral diseases.

Recent research showed that viral infections are more severe when contracted from a relative than from a genetically unrelated individual. This is because the virus has already gone through many generations and become adapted to reproduce in a genetic environment very similar to the new host's body. This also explains why populations with little genetic diversity, such as Native Americans, experienced high death rates from viral diseases introduced by Europeans, such as measles and smallpox. More diverse populations, such as Africans, are less susceptible.

## SUMMARY

Viruses are tiny parasitic particles that can be produced only inside living cells. They resemble cells in that they contain nucleic acid and protein molecules, and some are surrounded by membranous envelopes of lipid and protein. However, viruses do not share all the features of cellular organisms and do not belong in the kingdoms of living organisms. Viruses do not eat, grow, metabolize, make proteins, or reproduce by themselves, but they do evolve. Cells reproduce by dividing in two, but hundreds of viruses may be produced in a host cell after infection by a single virus particle. Many viruses can be crystallized.

1. A virus particle consists of a DNA or RNA molecule, comprising the viral genome, inside a protein capsid, which is sometimes enclosed in a membranous envelope.
2. Viral infection of a host cell begins when a virus particle's capsid or envelope proteins bind specifically to receptor proteins on the host cell's surface. The viral genome enters the cytoplasm.
   - In a lytic cycle, the virus immediately takes over the host cell's metabolic machinery, causing it to produce more viral nucleic acid and protein molecules. These are assembled into many new virus particles, which are freed by lysis of the host cell shortly after the original virus invaded.
   - In a lysogenic cycle, the viral genome "goes underground," becoming incorporated into the host's DNA, where it is replicated and passed along to the cell's progeny at each division. Eventually the viral genome may cause the production and release of many new virus particles as in the lytic cycle. Sometimes the virus takes along some of the host's genetic material and introduces it into the genome of the next host cell.
   - A third kind of cycle involves the gradual production and release of viral particles, leaving the host cell intact, at least for a while.
3. RNA viral genomes are reproduced by way of RNA or DNA intermediates, depending on the type of virus. Because cells do not contain RNA-copying enzymes, RNA viruses must encode the enzymes for replication of their genomes and must either carry these enzymes in their capsids from previous hosts or induce the host to make the enzyme as the first order of business following infection.
4. Viral disruption of host cells is responsible for many diseases, including some tumors, in plants and animals. Persistent virus infections may cause some cells to reduce their output of hormones or other chemical messengers.
5. Some plant diseases are caused by viroids, which are short, naked strands of RNA that do not code for protein. How they cause disease is unknown.
6. The evolutionary origin of viruses is not understood, although they are often very similar genetically to their hosts. Viruses play important roles in the evolution of their hosts by acting as selective pressures and by moving genes from one host to another. The rapid evolution of viruses enables them to stay ahead of their hosts' defenses.

## Self-Quiz

1. One reason that viruses are considered nonliving is that:
   a. they lack replicable nucleic acids
   b. their nucleic acids do not code for proteins
   c. they cannot make their own food molecules
   d. they cannot carry out their own reproduction
   e. they do not undergo genetic change (mutation) and so do not become adapted to changes in their environment
2. A virus always contains:
   a. DNA, RNA, and proteins
   b. nucleic acids and lipids
   c. proteins and nucleic acid
   d. DNA and proteins
   e. nucleic acids, proteins, lipids, and carbohydrates
3. The genetic component of a virus may be:
   a. single-stranded DNA
   b. single-stranded DNA or RNA
   c. single- or double-stranded DNA or single-stranded RNA
   d. single- or double-stranded DNA or RNA
   e. double-stranded DNA
4. Examine each statement below, and tell whether it is true of lytic viruses, lysogenic viruses, or both.
   _____ a. Many virus particles may be released from each host cell.
   _____ b. The host's DNA is broken down soon after the virus enters the cell.

*(Continued)*

_____ c. Part of the host's DNA may be carried to a new host cell by the virus.

_____ d. The viral DNA may be incorporated into the host's genome.

_____ e. Viral proteins are synthesized on host ribosomes.

5. True or False. The genomes of some RNA viruses are replicated by the formation of RNA templates, whereas the genomes of others are replicated by the formation of DNA templates.

6. A virus can infect only certain kinds of cells in certain host species because:
   a. it carries genes from previous hosts of that type
   b. it seeks the same type of host cell that produced it
   c. its envelope contains proteins of the same type as the correct host's
   d. its capsid or envelope proteins bind specifically to membrane proteins on those types of host cells
   e. all of the above

## Questions for Discussion

1. Why is it difficult to produce convincing evidence that viruses cause cancer in humans?

2. Geneticist Robert L. Sinsheimer has said that one reason experiments with recombinant DNA might be dangerous is that a phage could be used to carry a gene for a bacterial restriction endonuclease enzyme (Section 13–B), which destroys forms of DNA that are not specifically protected against it, into a eukaryotic cell. What might be the deleterious effects of such a transduction?

3. In what ways do viruses resemble living organisms, and what criteria of "life" do they lack?

4. One theory of the origin of viruses is that they began as stray pieces of host cell DNA or RNA, much like the "naked" viroids discussed in Section 23–E. Outline how such an escaped fragment of the cellular genome could have evolved into one of the complex modern viruses.

5. Why don't antibiotics work against virus infections? (Refer to Figure 10–16.)

## Suggested Readings

Beardsley, T. "Oravske kuru." *Scientific American,* August 1990. A survey of recent epidemics of slow brain infections and their causes.

Cohen, F. E., et al. "Structural clues to prion replication." *Science* 264:530, 1994.

Diener, T. O. "Viroids." *Scientific American,* January 1981. How the existence of viroids was established.

Hirsch, M. S., and J. C. Kaplan. "Antiviral therapy." *Scientific American,* April 1987. How drugs such as acyclovir (used against herpes infections) and AZT (used against AIDS) work.

Hogle, J. M., M. Chow, and D. J. Filman. "The structure of poliovirus." *Scientific American,* March 1987.

Oldstone, M. B. A. "Viral alteration of cell function." *Scientific American,* August 1989. Recounts the discovery of how viruses may alter cell metabolism without causing signs of infectious disease.

Prusiner, S. B. "Molecular biology of prion diseases." *Science* 252:1515, 1991.

Simons, K., H. Garoff, and A. Helenius. "How an animal virus gets into and out of its host cell." *Scientific American,* February 1982. The story of Semliki Forest virus.

Tiollais, P., and M.-A. Buendia. "Hepatitis B virus." *Scientific American,* April 1991.

Varmus, H. "Reverse transcription." *Scientific American,* September 1987.

Weiss, R. "The viral advantage." *Science News* 136:200, 1989. A chilling account of recent outbreaks of viral diseases.

Winkler, W. G., and K. Bögel. "Control of rabies in wildlife." *Scientific American,* June 1992.

# *Bacteria*

## O B J E C T I V E S

*When you have studied this chapter, you should be able to:*

1. Use the following terms correctly:
   peptidoglycans, capsule, pili
   binary fission
   aerobe, obligate anaerobe, facultative anaerobe
   spore
   autotroph, photoautotroph, chemoautotroph, heterotroph,
     saprobe, pathogen, symbiont
   nitrogen fixation, heterocyst

2. State the criterion used to place organisms in kingdom Prokaryotae; describe the distinguishing characteristics of prokaryotic cells and of archaeobacteria, eubacteria, and these special groups of eubacteria: Gram-positive and Gram-negative bacteria, actinomycetes, spirochetes, and cyanobacteria.

3. Describe the operation of bacterial flagella and how it changes during a behavior such as chemotaxis.

4. List and describe the four sources of genetic change known in prokaryotes; discuss the contributions of these sources and of reproductive rate and small size to the evolutionary success of prokaryotes.

5. State how photosynthesis in cyanobacteria and prochlorophytes differs from that of other bacteria.

6. Describe the metabolism of a sulfur bacterium and the conditions that promote or inhibit the growth of these bacteria.

7. Summarize the ecological significance of prokaryotes, including their roles in food production, nitrogen fixation, nitrification, and decomposition.

8. Explain what is meant by a "bloom" of cyanobacteria; list some factors that can cause such blooms, and describe measures to prevent them.

9. List the three main shapes of bacteria; give their Latin names.

10. Give some examples of the use of bacteria in food production.

11. Describe the difference between bacterial endotoxins and exotoxins, and state their role in disease.

12. Explain how drug resistance arises in bacterial populations, including the roles of natural selection and plasmid transfer (Sections 24–B and 24–F).

13. State what the normal microbiota of an animal is, and list three reasons why it is important to the animal.

14. Give evidence for the endosymbiont theory of the origin of mitochondria and plastids.

The bacteria are assigned to kingdom Prokaryotae on the basis of their key feature: the prokaryotic cell, which has no membrane-bounded nucleus. The kingdom's alternative name, Monera (mono = one), signifies that most bacteria function as single cells, even though many remain attached after cell division in characteristic formations. However, in some species the cells cooperate, with limited division of labor.

Fossil bacteria dating from about 3.5 billion years ago are the earliest signs of life on Earth. Cells recognizable as eukaryotes of the modern type did not appear until more than 1.5 billion years later.

Today, bacteria still outnumber other organisms: a handful of fertile soil contains billions of them. They flourish in almost every conceivable habitat, including our own bodies. But because of their microscopic size, bacteria were the last major group of organisms to be discovered. They were first seen in 1676 by Anton van Leeuwenhoek, an inquisitive Dutch linen draper. Leeuwenhoek built his own microscopes and examined samples from many sources, in which he found a great variety of microorganisms, including some of the larger bacteria.

About two centuries later, in the 1860s and 1870s, the classic studies of Louis Pasteur and Robert Koch established the role of bacteria in causing food spoilage and diseases. To most people, bacteria came to mean illness and decay, banes of human health and wealth. Indeed, the traditional emphasis of bacteriology is on identifying and controlling bacteria.

Today, however, we are coming to appreciate that bacteria play a broader range of roles in the environment. Bacteria change and recycle mineral nutrients, help clean up pollution, and control many organisms harmful to humans.

In fact, one of the most notable features of bacteria is their metabolic diversity: as a group, prokaryotes have a wider range of chemical abilities than eukaryotes do.

In this chapter we shall study how the structure, reproduction, and metabolism of bacteria are related to their ways of life. We shall also look at the main groups of bacteria and examine some of their roles in nature and in human life.

## KEY CONCEPTS

♦ Bacteria include free-living autotrophs and saprobes, as well as forms living in symbiotic relationships with other species, as parasites or mutualistic symbionts.

♦ Many bacteria cause disease, and some have changed the course of human history.

♦ Bacteria (and fungi) play important ecological roles as decomposers. Some bacteria fix nitrogen and so support all plant life on Earth. Plants, in turn, support virtually all animal life and fungi.

## 24–A    PROKARYOTIC CELLS

We have seen that the prokaryotic cells of bacteria differ in several ways from eukaryotic cells (Section 5–D and Table 5–1):

1. **Absence of a nucleus.** The definitive feature of prokaryotic cells is the absence of a nuclear envelope separating the genetic material from the cytoplasm. Nor do they have any of the other internal membranous organelles of eukaryotic cells, such as mitochondria, chloroplasts, endoplasmic reticulum, or Golgi complexes. However, some do have internal membranes, such as photosynthetic vesicles (see Figure 24–5).

2. **Size.** Prokaryotic cells are generally much smaller than eukaryotic cells, and the absence of most membrane-bounded organelles makes them much simpler in structure. They generally measure less than 1 $\mu$m across and about 1 to 10 $\mu$m long. However, the largest, at 500 $\mu$m (0.5 mm) long, dwarf most eukaryotic cells.

3. **Organization of the genetic material.** A prokaryote's genome consists of one large, circular DNA double helix associated with small amounts of RNA and non-histone proteins. This structure contains only about one thousandth as much DNA as the smallest eukaryotic cells. The region it occupies is sometimes called a **nuclear area.**

4. **Size of ribosomes.** Prokaryotes have smaller ribosomes, the sites of protein synthesis.

5. **Differences in external features.** Prokaryotes have distinctive extracellular structures, discussed next.

### Extracellular Structures

Bacteria produce various extracellular structures, which play a role in their interactions with the environment (Figure 24–1).

The cell wall protects the cell from injury and prevents the cell from taking up so much water by osmosis that it bursts (Section 4–D). Bacteria without cell walls can be produced by culturing them in isotonic solution, which is osmotically balanced so that they do not gain water, and using enzymes to digest the wall away. These bacteria do not divide, indicating that the cell wall is necessary to reproduction (except in the few groups of bacteria that naturally lack cell walls).

Prokaryotic cell walls contain **peptidoglycans,** polymers made up of amino sugars and various amino acids. Drugs of the penicillin family interfere with the synthesis of this part of the cell wall and are therefore toxic to bacteria but not to host organisms, which do not make peptidoglycans.

Some bacteria produce a **capsule** of polysaccharide or polypeptide outside the cell wall. *Streptococcus pneumoniae* causes pneumonia only when it forms a capsule, which apparently protects it from being engulfed and destroyed by the host animal's phagocytic cells (see Figure 9–1). The capsule material also forms a barrier against bacteriophages (viruses that kill bacteria), and sometimes even against disinfectant chemicals. In addition, a capsule holds water around the cell, a safeguard against drying out.

In nature, many prokaryotic cells produce a dense felt-like mat of capsule polysaccharides that enables them to stick to surfaces—soil particles, rocks in stream beds, or cells of host animals, for example. *Streptococcus mutans,* the chief agent of tooth decay, attaches to teeth better when it produces such a capsule. To do this, it needs sucrose as a raw material; other sugars won't do. This is why candy and other sucrose-rich foods are bad for our teeth.

Some bacteria produce hundreds of hollow protein strands called **pili** (singular: **pilus** = hair). Pili may serve to attach the bacterial cell to other structures. In *Neisseria gonorrhoeae,* only strains that produce pili can attach to host cells and cause gonorrhea. Special F (fertility) pili may

THE CELL :

EXTRACELLULAR STRUCTURES :

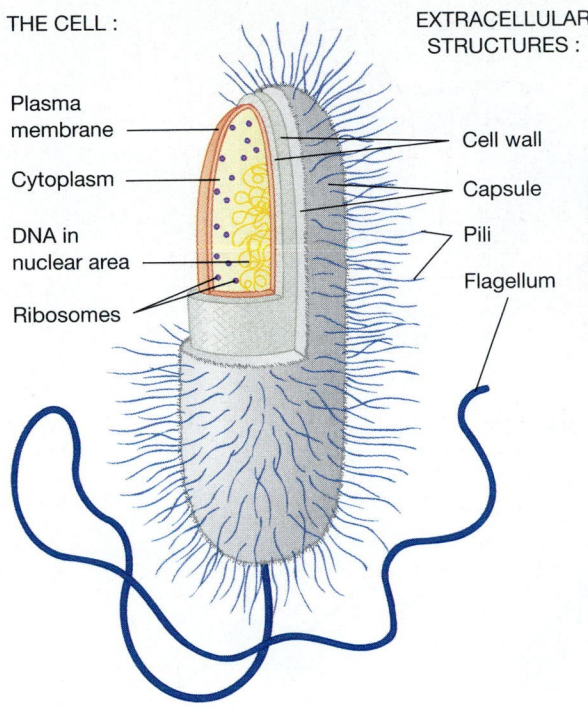

Plasma membrane

Cytoplasm

DNA in nuclear area

Ribosomes

Cell wall

Capsule

Pili

Flagellum

(a) Bacterial cell structure

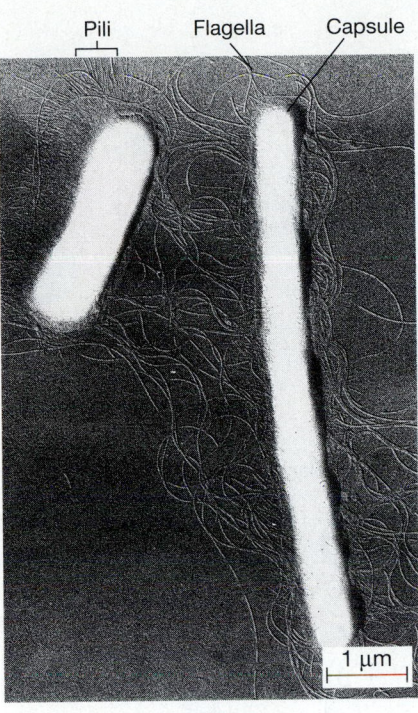

Pili    Flagella    Capsule

1 µm

(b) External structures (SEM)

**FIGURE 24–1** Bacterial cells. (a) Generalized internal and external structures. (b) Two bacteria surrounded by their capsules and many pili and flagella. (b, Biophoto Associates)

also link bacterial cells together before DNA passes from one to the other during the process of conjugation (Section 24–B).

Bacteria of many species move by rotating their long, thin flagella (Figure 24–2). A flagellum is composed of three strands of the globular protein flagellin, wound in a helix, which extends outside the plasma membrane and cell wall. (In contrast, the microtubules of eukaryotic flagella are made up of the globular protein tubulin, and the whole structure is enclosed in the plasma membrane.)

Because of their small size, bacteria swimming in water must overcome a drag comparable to humans trying to swim through honey. As the helical flagella rotate, the cell twists through water like a corkscrew through a cork. Energy to power the flagella comes from hydrogen ions ($H^+$) moving down their concentration gradient into the cell through channels in the plasma membrane.

Some bacteria have flagella located at one or both ends, or poles, of the cell. These bacteria move forward or backward depending on which direction their polar flagella rotate. Other bacteria have several to many lateral flagella (on the sides) and move in a more complex way. When these lateral flagella rotate counterclockwise, they all wind together into a single bundle, which acts like a propeller driving the cell through the water. However, the flagella soon reverse direction and the bundle flies apart, with each flagellum pulling in its own direction. The cell tumbles randomly until the flagella reverse and form a bundle again. The cell then swims in a new direction, depending on where it was aimed when it stopped tumbling. So these bacteria move in an erratic course of alternating directional runs and random tumbles (Figure 24–2c).

## Behavior

In nature, conditions are generally better for bacteria in some areas than in others. Bacteria respond to their environment by swimming toward more favorable areas or away from less favorable ones, moving along gradients of factors such as light, temperature, or chemicals.

Such directional movement in response to an environmental cue is called a **taxis** (plural: **taxes**). A taxis is named for the cue that elicits it, and whether the organism moves toward or away from this cue. For example, photosynthetic bacteria show **positive phototaxis** by swimming toward light.

Bacteria also exhibit **positive chemotaxis,** swimming toward attractant chemicals (generally nutrients such as amino acids and sugars), or **negative chemotaxis,** moving away from repellents such as toxic chemicals. Chemical receptor molecules in the plasma membrane detect moment-

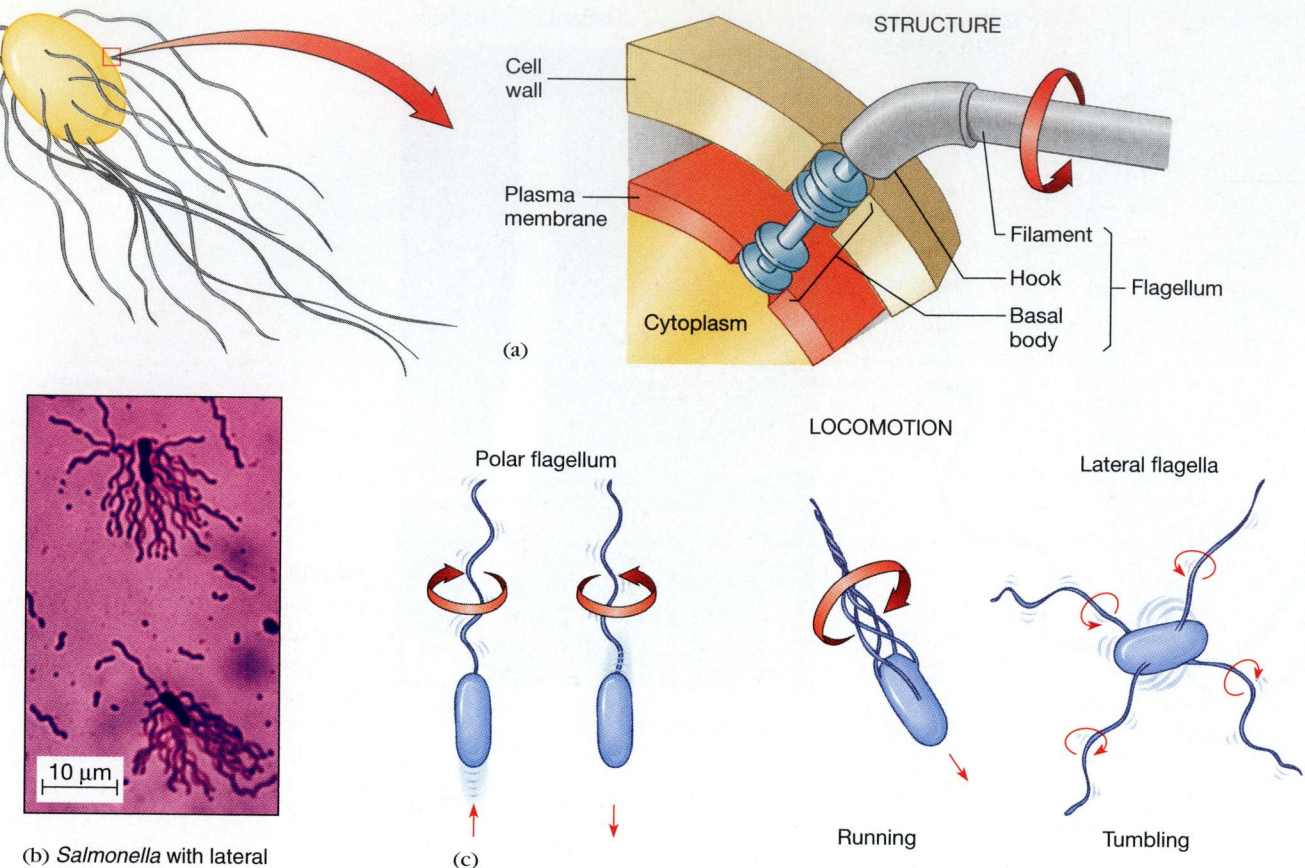

STRUCTURE

Cell wall

Plasma membrane

Cytoplasm

Filament

Hook

Basal body

Flagellum

(a)

10 µm

(b) *Salmonella* with lateral flagella

LOCOMOTION

Polar flagellum

Lateral flagella

Running

Tumbling

(c)

**FIGURE 24–2**  Bacterial flagella. (a) Structure. The basal body is inserted into the cell wall and plasma membrane. It connects to the filament by means of the hook. The basal body is thought to spin the filament like a propeller, at up to 50 revolutions per second. (b) *Salmonella typhosa* with many lateral flagella. (c) How bacteria swim using flagella.    (b, A. M. Siegelman / Visuals Unlimited)

to-moment changes in concentration around the cell and somehow control rotation of the flagella. When the cell is moving toward attractant molecules, tumbling is inhibited, and so the directional runs last longer (Figure 24–3). When the cell is going away from attractants, or toward repellents, tumbling increases. Because the cell spends more time swimming in favorable than in unfavorable directions, its overall progress is in an appropriate direction.

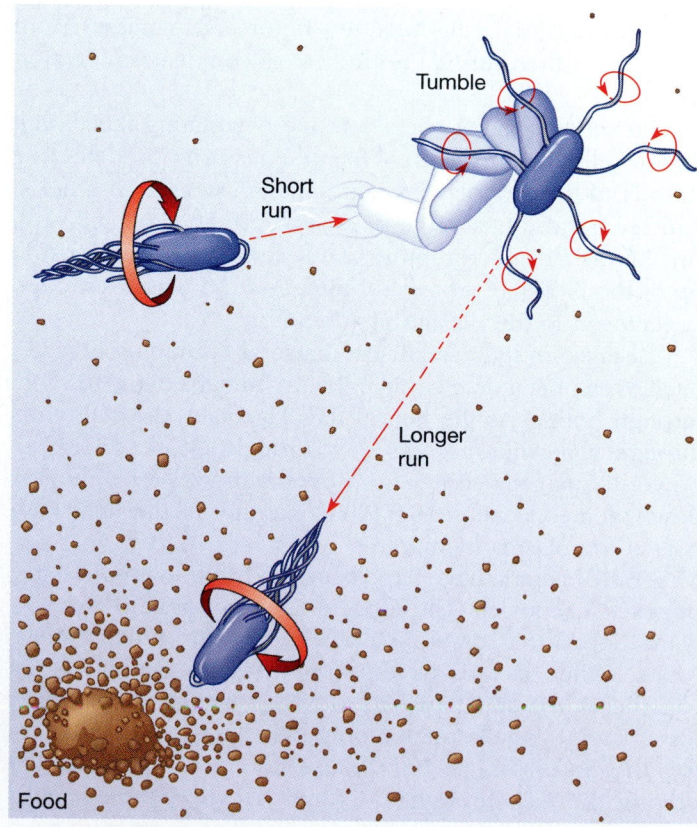

Tumble

Short run

Longer run

Food

**FIGURE 24–3**  Positive chemotaxis in bacteria. When the cell is moving up a concentration gradient toward food or other attractants, the periods of straight-line runs increase and tumbling decreases, so that the cell progresses toward the more favorable location.

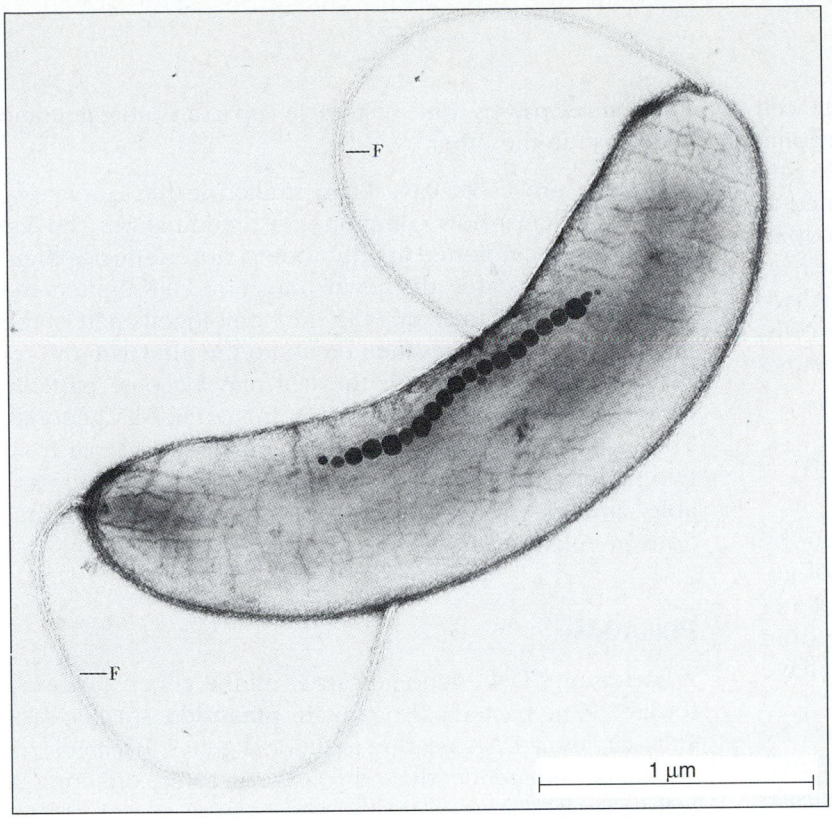

**FIGURE 24–4** A magnetotactic bacterium. The chain of dark (electron-dense) crystals of magnetite gives the bacterium a compass by which it orients in Earth's magnetic field. (R. P. Blakemore and R. B. Frankel, *Scientific American,* December 1981. Courtesy of the authors, publisher, and N. Blakemore)

Some bacteria find food-rich sediments by following a different cue: Earth's magnetic field. Such a **magnetotactic** bacterium contains a chain of magnetic iron ore crystals, which orient parallel to lines of magnetic force (Figure 24–4). Because these lines point toward the center of Earth, bacteria following then go downward until they reach the decaying bottom ooze.

## 24–B PROKARYOTE REPRODUCTION AND EVOLUTION

Most prokaryotes reproduce by **binary fission,** dividing into two cells of about equal size, each with a copy of the parent cell's genetic material (Figure 24–5). However, some bacteria reproduce by budding of smaller cells from the parent.

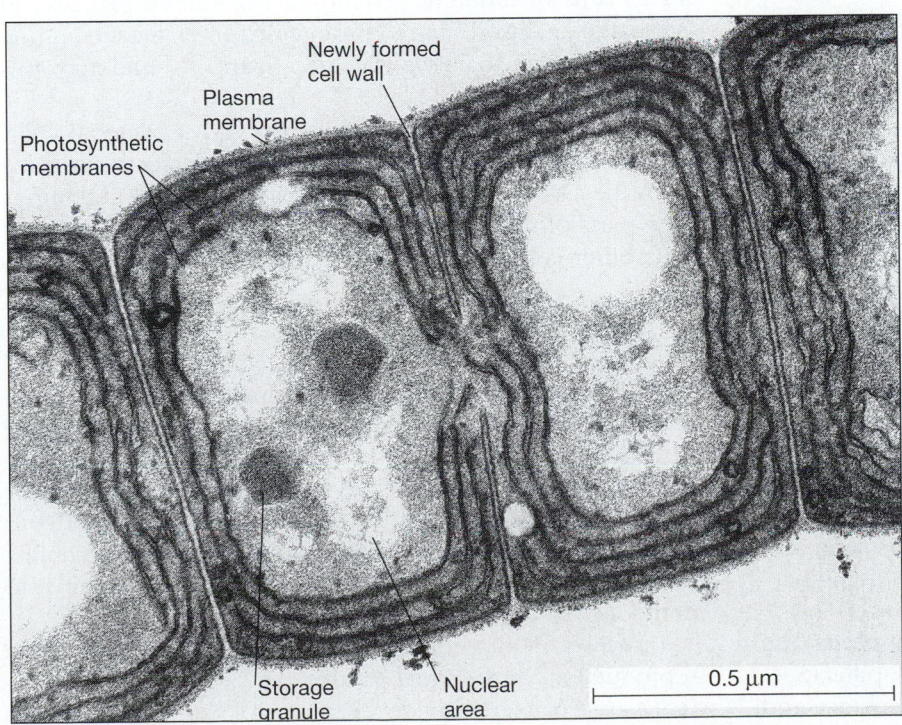

**FIGURE 24–5** Binary fission. A cell in a filament of the cyanobacterium *Oscillatoria* is dividing in two as a new cell wall forms at the edges and moves toward the center. Notice the several layers of photosynthetic membranes inside each cell. (Biophoto Associates)

A prokaryotic cell replicates its DNA well before cell division and may contain up to four copies of its genome if the rate of cell division does not keep up with the rate of DNA replication. The DNA is thought to be attached to the plasma membrane, and in a growing cell the DNA copies become separated as material is added between their attachment points. Eventually a cell containing at least two complete copies of the DNA divides into two new cells, each with at least one DNA molecule. Mitosis and meiosis do not occur in prokaryotes.

## Genetic Recombination

Bacteria do not reproduce sexually, as eukaryotes do, by fusion of two gametes with roughly equal amounts of genetic material. However, if we consider that the essence of sexual reproduction is combining genetic information from two individuals into one, then some prokaryotes do have sexual systems.

Genetic recombination in bacteria was first shown by Joshua Lederberg and Edward Tatum. They cultured two strains of *Escherichia coli;* one strain could not synthesize three amino acids needed for growth, whereas the other strain could not make two other needed amino acids nor the vitamin biotin. Each strain could grow only when supplied with the three substances it could not make. Lederberg and Tatum grew the two strains together in a broth that supplied all their needs. Then they spread some of the cell mixture on agar, a medium with a gelatin-like consistency, to which they added sugars and salts but no amino acids nor vitamins. A small fraction of the cells grew and reproduced; they must have been able to make all of the substances they needed and were therefore **recombinants,** containing genes from both original strains.

Genes can be transferred between bacteria by three means:

1. **Transduction.** A bacteriophage (virus) carries bacterial DNA from one bacterium to another. This may happen when a piece of DNA from an infected bacterium remains attached to the DNA of a lysogenic phage, which enters a new host cell (Section 23–B). Alternatively, a fragment of bacterial DNA rather than phage DNA becomes packaged into a phage protein coat and is carried to a new bacterial cell. This is the most common means of genetic recombination in bacteria, not surprisingly when you consider that a recent study of marine bacteria showed at least 15% to be infected with viruses.
2. **Transformation.** DNA from a broken cell is taken up, sometimes in surprisingly large pieces, by a living bacterium (Section 9–A).
3. **Conjugation.** Two cells of different **mating types** (the equivalent of sexes) become joined by an F pilus (Section 24–A). Later the cells come into direct contact, and

one cell passes some of its DNA (up to an entire genome copy) to the other.

These processes have been studied in the laboratory, but it is not clear how common they are in nature. The last two have been reported in only about a dozen species each.

In all three cases, the newly transferred DNA enters the recipient cell and lines up with the homologous part of the resident DNA. Enzymes then break up the old DNA and replace it with the new. Or the cell may become partially diploid, with two strands of DNA for part of its genome. The resulting recipient cell contains genetic material from two parents. However, the parents gave unequal, and variable, amounts of DNA, in contrast to the equal contributions in eukaryotic sex.

## Plasmids

A bacterium's DNA genome carries all the genes necessary for life. Some bacteria also contain **plasmids,** smaller, separate circles of DNA bearing additional genes. Plasmids are replicated independently and passed on to the offspring at cell division. They can also be passed from one bacterium to another by any of the three modes of gene transfer.

The first plasmid discovered was the F (fertility) particle of *Escherichia coli,* which codes for the proteins of the F pilus used in conjugation. The pilus then conveys a copy of the F particle from the donor to the recipient cell.

Of more practical importance are the various plasmids that carry genes for resistance to **antibiotics,** substances that destroy microorganisms. Some plasmids carry genes for resistance to several different antibiotics and can even transfer to cells of different genera—a frightening prospect in our battle against disease. Fortunately for us, most common disease-causing bacteria do not seem to undergo this process readily.

Plasmids are sometimes inserted into a bacterial cell's genome. Such incorporated plasmids can later be excised from the DNA, sometimes taking adjacent genes along as new parts of the plasmid. If the plasmid is later transferred to another cell, the captured genes go too.

## Prokaryote Evolution

New gene combinations are the raw material for natural selection. The transfer of genes from cell to cell plays a role in prokaryote evolution, but mutations may be more important. Mutations are generally rare—a given gene mutates perhaps once in every 10 thousand to 10 billion ($10^4$ to $10^{10}$) individuals. How can this provide enough variability to explain, say, the widespread evolution of resistance to antibiotics in the last few decades?

The secret of prokaryote evolution lies in their rapid reproduction. In the ideal environment of a laboratory culture, some bacteria divide every 20 to 30 minutes. At this rate, a single cell placed on the surface of a nutrient medium can produce a visible colony containing millions of cells in about two days. In nature, bacteria seldom find such excellent growing conditions; their growth and reproduction are retarded by cold, scarcity of food, predation, infection by viruses, and inhibition by their neighbors. Nevertheless, they still reproduce extremely rapidly.

With so many cells in a population, even rare mutations can produce many genetic variants. In other words, a huge population coupled with a high rate of population growth can produce a great deal of genetic variation among individual prokaryotes.

## 24–C  BACTERIAL METABOLISM AND WAYS OF LIFE

Bacteria live in a variety of habitats, as remote as the sea floor, icebergs, and hot springs, or as close as your mouth (Figure 24–6). This far-flung distribution is possible because prokaryotes are the metabolic wizards of the living world, with a range of chemical abilities eukaryotes cannot match.

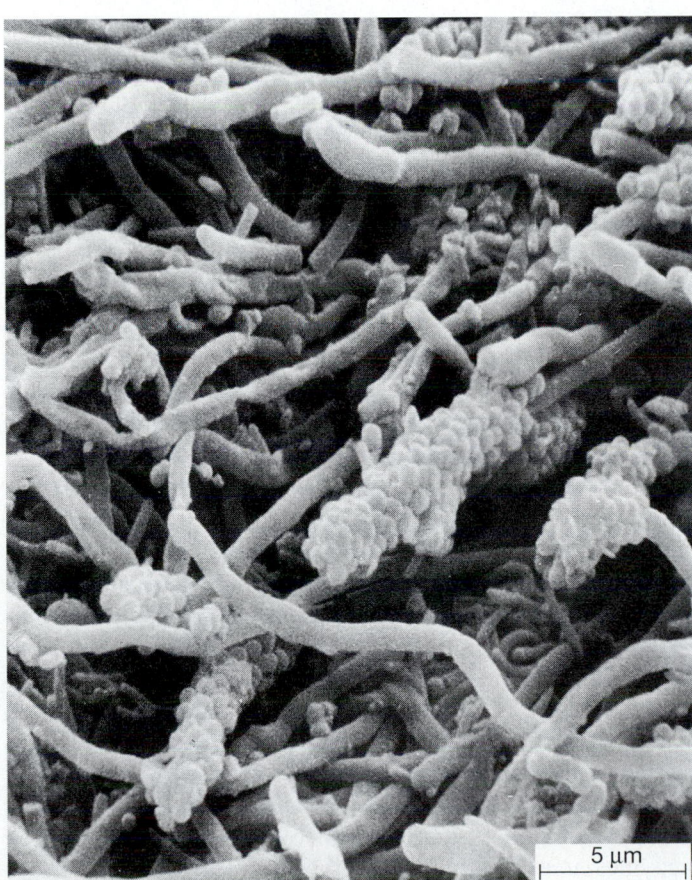

5 µm

**FIGURE 24–6**  Oral bacteria. Plaque from a human tooth that has not been brushed for three days, showing a dense growth of bacteria.  (Z. Skobe, Forsyth Dental Center/BPS)

For one thing, bacteria have more varied pathways of energy metabolism. Many species obtain some or all of their energy by **fermentation** (the breakdown of organic molecules, using organic molecules as the final electron acceptors), whereas others carry out respiration (using inorganic electron acceptors). **Aerobes** require $O_2$ for cellular respiration; **obligate anaerobes** are killed by exposure to $O_2$; **facultative anaerobes** can take it or leave it (an = not; aero = air; bios = life). The unusual process of **anaerobic respiration** is found only in some bacteria that can use nitrate ($NO_3^-$), sulfate ($SO_4^{2-}$), or iron ($Fe^{3+}$) instead of $O_2$ as electron acceptors.

Because of their rigid cell walls, bacteria cannot engulf large food particles. Rather, they absorb small molecules and ions that filter through their walls. In terms of nutrition, bacteria can be divided into two groups. **Autotrophs** can make their own food from inorganic compounds. **Heterotrophs** must absorb organic food molecules from other organisms, living or dead.

Many heterotrophic bacteria can synthesize any organic molecule they need starting with only one kind of organic carbon source, such as glucose or a fatty acid. Others cannot form all of the organic compounds they need, but have more complex nutritional requirements.

When conditions are unfavorable for growth, some bacteria form thick-walled, resting **spores** resistant to heat, drying, and ultraviolet radiation. Most bacterial spores are not means of reproduction, because no increase in cell number occurs. Rather, spore formation permits cells to survive adverse conditions and to disperse to new locations, where a new cell may germinate from the spore.

Bacteria undoubtedly owe their biological success to their varied metabolic capabilities, coupled with small size, rapid reproductive rate, and the ability to survive adverse conditions by forming resistant spores. These features permit bacteria to live in many habitats that are here today and gone tomorrow. A raindrop on a leaf evaporates in a few hours, but by then a bacterium has divided several times and its descendants have formed spores that can blow away to new homes.

As we discuss the metabolism of bacteria, notice how the chemical activities of the various species interact: one microbe's wastes are another's food and drink.

### Autotrophic Bacteria

Autotrophs are organisms that can use inorganic molecules to synthesize their food molecules. They need to absorb only raw materials such as carbon dioxide, mineral ions, and water (Figure 24–7).

**Photoautotrophs** use light as an energy source to make food. For example, the purple bacteria and green bacteria use light energy to drive primitive forms of photosynthesis, which differ from that of green plants in several ways. First, the photosynthesis of these bacteria does not obtain

**FIGURE 24–7** Autotrophic bacteria. The pastel bands of color in the runoff from this geyser are populations of photosynthetic and chemosynthetic bacteria adapted to different ranges of water temperature, which is near-boiling close to the source and progressively cooler at a distance.

electrons and hydrogen by splitting water, and so it is **anoxygenic:** not producing $O_2$. Various photosynthetic bacteria obtain hydrogen from hydrogen gas, hydrogen sulfide, fatty acids, or alcohols. Second, bacterial chlorophylls absorb, not visible light, but light in the longer-wavelength, near-infrared range (see Figure 8–2). Hence, these bacteria can photosynthesize in what we would consider darkness. Third, many photosynthetic bacteria are anaerobic. Fourth, these bacteria have only one photosystem: photosystem II in purple sulfur and nonsulfur bacteria, photosystem I in green sulfur bacteria.

These features make some photosynthetic bacteria, such as sulfur bacteria, ideally suited to life in the mud at the bottom of relatively stagnant bodies of water. This mud usually contains no $O_2$ because aerobic organisms nearer the surface use up what little $O_2$ dissolves in the water. Also, water and algae near the surface absorb most of the visible light, and only the longer wavelengths reach the bottom. Anaerobic mud usually contains hydrogen sulfide ($H_2S$), which smells like rotten eggs, produced by sulfide bacteria as they break down plant protein. The photosynthetic sulfur bacteria use this hydrogen sulfide (instead of water) as a source of hydrogen in photosynthesis:

$$CO_2 + 2\ H_2S \longrightarrow CH_2O + 2\ S + H_2O$$

Normally, sulfur bacteria use hydrogen sulfide as fast as sulfide bacteria produce it. If sewage is dumped into the water, however, the sulfide bacteria grow rapidly on nutrients in the sewage. Soon there are so many of them that they cloud the water. Then the photosynthetic sulfur bacteria grow poorly because they receive even less light than

usual. This leaves much hydrogen sulfide unused, and it rises to the surface, producing the stench of a sewage-polluted lake.

Cyanobacteria and prochlorophytes use the kind of photosynthesis found in eukaryotes. They contain chlorophyll *a* and therefore use visible light in photosynthesis. They have both photosystem I and photosystem II and therefore obtain hydrogen by splitting water and release $O_2$ as a byproduct (Chapter 8). The atmosphere of early Earth contained little $O_2$ until early cyanobacteria began producing it. Cyanobacteria and prochlorophytes, and their relatives the eukaryotic chloroplasts, produced virtually all the $O_2$ in the atmosphere today. Some cyanobacteria can also carry on photosynthesis using $H_2S$, like sulfur bacteria.

**Chemoautotrophy** (also called **chemosynthesis** or **chemolithotrophy**) is a form of autotrophic nutrition found only in various bacteria (Figure 24–8). They make food using energy, not from light, but from inorganic chemical reactions, involving the oxidation of hydrogen, ammonia, nitrites, sulfides, or iron. Chemoautotrophic bacteria are the sole food producers of some ecosystems near vents in the deep sea floor that spew forth warm, sulfide-laden water. The bacteria, in turn, may supply food to animals such as clams and worms (see Figure 24–20b).

Of practical importance are bacteria that oxidize sulfur compounds in low-grade metal ores. As a result of their activities, metals form soluble ions that leach out of the ore and can be collected from runoff water. Human miners have taken advantage of this phenomenon for about 3000 years, but only recently have they realized the crucial role of bacteria. Some copper, gold, and uranium mine operators now

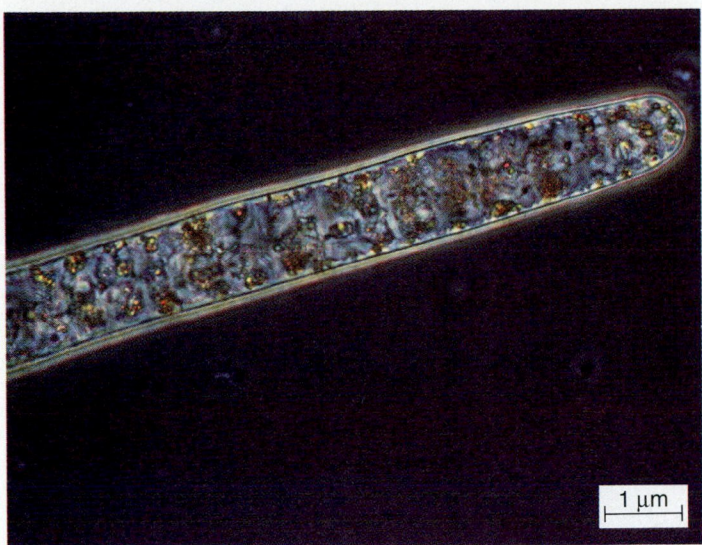

**FIGURE 24–8** A chemosynthetic bacterium. The relatively large cells of *Beggiatoa* grow in long filaments (note size bar). The cells obtain energy, not from light, but by oxidizing $H_2S$ to sulfur, which builds up in yellow particles visible in the cell. Some species of *Beggiatoa* are important in the culture of rice because they remove toxic $H_2S$ from the water of the rice paddies. (P. W. Johnson and J. McN. Sieburth, University of Rhode Island/BPS)

deliberately encourage bacterial leaching in order to recover metal from low-grade ores without using expensive fuels for smelting.

These same bacteria are responsible for much of the pollution from coal and other mines. "High-sulfur" coal (or oil) contains iron sulfides. Bacteria convert the sulfides to sulfur dioxide ($SO_2$), which dissolves in water and forms sulfuric acid ($H_2SO_4$). Toxic elements dissolve in the acidic water draining from a mine or from heaps of mining waste, causing pollution of rivers, lakes, and soil water in mining areas throughout the world.

**Nitrifying bacteria** are chemoautotrophs that oxidize ammonia ($NH_3$) or ammonium ($NH_4^+$) to nitrites ($NO_2^-$), and nitrites to nitrates ($NO_3^-$). These bacteria play a crucial role in the nitrogen cycle (Section 48–D).

## Heterotrophic Bacteria

Most bacteria are **heterotrophs,** using food made by other organisms. **Saprobes** are bacteria or fungi that live on dead organic material, such as dead animals or plants, feces, leaves, or bark. They secrete hydrolytic (digestive) enzymes into the food around them and absorb the resulting small organic molecules and inorganic ions.

Saprobic bacteria and fungi are the most important decomposing organisms. When a dead leaf drifts to the forest floor or an animal dies, fungal and bacterial spores floating in the air have already settled on it. The spores

quickly begin to grow and break down the dead organism, recycling it in the form of carbon dioxide, water, and minerals that autotrophs can use to produce more food.

Bacteria may live on unlikely foods. One kind lives on mixtures of aviation fuel and aluminum in airplane tanks; some increase the rate of corrosion in metal pipes, storage tanks, and bridges; others break down agricultural pesticides or herbicides, perhaps before the chemicals have killed the intended insects or weeds; and still others can degrade toxic chemicals such as dioxins, DDT, or polychlorinated biphenyls (PCBs) into less harmful compounds. Some of these organisms may have evolved recently in response to modern human activities, but it is equally likely that they have always been around in small numbers, living on other food sources, until we provided them with vast new areas of favorable habitat.

Many heterotrophic bacteria are parasites, obtaining food directly from the bodies of living organisms (Sections 24–F and 24–G).

## Nitrogen-Fixing Bacteria

**Nitrogen fixation** is the reduction of gaseous nitrogen ($N_2$) from the air to ammonia ($NH_3$). The only organisms that can perform this complex set of reactions are some bacteria and cyanobacteria, including both autotrophic and heterotrophic forms. All other life on Earth is directly or indirectly dependent on their activities. Green plants need nitrogen, from ammonium or nitrates, to produce amino acids for their proteins. (Plant-eating animals later use these amino acids to build their own proteins.) However, most of Earth's nitrogen exists as $N_2$ in the air, which plants cannot use. The nitrogen-fixing prokaryotes link the vast $N_2$ supply of the air to the rest of the living world by providing ammonia. This can be used by plants directly or it can be converted to nitrites and then to nitrates by chemoautotrophic nitrifying bacteria (discussed before) and then used by plants.

Because it takes a lot of energy to break the very stable triple covalent bond linking the two nitrogen atoms in $N_2$, nitrogen fixation uses about 12 ATP per molecule of $N_2$ fixed. The important nitrogen-fixing enzyme **nitrogenase** is readily inactivated by $O_2$ and cannot function in a cell that gives off $O_2$ during photosynthesis. Some filamentous cyanobacteria fix nitrogen inside specialized cells called heterocysts. These cells use photosystem I to supply energy for nitrogen fixation but lack photosystem II—the one that evolves oxygen (Section 8–E).

The nitrogen-fixing, photosynthetic cyanobacteria are remarkably self-sufficient; they need only carbon dioxide and some inorganic substances to meet their few metabolic needs. They may colonize new habitats, such as newly formed volcanic rock, before anything else can live there.

Nitrogen-fixing bacteria have great importance to agriculture. Members of the genus *Rhizobium* live in nodules

(a) Nodules on pea roots

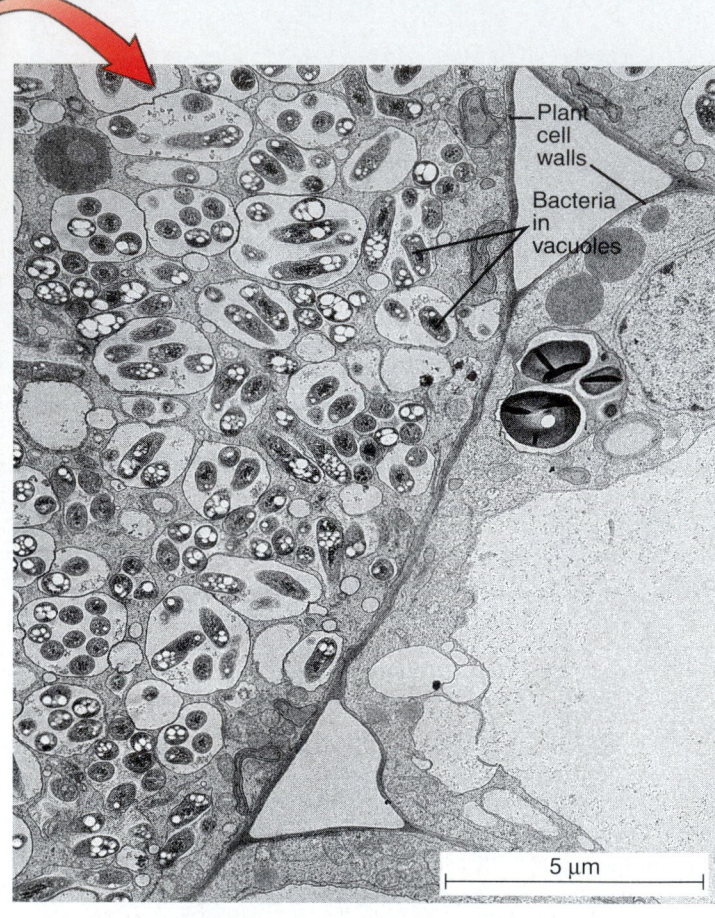

Plant cell walls

Bacteria in vacuoles

(b) Bacteria in cells (TEM)

**FIGURE 24–9**   Root nodules in legumes. (a) These swellings on the roots of a pea plant house nitrogen-fixing *Rhizobium* bacteria. (b) Thin section of cells from a soybean root nodule, showing an infected cell (left) and an uninfected cell (right). The cytoplasm of the infected cell contains many vacuoles, each containing several bacteria.   (a, Hugh Spencer/Photo Researchers; b, E. H. Newcomb and S. R. Tandon, University of Wisconsin, Madison/BPS)

on the roots of plants in the **legume** family (e.g., peas, beans, clover, alfalfa, vetch) (Figure 24–9). These bacteria use sugars produced by the legume's photosynthesis and supply the plant with ammonium. Growing legumes as part of crop rotation adds nitrogen compounds to the soil, improving its fertility.

Some crops that are not legumes, such as sugar cane, sorghum, and potatoes, also support nitrogen-fixing bacteria among their roots, although they do not form nodules. Several seed companies now sell cultured dried bacteria to be sown along with seeds of the crop to increase the yield.

Genetic engineers are working to make other crops able to house such nitrogen-fixing bacteria. This would reduce the need for expensive nitrogen fertilizers as well as improve the nutritional quality of plant protein. Even better would be developing a recently discovered species of bacteria that forms nodules on stems of legumes, where it carries out photosynthesis as well as nitrogen fixation. Such self-supporting bacteria would save their host plants energy, which could go into increased crop yields.

The yield of rice is boosted indirectly, by managing rice paddies to promote the growth of aquatic ferns that form symbioses with nitrogen-fixing cyanobacteria.

## 24–D   CLASSIFICATION OF PROKARYOTES

Biologists classify most organisms by morphology (structure), which usually provides good evidence of phylogeny (Chapter 22). However, many kinds of bacteria look too much alike to tell them apart by structure alone.

Using a biological definition, based on a shared gene pool, also presents problems. All the descendants of a single ancestor share a common gene pool, but we cannot hope to trace all the lines of descent in organisms that may have produced dozens of generations a week for 3.5 billion years. A further complication is that bacteria sometimes receive DNA from another individual, which need not be a close relative but can in fact be any other bacterium in the vicinity! On these grounds, some extremists claim that all the bacteria on Earth belong to just one species. However, the traditional, more practical view divides bacteria into many species based on similarities among individuals.

Bacteria have long been classified **phenetically,** that is, by phenotypic features such as structure and metabolism (Sections 24–A and 24–C). Bacterial cells come in three main shapes (Figure 24–10):

♦ **bacillus** (plural: **bacilli**), rod-shaped
♦ **coccus** (plural: **cocci**), spherical
♦ **spirillum** (plural: **spirilla**), spiral

Morphology and metabolism give us a useful way to identify unknown bacteria, a necessary first step in practical situations such as diagnosing a disease or finding the cause of food spoilage. However, such characters do not necessarily reflect evolutionary relationships.

Recently, researchers have been working out a phylogenetic classification for prokaryotes by comparing nucleotide sequences, particularly in ribosomal RNA molecules. This is

**FIGURE 24–10**  Three shapes of bacteria. This mixed population of symbiotic organisms from a cow's rumen (first stomach) contains cocci, bacilli, and spirilla as well as large eukaryotic yeast cells. (J. W. Costerton, University of Calgary)

probably more reliable than morphology or metabolism for sorting out evolutionary relationships among bacteria.

Drawing a phylogenetic tree based on these data has required moving some species from their traditional spots in the bacterial classification scheme. For example, traditional classification placed photosynthetic and nonphotosynthetic forms into separate groups. However, it turns out that there are at least four distinct lines of photosynthetic bacteria, some of which gave rise to lines of heterotrophic forms that lost the ability to make their own food. Likewise, several different lines of anaerobic ancestors apparently gave rise to aerobic forms independently.

There are too many groups of bacteria to discuss in this book, but we shall list the major lineages and some of their distinctive features.

## Archaeobacteria

Archaeobacteria (or archebacteria) resemble other prokaryotes in the absence of a nucleus and in the small sizes of their cells, genomes, and ribosomes. However, their ribosomes have a different shape. Strangely, in some details of protein synthesis the archaeobacteria seem closer to eukaryotes than to other prokaryotes.

In other respects, archaeobacteria are like nothing else living on Earth today. First, they have unusual kinds of lipids in their plasma membranes (Figure 24–11). They also have unique cell wall molecules, metabolic coenzymes, lipid-synthesis pathways, transfer RNAs, and RNA polymerase enzymes.

Basic biochemical characters such as these are usually highly conservative (Section 22–D). Therefore, according to many biologists, archaeobacteria diverged from other organisms very early in evolution, after the origin of the genetic code but before the machinery of metabolism, protein synthesis, and gene regulation had stabilized. Their name reflects this viewpoint (archaeo = ancient). By this reckoning, the archaeobacteria belong in a kingdom by themselves, and many biologists subscribe to this view.

Recently, however, new data and new interpretations have led other biologists to claim that archaeobacteria are so weird because the first ones were adapted to the peculiar environment of hot, acid springs. They point out that archaeobacteria are relative latecomers in the fossil record and contend that both archaeobacteria and eukaryotes evolved from eubacterial ancestors, or from an ancestor common to both.

There are three main types of archaeobacteria. The first group is thought to be ancestral to the other two (and possibly to all cells on Earth, depending on who's right about when archaeobacteria evolved).

1. **Thermoacidophiles**, as their name suggests, live in hot, acidic springs, growing best at a temperature of 70 to 75 °C and pH of 2 to 3. By contrast, the pH inside

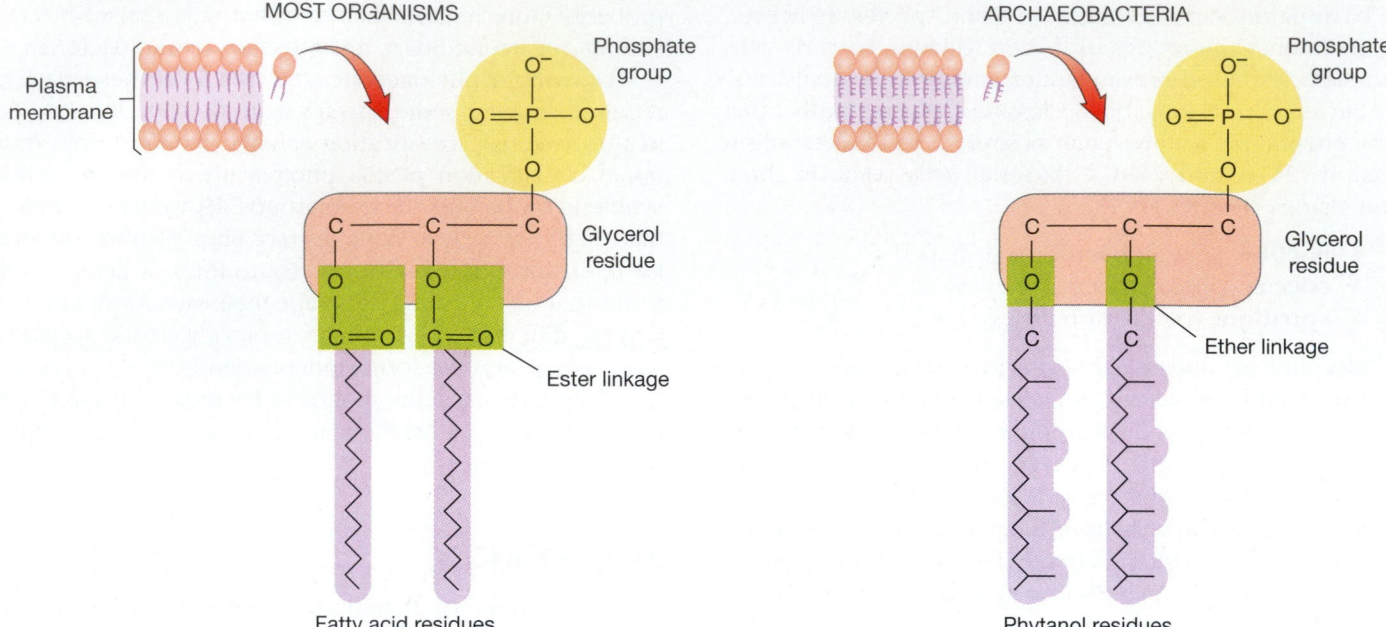

**FIGURE 24–11**   Differences between lipids in the plasma membranes of archaeobacteria and of other organisms. (Left) Other bacteria, and eukaryotes, have lipids made up of straight-chain fatty acids attached to glycerol by ester linkages. (Right) In contrast, the membrane lipids of archaeobacteria are made up of branched phytanol molecules attached to glycerol by ether linkages.

the cell is nearly neutral, and the thermoacidophiles use the energy of this pH gradient across the plasma membrane to power the active transport of substances into and out of the cell (Section 4–E). The autotroph *Sulfolobus* lives in hot sulfur springs, where it obtains energy by oxidizing sulfur. *Thermoplasma* is a heterotroph that inhabits smouldering coal tailings; it has no cell wall.

2. **Methanogens** produce methane ($CH_4$) as a by-product of the unique reaction they use to obtain energy:

$$CO_2 + 4 \ H_2 \longrightarrow CH_4 + 2 \ H_2O$$

Methanogens require carbon dioxide and hydrogen for this reaction, and they can use only three organic compounds for food: methanol ($H_3COH$), formate ($HCOO^-$), and acetate ($H_3C—COO^-$). These five raw materials are all small, simple molecules believed to have been abundant on early Earth. Because of this, and because methanogens are obligate anaerobes, living only in the absence of $O_2$, which was essentially absent in the early atmosphere, methanogens are thought to represent an ancient lineage. Today they live in a variety of oxygen-free habitats, such as at the bottoms of bogs, streams, lakes, and sewage treatment ponds, and in the digestive tracts of animals, especially in the rumens of animals such as sheep and cattle (Section 31–D). Here they live on simple molecules released by other bacteria, which break down more complex organic materials.

3. **Extreme halophiles** (halo = salt; phile = lover) inhabit extra-salty environments such as salt evaporation ponds, the edge of the ocean, Great Salt Lake, and the Dead Sea. Here they maintain a steep ionic gradient between the cell interior and the environment and use the gradient as an energy source powering the active transport of substances through the plasma membrane. Halophiles are aerobic heterotrophs, but they can supplement respiration by means of a light-driven $H^+$ pump in the plasma membrane, which supplies additional energy to the electrochemical gradient used in chemiosmotic ATP synthesis. The pump uses the purple pigment bacteriorhodopsin, similar to the rhodopsin in our eyes (*Essay: The Solar-Powered Purple Proton Pump,* Chapter 6). Large populations of extreme halophiles give salt evaporation ponds a red tint, due to a red sun-screening pigment also present in the cells.

## Eubacteria

The vast majority of prokaryotes falls into a second main evolutionary group, the eubacteria ("true bacteria"). Our traditional concept of bacteria is based on these familiar and widespread forms. Here we find photosynthetic and chemosynthetic autotrophs, as well as harmless saprobes, normal residents of humans or other eukaryotes, and forms that cause diseases. Indeed, most of the discussion of prokaryotes in this and other chapters of this book pertains to eubacteria.

Two main groups of eubacteria can be distinguished by means of the **Gram stain,** invented in 1884 by Danish microbiologist Hans Christian Gram. **Gram-negative** cells have an outer membrane covering their cell walls; **Gram-positive** cells do not, but tend to have much thicker, rigid walls (Figure 24–12). These features correlate well with other aspects of cell chemistry, and the traditional distinction between Gram-positive and Gram-negative bacteria does appear to reflect phylogeny. The difference between Gram-positive and Gram-negative bacteria also has medical importance (Section 24–F).

Because they do not have outer membranes, Gram-positive bacteria are more susceptible to most antibiotics and to **lysozyme** (lyso = unbinding; zyme = enzyme). This

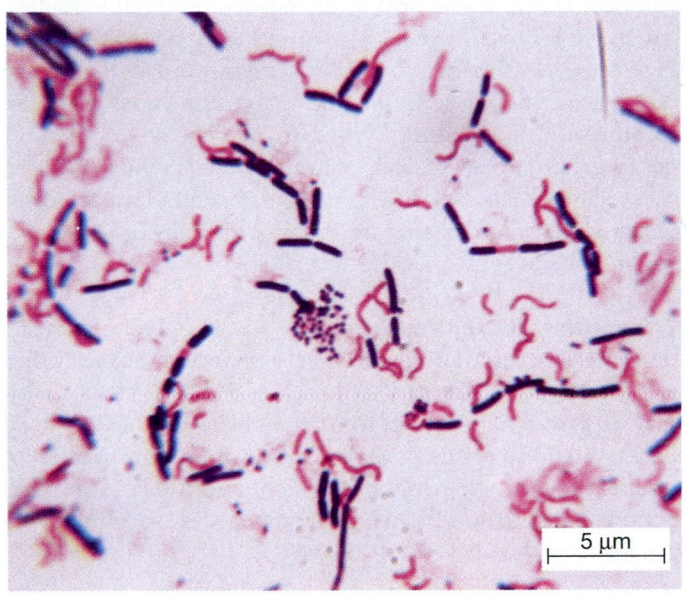

(a)

GRAM-NEGATIVE BACTERIA          GRAM-POSITIVE BACTERIA

Outer membrane
Cell wall peptidoglycans
Periplasmic space
Plasma membrane
Cytoplasm

(b)

**FIGURE 24–12** Gram-negative and Gram-positive bacteria. (a) A mixture of Gram-stained bacteria. The cells were smeared on a microscope slide and treated with crystal violet dye, then iodine solution, and then an alcohol bath, which washed the violet dye out of Gram-negative cells. Last, they were counterstained with a red dye such as safranin, which recolored the Gram-negative cells. Under a light microscope, Gram-positive cells appear purple, Gram-negative cells pink. (b) Comparison of external cell layers of Gram-negative and Gram-positive bacteria.    (a, Jack M. Bostrack/Visuals Unlimited)

enzyme, in nasal secretions and other body fluids, destroys bacterial cell walls. Lysozyme was discovered by Alexander Fleming when he noticed that bacterial cultures died after he had sneezed on them. Gram-negative cells are resistant to lysozyme because the outer membrane keeps such large molecules from reaching the underlying wall.

The new phylogenetic picture emerging from comparisons of macromolecules shows several main evolutionary lines of eubacteria.

***Gram-Positive Bacteria***    Almost all the Gram-positive bacteria belong to one major evolutionary line. Gram-positive bacteria tend to have especially thick and rigid cell walls. Table 24–1 lists many forms of importance to humans. This group also boasts the largest bacterium known: a resident of the gut of surgeonfish, it reaches a length of 0.5 mm, visible to the unaided eye.

**Mycoplasmas** have two notable features: they have lost their cell walls, and they are easily the smallest living cells, only 0.1 to 0.25 $\mu$m in diameter. These features permit them to pass through filters used to trap other bacteria. The lack of cell walls also makes them resistant to penicillin and other antibiotics that act by inhibiting cell wall formation. The mycoplasma genome contains enough DNA to code for about 750 proteins, which may be the minimum necessary for life. Mycoplasmas live as parasites in plant or animal cells. They cause many animal diseases, one kind of human pneumonia, some kidney stones and urinary infections, and they may be the most common cause of premature labor.

**Actinomycetes** produce branching, multicellular filaments that resemble fungi. Members of the genus *Streptomyces* are the source of many valuable antibiotics, which were discovered during a systematic screening of microbes to find agents that would kill Gram-negative bacteria. The first known actinomycete antibiotic, streptomycin, was announced in 1943; others include tetracycline, chloramphenicol, erythromycin, neomycin, and nystatin. Many antibiotics are produced by growing the bacteria in huge tanks, but some can now be synthesized artificially.

The unicellular relatives of actinomycetes include *Corynebacterium diphtheriae,* which causes diphtheria; *Mycobacterium tuberculosis,* which causes tuberculosis; and *M. leprae,* which causes leprosy. Tuberculosis kills more people than any other infectious disease: about 8 million new cases occur each year, and about 2.9 million fatalities. Leprosy, a disfiguring chronic infection, is also widespread; it is estimated to affect more than 20 million people. *M. leprae* grows remarkably slowly, dividing only once every 12 days; this explains why it takes so long for leprosy to appear after a person is exposed to it.

***Spirochetes***    Spirochetes are extremely thin, flexible, curved or spiral bacteria. The genus *Spirochaeta,* probably observed by van Leeuwenhoek as early as 1683, has mem-

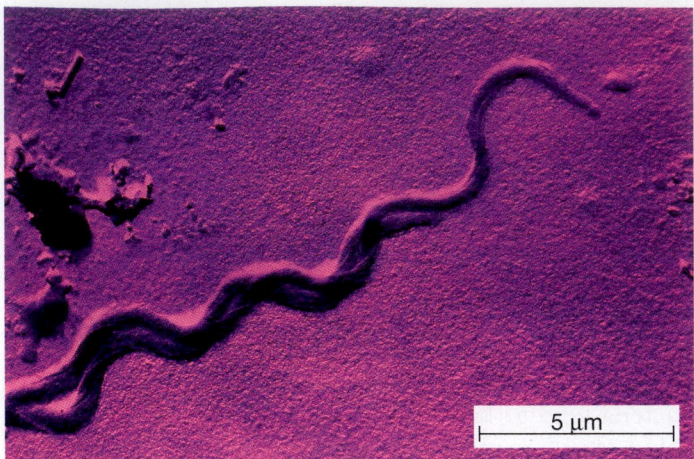

(a) The spirochete that causes syphilis (SEM)

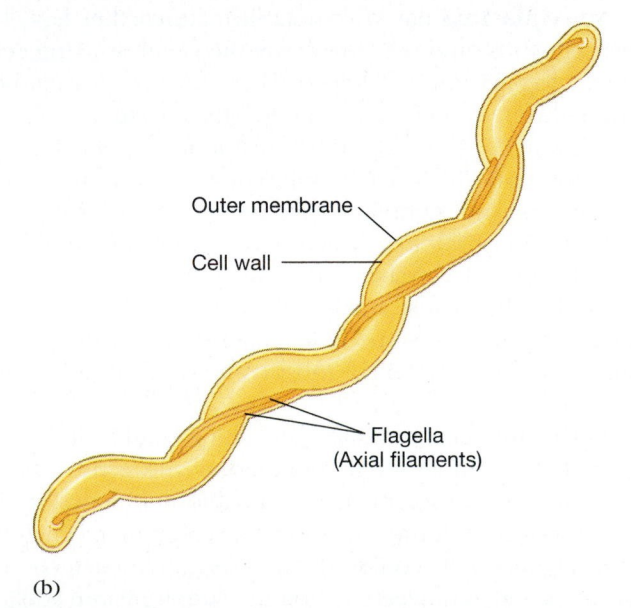

Outer membrane

Cell wall

Flagella
(Axial filaments)

(b)

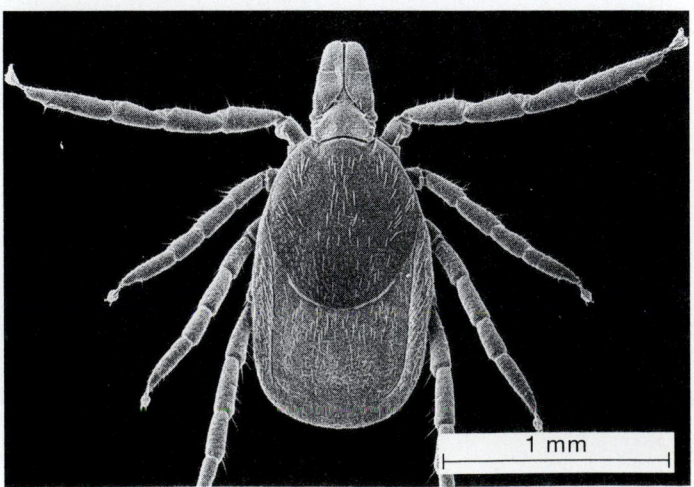

(c)

bers up to 500 $\mu$m long. Spirochete locomotion is unique: their flagella, often called axial filaments, lie in the periplasmic space, between the outer membrane and the cell wall (Figure 24–13). In this position, they enable the cell to swim well through viscous media or crawl against solid surfaces.

Parasitic spirochetes include *Treponema pallidum,* which causes the sexually transmitted disease syphilis, and the related *T. pertenue,* which causes yaws (a tropical eye disease). Lyme disease is caused by spirochetes transmitted by ticks in rural areas of the United States, Europe, and Australia. Many spirochetes live symbiotically with invertebrates such as clams and termites.

***Myxobacteria*** Myxobacteria surround themselves with a slimy polysaccharide secretion (myxo = mucus). Most are saprobes, living in soil and in decomposing organic matter. Here they move around by a mysterious gliding motion. They have no flagella, but fibrils inside the cell may be involved in this movement. They have the most complex reproduction of any prokaryotes: many cells come together and form structures called fruiting bodies, which contain spores (Figure 24–14). This primitive social behavior began at least 2 billion years ago, as shown by fossils of myxobacteria in several stages of the life cycle.

***Rickettsiae*** Rickettsiae are tiny, rod-like bacteria, only about 1 $\mu$m long, that usually live as parasites inside other cells. They are carried by arthropods such as ticks or insects, which transmit them to mammals by bites. Rickettsial diseases include typhus, one of the all-time great killers of humans, which is transmitted by lice, and Rocky Mountain spotted fever, carried by ticks. Most rickettsiae are relatively harmless to their arthropod hosts.

***Purple Bacteria*** One very large and diverse group of Gram-negative eubacteria contains photosynthetic purple sulfur and nonsulfur bacteria, as well as many heterotrophs that apparently evolved from photosynthetic ancestors (Table 24–1).

***Green Sulfur Bacteria*** The green sulfur bacteria represent still another line of evolution. They use $H_2S$, or sometimes $H_2$, as the hydrogen source in photosynthesis.

**FIGURE 24–13** Spirochetes. (a) *Treponema pallidum,* which causes syphilis. (b) A spirochete's flagella are attached at either end of the cell and lie between the cell wall and outer membrane. Each cell has at least two flagella and sometimes hundreds. (c) The pinhead-sized tick that transmits spirochetes responsible for Lyme disease. The disease begins with a rash spreading from a small red bump at the bite, accompanied by aches, chills, and fever. Left untreated, it later progresses to nerve disorders and arthritis.   (a, Science VU/Visuals Unlimited; c, Stanley F. Hayes, Willy Burgdorfer, and M. D. Corwin/Rocky Mountain Laboratories, Hamilton, MT)

**Major Lineages of Bacteria**

| Group | Characteristics | Examples |
|---|---|---|
| **Archaeobacteria:** | Membrane lipids with branched phytanols joined to glycerol by ether linkages; unique lipid synthesis, coenzymes, transfer RNAs, RNA polymerases, and cell wall chemistry | |
| Thermoacidophiles | Inhabit hot, acidic environments | *Sulfolobus, Thermoplasma* |
| Methanogens | Anaerobic, producing methane from $CO_2$ and $H_2$; unable to use complex organic compounds for food | *Methanobacterium* |
| Extreme halophiles | Aerobic photoheterotrophs, using light-driven bacteriorhodopsin $H^+$ pumps to supplement respiration for ATP synthesis; inhabit salt-rich environments | "Purple salt bacteria": *Halobacterium, Halococcus* |
| **Eubacteria:** | "True bacteria"; cell wall containing peptidoglycans and, in Gram-negative forms, surrounded by an outer membrane with lipopolysaccharide | |
| Gram-positive bacteria | Bacteria mostly having thick cell walls with many layers of peptidoglycans, no outer membrane; susceptible to antibiotics and lysozyme | *Staphylococcus epidermidis* on skin. *Streptococcus*: various species normal residents of intestine, mouth, and throat; other forms used to produce dairy products. *Lactobacillus*, also used in dairy products. |
| Spore-forming Gram-positive bacteria | Form heat-resistant spores | *Bacillus anthracis* (causes anthrax); *B. thuringiensis* (used for biological control of some insects). *Clostridium tetani* (causes tetanus);*C. botulinum* (botulism); *C. perfringens* (gas gangrene, food poisoning) |
| Mycoplasmas | Tiny, wall-less, many species normal residents of human mucous membranes | *Mycoplasma pneumoniae* (causes pneumonia) |
| Actinomycetes and relatives | Branching, filamentous prokaryotes; major bacteria in soil | *Streptomyces* (species produce many antibiotics); *Actinomyces* (in human respiratory and digestive tracts but may cause disease in other areas). Unicellular relatives: *Mycobacterium tuberculosis* (causes tuberculosis); *M. leprae* (leprosy); *Corynebacterium diphtheriae* (diphtheria) |
| Spirochetes | Spiral or curved, moving by means of axial filaments; free-living, parasites, or symbionts | *Treponema pallidum* (causes syphilis); *T. pertenue* (yaws); *Borrelia burgdorferi* (Lyme disease) |
| Myxobacteria | Social bacteria. Aerobic saprobes in soil or decaying matter; complex reproductive structures formed by congregation of many cells, releasing cysts enclosing resistant spores | *Stigmatella aurantiaca* |
| Rickettsiae | Small Gram-negative intracellular parasites carried by arthropods | *Rickettsia prowazekii* (causes typhus) |
| Purple bacteria and their relatives | Large, diverse group probably arising from photosynthetic ancestors but including many nonphotosynthetic forms | *Rhizobium* (nitrogen-fixing symbiont in legumes); *Escherichia coli* (in human intestine); *Yersinia* (plague bacterium, in fleas); photosynthetic purple sulfur and nonsulfur bacteria (*Chromatium, Rhodospirillum, Rhodomicrobium*) |
| Green sulfur bacteria | Use $H_2S$ or $H_2$ in photosynthesis | *Chlorobium, Chloroflexus* |
| Cyanobacteria | Mostly photosynthetic, using chlorophyll *a*, splitting water, and releasing $O_2$; phycobilin and carotenoid pigments also present; specialized photosynthetic membranes; some with gas vacuoles; some with nitrogen-fixing heterocysts | *Anabaena, Oscillatoria, Nostoc, Synechococcus, Gloeocapsa* |
| Prochlorophytes | Photosynthetic, using chlorophyll *a* and *b*, splitting water, releasing $O_2$; free-living or symbiotic | *Prochloron* |

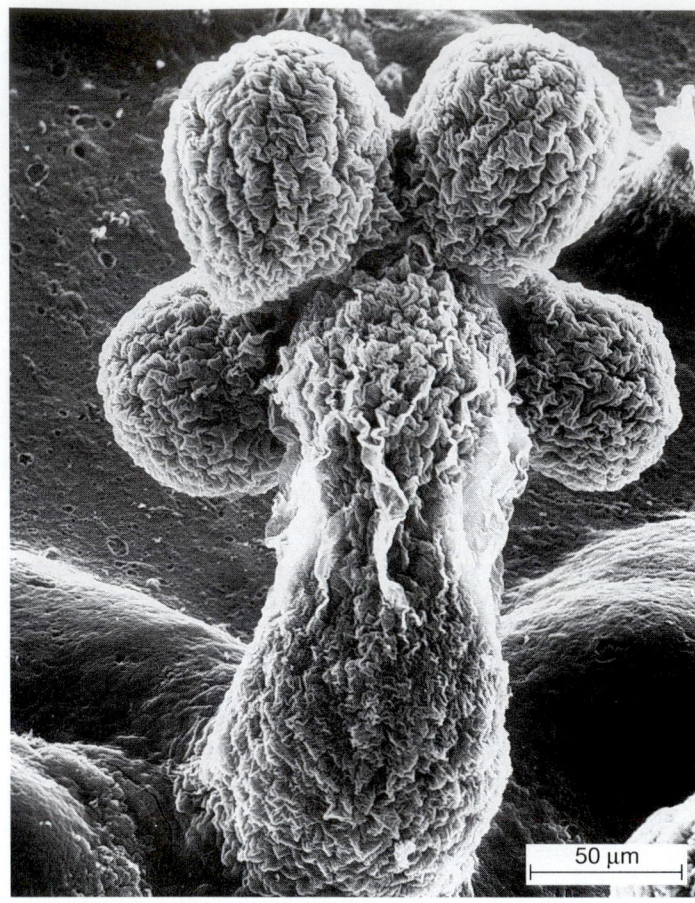

**FIGURE 24–14**  Myxobacteria. This fruiting body of *Stigmatella aurantiaca* consists of thousands of cells.    (Karen Stephens/BPS)

***Cyanobacteria***  Cyanobacteria (also called blue-green bacteria; cyano = blue) are nearly all photosynthetic. Like eukaryotic algae and plants, they split water and give off $O_2$ during photosynthesis (Section 24–C). Most are blue-green because they contain green chlorophyll *a* plus the blue phycobilin pigment phycocyanin. They often contain the red phycobilin, phycoerythrin, as well. Various carotenoid pigments contribute yellow and red tints.

Compared with other prokaryotes, cyanobacteria tend to have larger cells and more specialization of membranes devoted to particular functions. Many cyanobacteria contain protein-bounded gas vesicles that enable them to float near the surface in a body of water, where sunlight is plentiful. Many also have a gelatinous outer sheath that may contain light-screening pigments in addition to toxins poisonous to many animals. Unlike other prokaryotes, many cyanobacteria contain unsaturated fatty acids. No cyanobacteria have cilia or flagella, but some species exhibit a gliding movement. Some are anaerobic and some aerobic.

Cyanobacteria may form filamentous or clustered colonies (Figure 24–15). Some filamentous forms show division of labor within a colony: besides vegetative (photo-

synthetic) cells, there may be spore-producing cells, other cells specialized for attachment to the substrate, and thick-walled **heterocysts,** cells specialized for nitrogen fixation (Section 24–C).

Cyanobacteria are found in almost every moist environment, in the sea, in fresh water, and on land. In these habitats they produce oxygen as a by-product of photosynthesis, and some provide food for other organisms. Some tolerate extremes of temperature, pH, and salinity that would kill photosynthetic eukaryotes.

Population explosions ("blooms") of cyanobacteria are an unpleasant result of water pollution. These organisms thrive on phosphates and nitrates in untreated sewage. Because these cyanobacteria produce toxins and also encase themselves in slimy inedible sheaths, fish cannot control the

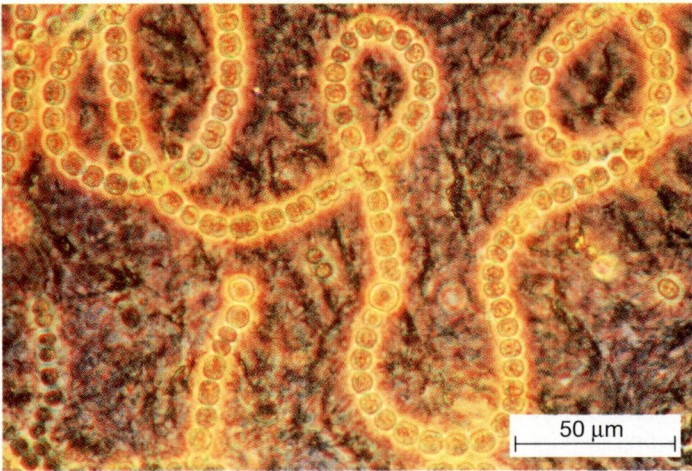

(a)

(b)

**FIGURE 24–15**  Cyanobacteria. (a) Filaments of *Nostoc,* a nitrogen-fixing form that enriches the soil in rice paddies. Numbers of these filaments often form mucilaginous masses on the soil surface. The larger cells are heterocysts. (b) *Merismopedia,* a colonial form.    (a, Sinclair Stammers/Science Photo Library/Photo Researchers; b, J. R. Waaland, University of Washington/BPS)

blooms by eating them. In fact, some fish are killed by cyanobacterial toxins. As cyanobacteria die, aerobic bacteria decompose them and use up all the oxygen in the water, and so any remaining fish suffocate.

***Prochlorophytes***   The prochlorophytes, discovered in the late 1970s, probably evolved from cyanobacteria (Figure 24–16). They have very similar ribosomal RNA sequences, and they also carry on water-splitting, oxygen-producing photosynthesis. However, unlike cyanobacteria, they contain two chlorophylls—*a* and *b*—which also occur in most chloroplasts. In addition, their thylakoids are often associated in twos, rather than occurring singly in the cytoplasm. For these reasons, it seems likely that some member(s) of the group gave rise to the chloroplasts of most photosynthetic eukaryotes (Section 24–H).

Prochlorophytes are also important ecologically: one tiny species, discovered in 1988, is one of the two most numerous photosynthetic organisms in the oceans (the other is a cyanobacterium). Ten gallons of sea water contain about 5 billion of these cells—almost as many as there are people on Earth!

## 24–E   BACTERIA AND FOOD

All food contains bacteria. Even though milk is sterile when it leaves a healthy cow, it contains several types of bacteria by the time it reaches the table. Pasteurization (heating) retards spoilage by reducing the population of these bacteria (originally, it was used to kill disease-causing bacteria, which are seldom a problem nowadays).

Milk spoils when *Streptococcus lactis* and species of *Lactobacillus* ferment lactose (milk sugar) to lactic acid. This lowers the milk's pH and coagulates the milk proteins. On the other hand, *S. lactis* is used to produce many cheeses, including cottage cheese, and species of *Lactobacillus* are used to produce yogurt. Fermenting bacteria are also used to make other foods, such as sauerkraut and pickles. Vinegar is made using bacteria such as *Acetobacter aceti* to oxidize ethyl alcohol to acetic acid in apple cider or wine.

"Food poisoning" comes from toxins produced by bacteria growing in food. The toxins made by the common bacteria *Staphylococcus aureus* and *Clostridium perfringens* are seldom lethal to healthy adults. On the other hand, the common anaerobic soil bacterium *Clostridium botulinum* produces the toxin that causes botulism, one of the most dreaded, but rare, forms of food poisoning. This toxin interferes with nerve activity, causing paralysis, and is fatal if the breathing muscles are paralyzed. The toxin is easily destroyed by heat, but the bacterium also produces endospores that must be heated for a long time to kill them (Figure 24–17). Because *Clostridium* is strictly anaerobic, it usually does not grow in fresh or frozen foods, but thrives in canned goods not heated enough to kill the spores dur-

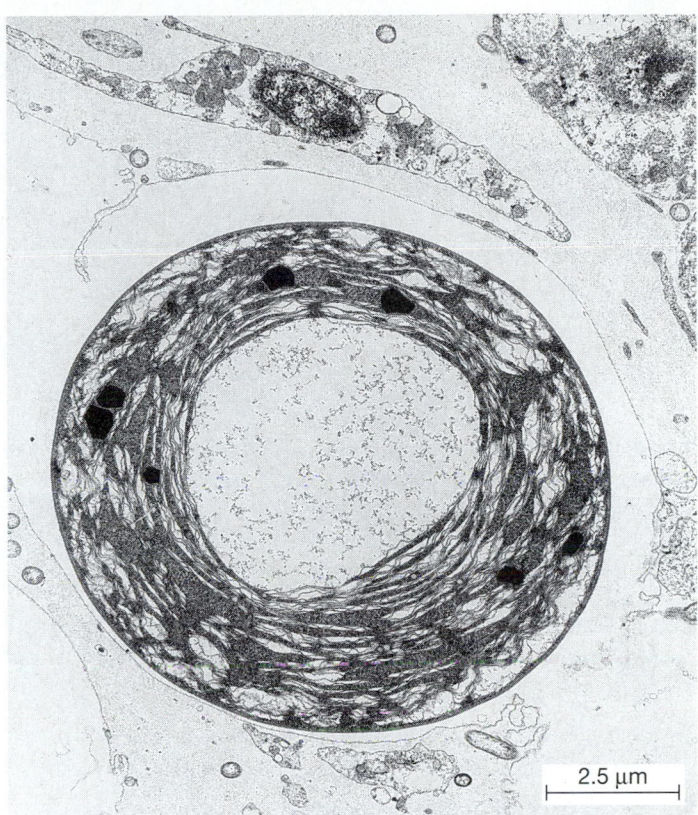

**FIGURE 24–16**   Prochlorophytes. Compare the layers of photosynthetic membranes in this round *Prochloron* cell with those of the cyanobacterium in Figure 24–5. Cells of the invertebrate host are seen around this photosynthetic symbiont.   (by T. D. Pugh, courtesy of E. H. Newcomb, University of Wisconsin)

ing processing. Most cases of botulism come from improperly home-canned foods. The most effective safeguard is to can only acid foods such as fruit and pickles, since *Clostridium* cannot survive at a low pH; vegetables such as beans and peas do not contain enough acid to kill the bacteria.

Food contaminated by populations of *Salmonella* bacteria may also cause illness (mainly diarrhea); pork, poultry, and eggs are common sources. It is not clear whether the trouble is caused by the living bacteria themselves or by their endotoxins (Section 24–F). Recently, ground beef contaminated with certain strains of *Escherichia coli* from the intestines of cattle has been blamed for some cases of serious illness, and even deaths. Simple safety rules minimize the chances of *Salmonella* and *E. coli* outbreaks:

1. Refrigerate food until time to cook or serve it; this retards bacterial growth.
2. Cook chicken, pork, and beef thoroughly; about one in three chickens contains these bacteria.
3. Wash cutting boards, counters, utensils, and hands after handling chicken and before preparing other foods, to avoid cross-contamination.

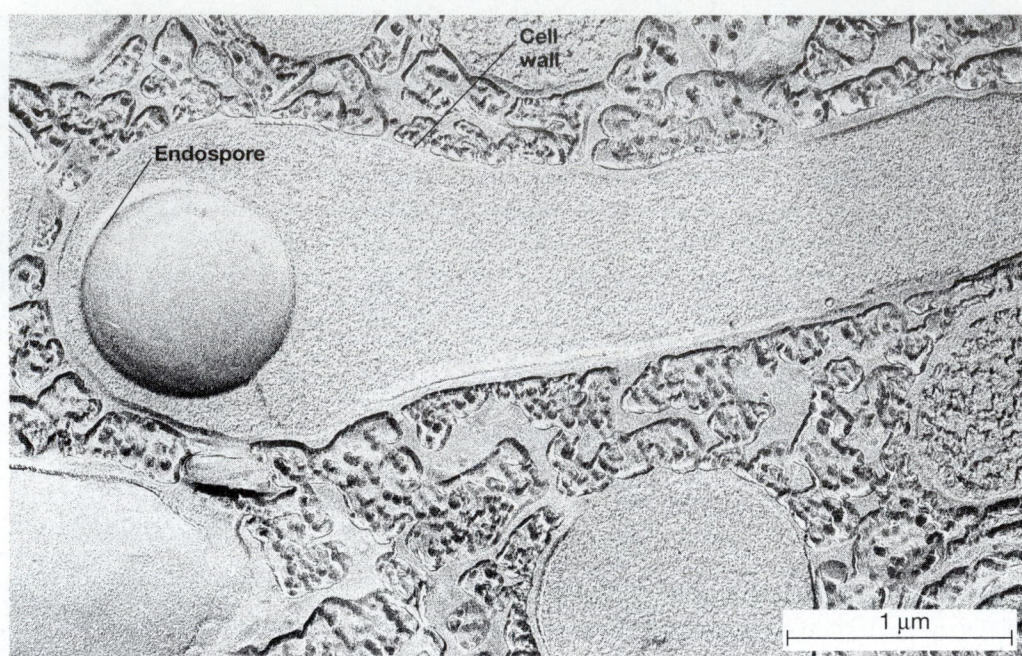

Cell wall

Endospore

1 µm

**FIGURE 24–17**   The heat-resistant endospore of *Clostridium botulinum.* An endospore forms inside the cell wall of the vegetative cell. Canned food must be sterilized under intense heat and pressure to kill these spores. (Freeze fracture)   (T. J. Beveridge, University of Guelph/BPS)

## 24–F   BACTERIAL DISEASES

A **pathogen** is an organism able to produce disease. Some bacteria are **opportunistic pathogens,** causing disease under some conditions but harmless under others. Pathogens are specific for particular host species and particular tissues because infections begin with the specific binding of certain bacterial and host cell-surface molecules (Figure 24–18).

Some pathogenic bacteria destroy host cells, but most cause disease by producing **toxins,** poisonous substances that damage the host's metabolism. There are two kinds of bacterial toxins: endotoxins and exotoxins. **Endotoxins** (endon = within) are lipopolysaccharides in the outer membranes of all Gram-negative bacteria, where they form a shield against entry by antibiotics. This explains why Gram-negative cells are not very susceptible to antibiotics. Because all endotoxins have very similar molecular structures, all Gram-negative bacterial infections produce the same symptoms: chills, fever, damage to the circulatory system, and in high concentrations, fatal shock.

The much less common **exotoxins** ("outside toxins") are proteins secreted from the bacterium and often carried throughout the host's body by the bloodstream. Exotoxins are produced by the bacteria that cause diphtheria, tetanus (lockjaw), and botulism, among others. Since exotoxins are proteins, they differ from each other in structure and in their effects on the host's body. For instance, the exotoxin of *Vibrio cholerae* causes the severe diarrhea of cholera, whereas that of *Yersinia pestis* causes the boils, swollen lymph nodes, bleeding under the skin, fever, and delirium of the plague.

## Control of Bacterial Disease

Pathogens are best controlled by preventing their spread. Victims of diseases such as diphtheria, scarlet fever, whooping cough, and tuberculosis may be quarantined to prevent the airborne bacteria from reaching others in cough or sneeze droplets. Vaccination prevents many bacterial (and viral) diseases from spreading by breaking the chain of infection (Section 34–G).

However, the most important preventive measures are hygiene and sanitation. In the last century, Ignaz Semmelweiss in Hungary and Oliver Wendell Holmes in America found that doctors and midwives could reduce maternal deaths following childbirth about tenfold by washing their hands and their instruments between patients. English surgeon Joseph Lister developed aseptic surgical techniques to reduce the risk of bacterial infections in patients undergoing surgery.

Even today, keeping food and water reasonably free of bacteria saves many more lives than antibiotics do. Cleanliness is especially important in protecting infants from respiratory and intestinal infections. Indeed, the increase in average life expectancy in most countries in the last 100 years can be ascribed almost entirely to better hygiene, which has decreased infant mortality.

Heating to 60 °C for 30 minutes destroys exotoxins, which are proteins, and kills most bacterial cells. For this reason pasteurization (heating) has proved both easy and effective in protecting against botulism in canned food, brucellosis and tuberculosis transmitted in milk, and dysentery caused by drinking water contaminated with human feces.

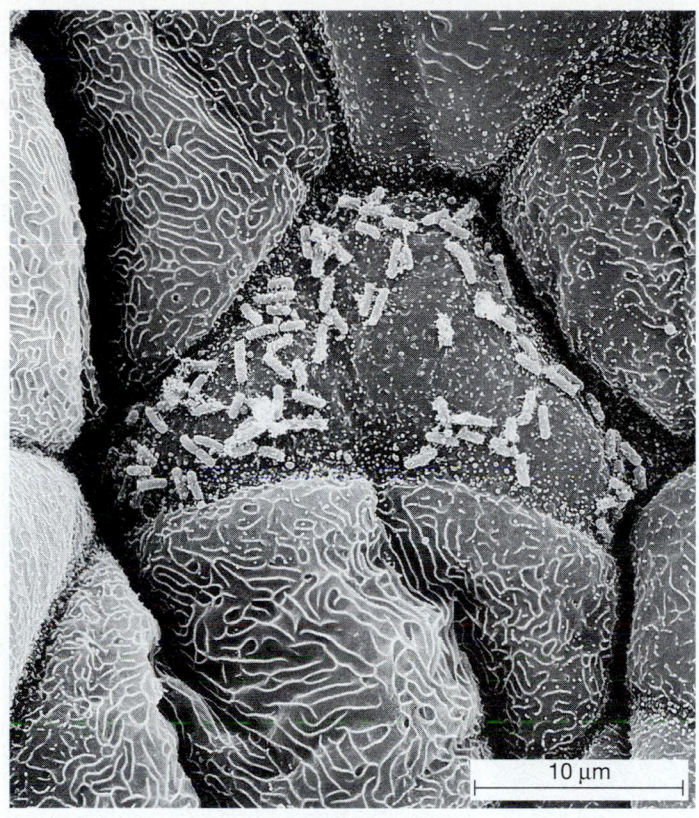

(a)

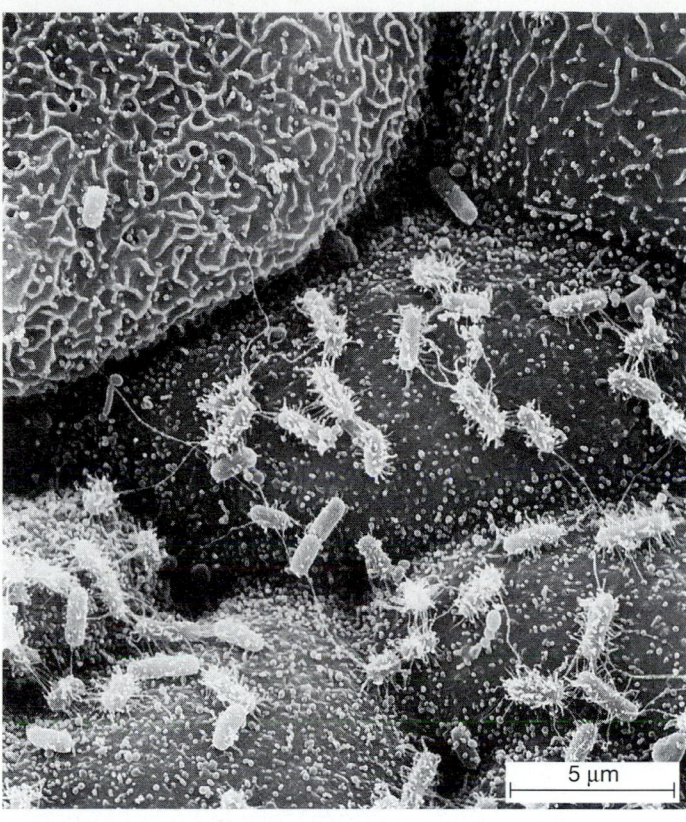

(b)

**FIGURE 24–18**   Selective binding of pathogenic bacteria. (a) Many rod-shaped *Escherichia coli* bacteria have migrated up the urinary tract and attached to a young cell, which has microvilli on its surface. The bacteria do not attach to the surrounding mature cells, which have microfolds on their surfaces. (b) A closeup shows the pili by which the bacteria bind selectively to carbohydrates on the eukaryotic cell's surface. (SEMs)   (Kazuhiko Fujita, Juntendo University School of Medicine, Tokyo)

Antibiotics have also saved thousands of lives since about 1940. During all wars before World War II, more lives were lost to disease (from unsanitary camps and infected wounds) than to enemy action.

## Drug Resistance in Bacteria

At their debut in the 1940s, antibiotics were hailed as miracle drugs that would soon eradicate most infectious diseases. However, as we saw in our study of evolution, selection has favored bacteria that happened to have mutations making them resistant to antibiotics. When a bacterial population is exposed to a particular drug, susceptible cells die, but resistant ones multiply rapidly and soon become widespread. For example, some strains of the bacteria that cause gonorrhea and tuberculosis can no longer be killed by any known drug, and both diseases are now spreading and posing serious public health problems. Even pathogens without genetic resistance to antibiotics may be protected if the body houses resistant nonpathogenic bacteria that can render the drug harmless before it can reach the pathogen.

Some drug-resistance mutations are well understood. Some bacteria, for instance, make penicillinases, enzymes that destroy penicillin. Others have enzymes that add a molecular group to the drug, covering the site where the drug would normally bind to the bacterium.

## 24–G   SYMBIOTIC BACTERIA

Many kinds of bacteria live with other organisms in an intimate symbiotic relationship, either parasitic, commensal, or mutualistic (Section 18–A). The partners in a symbiosis, especially in a mutualistic association, are **symbionts.** If the partners differ in size, as in the case of a plant or animal housing much smaller bacteria or protists, the larger organism is often referred to as the host.

Many symbiotic bacteria are part of an animal's **microbiota:** organisms that normally live on or within it. These

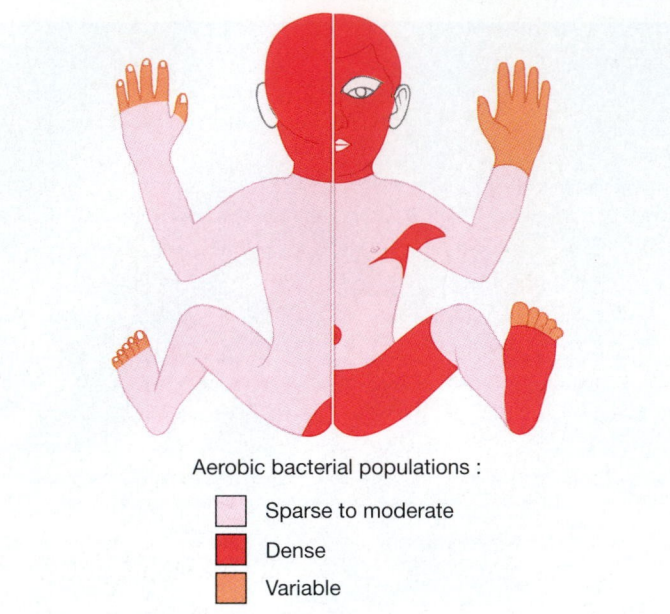

Aerobic bacterial populations :

▢ Sparse to moderate

■ Dense

■ Variable

**FIGURE 24–19**   Microbiota of the human skin.

are also called the animal's **flora** ("plants"), a holdover from the old classification of bacteria in the plant kingdom. They may be saprobes, living on such things as dead skin cells or digested food in the intestines, or parasites or symbionts absorbing nutrients from living tissue. The microbiota of human beings includes *Staphylococcus epidermidis* on the skin, *S. aureus* in the nostrils, and *Bacteroides fragilis* and *Escherichia coli* in the intestine (Figure 24–19).

An animal's microbiota benefit the host because these established populations prevent invading pathogens from settling in. (All organisms grow better without competition for food and space.) Foreign bacteria or fungi invade the skin, gut, or mucous membranes more readily if the resident bacteria have been removed by antibiotics, douching, or other means. Some bacteria also produce antibiotics that inhibit the growth of other organisms.

Beyond this, both humans and many other animals depend on their gut microbiota to make some vitamins. And many animals that use cellulose for food (e.g., cows and colobus monkeys) can digest it only because some of the bacteria in their stomachs secrete cellulose-digesting enzymes (Section 31–D).

On the other hand, members of an animal's microbiota may cause disease if they settle in the wrong places. *Escherichia coli* causes cystitis if it invades the urinary bladder, and *Staphylococcus aureus* can cause serious infections if it gets into a wound; this is one reason why surgeons wear gowns, gloves, and masks.

Other symbiotic bacteria include the nitrogen fixers in the root nodules of legumes (see Figure 24–9), cyanobac-

teria inhabiting some water ferns, and bioluminescent (light-producing) forms in ponyfish and flashlight fish (Figure 24–20a). Chemosynthetic bacteria live as symbionts of various invertebrate animals in anaerobic environments such as salt marshes, mangrove swamps, and deep-sea hydrothermal vents (Figure 24–20b). About 20 different kinds of bacteria reside in various strains of the protist *Paramecium* (Section 25–J).

One strange case is *"Methanobacillus omelianski,"* once thought to be a methanogen (Section 24–D) but now known

(a)

(b)

**FIGURE 24–20**   Hosts of symbiotic bacteria. (a) This flashlight fish from the Indian Ocean emits light thanks to bioluminescent bacterial symbionts housed in an organ below the eye. The fish can block the light by raising a black partition. By "blinking," the fish can lure prey, confuse predators, and communicate with other flashlight fish. (b) Near a thermal vent in the sea floor, chemosynthetic bacteria use hydrogen sulfide as an energy source to make their own food. These bacteria, in turn, support the animals seen here. Some of the bacteria live inside the huge white worms with red tips, in a mutualistic symbiosis.   (a, Ken Lucas/BPS; b, John B. Corliss)

to be a symbiosis between two look-alike species with different metabolic talents: an anaerobe that ferments ethanol and produces $H_2$, and its methanogen partner, which removes the potentially toxic $H_2$ by combining it with $CO_2$ to form $CH_4$.

Perhaps the most fascinating symbiosis is that between the anaerobic protist *Mixotricha paradoxa* and four different kinds of bacteria. One kind lives inside the cell and serves much as mitochondria do in aerobic eukaryotes, breaking down the protist's fermentation products. The others act in locomotion. The cell's surface is covered with up to half a million small spirochetes, interspersed with fewer, larger spirochetes of another species. Bacteria of the fourth species live in pockets on the cell's outer surface, where they anchor the small spirochetes (Figure 24–21). The small spirochetes act as organs of locomotion, propelling the whole assemblage through its environment, which is the gut of a termite: *M. paradoxa* itself is a symbiont, digesting the termite's meals of wood! The termite cannot produce wood-digesting enzymes, and without its symbionts it would starve.

The evolution of one symbiotic relationship has been documented. Bacteria infected and killed most of the *Amoeba proteus* cells in a laboratory culture. Over the course of about 200 generations, the surviving amoebas, also infected, became dependent on their bacterial residents. After about five years, nuclear transplantation experiments showed that the nuclei of infected amoebas could no longer function in cytoplasm lacking the bacteria. The inhabited amoebas have different sensitivity to temperature and antibiotics and grow more slowly, but they are less susceptible to fatal infection by other bacteria.

We can view mutualistic symbioses as yet another mode of evolution available to prokaryotes because it provides a way for distantly related species to combine their genomes into a single lineage. In many symbioses, the partners become genetically interdependent. For example, root nodules of legumes contain leghemoglobin, a pigment that binds $O_2$ and so prevents it from interfering with nitrogen fixation. The heme part of leghemoglobin is produced by the nodule bacteria, the globin protein by the leguminous plant.

Often one member of a symbiosis, usually the smaller one, loses genes for proteins similar to those made by its partner, and some of its genes move into the partner's genome. As a result, many symbiotic organisms are hard to grow without their partners because they normally rely on their partners for a variety of vital nutrients.

Most bacterial symbionts are enclosed in a vacuole composed of membrane made by the host cell (see Figure 24–9b). As the relationship becomes more intimate, this barrier may disappear. Symbionts not enclosed in such a membrane usually cannot be grown outside their hosts in laboratory cultures.

## 24–H   ORIGIN OF MITOCHONDRIA AND PLASTIDS

Plastids and mitochondria are eukaryotic organelles that evolved from prokaryotic symbionts living inside ancient eukaryotic cells. This **endosymbiont theory** (endon = within) was first proposed in 1918 and has been championed since the 1960s by Lynn Margulis. She and others have accumulated impressive supporting evidence:

1. Some prokaryotes live inside eukaryotic cells today (Figure 24–22).
2. Plastids and mitochondria are about the size of prokaryotic cells. They also contain circular DNA without histone proteins and reproduce by binary fission.
3. Plastids and mitochondria make some of their own proteins (although most are imported from the cytoplasm). As in prokaryotic protein synthesis, most of their polypeptide chains start with the modified amino acid formylmethionine rather than with methionine as in eukaryote cytoplasm.
4. Organelle and prokaryote ribosomes are the same size. Furthermore, chloroplast and prokaryote ribosomes can

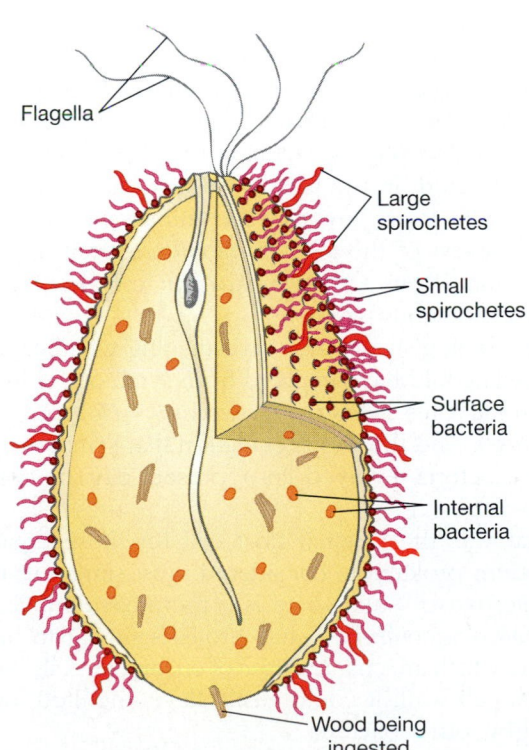

**FIGURE 24–21**  A symbiont's symbionts. The protist *Mixotricha paradoxa*. This symbiont in the gut of Australian termites has four species of symbiotic bacteria, which digest its food and provide locomotion as described in the text. The protist's four flagella do not move it, but serve as rudders.

Labels on figure:
Flagella
Large spirochetes
Small spirochetes
Surface bacteria
Internal bacteria
Wood being ingested

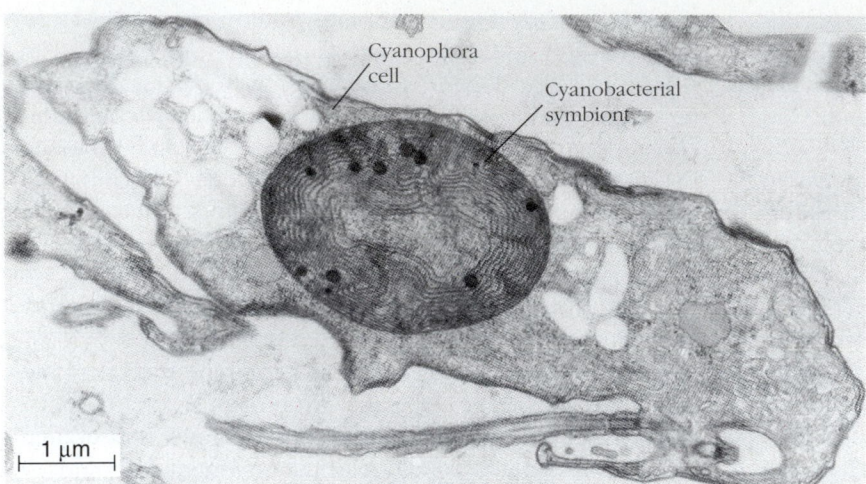

**FIGURE 24–22** The protist *Cyanophora* with a symbiotic cyanobacterium in its cytoplasm. The cyanobacterium produces food for itself and its host, and *Cyanophora* uses its flagellum to move to areas where light is available for its symbiont's photosynthesis. Notice the many layers of photosynthetic membranes in the cyanobacterial cell. (Biophoto Associates)

be hybridized: in laboratory experiments, small subunits from one can combine with large subunits from the other, forming hybrid ribosomes that carry out protein synthesis. In contrast, ribosomal subunits from eukaryotic cytoplasm do not hybridize with those from their own organelles nor from prokaryotes.

5. Many antibiotics block protein synthesis by the ribosomes of prokaryotes and organelles but not by the cytoplasmic ribosomes of eukaryotes.

6. Nucleotide sequencing shows great similarity among ribosomal and transfer RNAs of several cyanobacteria, red algae, and chloroplasts of various plants. Some of these molecules remained nearly unchanged during 1 to 2 billion years on separate evolutionary paths. These molecules are highly conserved because mutations that impair protein synthesis spell certain death.

The ancestors of mitochondria may have been accepted into host cells because they could detoxify $O_2$. Later, they evolved a way of exchanging ADP and ATP with the cytoplasm and so became the host cell's main power supply.

The evidence indicates that these mitochondrial ancestors moved into their eukaryotic host cells long before ancestral plastids arrived. First, all organisms with plastids also have mitochondria, but not vice versa. Second, mitochondria are less like free-living bacteria, reflecting a longer history as symbionts. For example, a significant amount of DNA has moved from mitochondria into the cell nucleus. Also, mitochondria make fewer of their proteins (only about 10%); the rest are imported from the cytoplasm. Chloroplasts still make most of their own membrane lipids, whereas mitochondria import most of theirs from the cytoplasm, molecule by molecule. (In contrast, nonsymbiotic eukaryotic organelles swap patches of membrane [see Figure 5–16].)

Which prokaryotes were ancestral to these organelles? Mitochondria must have come from aerobic bacteria, most likely a member of the purple nonsulfur group. Chloroplasts from different groups of algae have different structure and chemistry and may have descended from members of different lines of photosynthetic bacteria (Chapter 25). For example, the chloroplasts of red algae probably evolved from cyanobacteria, whereas those of green algae and plants resemble prochlorophytes. (In fact, the first prochlorophyte to be discovered, in the late 1970s, lives as a symbiont of tunicates, which are invertebrate animals [Section 29–J].) The chloroplast of the alga *Cryptomonas* even appears to be descended from a small *eukaryotic* cell that became a photosynthetic endosymbiont!

Some biologists are also investigating whether eukaryotic flagella could have evolved from symbiotic spirochete-like prokaryotes (*Essay: Eukaryotic Flagella: Senior Symbionts?*). And there is speculation that symbiotic Gram-positive bacteria evolved into present-day peroxisomes (Section 5–L).

What was the original host cell for these organelles? The modern prokaryote considered most similar to the cytoplasmic part of eukaryotes is a thermoacidophile, *Thermoplasma acidophilum*. It has proteins similar to histones and actin, which are characteristic of eukaryotic cells. Because it lacks a cell wall, it could easily have engulfed, or been invaded by, other bacteria.

The endosymbiont theory of eukaryote origins changes our traditional view of an evolutionary tree with repeated branchings that fan out from a common origin. Instead, the endosymbiont tree also has branches that fuse and grow on as one, often fanning out vigorously as the new partnership undergoes adaptive radiation (Figure 24–23).

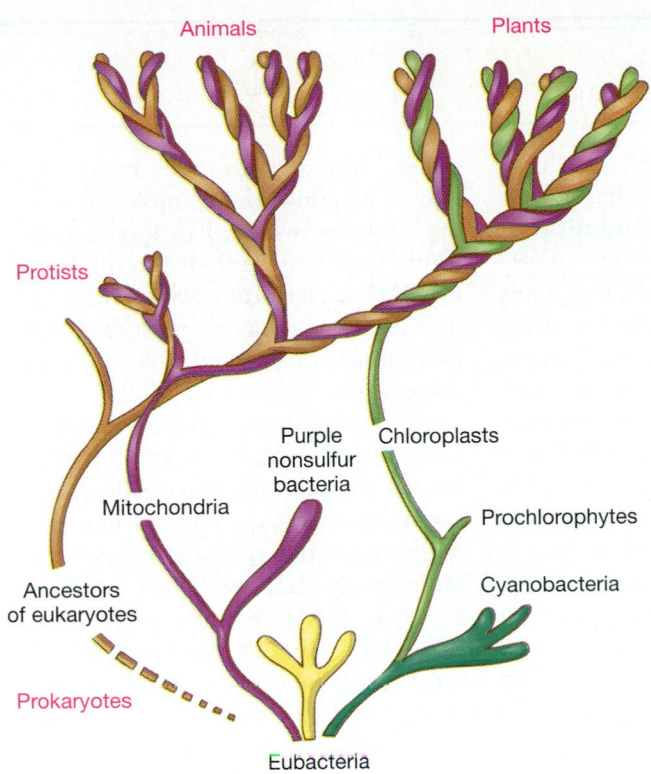

**FIGURE 24-23**  General scheme of evolution. Different lineages merged and grew as one when symbiotic bacteria evolved into organelles. The branching patterns shown are schematic and are not intended to represent actual taxa.

## SUMMARY

The kingdom Prokaryotae contains all the bacteria, although some biologists would erect a separate kingdom for archaebacteria. All are prokaryotic and most are unicellular in function, if not in structure.

Three billion years ago, bacteria were Earth's only inhabitants, living in communities of many different, interdependent species. Some were autotrophs, absorbing raw materials from the environment and making their own food; photoautotrophs used solar energy, whereas chemoautotrophs obtained energy from chemical reactions, mostly of sulfur or nitrogen compounds. Other bacteria lived as heterotrophs, exploiting their neighbors for food. When a cell died or eliminated its wastes, it provided raw materials for other bacteria, and so nutrients were recycled in the bacterial community.

Today, bacteria are still the most numerous and ubiquitous organisms on Earth, and they still play all of these roles. However, eukaryotes (themselves the descendants of multi-cell symbiotic alliances) have become the chief photosynthesizers and the most conspicuous heterotrophs wherever resources are abundant. Bacteria predominate where resources are scarce or where their unique metabolic abilities enable them to survive in conditions too harsh to support most eukaryotes.

1. Prokaryotic cells are smaller and less complex than eukaryotic cells. They lack membrane-bounded organelles, particularly a nucleus, but some have internal membranes. A prokaryote's genome consists of a single, circular DNA molecule with little associated protein.

2. Bacteria also differ from eukaryotes in their cell wall composition: peptidoglycans in eubacteria, and various other compounds in archaeobacteria. Gram-negative eubacteria also have an outer membrane around the cell wall, which confers resistance to many antibiotics and also contains lipopolysaccharide endotoxins that produce disease symptoms in host animals. Many bacteria also have polysaccharide capsules or protein pili by which they anchor to some surface. Capsules also offer some protection from viral invasion, predation, toxic chemicals, and desiccation.

3. Some bacteria have flagella, which rotate and move the cell through its liquid environment. These bacteria may respond to the environment by moving up or down gradients of light, temperature, chemicals, or magnetic force.

4. Most bacterial cells reproduce by binary fission.

5. Genetic recombination can occur by the transfer of a variable amount of DNA from a donor to a recipient bacterium, but the frequency and evolutionary importance of such events in nature is unknown.

6. Mutation and rapid reproduction, coupled with small size, metabolic diversity, and ability to survive adverse conditions by forming resistant spores, are believed to account for the evolutionary success of bacteria.

7. Bacteria obtain nutrients by absorption. Autotrophic bacteria make their own organic molecules from inorganic substances, using energy from light (photoautotrophs) or inorganic chemical reactions (chemoautotrophs). Heterotrophic bacteria must obtain their organic molecules from other organisms.

8. Purple and green bacteria carry out primitive forms of photosynthesis. They use bacterial chlorophyll rather than chlorophyll *a* and do not split water nor produce oxygen. Cyanobacteria and prochlorophytes have chlorophyll *a* and carry out water-splitting, oxygen-producing photosynthesis.

9. Saprobic bacteria (and fungi) play the vital ecological role of decomposers, breaking down organic material and releasing its components for recycling.

10. All life on Earth depends on nitrogen-fixing and nitrifying bacteria to provide the forms of nitrogen used in making proteins.

11. For practical purposes, bacteria are usually classified by morphology and metabolism. Nucleotide sequencing has revealed that these criteria do not necessarily reflect phylogeny.

12. Archaeobacteria include thermoacidophiles, methanogens, and extreme halophiles, all adapted to unusual habitats. They differ in many metabolic fundamentals from eubacteria, which include most prokaryotes alive today.

## ESSAY: *Eukaryotic Flagella: Senior Symbionts?*

Some biologists believe that the ancestors of mitochondria were not the first bacteria to merge with eukaryotic cells. Even earlier, symbiotic spirochete bacteria, with their unusual mode of locomotion, moved in and evolved into eukaryotic cilia and flagella (Figure 24–A). However, the evidence for this is less impressive and has not convinced most biologists.

The evolution of eukaryotic cilia and flagella is indeed a mystery. They differ from prokaryotic flagella in many ways (Table 24–A). Furthermore, we do not know of any forms intermediate between the two, which we would expect to find if prokaryotic flagella had evolved into the eukaryotic type. In fact, the two are so different that some biologists use the term "flagellum" only for the prokaryotic type. For eukaryotic flagella (and cilia, which have the same basic structure) they use the term **undulipodia** (undula = little wave; podia = feet).

One hypothesis holds that early eukaryotes had cytoskeletons made up of microtubules, whose components rearranged into the mitotic spindle during cell division. Eventually, some microtubules assembled into the more elaborate pattern found in centrioles and in undulipodia and their basal bodies (Section 5–Q).

In contrast, the endosymbiont hypothesis proposes that this pattern of microtubules is inherited from symbiotic spirochete ancestors that lived attached to the eukaryotic host cell's surface and moved the host around. Eventually the bacteria fused with the host and lost their separate identity. Only their locomotory structures persisted, along with the necessary genes. At some point, molecules of the proteins involved in locomotion found their way further into the host cytoplasm. Here they were put to a new use, parceling chromosomes into two groups during cell division, and mitosis was born.

This hypothesis arose because some single-celled eukaryotes and animal tissues today have spirochete symbionts that look and act like undulipodia. (Some of these hosts have undulipodia as well; see Figure 24–21.) Electron micrographs of some large spirochetes show structures resembling microtubules at the base of their flagella, but nothing like the specific arrays in undulipodia.

Bacteria of the genus *Spirochaeta* have two proteins similar to the tubulin in eukaryotic undulipodia. Both

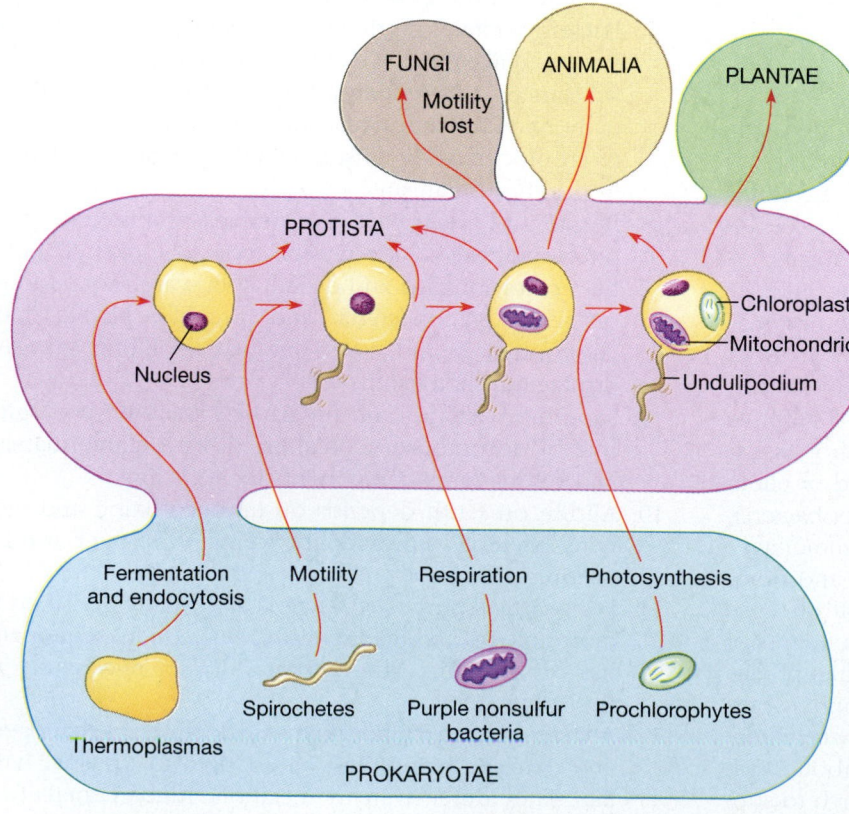

**FIGURE 24–A** Proposed symbiotic evolution of major eukaryotic lineages. An ancient cell evolved a nucleus, thereby becoming a eukaryote. Its descendants later formed symbiotic partnerships with bacteria, which eventually evolved into organelles: first undulipodia, then mitochondria, and later plastids. At each stage, some lineages evolved into forms with only those organelles, whereas others added new symbionts and evolved further diversity. This stepwise acquisition of organelles is sometimes called the serial endosymbiont theory.

spirochete proteins and tubulin share the unusual property of temperature-dependent polymerization: they join into tubules at 37 °C, but separate at 0 °C. Both spirochete proteins also react with antibodies (immune system proteins) prepared to bind specifically to tubulin (*How Do We Know? Making Monoclonal Antibodies,* Chapter 34).

In addition, the single-celled flagellated eukaryote *Chlamydomonas reinhardtii* appears to have a prokaryote-like DNA containing genes governing its locomotion. More than 200 of this organism's genes have been mapped by making crosses between normal and mutant individuals. Most of these genes map to linear chromosomes in the nucleus. However, 20 mutations that affect locomotion all occur in a separate linkage group with a circular map, about 7 million bases long, on the same order as prokaryote DNAs. This linkage group does not lie in the nucleus. Some researchers claim it occurs in the basal bodies of the two undulipodia, but others have found no sign of nucleic acid there.

Arguing against the idea of a spirochete origin of undulipodia is the finding that antibodies to the spirochete proteins do not also bind tubulin. (That is, the antibody reactivity is not reciprocal, contrary to what we would expect from homologous proteins.) Also, even though the spirochete proteins and tubulin behave similarly, there is scant correspondence in the nucleotide sequences encoding them. However, such a difference would not be surprising in distantly related molecules that have evolved independently for a billion years or so; in many proteins, only certain key positions must remain occupied by the same amino acids, while the rest may be changed with little effect on the protein's function except perhaps to fine-tune it.

Is it just coincidence that symbiotic spirochetes look like undulipodia at first glance or are these indeed distant relatives? We may never know for certain. However, there are a variety of primitive eukaryotes, and many symbiotic spirochetes, that have not yet been well studied. If undulipodia did have a symbiotic origin, some of these organisms may provide evidence of intermediate stages in this evolution.

**TABLE 24–A**

**Comparison of Prokaryotic and Eukaryotic Flagella**

| Feature | Prokaryotes | Eukaryotes |
| --- | --- | --- |
| Structure | Basal body "motor," hook, filament, not covered by membrane | 9 + 2 shaft covered by extension of plasma membrane; 9 + 3 basal body |
| Main protein | Flagellin | Tubulin |
| Motion | Rotation at base | Sliding force between tubules throughout shaft |
| Power source | $H^+$ gradient across plasma membrane | ATP |

13. Food poisoning is caused by exotoxins or pathogenic bacteria in food. We use some bacteria to produce foods such as yogurt, cheeses, and vinegar.

14. Bacterial diseases have played major, roles in human history. Immunization, improved hygiene, and antibiotics have reduced human and animal deaths from bacterial disease. One disadvantage of using antibiotics is that it selects for mutant bacteria and for transferable plasmids carrying genes for antibiotic resistance.

15. The small size of bacteria has permitted many of them to form especially intimate symbioses with other organisms, lending their unique metabolic abilities to fellow prokaryotes or to eukaryotic hosts. Some bacteria are pathogens, producing toxins that cause disease, and some are part of the normal microbiota of animals.

16. Chloroplasts and mitochondria evolved from endosymbiotic bacteria and still show many similarities to prokaryotic cells.

## Self-Quiz

1. An organism should be placed in kingdom Prokaryotae if:
   a. it consists of a single cell
   b. it has a cell wall
   c. it is surrounded by a capsule
   d. it lacks a nuclear envelope separating its genetic material from the cytoplasm
   e. it causes diseases
2. Prokaryotic cells differ from eukaryotic cells in:
   a. structure of genetic material
   b. possessing cell walls
   c. possessing vacuoles
   d. possessing ribosomes
   e. all of the above
3. Newly started rice paddies produce poor crops until they have established a flourishing population of cyanobacteria. This is probably because:
   a. the rice needs nitrogen fixed by the cyanobacteria
   b. the rice cannot compete with weeds, which are poisoned by toxins produced by the cyanobacteria
   c. the cyanobacteria use up surplus nutrients from sewage in the rice paddies
   d. the cyanobacteria provide plasmids carrying genes that increase fertility
   e. cyanobacteria form a protective coating on the rice plants
4. When you must take antibiotics for a long time, a doctor will frequently prescribe a combination of antibiotics instead of only one. This is because:

   a. The combination of drugs aids in singling out pathogenic bacteria without affecting others.
   b. Some of the antibiotics may encourage the growth of non-pathogenic bacteria, but these can be combatted by careful choice of a second antibiotic.
   c. Chances are very small that single bacteria resistant to both antibiotics are present.
   d. You may be allergic to one of the antibiotics but not to them all.
   e. It takes longer for a bacterial cell to develop resistance to two drugs to which it is exposed.
5. Bacteria are used in the production of:
   a. wine
   b. vinegar
   c. gelatin
   d. pasteurized milk
   e. marshmallows
6. List three differences between photosynthesis in sulfur bacteria and that in cyanobacteria.

Mark each of the following statements true or false.

_____ 7. A bloom of photosynthetic cyanobacteria can lead to shortage of oxygen in a lake or pond.
_____ 8. Many bacteria protect their hosts from pathogens by producing antibiotics.
_____ 9. All food containing bacteria is unsafe for consumption and should be thrown out.
_____ 10. Bacteria swim by lashing their flagella back and forth.

## Questions for Discussion

1. Why are fossil prokaryotes more difficult to find and study than fossils of other organisms?
2. What is the adaptive value to an organism of producing a toxin? What is the disadvantage?
3. Why is it advantageous for a bacterium to produce a capsule? Why do mutants lacking a capsule come to predominate in laboratory cultures?
4. *Escherichia coli* produces flagella only when its environment does not contain enough food. What is the advantage of this?
5. Many prokaryotes can make all the molecules they need when provided with a nutrient medium containing a supply of inorganic salts and one type of small organic molecule (e.g., a monosaccharide or fatty acid) as an organic carbon source. How can prokaryotes be so self-sufficient if their DNA is so small compared to even one of the many chromosomes found in eukaryotic cells?

6. Why did Lederberg and Tatum do their experiment (Section 24–B) with strains of bacteria deficient in three nutrients each, instead of just one each?
7. What are the biological roles of steroids? Why don't most bacteria need steroids?
8. Explain why symbiotic bacteria in the cells of photosynthetic diatoms (protists) fix more nitrogen at night than during the daytime.
9. Mitochondria and plastids have lost the cell walls and some of the genes of their bacterial ancestors, while other genes have moved into the host cell's nucleus. (The genes that remain in mitochondria code for polypeptide chains that are very insoluble in water.) What adaptive advantages would have selected for these changes?

## Suggested Readings

Brili, W. J. "Biological nitrogen fixation." *Scientific American,* March 1977.

Brock, T. *Biology of Microorganisms,* 6th ed. Englewood Cliffs, NJ: Prentice-Hall, 1990.

Costerton, J. W., G. G. Geesey, and K.-J. Cheng. "How bacteria stick." *Scientific American,* January 1978.

Fischetti, V. A. "Streptococcal M protein." *Scientific American,* June 1991. How bacteria that cause strep throat and rheumatic fever evade the body's defenses.

Fox, G. E., et al. "The phylogeny of prokaryotes." *Science* 209:457, 1980.

Kantor, F. S. "Disarming Lyme disease." *Scientific American,* September 1994.

Margulis, L. *Symbiosis in Cell Evolution.* San Francisco: W. H. Freeman, 1981. A leading advocate of the hypothesis of symbiotic origin of organelles presents evidence to support her case.

Rietschel, E. T., and H. Brade. "Bacterial endotoxins." *Scientific American,* August 1992.

*Scientific American. Industrial Microbiology.* September 1981. An issue devoted to the many ways humans have harnessed microbes.

Shapiro, J. A. "Bacteria as multicellular organisms." *Scientific American,* June 1988.

Thomsen, D. E. "Swimming for the good life." *Science News* 125: 298, 1984. How bacterial chemotaxis works.

Woese, C. "Archaebacteria." *Scientific American,* June 1981.

# Protista and the Origin of Multicellularity

## OBJECTIVES

*When you have studied this chapter, you should be able to:*

1. Define the following terms, and use them correctly:
   protozoa, algae, phytoplankton, colony, thallus
   sessile, motile, taxis
   cyst, spore, fruiting body, conjugation
2. State the criteria for placing organisms in the kingdom Protista.
3. List and describe the three main types of locomotion found among protists.
4. List or recognize distinguishing characteristics of members of the major phyla of protists: Foraminifera, Zoomastigina, Api-

complexa, Ciliophora, Dinoflagellata, Bacillariophyta, Phaeophyta, Rhodophyta, Chlorophyta, and Conjugaphyta.
5. Place each of the following in the correct phylum: *Amoeba,* foraminiferan, cellular slime mold, plasmodial slime mold, *Trypanosoma, Plasmodium, Paramecium,* water mold, dinoflagellate, *Euglena,* diatom, *Fucus,* kelp, red algae, *Chlamydomonas, Ulothrix, Ulva, Spirogyra,* desmids.
6. State how each of the following is of adaptive advantage to the algae that possess it: flagella, storage of food as oil, water-retaining cell-wall components, accessory photosynthetic pigment, holdfast, air bladder, sexual reproduction, asexual reproduction, zoospores.

---

**K**ingdom Protista is a catchall for any eukaryote that does not fit the definition of kingdoms Fungi, Plantae, or Animalia.[1] Besides eukaryotic cells, the only thing protists have in common is dependence on a watery environment: when they are actively growing, most protists inhabit fresh or salt water or the bodies of larger organisms. The terrestrial (land-dwelling) forms live in soil water or in wet leaf litter, wood, and similar damp places. Nearly all protists are aerobic.

The organisms in the protistan grab bag span a wide range of size and complexity. Here we find virtually all single-celled eukaryotes, some of them having the most elaborate cells of any organisms. Some protists are **syncytial** or **coenocytic:** having many nuclei lying in cytoplasm

[1]The term "protists" originally meant single-celled organisms, and the five-kingdom system popularized by Robert Whittaker put only single-celled eukaryotes without close multicellular relatives into kingdom Protista. With the trend to expand the kingdom's boundaries to include some phyla having multicellular members, some biologists favor changing its name to Protoctista (protos = very first; ktistos = establish). In this book we keep the kingdom's shorter name and expand its definition, as many biologists prefer.

that is not divided into cells (Section 14–D). Others form filaments or colonies of many similar, independent cells, with limited cooperation. The largest protists are enormous seaweeds, with simple body plans composed of several different kinds of specialized, interdependent cells.

Ancient protists radiated into many different habitats and ways of life, and some of them also founded the higher kingdoms. The many kinds of single-celled, mobile, heterotrophic protists that lack cell walls and ingest their food are all lumped under the term **protozoa** ("first animals"). They may live as predators on smaller organisms or as parasites or symbionts of larger ones. Photosynthetic protists, ranging from single cells to giant seaweeds, are collectively called **algae.** There are also some fungus-like forms, heterotrophs that have walls and live by absorbing nutrients as saprobes or parasites. In the old two-kingdom system, algae and fungus-like forms were classified as plants because they have cell walls, whereas protozoans were classified as animals.

The study of protists is important for several reasons. First, the kingdom contains many oddballs that do not conform to our general concept of eukaryotic cells; these give clues to the origin and early evolution of eukaryotes. Of more practical concern, algae carry out most of the photosynthesis in the oceans and so play a key role in global ecology. Some algae also make economically useful products. Finally, parasitic protozoa infect more than one fourth of the world's human population—more than 1 billion people. Protists also cause many diseases of other animals and of plants. We cannot even estimate the true cost of all these diseases in prevention, treatment, and lost productivity.

Authorities recognize two to four dozen phyla of protists, a lot more than the handful your authors learned as students and wrote about in the previous edition of this book. This longer list comes partly from splitting older taxa and partly from redefining the fungal and plant kingdoms more narrowly and expanding the boundaries of kingdom Protista to include the misfits. This chapter covers only the major phyla (which we arbitrarily define as those with more than 1500 species), plus a few smaller ones whose members have special evolutionary or economic significance. Appendix II lists the rest.

In this chapter we discuss two major steps in the evolution of life, the origin of eukaryotic cells and of multicellularity, and survey the diversity of kingdom Protista.

### KEY CONCEPTS

- The origin of large, complex eukaryotic cells, and especially of those with sexual reproduction, spurred the adaptive radiation of many kinds of unicellular protists.
- Additional radiations occurred in many protistan lineages that developed multicellular structures, including those that produced ancestral fungi, plants, and animals.
- The kingdom Protista contains unicellular, syncytial, colonial, and simple multicellular eukaryotes, all found in aquatic or very moist habitats.

- Protists include protozoa, which are predators, scavengers, parasites, or symbionts; algae, which are photosynthetic autotrophs, either free-living or symbiotic; and fungus-like forms, which are saprobes or parasites.
- Many protists cause diseases, often of devastating impact.

## 25–A  ORIGIN AND EVOLUTION OF EUKARYOTES

The evolution of eukaryotes, perhaps 1.8 billion years ago, was a tremendous advance (Section 5–E and *Essay: Evolution of Eukaryotic Cells,* Chapter 5). The remains of the thick outer coverings of dormant protistan cysts are common in the fossil record. However, fossils do not preserve the internal cell structure. To trace the evolution of eukaryotic cells, we must study living forms, which provide only a broad outline.

The hallmark of eukaryotic cells is the nucleus. However, the key to the early eukaryotes' success was probably that they did not have an external cell wall, like most bacteria, but instead an internal skeleton. By pulling and pinching the cell's flexible membrane, this cytoskeleton could have produced the nuclear envelope and other internal membranous organelles (see Figure 5–A). It also gave early eukaryotes a way to engulf food in bulk, by phagocytosis, and formed the tracks of a rapid internal mass transport system. With these features, the cell could grow much larger. In addition, some of the cytoskeleton's microtubules form the framework of flagella and cilia, the locomotory structures of many protists (Figure 25–1).

The cytoskeleton also played a crucial role in a new mode of cell division, mitosis, because it makes up the mitotic spindle. Mitosis divides up the eukaryotic cell's many linear chromosomes, which carry much more genetic information than a bacterium's single circular DNA molecule. We know little about the origin of mitosis.

Another eukaryotic innovation was sexual reproduction, with its two complementary processes, meiosis and fertilization. The divisions of meiosis undoubtedly evolved from mitosis, whereas the union at fertilization may have arisen from phagocytosis (Section 20–C). Crossing over and independent assortment during sexual reproduction produced a wide variety of new genetic combinations. No longer did selection operate only on clones, which often meant the death of entire genomes despite a favorable new allele. Instead, the various alleles of each gene could be expressed in many genetic backgrounds. Now the better alleles of different genes could be wedded, the worse weeded out. This genetic variation fostered the evolutionary success of early sexual eukaryotes and led to tremendous adaptive radiation.

Eukaryotes also typically have mitochondria and often chloroplasts. These organelles evolved from symbiotic bacteria that provided metabolic services their larger hosts could not perform for themselves (Section 24–H).

Both mitochondria and chloroplasts evolved more than once, from similar or different kinds of symbiotic bacteria residing in different protists. Possessing these organelles does not necessarily indicate kinship among eukaryotes. In fact, taken all together, protists have a diverse array of symbioses not found among members of the other kingdoms. Some were so successful that they helped spark a burst of adaptive radiation and became defining features of the resulting taxon, as we shall see.

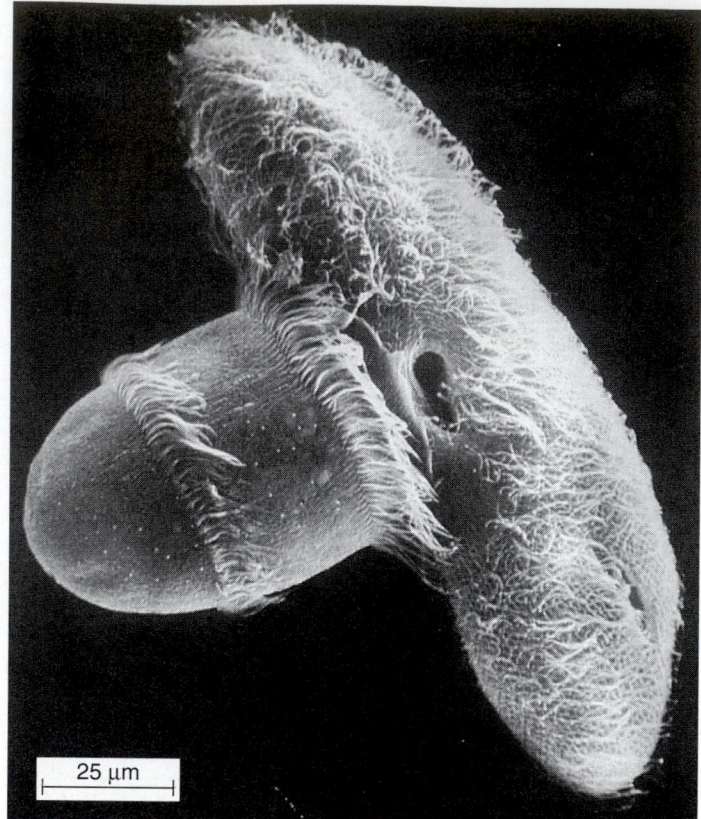

(a) *Didinium* attacks *Paramecium* (SEM)

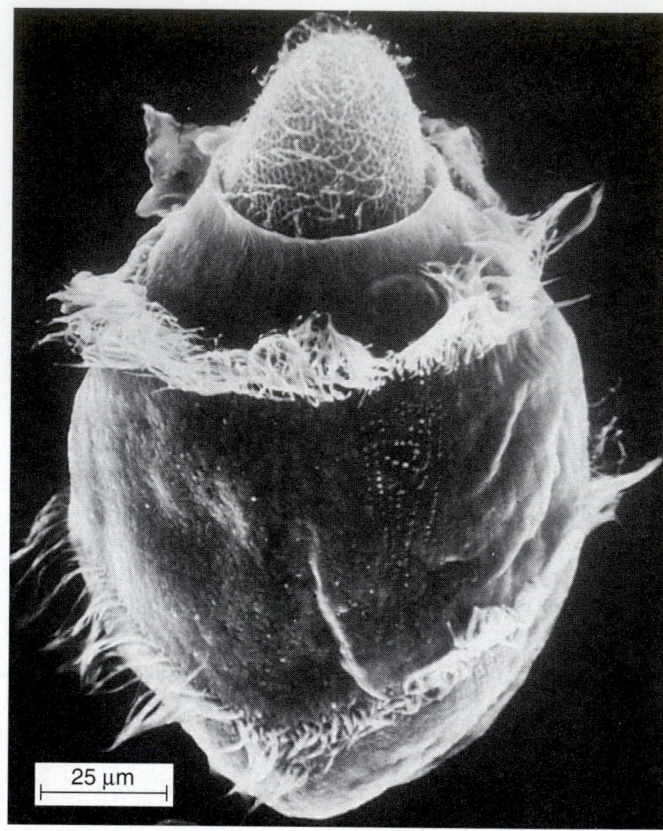

(b) All but the tip of the *Paramecium* engulfed

**FIGURE 25–1**   The cytoskeleton in action. Both these protozoans move by means of their many cilia, and both also ingest whole food particles, as *Didinium* demonstrates by engulfing *Paramecium*. In (a) the cavity halfway along the *Paramecium* is its gullet, through which it has ingested its last meal. (Phylum Ciliophora, Section 25–J)   (Biophoto Associates, courtesy of Drs. G. Antipa and E. Small)

We shall also see that many features we consider typical of eukaryotes are not found in them all. Various protists lack true mitosis, sexual reproduction, flagella, mitochondria, chloroplasts, or more than one of these.

The absence of a feature is said to be primitive, or ancestral, if the feature never evolved in the organism's ancestors. This condition is often accompanied by other primitive features. Such organisms seem to be frozen at some ancient stage of evolution. On the other hand, loss of a feature present in the ancestral line is secondary, or derived. A feature generally disappears if it is not needed in a new, specialized way of life, because individuals who lose it have lower material and energy costs and eventually outreproduce their peers. For example, secondary loss of structures is notoriously common among parasites, a category that includes many protists.

Much of the evolution of protists is obscure, for several reasons. First, it is often hard to tell whether the absence of various features is ancestral, derived, or a mixture of the two. Second, symbiotic mergers occur independently of the evolution of other features. Third, it is often not clear whether similar cell structures signify descent from a common ancestor or instead reflect either parallel or convergent evolution.

## On to Multicellularity!

Once upon a time all organisms were single cells. But wherever such life was abundant, there must have been strong selective pressure for increased size, for two reasons. The first was predation: a larger organism could eat more of its neighbors and be eaten by fewer. The second was economy of size: a larger organism has a lower metabolic rate, and hence a lower food requirement, per unit of mass (although it needs more *total* food). These advantages favor the evolution of larger size provided they offset the chief disadvantage: a longer generation time, which means that fewer descendants are produced in a given period.

A single cell cannot just become larger and larger; eventually its center is too far from the outside environment to

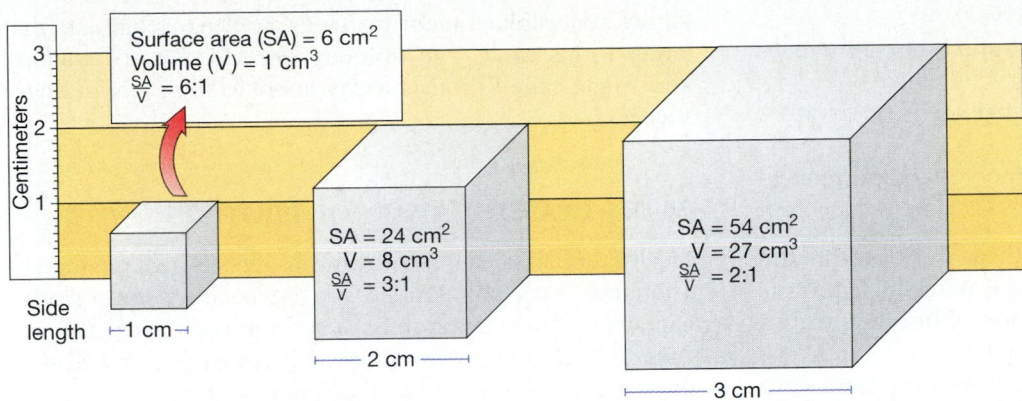

Surface area (SA) = 6 cm²
Volume (V) = 1 cm³
$\frac{SA}{V}$ = 6:1

SA = 24 cm²
V = 8 cm³
$\frac{SA}{V}$ = 3:1

SA = 54 cm²
V = 27 cm³
$\frac{SA}{V}$ = 2:1

Centimeters

Side length   ⊢1 cm⊣

⊢ 2 cm ⊣

⊢ 3 cm ⊣

**FIGURE 25–2**   Surface-area-to-volume ratios (SA/V). Surface area increases with the square of the object's linear dimension, whereas volume increases with the cube. Therefore, SA/V decreases as an object increases in size while retaining the same shape.

obtain the substances it needs fast enough. Also, as a cell increases in size while staying the same shape, its ratio of cell surface area to cytoplasmic volume decreases (Figure 25–2). Food, oxygen, and wastes are exchanged through the surface. Hence, a decrease in the surface-to-volume ratio also decreases the rate at which the cell takes in supplies per unit volume of cytoplasm.

Some cells evolved shapes with greater surface areas. We can view a eukaryotic cell as having a large, complex surface in separate but fusible pieces: the plasma membrane and transport vesicle membranes (Figure 25–3). Plant cells grew even larger by confining the metabolically active

cytoplasm and organelles to a thin surface layer, with the interior occupied by a large storage vacuole. But these strategies have limits. Eventually body size must be increased by an increase in cell number, with each cell's surface-to-volume ratio able to sustain its metabolism.

The trend toward multicellular bodies began with cells sticking together after cell division. This has occurred in many different lineages during the history of life. At the simplest level, some protists form **colonies** of many independent but cooperating cells. Each cell is small and exchanges substances with the environment, but the cells are still more or less identical (Figure 25–4). In true multicellular organisms, there is division of labor among different kinds of cells.

Every cell and every organism must carry out certain basic life functions:

♦ Feeding or making food
♦ Gas exchange

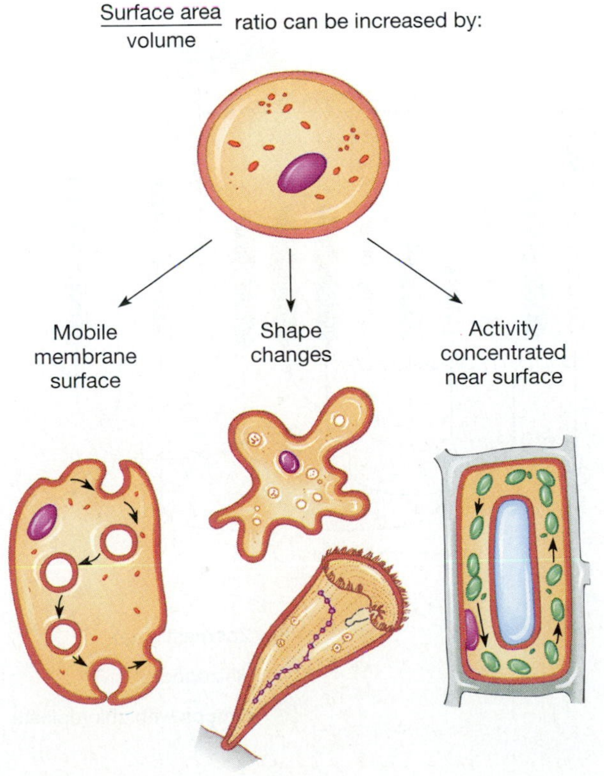

$\frac{Surface\ area}{volume}$ ratio can be increased by:

Mobile membrane surface

Shape changes

Activity concentrated near surface

**FIGURE 25–3**   Increasing a low surface-area-to-volume ratio.

**FIGURE 25–4**   Colonies of marine diatoms. (LM)   (Biophoto Associates)

- ◆ Waste removal
- ◆ Internal transport of food, gas, and other substances
- ◆ Sensing environmental stimuli
- ◆ Dispersal (locomotion, scattering seeds or larvae)
- ◆ Support and protection
- ◆ Coordination of all functions (nerves and hormones)
- ◆ Reproduction

Bacteria and most protists do all these things within the confines of a single cell. Indeed, some unicellular protists have a high degree of division of labor within their elaborate, specialized cytoplasm and organelles.

In multicellular organisms, each cell not only carries out most of these functions for itself, but also helps carry out one or more of them for the entire body. During the evolution of multicellular organisms, groups of cells formed increasingly specialized tissues and organs. The first sign of division of labor is usually the specialization of some cells for reproduction, whereas others remain **vegetative** (acquiring energy by feeding or photosynthesis). Other early specializations include structures for protection and anchorage.

We tend to view greater size and complexity as "progress" over yesteryear's smallness and simplicity. However, unicellular organisms do have some advantages over larger creatures. A single cell can live in a tiny space and needs only a little food before it is ready to reproduce. This allows unicellular organisms to exploit many habitats that larger forms can't. The ubiquity and diversity of bacteria and single-celled protists today attest to their evolutionary success.

## 25–B  CLASSIFICATION OF PROTISTS

Ideally, the classification of organisms reflects their phylogeny. Therefore, taxa are defined on the basis of conservative characters. Early taxonomists used morphological features such as locomotory structures, cell coverings, and distinctive organelles to classify unicellular protists.

Today many of their groups are being discarded and new ones erected on the basis of biochemistry and ultrastructure (fine detail revealed by electron microscopy). Vital chemicals and structures, such as photosynthetic pigments and flagella, tend to be conservative, and so they make good criteria for classifying protists. Progress is often frustratingly slow because of technical difficulties: many protists are difficult to grow in laboratory culture in order to get enough material to study their chemistry, morphology, and reproduction.

Nucleotide sequencing of ribosomal RNA is starting to reveal the relationships among different phyla of protists and other eukaryotic groups (Figure 25–5). Despite discrepancies among different studies, it is becoming obvious

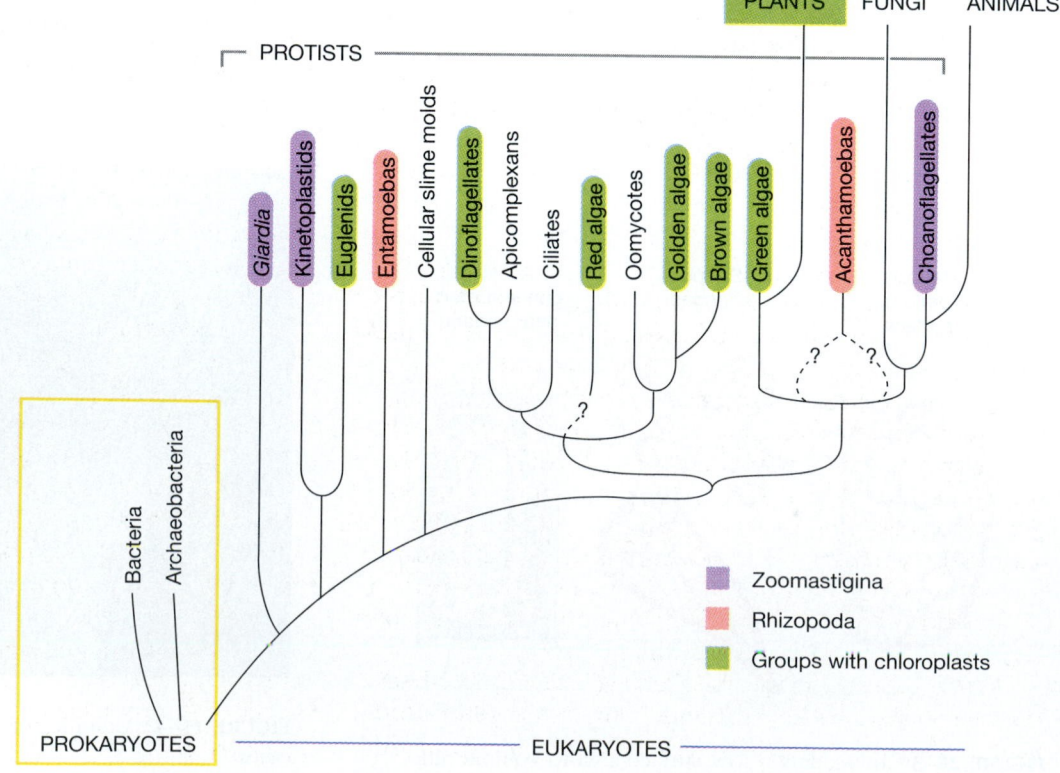

**FIGURE 25–5** Evolutionary relationships. This evolutionary tree, based on ribosomal RNA sequencing data, shows how protistan phyla are related to each other and to other kingdoms. Groups shown in green are photosynthetic forms. Note that acquisition of chloroplasts occurred independently in several eukaryote lineages. Forms shown in purple are classified as Zoomastigina, whereas those shown in pink are classified as Rhizopoda. According to this diagram, further subdivision of these phyla is warranted in order to produce a classification that reflects phylogeny.

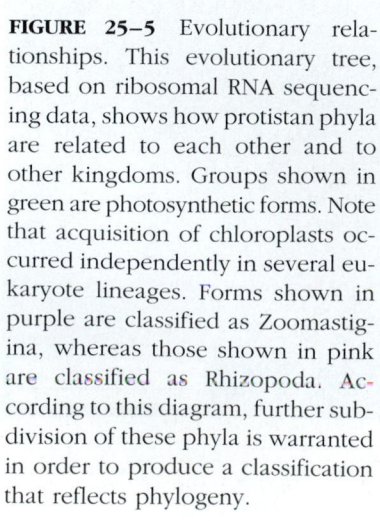

that many protistan phyla are only distantly related: their ribosomal RNAs differ more than those from different phyla of plants, animals, and fungi. The protists appear to be descendants of a number of pioneers with a wide variety of eukaryotic organization and of adaptations to specialized niches. Most lineages never produced multicellular descendants.

In an admittedly artificial classification scheme, the kingdom Protista is certainly the most artificial, and so in many ways the most unsatisfactory. There are many ways of defining and drawing the boundaries of the protistan, fungal, and plant kingdoms, and you will see some groups assigned differently in other books (including our previous edition). Good arguments can also be made for splitting the kingdom Protista, but we need more information before biologists can agree on how to do this.

For convenience, we group the protistan phyla into two main categories, heterotrophs and autotrophs. However, this is not a natural boundary. Autotrophic protists evolved from several unrelated groups of heterotrophs that acquired chloroplasts independently. Therefore, some autotrophic phyla are more closely related to heterotrophic phyla than to each other. In fact, several phyla contain both autotrophic and heterotrophic species, or even species with both modes of nutrition.

## 25–C  PHYSIOLOGY OF PROTISTS

The Protista are so heterogeneous that few generalizations can be made about their physiology (how they work). Some have a cell wall or a **test** (shell) that provides protection and support and may also keep the cell from bursting by limiting its osmotic water uptake from a hypotonic environment (Section 4–D). Many protozoa have neither, and they gain a lot of water by osmosis, especially those living in fresh water. These forms often have **contractile vacuoles,** which collect excess water and then contract and expel it from the cell (see Figure 25–14).

Protists exchange gases with the environment by diffusion through the plasma membrane. Although nearly all protists can carry on respiration, many can also live indefinitely using fermentation when oxygen is not available.

Under adverse conditions, many protists can enclose themselves in thick walls, forming dormant **cysts,** resistant to desiccation (drying out) and temperature changes. Here the cell survives until favorable conditions return. Some desert dwellers emerge from their cysts to feed and reproduce only for the few hours each year when water is available. Many protists are also dispersed as cysts, carried in dried mud on the foot of a migrating bird, in the fur of a shipboard rat, or in human clothing or goods.

Some protists are **sessile** ("sitting"), living attached to objects such as rocks or plants. Others are **planktonic,** floating passively in their watery homes. Many are **motile,** moving actively by means of cilia, flagella, or pseudopods or by flexing the cell body. Many protists have more than one means of locomotion.

In flagellar locomotion, wave-like bending motions pass from one end of the flagellum to the other. This pulls or pushes the cell through the water, depending on whether the flagellum is anterior or posterior (at the front or rear of the cell) (Figure 25–6). Some flagella move in one plane, others in a spiral.

Cilia have the same structure as flagella but are shorter and generally more numerous. Ciliary locomotion can be compared to rowing a boat, with each cilium moving in one plane on the power stroke but bending aside on its recovery stroke, offering less resistance to the water.

In amoeboid locomotion, amoebas or other cells form temporary extensions called **pseudopods** ("false feet"). A pseudopod is extruded, anchors at the tip, and the cytoplasm moves in that direction. The cytoskeleton's actin microfilaments play a role in this activity, in a poorly understood process. The formation of pseudopods may look random, but amoebas do direct their movements in response to food or other stimuli in the environment.

Protists feed in various ways. Heterotrophs ingest food by endocytosis (see Figure 25–1) or absorb small organic molecules from their environment. Autotrophs are photosynthetic. Many protists have more than one type of nutrition: some species can switch from photosynthesis to endocytosis to absorption, as conditions dictate.

Most protists can sense stimuli such as light, temperature, touch, gravity, and chemicals. Many have **photoreceptor** organelles containing a light-detecting pigment, rhodopsin, which also occurs in our own eyes. Many protists can detect objects that touch their membranes. They also have various membrane receptor proteins that detect chemicals in the environment by specific binding.

Like bacteria, protists respond to stimuli with directed behaviors called **taxes** (singular: **taxis**), moving toward or away from a stimulus. For instance, photosynthetic forms often move to areas where light is ample (although they avoid high light intensities, which would damage vital molecules).

In general, stimuli cause changes in the shapes of some membrane proteins and thereby change the flow of ions through the membrane. This in turn changes the activity of response mechanisms such as cilia or flagella.

The fine tips of flagella often serve as adhesive recognition organelles. Flagellated gametes may recognize and bind each other by the tips of their flagella before fusing at fertilization (see Figure 20–5). The tip of a cell's flagellum may also recognize and attach to a suitable substrate where it can settle.

Both asexual and sexual reproduction occur among protists, and most can do both. Asexual reproduction in unicellular forms usually occurs by dividing in two after a mitotic division of the nucleus. Many larger forms repro-

(a) FLAGELLAR MOTION
(Anterior flagellum)

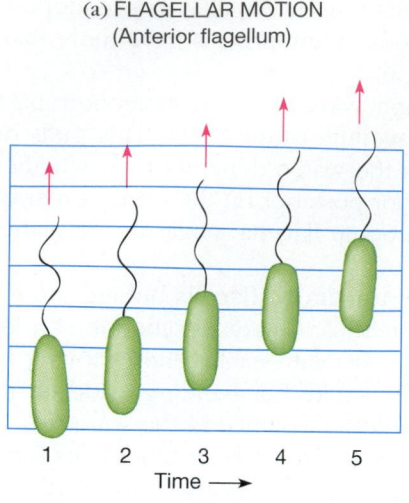

(b) CILIARY MOTION

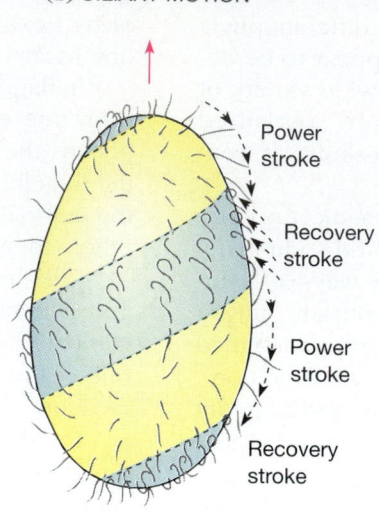

Power
stroke

Recovery
stroke

Power
stroke

Recovery
stroke

(c) AMOEBOID MOTION

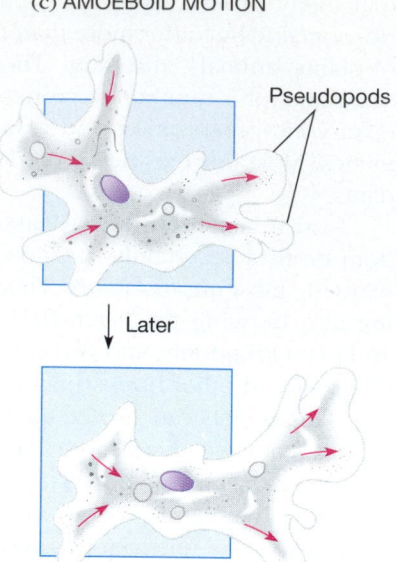

Pseudopods

Later

1    2    3    4    5
Time →

**FIGURE 25–6**    Protistan modes of locomotion. (a) Flagellar motion. Numbers 1 through 5 show the cell's position with respect to reference lines as its anterior flagellum undulates, pulling the cell forward through the water (arrows). (b) Ciliary motion. Cilia are held straight on the power stroke; as they push backward, the cell moves forward. By bending on the recovery stroke, the cilia avoid pushing the cell backward. The cell's many cilia move in a coordinated sequence. (c) Amoeboid motion. Streaming of the cytoplasm pushes out temporary pseudopods, and the rest of the cell follows. In the second drawing, the cell has moved with respect to the reference square.

duce by fragmenting into smaller pieces or by releasing asexual **spores:** single cells that serve as units of reproduction and dispersal. Protists tend to reproduce asexually when conditions favor feeding and growth and sexually when they do not. Sex often occurs in conjunction with forming a cyst, which enters dormancy until good times return. This is thought to be adaptive because the organisms use their "down time" for renovation of their genomes (Section 20–C).

## PROTOZOA

Protozoa are generally motile, unicellular or syncytial, wallless heterotrophic protists (Table 25–1). They may be freeliving predators or scavengers, ingesting other organisms or bits of organic matter, or parasites or mutualistic symbionts. Members of the first two phyla (plus Phylum Actinopoda, Appendix II) were formerly placed into the single phylum Sarcodina, based on their common feature: pseudopods.

### 25–D    RHIZOPODA: AMOEBAS

Phylum Rhizopoda ("root foot") contains only 200 species. However, no biology course would be complete without its members, the amoebas (Figure 25–7). The freshwater

*Amoeba proteus,* which you may have observed in the laboratory, is one of the most thoroughly studied protists. It is easy to rear on a diet of bacteria and is large enough to be used to study amoeboid locomotion and endocytosis. It is also used in experiments on nucleus-cytoplasm interactions, since its nucleus can be removed or transplanted by microsurgery to find out how the cytoplasm reacts.

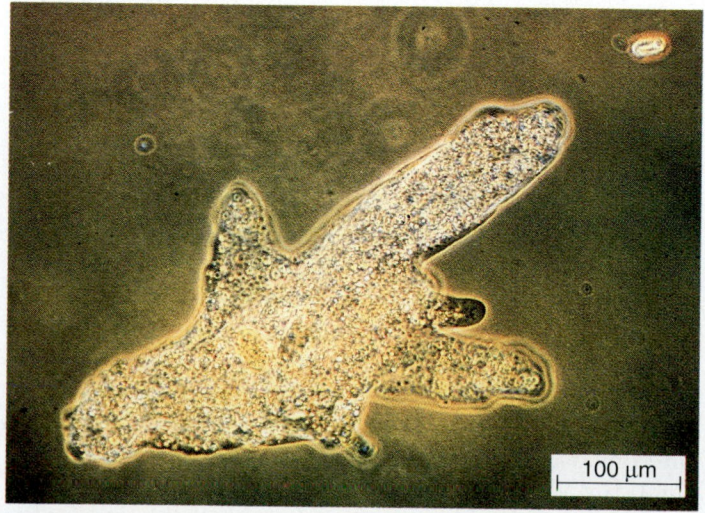

100 µm

**FIGURE 25–7**    An amoeba with several pseudopods. (LM)    (Biophoto Associates)

TABLE 25–1

**Important Phyla of Heterotrophic Protists**

| Phylum | Characteristics | Examples |
|---|---|---|
| Rhizopoda (~200 species) | Amoebas. Unicellular; naked or with test; pseudopods used in locomotion and feeding; no flagellated stages; no sexual reproduction; terrestrial, marine, freshwater, or parasitic | *Amoeba, Arcella, Chaos, Difflugia, Entamoeba* |
| Foraminifera (Granuloreticulosa) (~4000 species*) | Foraminiferans. Unicellular or syncytial; tests containing calcium carbonate, sand grains, etc.; thin pseudopods used mostly for feeding; some with sexual reproduction; predaceous; mostly marine | *Fusulina, Globigerina, Nodosaria, Textularia; Nummulites* (extinct form found in "nummulitic" limestone used to build Egyptian pyramids) |
| Acrasiomycota (~50 species) | Cellular slime molds. Unicellular feeding amoebas, congregating into a pseudoplasmodium before forming an asexual fruiting body with cellulose cell walls producing asexual, walled spores; freshwater and terrestrial | *Acrasia, Dictyostelium, Polysphondylium* |
| Myxomycota (~550 species) | Acellular slime molds. Mobile, coenocytic plasmodial feeding stage; fruiting body with cellulose walls, spores formed by meiosis; terrestrial | *Arcyria, Physarum, Stemonitis* |
| Zoomastigina (~1500 species) | Zooflagellates (flagellates without chloroplasts). Unicellular; no cell wall; one or more flagella; freshwater, many symbionts or parasites | *Trypanosoma, Trichomonas, Leishmania, Lophomonas, Giardia, Barbulonympha* |
| Apicomplexa (~2400 species) | Parasitic protists with apical complex used to penetrate host cell. Complex life cycle, including sex; flagellated male gametes | *Toxoplasma, Plasmodium* (causes malaria), *Monocystis, Gregarina* |
| Ciliophora (~8000 species) | Ciliates. Unicellular; two types of nuclei, complex sexual reproduction; pellicle; cilia used in locomotion and food collection; almost all with oral opening; usually have trichocysts; freshwater and marine, some parasites and symbionts | *Paramecium, Tetrahymena, Vorticella, Euplotes, Stentor* |
| Oomycota (~800 species) | "Water molds" and their terrestrial relatives. Coenocytic diploid hyphae with cellulose walls; unwalled spores with two unlike flagella, and walled airborne spores; oogamous sexual reproduction; aquatic and terrestrial saprobes and parasites | *Phytophthora infestans* (late blight of potatoes), *Plasmopara viticola* (downy mildew of grapes), *Albugo candida* (white rust on cabbage and other plants), *Peronospora parasitica* (downy mildew on crops), *Saprolegnia* (mold on living or dead animals in water) |

*Only numbers of living species given for each group.

Although members of the genus *Amoeba* are naked, many other rhizopods have tests, which they construct from bits of solid matter. Rhizopods use their pseudopods for both locomotion and feeding.

Rhizopods live worldwide, in soil, in salt or fresh water, and in the bodies of animals. Two species of *Entamoeba* live harmlessly in the human mouth and intestine, but a third, *E. histolytica,* causes devastating amoebic dysentery.

Amoebas never have cilia or flagella, nor meiosis and sex. They do form mitotic spindles, but their mitosis is unusual, with the nuclear envelope persisting during most or all of the process.

## 25–E FORAMINIFERA (GRANULORETICULOSA)

**Foraminiferans** have slender, granular pseudopods used more for food capture than for locomotion. These pseudopods poke out through holes in a test made up of calcium carbonate ($CaCO_3$) or of sand grains cemented by organic secretions. Outside the test, the pseudopods may branch and join, forming a net that traps and digests prey. These features are described by the phylum's two suggested names: Foraminifera (foramen = an opening; ferre = to carry) and Granuloreticulosa (granulum = little grain; reticulum = little net) (Figure 25–8).

250 µm

**FIGURE 25–8**  A foraminiferan. The thread-like pseudopods protruding through holes in the test have trapped several smaller organisms. (LM)   (Biophoto Associates)

Forams, as they are nicknamed, contain one to many nuclei. The larger forms may exceed a millimeter across, and one fossil species grew larger than 10 centimeters.

Nearly all forams are **marine** (ocean-dwelling). Most live in the sand or attached to other organisms, but planktonic forms occur in large numbers and provide food for many invertebrate animals. Some forams retain chloroplasts from algal prey, thereby gaining extra food from photosynthesis. Most shallow-water species contain symbiotic algae and respond to light by moving toward it.

Living species of forams are about 4000 strong, but many times this number are known from fossilized tests, which sank to the bottom after their occupants died. Millions of years' worth of this foraminiferan debris formed chalk rocks or limestone, such as England's famous white cliffs of Dover. Fossil forams are also common in deposits of oil, which forms from the remains of ancient floating algae. As an oil well is drilled, the bit passes through successive layers of ancient sediment, each containing abundant fossils of Foraminifera that lived during a particular geological period. These fossils provide clues to where oil is likely to be.

Some forams have never been seen to reproduce sexually, but others have both mitosis and meiosis in complex life cycles.

## 25–F   ACRASIOMYCOTA: CELLULAR SLIME MOLDS

Members of this phylum and the next are not closely related, but both combine amoeba- and fungus-like features. Because the fungus-like stage is more obvious, these organisms were first studied by mycologists (people who study

fungi), who gave them the common name "slime molds" and claimed them as fungi.

These organisms spend the active part of their lives in a mobile, naked, amoeba-like state, engulfing organic matter and bacteria. However, when the going gets tough, they form fungus-like reproductive structures called **fruiting bodies:** some cells form a stalk, and others become spores with cellulose walls. This cooperation and division of labor show a primitive degree of multicellularity. Indeed, sequencing of ribosomal RNA and of homeobox genes (Section 11–E) suggest that cellular slime molds lie close to the base of the phylogenetic tree that includes the three multicellular eukaryotic kingdoms (animals, fungi, and plants) as well as several phyla of protists.

Acrasiomycota (akrasia = bad mixture; mykes = fungus) are called cellular slime molds because the feeding stage is a uninucleate amoeba. These amoebas, and even the reproductive structures they form, occur widely in soil but are so small they are seldom observed. When their food supply is exhausted, some of the amoebas secrete a chemical that attracts others. The amoebas crawl along the chemical's concentration gradient and aggregate into a slug-like clump, a **pseudoplasmodium,** which may crawl around for a while. Then it stops and forms a fruiting body topped by spores with cellulose cell walls (Figure 25–9). After they are released, the spores germinate and form new amoebas. The formation of fruiting bodies by cellular slime molds provides a simple and useful model for the experimental study of how cells differentiate. Sexual reproduction is rare and very different from the asexual production of fruiting bodies.

Several species of cellular slime molds may occur in the same soil and form fruiting bodies in response to the same environmental cues. How do they manage to aggregate so that each cluster contains only members of the same species? In some cases, different species secrete different chemical attractants and therefore congregate at different centers. However, some species appear to have the same aggregation chemical, and so all the amoebas come together in the same place. Here they sort themselves out into species-specific clumps based on the compatibility of cell-surface molecules.

## 25–G   MYXOMYCOTA: PLASMODIAL SLIME MOLDS

Plasmodial (= acellular) slime molds are placed in phylum Myxomycota (myxa = mucus). The feeding stage is a **plasmodium,** an amoeba-like syncytium (Figure 25–10a). (Do not confuse this term with *Plasmodium,* the generic name of the protozoan that causes malaria.) A plasmodium usually forms by the fusion of many small plasmodia. The plasmodium streams around in soil, wood, dung, or decayed vegetation, engulfing bacteria or particles of food. When

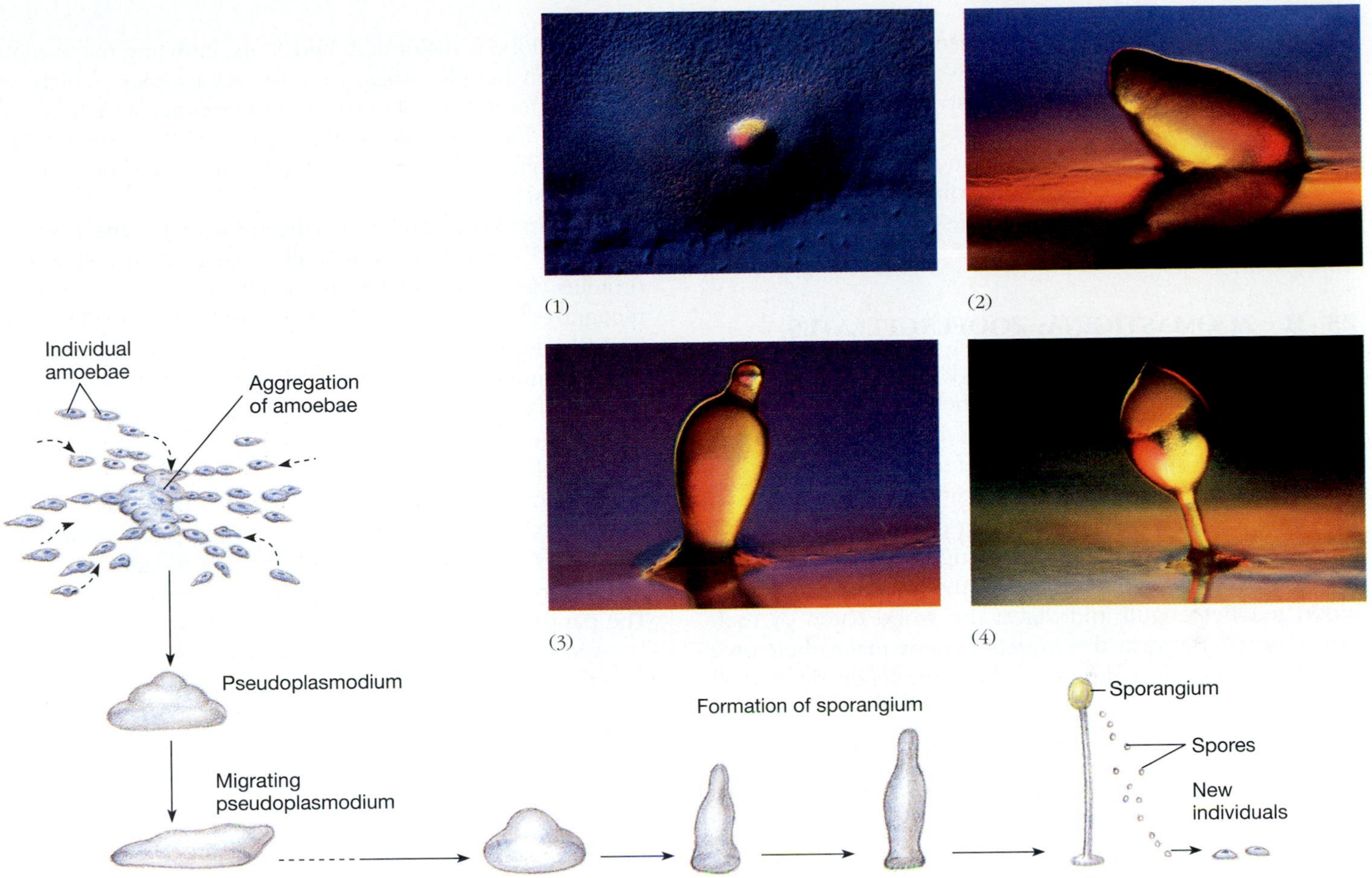

Individual
amoebae

Aggregation
of amoebae

(1)   (2)

(3)   (4)

Pseudoplasmodium

Formation of sporangium

Sporangium

Spores

Migrating
pseudoplasmodium

New
individuals

**FIGURE 25–9**   Reproduction in *Dictyostelium*, a cellular slime mold. *Photos:* (1) Individual amoeba-like cells congregate into a single mass, visible as a spot in the center of the frame. (2) The resulting slug-like pseudoplasmodium crawls about until it (3) settles in one spot and (4) forms a reproductive structure, a stalk topped by a rounded sporangium (spore-producing structure). (Carolina Biological Supply Company)

(a)   (b)

**FIGURE 25–10**   Acellular slime molds. (a) A plasmodium, the feeding stage of an acellular slime mold. (b) Developing fruiting bodies of another species.   (b, Biophoto Associates)

conditions become too dry, the plasmodium forms a fruiting body with cell walls (Figure 25–10b). Spores are produced by meiosis. Germinating spores release haploid amoebas, which may develop flagella. Two compatible amoebas or flagellated cells fuse and form a diploid cell. This develops into a young plasmodium as the diploid nucleus divides but the cytoplasm does not.

## 25–H  ZOOMASTIGINA: ZOOFLAGELLATES

Phylum Zoomastigina is a polyphyletic collection of heterotrophic protozoa with flagella, including the group that is probably ancestral to animals and possibly to fungi. (Zoo = animal; mastix = whip, a reference to the flagella.) Some zooflagellates are free-living, either freshwater or marine, whereas others are symbionts or parasites. The parasitic forms often have life cycles involving two hosts.

The zooflagellate symbionts in the guts of termites and wood roaches engulf and digest the wood eaten by their insect hosts. Because the insects cannot make their own wood-digesting enzymes, they are completely dependent on their symbionts.

Of great medical and veterinary importance are parasitic zooflagellates of the genus *Trypanosoma,* which live in the blood of vertebrates, mainly mammals (Figure 25–11). Human diseases caused by trypanosome infection include sleeping sickness (Africa) and Chagas' disease (South and Central America). Nagana is a trypanosome disease that kills 3 million cattle a year and makes cattle farming impossible in much of sub-Saharan Africa. Both sleeping sickness and nagana are transmitted to their mammalian hosts by bites of the tsetse fly, the host during part of the trypanosome life cycle.

In Latin America, Chagas' disease affects about 12 million people. It is transmitted by bloodsucking insects, which

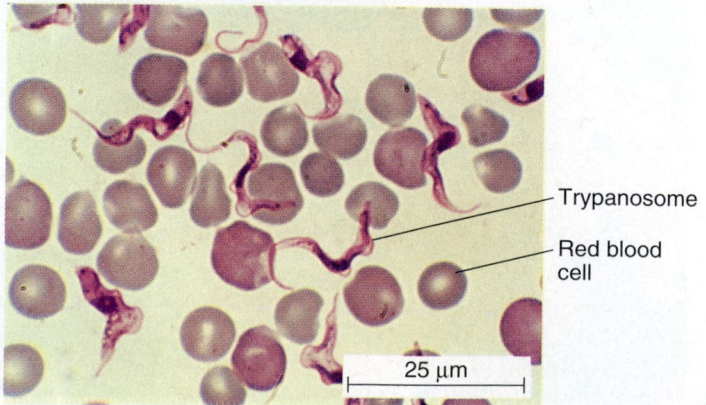

Trypanosome

Red blood cell

25 µm

**FIGURE 25–11**  Phylum Zoomastigina: trypanosomes. These parasites of vertebrate blood are seen here swarming among red blood cells. (LM)  (Biophoto Associates/Photo Researchers)

hide in adobe walls or thatched roofs, crawling out at night to feed on people, dogs, cats, or guinea pigs, which are raised for food. Only about 10% of the victims of Chagas' disease die during the acute first stages of the disease. The later, chronic phase may last years, damaging the heart or other muscles.

Trypanosomes are especially difficult for the body to conquer because they continually change their cell-surface proteins (antigens), which the host's immune system must recognize before it can kill the parasites. As quickly as the host develops a defense against a clone of parasites expressing one antigen, some parasite cells switch to expressing another of their many surface-antigen genes, and the immune system must start all over to combat what is essentially a new disease.

Leishmaniasis, caused by parasitic flagellates of the genus *Leishmania,* afflicts millions of people in Africa and Asia. These organisms cause ulcers on the skin and internal organs, and untreated cases are fatal within two years. The parasites are transmitted by sand flies.

Vaccines against all of these diseases will not be available for years, if ever. Meanwhile, the best preventive measure is to control the insect carriers necessary for the parasites to complete their life cycles. Researchers are also working on some promising drug treatments for Chagas' disease and leishmaniasis.

One of the first known protists was a species of the parasitic flagellate *Giardia,* discovered by the dedicated amateur microscopist Anton van Leeuwenhoek in his own feces during a bout of diarrhea. Expanding muskrat and beaver populations are carrying *Giardia* to ever more lakes and reservoirs in North America. These mammals are not affected by the parasites, but humans drinking contaminated water come down with giardiasis: diarrhea and cramps, fatigue, and weight loss. To avoid this, campers in back country should boil all water before they use it. *Giardia* is another parasite that is hard for the body to combat because of rapid mutation of its surface antigens.

***Choanoflagellates***  Choanoflagellates have a collar of microvilli around the base of the flagellum. This feature is also found in choanocytes, specialized cells typical of sponges, which are simple animals. Some choanoflagellates are solitary, whereas others are organized into colonies (Figure 25–12). These features suggest that sponges evolved from choanoflagellate ancestors. Ribosomal RNA nucleotide sequencing shows that these flagellates may be closely related to the ancestors of all animals and fungi.

## 25–I  APICOMPLEXA

Phylum Apicomplexa is the larger of two phyla created by splitting an older phylum, Sporozoa (the smaller is phylum Cnidosporidia, Appendix II).

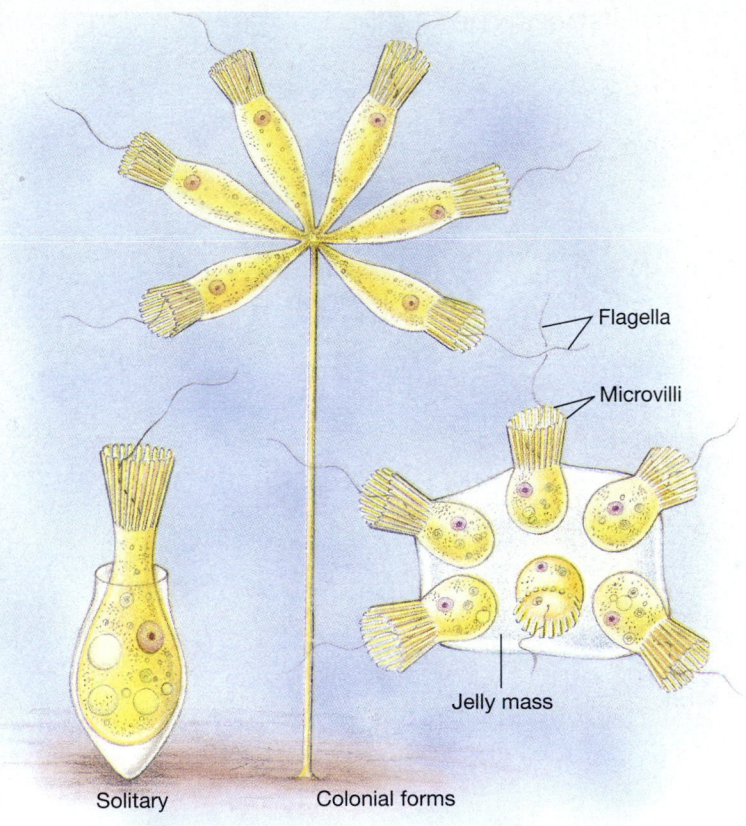

Flagella

Microvilli

Jelly mass

Solitary          Colonial forms

**FIGURE 25–12**   Choanoflagellates. The flagella move water currents through the collar of microvilli, which strain food out of the water.

All apicomplexans are parasites. In typical parasite fashion, most have complicated life cycles, often with two different host species.

The phylum is named for the "apical complex," a distinctive arrangement of organelles and cytoskeletal fibers at one end of the infective motile stage, used to enter host cells. This feature is thought to indicate that all apicomplexans share a common lineage.

The apicomplexan *Toxoplasma gondii* may well be the most common parasite in humans: half of all adults in the United States have probably been infected at some time. This protozoan invades body cells and usually causes mild symptoms (enlarged lymph nodes) that pass unnoticed. However, it can cause serious disease in a fetus, newborn, or a person debilitated by AIDS, leading to blindness, mental retardation, or death. *Toxoplasma* is usually spread in the feces of cats or in undercooked meat from infected pigs.

Human malaria, caused by four species of the genus *Plasmodium,* illustrates the complex life cycles typical of parasites. Malaria parasites require two different hosts to complete the life cycle: humans and female mosquitoes of the genus *Anopheles,* which transmit the disease from one person to another (Figure 25–13).

Malaria was attacked vigorously early in the twentieth century by draining stagnant fresh water where mosquitoes breed, by spraying houses with DDT to kill mosquitoes, and by treating victims with drugs to destroy the parasites. The disease was eradicated from much of its former range, including the southern United States and most of southern Europe, and islands such as Trinidad, Sicily, Taiwan, and Mauritius. It was also drastically lowered in most other areas. But just when victory seemed near, some parasites became resistant to the drugs, and mosquitoes to DDT, bringing a resurgence of malaria to many areas of Asia, Latin America, and Africa. About 270 million people are infected at any one time, and 2.5 million die each year, mainly young children. Most victims live in developing nations, and neither they nor their governments can afford new drugs, insecticides, and drainage programs.

In 1983, researchers cloned a gene for a cell-surface protein of the malaria sporozoite, hoping to develop a vaccine against this first infective stage. However, it turned out that many sporozoite (and merozoite) surface proteins decoy the immune system into futile attacks. Furthermore, sporozoites are vulnerable only on the short trip from the bite site to the liver, and they continuously shed their surface proteins and synthesize replacements; the body's defenses end up attacking the empty coats, not live parasites. As if this weren't enough, the parasite also eludes the immune system by rapid mutation of the merozoite-produced proteins that cause infected red blood cells to stick to the inner walls of blood vessels. In effect, this continuously creates a "new" disease. The parasite's quick-change adaptation works so well that most infected people develop only limited immunity to malaria, whereas a bout of most other diseases confers resistance to further infections by that pathogen. Vaccines combining proteins from several stages are having moderate success. Until we have an effective vaccine against malaria, the best means of keeping the disease from spreading even further seems to be biological control of mosquitoes by releasing predators, bacterial diseases, or sterile males.

## 25–J   CILIOPHORA: CILIATES

All protists with cilia belong to one highly successful lineage, placed in phylum Ciliophora ("eyelash bearers"). They have rows of cilia either all over the body or in specialized areas of the cell surface (see Figure 25–1).

Ciliates have a very complex organization. The cell covering, the **pellicle,** consists of two layers of membrane sandwiching a layer of vesicles between them, much like bubble wrap. The outermost layer of cytoplasm, the cortex, contains a network of protein fibers connecting the basal bodies of the cilia. It may also contain many **trichocysts,** barbed or poisoned thread-like organelles that can be discharged to the outside. Trichocysts serve for anchorage, defense, or prey capture.

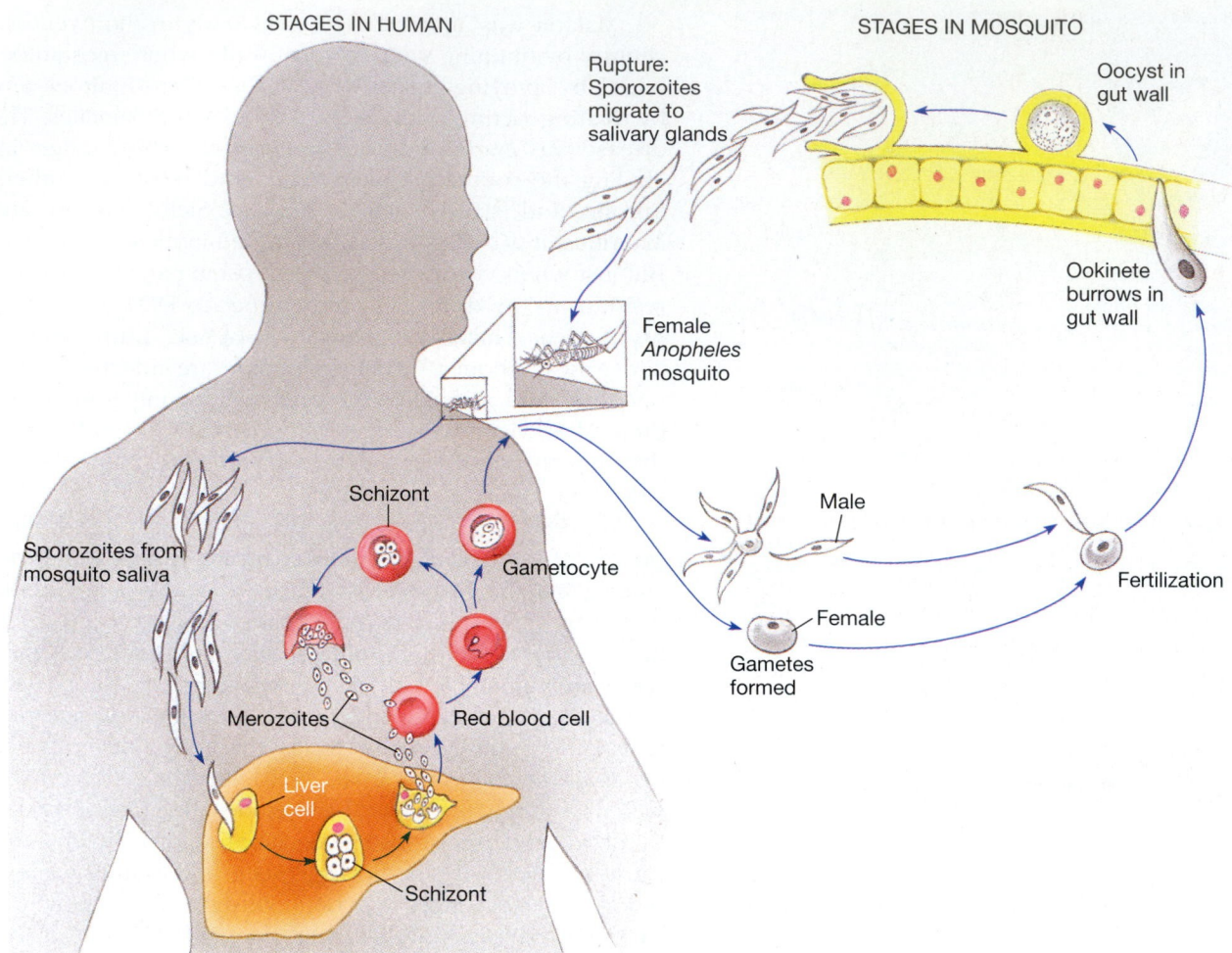

STAGES IN HUMAN

STAGES IN MOSQUITO

Rupture:
Sporozoites
migrate to
salivary glands

Oocyst in
gut wall

Ookinete
burrows in
gut wall

Female
*Anopheles*
mosquito

Schizont

Gametocyte

Male

Sporozoites from
mosquito saliva

Fertilization

Merozoites

Red blood cell

Female

Gametes
formed

Liver
cell

Schizont

**FIGURE 25–13** Life cycle of *Plasmodium vivax,* which causes human malaria. Parasites in the sporozoite stage enter the bloodstream with saliva from a female *Anopheles* mosquito as she prepares to take a blood meal. Within a few hours, sporozoites invade cells in the liver. For a week or two they grow, divide, and develop into schizonts, which eventually rupture, each releasing 20,000 or more merozoites.

Each merozoite enters a red blood cell and induces it to form protein knobs on its membrane. These knobs cause the red cell to stick to the inner surface of a blood vessel instead of circulating throughout the body. This protects the parasitized cell from destruction in the spleen.

The parasite repeats the sequence of development into schizonts and merozoites, with infection of even more red cells. Repeated cycles of rupture and invasion occur every two or three days, producing the symptoms of malaria: periodic bouts of shaking, chills, fever, and sweating.

Some of the parasites in the red blood cells develop into sexual gametocytes, which may be taken up by a mosquito when she bites a malaria victim. Fertilization occurs in the mosquito's gut, and the zygote develops into a stage that encysts in the gut wall and grows. After about 12 days, each cyst releases many sporozoites, which migrate to the salivary glands, ready to be injected into a new human host at the mosquito's next meal.

Most ciliates prey on bacteria, small animals, or fellow protists; some eat organic particles from the surrounding water, some are symbionts, and a few are parasites. Specialized cilia around the mouth region sweep food into a gullet. Here food enters a food vacuole, which fuses with a lysosome full of digestive enzymes. Digested food is ab-

sorbed into the surrounding cytoplasm, and undigested remains are discharged at a specific site on the cell surface. The contractile vacuoles, which discharge excess water, also expel their contents at specific sites.

A unique feature of ciliates is that they have two different kinds of nuclei. Each cell has one or more small, diploid **micronuclei** and a large **macronucleus,** containing hundreds of copies of the DNA. The macronucleus controls the cell's growth and metabolism, whereas the micronucleus is involved in sexual reproduction.

When a ciliate reproduces asexually, the micronucleus divides by mitosis, with the spindle forming inside the nuclear envelope. However, the macronucleus divides into two roughly equal parts without mitosis, a process that reduces its quality over many generations.

Sexual reproduction restores genetic vigor. During **conjugation** two cells come together and exchange haploid gamete nuclei resulting from meiosis of their micronuclei. After a long, complex process, both cells end up with genetically identical micronuclei, which divide and form new macronuclei. Meanwhile, the old macronuclei have degenerated.

*Paramecium* is the most familiar genus of freshwater ciliates (Figure 25–14). Some paramecia are hosts for other organisms. For example, *Chlorella*, a unicellular green alga, sometimes lives in the cytoplasm of *Paramecium bursaria*. In this symbiosis, the alga gains protection, and the ciliate has captive photosynthesis. Strains of paramecia containing **kappa particles** (Gram-negative bacteria with bacteriophages) are called "killers" because they produce a toxin lethal to paramecia that do not contain the particles. **Lambda particles,** found in other strains of paramecia, have been grown outside their ciliate hosts and have also been identified as Gram-negative bacteria.

Ciliates are sensitive to pollution and therefore are used as indicators of water quality.

## FUNGUS-LIKE PROTISTS

Three phyla in this chapter were once classified as fungi: the cellular and plasmodial slime molds (Sections 25–F, 25–G) and the Oomycota. There are also a handful of other, less well known fungus-like protistan phyla (Appendix II). The resemblance of all these groups to fungi seems to be a result of convergent evolution.

### 25–K  OOMYCOTA: WATER MOLDS

The Oomycota resemble fungi in having bodies made up of thread-like filaments called **hyphae** and in reproducing and dispersing by means of spores. Oomycotes live as saprobes or parasites. They feed by growing hyphae into a food source, releasing digestive enzymes, and absorbing the resulting small molecules. The vegetative hyphae are coenocytic, with diploid nuclei and cellulose walls. They form motile asexual spores, which swim through water by means of two unlike flagella, one hairy and forward-pointing, the other smooth and trailing.

These distinctive unlike flagella are also found in several other phyla of protists, including photosynthetic golden algae and brown algae (Sections 25–O, 25–P). Ribosomal RNA sequencing confirms that all these phyla are closely related. One theory holds that oomycotes evolved from coenocytic, diploid algae that lost their chloroplasts.

Some Oomycota are commonly called water molds because their hyphae form a white fuzz on aquarium fish or

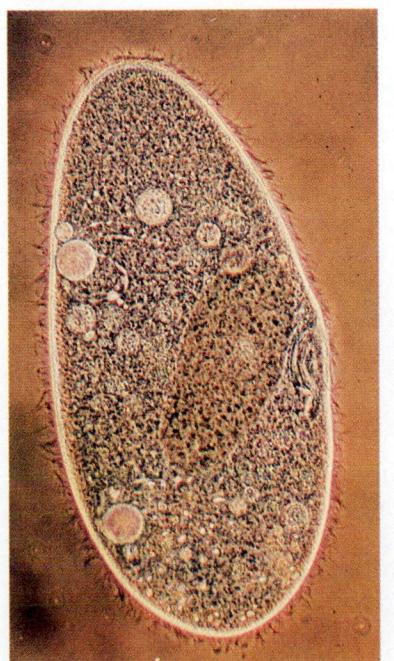

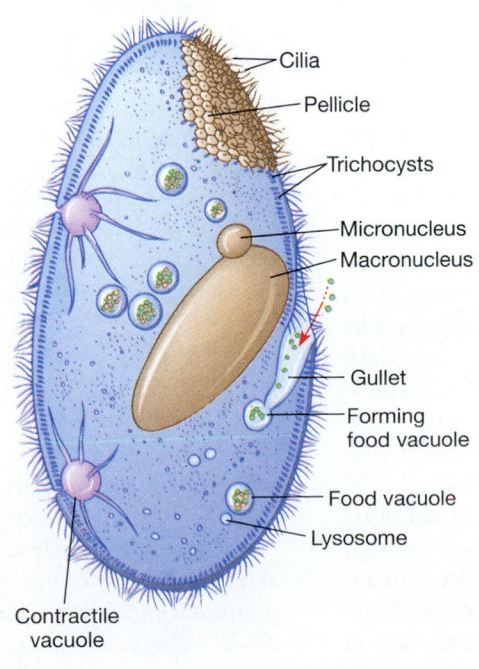

Cilia

Pellicle

Trichocysts

Micronucleus

Macronucleus

Gullet

Forming food vacuole

Food vacuole

Lysosome

Contractile vacuole

**FIGURE 25–14** *Paramecium*, a ciliate. The body is completely covered with cilia, used in locomotion and also to sweep food particles into the gullet. The cytoplasm at the end of the gullet engulfs food into vacuoles by phagocytosis. Lysosomes containing digestive enzymes fuse with the food vacuoles. Food is digested as the food vacuoles move around in the cytoplasm. Water taken up by osmosis is collected into two contractile vacuoles and expelled from the cell. Trichocysts are defensive, barbed threads that the cell may discharge at its surface when it is disturbed.  (a, Biophoto Associates)

(a)                                    (b)

on organic matter sitting in water. However, many oomy-cotes are terrestrial, including some very damaging plant parasites such as downy mildews and root-rotting forms. Some of these can disperse by means of airborne spores.

"Oomycota" means "egg fungi," referring to the large eggs formed in the female reproductive structures (Figure 25–15).

*Phytophthora infestans*, the parasitic oomycote that causes late blight of potatoes, was responsible for the Irish potato famine. The potato, native to South America, was brought to Europe in the sixteenth century. Potatoes require little labor and produce high yields of one of the most nutritious plant foods. By the nineteenth century, they were almost the only crop grown in Ireland. However, **mono-cultures** (plantings of only one crop) are especially susceptible to the rapid spread of disease, and in 1845 and 1846 late blight destroyed virtually all the Irish potato crop, leading to a devastating famine. From 1845 to 1851 a million people died in Ireland and a million and a half emigrated, mainly to the United States and Canada. The European potato shortage stimulated the agricultural economy of North America, where grain for export was produced in increasing quantities until the 1980s.

## ALGAE

The term algae embraces all photosynthetic protists. It refers to an aquatic, photosynthetic way of life, not to evolutionary kinship: many groups of algae arose independently from heterotrophs that gained photosynthetic symbionts, which evolved into chloroplasts. (In ecological discussions, the term algae may also include cyanobacteria, once known as blue-green algae.)

Most algae live in water, but some are terrestrial. The aquatic forms are ecologically important as the ultimate source of food for many protozoa and animals. They also produce an estimated 30 to 50% of the world's oxygen.

The photosynthetic way of life places a high selective premium on obtaining light. Because water absorbs light passing through it, algae must live reasonably near the water's surface. Various adaptations help maintain their favorable position. For instance, many algae attach to rocks in shallow water and so avoid being swept into the dark depths. Too much light can be as bad as too little because light damages some biological molecules. Algae that live on exposed surfaces generally have protective sunscreen pigments; motile forms may swim to dimmer depths.

Unicellular or filamentous algae, along with prokaryotic cyanobacteria and prochlorophytes, are often small enough to be members of the **phytoplankton** (phyto = plant; plankton = wanderer): swimming or floating photosynthetic organisms in bodies of water. Some use flagella to stay in the well-lit upper layer of water; others have projections that act as water wings, increasing their surface area and

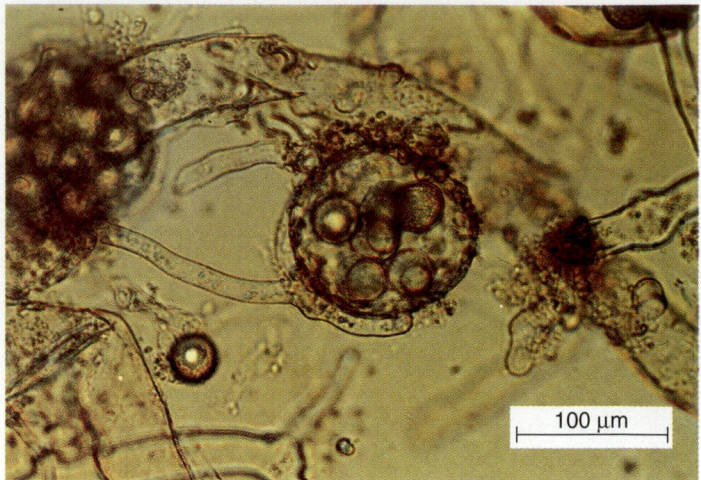

(a) Hyphae and sexual reproductive structures of the oomycote *Saprolegnia* (LM)

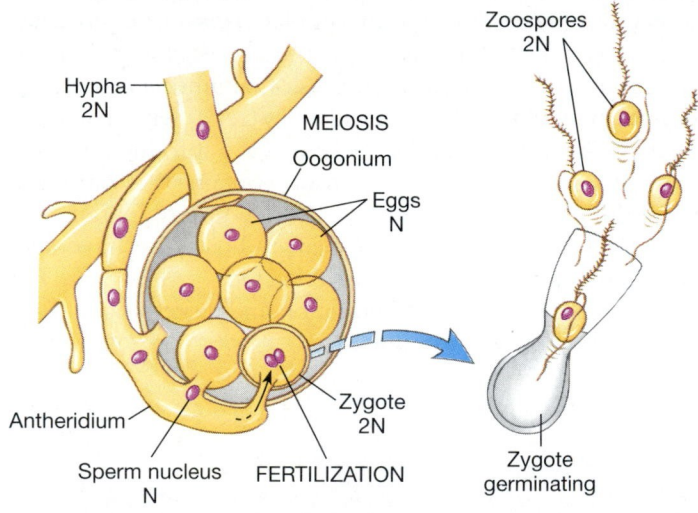

(b) Sexual reproductive structures

(c) Zoospores

**FIGURE 25–15**    The oomycote *Saprolegnia*. This common saprobe lives on organic matter in water. (a, b) Meiosis occurs in specialized sexual reproductive structures. The female oogonium produces many eggs, and the male antheridium produces sperm nuclei and releases them into an oogonium. The diploid zygotes become dormant for a time. (c) Upon germination, a zygote produces a short hypha with a sporangium that releases flagellated, diploid zoospores. Zoospores swim away and grow into new diploid mycelia.    (a, James W. Richardson, Visuals Unlimited)

so providing buoyancy; still others store their extra food as buoyant oil rather than dense starch. Many phytoplankton have more than one of these adaptations.

Algae are classified on the basis of conservative characters such as the type of cell wall, flagella, photosynthetic pigments, and the form in which food is stored (Table 25–2). There is strong selection against changes in these features because even a small change may affect their function so drastically that the organism cannot survive.

TABLE 25–2

**Important Phyla of Algae**

| Phylum | Characteristics | Examples |
|---|---|---|
| Dinoflagellata (Pyrrophyta) (~2000 species*) | Dinoflagellates. Mostly unicellular, some colonial; some with internal cell wall of cellulose plates; two flagella; chlorophylls *a* and *c* and carotenoids in photosynthetic forms; food stored as starch and oils; usually have trichocysts; mostly marine; free-living, photosynthetic or predatory, or symbiotic | *Gonyaulax, Noctiluca, Gymnodinium, Ceratium* |
| Euglenida (~800 species) | *Euglena* and its relatives. Mostly unicellular; pellicle of protein, no cell wall; usually two anterior flagella; photoreceptor, chlorophylls *a* and *b*, and carotenoids in photosynthetic forms; food stored as paramylon; no sexual reproduction; mostly freshwater, some parasitic | *Euglena, Peranema* |
| Bacillariophyta (~10,000 species*) | Diatoms. Mostly unicellular; diploid; cellulose and pectin cell walls containing silica; mostly nonmotile, a few species with flagellated male gametes; chlorophylls *a* and *c*, fucoxanthin and other carotenoids; food stored as oil and chrysolaminarin; sexual reproduction present; very abundant: marine, freshwater, terrestrial | *Asterionella, Fragilaria, Navicula, Pinnularia, Chaetoceros* |
| Chrysophyta (~1000 species) | Golden algae. Unicellular or colonial; cellulose and pectin cell walls; two unlike flagella in motile stages; chlorophylls *a* and *c*, fucoxanthin and other carotenoids; food stored as oil and chrysolaminarin; sexual reproduction present; mostly freshwater, some marine | *Dinobryon, Synura, Trentonia* |
| Phaeophyta (~900 species) | Brown algae. All multicellular, some very large and elaborate; sperm and motile spores with two unlike flagella; chlorophylls *a* and *c*, fucoxanthin and other carotenoids; store laminarin; almost all marine | *Fucus, Ascophyllum, Nereocystis, Laminaria* |
| Rhodophyta (~4100 species) | Red algae. Unicellular to multicellular; complex, sexual life cycles with no flagellated cells; chlorophyll *a*, phycobilins, carotenoids; chloroplast membranes not stacked; food stored as floridian starch in granules outside plastids; mostly marine, some freshwater, a few terrestrial | *Chondrus crispus* (Irish moss), *Polysiphonia, Porphyra* (nori), *Porphyridium* |
| Chlorophyta (~4300 species) | Green algae. Unicellular to multicellular; motile stages with two or more flagella; chlorophylls *a* and *b*, carotenoids; store starch in plastids; mostly freshwater, many marine, some terrestrial | *Chlamydomonas, Chlorella, Ulothrix, Oedogonium, Ulva* |
| Conjugaphyta (~10,700 species) | Conjugating green algae. Desmids (double cells) and filaments; no flagellated stages; isogamous; chlorophylls *a* and *b*, carotenoids; store starch in plastids; freshwater | Desmids, including *Cosmarium* and *Micrasterias*; filamentous forms such as *Mougeotia, Spirogyra, Zygnema* |

*Only numbers of living species given for each group.

Like cyanobacteria and prochlorophytes, all algae contain chlorophyll *a*. Algae are divided into phyla partly on the basis of their accessory photosynthetic pigments, such as chlorophyll *b* or *c* and various carotenoids (Section 8–B). Indeed, the names of some algal phyla are based on the characteristic colors of most of their members, imparted by their accessory pigments.

Characteristics such as structure and reproductive patterns tend to be more plastic, or able to change without endangering the alga's livelihood, and these variations may be adaptations to different habitats. Some algae once classified as separate organisms are now known to be members of the same species, which develop different body forms depending on the chemistry, depth, temperature, movement of the surrounding water, or stage of the life cycle.

## 25–L   DINOFLAGELLATA (PYRROPHYTA)

Dinoflagellates (deinos = whirling) are unicellular or colonial organisms with two flagella: one strap-shaped and wrapped around the middle of the cell in a characteristic groove, and the other a whiplash type beating at the rear (Figure 25–16). About half the species have cellulose "armor plating" under the plasma membrane. Most species have trichocysts.

About half of the living species are photosynthetic, capturing sunlight with chlorophylls *a* and *c* and various carotenoids and storing their food in the form of starch and oils. Dinoflagellates are common in the coastal marine phytoplankton. Some photosynthetic forms live as symbionts of larger organisms such as corals, sea anemones, flatworms, and clams.

Many dinoflagellates are colorless, that is, without chloroplasts and photosynthetic pigments and living as heterotrophic predators or parasites. A few predatory forms have an extremely complex light-detecting organelle, obviously evolved from a chloroplast, which can protrude and point in different directions in search of prey; it can even focus light with a moveable lens!

Dinoflagellates are primitive eukaryotes, in that their chromosomes have no histones and few other proteins. In addition, in most forms the chromosomes are always condensed. The kinetochores are embedded in the nuclear envelope, which remains intact when the nucleus divides. The chromosomes are segregated by their attachment to the envelope, and the stages of mitosis do not occur. Cell division varies greatly in this phylum, and only some members have spindle microtubules, which form outside the nucleus and make contact with chromosomes at points on the nuclear envelope. Most dinoflagellates are haploid, and the majority seem not to have discovered sex, although some reproduce sexually on rare occasions.

Several dozen species of dinoflagellates, including the largest, 2-millimeter-long *Noctiluca,* are bioluminescent: they give off light like a firefly. Alister Hardy described an encounter in the English Channel:

"I looked over the side to see a small shoal of fish, most likely mackerel, lit up by each individual being covered by a coat of fire; they were being chased this way and that by some much larger fish similarly aflame. On putting over a tow-net, which came up brilliantly illuminated, the sea was seen to be full of very small *Peridinium*-like dinoflagellates of the genus *Goniaulax*." [now spelled *Gonyaulax*]

Some species of dinoflagellates produce nerve poisons toxic to vertebrates. During population explosions of these algae, shellfish and fish may eat enough of them that the accumulated toxin kills them or poisons human seafood lovers. Since these organisms contain a reddish pigment, such population explosions may color the water red, a fitting warning signal known as a "red tide."

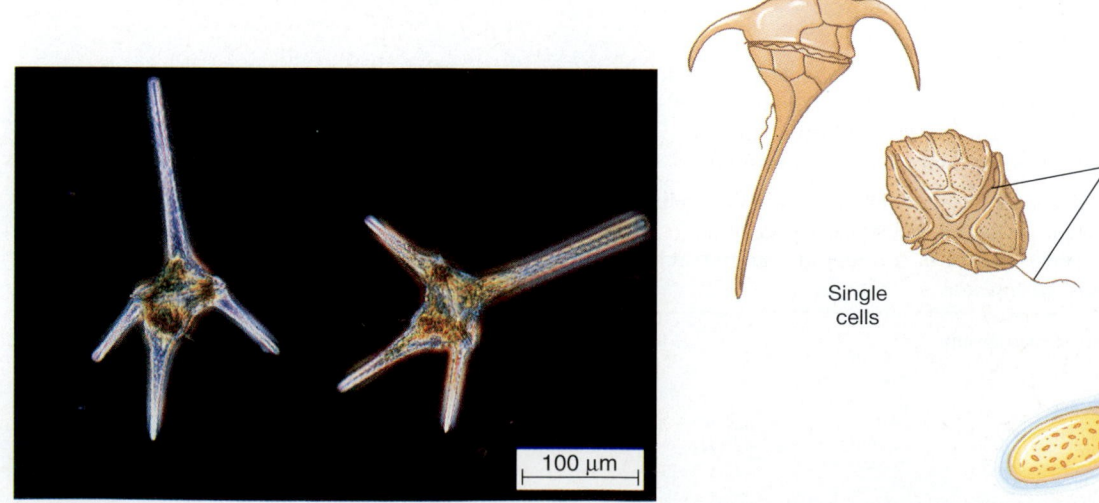

(a) Dinoflagellates: *Ceratium hirudinella* (LM)

(b) A variety of dinoflagellates

**FIGURE 25–16**   Dinoflagellates. (a) *Ceratium,* showing one of the two flagella of each cell. The long projections serve as water wings. (b) Some of the many other forms of dinoflagellates. (a, M. I. Walker/Science Source/Photo Researchers)

The phylum's alternative name, Pyrrophyta, combines the root phyton ("plant") with one meaning fire or red, apt references to bioluminescence and red color.

## 25–M    EUGLENIDA

The Euglenida are named after the familiar genus *Euglena* (eu = good; glene = eyeball). Most members of this group live in fresh water, being especially abundant in polluted habitats. Most euglenids have two flagella, at least one of them used for locomotion. Euglenids have no cell walls but may have an elastic or rigid transparent pellicle, made of protein, just beneath the plasma membrane. About one third are photosynthetic, with chloroplasts containing chlorophylls *a* and *b* and carotenoids. These photosynthetic forms have a swelling at the base of the locomotory flagellum that acts as a photoreceptor (light sensor). An eyespot containing red-orange pigment is thought to act as a shade that permits the sensor to detect the direction of light. The cytoplasm contains storage granules of the polysaccharide paramylon (Figure 25–17).

Euglenids reproduce asexually by dividing lengthwise into two. Sexual reproduction is unknown. The nuclei of members of the same species often contain varying amounts of DNA; the many small chromosomes remain condensed during interphase; and chromosome movements during mitosis are irregular, without the precise metaphase and anaphase seen in most other eukaryotes.

Because the euglenids include both autotrophs and heterotrophs, early biologists were torn between placing them in the plant or animal kingdoms. In fact, if *Euglena* is raised in the dark, where it cannot carry on photosynthesis, it loses its green color and becomes a heterotrophic predator. It now appears that the euglenids' closest relatives are parasitic flagellated protozoa, including *Trypanosoma,* and there are no close links to other algal groups. Early in evolution, some predatory euglenid ancestor probably gained chloroplasts either as prokaryotic symbionts or by saving the chloroplasts from its diet of green algae.

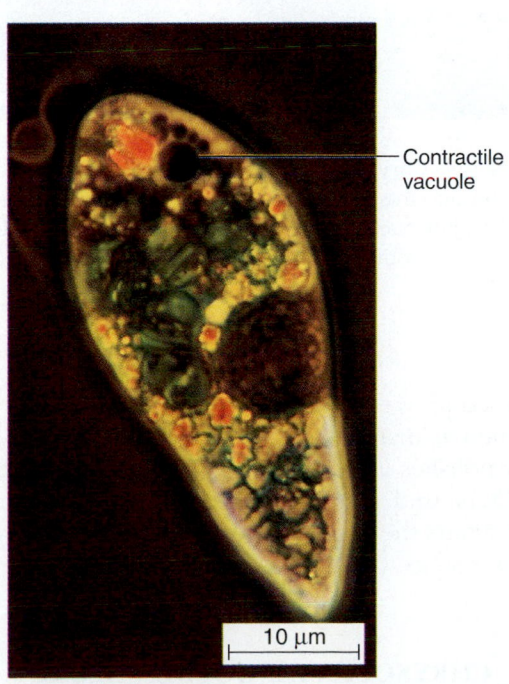

(a) Euglena (LM)

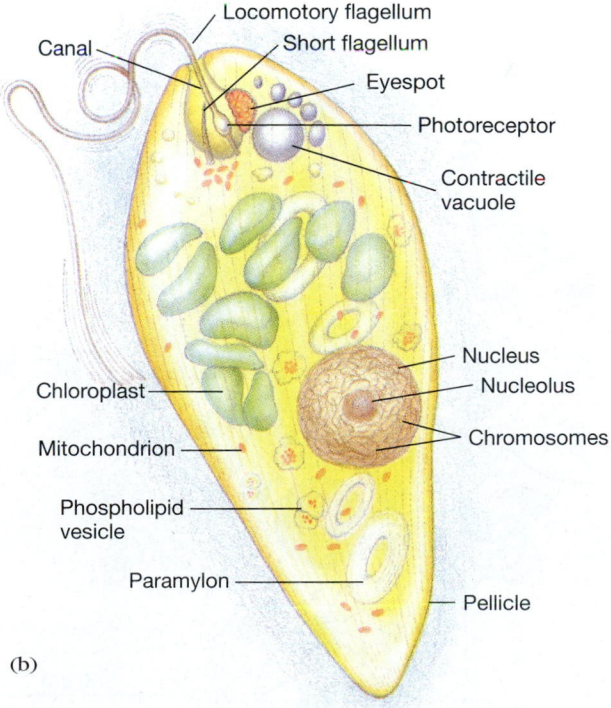

(b)

**FIGURE 25–17**  *Euglena.* The photoreceptor and eyespot detect the direction of light. Using its locomotory flagellum, the cell moves toward light, where its chloroplasts carry on photosynthesis, making the cell's food. Any surplus food is stored as paramylon. The contractile vacuole expels excess water taken in by osmosis.   (a, Biophoto Associates)

## 25–N  BACILLARIOPHYTA: DIATOMS

Phylum Bacillariophyta ("little stick plants") contains the diatoms, probably the most abundant aquatic eukaryotes, in numbers of individuals and of species. Living singly or in simple filaments or colonies, they occur just about anywhere there is water: floating in the marine or freshwater phytoplankton or attached to surfaces such as larger algae, rocks, wet wood, or rainforest plants. They are basically nonmotile, though some secrete slime and glide along slowly. Flagella occur only in the male gametes of a few genera. Although sexual reproduction is common, asexual reproduction predominates. Most unicellular algae are haploid, but diatoms are diploid.

Diatoms contain chlorophylls *a* and *c* and the accessory pigment fucoxanthin, a carotenoid that gives them their characteristic yellow-brown color. They store food as oil and the polysaccharide chrysolaminarin.

The most distinctive feature of diatoms is the intricately patterned cell wall, which is composed of two pieces that fit together like a box and its lid. The wall is impregnated with silica, the same material that makes up glass and sand (Figure 25–18). Diatoms are extremely efficient at removing dissolved silica from the surrounding water, and scarcity of silica often limits their population growth. Because silica does not decay, diatom walls accumulate on the ocean floor in enormous numbers. Some ancient deposits of these walls

**FIGURE 25–18**  Diatoms live in glass houses. The intricate patterns of pits and ridges in diatoms' walls were once used to test for aberrations in microscope optics. Victorian microscopists painstakingly arranged diatoms to form designs; this photo shows one such arrangement. (LM)  (Biophoto Associates)

**FIGURE 25–19**  Chrysophyta. In *Dinobryon,* the beating of flagella forces water into the transparent casing that surrounds the colony. The cells engulf bacteria from this water current, supplementing the food they produce during photosynthesis. (LM)  (Biological Photo Service)

were raised above sea level by geological activity and are now mined as diatomaceous earth, used as a fine abrasive in silver polishes and toothpaste, as the packing in air and water filters, and as a pesticide in organic gardening because it scours the protective waterproofing off insects and dehydrates slugs.

## 25–O  CHRYSOPHYTA: GOLDEN ALGAE

Chrysophytes are the "golden algae," most of which are photosynthetic. The phylum once included diatoms because these organisms share the same kind of photosynthetic chemistry—chlorophyll *a* and *c* and fucoxanthin pigments—and both store chrysolaminarin. However, the two groups were split because chrysophytes generally have two unlike flagella at some stage of the life cycle, marking them as relatives of the Oomycota.

Chrysophytes occur as single cells or as colonies of great diversity and complexity. Most are freshwater, a few marine. Some lake-dwellers supplement photosynthesis by eating bacteria, which are swept into the colony by the beating of flagella (Figure 25–19).

## 25–P PHAEOPHYTA: BROWN ALGAE

Members of the phylum Phaeophyta (phaios = dusky, brown) are all multicellular, ranging from microscopic filaments to Pacific kelps that may reach a length of 70 meters —the biggest and most complex algae. They contain chlorophylls *a* and *c*, and the carotenoid pigment fucoxanthin imparts the typical brown to olive-drab hue of most phaeophytes. These pigments also occur in the Chrysophyta, and both groups have similar food storage molecules as well. In addition, the flagellated cells of brown algae, both sperm and motile spores, have two unlike flagella similar in structure and placement to those of the golden algae. Therefore, phaeophytes are thought to have evolved from chrysophyte ancestors.

Brown algae are especially noticeable in cool, shallow waters along the seacoast in temperate and subpolar areas. There are a few small freshwater forms. Most brown algae grow attached to a solid surface.

The larger brown algae are interesting because of their tissue differentiation and their adaptations to life along the coast, where they may be pounded by the surf and then left high and dry when the tide goes out. Members of the genus *Fucus* are good examples. The algal body is a **thallus,** a multicellular structure that looks like a plant but has no vascular (transport) tissue. A foot or more long, the thallus grows on rocks in the intertidal zone, between high and low tide levels, attached by a specialized **holdfast** (Figure 25–20). A stalk-like **stipe** connects the holdfast to the flat photosynthetic **blades.** When the tide is in, the blades float

Reproductive structures

Air bladder

Blade

Holdfast | Stipe

*Fucus* submerged

(a)

(b)

**FIGURE 25–20** Brown algae. (a) *Fucus* shows the considerable differentiation of parts found in many large brown algae. Here the thallus is shown submerged at high tide. (b) Intertidal algae hang down from the rocks when the tide is out. Most of these algae are *Fucus,* some with air bladders and reproductive structures visible. The more slender plants (lower left) are *Ascophyllum,* another brown alga. (c) Air bladders keep the blades of kelp floating in the well-lit surface layers of the sea. (c, Tammy Peluso/Tom Stack & Associates)

(c)

near the surface of the water, close to the light, buoyed by gas-filled **air bladders.** Numerous swellings at the ends of the blades contain chambers where reproductive cells develop. The seaweed's cell walls are impregnated with a gummy substance, alginate, which helps retain water when the thallus is exposed to the air at low tide. Dark pigment shields the living tissue from bright sunlight until the tide returns.

*Fucus* has differentiated tissues. The tightly packed cells of the outer layer receive the most light and also have more chloroplasts than the larger, more loosely spaced cells of the next layer inside. A column of tough filaments in the center holds the thallus together. Strands of mucilage in this region hold water and release it gradually, replacing water lost from the outer layers to the air at low tide. Cells in the central column also transport food produced by the blade to the holdfast, which usually receives less light and so cannot carry on photosynthesis and feed itself.

Giant kelps such as *Laminaria* and *Nereocystis* can live with their holdfasts anchored in deeper water and still have parts that float at the surface, near the sunlight. These forms have cells resembling the phloem tissue that transports food from the leaves to the roots of plants, an example of convergent evolution.

The larger brown algae provide hiding places for many animals. They may also provide a surface for the attachment of smaller algae.

Many brown algae have economic importance. Algae tend to accumulate the nutrients nitrogen, potassium, and iodine in great quantities, and this makes them useful as fertilizer and as food for livestock or humans. Some kelps are processed to extract alginate, which is used as an ingredient in about half the ice cream produced in the United States because it gives a smooth texture and helps prevent formation of ice crystals. Alginate is also used in various drugs and cosmetics.

## 25–Q   RHODOPHYTA: RED ALGAE

Phylum Rhodophyta, or red algae, contains single cells as well as thalli that grow as filaments, branching structures, and broad flat plates or ruffles (Figure 25–21). Some of the more complex red algae grow up to a meter long, but most are small and delicate. All live attached to some surface: a rock, coral reef, animal shell, or even a larger alga. Many forms cement themselves to the surface and spread over it to form a crust; larger species attach to the surface by a holdfast.

The chloroplasts of red algae show strong evidence of descent from cyanobacteria (Section 24–D). The arrangement of photosynthetic membranes is similar (Figure 25–22); both have chlorophyll *a* as the only chlorophyll; and both contain the phycobilin accessory pigments phycocyanin and phycoerythrin. But whereas the blue phycocyanin predom-

(a)

(b)

**FIGURE 25–21**   Rhodophyta. (a) *Rhodymenia* exposed at low tide. (b) Divers in Japan washing *Gelidium,* from which agar is extracted.   (a, Biophoto Associates; b, W. Johansen/Visuals Unlimited)

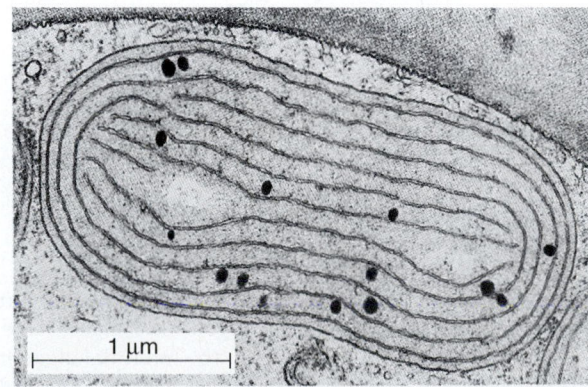

**FIGURE 25–22**   A chloroplast from a red alga. The photosynthetic thylakoids are single, like those of cyanobacterial cells, rather than piled in stacks, as in most other chloroplasts (compare Figures 24–22 and 8–7c). (TEM)   (Curt Pueschel)

inates in cyanobacteria (combining with green chlorophyll to give a blue-green hue), red phycoerythrin predominates in the red algae. Here it masks the green chlorophyll, giving most members of this group their characteristic reddish tints. Carotenoid accessory pigments are also present. Rhodophytes store food in a form called floridean starch.

No red alga has flagella, even in sperm cells. This seems to be their primary condition, unlike some other eukaryotes that have lost their flagella secondarily.

Rhodophytes have complex life cycles with many variations. Sexual reproduction is oogamous: a large female gamete is fertilized by a smaller male gamete. The male gamete is released near the female organ and must move down a long neck to reach the egg, but we do not know just how this occurs.

Most red algae are marine, with a few freshwater or terrestrial forms. They may be found far up the shore in the splash zone, where the ground is never covered by water even at high tide. At the opposite extreme, red algae grow at depths of almost 100 meters in clear seas, and one form thrives deeper than 260 meters near the Bahamas. These rhodophytes are probably the only organisms able to carry on photosynthesis at such depths, where little light penetrates.

Red algae are prominent residents of tropical coral reefs, where they play an important part in reef-building. The construction of reefs was once attributed solely to coral animals, which secrete calcium carbonate (lime) tubes around themselves (Section 28–E). However, many algae encrust themselves with calcium carbonate extracted from the sea, thereby adding material to the reef.

Some red algae are eaten by people in the Orient and in coastal areas of Europe. Components of the cell wall similar to alginate in brown algae are extracted from some forms. Carrageenan, a substance used in puddings, candies, and ice cream, comes from the red alga called Irish moss. Agar, also from a red alga, is used to solidify nutrient media for laboratory cultures of microorganisms.

## 25–R  CHLOROPHYTA: GREEN ALGAE

The ancestors of all plants were undoubtedly members of phylum Chlorophyta (green algae), named for their vivid green chloroplasts. Indeed, some taxonomists classify chlorophytes in the plant kingdom. Green algae show great diversity of form and live in a variety of habitats. Many are single-celled; others form simple or branched filaments, or hollow balls of cells, or broad, flat sheets (Figure 25–23).

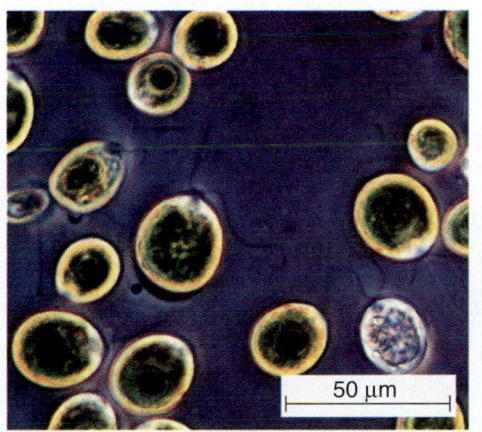

(a)

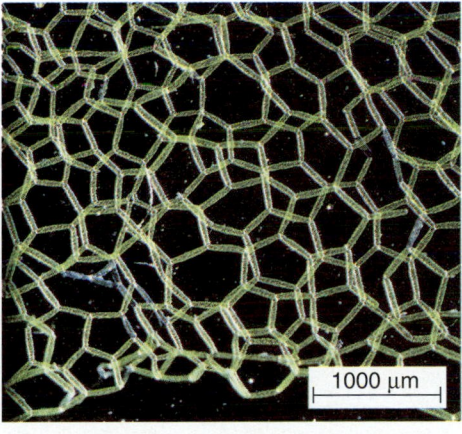

(b)

(c)

(d)

**FIGURE 25–23**  Phylum Chlorophyta. (a) *Chlamydomonas* may resemble the ancestral green alga. (b) *Hydrodictyon* (water net) grows as a tube of interconnected cells. (c) A calcium-secreting marine form. (d) *Ulva* (sea lettuce), a marine intertidal form.   (a, Biophoto Associates; b, Dwight R. Kuhn/DRK Photo; c, Steven Webster; d, J. M. Kingsbury)

There is no tissue differentiation, although some cells may be specialized as holdfasts or reproductive cells. Many species of green algae form coenocytes rather than colonies of individual cells.

Most Chlorophyta inhabit fresh water, both still and running, and some live on moist rocks, soil, and tree trunks on land. There are also many marine species, including some that contribute limy material to coral reefs.

Members of the genus *Chlamydomonas* (Figure 25–23a) may well resemble the ancestral green alga. With their cup-shaped chloroplasts and two similar anterior flagella, they look much like the individual cells in some colonial forms. They also bear a striking resemblance to the flagellated gametes and zoospores of the phylum's multicellular members. It is not hard to imagine the evolution of multicellular thalli from cells much like modern *Chlamydomonas,* which stuck together after dividing and suppressed the formation of flagella except during reproduction.

The chloroplasts of plants and chlorophytes contain chlorophylls *a* and *b,* as well as β-carotene and other carotenoids, and store food as starch. These features also occur in prochlorophyte bacteria, the presumed ancestors of chlorophyte and plant chloroplasts. One group of green algae, the charophytes, shows details of cell structure and division, peroxisomes, and photorespiration enzymes suggesting that some of its ancient members gave rise to plants.

Green algae show some important evolutionary trends. Many different lineages show a trend from unicellular to multicellular body forms. Another trend is in sexual reproduction, from the primitive condition of **isogamy,** the union of two similar gametes at fertilization, to the more advanced (and widespread) oogamy, the union of a large, immobile egg and a small, motile sperm (Section 20–C).

A third important trend is seen in the life cycles of some multicellular green algae. In the ancestral pattern, the thallus is haploid (N) and the zygote is the only diploid (2N) stage in the life cycle. The zygote often overwinters as a dormant, resistant cyst and then undergoes meiosis and produces four haploid spores, which develop into new haploid thalli.

$$\underset{\text{N}}{\text{thallus}} \xrightarrow{\text{(fertilization)}} \underset{\text{N}}{\text{gamete}} \longrightarrow \underset{\text{2N}}{\text{zygote}} \xrightarrow{\text{(meiosis)}} \underset{\text{N}}{\text{spore}} \longrightarrow \underset{\text{N}}{\text{thallus}}$$

In some lineages this pattern is changed by adding another stage to the life cycle: the zygote divides by mitosis and produces a multicellular diploid thallus (Figure 25–24). Many

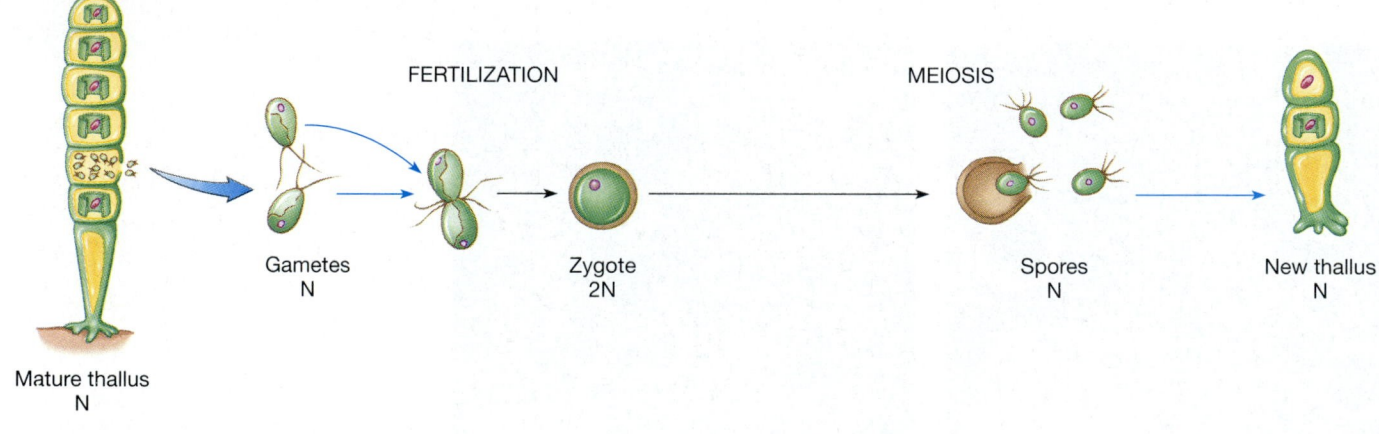

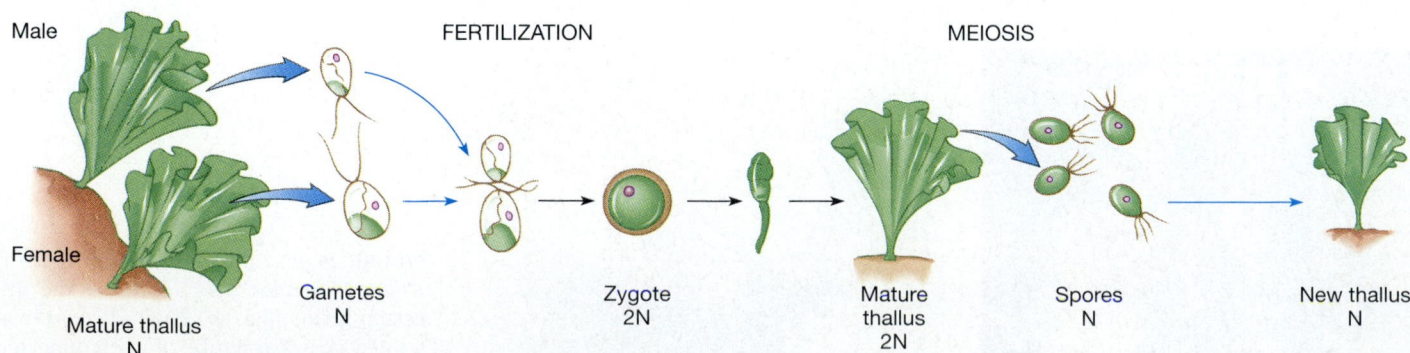

**FIGURE 25–24**   A comparison of two life cycles. (top row) In *Ulothrix* the thallus is haploid, and the zygote is the only diploid stage in the life cycle. (bottom row) In *Ulva* meiosis is deferred. Instead, the zygote divides and forms a new, multicellular diploid stage, which alternates with the multicellular haploid stage.

# ESSAY: *Phytoplankton: The Moveable Feast*

Phytoplankton are the microscopic, floating fodder that supports flocks of aquatic grazers, both protozoa and animals, in oceans, lakes, and ponds (Figure 25–A). Diatoms and dinoflagellates are the two main groups of eukaryotic marine phytoplankton, although both are outnumbered by prochlorophyte bacteria. In unpolluted fresh water, diatoms and green algae predominate, whereas pollution by fertilizer runoff and sewage permits cyanobacteria and euglenids to proliferate.

Unlike plants and attached algae, phytoplankton are not fixed in one place. They are carried around by waves and water currents, although they do control their depth in the water: by swimming with their flagella or by changing their buoyancy, these minute algae make daily vertical migrations, sinking away from intense light at midday, rising as the afternoon light fades, and sinking again in the dark, when their predators move to the surface.

Phytoplankton also have seasonal cycles of fluctuating population density. Their numbers are usually limited either by light intensity (and temperature) or by the concentration of nutrients, especially nitrates and phosphates. In winter in temperate climates, the phytoplankton population is small because there is not much light. In summer in temperate climates, and throughout the year in the tropics, nutrients are limiting.

As day length increases in the temperate zones in the spring, an enormous burst of reproduction occurs among the phytoplankton. However, soon all the dissolved nutrients are absorbed and then carried to the bottom as planktonic organisms die and sink. Thus, despite plentiful light, the phytoplankton population declines in midsummer from scarcity of nutrients.

Later in the year, water from the deeper layers of the lake or sea wells up, bringing nutrients back to the surface. Phytoplankton density is greatest in those areas of the ocean where currents bring nutrient-rich waters up from the deep throughout the year, or where rivers bring nutrients washed off the land. Larger phytoplankton populations, in turn, support more animal life.

For a time, marine biologists assumed that phytoplankton populations were sparse in much of the open ocean. This conclusion was based on two observations: first, samples of sea water contained few dissolved nutrients, and second, few organisms grew from these samples in the laboratory.

Recently, some areas of the ocean thought to be barren have turned out to have thriving populations of diatoms with symbiotic nitrogen-fixing bacteria. These do-it-yourself fertilizer microfactories support the growth of whole communities of microscopic life. Also, biologists have discovered that many phytoplankton failed to grow in the laboratory because their fragile cells were destroyed during collection or because they have exotic dietary needs. For example, dinoflagellates that cause red tide have heavy vitamin requirements, but vitamins are not normally supplied to cultures of autotrophic organisms.

When oil-storing phytoplankton die and sink to the bottom, they may be covered by sediments and water that exert enormous pressures. Dinoflagellates, diatoms, and chrysophytes that lived hundreds of millions of years ago made the fossil fuel oil we use today.

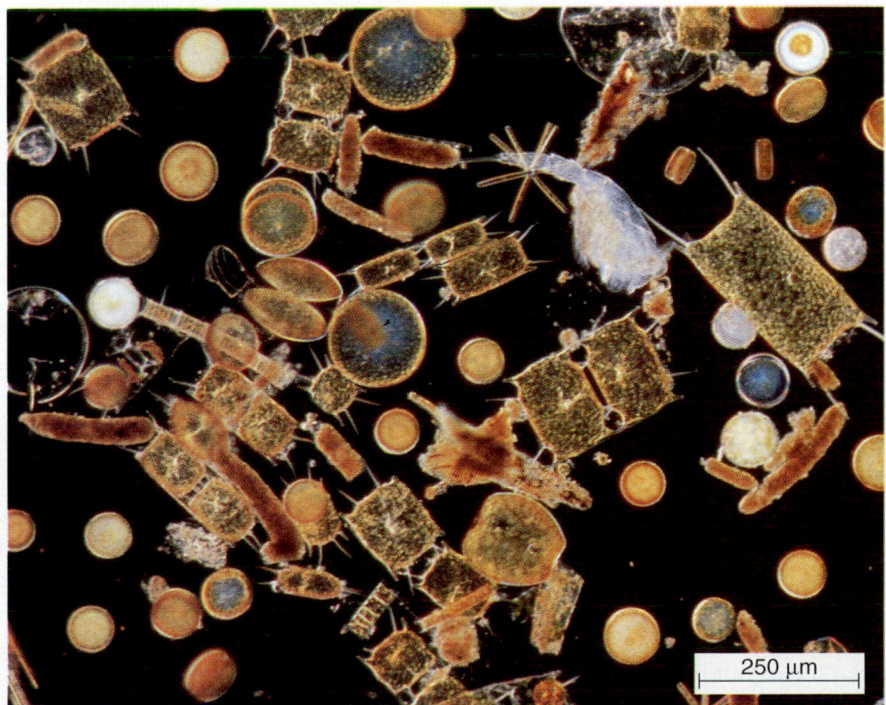

250 µm

**FIGURE 25–A**  Phytoplankton. This sample of seawater shows many planktonic diatoms and some of the small animals that feed on them. (LM)  (D. P. Wilson/Science Source/Photo Researchers)

of its cells later undergo meiosis, and so this diploid stage produces many spores, with a more varied range of genetic combinations than the single diploid zygote nucleus could have produced.

$$\text{thallus} \underset{N}{\longrightarrow} \text{gamete} \overset{\text{(fertilization)}}{\underset{N}{\longrightarrow}} \text{zygote} \overset{\text{(meiosis)}}{\underset{2N}{\longrightarrow}} \text{thallus} \underset{2N}{\longrightarrow} \text{spore} \underset{N}{\longrightarrow} \text{thallus}$$

This pattern adds a multicellular diploid stage to a life cycle that already had a multicellular haploid stage. Because the life cycle goes through these multicellular haploid and diploid stages in turn, this pattern is called **alternation of generations.**

This same basic life cycle also occurs in some brown algae. (Red algae have a variety of complex life cycles beyond the scope of this book.) The diploid and haploid thalli of some algae (including *Ulva*, Figures 25–23d and 25–24b) look alike, but in many species the two generations differ. This has caused confusion in algal taxonomy because in some cases the haploid and diploid forms of the same alga were named as different species, until life cycle studies revealed their secret.

## 25–S   CONJUGAPHYTA

Members of phylum Conjugaphyta were formerly classified as chlorophytes on the basis of their similar chloroplast pigmentation and food storage. However, they have differences considered fundamental enough to warrant a separate phylum. First, these algae have no flagellated stages. Second, sexual reproduction occurs by isogamous union of amoeboid gametes formed from the entire contents of vegetative cells in the process of conjugation, which gives the phylum its name (see Figure 20–6a).

All conjugaphyta live in fresh water. The phylum includes filamentous forms such as *Spirogyra* (see Figure 8–1b). Another large group is the desmids, each consisting of two mirror-image halves sharing a single nucleus housed in an isthmus between them (Figure 25–25).

## SUMMARY

The kingdom Protista contains unicellular and simple multicellular eukaryotes that live in aquatic or very moist terrestrial habitats or in the body fluids of other organisms. Most are aerobic. The kingdom originally contained only unicellular forms but has recently been expanded to embrace slime molds and oomycotes, formerly classified as fungi, and multicellular brown, red, and green algae, formerly classified as plants.

The origin of eukaryotic cells was a major leap in the evolution of life. It produced a complex and versatile basic pattern that could be modified in many ways, especially

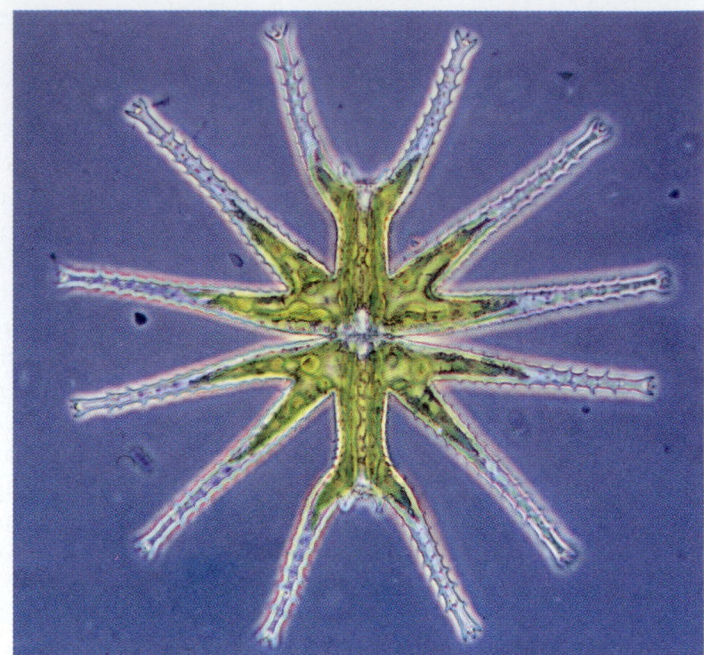

**FIGURE 25–25**   Conjugaphyta. *Micrasterias* is a desmid, consisting of a single cell with two mirror-image halves. (LM)   (Biophoto Associates)

under the direction of genetic variations produced by sexual reproduction. This led to the adaptive radiation of a wide range of unicellular protists suited to many modes of life: autotrophic and heterotrophic; motile, floating, and sessile; free-living, parasitic, and symbiotic. Many of these protists house an elaborate organization and division of labor within a single cell.

A number of unicellular protists took another big step: evolution into colonial and multicellular forms, including the ancestors of members of the animal, plant, and fungal kingdoms and some whose modern descendants are still classified as protists.

Tables 25–1 and 25–2 list the major phyla of protists and their definitive characteristics.

1. Eukaryotic cell structure permitted cells to become larger, thanks mostly to the cytoskeleton, which provides internal support as well as ways to obtain and transport materials in bulk by remodeling membranes as needed. Mitosis provided for accurate distribution of copies of a genome made up of many chromosomes, and sexual reproduction permitted reassortment in the genome.

2. Although early eukaryotes had more elaborate cell structure than bacteria, they lacked some metabolic talents found among prokaryotes. Many ancient eukaryotes set up symbioses with bacteria, especially those that could carry out aerobic respiration or photosynthesis. Some of these bacterial symbionts evolved into mitochondria and chloroplasts, respectively.

3. Continued selection for larger size favored some protists that remained together after cell division and formed multicellular structures. Each cell maintained a favorable ratio of surface area to volume, and all gained the advantage of the structure's larger size. At first these were colonies of undifferentiated cells, but in some of them cells eventually began to assume specialized roles, adding division of labor among cells, tissues, and organs to the division of labor among the organelles within each cell.

4. The classification of protists is changing rapidly as researchers gather new evidence about their biochemistry, ultrastructure, cell division, and life cycles. More importantly, the endosymbiont theory changes our fundamental picture of evolution, and this affects the classification of protists more than that of any other kingdom.

5. Protists may be naked or have cell walls or tests. Some float passively in their watery homes; others live attached to some surface; many are motile, using flagella, cilia, or pseudopods to move toward or away from various stimuli. Most unicellular and small multicellular protists can reproduce rapidly, and many of them survive unfavorable environmental conditions in the form of dormant cysts or spores.

6. Protozoa are unicellular or syncytial, heterotrophic, motile protists without cell walls, although many have a test or a pellicle. They may be free-living predators, eating other protists, bacteria, and small multicellular organisms; scavengers of organic debris; or symbionts or parasites of larger forms of life. The major phyla are characterized by conservative features, especially type of locomotion (by pseudopods, flagella, or cilia), cell division, and life cycle. Many have contractile vacuoles, which expel excess water absorbed by osmosis.

7. Some parasitic protozoa have an enormous impact on human health. They create economic burdens in the form of health care costs and lost labor. The major human diseases caused by protozoa are trypanosome infections (sleeping sickness and Chagas' disease), leishmaniasis, and malaria. Amoebic dysentery and giardiasis are also serious illnesses.

8. Fungus-like protists live as saprobes, digesting food externally and absorbing small molecules. The most important phylum, Oomycota, contains the organism that causes late blight of potatoes.

9. Algae live either free in the phytoplankton, or attached to moist or submerged surfaces, or as symbionts inside larger organisms. The clues to their phylogeny include such conservative characters as photosynthetic pigments; form of food storage; chemistry and form of the cell wall or other cell covering; and presence, type, and arrangement of flagella.

10. Algae have various adaptations that enable them to maintain a place in the sun: flagella, cell-wall projections, stored oil droplets, adhesive secretions, holdfasts, and air bladders. Algae living on land or between tide lines have moisture-retaining substances in their cell walls and produce dark pigments that screen out intense solar rays.

11. Aquatic algae produce most of the basic food supply in their environments and also contribute a healthy percentage of the world's oxygen. Their ability to accumulate essential minerals from seawater has also made them useful as food for livestock and humans. In addition, they produce various economically useful substances.

12. Different groups of algae predominate in different habitats. Red algae tend to live high up the shore or in the deepest layers of water penetrated by sunlight. Brown algae dominate shallow waters, particularly in colder seas. Green algae live less conspicuously in most marine environments and also flourish in freshwater and moist terrestrial habitats. Diatoms are abundant in both marine and freshwater phytoplankton, whereas dinoflagellates are mainly marine.

13. Green algae show trends from unicellular isogamous haploid forms to multicellular oogamous forms, and in some cases to alternation of diploid and haploid generations.

14. Similarities of photosynthetic chemistry, cell structure, and life cycle indicate that plants evolved from some green algae.

## Self-Quiz

1. An organism should be placed in kingdom Protista if it meets two criteria:
   a. _____
   b. _____
2. State whether each description below applies to locomotion by means of cilia, flagella, or pseudopods:
   _____ a. Cytoplasm flows out into temporary extrusions from the cell body.
   _____ b. Oar-like beating propels the cell through the water.
   _____ c. Wave-like undulations pull or push the cell through the water.

3. For each phylum listed below, indicate whether its members have cilia, flagella, pseudopods, or none of these:
   _____ a. Conjugaphyta
   _____ b. Zoomastigina
   _____ c. Rhodophyta
   _____ d. Phaeophyta
   _____ e. Foraminifera
4. A dinoflagellate would have which of the following sets of characteristics?
   a. a mouth, contractile vacuole, and mitochondria
   b. a macronucleus, a micronucleus, and a chloroplast

*(Continued)*

c. chloroplasts, one circular chromosome, and cilia

d. a contractile vacuole, pseudopods, and cilia

e. chloroplasts, mitochondria, and flagella

Give the names of the algal phyla whose members fit each description below.

_____ 5. May reach several meters in length; no unicellular forms known

_____ 6. Contain chlorophylls *a* and *b;* store food as starch; have flagella at some stage of the life cycle

_____ 7. Contain chlorophylls *a* and *c;* store food as laminarin

_____ 8. Chloroplasts contain chlorophyll *a* and phycobilins; internal chloroplast membranes occur singly

_____ 9. Unicellular, diploid, with cell walls containing silica

For each adaptation of algae on the left, pick its function from the list on the right. The functions may be used one or several times or not at all.

| *Adaptation* | *Function* |
|---|---|
| _____ 10. Accessory photosynthetic | a. maintains a favorable location |
| _____ 11. Flagella | b. obtains more nutrients |
| _____ 12. Holdfast | c. protects from desiccation |
| _____ 13. Air bladder | d. obtains more energy |
| _____ 14. Water-retaining cell-wall components | |

## Questions for Discussion

1. Can you think of selective pressures other than the ones discussed in this chapter that might have selected for evolution of multicellular organisms? What other advantages do multicellular organisms have?

2. How do the protists you studied in this chapter carry out each function listed on pages 521–522?

3. Explain why the bacterial symbionts of diatoms fix nitrogen mainly at night (*Hint:* see Section 24–C).

4. Nonmotile algae must be able to get gametes together for fertilization. How does this happen in green algae such as *Ulothrix* and *Ulva?* In red algae?

5. What modifications in the structure and reproduction of aquatic algae would be necessary for them to live permanently on land?

6. Are the cells specialized for the transport of food in kelps and land plants analogous or homologous? Kelps live in the sea; what selective pressures favored their evolution of food-transporting cells similar to those of land plants?

7. Algae lack roots; how do they obtain nutrients?

## Suggested Readings

Baker, D. *"Giardia!" National Wildlife,* August–September 1985.

Bold, H. C., and M. J. Wynne. *Introduction to the Algae,* 2d ed. Englewood Cliffs, NJ: Prentice-Hall, 1985.

Bonner, J. T. "Chemical signals of social amoebas." *Scientific American,* April 1983. How cellular slime molds aggregate.

Curtis, H. *The Marvellous Animals.* Garden City, NY: Natural History Press, 1968. A delightfully written introduction to the kingdom Protista.

Donelson, J. E., and M. J. Turner. "How the trypanosome changes its coat." *Scientific American,* February 1985.

Godson, G. N. "Molecular approaches to malaria vaccine." *Scientific American,* May 1985.

Jensen, W. A., and F. B. Salisbury. *Botany: An Ecological Approach,* 2d ed. Belmont, CA: Wadsworth, 1984. Chapters 17 and 18 treat algae in their ecological groupings rather than by taxonomy; an interesting change.

Laybourn-Parry, J. *A Functional Biology of Free-living Protozoa.* Berkeley: University of California Press, 1984.

Margulis, L., and K. V. Schwartz. *Five Kingdoms: An Illustrated Guide to the Phyla of Life on Earth,* 2d ed. San Francisco: W. H. Freeman, 1988. Covers all groups in this chapter as well as several more obscure ones.

# *Fungi*

## OBJECTIVES

*When you have studied this chapter, you should be able to:*

1. Define the terms saprobe, parasite, hypha, mycelium, haustorium, dikaryon, spore, sporangium, fruiting body, ascus, basidium.
2. List and describe the major characteristics of fungi, and tell how fungi differ from plants and animals.
3. List at least three ways in which fungi may spread.
4. Describe both sexual and asexual life cycles of a fungus such as *Rhizopus,* the black bread mold.
5. List the four divisions of fungi and the distinguishing characters of each. Sketch and label their characteristic sexual reproductive structures.
6. Place each of these organisms in the correct division: *Rhizopus* (black bread mold), yeast, *Penicillium, Neurospora* (pink bread mold), mushrooms, bracket fungi, *Epidermophyton* (organism causing athlete's foot).
7. Describe why the activity of fungi as decomposers is vital to life on Earth.
8. Explain why there are many more fungal diseases of plants than of animals.
9. Discuss methods of controlling fungi.
10. State what lichens and mycorrhizae are, and explain their ecological roles.
11. State how yeasts differ from other fungi, and list three products made using yeasts.
12. List or recognize ways in which fungi are economically beneficial and ways in which they are economically harmful.

---

The kingdom Fungi might well be nicknamed the undercover kingdom because its members typically live as masses of thin threads hidden in some food source. Only the reproductive structures surface at times, releasing tiny, dust-like spores into the air (Figure 26–1). So fungi are always around but seldom seen. Indeed, the largest living organism known is an underground fungus, lying beneath more than 100 hectares of ground in the Midwest and having a calculated mass of more than 500 tons. Sampled pieces of this fungus were shown to be part of the same organism by genetic analysis using the polymerase chain reaction (Section 13–D).

In Robert Whittaker's 1969 five-kingdom proposal, kingdom Fungi contained eukaryotic, mostly multicellular heterotrophs adapted to absorbing their food. Further study showed that most of these organisms share an additional suite of characters that has proven extremely successful. Most have a multicellular, thread-like body covered by a wall of the polysaccharide chitin. These organisms are

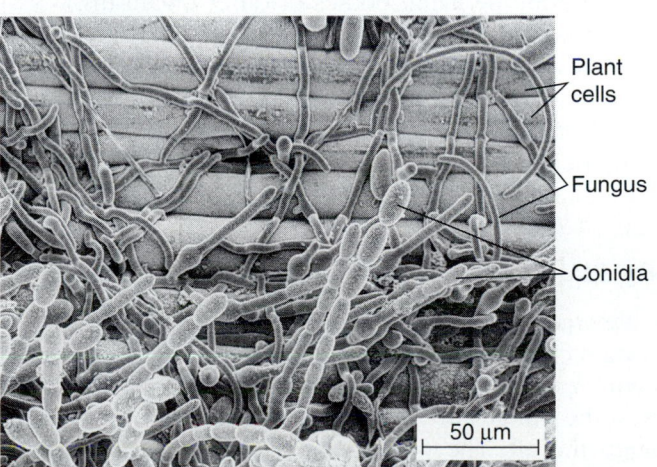

**FIGURE 26–1** The surface of a barley plant infected with powdery mildew. The fungus forms a tangle of thin threads, which penetrate the plant cells. Thicker threads thrust into the air and pinch in, forming asexual spores known as conidia. (Biophoto Associates)

mostly terrestrial, and so they have no flagellated, swimming stages: both sexual and asexual reproduction and dispersal occur by means of walled, usually airborne spores.

This large, flourishing lineage is not closely related to several smaller ones that Whittaker considered to be fungi. Hence, many biologists now favor using phylogeny, not lifestyle, to define kingdom Fungi. Moving oddball phyla such as Acrasiomycota, Myxomycota, and Oomycota to kingdom Protista (Sections 25–F, G, and K) also means that the remaining fungi are probably monophyletic.

All fungi are heterotrophs, living as saprobes, parasites, or partners in a mutualistic symbiosis. Nearly all are aerobic.

Fungi are vitally important in their ecological role as decomposers. When a dead leaf drifts to the forest floor or an animal dies of disease, fungal and bacterial spores floating in the air have already settled on it. These spores quickly germinate and begin to break down the dead organism, releasing small organic molecules that they can use as food, as well as minerals that may be absorbed by the decomposer or by nearby plants.

Fungi are also important in the human economy. Some fungi cause tremendous losses of crops or stored food every year, but fungi are also used to produce foods and medicines.

### KEY CONCEPTS

♦ Fungi are multicellular, mostly terrestrial eukaryotes with heterotrophic, absorptive nutrition, a thread-like mycelial body form with chitinous walls, and reproduction by walled, unflagellated spores.

♦ Fungi live as saprobes, parasites, or mutualistic symbionts. They are important in nature as decomposers, and they have tremendous importance on both sides of the ledger in human economies.

## 26–A   THE FUNGAL WAY OF LIFE

### Nutrition

Fungi cannot produce their own food by photosynthesis like plants nor ingest it like animals. They obtain food as heterotrophic bacteria do: they secrete digestive enzymes, which hydrolyze the organic matter around them into small organic molecules and minerals. The fungus absorbs these through its cell walls and plasma membranes. Fungi may be **saprobes,** which absorb food from dead organic matter; or symbionts, either mutualists or **parasites,** which obtain food from the living bodies of other organisms.

The area around a fungus is enriched by digestive products it does not absorb and by its metabolic wastes, which may include organic molecules useful to other organisms. Humans add fungal cultures to certain foods to increase their nutrient content.

### Body Plan

The absorptive lifestyle of fungi is intimately linked to two important characteristics: production of spores and mycelial growth. A **spore** is a tiny, usually haploid, cell that disperses the fungus to new habitats, usually by floating through the air. The production of many tiny spores increases the chance that at least a few will fall onto a suitable food source. The spore then germinates and grows into a thread-like **hypha,** which rapidly branches and forms a **mycelium,** a mass of many hyphae (Figure 26–2).

The mycelium, with its high surface-to-volume ratio, is well suited to absorbing food. A hypha grows from the tip

and releases chemicals that cause other hyphae to grow away from it. This reduces competition between hyphae and permits the fungus to spread and obtain food efficiently. Parasitic fungi absorb nutrients from the host's body fluids.

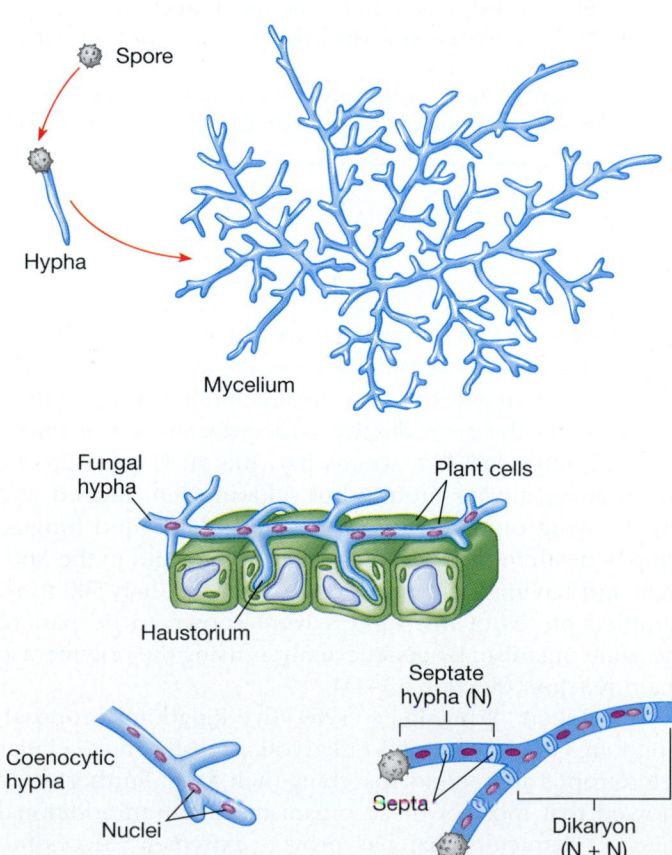

**FIGURE 26–2**   Some basic features of fungal anatomy, described in the text.

Parasites of plants may produce **haustoria** (singular: **haustorium**): specialized hyphae that penetrate the plant's cell walls and lie against the plasma membranes, where they can absorb food.

Some hyphae are **coenocytic,** with many nuclei lying in the same cytoplasm. Others are divided into compartments by cross-walls called **septa** (singular: **septum**). Each compartment contains one or more nuclei. The septa are often incomplete (perforated), permitting cytoplasm and organelles such as nuclei, ribosomes, and mitochondria to move between compartments. A **dikaryotic hypha,** or **dikaryon,** contains two (haploid) nuclei between each pair of adjacent septa; it is usually symbolized as N + N to distinguish it from uninucleate haploid (N) and diploid (2N) cells. A dikaryon generally forms as a preliminary to sexual reproduction: two hyphae fuse but their nuclei remain separate, so that each cell in the resulting new hyphae has one nucleus from each parent.

Fungi have rigid walls containing fibrils of chitin embedded in a matrix of other polysaccharides and short proteins. Chitin is a polymer of a nitrogen-containing glucose amine (see Figure 3–14). (Chitin is most familiar as the basic material in the external skeletons of arthropods—insects, spiders, crabs, and their kin.)

## Reproduction

Fungi may spread by **vegetative reproduction,** that is, by growth or fragmentation of the mycelium. Spores may be formed asexually or as a result of sexual processes. Spores are often produced on aerial structures that hold them away from the food source, out where an air current wafts them to a new home.

The parts of a fungus we normally see are reproductive structures. If you use a microscope to examine the green, white, pink, or black fuzz on moldy food or diseased plants, you will see that it consists of masses of hyphae tipped with strings or spherical clusters of thick-walled spores. Some spores form inside **sporangia,** which rupture and release the mature spores. Many species form **conidia,** asexual spores formed when the end of a hypha pinches into compartments, each containing a single nucleus (Figure 26–3). Mushrooms and cup fungi are **fruiting bodies:** large, complex, sexual reproductive structures, composed of many hyphae. Fruiting bodies are produced by vegetative (feeding) mycelia hidden in the food source.

Sexual reproduction occurs between compatible hyphae. In some species of fungi, this means mycelia of different **mating types,** the equivalent of sexes in other organisms. Individuals of different mating types are alike in structure but somewhat different in chemistry. Mating types are usually designated as + and −. Some fungi have more than two mating types: up to 20 are known in some species. Other species have no distinct mating types, and sexual reproduction may take place between any two mycelia, even different hyphae of the same mycelium.

Fungi do not produce and release gametes. Rather, sexual union occurs when two compatible hyphae grow together: the cytoplasm merges, and after a variable delay, nuclei derived from different hyphae fuse (fertilization) and then undergo meiosis. A mycelium may reproduce sexually with more than one neighbor it meets during its growth.

(a)

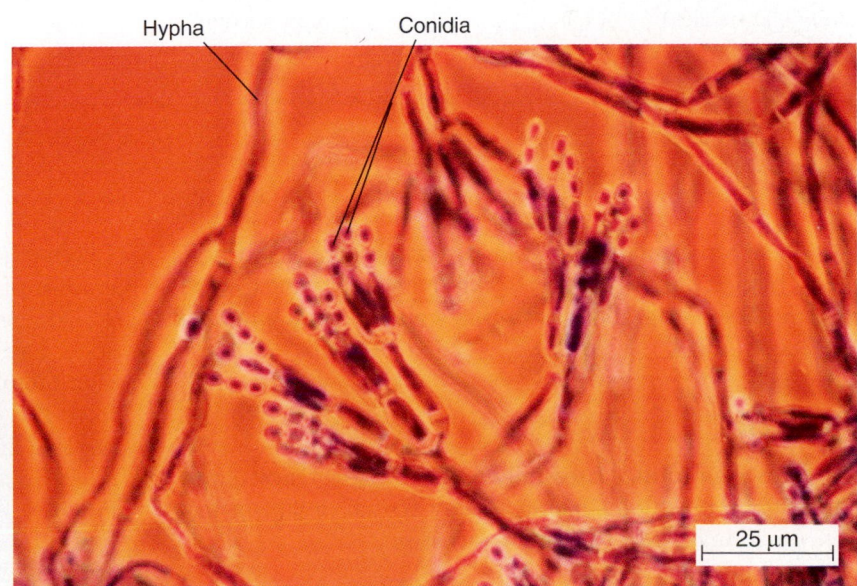

Hypha    Conidia

25 µm

(b) *Penicillium* with conidia

**FIGURE 26–3** *Growth and reproduction of Penicillium.* (a) Vegetative parts of a mycelium are growing within this orange, turning it to mush. The white fuzz consists of immature reproductive hyphae that have not yet formed spores. The oldest, most mature aerial hyphae in the center have formed conidia, which impart the green color seen here. (b) Hyphae and conidia of *Penicillium* as seen through the light microscope. (b, Biophoto Associates)

## Divisions of the Kingdom Fungi

| Kingdom Fungi | Body usually a mycelium composed of hyphae; cell walls containing chitin; nearly all aerobic; sexual and asexual reproduction by spores; no flagellated stages; mostly terrestrial, some marine; saprobes and parasitic or mutualistic symbionts |
|---|---|

| Division | Sexual Reproductive Structure | Characteristics | Examples |
|---|---|---|---|
| Zygomycota (~765 species) | Zygospore | Sexual reproduction by zygospores; coenocytic hyphae; asexual spores produced in sporangia | *Rhizopus* (black bread mold), *Entomophthora muscae* (parasite on the housefly), *Pilobolus* (a dung fungus) |
| Ascomycota (~28,650 species) | Ascus → Ascospores | "Sac fungi." Sexual reproduction by ascospores formed in sac-like asci; some unicellular (yeasts); hyphae of multicellular forms divided by perforated septa; short dikaryotic stage before sexual reproduction | *Neurospora crassa* (pink bread mold), *Saccharomyces cerevisiae* (bread and wine yeast), *Claviceps purpurea* (ergot disease of grasses), *Peziza* (a cup fungus), *Ceratocystis ulmi* (Dutch elm disease), *Cryphonectria parasitica* (chestnut blight fungus), *Aspergillus* (a common mold on foods; one species used to ferment beans for *shoyu*, soy sauce); *Penicillium* (various species used for production of penicillin and of Roquefort and Camembert cheeses) |
| Basidiomycota (~16,000 species) | Basidiospores Basidium | "Club fungi." Sexual reproduction by basidiospores borne by club-shaped basidia; septate hyphae; long dikaryotic stage in life cycle | Mushrooms and toadstools, bracket fungi, puffballs, rusts, and smuts. Mushrooms (poisonous *Amanita* species as well as *Agaricus campestris,* the mushroom sold in grocery stores); *Puccinia graminis* (wheat rust); *Ustilago* (smuts of corn, oat, wheat, etc.) |
| Deuteromycota (~17,000 species) | | "Imperfect fungi." No known sexual reproduction | *Microsporum gypseum* (causes ringworm of dogs), *Epidermophyton floccosum* (athlete's foot), *Botrytis cinerea* (spear rot of asparagus) |

Sexual reproduction often involves the secretion of hormones by one or both partners, with hyphae growing toward one another along the hormone's concentration gradient until they touch.

## Life Cycles

Many fungi reproduce asexually for many generations, and many also reproduce sexually. Every sexual organism has a life cycle involving two key processes:

$$\text{meiosis: } 2N \longrightarrow 4\ (N)$$

A diploid (2N) cell divides twice, forming four haploid (N) cells with new genetic assortments.

$$\text{fertilization: } N + N \longrightarrow 2N$$

Two haploid cells or nuclei (gametes) unite, forming a diploid cell (zygote) with a new gene combination.

Taken together, these processes reshuffle genes in each generation of sexual reproduction.

Many eukaryotes spend most of their lives in the diploid state, like humans, or in the haploid state, like the green alga *Ulothrix*. Many other algae, some animals, and all plants have life cycles of alternating generations of multicellular haploid and diploid individuals (see Figure 25–24).

Some fungi add a further wrinkle to life-cycle patterns. Fungi begin life by growing from haploid spores into haploid hyphae. Then, two compatible haploid hyphae meet and fuse. The resulting mycelium grows as a (N + N) dikaryon, with each cell containing two separate haploid nuclei, one derived from each parent hypha. After living together like this for a time, sexual reproductive structures are formed, and pairs of nuclei take the final plunge: they fuse, forming zygotes, the only truly diploid stage in the fungal life cycle.

## 26-B CLASSIFICATION OF FUNGI

Because fungi have a stringy body form and no flagella, it is assumed that they evolved from protists with similar features, perhaps filamentous Rhodophyta (red algae) or Conjugaphyta that lost their chloroplasts (Sections 25–Q and 25–S). On the other hand, ribosomal RNA sequence data indicate that fungi are more closely related to choanoflagellates and animals than to any extant algae (see Figure 25–5).

The oldest known fossils that are certainly true fungi are found in plant tissue from the Devonian period (about 370 million years ago). This is surprisingly recent when we consider the simple structure of most fungi, but not when we take into account that fungi are essentially land organisms. They probably could not have evolved extensively until the emergence of land plants, which are the major terrestrial food source for decomposers and the hosts of many symbiotic fungi.

Because fungi were originally placed in the plant kingdom, fungal taxonomy follows the rules for plants, which use the term **division** instead of phylum in the taxonomic hierarchy. The names of fungal divisions end in –mycota (mykes = fungus). The divisions are based on, and named after, their members' characteristic sexual reproductive structures (or lack thereof), shown in Table 26–1. These sexual structures are one of the few consistent and conservative features in each division.

It is generally thought that Zygomycota is the most primitive division of fungi and gave rise to the other two sexual divisions.

## 26-C ZYGOMYCOTA

Zygomycotes include many saprobes, some parasites, and some forms that live in mutualistic symbioses with plant roots (Section 26–I). Zygomycota have coenocytic hyphae, forming cross-walls only between the vegetative mycelium and reproductive structures. Asexual spores are formed inside sporangia. Sexual reproduction is characterized by formation of diploid zygote nuclei enclosed in thick-walled **zygospores.**

The common zygomycote *Rhizopus* (black bread mold) usually reproduces asexually by means of spores shed from sporangia on aerial hyphae. Sexual reproduction, a rare event, occurs only when hyphae of + and − mating types meet. Each forms a **gametangium** (gamete-producing structure), closed off from the rest of the mycelium by a cross-wall and containing many nuclei. The two gametangia then fuse, and nuclei of opposite mating types pair and fuse, forming many diploid zygote nuclei. The fused gametangia become covered with a thick protective wall enclosing many identical zygote nuclei; this resistant **zygospore** enters a period of dormancy. Meiosis occurs when the zygospore germinates. It then sends up an aerial hypha, which forms a sporangium and releases haploid spores (Figure 26–4).

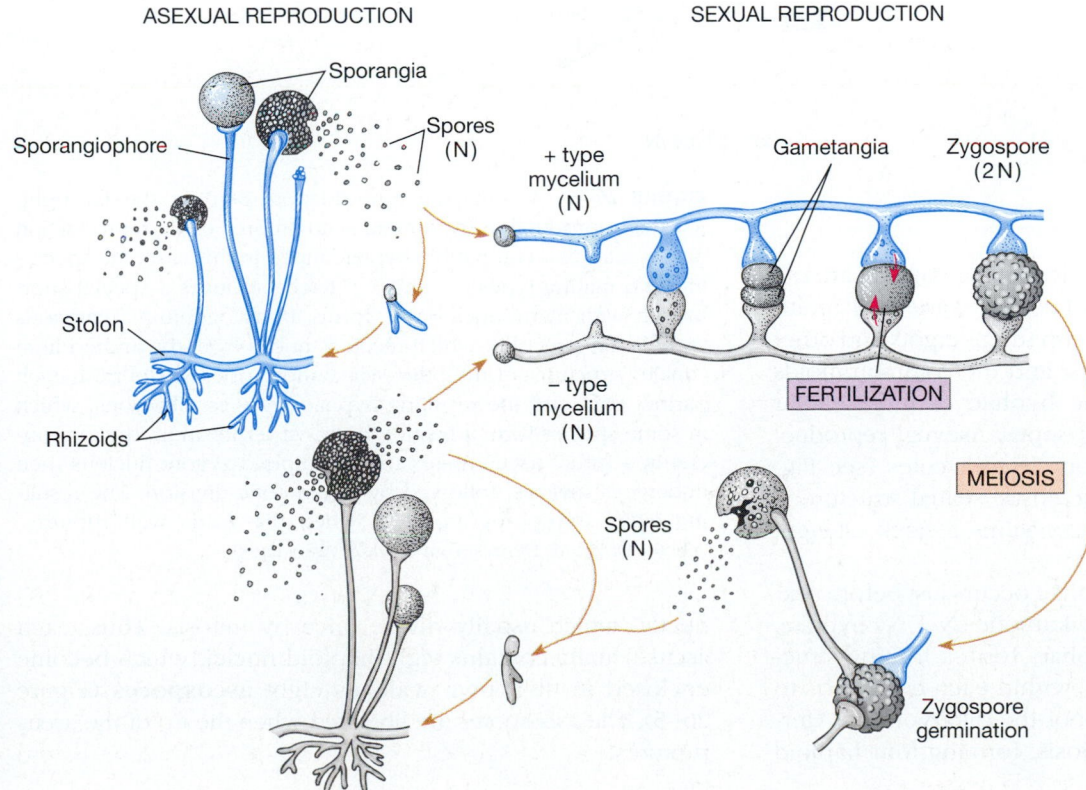

ASEXUAL REPRODUCTION    SEXUAL REPRODUCTION

**FIGURE 26–4** Growth and reproduction in *Rhizopus* (black bread mold). The vegetative hyphae are stolons, on the surface of the food source, and rhizoids, penetrating into the food. Sporangiophores are aerial hyphae bearing sporangia. Spores are shed into the air, which wafts them to new habitats. The mycelium is colorless; the common name of this organism comes from the masses of black spores it produces. The organism usually reproduces by this asexual process.

Sexual reproduction occurs between haploid (N) mycelia of opposite mating types (+ and −). The many zygotes in the zygospore are the only diploid (2N) stage.

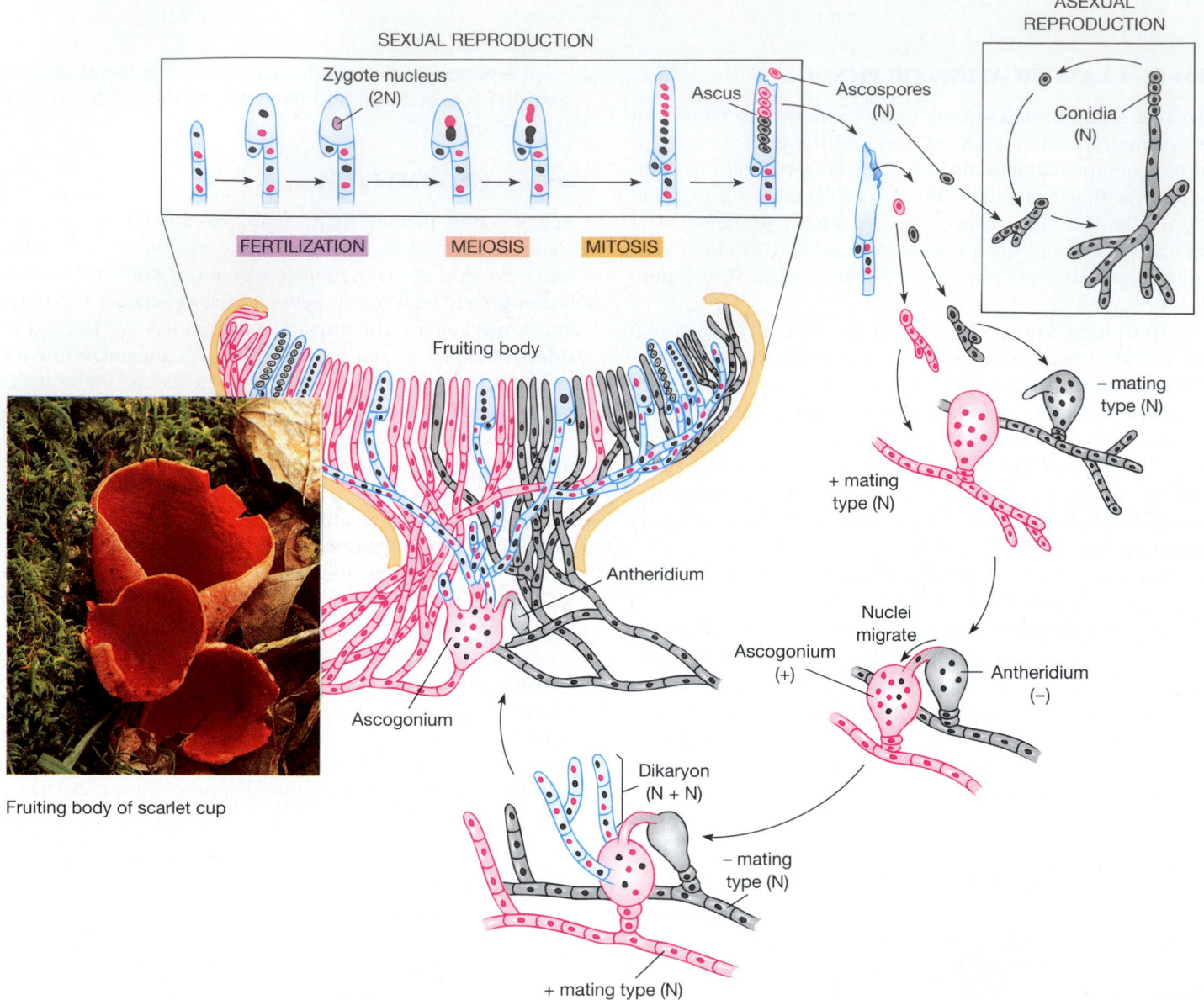

ASEXUAL
REPRODUCTION

Zygote nucleus
(2N)

Ascus

Ascospores
(N)

Conidia
(N)

**FERTILIZATION**      **MEIOSIS**      **MITOSIS**

Fruiting body

Antheridium

+ mating
type (N)

− mating
type (N)

Nuclei
migrate

Ascogonium
(+)

Antheridium
(−)

Ascogonium

Dikaryon
(N + N)

− mating
type (N)

Fruiting body of scarlet cup

+ mating type (N)

## 26–D ASCOMYCOTA

Among the more than 30,000 species of Ascomycota are the yeasts, which are unicellular, and a great variety of multicellular forms, including morels, cup fungi, ergots and other plant parasites, powdery mildews, and the common molds *Aspergillus* and *Penicillium*. The hyphae of multicellular forms are divided by perforated septa. Asexual reproduction by conidia is common among ascomycetes (see Figures 26–1 and 26–3). The characteristic sexual structure is an **ascus** ("sac"), which usually contains a stack of eight haploid **ascospores.**

The fusion of compatible hyphae occurs just before sexual reproduction. The resulting dikaryotic (N + N) hyphae, together with sterile haploid hyphae, form a fruiting structure typical of the species. Here, within each cell that is to become an ascus, the two nuclei of the dikaryon fuse (fertilization) and then undergo meiosis, forming four haploid

**FIGURE 26–5** Ascomycote reproduction. Start at the far right. Asexual reproduction by conidia is common. Sexual reproduction occurs after two compatible hyphae meet. In this case, the species has two mating types (+ and −). Each produces a special short branch with many nuclei: one forms an ascogonium, analogous to a female structure, which receives nuclei from the antheridium ("male" structure) of the other, via a short tube. Nuclei from each partner pair, and the resulting hyphae grow as dikaryons, which in some species form a fruiting body. At fertilization, the two nuclei in a future ascus unite, and the diploid zygote nucleus then undergoes meiosis, followed by one mitotic division. The resulting eight ascospores are shed when the ascus wall ruptures. (photo, David M. Dennis/Tom Stack & Associates)

nuclei, which usually divide once by mitosis. Thus, each ascus usually contains eight haploid nuclei, which become enclosed in their own walls as eight **ascospores** (Figure 26–5). The ascospores are liberated when the tip of the ascus ruptures.

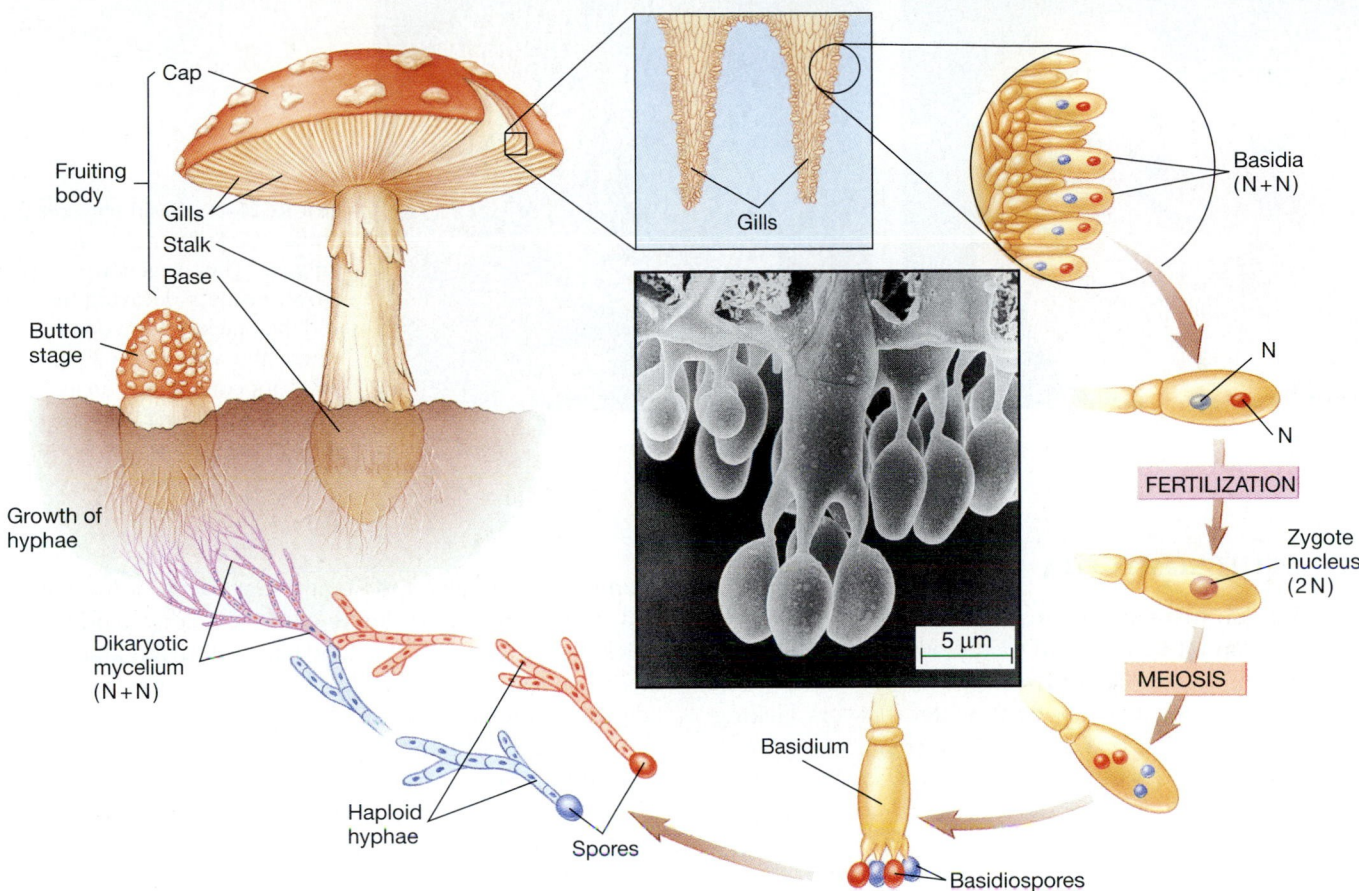

**FIGURE 26–6** Reproduction of a mushroom. A mushroom is the fruiting body of a basidiomycote, made up of masses of dikaryotic hyphae (N + N). The vegetative, feeding mycelium, also made up mostly of dikaryotic hyphae, remains underground. The mushroom's gills produce many basidia with basidiospores along the outer surfaces. Fertilization occurs in each basidium, followed by meiosis of the resulting zygote nucleus (2N), producing four haploid (N) basidiospores. These germinate into new haploid hyphae. (photo, Biophoto Associates)

## 26–E  BASIDIOMYCOTA

Like ascomycotes, basidiomycotes have perforated septa dividing their hyphae into compartments, and they begin life as haploid mycelia, growing from spores. However, compatible mycelia soon fuse, and basidiomycotes spend most of the life cycle in the dikaryotic (N + N) state.

The sexual reproductive structure is the **basidium** ("club"; plural: **basidia**), produced at the tip of a dikaryotic hypha. Within each basidium, the two haploid nuclei fuse, forming the only diploid nucleus in the life cycle. Meiosis follows immediately, forming four haploid nuclei. These move to the outer edge of the basidium, and here the cell wall forms finger-like extensions into which the nuclei move, forming **basidiospores** attached only by a delicate, easily broken stalk. Once released, the basidiospores are wafted away on the slightest air current.

Mushrooms are the most familiar basidiomycotes. In these fungi, a well-fed mycelium forms an underground mass of hyphae that differentiates into a bulbous base, a stalk, and a knob-like cap. Some morning after a heavy rain we awake to find that the hyphae of the stalk have swelled with moisture and elongated, lifting the cap above ground. The cap opens like a belated umbrella, and hundreds of basidia along the edges of the gills or pores beneath the cap shed their spores (Figure 26–6).

The Basidiomycota include rusts, smuts, mushrooms, puffballs, bracket fungi, and coral fungi (Figure 26–7).

(a)

(b)

**FIGURE 26-7** Basidiomycote fruiting bodies. (a) Sulfur polypore, a bracket fungus. The vegetative mycelium grows inside a decaying tree stump, and the brackets form on the surface, where the spores are shed into the air. (b) A coral fungus in a Malaysian rain forest. (a, Grant Heilman; b, George Loun/Visuals Unlimited)

## 26-F DEUTEROMYCOTA

Division Deuteromycota contains a rummage-sale collection of fungi that cannot be assigned to any other group because their sexual reproduction (if they have any) has never been observed. Borrowing from botany, where flowers lacking either male or female parts are termed "imperfect," deuteromycotes are called "imperfect fungi." They have septate hyphae and are thought to comprise mostly ascomycotes and perhaps some basidiomycotes that lost the ability to produce sexual structures. In the laboratory they have been shown to go through poorly understood **parasexual cycles,** in which hyphae fuse and eventually produce true-breeding offspring with new gene combinations.

Members of this group include the organisms that cause ringworm, athlete's foot, and other skin infections. Other deuteromycotes cause important diseases of crops, including root rots, strawberry leaf blight, and bitter rot of grapes.

## 26-G FUNGI AS DECOMPOSERS

All fungi are heterotrophic, so they must obtain their food from outside sources. The thousands of species of fungi have adapted to a variety of lifestyles. One of their ecological roles is to break down organic matter: dead animals and animal wastes, dead plants, and fallen leaves (Figure 26-8).

Fungi are not the only organisms that consume dead organisms. Large animals such as crows, vultures, coyotes, and hyenas consume a lot of carrion (dead animals). Many insects and other invertebrates also feed on dead plants and animals, but fungi and bacteria are the most important decomposers. Both are also being explored as ways to break down environmental pollutants such as chlorinated organic compounds.

Fungi and bacteria are adapted to different conditions. For instance, some fungi can grow in relatively high con-

centrations of salts and sugars that would kill most bacteria. Many fungi also tolerate extreme acidity; acid foods like pickles and jams (fruit is acid) are safe from attack by bacteria but not by fungi. Because fungi can absorb water from damp air, they can grow in environments where there is no liquid water. On the other hand, many bacteria tolerate anaerobic conditions better than most fungi do. Although fungi such as yeasts can survive under anaerobic conditions using fermentation, neither they nor other fungi can grow and reproduce without oxygen. This is why some items normally broken down by fungi do not decompose in landfills.

Unlike parasitic fungi, which cause many diseases, most saprobic fungi are beneficial to humans, and even those that form molds on food are usually harmless. However, some saprobic fungi cause considerable inconvenience by consuming such unlikely substances as leather, hair, wax,

**FIGURE 26-8** The fruiting bodies of bird's-nest fungi. The mycelium decomposes hard-to-digest lignin, a complex substance in wood. (Michael Fogden/DRK Photo)

cork, and polyvinyl plastics. Although their action in the compost heap is convenient, fungi are a nuisance when they break down telephone insulators, clothes, shoes, books, or rafters. During their long years of global supremacy, the Spanish and British navies lost more ships to wood-rotting fungi than to enemy action. It was quite common for the bottom of a ship to fall out in mid-ocean!

## 26–H FUNGAL DISEASES

Some fungi seem to have evolved from saprobes, living on dead matter, through intermediate forms living on almost-dead matter such as injured tissues, to parasites, which absorb nutrients from living organisms.

There are hundreds more fungal diseases of plants than of animals. Fungi can easily enter plants through injured areas or stomatal pores of leaves. Air spaces within leaves permit the fungus to obtain plenty of oxygen and still send haustoria into the leaf cells for food. When the well-fed mycelium is ready to reproduce, hyphae may grow out through the stomata, or a clump of sporangia may form inside the leaf, grow, and finally rupture the host's epidermis and release the spores into the air (Figure 26–9). Fungi spread rapidly to other plants in this way and may decimate crops or tree populations, especially during unusually wet seasons. On the other hand, these same features make fungi good candidates for biological control of weeds, provided researchers can find fungal parasites highly specific for the target plants.

Some pathogenic (disease-causing) fungi are spread by insects. Because insects can specifically search out new host plants of the correct species, this is a more efficient means of fungal dispersal than hit-or-miss air travel. For example, ambrosia beetles tunnel in trees, introducing fungi that grow in the wood and provide the beetles with food. Female *Sirex* wasps even have special fungus incubator pouches. These wasps inject fungal fragments into weakened pine trees along with their eggs, using egg-laying appendages that are several centimeters long. In this way the fungus bypasses the tree's natural defenses in the bark and grows in the wood, aided by toxic wasp mucus that inhibits the tree's production of defensive **fungicides** (fungus-killers). As the fungus grows it softens the wood and releases vitamins and sterols. The developing wasp larvae thrive on their diet of enriched wood pulp.

The fungus *Ustilago violacea* infects a roadside weed, catchfly, causing female plants to develop male rather than female parts, and plants of both sexes to produce fungal spores rather than pollen in the male parts. Insects attracted to the flowers then carry spores on to new hosts. In this case the fungus is not mutualistic with the insect but is parasitizing the symbiotic relationship between flower and pollinator.

Animals contain fewer air spaces than plants, making their bodies less hospitable environments for aerobic fungi.

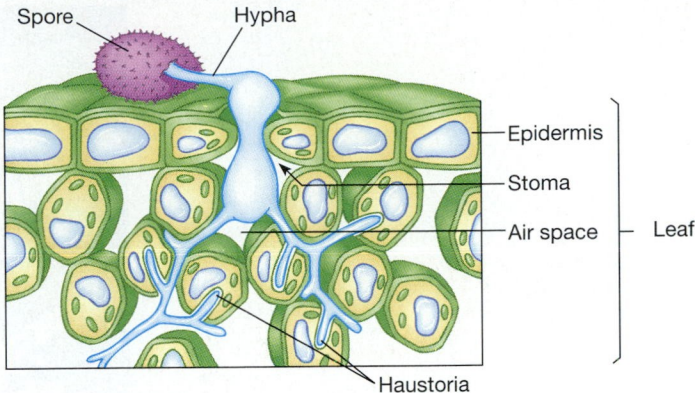

(a) Fungus invading leaf

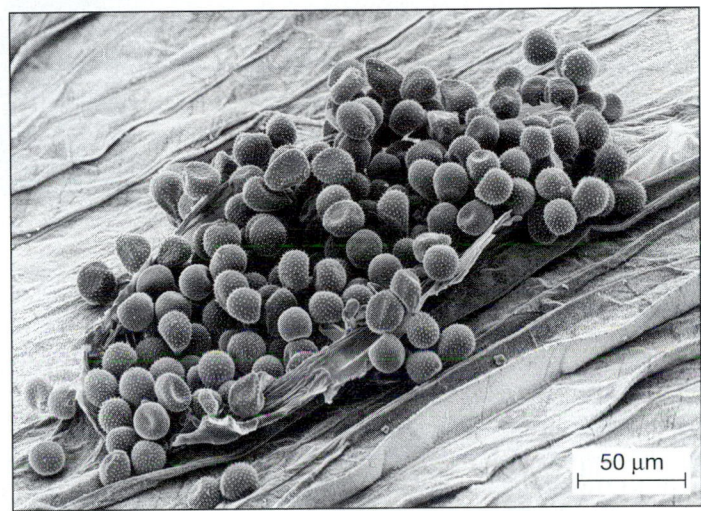

(b) New spores rupture epidermis (SEM)

**FIGURE 26–9** Fungal parasites of plants. (a) Fungi may enter plants through their stomata and grow through interior air spaces, sending haustoria into some of the internal photosynthetic cells. (b) A cluster of asexual spores of wheat rust bursts through the plant's epidermis and scatters, infecting other plants. (b, Biophoto Associates)

When fungi do invade animals, they attack areas exposed to the air, such as the skin, lungs, and mucous membranes (Figure 26–10). The body surface also tends to be slightly cooler, and most fungi do not grow well at body core temperature.

Ordinarily, the skin and mucous membranes provide a healthy animal with considerable defense against penetration by fungi. For example, human sweat contains fatty acids that serve as a natural fungicide. Sick or injured animals are more susceptible to fungi. When the immune system is severely damaged, a fungal infection may become **systemic,** spreading though the interior of the body, an extremely serious condition.

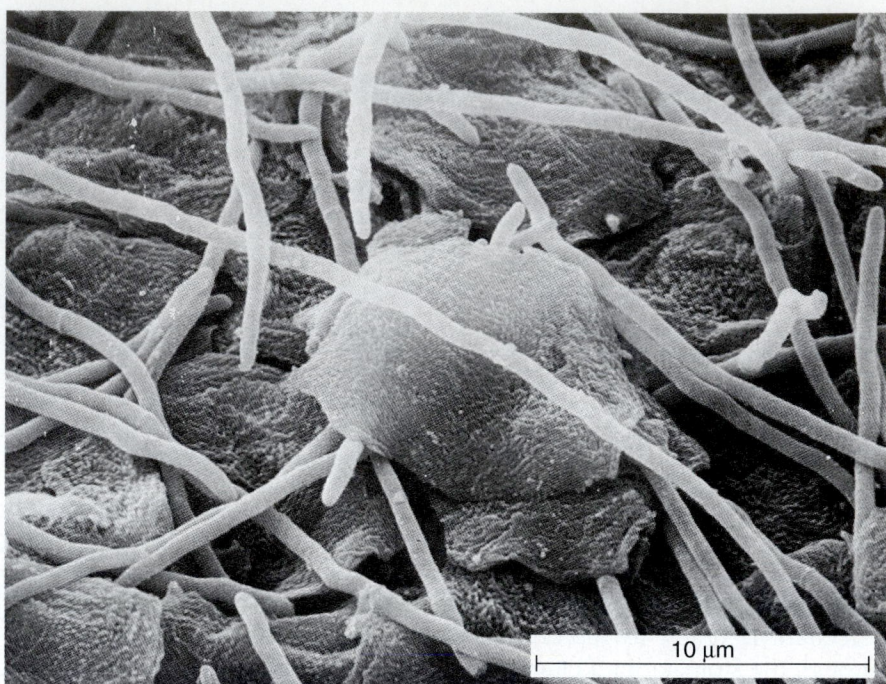

**FIGURE 26–10**   Fungal infection of human skin. Slender, tubular hyphae of *Trichophyton interdigitale* (Division Deuteromycota) grow over and under flat human epidermal cells.   (Biophoto Associates)

10 μm

## Treatment of Fungal Diseases

Fungal diseases of plants rank among the greatest threats to human welfare. Hundreds of millions of dollars are spent each year to protect crops by treatment with fungicides and to develop new fungicides and new strains of crops resistant to fungal attack. Even so, the average usefulness of a new agricultural fungicide is estimated at only four years because most fungi rapidly evolve resistance to chemical controls.

Chemicals that combat bacterial diseases are easier to develop than fungicides. This is mainly because it is much easier to duplicate growth conditions for parasitic bacteria in the laboratory. Even so, many agents that cure or prevent fungal diseases have been produced.

Many drugs are useful in destroying both fungi and bacteria. For instance, the sulfonamides (such as sulfanilamide) prevent the formation of folic acid, needed for synthesis of purines and hence of nucleic acids. Organisms unable to form nucleic acids eventually die.

A number of antibiotics (substances that destroy organisms) act by preventing protein synthesis. Some are specific, in that they block protein synthesis in prokaryotes but not in eukaryotes. These drugs can be used against bacteria but not against fungi. Other drugs, such as cycloheximide and puromycin, block protein synthesis in the cells of both parasite and host (Section 10–H). They can be used to kill parasites selectively because parasites metabolize and divide so much faster than host cells. Slowing the parasite down permits the host's natural defenses to prevail. The fungicide nystatin binds to the membranes of some fungi

and changes their permeability, killing them by leakage of ions. All of these drugs can be used to treat fungal infections of animals, but they are too expensive to be used for treating crops.

Besides these antibiotics, many more general antifungal agents are known. Sodium propionate is a fatty acid salt used by bakeries to inhibit growth of mold in bread. It is also used to treat fungal infections of the skin, as are sulfur, benzoic acid, salicylic acid (aspirin), gentian violet, and potassium permanganate.

## Prevention of Fungal Diseases

Some fungal diseases of animals can be controlled by prolonged medical treatment. However, once a fungus becomes established in a plant or an animal, or in wood, paper, or leather goods, it is virtually impossible to eradicate. Therefore, measures that prevent spores from germinating are the most effective defenses against fungi.

The discovery of chemical sprays that kill fungal spores in crops was a milestone in the battle between humans and fungi for food. In the 1860s most of the grape vines in France were torn up and grafted onto rootstocks of American vines to protect them from root-feeding aphids: small, soft-bodied insects that suck the juices of plants (see Figure 20–3). The aphids had almost destroyed the French wine industry when they were accidentally introduced into that country from North America. Unfortunately, the American vines also imported *Plasmopara viticola*, downy mildew of grapes (an oomycote, not a true fungus). This

**FIGURE 26–A** *Amanita muscaria,* fly agaric. (Biophoto Associates)

Many fungi are famous for producing toxic chemicals. Poisonous mushrooms such as the death angel or fly agaric spring immediately to mind (Figure 26–A). Because they kill those unwise enough to eat them, the obvious assumption is that the toxins are an adaptation against predation. But when we look at more cases we find that toxins play various roles in fungal survival.

For example, the toxins of some mushrooms are a means of supplementing a diet of nutrient-poor rotting wood. The potent toxins inactivate passing roundworms (Section 28–G) within 30 seconds, and then the fungal hyphae grow rapidly in through the worm's mouth. A day later, the worm has been digested and its nutrients, particularly nitrogen from proteins, have been absorbed into the mycelium of the carnivorous fungus.

The ascomycote *Claviceps purpurea* infects the flowers of rye and other cereals and produces a structure called an **ergot** where a seed would normally be found in the head of grain (Figure 26–B). The ergot produces many extremely toxic substances, and humans may be poisoned by eating bread made from infected rye. Ergotism caused the death of thousands in medieval Europe. It is also called St. Anthony's Fire, after a famous victim. Its symptoms include fever, convulsions, loss of reason, violent hallucinations, and dry gangrene, with fingers, toes, and limbs turning black and dropping off. From this list we can understand the source of the hideous demons tormenting St. Anthony in the surreal paintings of Hieronymus Bosch, and also ergotism's proposed role in the behavior of the witches of Salem. It is also not surprising to learn that ergot supplied the chemicals from which the hallucinogenic drug lysergic acid diethylamide (LSD) was first synthesized.

Although the substances produced by ergots are deadly in high concen-

trations, in small quantities they are valuable drugs, used to induce labor, control bleeding, ease migraine headache, and treat high blood pressure and varicose veins.

Many plant pathogens produce toxins that aid their growth by weakening or killing the host. However, the toxic alkaloids secreted by the fungus *Epichloe typhina* are not poisonous to the host grasses but to grazing animals. The fungus also makes

the grass unable to reproduce sexually, thereby preventing the production of offspring that might not be susceptible to the fungus. Instead, with no grazers eating it, the grass spreads by vigorous vegetative growth, filling a widening area with more host for the fungus.

In 1927 Alexander Fleming noticed that his bacterial cultures had been killed by the ascomycote *Penicillium notatum* (Figure 26–C). This led to the discovery of the world's most widely used and effective antibiotic, penicillin. The antibiotic presumably serves the fungus that produces it by reducing competition from bacteria; humans now use it for the same purpose.

So we see that fungal toxins not only serve to deter predators, but also catch animal prey as dietary supplements, make diseased hosts better places for fungi to live, and reduce competition from bacteria or other fungi.

**FIGURE 26–B** In the head of grain on the left, ergots have formed in place of some seeds. (Dennis Drenner)

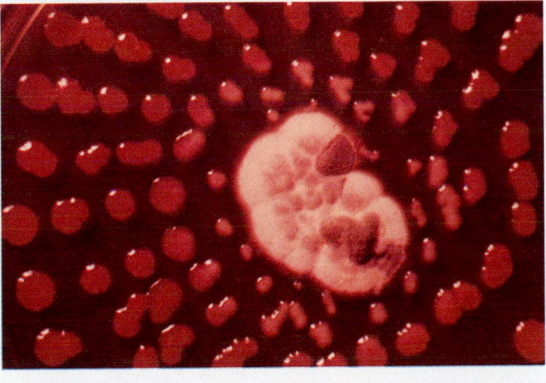

**FIGURE 26–C** Penicillin in action. The white fuzzy mycelium of *Penicillium* produces penicillin, which kills bacteria. Diffusion of penicillin from the fungal mycelium through the agar in the dish has killed the nearest colonies of bacteria (red dots). (Biophoto Associates)

disease almost wiped out the French grape harvest in wet years during the 1870s.

A preventive measure against downy mildew was discovered in 1882 by Pierre Millardet of the University of Bordeaux. A local vineyard owner was spraying his grapes with a vile-looking mixture of copper sulfate and lime to discourage passersby from eating the grapes; Millardet noticed that these vines were free of mildew. He made up a mixture of copper hydroxide and lime, which he called **Bordeaux mixture,** the first and one of the most successful fungicides ever produced by human ingenuity. Germinating spores of many parasitic fungi produce just enough acid to dissolve the very insoluble copper hydroxide. The copper kills the spore, and the host plant is not affected by the remaining undissolved copper.

Another way to fight fungal disease is by breeding resistant strains of plants, a never-ending struggle because fungi soon breed their own counterstrains and spread rapidly once more. Breeding resistant trees is especially difficult because they take so long to mature and reproduce; meanwhile they may have succumbed to fungi. The American chestnut, once a dominant tree throughout the eastern United States, was virtually eliminated by the chestnut blight fungus. Between 1900 and 1950, 3 to 4 billion trees died. Although surviving root systems send up new shoots, the fungus kills them, too, before they can reproduce. Researchers have had no luck breeding resistant trees but are now working on a promising therapy: a virus that makes the fungus less virulent.

## 26–I  SYMBIOTIC RELATIONSHIPS OF FUNGI

Although many fungi are harmful parasites, almost one third of known species form mutualistic symbioses, in which both partners benefit. Some of these permit the partnership to live in habitats where neither can survive alone.

### Mycorrhizae

Many fungi grow associated with plant roots in a symbiosis called a **mycorrhiza** (= "fungus root"). The fungal hyphae spread into the soil and use their superior absorptive ability to take up mineral nutrients. By using radioactive elements as tracers, researchers found that the fungi pass on some of these minerals to the plant roots and receive sugars and other nutrients from the plant. Apparently the fungi also supply growth hormones to the roots.

Most plants form mycorrhizal associations. The fungal partners are basidiomycotes, zygomycotes, or ascomycotes. When pine trees were introduced into new areas, such as Puerto Rico and Australia, they grew poorly until supplied with soil from pine forests, containing the appropriate mycorrhizal fungi; after this, they grew rapidly.

The most common mycorrhizae are **endomycorrhizae** (endon = within): the fungus grows in the root's outer layers and sends highly branched hyphae similar to haustoria inside the cell walls, where they lie pressed against the plant cells' plasma membranes. Such associations occur in the majority of plants, usually with zygomycote partners.

In **ectomycorrhizae** (ektos = outside), the fungus forms a sheath several cell layers thick around the outside of the root and also extends hyphae between the cells of the root's outer layers. The participants are usually woody plants and basidiomycotes, particularly some common forest mushrooms or puffballs. Some of these fungi are species-specific, but others can form associations with many different kinds of plants.

Many plants can also form mycorrhizae with more than one species of fungus, and which fungus gets there first may make the difference between life and death. Some fungi help protect their plant partners from acid rain caused by industrial pollution. Acid promotes two changes in the soil unfavorable to plants: leaching (washing away) of required nutrients, making them unavailable to plants; and increased solubility of toxic minerals such as zinc, copper, aluminum, and manganese, allowing them to be carried into roots. The right mycorrhizal fungus can absorb nutrients from depleted soil water and pass them to the plant and can also protect the plant from toxic substances in the soil. Let us hope that mycorrhizal fungi can adapt to local conditions as fast as the pathogenic and parasitic fungi which also invade an area, killing trees, in the wake of acid rain.

Some plants are even more dependent on fungi. Orchid seeds will not germinate unless infected by their fungal partners. Fungi extend throughout the body plus the outer layers of the seeds in members of the family Ericaceae: azaleas, rhododendrons, blueberries, and their kin. Substances secreted by these fungi are necessary to induce root growth.

### Lichens

A **lichen** looks like a plant (Figure 26–11). However, it is really an intimate symbiosis between two kinds of organisms: an ascomycote (or more rarely, basidiomycote) fungus and photosynthetic cells—either cyanobacteria or green algae (Sections 24–D and 25–R). The fungus obtains organic compounds from the photosynthetic partner, but it is unclear what it provides in return. Possibly it absorbs water and minerals.

The photosynthetic members of lichens are sometimes found living by themselves in nature, but the fungi are usually not, although in the laboratory they can be grown in pure culture (that is, with no other living organisms present). More than one sixth of known fungal species are lichenized. The association certainly extends the available habitat: a lichen may be found in areas where neither partner can grow alone.

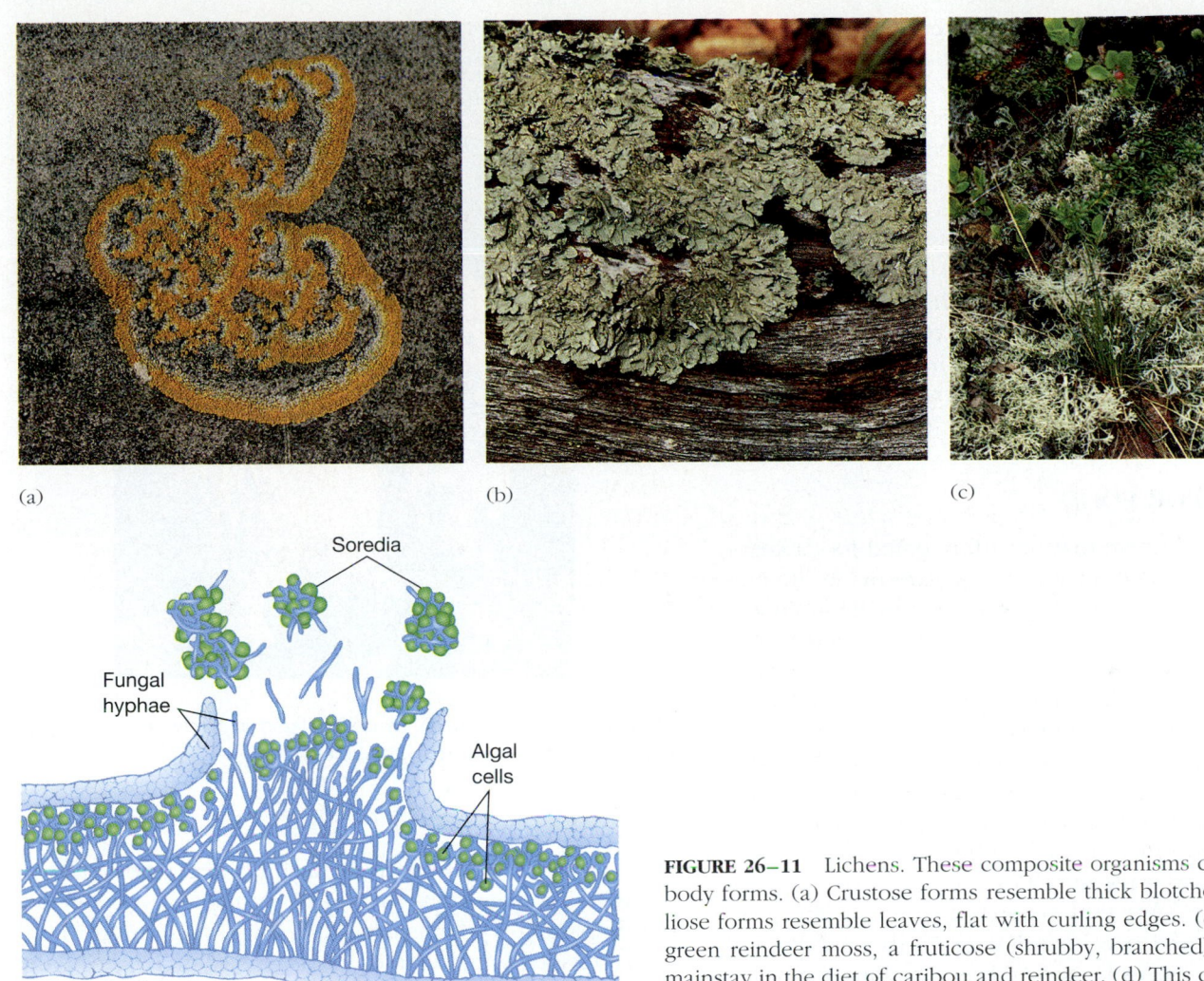

(a)

(b)

(c)

Soredia

Fungal
hyphae

Algal
cells

(d) Lichen structure and dispersal units

**FIGURE 26–11** Lichens. These composite organisms come in three main body forms. (a) Crustose forms resemble thick blotches of paint. (b) Foliose forms resemble leaves, flat with curling edges. (c) A clump of pale green reindeer moss, a fruticose (shrubby, branched) form, which is a mainstay in the diet of caribou and reindeer. (d) This cross section shows how the symbionts are arranged in the lichen body and how they disperse by soredia, clumps containing cells of both partners.   (c, Walter Sharp)

Lichens have very slow metabolism and growth, but they are extremely resistant to drought and cold. They are the most important autotrophs in the low-growing vegetation of the **tundra,** found in the Arctic and also at high altitudes on mountains. Lichens absorb minerals from the air and so can grow without soil, on stones or rocky ocean islands where nothing else survives. Archaeologists have estimated the age of the mysterious stone heads of Easter Island by measuring patches of lichens growing on them.

Because lichens draw their nutrients from the air, they are particularly sensitive to air pollution. Surveys in many parts of Europe showed that 40 to 100% of lichen species once present had disappeared from the vicinity of factories or other sources of pollution, and the survivors were in poor health. In the 1950s, European biologists found that the presence or absence of particular lichens indicates whether or not the air contains pollutants from burning coal. Dozens of lichens are now known to indicate particular pol-lutants. Lichens that die at low levels of air pollution provide early warning of declining air quality.

In other lichens, nutrients from the air are metabolized, and pollutants are bound to cell walls in concentrations that mimic those in the air. Analyzing heavy metals in the lichen indicates the types and concentrations of heavy metals in the air. This is more accurate than analyzing air samples because air quality fluctuates widely depending on factors such as wind direction. Pollution from radioactive fallout conta-minated reindeer moss (a lichen) during open-air atomic bomb tests in the 1950s and 1960s and again after the Cher-nobyl power plant accident in 1986. Radioactivity built to dangerous levels in caribou and reindeer, which eat the lichens, and in northern peoples who depend on these an-imals for much of their meat.

Reproduction is tricky for a composite "organism" like a lichen. Most reproduce by fragmentation, in which small pieces containing cells of both partners break off and blow away. In some, the fungus forms spores that blow away and find the appropriate photosynthetic partner only by luck.

Lichens are important in many ways. They are usually the first organisms to colonize bare rock, and they contribute to breaking down rock into soil. They are also the main food of most tundra animals.

Many lichens have crystalline deposits of unusual organic acids on their surfaces. Their function to the lichens is unknown, but it has been proposed that they protect the cells from high light intensities. Many of these chemicals are used as dyes in Harris tweed and in fabrics from the Orkney and Shetland Islands of northern Scotland.

## 26–J   FUNGI FOR FOOD

People have long known that most fermented foods keep longer than the foods from which they are made. Before refrigerators were invented, this was a compelling motive for making wine, beer, and cheeses using fungi (and sauerkraut, fermented sausages, and yogurt using bacteria). Fermented foods are also often more nutritious, flavorful, and digestible than their raw materials.

Beer is made by fermenting germinated grain (malt), usually barley, flavored with hops. The germinating plant embryo breaks down its starchy food supply to monosaccharides, which are then fermented by yeast.

Brewer's yeast, the unicellular ascomycote *Saccharomyces cerevisiae,* ferments sugar to alcohol in the production of wine and beer (see Figure 7–10). A different strain of the same yeast is used in bread-making. The carbon dioxide it gives off is trapped in the dough as bubbles that give bread its light texture.

English Stilton and French Roquefort, Brie, and Camembert cheeses all get their flavors from specific ascomycotes introduced as part of the production process.

In the Orient, soy sauce (*shoyu*) is traditionally made by fermenting boiled soybeans and wheat with the ascomycote *Aspergillus oryzae* for about a year. This produces a flavorful sauce rich in vitamins and amino acids, a valuable addition to a low-protein diet of rice. (Most modern soy sauce is produced by hydrolyzing soybeans with hydrochloric acid, and the result tastes very different.)

Fungi are also grown commercially as a source of vitamins, amino acids, enzymes, sterols (precursors of steroid hormones), and other organic compounds used in the food and drug industries.

Some fungi that cause diseases are the same as those used in food production. An interesting example is *Botrytis cinerea,* which causes spear rot in asparagus but is greatly valued in some wine-producing areas. Grapes infected with this "noble rot" ferment to produce highly esteemed Sauternes and Barsac in France and Beerenauslesen in Germany and in the eastern United States.

**FIGURE 26–12**   A morel, an edible fruiting body of an ascomycote.   (Matt Meadows/Peter Arnold, Inc.)

Growing the grocery store mushroom, a basidiomycote, is a million-dollar industry in many parts of the world. Several species of wild mushrooms, many of them mycorrhizal forest residents, are harvested as specialty foods in the United States and other countries. The morels and edible truffles are ascomycotes (Figure 26–12). Truffles are underground fruiting bodies of mycorrhizal fungi growing on tree roots. In France, truffle hounds and pigs are trained to hunt them by smell. These valuable fungi sold at $720 per pound in 1990.

## SUMMARY

Members of the kingdom Fungi are eukaryotic saprobes, parasites, and mutualists adapted to obtaining food by absorption. Fungi are ubiquitous on land. Here the saprobes decompose and recycle organic matter, and the parasites cause many diseases in plants and animals. Most land would probably be virtually devoid of vegetation were it not for the mutualistic mycorrhizal fungi of plants and, in harsher environments, the lichens.

Many fungi are economically harmful because they destroy seed, standing crops, and harvested food, as well as buildings and possessions. Fungi also cause disease in humans and other animals. On the other hand, the fungal antibiotic penicillin has probably saved millions of human lives. And what would these and other lives be without the flavor and nutrition fungi add to foods such as bread, cheeses, wine, and beer?

1. The thread-like mycelial bodies of fungi are adapted to penetrate food sources or host tissues. Most fungi are multicellular, with chitinous cell walls. However, one important group, the yeasts, are essentially unicellular.
2. Fungi reproduce and disperse mainly by walled, airborne, asexual or sexual spores.
3. Sexual reproductive structures are the basis for classifying the true fungi into three main divisions: Zygomycota, Ascomycota, and Basidiomycota. A fourth division, Deuteromycota, contains all the fungi for which no sexual reproduction is known. Table 26–1 lists the features of each division and illustrates the typical sexual reproductive structures of the first three.
4. In the life cycles of the three sexual divisions of fungi (Zygomycota, Ascomycota, and Basidiomycota), the mycelia are haploid (N) or dikaryons (N + N) formed by fusion of compatible haploid hyphae; the only truly diploid (2N) cells are the zygotes, which undergo meiosis and form four new haploid nuclei. The resulting new gene combinations become enclosed in spores and disperse to new food sources. New mycelia develop from these spores by mitosis of haploid nuclei.
5. One important ecological role of fungi is as decomposer organisms, which break down dead plants and animals and absorb the resulting small food molecules. Inevitably, however, some of these molecules escape the fungi. Minerals released by fungal breakdown are important to plants, which absorb and reuse these vital nutrients, often by way of their own mycorrhizal fungi.
6. Fungi are also important as parasites, causing diseases of both plants and animals. Most fungal diseases attack plants. Once established, a fungal disease of a plant is often impossible to combat, and fungal diseases of animals can be cured only with much difficulty. Thus, prevention of infection is important in combating fungal diseases.
7. Some fungi form mutualistic, mycorrhizal associations with the roots of higher plants, to which they supply minerals and from which they receive food. Most land plants require such fungal associates for normal growth and health.
8. Lichens are symbiotic associations between fungi and photosynthetic cells. They are important food producers in cold or barren areas and may also help to form soil from bare rock.
9. Fungi are useful to humans in the production of many edible fruiting bodies, of some fermented foods, and of drugs, antibiotics, and various organic chemicals.

## Self-Quiz

1. Commercial mushrooms are grown in straw and horse manure. These mushrooms are:
   a. autotrophic
   b. parasitic
   c. saprobic
   d. chemosynthetic
2. The organism below that would *not* have hyphae is:
   a. *Penicillium*
   b. black bread mold *(Rhizopus)*
   c. yeast
   d. commercial mushroom
3. The bracket fungi found on trees are:
   a. fruiting bodies of mycelia growing hidden in the tree trunk
   b. mycelia absorbing nutrients from the exposed surface of the wood
   c. sporangia
   d. lichens
4. Fungi are most widely spread by:
   a. airborne spores
   b. ingestion and subsequent deposition in the feces of animals
   c. migration of insects with spores or bits of hyphae stuck to them
   d. fragmentation of vegetative mycelia
   e. water currents

5. Fungi are placed into one of four divisions on the basis of:
   a. life cycle
   b. sexual reproductive structures
   c. mode of nutrition
   d. complexity of vegetative structures

Match each description below to all correct divisions:

_____ 6. No known sexual reproduction     a. Ascomycota
_____ 7. Septate hyphae     b. Basidiomycota
_____ 8. Sexual reproduction results in a sac     c. Zygomycota
        usually containing eight spores     d. Deuteromycota
_____ 9. Some members are parasites

10. Fungal pathogens invade a living host by:
    a. digesting away the epidermal layer with their powerful enzymes
    b. growing in through a break in the epidermis
    c. secreting hormone-like substances that cause the host's cells to accept them as part of the body
    d. growing hyphae under a cell of the host and prying it up as a lever would
11. Which is *not* a means of limiting the growth of fungi?
    a. spraying with compounds containing sulfur
    b. refrigeration
    c. dehydration
    d. scrupulous cleanliness
    e. humidification

12. In the mycorrhizal association between a pine tree and a fungus, the fungus:
    a. eventually depletes the tree's mineral supply
    b. secretes toxic materials that inhibit the growth of nearby trees
    c. absorbs nutrients from the soil
    d. converts nitrogen into a form the tree can use

## Questions for Discussion

1. How is the anatomy of fungi related to their way of life?
2. In what ways is a dikaryon similar to a diploid organism? How do the two differ?
3. Compared with unpolluted waters, fresh water heavily polluted with industrial wastes contains few fungi. How might this affect the life in a lake or stream?
4. The mycorrhizal fungus of pines cannot produce the polysaccharide-digesting enzymes characteristic of most soil microbes. What is the adaptive significance of this?
5. Pine trees with mycorrhizae can take up two or three times more phosphorus, nitrogen, and potassium than those without. Explain why they also take up more water and distribute carbohydrates faster.
6. Is it valid to conclude that *Penicillium* secretions aid the fungus by reducing competition from bacteria in nature because they do so in the laboratory? What experiments could you do to investigate this question?
7. Reread the discussion of the origin of multicellularity and cell specialization in Section 25–A, and apply it to what you have learned in your study of fungi.

## Suggested Readings

Bruemmer, F. "In praise of the lowly lichen." *International Wildlife,* November–December 1991. Fascinating account of the biology and uses of lichens.

Hansen, J. "Let them eat truffles." *Science 80,* December 1980. About the modern truffle industry.

Hudson, H. J. *Fungal Biology.* London: Edward Arnold Ltd., 1986. Ecological relationships of fungi.

*Industrial Microbiology. Scientific American,* September 1981 issue.

Jensen, W. A., and F. B. Salisbury. *Botany: An Ecological Approach,* 2d ed. Belmont, CA: Wadsworth, 1984. The chapters on fungi show how intriguing these organisms can be.

Kendrick, B. "Fungal symbioses and evolutionary innovation." In L. Margulis and R. Fester, eds. *Symbiosis as a Source of Evolutionary Innovation: Speciation and Morphogenesis.* Cambridge, MA: MIT Press, 1991.

Moore-Landecker, E. *Fundamentals of the Fungi,* 3d ed. Englewood Cliffs, NJ: Prentice-Hall, 1990. A comprehensive textbook with many interesting examples.

Newhouse, J. R. "Chestnut blight." *Scientific American,* July 1990.

Strobel, G. A. "Biological control of weeds." *Scientific American,* July 1991.

# The Plant Kingdom

## OBJECTIVES

*When you have studied this chapter, you should be able to:*

1. Define the following terms, and use them correctly:
   vascular tissue, cuticle, stomata
   antheridium, archegonium, sporangium
   rhizoid, rhizome, microphyll, megaphyll
   heterospory, pollen, pollination, seed
   stamen, carpel, fruit
2. Contrast water and land as environments for plant life.
3. List the problems faced by plants living on land that are not faced by algae living in water, and for each problem name the adaptation(s) allowing plants to survive on land.
4. List the characteristics of plants and explain why plants are believed to be descended from green algae.
5. List the adaptations of bryophytes that allow them to live on land and the characteristics that restrict them largely to moist land environments.
6. Trace the evolutionary advances shown by lycophytes (club mosses) and sphenophytes (horsetails), ferns, gymnosperms, and angiosperms.

7. Compare and contrast any of the groups mentioned in Objectives 5 and 6 with respect to structure, reproduction, and life cycle.
8. Given a plant from any of the groups listed in Objectives 5 and 6, name the group to which it belongs.
9. (a) Briefly explain what alternation of generations is;
   (b) define each of these stages in the plant life cycle, explain its relation to the other stages, and state whether it is diploid or haploid: spore, gametophyte, gamete, zygote, embryo, sporophyte;
   (c) point out when in the plant life cycle meiosis and fertilization occur;
   (d) trace the increasing dominance of the sporophyte and reduction of the gametophyte during evolution of vascular plants, giving examples, and discuss the selective value of this shift;
   (e) correctly identify any plant as a sporophyte or gametophyte, and point out or explain where the opposite generation would be found.

Plants led the last great wave of evolution, the colonization of land. These green pioneers emerged from the water about 420 million years ago: a remote age that nevertheless was nearly 90% of the way from the origin of life to today. By then, the water teemed with hungry animals and with algae competing for light and nutrients. But the beckoning sun lit a terrain inhabited only by patches of cyanobacteria, algae, and possibly fungi: literally a land of opportunity. All these factors selected for the hardy descendants of some multicellular green algae that moved onto the shore and later spread to drier inland areas. These early plants provided abundant food, which in turn stimulated the evolution of fungi and lured many kinds of animals onto land. The history of life on land is a tale of successive bursts of adaptive radiation by groups of plants and animals with new adaptations to terrestrial existence.

Plants probably evolved from green algae living at the edges of lakes or streams, where fluctuating water levels sometimes left them exposed. At other times, heavy rains might sweep them away. These conditions would select for several adaptations: structures anchoring the body in place; division of labor among cells, supplying unlit lower cells with food from photosynthetic ones above; a sizable body storing water against dry spells; and a lipid surface coat reducing evaporation of this water into the air.

These features can be considered **preadaptations** that made the move to land possible. They were selected for because of their advantages in the existing habitat. However, like many other adaptations, they happened to have more than one advantage, in this case permitting the organism to exploit the untapped resources of a new habitat. At first the ancestors of plants could probably survive only short dry spells, but with continued selection eventually some could live on dry land.

The overall trend in plant evolution is toward greater independence of water and hence the ability to live in drier habitats. This involved adaptations in body structure, reproduction, and dispersal. An important structural adaptation was the evolution of tissues that transport food and water long distances within the plant body and also have strong cell walls that provide physical support: these two functions permitted plants to evolve larger and more complex bodies, with greater division of labor (Figure 27–1). The reproduction of early plants was tied to periodic moisture because they had swimming sperm like those of their algal ancestors. Later, some plants evolved adaptations that freed them from dependence on water for reproduction.

From the human standpoint, the ecological and economic impact of plants is overwhelmingly beneficial. They are the direct or indirect source of food for humans and virtually all other forms of terrestrial life, as well as a major source of the world's oxygen. In addition, they make up much of the physical structure in which animals live and move. They also form and anchor soil, break the force of the wind, and moderate the water cycle. Humans use plants for many purposes. A few species do make pests of themselves, usually because humans have moved them to new areas, away from their natural enemies. Then we call them weeds and spend huge sums to eradicate them and make way for less competitive plants of more use to us.

This chapter examines the adaptations that permit plants to cope with the problems of life on land. It also surveys the major groups of living plants and their adaptations for life in increasingly difficult terrestrial environments, and it follows several trends in plant evolution.

**FIGURE 27–1** Representatives of many plant groups in the Olympic rain forest. The tiny mosses do not have the transport tissues found in the larger plants. Mosses and ferns disperse by means of dust-like asexual spores, whereas the trees and shrubs seen here produce seeds.

## KEY CONCEPTS

♦ Members of the plant kingdom are mostly photosynthetic and terrestrial, with cellulose-walled cells arranged into the tissues and organs of complex bodies and with sexual life cycles involving alternation of generations, reproduction by both spores and gametes, and development by way of embryos nurtured in the female parent's body.

♦ Because plants cannot move around, obtaining light and raw materials for photosynthesis and bringing gametes together for sexual reproduction have been dominant selective pressures in their evolution.

♦ Plants evolved from green algal ancestors that colonized land. The most successful plants have vascular tissues that transport food and water rapidly throughout the body and support it in the air, permitting division of labor between absorptive, supportive, and photosynthetic structures, and growth to a large size.

♦ The most successful vascular plants have pollen and seeds, adaptations that free their reproduction from dependence on environmental water.

♦ Food produced by plants made it possible for many heterotrophs to move onto land and supports their descendants today.

## 27-A   THE MOVE TO LAND

The main challenges facing plants are obtaining light, water, carbon dioxide, oxygen, and minerals for photosynthesis and other metabolic processes and bringing gametes together for fertilization.

Land offers many advantages to photosynthetic organisms that can function out of water (Table 27–1). More light is available on land because water itself absorbs much of the sun's energy. The concentrations of carbon dioxide for photosynthesis and oxygen for respiration are also much higher in air than in water. As an added bonus, the earliest plants faced little competition for these resources and little predation, although a few animals may have ventured out of their watery homes to feed at night.

The main disadvantage of moving out of water is that water itself becomes hard to obtain. On land, it no longer bathes the plant's entire body but lies below the soil surface, where it is too dark for photosynthesis. And once obtained, water is easily lost by evaporation from a plant's surface into the air.

Living on land presents other problems. Air is much less dense than water, and so it gives virtually no support to the plant body, in contrast to the buoyancy provided by water. Nor does air afford protection from the sun's ultraviolet light, which can damage nucleic acids. This exerted selection for the production of light-absorbing pigments or other shields that prevent damage to genetic material. Temperature changes are also wider and more rapid in air and soil than in water, and plants have adaptations that cope with this. Finally, water is no longer available for reproduction. Many algae rely on water for their flagellated sperm to swim to eggs and for spores or zygotes to disperse to new locations. On land, plants must carry out these functions without a constant supply of water.

### Origin of Plants

Plants' algal ancestors were probably one or more species of charophytes, a group of green algae. Modern charophytes share many features with plants. Only the first in this list is also found in other green algae:

1. Chloroplasts with chlorophylls *a* and *b* and β-carotene as the main photosynthetic pigments and starch as their main food store.
2. Chloroplasts with thylakoids arranged into granal stacks (see Figure 8–7) and with similar introns in the chloroplast DNA (Section 10–D).
3. Cellulose cell walls.
4. Cell division in which the new cells form a common partition from material in Golgi vesicles, starting at the center and moving toward the side walls. Then each cell builds a new wall on its side (see Figure 14–10). (In contrast, the new cell walls of many other green algae grow from the sides toward the center.)
5. Growth in length by the activity of **apical meristems,** regions of dividing cells at the tips of the body.
6. Phytochrome, a pigment that plays an important role in plants' response to the environment (Section 45–K).
7. Similar photorespiration (Section 8–H).
8. Oogamous sexual reproduction, in which large eggs are fertilized by small sperm.
9. Sperm with similar flagellar structure. (Flagellated sperm have been secondarily lost in most modern seed plants.)

Some charophytes also exhibit cell division in three planes, producing not filaments or sheets of cells (as in some other green algae) but three-dimensional bodies, with parenchyma cells filling the interior of the body (Section 5–R). Plants have elaborated this body type into complex structures with many kinds of specialized cells.

In addition, charophytes have some features apparently only a step away from those of plants. For example, after a charophyte egg cell forms, nearby cells grow and surround it with sterile (nonreproductive) cells. In plants, eggs and sperm are produced inside multicellular reproductive organs with walls composed of sterile cells. These organs could be derived from the algal condition by a change in timing that produced the sterile cells before gamete formation.

**T A B L E   2 7 – 1**

**Comparison of Water and Land as Habitats for Algae and Plants**

|  | Water | Land |
|---|---|---|
| Water | Close to each cell | Under land surface; evaporates quickly above surface |
| Minerals | Dissolved at low concentrations, close to each cell | On or under land surface |
| Gases | Dissolved at low concentrations, close to each cell | Plentiful in the air |
| Light | Cuts out some wavelengths, and lowers intensity | More light available, with potentially damaging levels of ultraviolet |
| Temperature | Changes slow, narrow range | Changes more rapid, extremes wider |
| Support | Provides buoyancy, support | Much less support for parts in air |
| Reproduction | Motile gametes swim | Water seldom available for swimming gametes |
| Dispersal | Water carries offspring to new locations | Water seldom available to carry offspring to new locations |

A general feature of most algae also seen in plants is large, fluid-filled vacuoles. A plant's vacuoles provide turgor pressure and so help support the body. They also serve as storage tanks releasing water to the cytoplasm during dry periods.

In addition to the features inherited from their algal ancestors, plants evolved new adaptations to life on land, both in body structure and in reproduction.

## Adaptations of the Plant Body

On land, the resources a plant needs are segregated: water and minerals lie at or below the soil surface, light and air above it. This division of resources selected for division of labor in the plant body. Underground structures serve as anchors and often absorb water and minerals for the entire plant. In most plants this absorption is aided by symbiotic mycorrhizal fungi (Section 26–I). The photosynthetic structures above ground produce enough food for all the plant's cells.

This division of labor depends on the ability to move substances between the two areas. Most plants have **vascular tissue,** pipe-like systems of cells that transport substances within the plant, especially between the food-making parts above ground and the water-absorbing parts below. Plants with vascular tissues, such as ferns, grass, and trees, are called **vascular plants** (Table 27–2).

There are two complementary vascular tissues. **Phloem** conducts food and other organic materials from the sites of synthesis to sites of use or storage. **Xylem** transports mainly water and minerals from the roots to the stems and leaves. Many of the cells in xylem tissue have secondary cell wall thickenings containing **lignin** (lignus = wood), a material that provides strength and support. A plant with xylem may be compared to a building whose plumbing pipes double as some of its supporting columns. This is particularly obvious in trees.

Plants without vascular tissue, such as mosses, cannot be more than a few centimeters tall because diffusion over longer distances is too slow to support the cells at either end of the line. Vascular tissue permitted plants to break through this height barrier. Phloem delivers food fast enough to support an extensive root system, which can absorb a lot of water. Xylem delivers this water so efficiently that the stems can grow tall and produce many leaves, which make food supporting even more root growth, and so on. Once vascular tissue evolved, plants quickly reached heights measured not in centimeters, but in meters and even tens of meters.

In vascular plants the aerial parts are covered by a waxy **cuticle,** which is virtually impermeable to water and reduces the plant's loss of water by evaporation. However, the cuticle is also impermeable to carbon dioxide and oxygen, which the plant must exchange with the air. Photosynthetic parts more than a few cells thick exchange gases

### TABLE 27–2

**Adaptations of Vascular Plants to Terrestrial Environments**

| Problem | Adaptation |
|---|---|
| 1. Obtaining water and mineral nutrients when they no longer surround the entire plant | Roots or other structures in contact with moisture; symbiosis with fungi |
| 2. Transporting water within the plant | Xylem ⎫ Vascular |
| 3. Transporting food from sites of manufacture to sites of use | Phloem ⎭ tissues |
| 4. Preventing evaporation from surfaces exposed to air | Cuticle |
| 5. Obtaining gases for photosynthesis and respiration | Stomata* |
| 6. Obtaining sunlight for photosynthesis | Leaves or green stems |
| 7. Supporting body in medium lacking buoyancy | Xylem |
| 8. Coordinating growth and response to environment | Hormones |
| 9. Getting gametes together without reliable supply of water for sperm | Wait for moisture (in seedless plants) Pollen (in seed plants) |
| 10. Dispersing new individuals to suitable locations | Airborne spores in seedless plants*; seeds |

*Also found in some nonvascular plants (some liverworts have air pores rather than stomata)

through tiny air pores called **stomata** ("mouths"; singular: **stoma**). Stomata are surrounded by pairs of **guard cells,** which can regulate the size of the opening (Figure 27–2). Some water is inevitably lost through the stomata, but much less than if water evaporated freely from the plant's entire aboveground surface.

Plants also have a variety of **hormones,** chemicals that coordinate their growth and response to environmental cues. Some kelps (large brown algae) also have hormones, but hormones play a much more important role in plants. Hormones make roots grow down into the soil and stems turn up toward the light. Hormones also initiate reproduction, dormancy, and the breaking of dormancy (Chapter 45).

## Adaptations of Reproduction

Plant life cycles feature **alternation of generations,** a pattern in which a multicellular haploid stage produces multicellular diploid offspring, which then produce new haploid forms. The haploid stage is called the **gametophyte** generation because it produces gametes: eggs and sperm. The gametes are formed by mitosis. At fertilization, two gametes unite and form a diploid zygote. The zygote develops into an embryo, which grows into an adult of the diploid gen-

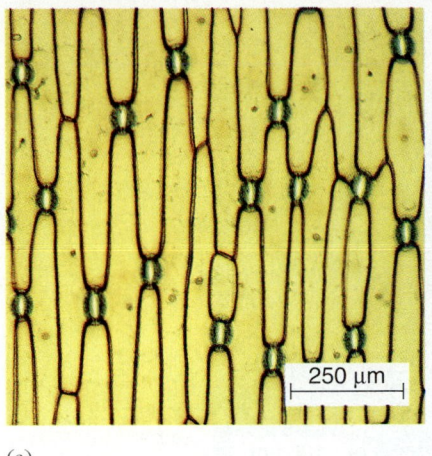 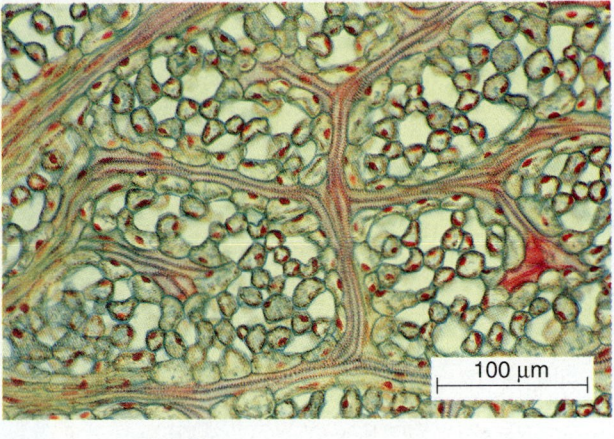

(a)  (b)

**FIGURE 27–2** (a) Pairs of green lip-like guard cells surround stomata ("mouths") on the surface of an iris leaf. (b) Veins of vascular tissue branch among photosynthetic mesophyll cells inside a leaf. The spiral-shaped red thickenings reinforce xylem cells, which transport water to the leaf. (LM) (James Mauseth)

eration. This is called a **sporophyte** because it produces spores: haploid reproductive cells formed by meiosis.

We can diagram the plant life cycle:

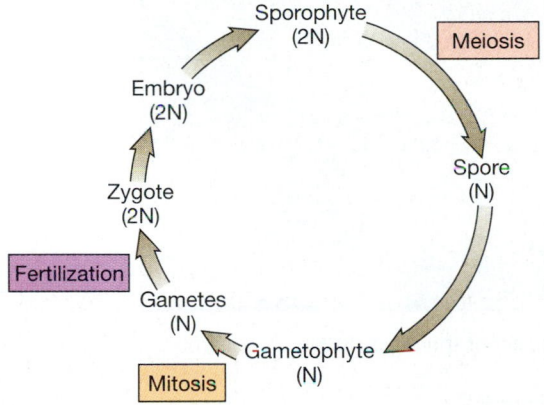

The plants' algal ancestors reproduced by means of flagellated, swimming spores and sperm. Because these cells require at least a film of moisture, they are of limited use on land. Furthermore, reproductive cells and young individuals are extremely susceptible to **desiccation** (drying out) because a small object has a lot of surface area exposed to evaporation compared with the small volume of water it can hold. All plants have multicellular reproductive organs, where the vulnerable developing reproductive cells— spores, sperm, and eggs, and later the zygote and embryo— are protected, at least for a time, by a surrounding jacket of sterile cells.

This chapter traces a few important trends in life cycles that occurred during plant evolution:

1. Gametophytes and sporophytes changed in their relative sizes and in their relationship to each other. The larger and longer-lived form is the **dominant generation.** In nonvascular plants the dominant generation is the

haploid gametophyte, such as a moss plant. The diploid sporophytes are generally smaller, growing on the gametophytes and to some extent dependent on them. The vascular plants show a trend in the opposite direction, with sporophytes, such as fern plants, increasingly dominant. In ferns and similar seedless plants, the small gametophyte and much larger sporophyte are independent. In seed-bearing plants, such as conifers and flowering plants, the sporophyte is dominant. The greatly reduced gametophyte grows within a cone or flower on the sporophyte and is dependent on it.

2. Some plants still retain flagellated sperm that must swim to the eggs, an ancestral feature. Others evolved a new structure, the pollen grain, carried to female reproductive structures by the wind or by animals. The pollen then produces sperm nuclei (Figure 27–3).

3. In the seedless plants, airborne spores are the main means of dispersal; in seed plants there is a new structure, the seed, which is better adapted for establishing a new individual in a new location.

Even though plants are nonmotile, they have means of bringing gametes together and of dispersing offspring to new locations. Flagellated sperm move under their own power, and plants release sperm only when moisture is available for swimming. Spores, pollen, and many seeds travel passively in air currents. These windborne structures are light and may have wings or parachutes that keep them aloft longer. They are also generally produced high on the plant, and so the breeze carries them farther before they alight. Many plants actually fling their spores or seeds into the air by means of specialized cell wall structures that make sudden, spring-like movements when they dry out to a critical threshold (Figure 27–4). Some plants have symbiotic associations with animals that carry their pollen or seeds. And finally, growth is a kind of slow movement by which many plants spread and occupy a wider area.

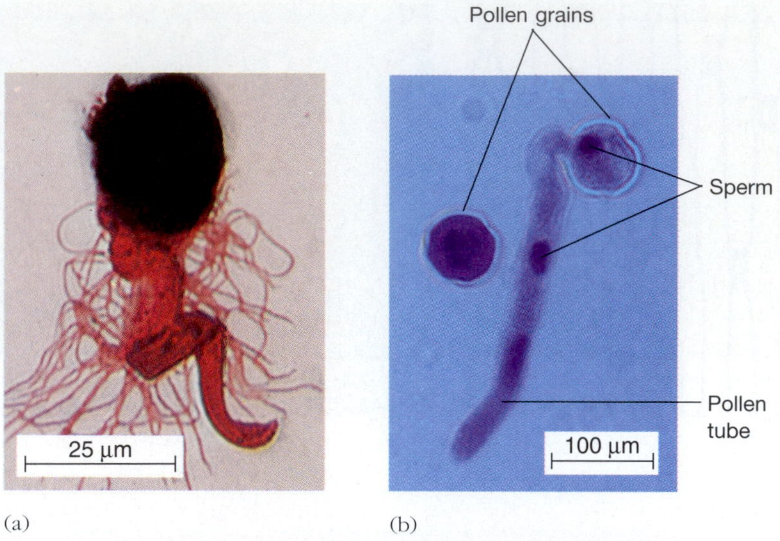

**FIGURE 27–3** Male gametes. (a) A fern sperm swims to an egg using its many flagella. (b) A pollen grain is carried to the vicinity of the female gametophyte by wind or animals. Here it grows a pollen tube, through which the sperm nuclei migrate to the egg. (LM)   (a, Carolina Biological Supply Company; b, Runk/Schoenberger from Grant Heilman)

(a)

(b)

Pollen grains

Sperm

Pollen tube

25 μm

100 μm

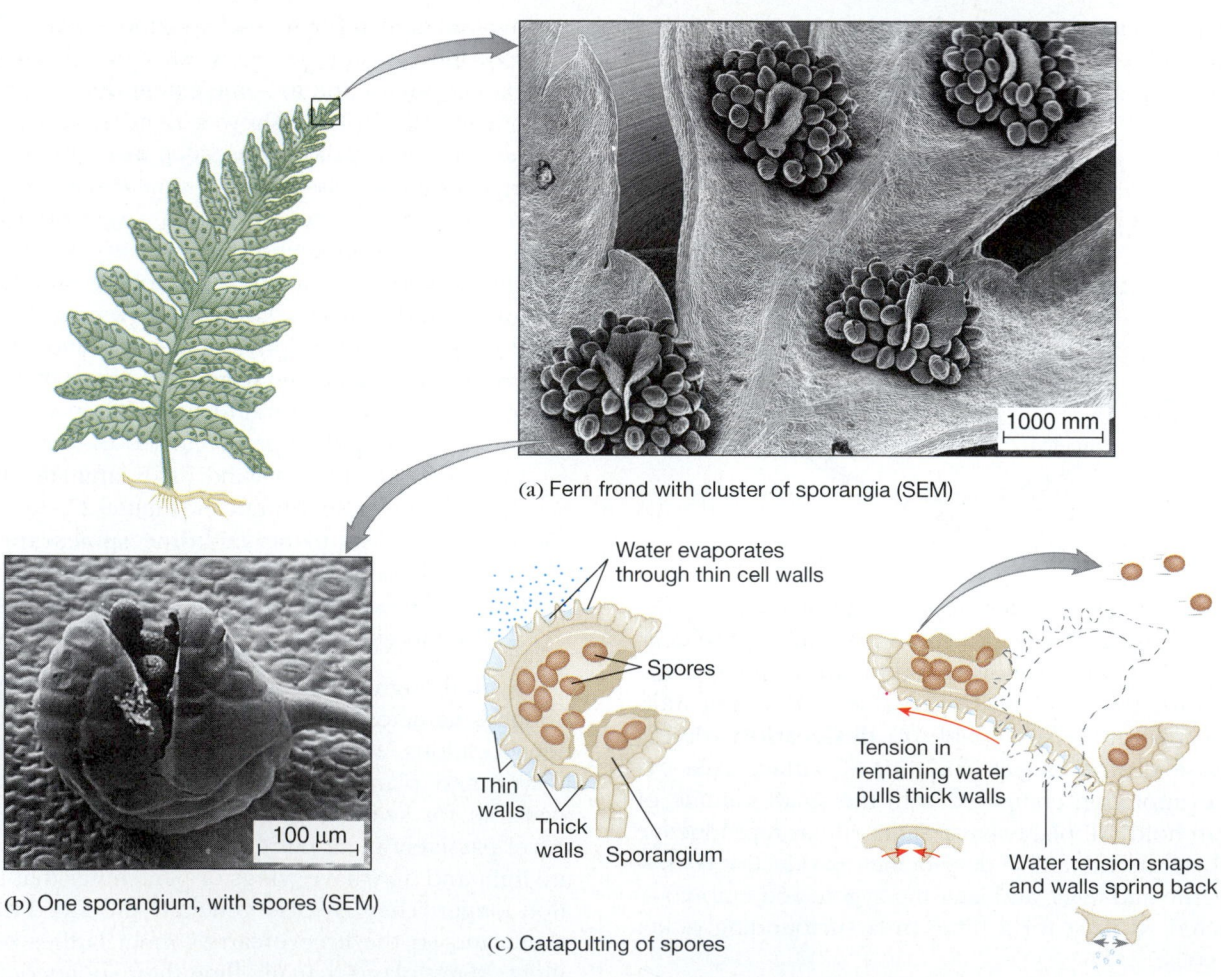

(a) Fern frond with cluster of sporangia (SEM)

1000 mm

(b) One sporangium, with spores (SEM)

100 μm

Water evaporates through thin cell walls

Spores

Thin walls

Thick walls

Sporangium

Tension in remaining water pulls thick walls

Water tension snaps and walls spring back

(c) Catapulting of spores

**FIGURE 27–4** Dispersal of fern spores. (a) Clusters of sporangia on the underside of a fern leaf. (b) Closeup of one sporangium, which has split open before releasing its spores. (c) The sporangium acts as a catapult because of a ring of cells with unevenly thickened cell walls. As water evaporates through the thin cell walls, the surface tension in the remaining water pulls the thick walls closer together, and the sporangium opens. Evaporation continues until the tension snaps like an overstretched rubber band, and the sporangium springs back to its original shape.   (a, b, Biophoto Associates)

Botanists think plants evolved from charophytes because the two share a long list of features (Section 27–A). However, this list lacks one major item. Whereas many other green algae, and all plants, have alternation of generations, charophytes do not: the thallus is haploid, and the diploid zygote divides by meiosis, releasing four haploid zoospores.

If plants evolved from such algae, they must have added the multicellular, diploid sporophyte stage to this ancestral life cycle. This could happen by postponing meiosis, with the zygote instead dividing by mitosis and forming a multicellular diploid body. Some of these diploid cells eventually divided by meiosis and formed spores.

Enter *Coleochaete*, a mat-like charophyte that grows attached to other plants or nonliving objects in shallow water. Its structure and habitat are very similar to fossil forms known from Upper Silurian and Lower Devonian times, about when plants were evolving. Like other charophytes, *Coleochaete* is ooga-mous, and sterile cells grow out from the thallus and cover the egg.

*Coleochaete* is the only green alga known to retain the zygotes in the parent's body after fertilization, and even after the parent's death (Figure 27–A). This is apparently an adaptation that keeps the zygote in the alga's shallow-water habitat during the winter instead of being washed into deep water. The flagellated spores are therefore saved a long, perilous swim back up in the spring.

The zygote of *Coleochaete* is also unusually large. It is thought to obtain food not only by its own photosynthesis but also by delivery from the surrounding parental cells. The zygote's DNA undergoes extra replications before meiosis, and each zygote produces 8 to 32 spores instead of the regulation four. This is thought to be an adaptation that gets the *Coleochaete* population off to a faster start in the spring.

From this discussion we see that the life cycle of *Coleochaete* has several features also found in plants: oogamous sexual reproduction, retention and nourishment of the diploid zygote in the haploid parental body, and delay of meiosis in the diploid stage, with each zygote eventually giving rise to more than four spores, containing a greater variety of genetic reassortments. Some species of *Coleochaete* also have multicellular antheridia, where sperm are produced. Indeed, *Coleochaete* seems poised on the brink of a life cycle like that of plants, with their embryonic development and true alternation of generations.

*Coleochaete* also makes two chemicals characteristic of plants but not of other algae. The zygote contains **sporopollenin,** a decay-resistant polymer found in the walls of plant spores and pollen grains. The thallus contains a polymer much like lignin, which probably serves as an antimicrobial agent, defending the alga against bacteria and fungi. Only minor mutations would permit the alga to produce lignin.

Because of all these features, botanists think *Coleochaete* is a good model of what the charophyte ancestors of plants must have been like.

(a)

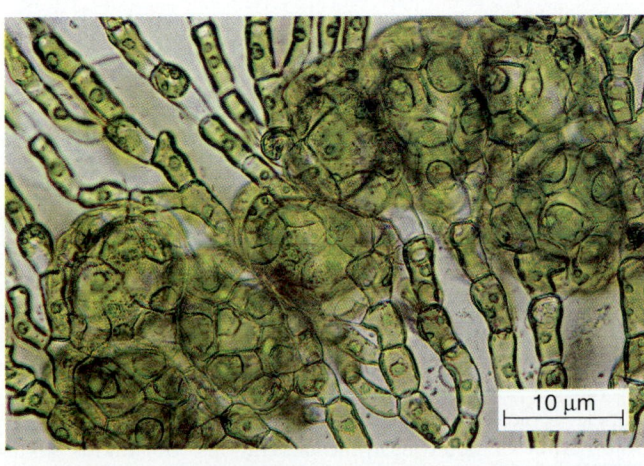

(b)

**FIGURE 27–A**   *Coleochaete.* (a) The mat-like thallus. (b) Closeup showing the large zygotes, which are retained in the thallus, surrounded and probably nourished by sterile cells.   (Linda Graham, University of Wisconsin, Madison)

## 27–B   CLASSIFICATION OF PLANTS

The boundaries of the plant kingdom have been changed many times. Many botanists now favor defining plants not just as multicellular eukaryotic autotrophs (as before) but instead as those with embryonic development and a high degree of organization. On this basis, multicellular algae no longer qualify as true plants but are placed in kingdom Protista (Sections 25–P to 25–S).

In its present incarnation, the plant kingdom contains multicellular, photosynthetic organisms with sexual reproduction, alternation of generations, and adaptations to life on land. However, some plants have secondarily evolved new adaptations to aquatic habitats or lost the ancestral characters of photosynthesis or sexual reproduction.

Plants have complex body plans with tissues and organs, including multicellular reproductive organs with one or more layers of sterile cells surrounding the reproductive tissues. The definitive feature of plants is development by way of an embryo retained in the female reproductive organ and nourished by the parent plant. This plant pregnancy is an adaptation to reproduction on land because it protects the tiny embryo from desiccation and provides life support until the young sporophyte can cope with the outside world.

Table 27–3 summarizes the characteristics of the plant kingdom, including many inherited from ancestral green algae (Section 27–A).

Botanists use the term **division,** equivalent to phylum, for the next categories below the kingdom level. The conservative characters used to place plants into divisions include morphology of both the vegetative body and reproductive structures. The members of a division also have very similar life cycles. All division names end in "–phyta" (= plants).

Fossil evidence supports and extends the classification scheme based on living plants. Early plants left many fossils. This is partly because plants evolved so recently, partly because much of their evolution occurred in ideal fossil-forming areas such as swamps and flood plains, and partly because they have large complex bodies with stiff, decay-resistant cell walls. Microscopic study of thin sections shows how tissues were organized. This supplements the information provided by features such as size and shapes of leaves, texture of bark, and so on.

We don't have enough evidence to be sure whether the plant kingdom is monophyletic or polyphyletic, descended from a number of different chlorophytes-turned-plants.

In several cases members of more than one division exhibit similar degrees of adaptation to life on land, in the plant body, reproductive structures, and life cycles. We often find it convenient to group all the similar divisions under a common term (which, however, has no taxonomic standing). The ones we will use are shown in *italics* in the first column of Table 27–4. There you can see that the divisions of plants fall into two main groups, nonvascular and vascular. Nonvascular plants are small and generally confined to moist habitats. By contrast, the evolution of vascular tissue permitted the vascular plants to become the dominant plants in Earth's terrestrial vegetation. Vascular plants are further subdivided into three levels with progressively more efficient adaptations to life on land.

## 27–C   NONVASCULAR PLANTS

Nonvascular plants fall into three divisions: Bryophyta, the mosses; Hepatophyta, the liverworts; and Anthocerotophyta, the hornworts. These plants have no true vascular tissues, although conducting cells with transport functions occur in some. Because they do not have vascular tissue, and because they reproduce by means of swimming sperm that require water to reach the eggs, nonvascular plants have remained small and generally confined to moist habitats. This is not to say that they are unsuccessful: their continued existence, more than 23,000 species strong, attests to their worldwide success in a variety of permanently or intermittently damp spots.

The (haploid) gametophyte generation is the dominant stage in the nonvascular plant life cycle, and it carries on most or all of the photosynthesis. The (diploid) sporophyte is smaller and shorter-lived; it grows attached to the gametophyte and remains generally dependent on it, not only as an embryo but through maturity.

The three divisions of nonvascular plants probably represent independent lineages: the earliest fossils of each group appear at markedly different times and already show their distinctive differences. It is not known whether each group evolved directly from green algae or whether some of them evolved from early plants. Some botanists think that mosses evolved from pre-vascular ancestors in which the conducting tissue was secondarily reduced. This view is

**T A B L E   2 7 – 3**
### Characteristics of Plants

1. Macroscopic (visible to the unaided eye); complex multicellular bodies with true parenchyma, distinct tissues and organs
2. Chlorophylls *a* and *b* and β-carotene (except a few secondarily nonphotosynthetic); food usually stored as starch
3. Growth by cell division in apical meristems
4. Cellulose cell walls laid down on either side of middle lamella after cell division
5. In vascular plants, aboveground surfaces covered by waxy cuticle, typically interrupted at intervals by stomata
6. Alternation of heteromorphic generations: oogamous sexual generation (gametophyte) alternates with asexual generation (sporophyte)
7. Multicellular reproductive structures producing spores and gametes
8. Embryo retained within protective jacket of sterile cells in (female) reproductive organ

**T A B L E   2 7 – 4**

## Summary of the Plant Kingdom

| Group | Common Name | Characteristics |
|---|---|---|
| **Nonvascular Plants** | | Small, no vascular tissue; gametophyte independent, dominant, photosynthetic, anchored by rhizoids; sporophyte dependent on it; flagellated sperm requiring water to reach egg; moist habitats |
| Division Bryophyta (~14,500 living species) | Mosses | Gametophyte usually with pointed leaf-like structures; sporophyte a stalk with capsule; stomata in sporophyte; conducting cells in both generations of some species; e.g., *Mnium, Polytrichum, Sphagnum* |
| Division Hepatophyta (~9000 living species) | Liverworts | Gametophyte with round-lobed leaf-like structures or flat and ribbon-like; characteristic oil bodies in each cell; e.g., *Marchantia, Riccia* |
| Division Anthocerotophyta (~200 living species) | Hornworts | Gametophyte mat-like; sporophyte long, cylindrical, horn-shaped, producing multicellular spores in some species; only one chloroplast per cell; e.g., *Anthoceros, Megaceros* |
| **Vascular Plants** | | Small to huge; vascular tissue in sporophyte, most with roots, stems, and leaves; sporophyte dominant, independent, usually photosynthetic; wide range of habitats |
| *Seedless Vascular Plants* | | Sporophytes mostly small to medium size, usually with rhizomes bearing small, slender roots; gametophytes free-living, independent of sporophyte; flagellated, swimming sperm; dispersal by airborne spores |
| Division Lycophyta (~1200 living species) | Club mosses, ground pine | Small, but some tree-like fossil forms; scale-like leaves with single vein; sporangia on top of leaves or in their axils; gametophytes autotrophic or heterotrophic; e.g., *Lycopodium, Selaginella, Isoetes* |
| Division Psilotophyta (3 living species) | Whisk fern | *Psilotum,* simplest living vascular plant: sporophyte short, with no leaves nor roots, sporangia at tops of aerial stems; gametophytes heterotrophic, with vascular tissue; also *Tmesipteris* |
| Division Sphenophyta (~25 living species) | Horsetails, *Equisetum* | Most short, some tree-like fossil and tropical forms; hollow, jointed stems with vertical ribs, some with slender branches; reduced leaves around joints; sporangia in terminal cones; gametophytes photosynthetic |
| Division Pterophyta (~12,000 living species) | Ferns | Most short, some tropical tree-like forms; large, many-veined leaves (fronds), often highly divided, coiled when young; sporangia on underside of fronds or on separate nongreen fertile fronds; gametophytes usually photosynthetic; e.g., *Adiantum, Osmunda, Polypodium* |
| *Gymnosperms* | | Medium to huge; woody xylem; heterosporous; sexual reproduction by airborne pollen; dispersal by seeds |
| Division Cycadophyta (~160 living species) | Cycads | Palm-like shrubs and small trees; tropical and subtropical; sexes separate, pollen and seeds borne in cones, sperm flagellated; e.g., *Cycas, Zamia* |
| Division Gnetophyta (~70 living species) | | Desert or tropical plants with variable body and leaf forms; vessels in xylem; e.g., *Ephedra, Gnetum, Welwitschia* |
| Division Ginkgophyta (1 living species) | *Ginkgo* | Tree with broad, fan-shaped leaves, smooth naked seeds, sexes separate |
| Division Coniferophyta (~550 living species) | Conifers | Shrubs and trees; needle- or scale-like leaves, mostly evergreen; especially abundant in dry or cold climates; most produce pollen and seeds in cones; e.g., *Pinus, Taxus* |
| *Angiosperms* | | |
| Division Anthophyta (~235,000 living species) | Flowering plants | Trees, shrubs, herbs; usually with vessels in xylem, most with broad leaves; sexual reproductive organs in flowers, ovule within closed carpels, pollen carried by wind or animals; double fertilization, producing triploid endosperm; dispersal by seeds, which develop inside fruits |

**FIGURE 27–5** A clump of moss. Sporophytes are tan stalks with capsules growing up from the "leafy" green gametophytes. (Runk/Schoenberger from Grant Heilman)

supported by the late appearance of mosses in the fossil record (in the Mississippian, about 335 million years ago) and by the belief that the absence of conducting cells in some mosses is a derived character.

## Bryophyta: Mosses

Division Bryophyta ("moss plants") contains the most familiar nonvascular plants and the largest number of species. The low-growing green moss plants are the dominant gametophyte generation. A moss sporophyte consists of a long thin stalk topped by a capsule, in which the spores develop (Figure 27–5).

Moss gametophytes have parts that look like the true roots, stems, and leaves of vascular plants but have somewhat different structures and functions. **Rhizoids** ("rootlike") are slender, unicellular, hair-like structures that anchor the plant but do not seem to absorb water or minerals for

**FIGURE 27–6** Life cycle of a moss. Sperm formed in the male gametophyte's antheridia may be splashed out by raindrops. The sperm swim through a film of water to the top of a female gametophyte and down the neck canal of an archegonium to the egg. The zygote develops in the archegonium first into an embryo and then a mature sporophyte, a stalk bearing a capsule. Cells inside the capsule undergo meiosis and form haploid spores, which are shed into the air when the operculum covering the tip of the capsule opens. A spore germinates into a thread-like protonema, which produces buds of new gametophytes.

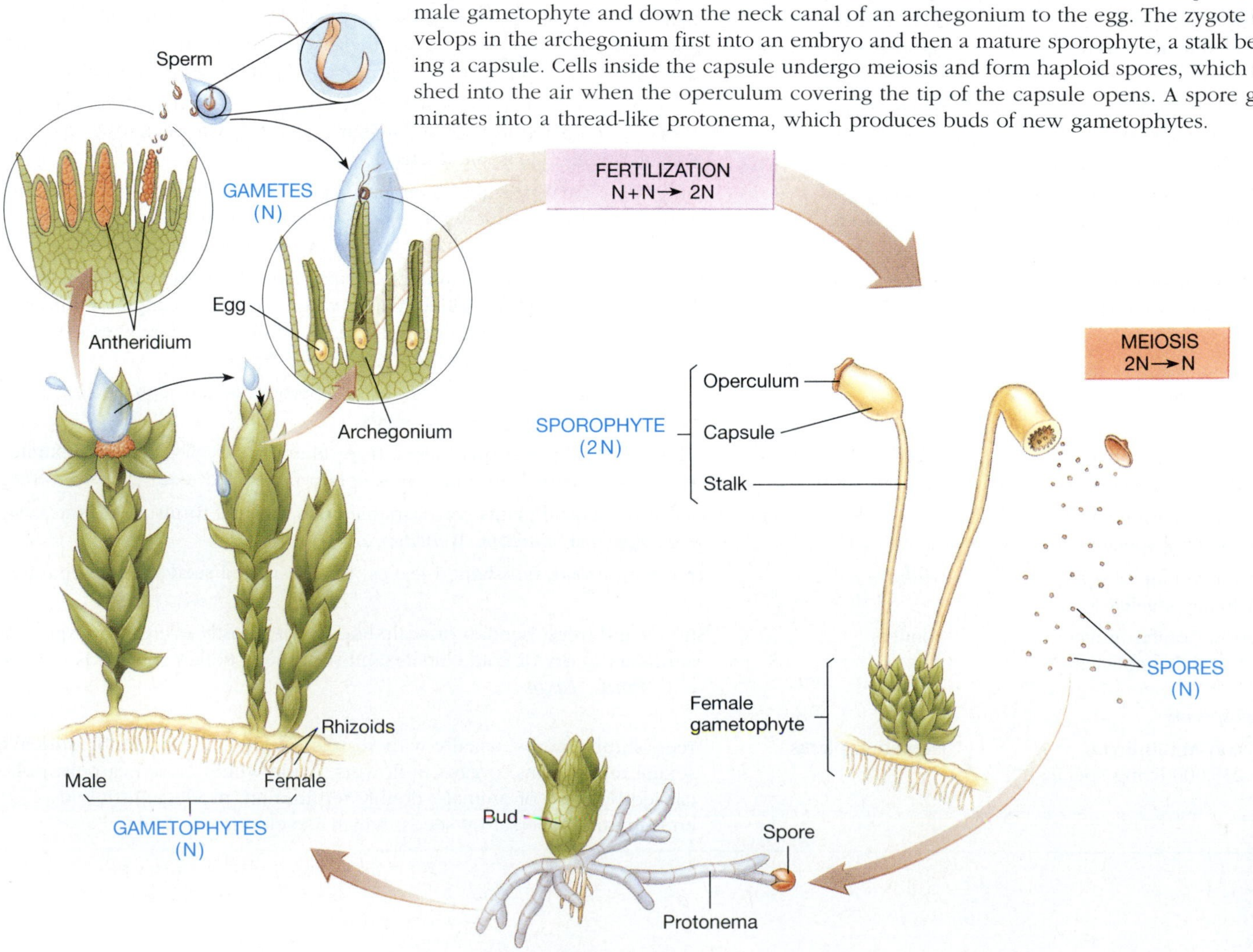

Sperm

GAMETES (N)

FERTILIZATION
N + N → 2N

Antheridium

Egg

Archegonium

SPOROPHYTE (2N)

Operculum

Capsule

Stalk

MEIOSIS
2N → N

SPORES (N)

Female gametophyte

Rhizoids

Male          Female

GAMETOPHYTES (N)

Bud

Spore

Protonema

it. Rather, this absorption occurs through the surfaces of the thin green leaf-like structures, which also carry on photosynthesis. Because these "leaves" are generally only one or two cells thick, stomata are not required for gas exchange, but stomata do occur in moss sporophytes.

In most bryophytes, cells similar to the conducting cells in xylem and phloem carry out slow transport of water and food in both gametophyte and sporophyte. However, these are far less efficient than true vascular tissue.

The key to the water relations of mosses lies not in their internal but their external structure, especially their habit of growing in clumps. This creates a network of tiny spaces that move rain or dew down through the clump by capillary action and hold it like a sponge. The water in these spaces can be absorbed by the leaf-like parts, and it also aids reproduction. Because water evaporates quickly in direct sunlight, many mosses can survive only in the shade, and their photosynthetic pigments and enzymes can function in dim light. However, some bryophytes can tolerate desiccation and survive in surprisingly hot or dry conditions.

Mosses play an important ecological role. Because they are small and have no roots, they can live on bare rock, and they help to break the rock surface into particles. Organic material is added as the mosses die. Eventually soil forms, and larger, vascular plants can grow.

***Life Cycle of Mosses***   Some species of moss have bisexual gametophytes, with both male and female organs; others have separate male and female plants. Male organs, **antheridia,** produce many flagellated sperm and shed them when there is a film of moisture to swim through. Female organs, **archegonia,** develop at the top of the female plant (or branch). Each is shaped like a bud vase, with a long neck and a single egg at the base (Figure 27–6). A sperm swims to the top of the female plant and down the long neck of an archegonium, attracted by minute amounts of a sucrose secretion. Here it fertilizes the egg, forming a diploid zygote.

From the archegonium, the zygote grows into a sporophyte, a long stalk with a capsule on top. The young sporophyte is green and carries on photosynthesis, but it generally receives water and minerals, and probably some of its food, from the female gametophyte.

Cells inside the sporophyte's capsule undergo meiosis, producing haploid spores, which are released into the air when the capsule opens. If a spore lands in a suitable place, it germinates into a **protonema** ("first thread"), growing along the surface of the ground and producing buds, which develop into a clump of new gametophytes.

## Hepatophyta: Liverworts

The gametophytes of most liverworts look like flattened mosses, with round-lobed leaves fancied to resemble the lobes of the liver: hence the name Hepatophyta ("liver plants").

Other liverworts have flat, ribbon-like bodies. These include *Marchantia,* the one most often studied in biology classes because it is easy to grow in greenhouses. Its gametophytes have a **dichotomous** pattern of branching: the growing tip of the plant forks into two equal branches, which then fork again, and so on. As the older parts of the plant die, the new forks become separate individuals. Another means of vegetative reproduction in *Marchantia* is by **gemmae,** tiny balls of vegetative cells that pinch off from the parent plant. They are produced inside cunning little cups that grow on the plant's upper surface (Figure 27–7). When raindrops splash into the cups, the gemmae bounce out, growing into new plants if they land in a suitable spot.

(a) *Marchantia* with gemmae cups

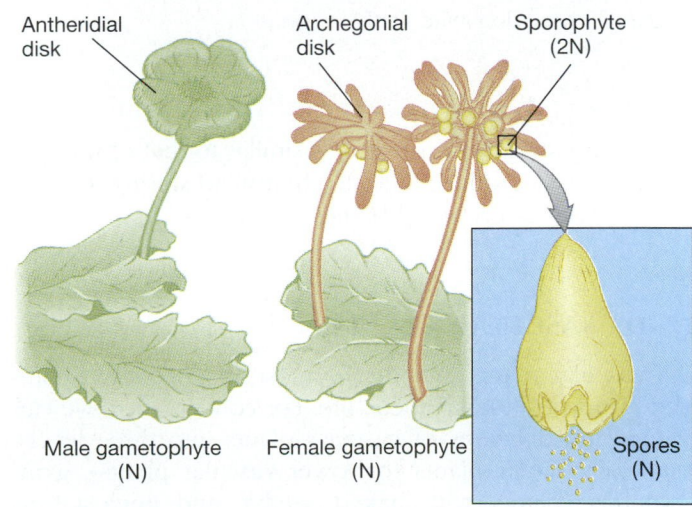

(b) Sexual reproductive structures

**FIGURE 27–7** The liverwort *Marchantia.* (a) Note the flat body, dichotomous branching pattern, and the little gemmae cups, asexual reproductive structures containing balls of cells that may be splashed out and grow into new plants. (b) The stalked antheridial and archegonial disks each bear several reproductive organs. After fertilization, the zygotes remain in place and develop into sporophytes.

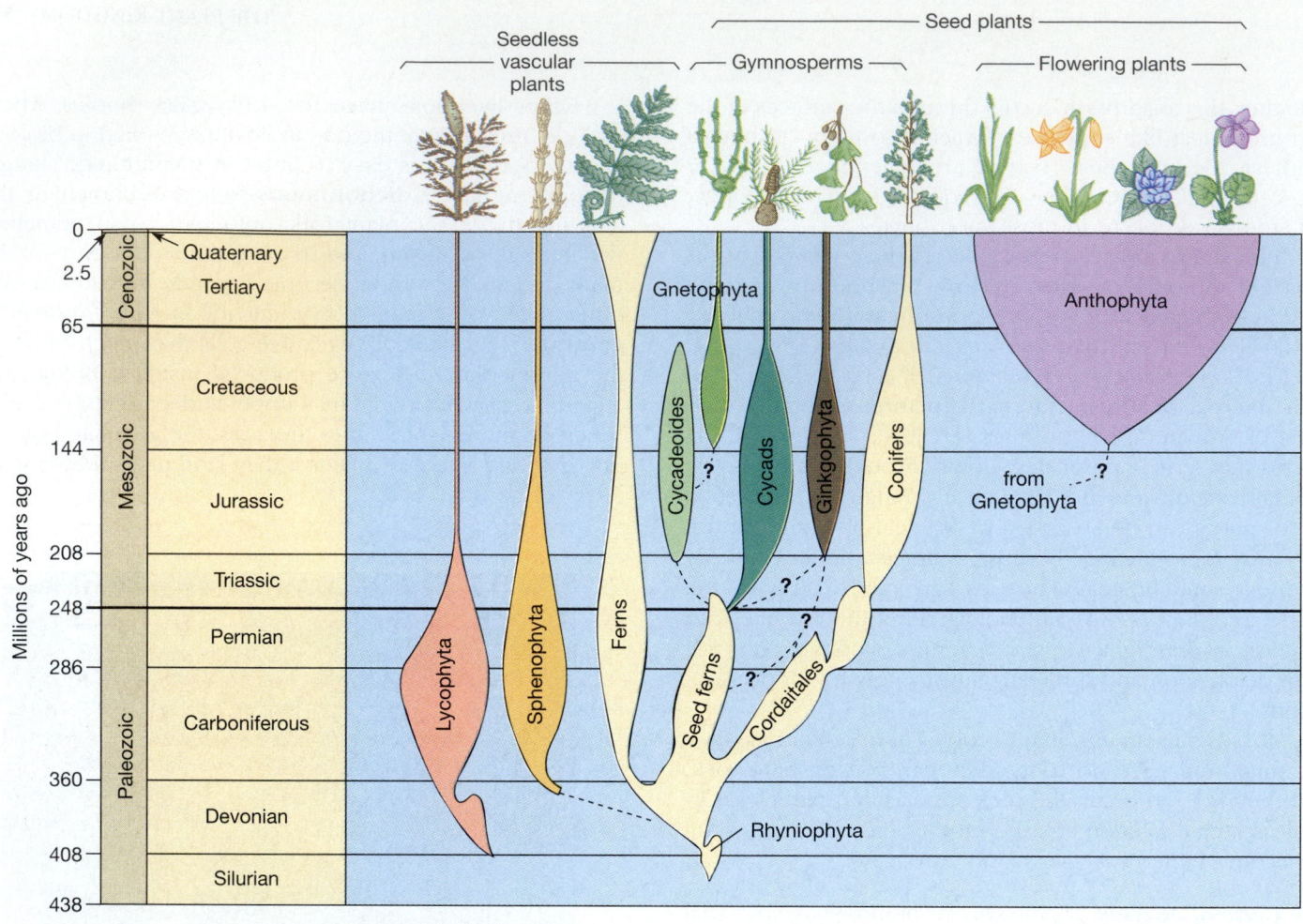

**FIGURE 27–8** Evolutionary replacement. The width of each colored area indicates the relative abundance through geological time. Seedless vascular plants evolved first but were largely replaced by gymnosperms during the Mesozoic Era. Flowering plants, most highly adapted to terrestrial life, now dominate Earth's vegetation.

The life cycle of liverworts is similar to that of mosses. Liverworts produce little starch, instead storing food reserves in characteristic oil bodies.

## 27–D VASCULAR PLANTS

The vascular plants include the most advanced and complex forms of photosynthetic life. For convenience, we can arrange the divisions of vascular plants into three levels: seedless vascular plants (or lower vascular plants), gymnosperms (plants with naked seeds), and angiosperms (flowering plants, with seeds inside fruits). The plants at each level show a similar grade of evolution in features such as life cycle, reproductive structures, and vascular tissue. The new adaptations we see from one level to the next signify major evolutionary advances in adaptation to life on land, conferring tremendous advantages over plants in the previous level(s).

Seedless vascular plants appeared in the fossil record first, gymnosperms later, and flowering plants most recently.

Each new group generated a wave of adaptive radiation, settling in habitats so challenging that earlier plants could not live there, and also outcompeting many earlier plants in much of their range. Overall, this resulted in **evolutionary replacement** of older by newer groups (Figure 27–8). Although older groups declined, both in numbers of species and in ecological importance, many still have living representatives. Studying these relics of bygone ages sheds light on the stages in the evolution of more advanced plants.

Vascular tissue was the first major evolutionary leap. By providing rapid long-distance transport and firm support, it permitted plants to grow taller in the competition for sunlight and their roots to obtain water deeper underground. Vascular tissue runs throughout a plant's roots, stems, and leaves—in fact, these three familiar plant parts are officially defined as containing vascular tissue. Vascular tissue also permitted more division of labor and specialization among these parts.

Fossil evidence shows that early vascular plants had no roots nor leaves. They consisted of a **rhizome,** an under-

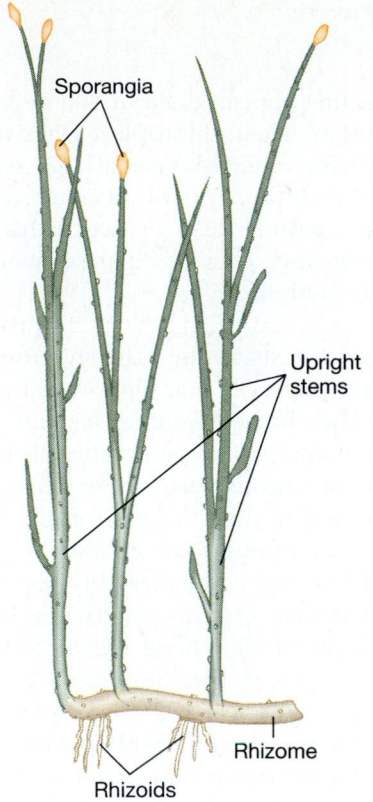

**FIGURE 27–9** Sporophyte of *Rhynia gwynne-vaughanii*. This primitive vascular plant lived about 400 million years ago (Lower Devonian). It had an underground stem, the rhizome, with rhizoids like those of bryophytes. Notice the dichotomous branching pattern of the upright photosynthetic stems. These bore sporangia, which produced spores. The plants stood about 18 cm tall.

ground stem, which produced aerial photosynthetic stems with dichotomous branching. The rhizome was anchored by rhizoids similar to those of bryophytes (Figure 27–9). These plants represent the sporophyte generation.

## 27–E SEEDLESS VASCULAR PLANTS

There are four divisions of seedless vascular plants: Lycophyta (club mosses), Psilotophyta (whisk fern), Sphenophyta (horsetails), and Pterophyta (ferns). The first three groups were once much more abundant, but seed plants crowded them out in most habitats, and only a few members survive today.

Sporophytes of all these groups show many similarities. All grow from an underground rhizome or from a trailing aboveground stem, and all but psilotophytes have small, slender roots.

However, these groups differ in their types of leaves. Some psilotophytes have no leaves. Lycophytes have **microphylls** ("small leaves"), each containing a single **vein,** a strand of vascular tissue that brings water to the leaf from the roots and carries away excess food made by the leaf. All other plants have **megaphylls,** which are larger and have many, often highly branched veins, which support the more delicate photosynthetic tissue. The two types of leaves are thought to have had different evolutionary origins (Figure 27–10).

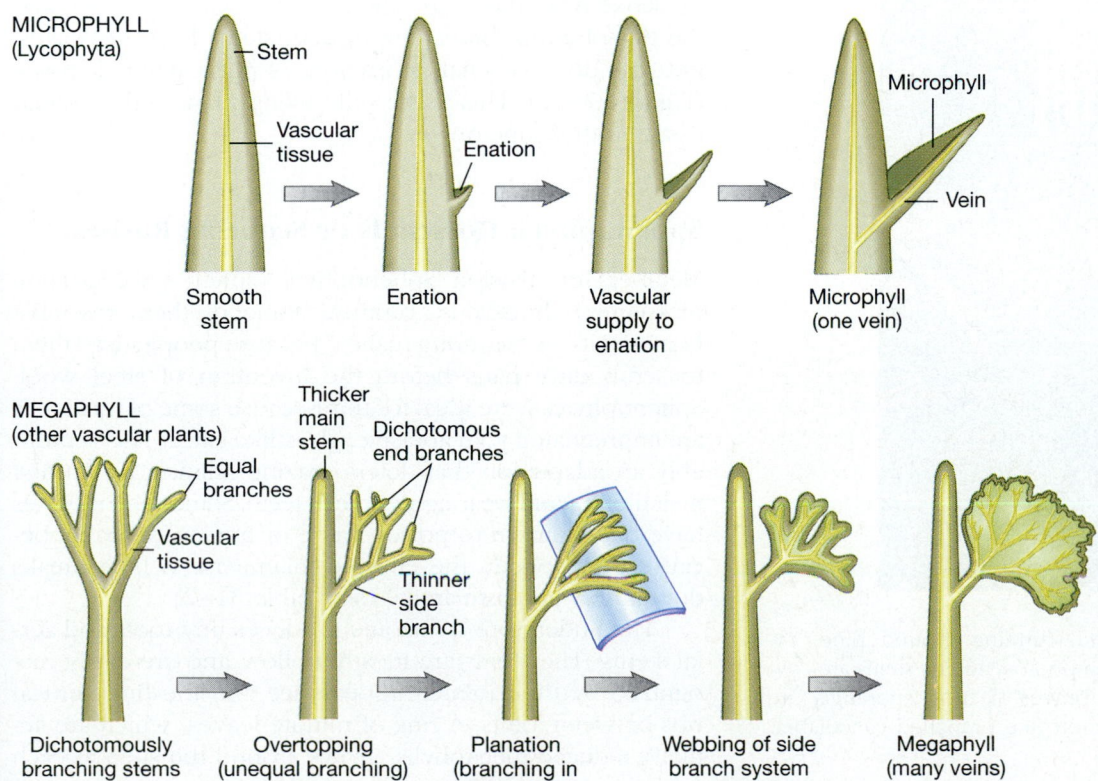

EVOLUTION OF LEAVES

MICROPHYLL (Lycophyta)

Smooth stem — Enation — Vascular supply to enation — Microphyll (one vein)

MEGAPHYLL (other vascular plants)

Dichotomously branching stems — Overtopping (unequal branching) — Planation (branching in same plane) — Webbing of side branch system — Megaphyll (many veins)

**FIGURE 27–10** Evolution of leaves. Microphylls began as enations, thin flaps increasing the stem's photosynthetic surface. They became larger and eventually received a single vein from the stem's vascular system.

Megaphylls derive from telomes (tele = far), the last few forks in a system of side branches. First, the stems at these forks changed from equal dichotomous branching to unequal branching, known as overtopping because it produces a dominant main stem and a smaller side branch. Next, the telomes began growing all in one plane and became connected by a web of photosynthetic tissue. Meanwhile, the vascular tissue became increasingly branched into a complex pattern of veins.

All of the seedless vascular plants also have similar life cycles (described under ferns). Their sexual reproduction still depends on water for flagellated, swimming sperm (see Figure 27–3a). Thus, they grow mainly in areas with frequent rain or dew. But all is not lost if conditions are too dry for sexual reproduction one year. The sporophytes are **perennial,** living for many years and releasing spores each year for another try. Moreover, these plants do not rely on sexual reproduction alone to produce new individuals. Many species show vigorous vegetative growth, with the rhizome sending up aerial parts as it spreads, forming large clumps of plants.

## Lycophyta: Club Mosses and Ground Pines

Division Lycophyta is named for the common genus *Lycopodium* ("wolf foot," a reference to their claw-shaped rhizomes). Many members of this division are commonly called "club mosses," but in fact they are vascular plants, not true mosses. Small, scale-like microphylls cover the sporophytes' aerial stems. **Sporangia** (singular: **sporangium**), the spore-producing structures, appear on the upper surfaces of leaves or in the leaf axils (the angles between the upper surfaces of leaves and the stem). In some species, the leaves bearing sporangia are green like the others, but in other species they are modified nongreen structures clustered near the tops of stems (Figure 27–11).

**FIGURE 27–11** Division Lycophyta: running ground pine, *Lycopodium complanatum.* The green photosynthetic stems are covered with tiny scale-like leaves. Leaves bearing sporangia are arranged in golden clusters held aloft like branched candelabra. (Gilbert S. Grant/Photo Researchers)

Lycophytes first appeared about 400 million years ago, in the late Silurian. Their chloroplast DNA resembles that of mosses and liverworts, whereas a large section of DNA is inverted in the chloroplasts of all other vascular plants. This difference confirms fossil evidence that the split between lycophytes and other vascular plants occurred very early in plant evolution.

Lycophytes reached their heyday as prominent members of the coal forests in the Carboniferous (see Figure 27–8). Many lycophytes of that time were trees more than 35 meters tall. They had efficient leaves and wood, but the amount of wood was small compared with their diameter, and the trunks were mostly bark. These trees did not branch until they reproduced, near the end of life. When the climate became drier during later geological periods, these enormous lycophytes died out, probably because in the new climate their vascular systems could not cope with the water requirements of such large plants. Some smaller lycophytes survived this period, and about 1200 species are still living, mostly in tropical and semitropical areas. All are less than half a meter high and probably always have been small. Modern lycophytes are inconspicuous, but they are quite common in fields and forests.

## Psilotophyta: Whisk Ferns

Members of division Psilotophyta are thought to be remnants of an ancient line. *Psilotum,* the simplest living vascular plant, has sporophytes very similar to the early vascular plant *Rhynia* described before (although no fossil evidence links the two). The rhizome bears no roots, and the dichotomous-branching upright stems have no leaves, but they do have small projections of photosynthetic tissue (Figure 27–12). This is the only living plant with vascular tissue in the gametophyte.

## Sphenophyta: Horsetails or Scouring Rushes

Members of division Sphenophyta (sphen = wedge) are nicknamed "horsetails" because some of them resemble horses' tails, or "scouring rushes" because people used them to scrub dirty pans before the invention of steel wool. Sphenophytes were ideal for this because some of their cells are impregnated with abrasive, glass-like silica. This is probably an adaptation that deters grazing animals by hurting their mouths and wearing away their teeth. Some sphenophytes have been known to poison cattle or horses, probably because they contain the enzyme thiaminase, which breaks down the vitamin thiamine (see Table 31–2).

The underground rhizome produces tiny roots and aerial stems. The stems are mostly hollow and are easily recognized by their jointed appearance and the fine vertical ribs between joints. A ring of minute leaves, which are actually reduced megaphylls, grows around the stem at each

**FIGURE 27–12** Whisk fern, *Psilotum nudum*. This intriguing plant has no leaves nor roots. Sporangia are the knobby structures near the tops of the dichotomously branching stems.    (David S. Addison/Visuals Unlimited)

joint. Some species have slender branches looking so much like pine needles that the plants are sometimes mistaken for young pine trees. Depending on the species, sporangia grow in clusters atop green vegetative stems or specialized nongreen **fertile shoots** (Figure 27–13).

Like lycophytes, sphenophytes flourished in the Carboniferous forests, with some woody members attaining heights of about 20 meters. However, only one genus, *Equisetum* ("horse bristle"), with about 25 species, survives today. Our remarks about ancient and modern lycophytes apply here too: the tree-size species did well as long as moisture was abundant, but became extinct when the climate grew dry. Some small forms persisted in the kinds of habitats to which they were already adapted, and they are common residents of ditches, railroad embankments, and moist woods. Most species of *Equisetum* grow in temperate regions.

## Pterophyta: Ferns

Division Pterophyta (pteron = feather) is named for the ferns' most noticeable feature: large, highly divided leaves with many veins. These leaves, or **fronds,** usually arise directly from a rhizome, which also has many small roots and may spread vegetatively underground. Tropical tree ferns

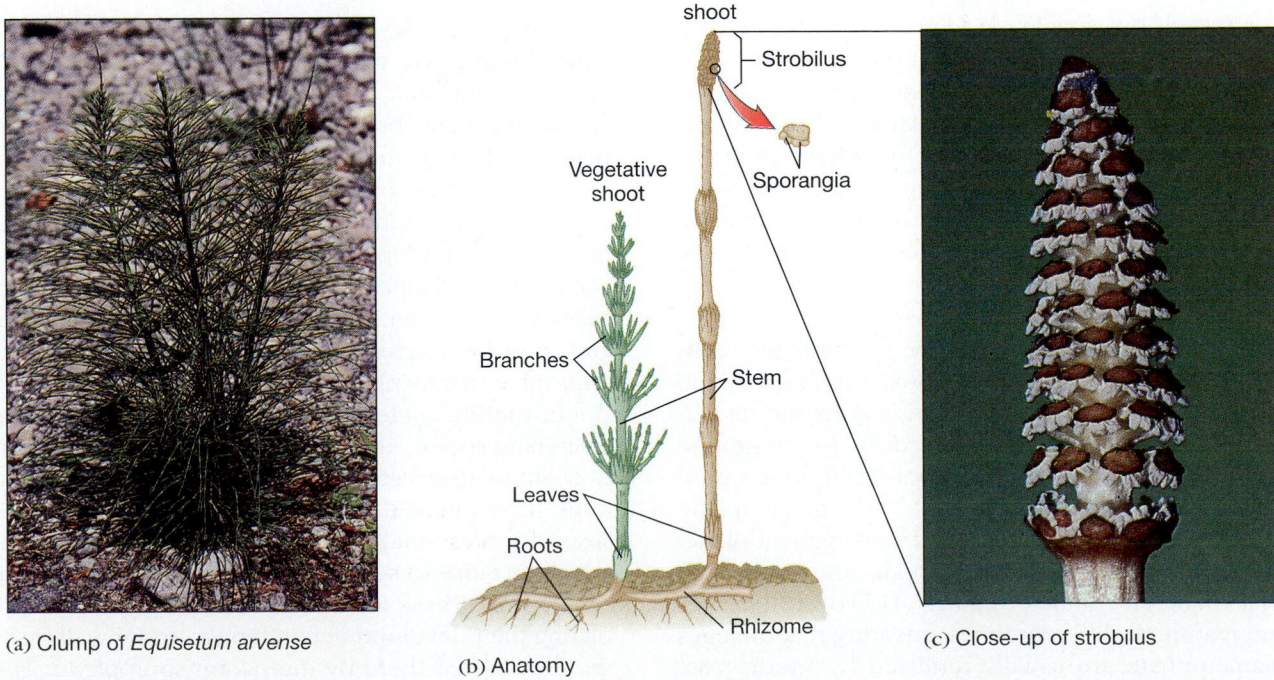

(a) Clump of *Equisetum arvense*

(b) Anatomy

(c) Close-up of strobilus

**FIGURE 27–13** Division Sphenophyta: *Equisetum arvense.* (a) A clump of vegetative shoots in a typical habitat: a gravel parking lot. (b) Anatomy of the vegetative and fertile shoots. (c) Closeup of a strobilus (cluster of reproductive structures) at the tip of a fertile shoot.

(a)                                   (b)                                   (c)

**FIGURE 27–14**  Ferns. (a) Tree ferns in a eucalyptus forest. (b) Fronds of most ferns develop as "fiddleheads," which unroll and expand to their final form. (c) A fern gametophyte. Note the photosynthetic cells with their green chloroplasts, and the colorless rhizoids.   (a, J. N. A. Lott/McMaster University/Biological Photo Service; c, Biophoto Associates)

may be quite tall, but species in temperate zones are seldom more than a meter high (Figure 27–14).

Ferns were abundant in the Carboniferous coal forests, and some living tree ferns are descendants of the only surviving trees from that time. However, ferns have probably never been a dominant part of the vegetation. Many became extinct at the same time as the lycophyte and sphenophyte trees, but the survivors have evolved extensively since then, producing many new species.

***Life Cycle of Ferns***  Fern plants are sporophytes. Cells within their sporangia undergo meiosis, forming haploid spores. These are shed into the air, serving as the dispersal stage in the life cycle. If the spore drifts to the ground in a suitably moist spot, it grows into a small, green photosynthetic gametophyte (Figure 27–14c). The gametophyte is anchored by rhizoids like those of a moss gametophyte. Most fern gametophytes produce both antheridia and archegonia (male and female organs). However, the eggs and sperm mature at different times, ensuring that the eggs of one gametophyte are usually fertilized by sperm from another, a system that enhances genetic variability. Fern sperm have flagella and must have moisture on the soil surface to swim through. The zygote remains within the ga-

metophyte and develops into a sporophyte embryo. The young sporophyte pushes a leaf up into the air and a root down into the soil. At first it depends on the gametophyte for nourishment, but it soon establishes itself as an independent plant (Figure 27–15).

The life cycle of ferns is basically similar to that of lycophytes, psilotophytes, and sphenophytes. In all four groups, the gametophyte and sporophyte live independently of each other. Gametophytes of sphenophytes and some lycophytes are green like those of ferns; in other lycophytes and in psilotophytes, however, the gametophytes are heterotrophic, receiving food from symbiotic fungi.

In the life cycle just described, the tiny gametophyte, swimming sperm, and young sporophyte are all vulnerable to death by desiccation or starvation. A few lycophytes and ferns have evolved **heterospory,** the production of two sizes of spores: small **microspores** and larger **megaspores,** which give rise to separate male and female gametophytes, respectively. These gametophytes show a trend toward completing their development within the spore wall, using only the food stored there by the parent sporophyte. Their role in the life cycle is strictly fertilization: the microspore wall opens only to release sperm from the enclosed male gametophyte, the megaspore wall only to admit them to the egg.

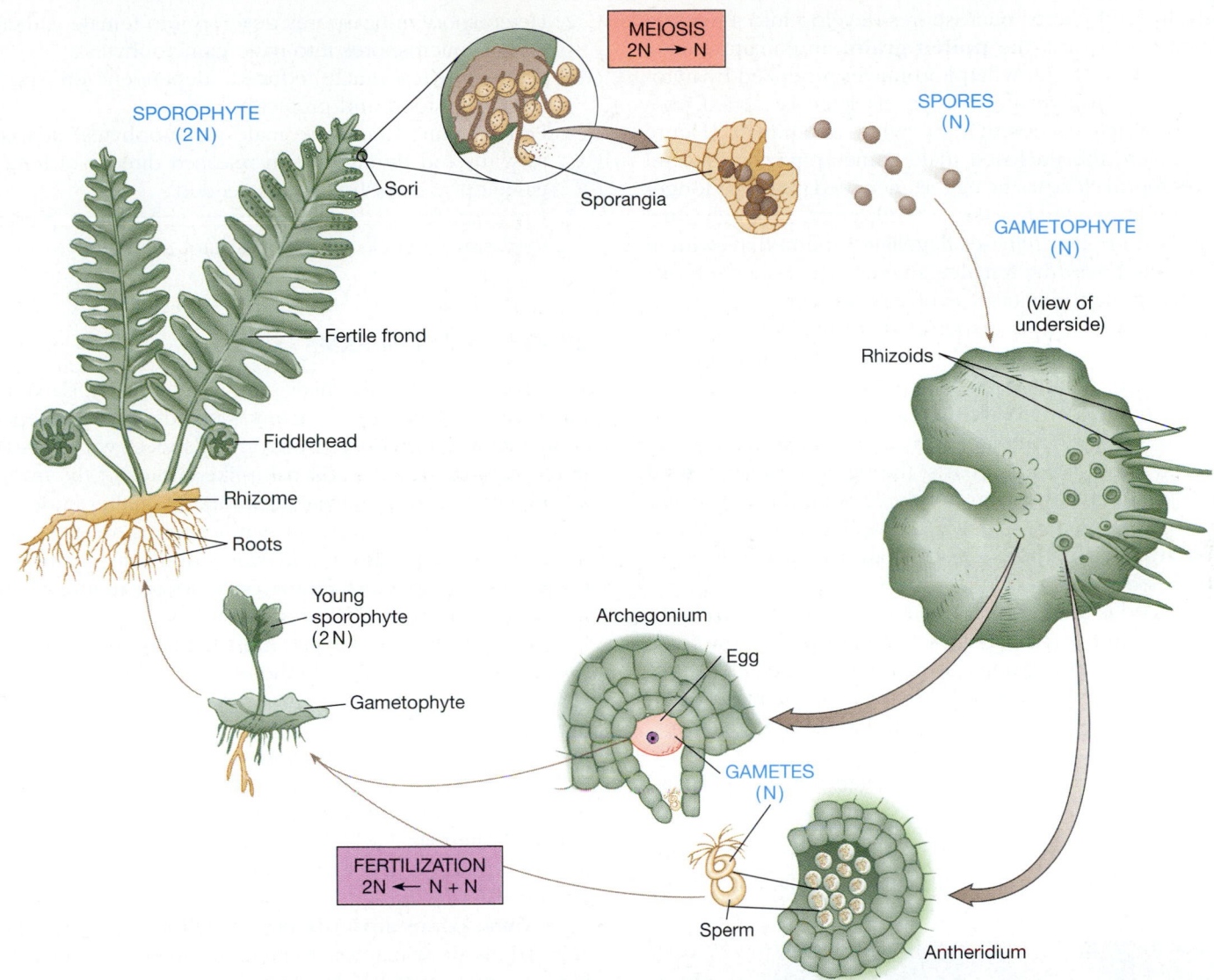

**FIGURE 27–15** Life cycle of a fern. The large sporophyte and tiny gametophyte are independent of each other. In the sporophyte shown here, sporangia grow on the underside of photosynthetic fronds in clusters called sori. Meiosis of cells inside the sporangia forms spores, which are flung into the air and then grow into tiny gametophytes only a few millimeters across. Male and female organs (antheridia and archegonia) develop on the underside of the gametophyte. The male gamete is a sperm with many flagella; it swims through environmental moisture to an archegonium and fertilizes the egg. The zygote develops into a new sporophyte.

## 27–F  EVOLUTION OF SEED PLANTS

The key advantage of the seed plant life cycle is that the tiny gametophytes, sperm, and embryonic sporophytes are no longer at the mercy of the environment. Instead, they live within the body of the large, established sporophyte, which provides all these stages with food, water, and protection from desiccation. Thus, seed plants' gametophytes can survive anywhere the sporophytes can, even though they are much smaller than the gametophytes of other plants.

The evolution of seeds began when some heterosporous plants retained their megaspores instead of releasing them into the air. There was strong selection for this because becoming part of the parent sporophyte's body, with access to its resources, greatly increased the survival of the megaspore and of the female gametophyte, zygote, and embryonic sporophyte developing in turn. In addition, nearby sterile tissue evolved into a new protective structure, the **ovule,** surrounding this reproductive organ.

Selection now favored microspores that landed not on the ground but on another sporophyte, near the megaspores. In seed plants, microspores develop into a new reproductive structure, the **pollen grain,** an immature male gametophyte with a few haploid nuclei produced by mitosis. This special air male delivery package is carried from one sporophyte to another by wind or animals (Figure 27–16). Here the enclosed male gametophyte grows and releases sperm close to the egg. Hence seed plants no longer rely on environmental water for fertilization, although some gymnosperm pollen releases flagellated sperm that swim in fluid secreted near the female gametophyte. So we see that seeds and pollen inevitably evolved together.

The seed is a new dispersal stage that gives the offspring a head start compared with the spores that disperse more primitive plants. (However, as we just saw, seeds are not equivalent to spores in the earlier life cycles.) Instead of a spore's single, haploid cell, a seed contains many cells, organized into three main parts: the outer, protective **seed coat,** which develops from the ovule wall; the multicellular diploid **embryo** of the new sporophyte generation; and a **food supply** that the seedling plant uses while its roots and leaves develop.

The seed is an extraordinary evolutionary invention for the survival and perpetuation of terrestrial plants. The fossil record shows many groups of plants with seeds. It is not clear whether seeds evolved just once or many times, in different lineages, in response to the same selective pressures.

We can summarize the general characteristics of living seed plants:

1. More efficient vascular tissue than living seedless plants. All gymnosperms, and many angiosperms, are woody.

2. Heterospory: megaspores develop into female gametophytes, microspores into male gametophytes.

3. Gametophytes much reduced, dependent on sporophytes for food and protection.

4. Pollen grains (immature male gametophytes) adapted to withstand drying; sperm released directly at female gametophyte; no free water needed.

5. Dispersal by seed: an embryonic sporophyte plus its food supply enclosed in a resistant seed coat.

## 27–G    GYMNOSPERMS

The redwoods of California's coastal mountains, Canadian pines and hemlocks, bald cypress standing knee-deep in Southern swamps, *Ginkgo* trees on city streets, wiry *Ephedra* in the western deserts, and palm-like cycads of the tropics are all members of the divisions lumped together under the term gymnosperm. Nearly all gymnosperms are trees, but there are also some shrubs and even woody vines. Although fewer than 750 species of these once dominant groups survive today, they cover large areas of land.

Conifers form the largest, most familiar group of gymnosperms. What is probably the oldest tree in the world, a 4900-year-old bristlecone pine (*Pinus longaeva*) in the mountains of eastern Nevada, is a conifer; so are the tallest tree, a coast redwood (*Sequoia sempervirens*) more than 100 meters high, and the tree with the greatest bulk, a giant sequoia (*Sequoiadendron giganteum*) nicknamed "General Sherman," which is more than 80 meters tall, 20 meters around at its base, and 3500 to 4000 years old (Figure 27–17).

Gymnosperms arose about 300 million years ago. Their adaptations allow many of them to live in environments too harsh for most of the plants discussed so far and have made gymnosperms a very successful group. Although individual species of gymnosperms died out during two major periods of extinction in the geological past, many have survived.

One major gymnosperm feature is wood, xylem tissue made up of heavily lignified cells. New wood is added each year, and so the plant grows in diameter as well as in height. Wood strengthens the stem and allows the plant to grow tall and compete for sunlight, and it also delivers water to the leaves efficiently. The extinct lycophyte and sphenophyte trees also had wood but lacked the ability to keep increasing the diameter of the trunk to the extent that gymnosperm and angiosperm trees do.

As a further adaptation to dry habitats, gymnosperm leaves produce a thick cuticle.

A major reason for gymnosperm success was the evolution of two new reproductive structures: wind-borne pollen, and seeds. "Gymnosperm" means "naked seed" (in contrast, the seeds of flowering plants are covered by an ovary). "Naked" does not mean "unprotected," as you know if you have ever pried seeds out of a pine cone. In fact,

**FIGURE 27–16**   Pine cones. Two green seed cones and a cluster of pollen cones, which are releasing a cloud of yellow pollen. (Runk/Schoenberger from Grant Heilman)

the seed represents a major advance in the plant's protection and support of its offspring (Section 27–F).

## Cycadophyta: Cycads

**Cycads** live mainly in tropical or semitropical regions. With their large, palm-like leaves they might easily be mistaken for palm trees, except that they have cones (Figure 27–18). Cycads have separate microsporangiate ("male") and megasporangiate ("female") plants, and the sperm are flagellated. Special roots growing at or above the soil surface house symbiotic nitrogen-fixing cyanobacteria.

The living cycads are mere remnants of a group that was once much more abundant. Their extinct relatives were a dominant group from the late Triassic through the Jurassic and into the early Cretaceous, and there is evidence that some dinosaurs ate cycad leaves and seeds. The stems and seeds of living cycads are eaten by some human populations, but some must be prepared specially to eliminate toxins.

## Gnetophyta

Division Gnetophyta may contain the closest living relatives of flowering plants. The living gnetophytes are alike in certain details of their vascular tissue, but otherwise are very different and thought to be only distantly related to each other.

Members of the genus *Gnetum* are woody vines with leaves that look like those of flowering plants. They live mainly in tropical rainforests.

*Welwitschia* is a bizarre resident of the driest deserts in southwestern Africa. It has an extremely long taproot, a woody trunk up to a meter wide but emerging only a few centimeters above the soil surface, and two enormous strap-

**FIGURE 27–17**  Gymnosperm giants. Mature giant sequoias are the largest of living plants, dwarfing the 6-foot man standing beside a young specimen (perhaps 50 years old) in the foreground.

(a)

(b)

**FIGURE 27–18**  Cycads. (a) Cycads growing in South Africa. (b) Closeup of a cycad cone.   (a, Walter H. Hodge/Peter Arnold, Inc.)

**FIGURE 27–19** *Ephedra* growing on the rim of the Grand Canyon. The leaves of this desert-adapted gymnosperm are much reduced; the green stems carry on photosynthesis.

like leaves that grow from the trunk throughout the plant's life of perhaps centuries.

The members of the genus *Ephedra* are small, wiry shrubs with jointed stems and scale-like leaves (Figure 27–19). They live in cool deserts in Europe, Asia, and the Americas. Native Americans reportedly used these shrubs for flour and tea and for a medicine that constricts blood vessels, which was useful for respiratory problems. These plants are the source of ephedrine, used to treat asthma.

## Ginkgophyta

*Ginkgo biloba* (Figure 27–20) is the sole surviving species of a group that flourished during the Mesozoic. Its vascular tissue is similar to that of conifers. However, it has broad, fan-shaped leaves and round, smooth seeds. The trees are either male or female, and the pollen releases flagellated sperm.

Wild *Ginkgo* trees may still exist in remote mountainous parts of China, but for centuries this species was known only from cultivated specimens in oriental gardens and temples. Today it is often planted in American and European cities because of its remarkable resistance to urban smog and to insect pests. Male trees are usually preferred because the females produce foul-smelling seeds. The seed kernel, however, has been used for food and drugs in Asia.

## Coniferophyta: Conifers and Their Relatives

The most familiar gymnosperms are the **conifers** ("cone-bearers"). Most have cone-like reproductive structures and needle-like or scale-like leaves with little surface area and thick cuticle (see Figure 27–16). Pines, firs, cedars, yews, hemlocks, junipers, larches, spruces, redwoods, and cypresses are all conifers. Many conifers grow better than flowering plants in poor or shallow soil, where nutrients are scarce, and in areas subject to regular cold or dry spells. This is probably one reason why conifers are so common.

The earliest known fossil conifers come from the late Carboniferous of 290 million years ago. The drought-resistant needles found in so many members of the group appeared during the Permian and were probably a major reason for the conifers' success during those times of global aridity. As we have seen, many other plants became extinct at this time.

**FIGURE 27–20** *Ginkgo.* (a) A male specimen. (b) A branch of a female tree, showing a seed and the distinctive fan-shaped leaves.

(a)

(b)

The discovery of a plant that controlled malaria had dramatic effects on human history. Malaria is one of the most devastating human diseases (Section 25–I). Carried from one person to another by blood-sucking female mosquitoes, it was endemic in southern Europe, Asia, and northern Africa during the Middle Ages. European travelers took malaria all over the world. Thousands died of it during the settlement of the United States. In nineteenth-century India, it killed about 2% of the population each year and made another 2% too ill to work. Even in this century, ten million cases were reported during an epidemic in the Soviet Union in the 1920s, and each year two million people worldwide still die from it.

The treatment for malaria came from Peru, where malaria was introduced by European settlers. In 1638 the Spanish Countess of Cinchon recovered from a bout of malaria after taking an extract of the bark of the tree now called *Cinchona* (Figure 27–B). Native Peruvians had recommended this to Jesuit missionaries as a cure for fever. The bark was soon imported to Europe, but many Protestants, such as English leader Oliver Cromwell, distrusted the Catholic Jesuits, refused their drug, and died of malaria. In 1852 the indefatigable Louis Pasteur discovered that quinine, the active principle in the bark, was a mixture of substances similar to strychnine and morphine. Not surprisingly, quinine is somewhat toxic; it can cause temporary ringing in the ears when taken, and large quantities can cause permanent hearing loss.

We have seen (Section 16–B) that some west Africans (and some southern European populations) had already evolved sickle cell hemoglobin, which affords some protection from malaria. Genetic resistance to malaria was a major reason west Africans were prized as slaves in the southern United States and the Caribbean, be-

**FIGURE 27–B** *Cinchona.*

cause quinine was expensive. During the nineteenth century, the cost of quinine for the British Army occupying India was more than $320 million each year (in 1990 dollars).

Tonic water is quinine dissolved in carbonated, sweetened water. Few who enjoy "gin and tonic" today realize that this drink was invented to make daily doses of bitter quinine palatable to people living in malarial areas. Quinine permitted northern Europeans, who had no natural defenses against the disease, to establish vast empires in tropical areas better defended by malaria than by any army.

Quinine also permitted Europeans to move about 20 million Indian and Chinese laborers to tropical areas where they would have died without quinine. The great European empires of the nineteenth century in Africa, Madagascar, Malaya, and Ceylon were based on plantations worked by the cheap labor of these immigrants, the ancestors of large modern populations in these areas.

Huge new industries were based on these population movements: sugar in the Indian Ocean and the Caribbean, tin and rubber in Malaysia,

and tea in India and Ceylon were all made possible by quinine. The drug also probably permitted the United States to win the Second World War in the Pacific against Japan. During that war, 25 million Allied troops traveled to areas where malaria was epidemic, including most of the Pacific, the Mediterranean, and northern Africa, areas where they could not have survived without quinine.

A German scientist, Paul Ehrlich, first treated malaria with an artificial substitute in 1881. His efforts to make quinine led to a wealth of artificial fertilizers, pharmaceuticals, and plastics that, in the 1920s, became the basis for the modern chemical industry.

However, most quinine is still extracted from the bark of *Cinchona* grown in plantations. The natural product is cheaper and pleasanter-tasting than the synthetic variety and remains one of the world's most important drugs. If one South American tree affected world history in all these ways, no wonder scientists are sure that in destroying unexplored tropical forest, we are depriving ourselves of products whose value we cannot even imagine.

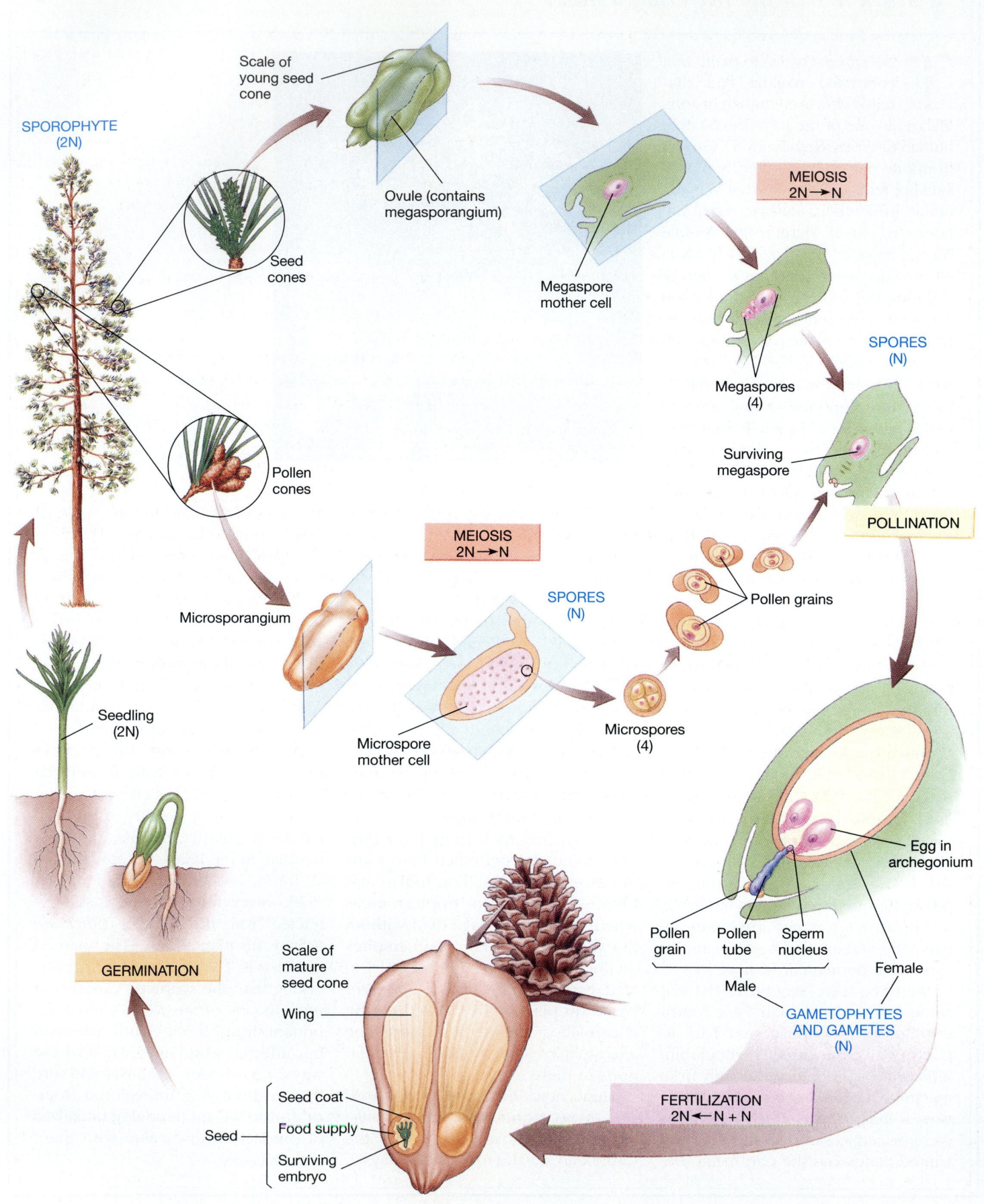

SPOROPHYTE
(2N)

Scale of young seed cone

Ovule (contains megasporangium)

Seed cones

Pollen cones

Megaspore mother cell

MEIOSIS
2N → N

Megaspores (4)

SPORES
(N)

Surviving megaspore

POLLINATION

Microsporangium

MEIOSIS
2N → N

SPORES
(N)

Microspore mother cell

Microspores (4)

Pollen grains

Seedling (2N)

GERMINATION

Scale of mature seed cone

Wing

Seed coat

Food supply

Seed

Surviving embryo

Pollen grain

Pollen tube

Sperm nucleus

Male

Egg in archegonium

Female

GAMETOPHYTES AND GAMETES
(N)

FERTILIZATION
2N ← N + N

***Life Cycle of Pines***    We can use pines, which are conifers, as an example of the gymnosperm life cycle (Figure 27–21). A mature pine tree produces two types of cones: seed (ovulate) cones and pollen (staminate) cones. The upper surface of each seed cone scale bears two ovules, each enclosing a **megasporangium.** Here a megaspore mother cell undergoes meiosis and forms four megaspores. Three disintegrate, and the fourth develops into a female gametophyte.

Similarly, cells in the pollen cone's **microsporangia** divide by meiosis and produce four haploid microspores apiece. Within the microspore walls the male gametophytes begin to develop and are released as pollen grains, each containing a few haploid nuclei. Pine pollen has flattened, wing-like structures adapted to wind dispersal.

When pollen is shed, the scales of the seed cones are spread apart, creating eddies that cause pollen to settle out of air currents. Because the cones and pollen of each species have a distinct size and shape, cones tend to trap mostly pollen of their own species. Pollen grains settle at the bases of the cone scales, near where the female gametophytes will develop. After this **pollination** (the receiving of pollen), the seed cone keeps growing and the scales close. The pollen grain resumes development into a mature male gametophyte. First, it grows a slender **pollen tube** toward the female gametophyte. Then one of the haploid nuclei divides by mitosis and forms two sperm nuclei.

Pollination occurs at about the time of meiosis in the seed cone. While the pollen tube grows, the surviving megaspore develops into a female gametophyte, containing two or three archegonia, each with an egg. **Fertilization** occurs when a sperm nucleus leaves the pollen tube and fuses with an egg nucleus. Several zygotes are produced in each female gametophyte, and the embryos compete until one crowds out the others. Meanwhile, the female gametophyte becomes a food supply by absorbing nutrients from the parent sporophyte, and the protective seed coat develops from the ovule wall, part of the parent sporophyte.

All of this takes a surprisingly long time. During its first spring, a seed cone forms and opens its scales for pollination. In pines, the male and female gametophytes take about a year to mature, and fertilization occurs in the seed cone's second spring. (This points up the fact that pollination and fertilization are two separate events, and these terms are not interchangeable.) In the seed cone's second summer, the embryo develops and its food supply is absorbed from the parent sporophyte. Not until late summer do the seed cone's scales reopen and the mature seeds spin away through the air on their thin, papery wings. Many conifers reproduce faster than pines, and their seed cones release seeds in the same year that they first appear.

## 27–H    ANGIOSPERMS (FLOWERING PLANTS)

Flowering plants include an estimated 235,000 species, six times the number of species of all other plants! They vary from duckweeds to dogwoods, onions to oak trees, and range in size from tiny *Wolffia,* about 1 millimeter long, to towering *Eucalyptus* trees that vie with redwoods for botanical height records. Flowering plants are represented in deserts, on mountain tops, and in polar regions, salt marshes, lakes, and streams. Their flowers borrow every hue of the rainbow.

Flowering plants are crucial to the existence and economy of human beings (Table 27–5). We may occasionally snack on pine seeds or spruce beer or sauté a batch of fern fiddleheads (the unrolling fronds) as a novel spring vegetable. However, almost all of our plant food comes from flowering plants, as does most of the food for domestic animals. And, although gymnosperms provide most utility lumber, angiosperm trees are used to make objects that require wood of more beauty or strength. Cotton and linen come from angiosperm plants, and they can be dyed with pigments extracted from roots and berries of other angiosperms. Flowering plants give us medicines and drugs, teas and spices. Finally, we spend millions of dollars each year buying and tending live plants.

### Anthophyta

The fossil record provides few answers about the origin of the Anthophyta. Botanists now think that the group is monophyletic, having evolved from gymnosperms. However, they do not agree about which group gave rise to them nor about the structure and habitat of the earliest forms.

The unique and definitive anthophyte feature is the flower (anthos = flower). The oldest identified fossil flower dates from the early Cretaceous, 110 million years ago, although fossil flower pollen and angiosperm leaves are known from 130 million years ago, and some even older fossils have been interpreted as angiosperms.

Flowering plants radiated into a variety of different forms during the Cretaceous, and by the mid-Cretaceous, 75 million years ago, two thirds of plant fossils were flowering plants. During the late Cretaceous, around 70 million years ago, there was much drying and cooling around the

**FIGURE 27–21**    Life cycle of pine. The mature sporophyte (pine tree) produces megaspores in the seed cones, microspores in the pollen cones. Microspores develop into pollen grains (immature male gametophytes), which are carried by the wind to seed cones. Here, each pollen grain completes development by growing a pollen tube, which releases sperm nuclei. Meanwhile, megaspores develop into female gametophytes within the seed cone. After fertilization, the zygote grows into the embryonic sporophyte, surrounded by a food supply and a protective seed coat. When the seed cone scales open, the winged, wind-blown seed is shed and serves as the dispersal unit in the life cycle.

(a)

(b)

(c)

(d)

(e)

(f)

**FIGURE 27–22**  The variety of flowering plants. (a) The prickly pear cactus receives little water in its arid home. (b) At the opposite extreme, water lilies live surrounded by water. (c) This rain forest in the Andes contains tall trees, shrubs with vividly colored flowers, and epiphytes—plants that grow on other plants—in this case, many bromeliads. (d) Wheat is a member of a major family, the grasses. (e) Black-eyed Susans belong to another major family, the composites, so called because the "flower" really consists of a group of many tiny flowers. (f) This snow plant, from the Sierra Nevada, cannot photosynthesize but obtains food from dead matter by way of a mycorrhizal fungus.

world. The dinosaurs, most of the cycads, and some conifers became extinct, while mammals and flowering plants started to dominate terrestrial habitats (Figure 27–22).

Why are flowering plants so successful? One theory holds that they are superior competitors. Thanks to more efficient vascular tissue, they can produce larger leaves, which carry out more photosynthesis, which in turn allows faster growth and reproduction. In addition, their seedlings can often survive with less light than gymnosperm seedlings require. Angiosperms also have short reproductive cycles, with greatly reduced gametophytes and rapid formation of seeds. Many families with ancestral features have large numbers of chromosomes, suggesting that polyploidy may have been a major factor in their evolution.

The speed of angiosperm growth and reproduction may account for their rapid evolution and for their phenomenal success in taking over vast areas once covered by coniferous forests, such as the middle Atlantic coast of the United States. Only in parts of the South, with nutrient-poor, sandy soil, and frequent fires, and in the Canadian north, with its shallow, rocky soil and long, frigid winters, do conifers still hold sway. Some botanists attribute this to the evergreen habit of most conifers, which enables them to use every sunbeam.

Another factor in angiosperms' success is their coevolution with animals. Many flowering plants have mutualistic symbioses with pollinating animals, and they may also rely on animals to disperse their seeds. Indeed, one theory holds that widespread dispersal of seeds by birds was crucial to the evolution of so many species of angiosperms.

However, many angiosperms rely not on animals but on wind for pollination, dispersal, or both. Wind-pollinated angiosperms, such as oak or maple trees or the grasses, generally have small, inconspicuous flowers (Figure 27–23). Both angiosperms and gymnosperms that rely on the wind for pollination must produce prodigious quantities of pollen to ensure that at least some of it reaches the female counterparts. This works best for plants living in vast stands of only one or a few species.

Animal-pollinated angiosperms gain an advantage over this hit-or-miss method. Aided by multiple sensory cues such as species-specific size, shape, color, and odor, pollinators easily learn to pick out and visit just one kind of flower. Hence, they carry pollen directly between flowers of the

**TABLE 27–5**

## Economic Importance of Plants

| | |
|---|---|
| Bryophytes | Sphagnum moss: Used as fuel (peat) in Ireland, Scotland, etc.; used as mulch and planting medium in gardening and nursery industries |
| Lycophytes | Formerly used as Christmas greens, the ground pine is now rare and protected in most areas<br>Waterproof spores once used to dust pills so they would not stick together in humid weather; highly inflammable, they were also used in fireworks |
| Ferns | Foliage used in florist industry; plants sold for house and garden |
| Gymnosperms | Lumber: Douglas fir, hemlock, spruce, various pine species, cedar, redwood<br>Turpentine: Distilled from pine trees<br>Pulp for paper: Various conifers<br>Christmas trees: Spruces, pines, firs, eastern red cedar<br>Landscape plants: Spruces, junipers, yews, cedars, cypress, hemlock, pines, cycads, *Ginkgo*<br>Gin: Sometimes flavored by redistilling spirits with juniper "berries" |
| Flowering plants | Food: Fruits, berries, seeds, nuts, grains, stalks, leaves, roots, tubers; extracted juices, syrups, fats<br>Textiles: Cotton, linen<br>Lumber: Oak, maple, ash, birch, poplar, walnut, cherry, pecan<br>Fuel: Wood, charcoal<br>Landscaping: Grass, oak, maple, magnolia, birch, and many other trees; flowering shrubs; annual and perennial herbaceous flowers<br>Beverages: Coffee; tea; fermentation of many angiosperm species to make beer, wine, and liquors<br>Drugs and medicines: Tobacco; aspirin (originally derived from willow bark); morphine, opium; marijuana; atropine (from *Belladonna* plant); digitalis (from foxglove); various tonics, from sassafras, dandelion, coltsfoot, etc.; quinine (from *Cinchona* bark) used to treat malaria |

(a)

(b)

**FIGURE 27–23**  Pollination of flowers. (a) The flowers of grasses are wind-pollinated. Pollen is released from the dangling golden anthers into the breeze. The tiny brush-like structures are the tips (stigmas) of female flower parts, which catch pollen from passing air currents. (b) Animal pollinators fly directly from flower to flower of the same species. Here, a lesser long-nosed bat is about to deliver a faceful of pollen to a cactus flower. The flower's pale color and sweet odor make it easy to find in the dark.   (a, Biophoto Associates; b, Merlin D. Tuttle/Bat Conservation International)

same species. This assures plants of specific and reliable pollination, even if the members of a species are relatively few and far between. This permitted evolution of many different kinds of flowering plants, perhaps specialized to exploit particular microenvironments in the habitat, or perhaps living widely scattered as a defense against the ravages of fungal diseases and leaf-eating insects. Flowering plants underwent tremendous adaptive radiation.

The most distinctive feature of angiosperms is the flower. Its parts are modified, highly specialized leaves. The two outermost sets of flower parts are the **sepals** and the **petals,** which are sterile but often large and brilliantly colored, attracting pollinators and therefore crucial to reproduction. The **stamens** are the "male" structures, tipped with **anthers,** which produce pollen. A "female" structure, or **carpel,** consists of three parts: the **stigma,** often covered with a sticky substance to which pollen grains adhere, the **style,** and the **ovary** (Figure 27–24).

A closed carpel is a unique and definitive feature of division Anthophyta. The ovary completely surrounds one or

**FIGURE 27–24**  Parts of a flower. Beginning with the sepal (lower left) and going clockwise, the labels progress from the outermost to innermost floral parts. Pollen produced in the anthers of the stamens lands on the sticky stigma of the carpel and grows a pollen tube down through the style to the female gametophyte inside the ovule.

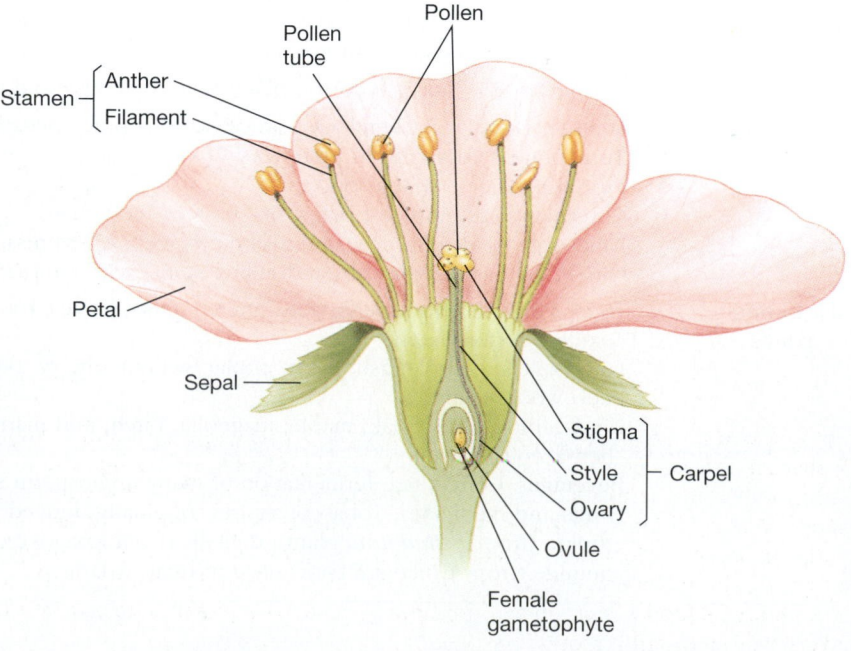

more ovules, in which a megaspore forms and develops into a tiny female gametophyte. After fertilization, the ovule matures and becomes the seed coat surrounding the embryonic plant. The term "angiosperm" ("hidden seed") comes from the fact that the seed grows within a **fruit,** the matured ovary surrounding the seed. Not all fruits are juicy and delicious; the pods of peas and milkweed are fruits, and so are pumpkins and peanut shells.

***Life Cycle of a Flowering Plant***    Like gymnosperms, angiosperms produce two kinds of spores. Each megaspore, produced in an ovule, develops into a female gametophyte. Microspores in the anthers develop into male gametophytes, the pollen grains.

Angiosperms have the smallest and simplest female gametophytes of any plant group. These vary considerably, but typically the female gametophyte contains several haploid nuclei, among them the **egg nucleus** and two **polar nuclei** (Figure 27–25).

The pollen grain, as in gymnosperms, is an immature male gametophyte, which is transported to the stigma by the wind or by an animal. Here it produces a pollen tube, which grows down the style and into an ovule in the ovary, where it releases two haploid sperm nuclei. These two nuclei take part in **double fertilization:** one sperm nucleus fertilizes the egg nucleus, producing a diploid zygote nucleus, and the other unites with the polar nuclei, forming a triploid (3N) **endosperm nucleus.** This divides rapidly

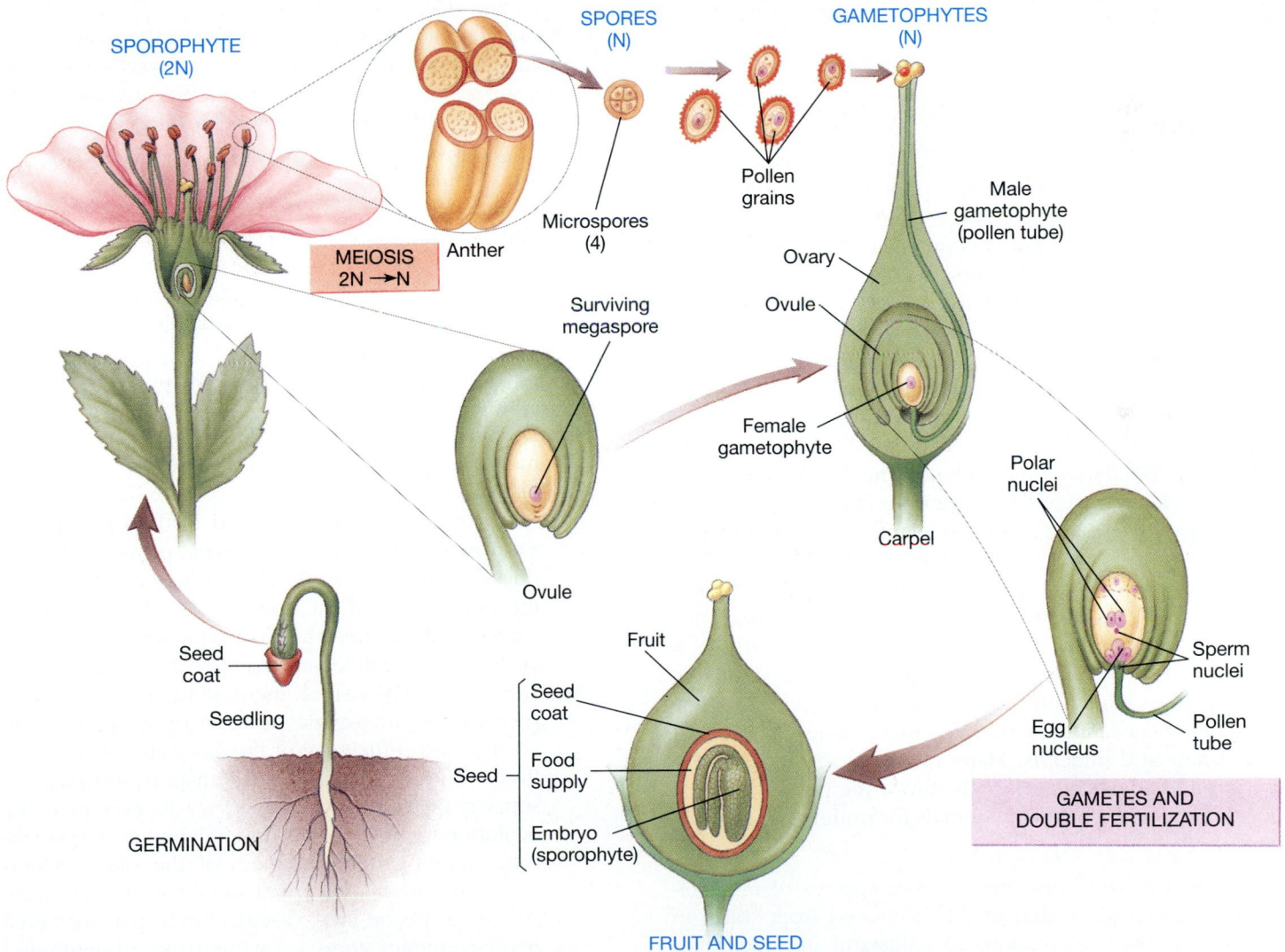

**FIGURE 27–25**   Life cycle of a flowering plant. Microspores produced in the anthers develop into pollen grains. A pollen grain germinates after being deposited on the stigma of the same or another flower, growing a pollen tube down through the style to an ovule in the ovary. Here a megaspore has formed and developed into a female gametophyte with several nuclei. In double fertilization, the pollen tube releases two sperm nuclei: one unites with the egg nucleus, forming a zygote; the other unites with the polar nuclei. The resulting endosperm nucleus gives rise to the embryo's food supply in the developing seed. The ovule develops into the seed coat, the ovary wall into a fruit around the seed.

and gives rise to a triploid nutritive tissue, the **endosperm.** As the zygote develops into an embryo, the endosperm absorbs food from the parent sporophyte, and the ovule wall develops into the seed coat. The ovary wall develops into the fruit. Recent evidence shows that double fertilization, but not endosperm formation, also occurs in the gymnosperm genus *Ephedra* (Section 27–G).

The food absorbed and stored by the endosperm is advantageous because it later supports the embryo as it grows into a new plant seedling. However, gymnosperms store food in their seeds without using an endosperm, and so the adaptive advantage of double fertilization to flowering plants is unclear.

Endosperm is invaluable from the human viewpoint. Humans rely on the calories stored in the endosperm of grains—rice, wheat, corn, oats, and so on—to meet our energy needs.

## SUMMARY

Plants are complex, multicellular, photosynthetic organisms that evolved from multicellular green algae and have adapted to life on land. The overall trend in plant evolution is toward greater independence of water, with increasing size and division of labor in the vegetative body, increasing parental care of small, vulnerable reproductive stages, and increasing use of media other than water for transport of male gametes and for dispersal. Advantageous new features spurred the adaptive radiation of new groups. These replaced many older forms and also extended the range of plant life into harsher environments. The four major phases in plant evolution were the move onto land and the evolution of vascular plants, of seed plants, and of flowering plants.

Plants changed the land by forming soil and by moderating the effects of sun, water, and wind. The food they produce also permitted the evolution of fungi and terrestrial animals and continues to support virtually all life on land. Even though plants make their own food, most rely on symbiotic relations with mycorrhizal fungi for most of their water and minerals. Many members of the most numerous and advanced group, the flowering plants, also rely on symbiotic relations with animals for pollination and dispersal.

1. Plants evolved from one or more species of multicellular green algae that gradually moved from water to land as their adaptations to withstand dry spells improved. The main advantages of this move were the decreased competition and greater intensity of solar energy available for photosynthesis. Competition among plants for light selects for rapid growth to a large size, with a large total photosynthetic surface spread out to the light.

2. The algal heritage of plants includes chloroplasts using chlorophylls *a* and *b* and carotenoids as photosynthetic pigments and storing food as starch; cellulose cell walls; large, water-filled vacuoles; patterns of cell division and growth; and oogamous sexual reproduction with eggs fertilized by flagellated, swimming sperm.

3. The alternation of multicellular haploid and diploid generations, universal in plant life cycles, is apparently a new adaptation not found in their algal ancestors.

4. The difficulties of life on land all boil down to the scarcity of water itself: a restricted, largely underground water supply, loss of body water by evaporation, loss of the physical support provided by water, and loss of the use of water as a transport medium bringing dissolved minerals close to every cell, carrying sperm to eggs, and dispersing spores.

5. Plants' adaptations to land include complex bodies with multicellular tissues and organs specialized for anchorage, support, photosynthesis, and reproduction. The female organs retain and protect the developing embryos. Dispersal occurs by waterproof spores or seeds.

6. In the vascular plants, vascular tissue transports water taken in by the roots, supports the stems and leaves in the air, and transports food from the photosynthetic parts to the roots. Leaves expose large surface areas for photosynthesis. A waxy cuticle retards evaporation from the aboveground parts, and stomata allow gases to enter with minimal water loss.

7. The adaptations of plants to life on land show several evolutionary trends:
   - Dominant generation. In nonvascular plants, the gametophyte became dominant, and the sporophyte became dependent on it.

     The vascular plants evolved in the opposite direction. The gametophyte became progressively reduced as it went from living independently to being dependent on the sporophyte. Meanwhile, the sporophyte became the dominant stage in the life cycle, and overall the size of sporophytes increased.
   - Larger size. The overall increase in size of vascular sporophytes was made possible by an increase in strength and efficiency of the vascular tissue.
   - Body complexity. The rapid transport and physical support provided by vascular tissue permitted the evolution of more division of labor and specialization among different parts of the plant. Water-absorbing roots and photosynthetic leaves were added to the original vascular body plan of aerial photosynthetic stems growing from underground stems (rhizomes). Later came woody xylem and bark.
   - Male delivery. Nonvascular plants and seedless vascular plants still have flagellated sperm, which are released when water is available for swimming to the egg. During the evolution of seed plants, the male gametophyte evolved into a waterproof pollen

grain, which travels to the female gametophyte and completes its development by growing a pollen tube and releasing sperm.

◆ Protection of embryo. The female gametophyte of all plants retains the egg and protects the zygote as it develops into an embryo. In higher plants, the female gametophyte itself is retained on or in the sporophyte parent, which contributes food and protective coatings to the seed, a new dispersal structure containing the embryo of the next sporophyte generation.

◆ Dispersal. In nonvascular plants and seedless vascular plants, the dispersal stage is an airborne spore, a small, single haploid reproductive cell. In seed plants the multicellular seeds replaced spores as the dispersal stage in the life cycle.

8. Plants are placed into divisions (the equivalent of phyla) on the basis of morphology of vegetative and reproductive structures. Members of each division also have very similar life cycles. Table 27–4 lists the divisions of living plants and their characteristics.

9. Living plants show four distinct grades of evolutionary adaptation to life on land, based on efficiency of transport within the body and on similarity of life cycles and reproductive structures:

◆ **Nonvascular plants:** mosses, liverworts, and hornworts. These plants have inefficient conducting cells, if any, and so are limited to small size and moist microhabitats, where they can absorb water through much of the body surface. Their sexual reproduction also depends on environmental moisture and low height, allowing swimming sperm to reach the eggs at the top of gametophyte plants. However, dispersal occurs by airborne spores, produced by sporophytes that grow as dependents of the dominant gametophytes.

◆ **Seedless vascular plants:** club mosses, whisk fern, horsetails, and ferns. In these plants the larger, dominant sporophyte generation has vascular tissue and is usually differentiated into roots, stems, and leaves. Dispersal is still by unicellular airborne spores. The gametophyte generation is small, independent, and usually nonvascular. These plants are still restricted to moist habitats by reliance on swimming sperm for sexual reproduction.

◆ **Gymnosperms:** cycads, gnetophytes, *Ginkgo,* and conifers. Compared with the surviving seedless vascular plants, the living gymnosperms generally have larger sporophytes, with more efficient vascular tissue. Their gametophytes are even smaller and are now dependent on the sporophyte, as are the young sporophyte embryos: all are held and nourished within the parent sporophyte's body. Airborne pollen replaces swimming sperm as the means of bringing gametes together, and multicellular, food-rich, airborne seeds replace unicellular spores as the dispersal stage. The seeds are naked.

◆ **Angiosperms:** flowering plants. Flowering plants are the most recently evolved group of plants, with the greatest number of species and widest range of sizes and habitats. They also have more efficient vascular tissue in their sporophytes, faster growth, even more reduced gametophytes (still dependent on the sporophyte), and more protection for delicate reproductive and dispersal stages. Their pollen is carried by wind or animals, the ovules grow inside closed carpels, the seeds are covered by a fruit developed from the ovary, and the seedling is nourished by food absorbed by an endosperm.

## Self-Quiz

1. What adaptation permits plants to reach large sizes?
   a. airborne pollen      d. diploidy
   b. seeds                e. stomata
   c. vascular tissue
2. An adaptation of plants that reduces evaporation of water from the body surface into the air is:
   a. roots                d. cuticle
   b. rhizoids             e. wood
   c. stomata
3. Mosses are adapted to live in a land environment in that:
   a. they have means of vegetative reproduction
   b. they have alternation of generations
   c. they have dependent sporophytes
   d. they hold water near their bodies
   e. they are no more than a few inches tall

4. An adaptation of gymnosperms not found in any of the more primitive groups of living plants is:
   a. growth exceeding a few feet high
   b. dominance of the sporophyte over the gametophyte generation
   c. protection of the gametophyte within the sporophyte body
   d. presence of true leaves
   e. production of reproductive structures at tips of the plant
5. A haploid reproductive cell that may divide and give rise to a new plant is called a(n):
   a. spore                c. zygote
   b. gamete               d. embryo

6. Which of the following stages in the life cycle of a plant are diploid?
   a. gamete
   b. zygote
   c. gametophyte
   d. sporophyte
   e. spore

7. Beginning with the zygote stage, arrange the terms in Question 6 in the order found in a plant's life cycle.

8. A seed encloses an individual of the next sporophyte generation in the form of a(n):
   a. spore
   b. gamete
   c. zygote
   d. embryo
   e. pollen grain

9. Where would you find the gametophyte of
   a. a moss?
   b. a fern?
   c. a pine tree?
   d. a lily?

10. Check the appropriate boxes to indicate which groups possess each feature:

| | Bryophytes | Ferns | Gymnosperms | Angiosperms |
|---|---|---|---|---|
| flagellated sperm | | | | |
| true roots | | | | |
| vascular tissue | | | | |
| pollen | | | | |
| protection of embryo within parent plant | | | | |
| protection of embryo within seed coat | | | | |
| photosynthetic gametophyte | | | | |
| photosynthetic sporophyte | | | | |

## Questions for Discussion

1. Parental care of the young was an adaptation that played a major part in the evolutionary success of birds and mammals. What parallels can you find in the plant kingdom?

2. Most conifers in temperate climates keep their leaves through the winter. Most angiosperm trees in the same environment are deciduous; that is, they drop their leaves each fall and produce a new set each spring. What are the advantages and disadvantages of being evergreen? Of being deciduous?

3. Why are there so many species of angiosperms? (You may be interested in reading Philip Regal's theory, listed in the references; do you agree with his argument?)

4. What energy expenditures must a plant make if it is pollinated by animals? What energy savings does the plant gain by having animal pollination? What is the adaptive advantage to a plant of being animal-pollinated rather than wind-pollinated?

5. List all possible combinations of animal or wind pollination and dispersal, and give examples of flowering plants with each combination.

6. What is the adaptive advantage of the fact that moss sperm are attracted to eggs by sucrose secretions, fern sperm by malic acid?

## Suggested Readings

Graham, L. E. "The origin of the life cycle of land plants." *American Scientist* 73:178, 1985. An argument for *Coleochaete* as a model of the plants' algal ancestors.

Hobhouse, H. *Seeds of Change: Five Plants that Transformed Mankind.* New York: Harper and Row, 1986. The story of quinine—and potatoes, tea, sugar cane, and cotton.

Mauseth, J. D. *Botany: An Introduction to Plant Biology,* 2d ed. Philadelphia: Saunders College Publishing, 1995. Chapters 22 through 25 present the evolution and current representatives of the various plant groups.

Niklas, K. J. "Aerodynamics of wind pollination." *Scientific American,* July 1987. Despite its daunting title, this article is more biology than physics.

Regal, P. J. "Ecology and evolution of flowering plant dominance." *Science* 196:622, 1977. A theory that angiosperms speciated so rapidly because of particular interactions with animals.

Simpson, B. B., and M. Conner-Ogorzaly. *Economic Botany: Plants in Our World.* New York: McGraw-Hill, 1986. A favorite botany book, full of fascinating stories, applications, and history.

Taylor, T. N., and E. L. Taylor. *The Biology and Evolution of Fossil Plants.* Englewood Cliffs, NJ: Prentice-Hall, 1993. An excellent, up-to-date text with enough background for beginners and breadth for serious students.

Wilkins, M. B. *Plantwatching: How Plants Remember, Tell Time, Form Relationships, and More.* New York: Facts on File, 1988. A survey of various plant groups and how plants live and work.

# Lower Invertebrates

## OBJECTIVES

*When you have studied this chapter, you should be able to:*

1. Use the following terms correctly:
   invertebrate
   plankton, zooplankton, filter feeder, parasite
   primitive, acoelomate, hermaphroditic
   gastrovascular cavity, pharynx
2. List or identify the characteristics of members of the kingdom Animalia and of phyla Porifera, Cnidaria, Platyhelminthes, Nematoda, and Rotifera.
3. When presented with any common member of the phyla in Objective 2, name the phylum to which it belongs.

4. Explain the adaptive advantages of the following characteristics: motile as well as sessile stages of the life cycle; larval stage; radial symmetry, bilateral symmetry, cephalization; internal digestive cavity.
5. Name the three germ layers, outline their general fate during embryonic development, and name the two major animal phyla whose members do not have three germ layers.
6. Define pseudocoel, explain its advantages, and name the two major phyla of pseudocoelomate animals.

---

The first animals appeared in Precambrian seas as multicellular organisms that ate their neighbors. This way of life led to the evolution of a vast diversity of species that populated the seas, then fresh water, and eventually land. A parade of the animal kingdom displays a seemingly endless variety of shapes, colors, and sizes, ranging from simple body plans to the most complex forms of life on Earth.

Kingdom Animalia is made up of multicellular, diploid, heterotrophic eukaryotes with oogamous sexual reproduction, embryonic development by way of a blastula, and ingestive nutrition.

Ingestive nutrition means that animals consume food in bulk: eat now, digest later. The entire living world is the animals' lunch box: bacteria, protists, plants, fungi, and the great variety of fellow animals, not to mention dead organisms and body wastes. And there are many specialized ways to exploit each of these. Add to this the fact that animals are not confined primarily either to water (like bacteria and protists) or to land (like fungi and plants), and it is no wonder that the animal kingdom contains more species than any other. More than a million species of living animals are known, and biologists estimate that at least another million exist but are yet to be described.

A number of the landmarks in animal evolution were new methods of obtaining food. Although many animals spend most of their lives anchored in place or moving very little, most of the great leaps forward in animal evolution were made by forms that actively seek their food. Sometimes these new adaptations were so different from ancestral characteristics that we classify their owners in a new phylum, class, or other taxon. The new chasing, catching, or feeding adaptations then often led to bursts of adaptive radiation.

Thus, the ancestral group in many taxa consists of free-living species, often predators of other animals. These then gave rise to groups containing less mobile species, such as tube-dwellers that wait for their food to come to them and parasites that live inside a host's body.

Depending on which groups are lumped together in taxa, the living members of the animal kingdom comprise up to 35 phyla, distinguished mainly by their body plans and embryology. The next three chapters present the major animal phyla, which we arbitrarily define as those with more than 1500 known species. The other phyla appear, briefly, in this chapter's essay.

One subphylum of the phylum Chordata contains all the **vertebrates,** animals with backbones, including fish,

**FIGURE 28–1** Marine invertebrates at home on a reef in the Pacific Ocean. A white nudibranch with black spots slides over a group of sea anemones. The blunt red spines belong to sea urchins. (Steven Webster)

birds, and mammals. The rest of the chordates, and the members of all other animal phyla, are **invertebrates,** animals without backbones (Figure 28–1). Vertebrates attract more of our attention because they cluster with us at the upper end of the scales of animal size and intelligence where invertebrates have few representatives. However, vertebrates comprise only about 5% of the known animal species. The other 95% are invertebrates, and the vast majority of these are insects.

Relatively few animal species affect human economies directly, but those that do are very important. Domesticated vertebrates supply most of our meat as well as milk, eggs, wool, leather, feathers, and fertilizer. We also consume animal protein from many aquatic vertebrates and invertebrates. Invertebrates include many disease vectors and crop pests as well as devastating parasites. On the other hand, they also include crucial crop pollinators and decomposers.

Animals are also major players in the ecology of every part of the world. One species—our own—is causing massive ecological changes worldwide and has the potential to destroy most of the species on this planet, including itself.

This chapter begins our survey of kingdom Animalia with a look at some general animal adaptations and structures. We then turn to the phyla of so-called lower invertebrates—animals with simple body plans we believe originated quite early in evolution.

## KEY CONCEPTS

- The animal kingdom contains the widest variety of multicellular organisms, including the most complex forms of life on Earth.
- Animals are multicellular, diploid heterotrophs with ingestive nutrition, oogamous sexual reproduction, embryonic development by way of a blastula, and motility during at least one stage of the life cycle. Some animals have secondarily lost one or more of these features.
- The first animals evolved in Precambrian seas. The few groups that successfully colonized fresh water and land evolved many adaptations of structure, function, and reproduction to these difficult environments.

- All but the simplest animals have a number of specialized tissues, organs, and organ systems.
- Most animals develop from three-layered embryos. The outer layer of cells develops into protective and sensory structures; the inner layer produces a digestive tube; and the middle layer forms the organs of motion, reproduction, excretion, and transport.
- Free-living predators made many of the major evolutionary advances among animals. In many groups adaptive radiation later produced members with other lifestyles, such as herbivory, filter feeding, or parasitism.

## 28–A   ANIMAL AND ENVIRONMENT

Animals evolved in the sea. Competition for food, or predation pressure, eventually drove many animals into new habitats. Some moved from bays into rivers as adaptations to the hypotonic environment of fresh water evolved (Sections 4–D and 35–A). Other animals, living on the shore, adapted to ever higher and drier sites and became **terrestrial** (land-dwelling).

The sea is a more stable and hospitable environment for life than either fresh water or land. For instance, most marine invertebrates have no problems of gaining or losing water by osmosis because the total salt concentration of the sea is very similar to that of a cell. (However, a cell maintains different *proportions* of the various salts present in sea water.) Although light and temperature vary, largely with depth, the temperature in any one area of the sea changes slowly, within fairly narrow limits. Furthermore, algae and protozoa thrive in areas of the sea with abundant nutrients, providing a constantly renewed source of food.

In contrast, fresh water makes up less than 5% of Earth's surface water and usually contains lower concentrations of nutrients than the ocean. It supports less life and many fewer species. Because fresh water is hypotonic to living cells, a freshwater animal must constantly expend energy to retain its salts and expel the excess water that enters its body by osmosis. Land is an even more difficult environment because water is often in short supply, and it readily evaporates into dry air, making desiccation a constant danger for many land animals.

Relatively few animals have invaded fresh water successfully, and only two groups, terrestrial arthropods (spiders, insects, and their kin) and higher vertebrates (reptiles, birds, and mammals), have really solved the problems of life on land. Many members of these groups move about freely even in the dehydrating conditions of warm, sunny days, when other terrestrial animals must retreat to cool, moist places, emerging only at night.

Despite its advantages, life in the sea poses difficulties. Marine animals depend on algae as the ultimate source of food, and these algae exist only near the surface, where there is enough light for photosynthesis. However, the ocean surface is constantly tossing about. Fish, which are vertebrates, can swim where they will despite water movement, but few invertebrates are strong enough swimmers to do this.

Invertebrates have two types of adaptations that permit them to cope with this problem. One is to be small enough to float around as members of the **zooplankton,** or floating animals, close to their food supply of other members of the plankton. The other is to live on the bottom, usually in water shallow enough for photosynthetic algae to live and supply food for other forms of life. A bottom-living animal may burrow in the sea floor, cling to rocks or other stable objects using structures such as claws or suckers, or be **sessile** ("sitting") on any available hard surface.

Most sessile animals are **filter feeders.** They may sit in a water current and trap passing food in a net of tentacles or mucus, or they may use cilia or muscles to create water currents past or through their bodies and filter out or seize any food the current brings.

A sessile organism needs protection from mobile predators. It may have active protection such as stingers or passive protection such as a thick shell, a burrow, or a coating of toxic mucus.

Sexual reproduction poses problems for sessile animals because they cannot move around and find a mate. Instead, most of them have behavioral adaptations in which all members of a species release their gametes into the water at the same time. This occurs in response to environmental cues, such as a particular water temperature and phase of the moon. The gametes then must find each other.

Most sessile or slow-moving marine invertebrates develop from a larva that is **motile,** able to move around. A **larva** (plural: **larvae**) is a juvenile stage of an animal differing from the adult in its structure and usually in habitat and diet. These differences reduce competition for food between the larva and its parents. In many species, the larva undergoes several changes of shape as it develops, and the different larval stages may have different names. In marine invertebrates, the larva is usually the dispersal stage. It swims or is swept away, often as part of the plankton, and eventually settles in a new location.

The larvae of marine invertebrates vary in size and shape, but they have certain features in common. Like other members of the plankton, most are transparent, making them difficult to see. In addition, all have adaptations that prevent them from sinking. Some remain afloat thanks to a large surface area. For example, they may be flat and thin, or have long tentacles or flaps. Others swim using flagella or cilia (Figure 28–2).

The disadvantage of larvae is their high mortality. Predators eat many, and many others do not end up in a suitable habitat. This has selected for the production of vast numbers of offspring in many invertebrates.

The danger of being carried into unfavorable habitats by currents is especially great in streams and rivers, and in most freshwater invertebrates the larval stage has been suppressed. Thus, marine snails produce larvae but freshwater snails do not. Instead of developing by way of a larva, these animals have direct development into a miniature adult.

Many sessile animals can reproduce asexually. **Fission** occurs by dividing into (usually) two parts. **Budding** is production of a smaller individual, which may later become independent, or may remain attached as part of a colony. Asexual reproduction avoids the problem of synchronizing gamete release and also permits an animal that has found a good spot to populate the area with a clone of individuals containing its own genes.

## 28–B    ANIMAL STRUCTURE

Animals generally have oogamous sexual reproduction, in which large eggs are fertilized by tiny sperm. The zygote develops into a solid ball of cells and then a hollow ball, the blastula, containing a cavity, the blastocoel. The embryo develops inside the egg or inside its mother's body until it is hatched or born.

Embryonic development, with cell differentiation and specialization, is the hallmark of higher eukaryotes (plants and animals). Animals could not have evolved without it because an animal cannot eat until the many cells of its feeding structures have formed.

In all but the simplest animals, cells organize themselves into tissues (Section 5–R). Tissues are arranged into organs,

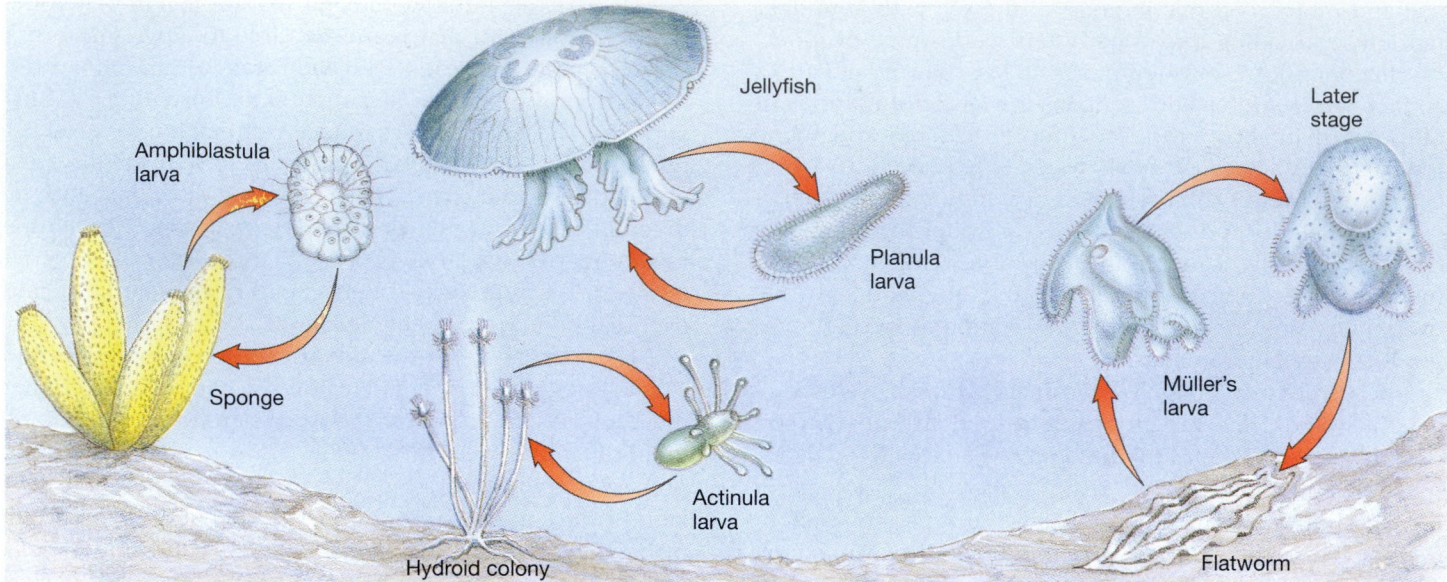

**FIGURE 28–2**   Some invertebrate adults and their larval stages. The larvae are shown enlarged with respect to the adults.

and organs into organ systems, such as the digestive, reproductive, nervous, and circulatory systems. Animals also have cells that move from place to place within the body, such as blood cells and some cells of the immune system.

The multicellular architecture of animals is strikingly different from that of plants, mainly because animal cells do not have cell walls cemented to their neighbors. The absence of cell walls permits more intimate contact between cell surfaces, greater flexibility of cells, tissues, and organs, and movement of cells within the body.

The absence of stiff cell walls also permitted the evolution of muscle tissue. All animals are motile in at least one stage of the life cycle, and animals with muscles can produce stronger, more rapid movements than organisms that depend solely on cilia and flagella. The evolution of muscles led to the evolution of a wide range of animal structures and behaviors used in hunting, capturing, and eating food; migration; escaping predators; finding and courting mates; and reproduction. Muscle tissue also speeds up internal functions such as digestion, circulation, and gas exchange.

The ingestive way of life exerts selective pressure for the ability to detect and obtain food with coordination among different parts of the body, and often for large size and superior strength as well. This favored the evolution of muscular, sensory, and nervous systems—features unique to animals.

## Body Symmetry

Animals tend to look symmetrical. Some invertebrates have body plans with **radial symmetry:** the generally cylindrical body has parts arranged around a central axis, and any

of several planes passing along this axis divides the body into mirror-image halves (Figure 28–3). Radial symmetry permits an animal to detect food or danger approaching from any side, an advantage to a sessile or passively drifting animal.

Animals with well-developed musculature can move and obtain food more efficiently if they have **bilateral symmetry,** with only one plane that divides the body into two mirror-image halves. Bilateral symmetry allows the evolution of a streamlined shape and concentration of the power of muscles and appendages into producing motion in one direction.

Along with the trend toward bilateral symmetry in animals comes the evolution of **cephalization** (kephale = head): the development of a head, with a concentration of sensory and nervous tissue that monitors the area the animal is entering.

## Body Layers

In most animals, the body plan boils down to a tube within a tube. The inner tube is the digestive system, which holds and processes ingested food. The outer tube deals with the environment. Here we find sense organs, which tell an animal what is going on in the world around it, and protective structures such as skin, shells, spines, and slime glands. Between these inner and outer body layers, most animals have a middle layer containing muscles used in locomotion, feeding, and defense against predators; structures for internal functions such as circulation and excretion; and reproductive organs.

This body plan is laid down early in development. Gastrulation converts the blastula into an embryo with three

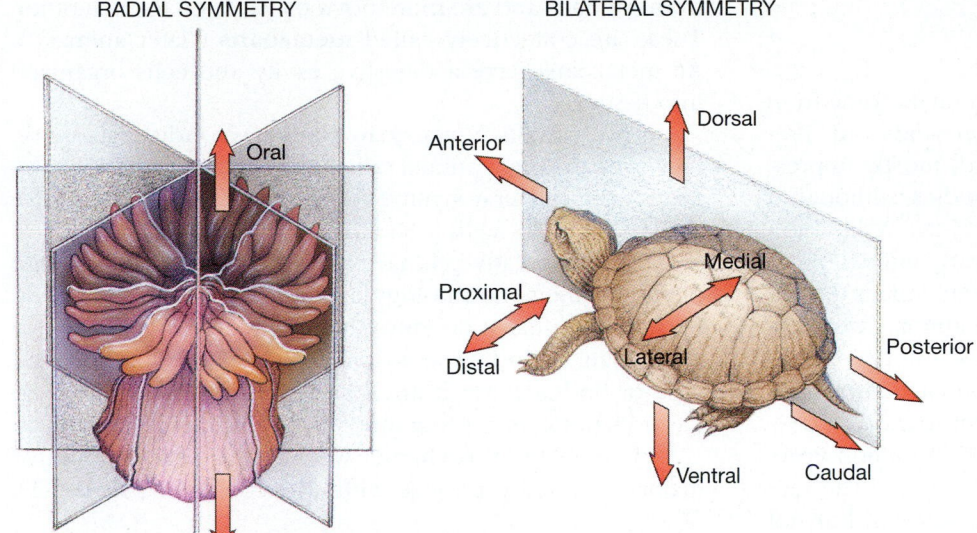

RADIAL SYMMETRY

Oral
Aboral

BILATERAL SYMMETRY

Dorsal
Anterior
Medial
Proximal
Distal
Lateral
Posterior
Ventral
Caudal

**Dictionary of Anatomical Directions**

**Oral:** of or toward the mouth
**Aboral:** away from the mouth
**Anterior:** of or toward the front
**Posterior:** of or toward the rear
**Dorsal:** of or toward the upper surface
**Ventral:** of or toward the lower surface
**Caudal:** toward the tail
**Lateral:** of the side
**Medial:** of or toward the midline
**Proximal:** of the part of an appendage nearer to the point of attachment to the body
**Distal:** of the part of an appendage farther from the body

**FIGURE 28–3** Radial and bilateral symmetry. In radial symmetry, a cut made in several planes along the central axis would cut the animal into mirror-image halves. A bilaterally symmetrical animal has only one plane of symmetry. The arrows indicate the anatomical directions defined in the dictionary panel at the far right.

germ layers of cells (Figure 28–4). The inner layer, the **endoderm,** develops into the lining that surrounds the digestive cavity, inside the body. The outer layer, the **ectoderm,** will form external structures. The third germ layer, the **mesoderm,** lies between these layers. (Endon = within; ektos = outside; mesos = middle; derma = skin.)

Animals such as sea anemones have only two well-developed cell layers, derived from the ectoderm and endoderm, with a membranous layer of mucus and a few cells between them in the **mesoglea** ("middle glue"). The body plan of these animals is better described as a sac within a sac because the digestive cavity has only one opening, the mouth, but no anus.

Animals with three body layers can be divided into three categories. In the primitive condition found in flatworms, there is no space between body layers, and so these animals are **acoelomate** (a = without; koilos = hollow). The most advanced animals are **coelomate.** They contain a **coelom,** a fluid-filled space that develops in the mesoderm (Figure 28–4). A coelom permits body organs to move independently, and this in turn permitted the evolution of rapid locomotion in coelomates such as insects and vertebrates (Section 29–A). Some lower invertebrates are **pseudocoelomates,** with a fluid-filled **pseudocoel** between the endoderm and mesoderm. A pseudocoel is the remains of the embryonic blastocoel, whereas a coelom opens up in the mesoderm later in development.

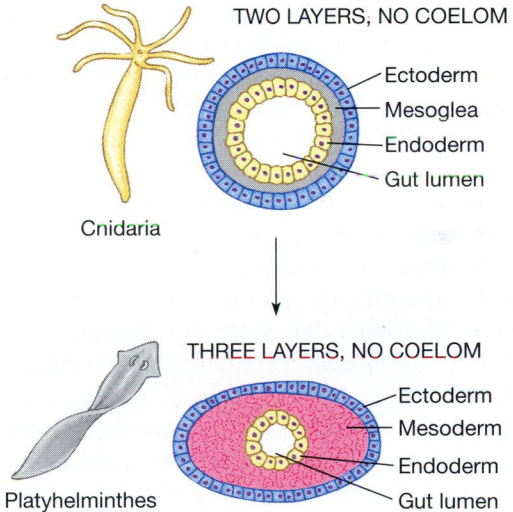

TWO LAYERS, NO COELOM
Cnidaria
Ectoderm, Mesoglea, Endoderm, Gut lumen

THREE LAYERS, NO COELOM
Platyhelminthes
Ectoderm, Mesoderm, Endoderm, Gut lumen

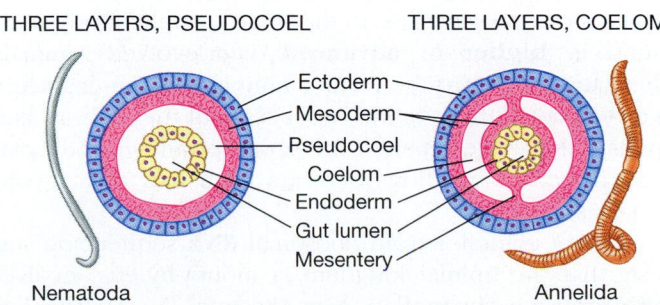

THREE LAYERS, PSEUDOCOEL — Nematoda
THREE LAYERS, COELOM — Annelida
Ectoderm, Mesoderm, Pseudocoel, Coelom, Endoderm, Gut lumen, Mesentery

**FIGURE 28–4** The four main body plans of animals. In animals with three layers and a coelom, part of the mesoderm forms membranous mesenteries, which suspend the internal organs in the coelom.

## 28–C ANIMAL EVOLUTION AND CLASSIFICATION

The oldest fossils generally accepted as animals come from late Precambrian rocks about 600 million years old. Precambrian animals left a poor fossil record, mostly impressions of tracks, burrows, and flat, soft bodies, although a few species had hard tubular shells or sheaths. Many biologists think that attempts to relate these early animal traces to still-living forms is stretching vague resemblances too far. In any case, almost all of the Precambrian animals were apparently evolutionary dead ends: they died out leaving no descendants, possibly in a mass extinction caused by a violent geological upheaval at the end of the period.

The Cambrian period dawned about 540 million years ago with an exuberant burst of animal radiation. The reasons for this "Cambrian explosion" are unknown, but not for lack of speculation: about 20 ideas have been proposed, such as changes in continents and ocean basins that opened more habitats; more nutrients in the oceans or oxygen in the atmosphere; and evolution of new predators or animal features such as collagen fibers in connective tissue and nerve cells.

Whatever the reason, increased competition and predation selected for an ever larger number of animals with hard shells, teeth, or spines, and these hard parts were preserved in the extensive Cambrian fossil record. In addition, the growing number and diversity of tracks and burrows in bottom sediments show that animals had new ways of moving, hunting, and burrowing.

Cambrian animals included the first recognizable members of most major phyla living today. However, these were outnumbered by fossils that were not members of extant phyla but were apparently evolutionary experiments long since extinct.

Many animals now alive have apparently changed little from their ancestors of hundreds of millions of years ago, and so they give us a good idea of what some ancient forms were like. We often refer to the simple body plans of some living animals as being **lower** or **primitive,** meaning that they evolved early, and theorize that animals with similar structure gave rise to the more complex body plans found in **higher** or **advanced** (later-evolved) animals. These terms refer to the relative timing of when characters evolved, not to how "good" or successful they are. In fact, animals that have retained the same, primitive body plan for hundreds of millions of years have had great evolutionary success.

Recent evidence from ribosomal RNA sequencing suggests that the animal kingdom is monophyletic, evolved from protistan choanoflagellates (Section 25–H). This surprised some biologists, because anatomical evidence suggested that only one major phylum, the sponges, evolved from choanoflagellates, and that other animals had different protistan ancestors. The other animal phyla differ greatly from sponges and are more obviously related to each other. These are collectively called **metazoans** ("later animals"). All metazoans have a digestive cavity and cells organized into tissues.

A few primitive metazoan phyla have radial symmetry. However, the main animal radiations occurred among metazoans with bilateral symmetry and cephalization, and later among creatures with a coelom (Figure 28–5).

The conservative characters used to divide animals into phyla are their morphology and embryonic development. Each phylum has a definitive suite of features: a distinctive body plan, including its symmetry, number of cell layers, type of body cavity (if any), and type of skeleton. It also has a characteristic program of embryonic development.

In the rest of this chapter we survey the most primitive major animal phyla: those without a coelom (Table 28–1).

## 28–D PORIFERA: SPONGES

Sponges are simple, sessile filter feeders shaped like tubes, cups, or crusts with porous walls and belonging to phylum Porifera ("pore-bearing"). Sponges grow singly or in colonies and range from a few millimeters to the size of a barrel. As filter feeders, sponges must live in water. They abound in all seas, usually in shallow water, on substrates such as rocks or the shells of other invertebrates. About 100 of the estimated 5000 species live in fresh water.

Sponges differ from metazoans in having several types of specialized cells but no tissues. Their cell layers do not correspond to the germ layers of other animals. In fact, their cells are remarkably independent. This was shown in 1907 by embryologist H. V. Wilson, who discovered that he could push a living sponge through fine silk, breaking it up into individual cells and cell debris. This would doom any other adult animal, but on standing in sea water, the sponge cells aggregated into larger masses, and after about three weeks they had formed a functional sponge.

The simplest sponge body plan is like a vase with porous sides. Some sponges are more complex, with the inner walls of the vase folded such that pores open into internal canals or chambers, which in turn lead to the central cavity. Sponges lack symmetry, although some do have a very regular shape.

A sponge's body has two layers of cells sandwiching a gelatinous matrix. This contains mobile amoeboid cells as well as a skeleton composed of a resilient fibrous protein, **spongin,** or of sharp **spicules** (little spikes) made of calcium carbonate or silica. The natural sponges sometimes used for bathing are spongin skeletons. Sponges are classified largely by the chemical composition and shape of the skeletal structures. The pores of a sponge are tunnels through **porocytes,** ring-shaped cells that span all three layers of the body wall.

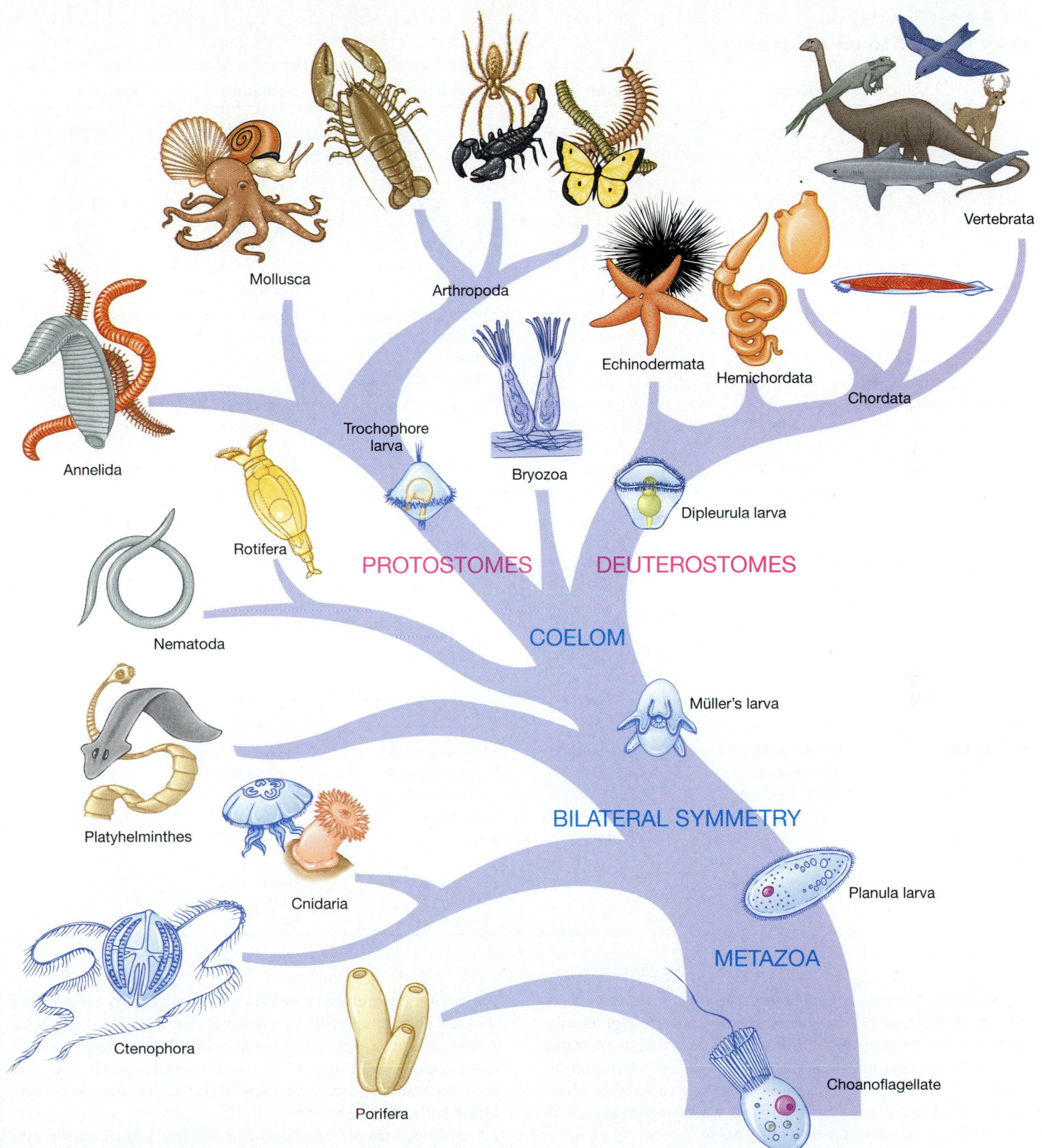

**FIGURE 28–5** Evolutionary relationships among major phyla of animals according to one widely held theory, based on the structure and development of modern animals.

**TABLE 28-1**

## Major Phyla of Lower Invertebrates

|  | Porifera | Cnidaria | Platyhelminthes | Nematoda | Rotifera |
|---|---|---|---|---|---|
| **Living species** | ~5000 | ~9000 | ~18,500 | ~12,000 | ~2000 |
| **Common names** | Sponges | Jellyfish, sea anemones, corals, *Hydra* | Flatworms: turbellarians, flukes, tapeworms | Roundworms | Wheel animals |
| **Unique features** | Collar cells<br>Pore cells<br>Skeleton of spongin or spicules<br>Body organized around water canals | Tentacles with nematocysts | Flattened body | Muscle cell extensions to nerves | Cilia around mouth |
|  |  |  |  | ——— Cuticle, adhesive glands ——— | |
| **Level of organization** | Specialized cells, no tissues | Tissues | ——————— Organs and organ systems ——————— | | |
| **Symmetry** | None | Radial | ——————— Bilateral, cephalization ——————— | | |
| **Germ layers** | None | Two, with mesoglea | Three, no coelom | ——— Three, pseudocoel ——— | |
| **Digestive cavity** | None | ——— Sac with mouth but no anus ——— | | ——— Tube with mouth and anus ——— | |
|  |  |  | ——————————— Pharynx ——————————— | | |
| **Way of life** | Sessile; solitary or colonial; filter feeding | Free-swimming or sessile; solitary or colonial; predatory | Free-living as predators or scavengers; or parasitic | Free-crawling or parasitic | Sessile or free-swimming, most solitary, few colonial |
| **Habitat:** | | | | | |
| Marine | Most | Most | Many | Many | Few |
| Freshwater | Few | Few | Some | Some | Most |
| Moist terrestrial | — | — | Few | Many | — |

Sponges have no muscle cells and therefore move very little. People assumed they were plants until the eighteenth-century discovery of flagella that propel a current of water into the body via the pores and out through the large opening at the top. This water current is the key to sponge physiology, serving for gas exchange and waste removal as well as carrying food particles into the body.

The flagella propelling this current are part of specialized cells dubbed **choanocytes** ("collar cells") because of the ring of microvilli encircling the base of the flagellum (Figure 28–6). Choanocytes are very similar to choanoflagellates, evidence that sponges probably evolved from com-plex colonial members of this protistan group (see Figure 25–12). The microvilli of choanocytes trap food particles swept into the sponge's body with the feeding current. Choanocytes then ingest the food by endocytosis and pass it on to amoeboid cells, which digest it and distribute it to other cells.

Sponges might be called the ocean's ultrafiltration system because they ingest very small particles. As little as 20% of their food consists of plankton large enough to be seen with a light microscope (that is, bigger than 0.4 $\mu$m). The rest of their food is smaller than this, mostly fine debris from dead organisms. Sponges' success is no doubt partly

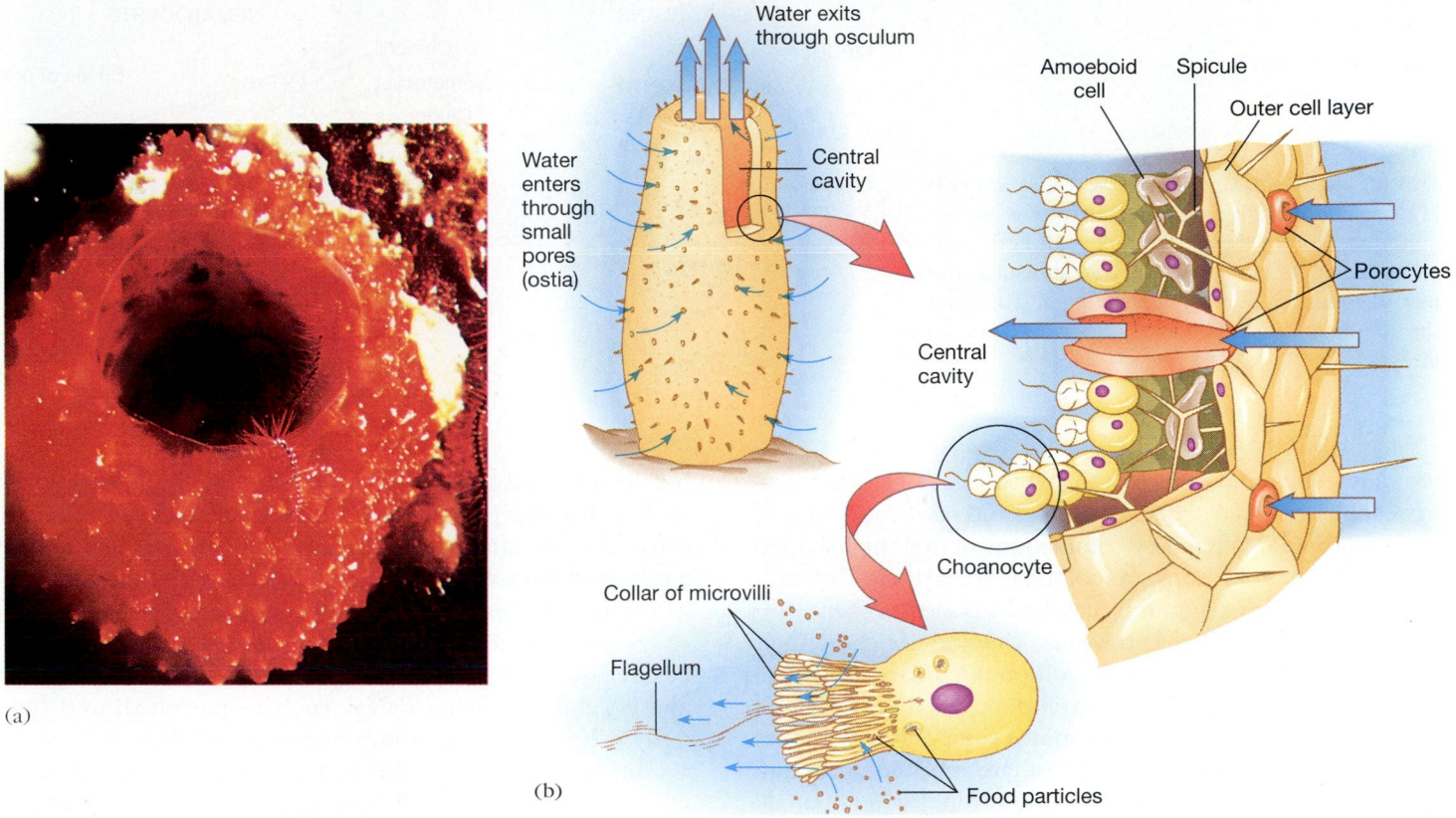

Water exits through osculum

Water enters through small pores (ostia)

Central cavity

Amoeboid cell

Spicule

Outer cell layer

Porocytes

Central cavity

Choanocyte

Collar of microvilli

Flagellum

Food particles

(a)

(b)

**FIGURE 28–6**    Sponges. (a) A solitary sponge, with a worm crawling over the edge of the hole where water exits. (b) The structure of a sponge.    (a, Steven Webster)

due to their ability to feed on these tiny particles, which many other animals cannot trap. Because they are sessile, sponges can live only in clear water, free of detritus that would bury them or clog their pores.

Some sort of defense is vital for sessile animals. More than half the species of sponges produce toxic chemicals, which prevent larvae of other sessile animals from settling on them and deter predators. The sharp, brittle spicules also make them noxious to predators. These defenses make sponges a safe haven for photosynthetic symbionts such as cyanobacteria or algae and for animal tenants of the central cavity.

Like other marine creatures, many sponges are brightly colored, either by the photosynthetic pigments of their symbionts or by their own pigments. Purple, yellow, and orange species abound. It is unclear whether this has some selective advantage or is merely incidental.

Most sponges are **hermaphroditic,** producing both sperm and eggs, but at different times. Hence, cross-fertilization is usual. Flagellated sperm leave the parent's body and enter another sponge with the feeding current. The zygote usually remains inside the parent until it develops into a ciliated, free-swimming larva, which lives in the plankton before settling to the bottom and forming a young sponge.

Sponges also reproduce asexually by budding of various kinds. Many freshwater and a few marine species form gemmules, little balls of amoeboid cells surrounded by skeletal spicules. Many freshwater sponges overwinter as gemmules.

## 28–E   CNIDARIA

Phylum Cnidaria contains hydras, sea anemones, corals, and jellyfish. The adults usually have radial symmetry, although the larvae and some adults have bilateral symmetry. These animals are metazoans, with cells forming three germ layers in the embryo and tissues in the adult. The endodermal and ectodermal body layers are well developed, but the mesoglea between them ranges from a noncellular membrane to masses of jelly-like substance containing scattered ectodermal cells. The gut is a blind sac, with one opening serving both to ingest food and to expel indigestible remains. This **gastrovascular** cavity (gastro = stomach; vascular = vessel) doubles as a circulatory system, distributing food around the body.

Cnidarians come in two essentially similar body forms: **medusa** and **polyp.** In both forms, a ring of

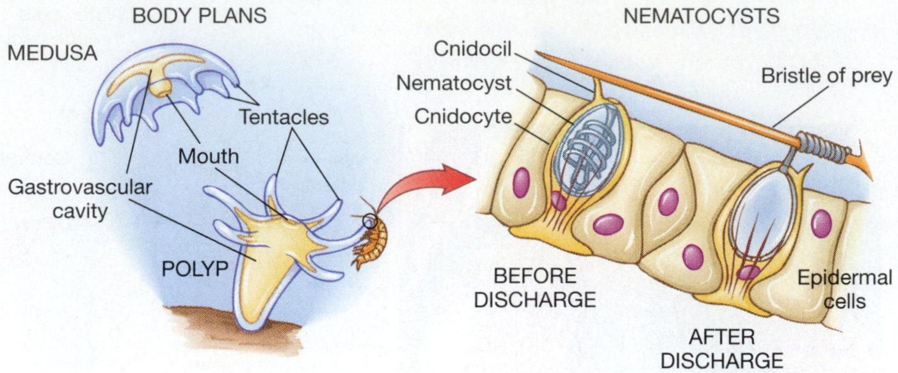

**FIGURE 28–7** Cnidarian features. There are two body plans: the free-floating medusa is an inverted version of the sessile polyp. In both, tentacles covered with nematocysts surround the mouth. The cnidocil acts as a trigger. Touching the cnidocil causes the nematocyst to eject its thread.

tentacles surrounds the mouth, which leads into the gastrovascular cavity. Polyps are sessile, and most medusas are free-swimming, moving by contractions of their bell-shaped bodies. Many cnidarians are colonial, consisting of numerous polyps and/or medusas attached to one another.

Most cnidarians are carnivores. They do not chase their prey but sit (or float) and wait. Although cnidarians are too weak to subdue animals with well-developed musculature, they have special weapons: **nematocysts,** giant organelles used in offense and defense. These are most concentrated on the tentacles. A nematocyst is the business end of a cell called a cnidocyte, which also usually possesses a trigger-like cnidocil. The cnidocil responds to touch or chemical stimuli, causing the cnidocyte to evert its thread-like nematocyst (Figure 28–7). The nematocyst may twist about

bristles on the body of the prey, entangling it, or may secrete a sticky or paralytic substance. Nematocyst toxins can cause a nasty sting, occasionally fatal to swimmers. Meat tenderizer rubbed on the sting is a good antidote because its proteolytic enzymes break down protein toxins.

Some predators are immune to the toxins of cnidarians. Leatherback sea turtles specialize in eating large jellyfish, and several reef-living fishes, such as parrotfish and butterfly fish, eat sea anemones and even corals.

In most cnidarians, the tentacles pull prey trapped by the nematocysts and stuff it into the mouth. Having nematocysts and a mouth and gastrovascular cavity permits cnidarians to eat much larger prey than a protozoan or sponge can manage. Digestion begins in the gastrovascular cavity. Small particles of food are engulfed from the cavity

**TABLE 28–2**

## Phylum Cnidaria and Its Classes

| Phylum Cnidaria (~9000 species) | Free-swimming medusas or sessile polyps; inner and outer tissue layers well developed, mouth surrounded by tentacles bearing nematocysts; gastrovascular cavity, no anus; predatory; solitary or colonial; radial symmetry; mostly marine. | | |
|---|---|---|---|
| **Class** | **Hydrozoa** | **Scyphozoa** | **Anthozoa** |
| | 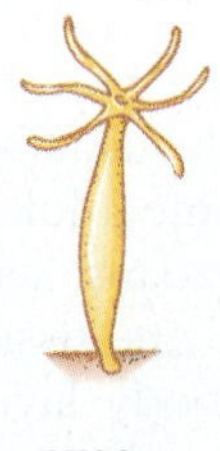 |  |  |
| **Number of living species** | ~2700 | ~200 | ~6100 |
| **Examples** | Portuguese man-of-war, *Hydra*, some corals, *Obelia*, *Tubularia* | Jellyfish, sea wasp, sea nettle | Sea anemone, most corals, sea fans, sea pansies |
| **Polyp** | Usually present, often colonial | Reduced | Present; sessile, often colonial |
| **Medusa** | Usually present | Well developed, large, free-swimming | Absent |
| **Habitat** | Marine, few freshwater | All marine | All marine |

by amoeboid cells, which complete digestion and distribute food to other cells.

Most cnidarians are marine, but there are a few freshwater forms.

Phylum Cnidaria has three classes: **Hydrozoa, Scyphozoa,** and **Anthozoa** (Table 28–2). Hydrozoa, the most primitive, is thought to have contained the ancestors of the other two.

Hydrozoan life cycles combine the polyp and medusa body forms in many different ways. Usually polyps, either alone or in colonies, are the feeding stage and reproduce asexually, giving rise to other polyps or to medusas. The medusas swim off and then reproduce sexually, giving rise to a new polyp generation (Figure 28–8). Many hydrozoans have a reduced medusa stage. Hence, it is not surprising that the medusa as well as the larva is absent in the commonly studied freshwater genus *Hydra. Hydra,* like most hydrozoans and scyphozoans, has separate sexes, whereas most anthozoans are hermaphroditic.

Scyphozoans, jellyfish, have a greatly reduced polyp stage and a prominent, planktonic medusa. Some species of Scyphozoa can swim quite actively by contraction of muscle fibers arranged around the bell. The almost transparent body is nearly invisible to prey, predator, and unwary swimmer alike (Figure 28–9).

Anthozoans, the sea anemones and corals, apparently lost the medusa stage during evolution, and a tiny, ciliated planula larva is the dispersal stage. Sea anemones are large anthozoans attached to rocks and other surfaces, mainly between high and low water marks around the ocean shore.

Most corals are tiny colonial polyps that build and inhabit a common skeleton of calcium carbonate. The calcium carbonate has formed rock layers, reefs, and islands in the warmer seas of the world (Section 47–H). In fact, Australia's Great Barrier Reef, more than 2000 km long, is the world's largest structure built by living organisms, a process that has taken an estimated one-half to one million years.

Corals contain symbiotic algae, which color the transparent polyps yellow or brown. The alga's photosynthesis often supplies more than half of the coral's food, and the coral's prey supplies nitrogen and phosphorus for both partners. Warmer-than-normal water temperatures during some years in the 1980s and 1990s are blamed for several episodes of widespread **coral bleaching,** when the polyps expelled their algae, leaving only the transparent animals in their white skeletons. This disrupts the corals' food supply and hence their growth and reproduction. It also drastically reduces the deposition of calcium carbonate, to only 4 to 30% of the normal rate, which may not keep pace with destruction of a reef by waves and human activity. Because a coral reef serves as a high-rise apartment complex for diverse plant and animal populations, coral bleaching threatens many forms of tropical marine life.

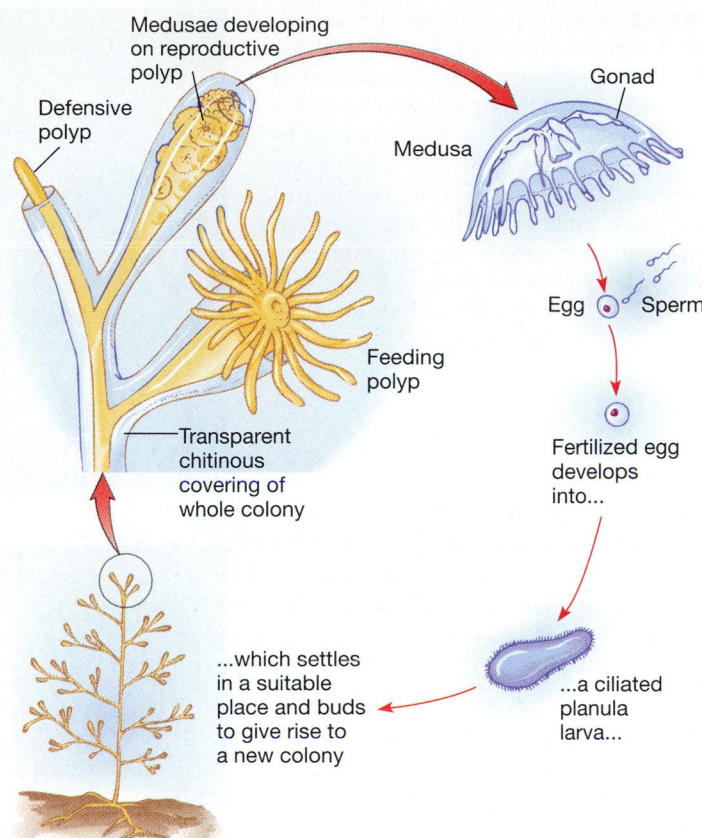

**FIGURE 28–8** The life cycle of a colonial member of the class Hydrozoa. The different individuals in the colony are specialized for different functions. Medusae bud off from reproductive polyps and swim away (or may remain attached to the colony). Gonads consist of clusters of developing gametes, which are released when mature.

## 28–F    PLATYHELMINTHES: FLATWORMS

Phylum Platyhelminthes contains acoelomate worms with flattened bodies (platys = flat; helmis = worm). They are thought to have evolved from cnidarian ancestors in which the ciliated larva developed into a new adult form with bilateral symmetry. This is an adaptation to a more active way of life, as are the beginnings of cephalization, with eyespots, chemical-sensing cells, and a "brain" (a cluster of nerve cells) at the anterior end. Flatworms also have true organs, including layers of muscles that are part of a well-developed middle layer of cells. This layer also contains an excretory system and reproductive structures. Most flatworms are hermaphroditic, some with complex reproductive systems.

Flatworms retain some ancestral characters: the mouth is the digestive tract's only opening, and there is no circulatory system. Instead, the digestive and excretory systems extend throughout the body, carrying food and collecting

**FIGURE 28–9** Representative cnidarians. (a) A siphonophore, a large colonial hydrozoan, which swims by contractions of its many bells. It traps prey, which include quite large fish, with nematocysts on its trailing tentacles. (b) A scyphozoan medusa. (c) A sea anemone, member of the class Anthozoa. The tentacles trap food and pass it to the mouth in the center of the ring of tentacles. (d) The tiny polyps of a coral of the class Anthozoa. (a, James A. King; b, Bruce Robison; c, Biophoto Associates; d, Steven Webster)

(a)  (b)  (c)  (d)

**FIGURE 28–10** The internal anatomy of a free-living planarian platyhelminth worm. The mouth is not at the animal's front end, but in the middle of the body. When the worm feeds, the pharynx is everted through the mouth. Both male and female reproductive organs are present.

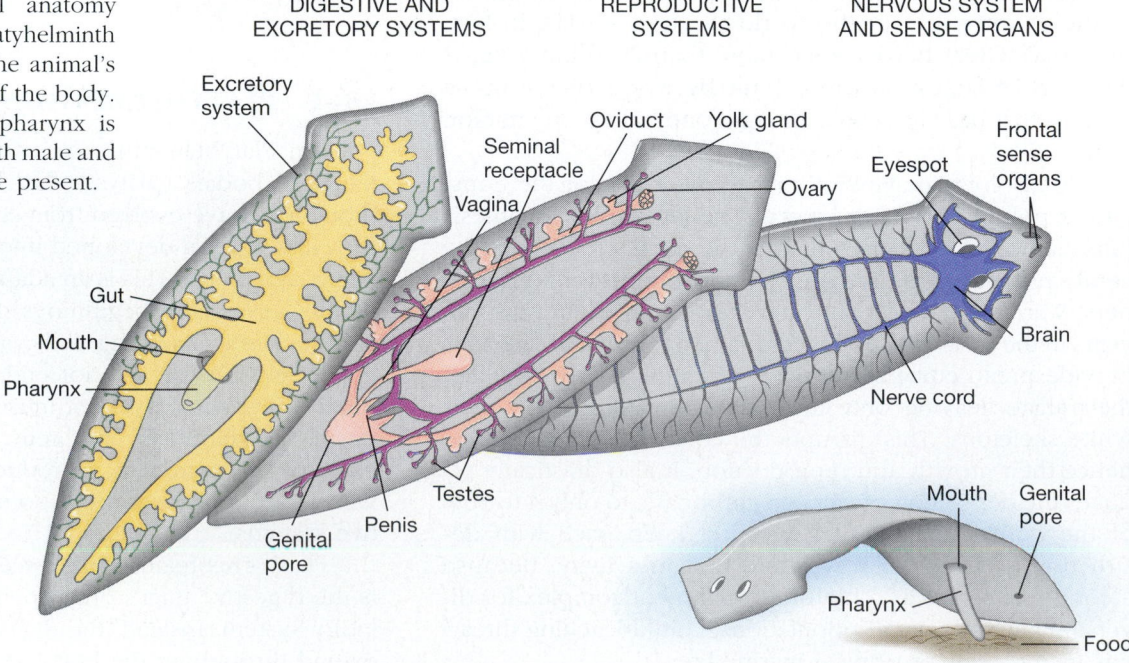

DIGESTIVE AND EXCRETORY SYSTEMS

Excretory system

Gut

Mouth

Pharynx

REPRODUCTIVE SYSTEMS

Seminal receptacle

Vagina

Oviduct

Yolk gland

Ovary

Testes

Penis

Genital pore

NERVOUS SYSTEM AND SENSE ORGANS

Eyespot

Frontal sense organs

Brain

Nerve cord

Mouth

Genital pore

Pharynx

Food

FEEDING

**TABLE 28–3**

## Phylum Platyhelminthes and Its Major Classes*

| Phylum Platyhelminthes (~12,700 species) | Flatworms. Free-living or parasitic; bilateral symmetry; three well-developed layers with organs; gut with mouth but no anus; marine or freshwater. | | |
| --- | --- | --- | --- |
| **Class** | **Turbellaria** | **Trematoda** | **Cestoda** |
| | | | |
| **Number of living species** | ~3000 | ~6000 | ~1500 |
| **Common names** | Free-living flatworms; planarians | Flukes | Tapeworms |
| **Characteristics** | Ciliated body surface | Two suckers attach to host | No head or digestive system; scolex attaches to host; proglottids produce eggs and break off after fertilization |
| **Way of life and habitat** | Predators and scavengers; most marine, some freshwater, a few terrestrial | Parasites, almost always of vertebrates; life cycle with intermediate host; larvae multiply asexually | Parasites of vertebrates; life cycle with one or more intermediate hosts |

*A fourth class, Monogenea, contains about 1100 species of fluke-like external parasites without intermediate hosts or larval reproduction.

wastes. Gas exchange occurs through the general body surface: the flattened body shape means that no cell is far from the external source of oxygen (Figure 28–10).

The phylum Platyhelminthes includes three major classes of worms. Two of these, the Trematoda (flukes) and Cestoda (tapeworms), contain only parasites and probably evolved from members of Class Turbellaria (Table 28–3).

## Turbellaria

Like the ancestral forms of many other phyla, turbellarians are mostly free-living predators or scavengers (Figure 28–11). They range from microscopic size to 60 cm long, although most are less than 1 cm. Cilia cover the entire body in small forms but only the ventral surface of larger ones. Turbellarians are often recognizable by light-sensing eyespots on the top of the head. Most are marine but a

**FIGURE 28–11** A colorful turbellarian. (Marjorie Banks/Norbert Wu)

few, like *Dugesia* (often studied in the laboratory), live in fresh water. The few terrestrial species spend most of their time in moist areas, under logs and leaf mold, for example. Small turbellarians move by secreting slimy mucus and gliding along it using ciliary action. The muscles are used in making turns or in curling up when disturbed. Larger forms move by muscle power.

Turbellarians feed on smaller organisms or on dead organic matter. Small bits of meat are used as bait to attract turbellarians for collection. The first part of the gut is a muscular bulb or tube, the **pharynx,** which is protruded through the mouth to feed.

Nearly all turbellarians are hermaphroditic. Sexual reproduction occurs by mutual copulation between two individuals, with internal fertilization. After embryonic development, the eggs of some species hatch as miniature adults, whereas others produce ciliated larvae that disperse in the plankton (see Figure 28–2). Some turbellarians are sometimes parthenogenetic, producing new individuals from unfertilized eggs, and in some species males are unknown. Many also reproduce asexually by budding or by dividing into two.

## Trematoda: Flukes

Flukes, members of the class Trematoda, are all parasites that live in other animals. They usually have adhesive organs or suckers near the mouth and a second sucker on the ventral surface. The pharynx sucks host tissue or body fluids in through the mouth and passes it into the two branches of the intestine.

Flukes produce huge numbers of offspring—perhaps releasing half a million eggs during a lifetime. Most flukes are hermaphroditic, and reproductive organs occupy most

**T A B L E   2 8 – 4**

**Some Parasitic Worms Common in Humans**

| Name | Symptoms | Means of Infection |
|---|---|---|
| **Platyhelminthes** | | |
| Chinese liver fluke | None in mild cases; destruction of liver, bile stones, and clogging of liver ducts in severe cases | Eating infected raw fish |
| Blood fluke (schistosomes) | Enlargement of liver and spleen<br>Urinary disorders<br>Bloated abdomen, wasted arms and legs | Drinking or wading barefoot in water containing infected person's urine. Infects about 300 million people in 70 nations. Not found where there are modern sewage disposal systems |
| Swimmer's itch | Itching after exposure of skin to infested water | Burrowing of fluke larvae of species that cannot successfully infect humans |
| Bladder worm *(Echinococcus)* (immature stage of a worm that lives in dogs when adult) | Cysts up to the size of an orange; symptoms depend on part of body invaded | Infected dogs licking people's hands or faces or contaminating drinking water |
| Pork, beef, and fish tapeworms | Immature worms: cysts<br>Adult worms may cause diarrhea, weight loss, perforation of intestine | Eating undercooked meat containing worm cysts |
| **Nematoda** | | |
| Pinworm | Anal itching | Females lay eggs around anal opening; hands may transfer eggs to mouth, maintaining infection in same person. Physical contact may also transfer to other people |
| Hookworm | Anemia, lethargy | Young worms burrow through skin (bare feet) from moist soil and grass contaminated by feces of infected humans |
| Filaroid worms | Lymphatic blockage, elephantiasis in extreme cases<br>Blindness | Bite of infected mosquitoes<br><br>Bite of infected blackflies |

of the body. Flukes have complex life cycles involving several different hosts. The host of the adult is the **primary host;** hosts of larval stages are **intermediate hosts.** Fluke larvae multiply asexually, and so a single fluke egg can generate many adults. Flukes' enormous reproductive capacity is an adaptation to a parasitic mode of life: the chances of any one egg or larva finding a proper host are minuscule.

It is never safe to bathe in a tropical pond because the water may contain larval blood flukes of the genus *Schistosoma,* the cause of the devastating disease schistosomiasis (bilharzia). These flukes afflict about 300 million people in more than 70 nations in the tropics, killing about 200,000 a year (Table 28–4). In some African tribes, the disease is a way of life: bloody urine, a symptom of schistosomiasis, is regarded as a sign of male puberty. Fluke eggs enter the water in human feces and urine. In fresh water, they hatch into ciliated larvae, which may be eaten by their intermediate host, a species of snail. Here they reproduce asexually. Cercaria larvae are released with the snail's feces and burrow in through the skin of any human in the water (Figure 28–12).

Economic development has worsened the problem by providing more host snail habitat: new irrigation ditches and reservoirs. In addition, money for research into all tropical diseases has declined drastically since many of the countries where they occur ceased to be colonies of wealthier nations. So far, the best prospect seems to be new, more effective drugs, which have achieved a reported 90% cure rate in China.

## Cestoda: Tapeworms

Tapeworms are very specialized parasites that live in the intestines of probably every species of vertebrate. Their adaptation to this way of life involved loss of the head, mouth, and digestive system. Surrounded by digested food, the worm simply absorbs it through the body surface, which is greatly expanded by microvillus-like projections. These worms are facultative anaerobes, adapted to their low-oxygen surroundings.

The anterior of the body usually has a **scolex,** bearing a circle of hooks, suckers, or both, which attach to the lin-

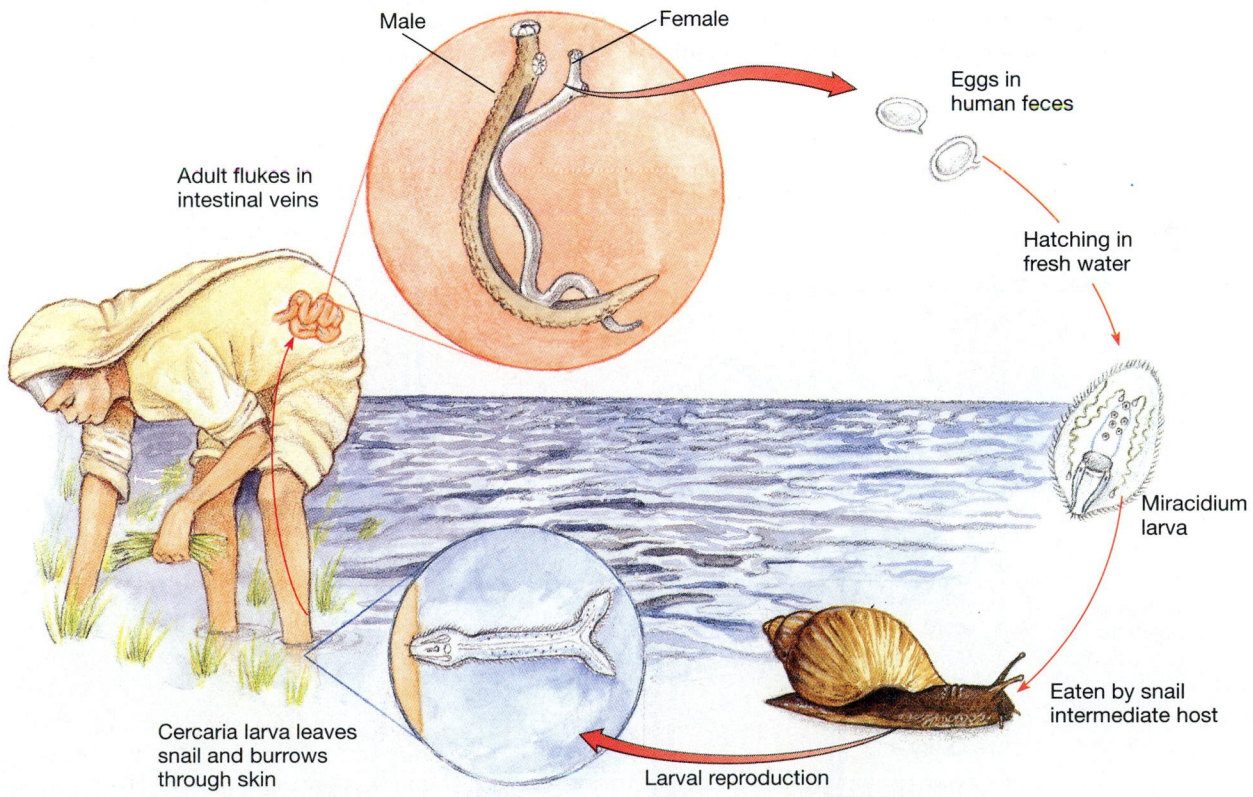

**FIGURE 28–12**   The life cycle of the fluke *Schistosoma mansoni,* the cause of schistosomiasis (bilharzia). Larvae eaten by a suitable aquatic snail may reproduce asexually and make several hundred offspring each. The motile cercaria discharged from the snail finds a human host, burrows through the skin, and migrates through the circulatory system to veins in the intestine. The female fits into a groove in the male's body. She lays eggs in small blood vessels and they eventually break into the intestinal lumen and leave in the feces. In fresh water, the eggs hatch into miracidium larvae.

ing of the host's intestine. Immediately behind the scolex is a growing region that constantly produces new body sections, called **proglottids,** by budding. The chain of proglottids may be a few centimeters to several meters long, depending on the species. Although the nervous, muscular, and excretory systems run throughout the chain, a mature proglottid's main features are its reproductive organs (Figure 28–13). It mates with itself or with another proglottid of the same or a different individual. Then, full of fertilized eggs, it breaks away from the rest of the worm and is passed out of the host with the feces. Some species of tapeworms produce nearly a million eggs per day. Depending on the species, either the egg or the hatched larva must be eaten by another host before developing further. Many cestodes must be eaten by several different host species in succession to complete their development.

The beef and pork tapeworms, *Taeniarhynchus saginatum* and *Taenia solium,* are the most common tapeworms infecting humans in the United States. However, the growing popularity of sushi is likely to raise the incidence of tapeworms from fish here, as in the Orient. Tapeworm eggs or larvae enter these hosts with the food. The larvae burrow through the host's gut wall, enter the blood, and travel to the muscles, where they form cysts. They infect humans that later eat this flesh if it has not been cooked thoroughly enough to kill the worms. Infected persons can also contaminate other foods if they do not wash their hands before preparing it. Dogs and cats may get tapeworms from eating wild animals such as rabbits and from ingesting fleas, which are intermediate hosts to several species.

## 28–G   NEMATODA: ROUNDWORMS

Nematodes are pseudocoelomates. The phylum is named for their long thin bodies with tapered ends (nema = thread). This shape is excellent for squirming into small spaces, be they between sand grains or within the tissues of other organisms, and roundworms have conserved it during their adaptive radiation into many habitats and many modes of nutrition. Most nematodes are less than a few millimeters long and therefore go unnoticed, even though they can be found by the millions in environments from deserts and forests to the ocean depths, from polar regions to the tropics. They are important inhabitants of marine and freshwater mud and of the soil; any handful of good soil contains thousands of tiny white or transparent roundworms. Nematodes may eat bacteria, burrow into roots or suck their juices from outside, capture smaller animals, or live as animal or plant parasites.

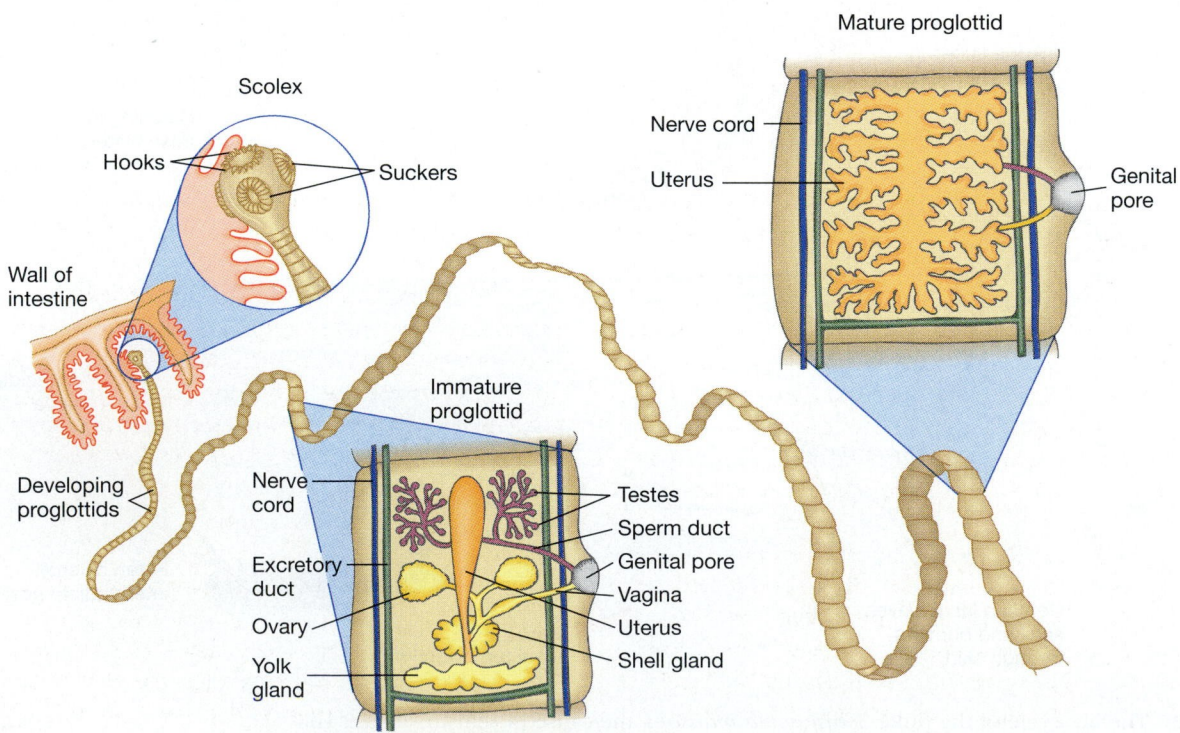

**FIGURE 28–13** *Anatomy of the pork tapeworm,* Taenia solium. *The hooks and suckers of the scolex attach to the wall of the host's digestive tract. Proglottids develop just behind the scolex. A mature proglottid is little more than a reproductive system containing both male and female organs, which may fertilize one another or mate with another proglottid. Ripe proglottids detach from the end of the worm and are shed in the host's feces.*

(a)

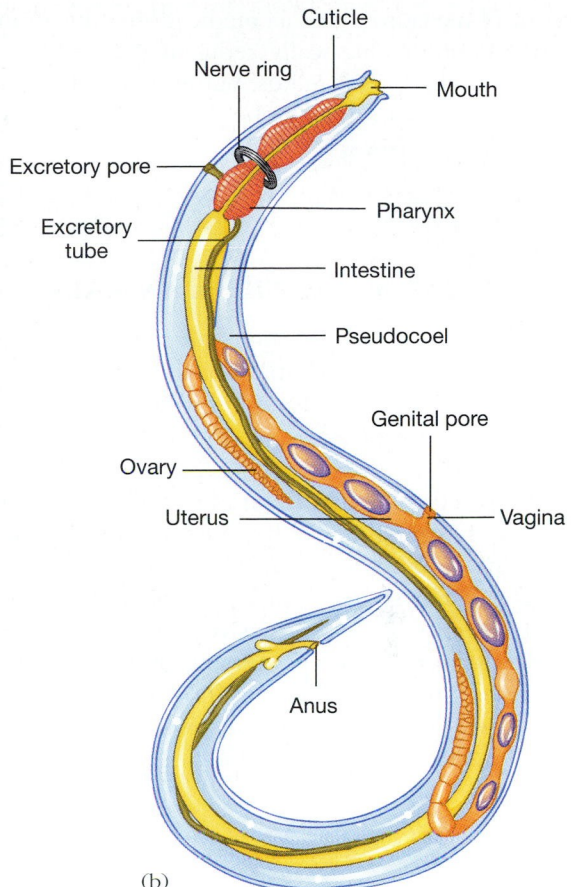

Cuticle
Nerve ring
Mouth
Excretory pore
Pharynx
Excretory tube
Intestine
Pseudocoel
Genital pore
Ovary
Uterus
Vagina
Anus

(b)

**FIGURE 28–14** Nematodes. (a) Scanning electron micrograph of a free-living nematode from the soil. (b) Internal anatomy of a female nematode. Advances over platyhelminths include a gut with an anus as well as a mouth, and a fluid-filled body cavity. (a, Biophoto Associates)

Nematodes have advanced over the flatworms in possessing a complete digestive tract in the form of a simple tube, with an anus as well as a mouth (Figure 28–14). Hence, they have a true "tube within a tube" construction. The pseudocoel lies between the digestive tube and the muscles and other organs derived from the mesoderm (see Figure 28–4). This fluid-filled cavity distributes food molecules and dissolved gases, a rudimentary but adequate arrangement for such small, thin-bodied animals. A tough cuticle covers the outside of the body, and an adhesive gland at the posterior tip provides temporary anchorage to nearby objects.

Roundworms do not have the layer of circular muscles that permit other worms to squeeze their bodies thinner and extend forward. Their muscles run only lengthwise and bend the body from side to side by alternate contractions, pulling against the elastic cuticle and the pressure of fluid in the pseudocoel. This results in inefficient thrashing unless the body has solid objects or a viscous fluid to push against. A peculiarity of nematodes' muscles is that they grow long thin extensions to nerve cells, rather than the nerve cells growing to the muscles, as in other animals.

Although hermaphroditic and parthenogenetic species exist, most nematodes have separate sexes. The male in-

troduces sperm into the body of the much larger female, and the eggs are fertilized internally. The sperm are unusual in that they do not have flagella—in fact, no nematode cells have cilia or flagella.

The largest nematodes are found among species that parasitize vertebrates, attaining lengths of several centimeters to more than a meter. About 50 species parasitize humans. *Trichinella spiralis* causes trichinosis, often contracted by eating undercooked, infected pork, although an outbreak in Canada was traced to undercooked bear meat. The human pinworm, *Enterobius vermicularis,* lives in the intestines of children throughout the world but does little damage.

Among the more harmful species are the intestinal roundworm *Ascaris lumbricoides* and the guinea worm. Filaroid worms include some that lodge in the lymph nodes and in extreme cases cause elephantiasis (enormous swelling of the leg) and some that cause river blindness in an estimated 40 million people in Africa and Latin America.

Hookworms enter the skin of people walking barefoot on soil contaminated with feces from infected humans. These worms caused the anemia, weakness, and fatigue—formerly attributed to laziness—of thousands of poor laborers in the southern United States. Early in this century, control measures such as medication, education, and modern plumbing drastically reduced the incidence of hookworm in the United States, but millions of people in warm, rural areas around the globe are still afflicted with these debilitating parasites.

## 28-H   ROTIFERA: WHEEL ANIMALS

The tiny, transparent rotifers are among the lesser-known invertebrates, but members of this phylum are very common and likely to be found in water collected from a pond or stream, a roof gutter, or a clump of moss. They can be recognized by the crown of cilia around the mouth, at the anterior end. Beating in rapid sequence, the cilia look like a wheel turning; hence the phylum name Rotifera ("wheel-bearing"). The body consists of a trunk containing the various organ systems, with a posterior foot, which often has one or more toes with adhesive glands (Figure 28–15). Many rotifers are no larger than some protists, but they are nevertheless multicellular, with all of the organs found in nematodes, including a pseudocoel and cuticle; in fact, the two phyla are believed to be closely related.

Most rotifers live as free-swimming individuals in fresh water, but some are sessile, and a few are colonial. Only about 50 species are marine. Some rotifers are symbiotic, some are parasites on other invertebrates, and some are predators, eating mostly protozoans and other rotifers.

A rotifer's cilia propel it through the water and also create a feeding current, which pulls small organisms and organic particles into the animal's mouth. Jaws in the pharynx grind the food and pass it on to the stomach and intestine.

Both sexual reproduction and parthenogenesis are common among rotifers, often in the same species. Males are unknown in some species. When present, they are much smaller than females and sexually mature at hatching. Females generally produce only one, comparatively large, egg at a time. During their lifespan of only a few weeks, they produce only 8 to 20 eggs. The eggs often remain within the mother's body, or attached by an external stalk, until hatching.

A peculiarity of rotifers is that the adults' cells never divide. Mitosis ceases following embryonic development, and the adult cannot grow or repair itself. Not surprisingly, rotifers are used in research designed to determine how mitosis is controlled.

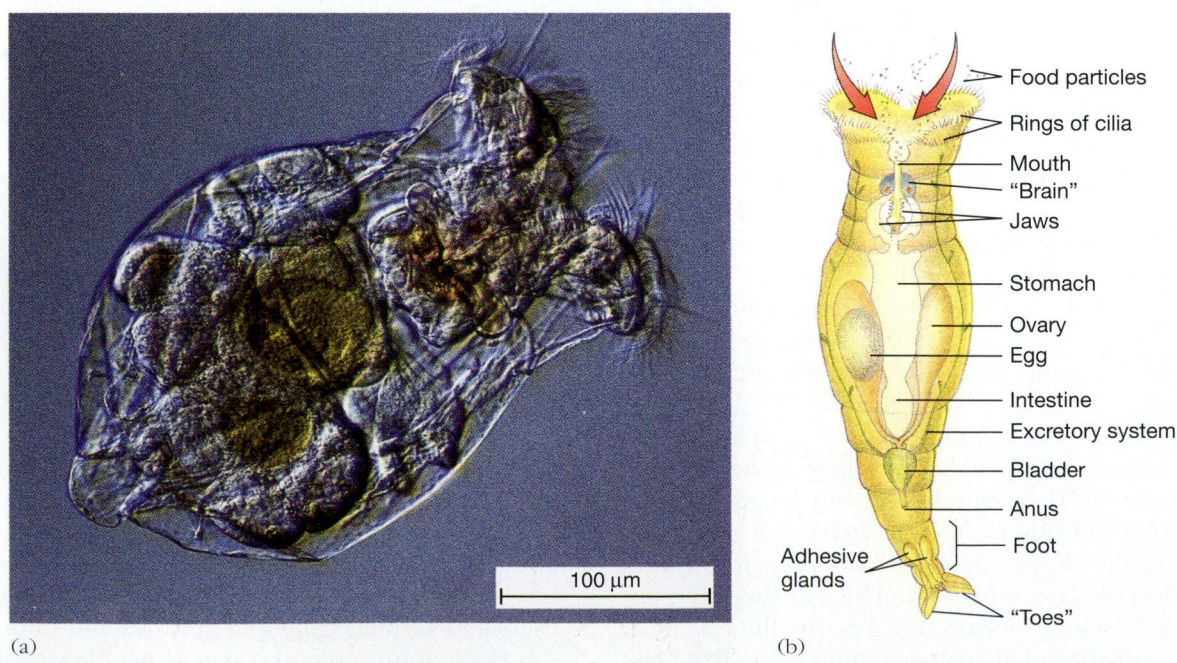

(a)                                                                 (b)

**FIGURE 28–15**   Rotifers. (a) A living rotifer (*Brachionus*). (b) Anatomy of a rotifer. Internal organs are easily seen through the transparent body wall of many forms. The "toes" are often provided with glands that secrete glue by which the rotifer can anchor itself to the substratum. Most rotifers are females, and most of their offspring are females.   (a, Robert Brons/BPS)

Chapters 28 to 30 discuss only the larger (in number of species) invertebrate phyla. Here we briefly introduce those invertebrate phyla with fewer than 1500 living species each.

## Phylum Ctenophora

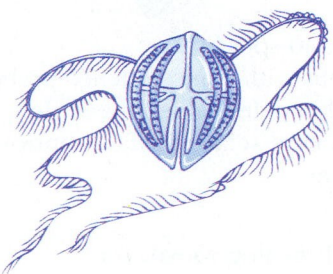

(~90 species)
Comb jellies. Once classified as Cnidaria, which they resemble, but only one ctenophore species has nematocysts; radial symmetry or secondary bilateral symmetry; free-swimming, predatory, with two tentacles and eight rows of ciliary combs for locomotion; marine.

## Phylum Mesozoa

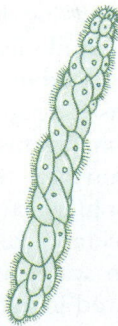

(~50 species)
Peculiar ciliated parasites of marine invertebrates; contain few, large cells and no organs.

## Phylum Nemertina (= Rhynchocoela)

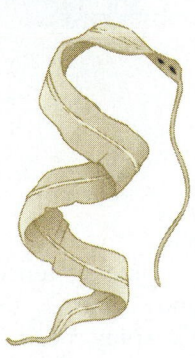

(~900 species)
Ribbon worms. Long, flattened acoelomate worms 0.5 mm to 30 m long; ciliated body surface; complex proboscis used to explore environment and catch prey; anus and circulatory system present; usually with separate sexes, external fertilization; most marine, a few freshwater or terrestrial.

## Phylum Gnathostomulida

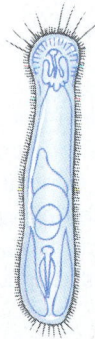

(~80 species)
Tiny, ciliated, acoelomate worms once classified with platyhelminths, which they resemble except in having only one cilium per body surface cell; live in marine sand or on surface of algae. First described in 1956.

## Phylum Gastrotricha

(~460 species)
Microscopic to 4 mm with slit-like pseudocoel spaces; body ciliated ventrally; adhesive tubes and spines on sides of body; cuticle with decorative patterns; marine and freshwater, among bottom particles.

## Phylum Kinorhyncha

(~100 species)
Tiny (up to a few mm), elongated worm-like body with spiny cuticle; move by anchoring snout and pulling body along; pseudocoel; anus present; marine, on muddy intertidal shores.

*(Continued)*

**E S S A Y** *(Continued)*

### Phylum Nematomorpha

(~230 species)

Hairworms: some members once believed to be horsehairs come to life in watering troughs. Thread-like, pseudocoelomate worms. Juveniles parasitic in arthropods; adults non-feeding, free-living in fresh water or damp soil.

### Phylum Acanthocephala

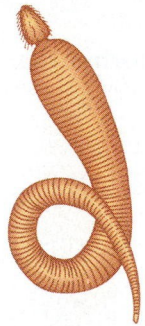

(~1000 species)

Small pseudocoelomate worms parasitic on vertebrates, with arthropod intermediate hosts; retractile proboscis with spines, no gut.

### Phylum Priapulida

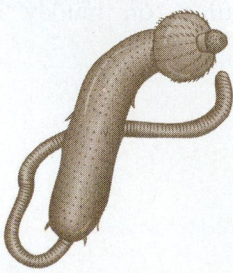

(13 species)

Cucumber-shaped animals, 1 to 200 mm long, with reduced coelom and proboscis; covered with spines; marine, bottom-burrowing predators.

### Phylum Loricifera

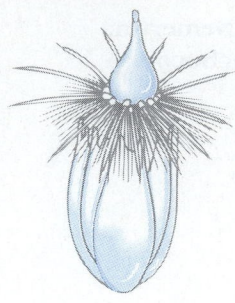

(~10 species)

Microscopic, clinging tenaciously to marine sand grains; telescoping mouth tube, spiny head, girdle of spiny cuticle plates around abdomen; pseudocoelomate; gut with mouth and anus, separate sexes. First species described in 1983.

### Phylum Sipuncula

(~320 species)

Plump, cylindrical coelomate worms with lobes or short tentacles around mouth used in deposit feeding; burrowing or inhabiting empty tubes or shells of other animals; possibly common ancestor of major deuterostome phyla; marine.

### Phylum Echiura

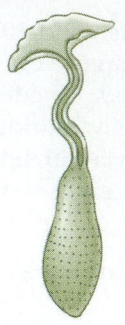

(~140 species)

Worms with large, flattened proboscis used in detritus feeding; coelom; pair of ventral setae (bristles); burrowing; marine.

### Phylum Pogonophora

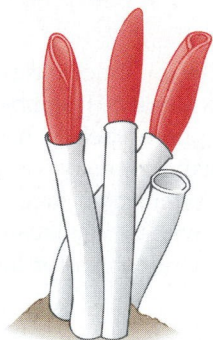

(~80 species)

Long (10 to 150 cm), sessile tube-living worms having long tentacles with microvilli and cilia; gut absent; food possibly absorbed through body surface; at least one, the largest, living near thermal vents in sea floor probably nourished by symbiotic chemoautotrophic bacteria that live in its body and metabolize $H_2S$ from water. Marine, in water at least 100 m deep; discovered in 1900.

## Phylum Tardigrada

(~375 species)
Water bears. Small (maximum 1 mm), segmented body covered with periodically shed cuticle; four pairs of clawed legs; stylus of mouth used to suck fluids from plants, rotifers, nematodes, or other tardigrades; form dormant stage able to survive extremely low temperatures and desiccation for long periods; most terrestrial, in water on moss and lichens, some marine or freshwater.

## Phylum Onychophora

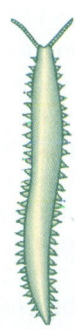

(~70 species)
Segmented, worm-like body 1.5 to 15 cm long, with thin cuticle; antennae and many pairs of conical, unjointed legs with claws; terrestrial carnivores. Living in humid areas south of Tropic of Cancer that were formerly part of Gondwana, the distribution of onychophorans provides strong evidence for theory of continental drift (*Essay: Continental Drift,* Chapter 19).

## Phylum Pentastomida

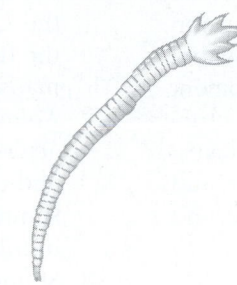

(~90 species)
Worm-like parasites in nasal passages or lungs of vertebrates; two pairs of leg-like anterior appendages with claws; chitinous cuticle; closely related to arthropods (Section 29–F).

## Phylum Phoronida

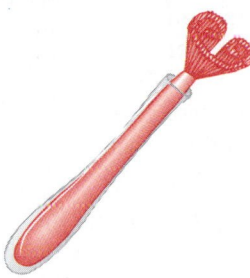

(15 species)
Lophophorate* worms living in chitinous tubes; marine.

## Phylum Entoprocta

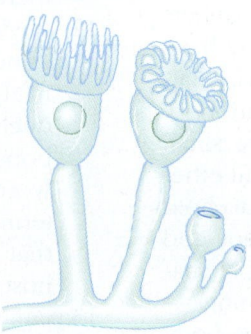

(~150 species)
Sessile lophophorates* with both mouth and anus opening within circle of tentacles; mainly sessile, attached by stalks, in colonies; almost all marine.

## Phylum Brachiopoda

(~300 species)
Lamp shells. Suspension feeders with lophophorate body enclosed by two unequal calcareous shells superficially resembling the shells of bivalve molluscs; attached to substratum by a stalk; marine. Remnants of a once-great phylum: 30,000 fossil species, mostly Paleozoic and early Mesozoic.

## Phylum Chaetognatha

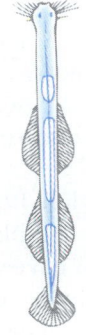

(~70 species)
Arrow worms. Arrow-like, active predators 6 to 70 mm long, with anterior chitinous grasping spines; using tail to swim, fins for stability; marine planktonic.

*Lophophorates are coelomate invertebrates with a crown of hollow tentacles (the lophophore) surrounding the mouth and used in filter feeding.

## SUMMARY

Animals are eating machines, adapted to detecting, moving toward, and ingesting organic material in bulk. In sessile forms only the larva moves around. Sexual reproduction permitted adaptive radiation into millions of species with a wide range of diets and habitats. Embryonic development permits new individuals to establish their basic body structures before they begin feeding and moving for themselves. In most animals, coordination among sensory, nervous, and muscular systems produces complex feeding, defensive, and reproductive behaviors.

1. The so-called lower invertebrates introduced in this chapter are believed to have evolved in the sea, where most forms still live. These animals are either planktonic or bottom-dwelling, with motile larvae. Some have adapted to life in fresh water, in damp terrestrial habitats, or in the watery interior of a host's body.

2. Sessile animals live as filter feeders or predatory trappers. Mating usually occurs by release of gametes into the water, dispersal by a motile larval stage. Local spreading by asexual reproduction or colony formation is common. Animals that can move around have a wider range of feeding methods.

3. Embryonic development occurs by way of a hollow, one-layered blastula. In most animals the embryo then develops three germ layers, which establish the basic body plan of an inner digestive tube within an outer sensory and protective tube, and between them a layer containing the body wall muscles, reproductive and excretory organs, and (in many higher forms) the circulatory system.

4. Members of phylum Porifera do not develop these three cell layers after the blastula stage. Sponges are primitive in that they have no true tissues, body symmetry, nor digestive cavity. Cnidarians have only two well-developed layers of cells and radial symmetry. Members of phyla Platyhelminthes, Nematoda, and Rotifera have three cell layers, true organs and organ systems, bilateral symmetry, and the beginnings of cephalization.

5. The most primitive animals, the sponges, take small food items directly into their cells. Members of all other major phyla have a space inside the body able to hold larger food, such as entire animals. In cnidarians and flatworms, this is a sac with only one opening. Nematodes and rotifers have a tube with two openings—mouth and anus—encased in an outer tube, the body proper.

6. There are few clues to the origin and early evolution of animals. Animals first appeared about 600 million years ago, in the Precambrian era. A rapid burst of adaptive radiation produced many new phyla at the start of the Cambrian, about 540 million years ago, including the first definite members of the phyla living today and many more long since extinct.

7. Animals are classified on the basis of conservative characters such as embryonic development, basic body plan, and definitive specialized cells and organs. Table 28–1 summarizes the characteristics of the major animal phyla in this chapter.

8. Sponges, phylum Porifera, exhibit a primitive level of organization, with specialized cells but no tissues or organs. They are sessile filter feeders, with flagellated choanocytes that draw water into the body and strain out food particles. Their cells show marked independence of one another.

9. Members of phylum Cnidaria have two well-developed tissue layers separated by mesoglea, but they lack most organs. Both sessile polyps and swimming medusas have the same body plan—a ring of tentacles bearing nematocysts surrounds the mouth, which leads into a blind sac, the gastrovascular cavity. *Hydra,* jellyfish, corals, and sea anemones belong to this group.

10. Platyhelminthes, or flatworms, show several advances, including bilateral symmetry, the beginnings of cephalization, and three well-developed tissue layers; there are several organ systems, but no circulatory system. The digestive system has only one opening, the mouth. Free-living flatworms and parasitic flukes and tapeworms belong to this phylum.

11. Nematoda (roundworms) and Rotifera ("wheel animals") possess complete digestive tracts, with both mouth and anus, and many other organ systems. A fluid-filled body cavity, the pseudocoel, may serve for transport, but there is no distinct circulatory system.

12. Phyla Platyhelminthes and Nematoda contain many parasitic worms, with greatly reduced digestive tracts (none in tapeworms) and expanded reproductive systems producing enormous numbers of eggs. Fluke larvae also boost their success by reproducing asexually. Parasitic worm life cycles often involve larval development in intermediate hosts, and there may be an inactive cyst stage that can survive until it is ingested by an appropriate host.

13. Parasitic flatworms and roundworms cause debilitating infections in hundreds of millions of humans living today.

## Self-Quiz

1. Jellyfish are members of phylum _____.
2. Sponges are members of phylum _____.
3. The most primitive phylum of animals with bilateral symmetry and true organs is:
   a. Cnidaria            c. Porifera
   b. Platyhelminthes     d. Nematoda
4. What is the advantage of having an internal digestive cavity?
5. The chief function of the larval stages of marine invertebrates is _____. It is generally advantageous for animals with larval stages to produce large numbers of offspring because _____.
6. Hookworms belong to phylum _____.
7. Phylum _____ contains filter feeders.
8. The embryonic ectoderm develops into _____.
9. Organisms with bilateral symmetry also have:
   a. rudimentary cephalization

b. a digestive tube with two openings
c. a well-developed circulatory system
d. two defined tissue layers

*Matching:* For each group listed below, pick out *all* the characteristics in the list (a through j) which pertain to that group.

| Group | Characteristic |
|---|---|
| _____ 10. Rotifera | a. gut with mouth and anus |
| _____ 11. Platyhelminthes | b. sessile members |
| _____ 12. Nematoda | c. parasitic members |
| _____ 13. Cnidaria | d. excretory system |
| _____ 14. Porifera | e. nematocysts |
| | f. no gut present |
| | g. choanocytes |
| | h. bilateral symmetry |
| | i. radial symmetry |
| | j. pseudocoel |

## Questions for Discussion

1. How do the organisms covered in this chapter survive without special organs for gas exchange and internal transport?
2. What selective pressures do you think have made it advantageous for flukes to produce hundreds of eggs at a time and for rotifers to produce only one or a few at once?
3. What are the advantages of parthenogenesis and of asexual reproduction? Why are they so common among the animals discussed in this chapter?
4. Paleontologist Steven Jay Gould contends that it was largely a matter of chance which Cambrian animals became extinct and which survived and founded the adaptive radiations in later geological periods. Do you agree?

## Suggested Readings

Barnes, R. D. *Invertebrate Zoology,* 5th ed. Philadelphia: Saunders College Publishing, 1987. An excellent text and reference book covering the invertebrates except insects.

Barth, R. H., and R. E. Broshears. *The Invertebrate World.* Philadelphia: Saunders College Publishing, 1982.

Brown, B. E., and J. C. Ogden. "Coral bleaching." *Scientific American,* January 1993.

Buchsbaum, R. *Animals Without Backbones,* 2d ed. Chicago: University of Chicago Press, 1976. A classic elementary textbook on invertebrates.

Childress, J. J., H. Felbeck, and G. N. Somero. "Symbiosis in the deep sea." *Scientific American,* May 1987. Animal adaptations to life in deep-sea hydrothermal vents.

Goreau, T. F., N. I. Goreau, and T. J. Goreau. "Corals and coral reefs." *Scientific American,* August 1979. Biology, growth, and ecology of reef-building corals.

Levinton, J. S. "The big bang of animal evolution." *Scientific American,* November 1992. Addresses the question, "Why have no new phyla appeared since the Cambrian Period?"

McMenamin, M. A. S. "The emergence of animals." *Scientific American,* April 1987.

Moore, J. "Parasites that change the behavior of their host." *Scientific American,* May 1984.

Richardson, J. R. "Brachiopods." *Scientific American,* September 1986. The biology of surviving members of a once-major animal phylum.

Wilson, H. V. "On some phenomena of coalescence and regeneration in sponges." *Journal of Experimental Zoology* 5:245, 1907. An early account by a pioneer embryologist; well worth reading for its elegant prose and clear thinking.

**29**

# Higher Invertebrates

## OBJECTIVES

*When you have studied this chapter you should be able to:*

1. State what a coelom is, and discuss its evolutionary importance.
2. List the distinguishing characteristics of annelids, bryozoans, molluscs, arthropods, echinoderms, hemichordates, chordates, urochordates, and cephalochordates.
3. Discuss reasons for the success of the annelids, molluscs, arthropods, echinoderms, and chordates.
4. When presented with an annelid, mollusc, arthropod, echinoderm, or chordate, name the phylum to which it belongs.

5. Name, describe, and give an example of each of the three classes of annelids.
6. Name, describe, and give examples of four classes of molluscs.
7. Name, describe, and give examples of each of five major classes of arthropods.
8. Describe the adaptations of insects to a terrestrial way of life.
9. Name, describe, and give examples of each of five major classes of echinoderms.

T he invertebrates in Chapter 28 are not familiar to most of us because they are largely marine. Even the ones abundant in fresh water and on land, such as rotifers and soil nematodes, are small and easily overlooked. By contrast, the invertebrates in this chapter show evolutionary advances that have allowed many of them to become larger. Some of these larger invertebrates are well represented in moist terrestrial habitats: earthworms, snails, slugs, centipedes, and millipedes abound in soil and in leaf litter. Spiders and insects are more fully terrestrial and may spend a lot of time out in the open (Figure 29–1).

Before we study these animals themselves, we will look more closely at their major evolutionary advance over the phyla we already met: the coelom. A **coelom** is a fluid-filled body cavity in the mesoderm (Section 28–B). The animals in this chapter are basically coelomate, although the coelom has secondarily become reduced in some groups.

The coelomate animals may be divided into two evolutionary lines, largely on the basis of their embryonic development, a feature that often provides evidence about relationships. One line contains the **protostomes,** including Annelida, Mollusca, and Arthropoda. The other, or **deuterostome,** line has three major phyla: Echinodermata, Hemichordata, and Chordata (including the vertebrates).

**FIGURE 29–1** Terrestrial invertebrates. This scorpion carries her young, defending them with her sting against other residents of a tropical dry forest in Costa Rica. Scorpions are members of phylum Arthropoda, class Arachnida.   (John Cancalosi/DRK Photo)

## KEY CONCEPTS

◆ The evolution of a coelom permitted the evolution of a circulatory system with a pumping heart.

◆ An efficient circulatory system opened the way for the evolution of a larger overall body size and larger, more compact internal organs.

◆ The enlarged, complex brain of many coelomates permits complex behavior patterns.

◆ The efficient circulatory system and well-developed nervous system of a coelomate animal carry supplies and information between distant parts of the body and permit considerable division of labor.

◆ Many coelomate animals have powerful muscles and varied modes of locomotion, which often permitted new ways of living and feeding.

## 29–A   THE COELOM

The origin of the coelom was one of the most important steps in animal evolution for several reasons. First, the coelom separates the muscles of the gut wall from the muscles of the body wall, and so food can move down the gut independent of muscle movements in the body wall. Second, the fluid in the coelom transports waste, food, and gases around the body. More important, the coelom provides space where a true circulatory system with blood vessels and a pumping heart can develop and function without being squeezed by other organs every time the animal moves.

Having a circulatory system permits animal size to increase because the blood transports oxygen and food rapidly to each cell and carries wastes away. Without it, cells must rely on diffusion for their oxygen supply. Because the rate of diffusion falls rapidly with distance, diffusion of oxygen can support respiration only in very small, flat, or sluggish animals. Similar constraints apply to the transport of food to the body cells and removal of their wastes. A circulatory system overcomes these limitations and permits the increases in size and metabolic rate seen in many higher animals.

A coelom also provides space where the **gonads** (ovaries and testes) can expand and allow eggs and sperm to accumulate over a period of time. Most coelomates have brief breeding seasons during which they produce large numbers of offspring at the time most favorable for their development.

Having a circulatory system permits the internal organs to evolve in various ways because the blood supplies their needs whatever their size and shape. In particular, the brain, which uses oxygen and food rapidly, can become larger. As a result, it can process more information, from more complex and specialized sense organs, and exert precise control over larger, more powerful muscles. Hence, coelomate animals have more complex behavior patterns than acoelomates.

## 29–B   PROTOSTOMES AND DEUTEROSTOMES

Living animals show several levels of body organization. Some biologists think that simple sponges probably gave rise to cnidarians. Flatworms, with three germ layers and bilateral symmetry, may have evolved from a cnidarian larva. Then came the evolution of fluid-filled body cavities—pseudocoels and coeloms—and dozens of new animal phyla evolved.

Nineteenth-century biologists divided the members of these new phyla into two groups on the basis of their embryonic development. Annelids, molluscs, and arthropods were grouped together as protostomes, and echinoderms and chordates as deuterostomes.

Older descriptions of the protostome-deuterostome division state that deuterostomes have the following characteristics:

1. Cleavage is indeterminate, meaning that each cell produced by the first few cleavage divisions is capable of developing into a complete embryo. It is indeterminate cleavage of the human zygote that makes identical twins possible.
2. Cleavage is radial, meaning that the cleavage furrows pass through the embryo at right angles to one another.
3. The coelom arises as a series of pouches in the mesoderm.
4. The embryonic blastopore becomes the anus in the adult.

Protostomes are said to have the opposite characteristics: cleavage is determinate and spiral, with cells that develop into only part of the embryo if they are separated and that form a spiral pattern in the embryo. The coelom is said to arise as a split in the mesoderm and the mouth from the blastopore.

Modern research has shown that many of these statements are wrong. We saw in Chapter 12 that many protostomes do not have spiral cleavage. For instance, the fruit fly *Drosophila* is a protostome with neither spiral nor radial cleavage. In humans and many other vertebrates the coelom arises as a split in the mesoderm, not as pouches, and the mouth and anus arise in many positions relative to the blastopore in members of both groups.

At a molecular level, there are more similarities than differences in development between protostomes and deuterostomes. Development is controlled by homeobox genes in both groups, and the position of the head is governed by the same transcription factor in amphibians (deuterostomes) as in the protostome *Drosophila* (Section 12–H).

Nevertheless, protostomes and deuterostomes do form distinct evolutionary lines. Echinoderms and chordates belong to one natural group, and their embryonic development is quite similar. But the protostomes are a heterogeneous group. It appears that this lineage gave rise to evolutionary experiments in many aspects of life: embryonic development, formation of the body cavity, and methods of locomotion.

# PROTOSTOMES

## 29–C  ANNELIDA: SEGMENTED WORMS

The annelids include earthworms, leeches, and a spectacular variety of marine worms (Table 29–1). The phylum name comes from their distinguishing feature: division of the body into many segments, resembling rings strung together (annellus = little ring). In addition, annelids have a coelom and most have a circulatory system, features that permitted annelids in general to become larger than other worms.

In an annelid, the space between the gut and body wall is divided into repeated segments **(metameres)** by partitions called **septa** (singular: **septum**). Each segment contains part of the fluid-filled coelom. Externally, each segment bears **setae** (singular: **seta**), bristles composed of the polysaccharide chitin (Figure 29–2a).

Most annelids also have a **closed circulatory system,** consisting of blood vessels with muscular walls, some of them enlarged as pumping "hearts" (Figure 29–2b). The

**T A B L E   2 9 – 1**

### Phylum Annelida and Its Major Classes

| | |
|---|---|
| **Phylum Annelida**<br>(~8900 species) | Segmented worms<br>Coelomate, with circulatory system (except Hirudinea)<br>Setae in most<br>Gut with mouth and anus<br>Gas exchange through general body surface and through gills in many<br>    polychaetes and a few oligochaetes and leeches |
| Class Polychaeta<br>(~5300 species)<br>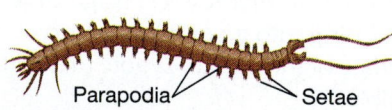 | Segmented worms with many setae per segment, usually borne on<br>    parapodia<br>Well-developed head<br>Tube-dwelling or free-moving<br>Development via trochophore larva<br>Mostly marine<br>E.g., *Aphrodite* (sea mouse), *Sabella* (fanworm), *Nereis* (ragworm),<br>    *Arenicola* (lugworm) |
| Class Oligochaeta<br>(~3100 species)<br> | Segmented worms with few setae per segment<br>Head reduced<br>Terrestrial and freshwater<br>E.g., *Lumbricus* (earthworm), *Megascolides* (giant Australian worm),<br>    *Tubifex* (often used to feed pet fish) |
| Class Hirudinea<br>(~500 species)<br> | Leeches<br>Body usually flattened, segmentation and coelom reduced, circulatory<br>    system and setae usually absent<br>Anterior and posterior suckers<br>Ectoparasites, predators, and scavengers<br>Mostly freshwater, some terrestrial<br>*Hirudo medicinalis* (medicinal leech) |

(a) BASIC BODY PLAN

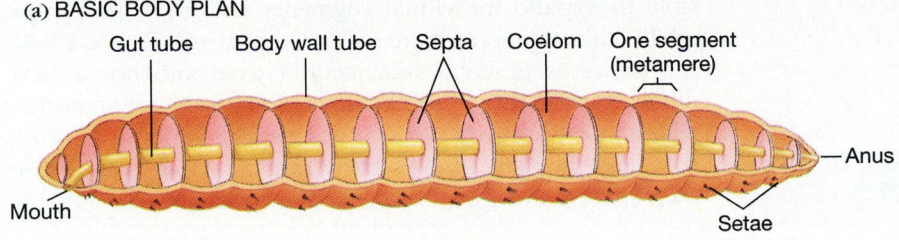

(b) EARTHWORM ANATOMY

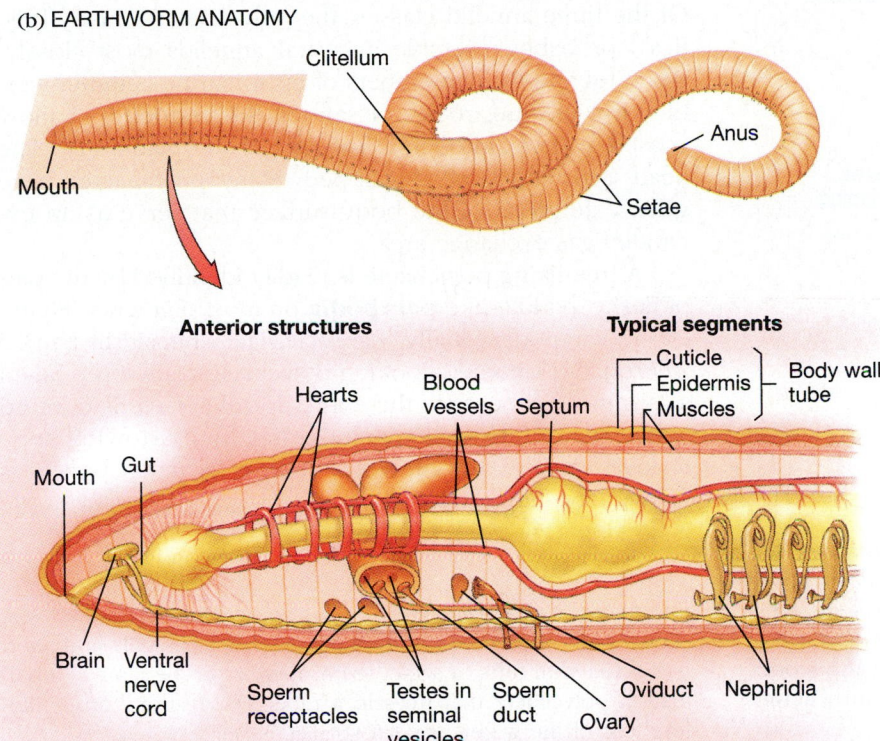

**Anterior structures**

**Typical segments**

**FIGURE 29–2** Annelid anatomy. (a) The tube-within-a-tube body plan, with the coelom divided into compartments by septa. (b) External and internal anatomy of an earthworm. The clitellum is an organ that secretes a cocoon around the eggs. The longitudinal section through the anterior segments of an earthworm shows that structures such as nerve ganglia and nephridia (excretory organs) are repeated in most body segments.

blood picks up digested food from the intestine and exchanges gases with the environment through the general body surface.

The main blood vessels and the ventral nerve cord run the length of the worm's body, giving off paired blood vessels and nerves in each segment. Most segments also contain a pair of tube-like **nephridia,** excretory organs that take in coelomic fluid through a funnel-like opening, resorb the solutes needed by the body, and expel the remaining wastes at the body surface. In this way they regulate the composition of the body fluid.

The body wall of each segment contains two sets of muscles, longitudinal muscles running lengthwise and circular muscles running around the segment. Contraction of the longitudinal muscles makes the segment shorter and wider; contraction of the circular muscles makes it longer and thinner. These muscles generate force by contracting against the enclosed coelomic fluid of each segment.

Thus, the fluid-filled segments serve as a **hydrostatic skeleton,** rather like a line of water-filled bags. Because fluid cannot be compressed, it provides a base that the muscles can work against. And because this skeleton is fluid, it changes shape as the muscles contract against it, permitting great flexibility. Annelids burrow and crawl by squeezing some segments into a short, wide shape and gripping the sides of the burrow with their setae. The segments forward of this point are squeezed long and thin, pushing through mud, sand, or soil. The front segments then widen and grip the side of the burrow, while the segments behind relax their hold and are pulled forward. Anyone who has dug for earthworms or lugworms will testify that this system permits them to burrow rapidly.

Segmentation is crucial to this form of locomotion. If the coelom were not partitioned by septa, muscular contraction in the middle of the body would push the coelomic fluid both forward and backward, and it would be impos-

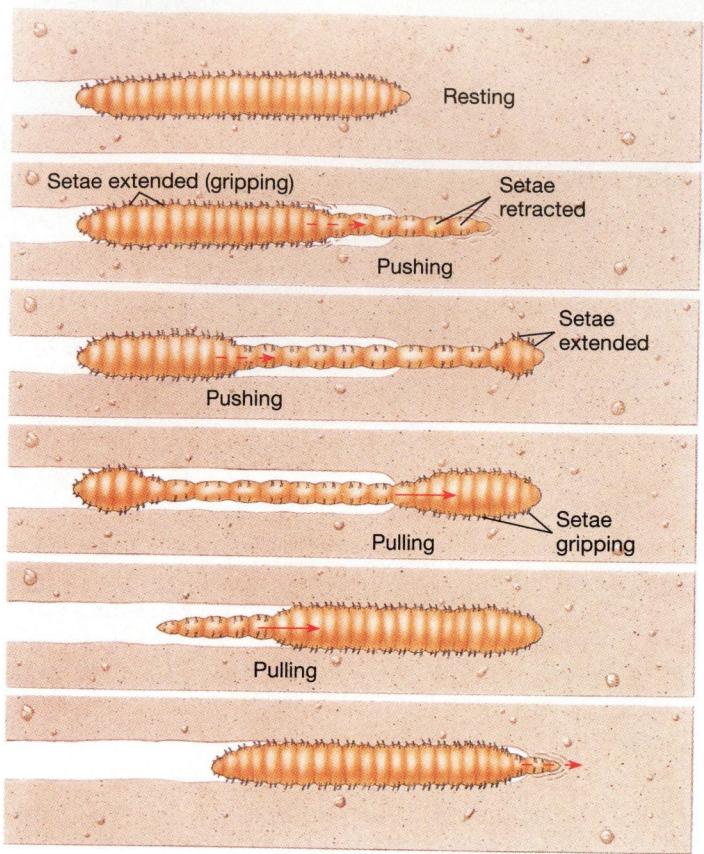

**FIGURE 29–3** An earthworm crawling, using its hydrostatic skeleton. Each segment acts as a closed sac of fluid surrounded by two sets of muscles. Alternate contraction of these muscles shortens and lengthens the segment. When the segment is short and wide, its setae are extended against the substrate, keeping the worm from slipping backwards. Each segment is extended in turn; at the end of the sequence, the worm has moved forward.

sible to expand individual segments and grip the burrow while other segments were extended forward (Figure 29–3).

Hence, segmental organization gave annelids a new mode of locomotion: unlike platyhelminths and nematodes, which mostly squeeze into existing spaces in bottom ooze, soil, leaf litter, or a host's body, annelids can create their own spaces, even in well-packed substrates.

## Polychaeta: Bristle Worms

Of the three annelid classes, the polychaetes ("many bristles") probably resemble ancestral annelids most closely. Beautiful and lively members of most marine communities, they live in mud, rocks, or sand at the bottom of shallow coastal waters, although a few species are planktonic. The head is usually well developed. Many polychaetes have **gills,** extensions of the body surface that serve as an expanded gas-exchange area.

A free-living polychaete is readily identified by the pair of fleshy, paddle-like **parapodia** on most segments (Figure 29–4). Parapodia usually bear setae. The whimsically named *Aphrodite* is a homely-looking polychaete commonly called a "sea mouse" because the entire dorsal surface is covered with a mat of fine setae that looks like fur. Burrowing polychaetes, such as the lugworm *Arenicola,* much prized as fishing bait, have reduced parapodia.

**FIGURE 29–4** Polychaetes. (a) A bristle worm from the Red Sea, showing the segmented body. Each segment has gills and short parapodia with tufts of setae. (b) A member of the family Serpulidae, a polychaete that lives in a tube. The feathery arms filter food out of the water.   (a, Jeff Rotman; b, Steven Webster)

(a)

(b)

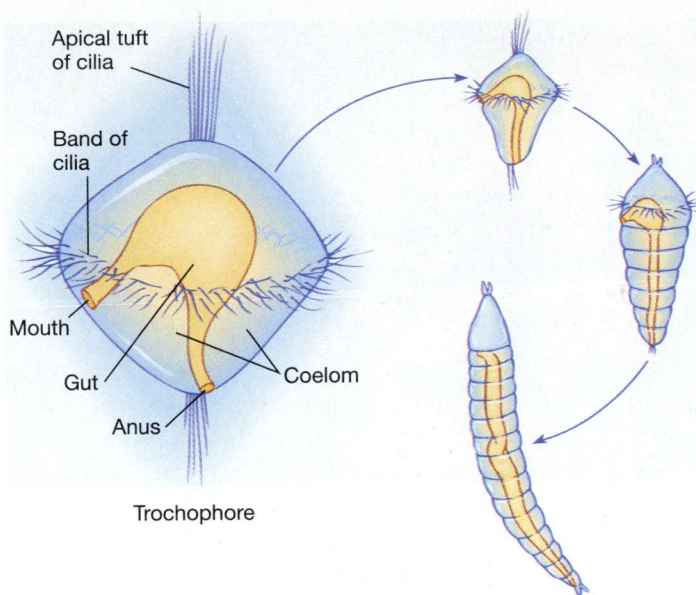

**FIGURE 29–5**  The planktonic trochophore larva of a polychaete (annelid) worm. The larva moves by means of its band of cilia. Many molluscs (Section 29–E) have a similar larva.

The habit of living in tubes evolved separately in a number of polychaete families (Figure 29–4b). The tubes are built by cementing sand grains and bits of shell together. The parapodia are usually reduced or absent, and feeding arms develop from the head. Tube-dwelling polychaetes are carnivores, feeding on other animals, or filter feeders, straining small organisms or food particles out of the surrounding water.

Polychaetes have separate sexes and external fertilization, and nearly all develop via a planktonic **trochophore** larva (Figure 29–5).

## Oligochaeta: Earthworms and Their Kin

Oligochaetes are freshwater, terrestrial, and marine annelids, most burrowing in mud or soil. Some measure less than half a millimeter, while the giant Australian earthworm grows to 3 m long and twice as thick as a man's thumb (at this size, its quick retreat into a burrow sounds like the gurgle of an emptying bathtub).

Oligochaetes have fewer setae than polychaetes (oligo = few). There are no parapodia, and the head is reduced, probably an adaptation to burrowing.

Most oligochaetes feed on dead organic matter, particularly plant material. Earthworms and other species ingest soil as they burrow, digest any organic matter, and void the rest as worm casts, which can often be found in a garden. Earthworms play a crucial role in loosening, aerating, and mixing soil. Researchers use earthworms as model animals in experiments to test for toxic pesticides or wastes that may make their way into soil.

Unlike the polychaetes, oligochaetes are hermaphroditic: each worm has both male and female organs, and fertilization is internal. During copulation, two animals touch each other, and each passes sperm to the other (see Figure 20–7). Fertilized individuals later lay eggs in a cocoon secreted by the clitellum, the most obvious structure on the outside of an earthworm.

## Hirudinea: Leeches

With only about 500 species, leeches comprise the smallest annelid class. Most live in fresh water, but some are marine, and a few live in humid jungles. Leeches range in length from about 1 to 20 cm. Unlike other annelids, they have a constant number of segments (34).

Leeches evolved new modes of locomotion in place of annelids' ancestral hydrostatic crawling and burrowing. Their derived characters include a flattened body with a sucker at each end, and loss of the setae and septa (and hence loss of internal segmentation). In most species, a series of small coelomic channels is all that is left of the circulatory system.

Using this modified body plan, some leeches move inchworm-fashion: they attach the posterior sucker to a surface, extend the body forward, attach the anterior sucker, and pull the rear up to join the front end. Others are swift and graceful swimmers, throwing their long bodies up and down into a series of curves.

Like oligochaetes, leeches are hermaphroditic. Leeches often stay with their eggs, arching over them and moving up and down, ventilating them with fresh, oxygen-bearing water. The young hatch as little leeches, not larvae, and often fasten onto the parent until they find their first meal (Figure 29–6).

**FIGURE 29–6**  Three leaf-eating leeches in a pond in New South Wales, Australia. The left-hand animal has a brood of young on its back.  (Kathie Atkinson/Oxford Scientific Films)

Many leeches prey on other invertebrates or scavenge dead matter, but the most famous are ectoparasites (external parasites) that suck the blood of animals such as fishes, turtles, amphibians (frogs and newts), water birds, or snails. Only a few species have "teeth" that can break through the tough skin of a mammal. These species secrete a local anesthetic that makes their bites painless and an anticoagulant that keeps the blood from clotting. Powerful sucking muscles remove blood quickly, and the leech swells to many times its normal size as it gorges. Opportunities to feed may be infrequent, and leeches can survive for months between meals.

In the last century, Western doctors often used the medicinal leech, *Hirudo medicinalis,* to bleed patients, a treatment for all manner of diseases. Indeed, "leech" was a nickname for doctors, and doctors kept aquaria with leeches that could be popped into containers in their medical bags when they went out on house calls. Leeches are still sometimes used to prevent bruising and blood pooling after surgery.

## 29–D  BRYOZOA: ECTOPROCTA

The Bryozoa, or "moss animals," are common inhabitants of fresh water and shallow seas. However, they are little known, probably for two reasons: first, the individual animals are small, most less than 0.5 mm long; second, individuals grow in colonies that look so much like seaweed that they are often mistaken for plants.

Bryozoans are lophophorates, animals with a filter-feeding **lophophore,** a crown of tentacles around the mouth. A lophophore differs from the feeding tentacles of cnidarians and annelids in that it is an extension of the body wall and contains part of the coelom. The lophophore is extended or retracted by changes in the hydrostatic pressure within the coelom. Cilia on the tentacles pass food to the mouth at their base and thence into the U-shaped digestive cavity (Figure 29–7).

Bryozoans have no gas exchange, circulatory, or excretory organs, probably because the animals are so small that all these processes can take place by diffusion.

All bryozoans are sessile. They live in colonies of several to many individuals, each enclosed in an **exoskeleton** (external skeleton) that also holds the colony together and helps attach it to the substrate. The exoskeleton is made largely of chitin, protein, and calcium carbonate.

Most bryozoans are hermaphroditic. Although some release eggs and sperm into the water, the vast majority have internal fertilization and brood their eggs in the body cavity. The lophophore and digestive tract may degenerate to make room for one large egg to develop. The embryo develops into a larva, which is released into the water and rapidly settles onto any suitable substrate. Here it founds a new colony, producing other individuals by budding.

(a)

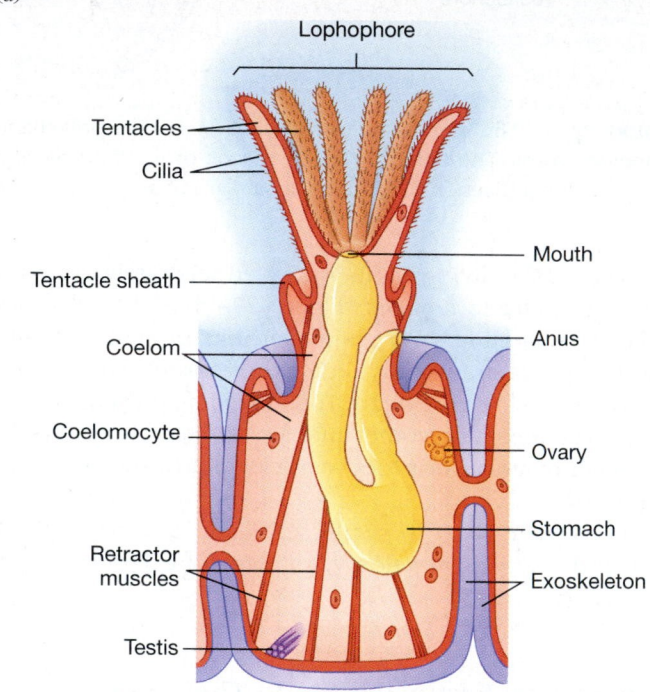

(b)

**FIGURE 29–7**  Bryozoa. (a) A colony of red bryozoans (*Iodictyum* sp.) near the Solomon Islands. Each dark spot marks the position of one zooid (individual) seen from the top. (b) Section of one zooid in a colony. The external layer consists of the epidermis plus the exoskeleton of chitin and calcium carbonate that it secretes. The zooid's tentacles trap food and pass it into the mouth, whence it proceeds through the digestive tube, undigested remains being voided through the anus. The hermaphroditic zooids contain reproductive organs of both sexes. When the zooid is disturbed, its retractor muscles pull the lophophore down into the protection of the exoskeleton.   (a, DRK Photo)

## 29–E  MOLLUSCA

Molluscs have soft bodies usually enclosed in hard shells (molluscus = soft). They include chitons, snails, slugs, octopuses, and bivalves such as clams and oysters (Table

**TABLE 29-2**

## Phylum Mollusca and Its Major Classes

| | |
|---|---|
| **Phylum Mollusca** (~57,000 species) | Body covered by a mantle, which may secrete a shell<br>Head and muscular foot usually present<br>Feeding by scraping radula<br>Bilaterally symmetrical with segmentation reduced<br>Gas exchange by gills or lining of the mantle cavity<br>Excretion by nephridia<br>Circulatory system with heart<br>Development via trochophore-like larva or direct |
| Class Amphineura (~500 species) 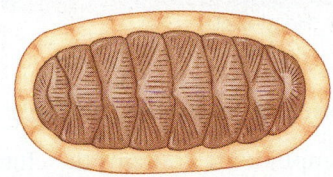 | Chitons<br>Shell of eight plates<br>Head reduced<br>One foot, used for locomotion<br>Marine<br>E.g., *Chaetopleura* (Atlantic chiton), *Chiton* |
| Class Gastropoda (~35,000 species)  | Snails, slugs, nudibranchs<br>Asymmetric body<br>Shell when present usually coiled<br>Marine, freshwater, or terrestrial<br>E.g., *Patella* (limpet), *Haliotis* (abalone), *Littorina* (periwinkle), *Strombus* (conch), *Buccinum* (whelk), *Aplysia* (sea hare), *Aeolis* (sea slug), *Lymnaea* (pond snail), *Planorbis* (ramshorn snail), *Helix* (escargot), *Limax* (slug) |
| Class Bivalvia (~20,000 species) 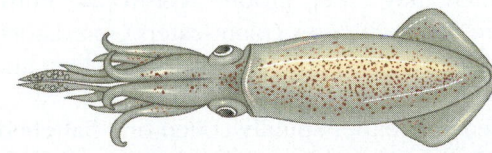 | Clams, oysters, scallops, mussels<br>Flattened shell with two valves<br>Head reduced<br>Mantle forms siphons<br>Paired gills<br>Most filter feeding<br>Marine and freshwater<br>E.g., *Tridacna* (giant clam), *Teredo* (shipworm), *Cardium* (cockle), *Pecten* (scallop), *Mytilus* (mussel), *Ostrea, Spondylus* (oyster), *Mya, Ensis* (razor clam), *Pinna* (pen shell) |
| Class Cephalopoda (~600 species) | Squids, octopuses, cuttlefish, *Nautilus*<br>Head surrounded by prehensile (grasping) tentacles, usually with suckers<br>Locomotion by jet propulsion using siphon made from mantle<br>Shell external, internal, or absent<br>Mouth with or without radula<br>Large eyes<br>Direct development<br>Marine<br>E.g., *Nautilus, Octopus, Argonauta* (paper nautilus), *Sepia* (cuttlefish), *Loligo* (squid), *Architeuthis* (giant squid) |

29–2). Data from ribosomal RNA sequencing supports the theory that molluscs and annelids share a common ancestor, which was partly segmented when the two lines diverged. Most members of the molluscan line lost their tendency toward segmentation, whereas other molluscs and the annelids evolved more pronounced segmentation.

The basic molluscan body plan consists of a muscular **head-foot,** with the part of the body containing the internal organs on top of it. This in turn is covered by a **mantle,** a flattened piece of tissue, which in many species secretes a calcareous shell—that is, one containing calcium carbonate. Between part of the mantle and body lies a **mantle cavity,** used for gas exchange and in some species also adapted for feeding and locomotion (Figure 29–8). The mouth contains a unique feeding structure, the rasp-like **radula,** covered with chitinous teeth used to scrape up food, such as algae on rocks. In many molluscs the radula has become adapted for other feeding methods or secondarily lost. As in annelids, the excretory organs are nephridia.

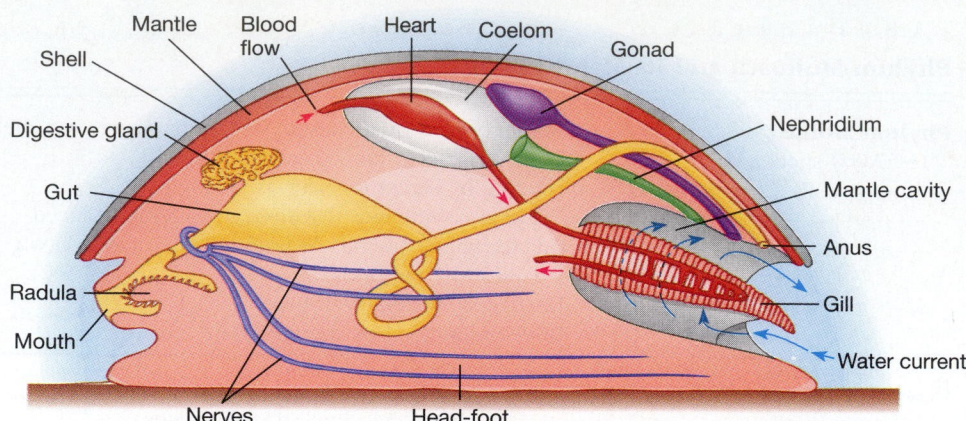

**FIGURE 29–8** A hypothetical ancestral mollusc, showing the general anatomy of members of this phylum. The muscles of the lower body form a continuous mass from which both the head and foot are formed.

Most molluscs have separate sexes. In general, gametes are shed into the water, where fertilization occurs, and the young develop via a trochophore larva. However, some species have **direct development:** the egg hatches as a miniature version of the adult, with no distinct larval stage. Land snails, nudibranchs, and some bivalves are hermaphroditic. Internal fertilization occurs in terrestrial snails (the sperm would dehydrate if released into the environment).

Molluscs are diverse and widespread. With nearly 57,000 described species, the phylum comes in a distant second to the arthropods, the animal phylum with the most species. Although most molluscs are marine, members of the phylum live in nearly every kind of habitat.

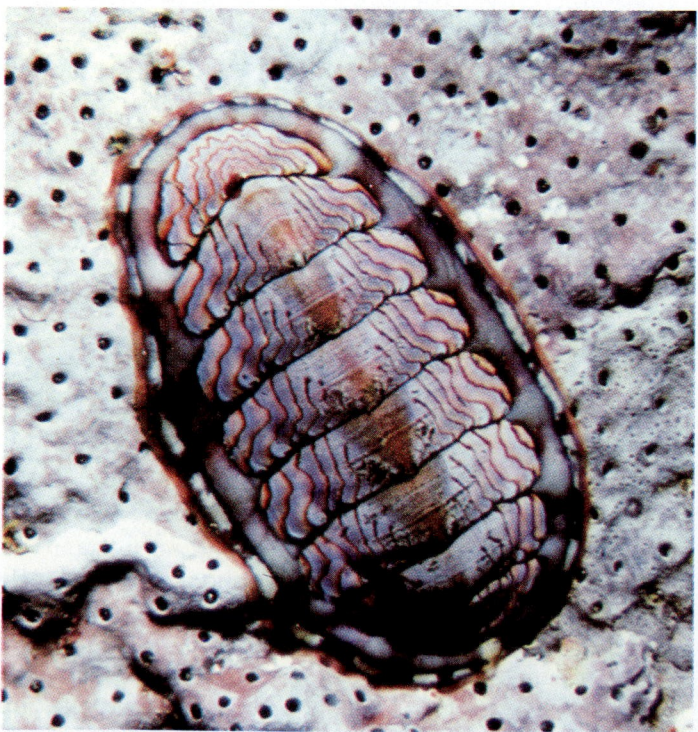

**FIGURE 29–9** A chiton grazing algae as it creeps over a rock on the seashore. Note the shell, composed of eight plates. (Steven Webster)

## Amphineura: Chitons and Their Relatives

The class Amphineura contains the chitons—oval, flattened creatures, with a tough mantle and eight crosswise plates (Figure 29–9). They are found in the intertidal zone and shallow water along the coast, firmly attached to rocks by the muscular foot. They move around very slowly, scraping algae off the rocks with the radula. When exposed to air at low tide, chitons, like many other molluscs, pull themselves firmly down onto a rock; the mantle covers the rest of the body tightly like a suction cup and prevents desiccation. Eggs are discharged directly into the sea, where they are fertilized and develop by way of a trochophore larva.

## Gastropoda

The class Gastropoda ("stomach foot") contains the most species of any molluscan class, including whelks, slugs, nudibranchs, limpets, and marine, freshwater, and terrestrial snails. Gastropods have well-defined heads, usually with eyes and tentacles, and most have an elongated, flattened foot by which they creep around. Gastropod nutrition is varied: there are herbivores (plant-eaters), predators, scavengers, deposit feeders, and suspension feeders, and even parasites.

A single shell, either spirally coiled or a flattened cone, is characteristic of the class, although it is reduced or absent in forms such as the terrestrial slugs (Figure 29–10). Shelled gastropods start life by secreting a tiny shell and then add more material around the opening, enlarging the shell as they grow. When in danger from predators or desiccation, some snails can withdraw entirely into the shell and close the "door": a horny operculum located on the upper surface of the foot.

All gastropods undergo **torsion,** or twisting, during their embryonic development: viewed from the top, most of the body behind the head twists through 180 degrees counterclockwise over the foot, bringing the anus and nephridia around to where they empty on top of the head! (Torsion is not the same as the spiral coiling of the shell,

(a)

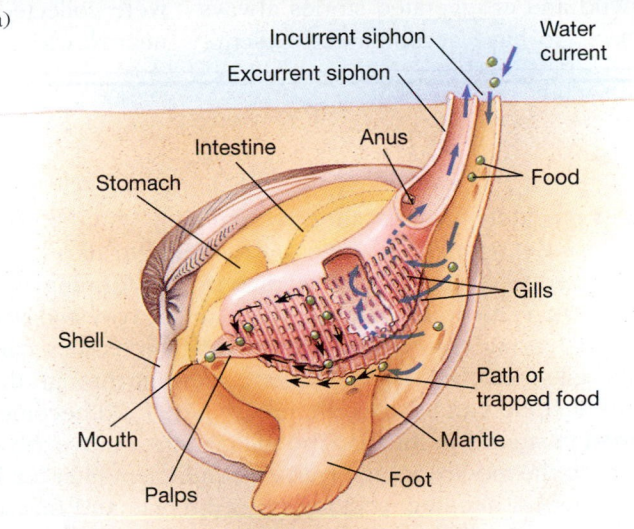

(b)

and the animals tend to crawl into the shade or under rocks during the heat of the day.

Some of the world's loveliest animals are nudibranchs, commonly known by the unlovely name of sea slugs. The mantle, shell, and gills of these gastropods have disappeared, leaving a naked body that is often brilliantly colored and covered by many projections that increase the surface area used for gas exchange (Figure 29–10b).

## Bivalvia

Bivalves are animals such as clams, oysters, and mussels, with the body flattened between the two valves of a hinged shell (Figure 29–11). Most bivalves can protect themselves

(a)

**FIGURE 29–10** Gastropods. (a) A snail. Note the head-foot and the coiled shell. The hole beside the shell opening leads to the lung cavity. (b) A nudibranch, a marine gastropod without a shell. The two orange-tipped tentacles bear sense organs that detect chemicals and other stimuli. The feathery yellow structures are gills. (a, Ken Lucas; b, David Fleetham/Natural Selection)

a separate and earlier evolutionary development.) Since emptying the body wastes onto the head is assumed to be disadvantageous, the evolutionary significance of torsion is much debated.

Most gastropods have gills inside the mantle cavity. However, in some forms the mantle cavity acts as a lung, and the animal obtains oxygen from air instead of from water. Some gastropods have both gills and lungs.

Like chitons, many intertidal gastropods, such as limpets and periwinkles, always return to the same "home" on a rock when the tide is out. Here they pull the shell tightly down against a groove they have worn in the rock, holding on by suction of the foot.

Land-living slugs and snails have adaptations of structure and behavior that conserve water. The mantle cavity, which must be kept moist, has a small external opening,

(b)

**FIGURE 29–11** Bivalves. (a) A marine bivalve. Note the edges of the mantle visible within the shell, and the long foot protruding out toward the left. (b) Cutaway view of a clam. The gills are built like large, flat sieves, which strain food out of the incoming water current and pass it, trapped in strands of mucus, along a groove at the bottom of the gill to the mouth. (a, Biophoto Associates)

## ESSAY: *Giant Squid*

Giant cephalopods appear as sea monsters in stories from many nations. Beasts similar to octopuses appear in a number of ancient Greek and Roman legends, but the most familiar mythical cephalopod is the "kraken" of Norwegian folklore. Scandinavians from the sixteenth century onwards described encounters in which "one of these Sea-Monsters will drown easily many great ships provided with many strong Mariners."

For years scientists were uncertain whether the kraken was fact or fiction. In the original 1735 version of his *Systema Naturae,* a catalogue of all the known animals, Linnaeus included the kraken under the name *Sepia microcosmos,* plainly believing that it was a giant squid. By the time the second edition was published, Linnaeus had apparently become convinced that this animal existed only in the minds of imaginative sailors and omitted it from his list. Wild and exaggerated stories always seem to have accompanied descriptions of the kraken, and for about 200 years most people believed that there was no such animal.

In 1861, however, a French naval vessel encountered a kraken near the Canary Islands, and her crew's description of the incident could not be dismissed as just another tall story. The animal lay on the surface long enough for them to make a detailed drawing, after which they harpooned and lassoed the beast, bringing a 20-kilogram portion of it on board the boat. The kraken had eight arms, each 2 m long, and two longer tentacles. It could open its beak to almost half a meter, and its weight was estimated at more than 1000 kg, or about 1 ton.

It is said that this incident prompted Jules Verne to describe the battle between a kraken and a submarine in his *Twenty Thousand Leagues Under the Sea.* Finally, at the end of the nineteenth century, some specimens and parts of giant squid were collected by fishermen, mainly near Newfoundland.

One species of giant squid is called *Architeuthis harveyii* after the Reverend Moses Harvey, an amateur naturalist, who lived in St. John's, Newfoundland, and recognized the importance of the fishermen's catches. In 1873 he acquired part of a tentacle that was 6 m long and had been cut off by a 12-year-old herring fisherman in a small boat when he was attacked by a squid. Finally, Harvey became the proud possessor of an intact *Architeuthis.* He wrote:

I stood on the shore of Logy Bay . . . and . . . thought of how I would astonish the savants, and confound the naturalists, and startle the world at large. I speedily completed a bargain with the fishermen, whom I astonished by offering 10 dols. to deliver the beast carefully at my house.

Next day, to my great satisfaction, a cart arrived at my door almost filled with the hideous, corpse-like creature, which I speedily stowed away in an outbuilding, in a huge vat filled with the strongest brine . . . A stream of daily visitors came to gaze in shuddering horror at the dead giant.

---

against predators or desiccation by retreating into the shell and closing it tightly, but some are too large to fit.

The edge of the mantle is drawn out, forming two **siphons,** one for water entering the mantle cavity and one for water leaving. Most bivalves are ciliary filter feeders: cilia on the gills beat and draw a water current in through the incurrent siphon and across the gills, where food becomes trapped in strands of mucus. The gill cilia then move these strands to the mouth. Some bivalves supplement filter feeding by deposit feeding, sucking up sediments and digesting the organic matter so obtained, or by housing photosynthetic symbionts in the mantle.

As one might expect of filter feeders, many bivalves live sedentary lives: clams buried in sand or mud, mussels attached to some substrate by adhesive threads. A few species are more mobile and can swim for brief periods by clapping the valves of their shells together; others burrow or creep around using the muscular foot.

Many bivalves are important as "shellfish" for human food, and some are specially cultivated for this purpose. Pearls are also cultivated, by placing a rough bit of foreign matter between the mantle and shell of a pearl oyster and waiting several years while the animal covers this core with layers of smooth inner shell material. Analysis of the adhesives by which bivalves attach to solid surfaces has led to the invention of better glues for underwater use, including surgery and dentistry.

A number of bivalves are very destructive. "Shipworms" are marine bivalves that use the edges of their shells to burrow into wood. They eat the sawdust produced, and their symbiotic, nitrogen-fixing gut bacteria digest it, providing both partners with food. One exotic species, the tiny European zebra clam (zebra mussel), has caused extensive damage throughout the Eastern United States, clogging cooling systems and water intake pipes and outcompeting native species in many freshwater habitats. The zebra clam

Harvey found that the tentacles were 8 m long and that the animal was 10 m long overall. Larger specimens have been reliably reported, but few have been captured intact or photographed to show the whole body, as this one was.

Harvey's specimens went to A. E. Verril, Professor of Zoology at Yale, who in 1879 published all the material he had accumulated on giant squid. He named two species of *Architeuthis*, and other people have described another ten. There is not, however, enough information for us to be sure how many species really exist. *Architeuthis* is plainly a deep sea creature with poor swimming ability; it comes to the surface only under exceptional circumstances. Few people who sight giant squid are equipped to capture and preserve them for proper study. In fact, sometimes quite the reverse: in World War II a British sailor was pulled off a life raft by one.

Giant squid are undoubtedly the largest and heaviest invertebrates ever described. The largest apparently authentic measurements come from a specimen stranded on a New Zealand beach in 1888. This animal was nearly 20 m in overall length, with tentacles of 16 m and a body nearly 2 m long. Members of other species with bodies almost twice as long have been measured, but they all had shorter tentacles. It has been estimated that *Architeuthis* may grow until the body is more than 4 m long.

The best-known predator of giant squid is the endangered toothed cachalot, or sperm whale, which may grow up to 20 m in length and would be considerably heavier than the largest reported squid. Captured sperm whales are often described as bearing the marks of titanic battles with giant squid: sucker scars and gouges from a huge beak on their skins, and tentacles and beaks in their stomachs. Moby Dick was a sperm whale, and the attacks of these voracious carnivores have occasionally sunk quite large ships.

A naturalist aboard the whale ship *Cachalot* in about 1875 left a dramatic account of a fight between a sperm whale and a giant squid:

I was leaning over the rail, gazing steadily at the surface of the sea, where the . . . tropical moon made a broad path . . . There was a violent commotion in the sea . . . and . . . a very large sperm whale was locked in deadly conflict with a cuttlefish, or squid, almost as large as himself, whose interminable tentacles seemed to enlace the whole of his great body. The head of the whale especially seemed a perfect network of writhing arms, naturally, I suppose, for it appeared as if the whale had the tail part of the mollusc in his jaws, and, in a business-like, methodical way, was sawing through it.

By the side of the columnar black head of the whale appeared the head of the great squid, as awful an object as one could well imagine even in a fevered dream. The eyes were remarkable for their size and blackness, . . . at least a foot in diameter.

All round the combatants were numerous sharks, like jackals round a lion, ready to share the feast, and apparently assisting in the destruction of the huge cephalopod.

was accidentally introduced into the Great Lakes in ballast water emptied from cargo ships, and the population has exploded—to densities of 750,000 per square meter in some areas.

## Cephalopoda

The exclusively marine class Cephalopoda contains the squids, octopuses, cuttlefish, and nautiluses. These are specialized for life as active predators and are among the largest, fastest, and most intelligent invertebrates.

Cephalopods appear quite closely related to gastropods, but the body has been rearranged and streamlined. The mouth is now in the middle of the foot, whose edges are drawn out into grasping arms and tentacles lined with rows of suckers. Most cephalopods swim by jerky jet propulsion. They take water into the mantle cavity and squirt it out through a siphon, which the animal can point in various directions to determine which way it will move. Some squids hold the swimming speed record for invertebrates at 40 km per hour! Octopuses are more sedentary, usually lying in wait for prey near the entrances of their dens in rock crevices.

The fossil record contains thousands of cephalopod species with straight or curved shells, but the only cephalopods with external shells today are those of the genus *Nautilus*, which have curled shells filled with gas. The cuttlebones we give pet birds to sharpen their bills and provide calcium are the internal shells of cuttlefish, containing gas-filled spaces that give the animal neutral buoyancy. Squids also have reduced, internal shells, and octopuses have none.

Cephalopods grasp their fish or invertebrate prey with their arms and tentacles and tear it apart with beak-like jaws. Some octopuses can inject poison into their prey through their beaks.

(a)

(b)

**FIGURE 29–12** Cephalopods. (a) A juvenile octopus swimming in the plankton of the Southern Ocean near Antarctica. (b) A nautilus.    (a, Minden Pictures)

Cephalopods have gills in the mantle cavity, and the forceful pumping of water into and out of the cavity during locomotion supplies them with plenty of oxygen. Unlike other molluscs, cephalopods have a closed circulatory system, which transports food and oxygen rapidly and supports their high metabolic rate (Figure 29–12).

Large eyes are the main sense organs. Cephalopod and vertebrate eyes show striking convergent evolution: both have a cornea, iris diaphragm, and lens. The brain's large size correlates with cephalopods' keen senses, complex be-

havior, control of many appendages, and ability to learn and remember. Behavioral studies have shown that octopuses are quite intelligent and can solve many problems.

Most cephalopods have pigment cells called **chromatophores** in their skin and can change color rapidly by expansion and contraction of different-colored chromatophores. This ability is used in camouflage, courtship, and other communication. Most cephalopods also have a large ink sac and when attacked eject a cloud of dark ink and shoot off in another direction. Ink from the genus *Sepia* was used for years to make sepia-colored writing ink.

The sexes are separate, and fertilization is internal. During copulation, the male uses a specialized tentacle to transfer packages of sperm, called spermatophores, from his own mantle cavity to the female's. The female then lays one or a string of fertilized eggs. Development is direct, and the young hatch as miniature versions of the adults. Female octopuses tend their eggs, removing debris and squirting water from the siphon over the nest until the eggs hatch.

## 29–F    ARTHROPODA

Arthropods and vertebrates are the only two animal groups with many members fully adapted to life on land. The oldest known land animal fossils are arthropods: centipedes and spider-like forms dating from 414 million years ago.

There are more than three times as many species of arthropods as of all other animal groups combined (Table 29–3). This spectacular success sprang from an evolutionary leap forward, elaboration of the cuticle covering the body into an exoskeleton much like a suit of armor. The exoskeleton is made up of layers of protein complexed with chitin, arranged in a set of plates joined by thin flexible hinges. This armor covers the body and the many jointed appendages for which the phylum is named (arthro = joint; pod = foot). This structure is strong enough to provide protection against attack and injury, while also permitting mobility. Compared with the calcareous exoskeletons of molluscs and corals, it is both lighter and more maneuverable.

In many species, wax in the exoskeleton's outer layer also prevents loss of water. This adaptation is vital to the survival of terrestrial arthropods. Since Roman times, people in Africa and Europe have mixed fine dust with stored grain to keep it free of insects. The dust abrades the soft cuticle between segments, destroying the insect's waterproofing, and it dries up and dies.

The arthropod exoskeleton does not grow with the animal. Instead, arthropods must **molt,** shedding their cuticles and developing new ones, as they grow. In many arthropods, successive molts take the animal through a series of larval stages. Some young arthropods look more and more like the adult with each molt, as new adult structures are added. Others develop through larval stages very different from the adult and with a totally different way of life,

**TABLE 29–3**

## Phylum Arthropoda and Its Major Classes

| | |
|---|---|
| **Phylum Arthropoda** | Segmented animals with a jointed exoskeleton containing chitin<br>Jointed appendages<br>Body cavity a hemocoel<br>Respiration through body surfaces or by gills or tracheae<br>Early embryo syncytial in many species<br>Marine, freshwater, and terrestrial forms |
| Class Arachnida<br>(~72,000 species)<br> | Spiders, scorpions, ticks, mites, harvestmen<br>Body having one or two main parts<br>Six pairs of appendages (chelicerae, pedipalps, four pairs of walking legs)<br>Mostly terrestrial |
| Class Crustacea<br>(~42,000 species)<br>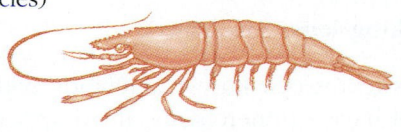 | Shrimps, krill, lobsters, crabs, barnacles, cladocerans, ostracods, copepods<br>Skeleton usually impregnated with $CaCO_3$<br>Body of two or three parts<br>Two pairs of antennae, chewing mouthparts, three or more pairs of legs<br>Development usually via nauplius larva<br>Mostly marine |
| Class Insecta<br>(~700,000 species)<br> | Insects (see Table 29–4)<br>Body divided into head, thorax, and abdomen<br>One pair of antennae; mouthparts modified for chewing, sucking, or lapping;<br>  usually with two pairs of wings and three pairs of legs<br>Breathing mainly by tracheae<br>Excretion by Malpighian tubules<br>Mostly terrestrial |
| Class Chilopoda<br>(~2000 species)<br> | Centipedes<br>Body with distinct head bearing large antennae and chewing mouthparts;<br>  appendages of first body segment modified as poison claws; remaining<br>  segments bearing a pair of walking legs each<br>Breathing by tracheae<br>Excretion by Malpighian tubules<br>Predaceous on insects<br>Terrestrial in damp areas, including houses |
| Class Diplopoda<br>(~7000 species)<br> | Millipedes<br>Body with distinct head bearing antennae and chewing mouthparts; most seg-<br>  ments of body grouped in pairs covered by a single skeletal plate, with each<br>  apparent segment bearing two pairs of walking legs and two pairs of spiracles<br>Breathing by tracheae<br>Excretion by Malpighian tubules<br>Scavengers on dead vegetable matter, or herbivores<br>Terrestrial in damp areas |

as in the case of caterpillars, which are larval stages of butterflies and moths.

Arthropods are thought to have evolved from annelid-like ancestors, because the basic body plan is a long segmented body. (Trilobites are fossil forms with this type of body plan.) The ancestral body plan has one pair of jointed appendages per segment, as in modern centipedes. The appendages became variously modified during evolution into specialized antennae, mouthparts, walking legs, claws, and swimming paddles. In many arthropods, the segments became grouped and form distinct parts of the body, such as the **head, thorax,** and **abdomen** in insects.

Arthropods have many different kinds of sense organs that detect stimuli in the external environment. In each case, the cuticle is modified such that stimuli can be transmitted to the underlying sensory cells.

The coelom of an arthropod is called a **hemocoel** (hemo = blood; coel = cavity) because the body fluid, or **hemolymph,** also serves as the blood. Arthropods have an **open circulatory system,** in which hemolymph flows through channels within and between body organs, instead of in discrete blood vessels.

The sexes are almost invariably separate. Internal fertilization is the rule in terrestrial species, and in many cases

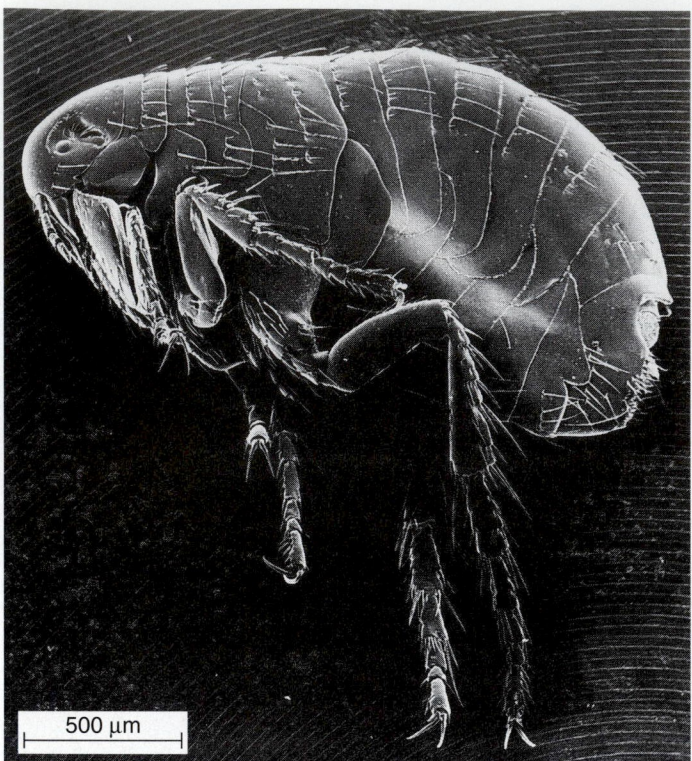

**FIGURE 29–13** Full chitin jacket. The jointed exoskeleton looks like a suit of armor in this scanning electron micrograph of a flea. Fleas are members of the class Insecta, order Siphonaptera (see Table 29–4). (Biophoto Associates)

the complex structure of the external genitalia ensures that only members of the same species can copulate with one another (Figure 29–13). Aquatic species usually release eggs and sperm into the water for external fertilization.

## Arachnida

Arachnids include the spiders, ticks, mites, and scorpions. The most familiar ones are predators or ectoparasites, although many mites eat plants or dead organic matter.

Arachnids have six pairs of jointed appendages (Figure 29–14a). Starting at the front, these pairs are:

1. **Chelicerae,** adapted for feeding by piercing the prey or host. There are often associated poison glands, which inject a substance that anesthetizes or kills the prey.
2. **Pedipalps,** adapted as sense organs detecting touch and chemicals and also helping to grasp the food. In male spiders, the pedipalps aid in sperm transfer to the female during mating. The pedipalps of scorpions end in a large pincer.
3–6. **Walking legs** (four pairs).

Spiders are the best known arachnids, with about 32,000 species. All have spinnerets, modified appendages on the abdomen, with which they spin silk. It is used to make the webs that trap their prey and the cocoons that protect their eggs. Silk is a fine, elastic protein extruded from the spinnerets as a liquid that hardens as it meets the air. All spiders are carnivorous, usually preying on insects, which they inject with a paralytic poison (Figure 29–14b).

Some ticks and mites transmit diseases. Ticks may carry Rocky Mountain spotted fever and Lyme disease, and chiggers (larval mites) may carry Asian scrub typhus. Mites themselves cause the itching of mange and scabies. Some mites are also serious plant pests (Figure 29–15). However, the vast majority of mites lives around us unnoticed because of their tiny size—a species that reaches a millimeter or two is a mighty mite.

**FIGURE 29–14** Spiders. (a) The appendages and silk-spinning apparatus of a spider. (b) A garden spider, member of the class Arachnida, wrapping an insect it has caught in silk before storing it for later use. (b, John Gerlach/ Animals Animals)

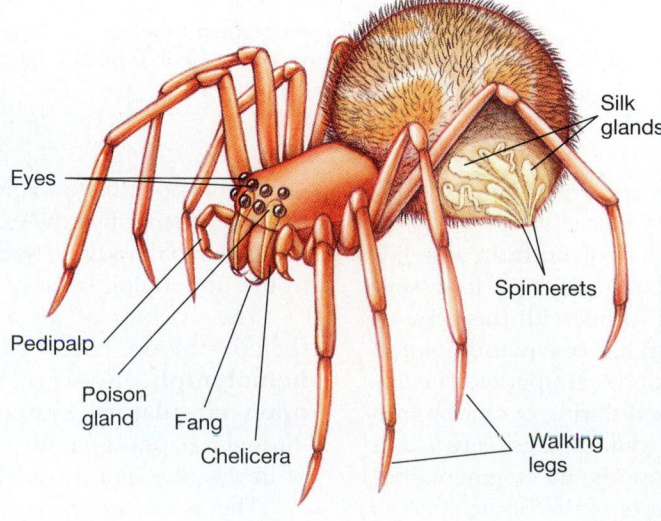

Eyes
Pedipalp
Poison gland
Fang
Chelicera
Silk glands
Spinnerets
Walking legs

(a)

(b)

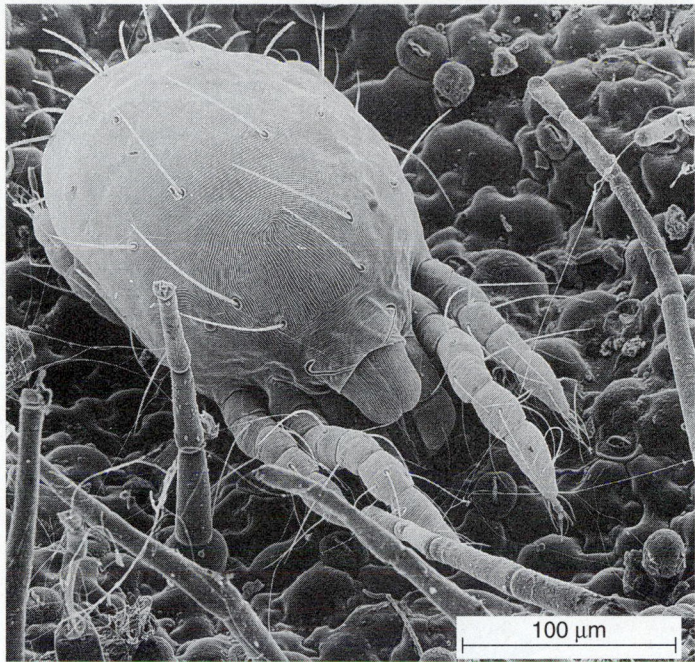

**FIGURE 29–15** Scanning electron micrograph of a red spider mite, miniature member of the arthropod class Arachnida. (Biophoto Associates)

## Crustacea

Crustaceans make up a large and diverse group of arthropods that includes lobsters, crabs, shrimps, crayfish, barnacles, wood lice, pill bugs, and water fleas. In general, crustaceans are aquatic with two-pronged appendages: two pairs of antennae, three pairs of feeding appendages used for food handling and chewing, and legs that may bear claws. However, many crustaceans do not conform to this description.

Some crustaceans crawl using their walking legs; others swim by the rhythmic beating of paddle-like appendages. In the tiny ostracods and water fleas, the main swimming appendages are the second pair of antennae! Typically, some anterior appendages are modified to grasp food and convey it to the mouth. However, filter feeding has evolved many times. Unlike the filter-feeding annelids, which use cilia for filters, these crustaceans use setae.

Crustaceans are the most prominent animals in plankton. Indeed, copepods are so abundant that there may well be more copepods than there are insects. Planktonic crustaceans are the main food of many vertebrates and include the shrimp-like "krill" upon which the biggest whales feed.

Barnacles are highly modified for a sessile way of life. Essentially, barnacles are attached to the substrate by their heads, and they shovel food into their mouths with their thin feathery legs (Figure 29–16). The body is protected by

(a)

(c)

(b)

**FIGURE 29–16** Crustacea. (a) Water fleas, *Daphnia* sp. live in ponds. (b) A barnacle. These odd crustaceans attach to a solid surface by their heads and secrete calcareous shells around themselves. They use their legs to shovel food into their mouths. The animal behind the barnacles is a red sea anemone. (c) A decapod: a crab. Eyes on stalks and the heavy calcareous skeleton are characteristic of the larger members of this class. (a, Harry Taylor/Animals Animals; b, Bruce Robison)

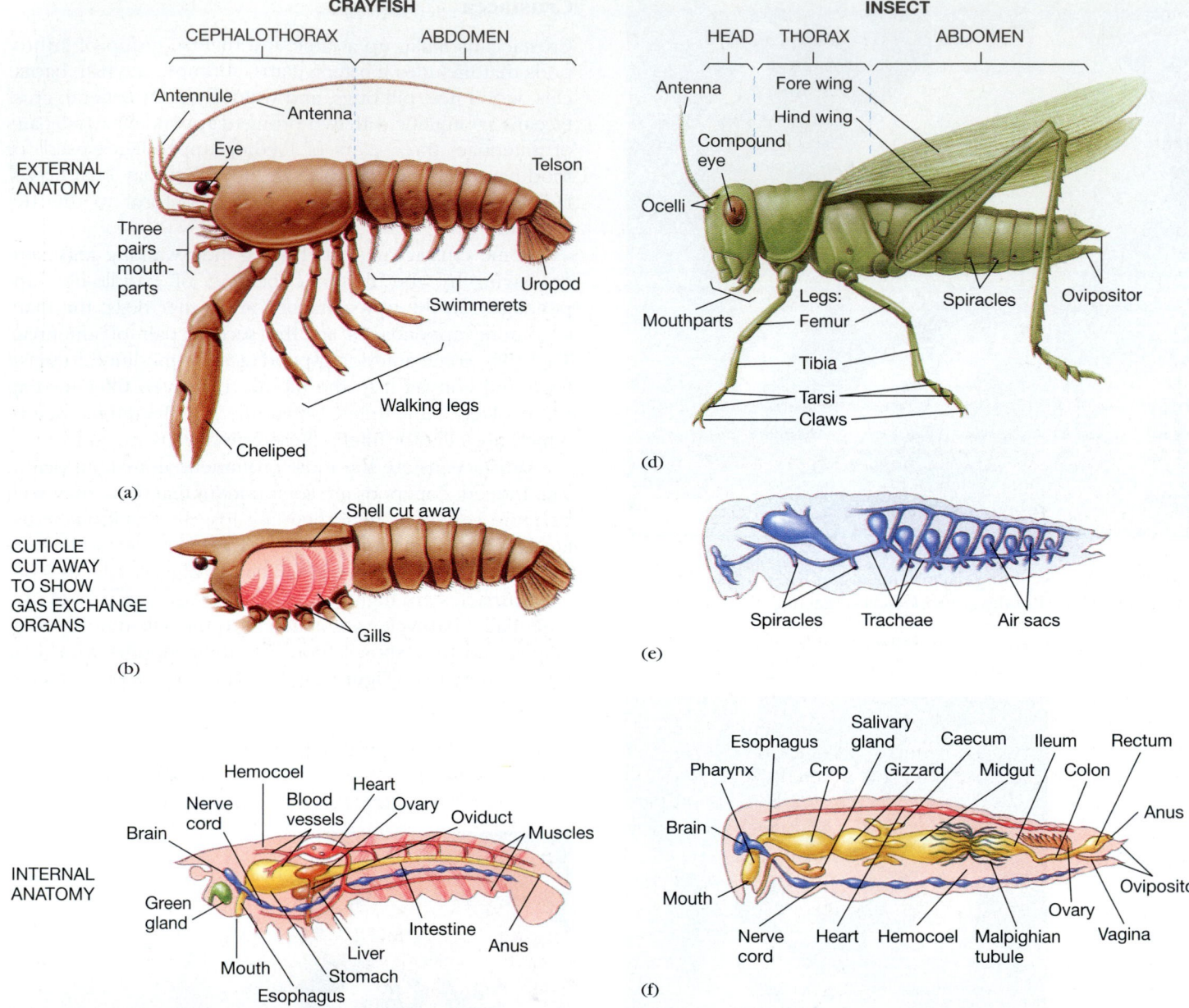

**CRAYFISH**

CEPHALOTHORAX    ABDOMEN

EXTERNAL ANATOMY

Antennule
Antenna
Eye
Telson
Three pairs mouthparts
Uropod
Swimmerets
Walking legs
Cheliped

(a)

CUTICLE CUT AWAY TO SHOW GAS EXCHANGE ORGANS

Shell cut away
Gills

(b)

INTERNAL ANATOMY

Hemocoel
Heart
Blood vessels
Ovary
Nerve cord
Oviduct
Brain
Muscles
Green gland
Intestine
Mouth
Anus
Esophagus
Liver
Stomach

(c)

**INSECT**

HEAD    THORAX    ABDOMEN

Antenna
Fore wing
Hind wing
Compound eye
Ocelli
Spiracles
Ovipositor
Mouthparts
Legs:
Femur
Tibia
Tarsi
Claws

(d)

Spiracles    Tracheae    Air sacs

(e)

Esophagus
Salivary gland
Caecum
Ileum
Rectum
Pharynx
Crop
Gizzard
Midgut
Colon
Brain
Anus
Mouth
Ovipositor
Nerve cord
Heart
Hemocoel
Malpighian tubule
Vagina
Ovary

(f)

**FIGURE 29–17**  Arthropod structure. (a, b, c) A crayfish, a typical decapod crustacean, shows ancestral characters such as paired appendages on each segment. (a) Segments are grouped into the cephalothorax and abdomen. (b) Gills lie under the cuticle of the cephalothorax. (c) Internal organs. The green gland is the excretory organ. Arthropods have an open circulatory system. Hemolymph enters the heart through slits in its walls and is pumped out through vessels that empty into the hemocoel.

(d, e, f) Insects have many derived characters. (d) The segments are grouped into three areas: head, thorax, abdomen. Each of the three thoracic segments bears a pair of legs and the last two typically also bear wings. In adult insects only the last abdominal segment has appendages. (e) The tracheal system. Air enters and leaves through spiracles at the body surface. (f) Internal anatomy. Like crayfish, insects have an open circulatory system and solid ventral nerve cord. Malpighian tubules are excretory organs. They empty into the gut. Fertilization is internal, and the female uses her ovipositor to place eggs in appropriate positions.

calcareous plates, which may or may not be cemented directly to the substrate. Barnacle eggs hatch into planktonic larvae, which attach to any solid substrate and grow very rapidly. A layer of barnacles ruins the streamlining of any ship, from a nuclear submarine to a fiberglass sailboat, and scraping barnacles from the bottom of a boat has been a chore for sailors since prehistoric times.

The decapod (ten-footed) crustaceans include the larger bottom-living and swimming forms such as shrimps, lobsters, crabs, hermit crabs, and crayfish. Most have a cuticle reinforced with calcium carbonate, making it harder and more brittle than that of most other arthropods.

### Insecta

Insects are, by far, the most successful terrestrial animals, with several evolutionary advances over their crustacean ancestors (Figure 29–17). It has been estimated that all the insects on Earth weigh many times as much as all the humans and that there are 200 million insects for every person alive. Nearly a million insect species have been described, and it is estimated that there are several million more. Insects, bats, and birds are the only living animal groups with members that can fly.

Claude Bernard, a nineteenth-century French physiologist, pointed out that an animal can move into new habitats only if it can regulate its internal environment in the face of inhospitable external conditions. Most invertebrates cannot maintain the salt compositions of their bodies in fresh water nor conserve body water when exposed to air. Insects won their freedom from the ocean by evolving the ability to tolerate fluctuations in salt and water content that would be fatal to most other organisms.

In insects, the segments became grouped into three main body areas: head, thorax, and abdomen. The thorax of most adult insects bears three pairs of walking legs and two pairs of wings. However, flies and mosquitoes (Order Diptera) have only one pair of wings, and fleas, lice, and silverfish have none. The head bears one pair of antennae, specialized mouthparts, and usually compound eyes (Figure 29–17).

Insects show several adaptations to life on land. Specialized excretory structures, the **Malpighian tubules,** produce nearly solid wastes, minimizing the loss of body water. The waxy cuticle retards loss of body water and also supports the body in the air. Wings develop as outgrowths from the cuticle, strengthened by rib-like structures called **veins.**

An open circulatory system is less efficient than closed circulation, and insects could not be as active as they are if they had not evolved an additional system for delivering oxygen to the tissues. Insects breathe by means of **tracheae,** air-filled tubes that branch throughout the body and carry oxygen close to all the cells. Air enters the tracheae via **spiracles,** tiny openings in the thorax and abdomen.

**FIGURE 29–18**  Insect development. A dragonfly emerges as an adult from the exoskeleton of its immature naiad stage in a Michigan pond.   (Barbara Gerlach/DRK Photo)

Insects also evolved sense organs that are effective on land, including eyes that permit color vision in some species. Depending on the species, receptors that detect chemicals, vibrations, or touch may be found on the antennae, mouthparts, and legs.

Because insects are basically terrestrial, they have internal fertilization. The eggs are laid with a waterproof covering, which protects them from dehydration. Females select a place to lay their eggs where the young will find food when they hatch. The young undergo a series of molts. The degree of resemblance between young and adult varies among different insect groups, and the higher insects pass through larval stages (such as caterpillars or maggots) that are completely different from the adult in appearance and way of life (Figure 29–18).

Why are most insects so small? They range in size from beetles only 0.1 mm long to tropical moths with a wingspan of 30 cm and Indonesian walking sticks about twice that length. The mechanics of an exoskeleton must impose some theoretical upper limit on the size of a flying insect. Nevertheless, modern insects do not reach this limit, and some extinct insects were much larger. This suggests that small size permits insects to specialize in habitats where they are not in direct competition with vertebrates, the other successful group of land animals. Leaf miners are insects that live within a living leaf. Other insects live inside the egg or the body of another insect or inside a single seed. In these

**TABLE 29-4**

## Twelve Major Orders of the Class Insecta

**Incomplete Metamorphosis**
**Development: Egg → Nymph → Adult**
At each molt, the nymph's wing pads increase in size; the last molt produces a (usually) winged adult. Nymphs and adults share the same habitat and way of life (except in Ephemeroptera and Odonata, which have aquatic nymphs, often called naiads).

**Complete Metamorphosis**
**Development: Egg → Larva → Pupa → Adult**
The larvae lack compound eyes and do not resemble adults. Larvae and adults usually have different modes of life (and often different habitats). Pupa is an immobile, nonfeeding stage in which the larva metamorphoses to the (usually) winged adult.

| Order | Characteristics and Examples | Order | Characteristics and Examples |
|---|---|---|---|
| Orthoptera (~23,000 species) 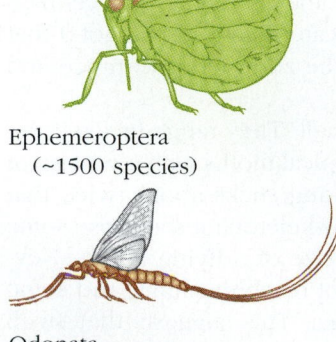 | Chewing mouthparts Forewings straight, leathery; hindwings membranous, folded under forewings at rest Grasshoppers, crickets, katydids, mantids, walking sticks, roaches | Neuroptera (~4600 species)  | Mouthparts sucking in larvae, chewing in adults Wings membranous, two pairs similar Lacewings, ant-lions, aphid-lions |
| Isoptera (~1800 species) | Chewing mouthparts Wings membranous, both pairs alike, or absent Social Termites | Coleoptera (~280,000 species)  | Chewing mouthparts Forewings hard, hindwings membranous and folded under forewings at rest Beetles, including weevils and fireflies |
| Hemiptera (~40,000 species) | Sucking mouthparts Forewings leathery at base, membranous at tip; hindwings membranous True bugs: e.g., chinch bugs, stink bugs, plant bugs, water boatmen, water striders, assassin bugs, bed bugs | Diptera (~85,000 species)  | Mouthparts variable in larvae, sucking or vestigial* in adults Only one pair of wings; hindwings reduced to knob-like halteres True flies: e.g., mosquitoes, gnats, midges, houseflies, horseflies, crane flies |
| Homoptera (~20,000 species) | Sucking mouthparts Wings membranous and held roof-like over the abdomen, or absent Cicadas, aphids, scale insects, leaf hoppers, spittle insects | Trichoptera (~4500 species) | Mouthparts chewing in aquatic larvae, vestigial* in adults Wings moth-like, covered with short hairs Caddis flies |
| Ephemeroptera (~1500 species) | Mouthparts chewing in naiads, vestigial* in adults Wings membranous, forewings large and triangular Mayflies | Lepidoptera (~110,000 species) 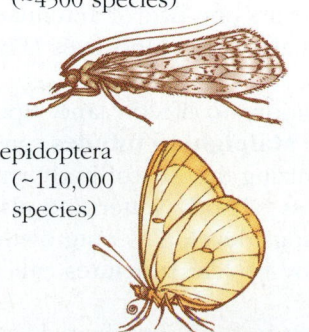 | Mouthparts chewing in larvae, sucking in adults Wings very large, covered with scales Butterflies, moths, skippers |
| Odonata (~5000 species) | Chewing mouthparts Wings membranous, long and narrow Dragonflies and damselflies | Hymenoptera (~100,000 species) 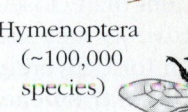 | Mouthparts both sucking and chewing Wings membranous when present Some forms social Wasps, bees, ants |

*Vestigial = much reduced in size and nonfunctional

**FIGURE 29–19** Beneficial insect: a ladybug feeding on aphids. Many predatory insects are valuable to humans in that they control populations of herbivorous insects that damage crops and ornamental plants.   (Biophoto Associates)

tiny habitats they face little competition, and they need little food to reach their adult size. The evolutionary tendency to small size has been an enormously successful one for the insects (Table 29–4).

***Humans and Insects*** Insects perform many roles that are vital to human life. Chief among these is pollination of flowering plants, especially many important fruit crops. Many insects also play major roles in recycling dead plants and animals and animal wastes and in controlling pests (Figure 29–19).

Several species of insects have been domesticated, including honeybees, which produce honey and wax, and lac insects, source of the main ingredient in shellac. Silk, from the cocoon of the silkworm pupa, has been a major product of China for centuries. Cochineal insects are the source of a bright red dye that the Aztecs and Incas, and their Spanish conquerors, prized nearly as much as gold and silver. The Spanish developed a lucrative trade in this dye, jealously guarding the secret of its insect source. The Star-spangled Banner and the red coats of the invading British Army were both colored with this dye, as were the paints depicting them on canvas. Today it is used in some cosmetics and is the only natural red dye approved by the U.S. Food and Drug Administration.

Crime investigators have also found uses for insects. A thirteenth-century Chinese law enforcement manual recounted how a murderer was caught when his sickle (the murder weapon) was the only one in the village to attract flies sensitive to the odor of decaying flesh. In New Zealand, the bodies of Asian insects in a load of confiscated marijuana provided the evidence to convict drug dealers on charges of importing the drug. Police have also found that the age of maggots, and the other kinds of insects present, can sometimes provide a remarkably accurate estimate of how long a corpse has been dead.

On the whole, people have devoted more time to killing insects than to praising them because the harm done by some is so staggering. Insects attack human beings directly with bites and stings. Much more important, insects transmit many diseases. Malaria, river blindness, and sleeping sickness carried by blood-sucking insects blind or kill millions of people a year. Insects probably do more damage indirectly, however, by transmitting plant diseases, such as Dutch elm disease and many viral diseases of crop plants, and by eating crops and killing trees. In the United States alone, the gypsy moth, tussock moth, southern pine beetle, and spruce budworm destroy enough forest trees every year to build nearly a million houses.

Insects destroy more than 10% of all crops grown in the United States, but the damage is even worse in the tropics, where hot weather throughout the year permits insects to grow and reproduce faster. In Kenya, officials estimate that insects destroy 75% of the nation's crops. A locust swarm in Africa may be 30 m deep along a front 1500 m long and will consume every fragment of plant material in its path (Figure 29–20).

**FIGURE 29–20** A swarm of locusts has just about stripped this field in Ethiopia, despite the efforts of the young farmer.   (Photo Researchers)

(a)

(b)

**FIGURE 29–21** Myriapods. (a) Chilopoda: a centipede cleaning its antennae. The antennae are larger than in millipedes; the legs are fewer but swifter. (b) Diplopoda: a millipede. The rounded head bears a pair of antennae. The remaining appendages visible here are the "thousand" legs. The legs move in coordinated waves, which progress along the body from one pair to the next. Despite the abundance of locomotory appendages, millipedes travel slowly, and when disturbed they curl up and secrete repellent chemicals.   (a, David Dennis/Tom Stack & Associates)

Pesticides have not solved the insect problem. This is partly because many insecticides are also dangerous to people, but mainly because pesticides act as selective pressures for the evolution of resistant strains of insects, which evolve too fast for expensive pesticide research to keep up. The list of pesticide-resistant insects nearly doubled between 1980 and 1990. Workers now direct much of their effort to using a combination of chemical and biological methods to control damaging outbreaks of insects. Biological controls include raising, sterilizing, and releasing large numbers of males (used for species in which females will mate only once), using sex attractant chemicals (pheromones) to attract males to traps instead of to females, breeding pest-resistant plants, and introducing predators and parasites of

pest insects. Yet despite the fact that human beings have waged war on insects since the two have existed together, human efforts have apparently not succeeded in exterminating even a single species of unwanted insects.

## Myriapods: Chilopoda and Diplopoda

The insects' nearest living relatives are the centipedes and millipedes, often lumped under the term myriapods ("many feet") (Figure 29–21). Like insects, myriapods have a single pair of antennae. Gas exchange occurs by a system of tracheae, excretion by Malpighian tubules. Despite these terrestrial adaptations, myriapods cannot tolerate hot, dry conditions as many insects do, but must remain in moist soil or leaf litter.

Centipedes belong to order Chilopoda, so named because the appendages on the first body segment are modified as poison claws (chele = claw). These are used to subdue prey, usually other invertebrates. Despite their common name, centipedes seldom have as many as a hundred legs. Some centipedes run very quickly, whereas others burrow through soil much like earthworms, using their legs for anchorage.

Millipedes ("thousand feet") appear to have even more legs than centipedes. This is because the body segments after the first four are fused in pairs, and so each of these apparent segments bears two pairs of legs: hence the class name Diplopoda ("double feet"). These legs move in a wave from the rear forward, and the animal crawls slowly. Millipedes eat decaying plants. When disturbed, they may secrete a repellent chemical and curl into a spiral.

## DEUTEROSTOMES

### 29–G    ECHINODERMATA

Sea stars, brittle stars, sea cucumbers, sea lilies, sea urchins, and sand dollars belong to the exclusively marine phylum Echinodermata (Table 29–5). The name Echinodermata ("spiny-skinned") refers to the calcareous spines and plates that form a skeleton just under the skin. Another distinctive feature is the **water vascular** system, a series of fluid-filled tubes used in gas exchange, feeding, and locomotion. Water enters the system via the sieve-like madreporite on the surface. Alterations of fluid pressure within this system permit an echinoderm to extend and retract its hollow, suction-cup **tube feet** (Figure 29–22). Some echinoderms use their many small tube feet for locomotion, and predatory forms may use their feet to grasp prey.

Most adult echinoderms show **pentaradial symmetry:** the body is divided into five parts around a central area where the mouth lies. This is a secondary adaptation to the echinoderms' slow-moving existence: the larvae are bilat-

**TABLE 29-5**

## Phylum Echinodermata and Its Classes

| **Phylum Echinodermata** (~5800 species) | Skeleton of calcareous plates under skin<br>Water vascular system with tube feet<br>Pentaradial symmetry in adults<br>Marine |
|---|---|

| | | | | |
|---|---|---|---|---|
| Class Crinoidea (~80 species) 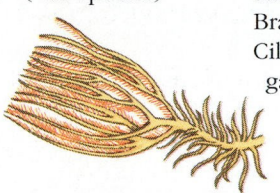 | Sea lilies.<br>Stalked or free<br>Branched arms with feeding grooves<br>Ciliated tube feet used for feeding and gas exchange | | Class Ophiuroidea (~2000 species) 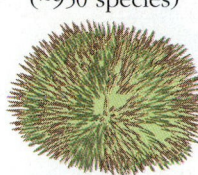 | Brittle stars.<br>Free-moving with thin arms marked off from disc and used for locomotion<br>Tube feet used as sensory organs and for feeding |
| Class Asteroidea (~1500 species)  | Sea stars.<br>Free-moving with arms merging into disc<br>Suctional tube feet used for locomotion and feeding | | Class Echinoidea (~950 species) | Sea urchins and sand dollars.<br>Free-moving with skeleton of fused plates<br>No arms<br>Suctional tube feet used for locomotion and gas exchange |

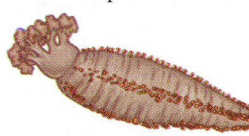

Class Holothuroidea (~900 species) — Sea cucumbers. Free-moving with mouth at one end of long body. Skeleton reduced. Some tube feet used for locomotion, others modified into tentacles around mouth

erally symmetrical. The absence of a head and an excretory system also reflect a sluggish lifestyle. The adults are primarily bottom-dwelling and slow-moving. Sea lilies are the only extant group of echinoderms with sessile members.

Most echinoderms have separate sexes. Gametes are usually shed into the water, where the eggs are fertilized and grow into extraordinarily beautiful, transparent larvae. In some species fewer, larger eggs are formed, and after external fertilization one parent broods them while they undergo direct development.

### Crinoidea: Feather Stars

Crinoids include the sea lilies and feather stars. Some are sessile, attached to the substrate by a stalk, and some are mobile, but all use their branched, feathery arms to trap food particles in a mucous web (Figure 29–23). Food is then carried to the mouth by ciliated tube feet. The tube feet also provide surfaces for gas exchange.

### Asteroidea: Sea Stars

The asteroids, or sea stars, are flattened, with a central disc from which radiate five or more arms. The mouth lies on the underside of the central disc. Asteroids move mainly by their tube feet, located on the underside of the body. The skeleton is organized as a series of plates that permit a certain amount of movement in the arms.

Most sea stars are carnivorous and feed on snails, crustaceans, bivalves, polychaetes, other echinoderms, and even fish, using their tube feet to grip their prey. Predators on bivalves can exert enough suction to pry open a narrow slit between the valves. The sea star then everts its stomach, which squeezes into the shell and digests the prey. The "crown of thorns" sea star feeds on cnidarian polyps and is notorious for the damage it does to coral reefs.

Sea stars are famous for their ability to regenerate a whole body from one arm that is still attached to part of the central disc. At times misguided fishermen have tried to destroy asteroids preying on mussel and oyster beds by hacking them into pieces and throwing them back into the

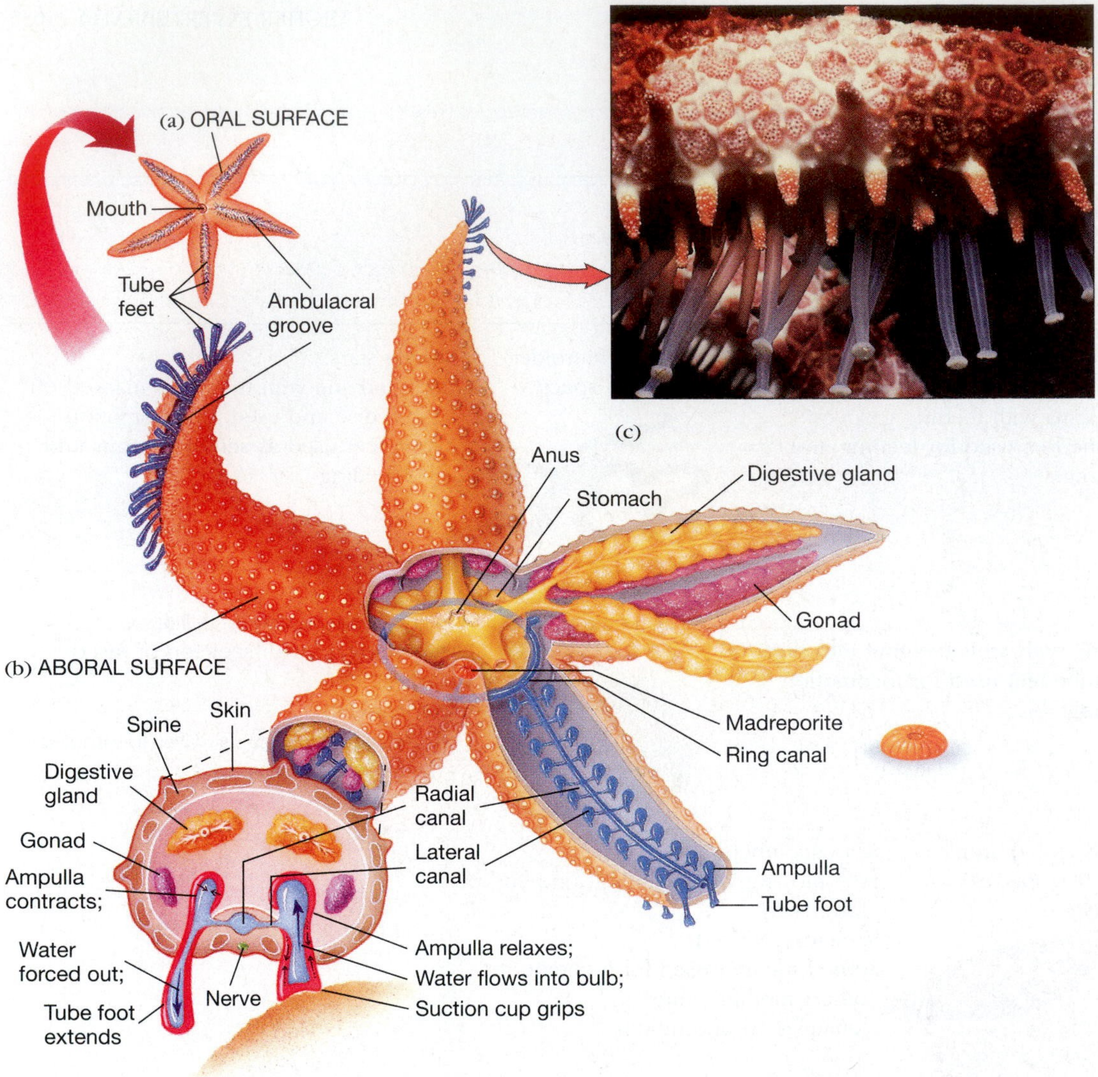

**(a) ORAL SURFACE**

Mouth

Tube
feet

Ambulacral
groove

**(b) ABORAL SURFACE**

Anus

Stomach

Digestive gland

Gonad

Madreporite

Ring canal

Spine

Skin

Digestive
gland

Gonad

Ampulla
contracts;

Water
forced out;

Tube foot
extends

Nerve

Radial
canal

Lateral
canal

Ampulla

Tube foot

Ampulla relaxes;

Water flows into bulb;

Suction cup grips

(c)

**FIGURE 29–22** Echinoderm anatomy. (a) The oral surface
(underside) of a sea star, showing the mouth and the ambulacral
grooves, bordered by tube feet, along each arm. (b) Aboral surface of a sea star. The top arm shows the upper surface. Proceeding clockwise, in the next arm the skin and calcareous plates
have been removed to show digestive structures and gonads.

The lower right arm shows the water-vascular system, and
the lower left arm is in cross section. When muscles in the ampulla contract, fluid is forced into the tube foot and it extends.
When muscles in the wall of the tube foot contract, they force
fluid back into the ampulla and radial canal. This lowers the fluid
pressure in the tube foot and it shortens. Low pressure between
the sucker of the foot and the substrate causes the foot to stick
like a suction cup. (c) Part of the arm of a crown of thorns sea
star from the Red Sea. Transparent tube feet hang down below
the arm.   (c, Jeff Rotman)

sea, a practice that merely increases the sea star population.

## Ophiuroidea: Brittle Stars

Ophiuroids, the brittle stars and serpent stars, look much
like asteroids except that their arms are sharply marked off
from the central disc. Ophiuroids also move differently from

**FIGURE 29–23** Class Crinoidea. A yellow crinoid attached to a
red coral in the Red Sea. (Jeff Rotman)

636

(a)

(b)

**FIGURE 29–24** More echinoderms: asteroid and ophiuroids. (a) A pink sea star on a rock displays the pentaradial symmetry characteristic of adult echinoderms. (b) Two brittle stars on a sponge in the ocean off Honduras. (b, Norbert Wu)

asteroids, by wriggling their jointed arms rather than using their tube feet. Some even use the arms to swim. The tube feet are used mainly as sense organs and for passing food to the mouth. Many ophiuroids are covered with spines, which give the arms a feathery appearance (Figure 29–24).

Most ophiuroids eat organic detritus from the sea floor and whatever small organisms they encounter. They have a very simple gut with a sac-like stomach and no intestine or anus.

## Echinoidea: Sea Urchins

Sea urchins and sand dollars, in the class Echinoidea, are tubby little echinoderms very different in appearance from the graceful crinoids and ophiuroids. They live mouth-down on the bottom of the sea, protected from intrusion by brittle calcareous spines, which easily penetrate human flesh and are difficult to remove (Figure 29–25a). The calcareous skeletal plates are fused into a sphere (for sand dollars, a flattened envelope), pierced only by holes for the mouth, anus, and tube feet. The mouth contains a chewing apparatus with a ring of large triangular teeth (Figure 29–25b). Echinoids can chew almost any sort of organic material they encounter. They walk either on their spines, which are hinged to the rest of the skeleton by a sort of universal joint, or on their tube feet. The tube feet are also used for gas exchange.

## Holothuroidea: Sea Cucumbers

Holothurians, or sea cucumbers, look more like large flabby sausages as they wash gently to and fro or lie partly buried on the sea floor. One would never guess that they are echinoderms, because their calcareous skeleton is reduced to microscopic plates scattered through the skin. Like echinoids,

(a)

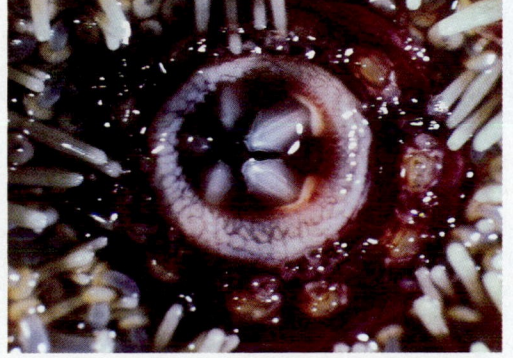

(b)

**FIGURE 29–25** Echinoids. (a) A sea urchin on a wreck. (b) Five large teeth, part of the feeding mechanism known as an "Aristotle's lantern," protrude from the mouth of a sea urchin. (Steven Webster)

holothurians lack arms. They are elongated, with the mouth at one end so that they are really lying on their sides. Holothurians can move slowly, either by using their tube feet or by wriggling. Some are quite competent burrowers. The mouth is surrounded by modified tube feet known as tentacles (Figure 29–26). In many species the tentacles secrete a net of mucus, which is used to catch small planktonic organisms for food. Other holothurians are detritus feeders, engulfing sand and mud from which they digest the organic matter before voiding the inorganic particles through the anus.

## 29–H   HEMICHORDATA: ACORN WORMS

Most hemichordates are worms that burrow in sediments at the bottom of shallow seas. A few species, the pterobranchs, are sessile and look like bryozoans. But whether burrowing or sessile, all hemichordates have a body divided into three parts: the **proboscis,** used for burrowing and feeding, a short **collar,** and a long **trunk** (Figure 29–27).

Hemichordates have features in common with both echinoderms and chordates. The transparent hemichordate larva is very similar to the larva of an asteroid echinoderm. Similarly, although they have no tube feet, pterobranchs have a hydrostatic system in their tentacles reminiscent of the water vascular system of echinoderms or of a lophophorate's system of tentacles. Hemichordates may have both a

**FIGURE 29–26**   A sea cucumber from the South Pacific.   (Brian Parker/Tom Stack)

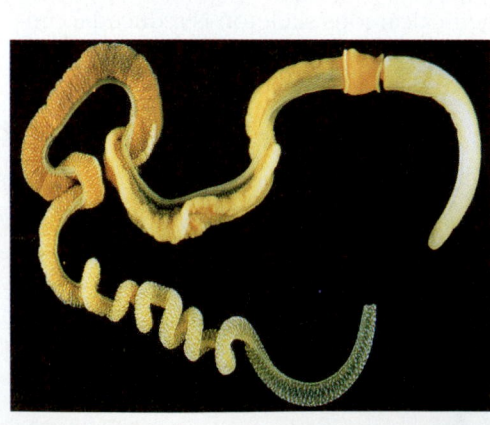

(a)

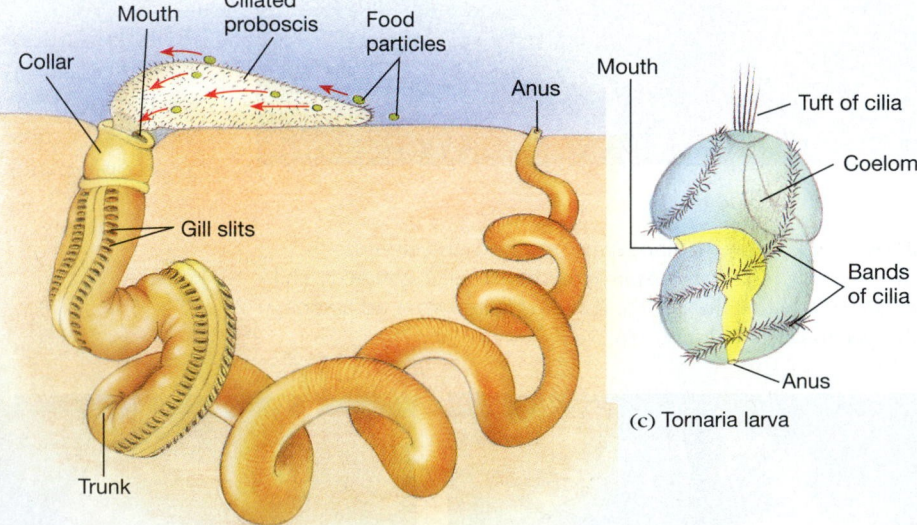

(b)

(c) Tornaria larva

**FIGURE 29–27**   Hemichordates. (a) *Saccoglossus kowaleskii.* (b) *Saccoglossus.* This species is up to 40 cm long, but other species may be as long as 150 cm. The body is flattened and the ciliated proboscis is conical. The proboscis is used mainly for burrowing, but its cilia convey water, food, and mucus to the mouth. Water is pushed out through the gill slits, and food particles pass to the digestive tract, which runs the length of the body. (c) The tornaria larva. (a, C. R. Wyttenbach, University of Kansas/BPS)

ventral nerve cord (as in most other invertebrates) and a dorsal nerve cord (a chordate feature). Hemichordates were once classified as chordates on the grounds that the connective tissue in the collar was a notochord, a definitive feature of chordates. It is now generally agreed that this is not so, but hemichordates do have one other chordate characteristic: gill slits leading from the pharynx to the exterior.

Although their relationship to the chordates is not as close as zoologists used to think, hemichordates do appear to represent some sort of evolutionary link between echinoderms and chordates. Anatomist Libby Hyman suggested hemichordates as a common ancestor of the two groups.

## 29–I  CHORDATA

The phylum Chordata contains two subphyla of invertebrates and a much larger subphylum containing all the vertebrates. Although we have a remarkable fossil record of the adaptive radiation of the vertebrates, details of the origin of the earliest chordates are lost, probably forever, owing to the poor Precambrian and Cambrian fossil record of this group. The only way to reconstruct early vertebrate history is to study the invertebrate groups most closely related to the vertebrates.

All chordates share several features (Figure 29–28). The first three listed here define the phylum:

1. **Notochord.** The name Chordata refers to the stiff, rod-like **notochord,** present in all chordates at some stage of life. In the invertebrate chordates, the notochord serves as a skeleton and prevents the body from shortening when the body muscles contract. During the em-

bryonic development of vertebrates, the notochord is surrounded or replaced by a column of vertebrae, which form the backbone.

2. **Pharyngeal gill slits.** At some time in their lives, all chordates have gill slits leading from the **pharynx,** the throat cavity behind the mouth, to the exterior.

3. **Dorsal tubular nerve cord.** Chordates have a hollow dorsal nerve cord, in contrast to most invertebrates, whose main nerve cord is solid and ventral.

4. **Segmented body.** Most chordates have more or less segmented bodies.

5. **Endoskeleton.** The chordate skeleton is internal (endo = within).

6. **Post-anal tail.** At some stage of life, a tail extends behind the anus at the posterior end of the body. (In non-chordates the anus is usually, although not always, at the end of the tail.)

These features comprise a particularly successful set of adaptations for a more active life than that of most invertebrates. Locomotion is by segmented blocks of muscle, the **myotomes,** pulling on the internal notochord. This allowed early chordates to swim quickly and efficiently by side-to-side wiggles of the body. As they swam forward, they took in food and water through the mouth and the water then left through the gill slits, having given up oxygen to the blood in the gills. Sense organs in the head detected where the animal was going and found food, and the brain and nervous system became well developed.

## 29–J  UROCHORDATA: TUNICATES

Urochordates, the sea squirts and their relatives, are an entirely marine subphylum of the chordates. As adults, these animals look nothing like any other chordate. They are classified as chordates because they have a tadpole-like larval stage with a notochord and a hollow dorsal nerve cord. However, the notochord disappears during metamorphosis to the adult form.

One class of tunicates contains sessile animals, the sea squirts, which are common around the coast. The other two classes contain planktonic forms that are rarely seen because, although common, they are fragile and transparent.

### Ascidiacea

Ascidians, or sea squirts, are sessile tunicates with the odd property of secreting a test (protective outer covering) of the polysaccharide cellulose, found nowhere else in the animal kingdom. Ascidians are filter feeders: they take in water through the mouth, filter it through a complicated basket derived from the pharyngeal gill slits of the larva, and push it out through a pore. Many live in colonies, which may share a common mouth and a common cellulose test.

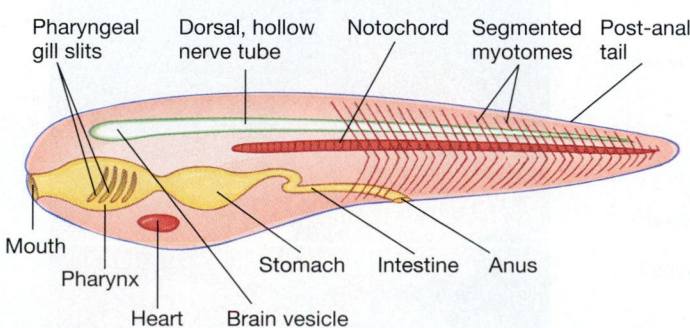

CHORDATE CHARACTERISTICS:

Pharyngeal gill slits    Dorsal, hollow nerve tube    Notochord    Segmented myotomes    Post-anal tail

Mouth

Pharynx

Heart    Brain vesicle    Stomach    Intestine    Anus

**FIGURE 29–28** Diagram of a generalized, primitive chordate, showing the features characteristic of chordates: the notochord, the dorsal hollow nerve tube above it, pharyngeal gill slits, segmentally arranged blocks of muscle (myotomes), and tail extending behind the anus.

The ascidians' motile, tadpole larva could almost have posed for our drawing of a generalized chordate (see Figure 29–27). The main differences are that the larva does not feed, so some have no mouths; and the only evidence of segmentation is the myotomes of the tail. Upon hatching, the larva swims to the surface with fish-like wriggles, using the action of its myotomes against its notochord. It drifts a short distance in the plankton, turns, swims down, and searches out a suitable surface on a rock or a dock piling. Here it attaches by adhesive projections on the tip of its nose and metamorphoses into an adult, losing its notochord and tail, while the gill slits expand tremendously (Figure 29–29).

Because the ascidian larva is the most primitive known chordate, it looks as if the typical chordate characteristics evolved, not as adaptations of an adult to its way of life, but in a larva. In 1928 Walter Garstang, a British zoologist, proposed that vertebrates evolved from a tadpole-like creature that failed to metamorphose but became sexually mature while still a larva. Most biologists today accept this idea. It is supported by the facts that a number of living animals develop some degree of sexual maturity while still larvae and that groups other than vertebrates also seem to have originated from young stages of their ancestors.

## 29–K    CEPHALOCHORDATA

Members of the subphylum Cephalochordata are commonly called amphioxus or lancelet. There are only 29 living species in two genera (*Asymmetron* and *Branchiostoma*). Whether they evolved from urochordates or from fish-like ancestors is a hotly debated question: they look like simple fish with too many gill slits, or like tunicate tadpoles

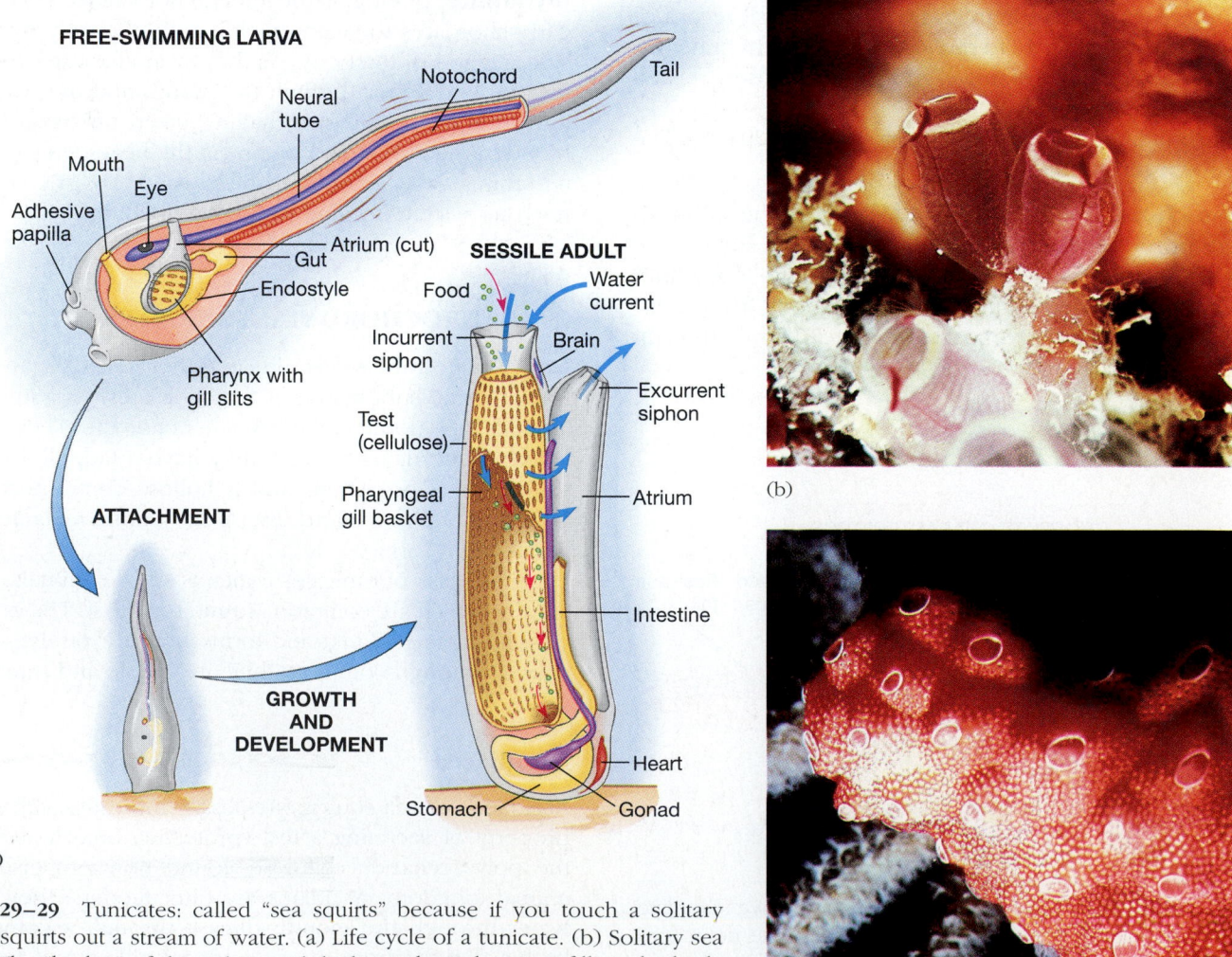

**FIGURE 29–29**  Tunicates: called "sea squirts" because if you touch a solitary adult, it squirts out a stream of water. (a) Life cycle of a tunicate. (b) Solitary sea squirts. The "basket" of thin white and thicker pale pink stripes filling the body is pharyngeal gill slits arranged to filter food out of the water. (c) Colonial ascidians.    (b, c, Steven Webster)

(a)

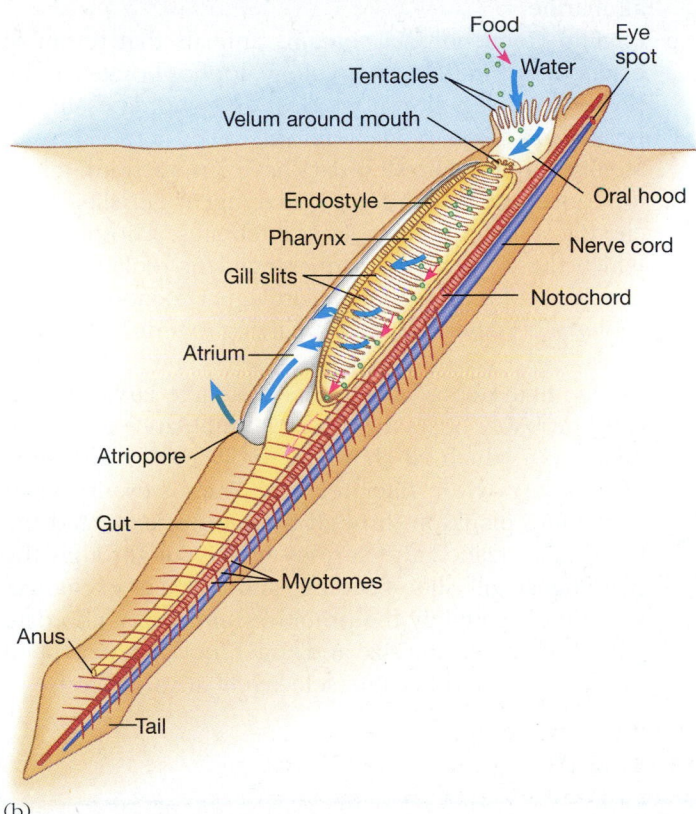

(b)

**FIGURE 29–30** Amphioxus. (a) A living animal. (b) Diagram to show the chordate characters of notochord, dorsal nerve cord, and pharyngeal gill slits. An amphioxus lies on its back, partly buried in sand or mud. The tentacles waft food and water into the mouth. Excess water is pushed out through the gill slits into an atrium and leaves the body via the atriopore. The notochord is a skeletal structure that stiffens the body.   (a, G. I. Bernard/Oxford Scientific Films)

with the gill system expanded into an enormous pharyngeal gill basket (Figure 29–30). The gill basket leaves little room for swimming muscles, and these animals swim poorly.

An amphioxus lives buried in the sand in shallow, warm oceans, with only its head end protruding. Two rings of tentacles around the entrance to the mouth screen out large particles and sense chemicals and touch. These tentacles edge the **oral hood,** extending like a vestibule before the mouth, and the membranous **velum,** which surrounds the mouth opening.

An amphioxus feeds like a urochordate. Cilia pull a current of water into the mouth. Any food in the current is filtered out by a mucous net secreted by the **endostyle,** a ventral groove in the pharynx. The endostyle also secretes iodine-containing molecules and is homologous to the vertebrate thyroid gland, which secretes iodine-containing hormones. Mucus containing trapped food particles passes down the pharynx into the gut, where the food is digested, and the water current is pushed out through the gill slits. The gill slits do not lead directly from the pharynx to the exterior as in fish, but open into a large **atrium,** which in turn has an opening to the exterior called the atriopore.

## 29–L   VERTEBRATA: A PREVIEW

Vertebrates differ from other chordates in having a vertebral column, which replaces the notochord to a greater or lesser extent. Cephalization is pronounced, with sense organs and nerves concentrated at the front end of the body, forming a very obvious head. In addition, all vertebrates have some sort of a liver; endocrine organs, which secrete hormones; and kidneys that differ completely from the various excretory organs of invertebrates.

Primitively, vertebrates were aquatic animals that moved as sharks do today, by sinusoidal swimming, with the segmented myotomes pulling against a dorsal notochord or vertebral column and pushing the paddle-like tail against the water. Early fish fossils lacked movable jaws, and it is likely that the original feeding method was a type of filter feeding like that of amphioxus. Food was probably filtered out of a water stream, which entered through the mouth, passed over the gills, and exited through the gill slits. Gills are used for gas exchange in modern fish, but feeding was probably their original function in chordates.

## SUMMARY

A coelom is a fluid-filled body cavity that originates as a space in the mesoderm during embryonic development. The coelom permitted the evolution of more efficient modes of digestion, circulation, and reproduction and thus led to an enormous burst of adaptive radiation that resulted in the spectacular variety of coelomate animals.

Coelomates include various protostomes (phyla Annelida, Mollusca, and Arthropoda) and the evolutionary line known as deuterostomes (phyla Echinodermata, Hemichordata, and Chordata).

1. The phylum Annelida includes the classes Polychaeta (bristle worms), which contains mostly marine forms; Oligochaeta, which contains mostly freshwater or terrestrial forms, including earthworms; and Hirudinea (leeches), which contains mostly freshwater forms. Polychaetes and oligochaetes are generally segmented. This segmental organization gives them a way to move and burrow rapidly, by means of a hydrostatic skeleton acting together with their setae. Leeches lose their segmentation during development and move by means of powerful body muscles.

2. The members of phylum Bryozoa are tiny aquatic animals living colonially within an exoskeleton attached to the substrate. The animals feed by means of a lophophore, a crown of tentacles around the mouth.

3. Most members of phylum Mollusca are unsegmented marine animals, but many gastropods are well adapted to terrestrial life. A mantle covers most of the body and usually secretes a calcareous shell; gas exchange occurs through gills in the mantle cavity or through the lining of the mantle cavity itself. Molluscs also have a muscular foot, which is used for locomotion or burrowing; cephalopods use their tentacles, derived from the foot, for gripping prey.

    The main molluscan classes are:
- Amphineura: chitons
- Gastropoda: snails, slugs, and nudibranchs
- Bivalvia: shellfish
- Cephalopoda: nautiluses, cuttlefish, squids, and octopuses

4. Phylum Arthropoda contains highly successful marine, freshwater, and terrestrial members. The arthropod body is segmented, with a protective chitinous exoskeleton and many jointed appendages performing a variety of specialized jobs. This body plan proved so versatile that the arthropods have undergone impressive adaptive radiation, with more species and individuals and a larger total mass than any other animal phylum.

    The main arthropod classes are:
- Arachnida: spiders, scorpions, ticks, and mites
- Crustacea: crabs, lobsters, shrimp, barnacles, copepods, pillbugs, and so on
- Insecta: insects
- Chilopoda: centipedes
- Diplopoda: millipedes

5. The relationship of echinoderms and chordates is deduced from the similarity of the embryos but is not evident in the very different adults. Adult echinoderms have pentaradial symmetry and a calcareous endoskeleton under the thin epidermis. Tube feet operated by pressure changes in a water vascular system are unique features of this group. The echinoderms are all marine.

6. Phylum Hemichordata contains animals that resemble echinoderms in their embryology, lophophorates in the possession of hollow tentacles, and chordates in the possession of pharyngeal gill slits.

7. Members of phylum Chordata have a notochord, a hollow dorsal nerve cord, and pharyngeal gill slits. Most chordates living today are vertebrates, chordates in which the notochord is surrounded or replaced by a vertebral column of bone or cartilage. Urochordates and cephalochordates are the only living invertebrate chordates.

8. Among living chordates, the tadpole-like larva of tunicates probably bears the closest resemblance to the animals from which all chordates evolved. The ancestors of chordates were filter feeders. They drew in water containing plankton through the mouth, extracted the food, and pushed the surplus water out through the pharyngeal gill slits. Feeding, and not gas exchange, was almost certainly the primitive function of chordate gills. With increasing size and larger food, greater speed to catch the food became selectively advantageous.

## Self-Quiz

1. The evolutionary importance of a coelom is that:
   a. it permitted animals to have a circulatory system and other internal organs that move
   b. it permitted animals to move onto land with an internal storage place for extra body fluid
   c. it provided the possibility of evolving a hard, protective exoskeleton
   d. it allowed organisms to have excretory systems
   e. it paved the way for evolution of locomotory appendages
2. The possession of a mantle, a trochophore-like larva, and a calcareous exoskeleton is characteristic of the:
   a. Annelida      d. Mollusca
   b. Bryozoa      e. Arthropoda
   c. Crustacea

Match (choose all correct lettered items for each description):

_____ 3. Part of the insect respiratory system
_____ 4. Excretory organ of an insect
_____ 5. Locomotory apparatus of a polychaete
_____ 6. Tongue-like rasping organ of some molluscs
_____ 7. Feeding apparatus of a spider
_____ 8. Feeding apparatus of bryozoans

a. chelicerae
b. hydrostatic skeleton
c. lophophore
d. Malpighian tubule
e. mantle
f. muscular foot
g. radula
h. setae
i. spiracle
j. suckers
k. trachea
l. walking legs

9. Which set of animals below all belong to the same class?
   a. crab, scorpion, lobster      d. snail, slug, scallop
   b. shrimp, barnacle, pillbug    e. clam, nautilus, oyster
   c. leech, tick, flea
10. Filter feeding is found in which of the following? (Choose all correct answers.)
   a. adult urochordates      d. chitons
   b. adult cephalochordates  e. bryozoans
   c. chelicerates

Match:

_____ 11. Orb-weaving spider      a. Amphineura
_____ 12. Sea urchin              b. Arachnida
_____ 13. Medicinal leech         c. Asteroidea
_____ 14. Chiton                  d. Bivalvia
_____ 15. Earthworm               e. Cephalopoda
_____ 16. Squid                   f. Crinoidea
_____ 17. Brittle star            g. Crustacea
_____ 18. Sea cucumber            h. Echinoidea
_____ 19. Mystery snail           i. Gastropoda
_____ 20. Crayfish                j. Hirudinea
                                    k. Holothuroidea
                                    l. Insecta
                                    m. Oligochaeta
                                    n. Ophiuroidea
                                    o. Polychaeta

21. The members of phylum Echinodermata do *not* show:
   a. calcareous skeleton
   b. pentaradial symmetry
   c. tube feet
   d. metameric segmentation
   e. slow locomotion
22. A long, cylindrical animal that has tube feet is a member of:
   a. phylum Chordata
   b. class Ophiuroidea
   c. subphylum Urochordata
   d. class Echinoidea
   e. class Holothuroidea
23. You would be most likely to find an adult urochordate:
   a. in a mountain stream
   b. in a large river such as the Mississippi
   c. preying on clams
   d. along Cape Cod, attached to the piling of a dock
24. Which of the following chordate characteristics contributes *least* to its efficiency of locomotion?
   a. myotomes
   b. pharyngeal gill slits
   c. notochord
   d. post-anal tail
   e. streamlined body shape

## Questions for Discussion

1. In several species of leeches, individuals fix their spermatophores (sperm packets) onto the external body wall of another leech. When many individuals are kept in a jar together, the largest individuals receive more spermatophores than do smaller individuals. What is the adaptive advantage of this behavior?
2. Why is the exoskeleton of an arthropod so much more effective as a waterproofing device for a terrestrial animal than is the calcareous exoskeleton (shell) of a mollusc? (Think of the difference between the shell of a lobster and the shell of a snail.)
3. It is thought that the main limit to the size of insects is the tracheal system. Terrestrial vertebrates with lungs can grow to much larger sizes. Why is this?
4. An early theory held that vertebrates might have originated from annelids (turn an annelid worm upside down and it looks quite like a fish). What similarities between the two groups led to such a theory?
5. Why is filter feeding such a common way of life among invertebrate animals?
6. Cephalization is pronounced in the vertebrates but not in the cephalochordates, urochordates, or echinoderms. What differences in selective pressures may have caused this difference in degree of cephalization?

## Suggested Readings

(See also references in Chapter 28.)

Barrington, E. J. W. *The Biology of the Hemichordates and Proto-chordata.* Edinburgh: Oliver and Boyd Ltd., 1965. A thorough discussion of the lower chordates and some of their invertebrate relatives.

Borror, D. J. et al. *An Introduction to the Study of Insects,* 6th ed. Philadelphia: Saunders College Publishing, 1989. A good introductory entomology text.

Cameron, J. N. "Molting in the blue crab." *Scientific American,* May 1985. The life and times of a favorite invertebrate delicacy.

Evans, H. E. *Life on a Little-known Planet.* New York: E. P. Dutton, 1978. An entertaining and enlightening account of our insect neighbors by a leading scientist.

Evans, H. E. *Insect Biology: A Textbook of Entomology.* Reading, MA: Addison-Wesley, 1984.

Gosline, J. M., and M. E. DeMont. "Jet-propelled swimming in squids." *Scientific American,* January 1985. An advanced version of the hydrostatic skeleton makes swift, efficient swimming possible.

Jackson, R. R. "A web-building jumping spider." *Scientific American,* September 1985. Covers structure, habits, ecology, and evolution of spiders.

Lane, F. W. *Kingdom of the Octopus*. New York: Sheridan House, 1965. This fascinating book on cephalopods contains Harvey's account of the acquisition of *Architeuthis*.

Moynihan, M. *Communication and Noncommunication by Cephalopods*. Indianapolis, IN: Indiana University Press, 1986. A delightful book with lovely black and white wash drawings.

Nichols, D. *Echinoderms,* 4th ed. London: Hutchinson, 1969. An authoritative and readable book on the echinoderms.

Roper, C. F. E., and K. J. Boss. "The giant squid." *Scientific American,* April 1982.

Ryker, L. C. "Acoustic and chemical signals in the life cycle of a beetle." *Scientific American,* June 1984. Adaptations of a major lumber industry pest.

Vollrath, F. "Spider webs and silks." *Scientific American,* March 1992.

# Vertebrates

### O B J E C T I V E S

*When you have studied this chapter, you should be able to:*

1. List characters that would permit you to distinguish members of the following vertebrate groups: jawless fish, Chondrichthyes, actinopterygians, amphibians, reptiles, turtles, diapsids, birds, mammals.

2. State the order in which the following groups appeared in the fossil record: invertebrates, jawless fish, Chondrichthyes, bony fishes, amphibians, reptiles, birds, mammals.

3. List and explain: (1) the selective pressures encountered by previously water-dwelling vertebrates in adapting to life on land; (2) adaptations that enable modern terrestrial vertebrates to live on land.

4. List the evolutionary advances made by each main group of vertebrates over its ancestors.

5. Outline human evolution.

W hen you say "animal," most people think of animals with backbones: fishes, amphibians, reptiles, birds, and mammals. These are **vertebrates,** chordates with backbones. As a chordate, a vertebrate has a notochord, pharyngeal gill slits, and a dorsal hollow nerve cord. During early vertebrate evolution the notochord became encased in bony or cartilaginous vertebrae, which form the axis of the skeleton. This permitted rapid, efficient locomotion by providing a strong but flexible rod for the segmental muscles to pull against in the body's sinuous swimming movements.

An animal that can move fast can be carnivorous, feeding on other animals. A carnivore can spend less time actually eating than a herbivore does because animal tissue is richer than plant material in energy and nutrients. As we shall see, many evolutionary advances by vertebrates were associated with an active carnivorous way of life.

Various other characters are also found in vertebrates: kidneys that differ markedly from the excretory organs of invertebrates; a ventral heart, closed blood vessels, and some degree of segmentation. Cephalization is pronounced; that is, sense organs and nerves are concentrated at the front end of the body, and so vertebrates have very obvious heads. The sense organs are protected by a skull, formed from skeletal elements at the anterior end of the spinal cord. A tail extending beyond the anus is an ancestral character found in most vertebrates.

Vertebrates range in size from tiny fish of less than a gram to 100,000-kg whales (Figure 30–1). But most vertebrates fall at the large end of the range of animal sizes, and this is why they are the organisms most threatened with extinction by spreading human populations. There are about 50,000 species of vertebrates today, and about ten times as many have existed during the 500 million years since vertebrates first evolved in the Cambrian.

**FIGURE 30–1** Fish on a coral reef. The evolutionary advances of early fish—in feeding and locomotion—permitted the evolution of all other vertebrates. (James T. Spencer)

**TABLE 30-1**

**Classification of Vertebrates**

| Traditional (Classes) | Common Name (Examples) | | Modern (Names Used in This Chapter) |
|---|---|---|---|
| Class Agnatha | Jawless fish (hagfish, lamprey) | | Jawless fish |
| Class Chondrichthyes | Cartilaginous fish (shark, ray) | | Cartilaginous fish, Chondrichthyes |
| Class Osteichthyes | Bony fish (trout, herring) | | Ray-finned fish, bony fish, Actinopterygians |
| | Coelocanth, lungfish (coelocanth) | | Lobe-finned fish, Sarcopterygians |
| Class Amphibia | Amphibians | (newt, salamander) | Urodeles |
| | | (frog, toad) | Anurans |
| | | (caecilians) | Apodans |
| Class Reptilia | Reptiles | (turtles) | Testudomorphs |
| | | (dinosaurs, lizards, snakes) | Diapsids |
| | | (crocodiles, alligators) | Crocodilians |
| Class Aves | Birds | | Birds, Aves |
| Class Mammalia | Mammals (monotremes, marsupials, placental mammals) | | Mammals |

After the first vertebrates evolved, the continents drifted together to form Pangaea and then broke up into Gondwana and Laurasia before forming the modern continents (*Essay: Continental Drift,* Chapter 19). Vertebrate evolution was strongly influenced by these movements and their effects on the climate on land. For instance, much of Pangaea lay near the equator, so early land vertebrates evolved in a warm climate. When Pangaea broke up, land masses moved north and south and cut the Arctic Ocean off from warm ocean currents. The temperature over much of the land fell, contributing to the extinction of the dinosaurs. Later, the formation of Central America, joining North and South America, permitted migrations that had drastic effects on the evolution of vertebrates throughout the Americas.

Increasing use of cladistic taxonomy is causing changes in vertebrate classification (Chapter 22). Traditionally, vertebrates have been divided into seven classes: three classes of fish plus the amphibians, reptiles, birds, and mammals (Table 30–1). But cladists argue that crocodiles are more closely related to birds than they are to turtles or snakes, so the reptiles should be divided into several classes. In this chapter we use both old and new names for groups of vertebrates, without trying to decide which of these groups should be considered classes.

**KEY CONCEPTS**

♦ In the vertebrates, the notochord was replaced by a column of vertebrae, permitting efficient swimming upon which the success of fish depends.

♦ Most vertebrate evolutionary advances have been made by carnivorous groups and involve improvements in locomotion and feeding.

♦ The radiation of terrestrial vertebrates was associated with evolution of the amniotic egg and the radiation of insects.

## 30-A  JAWLESS FISH: HAGFISH AND LAMPREYS

The earliest complete vertebrate fossils are of ostracoderms, fish without jaws from the Ordovician to Devonian Periods. Their modern relatives are the lampreys and hagfishes, long narrow creatures specialized as scavengers and parasites.

Hagfishes, which are all marine, live in warm waters of the Atlantic and Pacific. They attack dead, diseased, or disabled fish and also eat various invertebrates.

Lampreys have a sucking mouth and a rasping tongue covered with teeth, which are used to break the skin and suck the blood of bony fish. They do not usually kill these hosts (Figure 30–2). Unlike hagfishes, lampreys have a larval stage, an **ammocoete,** which lives for up to seven years as a filter feeder buried in the mud of a river or stream. The adult usually lives only a few years and dies after migrating upriver to spawn.

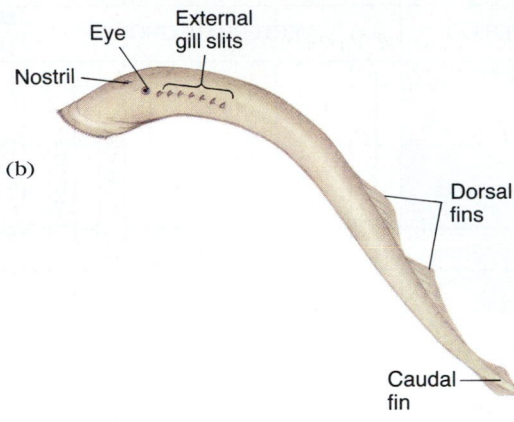

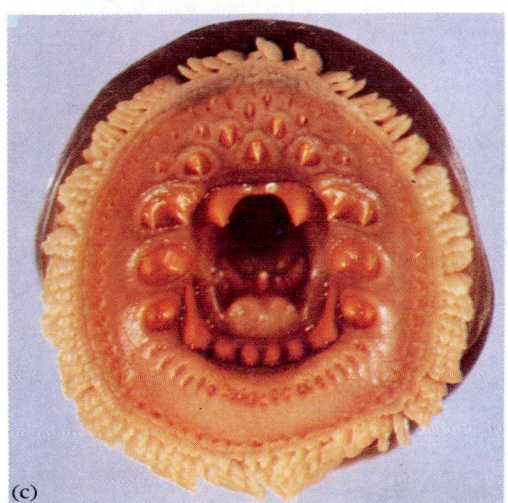

**FIGURE 30–2** Lampreys. (a) Two lampreys attached to a bony fish (carp) by their round sucking mouths. Note the row of circular gill slits. (b) Drawing to show the external gill slits and unpaired fins. (c) The round, suction-cup mouth, with many rasping teeth. (a, Tom Stack & Associates; c, Kiyoko Uehara)

Lampreys and hagfishes have no paired fins, jaws, or scales, and the skeleton is just the notochord, which persists throughout life, and various small cartilages. Cartilage is a flexible connective tissue typical of vertebrates. (It is found in the moveable parts of our noses and ears.) In the skeletons of most vertebrates, much of the cartilage is replaced during development by bone, a hard tissue containing calcium and phosphate salts. Bone must have evolved in early vertebrates: many ostracoderms were covered with bony armor. Unfortunately, neither lampreys nor hagfishes contain bone, so they provide no evidence about the origin of this useful tissue. However, lampreys do provide insight into other evolutionary advances.

### Early Fish Evolution

The mouth and pharynx of the ammocoete larva of a lamprey are very like those of an amphioxus. Both animals filter food from a current of water that enters the mouth and leaves through the gill slits. However, there is one important difference: in an amphioxus the water current is propelled by cilia, mainly on the gills, but in the lamprey larva it is moved by muscles in the walls of the gill pouches. Because water can be moved faster by muscles than by cilia, the ammocoete's feeding method allows more food to be gathered in a given period of time. As a result, an ammocoete can grow larger than an amphioxus. Ostracoderms undoubtedly fed by this type of muscle-powered filter feeding.

One of the most useful vertebrate organs is the kidney. Kidneys keep the composition of the body fluids constant by controlling water and salts that leave the body in urine. Without kidneys, vertebrates could not live in fresh water, where the body tends to gain water by osmosis, or on land, where the body loses water by evaporation. Lampreys and hagfishes have kidneys and so, presumably, did their ostracoderm ancestors. Although ostracoderms were marine, they were preadapted to move into fresh water when freshwater lakes began to spread in the Silurian. A diverse collection of marine and freshwater fish rapidly evolved. Figure 30–3 shows when the main vertebrate groups appeared in the fossil record.

The most successful fish groups were the cartilaginous fishes (including the sharks) and the bony fishes. Both had

| ERA | PERIOD | VERTEBRATE GROUP | Millions of years ago |
|---|---|---|---|
| Cenozoic | Quaternary | | 2 |
| Cenozoic | Tertiary | | |
| Mesozoic | Cretaceous | | 66 |
| Mesozoic | Jurassic | | 144 |
| Mesozoic | Triassic | | 208 |
| Paleozoic | Permian | | 245 |
| Paleozoic | | | 286 |
| Paleozoic | Carboniferous | | |
| Paleozoic | Devonian | Invasion of land | 360 |
| Paleozoic | Silurian | Invasion of fresh water | 408 |
| Paleozoic | Ordovician | | 438 |
| Paleozoic | | | 505 |
| Paleozoic | Cambrian | | |
| Paleozoic | | | 570 |

(Vertebrate groups shown as bars: Jawless fishes, Cartilaginous fishes, Bony fishes, Amphibians, Reptiles, Birds, Mammals)

**FIGURE 30–3** The vertebrate fossil record. The lines show when groups with modern descendants first appear in the fossil record.

two evolutionary advances over their ancestors. First, part of the gill skeleton moved forward and evolved into jaws, permitting these fish to bite and chew their food instead of sucking or filtering it. Second, both groups have two pairs of lateral fins: **pectoral** fins at the front and **pelvic** fins, as well as unpaired **dorsal, anal,** and **caudal** (tail) fins (Figure 30–4). Paired fins allow a fish to balance and maneuver in ways that are not possible without them. Both jaws

and better locomotion permit an active carnivorous lifestyle, permitting fish to catch and eat larger prey than their ancestors could.

## 30–B CARTILAGINOUS FISH: CHONDRICHTHYES

Cartilaginous fish get their name from their skeleton, which is composed entirely of cartilage. Although many Chondrichthyes are known from Paleozoic fossils, only about 700 species survive today: the sharks, dogfish, rays, skates, and ratfish. Almost all are marine, and few deserve their ferocious reputations. Many species are less than 15 cm long, and the largest living fish, the 10-m whale shark, is a filter feeder. Rays are flattened and swim by undulating their huge pectoral fins, often feeding on the bottom. Some rays have poison spines, which they use to defend themselves, on the back or tail. The electric ray repels intruders with an organ that can produce quite a powerful electric shock.

A shark's teeth are enlarged versions of the pointed **denticles** found all over its skin. Sharks are not picky eaters. One killed in the Adriatic Sea had in its stomach two raincoats, part of a horse, an automobile license plate, and a length of rope. The stomach of a small dogfish, such as you might dissect in the laboratory, is more likely to contain crustaceans and fish. Digestion takes place mainly in the stomach, and absorption occurs in the intestine, which, unlike that of most other vertebrates, is short and fat. The surface area for absorption is increased by a spiral valve, which runs down the inside of the intestine. A shark's liver is enormous in proportion to its body size. This is because sharks store lipids in their livers, and by altering the lipid content of the liver, they can slowly increase or decrease the buoyancy of the whole body, reducing the effort needed to stay at a particular depth in the water. We shall see that bony fish evolved a different buoyancy device—a gas-filled swimbladder.

CARTILAGINOUS FISH

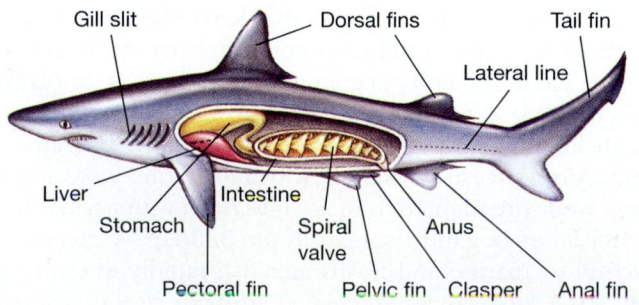

BONY FISH

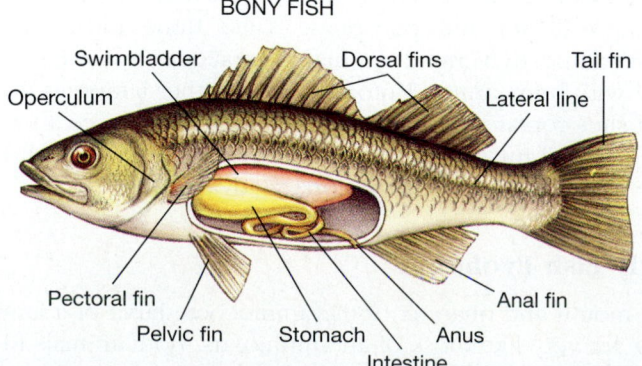

**FIGURE 30–4** Comparison of a shark and a bony fish. Note the asymmetry of the shark's tail fin. Compared with the shark's, the pelvic fins of the bony fish are placed far forward. The large, fat-filled liver of the shark and the swimbladder of the bony fish both provide buoyancy.

Sharks swim in primitive vertebrate fashion, by sinuous waves of the body, using their segmental muscles and jointed backbones. Sharks (and bony fish) that spend all of their time swimming depend on their constant forward motion to pass oxygen-rich water over their gills, where gas exchange occurs. If these active swimmers are restrained, they "drown" because they lack the pumping muscles in the head that allow less active fish to stay in one place and draw a water current across the gills.

Sharks have internal fertilization. Females produce a few large yolky eggs. The male passes sperm into the female's body where the eggs are fertilized, using his claspers, extensions of the pelvic fins (see Figure 30–4). In some species the large eggs are covered with a leathery coat before being deposited. In most species, however, the eggs and embryos remain in the oviduct, the tube leading from the ovary toward the exterior of the body, until they are fully developed baby sharks. Most sharks are **ovoviviparous**—that is, the embryo receives all its nourishment from the yolk of its egg, and the mother's body serves merely as an incubator. A few are **viviparous:** the embryos receive nourishment from the mother's bloodstream, with close contact between their blood vessels and those of the oviduct. The embryos of the sand tiger, a shark common in public aquariums, obtain nourishment by eating any smaller embryos! This prenatal cannibalism results in the birth of only two offspring at a time, one from each oviduct.

Sharks have efficient pressure and chemical sense organs. Large olfactory organs in their heads permit sharks to detect chemicals (notoriously blood) in the water. In addition, all kinds of fish possess **lateral line systems,** a line of pressure receptors along each side of the body, which are sensitive to vibrations or sound waves. In a shark's head, the lateral line receptors are modified to form electroreceptors. All animals generate electric fields, and experiments have shown that a dogfish shark can use its electroreceptors to detect an electrical potential of as little as 2 microvolts emitted by a worm that the shark can neither see nor smell. Once the olfactory and lateral line organs have guided the shark to the vicinity of its prey, the eyes guide the actual attack.

## 30–C  RAY-FINNED BONY FISH: ACTINOPTERYGIANS

Figure 30–5 shows the probable phylogenetic relationships of vertebrate groups with members that have survived to this day. Fish with skeletons of bone have such versatile anatomy and physiology that bony fish gave rise to the greatest diversity of any aquatic vertebrate group. Nearly all modern bony fish are among the 20,000 species of **actinopterygians,** the ray-finned bony fish. These have fins supported by rays of bone or cartilage, distinguishing them

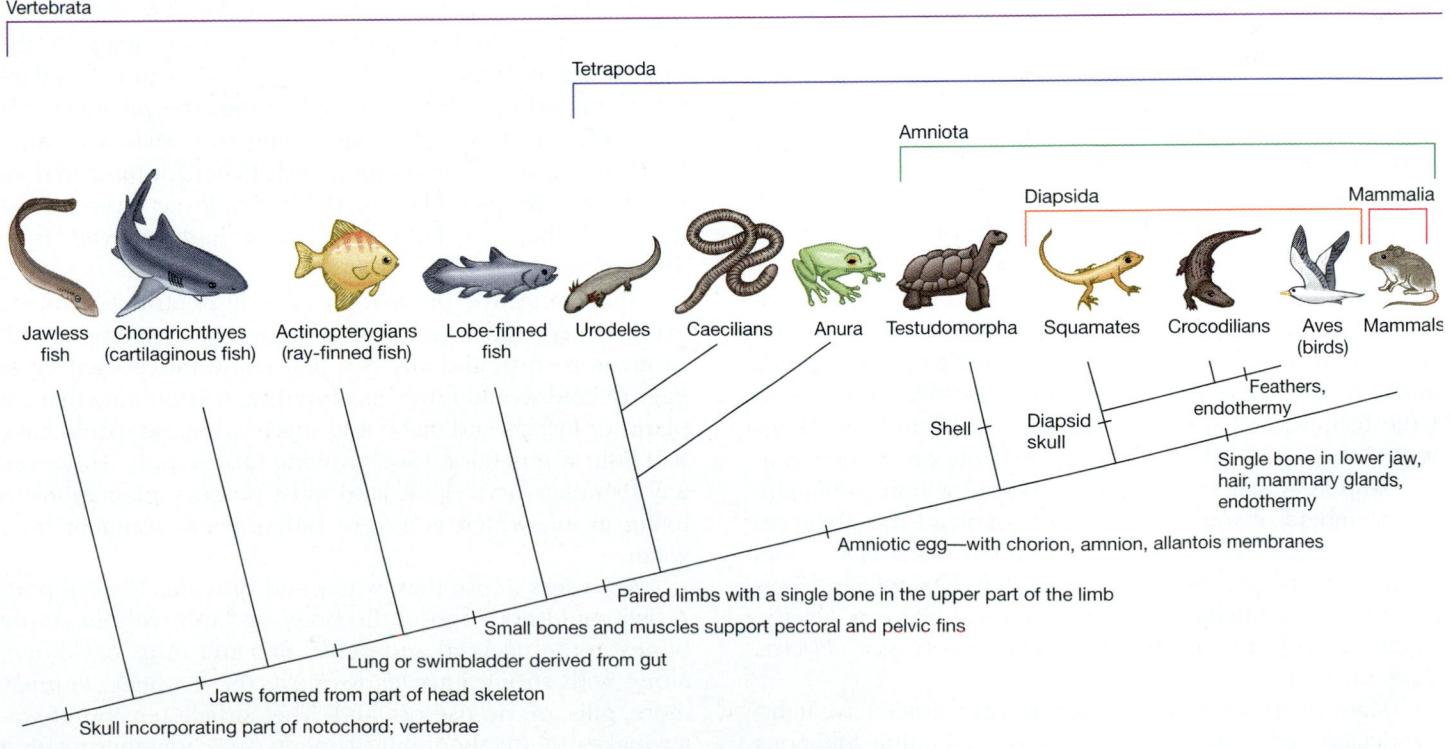

**FIGURE 30–5**  A cladogram showing the probable phylogenetic relationships of modern vertebrates.

from the other group, the **sarcopterygians,** lobe-finned fish, which gave rise to terrestrial vertebrates.

Among the actinopterygians are filter feeders like herrings; parrot-fishes, which crunch up coral; insectivores (insect-eaters) like trout; and predaceous carnivores like pike, barracuda, and blennies. Boxfish are practically spherical, moray eels are snake-like, a sea horse looks like a horse's head on a monkey's tail, and a stonefish looks like a lump of rock.

The success of bony fish is due largely to a few evolutionary innovations. One of the most important was the development of a **swimbladder,** a gas-filled sac formed as an outgrowth of the pharynx (see Figure 30–4). In water containing little oxygen, many fish come to the surface and gulp air into the swimbladder, which serves as an accessory breathing organ. In deeper water, a fish can alter the gas pressure in the bladder and hence its buoyancy, so that it can remain at any depth in the water with no muscular effort. A swimbladder is found in many but not all actinopterygian fish.

Most of the push in actinopterygian locomotion comes from the tail; the paired fins are usually used for fine control. The pelvic fins are often further forward and higher on the body than those of a shark.

The gill slits of bony fish do not open separately to the exterior, as do those of agnathan and cartilaginous fish, but are instead covered by a common **operculum.** Oxygen-laden water enters through the mouth, passes over the gills, and leaves through the gill slits, moved by muscles in the head and at the base of the operculum.

Reproduction is as varied as everything else about this group. Unlike Chondrichthyes, most bony fish have external fertilization and lay large numbers of small eggs which they abandon. Some build nests and care for the eggs. In pipefishes and sea horses, the males have brood pouches, where they incubate the young. Other species have internal fertilization, and the young develop within the female's body.

Many bony fish have spines on their dorsal fins or opercula. In some species, like the stonefish, rockfish, and scorpion fish, these spines are connected to poison glands and can inject poisons powerful enough to kill humans or large fish. Several species of puffer fish protect their eggs with a coating of jelly containing tetrodotoxin, a nerve poison so potent that less than a microgram will kill a human.

Members of four families have evolved the ability to produce electric discharges. *Electrophorus,* the electric eel of the Amazon, can generate up to 550 volts, which it uses for offense and defense. More commonly, fish use electric organs and electroreceptors for navigation (*Essay: Electric Fish,* Chapter 39).

Many deep-sea fishes are luminescent. Some have light-producing cells; others have organs containing luminous bacteria. Light flashes are probably used to signal the op-posite sex and to startle attackers. In the deep-sea angler fish, the luminous tip of a fin is used as a lure that attracts prey. In addition, many fish can change color by using tiny muscles in the skin to alter the size of different chromatophores (color cells).

## 30–D  LOBE-FINNED FISH: LUNGFISHES AND COELOCANTH

Paleontologists agree that terrestrial vertebrates evolved from lobe-finned fish (sarcopterygians), although debate rages over which sarcopterygian group was ancestral. Only five genera of lobe-finned fish survive today: four genera of lungfishes in oxygen-depleted freshwater swamps and the coelocanth, adapted to life in the deep ocean.

Lungfish have lungs, which supplement gill breathing. Lungs and swimbladders both develop as outgrowths of the embryonic gut. It is not difficult to imagine that ancient air-breathing lungfishes, using their fleshy pectoral and pelvic fins to move about on land, were the ancestors of the first terrestrial vertebrates. The fossil record shows that animals much like modern lungfish evolved in the Carboniferous era, about 300 million years ago. These creatures had fish-shaped bodies, short stubby legs, and no gills.

## 30–E  THE MOVE TO LAND

In the Cambrian, when vertebrates evolved, the land was an unappealing environment with no plant life except perhaps a film of bacteria and algae in moist areas. By the mid-Devonian some 200 million years later, when the adaptive radiation of fish was in full swing, the prospect was very different. Plants were spreading over swampy, tropical Gondwana, providing food and shelter for terrestrial invertebrates. By the end of the Devonian, insects were evolving, and the first land vertebrates had evolved from lobe-finned fishes.

Many selective pressures probably contributed to the evolution of land vertebrates. The seas were teeming with carnivorous fish, and any fish that could move itself or its eggs to land would lower its mortality. A vertebrate that ate plants or insects and that could survive on land would have had little competition for a growing food supply. However, any fish that survives on land must possess adaptations to living in air, which is a very different environment from water.

Air is less dense than water and provides less support. A fish could not support the body on land without sturdy bones in the pectoral and pelvic fins and in its backbone, along with strong muscles to move these bones. Furthermore, gills are no use on land. The surface tension of water makes the feathery gill filaments stick together when a fish comes out of water; a swimbladder, lung, or body surface that keeps its shape in air is necessary. In addition, sound, light, and chemicals travel differently in air than in water. The sense organs of an aquatic animal will work on land, but they are less than ideal.

Probably the biggest problem faced by an aquatic animal moving to land is dehydration. Waterproofing the body surface reduces the amount of water lost. However, water is still lost in the urine and feces and from areas that must be kept moist, such as the respiratory surface and mouth. Fish that can live on land for long periods breathe using internal lungs or swimbladders, which lose less water than external gills or skin. Their kidneys produce a concentrated urine, and the walls of the digestive tract absorb water into the body, thus reducing the amount lost in the feces.

## The First Tetrapods

Amphibians, reptiles, birds, and mammals are known as **tetrapods** ("four feet"). They are the land vertebrates. Only reptiles, birds, and mammals are fully terrestrial, however. Most amphibians—frogs, toads, salamanders, and newts—must return to water to reproduce.

Even though the adults of most amphibian species are terrestrial, most must still live in damp places. Even desert toads spend most of their time in burrows where the humidity is high. This is because gas exchange takes place partly through the skin, which must therefore be kept moist despite its tendency to lose water to the air. Because the skin is permeable, few amphibians can live in the sea, where they lose water by osmosis through any permeable body surface.

The earliest full fossil of a tetrapod is that of the late Devonian genus *Ichthyostega* (Figure 30–6). These animals had no gills but were fish-like in having scales and a tail supported by fin rays. Their adaptations to land included well-developed limbs, ribs, and limb girdles. In tetrapods, the function of the pectoral and pelvic girdles (shoulder and hip girdles) is to form a strut between the limbs and the spine, permitting the body to be lifted off the ground. Ribs perform a similar function, supporting and lifting the internal organs.

## 30–F   AMPHIBIANS: FROGS, SALAMANDERS, CAECILIANS

The two largest groups of amphibians are the urodeles (newts and their relatives) and the anurans (frogs and toads). Urodeles are the more generalized and show the transition from fish to tetrapod more clearly. Their limbs contain small bones and muscles, like those found at the base of lungfish fins.

Anurans (an = without; uro = tail) have specialized skeletons that give them their ability to jump long distances. The backbone is shortened, with many vertebrae fused together, and the limb girdles are firmly attached to the backbone.

Caecilians belong to a third, small group, the Apoda ("without feet"). They are legless, worm-like, tropical animals adapted to burrowing in leaf litter on the forest floor (Table 30–2).

Amphibians have a soft glandular skin, which is used for gas exchange in most species, despite the fact that most also have small lungs. The aquatic larvae have gills. Only caecilians have scales. The sense organs of amphibians are adapted to land in that the anterior end of the lateral line system has evolved into simple ears, which respond to sound (pressure) waves in air. Amphibians were also the first vertebrates to develop true tongues. In most frogs and toads the tongue is long and sticky and can be shot out rapidly to catch flies.

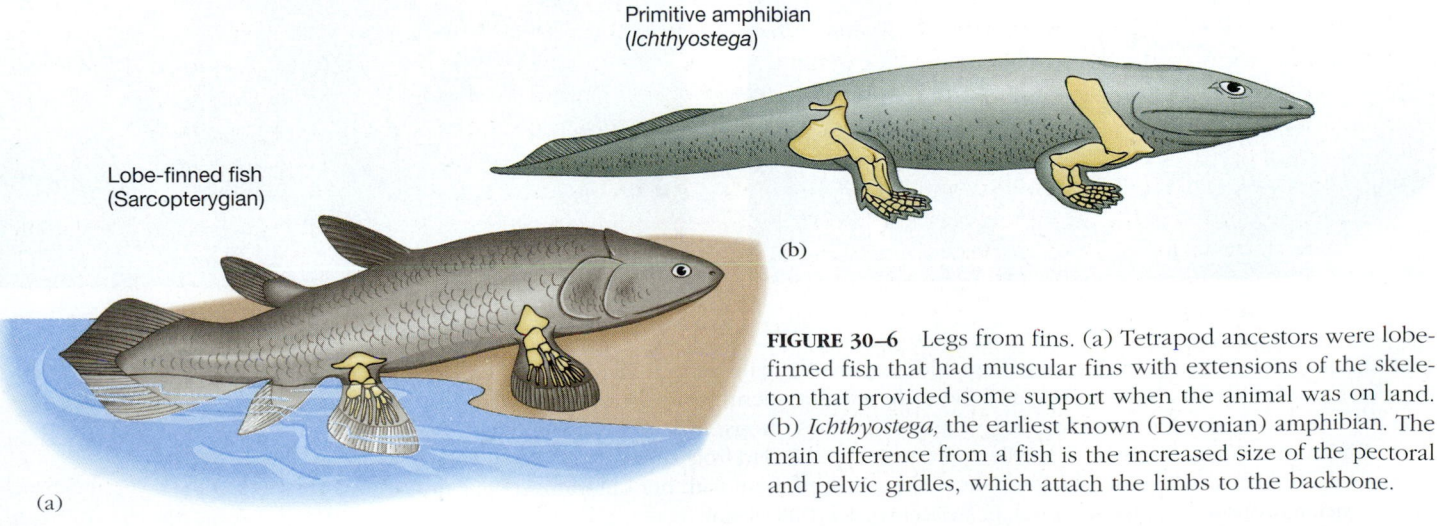

Primitive amphibian
(*Ichthyostega*)

Lobe-finned fish
(Sarcopterygian)

(b)

(a)

**FIGURE 30–6**   Legs from fins. (a) Tetrapod ancestors were lobe-finned fish that had muscular fins with extensions of the skeleton that provided some support when the animal was on land. (b) *Ichthyostega*, the earliest known (Devonian) amphibian. The main difference from a fish is the increased size of the pectoral and pelvic girdles, which attach the limbs to the backbone.

## TABLE 30-2
### Modern Amphibians

| Order | Habitats | Examples |
|-------|----------|----------|
| Urodela (~350 species) | Freshwater and moist terrestrial worldwide | Salamanders, mud puppies, newts, hellbender |
| Anura (~3500 species) | Freshwater, moist terrestrial; some desert and brackish water species | Frogs, toads |
| Apoda (Gymnophiona) (~170 species) | Moist tropical terrestrial | Caecilians (look like large segmented worms) |

Another amphibian adaptation to terrestrial life is the production of **vasopressin** (antidiuretic hormone), a hormone that increases the amount of water resorbed into the body from the urine before it is voided.

Despite their adaptations to land, amphibians are not fully emancipated from water. This is most obvious in reproduction. Amphibian eggs lack the membranes and shell that protect the eggs of other tetrapods from desiccation. Many lay their eggs in fresh water, but some species in all the major groups have evolved the ability to lay their eggs elsewhere (Figure 30-7). Some lay their eggs in damp places on land. In the frog *Pipa dorsalis,* the eggs develop in pouches on the mother's back. A male frog of the genus *Rhinoderma* carries the developing young in his vocal pouch, and the female *Rheobatrachus* carries the tadpoles in her stomach. The tadpoles release a hormone that inhibits secretion of stomach acid, thereby protecting themselves from digestion. In a number of species, the young develop in the female's oviduct and are born as miniature adults.

The eggs of most frogs and toads hatch as herbivorous tadpoles, whereas salamander larvae are carnivorous and look more like miniature adults. Some salamanders, such as one species of *Ambystoma*, never undergo metamorphosis but look like larvae even when they become breeding adults. (Such sexual maturity in an otherwise larval animal is known as **paedogenesis** or **paedomorphosis.**)

Amphibians are seen in large numbers only during the breeding season, when they congregate around ponds and streams. At mating sites, male frogs and toads produce mating calls that permit the female to recognize and approach a male of her own species. For the rest of the year they are secretive and silent, and little is known about their behavior. Many amphibians hibernate on land or under water during the winter; metabolism slows down, and the animal survives on energy stored in the body. Since they are poorly protected from desiccation, many amphibians also **estivate** in summer, reducing the metabolic rate to conserve water and other resources, and emerging only at night when it is cooler and more humid or on wet days.

(a)

(b)

**FIGURE 30-7**    Amphibians. (a) Frogs mating, with the smaller male on top. The female sheds her eggs into the damp moss beneath her, and the male releases sperm, which fertilize the eggs outside the female's body. Their bright color suggests that these frogs, like many others, have poisonous mucus in their skins. (b) A forest-living Peruvian poison dart frog carries two newly hatched tadpoles on her back to a pool of water, where she will feed them with her unfertilized eggs until they metamorphose.    (a, Richard Laval; b, Michael Fogden/DRK Photo)

Most amphibians are protected from predators by poisons in the mucus on their skin (and in the jelly surrounding their eggs). The disgust of a cat or dog that attacks a toad is dramatic evidence that the defense works. (Frog legs are skinned before they are served as human food.) Some tropical species are extremely poisonous and advertise themselves to potential predators by spectacular fluorescent green and red coloration. Colombian Indians tip poison darts with nerve poisons such as batrachotoxin, which they obtain by heating the frogs *Dendrobates* and *Phyllobates* over a fire, inducing the frogs to secrete quantities of poisonous mucus. A milligram or less of batrachotoxin will kill a human.

In the 1980s, biologists raised the alarm that amphibian species were disappearing from habitats all over the world. Further study in some locations showed that this was the result of large fluctuations in natural populations. In dry years, the numbers of some species fall dramatically and then rebound in population explosions in rainy years. In other cases many amphibians are extinct or endangered because their habitat has been destroyed. Humans are filling in wetlands all over the world (Section 47–G), and with the wetlands go the amphibians that breed there.

## 30–G    REPTILES

In the late Carboniferous, about 300 million years ago, life took a turn that caused a burst of tetrapod evolution: insects radiated widely into habitats on land. Many insects and plants evolved symbiotic relationships, which promoted the adaptive radiation of both groups. Probably for the first time in history there was enough animal life on land to support terrestrial vertebrate predators, and this permitted the explosive radiation of tetrapods. Early tetrapods were probably all carnivorous. This theory is supported by the fact that the adults of all living adult amphibian species are carnivorous and by the jaw shape and teeth of fossil amphibians and early reptiles.

Early reptiles were less than 20 cm long. We can imagine them as creatures looking much like salamanders, scurrying around snapping up insects. Their limbs were stronger and were tucked further under their bodies than those of amphibians, enabling them to move faster. Their jaws were more firmly attached to the skull, permitting them to subdue and eat larger prey, and their skins were waterproofed and scaly, minimizing the loss of water across the outer body surface. Without a moist skin, reptiles had to breathe entirely with their lungs.

### Amniotic Eggs

The derived character that distinguishes all later vertebrates from amphibians is the amniotic egg, with a number of features suited to life on land (Figure 30–8). The developing

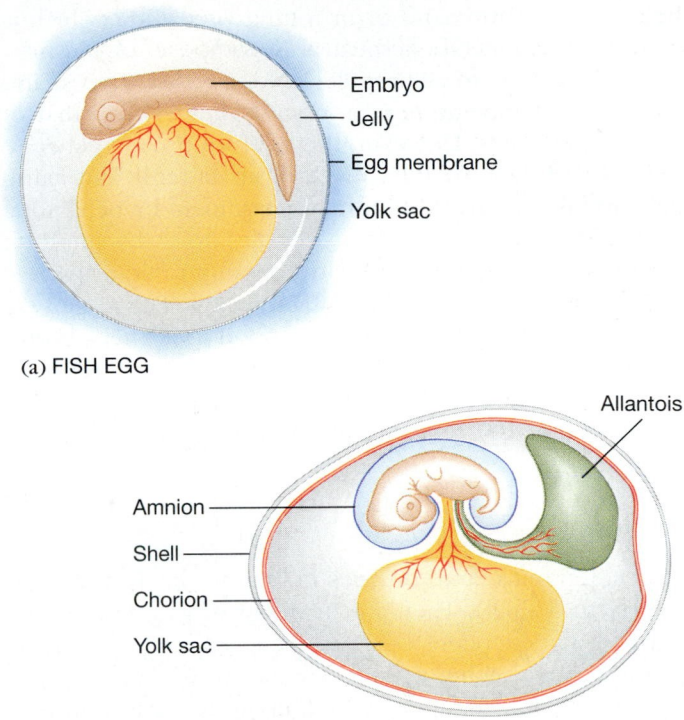

(a) FISH EGG

(b) AMNIOTIC EGG

**FIGURE 30–8** Eggs with and without embryonic membranes. (a) The egg of a bony fish or amphibian. The embryo obtains its food from the yolk sac and is protected and buoyed up by the jelly coat. (b) The amniotic egg of a reptile or bird, adapted to develop on land.

embryo is surrounded by a membrane, the **amnion,** enclosing **amniotic fluid,** which protects the embryo from dehydration and jolting. Two membranous sacs are attached to the embryo. One is the **yolk sac** (found also in fish embryos), containing yolk that the embryo uses for food. The other is the **allantois,** which stores the embryo's nitrogenous waste until hatching. Blood vessels grow out from the embryo through the membranes of the yolk sac and allantois until they come close to the surface of the egg, where they absorb oxygen from the environment and release carbon dioxide. The embryo, amnion, yolk sac, and allantois are all surrounded by a membrane, the **chorion,** which controls the overall permeability of the egg. The chorion, in turn, is surrounded by the outer **egg shell,** which may be leathery or hardened with calcium carbonate.

Because the egg is laid in a shell or the young develop within the mother's body, reptiles have internal fertilization, and the male has a penis (or even two). Since embryonic development proceeds faster at higher temperatures, it is selectively advantageous to keep the eggs warm. Reptiles have many adaptations that do this. Female pythons coil around the eggs and brood them for three months until they hatch. Other snakes, especially in colder climates, retain the embryos inside the body. Reptiles may also bury their eggs in warm sand or in rotting vegetation, which is warm from the metabolic heat of decomposer organisms.

It is incorrect to call reptiles cold-blooded. They maintain a body temperature considerably higher than that of their surroundings. However, they do not usually generate most of their heat by their metabolism as birds and mammals do but gain heat from the environment by behavioral adaptations. For instance, many reptiles lie in the sun for a period before they become warm enough to be active.

The skin of a reptile is dry and scaly. The scales are made up of keratin, a protein also found in feathers, horns, and hair. In crocodiles and tortoises the scales are replaced as they wear away. In lizards and snakes, the scales are shed and replaced several times a year during a molt.

## Turtles

Turtles, terrapins, and tortoises belong to the Testudomorpha (testu = shell). They are in many ways the most peculiar of all vertebrates because their anatomy is uniquely adapted to life within a shell (Figure 30–9). For example, they are the only tetrapods with limb girdles inside, instead of outside, the ribs. Some can withdraw their legs and heads completely into their shells; others cannot.

Most turtles are herbivorous, and there are species adapted to life on land, in fresh water, and in the sea. Many marine species are famous for their annual migrations to the beaches where they lay their eggs. Alone among tetrapod groups, no turtles provide parental care to the young. The female digs a nest, lays the eggs, and leaves the young to fend for themselves.

Many larger land turtles and all sea turtles are endangered by human activities, mostly habitat destruction and hunting. Part of the reason is that turtles are long-lived and take years to reach sexual maturity, so their populations do not grow rapidly once numbers decline. Some species have another characteristic that makes conservation difficult: the sex of an individual is determined by the temperature in the nest where the embryo develops. As a result, lack of one sex may frustrate efforts to breed turtles in captivity.

## Mesozoic Diapsids

Amniotes, animals with amniotic eggs, can be divided into three groups by the number of holes in the temporal region of the skull. In early amniotes, including modern turtles, the skull is **anapsid** (an = without), with no holes in this area behind the eye (Figure 30–10). Later amniotes evolved holes in this region, with bony arches between them, which provide better leverage and more area for attachment of the jaw muscles. A two-arched skull is the derived character of **diapsids** (di = two; apsid = arched), the largest amniote group, including dinosaurs, birds, snakes, and crocodiles. Mammals and their extinct reptile relatives are **synapsids,** with a single temporal arch.

Diapsids are the only animals with **Jacobson's organs,** chemical-sensing organs in the roof of the mouth. Snakes and lizards use these organs in social interactions. A snake's tongue is not poisonous; it flickers as it carries chemicals from the air or ground to the Jacobson's organs, detecting potential mates and enemies. In birds and crocodilians, the Jacobson's organs form in the embryo but later disappear.

The Mesozoic era, which lasted about 180 million years, is sometimes known as the "age of reptiles." Reptiles radiated into all the niches that terrestrial vertebrates occupy today. There were species that flew, walked, and swam. Some were insectivores, some voracious carnivores, and some placid herbivores of enormous size.

Within this adaptive radiation, several evolutionary trends can be seen, often with convergent evolution in sev-

(a)

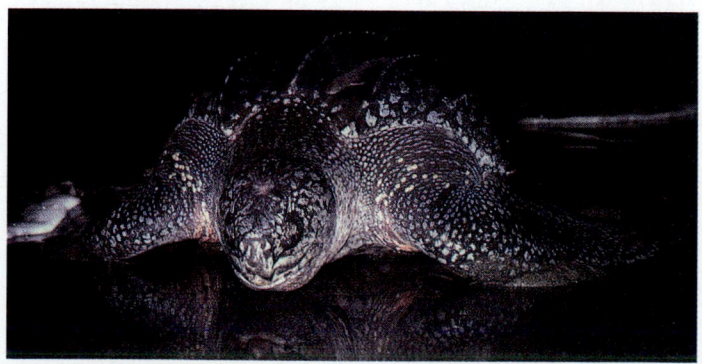
(b)

**FIGURE 30–9**    Turtles. (a) An American desert tortoise, member of a species endangered by destruction of its habitat in the southwestern United States. (b) A leatherback sea turtle with limbs modified into paddle-like flippers used for swimming. Note the streamlined shape of the shell. (a, Peter Brussard; b, Fundación Neotrópica)

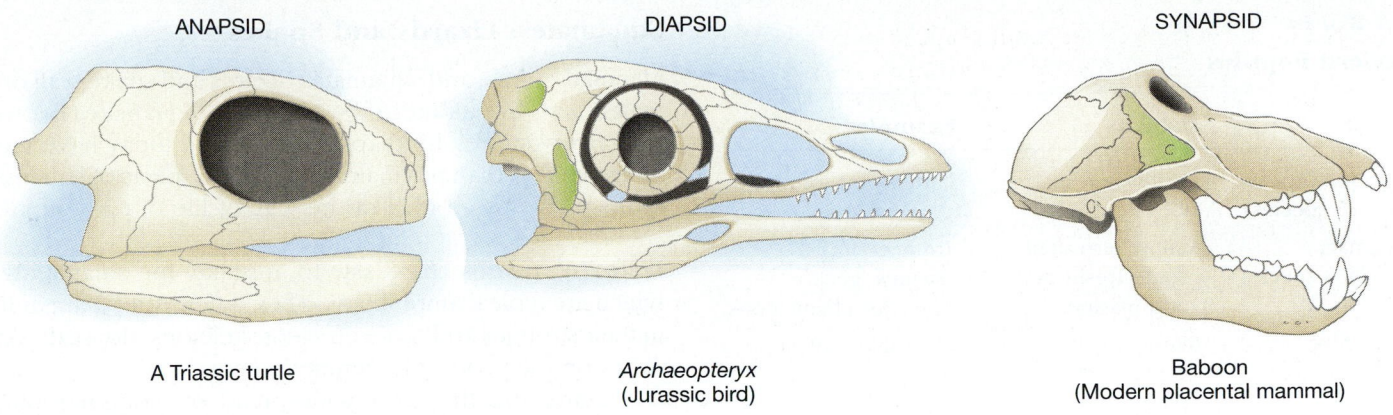

ANAPSID                    DIAPSID                    SYNAPSID

A Triassic turtle          *Archaeopteryx*            Baboon
                           (Jurassic bird)            (Modern placental mammal)

**FIGURE 30–10**  Vertebrate skulls showing the holes (green) and arches in the temporal region that give them their names.

eral groups. For instance, low-slung animals with long snouts adapted to eating fish evolved in several diapsid lines. Crocodilians are the only surviving examples. Flight evolved independently in pterodactyls (pterosaurs) and birds. Many lines became increasingly bipedal (bi = two; ped = foot), walking on their hind legs, using the tail for balance and as a weapon and using the shorter forelimbs for other activities. Modern birds are bipedal, and crocodilians have short forelimbs, suggesting bipedal ancestors.

***Dinosaurs***  Our ideas of dinosaur life have changed in recent decades. Many dinosaurs are now thought to have had behavior patterns at least as complex as those of modern birds and crocodilians, with social interactions and parental care of the young. Fossilized nests, eggs, footprints, and stomach contents all help to flesh out our picture of dinosaur life. Some biologists even think it may one day be possible to isolate dinosaur DNA and examine dinosaur genes.

Some dinosaurs were the largest animals that have ever lived on land. At 30 m long and 100,000 kg, *Diplodocus* and its relatives were 6 times as long and 20 times as heavy as elephants. Paleontologists once doubted that these animals could have supported their weight on land, and they were viewed as swamp-dwellers. Modern workers, however, conclude that these giants were terrestrial with amazing adaptations to their weight, including vertebrae built like the flying buttresses that hold up the walls of old cathedrals! Stomach contents, from a fossil found in Utah, show that these huge dinosaurs were herbivores, eating leaves and twigs. (The largest animals are all herbivores or filter feeders, for reasons we shall explore in Chapter 48.)

Many people have suggested that dinosaurs were **endotherms** ("warm-blooded," with the same type of temperature control as modern mammals). They point out that some dinosaurs were very active and lived in cold areas, a combination impossible without means of maintaining high body temperature. Living diapsids are no guide to dinosaur temperature regulation: birds are endotherms, but lizards and crocodilians are not. Many biologists, however, think that extinct reptiles remained warm and active by the same behavioral means as modern reptiles. Some dinosaurs had large sails on their backs, once interpreted as heat-collection devices (Figure 30–11). (Many butterflies today use their wings for this purpose.) However, recent calculations show that larger dinosaurs probably had difficulty getting rid of excess metabolic heat produced by rapid movement, and the sails probably served to dissipate rather than to collect heat.

*Stegosaurus*

*Kentrosaurus*

**FIGURE 30–11**  Absorbing or dispersing heat? Sails on the back of two dinosaurs.

## TABLE 30-3
### Modern Reptiles

| Group | Habitats | Examples |
|---|---|---|
| Testudomorpha (~225 species) | Freshwater, marine, terrestrial | Turtles and tortoises |
| Squamata (~5600 species) | Mainly terrestrial; some freshwater and marine | *Tuatara*, lizards, snakes, geckos, iguanas, chameleons |
| Crocodilia (21 species) | Mainly freshwater, some marine, most partly terrestrial | Crocodiles, alligators, caimans |

There is still some mystery about the extinction of many of the reptiles, for large numbers of them, including all the dinosaurs and all the flying reptiles, disappeared from the fossil record during a short time at the end of the Cretaceous Period (*Essay: Mesozoic Murder*).

## Squamates: Lizards and Snakes

Mammals, birds, and squamates (lizards and snakes) all originated and diversified in the Mesozoic. Squamates evolved into a very successful diapsid group, and today have more living species than the mammals. Some aspects of squamate biology give us an idea of the way of life of early reptiles.

***Lizards*** Lizards are easy to mistake for salamanders, which are typical amphibians. They differ from salamanders in their stronger and more efficient skeletons, dry scaly skin, claws on their toes, and amniotic eggs.

Lizards are the most widespread of modern reptiles. They are found in jungle treetops, grasslands, deserts, rivers, and sea coasts. They range in size from one- or two-gram geckos to the 100-kg Komodo dragon of Indonesia, which is reported to kill small pigs on occasion. The Malaysian flying dragon is a lizard that can glide from tree to tree, and a number of lizard species have lost their legs in a way that parallels the evolution of snakes. Most chameleons, with their rolling, turreted eyes, eat insects, as do a majority of the smaller lizards (Figure 30–12). Larger species will eat

**FIGURE 30–12** Squamates. (a) A lizard: a broad-headed skink broods her eggs. (b) Corn snake eating a mouse. (c) A snake skeleton. The many vertebrae, with their attached ribs free at the distal end, and the loosely hinged jaw contribute to the flexibility of the snake's body. (a, Florida Audubon Society; b, John Cancalosi/Tom Stack & Associates; c, Biophoto Associates)

(a)

(b)

(c)

mammals and birds when they can catch them, and a few species, like the basilisk (whose gaze, in legends, is death), eat vegetable matter and fruit.

***Snakes*** Snakes are legless squamates that probably evolved their specialized anatomy by parallel evolution from several groups of burrowing lizards. A few dangerous snakes, and ignorance and superstition about many harmless ones, have been responsible for giving all reptiles a bad name. In fact, snakes are not typical reptiles but have highly specialized anatomy. Most move by using their muscles to throw the body into curves. Scales on the ventral surface, or the curves of the body itself, provide traction. The group includes expert swimmers, burrowers, and climbers. The backbone is greatly elongated, and most of the vertebrae bear long, flexible ribs that hold the body in shape (see Figure 30–12).

Snake jaws are exceptionally mobile and loosely attached to the rest of the skull. The two halves of the lower jaw are joined in front only by elastic tissue, and the mouth can be opened to enormous size, permitting a snake to swallow prey much larger than its head. The front of the windpipe can be protruded so that breathing is not obstructed as prey passes slowly down the throat. The teeth curve backwards, preventing the prey from popping back out of the mouth. Some egg-eating snakes use sharp hemal arches on the neck vertebrae to break the shells of birds' eggs as they are swallowed. In some snake species, two or more teeth are specialized as hollow fangs connected to muscular salivary glands that produce venom and pump it through the fangs into the prey. Snake venoms generally contain agents that lyse blood cells and toxins that paralyze nerves and muscles.

Although snakes' ears have no external openings, they are well developed, responding mainly to vibrations of the ground detected through the lower jaw. Pit vipers and some boas also have heat-detecting organs on the head, which allow them to strike warm-blooded prey accurately on dark nights or in deep burrows.

The most primitive snakes retain traces of their presumed lizard ancestry, such as hindlimb girdles and rather immobile jaws. Most of these snakes belong to the New World boas and the Old World pythons, including the world's biggest snakes. The longest are the awe-inspiring South American anacondas, which may grow nearly 10 m long.

## Crocodilians

Crocodilians are the closest living relatives of the dinosaurs and of their descendants, the birds. Three groups of crocodilians survive today: crocodiles in Africa, Asia, and America; alligators and caimans in the southern United States, China, and Central America; and the gavial of Southeast Asia. All spend much of their time in water and have a special arrangement of their nostrils that permits them to breathe while the rest of the body is submerged (Figure 30–13).

All crocodilians are carnivorous. Females lay their eggs in nests and guard the nest and newly hatched young from their relatives and other predators. Fear of crocodilians and desire for their skins to make shoes and handbags have brought most species close to extinction. However, efforts to protect alligators in the southern United States have been spectacularly successful. Farming of alligators in the United States, and long-snouted, fish-eating gavials in northern India, seems likely to preserve these species from extinction and at the same time provides human income.

## 30–H   AVES: THE BIRDS

Birds can be simply defined as the only organisms with feathers. They are diapsids that evolved in the Mesozoic from theropods—bipedal, carnivorous dinosaurs with S-shaped necks. Birds differ from other theropods mainly in flight and endothermy. The evolution of feathers made both these advances possible.

**FIGURE 30–13**   An American alligator. You can see several characteristics of crocodilians in this photograph: conical teeth, hindlimbs that are longer than the forelimbs, and the sprawling gait. Some of the muscles that move a reptile's hindlimbs attach to the tail rather than to the hip girdle as in mammals, and the limb bones are attached in such a way that the limbs cannot be tucked under the body as they can be in mammals.   (Florida Audubon Society)

Feathers are made of beta-keratin, a protein related to the keratin of scales, hair, and horn, which is one of the strongest known materials, by weight. Feathers are flexible, and they are also excellent insulation (Figure 30–14). Colorful feathers are used in courtship displays and other types of social signaling. Feathers are molted and replaced periodically as they become too damaged to function.

Feathers form a flying surface that is a distinct improvement over the patagium, formed of skin and connective tissue, used by bats and the extinct flying reptiles. The trouble with a patagium is that a bad tear may put the wing out of order. With feathers, birds can suffer considerable damage to the wings without losing the ability to fly. A patagium also invariably stretches from forelimb to hindlimb. Birds use only the forelimbs for flight; the hindlimbs are free to be used for running or swimming, and nearly all birds have two different types of locomotion (flying and swimming, running and flying).

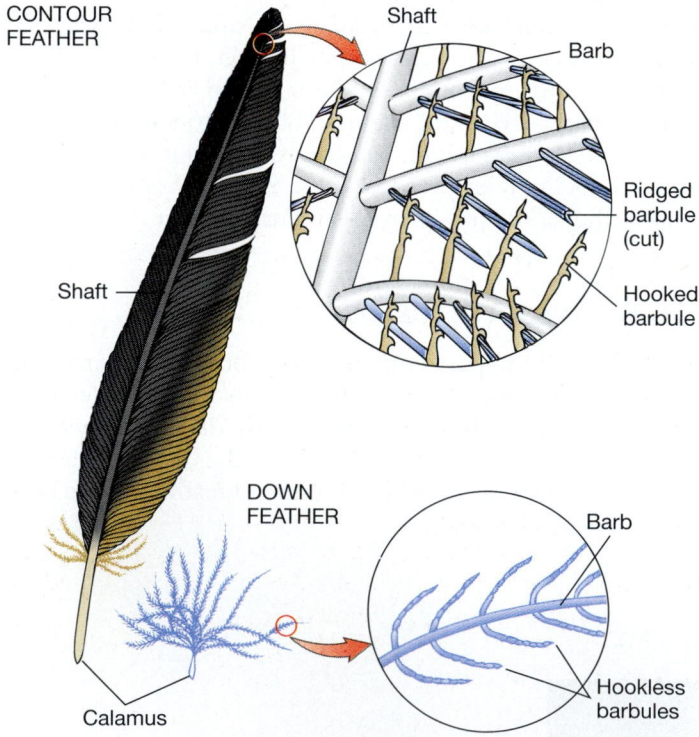

**FIGURE 30–14** Feathers. Contour feathers provide a smooth, streamlined surface for flight. When the bird preens, it draws the barbs through its bill from base to tip. As a result, the hooked barbules on one barb hook onto the ridged barbules on the next, linking the barbs into a smooth but flexible surface. Down feathers are the last word in insulation. Their long, unhooked barbules provide a mass of spaces where air is trapped and warmed by the body. They are the first feathers to develop on young birds, and they underlie the contour feathers on parts of the body of most adult birds.

Birds use metabolic heat to maintain their body temperature at a high level. The down feathers, and a layer of fat just under the skin, insulate the body against changes in temperature. Endothermy permits birds (and mammals) to be active whatever the external temperature, which is not possible for reptiles or amphibians.

Both flight and the generation of body heat require large amounts of energy, and so birds have to eat a lot. Birds that live on land usually eat seeds and fruit (finches, parrots, fowl, pheasants), insects (thrushes, swifts, woodpeckers, wrens, swallows), smaller vertebrates (owls, eagles, falcons, hawks), or carrion (vultures and crows). A large number of birds find most of their food in fresh water or the sea, in the form of invertebrates, algae, and fish. Some are wading birds, such as the sandpipers and herons, whereas others are capable of swimming and diving, with feet modified as paddles. Gulls, pelicans, ducks, geese, and cormorants fall into the latter category.

Many birds are strongly social, with complex behavior patterns. Because they must feed more often than most reptiles, both parents often collaborate in nest-building, incubating eggs, and feeding the young. This has led to extensive interaction between the sexes and has apparently resulted in complex courtship, monogamy, or reproduction in large colonies in many species.

All birds lay eggs, which are then usually incubated by the parents' body heat, although the Australian brush turkey buries her eggs in a warm pile of rotting vegetation. Because the egg is laid in a hard calcareous shell, fertilization is internal, although the male has no penis in most species.

## Bird Flight

Because the ability to fly limits the possible range of structural variation, bird anatomy is more conservative than that of other vertebrates. The whole skeleton is both light and strong, undoubtedly adaptations to flight. The center of gravity lies well forward, under the wings and above the legs. Most of the bones in birds' skeletons are fused to one another, strengthening the air frame, and only the legs, neck, bill, and wings are separately moveable.

Weight reduction can be seen in several other aspects of bird anatomy. Heavy teeth are replaced by a light horny bill (beak) and a gizzard, a grinding chamber in the gut. The tail vertebrae are reduced to a tiny bump behind the pelvic girdle, and the tail is made up of feathers. Many bones are hollow and reinforced by internal girder-like structures similar to those in an airplane wing.

The most important flight muscles are the pectorals, or breast muscles, which attach the base of the wings to the keel of an extended sternum, or breastbone. The lungs of a bird connect to air sacs that spread throughout the body and even into some of the hollow bones. The air sacs provide buoyancy and allow heat to be lost rapidly, particu-

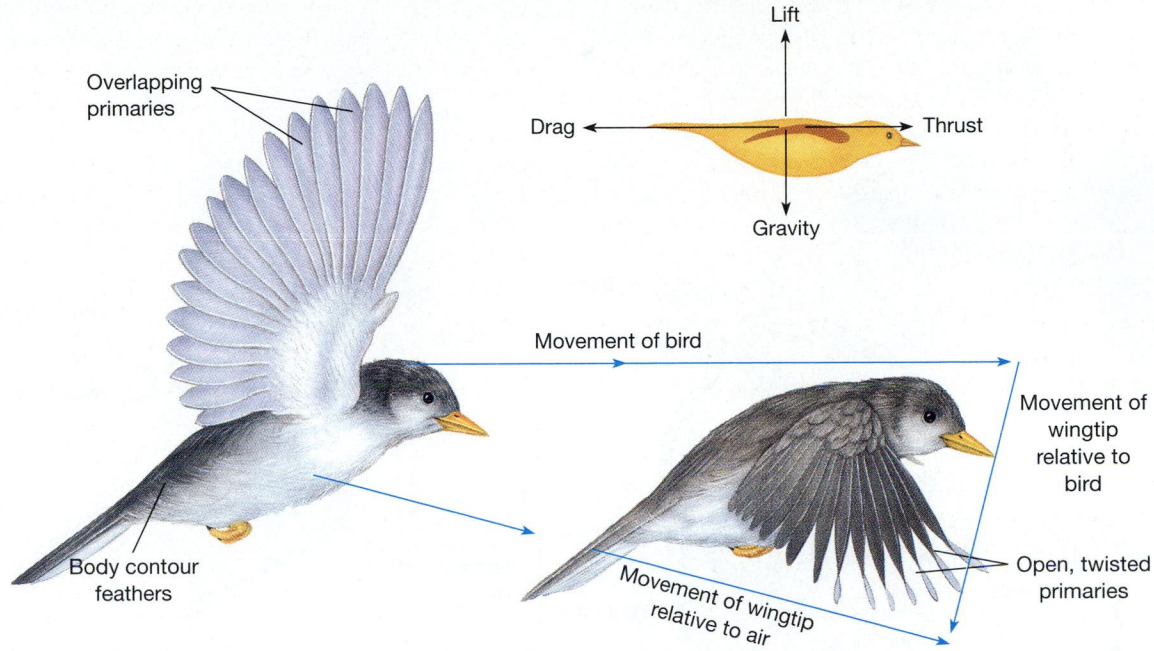

**FIGURE 30–15**   Bird flight, showing the movements and forces involved. The fixed wing of an airplane acts as an airfoil (lifting surface). A bird wing is both airfoil and propeller. The primaries separate and twist as the bird flaps its wing.

larly during flight, when the flight muscles and heart generate a lot of metabolic heat.

Birds are primarily visual animals, with acute color vision, reflecting their arboreal (tree-dwelling) origin. Arboreal animals tend to have the good vision needed for landing and jumping in trees.

Bird flight ranges from flapping flight, such as that of sparrows, robins, chickadees, and so forth, to the soaring flight of the hawks, vultures, and albatrosses (Figure 30–15). Soaring birds ride the thermal currents in the air, like a glider (though more efficiently), and flap their wings infrequently. The largest of these, with a wingspan of almost 8 m, is a fossil form from Argentina.

Not all birds can fly. The most highly specialized water birds are the penguins, which cannot fly because their wings as well as their feet are modified as paddles. The other birds that cannot fly are typified by ostriches, which rely on their running speed to escape predators.

## The Origin of Birds

The earliest known bird is *Archaeopteryx*, from fossils laid down in the Jurassic, 150 million years ago. *Archaeopteryx* is classified as a bird because it had feathers. Without them, it would be classified as a dinosaur. Its main characteristics are shown in Figure 30–16.

The discovery of the first *Archaeopteryx* in 1863 touched off a debate, which still flares up periodically, on the origin of bird flight and on the lifestyle and flying ability of *Archaeopteryx*. One theory holds that birds used their broad, feathered forelimbs at first as stabilizers when jumping among branches, and later as parachutes (as used by flying lizards, flying frogs, and flying squirrels). According to another theory, birds were bipedal insectivores. The broadening and flattening of the forelimbs would have allowed such animals to make sustained jumps into the air and so increased their insect-catching prowess. Still another theory holds that feathered wings evolved as "insect nets," used to scoop flying prey toward the mouth.

Feathers probably evolved from reptilian scales, although there is no trace of this evolution.

## 30–I   MAMMALS

Mammalian characteristics evolved in various lines of synapsid reptiles during the Permian. Mammals and dinosaurs both appeared during the Triassic, well before the first birds. By the late Mesozoic, all three modern groups of mammals had evolved—monotremes, marsupials, and placentals—as well as several groups that are now extinct.

A mammal can be defined as a vertebrate with only one bone on either side in the lower jaw. This was prob-

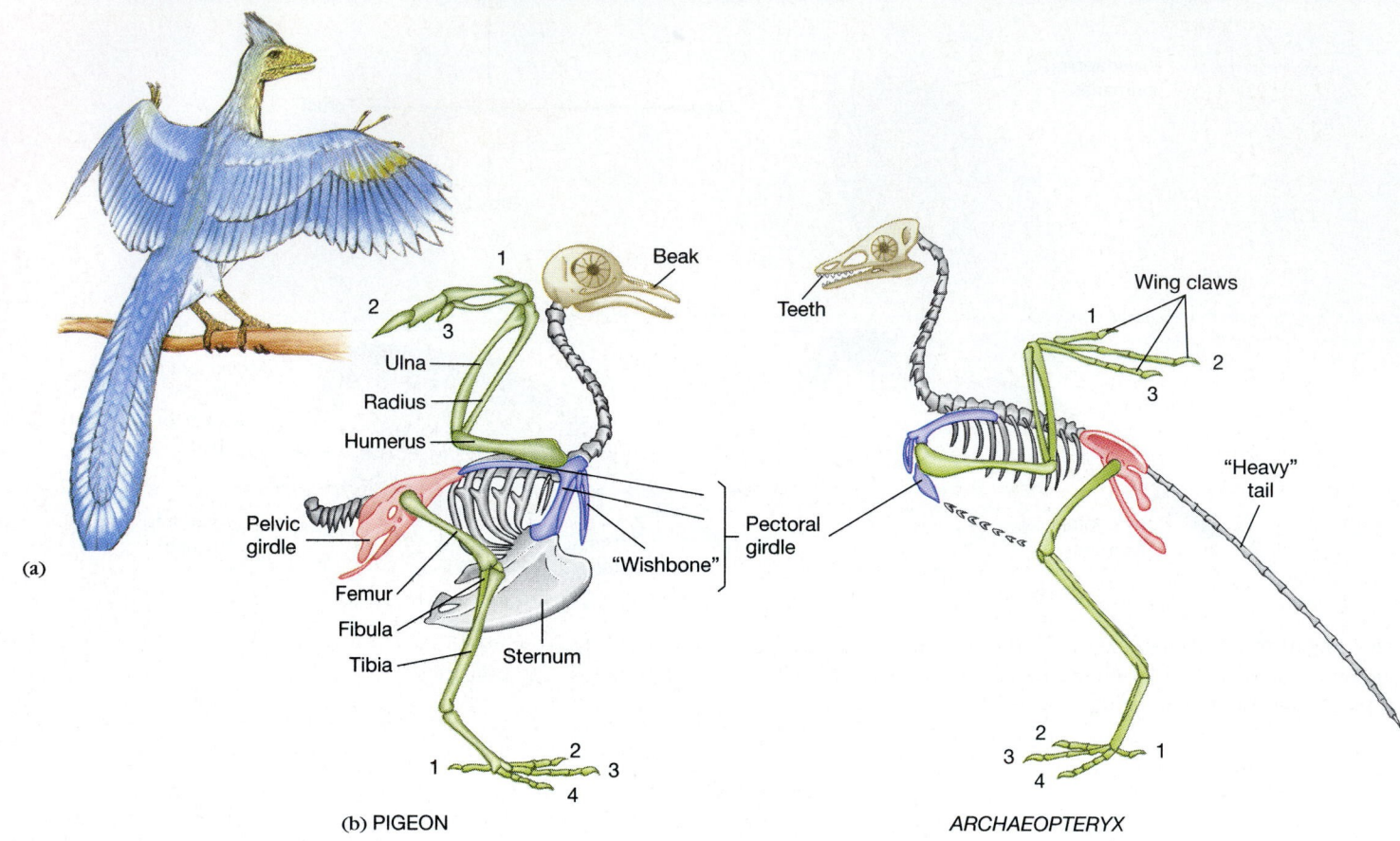

**FIGURE 30–16** *Archaeopteryx*. (a) Artist's impression of *Archaeopteryx*, which was the size of a crow. (b) *Archaeopteryx* skeleton compared with that of a modern bird (pigeon). Note the heavy tail of *Archaeopteryx*, the teeth, and the claws on the digits in the wing, features that have been lost in modern birds. The skeleton of a modern bird is much more compact and lightweight.

ably an adaptation that allowed a carnivore to grip a struggling victim more firmly, with less chance of dislocating the jaw. The other distinctive characters of modern mammals are the possession of hair, mammary glands, and endothermy.

Early mammal evolution is the history of the perfection of quadrupedal locomotion and of new adaptations to carnivory. Adaptations for fast quadrupedal locomotion include a narrow foot track, with the legs slung further under the body than in reptiles, and further evolution of the limb girdles, which in some fast mammals are highly mobile (Figure 30–17). (Part of the pectoral girdle of a cat moves back and forth several inches as the legs move, lengthening the stride.) Muscles that move the limbs became clustered at the top of the leg, making the lower leg slim and light and therefore enabling the animal to move its legs quickly. Muscles moving the hind leg backward became attached to a spur on the back of the pelvic girdle instead of to the tail. This freed the tail for other uses, for instance as a fly swatter, as a fifth hand, or for balance (Figure 30–18).

The evolution of specialized teeth is another advance seen in mammals. Whereas fish and reptiles have teeth that are all roughly the same size and more or less conical, early mammals rapidly evolved different kinds of teeth: chisel-like incisors used for cutting, pointed canines for gripping and tearing, and grindstone-like molars for crushing and breaking (Figure 30–19). Fossil teeth suggest that mammals had given rise to herbivores and omnivores (using both plants and animals for food) as well as carnivorous species by the end of the Mesozoic.

Early mammals were small, about the size of a small mouse or shrew. Their large eye sockets suggest that they were nocturnal, like owls and large-eyed modern primates. It is quite likely that endothermy was advantageous partly because it permitted mammals to be active during the cooler nights and so avoid competition with reptiles. During the Cenozoic, many larger species evolved, exploiting resources formerly monopolized by now-extinct reptiles.

Most mammals are viviparous. The egg is tiny with very little yolk, and the embryo is completely dependent on the

AMPHIBIAN

**FIGURE 30–17** Changes in the foot track during the evolution of tetrapods. The joint between limbs and limb girdles has evolved so that the legs can be tucked more vertically under the body, leading to more rapid locomotion.

DIAPSID REPTILE

MAMMAL

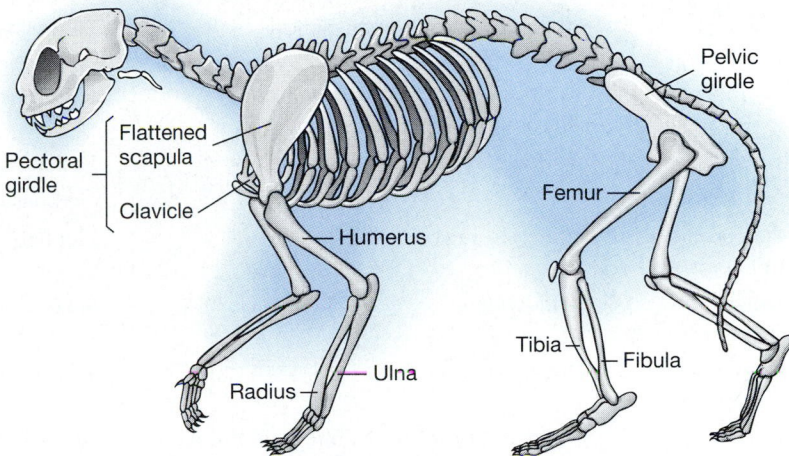

**FIGURE 30–18** The skeleton of a mammal (a cat). The scapula (shoulder blade) is only loosely attached to the backbone, so that it moves back and forth, giving the cat its long stride. Most of the muscles that move the hindlimb are attached to the pelvic girdle so that the tail, unlike that of a reptile, is not involved in locomotion. The limbs are slim and the foot track narrow, contributing to the rapid movement characteristic of mammals.

mother. Viviparity permits mammals to remain mobile while incubating embryos that, like bird embryos, must be warm to survive. While not all mammals are viviparous, all female mammals nourish the young after they are born with milk produced in mammary glands.

The mammalian integument, consisting of the skin and associated structures, has many different roles. First, it assists in temperature control: the outer layer of hair and underlying layer of fat provide insulation, and the sweat glands provide water for evaporative cooling. In many species, chemicals produced by glands in the skin play an important role in communication between individuals. Other integumentary structures include claws, nails, or hoofs; and horns or antlers, often underlain by a supporting bony core.

Although the earliest mammalian design is that of a fast, quadrupedal carnivore, many modern mammals are neither fast, quadrupedal, nor carnivorous. The success of mammals led to extensive adaptive radiation, during which herbivory, bipedalism, flight, and many other adaptations evolved (Table 30–4).

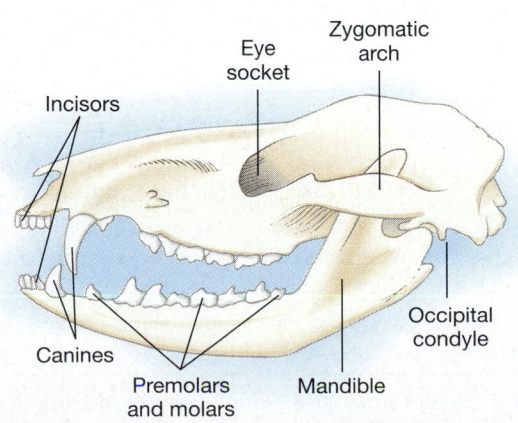

**FIGURE 30–19** The skull of a mammal, showing different kinds of teeth. The zygomatic arch protects the eye from being jogged by the jaw muscles. The backbone attaches to the skull at the occipital condyle.

**TABLE 30-4**

## Orders of Mammals

| Order | Distribution* | Examples |
|---|---|---|
| Monotremata (3 species) | Australian region | Egg-laying mammals: duck-billed platypus (*Ornithorhynchus*) and echidnas *(Tachyglossus, Zaglossus)* |
| Marsupialia (~260 species) | Australian and Neotropical (opossum has migrated to North America) | Mammals with marsupial pouches, e.g., opossum, kangaroo, koala |
| Edentata (29 species) | Neotropical (armadillo spreading into North America) | Sloths, anteaters, armadillos |
| Pholidota (7 species) | Ethiopian and Oriental regions | Pangolins |
| Lagomorpha (~65 species) | Worldwide (except Australia until introduced) | Rabbits, hares, pikas |
| Rodentia (~1750 species) | Worldwide | Rodents: mice, rats, voles, beavers, porcupines, guinea pigs, hamsters, jerboas, squirrels, gophers |
| Macroscelidea (19 species) | Ethiopian region | Elephant shrews |
| Insectivora (~390 species) | Worldwide | Small insect-eating mammals; e.g., hedgehogs, moles, shrews |
| Scandentia (16 species) | Oriental region | Tree shrews |
| Primates (~180 species) | Oriental, Ethiopian, Neotropical; humans now worldwide | Lemurs, monkeys, apes, humans; e.g., *Lemur, Tarsius* (spectral tarsier), *Macacus, Pongo* (orangutan), *Gorilla, Pan* (chimpanzee), *Homo* |
| Dermoptera (2 species) | Oriental region | Flying lemurs |
| Chiroptera (~920 species) | Worldwide | Bats |
| Carnivora (~235 species) | Worldwide | Dogs, cats, hyenas, bears, wolves, raccoons, pandas, weasels, otters, badgers, skunks, jackals, civets, mongooses |
| Tubulidentata (1 species) | Ethiopian region | Aardvark |
| Hyracoidea (7 species) | Ethiopian and southern Sahara region | Hyraxes |
| Sirenia (4 species) | Tropical and subtropical oceans and estuaries except eastern Pacific | Dugongs, manatees |
| Pinnipedia (~30 species) | Worldwide | Seals, earless seals, sea lions, walruses |
| Proboscidea (2 species) | Ethiopian and Oriental regions | Elephants |
| Perissodactyla (18 species) | Africa, Asia, Central and South America | Odd-toed ungulates; e.g., horses, zebras, asses, tapirs, rhinoceroses |
| Artiodactyla (~185 species) | Worldwide; introduced into Australia and New Zealand | Even-toed ungulates; e.g., pigs, hippopotamuses, camels, deer, giraffe, buffaloes, cattle, gazelles, goats, llamas, antelopes, sheep |
| Cetacea (78 species) | Worldwide in oceans, some rivers | Whales, dolphins, porpoises |

* The world (except Antarctica, which has no endemic species of mammals) is divided into **biogeographic regions,** including:
Nearctic: arctic and temperate regions of America
Neotropical: American tropics
Ethiopian: Africa and Arabia
Oriental: Tropical Asia and the adjacent islands of the Malay archipelago

(a)

(b)

**FIGURE 30–20** Monotremes. (a) An echidna, a spiny anteater. (b) A duck-billed platypus, found only in Australia and Tasmania. (a, Merlin D. Tuttle; b, Tom Stack & Associates)

## Monotremes

There are only three species of living monotremes: the duck-billed platypus and two kinds of spiny anteaters, also known as echidnas (Figure 30–20). Unlike other mammals, they lay eggs, which are much like reptile eggs, although the females feed the hatched young with milk from mammary glands in typical mammalian fashion. Monotremes thus provide evidence that viviparity and milk production did not necessarily evolve at the same time. Monotremes lay their eggs in burrows in the ground and incubate the eggs and young with the warmth of their bodies.

The Australian duck-billed platypus lives rather like a water rat on the edges of streams. It swims with its flattened tail and eats aquatic invertebrates that it detects with electroreceptors in its duck-like bill. Echidnas, found in Australia and New Guinea, dig out ant nests and trap the ants with their sticky tongues.

## Marsupials

Marsupials are mammals whose young are born at an early stage of development and finish developing in a pouch, or marsupium.

Marsupials are found only in Australia and America, although fossil evidence shows that they once lived in Europe. The only North American marsupial is the opossum, once restricted to South and Central America but now established in much of the United States and spreading northward into Canada. Well-known Australian marsupials include kangaroos and the koala.

Most mammals have a **placenta,** an organ that develops from maternal and fetal tissues during pregnancy and permits their blood systems to exchange substances. Most marsupials have no placenta. The eggs are yolky; the embryo absorbs food from this yolk and from fluid in the uterus as it develops, depending on the mother only for protection. The embryos are born as few as eight days after fertilization and crawl to the pouch. Here, the young attach their lips firmly to a teat, and muscles around the mother's mammary gland pump milk into them while they complete development (Figure 30–21).

## Placental Mammals

Most modern mammals are placental. The egg has hardly any yolk, and the mother's blood supplies the developing embryo with food and oxygen and removes its wastes. The young are born at various stages of development, from the furless, blind babies of a mouse, to a horse foal, which can walk within a few minutes of birth. Many mammalian species have developed social systems, which probably originated from the advantage gained when both parents cared for and fed the growing young.

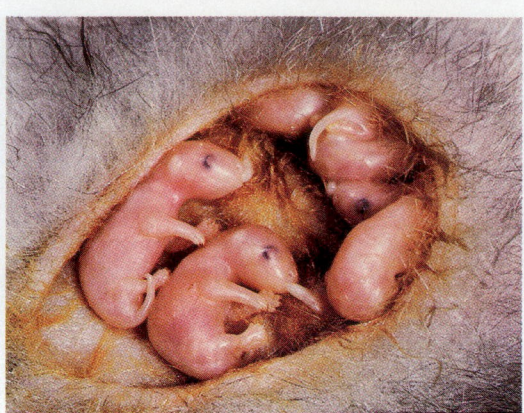

(b)

(b)

**FIGURE 30–21**  Marsupials. (a) A red kangaroo. (b) These young opossums are completing their development attached to a teat in the marsupium, or pouch, of their mother.   (a, Dave Watts/Tom Stack & Associates; b, Carolina Biological Supply Company)

Modern placental mammals are classified into a number of orders, with adaptations to many different ways of life. Their phylogeny is much disputed, partly because of convergent evolution between members of different groups. Here we consider the most important orders briefly, starting with the mammals that appear first in the fossil record and are thought to resemble their reptilian ancestors most closely.

Members of the order **Rodentia,** mammals with gnawing teeth, are the most successful modern mammals apart from *Homo sapiens* (Figure 30–22). One striking rodent characteristic is that they have nearly all remained small and reproduce rapidly. The **Lagomorpha** are similar but not related. Lagomorphs include the small, compact pikas and the rabbits and hares, with their long legs specialized for jumping.

Hedgehogs, shrews, and moles are members of order **Insectivora,** small active mammals that feed on insects and other invertebrates. Because their small bodies lose heat rapidly, insectivores must eat almost constantly to provide metabolic fuel, and many are active day and night.

**Primates** (lemurs, monkeys, and apes) retain many features of early mammals, with additional adaptations to arboreal life (Section 30–J). Most primates face extinction as a result of the destruction of their tropical forest homes by a fellow primate, *Homo sapiens.*

Bats (**Chiroptera**) and birds are the only vertebrates with true flapping flight (Figure 30–23). Bats' wings con-

sist of a web of skin stretched over long thin bones of the forelimb digits and attached also to the hindlimbs. Bat flight is slower but more maneuverable than that of most birds, enabling many species to live by catching insects. It is probably advantageous for bats to avoid competition with birds, which is why most species are nocturnal. Some have developed echolocation systems that permit them to fly and to catch food in the dark (*Essay: Moths and Bats,* Chapter

**FIGURE 30–22**  A mouse, member of the mammalian order Rodentia.   (Biophoto Associates)

**FIGURE 30–23** Red bats (*Lasiurus borealis*). (Merlin D. Tuttle/Photo Researchers)

38). Adaptive radiation has produced species that catch fish and frogs, blood-feeding vampires, pollen and nectar feeders, and large diurnal fruit-eaters with good vision. Although they are not well known, bats actually make up more than 20% of all mammal species.

The **Carnivora** are the most specialized terrestrial hunters (Figure 30–24). Their bodies are the ultimate development in the trend toward quadrupedal carnivory. Their behavior is complex and involves the ability to learn much of their hunting behavior. The sense of smell is usually well developed, and carnivores use odors, often produced by anal glands, in many of their social interactions as well as for hunting. Social and family behavior is common and often involves teaching the young, as well as protecting and feeding each other.

**Pinnipeds,** the seals, walruses, and sea lions, are undoubtedly descended from carnivore ancestors. They are less fully adapted to aquatic life than cetaceans (whales and their kin), since most must come on land to copulate and to bear their young. They feed almost exclusively on fish. The main exceptions are walruses, sad-eyed giants that gouge molluscs from rocks or sea-bottom sediments with their enormous canine teeth.

There are three orders of **ungulates,** mammals that walk on the tips of their toes. Most of them are hoofed mammals, divided according to whether they walk on an odd number of toes (horses, rhinoceroses, and tapirs) or an even number (deer, cattle, and so on). The only other ungulate group is the elephants, with two modern genera, one in Africa and the other in Asia.

Ungulates are the mammals most highly specialized for a herbivorous diet. In most species the flattened teeth used to crush and grind tough plant material are not well equipped to fight a potential predator. Most ungulates rely on running fast to escape their enemies, and they also tend to feed in herds, where every animal's eyes watch out for danger.

The main defenses of elephants (order **Proboscidea**) are their large size, their trunks, and their huge tusks. The weight and maneuverability of an elephant's trunk, developed from the nose and upper lip, make it a formidable opponent. In order to sustain their huge bodies, elephants must feed for as much as 18 hours a day (Figure 30–25).

The order **Perissodactyla** contains the odd-toed ungulates. These range from the jungle-dwelling tapirs to enormous rhinoceroses, which have a thick horny skin and very little hair. Horses, asses, and zebras of various species are widespread and are as well adapted to grazing in herds on

**FIGURE 30–24** Carnivores. A wolf pack with its kill. (Robert Winslow/Tom Stack & Associates)

**FIGURE 30–25** Order Proboscidea. A mother elephant pauses to graze while her calf nurses and other members of the herd forage nearby.

grassy plains as many of the even-toed ungulates (Figure 30–26).

Just as the Carnivora can be considered the ultimate in mammalian carnivores, so the order **Artiodactyla,** the even-toed ungulates, represents the peak of evolution for herbi-

vores. One reason for their evolutionary success was the development of a ruminant digestive system, a series of stomach compartments in which symbiotic bacteria break down plant cellulose (Section 31–D). In addition, most artiodactyls are keen of sense and fleet of foot. Their adap-

(a)

(b)

**FIGURE 30–26** Perissodactyls and artiodactyls. (a) Zebras form herds, where some individuals keep a lookout for predators while others graze. (b) An elk buck.

tive radiation is impressive: they have given rise to such diverse forms as pigs, hippopotamuses, and camels. Cattle, sheep, and deer often have horns or antlers, which probably evolved originally as defensive weapons. No artiodactyl will fight if it can help it because injuries to the delicate legs are usually fatal. Many artiodactyls, however, give a good account of themselves with horns and hoofs if they are cornered.

Whales belong to the order **Cetacea,** mammals adapted to living permanently in the sea. They can be divided broadly into the toothed and the baleen whales. Toothed whales, including dolphins and porpoises and the sperm, bottlenose, white, and killer whales, feed mainly on fish and large invertebrates.

Baleen whales are filter feeders and include the blue whale, the largest animal that has ever lived. Baleen whales feed on crustaceans in the plankton, which they engulf in huge mouthfuls. They then use their large tongues to push water and plankton against the sieve of whalebone or baleen plates lining the jaws, pushing out the water and retaining the plankton. Whales are intelligent, social animals, communicating with one another by sound. They can also hear echoes of their own voices that have bounced off other objects. This process of echolocation helps them identify food and other objects in the water.

## 30–J  HUMAN EVOLUTION

All the people on Earth today belong to the species *Homo sapiens,* in the mammalian order Primates. Primates can be divided into "lower" primates, including lemurs and lorises, and "higher" primates, including monkeys, apes, and humans. Primate radiation is the story of arboreal herbivores with excellent manual dexterity and vision and increasingly complex social behavior and communication. Humans and some other primates have become terrestrial, but some of our most distinctive features are inherited from arboreal ancestors.

Primate skeletons are unspecialized, looking much like those of primitive mammals. The main evolutionary trend is enormous expansion of the cerebral hemispheres of the brain, and particularly of the frontal lobes, which are responsible for intelligent behavior and muscular dexterity (Figure 30–27). The olfactory part of the brain is small, and the snout is reduced in size. This evolution of the nervous system is closely tied to the muscles and sense organs needed to live in trees. Well-developed eyes are almost essential to a bird or a monkey that must land on a branch. In most primates, both eyes face forward and therefore see the same thing; the two superimposed images provide **stereoscopic** (three-dimensional) vision. (Compare your depth perception using one instead of both eyes.)

Primates are also notable for their ability to grasp and manipulate tree branches and other objects with their feet.

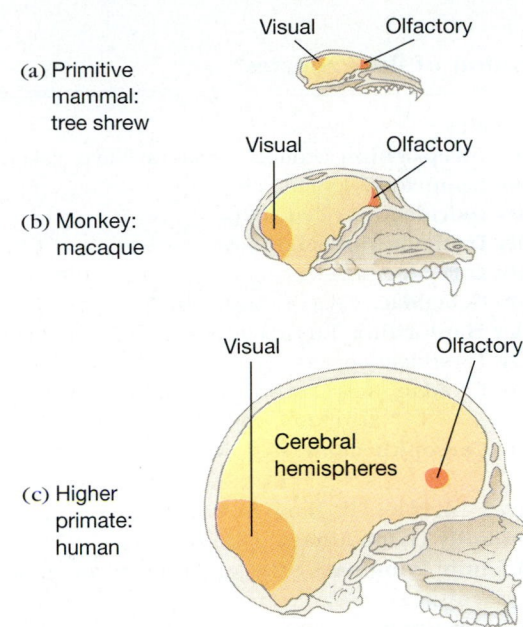

**FIGURE 30–27**  Expansion of the visual center and cerebral hemispheres of the primate brain. By comparison, the olfactory area has hardly enlarged as brain size increased.

They have five digits (fingers or toes) on each limb, usually with one digit at least somewhat opposable to the other four. The digits end in sensitive pads, often covered by flattened nails rather than the curved claws of other mammals (Figure 30–28).

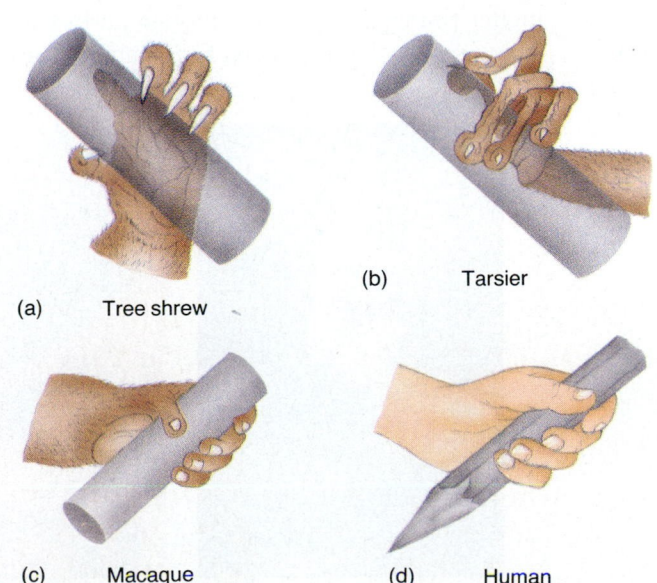

**FIGURE 30–28**  Hands of primates. This series of drawings shows the evolutionary trend from relatively immobile digits with claws to the human hand, with opposable thumb and with fingertips protected by nails.

## TABLE 30–5

### Classification of the Primates*

**Order Primates**
  **Suborder Strepsirhini:** Lemurs, lorises, aye-aye, galagos
    **Family Lemuridae:** lemurs
    **Family Indridae:** woolly lemurs
    **Family Daubentoniidae:** aye-aye
    **Family Lorisidae:** lorises
    **Family Galagidae:** galagos, bushbaby
  **Suborder Haplorhini:** Tarsiers, monkeys, apes, humans
    **Family Tarsiidae:** tarsiers
    **Family Cebidae:** New World (American) monkeys, including marmoset, capuchin
    **Family Cercopithecidae:** Old World monkeys, including macaques, baboons
    **Family Pongidae:** Gibbons, great apes
      **Subfamily Hylobatinae:** gibbons
      **Subfamily Ponginae:** great apes (orangutan, gorilla, chimpanzee)
    **Family Hominidae:** humans and prehumans
      *Australopithecus* (extinct prehumans)
      *Homo habilis, H. erectus, H. sapiens*

*After T. A. Vaughn. *Mammalogy*, 3d ed. Philadelphia: Saunders College Publishing, 1986.

## Haplorhini

The most primitive higher primate is the Indonesian tarsier, an arboreal nocturnal creature with huge eyes, fully stereoscopic vision, and nails instead of claws. In addition, the upper lip is free of the gums. This feature is a large part of the reason higher primates have such mobile and expressive faces. They use facial expressions for communication,

undoubtedly because vision is the dominant sense, rather than smell, which most mammals use to communicate.

Tarsiers, monkeys, apes, and humans make up the haplorhine ("half-nosed") primates (Table 30–5). The short nose gives the forward-looking eyes an unimpeded view of the world. As the snout became shorter, so did the jaws, and some of the teeth were lost. In humans this trend is still evident as variations among individuals in the production of wisdom teeth. Haplorhines have stereoscopic color vision, rounded heads, and an upright posture. Even quadrupedal monkeys sit upright for long periods, freeing their hands to manipulate objects and handle their young (Figure 30–29). In addition, some arboreal monkeys spend long periods in a vertical position as they swing through the trees by their arms, a type of locomotion called brachiation.

Monkeys are similar to most other mammals in their basically quadrupedal locomotion, in the proportions of their limbs, and in the compression of the rib cage from side to side. In contrast, the arms of apes are long compared with their hindlimbs, and the rib cage is flattened from front to back. Apes' tails are reduced to the few fused vertebrae of the coccyx.

There are only four genera of modern apes: gibbon, orangutan, gorilla, and chimpanzee. All live in the Old World, and their structure and behavior bridge the gap between monkeys and humans.

There are more similarities than differences between apes and **hominids** (humans and their immediate ancestors). In fact, by traditional standards of classification, we would all be placed in the same family. Nevertheless, the four genera of modern apes share some features not found in hominids. For instance, apes have powerful canine teeth, large incisors, and specialized feet, features not shared by humans. Gorillas and chimpanzees spend a lot of time on

(a)                               (b)

**FIGURE 30–29** Haplorhines. (a) A white-faced monkey from Central America. Like most other higher primates, this species can comfortably sit upright, freeing its hand to carry things. (b) An orangutan mother and infant in Borneo. (Peter Arnold)

the ground, walking on their hind feet and the knuckles of their hands. This allows them to use their fingers to carry objects such as food or stones.

## Human Characteristics

How did our distinctively human traits evolve? What changed our ancestors from arboreal animals into terrestrial creatures who could not live in trees if we tried? And where did the big brains and tool-using hands that permit our complex technological societies come from?

These characters appear to have originated with changes of climate in the Miocene. The climate became cooler, with increasing seasonal differences. This favored the growth of herbs and annual plants over trees, and fossil pollen grains show that many forested areas became open grasslands. So there was selection for spending more time on the ground and less in trees. This stage is represented by fossils such as *Ramapithecus* from Africa, Europe, and Asia. These ancestors of both apes and humans had teeth adapted to an omnivorous diet, including seeds, nuts, and small animals.

For reasons that we do not understand, spending more time on the ground led to complete bipedalism for hominids but not for the ancestors of apes. Once bipedalism developed, however, it may well have shaped later hominid evolution. No longer needed for locomotion, hominid hands coevolved with the expanding brain and became increasingly efficient at other tasks, such as collecting food and carrying it back to the rest of the group. This may well have increased survival of offspring. This in turn may have accelerated a trend already apparent in primate evolution— a relative lengthening of all the stages in life. This culminates in chimpanzees and humans, which have a long period of helpless infancy, late sexual maturity, a considerable interval between births, and survival beyond

reproductive age. These traits all select for social organization to raise offspring and language for communication. These trends culminated in humans in the ability to cooperate in the development of technology (tool use) for activities such as hunting and agriculture.

## The First Hominids

Molecular evidence suggests that African apes and hominids diverged much later than had previously been imagined. They appear to have had a common ancestor in the Pliocene, between 5 and 10 million years ago. Shortly thereafter, the first undoubted hominid fossils appear in Africa. The first specimen found, in 1924, was named *Australopithecus africanus.*

Australopithecines were presumably the ancestors of species of *Homo*. They had essentially human teeth, with small incisors and canines, but they were ape-like in having large, heavy jaws, and their brains were little larger than those of modern apes (Figure 30–30). It seems that these early hominids were largely vegetarian. There is no convincing evidence that they used tools, but we would not necessarily expect to find remains of the earliest tools. By analogy with the tools used by chimpanzees today, the first tools of hominids must have been "found" materials— unworked rocks, bones, sticks, large thorns, or lengths of vine—which either would not have been preserved or, if they were, would not bear noticeable marks of having been used as tools. By 2.5 million years ago, an advanced group of *A. africanus* was apparently making stone tools.

In 1979 a group of fossils, known as the Hadar australopithecines, was discovered in the Ethiopian desert. They belong to an earlier species, *Australopithecus afarensis,* and include the most nearly complete skeleton of this genus found so far. This specimen, nicknamed "Lucy," was a full-grown female that lived about 3.6 million years ago. She

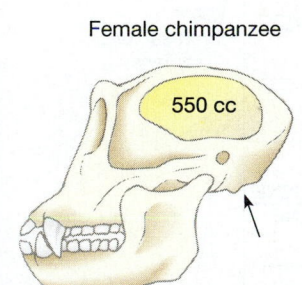
**Female chimpanzee**
550 cc

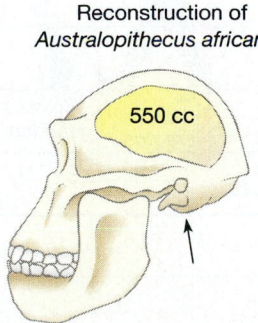
**Reconstruction of *Australopithecus africanus***
550 cc

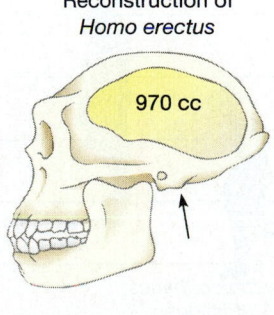
**Reconstruction of *Homo erectus***
970 cc

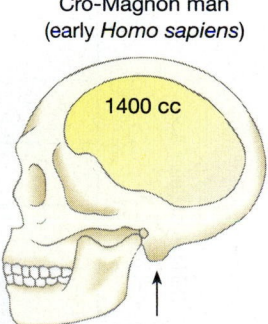
**Cro-Magnon man (early *Homo sapiens*)**
1400 cc

**FIGURE 30–30** Changes in the brain and skull from apes to modern humans. Note the increase in size of the braincase, change in the angle at which the skull joins the backbone (arrow) as hominids became more bipedal and upright, and decrease in relative length of the jaw as hominids changed from a herbivorous to an omnivorous diet.

stood upright and was about a meter tall, although males of her species were much larger. (Anatomical differences between the sexes decreased during the course of later evolution.) There is some debate about whether the Hadar australopithecines were completely ground-dwelling or spent considerable time in trees.

About 2 million years ago, the small-brained, vegetarian *Australopithecus robustus* coexisted with the first species assigned to the genus *Homo, H. habilis,* which had a larger brain (Figure 30–31). Piles of animal bones associated with this species show that meat had been added to plants as a regular part of the diet by this time, but whether it came from hunting or scavenging is unclear. These hominids also used crude stone tools, sometimes carrying the stones considerable distances from their sources.

*Homo habilis* lasted only a few hundred thousand years. By about 1.75 million years ago it had been replaced by a similar species, *H. erectus,* possibly descended from a population of *H. habilis. H. erectus* was a fully bipedal, tool-using hominid with an omnivorous diet. A virtually complete skeleton of *H. erectus* from Lake Turkana in Kenya was discovered in 1984. This fossil shows that *H. erectus* was taller than most *H. sapiens* populations, with little difference between the sexes and a hip joint that differs from that found in either *Australopithecus* or any other species of *Homo.* The skull was thick, with heavy jaws and teeth, brow ridges, and a low forehead. Some *H. erectus* bones are found in caves, suggesting the use of at least temporary home bases. Besides animal bones and quite advanced stone tools, some of the caves contain heaps of charcoal and charred bones, showing that fire had been domesticated and brought indoors by this time. Presumably this habit originated in the use of natural fires (started by lightning or spontaneous combustion) to keep warm, cook food, or split stones.

Such beginnings of culture contributed to the success of *Homo erectus.* The species spread widely, and emigrants from Africa colonized other, colder areas in Asia and later in Europe. Anatomical or physiological adaptations alone will not permit hominids to survive winters that are as cold as those of Central Europe and China. Behavioral adaptation or technological expertise is necessary. Plainly the prehuman brain of *H. erectus* could produce social and technological solutions—such as fire, clothing, stored food, and communal living in caves—to the problems of surviving cold winters. These solutions probably led to the development of some of the most important features of *H. sapiens.*

Neanderthals, which had human-sized brains but heavy skulls and teeth and powerful bodies, appeared about 130,000 years ago. They coexisted in Europe and Israel with *Homo sapiens* but disappeared about 50,000 years ago. Some experts even think that Neanderthals were merely a race of *H. sapiens.*

## Out of Africa

*Homo sapiens* probably evolved from *H. erectus* a few hundred thousand years ago. Analysis of mitochondrial DNA

**FIGURE 30–31**  Time line of human evolution. Note that several hominid species existed at the same time.

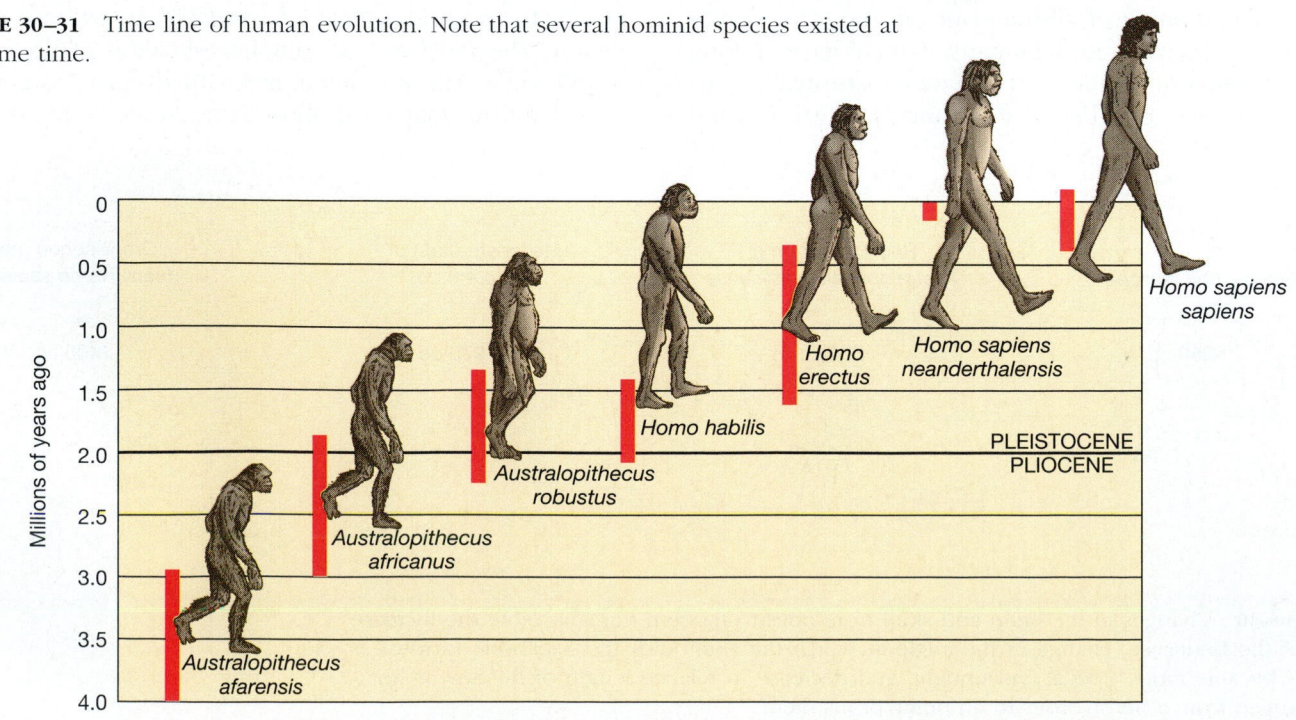

from human populations in many parts of the world suggests that all modern humans have a common female ancestor. (Recall that mitochondria are passed only through the female line, in egg cytoplasm, Section 12–A). Allan Wilson, who first proposed this theory in 1987, claimed that this "mitochondrial Eve" lived about 200,000 years ago in Africa. The date when "Eve" lived is based on a molecular clock (Section 22–E). Wilson's group calculated that it must have taken 140,000 to 280,000 years for mitochondrial DNA to have accumulated all the genetic differences found in different African populations.

This theory is hotly debated, with some paleontologists claiming that *Homo sapiens* originated from *H. erectus* populations in Asia. An African origin seems more likely, however, because the DNA of Africans in the southern half of the continent contains more genetic variation than that of people in any other part of the world. This suggests that non-African populations are descended from founders who carried only a sample of genes from parent African populations (Figure 30–32). Whatever the outcome of this debate, it seems clear that modern humans slowly spread around the world, either wiping out other hominid species as they colonized new areas or, possibly, interbreeding with them.

## Permanent Settlements

Permanent settlements in Eurasia may have been responses to the difficulty of a nomadic life during cold winters. Cooperative effort permitted the group to trap large game animals. During the summer, the main source of food would have been plant materials—seeds, fruits, leaves, nuts, roots, and berries—gathered as they were used or dried to preserve them for later. The use of fire opened up a new range of plant foods because cooking can remove volatile toxins and can also soften plants and make them more digestible. A settled hunter-gatherer society such as this must have been the precursor to the development of agriculture. It would demand social organization and communication among individuals that would be a strong selective pressure for the development of language, social rites, laws, and customs. Written symbols first appear in the archeological record about 40,000 years ago and probably reflect the origin of language at about this time. Settled living in villages produced the decorated tools, pots, and dwellings that began to appear in Eurasia about 40,000 years ago and in America about 15,000 years ago, after humans reached North America by crossing the Bering land bridge some 25,000 years ago.

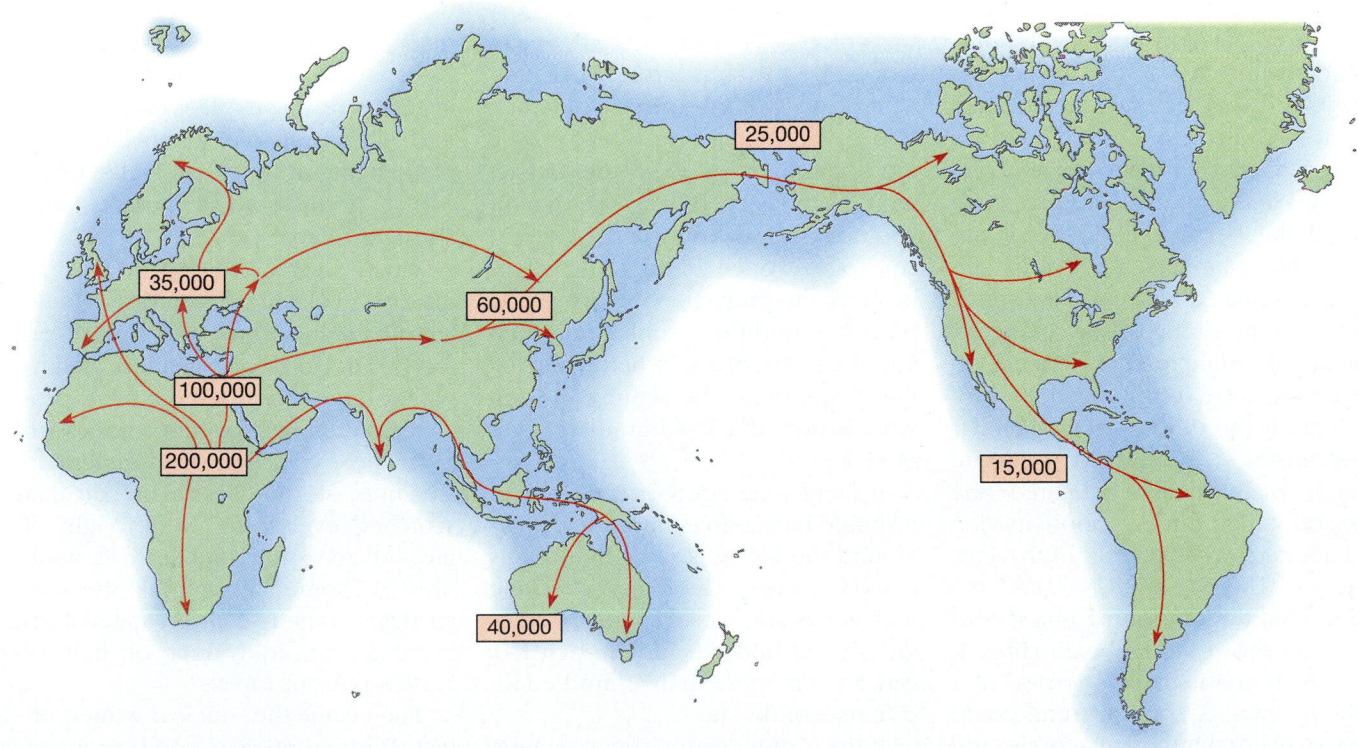

**FIGURE 30–32**   Out of Africa. The migrations of *Homo sapiens* populations according to the theory that all modern humans are descended from one African population. The figures show the age (in years) of the earliest human settlements known along the migration routes.

## ESSAY: *Mesozoic Murder*

Fans of "Jurassic Park" must be thankful that dinosaurs became extinct 65 million years ago, at the end of the Cretaceous Period. What toppled the fearsome Mesozoic monsters from their 150-million-year reign?

Early mammals were the first suspects. With warm, furry bodies suited to nocturnal activity, they could prey on dinosaur eggs under cover of darkness, as rats, wild pigs, and raccoons prey on turtle nests today. Early mammals could easily have been responsible for the demise of many lizards and marsupials along with the dinosaurs.

However, this does not explain enough. A large percentage of plants, and most marine life, perished at the same time as the dinosaurs. Paleontologists have long attributed these extinctions to changes in climate. Near the end of the Cretaceous, Earth became cooler and drier, with shifts in land forms and drying up of shallow inland seas. Perhaps the doomed species could not adapt quickly enough as new habitats replaced old. Falling temperatures might have tipped the balance in favor of endothermic mammals and birds, although some dinosaurs apparently inhabited the Arctic. Also, sex determination in some reptiles depends on the temperature during incubation of the egg. If the dinosaurs had similar mechanisms, prolonged climate change might have eliminated one sex and wiped out the population.

Either script—"Raiders of the Lost Egg" or "The Big Chill"—could explain what the dinosaur fossil record seems to show: gradual extinction of various dinosaurs over a period of 2 million years. A cooling trend could also explain why many corals and reef-forming bivalve molluscs perished at the same time. Such animals are notoriously sensitive to temperature.

### New Ideas

In 1978, physicist/Nobel laureate Luiz Alvarez, geologist Walter Alvarez (father and son), and Helen Michel analyzed a thin layer of clay laid down in Italy between marine limestones of the Cretaceous and the Triassic: at the so called K-T boundary. This clay contained 25 times the normal concentration of iridium, a metal rare in Earth's crust, but more common in deeper layers and in meteors.

Also present in the K-T clay layer are quartz crystals fractured in a way typical of crystals shattered by high-speed impacts at meteor craters and nuclear test sites. The clay also contains soot, as well as small pebbles apparently formed by rapid cooling of drops of molten rock.

The Alvarez group tested various hypotheses by computer simulation. Only one survived: the iridium must have come from a huge extraterrestrial object—a comet, asteroid, or meteor—about 10 km across. This Manhattan-sized chunk of rock slammed into Earth at a speed of 36,000 km/hour. The force of the impact would have been 10,000 times that of exploding the world's entire nuclear stockpile at once, enough to pulverize both the projectile and the rock it hit and to hurl the resulting dust high into the stratosphere, where it circled the globe, gradually settling to form what is now the K-T boundary layer of clay.

If a mega-meteor impact did occur, it should have left its impact on rocks around the globe, so this hypothesis could be tested. Geologists checking rock layers and ocean sediment cores did indeed find the iridium-rich K-T clay layer in more than a hundred locations worldwide.

If the iridium came from a meteor, where was the impact crater? One crater, under glacial deposits at Manson, Iowa, is the right age and occurs, suggestively, in quartz rock.

However, at 32 km across, it is much too small. Then crater-hunters heard of one in Mexico's Yucatán Peninsula in the right size range: 180 km across (Figure 30–A). The Alvarez team explored the area and concluded that this crater had been formed at about the right time: the end of the Cretaceous.

Meanwhile, other geologists proposed a different source of the iridium-rich clay: massive volcanic eruptions of molten rock from the deep mantle of Earth. They point to the Deccan Traps, immense formations of basalt rocks in India. These were formed by a series of eruptions spanning 3.5 million years, with the latest and largest at the K-T boundary. Altogether, they spewed forth at least 1.5 million km$^3$ of molten lava into the air, enough to cover the entire planet 3 m deep! The Deccan Trap basalt layers interleave with sedimentary deposits containing late Cretaceous dinosaur fossils. If, as often occurs, volcanic explosions preceded the outpouring of molten lava, this could account for the K-T clay layer, with its iridium, basalt pebbles, and shocked quartz.

Either a giant meteor impact or massive volcanic eruptions would have had disastrous, and similar, effects. The meteor scenario begins with a huge, incandescently hot projectile plunging into the sea floor with a terrific explosion. This generated tsunamis, storms, and a fireball that scorched everything within sight of ground zero and touched off wildfires. By some calculations, the soot particles in the K-T clay resulted from burning the equivalent of half of Earth's present forests.

Then came the "nuclear winter" effect. When a nuclear bomb or a volcano explodes it throws dust into the atmosphere that blocks out the sun and cools Earth's surface (Figure 30–B). Dust from a huge meteor

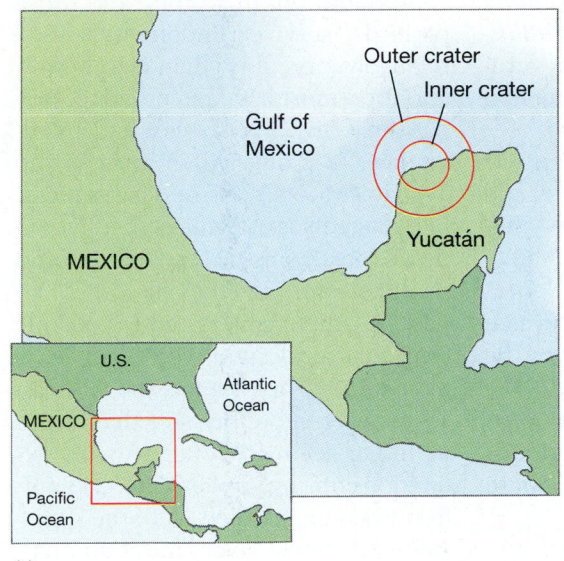

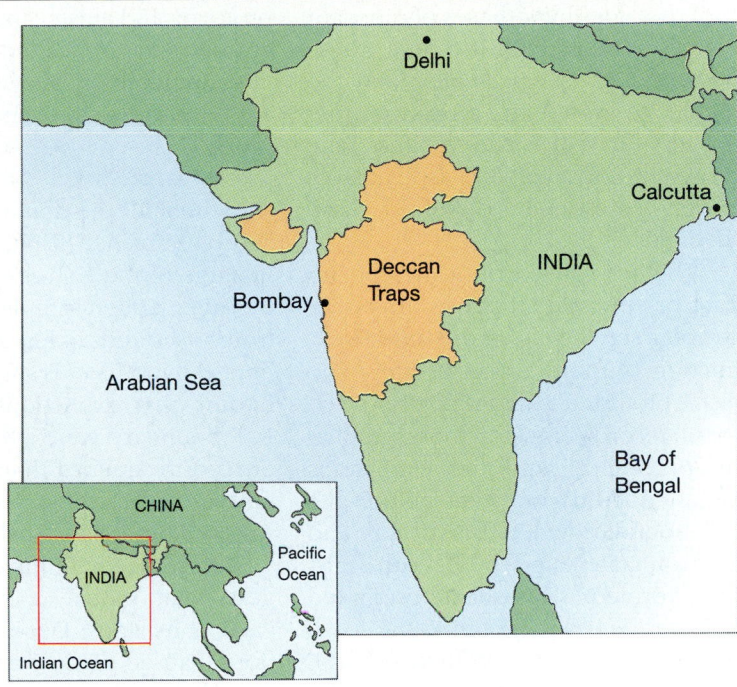

**FIGURE 30–A**  Sites of catastrophes that may have caused the K-T extinctions. (a) Site of impact of a meteor at the edge of the Yucatán peninsula. (b) The Deccan Trap formations, products of a series of volcanic eruptions in India.

would have circulated in the stratosphere, shrouding the planet in cold and darkness for at least a year, and more likely five to ten years. Plants cannot photosynthesize in the dark. Most would have perished before the dust, soot, and rock particles finally settled to Earth and formed the K-T clay layer.

But this environmental disaster was not over. The energy of the meteor impact would have catalyzed the reaction of atmospheric nitrogen and oxygen to form nitrous oxide, which dissolved in water and fell as nitric acid rain, killing plankton on the ocean surface. The acid rain would have dissolved the calcium carbonate shells of many species, as well as any exposed limestone, and this would have released carbon dioxide into the air. With most carbon-fixing photosynthetic organisms dead, carbon dioxide would have accumulated in the atmosphere, and the survivors of the cold, dark years now faced a longer period of global warming.

*(Continued)*

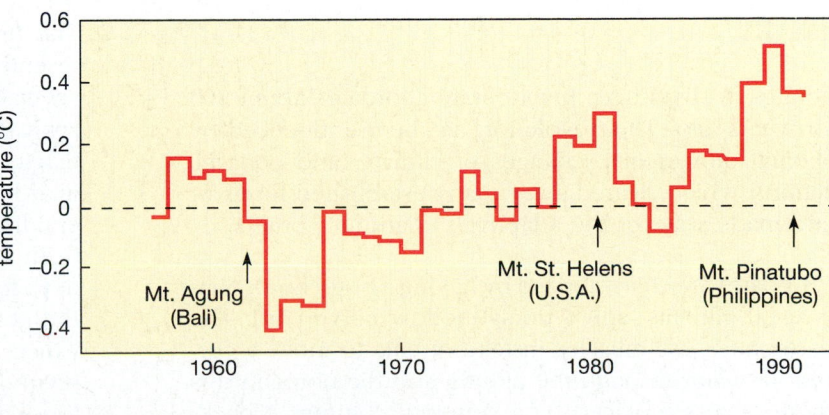

**FIGURE 30–B**  The nuclear winter effect: cold and dark caused by soot and dust in the atmosphere. (a) A few of the oil wells in Kuwait set afire by Iraqi troops during the Gulf War belch clouds of soot into the air, turning day to night. (b) The drop in temperature caused by dust spewed into the air by three recent volcanic eruptions.  (a, Steve McCurry, Magnum Photos)

**ESSAY** *(Continued)*

If the older hypotheses of dinosaur extinction explained too little, these newer ones seem to explain too much. We now have a mass murder mystery in which the victims were drowned, cremated, asphyxiated, frozen, poisoned, dissolved, and steamed!

Whether caused by a meteor impact or volcanic eruptions, this catastrophic script calls for the sudden extinction of many forms of life. "Too fast," protested paleontologists. "The fossil record shows a more gradual disappearance, and the extinctions began millions of years before the K-T boundary when the Yucatán and Manson craters and the Deccan Traps were formed." This debate prompted paleontologists to examine the fossil record more closely. With more data, it now appears that everyone is right, but the big picture is still not clear.

It is true that three quarters of dinosaur species were extinct before the end of the Cretaceous, but a higher proportion survived in North America, which contains 20 of the 26 dinosaur fossil beds known from this time. In the oceans, coral reef and bivalve extinctions also occurred gradually, ending before the K-T boundary. However, up to 90% of marine protozoa and algae vanished abruptly at the boundary, as did the *Nautilus*-like ammonites. The seas were virtually sterile for the next half-million years.

On land, fossil pollen deposits show that all the dominant trees and shrubs vanished within a few millimeters of the K-T clay layer. Above the layer, pollen is replaced by spores from vast stands of ferns. The Yucatán impact hypothesis is supported by the finding that extinctions right at the K-T boundary were greater in Western North America than in more distant areas such as Canada, South America, and New Zealand.

Taken together, this evidence suggests that global cooling, perhaps caused by early Deccan Trap eruptions, caused gradual extinctions, followed by more rapid extinctions caused by the catastrophes at the K-T boundary.

## Unanswered Questions

Many questions remain. What was the sequence of impacts, eruptions, and extinctions at the K-T boundary? Which organisms were killed by cold and dark, which by heat or acid rain? How long did the period of massive extinctions last?

Why did our ancestors, and those of birds, survive? Endotherms need relatively more food than ectotherms, and food must have been scarce. Perhaps insectivorous species had an advantage at a time when most of the available food consisted of species like maggots and termites that live on dead organisms. Many birds and mammals are insectivores.

Are other craters and volcanic eruptions associated with other mass extinctions? For instance, the Siberian Traps, even more massive than those in India, seem to coincide with extinctions during the Permian, 248 million years ago, which were the greatest extinctions since the Cambrian Period (and perhaps of all time). The Triassic extinctions, 200 million years ago, cleared the way for adaptive radiation of the dinosaurs. A large crater in Quebec and the Karoo flood basalt in Africa date from this time. Far from answering our original question, we have wound up with a host more.

By the close of the Cretaceous, 60 to 80% of species were gone, and the Mesozoic Era, the Age of Reptiles, came to an end. From its survivors came a new era, the Cenozoic, the Age of Mammals.

## SUMMARY

Vertebrates evolved from invertebrate chordates about 500 million years ago. Their evolution has been influenced by continental movements, changes of climate, and periodic extinctions, which cleared away many established forms of life and made way for the adaptive radiation of others.

1. Chordates are characterized by having a notochord, pharyngeal gill slits, and a dorsal hollow nerve cord. The vertebrates use muscles instead of cilia to move a current of water through the mouth, and the notochord is more or less replaced by a vertebral column. A backbone permits efficient fish-like aquatic locomotion, particularly useful to a carnivore. As in other animal groups, the main evolutionary advances of vertebrates were made by carnivores.

2. The first vertebrates were marine jawless fishes, represented by modern hagfish and lampreys, which are specialized as scavengers and external parasites. Early jawless fishes probably gave rise to the bony and cartilaginous fishes, which were the dominant forms of animal life in the later Paleozoic era. All fishes have lateral line sense organs. Bony fish, in particular, invaded fresh water, a move made possible by use of the kidneys to control the body's water balance.

3. Both cartilaginous and bony fish showed two major advances over the jawless fish, both adaptations to carnivory: first, jaws enabled them to snap and bite their food; and second, paired pectoral and pelvic fins provided balance while the body or tail gave the main impetus for locomotion, increasing their maneuverability.

4. Cartilaginous fish are mainly marine, with short intestines, a tendency to ovoviviparity.

5. The actinopterygian bony fish are a huge group that radiated widely. Bony fish evolved an operculum covering the gill openings, and many have a swimbladder, which provides buoyancy control and allows for gas exchange with the air.

6. Some lobe-finned fish evolved lungs, used for gas exchange in shallow, oxygen-poor water. Modern survivors of this group are the lungfish and lungless coelocanth.

7. The first terrestrial vertebrates, ichthyostegid amphibians, evolved from lobe-finned fish with sturdy paired fins able to support the body on land. Amphibians have remained in moist habitats, most in or near bodies of fresh water. Their eggs must be laid in water, and their thin, moist skin is still used as a major gas exchange surface, although they also have small lungs.

8. Fully terrestrial vertebrates appeared in the Devonian with evolution of the amniotic egg, characterized by three embryonic membranes that make the egg capable of developing on land. Many organ systems have been modified as adaptations to life on land. Amniotes use internal lungs to breathe air, and their skins and kidneys slow dehydration. The skeleton is modified to support the body in air. Radiation of insects on land during the Carboniferous provided a plentiful food source that fueled the adaptive radiation of reptiles.

9. Turtles are survivors of an ancient group adapted to life inside a shell. They are unique among living reptiles in providing no parental care. All sea turtles and many terrestrial species are endangered, mainly by habitat destruction.

10. Mesozoic diapsids included the dinosaurs and their relatives, which included the ancestors of birds, squamates, and crocodilians. Diapsids are characterized by their skull arches, Jacobson's organs, and a tendency to bipedalism (longer hindlimbs than forelimbs). They underwent wide adaptive radiation.

11. Squamates are the lizards and snakes, a very successful group. Most are terrestrial and carnivorous, but aquatic and herbivorous forms have also evolved. The snakes are highly specialized, carnivorous animals that lost their legs and evolved jaw mechanisms that permit many of them to eat large prey.

12. Crocodilians are the closest surviving relatives of dinosaurs and birds. Their long snouts are adapted to eating fish. All build nests and care for their young.

13. Both birds and mammals are endothermic, using rapid metabolism to generate body heat, which is retained by a layer of fat under the skin and an outer layer of feathers or hair. Parental care is well developed in both groups; parents warm their embryos and young by incubation and protect and teach them.

14. Birds are vertebrates with feathers that evolved from bipedal carnivorous dinosaurs in the Mesozoic. The origin of bird flight is much debated. Feathers provide lightweight insulation as well as strong flight surfaces. The structure of most birds is very similar, highly modified for flying, with a greatly reduced, lightened, and streamlined skeleton.

15. Mammalian characteristics, which are adaptations to quadrupedal carnivory, evolved in several synapsid lines that became extinct. Modern mammals are endothermic, with one bone in each side of the lower jaw, teeth of varying structure adapted to different tasks, hair, and mammary glands. The limbs are slung further under the body than those of reptiles, and muscles moving the hindlimbs attach to the pelvic girdle rather than to the tail, freeing the tail for various other uses. Surviving mammals—monotremes, marsupials, and placentals—probably had a common ancestor. Placentals, the most numerous, are viviparous. Marsupials, found mainly in Australia and South America, give birth to very immature young, which finish development in a pouch. Monotremes are the only surviving egg-laying mammals. All mammals nourish the young with milk produced by mammary glands. Mammals have undergone wide adaptive radiation.

16. The relationships of modern mammalian groups are disputed. The Carnivora are the most specialized carnivores, and Artiodactyla are the most specialized herbivores. Cetaceans are highly adapted to marine life. Rodents, the group with the most species, have gnawing teeth and are generally small. Chiroptera (bats), second in number of species, are also small and adapted to flapping flight.

17. Humans are primates, characterized by arboreal adaptations including grasping hands, stereoscopic vision, large brain, and communication by vision and sound.

## Self-Quiz

1. Which of the following features does *not* distinguish chordates from other invertebrates?
   a. a closed circulatory system
   b. a notochord
   c. pharyngeal gill slits
   d. dorsal, hollow nerve cord

2. Which of the following is *not* a vertebrate?
   a. an amphioxus
   b. a lamprey
   c. a shark
   d. a kangaroo
   e. a duck-billed platypus

3. The first vertebrates to be totally independent of permanent bodies of water during reproduction were:
   a. amphibians
   b. birds
   c. mammals
   d. reptiles

4. What is the most characteristic feature of Aves, found in no other living vertebrates?
   a. flapping flight
   b. eggs incubated by parents
   c. a beak (bill) instead of teeth
   d. feathers
   e. nest-building

5. Which of the following is *not* a diapsid?
   a. goose
   b. dinosaur
   c. rattlesnake
   d. alligator
   e. turtle

6. The biggest problem that vertebrates faced in the move from an aquatic to a terrestrial environment was:
   a. lack of food
   b. inability to find water
   c. dehydration
   d. decreased availability of oxygen

7. Compared with reptiles, which of the following evolutionary advances are seen in mammals and contribute to their success?
   a. specialized teeth
   b. the integument
   c. endothermy
   d. all of the above

8. Which of the following organisms does *not* produce an amniotic egg?
   a. snake          c. robin
   b. frog           d. platypus

9. Bats belong to the taxon:
   a. Aves
   b. Mammalia
   c. Dermoptera
   d. Diapsida

Match the characteristic with the type of animal.

10. _____ hard, dry, horny scales          a. bird
11. _____ hair, differentiated teeth        b. amphibian
12. _____ jaws, swimbladder                 c. reptile
13. _____ feathers, endothermy              d. bony fish
14. _____ tetrapod, lays eggs in water      e. mammal

15. A vertebrate that has a body covering of scales, glandless skin, no limbs, and internal fertilization and that suns itself to increase its body temperature in the morning belongs to the _____ .

## Questions for Discussion

1. We often talk as if *Homo sapiens* were the most highly evolved and specialized mammal. In what ways is this true, and in what ways is it not true?

2. Many mammals have adapted to a life permanently at sea. Why haven't they gone back to using gills for respiration?

3. What are the advantages of social and parental behavior that have made it profitable for organisms to spend some of their energy budgets in these activities? What are the drawbacks of such types of behavior?

4. Why do so few birds live as grazers on leaves and grass?

5. In Tables 30–1, 30–2, and 30–3, note the number of living species shown for each group listed. Can you account for the differences in species numbers among the various groups?

6. We tend to talk as though mammalian adaptations are amazingly successful. If this is so, why are there more species of diapsids than of mammals?

7. Birds are sensitive to environmental perturbations. Miners used to use the deaths of caged canaries as early warnings that air quality in the mine was deteriorating. The decline of many bird species today results from environmental deterioration. How, then, did birds survive the K-T extinctions?

## Suggested Readings

Alvarez, W., and F. Asaro. "An extraterrestrial impact." *Scientific American,* October 1990. The K-T extinctions.

Pilbeam, D. "The descent of hominoids and hominids." *Scientific American,* March 1984. A clear account of how thinking on the subject changed during the preceding five eventful years, summarizing a new emerging consensus and pointing out what we know and don't know about our human "roots."

Pough, T. J., J. B. Heiser, and W. N. McFarland. *Vertebrate Life,* 3d ed. New York: Macmillan Publishing, 1989. A readable general text on vertebrates, living and extinct.

Rismiller, P. D., and R. S. Seymour. "The echidna." *Scientific American,* February 1991.

Romer, A. S., and T. S. Parsons. *The Vertebrate Body,* 6th ed. Philadelphia: Saunders College Publishing, 1986. A standard text of the comparative anatomy of vertebrates, living and extinct.

Tattersall, I. "Evolution comes to life." *Scientific American,* August 1992. Behind the scenes: a scientist and artist make life-sized figures for a museum exhibit of ancient hominid life.

Vaughan, T. A. *Mammalogy,* 3d ed. Philadelphia: Saunders College Publishing, 1986. A useful reference book on mammals.

Wellnhofer, P. "Archaeopteryx." *Scientific American,* May 1990. Only six fossils of this species are known. So far, they do not resolve the controversy about the origin of bird flight.

Welty, J. C., and L. Baptista. *The Life of Birds,* 4th ed. Philadelphia: Saunders College Publishing, 1988. A readable account of all aspects of bird biology.

# Animal Biology

31

# Animal Nutrition and Digestion

**O B J E C T I V E S**

*When you have studied this chapter you should be able to:*

1. Name the major classes of macronutrients and micronutrients, and list the general functions of each class.
2. Explain the selective advantages of the following digestive adaptations found in animals: extracellular digestion, discontinuous feeding, digestive tract with mouth and anus, crop, gizzard, and caeca.
3. List the parts of the human digestive tract in order, and state what happens to food in each part of the tract.
4. List the functions of the mammalian liver, and explain the importance of this organ.

5. List the organs that secrete digestive enzymes in mammals, and state the type of substrate digested by the enzymes secreted by each organ.
6. List the parts of the gut in which various substances are absorbed, and which types of substances are absorbed in each part.
7. Describe the digestive adaptations of herbivorous and carnivorous mammals and of birds.
8. Discuss how symbiotic microorganisms contribute to the nutrition of their hosts.

---

A nimals are heterotrophs: they cannot make their own food from inorganic substances as plants do but must ingest organic molecules from the environment (Figure 31–1). Animals can be broadly divided into **herbivores,** which eat plants; **carnivores,** which eat animals; and **omnivores,** which eat both. Animals with each mode of nutrition have digestive systems suited to handling and digesting the type of food they ingest.

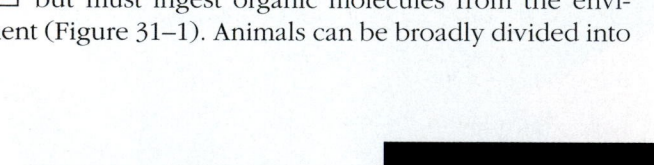

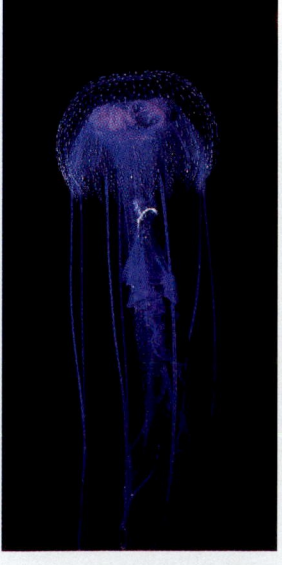

**FIGURE 31–1** Feeding. (a) A jellyfish in the Pacific pulls a white pipefish (in the center) caught in its sticky tentacles up to its mouth. (b) A pony mare on Assateague Island, Virginia, nurses her foal. (a, Robert Frerck/Odyssey)

Feeding is a necessary evil. The longer an animal spends feeding, the less time it has for activities that are more selectively advantageous, such as reproduction and raising its offspring. Thus, there is strong selective pressure for an animal to feed as rapidly as possible.

**Digestion** is the mechanical and chemical breakdown of food into small organic molecules that can be absorbed from the gut. In this chapter we shall consider what food animals need, how animals obtain food, and the role of the digestive system and liver in processing food so that it can be used by the body.

## KEY CONCEPTS

♦ Animals need large amounts of macronutrients (proteins, carbohydrates, and fats) and smaller amounts of micronutrients (vitamins and minerals) in their diets.

♦ Animal digestive systems have evolved into disassembly lines where food is broken down first mechanically,

and then chemically by digestive enzymes, after which it is distributed throughout the body.

♦ Most animals spend much of their feeding time in the pursuit of nutrients that are in short supply in their main food.

## 31–A  MACRONUTRIENTS

The nutrients any animal must ingest may be divided, for convenience, into **macronutrients**—nutrients that are needed in large quantities—and **micronutrients**—nutrients that are required in lesser amounts.

The macronutrients are fats, carbohydrates, and proteins. All three can serve as energy sources because they can be broken down into molecules that are respired to produce ATP. The amount of energy available from a given amount of a macronutrient is commonly measured as the number of Calories of heat that it will produce when fully oxidized (Table 31–1). All three classes of macronutrients also provide carbon atoms that can be used to form the body's own organic molecules.

The macronutrient that supplies most of the human population's energy needs is **starch,** the carbohydrate that plants commonly make and store. All the most important human food plants, such as grains, potatoes, and beans, contain high concentrations of starch. In the body, starch

and other carbohydrates are usually broken down to supply energy.

The body breaks proteins down into their amino acid monomers, most of which are used to build the body's own proteins.

Lipids are fats and oils. They are important components of all the membranes in every cell. Steroid hormones are also lipids.

Macronutrients can be stored until the body needs them for energy. Carbohydrates are stored as the polysaccharide glycogen, in muscle and liver, and fats are stored as fat. Proteins cannot be stored; excess protein is broken down to amino acids, which are deaminated—that is, their amino ($—NH_2$) groups are removed. The rest of the molecule is then processed as fat or carbohydrate (see Figure 7–12).

### Macronutrient Deficiencies

The most common dietary problem for people in industrialized countries is obesity. This is especially true in the United States, where food is very cheap by world standards. If more food is ingested than used, the excess is stored as fat. Overeating and lack of physical activity are the usual causes of obesity. Once obesity sets in, it may alter the body's metabolism in such a way as to perpetuate itself in a vicious cycle of overeating and inactivity.

Women have a greater tendency than men to store fat under the skin. This is biologically valuable, if currently unfashionable, because it provides food reserves that can carry not just the woman but also her unborn child or nursing infant through times of shortage. Without sufficient body fat, a woman ceases to ovulate. Hence, she cannot conceive a fetus that her body could not support. This is why women athletes with very little body fat may have difficulty becoming pregnant.

**TABLE  31–1**

### Energy Yields of Macronutrients

| Class of Macronutrient | Composition | Energy Yield |
|---|---|---|
| Carbohydrates | Polymers made of sugar monomers | 4 kilocalories/gram* |
| Lipids | Fatty acids + alcohols | 9 kilocalories/gram |
| Proteins | Polymers made of amino acid monomers | Varies: about 4 kilocalories/gram |

*A calorie is the heat needed to raise the temperature of one gram of water by 1 °C. The "Calories" in food are actually kilocalories, often indicated by a capital C. 1 kilocalorie (Kcal) = 1 Calorie = 1000 calories = 4.187 kilojoules.

It was long assumed that an animal's feeding habits were determined largely by its energy needs. However, it is now clear that many animals spend a lot of their feeding time and energy obtaining enough of a specific nutrient that is in short supply rather than merely consuming calories.

Probably the most common dietary problem for most animals is deficiency of protein, or of nitrogen compounds that can be converted into protein. This occurs when the diet contains too little protein or when the protein is not of high quality because it does not contain enough of some essential amino acid. **Essential amino acids** must be supplied in the diet because an animal cannot synthesize them from other amino acids. Essential amino acids may be used as they are or may serve as precursors of other amino acids.

### Protein Deficiency

One of the best-known human protein deficiency diseases is **kwashiorkor,** often found in African populations where the diet consists primarily of cornmeal. Cornmeal contains very little of the essential amino acids tryptophan and lysine. Most victims of kwashiorkor are growing children, who need more protein than adults. They show symptoms such as skin problems, failure to grow normally, lethargy, and edema (swelling due to fluid retained in parts of the body).

**Marasmus** is a common disease of human malnutrition. It occurs when the diet is deficient in both total calories and essential amino acids. Most victims are infants in poor families where the babies are not breast-fed or where there is not enough food for young children. A child suffering from marasmus has a bloated belly, shriveled skin, wide eyes, and an aged-looking face. This is the "face of poverty" we see in news reports on famine. Both kwashiorkor and marasmus cause permanent brain damage unless they are cured rapidly. Commentators say that "an entire generation of Somalis" has been lost to the widespread starvation of 1991 and 1992, not because an entire generation has died, but because many of those who survive will grow up to be mentally handicapped adults.

### Protein and Legumes

Some people eat only plants, either because of preference or tradition, more often because they are too poor to buy meat. It is more difficult to obtain adequate nutrition with a vegetarian diet than with a mixed diet because plant protein is relatively poor in the essential amino acids humans need (Figure 31–2). In addition, plant material contains less protein than meat does. Plants such as wheat and rice are about 10% protein. This is enough for an adult human, but not for a growing child. The protein content of plants varies from about 3% in leafy vegetables to 34% in soybeans.

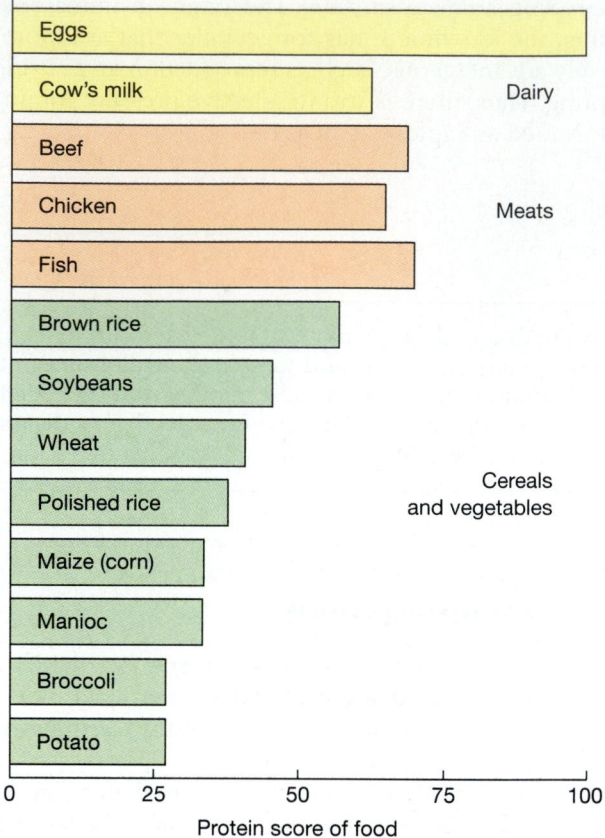

**FIGURE 31–2**   The essential amino acid content of various human foods. Other foods are given a "protein score" by which they are rated in comparison with eggs, which have an almost ideal amino acid content for human nutrition.   (U.S. Department of Agriculture)

Soybeans and other legumes house nitrogen-fixing bacteria in their roots. As a result, legumes have plenty of nitrogen and a higher protein content than other plants. This makes them a vital part of the human diet.

Soybeans, peanuts, beans, and peas are the most important leguminous human food plants worldwide. Soybeans are known in China as "poor man's meat," and they are used to make *tofu,* common in Japanese recipes. *Tofu* consists of protein concentrated from soybeans by boiling them with magnesium compounds. The Arab dish *hummus* is made from chickpeas, another legume, and peanuts are widely used in Indian dishes.

Legumes are often combined with starchy foods, which provide extra calories and amino acids. For instance, the combination of red beans and rice is common in the Southern United States. Costa Rica's national dish is *gallo pinto,* black beans and rice. Tortillas and beans are characteristic of Mexican cooking.

## Macronutrients and Health

Nowadays many people are interested in the relationship between diet and health. People may inherit genetic tendencies to develop fatal conditions such as cardiovascular disease or various cancers. However, environment also plays a role. It has been observed that immigrants tend to die of the causes of death common in their adopted countries. For instance, people of Japanese origin living in the United States are more likely to die of breast cancer and heart attacks than are Japanese people living in Japan. Researchers asked if the change to an American diet could account for the shift in the cause of death among these immigrants. The answer to this question is still not clear.

A compilation of dietary research prepared for the National Academy of Sciences tentatively concluded that Americans could reduce (or at least postpone) many causes of death by eating a higher proportion of unprocessed carbohydrates. These include raw or lightly cooked fruit and vegetables and unprocessed (or minimally processed) cereals such as wheat, oats, corn, rice, and the like. Increased consumption of these foods would raise the fiber content of the diet and would automatically reduce the amount of fat

and protein consumed. A typical Western diet almost certainly contains too much animal fat and too much protein for perfect health (Figure 31–3). As a bonus to the recommended change in diet, widespread adoption of plant rather than animal food would save much of the fuel and water now used in farming and would help to conserve land resources as well.

## Micronutrients

Micronutrients are the substances an organism must have in its diet in small quantities because it cannot make them for itself or because it cannot make them as fast as it needs them. Micronutrients can be divided into **vitamins,** which are organic compounds, and **minerals,** which are inorganic. Various deficiency diseases result from shortage of certain vitamins and minerals in the diet.

*Vitamins*   The vitamin requirements of various animals differ. For instance, many animals can make ascorbic acid from other molecules, and so vitamin C is not necessary in their diets, as it is in the human diet. Vitamins needed in the human diet are generally divided into two categories: water-soluble and fat-soluble. Water-soluble vitamins (Table 31–2) are coenzymes needed in metabolism. The water-soluble vitamins are easily excreted by the kidney.

The fat-soluble vitamins (Table 31–3) have various poorly understood functions. Fat-soluble vitamins that are not used immediately may be stored in fatty tissue. Because these vitamins are not soluble in water, the body's enzymes must process them before they can be excreted by the kidneys. As a result, some of them can accumulate to toxic levels if eaten in amounts larger than the body can use. Some modern "diets" recommend dangerous doses of certain fat-soluble vitamins. (In 1978, for instance, two people on high-vitamin diets died of vitamin A poisoning.) The key to avoiding such dangers is common sense and moderation. Almost anything can be poisonous if consumed in sufficient quantity.

Vitamin-rich foods include fresh fruits and vegetables and whole or enriched grain products, as well as fortified milk, liver, and cod-liver oil.

*Minerals*   We need some minerals in relatively large amounts. Sodium and potassium, for instance, are vital to the working of every nerve and muscle in the body (Table 31–4). Large quantities of these minerals (particularly sodium) are excreted in the urine every day. Sodium excretion is also a vital part of sweating, which is necessary to the regulation of body temperature in some mammals, including humans. Calcium is necessary for muscular activity and, with phosphorus, is needed in large amounts for bone formation.

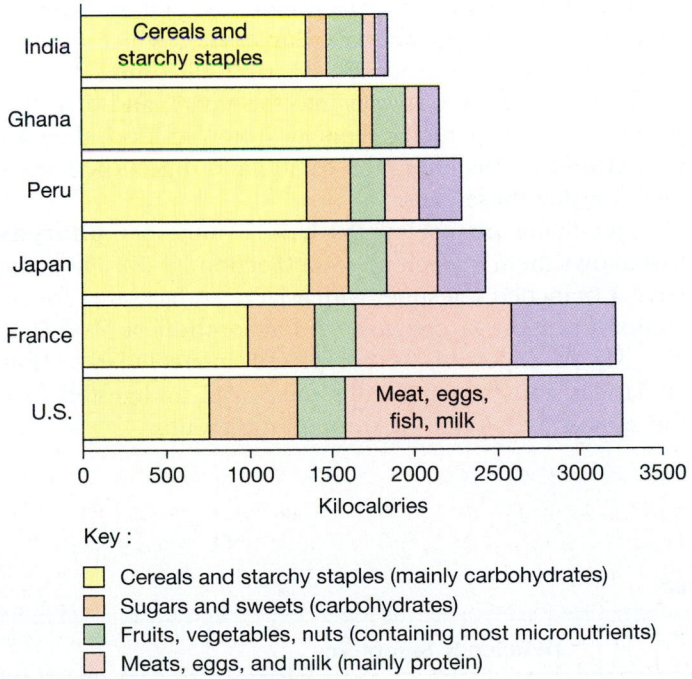

Key :

- ☐ Cereals and starchy staples (mainly carbohydrates)
- ☐ Sugars and sweets (carbohydrates)
- ☐ Fruits, vegetables, nuts (containing most micronutrients)
- ☐ Meats, eggs, and milk (mainly protein)
- ☐ Fats and oils (lipids)

**FIGURE 31–3**   The number of kilocalories of energy from different types of food in the average diet of various countries. Note the high fat and protein content of Western diets. The U.S. National Academy of Sciences recommends a daily intake of 2700 kilocalories for an active man weighing 70 kg. (The amount is less for women unless they are pregnant.)   (World Health Organization)

TABLE 31–2

**Functions and Deficiency Symptoms of Water-Soluble Vitamins**

| Vitamin | Metabolic Role | Deficiency Symptoms |
|---|---|---|
| Thiamine (B$_1$) | Coenzyme in carbohydrate metabolism | Beriberi, loss of appetite, fatigue |
| Riboflavin (B$_2$) | Part of FAD, coenzyme in respiration and protein metabolism | Inflammation and breakdown of skin, swollen tongue |
| Niacin | Part of NAD$^+$ and NADP$^+$, coenzymes in energy metabolism | Pellagra, fatigue |
| Pyridoxine (B$_6$) | Coenzyme in amino acid metabolism | Anemia, nerve problems |
| Pantothenic acid | Part of coenzyme A, in carbohydrate and fat metabolism | Similar to other B vitamins |
| Biotin | Coenzyme in addition of carboxyl groups | Rare; tiny amounts required |
| Folic acid | Coenzyme in formation of nucleotides and hemoglobin | Most common U.S. vitamin deficiency; causes some cases of spina bifida (birth deformity) |
| Cobalamin (B$_{12}$) | Coenzyme in formation of proteins and nucleic acids | Pernicious anemia |
| Ascorbic acid (C) | Helps build intercellular cement for bones, cartilage, skin | Scurvy, anemia, slow wound healing |

Other minerals are known as trace minerals (Table 31–5). Some of them are required in tiny amounts as cofactors of enzymes in various metabolic pathways. The functions of other trace minerals are unknown or poorly understood.

Foods rich in minerals include meats, seafood, milk and cheese, whole grains, nuts, legumes, and spinach.

## 31–B  DIGESTIVE SYSTEMS

**Digestion** is the hydrolysis of food macromolecules into monomers, particularly amino acids, monosaccharides, glycerol, and fatty acids. Digestion can be either **intracellular,** that is, in vacuoles within cells, or **extracellular,** in a digestive cavity in the animal's body.

Sponges have mainly intracellular digestion (Section 28–D). They are limited to food items small enough to be taken into a cell, and this often means that they must feed continuously.

Extracellular digestion becomes more specialized and efficient as we progress up the evolutionary scale. Cnidarians, such as *Hydra* and sea anemones, pull surprisingly large prey into the **gastrovascular cavity,** which doubles as a transport system (Section 28–E). Cells lining the cavity secrete digestive enzymes into the cavity, and digestion begins. As small particles separate from the food, they are picked up by cells lining the cavity, and digestion is completed inside these cells.

Free-living platyhelminths have a muscular **pharynx** that allows them to suck up food (Section 28–F). They also have a branched intestine, with a large surface area for secretion of digestive enzymes and absorption of food (Figure 31–4). Digestion is completed intracellularly. Both cnidarians and platyhelminths must void undigested food the same way it came in, through the mouth.

TABLE 31–3

**Functions and Deficiency Symptoms of Fat-Soluble Vitamins***

| Vitamin | Physiological Role | Deficiency Symptoms |
|---|---|---|
| A (retinol) | Part of visual pigments in eye | Night blindness, drying of mucous membranes |
| D (calciferol) | Increases absorption of calcium and phosphorus used to build bone | Rickets in children |
| E (tocopherol) | Protects blood cells, vitamins, and other molecules from oxidation | Lysis of red blood cells, anemia |
| K (menadione) | Needed in synthesis of prothrombin for blood clotting | Hemorrhage in newborns, who lack gut bacteria that produce vitamin K |

*Vitamins A, D, and K are toxic in large amounts.

**TABLE 31–4**

## Physiological Roles of Minerals Required in Large Amounts

| Mineral | Major Physiological Roles |
| --- | --- |
| Sodium (Na) | Major extracellular cation; active transport, osmotic and acid-base balance, nerve and muscle activity |
| Potassium (K) | Major intracellular cation; acid-base balance, nerve and muscle activity |
| Calcium (Ca) | Component of bones and teeth, messenger within and between cells; membrane permeability, blood clotting, nerve and muscle activity |
| Phosphorus (P) | Bone formation; part of nucleotides, energy metabolism |
| Magnesium (Mg) | Constituent of bones and teeth; carbohydrate and protein metabolism; protects against atherosclerosis |
| Chlorine (Cl) | Major extracellular anion; osmotic and acid-base balance; stomach acid |
| Sulfur (S) | Part of some amino acids; detoxification reactions |

**TABLE 31–5**

## Physiological Roles of Trace Minerals

| Mineral | Major Physiological Roles |
| --- | --- |
| Iron (Fe) | Component of heme group in hemoglobin, cytochromes |
| Copper (Cu) | Needed to make hemoglobin and bone; part of cytochromes |
| Iodine (I) | Component of thyroid hormone, which regulates cellular respiration |
| Manganese (Mn) | Needed in urea formation, protein metabolism, insulin regulation, glycolysis, citric acid cycle, bone formation |
| Cobalt (Co) | Component of vitamin $B_{12}$, required for red blood cell formation |
| Zinc (Zn) | Component of many enzymes and transcription factors; senses of smell and taste |
| Molybdenum (Mo) | Component of some enzymes |
| Fluorine (F) | Reduces tooth decay |
| Selenium (Se) | Needed in fat metabolism; antioxidant, may reduce risk of some cancers |
| Chromium (Cr) | Needed in glucose metabolism |
| Boron (B) | Needed for calcium metabolism and use of copper |

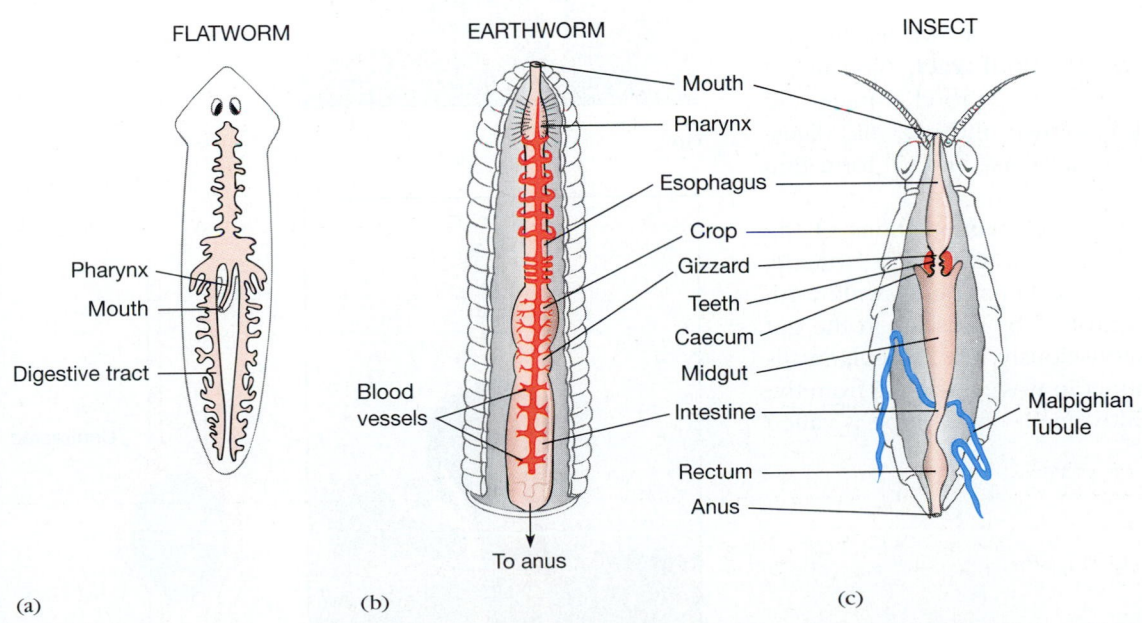

FLATWORM    EARTHWORM    INSECT

Pharynx
Mouth
Digestive tract

Blood vessels
To anus

Mouth
Pharynx
Esophagus
Crop
Gizzard
Teeth
Caecum
Midgut
Intestine
Rectum
Anus

Malpighian Tubule

(a)    (b)    (c)

**FIGURE 31–4**  Invertebrate digestive tracts. (a) A freshwater flatworm (platyhelminth) with a mouth but no anus. (b) Anterior part of an earthworm's gut. Most of the digestive tract consists of a long intestine, which extends to the anus. (c) An insect's digestive tract. Several caeca may be found just before the midgut. (Malpighian tubules are excretory structures. They discharge wastes into the gut.)

In all higher animals, the **digestive tract** is basically a tube, with a mouth for ingesting food and an anus for **egestion.** Such a tubular digestive system allows food processing to be organized as a one-way conveyor belt. Additional food can be taken in while previously eaten food is being digested, and different parts of the tube have specialized structures and functions (Figure 31–4b and c). The food is broken down step by step, and the nutrients released are absorbed farther down the tube.

Somewhere near the front end most animals have structures that break food into smaller parts—teeth in mammals, fish, and sharks; a bill in turtles and birds; a muscular **gizzard** containing stones or grinding edges in earthworms, insects, most birds, and crocodiles. A thin-walled **crop** often stores food and releases it in small amounts to the gizzard, where it is ground up.

The intestine is the next part of the digestive tract. The anterior part may be specialized as a **stomach** that stores food and secretes digestive enzymes, and following portions are increasingly specialized to absorb food molecules. The digestive tracts of many animals have **caeca** (singular: **caecum**), blind outpocketings that hold food destined for a longer stay in the intestine (Figure 31–4c). Many molluscs and arthropods have large digestive glands, where food is digested intracellularly.

## 31–C   DIGESTION

### The Human Digestive Tract

The digestive tract of humans is also known as the **gut, alimentary canal,** or **gastrointestinal tract.** Basically, it has five functions: (1) food intake, (2) food storage and transport, (3) mechanical breakdown and enzymic digestion of food, (4) absorption of nutrients, and (5) formation and evacuation of feces.

Food enters the human gut by manipulations of the mouth and associated structures such as the lips, tongue, and jaw muscles. Once a bite of food has been swallowed, the rest of its journey is controlled by muscles in the gut wall that we cannot control consciously. The muscular walls of the alimentary canal contract in waves, starting from the end closest to the mouth. Movement of this type is called **peristalsis** (Figure 31–5).

(a)

(b)

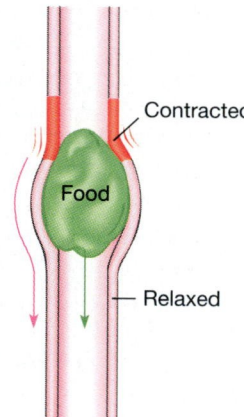

(c)

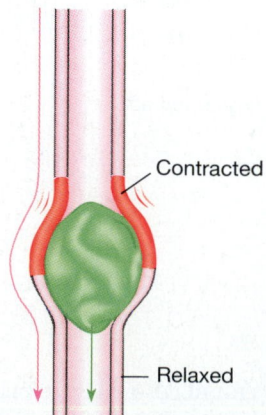

**FIGURE 31–5** Signs of peristalsis in an ostrich. The ostrich takes a mouthful of grass (a), which is formed into a bolus in the mouth. The bolus is swallowed into the esophagus, where it is pushed down to the stomach by peristalsis (b, c).

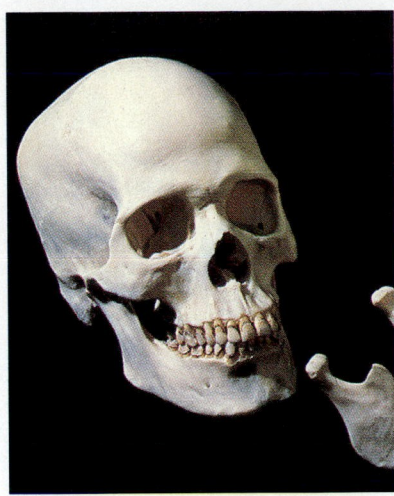

**FIGURE 31–6** A human skull, showing the teeth. Starting from the center front and working back, there are two chisel-shaped incisors, a bluntly pointed canine, two premolars, and two or three molars. (The third molars, "wisdom teeth," never grow in some people.) (Biophoto Associates)

Besides taking in food, the mouth begins the process of mechanical breakdown, using teeth of various shapes and sizes (Figure 31–6). The **incisors,** chisel-like teeth in the front of the mouth, cut bite-size pieces of food from a larger portion. The tongue pushes the food back to the **molars,** the millstones of the mouth, which mash the food into small pieces. Meanwhile, salivary glands secrete **saliva,** a fluid containing mucus and a starch-digesting enzyme. Saliva enters the mouth through ducts. It moistens and lubricates the food, and sticks it together in a bolus (ball) for swallowing.

As a bolus of food is swallowed, it passes through the pharynx, gliding over the epiglottis, a sort of trap door that prevents food from entering the trachea (windpipe), where it could cut off the air supply to the lungs. Food then drops into the **esophagus,** a long tube with thin muscular walls.

The **stomach** serves several functions. It is the widest part of the digestive tract, in keeping with its function as a holding chamber that stores food and releases it into the intestine in small servings. Having a stomach enables us to feed discontinuously and do other things between meals. The stomach's muscular walls churn the contents around. Glands in the stomach wall also release digestive enzymes and acid that kills most microorganisms that have been ingested.

From the stomach, food is released in spurts into the **duodenum,** the first part of the **small intestine** (Figure 31–7). A **sphincter,** a circular muscle, closes the stomach off from the intestine except when it relaxes briefly to permit a small amount of food to squirt through, propelled by the stomach's muscular contractions.

In the intestine, more digestive enzymes are added. Some are secreted by the lining of the intestine itself and some by the **pancreas,** which empties its digestive enzymes into the duodenum by way of a duct. Bile, produced in the liver and stored in the gall bladder, also enters the duodenum through a duct. Digestion continues in the small intestine, and nutrients are absorbed through its lining.

The leftovers pass into the **colon,** or **large intestine,** where millions of bacteria live and work. The large intestine absorbs water, minerals, and vitamin K (produced by intestinal bacteria) and pushes the remaining fecal matter into the **rectum,** where it is held until it is voided. Defecation, or expulsion of the feces from the body, depends on contraction of the walls of the rectum and relaxation of the anal sphincter, another circular muscle at the very end of the digestive tract.

## The Mammalian Liver

The liver's main function is to control the level of many substances in the blood, starting with substances absorbed in the intestine. Whereas most blood vessels empty into the larger vessels returning blood to the heart, vessels from the intestine lead to the **hepatic portal vein,** which carries blood to the liver. Before food molecules in the blood travel to the heart and thence to the rest of the body, the liver may change their concentration and chemical structure. For instance, the liver removes glucose from the blood under the influence of the hormone insulin and stores it as glycogen. When the level of glucose in the blood falls, the hormone glucagon causes the liver to break down glycogen and release glucose into the blood.

One vital role of the liver is detoxifying otherwise poisonous substances. The liver also synthesizes many blood proteins, such as albumins (Table 31–6). In addition, the liver converts nitrogenous wastes into urea, which can be excreted by the kidneys. With the kidneys, the liver is vital in regulating what the blood contains when it reaches all the other organs of the body. Because the liver is the body's major organ for making all these biochemical adjustments, severe liver damage or loss of the liver is rapidly fatal.

The liver also acts as a gland, supplying bile to the intestine. In adult vertebrates, this function is a minor one, but the liver originally arose as a digestive gland in lower chordates. Throughout the liver, a network of tubes collects **bile**—a solution of salts, cholesterol, fatty acids, and bilirubin (made when hemoglobin from red blood cells is broken down in the liver). Bile accumulates in the **gall blad-**

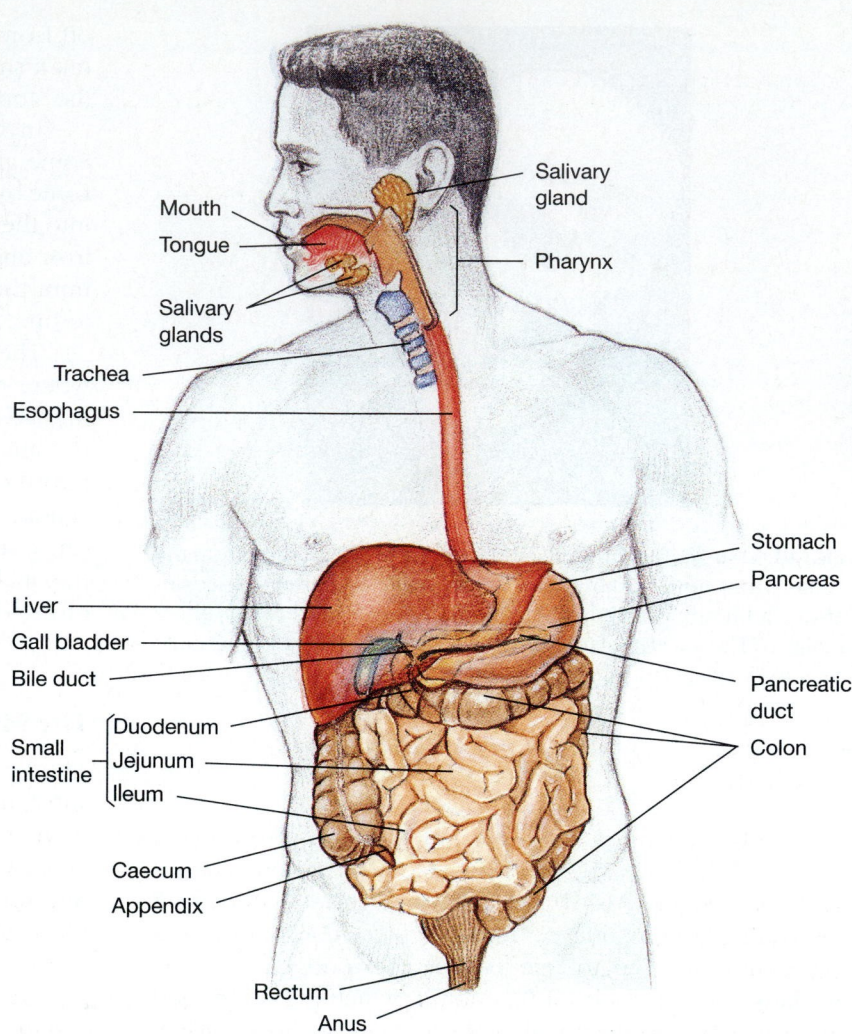

Salivary gland

Mouth

Tongue

Pharynx

Salivary glands

Trachea

Esophagus

Liver

Gall bladder

Bile duct

Small intestine { Duodenum  Jejunum  Ileum

Caecum

Appendix

Stomach

Pancreas

Pancreatic duct

Colon

Rectum

Anus

**FIGURE 31–7** The human digestive tract (color) and associated structures: the salivary glands, liver, gall bladder, and pancreas.

**der,** which empties into the small intestine by way of a duct. (Most gall stones form when cholesterol precipitates out of the bile in the gall bladder, but some are composed

of bilirubin.) Bile has two functions in the intestine. First, it acts as a detergent or emulsifier, breaking fat into small globules that can be attacked by digestive enzymes. Second, and more important, bile salts aid in the absorption of lipids; removal of the gall bladder sometimes causes temporary difficulty with lipid absorption.

**TABLE 31–6**

**Functions of the Liver in Regulation of Blood Composition**

Blood glucose ⟷ Liver glycogen

Amino acids
  Uptake for manufacture of blood proteins
    Clotting proteins
    Albumins
  Excess deaminated

Heme groups from hemoglobin of old red blood cells
  Broken down in liver
  Iron conserved
  Porphyrin rings excreted in bile

## Human Digestive Enzymes

Digestive enzymes are **hydrolases,** enzymes that break substances down by adding water molecules into the bonds between monomers. Digestive enzymes are produced by ductless glands lining the stomach and small intestine and by the pancreas and salivary glands, which are connected to the digestive tract by their ducts (Table 31–7). In addition, the entire gut is lined by millions of ductless mucous glands secreting mucus, which lubricates the food and prevents it from injuring the gut by friction.

**T A B L E   3 1 – 7**

## Some Mammalian Digestive Enzymes

| Origin | Enzyme | Hydrolytic Action: Substrate → Product |
|---|---|---|
| Salivary glands | α-Amylase | Starch, glycogen → maltose, oligosaccharides |
| Stomach | Pepsin (endopeptidase) | Proteins → polypeptides |
| Pancreas | Lipase | Triacylglycerols → fatty acids, monoacylglycerols, glycerol |
| | α-Amylase | As in saliva |
| | Trypsin } Endopeptidases<br>Chymotrypsin | Proteins → peptides |
| | Carboxypeptidase A and B (exopeptidases) | Short peptides → amino acids from carboxyl terminal |
| | Ribonuclease | RNA → nucleotides |
| | Deoxyribonuclease | DNA → nucleotides |
| Small intestine (lining) | Sucrase | Sucrose ⎫ |
| | Maltase | Maltose ⎬→ monosaccharides |
| | Lactase* | Lactose ⎭ |
| | Enteropeptidase | Trypsinogen → trypsin |
| | Aminopeptidases (exopeptidases) | Short peptides → amino acids from amino terminal |
| | Alkaline phosphatase | Compounds containing phosphate |

*Lactase digests lactose (milk sugar) in young animals. It is not found in most human adults except whites of European origin and some African cattle-herding tribes.

***Salivary Glands***   Each day the three pairs of salivary glands discharge more than one liter of saliva into their ducts, which lead to the mouth. Saliva contains an α-amylase. This is a **glycosidase,** an enzyme that hydrolyzes the glycosidic linkages that join monosaccharides together in starch. However, saliva is mainly mucus, important as lubrication. Without saliva, it is extremely difficult to swallow food, even with plenty of water to wash it down.

***The Stomach***   When the stomach contains food, some cells in its lining secrete hydrogen ions, and others secrete chloride ions. This produces a fluid containing hydrochloric acid (HCl) with a remarkably low pH of less than 2.0. Still other stomach cells secrete **pepsin,** a blanket name for a number of **proteases,** enzymes that hydrolyze peptide bonds in proteins. Only vertebrates produce pepsin, which works only at a very low pH. One of pepsin's important functions is to digest collagen, a major constituent of fibrous tissue in meat, which is soluble at the low pH in the stomach.

It is remarkable that the stomach does not digest itself, because the hydrochloric acid and pepsin it secretes can digest most flesh. The stomach protects itself by secreting acid- and enzyme-proof mucus, which coats the stomach wall. Even so, a stomach cell's life is short. About half a million cells die each minute, and all the cells lining the stomach are replaced every three days. Ulcers, often caused by the bacterium *Helicobacter pylori,* form when the stomach does not secrete enough mucus to protect itself from its own acid.

***The Pancreas***   Most digestive enzymes are produced in the pancreas and empty into the small intestine close to the stomach. Because pancreatic enzymes work best at a pH of 7 to 8, the first role of the pancreas is to secrete a concentrated solution of sodium bicarbonate, a buffer that neutralizes the hydrochloric acid entering the intestine from the stomach. Pancreatic juice also contains enzymes that hydrolyze all three types of macronutrient.

The pancreas synthesizes proteases in inactive forms. Occasionally this system breaks down, and the pancreas is digested by its own enzymes within a few hours in an attack of acute pancreatitis, which is usually fatal. Normally, the inactive enzymes become active only after they reach the intestine. Here, the inactive precursor **trypsinogen** is converted into the active protease **trypsin** by the enzyme **enteropeptidase** (enterokinase), secreted by the intestine. Trypsin itself activates the other pancreatic enzymes.

Proteases produced by the pancreas can be divided into **exopeptidases,** which cleave peptide bonds to free the terminal amino acids in a peptide, and **endopeptidases,** which cleave peptide bonds in the interior of a peptide chain. Endopeptidases start protein digestion by breaking a long polypeptide into shorter chains. Then exopeptidases attack the ends of short chains and break off amino acids one by one.

Cells lining the small intestine produce other digestive enzymes. Some of these stay inside the cell, where they digest small molecules absorbed from the intestinal lumen (the hollow inside a tube). Others are integral parts of the membrane lining the lumen. Like stomach cells, the cells lining the intestine are protected from digestion by mucus, but even so the enzymes do considerable damage, and the cells live only two or three days. Then they slough into the lumen, where their enzymes continue to operate. Intestinal enzymes digest mainly small molecules, many of them produced by stomach or pancreatic enzymes hydrolyzing larger food molecules.

*Control of Enzyme Secretion*    The secretion of digestive enzymes is regulated by the nervous system and by hormones. The hormonal interactions are complex. The hormone **gastrin** is secreted by cells in the stomach lining in response to either of two stimuli: peptides in the stomach or stretching of the stomach wall by food after a meal. Gastrin stimulates other stomach cells to secrete acid and pepsin. These digest the peptides and permit the stomach to get smaller, thereby removing the stimuli (peptides and stretching) that cause gastrin to be released. Hence, gastrin secretion stops.

Later, in response to acidic food entering the duodenum, cells around the gut lumen release another hormone, **secretin,** which carries messages to three other organs: it stimulates release of sodium bicarbonate by the pancreas, neutralizing stomach acid; it inhibits acid production by the stomach itself; and it causes the release of bile from the gall bladder in the liver. Meanwhile, another intestinal hormone, the peptide **cholecystokinin,** stimulates the pancreas to release digestive enzymes. A third intestinal hormone, **gastric inhibitory peptide,** inhibits peristalsis and secretion of stomach acid (Figure 31–8). These mechanisms ensure that digestive enzymes from the stomach and pancreas are present when needed, but not at other times.

## Absorption of Food

In a sense, the lumen of the digestive tract is a continuation of the external environment because its contents lie outside the body's cells. Only after digestion do small molecules released from the food pass through plasma membranes into living cells. So we can say that our food does not enter our bodies until some time after a meal.

Anything we swallow reaches the stomach quickly. However, the stomach absorbs only lipid-soluble substances such as alcohol and a few drugs. Nearly everything we ingest is absorbed into the body in the small intestine.

The small intestine, about 3 meters long, is highly adapted for absorption. Its lining is thrown into circular folds, and the surface of the lining forms finger-like exten-

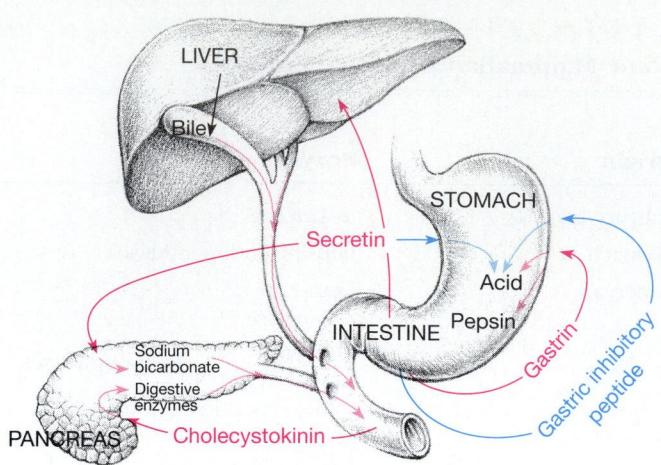

**FIGURE 31–8**    Hormonal control of some digestive secretions in humans. Red arrows show stimulatory pathways. Blue arrows indicate inhibition.

sions called **villi,** which make the intestinal lining look like velvet. The surface of each villus, in turn, is covered with a "brush border" of tiny **microvilli,** formed by extensions of the plasma membranes of cells lining the lumen. These folds on folds on folds enormously increase the surface area available to take up food molecules (Figure 31–9).

In this surface area is an array of transport molecules, which absorb food molecules from the lumen selectively. Food is prevented from entering the body indiscriminately by the tight junctions that seal the edges of adjacent cells to each other (see Figure 4–17). The absorptive layer of a villus is only one cell thick. Inside it lies a network of blood vessels and a lacteal, a vessel of the body's lymphatic drainage system (Section 33–E). Food absorbed from the intestinal lumen is passed on to these vessels, whence it is transported to other parts of the body.

Many molecules move out of the lumen by active transport or facilitated diffusion. Glucose and amino acids are taken up by active transport carriers powered by the high concentration of sodium ions from pancreatic juice and bile. The intestinal cells can also absorb di- and tripeptides and digest them intracellularly before passing them on to the bloodstream.

Long-chain fatty acid molecules enter cells of the intestinal lining by diffusing through the plasma membrane. Here they are converted mainly into fats and are then voided by exocytosis from the far end of the cell as tiny fat droplets called **chylomicrons.** Chylomicrons are coated by a layer of protein, which makes them water-soluble. Chylomicrons travel by way of lymphatic vessels that drain into blood vessels near the heart. Small fatty acids are absorbed directly into the bloodstream near the intestine without being converted into triacylglycerols.

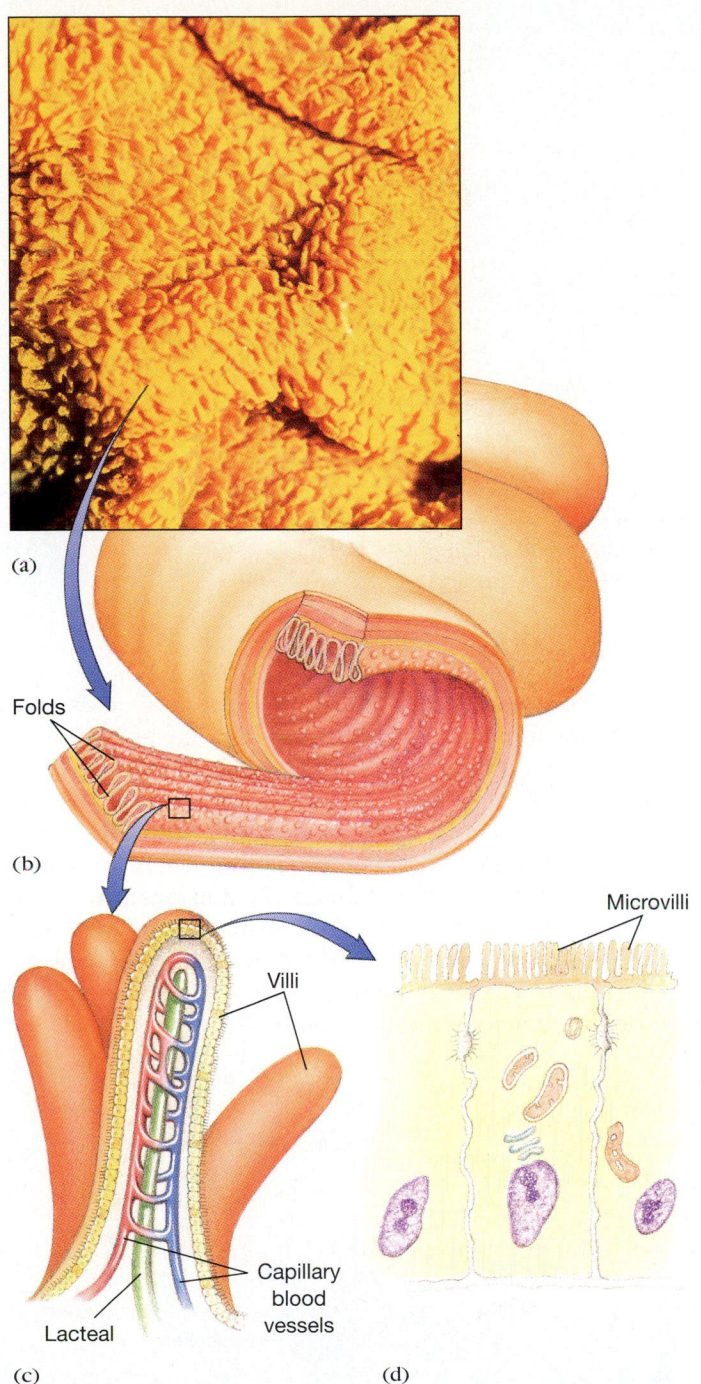

Microvilli

Villi

Capillary blood vessels

Lacteal

Folds

(a)

(b)

(c)    (d)

**FIGURE 31–9**   The large surface area of the small intestine. (a, b) The folded lining of the intestine. (c) Finger-like villi containing blood capillaries and lymphatic vessels line the folds. (d) The plasma membranes of cells covering the villi are folded into microvilli, further increasing the absorptive area facing the lumen. (a, Biophoto Associates)

***Absorption of Minerals and Vitamins***   In addition to all these food molecules, about 10 liters of fluid must be absorbed from the digestive tract into the blood every day.

Of this, about 1.5 liters consists of fluid we have drunk, and about 8.5 liters consists of the fluid secreted into the gut with digestive enzymes and mucus. Most of these 10 liters are absorbed in the small intestine.

Water is absorbed indirectly as a result of sodium transport. Sodium is actively transported out of the intestinal cells, not back into the lumen but into the extracellular fluid in the interior of the villi. Negatively charged chloride ions follow passively, attracted by the positively charged sodium ions. The accumulation of sodium and chloride ions lowers the osmotic potential outside the intestine compared with that in the lumen, and water moves from the lumen to the extracellular body fluid by osmosis.

Sodium, small amounts of other ions, vitamin K, and water are the main substances absorbed in the large intestine. This absorption ensures that only about 100 ml of water and small amounts of inorganic ions are lost in the feces every day. The feces are about three-fourths water and one-fourth solid matter. Of the solid matter, about 30% is made up of bacteria (from the intestine where they live), 15% is inorganic matter, 3% protein, 20% fat, and 30% undigested roughage.

## 31–D   FEEDING ADAPTATIONS

Most aquatic herbivores are filter feeders, straining tiny, floating plants out of large volumes of water. The filtering system usually traps the plants in mucus, which is then moved to the gut by cilia. Aquatic plants are supported by the surrounding water, so they have little tough supporting tissue and are easily digested.

Animals that eat terrestrial plants have special nutritional problems because these plants have tougher cell walls than aquatic plants. Breaking these thicker cell walls and digesting cellulose are major problems for terrestrial herbivores.

Most herbivorous insects have mouthparts adapted to breaking or piercing cell walls, and they actually feed on the cytoplasm. One reason herbivorous insects such as locusts and grasshoppers damage crops so heavily is that they use only a fraction of the food they eat. Cell walls and starch grains generally appear in the feces unchanged. Termites are an exception: their guts contain protistan symbionts that secrete cellulose-digesting enzymes. As a result, termites can live on wood, which is made up entirely of very tough cell walls—a food source unavailable to most other insects.

### Mammalian Herbivores

Herbivorous mammals have greatly enlarged molars that grind plant cell walls (Figure 31–10). But their most effective adaptation is a collection of symbiotic microorganisms

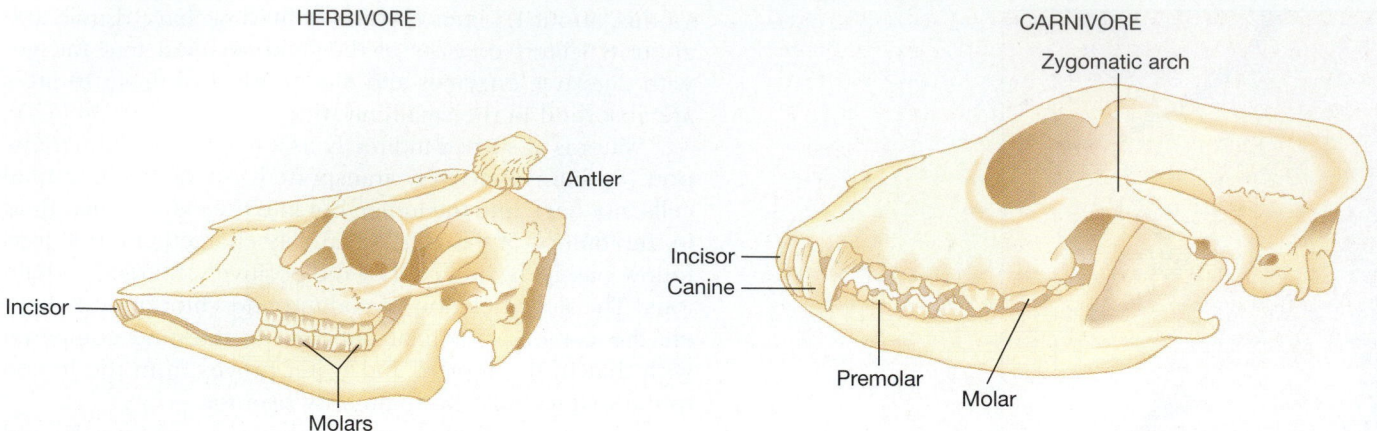

HERBIVORE

Antler

Incisor

Molars

CARNIVORE

Zygomatic arch

Incisor

Canine

Premolar

Molar

**FIGURE 31–10**    Teeth of a mammalian herbivore and carnivore. The herbivore's incisors are specialized to clip vegetation. The molars and premolars are flattened, and the jaws move sideways to grind plant food. The carnivore's incisors and canines are specialized for cutting and pulling. The saw-like molars and premolars are used to slice and tear as the jaws move mainly up and down.

in the gut, including cellulose-fermenting bacteria. These make various polysaccharide-splitting glycosidases and so can use polysaccharides as food under the anaerobic conditions in the stomach, breaking down food that their host cannot digest. Because the food must be exposed to the bacteria for a fairly long time, herbivores tend to have longer intestines than do omnivores or carnivores of similar size. (Some bacterial fermentation of food probably occurs in the intestines of all terrestrial vertebrates, including omnivores such as pigs, rats, and humans.)

Many vertebrate herbivores house their symbionts in a caecum, a blind sac set off to one side of the gut at the junction of the small and large intestines and used as a fermentation chamber. In rats and horses, the caecum and colon are enlarged to house fermentation. In humans, the caecum is much smaller, and part of it has become reduced to a vestigial structure, the appendix.

***Ruminants***    Digestion aided by symbiotic bacteria has reached its greatest complexity in the **ruminants.** In these mammals, parts of the esophagus and stomach form a large sac, the rumen, containing microorganisms in an alkaline fluid. Most ruminants are artiodactyls, the even-toed ungulates, including sheep, cattle, and deer. Ruminant digestion has also evolved in unrelated animals, including some marsupials, colobus monkeys, and sloths, and even in the hoatzin, an unusual leaf-eating bird.

A ruminant such as a sheep secretes more than 10 liters of saliva per day. This maintains an alkaline pH (about 8.5) in the first three of the stomach's four digestive chambers (Figure 31–11). Food descends first into the **rumen** and the **reticulum.** Here it is fermented by anaerobic bacteria and protozoans, making a bubbling noise that you can hear by putting your ear against the animal's side. The microor-

ganisms feed on sugars in the food, and fatty acids produced during fermentation are absorbed by the host through the rumen. Unchewed food floats to the surface and is regurgitated into the mouth as "cud," which is chewed some more. On its second descent the food bypasses the rumen and reticulum and enters the **omasum,** where it is mechanically churned. Finally it enters the **abomasum,** which corresponds to the stomach of other mammals. Here acid and digestive enzymes are secreted, and the ruminant recaptures many of the nutrients its symbionts have used by digesting the microorganisms themselves!

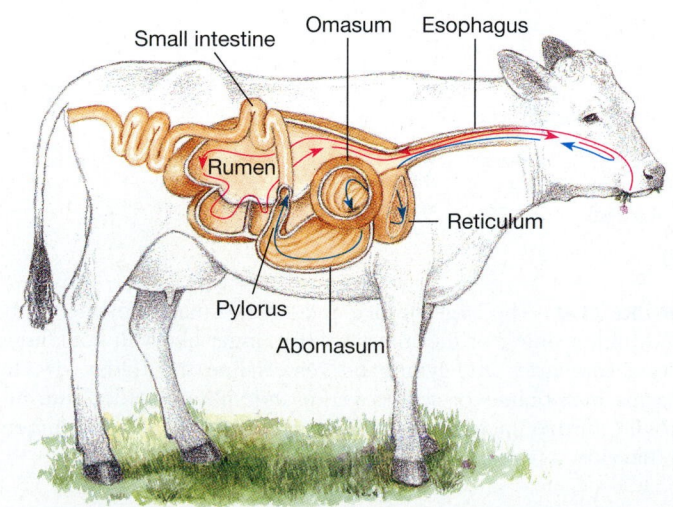

Small intestine    Omasum    Esophagus

Rumen

Reticulum

Pylorus

Abomasum

**FIGURE 31–11**    The digestive system of a cow with the wall cut away to show the four compartments in the stomach. Arrows show the direction in which the food travels—not straight through the stomach but also sideways and even backwards.

There are many advantages to having gut symbionts. Some microorganisms can synthesize amino acids using urea and ammonia; animal enzymes cannot do this. Thus, microorganisms are valuable when the diet is low in protein. In addition, symbionts synthesize many vitamins that can be used by the host. Herbivores such as baboons, which do not have ruminant digestion, must eat meat occasionally (grasshoppers, snakes, baby monkeys) to replenish their B vitamins. A ruminant needs few dietary vitamins except vitamin A (which can be made from β-carotene, a pigment common in plants) and vitamin D.

## Mammalian Carnivores

Herbivores eat food that is hard to digest. Carnivores, on the other hand, find their main nutritional problems in catching their food in the first place. Once it is caught, their worries are pretty well over. Since the food of carnivores consists of other animals, its composition is very similar to that of their own bodies, with little of the waste that results from eating plants with thick cell walls and a high water content.

Carnivorous mammals have teeth adapted to killing their prey and shredding it into bite-size pieces. The canine teeth are often elongated into formidable fangs that can inflict swift and extensive damage on the prey. The muscles and bones of the skull are powerful, enabling them to subdue their meals without damaging their jaws. The molars are modified so that they resemble the blades of short saws, adapted to shredding meat into chunks that can be swallowed. Extensive chewing is not necessary, since there are no thick cell walls to break. The strong stomach acid and powerful proteases make short work of the food, and the intestine is short compared with that of herbivores and omnivores of the same size.

## Adaptations of Birds

A bird's jaws are composed of a **beak** or **bill** made up of bone and **keratin,** a protein also found in hair and fingernails. The beak size and shape depend on the feeding habits of the species (Figure 31–12). Although their reptilian ancestors had teeth, modern birds do not. Instead, birds that eat hard food grind it in a muscular gizzard, the hindmost part of the stomach. The gizzard usually contains stones that the bird picks up and swallows. This arrangement moves weight from the head (where teeth would have been) closer to the center of gravity, an adaptation that makes flying more efficient.

The jaw muscles are particularly strong in carnivorous birds and in birds that eat large, hard seeds. Carnivores may also have powerful digestive enzymes capable of breaking down even "the bones and the beak" of their prey, although cormorants, hawks, and owls regurgitate bones, fur, scales, and feathers in characteristic pellets.

A special organ found particularly in seed-eating birds is the crop, a storage sac between the esophagus and the stomach. Birds use a lot of energy in flying and in maintaining a high body temperature. Storing food in the crop ensures an almost continuous supply of food to the stom-

**FIGURE 31–12**    Bird beaks. (a) The narrow, tubular beak of a hummingbird, adapted to sucking nectar from the nectaries deep inside a flower. (b) A parrot's bill is strong and curved, adapted to breaking open even large, hard-coated seeds.    (a, Robert A. Tyrrell/Oxford Scientific Films/Animals Animals)

ach, and so birds do not have to spend all their time feeding. Nevertheless, food takes only about an hour to pass through the body of a young bluejay, and a thrush fed blackberries will void the seeds less than an hour later.

Birds generally use their food more efficiently than do mammals. A three-week-old stork converts about one third of the mass of the fish it eats into stork. A young mammal converts only about one tenth of its food into body mass.

## 31–E   STORED FOOD

The body's carbohydrate stores (glycogen in the liver and muscles) would supply its energy needs for only about 12 hours if they were used alone. However, a human of normal weight can usually survive without food for about six weeks, using fat reserves for energy (Figure 31–13). A hibernating animal, with a lowered temperature and metabolic rate, can survive for months on the fat reserves that it built up by eating ravenously during the autumn.

Because a given weight of fat provides about twice as many calories as the same weight of carbohydrate or protein, energy is stored most compactly in the form of fat. Fat is stored in the fat cells of **adipose tissue,** a storage tissue found under the skin, between muscle fibers, in the breasts and buttocks, between folds of the intestines in the abdomen, and elsewhere in the body. Excess carbohydrate or protein in the body is converted into triacylglycerols and stored in the fat cells of adipose tissue. Fat is constantly exchanged between the bloodstream and adipose tissue: every molecule of fat in adipose tissue is replaced about every three weeks.

Muscular activity uses up more than half of human energy. Most goes to maintain our body temperature as the contraction of tiny muscles that we don't even notice releases heat and warms the body. This is why we use up more calories in winter, especially in a cold house. Most of our other daily calorie usage goes for functions such as urine formation and nervous system activity, whose energy requirements vary little from day to day.

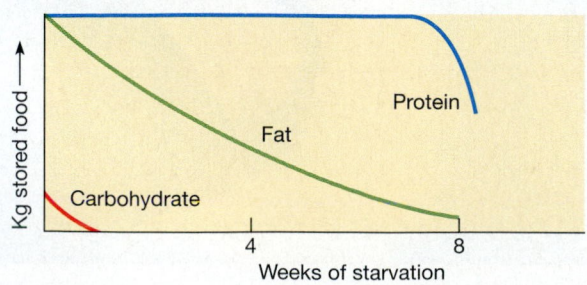

**FIGURE 31–13**   The fate of stored food in a starving human being whose initial body weight was 15% fat. (A human being of average weight takes weeks to die from starvation, although death from thirst occurs in a few days.)

## 31–F   REGULATION OF FEEDING

Feeding is governed by long-term and short-term controls. Long-term regulation ensures that enough stored food is maintained in the body. It can be altered by variations in the levels of hormones that ensure, for instance, that an animal builds up its fat reserves before its hungry young are born or hatched and before hibernation. Short-term regulation ensures that an animal eats regularly on a day-to-day basis, so that food passes through the gut more or less continuously.

The control of feeding appears to involve the interaction of a number of hormones and other chemicals produced in the gut and brain. These chemicals start or stop feeding and control the selection of food.

Habit has a major effect on short-term feeding. People accustomed to three meals a day become hungry and experience muscular contractions ("hunger pangs") of the stomach if they miss one meal. However, people living alone or working intensely so that they are not reminded of mealtimes frequently miss meals without feeling hunger. After a large meal, we stop eating because of several factors, including distension of the stomach and increased production of several chemicals by the feeding centers in the brain. On the other hand, people and animals whose food never reaches the stomach (because a tube has been inserted into the esophagus to divert food to the outside of the body instead of the stomach) also stop eating after awhile. This suggests that the quantity of food that has passed the mouth is monitored in some unknown way.

Recommended weights in life insurance tables were adjusted upward after studies showed that many people as much as 20% heavier than their "ideal" weights (by 1960s standards) lived longer than those of "ideal" weight. Research in the 1980s also showed that healthy people increase in weight as they age. People vary considerably in how much body fat they carry or should carry. If you have struggled to lose weight over and over and always seem to return to a weight somewhat heavier than your ideal weight, it is very likely that your long-term regulator is "set" to the higher weight and that, as a result, you are no less healthy than your slimmer friends. Those of us in that situation should probably give up fighting the battle of the bulge, exercise to maintain our health, and hope that seventeenth-century fashions in bodies will return during our lifetimes.

Why an animal eats what it does is even more complicated. For carnivores, things are reasonably simple because carnivores eat food that is generally nutritious and nontoxic. Carnivores rapidly learn to avoid food that tastes unpleasant, as you realize if you have ever seen a cat attack a toad.

Herbivores face a bigger problem because many plants contain toxic chemicals, and even more (including some of our common foods) are toxic in large amounts. Furthermore, few herbivores can eat just one kind of plant because no single plant species contains the complete mix of nutri-

**FIGURE 31–14**  A sheep feeding. Sheep are widely used in studies of feeding behavior and physiology in herbivores.

ents that animals need. A sheep presented with a field of grasses, most of which are new to it, adds only one new species to its diet at a time and eats very little of that (Figure 31–14). Presumably, this gives the long-term learning system time to "tell" the sheep whether the new plant is a good thing or not. Using this system, a herbivore can work up to a varied and nutritious diet.

Small, slow-moving herbivores, such as caterpillars, may eat only one plant in their lives—the one where they hatch from the egg. Special arrangements make up for the nutrient deficiencies of the food plant. For example, the caterpillar of a tiger swallowtail butterfly cannot obtain all the essential amino acids and sodium it requires from its diet of cherry tree leaves. However, the caterpillar hatches with extra supplies of these nutrients, deposited in the egg by its mother, who obtained them from plant nectar as an adult.

## Specific Appetites

One of the least-understood aspects of feeding is the development of an appetite for a specific substance. People who live on diets of fish, rice, or fruit, which are all very low in fat, develop a craving for fats and treat them as a delicacy. It appears that the craving for fat is due to the brain peptide galanin, produced by a group of appetite-controlling cells in the brain's hypothalamus. Other brain chemicals stimulate appetites for carbohydrates.

Appetites for minerals are also common. Children with calcium deficiencies have been known to eat the plaster from walls (which is mainly $CaCO_3$). Female red-cockaded woodpeckers seek out pieces of bone or shells during the breeding season, when they need extra calcium to make shells as they lay their eggs. Laborers in the tropics fre-

quently develop salt deficiency because of the volume of sweat they produce. They drink salt water or salty beer and find it delicious, although it tastes repulsive to anyone who is not short of sodium.

Sodium appetite has been the subject of considerable research. Many herbivores experience a constant shortage of sodium because many plants contain very little of this essential animal nutrient. Indeed, the size of an animal population may actually be limited by the availability of sodium. Isle Royale in Lake Superior has a population of moose, and measurements of sodium eaten and excreted show that when the population rises to its peak of about 1200 animals, the moose are using every speck of available sodium. The island could not support any more of these large herbivores.

Animals increase the sodium content of their diet in various ways. Aquatic plants generally contain more sodium than do the nearby land plants, and moose or elk knee-deep in a swamp, grazing on water weeds, are a common sight in the wilds of Canada. Colobus monkeys in India do the same thing, and African elephants will travel many miles from their usual feeding grounds to salt springs. "Salt licks" are areas where the soil contains more sodium than it does elsewhere, attracting many different herbivores in such places as Yellowstone Park and many game parks in Africa. Any provident farmer sets out cakes of salt for horses and cows.

Plainly, specific appetites are valuable because they lead animals to seek and eat food or soil containing the nutrients they need. Somehow, the deficiency of a particular nutrient changes the reaction of the sense organs or the brain to potential food so that items containing the missing nutrient taste or look much more appetizing to an animal that is short of that substance.

Scurvy is a disease that is hardly ever seen these days. It plagued sailors on long voyages before the end of the eighteenth century.

In the 1740s, Englishman George Anson sailed around the world in *H. M. S. Centurion* (Figure 31–A). He captured a fortune from Spanish possessions on the West Coast of America and from treasure galleons in the Pacific. Richard Walter and Pascoe Thomas wrote the story of this voyage, during which more than half the 400 men on board died of scurvy. Here is their account of the disease as the ship battled gales off Cape Horn, the southernmost tip of South America:

And now as it were to add the finishing stroke to our misfortunes, people began to be afflicted with the most terrible, obstinate and, at sea, incurable disease, the scurvy, which quickly made a most dreadful havoc among us, beginning at first to carry off two or three a day, but soon carrying off eight or ten; and as most of the living were very ill, and the little remainder who preserved their healths better, in a manner quite worn out with incessant labour, I have sometimes seen four or five dead bodies sewn up in their hammocks, others not, washing about the decks, for want of help to bury them in the sea.

Its symptoms are inconstant and its progress and effects irregular; for scarcely any two persons have the same complaints. . . . The common appearances are large discoloured spots over the whole surface of the body, swelled legs, putrid gums, and above all, an extraordinary lassitude of the whole body, especially after any exercise, however inconsiderable. And this lassitude at last degenerates into a proneness to swoon on the least exertion of strength, or even the least motion. . . . At other times, the whole body but more especially the legs, were subject to ulcers of the worst kind, attended with rotten bones, and such a luxuriancy of fungous flesh as yielded to no remedy.

A most extraordinary circumstance is that the scars of wounds which had been for many years healed, were forced open again by this violent distemper. One of the invalids on board the *Centurion,* who had been wounded about fifty years before at the Battle of the Boyne was cured soon after, and had continued well, yet in his being attacked by scurvy, his wounds broke out afresh, and appeared as if they had never been healed: Nay what is still more astonishing, the callous of a broken bone was found to be hereby dissolved, and the fracture seemed as it had never been consolidated. Indeed, the effects of this disease were in almost every instance wonderful; it was no uncommon thing for those who were able to walk the deck and do some kind of duty, to drop down dead in an instant on any endeavour to act with their utmost vigour.

Sad to relate, a cure for scurvy was already known. Various people had shown that scurvy was a nutritional deficiency that could be cured by eating raw fruit and vegetables. Since fresh vegetables would not survive long voyages in the days before refrigeration, this was not a very helpful observation, until it was found that acid fruit juices retained their anti-scurvy properties for long periods. By the 1820s, the British were known as "limeys" because their sailors consumed a daily ration of lime juice, but it was almost a century after Anson's voyage before this habit became widespread.

Scurvy is caused by dietary deficiency of ascorbic acid, vitamin C, which most vertebrates (but not humans or monkeys) can synthesize for themselves. Ascorbic acid is necessary to the synthesis of the amino acid hydroxyproline, which is required for connective tissue synthesis and repair. Without it, the body's connective tissue breaks down; hence the loosening of teeth in degenerating gums and the breakdown of scar tissue and blood vessels. Sufferers from scurvy usually died as a result of hemorrhage from broken blood vessels.

**FIGURE 31–A**  *H. M. S. Centurion*

## SUMMARY

An animal's food contains both materials and energy needed to build and run its body. Before these can be used, the body must break down large molecules into their constituent monomers, absorb them into the body, and distribute them to the cells.

1. Because animals are heterotrophs, their diet must contain the organic and inorganic substances they need for metabolism, growth, and energy. Animals obtain macronutrients (fats, carbohydrates, and proteins) and micronutrients (vitamins and minerals) in their food.

2. Lack of enough of any nutrient in the diet causes deficiency diseases such as marasmus, when the diet is low

in calories and essential amino acids, and scurvy, beriberi, pellagra, and anemia, caused by deficiencies of various micronutrients.

3. The function of digestion is to break food down into small molecules that can be absorbed into the body from the gut. In the vertebrates, digestive enzymes are synthesized in the salivary glands, pancreas, and lining of the stomach and small intestine.

4. Digested food is absorbed into the extracellular fluid and blood by diffusion, facilitated diffusion, and active transport through the enormous surface area of the small intestine.

5. Many animals, particularly herbivores, harbor symbiotic microorganisms in the alimentary canal. These symbionts secrete digestive enzymes and sometimes vitamins that the host cannot synthesize.

6. The liver plays a major role in controlling the fate of newly absorbed food molecules. It stores excess glucose as glycogen, synthesizes many blood proteins, and converts nitrogenous and other wastes into a form that can be excreted by the kidneys.

7. Excess carbohydrate or protein in the diet is converted into triacylglycerols and stored in the fat cells of adipose tissue.

8. Feeding is regulated by long-term and short-term control mechanisms that are not well understood. These controls ensure that the alimentary canal is efficiently occupied most of the time and that the animal maintains its body reserve of fat. All animals have poorly understood regulatory systems that control what, as well as how much, they eat.

## Self-Quiz

Matching: For each numbered phrase below choose the letter of the correct class of nutrient on the right. More than one letter may be correct; choose all that apply.

| | | |
|---|---|---|
| _____ 1. Inorganic nutrients | a. protein | |
| _____ 2. Macronutrient that cannot be stored in the body | b. carbohydrate | |
| | c. fat | |
| _____ 3. Absorbed in large intestine | d. water-soluble vitamins | |
| _____ 4. May be the source of energy for the body's metabolism | e. fat-soluble vitamins | |
| _____ 5. Source of material for cell membranes | f. minerals | |
| _____ 6. Digested by enzyme in saliva | | |
| _____ 7. Coenzymes for metabolic enzymes | | |

8. The main advantage of having a digestive tract with a mouth and anus is:
   a. it permits different parts of the gut to become specialized to perform different parts of the digestive process in turn
   b. it permits an animal without teeth to have a means of grinding its food
   c. it permits animals to eat a great deal at once and digest it while doing something else
   d. it permits animals to eat larger organisms as food
   e. it permits animals to eat food in larger chunks

9. In humans, digestion of food is completed in the:
   a. mouth
   b. stomach
   c. small intestine
   d. large intestine
   e. rectum

10. In humans, protein digestion is carried out by enzymes secreted by the:
    a. stomach, pancreas, and salivary glands
    b. liver, salivary glands, pancreas, and small intestine
    c. salivary glands, stomach, pancreas, and small intestine
    d. liver, stomach, pancreas, and small intestine
    e. stomach, small intestine, and pancreas

11. A portion of the stomach that has evolved extremely thickened muscular walls and is quite efficient at grinding hard food is called a(n):
    a. rumen
    b. gizzard
    c. crop
    d. omasum
    e. caecum

12. Which of the following is probably *not* an action of symbiotic microorganisms of the gut?
    a. use of the host's food for their own nutrition
    b. extracellular digestion
    c. respiration
    d. breakdown of substrates that the host cannot digest
    e. manufacture of vitamins needed by the host animal

13. Which of the following is *not* a function of the mammalian liver?
    a. secretion of digestive enzymes for export to the gut
    b. regulation of blood glucose and amino acid content
    c. production of the nitrogenous waste urea
    d. production of plasma proteins for the blood
    e. detoxification of poisonous substances

## Questions for Discussion

1. Herbivores can seldom survive by eating only one species of plant (for example, corn is low in the amino acids tryptophan and lysine; many plants contain too little sodium). This is probably no evolutionary accident. What's in it for the plant?

2. Good manners dictate that you take small bites of your food and chew it thoroughly, but there are also sound biological reasons for such behavior. What are they?

3. Why does it take longer to become hungry after a protein-rich meal than after a meal that is mostly carbohydrate?

4. Trace the fate of a piece of green pepper pizza through the human digestive tract. (Contents of pizza:   crust: carbohydrate and various B vitamins; cheese: protein, fat, calcium, phosphorus; tomato: vitamin C, potassium; pepper: vitamin A, iron, cellulose).

5. Some kinds of stress can upset an animal's normal nutrient balance. For example, infection increases the rate of utilization of vitamin C. How might the organism compensate for this disturbance? Will this invariably change the optimum dietary level?

## Suggested Readings

Gordon, M. S. *Animal Physiology: Principles and Adaptations,* 4th ed. New York: Macmillan Publishing Co., 1982. An excellent textbook on the comparative physiology of different animals.

Scrimshaw, N. S. "Iron deficiency." *Scientific American,* October 1991. A little-known health problem that affects both mental and physical ability.

Uvnäs-Moberg, K. "The gastrointestinal tract in growth and reproduction." *Scientific American,* July 1989. Hormonal regulation of the gut and its changes during pregnancy.

# Gas Exchange in Animals

## OBJECTIVES

*When you have studied this chapter you should be able to:*

1. Distinguish between ventilation and respiration.
2. Discuss the advantages and disadvantages of air and water as respiratory media.
3. Describe the relationship between metabolic rate, size (for endothermic animals), and environmental temperature (for ectothermic animals).
4. List and explain the characteristics of animals that use the general body surface for gas exchange.
5. Compare and contrast lungs and gills with respect to structure, function, and their advantages and disadvantages.
6. Explain how the countercurrent exchange mechanism works in the gills of a bony fish; state the importance of this adaptation to the fish.
7. Explain how the positive-pressure and negative-pressure mechanisms of breathing work, and state the advantages of a negative-pressure breathing mechanism.
8. Describe how birds' respiratory systems differ from mammals' in structure and operation. List three functions of birds' air sacs.

9. Describe the main differences in structure and function between insect tracheal systems and other respiratory systems; explain two ways ventilation of the tracheal system is enhanced in an active insect.
10. (a) State the function of respiratory pigments; (b) draw an oxygen dissociation curve for hemoglobin, label both axes, indicate the parts of the curve that are significant for binding and release of oxygen, and state how these are significant to the organism; (c) state the main factor governing whether hemoglobin binds or releases oxygen; (d) state how pH influences the curve; and (e) predict the change in shape and position of such a curve with a change in the animal's size or environment.
11. Compare and contrast the transport of oxygen and carbon dioxide in the blood.
12. Describe how hemoglobin, other blood proteins, and carbon dioxide buffer the blood.
13. Explain how breathing is regulated in the human body.
14. List two functions of the swimbladder of bony fishes.

---

**F**rom a climber struggling to the top of Mount Everest with an oxygen cylinder, to an invertebrate on a reef in the ocean, every animal is continually exchanging gases with its environment (Figure 32–1). Why do animals need a regular supply of oxygen? In Chapter 7 we saw that living cells obtain energy from the oxidation of food molecules, usually by cellular respiration. Respiration requires oxygen, which must be obtained from the environment, and produces carbon dioxide ($CO_2$), which must be expelled. Molecular oxygen ($O_2$) in high concentrations is dangerous to living tissues because it oxidizes not only food but also the body's own organic molecules. For this reason, little $O_2$ can be stored in the body, and $O_2$ must be obtained continuously from the environment. Our own bodies can survive for weeks without food and for days without water, but only for minutes without $O_2$.

**FIGURE 32–1** Respiratory surfaces. A white nudibranch crawls across a pink sponge in Puget Sound. The feathery dorsal extensions of the body are gills by which the animal exchanges carbon dioxide and oxygen with its environment. (Jeff Rotman)

Small animals can take in $O_2$ (and give off $CO_2$) by diffusion through plasma membranes of cells at the outer surfaces of their bodies. The $O_2$ does not have far to go to reach any cell in the body.

As animal size increases, this means of supplying $O_2$ falls behind the body's demand, for two reasons. First, the surface area available to take up $O_2$ does not increase as fast as the volume of interior cells that use it (see Figure 25–2). Second, the innermost cells are farther from the surface, and the time it takes $O_2$ to reach these cells by diffusion increases with the *square* of the distance. For instance, in the tissues $O_2$ travels an average of 0.1 mm in one second. To go ten times this distance (1 mm) would take a hundred times longer (100 seconds), and to go 10 mm would take 10,000 seconds, or nearly three hours—assuming (wrongly) that other cells along the way did not absorb any of the oxygen! The evolution of larger animals depended on more extensive surface areas for gas exchange, as well as transport systems that move substances rapidly over long distances throughout the body.

This chapter describes the most common arrangements for gas exchange in animals and some of the theoretical considerations that govern their working. In the next chapter we shall examine transport systems, which most animals use to move substances around inside the body.

### KEY CONCEPTS

- Gas exchange between cells and their environment occurs by diffusion through moist plasma membranes.
- In multicellular animals, every cell lies close to a gas exchange medium, which may be the external respiratory medium or extracellular fluid within the body.
- In large or active animals, gas exchange is speeded up by ventilation of a respiratory surface and by transport using respiratory pigments in a circulatory system.

- The main respiratory organs in animals are the body surface, gills, lungs, and tracheae.
- Respiratory pigments take up oxygen at the respiratory surface, where it is plentiful, and release it into the oxygen-poor fluid around the cells that use it.

## 32–A    SUPPLYING OXYGEN

Most of Earth's available $O_2$ is found in the air, which is about 21% $O_2$, and some is dissolved in water. Either water, air, or both may serve as an animal's **respiratory medium,** the immediate source of its $O_2$.

An animal obtains $O_2$ from the respiratory medium by diffusion through the body's thin, moist **respiratory surface,** either the general body surface or a specialized area. This process is **gas exchange.** It is often called respiration because the $O_2$ is used for cellular respiration, but the two are separate processes. Nor should respiration be confused with **ventilation,** the process of moving the respiratory medium so that the respiratory surface is constantly exposed to a fresh supply of $O_2$. **Breathing** is ventilation of a respiratory surface with a particular medium: air.

Ventilation is necessary because gas exchange occurs by the diffusion of $O_2$ and $CO_2$ down their concentration gradients. The steeper the concentration gradients of these gases across the respiratory surface, the faster gases will diffuse between the respiratory medium and the body fluids. At the respiratory surface $O_2$ leaves the respiratory medium, and $CO_2$ builds up. This reduces the concentration gradients of both gases and slows their diffusion through the respiratory surface. Ventilation brings a fresh supply of air or water, which renews the $O_2$ and removes the $CO_2$, permitting gas exchange to continue rapidly.

$O_2$ moves from the environment into the blood partly by diffusion and partly by **facilitated diffusion,** using a carrier protein (Section 4–E). The carrier is cytochrome P450, which speeds up diffusion, allowing the blood to pick up $O_2$ faster. Cytochrome P450 also oxidizes toxic substances, rendering them less harmful to the body. Although the liver is the main organ of detoxification, the lungs also contribute to this activity. The ready supply of $O_2$ at the lung surface makes it an ideal site for oxidizing foreign substances.

In all but the smallest animals, most cells lie some distance from the respiratory surface. These interior cells obtain $O_2$ from their immediate environment, the **extracellular fluid** that bathes every cell. They rely on the blood in the transport system to bring $O_2$ from the respiratory surface to the extracellular fluid and to carry their $CO_2$ waste away (Figure 32–2). This second gas exchange, from the blood to the extracellular fluid and cells, also occurs by diffusion.

## 32–B    PROBLEMS OF GAS EXCHANGE

Animals living in different environments face different problems with gas exchange. For instance, fish extract $O_2$ from water, which contains less $O_2$ than air does. Water pollu-

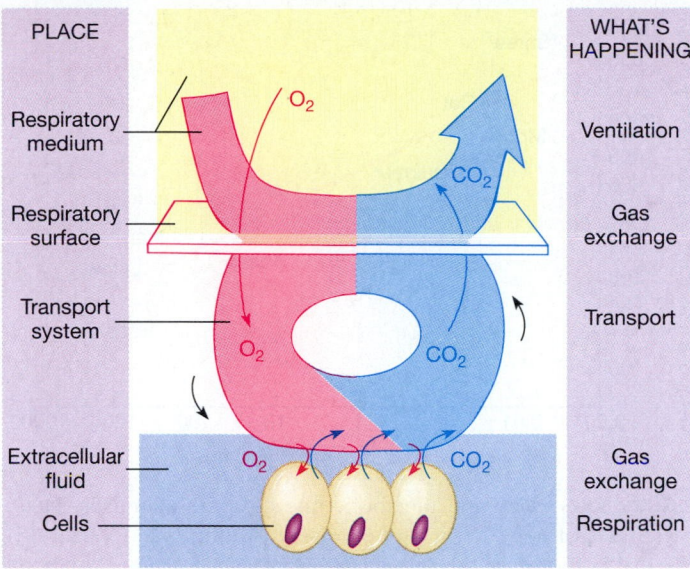

| PLACE | | WHAT'S HAPPENING |
|---|---|---|
| Respiratory medium | $O_2$ / $CO_2$ | Ventilation |
| Respiratory surface | | Gas exchange |
| Transport system | $O_2$ / $CO_2$ | Transport |
| Extracellular fluid | $O_2$ / $CO_2$ | Gas exchange |
| Cells | | Respiration |

**FIGURE 32–2** Summary of gas exchange. Gas is exchanged between the respiratory medium and body fluids at the respiratory surface (top) and between the body fluids and living cells (bottom). Oxygen enters the body fluids from the respiratory medium, whereas carbon dioxide enters the system as it is produced by body cells. $CO_2$ is picked up from the cells by the body fluids and leaves the body at the respiratory surface.

tion may reduce the available $O_2$ still further. This is why fishery managers monitor the dissolved $O_2$ content of lakes and rivers for signs that the $O_2$ level has fallen too low to support active species such as trout and bass. Likewise, air-breathing animals face oxygen shortages at high altitudes. Athletes entering events in mountain areas such as Denver, Mexico City, or the European Alps arrive several weeks early so they will have time to adapt to the low $O_2$ levels.

In fact, neither water nor air is an ideal respiratory medium at the best of times. Let us examine the two and see why.

The problem with water as a respiratory medium is that it contains comparatively little dissolved $O_2$, and the warmer or saltier the water, the less oxygen it can hold (Table 32–1). To obtain 1 liter of $O_2$, a fish must process about 100,000 liters of water!

Volume for volume, air is about a thousand times less dense than water, and so it takes less energy to move air to and from the respiratory surface. In addition, air contains up to 30 times more $O_2$, and $O_2$ molecules diffuse half a million times faster in air than in water. However, the problem with breathing air is that gas molecules must diffuse through plasma membranes in solution, and so respiratory surfaces must always be moist. Air-breathing animals pay a different price for $O_2$: the loss of body water by evaporation from their respiratory surfaces into the air.

TABLE 32–1

**Oxygen Content of Some Respiratory Media**

| Medium | Oxygen Content (mL/L) |
|---|---|
| Sea water at 5 °C | 6.4 |
| Fresh water at 5 °C | 9.0 |
| Fresh water at 25 °C | 5.8 |
| Air (any temperature) | 209.5 |

How fast an animal can absorb $O_2$ depends on how much $O_2$ is available in the environment and on the total area of its respiratory surface. All but the smallest animals have adaptations that increase the area of their respiratory surfaces. Once $O_2$ enters the body, its low solubility in water is still a problem because the body fluids that transport it to the cells are mainly water. We shall see how animals cope with this problem in Section 32–D.

$CO_2$ is nearly 30 times more soluble in water than $O_2$ is, and it is easily converted to highly soluble bicarbonate (Section 32–E). Hence, any respiratory system that can supply an animal's $O_2$ needs can also dispose of $CO_2$ fast enough.

How much $O_2$ does an animal need? This depends on its activity and on its size (that is, how many cells must be supplied with $O_2$). An active animal uses energy faster than a sluggish one and so must obtain $O_2$ for respiration at a greater rate (Table 32–2). This is why active fish, such as trout that catch insects for food, are the first to disappear from a polluted river containing little $O_2$. In addition, an animal's activity level and $O_2$ demand vary; for instance, human $O_2$ consumption is 15 to 20 times greater during exercise than at rest. The respiratory and transport systems must be able to supply more $O_2$ when an animal is active.

Biologists often use an animal's metabolic rate as a measure of its activity. The **metabolic rate** is the rate at which an organism uses energy in its metabolism, which supports all its activities such as muscle contraction and urine formation. Since most energy is released by cellular respiration, using $O_2$, the metabolic rate is usually measured as the volume of $O_2$ used per unit of body mass per unit of time. Thus:

$$\text{Metabolic rate} = O_2 \text{ used} \div \text{body mass}$$

Endothermic ("warm-blooded") animals, notably mammals and birds, have high metabolic rates because they produce their body heat by continually oxidizing food. These animals constantly lose heat to the environment through the body surface. The smaller the animal is, the higher its ratio of heat-losing surface area to heat-producing volume of cells. Therefore, small endothermic animals lose relatively more heat than do large ones and so have higher metabolic

## TABLE 32–2

### Metabolic Rates of Some Animals

| Animal | O₂ Consumption* (mm³/body mass/hr) |
|---|---|
| Sea anemone | 15 |
| Jellyfish | 5 |
| Flatworm | 75 |
| Earthworm | 60 |
| Snail | 250 |
| Octopus | 280 |
| Squid | 320 |
| Crab | 80 |
| Cockroach | 450 |
| Butterfly, resting | 600 |
| Butterfly, flying | 100,000 |
| Starfish | 400 |
| Sea urchin | 15 |
| Sea squirt | 5 |
| Goldfish | 420 |
| Trout | 225 |
| Frog | 55 |
| Iguana | 60 |
| Parakeet, resting | 4,500 |
| Parakeet, flying | 22,000 |
| Mouse, resting | 2,000 |
| Mouse, running | 20,000 |
| Dog | 360 |
| Human | 200 |

*At typical environmental temperatures

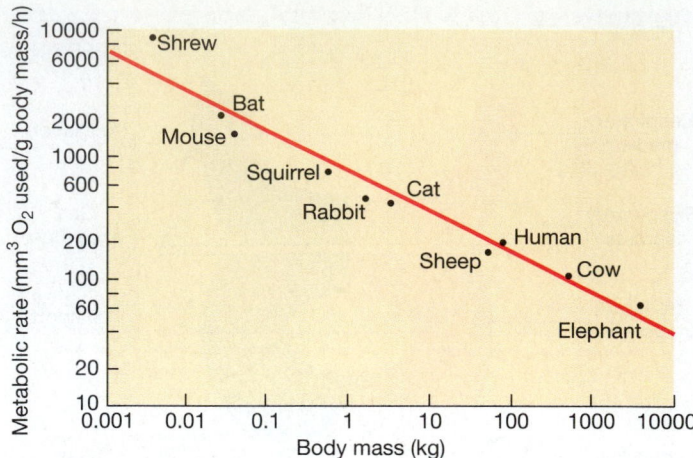

**FIGURE 32–3** Size and metabolic rate of some mammals. Metabolic rates of endotherms are inversely proportional to size. Notice that both axes in this graph have logarithmic scales. Both body mass and metabolic rate range over several orders of magnitude for the mammals indicated.

rates and use relatively more $O_2$ (Figure 32–3). However, a large endotherm still requires more *total* $O_2$ per unit of time than does a small one.

$$O_2 \text{ used} = \text{metabolic rate} \times \text{body mass}$$

**Ectotherms** are animals that get much of their body heat from the environment rather than by conserving metabolic heat. They include reptiles and amphibians, which often bask in the sun to raise their body temperature. Ectotherms may show a drastic change in $O_2$ consumption with changing environmental temperature (Figure 32–4). The higher the temperature, the faster the chemical reactions of metabolism occur, the higher the animal's metabolic rate, and the more $O_2$ it uses.

An animal that obtains its $O_2$ supply from water has a dual problem as the temperature rises. It needs more $O_2$ because of the rise in its metabolic rate, but there is less $O_2$ available because less $O_2$ is dissolved in water at higher temperatures (see Table 32–1). A few active fish, such as trout, can obtain enough $O_2$ only in very cold water. Thus, the discharge of waste heat (for instance, from a power plant) into a lake or stream may ruin the trout fishing.

## 32–C RESPIRATORY SURFACES AND VENTILATION

Despite the variety of animal life on Earth, there are only four main types of respiratory surface: the body surface, gills, lungs, and tracheae, all of which are thin-walled and moist (Figure 32–5). Animals that use water as a respiratory medium obtain $O_2$ through the general body surface or through gills, outward expansions of a local surface area. Some air-breathing animals obtain $O_2$ from moist air through

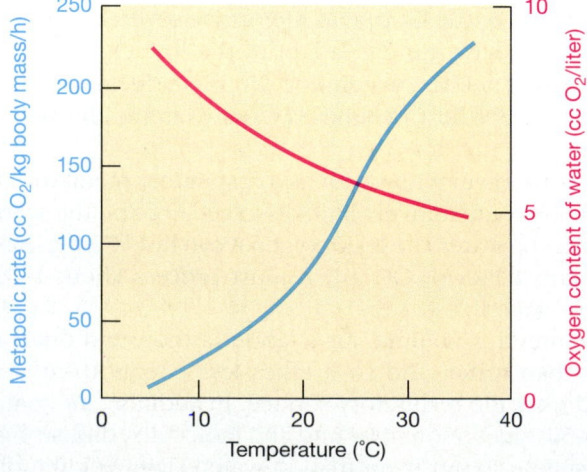

**FIGURE 32–4** How temperature affects the metabolic rates of ectotherms. Metabolic rates vary with the temperature of the environment (blue line). For an aquatic animal, higher temperatures decrease the amount of oxygen available, because less oxygen dissolves in warm water than in cold water (red line).

Comparison of Four Main Kinds of Respiratory Surfaces

| | Body surface | Gills | Lungs | Tracheal system |
|---|---|---|---|---|
| Body location | General, external | Local, external | Local, internal | General, internal |
| Respiratory medium | Water, moist air | Water | Air | Air |
| Use of transport system to carry $O_2$ | Yes or no | Yes | Yes | No |

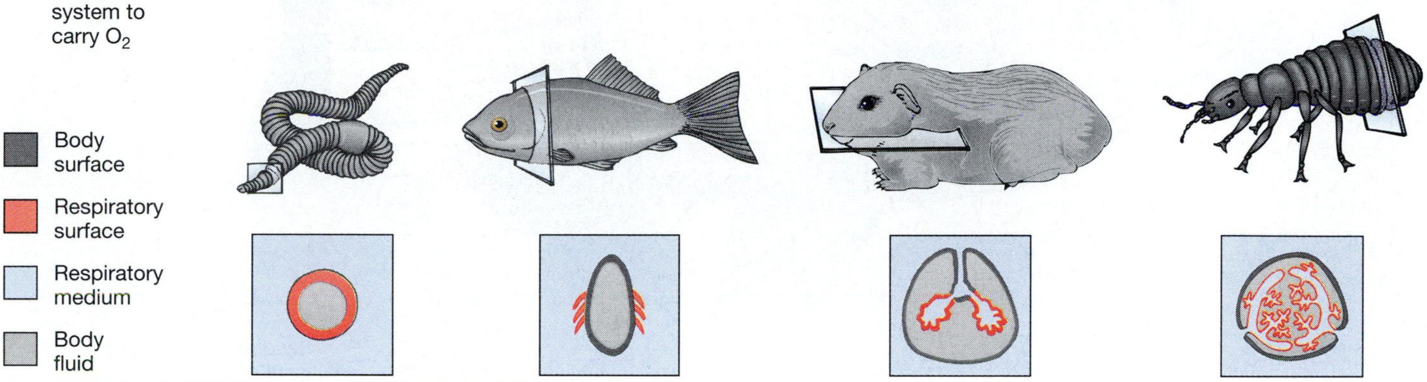

■ Body surface
■ Respiratory surface
■ Respiratory medium
■ Body fluid

**FIGURE 32–5**    The four main types of respiratory surfaces.

## The Body Surface

Animals such as sponges, cnidarians, and many worms obtain all their $O_2$ through the general body surface. These animals have four main features in common:

1. **High surface-to-volume ratio.** This is a result of small size or of a shape that increases the surface area. For example, sponges are essentially hollow cylinders with thin walls, which may be folded into complex shapes, and platyhelminths are flat.
2. **Moist body surface.** Because gas exchange occurs through moist membranes, animals that obtain $O_2$ from the air through the body surface must live in damp places, for example, under logs or in wet vegetation.
3. **Low metabolic rate.** These animals are relatively sluggish and so do not use $O_2$ at a great rate.
4. **Surface protection.** The thin moist body surface is protected from injury, often by a slimy mucus that makes it too slippery to be damaged by sharp objects.

For many animals in this category, no cell is far from the respiratory surface, and so $O_2$ can be distributed, and $CO_2$ eliminated, by diffusion. In the larger animals, such as earthworms, a transport system picks up $O_2$ at the body surface and carries it to cells throughout the body.

Gas exchange through the body surface is not restricted to small invertebrates. In some fish and reptiles, the body surface supplements gas exchange by the gills or lungs. Among vertebrates, this ability is most highly developed among amphibians with thin moist skin. Some amphibians obtain more than half of their $O_2$ through the body surface. Indeed, members of one group of salamanders lack lungs and carry out all of their gas exchange through the skin and membranes in the mouth.

## Gills

Animals such as fish and many aquatic arthropods, molluscs, and amphibian larvae carry out gas exchange through **gills,** feathery tissue outgrowths with thin membranes that are exposed to the water of the respiratory medium.

Unlike air, water is heavy for the amount of $O_2$ it contains. Therefore, animals with gills must use a large portion of their energy to push a current of water across the gill membranes. It would take even more energy to stop the water, reverse its direction, and push it back out of the gill area. Meanwhile, no new oxygen-carring water could enter. Therefore, in most aquatic animals, water passes in through one opening and out by another, bathing the gills in a constant stream of oxygen-laden water.

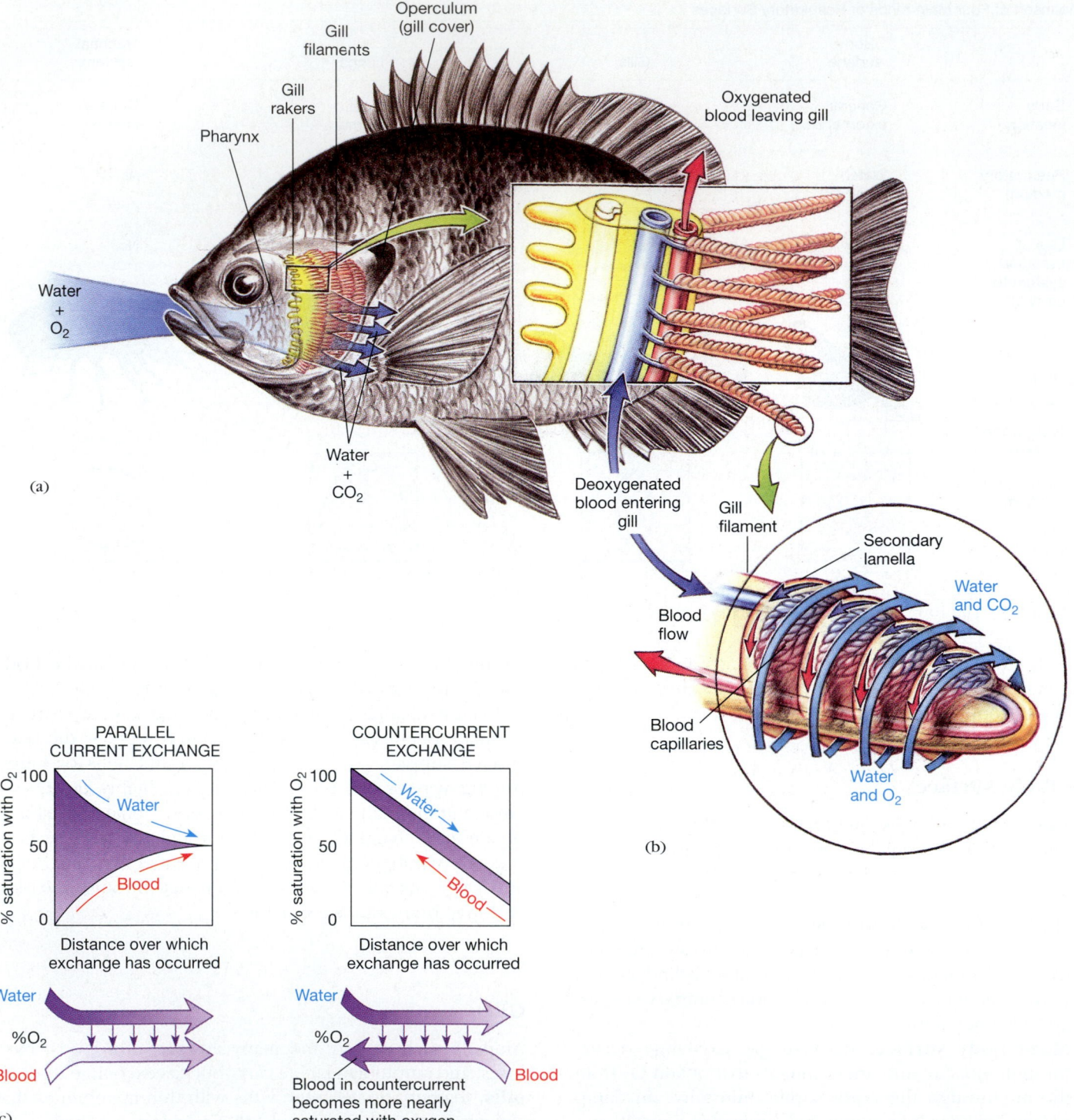

**FIGURE 32–6**  The gills of a fish. (a) A one-way current of water enters the mouth, flows past the respiratory surfaces of the gills, and exits behind the head. (b) Blood flows through the gill lamellae in the opposite direction. (c) This countercurrent flow permits the blood to extract more $O_2$ from the water than it could with parallel flow.

The gills of a bony fish are located behind the head (Figure 32–6). Water enters through the mouth, passes from the pharynx across the gills, and leaves via the opening behind the **operculum,** which covers the gills.

A fish's blood circulates through the gills in a pathway that allows it to remove $O_2$ from the water with maximum efficiency. In the gills, thin-walled blood vessels called **capillaries** are arranged so that blood and water flow in opposite directions, much like traffic in a two-lane road (Figure 32–6b). Such an arrangement, in which two fluids exchange substances (or heat) while moving in opposite directions, is called a **countercurrent exchange system.** It maintains a gradient of whatever the fluids are exchanging (in this case, $O_2$) at all times. As water enters the gill area, it flows past blood that has already picked up some $O_2$ and is about to leave the gills. However, this blood can hold still more $O_2$, which it picks up from the fresh, oxygen-rich water just entering. As the water moves on, it loses more and more $O_2$ to the blood, but it is passing blood that is less and less saturated with $O_2$. The water always contains more $O_2$ than the blood it encounters, and so the water keeps losing oxygen and the blood keeps gaining it (Figure 32–6c). The countercurrent mechanism is so efficient that a fish's gills may remove more than 80% of the oxygen from the water in the respiratory current.

Ventilation of the gills is sometimes used for other functions as well as for gas exchange. Many bivalve molluscs and the lower chordates (tunicates, amphioxus [Sections 29–J, 29–K]) filter food out of water drawn into the gill area, and gas exchange occurs at the same time. Squid and octopuses use the respiratory water current for locomotion. They can eject water from the siphon with considerable force, creating a jet-propulsion stream that moves them rapidly forward or backward depending on which way the siphon is aimed (Figure 32–7).

Gills function poorly in air because the surface tension of the water on the gill membranes causes the soft gill processes to stick together, reducing the surface area exposed to the air. Some fish can survive in air for a short time. While breathing air, a mudskipper uses mainly its skin for gas exchange, the electric eel uses its mouth, and some catfish use their stomachs. Instead of gills, terrestrial animals evolved internal respiratory surfaces less subject to desiccation and collapse.

## Lungs

**Lungs** are the invaginated respiratory surfaces of air-breathing vertebrates, gastropods, and some spiders. They connect to the outside air via narrow tubes, which minimize water loss from the respiratory surface of the lungs. The first vertebrates with lungs were fish living in warm, shallow water where $O_2$ was scarce. These primitive lungs were outpocketings of the pharynx, providing an expanded internal surface where gas was exchanged with air gulped through the mouth.

The problems faced by animals with lungs are different from those faced by animals with gills. Because air is

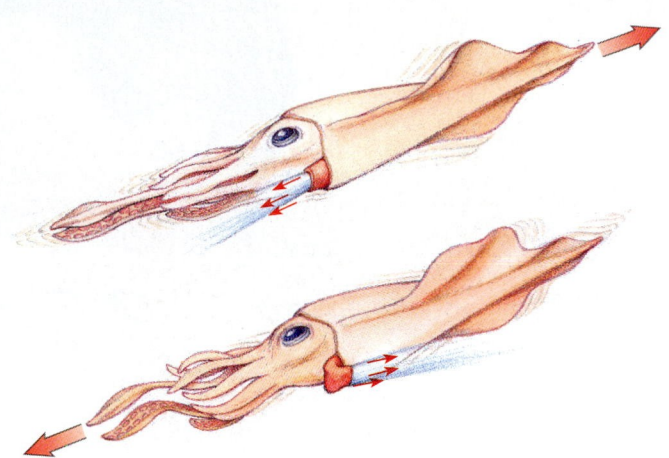

**FIGURE 32–7** Squid locomotion. The squid can move forward or backward depending on which way the siphon (orange) is aimed as it ejects water from the mantle cavity surrounding its gills.

much lighter and richer in $O_2$ than water is, air-breathing animals can ventilate the lungs by a **tidal** (in-and-out) flow of air rather than by the one-directional stream of water usual with gills. This makes it unnecessary to have separate entrance and exit openings at the body surface. The main disadvantage of breathing air is water loss from the respiratory surface.

As an example of a mammalian respiratory system, let us examine our own. Air enters the body through the nose or mouth. It is more healthful to breathe through the nose because its complex passages warm, moisten, and filter the incoming air. The nose also contains nerve endings that detect the odor of airborne molecules.

Air next passes through the pharynx, a common passageway for both air and food, and then the **larynx,** also known as the voice box or Adam's apple. The larynx contains the vocal cords, which vibrate and produce the sounds of the voice when air is forced between them. The larynx is a vestibule that admits air into the **trachea** (windpipe). The walls of the trachea contain rings of cartilage that hold the tube open. The trachea's inner surface is lined with cilia, which keep the air passages clear by sweeping foreign particles up into the pharynx, where they can be swallowed. (No spitting, please: esthetics apart, spitting can transmit disease, whereas swallowing destroys most pathogens on contact with stomach acid.) The posterior end of the trachea divides into two **bronchi,** which in turn divide into finer and finer tubes, the **bronchioles.**

The smallest bronchioles end in a myriad of tiny sacs, the **alveoli,** whose thin moist walls are the actual respiratory surfaces of the lungs. A vast network of capillaries surrounds the alveoli. Blood traveling in these capillaries picks up $O_2$ for transport to the rest of the body and gives up $CO_2$.

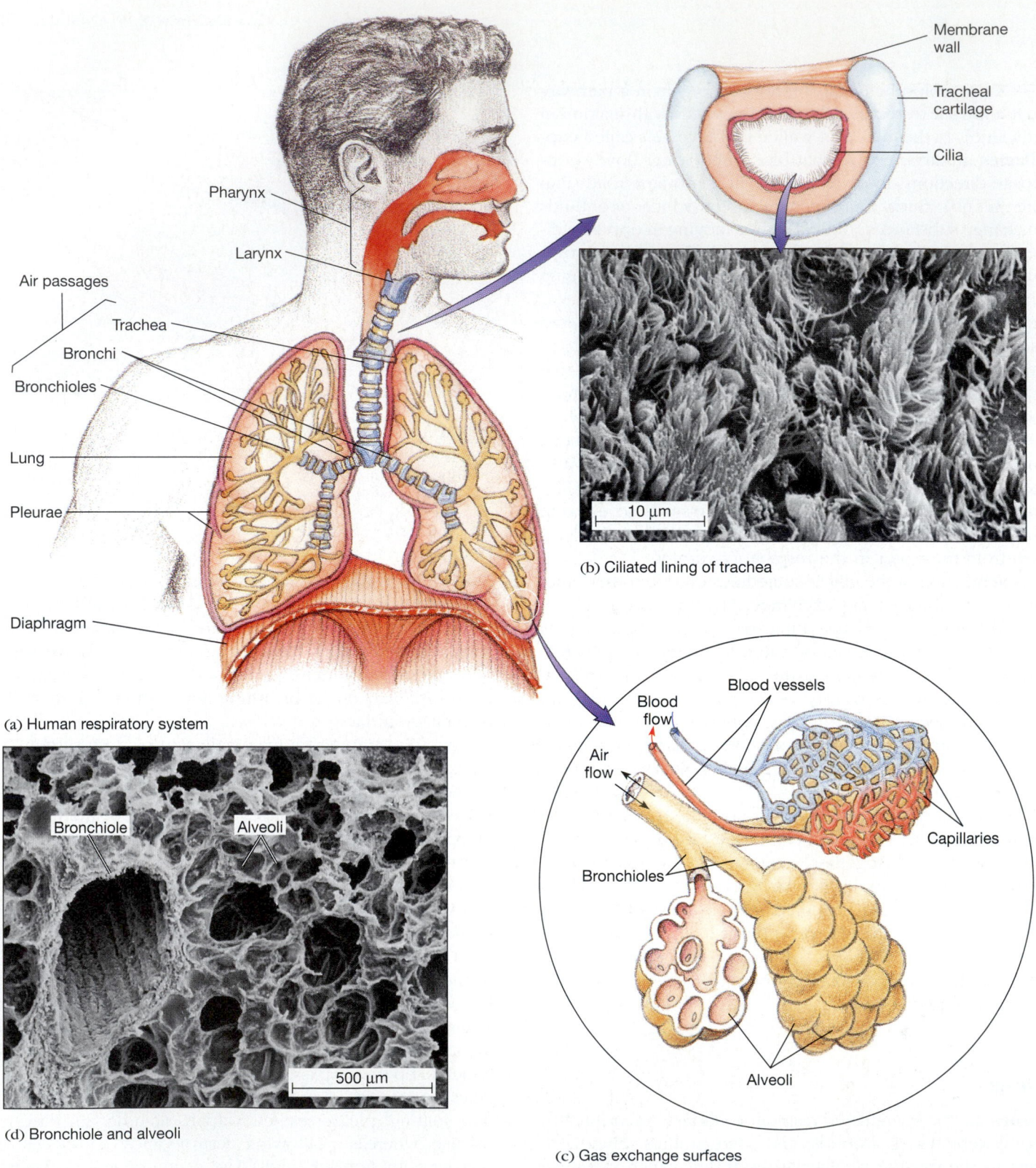

Membrane
wall

Tracheal
cartilage

Cilia

Pharynx

Larynx

Air passages

Trachea

Bronchi

Bronchioles

Lung

Pleurae

Diaphragm

(a) Human respiratory system

10 µm

(b) Ciliated lining of trachea

Blood vessels

Blood
flow

Air
flow

Bronchioles

Capillaries

Bronchiole

Alveoli

500 µm

Alveoli

(d) Bronchiole and alveoli

(c) Gas exchange surfaces

**FIGURE 32–8** The human respiratory system. (a) The anatomy of the respiratory system. Air travels from the nose or mouth to the lungs by way of the air passages: the trachea, bronchi, and bronchioles. The actual respiratory surfaces are the walls of the numerous alveoli in the lungs. (b) A scanning electron micrograph of the inside of the trachea. The globular structures are glands that secrete mucus. Most of the surface is covered with cilia, which move the mucus and any foreign particles it traps up into the throat. (c) The bronchioles end in tiny sac-like alveoli throughout the lungs. Each alveolus is surrounded by capillaries of the blood system. Gas exchange takes place across the moist surfaces of the alveoli, between air inside the alveoli and blood flowing through the capillaries. (d) Scanning electron micrograph of a section of lung tissue, showing part of a bronchiole surrounded by spongy alveoli. Note the vast alveolar surface area. (b, d, Biophoto Associates)

Most of the air in our lungs remains in the alveoli even when we exhale as hard as we can (Figure 32–8). This residual air continues to provide O$_2$, and it also keeps the walls of the alveoli apart so that they do not collapse. As a further protection against collapse, the alveoli produce **surfactants,** substances that lower the surface tension of fluids. The lungs' surfactants prevent the moist alveolar membranes from sticking together by surface tension. Infants born prematurely often cannot breathe well because they are not yet secreting these surfactants. The walls of their alveoli stick together like the sides of a wet plastic bag, instead of inflating like a dry bag full of air.

Two membranes that normally behave much like wet bags are the **pleurae,** one covering the lungs like a skin, the other lining the inside of the chest cavity. The space between the pleurae contains a lubricating fluid that holds them together, yet allows the lungs to slide freely. Whenever the chest cavity enlarges, the outer lung surfaces are pulled along, and so the lungs expand.

How does an air-breathing vertebrate ventilate its lungs? There are two main ways, positive pressure breathing and negative-pressure breathing. **Positive-pressure breathing** moves air by pushing on it, as when we blow up a balloon. This mechanism evolved first, as a modification of the ventilation movements of a fish. A frog breathes by a positive-pressure mechanism, using muscles to force air from the mouth into the lungs (Figure 32–9).

Mammals have **negative-pressure breathing,** which works by sucking air into a low-pressure area. We have a

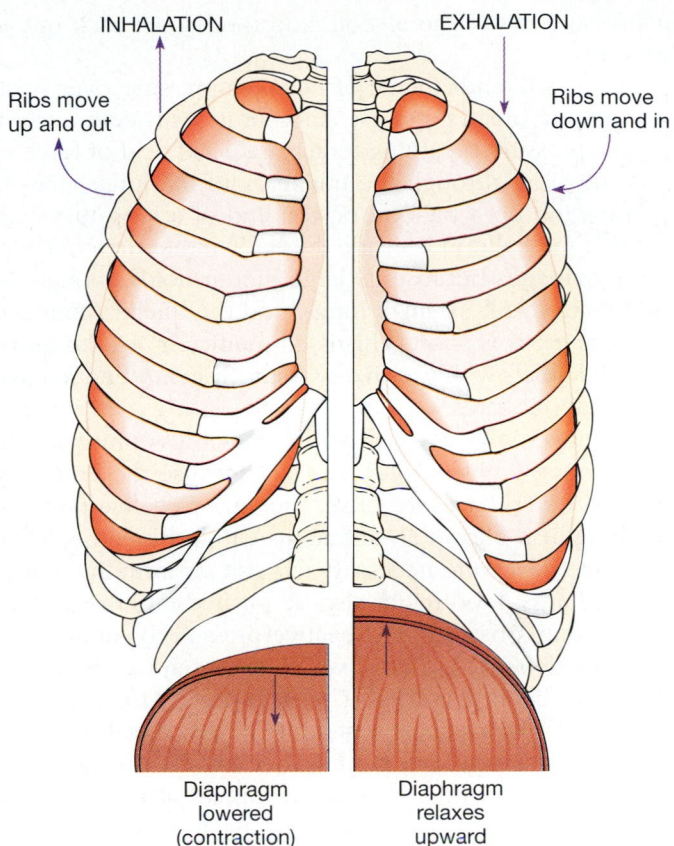

FIGURE 32–10 Negative-pressure breathing. During inhalation, the ribs are lifted up and out and the diaphragm is lowered. This increases the size of the chest cavity and decreases the pressure within it. Air rushes in through the nose, mouth, or both. Exhalation occurs when the breathing muscles relax, forcing air back out; it is a passive process.

respiratory muscle, the **diaphragm,** not found in other vertebrates. (Some reptiles have an incomplete diaphragm.) The dome-shaped diaphragm, beneath the lungs, separates the chest cavity from the abdominal cavity. When we inhale, the diaphragm contracts and flattens down toward the abdominal cavity. At the same time, the muscles between the ribs contract and lift the ribs outward (Figure 32–10). Both movements expand the chest cavity. This pulls the lungs to fill a larger volume and therefore lowers the air pressure inside the alveoli. Air then rushes into the nose, down the trachea, and into the lungs, moving down the air pressure gradient.

You can get an idea of how negative-pressure breathing works by closing your mouth and holding your nostrils shut while you expand your rib cage and lower your diaphragm; you will feel the partial vacuum created. Then remove your fingers from your nose and you can hear and feel the air rushing in as the pressure equalizes.

When we exhale, the rib muscles and diaphragm relax, decreasing the volume of the chest cavity, increasing the

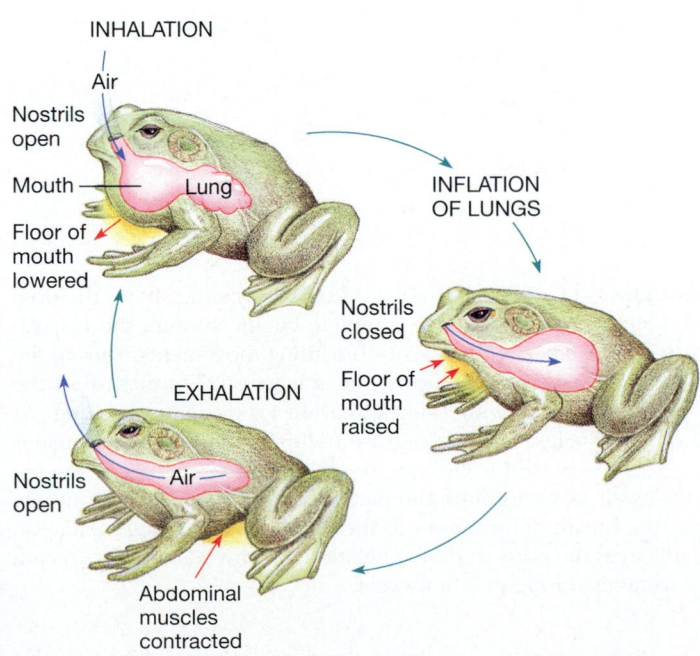

FIGURE 32–9 Positive-pressure breathing in a frog. Movements of the mouth push air into the lungs.

air pressure inside the alveoli, and forcing air back out of the lungs.

What is the advantage of negative-pressure over positive-pressure breathing? First, since air is easily compressed, positive-pressure breathing requires a good deal of force to push much air through the trachea into the lungs. This is like blowing up a balloon on the end of a long tube, except that the balloon is inside the body rather than outside. A frog manages because its large mouth holds enough air at a breath to fill its small lungs, and the trachea connecting these areas is short. This is an inefficient way to move air, and animals with positive-pressure systems tend to have low metabolic rates.

Negative-pressure breathing requires less muscular effort to move a given volume of air because the breathing muscles don't have to push the air. Negative-pressure breathing also allows an animal to eat and breathe at the same time. If it were necessary to push air from the mouth into the lungs, food might also be pushed into the trachea and cause an obstruction. Negative-pressure breathing creates a gentler stream of air, which is less apt to pull food along into the air passages. This is important for endothermic animals, which must sustain their high metabolic rates by eating and breathing more frequently than frogs do.

How big is a breath of air? A human at rest normally inhales about 500 mL (0.5 liter) of air at a time. The last 150 mL never gets beyond the air passages (the trachea, bronchi, and bronchioles). Only about 350 mL makes it to the alveoli of the lungs. Here it mixes with about 2300 mL of air remaining in the lungs after the previous exhalation. This means that only about one seventh of the air in the lungs is exchanged with each breath. The low turnover keeps the concentrations of gases in the alveoli fairly constant during both inhalation and exhalation. Although the concentration of $O_2$ is lower in the alveoli than in the atmosphere, there is still a steep concentration gradient between the alveolar air and the blood. Since a resting human takes about 12 breaths per minute, the alveoli receive a total of about 4200 mL of new air each minute.

During deep breathing, humans can take in about 1500 to 3000 mL of air instead of the normal 500 mL. It is also possible to exhale more forcefully than usual, expelling 1000 to 1500 mL. Even then, the lungs still contain 1000 to 1200 mL of residual air.

### Ventilation in Birds

The tidal air flow in the lungs of amphibians, reptiles, and mammals leaves a great deal of used air in the blind ends of the alveoli at each exhalation. The next breath of fresh air merely increases the proportion of $O_2$ and decreases the proportion of $CO_2$ in the lungs.

Birds have evolved more effective gas exchange, which supports the extremely high metabolic rates required for flapping flight. Flying at high altitudes, where $O_2$ is scarce, is especially demanding. Many birds spend considerable time flying at several thousand meters above sea level.

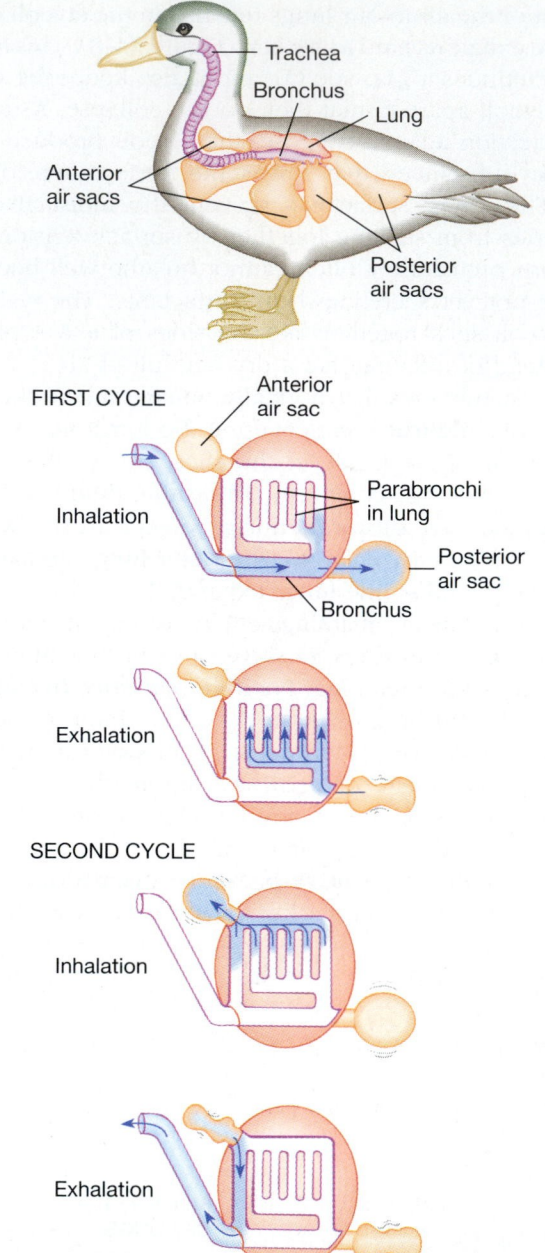

**FIGURE 32–11** Ventilation in a bird. Blue shading shows the flow of a single breath of air. A complete circuit through the respiratory system takes two cycles of breathing movements. During the first cycle, air is drawn down the trachea and bronchi into the posterior air sacs, with a small amount leaking into the lung. As the bird exhales, contraction of muscles in the body wall pushes air from the posterior sacs into the lungs, where gas exchange occurs as air flows through the parabronchi. During the second cycle, the breath of air moves to the anterior sacs during inhalation and leaves the body during exhalation. Meanwhile the next breath of air goes through its first cycle.

Birds' lungs do not have dead-end alveoli. Instead, the bronchial air passages divide into thousands of tiny parallel tubes, the **parabronchi,** open at both ends (Figure

32–11). The walls of the parabronchi are riddled with even smaller tubes, the **air capillaries,** the actual surfaces for gas exchange as air moves in one end of a parabronchus and out the other. This flow-through system constantly exposes the respiratory surfaces to fresh air instead of the partly used air in the lungs of other air-breathing vertebrates. Birds also have many **air sacs** in the body cavity and even extending into some of the larger bones. The air sacs do not contain respiratory surfaces, but they help to circulate air through the respiratory system.

A bird's respiratory system is arranged in such a way that it takes two cycles of inhalation and exhalation for each "breath" of air to complete the circuit. This pattern of air flow has two important features. First, the inhaled and exhaled air hardly mix, and so used air in the system does not appreciably reduce the $O_2$ concentration at the respiratory surface. Second, air flow in the lung is always in the same direction. This is a necessary prerequisite to countercurrent gas exchange like that in a fish's gills. To the surprise of most observers, however, birds do not have countercurrent exchange between blood and air. This was shown by reversing the direction of air flow through the lungs of a duck, a change that did not affect the efficiency of gas exchange. Birds apparently have "crosscurrent" gas exchange, with blood and air moving at right angles to each other. This has some, but not all, of the efficiency of a countercurrent system.

Gas exchange is certainly much more efficient in birds than in mammals. At rest, a bird and a mammal of equal body mass have nearly the same metabolic rate and therefore use the same amount of $O_2$. However, the bird obtains this much $O_2$ using lungs little more than half the size of the mammal's and breathing only about two thirds as much air per unit of time.

In addition to storing air and acting as bellows during the breathing cycle, a bird's air sacs help to lower its body density, a useful adaptation to flight and to floating in water. The air sacs also provide additional spaces where water can evaporate into the air and carry off excess heat produced during strenuous flight. You may have seen movies of male frigate birds or prairie chickens inflating air bladders in their necks as part of their courtship displays; this air comes from nearby air sacs. Some birds also use air from the sacs for singing.

## Tracheal Systems

Air-breathing vertebrates have lungs. The other major group of land animals, the terrestrial arthropods (insects, centipedes, millipedes, and some spiders), breathe air by a different system. They have **tracheae** (singular: **trachea**), air-filled tubes that extend throughout the body, starting from **spiracles,** tiny openings in the body surface. Animals with tracheae do not rely on the transport system to carry $O_2$

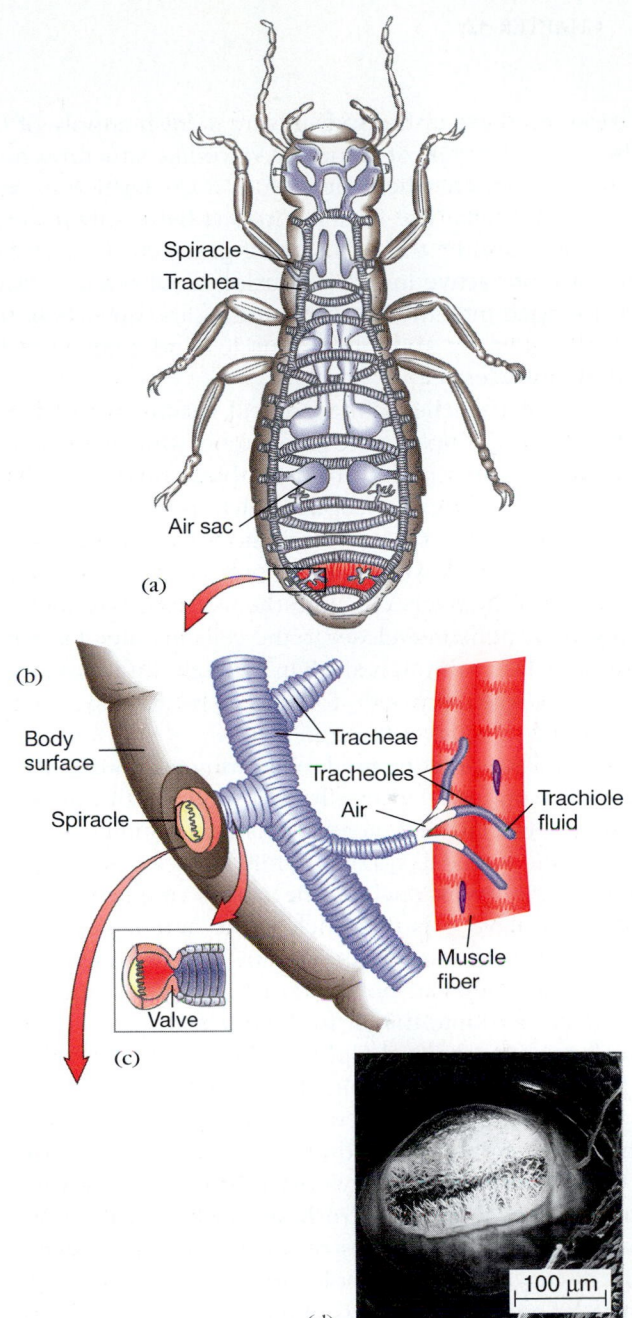

(a)

(b)

(c)

(d)

**FIGURE 32–12** The tracheal system of an insect. (a) Like the mammalian trachea, the large tracheal trunks have thickened rings that prevent their collapse. The fine endings of tracheoles lie close to each body cell. (b) Enlargement of one spiracle, branching tracheae, and the fluid-filled tracheoles where gas exchange with the tissues occurs. (c) A spiracle to show the valve that can be used to close it. (d) A scanning electron micrograph of one of a caterpillar's spiracles. The hair-like structures prevent dust from entering the trachea. (d, Biophoto Associates)

from the gas exchange surface to the body's cells. Instead, the tracheae branch into all parts of the body, ending close to every cell (Figure 32–12). Cells that are short of $O_2$ produce chemicals that stimulate nearby tracheae to grow branches toward them.

How do these arthropods ensure a fresh supply of $O_2$ in the tracheal system at all times? Valve-like structures near the spiracles open when the tissues' $CO_2$ level rises and close when it falls. The smallest insects have only to open these valves and let the diffusion of gases do the rest. For larger or more active insects, diffusion would be too slow. Air is pumped into and out of the tubes and air sacs of the tracheal system by abdominal muscles and sometimes by the flight muscles.

The main branches of the tracheal system are filled with air. However, the tips of the finest tubes (tracheoles) contain fluid, which is pulled from the body fluids into these fine tracheal endings by capillary action. If an insect becomes very active, $CO_2$ and other molecules accumulate in the tissue fluids, lowering their osmotic potential. Fluid moves out of the tracheoles into the tissues by osmosis, allowing air to penetrate closer to the cells that need it. Since $O_2$ diffuses faster through air than through fluid, this mechanism speeds delivery of $O_2$ to the cells by leaving a shorter distance for it to travel through fluid.

Most insects are terrestrial, but during the course of evolution some terrestrial insects have invaded freshwater habitats. These insects have adaptations for obtaining gaseous $O_2$ for their tracheal systems while they are submerged. Some aquatic forms, such as the naiads (immature forms) of mayflies, have gills that enclose the endings of the tracheal system. As long as the water in which they live is well aerated, they can obtain $O_2$ by diffusion through the gill surfaces and into the air in the tracheal system. A mosquito larva has a special siphon tube, which encloses the tracheal tubes and has a spiracle opening on the end. The larva floats head-down just under the surface of the water, with only its siphon protruding above the surface into the air. If it is disturbed, the larva dives below the surface, but it must return soon for a fresh supply of air. One of the most interesting adaptations of aquatic insects is found in some diving beetles that hold an air bubble under their wings and over the openings of their spiracles. Thus, when they dive, they carry air supplies in their own little scuba tanks. As the $O_2$ in the air bubbles is used up, more $O_2$ diffuses in from the surrounding water.

## 32–D   RESPIRATORY PIGMENTS

Salty water, such as body fluids, cannot carry much dissolved $O_2$. The evolution of large active animals using lots of oxygen depended on adaptations that vastly increase their blood's capacity to carry $O_2$. Most commonly, these animals have **respiratory pigments,** protein molecules containing metal atoms (iron or copper) that can bind $O_2$. Respiratory pigments bind $O_2$ in areas where it is abundant (at respiratory surfaces) and release it where it is scarcer (in tissues that use $O_2$ in respiration). The respiratory pigment hemoglobin in the red blood cells of a mammal carries 98% of the $O_2$ in the blood.

**Hemoglobin (Hb)** is a general name for a group of oxygen-carrying compounds that all share a common feature: a heme group with an iron atom at its center (Figure 32–13). It is this iron atom that actually binds the oxygen. When the pigment has bound $O_2$, it is said to be **oxygenated** (not oxidized) and is called **oxyhemoglobin** (HbO). Oxyhemoglobin is bright red, whereas deoxygenated hemoglobin is a darker, purplish red. Vertebrate hemoglobin consists of four polypeptide chains, each with a heme group attached.

Similar hemoglobin molecules are found in almost all vertebrates, so it is believed that all are descended from the hemoglobin of some common ancestor (*How Do We Know? Molecular Phylogeny of Globin Genes,* Chapter 22). Hemoglobins also occur in various invertebrates, including earthworms and some other annelids, and a few echinoderms, molluscs, and insects. Since these examples are scattered among various phyla, hemoglobins are thought to have evolved independently in various forms of life. Hemoglobins evolved by fairly modest changes in cytochromes containing heme groups; such cytochromes are found in all aerobic organisms.

Many arthropods and molluscs have copper-containing respiratory pigments called **hemocyanins** dissolved in their

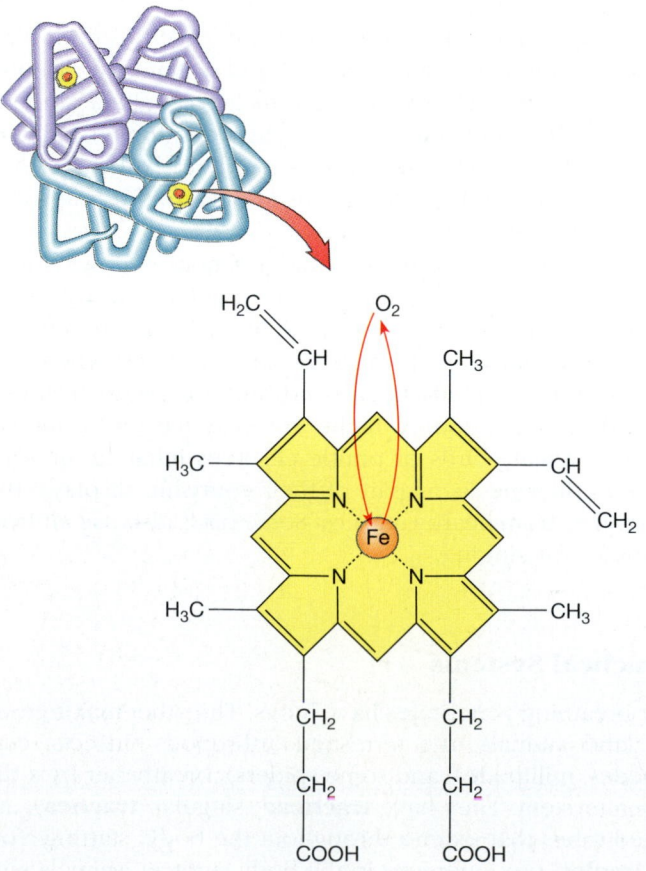

**FIGURE 32–13**   The heme group of a hemoglobin molecule. Oxygen attaches to the iron (Fe) atom. The structure around the iron atom is attached to the hemoglobin's protein chains.

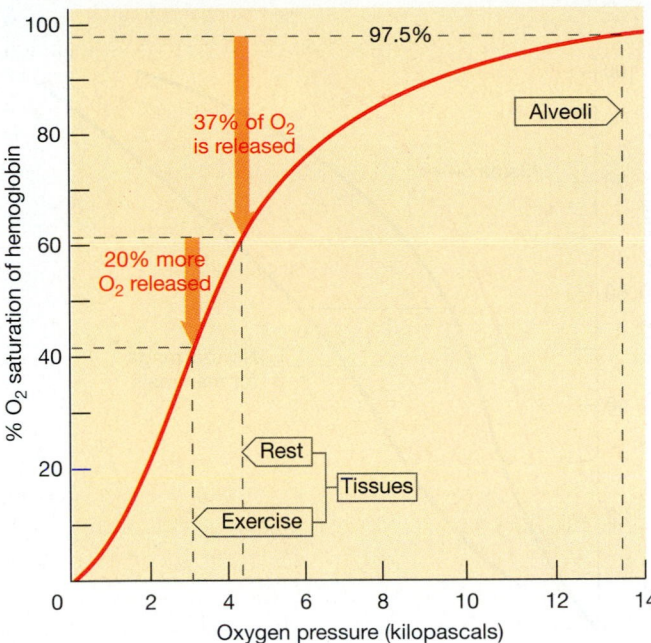

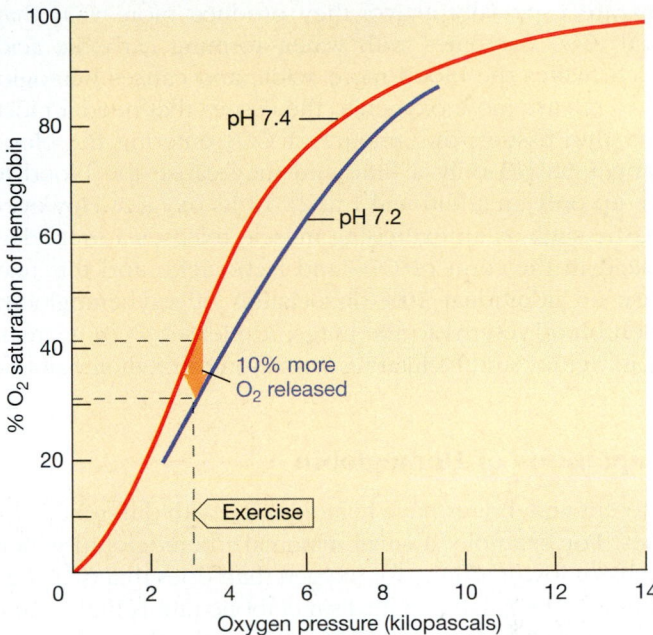

**FIGURE 32–14**  Oxygen dissociation curve for human hemoglobin. Blood passing through capillaries around the alveoli becomes about 97.5% saturated with $O_2$ (far right). Then, as it passes through the tissues, it releases $O_2$ (gold arrows). About 37% of the $O_2$ is given up in tissues at rest. In exercising muscle, which uses $O_2$ faster, the $O_2$ pressure is lower, and hemoglobin gives up about 20% more of its $O_2$ (that is, a total of about 57%).

**FIGURE 32–15**  Dissociation of $O_2$ under acidic conditions. The gold arrow starts where the one in Figure 32–14 left off: when the $CO_2$ and lactic acid released by exercising muscles lower the pH of the blood from 7.4 to 7.2, hemoglobin gives up another 10% of its $O_2$, providing the muscles with extra $O_2$.

blood plasma. Hemocyanin appears bluish when oxygen is bound to its copper atoms and colorless when it is deoxygenated.

## Oxygen Dissociation Curves

A respiratory pigment's ability to bind $O_2$ is important, but just as important is its ability to release $O_2$ to the tissues that need it. $O_2$ binds to hemoglobin by a reversible reaction:

$$Hb + O_2 \leftrightarrow HbO_2$$

From our study of chemical equilibrium (Section 6–B), we know that a surplus of the reactant $O_2$ will drive the reaction to the right. This is what happens as blood moves through the lungs: hemoglobin picks up $O_2$ until all its binding sites are filled. When the blood gets to the tissues, there is much less $O_2$, thanks to the body's respiring cells. Now the reaction is driven to the left; that is, the hemoglobin releases $O_2$. So we see that whether hemoglobin binds or releases $O_2$ depends on the $O_2$ concentration around the hemoglobin.

This relationship can be shown by a graph called an **oxygen dissociation curve;** Figure 32–14 shows the one for human hemoglobin. The horizontal axis gives the **partial pressure of oxygen** (the portion of the total air pressure due to $O_2$; also called **oxygen pressure** or $pO_2$ for

short). The vertical axis shows the percent saturation of hemoglobin. Hemoglobin is said to be **saturated** with $O_2$ when virtually all the iron atoms in all the hemoglobin molecules have bound $O_2$. In general, the higher the $O_2$ pressure, the more nearly saturated with $O_2$ the hemoglobin will be. In Figure 32–14 the vertical line at the far right shows that the oxygen pressure of air in the alveoli is about 13 kilopascals. At this oxygen pressure, hemoglobin is about 97.5% saturated (top of curve).

The $O_2$ pressure is much lower in the tissues. The steep slope in this part of the curve means that a fairly small drop in $O_2$ pressure as the tissues use more $O_2$ makes a big difference in how much $O_2$ the hemoglobin can hold: $O_2$ and hemoglobin dissociate, freeing $O_2$ to diffuse into the cells. For example, when the $O_2$ pressure in the tissue fluids is about 4.3 kilopascals, $O_2$ saturation is a little over 60%; hence, about 37% of the $O_2$ is given up to the tissue fluid. During exercise, muscle cells use more $O_2$ than normal, the $O_2$ pressure in the tissue fluid falls even lower, and the hemoglobin releases more $O_2$ than usual.

Notice that the $O_2$ dissociation curve for hemoglobin is S-shaped. This is because hemoglobin's four polypeptide-heme subunits bind $O_2$ cooperatively: after the first one has bound $O_2$, adding $O_2$ to another one is easier, and so on, until the hemoglobin molecule has a full complement of four $O_2$. Conversely, dissociation of the first $O_2$ makes it easier to shed the next one, and so on.

Increased acidity boosts the tendency of hemoglobin to release oxygen (blue curve in Figure 32–15). When the tissues are especially active, they produce more $CO_2$ than usual. $CO_2$ combines with water, forming carbonic acid, which makes the blood more acidic and causes hemoglobin to release more oxygen to the tissues that need it most. In resting tissues, the amount of $CO_2$ entering the blood changes its pH only a little and may cause the blood to give up only an additional 1 to 2% of its oxygen. However, a tissue such as an exercising muscle releases a great deal of acid in the form of $CO_2$ and lactic acid, and this may cause an additional 10% dissociation of oxyhemoglobin. When blood returns to the lungs, it releases $CO_2$, decreasing its acidity and facilitating formation of oxyhemoglobin.

## Adaptations of Hemoglobin

Different vertebrates have hemoglobins with different properties. For example, a small mammal's hemoglobin generally has a lower affinity for oxygen than does that of a large mammal. The small mammal's metabolic rate is higher, and so it uses oxygen more quickly. The smaller the animal, the more oxygen its hemoglobin will give up to the tissues at a particular oxygen pressure, and the further its dissociation curve is shifted to the right (Figure 32–16).

Animals native to high altitudes have hemoglobin with a high affinity for oxygen. The fundamental problem of gas exchange at high altitude is that the air pressure is lower than it is at sea level. Thus, although air at these high alti-

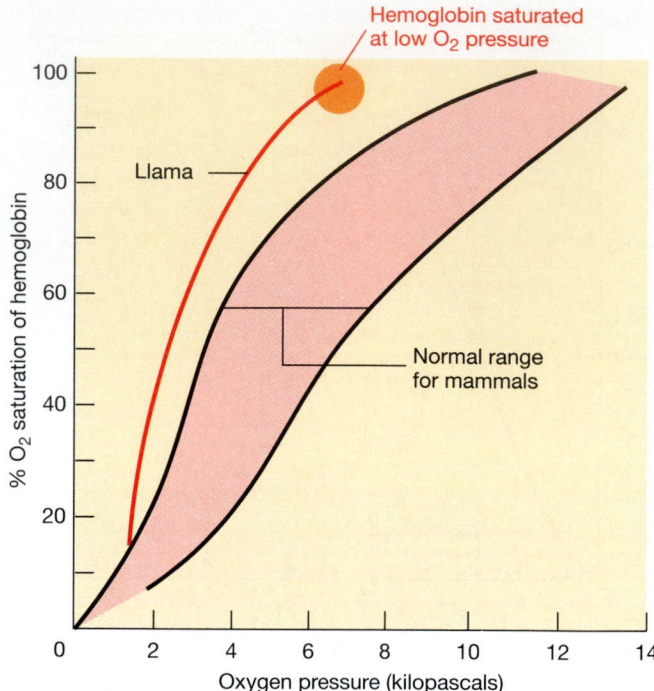

**FIGURE 32–17** Adaptation of llama hemoglobin to high altitudes. The llama's hemoglobin oxygen dissociation curve lies to the left of that of most other mammals, permitting the llama to obtain enough $O_2$ despite the low $O_2$ pressure at high altitudes.

tudes contains the same proportion of oxygen, there is less oxygen available because there is less air. The oxygen dissociation curve of a highland mammal lies to the left of that of a lowland species (Figure 32–17). High-altitude hemoglobin, with its higher affinity for oxygen, picks up and releases oxygen at the relatively lower oxygen pressures present in the alveoli and tissues at high altitudes.

An animal may have hemoglobin with different properties at different times in its life. Before birth, human fetuses produce several kinds of hemoglobin that adults do not. After birth, fetal hemoglobin is gradually replaced by adult hemoglobin. All of a fetus's oxygen comes from its mother's bloodstream. If fetal and adult hemoglobins had the same dissociation curves, the fetus would not be able to pick up very much of the oxygen released by the mother's blood. But a fetus's hemoglobin has a higher affinity for oxygen than the mother's, permitting it to pick up oxygen at oxygen pressures that cause the mother's hemoglobin to release oxygen (Figure 32–18).

## 32–E    CARBON DIOXIDE TRANSPORT

Carbon dioxide produced during respiration must be carried by the blood to the lungs, where it is excreted. Some $CO_2$ travels in solution, but most of it reacts with other substances in the blood. Hemoglobin and other blood proteins carry some $CO_2$. Instead of attaching to hemoglobin's iron

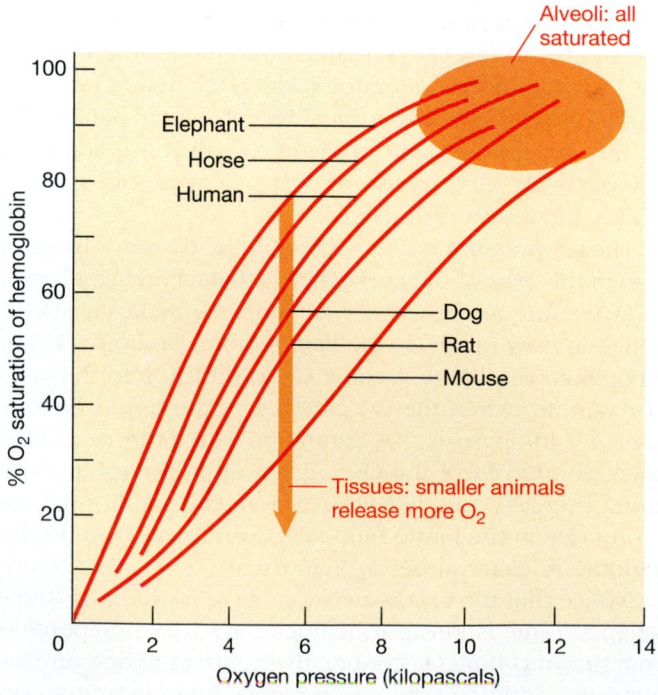

**FIGURE 32–16** Oxygen dissociation curves for mammals of different sizes. The smaller the animal, the farther its curve is shifted to the right, and so the more $O_2$ its hemoglobin gives up to tissues at a given $O_2$ pressure (gold arrow).

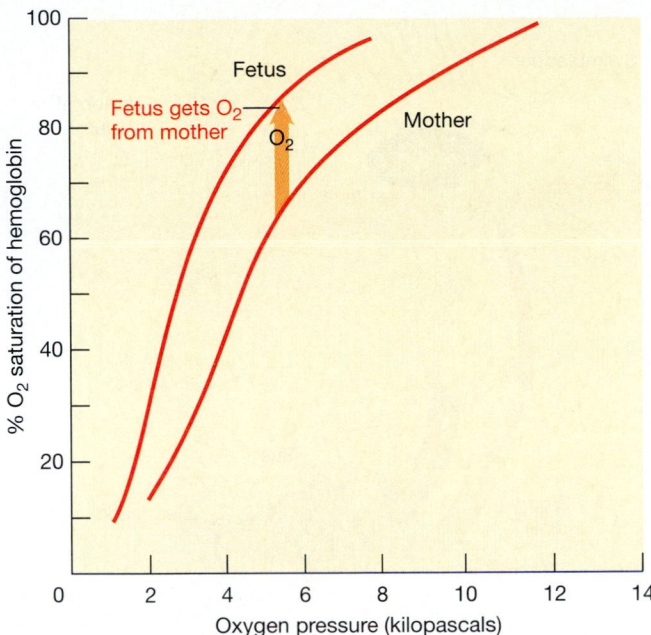

100
80
60
40
20

% O₂ saturation of hemoglobin

Fetus
Fetus gets O₂ from mother
O₂
Mother

0  2  4  6  8  10  12  14
Oxygen pressure (kilopascals)

**FIGURE 32–18** Oxygen dissociation curves for fetal and maternal hemoglobin in mammals. The fetal curve lies to the left of that for adult hemoglobin, meaning that fetal hemoglobin has a higher affinity for $O_2$. As the mother's hemoglobin gives up $O_2$ at normal tissue $O_2$ pressures, fetal hemoglobin is able to pick it up (arrow).

atoms, as oxygen does, $CO_2$ forms bonds with amino groups at the ends of the polypeptide chains. Oxygenated and deoxygenated forms of hemoglobin are allosteric. Deoxygenated hemoglobin takes up $CO_2$ more readily, whereas oxyhemoglobin releases $CO_2$ more easily. The allosteric change facilitates both uptake of $CO_2$ at low oxygen pressures (in the tissues) and release of $CO_2$ where oxygen levels are high (in the lungs).

Dissolved $CO_2$ makes a vital contribution to the blood's **buffering capacity**—its ability to absorb acids and bases with little change in pH. Most of the $CO_2$ in the blood is carried as bicarbonate ions dissolved in the blood plasma and inside the red blood cells. The enzyme **carbonic anhydrase** in red blood cells catalyzes the formation of carbonic acid from water and $CO_2$. Once formed, carbonic acid dissociates readily into hydrogen and bicarbonate ions:

$$H_2O + CO_2 \overset{\text{Carbonic anhydrase}}{\rightleftharpoons} H_2CO_3 \rightleftharpoons H^+ + HCO_3^-$$

Water  Carbon dioxide  Carbonic acid  Hydrogen ion  Bicarbonate ion

The hydrogen ions produced by this reaction can be neutralized by combining with hemoglobin. Hemoglobin has negative charges that attract positive potassium ions. When the blood becomes more acidic than usual, hydrogen ions displace the potassium ions, and the hemoglobin becomes **acid hemoglobin,** reducing the acidity of the blood. If the blood becomes too alkaline, acid hemoglobin dissociates,

releasing hydrogen ions and the negative hemoglobin ion, which takes up potassium. The increase in $H^+$ makes the blood more acidic, but the equilibrium of the overall reaction lies far in favor of acid hemoglobin.

When blood returns to the lungs, it gives up some of its $CO_2$ into the air in the lungs. This reduces the concentration of $CO_2$ in the blood, altering the equilibrium between $CO_2$ and carbonic acid. The equilibrium is restored by the dissociation of carbonic acid into water and more $CO_2$. This, in turn, alters the equilibrium of the equation, so that hydrogen and bicarbonate ions combine, forming more carbonic acid (colored arrows).

The blood gives up only about 7% of its $CO_2$ as it passes through the lungs. The rest is retained, mostly in the form of bicarbonate ions, which act as important blood **buffers,** substances that keep the pH from fluctuating. Therefore, although $CO_2$ is a waste product, its presence is essential in regulating the pH of the blood.

## 32–F  REGULATION OF VENTILATION

The oxygen and $CO_2$ pressures in our blood must be continually adjusted so that they stay within physiological limits. This adjustment is made by changes in the depth and rate of breathing. As blood passes through the capillaries of the lungs, the gases in the blood come into equilibrium with those in the alveoli. The faster and deeper we breathe, the greater is the percentage of oxygen in the alveolar air and the lower the percentage of $CO_2$, because the air in the lungs is replaced more completely and more often.

The breathing rate is controlled by the content of $CO_2$, $H^+$, and oxygen in the blood and cerebrospinal fluid (the fluid around the brain and spinal cord). The concentrations of these substances in the blood are monitored by receptor organs in some of the blood vessels (aorta and carotid arteries; see Figure 33–13). These receptors send nerve signals to the respiratory center in the medulla of the brain (see Figure 37–16). Cells in the brain also monitor the cerebrospinal fluid's content of $H^+$, which reflects the blood's $CO_2$ content. The brain responds to this information by sending nerve signals to the diaphragm and rib muscles, speeding or slowing the rate of breathing.

Perhaps surprisingly, the $CO_2$ level affects breathing more than the oxygen level does. Breathing is normally adjusted to keep the blood's $CO_2$ pressure at 5.3 kilopascals. If $CO_2$ drops below this level, breathing is inhibited. By purposely hyperventilating—that is, taking several deep breaths in swift succession—you can hold your breath longer. Swimmers often do this before an underwater dive. At each breath, however, the blood's $CO_2$ content goes down, and if it goes too far, you lose consciousness. This may seem annoying, but it prevents you from lowering the $CO_2$ pressure of the blood to a dangerous level.

Similarly, when you hold your breath, the $CO_2$ level in the blood rises, and if it goes above a certain level, you lose consciousness. Once this happens, the breathing reflexes take control again and cause you to inhale. It is impossible to kill yourself just by holding your breath.

Changes in the blood's oxygen pressure have little effect on the breathing rate until blood oxygen pressure falls below 8.0 kilopascals (from a normal level of 12.7 kilopascals). However, a lower-than-normal oxygen pressure makes the receptors more sensitive to a rise in $CO_2$ pressure.

Breathing is controlled less closely during sleep, and the response to changes in $CO_2$ pressure becomes more variable. Periods of **apnea** (cessation of breathing due to a drop in $CO_2$) occur several times each night, causing the person to wake up briefly and resume breathing. The frequency of apnea is increased by consumption of alcohol, antihistamines, and tranquilizers.

## 32–G  SWIMBLADDER PHYSIOLOGY

The "lungs" of some early air-breathing fishes evolved into the gas-filled **swimbladder,** or **air bladder,** found in many modern bony fishes. In most fishes, its chief function is to provide variable buoyancy by altering the density of the fish's body. The amount of air in the bladder can be regulated so that the fish can float at a particular depth, rise, or sink with little muscular effort.

The swimbladder arises as an outpocketing of the pharynx. Some fish retain a connection between the swimbladder and pharynx, and these fish can spit out air to increase their density, or swallow bubbles of air to fill the swimbladder. Other fish have no connection between the digestive tract and the swimbladder and move gas in the blood into and out of the swimbladder.

A swimbladder with a connection to the pharynx functions as an accessory breathing structure in some fish that live in water with low $O_2$ levels. Such fish come to the surface to gulp air into the swimbladder. Oxygen is then picked up by blood passing through the vessels in the walls of the swimbladder. Bowfins, garpikes, and many tropical freshwater species supplement their oxygen intake in this way.

The air in the swimbladder of a fish 2000 meters below the water surface is under about 20,000 kilopascals of pressure. Blood coming into the swimbladder carries oxygen at about 20 kilopascals (the same as the oxygen pressure in the water) and encounters air at 1000 times this pressure in the swimbladder. Under these conditions, the blood will pick up huge amounts of gas from the swimbladder, but none of this extra gas must be removed if the volume of the swimbladder is to remain constant. The volume of gas in the swimbladder is kept constant by a special arrangement of blood vessels in the swimbladder wall. The **rete mirabile** ("miraculous net") consists of a tremendous number of hairpin-shaped capillaries lying side by side. The parts of the capillaries going toward the swimbladder lie close to the parts coming back out (Figure 32–19). This

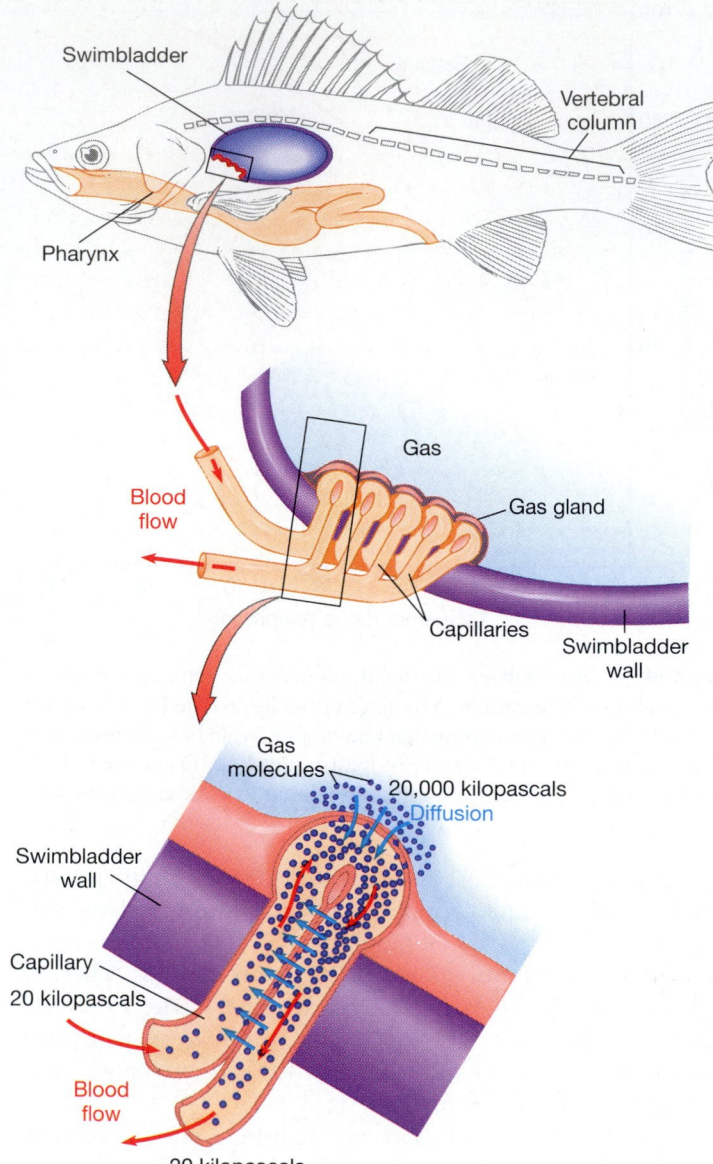

**FIGURE 32–19**  Countercurrent exchange in the rete mirabile in the swimbladder. At depths of 2000 to 3000 m, the gas pressure in the swimbladder is about 20,000 kilopascals. Blood entering the swimbladder is in equilibrium with the low gas pressure in the water, about 20 kilopascals. Gas is exchanged between the long incoming and outgoing portions of capillaries, so that the blood in the body is kept in equilibrium with the low gas pressure in the water, even though the blood comes into equilibrium with the much higher gas pressure inside the swimbladder as it flows through the walls of the swimbladder. Red arrows indicate the direction of blood flow; blue arrows indicate the movement of gases.

arrangement allows for countercurrent exchange: the outgoing blood gives up its extra gas to the incoming blood. So, by the time the incoming blood reaches the swimbladder, it is already in equilibrium with the gas there, and no more gas is removed.

## ESSAY: *Air Pollution and Smoking*

In Greece, six times as many people as usual die in Athens on days when the air is heavily polluted. The Hungarian government attributes 1 in 17 deaths to outdoor air pollution, and the American Lung Association blames 120,000 U.S. deaths each year on this cause. In the United States, Los Angeles, New York, Chicago, and Houston have the worst air pollution.

As we would expect, air pollution affects the parts of the body directly exposed to air—the air passages and the respiratory surfaces of the lungs. (Although skin is exposed to air, its surface is protected by layers of dead cells.) Most of those stricken down during air pollution peaks are elderly people whose lung function was already reduced by emphysema, asthma, bronchitis, pneumonia, or cardiovascular problems.

Meanwhile, air pollution also generates new cases of respiratory infections and lung damage in younger people, who may become air pollution fatalities years later. In Los Angeles, a study of the lungs of traffic and homicide victims aged 14 to 25 showed that 98% had **chronic bronchitis** (infection of the bronchioles), and most showed signs of other lung damage that would have developed into future respiratory and cardiovascular problems.

Most urban air pollution comes from motor vehicles and industry. Deaths on peak pollution days are correlated with the air's content of dust-sized particles of soot, sulfates, and other materials. These particulates are more toxic than the ozone that forms in smog, which is also linked to a spectrum of respiratory symptoms.

Except on air pollution alert days, the air inside a building is nearly always more polluted than the air outside. Indoor air pollutants include fungal spores and dust in air conditioning and heating ducts; chemical fumes from glue in chipboard and other building materials and from plastic in furniture and carpets; and tar and particulate matter in smoke from fires, cooking units, and furnaces. All these pollutants can damage our bodies.

Smoking cigarettes and marijuana is an even more serious source of damage because it introduces particulates and other pollutants directly into the air passages and lungs, regularly and at high concentrations.

Most of the particulate matter in smoke settles out in the lungs (Figure 32–A). **Tars** form a brown, sticky substance that can damage lung tissue. They also contain carcinogenic (cancer-causing) chemicals.

Nicotine in tobacco smoke paralyzes the cilia in the trachea and bronchi. This permits debris to accumulate in the air passages. Hence, smokers suffer excessive damage to the respiratory system from the effects of other substances in tobacco smoke and from other air pollutants. Two common results are chronic bronchitis and **emphysema,** destruction of the gas exchange surface of the lungs' alveoli. Both reduce the rate of gas exchange in the lungs and can lead to disabling shortage of breath. Bronchitis and other respiratory infections result partly from failure of paralyzed cilia to sweep away disease-causing bacteria and partly from smoke's inhibition of bacteria-fighting immune cells that reside in the lungs.

All smoke also contains carbon monoxide, produced by incomplete burning. This gas occurs in smokers' blood at 4 to 15 times the level found in nonsmokers (depending on how much the person smokes). Carbon monoxide combines with hemoglo-

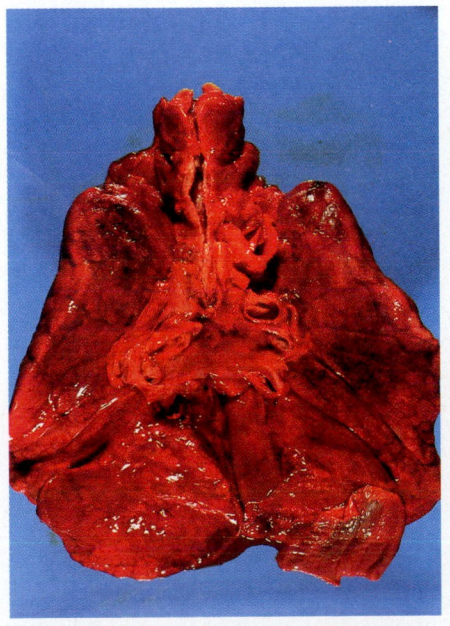
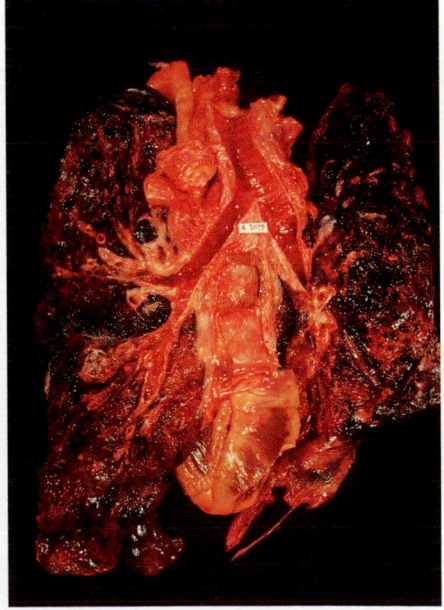

(a)    (b)

**FIGURE 32–A**  Effects of smoking on human lungs. (a) Normal human lungs. (b) Lungs of a cigarette smoker.  (Martin M Rotker/Taurus Photos Inc.)

*(Continued)*

**ESSAY** *(Continued)*

bin and reduces the amount of oxygen the blood can carry. Hemoglobin binds carbon monoxide very tightly, and so the gas lingers in the blood, robbing the body of oxygen, for as long as 6 hours after the cigarette is finished.

The effects of carbon monoxide can be measured after someone smokes just one cigarette. Smoking one cigarette also provides enough nicotine to cause blood vessels to constrict (become narrower). This cuts down the flow of blood and oxygen to the body. It also speeds up the heart rate and increases blood pressure. In addition, breathing smoke for short periods makes blood cells called platelets stickier and more apt to form damaging clots inside blood vessels. A steady smoking habit renews all these effects with each cigarette. This can eventually damage the circulatory system and lead to cardiovascular disease. In fact, smoking contributes to more deaths from cardiovascular disease than from cancer. Smoking damage to the circulatory system may have other consequences, as well. For instance, smokers are usually not eligible for procedures such as kidney or liver transplants. Transplant organs are in short supply and must be given to those with healthy circulatory systems and, therefore, a greater chance of recovering from the surgery.

Another effect of breathing polluted air is that the body is forced to detoxify large amounts of foreign substances. Tobacco smoke, other pollutants, or anesthetic gases in the lungs force cytochrome P450 (Section 32–A) to use up a lot of oxygen detoxifying these substances, instead of transferring this oxygen to the blood.

Pregnant women are especially affected by pollutants in the air. Not only do pollutants reduce the amount of oxygen transported from the lungs into the mother's bloodstream, but they also restrict the amount of oxy-

gen transferred to the fetus at the placenta, where capillaries from the mother and fetus intermingle. If the cytochrome P450 in the placenta is tied up detoxifying substances from the mother's blood, the fetus may lose a lot of its oxygen supply. This may account for the fact that nurses who work with anesthetics have 30% more miscarriages and babies born with birth defects than women in other occupations, and also for the fact that women who smoke tend to give birth to premature or small babies.

Fetuses are not the only nonsmokers affected by smoking. The smoke from an "idling" cigarette contains more tar, nicotine, and cadmium (a toxic metal) than does the smoke inhaled by a smoker. People inhaling cigarette smoke in the air around them are affected by the same kinds of changes as the smoker, but to a lesser degree. Children living with smokers have twice as many respiratory infections as children in nonsmoking homes. They also have more allergies, slower lung growth, and poorer lung function.

In 1919 doctors in a Boston hospital hastily called medical students to an autopsy to see a type of cancer so rare that they might never witness it again: lung cancer. Little did they suspect that smoking cigarettes, a habit adopted by many men around World War I, would soon change this. Today, lung cancer is the most common type of fatal cancer in the United States. (Few women smoked before World War II, and not until 1986 did lung cancer pass breast cancer as the most common cause of cancer death among women.)

Tobacco smoke contains many carcinogens. They cause several types of lung cancer, most of them highly malignant and rapidly fatal. Cells in the lungs of smokers show a number of chromosomal abnormalities. So do cells in the nearby larynx and esoph-

agus, which lie along smoke's path to the lungs, and even in more distant organs such as the kidneys and liver, which process and detoxify carcinogens absorbed into the blood. In addition, the radioactive breakdown products of the radioactive gas radon become attached to particles in smoke and lodge with them in the lung. These breakdown products also linger at higher levels in smoky air. Because smoking exacerbates the effect of radon in this way, about 85% of the cancers attributed to radon occur in smokers. Smokers are similarly much more likely to die of lung cancer caused by inhaling certain kinds of asbestos fibers than are nonsmokers. Studies have implicated smoking in various cancers other than lung cancer. The carcinogen benzene in cigarette smoke is blamed for about 550 cases of adult leukemia every year.

In summary, smoking is a major cause of chronic bronchitis, emphysema, heart disease, and cancer. Smokers can expect to live about ten years less than nonsmokers. About one in every 15 smokers can expect to die of lung cancer (some 40,000 people each year in the United States), and several times that number will die prematurely of cardiovascular disease. In fact, smoking tobacco causes greater health damage than any other drug, legal or illegal.

With all these disadvantages, why do so many people smoke? For reasons that are not fully understood, smoking is highly addictive. However, after learning of the adverse effects of smoking, more than 50 million Americans have managed to stop. In people who quit, the physiological damage done by smoking is gradually reversed. Ten years later, a person's chance of suffering adverse health effects from a previous smoking habit are hardly any higher than those of a lifelong nonsmoker.

# SUMMARY

All animals whose cells carry out respiration must obtain oxygen from their environments and expel $CO_2$. Every cell must be near a medium from which it can obtain oxygen and into which it can dispose of $CO_2$. Obtaining enough $O_2$ is the more difficult task because $O_2$ is much less water-soluble than $CO_2$, and $O_2$ must enter the body in solution. The rate of $O_2$ consumption increases with temperature (for ectothrems) and with metabolic rate. Metabolic rate increases with activity and, in warm-blooded animals, with increasing surface-to-volume ratio and therefore decreasing size.

1. Gas exchange with the environment can take place only by diffusion through a moist surface. Ventilation moves the respiratory medium past the respiratory surface and maintains the steep concentration gradient of gases required for rapid diffusion. Animals that breathe air instead of water use less energy for ventilation, since air has less mass and contains more oxygen. However, air-breathing animals lose body water from the thin, permeable respiratory surface.

2. The four main types of respiratory organs are the body surface, gills, lungs, and tracheal systems. Small or inactive animals can obtain all their oxygen by diffusion through the general body surface. Many have body shapes with a high surface-to-volume ratio. Larger or more active animals usually have part of the body specialized as an expanded respiratory surface. Gases are carried to and from this surface by way of a transport system. An exception is the tracheal system of insects, in which the respiratory surface itself branches throughout the interior of the body.

3. In many animals, the blood contains respiratory pigments that vastly increase its oxygen-carrying capacity. The oxygen saturation of the respiratory pigment hemoglobin depends on the oxygen pressures of the nearby fluid and air. An oxygen dissociation curve shows the oxygen pressures at which a pigment picks up and releases oxygen. A study of these curves shows that pigments are adapted to bind and release $O_2$ at the oxygen pressures in the animal's normal environment.

4. Carbon dioxide is transported mainly in the form of bicarbonate ion, which also acts as an important buffer that helps prevent wide fluctuations in the pH of the body fluids.

5. In vertebrates, the gas content of the blood is regulated by controlling the depth and rate of breathing and therefore the gas composition in the alveoli. Normally, carbon dioxide levels in the body determine the breathing rate.

6. The lungs of terrestrial vertebrates are homologous to the swimbladder of bony fishes. The swimbladder is a gas-filled organ that permits the fish to regulate its buoyancy; it may also be used as an accessory breathing organ.

# Self-Quiz

1. A disadvantage of using air as a respiratory medium is:
   a. it carries less oxygen than water does
   b. it increases the risk of desiccation
   c. oxygen diffuses faster in air than in water
   d. air contains nitrogen as well as oxygen
   e. air pressure changes more than water pressure with changes in temperature
2. The metabolic rate of an ectothermic animal increases:
   a. with increasing environmental temperature
   b. with decreasing environmental temperature
   c. with increase in size
   d. with decrease in muscular activity
   e. with increase in age
3. Which of the following is a handicap to using the general body surface as a respiratory surface?
   a. having a low surface-to-volume ratio
   b. plentiful supply of water in the environment
   c. low metabolic activity
   d. thin slimy mucus covering the skin
4. The countercurrent exchange mechanism in the gills of fish works by:
   a. running the respiratory medium and blood in the same direction
   b. running the respiratory medium in a direction perpendicular to that of the blood flow
   c. maintaining a gradient such that the respiratory medium always contains a concentration of oxygen higher than that of the blood
   d. maintaining a gradient such that the respiratory medium always contains an oxygen concentration lower than that of the blood
   e. keeping the oxygen concentration in the respiratory medium equal to that in the blood
5. The main difference between the insect tracheal system and most other types of respiratory systems is:
   a. tracheal systems do not rely on the blood to transport oxygen to the tissues
   b. insects do not ventilate their tracheal systems
   c. insects do not dispose of carbon dioxide via their tracheal systems
   d. insects do not rely on diffusion to exchange gases in their tracheal systems
   e. oxygen need not be in solution to cross the membranes in the tracheal systems of insects

*(Continued)*

6. The main factor that determines the saturation of hemoglobin with oxygen is:
   a. oxygen concentration in the blood
   b. carbon dioxide concentration in the blood
   c. pH of the blood
   d. hemoglobin concentration in the blood
   e. breathing rate

7. A cat's hemoglobin dissociation curve will lie further to the (left/right) than that of a bear.

8. As blood passes through the capillaries around the alveoli of the lungs, in which direction would you expect each of the following equations to go (i.e., toward the left or toward the right)?
   a. $Hb + O_2 \leftrightarrow HbO$
   b. $H^+ + Hb \leftrightarrow HHb$
   c. $CO_2 + H_2O \leftrightarrow H_2CO3$
   d. $H_2CO_3 \leftrightarrow H^+ + HCO_3^-$

9. Both hyperventilation and holding the breath can cause unconsciousness. Under normal circumstances, why does this occur?
   a. change in carbon dioxide levels in the blood
   b. change in oxygen levels in the blood
   c. loss of hemoglobin from red blood cells
   d. distress of the lungs
   e. excess dissociation of oxygen from hemoglobin

10. Which of the following is the main function of swimbladders in fish?
    a. adjustment of total body density
    b. gas exchange
    c. incubation of the eggs
    d. locomotion by jet-propulsion

## Questions for Discussion

1. Ice fish are a family of bony fishes that inhabit Antarctic waters. These fish have no hemoglobin in their blood. How do you think they are able to survive without this respiratory pigment that all other adult vertebrates possess? What characteristics would you expect a member of this family to show?

2. Many adult amphibians use gills for breathing. There are many reptiles, birds, and mammals that spend most or all of their time in water. Why has none of them evolved so that it retains the embryonic gills and uses them for gas exchange in the adult stage?

3. Animals with gill coverings must have openings to and from the gill area as well as muscles to draw a water current across the gills. Yet few animals have gills without some sort of covering. What is the advantage of a gill covering that has selected for evolution of the covering plus all these accessory arrangements?

4. Do you consider that the respiratory surfaces of your lungs are exposed to the environment? Why or why not?

5. Draw a hypothetical dissociation curve for frog hemoglobin, and on the same graph draw the curve you would expect for frog tadpole hemoglobin. Explain why you drew your curves as you did. (Tadpoles live in water and metamorphose into air-breathing frogs.)

6. Which would you expect to have larger swimbladders, marine fish or freshwater fish? Why?

7. Flounders, sole, and other fish that live on the bottom of the sea lack swimbladders. How can you account for this?

8. Mackerel and other fish that spend their entire lives swimming actively in the upper layers of the ocean also lack swimbladders. How can you account for this?

## Suggested Readings

Avery, M. E., N-S. Wang, and H. W. Taeusch, Jr. "The lung of the newborn infant." *Scientific American,* April 1973. Describes the physiological changes in the lung just before birth and efforts to speed these changes in premature infants.

Comroe, J. H., Jr. "The lung." *Scientific American,* February 1966. Anatomy and physiology of the human lung and respiratory tract, describing physiological measurement techniques and applications.

Feder, M. E., and W. W. Burggren. "Skin breathing in vertebrates." *Scientific American,* November 1985.

Hainsworth, F. R. *Animal Physiology.* Reading, MA: Addison-Wesley, 1981. A comparative physiology text with a good section on gas exchange.

Hock, R. J. "The physiology of high altitude." *Scientific American,* February 1970.

Schmidt-Nielsen, K. "How birds breathe." *Scientific American,* December 1971. Describes the experiments that elucidated the path of air movement in birds' respiratory systems.

Schmidt-Nielsen, K. *Animal Physiology: Adaptation and Environment,* 4th ed. New York: Cambridge University Press, 1990.

Stryer, L. *Biochemistry,* 3d ed. New York: W. H. Freeman, 1988. Chapter 7 covers the oxygen-transporting proteins myoglobin and hemoglobin.

# Animal Transport Systems

## O B J E C T I V E S

*When you have studied this chapter, you should be able to:*

1. Describe the transport system in a cnidarian, an earthworm, an insect, a fish, and a mammal.
2. Describe how an open circulatory system such as that of an insect differs from the closed circulatory system of an earthworm or a vertebrate.
3. Describe the circulation of amphibians and reptiles, and explain how their systems meet the needs of these animals. Describe the double circulation of birds and mammals and list its selective advantages over the single circulation of fishes.
4. Describe the structure and state the main functions of arteries, veins, capillaries, and the heart.
5. Describe the locations of valves in the circulatory system, and explain their structure and function.
6. Trace the flow of blood through the mammalian circulatory system, indicating the sites at which oxygen, carbon dioxide, and food molecules enter and leave the bloodstream, and using the correct names for the chambers of the heart and for the major arteries and veins labeled in Figures 33–12 and 33–13.

7. State why the hepatic portal system is important to the regulation of the blood's composition.
8. Describe the three levels of control of blood pressure in mammals.
9. Describe adjustments made in the circulation as the body's needs increase during exercise.
10. List the principal substances found in blood.
11. State the principal functions of red blood cells, white blood cells, and platelets.
12. List or recognize four functions of the lymphatic system.
13. Describe the mechanism and routes by which fluid moves between the circulatory and lymphatic systems.
14. Define ectothermy and endothermy, explain the differences between them, and describe how endothermic and ectothermic animals regulate the temperatures of their bodies.

---

The earliest organisms lived in the sea, which provided them with oxygen, carried away carbon dioxide and other wastes, and surrounded them with an environment of relatively constant temperature and chemical composition. The immediate environment of cells in higher animals is the **extracellular fluid (ECF).** The ECF may be likened to a tiny captive sea, because it performs the same functions inside the body that the sea outside performs for more primitive organisms. Cells obtain food and oxygen from this fluid and discharge their wastes into it. The body's transport system supplies the ECF with fresh food and oxygen and removes wastes from it, ensuring that the fluid remains an environment where cells can flourish (Figure 33–1).

Most animals have **vascular systems,** which transport fluids. A **circulatory system** is a vascular system in which

**FIGURE 33–1** Two moon jellies (*Aurelia*) in the Pacific. Every cell is bathed by fluid inside the body or by sea water. In higher animals, the extracellular fluid and circulatory system perform the same function of bathing the cells, supplying them with oxygen, and removing waste. (Robert Brons/BPS)

the fluid moves in a particular direction, usually because it is propelled by a muscular, pumping structure, generally called a heart.

Having an efficient transport system permits an animal's cells to obtain and use food and oxygen more rapidly even if the animal is quite large. It also supports a higher metabolic rate, permitting the animal to be more active than it otherwise could. For instance, the metabolic rate of a bird in flight, measured by its use of oxygen, is more than a thousand times that of a sea anemone with its inactive lifestyle.

In addition to transporting gases, food, and wastes, the circulatory system carries hormones, as well as molecules and cells that help to protect the body from disease (Chapter 34). Another important, and indeed inevitable, function performed by circulatory systems is the distribution of heat. Many animals can control their exchange of heat with the environment by adjustments within the circulatory system.

---

### KEY CONCEPTS

♦ Vascular systems transport substances and heat from one part of the body to another and between the animal's external environment and the extracellular fluid that surrounds every cell.

♦ Animals with faster, more direct transport systems can metabolize faster and lead more active lives than those with less efficient vascular systems.

---

## 33–A    TRANSPORT IN INVERTEBRATES

### Cnidaria

Cnidarians are slow-moving animals with low metabolic rates. They obtain most of their oxygen by diffusion through the body surface, and soluble wastes diffuse out in the opposite direction. Food is transported by the **gastrovascular cavity,** which doubles as a digestive (gastro) and transport (vascular) system.

In *Hydra* the gastrovascular cavity extends from the center of the body into each tentacle. In larger jellyfish, it extends into a series of canals that branch throughout the body (Figure 33–2). The fluid in the system is pushed in

defined pathways by cilia lining the canals. A cnidarian's metabolic needs increase when it swims or feeds, and the contraction of its muscle fibers during these activities also speeds the flow of fluid through the canals, automatically speeding delivery of food as the demand increases.

### Turbellarians

Turbellarians, which are free-living platyhelminths, have higher metabolic rates than cnidarians. However, because they have flattened bodies, the general body surface still provides enough surface area for gas exchange. There is no separate circulatory system or blood. Food is distributed by the digestive cavity, which branches throughout the body, providing a large surface area from which food can be absorbed. The excretory system also branches throughout the body and collects wastes (Figure 33–3).

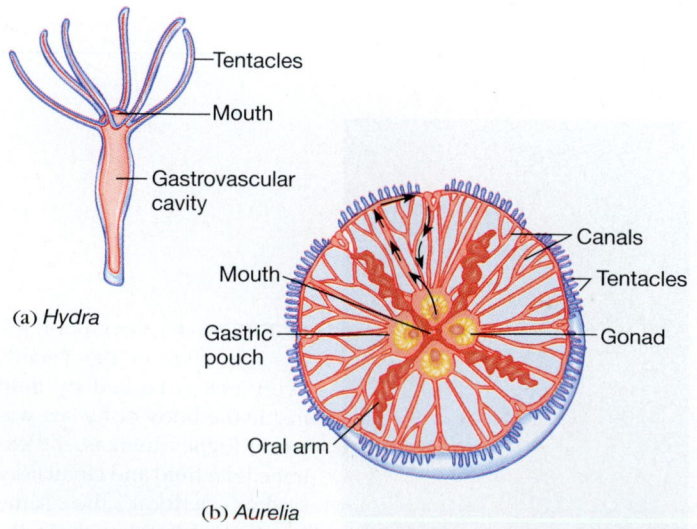

(a) *Hydra*

(b) *Aurelia*

**FIGURE 33–2**  Gastrovascular cavities of two cnidarians. Arrows show the movement of fluid in the jellyfish *Aurelia,* from the gastrovascular cavity to outlying parts of the body by way of a branching system of canals.

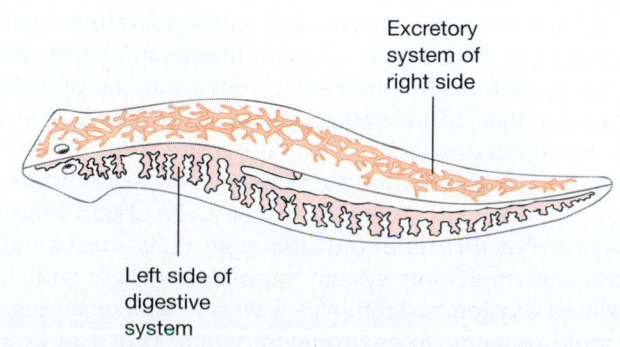

**FIGURE 33–3**  Transport in a turbellarian. The digestive and excretory systems branch throughout the body and perform their separate functions. Only half of each system is shown.

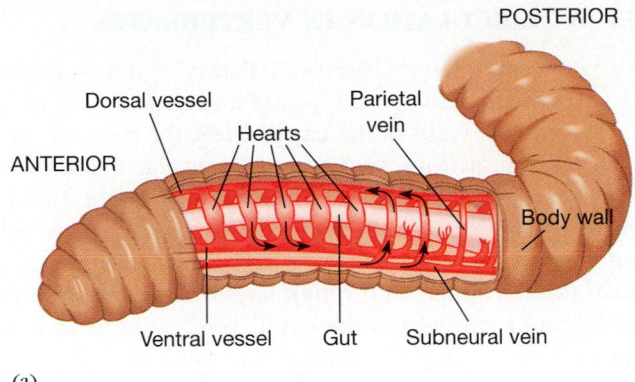

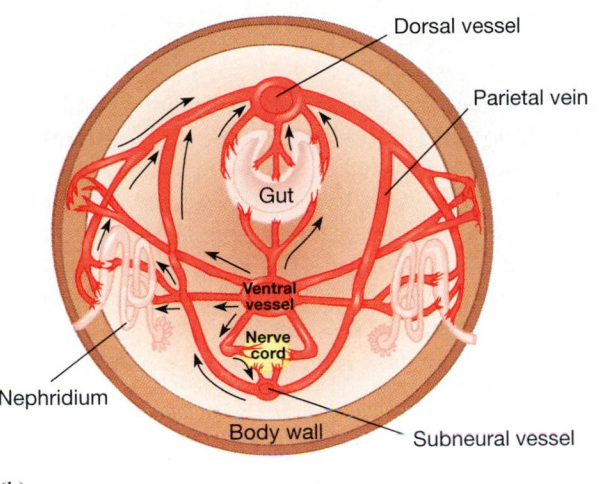

**FIGURE 33–4**  Circulatory system of an earthworm. (a) Arrangement of the major blood vessels in several segments. (b) Some of the blood vessels found in a single segment. Nephridia are excretory organs, which control the composition of the blood.

## Annelids

An annelid has a **coelom,** a fluid-filled cavity in which organs can move independently of one another. The coelom provides room for expansion of the dorsal blood vessel and pumping movements by the five pairs of lateral blood vessels enlarged as muscular "hearts." Muscles are more powerful than cilia, and so an earthworm's hearts provide faster circulation than the ciliated canals of cnidarians.

An earthworm has a simple **closed circulatory system,** in which the blood never leaves the vessels (Figure 33–4). The walls of the vessels are thin, and substances can diffuse between the blood and the ECF bathing the cells. The blood of earthworms contains a type of hemoglobin, which increases the amount of oxygen transported.

A circulatory system that moves the blood steadily at all times permits division of labor among organs and tissues. In platyhelminths, which have no circulatory system, the gut and excretory organs branch throughout the body, serving as their own circulatory systems. In an annelid, by

comparison, the gut is simple and unbranched, specialized only for digestion. The separate circulatory system carries the digested food to all of the body cells.

Division of labor among organs is a major reason for the physiological efficiency of higher animals. One example is the concentration of nerve cells in a brain. Here information can be processed more rapidly, and the animal can respond faster, than it could if messages had to travel around a nerve network scattered throughout the body. However, because nerve cells are easily damaged by temporary shortages of food and oxygen, a brain with many nerve cells can exist only when there is an efficient circulation to supply its needs rapidly.

## Insects

The insect circulatory system is of peculiar interest. Because many insects are much more active than other invertebrates, we would expect to find a very efficient circulatory system of closed vessels and a heart. However, in even a large dissected insect, such as a big cockroach, it is difficult to see any circulatory system at all. Surprisingly, most insects have **open circulatory systems,** with only one, open-ended blood vessel. The blood (**hemolymph**) is not distinct from the other body fluids, and the entire body cavity is a **hemocoel,** filled with hemolymph.

The hemolymph is moved by contractions of a long, thin-walled heart, lying just below the dorsal surface of the exoskeleton (Figure 33–5), and by movements of the body

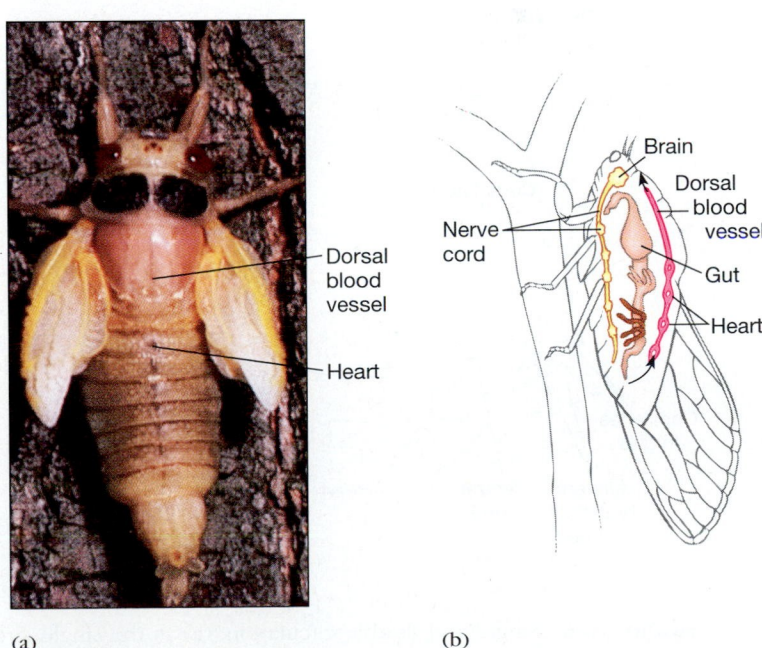

**FIGURE 33–5**  Insect circulatory systems. (a) The heart of this newly molted cicada can be seen just beneath the transparent exoskeleton of the dorsal body surface. (b) Circulation: blood from the hemocoel enters the heart through slits in its wall and is pumped out the anterior end of the dorsal blood vessel.

muscles. The heart pushes blood forward and out through a vessel that usually ends in the head, near the brain. In many insects, auxiliary hearts at the bases of appendages (wings, antennae, legs, and so forth) pump blood into channels that extend out into the appendages.

How can such an open circulatory system work efficiently enough to supply an insect's needs? The secret is that the circulatory system does not transport oxygen. The high blood pressure and rapid circulation of mammals are necessary because some mammalian organs, such as the brain, need vast amounts of oxygen. However, food, hormones, and wastes need not be moved with the same speed and efficiency. In an insect, oxygen reaches cells by way of the **tracheal system,** a series of air-filled tubes branching throughout the body. Other substances can travel more slowly via the hemolymph.

In a closed circulatory system, the blood moves in one direction along a defined route. In an open system, such as that of an insect, the pattern of flow is still predictable and generally one-way, rather than random sloshing back and forth. Also, without blood vessels, hemolymph bathes the tissues directly, with no vessel walls separating the two. Furthermore, although the heart provides the main impetus to the blood in a resting insect, the movement of other muscles speeds circulation in an active insect.

## 33–B  CIRCULATION IN VERTEBRATES

All vertebrates have closed circulatory systems. Exchange of substances between the blood and the ECF occurs only across the thin walls of the **capillaries,** the narrowest blood vessels. Contractions of a strong, muscular heart exert the pressure needed to force the blood through the fine tubes of the capillaries. The heart contains chambers, which fill with blood and then pump it out. Blood travels from the heart to the capillaries through larger blood vessels, the arteries, and returns from the capillaries to the heart through veins.

### Fish

In fish, the heart has an **atrium** (receiving chamber), which collects blood and then releases it into the **ventricle** (pumping chamber), with thicker, more muscular walls (see Figure 33–7). If the heart had only one chamber, blood flow would stop every time it contracted. Having two chambers means that blood from the body can fill the atrium while the ventricle pumps blood out into a short, muscular artery, the **ventral aorta** (plural: **aortae**). The ventral aorta carries blood to the gills, where the aorta branches into smaller

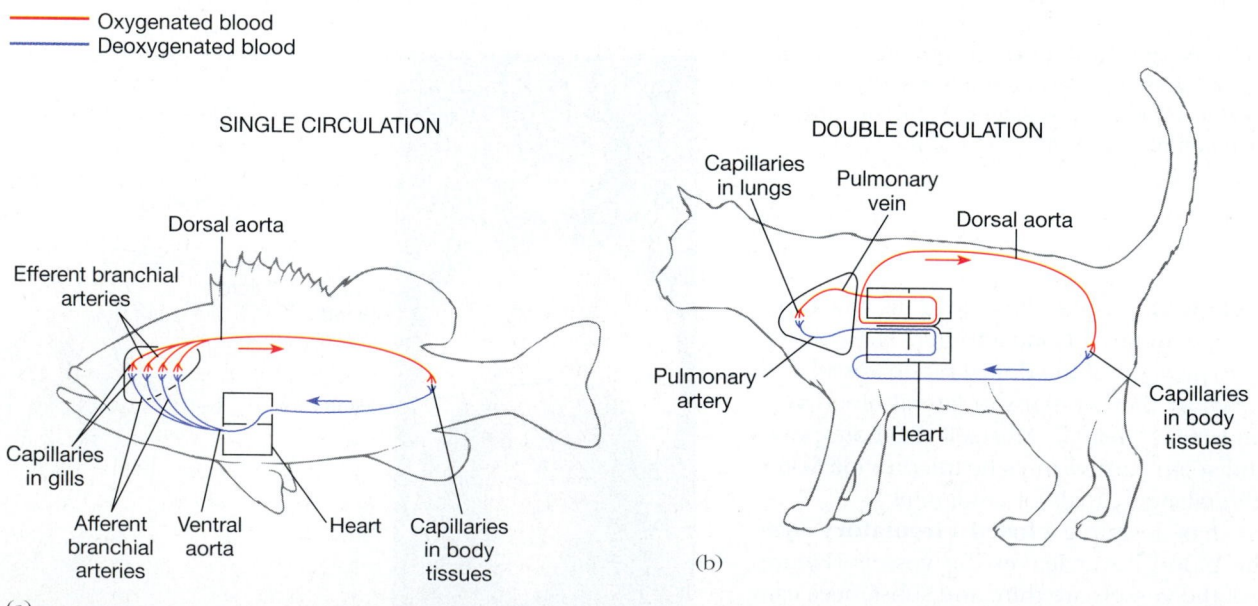

**FIGURE 33–6**  Single and double circulation. (a) In the single circulation of a fish, blood travels in one loop, passing through the heart once during each circuit of the body. Pressure in the dorsal aorta is low because the narrow gill capillaries slow the blood as it picks up oxygen. (b) In the double circulation of birds and mammals, blood must pass through the heart twice before it returns to the same point. Blood returning to the heart from the lungs is pumped at high pressure into the systemic (body) circulation.

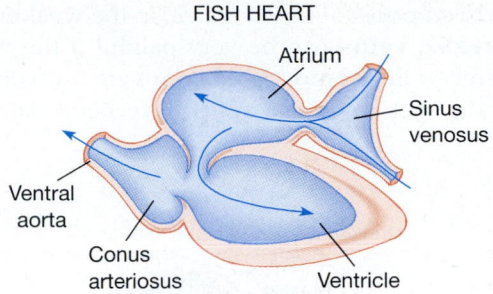

FISH HEART

Atrium

Sinus
venosus

Ventral
aorta

Conus
arteriosus

Ventricle

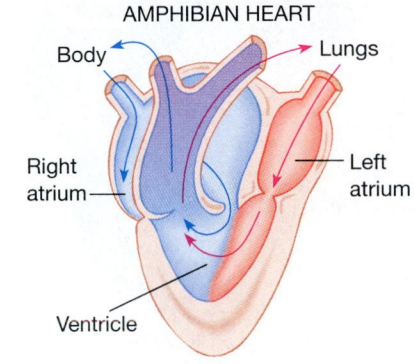

AMPHIBIAN HEART

Body

Lungs

Right
atrium

Left
atrium

Ventricle

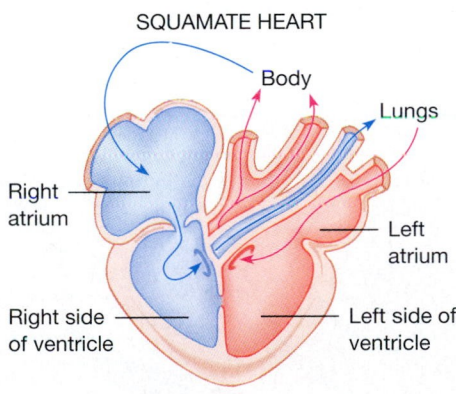

SQUAMATE HEART

Body

Lungs

Right
atrium

Left
atrium

Right side
of ventricle

Left side of
ventricle

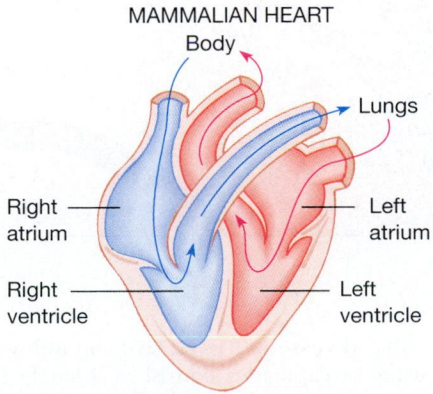

MAMMALIAN HEART

Body

Lungs

Right
atrium

Left
atrium

Right
ventricle

Left
ventricle

**FIGURE 33–7** Blood flow through the hearts of some vertebrates. Blue = deoxygenated blood; red = oxygenated blood. The amphibian heart has two atria but only one ventricle. In most reptiles, including all squamates, the ventricle is partly divided, and oxygenated and deoxygenated blood are largely separate. The heart of mammals and birds has two atria and two completely separate ventricles.

vessels and eventually into capillaries. Here the blood picks up oxygen. From the gill capillaries, blood flows into the **dorsal aorta,** whose branches distribute blood to the capillaries of all the body organs. Blood returns to the heart through the veins.

In this **single circulation,** blood passes through the heart only once in a complete circuit around the body (Figure 33–6). Such a system has the advantage that blood going to the body has already been oxygenated in the gills. A disadvantage is that the narrow gill capillaries resist passage of the blood, so that blood leaves the gills at a lower pressure than when it entered. No matter how hard the heart pumps, the blood in a fish's dorsal aorta travels relatively slowly and at low pressure because it has had to pass through the gill capillaries. This slows oxygen delivery to the cells and limits the metabolic rate that fish can attain.

## Amphibians and Reptiles

In tetrapods, the problem of low blood pressure resulting from the network of capillaries at the respiratory surface is overcome by a **double circulation,** which passes blood through the heart twice in each complete circuit of the body. Blood is first pumped from the heart to the lungs, where it is oxygenated. Blood then returns to the heart where the second pumping raises the blood pressure again before the blood travels to the rest of the body.

The hearts of birds and mammals are divided into two sides, each with an **atrium** (receiving chamber) and a **ventricle** (pumping chamber). The right side of the heart receives blood from the body and pumps it to the lungs. The left side receives blood returning from the lungs and pumps it to the body (Figure 33–7).

The hearts of most amphibians, squamates, and turtles are not fully divided into two. In amphibians, some blood from the lungs returns to the lungs instead of passing to the rest of the body each time the heart contracts. This is not so inefficient as it sounds because many amphibians absorb more oxygen through the skin than through the lungs or gills. Blood returning to the heart from the skin often contains more oxygen than that in the **pulmonary vein** from the lungs, and there would be little advantage in keeping the blood from these two sources separate.

Reptiles obtain nearly all their oxygen from the lungs. Blood returning from the lungs to the left atrium of the heart through the pulmonary vein is well oxygenated. Because there are two atria this blood does not mix with blood returning from the body to the right atrium. In most reptiles the ventricle is only partially divided. However, experiments show that valves in the ventricle ensure that there is little mixing of blood from the two atria. Thus, the heart is functionally, if not anatomically, divided. (Crocodilians are the exception, with completely divided atria and ventricles.)

In reptiles, blood travels from the heart to the body via paired dorsal aortae, one on each side of the body.

## Mammals and Birds

Mammals and birds have only one dorsal aorta, instead of the two smaller ones of their reptilian ancestors. This gives mammals and birds higher blood pressure, supporting their very high metabolic rates. This adaptation arose independently in diapsids and synapsids. Birds have retained the right, and mammals the left, of the reptilian paired aortae.

Both birds and mammals have a double circulation with the ventricles completely separated. This has two important effects. First, keeping oxygenated and deoxygenated blood separate in the heart ensures that blood reaching the body organs from the aorta contains as much oxygen as possible. Second, it ensures high blood pressure in the aorta. Higher blood pressure means faster circulation; oxygen and food reach the tissues faster, and waste is removed more rapidly.

## 33–C   HUMAN CIRCULATION

### Blood Vessels

Arteries, capillaries, and veins are the pipes through which blood travels to the tissues (Figure 33–8). **Arteries** are vessels that carry blood away from the heart. Their walls are muscular and highly elastic. The arteries branch and re-branch into smaller **arterioles,** which divide even further into capillaries.

Capillaries are so narrow that red blood cells must pass through them in single file (Figure 33–9). The capillary walls are only one cell thick, in keeping with their role as the sites where substances pass between the blood and the ECF. The far ends of capillaries rejoin, forming larger vessels, **venules,** which join into **veins,** blood vessels leading back to the heart. The walls of veins contain connective tissue and muscle, as do those of arteries, but veins are much less elastic and tend to have larger internal diameters.

Another important difference between veins and arteries is that veins in the lower body contain **valves,** flaps of tissue that help to keep blood flowing in one direction. These valves open under the pressure of blood going toward the heart and close when it begins to go backward under the pull of gravity (Figure 33–10).

When a vein's walls are weakened, blood may collect in the vein and distend it so that the valve flaps cannot meet. The valve cannot now prevent blood from flowing

backwards, so pools of blood collect in the weakened vein. Such **varicose veins** can be very painful if the weakness is in a large vein. **Hemorrhoids** are varicose veins in the walls of the rectum. These veins have been damaged by

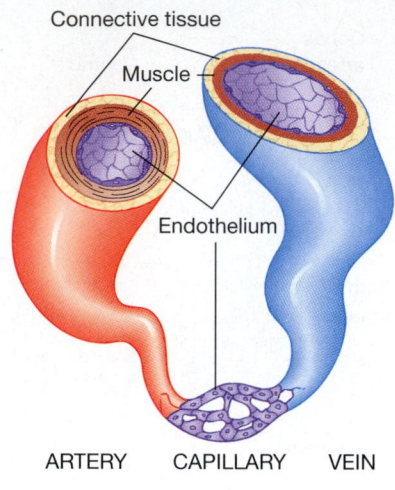

(a)

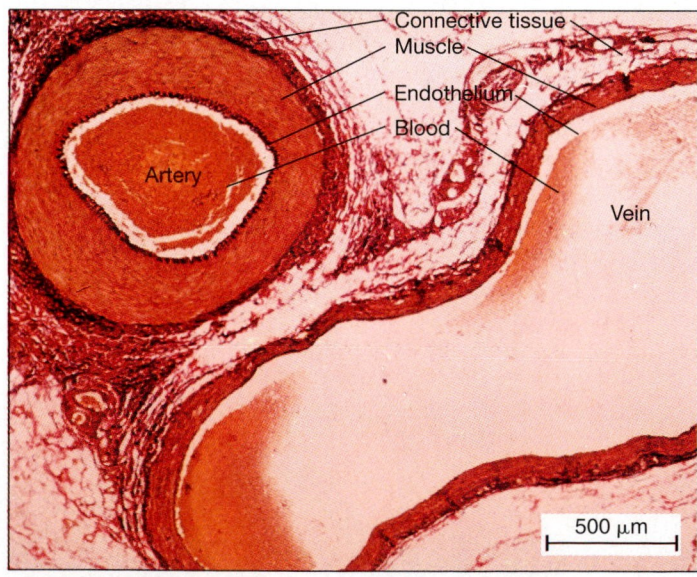

(b)

**FIGURE 33–8**   Blood vessels. (a) The layers in the walls of blood vessels. The walls of capillaries consist of a single layer of epithelium, called the endothelium. In veins and arteries the endothelium is surrounded by muscle and connective tissue. The walls of veins are thinner and flabbier than those of comparable arteries. (b) A thin-walled vein and a thick-walled artery.   (b, Biophoto Associates)

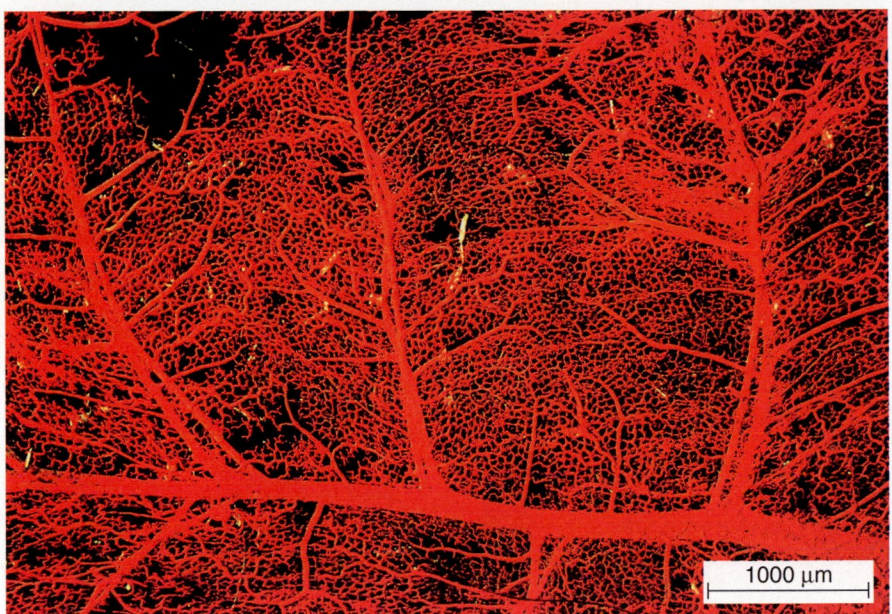

**FIGURE 33–9** A capillary bed. Blood flows from the bottom to the top of the picture. Across the bottom is an artery with three main pairs of smaller arteries leading up from it. The arteries branch into ever-smaller blood vessels, until the blood reaches the fine network of capillaries from which it flows into the veins at the top of the photo. (Biophoto Associates)

pressure resulting from conditions such as constipation or pregnancy.

The heart too has valves that direct the flow of blood in a one-way path. Valves between the atria and ventricles prevent backflow of the blood into the atria when the ventricles contract, and valves between the ventricles and arteries prevent blood from falling back into the heart when the ventricles relax after pumping the blood out.

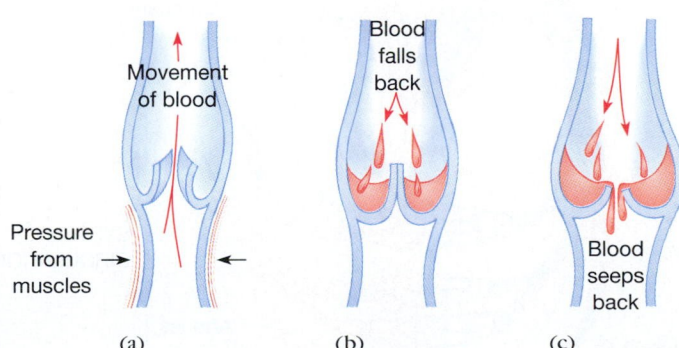

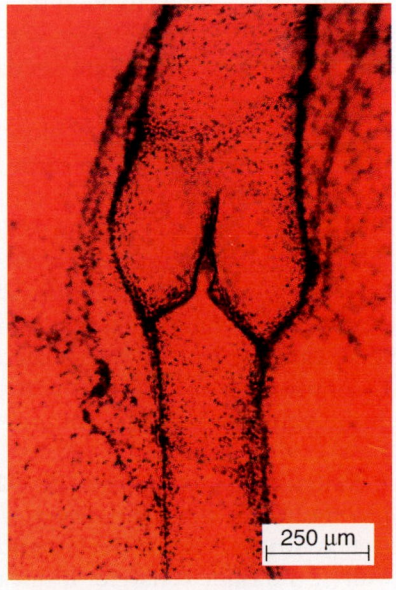

**FIGURE 33–10** Valves in veins. (a) When blood flows toward the heart, the valve opens and allows it to pass. (b) If blood moves in the reverse direction, it fills the cup-like flaps of the valve and presses the edges together, preventing backflow. (c) The walls of a varicose vein are weak and allow blood to collect and distend them so that the edges of the valve flaps cannot meet. Blood may then return through the valve, and circulation is impaired. (d) A valve in a vein. (d, John D. Cunningham/VU)

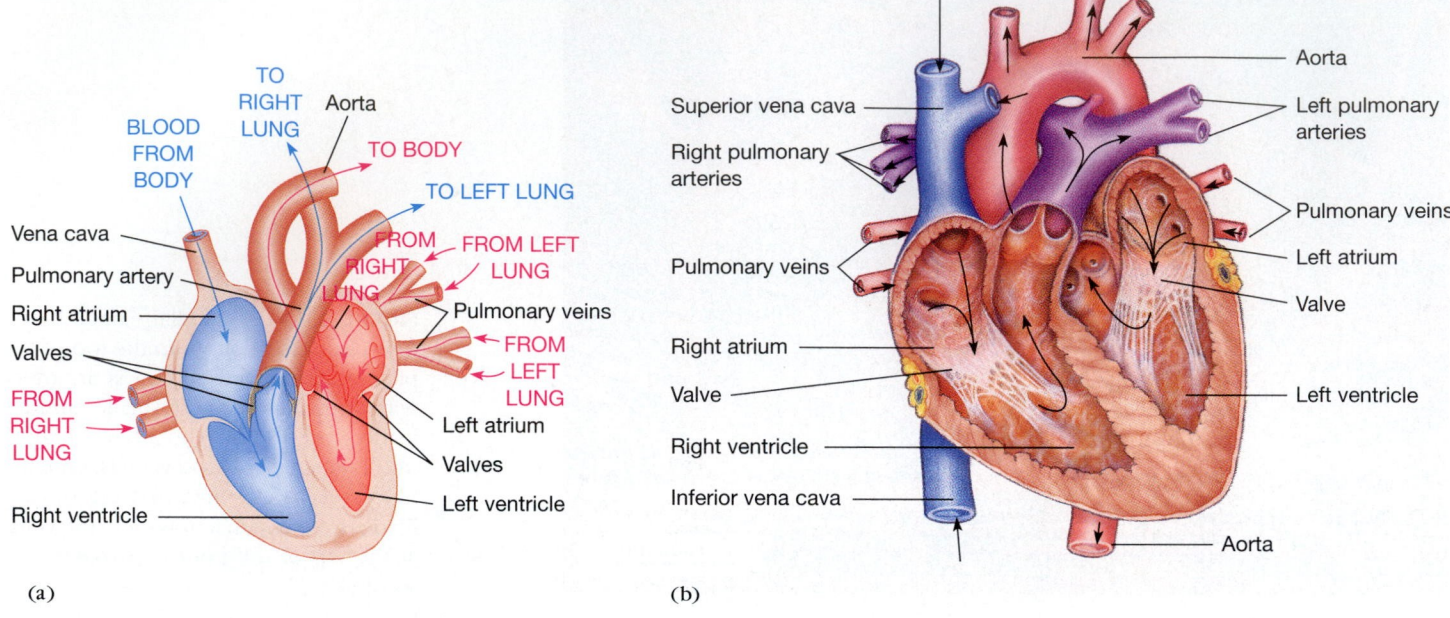

(a)                    (b)

**FIGURE 33–11**   The mammalian heart. (The heart's owner is facing us, so that the heart's right side is on our left.) (a) Basic structure of the heart, showing the major blood vessels. (b) The main vessels and the valves in more detail.

## The Circuit of Blood in the Body

Closed blood vessels, with valves preventing backflow, ensure that the blood of a vertebrate flows in only one direction and in definite channels. Blood returns to the heart from the body via two large veins, the **venae cavae.** It then flows through the right atrium and on into the right ventricle (Figure 33–11). Contraction of the right atrium sends more blood in to "top off" the ventricle.

When the right ventricle contracts, it pushes the blood through a valve into the **pulmonary artery,** which carries it to the lungs. In the lungs the blood flows through capillaries surrounding the air-filled alveoli. Blood picks up oxygen and loses carbon dioxide through the thin walls of the alveoli and lung capillaries. The freshly oxygenated blood then flows through venules and veins, which eventually join to form the pulmonary veins carrying freshly oxygenated blood back to the heart. This time the blood enters the left atrium and passes through the valve into the left ventricle (Figure 33–12).

When the left ventricle contracts, the oxygenated blood passes through a valve into the aorta, the main artery to the body. The wall of the left ventricle is much thicker and more muscular than the wall of the right ventricle. It must push the blood throughout the body, not just on the short journey to the lungs.

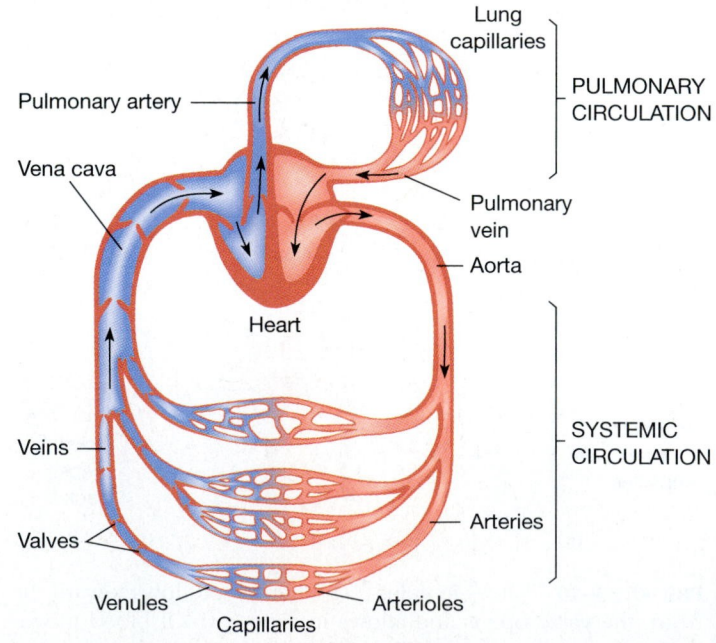

**FIGURE 33–12**   Outline of the circulation in a mammal. The heart and its blood vessels have been rearranged somewhat for clarity. Blue = deoxygenated blood. Red = oxygenated blood. Follow the direction of the arrows through a complete circuit, noting that the blood passes through the heart twice before it returns to the starting point.

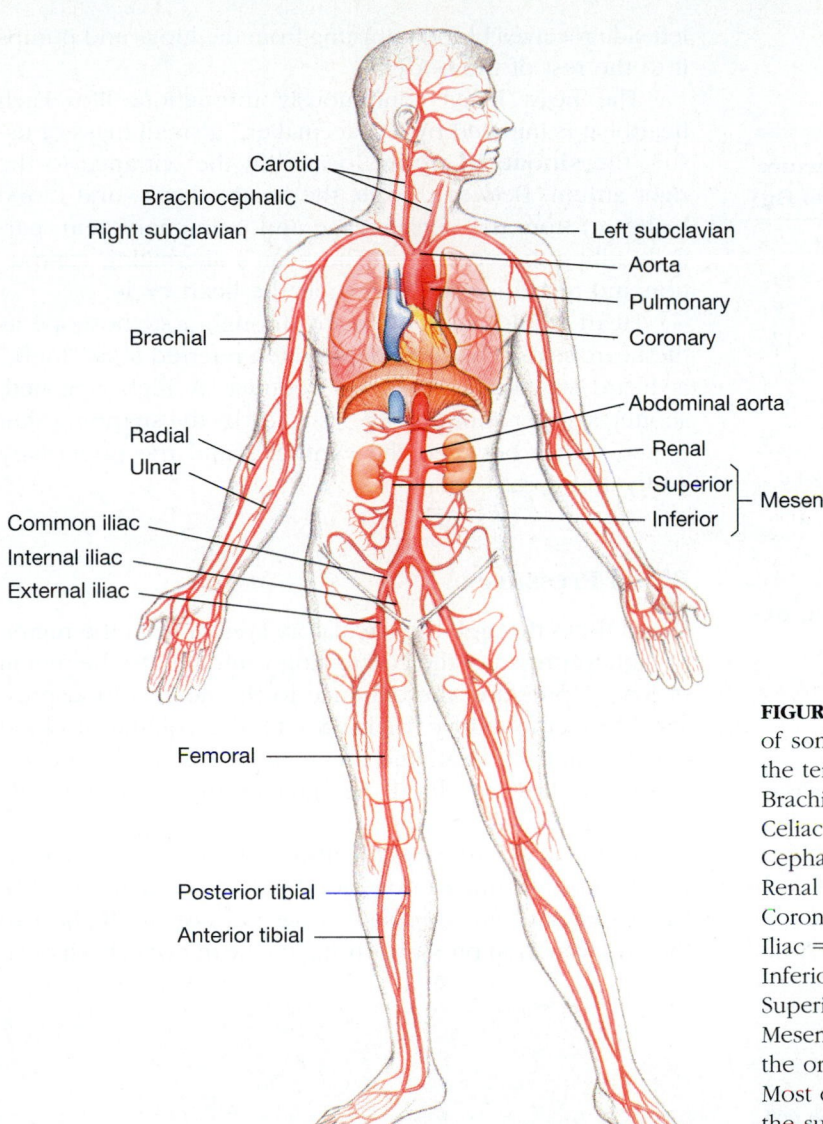

Carotid
Brachiocephalic
Right subclavian
Left subclavian
Aorta
Pulmonary
Coronary
Brachial
Abdominal aorta
Radial
Renal
Ulnar
Superior ⎤
         ⎬ Mesenteric
Inferior ⎦
Common iliac
Internal iliac
External iliac
Femoral
Posterior tibial
Anterior tibial

**FIGURE 33–13**   The major arteries of the human body. The names of some of the arteries will make more sense if you know what the terms mean.
Brachial = of the arm.
Celiac = abdominal (between diaphragm and hips).
Cephalic = of the head.
Renal = of the kidneys.
Coronary = of the heart.
Iliac = of the groin.
Inferior = lower.
Superior = upper.
Mesenteric = of mesenteries, sheets of connective tissue that hold the organs in place.
Most of the other names are also those of bones in the skeleton: the subclavian artery lies below the clavicle, or collar bone. The radius and ulna are the bones of the forearm; the femur and tibia are bones in the leg.

The aorta gives rise to many branch arteries, which take blood to the wall of the heart itself (**coronary arteries**), to the head (**carotid arteries**), and to the digestive system, kidneys, and other organs (Figure 33–13).

Blood leaving capillaries in the body enters venules and finally veins, all of which join into the venae cavae and empty into the right atrium of the heart.

As the ventricles push blood into the arteries, the artery muscles relax and the elastic walls expand to accommodate the blood flowing through them. As the blood passes, the artery walls contract and exert pressure on the blood. Blood pressure decreases steadily in the arterioles, capillaries, venules, and veins, mainly because the vessel walls exert frictional resistance to the movement of the blood (Table 33–1). The narrower the vessel, the greater the resistance,

and so blood pressure and flow rate drop almost to nothing in the capillary beds. The veins are flabby and do not help to push the blood back to the heart. In addition, blood in the veins below the heart must be returned against the pull of gravity. Thus, blood tends to collect in the veins.

Blood in the veins obviously does return to the heart, propelled mainly by the muscles of the body. When muscles contract, they squeeze against the veins, forcing blood along. Because muscular contraction is needed to push blood through the veins, it is more tiring to stand still than to walk for an equal period. Standing allows blood to collect in the veins of the feet and legs. The feet swell with stranded blood, and the body temporarily loses the use of blood that should be distributing oxygen and nutrients to other tissues. Studies have shown that students who jiggle

**T A B L E   3 3 – 1**

**Pressure, Volume, and Velocity of Blood in the Human Circulation***

|  | Volume (cc) | Velocity (cm/sec) | Pressure (mm Hg) |
|---|---|---|---|
| Aorta | 100 | 40 | 100 |
| Arteries | 325 | 40 – 10 | 100 – 40 |
| Arterioles | 50 | 10 – 0.1 | 40 – 25 |
| Capillaries | 250 | less than 0.1 | 25 – 12 |
| Venules | 300 | less than 0.3 | 12 – 8 |
| Veins | 2200 | 0.3 – 5 | 10 – 5 |
| Vena cava | 300 | 5 – 20 | 2 |

*From "The Venous System," by J. Edwin Wood. Copyright © 1968 by *Scientific American Inc.* All rights reserved.

their feet are more alert, and perform better on long exams, than their peers who sit still.

## The Heart Cycle

The heart of a bird or mammal is really two pumps working side by side. Each pump is made up of a thin-walled atrium, which receives blood from veins and pumps it into the adjoining ventricle, and a thick-walled ventricle, which pumps blood into arteries. The right side of the heart receives blood from the body and pumps it to the lungs; the left side receives blood returning from the lungs and pumps it to the rest of the body.

The heart beats continuously throughout life. Each heartbeat is initiated by a "pacemaker," a small mass of tissue, the **sinoatrial node,** located at the entrance to the right atrium. Between beats, the heart relaxes, and blood rushes in from the venae cavae and pulmonary vein, partially filling the ventricles. The heart's alternating contraction and relaxation are known as the heart cycle.

Heart sounds may be heard through a stethoscope as the heart beats. The first sound, often referred to as "lubb," is heard when the ventricles contract. A higher-pitched, shorter, sharper sound, "dup," is made by the snapping shut of the valves between the ventricles and the pulmonary artery and aorta.

## Blood Pressure

Blood flows through the circulatory system from the region of highest pressure, the contracting ventricles, to the region of lowest pressure, the entrance to the heart. Blood pressure is determined by the heart rate, the volume of blood expelled at each beat, and the resistance of the blood vessels to the flow of blood. The greater any of these is, the higher the blood pressure.

The brachial artery in the arm, just above the elbow, is usually used for measuring blood pressure (Figure 33–14). Blood pressure is expressed as the ratio of the highest to the lowest arterial pressure during the heart cycle, both mea-

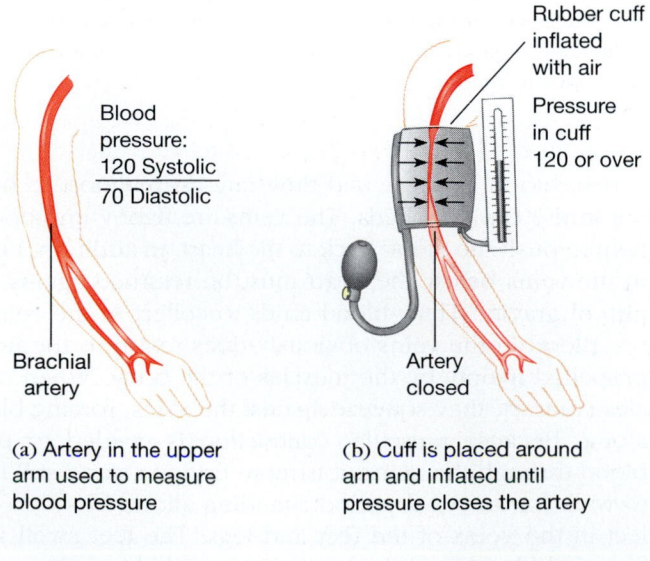
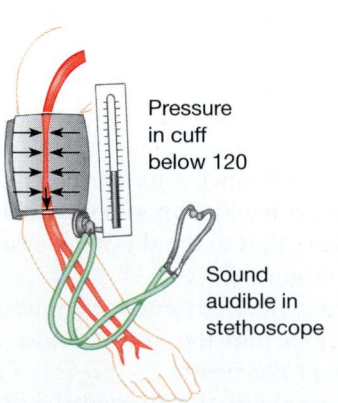
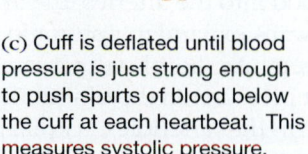
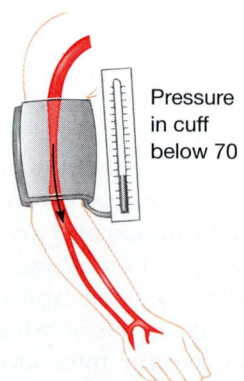

(a) Artery in the upper arm used to measure blood pressure

(b) Cuff is placed around arm and inflated until pressure closes the artery

(c) Cuff is deflated until blood pressure is just strong enough to push spurts of blood below the cuff at each heartbeat. This measures systolic pressure.

(d) Cuff is loosened until blood flows freely. Sounds disappear. Pressure now is diastolic pressure, between heart contractions.

**FIGURE 33–14** How human blood pressure is measured. The instrument used is a sphygmomanometer, an inflatable cuff attached to a gauge that measures air pressure in the cuff.

sured in millimeters of mercury (mm Hg). For instance, 120/80 is considered "average" blood pressure for a man of 20. The first figure, **systolic pressure,** indicates the force with which the left ventricle pushes blood. The second figure, **diastolic pressure,** measured when the heart is relaxed, indicates the resistance of the blood vessels: it is useful in diagnosing hardening of the arteries or strain on their walls. The mean pressure over the whole heart cycle is usually maintained at about 100 mm Hg.

The blood pressure of a normal adult seldom varies by more than 15% because the body has three main pressure control systems that keep the blood pressure constant. Fastest to act are nervous receptors that respond within seconds to correct abnormal pressure. Minutes later, changes in blood pressure activate hormonal controls. Still more slowly, within hours or days, the kidneys react by increasing the volume of fluid in the body when blood pressure falls and decreasing it when blood pressure rises.

1. **Baroreceptors.** The walls of the aorta and carotid arteries contain **baroreceptors,** pressure receptors that are stimulated when increased pressure stretches the artery wall. These baroreceptors send nervous signals to the brain, which in turn signals small blood vessels to dilate (relax), increasing the diameter of the blood vessels to accommodate the blood flow more easily. The brain also slows the heartbeat. Both of these measures lower the blood pressure.

   Similar fast-acting pressure receptors in the brain react when blood pressure in the brain becomes too low. Low blood pressure in the brain means slower delivery of food and oxygen and the risk of brain damage or death within a few minutes. If blood pressure falls, the brain's pressure receptors send strong signals to the heart and blood vessels to increase the blood pressure.

2. **Renin and angiotensin.** Within minutes after the nervous system responds to a change in blood pressure, hormonal systems are activated. The most important of these responds when low blood pressure reduces the flow of blood to the kidneys. This makes the kidneys secrete the enzyme **renin** into the blood. Renin hydrolyzes a blood protein into the hormone **angiotensin,** a small peptide containing only eight amino acids residues. Angiotensin travels in the blood, binding to receptors on the membranes of cells lining small blood vessels. This causes muscles in these blood vessels to contract, the blood vessels decreasing the diameter of and thereby increasing the blood pressure.

3. **Kidney-fluid system.** When arterial pressure rises above normal, the excess pressure causes the kidneys to excrete water and salt faster than the body takes them in. This decreases the volume of blood that the heart must pump and the arterial pressure falls. Conversely, when blood pressure falls below normal, excretion by the kidneys slows, retaining fluid in the blood and maintaining blood pressure. This system takes several days to correct abnormal blood pressure.

Baroreceptors and angiotensin are adapted mainly to prevent damage to the body when the blood pressure falls as a result of bleeding. They are stopgap measures that improve the situation until long-term regulation by the kidney-fluid system can correct changes in blood pressure.

About 20% of the U.S. population suffers from **essential hypertension,** high blood pressure of unknown origin. Hypertension must be controlled because it increases the likelihood of strokes, which can cause brain damage or death. Drugs that control hypertension include **diuretics,** which increase the volume of water excreted by the kidneys. In people with diseased kidneys that do not excrete salt normally, a diet very low in salt may help to control hypertension. Hypertension is poorly understood because we still do not understand how all the factors that control blood pressure interact.

## Cardiovascular Disease

**Cardiovascular diseases** are diseases of the heart and blood vessels. They include hypertension and atherosclerosis. In **atherosclerosis** the artery walls develop deposits of lipids that reduce the internal diameter of the blood vessel and also make the artery walls less elastic.

Cardiovascular diseases cause death in many ways, such as strokes and heart attacks. A stroke is a blood clot or rupture of a blood vessel in the brain, which cuts off the blood supply to part of the brain. A heart attack occurs when the blood supply to part of the muscle that makes up the heart fails. With their blood supply cut, the heart muscle cells stop contracting and may die. Even if the victim recovers from the heart attack, part of the heart muscle may have been killed and the heart permanently weakened.

Susceptibility to heart attacks is correlated with blood levels of HDLs and LDLs, lipoprotein molecules that transport cholesterol in the blood. Most cholesterol is carried by **low-density lipoproteins (LDLs);** some is carried by **high-density lipoproteins (HDLs).** Studies of people from many ethnic groups have shown that the risk of heart attack is greater the higher the LDL concentration and the lower the HDL concentration in the blood. Some people who have permanently high levels of HDL or low levels of LDL because of their genetic makeup apparently never die of heart attacks caused by atherosclerosis.

The factors correlated with high HDL levels are those long known to be associated with a low risk of heart attack: being female (a woman's HDL level is about 10% higher than a man's), being slim, exercising, not smoking, and consuming moderate amounts of fat and alcohol.

Various "risk factors" render people more liable to heart disease. Three genetic factors seem to make people particularly susceptible: having at least one parent with cardiovascular disease, being male, or carrying a gene that is more common in people of African than of European or Asian origin. For the foreseeable future, these genetic factors are beyond our control. The "big three" nongenetic risk factors for heart attack are high blood pressure (hypertension), high blood cholesterol levels, and smoking. The first two of these may also have a genetic basis, but they can often be decreased by some combination of diet, exercise, and medication. Smoking is a greater cause of heart disease than of cancers and is especially risky to women taking oral contraceptives.

Deaths from heart attacks have decreased in the United States since about 1970, a change that has occurred to a lesser degree in other developed nations. Most experts credit changes in lifestyle: less smoking and more aerobic exercise—walking, jogging, bicycling, swimming, and so on. We are also changing our dietary habits, monitoring our blood pressure, and becoming more aware of the early warning signs and symptoms of cardiovascular disease. In addition, there are fewer cases of rheumatic fever and heart disease caused by complications of bacterial infections, which were much more common before the discovery of antibiotics.

The factors that cause cardiovascular disease seem to have additive effects that are difficult to disentangle, and this is probably why it is impossible to give a single reason for the decline in heart disease. Reducing deaths from cardiovascular disease has not increased U.S. life expectancy, which is now the lowest among developed nations.

## Adjustment to Exercise

The circulatory system adjusts in various ways to changes in physiological conditions. These adjustments are usually controlled by **negative feedback,** whereby a change in some condition, such as blood pressure, stimulates compensating activity that brings the condition back to its normal range. Negative feedback systems ensure that the composition of the ECF remains almost constant. We shall consider, as an example, some of the circulatory system's responses to vigorous exercise.

As exercise begins, the nervous system sends impulses to the **adrenal glands** near the kidneys, causing them to release the hormone **epinephrine** (also called **adrenalin**) into the bloodstream. Epinephrine causes blood vessels in the skin and abdominal organs to constrict, decreasing the blood supply to these organs and sending blood in these areas into more active circulation (Table 33–2). This in effect increases the volume of blood available. Epinephrine also causes local **vasodilation,** or widening of the vessels, in the muscles and heart, increasing the blood supply to

**T A B L E   3 3 – 2**

### Changes in Blood Supply to Organs During Strenuous Exercise

|  | Normal (ml/min) | Exercise (ml/min) |
|---|---|---|
| Heart output | 5400 | 17,500 |
| Blood flow to: |  |  |
| Brain | 750 | 750 |
| Abdomen | 1400 | 600 |
| Kidneys | 1100 | 600 |
| Muscle | 850 | 12,500 |
| Skin | 450 | 1900 |
| Heart | 250 | 750 |

these organs. This tradeoff of blood supplies helps to maintain the blood pressure. There is not enough blood to fill the whole circulatory system in the dilated state.

The "stitch" you may feel in your left side when running is caused by blood pooling in the spleen, a blood-filtering organ behind the stomach. This blood exerts extra pressure for a short time before the circulatory system becomes adjusted to the change in activity. Besides shifting blood supply to various systems, epinephrine also stimulates faster breathing rate and heartbeat rate, speeding delivery of oxygen to the muscles and removal of wastes.

Exercising muscles produce more carbon dioxide and lactic acid than usual. These make the blood more acidic as it passes through the muscles, and an increase in acidity does three things: it makes the blood give up more of its oxygen in the muscles, it increases the dilation of blood vessels in the muscles, and it also stimulates the nervous system to increase the secretion of epinephrine.

Intense muscular activity also generates a great deal of heat. When the hypothalamus (part of the brain, Figure 37–16) becomes too warm, it sends nerve impulses that cause dilation of blood vessels in the skin. The resulting increase in blood flow to the body surface allows heat to be given off to the environment and accounts for the sweating and flushed face when we exercise vigorously.

These are only a few of the interactions involved in the body's adjustment to exercise, but they illustrate the complexity of the physiological mechanisms that adjust the body's vital functions to changes in its activity.

## 33–D   BLOOD

The familiar red fluid called **blood** is really a tissue made up of a liquid containing several types of cells (Table 33–3). About half the volume of blood is made up of a fluid called **plasma,** and the other half is blood cells. The plasma contains water, various salts, and a great variety of plasma pro-

**T A B L E   3 3 – 3**

## Main Components of the Blood

| | |
|---|---|
| **Water** | 45–54% vv* |
| **Salts** | |
| Sodium | 2400 mg/l |
| Potassium | 80 mg/l |
| Calcium | 80 mg/l |
| Magnesium | 28 mg/l |
| Chloride | 2600 mg/l |
| Bicarbonate | 1500 mg/l |
| **Plasma Proteins** | 7–9% wv† |
| **Blood Cells** | 40–50% wv |
| White cells (leukocytes) | $4.7–9.7 \times 10^3/\mu l$ |
| Red cells (erythrocytes) | $3.6–5.5 \times 10^6/\mu l$ |
| Platelets (thrombocytes) | |
| **Substances Transported by Blood** | |
| Sugars | |
| Amino acids | |
| Fatty acids, glycerol | |
| Hormones | |
| Nitrogenous wastes | |
| Carbon dioxide | |
| Oxygen | |

*vv means volume per volume; e.g., 12 ml per 100 ml is 12% vv.
†wv means weight per volume; e.g., 13 g per 100 ml is 13% wv.

teins. **Serum** is plasma from which the proteins involved in clotting have been removed.

Blood cells are produced in the bone marrow in the center of some long bones. They can be divided into three main groups: **white cells** or **leukocytes, red cells** or **erythrocytes,** and **platelets** or **thrombocytes.** Most of the many different types of white blood cells help protect the body from disease (Chapter 34).

Red blood cells are by far the most numerous cells in the blood. Their main function is oxygen transport. Mature mammalian red cells have no nuclei and contain little except hemoglobin, the protein that binds $O_2$. These red blood cells usually last for about four months. Then they break up, and white cells destroy their remains by phagocytosis. **Anemia** is a condition in which the blood contains fewer red blood cells or less hemoglobin than usual as a result of unusually slow production or fast destruction of red cells or hemoglobin. Anemia may be caused by a variety of diseases and nutrient deficiencies.

If the concentration of red cells in the blood falls, the resulting oxygen shortage causes kidney cells to secrete the hormone **erythropoietin** into the blood. This stimulates nucleated, dividing cells in the bone marrow to increase red blood cell production. The new cells boost the blood's oxygen-carrying capacity, which cuts off production of erythropoietin, an example of negative feedback control. Living at high altitudes and leading a physically active life are two other factors that increase red cell production, because both conditions increase the body's oxygen demand.

Platelets are important in blood clotting. They are not really cells but are parts of the cytoplasm pinched off from large cells in the bone marrow.

The blood also contains various salts, and it transports food molecules. The amount of water, salts, and nitrogenous waste in the blood is controlled largely by the kidneys (Chapter 35).

The liver is the organ most responsible for regulating the concentration of food molecules in the blood. Blood passing through capillaries in the wall of the intestine picks up food from the digestive tract and carries it directly through the hepatic portal vein to capillary beds in the liver (Figure 33–15). (A portal system is one in which blood passes through two separate capillary beds before it returns to the heart.) The liver removes excess glucose from the blood and stores it as the polysaccharide glycogen. When the level of blood glucose falls too low, the liver breaks down glycogen and releases glucose into the bloodstream. The liver also regulates the blood's amino acid content and makes plasma proteins.

## Blood Clotting

When the wall of a blood vessel is broken or damaged, clotting begins. The blood forms a clot, sealing the vessel wall and preventing loss of blood until the blood vessel re-

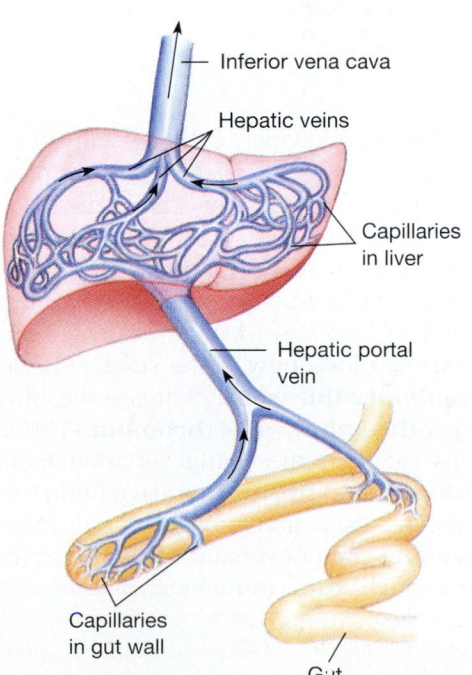

**FIGURE 33–15**  The hepatic portal system. Blood circulated to the digestive system passes through two capillary beds before returning to the heart. In the second capillary bed, the liver removes excess glucose and amino acids from the blood.

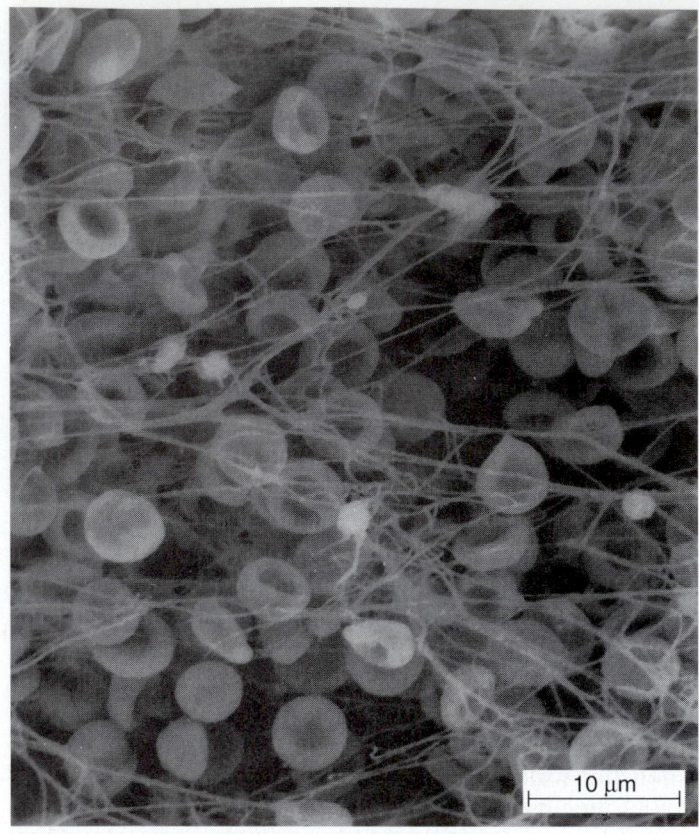

(a)

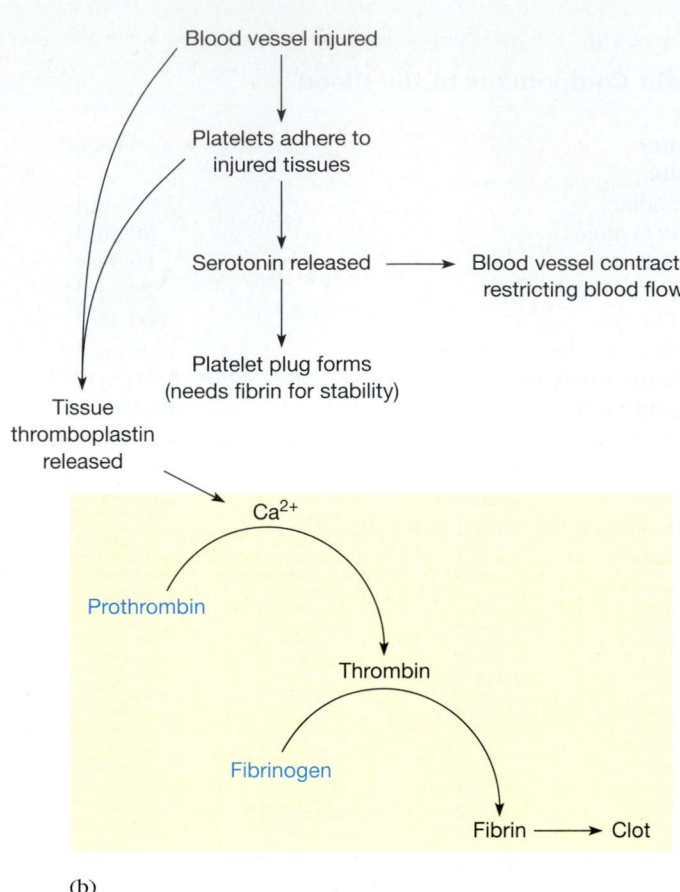

(b)

**FIGURE 33–16**  Blood clotting. (a) SEM of blood clotting showing fibrin strands with red blood cells (doughnut-shaped) and platelets trapped among them. (b) A simplified diagram of some of the reactions involved in the clotting of blood.   (a, R. P. Apkarian/BPS)

pairs itself. The injured cells of the vessel wall release substances that attract blood platelets. When the platelets come into contact with collagen fibers exposed by the injury, they disintegrate and form a temporary plug for the injured area (Figure 33–16). They also release two substances. The first is **serotonin,** which causes muscles in the blood vessel wall to contract and constrict the vessel, reducing blood loss by reducing blood flow in the vessel. Platelets also release **thromboplastin,** which changes one of the plasma proteins, **prothrombin,** into **thrombin.** Thrombin is an enzyme that catalyzes the change of another plasma protein, **fibrinogen,** into **fibrin.** Strands of fibrin form a meshwork around the disintegrated platelets. Still another plasma protein catalyzes the conversion of the loose fibrin meshwork into a tough, hard, permanent plug or clot.

## 33–E   THE LYMPHATIC SYSTEM

In many ways, the body's capillary beds are the most important parts of the circulatory system, for it is here that the exchange of substances between blood, ECF, and cells takes

place. Most substances, such as glucose and oxygen, leave the blood for the ECF by diffusing down the concentration gradient between the two fluids. Waste and carbon dioxide move from the ECF to the blood in the same manner. In addition, water and larger molecules, such as hormones and small proteins, enter and leave the blood either by moving through the spaces between the cells of the capillary walls or by passing through the cell itself by way of endocytotic vesicles. Large proteins cannot squeeze through spaces between cells, so they remain in the blood.

Water leaves capillaries under the pressure generated as blood is forced through a tube of small diameter. Toward the end of a capillary, so much water has been lost that the proteins left behind lower the blood's osmotic potential, and most of the water returns to the capillary. Overall, however, slightly more fluid leaves than enters the blood in the capillary beds.

This excess fluid, now in the ECF, is eventually collected and drained away through **lymphatics,** thin-walled vessels with valves that ensure one-way flow. The lymphatics eventually join to form the **thoracic duct** and the **right lymph duct,** which empty into veins near the heart (see Figure 34–5). Often these are the only lymph vessels large enough to be visible. The lymphatics perform several vital functions:

1. They drain excess water from the ECF back into the circulatory system.
2. They temporarily store fluids taken into the body. Some of the fluid absorbed from the digestive tract finds its way into the lymphatic system, which releases it gradually, so that the kidneys do not have to cope with sudden surges of fluid.
3. They carry large molecules, such as large proteins and hormones, into the bloodstream. Such molecules are too large to cross the wall of a capillary and so cannot reach the bloodstream directly.
4. Some food molecules, especially fats, move into the lymph rather than into the blood when they are absorbed from the intestine. The lymphatics form the main route by which such molecules reach the blood.
5. Lymph nodes occur in several areas of the body. These nodes are an important part of the body's defense against disease (see Chapter 34).

## 33–F  TEMPERATURE REGULATION

Heat spreads rapidly through any volume of water. Since cytoplasm, extracellular fluid, and blood are mostly water, heat spreads rapidly through an animal's body, and the circulation of the blood enhances this spread of heat. In fact, heat is one of the important things transported by blood.

The temperature of a cell determines the rate of its metabolic processes. An organism can grow faster and respond to the environment more rapidly if its cells are kept warm, and it would probably be advantageous for any animal to be able to control its body temperature. However, water has such a high heat capacity, and conducts heat so rapidly, that small aquatic animals such as small fish and amphibians cannot maintain much temperature difference between the body and the surrounding water. Their body temperature rises and falls with the water temperature. Air has lower heat capacity and conductivity, and most terrestrial vertebrates have evolved adaptations giving them greater control of their body temperatures.

We usually think of all animals except mammals and birds as "cold-blooded," but in fact many other vertebrates, and some invertebrates, also have adaptations that permit them to maintain body temperatures different from those of the environment. This is called **thermoregulation,** controlling the body temperature.

### Acclimation

The vast majority of reptiles, amphibians, fish, and invertebrates can tolerate body temperatures in the range of about 0 to 35 °C. However, a rapid change of 30 °C will inactivate most enzymes and kill the animal. When the temperature changes slowly, many animals can undergo **acclimation,** adjustments to the change that permit them to remain active. For instance, if a goldfish has been living in water at 30 °C, its nerve impulses will be blocked at temperatures below 10 °C, but after the same fish has spent some time in water at 15 °C, its nerves will function at temperatures down to 1 °C: it has become acclimated to the lower temperature.

In many animals, acclimation is due to synthesis of new enzymes that function at the new temperature. In a number of fish and aquatic invertebrates, this changeover from one set of enzymes to another occurs regularly as the water temperature slowly changes with the seasons.

### Endotherms and Ectotherms

All animals produce **metabolic heat** from the chemical reactions in their bodies and exchange heat with their environments (Figure 33–17). Most land animals regulate their body temperatures by controlling the rate at which metabolic heat is produced and dispersed. **Ectotherms** absorb heat from outside the body (ecto = outside), using behavioral adaptations for thermoregulation. For instance, they move into areas where the environmental temperature is appropriate. **Endotherms,** mainly birds and mammals, depend more on physiological adaptations and insulation to control body temperature by regulating loss of heat generated inside the body.

The distinction between ectothermy and endothermy is seldom clear-cut: most animals fall somewhere between the two. Thus, desert rodents, which are endotherms, escape the midday heat by burrowing, a behavioral adaptation. Moths, which are primarily ectotherms, fly poorly unless their wing muscles are at least at 35 °C, and they warm the muscles by physiological means, contracting them rapidly (fluttering their wings) before they take to the air. Behavioral and physiological thermoregulation, then, are found in both ectotherms and endotherms.

### Torpor and Hibernation

Thermoregulation uses considerable energy. There is no advantage in being warm and active when neither food nor a mate is available, and many animals, both ectotherms and endotherms, maintain high body temperatures only at certain times of the day or year. At other times, they conserve energy by allowing the body temperature to vary with the temperature of the surroundings.

A period of daily torpor or inactivity is common in terrestrial invertebrates, amphibians, and reptiles. It is also common in small birds and mammals, which have a high ratio of heat-losing body surface to heat-producing body volume. Small mammals use more than 80% of their food calories merely to maintain their body temperatures; a shrew

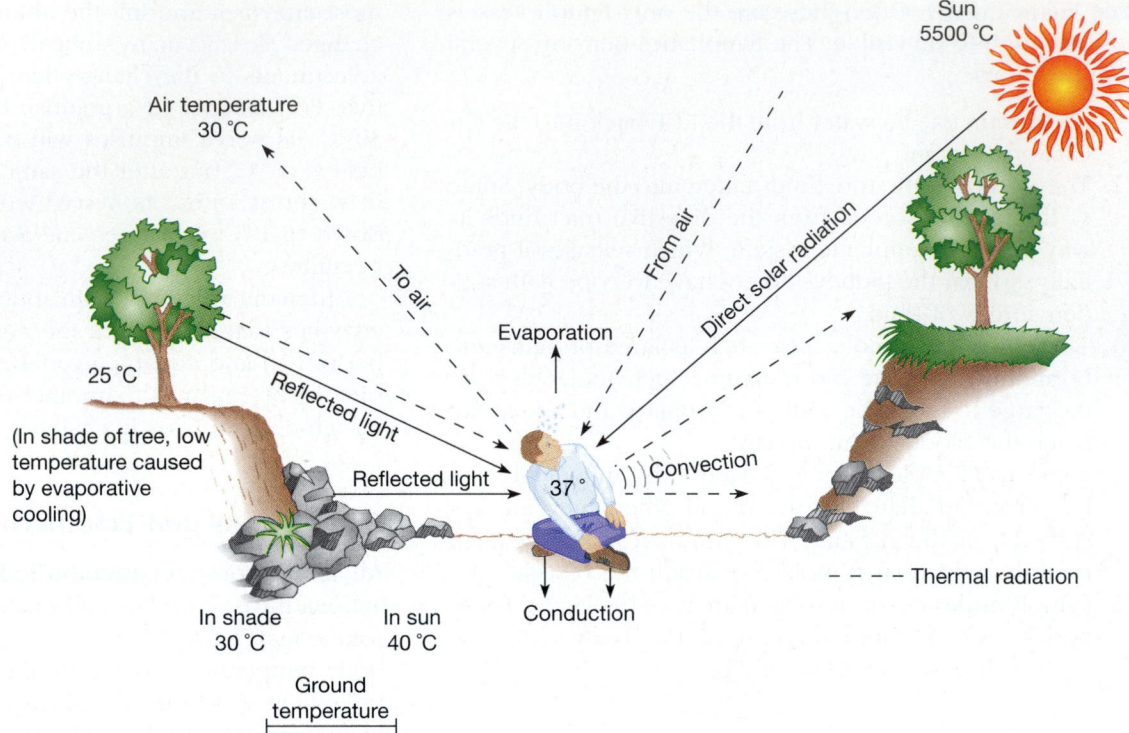

**FIGURE 33–17** Heat exchange between a mammal and its environment when there is no wind.

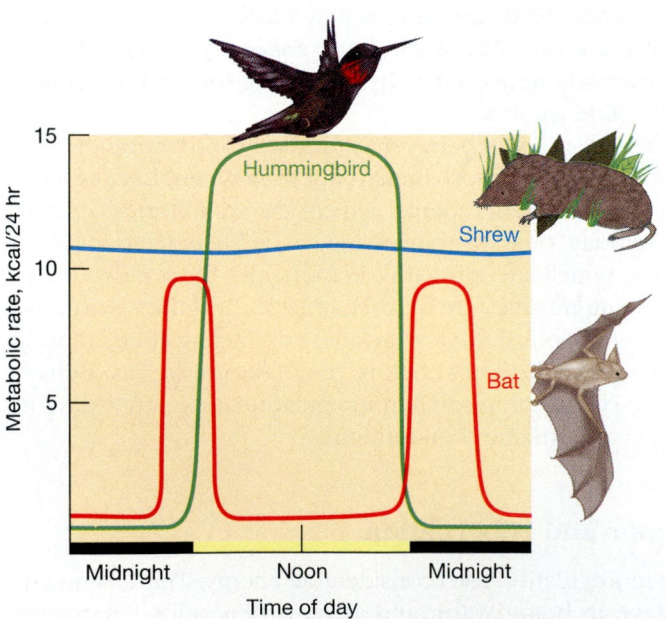

**FIGURE 33–18** Daily fluctuations in metabolic rates of three small endotherms. The hummingbird and insectivorous bat use less than half as much food per day as a shrew of the same size because they are active only when their food is available. A hummingbird's food (nectar from flowers) is available only during the day, a bat's (nocturnal moths) only at night. At other times, the bat and hummingbird become torpid, allowing their body temperatures and metabolic rates to fall.

or a bat is never more than a few hours away from starving to death. A period of torpor, when its body temperature falls, relieves the constant drain on the animal's energy resources (Figure 33–18). Roadrunners, predatory birds of Mexico and the southwestern United States, permit their body temperature to fall on cold nights, again saving energy.

Some vertebrates **hibernate,** avoiding activity in winter (hibernus = winter). The heart rate, metabolic rate, and body temperature decrease. The body thermostat resets to a lower level, and the body temperature of a hibernating mammal falls almost to that of the environment. These changes reduce the animal's energy expenditure and therefore its need for food at a time of year when food is scarce. With so many activities at low levels, the animal produces much less metabolic heat than when it is active. Even though its body temperature falls, it must produce some heat. This happens in brown fat tissue, whose color results from its many mitochondria. These produce heat by carrying out electron transport and releasing energy as heat rather than trapping it in ATP (Section 6–F).

During hot or dry seasons, some animals undergo a similar dormancy, but with a lesser drop in body temperature. This type of seasonal dormancy is called **estivation** (aestes = summer).

## Behavioral Thermoregulation

A dog lying in the shade on a hot day and bees shivering in their hive on a cold one both exemplify behavior patterns that control body temperature. The dog avoids being further warmed by the radiant heat of the sun; the bees produce metabolic heat by muscle movement. The bees also close the entrance to the hive and huddle together, slowing the loss of the heat generated by their metabolism.

Behavior that decreases the loss of metabolic heat to the environment can help terrestrial animals keep their bodies above environmental temperatures, but this is not practical for most aquatic animals with gills. These animals obtain oxygen from a current of water rather than from air, and water has a heat-absorbing capacity thousands of times greater than that of air. As blood in the gills picks up oxygen from the water current, it also loses metabolic heat that it absorbed in the interior of the body. Hence, the body's metabolic heat is lost almost as fast as it is generated. Only a few large fish, such as tunas, can raise their body temperature above that of the surrounding water; these fish have countercurrent heat exchangers that work on the same principle as those of endotherms, discussed later.

The vast majority of terrestrial ectotherms regulate body temperature by moving into areas that have favorable temperatures. Many earthworms, reptiles, and arthropods retire to burrows during the hottest or coldest parts of the day; this is advantageous because the soil changes temperature more slowly than air does. Millipedes, houseflies, and crabs escape the heat of the sun by avoiding light.

For most animals, avoiding extreme heat is only an occasional problem. The muscles of most vertebrates and terrestrial arthropods work most efficiently at the surprisingly high temperature of 35 to 40 °C. Because temperatures on Earth are usually lower than this, most terrestrial animals, rather than avoiding heat, need to capture as much environmental heat as possible. Living in warmer areas is the simplest way to do this, and in fact there are many more species of ectotherms of all sorts in the tropics than in temperate or cold climates.

In cold areas, basking in the sun is the most common behavioral adaptation to capturing environmental heat. Even though the air may be very cold, an animal can often absorb much radiant heat by sitting directly in the sun's rays, especially on dark-colored surfaces. Many lizards adjust their body shape and orientation while basking, altering the surface area exposed to the sun and therefore the rate of heat gain. By moving its ribs, a horned lizard can increase the body surface exposed to the sun more than sixfold (Figure 33–19). Many lizards can also change color, and the skin absorbs more heat when the color is darker than when it is lighter. Many ectotherms are fairly dark in color, and many mammals in colder areas have black-tipped noses, ears, and

**FIGURE 33–19** A Texas horned lizard, capable of moving its ribs so as to expand the body surface for heat absorption. This ability is highly developed in squamates but not in other reptiles such as turtles and crocodilians.  (Florida Audubon Society)

paws. Dark-colored areas absorb radiant heat more readily, but they also lose heat more rapidly, and these animals often have behavior patterns—such as curling up to sleep or moving about only when the sun is out—that reduce heat loss through their dark extremities.

The insects that pollinate an Arctic species of poppy, *Papaver radicatum,* bask in the open flower and may raise their temperatures to 35 °C, about 20 °C above the air temperature. The poppy flower itself turns and faces the sun, an adaptation that encourages the visits of the insects—and also speeds up development of the poppy's seeds.

## Physiological Thermoregulation

Some ectotherms use physiological mechanisms to adjust the rate at which they gain and lose heat. Large lizards, such as the Galápagos marine iguana, can raise their body temperature rapidly in the morning and cool down slowly in the evening by controlling dilation of blood vessels in the skin. In the morning, when heating up rapidly is advantageous, blood vessels in the skin dilate, speeding absorption of heat from the sun. In the evening, when the temperature falls, blood vessels near the surface constrict, minimizing heat loss to the cool air.

Birds and mammals maintain high body temperatures, from 35 to 42 °C, depending on the species. They do this physiologically, by producing a great deal of metabolic heat and regulating the rate at which this heat is lost to the environment. When the environment is too cold, for example, the brain induces shivering, muscular movement that produces metabolic heat. When the environment is too warm,

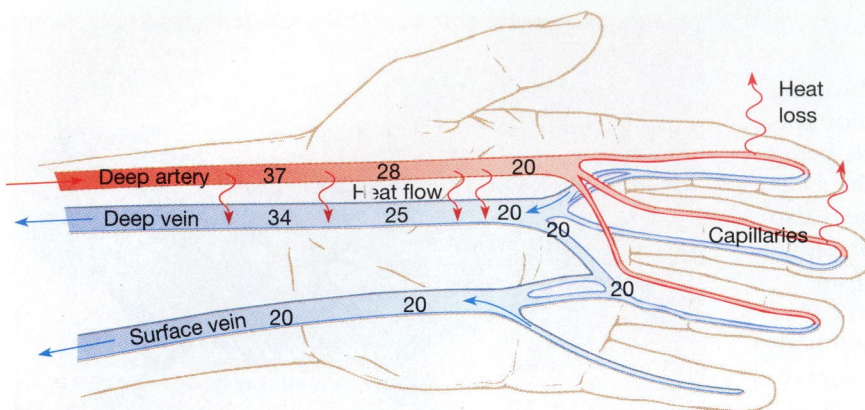

**FIGURE 33–20** Countercurrent heat exchanger between an artery and a vein deep in the human arm. Heat is lost in the capillaries of the hand. As the blood runs through the adjacent deep vein and artery, it exchanges heat so that the arterial blood is cooled by the time it reaches the hand, and venous blood is warmed as it returns to the body. When heat conservation is unnecessary, the heat exchanger can be bypassed, and most of the blood is returned to the body via a surface vein where its temperature does not change. The nervous system controls the degree of constriction of the two veins and so controls heat loss.

the animal becomes less active, and it may rest in the shade or in a burrow.

Layers of fat plus a body covering of fur or feathers insulate the interior of an endotherm's body from the environment and decrease heat exchange; such insulation is not found in ectotherms. Cold-climate endotherms, such as whales, seals, and polar bears, tend to have thicker layers of insulation than their relatives in temperate areas. At the other extreme, a camel living in a hot desert is protected against gaining too much heat from the sun by the thick layer of fur and fat insulating its back.

Several physiological mechanisms enable an endotherm to decrease its heat loss when its body temperature falls. Because heat is lost through the body surface, blood coming to the skin from the interior of the body is usually warmer than the skin. When more blood passes through the skin, more heat is lost from the body. When the body is too cool, nervous and hormonal signals constrict the surface blood vessels and decrease the blood flow to the skin, reducing heat loss.

Extremities, such as the ears, nose, and legs, are often slim and streamlined, with little insulating fat. Endotherms lose considerable heat from the blood in these parts of the body. To take extreme examples of this problem, consider how much heat must be lost through the legs of a bird fishing in an icy mountain stream, or through the paws of an Arctic fox trotting across a snowfield.

Endotherms minimize such heat loss by means of **countercurrent heat exchangers,** arrangements in which

the blood vessels entering and leaving the ear or leg run next to each other. This allows blood on its way to the limb to give up heat to blood returning to the body. By the time outgoing blood reaches the limb it has been cooled and has little heat left to lose to the environment. Its heat has been given up to blood returning to the body, and this blood has been warmed almost to body core temperature (Figure 33–20).

When endotherms need to cool themselves, their bodies radiate heat by increasing the supply of warm blood to the skin and by rearranging the fur or feathers so that more heat escapes. Most also employ **evaporative cooling** as a second line of heat control: as water evaporates from the skin or respiratory tract, the heat needed to vaporize the water is removed from the body. Humans and some other mammals have sweat glands in the skin that permit evaporative cooling; many carnivores, ungulates, and primates pant, increasing evaporation from the respiratory tract. The metabolic heat generated by a flying bird is tremendous, and on its way to and from the lungs, air passes through air sacs, which provide extra surfaces for evaporative cooling (see Figure 32–11).

Such physiological and behavioral mechanisms enable birds and mammals to maintain a high, constant body temperature. These endotherms can move about when the environmental temperature is so low or so high that other animals cannot be active. It is not surprising that the coldest regions of the world are populated only by mammals and birds, such as whales, polar bears, penguins, and seals.

# ESSAY: *Adaptations of Diving Mammals*

Many mammals dive to catch food or escape predators. Diving presents several problems. First, a mammal must breathe air, but its trips to the surface may be infrequent (Table 33–A). Second, water pressure increases with depth. As pressures increases, gas in the lungs is compressed and forced into solution in the blood. When the animal surfaces, the dissolved gas, mainly nitrogen (air is 79% $N_2$), comes out of solution and forms bubbles that may block blood vessels, causing Caisson disease, or "the bends." Third, water absorbs heat faster than air, threatening a warm-blooded animal with death from **hypothermia:** chilling of the body. The adaptations of diving animals that permit them to overcome these problems are exaggerations of features found in other animals.

Some mammals show **bradycardia,** slowing of the heart rate, when they dive. A diving seal's heart rate drops from 150 to 10 beats per minute. Humans also experience bradycardia when they dive, and fish show it when removed from water. Bradycardia saves the body energy and oxygen. In addition, blood vessels constrict, reducing the circulation to many organs. This conserves oxygen for use by the brain, which must not be allowed to run short of oxygen.

Although circulation to the head is maintained, the respiratory center in the brain of a diving animal tolerates high levels of carbon dioxide, and so the animal can go longer between breaths. Other body organs carry out fermentation rather than respiration, and most of the carbon dioxide and lactic acid they produce is retained in the tissues. When the animal surfaces, a surge of metabolic wastes enters the bloodstream and is excreted.

Many diving mammals can store extra oxygen for use during a dive. They have more red blood cells, and their muscles contain extra quantities of myoglobin, an oxygen-storage pigment. Seals carry most of their oxygen in the blood, whereas whales store more oxygen in myoglobin.

The lungs do not carry more oxygen than usual during a dive. In fact, seals exhale as they dive, and whales have smaller lungs for their size than other mammals. These are adaptations that prevent Caisson disease: the less air there is in the lungs, the less nitrogen dissolves in the blood during a dive. Furthermore, as a mammal dives, the pressure on the lungs forces much of the air into the air passages, whose walls are impermeable to gases. This air cannot dissolve in the blood. However, some nitrogen inevitably does enter the blood. The tissues of dolphins have been found to tolerate levels of dissolved nitrogen that would be dangerous to a human diver. An important behavioral adaptation to diving is slow resurfacing, so that the nitrogen dissolved in the blood comes out of solution gradually and returns to the lungs instead of forming gas bubbles that block blood vessels.

Diving mammals reduce heat loss by reducing the flow of blood to the skin during a dive. Their bodies are often shaped with a low surface-to-volume ratio, and the blood vessels to their appendages are arranged as countercurrent heat exchangers (see Figure 33–20).

On its return to the surface, a diving mammal must spend time breathing to expel carbon dioxide and replenish its oxygen supply before it can dive again (Figure 33–A).

**FIGURE 33–A** An Alaskan harp seal greets her white-coated (camouflaged) offspring as she surfaces to breathe through a hole she keeps clear of ice. (Fred Bruemmer/Peter Arnold)

### TABLE 33–A

**Duration and Depth of Diving for Some Mammals**

| Animal | Duration (minutes) | Depth (meters) |
|---|---|---|
| Beaver | 15 | shallow |
| Muskrat | 12 | shallow |
| Walrus | 10 | 80 |
| Gray seal | 20 | 100 |
| Bottle-nosed whale | 120 | unknown |
| Blue whale | 49 | 100 |
| Most people | 1 | shallow |
| Trained skin divers | 2.5 | 20 |

## SUMMARY

Most animals have transport systems that move molecules and heat within the body. The transport system's main function is to maintain the relatively constant composition of the ECF, with which every cell exchanges food, gases, and waste. The more efficient its transport system, the more active an animal can be.

1. In lower invertebrates such as cnidarians and platyhelminths, food is carried by the digestive system, and there is no separate circulatory system.

2. Animals with coeloms have true circulatory systems. An open circulatory system has few blood vessels, and blood bathes the cells directly. The closed circulatory systems of many invertebrates and of all vertebrates have blood vessels through which blood is pumped by the heart.

3. The circulatory systems of vertebrates show an evolutionary trend from single to double circulation. The double circulation of birds and mammals achieves complete separation of oxygenated and deoxygenated blood as well as high blood pressure in the body's capillaries.

4. Blood is pumped by the heart through a set of pipes, the blood vessels. Blood flows through the circuit from the region of high pressure, the contracting ventricles of the heart, through the vessels at progressively lower pressure, until it returns to the heart. Valves in the veins and the heart prevent backflow. Blood pressure and blood supply in various parts of the body may be regulated by dilation and contraction of arterioles and capillaries and by changes in heartbeat rate and volume of blood pumped by the heart at each stroke.

5. The circulatory system responds to changes in the body's activities so that the body's new needs are met. During exercise, the amount of blood flowing and the rate of flow are increased, and more blood is diverted to active muscles.

6. Blood is a tissue consisting of a watery matrix containing salts, proteins, and blood cells. White blood cells defend the body from disease and phagocytose dead cells and other debris. Red blood cells carry oxygen bound to their hemoglobin, a protein that combines with oxygen in the lungs and releases it in the capillaries of the body tissues. Blood platelets play an important role in the clotting of blood. Clotting helps to plug vessel walls after injury, preventing loss of fluids and entry of pathogens.

7. The lymphatic system consists of vessels that collect ECF, proteins, and digested fats, and empty them into the venous system.

8. Living cells can function only at certain temperatures. Animals cope with temperature changes either by tolerating them or by making behavioral or physiological responses that help to maintain their bodies within a suitable temperature range. Ectotherms have more behavioral than physiological adaptations for thermoregulation. They may move into hotter or colder areas, or they may sunbathe, burrow, or huddle together to maintain a favorable body temperature.

9. Torpor, hibernation, and estivation are temporary reductions in metabolic rate and body temperature that allow many animals to conserve energy.

10. Thermoregulation by physiological mechanisms is most apparent in endotherms. They produce large quantities of metabolic heat and regulate its escape into the environment by such means as insulation, alteration of the blood supply to the skin, regulation of the temperature of blood reaching the extremities, and evaporation of water from the body surface.

## SELF-QUIZ

1. Place an "X" in the boxes to indicate which transport systems exhibit the feature mentioned:

|  | Cnidarian | Earthworm | Insect | Fish | Mammal |
|---|---|---|---|---|---|
| food transport |  |  |  |  |  |
| oxygen transport |  |  |  |  |  |
| high pressure fluid picks up food |  |  |  |  |  |
| muscular circulatory pump(s) |  |  |  |  |  |

2. Which of the following is *not* a difference between the circulatory systems of mammals and those of insects?
   a. Substances must cross a vessel wall to reach the cells of mammals, but the blood of insects is in direct contact with their cells.
   b. Blood moves randomly in an insect, but in a definite path in a mammal.
   c. Insects do not transport oxygen in their transport system, whereas mammals do.
   d. Mammals have much higher blood pressure than do insects.
3. Select *two* advantages of a double circulation over a single circulation.
   a. In the double circulation, all the blood going to the tissues is oxygenated, whereas in the single circulation it is not.
   b. In the double circulation, the blood can transport more types of substances.
   c. In the double circulation, the blood is at higher pressure when it enters the body tissues.
   d. In a double circulation, the blood travels around the body faster.
   e. In a double circulation, there are twice as many blood vessels servicing the body tissues.
4. The greatest amount of oxygen will be lost from the blood while it is traveling through:
   a. the capillaries around the alveoli
   b. the left atrium of the heart

c. the arteries
d. the capillaries in the body
e. the veins

5. If you were asked to dissect an animal so as to reveal a valve, all of the following places would be good to try except:
   a. the opening between the right atrium and the right ventricle
   b. the fork where the pulmonary artery splits and one branch goes to each lung
   c. the base of the aorta where it leaves the left ventricle
   d. a vein in the arm
   e. a lymph vessel that empties into the thoracic duct
6. Trace the path of a fat molecule from the time it leaves the intestine until it is deposited in the fatty tissue of the body. Name, in order, all the structures it passes through on the way. (Give the shortest possible route.)
7. Which of the following organs will receive a decreased flow of blood during strenuous exercise?
   a. brain      d. heart
   b. skin       e. lungs
   c. liver
8. State whether each phrase below is characteristic of endotherms, ectotherms, or both.
   a. Regulation of body heat is mainly by moving to locations with favorable temperatures.
   b. Heat is generated by the body's metabolism.
   c. An insulating body covering reduces heat exchange with the environment.

## Questions for Discussion

1. List some forces, besides contraction of the heart, that may move fluids in the bodies of animals.
2. Explain how the open transport system of a butterfly can deliver the oxygen and sugar needed to sustain the high metabolic rate of the flight muscles as the butterfly flits from flower to flower.
3. What restrictions in size and activity are faced by animals that possess an open circulation combined with a tracheal system?
4. Can you think of any reasons why cephalopods (such as squid and octopus) are the only molluscs with closed circulatory systems, and why other molluscs (snails, clams, and chitons, for example) manage well with open systems?

5. Birds and mammals have four-chambered hearts, and most keep their body temperatures high and stable. In what way might these two characteristics be linked?
6. Arteries usually lie deep in the body, whereas veins lie near the surface. What is the advantage of this arrangement?
7. Misinformed people often define arteries as blood vessels that contain oxygenated blood, and veins as vessels that contain deoxygenated blood. What is wrong with these definitions?
8. Explain why the body core of an alligator basking in the sun warms up faster when the animal is moving than when it lies still.
9. Many mammals hibernate, but very few birds do. Why not?

## Suggested Readings

Brown, M. S., and J. L. Goldstein. "How LDL receptors influence cholesterol and atherosclerosis." *Scientific American,* November 1984.

Gordon, M. S. *Animal Physiology: Principles and Adaptations,* 4th ed. New York: Macmillan, 1982. A well-written modern comparative physiology; discusses the changes in physiological systems in different animals.

Kanwisher, J. W., and S. H. Ridgway. "The physiological ecology of whales and porpoises." *Scientific American,* June 1983.

Lawn, R. M. "Lipoprotein (a) in heart disease." *Scientific American,* June 1992.

Schmidt-Nielsen, K. *Animal Physiology: Adaptation and Environment,* 4th ed. New York: Cambridge University Press, 1990. An excellent book on environmental physiology by its elder statesman.

Zucker, M. B. "The functioning of blood platelets." *Scientific American,* June 1980.

C H A P T E R

34

# Defenses Against Disease

## O B J E C T I V E S

*When you have studied this chapter, you should be able to:*

1. Use the following terms correctly:
   antigen, antibody
   lymph node
   T cell (T lymphocyte), B cell (B lymphocyte), T helper cell, T killer cell
   phagocyte, macrophage, memory cell, plasma cell
2. Describe four nonspecific defenses of animals against disease.
3. List three characteristics of an immune response.
4. Describe the difference between cellular and humoral immune responses.

5. Describe an example of a cellular immune response.
6. Describe the steps by which bacteria are destroyed during a primary humoral immune response.
7. List three roles played by lymphocytes in the immune system.
8. Explain how vaccination protects against a serious case of a disease.
9. Explain why skin grafts and transplanted organs may be rejected by the recipient's body, and why a second skin graft from the same donor is rejected faster than the first.
10. Describe how an allergic reaction comes about.

---

A living body is an ideal incubation chamber, providing food, shelter, and just the right combination of water, minerals, and temperature for cells. It is no wonder that many small organisms have adapted to life inside the bodies of their larger neighbors. Some, like symbiotic bacteria in digestive tracts, cause their hosts no trouble or are beneficial. Other organisms, called **pathogens,** cause disease.

The twentieth century witnessed a decline in deaths from infectious diseases, largely as a result of worldwide vaccination programs, improved hygiene, and antibiotics. Vaccination has saved millions from infectious diseases such as poliomyelitis, measles, whooping cough, mumps, and influenza (flu). Improved hygiene and clean water supplies made people less likely to become infected with bacteria, protistan parasites, and fungi. When infections did occur, antibiotics often cured them.

As a biologist might have predicted, however, pathogens evolve resistance to antibiotics. The problem is especially acute in the United States, where more antibiotics are used than in any other country. In addition, antibiotics are used in massive amounts in U.S. livestock destined for food: the animals provide a breeding ground for antibiotic-resistant pathogens. As a result, deaths from infections are now rising rapidly in the United States and are

rising, although less rapidly, in other countries. Blood poisoning (septicemia), which caused 3500 deaths in the United States in 1970, caused 20,000 deaths in 1992.

The biggest killer among infectious diseases worldwide is tuberculosis (TB), a bacterial infection that causes ulcers in the lungs. Hundreds of millions of people every year also develop each of the most devastating diseases on the worldwide health list: acquired immunodeficiency syndrome (AIDS), leprosy (Chapter 24), trypanosome infections, leishmaniasis, and malaria (Chapter 25), and schistosomiasis and filariasis (Chapter 28). Disease has been, and still is, a potent selective force in human evolution.

Probably all organisms have defenses against pathogens, but here we consider only the defenses of animals. Some defenses are nonspecific; that is, they protect the animal from many different pathogens. For instance, most animals contain **phagocytes,** wandering cells that engulf and destroy any pathogens and debris they encounter. Other defenses are specific, working against particular pathogens. These include the reactions of the **immune systems** of vertebrates, whose cells can mount an attack specific to each pathogen. When the disease has been overcome, cells of the immune system "remember" the pathogen and combat it more effectively in the future.

◆  Animals defend themselves against disease by nonspecific mechanisms and by immune reactions, which recognize foreign substances in the body and destroy them.

◆  In vertebrates, the identity of a pathogen attacked by the immune system is stored in memory cells, which enable the animal to respond more effectively to the pathogen if it is encountered again.

◆  The distinguishing features of vertebrate immune responses are specificity, recognition of foreignness, and memory.

◆  Failure of the immune system leads to devastating diseases, either because the immune system fails to defend the body from foreign organisms or because it attacks its own body.

## 34–A  NONSPECIFIC DEFENSES

Animals defend themselves against pathogens by preventing foreign organisms and objects from entering the body in the first place and by destroying them if they do enter.

### Skin and Mucous Membranes

The body surfaces in contact with the outside world are the skin and the mucous membranes. These surfaces are barriers that prevent most pathogens from entering the body. In addition, millions of bacteria live on these surfaces. These resident bacteria produce substances that protect their homes—our skin and mucous membranes—from foreign organisms.

**Skin** is the human body's largest organ and its main protection against infection (Figure 34–1). The skin's importance can be gauged from the fact that whether a burn victim lives or dies depends almost entirely on how fast damaged skin can be replaced.

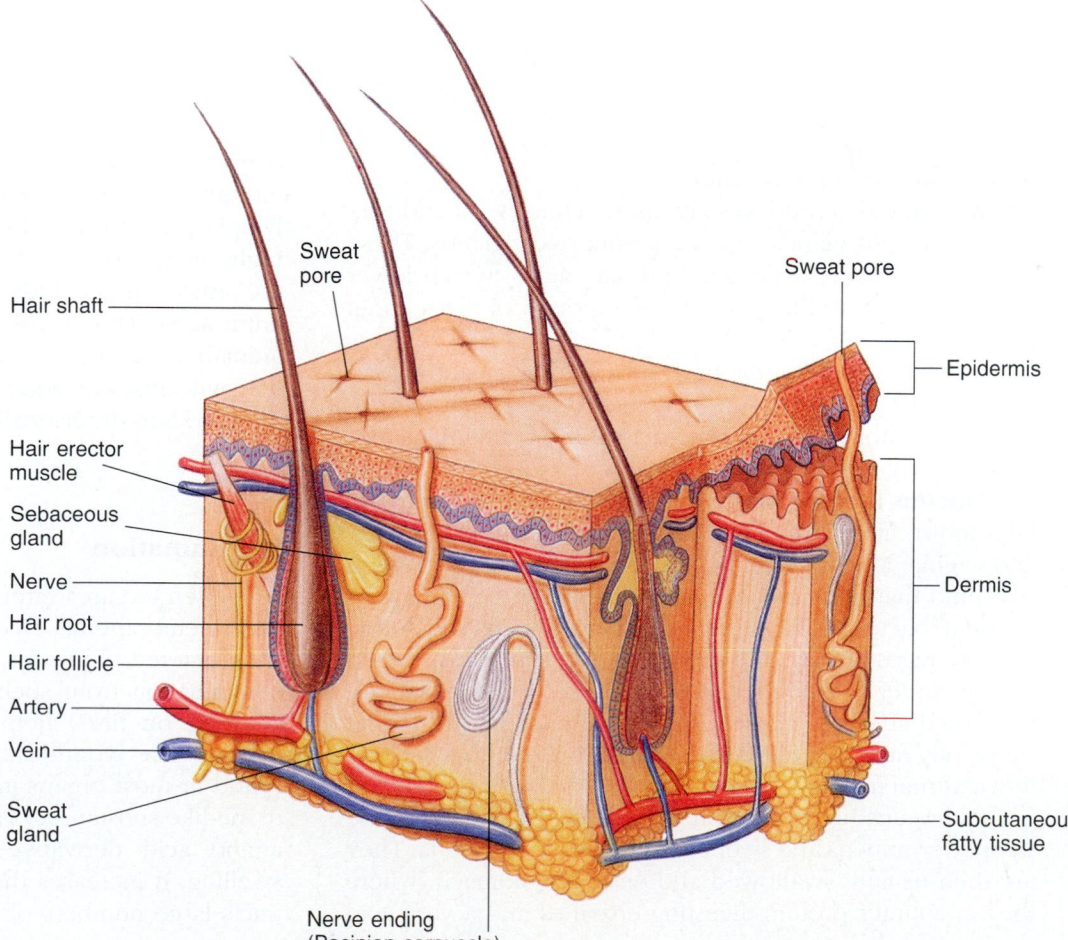

**FIGURE 34–1**  A section through human skin. The cells in the outer layer of the epidermis constantly die and slough off. They are replaced by division of cells in the basal layer. Chemical defenses of the skin include acids in the oily secretion of sebaceous glands and in sweat; these acids combat fungi and bacteria. Evaporation of sweat cools the body. The hair erector muscles can raise and lower the hairs, altering the air circulation next to the skin and contributing to thermoregulation.

Sweat pore

Sweat pore

Hair shaft

Epidermis

Hair erector muscle

Sebaceous gland

Nerve

Dermis

Hair root

Hair follicle

Artery

Vein

Sweat gland

Subcutaneous fatty tissue

Nerve ending (Pacinian corpuscle)

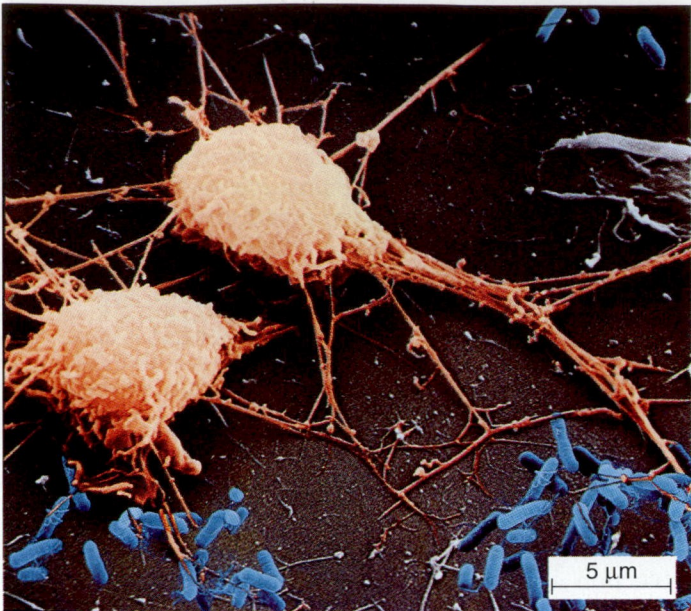

5 μm

**FIGURE 34–2**   Human phagocytes. Macrophages (brown) attacking *Escherichia coli* bacteria (blue-green). (Manfred Kage/Peter Arnold)

**Epidermal cells** at the skin's surface are constantly dying and sloughing off. This process becomes wholesale when a sunburned nose peels. The lost cells are replaced by division of cells beneath them. In this way the body soon repairs minor damage to the skin and maintains a constantly renewed barrier against infection.

The skin also produces chemical defenses: oil and wax from sebaceous glands and sweat from sweat glands. These secretions contain lactic acid and fatty acids, which lower the pH enough to kill or inhibit the growth of many fungi and bacteria.

Besides protecting the body against disease and injury, the skin plays vital roles in regulation of temperature, production of cholesterol, and the detection of stimuli such as pressure and temperature.

**Mucous membranes** cover body surfaces that must be kept moist, lining the eyelids, nose, digestive tract, urethra, and vagina. Mucous membranes secrete **mucus,** a protein-rich fluid that traps microorganisms and that contains bactericidal (bacteria-killing) enzymes such as **lysozyme,** found in tears, nasal mucus, and saliva. Lysozyme was discovered by Alexander Fleming when he noticed that bacterial cultures died after he sneezed on them. Douching the vagina frequently may lead to infection because it removes the bactericidal mucus.

Pathogens in the respiratory tract are likely to be trapped by mucus and swept into the pharynx by cilia. They are then usually swallowed and enter the stomach, where they encounter protein-digesting enzymes and a very acid

environment. Many fungi and bacteria that survive this and reach the large intestine are attacked by antibiotics secreted by gut symbionts.

Despite these defenses, some pathogens do enter the body, usually by crossing mucous membranes, which are thinner and more vulnerable than the dry, oily skin.

## Phagocytes

Phagocytes are the most widespread cells protecting animals from infection. They travel throughout the body, destroying any pathogens they encounter and engulfing and recycling debris and dead body cells. The coelomic fluid of most invertebrates contains phagocytes called **coelomocytes.** Poisons such as heavy metals and PCBs (polychlorinated biphenyls) disable these cells. Environmental toxicologists are using this fact to develop a **bioassay** (a test using organisms) for the success of cleanup efforts at toxic waste dumps. Earthworms are exposed to the poisons and then tested to see how effectively their coelomocytes destroy bacteria.

Vertebrates contain **macrophages** ("big eaters"), phagocytes that circulate in the extracellular fluid (Figure 34–2). Organs such as the lungs and brain also contain resident macrophages. In infected tissue, **neutrophils,** another group of white blood cells, also become phagocytic and engulf pathogens and dead tissue (Figure 34–3).

In the early 1970s, researchers studying immune responses discovered that humans and mice with tumors contained lymphocytes (white blood cells) that attacked those tumors. As controls for the experiments, investigators used lymphocytes from individuals without tumors. Embarrassingly, many of these "control" lymphocytes also attacked the tumors under study and indeed a wide range of different tumors. This surprising result led to the discovery that animals contain a special population of white blood cells that will attack a wide range of tumor cells. These lymphocytes are dramatically called **natural killer cells.**

## Inflammation

A swollen red area often develops where the skin or a mucous membrane has been cut or punctured and pathogens have entered the body. A pimple or a boil is the result of inflammation from such a local infection. **Inflammation** ("setting on fire") helps to fight infection and heal the wound. The wound attracts large **mast cells,** which are found in most organs and which contain a variety of hormone-like substances. The mast cells release **histamine,** an amino acid derivative that causes heat, redness, and swelling. It increases the blood supply to the area and attracts large numbers of phagocytes, which engulf bacteria

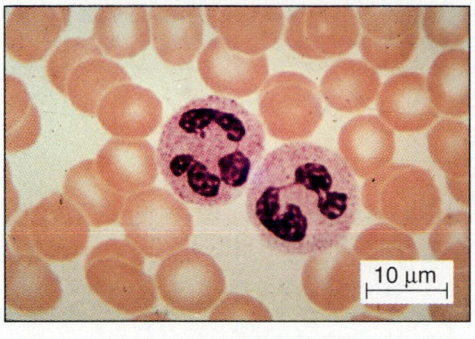
(a)

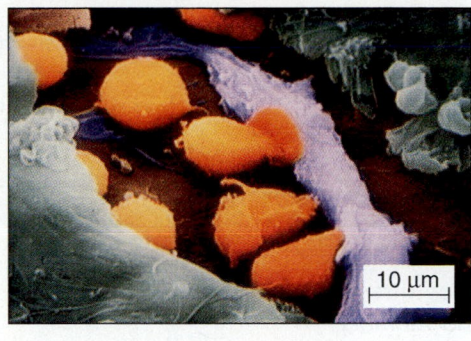
(b)

**FIGURE 34–3**   Phagocytic cells. (a) Two neutrophils in a group of red blood cells. This photograph shows the large, variably shaped nucleus characteristic of a neutrophil. Because the nucleus is so variable in shape, neutrophils are also known as polymorphonuclear granulocytes. (b) Two macrophages in a scanning electron micrograph of part of the human intestine.   (Biophoto Associates)

and dying body cells. Eventually the phagocytes die and may accumulate as **pus.**

***Fever***   An infected area often feels warm to the touch because heat is one of the body's ways of fighting pathogens. Normally, the brain keeps the human body at about 37 °C (98.5 °F). However, when the body is infected, some white blood cells respond by releasing hormones that act as **pyrogens** ("fire-producers"). If enough pyrogens reach the brain, the body's thermostat is reset to a higher temperature, allowing the body temperature to rise, producing a **fever.**

Very high fevers are dangerous and must be reduced. But a study of children with influenza ("flu") showed that a few degrees of fever helps the body fight infection. In this study, children treated with aspirin to reduce their fevers were compared with those whose temperatures were permitted to rise naturally. The children whose fever was kept down were ill longer, and had more serious symptoms, than those with fevers.

Cells metabolize faster at higher temperatures, so fever increases the rate at which cells fight infection. In addition, many bacteria require more iron in order to reproduce at higher temperatures. Pyrogens not only raise the body temperature but also reduce the concentration of iron in the blood, slowing bacterial reproduction even further.

Ectotherms cannot alter their temperature physiologically as endotherms can, but they too may raise their body temperature while fighting infection. Researchers kept desert iguanas in an enclosure with sunny and shady areas and infected some of the reptiles with bacteria. They found that uninfected iguanas kept their body temperatures between 37 and 41 °C. However, infected iguanas spent more of their time in the sun and maintained temperatures between 39 and 43 °C. Once the infection was overcome, the iguanas resumed their normal behavior.

## Complement

The **complement reactions** complement (round out) the pathogen-destroying effects of inflammation. **Complement** consists of about 20 kinds of proteins, produced mainly by the liver, which circulate in the body fluids in inactive form. The complement system is activated either by an immune reaction to a pathogen or by the polysaccharide coats on some bacteria. In the latter case, the reaction activates enzymes that destroy the bacterial coat. Complement proteins may also bind to a pathogen, stimulating a macrophage to engulf it.

The complement reactions, then, enhance phagocytosis as part of an immune response and induce phagocytosis in other cases. They are an example of the many ways in which nonspecific responses work with specific immune reactions to rid the body of pathogens.

## Interferon

Fever increases the production of glycoproteins that are called **interferons** because they interfere with virus replication. A cell infected with viruses rapidly produces interferons. These cannot save the infected cell, but they bind to cell-surface receptors on nearby cells. Within minutes, these cells produce a variety of antiviral proteins, including an enzyme that cleaves viral messenger RNA, thereby preventing synthesis of viral proteins.

Different body cells produce different interferons, all capable of defending against many viruses. This led researchers to hope that interferons might be useful in treating cancers caused by viruses. However, experiments using interferon against cancer have proved disappointing. This may be because the main role of interferons is not actually to defend against viruses directly but to organize and enhance the immune system's response.

## 34–B  OVERVIEW OF IMMUNE RESPONSES

Few people suffer twice from diseases such as measles, chickenpox, and mumps. The body's first encounter with the pathogens that cause these diseases equips it to get rid of the same kinds of pathogens when they next invade. A response that is bigger and better the second time around tells us that this reaction is produced by the immune system.

Immune responses can be identified by three features: specificity, recognition of foreignness, and memory:

1. **Specificity.** If you have had measles, you are immune to further attacks of measles but not to rubella (German measles) or mumps. In other words, the immune response to measles is specific for the measles virus.
2. **Recognition of foreignness.** The immune system attacks measles viruses, but it does not attack its own cells. This shows that the body can distinguish the measles virus as foreign, that is, not a normal part of the body.
3. **Memory.** The immune response to a bout of measles confers **immunity,** protection against measles viruses encountered later in life. This shows that immunity involves memory: the body "remembers" that it has previously encountered this type of virus. To be protected against other diseases, an individual must build up specific immunological memory for each one.

The role of the immune system is to recognize and destroy foreign antigens that invade the body. An **antigen** is any substance that can stimulate the body to mount an immune response against it. The most common antigens are proteins and polysaccharides from other organisms, such as toxins produced by bacteria and the coats of viruses. Many substances are antigens for one person but not for another: the glycoproteins on your liver cells are not antigens to your own body, but they would act as antigens if injected into another person.

An immune reaction is specific because it starts when an antigen binds a receptor produced by a lymphocyte. **Lymphocytes** are white blood cells, each of which produces only one kind of antigen-binding receptor, which can bind only one kind (or a few very similar kinds) of antigen. Different lymphocytes produce different receptors. So among all the lymphocytes in the body, there are receptors that can bind almost any antigen.

The binding of antigen to receptor stimulates the lymphocyte to reproduce, forming a clone of cells that differentiate into two cell types. One type fights the pathogen that formed the antigen. A smaller number are long-lived **memory cells,** which survive for years. Some lymphocytes produce receptor-like molecules called **antibodies,** some of which are secreted into the body fluids. Here they can bind and neutralize antigens.

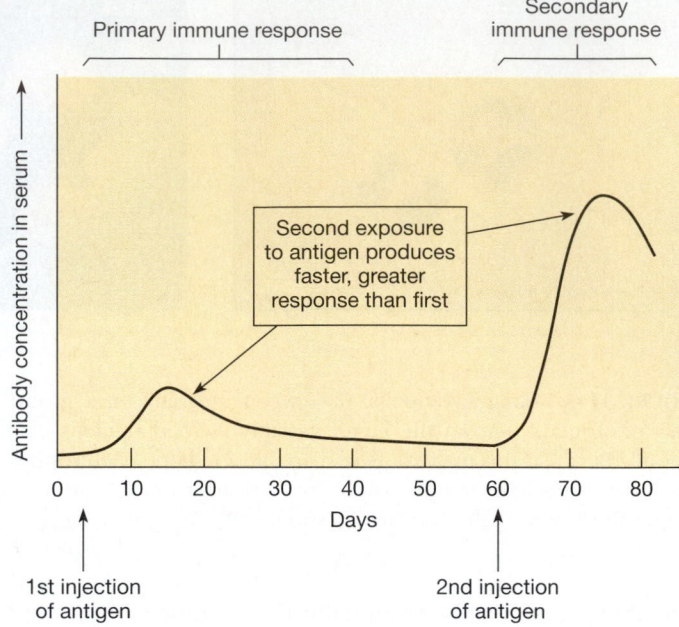

**FIGURE 34–4**  Primary and secondary immune responses. The graph shows the amount of antibody to a specific antigen detected in the blood of a rabbit. Arrows below the graph indicate the times of the first and second injections of antigen. The second time the antigen is injected, the rabbit produces the specific antibody more rapidly and in greater amounts.

An immune response to the body's first encounter with a foreign antigen is called a **primary immune response.** During a primary response, the antigen eventually disappears from the blood, bound by antibody or engulfed by phagocytes. The lymphocytes that secrete antibody also die. However, the memory cells remain. If the same antigen enters the body again, the memory cells permit the immune system to mount a **secondary immune response,** faster and more extensive than the primary response (Figure 34–4). The secondary immune response quickly eliminates the antigen again.

It is essential that the immune system should not attack the body's own cells. How does it distinguish between a foreign antigen and parts of its own body? Body cells are naturally "labeled" with glycoproteins that identify the cell as belonging to a particular tissue in a particular individual. In mammals, many of these cell-surface proteins are encoded by genes of the **major histocompatibility complex (MHC).**

The body is said to have **tolerance** for its own antigens. Tolerance develops during embryonic life. As a fetus or newborn, an animal destroys any lymphocytes that produce receptors that bind the body's own antigens. The earlier in life an animal encounters an antigen, the more likely it is to become tolerant to it. The immune system can distinguish thousands (probably millions) of different foreign antigens by their lack of "self" MHC proteins. Cells trans-

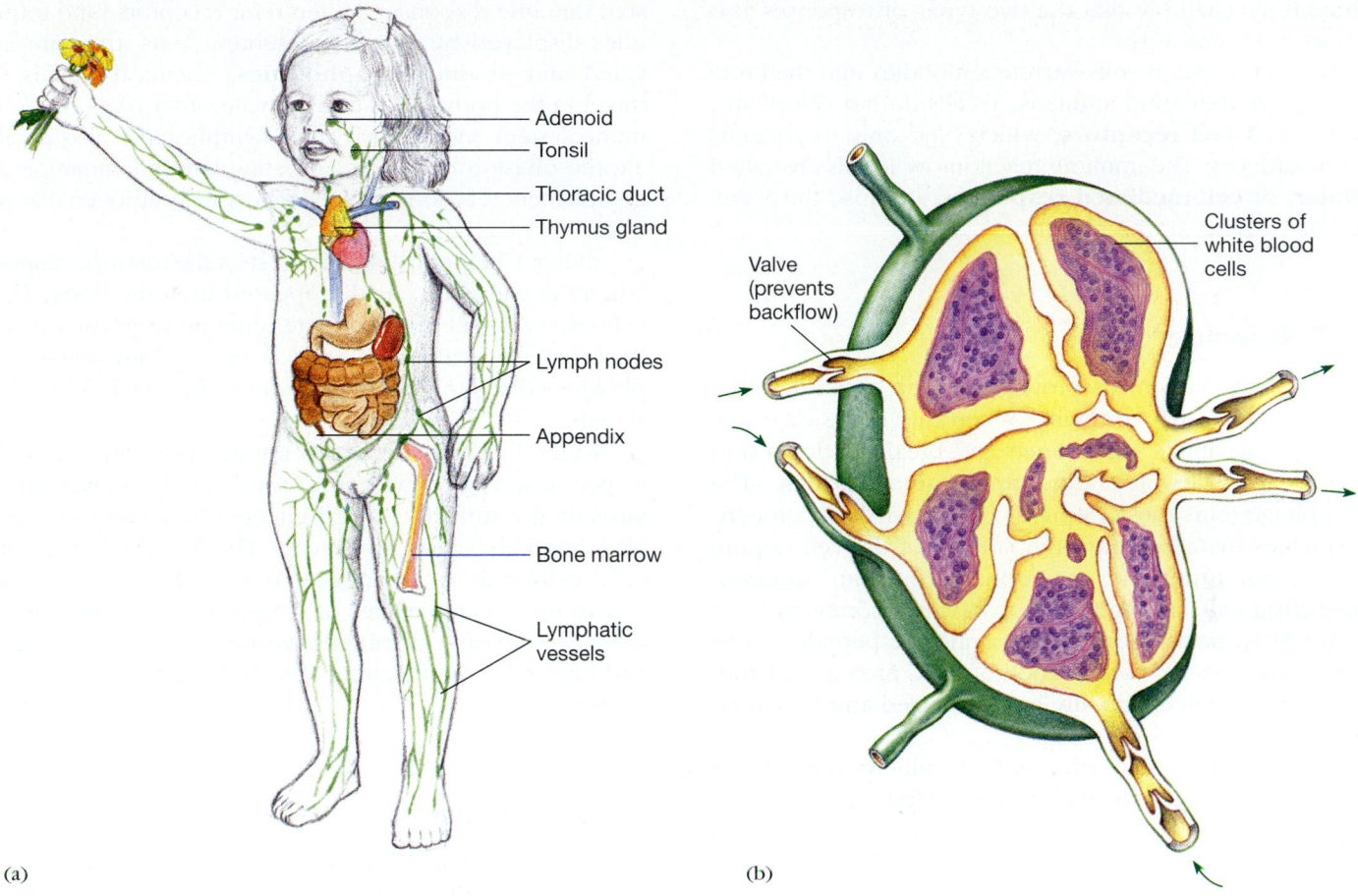

Adenoid
Tonsil
Thoracic duct
Thymus gland
Lymph nodes
Appendix
Bone marrow
Lymphatic vessels

Valve (prevents backflow)
Clusters of white blood cells

(a)

(b)

**FIGURE 34–5**    The human immune system. (a) Location of the major lymphatic vessels and lymph nodes in the human body. (b) Enlarged view of a lymph node.

planted from another person carry foreign collections of MHC proteins, so the immune system also reacts to them as foreign antigens.

## 34–C    THE IMMUNE SYSTEM

Specific responses to foreign antigens are not confined to vertebrates. In insects, injury and bacterial infection cause the animal to produce specific proteins in the hemolymph. Some of these proteins are bactericidal; others form clot-like networks that reduce bleeding. Some snails produce proteins that bind and immobilize invading fluke (platy-helminth) larvae. However, specific responses are most highly developed in the immune systems of vertebrates and reach their greatest complexity in placental mammals.

The immune system is not a set of organs like the di-gestive or respiratory system. It is made up of blood cells that are basically mobile and that are produced and housed and are at work at many sites around the body.

In Chapter 33 we saw that fluid and cells seep out of the blood in the body's capillary beds and join the extra-cellular fluid, or **lymph,** that surrounds all cells. Lymph drains slowly into thin-walled lymphatic vessels, which drain back into the blood via the thoracic duct near the heart (Figure 34–5). At various places, lymph traveling toward the heart passes through **lymph nodes** (sometimes called lymph glands), which contain a mesh lined with white blood cells. Lymph nodes filter pathogens out of the lymph for attack by the cells of the immune system. Tonsils and ade-noids are lymph nodes in the throat and nose, respectively. The spleen acts as a similar filter for the blood.

## 34–D    CELLULAR AND HUMORAL RESPONSES

There are two main kinds of lymphocytes, **T lymphocytes** (or **T cells**) and **B lymphocytes** (or **B cells**). B lympho-cytes are responsible for humoral immune responses, and T lymphocytes are responsible for cellular responses, although we shall see that the two types of responses also interact in several ways.

Whereas some B cells secrete antibodies into the body fluids where they bind antigens, T cells do not. T cell surfaces bear **T-cell receptors,** which bind only to antigens on cell surfaces. The immune reactions of T cells are called **cellular,** or **cell-mediated responses,** because the T cell itself participates in the reaction.

## T Cell Response

A macrophage recognizes a foreign antigen because it does not have "self" MHC proteins. When this is the case, the macrophage engulfs the antigen and breaks it down into smaller molecules, including **antigenic peptides.** The macrophage joins these peptides to its own MHC proteins and pushes them out onto the surface of the cell (Figure 34–6). This turns the macrophage into an **antigen-presenting cell,** distinguished by having a combination of its own MHC protein and foreign antigenic peptides on its surface. The specific combination of MHC protein and antigenic peptide will eventually be recognized and bound by a T-cell receptor.

There are two main groups of T cells: **helper (CD4)** and **killer (CD8** or **cytotoxic)** cells. Helper cells stop and

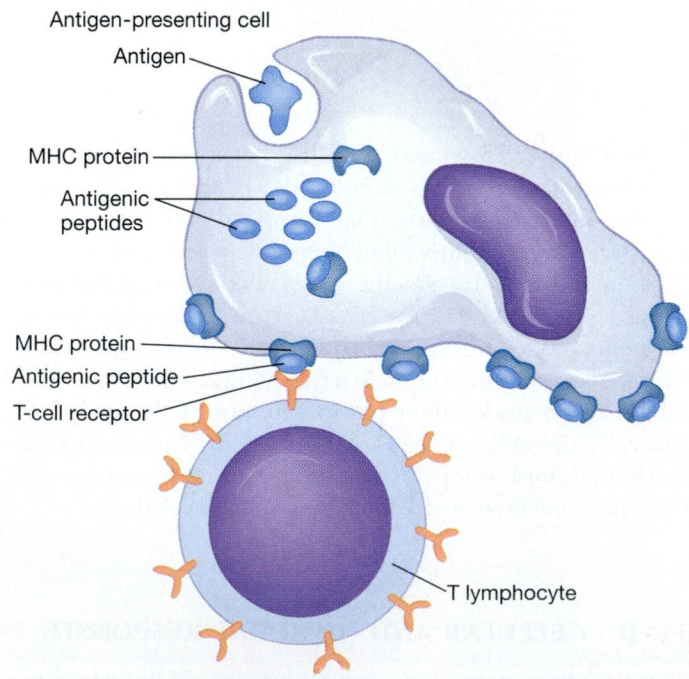

**FIGURE 34–6** How an antigen-presenting macrophage presents antigen to a T cell during the cellular immune response.

Antigen-presenting cell
Antigen
MHC protein
Antigenic peptides
MHC protein
Antigenic peptide
T-cell receptor
T lymphocyte

start immune responses. When their receptors bind to peptides displayed by antigen-presenting cells, they are activated and produce **lymphokines,** chemical signals that travel in the body fluids and activate other parts of the immune system, including B cells. Lymphokines also accelerate the division of T cells and stimulate inflammation and complement reactions, which destroy the pathogen that produced the antigen.

Other T helper cells work to stop the immune response when the pathogen has disappeared from the body. These cells also prevent inappropriate immune responses such as those that cause allergies (Section 34–H). They release lymphokines that suppress the activities of other T cells and of B cells.

Killer T cells react not to antigen-presenting cells but to peptides originating within a cell, such as fragments of virus on the surface of a cell infected by a virus, or mutant MHC markers on a cancer cell. The T-cell receptor on a killer cell binds to the antigens of the infected or damaged cell and inserts molecules of the protein **perforin** into the other cell's plasma membrane. Perforin molecules aggregate and form pores in the membrane so large that the cell leaks to death.

## B Cell Response

B lymphocytes also have specific receptors on their surfaces but unlike T cells, these are antibodies, which bind antigens such as toxin molecules in the blood or antigens on a pathogen's surface, not those attached to body cells. B cells can also secrete their antibodies into the body fluid as **circulating antibodies.** B cell responses are called **humoral immune responses** because they involve free antibodies in the body fluids. (Fluids were called "humors" in medieval medicine.) Humoral responses are the body's main defense against pathogenic bacteria.

A B cell response starts when an antigen binds to an antibody on the surface of a B cell, usually in a lymph node. Swollen, aching lymph nodes under the jaw are a sure sign that the body is mounting a humoral response to bacteria in the head, often as a result of an infected throat or an abscess under a tooth.

The bacterial antigen activates any B cells it binds, and the activated B cells divide to form a clone. This process is sometimes called **clonal selection** because the antigen "selects" which lymphocytes will divide to form clones. Clonal selection permits the body to produce large quantities of antibody only when needed. It also permits the immune system to remember antigens it has encountered before because some members of the clone will become long-lived memory cells. Most cells of a clone differentiate into **plasma cells,** which secrete many copies of the clone's antibody molecule. Plasma cells develop large areas of rough endoplasmic reticulum where antibody is synthesized and packaged for secretion. Whereas the original B cell had copies of its antibody on its surface but secreted very few of them, a plasma cell may spew out 10 million copies of its antibody every day for four or five days until it dies. The circulating antibodies travel in the lymph and blood, binding the bacterial antigen wherever they encounter it.

Antibody binding may neutralize a bacterium by preventing it from reproducing and preventing its antigens from binding and damaging body cells. But its main role is to label the bacterium for destruction by a phagocyte, often aided by the complement system.

## Lymphocyte Interactions

B and T cells complement each other. B cell antibodies respond swiftly to toxin molecules or to the outer surface of a pathogen, which a T cell cannot recognize. On the other hand, T cells respond to abnormal body cells, such as cancerous cells or those infected by viruses. These cells would not bind antibody.

Besides responding to different situations, T and B cells exchange information. B cells may be activated not by free antigen but by binding to the antigenic peptide attached to a T-cell receptor (Figure 34–7). When this happens, the T cell passes lymphokines to the B cell, speeding cell division and antibody production. At the same time, the B cell may influence the T cell's activities by transferring chemical messengers in the other direction.

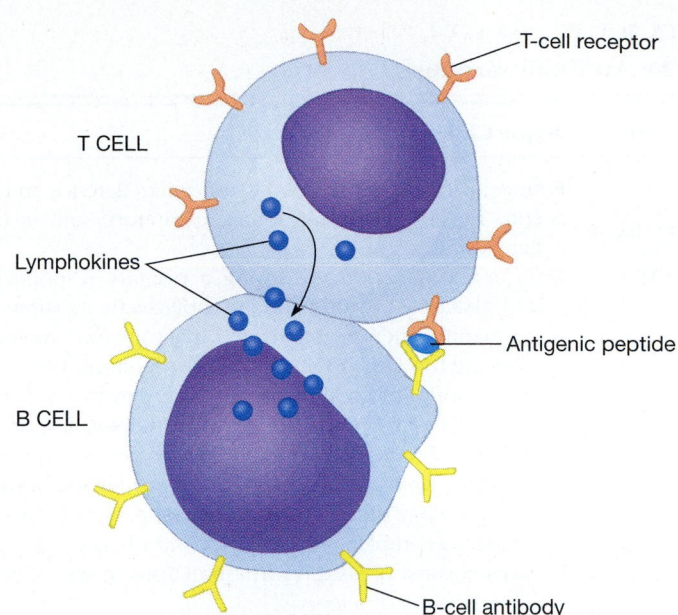

FIGURE 34–7    A T cell and a B cell exchange information. Antibody on the surface of the B cell binds an antigen attached to a T-cell receptor. This stimulates the T cell to transfer lymphokines to the B cell, activating the B cell.

The **constant region** of an antibody, the stem of the Y, is one of only five possible types. Antibodies are classified by this region into group A, M, G, E, or D, with each group having its own general biological properties (Table

## 34–E    ANTIBODIES

The body produces millions of different antibodies, but all belong to a group of proteins called **immunoglobulins.** (The terms "immunoglobulin" and "antibody" can be used interchangeably.) Each antibody molecule is made up of four peptide chains: two identical heavy chains and two identical light chains, all joined by sulfur bridges into a Y-shaped structure (Figure 34–8). Each chain has two parts, a variable region and a constant region, which are joined by a short "joining" segment. These parts are responsible for the two functions of the antibody: binding a specific antigen and assisting in its destruction.

Antigens bind to the tips of the Y's arms, the **variable region,** so called because it comes in thousands of different forms in different B cell clones. An antibody binds to an antigen having a complementary three-dimensional shape and distribution of electrical charges; no covalent bonds or chemical reactions are involved. In this respect, antigen-antibody binding resembles the binding of an enzyme to its substrate.

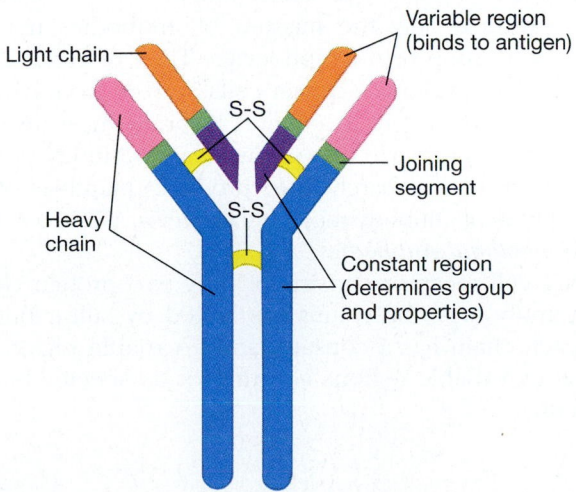

FIGURE 34–8    Antibody structure. Each antibody (immunoglobulin) molecule consists of two identical light and two identical heavy peptide chains joined by disulfide bonds (yellow). Part of each chain is variable and part is constant.

**TABLE 34–1**
**The Antibody Groups**

| Group | Major Characteristics |
|---|---|
| A | Found in mucus secretions, tears, milk; defends mucous membranes in intestinal, respiratory, and urogenital tracts |
| M | Main antibody produced during a primary response; sticks bacteria together and immobilizes them; stimulates complement reactions and macrophages; molecules are too large to leave blood and lymph vessels |
| G | Main antibody in blood; chief antibody produced during a secondary immune response; the smallest antibody molecule, it can leave blood vessels easily, cross the placenta, and defend tissue spaces and body surfaces; combats microorganisms and their toxins; stimulates complement reactions and phagocytes |
| E | Effective against parasitic worm infections; responsible for symptoms of allergy |
| D | Rare; found on the surfaces of lymphocytes; function unknown |

34–1). For instance, group G antibodies combine with bacteria and viruses in the blood. The group G constant region attracts macrophages that engulf and destroy the pathogen. It also enables a G antibody to cross the placenta. Hence, antibodies produced by the mother's body can enter the fetus and protect it from disease.

Because an antibody is so specific for its antigen, antibodies are widely used in research and in medical tests to detect the presence of particular molecules and to immobilize specific antigens. However, until the 1970s researchers could produce only the mixture of antibodies normally made by a group of B lymphocytes. Then researchers discovered that myelomas, cancer cells derived from lymphocytes, live indefinitely in culture. Hybridizing a myeloma cell with a lymphocyte that produces an antibody of interest now permits researchers to produce quantities of any desired type of antibody molecule (*How Do We Know? Making Monoclonal Antibodies*).

A T-cell receptor consists of only two protein chains. As in antibodies, the chains are linked by sulfur bridges, and each chain has a constant and a variable region. The receptor's variable regions account for its specific binding properties.

## Antibody Genetics

The development of B cells is particularly interesting because these cells produce thousands of antibodies using relatively few genes. During B cell differentiation, segments of antibody genes are physically rearranged in the genome and spliced together before they are transcribed. This extraordinary system uses an enzyme unique to the nuclei of immature B cells: terminal deoxynucleotide transferase, which cuts and splices DNA. In mice, immunologists have analyzed about 250 genes for variable regions and 8 genes for constant regions, as well as about 12 genes for joining segments that lie between the two. Combined in various ways in different B cells, these genes permit the body's B cell population to produce hundreds of thousands of different antibody molecules from fewer than 300 genes. Even more variation comes from the B cell's ability to insert extra nucleotides in the joins between DNA fragments as it splices them together.

Further variation comes into action once B cells have been exposed to antigen and cells that bind the antigen have been selected to proliferate. In a growing clone, the immunoglobulin genes undergo small mutations. As a result, some of these B cells become fine-tuned, with even greater specificity to the particular antigen present and hence greater effectiveness against it.

Another set of genes provides a similar "kit" of DNA segments that can be spliced in different ways to produce a great variety of T-cell receptors.

***Antibody Isotypes*** After it is activated, a B cell (plasma cell) has the remarkable ability to switch from making antibodies of one group (usually group M) to making antibodies of another group (G, A, or E). This switching involves rearranging the DNA so that the gene for a different constant region is expressed. (The constant region determines the antibody's group.) The antibody still has the same variable region and hence binds the same antigens. The two different antibodies are known as **isotypes.**

The advantage of isotype switching is that different isotypes act in different places while still attacking the same pathogen. Thus, if it is stimulated by a bacterial infection, a plasma cell can first make millions of antibodies of group M, which will inactivate bacteria in the blood and stimulate complement reactions to destroy them. Then it can switch to the group A isotype so that the body secretes lots of antibody-laden mucus, preventing any more of that type of bacteria from entering the body.

## 34–F IMMUNE CELL DEVELOPMENT

Because the white blood cells of the immune system travel around the body, and some of them change from one form to another, it has proved difficult to discover where they come from.

Modern understanding of blood cells started with studies of radiation damage from the atomic bombing of Hiroshima and Nagasaki in 1945. Many people exposed to radiation in the attacks died about two weeks later from internal bleeding and infection. Radiation had destroyed the precursors of blood cells, and without the cells that fight infection and permit blood clotting, these people died.

Experiments on mice showed that radiation damage could be reversed by injecting bone marrow from a genetically compatible donor. **Bone marrow** is soft, pulpy tissue found inside many bones in reptiles, birds, and mammals. It became obvious that even a small sample of bone marrow contained cells capable of reproducing and of differentiating into all the blood cells that the body needs, including red blood cells, platelets, phagocytes, and lymphocytes. These bone marrow cells were named **hematopoietic** ("blood-producing") **stem cells.**

In the 1960s, researchers found that stem cells gave rise to T and B lymphocytes. B cells were discovered in chickens, where they differentiate in an organ called the bursa of Fabricius. In mammals, B cells differentiate in the bone marrow. T cells differentiate from stems cells that migrate into the thymus gland in the neck. Interestingly, the thymus gland disappears in many adult mammals. It appears to be needed only in the embryo and very young animals, to get the T cell populations started.

## B Cell Differentiation

The signals that tell a mammalian stem cell to become a B cell rather than some other type of blood cell appear to be produced by large **stromal cells** in the bone marrow shortly before birth. The signals cause the stem cells to shuffle and express the genes coding for heavy immunoglobulin chains. The light-chain genes are rearranged and expressed later. When both have been expressed, the antibodies move to the cell surface. The now-mature B cell enters the circulation, which carries it to lymph nodes and other parts of the body (Figure 34–9).

More than half of the pre-B stem cells die in the bone marrow as the body develops immunological tolerance to its own antigens. By chance, some B cells produce antibodies that react with the body's own antigens. If its antibodies bind tightly enough to neighboring cells, the cell commits suicide by digesting its own DNA (apoptosis, or programmed cell death, Section 12–G). This ensures that the adult does not contain B cells that could destroy its own body cells.

## T Cell Differentiation

Early in embryonic development, pre-T stem cells migrate in waves to the thymus. Each wave of cells develops a different type of T-cell receptor on its surface and migrates to a different part of the body. For instance, the first two waves of cells populate the skin and the reproductive organs. The third wave consists of T helper and killer T cells, which are produced throughout life and which end up mainly in the spleen and digestive tract.

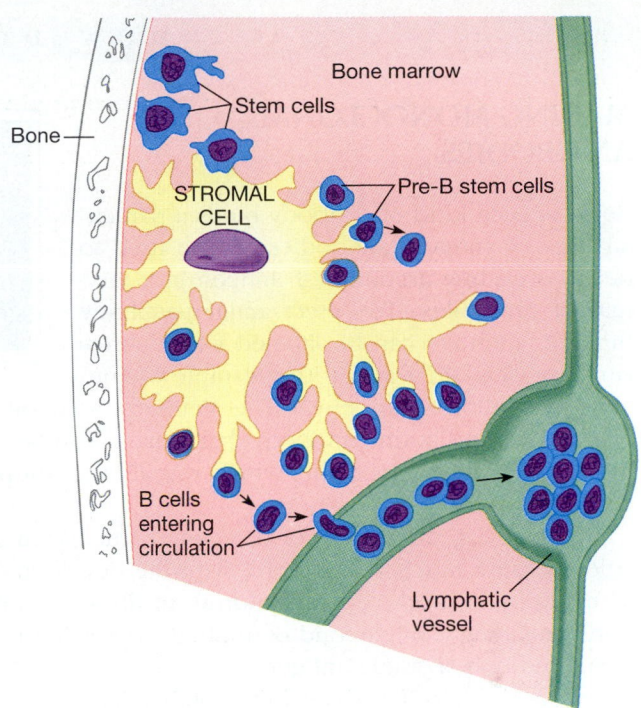

**FIGURE 34–9** Differentiation of B lymphocytes in the bone marrow. A hematopoietic (blood-forming) stem cell may either divide and give rise to new stem cells or differentiate to produce one of several types of blood cell that enter the circulation via a lymphatic.

T cells, like B cells, have to survive selection processes. The first tests their ability to recognize antigens on other cells. Stem T cells in the thymus die if they cannot bind to any "self" MHC antigens on cells in the thymus. The next selection step is like that for B cells: the destruction of T cells that bind too tightly to the body's own antigens. Only T cells that survive both tests leave the thymus to take up residence elsewhere in the body.

## 34–G MEDICAL ASPECTS OF IMMUNE RESPONSES

### Organ Transplants

"Histocompatibility" means compatibility between tissues and refers to the fact that organs can be transplanted successfully from one animal to another only if the MHC antigens of the two animals are compatible. MHC compatibility between two individuals is extremely unlikely because there are so many genes in the MHC complex that the number of possible combinations of antigens in a population is

# MAKING MONOCLONAL ANTIBODIES

Because each kind of antibody binds a particular kind of antigen, antibodies can be used to detect even tiny amounts of antigen in a mixture of molecules. To detect antigen-antibody binding, antibody can be labeled with fluorescent dye. It is then added to a sample of antigens and allowed to bind. The antibody solution is washed off, and bound antibody that remains can be detected with ultraviolet light, which reveals the fluorescence.

In the 1970s, this means of detecting particular molecules became much more useful with the development of methods to produce **monoclonal antibody,** a solution containing a single kind of antibody that will, therefore, bind only a single antigen.

First a mouse is injected with the pure antigen. This induces it to produce large numbers of plasma cells (acti-

vated B lymphocytes) that secrete antibody specific for the antigen. Then plasma cells are isolated from a sample of the mouse's blood. Most of these will be secreting the desired antibody. However, plasma cells normally die within a few days. To keep them alive and secreting antibody, they are mixed with cultured myeloma cells, derived from mouse B lymphocyte tumors, which live indefinitely in culture.

Some of the mouse plasma cells fuse with myeloma cells, forming **hybridoma cells.** The myeloma cells contain a mutation that prevents them surviving without a particular nutrient, which the plasma cells supply. By placing a mixture of myeloma and plasma cells in a medium lacking the nutrient, immunologists select for hybridomas, which combine the myeloma's immortality with the plasma cell's ability to provide the nutrient (Figure 34–A).

Each hybridoma forms a clone of cells, which is tested to see if it produces antibody to the desired antigen. If it

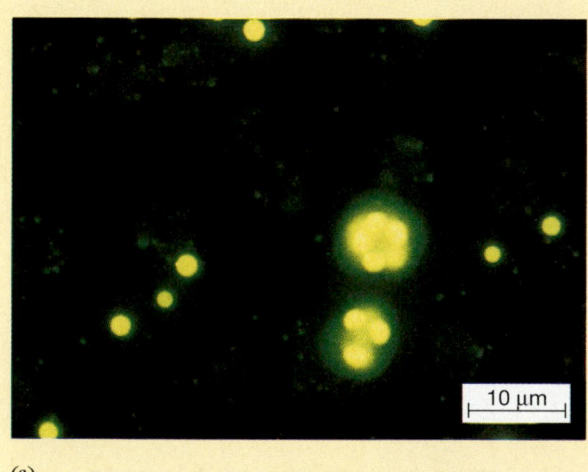

(a)

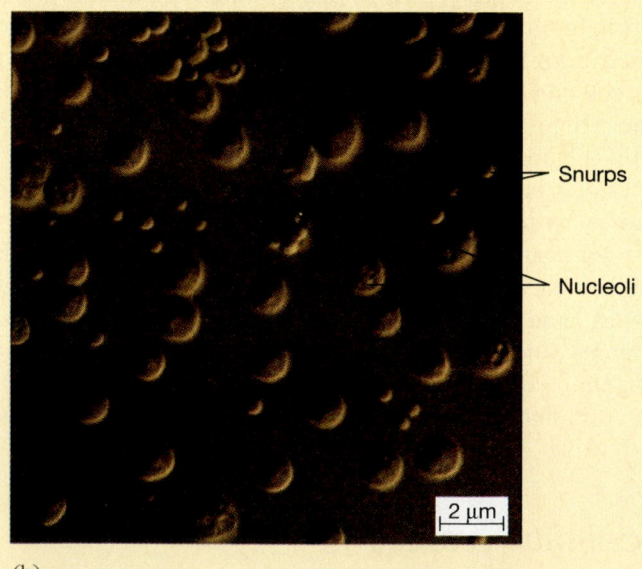

— Snurps

— Nucleoli

2 μm

(b)

**FIGURE 34–A**  Staining with monoclonal antibody. (a) Contents of nuclei from toad oocytes. You cannot tell by looking at them what these blobs are. (b) The same preparation stained with monoclonal antibody to a protein found in spliceosomes—the structures that splice pre-RNAs. From the size, we know that these are not whole spliceosomes, but snurps, the subunits of spliceosomes. Similar staining shows that the larger blobs in (a) are nucleoli—the sites of ribosome manufacture.   (J. G. Gall and Z. Wu)

does, it is isolated and cultured for large-scale production of the antibody (Figure 34–B).

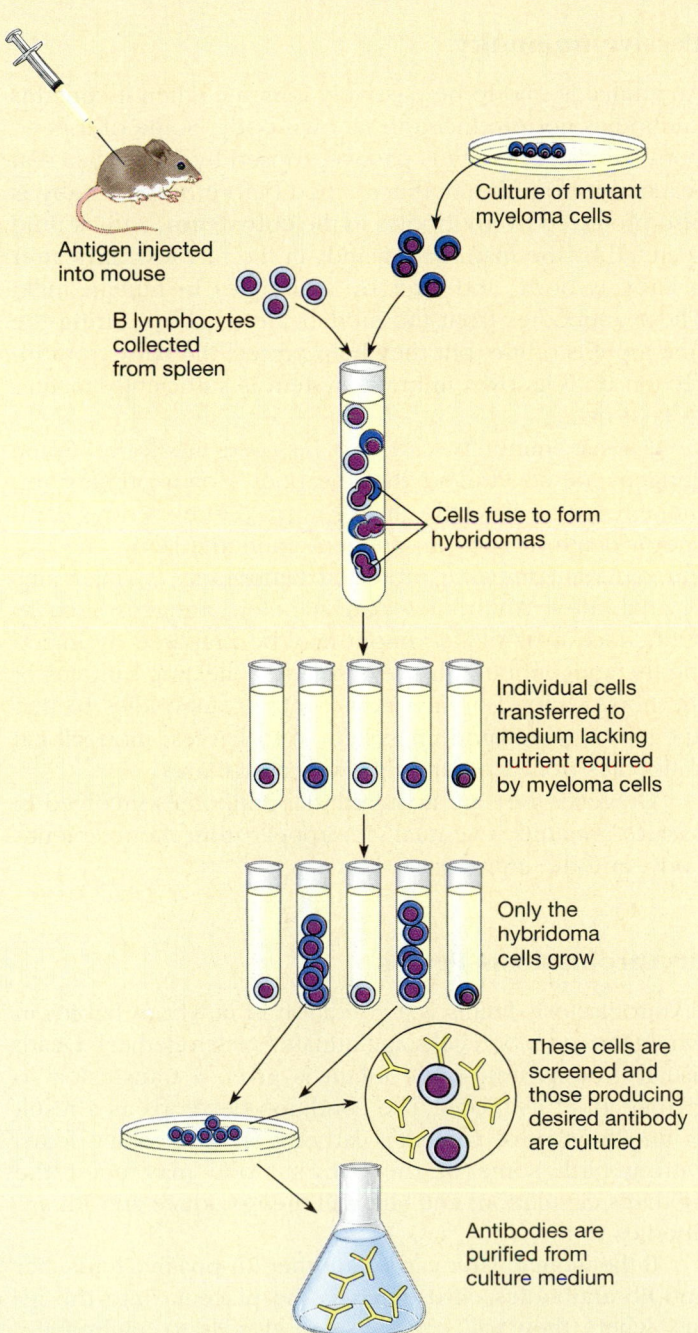

Antigen injected into mouse

B lymphocytes collected from spleen

Culture of mutant myeloma cells

Cells fuse to form hybridomas

Individual cells transferred to medium lacking nutrient required by myeloma cells

Only the hybridoma cells grow

These cells are screened and those producing desired antibody are cultured

Antibodies are purified from culture medium

**FIGURE 34–B**  Production of monoclonal antibodies.

almost infinite. Identical twins have identical MHC antigens, but even close relatives such as nonidentical siblings, or parent and child, often do not show antigen matches close enough to permit a successful transplant.

Because cells from different individuals generally display different MHC antigens on their surfaces, transplanted organs provoke cellular immune responses. Immunologists often study cellular responses using skin grafts, which are easy to work with and do not harm the recipient. If skin is transplanted from one mouse to another, the graft is quickly invaded by blood vessels, and it looks healthy for several days. Within nine days, however, the blood supply to the graft diminishes, and the graft is infiltrated by T cells. The graft begins to wither, and within a day or two it sloughs off. A second graft from the same donor to any part of the same recipient is rejected faster than the first one, showing that the body "remembers" the graft antigens.

Medical researchers are primarily interested in preventing rejection of transplanted tissue. One way of doing this is to use **immunosuppressant** drugs, which suppress the body's immune responses. Such drugs are invariably used after heart, liver, or kidney transplant operations. Most immunosuppressant drugs work by suppressing lymphocyte cell division. Although they suppress graft rejection, these drugs also prevent the immune system's normal activity, so transplantees are susceptible to infections that would not be dangerous to those with functioning immune systems.

The other, and safer, way to minimize graft rejection is to match the MHC antigens of donor and recipient as closely as possible.

## Adaptive Value of MHC Antigens

Animals all the way back to sponges reject foreign tissue grafts, which suggests that there is a strong selective advantage to having different antigens from everyone else in the population. There are various hypotheses about how this system evolved and why it has been retained in all animals.

One hypothesis comes from the fact that T cells produce a cellular response only to cells that display both self and foreign antigens. This being the case, there is selection for a pathogen to produce an antigen that imitates the body's own antigens well enough to fool the T cell into accepting it as a legitimate part of the body.

A virus with such an adaptation would sweep through the host population, surviving and reproducing in every host having the same antigen. Who would be spared? Only individuals with antigens different from those produced by the virus: their T cells would not accept the viral antigen but would destroy it. Hence, these individuals would enjoy a strong selective advantage from having different antigens, and the population would evolve in such a way that the antigen diversity of its members increased.

Furthermore, the more antigenic variety there is in a host population, the less selective advantage there is for viruses to evolve an antigen mimicking one host antigen, because relatively few potential hosts will have that antigen.

This theory also accounts partly for variations in how sick different individuals become when infected by the same strain of influenza or other disease.

## Vaccination

Vaccination induces the immune system to mount a primary immune response against a disease and to produce memory cells, ready to trigger a secondary response at the body's first real battle against that disease antigen.

Arabic and Chinese manuscripts more than a thousand years old refer to vaccination against smallpox. Lady Mary Wortley Montagu, wife of the British ambassador to Turkey, introduced this ancient custom into England in 1718. She had her children vaccinated by rubbing part of the scab from a healed smallpox sore into a small wound in the skin. This **vaccination** introduced a few live smallpox viruses into the body, stimulating a primary immune response and thereby conferring immunity to smallpox in later life. The snag was that vaccination with even a small amount of live virus sometimes caused a case of smallpox, which could be fatal.

English physician Edward Jenner found a way around this problem in 1796. Jenner noticed that dairy workers who had caught the relatively mild disease cowpox from cows seemed to be immune to smallpox. He found that rubbing pus from cowpox sores into scratches in the skin prevented people from coming down with smallpox later. In this case, the antigens of smallpox and cowpox are so similar that the same antibodies work against both.

Almost a century later, Louis Pasteur found a safer way to prepare vaccines by making them from pathogens disabled by heat and other treatments. This rendered the pathogens unable to cause disease but left their antigens intact as vaccines that could stimulate a primary immune response.

Smallpox was not only the first disease to be prevented by vaccination but also the first disease to be officially declared wiped out by human efforts. The last known outbreaks of smallpox occurred in India and Africa in the late 1970s. International vaccination programs had greatly reduced the number of smallpox cases. The final conquest came after health officials adopted a different strategy: searching out pockets of infection (people were rewarded for each case they reported), quarantining the victims, and vaccinating their friends and relations.

Poliomyelitis may join smallpox on the extinct list. It has dropped from 58,000 cases in the United States in 1952 to 10 in the Western Hemisphere in 1990. Once as feared as AIDS is today, this crippling and killing disease is now seldom seen.

Nowadays, we have vaccines for a number of bacterial and viral diseases. However, several important infectious diseases remain without effective vaccines, including malaria, trypanosome infections, and AIDS. All these diseases are caused by pathogens whose cell-surface antigens change frequently. As a result, no single antibody is effective against very many cases.

## Passive Immunity

An animal is said to be passively immune when it contains antibodies not produced in its own body. Some of a newborn baby's immunity is passive, caused by antibodies that reached it from the mother's blood before birth. A baby is also protected by antibodies in the **colostrum,** a thick fluid secreted by the mammary glands in the first few days after a baby is born, and later by antibodies in human milk. These antibodies from the mother are used up during the first months of life, but they help protect the baby from infection until its own immune system is sufficiently mature to take over.

Passive immunity can also be used medically. Some antigens are so virulent that the body's own primary immune response has little chance of preventing serious damage or death. If such an antigen enters the body, the victim can sometimes be protected temporarily by injections of antibodies. Antibodies against potent antigens such as tetanus toxin or snake venom may be prepared by injecting the antigen into a horse and later collecting samples of the horse's blood, which now contains antibodies to that antigen. A more modern system is to harvest monoclonal antibodies from laboratory-grown cell cultures.

However they are acquired, the antibodies involved in passive immunity eventually disappear from the recipient's body, and the immunity is lost.

## Erythroblastosis Fetalis

Erythroblastosis fetalis is a condition in newborn babies in which the red blood cells agglutinate (stick together). Death usually results unless the infant is given a transfusion to change all of its blood. The condition develops as a result of an Rh-negative mother's carrying an Rh-positive fetus. During birth some of the baby's blood may enter the mother's circulation, and she will then produce anti-Rh antibodies to it.

If the woman later carries another Rh-positive fetus, her anti-Rh antibodies diffuse across the placenta into the fetus, where they will cause its blood cells to agglutinate. Since Rh-negative is a recessive genetic trait, and about 85% of humans are Rh-positive, an Rh-negative woman is quite

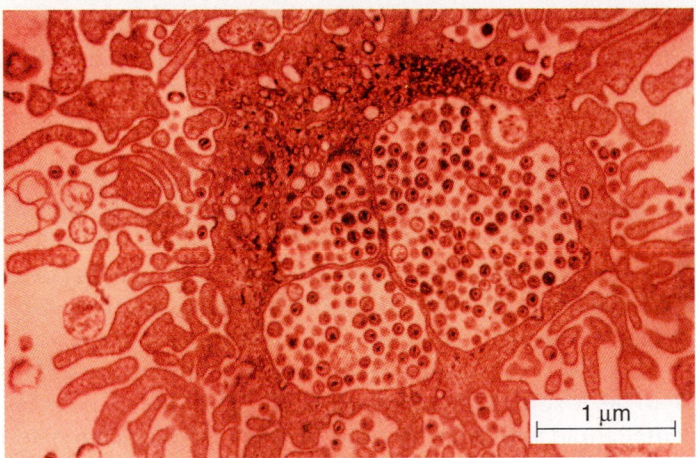

**FIGURE 34–10**  Blood cells infected with HIV. Newly formed HIV can also be seen budding off the edges of the cell.  (Hans Gelberblom/VU)

likely to marry an Rh-positive man and bear Rh-positive babies. Leakage between the fetal and maternal circulations is not inevitable, however, and many Rh-negative women bear several Rh-positive children without problems. Nowadays, an Rh-negative mother may receive injections of anti-Rh antibody after delivery, to remove any Rh factor that may have entered her blood from the blood of the fetus and so keep the mother's body from manufacturing antibodies that could damage future fetuses.

## AIDS

**Acquired immune deficiency syndrome (AIDS)** is an immune disorder first identified in 1981 and characterized by a decline in the number of T helper cells. Because these cells activate both B and T lymphocytes, their loss eventually disables the immune system (Figure 34–10). Most individuals with AIDS die of infections or cancers that are no threat to people with healthy immune systems.

In 1984, French scientists identified the pathogen that causes AIDS, **human immunodeficiency virus (HIV),** a retrovirus (Section 23–C). How HIV kills T helper cells is unclear. One peculiar feature is that both infected and uninfected T helper cells die. Recent evidence suggests that HIV activates uninfected T helper cells, but instead of normal cell division, this activation switches on the cell-death program, resulting in apoptosis.

Someone newly infected with HIV mounts a vigorous immune response, which may cause flu-like symptoms for several weeks and destroy most of the viruses. Antibodies to HIV can later be detected in the blood. Possibly, some people's immune systems destroy all viruses at this stage. Usually, however, the person is permanently infected (Figure 34–11). Small numbers of viruses elude the immune system and continue to replicate slowly in several types of cells for up to 12 years. Eventually, the viruses reduce immune responses to the point where various tumors and infections take hold in the body, and the person develops full-blown AIDS. As far as we know, AIDS is always fatal, usually several years after symptoms first appear.

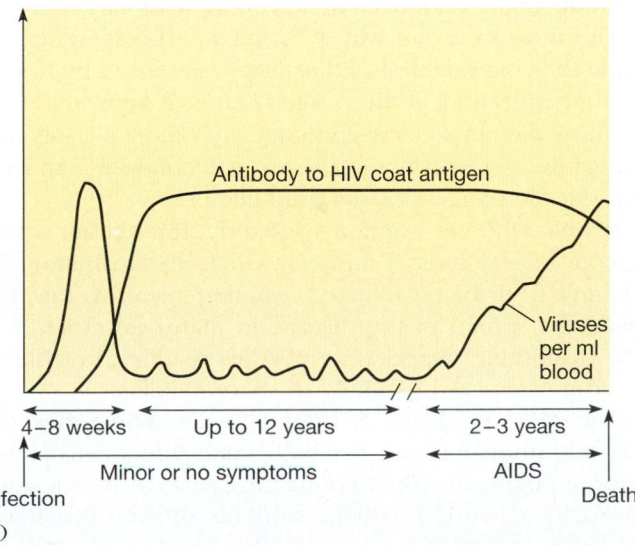

(a)

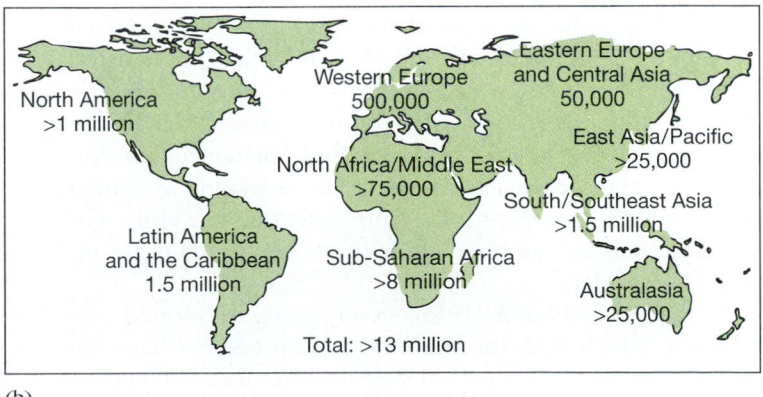

(b)

**FIGURE 34–11**  AIDS. (a) Progress of HIV infection to AIDS with signs of the various stages found in the blood. (b) The estimated incidence of HIV infection in various parts of the world in 1993.

HIV is transmitted when body fluids—usually blood or semen—containing the virus pass from one person to another. The virus can pass from one person to another via semen during anal or vaginal intercourse. Blood-to-blood transmission may occur when the same hypodermic needle is used to inject drugs or vaccines into more than one person. It can also occur when someone receives a transfusion of infected blood or blood products, or even when one person's blood accidentally enters another person through a lesion, such as a cut or ulcer. Similarly, blood vessels break during childbirth, allowing blood from an infected mother to reach her baby. Since 1985, blood for transfusions has been screened for antibodies to HIV in most countries and is largely safe.

HIV disintegrates rapidly outside the body so is not spread by casual contact, hugging, changing diapers, using the same toilet seats, or even sharing toothbrushes. Doctors, nurses, friends, and family members who live with people who have AIDS almost never catch the disease. The virus has been found in urine, tears, saliva, breast milk, and vaginal secretions, but it seems not to be transmitted by these fluids unless they get into a cut. The virus apparently does not cross intact mucous membranes (in the mouth, vagina, or rectum) or skin. However, there are often gaps in mucous membranes that have been stretched or that have been injured by ulcers or by fungus, herpes, or gonorrhea infections. These gaps in the membrane form a route by which the virus can pass from one person to another.

In Western nations, infection from a man's semen was the first method of transmission recognized, particularly between male homosexuals. AIDS has spread to heterosexual men and women and to babies, mainly by way of intravenous drug users who share needles. Infected women can pass the virus to their sexual partners, especially during menstruation.

In Africa the virus is transmitted mainly by sexual promiscuity, genital mutilation of girls with contaminated surgical instruments, vaccination of many people with the same needle, and transfusion of unscreened blood into malaria patients. Africa also has a high incidence of sexually transmitted diseases. These produce lesions in genital areas, which in turn provide entry sites for the virus. For all these reasons, as many African women as men are infected with AIDS.

Drugs to combat AIDS are slowly being developed. Antibiotics, which help the immune system combat bacterial infections, have no effect on viruses. The drug zidovudine (also called azidothymidine [AZT]) resembles the nucleotide thymidine and prolongs the life of some people with AIDS, although only by a matter of months. However, AZT-resistant forms of the virus have already evolved.

There are three potential defenses against AIDS or any other viral disease: vaccination, immune responses, and prevention. It is not easy to produce a vaccine to the AIDS virus because the virus evolves rapidly and changes its antigens frequently. The immune system is little help because HIV disables it. This leaves us with one real line of defense: prevention.

Some people long infected with HIV have not yet developed AIDS. Some of these people, presumably, already have immune systems that defend them against the virus. This is natural selection in action, and we can see that even if a cure for AIDS is never found, a human population resistant to the virus will probably evolve eventually.

This is no consolation to those now suffering from AIDS or to health officials overwhelmed by the scope of the problem. People with AIDS require frequent hospitalization and treatment, and their average hospital stay is twice as long as that of other patients. The psychological strain on those caring for so many patients with no hope of recovery is another inestimable burden. The U.S. Public Health Service estimates that about 2 million people in the United States are infected with HIV. In parts of West Africa, one quarter of the population is infected, and millions of children have been orphaned by AIDS.

Most cases of AIDS result from sex with an infected partner or injection from a hypodermic needle that has not been sterilized after being used by someone infected by HIV. Therefore, no one is completely safe from AIDS if, during the last 15 years, he or she has had sex with anyone who has had sex with a third person, shared a needle to inject drugs, or received a medical injection by an improperly sterilized needle. Chastity, lifelong monogamy, and a clean needle for each injection are the only ways to prevent the spread of AIDS. Even these practices would not stop AIDS for many years (because of the risk from blood transfusions and from individuals already infected but not yet producing antibodies that screening tests can detect).

The most effective way to combat AIDS is educating people to avoid infection. Education was started by homosexual organizations in the United States. It aims to (1) explain how the virus is transmitted, (2) promote chastity and monogamy, (3) encourage the use of condoms, and (4) point out the dangers of sharing needles.

People who use condoms properly, throughout sexual intercourse, are almost completely safe from transmitting the virus and from being infected. Another measure that has slowed the spread of the disease in many countries is to make disposable hypodermic needles readily available so that drug users are less likely to share needles.

The AIDS epidemic is bound to get worse, because AIDS education has not yet had much effect on the behavior of people at risk of contracting AIDS. A recent study of sexually active U.S. college students showed that fewer than half were monogamous and that condom use had increased only slightly in the previous five years.

## 34–H   IMMUNE SYSTEM MALFUNCTIONS

When something goes wrong with the immune system, as in the case of AIDS, the consequences are often fatal.

For instance, if the thymus gland is abnormal, T lymphocytes fail to develop. A baby without T lymphocytes fails to produce B cell clones to combat invading organisms and so is usually killed by the first pathogens it encounters. A few such babies have been saved by keeping them in sterile environments or by transplants of bone marrow cells and thymus tissue, which may permit them to make antibodies.

### Autoimmune Diseases

**Autoimmunity** is a condition in which the body mounts immune responses against some of its own antigens, sometimes because it has been invaded by a foreign antigen very similar to one of the body's own antigens. In this case, the immune response destroys the body's similar protein as well as the foreign antigen. Table 34–2 shows some of the diseases this causes.

Rheumatic fever starts when the body forms antibodies against streptococcal bacteria. These antibodies also attack the body's own proteins, particularly those in the heart valves, permanently weakening the heart. Because of the danger of developing rheumatic fever, "strep throat" should be seen as a serious disease that should be treated quickly with antibiotics.

Some autoimmune diseases occur almost exclusively in people with particular MHC antigens—presumably ones

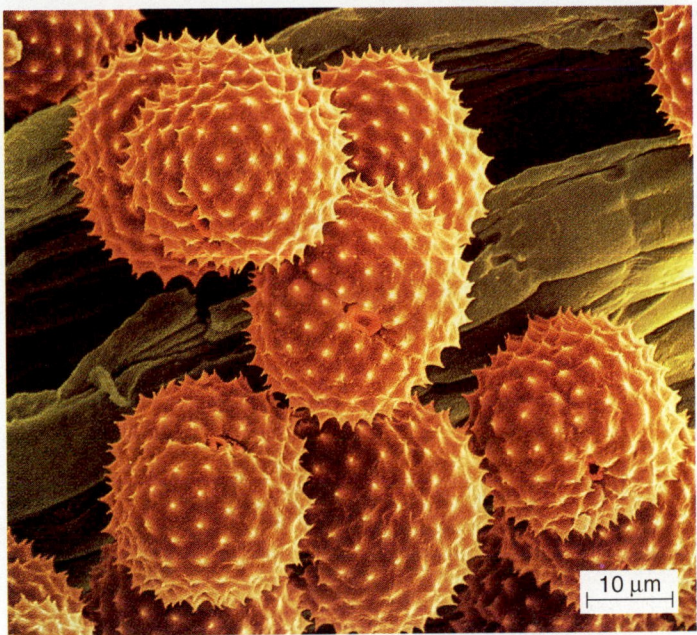

**FIGURE 34–12** Ragweed. The pollen of this summer-flowering weed is a common cause of allergy.   (David Scharf/Peter Arnold)

that are easily mistaken for foreign by the body's immune cells. Thus, people with the genes for these MHC antigens have an inherited risk of developing the corresponding autoimmune disease.

### Allergies

About 10% of the human population suffers from allergies: inappropriate immune responses to harmless substances in the environment or in food or medicine—for example, milk, chocolate, ragweed pollen, penicillin, cat saliva, or the feces of mites in house dust (Figure 34–12). Generally the first exposure to the allergy-producing antigen (an **allergen**) produces no symptoms, but it **sensitizes** the body by evoking a primary immune response. The next encounter with the allergen evokes a secondary response known as an allergic, or **hypersensitivity,** response, which may be so violent as to cause tissue damage.

Allergic reactions are due to group E antibodies. The normal role of E antibodies is not to cause allergy but to protect the body from infection by platyhelminth parasites (tapeworms and flukes). In most people, the first time an allergen enters the body, T helper cells recognize it as harmless and prevent B cells from responding to it. In a person who will produce an allergic reaction, however, something goes wrong. T cells induce B cells to secrete group E

### TABLE 34–2
### Autoimmune Diseases

| Disease | Site(s) Attacked | By |
|---|---|---|
| Graves' disease | Thyroid gland | Antibodies |
| Myasthenia gravis | Neuromuscular junctions | Antibodies |
| Rheumatic fever | Heart valves | Antibodies |
| Systemic lupus erythematosus | Joints, skin, kidneys, etc. | Antibodies |
| Psoriasis | Skin | T cells |
| Rheumatoid arthritis | Joints | T cells |
| Multiple sclerosis | Myelin sheath around nerve cells | T cells |
| Insulin-dependent diabetes mellitus | Insulin-making cells of pancreas | T cells |
| Reiter's syndrome | Eyes, joints, genital tract | T cells |

10 μm

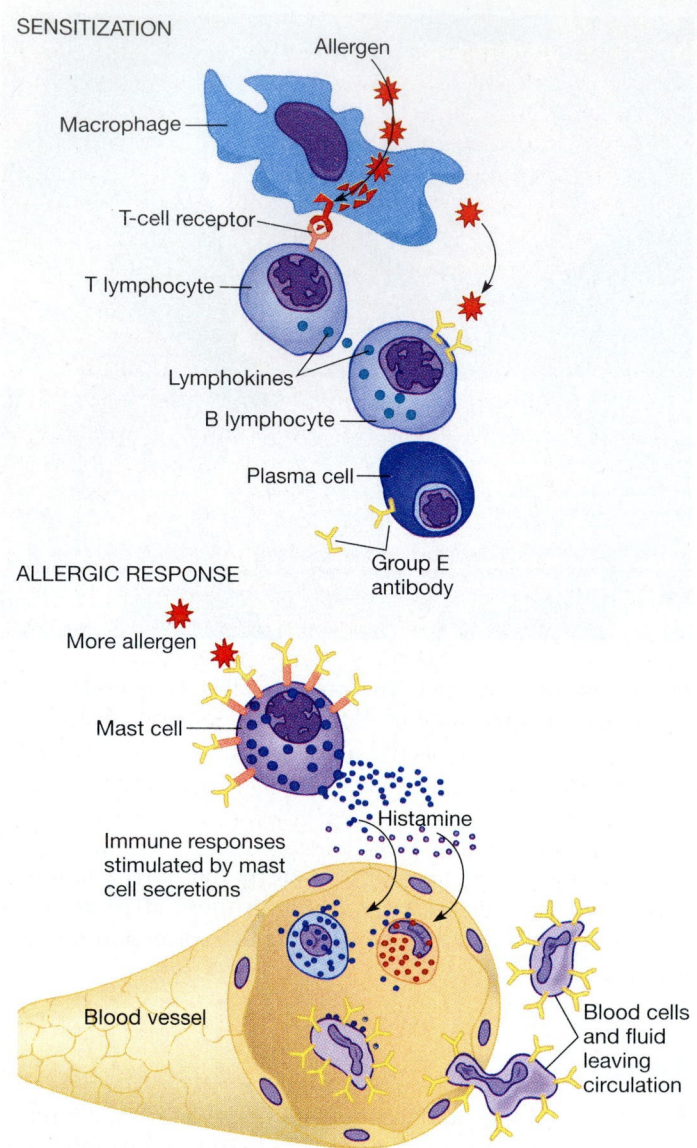

SENSITIZATION

Allergen

Macrophage

T-cell receptor

T lymphocyte

Lymphokines

B lymphocyte

Plasma cell

Group E
antibody

ALLERGIC RESPONSE

More allergen

Mast cell

Immune responses
stimulated by mast
cell secretions

Histamine

Blood vessel

Blood cells
and fluid
leaving
circulation

**FIGURE 34–13** Simplified diagram of an allergic reaction. Sensitization to the allergen causes production of group E antibodies. These bind mast cells, which cause an allergic reaction if the allergen is encountered again.

antibodies, which bind to receptors on mast cells (Figure 34–13). When the same allergen next enters the body and reaches such a bound mast cell, the antigen binds to the E antibody, and the mast cell releases granules of histamine by exocytosis so rapidly that it appears to be falling apart.

Histamine causes a hypersensitivity reaction. It makes blood vessels dilate and increases the permeability of capillary walls, so that fluid and cells escape and swell the tissues. In extreme cases, **anaphylactic shock** sometimes occurs in people allergic to such things as penicillin or insect stings: the muscles of the bronchiole walls contract, con-

stricting the air passages to the lungs, and the capillaries dilate. Injections of the hormone epinephrine can be used to open the air passages and prevent possible death from suffocation.

Allergies tend to run in families, suggesting that they have a genetic component. Studies have also shown that breast-fed infants are less prone to develop some kinds of allergies in later life than those fed on baby formula. For unknown reasons, allergy sufferers are apparently less likely to develop tumors than are other members of the population.

Allergies are sometimes treated by attempting to induce tolerance to the allergen(s). Small amounts of allergen are repeatedly injected, in the hope that the body will cease to react to it. This treatment does not always work, probably because the body usually develops tolerance as an embryo, and the adult has less ability to develop it.

## SUMMARY

All animals have several lines of defense against diseases caused by pathogens. Some of these defenses are nonspecific, and some are directed specifically against one or a few very similar pathogens.

1. Skin and mucous membranes form continuously renewed physical barriers to the entry of pathogens. Both also secrete substances that kill pathogens.
2. When pathogens do invade the body, they are attacked by nonspecific mechanisms such as interferon (against viruses), inflammation, fever, complement reactions, phagocytes, and natural killer cells. Vertebrates, and to a lesser extent invertebrates, are also protected by much more specific immune responses.
3. Vertebrate immune responses are characterized by specificity, by the recognition of antigens either as part of the body or as foreign, and by formation of memory cells, which store receptors specific for foreign antigens that the body has encountered. As a result, the second or later reaction to an antigen is faster and more extensive than the first.
4. The immune system is made up of mobile blood cells, particularly phagocytes and lymphocytes. These cells travel in the blood and extracellular fluid, collecting in antigen-filtering organs such as lymph nodes and the spleen.
5. Macrophages sometimes act as antigen-presenting cells. They engulf foreign antigens and break them down to smaller antigenic peptides, which they display on their cell surfaces attached to their own MHC proteins.
6. A T helper (CD4) lymphocyte bears a T-cell receptor that will bind to a particular combination of self-MHC protein and antigenic peptide displayed on the surface

of an antigen-presenting cell. Binding activates the T lymphocyte to form a clone of cells bearing the same receptor. Some of these become memory cells for the antigenic peptide. The rest produce lymphokines, which stimulate division of T and B cells and cause inflammation and complement reactions, which help destroy the pathogen. Other T helper cells stop the immune reaction when the pathogen is destroyed.

7. Killer (CD8) T cells have receptors that bind to inappropriate MHC antigens on body cells. Bound killer cells secrete proteins that destroy the nonself cell by puncturing its membrane. Cells destroyed in this way include cells that have mutated into cancer cells, cells infected with viruses, and cells from another individual.

8. The surface receptors on B cells are antibodies. B cells are activated when these receptors bind antigens that are free in the body, or part of a pathogen's surface, or presented by a T cell. This stimulates the B cell to form a clone, some of whose members become memory cells for the antigen. Most of the cells differentiate into plasma cells, which secrete large quantities of their antibody. Antibody binds the appropriate antigen, stimulating its phagocytosis.

9. An antibody (immunoglobulin) is a Y-shaped protein. It has a variable region, which binds complementary antigen, and a constant region, which determines features such as whether the molecule is small enough to be secreted in mucus or to cross the placenta.

10. The body makes thousands of different antibodies from a few hundred antibody genes, which are rearranged and spliced in various ways in differentiating B cells. Mutations in plasma cells fine-tune these antibodies to the specific antigen. A plasma cell may also switch from making one isotype of its antibody to making another.

11. Blood cells form in bone marrow. T cells undergo later differentiation in the thymus gland. At least half of all lymphocytes undergo apoptosis: those that bind to the body's own antigens are destroyed as tolerance to the body's own proteins develops in the embryo.

12. Compatibility of MHC antigens determines whether cells transplanted from another individual will be rejected and destroyed by a cellular immune response. Immunosuppressant drugs help to prevent such rejection.

13. Vaccination stimulates a primary immune response to a pathogenic antigen so that the body responds with the more effective secondary response if it later encounters the pathogen.

14. AIDS is caused by HIV, a virus that destroys mainly T helper cells, thereby disabling the immune system. HIV is transmitted in body fluids containing the virus. It is incurable, but its spread can be prevented by chastity and by the use of condoms and clean needles.

15. Autoimmune diseases occur when the immune system attacks components of the body. They result from a failure of immunological tolerance or an inability to distinquish self from very similar foreign antigens.

16. Allergic reactions occur when an allergen induces mast cells bound to E-group immunoglobulins to release histamine.

## Self-Quiz

1. Nonspecific defense mechanisms include all of the following *except:*
   a. complement reactions
   b. phagocytosis
   c. the skin
   d. production of antibodies
   e. production of interferon

2. Interferon:
   a. acts as a pyrogen
   b. stimulates cells to produce antiviral substances
   c. stimulates B lymphocytes to produce lymphokines
   d. all of the above

3. In cellular immunity, _____ are activated and divide repeatedly to produce a clone.
   a. macrophages       c. B lymphocytes
   b. T lymphocytes     d. antigens

4. Molecules that bind circulating antigens are called:
   a. allergens         c. antibodies
   b. histamines        d. interferons

5. An allergic reaction involves:
   a. histocompatibility antigens
   b. natural killer cells
   c. mast cells
   d. interferon

6. The specificity of an antibody for an antigen resides in the:
   a. variable region of its heavy chain
   b. variable region of its light chain
   c. constant region of its heavy chain
   d. constant region of its light chain
   e. variable region of both chains

7. In contrast to the B lymphocytes, the T lymphocytes:
   a. secrete antibodies directed against a specific antigen
   b. have no effect on the proliferation of B lymphocytes
   c. bind only antigens on cell surfaces
   d. do not differentiate into memory cells

8. You need a skin graft. In order to decrease the risk of graft rejection you and the donor should have:
   a. the same blood type
   b. the same or similar histocompatibility antigens
   c. your thymus removed
   d. different genes in the MHC group

*(Continued)*

9. A person's bone marrow is destroyed by massive doses of radiation released from a meltdown at a nuclear power plant. Which of the following can be direct or indirect effects resulting from the radiation damage to the bone marrow?
   a. inability to produce erythrocytes
   b. lack of protection against pathogens
   c. lack of rejection of a skin graft from another person
   d. all of the above

10. The _____ are responsible for the fact that a secondary response to an antigen is faster than the primary response to the same antigen.

   a. T lymphocytes    c. macrophages
   b. B lymphocytes    d. memory cells

11. Vaccination protects the body against catching a disease because:
   a. it provides antibodies synthesized by another animal
   b. it makes the disease organism histocompatible with the body
   c. it produces an enlarged clone of memory cells against that disease
   d. it builds up an immunological tolerance for the disease antigen
   e. it releases large amounts of nonspecific defensive secretions

## Questions for Discussion

1. What might be the selective advantage of a baby's being born before its immune system has matured?

2. Breast cancer nearly always develops in women past childbearing age. Does this mean that natural selection cannot increase the immune response to tumors of the breast?

3. The blood of people who use large amounts of opiates (heroin, opium, codeine) contains fewer T cells than the blood of nonusers. What characteristics would you expect users to show as a result?

4. Recently, researchers cloned the genes for antigens that should make it possible to produce a vaccine against sporozoites of *Plasmodium vivax,* which causes malaria. Referring to the information on the life cycle of this pathogen (Figure 25–13), why do you think health officials feel that a vaccine against this stage of the life cycle can have only limited success in preventing this major disease?

5. If a mouse (#1) that has rejected a skin graft from mouse #2 is then given a new graft from mouse #3, would you expect the rejection of this new graft to follow the same pattern as would be seen with a second graft from mouse #2?

## Suggested Readings

Ada, G. L., and G. Nossal. "The clonal-selection theory." *Scientific American,* August 1987. The history of experiments showing how the body responds to antigen by producing specific antibodies.

Anderson, R. M., and R. M. May. "Understanding the AIDS pandemic." *Scientific American,* May 1992. How AIDS is spreading in different parts of the world and what is likely to happen next.

Cohen, I. R. "The self, the world and autoimmunity." *Scientific American,* August 1988. A discussion of autoimmune diseases and the possibility that the immune system might be induced to control them.

Grossman, C. J. "Interactions between the gonadal steroids and the immune system." *Science* 227:257, 1985. Interesting discussion of the recent discovery that immune responses are at least partly controlled by sex hormones—possibly accounting in part for the greater longevity of women than men.

*Scientific American,* October 1988. What Science Knows about AIDS. An entire issue devoted to the origin, spread, prevention, treatment, and social significance of the disease.

*Scientific American,* September 1993. An entire issue devoted to the immune system, including articles on AIDS, autoimmune diseases, and allergies.

Tizard, I. R. *Immunology: An Introduction,* 3d ed. Philadelphia: Saunders College Publishing, 1992. A fairly short textbook on the subject.

# Excretion

## OBJECTIVES

*When you have studied this chapter, you should be able to:*

1. (a) List three substances excreted by the human body; (b) name the organ(s) that excrete(s) each substance; (c) state the difference between excretion and egestion.
2. (a) State the origin of nitrogenous wastes in animal metabolism; (b) name three common nitrogenous wastes considered in this chapter; (c) state the advantages and disadvantages of each substance as an excretory end product, and relate these advantages and disadvantages to the animal's habitat; (d) name the animal groups commonly associated with each of these excretory end products; (e) use your knowledge of parts (c) and (d) to predict the main nitrogenous waste excreted by an animal.
3. Explain the relationship between nitrogenous waste formation, osmoregulation, energy expenditure, and habitat as factors in body fluid regulation. Interpret Figures 35–4 and 35–15 (in Self-Quiz) as they relate to this problem.
4. (a) List two problems encountered by a freshwater animal in maintaining homeostasis, and the adaptations enabling freshwater bony fish to cope with these problems; (b) list the problems encountered by animals in maintaining homeostasis in a

hypertonic environment, and the adaptations of (1) marine Chondrichthyes, (2) marine bony fish, (3) marine birds, and (4) marine mammals in meeting these problems.
5. List three steps common to the action of all excretory organs.
6. Name the excretory structures found in the following forms: planarians, earthworms, vertebrates, and insects; briefly explain how each works.
7. Define the terms resorption and secretion, and be able to explain kidney function in terms of these activities.
8. Draw or identify, and give the functions of, the following parts of a nephron and its associated structures: nephric capsule, glomerulus, renal artery, renal vein, proximal convoluted tubule, distal convoluted tubule, loop of Henle, collecting duct.
9. Explain how the combined functions of the loops of Henle and the collecting ducts concentrate the glomerular filtrate.
10. Describe the effects of vasopressin, aldosterone, renin, angiotensin, blood pressure, sweating, and excessive bleeding on the volume and composition of the urine, as these are outlined in this chapter.

---

An animal's body fluids must be maintained as a medium in which its cells can live. This requires regulation of temperature, pH, and the amounts and proportions of salts and water in the body fluids. Of the three main environments—land, fresh water, and the ocean—it is easiest for animals that live in the sea to maintain **homeostasis** ("same-standing") of their body fluids. The ocean has a stable salt composition and pH, and its high thermal capacity means that its temperature changes little and slowly. Fresh water and land lack these properties.

Life began in the sea. Most invertebrates and algae are marine, and the body fluids of invertebrates have a composition similar to that of sea water. As animals invaded fresh water and land, they evolved new adaptations that maintain body fluid homeostasis. For instance, fresh water contains few salts. Early vertebrates invading fresh water evolved the ability to live with body fluids with a low salt concentration. Today, vertebrate body fluids resemble sea water in salt composition but are only about one-third as concentrated. Kidneys—excretory organs found only in vertebrates—enable freshwater fish to excrete excess water that enters the body by osmosis.

All the body's fluid compartments are connected. Blood and extracellular fluid (ECF, which surrounds all cells) exchange substances as the blood passes through the capillaries of the circulatory system. Blood is also the source of fluid and solutes in the lymph and in the **cerebrospinal fluid,** which bathes the brain and spinal cord. Both of these fluids drain back into the veins and return to the heart. Because all the body's fluid compartments connect with one another, an animal can regulate the composition of all its body fluids by controlling the content of any one of them. In the vertebrate body, the liver and kidneys are the most important organs that monitor and adjust the composition of one fluid, the blood compartment, and thus keep the composition of all the body fluids constant. The liver regulates the blood's content of food molecules. The kidneys dispose of nitrogenous and other wastes and regulate salts and water.

Homeostasis might be easier if an organism were a self-contained system, but every organism must constantly take in substances from its environment, use them in the chemical reactions of metabolism, and discharge the resulting waste products back into the environment (Figure 35–1). The task of maintaining the constant composition of the fluid surrounding the cells is further complicated by the fact that wastes and substances needed by the body are mingled in the body fluids. The excretory system must remove waste, especially toxic nitrogenous waste, while maintaining the body fluids' pH, water, and salt balance.

**FIGURE 35–1**    The daily exchange of substances between the human body and its environment.

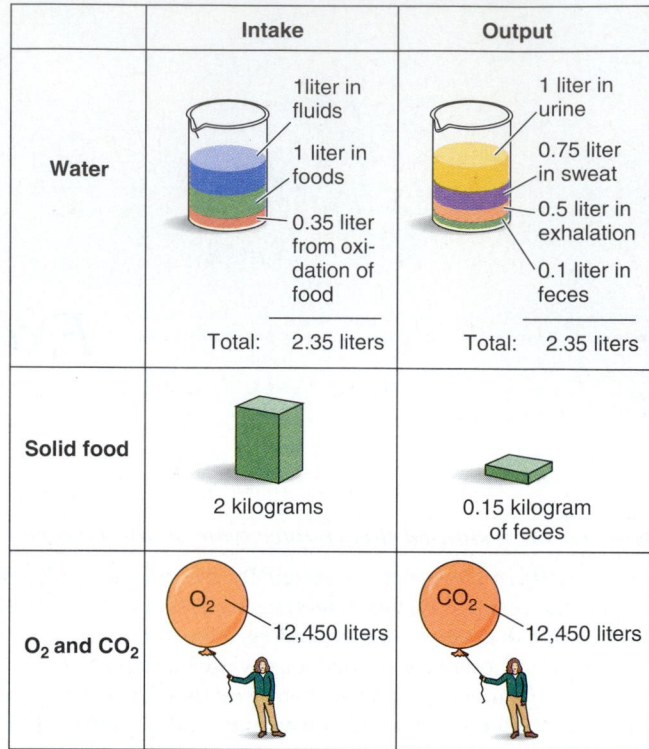

**Amount of Vital Substances Exchanged Daily**

| | Intake | Output |
|---|---|---|
| Water | 1 liter in fluids<br>1 liter in foods<br>0.35 liter from oxidation of food<br><br>Total:    2.35 liters | 1 liter in urine<br>0.75 liter in sweat<br>0.5 liter in exhalation<br>0.1 liter in feces<br>Total:    2.35 liters |
| Solid food | 2 kilograms | 0.15 kilogram of feces |
| $O_2$ and $CO_2$ | $O_2$    12,450 liters | $CO_2$    12,450 liters |

| Substances excreted | Excretory organs |
|---|---|
| Nitrogenous wastes | Kidneys, Skin (small amount in sweat) |
| Water | Kidneys, Skin, Lungs |
| Salts | Kidneys, Skin (in sweat) |
| $CO_2$ | Lungs |

---

### KEY CONCEPTS

♦ Living cells require an almost constant chemical composition in the extracellular fluid which is their immediate environment.

♦ An animal's body fluids are in contact with one another. Regulating the content of any one body fluid (usually the blood) therefore maintains homeostasis in all of them.

♦ An animal's excretory organs must rid the body of nitrogenous wastes while keeping salts and water in proper balance.

♦ The pressure of body fluids is determined by their volume, so regulating fluid volume controls blood pressure.

---

## 35–A   SUBSTANCES EXCRETED

The chief wastes produced by the body are carbon dioxide and water from breakdown of organic molecules and nitrogenous wastes from breakdown of proteins. Carbon dioxide is excreted through the body's respiratory surfaces.

Excretory organs such as kidneys have two major functions: removing nitrogenous wastes and regulating the

body's salt and water content. They also control the body's content of substances such as drugs and hormones. Onions, garlic, and some spices have volatile components that leave the body through the lungs. Other components of the same spices are excreted through the kidneys. Penicillin and other drugs are removed from the system primarily by the kidneys. The kidneys, liver, and lungs carry out **detoxifica-**

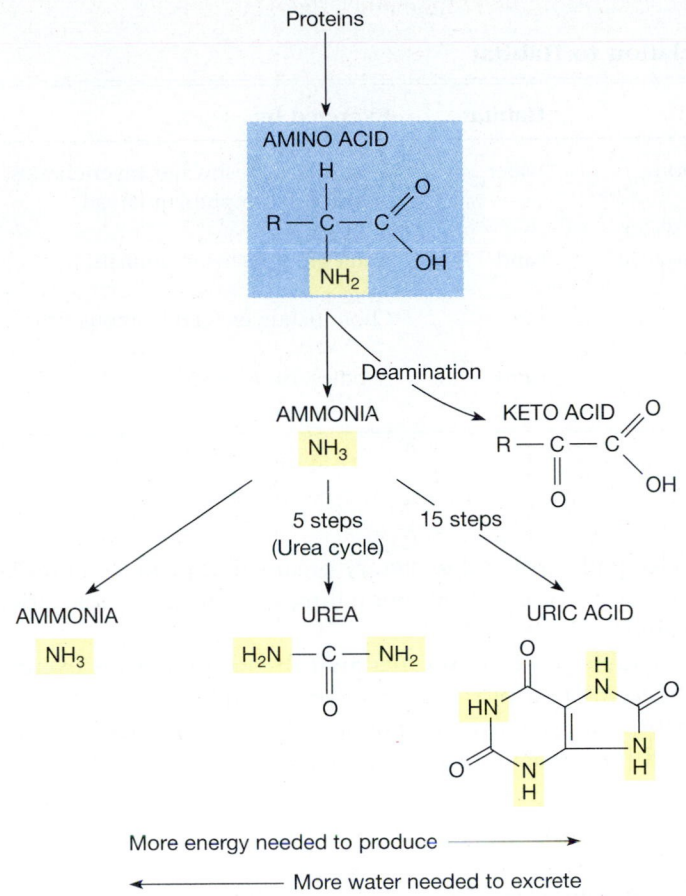

More energy needed to produce ──────────────▶

◀────────── More water needed to excrete

**FIGURE 35–2** Formation of nitrogenous wastes by deamination of excess amino acids from protein digestion. The first products of deamination are ammonia and keto acids. In many animals ammonia is converted to urea by the urea cycle (Figure 35–3). Others convert ammonia to uric acid. Production of urea and uric acid uses more energy but saves water.

**tion,** altering substances to forms that are not poisonous to the body.

Although the kidneys control the salt composition of the blood, some salts are also excreted through the skin in sweat, and some leave with the feces. When a lot of water and salts are lost through the skin as sweat, the kidneys form less urine. A person working in the desert may lose

more than 1 liter of sweat an hour, and this produces a loss of 10 to 30 grams of sodium chloride per day. This loss of salt causes no immediate problem because water is also lost, and so the salt concentration of the body fluids remains the same. However, drinking a lot of water after such heavy sweating dilutes the ECF and leads to muscle cramps —an example of the importance of maintaining the composition of the body fluids. This is one reason athletes drink "Gatorade" or a similar solution of salts and sugars after exercise instead of plain water.

Undigested food from the gut does not appear on our list of substances excreted by the body. Food that passes down the digestive tract and out the anus is not excreted but **egested**—that is, it travels through and is expelled from the body without ever passing through a plasma membrane to become a part of the body. The term "excretion" is correctly applied only to substances that must cross plasma membranes to leave the body.

## Nitrogenous Wastes

Nitrogenous wastes are a by-product of the breakdown of proteins. First, proteins are hydrolyzed to amino acids. This occurs in the digestive tract as food proteins are digested and in the body cells, where proteins are constantly made and destroyed. Fats and carbohydrates are stored in the body for future use, but the body cannot store proteins or amino acids. Amino acids that the body cannot use immediately are **deaminated;** that is, their amino ($-NH_2$) groups are removed. The remaining organic acids can be used for energy or converted into carbohydrate or fat and stored. Each $-NH_2$ group removed picks up another hydrogen atom and becomes $NH_3$, ammonia (Figure 35–2).

Ammonia is the first metabolic breakdown product of amino acids, and it can be produced using little energy. However, it is toxic except in a very dilute solution. Many aquatic animals dissolve their ammonia in large amounts of water from the environment and excrete this dilute solution. Thus, excreting ammonia saves energy but uses a lot of water.

Many land animals, such as adult amphibians and mammals, conserve water by converting ammonia to urea (Figure 35–3). For an animal with a limited supply of water, it is worth using the extra energy needed to produce urea be-

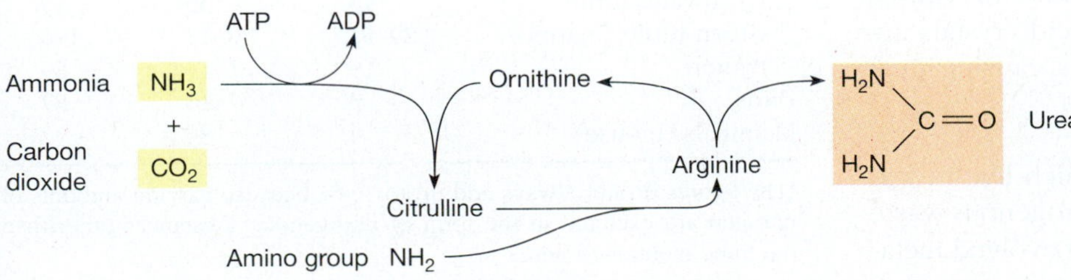

**FIGURE 35–3** The urea or ornithine cycle. This is the pathway by which mammals and adult amphibians convert ammonia and carbon dioxide (yellow) into urea (orange).

## TABLE 35-1

### Advantages and Disadvantages of Nitrogenous Wastes in Relation to Habitat

| Waste | Advantages | Disadvantages | Habitat | Excreted by |
|---|---|---|---|---|
| Ammonia | Production requires little energy | Toxic in concentrated solution | Water | Marine and freshwater invertebrates, bony fish, amphibian larvae |
| | | Must be excreted in lots of water | | |
| Urea | Less toxic than ammonia | Requires some energy to produce | Land | Adult amphibians, mammals |
| | Less water needed to excrete it | | Sea | Chondrichthyes (cartilaginous fish) |
| Uric acid | Very little water excreted with it | Requires considerable energy to produce | Land | Reptiles, birds, insects |

cause urea is less toxic than ammonia. Hence, it can accumulate in higher concentrations without damaging the tissues and can be excreted in a more concentrated form, using less water.

In mammals, most urea is formed by the liver, which takes excess amino acids out of the blood, deaminates them, and incorporates their nitrogen into urea molecules. The brain and kidneys also form urea, but in lesser amounts.

Other land animals, notably reptiles, birds, and insects, incorporate their ammonia into **uric acid** (Table 35–1). Uric acid synthesis takes about 15 enzymes and requires a great deal of energy. This investment of energy may be worthwhile as a water conservation measure, however, because uric acid can be excreted in almost solid form. In fact it is so insoluble that it precipitates spontaneously out of a concentrated solution.

Since the kidneys can handle nitrogenous waste only in solution, birds and reptiles, which excrete uric acid, pass a dilute solution of uric acid from the kidneys into the **cloaca,** a common reservoir at the end of the urinary, digestive, and reproductive tracts. Here most of the water is resorbed, and uric acid crystals precipitate and mix with the feces so that the two are voided together. A similar arrangement is found in insects.

Although most mammals do not excrete uric acid, it appears in the urine of humans, the great apes, and Dalmatian dogs in small quantities. However, in mammals, uric acid is produced by the breakdown of adenine and guanine rather than by the protein-breakdown pathway used by birds and reptiles. Humans with certain metabolic disorders produce more uric acid than normal, and this causes the disease known as gout. In gout, uric acid crystals accumulate in some joints, especially in the toes, causing great pain.

**Wastes and Habitats**    Lower animals, which live in water or moist habitats, excrete most of their nitrogenous waste as ammonia. Some groups of higher animals evolved metabolic pathways and excretory organs that produce and excrete urea or uric acid, permitting them to move into drier habitats.

Table 35–2 shows that animals do not always excrete the form of nitrogenous waste predicted by their taxonomy; habitat also counts. For example, most reptiles excrete uric acid, but aquatic turtles also excrete a good deal of am-

## TABLE 35-2

### Proportion of Nitrogenous Waste Excreted as Ammonia, Urea, and Uric Acid by Various Animals

| Animal | Percentage* of Waste Nitrogen Excreted as: | | |
|---|---|---|---|
| | Ammonia | Urea | Uric Acid |
| Protozoans | >98 | | |
| Cnidarians | 52 | 4 | |
| Earthworm | 72 | 5 | 2 |
| Molluscs | | | |
|   Squid | 67 | 2 | 2 |
|   Terrestrial snail | 22 | 17 | 7 |
| Arthropods | | | |
|   Marine crustacean | 87 | | |
|   Terrestrial insect | 0 | | 92 |
|   Aquatic insect | 90 | | |
| Frog | | | |
|   Tadpole | 78 | 20 | |
|   Adult | 3–38 | 62–88 | <1 |
| Reptiles | | | |
|   Freshwater turtle | 15 | 39 | 19 |
|   Green turtle (marine) | 29–51 | 0–12 | 1–6 |
|   Python | 9 | | 89 |
| Birds | 3 | 10 | 87 |
| Mammals (mouse) | 4 | 84 | 2 |

*The figures do not always add up to 100% because varying amounts of nitrogen are excreted in the form of nitrogenous substances other than the three mentioned here.

monia and urea. Also, some animals that could produce urea do not.

Conversely, most fish, although surrounded by water, do excrete a little urea. Lungfishes live in freshwater ponds and excrete ammonia. When the water dries up, a lungfish buries itself in the mud, reduces its metabolic rate, and accumulates nitrogenous wastes in the form of urea. When it rains and another pool of water collects, the lungfish excretes a lot of urea and then returns to excreting ammonia while water remains plentiful.

Why do reptiles and birds excrete uric acid, whereas mammals turn their nitrogenous wastes into urea? One might expect all three groups to have the same excretory end product because they have fairly similar habitat ranges, as well as common ancestors. The production of urea or uric acid correlates with the reproduction of these vertebrates.

Birds and reptiles lay their eggs on land, and the egg shells are not very permeable to water. The water in the egg must last the embryo until it hatches. Thus, if the embryo produced ammonia, or even urea, the concentration of nitrogenous wastes might become toxic before the embryo hatched. Uric acid, however, precipitates from solution and sits as a mass of solid crystals, leaving the water in the egg free for other uses.

On the other hand, the body fluids of a mammalian embryo make contact with its mother's fluids via the placenta, and so it can rely on her system to dispose of its wastes. Any water available to the mother can also be used by the mammalian embryo. Thus, a mammalian embryo does not have to spend the extra energy needed to produce uric acid rather than urea. Mammals appear to have lost the metabolic pathway by which birds and reptiles produce uric acid.

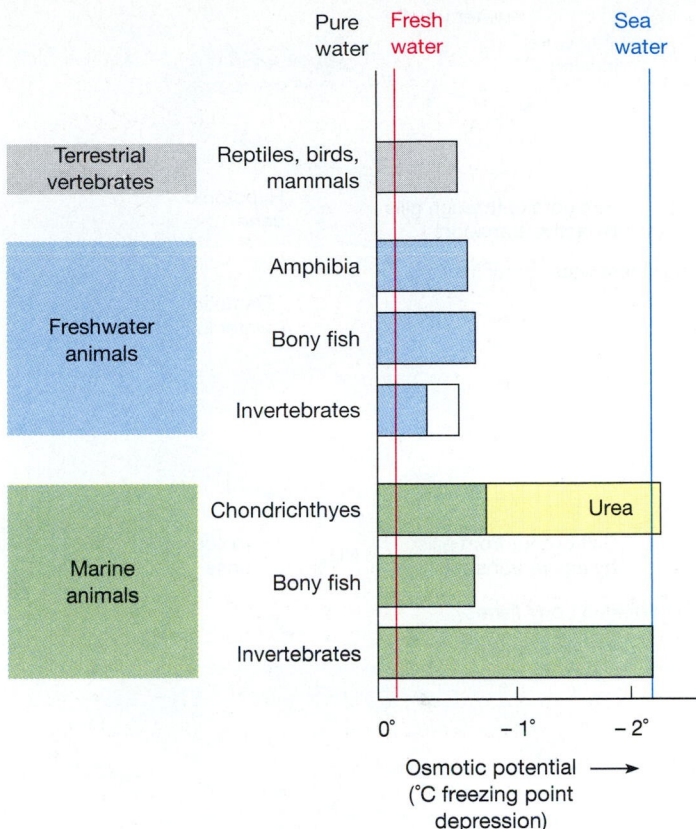

**FIGURE 35–4**  Osmotic potentials of animal body fluids compared with those of sea water and fresh water. Water tends to move toward the lower (more negative) osmotic potential. Hence, among the marine groups shown here, bony fish tend to lose water to the sea, whereas marine invertebrates and Chondrichthyes (cartilaginous fish) are in approximate osmotic balance with their surroundings. All freshwater animals tend to absorb water from the environment by osmosis.

## 35–B  OSMOREGULATION

The problem of expelling nitrogenous waste is tied to the problem of **osmoregulation**—regulation of the fluids' osmotic properties, that is, the balance between salts and water. In general, the body fluids of marine invertebrates are **isotonic** (iso = same; tonic = tension, stretching) with sea water, and no net gain or loss of water occurs across the membranes separating the two. Freshwater invertebrates and most vertebrates have body fluids that are **hypotonic** (hypo = under) to sea water but are **hypertonic** (hyper = above) to fresh water (Figure 35–4). Therefore, these animals tend to lose water by osmosis if they are placed in sea water and to gain water when in fresh water (see Section 4–D). How do vertebrates live with these osmotic problems?

***Bony Fish***  Freshwater fish secrete mucus, which retards passage of water and salts through the body surface. A freshwater bony fish does not drink water, but water inevitably enters the body along with oxygen through the permeable gill membranes. Such a fish expels water by producing a lot of very dilute urine, but it loses salts both via the urine and via diffusion through the gill membranes. This is counteracted by special cells in the gills that take up salt from the environment by active transport. Freshwater fish also take in salts as part of their food.

In a sense, although marine bony fish are surrounded by water, they actually live in a physiological desert because their bodies tend to lose water to the hypertonic environment. These fish have evolved adaptations that conserve water, enabling them to survive in this desert. A marine bony fish must rid itself of excess salts, and its gill cells actively transport salts out of the body. The marine fish loses water through the gills because its body fluids are less concentrated than sea water. It also loses water in its urine, which is approximately isotonic with its body fluids. A ma-

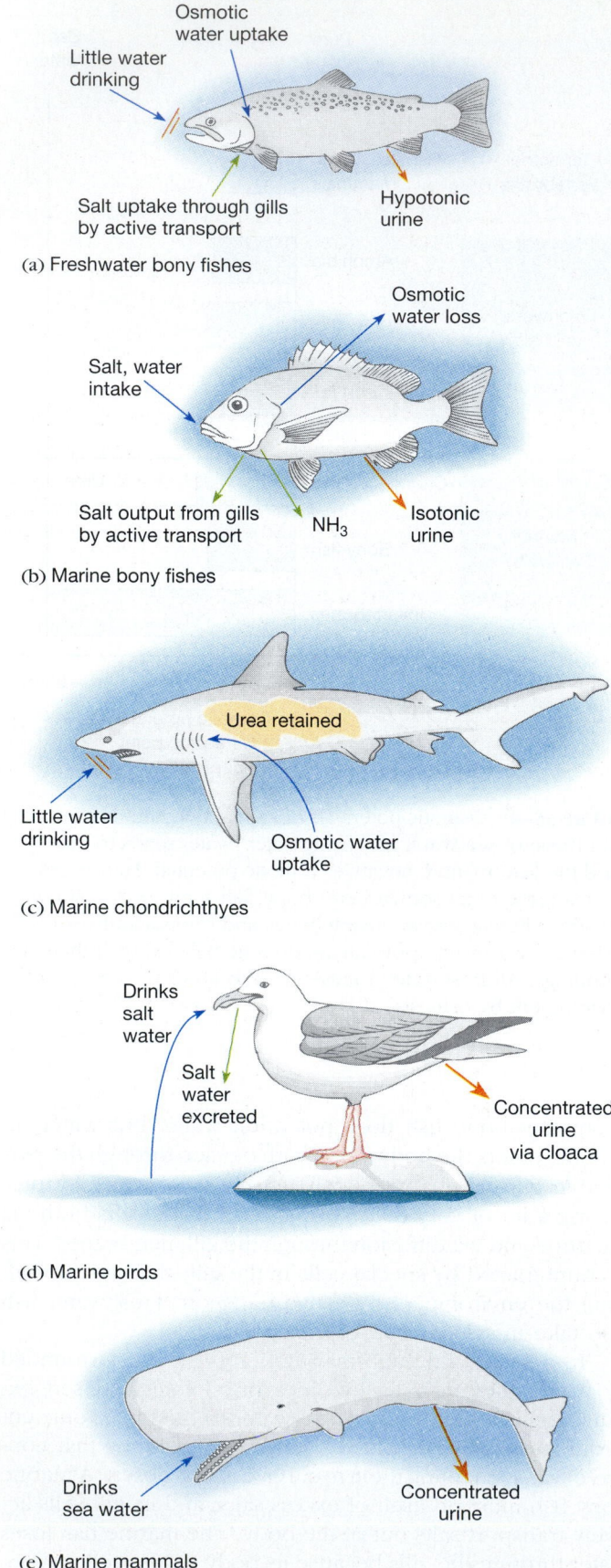

Little water drinking

Osmotic water uptake

Salt uptake through gills by active transport

Hypotonic urine

(a) Freshwater bony fishes

Salt, water intake

Osmotic water loss

Salt output from gills by active transport    NH₃

Isotonic urine

(b) Marine bony fishes

Urea retained

Little water drinking

Osmotic water uptake

(c) Marine chondrichthyes

Drinks salt water

Salt water excreted

Concentrated urine via cloaca

(d) Marine birds

Drinks sea water

Concentrated urine

(e) Marine mammals

**FIGURE 35–5** Osmotic adaptations of vertebrates.

rine fish makes up for these losses by drinking sea water and excreting much of the salt through its gills, again by active transport (Figure 35–4).

***Chondrichthyes*** Cartilaginous fishes (sharks, skates, and rays) have an unusual method of coping with their marine environment. Like most other vertebrates, they have body fluids with a salt concentration about one-third that of sea water, but they also produce a lot of urea and retain much of it in their body fluids. Their tissues are adapted to functioning at high levels of urea, which would be toxic to most other organisms. The combination of salts and urea lowers the osmotic potential of the body fluids to slightly below that of sea water, and so these fish actually gain some water from the sea through their gills by osmosis. This water can be used for excretion.

***Marine Birds*** Birds that live at sea without access to fresh drinking water acquire water in a manner similar to that of a bony fish: they drink sea water, and a **salt gland** (or **nasal gland**) in the head excretes a very concentrated salt solution. This drips out of the nares (nostrils). Birds also excrete uric acid in very concentrated form; their kidneys and cloaca conserve as much water as possible.

***Mammals*** Marine mammals, such as whales and porpoises, take in sea water with their food. Their kidneys can produce a urine several times as concentrated as sea water. This is especially important for the carnivorous marine mammals because their high-protein diet yields much urea that must be excreted.

The kidneys of some land vertebrates can also produce highly concentrated urine. Laboratory rats can live indefinitely when all they are given to drink is sea water. Sea water is too concentrated to support human life. Although the human kidney can produce urine that is slightly more concentrated than sea water, this is not enough to offset the body's other water losses through the lungs and skin. Furthermore, magnesium and sulfate in sea water may cause diarrhea, increasing water loss in the feces. For every swallow of sea water, even more precious body water must be used to excrete the salts taken in. Thus, humans lost at sea are indeed surrounded by "water, water everywhere, nor any drop to drink."

If shipwrecked sailors had read this far, they might recall that bony fish have body fluids only one-third as concentrated as sea water and think that eating fish would be easier on the kidneys. This is of little help, however, because fish is high in protein, and so the body must produce a lot of urea, again requiring more water for urine production. It is, however, possible to improve the osmotic situation by drinking the dilute body fluids squeezed from bony fish. In addition, shipwrecked sailors can eat algae, which contain more carbohydrate and less protein than fish. It usually takes only a few days for a human to die of thirst without drinking at all, but people have survived at sea with no fresh water for more than two months by eating a low-protein diet and drinking any available hypotonic fluids (human urine may be one of these).

## 35–C    EXCRETORY ORGANS

Any excretory structure does three things:

1. It receives fluids from somewhere inside the body, usually from the blood or from spaces between organs.
2. It modifies the composition of this fluid by resorbing substances the body needs to keep and transporting waste substances into the excretory product.
3. It provides some means of expelling the excretory product from the body.

These features are present even in the contractile vacuoles of single-celled protozoans (see Figure 25–14).

During excretion, an organism expends metabolic energy. First, it uses energy in the breakdown of proteins, and often in the formation of urea or uric acid. Second, energy is used by active transport mechanisms, such as the sodium pump, that help to modify fluids collected from the body into final excretory products. Although human kidneys make up less than 0.5% of the body mass, they use 7.2% of the oxygen consumed by the body. Pumping blood from the heart to the kidneys takes another 2.7% of the body's total oxygen consumption, so that about 10% of the human body's energy is spent just moving blood to the kidneys and cleansing it.

## Invertebrates

Jellyfish, corals, and other marine Cnidaria live in an isotonic environment and lose most of their wastes by diffusion. Freshwater flatworms (phylum Platyhelminthes), on the other hand, void excess water by way of organized, multicellular excretory systems (Figure 35–6). Body fluids are collected into **flame cells,** so called because the fluid is pulled in by the beating of cilia, which resembles the flickering of a candle. The fluid then passes through a series of tubules until it reaches an excretory pore at the body surface.

The functional unit of excretion in an earthworm is the **nephridium** (Figure 35–7). The **nephrostome,** a ciliated funnel, opens into the coelom and draws coelomic fluid into the long, thin, coiled tubule of the nephridium. As fluid flows through the nephridium, substances that the body needs are reclaimed and passed into surrounding capillaries of the circulatory system. Fluid that is expelled from the body through the **nephridiopore** contains water, nitrogenous wastes, and any salts that have not been resorbed.

Like earthworms, most other invertebrates have nephridia, but insects have a completely different system: long, slender **Malpighian tubules,** attached at one or both ends to the gut. Nitrogenous wastes from the body fluid are transformed into uric acid, which moves down the Malpighian tubule into the gut. Cells in the lining of the

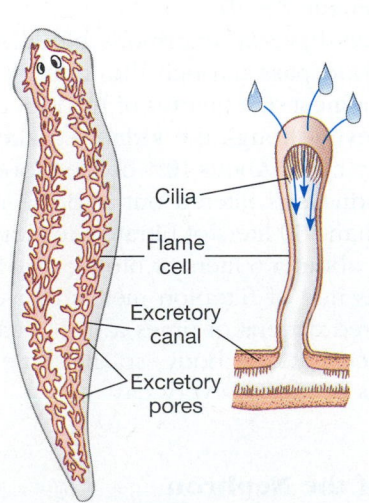

**FIGURE 35–6** Excretory system of a planarian (Platyhelminthes). Flame cells collect body fluid and push it along a system of excretory ducts to pores on the surface of the body. The system serves primarily in osmoregulation, removing excess water.

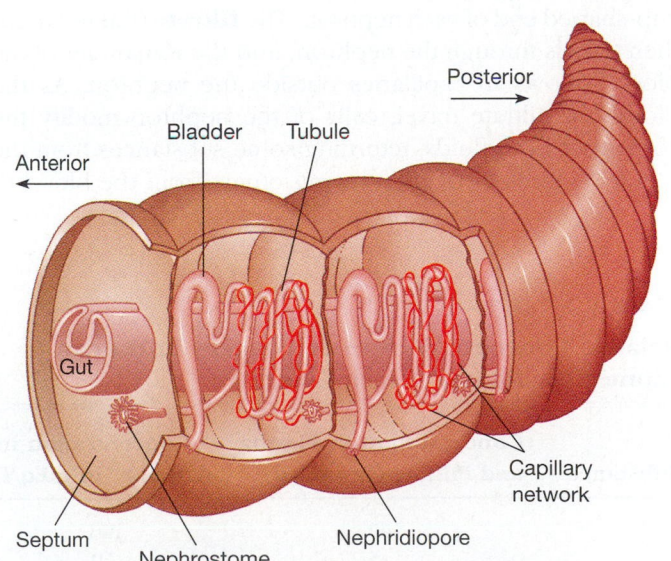

**FIGURE 35–7** Earthworm nephridia. The funnel-shaped opening of the nephrostome collects fluid from the coelom. The rest of the nephridium lies in the next segment back. Capillaries of the circulatory system intertwined with the tubule resorb some substances before the fluid reaches the bladder, from which the urine is discharged through the nephridiopore. Each segment, except a few in the anterior end, contains a pair of nephridia.

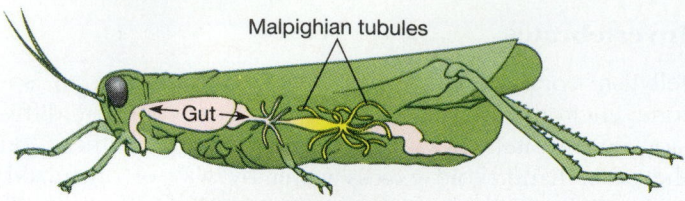

**FIGURE 35–8**  Malpighian tubules of an insect. The tubules discharge uric acid into the gut, whence it is voided with the feces.

rectum resorb water from the combined wastes (uric acid and undigested food) before they are eliminated as fairly dry fecal pellets (Figure 35–8).

## 35–D  KIDNEYS

Every vertebrate has a pair of kidneys. In lower vertebrates, the kidneys are long thin organs extending along either side of the backbone, outside the **peritoneum,** the membrane lining the abdominal cavity. Mammalian kidneys are the only short "kidney-shaped" kidneys.

The functional units of vertebrate kidneys are **nephrons,** which resemble the nephridia of invertebrates. In humans, each kidney is made up of more than a million nephrons.

The **renal arteries** carry blood from the dorsal aorta into the kidneys. Here fluid filters out of capillaries into the cup-shaped end of each nephron. The **filtrate** (filtered fluid) then travels through the nephron, and the remainder of the blood follows in capillaries outside the nephron. As the blood and filtrate travel, cells of the nephron modify the content of both fluids, returning some substances from the filtrate to the blood, and moving others from the blood to

### T A B L E   3 5 – 3

**Relative Concentrations of Substances in Human Glomerular Filtrate and in Urine**

| Substance | Concentration in Plasma and Filtrate (in mEq/l) | Concentration in Urine (in mEq/l) |
|---|---|---|
| $Na^+$ | 142 | 128 |
| $K^+$ | 5 | 60 |
| $Ca^{2+}$ | 4 | 4.8 |
| $Mg^{2+}$ | 3 | 15 |
| $Cl^-$ | 103 | 134 |
| $HCO_3^-$ | 28 | 14 |
| Creatinine | 1.1 | 196 |
| Glucose | varies | 0 |

*Glomerular filtrate forms at a rate of about 125 ml/min; urine at 1 ml/min.

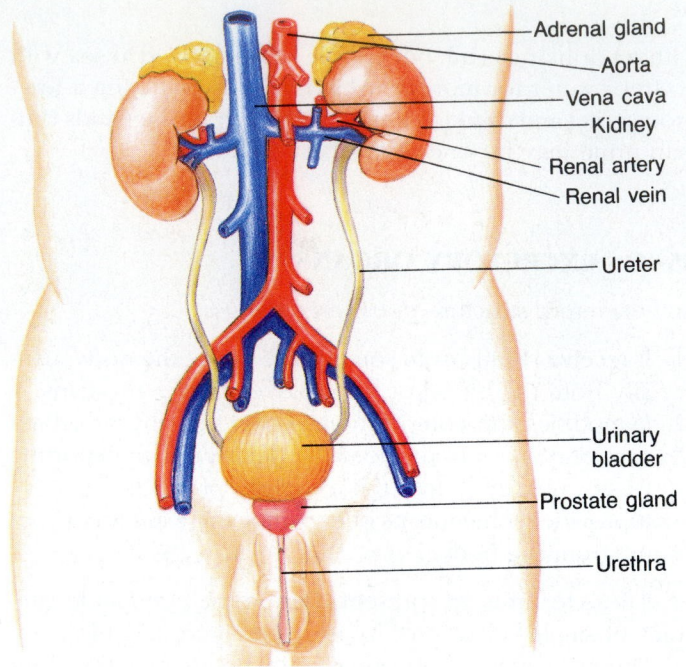

**FIGURE 35–9**  The human urinary system and associated blood vessels. The kidneys lie in the small of the back, behind the peritoneum and against the spinal column.

the urine-to-be (Table 35–3). The purified blood leaves the kidneys via the **renal vein,** whereas the urine finally passes down a collecting duct, leaves the kidney, flows down a **ureter,** and is stored in the **urinary bladder.** Water may be resorbed from the bladder under some hormonal conditions. Eventually the urine is expelled from the body via the **urethra** (Figure 35–9).

The kidneys have an enormous blood supply. About 1200 ml of blood pass through the kidneys each minute, amounting to almost one quarter of the blood pumped out by the heart, even though the kidneys make up less than 1% of the body mass. About 10% of the blood volume that reaches the kidneys is filtered out into the nephron, producing more than 150 liters of filtrate each day. The human body contains about 5.6 liters of blood, of which 3 liters is plasma, so this rate of filtration means that every drop of plasma is filtered dozens of times a day. Nearly all the filtrate is resorbed into the body, so that little more than 1 liter of urine is produced every day.

### Functions of the Nephron

Let us follow the process of urine formation in a nephron. The nephron's cup-shaped **nephric capsule** (also called a Bowman's capsule) surrounds a knot of capillaries called a **glomerulus** (Figure 35–10). Because the blood is under pressure and the capillary walls are permeable, some of the

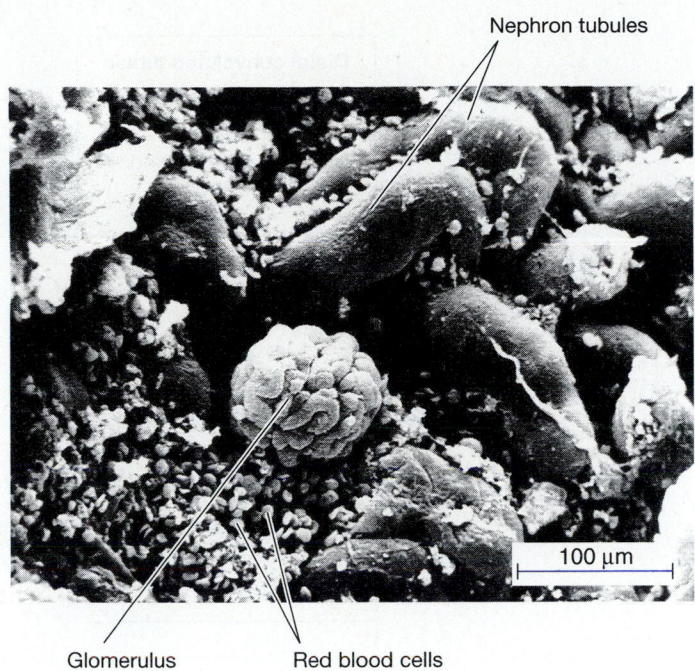

**FIGURE 35–10**  Scanning electron micrograph of the kidney cortex. The red blood cells visible in the picture have leaked from blood vessels broken during preparation of the specimen.  (Biophoto Associates)

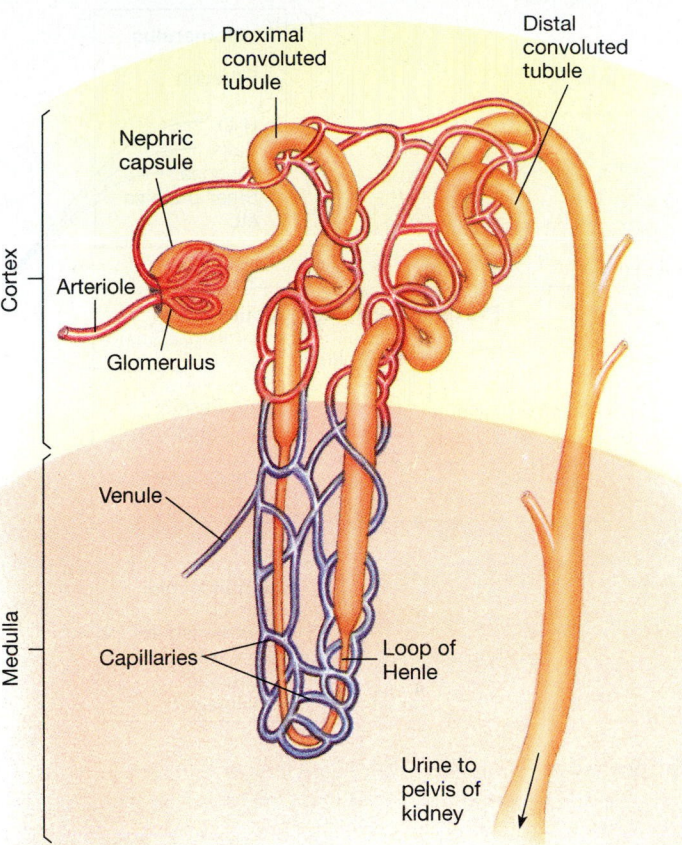

**FIGURE 35–11**  A nephron and its associated blood supply. (Figure 35–13 shows the orientation of the nephron within the kidney.) A human has about 2 million nephrons. Blood pressure pushes fluid out of the blood in the glomerulus and into the nephric capsule. The filtrate so formed travels through the nephron tubule while substances are exchanged among the filtrate, the extracellular fluid surrounding the tubule, and the blood in the nearby capillaries. The fluid that reaches the collecting duct is urine, which flows through the pelvis of the kidney and down the ureter to the urinary bladder.

fluid from the blood filters into the capsule along with solute molecules up to the size of small proteins. Left behind are large proteins and cells, which are too large to pass through the filter, along with the rest of the blood fluid (plasma).

From the capsule, the filtrate passes into the nephron tubule. Meanwhile, the rest of the blood follows along in capillaries outside the nephron. The cells of the tubule modify both fluids, **resorbing** some substances from the filtrate and returning them to the ECF, from which they reenter the blood, and **secreting** others from the ECF and blood into the forming urine.

The nephron tubule has four main parts: the **proximal convoluted tubule,** the U-shaped **loop of Henle,** the **distal convoluted tubule,** and the **collecting duct.**

In the proximal convoluted tubule, a considerable amount of resorption takes place (Figure 35–11). Small proteins, glucose, and ions such as sodium, magnesium, potassium, and calcium are returned to the blood by active transport. Negatively charged chloride ions follow passively, attracted by the positively charged ions. Water follows these solutes passively by osmosis. About 65% of the salt and water in the filtrate returns to the blood from the proximal convoluted tubule.

Normally, all the glucose in the filtrate also returns to the blood in this part of the nephron. However, glucose sometimes appears in the urine in conditions such as diabetes. This happens if there is so much glucose in the filtrate that active transport carriers, working at top speed, cannot remove all the glucose from the filtrate as it passes through the proximal convoluted tubule.

In many mammals and birds the loops of Henle are quite long, and their activities (discussed later) permit the production of relatively concentrated urine. From the loop of Henle, the fluid in the nephron passes to the distal convoluted tubule, another area of resorption and the chief area where secretion into the tubule occurs. The main substances secreted are potassium and hydrogen ions, by active transport, and ammonia, by diffusion.

Secretion of hydrogen ions is a means of regulating the pH of the blood. When the blood becomes too acidic, more hydrogen ions and fewer potassium ions are secreted. More sodium is also resorbed in the distal tubule, being exchanged for hydrogen ions to preserve the balance of electric charges. Various drugs, such as penicillin, are also se-

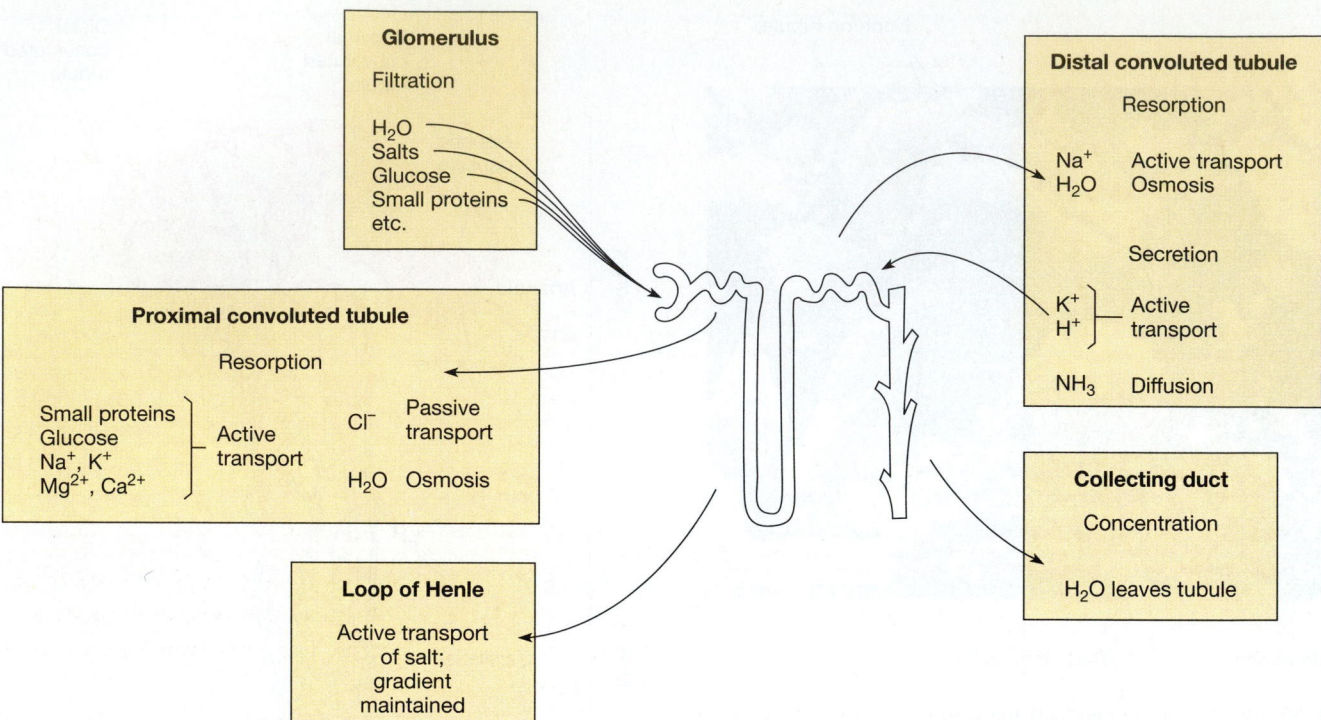

**FIGURE 35–12** Summary of the activities in different parts of a nephron.

creted into the urine in this area. The collecting duct, along with the loop of Henle, plays a vital role in water balance (Figure 35–12).

***Urine Concentration*** Cells move water indirectly by actively transporting solutes, with water following passively by osmosis (Section 4–D). But to produce a concentrated urine, the kidneys must be able to separate the movement of solutes from the osmotic movement of water. Birds and mammals can produce urine more concentrated than blood thanks to a specialized area, the loop of Henle, found in some of their nephrons. The actual concentration of urine takes place in the nephrons' collecting ducts, but the process depends on the ability of the loops of Henle to produce a concentration gradient of salt in the surrounding extracellular fluid.

To understand how this works, you must know that the loops of Henle and collecting ducts lie intermingled in the **medulla** of the kidney, whereas the rest of the nephron lies in the **cortex,** the outer layer of the kidney (Figure 35–13). The ECF in the medulla contains an osmotic gradient, with solutes more concentrated deeper within the kidney. The gradient contains two kinds of solutes: salt ions (mostly sodium and chloride) and urea.

Filtrate enters the loop of Henle from the proximal convoluted tubule. Most of its salt and water have returned to the blood, and the filtrate is isotonic to the ECF and blood around the nephron at this point. The walls of the descending limb of the loop of Henle are more permeable to water than to solutes, and water leaves the tubule by osmosis. As it descends, the filtrate becomes more concentrated but is entering a region of ever-higher solute concentration, and it continues to give up water to the ECF until the two have equal but now higher solute concentrations.

At the bend of the loop, the tubule walls become impermeable to water but much more permeable to salt. Because the high solute concentration in this part of the medulla is only partly salt (the rest is urea), the filtrate's salt concentration is higher than that of the ECF, and so salt diffuses down its concentration gradient, from the filtrate out into the ECF of the medulla.

Farther up, cells in the thick part of the ascending loop of Henle actively transport salt out of the filtrate. As the fluid moves upward, less salt remains to be exported. This is why a gradient of salt forms, with a higher concentration near the bottom of the loop. The ascending part of the loop of Henle is relatively impermeable to water, so that water cannot follow the ions out on the upward journey.

Having lost so much salt, the filtrate reentering the cortex in the distal convoluted tubule is now hypotonic to the surrounding fluid, and the active export of additional salts makes it even more so.

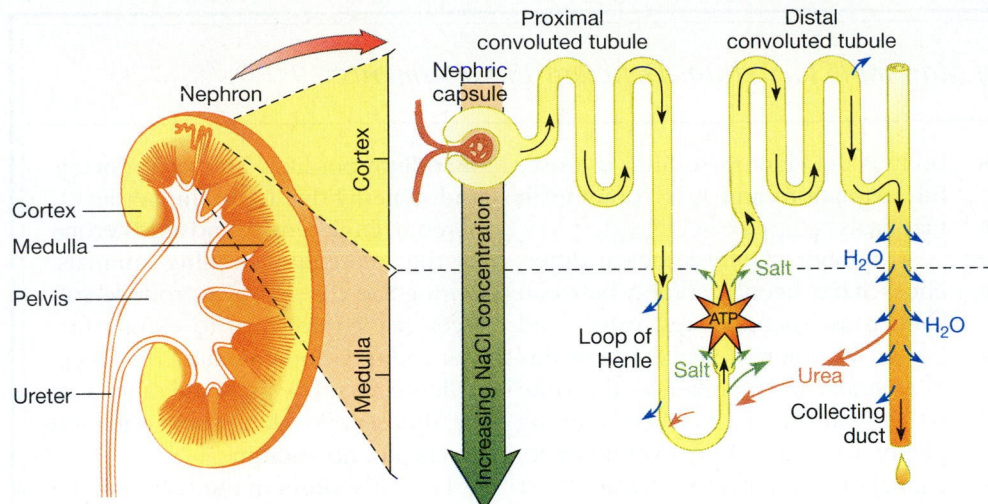

**FIGURE 35–13** How urine is concentrated. The loops of Henle and collecting ducts lie in the medulla of the kidney. Here the loops of Henle maintain a high sodium chloride gradient in the extracellular fluid by active transport of salt into the ECF. As urine passes down the collecting duct, it loses water by osmosis to the ECF, which has a lower osmotic potential. Urine also loses urea, which further lowers the osmotic potential in the inner medulla ECF.

Now the hypotonic filtrate begins its second descent through the medulla, in the collecting duct. Here the final urine concentration is determined, according to whether the body needs to conserve water by reclaiming it from the filtrate. If not, the filtrate passes through the duct with little change, and on through the ureter to the bladder.

When the body needs to save water, the hormone vasopressin is secreted into the blood. It acts on the walls of the collecting ducts to increase their permeability to water. Filtrate descending the collecting duct then loses most of its water to the highly concentrated ECF in the medulla by osmosis. Near the end of the duct, so much water has been lost that urea is now quite concentrated, and some of it diffuses out of the duct, down its concentration gradient, providing part of the solute gradient in the inner medulla.

Some urea diffuses back into the filtrate in the lower loop of Henle and in this way becomes trapped, circulating back to the collecting duct. Reusing the same urea in this way multiplies its effect.

In summary, the loops of Henle and collecting ducts together provide a way to concentrate urine, by creating an osmotic gradient using two solutes, salt and urea. The salt gradient comes from active transport by the thick part of the ascending loop of Henle, the urea gradient from the lower part of the collecting duct. The combined gradient is used to move water separately from solutes, permitting production of a concentrated urine.

## 35–E  REGULATION OF KIDNEY FUNCTION

Kidney function is under the minute-by-minute control of many systems that work together and ensure that the composition of the blood, and therefore of the body's ECF, remains constant. For instance, the filtration rate remains nearly constant because cells in the kidneys detect changes in blood pressure and adjust contraction of muscles in the walls of the renal arterioles, maintaining the proper blood pressure in the glomeruli.

On the other hand, the rate of urine formation and the composition of urine change dramatically under different circumstances. Drinking a lot of water dilutes the blood but raises the filtration rate only slightly. The major change in this case is a drop in the rate of water resorption, so that the excess water is excreted from the body.

Urine composition and the rate of urine formation are largely regulated by the hormones vasopressin, angiotensin, and aldosterone. These hormones work in three main ways:

1. Vasopressin increases the permeability of membranes to water.
2. Angiotensin increases contraction of tiny muscles in blood vessel walls, thereby constricting arterioles and raising blood pressure. It also causes secretion of aldosterone.
3. Aldosterone boosts the active transport of sodium.

The concentration of sodium is the most important factor determining the pressure, volume, and concentration of the body fluids. This is because sodium accounts for about 90% of all the cations outside cells. Since water moves passively by osmosis, the volume of water in the blood and ECF is determined by their sodium content.

The amount of sodium in the body is determined by the balance between intake in the diet and loss in the urine. (Sodium is also lost in the feces and in sweat, but less control can be exerted over these.) Vertebrates have nervous and hormonal systems that permit them to regulate how much sodium they eat and how much leaves the body. These mechanisms permit survival in a much greater range of habitats than would otherwise be possible (*Essay: Adaptations of Mammals to Sodium-Deficient Environments*).

Many arid and mountainous areas contain little sodium. Although no animal can survive with much less than the usual amount of sodium in its body, many mammals (and some insects) have adaptations that permit them to survive in such areas.

These adaptations may be divided into those that increase the intake of sodium and those that reduce its loss. Herbivores, including reindeer, elephants, gorillas, elk, moose, sheep, and kangaroos, have been seen eating soil, drinking sea water, or consuming other improbable items of diet whose only redeeming value appears to be their high sodium content. Plainly an appetite for salt when the body is deficient in sodium is a useful adaptation, and it is common in mammals.

A number of physiological differences have been recorded between kangaroos, sheep, foxes, cattle, and rabbits in sodium-deficient mountains of Australia and members of the same species on the coast, where there is plenty of sodium. Whenever possible, animals in mountainous areas select food plants containing sodium and also hold sodium loss from the body to a minimum.

For instance, the urine of mountain animals was found to contain virtually no sodium, whereas urine of kangaroos and wombats on the coast contained up to 300 milliequivalents of sodium per liter. This difference is undoubtedly due to the higher levels of renin, angiotensin, and aldosterone in the mountain-dwelling animals. During the day, rabbits produce soft feces, which they eat to extract further sodium (and vitamins). The fecal pellets egested by highland rabbits after the second digestion contained practically no sodium.

Loss of sodium in the feces can be reduced by resorbing it from the gut. Sodium-deficient cattle, deer, sheep, and impalas also reduce sodium loss by changes in the composition of their saliva (Figure 35–A). These ruminants produce many liters of saliva a day. This usually contains a high concentration of sodium bicarbonate, which produces an alkaline environment for symbiotic microorganisms in the stomach. Some of this sodium bicarbonate is lost with the feces.

Aldosterone decreases the amount of sodium in the saliva and causes potassium to be secreted in its place. Although the secretion of large quantities of sodium in the saliva every day would seem to impose a sodium-supply problem on ruminants, the effect may, in fact, be the other way around. Sheep survive temporary sodium shortages better than do non-ruminants such as humans or foxes. This is probably because the ruminant stomach provides a large store of sodium that can be steadily replaced by potassium and used for more vital functions in the body during times of sodium shortage.

**FIGURE 35–A**   Impalas eating salt-rich soil at the edge of a waterhole on the African savanna.

## Vasopressin (ADH)

**Vasopressin (antidiuretic hormone, ADH)** is a hormone released from the posterior pituitary gland in the brain. It conserves body water by increasing the permeability of the collecting ducts to water, thereby boosting resorption of water by the nephron tubules. The body may detect a decrease in its water content in one of two ways: either as a reduction in blood volume (for example, caused by severe bleeding) or as an increase in the concentration of the blood due to loss of water (for instance, from sweating). Either way, loss of water from the body stimulates vasopressin secretion and so slows the loss of water via the urine.

In the absence of vasopressin, the walls of the ducts are practically impermeable to water, and very little water is resorbed from the urine. Lack of vasopressin results in the disease known as diabetes insipidus, characterized by thirst and the production of large quantities of dilute urine. It is much less common than diabetes mellitus, characterized by sugar in the urine.

## Aldosterone, Renin, and Angiotensin

**Aldosterone** is one of several steroid hormones secreted by the cortex of the **adrenal gland,** attached to the kidney (see Figure 35–9). It promotes resorption of sodium by the distal convoluted tubule, retaining sodium in the body. If the sodium content of blood reaching the adrenal cortex falls, aldosterone secretion increases. This leads to resorption of more sodium from the filtrate, conserving the body's salt supply and decreasing the loss of sodium in the urine (Figure 35–14).

Aldosterone secretion also increases when the level of angiotensin in the blood rises. In Section 33–C we saw that one of the three main methods of controlling blood pressure is the renin-angiotensin system. If kidney cells detect a decrease in sodium or potassium in the blood, or a reduction in blood pressure, they secrete the enzyme **renin,** which converts a plasma protein (angiotensinogen) into the hormone **angiotensin.** Angiotensin causes contraction of muscles in small blood vessels all over the body. This constricts the vessels, raising the blood pressure. It is advantageous for this detection system to lie in the kidney because kidney function depends on filtration, which is driven by blood pressure. So it is important that the kidneys should detect and correct a falling blood pressure rapidly.

The overall effect of the renin-angiotensin system is to increase blood pressure and sodium retention whenever the sodium content of the blood falls. This maintains the blood pressure directly, and it also indirectly increases the volume of the body fluids.

## SUMMARY

Animals have evolved various specialized mechanisms that maintain homeostasis of their body fluids in the face of the continuous flow of materials between the external environment, the extracellular fluid, and the interiors of cells.

1. Animals must have adaptations that allow them to collect and dispose of nitrogenous wastes before they reach toxic levels while maintaining the salt and water (osmotic) balance in the body fluids. This is difficult when salt or water is in short supply in the environment and must be conserved.
2. The main nitrogenous wastes are ammonia, urea, and uric acid. Ammonia requires the least energy to produce but is the most toxic and therefore requires the most water to excrete. Uric acid takes the most energy to produce but requires little water to excrete. Its production is, therefore, worthwhile for terrestrial animals that have very little water available to them, such as the embryos of birds and reptiles.
3. Animals maintain osmotic homeostasis by varying their intake and output of salts and water.
4. Excretory organs collect body fluid, modify its composition, and void it from the body. They thereby control the composition of the fluid that remains in the body.
5. The function of the vertebrate kidney is to maintain the composition of all of the body's fluids within the narrow limits required by the body's cells. The basic unit of the kidney is the nephron, a long thin tube closely associated with a capillary bed of the circulatory system. Blood plasma is filtered, under hydrostatic pressure, into one end of this tubule. As the filtrate passes through the nephron, substances needed by the body are resorbed through the cells of the nephron tubule into the extracellular fluid and then into the capillaries,

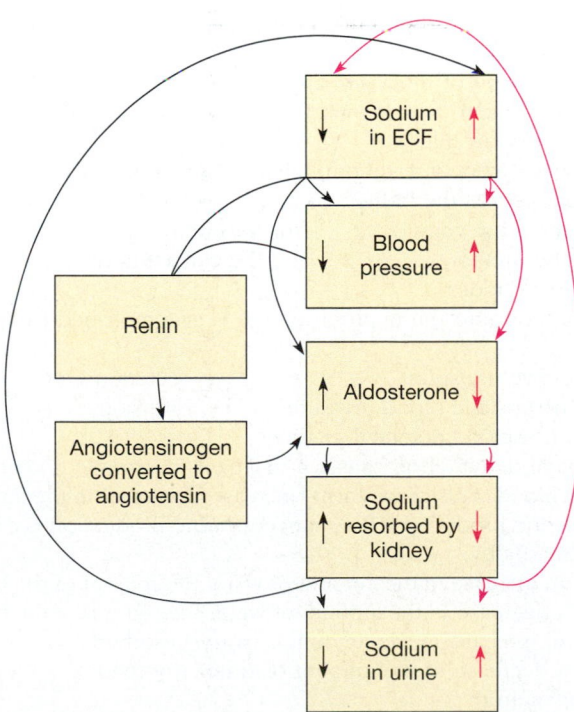

**FIGURE 35–14** Interactions among aldosterone, renin, and angiotensin regulate sodium in urine and hence in the ECF, which in turn determines blood pressure. Curved arrows indicate events that tend to return low sodium levels to normal and those that lower the sodium levels when there is too much sodium.

either by diffusion or by active transport. Some unneeded substances are secreted from the blood into the filtrate. After being changed in these ways during its passage through the tubule, the filtrate is collected as urine.

6. Homeostatic mechanisms under hormonal control regulate the amount and composition of the urine produced.

## Self-Quiz

1. Which of the following pairs does not indicate an organ and a substance it excretes?
   a. lungs/$CO_2$
   b. kidneys/nitrogenous wastes
   c. liver/nitrogenous wastes
   d. skin/salts

2. One advantage of excreting nitrogenous waste as uric acid as opposed to urea is that:
   a. uric acid requires less energy than urea to produce
   b. uric acid can be excreted in concentrated form, thereby conserving water
   c. uric acid is harmless to the kidneys and is excreted easily
   d. the formation of uric acid requires a great deal of energy
   e. uric acid is the first metabolic breakdown product of amino acids

3. All of the following animals excrete ammonia except:
   a. sharks
   b. amphibian tadpoles
   c. earthworms
   d. lobsters

4. Marine bony fish maintain water balance by:
   a. producing hypotonic urine in large quantities
   b. actively transporting salt from their gills
   c. drinking salt water and removing salt from their gills by active transport
   d. excreting uric acid

5. The osmotic potential of sea water is lower than that of the body fluids of marine bony fish. Because bony fish are _____ to their environment, they produce _____ .
   a. hypotonic/large amounts of urine
   b. hypertonic/very little urine
   c. hypotonic/very little urine
   d. hypertonic/copious amounts of urine

6. Salmon have gills that are more permeable to water than to salts. Salmon hatch in freshwater streams and then migrate to the ocean. Once they reach the ocean, you would expect the rate of uptake of water into their bodies through the gills to:
   a. increase
   b. decrease
   c. remain the same

7. The pack rat, a rodent, often goes for long periods without drinking, eats leaves of juicy plants, and moves about in the open only in the evening and at night. From these habits, you can guess that it lives in:
   a. desert areas
   b. the Arctic tundra
   c. the woodlands of the eastern United States

8. The main nitrogenous waste substance excreted by the pack rat will probably be:
   a. ammonia
   b. urea
   c. uric acid

Match the term with its definition.

_____  9. Regulates $Na^+$ resorption
_____ 10. Removal of undigested food materials from the body
_____ 11. Removal of $H^+$ and $K^+$ from body fluids to form urine
_____ 12. Removal of nitrogenous wastes from the body
_____ 13. Regulates permeability of collecting duct

14. The primary site of resorption of glucose and amino acids is the:
   a. glomerulus
   b. collecting duct
   c. proximal convoluted tubule
   d. loop of Henle

15. Which of the following activities does *not* require energy?
   a. resorption of glucose
   b. secretion of hydrogen and potassium ions
   c. movement of salt ions from the filtrate in the ascending loop of Henle
   d. movement of water through the walls of the collecting ducts

16. Urine leaves the kidney via:
   a. the renal vein
   b. the urethra
   c. the bladder
   d. the ureter
   e. the collecting duct

17. Filtration into the nephron tubule is accomplished by means of:
   a. active transport
   b. hydrostatic blood pressure
   c. an osmotic potential gradient
   d. secretion
   e. diffusion

18. Severe dehydration causes a decrease in osmotic potential in the blood. This causes a(n) (increase/decrease) in the amount of urine produced. This change in urine production is caused primarily by:
   a. an increase in the amount of water filtered out of the blood
   b. a decrease in the amount of water filtered out of the blood
   c. an increase in the amount of water resorbed
   d. a decrease in the amount of water resorbed

19. Angiotensin:
   a. decreases aldosterone secretion
   b. increases blood pressure by causing constriction of blood vessels
   c. converts renin to reninogen
   d. stimulates red blood cell production

## Questions for Discussion

1. Why does an increase in aldosterone secretion increase the volume of extracellular fluid and increase blood pressure?
2. If we divide animals into marine and freshwater invertebrates, marine and freshwater bony fish, cartilaginous fish (Chondrichthyes), amphibians, and terrestrial vertebrates, which animals are in the following osmotic situations?
   a. approximately in osmotic equilibrium with their environment
   b. must have adaptations against loss of body water to the environment
   c. must have adaptations to prevent gain of excess water from the environment
   d. in danger of taking up too many salts from the environment
   e. must have adaptations to prevent loss of salts to the environment
3. Many fish lack a urinary bladder, but urinary bladders are found in amphibians and all higher vertebrates. What is the advantage of having a urinary bladder?
4. Would you expect a dolphin to have longer or shorter loops of Henle than those in your kidneys?
5. Certain reptiles have kidneys with no glomeruli in the nephric capsules. What effect would this have on urine formation? What type of habitat are such animals adapted to?
6. Why are animals that eat a lot of protein better able to produce concentrated urine than those with low-protein diets?
7. Study Figure 35–15. The crab *Carcinus* inhabits estuaries (mouths of rivers where they join the sea). Salinity of sea water is about 35 grams per liter.

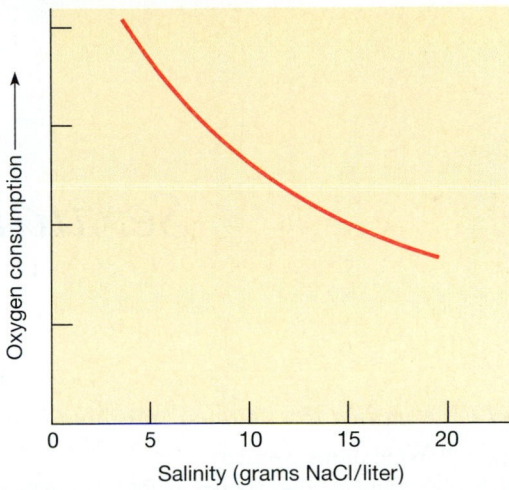

**FIGURE 35–15**  Oxygen consumption of an estuarine crab, *Carcinus,* as a function of the salinity in the external environment.

   a. Would the crab expend more energy resting on the bottom during high tide or low tide? (The salinity would be higher during high tide, when ocean water pushes into the mouth of the river.)
   b. What do you think it is using this energy for?
   c. What is the main nitrogenous substance it excretes?
8. What changes in physiology must a fish such as a salmon undergo as it returns from the sea to spawn in the stream where it hatched?

## Suggested Readings

Baldwin, E. *An Introduction to Comparative Biochemistry,* 4th ed. New York: Cambridge University Press, 1964. A short and entertaining book with emphasis on osmoregulatory problems and their solutions by animals.

Ganong, W. F. "Renal function." *Review of Medical Physiology,* 15th ed. Los Altos, California: Lange Medical Publications, 1991.

Scoggins, B. A., et al. "The physiological and morphological response of mammals to changes in their sodium status." *Memoirs of the Society of Endocrinology* 18:577–601, 1972. An advanced discussion of how hormonal control permits mammals to regulate their sodium content by regulating the amount of sodium excreted by the kidneys and by other mechanisms.

Smith, H. W. *From Fish to Philosopher.* Garden City, New York: Doubleday, 1961. Evolutionary history of vertebrate internal homeostasis delightfully written by an eminent renal physiologist.

◇ 36 ◇

# Sexual Reproduction

## OBJECTIVES

*When you have studied this chapter, you should be able to:*

1. Use the following terms correctly:
   fertilization, gamete, egg, sperm, zygote, copulation
   secondary sexual characteristics
   vulva, clitoris, ovulation, endometrium, cervix
   scrotum, semen
   orgasm
   prolactin, lactation
   contraception
   placenta, amnion, allantois, chorion
2. List the organs in the human female and male reproductive tracts, give their functions, and explain how and where fertilization occurs.

3. Name the most important androgen and the organ in the male primarily responsible for its secretion.
4. Name the five hormones that interact in controlling the female menstrual cycle, and explain what hormonal events lead to the release of an egg from the ovary.
5. Give the roles of human chorionic gonadotropin and oxytocin in pregnancy, and name the source of each.
6. State how "the pill," IUD, diaphragm, condom, rhythm, induced abortion, and sterilization of males or females operate as birth-control methods. State the part of the reproductive process with which each method interferes.

---

**A**nimals have evolved many adaptations for obtaining food, exchanging gases, regulating body temperature, and retaining body water and salts. These adaptations allow some animals to survive in a variety of challenging environments. But the true measure of evolutionary success is reproduction. Many animals can reproduce asexually (Chapter 20), but sexual reproduction is much more common.

The crucial event in sexual reproduction is **fertilization,** the union of two haploid reproductive cells, the **gametes,** forming a diploid **zygote.** In animals, the gametes are different from each other. The female gamete is a large, nonmotile **egg,** swollen with nutrients and other molecules needed to support embryonic development. The male gamete is a motile **sperm,** which actively moves toward the egg. After fertilization, the animal zygote undergoes a period of embryonic development before it is born or hatched, ready to carry out the various activities of life.

In most animal species, the two gametes that unite at fertilization come from different parents. The parents' anatomy, physiology, and behavior must be coordinated so that their gametes are produced at the same time and brought together in the same place (Figure 36–1). Hormones often play an important role in this coordination.

**FIGURE 36–1** Copulation. A male "ladybug" (actually a beetle) introduces sperm into the body of a female, a prelude to internal fertilization. The female later lays eggs with waterproof shells, which develop externally without further parental attention. (Peter J. Bryant, University of California at Irvine/BPS)

KEY CONCEPTS

♦ Sexual reproduction involves (1) getting sperm and egg together for fertilization and (2) ensuring that the fertilized egg has a suitable environment where it can develop until the young animal is ready to survive on its own. In animals, fertilization and development may be external or internal, or fertilization may be internal, followed by external development.

♦ Human fertilization and embryonic development are both internal. The male reproductive tract produces sperm, and the penis inserts them into the female tract. The female reproductive system produces eggs and provides the sites of fertilization and embryonic development.

♦ Hormones coordinate production of gametes in both sexes. In females, hormones maintain pregnancy and, after childbirth, stimulate milk production.

♦ Understanding how pregnancy occurs can help people avoid unwanted pregnancy and enhance the chances of beginning a desired pregnancy.

## 36–A  REPRODUCTIVE PATTERNS

Animal reproduction falls into three main patterns, depending on whether fertilization and embryonic development occur within or outside the body of the female parent (Table 36–1):

1. External fertilization and development
2. Internal fertilization followed by external development
3. Internal fertilization and development

### External Fertilization and Development

In general, the number of offspring that animals produce at one time is related to their reproductive pattern. In animals with external fertilization and development, mortality of gametes, zygotes, embryos, and young larvae or hatch-lings is extremely high. These animals produce so many young that at least some are likely to survive. The eggs are usually very small and are produced by the hundreds or even the millions in extreme cases such as herring (bony fish).

Eggs and sperm are released from the parents' bodies into the surrounding water, where they must find each other. For this to succeed, male and female must be in the same place at the same time and release their gametes together. Hormonal conditions and environmental triggers interact to produce the necessary coordination.

For instance, the hormone cycles of some marine worms ensure that they come into breeding condition once a month for several months. The moon's cycle then triggers males and females in breeding condition to come to the surface of the sea during the hours after the full moon rises. Females release their eggs, and the eggs phosphoresce for a

TABLE 36–1

**Some Examples of Animals Showing the Three Main Reproductive Patterns**

| External Fertilization and Development | Internal Fertilization, External Development | Internal Fertilization and Development |
|---|---|---|
| Cnidarians | Flatworms | |
| | Roundworms | |
| Polychaete annelids | Earthworms; leeches | |
| Chitons, bivalves | Most snails; cephalopods | |
| Many aquatic arthropods | Many aquatic arthropods | |
| Echinoderms | Terrestrial arthropods: spiders, insects, centipedes, millipedes | |
| Most bony fish | Skates (cartilaginous fish) | Sharks; some bony fishes |
| Most amphibians | Most reptiles | Some snakes |
| | Birds | |
| | Egg-laying mammals | Placental mammals |

(a)

(b)

**FIGURE 36–2**  External fertilization. (a) The yellow material draping these brown tube sponges consists of masses of eggs, extruded from the body to await fertilization by sperm shed into the surrounding water. (b) Fertilization is also external in frogs, but it is preceded by courtship behavior, in which the female approaches the male in response to his mating song. In the mating embrace, called amplexus, the male frog (top) uses his forelimbs to grasp the female, whose body is swollen with masses of eggs. The male sheds sperm over the eggs as they are squeezed out through the female's cloaca.   (a, S. K. Webster, Monterey Bay Aquarium/BPS; b, Biophoto Associates)

few minutes, lighting up the sea with a swirl of greenish light. Males swim toward the light, releasing their sperm as they reach it. The combination of hormonal conditioning, with the behavior triggered by the moon and by the phosphorescent eggs, ensures that sperm and eggs come together for fertilization.

Why is this behavior, and the reproductive behavior of many other marine animals and plants, linked to the lunar cycle? The reason is thought to be that the height of the tides is also related to the moon. Offspring produced at a particular phase in the tide cycle have a better chance of being dispersed in favorable ways, perhaps to new beaches.

In the case of these worms, males and females do not interact much, if at all, with one another. The same is true of many other invertebrates. On the other hand, many species of aquatic vertebrates with external fertilization and development rely on behavioral interactions between the sexes to ensure that potential mates recognize each other and that the female releases her eggs only when the male is ready to fertilize them. This is true of most bony fishes and amphibians (Figure 36–2).

## Internal Fertilization with External Development

Internal fertilization has several advantages. First, the female reproductive tract provides a space where sperm and egg can get together without danger of being eaten or washed away. Second, internal fertilization permits the sperm access to the egg before it is actually prepared for laying.

Then, as the fertilized egg passes through the female reproductive tract to the exterior, it can be surrounded with impermeable membranes or shells that will protect the developing embryo from desiccation.

**Courtship behavior** brings male and female of the same species together for fertilization. It is found in some animals with external fertilization, such as amphibians. In animals with internal fertilization it is essential. Courtship allows two potential parents to recognize one another as members of the same species but of the opposite sex and to approach close enough for fertilization. Preludes to courtship include various forms of "advertising": distinctive coloration; the songs of male frogs, birds, and crickets; odors wafted into the air by female moths and other insects; and so on.

Male squids and spiders use some of their many appendages to pass a packet of sperm, called a **spermatophore,** to the female. In other invertebrates, sharks, reptiles, birds, and mammals, sperm is transferred by **copulation,** direct passage of sperm from inside the male's body to inside the female's body. In some animals, but not in all, this transfer involves an organ such as a penis. Some species can copulate without this. In most species of birds, for instance, the males have no penis, and birds of these species mate by placing their cloacal openings tightly together. Male reptiles have at least one penis (sometimes two) and presumably so did the ancestors of modern birds. Loss of the penis in male birds and of one of the two ovaries in females are undoubtedly weight-saving adaptations.

In animals with internal fertilization and external development, one or both parents often guard the eggs after they are laid, reducing the mortality of embryos and young. These animals tend to produce relatively few eggs at a time, from several hundred for some insects and spiders down to one for a penguin.

## Internal Fertilization and Development

If development is internal, fertilization must be too. Since the space in a female's body is limited, animals with internal development produce the fewest offspring of the three groups, perhaps a few dozen for a snake and usually fewer than ten for a mammal.

Internal development gives an embryo many advantages. The mother's body provides moisture and the right chemical conditions and, for the mammalian embryo, warmth as well. Because the mother carries the embryo everywhere she goes, it is not vulnerable to the predators that plunder the egg masses or nests of animals that develop externally.

In most species other than mammals, internal development simply means retaining the egg inside the body instead of laying it. As with animals that develop externally, the food stored in the egg must support the entire embryonic development. In mammals, however, the embryo shares whatever food its mother takes in during her pregnancy. Such a continuously renewed food supply may permit the embryo to reach a larger size before birth.

## 36–B  HUMAN REPRODUCTIVE ORGANS

Human reproductive tracts are very similar to those of other vertebrates, except that only mammals have a uterus, the part of the female tract where embryos develop.

The human reproductive organs can be roughly divided into the internal organs and the external or accessory sex organs. In addition, each sex has **secondary sexual characteristics,** such as enlarged breasts in the human female and a higher metabolic rate in the human male, which are not part of the actual reproductive apparatus.

### Female Reproductive Organs

Figure 36–3 shows the external sex organs of a woman, collectively known as the **vulva.** The **labia majora** and **labia minora** ("major and minor lips") cover and protect the urinary and genital openings and the surrounding tissue. Note that the openings of the urethra, from the urinary tract, and of the vagina, from the reproductive tract, are separate. Among the vertebrates, only females of placental mammals show this separation.

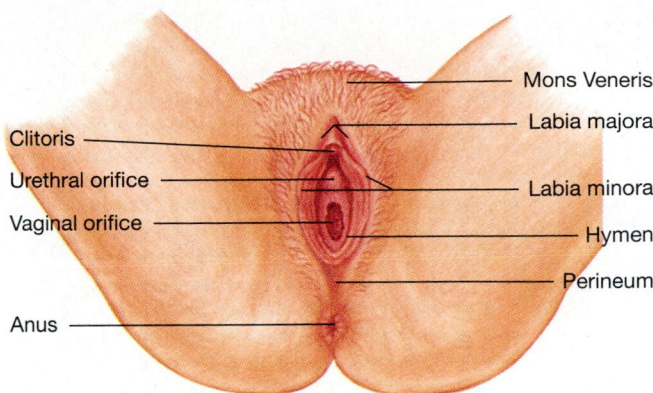

**FIGURE 36–3**  Vulva of a woman. Note the three separate openings characteristic of females of higher mammals—the urethral orifice from the urinary tract, vaginal orifice from the genital tract, and anus from the digestive tract. The mons Veneris ("mountain of Venus") is a fatty pad over the pubic bones. The labia majora and minora and the hymen play protective roles.

Girls are born with a membrane, the **hymen,** covering the vaginal orifice. One or more slits in the hymen allow the menstrual flow to pass. The hymen is absent in many women and unusually tough in others. During embryonic development, the **clitoris** arises from the same structure that develops into the penis in a male. Like its male counterpart, the clitoris produces a pleasurable sensation when it is stimulated by touch, and indeed this seems to be its only function.

The internal female reproductive organs consist of the ovaries and oviducts (fallopian tubes), uterus, and vagina (Figure 36–4).

The two **ovaries** are the female gonads, or gamete-producing organs. The ovaries produce eggs. All the **oocytes,** immature eggs, form during embryonic development. A girl is born with about 1 million oocytes, but more than two thirds die before she reaches puberty, and most of the others die during the next several decades. Only 400 to 500 are ovulated before the woman's reproductive life ends at menopause.

An ovary contains many **follicles,** each consisting of an immature egg surrounded by nutritive follicle cells (Figure 36–5). At **ovulation,** the wall of a mature follicle ruptures, and the egg pops out into the coelom. This release of a ripe egg stimulates the finger-like endings of the **oviduct** to reach and grasp the egg. The beating of cilia lining the tube draws the egg into the oviduct and on toward the uterus. Fertilization usually occurs about halfway down the oviduct. An unfertilized egg disintegrates as it passes through the uterus about 72 hours after ovulation.

The **uterus** is a strong, elastic organ whose main function is to hold a developing embryo and expel it during

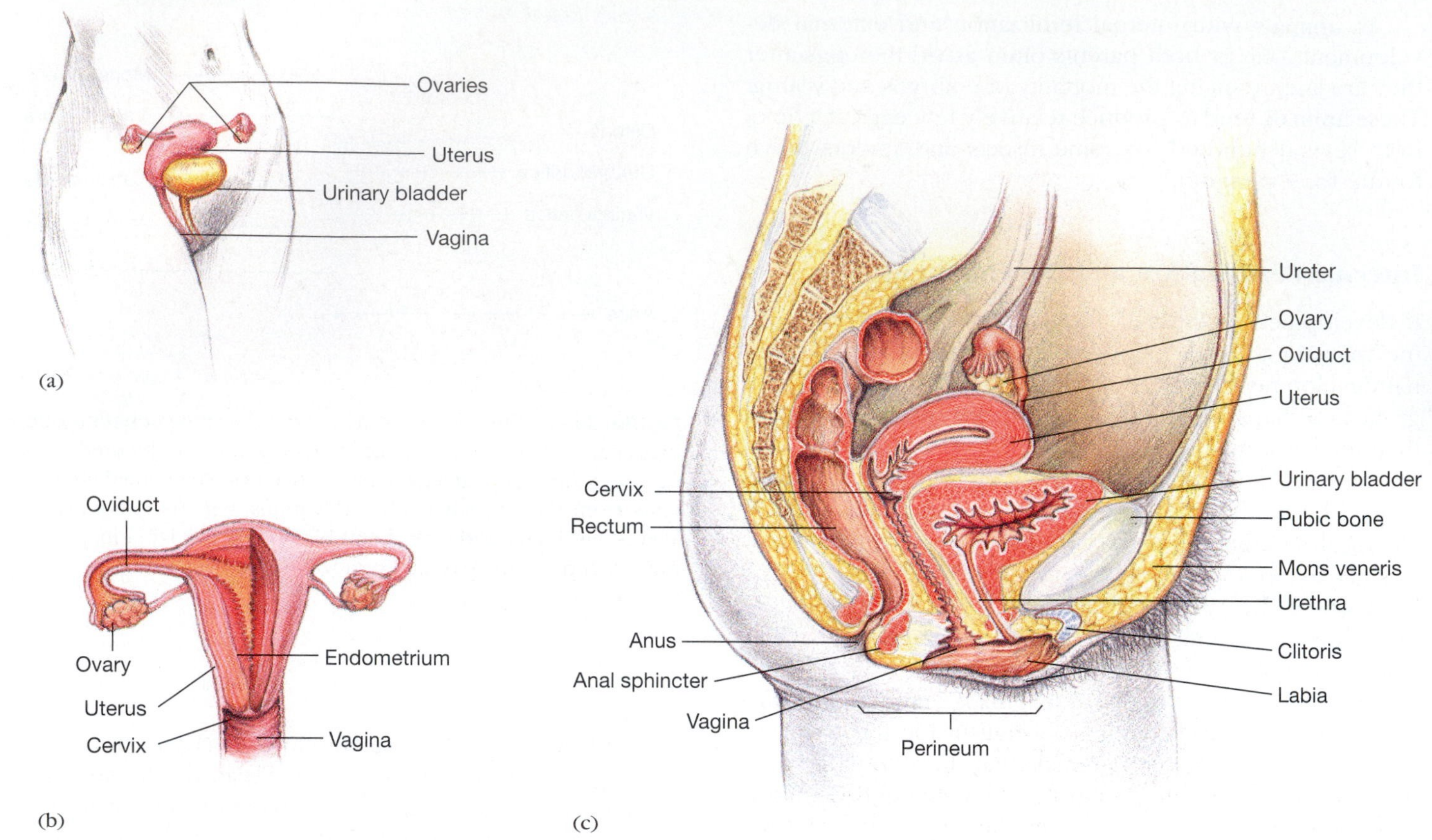

Ovaries
Uterus
Urinary bladder
Vagina

(a)

Oviduct
Ovary
Uterus
Cervix
Endometrium
Vagina

(b)

Ureter
Ovary
Oviduct
Uterus
Cervix
Rectum
Urinary bladder
Pubic bone
Mons veneris
Urethra
Anus
Anal sphincter
Clitoris
Vagina
Labia
Perineum

(c)

**FIGURE 36–4**  The internal reproductive organs of a woman. (a) The position of the reproductive tract. (b) Closeup of the reproductive organs, with one side cut open to show the pathway taken by the egg from the ovary to the vagina. (c) The relationship of the reproductive tract to other lower abdominal structures.

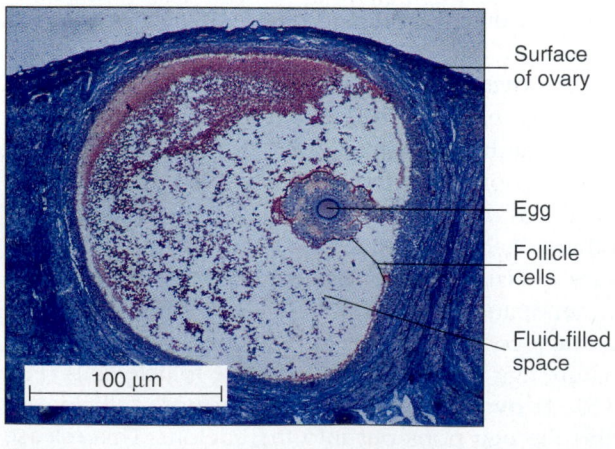

Surface of ovary

Egg

Follicle cells

Fluid-filled space

100 μm

Mature ovarian follicle (LM)

**FIGURE 36–5**  Cross section of a mature ovarian follicle of a rabbit. A cavity forms in the maturing follicle and swells with fluid, so that the mature follicle balloons out on the surface of the ovary and eventually ruptures, expelling the egg into the abdominal cavity.   (Biophoto Associates)

childbirth. The nonpregnant uterus is about the size and shape of a pear. Its walls contain powerful smooth muscles that contract and push the baby out during childbirth. The inner lining of the uterus, the **endometrium,** provides a suitable environment for the growth of the early embryo and later forms the maternal portion of the placenta (Section 36–E). The external opening of the uterus is the **cervix,** made up largely of the biggest, most powerful sphincter muscle in the body. (A sphincter is a circular muscle whose contraction closes a tube; other sphincters include those around the mouth, the anus, and the opening of the urinary bladder.) The strength of the cervical muscle is necessary to hold about 7 kg of fetus and fluid in the uterus against the pull of gravity during pregnancy. The cervix protrudes into the upper end of the vagina.

The **vagina** is the receptacle for the penis during copulation and is the pathway to the exterior for the baby during childbirth. In accordance with these functions, it has

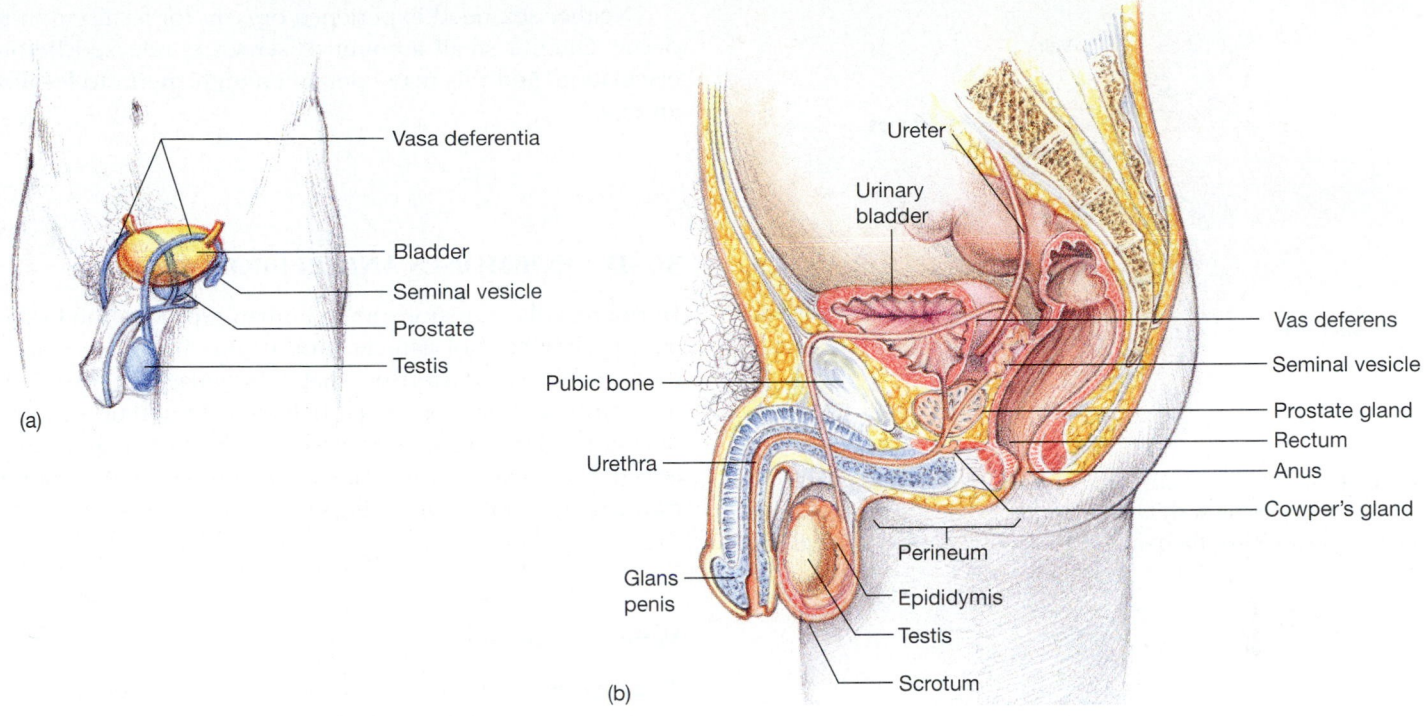

**FIGURE 36-6**   Male reproductive tract. (a) Position of the reproductive organs in a human male. Sperm are formed in the testes and travel through the vasa deferentia to the urethra, leaving the body through the opening of the urethra at the tip of the penis. (b) Cross section through the pelvic region of a man. Notice that the urethra and vasa deferentia pass through the prostate. Enlargement of the prostate can constrict this passageway and interfere with urination.

extremely elastic walls. The muscles in the vaginal wall are much thinner and weaker than those of the uterus.

## Male Reproductive Organs

The gonads of a human male are **testes** (singular: **testis**). They produce spermatozoa, or sperm, from the time of sexual maturity, at puberty, until death. Sperm production in mammals requires a lower temperature than that of the body. Mammalian testes usually lie in the cooler **scrotum,** a sac outside the body cavity. The testes form in the abdominal cavity during embryonic development and descend into the scrotum through a canal in the abdominal wall before birth. The presence of this canal between the abdominal cavity and the scrotum is one reason more men than women experience inguinal hernias, which are splits in the sheet of muscle of the abdominal wall at this point.

Sperm develop from cells lining the **seminiferous tubules** of the testis (Section 14–G). Muscular action in the seminiferous tubules carries mature sperm to the **epididymis,** where they are stored and their final maturation

occurs. During sexual stimulation, the sperm move through the **vas deferens** (plural: **vasa deferentia**) by contraction of its walls, until they reach the **seminal vesicles.** The sperm then move into the urethra, where they are joined by secretions from the seminal vesicles and the **prostate** and **Cowper's glands.** The sperm and their attendant secretions, collectively called **semen,** leave the penis via the urethra (Figure 36–6). Urine leaves the male's body by the same route, although the two fluids cannot pass through the urethra at the same time.

## 36–C   PHYSIOLOGY OF SEXUAL INTERCOURSE

During copulation (often called sexual intercourse in humans), the male's penis introduces sperm into the female's body. Before the penis can enter the vagina, it must become at least partly erect under the influence of sexual stimulation. Sexual stimulation can be brought about by any of the senses, usually most effectively by touch. The external genitals, and especially the glans of the penis, are the most sexually sensitive areas.

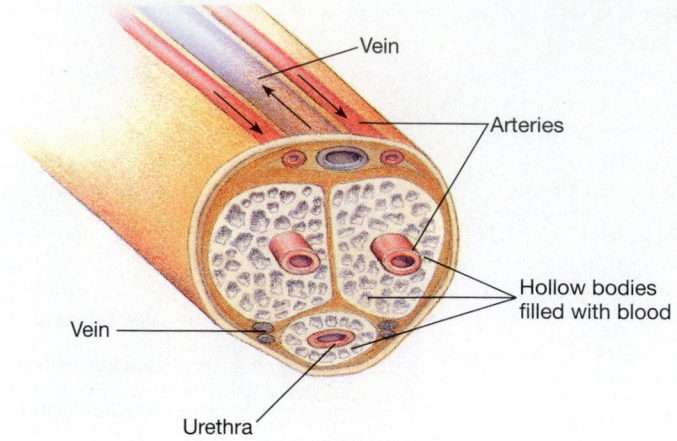

**FIGURE 36–7**  Cross section of a human penis. During erection, the hollow bodies fill with blood, partly blocking the veins so that little blood can leave the penis. The engorged penis is large and stiff.

The penis consists of three cylinders of spongy tissue that extend into the body (Figure 36–7). The cells within the cylinders contain many hollow bodies, which are usually empty and collapsed. During sexual stimulation, about ten times the normal volume of blood is carried from the arteries into these hollow bodies. The blood filling the spaces presses against the outsides of the veins, narrowing them so that the flow of blood leaving the penis is restricted; the penis thus enlarges and becomes rigid. Sexual stimulation also increases the muscular contractions that move the sperm.

In females, sexual arousal can also result from many different stimuli. The clitoris is the female organ most sensitive to touch. In a sexually aroused woman, the vulva becomes swollen because of an increased blood supply, and the walls of the vagina secrete fluid, which acts as a lubricant for entrance of the penis. This lubrication is caused by contraction of muscles in the walls of blood vessels around the vagina, forcing fluid out of the enlarged vessels.

When the erect penis has been inserted into the vagina, stimulation by the movement of the genitals against each other may result in **orgasm.** Orgasm is characterized by an increase in heart rate and blood pressure, engorging of various tissues with blood, and faster and deeper breathing, which finally result in an explosive burst of involuntary muscular contractions. In men, orgasm is accompanied by **ejaculation,** the forceful ejection of semen, propelled by peristaltic waves of contraction of the muscles in the sperm ducts. In women, orgasm is characterized by rhythmic spasms of the muscles surrounding the vagina.

Neither sex need experience orgasm for fertilization to occur. Often a small amount of semen is released before ejaculation, and this may contain enough sperm to fertilize an egg.

## 36–D  HORMONES AND REPRODUCTION

Hormones play a major role in human reproduction (Table 36–2). They control gamete production. They also control the superficial characteristics that differentiate the sexes, and they are necessary for sexual behavior. In addition to producing the hormones appropriate to their own sex, members of both sexes produce small quantities of the hormones characteristic of the opposite sex.

### Male Hormones

**Androgens** are hormones with masculinizing effects. The most important is **testosterone,** a steroid hormone secreted primarily by the testes. Some testosterone and other androgens are secreted by the adrenal glands, which lie on top of the kidneys (see Figure 40–4).

A boy's testes remain dormant until the onset of puberty, at the age of 10 to 14. From then on, the pituitary gland, located beneath the brain, releases **luteinizing hormone (LH),** a gonadotropic ("gonad-feeding") hormone into the circulation. Another gonadotropic pituitary hormone, **follicle-stimulating hormone (FSH),** also helps to regulate spermatogenesis. (These two hormones play important roles in females as well, and their names come from their effects on the ovaries.) One effect of FSH on the testes is to stimulate production of **androgen-binding protein** (ABP). This protein has a high affinity for testosterone and holds that hormone within the tubules of the testis, where it promotes spermatogenesis, along with FSH.

In addition to its direct effect on spermatogenesis, testosterone stimulates development of the secondary sexual characteristics of a sexually mature male. These include deepening of the voice, development of male musculature, and growth of hair on the face and other parts of the body.

In medieval times, when music written for boys' soprano voices was popular, some nobles maintained choirs of *castrati,* men castrated (by removing their testes) when they were boys. Lacking testosterone, *castrati* kept the high voices of childhood into adult life. Eunuchs, once used as male attendants in harems, were also castrated men. Men isolated from most sexual stimulation produce less testosterone and as a result have slower beard growth than at other times in their lives.

## Female Hormones

Feminizing hormones control women's secondary sexual characteristics and menstrual cycles. As in the male, the pituitary hormones FSH and LH are secreted at puberty, triggering the development of the secondary sexual characteristics and the onset of menstrual cycles. The hormonal progress of the menstrual cycle determines a woman's fertility.

The secondary sexual characteristics of human females include breasts and the rounded contours imparted by a thick, widespread layer of subcutaneous ("under skin") fat. This fat provides food during periods of starvation and insulation against cold, giving women (and the unborn babies or nursing infants that depend on the mother's body for food) a survival edge over men. However, following current fashions, many human females are unenthusiastic about their subcutaneous fat.

***The Menstrual Cycle***    At puberty (10 to 14 years of age), the pituitary gland starts a series of hormonal cycles that periodically render a woman fertile (capable of becoming pregnant) until the cycles cease at menopause, some 30 to 40 years later. These hormonal changes and their effects on the body are **menstrual cycles.**

Hormone levels in the blood change during the menstrual cycle, preparing the body for a potential pregnancy by causing ovulation and thickening the lining of the uterus, ready for implantation of a fertilized egg. If fertilization does not occur, the uterus sheds its lining during a **menstrual period,** and the cycle of preparation for pregnancy starts again. Changing hormone levels alter cell division, metabolism, and secretion in the reproductive organs, and these in turn cause changes in hormone synthesis. As long as the pituitary remains active, each change in hormonal and reproductive systems causes the next step in the cycle, so the cycles are endlessly repeated until menopause unless interrupted by pregnancy. Let us first consider the cyclical changes that occur in the uterus.

Human menstrual cycles are notoriously variable, but the average cycle lasts 28 days. The days are numbered from the first day of the menstrual period. Menstrual bleeding persists for a few days, and then the endometrium of the uterus thickens during the **proliferative phase** of the cycle. Next comes the **secretory phase,** about two weeks long, during which the endometrium continues to thicken and secretes a fluid rich in glycogen, creating an appropriate environment for an embryo if one is present. If an embryo has not implanted in the uterus by the end of the secretory phase, the newly added outer layer of endometrium sloughs off in another menstrual period, marking day one of the next cycle.

Meanwhile, changes in the ovaries ensure that an egg is produced and may be fertilized at a time when the uterus is ready to receive it. Ovarian changes start with the follicular phase, during which several follicles in the ovaries start to grow. Only one of these usually matures, while the others degenerate. The maturing follicle becomes large and filled with fluid, forming a bulge near the surface of the ovary. The follicular phase ends with ovulation, when the follicle bursts, releasing its egg. The remains of the burst follicle develop into a **corpus luteum** ("yellow body"), a

### TABLE 36-2

### Hormones Involved in Human Reproduction

| Source | Hormone | Role |
|---|---|---|
| Hypothalamus | Gonadotropin-releasing hormone (GnRH) | Induces secretion of FSH and LH |
| Pituitary | Follicle-stimulating hormone (FSH) | Production of gametes in both sexes |
| | Luteinizing hormone (LH) | Secretion of sex hormones by gonads in both sexes |
| | | Ovulation in females |
| | Oxytocin | Letdown of milk from mammary glands to nipple |
| | Prolactin | Growth of mammary glands and milk production (lactation) |
| | | Inhibition of FSH and LH |
| Testes | Testosterone | Sperm formation, with FSH |
| | | Development of male secondary sexual characteristics |
| Ovaries | Estrogen | Development of female secondary sexual characteristics |
| | | Preparation of uterine lining for implantation of embryo |
| | Progesterone | Preparation of uterine lining for implantation of embryo |
| | | Enlargement of breasts for lactation |
| Placenta | Human chorionic gonadotropin (HCG) | Maintenance of pregnancy |
| | Oxytocin | Contractions of uterus during childbirth |

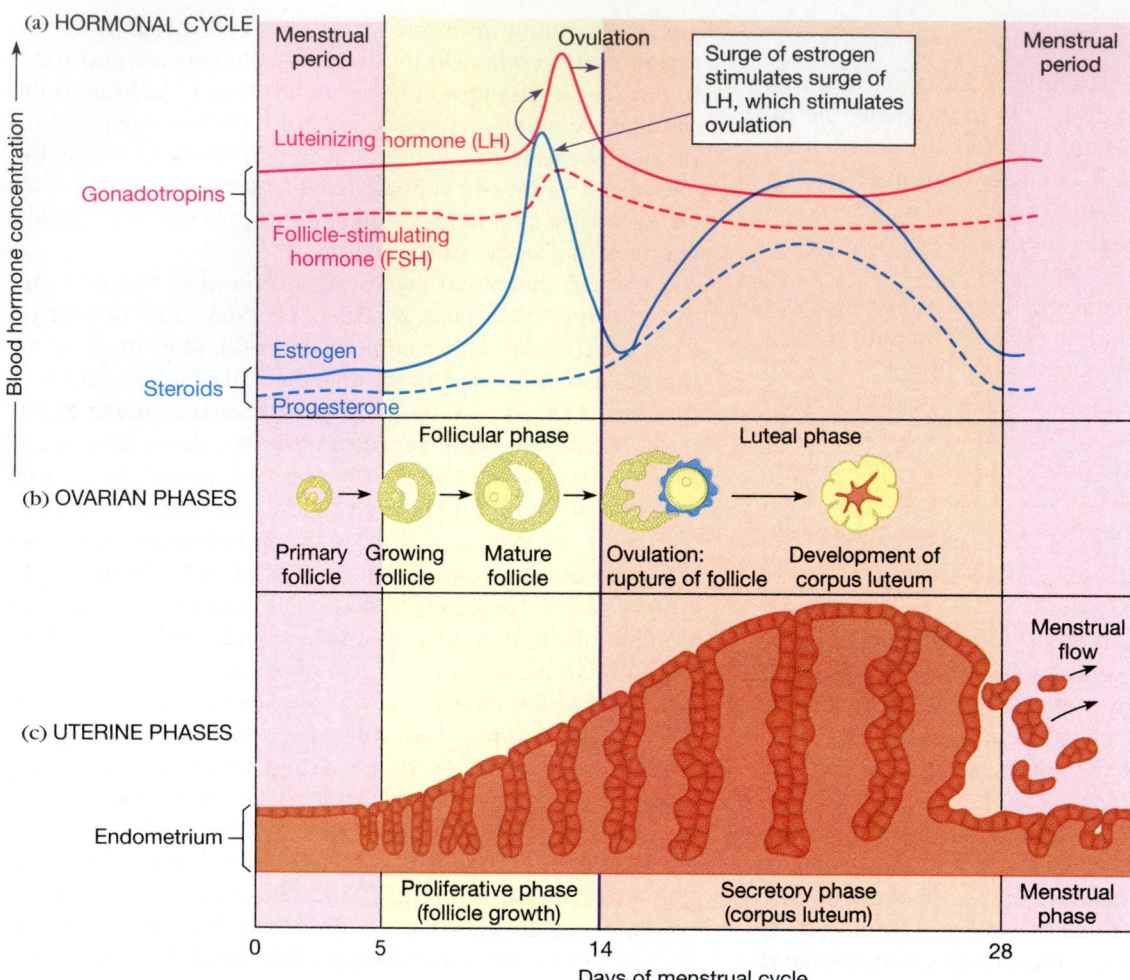

(a) HORMONAL CYCLE

Blood hormone concentration

Menstrual period

Ovulation

Surge of estrogen stimulates surge of LH, which stimulates ovulation

Menstrual period

Luteinizing hormone (LH)

Gonadotropins

Follicle-stimulating hormone (FSH)

Estrogen

Steroids

Progesterone

(b) OVARIAN PHASES

Follicular phase

Luteal phase

Primary follicle — Growing follicle — Mature follicle — Ovulation: rupture of follicle — Development of corpus luteum

(c) UTERINE PHASES

Menstrual flow

Endometrium

Proliferative phase (follicle growth)

Secretory phase (corpus luteum)

Menstrual phase

0    5    14    28

Days of menstrual cycle

**FIGURE 36–8** Events of the menstrual cycle. (a) Levels of sex hormones in the bloodstream during the phases of one menstrual cycle in which pregnancy does not occur. (b) Development of an ovarian follicle during the cycle. At ovulation the follicle ruptures, releasing the oocyte. The remaining follicle walls develop into a corpus luteum, which secretes progesterone and then degenerates. (c) Phases of development of the endometrium. The endometrium thickens during the proliferative phase, developing long narrow glands and a rich blood supply. In the secretory phase, the glands become wide and twisted, secreting a glycogen-rich nutritive material. When no embryo implants, the new outer layers of the endometrium disintegrate, and the blood vessels rupture, producing the menstrual flow.

tissue that secretes hormones during the final, luteal, phase of the cycle. At the end of the cycle, the corpus luteum degenerates. Then new follicles start to grow, marking the beginning of the next cycle.

Five hormones are involved in coordinating these changes in the reproductive organs, three secreted by the brain and two by the ovaries. At the beginning of the menstrual cycle, the hypothalamus in the brain secretes **go-nadotropin-releasing hormone (GnRH),** which induces the nearby pituitary gland to secrete small quantities of FSH and LH. Immature follicle cells have receptors for follicle-stimulating hormone but not for LH. They respond to FSH

by growing and secreting increasing quantities of the steroid hormone **estrogen.** This has a feedback effect on the brain. Whereas a low concentration of estrogen in the blood keeps secretion of pituitary hormones at a low level, a high concentration of estrogen increases GnRH secretion by the hypothalamus and therefore causes more rapid secretion of LH and FSH (Figure 36–8).

Follicles develop LH receptors as they mature, and LH has a positive feedback effect on the follicle: it causes final maturation and ovulation about a day after the surge in LH concentration, on about the fourteenth day of the menstrual cycle.

Still under the influence of luteinizing hormone, the cells of the ruptured follicle grow and form a corpus luteum, which secretes more estrogen and yet another steroid hormone, **progesterone.** Increasing levels of progesterone and estrogen exert negative feedback on the pituitary, inhibiting secretion of LH and FSH. The corpus luteum requires LH in order to function, so as LH levels fall, the corpus luteum begins to degenerate, and progesterone and estrogen secretion decline sharply. This removes their inhibitory effect on the pituitary, which then begins to secrete enough FSH to stimulate growth of follicles in the next cycle.

Changes in the ovary are regulated by hormones from both the brain and the ovary. In contrast, changes in the uterus are affected only by the ovarian hormones, progesterone and estrogen. Increasing estrogen concentration causes the endometrium to thicken during the proliferative phase of the cycle. After ovulation, both estrogen and progesterone levels are high. This increases the blood supply to the uterus and causes glands in the endometrium to secrete nutrient fluid that can be absorbed by an embryo even before it implants. When the corpus luteum degenerates, causing the level of ovarian hormones to fall, arteries to the uterus constrict, depriving the endometrium of blood. The surface of the endometrium sloughs off in a menstrual period.

Ovarian hormones affect other parts of the body as well as the uterus and ovaries. They are responsible for secondary sexual characteristics. They cause deposition of fat in the breasts and hips, water retention, and sometimes changes in behavior. Some women find themselves inefficient or bad-tempered immediately before the menstrual period starts, probably as a result of the high levels of progesterone in the blood at this time. Others experience pain ("cramps") during menstrual periods and occasionally at the time of ovulation. Menstrual cramps are thought to be due to the accumulation of local lipid hormones called prostaglandins in the uterus (Section 40–E).

## Hormones of Pregnancy

Menstrual cycles are interrupted if an egg is fertilized and pregnancy ensues. At the beginning of pregnancy, the embryo and the lining of the uterus jointly form a special organ, the placenta, which provides for nourishment of the embryo. The placenta also secretes **human chorionic gonadotropin (HCG),** a hormone related to LH. Elevated HCG, like elevated LH, promotes production of progesterone, which inhibits production of FSH and LH by the pituitary gland. This prevents ovulation and thus prevents formation of any new embryo while the first one is developing. If the embryo is abnormal, or dies, secretion of HCG by the placenta stops and the HCG level in the bloodstream falls, permitting the body to abort or miscarry the pregnancy. As many as one third of human embryos that implant abort in this manner. If a hormone from the mother's body maintained pregnancy, there would be no way to discard dead and abnormal embryos.

By the same reasoning, it is not surprising that hormones produced by the fetus determine when birth shall occur. It is important that the baby be born when it is mature and not when the mother feels like it. Researchers studying sheep have found recently that the hypothalamus in the fetus's brain signals the pituitary to secrete a releasing hormone (ACTH), which in turn causes the fetal adrenal glands to secrete the steroid hormone cortisol. The fetal hormones diffuse across the placenta and cause the placenta and uterus to secrete the hormone **oxytocin.** Oxytocin stimulates the muscles of the uterus to contract and cause birth.

The placenta is expelled within an hour after the baby is born. Many mammalian mothers, including herbivores, eat their placentas. This behavior may be advantageous because it prevents loss of valuable nutrients. It also destroys evidence that might reveal the existence of a vulnerable newborn to potential predators. One important result of this behavior is that hormones ingested in the placenta trigger maternal behavior. Whether women would make better mothers if they ate their placentas has not, as far as we know, been tested!

Expulsion of the placenta during childbirth causes the mother's pituitary to secrete **prolactin,** a hormone that initiates lactation (milk production) by the mammary glands (Figure 36–9). Once this has occurred, milk production is thereafter on a supply-and-demand basis. Suckling by the

**FIGURE 36–9** A baboon nursing an infant at a national park in Botswana. (Frans Lanting/Minden Pictures)

baby is the stimulus for secretion of more prolactin, and the more a baby suckles, the more milk is produced about 24 hours later. Suckling also induces secretion of oxytocin by the lactating woman's pituitary gland. This hormone in turn causes milk letdown: the release of milk from mammary glands in the breast into ducts leading to the nipple.

Prolactin also inhibits the release of LH from the pituitary and counters the effects of FSH and LH on the ovarian follicles. Hence, in nursing mothers the maturation of subsequent eggs, production of estrogen and progesterone, and resumption of the menstrual cycles are suppressed, particularly when the baby suckles frequently. This is adaptive because studies show that baby and mother are likely to be healthier if pregnancies are no closer than two and a half years apart. Breast feeding is not a reliable means of birth control, but it helps to space pregnancies where artificial birth control is not available.

The habit of feeding babies artificial "formula" from a bottle has spread throughout the world in the twentieth century. The reasons are probably that bottle-feeding is expensive, and therefore fashionable, and that bottle-feeding prevents the change in breast shape that breast-feeding causes, particularly in women with large breasts.

Breast-feeding, however, is pleasurable for the mother and is better for the health of both mother and child. Breast-feeding transfers energy from mother to baby, using up the fat deposits that a woman accumulates during pregnancy. The hormonal changes of lactation also increase a woman's muscle tone and, for reasons that are not understood, make her less likely to develop breast cancer later in life.

Breast-fed babies have less chance of developing colic (painful upset stomach) and asthma than bottle-fed babies. In addition, they are less susceptible to colds and other infections because milk contains antibodies from the mother's immune system, which protect the baby from infection while the baby's immune system is still immature (Section 34–G). Breast-feeding is also the best way to ensure that the baby receives adequate nutrition and avoids infections from unsterilized formula and bottles. Health authorities encourage breast-feeding because breast-fed infants have been found to require fewer doctor visits than those fed formula.

## 36–E   FERTILIZATION AND IMPLANTATION

Sperm released into the vagina during ejaculation swim through the cervix and uterus into the oviduct, where fertilization occurs. (This description of pregnancy applies primarily to humans, but the situation is similar in other mammals.)

As a sperm penetrates the egg cell membrane, a rapid electrical reaction, followed by a slower chemical change, runs through the membrane. After this, the egg cannot be penetrated by another sperm. Although an egg is fertilized by only one sperm, a male's semen must contain many millions of sperm per milliliter if it is to be capable of fertilization. The excess sperm are apparently needed to penetrate the barriers that surround the egg.

A man with a "low sperm count" may be infertile even though his semen contains millions of sperm. The important quantity is the concentration of sperm, not the total volume of fluid ejaculated. The average sperm counts of men have declined since about 1950, and the same is true in most other vertebrates that have been studied, from alligators to cheetahs. This is believed to be the reason for the relative infertility of a number of endangered species. Biologists speculate that a change in some environmental factor must be the reason for the decline in sperm count, but no one cause has been identified.

During evolution, selection has strongly favored the birth of only one human baby at a time. The energy available from the mother, for nourishing the unborn child and for milk production and care after the child is born, is limited. Babies borne singly tend to be larger and healthier, with a better chance of survival, than those with wombmates. However, multiple births result from a small percentage of pregnancies. Nonidentical twins, also called fraternal twins or dizygotic ("two-zygote") twins, result when two eggs are ovulated at the same time and both are fertilized and implant in the uterine wall. Because the resulting embryos come from different eggs and different sperm, they are no more alike than any other children of the same parents, except that they are the same age.

Identical twins are produced when the mass of cells formed by cell division of the zygote separates into two groups in the first week after fertilization. Each group of cells develops into a separate embryo. Because these cells contain identical genes, the resulting embryos are genetically identical and so must be of the same sex. Identical twins are also called monozygotic twins because they originate from a single zygote.

The number of multiple births has increased with the use of modern fertility drugs to stimulate ovulation at a predictable time. These drugs are used to increase the fertility of couples who have various problems in trying to conceive a child and also to obtain eggs for test-tube fertilization. The disadvantage is that these drugs may cause many ovarian follicles to mature at the same time, so that more than one egg is released. As many as eight babies at once have been born to women treated with these drugs.

As a fertilized egg moves slowly down into the uterus, it is already a morula undergoing the cell divisions of cleavage (Section 12–B). Because the morula is not attached to the mother, modern medicine and animal husbandry can intervene in the reproductive process at this time (*Essay: Test Tube Babies and Surrogate Mothers*). About a week after fertilization, the embryo starts to implant in the endometrium of the uterus, establishing an intimate anatomical and physiological link between the mother and embryo that lasts throughout the pregnancy.

**E S S A Y :** *Test Tube Babies and Surrogate Mothers*

During the brief period between fertilization and implantation, the early embryo lives free from attachment to the mother. Modern medicine and animal husbandry take advantage of this phase in development to intervene in the reproductive process. Eggs or developing morulas can be removed from the ovaries or oviducts, treated in the laboratory, and then returned to the uterus without damage, if they are provided with a suitable fluid environment during their stay in the outside world.

"Test tube" babies do not develop in test tubes, but merely undergo fertilization and a few rounds of cell division outside a woman's body. The more scientific term for this is *in vitro* ("in glass") fertilization. If a woman has blocked oviducts, or if her vagina produces spermicidal secretions (as in one type of autoimmune disease), or if her husband's sperm count is low, fertilization cannot occur naturally in her body. However, a doctor can remove eggs from her body, add the husband's sperm in a laboratory dish, check to be sure that development has begun, and then introduce one or more morulas to the uterus.

*In vitro* fertilization has a very low chance of resulting in pregnancy (possibly because infertility may be caused by one member of the couple producing a large proportion of abnormal gametes). Indeed, about half of all "fertility clinics" have yet to achieve a single pregnancy. To increase the chances of success, doctors usually treat the prospective mother with fertility drugs, obtain several eggs, and transfer more than one embryo to the uterus. Hence, when successful, this technique results in a high proportion of multiple births.

Once the embryo is outside the female's body, it can just as easily be inserted in the uterus of a different female, provided her hormones are in the proper phase of the reproductive cycle for implantation to occur. Human babies and the young of many other species have been born to such surrogate mothers. This permits a woman who has a normal uterus but damaged ovaries to bear a child (which of course will have the genes of the woman who donated the egg).

However, the greatest use for embryo transplants is not for humans but for other mammals. With it, a breeder can obtain many embryos from a genetically superior cow or mare, for instance, and use inferior stock as surrogate mothers, vastly increasing the rate of production of desirable young. It also turns out that female mammals can successfully gestate embryos of closely related species. This allows zookeepers to produce more young of endangered species, using surrogate mothers of species that are common and easily bred in captivity (Figure 36–A).

**FIGURE 36–A** A newborn bongo and its surrogate mother, an eland. As a young embryo, the bongo was transplanted into the eland's uterus, where it implanted and developed. Bongos are a rare and elusive species inhabiting dense forests in Africa. The dark color and arched back are adaptations to this habitat. Larger and more common elands, members of the same genus, inhabit open areas.   (Cincinnati Zoo)

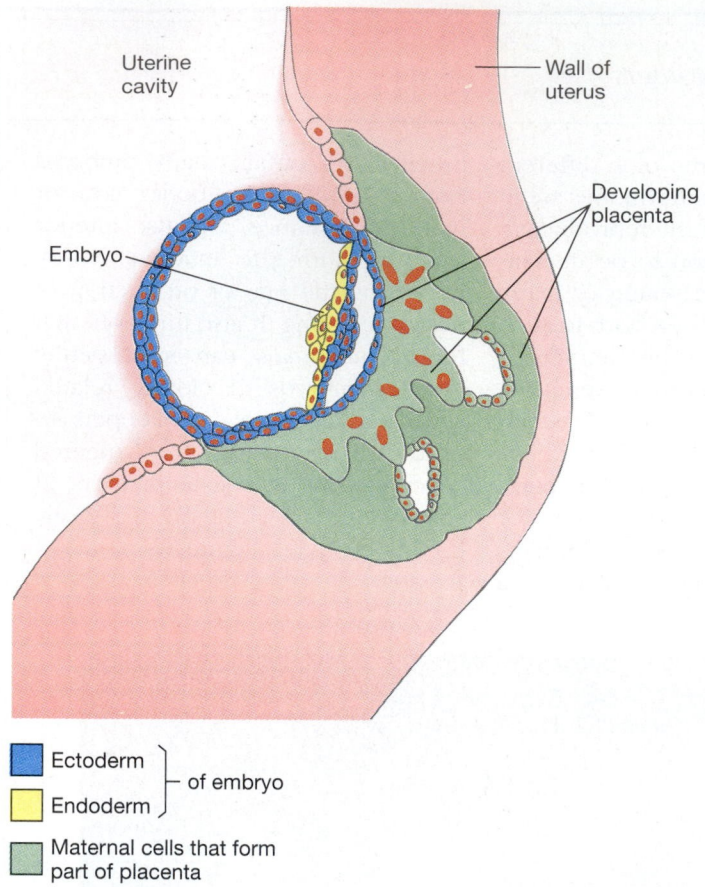

Ectoderm ⎫ of embryo
Endoderm ⎬

Maternal cells that form part of placenta

**FIGURE 36–10**   Implantation: an eight-day human embryo. Most of the morula will form extra-embryonic membranes and part of the placenta. The embryo develops from a few cells of ectoderm and endoderm. (The mesoderm has not yet differentiated.)

When a morula reaches the wall of the uterus, it consists of a little ball of cells, which will develop into the embryo proper, and a sphere of cells surrounding this future embryo (Figure 36–10). Some of these outer cells invade the wall of the uterus, anchoring the embryo and beginning the development of the most important organ of pregnancy, the placenta. Other cells in this outer sphere give rise to the three **extra-embryonic membranes**—the amnion, chorion, and allantois—parts of which also form part of the placenta.

At first the placenta is tiny, but by the time of birth it will develop into an organ that looks like a raw hamburger, about the size of a stack of four dinner plates. The membranes joining the embryo to the placenta develop into a cord, the **umbilicus,** which grows thicker and longer as development proceeds (Figure 36–11). The umbilical arteries and vein, within this cord, carry blood from the fetus to the placenta and back.

In the placenta, blood capillaries of the mother and fetus lie intermingled, and the maternal and fetal bloodstreams exchange substances by way of the surrounding extracel-

lular fluid. Here the fetal blood picks up food and oxygen and gives up wastes. Other substances present in the mother's blood may also enter the fetal circulation and may affect development. Some may interfere with differentiation and cause birth defects. Many drugs are known or suspected to harm developing fetuses, and doctors urge pregnant women to avoid smoking, drinking alcohol, and taking medication.

The three extra-embryonic membranes—the amnion, chorion, and allantois—are important in the development of all terrestrial vertebrates—reptiles, birds, and mammals (Figure 36–12). The amnion is a fluid-filled sac around the embryo, cushioning it against bumps. This is the sac that bursts when the "waters break" during childbirth. Because the amniotic fluid contains some cells sloughed from the embryo, drawing fluid from the sac into a hypodermic needle (amniocentesis) permits a geneticist to check for chromosomal defects in the embryo (Section 16–J).

The allantois is a sac connected to the embryo's gut. In reptiles and birds it stores the embryo's waste products until the egg hatches. In humans it becomes riddled with blood vessels and makes up most of the embryonic side of the placenta.

The third membrane, the chorion, which surrounds the fetus outside the amnion, also makes up a part of the placenta. The chorion secretes human chorionic gonadotropin (HCG), the hormone that maintains pregnancy and can be detected in pregnancy tests.

During the early stages of development, the embryos of all vertebrates look very much alike, reflecting their common evolutionary histories. Not until it acquires uniquely human characteristics, about nine weeks after fertilization, is the human embryo called a fetus (not until birth is it called a baby).

## Birth

In humans, the date of birth averages about 270 days after conception, but there is much variation in the time a baby takes to develop. The process in which uterine contractions expel the baby and the placenta is called **labor** and can be divided into three main stages. The first stage is **dilation,** which usually lasts from 2 to 20 hours and ends when the cervix of the uterus is fully open, or dilated. The second stage, **expulsion,** which lasts from about 2 to 100 minutes, begins with **full crowning,** the appearance of the baby's head in the cervix, and continues while the baby is pushed, head first, down through the vagina into the outside world, where it draws its first breath.

The third, or **placental,** stage begins when the baby is born. The uterus continues to contract while the umbilical cord is clamped, and some 5 to 45 minutes after the baby is born the uterus expels the placenta. The umbilical cord can now be severed, and the baby's independent existence begins.

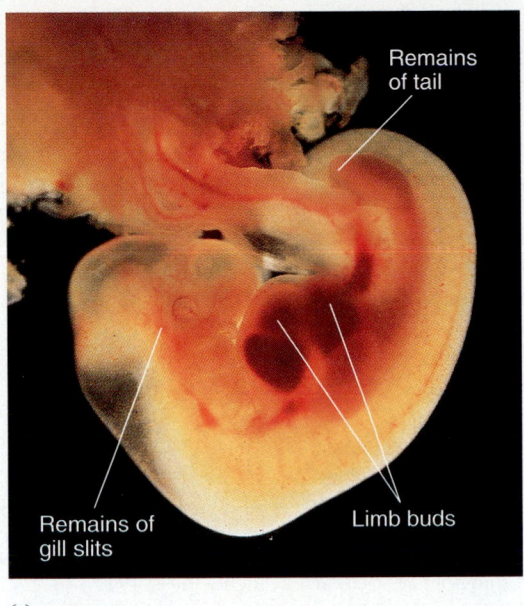

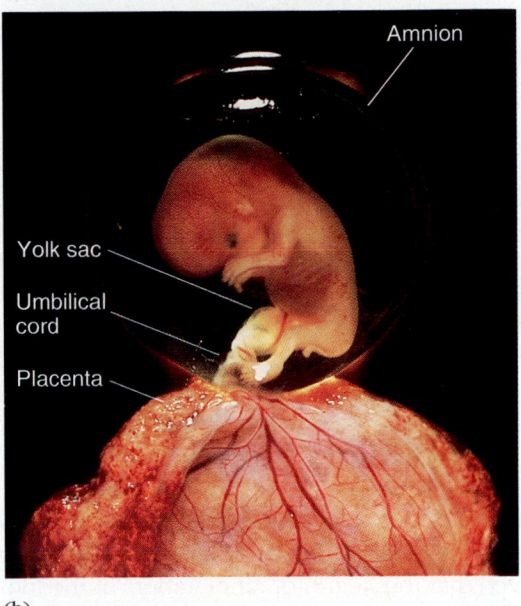

**FIGURE 36–11** Human embryos. (a) In a five-week embryo, buds that will develop into an eye, a forelimb, and a hindlimb are visible. The gill slits and tail, ancestral remnants, have almost disappeared. (b) In this two-month embryo, most of the bones are formed and the ears are forming. Part of the placenta is visible at right, attached to the embryo by the twisted umbilical cord in which blood vessels can be seen. The transparent amniotic sac, filled with amniotic fluid, surrounds the embryo. The circular structure between embryo and placenta is the remains of the yolk sac, which nourishes the very early embryo. (a, Lennart Nilsson: b, CNRI/Phototake)

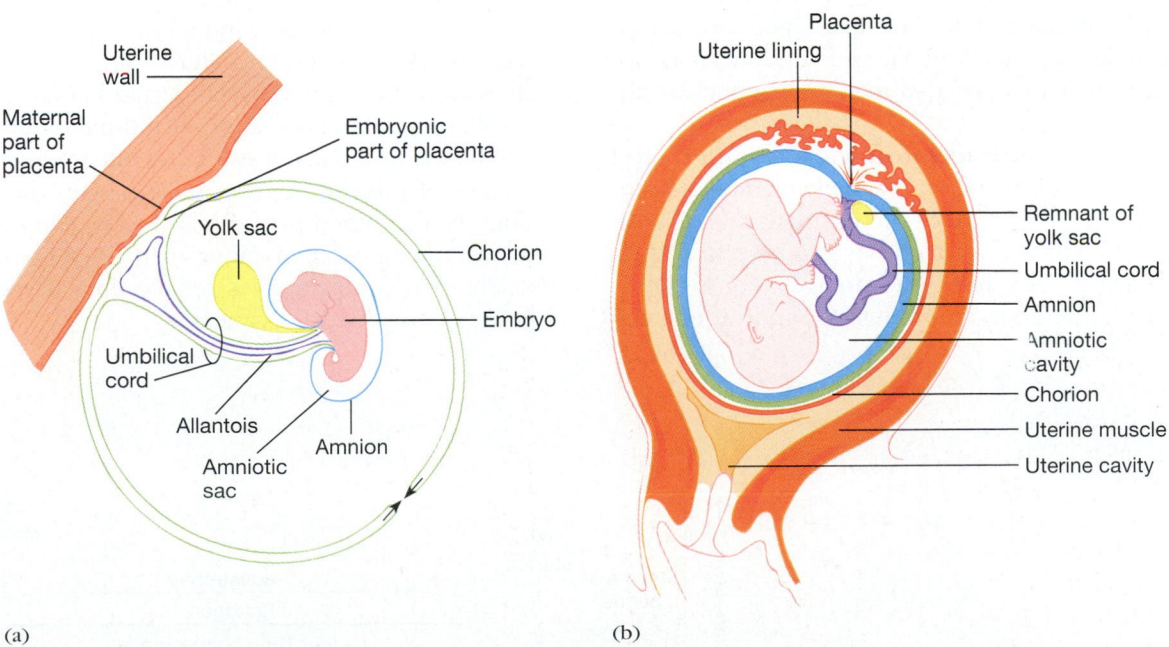

**FIGURE 36–12** The life support system of the human embryo. (a) Arrangement of the extra-embryonic membranes to form the umbilical cord and placenta. (b) A fetus in the uterus shortly before birth.

## 36–F    BIRTH CONTROL

The reproduction of humans and other animals is regulated so that the number of offspring bears some relation to the parents' ability to raise them. Rabbit embryos die when their mother is stressed by overcrowding, and her body resorbs the fetuses. Rats living in crowded conditions have lower fertility and a high rate of abortion. Birds abandon eggs or nestlings if food becomes scarce. Carnivorous birds with uncertain food supplies usually feed the youngest, and weakest, nestling last. If food is abundant, this nestling may receive enough to live; if not, its nestmates will receive all the food, and the weakling will be the first to die. In some species, the larger nestlings even kill their siblings if food is scarce. Thus, valuable energy has not been spent on offspring that are not likely to survive.

Controlling the number of children and spacing their births is just as adaptive in human populations. One study in Kenya showed that the health of all members of the family was improved if a woman had her first child after the age of 20 and had no more than three children, spaced two to three years apart. Studies in developed countries produce the same conclusion. Babies born to teenagers are disadvantaged compared with those born to older mothers. They are usually smaller than the optimal birth weight of eight pounds, have a greater risk of ill health and early death, and are the babies most likely to be born into poverty. The mother's health is better, and the educational achievement of each child is greater, if children are spaced more than two years apart.

Historically, the most common methods of human population control have been abstention from intercourse, abortion, prolonged breast-feeding, late marriage, and infanticide. Sponges placed in the vagina, and condoms made of leather or pigs' bladders, are also recorded in ancient Roman writings.

Strictly speaking, **contraception** refers to birth control methods that prevent fertilization. The only methods of contraception that are 100% reliable are abstention and sterilization. Artificial birth control methods can be safe and effective, but each of them has disadvantages. For instance, the most reliable methods are expensive and require the

**T A B L E   3 6 – 3**

**Contraceptive Use in the United States***

| Method | Estimated % Use | % Accidental Pregnancy in One Year of Use[†] |
|---|---|---|
| Male sterilization | 15 | 0.15 |
| Female sterilization | 19 | 0.4 |
| Oral contraceptive pill | 32 | 3 |
| Condom | 17 | 12 |
| Diaphragm + spermicide | 5 | 18 |
| "Rhythm" (periodic abstinence) | 4 | 20 |
| IUD | 3 | 6 |
| Contraceptive sponge | 3 | 18 |
| Vaginal foams, jellies | 2 | 21 |
| Norplant | 1 | less than 1 |

* Data from *Developing New Contraceptives: Obstacles and Opportunities.* Washington, DC: National Academy Press, 1990.

[†]About 89% of women using no contraceptive become pregnant within one year.

ongoing services of health-care professionals, so they are not readily available to everyone.

*Pills*    Birth control pills (**oral contraceptives**) are very reliable contraceptives if they are properly used (Table 36–3). "The pill" contains synthetic estrogen and progesterone that prevent ovulation (Figure 36–13). Pregnancy is avoided because there is no egg to be fertilized.

The pill does not reliably prevent ovulation and pregnancy until it has been taken regularly for at least two weeks. Women who have intercourse during this period, or who experience condom failure, can often protect themselves from pregnancy by taking additional contraceptive pills within 72 hours of unprotected intercourse.

The medical risks of using the pill are largely the same as those of pregnancy. The pill increases the risk of such things as high blood pressure and excessive blood clotting, mainly among women older than 35 who smoke. Conversely, oral contraceptives reduce their users' chances of getting certain kinds of cancer.

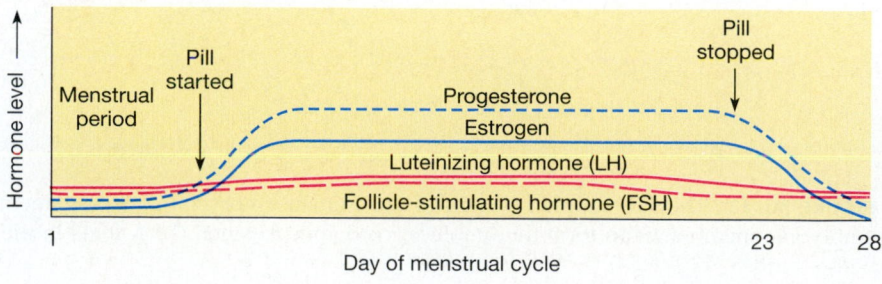

**FIGURE  36–13** Blood levels of reproductive hormones in a woman taking the combination contraceptive pill. Compare the shapes of these curves with those in Figure 36–8(a).

*Implants* Women who wish to postpone pregnancy for any length of time often choose implants of hormones that prevent ovulation for contraception. With Norplant, a set of matchstick-sized implants is placed under the skin of the upper arm. When it is removed, full fertility is restored. Because implants last for years, the user does not have to remember to take a pill every day. Implants are also generally cheaper to use than pills.

*Injections* There are several contraceptives that are injected and prevent conception for around three months. They have been used with considerable success in a number of developing countries but are not yet widely available in the United States.

*Condoms* The condom is the contraceptive device most commonly used by men. A condom is rolled onto the erect penis shortly before intercourse. It catches the semen so that sperm do not enter the female reproductive tract. Condoms are important as the only form of birth control that can reduce or prevent the spread of sexually transmitted diseases (STDs), such as AIDS, herpes, gonorrhea, and syphilis, from one partner to the other *(Essay: Sexually Transmitted Diseases)*. For this reason, improved materials for condoms and condoms that can be used by women are being tested. Research is also proceeding into other contraceptives that a man can use, including antifertility vaccines, reversible sterilization, and antisperm drugs.

*Barriers and Spermicides* Another contraceptive is a rubber diaphragm, which is smeared with a spermicidal (sperm-killing) jelly or cream and then inserted into the vagina each time it is used. It works by blocking the entrance to the uterus so that sperm cannot reach the egg. The cervical cap works in the same way but can be left in place for up to two days after intercourse instead of one. A similar device is a foam sponge soaked with spermicide, which has the advantage that one size fits all. Using a barrier plus spermicide protects women against some STDs but not against AIDS.

*Rhythm Method* The rhythm method consists of temporary abstinence, avoiding intercourse during the woman's "fertile period," the part of the menstrual cycle when there is an egg present to be fertilized. The difficulty is in determining when ovulation will occur, because menstrual cycles are so variable.

*Intrauterine Devices (IUDs)* IUDs are small, usually plastic objects inserted into the uterus by a doctor. IUDs are not true contraceptives. They somehow act on the uterine lining so that the already developing embryo cannot implant. IUDs are as effective as contraceptive pills in preventing pregnancy. On the other hand, a disquieting number of IUDs become deeply imbedded in the wall of the uterus, causing dangerous abdominal infections and requiring surgery to remove the device. Such infections may result in infertility.

*Sterilization* **Sterilization** is a more or less permanent change that prevents an animal from reproducing sexually. The most common sterilization operation for men is **vasectomy,** that is, severing and tying off the vasa deferentia. This is a simple operation, usually performed under local anesthesia (Figure 36–14). Afterwards, sperm are still

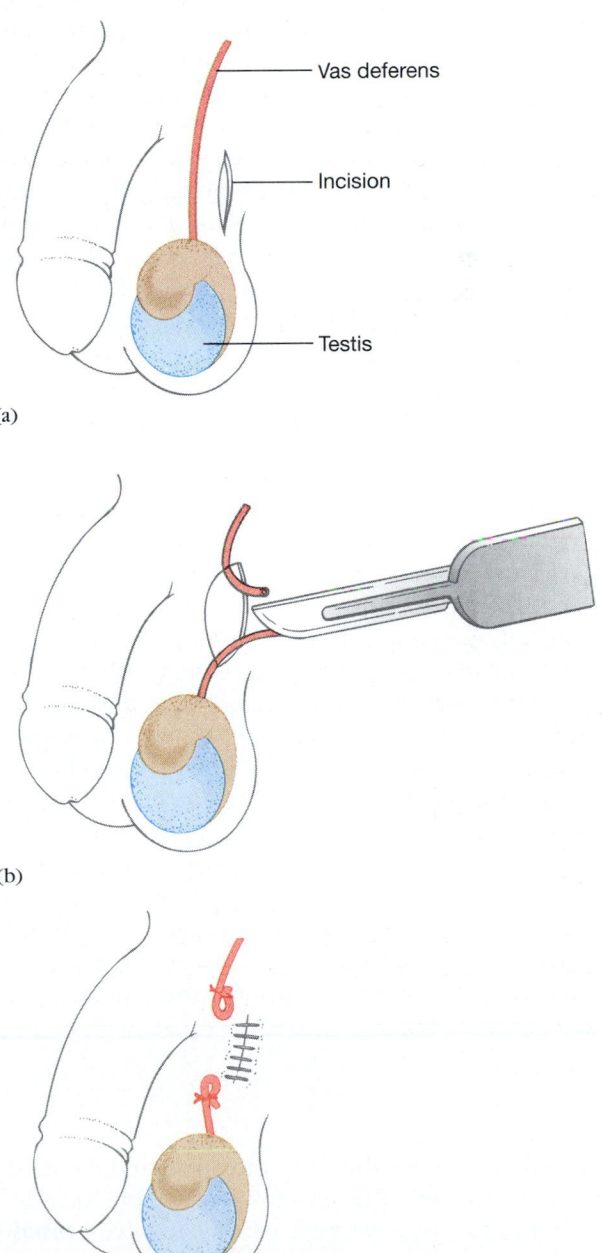

(a)

(b)

(c)

**FIGURE 36–14** Vasectomy, the surgical procedure in which the vas deferens is severed so that sperm can no longer enter the semen.

## ESSAY: *Sexually Transmitted Diseases*

Sexually transmitted diseases (STDs) are transmitted from one person to another primarily by contact between the genital organs during sexual activity. A dozen or more such diseases occur in humans, caused by pathogenic bacteria or viruses. The incidence of major STDs declined earlier in the century, but they are now at epidemic rates in the United States.

Historically, the most famous sexually transmitted disease was syphilis, caused by a spirochete bacterium. It is not clear whether this disease spread to Europe from the Americas with Christopher Columbus's crew or whether the spirochete was already in Europe but mutated to a more virulent form at about that time.

The first symptom of syphilis is a hard, painful ulcer called a chancre, which develops at the site of infection. The chancre soon disappears, to be followed some months later by the secondary stage of the disease: generalized infection of the skin and other organs. The disease then enters a latent period, which may last throughout life. However, in some people the disease enters a tertiary stage, involving damage to the nervous and circulatory systems, and sometimes death. Nervous degeneration caused by syphilis is believed to have been responsible for the strange behavior of England's King Henry VIII, and of mobster Al Capone, the

insanity of the composer Smetana, and the blindness of the composer Frederick Delius.

Syphilis is not very common because, even if it is not treated, it is not very contagious. Like herpes infections, syphilis may become **latent,** producing no symptoms. Syphilis cannot be transmitted to another person when it is in its latent stages, which is most of the time. Furthermore, the chance of catching syphilis from a single sexual encounter with someone who has an active case is only about 1 in 40. Syphilis can also be spread by means other than sexual contact, but only a very small percentage of syphilis cases begin this way.

Gonorrhea, one of the most widespread STDs, is somewhat more contagious than syphilis. It too is caused by a bacterium (Figure 36–B). In men the disease is readily detected because it produces a discharge of pus from the penis and a burning sensation during urination. Most infected men therefore seek treatment. The disease is widespread because it produces few or no symptoms in women, in whom the cervix is the area usually infected. In both sexes gonorrhea can cause sterility because the infection can produce scars that block the oviducts or the vasa deferentia.

Gonorrhea has usually been treated with penicillin. Predictably, penicillin-resistant gonorrhea bacteria

have evolved, as only those that happened to have mutations conferring resistance to penicillin survived and reproduced. There are now some strains of the disease that are incurable because they are resistant to all known antibiotics.

Infections caused by the bacterium *Chlamydia trachomatis* are also widespread: about 4 million cases occurred in the United States in 1987. These infections are so common because women usually experience mild or no symptoms. Hence, many cases go undetected and untreated and are unwittingly passed on to new sexual partners. In men, symptoms are much like those of gonorrhea and are sometimes mistaken for it, a problem compounded by the fact that both infections are often present. Unlike gonorrhea, *Chlamydia* is not susceptible to penicillin, and so *Chlamydia* infections cannot be cured until properly diagnosed. As with gonorrhea, infections can cause scarring of the reproductive tissues, especially in women with untreated infections. *Chlamydia* infections are believed to be the most common preventable cause of female infertility and of life-threatening tubal pregnancies in the United States today.

STDs caused by viruses cannot be cured by antibiotics. Herpes simplex virus Type 2 causes infectious genital blisters, much like the cold sores and fever blisters on the lips caused by a

produced but are resorbed into the body, and the fluid ejaculated contains only the secretions of various glands.

Sterilization of a woman usually involves **tubal ligation,** the cutting and tying off of the oviducts. This operation can also now be performed under local anesthesia. After tubal ligation the ovaries continue to function, but sperm cannot reach the eggs, and thus fertilization cannot occur. Sterilization that can be reversed without medical assistance

is under development and is undoubtedly the contraceptive technique of the future.

***Abortion***   Induced abortion is one of the oldest human birth control methods. Nowadays, the number of abortions is inversely related to the availability of family planning services. In countries where family planning services are cheap and available, abortion rates are lower than elsewhere.

related virus. The infection begins with burning, itching, or numbness, followed by flu-like symptoms. Between 10 and 20 days after infection, blisters develop on the infected area. The blisters rupture, leaving painful ulcers. These heal, and the virus becomes dormant but may reactivate at any time. Herpes is infectious only when it is active, but only a fraction of active cases produce noticeable blisters.

Herpes causes little permanent damage in adults. It is usually more painful for women than for men, and there is evidence that infected women are more likely than others to develop cancer of the cervix, making it especially important to obtain annual Pap smears and detect this cancer in its early stages.

The greatest danger posed by herpes lies in the high death rate (about one in three) among those newborn babies who catch the disease from their mothers during childbirth. (Most women with genital herpes, however, do not communicate the disease to their infants.)

Human papilloma virus (HPV) infections are the most common STD in the United States. Only a minority of cases produce obvious symptoms: genital warts. Several strains of the virus that do not usually produce warts are suspected to be contributing factors to cancers of the cervix and penis.

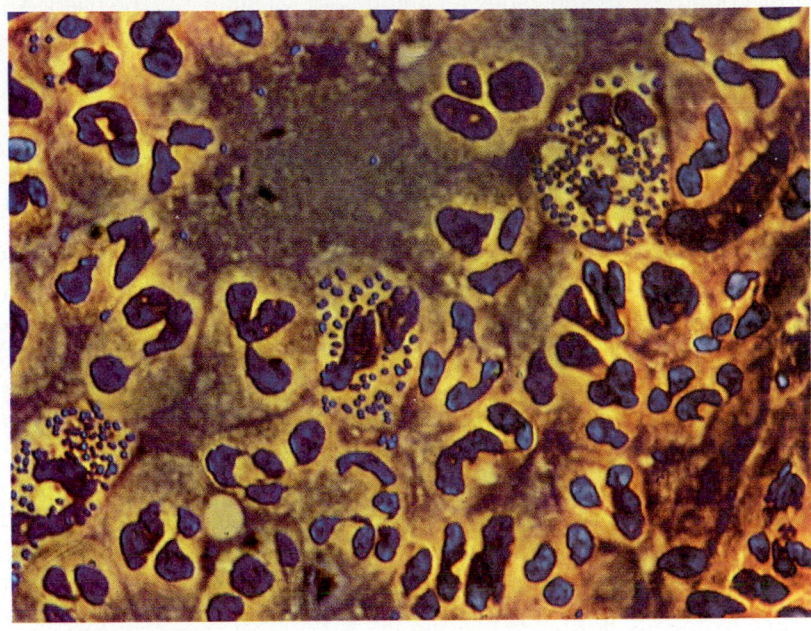

**FIGURE 36–B** Human cells infected with the bacteria that cause gonorrhea. Here, the stained bacteria appear as many tiny blue dots in the cytoplasm of three cells (lower left, center, and upper right). (Carolina Biological Supply Company, courtesy of Dr. Cecil Fox, Centers for Disease Control)

Acquired immune deficiency syndrome (AIDS) is an STD first recognized in the 1980s (Section 34–G). It infects new hosts most easily through ulcers caused by other STDs, and because it impairs the body's ability to fight other infections it promotes the spread of these companion STDs.

Various other sexually transmitted diseases are caused by viruses, bacteria, yeasts, and protists. It is extremely difficult to stop the spread of these diseases because victims often pass a disease on to others before they learn that they themselves are infected. The growing resistance to antibiotics among pathogens also makes treatment increasingly difficult.

Abortion was largely accepted, and fairly common, in the United States and Europe until the early nineteenth century. Before this time, a woman had up to a one-third chance of dying when she gave birth to a child, whereas abortions were less risky for her, and so abortions often saved women's lives. After midwives found that washing their hands and clothes reduced the spread of infection and improved their patients' survival rates, childbirth became less dangerous to women than nineteenth-century abortions, and doctors started to oppose abortion. Most religious and ethical opposition did not develop until some time later. The situation has changed since World War II, now that the danger of abortion to a woman is again less than the risk of a completed pregnancy. In 1988 about half the world's people lived in countries where legal abortions were freely available. However, in the United States about 83% of women do not have access to legal abortion because so few clinics perform the procedure.

There are several widely used methods of abortion. The French "morning-after" pill, RU-486, is available and popular in most countries. In the older techniques of vacuum curettage, and dilation and curettage ("D and C"), the fetus is sucked or scraped out of the uterus, usually under local anesthesia. ("D and C" is used for diagnosis and treatment of a number of uterine disorders as well as for aborting pregnancies.) These methods are used early in pregnancy. After the fifteenth week of pregnancy, saline injection can be used. An injection of salt solution kills the fetus, and the uterus subsequently expels the fetus and the placenta.

## SUMMARY

Most animals can reproduce sexually. This usually involves the coordination of the anatomy, physiology, and behavior of two parents, bringing small, motile sperm into position to fertilize the large, nonmotile egg.

1. Many aquatic species shed their gametes into the water, and both fertilization and development occur externally. Internal fertilization with external development permits fertilization to occur before the egg is surrounded with the impenetrable layers that will protect it as it develops outside the mother's body. Internal fertilization and development provide more protection of the developing embryo.

2. In vertebrates, males produce sperm in the testes, and females produce eggs in the ovaries. Sexual maturation, maturation of gametes, and the female reproductive cycle are controlled by hormones. In mammals, hormones secreted by the placenta control pregnancy, and hormones also control birth and lactation.

3. In mammals, the male's testes lie in the scrotum. Semen forms as glandular secretions join the sperm in the tubes leaving the testes. The penis is composed of spongy tissue that becomes erect by being engorged with blood during sexual stimulation; it can then be introduced into the vagina.

4. Sexual stimulation may eventually result in orgasm by both sexes and in ejaculation of semen by the male.

5. When a mature egg is released from the ovary, it travels through an oviduct, where it may be fertilized by a sperm.

6. A fertilized, developing egg descends into the uterus, where it implants in the uterine wall. Part of the embryo develops into the new individual, and part forms the embryonic part of the placenta and the extra-embryonic membranes surrounding the embryo.

7. The fetus cannot survive outside the uterus until about six months after fertilization. Birth occurs at about nine months and involves powerful uterine contractions that expel the fetus and placenta from the uterus through the vagina.

8. Humans commonly control their reproduction by techniques that either prevent fertilization or prevent a developing embryo from completing its growth in the uterus.

## Self-Quiz

Associate the reproductive organs on the right with the description on the left:

_____ 1. Tube for conducting sperm
_____ 2. Receptacle for penis
_____ 3. Production of seminal secretions
_____ 4. Conducts eggs
_____ 5. Holds baby in uterus
_____ 6. Produces sperm
_____ 7. Prepares nutritive lining for embryo

a. cervix
b. oviduct
c. Cowper's gland
d. ovary
e. prostate gland
f. testis
g. urethra
h. uterus
i. vagina
j. vas deferens

8. From the preceding list of reproductive organs and passages, construct a (correct) route for the passage of sperm from the site of production to the site of fertilization.

For each of the following birth control methods, choose the correct means of interference with reproduction:

_____ 9. Diaphragm and jelly
_____ 10. "The pill"
_____ 11. Vasectomy
_____ 12. IUD
_____ 13. Tubal ligation
_____ 14. Induced abortion
_____ 15. Condom

a. prevents fertilization of egg
b. prevents embryo implantation
c. prevents implanted embryo from developing
d. prevents ovulation
e. prevents sperm formation
f. prevents release of sperm into seminal fluid

16. In male mammals, luteinizing hormone stimulates:
    a. development of the seminiferous tubules
    b. production of testosterone
    c. release of FSH from the testes
    d. release of oxytocin by the posterior pituitary
17. In a typical human 28-day menstrual cycle, ovulation is most likely to occur:
    a. between days 1–5
    b. around day 14
    c. during the last five days of the cycle
    d. between days 5–10
18. Human chorionic gonadotropin is produced by the:
    a. corpus luteum    c. pituitary
    b. placenta    d. hypothalamus
19. The prostate gland:
    a. plays a major role in the production of sperm
    b. secretes FSH and LH in the male
    c. secretes testosterone
    d. secretes fluids in which sperm are transported
20. The level of progesterone is highest:
    a. at the time of ovulation
    b. at the time of menstruation
    c. during the growth of the follicle
    d. during the last half of the menstrual cycle

## Questions for Discussion

1. Does vasectomy affect male potency?
2. Does abortion affect a woman's subsequent fertility?
3. Can a woman become pregnant the first time she has sexual intercourse?
4. Why do you think the venereal disease gonorrhea is now epidemic in the United States and Western Europe?
5. Government regulations require that any hormonal contraceptive for males result in a sperm count of zero before it can be approved. In experiments with male animals, prospective hormonal contraceptives have produced complete infertility even though some sperm were present. Why does the "zero" requirement exist? Do you think it is justified?
6. In 1994 a researcher announced that he expects to be able to use eggs from aborted female fetuses to produce human zygotes by *in vitro* fertilization before 2000. Should this be permitted?

## Suggested Readings

Aral, S. O., and K. K. Holmes. "Sexually transmitted diseases in the AIDS era." *Scientific American,* February 1991.

Frisch, R. E. "Fatness and fertility." *Scientific American,* March 1988. The reasons women with very little body fat often have difficulty becoming pregnant.

Short, R. V. "Breast feeding." *Scientific American,* April 1984. A discussion of the effect of nursing on maternal and infant health and on population growth in various human cultures.

Silber, S. J. *How to Get Pregnant.* New York: Warner Books, 1980. A fertility specialist describes normal human reproduction and gives advice to infertile couples. Very readable and informative.

Ulmann, A., G. Teutsch, and D. Philibert. "RU486." *Scientific American,* June 1990. The story of the controversial French abortion pill.

Wassarman, P. M. "Fertilization in mammals." *Scientific American,* December 1988. An interesting account of this remarkable process.

Winikoff, B., and S. Wymelenberg. *The Contraceptive Handbook.* Yonkers, NY: Consumer Reports Books, 1992. A detailed guide to birth control methods available in the United States.

# Nervous Systems

## O B J E C T I V E S

*When you have studied this chapter, you should be able to:*

1. Define and use these terms:
   sensory and motor neuron, interneuron, effector
   depolarization, hyperpolarization, summation
   central nervous system, peripheral nervous system,
     autonomic nervous system
   brain, spinal cord, nerve, spinal nerve, cranial nerve
   cephalization, ganglion
2. Describe the basic structure and function of neurons.
3. Describe the resting potential of a neuron and explain how it is maintained.
4. Contrast the properties of local potentials (such as postsynaptic potentials) with those of action potentials.
5. Draw a graph of the potential changes that occur during an action potential and relate them to the flow of sodium and potassium ions through the axon's membrane. Explain how an action potential spreads down the length of an unmyelinated axon.
6. Describe the myelin sheath, and explain its effect on impulse conduction.

7. Draw a model chemical synapse with its principal components, explain the function of each component, and explain how information is transmitted to the postsynaptic cell and how the signal is stopped.
8. Explain the difference between an excitatory and an inhibitory synapse in terms of effect on the postsynaptic membrane.
9. Describe how the brain of a fish and the brain of a human differ in the relative sizes and functions of these structures: medulla, cerebellum, thalamus, hypothalamus, and cerebral hemispheres.
10. Draw a labeled diagram of a simple reflex arc, describe how it works, and explain the adaptive advantage of reflexes.
11. Outline the major structural and functional differences between the parasympathetic and sympathetic nervous systems.
12. Describe the vertebrate reticular formation, and explain the relationship of the reticular activating system to arousal.

---

Digestion, respiration, circulation, excretion, and reproduction may be studied separately, but in a living animal they work together, coordinated by the nervous and hormonal systems. We have already seen several examples of this coordination, such as the control of breathing and of body fluid volume.

The remaining chapters on animal biology deal with the nervous system and with the sense organs, muscles, and glands that work with it in coordinating an animal's activities. Various **receptors,** in the sense organs, detect **stimuli** -changes in the body's internal or external environ-

ment, such as blood chemistry, sound, or light. This information is converted into electrical impulses, the form in which **neurons**—nerve cells—carry information through the nervous system (Figure 37–1).

In an animal's nervous system, neurons are arranged to carry messages in pathways. **Sensory neurons** receive information from receptors in the sense organs and transmit it to **interneurons,** neurons that relay messages between neurons. Often a nervous pathway involves many interneurons, receiving and processing information from various parts of the body and sending out instructions for ap-

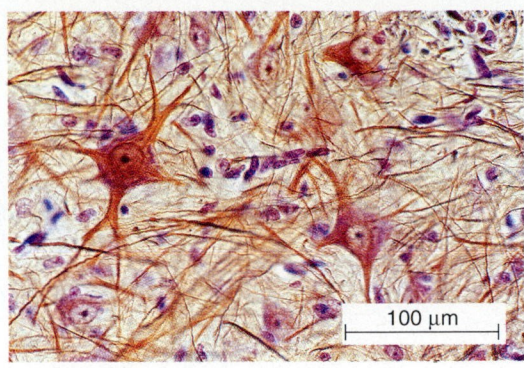

**FIGURE 37–1** Neurons. This section of nervous tissue shows (purple) neurons surrounded by other cells and extracellular fibers (LM). (Biophoto Associates)

propriate responses. **Motor neurons** convey these instructions to the body's **effectors,** the organs that carry out responses. The most common effectors are glands, which secrete hormones, digestive enzymes, and so on, and muscles, which contract and move parts of the body. The responses made by effectors do two main things: they maintain internal homeostasis, and they make appropriate behavioral responses to the external environment.

The vertebrate nervous system has two main divisions. The **central nervous system** consists of the brain and spinal cord, containing most of the body's interneurons and hence its information-processing systems. The **peripheral nervous system** consists of the nerves that carry sensory and motor information between the central nervous system and other parts of the body.

KEY CONCEPTS

◆ The nervous and hormonal systems coordinate the body's physiology and behavior. Nervous responses are generally rapid. Hormonal responses are slower and longer-lasting.

◆ Information progresses through a nervous system in three main steps:

1. Sensory reception, collecting information from stimuli outside and inside the body

2. Central processing of this information in the nervous system, initiating and relaying an appropriate response

3. Conveying instructions for the response to the body's effectors: muscles and glands

◆ Some sensory information and motor activities are stored as memories, changes in the nervous system based on past experience and available for future reference.

## 37–A CELLS OF THE NERVOUS SYSTEM

### Neurons

Neurons are the cells that transmit messages in the nervous system. Every neuron has a **cell body,** or **soma,** containing the nucleus and most of the cell's organelles. A neuron also has long, thin extensions, or processes, of two types. The neuron's many branching **dendrites** receive information from other cells or from the environment. A typical neuron also has one **axon,** the process that carries information to other cells (Figure 37–2). The end of an axon divides into many terminals, which make connections to other neurons (or, if it is a motor neuron, to muscle or gland cells).

A dendrite's diameter varies along its length, but an axon has a uniform thickness. Diameter determines electrical resistance, and so this difference in shape gives dendrites and axons different functions: the strength and speed of electrical signals in dendrites vary, whereas an axon conveys signals of uniform strength and speed.

A neuron may make connections to many others and so may receive information from, or send information to, many parts of the body. In the vertebrate brain, the dendrites and soma of a single cell may receive thousands of connections.

Some neurons in the central nervous system have no axons, and the dendrites of neighboring cells exchange information. Most neurons have short axons and communicate with near neighbors. However, some neurons have much longer axons, such as the axons of some sensory and motor neurons extending from the spinal cord to the tips of the toes. The longest individual cells in any animal are some of its neurons.

### Glial Cells

In addition to neurons, the nervous system contains **neuroglia** or **glial cells** (glia = glue). The human nervous system contains about ten times more glial cells than neurons. Glial cells support neurons and convey food and other molecules to them. One type, the **microglia,** act as scavengers of dying neurons and other debris in the central nervous system.

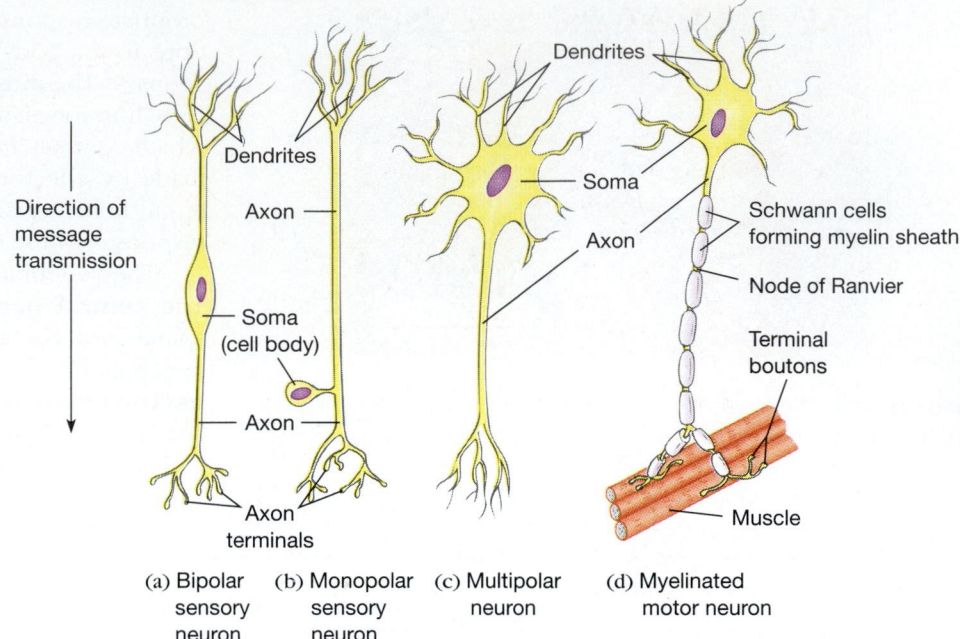

**FIGURE 37–2** Neuron structure. Neurons transmit nerve impulses from dendrites and cell body (top) to axon terminals (bottom). (a) A bipolar neuron has two main branches, on opposite sides of the cell body. (b) A monopolar sensory neuron, such as might carry impulses from the skin to the central nervous system. (c) A multipolar neuron has a more or less central cell body, many dendrites, and a branched axon. (d) A motor neuron to a vertebrate muscle. This one has an axon surrounded by fatty myelin insulating material (see Figure 37–8).

During embryonic development of the brain, star-shaped **astrocytes** serve as a scaffold for migration of neurons and for axon growth. Later in life, these glial cells also produce nerve growth factor and extracellular fibers, which help maintain neurons and promote growth of their processes. Astrocytes also induce cells in the walls of the brain's capillaries to form tight junctions, which prevent small, water-soluble molecules from seeping out of the capillaries into the brain. This **blood-brain barrier** helps maintain the constant composition of the **cerebrospinal fluid** (the extracellular fluid of the brain and spinal cord).

Other glial cells wrap layers of their plasma membranes around the axons of some neurons, forming a fatty **myelin sheath** that functions like insulation around an electric wire (see Figures 37–2 and 37–8). Myelin-forming cells are the **oligodendroglia** in the central nervous system and **Schwann cells** in the peripheral nervous system.

## 37–B   OVERVIEW: HOW NEURONS WORK

A characteristic of neurons is **electrical excitability,** the ability to generate and transmit an electrical impulse. A neuron can do this because ions are distributed unequally on the two sides (inside and outside) of its plasma membrane. These ions are poised to spurt through the membrane, down their concentration gradients, when the membrane's permeability changes briefly and allows them to pass. The result is an electrical **nerve impulse,** the form in which information travels down the neuron's axon.

Information is transmitted from one neuron to the next in three steps (Table 37–1):

1. The neuron's dendrites and cell body receive and "add up" incoming information.
2. If the neuron receives enough input, its axon fires a nerve impulse, a self-propagating electrical signal that travels to the axon terminals.
3. The axon terminals release chemicals that carry the signal to the next neuron(s) in the nerve pathway. These chemicals are the incoming information for the next neuron(s), which may in turn go through these three steps.

Nerve impulses can also be generated by applying electrical impulses to neurons, a common experimental technique.

## TABLE 37–1

**Neuron Structure and Function**

| | Structure | Function |
|---|---|---|
| | Dendrites and soma | Receiving and adding up information |
| | Axon | Transmitting electrical nerve impulse |
| | Axon terminals | Releasing transmitter chemicals to next cell(s) |

## 37–C ELECTRICAL PROPERTIES OF NEURONS

### Resting Potential

The asymmetric distribution of ions across a neuron's plasma membrane results in an **electrical potential difference,** or **membrane potential,** across the membrane (Section 4–E). A. L. Hodgkin and A. F. Huxley received a Nobel Prize for showing the role of the membrane potential in neuron function. They worked with the giant axons of squids, whose large diameter makes them easier to study than the thin axons of vertebrates (Table 37–2).

The membrane potential depends on two factors: active transport and the membrane's permeability to ions. Active transport is carried out by the membrane's **sodium-potassium pumps,** which use energy from ATP to pump sodium ions ($Na^+$) out of the cell and potassium ions ($K^+$) in.

The magnitude of the membrane potential depends on the membrane's permeability to ions. The lipid layers are largely impermeable to ions, but some membrane proteins form channels that let ions through. There are both $Na^+$ and $K^+$ channels that open and close, changing the membrane's permeability to ions. In a neuron at rest, these channels are usually closed, but there are also potassium leak channels, which permit $K^+$ ions to escape and a much smaller amount of $Na^+$ ions to enter (Figure 37–3). As a result, a net positive charge builds up outside the membrane. Because like charges repel one another, this slows the exodus of $K^+$ until it equals the $Na^+$ influx.

In a typical neuron the balance point, the membrane's **resting potential,** is about −70 millivolts, the inside of the neuron being negatively charged with respect to the outside. This difference in electrical charges on either side means that the membrane is **polarized.**

Chloride ions ($Cl^-$) distribute themselves passively through other channels in response to the membrane potential. They are more concentrated outside than inside the cell, attracted by the net positive charge outside and repelled by the net negative charge of proteins and other organic molecules confined inside. The distribution of $Cl^-$ ions decreases the membrane potential but does not cancel it entirely.

The resting potential has two sources of electrical potential energy: the stockpiles of $Na^+$ outside the cell and of $K^+$ inside. This potential energy can be converted into the energy of a moving electric current if ion channels open and permit some of the ions to flow back through the membrane, down their concentration gradients (Section 6–A). This is how a neuron works.

### Receiving Information: Local Potentials

When a neuron's cell body or dendrites receive a stimulus, some nearby ion channels open briefly and ions spurt through, causing a **local potential:** a disturbance in the

TABLE 37–2

### Concentrations of Ions in the Giant Axon of a Squid and in the Surrounding Fluid

| Ion | Concentrations (in millimoles) | | |
|---|---|---|---|
| | Sea Water | Squid Blood | Squid Axon |
| $K^+$ | 10 | 20 | 400 |
| $Na^+$ | 450 | 450 | 50 |
| $Cl^-$ | 550 | 550 | 100 |
| $Ca^{2+}$ | 10 | 10 | 0.5 |
| $Mg^{2+}$ | 55 | 55 | 10 |
| Organic anions | — | — | 400 |

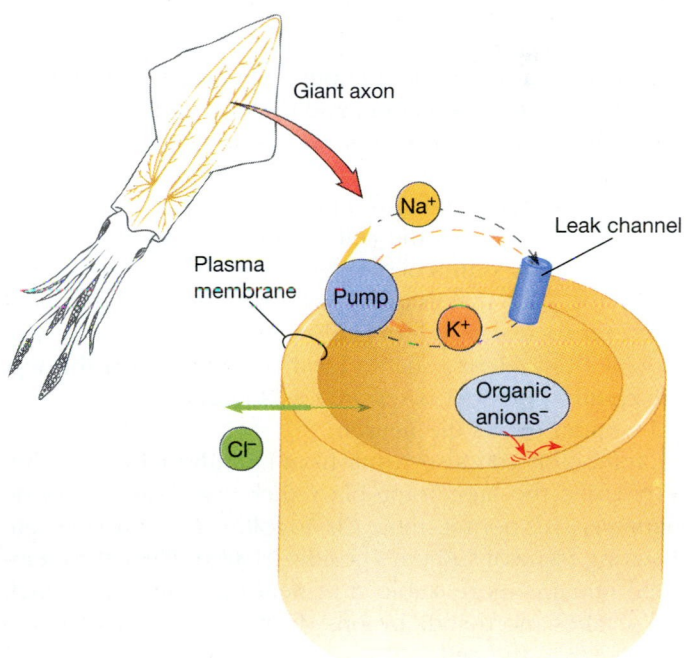

**FIGURE 37–3** Source of the resting potential. Ion concentrations in Table 37–2 were measured by analyzing squid blood and the cytoplasm squeezed out of the giant axons (color), which are up to 1 mm thick. Each ion is named inside or outside of the axonal membrane, depending on where its concentration is higher. The leak channel lets much more potassium out than sodium in.

resting potential in a small area of the membrane. A local potential is proportional to the intensity of the stimulus. If sodium channels open, $Na^+$ entering the neuron causes **depolarization:** a decreased difference in electrical charge across the membrane, making the membrane potential less negative. Depolarization is **excitatory,** making the neuron

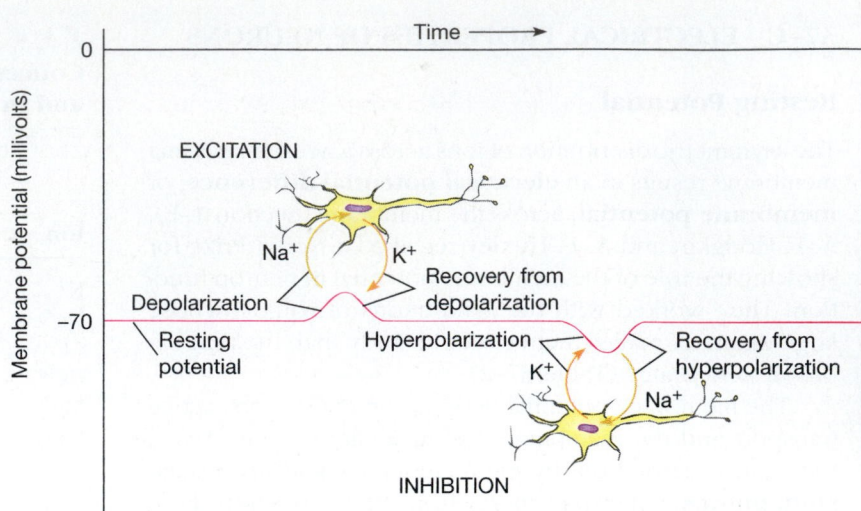

**FIGURE 37–4** Local potentials. The red graph line shows membrane potential changes during weak depolarization (excitation) and hyperpolarization (inhibition). The cell diagrams show the opposite flows of Na⁺ and K⁺ through the membrane as the resting potential is disturbed and then restored.

more likely to fire a nerve impulse (Figure 37–4). But the sodium channels soon close, and the flow of K⁺ in the opposite direction increases briefly, restoring the resting potential.

In contrast, if a stimulus initially opens potassium channels, more K⁺ leaves the cell, producing **hyperpolarization:** a greater difference in electrical charge across the membrane. This makes the membrane potential more negative and so makes it less likely that a nerve impulse will be generated. Hence such a local potential is **inhibitory.** The resting potential is restored as sodium channels open and sodium enters the neuron.

Inhibition can also result from stimuli that do not hyperpolarize the membrane but rather open more chloride channels. This permits more Cl⁻ to follow Na⁺ into the cell, offsetting some of sodium's positive charge. Thus, the membrane remains more stable at or near the resting potential, and it takes more sodium ions (from a stronger stimulus) to depolarize the cell.

Some channels are **electrically gated,** opening in response to changes in membrane potential, whereas **chemically gated** channels open when they bind certain chemicals. A change in potential at one point on the membrane causes electrically gated channels nearby to open. Here, in turn, the membrane potential changes as ions flow through, and so on.

In this way a local potential spreads along the membrane from the point of stimulation, much as ripples spread when you throw a stone into a pond. However, a local potential dissipates within a short distance, and its information is lost—unless it is reinforced by other local potentials.

Local potentials in a neuron's dendrites or cell body undergo **summation** (adding together) in space or time. **Spatial summation** occurs when local potential changes overlap in space on the membrane surface. **Temporal summation** occurs when there are two or more stimula-

tions very close in time in the same area of membrane, so that the later stimulations add their effect before the first local potential has completely dissipated (Figure 37–5). If inhibitory local potentials are occurring, it takes more excitatory input to counteract them.

## Transmitting Information: Action Potentials

Summation of excitatory local potentials in a neuron's dendrites and cell body may cause the axon to fire a nerve impulse. Each axon has a **threshold,** a level of depolarization that must be exceeded before it will fire. Once this threshold is reached, the resulting depolarization is **self-propagating:** it causes an above-threshold depolarization in nearby parts of the membrane, and so on, all the way to the axon terminals.

This self-propagating wave of depolarization down an axon is an **action potential:** a nerve impulse. Because a neuron fires an action potential maximally or not at all, an action potential is an "all-or-none" response (Table 37–3).

In an action potential, a threshold depolarization opens electrically gated sodium channels, and Na⁺ rushes into the axon. This depolarizes the membrane further, which opens even more sodium channels, until the inside is positive relative to the outside. So the inrush of Na⁺ is a self-reinforcing process:

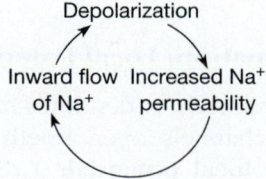

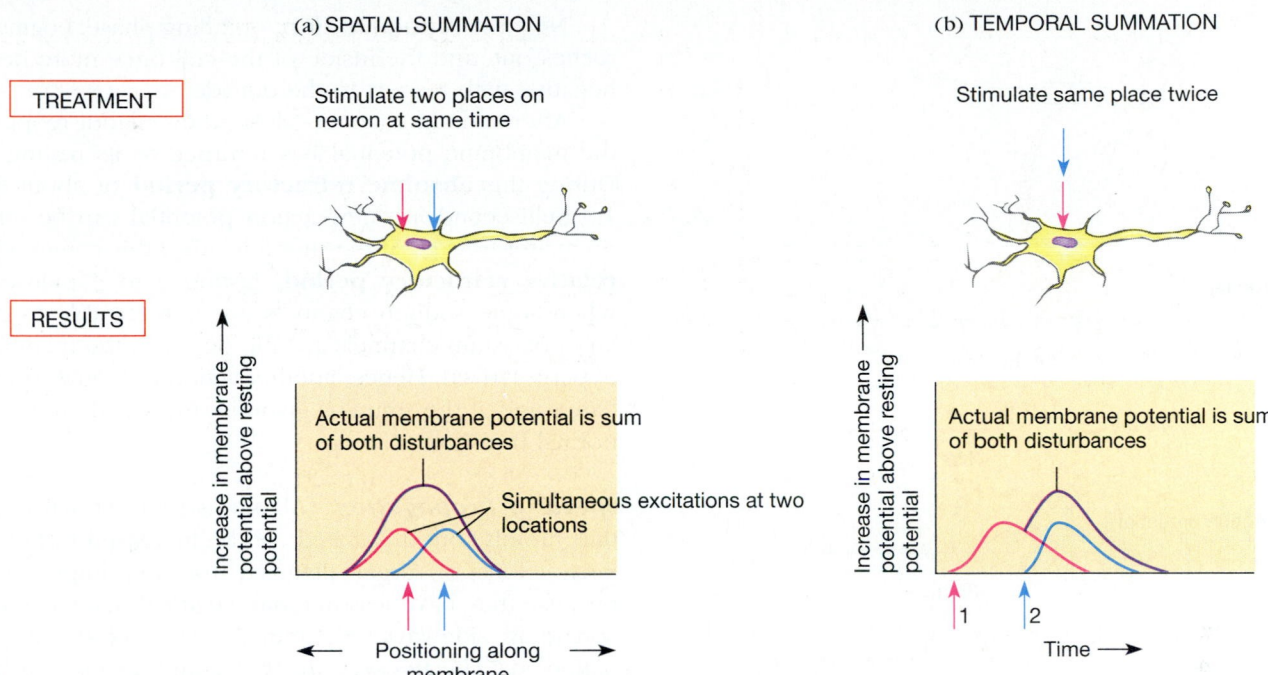

**FIGURE 37–5** Summation of separate excitatory impulses. (a) Spatial summation of two simultaneous depolarizations at nearby points on the membrane (colored arrows). (b) Temporal summation of two successive stimuli to the same point. Pulse 1 has not died out before pulse 2 adds to it.

## T A B L E   3 7 – 3

### Comparison of Local Potentials and Action Potentials

| Local Potential | Action Potential |
|---|---|
| Graded response | All-or-none response |
| Evoked by stimuli of any intensity | Evoked only by stimuli of at least threshold intensity |
| Amplitude proportional to stimulus intensity | Amplitude of same intensity for a given neuron, regardless of stimulus intensity |
| Decays rapidly as it spreads | Travels entire length of axon with no change in amplitude |
| Lasts for duration of stimulus | Lasts until it reaches the axon terminals (same time, depending on individual neuron) |
| May sum with nearby potentials | No summation |
| No refractory period; undergoes temporal summation with any previous responses still active | Refractory period during which no new action potential can be initiated |

An action potential is self-propagating because the flow of $Na^+$ into the axon and sideways to still-resting areas is great enough to cause threshold depolarization in these areas too: a wave of depolarization sweeps along the axon (Figure 37–6).

After the action potential has passed, the sodium channels close, returning $Na^+$ permeability to the resting level. Meanwhile, potassium channels have been opening, but more slowly, and the increased outflow of $K^+$ restores the membrane's resting potential. Then the potassium channels close and return the $K^+$ permeability to its resting value.

Only a negligible fraction of all the ions present has crossed the membrane, and the resting potential returns to normal even before the sodium pump transports them back. In fact, a neuron may respond to thousands of stimulations even when the sodium pump is stopped by chemical inhibitors. However, the pump is necessary to maintain the ion distribution in the long run.

Like a "yes" or "no" answer in a game of "Twenty Questions," a single all-or-nothing action potential provides little information. A neuron conveys information by its frequency of action potentials. It has an intrinsic rate of firing even when it is receiving no particular input. Excitation increases its firing rate, whereas inhibition decreases it. This permits a neuron firing all-or-nothing action potentials to send a larger amount of information.

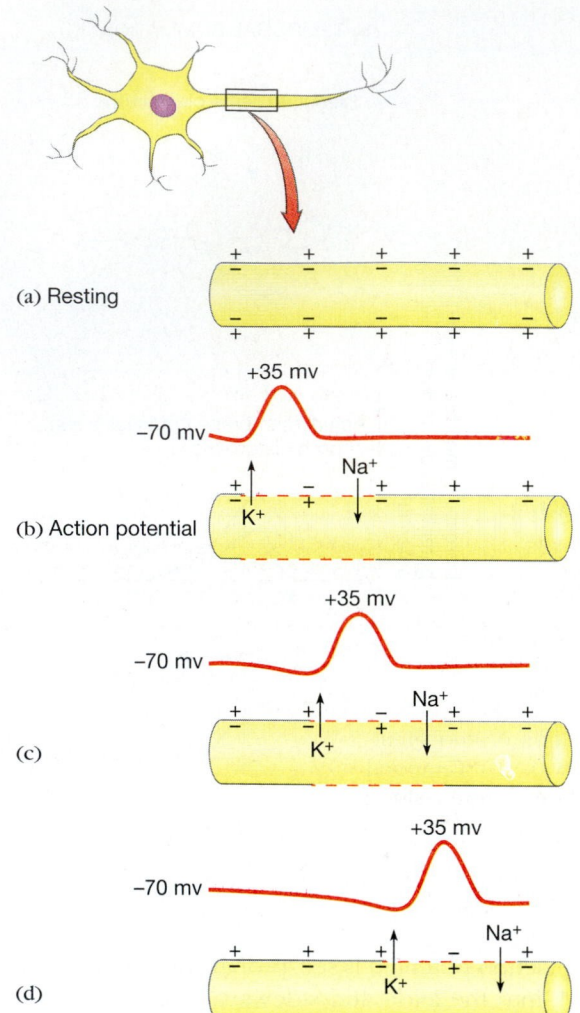

(a) Resting

+35 mv

−70 mv

Na+

K+

(b) Action potential

+35 mv

−70 mv

Na+

K+

(c)

+35 mv

−70 mv

Na+

K+

(d)

**FIGURE 37–6** Spread of an action potential along an axon. (a) Resting state, with the inside of the membrane negative with respect to the outside. (b, c, d) Successive stages in conduction of the impulse along the axon from left to right.

***Recording an Action Potential*** An action potential can be recorded by placing one electrode (electric current conductor) inside the axon membrane and another outside (Figure 37–7). The electrodes are connected to an oscilloscope, an instrument that records the difference in electrical charges at the two electrodes and thus measures the electrical potential difference across the membrane.

In a resting neuron, the inside electrode reads about 70 millivolts more negative than the outside electrode, for a baseline membrane potential of −70 millivolts. During the rising phase of the action potential, the inrush of Na+ depolarizes the membrane. The potential difference across the membrane decreases to zero and reverses briefly, peaking as the inside becomes about 35 millivolts positive with respect to the outside.

Now the **repolarization,** or falling phase, begins as K+ rushes out, and the inside of the cell once more becomes negative with respect to the outside.

After sodium channels close, they cannot reopen until the membrane potential has returned to its resting value. During this **absolute refractory period** of about 0.25 to 1.0 millisecond, no new action potential can be initiated, no matter how great the stimulus. After this comes a longer **relative refractory period,** lasting 2 to 3 milliseconds, when some sodium channels are not yet able to reopen and potassium channels are still open, so the membrane is hyperpolarized. Hence, another action potential can be initiated only if the stimulus given is stronger than the axon's normal threshold stimulus.

***Speed of Propagation*** All axons transmit action potentials rapidly, but this speed can be increased in two ways. First, axons of larger diameter transmit impulses faster because they have less internal electrical resistance to Na+ spreading sideways. All animals have axons of various diameters. The thickest are the "giant" axons involved in escape reactions in invertebrates (such as the squid axons studied by Hodgkin and Huxley). When you swat at a cockroach and miss, you are foiled by such giant fibers. Sensory neurons in the cerci (the roach's twin "tailpipes") detect the air currents from your downstroke and alert giant interneurons that run from the base of the cerci to the brain (see Figure 37–13). Side branches carry the signal to the legs along the way, stimulating the first leap.

The second adaptation, found mainly among vertebrates, is a myelin sheath around the axon. Each section of the sheath is formed by a single glial cell, which produces a great excess of plasma membrane that wraps many times around the axon (Figure 37–8). These membrane layers have an especially high lipid content, making them much like insulation around an electrical wire. The naked axon is exposed between the glial cells in short gaps called **nodes of Ranvier.**

In axons without myelin sheaths, the speed of impulse conduction is limited by how fast the membrane's ion channels can open and close and by how quickly ions flowing to neighboring areas cause the membrane potential to rise there. Without a myelin sheath, flow of K+ ions out through the membrane slows this rise in potential.

In myelinated axons, the tight, fatty sheath prevents ions from flowing through the membrane. Thus the membrane potential from that point to the next node of Ranvier rises nearly instantaneously. Therefore the electrical impulse appears to jump from one node of Ranvier to the next as a much faster electric current, passing through the fluid inside the axon and outside the myelin sheath. (In this situation, the ion-rich axon is analogous to a pipe tightly packed with marbles: if you force a marble [ion] in one end, a marble at the far end pops out almost instantly.) When the current reaches a node of Ranvier, the membrane potential

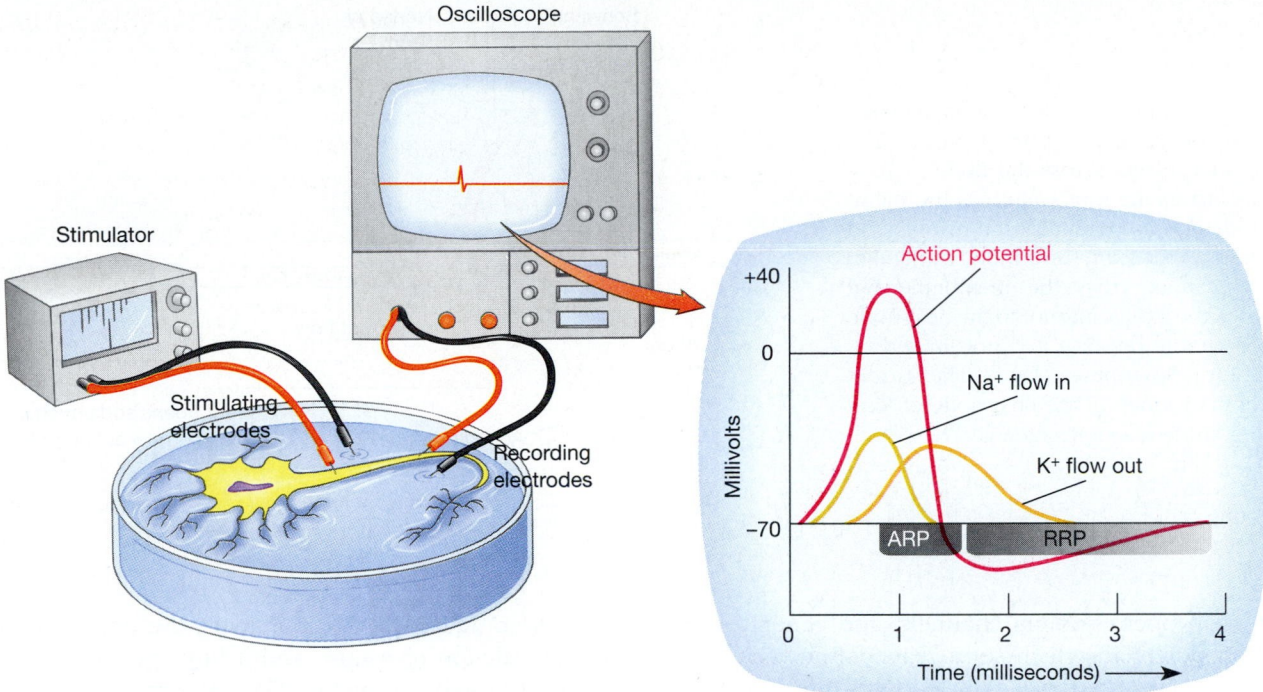

**FIGURE 37–7** Recording an action potential. Stimulating electrodes (one inside and one outside the membrane) deliver an electric shock to the axon of a neuron. This generates an action potential, which is detected as it sweeps past the recording electrodes. An oscilloscope displays the difference in electrical potentials at the two recording electrodes over time (red curve in the graph). Superimposed on the graph are the flows of sodium and potassium. During the absolute refractory period (ARP), no new impulse can be generated. During the relative refractory period (RRP), a stronger-than-threshold stimulus can start another action potential.

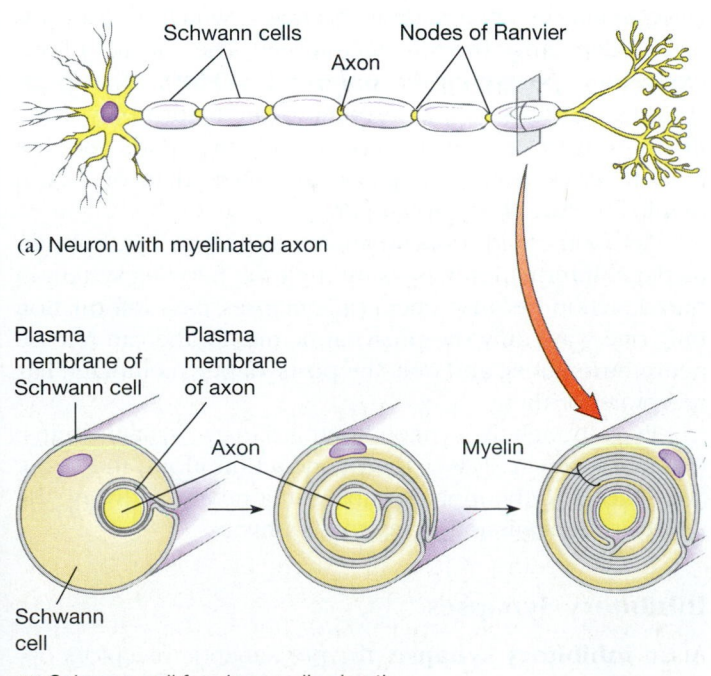

(a) Neuron with myelinated axon

(b) Schwann cell forming myelin sheath

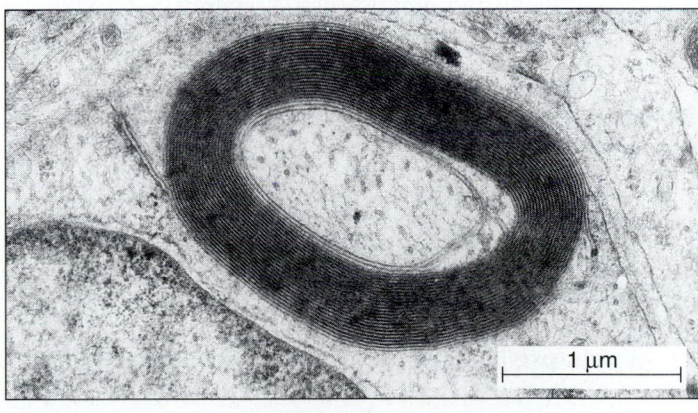

(c)

**FIGURE 37–8** The myelin sheath. (a) In a motor neuron, each section of the sheath is formed from one Schwann cell. The naked axon is exposed only at the nodes of Ranvier. (b) A Schwann cell forms a myelin sheath by wrapping several layers of its plasma membrane around an axon. (c) Cross section through a myelin sheath (TEM). (c, Biophoto Associates)

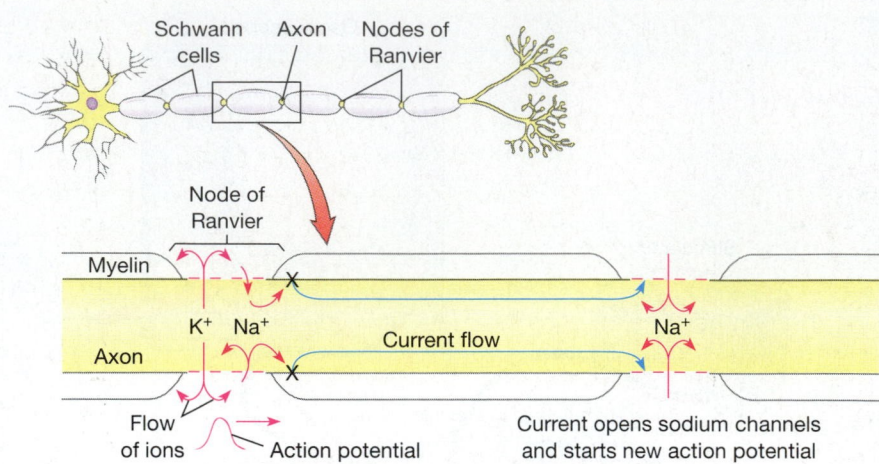

**FIGURE 37–9** Conduction along a myelinated axon. Red and blue arrows = flow of current. The action potential travels across the node of Ranvier only as fast as the ions can cross the membrane and change its potential (red arrows and curve). In parts of the axon surrounded by myelin, the current cannot cross the membrane but spreads down the axon's interior to the next node much more rapidly because it is not limited by the speed of ion flow (blue arrows). The current opens sodium channels at the next node of Ranvier, starting another action potential.

builds quickly, opens sodium channels, and generates another action potential, which soon reaches the other side of the node, where current again leaps to the next node (Figure 37–9).

This rapid conduction of action potentials is **saltatory** ("jumping") **conduction.** Although some unmyelinated giant axons in squids conduct action potentials at speeds of nearly 20 m per second, myelinated mammalian axons of much smaller diameter may transmit impulses at up to 100 m per second. Saltatory conduction saves time, space, and energy.

Bundles, or tracts, of myelinated axons occur in various parts of the nervous system. This **white matter** owes its color to the lipid in the myelin sheaths. **Gray matter** consists of unmyelinated nervous tissue, mainly cell bodies.

## 37–D   SYNAPTIC TRANSMISSION

When an action potential reaches the axon terminals, the message is passed to other cells. Having many axon terminals allows a neuron to pass information to several other cells and to make more connections with some cells than with others, thereby varying the intensity of the signal each receives.

A **synapse** is a junction where the **presynaptic membrane** of an axon terminal lies close to the **postsynaptic membrane** of the next cell in line, either another neuron or part of a muscle or gland. Between the two membranes lies the **synaptic cleft,** only about 20 nm wide.

Most synapses are chemical synapses, where the signal is carried between cells by chemical messengers (Figure 37–10). The ends of axon terminals form enlarged **synaptic knobs** or **boutons** containing membrane-bounded vesicles (sacs) of **neurotransmitter** molecules (or simply **transmitters**).

As an action potential reaches the end of an axon, it opens calcium channels, and a tiny spurt of calcium ions enters the axon terminal. The calcium acts as a "second messenger," causing some neurotransmitter vesicles to fuse with the plasma membrane and discharge their contents into the synaptic cleft.

Transmitter molecules cross the cleft and bind to receptor molecules in the postsynaptic membrane. Many postsynaptic receptors are associated with chemically gated ion channels, which open when neurotransmitters bind to the receptor. Ions flowing through the channel cause a local potential in the postsynaptic membrane.

Chemical synapses may be excitatory or inhibitory, depending on the postsynaptic receptors. Receptors at excitatory synapses open sodium channels, which let $Na^+$ enter and depolarize the postsynaptic cell. The result is a local **excitatory postsynaptic potential (EPSP).** The postsynaptic cell fires an action potential in its turn only if the EPSPs sum to exceed its own axon's threshold. The cell must receive excitatory signals from more than one axon terminal before this can happen.

An axon could transmit an action potential in both directions. Information passes through the nervous system in one direction because chemical synapses pass information only one way: only the presynaptic membrane can release neurotransmitters, and only the postsynaptic membrane has receptors for them.

Even though the synaptic cleft is narrow, synaptic transmission is much slower than conduction along the axon. So, in general, the more synapses in a neural pathway, the slower is transmission along that pathway.

### Inhibitory Synapses

At an **inhibitory synapse,** the postsynaptic receptors operate different kinds of ion channels when they bind trans-

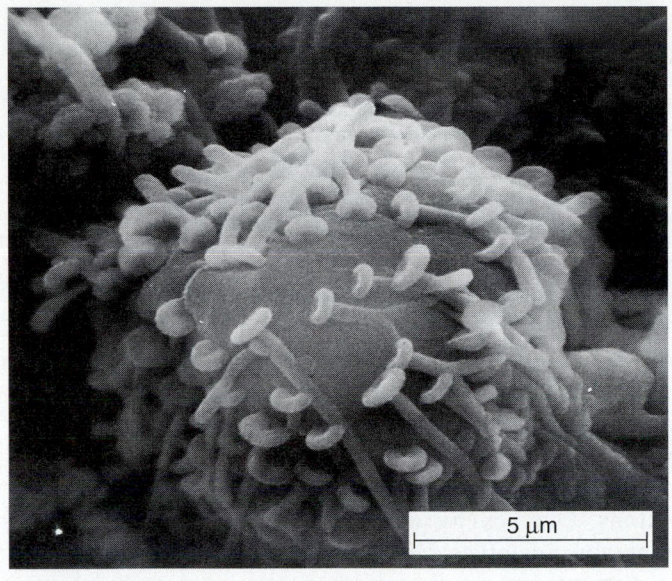

(a)

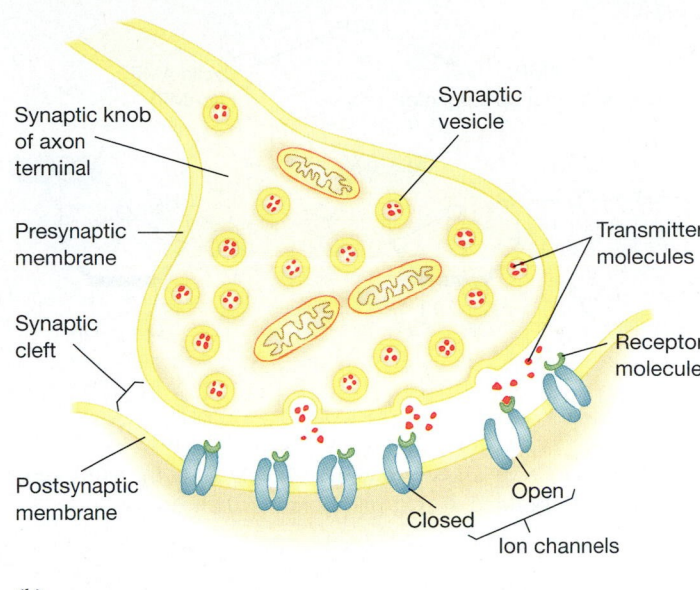

(b)

**FIGURE 37–10**   Synaptic transmission. (a) A neuron cell body receives messages from many axon terminals. (b) Closeup of one axon terminal and the postsynaptic membrane. When an action potential arrives, synaptic vesicles in the axon terminal fuse with the presynaptic membrane. Neurotransmitter molecules discharged into the synaptic cleft attach to receptor molecules in the postsynaptic membrane and cause opening of ion channels, increasing the flow of certain ions through the membrane.   (a, E. R. Lewis )

mitter molecules. Some open potassium channels (causing hyperpolarization of the postsynaptic membrane), whereas others open chloride channels, counteracting any excitatory sodium flow (Section 37–C). Either event results in a local **inhibitory postsynaptic potential (IPSP).** When inhibitory synapses are active, a greater excitatory postsynaptic potential is needed to produce a threshold depolarization and fire an action potential.

Inhibitory synapses are widespread in the brain. They serve to brake activity in the nervous system. Also, inhibitory neurons tune the activity of excitatory networks. For example, inhibitory synapses in sense organs suppress some incoming information and so sharpen the ability to pick out important features of stimuli. Thus, the eye is sensitive to motion largely because inhibitory synapses suppress information from receptors that are not detecting a change, making signals from receptors that do see a change relatively stronger.

## Electrical Synapses

Some neurons communicate by electric currents rather than by transmitter molecules. In an **electrical synapse,** the two cells are connected by gap junctions, and electric current flows from one cell to the other by way of these cytoplasm-to-cytoplasm connections (Section 4–H). In contrast to chemical synapses, electrical synapses work rapidly and in both directions. Electrical synapses occur between some neurons involved in the sense of equilibrium (balance).

Some synapses are mixed, containing both fast-acting gap junctions and slower chemical transmission, which prolongs the message. An example occurs at synapses between sensory and motor neurons involved in the escape behavior of goldfish.

If electrical synapses are so efficient, why do animals have chemical synapses at all? The reason seems to be that chemical synapses permit more complex and varied information processing. Electrical synapses can only be excitatory, but chemical synapses can be either excitatory or inhibitory. By activating different combinations of the excitatory and inhibitory input coming to a particular neuron, it is possible to block, enhance, shorten, or prolong the neuron's response to a signal. This makes it possible for the nervous system as a whole to adjust the intensity of the animal's response, or to make a different response, depending on circumstances.

## Changing Neuron Sensitivity

Some synaptic receptors do not operate ion channels directly. Instead they act by way of a series of intermediaries. Often the first of these are G proteins, so called because they bind the nucleotide GTP (Section 40–C). The G pro-

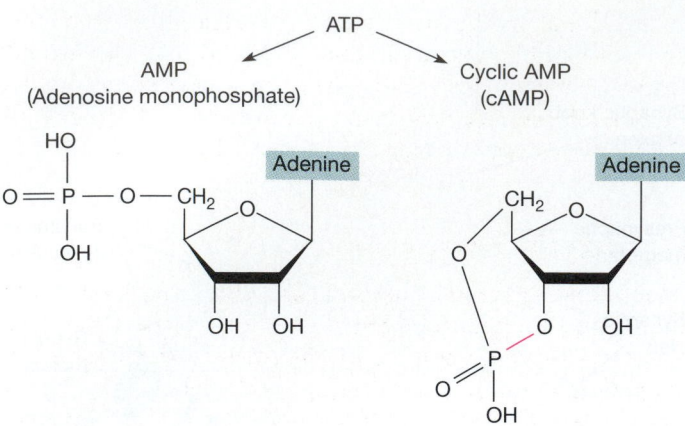

**FIGURE 37–11**    AMP and cyclic AMP. In cyclic AMP the phosphate and sugar are bonded at two places rather than one.

teins in turn regulate either ion channels or the enzyme adenylyl cyclase, which converts ATP to cyclic adenosine monophosphate—**cyclic AMP, or cAMP** for short (Figure 37–11). Cyclic AMP, in turn, acts as a second messenger inside the cell, regulating ion channels or else regulating enzymes that in turn operate on ion channels. The end result is a change in the membrane's permeability.

Not surprisingly, this roundabout series of steps produces a relatively slow response to the binding of a transmitter to a receptor. This system is not part of the transmission of messages to other cells. Instead, it changes the cell's own resting membrane potential and therefore its excitability, making it either more or less likely to respond to subsequent stimuli. This mechanism changes the nervous system's sensitivity to the same stimulus over time.

Once cyclic AMP has activated its target enzymes, it is broken down by another enzyme. Caffeine in coffee and tea, and theophylline in tea, are believed to act as stimulants by inhibiting the enzyme that degrades cyclic AMP.

Cyclic GMP is another second messenger in some neurons.

## 37–E    NEUROTRANSMITTERS

About 50 neurotransmitters have been identified. Each occurs in specific sets of neurons, and each neuron typically makes and releases one kind of transmitter. However, some neurons release neuropeptides (#5 in the following list) in addition to another transmitter. Some transmitters can be either excitatory or inhibitory, depending on which receptors occur at the synapse where they are released. In this way, the firing of a single neuron can excite some postsynaptic cells and inhibit others. Other transmitters appear to occur only at excitatory or at inhibitory synapses.

Some important transmitters are:

1. **Acetylcholine,** which occurs at many synapses in the brain and in other parts of the nervous system and also at **neuromuscular junctions,** synapses between neurons and skeletal muscles (muscles that move parts of the skeleton and are under our conscious control) (Figure 37–12).

2. **Norepinephrine,** important in the body's response to stress (Section 37–K).

3. **Dopamine,** the transmitter for a small group of neurons concerned with muscular activity. Parkinson's disease, with its bursts of uncontrollable muscular movement, is sometimes caused by lack of this transmitter and is often treated with l-dopa, the substrate of the enzyme that makes dopamine. Dopamine also occurs in areas of the brain involved with pleasure-seeking behaviors, motivation, and emotions. Overactivity of dopamine is linked to schizophrenia, but scientists don't yet know why.

4. **Serotonin,** produced by a group of cells in the medulla, the "stem" of the brain (see Figure 37–16). Serotonin seems to be concerned with functions such as sensory perception, regulation of body temperature, sleep,

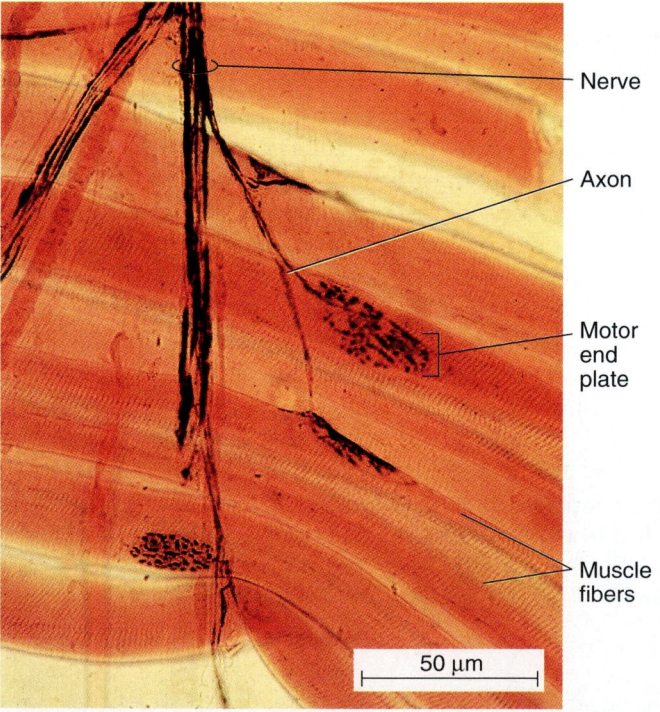

**FIGURE 37–12**    Neuromuscular junctions. These sites are used to study the action of the transmitter acetylcholine. A nerve is composed of the axons of many neurons. Near the muscle, the nerve divides into individual axons. Each axon may divide into several branches, each of which ends in a cluster of axon terminals called a motor end plate.    (Biophoto Associates)

consciousness, and emotions. Deficits of serotonin are linked to depression, anxiety, and aggression; excess can cause nausea.

5. **Neuropeptides,** the largest group of neurotransmitters, in both variety and molecular size: each contains 2 to 39 amino acids. Some neuropeptides are identical to local chemical messengers in other body areas (Section 40–E) or to hormones released from the pituitary gland, hypothalamus (see Figure 37–16), or gut. Indeed, they tend to diffuse out of synapses and act on any neuron in the local area. Their roles in the nervous system are poorly understood.

The larger **endorphins** and smaller **enkephalins** have been called the brain's "natural morphine." These neuropeptides were discovered after researchers found that morphine binds strongly to certain brain receptors. Reasoning that these receptors had not evolved as specific binders of morphine, which is not normally found in the body, the researchers looked for some naturally occurring brain transmitter at these synapses. Enkephalins consist of five amino acid residues each. The amino acid at one end has binding properties similar to those of morphine and presumably has the same effect on the neurons involved. This explains why the brain is so sensitive to morphine, opium, and related drugs.

Enkephalins occur in neurons that process information relating to emotion, mood, and pain. They apparently act by suppressing the response to other neurotransmitters. Endorphins cause the "runner's high" experienced by people who run or perform other intense exercise to the point of pain.

The neuropeptide bradykinin, made up of nine amino acid residues, elicits sensations of pain by binding to receptors in pain-signaling neurons. It also appears to be responsible for symptoms of the common cold. Drugs that bind bradykinin receptors without eliciting pain signals are being tested for treatment of pain and colds.

6. **GABA (gamma-aminobutyric acid),** an amino acid. Studies of this inhibitory transmitter, found throughout the brain, suggest that as much as 90% of the brain may be devoted to inhibition.

7. **Glutamate,** also an amino acid. This excitatory transmitter occurs at more of the brain's synapses than any other. One kind of glutamate receptor opens calcium channels into the neuron. The entering $Ca^{2+}$ ions bind to the regulatory protein calmodulin, which in turn binds and activates the enzyme that produces nitric oxide.

8. **Nitric oxide (NO),** a gas. NO has long been known as a toxic component of cigarette smoke and smog. From 1987 to 1992, biologists were excited to learn that it also plays physiological roles in many parts of the body, including certain neurons. Because it is a gas, it is not stored and transmitted like other neurotransmitters. It is made as called for by the arrival of glutamate and diffuses out into neighboring cells, where it activates enzymes that make the second messenger cyclic GMP.

At normal levels, NO appears to aid learning and memory: rats treated with inhibitors of NO production cannot learn their way through a maze. In males, certain pelvic nerves release NO, which causes erection of the penis by relaxing muscles in its blood vessel walls.

## Clearing the Synapse

After transmitter molecules have acted on the postsynaptic membrane, they must be removed or destroyed. Otherwise, their action would continue indefinitely, and all useful information would be lost. Norepinephrine is resorbed by the presynaptic membrane and reused. GABA and glutamate are absorbed by astrocytes and converted to their common precursor, the amino acid glutamine, which is recycled to neurons. Nitric oxide reacts with water and oxygen within 5 to 10 seconds, forming nitrites and nitrates.

Acetylcholine is broken down by the enzyme **acetylcholinesterase.** Many insecticides and nerve gases, such as the organophosphates, are **anticholinesterases,** inhibitors of this enzyme. Hence, they permit acetylcholine in the synapses to keep on stimulating the postsynaptic membranes. This causes contraction of the muscles in uncontrollable spasms and, eventually, death. (Such substances must be used with care. They are toxic to all organisms, including people and pets, that use acetylcholine as a transmitter.)

## Neurotransmitters and Brain Function

Variations in levels of neurotransmitters can affect people's moods and abilities. One form of depression results from having too many receptors for acetylcholine, so that postsynaptic neurons overreact to it. Depression can also result from deficiencies in norepinephrine, dopamine, or serotonin, whereas excesses of these transmitters may result in manic behavior. In fact, anything that affects postsynaptic receptors, or the enzymes that make or destroy transmitters, can lead to malfunction of the relevant neural pathways. This is the basis of many mental illnesses and is also how some drugs affect the nervous system. Understanding these interactions gives researchers clues for developing drugs to treat pain and some mental illnesses.

Excessive levels of NO kill nearby neurons (although those that produce it are somehow protected). Lethal concentrations of NO are produced after a stroke, when neurons in the brain are killed by blockage of blood vessels that bring them food and oxygen: dying neurons release glutamate, which stimulates synthesis of excess NO, which kills cells and causes them in turn to release more glutamate. This vicious spiral kills neurons in a widening area around the original site of damage. Emergency treatment started within 12 hours after the stroke can now rescue at-risk neurons.

*Text continues on page 806*

E S S A Y : *Drugs and Neurons*

Many drugs affect the nervous system, often by changing activity at synapses (Table 37–A). However, some have more general effects. For example, caffeine (in coffee, tea, cola, or chocolate) increases the metabolic rate of neurons, thereby increasing alertness. Alcohol is a depressant, probably acting on all neurons and affecting many functions such as alertness, coordination, judgment, memory, and mood. Alcohol also kills neurons faster than they would otherwise die—about 10,000 extra neurons per ounce of alcohol consumed. This probably accounts for the mental deterioration seen in some alcoholics.

Opium, from the seedpod of a poppy, has been used as a drug since ancient times, not only because it is the most effective pain-killer known, but also because it induces a state of euphoria. Addiction to opiates (opium and related compounds) has been a social problem in the United States ever since the Civil War, when they were used as pain-killers. Despite an intense search for a nonaddictive opiate, all known opium derivatives—including morphine, Demerol, methadone, codeine, and heroin—eventually produce addiction in many people who take them. They are therefore most useful as pain-killers for the terminally ill, for whom pain relief outweighs possible addiction.

Opiates bind to some postsynaptic receptors in the brain and block the binding of neurotransmitters. This prevents the transmission of messages along a tract of nerves that normally carry pain signals from the body to the brain. (Pain is a useful signal, alerting the brain to move an injured part of the body away from the source of the damage.) Opiates also depress the immune system, especially killer cells, which fight cancer and virus infections.

Marijuana contains tetrahydro-cannabinol (THC). This molecule interferes with short-term memory by binding to receptors in the hippocampus (Section 37–L). It also binds to receptors in the cerebral cortex (see Figure 37–16), where it impairs thought and reasoning, heightens sensory perception, changes the perception of time, and produces mild euphoria, relaxation, and relief from anxiety. Marijuana also lowers the levels of sex hormones and suppresses the immune system, both functions involving chemicals made in the nervous system.

Marijuana has several therapeutic uses: relieving fluid pressure in the eyes of glaucoma patients, countering nausea from cancer chemotherapy, relieving pain and high blood pressure, treating asthma, and reducing the frequency of epileptic seizures. However, because of the accompanying euphoria, the U.S. Food and Drug Administration has approved only a less-effective synthetic version of THC to treat chemotherapy nausea and weight loss accompanying AIDS (THC stimulates a craving for glucose).

Because THC is hydrophobic, researchers thought it might affect the lipid part of neuron membranes, but this idea was disproved when a receptor cloned in 1990 turned out to be specific for cannabinoids (THC and related molecules). In 1992, the natural brain transmitter that binds to this receptor was found and named anandamide (from the Sanskrit word for bliss).

At about the same time, the first opioid receptor was cloned. Now that researchers can study opioid and cannabinoid receptors directly and know the structures of the molecules that bind to each, they hope to design therapeutic drugs that do not also make patients "high."

Many other drugs also act on synapses in the brain. For example, it takes less of the inhibitory transmitter GABA to open chloride channels if the GABA receptor has bound barbiturates at other sites. Because barbiturates strengthen GABA's inhibitory input, they are useful for treating insomnia and anxiety and for preventing the muscular convulsions of epileptic seizures. Benzodiazepines (such as Valium and Librium) work as tranquilizers because they have similar effects when they bind to yet another site on these GABA receptors.

LSD and mescaline are thought to produce hallucinations by binding with receptors for the neurotransmitter serotonin, a molecule they resemble. Cocaine and amphetamines (Benzadrine, Dexedrine, and "speed") mimic or enhance the effects of dopamine and norepinephrine. Cocaine binds to molecules that normally remove the transmitter dopamine from synapses. Hence dopamine keeps firing the synapses after they should have stopped. At low doses this stimulates behavior and elevates mood, but overdoses are fatal. Most drugs of abuse are thought to act by ultimately stimulating a "reward pathway" operated by dopamine, although some of them work less directly than cocaine.

Addiction to drugs has two components, one physiological, the other psychological. People who stop using addictive drugs go through a period of withdrawal, with unpleasant physical reactions such as shaking, vomiting, hallucinations, pounding heart, and pain. But after their bodies have adjusted to working without the drug and returned to normal, many people still have a psychological craving for the drug.

Drug-taking by humans has parallels among other animals. Animals as different as birds, rabbits, deer, and elephants have been observed to seek out and eat such things as fermented fruit (containing alcohol) or intoxicating mushrooms.

TABLE 37–A

## Modes of Action of Some Drugs That Affect the Nervous System

| Mode of Action | Drugs Acting at Acetylcholine Synapses | Drugs Acting at Catecholamine (Epinephrine or Dopamine) Synapses |
|---|---|---|
| Inhibits manufacture of neurotransmitter | Hemicholinium: blocks neuron's uptake of choline to make acetylcholine | Alphamethylparatyrosine (AMPT), a sedative |
| Decreases stores of neurotransmitter | | Reserpine (sedative or depressant): used for epilepsy or hypertension<br>Tetrabenazine: blocks storage of neurotransmitter in vesicles |
| Decreases level of molecules that deactivate neurotransmitter | Diisopropyl fluorophosphate: insecticide; chemical warfare; used to treat glaucoma<br>Physostigmine<br>Neostigmine: boosts muscular stimulation in myasthenia gravis (weakness due to receptor shortage); antidote for atropine poisoning | Nialamide (antidepressant) |
| Enhances release of neurotransmitter | | Amphetamines (release of norepinephrine): CNS stimulant |
| Inhibits release of neurotransmitter | Botulin toxin (botulism poisoning) | Alphahydroxybutyrate |
| Activates receptor site | Pilocarpine (from a plant)<br>Carbachol<br>Muscarine (from fly agaric mushroom): at synapses with smooth muscle and heart muscle, glands, some CNS sites<br>Nicotine (from tobacco): at synapses with skeletal muscle, some CNS sites | Norepinephrine, epinephrine<br>Isoproterenol (asthma treatment): stimulates heart, relaxes bronchi<br>Mescaline |
| Blocks receptor sites | Atropine (from plants such as deadly nightshade): antidote for anticholinesterase poisoning<br>Scopolamine (from many plants): with morphine, induces "twilight sleep" for childbirth<br>Curare: used in poison darts and in research and surgery to immobilize skeletal muscles | Phentolamine: blocks epinephrine's constriction of blood vessels<br>Phenothiazines: decrease schizophrenia symptoms by blocking dopamine receptors<br>Propranolol<br>Chlorpromazine: used to treat schizophrenia or bad drug trips<br>Haloperidol |
| Inhibits resorption of neurotransmitter | | Imipramine: antidepressant<br>Nortriptyline<br>Cocaine: CNS stimulant<br>Amphetamines<br>Tricyclics (block reuptake of norepinephrine) |

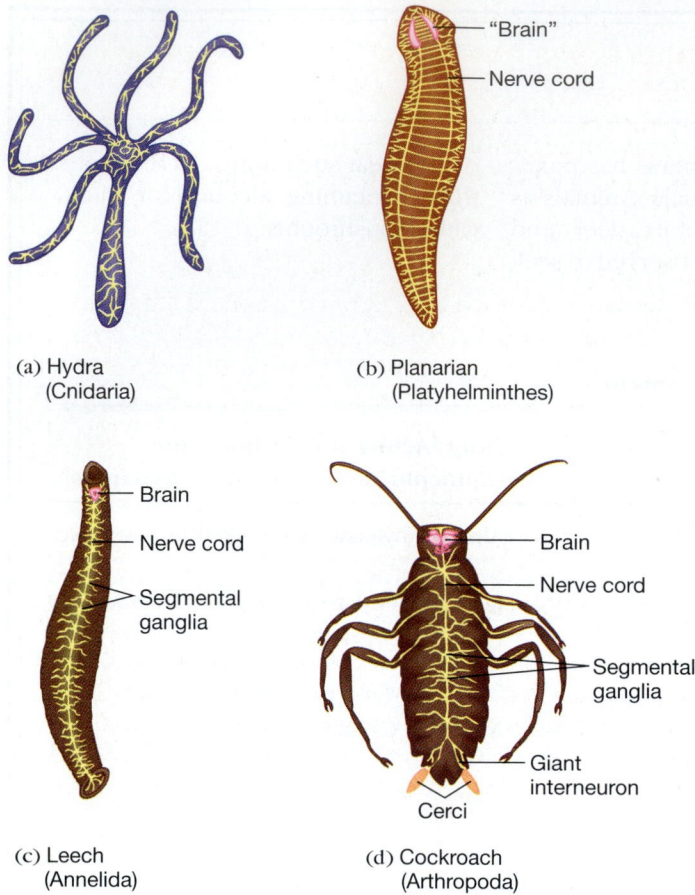

(a) Hydra
(Cnidaria)

(b) Planarian
(Platyhelminthes)
— "Brain"
— Nerve cord

(c) Leech
(Annelida)
— Brain
— Nerve cord
— Segmental ganglia

(d) Cockroach
(Arthropoda)
— Brain
— Nerve cord
— Segmental ganglia
— Giant interneuron
Cerci

**FIGURE 37–13**  Nervous systems of some invertebrates. Whereas Cnidaria (a) have a diffuse nerve net, the nervous tissue of higher invertebrates (b, c, d) is arranged in well-defined systems. The nerve cords are ventral and solid. In (d), note the giant axon involved in the cockroach's escape reaction.

## 37–F  ORGANIZATION OF NEURONS INTO NERVOUS SYSTEMS

In an animal's nervous system, neurons are "wired" to each other at synapses, forming nervous pathways. A neuron may synapse with many others and have excitatory synapses with some and inhibitory synapses with others. The number, arrangement, and synaptic connections of neurons determine how the animal responds to stimuli and the kinds of behavior it can perform.

During the course of evolution, nervous systems became increasingly complex as animals became larger and more mobile. Both trends call for more neurons. A tiny, sluggish, parasitic roundworm may have as few as 160 neurons and a correspondingly small range of behaviors. An octopus has more than a billion neurons, which give it precise control over its eight tentacles and considerable ability to learn new behavior patterns. A human is estimated to have about 100 billion neurons.

Large numbers of neurons are not enough; their organization is also crucial. The simple nerve nets of cnidarians have neurons scattered throughout the body rather than arranged in bundles (Figure 37–13). Many neurons have processes that are not differentiated as axons and dendrites but instead both send and receive information, and so the cell passes messages in any direction. This arrangement serves the needs of a radially symmetrical animal, whose food or enemies may approach from any direction. A cnidarian's reaction to most stimuli is generalized because each neuron transmits impulses to all of its neighbors. Surprisingly, even this simple nerve net is capable of learning. For example, sea anemones can learn to associate two stimuli and to anticipate a second stimulus that always follows the first (Section 41–E).

Platyhelminths (flatworms) show the beginnings of some important trends:

1. **Consolidation.** Instead of the diffuse nerve net of cnidarians, some neurons are arranged together in nerve cords running the length of the body, with cross-connections between them. This permits faster, more direct processing of information collected from outlying areas and quick signaling of actions to be taken by effectors.

2. **Specialization.** Cells have distinct roles as sensory or motor neurons or as interneurons. The patterns of synapses are more precise, so that incoming information is passed to specific neurons.

3. **Cephalization (formation of a head).** The evolution of bilateral symmetry gave the body a specific axis for movement and hence a definite front end. A head bears the major sense organs, such as receptors for light, touch, and chemicals, which detect what is happening in the outside world; the animal's leading end can sample the new environment for food or safety as it moves. The closer the "decision-making" neurons are to these sense organs, the faster they can signal appropriate reactions to the muscles. Consequently, nervous tissue became concentrated in the head, with neurons forming clusters called **ganglia** (singular: **ganglion**). The ganglia in the head are generally called a **brain,** the body's main nervous control center.

In annelid worms, arthropods, and other higher animals, ganglia also occur along the nerve cord and govern particular parts of the body. For example, annelids and arthropods have ganglia in each body segment. The nerve cord in these animals is single, running down the body's ventral midline.

Some invertebrate nervous systems have relatively few neurons, arranged in much the same pattern, and performing the same functions, from one individual to the next. This makes it possible to replicate experiments on the same neurons in many individuals. In a few species, such as the roundworm *Caenorhabditis,* medicinal leeches, nudibranchs (sea slugs), slugs, and crayfish, we now have maps showing exactly which neurons do what. In a leech, for instance, one particular pair of motor neurons everts the penis, and four neurons in each ganglion sense "pain." Researchers use these animals to examine basic principles of synaptic wiring and nervous system organization and their effects on behavior. However, it takes a lot of effort to understand even these simple nervous systems.

We are nowhere near ready to tackle such a cell-by-cell analysis of the vastly more complex nervous systems of higher animals. Instead, we study the nervous systems of vertebrates mostly by determining the functions of groups of cells in various areas.

## 37–G  THE VERTEBRATE NERVOUS SYSTEM

The vertebrate nervous system can be divided into the central and peripheral nervous systems. The **central nervous system** consists of the brain and spinal cord. The vast majority of neuron cell bodies, and most dendrites and axons, lie in the central nervous system and belong to interneurons. As these neurons relay information from one to another, sensory information is **integrated**—compared, changed, added up, or suppressed, determining a response appropriate for existing conditions. For example, your response to a plate of pastries depends on how full your stomach is, as well as how delicious the pastries look and smell. The central nervous system is also responsible for **association,** the channeling of sensory input into appropriate motor pathways.

In the human nervous system, some of these activities go on automatically, without our being aware of them. Others are carried out by "higher centers," parts of the nervous system where we are aware of sensory stimuli and of our own thoughts and emotions and may deliberately choose one response over another. We have no way of knowing whether other animals also have distinctly conscious nervous functions.

The brain is protected by the skull, the spinal cord by the vertebral column. Inside these bony coverings, the brain and spinal cord are covered by three layers of membranes, the **meninges.**

Within the meninges, the cerebrospinal fluid bathes the central nervous system and cushions it from jarring. This fluid, exuded from blood vessels in the meninges, is similar to extracellular fluid, but the blood-brain barrier excludes some substances that move freely out of capillaries elsewhere in the body. Disorders of the central nervous system can sometimes be diagnosed by doing a "spinal tap" to withdraw a sample of the fluid around the spinal cord. Inflammation of the meninges (meningitis) is often a serious condition because the meninges are so close to vital nervous tissues.

The **peripheral nervous system** connects the central nervous system with the rest of the body, including sense organs that in turn are in contact with the outside world. It consists of **nerves,** bundles of axons covered with sheaths of connective tissue, and **peripheral ganglia,** clusters of neuron cell bodies outside the central nervous system (Figure 37–14).

The peripheral nervous system has two parts, which play different roles. The **somatic nervous system** contains both motor and sensory neurons. It controls the activity of the skeletal muscles, such as those we use to smile, run, sing, or draw. The **autonomic nervous system** contains only motor neurons. It controls muscles and glands that usually operate without our conscious control. These effectors control things like blood pressure and the movement of food in the gut.

## 37–H  THE VERTEBRATE BRAIN

The central nervous system develops from a hollow tube formed during early embryonic development. The walls of the front part of the tube enlarge in a series of bulges, with three main parts—the forebrain, midbrain, and hindbrain—followed by the long, straight spinal cord (see Figure 12–12).

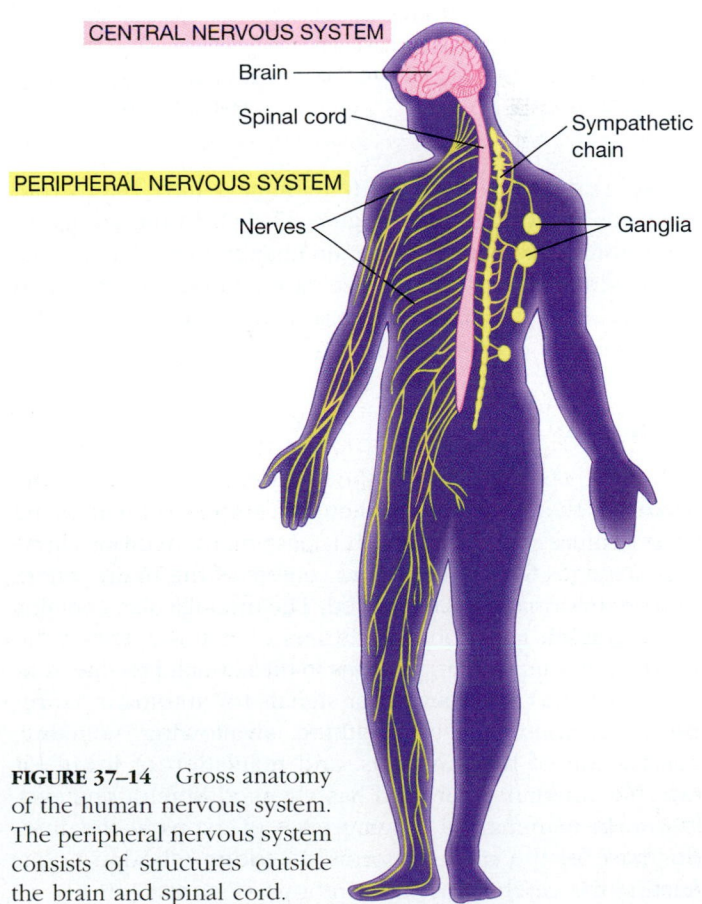

**CENTRAL NERVOUS SYSTEM**

Brain
Spinal cord
Sympathetic chain

**PERIPHERAL NERVOUS SYSTEM**

Nerves
Ganglia

**FIGURE 37–14**  Gross anatomy of the human nervous system. The peripheral nervous system consists of structures outside the brain and spinal cord.

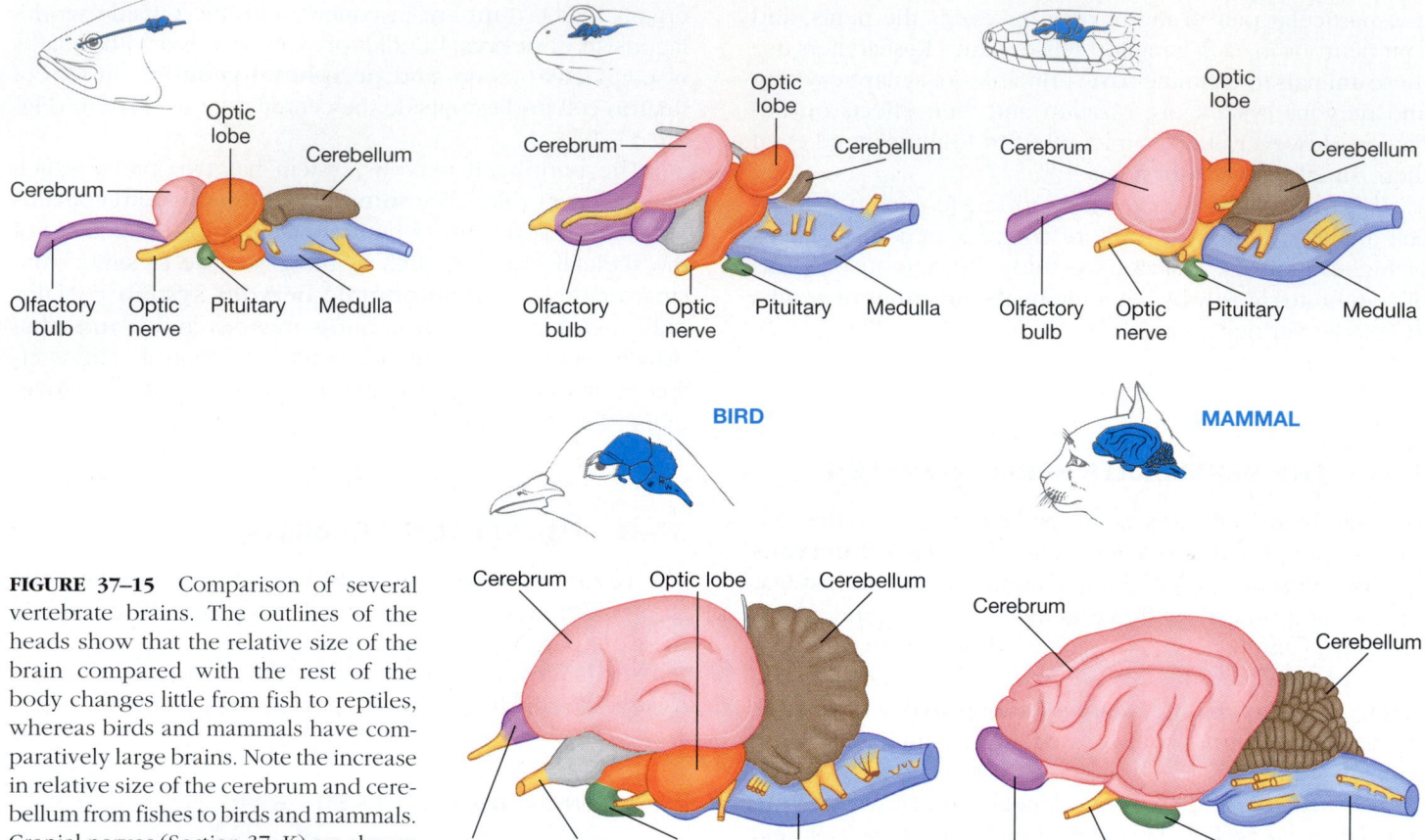

**FISH**

**AMPHIBIAN**

**REPTILE**

**BIRD**

**MAMMAL**

**FIGURE 37–15** Comparison of several vertebrate brains. The outlines of the heads show that the relative size of the brain compared with the rest of the body changes little from fish to reptiles, whereas birds and mammals have comparatively large brains. Note the increase in relative size of the cerebrum and cerebellum from fishes to birds and mammals. Cranial nerves (Section 37–K) are shown in yellow.

All vertebrates, from fish to mammals, have brains with the same basic structures (Figure 37–15). In the course of evolution, some parts of the brain changed very little, while others virtually exploded in size. Some brain areas retained their ancestral functions, whereas others took on new roles as vertebrates evolved (Table 37–4).

## Hindbrain

The most obvious part of the hindbrain of a fish is the **medulla,** the enlargement where the spinal cord enters the brain (Figure 37–15). Through it pass many neurons carrying messages to or from higher centers of the brain, where sensory information is integrated. The medulla also contains many **nuclei:** ganglion-like clusters of neurons within the central nervous system. Neurons in these nuclei receive sensory input and send out motor signals for automatic, or **reflex,** functions such as breathing, swallowing, vomiting, constriction of blood vessels, and regulation of heartbeat rate. The medulla's function has changed little during evolution. In mammals, a bulging mass of axons at the anterior base of the medulla forms the **pons** ("bridge") connecting the cerebellum and cerebrum (Figure 37–16).

The **cerebellum** is an outgrowth of the medulla. During evolution it has grown noticeably and taken on much of the central control of balance and movement. It is particularly large in animals with considerable muscular dexterity, such as birds.

## Midbrain

During evolution, the midbrain has changed more in function than in size or structure. In fish and amphibians it is the principal area for association of sensory input with appropriate motor output. In these lower vertebrates, a major part of the midbrain is the **optic lobe,** which receives signals from the **optic nerves,** carrying visual information from the eyes.

In mammals, the analysis of vision has moved out of the midbrain and into part of the forebrain. The midbrain of mammals consists of the **anterior colliculi,** which control reflexes of the eyelids and the iris of the eye, and the **posterior colliculi,** which analyze and relay information coming from the ear via the auditory nerve.

**TABLE 37-4**

## Functions of Major Parts of the Vertebrate Brain

| Derivation | Name | Function |
|---|---|---|
| Hindbrain | Medulla and pons | Passage of messages between brain and spinal cord; control of visceral reflexes |
| | Cerebellum | Coordination of equilibrium and movement |
| Midbrain | Optic lobe (in lower vertebrates) | Association of sensory and motor pathways |
| | Anterior colliculi (mammals) | Reflexes of iris and eyelid |
| | Posterior colliculi (mammals) | Receives sensory information from ear |
| Forebrain Diencephalon | Thalamus | Relays olfactory messages to midbrain (in fishes) Area of sensory integration (in higher vertebrates) |
| | Hypothalamus | Controls emotional states and drives (pleasure, pain, thirst, sex, rage); secretes hormones (Section 40–D) |
| | Posterior pituitary | Releases hormones produced in hypothalamus (Section 40–D) |
| Telencephalon | Olfactory bulb | Receives olfactory information (most important telencephalon area in fishes) |
| | Corpus striatum | Complex behavior patterns (in birds) |
| | Cerebrum (cerebral cortex) | Well developed only in mammals. Sensory and motor association, visual and auditory processing, seat of "intelligence," memory storage, and, in humans, of conscious thought and ability to use language, both written and spoken |

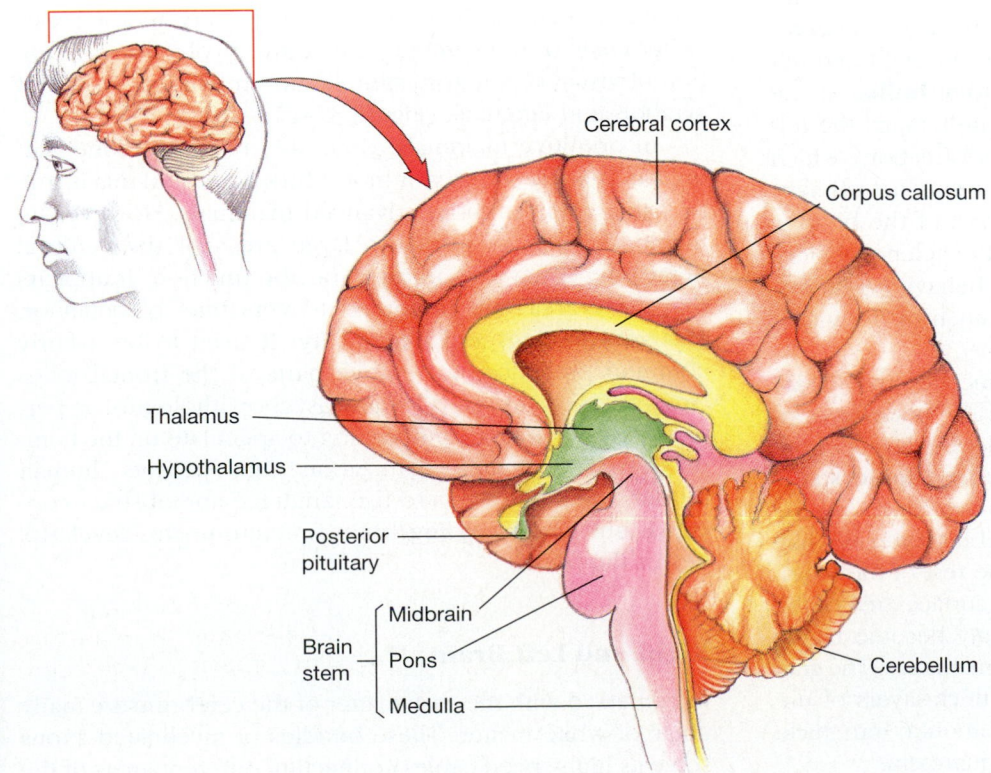

**FIGURE 37–16** The human brain. Only the major structures mentioned in the text are shown. The diencephalon is shown in green.

## Forebrain

The forebrain has changed a great deal during vertebrate evolution. It has two major parts, the diencephalon and the telencephalon. Lying just in front of the midbrain, the **diencephalon** contains the thalamus, the hypothalamus, and the posterior lobe of the pituitary gland (Figure 37–16). In fishes, the **thalamus** relays information to the midbrain from the **olfactory** (sense of smell) organs. However, in other vertebrates it is one of the centers that integrate all sensory information.

Immediately below the thalamus lies the **hypothalamus,** the part of the brain where the nervous and hormonal systems interact (Section 40–D). This area contains many nuclei that control homeostatic functions: regulation of body temperature, growth, sexual drive and maturity, hunger, thirst, and salt and water balance. This is the only part of the brain without a blood-brain barrier, and cells in the hypothalamus detect molecules in the blood directly.

Researchers have found that some nuclei in the hypothalamus are different sizes in male and female birds, rats, monkeys, and humans (and probably in other species not yet examined). In homosexual men, these nuclei are closer to the size found in women. Study of rat pups showed that the size differences in these hypothalamic nuclei depend on the presence of androgens during the five days after birth. The significance of these sex differences in the hypothalamus is not yet clear.

While the diencephalon slowly expanded to handle increased sensory input, the **telencephalon,** the anterior part of the forebrain, grew astoundingly in both size and complexity during vertebrate evolution. It consists of the **cerebrum,** which is divided by a fissure into the right and left **cerebral hemispheres,** and the **olfactory bulbs,** at the anterior end of the brain. In fish and amphibians the telencephalon handles mostly olfactory information, which plays a major role in the lives of these aquatic animals. However, in birds the most important part of the brain is the **corpus striatum,** at the base of the telencephalon, which is responsible for their complex behavior patterns. The corpus striatum of birds occupies much of the inside of the cerebrum.

In mammals, there is a progressive increase in the size and importance of the **cerebral cortex,** the outer layer of the cerebral hemispheres. The original, deeper layers of the hemispheres, the **hippocampus** and other **limbic structures,** regulate emotional state and short-term memory (see Figure 37–22). Above these areas, the cerebral cortex lies like the cap of a wrinkled mushroom over the rest of the brain. This expansion has greatly increased the surface area of the cerebral cortex, which has correspondingly become highly folded and convoluted in more intelligent mammals. The gray matter of the cerebrum is composed of thick layers of unmyelinated cells. In humans, this layer is about 6 mm thick, with a total surface area of about 1800 square cm.

## Mapping the Brain

How do we know which parts of the brain do what? Investigators have used several different methods. One traditional method is to catalog what functions are lost when parts of the brain are destroyed by accidents, strokes, or surgery.

Because the brain has no pain receptors, brain surgeons can use local anesthesia when they open the skull to operate, leaving the patient conscious and able to report the sensations or memories elicited by stimulating the brain with electricity. In the 1940s, Wilder G. Penfield used this technique to explore the cortical surface in more than 1000 patients during brain surgery.

A newer method uses radioactive tracers to detect which areas of the brain are active during a particular mental task. Neurons in the most active areas take up the most radioactive glucose analogues: molecules enough like glucose for neurons to absorb them but not to break them down for cellular respiration. In experimental animals, individual radioactive cells can be identified by slicing the brain into thin sections and examining them with a microscope.

In humans, researchers can narrow activity down to small areas of the brain by using positron emission tomography (PET). A mildly radioactive substance is injected into the blood, and brain scans show local increases in radiation as more blood flows to areas that become active while the subject performs certain mental tasks.

It turns out that many functions involve cells in several areas of the brain. However, certain areas tend to be "in charge" of certain activities. For example, the cerebral cortex has primary sensory areas and primary motor areas. Other areas of the cerebral cortex are involved in perception of visual or auditory stimuli and, in humans, in use of symbols and language (Figure 37–17).

In primitive mammals, each area of the cerebral cortex has specific sensory or motor functions, and this is true to some extent in more advanced mammals. However, in mammals such as primates, large areas of the cerebral hemispheres have no known specific function. It appears that these areas are important to versatility of behavior, abstract thinking, and personality. It used to be a fairly common procedure to remove parts of the frontal lobes (frontal lobotomy) to alleviate psychopathologies or personality disorders. It is intriguing to speculate on the complexity of cellular organization that permits human thoughts and allows us to use language and abstract concepts, but understanding brain function at this level still lies in the future.

## Right and Left Brain

Interspersed with the gray matter of the cerebrum are many tracts of white matter. These bundles of myelinated axons serve as high-speed cables connecting different areas of the

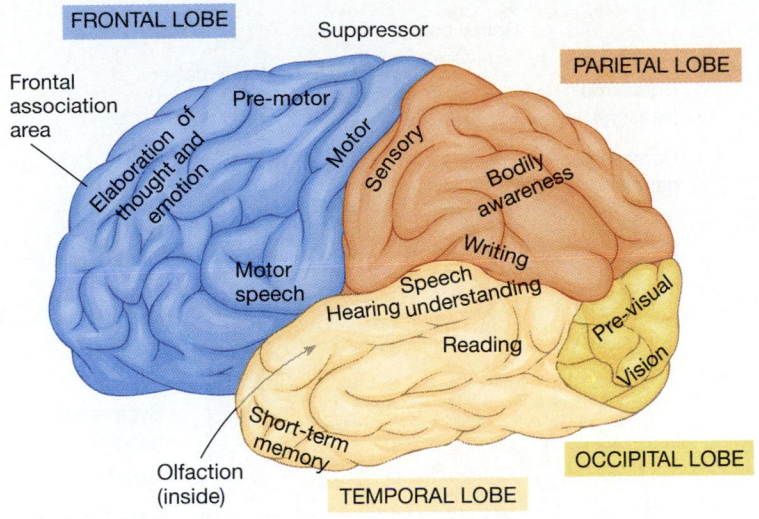

(a)

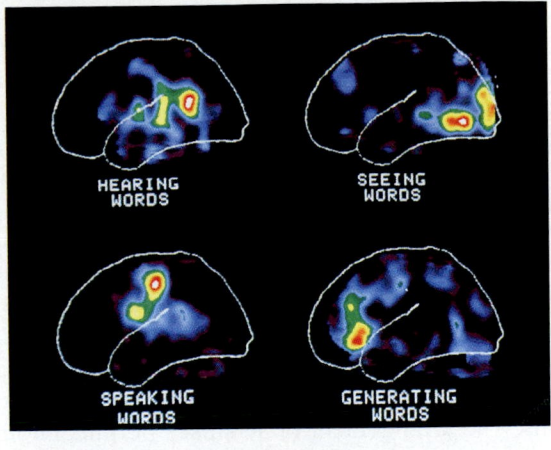

(b) PET scans of brain

**FIGURE 37–17** Functional map of the human cerebral cortex. (a) The functions assigned to several areas are written on this view of the brain's surface, as seen from the left side. (b) PET scans showing differences in brain activity while a person is hearing, seeing, speaking, or generating a word. (b, Marcus Raichle, Department of Neurology/Washington University School of Medicine)

brain. The right and left cerebral hemispheres are linked by way of a large tract of myelinated axons, the **corpus callosum** (see Figure 37–16). Its function is to tell the right half of the brain what the left half is doing (and vice versa).

Roger W. Sperry and his co-workers showed that the two cerebral hemispheres can operate somewhat separately; for this work, Sperry shared in a 1981 Nobel Prize. To understand his experiments, you must know that each half of the brain controls structures on the opposite side of the body (the nerve tracts cross over at lower nervous centers). Sperry severed the corpus callosum in experimental animals and also worked with people who had undergone similar "split-brain" surgery to treat epilepsy. When tasks are presented to the left eye and hand, which are controlled by the right side of the cerebral cortex, a split-brain subject may show no sign of recognizing them, even though the right eye and hand, and the left cerebral cortex, had previously mastered the task.

Furthermore, the two sides of the cerebral cortex do not have exactly the same functions. The left side is usually dominant, and it seems to have a strong tendency to assume control of logical thought and mathematical ability and also of language, whether written, spoken, or the sign language of the deaf. It also excels at motor control needed for speech. A person may be able to perform a task set for the right half of the brain correctly, but not be able to speak or write about it, whereas the left half of the brain can both learn to perform the task and generate a verbal explanation of what it is doing. Interestingly, Doreen Kimura found

that the two sexes differ in which parts of the left hemisphere control speech.

Kimura also studied left and right brain functions in people who had not undergone split-brain surgery. When she simultaneously played two different series of numbers into the left and right ears, people heard better with the right ear (left brain). By contrast, the left ear (right brain) showed more ability to recognize a tune.

The right side of the cortex seems to be involved in spatial relationships, musical and artistic activity, and expression of emotions. People with damage to this side of the brain may show emotional "flatness." Because the right side of the brain has so little ability to express itself in speech, it is hard to find out exactly what abilities it does have.

An interesting experiment using PET on people with intact brains supported this idea of a difference between the right and left hemispheres. Asked to remember a series of tones, some people showed more activity in the right hemisphere, others in the left. It turned out that people who used the right hemisphere had memorized the notes as a melody. Those who used the left hemisphere had mentally envisioned the notes on a music staff, and these people showed activity in the visual areas as well as the auditory areas of the brain.

Intriguing as all this is, attempts to divide functions neatly between the two halves of the brain are too simplistic. To produce normal functioning in most activities, both sides must work together, each contributing expertise in certain aspects of the activity.

## Sensory Analysis

In mammals, the various sensory systems relay information to the cerebral cortex for integration and decision making. The processing of sensory information has been particularly well studied in the visual system. In 1981 David Hubel and Torsten Wiesel shared a Nobel Prize with Sperry for their work on the visual system.

The organization of neurons in the visual system shows two general features. First, information from each area in the retina of the eye ends up in a corresponding location in the visual cortex, the posterior part of the cerebral cortex (see Figure 37–17a). Second, information reaches the visual cortex by passing through several consecutive layers of neurons. The arrangement of synapses permits information to be processed in increasingly complex ways at successive layers.

Cells in the first layer detect (that is, fire most rapidly in response to) small spots of light or of dark, with contrasting surroundings, in their area of the retina. Higher-order cells detect larger circles or ovals, still higher cells edges or corners of a certain size and orientation, and then moving edges (Figure 37–18). Other, even higher-order cells

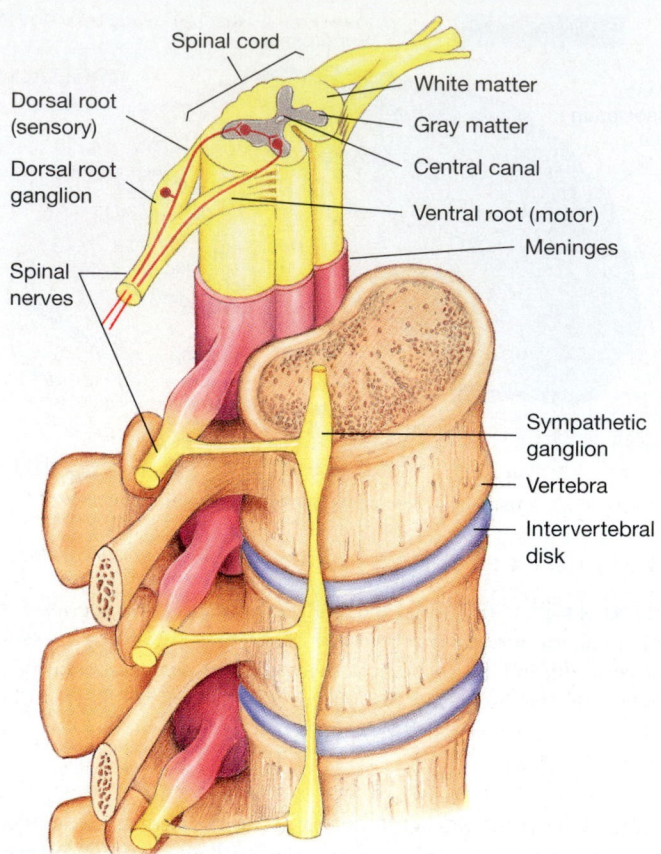

**FIGURE 37–19**   The human spinal cord. The vertebral column and the meninges surround and protect the cord. The paired spinal nerves protrude through spaces between the vertebrae. Each spinal nerve has a dorsal and a ventral root, which join a short distance from the cord. The cord contains both gray matter, containing cell bodies, and white matter, containing bundles of myelinated axons. Its fluid-filled central canal is an extension of the brain's.

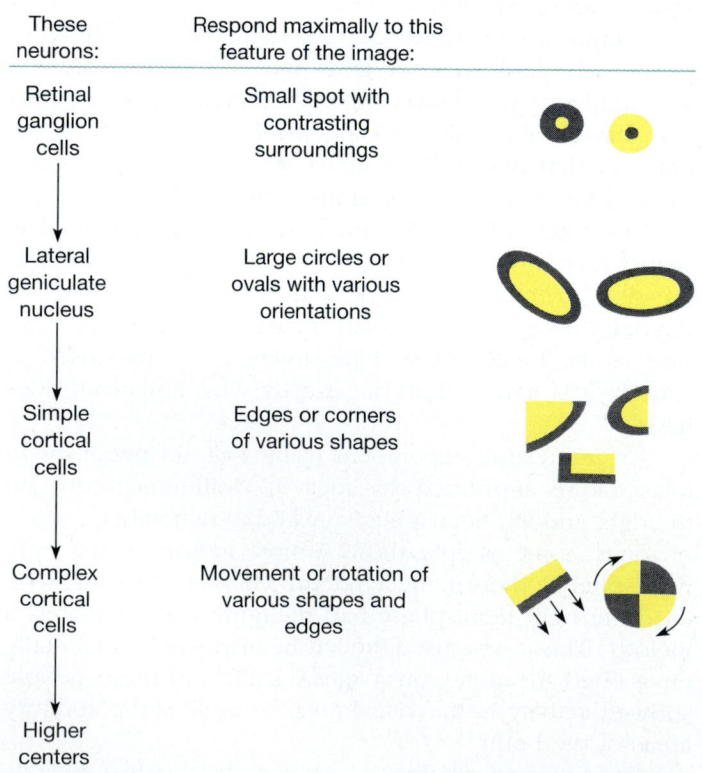

| These neurons: | Respond maximally to this feature of the image: | |
| --- | --- | --- |
| Retinal ganglion cells | Small spot with contrasting surroundings | |
| Lateral geniculate nucleus | Large circles or ovals with various orientations | |
| Simple cortical cells | Edges or corners of various shapes | |
| Complex cortical cells | Movement or rotation of various shapes and edges | |
| Higher centers | | |

**FIGURE 37–18**   Levels of visual processing. Information passes from ganglion cells in the retina to layers of other neurons in the brain. At each level, synaptic input is organized so that neurons detect more inclusive information.

in the visual cortex respond to more abstract features of visual stimuli and allow recognition of various kinds of visual patterns. For example, monkeys (and presumably humans) have "face cells," which fire strongly when the stimulus is a face but which do not respond to other stimuli, including pictures containing parts of faces scrambled out of their normal arrangement. This hierarchical arrangement of cells allows progressive extraction of more and more complex features of the image received by the eye.

Information on different aspects of an image, such as movement, color, and shape, travels from the eye to the brain via different sets of neurons, a phenomenon called divergence. Eventually, these different components converge again in a layer of neurons that receive input from the separate tracks. Just how the brain interprets the shape, motion, and color we see is still a mystery.

An animal's sensory system may have specific arrangements of synapses that make it "hard-wired" to extract particular features from the masses of stimuli in its environment. For instance, a frog's visual system is specialized to pick out small, dark moving objects—which in the frog's world tend to be juicy flies—and large darkenings of the visual field—danger! These "feature detectors" save time and energy by permitting the frog to react quickly to events significant to itself, while ignoring other aspects of its visual environment.

## Motor Pathways

In the cerebral cortex, sensory areas pass information to motor areas by way of association neurons in the deeper layers of the cortex. Motor areas send out commands for movements under conscious control. Axons may pass directly from the motor cortex to motor neurons in the spinal cord. Or the message may go less directly, crossing anywhere from three to thousands of synapses, allowing it to be adjusted on its outward path.

The cerebellum plays an important role in coordinating movements, especially rapid movements such as running or typing. It receives sensory information about where each part of the body is, makes moment-by-moment predictions of where each part will be next, and sends out instructions for fine adjustments so that actions are carried out smoothly. People with damaged cerebellums move jerkily, and their motions invariably overshoot the intended final position. For instance, a hand may reach too far to pick up a desired object, and then the next movement, intended to compensate, will bring the hand too far back.

## 37–I  THE SPINAL CORD

The spinal cord extends from the base of the hindbrain to the end of the vertebral column (Figure 37–19). It is a relay system carrying information between the brain and the peripheral nervous system. It is also the seat of **spinal reflexes,** which allow the body to make quick responses.

## Reflex Arcs

Our responses to certain stimuli are simple, unvarying, and quick. For instance, it is hard to hold your lower leg still when the doctor hits you below the knee with a little rubber mallet. The knee jerk, and the rapid withdrawal of a hand from a flame, are controlled by **reflex arcs,** pathways of a few neurons each, under little conscious control from higher brain centers. Reflex arcs contain sensory and motor neurons and usually one or more interneurons between them (Figure 37–20). A reflex pathway saves time because it has few synapses. In addition, the message need not make the longer trip to areas of consciousness in the brain and back before an appropriate motor response begins. By the time you are conscious of feeling the mallet below the knee, your foot has already begun to swing out.

Functions involved in physiological homeostasis, such as heartbeat rate, breathing, dilation of the pupils, and digestion, are controlled by reflexes of the autonomic nervous system. More complicated activities, such as posture control, locomotion, sexual behavior, and defensive responses, which are governed mainly by the somatic nervous system, also involve many reflex arcs.

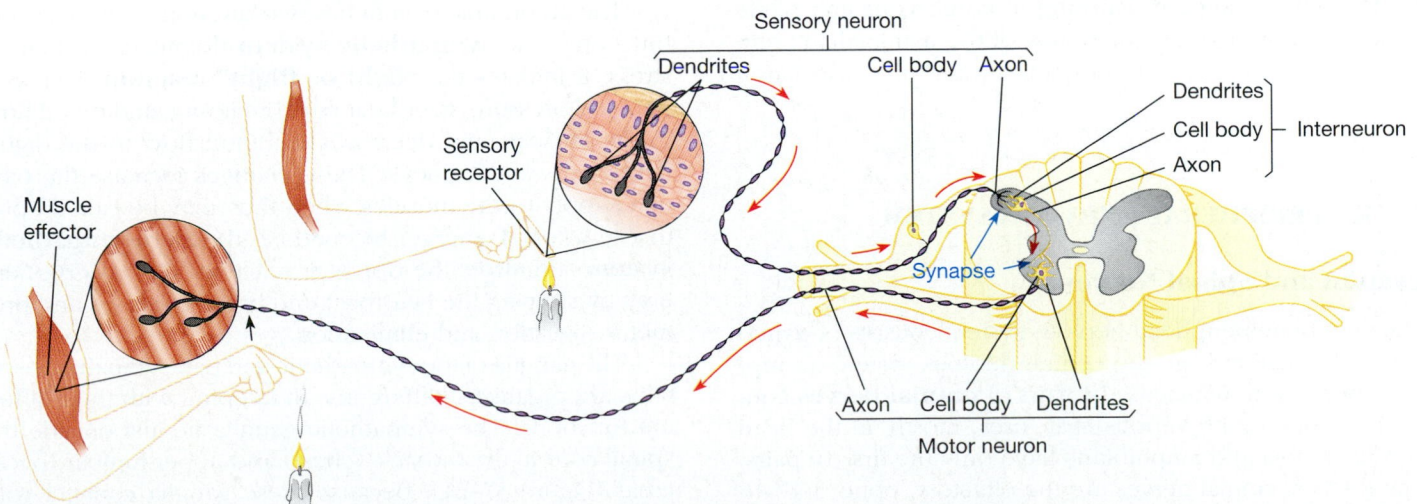

**FIGURE 37–20**  A simple reflex arc. Red arrows show the path of information. The arc shown here contains two synapses (blue arrows) and three neurons: sensory neuron, interneuron, and motor neuron.

A reflex may be quite complex, using many neurons and moving many muscles. For example, if you raise your arm to protect your face, muscles in your back must also adjust their contraction or you would lose your balance.

The spinal cord plays an important role in integrating reflex behavior. For instance, a "spinal" animal, one whose brain has been destroyed or removed, still shows reflexes. If a piece of acid-soaked paper is touched to the back of a spinal frog, one leg will come up and kick it away; the behavior is repeated no matter how many times the paper is placed on the skin. This response, involving the coordinated action of many muscles, clearly demonstrates one of the chief characteristics of a reflex: unvarying repetition. A frog with an intact brain might make the response two or three times, but eventually the higher centers would intervene and the frog would do something else—perhaps hop away.

## 37–J   PLASTICITY IN THE NERVOUS SYSTEM

The opposite of a reflex behavior pattern is a **plastic** behavior pattern, subject to modification by many neurons. For instance, the interaction of neurons at thousands of different synapses determines whether you immediately leap out of bed when the alarm clock rings, and for this reason the response may not follow hard upon the stimulus.

**Plasticity** is the ability to learn, that is, to change the interactions between various parts of the neural circuitry in the central nervous system. If a frog's eyes are surgically rotated 180° in their sockets, the frog will perceive its prey to be 180° from where it really is, and the frog will try to catch it by jumping in the wrong direction. A mammal, which has more plastic behavior, can compensate for such a shift in its visual world in a relatively short time. You have made such a change in learning to comb your hair while looking in a mirror. The flexibility of the mammalian brain is thought to be due to the organization of the association neurons.

## 37–K   PERIPHERAL NERVOUS SYSTEM

### Cranial and Spinal Nerves

The vertebrate peripheral nervous system consists of paired nerves branching from the central nervous system. In reptiles, birds, and mammals, 12 pairs of **cranial nerves** connect the brain with various structures, mostly in the head and neck; fish and amphibians have only the first 10 pairs. The thickest cranial nerves are the olfactory, optic, and auditory, which carry only sensory information coming to the brain from the major sense organs—the nose, the retinas of the eyes, and the ears. The other cranial nerves carry both motor and sensory information to and from the tongue, muscles of the eyes and face, and so on (see Figure 37–15).

The longest cranial nerve is the tenth, the vagus, which serves many internal organs of the chest and upper abdomen and is part of the autonomic system (see Figure 37–21).

Humans have 31 pairs of **spinal nerves,** which branch out from the spinal cord between adjacent vertebrae. Each spinal nerve leaves the spinal cord in two parts: a **ventral root,** containing the axons of motor neurons, and a **dorsal root,** containing sensory neurons. Along each dorsal root lies a **dorsal root ganglion,** containing the sensory neurons' cell bodies. This is one of the few places where neuron cell bodies are found outside the central nervous system. The sensory neurons' axons continue into the spinal cord and synapse with other neurons (see Figures 37–19 and 37–20).

The dorsal and ventral roots join outside the cord and run a short way together as a spinal nerve. The nerve soon splits into three branches, each containing both sensory and motor fibers. One branch serves the skin and muscles of the back, another those of the front of the body, and the third the internal organs.

Each spinal nerve serves its own segment of the body and also overlaps adjoining segments. So, if a spinal nerve is cut, the corresponding part of the body does not completely lose sensation and the ability to move.

### The Autonomic Nervous System

The autonomic nervous system, found in all vertebrates, governs most of the body's homeostasis: it regulates the heartbeat and controls contraction of the muscles in the walls of blood vessels and of the digestive, urinary, and reproductive tracts. Autonomic nerves also stimulate glands to secrete mucus, tears, and digestive enzymes.

The autonomic system has two divisions, with different functions. The **sympathetic system** dominates in time of stress. It initiates the **"fight or flight" reaction:** increases in blood pressure, heartbeat rate, breathing, and blood flow to the muscles and decreases in blood flow to the digestive organs and kidneys. These changes increase the oxygen supply to the muscles when they may be called upon to use a lot of energy. In contrast, the **parasympathetic system** stimulates the opposite reactions. It conserves energy by slowing the heartbeat and breathing rates and promotes digestion and elimination.

The ganglia of the sympathetic and parasympathetic systems are organized differently, in keeping with their different functions. The sympathetic ganglia lie just outside the spinal cord in the thoracic (chest) and upper lumbar (back) areas (Figure 37–21). Because these ganglia connect with the spinal cord and with each other, stimuli are quickly transmitted to all parts of the system, galvanizing the entire body for action in time of stress or danger.

In contrast, parasympathetic ganglia lie near their effector organs. They connect with the central nervous sys-

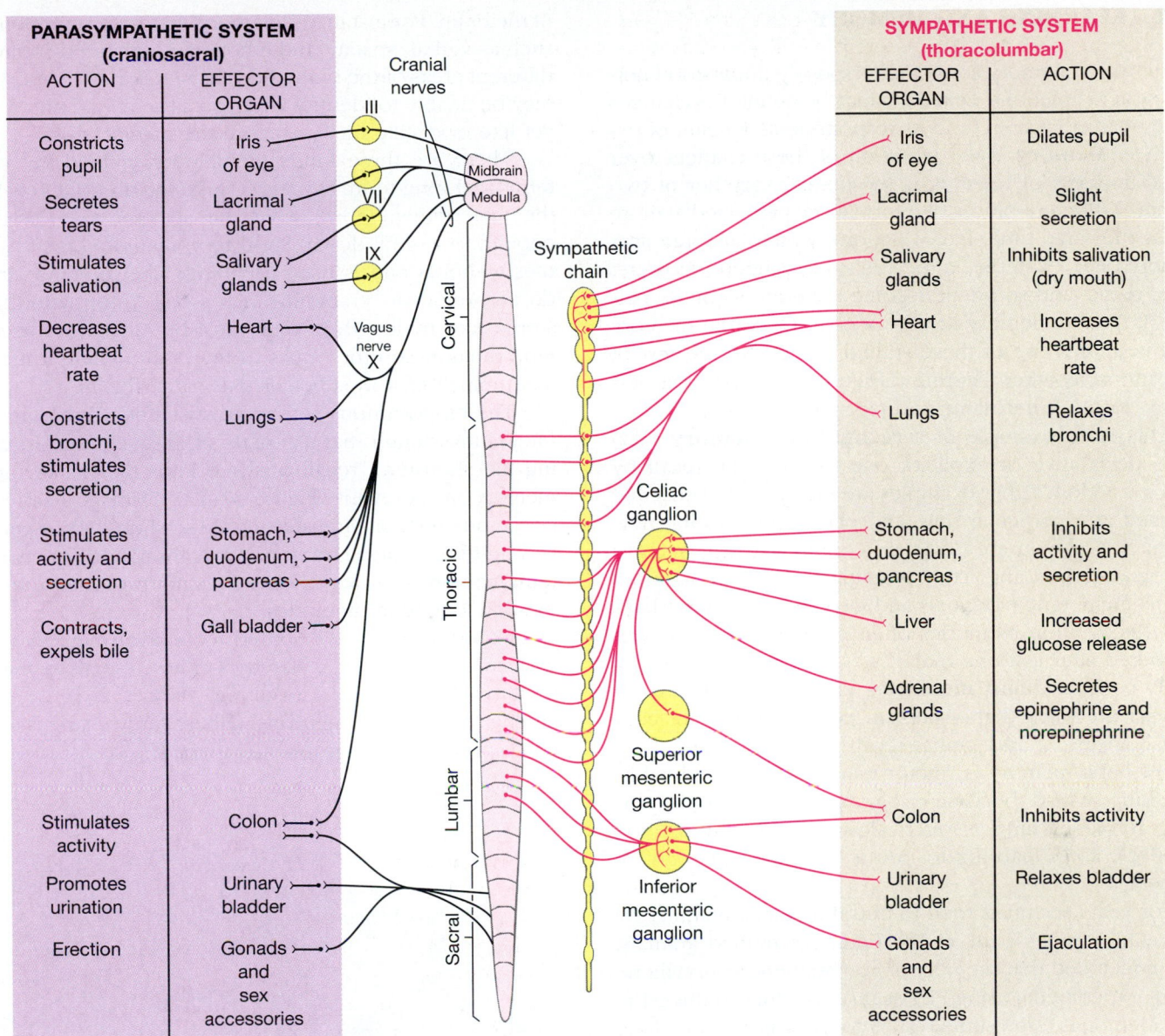

**PARASYMPATHETIC SYSTEM**
**(craniosacral)**

| ACTION | EFFECTOR ORGAN |
|---|---|
| Constricts pupil | Iris of eye |
| Secretes tears | Lacrimal gland |
| Stimulates salivation | Salivary glands |
| Decreases heartbeat rate | Heart |
| Constricts bronchi, stimulates secretion | Lungs |
| Stimulates activity and secretion | Stomach, duodenum, pancreas |
| Contracts, expels bile | Gall bladder |
| Stimulates activity | Colon |
| Promotes urination | Urinary bladder |
| Erection | Gonads and sex accessories |

**SYMPATHETIC SYSTEM**
**(thoracolumbar)**

| EFFECTOR ORGAN | ACTION |
|---|---|
| Iris of eye | Dilates pupil |
| Lacrimal gland | Slight secretion |
| Salivary glands | Inhibits salivation (dry mouth) |
| Heart | Increases heartbeat rate |
| Lungs | Relaxes bronchi |
| Stomach, duodenum, pancreas | Inhibits activity and secretion |
| Liver | Increased glucose release |
| Adrenal glands | Secretes epinephrine and norepinephrine |
| Colon | Inhibits activity |
| Urinary bladder | Relaxes bladder |
| Gonads and sex accessories | Ejaculation |

Labels in figure: Cranial nerves; III; VII; IX; Vagus nerve X; Midbrain; Medulla; Sympathetic chain; Cervical; Thoracic; Lumbar; Sacral; Celiac ganglion; Superior mesenteric ganglion; Inferior mesenteric ganglion

**FIGURE 37–21** The autonomic nervous system. This highly schematic and simplified drawing shows the general layout of the parasympathetic and sympathetic systems and how each affects the major organs it innervates. Parasympathetic ganglia lie near the target organs, and the postganglionic neurons lie within the target organs. In contrast, the sympathetic ganglia lie just outside the spinal cord and are connected to one another in the sympathetic chain.

tem by way of some of the cranial nerves and sacral (pelvic area) spinal nerves. The parasympathetic ganglia do not connect with each other, and thus each effector is controlled independently.

The sympathetic and parasympathetic systems also differ in the chemical transmitters released at the synapse with the effector: norepinephrine in most of the sympathetic system, acetylcholine in the parasympathetic and in sympathetic neurons to sweat glands and to blood vessels in muscles. However, both use acetylcholine in synapses in the ganglia.

Although the autonomic nervous system can carry out its tasks automatically, it is not completely independent of the animal's voluntary control. For example, it is possible to decide to stop breathing for a short time. Humans and animals can also be trained to change their heartbeat rates, blood pressures, and digestive reflexes voluntarily. However, any voluntary control that endangers life quickly disturbs homeostasis of the brain tissue, resulting in unconsciousness. Then the autonomic system takes over again and restores normal functions.

## 37–L   LEARNING AND MEMORY

Learning and memory are complex, poorly understood nervous system functions. **Learning** may be defined as changes in the nervous system (and its responses) as a result of experience. **Memory** is the retention of these changes over time. The basis of learning is the linking together of two (or more) different pieces of information, perceived at more or less the same time, into a pattern, which is stored as a memory. For example, a person's face, figure, posture, voice, name, and so on all become linked to form the pattern that is the memory of that person. Once the memory is formed, any one of these stimuli can evoke the rest of the pattern, as when hearing a friend's voice on tape permits us to recall her name and face.

This is an example of a **recognition memory** (also called declarative or explicit), one that we can recall by conscious effort. Other examples are memory of events that occurred at a particular time and place and learning new information as you are doing now: linking words such as memory, neurons, and action potential to older information, such as membrane potentials and ion channels, from Chapter 4. Recognition memories often can be formed and remembered after just one trial.

By contrast, **habit memories** (also called reflexive or implicit) are ones that a person (or animal) is not consciously aware of and cannot recall at will. One important kind of habit memory is "motor memory"—learned habits and skills, which become programmed into neural circuits. Habit learning is often slow, requiring repetition (practice), as in learning to ride a bicycle or play an instrument.

For years scientists tried to find the seat of memory by destroying various parts of the brains of trained animals. They concluded that memory is nowhere and everywhere. Recent experiments using PET scans give clues to this paradox. Memories are formed as information passes along neural circuits through many areas of the forebrain. Destroying any of these areas or cutting their connections impairs formation of new memories but may leave existing ones intact.

Existing memories are probably stored as patterns of altered synapses, also involving networks of neurons traversing many areas of the brain. Recognition memories appear to be stored in sensory analysis areas of the cerebral cortex, where the memory-forming circuits began. When cells in these areas receive a stimulus that is part of the memory, they trigger recall of the entire memory pattern. Habit memories are stored in deeper layers of the brain, often involving changes in both sensory and motor pathways. Therefore, they can be stored as part of unconscious reflex circuits.

These findings fit in with what is known about memory from studying people afflicted with amnesia (loss of memory) as a result of disease or damage to certain areas of the brain. For instance, "memory" has many components, such as verbal, spatial, and emotional, apparently stored in different areas of the brain. People with partial memory loss may be unable to identify a person as someone they know, yet like or dislike him as before the memory loss.

There are three kinds of memory: immediate, short-term, and long-term. When we take lecture notes, we use the first, remembering what the speaker said just long enough to write it down. Short-term memory lasts for minutes to hours and is used for things like remembering to do an errand or "cramming" for a test. Information to be stored for any length of time must be transferred to long-term memory in other areas of the brain, where it may remain, much of it in subconscious form, for life.

The hippocampus is a grand switchboard important in integrating stimuli that are parts of an event and in evoking the short-term recall of these links that make up the memory of the event (Figure 37–22). Damage to the hippocampus prevents transferring these short-term memories to long-term storage and so prevents formation of new long-term memories. However, this does not prevent retrieval of existing long-term memories.

Experimenters have started to unravel memory at the molecular level. Some researchers train sea snails to respond to stimuli and trace the resulting changes in circuits containing relatively few neurons. Others study changes in certain hippocampal neurons of mammals.

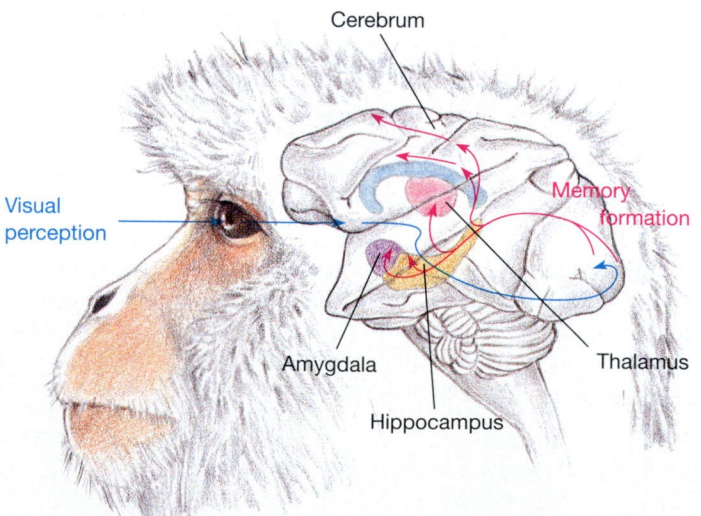

**FIGURE 37–22**   Memory formation. Blue arrows show the visual pathway from the eye of a macaque monkey to the visual cortex in the occipital lobe. From there, signals travel to many other parts of the brain that participate in remembering the stimulus (red arrows). The hippocampus in the deeper areas of the brain is vital to memory formation. If the amygdala is damaged, learning occurs more slowly, and its links to the hypothalamus appear to give memories their emotional content.

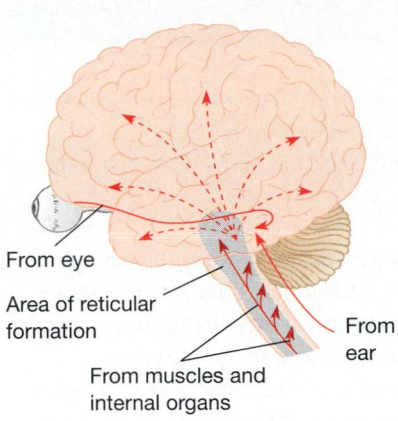

From eye

Area of reticular
formation

From muscles and
internal organs

From
ear

(a)

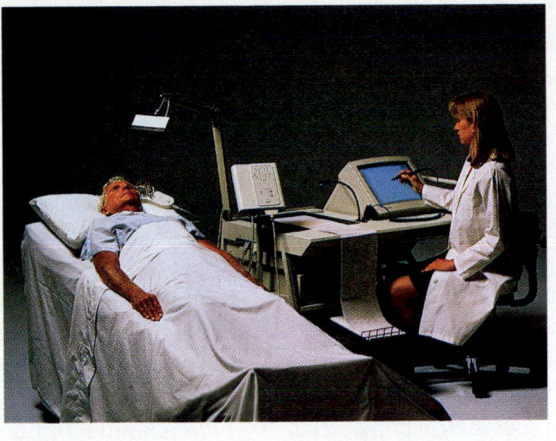

(b)

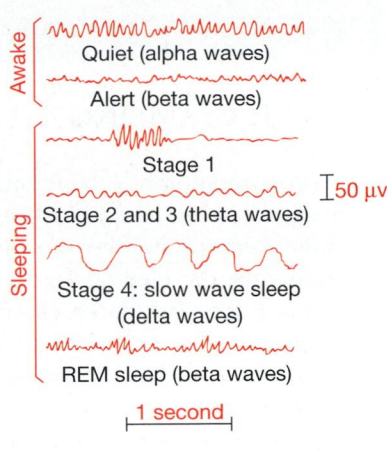

Awake

Quiet (alpha waves)

Alert (beta waves)

Stage 1

Stage 2 and 3 (theta waves)

50 µv

Stage 4: slow wave sleep
(delta waves)

Sleeping

REM sleep (beta waves)

1 second

(c) Electroencephalograms (EEG's)

**FIGURE 37–23** Levels of brain activity. (a) The reticular formation. Many sensory pathways provide input to the reticular formation (solid arrows). The reticular activating system (dashed arrows) in turn communicates with association areas throughout the cortex, stimulating wakefulness and arousal. (b) Recording the brain's electrical activity. Electrodes attached to the scalp detect the brain's electrical signals and can be recorded in an electroencephalogram (EEG). (c) EEGs. Desynchronized beta waves occur in the alert, conscious brain and in REM sleep (bottom recording in each group).   (b, Nicolet Biomedical)

It now appears that two signals are linked and form a memory when they cooperate in activating second messengers inside a neuron. This initiates a cascade of changes in enzyme activity, ion flow, and transmitter release that neither signal could have evoked alone. Short-term memories involve changes that boost the amount of transmitter released, making the postsynaptic neuron more apt to fire. For example, in the hippocampus, nitric oxide produced in postsynaptic cells diffuses back and primes presynaptic cells that have just fired, so that they will release more transmitter in the future. In the bigger picture, this makes it more likely that a given stimulus will activate the circuit of neurons the next time it occurs. This form of learning, **long-term potentiation,** involves the prolonged facilitation of synaptic transmission.

Formation of long-term memory involves activation of certain genes and hence the making of new RNA and proteins, as well as movement of proteins in specific branches of neurons and changes in synaptic structure. Synaptic changes do not occur in the entire neuron but only in particular branches, making those synapses respond more efficiently to specific stimulus patterns. For the first hours or days, these changes are reversible, but later they become more or less permanent. Because only some branches are involved, researchers speculate that one neuron can be part of more than one memory, as different branches strengthen their connections to various other neurons.

In summary, learning involves a cascade of chemical reactions. It first boosts the firing of particular neurons and

then may go on to make longer-lasting changes that remodel the neural circuit so that its increased response to specific combinations of stimuli becomes built-in. Researchers are still investigating and arguing about the details. However, it does appear that recognition and habit memories involve different kinds of neural circuits and chemical changes.

## 37–M  ATTENTION AND SLEEP

### Reticular Formation

The general level of alertness is controlled by the brain's **reticular formation,** named for the net-like pattern formed by its many small neurons in an area extending from the medulla into the midbrain (Figure 37–23). Its cells receive input from all types of sensory neurons entering the brain. Its output goes back down the spinal cord, where it amplifies or reduces incoming sensory signals (much like a volume control on a radio). This adjusts the nervous system's sensitivity to stimuli.

The **reticular activating system** consists of diffuse pathways extending from the reticular formation to the cerebral cortex. It acts as a filter determining which sensory information reaches the level of consciousness in the cortex. This system receives stimuli from all kinds of sense organs but does not keep the different senses separate: it simply stimulates the cortex to alertness and attention. When we

## PATCH CLAMPING

Neuron membranes have many kinds of ion channels, some gated electrically, others chemically. Recordings like the one in Figure 37–7 summarize the electrical activity of many channels at once. The patch clamp technique gives researchers a way to isolate and study a small patch of membrane, containing only one or a few channels.

First, enzymes are used to clean extracellular material off the neuron's plasma membrane. Then the tip of an extremely fine pipette is placed against the membrane, and gentle suction is applied. The rim of the pipette seals around a patch of membrane, isolating it physically and electrically from the rest of the cell surface. An electrode immersed in the solution inside the pipette records changes in current flow as experimental chemical solutions or voltages are applied. Because the patch is so small, this measures the activity of individual channels in response to the applied stimuli (Figure 37–A).

The pipette can also be used to remove the patch of membrane from the cell. Then whatever channels it contains can be studied by applying test chemicals to either side of the membrane. Researchers can also manipulate the pipette to open a hole in the membrane and do whole-cell recording from neurons too small to impale with conventional electrodes. This procedure is often used to study neurons in slices of brain tissue while they continue to interact with their neighbors.

Researchers have found many uses for patch clamping:

1. To measure current flow through the membrane with experimental changes in electrical potential, chemical environment, or concentrations of transmitters or drugs.
2. To measure the width of ion channels. For example, a sodium channel permits any positively charged ion to pass but is used mostly by sodium ions because they are small and numerous compared with other positively charged ions outside the cell. By using solutions of larger ions, researchers have been able to

**FIGURE 37–A** Patch clamping. *Preparation:* The tip of a micropipette is used to create a seal around a tiny patch of membrane. Electrodes in the pipette and in some other part of the cell record activity in the channels enclosed by the pipette tip. The pipette can also be used to do whole-cell recordings or to detach a patch of membrane from the cell.

*Recordings:* (top) This recording shows that channels have two distinct states: open or closed, but they vary in how long they stay open. When the channel is closed, the current (ion flow) is zero. (bottom) If the patch contains more than one channel, the membrane potential changes in discrete steps as channels open or close. In a recording from an entire neuron, these tiny steps merge into a smooth-looking overall curve (see Figure 37–7).

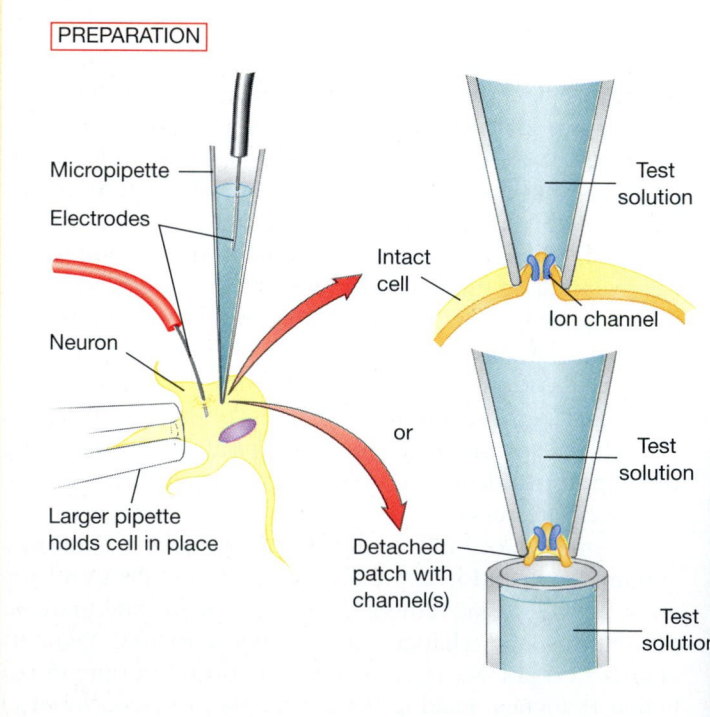

PREPARATION

Micropipette

Electrodes

Neuron

Larger pipette
holds cell in place

Intact
cell

Test
solution

Ion channel

or

Test
solution

Detached
patch with
channel(s)

Test
solution

find out which ones are too big to squeeze through certain channels. They have also investigated how mutations in channel proteins change the tunnel size.

3. To distinguish the effects of cytoplasmic substances. Solutions of known composition can be applied to the cytoplasmic side of detached membrane patches. In 1985, researchers using this procedure found that the second messenger operating sodium channels in the eye's light receptor cells (rods and cones) is not calcium ions, as previously supposed, but the nucleotide cyclic GMP.

By permitting the study of such details, the patch clamp technique has proven so useful that its inventors, Bert Sakmann and Erwin Neher, received a 1991 Nobel Prize.

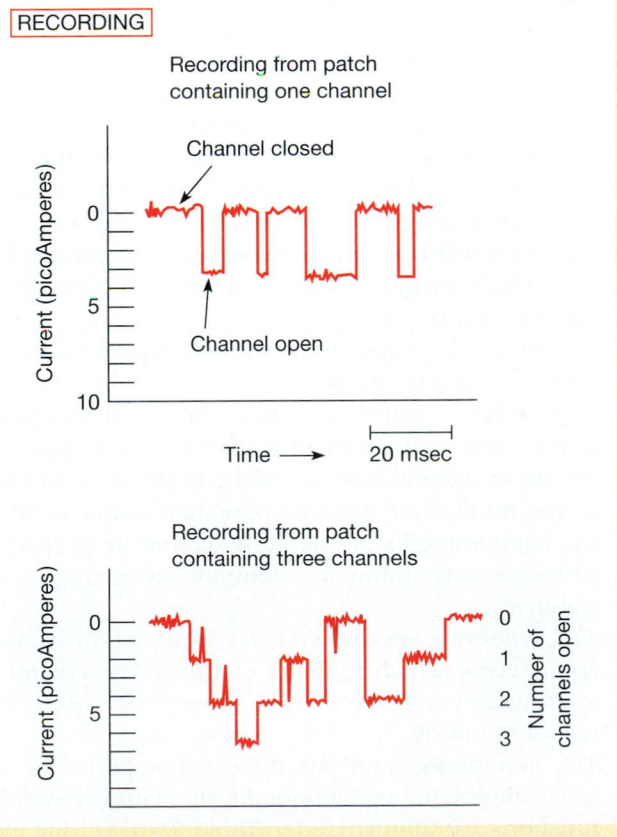

RECORDING

Recording from patch containing one channel

Channel closed

Channel open

Current (picoAmperes)

Time ⟶    20 msec

Recording from patch containing three channels

Current (picoAmperes)

Number of channels open

wake up suddenly, not knowing what woke us but certain that something did, it is because the reticular activating system flashed an "all points" arousal message throughout the cerebral cortex in response to some stimulus.

The brain's electrical activity can be detected by way of electrodes strapped to the scalp. Electroencephalograms (EEGs) recorded in this way show that at rest with the eyes shut, the conscious brain produces a more or less regular, synchronized pattern, the alpha rhythm. This regular rhythm is broken by mental concentration or arousal by sensory stimuli, for example from the reticular activating system. This results in a desynchronized, irregular pattern (beta waves) known as an arousal or alerting response.

## Sleep

Sleep is a nervous function we still know little about. It is produced partly by absence of stimuli transmitted to the cerebral cortex by the reticular system and partly by activity in parts of the thalamus, hypothalamus, reticular formation, and basal forebrain. Electrical stimulation of these areas puts experimental subjects to sleep promptly.

Most mammals and birds sleep, although many of them keep more reflexes active than humans do and remain upright. Reptiles, amphibians, and fishes also show periods when they are very unresponsive to stimuli. Sleep is undoubtedly related to circadian (daily) rhythms, which all animals display (Section 40–G). Indeed, the daily cycle of body temperature is synchronized with sleep, and most people wake spontaneously when the body temperature begins to rise.

We still have no idea why we must sleep each night nor why sleeping has such a profound effect on our temper, efficiency, ability to learn, and emotional stability. Nor do we know why some individuals need more sleep than others. Guinea pigs and humans with brain damage have lived for years without sleep, so sleep is obviously not necessary to survival. Since a sleeping animal is highly vulnerable to predators, sleep must have some powerful counteracting selective advantage.

All we can do now is describe the changes that occur during sleep. Heartbeat rate and blood pressure drop, and breathing becomes more shallow. Body temperature drops slightly, but the temperature of the big toes rises about 5 °C. More important, the brain's electrical activity changes (Figure 37–23). The alpha rhythm ceases at the onset of sleep, and brain activity then cycles through four different stages of sleep, from light to deep and back, followed by a period of **REM** or **paradoxical sleep,** which occurs every 80 to 120 minutes.

REM stands for "rapid eye movements." During REM sleep the reticular activating system arouses the brain, producing desynchronized activity resembling the waking state. Paradoxically, though, motor activity is strongly inhibited; the muscles are more relaxed, and the sleeper harder to awaken, than at any other stage. People awakened from REM sleep nearly always recall dreams, although dreaming can also occur during the other stages of sleep. We spend one fourth of our sleep time in this stage, spread among four to six periods each night.

REM sleep apparently helps consolidate memories to longer-term storage: research subjects who learned a task one evening performed it better the next morning after a good night's sleep or after being deprived of non-REM sleep, but did no better if deprived of REM sleep. REM sleep seems to be the sleep stage most crucial to our psychological well-being; people deprived of REM sleep become extremely tired, and they compensate by increasing REM sleep on subsequent nights. Alcohol, barbiturates, and morphine all decrease REM sleep.

The concentrations of various neurotransmitters in the brain are different in waking and in different kinds of sleep. Norepinephrine, dopamine, and acetylcholine occur at higher levels during wakefulness, whereas serotonin levels rise in the brain during sleep. Cause and effect are difficult to distinguish here: does an increase in serotonin induce sleep, or does sleep increase the activity of neurons that produce serotonin?

It seems likely that sleep permits the restoration of biochemical functions depleted by the day's activity and also permits processing and reorganization of information in the nervous system, but how and why these things happen will probably take a long time to discover.

## SUMMARY

An animal's nervous system is a major communication link, receiving and integrating sensory signals from receptors in various parts of the body and sending instructions for appropriate responses to muscles and glands, the body's effectors. All this information is processed by neurons. Messages travel the length of a neuron in the form of electrical impulses in its membrane and between neurons usually in the form of chemical transmitters.

In a few simple invertebrate nervous systems, it is possible to dissect out individual neurons and determine their exact role. Most nervous systems are too complex for such a direct approach. However, modern techniques permit researchers to locate cells and cell tracts that use particular neurotransmitters or synaptic receptors and to follow the pathways of information through the brain. We finally see glimmerings of the intricate wiring of this massively complex onboard computer. We can now make minor adjustments in structure or chemistry that improve its function, but we still cannot make true repairs.

1. Although neurons vary in size, location, and connections, they all have similar basic structure and function (see Table 37–1). The more numerous glial cells provide physical and life support services to neurons.

2. A resting neuron has two sources of stored electrochemical energy: a $Na^+$ gradient (higher concentration outside the cell) and a $K^+$ gradient (higher concentration inside). This asymmetrical ion distribution is maintained, in the long run, by the sodium-potassium pump. The differential permeability of the membrane to various ions results in a resting potential across the membrane.

3. Information passes along a neuron in the form of brief disturbances in the membrane potential, which occur when stimuli open membrane channels and permit ions to spurt through.

4. Local potentials in a neuron's dendrites and cell body add together. If the sum exceeds the threshold level, the axon fires an action potential.

5. An action potential travels down an axon faster if the axon has a large diameter or if it is electrically insulated by a myelin sheath.

6. When information reaches an axon terminal, it is usually transmitted as a chemical that crosses the synaptic cleft to the membrane of the next cell and disturbs its electrical balance in turn. Depending on the nature of the postsynaptic receptor, the binding of transmitter may produce excitatory (depolarizing) or inhibitory (hyperpolarizing) postsynaptic potentials. Or it may instead activate a second messenger system inside the cell, which changes the postsynaptic cell's sensitivity to further stimulation.

7. An electrical synapse passes action potentials in either direction via gap junctions.

8. During the evolution of animals, progressive cephalization resulted in the formation of a brain and major sense organs at the anterior end of the body. A vast increase in the number of neurons permitted better control of the many muscles in a large body and more flexibility of response to stimuli in a complex and changing environment.

9. The vertebrate nervous system consists of the brain and spinal cord, which together comprise the central nervous system, and the peripheral nervous system in the rest of the body.

10. The vertebrate brain has three major parts, the forebrain, midbrain, and hindbrain. Its main divisions and functions are summarized in Table 37–4. During evolution, the brain increased in size and complexity. Some parts retained their primitive functions, while others

took on new roles as body structure and behavior became more complex and as intelligence increased.

11. The function of the brain is to "make decisions." Using information coded as patterns of action potentials coming from the external or internal environment via the sense organs, the brain produces a set of directions coded as another set of action potentials that cause the effector organs to respond.

12. Information passes through various levels of organization in both sensory and motor areas as the brain analyzes and integrates sensory input and initiates appropriate responses. Each of these areas contains layers of interneurons with their synapses arranged in a hierarchy of information processing.

13. The spinal cord is primarily a relay system connecting the brain with the body by way of peripheral nerves. Some of its neurons mediate spinal reflexes, which save time and energy by performing rapid defensive reactions or routine actions without involving higher centers in the brain.

14. The peripheral nervous system consists of the somatic and autonomic nervous systems. The somatic system carries sensory information from the body and also serves the muscles under voluntary control, permitting us to react consciously to the outside world.

15. The autonomic system carries motor impulses to the muscles and glands of the internal organs, under little conscious control; it copes mainly with homeostatic mechanisms within the body.

16. A memory is a stored pattern of information, in the form of temporary or permanent changes in some of the synapses in neuron circuits that pass through many parts of the brain. Here it is available for future reference because it makes particular synapses more likely to fire when a stimulus that is part of the memory pattern is encountered again.

17. The reticular formation and reticular activating system are responsible for filtering sensory information and determining general levels of lethargy or liveliness.

## Self-Quiz

1. Action potentials travel in a neuron's:
   a. dendrites
   b. potassium channels
   c. myelin sheath
   d. axon
   e. cell body

2. Tell whether each of the following is true of an excitatory local potential, an action potential, or both.
   _____ a. It has a threshold.
   _____ b. It is a graded depolarization.
   _____ c. It has a refractory period.
   _____ d. It involves movement of Na$^+$ into the neuron.
   _____ e. It undergoes summation.

3. The myelin sheath around the axons of some vertebrate neurons:
   a. is rich in lipids because it is formed by many layers of membranes
   b. is a secretory product of glial cells
   c. is produced inside the axon and extruded out through the membrane
   d. is continuous all along the length of the axon
   e. secretes neurotransmitter substances for release at the synaptic boutons

4. Write a short sentence describing the function and importance of each of the following components of a synapse:
   a. neurotransmitter substance
   b. neurotransmitter vesicle
   c. receptor molecules
   d. enzymes that destroy neurotransmitter

5. In an inhibitory synapse:
   a. information travels from the postsynaptic to the presynaptic cell
   b. the neurotransmitter used is different from the neurotransmitter used in excitatory synapses
   c. there are no receptor molecules on the postsynaptic membrane
   d. the postsynaptic receptor molecules may cause hyperpolarization of the membrane rather than depolarization when they bind neurotransmitter
   e. the stimulus is transmitted electrically, not chemically

Matching: From the list of brain structures a through e, select the part that performs the functions listed in questions 6 through 10.
   a. cerebrum
   b. cerebellum
   c. hypothalamus
   d. medulla
   e. thalamus

6. _____ Regulatory control of deep body temperature, osmoregulation, thirst, and hunger
7. _____ Coordination of movement
8. _____ Visual and auditory processing and initiation of voluntary movements
9. _____ Conscious thought, intelligence, and memory
10. _____ Communication between spinal cord and brain
11. In looking at the evolution of the vertebrate brain from fish to humans, the greatest increases in size are seen in the:
   a. medulla and cerebellum
   b. cerebellum and optic lobes
   c. optic lobes and cerebral hemispheres
   d. cerebral hemispheres and cerebellum
   e. medulla and thalamus

12. Choose the letter(s) indicating the location of these features in the diagram below:

_____ a. An axon
_____ b. A synapse
_____ c. An interneuron
_____ d. Site of local potentials
_____ e. Vesicles containing acetylcholine
_____ f. Myelin

13. Indicate whether each feature is true of the parasympathetic or sympathetic nervous system or of both:

_____ a. ganglia lie outside the central nervous system
_____ b. arises from thoracic and lumbar segments of the spinal cord
_____ c. has direct connections between adjacent ganglia without passing through the CNS
_____ d. contains only motor neurons
_____ e. raises blood pressure, heartbeat rate, and breathing rate
_____ f. releases acetylcholine at the effectors

## Questions for Discussion

1. After an action potential arrives at the presynaptic membrane, what factors determine whether the postsynaptic cell will fire an action potential?
2. Why don't all the neurons in an animal's nervous system have either giant or myelinated axons? What is the selective advantage of having some neurons with axons that conduct impulses more slowly?
3. Tetrodotoxin is a chemical produced by puffer fish that interferes with the opening of sodium channels involved in transmission of an action potential. Why is tetrodotoxin a powerful poison?
4. Multiple sclerosis (MS) is characterized by patchy destruction of myelin. What symptoms would you expect this to produce?
5. Explain why cocaine addicts have decreased numbers of dopamine receptors.
6. Many neurons have synaptic arrangements in which the presynaptic membrane of a *dendrite* of one cell releases neurotransmitter that is received by the dendrite or soma of another cell. Neurons that process information about vision and smell contain many dendrite-to-dendrite reciprocal synapses: two synapses occur side by side, one sending information from cell A to cell B, the other from cell B to cell A. Such synapses release neurotransmitter in response to local potentials not large enough to initiate action potentials. How would such arrangements affect the functions of the two neurons? How would this affect the overall working of the nervous system?
7. What possible disadvantages are there in the evolutionary trend toward cephalization of the nervous system? What advantages?
8. During embryonic development of mammals, the nervous system forms all the neurons an animal will ever have. After birth, neurons never divide to form new neurons. What is the adaptive advantage of this? What are some drawbacks?
9. Studies using PET show that the human visual cortex uses increasing amounts of energy under the following series of conditions: eyes closed; looking at diffuse white light; looking at a checkerboard pattern; looking out the window at a park. Explain these findings using the information on visual processing in Section 37–H.

## Suggested Readings

Alkon, D. L. "Learning in a marine snail." *Scientific American,* July 1983. Extensive experiments identified the roles played by wiring and by chemical changes in neurons when an animal with a simple nervous system learns a new response.

Culotta, E., and D. E. Koshland, Jr. "NO news is good news." *Science* 258:1862, 1992. Describes the exciting work that earned NO *Science* magazine's award for "molecule of the year" among a field of outstanding contenders.

Franklin, J. *Molecules of the Mind.* New York: Atheneum, 1987. A well-informed journalist's thought-provoking look at the neurochemical revolution, mental illness, and society's attitudes about both.

Gottlieb, D. I. "GABAergic neurons." *Scientific American,* February 1988. A readable description of the discovery and roles of an important group of inhibitory neurons.

Julien, R. M. *A Primer of Drug Action,* 5th ed. New York, W. H. Freeman, 1988.

Kimelberg, H. K., and M. D. Norenberg. "Astrocytes." *Scientific American,* April 1989. The many functions of a major kind of glial cell in health and disease.

Lent, C. M., and M. H. Dickinson. "The neurobiology of feeding in leeches." *Scientific American,* June 1988. An interesting and readable case study of the nervous system's role in a vital activity. Outlines experiments that detected the nervous pathway responsible for feeding.

Melzack, R. "The tragedy of needless pain." *Scientific American,* February 1990. A claim that morphine used only to control pain is not addictive and should be prescribed long-term to keep patients with nonterminal illness pain-free.

Mishkin, M., and T. Appenzeller. "The anatomy of memory." *Scientific American,* June 1987.

*Scientific American.* "Mind and Brain," September 1992 issue. A collection of articles about brain structure and function.

Springer, S. P., and G. Deutsch. *Left Brain, Right Brain.* San Francisco: W. H. Freeman, 1989.

# Sense Organs

### O B J E C T I V E S

*When you have studied this chapter, you should be able to:*

1. Write a short paragraph discussing the general importance of sense organs to an animal.
2. Name the five general types of receptor cells, and name some sense organs in which each is found.
3. Explain how information about the intensity and type of a stimulus is transmitted in the nervous system.
4. Define sensory adaptation; state the difference between rapidly adapting and slowly adapting receptors, and explain the advantages of each.
5. State what a proprioceptor is, and briefly describe the importance of the muscle spindle to the body.
6. Describe the roles of mechanoreceptors in the inner ear; explain how the ear recognizes differences in pitch, tone, and volume of sounds.
7. Sketch and label the eyes of a vertebrate and of an insect.
8. Name the two types of vertebrate photoreceptors, state their function, and describe the role of rhodopsin in vision.

---

*I*'*ll be in* touch *later if* I feel *better, but I can't bear the* sight *or* smell *of hotdogs while my stomach* aches *like this.* That sentence contains five references to senses—of pain, sight, smell, and touch. Indeed, we seldom discuss anything for long without referring to our senses because our world, both around us and inside our bodies, is what our senses tell us it is.

We often speak of having five senses. In fact, we have more than a dozen different types of sense organs, which monitor conditions outside and inside our bodies. Internal sense organs detect and report changes in conditions such as body temperature, osmotic relationships, and pH. This information is used to maintain homeostasis. External sense organs report sights, sounds, and chemicals in the outside world, information used in feeding, in finding mates, in avoiding enemies, and in making other adaptive responses to the environment (Figure 38–1). Information collected by sense organs is passed to the nervous system, which determines and initiates an appropriate response.

Our studies of other animals' sense organs are biased by our own senses, which give us only one of many possible perceptions of the world. Humans are creatures of vision and, to a lesser extent, of hearing. It is hard for us to empathize with, say, an earthworm, which perceives neither line nor color.

Many animals have sense organs far more sensitive than ours, and some rspond to stimuli that we cannot detect at

**FIGURE 38–1** Sensing the environment. The sense organs of most animals are concentrated in the head. The head of a lion bears organs that sense mechanical stimuli (ears and whiskers), light (eyes), heat (tongue), and chemicals (nose and tongue).

all. For instance, the light in the sky is polarized in different directions relative to the sun and to the observer. Many invertebrates (and some vertebrates) can detect polarized light, using this ability for determining compass direction and for navigation. The electric fish *Gymnotus* has receptors that are sensitive to electrical impulses and can distinguish neighbor from stranger by the distinctive electrical pulses each fish emits. Bats emit ultrasonic squeaks and detect the faint echoes with such acute hearing that they can avoid telephone wires and catch flying moths on the darkest nights.

We cannot imagine what it would be like to have such senses. In terms of perception, we live in a different world from that of the worm burrowing in the flower bed or the pigeon strutting on the sidewalk. Nevertheless, there are many similarities in the types of sense organs found among different animals and in how these organs work.

### KEY CONCEPTS

♦ Sense organs gather information about changes inside the animal's body and in its external environment and pass this information to the nervous system.

♦ The particular collection of sense organs in each animal species gives its members a unique perception of their bodies and of their environment and are part of their adaptation to a particular way of life.

## 38–A SENSE ORGANS AND THEIR FUNCTIONS

The key elements of sense organs are **receptors,** structures that change in some way when they encounter an appropriate stimulus. In some sense organs, such as the vertebrate nose, the receptor consists merely of the dendrites of a sensory neuron. In others, such as the retina of the eye and taste buds in the tongue, the receptor is a separate, non-nervous cell close to the sensory neuron's dendrites. Either way, receptor cells respond to stimuli by producing electrical activity in sensory neurons of the nervous system.

A **stimulus** is some form of energy. Various receptors respond to energy in the form of pressure, light, electrical current, chemical changes, osmotic potential, magnetic fields, and heat. All of these stimuli will cause changes in the membrane of almost any cell. Even in unicellular organisms, however, there is specialization, with different areas of the cell peculiarly sensitive to one stimulus rather than another. In higher animals, each receptor cell is specialized so that it reacts maximally to a particular form of energy.

Many sense organs are incredibly sensitive. Receptor cells in our eyes can detect a single photon (the fundamental unit of light), and a male gypsy moth can detect a single molecule of the female's sexual attractant that comes in contact with his antenna!

Most sense organs are geared to alert the nervous system to *changes* in stimuli. Our eyes send particularly strong signals to the brain when they detect movement, a sudden shadow or a flash of light, and edges between light and dark areas (Figure 38–2). This is adaptive. Detecting movements of predators or prey, and outlines of objects in the surroundings, is more important to an animal than detecting a constant stimulus.

An animal responds to a stimulus in three steps:

1. Receptor cells are changed by the energy of a stimulus in such a way that the stimulus energy is converted into electrical energy, which can travel in the nervous system.
2. The nervous system may process information it receives from sense organs. It then passes signals for appropriate responses to the effectors.
3. Effectors are usually muscles and glands. Muscles contract, or glands secrete chemicals, producing suitable

**FIGURE 38–2** Sense organ sensitivity to changes and edges can produce optical illusion. Each stripe in this series is of uniform darkness. However, the human visual system is set up in such a way that it accentuates edges. Hence, it misinterprets these stripes, perceiving each one as having a darker edge where it meets its lighter neighbor.

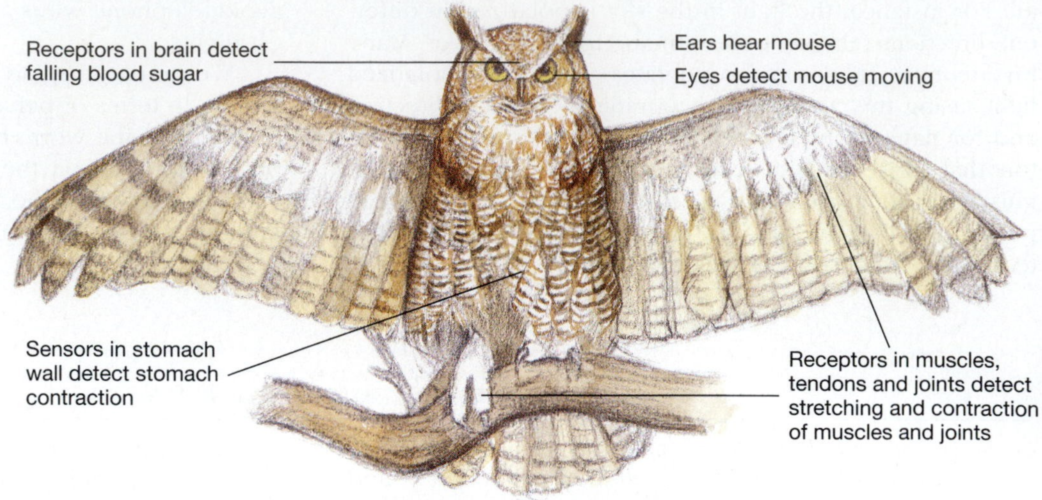

Receptors in brain detect
falling blood sugar

Ears hear mouse

Eyes detect mouse moving

**FIGURE 38–3**   Sense organs respond to change. This diagram indicates a few of the many sensory receptors responding to changes inside and outside the body of an owl about to start the night's hunting.

Sensors in stomach
wall detect stomach
contraction

Receptors in muscles,
tendons and joints detect
stretching and contraction
of muscles and joints

responses to the information passed to the nervous system from the sense organs (Figure 38–3).

"What about when I just sit and listen to music?" you may ask. "The music goes in (step 1), and the nervous system processes it (step 2), but no response occurs—I just enjoy." Not so; tiny muscles in your ears react to noises and adjust the ears' sensitivity, depending on the loudness of the music. In addition, depending on the type of music, your heart may beat faster or slower, and your other muscles may become more tense or relaxed. Even stopping a natural urge to tap your feet to the beat is a response.

## Stimuli

A sense organ reacts to a stimulus, which is some form of energy. Receptors may be classified by the form of energy to which they respond (Table 38–1):

- **Mechanoreceptors** detect mechanical energy in the form of movement, pressure, or tension (pull or stretch).
- **Photoreceptors** detect light energy.
- **Thermoreceptors** detect heat energy.
- **Chemoreceptors** detect chemicals.
- **Electroreceptors** detect electrical energy.
- **Magnetoreceptors** are recently discovered receptors that respond to Earth's magnetic field. Sea turtles, pigeons, dolphins, bees, monarch butterflies (and some bacteria) have these receptors. There will probably prove to be many more organisms containing particles of magnetite (iron oxide) that work like magnetic compasses, changing direction in response to Earth's magnetic field.

A receptor's most important role is to convert the energy of a stimulus into electrical energy, the only form of energy that can be transmitted by the nervous system. The receptor produces a local electrical potential, a **receptor potential,** with a magnitude proportional to the intensity

of the stimulus. If the receptor potential reaches the sensory neuron's threshold, it causes the neuron's axon to fire action potentials.

In 1950 Bernard Katz recorded receptor potentials near the dendrites of a sensory neuron of a vertebrate **muscle spindle,** a sense organ that detects stretch or contraction in the surrounding muscle. A receptor potential large enough to reach the threshold induced the neuron to fire action potentials. Figure 38–4 shows that the more the muscle was stretched, the higher the frequency of action potentials. Hence, the sensory neuron conveyed information about the extent of muscle contraction to the central nervous system.

Many sense organs filter incoming energy so that only part of the stimulus reaches the receptor cell. The human eye, for instance, contains various **accessory structures** that act as filters (see Figure 38–15). One of these, the iris, controls the size of the pupil and so changes the amount of light entering the eye. The lens adjusts the focus of the light, and it also filters out much of the ultraviolet light. Animals such as honeybees, which do not have ultraviolet fil-

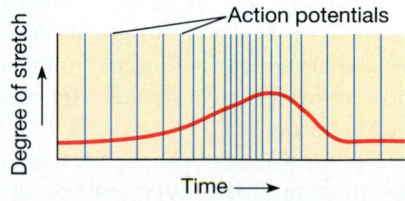

FIGURE 38–4   Recording from the sensory neuron of a muscle spindle, a stretch receptor in vertebrate muscle. The blue lines represent action potentials in the sensory neuron's axon. The red line shows the degree of stretch in the muscle. As the muscle is stretched, the frequency of sensory impulses increases and then falls again as the stretch subsides.

## TABLE 38-1

**Classification of Some Receptors by Stimuli**

| General Name | Examples | Effective Stimulus |
|---|---|---|
| Mechanoreceptors | Pacinian corpuscles (vertebrate connective tissue) | Deep touch and vibration |
| | Meissner's corpuscles | Touch on skin |
| | Proprioceptors | Position of parts of body |
| |   Joint receptors | Angular movement of joint |
| |   Golgi tendon organs | Stretch of tendon |
| |   Muscle spindles | Degree of muscle contraction |
| | Statocysts in invertebrates | Gravity |
| | Hair cells of vertebrates | |
| |   Lateral line organs in fish | Pressure waves and currents in water |
| |   Utriculus and sacculus | Gravity; linear acceleration |
| |   Semicircular canals | Angular acceleration |
| |   Cochlea | Airborne sound waves |
| Photoreceptors | Ommatidia of arthropods | Light |
| | Rods and cones of vertebrate retina | Light |
| Thermoreceptors | Pit organs of pit vipers; labial organs of boas; nerve endings in the tongue and skin of mammals | Increasing and decreasing infrared radiation |
| | Krause's end bulbs | Cold on skin |
| Chemoreceptors | Taste buds and olfactory organs of vertebrates | Chemistry of molecules in air or water |
| | Chemoreceptors of invertebrates | Chemistry of molecules |
| Electroreceptors | Organs in the skin of some fish; bill of platypus | Electric currents in surrounding water |
| Magnetoreceptors | Receptors containing magnetite in brains of sea turtles and some birds | Earth's magnetic field |

ters in their eyes, can perceive and react to ultraviolet light that we cannot see.

## 38–B  CODING AND INTERPRETING MESSAGES

Like computers, neurons convey information in digital form. Digital (di = two) devices have only two positions, on and off, and all the information they convey must be "written" in this two-word "code." Yet nerves convey information about not just the existence but also the size of a stimulus received by a sense organ. The nervous system encodes stimulus strength in three ways:

1. **Frequency of action potentials.** A strong stimulus causes more frequent action potentials than a weak stimulus. Thus a strong stimulus produces a higher frequency of action potentials (more action potentials per second).
2. **Duration of a burst of action potentials.** A weak stimulus may give rise to a short burst of pulses in the neuron, a strong stimulus to a longer burst.
3. **Number and kinds of neurons firing.** The threshold needed to initiate a nerve impulse varies from one neuron to another. Thus, a weak stimulus will cause only a few neurons to fire, whereas a stronger stimulus will fire these neurons, plus others with higher thresholds.

### Adaptation

If a constant stimulus is applied to a receptor for any length of time, the frequency of action potentials in the sensory neuron decreases as time passes. This diminishing response to a constant stimulus is called **adaptation.** Adaptation may result from the receptor's producing a smaller receptor potential with time, or from the sensory neuron's becoming less responsive to stimulation, or both.

Both receptors and sensory neurons may adapt slowly or rapidly to a constant stimulus, and there are advantages to possessing cells with both kinds of responses. A rapidly adapting receptor allows an animal to ignore an unchanging stimulus. For example, touch receptors in the skin adapt rapidly to a constant stimulus and fire rapidly again only if the stimulus changes. So, when we dress in the morning we notice the contact between our clothing and skin only briefly, and we are not distracted by the feel of our clothes all day. However, a fly crawling down the neck causes our touch receptors to respond immediately.

Receptors that adapt rapidly detect stimuli that usually change rapidly. Visual and sound receptors adapt rapidly, and animals are consequently much more sensitive to changing sights and sounds than to constant ones.

A slowly adapting receptor produces action potentials in the sensory neuron for as long as the stimulus lasts, continuing to provide information about stimulus intensity. Examples in humans include receptors for pain and cold.

Adaptation occurs not only in receptors and sensory neurons but also in the central nervous system. Indeed, adaptation at all levels in the nervous system is important in determining an animal's reaction to a particular stimulus.

## 38–C  MECHANORECEPTORS

Animals have many kinds of mechanoreceptors, which respond to many kinds of movement, including changes in pressure. Stretching of the receptor membrane causes the receptor potential.

Some mechanoreceptors adapt rapidly to a constant stimulus. These are admirably suited to detect changes in pressure—that is, movement—in their surroundings. Examples are the receptors at the bases of tactile hairs, which are sensitive to touch (Figure 38–5). In humans, rapidly adapting touch receptors are concentrated on the tongue, lips, face, and fingertips, reflecting the importance of the head and of the manipulative fingers. Other mechanoreceptors adapt slowly. The pressure receptors that constantly monitor blood pressure in the large arteries adapt slowly, and their rate of firing accurately reflects the blood pressure.

**FIGURE 38–5** Hairs on the back of a caterpillar have receptors at their bases that detect movement of the hairs. (Biophoto Associates)

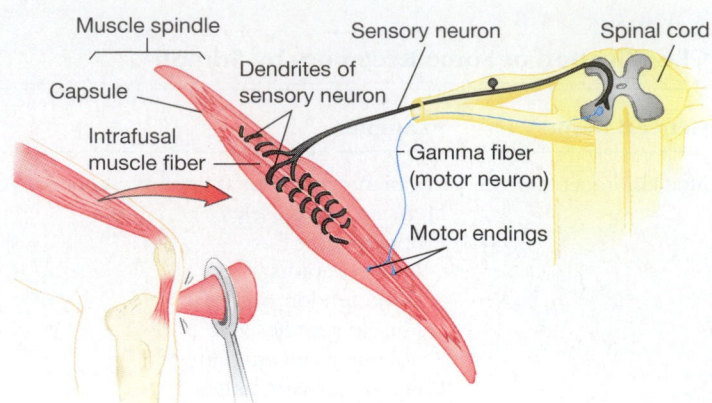

**FIGURE 38–6** A vertebrate muscle spindle. A muscle may contain hundreds of spindles. When the muscle stretches, it stretches the intrafusal fibers, producing a receptor potential. This stimulates the dendrites of the associated sensory neuron and initiates impulses that travel to the central nervous system. The intrafusal muscle fibers contract when stimulated by the gamma fiber. This changes the spindle's sensitivity to stretching of the muscle.

## Proprioceptors

Think of the enormous amount of information needed to control a dive into a swimming pool. The central nervous system receives information on the speed of falling and on the orientation of feet, hands, and head. Much of the necessary information comes from **proprioceptors,** mechanoreceptors in the joints and muscles that continually monitor the position and movements of parts of the body.

Vertebrates have three main types of proprioceptors: **joint receptors,** which detect angular movement in the ligaments that hold the bones of a joint together; **Golgi tendon organs,** which determine stretch in the tendons that hold muscles to bones; and **muscle spindles,** which detect muscle movement. Let us consider muscle spindles as an example of proprioceptors.

***Muscle Spindles***   A muscle spindle consists of several modified **intrafusal muscle fibers** and their sensory nerve endings, separated from the surrounding ordinary muscle by a capsule of connective tissue (Figure 38–6). The dendrites of sensory neurons lie in contact with the intrafusal fibers. When a muscle stretches, it also stretches the intrafusal muscle fibers, and the sensory neurons send information about the degree of stretch to the central nervous system.

The sensory neurons of a muscle spindle form part of a reflex arc that sends motor signals back to the same muscle and cause it to return to its previous state of contraction. In this way, the body constantly monitors and controls the contraction of muscles, such as those responsible for posture, so that we can sit or stand in one position without twitching or flopping.

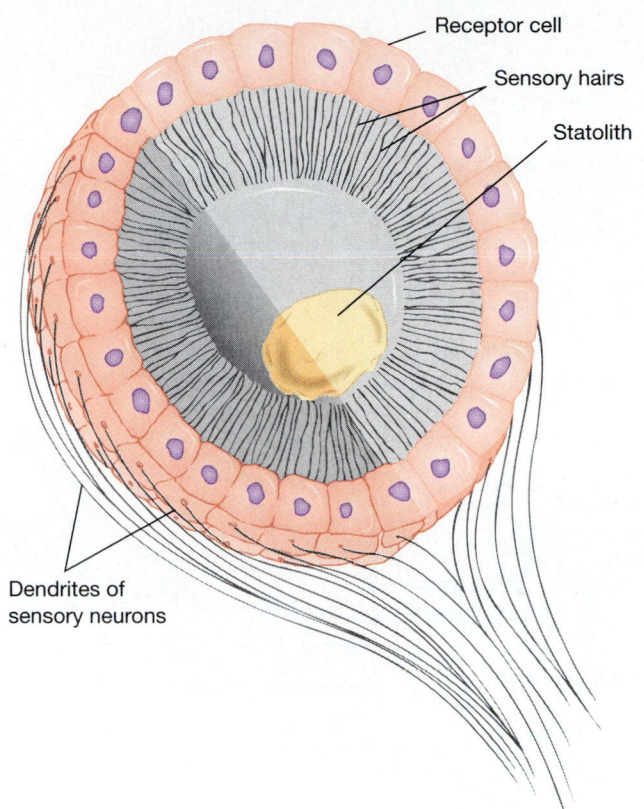

**FIGURE 38–7** The statocyst of a crustacean. The statolith is a particle, and it moves around under the influence of gravity. Wherever it comes to rest, it stimulates the underlying sensory hairs of receptor cells of the statocyst lining.

The muscle spindle itself receives signals from motor neurons (gamma fibers) that cause parts of the spindle to contract. This changes the sensitivity of the spindle to the degree of muscle stretch, so that sometimes only large changes in stretch are detected, whereas at other times even small changes are signaled. The central nervous system also controls the sensitivity of many sense organs.

## Statocysts

Many invertebrates have gravity receptors called **statocysts.** Each consists of a cavity lined with fine hair-like sensory structures and containing dense particles such as grains of calcium carbonate or sand (Figure 38–7). The particles are pulled downward by gravity and stimulate the sensitive cells on which they lie. By detecting which sensory cells are firing, the animal knows where "down" is and so determines how its body is oriented.

A classic experiment on statocysts was done by Hans Kreidl in 1893 using shrimp, whose statocysts are chambers at the bases of the antennae. The lining of the chambers is part of the exoskeleton, and so when a shrimp molts, it loses the lining of the statocyst as well as the sand grains inside the chamber. The shrimp must then pick up more grains of sand with its claws and place them in its statocyst chambers. Kreidl waited until his shrimp molted and then placed iron filings in their tank instead of sand. The shrimp refilled their statocysts with these filings and swam around quite normally. Kreidl then overcame the force of gravity by holding magnets above the shrimp. The filings were attracted by the magnets. The shrimp, responding to the pressure of the filings, promptly began to swim upside down in response to their new perception of "gravity."

## Hair Cells of Vertebrates

Vertebrates have remarkable mechanoreceptors called **hair cells,** each bearing a tuft of large microvilli of graduated lengths (Figure 38–8). Bending this tuft opens gated ion channels in the receptor cell membrane and allows ions to flow through the channels. This produces a receptor potential in the hair cell and then in the dendrites of a sensory neuron near the base of the hair cell. Typically, when the hair bundle is bent in one direction, the sensory neuron fires action potentials more frequently; when it is bent

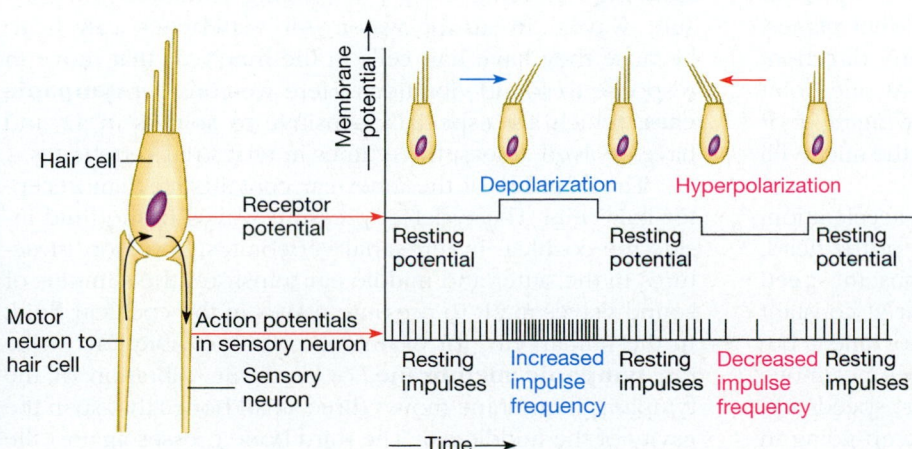

**FIGURE 38–8** A vertebrate hair cell responds to deflection of the hair bundle. The resting cell produces a train of impulses (action potentials) in the sensory neuron. When the hair bundle is bent one way (blue arrow), the cell produces a depolarizing receptor potential, which increases the impulse frequency in the sensory neuron. When the hair bundle is bent in the opposite direction (red arrow), the resulting hyperpolarization of the cell reduces impulse frequency in the sensory neuron. The impulse frequency informs the central neurons in the brain which way the hair bundle is bent.

in the other direction, the sensory neuron fires less frequently. Bending in either direction thus changes the pattern of action potentials sent to the central nervous system. Several kinds of organs contain hair cells.

***Lateral Line Organs***   The **lateral line organs** of fish and larval amphibians contain hair cell pressure receptors. The lateral line consists of a row of water-filled canals or tunnels extending along the animal's side and onto its head (Figure 38–9). In the canals are clusters of hair cells, with their microvilli tufts embedded in a gelatinous substance to form clumps called **cupulae** (singular: **cupula**). When the water in a lateral line canal moves, it pushes a cupula and bends the tuft, generating a signal in the sensory neurons. The lateral line organ is a remarkably sensitive system by which the animal can detect such things as water currents, the movements of other animals in the water, and pressure waves bouncing off stationary objects nearby.

***Labyrinth of the Ear***   We usually think of the ear as an organ of hearing, but much of the vertebrate ear acts like a statocyst, permitting the animal to monitor its orientation and movement. These are the roles of the **labyrinth,** the organ of balance in the inner ears of all vertebrates (Figure 38–10). Many zoologists believe the labyrinth evolved from the anterior end of the lateral line organ. Hair cells in different parts of the labyrinth detect sounds, the direction of gravity, and acceleration of the head. **Acceleration** is a change in the speed or direction of motion.

Two chambers, the **sacculus** and **utriculus,** contain hair cells that detect both gravity and linear acceleration (changes in speed when the body is moving in a straight line). Each chamber contains **otoliths,** crystals of calcium carbonate embedded in a gelatinous secretion atop a tuft of hair cells (Figure 38–11). These particles shift in response to gravity, stimulating the hair cells. Because the tufts in the two chambers lie in different planes, the animal can tell the direction of gravity when the head is in any position.

The three **semicircular canals** detect angular acceleration (rotation, or turning). They are the sense organs of equilibrium. Since the canals lie in three different planes, they can detect acceleration of the head in any direction. Each canal is a hollow ring filled with fluid. At one point in the ring is a cupula, which operates like a swinging door across the canal. Embedded in each cupula are the microvilli tufts of hair cell mechanoreceptors.

The semicircular canals can detect only acceleration, changes in the speed or direction of rotation of the head. They do not react if the head is rotating at a constant speed or moving in a straight line. Traveling in a car at constant speed in one direction gives no sense of movement because the fluid in the semicircular canals does not move with respect to the canals. When the car turns, speeds up, or slows down, however, the fluid tends to keep going in

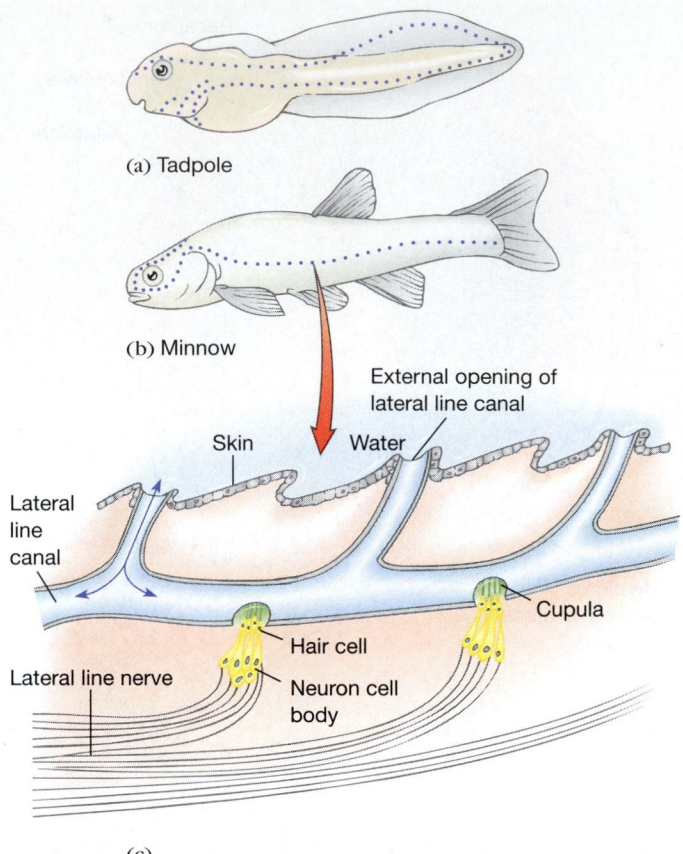

**FIGURE 38–9**   The lateral line system. (a, b) Colored dots show the distribution of openings of the lateral line system. (c) Longitudinal section through the lateral line organ. The lateral line canal connects with the water outside the fish via openings in the skin. The actual pressure receptors are hair cells embedded in cupulae lining the canal. Dendrites from sensory neurons synapse with the hair cell and detect their depolarization.

its original direction, and so it pushes the cupula into a new position, giving a sensation of movement.

***Hearing***   Hearing involves detecting vibrations, or pressure waves, in air or water. All vertebrates can hear, because they have hair cells in the inner ear that move in response to sound vibrations. Here we consider **tympanic ears,** which are especially sensitive to sounds in air and have evolved at least three times in terrestrial vertebrates.

The **cochlea** of the inner ear contains mechanoreceptor hair cells. These detect pressure waves in the fluid inside the cochlea. In terrestrial vertebrates, accessory structures in the outer and middle ear transform the stimulus of sound waves in air to pressure waves in the cochlear fluid. In the human ear, for example, air-borne vibrations strike the **tympanic membrane,** or eardrum. Vibration of the tympanic membrane moves three small bones that span the cavity of the middle ear. The third bone presses against the

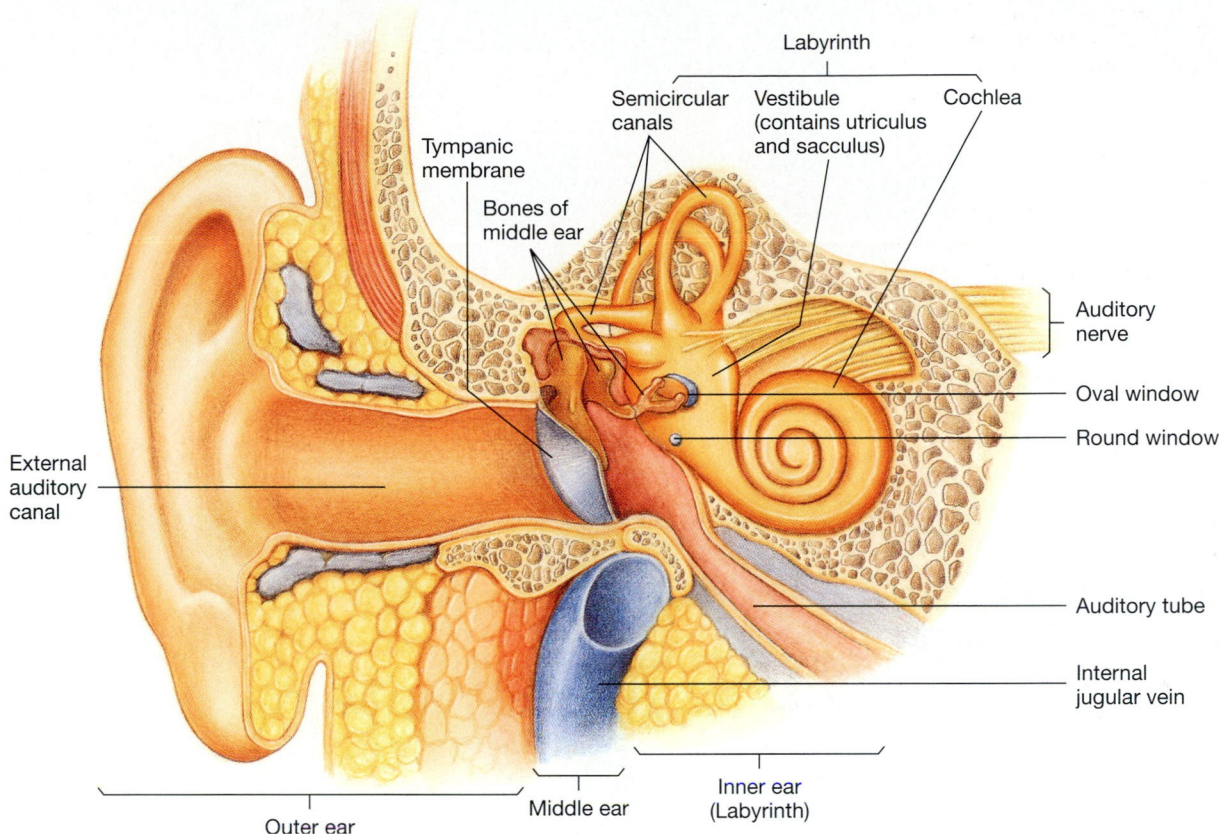

**FIGURE 38–10**   The human ear. Sound waves enter the outer ear, causing vibration of the tympanic membrane, which transmits the vibrations to the three bones in the air-filled middle ear. This, in turn, causes vibrations of the fluid in the cochlea in the inner ear. The inner ear also contains the vestibule, where gravity and acceleration are detected; and the semicircular canals, where rotation is detected.

oval window of the cochlea. Vibration of the oval window in turn moves the fluid inside the cochlea.

The snail-shaped cochlea is divided into fluid-filled canals by membranes that stretch across the inside of the cochlea along its entire length. The **organ of Corti** consists of hair cells sandwiched between the **basilar membrane** and the **tectorial membrane.** These hair cells synapse with neurons in the cochlear branch of the auditory nerve (Figure 38–12).

**FIGURE 38–11**   Labyrinth of the human ear. (a) Sensory cells in the utriculus and sacculus have their tips embedded in a gelatinous material. Particles of calcium carbonate (otoliths) lying on the gelatinous material respond to gravity, deflecting the hair cells as they do so. This signal is passed to the associated sensory neurons. (b) A semicircular canal contains a cupula in an expanded chamber. Here, hair cells embedded in gelatinous material are bent by movement of fluid in the canal. Bending the hair cells one way increases the rate of action potentials in the neuron; bending the opposite way decreases the rate of action potentials.

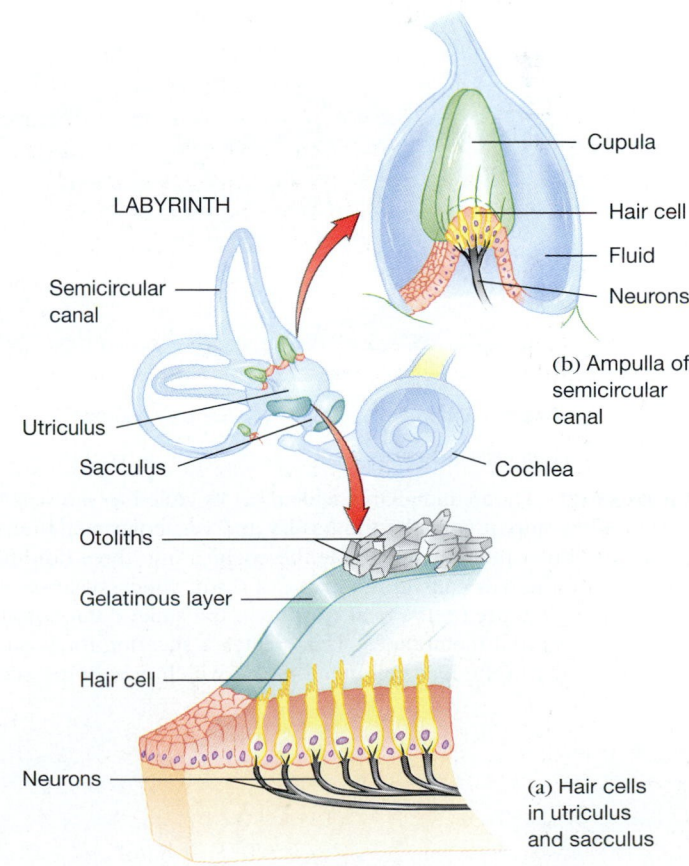

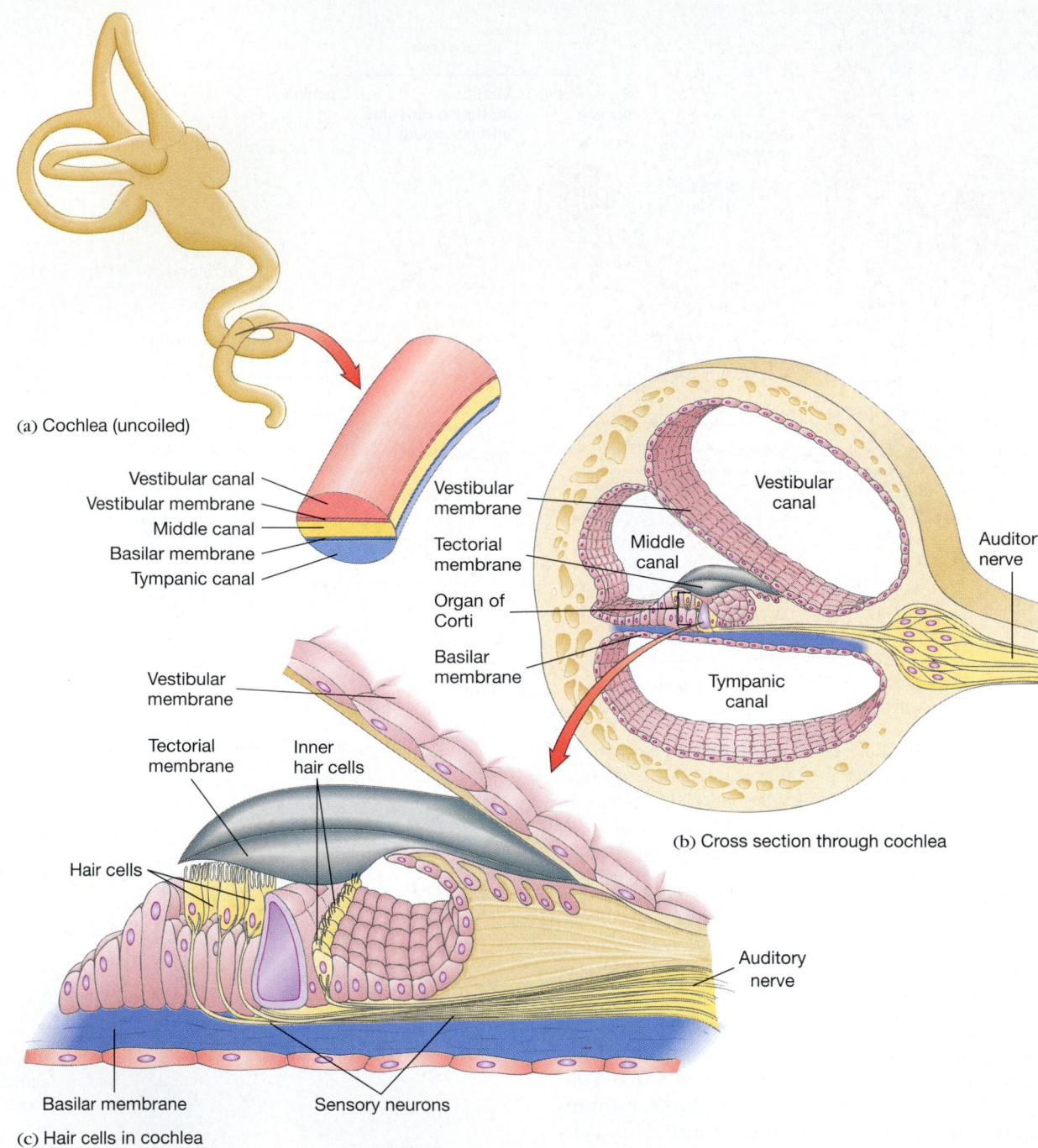

(a) Cochlea (uncoiled)

Vestibular canal
Vestibular membrane
Middle canal
Basilar membrane
Tympanic canal

Vestibular membrane
Tectorial membrane
Organ of Corti
Basilar membrane

Vestibular canal
Auditory nerve
Middle canal
Tympanic canal

(b) Cross section through cochlea

Vestibular membrane
Tectorial membrane
Inner hair cells
Hair cells
Basilar membrane
Sensory neurons
Auditory nerve

(c) Hair cells in cochlea

**FIGURE 38–12** The mammalian cochlea. (a) Uncoiled to show its basic structure, a tapering tube divided into compartments by the basilar and vestibular membranes. (b) Cross section. The basilar and vestibular membranes divide the cochlea into three fluid-filled canals. The actual sensory receptors are the hair cells in the organ of Corti, which synapse with sensory neurons of the auditory nerve. Pressure on the oval window is transmitted through the cochlear fluid, distorting the basilar and tectorial membranes. This creates a shearing force on the hair cells, producing a receptor potential in the hair cell membranes, which may be passed on to the sensory neurons.

Pressure on the oval window moves the fluid in the vestibular canal, which pushes on the other fluid-filled canals. This distorts the basilar membrane so that the hair cells of the organ of Corti are moved, initiating receptor potentials, which may increase or decrease the rate of action potentials in the associated neurons.

The entire cochlea is shaped like a snail shell, a coiled tube of gradually changing diameter. This shape permits the nervous system to tell the pitch of a sound—how high or low it is. Pitch is a function of frequency, the number of vibrations per second, or hertz (Hz). Low-frequency vibrations produce a sensation of low pitch, and high-frequency vibrations produce a sensation of high pitch.

At frequencies below 60 Hz, the entire basilar membrane moves, firing the sensory neurons in synchrony with the rhythm of the sound and so conveying information about its pitch. Frequencies above 60 Hz cause the basilar membrane to vibrate unequally along its length. Each frequency produces a maximum vibration at one point, where the width of the organ of Corti is tuned to the frequency. Hence, the brain "knows" the pitch of the sound because it knows the location of the sensory neurons that are firing. The cells that vibrate most inhibit firing of neighboring cells, so that the brain receives a clear sensation of the true pitch.

Besides pitch, we distinguish two other characteristics of a sound: volume and tone quality. The volume (loudness) of a sound is a function of the amplitude of its vibrations (the height of the sound waves) and is measured in units called decibels (Table 38–2). High-amplitude vibrations in air produce large oscillations of the cochlear fluid and basilar membrane, producing more intense stimulation of the hair cells and a greater frequency of action potentials in the auditory nerve.

The **tone quality** of a sound is a function of the frequency and amplitude (height, size) of its harmonics. **Harmonics** are vibrations at integer multiples of the main pitch of the sound. If a violin, a piano, and a clarinet play a note at the same pitch and volume, each produces a characteristic tone quality because of the differences in the loudness of its various harmonics. Each harmonic pattern stimulates a different combination of hair cells in various regions of the cochlea.

## Noise and Hearing Loss

Noise pollution is one of the prices we pay for modern living. It is a day-to-day irritation, and it damages our hearing by destroying hair cells in the cochlea.

A baby is born with up to 20,000 hair cells in each cochlea—not very many when you consider that this is a lifetime's supply, since damaged or killed hair cells are never replaced. Loud noise damages the hair cells' microvilli by disrupting their cytoskeletons, a core of actin filaments, so that the microvilli flop like wet noodles. Noise also causes the microvilli surfaces to develop blister-like vesicles, which rupture as the noise continues.

Hearing loss caused by noise is insidious because it is painless and so gradual that we do not notice it. At first, sounds must simply be louder (or closer) in order to be heard. The first hair cells to die are those that detect high pitches, above the range of the human voice, and so people can still hear conversations. Then the higher-pitched speech sounds fade out, and people can no longer distinguish between similar words. This often gives rise to complaints that speakers are mumbling.

Hearing loss is not an inevitable part of aging. It is the result of a lifetime of noise. The elders of a Sudanese tribe living in a quiet Stone Age–like culture hear better at 80 than most Americans do at 30.

Tests of more than 4000 students entering college in the 1980s showed that more than one in three already had measurable loss of hearing for high-pitched sounds. These were mostly not serious yet, but today's young people can expect to suffer greater hearing loss by the time they reach their 50s and 60s than we see in people now that age who, on the whole, have lived in quieter environments for most of their lives.

**TABLE 38–2**

**Comparison of the Average Intensities of Various Sounds**

| Source | Intensity (in decibels) |
|---|---|
| Jet aircraft at takeoff | 145 |
| **Pain occurs** | **140** |
| Air hammer | 130 |
| Unmuffled motorcycle, car, or truck | 110 |
| Subway train | 100 |
| Gasoline-powered lawn mower | 96 |
| Food blender | 93 |
| Heavy truck 15 meters away | 90 |
| Heavy city traffic | 90 |
| Vacuum cleaner | 85 |
| **Hearing loss occurs with long exposure** | **85** |
| Garbage disposal unit | 80 |
| Dishwasher | 65 |
| Window air conditioner | 60 |
| Normal speech | 60 |

Prolonged exposure to 85-decibel noise damages the hearing. But noise well below this level has psychological effects. Someone accustomed to sleeping in a 50-decibel city apartment where sirens, traffic noises, and music continue throughout the night may be unable to sleep in a country bedroom where the only noise is the 10-decibel sound of leaves rustling outside the window. More commonly, honking horns and other people's radios produce stress reactions such as irritation or anger, raised blood pressure, and a short attention span.

More than 1000 years ago, Julius Caesar banned iron-wheeled chariots from the Roman cobblestones during the night. Mufflers were added to early automobiles to prevent them from frightening horses. Noise is the most widespread occupational hazard, and regulations designed to control noise have proliferated throughout the world as noise has increased. People who work with noisy equipment must wear ear protectors, and machinery is designed to be as quiet as possible. However, many workers do not realize the damage noise can do and do not use this protective equipment properly.

## 38-D   PHOTORECEPTORS

Most animals have photoreceptors that transduce light energy using **pigments,** colored molecules that undergo chemical changes when they are struck by light. In platyhelminths and many annelids, photoreceptors are merely **eyespots** that detect light as opposed to darkness (Figure 38–13). **Eyes** detect more detail. In an eye, light from the visual field projects an image onto light-sensitive receptor cells.

There are three basic types of eyes (Figure 38–14). The simplest image-forming eye, found in invertebrates including some polychaete worms and molluscs, is analogous to a pinhole camera. Relatively little light can enter the eye through the tiny aperture. A second, larger type of eye found

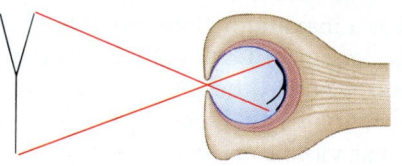

(a) Most molluscs

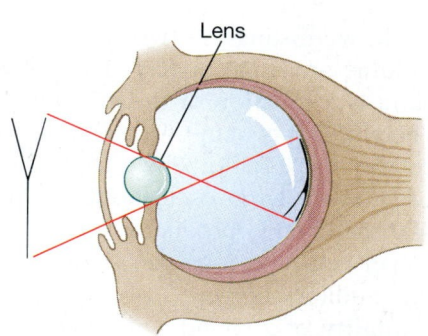

(b) Most cephalopods and vertebrates

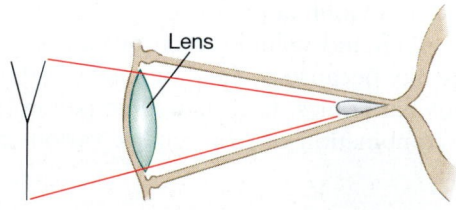

(c) Most arthropods

**FIGURE 38–14**   The three main types of eyes, showing the position of the receptor cells. An object in front of the eye produces an image, shown in black inside the eye. (a) The "pinhole camera" type of eye, without a lens, found in some molluscs. (b) A vertebrate eye, with an adjustable lens that can change the focus. Some molluscs, such as some bivalves and cephalopods, have a very similar eye. (c) A crustacean eye, containing a lens but only one or two receptor cells. The compound eyes of some insects may contain thousands of these individual units.

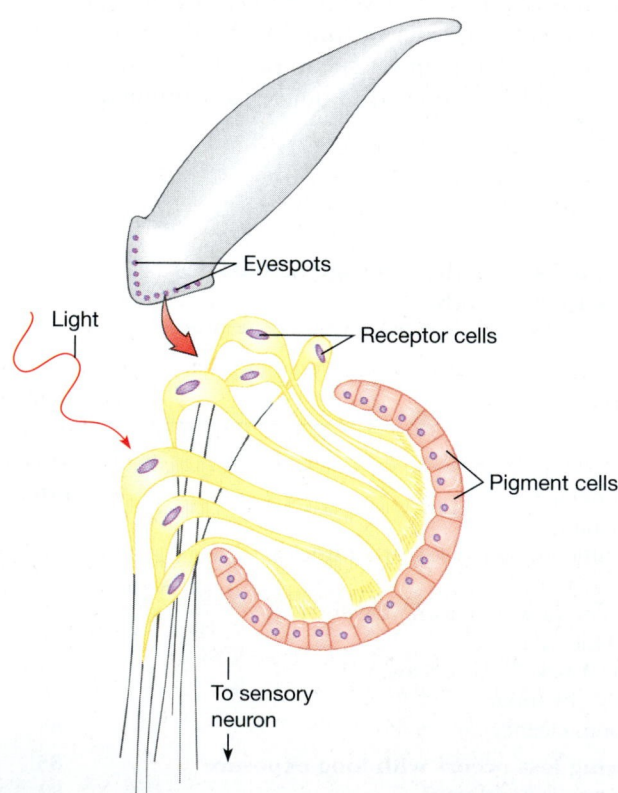

**FIGURE 38–13**   Eyespots. Like many other invertebrates, this planarian worm has many eyespots. In an eyespot, one or more receptor cells are covered by one or several pigment cells where light is absorbed, initiating changes that cause the plasma membrane of the receptor to depolarize.

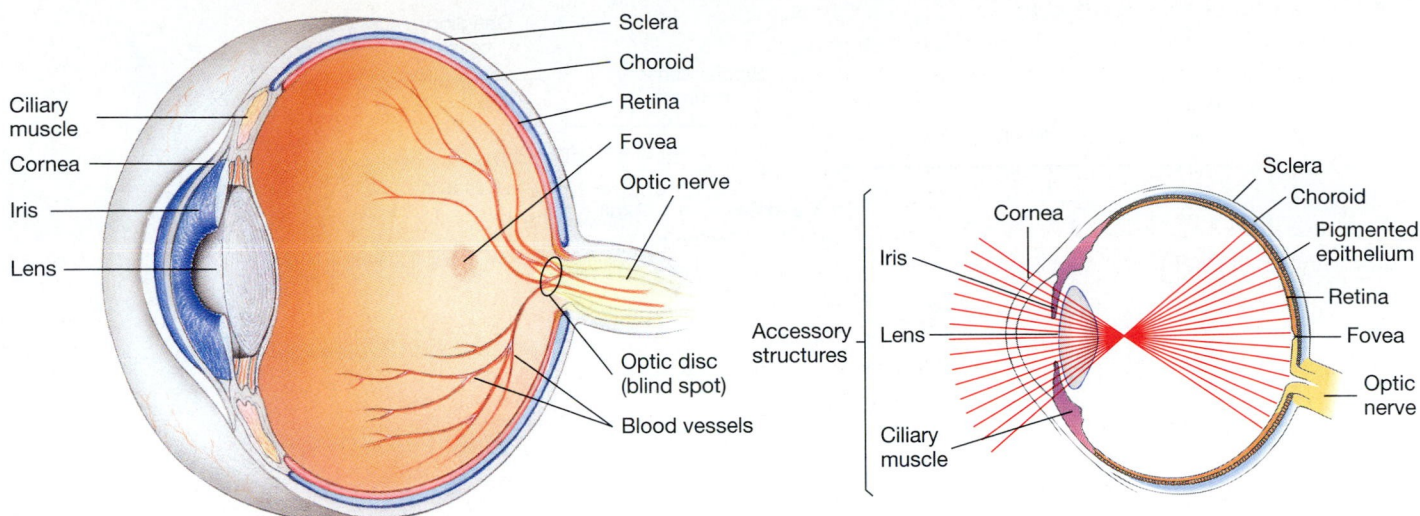

**FIGURE 38–15**  The human eye. *Left* The lateral half of the right eye, which has been sectioned top to bottom. *Right* Schematic view of the eye sectioned through its equator and viewed from above. The retina contains the receptor cells and their associated neurons, which detect light (lines) and send messages to the brain via the optic nerve. The choroid layer contains blood vessels, and the layer of dark pigment epithelium absorbs light that has passed through the retina, thereby preventing it from reflecting back to the receptors and blurring the image.

in vertebrates and cephalopods (octopuses and squids) contains a **lens,** which can focus incoming light. The third type of eye is the multifaceted compound eye of some arthropods. Such an eye contains many closely packed individual units, the ommatidia, each receiving light from a narrow area of the visual field.

In the eyes of vertebrates, the curved **cornea** focuses light entering the eye (Figure 38–15). The focus is adjusted by the **lens,** which is somewhat elastic and changes shape when it is pulled by muscles around its edges. The **iris** is a diaphragm that controls the size of its central opening, the **pupil,** and so regulates the amount of light entering the eye. The **retina** is a delicate layer containing the photoreceptors and their associated neurons. Visual information passes to the brain by way of the **optic nerve.**

The actual photoreceptors of the retina are specialized receptor cells called **rods** and **cones** (Figure 38–16). Each receptor synapses with the dendrites of a sensory neuron. Rods can detect very dim light, but they produce poorly defined images because the sensory neurons receive input from many rods. In bright light, we use the cones and their associated neurons. These neurons tend to keep information from different cones separated, giving good resolution for detail. The cone system is also much better than the rod system at interpreting color. Cones are especially densely packed in an area of the retina called the **fovea.** We tend to move our eyes so that the image of the object we want to see most clearly falls on the fovea. The fovea contains only cones, and most of the other cones of the retina are

nearby. Rods are most concentrated in a ring about 20° from the fovea. This explains why you can see faint stars at night out of the corner of your eye, but not when you move your eyes to look straight at them. These dim stars seem to disappear because the cones in the fovea cannot detect them.

## Chemistry of Vision

The light-absorbing molecule in animals' eyes is **retinal.** The raw material for retinal is vitamin A, which in turn is formed by breaking down $\beta$-carotene, a common plant carotenoid. Animals cannot make carotenoids but must obtain these precursors of visual pigments directly or indirectly from plants.

A visual pigment consists of a retinal molecule nestled in a membrane protein called an **opsin.** Different pigments have slightly different opsins, and this makes the pigments sensitive to different wavelengths of light. The pigments in our eyes absorb wavelengths of about 400 to 750 nanometers, which are therefore called visible light.

Retinal exists in *cis* and *trans* isomers. In rods, the *cis* isomer combines with opsin, forming the visual pigment **rhodopsin.** When light strikes rhodopsin, it causes retinal to change to the *trans* isomer. This change in shape shifts rhodopsin into an enzymatically active form, which starts a chain of chemical events that changes the firing rate in the sensory neuron that synapses with the rod. Retinal detaches from opsin and is converted back to the *cis* form. It can

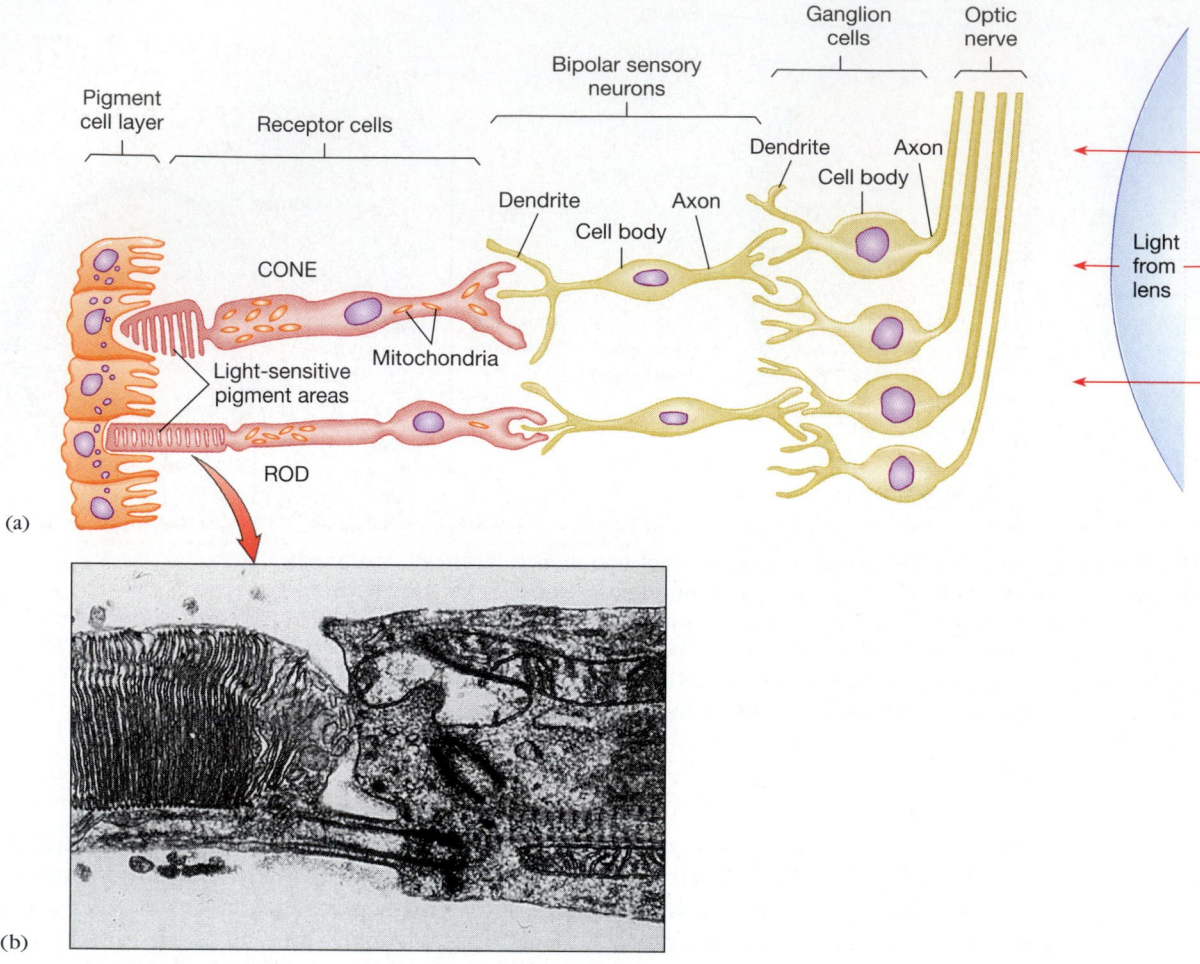

(a)

(b)

**FIGURE 38–16**  Rods and cones in a vertebrate. (a) Pigment-filled rods and cones synapse with bipolar neurons, which synapse with ganglion cells, neurons whose cell bodies lie in the retina and whose axons form the beginning of the optic nerve. Note that light entering through the lens must pass through several layers of cells before it reaches the receptors. (b) Electron micrograph of part of a rod cell. Mitochondria are plentiful in the cell body. The light-sensitive area of the cell consists of stacks of membranous discs bearing the photosensitive pigments.  (Biophoto Associates)

then recombine with opsin to form rhodopsin again. This completes the sequence of reactions called the **visual cycle** (Figure 38–17).

Cones have a similar system but require more light energy in order to produce a response. Because a series of chemical reactions occurs between absorption of light and production of a receptor potential, rods and cones respond much more slowly to stimuli than mechanoreceptors such as hair cells in the ear.

All vertebrates seem to have at least a sprinkling of cone cells in their retinas, but many of them do not re-spond to different colors (that is, they are in effect color-blind). An animal with color vision has two or more types of cone cells, each producing a different pigment. Humans and other primates (monkeys, apes) have three types. The different pigments absorb light of different wavelengths—red, green, and blue—most strongly. Therefore, each kind of cone is most sensitive to a particular range of wavelengths. The visual system detects color by comparing the levels of excitation in cones containing different pigments. Various degrees of color blindness result if one or more cone pigments is missing.

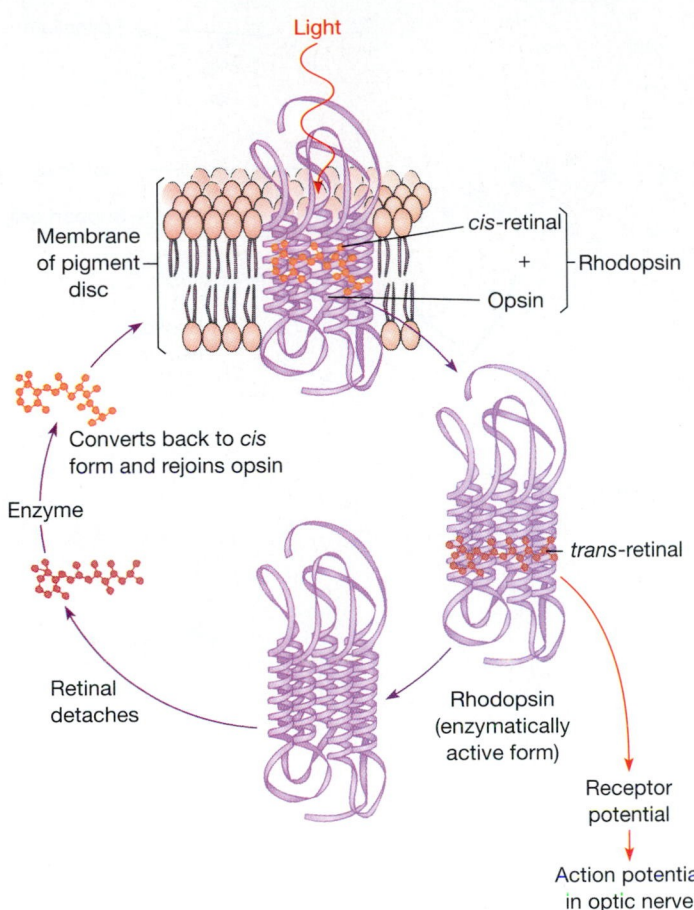

**FIGURE 38–17**   The visual cycle in a rod. Light causes retinal to change shape, converting rhodopsin to an enzymatically active form. It catalyzes the first reaction in a series that leads to a signal in the sensory neuron. Retinal detaches from opsin, and an enzyme converts it back to the *cis* isomer, which recombines with the protein opsin to form the visual pigment rhodopsin, ready to detect more light.

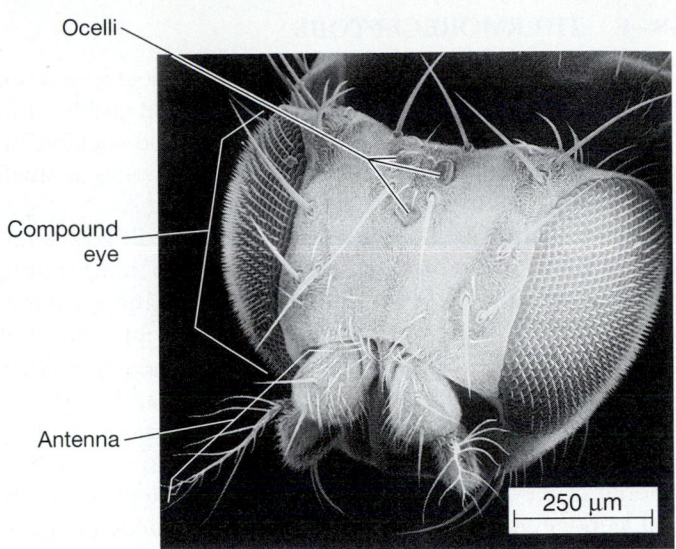

**FIGURE 38–18**   A fly's big wraparound compound eyes allow it to view most of its surroundings at once and to detect slight movements as new patterns of ommatidia become excited. Numerous hair-like pressure receptors, which respond to air movement or touch, are also shown in this picture. Flies also have single eyes (ocelli), each similar to one ommatidium of a compound eye. (The Johns Hopkins Applied Physics Laboratory)

are active in dim light, such as moths and lobsters, the pigment can move out of the way so that it does not block light from passing between ommatidia. The function of these compound eyes is not well understood.

## Compound Eyes

The compound eyes of many arthropods may be composed of as many as 20,000 units, the **ommatidia** (Figure 38–18). Each ommatidium has a lens-like structure that focuses incoming light rays onto a receptor called a **rhabdom,** made up of parts of the membranes of several cells. In the eyes of bees, flies, and many other diurnal insects, opaque walls around the receptor cut out light coming in from the side. Hence, each receptor detects only a narrow beam of light parallel, or nearly parallel, to the long axis of the ommatidium (Figure 38–19). Such an eye appears to present a rather crude mosaic picture of the world. Its main advantages are that it permits an animal to see things very close to its eyes and to detect rapid movement with a sensitivity about five times that of the human eye. In arthropods that

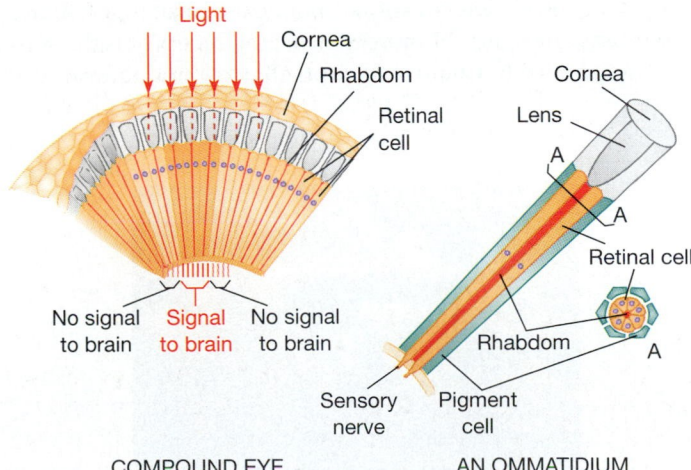

**FIGURE 38–19**   The compound eye of insects and some other arthropods is composed of many units, called ommatidia. Each ommatidium consists of a cornea, a lens, and a light-sensitive rhabdom surrounded by retinal cells, which transmit the sensory stimulus. Pigment cells around each ommatidium prevent light from passing into surrounding ommatidia. Only light rays that are parallel to the rhabdom will excite the retinal cells.

## 38–E   THERMORECEPTORS

Most animals have thermoreceptors, which detect heat, or infrared radiation, with wavelengths longer than visible light. Thermoreceptors may be very sensitive. Blood-sucking insects and ticks can distinguish temperature changes as small as 0.5 °C in their search for warm-blooded hosts.

Since temperature influences the rate of all biochemical reactions, every living cell responds to temperature changes, and this makes it difficult to locate the receptors responsible for behavior patterns such as moving toward or away from heat. The most thoroughly studied thermoreceptors are the pits in the faces of pit vipers and the labial organs of some boas (Figure 38–20). These snakes use their heat-detecting organs to locate warm-blooded prey, such as rodents in dark burrows. A pit viper can detect the body heat generated by a small mammal at distances of up to about 1 meter. There are pits on both sides of the head, and when the snake moves its head until the temperatures detected on each side are the same, it is facing its prey directly and centered to strike.

Mammals' thermoreceptors consist of free nerve endings scattered over the surface of the body, particularly on the tongue. There are also thermoreceptors that detect internal body temperature in the hypothalamus of the brain. Information from internal and external thermoreceptors is integrated in the hypothalamus to produce appropriate responses such as shivering or sweating.

## 38–F   CHEMORECEPTORS

Chemoreceptors detect various chemicals. In all invertebrate and most vertebrate chemoreceptors, the receptor is the sensory neuron itself. Chemoreceptors that are not neurons oc-

**FIGURE 38–20**   A pit viper.   (Joe McDonald/VU)

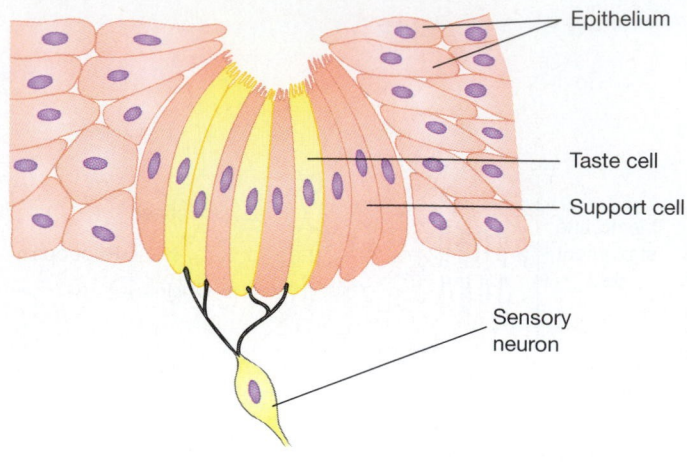

(a) Taste bud

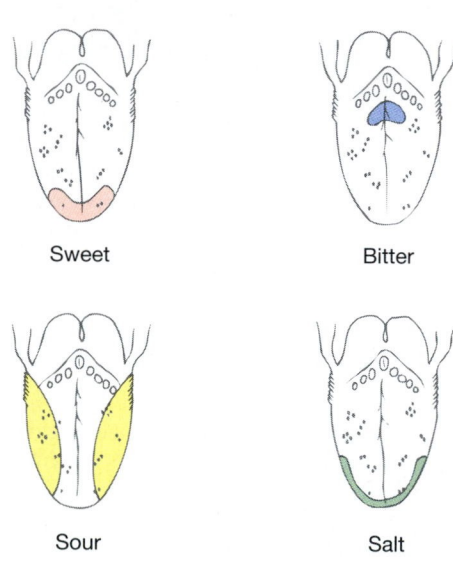

(b) Tongue

**FIGURE 38–21**   Chemoreceptors involved in the sense of taste in humans. (a) A taste bud in the tongue. (b) The human tongue, showing the distribution of taste buds sensitive to sweet (front), bitter (back), sour (sides), and salt (front and sides).

cur only in vertebrate taste organs (and possibly in some chemoreceptors in the circulatory system).

Our bodies have many internal receptors for chemicals such as nutrients, oxygen, carbon dioxide, hormones, and neurotransmitters. Internal chemical reception is covered in other chapters; here we consider only receptors that detect chemicals in the external environment.

External chemoreceptors in humans are responsible for the senses of smell (**olfaction**) and taste (**gustation**). These receptors and their nervous connections are much simpler than those for vision or hearing.

The vertebrate tongue is covered with numerous bumps, or **papillae.** In the grooves around the sides of a papilla lie the clusters of cells known as taste buds. A taste

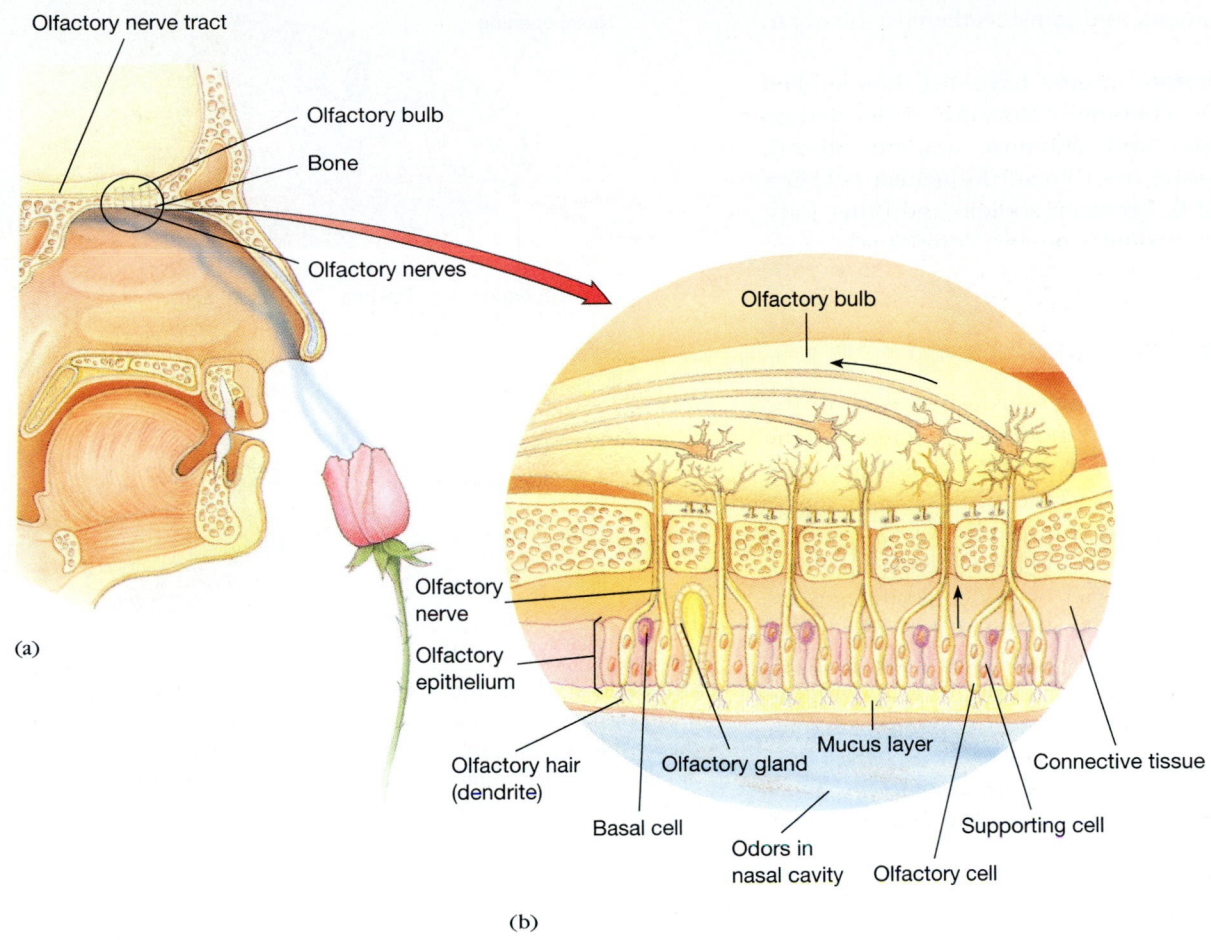

Olfactory nerve tract

Olfactory bulb

Bone

Olfactory nerves

(a)

Olfactory bulb

Olfactory nerve

Olfactory epithelium

Olfactory hair (dendrite)

Olfactory gland

Basal cell

Odors in nasal cavity

Olfactory cell

Mucus layer

Connective tissue

Supporting cell

(b)

**FIGURE 38–22**   Olfaction. (a) Olfactory receptors lie in the top of the nasal cavity in a region of spongy tissue. Air reaches this area through a winding passageway (black arrows). (b) Chemoreceptors in the nose are free dendrites of neurons in the olfactory epithelium.

bud consists of sensory hair cells with tips that project to the edge of the epithelium, and supporting nonsensory cells (Figure 38–21). A single bipolar neuron sends branches of its dendrite to all of the hair cells in several taste buds.

The number of possible tastes is believed to be only four: sweet, sour, bitter, and salt. The "taste" of more complicated flavors is in fact a combination of olfactory and gustatory sensations. The important role of smell in the sensation of "taste" is evident during a bad cold: when you have a stopped-up nose, food has little taste. The "hot" sensation of foods like chili peppers is detected by pain receptors, not chemoreceptors.

Most odors are probably mixtures of primary olfactory stimulants, sometimes described by words such as "musky," "floral," "pungent," "camphoraceous," and "putrid." Some researchers on the human sense of smell distinguish six primary odors, others more than 50. We do not know how many different kinds of olfactory receptors exist, but certainly we can distinguish many more odors than there are receptor types.

A vertebrate's olfactory receptors are nerve endings in the upper reaches of the nasal cavity; a dog has up to 38 million such endings per square centimeter. The dendrite of the olfactory sensory neuron extends beyond the epithelial cells in the nose, and its end is fringed with processes, where the initial chemoreception takes place (Figure 38–22). Since the olfactory epithelium of the nose is not on the main pathway to the lungs, air must travel a circuitous route to reach the olfactory endings. Sniffing aids the sense of smell by moving the air more rapidly into the upper recesses of the nose, where the olfactory endings lie. Here, contact between the olfactory chemoreceptor and the appropriate chemical creates a receptor potential. Presumably, the chemical combines with a specialized macromolecule on the surface of the neuron, and this binding makes the macromolecule change its shape and render the cell leaky to sodium.

Some snakes and lizards have specialized areas of chemoreception in the **organ of Jacobson** in the roof of the mouth. The animal's tongue flicks out, gathers chemicals from the environment, and transfers them to the organ (Figure 38–23).

Among invertebrates, insects have the best-studied chemical senses. Insect chemoreceptors may be located on the mouthparts, legs, and antennae (Figure 38–24). Chemoreceptors of some insects can distinguish between sugars and amino acids, between sodium and other ions, and between many very similar organic compounds.

## 38–G   ELECTRORECEPTORS

In a number of bony and cartilaginous fishes, parts of the lateral line organ have become modified to detect electric currents in the surrounding water. Every living organism generates weak electrical fields, and the ability to detect these fields may permit a fish to capture prey or to avoid predators. Such an ability is especially valuable in turbulent, deep, or murky water, where vision and olfaction are of little use. Some fish generate electrical fields and then use their electroreceptors to detect how surrounding objects distort the field; this allows the fish to navigate in the muddy rivers where they live (*Essay: Electric Fish,* Chapter 39).

Among mammals, the platypus uses receptors in its duck-like bill to detect weak electric currents from the small aquatic animals that are its prey.

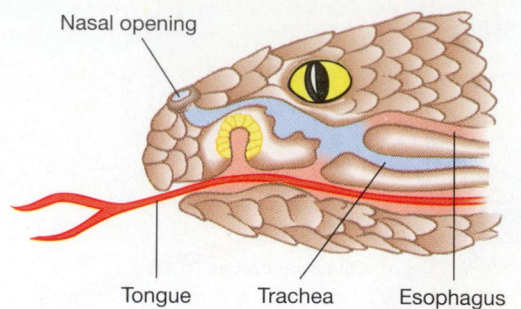

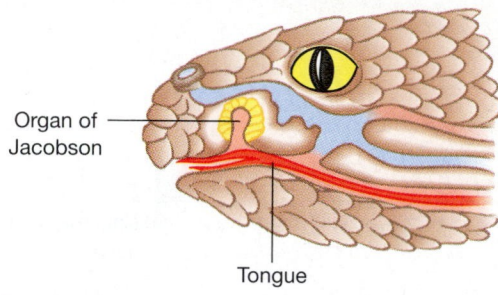

**FIGURE 38–23**   Chemoreception in the organ of Jacobson of a lizard. A lizard or snake shoots out its tongue to pick up chemicals in the environment. The tongue is then drawn back into the mouth, and the chemicals enter the organ of Jacobson, which is lined with chemoreceptor cells.

**FIGURE 38–24**   Chemoreceptors on a butterfly's proboscis. (a) Scanning electron micrograph of the tip of the proboscis of a black swallowtail butterfly. The button-like structures are chemoreceptors. (b) Closeup of one chemoreceptor. (c) A schematic cross section of the chemoreceptor. The hair-like dendrites of sensory neurons are the actual receptors.

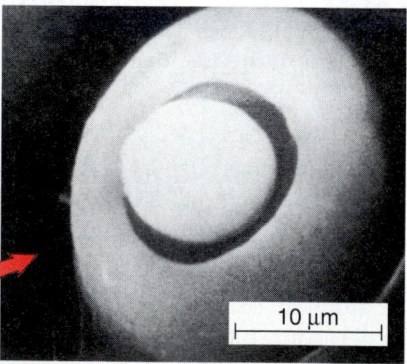

(b)

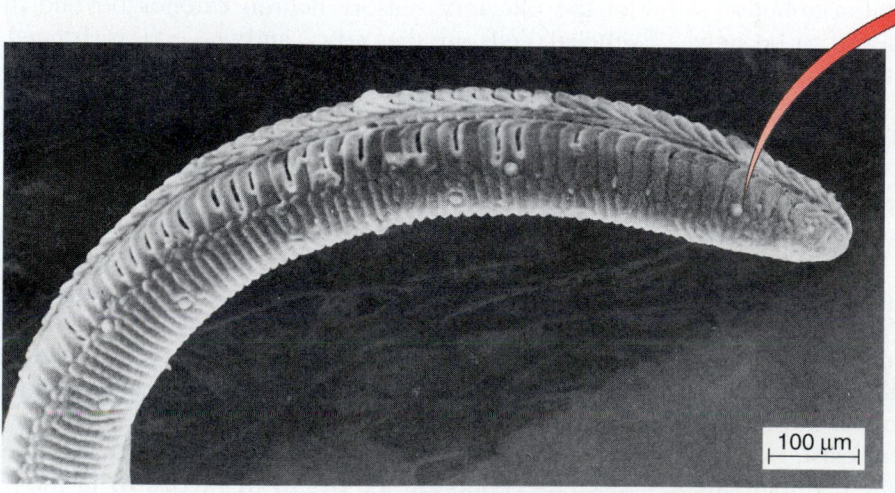

(a)

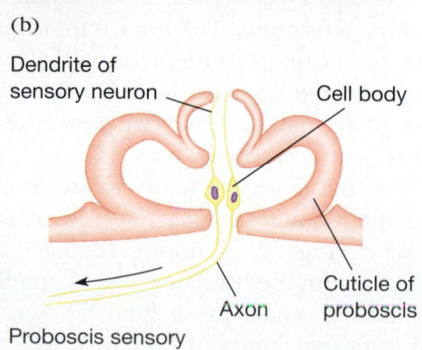

(c)

Sense organs are vital in the perennial battle between predators and their prey. Nocturnal bats and some of the moths they eat provide one of the best-studied examples of the coevolution of sense organs.

It is a warm spring night. As a bat awakes, the receptors in her empty stomach stimulate her to go in search of food. As she flies, she utters four or five short chirps per second, far too high-pitched for us to hear. The bat then listens for the echo of each chirp. Since the echoes are very faint, it seems strange that the bat is not deafened to them by the cry she made a split second earlier. But every time she makes a cry, an accessory structure moves in her middle ear, and the three bones that conduct sound from the tympanic membrane to the inner ear slide aside so that they no longer transmit sound to the cochlea. When the cry is finished, the three bones slide back into place, and the bat can hear the echo of her cry bouncing off nearby objects. Since she can tell both the time elapsed between uttering the cry and hearing its echo, and the direction of the echo, the bat can use the echoes of her cries as a sonar system to avoid obstacles and to detect other moving objects, an ability called echolocation. Eventually, the bat hears echoes she identifies as coming from a fat noctuid moth.

The moth is looking for a mate, using chemoreceptors on his antennae to detect her scent. He also has simple ears, with only two receptor cells apiece, but the cells are sensitive to the high-pitched sound of a bat. One receptor is sensitive to faint sounds, and the other responds only to loud sounds. In effect, the first receptor detects distant sounds and the second detects nearby sounds. As the bat approaches, her cries stimulate the moth's long-distance auditory cells more and more strongly.

If the bat approaches from the moth's left, his left-hand long-distance cell receives a stronger stimulus. The moth's defensive reaction is to turn as he flies, until the stimulus intensity is the same in both ears and he is flying either directly away from or directly toward the bat. He can tell whether the bat is ahead of him or behind him because his ears are located at the rear of his thorax, just in front of his abdomen, and the sound coming from certain directions is deflected as the wings pass over the ears during flight.

The information about differences in sound intensity, combined with information about wing position, gives the moth an accurate indication of where the bat is (Figure 38–A). If the bat is still far away, the moth may escape by turning and flying away. However, when the bat is very close, the "loud" cells in the moth's ear begin to fire. This signals the moth's brain that immediate evasive action is necessary. He may execute a loop, or he may close his wings in a crash dive. Sometimes these tactics succeed, sometimes not. The battle of the sense organs continues night after night, with survival as the stake, a strong selective pressure for sensitive and efficient sense organs.

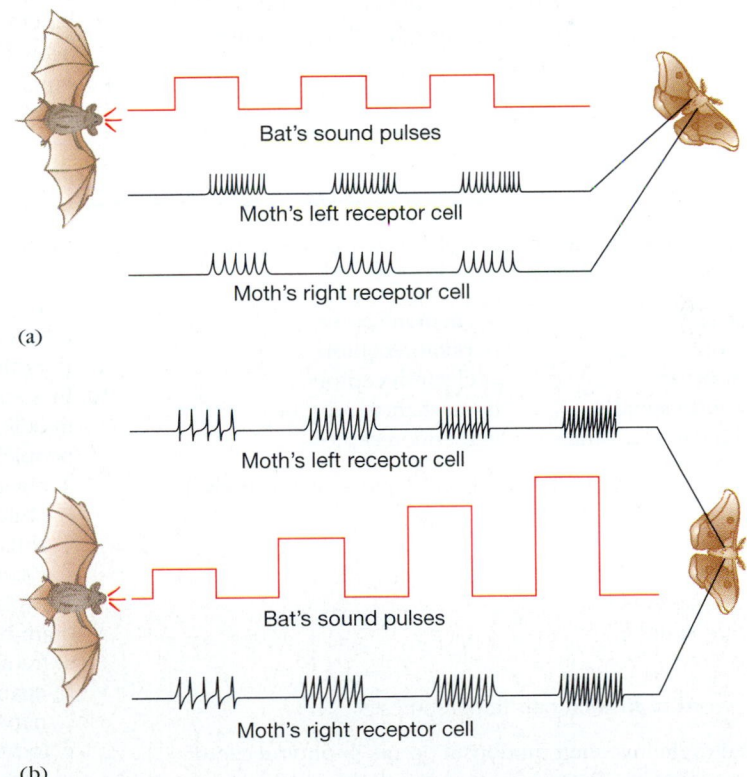

(a)

(b)

**FIGURE 38–A** Interactions between a bat and the moth it pursues. The nerve impulses recorded from the moth's right and left receptor cells are shown with the pulses of sound produced by the bat. (a) When a bat approaches from the left, the receptor on the left of the moth receives more intense stimulation and fires at a more rapid rate than the receptor on the right. (b) As the bat approaches the moth directly from the rear, the sounds emitted by the bat become louder (from the moth's point of view), and the moth's receptor cells fire at a greater rate.

## SUMMARY

An animal must be able to detect changes inside its body and in the world around it in order to produce appropriate physiological and behavioral responses. The particular collection of sense organs in each animal species gives its members an appropriate perception of their bodies and of their environment.

1. Sense organs are collections of cells specialized to react to particular forms of energy by producing local receptor potentials that change the rate of impulses carried to the nervous system by sensory neurons. Sense organs may detect mechanical stimuli, light, heat, chemical changes, electric current, or Earth's magnetic field, as summarized in Table 38–1.
2. The actual receptors in sense organs may be the dendrites of sensory neurons or specialized non-nervous cells.
3. Adaptation at the level of the receptor, sensory neuron, or central nervous system changes the signal transmitted or received in the presence of a constant stimulus. The central nervous system usually has feedback control of the sensitivity of sense organs.

4. Mechanoreceptors detect mechanical distortions such as pressure, touch, sound, muscle stretch, and movement of joints. Hair cells are vertebrate mechanoreceptors found in lateral line organs and the labyrinth and cochlea of the inner ear. Movement of the hair bundle depolarizes or hyperpolarizes the hair cell membrane, and this potential change is transmitted to dendrites of a sensory neuron that synapses with the hair cell.
5. Photoreceptors in eyespots, eyes, and ommatidia detect light by means of pigments whose structure is changed by electromagnetic radiation of appropriate wavelength. This photochemical reaction causes a receptor potential in an adjacent sensory neuron.
6. Thermoreceptors are sensory neurons that respond directly to infrared radiation.
7. Chemoreceptors are usually dendrites of sensory neurons, although the taste buds, and possibly some internal chemoreceptors, of vertebrates contain non-nervous receptor cells.
8. Electroreceptors are found in some bony and cartilaginous fishes and in the platypus; they detect electric currents generated by organisms.

## Self-Quiz

Match the type of receptor on the right with the sense organs listed on the left:

| | |
|---|---|
| _____ 1. Nose | a. chemoreceptors |
| _____ 2. Retina | b. photoreceptors |
| _____ 3. Pit organ | c. electroreceptors |
| _____ 4. Muscle spindle | d. mechanoreceptors |
| _____ 5. Semicircular canals | e. thermoreceptors |

6. A receptor that detects the position of parts of the body is called:
   a. a proprioceptor
   b. a hair cell
   c. a mechanoreceptor
   d. a muscle spindle
   e. a statocyst

Choose one word from each pair in parentheses:

7. The visual cycle involving rhodopsin occurs in photoreceptors called (cones/rods). These receptors and their associated neurons are primarily used for vision in (bright/dim) light, and they are (good/poor) discriminators of color.
8. In the ear, airborne sound waves first strike the:
   a. basilar membrane
   b. tympanic membrane
   c. tectorial membrane
   d. vestibular membrane

9. In vertebrates, thermoreceptors function in:
   a. location of warm-blooded prey
   b. regulation of body temperature
   c. detection of electric currents in water
   d. both a and b
10. In some people, calcium deposits form on the bones of the middle ear. Which of the following would be affected in these people?
    a. ability to detect pitch
    b. ability to maintain equilibrium
    c. ability to detect gravity
    d. changes in the position of muscles
11. The pitch of a sound arriving at the ear is recognized in the brain by the:
    a. frequency of action potentials in the auditory nerve
    b. change in amplitude of the action potentials in the auditory nerve
    c. location of sensory neurons firing action potentials in the auditory nerve
    d. frequency and amplitude of action potentials in the auditory nerve
12. Compared with the retina of a diurnal animal, the retina of a nocturnal animal would have:
    a. a larger proportion of cones to rods
    b. a larger fovea
    c. a greater amount of rhodopsin
    d. all of the above

## Questions for Discussion

1. Why do we say that sense organs detect changes in an animal's internal or external environment rather than just detecting the state of the environment?
2. In humans, the ability to taste phenylthiourea (PTU) is inherited as a dominant trait. Can the theory of taste presented in this chapter account for the genetic evidence?
3. Why do some people like certain foods that others do not?
4. Why does vitamin A deficiency result in "night blindness?"
5. What visual problems would be experienced by people whose eyes lacked rods? Whose eyes lacked cones?
6. Thermoreceptors of mammals are particularly concentrated on the tongue. These receptors keep us from burning our mouths with hot food, but cooking is a recent invention (in evolutionary terms), practiced only by humans. What is the advantage to a wild mammal of having so many thermoreceptors on the tongue?

## Suggested Readings

Hudspeth, A. J. "The hair cells of the inner ear." *Scientific American,* January 1983.

Levine, J. S., and E. F. MacNichol. "Color vision in fishes." *Scientific American,* February 1982. Fish provide an excellent example of the effect of the environment on an animal's sense organs and what they detect.

Koretz, J. F., and G. H. Handelman. "How the human eye focuses." *Scientific American,* July 1988.

Parker, D. E. "The vestibular apparatus." *Scientific American,* November 1980. Describes how the labyrinth interacts with other sense organs in detecting position and motion, and how experiments in space stations can extend our knowledge.

Renouf, D. "Sensory function in the harbor seal." *Scientific American,* April 1989. Seals face problems because they must be able to use their senses both on land and under water. Interesting adaptations permit this transition.

# Muscles and Skeletons

## OBJECTIVES

*When you have studied this chapter, you should be able to:*

1. Define these terms and use them correctly:
   antagonistic muscles, extensor, flexor, reciprocal inhibition, joint, ligament, tendon
2. Name the three types of muscle found in the vertebrate body, and describe the location, cellular organization, and nerve connections of each type.
3. Describe how electrical impulses are conducted through the heart.
4. Draw a sarcomere from a skeletal myofibril, name the main proteins in its thick and thin filaments, describe their interaction during muscle contraction, and state the role of calcium and ATP in contraction.
5. Describe the sequence of chemical and cellular events from the time a nerve impulse arrives at a neuromuscular junction in a skeletal muscle, through contraction of the muscle fiber, to subsequent relaxation of the fiber.
6. Explain how the duration and strength of contraction of a skeletal muscle are controlled.
7. Define tetanic contraction and fatigue in skeletal muscle.
8. List three functions of the skeleton as a whole and two nonskeletal functions of bones.
9. Describe the structure of cartilage and of bone.
10. Describe the role of bone in maintaining the body's circulating calcium supply and the two ways in which bone releases calcium into the bloodstream.

An animal detects things with its sense organs, processes the information in its nervous system, and reacts with its **effectors:** cells, tissues, or organs specialized to respond to stimuli detected by other cells. Muscles and glands are the most widespread effectors, but electric organs and light-emitting organs also respond to information gathered by the sense organs. Glands are considered in Chapter 40. Here we concentrate on muscles.

Muscles react to signals from motor neurons by contracting. We usually think of muscle contraction as moving some part of the body, but muscle contraction also helps to maintain the body's homeostasis by movements such as circulating blood and pushing food down the digestive tract. Muscle contraction does not always result in movement. Sometimes it prevents movement, as when the continuous contraction of a scallop's adductor muscle (the part people eat) keeps the shell closed, and a would-be predator out. In all these cases, muscle contraction performs adaptive responses to conditions in the animal's internal or external environment.

**FIGURE 39–1** The epitome of fast and graceful locomotion, a cheetah can use its muscles and skeletons to reach speeds of more than 40 km/h. (Erwin and Peggy Bauer/Natural Selection)

The force of muscle contraction cannot do useful work without something to pull or push against such as the scallop's shell, the bones of the human body, blood in the heart, or food in the gut. An animal's skeleton provides anchorage for its muscles and a system of levers and pivots for them to pull against, permitting them to perform useful work in moving body parts or in locomotion (Figure 39–1). Even the muscles of the internal organs, which are not attached to the skeleton, have more or less firm attachments to other structures, which serve the same sort of "skeletal" functions.

The muscular and skeletal systems, working together, lead an animal into temptation—in the form of food or mates—and deliver it from evil—such as predators or too much sun. The rigidity of a skeleton also provides support for the body and protects delicate parts such as the brain and other soft organs.

## KEY CONCEPTS

- An animal's effectors—mostly muscles and glands—"do something" to maintain internal homeostasis or make suitable responses to external events.
- Muscles and skeletons work together when moving parts of the body or holding something steady: the muscle provides a force, and the skeleton provides an object for the force to work against.

- In addition to their role in movement, skeletons provide support and protection to other organs.
- Electrical excitation of muscle membranes leads to the mechanical events of contraction.
- In muscle contraction, the heads of myosin molecules in thick filaments attach to actin molecules in thin filaments and push them with a ratchet-like motion.

## 39–A  MUSCLE TISSUE

Muscle tissue has two distinguishing properties: electrical excitability and contractility. Excitability is also a characteristic of neurons. In both muscle and nervous tissue, excitability is due to energy stored in an electrical potential difference across the membrane. The excitable muscle membrane depolarizes in response to a chemical transmitter released at a **neuromuscular junction,** the synapse between a neuron and a muscle. This excitation in the membrane initiates contraction in the muscle (Figure 39–2). Muscle cells are packed with contractile proteins, which interact in such a way that they shorten the cell.

Vertebrates have three main types of muscle (Figure 39–3):

1. **Smooth muscle** lines the walls of many internal organs.
2. **Cardiac muscle** makes up the walls of the heart.
3. **Skeletal** or **striated muscle** is responsible for locomotion and change of position. It is generally the only type of muscle under the animal's voluntary control.

### Smooth Muscle

Smooth muscle is made up of sheets of muscle cells. Neurons can often be seen between the muscle cells, but these end in fine branches rather than in the enlarged synaptic knobs of neuron terminals on striated muscle.

Smooth muscle occurs in many internal organs, including the walls of the arteries, veins, digestive tract, urinary bladder, and reproductive organs. The usual role of smooth muscle is to exert pressure on the contents of these organs: it constricts blood vessels, moves food along the gut, and expels urine from the bladder, semen from the seminal vesicles, or a baby from the uterus.

Some smooth muscles contract spontaneously, but others must be stimulated by nerves or hormones. Most smooth muscle is innervated (supplied with nerves) by the autonomic nervous system. Most neurons of the sympathetic

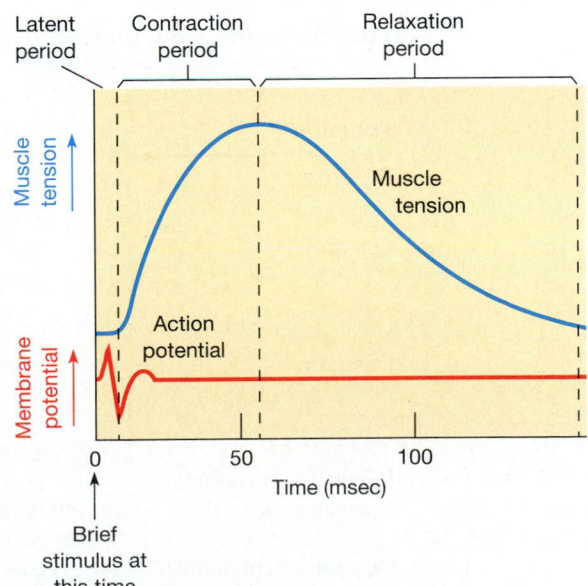

**FIGURE 39–2** Electrical excitation and contraction recorded from a skeletal muscle fiber. When a skeletal muscle is electrically stimulated, it responds by firing an action potential (electrical activity) and by contracting with a rapid twitch (mechanical activity).

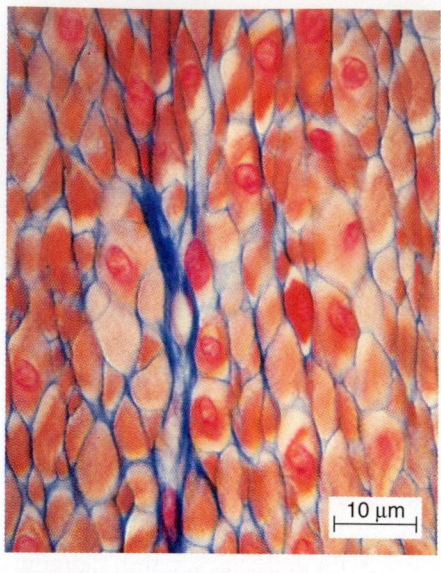
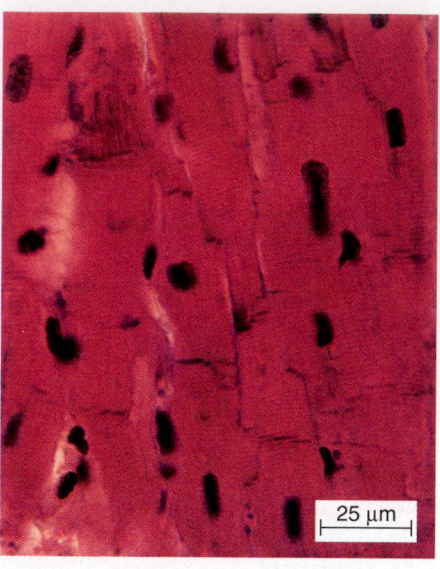
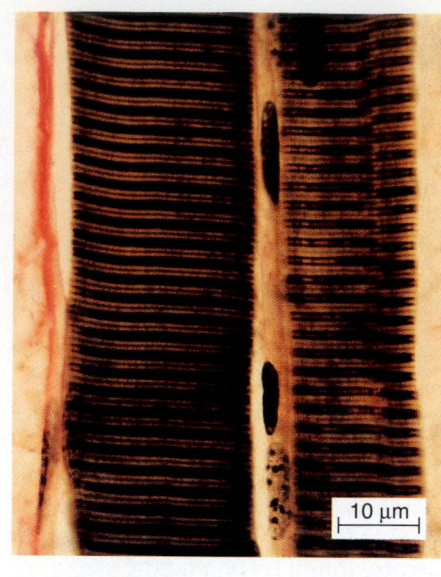

(a)       (b)       (c)

**FIGURE 39–3** Types of muscle. (a) Smooth muscle consists of individual cells whose nuclei are their most obvious feature. (b) Cardiac muscle also consists of individual cells with highly visible nuclei. The contractile proteins are arranged in arrays that give the cells a faintly striped appearance. (c) Skeletal (striated) muscle. During development, individual cells fuse into syncytia, forming long, slender muscle fibers, each with many nuclei lying just within the membrane. Parts of three fibers can be seen, with two dark oblong nuclei in the light area between two fibers. The rest of the fibers are filled with regularly arranged contractile proteins that give the entire fiber its striped ("striated") appearance. (Biophoto Associates)

**TABLE 39–1**

## Functions and Distribution of Muscle Types in Human Organs

| Function | Structures | Muscle Type |
|---|---|---|
| Circulation | Heart | Cardiac |
| | Walls of arteries, veins | Smooth |
| Excretion | Walls of renal pelvis, ureter, bladder | Smooth |
| | Internal sphincter between bladder and urethra | Smooth |
| | External sphincter near exit from body | Striated |
| Digestion | Tongue, muscles of jaw and pharynx | Striated |
| | Walls of esophagus, stomach, intestines | Smooth |
| | Internal sphincter of anus | Smooth |
| | External sphincter of anus | Striated |
| Ventilation (breathing) | Diaphragm, intercostal muscles | Striated |
| Ejaculation | Walls of genital ducts | Smooth |
| | Muscles at base of penis | Striated |
| Parturition (childbirth) | Uterine wall, cervix | Smooth |
| | Abdominal muscles, diaphragm | Striated |
| Heat production | Skeletal muscles (exercise, shivering) | Striated |
| Maintenance of posture Change of position Locomotion | Skeletal muscles | Striated |

nervous system release the neurotransmitter norepinephrine, whereas those of the parasympathetic system release acetylcholine at neuromuscular junctions. Each transmitter excites contraction of smooth muscle in some organs and inhibits it in others, and the two generally have opposite effects on the smooth muscle in any particular organ (see Figure 37–21). Whether acetylcholine or norepinephrine causes excitation or inhibition in a particular smooth muscle depends on the type of receptor for that transmitter made by the muscle.

A single sympathetic neuron may activate many smooth muscle cells because the neurons have many branches and release neurotransmitter from vesicles scattered along the end of the axon. From these vesicles the transmitter diffuses to the muscle and stimulates the muscle cells. Because norepinephrine is similar to epinephrine, it is not surprising that smooth muscle also reacts to the hormone epinephrine arriving from the adrenal glands by way of the bloodstream.

A smooth muscle cell's electrical activity is not an all-or-none phenomenon like that of a neuron, but is a graded response. Smooth muscle produces a gradual contraction of variable force, depending on how much the cell is depolarized. Contraction of smooth muscle is often sustained for long periods of time without fatigue.

## Cardiac Muscle

Cardiac muscle occurs only in the vertebrate heart. Here, the individual muscle cells are arranged in long columns of fibers (Table 39–1).

Cardiac muscle illustrates the similarity between nerve and muscle. Some muscle cells in the heart are so specialized to conduct electrical impulses that they have lost their contractile proteins. These **conducting cells** behave much more like neurons than like muscle cells.

The conducting cells of the heart are part of the mechanism that controls the heartbeat. Each beat starts with spontaneous electrical activity of the heart's pacemaker, the **sinoatrial node** in the wall of the right atrium (Figure 39–4). The pacemaker cells fire in a rhythmic pattern because they are unusually leaky to sodium ions, and so they depolarize without any outside stimulus and reach the threshold to fire an action potential. This resets them, and the process repeats itself.

An impulse generated at the sinoatrial node spreads to all parts of the atrium and then to the **atrioventricular node,** on the partition between the two ventricles. From here the impulse spreads rapidly through the ventricular walls, triggering simultaneous contraction throughout the ventricles.

There are no nerves to carry electrical impulses from the pacemaker to the rest of the heart. Depolarization travels through the membranes of the heart muscle cells. An impulse travels from one muscle cell to another electrically,

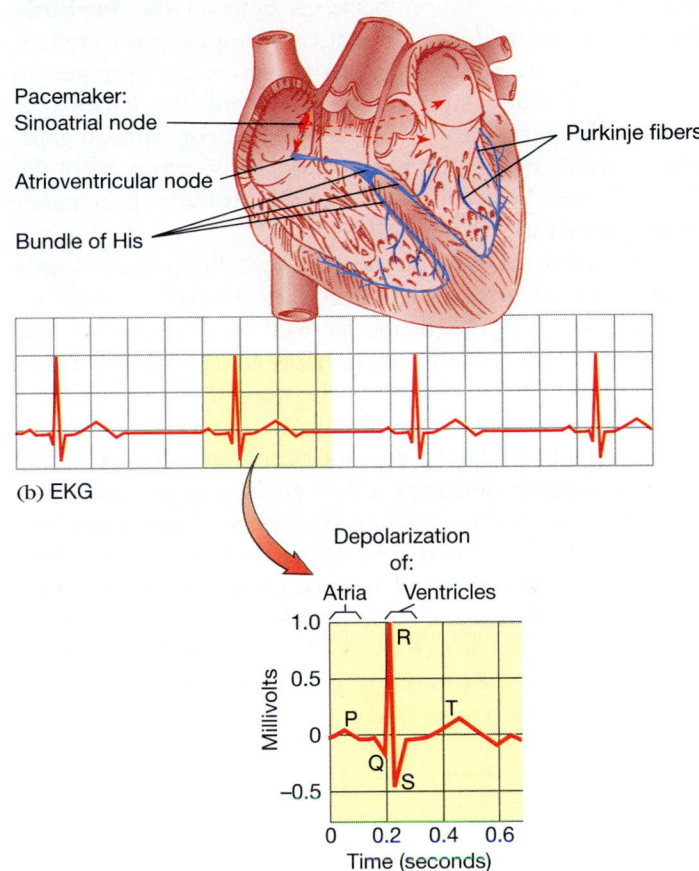

(a) Electrical conducting system

Pacemaker: Sinoatrial node

Atrioventricular node

Bundle of His

Purkinje fibers

(b) EKG

Depolarization of:

Atria   Ventricles

(c) One heartbeat cycle

**FIGURE 39–4** Electrical activity in the human heart. (a) The electrical conduction system. The sinoatrial node in the right atrium initiates each heartbeat. Electrical activity spreads from this pacemaker through the walls of the atria and to the atrioventricular node, and thence through the ventricular walls. (b) An electrocardiogram (EKG). Electrical excitation in the heart can be detected in all body fluids and monitored by electrodes attached to the skin. An EKG is used to determine whether the heart's electrical activity is normal. (c) One heartbeat cycle enlarged from an EKG. The P wave is produced by depolarization of the atria; Q, R, and S are produced by depolarization of the ventricles. T is a result of ventricular repolarization.

rather than by way of chemical transmitters like those at most synapses.

Any cardiac muscle cell can pass its electrical activity (which is much like an action potential in an axon) to any adjacent cell, but some cells are specialized so that transmission between them is particularly fast. These cells are joined by gap junctions, which contain pores through which ions, and therefore electric current, can pass. There is very little electrical resistance between cells linked by gap junctions, and the depolarization travels rapidly from one cell to the next.

Gap junctions occur between many cells in the heart. They are most common between cells in the **Purkinje fibers,** the conducting cells that carry impulses throughout the walls of the ventricles. Gap junctions are arranged in such a way that a single impulse leaving the pacemaker reaches different parts of the heart at precise times regardless of distance from the pacemaker. Thus, two parts of the ventricles that lie at different distances from the pacemaker both contract at the same time.

The main nerve to the heart is the tenth cranial **vagus nerve.** It contains branches of the sympathetic system, which speed up the heart, as well as branches of the parasympathetic system, which slow the heart rate. However, if all of the nerves from the central nervous system to the heart are cut, a vertebrate can survive in an apparently normal condition, and its heartbeat alters with the body's changing demands just as it does in an intact animal. This remarkable situation is possible for two reasons. First, even isolated heart cells contract spontaneously. Second, the rate and force of heartbeats are governed partly by hormones carried in the bloodstream and partly by reflexes in a system of nerves that lie completely within the heart and work even though they are not in contact with the rest of the nervous system. (As a result, a transplanted

heart beats normally in the recipient even though all nerves to the heart have been severed.)

Contraction in cardiac muscle is basically similar to that in other muscles, but there are some differences in detail. For instance, cardiac muscle must be highly resistant to fatigue if it is to beat regularly throughout an animal's lifetime. Not surprisingly, cardiac muscle has abundant mitochondria and a relatively enormous blood supply, which brings it adequate oxygen.

## Skeletal Muscle

The muscles that attach to, and move, the skeleton of a vertebrate are called skeletal or striated (striped) or voluntary muscles. Most of the research on vertebrate muscle has been done on this readily available tissue.

A skeletal muscle, such as the deltoid in the upper arm and shoulder, is made up of many **muscle fibers,** each running the entire length of the muscle. In contrast to smooth and cardiac muscle, which are made up of cells, each skeletal muscle fiber is a multinucleate syncytium formed by the fusion of many embryonic cells, the **myoblasts.** The plasma membranes of a group of these cells

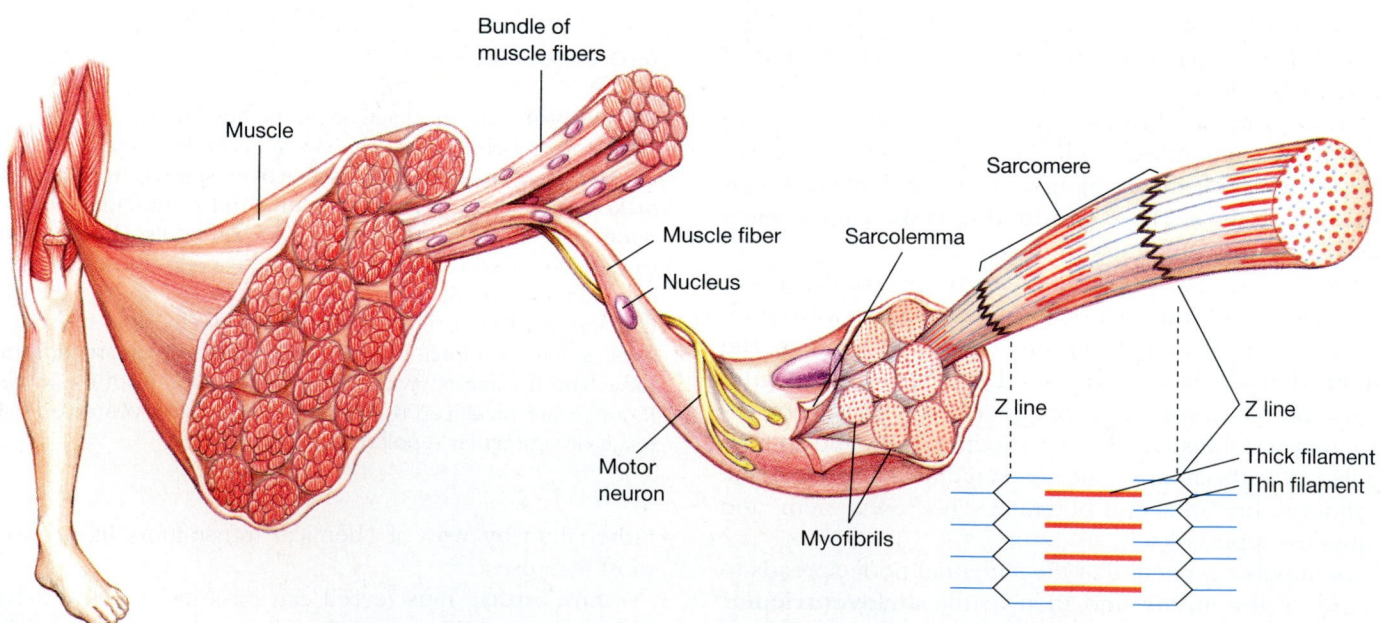

**FIGURE 39–5** Anatomy of a vertebrate skeletal muscle. This diagram shows serial enlargements of the muscle components, from the intact muscle to the pattern of contractile protein filaments that gives the muscle its striped appearance. The muscle is made up of bundles of muscle fibers. Each fiber is a syncytium in which many myofibrils and nuclei are enclosed within a common sarcolemma. Each myofibril contains many sarcomeres (the area between one Z line and the next), arranged end to end in single file.

coalesce to form a continuous membrane, the **sarcolemma,** around the whole muscle fiber. The many nuclei lie just under the sarcolemma (Figure 39–5).

Each muscle fiber consists of a bundle of **myofibrils.** Each myofibril is made up of units called **sarcomeres,** arranged end to end along the length of the fiber. A sarcomere is the part of the myofibril between two adjacent **Z lines,** which are protein-containing structures extending across the myofibril. Each sarcomere contains a well-developed cytoskeleton, consisting of a precise arrangement of two kinds of protein filaments. Attached to each side of a Z line are **thin filaments,** which extend less than halfway to the center of the sarcomere. Their free ends overlap **thick filaments,** which are centered in the sarcomere. Viewed with a microscope, the Z lines and the areas where thick and thin filaments overlap appear dark. The sarcomeres of neighboring myofibrils are lined up side by side, so that their light and dark areas produce a visible striped pattern extending across the muscle fiber.

## 39–B MUSCLE CONTRACTION

In the 1950s, A. F. Huxley and R. Niedegarde found that the Z lines of skeletal muscle fibers move closer together when a muscle contracts. H. E. Huxley and J. Hanson suggested that this was because each myofibril is made up of filaments that slide along each other and mesh together more closely as the muscle contracts. In this **sliding filament** mechanism of muscle contraction, the filaments stay the same length, but the free ends of the thin filaments move closer to the center of the sarcomere. The Z lines of each sarcomere move closer together because the other ends of the thin filaments are attached to the Z lines. This shortens the sarcomeres and therefore the whole muscle (Figure 39–6). Meanwhile, because the muscle's volume does not change, it also becomes thicker.

Thin filaments are made up of many molecules of the globular protein **actin,** with smaller amounts of the proteins **troponin** and **tropomyosin** (Figure 39–7). A thick filament contains hundreds of molecules of the protein **myosin,** each shaped like a golf club with a double head.

How do the filaments slide past each other? The heads of the myosin molecules in the thick filaments are the active part of the system. They attach to the actin in the thin filaments. The myosin heads swivel, pushing the thin filaments toward the center of the sarcomere. They then detach, reattach farther along the thin filaments, and push them still further. As the myosin heads complete one swiveling cycle, the sarcomere is shortened by about 1% of its length. Because there are many myosin-to-actin attachments and they do not all move at the same instant, the thin fil-

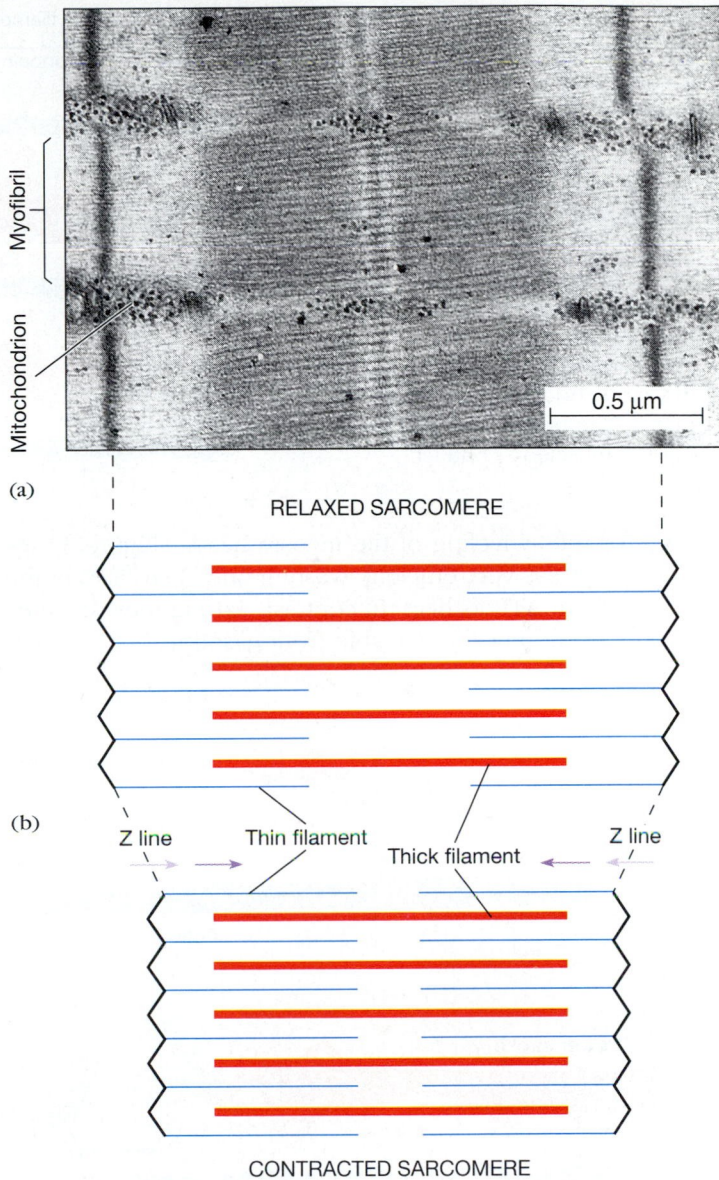

**FIGURE 39–6** The sliding filament model of muscle contraction. (a) An electron micrograph of parts of three relaxed skeletal myofibrils. (b) Diagram of the filaments of contractile proteins responsible for the striated appearance of myofibrils. The Z lines are strands of connective fiber to which the thin filaments are attached. The thick filaments lie between the thin filaments. (c) A contracted sarcomere. A muscle contracts when the thick and thin filaments slide past each other, reducing the distance between adjacent Z lines. (a, Biophoto Associates)

aments cannot slip back while individual myosin heads are detached.

Muscle contraction is powered by ATP, produced mainly in the muscle's many mitochondria. ATP binds to the heads of myosin molecules and is broken down to provide en-

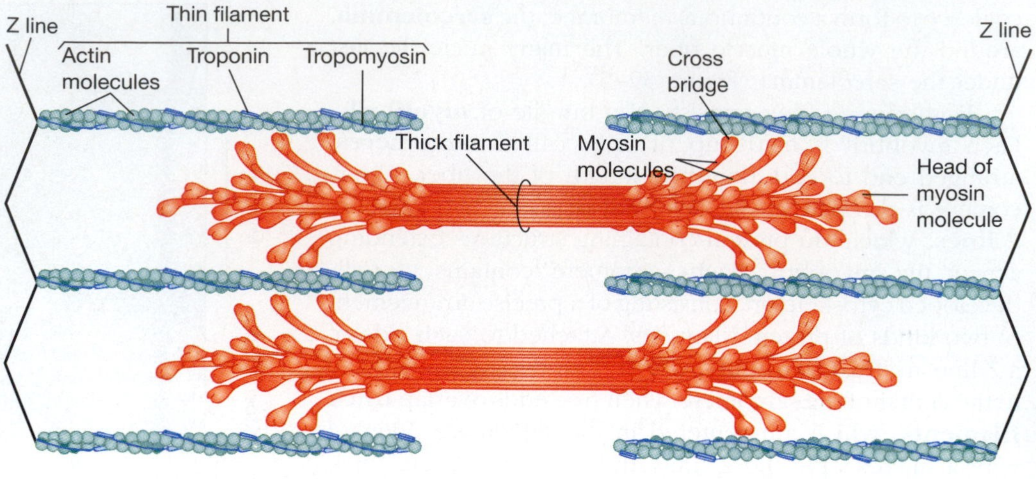

Z line    Thin filament    Cross    Z line
Actin    Troponin    Tropomyosin    bridge
molecules    Myosin    Head of
molecules    myosin
Thick filament    molecule

**FIGURE 39–7**  Molecular structure of thick and thin filaments. The filaments interact chemically when the heads of myosin molecules that make up the thick filaments attach to actin molecules in the thin filaments, forming cross bridges between the filaments. The heads then swivel, causing the muscle to contract.

ergy for the swiveling of the myosin heads (Figure 39–8). This process is very efficient, wasting only 30 to 50% of the energy from ATP as heat. In contrast, car engines waste 80 to 90% of the energy available from gasoline.

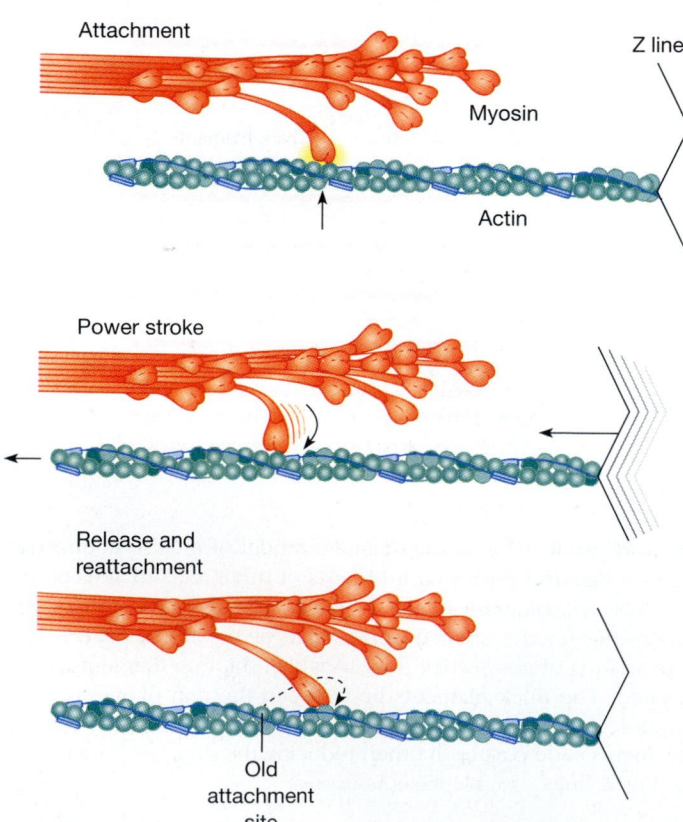

Attachment    Z line
Myosin
Actin

Power stroke

Release and
reattachment

Old
attachment
site

**FIGURE 39–8**  How filaments slide during muscle contraction. The head of a myosin molecule in a thick filament attaches to an actin molecule in a thin filament and swivels, moving the Z line closer to the thick filament. The myosin head then detaches and forms a new attachment to another actin and repeats the process.

Curiously, muscles need ATP to relax as well as to contract. Myosin heads cannot detach from the thin filaments until they have bound new ATP molecules. When an animal dies, its muscles soon run out of ATP and lose the ability to contract or relax. They become rigidly locked in whatever position they occupied when the ATP was used up, a phenomenon called **rigor mortis.**

A resting muscle's mitochondria can make ATP faster than it is used. Some of the ATP transfers phosphate groups to molecules of creatine, forming **phosphorylcreatine.** When the muscle becomes active, the phosphate can be transferred back to ADP, forming ATP. In this way muscle stores energy for later use, and an active muscle's energy supply is somewhat greater than its current rate of respiration plus glycolysis.

### Control of Contraction

A muscle usually contains ATP, and yet it contracts only some of the time. Between contractions, tropomyosin blocks the actin molecules' binding sites for myosin heads. Then myosin cannot bind to actin and cause contraction.

A muscle fiber contracts only when calcium ions ($Ca^{2+}$) are present to unblock the actin binding sites. Calcium ions are stored inside membranes that make up the **sarcoplasmic reticulum** (which resembles the endoplasmic reticulum of other cells). When a muscle fiber is activated, these ions rush out of the sarcoplasmic reticulum and then are pumped back in. A muscle fiber contracts if its cytoplasm contains enough $Ca^{2+}$ and ATP.

A skeletal muscle fiber is normally activated by the arrival of an action potential in its motor neuron. The axon terminals of a single motor neuron may form neuromuscular junctions with several to more than a hundred muscle fibers. The axon terminals release the neurotransmitter acetylcholine, which crosses the neuromuscular junction, binds to receptors on the sarcolemma, and causes the sarcolemma to depolarize. Extending from the sarcolemma into the muscle fiber is a system of membranes, the **transverse tubules (T-tubules).** Depolarization of the sarcolemma spreads along the T-tubules to the sarcoplasmic reticulum (Figure 39–9). The electrical signal opens calcium channels in the sarcoplasmic reticulum, and this releases calcium ions into the cytoplasm. Here they bind to troponin. This causes tropomyosin to move, exposing the sites on actin molecules that bind to myosin. Myosin heads attach to the actin and swivel, so that the muscle contracts.

Calcium is continually pumped back into the sarcoplasmic reticulum by active transport. When most of the calcium ions have unbound from troponin, tropomyosin moves out and blocks the binding sites on the actin again, and the fiber relaxes.

In smooth muscle of vertebrates, calcium controls contraction by a somewhat different mechanism. Here, the heads of myosin molecules have small regulatory protein chains that inhibit the breakdown of ATP even though actin binding sites are available to the myosin heads. When calcium ions are released around the contractile proteins, they activate a kinase enzyme that phosphorylates these regulatory chains, thereby removing their inhibitory effects.

## Graded Contraction

Muscles do not always contract or relax completely. They contract to different degrees as the situation demands. The mechanisms producing this graded response differ among animals.

In arthropods, some muscles receive inhibitory synapses as well as the more usual excitatory synapses, and the balance between inhibitory and excitatory impulses determines the degree of contraction.

Most invertebrate skeletal muscle fibers can produce graded responses. A series of impulses in the motor nerve produces ever greater depolarization of the muscle membrane, which causes slowly increasing contraction of the muscle fiber with time. Muscle fibers of this type, known as "slow" fibers, are also found in vertebrates, where they produce graded contractions in a similar manner. They are common in fish, birds, and reptiles.

By contrast, most mammalian skeletal muscle fibers are of the "fast" or "twitch" variety. They cannot produce a graded contraction, but respond to stimulation in an all-or-nothing manner, after which they relax. The main advantage of this type of muscle is that it contracts faster than "slow" muscle and so permits rapid movement. The flight muscles of some insects contract even more rapidly than mammalian "fast" muscles.

Having fast fibers makes it more complicated to produce a graded muscle response such as picking up a loaded tray. Each fast fiber is innervated by a branch of a motor

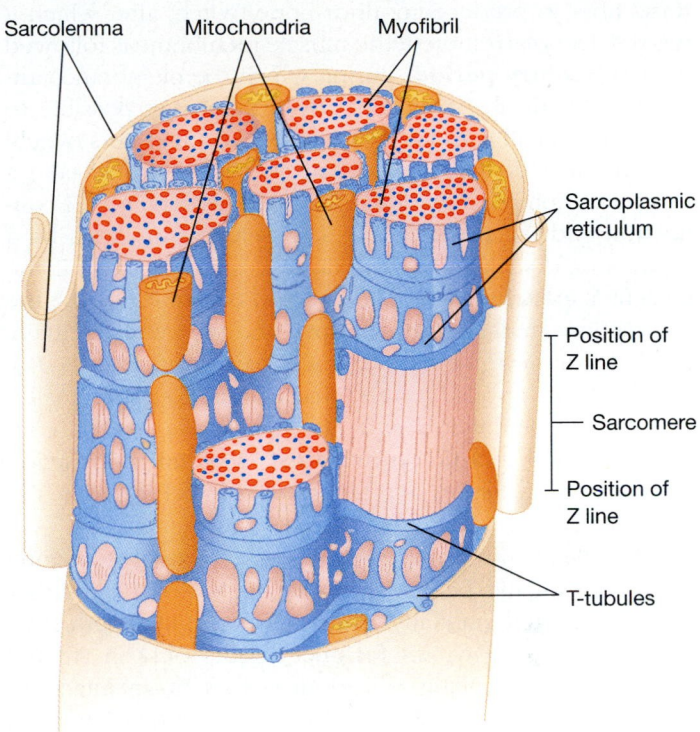

Sarcolemma   Mitochondria   Myofibril

Sarcoplasmic reticulum

Position of Z line

Sarcomere

Position of Z line

T-tubules

**FIGURE 39–9** T system and sarcoplasmic reticulum in part of a muscle fiber. The sarcoplasmic reticulum is a reservoir containing calcium ions. When the sarcolemma is depolarized, the depolarization spreads down the membranes of the T-tubules to the sarcoplasmic reticulum, which releases calcium ions into the myofibril.

neuron. The neuron and the muscle fibers it innervates are known as a **motor unit.** Since each skeletal muscle is made up of hundreds of motor units, the degree of contraction of the whole muscle can be altered by controlling how many motor units are active at any one time. As our hands and arms take the weight of the tray, more and more motor units are brought into action, until enough muscle fibers are contracted to lift the tray.

Contractions that last for a long time (while you walk down a corridor with the tray) are possible because the motor units fire in turn. For example, in the posture muscles, which work for long periods of sitting or standing, many fibers are active at any one time. These then relax while others contract. The result is that the entire muscle remains partially contracted, but no one fiber need stay contracted until it exhausts its supply of ATP.

The disadvantage of this system is that it requires more motor neurons than does a slow muscle system, since many different motor units are necessary. This is one reason that mammals have so many more central and peripheral neurons than do invertebrates or lower vertebrates.

## Tetanus and Fatigue

A single impulse from its motor neuron causes a vertebrate "fast" fiber to produce an all-or-none twitch, after which it relaxes. Depolarization of the muscle membrane is followed by a **refractory period,** during which the membrane cannot be electrically stimulated. How then, is it possible to carry a tray at all without spilling things as the fibers twitch?

The answer is that electrical events in muscle take a short time compared with the mechanical sequence of contraction and relaxation. So, if a muscle is stimulated by a rapid series of electrical impulses, it never relaxes. If the muscle is electrically stimulated again before it relaxes, the contractions undergo **summation,** and the muscle actually produces more force than a single twitch (Figure 39–10). As stimuli are applied to the muscle at smaller and smaller intervals, the individual twitches sum into a smooth contraction called **tetanic contraction.** A tetanic contraction is smooth and steady, as well as more forceful than a single twitch.

A skeletal muscle that is stimulated repeatedly for a long period eventually becomes unable to respond to further stimulations and gradually returns to its resting length. This phenomenon, known as **fatigue,** occurs only in striated muscle. Complete fatigue seldom occurs in an intact organism, but it is easily induced in a muscle that has been removed from the body.

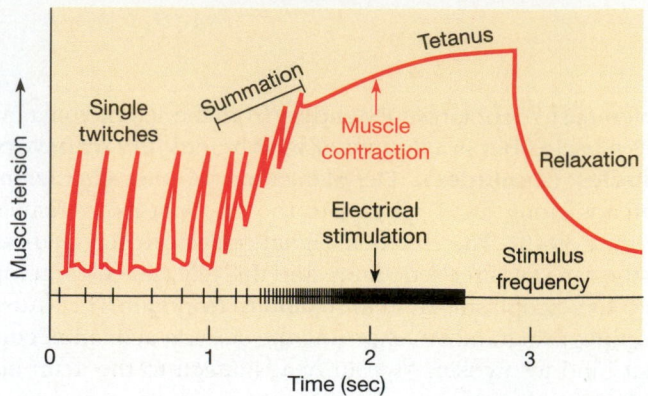

**FIGURE 39–10** Tetanic contraction. The muscle's motor neuron is stimulated electrically at increasing frequency, as shown along the lower line. When the rate of stimulation is low, the muscle contracts in individual twitches. As the stimulus frequency increases, the muscle does not have time to relax fully between stimuli. Summation occurs, finally resulting in a smooth, sustained tetanic contraction.

## 39–C  HOW MUSCLES AND SKELETONS INTERACT

Muscles can move parts of the body only because they work against skeletons. A **skeleton** may be defined as anything on which a muscle exerts force. Some structures that act as

(a)

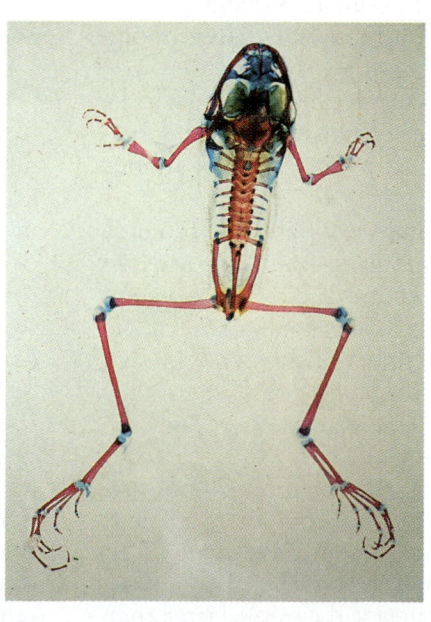

(b)

**FIGURE 39–11**  Skeletons. (a) The thin, tough exoskeleton of an insect or other arthropod provides a lot of strength for its weight and also supports and protects soft body parts. However, it cannot grow. Hence, the animal must molt its exoskeleton several times during its life, as this 17-year cicada is doing. (b) The endoskeleton of a frog consists of bone (stained red) and cartilage (blue). The bones grow as the animal grows, and bone replaces most of the cartilage at the ends of the bones.   (b, L. L. Sadler)

skeletons are not usually thought of as skeletal. For instance, the muscles that surround a blood vessel work against the walls of the vessel, and against the blood itself, when they contract and reduce the diameter of the vessel. When the muscles of the uterus contract during childbirth, they could do no useful work if there were no baby to push against.

Muscles that cause movement inside the body shorten so little, and exert so little force when they contract, that their attachment to nearby cells gives them enough leverage to operate. The muscles used in locomotion, on the other hand, may contract forcefully, and they usually exert their force on what we generally think of as a skeleton: a hydrostatic skeleton, such as the fluid-filled cavities of cnidarians or annelids; an exoskeleton, such as the cuticle of an insect or shell of a snail; or an endoskeleton, such as the bones of a vertebrate (Figure 39–11).

## Antagonistic Muscles

Movements are generally controlled by the action of two sets of **antagonistic muscles**—muscles with opposite effects. In systems without hard skeletons, such as soft-bodied invertebrates and the internal organs of vertebrates, muscles are arranged in circular and longitudinal sheets. Longitudinal muscles run lengthwise along the body (or organ) and make it shorter and wider when they contract. Contraction of circular muscles, which run around the body (or organ), makes it longer and narrower. The antagonistic actions of longitudinal and circular muscles against the contents of the alimentary canal move food down the gut in most animals. Similar forces exerted against the fluid-filled coelom of an earthworm permit the worm to move (see Figure 29–3).

Muscles attached to hard skeletons, such as the exoskeletons of arthropods or molluscs or the endoskeletons of vertebrates, are often arranged in bundles rather than sheets. A skeletal muscle is attached to the skeleton either directly or by way of a **tendon.** In some cases—as with the muscles that move our fingers—a tendon may be almost as long as the muscle.

Antagonistic muscles run parallel across a joint in the skeleton. The muscle that causes the joint to stretch out is called an **extensor,** and its antagonist, which causes the joint to close up, is called a **flexor.** The biceps is the main flexor across the human elbow joint, and the triceps is its antagonistic extensor (Figure 39–12).

Exoskeletons and endoskeletons may differ in the way muscles attach. In vertebrates most skeletal muscles are attached at a single point. However, the greater relative surface area of an invertebrate exoskeleton allows the attachment of muscle fibers all along the skeleton of a particular limb. The great strength of an insect relative to its size results partly from the variety of different types of muscle attachment possible with an exoskeleton.

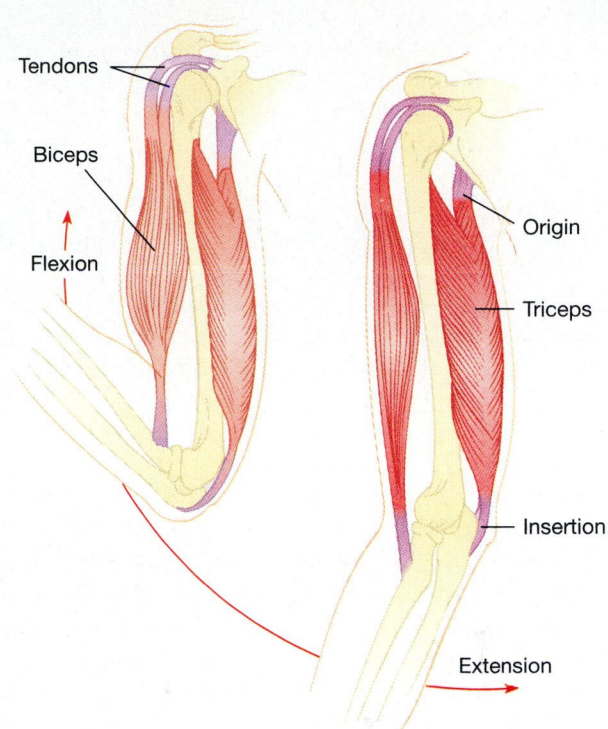

**FIGURE 39–12** Skeletal muscles in the human upper arm. Each muscle is attached to the skeleton by tendons at its insertion (the end attached to the bone that moves) and origin (the end attached to the less mobile bone).

## Reciprocal Inhibition

Locomotory muscles are so powerful that if a flexor and its antagonistic extensor were both to contract strongly at the same time, they could break a bone. This is prevented by **reciprocal inhibition** involving a reflex arc from each muscle to its antagonist. Proprioceptors in the muscle detect how far it is contracted (Section 38–C). Their sensory neurons send this information to interneurons in the spinal cord, which in turn inhibit the firing of motor neurons to the antagonistic muscle. Thus, any stimulus that causes a muscle to contract also inhibits the contraction of its antagonist, and the joint moves smoothly.

This is the simplest reflex involved in locomotion. Much more complex reflexes come into play when we walk. For instance, reflexes from the opposite limb ensure that one leg moves after another. Such reflexes are even more complicated in animals that have many legs or that use their tails for balance.

## 39–D THE VERTEBRATE SKELETON

The vertebrate skeleton has two major portions. The **axial** skeleton runs along the axis of the body and includes the skull, backbone, ribs, and tail. The **appendicular** (limb)

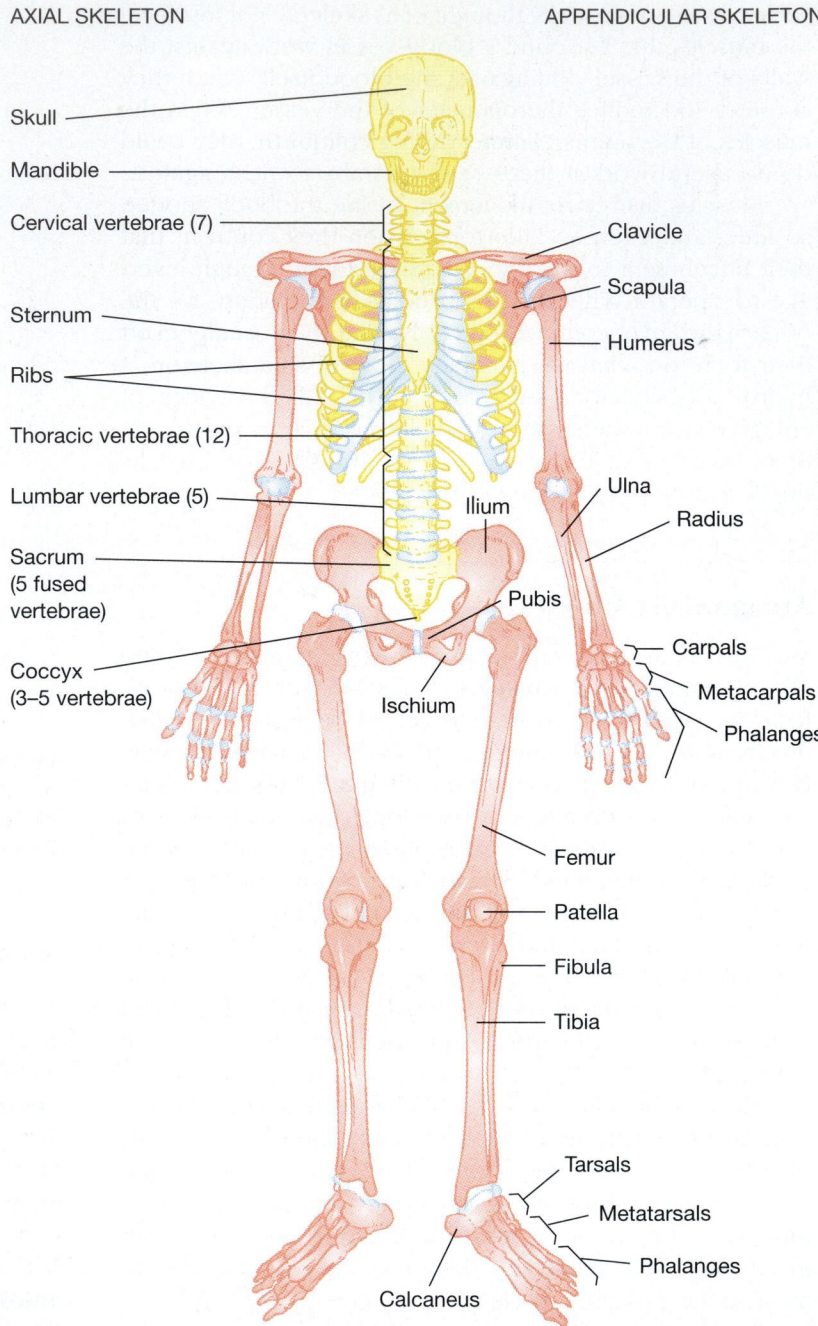

AXIAL SKELETON                    APPENDICULAR SKELETON

Skull

Mandible

Cervical vertebrae (7)                                    Clavicle

                                                          Scapula

Sternum                                                   Humerus

Ribs

Thoracic vertebrae (12)

Lumbar vertebrae (5)                                      Ulna
                                        Ilium             Radius

Sacrum
(5 fused                                Pubis
vertebrae)                                                Carpals

                                                          Metacarpals

Coccyx                                                    Phalanges
(3–5 vertebrae)              Ischium

                                                          Femur

                                                          Patella

                                                          Fibula

                                                          Tibia

                                                          Tarsals

                                                          Metatarsals

                                                          Phalanges
                     Calcaneus

**FIGURE 39–13**   Some major bones of the human skeleton.

skeleton includes the bones in the limbs and in the **limb girdles** that attach the limbs to the backbone. The **pectoral girdle** includes the collarbones and shoulder blades, and the **pelvic girdle** includes the large fused hip bones (ilium, ischium, and pubis) (Figure 39–13).

The bones and cartilages in our bodies anchor various muscles and permit them to move the body, but they serve other functions as well. First, they provide a framework that supports the body against the pull of gravity. Land vertebrates have skeletons more substantial than those of vertebrates that live supported by water. Parts of the skeleton also protect internal organs. The skull protects the brain and

major sense organs, the backbone surrounds and protects the spinal cord, and the pectoral girdle and ribs protect the heart and lungs.

## Vertebrate Joints

Joints between bones in the skeleton are of many types and degrees of rigidity. At one extreme are **sutures,** wiggly lines in the skull where bones meet at interlocking projections that hold them tightly together. However, most bones are joined by ligaments. An extremely flexible joint is the one

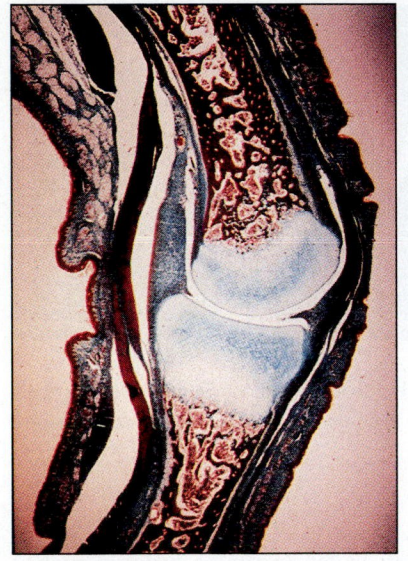

(a) Finger joint (LM)

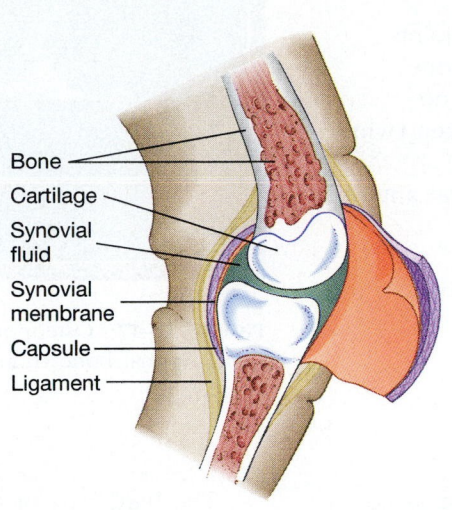

Bone
Cartilage
Synovial fluid
Synovial membrane
Capsule
Ligament

(b)

**FIGURE 39–14** Structure of a joint. (a) This photograph of the finger joint of a human child shows the large, pale areas of cartilage at the ends of the bones that form the actual joint. In an adult, the cartilages would be smaller, but would still be present to cushion the ends of the bones. The space between the bones is filled with fluid, kept in place by the surrounding membrane (b). A tough, fibrous capsule surrounds the membrane. Ligaments bind the bones together and limit the movement between them.  (a, Biophoto Associates)

between the two bones in the front of the female hip girdle, which are held loosely together by a ligament that stretches and allows the bones to separate during childbirth.

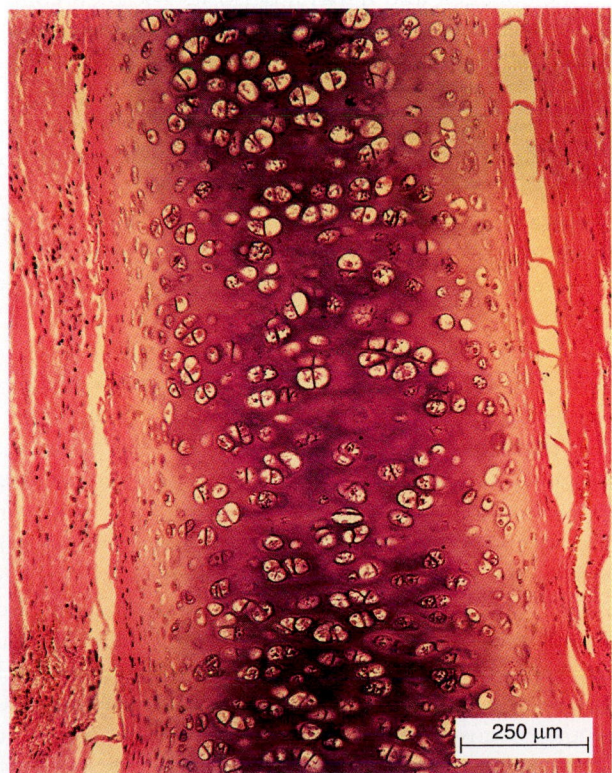

Cartilage (LM)

250 µm

**FIGURE 39–15** Cartilage is composed of scattered cells (white, with darker nuclei) surrounded by much extracellular material. (Biophoto Associates)

The joints of the limbs must permit smooth movement in various directions. The ends of the bones in these joints are covered by a smooth elastic sheet of connective tissue, and the whole joint is surrounded by a sac filled with **synovial fluid,** which lubricates the joint (Figure 39–14).

## 39–E  CONNECTIVE TISSUE

Bones, cartilage, tendons, and ligaments are all made up of various types of **connective tissue,** composed mainly of a matrix of substances secreted by scattered cells. This extracellular (outside cells) matrix contains fibers of a tough, elastic protein, **collagen** (see Figure 3–21c). Collagen is the most abundant protein in mammals, accounting for about one fourth of all the body's protein (it is also the main ingredient of gelatin). Cartilage has a great deal of a firm extracellular jelly surrounding the collagen fibers and cells (Figure 39–15), whereas tendons and ligaments consist mostly of fibers, with very few cells. Bone is hard and brittle because it has mineral deposits in addition to fibers.

### Cartilage

Unlike bone, cartilage receives no blood supply. Its scattered cells rely on diffusion of nutrients from capillaries in surrounding tissues. Cartilage makes up the entire skeleton in members of the Chondrichthyes (sharks, skates, and rays) and in all early vertebrate embryos. During embryonic development of vertebrates with bony skeletons, minerals are deposited in most of this embryonic cartilage, a blood supply and other typical features of bone develop, and much of the cartilage is replaced by bone. Cartilage persists in the adult only in areas where flexibility is necessary. These areas include:

1. The ends of bones where they form synovial joints
2. The discs between the vertebrae in the backbone
3. The ends of ribs where they join the breastbone
4. The rings that thicken the walls of the trachea (windpipe) and keep it from collapsing
5. The larynx ("voice box") in the throat, at the anterior end of the trachea
6. The external ear
7. The eustachian tube, which connects the throat to the middle ear
8. The tip of the nose

## Bone

Although the mineral matter of a dried human skeleton weighs only about 5 kg, the skeleton of a living human being is much heavier because bone is a living tissue that contains cells, blood vessels, nerves, fluid, and fat deposits.

Bone varies in structure depending upon its position and function in the body. At one extreme is **spongy bone,** composed of an irregular network of mineralized bars, and at the other is **compact bone,** composed of tubular units called **Haversian systems** (Figure 39–16).

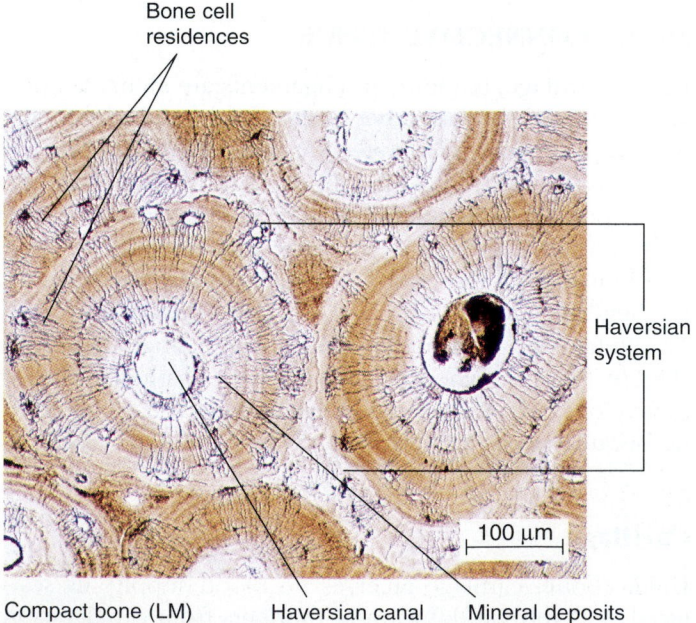

Bone cell residences

Haversian system

100 μm

Compact bone (LM)   Haversian canal   Mineral deposits

**FIGURE 39–16** A cross section of compact bone. Each set of concentric rings is a Haversian system, surrounding a central channel, the Haversian canal. In the living bone, the Haversian canals contain blood vessels and nerves. The many tan rings around each Haversian canal are mineral deposits, laid down by living cells that inhabit the small, dark, spider-shaped spaces. (Biophoto Associates)

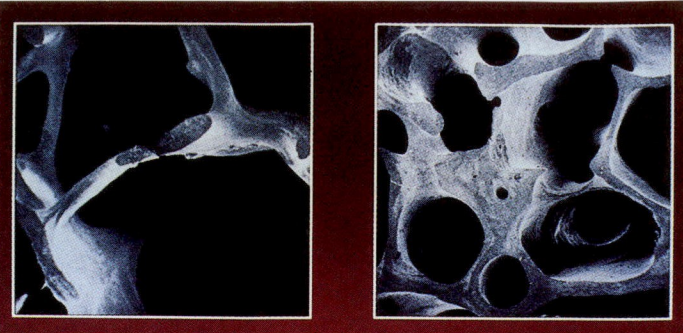

**FIGURE 39–17** Osteoporosis. Bone weakened by osteoporosis (left). Normal bone (right). (Courtesy of Wyeth-Ayerst Laboratories, Philadelphia)

The hard part of bone is made up of organic matter and inorganic salts. Most of the organic matter is collagen fibers, which give bone most of its ability to withstand tension (pull). The inorganic salts are mainly a complex phosphate of calcium and smaller amounts of other ions. These give bone its ability to withstand compression and side-slippage.

Distributed throughout a bone are the cells that lay down the collagen and mineral deposits and then continue to live within the bony walls they have constructed, connected to other cells only by long cytoplasmic extensions that form tight junctions. These cells are responsible for repair and replacement of broken bone and for the formation of calluses, which develop at points of pressure on a bone.

Repair and replacement of bone are the subjects of considerable research. One reason is that bone tends to degenerate with age, and this is an increasingly common medical problem as people live longer. For instance, in the United States more than 600,000 bone fractures a year are attributed to osteoporosis, and the resulting hemorrhage and trauma lead to many deaths.

**Osteoporosis** is the condition in which bones have lost so much inorganic matter that they become brittle and easily broken (Figure 39–17). The disorder has at least two distinct forms. Postmenopausal osteoporosis, involving loss of bone in the spine and forearms, afflicts women 50 to 65 years old. A contributing factor is the reduction in the body's level of the hormone estrogen at the time of menopause. Senile osteoporosis occurs in people of both sexes aged 75 and older and involves loss of bone in the hips and legs. The most effective way to prevent osteoporosis is for young children to consume enough calcium in the diet in the form of milk and other dairy products. In older people, osteoporosis can be counteracted by exercise, which stimulates the addition of material to bone.

Bones grow throughout life as the body increases in size and strength, and as they grow they are also remod-

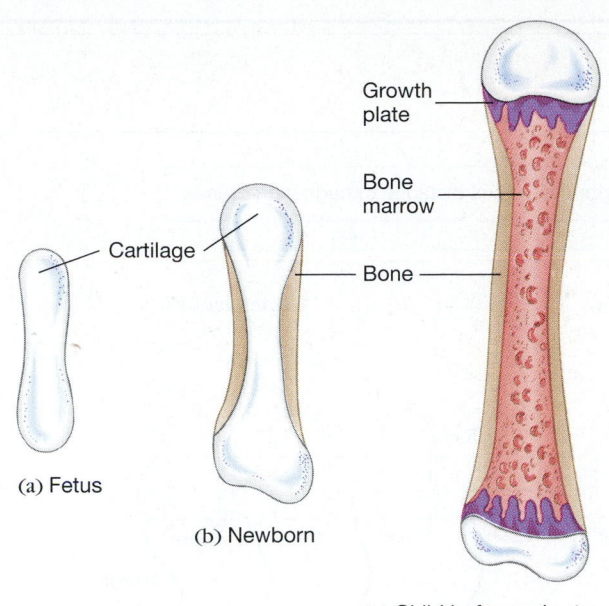

FIGURE 39–18 How a vertebrate long bone grows. The cartilage of the fetus is steadily replaced by bone surrounding a central core of bone marrow, except in the growth plate at each end of the bone where new cartilage is added until puberty.

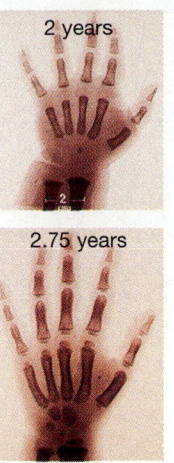
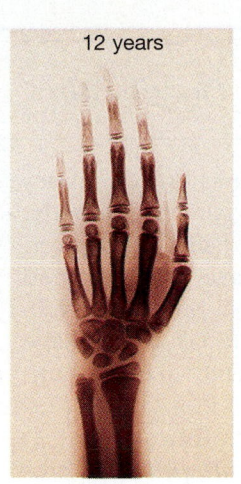
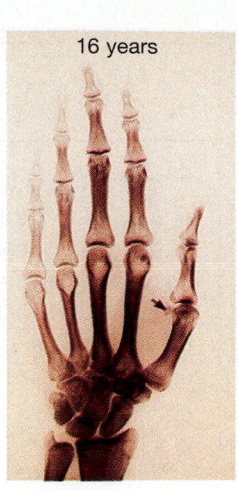

FIGURE 39–19 X rays showing growth of a human hand: at 2 years (top left), 2 years 9 months (bottom left), 12 years (center), and 16 years (right). Note that at 2 years of age, there appear to be gaps between the finger bones. These gaps contain the cartilaginous ends of the bones and the finger joints, which show up only faintly in an X ray. In the hand of a 12-year-old (center), the cartilage on either side of the joints has been almost completely replaced by bone, leaving only a thin cartilage cap over the end of the bone at the joint. In the fully developed hand (right), the ends of the bones are enlarged into knobs. The arrow points to a small spur of bone on the thumb. Many clinical conditions, from injury to arthritis, can cause such abnormal deposition of minerals in the body. When such depositions occur in joints, they may interfere with movement. (Biophoto Associates)

eled. New material is added to the outer surfaces and the ends, and old material is destroyed to enlarge the internal cavities that house the soft tissue of the bone marrow (Figure 39–18). The differentiation of bone is controlled by a number of local hormones, produced in juvenile animals while they are growing and also in response to physical stress, fractures, and other stimuli that indicate the need for more bone growth. **Bone growth protein,** for instance, is a hormone that occurs in the extracellular material of bone. When released, by a fracture or similar trauma, it induces undifferentiated cells to differentiate into bone-forming cells. The amount of bone growth protein declines with age, contributing to the increasing fragility of bone.

**Growth hormone,** secreted by the anterior pituitary gland, is also essential to normal bone growth. Its effects depend on age, changing at puberty. Before puberty, growth hormone stimulates growth and division of cells that lay down cartilage in the growth plates found in the long bones in arms and legs (Figure 39–19). As a result, it causes elongation of the long bones. Growth hormone deficiency leads to dwarfism in children but has no effect on adults. Excess growth hormone in children causes gigantism, extreme growth of the long bones, producing a tall individual with extra-long arms and legs. At puberty, the growth plates of the long bones "close," becoming unresponsive to growth hormone. In contrast, growth plates in the hands, feet, lower jaw, and skull do not close. As a result, excessive growth hormone in adults does not affect the long bones but causes **acromegaly,** excessive growth of fingers, toes, and skull bones.

*Nonskeletal Functions of Bones* In the centers of many bones are cavities filled with blood vessels and **bone marrow,** a soft tissue with a number of functions. Bone marrow is the single most important tissue of the immune system, because it is the site where all the body's white blood cells are produced, as well as red blood cells and platelets. Some bone marrow also acts as a fat depot.

Bone is involved in regulating the concentration of calcium ions in the blood. The calcium phosphate in bone is in equilibrium with that in the surrounding extracellular fluid (ECF). Thus, if the calcium level in the ECF rises, some of it is deposited in bone, and if the calcium concentration in the ECF falls, calcium from the bones dissolves into the fluid. The chemical equilibrium between solution and deposition determines the general level of calcium in the ECF and in the blood.

Fine tuning of the calcium level in the body fluids is under the control of hormones. **Parathyroid hormone** is secreted by the parathyroid glands, which lie behind the thyroid gland in the neck (see Figure 40–4). This hormone

## ESSAY: *Electric Fish*

The usual function of muscles is to move parts of the body. In some fishes, however, skeletal muscles have been modified for another function: producing electrical discharges. Electric organs have evolved independently in some cartilaginous fishes and in several different families of bony fishes.

The electric ray *Torpedo* can produce a powerful electric shock. Its electric organ consists of a row of parallel muscle fibers, called **electroplaques,** that have lost their contractile proteins and consist of little more than large neuromuscular junctions (Figure 39–A). Each electroplaque is innervated on the same side as all the others by a motor neuron. The electric organ works much like a battery with the electroplaques coupled in series. The motor neurons fire synchronously, and the depolarizations of the electroplaques sum to produce voltages up to 50 volts with several amperes of current. *Torpedo* uses its electrical discharges to stun its prey or to discourage invaders.

Cartilaginous fish have receptors called ampullae of Lorenzini embedded in a gelatinous substance at the bottoms of pits in the lateral line system. Dogfish sharks can use their ampullae to find prey, such as worms that they cannot see or smell, from a distance of many centimeters.

Members of some families of freshwater bony fish have electric organs that produce weak electrical discharges. H. W. Lissman found by behavioral experiments that these fish

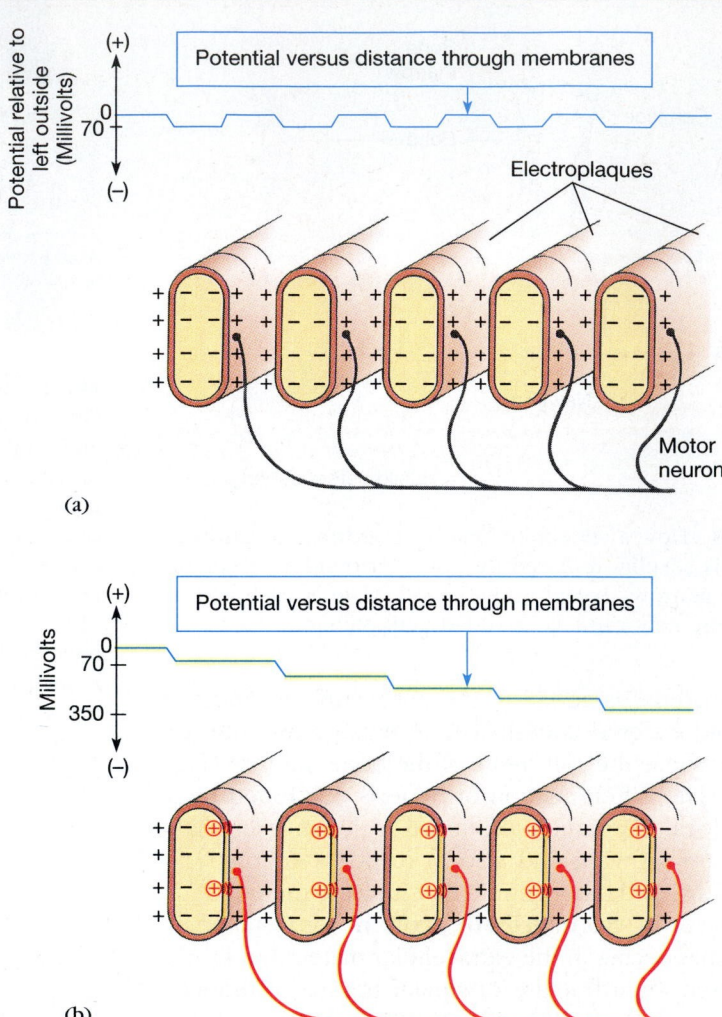

**FIGURE 39–A** Electroplaques of an electric ray. The graphs above the diagrams show potential versus distance from left to right through the membranes. (a) At rest, electrical potentials on the right- and left-hand membranes of an electroplaque cancel each other, for a total of zero potential across all the membranes shown. (b) When the motor neurons fire, they depolarize the right-hand side of each muscle cell to zero. Each of the left-hand membranes has the normal resting potential of −70 millivolts across it; with the right-hand membranes depolarized, these sum (as in a battery) to create a total potential across all the membranes of −350 millivolts in the five electroplaques shown.

stimulates the cells in bone to remove additional calcium from the bone and release it into the blood. It also causes the kidneys to reclaim more calcium from the forming urine. Third, it stimulates cells in the kidneys to convert vitamin

D to an active form, which promotes absorption of calcium from food in the intestine. A drop in the blood calcium level stimulates the parathyroid glands to produce this hormone. Calcium is then released from the bone, resorbed

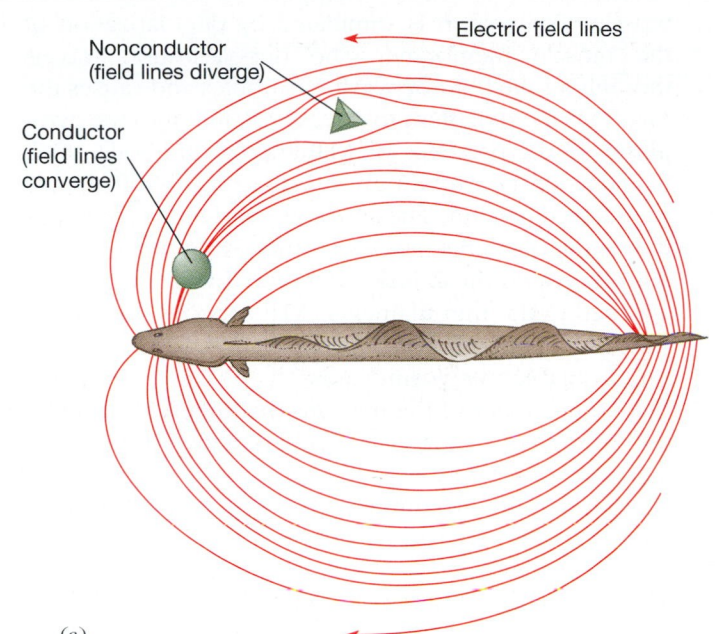

Electric field lines

Nonconductor
(field lines diverge)

Conductor
(field lines
converge)

(a)

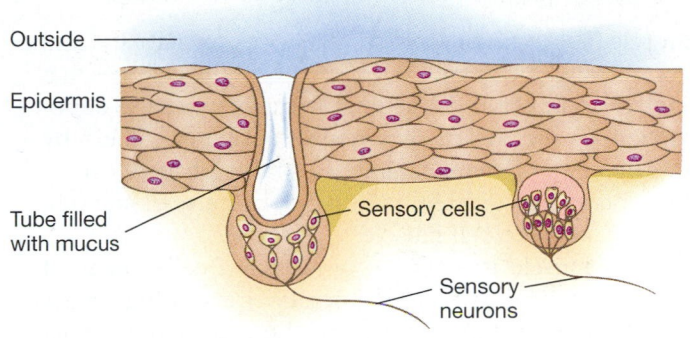

Outside

Epidermis

Tube filled
with mucus

Sensory cells

Sensory
neurons

(b)

**FIGURE 39–B** (a) The electric fish *Gymnarchus* generates an electric field with its tail and detects the field with electroreceptors, which are most concentrated in the head. The field is distorted by nearby objects. (b) The two types of electroreceptors in the skin of *Gymnarchus*.

use their electric organs for navigation. Several times a second, they emit electrical discharges lasting only a few milliseconds. Each discharge creates an electric field around the fish, and the fish's electroreceptors detect the shape of the field (Figure 39–B). The field is distorted by any object in the electric field with an electrical conductivity different from the surrounding water. The fish can detect such objects with phenomenal sensitivity. By rewarding a fish with food, Lissman showed that it could distinguish between glass and metal rods 2 mm in diameter and detect currents of about $3 \times 10^{-15}$ amp.

In the turbulent muddy waters where they usually live, sight and smell are of little use to these weakly electric fish. Their electroreceptors, in conjunction with their electric organs, provide them with a navigation system whereby they can move around freely and find food.

Recent research shows that weakly electric fish such as *Gymnarchus* (Figure 39–C) recognize individual members of their species by characteristics of the electric signals they produce, just as we recognize other people by their faces or voices.

**FIGURE 39–C** *Gymnarchus.* (Visuals Unlimited)

from the urine, and absorbed from food to make up the deficit.

A rise in the blood calcium level is counteracted by the hormone **calcitonin,** secreted by the thyroid gland. Calci-tonin inhibits the resorption of calcium from the bone deposits. This permits the opposite process, deposition of new mineral material in the bone, to bring the calcium level back down to normal.

## SUMMARY

Most of the movement in an animal's body is due to the action of its muscles. A muscle works by shortening so that it pulls against the skeleton or against adjacent tissues.

1. Muscle has two properties: electrical excitability and contractility. Electrical excitation causes the changes that initiate contraction. Some muscle membranes are depolarized by the action of a neurotransmitter, others by transmission of electrical impulses from adjacent cells.
2. Vertebrates have three main types of muscle: smooth, cardiac, and skeletal muscle.
3. Smooth muscle is made up of sheets of cells, many of them in the walls of internal organs. Depending on the organ where the muscle occurs, the cells may contract spontaneously, in response to hormones, or in response to norepinephrine released by neurons in the sympathetic nervous system or acetylcholine released by parasympathetic neurons. Smooth muscle contraction is a graded process, not an all-or-none response.
4. Cardiac muscle makes up the heart and consists of long columns of cells, many of them connected by gap junctions which conduct electrical activity directly from one cell to the next. Some heart cells have lost the ability to contract and serve only to conduct electrical activity through the heart muscle from the pacemaker in the sinoatrial node. This conducting system ensures that different parts of the heart contract at the right time. Cardiac muscle cells contract spontaneously, as well as in response to nerve impulses and hormones.
5. Vertebrate skeletal muscle is made up of syncytial fibers formed by the fusion of embryonic myoblast cells. The orderly arrangement of contractile proteins within a fiber gives the muscle a striped appearance. Fibers are innervated by motor neurons whose action potentials cause the release of neurotransmitters at neuromuscular junctions.
6. Muscle contraction has been most studied in vertebrate skeletal muscle, although it appears to be similar in all muscle. Contraction is stimulated by depolarization of the muscle membrane. The depolarization travels throughout the fiber by way of T-tubules and causes the sarcoplasmic reticulum to release calcium into the cytoplasm. In the presence of ATP and calcium ions, thick filaments of the protein myosin attach to thin filaments of the protein actin. The swiveling of the myosin heads moves the filaments past each other, shortening the distance between the Z lines of the sarcomere, which are attached to the thin filaments. ATP supplies the energy for contraction, whereas calcium makes actin's binding sites available to myosin heads.
7. The arrangement of the neurons that innervate a skeletal muscle consisting of fast fibers allows for graded muscle contractions.
8. Smooth, sustained contractions are brought about by trains of closely spaced action potentials from the motor neuron(s) active at any one time.
9. The vertebrate skeleton supports the body, protects internal organs, and permits the locomotory muscles to move the animal. Pairs of antagonistic muscles move bones back and forth at joints. Reciprocal inhibition prevents simultaneous strong contraction by both members of a pair of antagonistic muscles.
10. The vertebrate skeleton is made up of relatively flexible cartilage and harder bone. Cavities in the bone are filled with bone marrow, which stores fat and forms blood cells. Tiny spaces in the bony material itself house single cells that deposit or release calcium.
11. Bones store or release calcium to maintain equilibrium with the body fluids. Hormones maintain a fine tuning of the blood/bone calcium balance.

## Self-Quiz

For each phrase in questions 1 through 5, give the type(s) of muscle that show the characteristic.

_____ 1. Syncytial

_____ 2. Innervated by autonomic nervous system

_____ 3. Can contract without nervous stimulation

_____ 4. Found in the walls of the aorta

_____ 5. Unicellular organization

a. Cardiac

b. Skeletal

c. Smooth

6. Arrange these structures in the order in which an electrical impulse spreads during a heartbeat cycle:
   a. sinoatrial node          c. Purkinje fibers
   b. atrioventricular node    d. vagus nerve
7. ATP is required for:
   a. release of calcium ions from the T-tubules
   b. release of cross bridges
   c. removal of tropomyosin from the myosin binding site on the actin filaments
   d. binding of myosin to troponin to form tropomyosin
   e. release of calcium ions from the sarcoplasmic reticulum

8. Indicate which of the following items would be found at each of the places indicated on the diagram:

_____ actin
_____ myosin
_____ troponin
_____ calcium
_____ tropomyosin
_____ Z line
_____ mitochondria

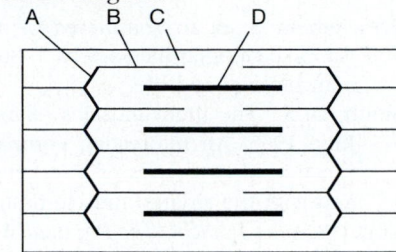

9. Contraction of a skeletal muscle fiber begins when the sarcolemma binds _____ released from a motor neuron. Electrical excitation spreads through the sarcolemma and into T-system tubules and sarcoplasmic reticulum. _____ is released into the cytoplasm and binds with _____, causing _____ to move and uncover binding sites on _____, which then bind to _____.

a. acetylcholine      d. calcitonin      g. norepinephrine
b. actin              e. calcium         h. tropomyosin
c. ATP                f. myosin          i. troponin

10. The extent of contraction of a mammalian skeletal muscle is controlled by:
    a. interaction of excitatory and inhibitory nervous input to individual fibers
    b. contraction of some sarcomeres in a muscle fiber but not others
    c. contraction of some entire muscle fibers while others remain relaxed
    d. reciprocal inhibition
    e. all of the above

11. The condition in which a living muscle is unable to respond to additional stimulations is referred to as:
    a. fatigue          c. tetanus
    b. rigor mortis     d. summation

12. All the following are functions of the vertebrate endoskeleton except:
    a. protection
    b. locomotion
    c. storage of calcium and phosphorus
    d. temperature regulation
    e. release of calcium and phosphorus to the blood

13. Which of the following is *not* a role of bones?
    a. storing calcium when blood calcium levels are too high
    b. removal of calcium deposits carried out by living bone cells in response to a hormone
    c. production of red blood cells
    d. production of white blood cells
    e. production of hormones to regulate calcium levels in body fluids
    f. storage of fat
    g. providing attachment sites for muscles
    h. protection of internal organs

14. Both bones and cartilage:
    a. contain collagen      c. store fat
    b. contain blood vessels d. all of the above

15. When the level of calcium in the blood drops below a certain point, calcium is released from:
    a. sarcoplasmic reticulum
    b. bone under the influence of calcitonin
    c. bone under the influence of parathyroid hormone
    d. cells in the bone marrow
    e. T-tubules

## Questions for Discussion

1. If you are a murder mystery fan, you know that it takes a variable length of time for rigor mortis to set in after death. What are some reasons for this variation?
2. Rigor mortis lasts for several hours and then disappears. Why does it eventually go away?
3. What are some advantages of having muscles attached to bones via tendons rather than directly?
4. Comment on the biological validity of the sayings "I can feel it in my bones" and "dry as a bone."
5. The disease rickets is characterized by bending of the bones due to lack of calcium deposits to keep them stiffened into the proper shape. Why is it advantageous for the bones to give up these calcium deposits even though doing so results in permanent skeletal deformity?

6. Would you expect levels of parathyroid hormone or of calcitonin to be elevated in a victim of rickets? In a pregnant woman? In her fetus?
7. Sharks and rays have skeletons composed entirely of cartilage. How might these vertebrates regulate the level of calcium in their body fluids?
8. Pumiliotoxin B, produced by dart-poison frogs, facilitates the release of $Ca^{2+}$ from muscle storage areas and inhibits its return. What symptoms would you expect to see in an animal dying after being shot with a dart tipped with this poison?
9. Why does a person breathe heavily and sweat after 15 minutes of aerobic dancing?

## Suggested Readings

Caplan, A. I. "Cartilage." *Scientific American,* October 1984.

Huxley, H. E. "Mechanism of muscle contraction." *Scientific American,* December 1965. Describes experimental techniques used to isolate muscle proteins and study their structure. Many photographs of the isolated protein fibers.

McMahon, T. A. *Muscles, Reflexes, and Locomotion.* Princeton, N.J.: Princeton University Press, 1984. A delightful advanced text for the mechanically minded.

Murray, J. M., and A. Weber. "The cooperative action of muscle proteins." *Scientific American,* February 1974. Recounts the experimental unraveling of the roles of the four muscle proteins, ATP, and calcium in muscle contraction.

Schiefelbein, S., et al. *The Incredible Machine.* Washington, D.C.: National Geographic Society, 1986. Fascinating account of human biology and life.

Smith, D. S. "The flight muscles of insects." *Scientific American,* June 1965. An interesting comparison of the structure and function of two types of insect flight muscle with each other and with the striated muscle tissue of vertebrates.

Vogel, S. *Life's Devices: The Physical World of Animals and Plants.* Princeton, NJ: Princeton University Press, 1989. An entertaining book on how organisms adapt to the physical forces of the world they live in.

C H A P T E R

40

# Animal Hormones and Chemical Regulation

## O B J E C T I V E S

*When you have studied this chapter, you should be able to:*

1. State what a hormone is, and explain how hormones may be identified and their action determined.
2. Describe the process of feedback control of hormone production, and diagram it using a specific example.
3. Describe what is known about how lipid-soluble and water-soluble hormones and cytokines affect their target organs.
4. Explain the differences in the control exerted by the nervous system and by the endocrine system.
5. Describe the role of cyclic AMP, kinases, and phosphatases in a cell's response to a hormone.
6. Describe the "fight or flight" response, listing six physiological changes that occur, and explaining the role of the nervous and endocrine systems in the response; discuss the roles played by epinephrine and norepinephrine in the response.
7. Describe the relationship between the hypothalamus and both the anterior and the posterior pituitary gland.
8. Explain the role of the pituitary gland in the body's endocrine system.
9. Give two examples each of hormonal responses that maintain homeostasis and of hormonal responses to conditions in the animal's environment.
10. Define circadian rhythm, give an example of a biological clock, and tell what is known about their properties.
11. Define pheromone, and explain the biological roles of pheromones.

---

Wherever there is division of labor, there must also be an exchange of information to coordinate the work of different units. On a construction site, dozens of people may perform different jobs. They must know what their co-workers are doing and interact appropriately to build the structure properly. In any multicellular organism, the jobs to be done are divided among many types of cells. Like the workers on a construction site, these cells must exchange information and work together if the organism is to survive. In animals, the nervous system and the body's chemical messengers coordinate the activities of cells.

We can define a **chemical messenger** as a substance produced by one cell that affects other cells. Any cell that it affects is known as the messenger's **target cell.** The first chemical messengers to be discovered were named hormones ("to excite"). As new hormones were discovered, the definition of a hormone expanded to include them, until today it is impossible to write an accurate definition that includes all the known hormones and hormone-like messengers. For most purposes, however, a **hormone** can be defined as a substance secreted into the blood in tiny amounts by specialized cells or glands and carried to other parts of the body, where it interacts with receptors on its target cells, producing specific responses. Substances that act as messengers but don't fit this definition we simply call chemical messengers.

Hormones are generally produced by **endocrine** cells. A typical endocrine cell responds to a chemical signal, often a hormone, by secreting its own hormone. Most endocrine cells are housed in **endocrine glands,** which have no ducts, unlike the glands with ducts that produce substances such as tears or digestive enzymes.

Some hormones are produced not by endocrine glands but by neurons, indicating the close relationship between endocrine and nervous systems. For instance, some neurons respond to electrical stimulation by releasing neurohormones from their axons into the bloodstream. Hormones released in this way include oxytocin, which induces labor during childbirth.

The functions of hormones and other chemical messengers fall into three main groups. First, many hormones maintain homeostasis, acting to stabilize the level of some substance in the body fluids. Second, some chemical messengers control cell division and differentiation, particularly during embryonic development. Third, other hormones play a part in adaptive responses to events outside the body. For instance, in many animals, environmental cues stimulate production of sex hormones, which bring the animal into breeding condition at a season when conditions favor the survival of its young (Figure 40–1).

The most obvious interactions between the nervous and endocrine systems occur in response to environmental stimuli. An animal can detect environmental events only through its sense organs, and the nervous system must convey this

FIGURE 40–1   Reproductive behavior in many animals results from hormonal responses to environmental cues such as daylength. Here, a white tern in Hawaii feeds a juvenile in spring.   (Frans Lanting/Minden Pictures)

information to the glands that produce hormones. In vertebrates, the hypothalamus in the brain (part of the nervous system) communicates with the pituitary gland (part of the endocrine system), which in turn regulates most other endocrine glands.

Various chemical messengers occur throughout the body, but each cell responds only to some of the messages. This is because different kinds of cells have receptors for different messenger molecules. Furthermore, the changes caused by a messenger depend upon which genes and metabolic pathways are active within each kind of cell.

## KEY CONCEPTS

- Neurons and chemical messengers coordinate activities within an animal's body.
- Chemical messengers are involved in maintaining homeostasis within the body, in growth and differentiation, and in the body's response to outside stimuli.
- The body maintains homeostasis of various chemical components by negative feedback control, using hormones as messengers to raise or lower the levels of other chemicals, including other hormones.

- The need to secrete a hormone is determined, in many cases, by the nervous system, which stimulates secretion by some glands.
- Only those cells with receptors for a hormone can respond to that hormone, and the type of response depends on the cell's biochemistry.

## 40–A  CHEMICAL MESSENGERS

We now know that animals have dozens of chemical messengers, including:

1. **Hormones,** produced by neurons and by endocrine cells in endocrine organs or in other organs such as the stomach and gonads (ovaries and testes).

2. **Cytokines,** chemical messengers with local effects, usually acting only on neighboring cells because they are rapidly removed from the extracellular fluid. These local messengers include interleukins, which activate lymphocytes in the immune system; growth factors, which stimulate growth and cell division by particular tissues; prostaglandins, which are lipids with a variety of effects;

and short-range neuroregulators, which are released into a local area by neurons.

3. **Pheromones,** chemical signals produced by one animal that affect the behavior of other individuals of the same species.

## 40–B HORMONES

How do we know that a hormone exists and what it does? Most hormones have been discovered by removing the relevant endocrine gland and observing how this affects the animal. Early experiments on metamorphosis illustrate the types of studies that led to the discovery of most hormones.

**Metamorphosis** is a more or less abrupt alteration in an animal's anatomy and physiology as it changes from a larva, such as a caterpillar, maggot, or tadpole, into a very different adult. The term is used to cover a number of rather different phenomena, but all of them are controlled by hormones. The best-known examples are those involving rapid change, as when an insect caterpillar or nymph metamorphoses into a pupa or adult, or an amphibian tadpole into a frog.

### Insect Molting and Metamorphosis

Major changes from larval to adult form occur in many insects. Such alterations in an insect's structure can occur only when it molts its cuticle and grows a new one. Molting and metamorphosis are therefore intimately related, although the function of molting is primarily to permit an increase in size, and the function of metamorphosis is to produce a change in body shape.

If a ligature is tied tightly around an insect's body before a critical stage in development, only the front half of the body undergoes metamorphosis. This shows that the process is controlled by a hormone produced in the front of the body.

In the 1930s and 1940s, English insect physiologist Vincent Wigglesworth studied this problem in some detail, working with the blood-sucking bug *Rhodnius prolixus.* Wigglesworth found that the molting hormone, **ecdysone,** was produced by prothoracic glands in the thorax, just behind the head (Figure 40–2). The stimulus for secretion of ecdysone turned out to be distension of the abdomen. Thus, in real life, a nymph molts shortly after a large meal of blood. The meal distends the abdomen and also provides enough food to last the insect until the molt is completed and it can feed again.

Although *Rhodnius* undergoes five molts as it develops, metamorphosis occurs only during the last molt. Why is this? In an ingenious experiment, Wigglesworth diluted the blood of an early nymph by adding blood from an adult and found that the early nymph underwent premature par-

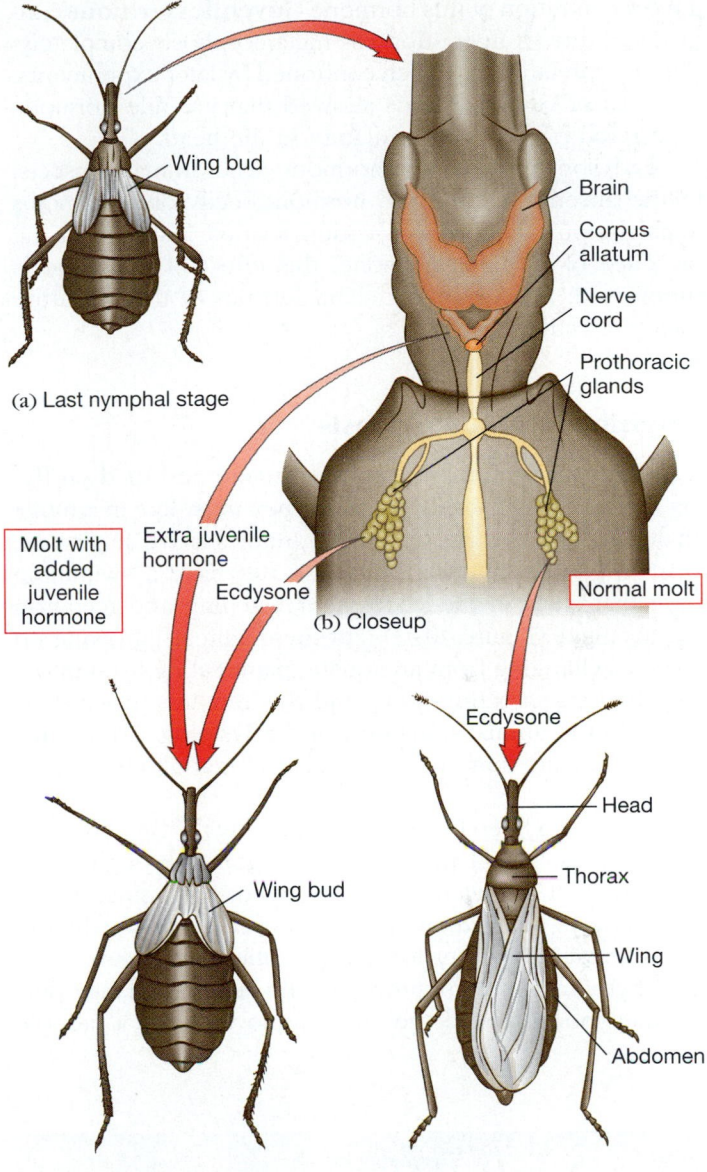

**FIGURE 40–2** Metamorphosis in *Rhodnius prolixus*. (a) The nymphal stages resemble the adult, but without wings. (b) Head and part of the thorax cut open to show the corpus allatum, where juvenile hormone is produced, and the prothoracic glands, where ecdysone is produced. (c) If the last nymphal stage is experimentally exposed to both juvenile hormone and ecdysone, it molts into a giant nymph. (d) Normally the last molt exposes the insect only to ecdysone. In the absence of juvenile hormone, the last nymphal stage metamorphoses into an adult.

tial metamorphosis. He interpreted this to mean that the blood of an early nymph normally contains a hormone that inhibits metamorphosis during the early molts but disappears at the fifth molt so that metamorphosis then takes place. By diluting the early nymph's blood, he had lowered the concentration of this hormone (**juvenile hormone**) until it could no longer suppress metamorphosis completely. This interpretation has been confirmed by later experiments. Removal of various glands showed that juvenile hormone is secreted by the corpus allatum in the head.

Ecdysone and juvenile hormone occur in many insects. In the absence of juvenile hormone, ecdysone promotes transcription of genes that code for adult features. In the presence of juvenile hormone, this effect of ecdysone is suppressed, and genes for characteristics of the immature stage are active.

## Amphibian Metamorphosis

As adults, many amphibians live on land and feed on flying insects. As embryos, however, they must live in a moist environment, because the eggs cannot survive in dry air. Most frogs that lay their eggs in water have a swimming larva that stays in the water after hatching and feeds on aquatic algae (Figure 40–3). For these animals, growing up involves changing from an aquatic plant-eating form into a terrestrial insect-eating form, and this requires many alterations in the animal's anatomy and physiology. Hormonal changes initiate transcription of genes for adult characteristics.

It has long been known that removing the thyroid gland from a tadpole will prevent it from going through metamorphosis. If the gland is reimplanted, the animal metamorphoses. The hormone thyroxin, secreted by the thyroid gland, must be present for metamorphosis to occur.

Thyroxin also acts directly on tissues that change during metamorphosis. If the tail is removed from a tadpole and placed in a bath containing thyroxin, the white blood cells in the tail will digest it so that it gets smaller, just as in the intact tadpole. In the control experiment, in which the bath contains no thyroxin, the tail is unchanged.

Thyroxin affects only tissues that are in an appropriate state. If thyroxin is injected into a fairly mature tadpole, the tadpole will metamorphose prematurely. However, the same injection has no effect on a very young tadpole because a tissue cannot react to a hormone until it has reached a particular stage in its own development. As an interesting corollary, in *Necturus maculosus,* a salamander that never undergoes metamorphosis but becomes sexually mature as a larva, lack of metamorphosis is not due to the absence of thyroxin. The tissues of *Necturus* will not react to the hormone at any stage in its life, no matter how much of it is injected into the animal.

Experiments such as these have led to the discovery of dozens of vertebrate hormones. Some of the most important of these are listed in Tables 40–1 and 40–2. Figure 40–4 shows the locations of the main endocrine glands of humans.

## Feedback Control of Secretion

Many of the hormones listed in Tables 40–1 and 40–2 help control the composition of body fluids or the rate of metabolism. For these homeostatic mechanisms to function, the hormones must be secreted only when they are needed. Their secretion is under feedback control, a process characteristic of the endocrine system. An endocrine cell produces and secretes a particular hormone, and the effects of that hormone feed back to the endocrine cell, altering the

(a)

(b)

**FIGURE 40–3** Metamorphosis in a frog. (a) A tadpole. (b) Partway through metamorphosis. The legs are fully developed, but the tail has not yet been resorbed. (Oxford Scientific Films/Animals Animals)

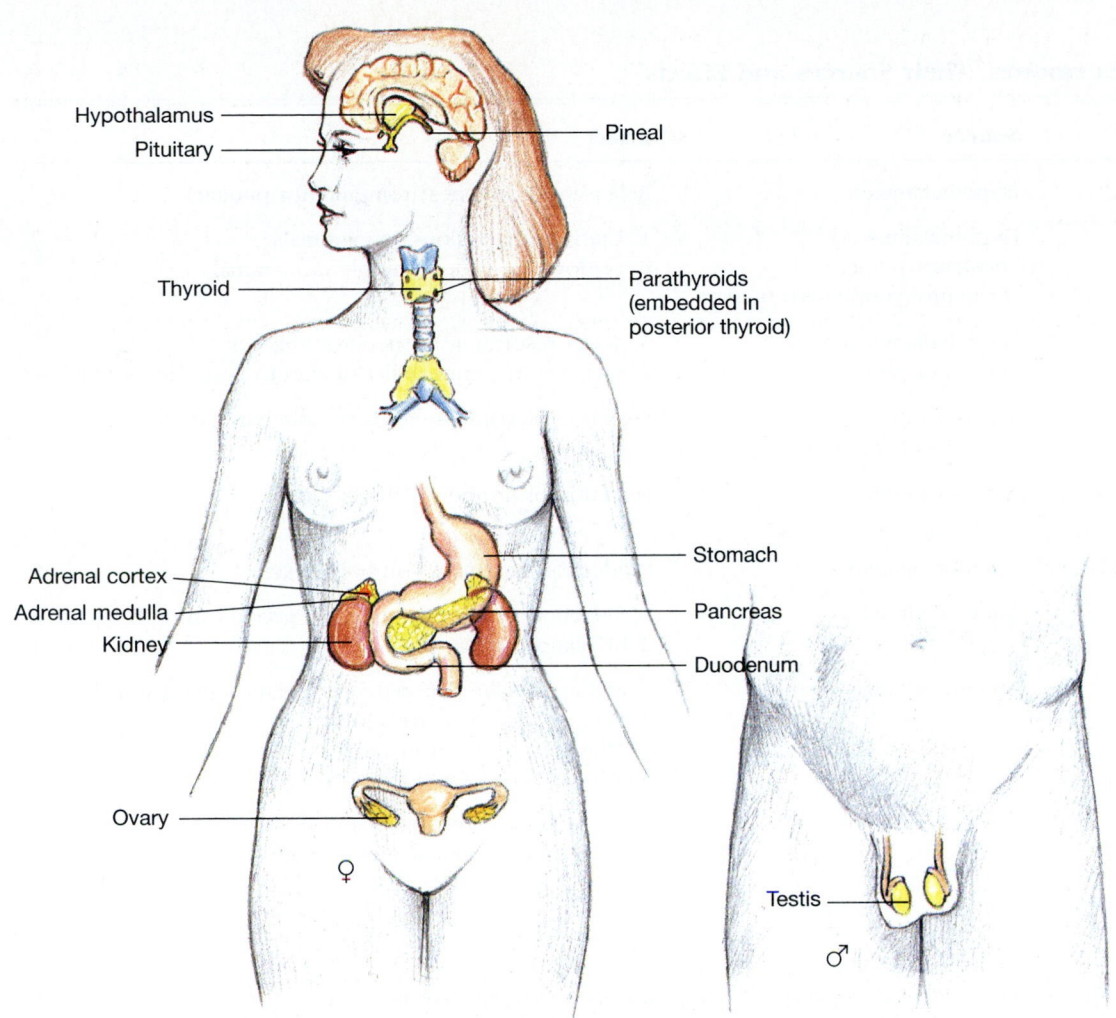

Hypothalamus

Pituitary

Pineal

Thyroid

Parathyroids
(embedded in
posterior thyroid)

Stomach

Adrenal cortex

Adrenal medulla

Pancreas

Kidney

Duodenum

Ovary

♀

Testis

♂

**FIGURE 40–4**   The locations of the major endocrine glands in humans.

rate at which it secretes the hormone. In most cases, feedback regulation involves **negative feedback,** although some cases of **positive feedback** are known.

The thermostat that turns a water heater on and off controls a negative feedback system. The thermostat responds to a drop in water temperature by turning the heater on. When the thermostat detects that the water temperature has risen to the set level, it turns the heater off. As a biological example, a drop in the level of calcium in the blood causes secretion of parathyroid hormone by the parathyroid glands (Figure 40–5, page 870). Parathyroid hormone stimulates release of calcium from the bones, decreases excretion of calcium by the kidneys, and increases absorption of calcium from food in the intestine. These actions raise the blood calcium concentration back to normal within a few hours. The rise in calcium, in turn, decreases the secretion of parathyroid hormone. This negative feedback loop is one of the control systems ensuring that the level of calcium in the blood remains within certain limits.

A hormonal feedback loop may involve two or more hormones, instead of a hormone and some other substance (such as calcium). This is how hormones secreted by the pituitary gland control the release of other hormones. For instance, a male vertebrate must have the right levels of both testosterone and luteinizing hormone (LH) in the blood for the testes to produce sperm and function properly. LH stimulates testosterone secretion, but testosterone inhibits the secretion of LH. So, if the level of testosterone in the bloodstream rises, it inhibits the secretion of LH, and less testosterone is produced until the testosterone level falls low enough that the inhibition of LH is turned off. LH secretion then rises again, stimulating the secretion of more testosterone. Thus, a rise or fall in the level of either hormone is automatically corrected by way of the other (Figure 40–5b).

Feedback control loops act at different speeds, depending on the number of hormones involved and the time that each takes to act. The system that controls the

TABLE 40-1
**Some Vertebrate Pituitary Hormones, Their Sources and Effects***

| Hormone | Source | Effect |
|---|---|---|
| Releasing factors (various) | Hypothalamus | Release of hormones from anterior pituitary |
| Oxytocin | Hypothalamus via posterior pituitary (=neurohypophysis); uterus | 1. Uterine contractions in mammals<br>2. Letdown of milk to nipple in mammals |
| Vasopressin (= ADH, antidiuretic hormone) | Hypothalamus via posterior pituitary | 1. Water resorption by nephron tubules<br>2. Increase in permeability of skin to water in amphibians |
| Adrenocorticotropic hormone (ACTH) | Anterior pituitary (= adenohypophysis) | Secretion of corticosteroids by adrenal cortex |
| Thyrotropin (TSH, thyroid-stimulating hormone) | Anterior pituitary | Secretion of hormones by thyroid |
| Follicle-stimulating hormone (FSH) | Anterior pituitary | Production of gametes in both sexes |
| Luteinizing hormone (LH) | Anterior pituitary | 1. Secretion of sex hormones by gonads in both sexes<br>2. Ovulation in females |
| Prolactin (= LTH, luteotropic hormone) | Anterior pituitary | 1. Mammary gland growth and lactation in mammals<br>2. Maintenance of corpus luteum in mammals<br>3. Migration to water in amphibians<br>4. Reproductive functions in birds |
| Somatotropin (growth hormone) | Anterior pituitary | 1. Body growth in reptiles and mammals<br>2. Increased blood sugar in mammals |

*All pituitary hormones are proteinaceous.

secretion of digestive enzymes in the stomach and intestine acts very rapidly in response to food in the stomach, whereas the feedback loop that controls fertility in women acts much more slowly; the menstrual cycle, one complete pass through a feedback control loop, takes about 28 days to complete.

## Chemistry of Hormones

Hormones generally fall into one of two chemical classes: proteins (including amino acids, polypeptides, and glycoproteins) and lipids of some sort.

Most protein hormones and messengers are water-soluble. These cannot cross the target cell's plasma membrane but bind to receptors on the cell surface. Others, such as thyroid hormones, are lipid-soluble. Steroids and other lipid messengers are also lipid-soluble. These lipid-soluble messengers are transported in the bloodstream by binding to protein carrier molecules. Once released from its carrier molecule, a lipid-soluble messenger passes through the plasma membrane and binds to a receptor inside the target cell.

Another difference between water- and lipid-soluble messengers is how long they generally stay in the body flu-ids. Water-soluble messengers are broken down or removed from the extracellular fluid rapidly. Some survive only for fractions of a second, others for up to a few minutes. Some lipid-soluble messengers, such as prostaglandins, are also short-lived, but many survive for much longer. Steroid hormones persist in the blood for hours, thyroid hormones for days.

The chemical nature of the hormone(s) secreted by a particular cell depends on the cell's embryonic and evolutionary origin. For instance, endocrine glands that develop from the embryonic endoderm produce protein or polypeptide hormones. Among endodermal glands are the thyroid, which evolved from part of the gut in early chordates, the associated parathyroids, and glands in the stomach wall and in the pancreas. The other glands that secrete polypeptide hormones are derived from nervous tissue in the embryo. These include the pituitary and hypothalamus in the vertebrate brain.

Related to the polypeptide hormones are the **catecholamines**—epinephrine and norepinephrine—from the medulla of the adrenal glands, at the anterior ends of the kidneys. Amines, as their name implies, are amino acid derivatives. One of the catecholamines, norepinephrine, is identical with the neurotransmitter produced by sympathetic neurons that synapse with effectors.

**T A B L E   4 0 – 2**

## Some Vertebrate Hormones, Their Sources, Chemical Nature, and Main Effects[*]

| Hormone | Source | Chemical Nature | Effect |
| --- | --- | --- | --- |
| Thyroxin | Thyroid | Iodinated amino acid derivative | 1. Stimulation of growth and metabolism<br>2. Metamorphosis in amphibians |
| Calcitonin | Thyroid | Polypeptide | Decrease in blood calcium by suppressing its resorption from bones |
| Parathyroid hormone | Parathyroids | Polypeptide | Increase in blood calcium |
| Insulin | Pancreas | Polypeptide | Decrease in blood sugar |
| Glucagon | Pancreas | Polypeptide | Increase in blood sugar |
| Gastrin | Stomach | Polypeptide | Secretion of HCl by stomach |
| Epinephrine (adrenalin) | Adrenal medulla | Catecholamine | 1. Dilation of blood vessels<br>2. Increase in blood pressure<br>3. Increase in blood sugar |
| Norepinephrine (noradrenalin) | Adrenal medulla | Catecholamine | 1. Same as epinephrine<br>2. Also serves as a neurotransmitter |
| Corticosterone, cortisol, etc. | Adrenal cortex | Steroids | Metabolism of carbohydrate, protein, fat |
| Aldosterone | Adrenal cortex | Steroid | 1. $Na^+/K^+$ retention by kidney<br>2. Sex drive |
| Chorionic gonadotropin | Placenta | Glycoprotein | Maintenance of body functions necessary for pregnancy in mammals |
| Progesterone | Corpus luteum of ovary | Steroid | 1. Maintenance of uterine endometrium in mammals<br>2. Enlargement of breasts during pregnancy in mammals |
| Estrogens (estradiol, estrone, etc.) | Ovary | Steroids | Sexual maturity in female mammals (affects uterus, vagina, mammary glands, skeleton, metabolism, etc.) |
| Testosterone | Testis | Steroid | Sexual maturity in male mammals; necessary for sperm production (affects voice, musculature, skeleton, metabolism, etc.) |
| Melatonin | Pineal | Amine | Involved in light-regulated control of reproduction |

[*]Unless a particular class of vertebrates is specified, the action applies to members of all classes.

A number of hormones are lipids, synthesized from acetyl CoA or cholesterol. Invertebrate hormones in this class include juvenile hormone of insects. Vertebrate lipid hormones are the prostaglandins and steroid hormones. Steroid hormones include the cortical steroids—cortisol and its relatives—which are synthesized in the cortex of the adrenal glands, as well as the sex hormones (see Figure 3–8). Steroid hormones are made in endocrine glands derived from the embryonic mesoderm.

Although we speak of "testosterone" or "estrogen" as if each were a specific molecule, there are actually dozens of slightly different versions of these and many other hormones. Different estrogens occur in different species and also within individuals of one species. Each molecule has somewhat different effects, although we lump them all together and call them "testosterone" or "estrogen" because they are similar chemically.

## 40–C  PATHWAYS WITHIN CELLS

Hormones are released from endocrine cells into the extracellular fluid and diffuse into the bloodstream, which carries them throughout the body. However, hormones are specific, influencing only their target cells, those cells with receptors to bind the hormone.

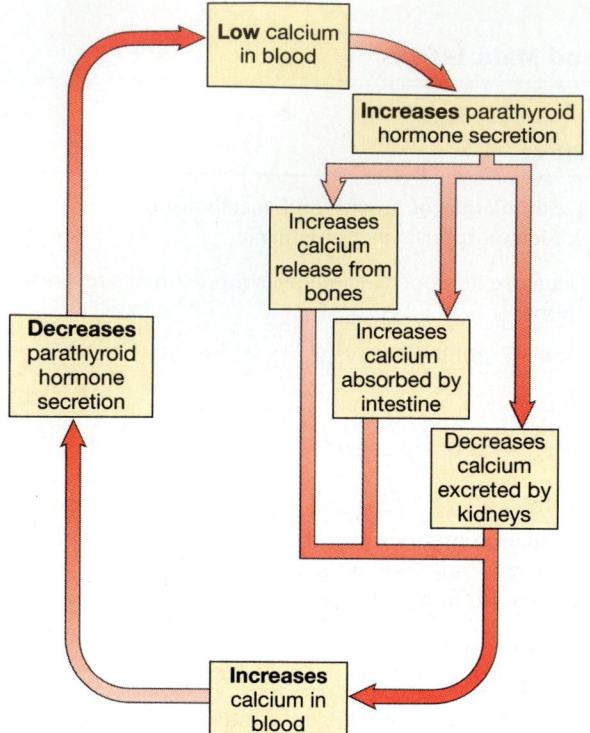

(a) CONTROL OF CALCIUM IN BLOOD PLASMA

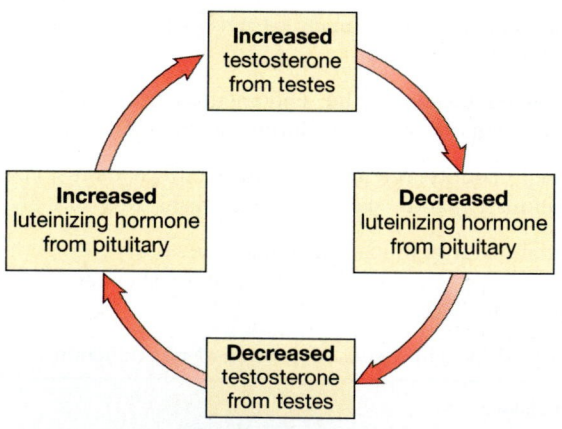

(b) CONTROL OF MALE HORMONE

**FIGURE 40–5**    Two examples of negative feedback control: (a) between parathyroid hormone and calcium levels in the blood; (b) between two hormones. In both cases, each substance serves to regulate the level of the other in the body fluids.

For instance, although insulin travels throughout the body, only certain types of cells, such as muscle and liver, have insulin receptors and respond to insulin by taking up glucose. Kidney and brain cells, in contrast, get all the glucose they need from the extracellular fluid and do not have the receptors to respond to the presence of insulin. The re-

action of the target tissue to a hormone differs at different times. For example, injection of the hormone prolactin into a sexually immature female will not cause lactation because her mammary tissue has not yet produced receptors for the hormone.

A hormone delivers its message to the target cell by binding to a receptor, thereby changing the receptor's shape. Other molecules in the cell are already set up in such a way that the receptor's new shape initiates changes in the cell, such as changes in permeability, enzyme activity, or gene transcription. All of these changes involve large numbers of ions or molecules, compared with the number of hormone molecules the cell received. In effect, the altered receptor **amplifies** the signal of a tiny amount of hormone into a much larger response by the cell. For instance, the binding of one molecule of the hormone glucagon to its receptor leads to formation of more than a million molecules of glucose-1-phosphate from glycogen in a liver or muscle cell.

Research in the 1990s is beginning to unravel the pathways by which binding of a chemical messenger to its receptor produces an appropriate reaction by the cell. These pathways are of two main types, depending on whether or not the chemical messenger can cross the plasma membrane and enter the cell. If it can, it binds to receptors in the cell. If it cannot, it binds to receptors on the cell surface.

### Intracellular Receptors

Steroid and thyroid hormones cross the plasma membrane and enter a cell, where they bind with **nuclear hormone receptors.** These are protein transcription factors that can bind to DNA and initiate transcription when they are also bound to a specific hormone (Figure 40–6). All known nuclear hormone receptors contain zinc fingers and pair up to form dimers that bind to DNA (see Figure 11–10).

A cell may respond to a steroid or thyroid hormone in several stages, amplifying the hormone's effect. In the case of the insect molting hormone, ecdysone, the ecdysone-receptor complex activates transcription of a few genes, some of which code for transcription factors. These cause a secondary response by initiating RNA synthesis at as many as 100 more sites on the DNA.

If each steroid hormone binds a specific receptor, how can it have different effects on different cells? For instance, estrogen causes division of cells in the breast, whereas in bone it does not cause cell division but induces cells to secrete calcium into the extracellular spaces. This involves turning on different genes in the two tissues. The answer is that more than one transcription factor is needed to turn on most eukaryotic genes. Although the nuclear hormone receptor for estrogen is the same in bone and breast cells, other transcription factors are different.

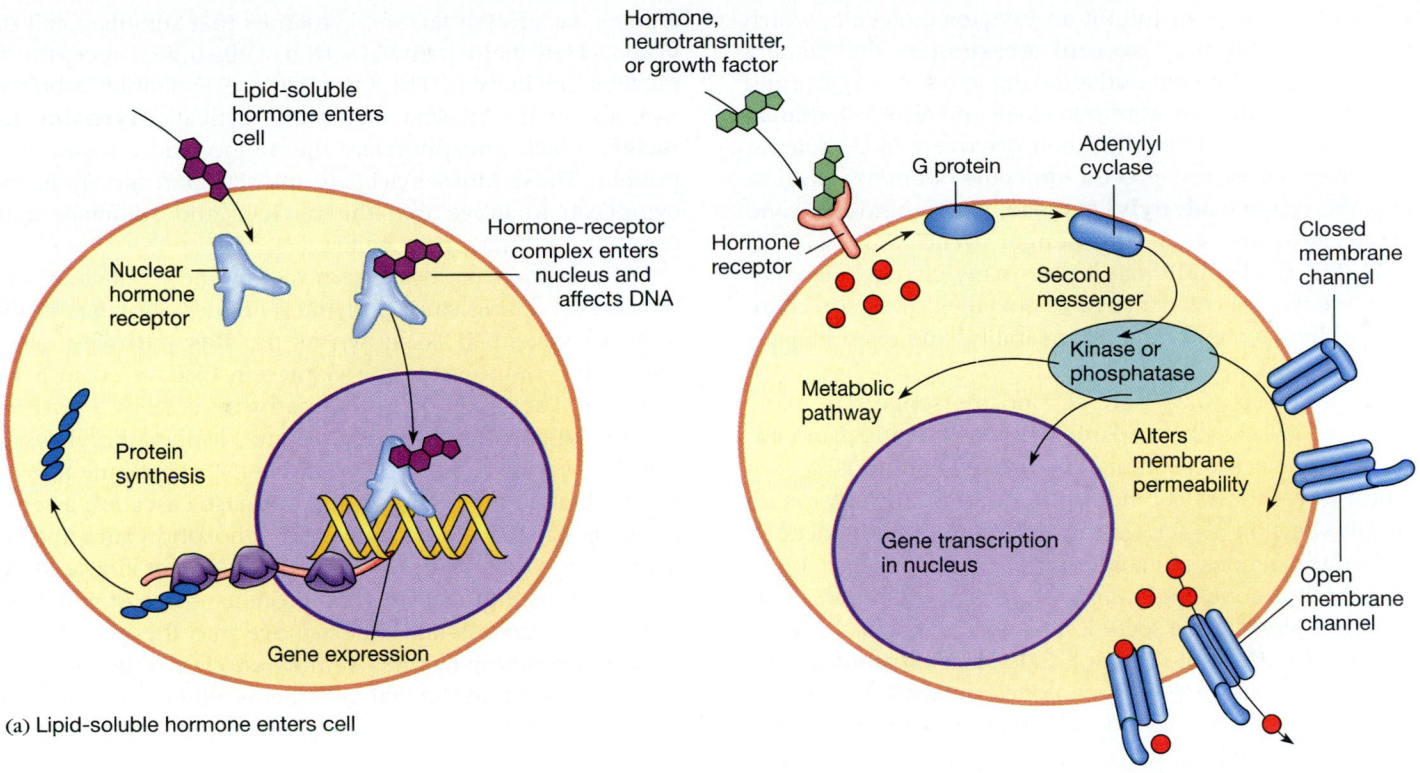

**FIGURE 40–6**  Two main types of hormone action. (a) Lipid-soluble hormones enter the cell and bind nuclear hormone receptors, which carry them to the nucleus, where they act as transcription factors. (b) Water-soluble hormones (and other messengers) bind cell-surface receptors, triggering reactions by second messengers that alter cell metabolism or membrane permeability.

***Hormone Pollution***   Recent research shows that environmental chemicals may mimic the effects of steroid hormones by binding to intracellular receptors. As a result, these chemicals may cause changes in development and physiology even at very low concentrations. The culprits include TCDD, a member of the dioxin group of chemicals; DDT, kepone, dieldrin, and other banned or restricted pesticides; endosulfan, the pesticide most widely used in the United States in the early 1990s; and some PCBs, plastics, and breakdown products of detergents.

French and British researchers have found that much fresh water is polluted with estrogen. Estrogen reaches waterways from the urine of women taking contraceptives and other hormone pills and of cattle fed hormones to increase milk and meat production. Some researchers blame environmental estrogen and related chemicals for decreases in sperm count and increases in rates of testicular cancer in men since the 1940s. This period also saw increases in male babies born with penis defects and undescended testicles (which are unable to produce sperm in later life). Many species of wildlife are similarly affected, including salmon,

sturgeon, trout, carp, alligators, otters, mink, and the Florida panther. Reproduction in some local populations of these species has declined precipitously.

## Cell-Surface Receptors

Many hormones, neurotransmitters, and growth factors cannot cross the target cell's membrane but bind to receptors on the cell surface. This changes the receptor's shape, setting off chemical pathways within the cell that amplify this initial signal.

More than 100 cell-surface receptors communicate with the cell interior by way of a single class of proteins, called **G proteins** because they bind guanine nucleotides. G proteins act as signal transducers for receptors that bind many hormones and neurotransmitters, as well as for photoreceptors in the retina and chemoreceptors in the nose. The G proteins lie at the plasma membrane's inner surface, where they briefly bind to the inner end of an activated receptor and themselves become activated. Then they move away and stimulate or inhibit an effector molecule, which in turn sends out many **second messengers** that amplify the signal, spreading out and affecting various enzymes and ion channels in the cell. Martin Rodbell and Alfred G. Gilman shared a 1994 Nobel Prize for their discovery of G proteins.

In many cases, the effector molecule bound by a G protein is the enzyme **adenylyl cyclase,** which binds ATP and converts it to the second messenger cyclic AMP (see Figure 37–11). Cyclic AMP may cause many changes depending on the type of cell: it activates enzymes, alters ion channels and hence membrane permeability, and even initiates protein synthesis.

Calcium ions ($Ca^{2+}$) are second messengers in many pathways. This has been particularly well studied in cells activated by electrical stimuli. The triggering of muscle contraction in response to binding of a neurotransmitter is an example (Section 39–C). Calcium is also an intermediate in the chemical actions of many hormones. For instance, binding of the hormone angiotensin to cells in the adrenal cortex triggers aldosterone secretion by way of events that raise the level of $Ca^{2+}$ in the cytosol. Calcium then alters the activities of enzymes involved in aldosterone production.

Binding of a hormone to a cell-surface receptor often causes changes in levels of both cyclic AMP and $Ca^{2+}$, which usually work together in the response and can regulate each other's concentration in the cell. In many cases, they combine forces to control **kinases,** enzymes that phosphorylate proteins by transferring a phosphate group from ATP. Many proteins are active as enzymes or transcription factors only when they are phosphorylated. Hence, many cell functions are altered by changing the balance between kinases and **phosphatases,** enzymes that carry out the opposite reaction, removing the phosphate group. A chemical messenger that stimulates kinase or phosphatase activity can have different effects depending on which proteins occur in its target cells.

In the 1960s, Earl Sutherland discovered the first kinase-phosphatase pathway while studying how the hormone glucagon causes conversion of glycogen to glucose in mammalian liver cells. When glucagon is bound to its cell-surface receptor, the receptor activates a G protein, which in turn activates adenylyl cyclase to convert ATP into cyclic AMP. Sutherland coined the term second messenger to describe the role of cyclic AMP in this pathway. Cyclic AMP regulates kinases that control the metabolic pathway from glycogen to glucose. A single bound receptor can activate many molecules of G protein, which in turn stimulates production of many molecules of cyclic AMP, thereby amplifying the response to a single molecule of glucagon.

***Two Pathways to the Nucleus*** Some water-soluble chemical messengers affect gene transcription in the nucleus even though they never enter the cell. Obvious examples are growth factors, cytokines that stimulate cell division. Two main pathways from cell-surface receptor to nucleus are known. The first of these is simple: a bound receptor in the plasma membrane activates **tyrosine kinases,** which phosphorylate the amino acid tyrosine in a protein. These kinases activate transcription factors in the cytoplasm to move into the nucleus and stimulate transcription of appropriate genes.

The second pathway from membrane to nucleus was discovered during studies of interleukins, cytokines in the immune system. It is known as the **Ras pathway,** since one of its components is a G protein discovered in a **ra**t **s**arcoma. The receptor in this pathway is itself a tyrosine kinase. When bound by its cytokine, it phosphorylates itself! This causes the Ras G protein in the membrane to bind the nucleotide GTP, stimulating a "kinase cascade": a series of enzyme reactions, each of which phosphorylates and activates the next enzyme of the series. The last kinase in the cascade phosphorylates several proteins, some of which act in the cytoplasm, while others move into the nucleus and act as transcription factors. We now know that the Ras pathway is involved in the action of many other cytokines that stimulate cell division, including several growth factors.

## 40–D    HORMONAL AND NERVOUS CONTROL

It usually takes longer for a hormone to act on its target cells than for the nervous system to activate an effector. This is because it takes longer for a hormone to reach its target tissue and because the hormone causes slower responses in the target tissue.

**FIGURE 40–7**   A mare suckling her foal, behavior that requires the hormone-triggered letdown reflex to release milk from the mammary glands.   (Biophoto Associates)

By having both nervous and endocrine control systems, the body is equipped to cope with a variety of occasions. The nervous system enables an animal to escape from an enemy in a fraction of a second, and some hormones also act relatively swiftly. For example, suckling by young mammals causes milk to let down (Figure 40–7). Suckling of the nipple sends sensory signals to the hypothalamus in the mother's brain. Cells in the hypothalamus extend into the pituitary, where they release the hormone oxytocin into the blood. Oxytocin reaches the mammary glands by way of the bloodstream and stimulates smooth muscles there to contract, pushing milk toward the nipple. This whole sequence of events takes less than a minute. Other hormones have much longer-lasting effects, like those that keep the body pregnant for many months.

Hormones are more suitable than nerves for controlling long-term changes involving many different organs, whereas nerves are more suitable for rapid reactions involving relatively few organs. Often the two systems acting together can control a situation more efficiently than either of them acting alone, as in the next example.

## Fight or Flight

A hot, red face, perspiring hands, and a rapidly beating heart commonly precede a stage appearance or an important exam. These symptoms are part of the "fight or flight" reaction, which prepares the body to meet stress or danger. We shall mention only a few aspects of this complex reaction here.

When a vertebrate senses danger or stress, the central nervous system stimulates the adrenal medulla to release the hormones epinephrine and norepinephrine into the bloodstream (Figure 40–8). In addition, much of the sympathetic nervous system is activated, releasing more norepinephrine at many target organs. Epinephrine and norepinephrine cause the heart to beat faster and to increase

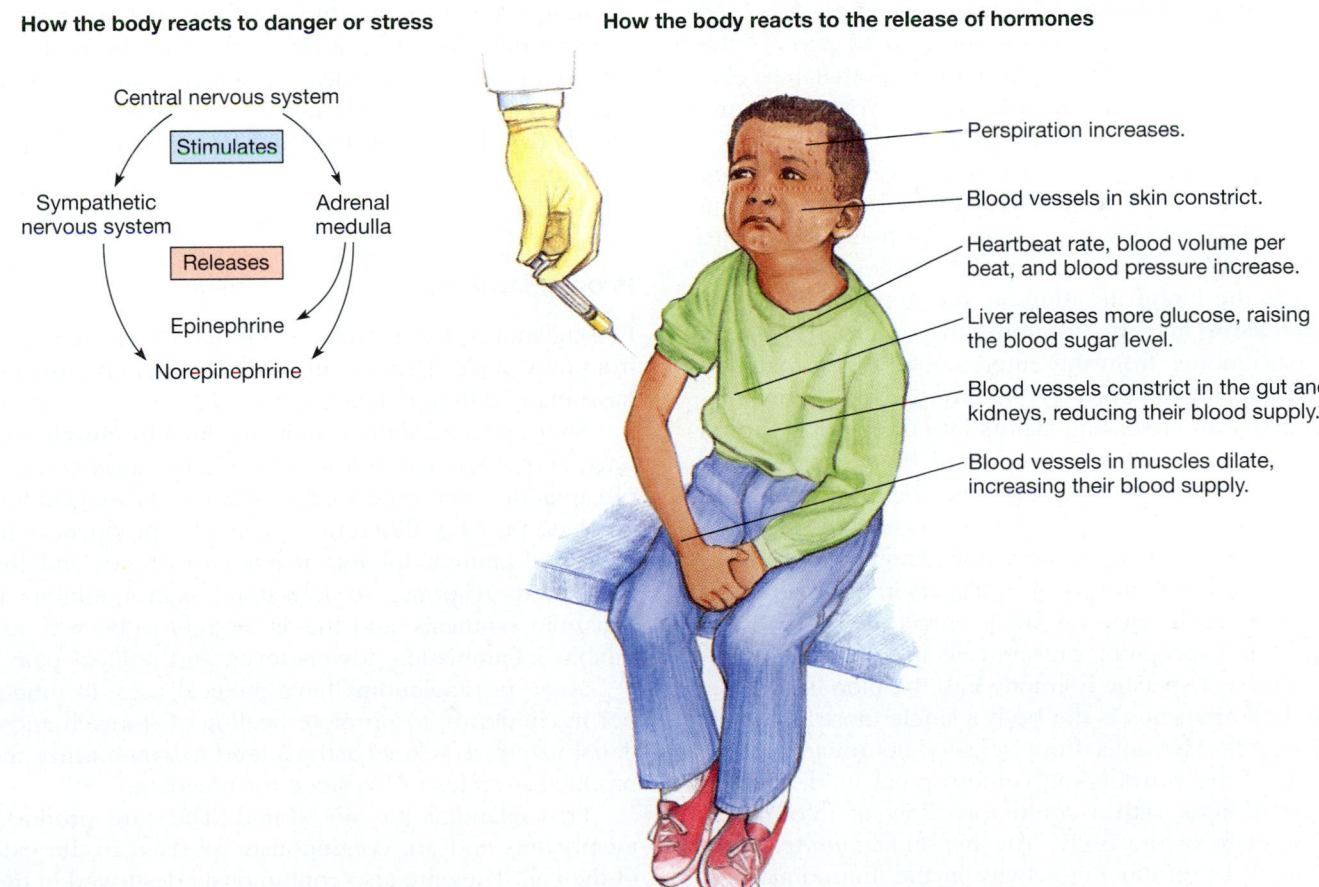

**How the body reacts to danger or stress**

Central nervous system

Stimulates

Sympathetic nervous system — Adrenal medulla

Releases

Epinephrine

Norepinephrine

**How the body reacts to the release of hormones**

- Perspiration increases.
- Blood vessels in skin constrict.
- Heartbeat rate, blood volume per beat, and blood pressure increase.
- Liver releases more glucose, raising the blood sugar level.
- Blood vessels constrict in the gut and kidneys, reducing their blood supply.
- Blood vessels in muscles dilate, increasing their blood supply.

**FIGURE 40–8** Fight or flight. The body's reaction to stress is mediated by both the nervous system and hormones.

the volume of blood pumped per stroke. This raises the blood pressure and circulates the blood more rapidly. In addition, the liver releases extra glucose into the bloodstream, raising the blood sugar level. Blood vessels in the muscles dilate, increasing the muscles' blood supply and preparing them for action by supplying them with extra oxygen and glucose. At the same time, constriction of the vessels supplying the kidneys and digestive tract reduces their supply of blood.

Nervous control evokes these reactions very rapidly in time of danger. Hormones provide a backup that can maintain the response for a long period. This explains why the state of "nervous energy" persists even after the performance is over or the exam paper completed.

## The Hypothalamus-Pituitary Connection

Most of the endocrine glands in the vertebrate body are controlled, directly or indirectly, by the brain. The nervous and endocrine systems interact by way of the connections between the hypothalamus in the brain and the pituitary gland. The **hypothalamus** is a small area of the forebrain (see Figure 37–16). It receives input from all parts of the brain. Stimulation of various cells in the hypothalamus elicits sensations and behaviors such as sex drive, pleasure, rage, fear, satiation, hunger, and thirst.

The hypothalamus synthesizes some hormones that are not released into the brain's blood supply but travel down the axons of secretory neurons from the hypothalamus to the posterior lobe of the pituitary gland, where they are released into the blood. In addition, the hypothalamus produces **releasing factors,** hormones that control the release of other hormones from the anterior lobe of the pituitary gland (Figure 40–9). Secretory neurons in the hypothalamus discharge the releasing factors into the pituitary portal system. A portal system is a series of blood vessels that carry blood between two capillary beds before it returns to the heart. In this case, the portal system carries the releasing factors from the capillaries in the hypothalamus through veins that branch into a second set of capillaries in the anterior pituitary. Here, each releasing factor enters the extracellular fluid and binds receptors, causing cells in the anterior pituitary to release a specific hormone into the bloodstream.

The hypothalamus is the body's single most important control center. Messages from sensory neurons, and the chemistry of the surrounding cerebrospinal fluid, provide the hypothalamus with a continuous flow of information about the state of the body. The hypothalamus reacts to these stimuli by producing activity in the autonomic nervous system, by initiating behaviors such as feeding, and by controlling the pituitary's secretion of hormones.

The posterior pituitary gland, or neurohypophysis, releases polypeptide hormones manufactured in the hypothalamus. The best-known of these are vasopressin, a water-conserving hormone in terrestrial vertebrates, and oxytocin, which induces muscular contractions by the uterus and by the ducts of the mammary glands in mammals. The posterior pituitary releases a number of other hormones, but little is known about them.

Many of the hormones secreted by the anterior pituitary (adenohypophysis) induce other endocrine glands in various parts of the body to secrete their particular hormones.

All of the anterior pituitary hormones are proteinaceous. In addition to hormones involved in growth and sexual reproduction, listed in Table 40–1, they include lipotropin, which mobilizes stored fats from fat cells in the adipose tissue. Melanocyte-stimulating hormone (MSH) causes pigment dispersal in the color cells in the skins of vertebrates that can change their color and affects sexual maturity and general health in birds and mammals.

## 40–E    CYTOKINES

Hormones travel in the bloodstream, and so they reach, even though they do not affect, most of the body's cells. By contrast, many chemical messengers are secreted into the extracellular fluid and absorbed by neighboring cells so rapidly that they never reach the bloodstream in significant amounts.

## Prostaglandins

Prostaglandins are a group of about 20 molecules made from fatty acids. They occur in most vertebrate tissues and have many different actions.

Some prostaglandins stimulate smooth muscle to contract, and others cause it to relax. Some cause constriction of capillaries, and others cause dilation. Prostaglandins are involved in many different aspects of reproduction and in menstrual cramps, allergic reactions to food, and the inflammatory response to infection. Aspirin inhibits prostaglandin synthesis, and this is thought to be why aspirin inhibits inflammation, lowers fever, and reduces pain.

Some prostaglandins have medical uses: to induce labor in childbirth, to promote healing of stomach and duodenal ulcers, to relieve asthma, and to synchronize the reproductive cycles of livestock for breeding.

Prostaglandins are not stored. They are produced in membranes and are continuously released to the exterior of the cell. They are also continuously destroyed in the extracellular fluid. When a cell is activated by changes in its environment, however, it may increase the rate of prostaglandin synthesis and release enough prostaglandin so that the messenger influences both the cell that produced it and its neighbors.

## Neuroregulators

Many neurotransmitters are deactivated as soon as they have crossed the synapse to the postsynaptic cell, and so they never escape from the synapse. If all signals in the nervous system reached only specific postsynaptic cells in this way, very few neurotransmitters would be needed because there would be no chance of confusion. But in fact more than 30 different neurotransmitters have already been identified in the vertebrate brain alone. Biologists are beginning to realize that many neurotransmitters diffuse out of synapses and so come into contact with many different cells, some of which have receptors for them. Neurotransmitters that behave in this way act as local hormones and may be distinguished by the special name of **neuroregulators.**

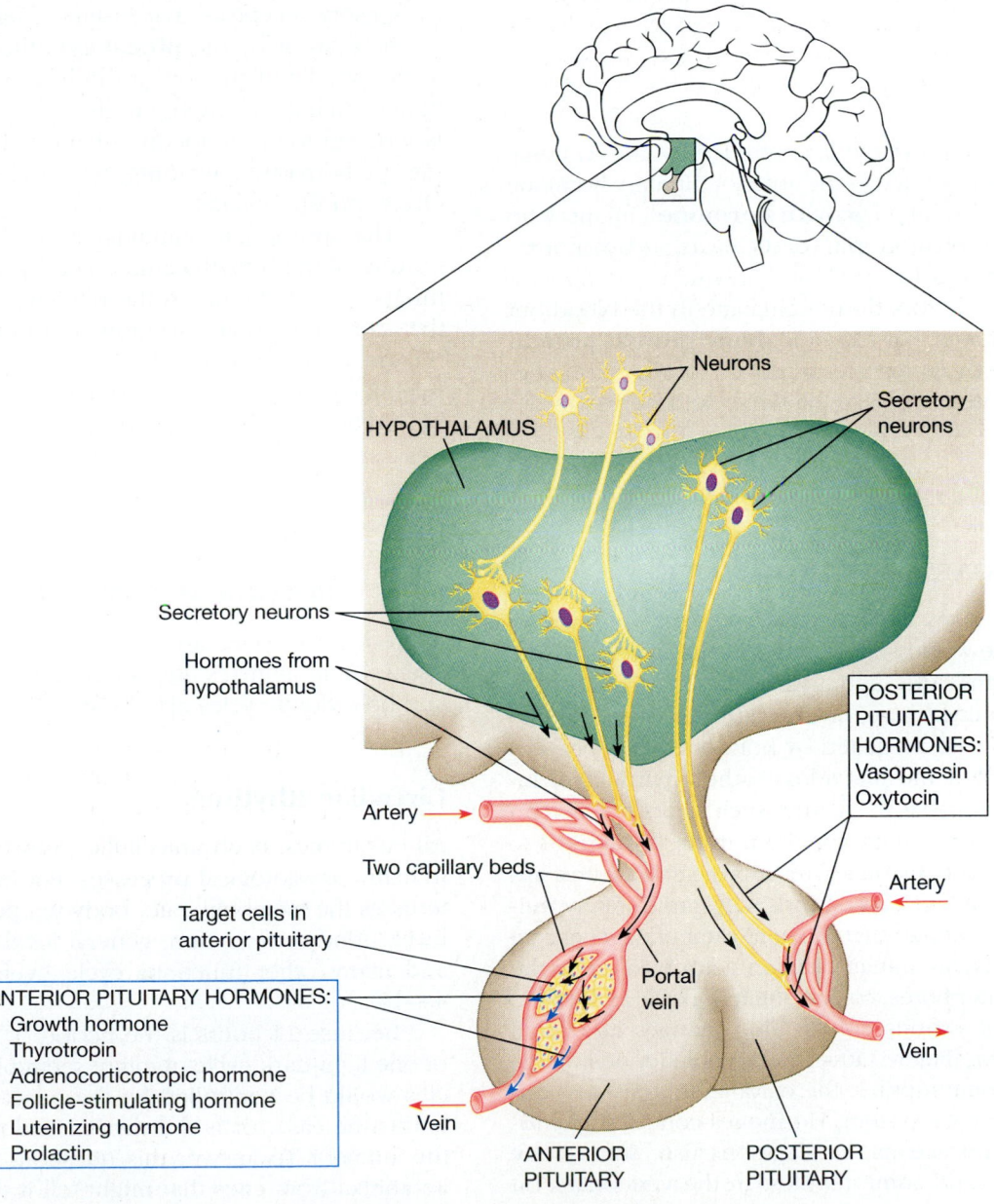

**FIGURE 40–9**    The interrelations of hypothalamus and pituitary in the vertebrate brain. The left part of the diagram shows secretory neurons in the hypothalamus that release hormones into the pituitary portal system. These hormones stimulate target cells in the anterior pituitary to release pituitary hormones into the bloodstream. At the right are other secretory neurons with cell bodies in the hypothalamus and axons extending to the posterior pituitary, where they release hormones into the bloodstream. Most of the hormones shown here are described in Table 40–1.

Sleep is at least partly controlled by a neuroregulator. If cerebrospinal fluid is extracted from a sleeping animal and injected into one that is wide awake, the second animal promptly goes to sleep. The fluid contains a sleep neuroregulator. Sleep is an ideal candidate for neuroregulation because it requires the suppression of many nerve cells. A neuroregulator can cause this suppression by affecting all the neurons that bear receptors for the sleep neuroregulator. The neurons need not be completely "turned off" by the neuroregulator but may remain responsive to other neurotransmitters.

## Growth Factors

In many tissues, cell division is controlled by conventional circulating hormones, such as the anterior pituitary hormone **somatotropin** (also called **growth hormone**). Infants who produce too little somatotropin (or its releasing factor or receptors) become dwarfs.

Other growth factors do not circulate in the blood but act as local cytokines. For example, **bone growth protein** and **nerve growth factor** are produced locally during development and are necessary for bones and nerves to develop normally.

## 40–F  HORMONES AND ANIMAL LIFE

Animals exhibit many adaptive responses to information about the outside world sent to the nervous system by way of the sense organs. Although most of these reactions, such as eating or running away, are brought about by muscle contraction, some are mediated by hormones. It is not surprising that most of the reactions to the environment involving hormones are slow changes such as occur when an animal comes into breeding condition in the spring.

Other hormonally mediated reactions, such as those involved in color change, are surprisingly rapid. Many animals can change color so that they are camouflaged against their backgrounds. An animal changes color by altering the size of its chromatophores, color-containing cells in the skin, which may be of various colors. This changes its overall color and pattern (Figure 40–10). In cephalopods, some bony fish, and some reptiles, the chromatophores are controlled by the nervous system. Hormones control the chromatophores in crustaceans, cartilaginous fish, some bony fish, amphibians, and some reptiles. In the vertebrates on the list, the pattern of light reaching the retina of the eye controls the release of melanocyte-stimulating hormone (MSH), which in turn changes the size of the chromatophores.

## Environmental Control of Reproduction

There is selective pressure for animals to reproduce when conditions favor survival of their offspring. For most animals this means birth or hatching in the spring, when warm weather and plentiful food offer the best possible conditions. Breeding often involves dramatic changes in an animal's anatomy, physiology, and behavior.

Animals come into breeding condition in response to environmental cues such as daylength, temperature, rainfall, food, and so on. These external stimuli are detected by sensory receptors. For instance, light may be detected by the eyes or by the **pineal eye,** the "third eye" that lies in the middle of the top of the head in some lower vertebrates. (In higher vertebrates the eye is gone, and the pineal is reduced to an endocrine gland within the head, secreting the hormone melatonin, which has poorly understood effects on the gonads.)

The appropriate stimulus causes hormone production by way of the hypothalamus. The hypothalamus does two things: first, it stimulates the pituitary to release hormones that cause the gonads to grow and produce sex hormones (Figure 40–11). Second, the sex hormones feed back to the hypothalamus, which then initiates reproductive behavior by sending out the appropriate signals in the nervous system.

## 40–G  BIOLOGICAL RHYTHMS

Reproductive cycles are examples of rhythmic or cyclical events in an animal's life. A number of other physiological and behavioral cycles also exist.

## Circadian Rhythms

All eukaryotes, even unicellular protists, show daily cycles in many physiological processes. For instance, in most vertebrates the metabolic rate, body temperature, blood sugar level, urine composition, general level of nervous activity, and many other functions cycle every 24 hours (Figure 40–12).

Because 24 hours is the period of Earth's rotation and of one light-dark cycle, it might seem obvious that daily cycles would be controlled by the onset of light or of dark. Is this the case, or is the rhythm endogenous ("built into" the animal)? To answer this question, investigators isolate an animal from cues that might tell it the time of day, and see whether the daily rhythm persists.

When the animal is kept in darkness with constant temperature and humidity, it still maintains these rhythms, but the cycles are no longer exactly 24 hours (Figure 40–13).

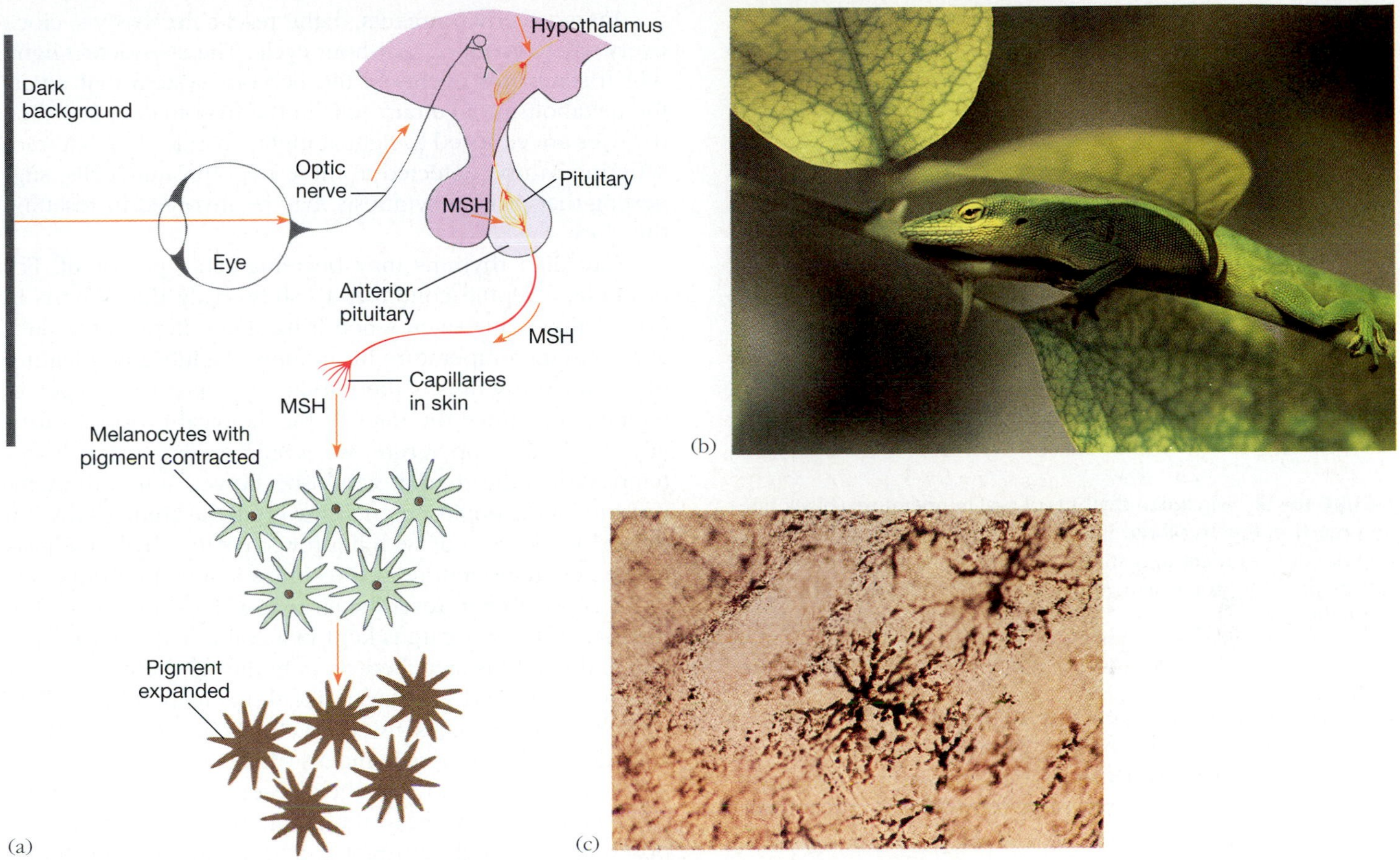

(a)

(b)

(c)

**FIGURE 40–10**   Color change. (a) The mechanism of color change in a frog placed on a dark background. The tone of the background is perceived via the eyes, which initiate a chain of nervous and hormonal events. Finally, MSH (melanocyte-stimulating hormone) from the anterior pituitary stimulates the pigments in melanocytes in the skin to spread out in the cell, darkening the skin. Arrows show the path of information flow. (b) This chameleon has turned bright green (from its usual brown color), matching the sunlit leaves around it. (c) Chromatophores (color cells) in a frog's skin. Melanocytes are one type of chromatophore.   (b, Steve Simonsen/Natural Selection; c, Biophoto Associates)

(a)

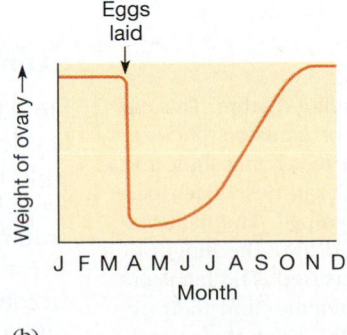

(b)

**FIGURE 40–11**   Hormonal responses to the environment: toad reproduction. Ovarian growth and egg laying are under hormonal control. (a) A toad (*Bufo bufo*) laying eggs in the spring. In amplexus, the male clasps her and sheds sperm on the eggs as they are laid. (b) The weight of the ovaries of *B. bufo* at different months of the year.   (a, Robert Brons/BPS)

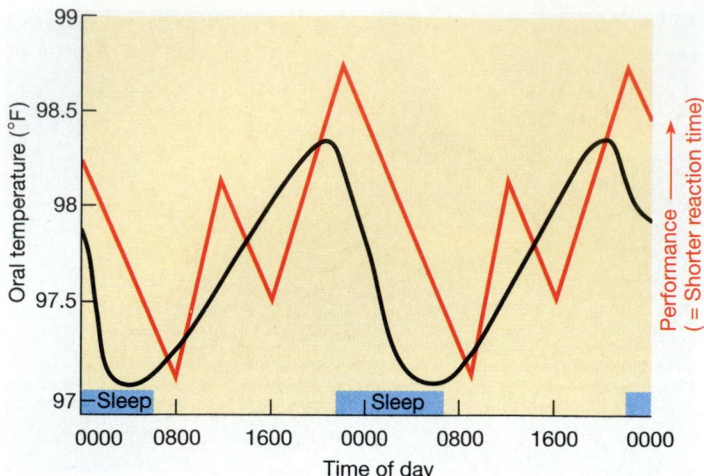

**FIGURE 40–12** Circadian rhythms of oral temperature (black line) and reaction time (colored line) in humans. (Reaction time is the time it takes to react to a stimulus when you are told to react as fast as possible, for instance, to press a button when you see a red light.)

Because the rhythms repeat approximately, but not exactly, every 24 hours in the absence of external cues, they are called **circadian rhythms** (circa = about; dies = day). These endogenous rhythms must be related to an internal "biological clock."

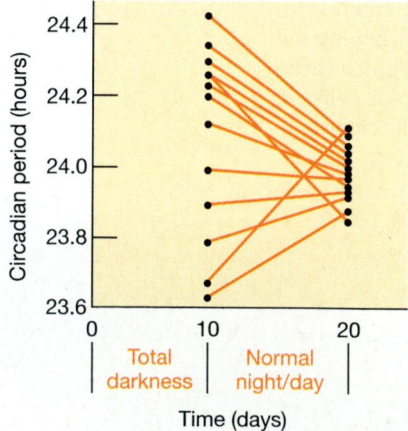

**FIGURE 40–13** Effect of light cycles on circadian rhythm. This experiment showed that the circadian period of hamsters is closer to 24 hours when the animals are exposed to normal light and dark cycles than when they are kept in total darkness. Each pair of dots joined by a line represents a single animal. The hamsters were kept in total darkness for ten days, and then the length of the daily cycle of sleep and activity was measured. The hamsters were then returned to a normal day/night regime, and their cycles were measured again ten days later. At this time, the animals were closer to a 24-hour cycle than when they were in total darkness. (After D. Pittendrigh and S. Daan, *Science* 186:548, 1974.)

In most environments, light resets the body's clock every day to produce a 24-hour cycle. The eyes detect light, which stimulates events in the nervous system that adjust the metabolism of certain cells in the hypothalamus. When the eyes are exposed to light at night, changes in DNA transcription can be detected in these hypothalamic cells, suggesting that protein synthesis may be involved in resetting the clock.

Circadian rhythms may be remarkably persistent. For example, a reptile continues to show daily fluctuations in body temperature even when it has been kept in the dark at a constant temperature for months. An intriguing feature of circadian rhythms is their ability to adjust to changes in temperature. Since the rates of biochemical processes usually vary with temperature, we would expect that at lower temperatures the rhythm would be slower. When the temperature of the environment is lowered, the animal's rhythm does slow down, but within a few days the rhythm adjusts to the new temperature and resumes its natural frequency.

It is possible to reset the circadian clock artificially. For instance, if a vertebrate is kept in a controlled environment where the light is switched on only during the night, night becomes day, and vice versa, for the animal, and its daily rhythm soon becomes reset exactly 12 hours later. Many zoos do this to nocturnal animals (those that are usually active only at night) so that they become active during the day when visitors want to see them. These animals are exposed to bright light at night and to a dim red light during the day. Their internal clocks are set so that the animals act as though it is night during what is really daytime.

What is the selective advantage of circadian rhythms? Night and day always follow each other on a 24-hour cycle, and so there is a perfectly good stimulus available to trigger daily cycles. Why should an organism also have an internal cycle of its own? The answer is probably that an animal's internal rhythm permits it to anticipate regular daily events before environmental cues appear. This ability is valuable. For instance, it permits a bat to start hunting at the time of day when its insect prey will also be active, without wasting the energy it would take to check on the light level or insect activity in the area every few hours.

## Annual Rhythms

In addition to circadian rhythms, some animals have yearly cycles. Examples have been found among both vertebrates and invertebrates. Many mammals continue to hibernate at roughly the right time even if they are deprived of environmental cues that could tell them the time of year. Deer kept under constant conditions continue to grow, and later to shed, their antlers at the same time every year (Figure 40–14). Several species of birds start migratory behavior at the same time of year even if they are deprived of environmental cues. In this last case, the adaptive advantage of

**FIGURE 40–14** Annual rhythms. Each year, a male moose produces antlers and later sheds them, growing another set the next year.

the annual rhythm is probably that it permits a bird to return north for the breeding season even though it receives few cues of seasonal change in its relatively constant tropical wintering grounds.

## Biological Clocks

Organisms plainly have biological clocks that keep track of time independently of environmental stimuli. Such a clock is very valuable. It provides an animal with a time sense that allows it to tell the season of the year by detecting the length of time between sunrise and sunset. It also permits animals to navigate using the sun as a compass. (The clock is necessary because the sun is at different points on the compass at different times of the day.)

The vertebrate body appears to contain many different clocks, each with its own endogenous rhythm. Biological clocks have been studied in some detail in the reproductive cycle of female white rats. The cycle starts with the release of LH from the pituitary, an event that must be due to the rat's biological clock since it does not depend on environmental factors. The clock is set so that LH is released only between certain hours of the day. In addition, LH release can occur only on days when the level of estrogen in the blood is sufficiently high. The interaction between the animal's clock and the estrogen level in the blood controls the reproductive cycle and ensures that females come into breeding condition every four or five days, depending on the strain of rat (in the five-day strain the level of estrogen in the blood builds up more slowly).

These studies with rats provided clues that the master clock of vertebrates is physically located in neurons in the hypothalamus. If part of the hypothalamus is destroyed, a female rat comes into permanent breeding condition, although the pituitary and the gonads produce their hormones as usual (but not in the normal time sequence). This part of the hypothalamus receives input from the eyes and passes this information along to the pineal gland, the location of another important clock.

***Melatonin***    Biological clocks are reset by the effect of light on the synthesis of melatonin by the pineal gland. The activity of the enzyme **acetyltransferase,** needed to synthesize melatonin, increases in the dark, and so the amount of melatonin in the blood rises at night and starts to fall with daybreak.

Acetyltransferase appears to be a very ancient enzyme, which probably explains why all organisms studied have biological clocks. It has been found not only in vertebrates but also in photosynthetic bacteria, where it increases the activity of photosynthetic pigments in dim light, thereby boosting the rate of photosynthesis.

## 40–H    PHEROMONES

Whereas a hormone carries information within the body, a **pheromone** is a chemical that travels outside the body, carrying information to other members of the same species. The first pheromones described were sex attractants from insects (Figure 40–15). In many species of moths, beetles, cockroaches, and flies, the female releases a chemical that attracts the male. He finds his mate by flying or crawling up the odor gradient toward her.

Many vertebrates, and particularly mammals, use pheromones in urine or feces, or from special scent glands, to mark trails and territories. When a dog urinates on a fire

**FIGURE 40–15** Pheromones: a July mating swarm of ladybugs. The beetles are attracted to the swarm by pheromones released by ladybugs in breeding condition.   (Mary Clay/Tom Stack & Associates).

## ESSAY: *Our Daily Spread*

An American will tell you that the normal adult human body temperature is 98.6 °F, or 37 °C. European fever thermometers give normal temperature as 98.4 °F. In fact, "normal" body temperature is anywhere from 36 to 37.6 °C (96.8 to 99.5 °F). If you take your temperature first thing in the morning, it is likely to be at the low end of this range, and if you keep taking it every hour or two, it will probably show the pattern in Figure 40–12, reaching a high point late in the evening.

Many other physiological factors also vary during the day, including blood pressure, heartbeat rate, excretion of various ions in the urine, secretion of different hormones, alertness and reaction time, and the ability to detect faint sounds. Each factor has its own predictable daily pattern, which may differ from the patterns of other factors (Figure 40–A). For example, one hormone may peak during the middle of the night and another first thing in the morning, whereas body temperature peaks shortly before bedtime.

Are these regular daily patterns in human body functions triggered by external cues, or are they endogenous? To distinguish between these two possibilities, experimenters have put people into isolation in caves, underground apartments, or windowless rooms for periods of several weeks. The subjects have no clocks or watches, no sunlight, no radio or television—nothing to tell them what time of day it is. They eat, sleep, read, or exercise when they want to. Under these conditions, people drift from their almost exactly 24-hour pattern of normal daily life into a new rhythm that depends on the person, frequently about 25 hours.

Ignorance of circadian rhythms can be hazardous to our health. For instance, blood pressure tends to be low in the morning and to rise during the day. People who always visit the doctor in the morning may have blood pressure readings that fall in the normal range, but their blood pressure may rise dangerously in the afternoons and evenings. These people will not receive the treatment needed to control their blood pressure.

Our circadian rhythms probably make themselves felt most acutely when we try to reset our internal clocks. People who have traveled by airplane across several time zones experience "jet lag," and people who work rotating shifts, such as airline flight crews, air traffic controllers, po-lice, and military and hospital personnel, have similar problems. The trouble is that some rhythms reset to the "new time" more quickly than others, and so the body goes through a period when its rhythms are out of synchronization with one another. This can have serious consequences. The accident at the Three Mile Island nuclear power station in 1979 would probably have been quite minor if the people on duty had been more alert. Their errors in handling the emergency may well have arisen from the shift rotation schedule followed at the plant. In one study, more than half of rotating-shift workers reported falling asleep on the job, including truck drivers and nuclear power plant operators. Some people have also developed ulcers because of the continuous stress of having desynchronized body rhythms.

The study of circadian rhythms has produced information that can help in the scheduling of shift rotations for the 20 to 30 million people in the United States who work at such jobs. For example, it is easier to reset the internal clock to a later time than to an earlier time. This explains why we adjust more easily to travel from east to west than in the opposite direction. It also suggests that rotating

---

hydrant, he is depositing a pheromone that tells other dogs that the hydrant is part of his territory. Pheromones also accelerate reproductive maturity in a number of species and permit members of one sex to distinguish which members of the opposite sex are in breeding condition.

The pheromones used to mark territories or attract a mate produce immediate effects on the nervous system, physiology, and behavior of the animal that receives the pheromone. There are other pheromones that act more slowly and have longer-lasting effects. For instance, if many female mice are caged together, the periods of estrus (sexual receptivity) of all of them are interrupted by periods in which the females are infertile because the normal estrous cycle does not occur.

Similarly, the odor of a strange male will terminate the pregnancy of a newly fertilized female mouse. In this case, a pheromone in the male's urine is received via the female's olfactory organs and triggers nervous activity in her hypothalamus. The hypothalamus causes the release of pituitary hormones that reduce the ovaries' output of steroids needed to permit implantation and pregnancy. The newly fertilized eggs cannot implant in the uterus, and the pregnancy aborts.

Pheromones may also synchronize reproduction within a group. In desert locusts, any sexually mature male pro-

shifts should be scheduled so that people work the day, evening, and night shifts in that order, rather than switching to nights and then evenings after the day shift. Furthermore, since it takes up to a week for a person to adjust to the new schedule, workers should be left on each shift for as long as possible. Workers at a mineral mine in Utah were switched from spending a week on each shift and then rotating to an earlier one, to a new schedule, on which they worked three weeks at each shift and then rotated to a later one. They reported increased job satisfaction and better health on the new schedule, and no one fell asleep during work hours any more. The employer gained, too, in less employee turnover and in increased output.

Related research shows that people tend to wake up when their body temperature begins to rise, regardless of how long they have slept or how long they were awake before going to sleep. This explains why people whose travel or work schedules force them to go to sleep at unusual times may wake up exhausted: their internal alarm, which is linked to deep body temperature, wakes them up when the temperature begins to rise even though they have not had enough rest to recover from previous waking periods.

Researchers are experimenting with ways to reset the human biological clock using meals, bright lights, or exercise at specific times of the day. The latest development is a melatonin pill designed to help reduce jet lag and fatigue from shift work. The pill speeds up the body's adjustment to clock-shifting.

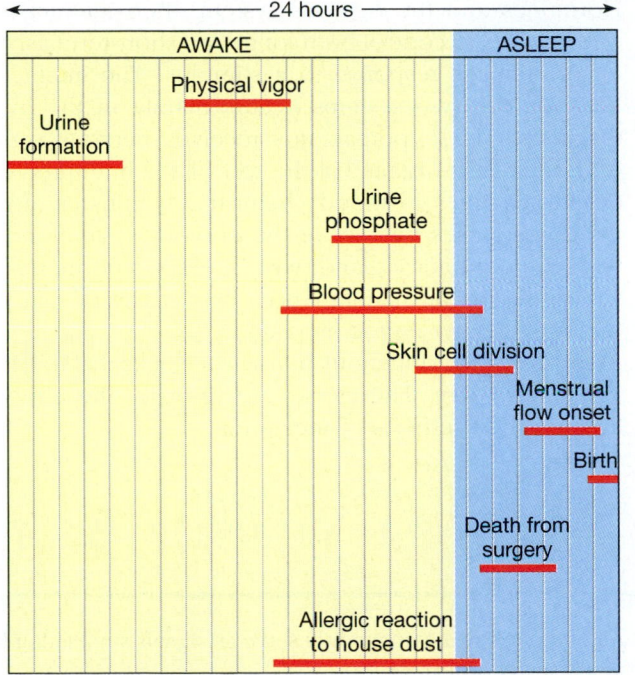

**FIGURE 40–A** Human circadian rhythms. Activities of the human body are not uniform throughout the day: different factors or events tend to be higher or more frequent at different times of the day in most people. The 24-hour time span shown begins, at the left, at the time when people get up in the morning (this varies from person to person so no clock times are given). The colored bars show the periods during the day when the high value of each function occurs. For example, urine formation is highest soon after rising, and cell division in the skin is highest just after going to sleep. Likewise, babies are not born uniformly throughout the day; there is an excess of births in the hours just before the mother would normally get up.

duces a pheromone that speeds up sexual development in immature members of both sexes. This system ensures that most of the locusts in the area reproduce at the same time: so many young locusts hatch at once that predators can kill only a small fraction of them. Hence, many more survive than would be the case if they hatched over a longer period and gave predators time to eat one brood after another.

The best evidence for human reproductive pheromones comes from anecdotal evidence that when numbers of women live together their menstrual cycles eventually synchronize. Other than this, there is no convincing evidence for human pheromones, despite the large number of reports on the subject that appear periodically.

In sexual systems involving pheromones, a single chemical determines whether an animal will reproduce. For this reason, a slight change in the pheromone might have profound evolutionary effects. For example, there are two pheromone "races" of the European corn borer moth. Males in Italy, the Netherlands, and the eastern United States respond to a pheromone mixture containing 96% *trans*-11-tetradecenyl acetate and 4% of the isomer *cis*-11-tetradecenyl acetate. Elsewhere in Europe and North America, males are attracted to the same compounds but in the reverse proportions. Both forms of the insect occur together in Pennsylvania, where they do not interbreed; they can therefore be regarded as separate species. There is a good chance that these species arose in the same area in Europe, before both were introduced accidentally into North America. Speciation would have required only two independent mutations, one altering a *trans* receptor protein to a *cis* receptor in the male and the second causing females to produce predominantly the *cis* instead of the *trans* isomer of the pheromone.

In the insect societies of bees, ants, and termites, pheromones organize not just the reproduction but also the behavior and social structure of a colony (Section 41–J).

## SUMMARY

Animals have three main types of chemical messengers: hormones, produced by endocrine glands or secretory neurons and carried throughout the body by the blood; cytokines such as prostaglandins, growth factors, and neuroregulators, which usually act near the cells that produce them; and pheromones, which carry information between individuals of the same species.

1. Hormones are involved both in homeostatic mechanisms within the body and in many of an animal's responses to its environment, such as reproductive cycles and fight or flight responses.
2. Each chemical messenger affects only target cells that carry receptors for the messenger. In addition, the same messenger may have different effects on cells containing different proteins and metabolic pathways. The effects triggered by receptors amplify the effect of the hormone.
3. The nervous and hormonal systems carry messages that travel between an animal's cells and coordinate their activities. Hormones may act on a wide range of cells over a long time; neurons affect specific cells briefly.
4. Lipid-soluble steroid and thyroid hormones enter cells, where they bind with nuclear hormone receptors, forming transcription factors that enter the nucleus and bind to DNA.
5. Water-soluble hormones bind to cell-surface receptors, activating signal pathways inside the cell, many of which involve calcium ions, G proteins, cyclic AMP, kinases, and phosphatases. These pathways affect protein activity in the cytoplasm and may produce transcription factors that control RNA synthesis and cell division.
6. Hormones and the nervous system often interact with one another to control both long- and short-term aspects of an animal's response to a stimulus. The interaction between the two systems occurs mainly in the hypothalamus. The hypothalamus receives nervous signals from the sense organs via the rest of the brain and also detects changes in blood chemistry. It initiates appropriate responses by way of the nervous system and by way of the pituitary gland, which releases hormones responsible for the maintenance and activity of many of the body's other endocrine glands.
7. Animals have endogenous biological clocks that tell them the time of day. The clocks are reset by the effect of light on the synthesis of melatonin.

## Self-Quiz

1. Which of the following techniques would be *least* likely to help elucidate the function of a hormone produced by an endocrine gland?
   a. removing the gland and analyzing what functions are lost
   b. transplanting the gland into an animal without one
   c. transfusing blood from an animal lacking the gland into an animal that has the gland and observing its effects
   d. observing effects of gland extract on various tissues grown in culture
   e. observing the condition of animals with tumors that causes the gland to be overactive.
2. Under normal conditions, the stimulus that initiates metamorphosis is:
   a. concentration of certain hormones
   b. state of nutrition
   c. age of the larva
   d. decrease in the population of adults of the species in the vicinity
   e. change of season

3. All of the following commonly serve as signals stimulating hormone secretion *except:*
   a. conditions outside the body
   b. rising levels of another hormone
   c. rising levels of the hormone in question
   d. falling levels of the hormone in question
   e. falling levels of another hormone
4. Hormones are known to cause all the following changes in target cells *except:*
   a. changes in genotype
   b. changes in permeability
   c. changes in metabolic rate
   d. increase in cyclic AMP concentration
   e. synthesis of different messenger RNAs and proteins
5. An advantage to having the endocrine system as well as the nervous system involved in the "fight or flight" response is:
   a. the endocrine system responds faster
   b. the endocrine response usually lasts longer
   c. the endocrine system is tuned more precisely to the degree of need

d. the endocrine system affects only the target organs whose response is needed to meet the emergency

e. response by the endocrine system frees the nervous system to think of a way out of the situation instead of simply maintaining the body in an alert state

6. Information from internal or external sensory receptors may initiate a response from the endocrine system by passing through the part of the brain known as the _____. This area connects with the anterior pituitary via _____ and with the posterior pituitary via _____. The pituitary stimulates various glands to secrete hormones by _____.

7. Thyroid stimulating hormone (TSH) is released from the anterior pituitary in response to low levels of the hormone thyroxin in the blood. Higher levels of thyroxin cause a lowering of the release of TSH. This regulation of levels of TSH and thyroxin is called _____. Receptor proteins for TSH exist only on cells in the thyroid gland. The thyroid is thus designated the _____ of TSH.

8. TSH is proteinaceous. Thus we would expect it to affect the cells of the thyroid by:

a. inducing transcription of DNA that codes for thyroxin-producing enzymes

b. accelerating release of already-formed thyroxin

9. Using the secretion of thyroxin as an example, diagram the feedback control system for hormone production.

## Questions for Discussion

1. Normal thyroxin levels are about 100 units. A patient has only 80 units in the bloodstream. Normal therapy for this situation is to inject the extra 20 units. Why might this therapy not be effective?

2. It's spring! Time to go on a diet so you won't bulge too much at the beach this summer. If you go on a low-sugar diet, your body will metabolize proteins and polysaccharides to sugar. Explain how the endocrine glands and hormones accomplish this, including what will happen to the glands and hormones when sugar levels return to normal.

3. Draw a diagram to show the glands and hormones that control growth. If the growth hormone has no known feedback mechanism, what sort of control mechanism might determine when growth stops?

## Suggested Readings

Berridge, M. J. "The molecular basis of communication within the cell." *Scientific American,* October 1985. Second-messenger pathways and some of the changes they initiate in cells.

Coleman, R. M. *Wide Awake at 3:00 A.M.: By Choice or by Chance?* New York: W. H. Freeman, 1986. An entertaining account of research on human biological clocks and recommendations for coping with time shifts.

Gwinner, E. "Internal rhythms in bird migration." *Scientific American,* April 1986. A biological clock with a period of about a year helps determine when birds begin and end their migratory flights and also helps them navigate.

Linder, M. E., and A. G. Gilman. "G Proteins." *Scientific American,* July 1992. Samples some effects of these ubiquitous signaling molecules.

Moore-Ede, M. C., F. M. Sulzman, and C. A. Fuller. *The Clocks That Time Us: Physiology of the Circadian Timing System.* Cambridge, MA: Harvard University Press, 1982.

# Animal Behavior

*When you have studied this chapter, you should be able to:*

1. Explain the theoretical difference between innate and learned behavior, and give two examples of each.
2. Explain what is meant by stereotyped behavior, and give two examples. Describe the neural characteristics of stereotyped behavior.
3. List the selective advantages of innate and learned behavior and of stereotyped behavior.
4. Give examples of motivation or drive, sign stimuli, and supernormal stimuli.
5. Describe the characteristics of territorial behavior, and give an example of such behavior.

6. Describe conflict behavior, and explain why it is thought to have played a role in the evolution of threat displays and courtship behavior.
7. Distinguish among habituation, conditioning, trial-and-error learning, insight learning, and imprinting.
8. Summarize what is known about migration and homing in animals.
9. Compare and contrast the societies of honeybees and vertebrates.
10. Describe the functions of threat and appeasement behavior in the maintenance of a dominance hierarchy.

W e are prone to conclude that a dog is "ashamed" when it puts its tail between its legs and sneaks into a corner after a scolding and "happy" when it wags its tail. This type of thinking is called **anthropomorphism,** ascribing human emotions to animals.

At the opposite extreme, we may say that a bird sings from instinct because it is incapable of behaving intelligently. The view of human behavior as intelligent and that of other animals as instinctive is reflected in the tendency to ascribe actions of which we are ashamed to "animal instincts." Neither of these approaches to animal behavior gives much useful insight into why animals behave as they do. Recent research has attempted to study animal behavior with as little bias as possible from our human prejudices (Figure 41–1).

In many ways, natural selection acts more directly on behavior than on anything else. Dozens of different behavior patterns may distinguish an individual that reproduces from one that does not, and all the adaptations of an animal's anatomy and physiology are useless if the animal does not feed itself, escape predators, and find a mate.

In this chapter we shall consider how behaviorists may try to disentangle genetic and environmental influences on behavior and the kinds of selective pressures that have produced the varied behavioral repertoires of different animals.

**FIGURE 41–1** Why do animals behave as they do? Is this African elephant taking a mud bath for fun, to cool itself, because mud kills skin parasites, or what? (Mark Phillips/Photo Researchers)

## KEY CONCEPTS

♦ An animal's genes determine the structure and function of its sense organs, nervous system, and effectors and the range of behavior patterns that these systems can develop in response to environmental stimuli.

♦ An animal's behavior may be innate, learned, or most often a mixture of the two.

♦ Behaviors that must be produced perfectly the first time they are performed are usually innate. Learning produces behavior that must be flexible to meet local or changeable conditions.

♦ Many behavior patterns develop from interactions of the developing nervous system with environmental factors normally encountered during development.

### 41–A  IMMEDIATE AND ULTIMATE CAUSES

A frog is sitting in the grass when a fly buzzes past. Zip! The frog's tongue flicks out and pulls the fly into the frog's mouth. How and why does the frog do this? The question "how" can be answered by describing the frog's sensory, nervous, and muscular systems. The stimulus of a moving fly before the eyes sends impulses along sensory neurons to the central nervous system, which in turn activates and directs the tongue muscles used in the simple reflex that catches the fly.

The question of "why" the frog catches the fly can be answered on two levels. The immediate reason is that this behavior pattern results from a nervous reflex activated by seeing a fly. The longer-term answer is that the behavior pattern exists because it has been selected for during the course of evolution.

The behavior patterns we see today result from three main selective pressures:

1. Ultimately, an animal's behavior patterns will be selected for as they contribute to its reproductive success. It is occasionally possible to see why a behavior pattern has evolved by showing that not performing the behavior is selected against. For instance, many birds remove the empty egg shell from the nest after a young bird has hatched (Figure 41–2). Niko Tinbergen showed that adult black-headed gulls that did not remove the shells lost more chicks to predators; the white inside of the egg

shell allowed certain predators to discover the otherwise camouflaged nest and chicks.

2. Behavior patterns solve immediate problems. Hungry animals must feed, and hunted animals must escape predators, if they are to survive and reproduce.

3. Behavioral adaptations permit an animal to detect sights, sounds, and other environmental stimuli that are important to survival or reproduction and then to carry out behavior patterns appropriate to those stimuli (Figure 41–3).

### Genes and Environment

Most genes that influence behavior operate indirectly, by controlling anatomy and physiology. In laboratory populations of the fruit fly *Drosophila,* single genes have been identified that cause a fly to do such things as court members of the same sex, copulate for much longer than normal, follow a 19-hour rhythm of activity instead of the usual 24-hour cycle, and fail to learn a simple association between a stimulus and subsequent electric shock. These behaviors

**FIGURE 41–3**  Reacting to a stimulus. *Helix pomatia* (which may reach the dinner table as escargot) reacts to ants crawling around it by secreting bubbly mucus, which traps the ants. (Hans Pfletschinger/ Peter Arnold)

**FIGURE 41–2**  Adaptive behavior. A black-headed gull removing an egg shell.

result from mutations that cause some abnormality in the flies' sense organs, nervous system, or muscles.

Environment affects behavior in two main ways. In the short run, animals perform many behavior patterns only when they are induced to do so by environmental stimuli. In the long run, the environment influences gene expression in the development of many behavior patterns.

## 41–B  DEVELOPMENT OF BEHAVIOR

An animal's behavior, like its anatomy and physiology, forms during its development, through the interaction between its genes and its environment.

There are often critical periods when a particular environmental influence must be present if a particular behavior pattern is to appear. An example occurs during **imprinting** of young animals. Goslings (young geese) and ducklings learn to follow their parents, and to respond to their parents' signals, during a critical period after they hatch. Konrad Lorenz found that young birds would follow him as if he were their mother if they saw him, rather than their mother, during the critical period. Many animals learn what their future mates will look like by a similar process of sexual imprinting during a critical period. Lorenz had a tame jackdaw (Figure 41–4) that unfortunately became sexually imprinted on him before he understood how the process worked. It caused Lorenz great inconvenience by stuffing regurgitated worms into his ear during its "courtship feeding."

The interactions between genes and environment in the development of behavior have been studied intensively in the case of bird song. Peter Marler and Masakazu Konishi showed that male white-crowned sparrows reared in isolation sing only the inherited song of their species, whereas in the wild the sparrows learn the dialect of their own lo-cal population by listening to adult birds sing. To compli-cate matters, a bird must hear the dialect during a critical period when it is about three months old if it is to produce the dialect when it first begins to sing at the age of one year. And even if it has heard the dialect during the criti-cal period, it will never sing this dialect correctly if it is deafened before it has also sung the dialect. Once a bird has sung the full dialect song, however, deafening has no effect on its further performance (Figure 41-5). This exam-ple shows that many factors may be involved in the de-velopment of a normal adult behavior pattern.

One further conclusion from this work is that an ani-mal inherits a tendency to learn some behavior patterns but not others. White-crowned sparrows learn the dialects of their own species, but exposing them to the songs of other (even closely related) species during the critical period does not make them learn the songs of these other species. Birds treated in this way end up with a song like that of the com-pletely isolated male.

The environmental stimulus necessary to the normal de-velopment of a behavior pattern may be very precise, or it may be rather general. Rats or mice that have been picked up and returned to the nest once or twice a week in their youth mature more quickly, in behavioral terms, than do those that are never handled. This is a rather unspecific stimulus to the maturation of behavior.

## 41–C  INSTINCT VERSUS LEARNING

The idea that a behavior pattern is either instinctive or learned (or, more often, a combination of the two) is com-mon among behaviorists and in everyday life. **Instinctive, or innate** (inborn), behavior is genetically programmed into the nervous system and is difficult to alter. Learned behav-ior is acquired or disappears as a result of experience.

**Experimental psychologists** study behavior in con-trolled laboratory environments where animals are pre-sented with specific tasks. These researchers have demon-strated that most animals are capable of learning many things. On the other hand, **field behaviorists** have shown that members of the same species tend to show identical behavior patterns in the wild, suggesting that much of be-havior is instinctive.

Instinct is defined by negatives: instinctive behavior de-velops without the animal's having to learn it. Such nega-tive definitions are notoriously difficult to use. Furthermore, the only possible experiment to determine whether or not a behavior pattern is instinctive is to deprive the develop-ing animal of as many environmental stimuli as possible and see if the behavior pattern still appears. Even if the pat-tern does appear under such circumstances, it may still not be instinctive. The experimenter might merely have failed to remove the stimuli that permit the animal to learn the behavior.

**FIGURE 41–4**  A jackdaw.  (John Cancalosi/Tom Stack & Associates)

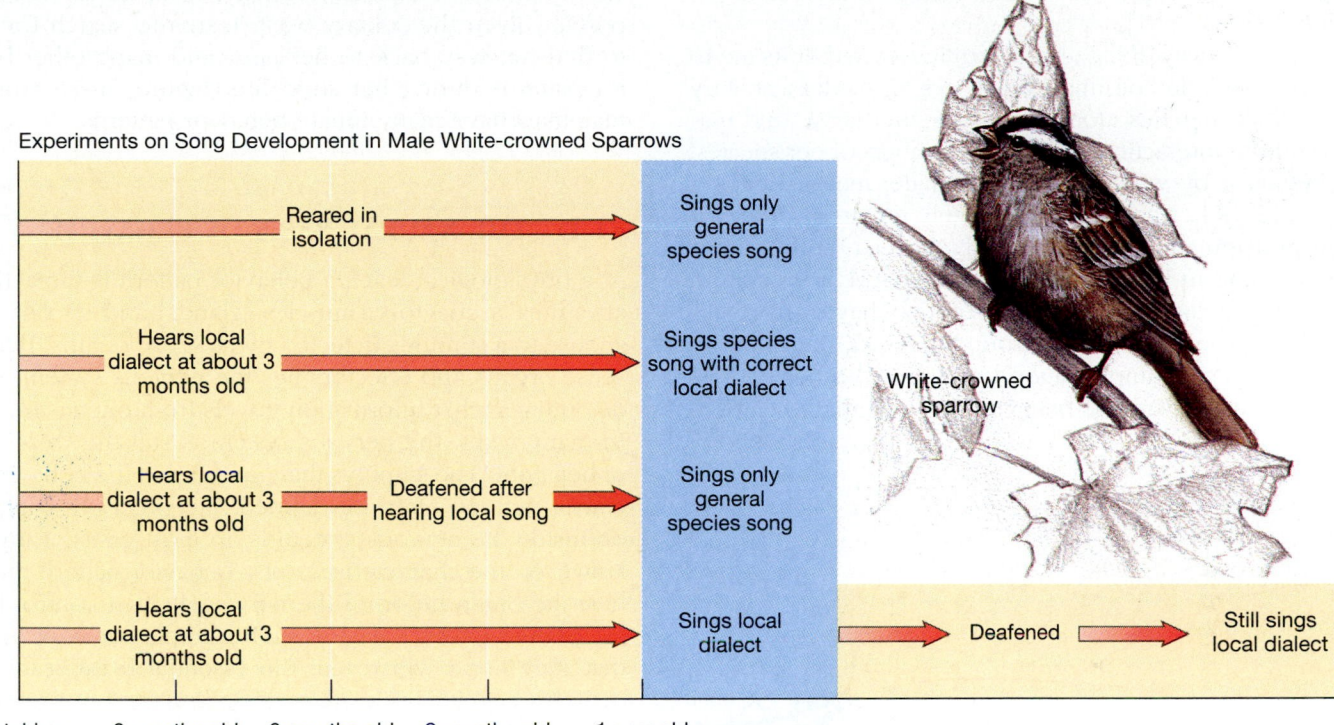

Experiments on Song Development in Male White-crowned Sparrows

| | | |
|---|---|---|
| Reared in isolation → | Sings only general species song | |
| Hears local dialect at about 3 months old → | Sings species song with correct local dialect | |
| Hears local dialect at about 3 months old → Deafened after hearing local song → | Sings only general species song | |
| Hears local dialect at about 3 months old → | Sings local dialect | → Deafened → Still sings local dialect |

Hatching    3 months old    6 months old    9 months old    1 year old

White-crowned sparrow

**FIGURE 41–5** Development of behavior. Summary of Marler and Konishi's findings on song development in white-crowned sparrows.

## Adaptive Value of Learning and Instinct

Behavior patterns that are arguably innate have a more restricted range of potential development than those that are learned. Each end of this continuum has some selective advantages.

As an example, kittiwakes are sea birds that nest on narrow ledges. The chicks keep still from the moment they hatch, whereas related herring gull chicks move around. This innate behavior (or nonbehavior) of kittiwakes is clearly adaptive, because a false step means death to a kittiwake chick. There is no room for learning.

Innate behavior also saves energy. There is clearly a selective advantage to the animal that does not have to waste energy learning responses that are sure to be required frequently and without variation.

With these advantages of innate behavior, why are so many behavior patterns learned? In particular, why are vital behavior patterns, such as recognizing a potential mate or learning to fly, so often partly learned? Learning gives an animal the flexibility to adapt to a changing environment by acquiring new behavior patterns as they become appropriate, or by responding in new ways to old stimuli. For animals such as vertebrates, which have relatively long lifespans and so experience changing environmental conditions, this flexibility often means the difference between life and death.

Learning is also necessary whenever a stimulus differs for individual members of a species. For instance, every mobile animal with a home base must learn to find that home. No amount of genetic programming will permit a crab to find its own burrow among all the holes in a beach (Figure 41–6). In addition, many social animals live in environments where the relationships between individuals

**FIGURE 41–6** Invertebrate learning. This ghost crab learns to find its home and also learns feeding spots on the beach where it lives. It will dig down into a loggerhead turtle nest several feet below the surface of the sand and return, night after night, to eat the eggs. (George Harrison/ U.S. Fish & Wildlife Service)

change constantly, and such relationships must almost always be learned.

A species' way of life also determines whether its members evolve learned or innate behaviors. Consider a solitary wasp, which hatches alone, develops as a larva, and matures without interacting with other members of her species. The behavior by which she finds a male, mates, builds a nest, and lays her eggs must be largely innate in order for her to perform each action perfectly the first, and perhaps the only, time in her life. On the other hand, a social animal such as a cat can learn much of its behavior from observing other members of its group. It would, however, be an enormous oversimplification to say that the behavior of an insect is innate and the behavior of a mammal is learned.

In any group of animals, both types of behavior are important. Even the solitary wasp learns to search for food, to find her way back to her nest, and many other behavior patterns during her short life (Figure 41–7). Similarly, mammals have many innate behavior patterns.

## 41–D   THE NEURAL BASIS OF BEHAVIOR

At a physiological level, a behavior pattern is the action of an animal's effectors (muscles, glands, and so on) in response to a stimulus detected by its sense organs. Between sense organs and effectors lies the nervous system, which determines what information travels from one to the other. In many ways, the nervous system is still the "black box" of behavior. The stimulus that goes in and the behavior that comes out can often be defined, but precisely what goes on inside the nervous system is, in most cases, a mystery. However, the characteristics of a behavior pattern must reflect the organization of the nerve cells that control it.

Many workers have looked for simple behavior patterns that they hope will reveal the essential features of more complex activities. We are now finding out how certain individual neurons function in locomotion and escape reactions of invertebrates such as cockroaches, crayfish, sea slugs, and leeches, but the picture is still far from complete.

Reflexes and more complex behavior patterns share a number of properties that result from the way neurons operate. For instance, both show **latency,** a time delay between stimulus and response (Figure 41–8). Latency is due to the time necessary for reception by the sense organ, conduction through the nervous system, and excitation of the effector.

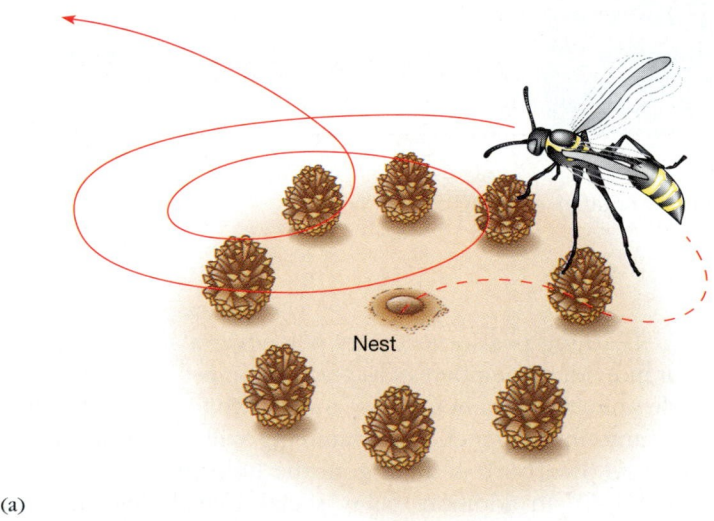

(a)

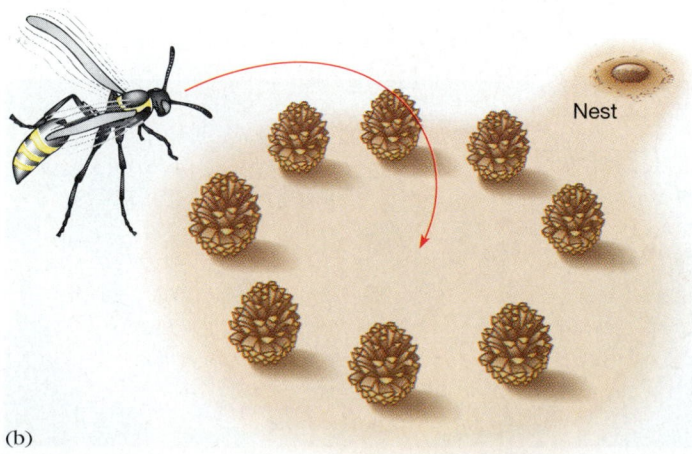

(b)

**FIGURE 41–7**   Learning by a solitary insect. A digger wasp finds her nest by visual landmarks. (a) The wasp makes an orientation flight over a nest entrance that the investigator has surrounded by pine cones. (b) She returns to the center of the ring of cones, which the investigator has moved in her absence.

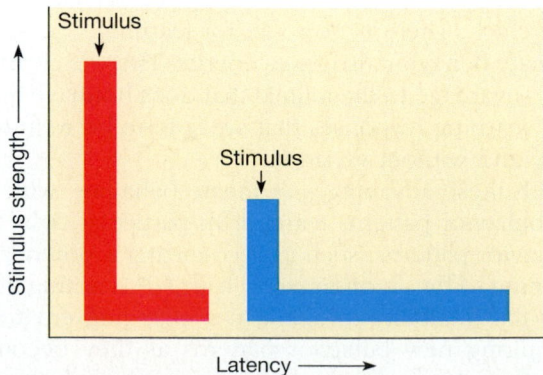

**FIGURE 41–8**   Relationship between stimulus strength and latency in the flexion reflex. Latency is the time that elapses between stimulus and response. If the dog's foot is pricked by a pin, the stronger the prick, the faster the dog withdraws its leg.

(a)

(b)

**FIGURE 41–9**    Stereotyped behavior. (a) Walking is a stereotyped behavior. These giraffes need little feedback from their sense organs to stroll across the savanna. (b) Landing is *not* a stereotyped behavior. This blue tit needs all the information it gets from its superb binocular vision, from proprioceptors in the wing muscles that detect air pressure against the wings, and many other sense organs, to stop flying and come to rest on this branch.    (b, Biophoto Associates)

## Stereotyped Behavior

A striking feature of the behavior of any animal is its repertoire of **stereotyped behaviors**—actions that use many muscles in a precisely timed sequence and that are always performed in an essentially identical pattern. Reflexes are the simplest examples, but more complicated activities such as locomotion, sound production, breathing, and feeding also fall into this category. For instance, a cockroach or cricket will leap forward in a standard escape reaction when receptors on its abdomen are stimulated by a puff of air. This is an example of an innate stereotyped behavior, sometimes known as a **fixed action pattern.** Other stereotyped behaviors are learned.

An example of a learned stereotyped behavior is a rat's pressing a lever for food. Each rat presses the lever with a characteristic gesture. One uses a fist, another one a toe, and each uses its own gesture time after time. Similarly, how you hold your pen, walk, play the piano, or ride a bicycle is a learned stereotyped behavior that is conservative and unique to you (Figure 41–9).

Studies of invertebrates suggest that stereotyped behaviors differ from other behavior in two ways: they are controlled by very few neurons in the central nervous system, and they can occur without feedback from the sense organs, although such feedback is available and is often used. For example, the fixed action patterns by which a crayfish flexes its abdomen when it swims are induced by the activity of single, identifiable neurons in its central nervous system. At least some fixed action patterns are "hardwired" into the nervous system: they are always produced in identical fashion because only one or a few control cells trigger all of the motor neurons for the entire behavior pattern.

By contrast, consider what happens when you pick up a glass of water. This behavior, which is not stereotyped, is a series of interactions between sense organs and muscles. The sense organs signal how much the water is slopping about and how far and how fast the glass is rising. These sensory messages reach neurons in the central nervous system, and hundreds of motor neurons respond, controlling the muscles in the arm so that the glass rises steadily and the water does not spill. Hundreds of central neurons and continual feedback from the sense organs are necessary for this behavior pattern.

The selective advantage of behavior patterns programmed into the nervous system is probably that they reduce the number of neurons used in a relatively complex task that must be performed perfectly and often. Some stereotyped behaviors, such as escape movements, must be performed perfectly to work at all. Others save energy because they ensure that the muscular movements of, say, writing or feeding need not be worked out with sensory feedback every time they are performed. All animals seem to be able to program behavior patterns into the nervous system during development and, in many cases, in later life.

## Sign Stimuli

What sorts of stimuli trigger behavior patterns? In the 1940s, Niko Tinbergen was studying male three-spined sticklebacks that sported the red belly characteristic of these fish in

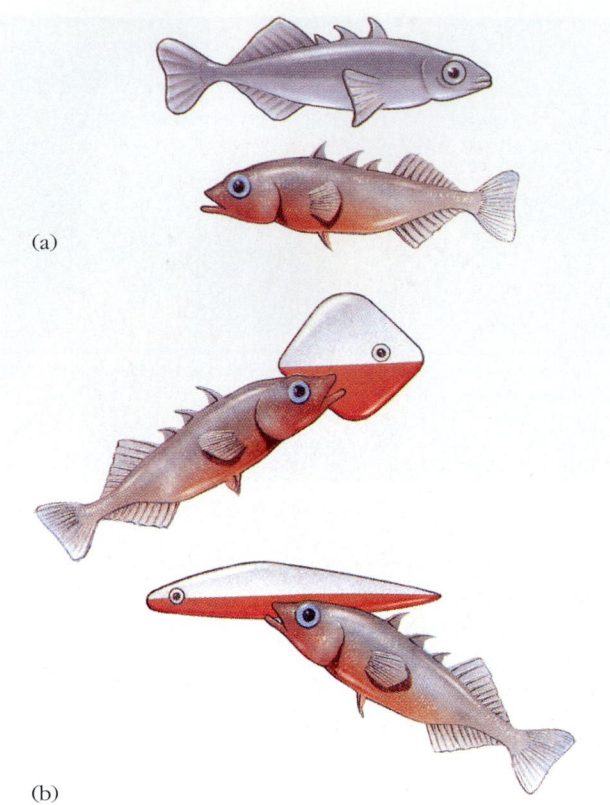

**FIGURE 41–10** The sign stimulus for attack by a male stickleback. Niko Tinbergen found that male sticklebacks in breeding condition do not attack the life-like model lacking a red belly (a), but they will attack either of the crude models with a red undersurface. (b) The eye is also necessary for the model to act as a stimulus. The presence of an eye is often necessary if an animal is to identify an object as another animal.

breeding condition. Every time a red truck drove past a nearby window, the fish made frantic attempts to swim through the glass of their tanks toward the truck, as if they would attack it. A male stickleback in breeding condition will also attack other breeding males.

What stimulus provokes this attack behavior? Tinbergen presented various models to the sticklebacks to find out. When he showed wooden models of sticklebacks to males in reproductive condition, they attacked crude models with an eye and a red belly in preference to life-like models without the red belly (Figure 41–10). The red belly of the male stickleback in reproductive condition is thus the sign stimulus that triggers the fixed action pattern of attack by another breeding male. A **sign stimulus** (also known as a **releaser**) is that portion of the total stimulus that releases a particular behavior pattern.

An interesting extrapolation of the theory of how sign stimuli work is that it is possible to produce a **supernormal stimulus,** which provokes a behavior pattern more effectively than does the normal stimulus. For instance, her-

ring gull chicks peck at the stimulus provided by a red spot on the parent's bill. This induces the parent to regurgitate fish to feed the chick. When models of the stimulus were tested, it was found that a bar with big red and white stripes provoked more pecks from young chicks than did a realistic model of the bill (Figure 41–11). The bar was a supernormal stimulus for this innate fixed action pattern.

## Drive and Motivation

A particular stimulus may evoke different responses in the same animal at different times. For example, an animal that sees food will eat if it is hungry but may ignore food if it has just eaten. Something inside the animal, which we may call **motivation** or **drive,** is different at these two times. Since different behaviors are appropriate at different times even when the stimulus is the same, variations in motivation help to ensure that an animal's behavior changes to fulfill its needs.

In the vertebrate brain, the hypothalamus appears to control motivation. Attack, escape, and sexual behavior can all be evoked by electrical stimulation of parts of the hypothalamus. The hypothalamus does not act alone to determine motivation. Its activities are modified by input from other parts of the brain and by hormone levels. For instance, only when they are in breeding condition, with high levels of the hormone testosterone, do male sticklebacks attack other fish with red bellies.

## 41–E   LEARNING

Learning produces adaptive changes in an individual's behavior as a result of its experiences. It occurs in so many different ways that we have to classify them somehow, although there is no evidence that the categories used here bear any relationship to the physiological basis of learning, which is not well understood.

**Habituation** is the loss of old responses. Animals may learn not to respond to stimuli that occur often and are unimportant to them. Young animals often show alarm behavior at a variety of stimuli, most of which they rapidly learn to ignore. For example, moving objects overhead signal danger because they may be hawks or eagles hunting. Young birds often react to airplanes or falling leaves by crouching and holding still until they learn that these objects do not pose a threat.

**Conditioned reflexes** are behavior patterns evoked by a previously neutral stimulus that an animal has learned to associate with the stimulus that normally elicits the reflex. Russian physiologist Ivan Pavlov showed that there is a reflex that causes hungry dogs to secrete saliva when they see food. Pavlov rang a bell when he showed food to the dogs, and after many trials the dogs would salivate when

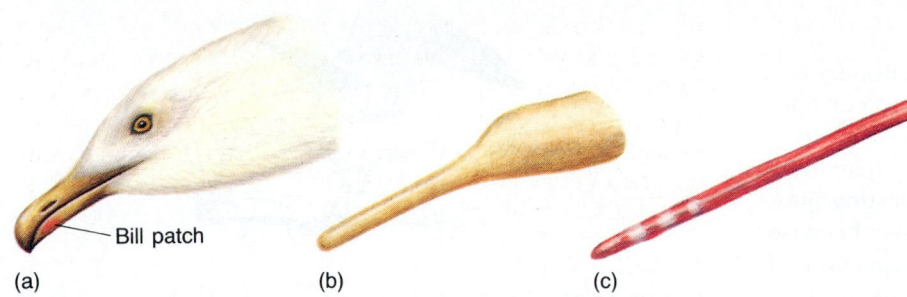

(a)     Bill patch       (b)          (c)

**FIGURE 41–11** Models used in testing for sign stimuli. Young herring gull chicks peck at a colored patch on the parent's bill. This induces the parent to feed the chick. (a) A life-like (though flat) model of the parent's head releases fewer pecks by a newly hatched chick than (b) a model in which the bill is longer and thinner than normal. This model is less effective than (c), a model that is long and thin and emphasizes the contrast between bill color and bill patch.

the bell was rung even though he stopped showing them food. The dogs had learned to respond to a previously neutral stimulus, the bell, to which they had not previously responded. Pavlov called the bell the conditioned stimulus. Fruit flies, and even their larvae, also show this kind of learning, for instance, when they have been taught to associate an odor with an electric shock.

**Trial-and-error learning** is what its name implies. An animal's spontaneous movements may by chance produce a reward, and the animal learns by trial and error to repeat that behavior pattern. The reward may often be the pleasure of performing an action more accurately than before. Trial and error is probably the most appropriate category for the learning of new motor skills (Figure 41–12). Young mammals and birds perfect their prey-catching movements, and humans learn to ride a bicycle, by a trial-and-error form of practice.

All these types of learning are varieties of **associative learning.** Reinforcement (reward or punishment) is a central feature of associative learning. Another characteristic of associative learning is that it improves with repetition.

**Latent learning** occurs without any obvious reward or punishment. It is learning that produces no obvious behavior at the time it occurs. This often happens during exploratory behavior. A recently fed animal may give no sign

that it has noticed a new food source until it later returns to feed there.

**Insight learning** is a form of reasoning that draws on the results of past experiences to arrive at the solution of a novel problem. The classic example of insight in animals came from the work of Wolfgang Kohler on chimpanzees. Presented with a bunch of bananas too high to reach, a chimpanzee would pile up boxes to make a stand from which it could reach the bananas (Figure 41–13). Reasoning of this sort has been shown in many mammals and in some birds, although it is often difficult to distinguish from other forms of learning.

**FIGURE 41–13** Insight learning. In Kohler's experiment, a chimpanzee was left in a room with a number of boxes and a bunch of bananas hanging from the ceiling. After a period (perhaps of thought?) the chimpanzee piled the boxes on top of one another, climbed on them, and reached the bananas.

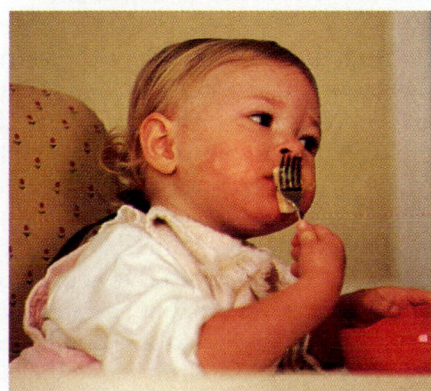

**FIGURE 41–12** Trial-and-error learning. Animals, especially juveniles, learn many things by trial and error.

## 41–F  TERRITORIAL BEHAVIOR

Now that we have seen something of the evolutionary origins and physical basis of behavior, we go on to consider particular behavior patterns that illustrate these ideas.

Many animals defend **territories,** areas where they have a monopoly on resources such as food or nesting sites. Holding a territory is evolutionarily advantageous because the resources contribute to the successful production of young. A territory holder attacks and drives away other members of the same species (Figure 41–14). It is to an animal's advantage to defend the territory with a minimum of attack behavior, since every attack carries the risk that the attacker will be injured or spotted by predators. Animals have evolved several features that minimize damage during territorial confrontations. For instance, real fighting is infrequent because there are "rules" about who wins encounters between two individuals, based on body language.

Consider a male thrush defending a territory before the female arrives in the spring. The male is most aggressive near the center of his territory. As he moves toward the boundary, his attacks on a trespassing neighbor become less violent, until he reaches a point at which he is as likely to retreat as to attack when he sees another male thrush. This point marks the boundary of his territory. When two neighbors meet at the boundary between their territories, they both act as if they have conflicting retreat and attack motivations. These tendencies are manifested as conflict behaviors.

**Conflict behavior** usually contains elements of two conflicting tendencies (in this case, movements toward retreat and toward attack), as well as movements apparently unrelated to the issue at hand, such as pecking at the ground (Figure 41–15) or preening: maintenance of the feathers by oiling, and smoothing them with the bill.

**FIGURE 41–15**  Conflict behavior. A gull involved in a territorial clash violently pulls up a clump of grass. The bird acts as if it were caught in a conflict between tendencies to attack and to flee. Instead of doing either, it engages in apparently irrelevant "displacement activity"—pulling up grass. A more placid form of grass-pulling is part of its nest-building behavior.

In many species, patterns of conflict behavior appear to have evolved into ritualized **threat displays** that are directed toward intruders (Figure 41–16). Threat is obviously more advantageous than actual fighting because it does not injure the animal. In the case of a mutual threat display between two animals, which is effectively a ritualized fight, an experienced observer can predict which animal will win by deciding which animal incorporates more attack movements in its display. The loser will eventually move away from the winner.

## 41–G  CONFLICT AND COURTSHIP

Most animals, even those that live in social groups, maintain a minimum **individual distance** from one another (Figure 41–17). For example, swallows sitting on a telephone wire are always a certain minimum distance apart—determined by the reach of the neighbor's bill. The invasion of individual distance is perceived as a threat, and the invading animal is usually attacked.

The conflicting tendencies to attack and to permit another animal to come close enough to mate are often evident in **courtship behavior,** the behavior patterns that precede mating in most animals. The courtship displays of many species seem to have evolved from such conflict behavior.

In a well-studied example of courtship behavior, the male black-headed gull attracts a female to his territory. She alights near him, and both gulls adopt a series of postures that resemble, but are slightly different from, the characteristic threat display of the species. If neither bird attacks the other, both display appeasement gestures, which imply that the hostility between them has lessened. Eventually the female flies off, but she may return many times, and each bird will display fewer threatening and more appeasement gestures with each visit. Eventually the greeting ceremony ceases entirely, and the male feeds the female. After this,

**FIGURE 41–14**  Territorial behavior. Two pairs of blue-footed boobies confront each other at the boundary between their territories.    (Fred Bavendam/Peter Arnold)

**FIGURE 41–16**   The threat display of a spike-headed katydid (*Copiphora* sp.) in the Amazon rain forest.   (Michael Fodgen/DRK)

copulation can occur, and a permanent pair bond forms (Figure 41–18).

In their initial encounters, the birds are displaying behavior that reveals three conflicting tendencies—to attack, to flee, and to stay together. The behavior patterns that result from the conflict have evolved into an elaborate courtship ritual.

**FIGURE 41–17**   Individual distance. These mallard ducks maintain a minimum distance from each other as they rest on a log.

## 41–H   MIGRATION AND NAVIGATION

Many animals have remarkable abilities to find their way over hundreds of miles of land and sea. Homing animals can do many things that we cannot do ourselves. A Manx shearwater (a sea bird), which had never been more than 10 miles from home, was removed from her nest on an island off the coast of Wales, flown to Boston, and released. She was back on her nest before the letter announcing her release reached the observers in Wales. To perform an equivalent feat, such as sailing from Boston to Wales across the Atlantic, a human would have to spend hours learning to use navigation instruments to cross the ocean and would still need a map to find the nest on the other side. Birds, caribou, monarch butterflies, sea turtles, fish, and salamanders all perform equivalent journeys without mechanical aids and with little or no learning (Figure 41–19).

Oblique–long call intimidates other males and attracts unmated females.

① ⑤

④

An unmated female lands and they greet each other with appeasement gestures.

②

She pecks him and flies off; he adopts a submissive posture and then tries again.

③

He turns his head away with bill down in another appeasement gesture. But she does not reciprocate; her head still points forward and her wings are raised in an aggressive posture.

⑥

This time with more success: both birds face away and copulation and nest building can now proceed without further hostility.

(a)

(b)

**FIGURE 41–18** Courtship in the black-headed gull. (a) The sequence of courtship behavior. Follow the arrows around from the top. (b) A gull uttering the long call by which it claims a territory and, if it lacks a mate, attracts unmated females to begin courtship. (Notice the red bill patch, the sign stimulus that prompts gull chicks to beg for food.)

Many animals, including horses, dogs, cats, and humans, orient themselves by landmarks that they learn and recognize visually. Many animals can also find their way around by using their chemoreceptors. Dogs can follow long and complicated scent trails in unfamiliar territory. Moths find mates and ants find their nests by following odor gradients. A dramatic case is that of the salmon, which hatches in a freshwater stream and matures hundreds of miles away in the ocean. Years later, when the time comes to spawn, each salmon "smells" its way back to the very stream in which it hatched.

## Compass Course

Many animals can move in a specific compass direction using the sun. Since the sun appears to move from east to west during the day, an animal using the sun as a refer-

(a)

(b)

**FIGURE 41–19** Migration. (a) Caribou in the Brooks Range, Alaska. Caribou migrate from the coastal plain to the uplands in spring and back again in the fall. (b) Monarch butterflies resting at Cape Cod in the lee of a hedge. (a Johnny Johnson/Natural Selection)

ence to maintain a compass direction must know the time of day. It must also be able to compensate for the sun's apparent movement throughout the daylight hours. This involves using the internal clock that controls circadian rhythms (Section 40–G).

Researchers found that "clock-shifting" salamanders, by resetting their internal clocks, disrupted navigation. The salamanders were exposed to artificial daylight that began and ended 6 hours (one fourth of a day) later than natural daylight. They then oriented themselves in a compass direction that was 90° (one fourth of a circle) clockwise away from the correct direction for migration to breeding ponds (Figure 41–20). These experiments confirmed that by combining information from the position of the sun and their internal clocks, salamanders can point themselves in a particular compass direction.

Birds have the same ability to use the sun as a compass, and birds that migrate at night can also use star patterns to find a compass direction. Navigation in birds has been studied intensively in pigeons, which have long been used to carry messages because they will fly home from wherever they are released. Studies of homing pigeons have not, however, produced a clear picture of how they navigate because pigeon behavior is complicated, and the birds can find their way home using any one of several cues.

Learning plays a part in homing behavior, and experienced birds released in an unfamiliar place reach home faster than naive ones. Inexperienced birds use a simple hierarchy of navigational cues. If the sun is shining, they use a sun compass, so that if they have been clock-shifted they fly off in the wrong direction. Experienced birds are less likely to be fooled by a clock shift. They appear to cross-check the information from their sun compass with other cues, such as odor or the magnetic field, which give conflicting information if the birds have been clock-shifted. An experienced bird solves this dilemma by going to sleep in the nearest tree until the effect of the clock-shift wears off. It then flies straight home.

The navigation of sea turtles is less obscured by learning than that of pigeons because these animals perform their most impressive navigation immediately after hatching and alone. American loggerhead turtles hatch on beaches in the southeastern United States and Caribbean Islands. Then they swim across miles of the Atlantic Ocean to the Sargasso Sea, an area full of floating seaweed and invertebrates. Here the turtles feed and mature. Years later, they mate, and the females swim back to the beaches to lay their eggs. Biologists have argued for years over how the turtles navigate on this remarkable trip, which requires both a map sense and changes of compass direction.

## Map Sense

In order to get from A to B using a compass you have to know whether A is north, south, east, or west of B. This is called a **map sense** because it means that you must know the relative positions of A and B.

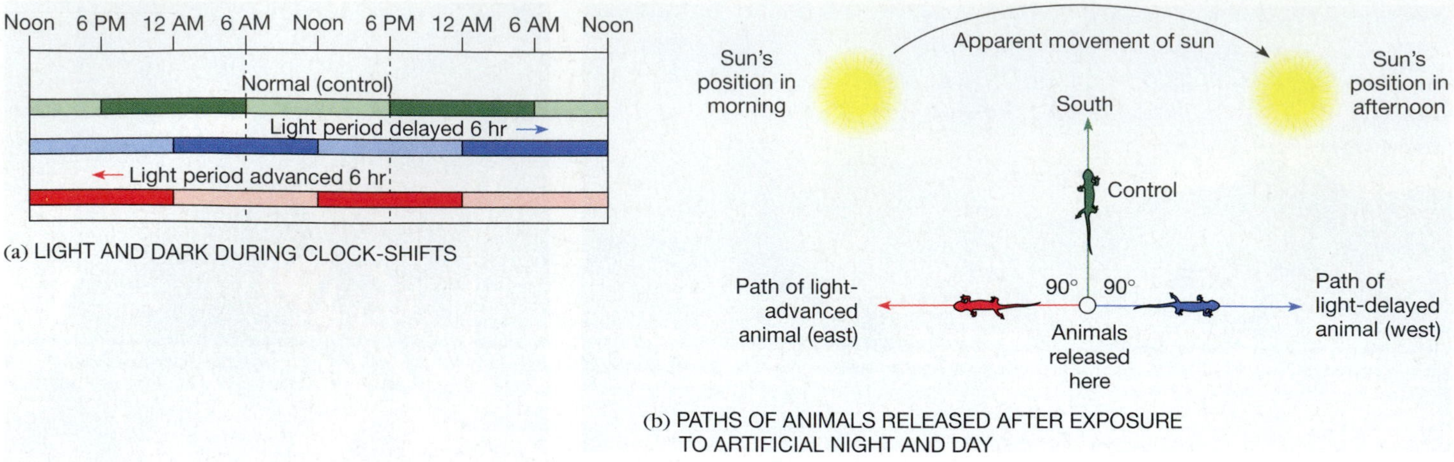

(a) LIGHT AND DARK DURING CLOCK-SHIFTS

Noon　6 PM　12 AM　6 AM　Noon　6 PM　12 AM　6 AM　Noon

Normal (control)

Light period delayed 6 hr →

← Light period advanced 6 hr

Apparent movement of sun

Sun's position in morning

Sun's position in afternoon

South

Control

Path of light-advanced animal (east)

90°　90°

Animals released here

Path of light-delayed animal (west)

(b) PATHS OF ANIMALS RELEASED AFTER EXPOSURE TO ARTIFICIAL NIGHT AND DAY

**FIGURE 41–20**　Effect of clock-shifting on migrating salamanders in the Northern Hemisphere. The animals were captured as they migrated south and exposed to artificial light and dark (a), which reflected normal night and day or a "day" in which "daylight" was advanced or delayed by six hours. (b) When they are released, the control animal sets off south as expected. The experimental animals go east or west, which is 90° away from south on the compass. The light-delayed animal's internal clock tells it that the time is 8 AM when the animal is released at 2 PM. To travel south, the animal keeps the sun on its left and therefore actually sets off to the west. The light-advanced animal perceives the time to be six hours later than it is. It keeps the sun on its right and heads east.　(Adler, K. *Photochem. Photobiol.* 23:288, 1976.)

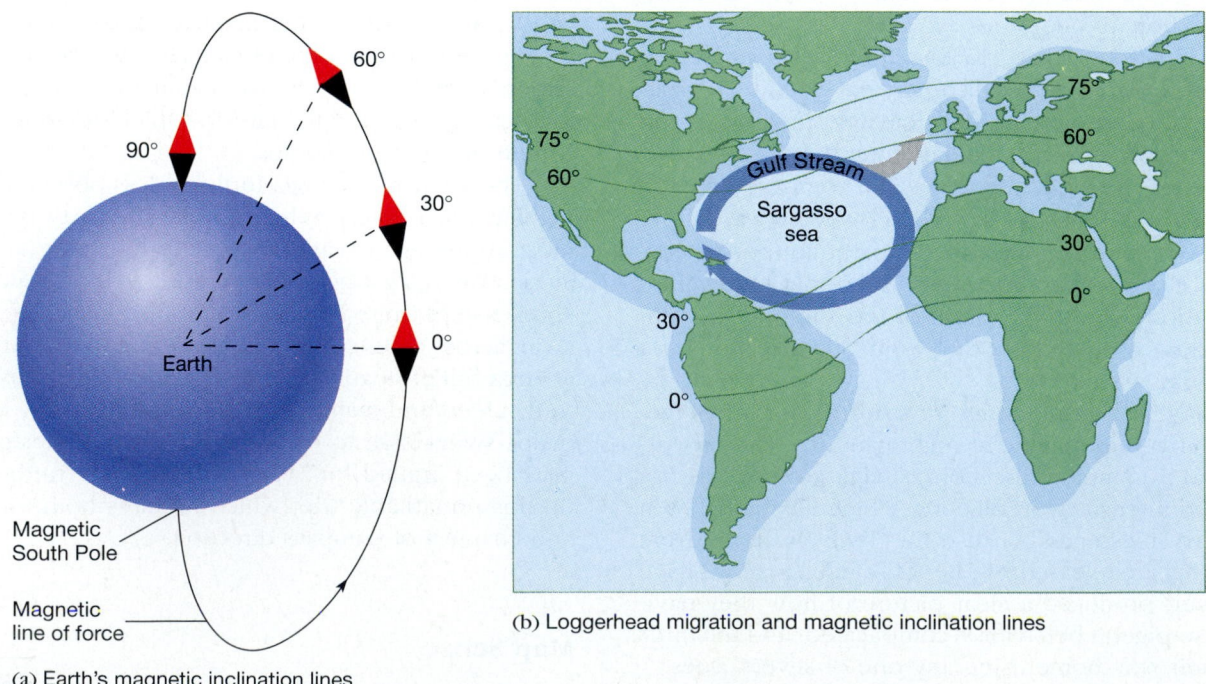

60°

90°

30°

0°

Earth

Magnetic South Pole

Magnetic line of force

(a) Earth's magnetic inclination lines

75°

60°

75°

60°

Gulf Stream

Sargasso sea

30°

0°

30°

0°

(b) Loggerhead migration and magnetic inclination lines

**FIGURE 41–21**　Navigation by loggerhead turtles, *Caretta caretta,* in the Atlantic. (a) Magnetic lines of force around Earth impart different degrees of inclination to a compass needle (red and black diamond) at different latitudes. (b) Magnetic inclination over the Atlantic and the Gulf Stream (pale blue), which hatchling turtles use to reach the Sargasso Sea.

Humans describe position on a map by using **latitude,** distance north or south of the equator, and **longitude,** distance east or west of a line from the North Pole to the South Pole that passes, arbitrarily, through Greenwich, England. Like many other animals, we can calculate our approximate longitude, but not our latitude, from the time of day and the position of the sun. For instance, we can tell whether we are in West Virginia or Utah by the time the sun rises. But this does not permit us to tell whether we are in Wisconsin or Mississippi, which straddle the same line of longitude but at different latitudes.

Many animals can detect magnetic fields using poorly understood magnetic sense organs in their heads, and biologists have long suggested that animals might use this sense to determine latitude. Earth acts as if it contains a bar magnet oriented between the magnetic North and South Poles. As a result, magnetic lines of force travel around Earth from the South to the North Pole (Figure 41–21a). A compass needle, or a rod of magnetite in an animal's brain, orients parallel to such a line of force. As a result, it lies parallel to Earth's surface at the magnetic Equator and sticks up farther and farther from the surface as you travel north and south. At the Poles it lies perpendicular to Earth's surface. Thus, the degree of **inclination** of the needle or rod to Earth's surface is a measure of latitude.

## Loggerhead Navigation

Marine biologist Kenneth Lohmann suspected that turtles used Earth's magnetic field to determine their latitude in the Atlantic. To find out if the turtles could detect the magnetic field, Lohmann tethered turtle hatchlings in a harness attached to an arm in the middle of a circular tank of sea water (Figure 41–22). The arm was attached to a computer so that the turtles could swim for hours while the computer made a continuous record of their movements. Lohmann then surrounded the tank with a system of coils that could be used to create different magnetic fields around the turtles.

Newly hatched loggerhead turtles swim toward light no matter what the inclination of the magnetic field around them, so Lohmann tested them in the dark. When he varied the inclination of the magnetic field, he found that at 57° inclination or less the hatchlings swam northeast, but at 60° inclination they made a right-angle turn until they were swimming south.

A genetic tendency to turn south when they reach 60° inclination makes sense if you look at the lines of magnetic inclination over the Atlantic (Figure 41–21b). Such a turn ensures that the turtles end up in the Sargasso Sea and not in the cold waters of the North Atlantic.

A tentative model of the navigation of loggerheads hatching on Atlantic beaches goes like this: loggerheads always hatch at night around the time of a full moon and

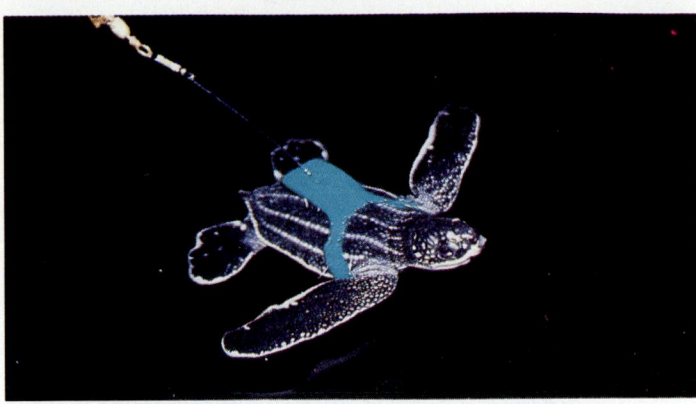

**FIGURE 41–22** Studying magnetic navigation. A turtle hatchling is harnessed so that a computer records the direction in which it swims, even in the dark. This is a leatherback turtle (*Dermochelys coriacea*), not a loggerhead (*Caretta caretta*).    (Kenneth Lohmann)

cross the beach toward the light of the moon, which rises in the east. When they reach the ocean, they stop responding to light and swim directly into the waves, which keeps them swimming away from land. At sea, wave patterns become confused. At some point the turtles switch from swimming into waves to swimming northeast. They could find this direction by the sun's position during the day, but Lohmann suspects they use their magnetic sense, which can act as a compass at night as well as during the day.

Within about 50 miles of the coast, the turtles reach the Gulf Stream, a wide current of warm water that flows north from the Caribbean and across the Atlantic Ocean to Europe. The current carries the turtles northeast no matter what direction they swim because it travels faster than a turtle normally swims. The Gulf Stream forks in mid-Atlantic, with the south fork flowing to the Sargasso Sea. By turning south when they reach 60° magnetic inclination, the turtles end up in the Sargasso Sea. Here, their motivation to migrate is lost, and they remain for years.

Eventually, the turtles reach sexual maturity. The hormonal changes involved presumably trigger the motivation to migrate back to the nesting beaches. Magnetic navigation alone would permit loggerheads to reach the nesting area, but females sometimes return year after year to the same beach to lay eggs, and it seems unlikely that magnetic navigation is precise enough to permit this behavior. Biologists suspect that their sense of smell guides the females the last miles to the beach.

## 41–I   DECISION-MAKING

We do not know whether other animals consciously consider their actions. Nevertheless, all behavior involves deciding among alternatives: to migrate this winter or stay

**FIGURE 41–23** A black-capped chickadee. This species was used in the first studies of "cost-benefit analysis." Here, an adult returns to its nest hole after a successful foraging trip.    (Uve Hublitz/Cornell Laboratory of Ornithology)

home, to continue feeding on this lawn or move on to the next, to fight for a territory or wait until one falls vacant.

## Foraging Behavior

The first studies on how animals make these decisions involved foraging (food-finding) behavior. Consider a chickadee eating insects in a patch of forest (Figure 41–23). When will it leave that patch and fly to the next? If it left every patch as soon as it arrived, it would spend all its time moving and find little food. But if it stayed in one patch all day, it would use up the food there and also take in little food that day.

There is selection for the bird to perform some sort of cost-benefit analysis. The costs of moving include the energy used to find and fly to the next patch and the risk that the next patch will contain little food. The benefit of moving is the chance of finding more food than in the patch the bird has already picked over. The result of this cost-benefit analysis will vary. The bird will get more to eat if it spends longer in patches with more insects. So selection will favor flexible decision-making that changes depending on the circumstances.

Computer models predict that a chickadee will find food most rapidly if it leaves every patch (no matter how much food it contains) as soon as its rate of feeding begins to slow down. When the bird enters a patch, it will first find all the obvious insects. After this, it will take more time to find each additional insect. It should leave at this point. Studies show that chickadees in a forest and parasitoids searching for a host behave very much as this model predicts. The chickadee's behavior can be described by saying that it remembers how long it takes to find an insect. When that time has increased for each of a few insects, it moves to a new food patch.

Behaviorists are beginning to apply cost-benefit analysis to more complicated decisions. For instance, animals tend to perform altruistic behavior when the cost is slight and the potential benefit to their kin is high. Territorial behavior may vary with the probable costs and benefits of attempting to raise a family at that time and in that place.

## Bird Reproduction

At each stage of reproductive behavior (courting, copulating, incubating eggs, feeding chicks), adult birds can decide to invest the necessary energy in the next stage or to abandon their reproductive efforts at the point they have reached. The adaptiveness of this series of decisions has been studied in detail using tawny owls.

Most tawny owl pairs that reproduce successfully defend territories year round (Figure 41–24). One study over many years examined a habitat containing 25 territories, each with its pair of owls. In a typical year, eight of the 25 pairs did not reproduce at all, another nine abandoned the eggs they had laid, and two more pairs allowed their chicks to starve to death. The remaining six pairs raised 18 young among them.

It appears that at each stage of reproduction the adults in effect perform a cost-benefit analysis, and that environmental cues determine which decision will be made. Tawny owls feed mainly on mice, and the number of mice in the territory at the time appears to be the main factor in each

**FIGURE 41–24** Tawny owls.    (Andy Rouse/DRK Photo)

decision. This is adaptive because if there are not enough mice to raise the young, any mice invested in the attempt will be wasted. The adults must eat to keep themselves alive whether they raise young or not. If they eat surplus mice themselves, they improve their chances of surviving to future years, when mice may be plentiful enough to raise young to maturity.

## 41–J    SOCIAL BEHAVIOR (SOCIOBIOLOGY)

Some animals have little contact with members of their own kind, but in many species, individuals do interact to some degree, either for a short time or throughout life. Social behavior is the cooperative interaction of members of the same species.

It is often difficult to tell whether particular groups of animals are truly social. For instance, gray herons spend much of their lives alone or with their mates. Occasionally, however, larger numbers of these birds spend days, and even weeks, together feeding. Is this feeding group social? Do the birds communicate with each other or help each other with the fishing? It turns out that the herons are all in the same place only because each of them has found food there. They do not interact any more than human diners at separate tables in a restaurant do, although they may find food because they have watched others feed there. Thus, it is not safe to assume that a group of animals constitutes a society. A group of animals is truly social only when there is considerable cooperation and communication among individuals. The individual's success is increased because members of the group cooperate in hunting for food or in defense, by taking turns at sentinel duty to warn of approaching predators and by working together to ward off the predator. There is often increased access to resources such as food or burrows, nests or other structures made by group efforts. Examples of extremely social animals are honeybees, humans, apes, and wolves, which form cooperative, long-lived societies upon which the individual's very life depends.

## Communication

All animals that live in societies communicate with other individuals. Among human means of communication are sound and hearing, used when we speak, clap, cheer, or laugh. Similarly, we use visual stimuli and vision in sign language, advertising posters, "body language," and choice of clothing. Birds, like humans, have highly developed vision. They communicate by movement and color, as well as by sound. Communication by sound can be highly elaborate even if it does not involve language, which is the most important aspect of our own sound communication.

Because our sense of smell is poor, we pay little attention to the chemical communication so common in other animals. Many mammals, such as dogs, mark their territories, determine another animal's mood, find their mates and food, and, for all we know, communicate in many other ways, by scent. A **pheromone** is a chemical that affects the behavior of another member of the species (Section 40-H).

## Honeybee Societies

Many insects are more or less social, but honeybee and ant societies are the most elaborate and widely studied. The unit of social organization is a family of related individuals. A society of honeybees typically consists of a reproductive female—the queen—and her daughters (and sometimes her sons).

A honeybee hive may contain 80,000 individuals, each with its own job. The queen lays eggs, drones (males) produce sperm needed to fertilize the eggs of new queens, and workers (sterile females) tend larvae, clean the hive, and forage for food (Figure 41–25). The tasks a bee performs,

(a)

(b)

**FIGURE 41–25** The life and times of honeybees. (a) Workers tending larvae (white objects). (b) A swarm of honeybees on a tree. (a, Robert and Linda Mitchell; b, D. Cavagnaro/DRK Photo)

# ESSAY: *Chimpanzee Societies*

The social structure of chimpanzees differs from that of most other social animals. This is of interest not only in itself, but also because chimpanzees are the nearest living relatives of humans. Chimpanzee behavior has some interesting parallels with, and differences from, human behavior.

Chimpanzees live in groups with 50 or more members, occupying territories of 10 to 30 square kilometers of tropical rain forest in Africa. A chimpanzee spends more than three fourths of its feeding time eating fruits, preferring figs, which are rich in protein. Chimpanzees also eat various other plant parts and insects, and they hunt mammals, including monkeys (which also reduces competition for fruit!).

In one study, the territory of a group contained 100 different species of trees, but only ten of these species bore edible fruit. Tropical trees produce fruit at unpredictable intervals, and the crop is ripe only for a short time. Hence, chimpanzees spend much of their time searching for trees with fruit. They must also fend off fruit-eating bats, birds, and monkeys, which are smaller and more agile in the treetops. Chimpanzees walk on the ground and then climb the fruiting trees, rather than moving through the branches of the forest canopy as monkeys do.

Few trees bear enough fruit to feed many chimpanzees at once. Hence, when chimpanzees travel in search of food, they go alone or in groups of three or four animals. By scattering in different directions, each little group manages to find enough food while minimizing the energy spent traveling. If a group finds a tree with more than enough fruit to feed its members, the males give loud cries, which soon attract other chimpanzees to feed on the tree, too.

Chimpanzees spend their "resting" time in larger groups. Much of this time is spent grooming: searching another animal's fur and picking lice off the skin. This closeness is thought to promote social bonds, but it also keeps the animals healthy by removing parasites that might carry disease. A grooming animal mostly picks lice off parts of the body that the animal being groomed cannot see or reach easily itself—the head, neck, and back.

In most social animals, the group's nucleus is the females, but in chimpanzees it is the males. Females leave home and join new groups on reaching maturity (Figure 41-A). Males remain in the community of their birth. Thus, a troop's males are related to each other, whereas its females may or may not be related. As males mature, they become the troop's defenders and the fathers of the new offspring. Hence, the territory is, in effect, held and passed down through the male line.

As a result of this social structure, it is to a male chimpanzee's advantage to promote the welfare not only of himself, but also of other members of the troop. The other adult males are his relatives, and the group's young are his own offspring or the offspring of his relatives. By helping any of these individuals, who share many of the same genes as his own, the male contributes to the welfare of many copies of his own genes. Hence, a male's calling when he finds an abundant food source, which attracts other troop members, is selected for.

An adult female, on the other hand, can best ensure her own evolutionary success by looking out for herself and her own offspring. A chimpanzee's rank in the troop, and therefore its access to food, mates, and assistance, is determined largely by its mother's status. It is therefore selectively advantageous for a female to maintain or increase her rank by behavior that commands the "respect" of other individuals, such as finding food efficiently and defending juveniles effectively.

The social structure of chimpanzees probably explains why males tend to spend more time with other males, females with other females and their offspring. It also accounts for the fact that male chimpanzees do not compete intensely for opportunities to copulate with females in estrus and so become fathers of the next generation. Males often ignore copulations occurring only a few meters away, and an estrous female may copulate with all the males in a party within a

---

the number of queens produced, and the founding of a new hive are all organized by pheromones.

A honeybee queen takes one mating flight, during which she mates with one to several males. She stores the sperm and uses them to fertilize the thousands of eggs she lays during her life of seven years or more. Her eggs hatch into larvae that are cared for by the workers. The diet fed to a larva determines whether it develops into a queen or a worker. A new worker usually first serves as a nurse, preparing cells for eggs and feeding the larvae after they hatch. After about two weeks the worker becomes a housebee, cleaning, secreting wax for the honeycomb, and guard-

short time. However, a dominant male sometimes takes an estrous female "on safari" apart from the group, thereby excluding other males from mating during her receptive period.

Chimpanzees have ten times more friendly than aggressive interactions with members of their own community. They give preferential treatment to individuals who were early associates, such as older siblings and their mother's mother and sisters. This is learned behavior, as is shown by the fact that early associates are favored even if they are not genetically related, for example in the case of adopted infants and unrelated females favored by their mother.

Chimpanzees are generally hostile to members of other groups. Groups of females have been observed to drive away "foreign" females. However, male aggression is much more frequent and more brutal. A community's males patrol the boundaries of their territory and keep members of other groups out. Males encountering a foreign female and her young have been observed to kill the infant. This behavior is selectively advantageous to the male on two counts: first, it eliminates competition from unrelated genes. Second, having lost her infant, a female quickly comes into estrus, and so the murderer or his relatives may be able to mate with her and father her next offspring if she remains in their territory. Hence, natural selection favors perpetuation of genes that make a male more likely to kill young born into other groups and of genes that make a female more likely to stay well within the boundaries of the territory protected by the males of her community. Males of one community have even been known to kill the males of a smaller group and take over their territory—a form of warfare.

These observations of chimpanzee behavior give much food for thought about our own behavior and its genetic consequences.

**FIGURE 41–A** A juvenile chimpanzee.

ing the hive. After this she forages outside the hive for the remaining five or six weeks of her life.

As we would expect from such a complex society, honeybees communicate extensively. Karl von Frisch found that foraging bees returning from a successful trip "dance" on the honeycomb, recruiting other bees to harvest a new good food source and indicating where to find it. Pheromones permit bees to identify their own hive and serve as alarm signals. A pheromone produced by the queen prevents the workers from producing any more queens and ensures that all the female larvae are fed so that they develop as workers. This continues until the hive is overcrowded, when the queen stops producing that particular pheromone, and the workers start to raise new queens.

Eventually the old queen may leave, for only one queen can survive in a hive. As she leaves, the queen secretes a swarming pheromone, which attracts many of the workers and keeps them with her. The swarm lands somewhere and may remain several days while scouts search for a new site. The scouts return to "dance" a description of the location of a possible new nest. The intensity of her dance conveys the scout's impression of the site's merits. Other workers go to inspect the sites, and finally a consensus emerges when all the scouts are dancing for one site. The swarm then flies to the new site and settles in. This method of making a decision impresses us by its resemblance to the way humans sometimes cooperate.

## Vertebrate Societies

Like insect societies, most vertebrate societies consist of genetically related individuals. Unlike insect societies, however, all members of the vertebrate society are fertile, and competition to reproduce is important in determining the social system. A typical vertebrate society consists of related females and their offspring plus unrelated males who join the society for shorter or longer periods. An interesting exception is chimpanzees: here a territory belongs to a group of related males, and females born in other families move into the area (*Essay: Chimpanzee Societies*).

Most vertebrates learn a much greater proportion of their behavior than do insects, and this difference affects social behavior. Young vertebrates learn hunting skills or seasonal location of food and water from older individuals. Juveniles also usually play with each other and with adults. Play behavior contributes to habituation and perfects motor skills through trial-and-error learning. Members of vertebrate societies can also identify each member of the group individually, whereas insects probably cannot.

In most vertebrate societies, there is a dominance hierarchy, or "pecking order," which ensures that dominant individuals have first choice of desirable but limited resources such as food, shelter, or mates. An individual's role in the society is largely determined by its position in the hierarchy, and so ability has more effect on an individual's role in a vertebrate society than in an insect society. Individuals usually fall in rank as the result of age or disability. In one baboon troop, the top male changed five times in two years. Position in the hierarchy is not determined solely by an individual's size or fighting ability, however. In many species (probably most primates), having a mother of high status gives one an initial boost up the social ladder.

Threat displays and related behavior patterns are important in maintaining dominance hierarchies. A dominant individual displaces a subordinate by threat behavior. A sub-ordinate responds with appeasement gestures, which inhibit other animals from attacking. Appeasement gestures show submission, often by turning away weapons (teeth or beak) or by presenting the vulnerable throat or belly to the dominant animal.

The evolutionary advantage of a social hierarchy is probably that it reduces the harmful effects, such as injury from fights, of the inevitable competition between related individuals living in the same area. The hierarchy also ensures that at times when resources such as food are in short supply, some individuals will get all the food they need instead of the whole group becoming half-starved and likely to die, as happens with honeybees. When members of a society are related, such behavior will be selected for because individuals carry many of the same genes, and an individual that starves while a relative lives to reproduce is actually contributing to the survival of many of his or her own genes in future generations.

## SUMMARY

An animal's genes interact with its environment in the development of sensory, nervous, and muscular systems that produce adaptive behavior. Behavior may be hard-wired into the nervous system, programmed by local or changing environmental conditions, or a combination of the two.

1. The immediate reason that an animal behaves in a particular way is that it has been exposed to environmental stimuli that induce the behavior pattern while the animal was in the appropriate physiological state. Ultimately, behavior patterns that must be produced perfectly at the first exposure to the stimulus are usually innate. So are those that must be performed in the same way by every member of the species.
2. The genes that an animal inherits determine the range of behavior patterns it can develop. In addition, most behavior patterns, innate or learned, develop normally only if the animal is exposed to the appropriate environmental conditions.
3. Learning requires time and energy and is reserved for behavior that must be flexible in meeting local or changing conditions.
4. Many behavior patterns, both innate and learned, become programmed into the nervous system. These stereotyped behaviors may be triggered by sign stimuli and controlled by a small number of neurons with minimal sensory feedback.
5. Animals are always exposed to a variety of stimuli, which may or may not evoke a response, depending on factors such as the animal's physiological state. Conflict behavior, frequently seen in courtship and territorial displays, is one possible outcome of mutually exclusive behavioral tendencies.

6. Many animals have behavior patterns that involve the ability to navigate across land and sea. It appears that the most important senses used for long-distance navigation are a combination of the biological clock with the ability to detect the direction of the sun, which permits an animal to move in a constant compass direction, and magnetic sense organs, which can be used to determine both compass direction and position on a "map."

7. Animals make adaptive decisions about which behaviors to perform, using cues that help them maximize the benefit and minimize the cost of the behavior.

8. Most animals seldom cooperate with other members of their own species, but true societies have evolved in some species of insects and of vertebrates. A society usually consists of genetically related individuals, whose cooperation enhances the survival of genes they share. Communication between individuals is most highly developed in social animals.

9. Vertebrate societies are characterized by hierarchies that determine an individual's access to limited resources. The society provides an individual with protection and the experience of older members, and its members cooperate in various aspects of their lives.

## Self-Quiz

From the list of types of behavior patterns below, choose the one exemplified by each of the following situations in questions 1–8.

a. appeasement
b. dominance
c. imprinting
d. insight
e. conditioned reflex
f. stereotyped behavior
g. territoriality
h. threat
i. latent learning
j. trial and error

_____1. A male cardinal attacks any other male cardinal that tries to come into your yard.

_____2. A puppy rolls on its back when a strange adult dog growls at it.

_____3. A cat meeting a strange (and not overly large or fierce) dog arches its back, fluffs its fur, and hisses.

_____4. A student in a typing class makes fewer errors on the tenth homework assignment than on the first.

_____5. Your signature looks much the same every time you write it.

_____6. Newly hatched ducklings follow a windup toy as if it were their mother.

_____7. One chicken pecks any other member of the flock that gets close, but no other chicken pecks it back.

_____8. A skilled musician can play a tune she or he has never heard before, after someone hums a few bars.

9. Courtship behavior is said to show conflict because:
   a. the two mates fight frequently
   b. the mates cannot immediately agree on a nest site
   c. the mates are sexually attracted to each other but do not normally permit another animal to get as close as copulation requires
   d. the mates must choose each other from a large number of members of the opposite sex
   e. the mates are in competition for food in a territory of limited size

10. Which of the following is *not* true of stereotyped behaviors?
   a. They may be triggered by sign stimuli.
   b. They are initiated by one or a few neurons.
   c. They exhibit latency.
   d. They can be learned or innate.
   e. None of the above

## Questions for Discussion

1. Despite everything you have read in this chapter, you probably still think that most of an insect's behavior is innate and that most of a human's or chimpanzee's is learned. Can you justify this position?

2. What are some of the probable selective advantages to defending a territory, an activity that consumes time and energy and increases the risk of injury?

3. William Dilger studied the nest-building behavior of parakeets of the genus *Agapornis*. He crossed members of a species that carries nest-building material in its beak with members of a species that carries nesting material tucked under the tail feathers and observed the behavior of the hybrid offspring. These offspring showed hybrid behavior and usually dropped the material whether they carried it in their beaks or their feathers. One particular bird tried to build a nest 48 times and failed. On the next try it was successful. What does this tell you about whether nest-building behavior in this genus is innate or learned?

4. Octopuses are solitary animals as adults, so researchers were surprised to find that they can learn by imitation. Italian scientists placed a red and a white ball in the tank with each octopus and trained some of the octopuses to attack the red and others to attack the white ball by rewarding them with food. Then untrained octopuses were positioned in nearby tanks where they could see the trained octopuses performing. The untrained octopuses were then given red and white balls to see if they would attack the same colored ball as the octopuses they had watched. The observer octopuses not only chose the right ball but they learned more quickly and made fewer mistakes when they learned by observation than did the octopuses that had been trained by trial and error by humans. Should this learning-by-observation be classified as insight or latent learning? What might its adaptive advantage be?

## Suggested Readings

Fitzgerald, G. J. "The reproductive behavior of the stickleback." *Scientific American,* April 1993.

Ghiglieri, M. P. "The social ecology of chimpanzees." *Scientific American,* June 1985. A fascinating study of our nearest relatives.

Gould, J. L., and P. Marler. "Learning by instinct." *Scientific American,* January 1987. An excellent account of how the innate tendencies of different animals to learn different kinds of information is adaptive for the species' way of life.

Gwinner, E. "Internal rhythms in bird migration." *Scientific American,* April 1986. A biological clock is involved in migration and navigation.

Krebs, J. R., and R. H. McCleery. *Behavioural Ecology: An Evolutionary Approach,* 2d ed. Oxford: Blackwell Scientific Publications, 1984. Contains an analysis of predicted and observed decision-making during foraging behavior.

Lohmann, K. J. "How sea turtles navigate." *Scientific American,* January 1992.

Lorenz, K. Z. *King Solomon's Ring.* London: Methuen, 1942. Delightfully written autobiographical account of life with animals.

Tinbergen, N. *The Animal in its World.* Cambridge, MA: Harvard University Press, 1972. Selections from Tinbergen's work, including experiments and general papers. A very readable description of the types of experiments you might perform if you became a behaviorist.

**PART SIX**

*Plant Biology*

CHAPTER

42

# Structure and Growth of Vascular Plants

## OBJECTIVES

*When you have studied this chapter, you should be able to:*

1. Explain how the basic structure (roots, stems, and leaves) and growth pattern of vascular plants are adapted to their functions.
2. List the three tissue systems in a vascular plant, and state their locations and functions.
3. List the three types of plant cells based on cell wall structure, and describe their functions.
4. List or recognize differences in structure and growth between monocotyledons and dicotyledons.
5. List or point out the parts of a bean seed and of a kernel of corn; state the function or future fate of each part.
6. List four functions of the root system, and compare the advantages and disadvantages of fibrous roots versus taproots in accomplishing these functions.
7. List the functions of the stem and leaves of a vascular plant.
8. State what is meant by primary and secondary growth in vascular plants, and describe the sequence of events in primary growth of a root or stem and in secondary growth of a stem.
9. Define meristem, and state the role of each of the following in the growth of a plant: apical meristem, zone of elongation,

zone of maturation, root cap, terminal or apical bud, axillary bud, intercalary meristem, pericycle, vascular cambium, cork cambium.

10. Sketch a cross section of a leaf and of a mature root and stem in which primary growth is complete, and label the following structures, if present: epidermis, root hairs, cuticle, stomata, guard cells, endodermis, pericycle, vascular tissue, pith, cortex, palisade and spongy mesophyll, air spaces; give the function of each part.
11. Briefly contrast the way branches are formed in roots versus stems.
12. Distinguish between simple and compound leaves and between alternate, opposite, and whorled arrangements of leaves; define node and internode.
13. Explain how leaf structure represents a compromise between the plant's requirements for water and for sunlight and carbon dioxide.
14. Explain how to tell the age of a twig, and identify the external features of a twig, including: bud scales, lateral or axillary bud, terminal bud, leaf scars, bud scale scars, lenticels.

---

**M**ost vascular plants are **autotrophic,** that is, nutritionally self-sufficient: they make all their organic molecules using carbon dioxide from the atmosphere, and water and minerals from the soil. This process is driven by solar energy captured during photosynthesis. A plant's structure and physiology are adapted to an autotrophic way of life.

First and foremost, plants generally spread a large surface area to intercept the energy of sunlight. Most plants produce this surface not in one vast sheet but in an array of modules, the leaves. The broad, flat shape of most leaves

provides a lot of surface area for very little weight. Usually the leaves are supported in the light by stems, which also deliver the water and minerals used in photosynthesis (Figure 42–1). A plant's stems and leaves together comprise the **shoot system.**

Besides light, leaves also absorb carbon dioxide, a gas in short supply in the air. Air enters through pores in the leaf surface, but at the same time a great deal of water vapor escapes. So plants require a lot of water, which is absorbed through the large surface area of another set of modules, the roots, which together comprise the **root system.**

Plants do not live alone but are surrounded by other organisms. The leaves of neighbors may shade a plant and so reduce the amount of energy it captures. Animals eat some of its leaves, and fungi may invade and damage others. A plant grows throughout its life, producing new sections of stem with new leaves. This replaces lost or damaged leaves and keeps up with the competition for light. Keeping pace with the growing shoot system, roots spread into new sections of soil and compete for water and mineral nutrients. Plants are said to have **indeterminate growth:** in a favorable environment, a plant becomes larger and adds new parts throughout its life.

In this chapter we follow the growth of plants and their main vegetative parts—roots, stems, and leaves—from seed to the formation of mature organs. Then we see how some plants grow in thickness.

**FIGURE 42–1** Solar panels. These leaves of Japanese maple grow in a pattern that intercepts sunlight with little overlap and shading of each other.

## KEY CONCEPTS

♦ A plant grows as dividing cells in its meristems lay down new cells, which then enlarge, differentiate, and mature as specialized cells in new plant tissues.

♦ The vascular plant body consists of roots, stems, and leaves. Each of these organs contains three tissue systems:

1. The dermal tissue system forms an outer protective layer.
2. The vascular tissue system transports water and food in the plant body.
3. The ground tissue system fills in between the other two.

## 42–A HOW PLANTS GROW

Most of a plant's body consists of mature, differentiated cells, which may have specialized functions. Normally, these cells will not divide again. But the plant can grow indefinitely thanks to **meristems,** tissues with cells that remain able to divide and produce new cells. Some of these cells differentiate and become new parts of the plant; others remain meristematic.

The growth of plants can be divided into two aspects, resulting from the activities of different meristems. **Primary growth** is growth in length of the shoots and roots, production of new **laterals** (root and shoot branches), and production of leaves (Figure 42–2). A root or shoot grows from an **apical meristem,** a group of dividing cells at its tip, or apex. The cells laid down during this primary growth enlarge and differentiate, making up the primary plant body.

**Secondary growth** adds new tissue to the primary plant body. This accounts for almost all the growth in girth, or thickness, of the stems and roots produced by primary growth. The meristems that produce secondary growth lie under the surface of stems and roots. Some of the secondary tissues resulting from secondary growth strengthen the plant and support new primary growth, both in height and in the spread of the branches. Although all vascular plants exhibit primary growth, secondary growth varies a lot: some species have none, whereas others produce extensive secondary growth.

A plant's structure depends on which parts grow, how long and how thick they become, where laterals arise, and how much the laterals grow with respect to the main shoot or root. Ultimately, a plant's size and shape reflect the activities of its meristems, which lay down the cells for each part.

## 42–B PLANT TISSUES AND CELLS

Cells in the apical meristems of root and stem tips divide and produce more cells. Cells farther back from the tip become organized into three **primary meristems,** tissues

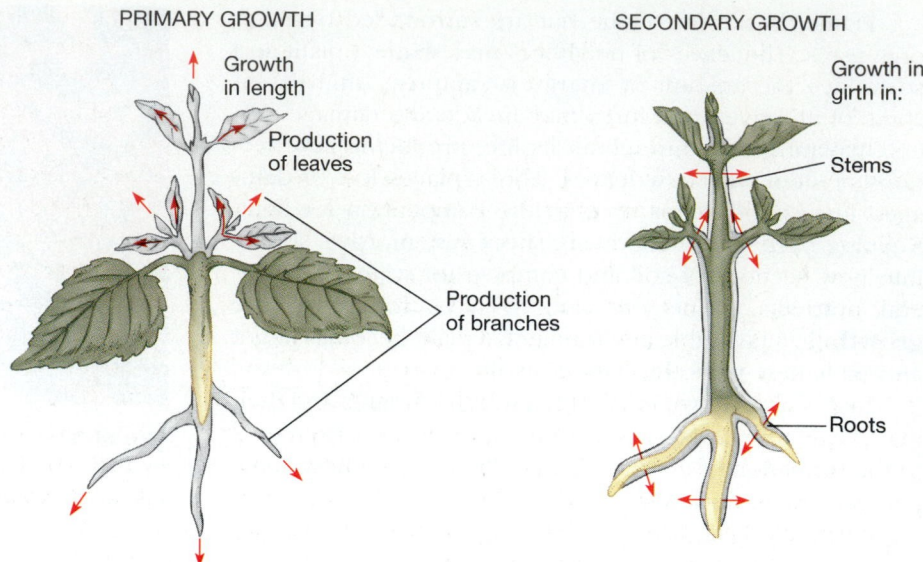

PRIMARY GROWTH

Growth
in length

Production
of leaves

Production
of branches

SECONDARY GROWTH

Growth in
girth in:

Stems

Roots

**FIGURE 42–2** Primary and secondary growth. Primary growth results from division and enlargement of cells in meristems at the tips of the plant (arrows). Primary growth produces the plant's primary tissues, which comprise the primary plant body. Secondary growth produces growth in girth (arrows).

containing cells that are continuing to divide and starting to differentiate into the tissue systems of the primary plant body:

| Primary Meristem | Produces | Primary Tissue System |
|---|---|---|
| Protoderm | ⟶ | Dermal tissue system |
| Procambium | ⟶ | Vascular tissue system |
| Ground meristem | ⟶ | Ground tissue system |

These three tissue systems are continuous throughout the plant body, although their arrangement differs somewhat in the roots, stems, and leaves (Figure 42–3):

1. The **dermal** ("skin") **tissue system** covers and protects the outside of the plant. This layer contains close-fitting cells that often form structures specialized for defense or for obtaining materials from the environment.

2. The **vascular tissue system** transports substances over long distances within the plant. There are two types of vascular tissue: **xylem** conducts water and minerals absorbed by the roots up to the stems and leaves, and **phloem** conducts food from sites of production to sites of use or storage. Conducting cells in the xylem and phloem are long and narrow, like sections of a pipe.

3. The **ground tissue system** makes up the bulk of the primary plant body, filling the space between the dermal and vascular tissues.

## Cell Types

In addition to their origin and location in a particular tissue, plant cells may be characterized as one of three types, based on cell wall structure. **Parenchyma** cells have rela-

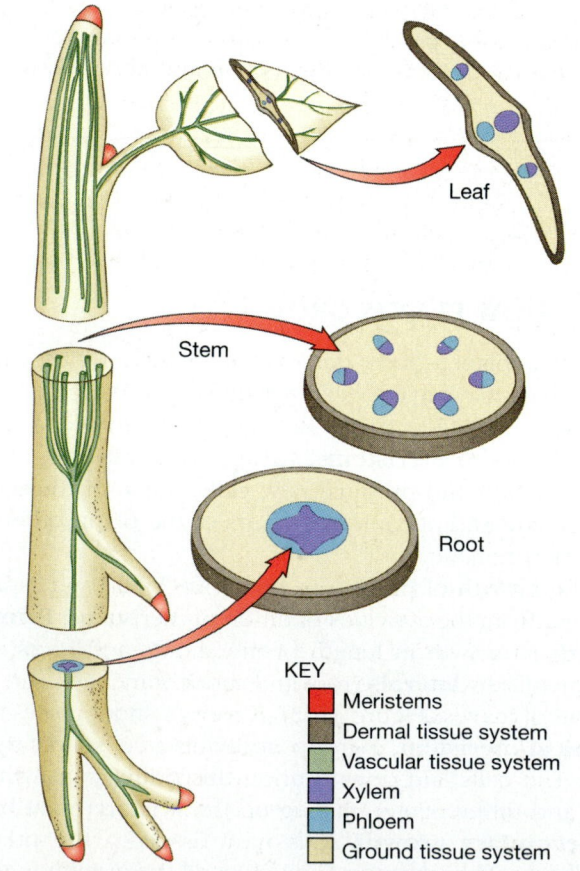

Leaf

Stem

Root

KEY

■ Meristems
■ Dermal tissue system
■ Vascular tissue system
■ Xylem
■ Phloem
■ Ground tissue system

**FIGURE 42–3** Plant tissue systems. This schematic diagram shows the distribution of the three tissue systems in the primary plant body.

tively thin walls, with no secondary thickening material laid down after the primary cell wall of cellulose is formed. These cells are rather rounded in cross section, with many spaces between cells, and they often contain many plastids: photosynthetic chloroplasts, yellow or orange chromoplasts, or starch-storing amyloplasts. The thin walls of parenchyma cells readily permit passage of light and chemicals, and these cells may carry out tasks such as photosynthesis or secretion.

Masses of parenchyma cells make up plants' soft parts, including most of the ground tissue system filling leaves, flower parts, fruits, and young roots and stems. Tight-fitting parenchyma cells also predominate in the dermal tissue system covering the primary plant body. Vascular tissue also contains some parenchyma cells. These include the specialized food-conducting cells of phloem, the **sieve tube members,** which have perforated sieve plates permitting rapid passage of food from cell to cell (Figure 42–4).

**Collenchyma** cells are column-shaped, with secondary thickening in ridges running the length of the cell wall and giving it added strength. These cells form cylinders or rope-like bundles just interior to the dermal tissue system of stems and leaf stalks and along leaf veins. The ribs of celery are rich in collenchyma, which gives them notable strength.

**Sclerenchyma** cells have varying amounts of secondary thickening in their walls and hence provide mechanical strength, support, and hardness (skleros = hard). Those with the thickest walls die after reaching full size and depositing the secondary thickening. For example, most cells in the wood used to make your pencils, paper, and desk died long before the trees were cut. Sclerenchyma includes the pipe-like cells that conduct water in xylem: long, thin **tracheids** and shorter, wider **vessel elements.** Other sclerenchyma cells are long, thin, nonconducting **fibers** in xylem, as well as rather cube-shaped **sclereids,** which give pears their gritty texture and cherry pits and thorns their hardness.

## 42–C  DICOTS AND MONOCOTS

There are two classes of flowering plants, commonly called dicots and monocots, short for **dicotyledons** and **monocotyledons. Cotyledons** are food-absorbing structures developed by the embryo. They are sometimes called "seed leaves" because they somewhat resemble true leaves in their nutritive role, flat shape, and development of chloroplasts upon exposure to light. Monocot embryos typically have one cotyledon, a condition derived from dicot ancestors by suppression of one of the two cotyledons. Monocots can usually be distinguished from dicots by their parallel leaf veins. In general, dicots and monocots also differ in several other characters; Table 42–1 lists those that are easily observed. However, there are many exceptions, in which a

species of dicot or monocot does not have one or more of the features listed for its class.

More than 70% of flowering plant species are dicots, including magnolias, oaks, maples, legumes, roses, mints, squashes, daisies, cactuses, violets, buttercups, and poppies. Almost all angiosperm trees and shrubs are dicots, as well as a wide variety of herbs: plants with only primary tissues and no woody secondary growth.

Most monocots are herbaceous: grasses, orchids, lilies, onions, irises, and flowering bulb plants such as tulips, daffodils, and crocuses. There are a few monocot trees, such as palms and Joshua trees. Although monocots comprise fewer species, they have tremendous economic importance because the seeds of the grasses known as cereals (rice, wheat, corn, oats, barley, rye, and so on) are the staples in the diets of most human societies.

**TABLE 42–1**

**Comparison of Dicotyledons and Monocotyledons**

| Characteristic | Dicotyledons | Monocotyledons |
|---|---|---|
| Cotyledon | Two | One |
| Flower parts | Five, four, their multiples, or irregular | Three, multiples of three, or irregular |
| Leaf veins | Usually netted or fan-like | Usually parallel |
| Primary vascular bundles in stems | Single ring | Scattered |
| Secondary growth and wood | Present in many, including some annuals | Absent in most |

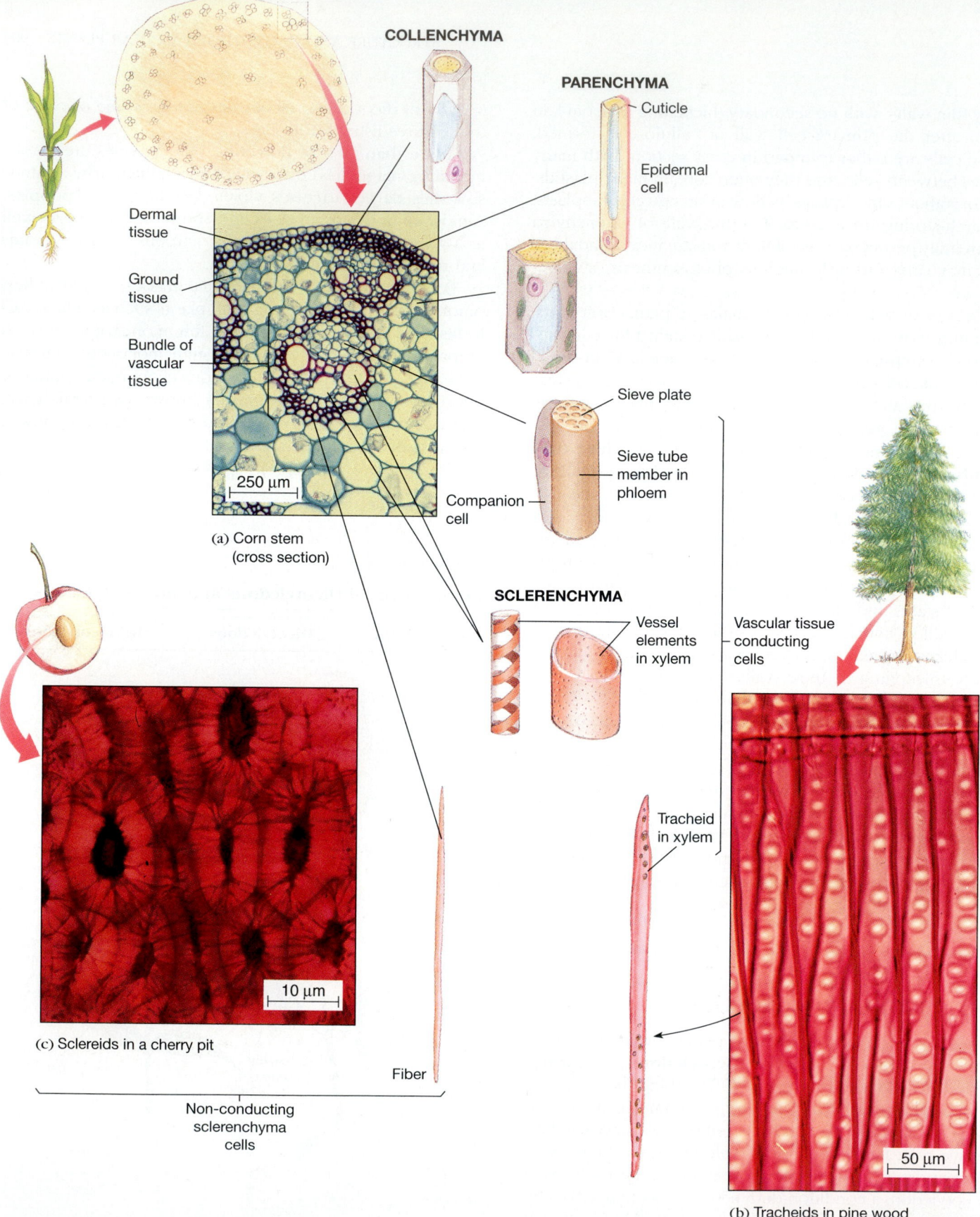

**COLLENCHYMA**

**PARENCHYMA**

Cuticle

Epidermal cell

Dermal tissue

Ground tissue

Bundle of vascular tissue

Sieve plate

Sieve tube member in phloem

Companion cell

250 µm

**(a) Corn stem (cross section)**

**SCLERENCHYMA**

Vessel elements in xylem

Vascular tissue conducting cells

Tracheid in xylem

10 µm

**(c) Sclereids in a cherry pit**

Fiber

Non-conducting sclerenchyma cells

50 µm

**(b) Tracheids in pine wood**

**FIGURE 42–4** Plant cell types. (a) Cross section of a corn stem, showing parts of the dermal, ground, and vascular tissue systems, and cells typical of each. (b) Tracheids are the kind of xylem conducting cell found in nonflowering vascular plants and also occur in some flowering plants. (c) Sclereids in a section of a cherry pit.   (a, James Mauseth; b, John Cunningham/Visuals Unlimited; c, Dennis Drenner)

## 42–D   SEEDS

A seed is a reproductive package containing a plant embryo in an arrested stage of development, along with its food supply, both wrapped within a protective **seed coat** (Figure 42–5). In the seeds of flowering plants, a special tissue, the endosperm, absorbs nutrients from the parent plant. The stored food is used by the germinating embryo until the seedling becomes self-sufficient. We begin our study of plant growth by examining two familiar seeds: the bean, a dicot; and a kernel of corn, a monocot.

The "skin" of a bean is the seed coat, composed of tightly packed sclerenchyma cells. By peeling this coat off, we can see the bean embryo within. The two "halves" of a bean are the greatly enlarged cotyledons, which have digested, absorbed, and stored most of the nutritive en-

dosperm. The seeds of many other dicots have some food stored in the cotyledons, but much remains in the endosperm surrounding the embryo.

If we separate the cotyledons of a bean, we can see that they are part of the embryo, attached to its tiny main axis. The **hypocotyl** is the part of the axis just below the cotyledons' place of attachment. Farther down is the embryonic root (**radicle**). Above the cotyledons is the **epicotyl** (or **plumule**), including the future shoot tip and the first pair of leaves. The bean embryo has two apical meristems, one at the tip of the stem in the epicotyl, between the first leaves, and one near the tip of the embryonic root.

The corn kernel is convenient for studying the structure of a monocot seed. A kernel of corn is actually a one-seeded fruit. Its "skin" contains both the seed coat and the

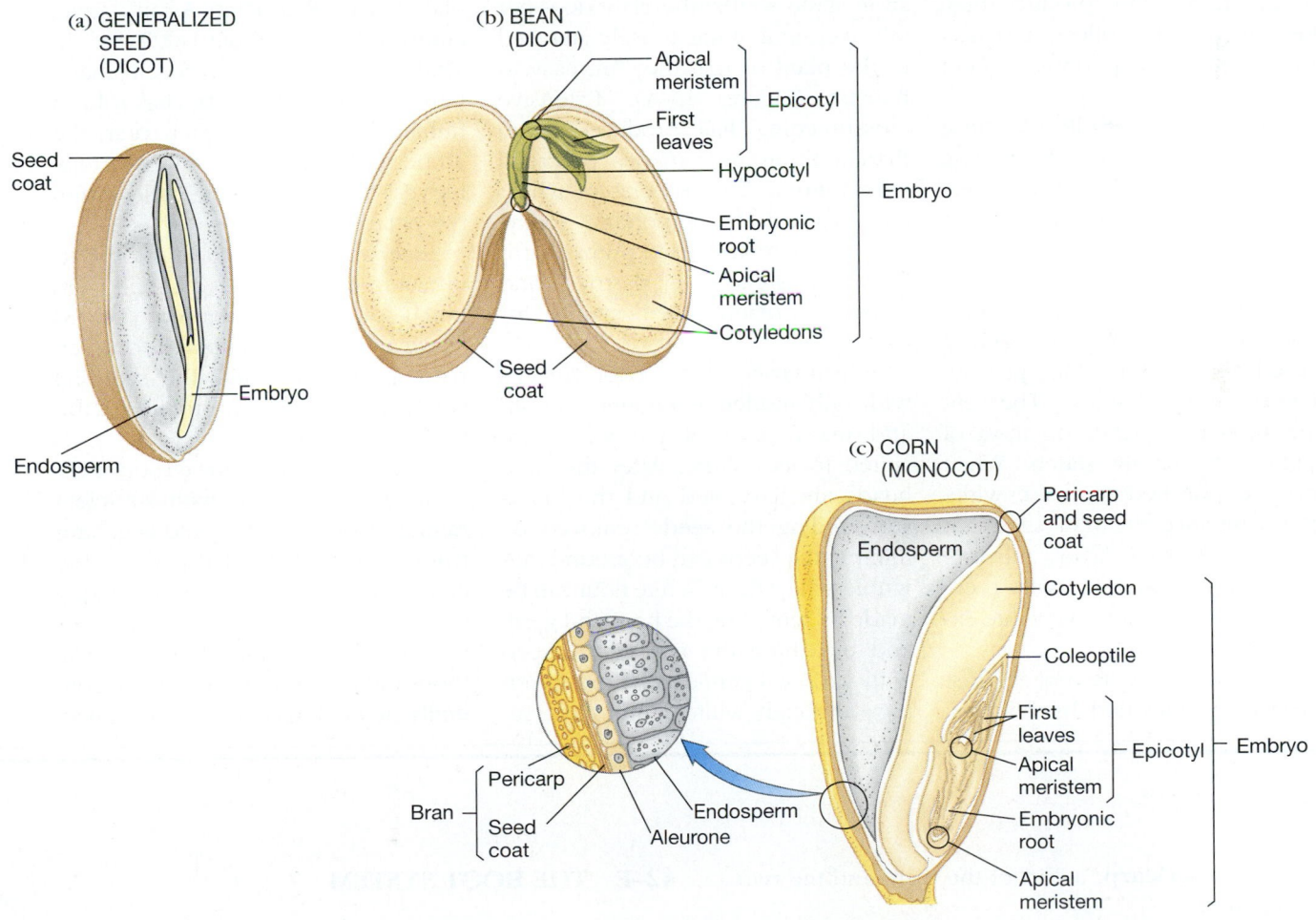

**FIGURE 42–5**   Seeds. (a) Dicot seeds have two cotyledons, and in many the endosperm surrounds the embryo. (b) A bean seed, a dicot seed in which the endosperm's stored food has been absorbed into the cotyledons, and the embryo now fills the seed coat. The apical meristem of the shoot is hidden between the first leaves. (c) A kernel of corn contains a monocot seed. The embryo has one cotyledon, and the nutritive endosperm lies outside the embryo. The aleurone layer stores proteins and fats. It also digests the endosperm during germination, providing the embryo with energy for growth.

# ESSAY: *Grains*

The vast majority of human food comes from only four plant species: wheat, rice, corn (maize), and potatoes. The first three of these are **grains,** or cereals, fruits from monocots of the grass family. The grains we grow for food also include barley, rye, and oats.

Grains have been cultivated since the beginning of agriculture, and today they occupy more than 70% of the world's farmland. It is no wonder that civilizations became so dependent on grains: they produce high yields, are easy to collect, and may be stored for long periods without spoiling.

Grains are easy to collect because they grow in clusters, developing from the ovaries of the many flowers in a compound flower head. Each fruit (grain) contains a seed, with a layer of stored protein and fat (aleurone) just inside the seed coat (see Figure 42–5c). The embryo (germ) of the seed also contains fats, protein, minerals, and vitamins. The endosperm, which makes up most of the grain, is mainly starch. When grains are polished to make white flour or white rice, the bran (seed coat plus fruit), aleurone layer, and embryo are removed and with them most of the grain's nutritive value except its starch.

Wheat and barley, two of the first plants to be cultivated by humans, were the staple foods of early empires in the Middle East. Today, wheat is the most widely grown plant, and barley is used mainly as animal food and as the source of malt for making beer.

## Wheat

Today's wheat bears little resemblance to its wild ancestors. Over the years, farmers and plant breeders have selected for varieties with particular genetic characteristics: resistance to disease, rapid germination, short stalks so that the plant does not fall over, and grains loosely attached to the plant so that they are easy to harvest (Figure 42–A). Extensive crossbreeding has produced more than a thousand varieties of bread wheat alone. Two species of wheat are particularly important today: *Triticum aestivum,* used primarily for flour for bread and pastries, and *Triticum durum,* used mainly for pasta.

Wheat grows best in cool climates with only moderate amounts of rain, and many parts of the world are suited to its culture. After the seed heads are harvested and the bracts surrounding the seeds removed by milling, the seeds can be ground into whole-wheat flour. White flour can be made by removing the bran and germ and bleaching the rest of the seed with various chemicals before grinding. Although white flours have su-

perior baking and keeping qualities, they are less nutritious than whole wheat. Therefore, many countries require producers to enrich white flour with vitamins and minerals, expensively replacing some of the nutrients removed during processing.

## Rice

Although rice is not planted as widely as wheat, it feeds more people, being the basic food of half of humankind, particularly in heavily populated Asia. Rice is an Asian native, cultivated since about 4000 B.C. in Thailand, and it is revered in many parts of the East for its vital role in human life. In Japan each year, the emperor himself joins in the ritual harvest of rice on a paddy field within the palace grounds.

Rice was introduced into the Carolinas in 1647 but was banned from much of the southeastern United States during the campaign to eliminate malaria, because the standing water in rice paddies provides habitat for mosquitoes.

Much rice is now grown on well-drained soil in areas with sufficient rainfall. However, the yield is greater from traditional flooded paddies because paddies contain nitrogen-fixing bacteria, which promote heavy yields of rice without added fertilizer. The thousands of varieties of rice are generally divided into two major groups.

---

closely attached **pericarp,** a part of the fruit, and the two peel off together.

In the corn seed, much of the nutritive endosperm tissue remains outside the embryo, whereas in beans it has mostly been absorbed into the cotyledons. The embryonic shoot (epicotyl) has several developing leaves, wrapped above the apical meristem. A tough tubular leaf, the **coleoptile,** covers the more delicate photosynthetic leaves and apical meristem.

## 42–E   THE ROOT SYSTEM

### Primary Growth of Roots: Growth in Length

As a seed germinates, the first part of the embryo to start growing is the embryonic root. In the root apical meristem lies a central **quiescent zone,** where little or no cell division occurs (Figure 42–6). All around this region are the meristematic cells, which divide and produce more cells.

(a)

(b)

**FIGURE 42–A** Grains: (a) wheat and (b) rice, ready for harvest. (a, Biophoto Associates; b, Grant Heilman/Grant Heilman Photography)

Japonica types have short grains and are sticky when cooked; indica types have long grains and are drier.

When it is prepared industrially, brown rice is whitened by removing its outer layers. As with wheat, this removes much of the nutritional value and leads to malnutrition among people whose diet consists almost entirely of white rice. The most widespread deficiency disease is **beriberi,** caused by a lack of B vitamins and leading to blindness and death.

### Corn (Maize)

Corn, native to Mexico, was first cultivated there and then spread north and south, becoming the most important cultivated plant of prehistoric America. Native North Americans were introduced to corn and beans by traders who brought seeds north from Mexico. Corn is not as good a human food as wheat. Like rice, it has a lower protein content, and it also contains too little of the amino acids tryptophan and lysine and of the B vitamin niacin.

Deficiencies of these nutrients cause the disease **pellagra,** characterized by weakness so acute it is often diagnosed as mental deficiency. Pellagra was common in the United States among the rural poor until the 1950s, and it is still occasionally seen. Plant breeders continue the search for high-vitamin corn, with some recent success. In Central and South America, where corn is a staple food, it is usually eaten with legumes such as beans, which make up for its nutritional deficiencies (and corn supplies nutrients lacking in beans).

In the United States, about 80% of the corn grown is fed directly to animals, but in Africa and Central and South America nearly all of it is used directly as human food. It is also used in some 200 products ranging from ethanol and cough syrups to tires. Because it grows so well in the United States, researchers are constantly looking for new uses for this plant.

---

Cells produced toward the tip of the root become part of the **root cap,** a thimble of cells protecting the growing root tip. Cells in the root cap perceive the direction of gravity and respond in such a way that the root grows downward. The root cap's outer cells secrete slime, a polysaccharide material that lubricates the root's passage between the rough soil particles. Even so, as the root cap is pushed through the soil, the outermost cells are scraped off. The lost cells are replaced by continuous cell division in the apical meristem.

New cells produced on the other side of the apical meristem develop into part of the root proper. These cells become arranged in definite strands. Above this area is the **zone of elongation,** where the cells of the primary meristems continue to divide more slowly and grow in length, pushing the root cap and apical meristem through the soil. This is the actual force of growth in the root.

Above the zone of elongation, cells in the **zone of maturation** have reached full size and are differentiating into specialized cells. These mature cells comprise the root's **pri-**

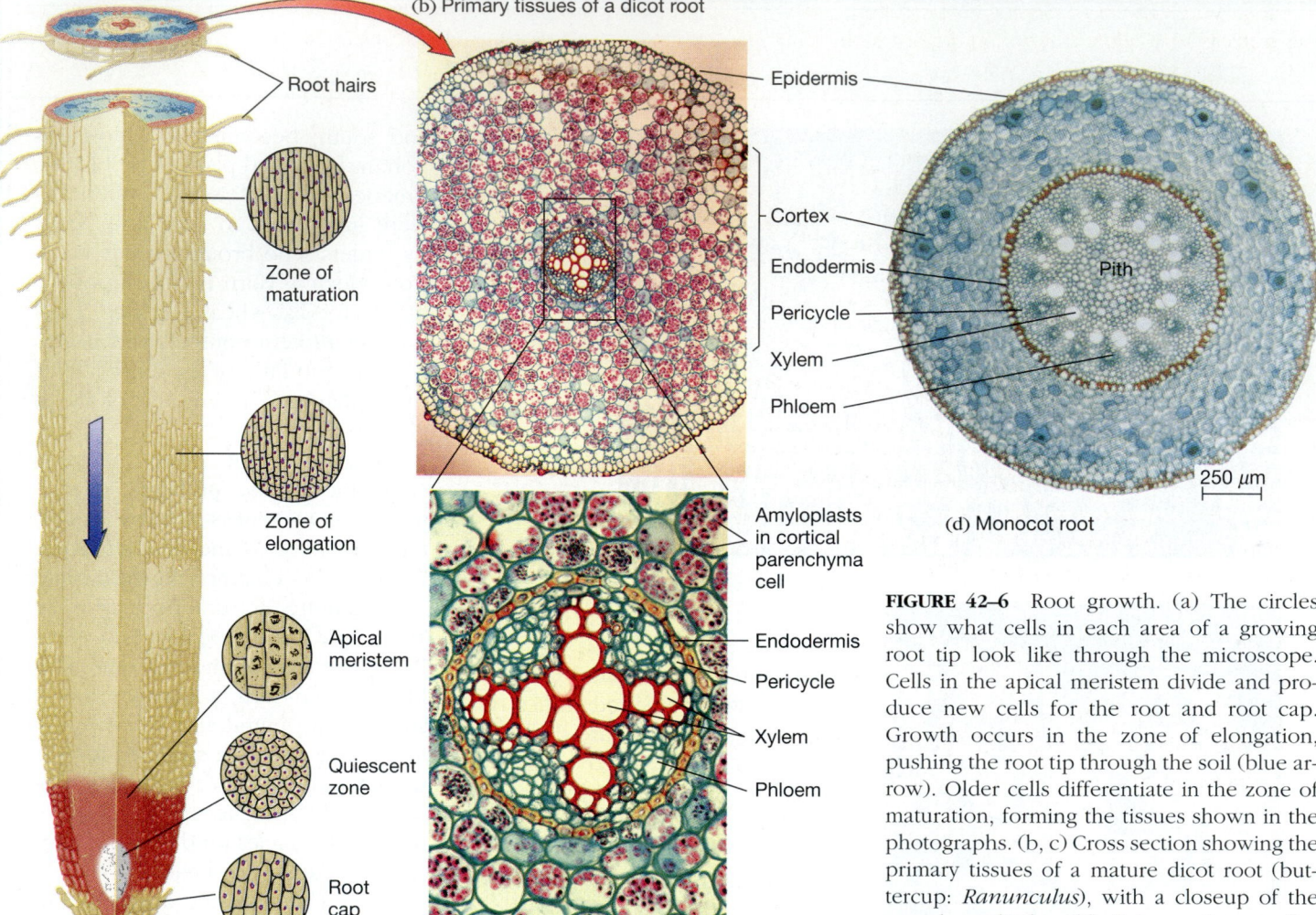

(b) Primary tissues of a dicot root

Root hairs

Zone of maturation

Zone of elongation

Apical meristem

Quiescent zone

Root cap

(a) Growing root tip

Epidermis

Cortex

Endodermis

Pericycle

Xylem

Phloem

Pith

250 μm

(d) Monocot root

Amyloplasts in cortical parenchyma cell

Endodermis

Pericycle

Xylem

Phloem

(c) Closeup of vascular cylinder

**FIGURE 42–6** Root growth. (a) The circles show what cells in each area of a growing root tip look like through the microscope. Cells in the apical meristem divide and produce new cells for the root and root cap. Growth occurs in the zone of elongation, pushing the root tip through the soil (blue arrow). Older cells differentiate in the zone of maturation, forming the tissues shown in the photographs. (b, c) Cross section showing the primary tissues of a mature dicot root (buttercup: *Ranunculus*), with a closeup of the vascular cylinder. (d) Primary tissues of a monocot root (corn: *Zea*). (b, c, Ed Reschke; d, Dennis Drenner)

**mary tissues,** that is, the tissues laid down by the activity of the apical meristem. In this zone grow **root hairs,** fine extensions of some cells at the root surface. Root hairs grow out among the soil particles, anchoring the plant in the soil and increasing the surface area for absorption of water and minerals. It is important that root hairs form in the zone of maturation rather than in the zone of elongation. If root hairs were formed where cells were still elongating and pushing through the soil, the delicate extensions growing out sideways between soil particles would be pulled off.

The apical meristem continues to divide and produce new cells, which follow the same sequence of division, growth, and development. So, at any one time, cells in different parts of the root tip are of different ages and are therefore at different stages of development: the farther back from the root tip, the older the cells.

## Primary Structure of Roots

At maturity, the cells laid down by the activity of the apical meristem form the root's primary tissues. The dermal tissue system consists of the outermost layer of tissue, the **epidermis.** Epidermal cells fit together tightly and form a protective covering, one cell thick, around the root. Some of them produce root hairs.

The bulk of the young root consists of the ground tissue system (Figure 42–6). The **cortex,** just interior to the epidermis, is made up of many parenchyma cells, which may contain starch-storing **amyloplasts.** There are many spaces between the cells of the cortex.

The innermost layer of the cortex is a single layer of cells, the **endodermis** ("inner skin"). It has a special structure that helps control the movement of substances between the cortex and the root's interior (Section 44–C).

Just within the endodermis is an irregular layer of cells, the **pericycle.** Cells in the pericycle are relatively undifferentiated and remain meristematic.

Interior to the pericycle are the vascular tissues, the root's xylem and phloem, which transport substances to and from the shoot system. Conducting cells of the phloem differentiate early and can then bring in food used in the growth and development of cells near the root tip. Before transport cells in the xylem become functional, they grow to their final size and die, leaving a hollow cell wall, which conducts water upward.

We shall examine the structure and function of vascular tissues in the next chapter. For now, note that they usually form a central core in dicot roots. In cross section, the xylem commonly forms a central "star," with the phloem in pockets between the arms of the xylem star. A cross section of a monocot root usually has many more "arms" of xylem (corn has up to 20), and in some the center of the star is not vascular tissue but a core of parenchyma, the **pith.**

## Primary Growth of Roots: Production of Laterals

The first root of a bean seedling grows quickly and soon begins to produce side branches, or laterals. Cells in the pericycle, which remained meristematic when their neighbors differentiated, divide and form a new apical meristem.

As the innermost cells elongate, they push the new meristem out through the endodermis, cortex, and epidermis and on into the soil (Figure 42–7a). This new branch root follows the same pattern of growth outlined before and becomes a lateral root with the same kinds of primary tissues. The new roots, in turn, may give rise to more branch roots. This produces a branching root system that establishes the plant firmly and absorbs water and minerals from an increasing volume of soil.

Some roots do not arise from existing roots. **Adventitious roots** are formed by meristems growing from other parts of the plant. For example, a corn plant's prop roots arise adventitiously from the base of the stem and provide support once they reach the ground (Figure 42–7b). Many climbing plants form tiny roots on the undersides of their stems as they grow up tree trunks or brick walls. African violets can form roots from the undersides of leaves placed on moist soil. Cuttings of many plants form adventitious roots if the stems are placed in water or moist soil.

There are two basic types of root systems. In a **taproot system,** typical of many dicots, the embryonic root develops into one main root, the **taproot,** by far the plant's longest and thickest root, with distinctly shorter and thinner side branches. Because its growth is concentrated into one axis, a taproot may grow quite deep. The taproot of an old grapevine may extend 15 meters down, reaching a reliable supply of water far beneath the soil surface. Many

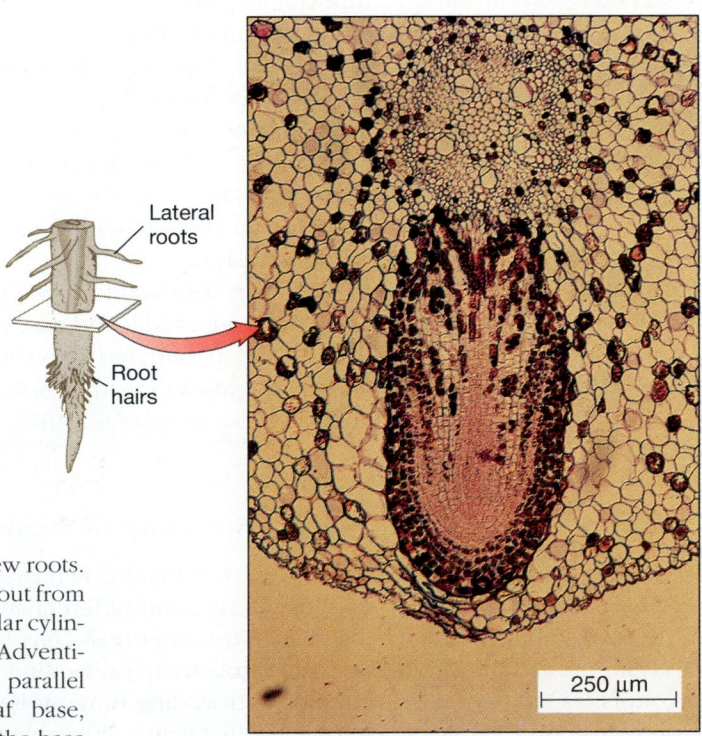

**FIGURE 42–7** Formation of new roots. (a) A new lateral root growing out from the pericycle around the vascular cylinder of an existing root. (b) Adventitious roots in corn. Note the parallel veins in the lowermost leaf base, which forms a sheath around the base of the stem.   (a, Biophoto Associates)

Lateral roots

Root hairs

250 μm

(a) New lateral root (LM)

(b) Adventitious roots

**FIGURE 42–8** Two types of root systems. Left, a taproot system; right, a fibrous root system.

kinds of trees have taproots when young and develop a more branching root system later.

In a **fibrous root system,** typical of most monocots and of some dicots, the embryonic root dies soon after germination, and root primordia at its base produce many roots of about equal size, each with smaller laterals (Figure 42–8). These roots branch throughout a large volume of soil, from which they absorb water and minerals. Plants with fibrous root systems, such as grasses, are good at holding soil in place and therefore controlling soil erosion.

### Functions of Roots

We have seen four main functions of roots:

1. **Roots anchor the plant in the soil.** Taproots are especially good anchors, since they are long and so thick and tough that it is hard to break them. Fibrous roots are good at holding soil in place.
2. **Roots absorb water and minerals from the soil.** Fibrous root systems often expose more absorptive surface area to the soil than do taproot systems, but taproots may grow deep enough to reach a more reliable supply of water.
3. **Roots transport water and minerals up to the shoot system.** The xylem serves this function. The endodermal layer exerts some control over the passage of substances into or out of the vascular tissue in the root interior.
4. **Roots store food.** The root cortex is the primary food storage area in some plants. Carrots, radishes, and beets are familiar examples of massive food-storing taproots. This stored food is why we eat them. Likewise, sweet

potatoes are food-swollen portions of fibrous root systems. In many other kinds of plants, the root stores only a small supply of food for itself.

## 42–F   STEMS

### Primary Growth of Stems: Growth in Length

Once a bean seedling has established roots, the shoot begins to grow. First, the hypocotyl forms a loop that pushes up through the soil and then straightens, pulling the cotyledons into the air (Figure 42–9). Meanwhile, the delicate apical meristem remains enclosed between the cotyledons. Once the cotyledons are above ground, they spread apart, revealing the epicotyl. The first pair of leaves, which were fully formed but tiny in the bean embryo, have already begun to expand. When they are exposed to sunlight they become green and expand more rapidly. Eventually, their photosynthesis produces enough food to supply the plant. Until then, the bean seedling continues to use the food stored in the cotyledons. In the light, the cotyledons too become green and carry out some photosynthesis. Eventually, the cotyledons shrivel as their food supply is used up, and they fall off.

The shoot's apical meristem consists of a little mound of rapidly dividing cells at the tip of the stem axis, between the first foliage leaves. It is similar to the apical meristem of the root tip, but it has no quiescent zone nor any structure similar to the root cap. The cells laid down by the apical meristem continue to divide in the primary meristems of the zone of elongation. This, rather than division in the apical meristem itself, produces most of the new cells during the stem's primary growth.

Growth in both length and diameter by cells in the shoot may take longer than in the root. Hence, growth can be detected farther behind the growing tip of the shoot's apical meristem.

When a corn seed grows, it too establishes roots first. Then the coleoptile lengthens, pushing up through the soil. Once it has penetrated into the light, the enclosed leaves and stem expand greatly, rupture the tip of the coleoptile, and grow out (Figure 42–10).

### Primary Structure of Stems

As in roots, cells laid down by the shoot's apical meristem divide, enlarge, and differentiate into the primary tissues of the stem. The outermost cells form the epidermis. Its cells fit snugly together, preventing both loss of moisture and damage by invading fungi or insects. These epidermal cells have a further water-conserving adaptation: they secrete a layer of waxy substances, the **cuticle,** which forms a wa-

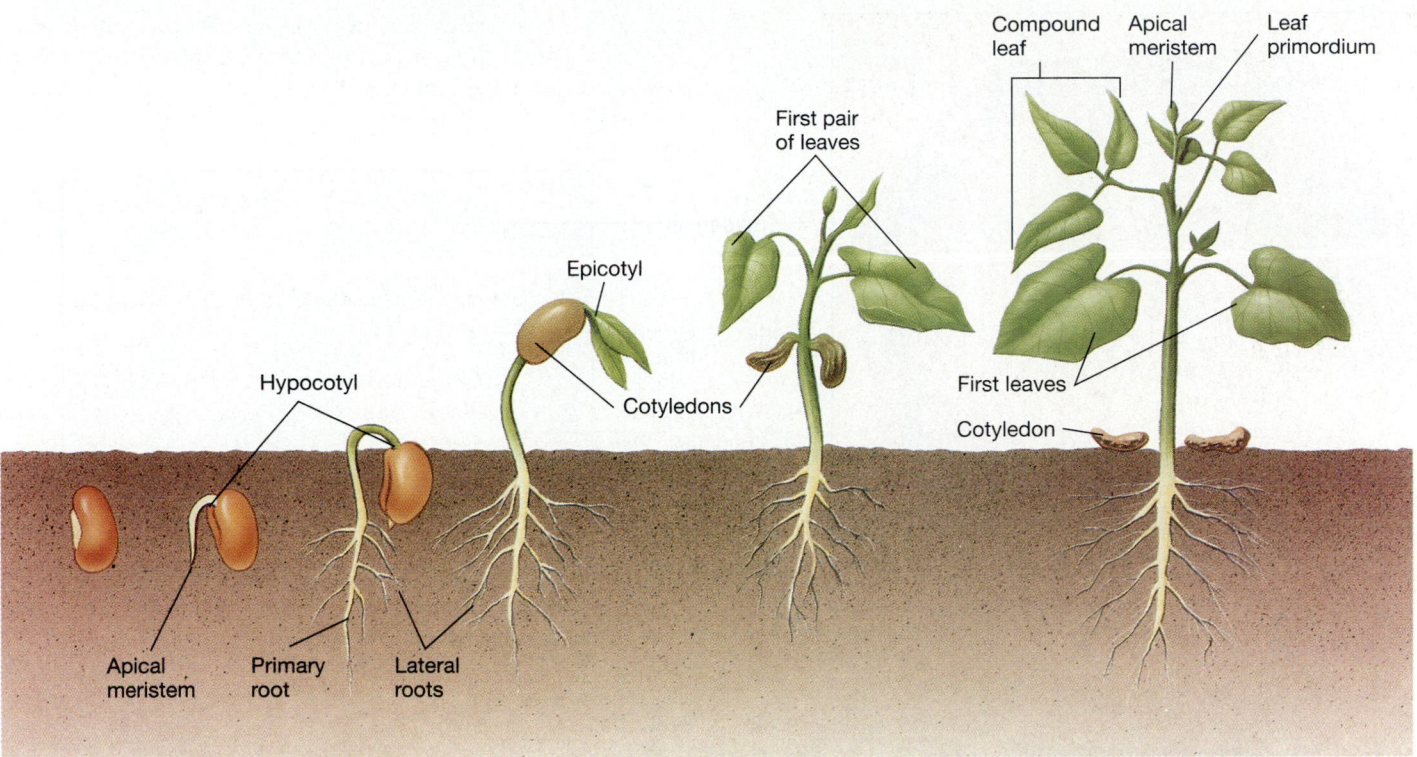

**FIGURE 42–9** Growth of a bean seed into a young plant. The root system starts to develop first, and then the shoot system grows and begins to carry out photosynthesis.

terproof covering over the surface of the stem. The cuticle of the stems is continuous with that covering the leaves.

Most of the primary tissue in the stem is ground tissue in the cortex and pith. The cortex lies interior to the epidermis. Its outer layer often contains collenchyma in cylinders or in rope-like bundles, particularly in stems with ribs or angles. However, most of the cortex consists of parenchyma cells: some contain chloroplasts and carry out photosynthesis, and some may store starch in amyloplasts.

In dicots, the center of the stem is occupied by pith, also composed of parenchyma cells. (Recall that the center of dicot roots is usually occupied by the cylinder of vascular tissue.) A stem grows as cells in the pith take up water and expand against the resistance of other, more rigid cells. The resulting turgor also provides support. As the stem continues to grow, the rapid elongation of cells may pull the pith apart, and the stem becomes hollow (see Figure 5–26).

In contrast to the pattern in roots, the stem's primary vascular tissue system forms discrete **vascular bundles:** strands of vascular tissue, each containing both xylem and phloem. In cross section, a dicot stem shows vascular bundles in a ring between the cortex and pith. Monocot stems have many small vascular bundles scattered throughout the

**FIGURE 42–10** Growth of a corn seedling. After the embryonic root grows down as the seedling's primary root, the pointed coleoptile grows upward. The coleoptile breaks through the soil surface, and light stimulates the foliage leaves within to expand, rupture the coleoptile, and spread to commence photosynthesis. (Runk/Schoenberger, from Grant Heilman)

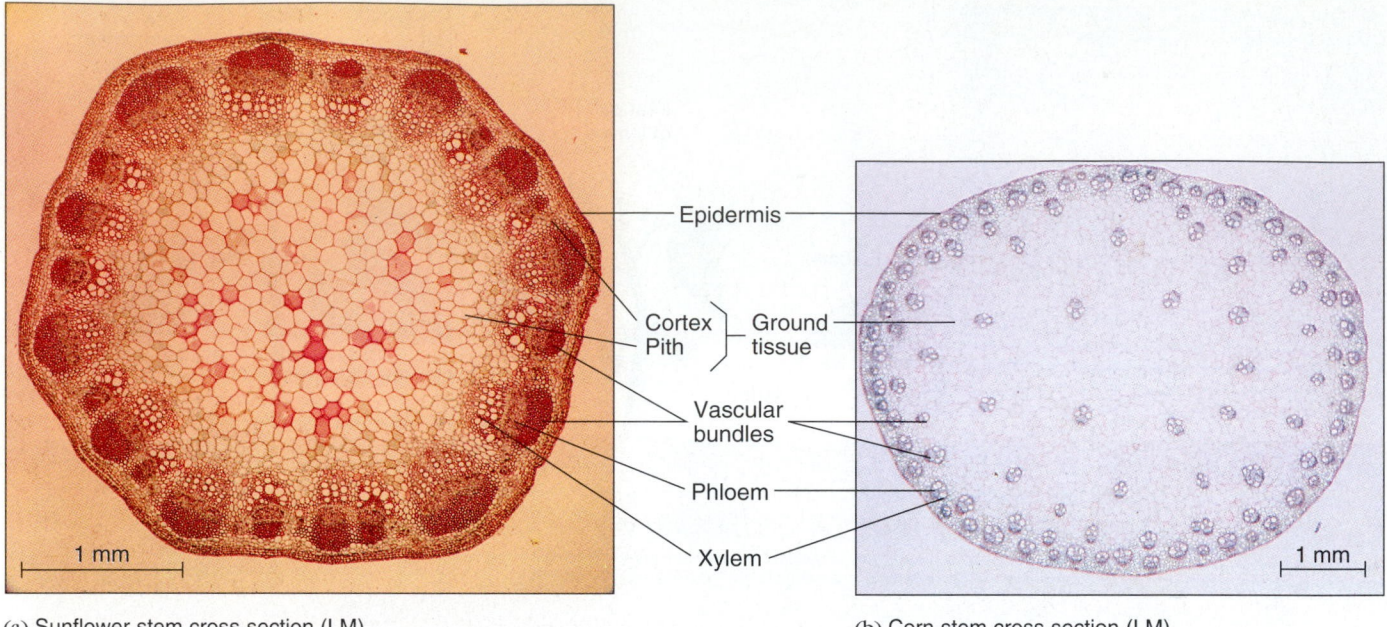

(a) Sunflower stem cross section (LM)

(b) Corn stem cross section (LM)

**FIGURE 42–11**   Tissues of mature primary stems. (a) Cross section of a dicot stem: sunflower (*Helianthus*). Most of the stem consists of pith, which lies inside the ring of vascular bundles. (b) Cross section of a monocot stem: corn.   (a, Runk/Schoenberger, from Grant Heilman; b, Dennis Drenner)

stem (Figure 42–11). Therefore, the ground tissue has no distinct pith and cortex regions. Generally, in both dicots and monocots the xylem lies in the part of the bundle toward the inside of the stem, the phloem toward the outside.

Compared with the soil around the roots, air provides little support to a plant's stems and leaves. The vascular tissue system in the shoot contains a lot of sclerenchyma, which provides strength. Each vascular bundle is partly surrounded by thick-walled, dead **fiber** cells, and the walls of the conducting cells in the xylem may also be especially thick. The strong-walled sclerenchyma and collenchyma, plus the turgor pressure of the fluid inside the parenchyma cells of the ground tissue system, all contribute to supporting the stem in the air.

Stems usually lack the endodermis and pericycle found in roots.

### Primary Growth of Stems: Production of Laterals

As the shoot's apical meristem grows, it lays down not only the cells that divide and form the primary tissues of the stem but also **leaf primordia,** which develop into leaves (Figure 42–12). **Nodes** are the parts of stems where leaves are attached; **internodes** are the stretches of stem between one node and the next. The angle between the base of the leaf and the stem is the leaf's **axil** ("armpit"). Cell division

in the axil produces a small group of meristematic cells, an **axillary** or **lateral bud,** which remains dormant for a time.

A new shoot branch arises when an axillary bud becomes active and begins to grow as the apical meristem of a new stem. These branches produce leaves with axillary buds, just as the main stem does.

While some axillary buds produce new branches, others remain dormant. Picture what the plant would look like if all the axillary buds grew: the plant would soon have many more branches than it could support, and the leaves would be too crowded. Normally, most of a plant's axillary buds are repressed by auxin, a hormone from the apical meristems, in the phenomenon of **apical dominance** (Section 45–H).

### Functions of Stems

We have seen four main functions of stems:

1. **Stems support structures of the shoot system.** The thick cell walls of the collenchyma and vascular tissue of stems form a tough framework that is often difficult to break. In addition, the parenchyma cells in the pith and cortex contribute turgor, which keeps the stem firm. With this tough, resilient structure the stem can hold the leaves up to the sunlight for photosynthesis and can bend in the wind without breaking.

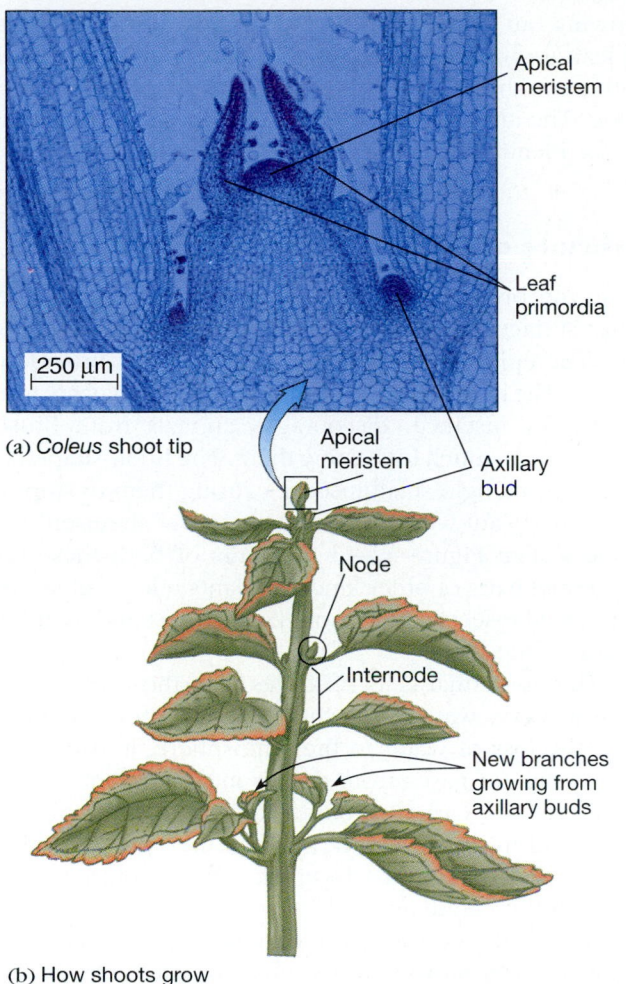

(a) *Coleus* shoot tip

(b) How shoots grow

**FIGURE 42–12**    Growth of a shoot tip. (a) Longitudinal section of a *Coleus* shoot tip. The apical meristem is flanked by new leaf primordia. Axillary buds have formed in the axils of the next older pair of leaves, which are still differentiating. (b) Growth of the apical meristem lengthens the shoot, and growth of the axillary buds produces new shoot branches.    (a, Dennis Drenner)

2. **Stems transport substances between the roots and leaves.** The vascular tissue system of stems is continuous with that of the roots and leaves. The xylem transports water and minerals taken in by the roots up to the leaves and to the living cells of the stem, and the phloem carries food from the leaves or from storage areas to the living and growing parts of the plant, some of which cannot carry on photosynthesis—for instance, roots and flowers.

3. **Stems produce food.** Some stems are green and photosynthetic. In most plants they supplement photosynthesis carried out in the leaves, but in plants such as cacti, stems are the main photosynthetic organs, and the leaves are modified as spines (see Figure 22–5).

4. **Stems store substances.** Some stems contain many amyloplasts. A potato is an underground shoot with a large stem specialized for storage. Its eyes are axillary buds. Other stems store lesser amounts of food. Some stems, such as those of some cacti, also store a lot of water.

## 42–G    LEAVES

A leaf begins as a leaf primordium laid down by the activity of an apical meristem. When mature, the leaf usually has two main parts. Vascular tissue with many supporting sclerenchyma cells makes up the **petiole,** or stalk, and its continuation, the midrib of the leaf. The photosynthetic **blade,** which is usually broad and flattened, consists of photosynthetic tissue supported by strands of vascular tissue, the **veins** (see Figure 42–13).

A dicot leaf primordium forms a slender rod of meristematic cells, which thickens and lays down the midrib, followed by thin, flat wings on either side, which develop into the blade. The shape of mature leaves varies a lot and is often a distinctive character used to identify plants. Depending on species, this shape may develop through differential division and growth of cells in different areas of the blade, selective cell death, or a combination of these. This pattern of leaf development is genetically determined, although in some species the environment influences which genetic development program occurs. For example, the leaves of the water buttercup have different shapes depending on whether they grow under water or above the surface. Unlike roots and stems, which have meristems that continue to produce new growth, dicot leaves have determinate growth: they stop growing at a genetically predetermined size.

In a **compound leaf,** the blade consists of several discrete parts, or **leaflets,** which resemble separate leaves. This appears to be an adaptation that minimizes damage to large leaves from wind, insects, or fungi, by splitting the blade into smaller, independent modules. Compound leaves can be distinguished from **simple leaves,** with undivided blades, by noting the position of axillary buds: these occur at junctions where the base of a leaf's petiole joins a stem, whereas no buds occur at the bases of leaflets (Figure 42–13).

Our bean plant has both simple and compound leaves. Each leaf in the first pair is simple and heart-shaped, with an axillary bud in its axil. However, all the leaves formed later are compound leaves, with three leaflets each (see Figure 42–9).

The bean plant also shows two different types of leaf arrangement. The first leaves are **opposite;** that is, there are two leaves at the node at a 180° angle to each other. The compound leaves formed later are **alternate,** with only one leaf per node and with the leaves at successive nodes

(a) Simple leaves          (b) Compound leaves

**FIGURE 42–13**   Leaf structure. The leaf's broad, flat blade contains photosynthetic tissue stretched on a supporting framework of highly branched veins. Simple leaves (a) can be told from compound leaves (b) by the position of the axillary bud: at the base of the petiole, where the leaf joins the stem.

growing out from the stem at different angles. A third type of leaf arrangement, not seen in bean plants, is **whorled,** with more than two leaves spaced around the stem at each node. The arrangement of leaves is often a helpful character for identifying plants (Figure 42–14).

## Structure of Leaves

As in the primary structure of the root and stem, a leaf's outer surface, both top and bottom, is covered by epidermis. The epidermis may have structures that provide protection. The itchy juice secreted by epidermal hairs of tomato plants, for instance, discourages animals from brushing against them, much less eating them. The hook-shaped hairs of beans entangle small insects, causing them to stop feeding on the plant while they struggle to free themselves. The nettle leaf in Figure 42–15 has hairs of both these types. Epidermal hairs of other kinds of plants release glue, which traps small insects and immobilizes the feet and mouthparts of larger ones.

The epidermal cells of leaves, like those of stems, secrete a waxy, waterproof cuticle. Although the cuticle retards the loss of water to the atmosphere, it also impedes the passage of gases from the air into the plant. Since the leaves require carbon dioxide from the air for photosynthesis and must release the oxygen they produce, the secretion of a continuous layer of cuticle would solve one problem but create another.

In fact, the cuticle is not continuous but is interrupted at intervals by **stomata** ("mouths"; singular: **stoma**). Each stoma is a pore between a pair of lip-like **guard cells,** which control the size of the pore (Figure 42–16). When

**FIGURE 42–14**   Arrangements of leaves.          (a) Opposite          (b) Alternate          (c) Whorled

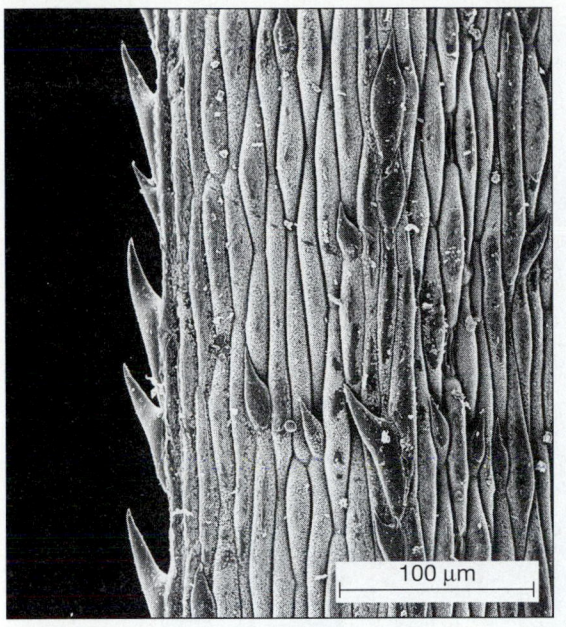

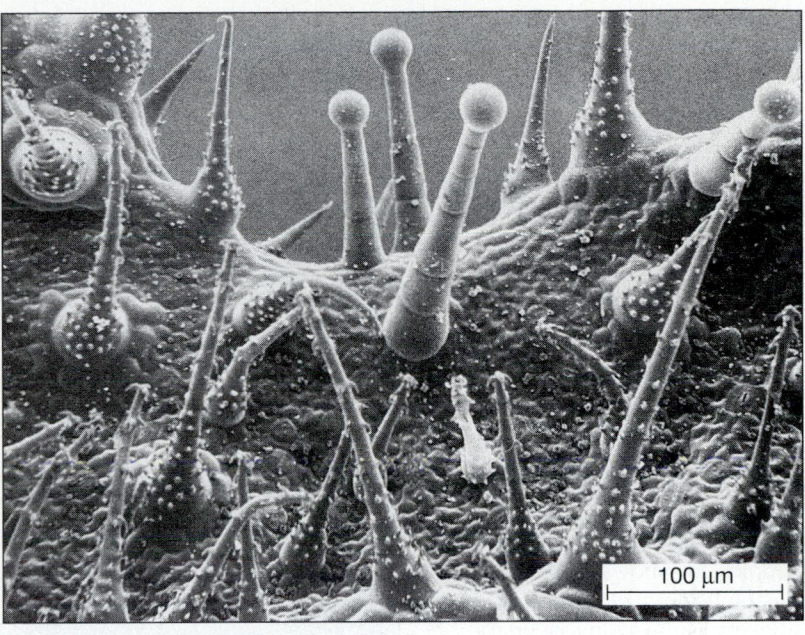

(a)                                                                                (b)

**FIGURE 42–15**   Leaf defenses. (a) SEM of a grass leaf with saw-like teeth, which can inflict a painful cut. The white flakes of wax covering the epidermis are part of the cuticle. (b) This South American nettle leaf has two kinds of epidermal hairs: one kind is pointed, with hooked ends, and the other kind secretes droplets of irritating fluid.   (Biophoto Associates)

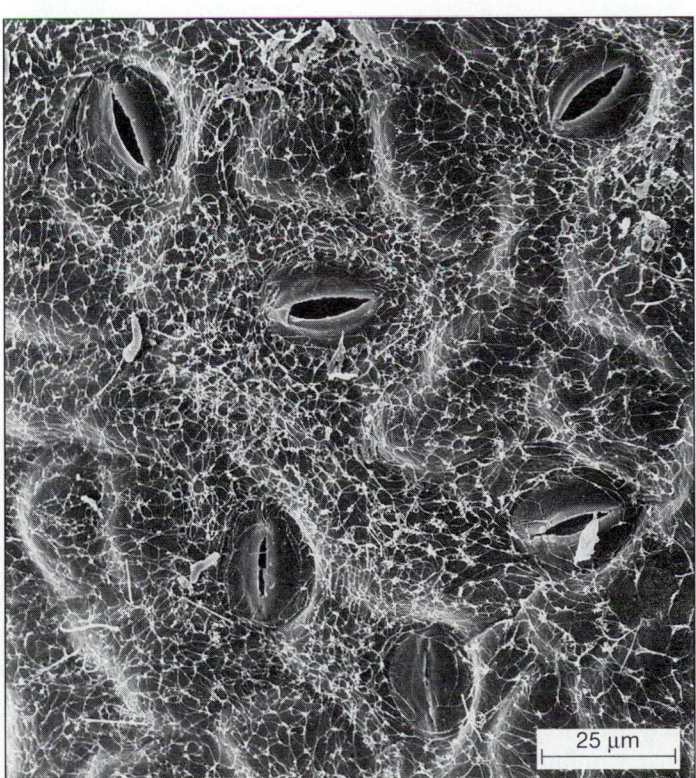

**FIGURE 42–16**   Guard cells and stomata. The waxy undersurface of a rose leaf, showing stomata varying from closed to well opened. Each stoma is surrounded by a pair of lip-like guard cells. (SEM) (Biophoto Associates)

the stoma is open, more carbon dioxide can enter the plant and more oxygen exits, but also more water vapor can escape into the atmosphere. Closing the stoma reduces the exchange of these gases between the leaf's interior and the atmosphere. Most dicot leaves have stomata only in the lower epidermis.

Stomata occur not only in leaves, but also in the epidermal layers of stems and flower parts and in aerial roots. All parts of a plant require oxygen for respiration, and some also require carbon dioxide for photosynthesis.

Sandwiched between the epidermal layers of dicot leaves are two layers of **mesophyll,** the photosynthetic ground tissue that makes up the bulk of cells in the leaf interior (Figure 42–17). The cells in these layers are modified parenchyma, with many chloroplasts.

Just beneath the upper epidermis is the **palisade mesophyll:** long, thin cells standing on end, perpendicular to the upper epidermis. There may be one or several layers of palisade cells, depending on the species. Below the palisade mesophyll is the **spongy mesophyll,** so named because of the extensive network of air spaces among the cells. The air spaces in both the palisade and spongy layers open to the atmosphere via the stomata, allowing gases to circulate to the photosynthetic cells inside the leaf.

The leaf's veins are strands of vascular tissue running through the mesophyll. The main vein, in the midrib, is continuous with one of the stem's vascular bundles and with the system of smaller veins in the leaf. The vascular

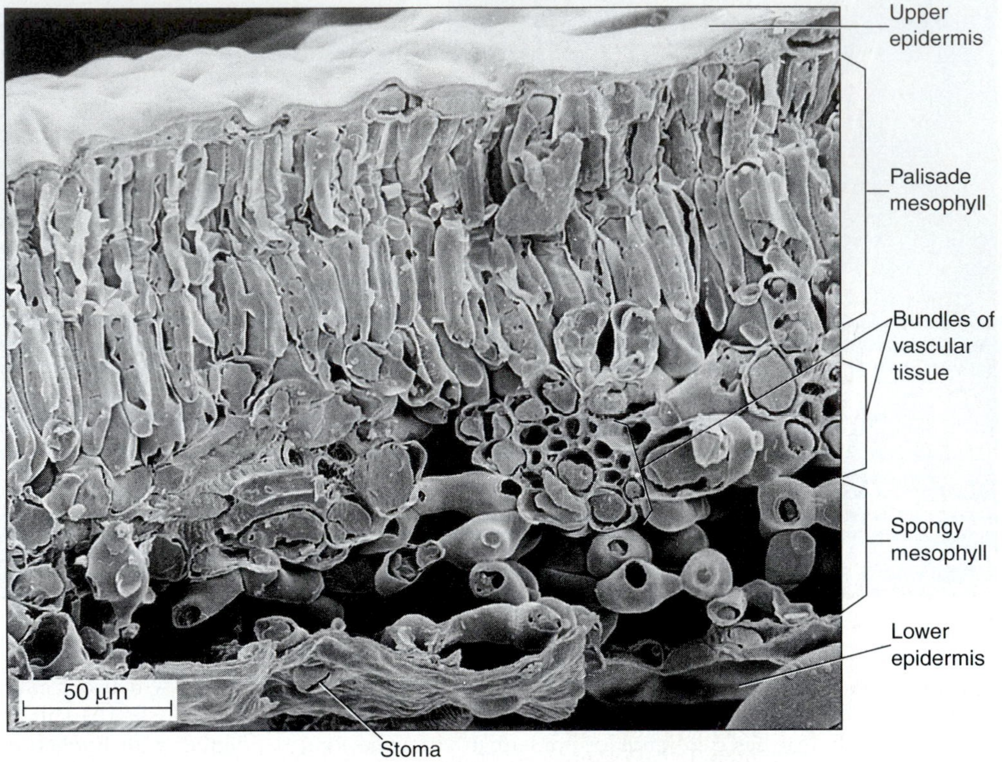

**FIGURE 42–17** Leaf tissues. This SEM of an apple leaf shows the typical structure of upper and lower epidermal layers sandwiching layers of palisade and spongy mesophyll. Veins of vascular tissue run among these photosynthetic cells. (Long Ashton Research Station, University of Bristol)

tissue conducts water to the leaves, carries away the products of photosynthesis, and provides a framework stretching the delicate photosynthetic tissues out where they can intercept the rays of the sun and the gases in the atmosphere (Figure 42–17). The pattern of leaf veins, or **venation,** varies among plant species.

A somewhat different arrangement of vascular tissue and mesophyll occurs in leaves that carry out $C_4$ photosynthesis (see Figure 8–16).

In monocot leaves, the principal veins of vascular tissue run parallel to each other, rather than in the fan-like or net-like pattern common in dicots. Monocot leaves often stand more or less upright, the mesophyll usually does not have distinct palisade and spongy layers, and stomata are plentiful in both epidermal layers. In many monocots, the leaves are long and narrow or tapering, such as the leaves of grass or onions. These leaves do not have a petiole, but the base of the leaf forms a sheath wrapped around the stem.

Grasses, lilies, and some other monocots have growth areas besides the apical meristems. **Intercalary meristems** at the bases of leaves produce new cells, and this permits the leaves to keep growing after their tips have been eaten by grazing animals or clipped by a lawn mower. This mode of growth keeps the developing cells that need the most nutrients close to the roots that supply some of these needs. These young, succulent cells have not yet toughened to maturity, but their basal position protects them to some extent from being eaten by hungry grazing animals, scorched by the sun, or desiccated by wind.

## 42–H    SECONDARY GROWTH

### Secondary Growth of Stems

Secondary growth is growth in the diameter of stems and roots by the addition of new cells. It is most familiar in woody, **perennial** (living for many years) dicots and gymnosperms. It is also found to some extent in nonwoody plants such as alfalfa, sunflowers, and some **annuals** (plants that live just one season).

The cells that form secondary tissues are produced by **lateral meristems.** One of these is the **vascular cambium,** which lays down cells that become the **secondary vascular tissues.** In the stem, cells between the primary xylem and primary phloem in the vascular bundles have retained their meristematic character, and they form part of the vascular cambium. Additional cells in the ground tissue between bundles become meristematic again. Hence, the vascular cambium can be seen in a cross section of the stem as a continuous ring of tissue, with the xylem and pith on the inside and the phloem, cortex, and epidermis on the outside (Figure 42–18b).

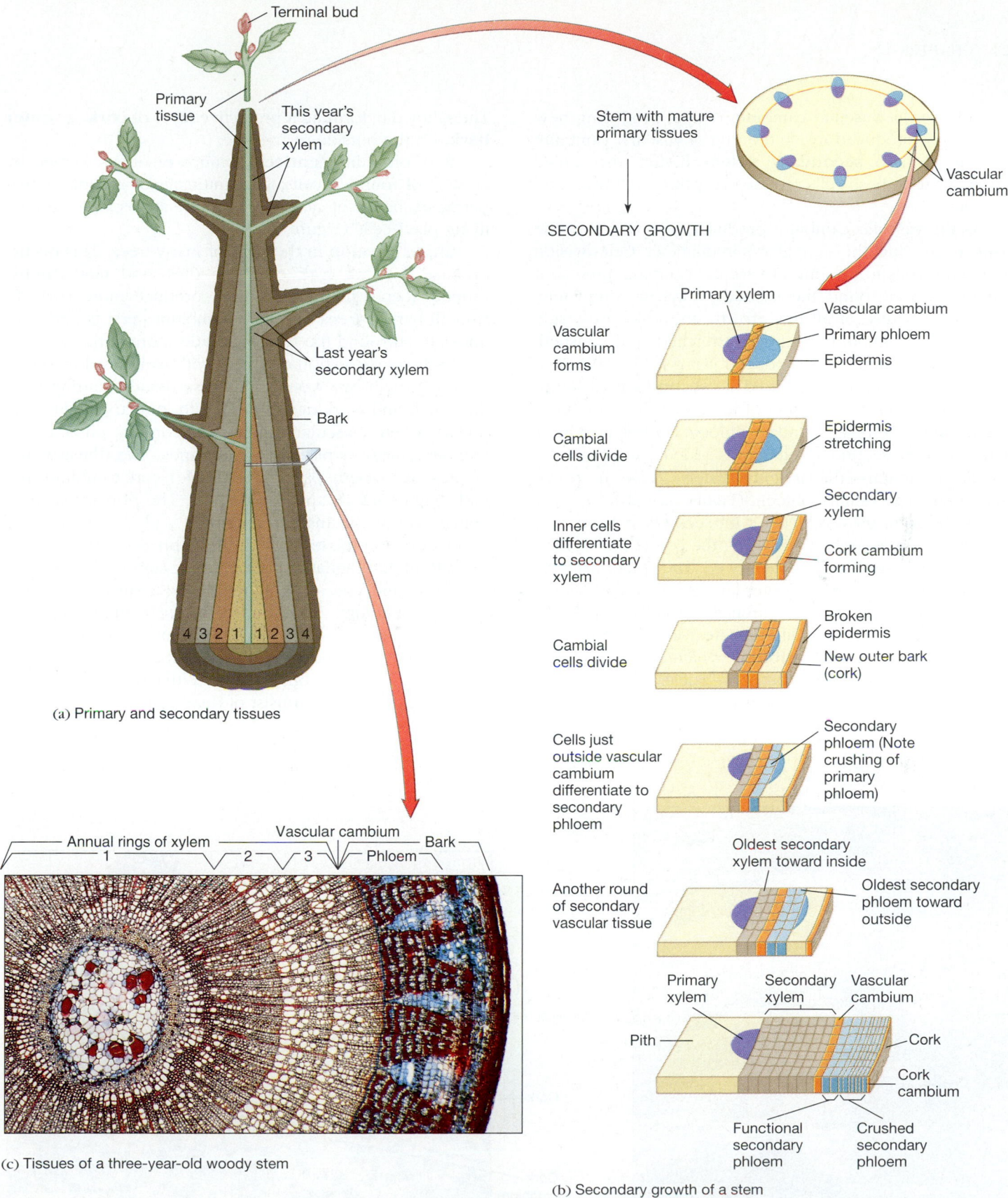

Terminal bud

Primary tissue

This year's secondary xylem

Last year's secondary xylem

Bark

4 3 2 1 1 2 3 4

(a) Primary and secondary tissues

Stem with mature primary tissues

SECONDARY GROWTH

Vascular cambium

**Vascular cambium forms**

Primary xylem

Vascular cambium

Primary phloem

Epidermis

**Cambial cells divide**

Epidermis stretching

**Inner cells differentiate to secondary xylem**

Secondary xylem

Cork cambium forming

**Cambial cells divide**

Broken epidermis

New outer bark (cork)

**Cells just outside vascular cambium differentiate to secondary phloem**

Secondary phloem (Note crushing of primary phloem)

**Another round of secondary vascular tissue**

Oldest secondary xylem toward inside

Oldest secondary phloem toward outside

Primary xylem

Secondary xylem

Vascular cambium

Pith

Cork

Cork cambium

Functional secondary phloem

Crushed secondary phloem

(b) Secondary growth of a stem

Annual rings of xylem

Vascular cambium

Bark

1    2    3    Phloem

(c) Tissues of a three-year-old woody stem

**FIGURE 42–18** *Secondary growth of a woody stem.* (a) Longitudinal section through a four-year-old tree, showing the primary plant body (green) and secondary growth for the first through fourth years. (b) Top: Cross section of a stem with mature primary tissues, showing the position of the vascular cambium. The series of closeup views shows the steps in the formation of secondary tissues: secondary xylem, secondary phloem, and cork. (c) Cross section of a three-year-old stem of basswood (*Tilia*). The secondary xylem forms distinct annual rings, each with larger cells on the inside, where more rapid growth occurs in spring. Cells produced later in the season are smaller. The phloem is part of the bark and can be removed by peeling the bark from the twig.   (c, Bruce Iverson)

Cells in the vascular cambium divide, producing new cells. Those produced inside the ring of vascular cambium differentiate into **secondary xylem** tissue, also called **wood.** Most of these cells produce very thick cell walls and then die.

As the vascular cambium produces new wood on the inside of the ring, the stem grows in diameter. Cell division adds more cells to the ring of vascular cambium, and so it continues to surround the secondary xylem completely. However, because of all this growth, the phloem outside the vascular cambium becomes stretched and crushed. Meanwhile, though, cells produced just outside the vascular cambium have differentiated as **secondary phloem,** which takes over the transport of food. As more secondary xylem forms, the first secondary phloem is destroyed in its turn, and more secondary phloem is added just outside the vascular cambium—that is, immediately *inside* the previously formed secondary phloem (Figure 42–18b).

As the stem grows thicker with secondary xylem and phloem, the outer primary tissues—the cortex and epidermis—are also stretched and destroyed. The epidermis is replaced by a new protective outer layer of secondary tissue. Depending on the species, parenchyma cells in the epidermis, cortex, or phloem differentiate, forming another lateral meristem, the **cork cambium,** which divides and produces new cells to the outside. These cells become impregnated with **suberin,** a waterproof waxy material.

Then they die, forming a protective layer of **cork,** or **outer bark,** on the outside of the tree.

Bark often has **lenticels,** bumps or ridges formed by clusters of rounded cells, with intercellular air spaces that permit exchange of gases between the atmosphere and the living plant cells (Figure 42–19).

Cracks develop in the bark of many trees. This occurs because the first cork cambium dies, and new cambia formed deeper in the bark then produce more cork. In smooth-barked trees, the cork cambium persists and continues to surround the trunk and add more bark.

In the cross section of a stem with well-developed secondary growth, we would find these tissues, starting from the center and working outward: pith, primary xylem, secondary xylem, vascular cambium, secondary phloem, the crushed remains of primary phloem and cortex (these would be present but perhaps not identifiable), cork cambium, and cork (Figure 42–18 and Table 42–2). The primary layer of epidermis has by this time ruptured and sloughed away. When you peel the bark off a tree, it breaks at the layer of delicate undifferentiated cells in the region of the vascular cambium, and you are holding all the tissues outside the vascular cambium. The trunk of a tree is almost entirely secondary xylem, with a slender column of pith and primary xylem in the center (Figure 42–18a).

From this discussion, we can see that most of the tree's wood and outer bark consist of dead cells. In essence, the

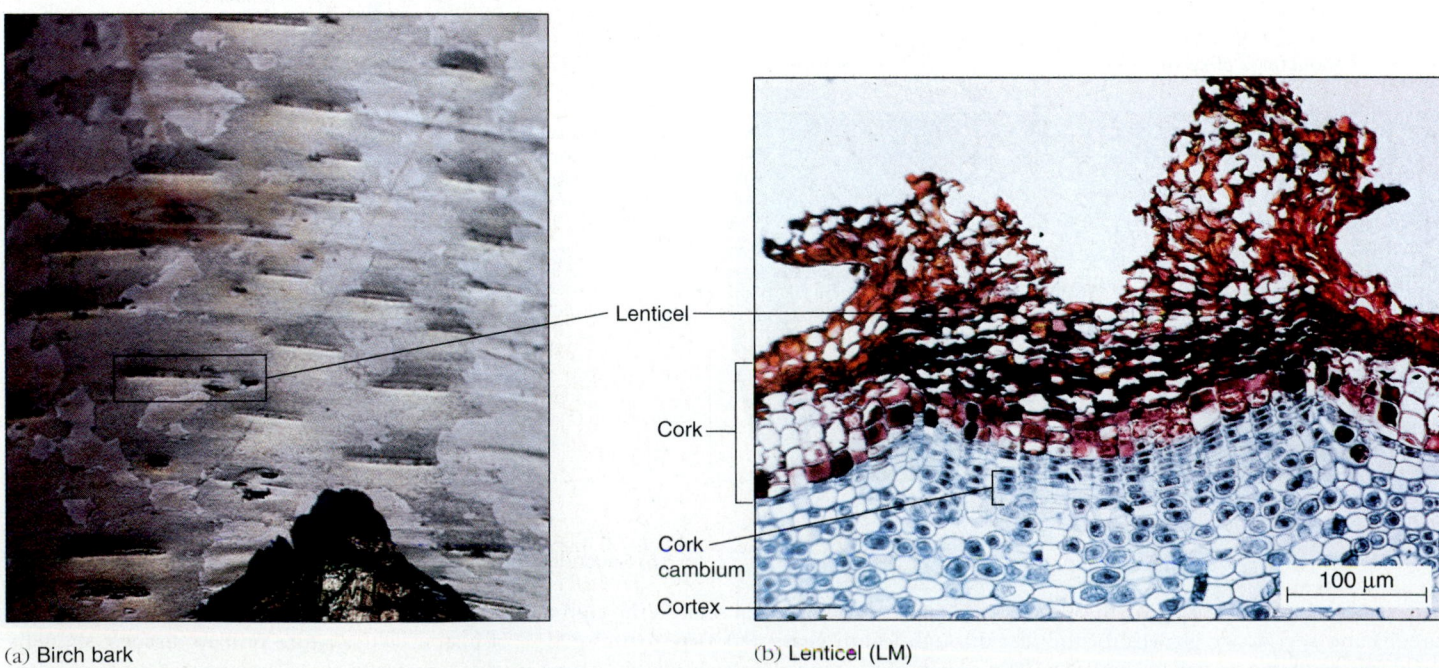

(a) Birch bark

(b) Lenticel (LM)

Lenticel

Cork

Cork cambium

Cortex

100 μm

**FIGURE 42–19** Lenticels. (a) The low ridges in the bark of a birch tree are lenticels. (b) A lenticel in the bark of an elder (*Sambucus*) twig.   (b, Carolina Biological Supply Company)

**T A B L E   4 2 – 2**

**Summary of Primary and Secondary Growth of a Woody Dicot Stem**

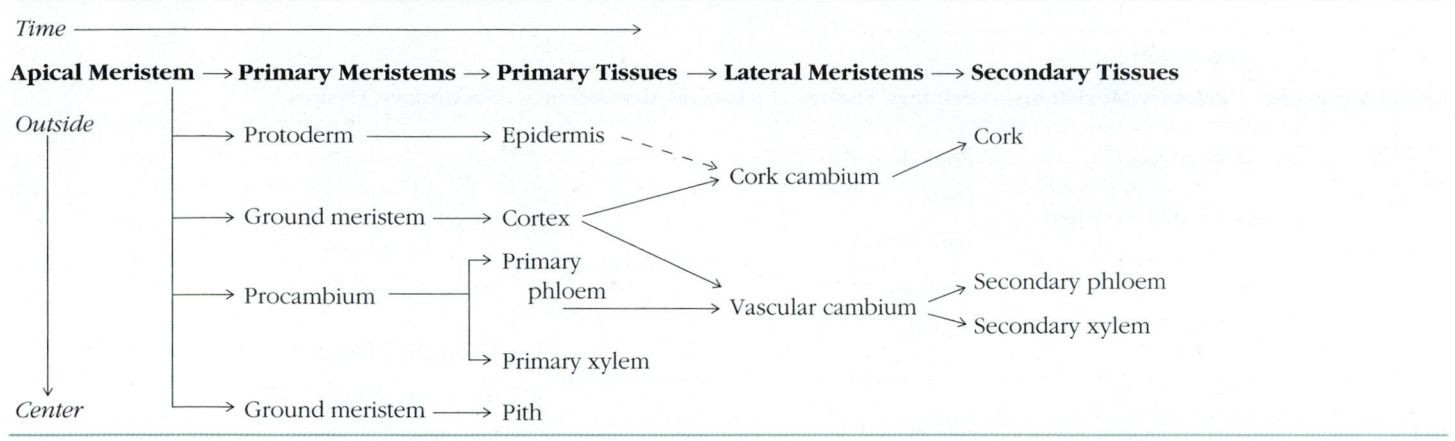

*Time* ⟶

**Apical Meristem ⟶ Primary Meristems ⟶ Primary Tissues ⟶ Lateral Meristems ⟶ Secondary Tissues**

*Outside*

Protoderm ⟶ Epidermis

Cork cambium ⟶ Cork

Ground meristem ⟶ Cortex

Procambium ⟶ Primary phloem

Vascular cambium ⟶ Secondary phloem / Secondary xylem

Primary xylem

*Center*

Ground meristem ⟶ Pith

---

living cells of the vascular and cork cambia keep growing a new tree outside the dead xylem core of its former self and inside its old, dead "skin"!

Most monocots are either herbaceous (soft-bodied, nonwoody) annuals or perennials with leaves that die back to the ground after each growing season. Except for *Joshua* trees and other yuccas, and dragon trees (which have an anomalous type of secondary growth), most monocots lack secondary growth and secondary tissue. Even the tallest monocots, palm trees, grow in diameter without secondary growth: they have "primary thickening meristems," close behind the apical meristem of the trunk.

## Structure of Twigs

The finer branches of a woody plant are the **twigs** that bear the leaves and the flowers and fruits in season. In woody plants living in areas with seasonal climates, the apical and axillary meristems pass the season of dormancy as **terminal** and **lateral buds,** respectively. Usually each bud is surrounded and protected by a cluster of **bud scales:** tough, modified leaves impregnated with wax (Figure 42–20).

The bud contains a very short stem with rudimentary leaves, flowers, or both, formed during the previous year. When the bud breaks dormancy, the bud scales are shed as these leaves and flowers enlarge tremendously. At the same time, the cells in the internodes of the new length of stem elongate, and the leaves spread in the sunlight.

The stem soon completes its primary growth and forms vascular and cork cambia, which in turn produce secondary xylem and phloem, and a layer of bark. Meanwhile, the apical meristem at the end of the twig may produce new stem growth and new leaves in addition to those originally con-

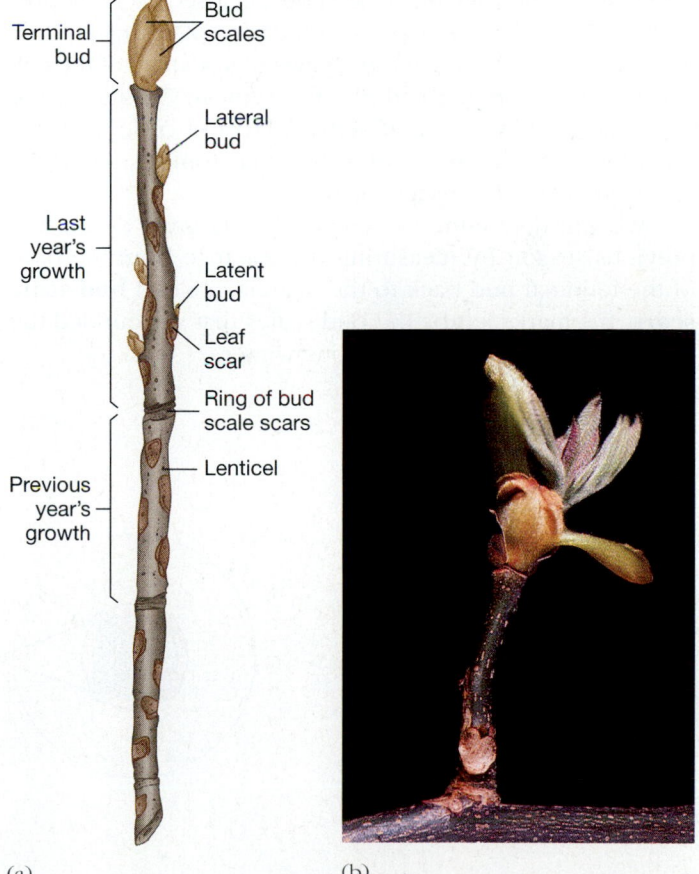

(a)    (b)

**FIGURE 42–20**    (a) External features of a hickory (*Carya*) twig. (b) A hickory twig breaking dormancy in the spring. The leaves expanding from within the terminal bud push the bud scales aside. Note examples of some other features shown in (a).

**TABLE 42–3**

**Summary of Primary and Secondary Growth of a Woody Dicot Root**

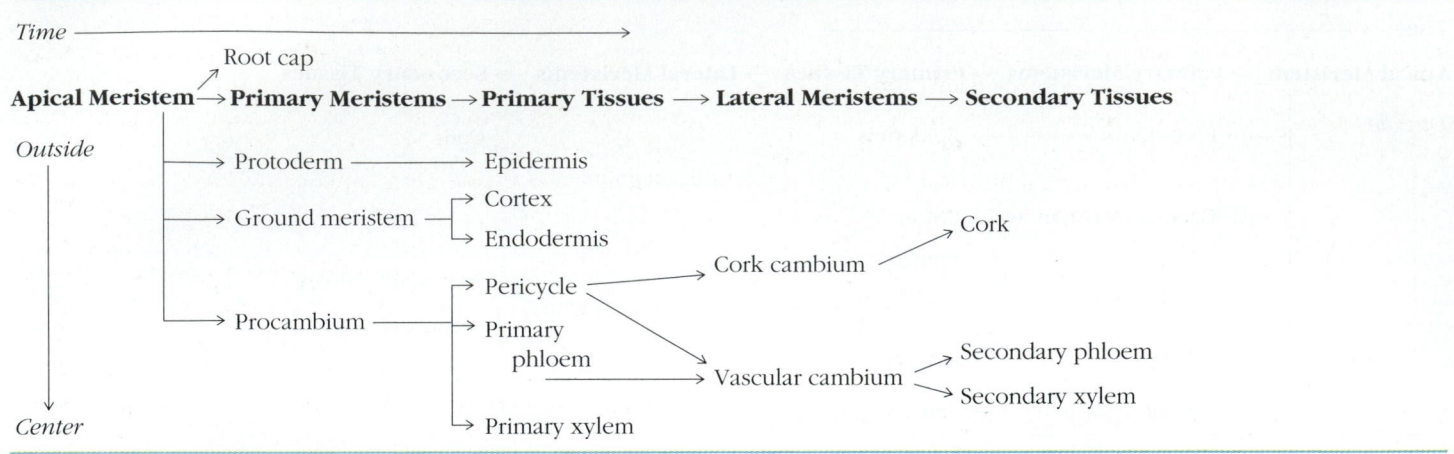

tained within the bud. Axillary buds occur in all the leaf axils. Some of these axillary buds may grow and form lateral twigs, but most remain dormant. At the end of the growing season, the buds become enclosed in their protective armor of bud scales. Near the base of each leaf, now grown old, cells divide and enzymes digest some of the cell wall material, forming a zone of weakness. Eventually the leaf falls off, leaving a **leaf scar.** A layer of corky material seals the leaf scar, protecting the twig from loss of water and from attack by insects or pathogens.

We can determine how much a twig grew during the previous season by measuring the distance from the base of the terminal bud back to the nearest circle of **bud scale scars,** the marks left by the bud scales that surrounded the terminal bud during the previous winter.

## Secondary Growth of Roots

Secondary growth in roots is similar to that in stems: vascular cambium forms between the primary xylem and primary phloem (Table 42–3). However, owing to the arrangement of the root's primary vascular tissues, these areas of vascular cambium are roughly U-shaped in cross section (Figure 42–21a). The vascular cambium is continued between these arcs by cells of the pericycle, producing a continuous layer with a complex shape (Figure 42–21b). Quite soon, however, the vascular cambium assumes a more circular cross section because secondary xylem is added more rapidly to the areas between the arms of primary xylem.

A tree's main roots are large and woody, tapering off rapidly in diameter near the trunk and then running for some distance with little change in diameter. Most trees that

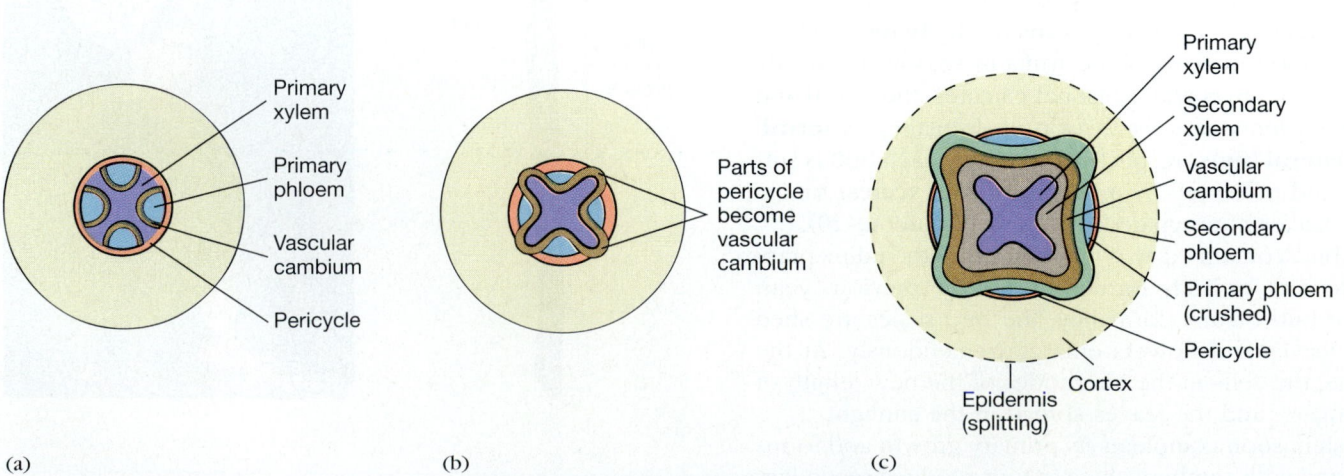

**FIGURE 42–21**  Beginnings of secondary growth in a root. (a) Vascular cambium forms between the primary xylem and primary phloem. (b) Parts of the pericycle become vascular cambium, forming a continuous layer. (c) Vascular cambium lays down secondary xylem unevenly, and the vascular cambium quickly assumes a more circular outline.

have taproots as seedlings or saplings develop a more branching root system later. Typically, many large roots spread in all directions not far below the soil surface. These roots provide support and serve as transport conduits. The longest roots might be up to twice as long as the longest stem, but most large roots are about the same length as the shoot system. Encased in their tough bark, these roots cannot absorb water and minerals from the soil, a task performed by young roots at the far ends of the root system. About 80% of the roots of even a large tree lie in the top 20 cm of soil.

## SUMMARY

A vascular plant grows in one place, making its own food and competing with other plants for sunlight, water, and minerals. The plant body consists of roots, stems, and leaves, all containing vascular tissue, which conducts water, minerals, and food rapidly from one part of the body to another. This body plan is adapted to obtaining water and sunlight efficiently. The plant's indeterminate growth pattern enables it to produce new organs throughout its life; if it is successful in exploiting its environment, it can continue to grow and obtain more resources.

1. A seed consists of a plant embryo with a rudimentary root, stem, and leaves, plus its food supply, both enclosed in a protective seed coat.
2. Early in life, all cells of the plant can divide. Later, most cells mature, specialize, and lose their capacity to divide. Cells that remain meristematic occur at specific locations. Cells laid down by division in meristematic tissue enlarge and differentiate, forming the dermal, ground, and vascular tissue systems.
3. Primary growth is principally growth in length and production of new root and shoot branches.
4. The tips of roots and stems grow longer through the activity of apical meristems. Meristematic cells divide, and newly formed cells elongate and then differentiate into specialized cells of the tissue systems in the mature primary root or stem. Cells in different parts of a root or stem tip are of different ages and so are at different stages of development: the farther from the apical meristem, the older the cells.
5. Lateral roots arise from the pericycle tissue inside the mature root. This produces a new apical meristem of a branch root. Stem branches arise from axillary buds, located at leaf nodes. An axillary meristem thus becomes the apical meristem of a new branch.
6. Leaves are the chief food-producing organs in most plants. The veins are the leaf's vascular tissue system, which supports the photosynthetic tissue and carries away food made in the leaf. They also provide the leaf with water, most of which escapes through the stomata into the air.
7. Secondary growth is growth in girth by production of secondary tissues. Vascular cambium, a lateral meristem, produces secondary vascular tissues. Secondary xylem (wood) adds strength and increases the capacity for conducting water to the leaves; secondary phloem replaces primary phloem, or older secondary phloem, destroyed as woody tissues grow. The cork cambium, another lateral meristem, produces secondary dermal tissue (cork), which replaces the epidermis that has been destroyed by expansion of the tissues interior to it.
8. Dicotyledons and monocotyledons differ in the number of cotyledons found in the embryo (two versus one, respectively), in the arrangement of vascular tissue, especially in the stems and leaves, and often in the arrangement of leaf stomata. In addition, very few monocots have secondary growth, whereas secondary growth is found even in some annual dicots.

## Self-Quiz

1. Sclerenchyma cells do *not* function in:
   a. transport of food
   b. transport of water
   c. support
   d. forming protective structures
   e. strengthening stems
2. One difference between a bean seed and a kernel of corn is that:
   a. a bean seed has a seed coat but a kernel of corn does not
   b. only the bean has two cotyledons; the corn has one
   c. the bean contains stored food; the corn does not
   d. the bean embryo has leaves; the corn embryo lacks leaves
   e. the bean embryo has two apical meristems; the corn embryo has only one
3. Primary growth of a tree:
   a. occurs through the activities of apical meristems
   b. occurs through the activity of a vascular cambium
   c. occurs through the activity of the root cap
   d. occurs only in the first year of the tree's life
   e. occurs in stems, but generally not in roots
4. Arrange the following events during the growth of a shoot in proper order:
   a. cell division
   b. cell maturation
   c. cell elongation
5. A fibrous root system is apt to perform which function better than a taproot system?
   a. absorption          c. food storage
   b. anchorage           d. transport          *(Continued)*

6. Secondary xylem and phloem are laid down by:
   a. apical meristems          d. cork cambium
   b. axillary meristems        e. intercalary meristems
   c. vascular cambium
7. One year's growth in length of a twig is the distance between successive:
   a. rings of bud scale scars
   b. leaf scars
   c. axillary buds
   d. branches
   e. any of the above
8. For each characteristic listed below, tell whether it is characteristic of dicotyledons, monocotyledons, or both:
   _____ a. stomata on both leaf surfaces
   _____ b. roots with root caps
   _____ c. parallel venation in leaves
   _____ d. many vascular bundles scattered throughout stem cross section
   _____ e. vascular cambium
9. Bud and Rose visited a forest on their honeymoon. Bud selected a tree 10 m tall and 30 cm in diameter. He carved their initials into its bark 1.5 m above ground level. On their tenth anniversary, Bud and Rose return to the forest; the tree is now 12 m tall and 33 cm in diameter. Their initials are now:
   a. 1.5 m above ground level
   b. 2 m above ground level
   c. 3.5 m above ground level

Match each function listed below with the letter indicating the corresponding structure in the diagram.

_____ 10. Stores food
_____ 11. Transports water to leaves
_____ 12. Produces lateral root
_____ 13. Carries food to be stored
_____ 14. Produces wood

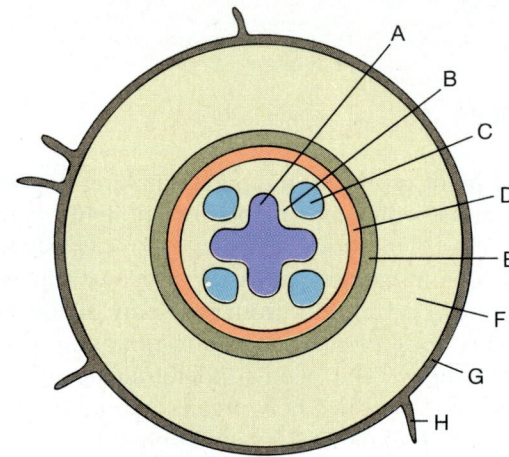

## Questions for Discussion

1. What is the advantage of the arrangement of xylem in a star-shaped pattern in the root, rather than its being internal to the phloem as it is in the shoot?
2. Since the cotyledons of a bean seedling become photosynthetic, one might expect them to be valuable organs of the plant even after their stored food is used up. Why might it be adaptively advantageous to the plant to shed them soon after it becomes well established?
3. Leaf structure varies from one species to another, and it is often closely correlated with the habitat of the plant. In what type of habitat would you expect to find each of the following modifications of leaf structure?
   a. more than one cell layer in the epidermis
   b. extra thick layers of cuticle
   c. little or no cuticle
   d. little or no cuticle; little or no xylem; no stomata
   e. large air spaces in the mesophyll; stomata in the upper epidermis instead of in the lower epidermis

## Suggested Readings

Blackmore, S. and E. Tootill. *The Facts on File Dictionary of Botany.* Aylesbury, U.K.: Market House Books Ltd., 1984. A small, very useful dictionary of botanical terms.

Esau, K. *Plant Anatomy,* 2d ed. New York: John Wiley and Sons, 1965. A reliable reference on plant structure, illustrated with numerous drawings and photomicrographs.

Kaplan, D. R. "The development of palm leaves." *Scientific American,* July 1983.

Mauseth, J. D. *Botany: An Introduction to Plant Biology,* 2d ed. Philadelphia: Saunders College Publishing, 1995.

Prance, G., and K. Sandved. *Leaves.* New York: Crown, 1985.

Sandved, K., G. T. Prance, and A. E. Prance. *Bark: The Formation, Characteristics, and Uses of Bark around the World.* Portland, OR: Timber Press, 1993.

Shigo, A. L. "Compartmentalization of decay in trees." *Scientific American,* April 1985. How the structure of wood and activities of its cells protect against disease.

# Transport in Vascular Plants

## OBJECTIVES

*When you have studied this chapter, you should be able to:*

1. Name the two conducting tissues in plants, give their functions, and list substances transported by each.
2. Describe a girdling experiment, and explain what it shows about the functions of conducting tissues in trees.
3. Sketch or describe the adaptations of tracheids, vessel elements, sieve cells, and sieve tube members to their roles in transport.
4. Outline the path by which water moves from the soil through a plant to the atmosphere, naming the molecular mechanisms and plant structures involved.
5. Explain the mechanism of xylem transport by (1) root pressure and (2) transpiration pull; describe experiments demonstrating each mechanism; discuss the relative importance of each mechanism in the total conduction that takes place in

the xylem of plants; predict how changes in the environment would affect each process.
6. Sketch two guard cells and the stoma between them and describe the mechanism by which the stoma is opened; state the advantage to the plant of the ability to open and close the stomata.
7. Sketch or identify these features of a woody stem, and state the function of each: growth ring, spring wood, summer wood, heartwood, sapwood, secondary xylem, bark, rays, and the approximate position in which you would find vascular cambium and secondary phloem.
8. Explain how transport in phloem may occur by mass flow, and predict how changing relevant conditions will modify transport in the phloem.

---

The root endings of an oak tree may be tens of meters from its leaves. The roots and leaves are connected by the vascular tissues, which transport substances throughout the plant body. Vascular plants actually have two transport systems, which carry different mixtures of substances in a complementary division of labor: **xylem** conducts water and minerals from the roots to the leaves, and **phloem** transports food from the leaves to the roots, as well as to growing buds, flowers, and fruits.

This dual transport system permitted the evolution of plants with specialized parts performing different functions,

such as absorption of water, collection of energy, and reproduction. Vascular tissue also provides strength and support, as seen most clearly in the wood of roots and stems and in the veins of leaves, which consist of vascular tissues. This support permits a tree to exploit a much larger volume of soil and air, an advantage in its competition with other plants.

How do materials move hundreds of feet to the various parts of a tall tree? To answer this question, we must study the structure of the vascular tissues, xylem and phloem, and the activities of cells near them.

## KEY CONCEPTS

- Transport in xylem and phloem follows a fluid pressure gradient.
- Water and minerals move upward in plants through xylem. The main force is the pull exerted when water evaporates from the leaves via the stomata.
- Food moves through phloem from leaves to sites of use or storage. Phloem transport follows a gradient of pressure built up by active transport of solutes and osmotic movement of water.

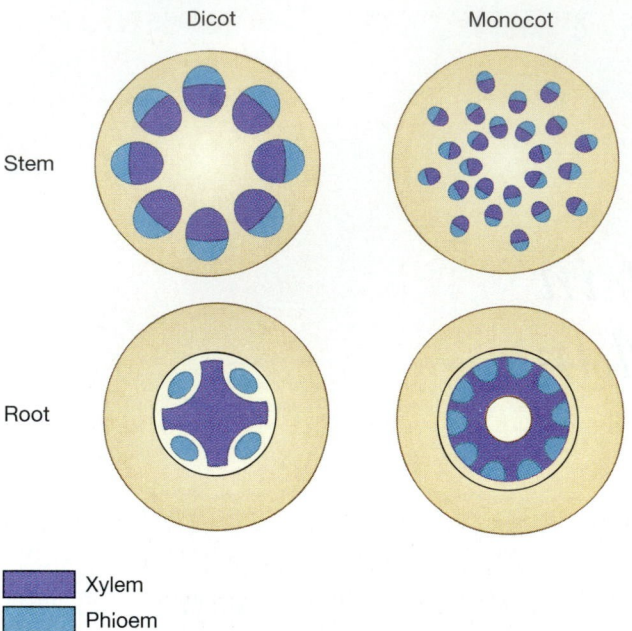

Dicot      Monocot

Stem

Root

◼ Xylem
◼ Phloem

**FIGURE 43–1** Arrangement of the vascular tissues—xylem and phloem—in stems and roots of monocots and dicots. Xylem generally lies more toward the interior of a root or stem, phloem more toward the exterior.

## 43–A  FUNCTIONS OF XYLEM AND PHLOEM

The arrangement of vascular tissue differs among plant groups and among different parts of a plant. In both roots and stems of most vascular plants, the xylem lies more toward the interior of the plant, the phloem more toward the exterior (Figure 43–1). How do we know that xylem and phloem are transporting tissues, and what materials are transported by each?

A white flower can be colored by placing its cut stem in a solution of food coloring and waiting for the dye to rise into the flower. If the stem is held up to the light, strands of color can be seen inside. Viewed with a microscope, a section of the stem shows that the dye is inside large cells in the xylem tissue but not in other cells. This is evidence that xylem transports water upward in a plant.

In 1679 Italian scientist Marcello Malpighi performed an experiment that showed the functions of xylem and phloem. He peeled off the bark in a complete ring around the trunk of a tree, a procedure known as **girdling.** This removes the phloem, which makes up the inner bark, but leaves the secondary xylem, or **wood,** intact. After this treatment, Malpighi found that a swelling appeared in the bark just above the girdled area. Fluid exuded from this swelling was sweet (we now know that it contained sucrose). The leaves showed no effects for days or months, but eventually they wilted and then died, and the entire tree was soon dead (Figure 43–2).

(a)    Bark removed    (b)    Sweet fluid exudes from girdle    (c)    Leaves wilt; tree dies

**FIGURE 43–2** A girdling experiment. (a) The bark, which includes the phloem, is removed in a complete ring around a tree. (b) A sugary fluid exudes from the phloem above the girdle, but the leaves remain green for a time, supplied with water through the intact xylem. (c) The wilting and death of the leaves signals death of the roots by starvation, cut off from the food normally supplied by the phloem.

From these observations, Malpighi concluded that phloem transports food, such as the sugar in the liquid exuded from the bark, to the roots. Without this supply of food, the roots died after they had used the food already stored below the girdle. Other girdling experiments have since shown that phloem also conducts food to growing buds, flowers, and fruits.

Leaves deprived of water wilt and die within hours. Because the leaves of the girdled tree remained healthy for much longer than this, Malpighi concluded that xylem was transporting water to the leaves.

## 43–B    STRUCTURE OF XYLEM

Xylem is called a complex tissue because it contains many different types of cells. Only some of these conduct the mixture of water and solutes called **sap.** Both conducting and nonconducting cells may strengthen the plant body.

Xylem structure shows many adaptations that enhance its dual functions of transport and support. After being formed by cell division, most plant cells lay down **primary cell walls** of cellulose and other polysaccharides. Later, the cells that will become xylem conducting cells produce additional strengthening material between the primary cell wall and plasma membrane. These **secondary thickenings** consist of cellulose and **lignin,** a tough, complex organic compound that makes wood woody. They vary from disconnected rings to extensive **secondary cell walls,** which cover the cell almost completely (Figure 43–3).

An important feature of xylem conducting cells is that they die once their cell walls are complete. The cell contents disintegrate, leaving a strong, hollow cylinder filled with sap. Many conducting cells may be stacked on top of one another. Sap may travel up for many centimeters in a more or less straight line within one of these "pipes," then flow laterally into another stack of conducting cells, and so on.

The xylem conducting cells of almost all nonflowering vascular plants are **tracheids:** long, extremely thin cells with slanting end walls. The wood of a gymnosperm such as a pine is almost all tracheids (Figures 43–4 and 43–5c). Heavy secondary cell walls slow the passage of sap from one tracheid to the next, except in thin areas, the **pits.** A pit is not a hole in the wall, but an area where the primary walls of neighboring cells are thin and little or no secondary cell wall material has been added. Pits occur both in the side walls and in the end walls of tracheids. This permits sap to move **laterally** (sideways) as well as upward in the plant and permits adjustment of the water supply to different sides of the tree.

A major evolutionary advance of angiosperms (flowering plants) was further specialization of cells in the xylem. Some conducting cells literally made an evolutionary breakthrough: their last act in life is to digest parts of their end walls, forming real holes, not just thin pit areas as in the

Early development          Late development

(a)

(b) Longitudinal section through a vascular bundle in corn (LM)

**FIGURE 43–3** Secondary thickenings in xylem conducting cells. (a) In a new length of stem, the first xylem cells lay down thickenings (pink) in disconnected rings or spirals, which permit the cell to elongate by stretching of the primary cell wall between them. Xylem cells formed after the stem reaches its final length lay down more extensive thickenings, which cannot be stretched, and the cell is fixed at this size. (b) Ring and spiral thickenings in xylem vessel elements.    (photo, Runk/Schoenberger from Grant Heilman)

side walls. The holes allow sap to flow faster from one cell to the next. In the ancestral condition, the end walls are perforated, as in magnolias; in derived forms such as oaks, they are entirely absent, and the cells are like sections of pipe.

These angiosperm conducting cells are also shorter and wider than tracheids. Their greater diameter reduces the cell walls' frictional drag, and so sap can travel faster. The hollow xylem tubes of angiosperms (and of a few gymnosperms) are **vessels,** and the individual cells in the vessels are **vessel elements,** or **vessel members** (Figure 43–4).

The wood of an angiosperm such as an oak contains many vessels, but most of it consists of other types of cells

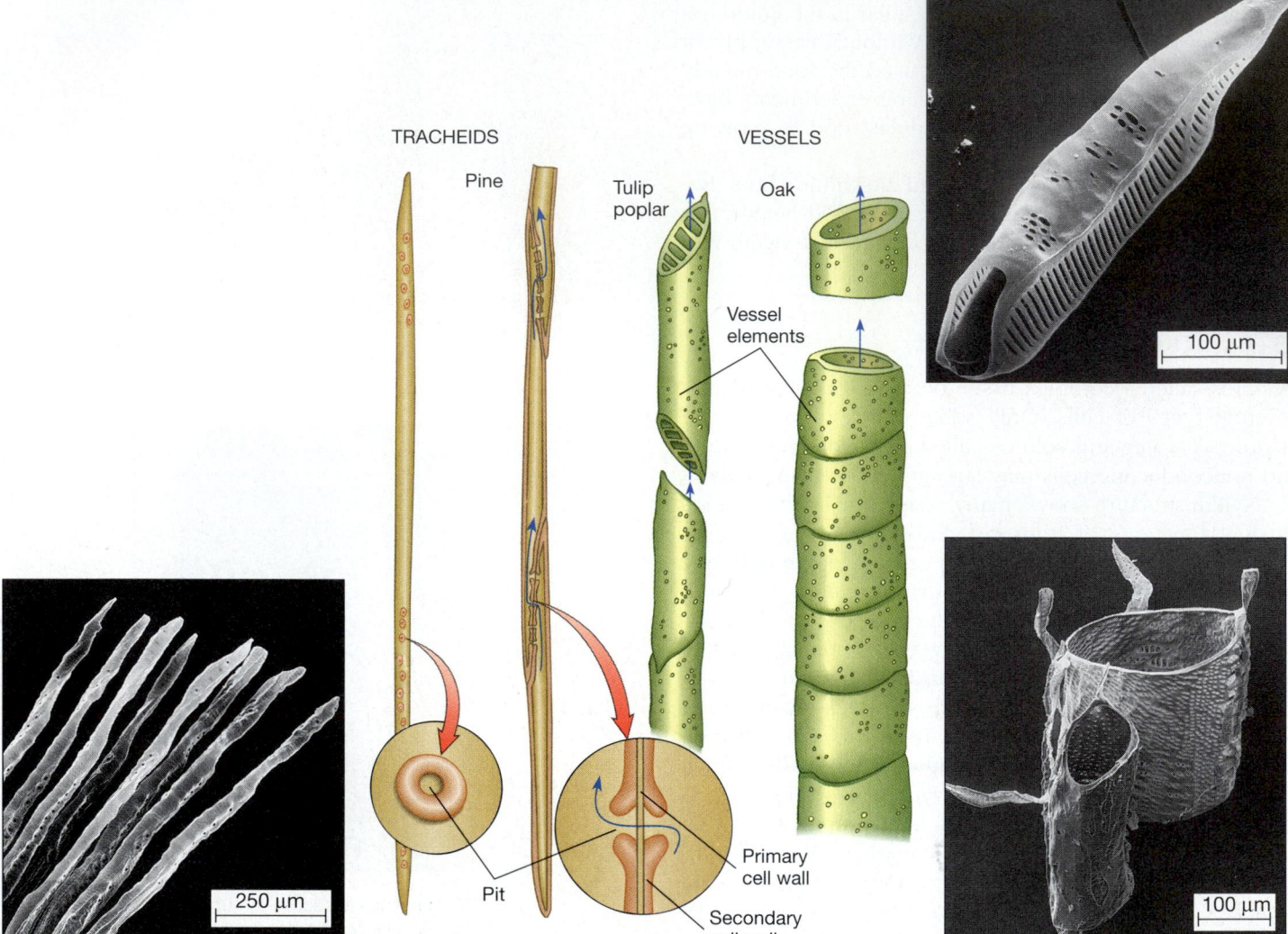

**FIGURE 43–4** Conducting cells in xylem. Blue arrows show the path of sap. Tracheids occur in all vascular plants, including angiosperms. They average 4 mm long, and sap flows unimpeded through the hollow interior. Gaps in the secondary cell wall occur in the pits, found at both ends and in the side walls, but sap moving from one tracheid to the next must still cross the thin primary cell wall. Vessel elements probably evolved from tracheids by becoming shorter and wider, with end walls perforated or absent. Sap moves freely from each cell to the one above. The bottom right photo shows the smooth primary cell wall around the secondary wall and the difference in diameter between vessel elements in summer wood (front) and spring wood (behind). (photos, N.C. Brown Center for Ultrastructure Studies/S.U.N.Y. College of Environmental Science and Forestry)

(Figure 43–5d). Some are **fibers**—long, thin cells with thick cell walls that help support the tree long after the cell contents die (see Figure 42–4). It is not hard to imagine that these cells evolved from tracheids by an exaggeration of the secondary cell wall thickenings, whereas vessel members evolved in another direction and lost parts of their cell walls completely. Angiosperm xylem may also contain irregularly shaped **sclereids,** or stone cells, similar to fibers in that they secrete a thick secondary cell wall and then

die. The wood of many angiosperms contains tracheids as well as vessels.

All the xylem cells discussed so far are different types of sclerenchyma: thick-walled, usually dead, strengthening and supporting cells.

Xylem also contains living parenchyma cells, with only thin primary cell walls. In both gymnosperm and angiosperm wood, many of the parenchyma cells are arranged in **rays** running out from the center of the tree. Ray cells

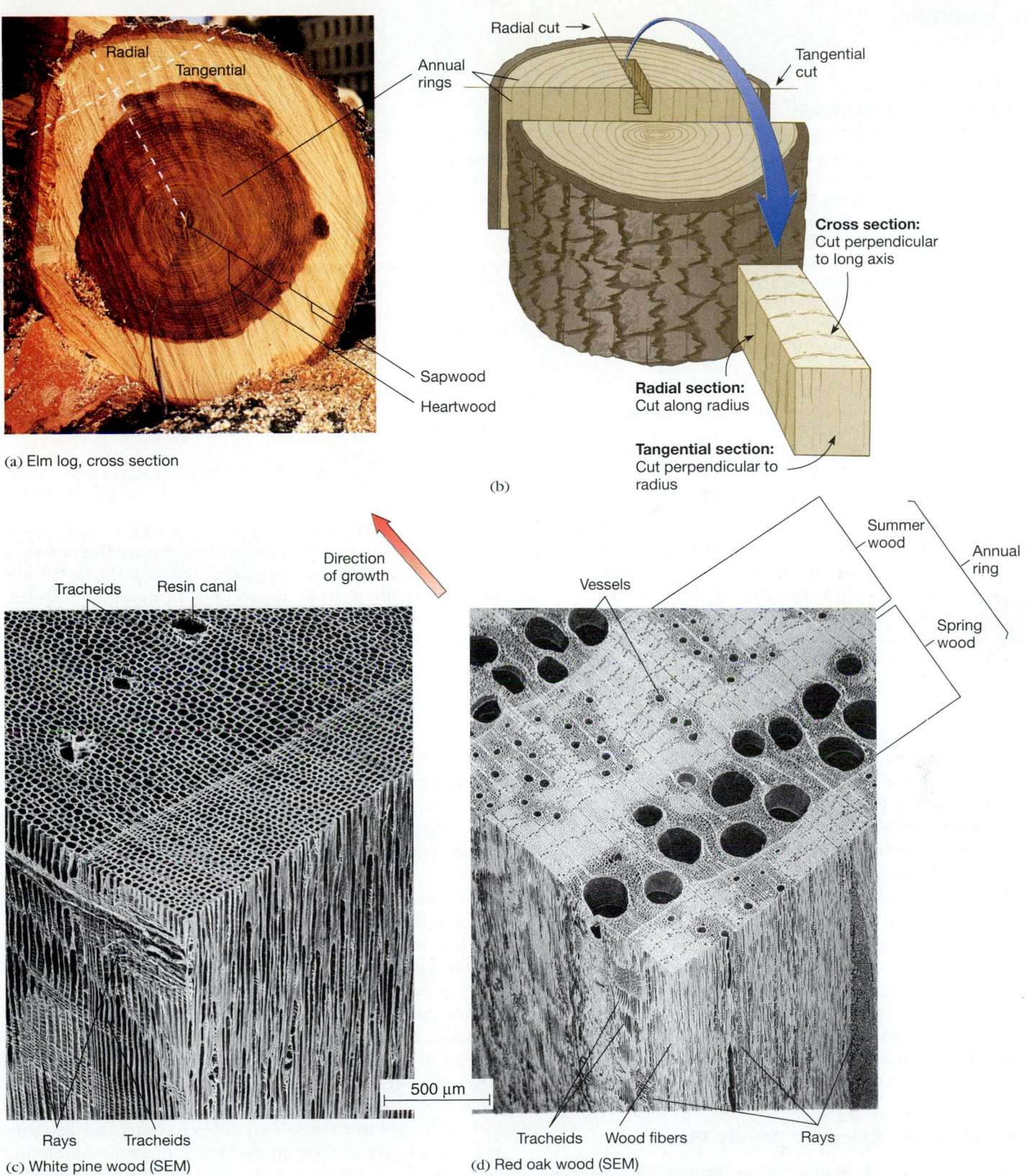

(a) Elm log, cross section

Radial cut →

Annual rings

Tangential cut

**Cross section:** Cut perpendicular to long axis

**Radial section:** Cut along radius

**Tangential section:** Cut perpendicular to radius

Sapwood

Heartwood

(b)

Direction of growth

Tracheids    Resin canal

Summer wood

Annual ring

Vessels

Spring wood

500 µm

Rays    Tracheids

(c) White pine wood (SEM)

Tracheids    Wood fibers    Rays

(d) Red oak wood (SEM)

**FIGURE 43–5** Structure of secondary xylem. (a) Cross section of an elm tree trunk, showing how radial and tangential cuts are made. (b) A log cut to produce blocks like those in (c) and (d). (c) Secondary xylem (wood) of a white pine. The tracheids conduct sap, and the resin canals produce antimicrobial secretions. (d) Secondary xylem of red oak, an angiosperm. The vessels conduct sap, and the fibers provide strength. Note that ray cells in both kinds of wood are elongated in a lateral direction. They conduct materials along a radius of the tree trunk. (c, d, Courtesy of Dr. Wilfred A. Côté, Jr.)

933

TABLE 43-1

## Cell Types in Xylem Tissue

| Cell Type | Description | Function |
| --- | --- | --- |
| Tracheids | Long, thin, thick-walled, dead; thin areas (pits) in walls let contents into adjoining cells | Transport of sap; support |
| Vessel elements | Shorter and wider than tracheids, thick walls with holes, which let contents into next cell; dead | Transport of sap; support |
| Fiber and sclereid cells | Fibers long, thin, thick-walled, often dead; sclereids variable shape, thick-walled, dead | Strengthening |
| Parenchyma | Relatively thin-walled, relatively unspecialized, living cells | Lateral conduction; starch storage; resin secretion |

may conduct materials laterally, out to the living tissues in the vascular cambium, phloem, and cork cambium. They may also serve as food storage depots (Table 43–1).

In pine wood, other living parenchyma cells form the walls of resin canals and secrete **resin.** This sticky, pungent fluid repels wood-boring insects and inhibits the growth of certain pathogenic (disease-causing) organisms. Hence, resin helps protect an injured tree from disease. Resin may be collected and distilled to make turpentine. It is also the raw material for pitch and tars, used to protect wood from rotting.

The wood of gymnosperms is commonly called softwood, whereas that of flowering trees is called hardwood. The distinction is one of taxonomy and cell structure, not physical hardness, because hardwoods such as poplar and basswood are actually softer than softwoods such as hemlock or yellow pine. Generally, softwoods have lower specific gravity. Softwood is 90% or more tracheids, the remainder being rays, usually only one cell wide. Hardwoods contain mostly fibers, but their distinguishing features are the vessels. Most hardwoods have higher, wider rays, occupying up to 30% of the wood's volume. In some hardwoods the rays and vessels are visible without magnification.

### Changes in the Xylem of Woody Plants

The xylem of a woody plant changes throughout its life. As we saw in Chapter 42, the vascular cambium adds new xylem outside the existing xylem. Trees in uniform climates grow continuously and do not form the **growth rings** seen in climates where good growing conditions alternate with cold winters or dry seasons. In temperate climates, a new ring is added each year, and these **annual rings** are distinctly visible, especially in hardwoods (Figure 43–5a).

The inner part of an annual ring, the **spring wood,** is light in color. The rains of spring (or of the tropical wet season) usually provide a great deal of moisture, stimulating production of a relatively high level of the growth hormone auxin. The new cells in the xylem grow rapidly and reach a large size before they produce the thick secondary cell wall and die. With less moisture available in the summer (or dry season), auxin levels decrease. Tracheids and vessel members grow more slowly and so are smaller when they produce their thick secondary walls and die. **Summer wood** appears darker than spring wood because cell walls occupy a greater proportion of its area (Figure 43–5). Annual rings can be counted to estimate the age of a tree or branch because there is usually one ring for each year of age.

A tree's wood is a record of its life story. Variations in cell size, thickness of growth rings, and so on indicate rainfall, fire, insect plagues, or the death of a neighbor. Annual growth rings are sometimes used to date archaeological finds or to determine the climate at some time in history. A wide growth ring reflects a good growing season, whereas a narrow band is formed in a poor year. By comparing the pattern of tree rings in building timbers with the rings in large, old trees in the same area, archaeologists can determine when the structures were built (Figure 43–6). Chemical analysis of tree rings from different years can also provide information on the area's history of air pollution and acid rain.

The latest several years' xylem is called **sapwood** because it actually conducts the sap in trees. Interior to the sapwood is the **heartwood,** older xylem that no longer conducts sap. When conducting cells cease to function, nearby parenchyma cells may block them off by secreting gums, resins, and dark, aromatic phenolic compounds, which are resistant to decay. These parenchyma cells also

**FIGURE 43–6** The Cliff Palace at Mesa Verde, Colorado, built between AD 1200 and 1300. The pattern of annual rings in the building timbers suggests that a long drought forced the inhabitants to abandon their cliff town.    (Allan O. Baer, Jr.)

die, and the heartwood contains no living cells. Sometimes it rots away, leaving a hollow tree that is still alive because its sapwood and phloem still function.

## 43–C  TRANSPORT IN XYLEM

Tracheids and vessels, the conducting cells of xylem, are hollow tubes stacked in columns, an advantage in the transport of sap. Sap is mostly water, and to see how it moves upward through these pipelines of dead cell walls we must briefly review the behavior of water.

**Water potential** is a measure of the energy of water, and it depends mainly on two factors: the osmotic potential, due to the presence of solutes, and the hydrostatic or turgor pressure of water in the cells pushing outward against the cell walls (Section 4–D):

water potential = osmotic potential + turgor pressure

The osmotic potential is zero for pure water and negative for water containing solutes. A living cell's turgor pressure is usually positive. Thus, the negative osmotic potential of a cell's cytoplasm results in a tendency for water to move into the cell, and a positive turgor pressure tends to push water outward. When the two are in balance they cancel out, and the water potential is zero, for no net movement of water into or out of the cell. The osmotic potential, and hence the water potential, can be lowered (made more negative) either by adding solutes or by removing some of the solvent, water. Water tends to move down the gradient, into areas with lower water potential.

If you touch the tip of a thin glass tube to the surface of water, water rises into the tube quickly by capillary action. This happens because water molecules are attracted to the charged walls of the tube (adhesion), and these molecules then pull others up into the center of the tube (cohesion). The narrower the bore of the tube, the higher water rises (Figure 43–7).

Water should also move in this way up the cell walls in the xylem tubes. However, calculations indicate that water moving by capillary action in the slenderest tracheids can reach a maximum height of about five feet, not high enough to account for the rise of sap in many plants.

### Root Pressure

If sap cannot climb on its own, it must be either pushed by some force from the bottom of the plant or pulled from the top. A push from below can be shown in some plants by cutting the plant off near the ground and sealing the stump into a glass tube: sap rises into the tube (Figure 43–8a). The force of the **root pressure** pushing sap upward can be calculated from the final height of the fluid in the tube.

Root pressure depends on active transport by living root cells, which move ions from the soil solution toward the xylem in the center of the root. The accumulation of ions lowers the water potential of the sap in the xylem, and water follows, building hydrostatic pressure in the xylem vessels. The endodermal layer surrounding the root's vascular tissues prevents the sap from oozing back to the soil, so the sap has nowhere to go but up.

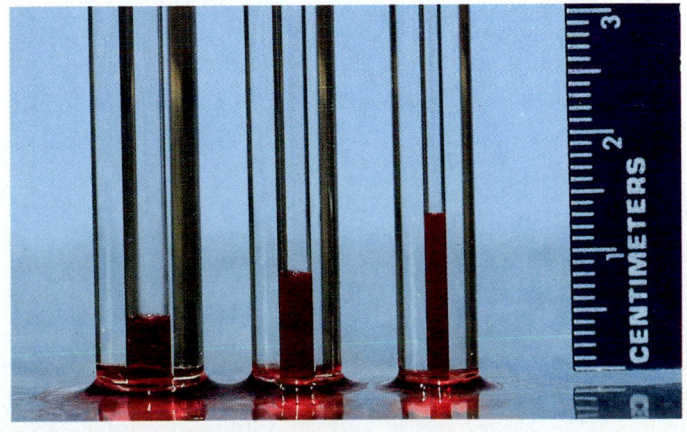

**FIGURE 43–7** Capillary action. When the end of a glass capillary tube is touched to water (dyed red here), water molecules are attracted to the walls of the tube, and these molecules pull others up into the center of the tube. The smaller the bore of the tube, the higher water rises.

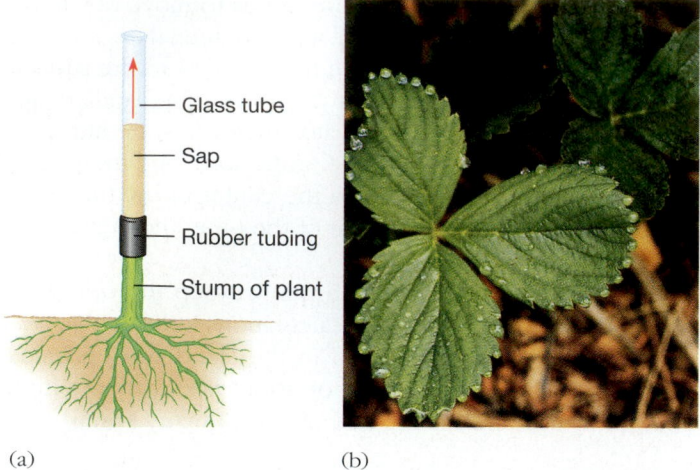

(a)                              (b)

**FIGURE 43–8**  Demonstration of root pressure. (a) A glass tube is sealed to a freshly decapitated plant. Sap rises into the tube from the roots. (b) Guttation in a strawberry plant. Guttation occurs in some short plants when ample water and oxygen are available in the soil and the air is too humid to permit evaporation of water as fast as it is pushed up from the roots. Sap is exuded from the tips of xylem columns in the leaves.

Because roots use a lot of energy for active transport, they must have an adequate oxygen supply. When the soil contains ample water and oxygen, root pressure may be high. If, at the same time, the air is very humid, water does not evaporate readily from the leaves, and sap forced up from the roots may be exuded from the tips of the leaves. This **guttation** occurs only in rather short plants, where the leaf tips are comparatively close to the root endings, the source of the root pressure (Figure 43–8b).

However, the greatest root pressures measured can push sap less than a meter, not far enough to reach the tops of many plants. Thus, root pressure alone can account for xylem conduction only in certain short plants growing in certain environmental conditions. Root pressure has not been found at all in gymnosperms, which include the tallest trees.

## Transpiration Pull

This evidence forces us to look at another mechanism for xylem transport, the pull from above. In 1727 English clergyman Stephen Hales demonstrated that **transpiration,** the evaporation of water from leaves, can pull sap up through the xylem of a plant (*How Do We Know? Transpiration Pull*). In the 1890s Henry Dixon and John Joly did similar experiments. They argued convincingly for this mechanism and also pointed out that the cohesion of water molecules to each other is vital to the rise of sap in tall trees.

***Anatomical Basis of Transpiration Pull***   The chain of events that moves sap upward depends on the structure of plants and the properties of water. The driving force starts at the top of the plant and works down (Figure 43–9):

1. Transpiration occurs: water evaporates from the porous walls of leaf cells into the air spaces inside the leaves and then diffuses out of the leaf into the atmosphere by way of the stomata.
2. The loss of water by evaporation lowers the water potential in the spaces between the cellulose fibers of the cell walls. Water is quickly replaced by capillary movement of water from neighboring cell walls, which in turn pulls the water from other neighbors behind it, and so on, until the replacement water comes from the end of a xylem veinlet.

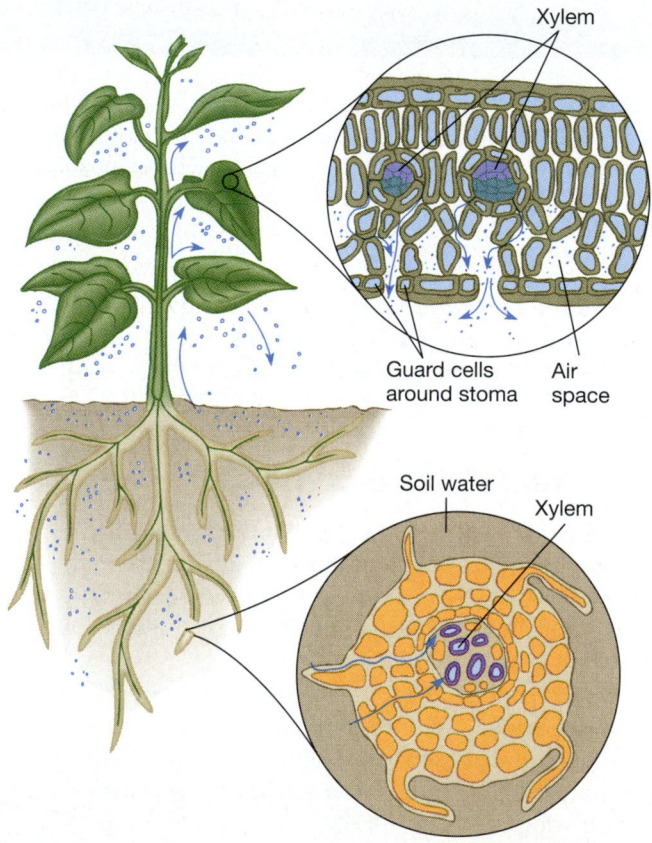

**FIGURE 43–9**  The transpiration stream. Movement of sap through a plant depends on cohesion of water molecules as they follow the water potential gradient. Evaporation of water from the leaves lowers the water potential in the cell walls lining the air spaces. This starts a chain of movement of water from cell walls nearer xylem veinlets. Eventually this water is replaced by water from the xylem, which coheres to water behind it, and the whole column of sap in the xylem, all the way to the roots, creeps up. The roots replace water by taking in more from the soil, provided the water potential in the soil is higher than that in the roots.

3. Because of the strong cohesion of water, pulling water out of the top of a xylem column is like pulling on a rope of water all the way down the thin xylem pipeline to its ends in the roots. All the water in the xylem moves up a bit. Eventually the pull reaches root cells, which take in soil water to replace that lost from the leaves.

Hence, a continuous "transpiration stream" moves from the plant's roots, up through the xylem, and out through the stomata. Water lost through the leaves is continuously replaced as soil water is taken up through the surfaces of root cells. Because transpiration sets this process in motion, and the cohesion of water molecules allows it to continue, this is called the **transpiration pull–water cohesion** mechanism of xylem transport.

Transpiration seems to be a necessary evil, the price plants pay for having leaves, which are very efficient structures for obtaining sunlight and carbon dioxide. The cuticle, guard cells, and internal air spaces seem to have evolved as a means of limiting water loss through the leaves.

### Cohesive Strength of Water

Water follows a pressure gradient from the soil, where there may be a low positive pressure, to the plant, where pressure rapidly decreases and usually becomes a negative water potential, because of transpiration pull from the leaves. When a plant is transpiring rapidly, water in the xylem is under **tension,** that is, a pull or negative force. Tensions in large trees may "stretch" the water in the xylem so much that it sucks the walls of the xylem in, and the tree's circumference shrinks measurably.

The cohesive strength that keeps water molecules together must be able to withstand tensions increasing by one bar (1000 kg/cm-sec$^2$) for every 10.2 meters in height. In addition, the walls of the xylem conducting cells exert a frictional resistance that at least doubles the pull required to raise the water column. When transpiration is especially rapid, or when little water is available from the soil, the tension on the water in the xylem column increases even more.

Because water in the xylem conducting cells is under tension, it is difficult to measure the actual cohesive strength that keeps water molecules together as water is pulled up. Cutting or puncturing the xylem to take internal measurements lets the water column snap apart, just as the two ends of a stretched elastic snap when it is cut. Air rushes into the xylem column, breaking the cohesion of the water and preventing further movement of sap by transpiration pull through the affected tracheids or vessels.

If you cut flowers on a hot day, their xylem water columns may snap in this way. When you then put the flowers in water, the water in the vase cannot pass the air in the xylem and connect to the sap, and so the flower may wilt (Figure 43–10). To re-establish a continuous stream of water in the xylem, you should cut off a few more inches of stem, using a sharp blade so that the xylem vessels are cut cleanly and remain open, rather than being squashed shut.

### Rate of Transpiration

Stephen Hales estimated that, weight for weight, a plant's daily water intake is 17 times that of a human being. This huge intake is required because the plant transpires so much water to the atmosphere. It has been estimated that plants lose 100 to 400 molecules of water for each molecule of $CO_2$ taken up. The rate of transpiration for any particular plant depends on many factors. Structural features such as density of stomata, thickness of cuticle, and geometry of leaves with respect to each other and to the ground all have an effect.

Environmental factors also affect transpiration:

1. **Availability of soil water** determines how much water a plant can lose by transpiration and replace readily from the soil.

2. **Sunlight** stimulates photosynthesis and causes the opening of the stomata and the faster uptake of carbon dioxide. It also increases leaf temperature. Hence, the plant loses water more rapidly on bright days.

1: Placed in water immediately

2: Left for 1/2 hour then 5 cm cut off before placing stem in water

3: Left for 1/2 hour then placed in water

**FIGURE 43–10** The importance of a continuous column of fluid in the xylem. Three stems were cut from the same plant and photographed one hour later, after receiving the treatments described below each.

# TRANSPIRATION PULL

Clergyman Stephen Hales (1677–1761) was an energetic disciple of Sir Isaac Newton, seeking to measure the forces acting in living systems, as Newton had for planets and their moons. Hales was the first to measure blood pressure, to compare blood pressures in animals of different sizes, and to show that the force of blood pressure was insufficient to drive muscle contraction, as some of his contemporaries claimed. Hales also founded the science of plant physiology with his measurements of the forces that cause sap to rise in plants.

In the summer of 1724, Hales measured how much water was "imbibed and perspired" by a 1-meter-tall sunflower plant over a 15-day period. On a warm, dry day the plant lost 0.9 liter of water by transpiration in 12 hours. On dry nights, this rate decreased to less than 0.1 liter in 12 hours, and on nights with dew, the plant gained water by absorbing the dew through its leaves. These changes supported the idea that "perspiration" was the force moving water through the plant.

In another experiment, Hales cut matched pairs of leafy branches and removed the leaves from one of each pair.

He then set each branch in a container with a measured amount of water and observed that the amount of water removed from the container was roughly proportional to the area of leaf surface on the branch (Figure 43–A). Hales decided that some activity of the leaves caused sap to rise, rather than capillary movement in the stem, as suggested by Malpighi and others.

Hales also found that for water to move through branches, the leaves must be dry and exposed to air. A branch with its leaves immersed in water could not "perspire," and almost no water moved through the branch (Figure 43–B).

In yet another experiment, Hales dug a hole to expose a root of a pear tree and measured how strongly the tree could pull on water (Figure 43–C). The pull was stronger on bright, sunny, dry days than on cloudy or damp days, and it slackened at night. These were exactly the results expected if evaporation of water from the leaves were indeed the force pulling water up through the tree. By

**FIGURE 43–A**  How Stephen Hales determined that water movement depends on the leaves. The amount of water removed by each branch in the top row was proportional to the number of leaves. Similar but leafless branches, in the bottom row, did not remove water from the containers.

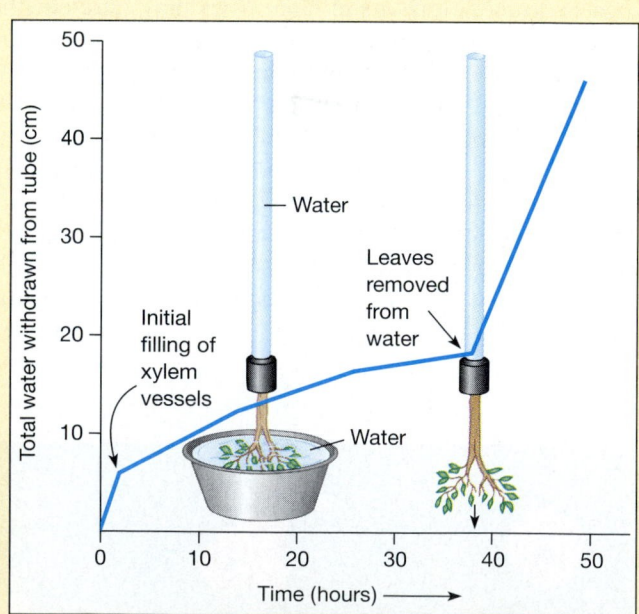

**FIGURE 43–B**  In this experiment, Hales showed that water moves through a plant much faster when its leaves are dry. A glass tube 2 meters long was attached to a branch and filled with water. Despite the pull of gravity on the water in the tube, little water moved through the branch when the leaves were immersed in water. When the leaves were dry, more water was pulled out of the tube.

using many different, simple approaches to test the hypothesis of transpiration pull, the ingenious and energetic Hales established this important principle of plant physiology.

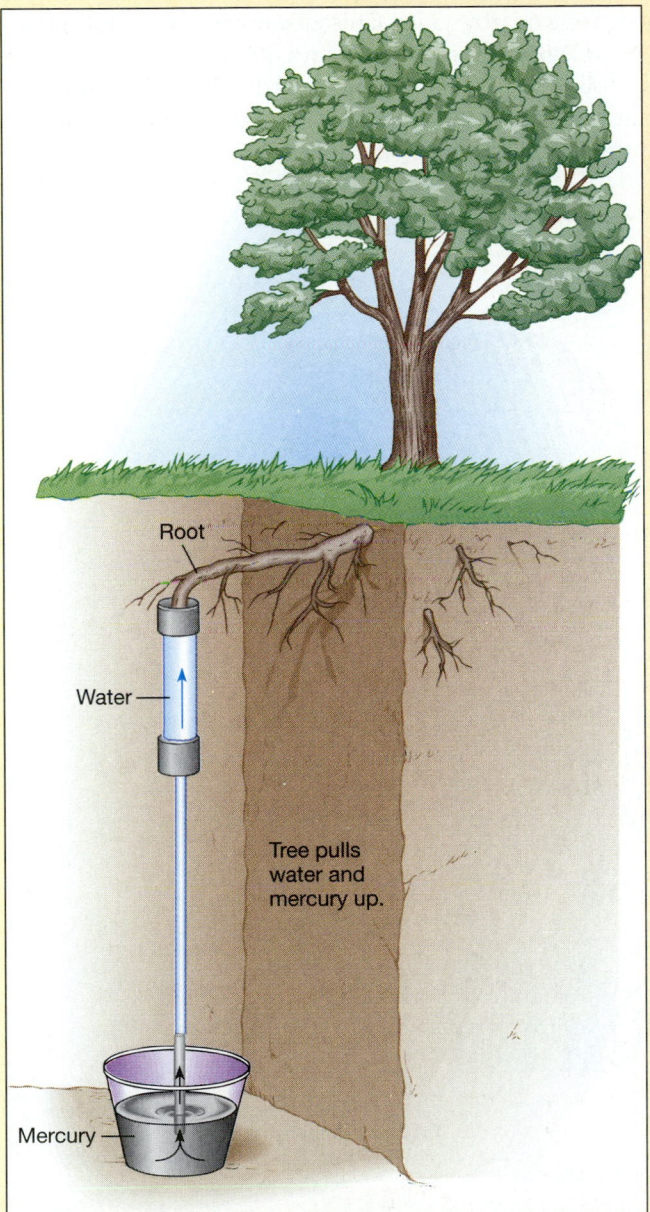

**FIGURE 43–C** Measuring transpiration pull. Hales attached a tube filled with water to the cut root of a pear tree. He placed a vessel of mercury in contact with the water and measured how far the mercury was pulled up into the tube as the tree withdrew water.

3. **Low humidity** increases transpiration because the concentration gradient between the leaf interior and the air is steeper: water diffuses from the leaf to the air at a greater rate.
4. **High air temperature** also increases evaporation. Furthermore, the higher the temperature, the more water vapor the air can gain before it becomes saturated.
5. **Wind speed** determines whether the water vapor transpired by a leaf remains nearby, forming a layer of saturated air that retards further evaporation from the leaf. If there is a breeze, this layer of still, saturated air blows away, and the leaf loses more water.

Transpiration can be regulated according to environmental conditions. This regulation is performed largely by the guard cells in the leaf epidermis, which open and close the stomata.

***Operation of Guard Cells***    Each stoma is surrounded by two guard cells. The ability of a pair of guard cells to control the size of the stomatal opening depends on their peculiar structure: their cell walls are thicker next to the stomatal opening than elsewhere (Figure 43–11).

Guard cells open the stoma by accumulating solutes, so that more water enters by osmosis and the guard cells

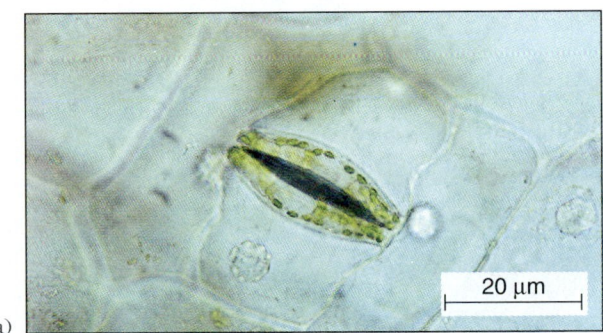

(a)

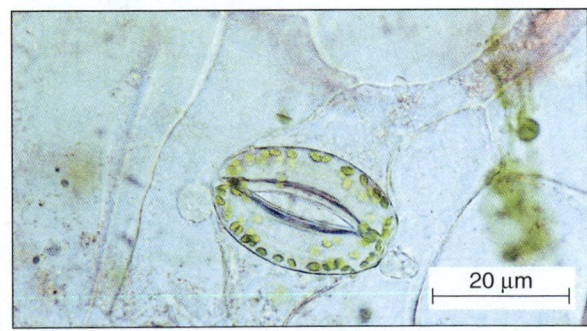

(b)

**FIGURE 43–11** Guard cells and stoma. (a) The epidermis of a *Zebrina* leaf. The two lip-like guard cells are the only cells with chloroplasts (green). In this view, the stoma between them is closed. (b) The guard cells have accumulated solutes, absorbed water, and expanded. Their thin outer walls bulge more than the thick walls bordering the stomatal opening, and as a result the stoma opens.    (Dwight R. Kuhn)

swell and become turgid. Because of the uneven thickness of their cell walls, the guard cells do not swell evenly all around. The thicker parts of the cell wall expand less than thinner parts, and so the guard cells assume a curved shape, with an opening between them. To visualize this better, think of the guard cells as toy balloons that bend when you blow them up because of a difference in the thickness in the rubber: the thicker parts of the balloon expand less than thinner parts.

Guard cells are the only epidermal cells with chloroplasts. ATP produced by these chloroplasts provides energy for active transport of solutes, mostly potassium ions, into the guard cells. This lowers the cells' osmotic potential, and hence their water potential. Water enters by osmosis, swelling the cells and opening the stomata.

The stomata are opened when the carbon dioxide concentration in the guard cells is low. This normally happens in the daytime, when photosynthesis is using up carbon dioxide, but it also occurs in the dark if carbon dioxide levels in the air are lowered artificially. We do not know how the low carbon dioxide level triggers the active transport of potassium into the guard cells.

Sometimes water is lost by transpiration faster than the xylem can replace it. When the water potential in the leaves becomes too low, the hormone abscisic acid is quickly released from nearby cells. This hormone causes the guard cells to release potassium ions and other solutes rapidly. As water follows by osmosis, the guard cells shrink, and the stomata close until the xylem's delivery of water catches up with the requirements of the leaves.

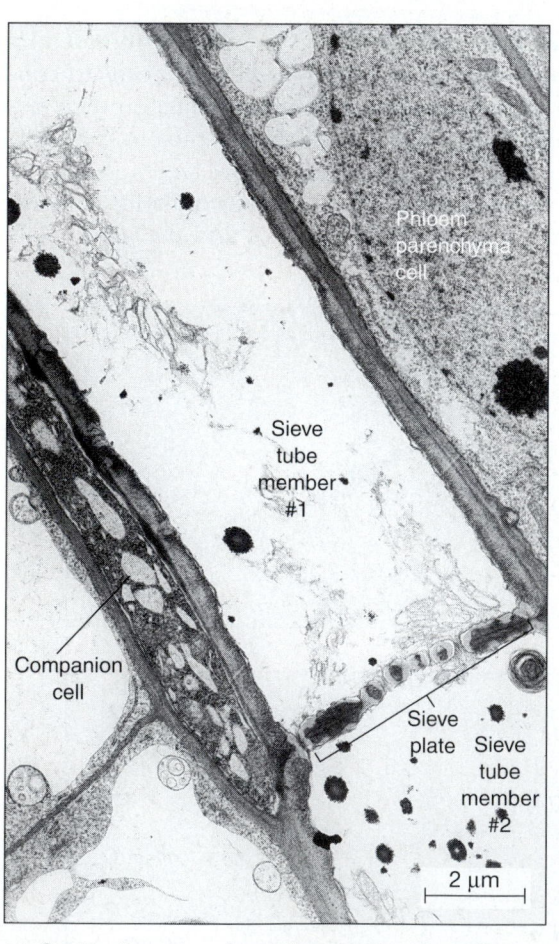

(a) Section through a sieve tube, showing sieve plate

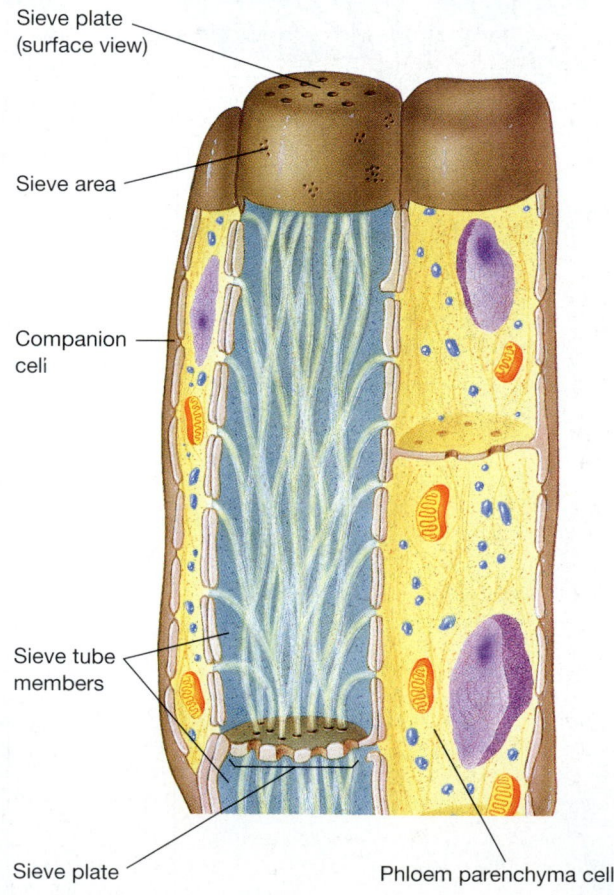

(b) Cell types found in phloem

**FIGURE 43–12**  Cell types in phloem of angiosperms. (a) A longitudinal section showing two sieve tube members, the sieve plate between them, a companion cell (of sieve tube member #1), and parts of some phloem parenchyma cells. (b) Materials pass up and down from one sieve tube member to the next through sieve plates, and sideways to neighboring sieve tube members through sieve areas. These phloem conducting cells contain living cytoplasm, in which substances are transported, but lack nuclei. Life support is provided by the nuclei of neighboring companion cells. Phloem parenchyma cells also contain nuclei and cytoplasm. Many plasmodesmata connect the cells.   (a, Biophoto Associates)

## 43–D   PHLOEM STRUCTURE

Phloem transports food and other organic compounds throughout the plant from sites of production to sites of use or storage. Most of the food moves in the form of the disaccharide sucrose (table sugar). Phloem may also contain minerals taken back into the plant from dying leaves. Phloem usually lies near the xylem (see Figure 43–1).

Like xylem, phloem is a complex tissue containing many types of cells. In nonflowering vascular plants, including gymnosperms, the phloem conducting cells are long, narrow **sieve cells.** Angiosperms evolved more specialized phloem conducting cells: shorter, wider **sieve tube members,** arranged in long columns called **sieve tubes.** The walls of both sieve cells and sieve tube members have special **sieve areas,** with many pores that permit exchange of substances between neighboring cells. Sieve areas in the end walls of sieve tube members have evolved as **sieve plates** with rather large pores (Figure 43–12).

Phloem conducting cells, unlike those of xylem, contain living cytoplasm. Before a cell becomes able to conduct, it loses its nucleus and most of its other organelles. In addition, there is a great enlargement of its **plasmodesmata,** strands of cytoplasm that pass through the sieve areas and sieve plates and connect neighboring cells. Strands of a material called **P-protein** pass through the cytoplasm both within the cell and in the plasmodesmata between cells.

After a conducting cell has lost most of its own metabolic machinery, it is thought to be maintained by the metabolism of an adjacent intact cell: a parenchyma cell in nonflowering vascular plants, or a **companion cell** in angiosperms. The sieve tube member and its companion cell are the offspring of a single cell that divided unequally (Figures 43–12 and 43–13).

The phloem of woody plants contains ray parenchyma cells that are extensions of the xylem rays, conducting materials laterally and storing food. Division of these cells helps the phloem expand with the growth of the internal woody rings of xylem.

Phloem also contains nonliving **phloem fibers,** similar to the fibers in xylem. These strengthening fibers are the only cells in the phloem with secondary walls.

## 43–E   TRANSPORT IN PHLOEM

When a phloem sieve tube is injured, plugs of protein and polysaccharide rapidly seal the sieve plates and stop transport. This reaction, like the clotting of blood in an animal, prevents loss of food molecules through the wound. Because of this sensitivity, transport in phloem cannot be studied directly, and so investigators have devised several indirect ways to observe phloem transport.

One often-used method is tracing the movement of radioactive materials applied to a plant's roots or leaves. A

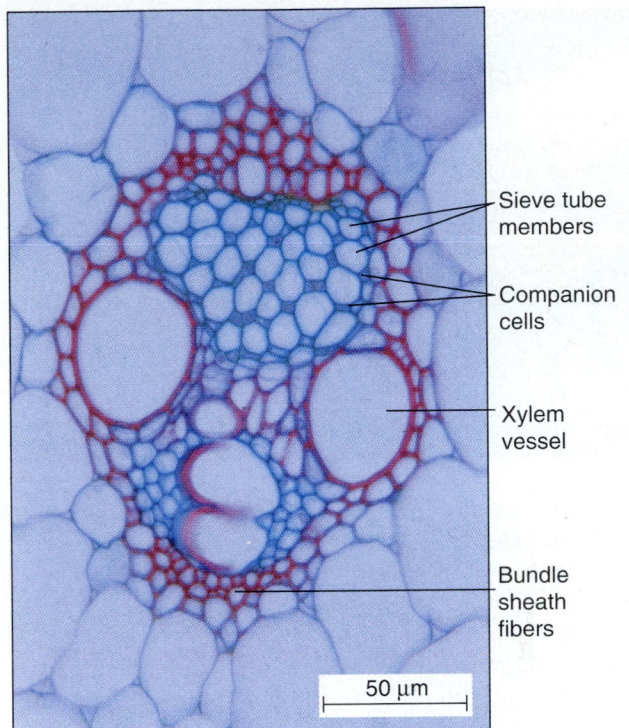

**FIGURE 43–13**   Cross section through a vascular bundle of a corn stem. The bundle is surrounded by a sheath of fiber cells. The xylem vessels conduct sap, and the sieve tube members of the phloem conduct food and other substances.   (Dennis Drenner)

substance moving in a large tree may be traced with a Geiger counter. For small plants, the usual method is autoradiography: the plant is given time to take up and transport the radioactive substance and is then pressed flat and placed on a sheet of photographic film. Particles emitted by radioactive decay expose (blacken) the film. The plant thus takes its own radioactivity portrait, which shows where the substance has been transported, by what route, and how fast (see Figure 43–16).

Phloem transport can also be studied using aphids, tiny insects that feed on plant fluids (Figure 43–14). An aphid's mouthparts form a long tube, which the aphid inserts into a plant in such a way that it taps a single sieve tube member. Because the cell's contents are under pressure, fluid from the phloem oozes into the aphid's mouth and on through the gut, often with such force that a drop of sweet fluid, called "honeydew," emerges from the anus. Because passing through the aphid changes the fluid's composition, researchers seldom work with intact aphids. After an aphid starts to feed, it is anesthetized with carbon dioxide and severed from its mouthparts. The fluid exuding from the phloem through the mouthparts can then be collected and analyzed. By using several aphids on different parts of the plant, an investigator can introduce substances at one point and study their speed and direction of travel.

**FIGURE 43-14** Aphids. These insects suck the fluid from individual phloem transport cells and so are useful tools in studying how the phloem works.   (Gill Renard)

Other methods of studying phloem include tracing dyes that move through the phloem, using chemicals that inhibit phloem transport, tracing a pulse of heat, and studying the microscopic structure of the tissue.

Any theory of phloem transport must explain several observations. For example, the contents of the phloem conducting cells are under pressure. A huge volume of fluid passes through each cell in a short time, moving at speeds of 30 to 200 centimeters or more per hour. The direction of flow in a particular cell may reverse at times, and neighboring cells may conduct in opposite directions at the same time. Killing the cells stops transport in that part of the plant.

Over the years, several hypotheses of phloem transport have been advanced. By far the most widely accepted is the **mass flow** or **pressure flow** hypothesis, suggested by Ernst Munch in 1926. Munch proposed a physical model that behaved very much like the phloem system in a living plant (Figure 43-15). Two chambers were connected by a tube, and each chamber had a membrane permeable to water but not to sucrose. One chamber was filled with plain water, the other with a sucrose solution. When the two chambers were immersed in plain water, water entered the chamber with the sucrose solution by osmosis because of the lower osmotic (and water) potential in this area. This increased the pressure in this chamber, forcing the solution toward the other chamber, taking some of the sucrose along. As sucrose solution arrived at the second chamber, water was forced out through the membrane into the water bath. The flow continued until sucrose became equally concentrated in the two chambers, giving them equal water potentials, and then stopped.

If sucrose could be continuously added to the first chamber, and removed from the second chamber as it arrived, the water potential gradient would be maintained, and the flow would continue indefinitely. Such conditions

are indeed found in the phloem system. In a living plant, some areas are **sucrose sources,** continually making sugar (in leaves or green stems) or releasing it by breaking down stored starch (in roots or stems). Other areas are **sucrose sinks,** where sucrose is consumed as it arrives. This could be any tissue short of energy, or food storage tissues converting sugar molecules into starch granules, which have less osmotic activity. Sucrose is the main solute in the phloem, and the phloem solution always moves down a sucrose gradient, regardless of the concentration gradients of other substances present.

In Munch's model, the water bath represents the xylem, the source of the water that enters the phloem solution by

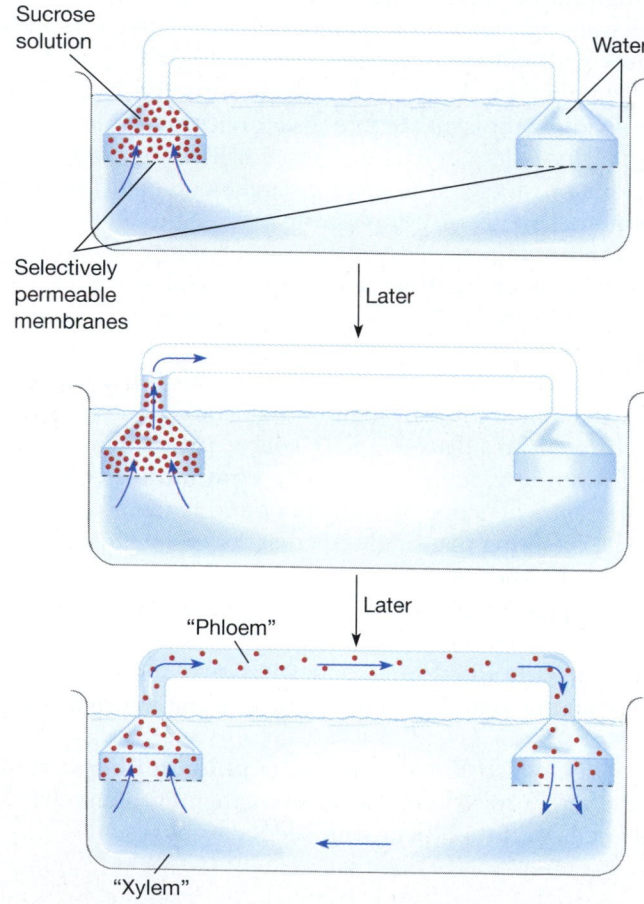

**FIGURE 43-15** Munch's model of mass flow in the phloem. The selectively permeable membrane is permeable to water (blue arrows) but not to sucrose (red dots). Sucrose solution is placed in the left-hand chamber, plain water in the right. The tube is then immersed with its ends in a water bath. Water enters the sucrose solution by osmosis through the membrane, and the solution rises into the tube and flows across to the right-hand chamber, where water is eventually forced out through the membrane. This system stops when the sucrose concentration becomes equal in the two chambers, but in a living plant some areas constantly produce sucrose as others remove it, and movement continues.

osmosis and forces the solution through the phloem to whatever tissues are removing the dissolved nutrients. The water would then be forced out of the phloem by the pressure of the more concentrated fluid moving behind it and taken back into the xylem.

Sugar enters the phloem conducting cells in the smallest leaf veins. These conducting cells are surrounded by companion cells or by special **transfer cells,** with highly infolded cell walls that increase the surface area of the plasma membrane. The many active transport proteins in this expanded surface move sugar into the phloem conducting cells, at concentrations of 10 to 25%. There are no further membranes to cross because strands of cytoplasm connect the conducting cells through the sieve areas or sieve plates. In tissues actively using or storing energy, sucrose is removed from the phloem, again often by transfer cells, and stored or metabolized.

The mass flow theory agrees well with the observed speed and pressure of the phloem contents. It can also account for changes in the direction of flow. Flow proceeds from higher to lower pressure areas. The lowest pressures occur in tissues withdrawing nutrients most actively, and this may change as different parts of the plant undergo spurts of growth. At one time, the roots may be most active as they store food, but later growing buds or fruits may begin to withdraw nutrients faster, reversing the flow. In this case, the food storage tissues may even switch roles, from sucrose sinks to sucrose sources, as they break down starch reserves and release sucrose to support growth or reproduction in other parts of the plant. Similarly, demands from two areas of the plant could account for the simultaneous flow of solution in different directions in different parts of the phloem system.

## 43–F DISTRIBUTION OF SUBSTANCES

The two transport tissues, xylem and phloem, between them carry the various substances that must be moved from one part of the plant to another. Water and minerals taken up by the roots are transported by xylem and may be removed by living cells in any part of the plant. The movement of minerals in the xylem is not always correlated with the movement of water.

Normally, the vast bulk of water moved to the leaves is lost to the atmosphere. However, some water becomes part of the cytoplasm, some serves as a raw material for photosynthesis, and some moves into the phloem (this water eventually returns to the xylem).

Some minerals also eventually make their way into the phloem. As leaves grow old, their minerals may be transported back into the plant for use in younger leaves or for storage over the winter (Figure 43–16). Phosphorus, potassium, and nitrogen especially are reclaimed, and sometimes iron as well. Calcium is a mineral that does not move once the xylem has delivered it to the leaves.

Minerals moved out of the leaves through the phloem may eventually make their way back into the xylem, and sucrose moved to storage areas by the phloem may appear in dilute solutions in the xylem when the sap "rises" in the plant the following spring. Thus, it appears that plants may have a slight and very slow "circulation" in their vascular systems, not just one-way transport.

The main substance transported by the phloem is sucrose, in concentrations of up to 25%. Other sugars and small polysaccharides are also transported. When sugars arrive in the roots, some of them are used by the root cells in their respiration or are stored as food reserves. Others are combined with ammonium ($NH_4^+$) or nitrate ($NO_3^-$)

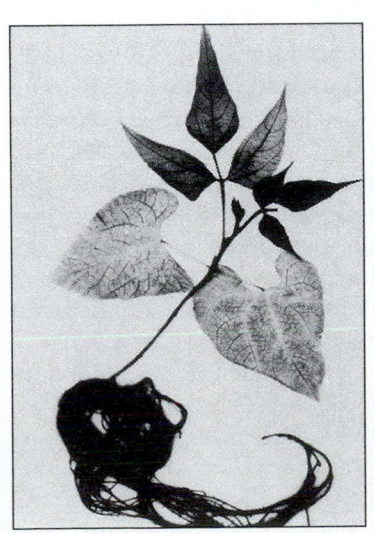

(a)          (b)

**FIGURE 43–16** Autoradiographs showing movement of phosphorus from older to younger plant parts. The roots of bean plants were placed in a solution containing radioactive phosphorus ($^{32}P$). After plant (a) had had time to take up and transport the phosphorus, it was pressed and autoradiographed. The darkest areas represent the parts of the plant with the most concentrated radioactive substance: the roots and youngest leaves. Plant (b) was removed from the solution containing $^{32}P$ at the same time as plant (a) but was allowed to continue growing for a time in nonradioactive solution before it was autoradiographed. Note that the radioactive phosphorus has been moved to the younger leaves, which had not yet formed when the plant was in the radioactive solution. (Susann and Orlin Biddulph)

ions taken in through the roots, forming amino acids. These amino acids are then transported to other parts of the plant by the xylem or the phloem. Phloem also carries nucleotides, some hormones, and various other organic compounds.

## Practical Applications

Transport in the vascular tissues of plants has more than academic interest. Knowing how various substances are transported in plants can be helpful in managing plants and plant pests. For example, leaves can absorb some substances placed on their surfaces in solution. These are moved into the phloem, which transports them throughout the plant. This is the basis of **foliar feeding,** fertilizing with a nutrient solution sprayed onto the leaves. Other substances must be worked into the soil because they are absorbed only by roots.

**Systemic pesticides,** watered into the soil or sprayed onto the leaves, are absorbed and distributed throughout the plant. Systemic pesticides can be used to combat both sucking insects (such as aphids) and internal pests (such as leaf miners, insect larvae that live entirely inside leaves and never come into contact with sprays applied externally). **Topical pesticides** are not absorbed into the plant, but remain on the leaf surfaces and protect the plant by killing leaf-chewing insects or invading fungi.

## 43–G   ROOT GRAFTS: TRANSPORT BETWEEN TREES

Roots, like stems, grow in diameter by adding secondary vascular tissue. If two roots grow near each other, their secondary tissues may touch and eventually merge, forming a **root graft.** The two roots can exchange materials through the graft. If the roots are from two different trees (of the same species), each tree can then affect the health and well-being of the other.

The effects of root grafting were observed during the droughts of the 1950s and 1960s. In plantations of red pine that had been started in the 1930s, clumps of trees in poor, shallow soil were dying because there was not enough water in the soil to support them. Surprisingly, all of the trees in a clump were dying at once, rather than the weaker or more spindly trees dying first, followed by the larger, healthier trees. Researchers found that in these clumps, root grafts formed an underground network linking the vascular systems of the trees to each other, and the available water supply was shared among the trees. This was not a conscious act of altruism, of course; the movement of sap in the xylem, as usual, was merely following water potential gradients, and moved throughout the clump of trees. When the entire soil volume had been depleted of its water, all of the trees died at the same time.

Similarly, when these plantations were thinned, foresters girdled the trees that they wished to remove. This was supposed to kill the unwanted trees. However, thanks to the root grafts, the roots of the girdled trees were not killed by being disconnected from the phloem of their own shoot systems. Food produced by their neighbors was transported through the phloem to the root grafts, where some of the food followed the pressure gradient and ended up in the girdled trees' roots.

Girdling will kill trees of many species even when root grafts are present, because fungal diseases and wood-boring insects invade through the wounded tissue of the girdle. However, in red pines and other trees that produce protective resins, the exposed wood is protected from these pests, and the wound soon heals over. The tree continues to produce new wood above the girdle and continues to grow new branches and needles. At last, after a decade or more, the tree becomes top-heavy and breaks at its weak point, the girdle, during a high wind or after a heavy snowfall.

## SUMMARY

In vascular plants, the task of transport is divided between two complementary vascular tissues: xylem moves water and minerals from the roots to the shoot system, and phloem moves food and other materials from leaves to roots or other organs that require food. In both, the transported solution moves down a pressure gradient, from a higher- to a lower-pressure area. Phloem contents are pushed by a positive pressure generated by osmosis. A similar mechanism may move substances in xylem. However, most often xylem transport results from a negative pressure—a pull—generated by evaporation of water from the leaves.

1. The conducting cells of xylem and phloem show structural adaptations for their rapid transport of substances. Xylem conducting cells are dead, hollow tubes reinforced with secondary wall thickenings. Tracheids, found in the xylem of both gymnosperms and angiosperms, have thin pits in their walls. Vessel elements, found only in angiosperm xylem, also have pits and may have their end walls perforated or absent. Conducting cells in the phloem have sieve areas in their cell walls, and the sieve tube members of angiosperms have well-developed sieve plates in their end walls.

2. Both xylem and phloem also contain thick-walled, non-living fiber cells, which provide strength, and living parenchyma cells, which may store food. Some parenchyma cells in the rays of woody plants conduct substances laterally through the secondary xylem and phloem. Companion cells of phloem are also parenchyma.

3. Root pressure pushes sap up a short distance through the xylem in the stems of some plants. However, in general the driving force of upward transport in xylem is the transpiration pull of water evaporating from the leaves at the top of the water column. Water follows a water potential gradient from the soil, through the roots and stem, and out through the stomata of the leaves. The cohesion of water and its adhesion to the walls of the xylem conducting cells is crucial to the transport of water to the top of tall trees.

4. Transpiration is controlled to some extent by paired guard cells. As the osmotic status of guard cells changes, the stoma between them is opened and closed. Transpiration is affected by the plant's anatomy and by environmental factors such as availability of soil water, intensity of sunlight, humidity, temperature, and wind.

5. The conducting cells of phloem are connected by way of porous sieve areas and by sieve plates in angiosperms. Hence, these cells form continuous tubes filled with living cytoplasm.

6. According to the mass flow theory, a high concentration of sugar transported into the phloem conducting cells of leaves creates a low osmotic potential. Water moves from the xylem into the phloem down the water potential gradient, generating hydrostatic pressure in the phloem. The contents of phloem conducting cells are pushed down a sucrose gradient from source to sink, moving from a high-sucrose, high-pressure area to a lower-sucrose, lower-pressure area.

7. Understanding plant transport mechanisms is important in planning fertilization, pest control, and hormone treatment programs in modern agriculture.

## Self-Quiz

1. The main solute transported by phloem is:
   a. glucose
   b. potassium ions
   c. sucrose
   d. starch
   e. amino acids

2. The girdling experiments performed by Malpighi supported the theory that:
   a. water moves in a tree by the root pressure mechanism
   b. water moves in a tree by a transpiration-cohesion mechanism
   c. xylem is primarily responsible for conducting water from the roots to the leaves
   d. phloem is primarily responsible for conducting water from the roots to the leaves

3. List two functions of xylem tissue, and describe at least one adaptation of cells found in the xylem that contributes to the performance of each function.

4. Movement of water up through a tree trunk depends on:
   a. capillary action
   b. exclusion of air molecules from the sap solution
   c. active transport into sieve cells
   d. attraction between water molecules
   e. low osmotic potential in the sap

5. Would root pressure increase, decrease, or remain the same under each of the following conditions?
   _____ a. high humidity
   _____ b. watering dried-out soil
   _____ c. darkness

6. Would the rate of transpiration increase, decrease, or remain the same under the following conditions?
   _____ a. high humidity
   _____ b. increased turgor pressure in the guard cells
   _____ c. increased light
   _____ d. increased wind

7. Xylem that is not conducting water is called:
   a. heartwood
   b. sapwood
   c. spring wood
   d. summer wood
   e. rays

8. Which of the following is *not* necessary to the operation of the mass flow theory as it is understood at present?
   a. ATP
   b. root pressure
   c. intact membranes in conducting cells
   d. different osmotic potential in different parts of the plant
   e. constant production or release of sugar molecules

## Questions for Discussion

1. In 1936 Bruno Huber performed an experiment in which he inserted thin heating wires into the xylem of a tree. He placed a thermocouple (a sensitive heat-detecting device) farther up the stem and timed how long it took before the heated sap passed the thermocouple. Huber found that the sap moved slowly at night. In the morning, the sap movement speeded up first in the twigs; later, the sap began to rise more quickly in the trunk further down the tree. Do these results support the root pressure or the transpiration pull theory of xylem transport? Justify your answer.

2. What is the best time of day to cut flowers from your garden for a centerpiece? Justify your answer.

3. We have seen in this chapter that cohesion of water keeps a column of water traveling up through the xylem of a tall tree. How did the water reach the top in the first place?

4. Why should the ground around evergreen plants be watered thoroughly before the ground freezes for the winter?

5. Contrast the conditions affecting transpiration experienced by a houseplant in winter versus in summer.

*(Continued)*

6. What is the advantage of bean leaves having more stomata in the lower epidermis than in the upper epidermis? Why are the stomata of grass leaves located in both epidermal layers?

7. What does wilting indicate about the movement of water in a plant? What does it indicate about the water content of the soil?

8. The virus disease known as "beet yellow" is transmitted from plant to plant by aphids. Why does the disease spread through the plant rapidly and kill it quickly?

9. Explain why the large brown algae known as kelps have a system of phloem-like "sieve filaments" that transport carbohydrates from the alga's blades (photosynthetic parts) to the stipe (stalk) and holdfast, but no xylem-like tissue.

10. Predict what will happen when foresters try to thin a red pine planting in which there are extensive root grafts by injecting poisons into the unwanted trees.

## Suggested Readings

Baes, C. F., and S. B. McLaughlin. "Trace elements in tree rings: evidence of recent and historical air pollution." *Science* 224:494, 1984.

Biddulph, S., and O. Biddulph. "The circulatory system of plants." *Scientific American,* February 1959. A highly interesting account of experiments showing how various substances move in plants.

Cohen, I. B. "Stephen Hales." *Scientific American,* May 1976. A biographical sketch of the life and experimental work of an energetic pioneer in plant physiology.

Jensen, W. A., and F. B. Salisbury. *Botany: An Ecological Approach,* 2d ed. Belmont, CA: Wadsworth, 1984. Especially good presentation of methods for measuring water cohesion and modern experiments on transport.

Shigo, A. L. "Compartmentalization of decay in trees." *Scientific American,* April 1985. How the structure of wood and activities of its cells protect against disease.

Stone, E. L., J. E. Stone, and R. C. McKittrick. "Root grafting in pine trees." *New York's Food and Life Sciences Quarterly* 6(2):19, 1973.

# Soil, Roots, and Plant Nutrition

## OBJECTIVES

*When you have studied this chapter, you should be able to:*

1. List or recognize the macronutrient and micronutrient elements required by plants.
2. Explain why a plant might show symptoms of nutrient deficiency even though the nutrient is present in the soil.
3. List the main components of soil, and explain their roles in plant nutrition.
4. Discuss several ways in which minerals become available to plants and ways in which they become unavailable.
5. Define the terms apoplast, symplast, epidermis, root hair, cortex, endodermis, and Casparian strip, and explain the role of each in the uptake of water and minerals by the plant.
6. Define the terms mycorrhiza and epiphyte; explain the nutritional adaptations of each.
7. Explain how carnivorous plants are adapted to capturing animals and why these adaptations are advantageous to such plants.

M ost plants are autotrophs; they take inorganic substances from the environment and use them to make organic food molecules. All living organisms, including plants, require the same major nutrients. Land plants obtain carbon, from carbon dioxide in the air, through their stomata (Section 42–G). However, it is their roots that absorb the wide array of mineral nutrients and the huge amounts of water they typically use. Taken all together, a plant's roots contain a vast total surface area of plasma membranes that absorb water and minerals from the soil.

In this chapter we study the intimate relationship between roots and the soil they live in. We examine how roots take in water and minerals and how plants use these mineral nutrients. We shall also study special nutritional adaptations that allow some plants to live in poor soils or with no soil at all.

## KEY CONCEPTS

♦ Life on land depends on healthy soil, the source of most nutrients required for plant growth. Plants, in turn, feed land-dwelling animals and microorganisms.

♦ Roots absorb water and minerals from the soil solution through a large surface area of selectively permeable plasma membranes.

## 44–A PLANT NUTRIENTS

Plants obtain carbon, hydrogen, and oxygen from carbon dioxide and water during photosynthesis. Plants also require the elements nitrogen (N), phosphorus (P), potassium (K), calcium (Ca), magnesium (Mg), sulfur (S), and iron (Fe) in order to grow well. This list of plant nutrients is easy to remember using a phrase containing the elements' chemical symbols: C HOPK'NS CaFe, Mighty good.

All of these nutrients except iron are required in relatively large amounts and so are called **macronutrients.** Calcium and magnesium are plentiful in many soils. Farmers have long fertilized their soil by adding the other elements in either organic or inorganic form. Nitrogen, phosphorus, and potassium are required in the greatest quantities, and commercial fertilizers are rated by the percentage of each of these elements they contain. For example, a "5-10-5" fertilizer contains 5% nitrogen, 10% phosphorus, and 5% potassium, by weight. (The other 80% is inert materials.)

Plants also require many **micronutrients,** minerals used in small amounts: iron, boron, zinc, manganese, chlorine, molybdenum, copper, and nickel. For instance, whereas nitrogen is applied to the soil at the rate of several hundred kilograms per hectare per year, the treatment for molybdenum-deficient soils in Australia is 0.14 kg of $MoO_3$ per hectare, applied once every 10 years. (A hectare [ha] is 10,000 square meters: about 2.5 acres.)

Many minerals are vital components of important biological molecules (Table 44–1). An inadequate supply of any one of several minerals may result in rather general deficiency symptoms, such as **chlorosis** (paleness) and poor growth (Figure 44–1).

On the other hand, deficiency symptoms of some nutrients may be quite specific. For example, one role of zinc is in production of the plant hormone **auxin,** which causes elongation or enlargement of cells during growth. If a plant lacks zinc, its cells do not grow to full size, the leaves are small, the cells in the internodes of stems, between leaves, do not elongate, and the plant grows in the form of a rosette: a flattened, rose-like shape. (Many plants, such as dandelions and sundew [see Figure 44–11], normally grow as rosettes even when well nourished.)

Some plants have special adaptations that require additional nutrients. For example, cobalt is a component of vitamin $B_{12}$, which is used by the symbiotic root nodule bacteria of legumes (pea family) in the process of **nitrogen fixation:** conversion of gaseous nitrogen ($N_2$) to ammonia ($NH_3$). Similarly, some grasses require silicon, which they incorporate into an abrasive material that both strengthens their stems and wears down the teeth of grazing animals, probably exerting selective pressure to look elsewhere for a meal.

## Determining Nutrient Requirements

Nutrient requirements are usually determined by growing plants in a solution of water containing all the known nutrients. If plants grow poorly under such conditions, then it is assumed that yet another nutrient must be supplied for normal growth.

In practice, there are several difficulties in determining the nutritional requirements of plants. For instance, the proportions of various nutrients to each other are important. In addition, the presence of some nonessential elements reduces the toxicity of others: if silicon is provided, a plant can tolerate higher levels of magnesium without developing symptoms of magnesium poisoning, but silicon itself is not a necessary nutrient for most plants. Plants can also sometimes use one nutrient in place of another that is in short supply. Thus, the ions $Ca^{2+}$, $Mg^{2+}$, and $Mn^{2+}$ can substitute for one another in some reactions.

Some nutrients are required in such small amounts that they are almost always present in sufficient quantities as impurities in the chemicals used to make up nutrient test solutions. The requirement for these nutrients was detected only after plants failed to thrive in test solutions concocted from extra-pure salts and water.

Yet another complication is that many nutrients are quite mobile in plants and can be moved from an older to a younger part of the plant as required. Nutritional studies generally use seedlings started under good nutritional conditions. These seedlings may have stored nutrients that permit them to grow in a seemingly normal fashion even after they are moved to a solution that lacks essential nutrients. To take an extreme example, a seed contains all the nickel the plant will require in its life—only 10 to 100 parts per billion. Nickel was found to be a necessary nutrient only in the second and third generation of plants given no nickel.

(a)

(b)

**FIGURE 44–1** Nutrient deficiencies in plants. (a) Corn plants grown for about 8 weeks in sand plus various nutrient solutions. The control plant (back row, far right) received all required nutrients. Each of the others received all nutrients except the one on its label. The poorest growth occurred in the plants not receiving the nutrients required in the largest amounts (front row). (b) Chlorosis in an allamanda vine. In this case the pale color was caused by magnesium deficiency, common in soils of the eastern United States and easily cured by applying Epsom Salts. (a, Dr. W. Shaw Reid, Soil, Crop and Atmospheric Science, Cornell University, Ithaca, NY)

**T A B L E   4 4 – 1**

## Roles of Mineral Nutrients in Plants

| Nutrient | Absorbed Form | Kg/Ha Removed by Wheat Crop in One Season | Role in Plant | Deficiency Symptoms |
|---|---|---|---|---|
| **Macronutrients:** | | | | |
| Nitrogen | $NO_3^-$, $NH_4^+$ | 84.5 | Component of proteins, nucleic acids, chlorophyll, some hormones, secondary chemicals (Section 18–C) | Mild: older leaves; yellow, purplish leaf veins, stems Severe: stunting |
| Phosphorus | $H_2PO_4^-$ | 15.6 | Component of nucleic acids, ATP, phospholipids, activated sugars, coenzymes | Various, general; dark color; loss of older leaves; stunting; slow maturation |
| Potassium | $K^+$ | 46.7 | Not well understood; important in membrane potentials, opening of stomata, osmotic balance; activates many enzymes | Various, general; mottled chlorosis and dead spots, starting in older leaves; weak stems |
| Sulfur | $SO_4^{2-}$ | 12.2 | Component of some amino acids and coenzyme A | Chlorosis, poor roots |
| Calcium | $Ca^{2+}$ | 13.3 | Component of middle lamella between cell walls; ties up oxalic acid and other waste products as insoluble salts; membrane function | Meristem death; abnormal cell division; deformed tissues; breakdown of membrane structure |
| Magnesium | $Mg^{2+}$ | 8.9 | Component of chlorophyll; cofactor of many metabolic enzymes | Chlorosis, appearing first in older leaves |
| **Micronutrients:** | | | | |
| Iron | $Fe^{3+}$ | 0.8 | Required for chlorophyll production; part of cytochromes and ferredoxins (electron transport) and of some enzymes | Chlorosis, appearing first in youngest leaves |
| Boron | $HB_4O_7^-$ | 0.3 | Mostly unknown; nucleic acid synthesis; pollen germination; carbohydrate transport | Various; thick dark leaves; malformations; cell division and flowering inhibited; meristem death |
| Zinc | $Zn^{2+}$ | 0.2 | Synthesis of tryptophan (precursor of auxin); component of some enzymes | Small, puckered leaves, short internodes |
| Manganese | $Mn^{2+}$ | 0.6 | Activates citric acid cycle enzymes; splits water during photosynthesis | Mottled chlorosis |
| Chlorine | $Cl^-$ | | Splits water during photosynthesis | Small, wilted leaves with chlorosis and dead spots; slow growth; thick, stunted roots |
| Molybdenum | $HMoO_4^-$ | | Part of enzymes for nitrate reduction and nitrogen fixation | Same as nitrogen deficiency |
| Copper | $Cu^{2+}$ | 0.03 | Component of cytochrome oxidase (respiration), ribulose bisphosphate carboxylase (photosynthesis) | Dark misshapen leaves with dead spots |
| Nickel | $Ni^{2+}$ | Not measured | Cofactor for urease enzyme; required for iron absorption | |

Nutrient deficiencies of plants may be due to low concentrations of nutrients in the soil, but deficiencies may also occur in soils where the nutrient is present but unavailable to plants.

## 44–B   SOIL

The minerals used by plants come ultimately from the soil's rock particles, formed as the underlying rock breaks down. The type of rock determines the presence of many minerals in the soil (this does not include nitrogen, discussed shortly).

Rock is broken down into particles by weathering. **Physical weathering** occurs through mechanical forces such as alternate expansion and contraction of the rock during hot days and cool nights, or wind, rain, water runoff, expansion of forming ice, and root growth. **Chemical weathering** breaks down rock particles by chemical reactions. For example, oxygen from the air oxidizes minerals at the rock surface. Acid, falling in rain or snow or excreted by organisms, dissolves mineral ions away from the particle surfaces.

Soils can be classified by the average size of their rock particles. The finest particles give rise to **clay** soils, larger particles to **silt,** and still larger particles to **sand.**

The size of soil particles influences the soil's capacity to hold **soil water.** Water is held in soil by adhesion to the charged surface of each particle, forming a thin film, and by capillary action, in which wedges of water fill the narrow spaces between soil particles by cohesion to water molecules on the particle surfaces (Section 2–G). Rainfall exceeding the soil's water-holding capacity drains away, carrying dissolved nutrients with it. This is called **leaching.**

The smaller the soil particles, the more water the soil can hold, since smaller particles have a greater collective surface area and more small-sized spaces between them than an equal volume of larger particles (Figure 44–2). In fact, clay soils, with the smallest particle size, tend to hold water so tightly that plants cannot withdraw much of it, and there may not be room for as much air as plant roots require. The large particles in sand, on the other hand, admit plenty of air but hold onto so little water that the soil dries out rapidly.

Most plants grow best in **loam,** soil with a mixture of particle sizes. Here, clay particles hold water, and larger particles permit drainage and penetration of air.

Fertile soil contains a great deal of organic matter. Manure, dead leaves, and bits of wood act like sponges, swelling as they soak up water when it rains, and then releasing it slowly. This alternate swelling and shrinking helps loosen the soil, allowing roots to grow through it easily. More important, the soil's organic matter is a reservoir of nutrients, which are released slowly by decomposition.

Living organisms are a vital component of soil. Soil residents include microorganisms such as bacteria, fungi, al-

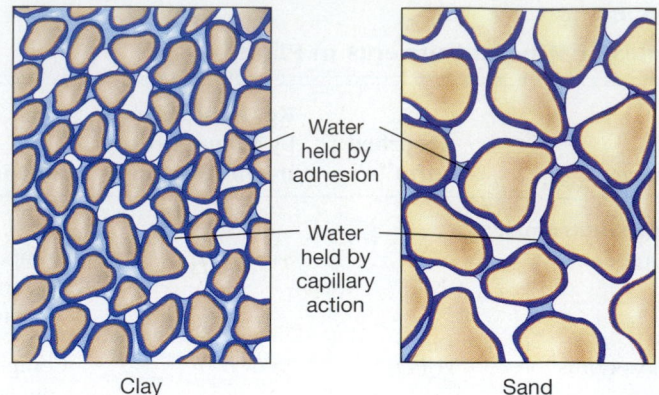

**FIGURE 44–2**   Soil particle size and water capacity. Water is held as a surface film adhering to the charged surfaces of individual soil particles and as capillary wedges in small spaces (up to 50 $\mu$m) between soil particles. Because small particles have a larger surface area/volume ratio and more small spaces between them, clay soils hold more water than sandy soils.

gae, and protozoans, and larger organisms, such as worms and insects, in addition to the roots of plants. Many soil organisms break down organic matter, slowly releasing minerals that plants can absorb. Soil organisms also affect conditions such as pH and the oxidation state of minerals.

Nitrogen-fixing bacteria convert nitrogen gas ($N_2$) dissolved in the soil water to ammonium ($NH_4^+$). Plants can use this ammonium, but not the more plentiful nitrogen gas, to make amino acids. If plants do not absorb ammonium quickly, nitrifying bacteria oxidize it to nitrite ($NO_2^-$) or nitrate ($NO_3^-$), which plants can also use, but which are easily leached out of the soil by rain. Plants of the legume family (peas, beans, clover, alfalfa) avert this problem by housing nitrogen-fixing bacteria in their root nodules (see Figure 24–9). Thanks to these symbionts, legumes have enough nitrogen to produce more protein than other plants do. They are planted as part of crop rotation programs and plowed under at the end of the season to increase the soil's nitrogen content.

Oxygen in the soil is important because most living things, including plant roots, require oxygen for respiration. The burrowing of organisms such as worms and insects mixes and breaks up the soil, allowing air to penetrate.

In waterlogged soil, only anaerobic organisms can make a good living. Some plants, such as mangroves and rice, do thrive in waterlogged soil, where their roots carry on alcoholic fermentation (Section 7–F). In general, however, anaerobic organisms cause changes unfavorable to the growth of most plants. This makes it doubly important for plants to live in well-aerated soil.

Digging a hole to plant a large tree, we notice characteristic layers of soil structure (Figure 44–3). On top is the litter layer, or mulch: dead leaves and other nonliving organic matter, which provides insulation from the sun's heat and reduces evaporation of soil water. Most soil organisms

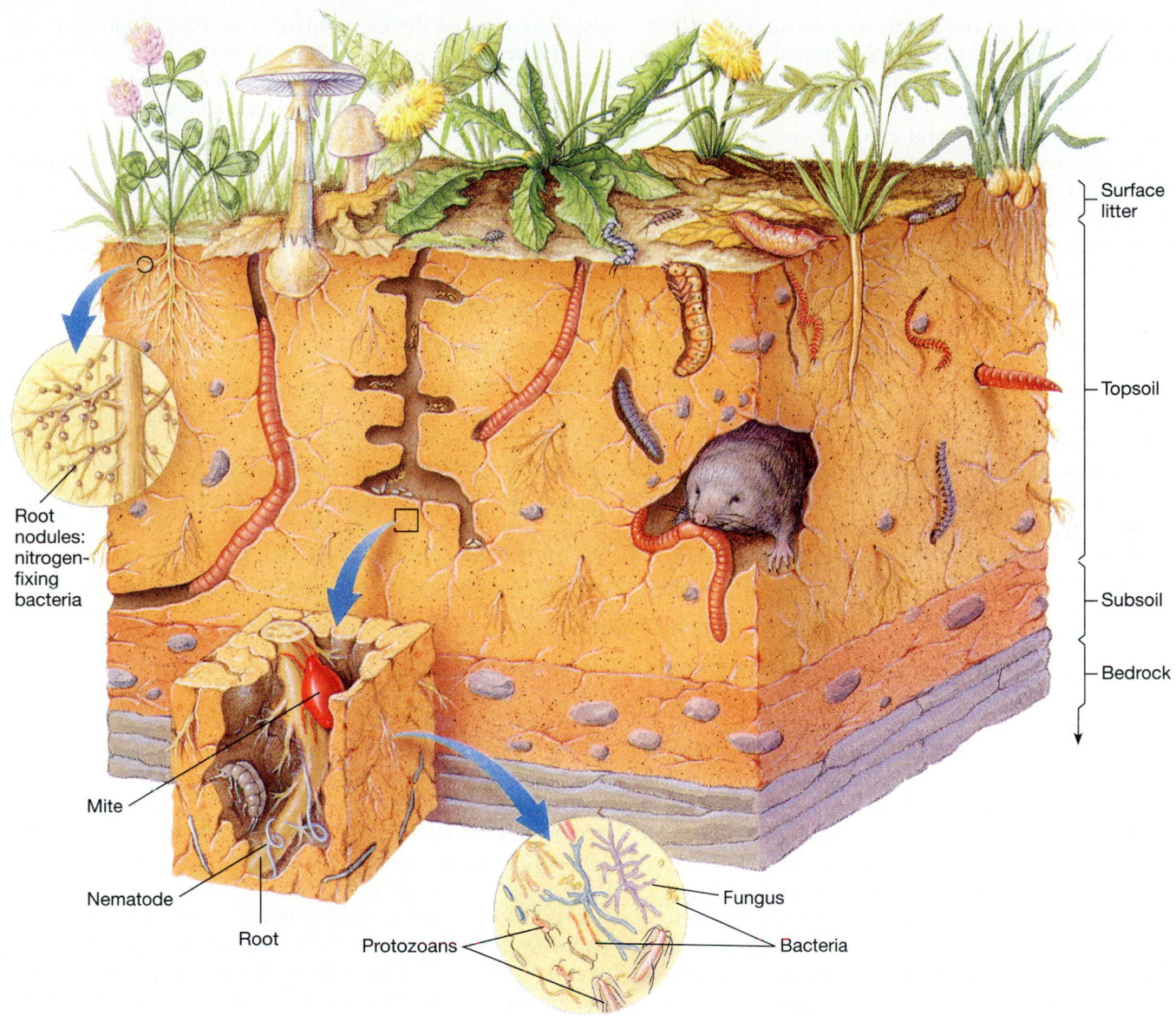

Root nodules: nitrogen-fixing bacteria

Mite

Nematode

Root

Protozoans

Fungus

Bacteria

Surface litter

Topsoil

Subsoil

Bedrock

**FIGURE 44–3**    Soil structure. Living organisms contribute to the breakdown of rock to finer particles. They also add organic matter and aerate the mixture by growing or burrowing through it. Productive topsoil contains many different kinds of organisms. Subsoil consists largely of rock particles and contains few organisms.

live in this layer or just below, in the **topsoil,** a loose layer containing rock particles and **humus** (partly decayed organic matter). Farther down, the **subsoil** consists largely of tightly packed rock particles, with much less organic matter and air than topsoil and hence few roots or other organisms. Under the subsoil, often far below the surface, lies solid bedrock, which breaks down very slowly into new soil particles.

Farmers and home gardeners alike improve the quality of their topsoil by mixing in organic matter such as manure or dead leaves. They may also create or add to the litter layer by spreading mulch on the surface to retard evaporation and weed growth and to add organic matter to the soil as the mulch decomposes.

## The Soil Solution

Soil water contains various dissolved molecules and ions. However, some minerals become tightly bound to soil particles. Bits of clay or organic matter have negatively charged surfaces and so bind most of the soil's positively charged ions. Only ions that stay dissolved in the soil solution are actually available to the roots of plants.

Acidic soil is low in nutrients because its $H^+$ displaces other positively charged ions from the surfaces of soil particles. The displaced mineral ions are then readily leached out of the soil. In addition, acidic soil water may dissolve so much aluminum and manganese that they reach toxic levels.

Some mineral nutrients are more soluble in acid, others in alkaline soil water. For example, iron is generally more available in acidic soil, calcium and nitrogen in neutral soil. In alkaline soil, as the pH rises above 7, the availability of manganese, boron, and phosphorus decreases rapidly. So soil pH determines the mix of nutrients that are dissolved and hence available to plants. Most plants do best at a soil pH of 6.0 to 7.5, although many plants are adapted to growing at higher or lower pH ranges.

Soil acidity affects more than plants. Researchers in the Netherlands investigated why great tits (birds related to chickadees) were laying eggs with thin, porous shells. They found that in areas where acid rain has produced acidic soils, the soil's calcium content had dropped to less than 10% of normal levels. This led to decreases in populations of snails, which require calcium for their shells, and snail-eating birds such as great tits no longer consumed enough calcium to produce normal eggshells. Some tits had taken to visiting farms and picnic areas, where they obtained calcium by eating the discarded shells of chicken eggs.

## Soil Treatments

Various treatments are used to improve the condition of soil for growing plants.

*Tilling*    One of the oldest agricultural practices is **tilling** the soil, turning it over before planting seeds and later around the growing plants. Tilling has two main purposes. First, it mixes nutrients and loosens soil particles, making root penetration easier. Second, it deliberately disturbs weeds and damages their root systems, thereby giving a competitive advantage to crop plants, which are left untouched.

Tilling also introduces more oxygen into the soil. This stimulates growth because roots require oxygen to take up minerals from the soil solution. The extra oxygen also permits aerobic microorganisms to break down organic material in the soil faster. This does not necessarily benefit the plants. If they are already taking up nutrients as fast as they can use them, releasing more into the soil solution may promote leaching.

Soil particles loosened by tilling are especially susceptible to erosion. The practice of **no-till planting** decreases this problem, and tilling is being abandoned by many farmers for this reason. Instead of plowing fields after the harvest, farmers leave the dying roots of the crop to hold the soil in place. The new crop is planted by machines that drill holes and place seeds in them. Weeds are sprayed with herbicides instead of being plowed under. In the midwestern United States, no-till methods have cut soil erosion from its former annual level of 22 to 33 tonnes per hectare to a current estimate of 2.25 to 4.5 tonnes per hectare.

*Fertilization*    Fertilization is another common agricultural practice. Of the mineral elements required by plants, nitro-

gen is most often deficient. Plants use a lot of nitrogen for their proteins and nucleic acids. Most of the world's nitrogen exists in the form of nitrogen gas in the air, which plants cannot use unless they have nitrogen-fixing symbionts (Section 24–C). Plants can absorb nitrogen only in the form of nitrate or ammonium. Nitrogen may be applied to the soil as a salt of either of these, as ammonia gas, or in a form such as urea ($H_2N$—CO—$NH_2$), which is broken down by microorganisms, releasing ammonia. Ammonia reacts with soil water, producing ammonium:

$$H_2O \; + \; NH_3 \; \longrightarrow \; NH_4^+ \; + \; OH^-$$
water + ammonia    $\longrightarrow$    ammonium + hydroxide

Any form of nitrogen applied is also available to soil microorganisms, which may convert the nitrogen into forms that plants cannot use. Nitrogen fixation by other soil organisms usually does not completely offset this effect. Another problem is that negatively charged nitrate leaches away easily because it is not attracted to soil particles as positively charged minerals are.

Applying positively charged ammonium also has drawbacks. Ammonium is taken up readily by plants and combined with carbohydrates to form amino acids, whereas nitrates must first be converted to ammonium. Ammonium may therefore give plants too much nitrogen too fast. They may outgrow their carbohydrate supply if photosynthesis cannot keep up with protein synthesis. Because of all of these considerations, nitrogen fertilizers must usually be applied several times during the growing season to provide a steady supply of nitrogen.

Does organic matter make better fertilizer than inorganic products? Plants can take up nutrients only in certain chemical forms. Inorganic fertilizers usually provide these forms, bypassing the steps of microbial digestion required to make organic fertilizers available to plants. Their nutrient content can also be controlled more precisely, and they are not so heavy, and therefore expensive, to spread.

For these reasons, chemical fertilizers are more often used in modern agriculture. Their main disadvantage is that they may leach rapidly out of the soil after each treatment. This wastes money and may pollute water supplies. Too much fertilizer may also make the soil solution so concentrated that plants lose water by osmosis. Such "fertilizer burn" may dehydrate and kill the plant. So despite their convenience, the use of inorganic fertilizers is decreasing in the United States and in many other countries.

Organic fertilizers do not provide better nutrition. However, they are superior because their minerals are released gradually, giving plants a sustained source of nutrients; they help hold water in the soil; and they improve soil texture.

An ongoing study is comparing the effects of inorganic and organic fertilizers on two adjacent farms in Washington state. One farm, first cultivated in 1908, has been conventionally managed since 1948, with inorganic fertilizer and pesticides applied as recommended by the state. Crops of winter wheat and spring peas are grown in succession in each field. The "organic" farm next door has received no inorganic fertilizer since it was first plowed in 1909. It too grows wheat and peas, but every third year, each field grows peas as **green manure,** a crop that is not harvested but left lying on the soil to add organic matter.

The organic farm now has 16 centimeters (6 inches) more topsoil, more soil organisms and organic matter, more water storage capacity, and less rain-shedding top crust. Its wheat yield is 8% less per hectare, and it harvests only two thirds of its land because its fields are each devoted to green manure one third of the time. Nevertheless it provides similar profits because the conventional farm has higher expenses: about one third of its total production goes to cover the cost of fertilizer and pesticides.

***Adjusting Soil pH*** Soil pH can be adjusted to suit the intended crop. Lime is applied to acidic soils to raise the pH, whereas organic matter, sulfur, or ammonium sulfate is applied to alkaline soils. These treatments have other effects besides changing the pH: lime provides calcium, ammonium sulfate contributes nitrogen and sulfur, and organic matter provides many nutrients.

Changing the pH may also affect the solubility of various minerals and hence their availability to plants. If an acidic soil's pH is raised too much, iron may precipitate out of the soil solution. One way around this is to apply **chelated iron**—iron bound to organic molecules so that it cannot be precipitated—along with lime. Lime also decreases the solubility of copper, manganese, and zinc, but it increases the solubility of some other minerals, notably phosphorus and molybdenum. So it is important to determine the pH of the soil carefully and then plan exactly how much of a chemical to apply to correct the situation.

## 44–C  ABSORPTION BY THE ROOTS

A plant's young roots, near the ends of the root system, absorb water and minerals from the soil solution. The external surface area of these roots is increased by root hairs, which are epidermal cells with extensions that grow out between soil particles (Figure 44–4).

But the epidermis is only part of a root's absorptive surface. The soil solution also moves freely into the root cortex via spaces between the cellulose fibers in cell walls. These cell wall spaces form a continuous functional unit, the **apoplast** (apo = apart from; plast = living material). The apoplast in the epidermis and cortex is virtually continuous with the external soil solution. Hence, a root can take up water and minerals through the entire surface area of the plasma membranes of cells in its epidermis and cortex. This greatly increases the surface area of contact between the soil solution and living root cells.

At the inner edge of the cortex, the soil solution in the apoplast reaches a barrier. The cells of the root's endodermis form a layer like a brick wall, separating the apoplast

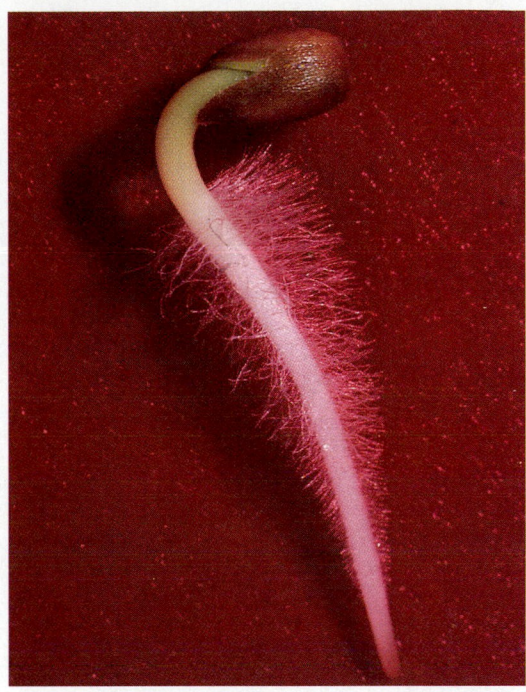

**FIGURE 44–4**  Root hairs. The primary root of this radish seedling has many root hairs, which provide anchorage among soil particles and increase the surface area available to take up water and minerals from the soil.

of the epidermis and cortex from that in the root's interior. A **Casparian strip,** made of an impermeable waxy material, **suberin,** runs right around each cell in the endodermis, as shown in Figure 44–5b. This material blocks the cell wall spaces and butts against the Casparian strips of the adjacent endodermal cells. Because the Casparian strips are impermeable to water and dissolved solutes, no substances can pass through the porous cell walls from the outside of the endodermis to the inside, or vice versa.

All substances that travel through the endodermis, therefore, must pass through the cytoplasm of the endodermal cells themselves. Thus, the endodermis lives up to the translation of its name, "inner skin," by serving as a selective barrier determining what substances can enter the vascular tissue and be transported to the rest of the plant.

Some substances from the soil solution are taken in through the plasma membranes of endodermal cells, but most are absorbed through the membranes of cells in the epidermis and cortex. All of these living cells form a second functional unit, the **symplast** (sym = together with; plast = living matter): a continuous system of cytoplasm extending throughout the root, connected by strands of cytoplasm, the **plasmodesmata.** Plasmodesmata are estimated to occupy about 1% of the total surface area of the root cells.

In 1938, Alden Crafts and Theodore Broyer proposed the symplast theory of mineral uptake by plant roots. According to this theory, all the cells in the epidermis and cor-

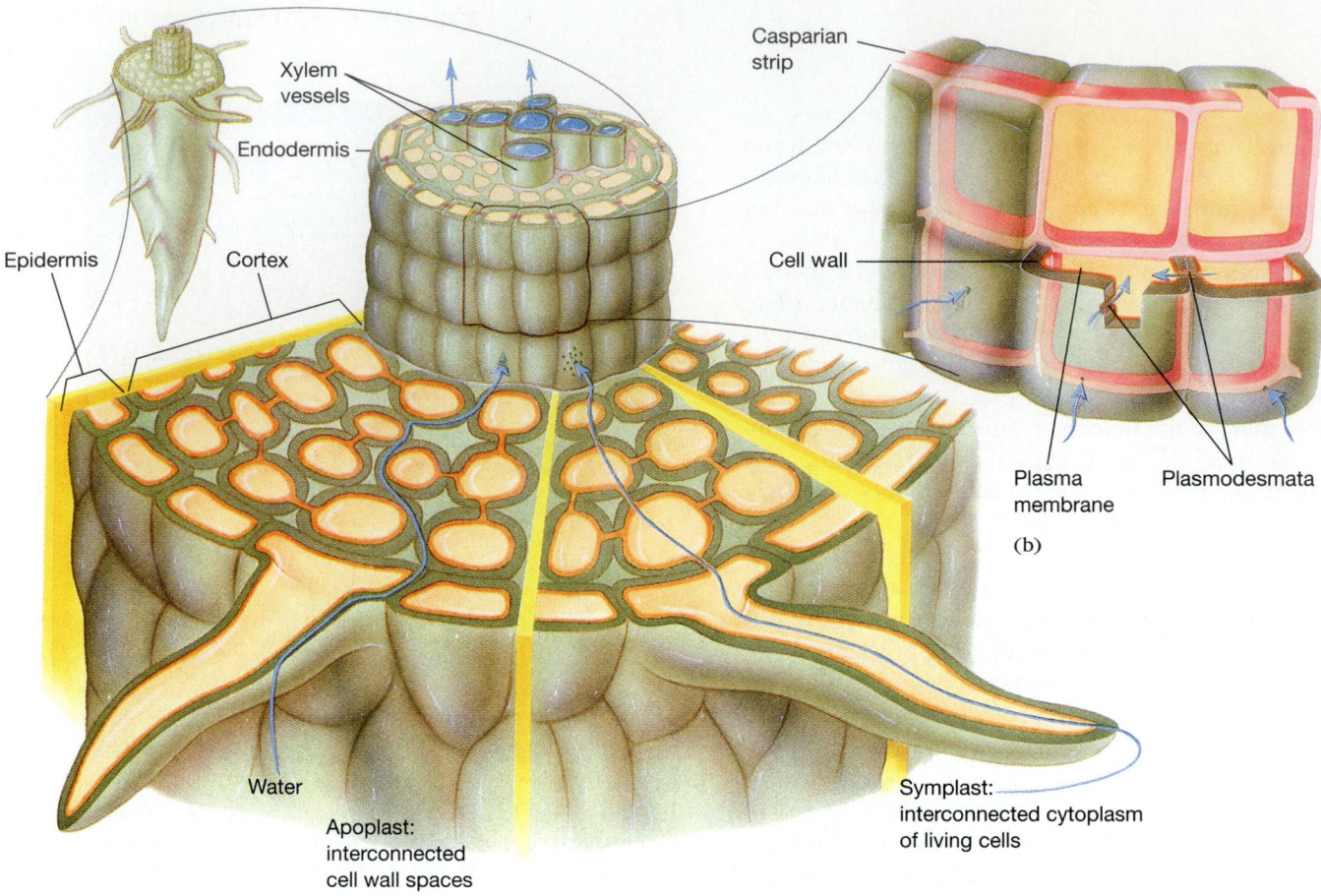

Xylem vessels

Endodermis

Casparian strip

Epidermis        Cortex

Cell wall

Plasma membrane        Plasmodesmata

(b)

Water

Apoplast: interconnected cell wall spaces

Symplast: interconnected cytoplasm of living cells

(a)

**FIGURE 44–5**  Uptake of substances from the soil solution by roots. (a) The apoplast consists of interconnected porous cell walls, which form a continuous pathway from soil to the outer walls of the endodermis. The symplast consists of all the living cytoplasm in the root cells. Water and minerals are taken in through a cell's selectively permeable plasma membrane and moved to other cells via plasmodesmata. (b) Structure of the endodermis. The watertight Casparian strips impregnate the cell walls in a continuous band. They prevent passage of the soil solution from the exterior to the interior of the root by way of spaces in or between the endodermal cell walls. Any substance that reaches the root's interior has passed through the plasma membrane and living cytoplasm of (at least) an endodermal cell (arrows).

tex, as well as the outward-facing surfaces of the endodermal cells, absorb minerals from the soil solution in the apoplast. Minerals then travel through the symplast, which continues right through the endodermis and into the living cells around the xylem conducting elements.

Thus, minerals move toward the root interior through both the apoplast and the symplast, although eventually they must pass through the symplast if they are to reach the xylem and, from it, the rest of the plant. The apoplast provides for rapid movement of materials, whereas the living symplast, with its large surface area of selectively permeable membranes, exercises some control over the kinds and amounts of substances taken in (Figure 44–5).

Some mineral ions, such as calcium and magnesium, enter root cells by diffusing through the plasma membranes. However, most are taken in by active transport, including potassium, nitrate, phosphate, and sulfate. Active transport uses the energy of ATP, and it stops in the absence of oxygen, which is required to make ATP during cellular respiration.

Once minerals have passed through the endodermis, they must leave the symplast, either by diffusion or by active transport, and move into the dead conducting elements of the xylem. From here they are transported upward to the rest of the plant.

The movement of minerals into the root cells decreases the cells' osmotic potential and hence their water potential (Section 43–C). The water potential may be further lowered by transpiration pull originating in the shoot system. Water moves into the root by osmosis, following the water potential gradient.

Although plants take up substances selectively, they cannot totally exclude substances that are unnecessary, or even toxic. This can pose problems. For example, nutrient-rich sludge from sewage treatment plants is often recycled by using it as fertilizer. This sludge must be free of toxic substances, because plants will take them up along with nutrients. Wastes from industrial towns often contain high levels of toxic elements, such as antimony, cadmium, and tin, as well as toxic levels of micronutrients such as selenium. Plants fertilized with these wastes may take up so much of these substances that they become unsafe for human consumption.

The ability of some plants to absorb toxins can be used to clean up polluted soil. In California, a wild mustard from Pakistan is planted on soil polluted with excessive selenium. The plant removes selenium from the soil and can then be fed to cattle, replacing the selenium supplement often added to their fodder.

## 44–D  NUTRITIONAL ADAPTATIONS

Many plants live in habitats where nutrients are scarce, and they have adaptations that boost their ability to obtain nutrients. We already saw that legumes and some other plants house nitrogen-fixing bacteria in root nodules. We shall examine a few other adaptations briefly.

### Mycorrhizae

The roots of most plants form **mycorrhizae,** symbiotic associations with some soil fungi (Section 26–I). Fungi have superior ability to absorb minerals from the soil. The hyphae of mycorrhizal fungi extend out into the soil around the root, where they absorb minerals, and also grow into the root itself. Here they pass along some of the minerals, especially phosphorus, which is not very mobile in the soil and is therefore hard to obtain. In return, the fungi receive organic molecules made by the plant. Most mycorrhizal fungi cannot grow without plant partners and are totally de-

pendent on the plant for food. The plants also do much better with fungal partners (Figure 44–6).

Fungi often produce plant growth hormones as well, and some plants rely on these hormones for normal growth. In nature, many orchid seeds cannot even germinate until they are infected by a fungus, and the young seedling grows as a parasite of the fungus until the plant becomes established.

The vast majority of mycorrhizae are **endomycorrhizae** (endon = within). The fungus germinates in the soil but does not survive unless it quickly allies itself with a plant root. It secretes growth-stimulating plant hormones, which boost its chances of success. After mutual recognition by the two partners, the fungus grows between the root's epidermal cells and penetrates the walls of cells in the cortex. Meanwhile, other hyphae grow out 70 cm or more into the soil, where they absorb minerals. These are transported into the root, where some are passed on to the plant through arbuscules ("little trees"), highly branched hyphae growing between the cell walls and plasma membranes of the cortical cells. These fungi also have many lipid-storing vesicles, and so endomycorrhizae are also called **vesicular-arbuscular mycorrhizae** (Figure 44–7).

Trees are the best-studied higher plants that form mycorrhizae. Trees by definition live in habitats where there is enough rainfall to support large plants, and this means that there is also enough rainfall to leach many nutrients out of the soil. Most trees in temperate forests form **ectomycorrhizae** (ektos = outside): fungal hyphae wrap a thick blanket around root endings in the humus layer and grow between but not inside cell walls in the root's outer layer.

**FIGURE 44–6**  Which plants have fungi? The two orange seedlings on the left were grown for six months with a mycorrhizal fungus, the two on the right without the fungus. All the plants were given nitrogen-containing fertilizer (calcium nitrate to the left-hand plant and ammonium nitrate to the right-hand plant in each pair). However, only the plants with mycorrhizae absorbed the nitrogen and grew well.   (Dr. J. Menge, University of California, Riverside)

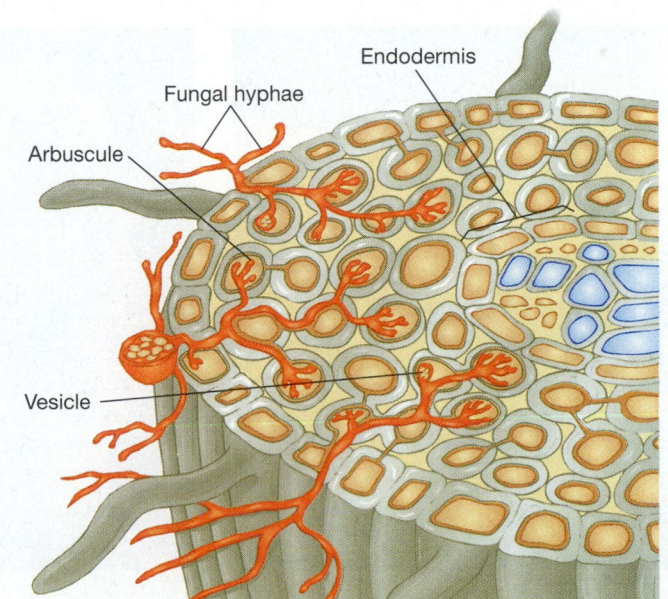

**FIGURE 44–7**  Endomycorrhizae. Fungal hyphae grow out into the soil and absorb minerals, which are passed on to plant roots via arbuscular hyphae penetrating cells in the root's cortex.

**FIGURE 44–8** Indian pipes. These flowering plants do not carry out photosynthesis but obtain food from trees by way of mycorrhizal fungi associated with both the tree roots and the Indian pipes.

The classic study of mycorrhizae was done by A. B. Hatch. Comparing pine seedlings that had mycorrhizae with those that did not, Hatch found that the seedlings with mycorrhizae absorbed nitrogen, phosphorus, and potassium more rapidly and also grew much more rapidly. Some mycorrhizal fungi are also believed to increase the solubility of soil nutrients by exuding substances into the soil, or to increase the rate of conversion of nutrients into forms that the plant can use. In return, the fungus receives organic compounds such as B vitamins and amino acids.

Some nonphotosynthetic plants of the forest floor rely entirely on their mycorrhizal fungus for food. The extreme case is that of Indian pipes (Figure 44–8). Experiments have traced nutrients from nearby trees into the Indian pipe by way of their mutual mycorrhizal fungus.

## Carnivorous Plants

Many plants can take up nutrients through the leaf surfaces. This ability is carried to fascinating lengths by carnivorous plants, which not only absorb organic food molecules through the leaves, but first use the leaves to capture and digest animal prey.

Carnivorous plants inhabit the wet, acidic soils of bogs and swamps. Acidity retards the growth of bacteria that release nutrients from dead organic matter, and it also makes the nutrients that are released very soluble, so that they are easily leached away. Thus, some plants, particularly in acid bogs, cannot obtain an adequate supply of nutrients through their roots. Some of these plants have turned the tables on some members of the animal kingdom by becoming carnivorous. The bodies of animals contain relatively high concentrations of protein, a good source of nitrogen, as well as minerals that plants can use. Carnivorous plants, like most other plants, produce food by photosynthesis, and indeed they experience little competition for sunlight because few noncarnivorous plants can survive in their habitats.

There are three basic types of carnivorous plants: pitfall traps, such as pitcher plants; flypaper traps, such as sun-

(a)                                        (b)

**FIGURE 44–9** Pitcher plants. (a) The leaves of pitcher plants are tubular. (b) Nectar-secreting glands on the lip of the pitcher attract insects. Downward-pointing hairs inside the pitcher cause prey to skid into the pitcher and prevent them from crawling back out. Instead, they plunge to a pool of digestive juices below.   (a, Milton H. Tierney, Jr./ Visuals Unlimited; b, Runk/Schoenberger from Grant Heilman)

**FIGURE 44–10**   Venus's flytrap. This endangered carnivorous plant is native only to the Carolinas. The red lining of the bilobed leaves attracts insects. Sensitive hairs on the inner surface of the leaves respond to an insect's jostling by starting an action potential like a nerve impulse. This causes rapid folding of the leaf, imprisoning the prey. Hairs along the edges of the leaf form a cage around the prey, as in the leaf at bottom right. Glands on the trap's inner surface secrete enzymes that digest the insect's soft parts. The rest of the insect blows away when the trap re-opens, ready for another victim.   (Carolina Biological Supply Company)

dews; and active traps, such as the Venus's flytrap. All have certain features in common. The traps are modified leaves, supplied with nectar glands that exude substances attractive to insects. In most other plants, nectar glands are confined to the flowers (which also consist of modified leaves). The trapping leaves also have glands that produce digestive enzymes. A third feature in common is the modification of leaf hairs in ways that aid in capturing prey (Figures 44–9, 44–10, and 44–11).

Once the trapped insect has been digested, the leaves absorb nitrogen and minerals from its body. A carnivorous plant absorbs most of its minerals through the leaves. The roots are small and serve mainly to anchor the plant and to absorb water.

Sundew and flytrap plants show growth adaptations as well as structural adaptations. When an insect is caught, certain cells expand rapidly in a manner that provides a more secure grip on the prey. After the insect has been digested, these plants "reset" their traps by another round of rapid differential growth. Because cells can grow only a limited amount, the leaves of these plants can capture only a few insects before their cells have grown too much to be of further use. As some leaves become too large, they are shed and replaced by new ones. Pitcher plants, however, can reuse the same pitchers over and over, and can even trap many insects at once in each leaf.

(a)

(b)

**FIGURE 44–11**   Sundew, a flypaper-type carnivorous plant. (a) Each leaf is covered with hairs that secrete glistening, sticky droplets attractive to insects. (b) Once an insect becomes entangled, nearby hairs grow toward it and hold it more firmly. Again, digestive enzymes are secreted and the digestion products absorbed.   (Carolina Biological Supply Company)

How did carnivorous plants evolve? Their adaptations do not include any real innovations. Rather, several common plant features have been modified or exaggerated, forming a suite of characters that permit carnivorous plants to live in acid bogs, where few other plants can. Reduced competition for resources would have been a major selective pressure for the evolution of the specializations of carnivorous plants. In fact, this is just one example of a general pattern seen among all organisms adapted to living in physiologically challenging environments, whether acid bogs, deserts, polar areas, mountaintops, or the deep sea.

## Some Other Nutritional Adaptations

Some plants inhabit soils with high concentrations of heavy metals, such as the slag heaps of mines. In Wales, a species

**FIGURE 44–13** Mistletoe on a host tree. The host branch has been split open to show the mistletoe's root-like haustoria (a term borrowed from parasitic fungi) within. The mistletoe's leathery leaves have a low transpiration rate, and the plant grows slowly. These adaptations reduce the stress it exerts on the host, which must live if the mistletoe itself is to survive. (Carolina Biological Supply Company)

**FIGURE 44–12** An epiphytic orchid growing on the trunk of a tree in a rain forest in Thailand. (Paul Feeny)

of the bent grass *Agrostis* grows well on soil containing tailings from copper mines. This grass tolerates levels of copper in its body that would kill other plants. Such tolerance has a double benefit: not only does the plant thrive where other species would perish, so that it has little competition from other plants, but also its copper content makes it toxic to herbivorous animals.

Copper-tolerant plants are immune only to copper, not to all heavy metals. This strongly suggests that the tolerance is genetic and involves the synthesis of a chelating agent that inactivates copper within the plant.

Prospectors make use of plants' abilities to tolerate high concentrations of heavy metals. Certain types of vegetation indicate likely places for mining particular metals.

Another group of plants with special nutritional adaptations are **epiphytes,** plants that grow on other plants. Some epiphytes have pockets at the bases of their leaves that catch water and minerals from the air. Others grow on trees whose rough bark catches dust and organic debris and forms a shallow layer of soil (Figure 44–12).

An interesting nutritional adaptation is seen in mistletoes, which make their own food by photosynthesis but parasitize their tree hosts for water and minerals. The roots of mistletoes invade the vascular tissue of the host tree and extract sap from it (Figure 44–13). Mistletoe is so well adapted to this way of life that if the host tree is girdled (Section 43–A), the tree's leaves die for lack of nutrients and water before the mistletoe does!

## 44–E  FOOD STORAGE

Nutrition involves not only obtaining and using nutrients but also storing them for use in future growth and reproduction. Most plants do not store much food in the leaves, where it is made, but move it to other parts of the plant.

Food storage organs include roots, such as those of radishes, carrots, and rutabagas; underground stems, such as those of potatoes; and sometimes even underground leaves, such as those of onions and other bulbs (Figure 44–14). One adaptive advantage of underground storage may be that it hides the food from hungry herbivores. In addition, underground storage organs are less vulnerable to freezing or desiccation.

During reproduction, stored food is transported to the developing seeds and fruits. The food in a seed nourishes the young embryo as it germinates, before it can make its own food. In some cases, storage of food in fruits is an adaptation to seed dispersal by animals: the animal is attracted by the nutritious fruit, eats it, and later deposits the seeds elsewhere, in a little pile of organic fertilizer.

Plants store inorganic nutrients as well as organic food. Each autumn, many proteins and pigments in a tree's leaves are digested, and nutrients such as nitrogen, phosphorus, and magnesium are moved back into the trunk before the leaves are shed. Some elements, such as calcium, stay in the leaves and must be recycled by the action of soil microorganisms.

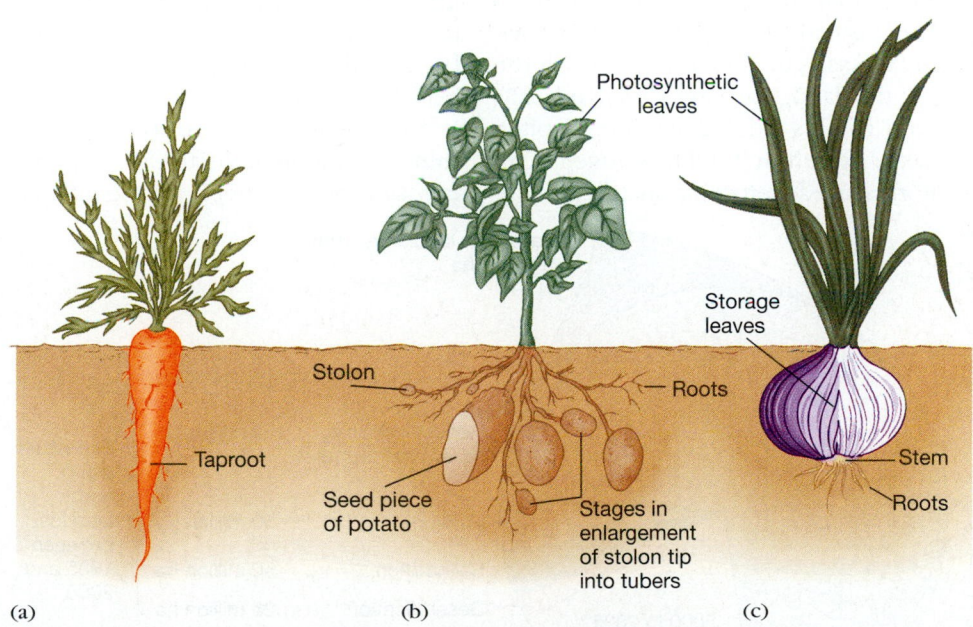

(a)　　　　　　　　　(b)　　　　　　　　　(c)

**FIGURE 44–14**  Underground food storage. (a) Many plants, such as carrots, store food in enlarged taproots. (b) A potato is an underground storage stem (tuber). Potato plants are grown from "seed" pieces of potatoes, each containing an "eye" (bud) that sprouts and forms the stems and leaves of the shoot system. As food is produced it is stored in modified underground stems, the stolons, whose tips swell and form tubers. (c) An onion is composed of modified, fleshy food-storing leaves formed underground at the bases of the green photosynthetic leaves.

Soil destruction is one of the most devastating environmental and agricultural problems that we face. Without fertile soil, we cannot grow the crops needed to feed our burgeoning human population.

**Soil erosion** is the movement of soil from one location, usually agricultural land, to somewhere else, usually a lake or river. Every year, an average of 25 to 30 tons of soil erodes from each hectare of American agricultural land. This soil is moved by the force of wind and water, often after being loosened by farm machinery or animals (Figure 44–A). In the United States, soil loss has reduced the productivity of about 70% of all farmland, and many areas have lost essentially all their soil. The soil of an area is considered destroyed when crop plants will no longer grow there.

Besides this agricultural loss, soil destruction causes other problems. For instance, soil washing into streams and rivers carries fertilizer and pesticides with it and so contributes to water pollution. The soil particles may also block out light required by photosynthetic organisms in streams and lakes.

Erosion is not the only cause of soil destruction (Figure 44–B). Soil in many parts of the world has become so salty that plants can no longer grow in it, especially in arid regions, including the semidesert and shrubland of the western United States. **Arid regions** are areas where more water leaves the soil by evaporation than reaches it through rainfall. (Water moves up from the water table underground, drawn by plant transpiration and by capillary action.) When water evaporates, the salts dissolved in it are left behind, so the soil of arid land contains high concentrations of salts. Because arid land receives less than 25 centimeters of rain a year, most agriculture in arid areas is irrigated. Irrigation water contains salts, and so when it evaporates, even more

(a)                                    (b)

**FIGURE 44–A** Soil erosion. (a) Animals erode soil. The sharp hoofs of these Cape buffalo in Africa loosen the soil surface, making it more likely to be blown away by wind or washed off by sudden hard rains. (b) Plants hold soil in place. Strong winds in Death Valley have blown away a foot or more of soil not held by plant roots.

salts are added to the soil. Eventually, the soil may become so salty that plants cannot grow in it. The accumulation of salts in the soil is called **salinization.**

The usual way to grow crops on salty soil is to irrigate with much more water than the plants can absorb, so that instead of evaporating and leaving the salts behind, the water will run off the land, carrying dissolved salts with it. This salt-laden water is often toxic. Kesterton Wildlife Refuge in California used to receive such high con-

centrations of the element selenium from the irrigated land of the San Joaquin Valley that many birds and plants in the refuge died. Selenium occurs naturally in the soil in low, harmless concentrations. Evaporation from arid land has concentrated it to a dangerous level. Scientists are attempting to clean up the toxic soil in Kesterton by growing plants and fungi that absorb selenium.

Saline soil can be restored to usefulness without intensive care by planting salt-tolerant trees. These absorb

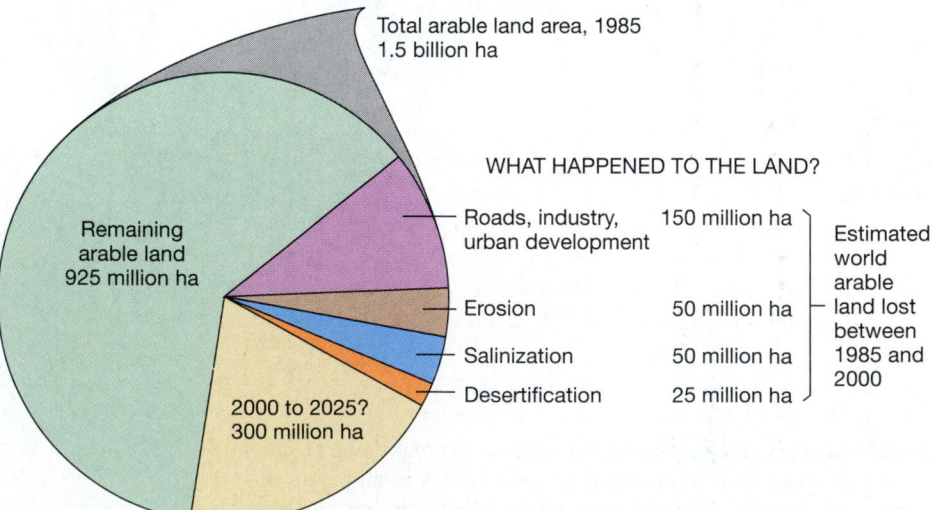

**FIGURE 44–B** Disappearing cropland. Worldwide, the area of cropland in 1985 was 1.5 billion hectares, represented by the whole circle. Experts estimate losses from 1985 to 2000, and from 2000 to 2025, as shown. This does not include 7 million hectares of rangeland lost to desertification each year.

salts and hold the soil in place. Dead leaves and twigs add organic matter to the soil. Australia has organized its Girl Scouts and other groups to plant more than a billion trees on land turned to saline desert by intensive agriculture. Tree seedlings survive best if they are watered and protected from animals for the first few years, so soil reclamation projects are most successful if residents of the area undertake to care for the trees.

Soil destruction is sometimes called **desertification,** the formation of desert from previously productive land as a result of human activities. The name arose because many of the world's deserts have been produced or enlarged in this way (Figure 44–C). Each year, desertification adds to them an estimated million hectares (nearly 4000 square miles). In pre-historic times, the Sahara Desert was less than one fifth its present size.

The problem of soil destruction is largely economic: an individual farmer profits for a short period by farming practices that destroy the soil. These practices include planting crops but not windbreak trees, irrigating arid land, or leaving land without plant cover between crops. Because the problem is economic, the solution usually involves altering government regulations and subsidies. For instance, in 1985 the United States passed a soil conservation act that rewards farmers for planting trees and grass on previously plowed land that erodes easily (such as that on steep hillsides). This program has greatly reduced the rate of soil erosion.

The desert created by destroying soil can be made to bloom again, but this is a long, slow process. The basic solution is to keep the soil covered with plants at all times and let some of this plant material decay into the soil.

If only rock or gravel remains when the soil is gone, soil will take thousands of years to form again. If sand and clay remain, the outlook is brighter. Both can be made into fertile soil by adding organic matter. This can be done intensively by planting green manure, plants that will grow in nutrient-poor soil and that are then plowed back into the soil to decompose. After only a few years of this treatment, soil will regain its fertility.

Governments around the world are beginning to realize this and to design economic incentives that will encourage farmers to conserve and restore the soil.

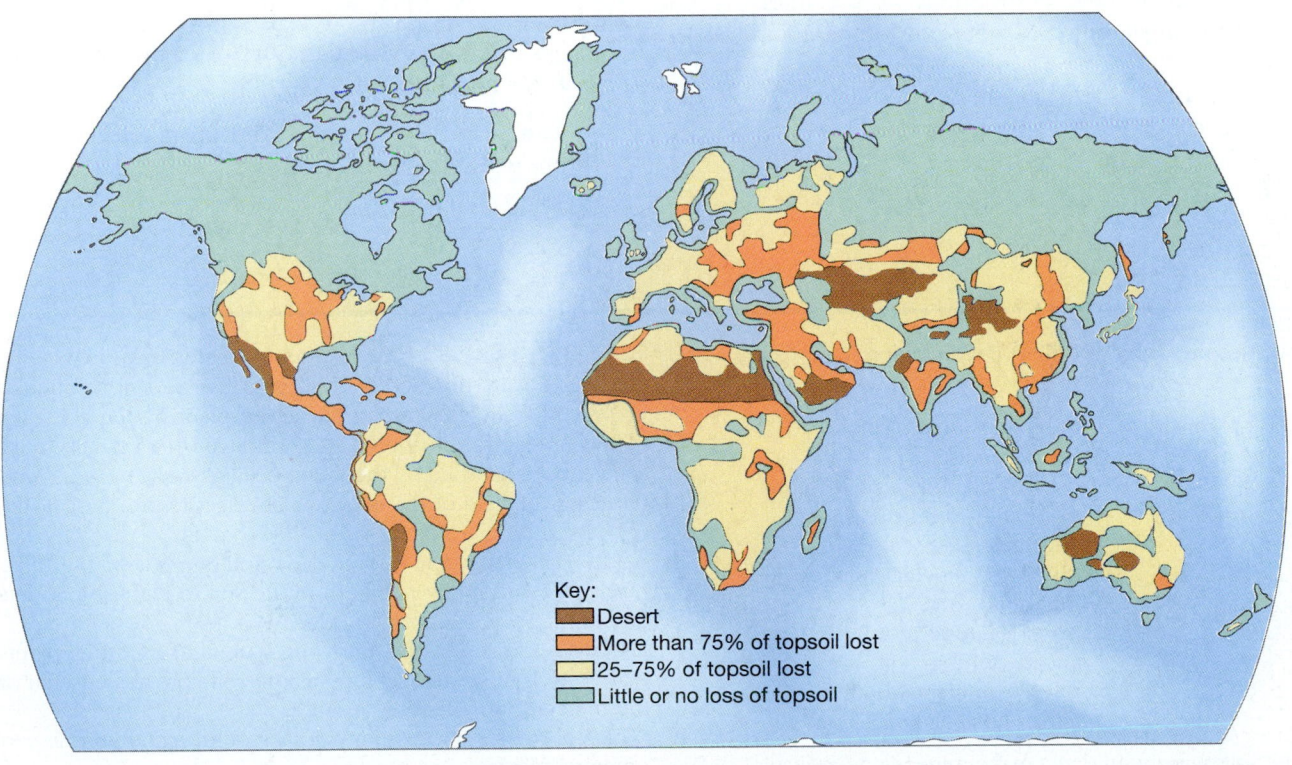

Key:
- Desert
- More than 75% of topsoil lost
- 25–75% of topsoil lost
- Little or no loss of topsoil

**FIGURE 44–C** Soil destruction around the world. (a) The worst soil loss (orange) occurs in arid areas, which lie on either side of the world's large deserts (tan). The yellow area in the Northern Hemisphere shows that much topsoil has also been lost from the world's most important food-producing regions in temperate parts of North America and Europe.

## SUMMARY

Plants obtain nutrients from carbon dioxide in the air, from water, and from substances dissolved in water. Most species take up most of their water and minerals from the soil.

1. Mineral nutrients are divided into two groups, macronutrients and micronutrients, depending on the quantities required by plants. Some nutrients are required only by plants with special adaptations.

2. The nutritional requirements of plants are difficult to assess because some micronutrients are used in very small amounts. Also, different concentrations or proportions of nutrients, and the ability of plants to use one nutrient in place of another, may change the outcome of experiments. Nutrient deficiencies of plants may result from lack of nutrients in the soil, a soil pH that makes nutrients unavailable to plants, or unfavorable proportions of one nutrient to another.

3. Soil is formed from rock broken down by the combined activities of physical and chemical weathering and from the remains of living organisms. The nature of the soil is determined mostly by the climate. The source rock, the size of the rock particles, organic matter, soil organisms, and oxygen are also important in determining the soil quality. All of these components of soil interact and affect the availability of water and minerals in the soil solution.

4. Agricultural practices such as tilling, fertilizing, or spreading lime can improve the soil to meet the requirements of crop plants.

5. The soil solution moves freely into the outer region of the apoplast: the cell wall spaces of the root's epidermis and cortex. The plasma membranes of these cells provide a large surface area for the uptake of minerals from the soil solution into the symplast, mostly by active transport. Water follows by osmosis. The endodermis forms a living, selective barrier between the soil solution and the rest of the plant. All water and minerals must move through these living endodermal cells before reaching the xylem, which transports them to the rest of the plant.

6. Plants growing in nutrient-poor conditions may show a variety of adaptations that enable them to acquire a better supply of nutrients.

   • Most plants form mycorrhizal associations with soil fungi, enabling them to take advantage of the fungus's superior ability to absorb nutrients and convert them into usable form. The fungus grows around or into the root and penetrates between, and sometimes into, the cells of the root cortex.
   • Carnivorous plants have leaves adapted to capturing, digesting, and absorbing the nutrients in the bodies of insects.
   • Epiphytes capture rain or dust from the air with their leaves.
   • Mistletoes tap the vascular systems of the host plants for water and nutrients.

7. Plants can store both organic and inorganic nutrients for future use in growth or reproduction.

## Self-Quiz

Match the following components of the soil with their role in plant nutrition:

_____ 1. Living organism
_____ 2. Organic matter
_____ 3. Oxygen
_____ 4. Rock particles
_____ 5. Water

a. ultimate source of most soil minerals
b. used in breakdown of organic molecules to release energy and minerals
c. dissolves minerals and carries them into roots
d. provides food used by fungi and bacteria in soil
e. releases minerals bound in organic molecules

6. Which of the following would *not* make minerals more available to plants?
   a. increasing the rainfall in a wet, forested area
   b. raising the pH of a very acidic soil
   c. tilling a packed-down or waterlogged soil
   d. introducing mycorrhizal fungi

7. The cell wall spaces of the root are collectively known as the (a) _____. Its function is (b) _____. The soil solution is prevented from penetrating the entire plant by the presence of the (c) _____ in the (d) _____ layer of root cells. Substances are absorbed by the living cells of the (e) _____ and (f) _____ layers of the root, which together are part of the (g) _____.

8. Aside from hydrogen, supplied by splitting water, the three nutrients required by plants from the soil in highest amounts are _____, _____, and _____.

9. List two reasons why the following statement is untrue: plants will be well nourished as long as the soil contains enough of the mineral nutrients they require.

10. Which of the following adaptations would *not* be seen in carnivorous plants?
    a. extensive root system
    b. nectar glands
    c. glands that secrete digestive enzymes
    d. trapping hairs
    e. specialized leaf shape

## Questions for Discussion

1. Determining what nutrients plants require would be much easier if a healthy plant were analyzed for its chemical makeup. However, this gives a less accurate picture of plant requirements than the more laborious method described in Section 44–A. Why is chemical analysis of the plant not a good indication of its nutritional requirements?

2. Why is vegetation often sparse in soil that contains many pebbles and boulders?

3. Plants may supply food to their mycorrhizal fungi. Why might it be advantageous to the plant to release organic compounds that stimulate the growth of nonsymbiotic soil microorganisms?

4. Boron-deficient soils are improved by applying sodium tetraborate at a recommended rate of 22 to 55 kg per hectare. One study showed that wheat plants took up only 0.3 kg of boron per hectare in one growing season. If these values are typical, what happens to the rest of the boron applied to the soil?

5. Examine Table 44–1 and see if you can guess why potassium and chlorine are highly mobile in the plant, whereas calcium, iron, manganese, boron, zinc, and molybdenum are mostly immobile.

6. From looking at Table 44–1, can you tell why the symptoms of molybdenum and nitrogen deficiency are alike?

7. Indian pipes (Figure 44–8) are flowering plants that presumably evolved from photosynthetic ancestors growing on the forest floor. Outline how their peculiar nutritional adaptations might have evolved in this habitat.

8. Design an experiment to determine how carnivorous plants obtain mineral nutrients other than nitrogen.

## Suggested Readings

Epstein, E. *Mineral Nutrition of Plants: Principles and Perspectives.* New York: John Wiley and Sons, 1972.

Epstein, E. "Roots." *Scientific American,* May 1973. Compares the form of root systems in different types of plants as well as how roots absorb minerals from the soil.

Gershuny, G. *Start with the Soil.* Emmaus, PA: Rodale Press, 1993. An organic gardener's guide to understanding and managing the soil in lawn and garden.

Heslop-Harrison, Y. "Carnivorous plants." *Scientific American,* February 1978. Briefly describes the types of carnivorous plants and presents experimental and photographic evidence on how prey is captured and digested.

Lloyd, F. E. *The Carnivorous Plants.* Waltham, MA: Chronica Botanica, 1942. The classic work on carnivorous plants.

Smith, S. E. "Physiological interactions between symbionts in vesicular-arbuscular mycorrhizal plants." *Ann. Rev. Plant Physiol. Plant Mol. Biol.* 39:221–244, 1988.

C H A P T E R

45

# Regulation and Response in Plants

## OBJECTIVES

*When you have studied this chapter, you should be able to:*

1. Name or recognize the five kinds of plant hormones (auxin, cytokinins, gibberellins, abscisic acid, and ethylene), list parts of the plant where each is produced, and discuss the roles of plant hormones.
2. Explain auxin's role in apical dominance, in phototropism of coleoptiles, and in gravitropism, and list or recognize at least two of its other roles.
3. List or recognize two main effects each for cytokinins, gibberellins, abscisic acid, ethylene, and phytochrome.
4. List five or more factors that may influence how a plant responds to application of a particular hormone.

5. Contrast tropisms, nastic responses, and electrical responses of plants, and give an example of each.
6. Explain how water shortage, heat, cold, injury, and attack by fungi or insects cause stress to plants, how plants respond to each, and why the responses to different stresses are similar to each other.
7. Explain what is meant by the terms short-day, long-day, and day-neutral plants; give the evidence that photoperiodic flowering responses are controlled by changes in phytochrome.

 plant grows through the activity of its meristems. Meristematic cells divide and produce cells that grow, differentiate, and mature (Chapter 42). How is this growth controlled? Why does one cell differentiate into an epidermal cell and another become part of a xylem vessel? Why do roots grow downward and shoots upward? What determines whether a lateral bud forms a new branch this year, next year, or never? Why do all the cherry trees in an orchard bloom at the same time in the early spring, and why do their fruits ripen over a short season in early summer? Why do all the thistles in a field bloom at once in the summer? And why do coffee trees bear flowers, un-

ripe fruits, and ripe fruits simultaneously during every season of the year? All of these events, and many other responses of plants, are under the control of plant hormones.

As in any other organism, a plant's response to internal and environmental stimuli entails three steps: perception, transduction, and response. In general, perception occurs when a stimulus interacts with its receptor. This somehow initiates transduction, conversion of the stimulus into a form that influences events in the cell, often by way of second messengers inside the cell. The response usually involves gene activation, protein synthesis, or other changes in metabolism.

## KEY CONCEPTS

♦ Every aspect of the production, differentiation, growth, and maturation of a plant is regulated by chemicals, the plant hormones.

♦ Environmental stimuli affect the production and distribution of hormones. The hormones, in turn, govern responses by which the plant adapts to these external factors.

♦ Our knowledge of chemicals that regulate plant growth affects what ends up on our dinner tables.

964

## 45–A  PLANT HORMONES

**Hormones** are chemical messengers produced in one part of an organism and affecting other parts in a manner out of all proportion to their very small concentrations. Unlike animals, plants do not have distinct endocrine organs but produce hormones in organs that have other functions as well (Figure 45–1). The types of hormones produced in an organ may change at different stages of development.

We know less about plant hormones than about animal hormones. However, current research is showing that plant hormones work in much the same way as animal hormones but of course produce outcomes that are adaptive to plants' very different way of life.

Hormones bind to receptor molecules in the plasma membrane or cytoplasm of their target cells. This triggers a response, which depends on the type and developmental stage of the target tissue and on the hormone's concentration. A hormone may change enzyme activity, membrane permeability, gene expression, or a combination of these. In each case, the presence of a small amount of hormone is amplified because it can cause the catalysis of many chemical reactions, the flow of many ions through the membrane, or the synthesis of many new protein molecules.

A cell's differentiation and metabolic activity are determined by the hormones and other chemicals it contains, which in turn are affected by the rest of the plant. Hormonal messages passing between different parts of the plant carry information about the state of affairs within the body and regulate growth of new parts. Thus, the root system remains in physiological balance with the shoot system, and the position of a branch or leaf or root determines where others are added.

Plants also respond to stimuli in the physical environment, such as light, temperature, wind, water, the pull of gravity, the change of seasons, and invasion by pathogens. These responses, too, are mediated by hormones and other chemical messengers.

The traditional list of plant hormones has five major entries: auxin, cytokinins, and gibberellins, which induce cell division or cell growth; and abscisic acid and ethylene. Each plays a leading role in certain activities (Table 45–1). However, many plant responses are governed by combinations or sequences of two or more hormones.

More recently, investigators have identified many other chemical messengers with regulatory roles in plants, often as partners or intermediaries for some of the five major hormones.

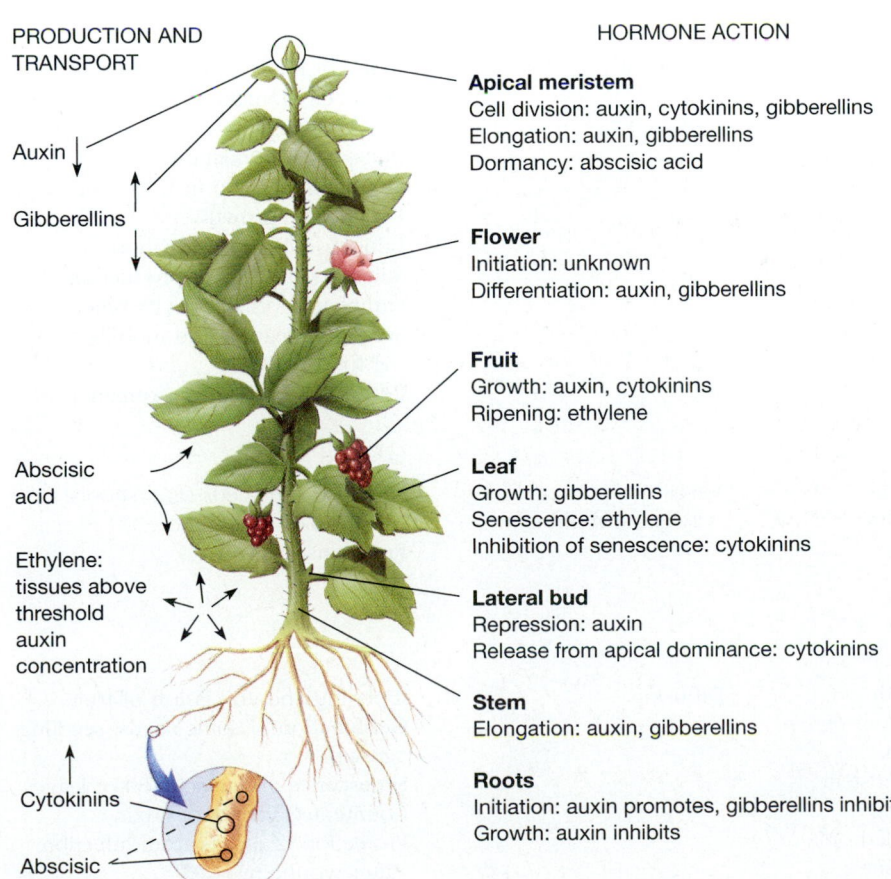

PRODUCTION AND TRANSPORT

Auxin

Gibberellins

Abscisic acid

Ethylene: tissues above threshold auxin concentration

Cytokinins

Abscisic acid

HORMONE ACTION

**Apical meristem**
Cell division: auxin, cytokinins, gibberellins
Elongation: auxin, gibberellins
Dormancy: abscisic acid

**Flower**
Initiation: unknown
Differentiation: auxin, gibberellins

**Fruit**
Growth: auxin, cytokinins
Ripening: ethylene

**Leaf**
Growth: gibberellins
Senescence: ethylene
Inhibition of senescence: cytokinins

**Lateral bud**
Repression: auxin
Release from apical dominance: cytokinins

**Stem**
Elongation: auxin, gibberellins

**Roots**
Initiation: auxin promotes, gibberellins inhibit
Growth: auxin inhibits

**FIGURE 45–1** Sites of hormone production in a plant. Auxin is produced in growing regions of shoots, including apical buds and young leaves. Gibberellins are produced in young leaves. Cytokinins are produced in the apical meristems of roots. Abscisic acid is produced in older leaves and in roots. Ethylene is produced in many areas when the concentration of auxin exceeds a certain threshold. Arrows indicate direction in which each hormone is transported.

**TABLE 45-1**

## The Major Plant Hormones

| Hormone | Produced in | Transported by | Actions |
|---|---|---|---|
| Auxin Indoleacetic acid | Young tissues: shoot apical meristem, young leaves, coleoptile tip, developing fruits and seeds | Slow cell-by-cell movement toward roots | Meristem growth, cell division and enlargement<br>Apical dominance<br>Xylem differentiation<br>Sex differentiation in flower parts<br>Fruit development and growth<br>Formation of new roots<br>Inhibition of root cell elongation<br>Stimulation of stem cell elongation<br>Promotion of ethylene production<br>Gravitropism: positive in roots, negative in shoots<br>Positive phototropism in coleoptiles<br>Maintenance of tissue polarity |
| Cytokinins Zeatin | Roots | Xylem | Cell division<br>Growth, including release of lateral buds from auxin inhibition<br>Production of fruits and seeds<br>Inhibition of dormancy<br>Inhibition of senescence |
| Gibberellins Gibberellic acid | Embryo<br>Young leaves, buds, upper stem | Unknown | Cell division in apical meristem<br>Stem cell elongation<br>Leaf growth<br>Phloem differentiation<br>Sex differentiation in flower parts<br>Pollen tube growth<br>Inhibition of root formation<br>Stimulation of auxin production<br>Stimulation of fruit development<br>Seed germination and mobilization of food reserves<br>Release of buds from dormancy |
| Abscisic acid | Cells containing plastids: mature leaves, roots | Vascular tissue (xylem and phloem) | Dormancy in seeds (and shoots?)<br>Initiation of senescence<br>Reactions to stress |
| Ethylene | Plant parts with high auxin concentration, especially meristem areas, ripening fruits, wounded or senescing tissue | Diffusion | Ripening and abscission of fruit<br>Germination of some seeds; seedling growth<br>Senescence and abscission of leaves<br>Counteracts effects of auxin<br>Protection of leaves from infection after wounding |

## 45–B  AUXIN

The first plant hormone to be discovered was **auxin,** which was isolated and identified as **indoleacetic acid** (Figure 45–2).

Auxin plays a role in many aspects of plant development—cell division, elongation, and differentiation—and has many different effects. For example, in shoots auxin stimulates meristematic growth and cell division but inhibits the growth of lateral buds (apical dominance, Section 45–H). Auxin also induces production of xylem, differentiation of sexual parts in flowers, and development of fruits. Although auxin stimulates formation of new roots, it inhibits elongation of root cells except at very low concentrations. High concentrations of auxin promote production of ethylene, another plant hormone.

Auxin is produced in the tips of coleoptiles and in growing apical regions of shoots, including young leaves. From here it is transported primarily toward the roots, not rapidly through the vascular tissues but slowly from one cell to another: auxin diffuses into the top of a cell and is ejected by proteins in the membrane at the bottom, using energy from ATP. It then diffuses into the next cell below, and so on. The production and polar transport of auxin helps organize the top and bottom directions in the polarity of the plant body.

Auxin also mediates many responses to the environment. It plays a role in the bending of stems and leaves toward light and in the growth of shoots upward and roots downward.

Auxin also mediates the production of secondary xylem. More auxin is produced in seasons with good growing conditions, and this results in wider growth rings (Section 43–B). Auxin also induces production of more wood in response to mechanical disturbance, such as wind. Trees grown in greenhouses or held up by guy wires do not toss in the wind. Compared with trees grown outdoors with no support, they are rather spindly. However, if greenhouse trees are periodically shaken, they produce additional xylem, which thickens and strengthens the tree trunk.

Auxin receptors occur on the cell surface and in the cytoplasm. Like animal steroid hormones, auxin inside the cell is a signal molecule stimulating transcription. Within five minutes of its arrival in a meristematic cell, genes coding for transcription factors are expressed. Some of these transcription factors are among the shortest-lived proteins known, surviving for less than 10 minutes. During this time, they activate other genes, resulting in a short burst of cell division that then stops unless more auxin molecules arrive.

When auxin binds to surface receptors, it stimulates cell growth. Auxin binding activates these receptors to export hydrogen ions ($H^+$) from the cell. Here this acid breaks some cross-links between the cellulose fibers of the cell

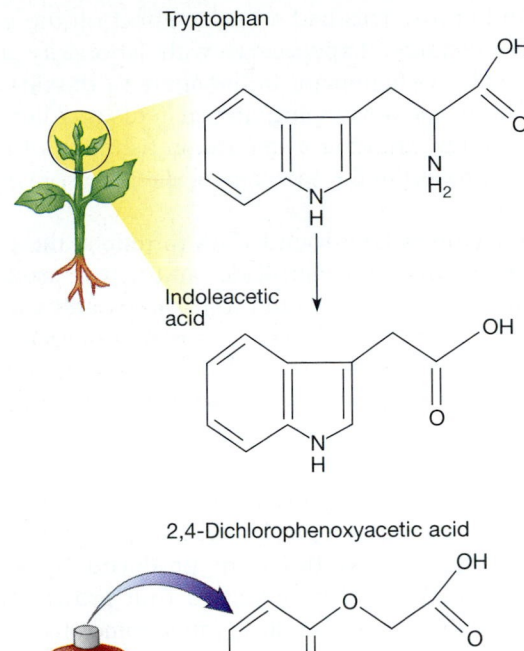

**FIGURE 45–2**  Auxin. Indoleacetic acid is synthesized from the amino acid tryptophan. 2,4-Dichlorophenoxyacetic acid (2,4-D) is a synthetic auxin used to kill weeds.

wall and activates enzymes that hydrolyze others. This loosening lowers the wall pressure and permits the cell's internal turgor pressure to push the wall outward. Turgor is maintained as more water enters the cell by osmosis. This expansion starts in the first half hour after auxin is supplied. Afterward, auxin continues to speed the synthesis of growth-promoting proteins.

Auxin also increases the concentration of calcium ions in the cytoplasm of some cells. By analogy with animals and microorganisms, this is thought to occur by a mechanism using G proteins to liberate calcium as a second messenger, which in turn activates kinase enzymes.

Synthetic auxins, such as 2,4-D, are used as weed killers in lawns (see Figure 45–2). Dicots, which include broad-leaved weeds, are much more sensitive to low levels of auxins than are grasses and other monocots. Hence, applying a low dosage of synthetic auxins to a lawn can cause the dicots to grow themselves to death, meanwhile becoming grossly deformed. Monocots are not visibly harmed. To be most effective, these herbicides must be applied when the weeds are actively growing and therefore sensitive to extra doses of hormones.

During the Vietnam conflict, American airplanes sprayed Agent Orange, a herbicide containing synthetic auxins, on forests and crops. This had a drastic effect on the ecology of the countryside. Experiments with laboratory animals showed that a contaminant in the spray (a dioxin) causes malformations of developing animal fetuses. After dioxin was detected in drinking water and in fish, one of the few sources of protein in the Vietnamese diet, the spraying was abandoned.

Manufacturers later found ways to reduce the concentration of dioxin in the herbicide, and it was used in the United States to kill unwanted forest trees and to control weeds on grazing ranges. However, further research showed that some dioxins are carcinogens, even in minute concentrations. After dioxin was found in beef fat and in human milk, the herbicide was banned.

Other synthetic auxins are sprayed on pear and apple trees to keep them from dropping fruits before they ripen or applied to cuttings of plant shoots to promote root formation. Some seedless fruits are produced by spraying plants with auxin, which stimulates fruit growth but not seed maturation. Other seedless fruits come from genetic strains especially developed to have naturally high auxin levels.

## 45–C    CYTOKININS

The first known cytokinin was a component of coconut milk. This compound proved difficult to isolate, but whole coconut milk became a standard ingredient in plant tissue cultures because it stimulates cell division. Without cytokinins, cells grow larger but do not divide.

Cytokinins were eventually isolated from coconut milk and from other sources. They are variations on the nitrogenous base adenine (see Table 45–1). Their action is not well understood. They are part of certain transfer RNAs and seem to be necessary for binding to ribosomes. Cytokinins also seem to participate in some change that occurs after DNA replication and initiates cell division.

Cytokinins interact with auxin in growth, including production of fruits, and they stimulate seed production. They also prevent the onset of dormancy and slow the aging of cut leaves or fruits. Holly that is cut early for the holiday season is kept fresh and green by spraying it with cytokinins. However, cytokinins are not used on food crops because nucleotide-like molecules are thought to be possible carcinogens.

Cytokinins are produced in the apical meristems of roots and transported through the xylem to the leaves and the shoot meristems, exactly the opposite of auxin, which is produced in shoot tips and transported downward. The amounts of auxin and cytokinins in a plant are proportional to the number of shoot and root endings producing them, and this exchange of messages keeps the development of shoots and roots in balance.

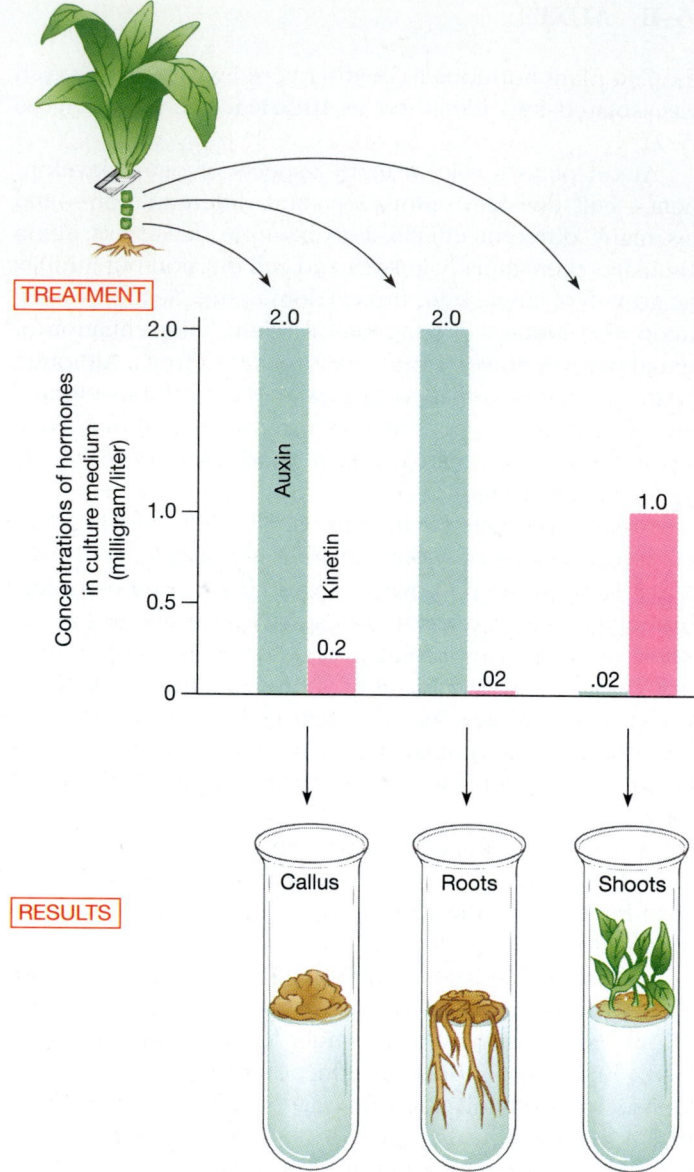

**FIGURE 45–3**    Differentiation of cultured plant tissue. Plugs of pith from a tobacco stem are grown on nutrient media containing auxin and kinetin, an artificial cytokinin. Depending on the concentrations supplied, these explants may grow as a lump of undifferentiated tissue (callus) or develop roots or shoots.

Experiments in tissue culture suggest that differentiation is largely determined by the relative concentrations of cytokinins and auxin (Figure 45–3). To grow an entire plant from cultured cells, researchers must apply hormones at certain concentrations, in particular ratios to each other, and in the correct sequence. Presumably, similar hormonal interactions in intact plants govern the switching on and off of genes, resulting in the differentiation of the various tissues and organs.

## 45–D   GIBBERELLINS

**Gibberellins** are an important group of plant hormones, comprising more than 80 related compounds. The first one known, discovered in the 1920s by Eiichi Kurosawa in Japan, was actually produced by a fungus, *Gibberella fujikoroi,* which causes "foolish seedling" disease of rice. Rice seedlings with the disease grow abnormally rapidly but are spindly and unhealthy and seldom yield fruit. The substance responsible for the rapid growth was isolated from cultures of the fungus and named **gibberellic acid** (see Table 45–1). Other gibberellins are produced not only by fungi but also by vascular plants and by brown and green algae.

Gibberellins are produced in young leaves, buds, and upper stems. From here they are distributed throughout the plant, and they affect virtually every part. Cell-surface gibberellin receptors have been identified.

One important role of gibberellins in plants is elongation in stem cells. Dwarf or miniature strains of plants are often genetic mutants that do not produce gibberellins but grow as tall as normal varieties if treated with gibberellins at the proper concentrations. Giving extra gibberellins to normal plants has little effect on their height (Figure 45–4).

**FIGURE 45–4**   Effect of gibberellin on dwarf and normal strains of corn plants. Left to right: Dwarf plant, untreated; dwarf plant treated with gibberellin; normal plant treated with gibberellin; normal plant, untreated. The dwarf plant has responded to gibberellin much more dramatically than the normal plant. The dwarf plant is homozygous recessive for a gene required for gibberellin production.   (B. O. Phinney)

Other effects of gibberellins include stimulation of leaf growth, especially in monocots, inhibition of root formation, stimulation of cell division at the stem apex, and stimulation of the development of flower parts. Gibberellins in germinating seeds turn on the genes for amylase, which breaks down stored starch into the sugars needed for growth. In conjunction with sugars, gibberellins also stimulate differentiation of phloem tissue. Gibberellins also seem to stimulate the production of auxin, as well as making plants more responsive to auxin treatment.

Gibberellins are also involved in the responses of plants to their environment. Some plants that require a cold period or exposure to light in order to germinate or to flower will make these responses instead after gibberellin treatment. However, flowering occurs in a different sequence, depending on whether cold or gibberellin was the initial stimulus. Cold-treated plants form flower buds before the flower stalk elongates, whereas gibberellin-treated plants form elongated stems, which can then form flower buds without cold treatment.

Gibberellins are used commercially in the wine and beer industries. They are sprayed in vineyards to elongate the stalks of the fruit cluster, giving each grape more space to expand. They are also used to speed germination of barley seeds, which are fermented to produce beer. Despite this head start, gibberellins are not applied to other crops because they often make the seedlings spindly and weak.

## 45–E   ABSCISIC ACID

Auxin, cytokinins, and gibberellins have many growth-stimulating roles. In contrast, **abscisic acid** is often called the "stress hormone" because of its roles in helping the plant cope with adverse environmental conditions. For example, roots growing in drying soil produce abscisic acid, which boosts uptake of ions, and therefore of water, and induces formation of branch roots, enabling the plant to obtain more water. It also moves in the transpiration stream to the leaves. Here its binding to receptors in the plasma membranes of guard cells opens ion channels and activates $H^+$ pumps. The guard cells lose ions, and therefore the stomata close. This root-to-shoot message reduces the plant's loss of water by transpiration during periods of water shortage even before the leaves start losing too much water (Section 43–C).

Abscisic acid was discovered almost simultaneously by P. F. Wareing, working with sycamore leaves, and Frederick Addicott, working with cotton bolls, which are seed pods containing the fibers used to make cotton thread and fabrics. In both cases, the substance isolated was found to accumulate in tissue as it aged and to counteract the effects of auxin application. This hormone was named **abscisic acid** (see Table 45–1).

Abscisic acid produced by mature leaves is thought to play a role in inducing dormancy in shoots. At the end of the growing season, its concentration in twigs is high compared with the levels of gibberellins and cytokinins. This induces the apical meristem to stop dividing. The newest leaf primordia form into bud scales around the tip instead of becoming normal photosynthetic leaves. The bud scales protect the delicate apical meristem from freezing or drying out in the cold of winter. In spring, overwintering twigs are released from dormancy by destruction of abscisic acid or production of substances that counteract it (Figure 45–5).

Abscisic acid also keeps some seeds dormant until their metabolism destroys enough of it to break dormancy, or until they produce enough of a stimulatory hormone, usually gibberellin. Dormancy prevents the embryo from growing into a tender seedling while cold or dry weather prevails. Earlier, as the embryo itself develops from the zygote, abscisic acid promotes the embryo's growth, synthesis of storage proteins, and drying as the seed enters dormancy.

Both abscisic acid and gibberellins are derived from a common precursor (mevalonic acid), by two different metabolic pathways. The path taken from this branch point determines many aspects of growth because the two hormones have antagonistic effects on stem growth, on dormancy, and on flowering in some short-day and long-day plants (Section 45–J).

## 45–F    ETHYLENE

Unlike the other plant hormones, **ethylene** is a gas (see Table 45–1). Its most notable effects are promotion of fruit ripening and plant senescence, including abscission (dropping) of fruits and leaves (Section 45–L).

Nineteenth-century growers of greenhouse plants knew that something in the gas used in gas lighting caused blossoms to wither prematurely. Mango and pineapple growers lit fires near the trees because something in the smoke synchronized flowering and fruit ripening. Eventually, ethylene was shown to be the gas responsible for these effects.

The production and effects of ethylene are closely tied to auxin. After auxin exceeds a certain level, it stimulates production of ethylene, which in turn counteracts the effect of auxin. Thus, production of ethylene by lateral buds as they receive auxin seems to play some part in repressing their growth.

After fruit has developed in response to stimulation by auxin, gibberellins, and cytokinins, ethylene promotes ripening. The level of auxin builds up and then drops; this is followed by production of ethylene. The ethylene receptor is believed to lie in the plasma membrane. Its binding activates a transcription factor to turn on genes for enzymes made only in fruits. Some convert starch and acids of the unripe fruit to sugars, making it sweeter; some syn-

thesize aromatic molecules, giving it flavor; and pectinase breaks down pectins in the cell walls, making it softer.

Some fruits, such as apples, bananas, and tomatoes, mature slowly and then undergo a surge of metabolism and ripening called the **climacteric.** This occurs because in these fruits the release of ethylene has a positive feedback effect: the more ethylene produced, the more the fruit is stimulated to produce additional ethylene. Hence, the fruit ripens all at once, and if there are many fruits in an area, the first to begin ripening stimulates ripening of its neighbors. By the same token, over-ripening is also contagious, and it is quite true that "one bad apple spoils all the good ones." Apples are now stored in refrigerated, airtight rooms in an atmosphere enriched with carbon dioxide, which inhibits the action of ethylene.

Fruits such as cherries, lemons, and oranges do not undergo the climacteric burst but have a steady ethylene concentration. These fruits tend to keep better during shipping and storage. The green fungus that often grows on oranges is *Penicillium.* It produces ethylene, which causes over-ripening in other oranges nearby, giving the fungus easy access to their soft, sweet interior.

Ethylene finds uses in the production of tropical fruits for far-away markets. Rather than letting the fruit ripen on the trees and risk loss by over-ripening in transit, growers pick fruits such as bananas, citrus fruits, and pineapples when they are still green and ripen them by applying ethylene after they reach their destination.

Because ethylene is a gas, it travels in the air as well as within the plant, and it affects not only the plant that produced it but others as well. For example, ethylene given off by apples inhibits the sprouting of potatoes stored in the same bin.

Ethylene is also formed by vegetative parts of the plant as a reaction to invading fungi. This ethylene activates genes for antifungal enzymes, including chitinase, which breaks down chitin in the fungal walls.

## 45–G    OTHER CHEMICAL REGULATORS

In addition to the five major plant hormones, with their wide-ranging effects, researchers have recently been discovering substances in plants that play more limited roles.

### Polyamines

Polyamines are molecules with many amino groups ($—NH_2$). Both polyamines and ethylene are produced from a derivative of the amino acid methionine, by alternative pathways. Polyamines inhibit ethylene production and counteract some of its effects. For example, tomatoes that produce higher-than-average concentrations of polyamines

(a) Entering dormancy

(b) Breaking dormancy

**FIGURE 45–5**   Azalea buds. As dormancy sets in, new leaves remain small and develop as tough, waxy bud scales containing little water and protecting the terminal bud, with its reduced water content, from further water loss.

ripen slowly and therefore have an extended shelf life. Treating normal tomatoes with polyamines has the same effect.

Polyamines are also produced when plants are stressed and are thought to give some protection from desiccation, salinity, and some pollutants.

## Polypeptides

Polypeptides have long been known to play prominent roles as hormones in animals, but the first polypeptide known to act as a plant hormone was discovered in 1991 by Clarence Ryan and colleagues after a 20-year search. Named **systemin** because it is transported throughout the plant, it consists of 18 amino acid residues. As little as one part in 10 trillion can trigger a stress reaction that produces proteinase inhibitors, molecules that may kill leaf-eating insects by preventing them from digesting proteins in their food. Researchers expect to find more polypeptide hormones in plants soon.

## Oligosaccharins

**Oligosaccharins** (oligo = few) consist of sugar residues in short branched chains. Such carbohydrates are challenging to analyze, especially because they occur at extremely low concentrations. Techniques to detect and study these substances were developed only recently, and so we still know little about them.

Oligosaccharins are released when enzymes cleave larger, complex polysaccharides in the cell wall matrix (the material in which the wall's cellulose fibers are embedded). Each oligosaccharin is liberated by a different enzyme, interacts with a membrane receptor in a hormone-like manner, and mediates a particular response by affecting gene expression.

Some oligosaccharins are produced in wounded tissue and help to defend the plant by inducing localized production of proteinase inhibitors (whereas systemin induces this throughout the shoot). Others cause production of enzymes that make **phytoalexins**, pathogen-killing antibiotics. Some may also act as second messengers, mediating a specific effect of one of the five major plant hormones. For example, auxin stimulates the production of one oligosaccharin, which in turn counteracts auxin-stimulated cell elongation, apparently serving as part of negative feedback regulation of auxin's effect.

## Salicylic Acid

**Salicylic acid** is named for the willow genus, *Salix*, from which it was first extracted. Aspirin (acetylsalicylic acid) is derived from salicylic acid. Recently, salicylic acid was also found to be a chemical regulator that helps defend plants against attack by pests or pathogens. When part of a plant is attacked, salicylic acid spreads throughout the plant. Once bound to its receptor inside the plant's cells, salicylic acid switches on the genes for a **systemic acquired resistance**

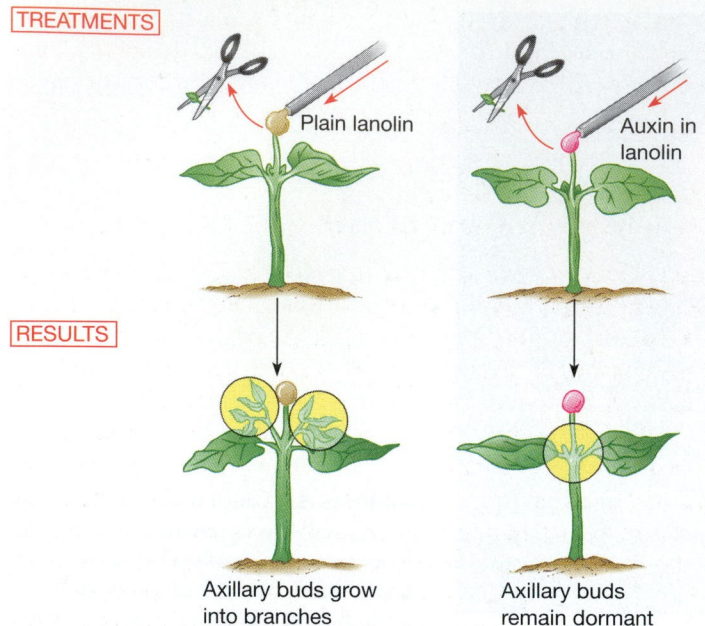

TREATMENTS

Plain lanolin

Auxin in lanolin

RESULTS

Axillary buds grow
into branches

Axillary buds
remain dormant

**FIGURE 45–6**  Demonstration that apical dominance is mediated by auxin. (Left) Removal of the apical bud permits growth of axillary buds, but if the apical bud is replaced with an auxin solution (right), the axillary buds are still inhibited from growing. (Why is plain lanolin applied to the plant on the left?)

## 45–H  APICAL DOMINANCE

Many types of plants exhibit **apical dominance,** in which the growing apical bud represses growth of lateral buds on the same stem, causing them to remain dormant. This is due primarily to auxin produced in the leaf primordia of the apical bud and transported toward the base of the plant. If the apical bud is cut off, the lateral buds develop into new branches. This can be prevented by placing auxin on the cut stump immediately after the apical bud is removed (Figure 45–6).

How can the apical meristem continue to grow in the presence of its locally produced auxin? A clue to the answer can be found by delaying the application of auxin in the experiment shown in Figure 45–6: once the lateral buds break dormancy and begin to grow as the apical meristems of new lateral branches, applying auxin cannot make them dormant again. Thus, the hormone's effect must depend on the physiological state of the tissue that receives it. If the meristem is active, auxin promotes its growth; if it is dormant, continued exposure to auxin keeps it dormant.

Other plant hormones may interact with auxin in apical dominance. Auxin stimulates lateral buds to produce ethylene, which in turn seems to play some part in inhibiting the buds' growth. In contrast, cytokinins may help to overcome apical dominance. Because cytokinins come from the roots, it has been proposed that a plant's lower buds break dormancy before those nearer the shoot's apical meristem because they receive more cytokinins and less auxin.

It is also known that food and water are transported preferentially into areas of the plant where auxin concentration is high. Removing the apical bud (and its auxin) may permit transport of more food to the lateral buds, and as

**response,** in which the cells produce proteins that fight infection and promote wound-healing.

Salicylic acid also causes plants such as voodoo lilies to produce a burst of heat, which volatilizes putrescent-smelling molecules that attract pollinating flies.

(a)

(b)

**FIGURE 45–7**  Tree shapes. (a) Many conifers show strong apical dominance. The main trunk grows much faster than the branches, and the terminal bud represses the growth of the nearest branches more than those farther away. (b) By contrast, very weak apical dominance results in a rounded, spreading shape, with many large branches.

they grow and produce their own auxin, they receive more nutrients.

It is hard to design experiments to show how each of these factors contributes to inhibition or growth of lateral buds.

A plant with strong apical dominance has a distinct main axis with few side branches, whereas a plant with weak apical dominance has many branches and a bushy appearance. Gardeners often pinch out apical buds to make their plants grow into compact, bushy shapes.

The degree of apical dominance varies a great deal among plants (Figure 45–7). This variation is partly genetic and partly a response to the environment. A beech tree in a forest is tall and slender, with short, thin branches, whereas another, growing in the open, is wide and spreading, and a third growing at the edge of the wood has stout, spreading branches on the open side and small, thin branches on the shady side. These variations in shape suggest that light may somehow counteract apical dominance; this allows the tree to grow new branches into any openings in the forest canopy.

## 45–I   RESPONSE TO THE ENVIRONMENT

A plant's growth and differentiation are governed by the interactions of hormones, produced in various parts of the plant and transported throughout its body in varying concentrations. Environmental factors influence hormone production and distribution. This results in adaptive adjustments in growth, as when light shining on a lateral bud helps to release it from apical dominance: the bud grows into a new branch, with leaves that use the light for photosynthesis and make more food for the plant. In addition to light, plants can respond to gravity, touch, heat, cold, oxygen shortage, and attack by insects and pathogenic fungi.

## Tropisms: Oriented Growth

**Tropisms** are growth responses oriented with respect to the direction of environmental stimuli. For example, **gravitropism** is the response of a plant to gravity. **Thigmotropism** is a response to contact, as when climbing plants wrap themselves around a supporting object.

Botanists once thought that roots exhibit **hydrotropism,** growing toward water. However, it now appears that roots simply grow more when they receive more water, and so a root that happens to be growing toward a source of water grows faster. Roots do not turn in the direction of water, as would be required for a true tropic response.

Tropisms orient a plant's parts toward resources. The first root of a germinating seed always grows downward, no matter what the position of the seed in the soil (Figure 45–8). This **positive gravitropism** ensures that the root anchors the plant. It also gives the best chance of obtaining water and nutrients.

Unlike roots, most shoots show **negative gravitropism,** growing away from the pull of gravity. Hence, a seedling's shoot grows upward and reaches the soil surface by the shortest route, conserving the seed's stored energy. Once the shoot reaches light, it also shows **positive phototropism,** growing toward light. If light shines mainly from one side, a plant bends in that direction (Figure 45–8b).

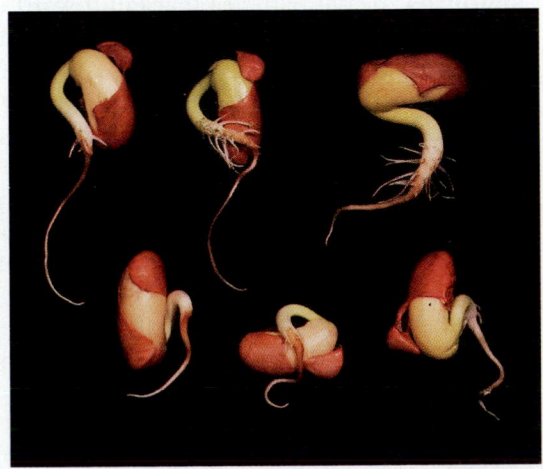

(a) Positive gravitropism in roots

(b) Positive phototropism in shoots

**FIGURE 45–8**   Tropisms. (a) Positive gravitropism: these bean seeds were germinated in the dark in several positions, but the roots always grew downward. (b) These houseplants show positive phototropism, growing sideways toward the window, the primary source of light, and positioning the leaves perpendicular to the sun's rays.   (b, Dennis Drenner)

Both negative gravitropism and positive phototropism in shoots are mediated by auxin.

***Phototropism***    Auxin was first detected because of its role in phototropism in oat seedlings (*How Do We Know? The Discovery of Auxin*). A tubular leaf, the **coleoptile,** sheaths the younger leaves of the oat seedling and other grasses and cereals (Figure 45–9). In the dark, a coleoptile grows upward (negative gravitropism). Light stimulates expansion of the leaves within the coleoptile. They rupture the coleoptile and grow out into the light. If the coleoptile is lit from one side, it shows positive phototropism and bends toward the light.

Charles Darwin and his son Francis studied this phototropic response. The coleoptile bends in the zone of elongation, but the Darwins found that covering this area did not affect phototropism. However, when they removed the coleoptile tip or covered it with an opaque cap, phototropic bending did not occur. They concluded that this response was controlled by the tip (Figure 45–10).

In order to respond to light, the plant must first detect it. Because light in the blue and ultraviolet range (440 to 480 nm) evokes the strongest phototropic response, a yellow or orange pigment must be the receptor. There is some evidence suggesting that this is a flavoprotein, which consists of a protein plus the colored, light-absorbing coenzyme riboflavin, a vitamin.

The coleoptile bends because cells on the shaded side of the zone of elongation lengthen more than those on the lighted side, until the tip is receiving light uniformly on all sides (see Figure 45–A). The coleoptile then grows straight toward the light source. So plants actually grow toward light by growing away from shade.

These results suggest that phototropic bending occurs because there is a higher auxin concentration on the shaded side than on the illuminated side of the coleoptile, producing greater elongation on the shaded side. Thus, auxin must either be produced in greater amounts on the shaded side, destroyed on the lighted side, or transported to the shaded side in response to the stimulus of light falling on the lighted side. By applying radioactively labeled auxin to coleoptiles, investigators established that auxin is transported from the lighted to the shaded side, where it stimulates increased cell growth. This also occurs in other tissues without much pigmentation.

Phototropism in green stems appears to be different. The shaded side grows about as much as it otherwise would, and the lighted side grows more slowly. Probably light causes the cells on this side to produce a growth inhibitor. Several inhibitory substances become more concentrated on the lighted side soon after illumination begins.

***Gravitropism***    Research shows that upward growth of coleoptiles and of stems (negative gravitropism) and downward growth of roots (positive gravitropism) all involve different response mechanisms, not yet fully worked out. However, in all three, gravity is perceived by means of **statoliths,** heavy particles that fall toward the bottoms of cells. These are usually amyloplasts, containing many starch grains of high density, but in starchless mutants other dense organelles apparently substitute in this role.

In roots, gravity is detected by means of statoliths in root cap cells: in most species, removing the root cap abolishes gravitropism. However, the root actually bends farther back, in the zone of elongation (see Figure 42-6). In a stem or coleoptile, gravity perception and the bending response occur in the same section of the organ.

Differential transport of auxin is required for gravitropism: applying inhibitors of auxin transport also inhibits bending. In roots, auxin is transported down the center to the root cap and then laterally back toward the zone of elongation. More auxin is transported to the lower side of both roots and shoots. In shoots, this auxin distribution causes cells on the lower side to elongate more, and the stem or coleoptile bends upward. In contrast, root cell growth is inhibited by all but the lowest levels of auxin. The uppermost cells elongate more because their lower auxin concentration inhibits their growth less (Figure 45–11).

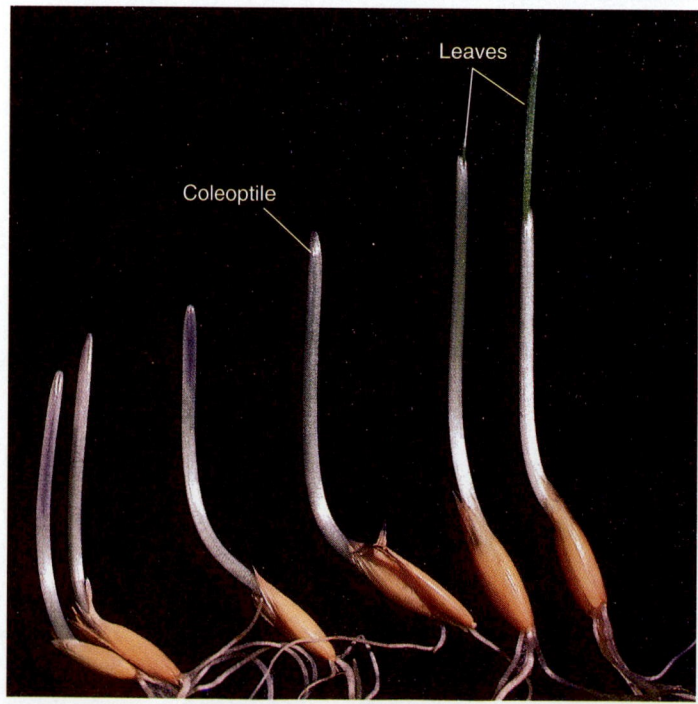

**FIGURE 45–9**  Oat seedlings. The coleoptile, a tubular outer leaf, is negatively gravitropic, growing upward, and positively phototropic, growing toward light. Illumination of the coleoptile stimulates the leaves within it to expand, rupturing its tip. The inner leaves then continue to grow out into the light where the coleoptile has positioned them, as seen in the two seedlings at the right.

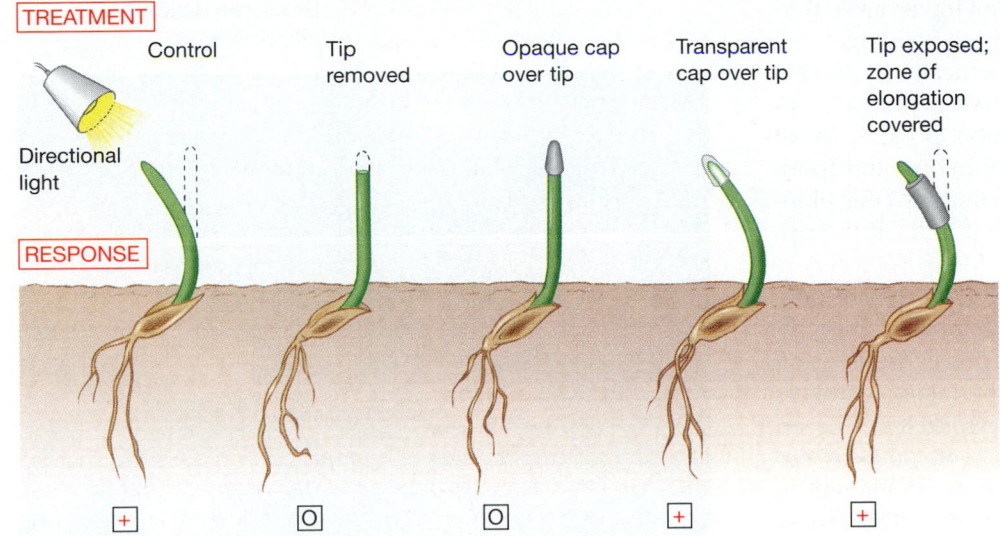

**TREATMENT**

Control — Tip removed — Opaque cap over tip — Transparent cap over tip — Tip exposed; zone of elongation covered

Directional light

**RESPONSE**

FIGURE 45–10 Phototropism in the coleoptiles of oat seedlings. These experiments by the Darwins showed that the coleoptile tip controls the bending response in the zone of elongation. The response to each treatment is scored as + if bending occurs, 0 if it does not.

Hence, the root bends until it is growing straight down and the auxin concentration equalizes.

Auxin is not the whole story of gravitropism because bending is not tightly correlated with auxin concentrations. Experiments suggest that gravitropism also involves an inhibitor, probably calcium ions (and possibly abscisic acid

in some species). In gravity-stimulated plants, calcium ions accumulate on the upper side of stems and the lower sides of roots. Calcium is known to inhibit cell elongation, and when calcium is applied to one side of a vertical root, the root bends to that side instead of growing straight down. Furthermore, applying a calcium-binding substance, which

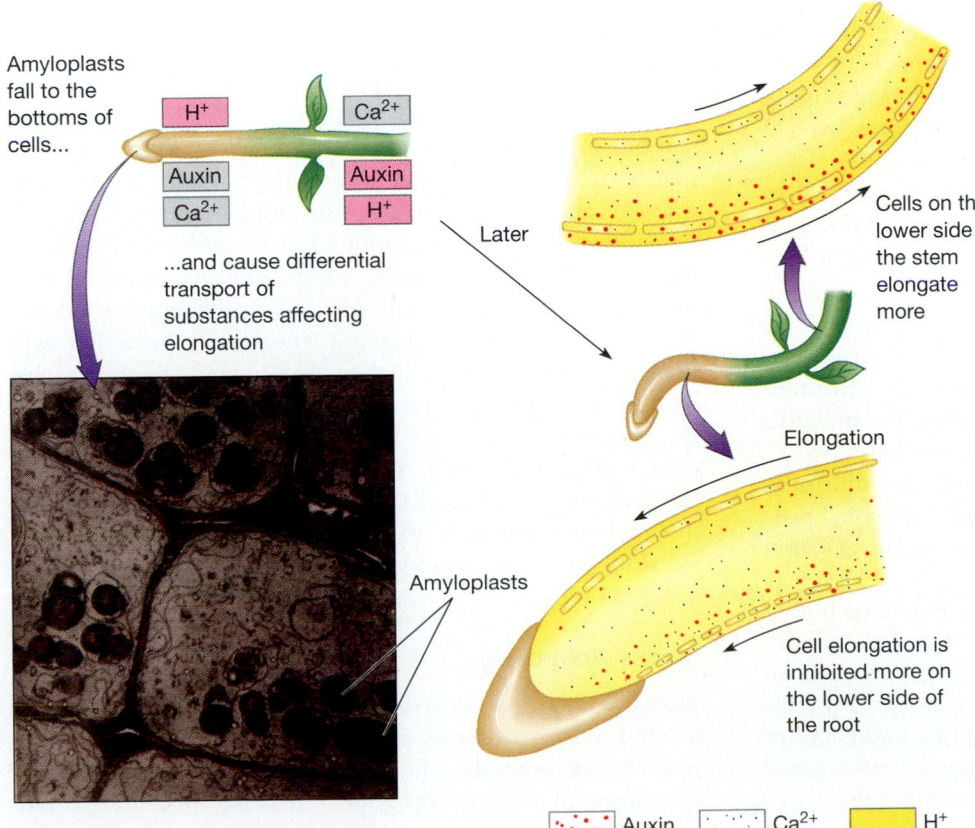

Amyloplasts fall to the bottoms of cells...

$H^+$   $Ca^{2+}$
Auxin   Auxin
$Ca^{2+}$   $H^+$

...and cause differential transport of substances affecting elongation

Later

Cells on the lower side of the stem elongate more

Elongation

Amyloplasts

Cell elongation is inhibited more on the lower side of the root

Auxin   $Ca^{2+}$   $H^+$

FIGURE 45–11 Gravitropism. (left) Statoliths in root cap and stem cells fall downward, stimulating differential transport of auxin and of calcium and hydrogen ions. These substances are distributed differently in roots and stems: here each is named on the side where it is more concentrated. Pink = promotion of elongation; gray = inhibition. (right) These differences in concentration result in differential elongation of cells on the upper and lower sides of roots and stems. Each bends in the appropriate direction. (photo, Barrie Juniper)

ties up the ions and keeps them from moving around, permits growth to continue but not gravitropic bending.

$H^+$ ions are also distributed asymmetrically in gravity-stimulated plants. They accumulate more on the side opposite the higher calcium ion concentration, possibly by an electrically driven counterflow to calcium ion transport. Growth promoted by $H^+$ may help explain why cells elongate more on the lower side of the shoot and on the upper side of the root.

One important missing piece of the gravitropism puzzle is how statoliths change the distribution of growth-mediating substances to the upper and lower sides of an organ. Some hypotheses propose that the falling statoliths exert pressure on components of the cytoskeleton or on membranes of calcium-storing compartments such as endoplasmic reticulum. This in turn causes some change in the pumping of both auxin and calcium ions. The movements of $Ca^{2+}$ and auxin are linked, going in the same direction (downward) in roots and in opposite directions in shoots.

## Nastic Responses

In contrast to tropisms, **nastic responses** are stereotyped responses that do not involve growth nor depend on the direction of a stimulus. For example, crocuses and tulips show **photonastic responses,** opening in the light, regardless of its direction, and closing in the shade (including heavy cloud cover) or after dark. (This is not the same as the regular, endogenous rhythmic responses of some other flowers, which open in the morning and close at night and will keep doing this at the appropriate times even if transferred to continuous dark.)

**Thigmonastic responses** occur when plants are touched or jostled by wind. This increases the expression of some genes, including one for the calcium-binding protein calmodulin, and slows the plant's growth.

*Electrical Responses*    Some plants produce electrical responses when they are touched. These responses are mediated not by hormones but by electrical currents similar to those in the nerves and muscles of animals (Chapter 37).

A notable example is the rapid closure and drooping of the leaves of *Mimosa pudica,* the "sensitive plant," possibly an adaptation protecting the plant from being eaten (Figure 45–12).

In these plants, certain cells in the phloem have highly negative membrane potentials. Stimuli such as touch, light, or certain chemicals, when sufficiently strong, can elicit an all-or-nothing action potential. This electrical signal travels from cell to cell by way of plasmodesmata. It spreads rather slowly, at about 1 to 2 cm/sec, rather than at tens of meters per second as in animals.

**FIGURE 45–12**  A sensitive *Mimosa* in the Big Thicket of Texas. As the finger strokes from left to right, the leaflets fold quickly, suddenly looking limp and wilted. The extent of the plant's response varies with the intensity of the stimulus.

The effectors in these plants are specific motor cells, which respond to the electrical signals by sudden membrane and osmotic changes. Potassium ions rush out of the motor cells, followed by water, and the cells collapse as they lose their internal turgor pressure. The motor cells are located at hinge regions, and their collapse results in rapid folding of the leaflets.

The closing of the Venus's flytrap around its insect prey also begins as an action potential evoked by a threshold stimulus of touch applied to trigger hairs on the trap's inner surface. This evokes rapid movement in an "acid growth" response by cells on the outside of the leaf's two lobes. These cells spend about 30% of their total ATP pumping out a burst of $H^+$ ions, which loosens their cell wall fibers, permitting the cells to expand rapidly under the force of their outward turgor pressure. The rapid, directed growth pushes the two lobes of the leaf together, and the trap snaps shut. If the leaves are treated with buffer solutions before the trigger hairs are touched, the action potential still occurs, but the release of $H^+$ does not acidify the cell walls. No growth occurs, and the trap does not close. Because this response involves irreversible growth, it is not, strictly speaking, a nastic response.

## Stress Responses

Plants experience many kinds of stressful conditions, such as cold, water shortage, air pollution, or attack by fungi or herbivorous animals. These sorts of stresses evoke similar responses in plants, apparently because they are all perceived by the same means: a stress-monitoring protein in the plasma membrane, very similar to those in nerve cells.

Earlier we saw that abscisic acid is sometimes called the stress hormone because various stresses increase its production. Abscisic acid in turn operates on a master gene, which causes expression of many other genes whose products help the plant cope not only with the stress that elicited it, but with others as well. For example, gardeners have long practiced cold-hardening of young plants started in greenhouses: exposing them to cold and wind for increasing periods before transplanting them outdoors in the spring. This makes the plants more resistant not only to mild, late frosts but also to drought and fungi. Plants can be cold-hardened even at normal temperatures by treatment with abscisic acid.

**Heat and Cold**  Extreme temperatures are potentially deadly to plants. On very hot days, a plant faces two hazards. First, heat increases the loss of water by transpiration. Closing the stomata conserves water but sacrifices the evaporative cooling provided by transpiration, which usually keeps leaves 3 to 5 °C below the air temperature. When the stomata close, the leaf temperature rises, increasing the likelihood of the second hazard: protein denaturation, in which enzymes and other proteins lose their three-dimensional structure and therefore cease to function properly. This of course shuts down metabolism and causes death.

Plants may survive temperatures above 40 °C by producing **heat shock proteins.** Messenger RNAs for these proteins appear about 3 to 5 minutes after leaves reach the critical temperature, and the proteins themselves appear in 10 to 12 minutes. Some heat shock proteins are **chaperonins,** which help fold other proteins into shapes better able to resist denaturation caused by heat or osmotic stress. Similar heat shock proteins are found in animals and in microorganisms.

Plants respond to seasonal temperature changes by producing different mixes of lipid-synthesizing enzymes. These enzymes in turn produce membrane lipids containing longer-chain, more saturated fatty acids in warm weather and fatty acids with shorter chains and more double bonds in cooler seasons (Section 4-B). Without these adjustments in lipid content, cell membranes would melt in summer and solidify in winter.

At subfreezing temperatures plant cells, with their big, water-filled vacuoles, may be killed as ice crystals form and rupture membranes. In areas with cold winters, plants generally survive by eliminating succulent parts, which by definition contain a lot of water and are therefore susceptible to ice formation. Most plants pass the winter in a dormant condition: as seeds, which have very low water content; as underground roots, stems, or bulbs; or as woody aboveground stems of trees or shrubs, with tender meristems insulated by bud scales and bark. Evergreens keep their leaves

and needles through the winter; these are covered by especially thick cuticle.

Many plants require a period of cold before they break dormancy and resume growth. Examples include germination of some seeds, blooming of spring bulbs, and breaking of dormancy in overwintering twigs of some woody plants. In all these cases, the requirement for a cold period delays growth until spring, when growing conditions are likely to be favorable. Many of these plant species do not grow well in climates where the winters are not cold enough to provide the necessary stimulus.

**Wound Reactions**  When a plant is wounded, the injury induces the formation of enzymes that quickly produce a burst of ethylene. This in turn activates the division of wounded cells, and the damage is soon repaired. For example, the corky scar tissue on potatoes is formed after a patch of skin has been scraped off, perhaps when the potato was dug up at harvest.

Earlier we saw that plants attacked by fungi or insects may produce protective substances (Section 45–G). Phytoalexins kill invading fungi, leaving spots of dead, brown, fungus-killed tissue surrounded by now-protected cells. (Some fungi, of course, are not susceptible and spread throughout the plant, killing it.)

In some plants, wounding causes the production of volatile jasmonates from lipids of disrupted membranes. This is part of the common pathway to the synthesis of proteinase inhibitors, which can be activated by either the polypeptide systemin or a local oligosaccharin (or less effectively by abscisic acid). Jasmonates shut down growth and promote the production of proteins that stabilize the plant against stress and the production of toxins that deter feeding by insects. They also boost formation of ethylene, which in turn initiates wound healing or, in the case of severely damaged leaves, abscission (Section 45–L).

Sagebrush is a plant that produces a lot of jasmonates even without being injured, and these volatile compounds cause stress responses in neighboring plants. For example, tomato seedlings sealed into airtight jars with sprigs of sagebrush produce proteinase inhibitors, which prevent insects from digesting the plant's proteins (Figure 45–13).

Producing jasmonates can backfire: they also function as sex pheromones for some insects and therefore attract unwanted dinner guests to already beleaguered plants.

## 45–J  FLOWERING

To many people, the parade of flowers blooming one after the other symbolizes the passing of the seasons. The fragrance of roses ushers in the summer, and the rich hues of goldenrod and purple asters herald the coming of autumn.

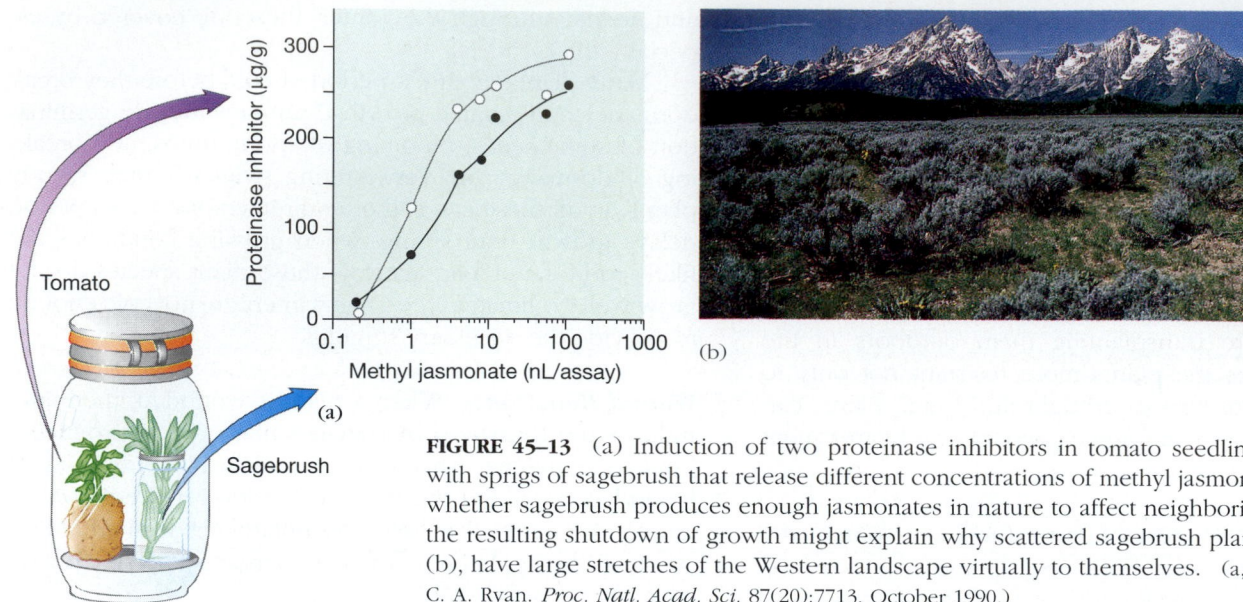

**FIGURE 45–13** (a) Induction of two proteinase inhibitors in tomato seedlings sealed into a jar with sprigs of sagebrush that release different concentrations of methyl jasmonate. It is not known whether sagebrush produces enough jasmonates in nature to affect neighboring plants. If it does, the resulting shutdown of growth might explain why scattered sagebrush plants, such as those in (b), have large stretches of the Western landscape virtually to themselves. (a, after F. E. Farmer and C. A. Ryan. *Proc. Natl. Acad. Sci.* 87(20):7713, October 1990.)

The predictable flowering of many plants depends on environmental cues from Earth's seasonal cycles. Some familiar plants, such as carrots and parsley, flower the second year after the seed began to grow—following a period of cold in the winter. Such plants can be forced to flower by placing them in the cold for several weeks or by applying gibberellins. Evidently, a cold period somehow triggers a hormonal change in these plants, switching them from the vegetative to the reproductive phase.

Unlike animals, plants do not have permanent reproductive organs ready to be activated by hormonal signals. Rather, hormonal changes cause some shoot meristems to differentiate as flowers instead of as vegetative shoots.

## Photoperiodism: Responses to the Length of Night and Day

Outside tropical regions, the daily period of light becomes noticeably longer in spring and shorter again in autumn. The flowering of many plants depends at least partly on exposure to light or dark periods of a certain length, a response called **photoperiodism.** This was recognized in 1920, when Wightman Garner and Henry Allard of the U.S. Department of Agriculture studied two flowering problems:

1. "Maryland mammoth" tobacco, a mutant variety, grew more than 3 m tall in the field but did not flower. However, cuttings rooted and grown in greenhouses during the winter flowered even at much smaller sizes.
2. "Biloxi" soybeans all flowered and produced beans at the same time in early September even when farmers tried to stagger the harvest by planting them at different times.

In each case, all of the plants flowered at the same time, regardless of size or age. Hence, their flowering must be determined by some cue in the environment. Garner and Allard found that different temperatures and light intensities had no effect on growth and flowering. They turned reluctantly to another hypothesis, that plants could respond to variations in daylength—that is, to different photoperiods. This seemed unlikely because it meant that plants would have to be able to measure time.

Garner and Allard artificially changed the daylength, lighting the greenhouse at night to give plants longer days, or placing plants in dark cupboards to give shorter days. And, sure enough, these strains of tobacco and soybeans would flower only if exposed to light for less than a certain maximum amount of time each day! Accordingly, these were dubbed **short-day plants.**

Thanks to a later experiment, we now know that these plants should really be called "long-night" plants. In 1938 K. C. Hamner and J. Bonner found that cocklebur, a short-day plant, would flower when it received more than 8.5 hours of darkness, regardless of the length of the light period. (They exposed plants to nonstandard "days" longer or shorter than 24 hours.) So the true requirement for flowering is a certain minimum period of uninterrupted darkness, the **critical night length** (which in nature occurs if the period of light—the **critical daylength**—is short enough). Interrupting the dark period with a flash of light prevents short-day plants from flowering, even if the critical daylength is not exceeded. This is why, if you own a Christmas cactus or poinsettia that you wish to flower in December, you must put the plant in a place kept dark from sunset to sunrise.

Later experiments showed that some types of plants are **long-day plants,** requiring a certain minimum length of

HOURS

TREATMENTS

RESPONSES

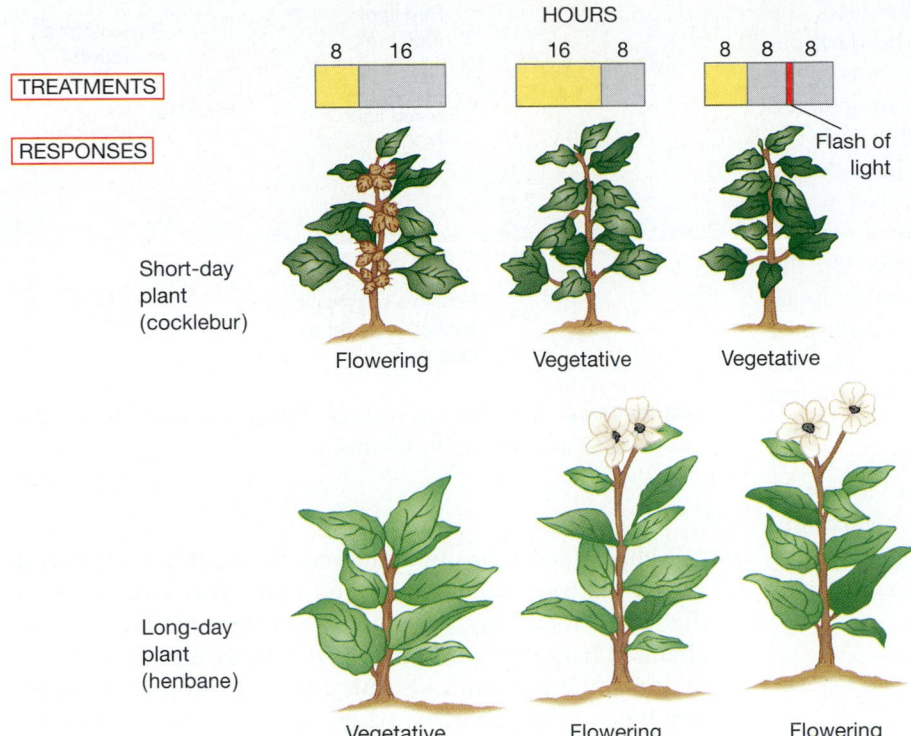

Flash of light

Short-day plant (cocklebur)

Flowering      Vegetative      Vegetative

Long-day plant (henbane)

Vegetative      Flowering      Flowering

**FIGURE 45–14** Demonstration that photoperiodic flowering depends on night length. (left) Short daylength (yellow) coupled with long night length (gray) induces flowering in short-day plants, whereas long-day plants remain vegetative. (center) With the opposite treatment of long days and short nights, short-day plants remain vegetative and long-day plants flower. (right) When the first treatment is varied by interrupting the dark period with a flash of red light, the plants respond as if to long days.

photoperiod (or less than a certain maximum length of dark) for flowering (Figure 45–14). Examples are spinach, radish, many cereals, and black-eyed Susan. (In some long-day plants, treatment with gibberellins can substitute for the proper light/dark regimen.)

To recapitulate: short-day and long-day plants are defined by whether the critical length of the photoperiod is a *maximum* (for short-day plants) or a *minimum* (for long-day plants), not by the actual length of critical light and dark periods.

In many widespread plant species, the critical photoperiod varies among local populations, being longer for those living farther north. Agriculturists planning to introduce new strains of crops into an area must take this into account, to ensure that seeds and fruit will have enough time to ripen and be harvested before the onset of winter.

There are also many **day-neutral** plants, such as tomatoes and cucumbers, which begin to flower at a certain stage of growth and continue to flower and bear fruit until they freeze to death.

In many plants, factors other than daylength influence flowering. A plant will often not flower until it has reached a certain stage of maturity, and increasing maturity may make its photoperiod requirements less exacting. Many plants require a particular light/dark regime but produce more flowers, or flower for a longer time, when they are also given the proper temperature, or day/night temperature changes, or moisture, or a sequence of short days after long days, or vice versa. Usually the light and dark periods must also be correct for several days running; most

kinds of plants do not respond to a single light/dark sequence. Cocklebur is a popular short-day research subject because a single dark period of the proper length induces it to flower.

The florist industry profits from studies of these factors. We can now obtain chrysanthemum plants in full bloom for Mother's Day and send carnations to our Valentines.

In addition to flower initiation, photoperiodism is also involved in the production of abscisic acid in the fall, leading to storage of the special food reserves of seeds and to formation of winter buds and general twig dormancy.

## 45–K  PHYTOCHROME

How do plants measure the length of photoperiods? Any timing mechanism must involve the biological clock found in all eukaryotes (Section 40–G). Measuring a period of light (or dark) must involve the interaction of this clock with pigment molecules, the receptors that undergo photochemical reactions when they absorb light, initiating chemical changes in the cell.

To discover which pigment is the receptor for photoperiodic control of flowering, investigators had to find out which wavelengths of light affect flowering and then isolate a pigment that absorbs those wavelengths. By using different wavelengths of light to interrupt the dark period of short-day plants, researchers found that red light of about 660 nm inhibited flowering most effectively.

Curiously, when far-red light (730 nm) was used, the plants acted as though no light were given in the dark period. Furthermore, when a flash of far-red light was given immediately after a flash of red light, the plants again acted as though they had received no light during the dark period! Given alternate flashes of red and far-red light during their dark period, the plants responded to the last wavelength, flowering when far-red was given last and remaining vegetative when red was last (Figure 45–15). On the other hand, long-day plants held on short-day photoperiods could be made to flower if they were given a flash of red light during the dark period (see Figure 45–14). This could be reversed by far-red light and re-reversed by red light, as just described.

The pigment that absorbs red and far-red light is the blue-green molecule **phytochrome.** It consists of a protein and a much smaller light-absorbing pigment unit very similar to the phycocyanin pigment of some algae. Phytochrome is so sensitive that a bright moon or a single match gives enough light to affect flowering in some plants.

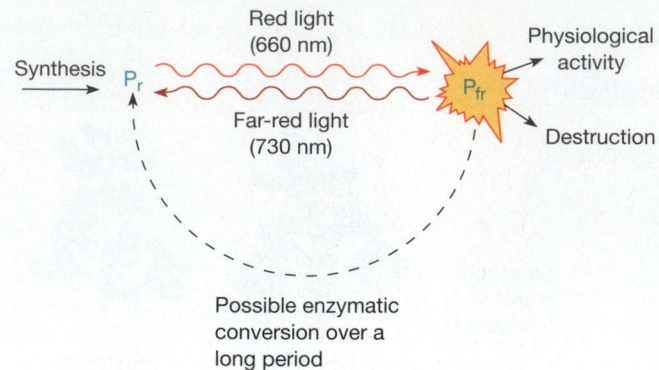

**FIGURE 45–16** The interconversion of phytochrome between the inactive ($P_r$) and active ($P_{fr}$) forms.

Phytochrome has two different forms. It is synthesized as **$P_r$,** the form that absorbs red light. This changes it to the other form, **$P_{fr}$,** which absorbs far-red light and is changed back to $P_r$ (Figure 45–16). $P_{fr}$ is the active form, in which the protein folds into a more hydrophobic shape and binds to membranes, where it triggers further physiological changes in the plant, including changes in membrane potential. So phytochrome works like a switch that is turned on and off by red and far-red light. Flowering apparently depends on the presence of $P_{fr}$ for a certain length of time during the right part of the plant's daily cycle.

Sunlight contains both red and far-red wavelengths, and at dusk about half of the plant's phytochrome is left in the $P_{fr}$ form. Short-day plants appear to require relatively high levels of $P_{fr}$ during the early night, because far-red light (converting it back to $P_r$) at the beginning of the dark period inhibits flowering. As the night progresses, $P_{fr}$ is destroyed, and the dark period must continue for a certain time thereafter: a dose of red light late in the dark period (converting $P_r$ to $P_{fr}$) also inhibits flowering. So flowering in short-day plants seems to rely on the proper sequence of two processes that occur during the dark period, the first requiring higher $P_{fr}$ levels than the second.

The flower-initiating sequence probably involves the interaction of phytochrome with metabolic reactions controlled by the plant's "internal clock": endogenous circadian rhythms (circa = around; dies = a day), which change in a pattern that repeats roughly every 24 hours. These rhythms continue for a time even when the plant is kept in constant darkness. In nature, the rhythm repeats almost exactly every 24 hours because the clock is reset daily by external cues such as sunrise and sunset. Illumination during particular parts of the dark period can apparently "reset" the clock.

In photoperiodic responses, light is detected by means of phytochrome in young leaves, but the plant's response, initiation of flower buds, occurs in the apical meristems. Hormones carry the message between these two sites: some kinds of plants flower after even a single leaf receives the

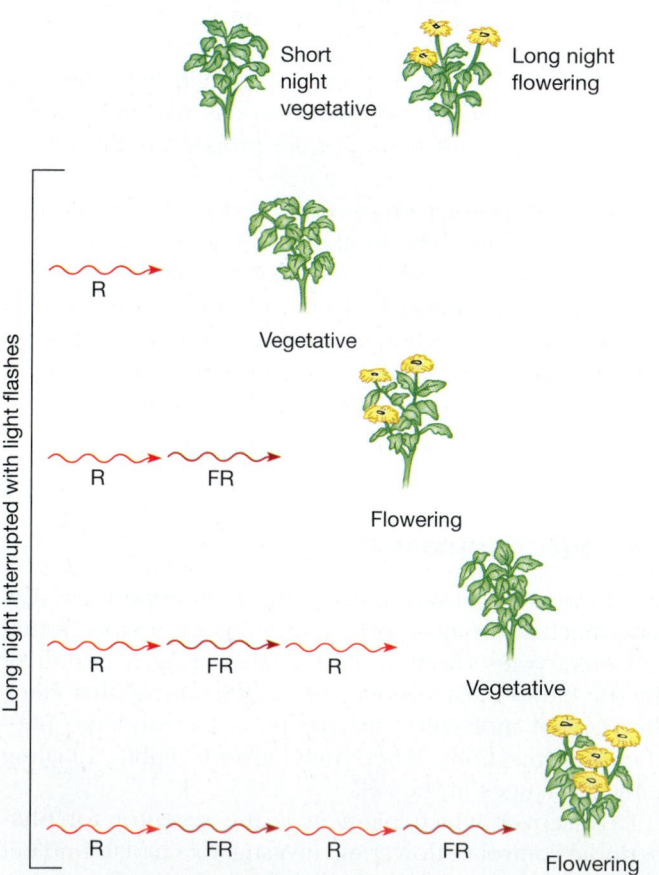

**FIGURE 45–15** Response of a short-day plant when its critical dark period is interrupted with flashes of light. Given a series of flashes, the plant responds to the last flash received, remaining vegetative if it was red light (R), flowering if it was far red (FR).

| TREATMENT | RESPONSE |
|---|---|

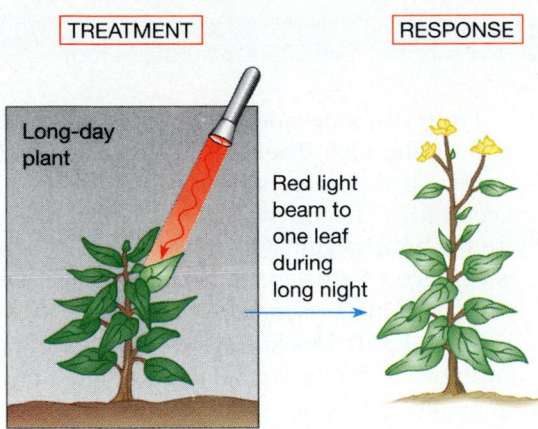

**FIGURE 45–17** Flowering of a long-day plant given short-day treatment. A single leaf of the plant is given a flash of red light during the dark period. Hormones produced in this leaf are transported to the apical bud and induce flowering.

appropriate light regime (Figure 45–17). Producing this response takes some time. If the leaf is removed right after receiving an inducing photoperiod, the plant does not flower, but if it is removed some hours later, flowering occurs.

A hormonal signal can also be demonstrated by grafting experiments, in which parts of two plants are cut off and the plants are then joined so as to connect their vascular tissues. If a plant that received its proper flowering stimulus is grafted to a plant that did not, the second plant will flower. This works even if the graft is made between a long-day and a short-day plant, provided that the first plant has been exposed to the proper photoperiod to stimulate its flowering (Figure 45–18). A single induced leaf grafted onto a noninduced plant is sometimes enough to induce flowering.

Despite decades of research, we still do not know what internal chemical changes initiate flowering. Researchers have variously guessed that it will turn out to be:

1. Ethylene.
2. Conversion of gibberellins to one of the less common forms.
3. A particular combination or sequence of hormones.
4. A yet-undiscovered flowering hormone, which has already been named "florigen"!

## Phytochrome and the Growth of Seeds

Phytochrome is the receptor not only for photoperiodic flowering but also for seeds that require light for germination. These include "Grand Rapids" lettuce and many kinds of weeds. As expected from a response mediated by phytochrome, red light is most effective because it switches phytochrome to the active $P_{fr}$ form, whereas far-red light inhibits germination.

| TREATMENT | RESPONSE |
|---|---|
| To first plant | Treated plant   Grafted to |

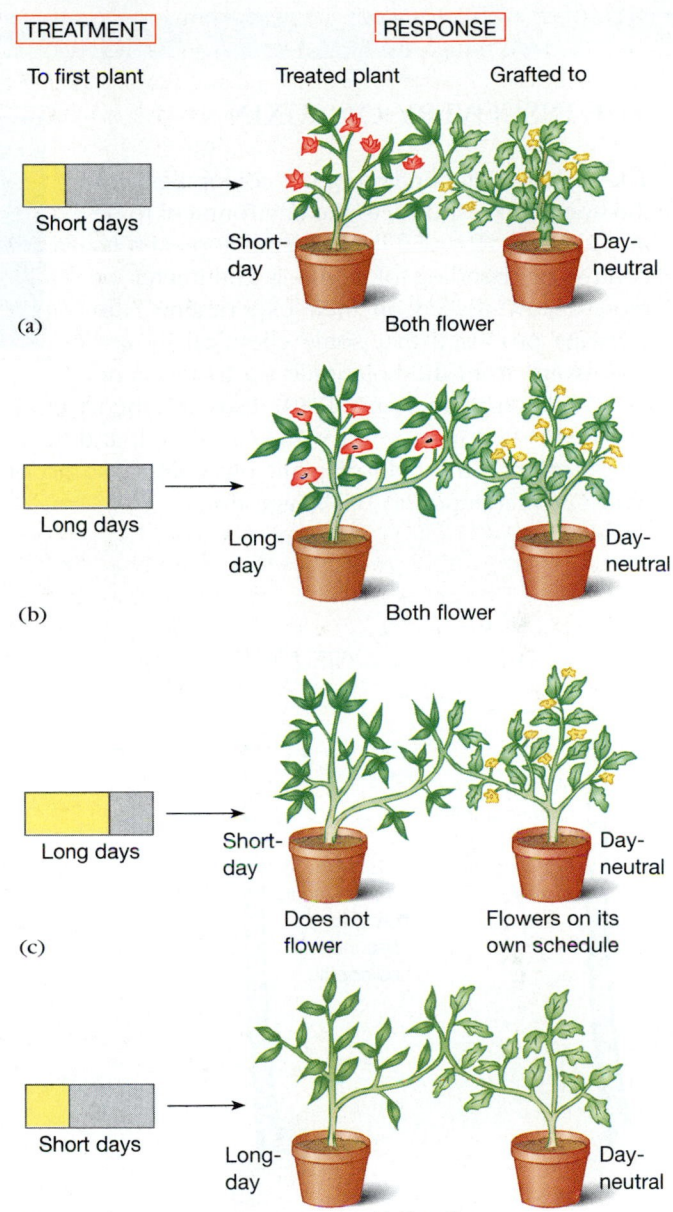

**FIGURE 45–18** Flowering of day-neutral plants grafted to photoperiodic plants. (a, b) Short-day and long-day plants given inducing photoperiods also induce flowering in day-neutral graft partners. (c) A short-day plant given long days does not flower. The day-neutral graft partner flowers on its own schedule. (d) However, a long-day plant given short days not only does not flower, but also apparently keeps a day-neutral graft partner from flowering when it otherwise would. This suggests that an inhibitor normally present in long-day plants must be removed before they will flower.

In nature, this response keeps seeds from germinating when they are buried too deeply in the soil for light to penetrate. If these seeds grew, the young seedlings might exhaust their food supply before their shoots could reach light

# THE DISCOVERY OF AUXIN

The phototropic bending of oat coleoptiles was the first plant response to the environment to be investigated. How did the coleoptile tip exert its control on bending of tissue a centimeter or more below? Based on their experiments, the Darwins proposed that some chemical moved downward from the coleoptile tip to the zone of elongation. At the time (1880), Darwin's theory of evolution was famous but not widely accepted, and he was an outsider to the field of plant physiology, so no one paid much attention to this suggestion.

From 1910 on, various researchers found support for this idea. Their experiments all involved removing the coleoptile tip. Phototropism still occurred if the tip was then put back on the stump or if gelatin (a water-based substance) was placed between tip and stump. However, a sliver of mica or metal foil, or a dab of cocoa butter (a lipid), blocked phototropic bending. In addition, if the tip was placed on just one side of the stump, the coleoptile bent toward the opposite side, even without the stimulus of light. And, when the tip was simply removed, the coleoptile grew less than intact controls.

Altogether, these experiments make a good case for the existence of a growth-stimulating, water-soluble chemical produced in the tip. However, they did not rule out the possibility that the response is evoked by electrical signals passed from the tip. The definitive test would

Agar + auxin

Leaf

Agar + auxin

Agar block placed under severed tip for several hours

Tip removed, block placed on side of second coleoptile

Hormone diffuses; coleoptile bends away from side of auxin application

Greater concentration of auxin on left stimulates more elongation; coleoptile bends toward right

**FIGURE 45–A** Went's experiment. The presence of a chemical growth mediator was shown by permitting the substance to diffuse into an agar block and then placing the block on one side of a decapitated coleoptile, where it induced bending.

and begin photosynthesis. The seeds have a better chance of survival by remaining dormant until the soil is later disturbed, bringing them into the light, which then stimulates their germination.

Even if light-sensitive seeds fall on the soil surface, chlorophyll in the green leaves of overhanging plants absorbs the red wavelengths while letting far-red through, and this keeps the seeds' phytochrome in the physiologically

be to isolate this chemical in pure form, apply it to one side of decapitated coleoptiles, and demonstrate bending like that seen in phototropism. However, technology was not yet good enough to isolate chemicals that occur in such tiny amounts.

Failing that, the next best thing is a **bioassay,** a test that detects a chemical and measures its relative concentration by observing the effect of unpurified extracts on living organisms or cells. In 1926 Frits Went, then a graduate student in The Netherlands, devised such a bioassay and confirmed the chemical's existence.

Went had the idea of making the transfer of the chemical from tip to stump a two-step process, first from the tip into a block of gelatin-like agar, then from this agar to the coleoptile stump (Figure 45–A). He first allowed a severed coleoptile tip to sit on a block of agar for a while; when he placed this block on one side of a freshly decapitated coleoptile, the coleoptile grew and bent away from that side. Furthermore, if he placed two coleoptile tips instead of just one on an agar block, the block produced twice as much bending when it was later placed to one side of a headless coleoptile. When he split the tip, the half from the lighted side produced less bending than its counterpart from the shaded side (Figure 45–B). These results could be explained only by the presence of a chemical messenger. The chemical responsible for the phototropic response was named auxin (auxein = to grow).

Bending occurs only when different concentrations of auxin are applied to different areas of the coleoptile sheath. If the concentration of auxin is uniform across the stump, the entire coleoptile grows straight up. Up to a certain point, the greater the concentration of auxin applied, the greater the elongation of the coleoptile.

Auxin was finally isolated, not from coleoptiles or other plant parts, but by screening various natural sources for substances that cause coleoptile bending. It turned out that one of the best sources was right under the researchers' noses (well, actually a bit farther down): human urine. Auxin was purified by separating urine into various fractions, using bioassays at each stage to test for curvature of coleoptiles, and was identified at last as indoleacetic acid.

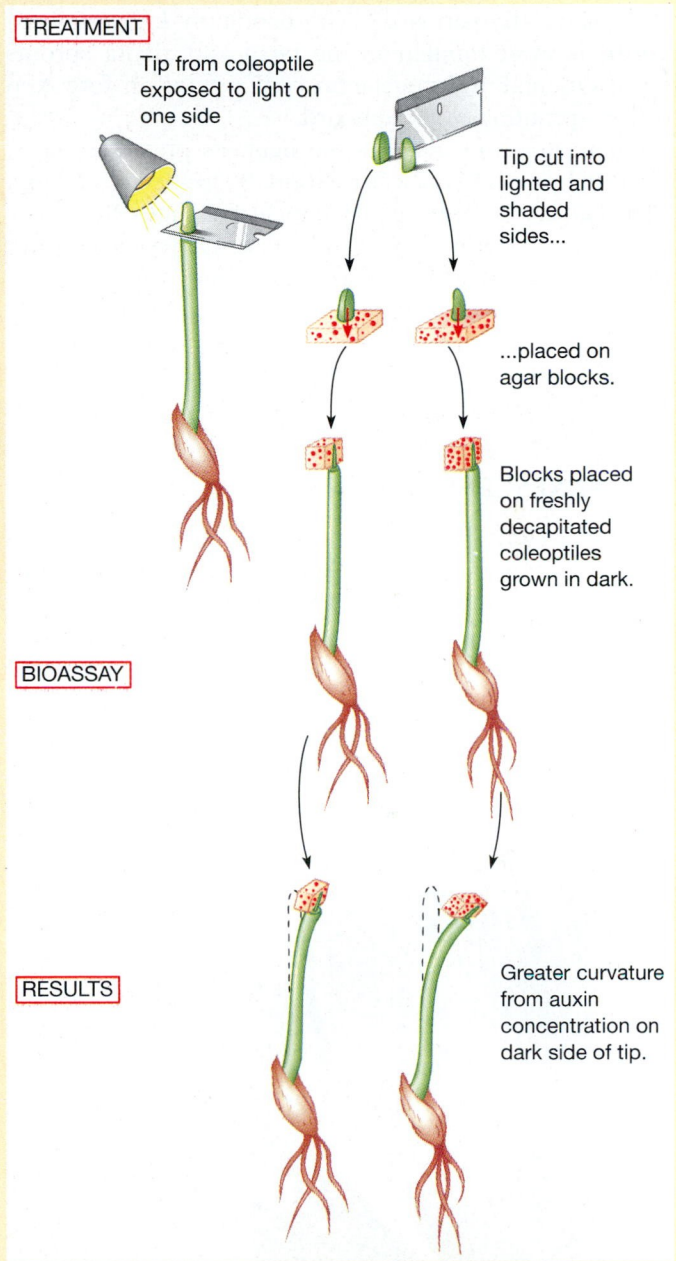

FIGURE 45–B   Bioassay comparing auxin concentrations in the two sides of a coleoptile tip lit from one side.

inactive $P_r$ form. Again, it is advantageous for the seeds to remain dormant until other plants no longer shade them. When the forest sheds its leaves or an overhanging tree dies, the seeds receive more sunlight, and the red wavelengths switch much of their phytochrome to the $P_{fr}$ form, initiating germination.

Phytochrome plays a further role in the development of the young shoot. While the shoot is growing in the dark, it elongates rapidly, an adaptation increasing its chances of reaching the light before its food supply runs out. The stem is long, spindly, and unpigmented (white or pale yellow), and the leaves remain small. This condition, known as **etiolation,** is most familiar in the bean and alfalfa sprouts used in Oriental and vegetarian cooking and in forgotten potatoes sprouting in a cupboard.

When the shoot reaches the light, its phytochrome is converted from the synthesized form, $P_r$, to the active form, $P_{fr}$. This initiates a series of physiological changes: the stem's rate of elongation decreases, and the stem becomes thicker and sturdier; the leaves expand; and genes for photosynthetic pigments and enzymes are turned on as undifferentiated plastids develop into chloroplasts, giving the stem and leaves a healthy green color.

## 45–L  SENESCENCE

Flowering, setting of seeds and fruit, and germination of seeds are events in a plant's life that occur predictably in response to environmental or genetic factors. **Senescence,** the process of aging that makes all or part of the plant more susceptible to death by factors such as dehydration or infection, is also an integral part of its life cycle. A wheat plant turns yellow, dries up, and dies after it has set seed; a plum tree drops its fruits during a short period in early summer and loses all its leaves during the fall.

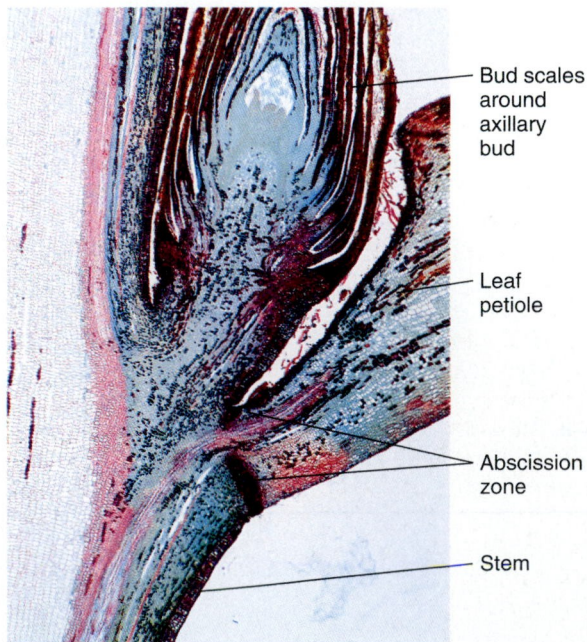

Bud scales around axillary bud

Leaf petiole

Abscission zone

Stem

**FIGURE 45–19**  Leaf abscission. The attachment of a leaf weakens as the abscission zone forms at the base of the petiole (stalk). The leaf eventually separates from the stem at this area of weakness and falls off the plant.  (James Mauseth)

Senescence is under hormonal control. It is thought to be initiated by abscisic acid, and ethylene produced in senescing tissue then plays an important role. In annual plants, the hormonal changes that initiate senescence of the whole plant often seem to be triggered by the setting of seed.

In perennial plants, the leaves, fruits, and withered flower parts senesce and drop off each year, a process known as **abscission.** Although its name suggests a link with abscisic acid, abscission is promoted mainly by ethylene. Auxin applied after the process begins may speed abscission. However, auxin produced in a healthy, active leaf inhibits abscission, as do cytokinins arriving from the roots.

The senescence and abscission of older leaves occur regularly as new ones grow above. Abscission may also follow stress or injury to a leaf, and wholesale abscission is a response of many plants to the onset of winter. It is thought that a rise in ethylene during the autumn suppresses auxin production, permitting abscission.

Before abscission, several changes take place. Enzymes that degrade the leaf tissue become active, and nutrients may be withdrawn from the leaf. Another change is cell divisions in the **abscission zone,** where the leaf joins the stem. In this region, ethylene stimulates the production and release of the enzyme cellulase, which breaks down the cell walls, thereby creating a zone of weakness (Figure 45–19). Finally, the leaf's weight, perhaps assisted by the wind, causes this zone to break apart, and the leaf drifts away. The leaf scar is quickly healed by the deposition of a corky, waterproof seal that prevents the loss of water or entry of pathogens (disease-causing organisms).

## SUMMARY

Hormones carry messages between different parts of a plant, coordinating growth and differentiation. They are also part of transduction pathways by which the plant responds to information received from the environment, such as direction of light, gravity, or prevailing winds, and changes in daylength and temperature that signal changes in the seasons.

1. Five groups of major plant hormones are known. Auxin, cytokinins, and gibberellins are generally growth-promoting; abscisic acid often prepares the plant to survive stressful conditions, and ethylene promotes ripening and abscission of fruits and senescence and abscission of leaves. Table 45–1 lists the roles of these hormones.

2. Interactions among hormones govern a plant's growth, so that different parts, such as the leaves and roots, remain in anatomical and physiological balance with each other.

3. How a plant responds to a particular hormone depends on the target tissue receiving the hormone, including

its age and physiological state; the concentration of that hormone and of others; and environmental factors such as temperature, light, or photoperiod.

4. In addition to the five major hormones, recent research has shown that plants have many more kinds of chemical messengers, including polyamines, polypeptides, oligosaccharins, and salicylic acid.

5. Plant and animal hormonal systems use many of the same second messengers to alter membrane permeability, enzyme activity, and gene expression.

6. Auxin, synthesized in shoot tips and transported toward the roots, establishes the polarity of the plant body and interacts with cytokinins from root tips in coordinating shoot and root growth and differentiation of new parts.

7. Auxin from the apical buds of shoots also tends to repress growth of lateral buds into new branches.

8. Tropisms are growth responses to the direction of stimuli such as light, gravity, or physical contact. Light is detected by a pigment, possibly a flavoprotein, and gravity is detected by pressure from dense amyloplasts resting on other cell components. Following poorly understood processes of transduction, differences in auxin distribution produce differential elongation in cells on opposite sides of a growing organ, resulting in positive phototropism of shoots and gravitropism of shoots and roots. Other substances may also play a role in these poorly understood responses.

9. Plants may also produce nondirectional nastic responses to stimuli such as light and touch.

10. Although hormones mediate most plant responses to environmental stimuli, some plants respond to touch by means of electrical action potentials.

11. A plant's response to stressful environmental conditions is started by a general stress-detecting protein and mediated by abscisic acid. This activates a set of changes that enable the plant to cope better with many stresses.

12. Plants respond to warmer weather by producing longer, more saturated membrane lipids and to very hot weather by producing heat shock proteins, some of which protect other proteins against denaturation.

13. Plants living in climates with regular cold or dry seasons have mechanisms that sense the onset of the season; eliminate tender, succulent leaves; and induce dormancy in the seeds, woody stems, and underground structures that carry the plant through the adverse period.

14. Wounded cells produce ethylene, which stimulates production of protective chemicals and also activates cell division in repair or abscission of the damaged part.

15. Plants can be classified as short-day, long-day, or day-neutral according to their flowering behavior. Short-day plants require periods of dark longer than a certain minimum, and long-day plants require dark periods shorter than a certain maximum. The flowering of day-neutral plants is governed by maturity. The hormonal mechanism of flowering is not yet understood.

16. Photoperiodism seems to involve interaction of phytochrome with metabolic reactions controlled by the plant's biological clock.

17. Phytochrome also mediates light-induced expression of genes needed for photosynthetic molecules, leaf expansion, and decreased elongation in the shoots of germinating seeds as they reach light.

## Self-Quiz

Match the hormones listed below to their effects.

_____ 1. Promotes ripening of fruits
_____ 2. Initiates cell division in tissue culture
_____ 3. High concentrations stimulate ethylene production
_____ 4. Substitutes for cold period in the flowering of some plants
_____ 5. Responsible for phototropic response in coleoptiles
_____ 6. Counteracts the effects of auxin
_____ 7. Promotes onset of dormancy

a. abscisic acid
b. auxin
c. cytokinin
d. ethylene
e. gibberellin

8. In phototropism, auxin:
   a. stimulates differential growth of cells on different sides of the plant
   b. promotes growth of cells
   c. inhibits growth of cells
   d. inhibits cell division
   e. absorbs stimuli and signals the direction of light

9. The folding of the leaflets of sensitive *Mimosa* plants is classified as a nastic response because:
   a. it releases distasteful chemicals
   b. it involves action potentials
   c. it occurs the same way no matter which side it was touched from
   d. it is mediated by abscisic acid
   e. it involves differential growth on the two sides of the petiole

10. The flowering of certain plants only under "short-day" conditions is an example of:
   a. apical dominance
   b. positive phototropism
   c. negative phototropism
   d. photoperiodism
   e. photonasty

*(Continued)*

11. In a short-day plant growing in a home garden, phytochrome is normally switched from one form to the other by:
    a. pure red and far-red light
    b. sunlight
    c. activation by gibberellin or abscisic acid
    d. different electrical potentials in the plant's plasma membranes
    e. measuring the length of the dark period between light periods

12. A long-day plant is one that:
    a. requires more than 12 hours of light in order to flower
    b. increases in height when it flowers
    c. requires a certain minimum length of photoperiod in order to flower
    d. is not affected by temperature in its flowering response
    e. will not flower if its dark period is interrupted by a flash of light

13. True or False?  In nature, short-day and long-day plants living in the same area cannot bloom at the same time.

## Questions for Discussion

1. Explain why some plants are dwarfs even though they produce normal levels of gibberellins and receive adequate water and nutrients.

2. The fungus that causes the "foolish seedling" disease in rice appears not to need the gibberellin it produces for its own growth. Why might it be selectively advantageous for the fungus to secrete this substance?

3. Why are apples, oranges, and grapefruit sold in plastic bags with holes in them rather than in unperforated bags? Why does it often turn out that produce packaged in market trays with clear wrap is too soft on the underside, which you could not see when you picked it out in the store?

4. Many plants of the forest floor produce seeds that germinate only in the early spring, before the canopy leafs out. Propose an explanation for the timing of this germination, and explain its adaptive advantage.

5. In some species of trees, individuals growing near streetlights become dormant later in the fall than do other individuals. How could you account for this?

6. Explain why exposure to a flash of light during the dark period can change the plant's subsequent flowering response (or lack of it), but interrupting the light period with an interval of dark has no effect.

7. Oligosaccharins originate from the cell wall rather than from structures inside the cytoplasm. What is the adaptive value to the plant of having regulatory molecules with such a source?

8. Chlorophyll absorbs red light of wavelengths near those absorbed by phytochrome ($P_r$) when it is converted to the active $P_{fr}$ form. What is the adaptive advantage of producing phytochrome as a light detector rather than chlorophyll?

9. Flower-inducing hormones move from a photoperiodically induced leaf to the apical meristem via the phloem. Explain why flowering is inhibited when an uninduced leaf is present between the induced leaf and the apical meristem.

10. In at least a dozen species of Amazon rain forest trees, roots grow out of the soil and up the trunks of other trees. What is the advantage of this reversal from the usual direction of plant root growth?

## Suggested Readings

Albersheim, P., and A. G. Darvill. "Oligosaccharins." *Scientific American,* September 1985.

Fosket, D. E. *Plant Growth and Development: A Molecular Approach*. New York: Academic Press, 1994. Emphasizes molecular and genetic mechanisms of plant development.

Galston, A. W. *Life Processes of Plants*. New York: Scientific American Library, 1994. An eminent senior researcher's beautifully written and illustrated guide through the experimental methods used to discover how plants work, relating their physiology to how they live in nature.

Overbeek, J. van. "The control of plant growth." *Scientific American,* July 1968. A brief history of the discovery of the plant hormones (except ethylene).

Salisbury, F. B., and C. W. Ross. *Plant Physiology,* 4th ed. Belmont, CA: Wadsworth, 1992. Outstanding for its clearly written, careful assessment of the evidence behind our understanding of how plants work.

Taiz, L., and E. Zeiger. *Plant Physiology*. Redwood City, CA: Benjamin/Cummings, 1991.

# Reproduction in Flowering Plants

## O B J E C T I V E S

*When you have studied this chapter, you should be able to:*

1. Define the following terms, and use them properly:
   stamen, anther, pollen, pollen tube
   carpel, pistil, style, stigma, ovary, ovule, embryo sac
   endosperm, seed coat, fruit
   pollination, fertilization, germination
2. Name the parts of a flower, and state the role of each in the plant's reproduction.
3. Explain what is meant by the terms microspore, megaspore, and male and female gametophytes, and tell where each is found in flowering plants.

4. Explain how pollination and fertilization occur in flowering plants.
5. Name the three main parts of a seed, and explain how each develops.
6. List four factors that may be required for germination of a seed.
7. Explain the advantage to a plant of asexual, or vegetative, reproduction; list some ways in which plants may propagate asexually; explain why humans use vegetative propagation of plants, and name some ways in which plants can be propagated by human manipulation.

---

P lants acquire resources from their environment as they grow (Chapters 42 to 44). Ultimately, natural selection ensures that the plant's resources are channeled into reproduction.

Two kinds of reproduction occur among flowering plants. **Vegetative reproduction** is an extension of the kinds of growth we saw in Chapter 42. It forms new individuals with genomes identical to the parent's, thus perpetuating gene combinations that are well adapted to the local environment. Individuals with the favorable genetic combination spread through the area around the parent plant.

**Sexual reproduction** involves more complex events: differentiation and growth of new groups of structures making up the flowers, production and fertilization of gametes, and development of embryos, seeds, and fruits. Sexual reproduction has two main advantages. First, it forms new genetic combinations in each generation. Second, it produces seeds, which can be dispersed over a wide area and which are protected against adverse environmental conditions that might kill the parent plant.

In Chapter 27 we studied the basic plant life cycle, in which the diploid (2N) sporophyte generation alternates with the haploid (N) gametophyte generation. We can diagram the life cycle for flowering plants:

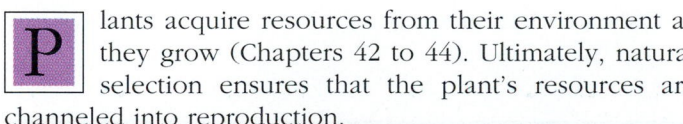

(You may wish to refer back to this diagram as you read.)

In flowering plants, the sporophyte generation is dominant, the gametophytes very much reduced. The familiar plants of garden, field, and forest are sporophytes. The male gametophyte consists of a pollen grain and the tube that grows from it, and the female gametophyte, the embryo sac, is hidden within the ovule.

Modern human society depends on our ability to grow plants for food and for many other uses. In this we employ both sexual and vegetative reproduction. Plant breeders manipulate the sexual reproduction of economically important plants in order to produce individuals with more desirable combinations of genetic features. Once such a set of features is achieved, vegetative propagation can be used to increase the number of plants available to farmers and gardeners.

## KEY CONCEPTS

♦ Sexual reproduction in flowering plants has two important outcomes:
  1. It produces new genetic combinations, the raw material for evolution by natural selection.
  2. It produces seeds, dispersal units by which offspring are spread to new areas some distance from the parent plant.

♦ Vegetative reproduction perpetuates combinations of genes suited to the local environment and allows plants with these combinations to spread within this local area.
♦ Both kinds of reproduction are important in human attempts to improve the strains of plants that we grow for our own use.

## 46–A    FLOWERS

Flowers are sexual reproductive structures, produced in response to hormonal changes in the plant. Some plants flower at a certain stage of maturity, whereas others are induced to flower by certain environmental stimuli, usually the length of the daily period of light or darkness the plant receives (Section 45–J). These responses may be modified by temperature, moisture, or other factors. Desert plants may respond to a heavy rain by flowering.

The new hormone balance makes some shoot meristems differentiate and develop into an abbreviated shoot with very short internodes and a cluster of many highly modified leaves, the flower parts. A typical flower has four types of modified leaves. Like vegetative leaves, they mature and differentiate in order:

1. **Sepals,** the outermost, basal leaves, develop first and protect the other parts maturing within the flower bud. They are often green.

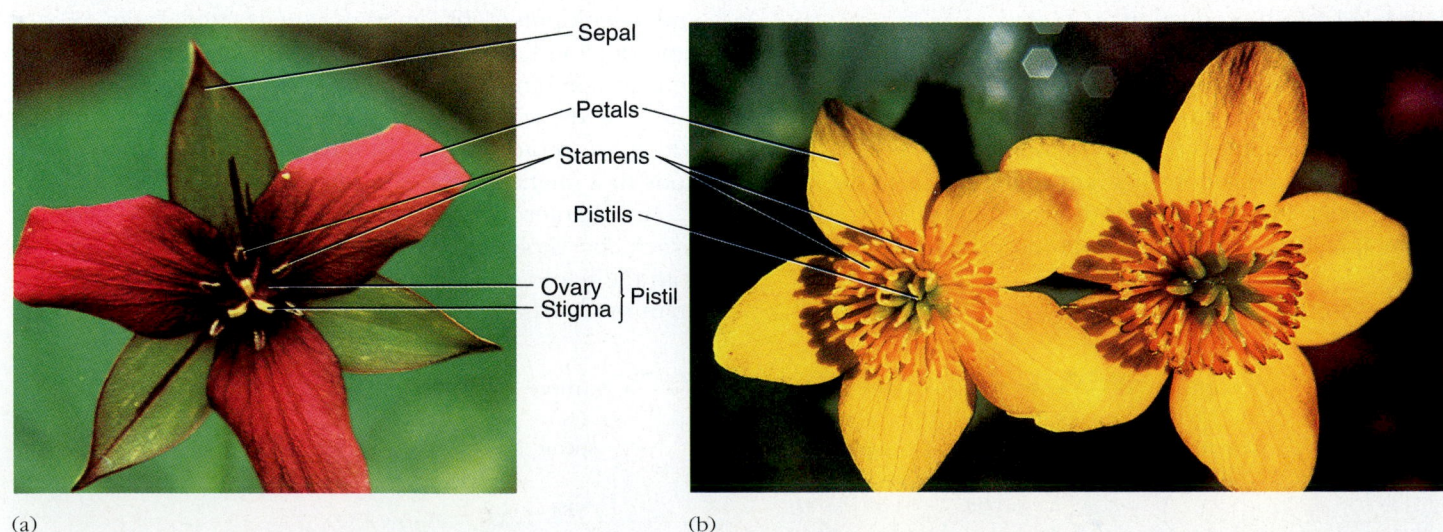

(a)                                                        (b)

**FIGURE 46–1**  Flower structure. (a) *Trillium,* with its flower parts arranged in threes, is a monocot. The three arms of the stigma and six lobes of the ovary indicate that the compound pistil forms from more than one carpel. (b) Marsh marigolds are dicots (note the five petals). A wreath of stamens surrounds a central group of simple pistils (green), each consisting of a single carpel.

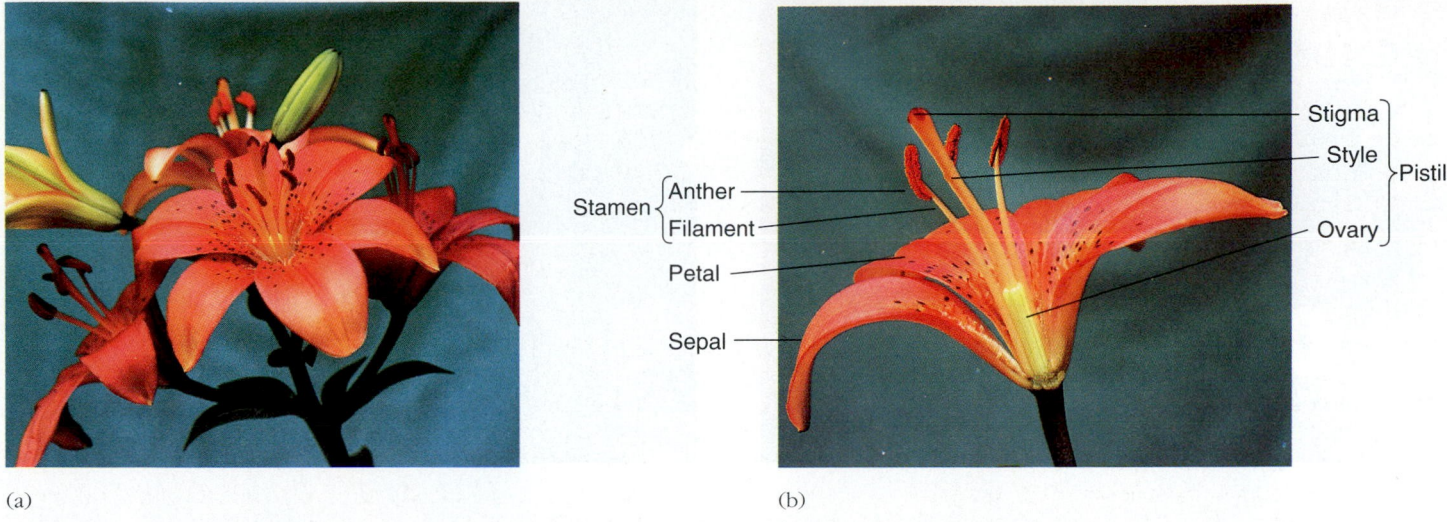

(a)

(b)

**FIGURE 46–2**   Structure of a lily flower. (a) The three petals and three sepals look almost alike. Six stamens surround the pistil. (b) Some sepals, petals, and stamens removed to show how the parts attach to the flower stalk. Each lobe of the ovary arises from one of the six carpels fused to form a compound pistil.

2. **Petals,** next inside the sepals, are often large and showy, with bright colors and patterns that attract animal pollinators.
3. **Stamens,** the pollen-producing ("male") parts, have a stalk, or **filament,** bearing an **anther,** a chamber where pollen grains develop.
4. **Carpels,** in the center of the flower, are modified leaves enclosing female reproductive organs. A flower may have one carpel or several, which may be separate or fused into a single structure. The term **pistil** refers either to a single, independent carpel **(simple pistil)** or to the structure formed by the fusion of several carpels **(compound**

**pistil)** (Figure 46–1). Each carpel has three parts. The **stigma,** at the tip, receives pollen. The stalk-like **style** connects the stigma to the third part, the **ovary,** which encloses one or more **ovules** (Figure 46–2).

Many variations on these typical flower parts are found among the 235,000 species of flowering plants. For example, in lilies the sepals and petals look almost exactly alike (Figure 46–2). In wind-pollinated plants, the sepals and petals are often reduced in size or absent, allowing greater exposure of the stamens as they shed pollen and of the stigmas as they receive it (Figure 46–3). Several plants

(a)

(b)

**FIGURE 46–3**   Wind pollination. Corn plants bear separate clusters of female and male flowers. (a) The pendulous anthers, clustered at the top of the plant, swing in the breeze and shed pollen. (b) Corn silk is the long styles of female flowers in the ears.

(a)

(b)

**FIGURE 46–4**   Nonflower "flowers." (a) Showy pink bracts (leaf-like structures below flowers) surround the small, inconspicuous flowers of bougainvillea. (b) The "petals" of a dogwood are also bracts surrounding a cluster of inconspicuous flowers.

produce separate staminate (male) and ovulate (female) flowers (for example, corn and members of the squash family, including cucumbers and pumpkins). Still others have separate male and female plants, as in spinach, willows, and some hollies. There are also many plants in which structures near the flowers act as parts of the "flower." The "petals" of poinsettias, dogwood, and bougainvillea, for example, are really modified leaves around clusters of small, inconspicuous flowers (Figure 46–4).

## 46–B   POLLEN

In plants meiosis gives rise to haploid **spores,** rather than to gametes as in animals. The haploid spores grow into haploid **gametophytes,** which in turn produce the gametes that take part in fertilization (page 987). Flowering plants produce spores of two sizes, microspores and megaspores, which give rise to male and female gametophytes, respectively.

Inside the anthers are chambers containing **microsporocytes (microspore mother cells).** These cells divide by meiosis, each forming four haploid cells, the **microspores** (Figure 46–5).

Each microspore begins to develop into a male gametophyte. The haploid nucleus divides once by mitosis, followed by unequal cytokinesis, forming two cells. The larger one, the vegetative cell, will eventually produce the pollen tube. The much smaller generative cell contains no mitochondria nor plastids, and therefore neither do the sperm it produces later. In 90% of angiosperm species, these organelles are inherited only from the female parent.

This immature male gametophyte becomes extremely dehydrated and metabolically inactive. It is enclosed in a protective wall, forming a **pollen grain,** which is shed when the anther splits open.

Just as leaf and flower structures vary among plants, so too do the shape and pattern of the pollen grain wall. In fact, experts can easily place a particular pollen grain into the proper genus (and sometimes species) by its distinctive cell wall pattern (Figure 46–6). The wall is resistant to strong acids and bases and to intense heat; indeed, pollen grains may last for millions of years, preserved in rock formations or peat deposits. The history of the vegetation in an area can be traced by examining pollen fossilized in rocks or deposited in layers of soil or aquatic sediments.

### Pollination

**Pollination** is the transfer of pollen from the anther, where it forms, to the stigma of the carpel. Pollen may simply fall from the anther onto a stigma in the same flower, resulting in **self-pollination.** Some flowers, such as peas and their relatives, are so constructed that the stamens and carpels are completely enclosed within the petals, resulting in a high percentage of self-pollination (see Figure 15–B).

**Cross-pollination,** the transfer of pollen to another individual of the same species, gives more genetic variety. This is often an evolutionary advantage, and many plants have adaptations that ensure cross-pollination. For example, a flower's carpels may mature only after its anthers have shed their pollen. The existence of separate staminate and ovulate plants or flowers is probably due to selective pressure for cross-pollination.

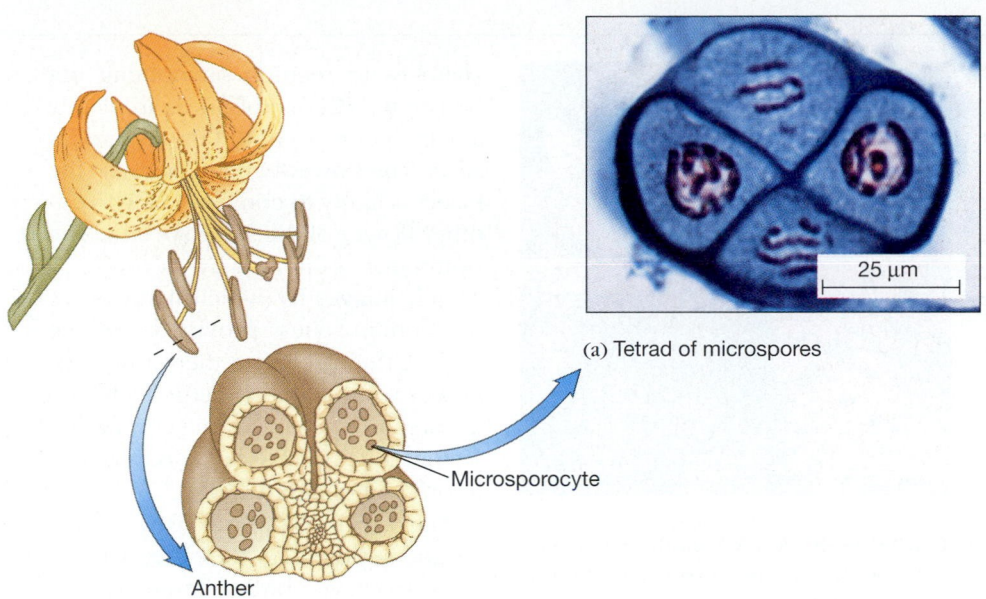

(a) Tetrad of microspores

25 µm

Microsporocyte

Anther

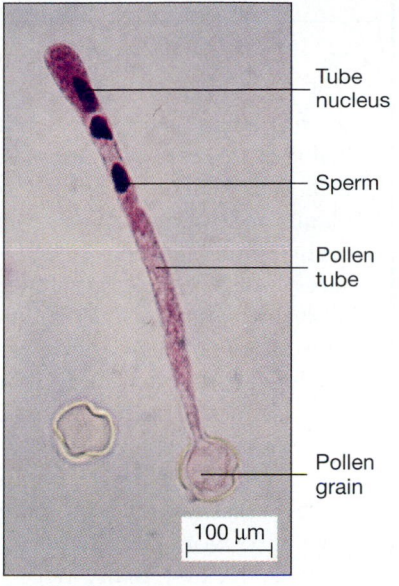

Tube nucleus

Sperm

Pollen tube

Pollen grain

100 µm

(b) Male gametophyte

**FIGURE 46–5**   Formation of pollen. Chambers in the anthers contain microsporocytes. Each of these cells undergoes meiosis, producing a tetrad of microspores (a). Each microspore develops into a pollen grain. (b) A germinated pollen grain. The vegetative cell, containing the tube nucleus, produces the tube, which grows to the ovule and releases two sperm.   (a, Carolina Biological Supply Company; b, Ed Reschke/Peter Arnold)

Pollen cannot move on its own; plants rely on wind or animals for pollination. From a plant's point of view, pollination by animals has some advantages over pollination by wind. Wind pollination wastes a lot of the energy invested in pollen production because much of the pollen never reaches another flower. Wind pollination is particularly inefficient for a plant that does not live in dense populations. Although pollen may be blown for miles, if few other

*Text continues on page 994*

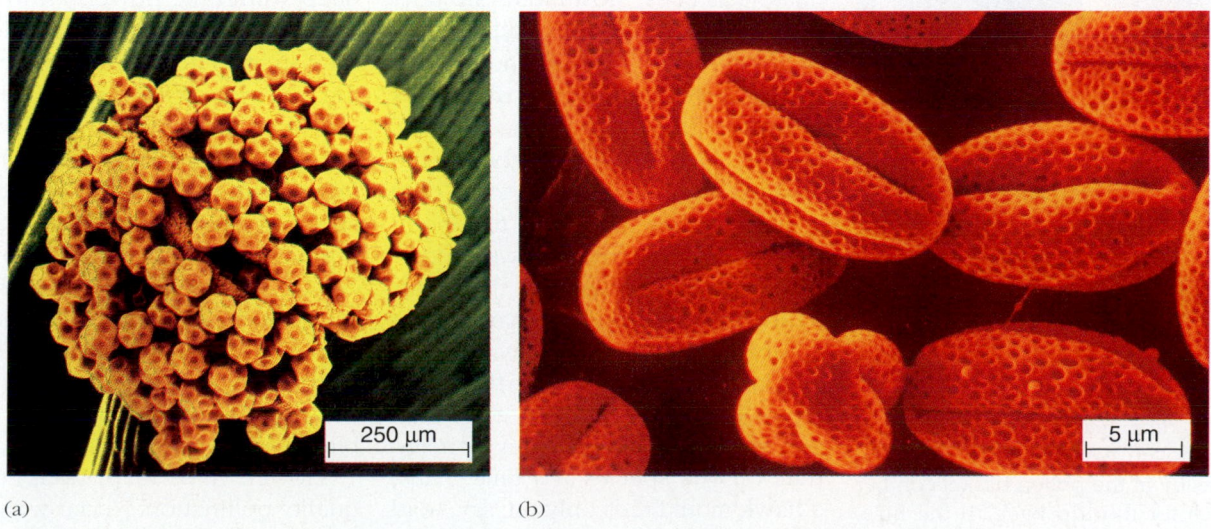

250 µm

(a)

5 µm

(b)

**46–6**   Pollen grains. (a) A ruptured chickweed anther releases its mature pollen grains. (b) Lemon pollen. The pollen tube grows out through a weak area in the wall of the pollen grain: one of the pores in chickweed and one of the furrows in lemon.   (a, Dr. Jeremy Burgess/Photo Researchers/ Science Photo Library; b, G. Shih and R. Kessel/Visuals Unlimited)

Much of the evolutionary success of flowering plants can be attributed to the fact that they evolved after animal life was well established on land. Flowering plants and animals have coevolved, each exerting strong selective pressures and shaping the evolution of the other in many ways. Pollination systems offer many fascinating examples.

The most important pollinating animals are the bees (Figure 46–A). A flower enjoys several advantages in being pollinated by bees. Bees are widely distributed and numerous. Bees also work very hard at visiting flowers because many bees depend entirely on the food they obtain from flowers, both to nourish themselves and to feed the larvae. The behavior of bees also makes them highly desirable pollinators. Bees quickly learn to tell different types of flowers apart, and they are faithful to one kind of flower for long periods. Bees are also

**FIGURE 46–B** A hawk moth sucking nectar from a cardinal flower through its tubular proboscis. (Sherman Thomson/ Visuals Unlimited)

available throughout the growing season, and they can remain active even at low air temperatures, which immobilize most other insects.

Various butterflies and moths (order Lepidoptera) are important flower pollinators in all parts of the world (Figure 46–B). Because these insects use nectar only as a supplementary food for their short-lived adult stage, however, they are not as effective pollinators as bees. Most moths are nocturnal, and the flowers that depend on them for pollination tend to have strong scents as well as pale colors visible in dim light. Some flowers, such as *Nicotiana* (a member of the tobacco family), produce scent only at night, when the moths that pollinate them are active. Flowers pollinated by butterflies, on the other hand, are more likely to have bright colors that stand out by day.

Although many butterflies and moths feed on nectar from more than one species of flower, they concentrate on one species at a time. Thus, a hawk moth feeds only on, say, toad flax for as many as five days, and then

switches to feeding on nothing but bedstraw. This faithfulness is plainly advantageous to both flowers and insects. The flower benefits because the insect is likely to convey pollen to another flower of the same species. The pollinator benefits by forming a search image, by which it "keys in" on certain cues provided by the flower; the insect can then find more flowers of that species efficiently and ignore the cues from competing "restaurants" (just as some people key in on the Golden Arches!).

Many species of birds feed on nectar and supplement their diet with insects. However, birds often pierce the sides of tubular flowers and so obtain the nectar without picking up pollen, a situation that is disadvantageous to the flower. This may be one factor that selected for flowers shaped in such a way that birds can reach the nectar more conveniently from a position where they also brush against the pollen.

The 300 species of hummingbirds are the largest group of bird pollinators. They nearly always feed while in flight, hovering in front of a flower and using their long bills and tube-like tongues to suck up the nectar deep within the flower. Flowers pollinated by hummingbirds usually have long stigmas that pick up pollen from the bird's head (Figure 46–C). In tropical areas particularly, the length of the bird's bill and the depth of the flower trumpet dictate considerable specificity, making any one species of hummingbird able to feed only on certain species of flowers. Most flowers growing at high elevations in the tropics are bird-pollinated. The frequent rains at these altitudes hinder the flight of insects but not that of birds.

Flowers have adaptations ensuring quality pollination. Nectar with a high concentration of sugar often also has

**FIGURE 46–A** A bumblebee foraging on cherry blossoms. As the bee sips nectar, her head rubs against the anthers and picks up pollen, which she will add to the load already in the pollen baskets on her hind legs for the trip back to the hive. (Runk/Schoenberger from Grant Heilman)

(a)

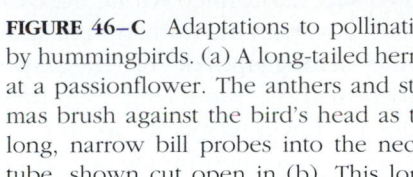

(b)

**FIGURE 46–C** Adaptations to pollination by hummingbirds. (a) A long-tailed hermit at a passionflower. The anthers and stigmas brush against the bird's head as the long, narrow bill probes into the nectar tube, shown cut open in (b). This long, narrow tube makes the nectar inaccessible to insects. (a, Michael Fogden/Animals Animals; b, Dan Cheatham/DRK Photo)

**FIGURE 46–D** Bee-fooled. A male long-horned bee copulates with part of an orchid flower that bears a remarkable resemblance to the abdomen of a female bee. (David Thompson/Animals Animals)

a high concentration of amino acids and some lipids, and this nutritious nectar tends also to contain alkaloids (see Figure 18–6). Alkaloids discourage butterflies, which find these bitter substances highly distasteful, but are tolerated by bees, the more efficient pollinators.

Other adaptations ensure not only that pollinators are faithful to one species but also that they visit the flowers frequently. For instance, flowers produce nectar slowly, making it necessary for the pollinator to visit many flowers and to revisit each flower again and again. Another adaptation that ensures frequent visits is the flower's distinctiveness: if the flower looks and smells different from others nearby, the pollinator can find the flower easily and discover more flowers of the same species quickly, so that it does not waste time and energy hunting around.

There is also strong selection for different plant species to bloom at different times, so that each receives the attentions of pollinators in its turn, rather than all competing during a brief period. Staggered blooming also gives pollinators a steady food supply throughout the growing season.

Although most flower-pollinator relationships are mutually beneficial, some plants have adaptations that secure animals' services without paying a reward. In some orchids, part of the flower resembles the rear end of a female bee (Figure 46–D). The flower may even emit the same chemicals used by the bee as her sex-attractant pheromone. Deluded male bees mistakenly copulate with the flower and pick up pollen, which they carry to other orchids of the same species as they repeat their mistake. This adaptation ensures that the flowers are visited frequently and faithfully.

The hairy red petals and rotting-meat stench of carrion flowers mimic dead mammals well enough to fool a female blowfly looking for a place to lay her eggs (Figure 46–E). The fly travels from flower to flower, transferring pollen as she leaves clusters of eggs. Since her larvae cannot survive without animal protein, this arrangement boosts the plant's evolutionary success but lowers the fly's.

**FIGURE 46–E** A fly on a carrion flower.

members of the species live within this area, there is a good chance that no pollen will reach the stigmas of their flowers. By contrast, an animal that visits only one kind of flower carries pollen directly from one flower to another of the same species, even if these flowers are widely scattered. Many flowers have evolved structures such that only one species of animal can pollinate them, and these flowers enjoy highly specific transfer of pollen from one individual to another.

Animal pollinators are attracted by some type of reward, usually a sweet nectar. The reward is made easier for the animal to find by an attention-catching "advertisement," such as the flower's odor, shape, or color—preferably all three. The reward is so located that the animal cannot reach it without picking up pollen at the same time. All of this has a cost: the animal-pollinated flower must invest energy in making its nectar and its large, showy petals, even though it need not make the prodigious amounts of pollen required for wind pollination.

Animals that serve as pollinators include insects—bees, butterflies, moths, wasps, flies, and beetles—and vertebrates such as birds, bats, and even a South African mouse!

## Pollen Maturation

A pollen grain completes development into a mature male gametophyte after it has landed on a compatible stigma, where sticky secretions may hold it in place. From the stigma the pollen absorbs water, nutrients, and possibly other substances that stimulate or inhibit its further development. Some pollen can be germinated in the laboratory in a sugar solution, but many kinds cannot. This suggests that in nature the stigma produces additional chemicals required for pollen tube growth. As water enters, the pollen grain swells and ruptures at one of the pores or furrows that form weak areas in its wall, and the **pollen tube** grows out.

The pollen grain wall contains glycoproteins that must be compatible with proteins in the stigma if the pollen is to grow. Hence, pollen will usually not germinate on the stigmas of flowers of a different species.

An estimated one half of the species of flowering plants also have a system of self-incompatibility genes, which prevent self-fertilization. Depending on the species, these genes prevent pollen from germinating or inhibit pollen tube growth if the pollen expresses one of the same alleles present in the carpel. This assures that the flower is cross-fertilized and so maintains genetic diversity in the population. In some species studied, self-incompatibility is controlled by a single genetic locus, which may have 40 or more alleles in the plant population. Self-incompatibility is not well understood. It appears to be complex and to vary among species.

If the pollen and stigma are compatible, the pollen tube grows down the style toward the ovule(s) in the ovary. The style contains complex macromolecules that provide nutri-

tion and guidance to the growing pollen tube. Many pollen grains may land and produce pollen tubes in the same style.

During the pollen tube's growth, the generative cell divides by mitosis and differentiates, forming two sperm cells. The pollen tube grows into the ovule through a tiny pore, the **micropyle,** and releases the sperm (Figure 46–7). The micropyle then closes, preventing the entry of any more pollen tubes. If there are more ovules, other pollen tubes enter through their micropyles. The traffic problem that must exist in the style of a cantaloupe flower is fearful to contemplate!

## 46–C    PREPARATION OF THE OVULE

Before the pollen tube arrives at the micropyle, a female gametophyte, or **embryo sac,** has formed within the ovule. Here again, the process begins with the formation of spores by meiosis. In the ovule, a **megasporocyte (megaspore mother cell)** undergoes meiosis, producing four **megaspores.** In most species of flowering plants, as in female animals, three of the four cells formed by meiosis disintegrate. Only one megaspore survives.

This megaspore enlarges greatly by absorbing nutrients, including the remains of the other megaspores. Its haploid nucleus typically undergoes three mitotic divisions, and the embryo sac has eight haploid nuclei distributed among seven cells: three at the end near the micropyle, three at the opposite end, and one in the center containing two haploid **polar nuclei** (Figure 46–8). This female gametophyte is now ready to be fertilized.

## 46–D    FERTILIZATION

Fertilization may take place as little as an hour after pollination, as in barley, or as much as several months later. For example, witch hazel flowers are pollinated in the late fall but not until the following spring do the sperm arrive at the micropyle. In some oaks the process takes a year or longer.

The sexual reproduction of flowering plants involves **double fertilization.** One sperm nucleus fertilizes the egg nucleus, forming the zygote. The other sperm nucleus fuses with the two polar nuclei, forming an **endosperm** nucleus that is typically triploid (3N, where N is the number of chromosomes in a haploid nucleus). The adaptive value of this second fusion is unclear, although the endosperm tissue that arises from this nucleus has a very important role, as we shall see shortly.

Dandelions, hawkweeds, and many grasses reproduce by the process of **agamospermy,** in which seeds develop from unfertilized ovules. In some of these plants pollination occurs, but it is only a stimulus to the ovule to de-

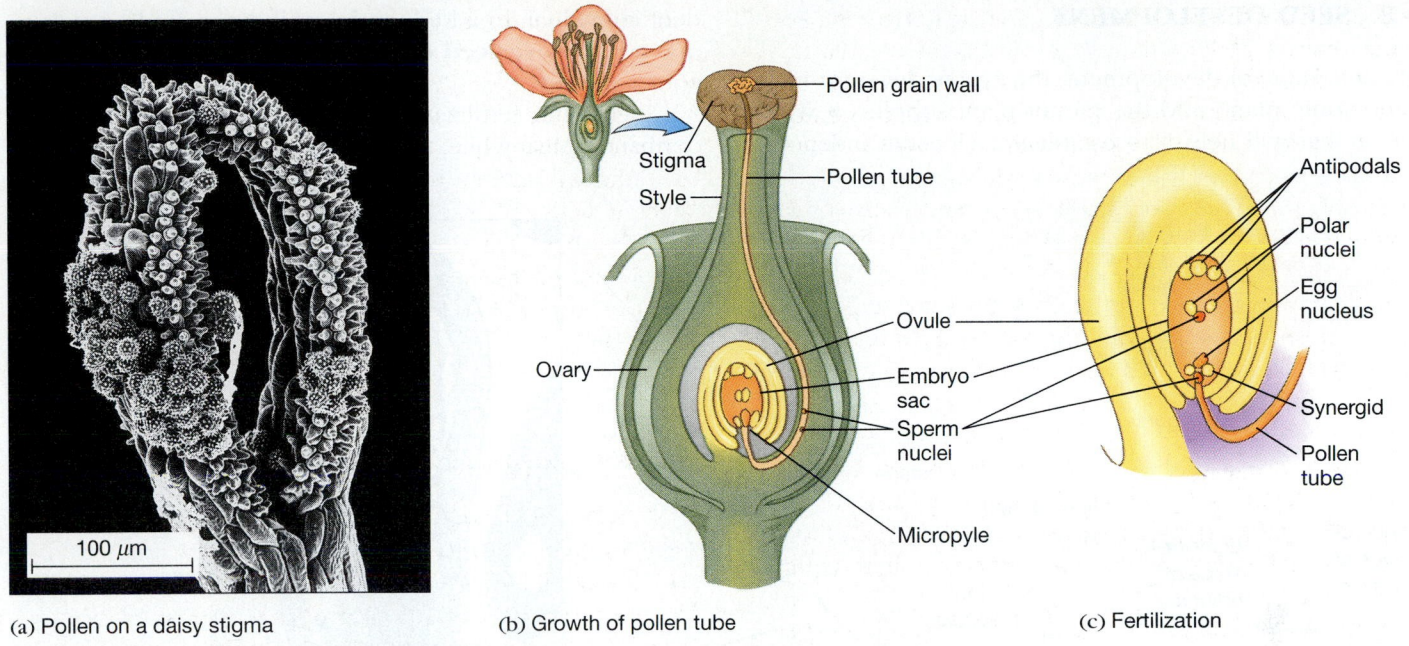

(a) Pollen on a daisy stigma    (b) Growth of pollen tube    (c) Fertilization

**FIGURE 46–7** Pollination and fertilization. (a) Pollen on a stigma from a daisy. (b) A pollen tube containing two sperm grows through the micropyle into the ovule. (c) In the embryo sac, the two sperm will take part in double fertilization. (a, Biophoto Associates)

velop into a seed, and the pollen contributes no genetic material to the offspring. In other species, agamospermy is an emergency method of reproduction when pollination does not occur.

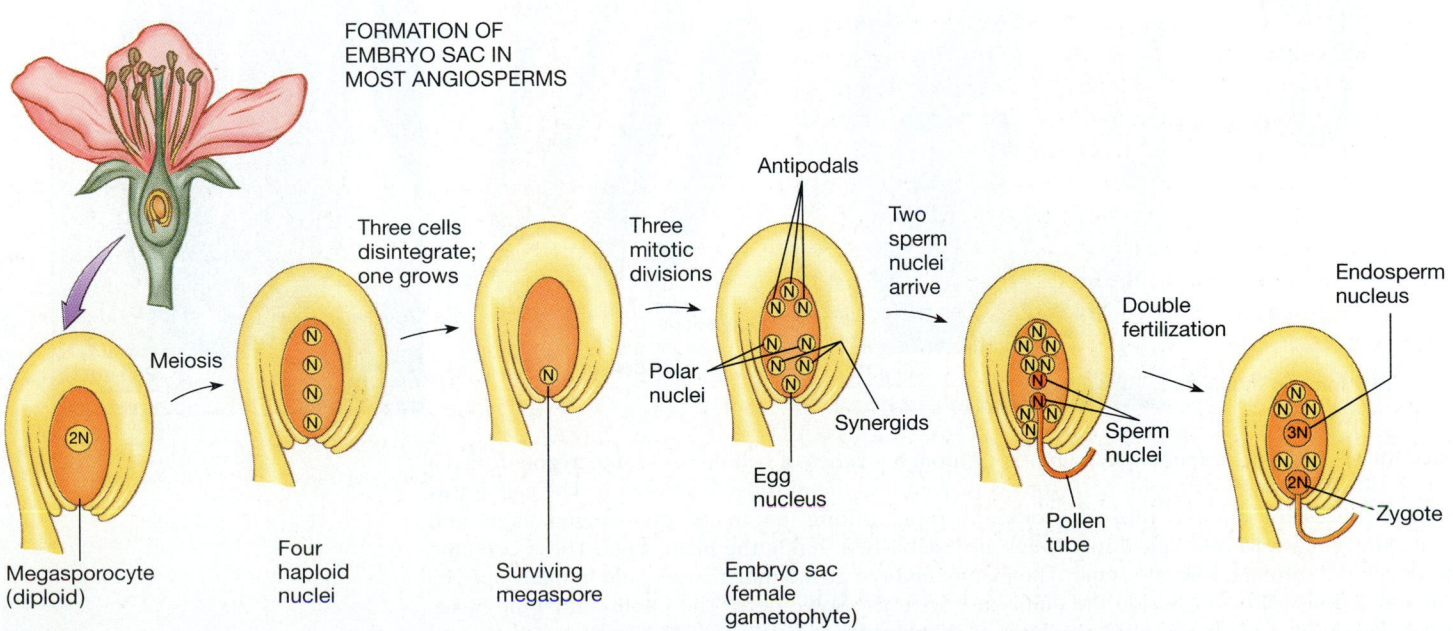

FORMATION OF EMBRYO SAC IN MOST ANGIOSPERMS

**FIGURE 46–8** Development of a megasporocyte into an embryo sac (female gametophyte). Most angiosperms show this pattern of nuclear divisions and rearrangements, but several variations occur. Double fertilization by two sperm nuclei forms a diploid (2N) zygote and triploid (3N) endosperm nucleus.

## 46–E   SEED DEVELOPMENT

In the next stage of development, the zygote develops into an embryonic plant, and the parent plant supplies it with nutrients that will help it to establish itself as an indepen-dent individual. In addition, the wall of the ovule develops into a protective **seed coat,** and the wall of the ovary becomes a **fruit.**

Right after fertilization, the zygote enters a period of dormancy. Meanwhile, the endosperm nucleus becomes ac-

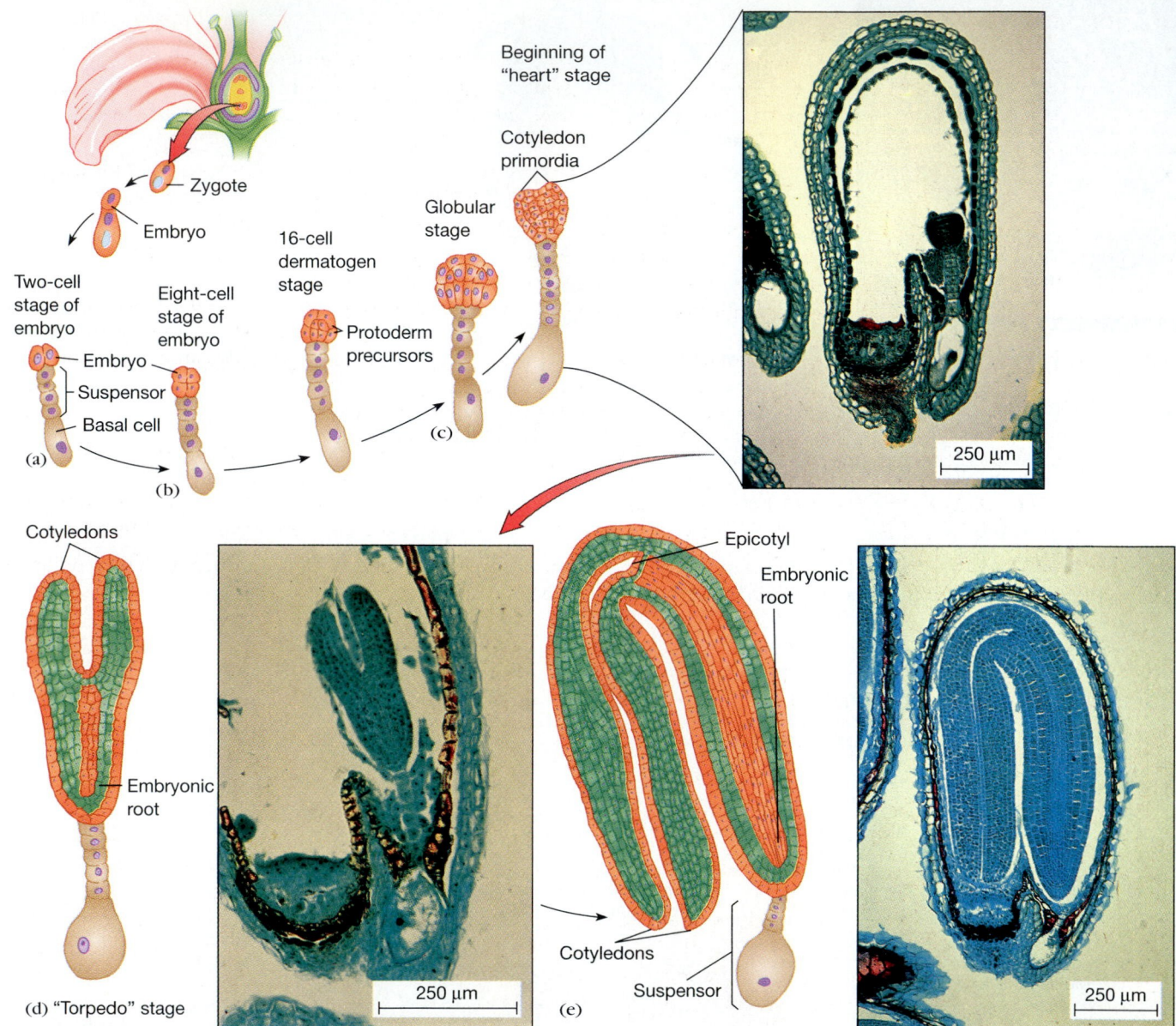

**FIGURE 46–9**   Development of the embryo. Through a series of cell divisions, the zygote forms a string of cells (suspensor); one end (top) is pushed into the nutritive endosperm. The cell at this end produces the embryo proper. Protoderm forms during the 16-cell dermatogen stage, and cotyledon primordia and apical meristems are established during the heart stage. These continue to develop during the torpedo stage. The mature embryo grows much larger, and in many species the cotyledons fold over beside the embryonic root, as shown here. The photos show the heart, torpedo, and mature stages of the embryo of shepherd's purse *(Capsella bursa-pastoris).* Around the embryo lie patches of endosperm, and encircling both is the wall of the ovule, which develops into the seed coat.   (photos, Biophoto Associates)

tive, dividing many times to form endosperm tissue, which enlarges and absorbs food from the parent plant. When the zygote breaks dormancy, the developing embryo grows by cell division and absorption of some of this food supply.

## Development of the Embryo

Embryonic development establishes the axis of the plant, with a root and a shoot meristem at either end and the precursors of the dermal, vascular, and ground tissues. In the first stage of embryonic development, the division of the zygote forms a line of cells, the **suspensor.** These cells elongate, pushing those at the end farthest from the micropyle into the nutrient-rich endosperm. Soon the terminal cell at this end starts to develop into the embryo, dividing into two cells, then four, eight, and sixteen (Figure 46–9). The suspensor disintegrates later as the plant embryo develops.

The first signs of tissue differentiation appear during the 16-cell dermatogen stage (derma = skin; genes = born). The embryo now consists of eight surface cells completely covering eight more in the interior. The surface cells are precursors of the protoderm, which produces the epidermis. During the following globular stage, these cells continue to divide and soon begin to secrete waxes characteristic of epidermal cells. The internal cells also divide in patterns that form the precursors of the other tissues.

During the transition from the spherical globular stage to the heart stage, the polarity of the plant axis is established. The heart shape is formed as the primordia of the cotyledons emerge by rapid, local cell division. The root and shoot apical meristems also form. Procambium, which will give rise to the vascular tissues, is present in the late heart stage.

The embryo continues to enlarge and elongate, forming the torpedo stage, with a distinguishable hypocotyl and embryonic root, where vascular tissue begins to differentiate. The cotyledon(s), often the largest part of the embryo, may continue to elongate. In many dicots they eventually fold over beside the root (Figure 46–9).

The genetic basis of plant development has been studied using *Arabidopsis thaliana,* a small, rapidly growing member of the mustard family with a small genome and short life cycle. This has become the workhorse of molecular embryology in plants, as the fruit fly *Drosophila* and the roundworm *Caenorhabditis* are in animals. By inducing mutations and screening for those affecting the embryo, researchers have found genes responsible for development of major parts of the plant axis, for size of organs, and for the pattern of tissue differentiation. For example, an embryo without functional genes specifying major parts of the plant axis lacks the relevant parts:

| Mutated Gene | Parts Missing in Embryo |
| --- | --- |
| *gurke* | Cotyledons and shoot apical meristem |
| *fackel* | Hypocotyl |
| *monopteros* | Root and root apical meristem |
| *gnom* | Roots and cotyledons |

Whatever parts are absent, those that do form have epidermal, ground, and vascular tissue. On the other hand, mutant plants with all these parts but without functional *knolle* and *keule* genes lack dermal tissue.

Researchers think these six genes will all prove to be homeobox genes, coding for transcription factors whose concentration gradients control the expression of the sets of genes required for development of the various organs and tissue types.

As the embryo grows, the endosperm continues to absorb food from the parent plant. The endosperm may persist as a food supply for the embryo or may be completely absorbed into the embryo's cotyledons as the seed matures, as in the bean seed (see Figure 42–5). Depending on species, the endosperm or cotyledons may also synthesize storage proteins during this time.

The wall of the ovule, which is part of the parent plant, becomes larger as the embryo grows and usually hardens, forming the protective seed coat as the seed matures.

## 46–F FRUIT GROWTH

Outside the seed coat, the wall of the ovary also enlarges and absorbs more nutrients, developing into a fruit (Figure 46–10). Fruit growth is initiated when the pollen tube releases tiny amounts of the hormones auxin and gibberellin (Chapter 45). Soon the developing seed begins to produce its own hormones, which continue to stimulate growth of the fruit. These activities enhance the transport of cytokinins, hormones that stimulate cell division, from the parent plant to the growing fruit. Together these three hormones promote fruit growth partly by cell division but mostly by cell enlargement. If developing seeds are removed, the surrounding part of the fruit does not develop.

Most types of plants do not set seed and develop fruits unless their flowers have been pollinated and fertilized. In some species, however, spraying the flowers with the proper concentration of auxin or auxin plus gibberellin stimulates development of seedless fruits. Some fruits, such as cultivated strains of bananas and pineapples, develop naturally without fertilization and are therefore seedless. Humans increase the populations of these economically desirable plants by vegetative propagation, and so the plants have become successful through artificial selection.

**FIGURE 46–10** Development: flower to fruit. (a) A winter aconite flower. (b) The ovaries enlarging as pod-like green fruits around the enclosed seeds. One developing fruit is dissected to show the white seeds inside.

(a)          (b)

## Four Main Types of Fruits

| Type | Characteristics | Example | Flower | Fruit |
|---|---|---|---|---|
| Simple | Develops from a single ovary, formed from a single carpel or from two or more fused carpels (= compound pistil) | Cucumber | | |
| Aggregate | Develops from fusion of many separate carpels in a single flower | Raspberry | | |
| Multiple | Develops from ovaries of many flowers grouped on same stalk | Pineapple | | |
| Accessory | Develops from ovaries plus nearby tissue | Strawberry | | |

(a)

(d)

(b)

Growth

Sepals

Floral
tube

Ovary

Seed coat

Embryo

(c) Apple

(e)

Receptacle

Achenes

Sepals

(f) Strawberry

**FIGURE 46–11** Accessory fruits. (a, b, c) Development of an apple blossom into an apple. The embryo develops inside the ovule, which itself develops into a seed coat. Meanwhile, the ovary wall enlarges greatly as part of the fruit. The outer part of an apple actually develops from parts of the floral tube around the base of the ovary in the apple blossom. Depending on the variety of apple, it takes the photosynthetic output of 15 to 35 leaves to support the growth of one apple. (d, e, f) Development of a strawberry. The flesh is the enlarged receptacle, studded with the tiny fruits (achenes), which contain the seeds. (d, Phil Gates/BPS)

Fruits may be classified as simple, aggregate, or multiple, according to the number of pistils and of flowers that contribute to their structure (Table 46–1). A fourth category is accessory fruits, in which other parts of the plant near the flower develop into fruit-like structures. For example, the outer part of an apple develops from the floral tube, which surrounds the base of the ovary in the apple blossom. A strawberry is really an enlarged **receptacle,** the area of the flower stalk that holds the flower parts. Its "seeds" are in botanical fact the fruits of the strawberry plant, each having arisen from a separate carpel of the flower (Figure 46–11).

Another way to classify fruits is according to whether they are fleshy or dry. Those just discussed are fleshy, with

(a)

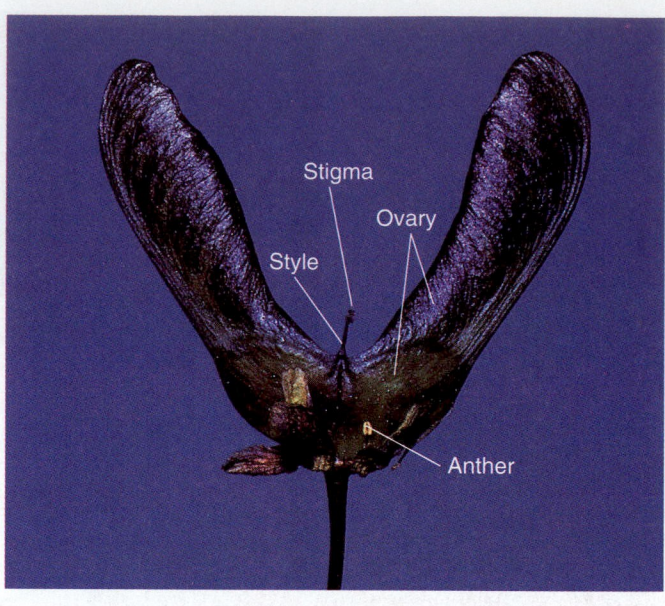

Stigma

Ovary

Style

Anther

(b)

**FIGURE 46-12** Wind dispersal. (a) After the fibers attached to milkweed seeds dry out, they spring apart because of repulsion among their like electrical charges. Tufts of fibers then function like parachutes as the seed wafts away—in fact, the fibers were used to make parachutes in World War II. (b) The "wings" of maple fruits, reminiscent of aircraft propellers, are enormous outgrowths of the ovary wall. The seeds are enclosed in the green swellings at their bases. The other flower parts are still visible at the base of the extremely enlarged ovary.

much thick, moist tissue. Dry fruits are often rather minimal protective layers that lose water and become papery or hard or brittle. When the seeds are ripe, the fruit may rupture along its lines of weakness and release the seeds. Dry fruits include the mature pods of peas, beans, and milkweeds and the capsules of poppies.

**FIGURE 46-13** Adaptations to animals. These porcelainberries change from white to deep blue by way of several intermediate shades, signaling herbivorous animals that they are ready to be eaten.

## 46-G DISPERSAL OF SEEDS AND FRUIT

Once mature, the seeds are ready for dispersal. In most cases it is advantageous for the seeds to grow well away from the parent. First, this distributes the population of the plant's descendants over a wider area. Second, it avoids competition for light, water, and soil minerals between the parent plant and its offspring.

Wind and animals are the usual agents of dispersal. Small, lightweight milkweed seeds and dandelion fruits have parachute-like tufts of fiber that enable them to disperse by floating through the air (Figure 46-12). The thin, flat wings of maple fruits whirl in the breeze like the blades of a helicopter. Larger seeds will have a distinct advantage when they start growing because they contain more food for the embryo, but wind cannot waft a large seed far from the parent plant.

Animals, on the other hand, may carry even a large seed quite a distance, and many plants invest a lot of energy in adaptations that promote dispersal by animals. Most commonly, the seeds are protected in indigestible seed coats and surrounded with a tasty, nutritious fruit that an animal will eat. The seeds then pass unharmed through the animal and are deposited with a small pile of organic fertilizer. In fact, the seeds of some plants will not grow unless their seed coats have been eroded somewhat by an animal's digestive enzymes.

**FIGURE 46–14** A herby hooker. This scanning electron micrograph shows a goosegrass flower with a pair of developing fruits. These fruits disperse by hooking onto an animal or onto human clothing when we walk through the meadow in late summer. The stigmas can still be seen (top right). (Biophoto Associates)

Fruits are usually protected from being eaten before the seeds have matured. Unripe fruit is often distasteful and may even contain toxic chemicals. As the fruit ripens, its chemical composition changes, and it becomes tastier. The fruit may also change color, a visual signal that it is now ready to be eaten (Figure 46–13).

Several thousand species of plants produce seeds with elaiosomes, nutritious fat bodies attached to the outside of the seed. Ants foraging for food carry the seed home, where the colony feasts on the elaiosome. The seed is discarded unharmed in the enriched, loose soil or rotting log inhabited by the ant colony.

Some fruits or seeds have hook-like extensions that attach to the feathers, fur, or clothing of passing animals, which give the seed a free ride to a new home (Figure 46–14). These hitchhikers gain the use of the animal's mobility for a small energy investment in the production of hooks.

## Seeds and Seed Predators

Because seeds contain the food supply for an embryonic plant, they also make good food for animals. Hence, there is strong selection for adaptations that protect the seeds

from predation. The types of adaptations that are effective depend on the main types of predators eating the seeds. An undiscriminating predator gobbles up every seed it finds. Plants with this type of predator usually produce many small seeds. This gives a good chance for some of the seeds to escape notice, so not all are eaten.

A discriminating predator maximizes its food intake for the energy it expends. It may attack plants that have the most seeds in a fruit or the seeds that are largest or easiest to chew. Plants attacked by such predators usually enclose their seeds in a hard covering, such as a nutshell, that discourages the predator (Figure 46–15). Producing smaller seeds works only if the seeds become smaller than those of another species to which the seed predator might switch.

Another adaptation is **seed masting,** the simultaneous release of seeds by all the plants of the same species in an area, at intervals of two years or longer. This makes seeds available to predators for a minimum time period. Beech trees and some oaks do this, but the most impressive examples are bamboos. Part of a bamboo stand in India was collected and sent to botanical gardens in the United States and Britain at the beginning of the nineteenth century. The plants grew vegetatively for the next 130 years, and then the bamboo stands on all three continents produced seeds in the same year! The advantage of seed masting to the plant is that most of the time there are no seeds to support the growth of large populations of seed-eating animals. When seeds are finally shed, there are so many that a small population of seed predators cannot eat them all, and some seeds escape to produce the next generation of plants. In addition, seed production in these bamboos triggered pro-

**FIGURE 46–15** A defensive arsenal. A large, nutritious horse chestnut seed (left) with a tough shell, which originally was covered by a prickly husk (right), a deterrent to prospective seed predators. (Biophoto Associates)

grammed death of the parent plants, a process also characteristic of annual and biennial plants (those living only one or two years, respectively). This left very little food for bamboo-eating herbivores such as pandas.

Some large seeds, such as nuts and acorns, are actually dispersed by would-be predators. Squirrels and some birds collect these large, nutritious seeds and hide them for later use. However, they do not return to all of their caches. The forgotten seeds sprout the next spring and may grow into new trees. In the long run, this predation behavior benefits the tree, which has gained some surviving offspring at the expense of others. Researchers attribute the wide distribution of oak trees to the industry of jays, which may fly several kilometers to bury acorns in soft soil or under moist leaves (squirrels have much smaller home ranges). In one study, 50 jays spent September hiding 150,000 acorns! The diligence of jays is thought to account for the rapid spread of oak trees north, following the retreating glaciers, after the last ice age—dispersal the heavy acorns could never have accomplished by themselves.

## 46–H  GERMINATION

The embryo is capable of germinating once it has formed the organ primordia, which will develop into its first root, leaves, and stem. However, in an intact plant this does not happen. At first this is because of high levels of the hormone abscisic acid, which promotes dormancy. The synthesis of RNA and proteins is largely suspended, and metabolism and respiration plummet. Some of the last genes expressed may be involved in the synthesis of dehydrins, responsible for the loss of water until it makes up less than 5% of the total mass of the dormant seed. The combination of abscisic acid and dryness inhibits germination.

In many kinds of seeds, dryness seems to be the main factor assuring the seed's **viability,** or ability to break dormancy and grow into a new plant after an extended period of time. Many commercial seed suppliers now dry their seeds thoroughly and wrap them in moisture-proof foil packets.

Seed viability varies among species and among individuals within a species. The viability record is held by water lily seeds found in a peat bed in Manchuria and dated at more than 1000 years old by radioactive isotope methods (*How Do We Know? Isotopes,* Chapter 2). Seeds from even older deposits have been grown, but no dating was done to prove that the seeds were as old as the layer in which they were found. Seeds of a few species stored as museum specimens for over a century have also been grown successfully.

In contrast, sugar maple seeds live less than a week. These seeds do not have a dormant period, and their viability is better if they do not dry out much.

In order to **germinate,** or begin growing into a new plant, seeds must be supplied with water. However, the seed coat is often so thick or impermeable that water cannot enter until the seed has had some special treatment, such as partial digestion of the seed coat by animals or decomposer organisms, or abrasion by soil particles. Most seeds also require oxygen, and many require particular temperature and light conditions or the leaching of inhibitory substances. Furthermore, germination requirements may vary not only from species to species, but also from one individual to another. Thus, a plant does not have all of its seeds "in one basket"; germination may spread over months or years, and at least some seeds are likely to find conditions that favor their survival.

During germination, a seed absorbs water, its cells expand, and its gibberellin content may increase. Applying suitable concentrations of gibberellins can often overcome a seed's special light or temperature requirements. For example, lettuce seeds normally induced to germinate by exposure to red light can be germinated by applying gibberellins instead. Likewise, gibberellins can substitute for the cold treatment **(vernalization)** required by seeds such as some types of rye and wheat. Normally, these stimuli induce production of gibberellins. Seedlings also produce ethylene, which promotes thickening of the hypocotyl as it pushes up through the soil. In some species, other plant hormones are needed for germination.

Germination has been studied best in barley. The barley embryo secretes gibberellin, which moves out to the **aleurone,** a tissue layer between the seed coat and the endosperm (Figure 46–16). Here it induces secretion of amylase and other hydrolytic enzymes into the endosperm, where they break down starch and other stored food, mak-

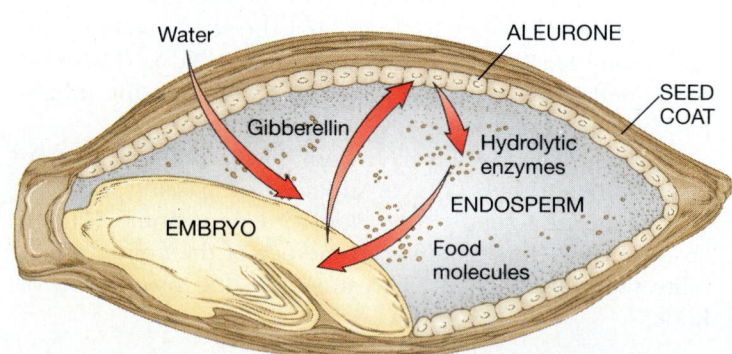

**FIGURE 46–16**  Germination. A barley seed breaks dormancy after absorbing water (top left). Arrows show the flow of substances as the food stored in the aleurone and endosperm is mobilized and used during the embryo's growth into a seedling.

(a)                              (b)

**FIGURE 46–17** Plant breeding. (a) Corn plants with striped bags over their anthers (top) to collect pollen and plain bags over the stigmas (lower on the plants) to prevent pollination until the breeder brings the desired pollen to the future ear of corn. (b) Almost 20 years of selective breeding went into developing "Bush Baby," a type of watermelon with desirable flavor, color, and disease resistance that also grows on space-saving bushy vines. (Most watermelon vines sprawl over several meters.)  (b, W. Atlee Burpee Co.)

ing it available for absorption by the developing embryo. As the embryo uses this food, the dry weight of the seed decreases until the embryo begins to make its own food through photosynthesis. The seedling then grows as described in Chapter 42.

## 46–I  BREEDING PROGRAMS

Probably since the beginning of agriculture, humans have bred plants selectively to improve the features that make them useful to us. For thousands of years, the only method available was to collect seeds from plants that produce well in the field and plant these for the next generation. This eventually produced strains of cultivated plants strikingly different from their wild ancestors, especially in improved food yield.

More control of the process can be gained by cross-pollinating plants selected for different desirable traits, in an attempt to obtain offspring with all the desired features in the same plant. Our modern knowledge of genetics now guides the selection of parental strains for such breeding programs.

The genetics of some crop plants are fairly well understood, especially for major crops like tomatoes, wheat, and corn. The popular strains of hybrid sweet corn are produced by crosses between a number of strains, each bred for particular traits. It may take several generations of crosses to produce the seed used to grow hybrid sweet corn, and

these crosses must be made anew each year in a continuous breeding program because the hybrids do not breed true. So seed for hybrid corn must be purchased from breeders each year; planting seeds saved from last year's hybrid ears results in a motley assortment of traits rather than the desired genetic uniformity.

Corn is a fairly easy crop to use in genetics programs. Because male and female flowers grow in large, separate clusters, it is easy to carry out a desired pollination (Figure 46–17). In addition, corn has only ten pairs of chromosomes, with a relatively limited number of possible combinations. And, since corn is an annual plant, the success of a cross can be judged within a year. Finally, there are millions of individuals from which desirable new traits can be chosen.

Other plants pose difficulties. For example, a breeder of apples must cross-pollinate blossoms of two varieties individually by hand. Later, the seeds must be cut from the fruit and grown in greenhouses, where the seedlings are inoculated with major diseases of apples in order to select for those with genetic resistance. The surviving seedlings take five to ten years to mature and bear fruit, which must then pass color, texture, and taste tests. It took 21 years to bring Empire apples from seed to market, and another 20 for this variety to become one of the most popular available.

Even in annual plants, it may take many years to develop new varieties if the genetics of the species has not been worked out well (Figure 46–17b).

Breeders have also realized that modern crops have low genetic diversity compared with their wild relatives. Researchers attempting to improve crops are traveling to the areas where crops were first domesticated, seeking genes for resistance to pests, diseases, and environmental stresses and for improved nutritional value. Under the 1991 Global Plant Genetic Resource Initiative, agricultural researchers and the governments of developing nations will cooperate in trying to identify such genetic resources among wild plants and locally grown crops, incorporate new genes into crop plants, and preserve habitats so that wild populations and their genes will continue to be available.

Plant breeders are turning more and more to genetic technology to produce plant varieties with desirable traits (Chapter 13). This permits them to introduce desired genes not only from other individuals of the same or closely related species, but also from other kinds of plants and even from bacteria or animals.

**FIGURE 46–18**  Vegetative reproduction. A "hens and chicks" plant surrounded by many smaller offspring (the "chicks").

## 46–J   VEGETATIVE REPRODUCTION

Sexually produced offspring often receive new genetic combinations not well suited to the environment, and even seeds with favorable genetic combinations have high mortality from predation, disease, and dispersal to unfavorable habitats. Thus, it is advantageous for plants that are well adapted to their environments to spread by vegetative reproduction and cover the surrounding area with a clone of genetically identical individuals (Figure 46–18).

Vegetative propagation may be by means of rhizomes (horizontal ground-level stems) or roots that send up new shoot crowns, with attached roots, at intervals; by **stolons,** which produce the roots and shoot of a new individual where they touch the ground, as in strawberries (see Figure 20–2); or by **layering,** in which a woody branch may produce new roots, and hence a potential new individual, at a point where it touches moist soil, as in blackberries and raspberries.

Vegetative propagation is often desirable from the human as well as from the plant point of view, and we use it to reproduce many economically important plants. Home gardeners root cuttings of *Coleus,* geranium, or ivy shoots by placing them in water or set leaves of African violets or jade plants on moist soil until they grow roots. Some plants can be rooted more successfully with the use of commercial plant hormone treatments. Many plants reproduce vegetative parts underground. People help spread plants such as daffodils and onions by digging up the bulbs when the tops die back and separating and replanting those that have multiplied, giving each more room to grow.

Potatoes are an important crop propagated by vegetative means. A potato is an underground stem, or **tuber.** Farmers get many offspring from a good potato plant by digging up "seed potatoes" (tubers of good quality), cutting them into pieces, each with an "eye" (bud), and replanting them. The seeds produced by potato flowers usually have inferior genetic combinations. However, breeders do grow plants from these seeds in an attempt to produce new strains with desirable traits, such as resistance to certain diseases and pests and the ability to form tubers when grown in tropical climates.

**Grafting** is another means of artificial vegetative propagation. A **scion,** a twig or bud of a desirable plant, is attached to a **stock,** the root system or stem of another plant, from which a twig or bud similar to the scion has just been removed (Figure 46–19). Scion and stock are then wrapped with their vascular cambia as close together as possible. The cut surface of each plant produces undifferentiated callus cells, which soon produce new cambial strands that form a vascular connection. Both stock and scion retain their genetic identity, but they interact physiologically by the exchange of nutrients and hormones.

Grafts work only between plants of the same or closely related species. The method is used to produce fruit trees, grape vines, or rose bushes combining the desirable fruits or flowers of the scion with a sturdy, pest-resistant rootstock. For example, all the Red Delicious, Golden Delicious, and McIntosh apple trees now in existence are derived, by grafting, from single fence-row "volunteer" trees that happened to have desirable fruits.

Dwarf fruit trees are produced by grafting scions onto rootstocks of related species. Thus, dwarf pear trees consist of pear scions grafted onto quince roots. Apple growers are rapidly switching from plantings of 100 full-size trees per hectare to anywhere from 1300 to 2200 dwarf trees,

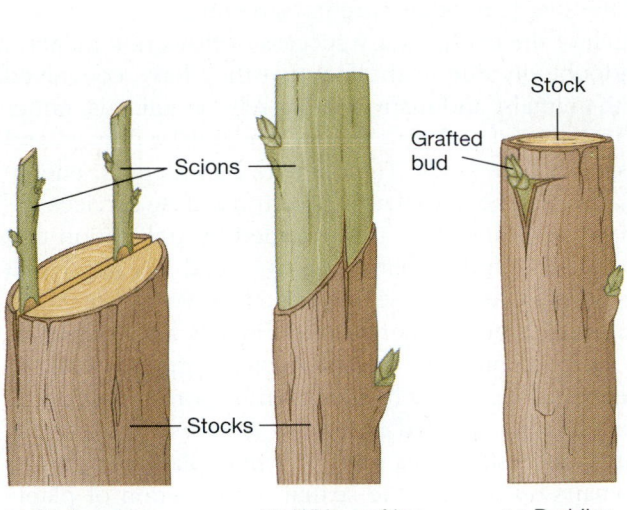

(a) Cleft grafting    (b) Whip grafting    (c) Budding    (d) A successful graft

**FIGURE 46–19** Grafting. (a, b, c) Three grafting techniques. (d) A successful graft in an apple tree: scion and stock formed a bulging callus as they grew together and sealed themselves into a single unit. Eventually the callus will disappear as new tissue grows smoothly around it.

which permit more light to reach fruit on the lower branches and are easier to harvest. For home gardeners, plant nurseries now offer small, manageable apple and pear trees with five or six grafted branches, each bearing fruits of a different variety.

Plants are also propagated vegetatively from cells grown in laboratory culture. The culture medium contains high levels of plant hormones, which produce high rates of mutation. Cells from plants with many desirable qualities multiply and mutate in laboratory culture and are then grown into whole plants. These plants are very similar genetically —more so than sexually produced offspring—but they may differ in important traits such as resistance to drought or to particular diseases. This method allows plant breeders to "fine-tune" the genetic makeup of crop plants.

Tissue culture also gives a way to produce more replicas of a desirable plant quickly. Only a small lump of cells is needed to start each new plant, rather than a large, leafy cutting. Cells kept in a flask or two in the laboratory can substitute for acres of plants formerly kept as sources for cuttings. Many plants, such as chrysanthemums and newly bred varieties of potatoes, are now grown commercially in laboratory culture from clumps of cells. Laboratory culture is also used to grow genetically engineered cells into complete plants (Figure 46–20).

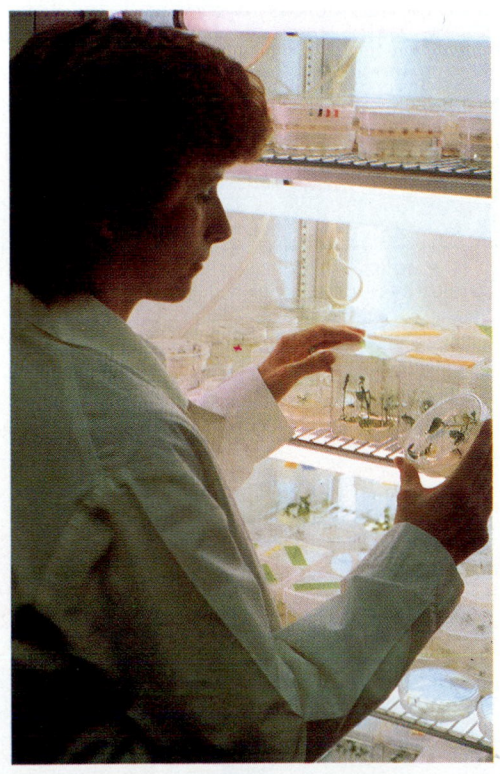

**FIGURE 46–20** Plantlets grown in tissue culture from cells containing foreign genes. The cultures are kept in a growth chamber where the environment is carefully controlled. (Calgene)

## SUMMARY

Plants flower in response to specific cues that differ greatly among species, and each species of angiosperm has its own distinctive flower structure. In all this diversity, however, we can find a basic unity in flower structure and function.

1. A flower is an abbreviated shoot, in which all the cells differentiate into parts of the flower stalk or its modified leaves, the flower parts. Flowers typically have four kinds of flower parts: protective sepals, petals, male stamens, and female carpels.

2. Meiosis occurs in certain cells in the anthers (part of the stamens) and in the ovules (enclosed by the ovary in the base of the carpels). The resulting haploid microspores and megaspores develop into the haploid male and female gametophytes— the pollen grains and embryo sacs, respectively.

3. Pollen grains are transferred from the anthers to the stigma of a pistil by wind or by animals. Here the pollen grain absorbs moisture and nutrients and germinates, growing a pollen tube down the style to the embryo sac in the ovary. Here the pollen tube releases two sperm.

4. Double fertilization forms a zygote and an endosperm nucleus in the embryo sac.

5. The endosperm nucleus divides and develops into the endosperm, which absorbs food from the parent plant, and the zygote soon develops into the embryo of a new plant. The parent plant, besides contributing food to the new embryo, also protects it and its food supply in a seed coat derived from the wall of the ovule, which in turn is surrounded by the fruit, derived from the wall of the ovary, another parental structure.

6. Much of the evolutionary success of flowering plants is undoubtedly due to the fact that they have coevolved with animals, and many species rely on animals, rather than on wind and water, to pollinate their flowers and disperse their seeds. They devote considerable energy to attracting animals that will perform these services appropriately, and they are rewarded by pollination that is efficient and specific and by seed dispersal that distributes even large seeds over a relatively wide range.

7. Many seeds enter a period of dormancy as the embryo, its food supply, and the seed coat complete development and become dehydrated. In response to environmental cues, the seed eventually absorbs water, germinates, and establishes itself as a new plant.

8. Humans manipulate the sexual reproduction of plants to produce individuals with desirable new combinations of genetic characters.

9. In many cases, the genetic combinations in sexually produced offspring are less desirable than those of the parents, from either the human or the plant point of view. Many plants have some means of asexual reproduction, which perpetuates a particularly favorable combination of genes unchanged, in addition to, or instead of, sexual reproduction. Humans propagate many desirable plants vegetatively by artificial means such as dividing, rooting, grafting, or growing plants from cells in tissue culture.

## Self-Quiz

Label the structures numbered in the diagram below:

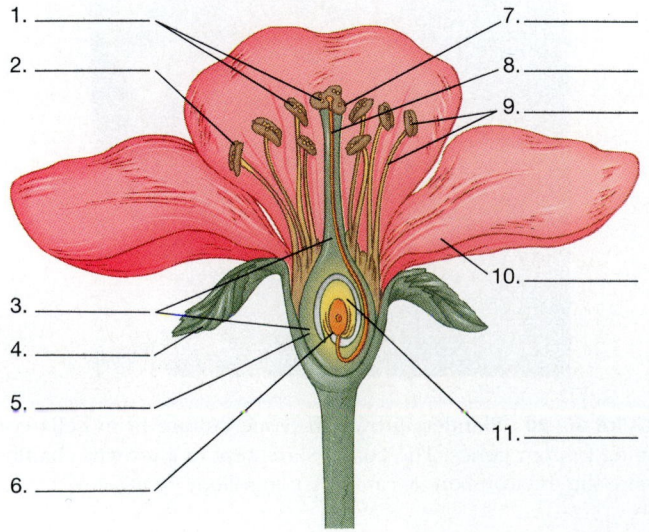

1. _____
2. _____
3. _____
4. _____
5. _____
6. _____
7. _____
8. _____
9. _____
10. _____
11. _____

For each of the following descriptions, give the name and number of the structure in the diagram:

_____ 12. Site of pollen production
_____ 13. Female gametophyte
_____ 14. Protective flower part
_____ 15. Develops into the seed coat
_____ 16. Contains structure that gives rise to endosperm
_____ 17. Develops into the fruit

18. True or False: The terms pollination and fertilization can be used interchangeably.

19. Which of the following is *not* required for seed germination?
    a. certain temperature conditions
    b. oxygen
    c. water
    d. light
    e. none of the above

20. Grafting is used to propagate plants because:
    a. it is faster than growing seeds
    b. it maintains a desired set of genetic characteristics
    c. it combines the genetic characteristics of two desirable strains of plants
    d. healthy plants will graft by themselves, so that they reproduce profusely
    e. a plant can produce many more scions than seeds

## Questions for Discussion

1. Why do banana plants put so much energy into producing fruits that contain no seeds?
2. Plants given large amounts of fertilizer, especially fertilizer containing much nitrogen, often flower poorly or not at all and do not accumulate food reserves; instead they engage in vigorous vegetative growth. Is there an adaptive advantage to this?
3. Some plants, such as dandelions and hawkweeds, have lost the ability to reproduce sexually but still produce flowers and set seed by development of the ovule without meiosis or fer tilization. What is the advantage of this system over vegetative reproduction?
4. Pollen is produced at the tips of the stamens, whereas ovaries lie at the bases of the carpels. What is the adaptive advantage of these differences in position?
5. What are some advantages to the plant in having its flower parts differentiate in sequence rather than all at once?
6. If pollen fertilizes an egg in another plant of the same clone, is this considered self-pollination or cross-pollination?

## Suggested Readings

Barrett, S. C. H. "Mimicry in plants." *Scientific American,* September 1987. Some plants cheat pollinators or farmers by resembling other species so closely that the animals are duped into helping the plant without reaping the expected reward.

Barrett, S. C. H. "Waterweed invasions." *Scientific American,* October 1989. How two aquatic plants have taken over waterways worldwide by vegetative reproduction.

Faegri, K., and L. van der Pijl. *The Principles of Pollination Ecology,* 3d ed. New York: Pergamon Press, 1979.

Handel, S. N., and A. J. Beattie. "Seed dispersal by ants." *Scientific American,* August 1990.

Heinrich, B. "The energetics of the bumblebee." *Scientific American,* April 1973. The relationships between bumblebees and the flowers they pollinate, viewed in terms of the influence of energy expenditure on evolution of adaptations.

Meeuse, B., and S. Morris. *The Sex Life of Flowers.* New York: Facts on File, 1984.

Miller, J. A. "Somaclonal variation." *Science News* 128:120, 1985. How new varieties of plants are grown from cells that mutate in tissue culture.

# The Biosphere

## OBJECTIVES

*When you have studied this chapter, you should be able to:*

1. Explain how and why tropical and temperate climates are different, why deserts are found at latitudes 20 to 30 ° north and south, why precipitation is usually higher on the west side of a continent, and how a rain shadow forms.
2. Name the two main factors that determine the distribution of biomes and explain why they do so.
3. Explain why widely separated areas with similar climate usually contain species with similar adaptations.
4. State the conditions under which you would expect to find each of the following biomes, and list the type(s) of plants characteristic of each: tropical rain forest, tropical savanna,

desert, temperate forest, temperate shrubland, temperate grassland, taiga, and tundra.
5. Describe the differences between oligotrophic and eutrophic lakes.
6. Explain the ecological importance of wetlands.
7. List three types of intertidal habitat and the types of organisms found in each.
8. Describe the conditions under which a coral reef may form.
9. List the main factors that determine the distribution of life in the oceans; describe the ocean surface and ocean bottom communities, and tell how organisms in these communities acquire food.

L ife on Earth requires a moderate temperature, water, a source of energy, and various chemical nutrients from the soil, water, and air. Suitable combinations of the things organisms need are found only in the **biosphere,** a narrow layer around Earth's surface that is the only place we know of where life can exist. The biosphere extends about 8 km up into the atmosphere (where insects and the spores of bacteria and plants have been found) and as much as 10 km down into the ocean.

Organisms do not live in isolation. They depend upon interactions with their natural environment and with each other. The study of these interactions is **ecology** (oikos = house), a term coined in 1869 by German zoologist Ernst Haeckel. Ecologists study the patterns of distribution and abundance of organisms in nature, how these patterns are maintained in the short run, and how they change during the course of evolution.

Organisms are not scattered haphazardly through the biosphere because each species needs particular conditions. **Communities** are collections of species living in the same area at the same time. For instance, a particular community of organisms lives on a coastal sand dune (Figure 47–1). A community of organisms together with its physical environment is known as an **ecosystem,** a term coined by A. J. Tansley in 1935. An ecosystem includes **abiotic** (non-

living) factors such as sunlight, temperature, water, and soil and **biotic** factors, all the organisms in the ecosystem.

**FIGURE 47–1** The community of plants on a coastal sand dune. The tall grasses are sea oats, protected in the southeastern United States because their roots prevent erosion and hold the dune in place.

The science of ecology grew out of natural history—the observation and description of organisms in nature. As Western naturalists explored the world, they discovered two main patterns. First, every new area explored contains species not previously known to science. Second, organisms live in only a small number of types of communities, which have similar characteristics wherever they are found. Thus, shrubland is found on the coasts of both California and the Mediterranean Sea. The species living in these two places are different, but convergent evolution in the two areas has produced plants of similar heights, spacing, and even chemistry and animals of similar size, types, and habits.

The world contains only a limited number of community types, dictated by the physical conditions. For instance, tropical forest is found in several parts of the world and is one example of a **biome,** a type of community recognized by the characteristic structure of its dominant vegetation, which in turn is determined by climate. Other biomes—desert, grassland, or tundra—occur in parts of the world with particular climates and also look much the same wherever in the world they occur.

F. E. Clements and V. E. Shefford introduced the term biome in 1939 and applied it only to communities on land. But the distribution of aquatic communities is also determined by the physical environment. For instance, plants and algae live only near the water surface because deep in the water there is not enough light for photosynthesis.

### KEY CONCEPTS

♦ The organisms on Earth make up only a small number of different community types, each of which is found in several parts of the world.

♦ The biome of an area of land is determined largely by the pattern of temperature and precipitation. Areas with high rainfall and temperature support more and larger plants than areas with little rainfall or low temperatures.

♦ The community found in a particular aquatic location depends largely on light, water temperature, salinity and currents, and the type of bottom in the area.

## 47–A CLIMATE AND VEGETATION

Climate is the main factor determining the type of soil in an area (Section 44–B). The climate and the soil, in turn, determine the plants that can grow, and the plant life and climate determine the types of animals that can inhabit the area.

### Temperature

Climate depends on solar energy, which warms Earth's surface, evaporates water that will fall as rain or snow, and drives the winds. The major climatic variations are determined by the position and orientation of the Earth in relation to the sun. The **solar energy flux,** the rate at which energy reaches a given area of ground, depends on the angle of the sunlight. Near the equator, the sun is almost directly overhead year round. As a result, a ray of sunlight is concentrated in a small area, and sunlight reaches the ground by a shorter path through the atmosphere than in other parts of the world (Figure 47–2a). As a result, tropical areas have little seasonal difference in daylength and fairly high temperatures all year. For instance, in coastal Ecuador on the equator, the temperature remains between 22 and 28 °C throughout the year.

North and south of the equator, in areas with subtropical and temperate climates, the sun is lower in the sky, so a given amount of sunlight is spread over a larger area. In addition, sunlight passes through more atmosphere, where it is absorbed and deflected, before it reaches the ground. Because Earth is tilted on its axis, both daylength and the angle of the sun vary more at higher latitudes, so the temperature differences that mark the seasons also increase with latitude (Figure 47–2b). Thus, in Pittsburgh, Pennsylvania, which lies due north of coastal Ecuador, the temperature changes by nearly 40 °C during the year, from about −13 °C in January to about 25 °C in July.

Only about half the solar radiation reaching the outer atmosphere actually penetrates to Earth's surface. The rest is either reflected back into space or absorbed by the mixture of gases, dust, and solid particles that make up the atmosphere. Most of the shortwave, high-energy ultraviolet radiation is absorbed by the **ozone layer,** about 25 km up in the atmosphere. Much of the solar radiation that does reach Earth's surface is radiated out again as infrared heat energy. Because water and $CO_2$ in the atmosphere are good absorbers of infrared energy, however, the atmosphere acts as an insulator. Instead of traveling directly back out into space, much of the heat energy is absorbed by the atmosphere and reradiated back to Earth's surface. This is why

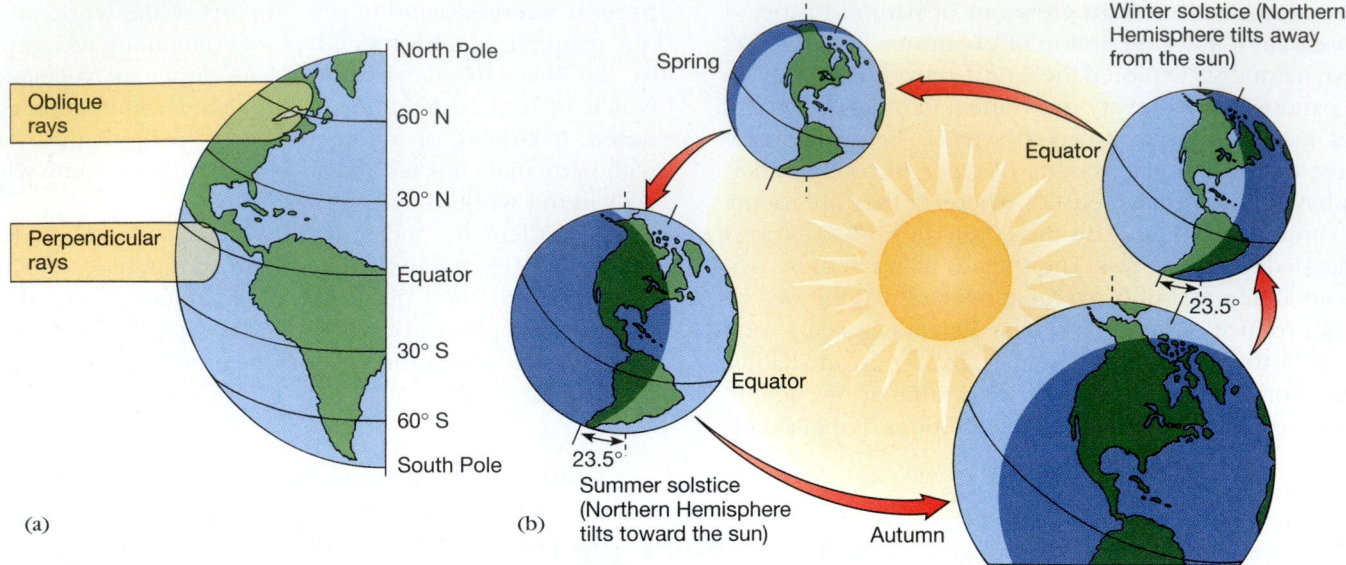

**FIGURE 47–2** Vertical versus oblique rays of sunlight. (a) A beam of sunshine striking Earth farther away from the equator is spread over a wider area. It is therefore less intense at any one point than a similar beam near the equator, which strikes Earth vertically. (b) Seasons occur outside the tropics because Earth's axis of rotation is tilted by 23.5° relative to its orbit around the sun. Winter occurs in the Northern Hemisphere when the North Pole is tilted away from the sun; summer occurs when it is tilted toward the sun.

cloudy or humid nights are warmer than clear, starlit nights. The insulating effect of the atmosphere is known as the **greenhouse effect** because it resembles the effect of the glass walls of a greenhouse.

The layer of the atmosphere that lies closest to Earth's surface is the **troposphere,** which contains the air we breathe. Two properties of air are particularly important to the weather:

1. Hot air is less dense than cold air, so hot air rises and cold air sinks. As air rises it expands and cools. When air descends, it is compressed, transforming energy into heat, and the temperature increases. As a result, temperature rises as we come down a mountain.

2. Warm air can hold more water vapor (a gas) than cold air can. This means that if warm air is cooled, some of its water vapor molecules condense into liquid water or solid ice, which form fog, rain, dew, or snow.

## Wind

Warm air at the equator rises, leaving low pressure and little wind near the surface, an area known to sailors as the "doldrums." As this air moves north and south from the equator at high altitude, it expands and cools until much of it sinks to the ground again at about latitudes 30°N and 30°S. This sinking air is compressed, forming regions of high pressure. One of these high-pressure zones usually lies

over the Pacific off the coast of California and causes the dry, stable climate of that state. Some of the descending air is forced back toward the equator as it reaches Earth's surface, creating the steady trade winds and completing the tropical circulation cell (Figure 47–3). At the poles, air falls and rises in cells similar to those of the tropics.

Between the tropical and polar cells, west winds prevail, but the winds are variable. At Earth's surface, some descending tropical air is deflected toward the poles. Eventually it meets cold dry air from the pole. These two bodies of air meet at the **polar front.** The behavior of this front in the Northern Hemisphere dominates the weather over most of Europe and North America. When cold polar air pushes under warm moist tropical air, a cold front moves in. When the warm air pushes the polar air back and rises over it, a warm front forms. In both cases, precipitation usually occurs.

The position of the polar front is strongly influenced by the position of the jet stream, which blows between the polar and tropical circulation cells high in the atmosphere. Here, the prevailing westerly wind blows at up to 320 km an hour. In the Northern Hemisphere, the jet stream tends to move south in winter and north in summer, affecting the position of the air masses beneath it.

Because of Earth's rotation, the prevailing winds making up the major circulation cells do not blow directly north and south relative to Earth's surface. They are deflected by the **Coriolis Effect** so that they curve to the east in the

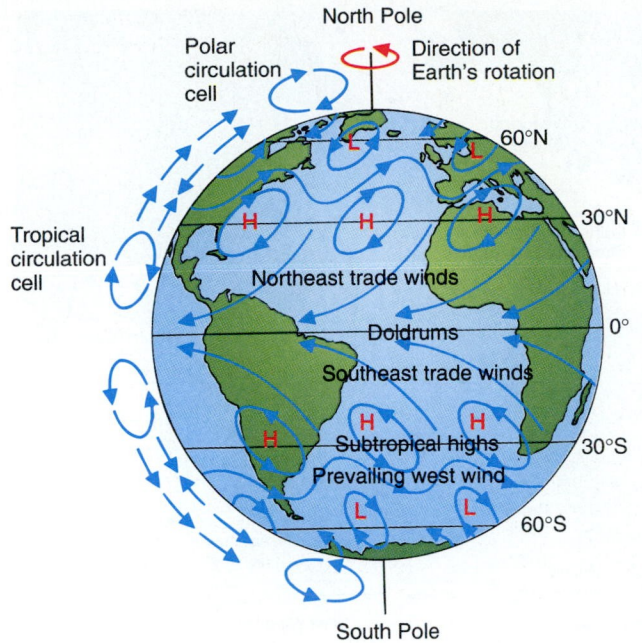

**FIGURE 47–3**  Prevailing patterns of wind circulation in the troposphere.

Northern Hemisphere. Figure 47–4 shows how the Coriolis Effect works.

## Ocean Currents

The oceans influence climate because they cover more than two thirds of Earth's surface and because water has a much higher specific heat than air or land. The **specific heat** of a substance is the energy required to raise the temperature of a given mass of the substance by 1 °C. The specific heat of water is more than 4186 J/kg, whereas that of most rocks is less than 500 J/kg. Because of its high specific heat, water gains and loses heat much more slowly than land or air. One result is that land in the middle of continents experiences greater extremes of temperature than land near the moderating effect of the ocean. For example, the coldest part of the Northern Hemisphere in winter is not the North Pole but part of Siberia.

Another effect of water's high specific heat is that warm ocean currents carry large quantities of heat from one part of the world to another. Surface waters move in the direction of the prevailing wind. Like wind, they are deflected by the Coriolis Effect. Therefore, ocean currents converge at the Equator, where they flow until stopped by land (Figure 47–5). Here they are deflected north and south, carrying enormous amounts of heat to higher latitudes and rotating clockwise in the Northern Hemisphere and counterclockwise in the Southern Hemisphere. As a result, water on the west sides of continents is cooler than that on the east sides. For instance, the west coast of the United States is bathed by the cold California Current, while the

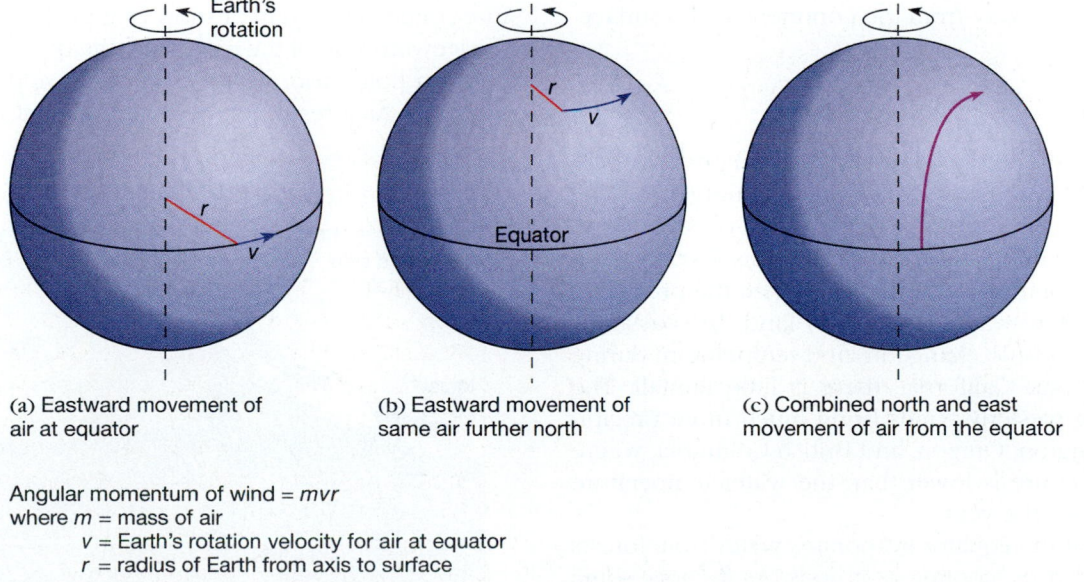

Angular momentum of wind = $mvr$
where $m$ = mass of air
$v$ = Earth's rotation velocity for air at equator
$r$ = radius of Earth from axis to surface

**FIGURE 47–4**  The Coriolis Effect, wind deflection caused by the conservation of angular momentum. (a) Air moving north from the equator also moves east at the speed of Earth's rotation. It has an angular momentum given by $mvr$. (b) As air moves north, its angular momentum remains the same, but Earth's radius from its axis of rotation ($r$) decreases. Since the air's mass ($m$) does not change, its angular velocity with respect to Earth's axis ($v$) increases. (c) Because of Earth's rotation, the wind travels northeast rather than due north.

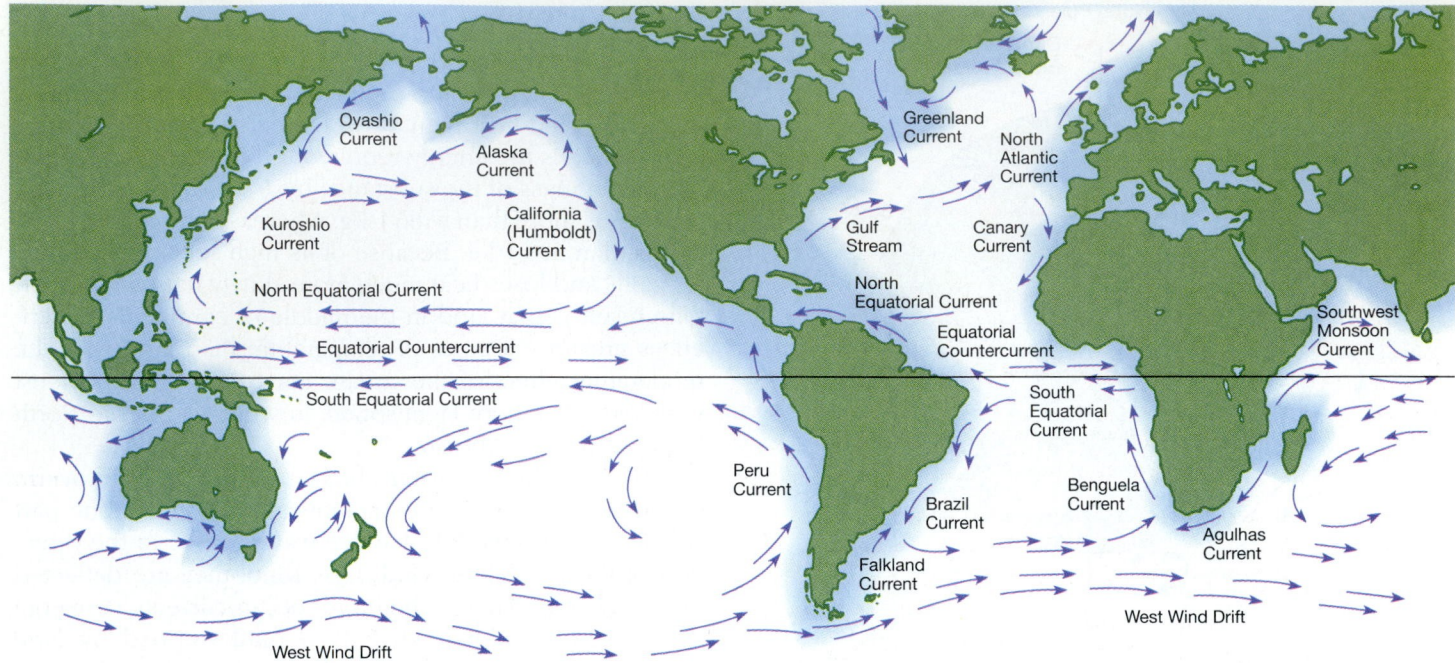

**FIGURE 47–5**   Ocean currents. Currents are driven by the prevailing wind and move in the same direction until they are deflected by bumping into continents.

east coast is exposed to the warm Gulf Stream. The cooling effect on the west side is increased in many places by the upwelling of cold water from the ocean depths to replace water blown away from the continent at the surface.

## Precipitation

Besides temperature, the other important component of climate is moisture, and this also depends on energy from the sun. The air picks up moisture as it travels over the ocean. When it reaches land, it releases moisture as it cools. As a result, precipitation on land is highest where the prevailing wind blows from warm ocean to cool land. In coastal areas where there is little change in land temperature during the year, as in Baja California, there is little rainfall. The same prevailing westerly winds bring much more precipitation to Washington, Oregon, and British Columbia, where the land temperature is lower than the water temperature for much more of the year.

Air heated at the equator evaporates water from forests and oceans and rises, cooling as it goes. As it cools, some of its moisture condenses. The result is the steamy rains of tropical jungles. The air moves on, high in the atmosphere, both north and south from the equator, until it sinks to Earth at about 30°N and 30°S, as described earlier. The descent of this dry air creates the world's great deserts. The Sahara and the Australian deserts are at these latitudes.

Mountain ranges affect precipitation by altering the temperature of the air. As air rises up the windward side of a mountain it expands and cools, and much of its moisture condenses as rain or snow (Figure 47–6). Coming down the leeward side of the mountain, the dry, sinking air warms and can hold more water vapor, so it seldom releases any moisture. As a result, the leeward sides of mountains lie in

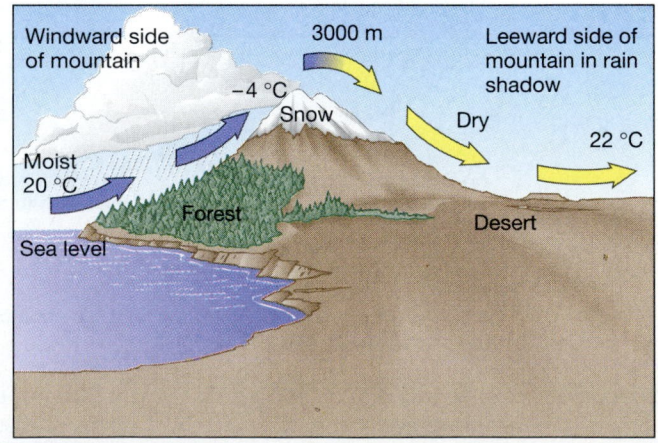

**FIGURE 47–6**   How a rain shadow forms on the leeward side of a mountain. Rising over a mountain causes air to cool, releasing any moisture it contains.

**rain shadows.** Rain shadows from prevailing west winds are the main reason for the large areas of desert and grassland found on the eastern sides of the Sierra Nevada, the Cascade Range, and the Rocky Mountains.

## Effects on Vegetation

One of the earliest naturalists to explore the effect of climate on vegetation was C. Hart Merriam, who later founded the forerunner of the U.S. Fish and Wildlife Service. While conducting a survey of part of Arizona in 1889, Merriam was impressed by the changing vegetation he encountered as he ascended the 4000-m San Francisco Mountains. From cactus desert at the base of the mountains, he climbed through successive belts of grassland, oak scrub, pines, Douglas fir, and spruce, finally reaching alpine turf above the tree line at 3500 m. The sequence of vegetation zones resembled in many ways the sequence encountered with increasing latitude at sea level. Since temperature decreases with both increasing altitude and increasing latitude, Merriam concluded that zones of vegetation are bounded by the limits of the plants' temperature tolerance.

Merriam's zones correlate quite well with temperature in parts of the western United States, but the classification breaks down in many other areas, where great differences in vegetation may occur within regions of similar temperature. The reason is that Merriam had omitted to consider the availability of water. By considering the amount and seasonal distribution of precipitation in a region, as well as temperature variations, it is possible to predict fairly well the height and kinds of vegetation that will be found there. Other factors influence vegetation structure, but temperature and moisture are by far the most important.

## BIOMES

### Convergent Evolution

Each biome, such as tropical forest or grassland, has vegetation with a characteristic structure wherever it is found. There are also many similar species of animals and decomposers (Figure 47–7). But the species of organisms that inhabit a biome in different areas are not the same and of-

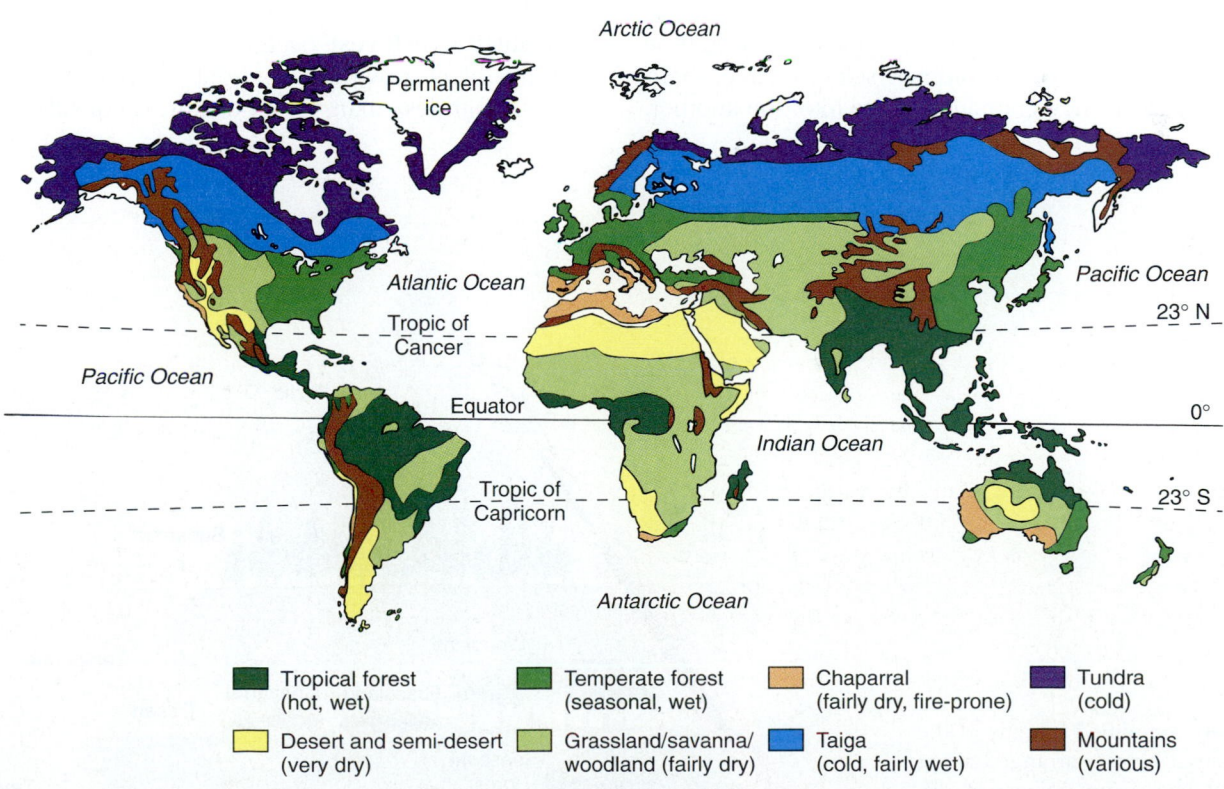

**FIGURE 47–7** World biomes. This map shows the major biomes, simplified to emphasize the overall pattern. The order of biomes northward from the equator is mirrored by the same biomes at similar latitudes south of the equator. Oceans, land masses, and mountain ranges affect climate, and therefore vegetation, making the map more complicated than it would be otherwise.

ten have very different evolutionary origins. For instance, deserts are inhabited by succulents, plants with water-storing stems and leaves that flower and set seed very rapidly when it rains. In the Sahara Desert in Africa, most of these plants are members of the family Euphorbiaceae; in America, they are members of the Cactaceae. Similarly, large grazing mammals have evolved in the grassland biome in many parts of the world. Kangaroos, horses, and antelopes have different evolutionary origins and are found in different parts of the world. Yet all have adaptations of the digestive tract that permit them to chew and digest cellulose-laden grasses, and all rely on keen senses and rapid locomotion to escape predators.

The structural similarity of the vegetation and the similar adaptations of the biomes' animals are the products of **convergent evolution.** This is the process whereby organisms evolve similar adaptations as a result of living in environments with similar selective pressures. In this case, the selective pressures are mainly those of the physical environment: temperature variations and the availability of water, which determine the adaptations of plants living in the area. The characteristics of the plants, in turn, determine the adaptations of the animals that feed on them.

## Classifying Biomes

Ecologists differ as to the number of biomes that should be recognized. This is because biomes seldom have sharp boundaries. Instead they gradually merge into one another,

forming **ecoclines,** gradients of changing community types along gradients of changing climate and soil type.

In temperate and tropical areas with similar temperatures, biomes can be arranged along gradients of increasing dryness (Figure 47–8). This is because different kinds of plants need different amounts of moisture, and this requirement varies with temperature. Plants lose water by evapotranspiration: evaporation from the surface and transpiration through the stomata. For every 10 °C rise in temperature, evapotranspiration approximately doubles. Thus, plants such as trees, which lose large quantities of water by evapotranspiration, can survive high temperatures only if the rainfall is also heavy. Sugar maple trees are found in the northeastern United States, where there is more than 100 cm of precipitation a year, but not in the center of the country, where temperatures are similar but the rainfall is less than 100 cm. Within areas with similar temperatures, progressively lighter rainfall supports communities dominated by small trees, shrubs, grasses, and finally scattered cacti or other desert plants. In extreme cases, there is so little rainfall that plants cannot grow at all.

## 47–B  TROPICAL BIOMES

### Tropical Forest

**Tropical rain forest** occurs where a warm moist climate permits plant growth throughout the year. In such areas, a month with less than 10 cm of rain is considered dry, and

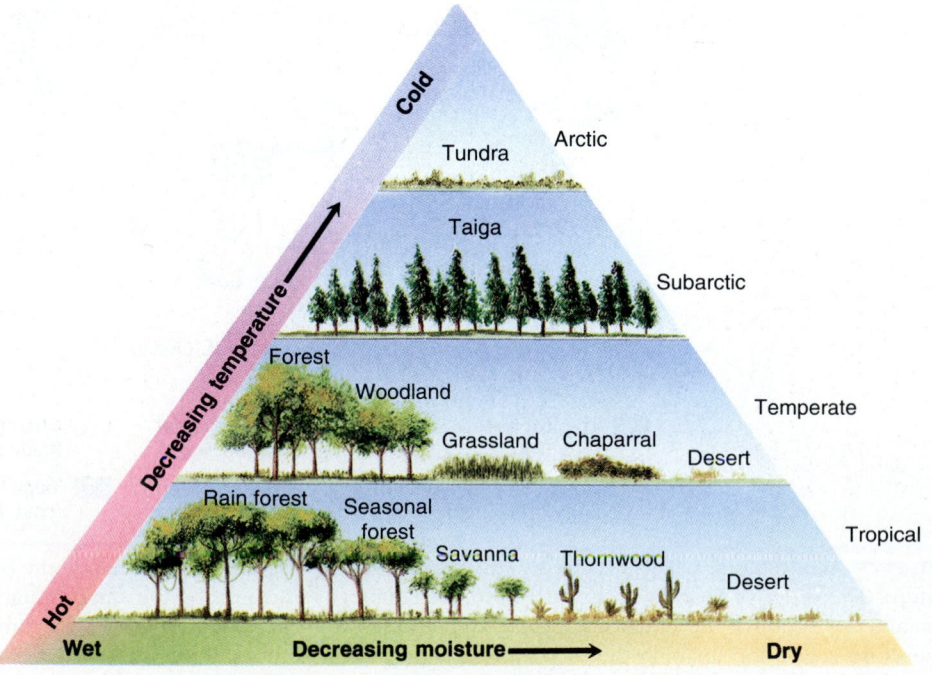

**FIGURE 47–8** Simplified scheme of the major terrestrial biomes arranged along ecoclines of increasing dryness at different latitudes. The graph illustrates the dominant influence of temperature and precipitation on the structure of plant communities.

annual precipitation may exceed 400 cm. The mean monthly temperature in a rain forest in Java was measured as varying between 24.3 °C in February and 25.3 °C in October; the temperature in the forest canopy varied by only 9 °C during any 24-hour period.

Soil in rain forests is thin and often waterlogged. The high temperature and moisture are ideal for decomposer organisms that break down organic matter. Here, a fallen leaf may decompose in two months, a process that takes one to seven years in a temperate forest. The minerals released by decomposition are rapidly taken up again by plants or leached away by water that percolates down through the soil. As a result, almost all the forest's nutrients are inside the bodies of living organisms, not in the soil.

Tropical rain forest is the richest of all biomes, in that it has the greatest diversity of species for a given area. A hectare of forest may contain more than 100 species of trees, compared with three to ten species per hectare in temperate forest. Although tropical rain forests cover only 7% of Earth's land surface, they contain more than half of all animal and plant species.

The dominant plants are tall trees with slender trunks that branch only near the top, covering the forest with a **canopy** of leathery evergreen leaves that shed water rapidly. To obtain minerals from the shallow soil, the trees must have shallow root systems. The thin soil provides little anchorage, and wide buttress roots support the bases of many trees. Mycorrhizal associations permit the trees to obtain some of their nutrients in the form of organic matter in the leaf litter, avoiding the nutrient loss that occurs when organic matter decomposes into inorganic compounds that

can leach out of the soil. These forests recapture nutrients so efficiently that water running off the forest floor is often as free of inorganic solutes as distilled water.

Thick "jungle" occurs only in open areas, along river banks or in clearings formed when trees die, where sunlight can reach the ground. Farther into the forest, the canopy permits little light to reach the ground. The forest floor is fairly open because few plants can grow here. Instead, many smaller plants use the trees as supports that hold their leaves up where there is enough light for photosynthesis. **Lianas** are large woody climbers that use trees to support their rapidly growing flexible shoots. The trees also provide surfaces on which grow many **epiphytes,** plants growing with their roots anchored on other plants. Epiphytes include a great variety of orchids, bromeliads, and ferns. Epiphytes disperse by wind-borne spores (ferns), tiny dust-like seeds (orchids), or berries (cacti, bromeliads) that are eaten by birds and later deposited on new tree branches in the birds' droppings.

Reflecting the lack of seasons, trees flower at various intervals. Different species may flower every few months, every 9 or even every 32 months, instead of every 12 months as in nontropical regions. As a result, there are some trees in flower and some in fruit throughout the year, providing a constant supply of food for pollinating animals and for fruit-eating animals such as monkeys and bats. The animal life of rain forests is exceedingly rich. Birds, butterflies, beetles, and frogs exhibit an almost bewildering diversity of striking color patterns. Since most of the plant food is high in the canopy, most of the animals also live in the trees (Figure 47–9).

(a)  (b)  (c)

**FIGURE 47–9**   Life in a tropical rain forest. (a) An orchid growing as an epiphyte on the trunk of a mahogany tree. (b) A South American macaw, member of the parrot family common in rain forests throughout the world. (c) A margay, an endangered arboreal carnivore from Central America, on a branch with a bromeliad.   (c, Fundácion Neotrópica)

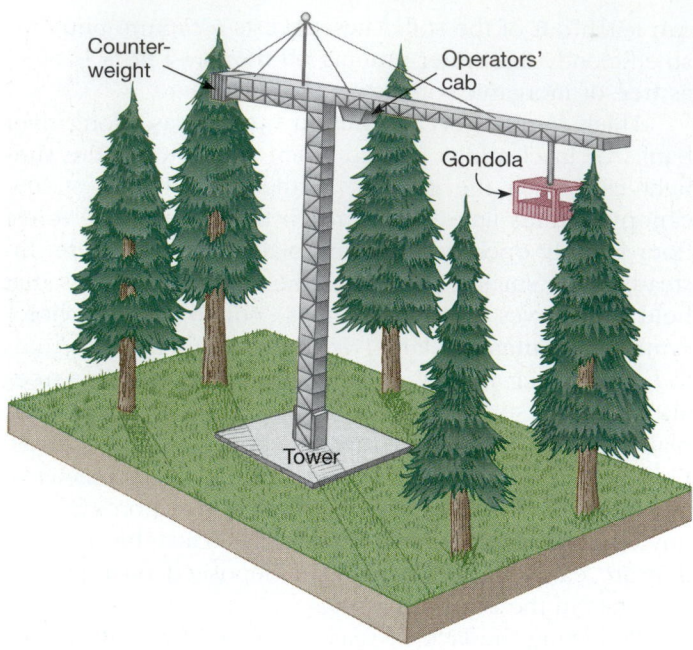

Counter-
weight

Operators'
cab

Gondola

Tower

**FIGURE 47–10** A crane developed for working in the canopy of forest in the Pacific northwest. (USDA Forest Service)

Methods for working in the high canopy are still being developed, and this is one reason we know so little about the species that inhabit tropical rain forests. Figure 47–10 shows a crane developed for studying the canopy of tall forests in the northwestern United States. Scientists hope that such cranes will also be used in the tropics because they provide access to much greater areas of canopy than the rope-walkway techniques usually used at the moment.

As we move north and south from the equator, we encounter **tropical seasonal forest.** Even though the temperature is still between 20 and 30 °C year round and precipitation averages about 200 cm a year, rainfall is concentrated during part of the year, and there is a definite dry season. In areas with longer dry seasons, the forest contains more and more **deciduous trees,** trees that lose their leaves for part of the year. They thereby avoid evapotranspiration at times when lost water cannot be replaced from the dry soil. Tropical seasonal forests include the monsoon forests of India and Southeast Asia.

Tropical forests of all types are being cut down rapidly as a result of the growth of human populations in these biomes. The loss of tropical forest is a pressing world problem (Section 50–I).

## Laterization

Because warmth and moisture speed decomposition, tropical soils tend to be highly weathered. As a result, the chemistry of the underlying rock has a greater effect than it does on most soils. A unique problem arises when agriculture is practiced in tropical areas where the rock is rich in **laterite,** particles containing iron or aluminum oxides. When lateritic soil dries out or is exposed to air, it **laterizes,** hardening into a substance that plant roots have difficulty penetrating.

Deforestation promotes laterization. In parts of Cameroon, West Africa, laterized soil has formed to a depth of two meters in less than 100 years. Crops cannot be grown on highly laterized soil, and even the native forest is slow to re-establish itself. Natives of many lateritic areas practice **swidden,** or **slash-and-burn,** agriculture. An area of forest is cut or burned and the land cultivated for a few years. Except in young, nutrient-rich volcanic soils such as some in Indonesia and Central America, the shallow tropical soil is then exhausted, and the farmer moves on to another area, leaving the cultivated patch to be overrun by forest plants again, allowing its soil to regenerate.

Swidden agriculture does not work unless soil is left uncultivated for several years. And it does not suit modern farming methods, which require cultivating large areas using heavy machinery. The failure to understand tropical soils and laterization has caused some spectacular agricultural disasters, including the loss of millions of American dollars invested in failed banana farms and similar projects in what was once Brazilian rain forest.

## Tropical Savanna and Tropical Thornwood

**Tropical savanna** consists of grassland dotted with scattered small trees or shrubs, such as acacias (Figure 47–11). It extends over large areas, often in the interiors of continents, where rainfall averages about 100 cm a year, which is too little to support many trees at average temperatures of about 25 °C. It also occurs where growth of trees is prevented by recurrent fires. Some savannas are entirely grassland, whereas others contain many small trees.

The proportion of trees in a savanna reflects competition between trees and grasses for water. Where rainfall is light, the roots of grasses absorb all of it during the wet season and survive the dry season by dying back to the roots. As rainfall increases, grasses are unable to absorb it all, leaving enough water in the soil for scattered trees to survive. Where rainfall is sufficient to support many trees, the shade of the tree canopy inhibits the growth of grasses, which are sun-lovers, and the competitive relationship is reversed.

Savannas are most extensive in Africa, where they support a rich variety of grazing mammals, such as zebras, wildebeest, and gazelles. The spectacular migrations of some of these species are related to shifting patterns of local rainfall that permit the growth of the young, nutritious foliage of grasses in different areas at different times of the year.

(a)

(b)

**FIGURE 47–11** African savanna. (a) The wet season: A lion strides across the grass between scattered small, thorny acacia trees. (b) An elephant herd.

**Tropical thornwood** occurs in many regions wetter than savanna but too dry to support forest and with at least a short rainy season each year. Spiny acacias and other drought-resistant trees of the pea family often dominate thornwoods of the Americas and Africa. Many of the plants in a tropical thornwood lose their small leaves during the long dry season, and grow and reproduce only during the wet season.

## 47–C  DESERT

**Deserts** occur in regions having less than about 20 cm of rain each year. In some deserts, there are great changes in temperature with the seasons, and even from day to night. The atmosphere over a desert is a poor insulator because it contains little water vapor. As a result, the ground loses heat rapidly at sunset and nights may be cold, although the days are often very hot.

Desert areas with less than 2 cm of rain a year support little life of any kind, and the terrain is mainly rocks and sand. Less extreme areas have highly specialized plants, many of them small woody shrubs that shed their leaves during the dry season. Desert animals have adaptations that restrict the loss of water through their skin and lungs and in their urine and feces. Many of them are nocturnal, avoiding water loss to the hot daytime air by spending the day in burrows in the cooler soil.

**Temperate (cold) desert** is found in dry regions in middle to high latitudes. It is especially common in the interiors of continents and in the rain shadows of mountain ranges, including much of the Great Basin east of the Cascade Mountains and northern Sierra Nevada in the western

United States (Figure 47–12). The species diversity is low: the landscape is dominated by a few species of low-growing shrubs such as sagebrush, interspersed with perennial grasses (Figure 47–13). Plant growth and reproduction is concentrated in the spring after the brief rains. The plants produce large quantities of seeds, and animals of the desert include many seed-eating birds, rodents, and insects. Typical animals include jack rabbits, sage grouse, pocket mice, and kangaroo rats. Temperate deserts also occur in central Asia, South America, and Australia.

Typical **hot deserts** are found around latitudes 20 to 30° north and south, where dry air from the equator sinks from the upper atmosphere. Hot deserts receive most of their rainfall in summer, and the temperature varies less than in a temperate desert. As a result, more plant species can survive, and the vegetation is more varied and structurally diverse. It includes many annuals that grow, bloom, and set seed in the few days when water is available, as well as succulents, cacti and other plants that store water in their tissues. The Sahara Desert, stretching across north Africa, is the world's largest hot desert. Hot deserts also occur in southwestern North America, the west coast of South America, and central Australia.

## 47–D  TEMPERATE BIOMES

North and south of the tropics and their adjacent deserts lie the world's **temperate regions,** so called because their climate typically has moderate temperatures (although Minnesotans may not think so as they struggle to start a car in February).

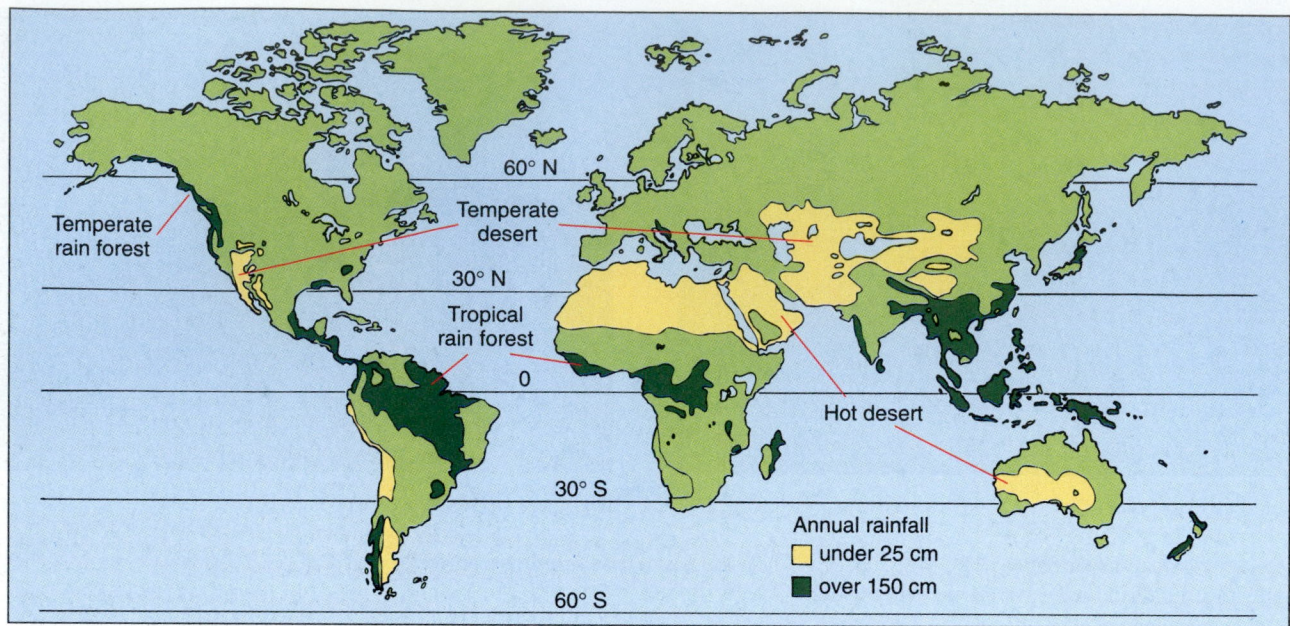

**FIGURE 47–12** Wet and dry areas of the world: the distribution of deserts and rain forests. Temperate deserts lie in the rain shadows of mountain ranges. Temperate rain forests are found on the western coasts of continents where the temperature of the land is lower than that of the ocean for much of the year. Prevailing westerly winds drop large quantities of moisture from the ocean on the cool coast.

## Temperate Shrubland

The temperate shrubland biome is best represented by the **chaparral** communities found in all five areas of the world with a Mediterranean climate: the Mediterranean region, southern Australia, the southern tip of Africa, and coastal Chile and California. These are areas on the west sides of continents with little or no rain in summer. The shrubs are mainly angiosperms up to about 5 m tall, with evergreen leaves. They are often distinctly aromatic. Not surprisingly,

many herbs we use in cooking, such as bay, sage, marjoram, and rosemary are natives of this biome. The leaves contain volatile organic compounds that catch fire easily. Fires are frequent and pose a constant threat to residents of Santa Barbara and other cities in this biome. After fires, the shrubs regrow from tissues near the ground, which are resistant to fire.

The shrubs and the biome's annual plants grow mainly in early spring after the winter rain. The annuals set seed and die before the dry summer, providing food for many

(a)

(b)

**FIGURE 47–13** Temperate desert in the Great Basin of western Arizona. (a) The woody shrubs are green because rain has fallen recently, inducing them to grow leaves. In dry periods, they lose their leaves and so lose little water by transpiration. (b) Desert predator: this handsome Sonora mountain king snake is largely nocturnal, feeding on lizards and rodents. It is protected in Arizona. (b, David Campbell)

seed-eating rodents and insects. Many of the shrubs produce fruits that are dispersed by birds. Olives are an example important in many human economies. The fruits ripen during the hot summer and are available to the many migratory birds that pass through the region in autumn.

## Temperate Forest

The **temperate forest** biome occurs in temperate regions with rainfall of 100 to 300 cm per year. The seasonal distribution of precipitation, temperature variations, and the nature of the soil determine the ratio of deciduous to evergreen species and the spacing and heights of the trees. Temperate forest falls into three major categories: deciduous, evergreen, and rain forests.

**Temperate deciduous forests** occur in moderately humid inland climates where precipitation occurs throughout the year, but where winters are cold and plants photosynthesize and grow only during the warm summers. Broadleaved deciduous trees, such as beeches, oaks, hickories, and maples, dominate this kind of forest. The soil is often rich in minerals and organic matter because decomposition is relatively slow.

There are more layers in a temperate than in a tropical forest. The canopy trees absorb only about 40% of the sunlight reaching them and the remaining light supports photosynthesis in other layers of leaves. Below the canopy grows an **understory** of smaller trees. Less than 10% of the initial sunlight may reach the third level down, the shrubs. Beneath the shrubs there is usually a layer of low-growing, nonwoody herbaceous plants (herbs) that receive less than 5% of the original sunlight striking the forest. Shade-adapted mosses and creeping herbs may provide yet another layer of vegetation close to the ground. Vertical structure continues down into the soil, where the roots of different plants extend to different depths.

Mammals of North American deciduous forests include white-tailed deer, chipmunks, squirrels, and foxes. Wolves, black bears, bobcats, and mountain lions roamed widely until they were largely eliminated by human activities. As winter draws near, many of the birds migrate south, and many of the mammals hibernate. In the spring, plants such as trilliums, violets, and Solomon's seal produce their leaves and flower before the tree canopy leafs out and reduces the amount of sunlight reaching the forest floor.

**Temperate evergreen forests** occur where poor soils, droughts, and forest fires favor gymnosperms or broadleaved evergreens over deciduous trees. In the United States, temperate evergreen forests include impressive stands of ponderosa and other pines in the west, as well as the pine forests of the southern states. These are now prime areas for commercial timber operations. Elsewhere in the world, temperate evergreen forests occur in eastern Asia, in southern Chile, in New Zealand, and in Australia, where forests are dominated by various species of eucalyptus.

**FIGURE 47–14**   A coast redwood tree, native to the coastal western United States.   (Paul Feeny)

**Temperate rain forests** occur in cool climates near the sea with summer cloudiness or fog and abundant rainfall of 250 to 350 cm in winter. They include the forests of giant trees along the Pacific coast of North America, from the mixed coniferous forest of Washington's Olympic Peninsula to the coastal redwood forests of Oregon and northern California. Although there is little rainfall in California in summer, the foliage of redwoods can absorb water from the frequent fogs (Figure 47–14).

**Temperate woodland** occurs in cool-winter areas too dry to support forests, yet with enough moisture to support scattered small trees as well as grasses. Woodlands of piñon pine and juniper cover extensive areas of the American west. Woodlands containing small oak trees are common in central California, and extensive evergreen oak and oak-pine woodlands occur in the southwestern states and in Mexico.

## Temperate Grassland

Early visitors to the American West were most impressed not by the forest but by the prairie with its burrowing prairie dogs and ground squirrels and large grazing mammals such

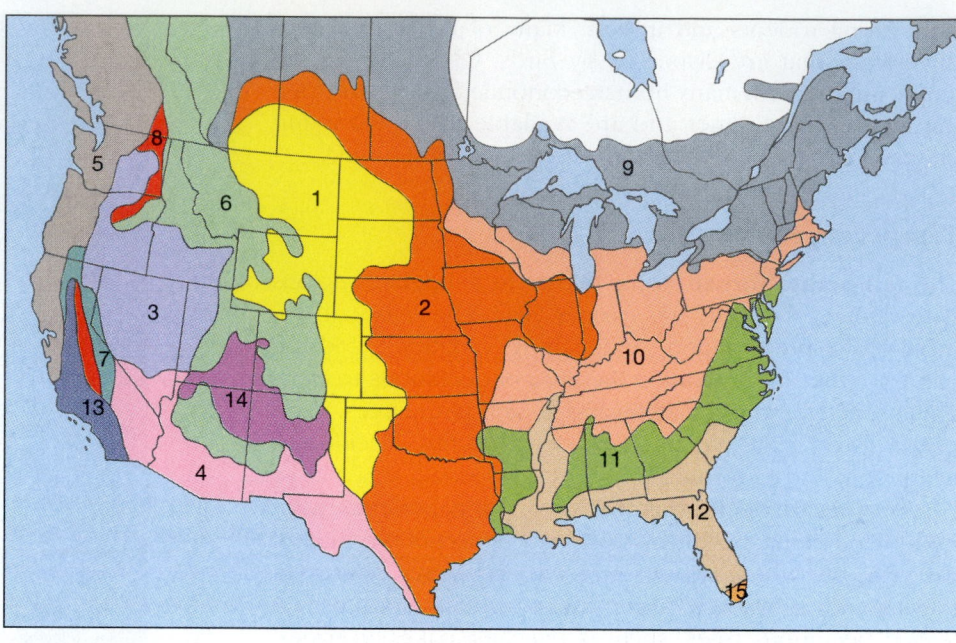

Key:

**GRASSLAND**

1 Short-grass prairie

2 Tall-grass prairie

**DESERT**

3 Cold desert

4 Hot desert

**FOREST**

5 Pacific coast rain forest

6 Rocky Mountain forest

7 Sierra-Cascade forest

8 Mountain forest

9 Northern mixed forest

10 Eastern mixed forest

11 Southeastern mixed forest

12 Eastern coastal forest

**SHRUBLAND AND WOODLAND**

13 Chaparral

14 Pinyon-juniper woodland

**SUBTROPICAL**

15 True subtropical

**FIGURE 47-15** Biomes of the United States. Much of the interior of the continent is covered by prairie: temperate grassland. In the rain shadow of the Rocky Mountains, there is little precipitation and short-grass prairie is found. Farther east, where there is more rainfall, luxuriant tall-grass prairie once covered almost one quarter of the continent with grasses and wildflowers up to 2 m tall.

as bison and pronghorns. There are fewer bird species in temperate grasslands than in deciduous forests, probably because the vegetation has little structural diversity. Prairie is **temperate grassland,** which covers large areas in the interiors of continents where there is 25 to 100 cm of precipitation a year (Figure 47–15). Scattered shrubs may occur, often in depressions or watercourses where extra water is available. The wetter parts of this biome could support woodland, but tree growth is prevented by recurrent fires, usually a result of lightning striking the dried-up grass in summer. Native Americans set fires if natural fires did not occur often enough to provide grass for large bison herds.

Although grassland vegetation forms only a single layer, many plant species may be present. Rich deep soil underlies much temperate grassland because dead vegetation is added to the soil faster than it decomposes. These regions of deep soil, including the midwestern United States, the Asian steppe, and Ukraine, have become prime areas for farming. As a result of its agricultural value, prairie has suffered more complete destruction than any other biome in North America. Some types of prairie have been so com-

pletely eliminated that ecologists are not even sure what plants and animals lived there. Conservation groups are undertaking the restoration of partly destroyed prairie in several parts of the United States (Figure 47–16).

## Fire-Maintained Communities

Fires occasionally sweep through large areas of grassland and temperate forest, burning trees and destroying entire communities of animals and plants. We have seen that in some biomes fire occurs often enough to determine the nature of the dominant vegetation. Such communities include some pine forests, chaparral, and temperate grassland. Grasses readily regenerate after fires that would kill trees; thus, recurrent fires may prevent grassland or savanna from turning into woodland.

Seedlings and saplings of deciduous trees are especially susceptible to fire, whereas many gymnosperms are adapted to survive, and even to exploit, fires. In some pines, for example, the cones open and their seeds germinate only when exposed to temperatures of several hundred degrees. This

(a)

(b)

(c)

**FIGURE 47–16** Prairie. (a) Spring in a Wisconsin prairie. (b) A wind storm ruffles the South Dakota prairie in summer. (c) A bison. (a, Annie Griffiths Belt/DRK Photo; b, Tom Bean/ DRK Photo)

ensures that the seeds germinate in areas that have just been burned. If fires are prevented in a fire-adapted pine forest, deciduous trees may become established. In addition, dead wood and litter build up on the ground, adding extra fuel. When a fire eventually does occur, it is more severe than usual, destroying not only any deciduous colonizers but also the pines and other species.

Odd though it may seem at first, frequent burning is essential for the preservation of many natural communities. This is the reason that the U.S. Park Service adopted the policy of letting fires in National Parks burn if they do not endanger human life or property. This policy caused an outcry when fires burned millions of hectares in Yellowstone National Park in 1988, because most people did not understand the ecology of fire-adapted communities. These fires became an opportunity for visitors to learn about the changes in an ecosystem following a fire. In 1989 the park had more visitors than ever before—come to survey the damage and to watch pioneer species colonizing the blackened landscape.

## 47–E TAIGA

The **taiga**, or **boreal forest**, biome stretches in a giant circle through Canada and Siberia. It is dominated by conifers — spruces, pines, and firs — that can survive extreme cold in winter. Trees tend to be farther apart than those in a deciduous forest because less water is available from the cold soil, and light penetrating to the forest floor supports a ground cover of shrubs. The forest is occasionally interrupted by extensive areas of bog, or "muskeg," in poorly drained areas.

Much of the precipitation in the taiga falls as snow, and in the winter many of the animals grow white fur or plumage that blends with the background. Animals of the North

American taiga include moose, wolverines, wolves, lynx, spruce grouse, ptarmigan, gray jays, crossbills, and snowshoe hare (Figure 47–17).

(a)

(b)

**FIGURE 47–17** Taiga. (a) A bog in the taiga of northern Ontario in summer. (b) A ptarmigan that is beginning to lose its white plumage in early spring. (a, Paul Feeny; b, Robert and Jean Pollock/BPS)

(a)

(b)

**FIGURE 47–18** Tundra. (a) The treeless landscape of the Arctic National Wildlife Refuge, Alaska. (b) A polar bear in the snow, Manitoba. (a, Tom Bean/DRK Photo; b, Art Wolfe)

## 47–F TUNDRA

The tundra, a treeless biome, occurs far north in the arctic regions, where winters are too cold and dry to permit the growth of trees (Figure 47–18). In many areas the deeper layers of soil remain frozen, as **permafrost,** throughout the year, and only the surface thaws during the summer. Because the ground is so cold, decomposition is slow, so the soil is shallow and plant growth slow. As a result, tundra takes a long time to recover when it is destroyed. This is why conservationists are so concerned about the effects of oil spills and oil industry traffic on tundra wildlife.

The species diversity of plants is low, dominated by sedges, grasses, mosses, lichens, and dwarf woody shrubs. Bogs are common because the permafrost prevents water from draining away. The largest animals of the tundra are muskoxen and caribou in North America and reindeer in Greenland, Europe, and Asia. Hordes of mosquitoes and deerflies breed in the wet spots during the brief arctic summer. These insects contribute to the food available for a variety of birds, including various plovers, sandpipers, and horned larks, which nest in the tundra.

Neither taiga nor tundra occurs at sea level in the Southern Hemisphere because the continents do not extend far enough south. Antarctica harbors only a few forms of life around its edges.

### Alpine Biomes

A variety of **alpine grasslands, alpine shrublands,** and **alpine semideserts** are found on high mountains, between the timberline and higher regions where nothing can live (Figure 47–19). These resemble arctic tundra in many ways.

However, nights are cool throughout the year in alpine areas, whereas they are warm during the brief summer in arctic regions. Also, alpine systems have little or no permafrost. In high mountains throughout the world, alpine meadows cover extensive areas above the timberline. These meadows are dominated by sedges and grasses, interspersed with shrubs and low-growing flowering herbs. Many plants cultivated in rock gardens, such as gentians, saxifrages, and edelweiss, are alpine species. Alpine meadows in North America are inhabited by mountain sheep, mountain goats, grizzly bears, marmots, and pikas (small relatives of rabbits). Many of the larger animals migrate to lower eleva-

**FIGURE 47–19** Alpine biomes: the timberline on a mountain in the northwestern United States. Above the timberline, the vegetation resembles arctic tundra. (Richard Feeny)

tions during the winter, and all organisms, like those of the tundra, are adapted to take advantage of the short growing season.

Alpine grasslands also occur above the timberline on tropical mountains. The **paramo** of the South American Andes is partly alpine grassland. Communities of similar structure, but widely different in evolutionary origin, occur in the alpine zones of African mountains, in New Zealand, and on sub-Antarctic islands.

## AQUATIC COMMUNITIES

Strictly speaking, the term biome refers only to communities on land. However, there are also many different kinds of aquatic communities, both marine and freshwater, which, like biomes, are similar wherever in the world they occur.

As on land, environmental conditions influence the distribution of organisms in water. Temperature, nutrient supply, light intensity, and salinity (salt concentration) determine what can live where. Shortage of water is not a problem, but in some areas a shortage of dissolved mineral nutrients or excess of sodium limits plant life. The types of organisms found in an aquatic community also depend upon the type of bottom (mud, sand, or rock) and upon water motion: currents and waves.

## 47–G  FRESHWATER COMMUNITIES

Only about 3% of the water on Earth is fresh water, and most of that is locked up in glaciers and polar ice caps. Less than 0.01% of Earth's water occurs in the atmosphere, rivers, and lakes where it is easy to get at, but this available fresh water is vital to human welfare and economy. Thus, the world's large temperate-region lakes, such as Lake Baikal in Siberia and the Great Lakes of North America, are essential sources of fresh water for domestic use, agriculture, and industry. The pollution of these and other freshwater ecosystems threatens human welfare more directly than most other forms of environmental degradation.

Rooted aquatic plants such as water lilies and rushes grow in shallow water around the edge of a lake or river. Much of the food supply comes from photosynthesis in this **littoral zone** (Figure 47–20). Fish, amphibians, insects and other arthropods, snails, and worms live and feed among the plants. In the lake's open surface waters live microscopic phytoplankton and larger floating plants, all of which need abundant light. Fish and arthropods that need much oxygen also inhabit the surface waters. Deep in the lake, the main biological process is decomposition, and the chief input of energy is **detritus** (dead organic matter) falling from above. This feeds decomposers, invertebrates, and fish.

As sunlight passes down through the water, some is used in photosynthesis and some is absorbed by the water

(a)

(b)

**FIGURE 47–20**  The littoral zone in fresh water, where plants grow rooted in shallow water. Many of these plants are monocots, a group that contains many members with roots adapted to wet soil containing little oxygen. (a) A crinum lily by a river in Central America. (b) Pickerel rush in a North American lake, with a butterfly laying eggs on its leaves.  (Steve Bisson)

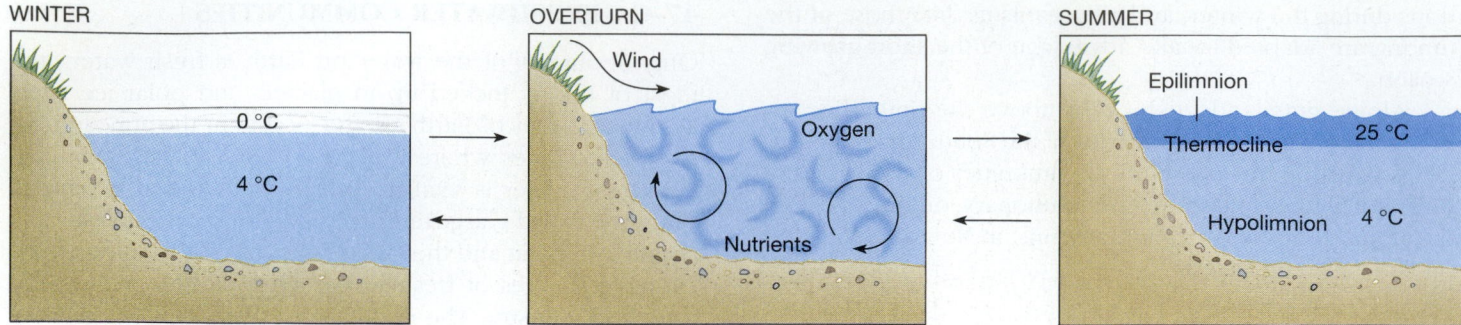

**FIGURE 47-21**    Seasonal stratification of temperature and overturn in a temperate dimictic lake.

itself. So, as light passes deeper into the water, it becomes dimmer. In deep lakes there is a **compensation depth** where the available light is just bright enough for green plants to eke out a living: their photosynthesis (production of food and oxygen) exactly offsets their respiration (use of food and oxygen). Above the compensation depth plants produce more oxygen than they use, and so extra oxygen is available for the respiration of other organisms. Below it there is not enough photosynthesis to offset respiration, and any available oxygen must come down from the water above.

The amount of oxygen dissolved in the water affects nearly every aspect of the lake, including which animals and plants can live where and the solubility of many inorganic nutrients. Oxygen enters the water from photosynthesis and by dissolving into the water from the air. Oxygen leaves the water when it is used in respiration or diffuses into the air above. The cooler the water, the more dissolved oxygen it can hold. A lot of oxygen is used by the bacteria that decompose detritus. Many desirable fish, such as trout, can survive only in waters containing a lot of oxygen.

## Lake Overturns

Temperature, as well as oxygen, affects life in a lake. Most temperate lakes are **dimictic** ("two mixings"): seasonal temperature changes mix water from the top and bottom of the lake twice a year during the **spring** and **fall overturns.**

Water is at its maximum density at 4 °C, so water at this temperature sinks below water that is either warmer or colder. In winter, the lake's surface may be covered with ice at 0 °C, but below the ice the water remains unfrozen, and fish and other organisms survive. In spring, the icy surface waters are warmed by the sun. When they reach 4 °C, they sink below the surface (Figure 47–21). Wind over the surface creates currents that accelerate the mixing of surface and deeper waters. This spring overturn brings nutrients to the surface from the layer of decomposition at the

bottom of the lake, and it carries oxygen from the surface down into deeper waters.

During the summer, surface waters warmed by the sun become less dense until they no longer mix with deeper layers. The lake becomes thermally stratified, with an upper layer **(epilimnion)** of warm water and a deeper layer **(hypolimnion)** of water at about 4 °C. Between the two lies a narrow **thermocline** where the temperature drops rapidly. On a calm day in early summer, water in the epilimnion may be warm enough for swimming but if you let your feet drop down they encounter the bone-chilling water of the hypolimnion.

During the summer, the depth of the epilimnion increases to as much as 20 m. In fall, the thermocline rises again as the temperatures of air and surface waters fall, eventually leading to the fall overturn as cooling, denser water sinks below the warmer layers under it. Once again, vertical currents bring nutrients to the surface and oxygen to the lake's bottom. People who fish for trout know that spring and fall are the only times when various species swim in the surface waters. Trout need richly oxygenated water, and since less oxygen can dissolve in warm than in cold water, these fish spend the summer deep in the hypolimnion.

Dimictic lakes, which turn over twice a year, are characteristic of temperate regions. In the arctic and parts of the tropics, **monomictic** lakes occur. Here the water reaches 4 °C only once a year, in midsummer or midwinter, and there is thus only one overturn.

Lakes can be divided into categories based on how much plant life they support. **Oligotrophic** ("few food") lakes are low in nutrients such as phosphorus, calcium, and nitrogen, so they support little plant growth and contain few organisms. Oligotrophic lakes are usually deep, with steep sides and narrow littoral zones. Their water is usually very clear, and the deep waters always contain oxygen (Figure 47–22). **Eutrophic** ("good food") lakes are rich in nutrients and organisms and are usually shallow. Such lakes contain little oxygen because decomposer organisms rapidly use it up metabolizing the organic matter produced by the lake's many other residents.

**FIGURE 47–22** Mount Hood towers over an oligotrophic lake in the Pacific northwest. (Richard Feeny)

In the normal course of events, a lake ages as it is steadily filled in with soil particles and organic matter, becoming more eutrophic as it ages. Natural eutrophication takes thousands or millions of years, but the process may be compressed into very few years if the lake becomes polluted. When nutrients wash into a lake in sewage or in chemical fertilizer runoff, they speed plant growth and hence eutrophication.

The main differences between rivers and lakes in the same climate are that rivers are usually shallower, with stronger currents. Where the current is strong, only organisms that can anchor themselves or swim against it can survive. A shallow, rapidly flowing river is usually well oxygenated because it has a large surface area to absorb oxygen from the air.

## Freshwater Wetlands

One tenth of the continental United States was once wetland (Figure 47–23). Today, freshwater wetlands are among the most threatened of habitats. They range from peat bogs and prairie potholes in the midwestern United States, where waterfowl nest or feed on migration, to the Pantanal in Brazil, the world's largest freshwater wetland, covering an area greater than Florida and home to endangered jaguars, anteaters, alligator-like caiman, and waterfowl.

Wetlands serve unique ecological functions. First, they are home to hundreds of species of plants and animals that cannot survive elsewhere (Figure 47–24). Second, wetlands provide flood control by absorbing large amounts of water

Key:
- ■ Prairie pothole region
- • Freshwater marshes
- ■ Areas containing many swamps

**FIGURE 47–23** The distribution of freshwater wetlands in North America.

**FIGURE 47–24** Animals of American freshwater wetlands. (a) An egret and a turtle stalk dinner in water covered with floating duckweed. (b) Common moorhens. (b, Steve Bisson)

(a)

(b)

when it rains or when snow thaws and rivers rise. Third, wetland plants absorb enormous quantities of pollutants, including mineral nutrients, preventing them from washing out into rivers and eventually to the sea.

Wetlands vary in the amount of water they contain. **Marshes** contain low-growing plants, such as pickerel weed and cattails, which stick up out of the water, and areas of open water. The Everglades is the largest freshwater marsh in the United States *(Essay: The Everglades)*.

**Swamps** are wetlands dominated by trees and shrubs that can survive floods. In northeastern swamps, red maples are the usual trees. Throughout the Southeast, bald cypress,

oaks, and gums stand in water year round. Big Cypress National Preserve in western Florida contains immense bald cypresses, festooned with orchids and other epiphytes (Figure 47–25).

**Bogs** and **fens** are wetlands found in colder parts of the world, where decomposition is so slow that dead plants are not completely decomposed into nutrients that living plants can absorb. Instead, the dead plants accumulate as peat. Northern Maine, Minnesota, Canada, and Alaska contain about 35% of the world's peat bogs. These areas are very acidic and low in available nutrients, and they contain specialized plants, such as sphagnum moss, heathers,

**FIGURE 47–25** Cypress swamp in the southeastern United States. (a) Bald cypresses (*Taxodium distichum*). These unusual gymnosperms thrive in water-logged soil and are deciduous, losing their leaves in winter. (b) Wood storks with a juvenile on their nest at the top of a bald cypress. Wood storks totter on the brink of extinction because their native Florida swamps have been drained. In 1993, biologists first induced wood storks to nest on artificial platforms in Georgia; this program may save the species. (Steve Bisson)

(a)

(b)

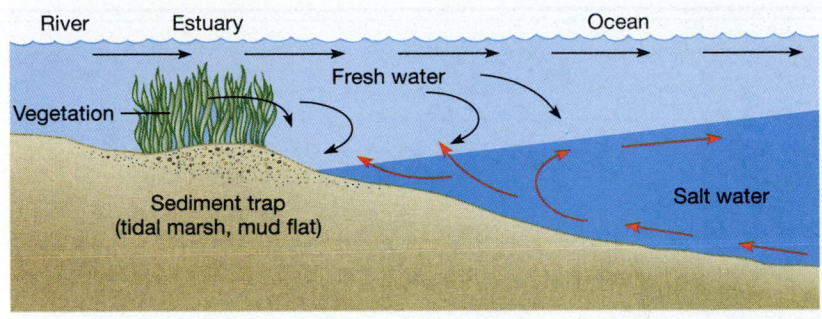

**FIGURE 47–26** Formation of an estuarine marsh. Fresh water and salt water meet in an estuary and create turbulence that deposits nutrients and sediment to form rich, muddy soil in which marsh plants grow.

pitcher plants, and black spruce, adapted to these unusual conditions.

## 47–H  MARINE COMMUNITIES

### Intertidal Communities

Along the seacoasts, many kinds of plants and animals thrive in the **intertidal zone,** the area between the high and low tide marks, where they are submerged for part of the day. There are three main types of intertidal zones: muddy, sandy, and rocky shores, which support very different communities.

**Mudflats** occur where the water moves slowly enough to deposit a sediment of small particles. Algae cover the particles and provide food for a multitude of burrowing molluscs, worms, and crustaceans.

**Sandy beaches** are less stable than mudflats, for sand shifts constantly and dries out faster than mud when the tide is out. Most of the tiny protists, worms, and crustaceans that live between the sand grains eat marine plankton stranded when the tide goes out or algae attached to the sand grains. A wide variety of shore birds feeds on the invertebrate inhabitants.

Neither muddy nor sandy shores provide much foothold for sessile animals or anchored seaweed. These are much more common on **rocky shores,** which support a wider variety of organisms. Since the water crashes onto the rocks, motile animals such as crustaceans anchor themselves firmly to rocks or seaweeds by their legs or hide in crevices. Few vertebrates live in the intertidal zone, although a number of birds come in at low tide to scavenge or to hunt invertebrates.

The **subtidal zone** occupies the **continental shelves** —the edges of continents—extending from the low-tide mark to a depth of about 200 m. Here, temperature fluctuates less, and wave movement is less violent than in the intertidal zone. Mineral nutrients are also readily available, washed from the land by rivers. Continental shelves are among the most productive and densely populated areas on Earth.

### Coastal Wetlands

In some parts of the world, the land rises steeply out of the sea. In others, the coast is flat, and the ocean's tide may rise and fall far up a river or in a coastal marsh. These areas where the ocean penetrates the land are the site of **coastal wetlands.** Coastal wetlands include **mangrove swamp,** found in tropical and subtropical regions, and **salt marsh,** its temperate equivalent.

The edges of the sea are the hatcheries and nurseries of many important species of marine animals such as shrimp, flounders, and other fish. Coastal wetlands and estuaries also serve as nesting, feeding, and resting spots for migratory waterfowl, and they reduce erosion and flooding inland. Recognizing these important but indirect contributions to human food, fun, and safety, ecologists are alarmed by the draining and filling of these areas to build towns, marinas, and resorts.

Most coastal wetlands form in estuaries. An **estuary** is a body of water where a river runs into the sea and salt water mixes with fresh water from the river. Because fresh water is less dense than salt water, river water flows out through an estuary over a wedge of salt water. Where the two bodies of water meet, currents cause sediment to fall to the bottom, slowly filling the estuary (Figure 47–26). Where the water is shallow enough, rooted plants start to grow in the bottom sediment, and a marsh forms.

Most estuaries, like Chesapeake Bay, are **drowned valleys,** river valleys that filled with water because the sea level rose or the land subsided. After an estuary has formed, the former coastal sand dunes become **barrier islands,** islands off the coast, like those found along most of the East and Gulf Coasts of the United States. Barrier islands are battered on the ocean side by wind, waves, and sand, but on the estuary side water moves more slowly and is quite shallow. As a result, barrier islands develop distinctive vegeta-

## ESSAY: *The Everglades*

The southern tip of Florida lies in tropical seasonal forest, the rest of the state in temperate evergreen forest. But Florida also contains communities found nowhere else in the world because of a peculiarity of its geology: gravel saturated with ground water lies close below the surface in many areas, and here forest is replaced by marshes and swamps.

Almost the whole of southern Florida was once freshwater wetland: the Florida Everglades. This was essentially a huge, slow-flowing, wide, shallow river that flowed from Lake Okeechobee to Florida Bay (Figure 47–A). In some places it flowed a foot or two beneath ground level (forming the Biscayne Aquifer); in others it flowed on the surface, partly filled with sawgrass and other water-loving plants (Figure 47–B). Near the coast, salt-tolerant mangrove trees grow in a dense tangle. Their prop roots are the nurseries where juvenile fish of more than 40 species grow to maturity. In the dry season, from December to April, animals collect in huge numbers around the remaining pools of water where herons, egrets, wood storks, roseate spoonbills, and anhingas catch fish. The Everglades were once home to hundreds of species of plants and animals. Hundreds of them are now extinct, and another 400 are endangered.

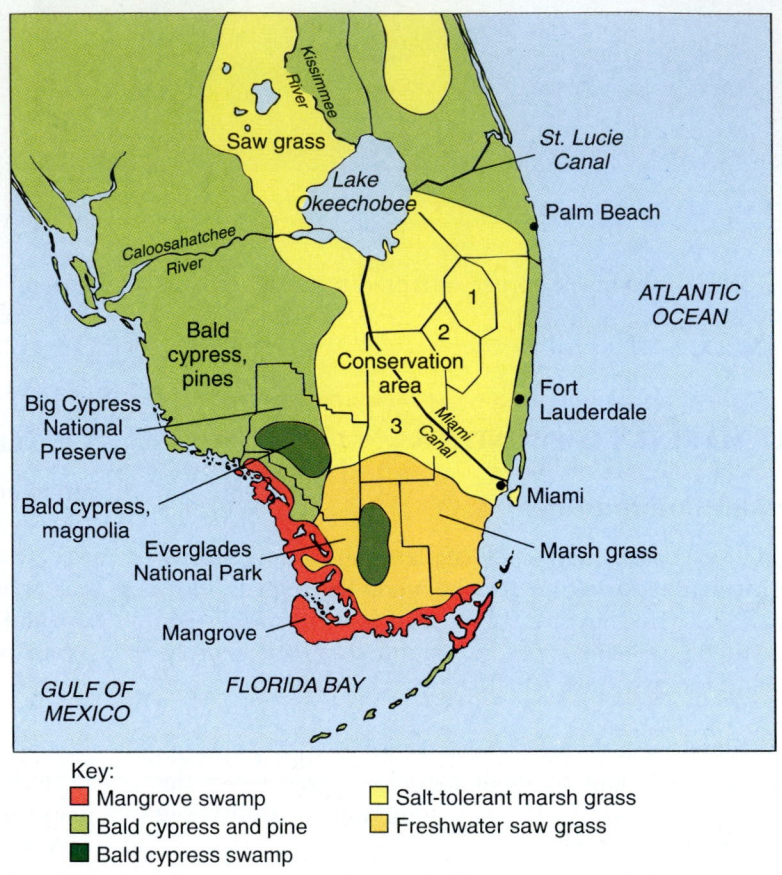

**Key:**
- ■ (red) Mangrove swamp
- ■ (light green) Bald cypress and pine
- ■ (dark green) Bald cypress swamp
- ■ (yellow) Salt-tolerant marsh grass
- ■ (gold) Freshwater saw grass

**FIGURE 47–A**  The Everglades: a wetland of saw grass, interspersed with hammocks of higher ground covered with trees and shrubs.

Agriculture and diversion of its water supply have destroyed the Everglades. Much of Florida's fresh water now flows, not to Florida Bay but into impoundments that supply water for agriculture and to cities such as Miami. More than 50 years ago, settlers started to drain the land around Lake

tion, with tough plants that can tolerate salty wind on the seaward side and marsh on the other side.

***Salt Marsh***  Salt marsh dominates much of the flat shoreline of the Gulf of Mexico and Atlantic Coast of the United States. Here, rivers such as the Mississippi deposit their load of mud and nutrients in estuaries. Smooth cordgrass (*Spartina alterniflora*) covers hundreds of thousands of hectares of marsh, with scattered stands of other plants on higher, drier ground (Figure 47–27).

Cordgrass is the backbone of an ecosystem that is unusual in two ways. First, it contains few species of plants. Few other flowering plants can tolerate wet, salty mud and sand, but cordgrass thrives. Second, few animals can eat cordgrass because its leaves contain quantities of glass-like silica.

ter supply. This coalition eventually led to the Kissimmee River restoration project, which includes plans to turn thousands of hectares of farmland near the lake back into marsh, which will absorb much of the fertilizer and pesticide pollution and send clean water down to the conservation areas. This land is not much loss to the sugar corporations because farming has depleted the soil's fertility to the point that agricultural experts estimate that it will be of little use for farmland after 2000.

**FIGURE 47–B** The Florida Everglades. Wide areas of wetland where saw grass grows are punctuated by hammocks, mounds of higher ground with trees and shrubs.

Okeechobee to grow crops on soil that was fertile because it contained the partly decomposed remains of sawgrass. A dike now rings the lake, preventing it from flowing south on its natural course. In times of drought, Everglades water is contained in the lake and in three "conservation areas." The Everglades National Park, which is all that remains of the Everglades, receives water only when it is released from a conservation area. As a result, the park is chronically short of water.

Sugar cane farms around Lake Okeechobee use large amounts of water and fertilizer. This deprives the Everglades of much of its water sup-

ply and pollutes the rest with nutrients in runoff from cane fields. The polluted water supports the growth of huge stands of European cattails, which crowd out native water plants (Figure 47–C).

Marjory Stoneham Douglas started the Friends of the Everglades to fight this situation with the rallying cry, "Sugar does not belong in the Everglades." Politics makes strange bedfellows, and Douglas's pleas were eventually heard not just by environmentalists but also by the South Florida Water Management District, which awoke to the fact that farm fertilizer was polluting the Biscayne Aquifer and threatening Miami's wa-

**FIGURE 47–C** Introduced European cattails (*Typha* spp.) choke what was once open water in many parts of the Everglades. (Steve Bisson)

Energy and nutrients from cordgrass pass to other organisms indirectly. Every year, the dead stalks of last year's cordgrass break off and wash up high into the marsh, where they are broken down by decomposers. The tide then washes the decomposing detritus toward the many creeks that meander through the marsh. The detritus provides nutrients for algae and food for crabs, mussels, worms, snails, and clams that live in the mud and among the cordgrass

stems. At the edges of creeks, oysters feed on the detritus and algae they filter out of the water. All these invertebrates, in turn, feed human collectors, raccoons, marsh rats, and a host of birds.

The marsh is the nursery where many animals find food and protection while they are small. For instance, southern commercial shrimp live in the ocean as adults. They lay their eggs on the bottom, and the young move in with the

**FIGURE 47–27** Springtime at low tide in a salt marsh. This year's young cordgrass is tall and green where nutrients are plentiful at the edges of creeks in the foreground. The sandy patch in the middle is high marsh, which is seldom flushed by the tide and, therefore, so salty that even cordgrass cannot grow on it. The decaying brown stalks of last year's cordgrass are visible, especially in the background. (Thom Smith)

tidal currents until they reach the marsh, where they grow to adulthood. As they grow to maturity, these animals migrate down the creeks and out to sea, where they may end up as seafood on the table or as food for ocean fish such as swordfish, snapper, grouper, and tuna.

## Coral Reefs

Coral reefs are restricted to warm oceans, where the water temperature seldom falls below 21 °C. Corals are cnidarians that live in symbiotic association with photosynthetic protists. The reef itself is made up of calcareous material, secreted by the coral animals and by red and green algae.

Since photosynthetic organisms are so important to their formation, coral reefs are found only in clear, shallow water where there is enough light for photosynthesis.

Most of a coral reef is submerged, although its top may be exposed at low tide. A reef acts physically like a rocky shore in providing anchorage for algae and sessile animals. A great variety of fish and swimming invertebrates finds shelter within the reef's crevices (Figure 47–28). Today, many coral reefs are threatened by boats dropping anchors, fishing nets dragged across their surfaces, the dumping of wastes from nearby tourist areas, and drilling for underlying oil deposits. Phosphorus pollution from waste water is a major threat. It increases the growth of algae on top of the reef, reducing the amount of light that reaches the photosynthetic algae of the reef itself. Countries such as Australia, and the American and British Virgin Islands in the Caribbean, enforce protective laws because a coral reef, once destroyed, takes many years to regrow.

## The Open Ocean

The open ocean can be divided into the top 100 m or so, where photosynthesis can occur, the deep ocean, and the ocean floor. In the surface waters live the **plankton:** drifting protists, plants, and animals not powerful enough to swim against currents. They can, however, control their vertical position in the water, thereby moving from one current to another, for currents flow in different directions at different depths in many parts of the ocean. Fish and similar large animals in the ocean make up the **nekton,** creatures that can swim independent of currents. These animals

**FIGURE 47–28** A coral reef. (Jeffrey Rotman)

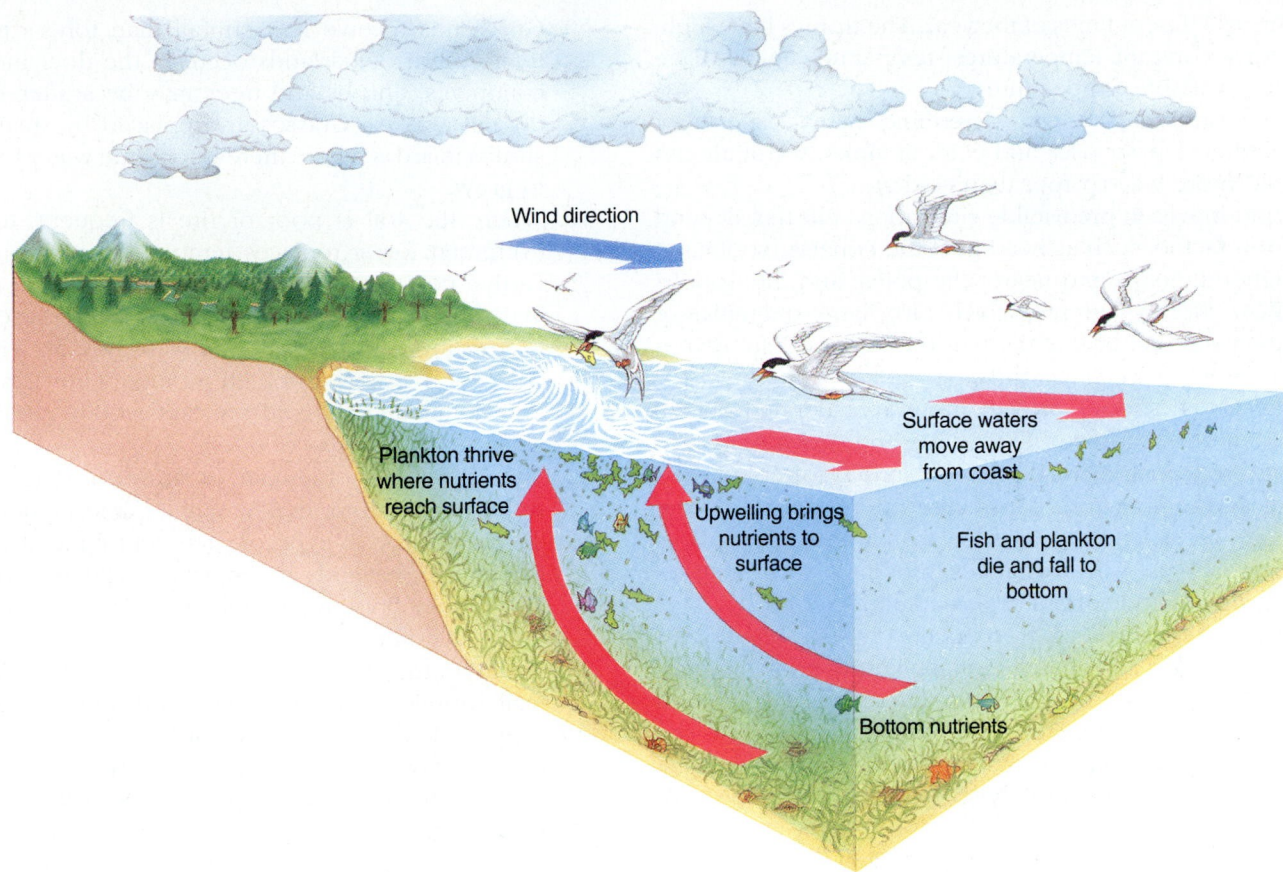

**FIGURE 47–29**   An area of upwelling. Water rising from the bottom of the ocean brings nutrients to the surface. This upwelling is caused by prevailing wind blowing surface water away from the land. Water from the bottom rises to the surface and takes its place.

feed mainly on plankton or on each other. Because mineral nutrients are scarce in many areas of the open ocean, the world's major fisheries lie over continental shelves, which receive minerals washed down rivers, as well as wherever upwelling currents carry minerals up from the bottom (Figure 47–29).

Seventy-five percent of the ocean's water lies more than 1000 m deep. Diving techniques permit sampling at depths of more than 6000 m and reveal fascinating communities on the ocean floor—communities that contain no plants. The **benthos** (depths of the sea), the community of the ocean floor, includes decomposer bacteria that live on dead organisms falling from the surface layers above them or on dead members of the deep-sea community. Other benthic communities are found around hydrothermal vents in the sea floor, where heat and minerals from Earth's interior support chemosynthetic bacteria and heterotrophs that feed upon them. Larger members of the benthos also feed on falling carcasses or on the decomposers, and filter-feeders strain food out of the water.

## SUMMARY

A worldwide survey of the distribution of organisms reveals that different areas of the world are inhabited by different species of plants and animals. However, terrestrial communities in different parts of the world can be divided into a fairly small number of biomes, each with a characteristic array of plant life. These biomes occur wherever on Earth a suitable climate exists.

1. Temperature and the amount and pattern of precipitation are the climatic factors determining what kinds of organisms can live in an area. Climate determines soil type, soil and climate determine the type of vegetation, and the type of vegetation determines the animal life of the area.
2. Climate depends on solar energy, which warms Earth's surface, evaporates water that will fall as precipitation, and drives the winds. Climate is determined mainly by the Earth's position and orientation in relation to the

sun at different times of the year. The tropics have high, nearly constant temperatures; temperate regions have more variable temperatures.

3. Two properties of air determine many aspects of weather: hot air rises and cold air sinks; warm air can hold more water vapor than cold air.

4. Wind moves in predictable circulation cells that depend upon factors such as heating at the equator, cooling at higher altitudes and nearer the poles, and the Coriolis Effect. Ocean currents, which carry large quantities of heat from one part of the world to another, are driven largely by prevailing winds.

5. The actual species present in an area depend on the evolutionary history, but the type and structure of vegetation depend mainly upon the annual pattern of rainfall and temperature. The higher the temperature, the more precipitation is needed to support large plants such as trees. Similar changes in biomes occur with increasing altitude and increasing latitude.

6. The richest biome is tropical rain forest, where high temperatures and rainfall permit plants to grow throughout the year. Most of the plant and animal life is found in the canopy among the evergreen leaves of trees. The soil is poor because decomposition is so rapid that most of the nutrients are rapidly locked up again in the bodies of living organisms.

7. In temperate deciduous forest, the soil is much richer in nutrients because the trees lose their leaves in the fall, creating a litter layer that decomposes and releases nutrients only slowly. Deciduous forest is an important biome of North America, Europe, and Asia in areas with warm, moist summers and cold winters.

8. Deserts have very little rainfall. Their plant life is mainly annuals with very short growing seasons, succulent perennials adapted to the low rainfall, and small-leaved deciduous woody shrubs.

9. Grasslands receive less rainfall than forests but more than deserts. Grasslands occur in the drier interiors of continents. Shrubs and trees may be scattered among the tall grasses. Grasses are replaced by small woody shrubs in areas where there is too little water for grasses to grow.

10. Where the soil is poor or fire is frequent, temperate evergreen forest replaces temperate deciduous forest. Farther north, both are replaced by taiga, a biome dominated by conifers adapted to growing in sparse soil and to resisting extreme cold and water loss in winter.

11. North of the taiga and high on mountains lies the tundra, dominated by cold-resistant woody shrubs or by grasses and lichens.

12. The distribution of aquatic organisms is determined by water temperature, depth, salinity, and motion and by the availability of light, oxygen, and minerals.

13. Freshwater wetlands are important wildlife habitats that also reduce flooding and absorb excess nutrients from polluted water.

14. Shallow areas near the coast and coastal wetlands are well supplied with both light and minerals, and they support dense communities of life.

15. Coral reefs are specialized, highly diverse communities found only in shallow tropical ocean waters.

16. In the open ocean, the availability of light for photosynthesis restricts plankton to the upper layers of the water, but scarcity of nutrients in these layers may limit the numbers of organisms. Larger nektonic organisms are found primarily where planktonic food is abundant. Dead organisms from the surface layers of the ocean supply food for a benthic community of bacteria and other organisms that live on the sea floor.

## Self-Quiz

1. Which of the following has a vegetation structure with only one level?
   a. tropical rain forest    d. shrubland
   b. taiga                   e. desert
   c. grassland

2. Which of the following communities would have trees?
   a. taiga                   d. tundra
   b. intertidal zone         e. plankton
   c. shrubland

3. A biome with high temperature, high rainfall, and poor soil is:
   a. shrubland               d. tropical rain forest
   b. coral reef              e. temperate evergreen forest
   c. semidesert scrub

4. Which of the following communities has no living green plants?
   a. a rocky shore           d. the deep ocean floor
   b. the plankton            e. a coral reef
   c. a mud flat

5. Compared with a eutrophic lake, an oligotrophic lake contains a greater concentration of:
   a. organic matter          d. bacteria
   b. plants                  e. mineral nutrients
   c. oxygen                  f. coastal wetland

6. The American prairies and the Asian steppes do not have the same species of grasses because _____. However, both are inhabited primarily by grasses because _____.

7. The biome characterized by the greatest amount of species diversity but having no dominant species and mineral-poor soil is:
   a. tropical rain forest
   b. tundra
   c. taiga
   d. tropical savanna

8. The two factors most important in determining the distribution of organisms are:
   a. soil type and temperature
   b. soil type and amount of rainfall
   c. temperature and amount of rainfall
   d. soil type and climate

9. The factors affecting the distribution of organisms in marine communities include:
   a. temperature
   b. intensity of sunlight
   c. availability of minerals
   d. salinity
   e. all of the above

10. Among marine communities the greatest number and diversity of organisms are found in:
    a. the intertidal zone
    b. estuaries and coastal wetlands
    c. the open ocean
    d. coral reefs

11. The biome characterized by large grazing animals and scattered shrubs and small trees is:
    a. tropical savanna
    b. taiga
    c. cold desert
    d. temperate grassland

12. Barnacles and algae with a holdfast and blade able to tolerate short periods out of water would be found in the:
    a. muddy intertidal zone
    b. open oceans
    c. sandy intertidal zone
    d. rocky intertidal zone

## Questions for Discussion

1. What biome do you live in?
2. The 30°N latitude line runs through southern Louisiana and northern Florida as well as through desert country in Mexico and Texas. Why is the area in Louisiana and Florida not desert like the area in Mexico and Texas?
3. Why is it proving difficult to carry out large-scale "agribusiness" farming in vast tracts of land cleared of their tropical rain forest vegetation?
4. Why is there less variation in size of vegetation in the tundra than in tropical regions?

## Suggested Readings

*BioScience* 39 (10), 1989. "Yellowstone Fires." An entire issue devoted to the effect of the 1988 fires on the ecology of Yellowstone National Park and the controversy over the extent to which fires should be extinguished or left to burn naturally.

Brewer, R. *The Science of Ecology*, 2d ed. Philadelphia: Saunders College Publishing, 1993. A general textbook on ecology, including ecosystems.

Colinvaux, C. "The past and future Amazon." *Scientific American*, May 1989. Colinvaux argues that the enormous diversity of species in the Amazon Basin is partly a result of frequent disturbances (of climate and geology). We can, therefore, be reasonably optimistic that the Amazon will survive recent human disturbances—as long as these are not too destructive.

Dolan, R., and H. Lins. "Beaches and barrier islands." *Scientific American*, July 1987. A description of why building seawalls and groins to protect East Coast beaches, wetlands, and barrier islands does not work. Argues that the best way to preserve coastal features is to let nature take its course.

Goulding, M. "Flooded forests of the Amazon." *Scientific American*, March 1993. The amazing ecosystems of the floodplain of the Amazon River.

Mitsch, W. J., and J. G. Gosselink. *Wetlands*. New York: Van Nostrand Reinhold, 1986. An analysis of wetlands ecosystems and the steps necessary to preserve them.

Perry, D. R. "The canopy of the tropical rain forest." *Scientific American*, November 1984. Its inventor tells the story of a climbing method that has made it possible to do research in the once-inaccessible rain forest canopy.

# Ecosystems and How They Change

**O B J E C T I V E S**

*When you have studied this chapter, you should be able to:*

1. Define the following words and use them in context:
   ecosystem, community
   producer, decomposer, consumer, herbivore, carnivore, detritivore
   food chain, food web, trophic level, biomass, productivity
   climax community, fugitive species, keystone species
2. Outline energy flow through an ecosystem and give a rough estimate of the energy loss between one trophic level and the next.

3. Explain the difference between gross and net primary productivity, and discuss factors that affect them.
4. Describe simplified nutrient cycles for carbon, nitrogen, phosphorus, and sulfur, and point out the important differences among them. Distinguish between atmospheric and sedimentary cycles.
5. Distinguish between primary and secondary succession, outline an example of each, and explain why succession occurs.
6. Explain what a population's niche is.

A tropical rain forest or a desert may cover a huge area. For convenience, ecologists usually study ecosystems, smaller units such as a hillside, a lake, or a field. An **ecosystem** consists of the **community** of all the organisms in the area, along with their physical environment: water, minerals, sunlight, air, and so on. We usually treat an ecosystem as an isolated unit, but in fact things invariably move from one ecosystem to another, as when leaves blow from a forest into a lake, or birds migrate between their summer and winter homes.

Not all ecosystems are natural: a space station, an aquarium, and a pot of houseplants are artificial ecosystems. A farm is often considered as an ecosystem because farmers must recognize the interactions among crop plants, fertilizers, pesticides, soil, climate, and the natural plant and animal life in order to manage the farm effectively (Figure 48–1).

Organisms in a community interact with each other and with their abiotic (nonliving) environment. These interactions can be viewed on two different time scales. Ecologists study what is happening here and now: plant growth, animals eating plants, and so on. But over the long term, every environmental event affecting organisms may also be a selective force that shapes their evolution. Each time an owl catches a mouse, it not only feeds itself and reduces

**FIGURE 48–1** Irrigating cow pasture on a farm, an artificial ecosystem.

the number of mice but also selects against a set of mouse genes that are not effective at avoiding capture by owls.

A community has group properties not exhibited by its populations of species of organisms. These include the community structure, such as the patchiness of the distribution of various species, and the number and relative abundances of species it contains. Communities also have a degree of permanence not shared by populations. Individual species may come and go, but the overall structure and function of the community is often not affected.

In this chapter we discuss the flow of energy and cycling of nutrients through ecosystems. Then we consider what happens to an ecosystem that is disturbed or destroyed, as it is when struck by fires, volcanic eruptions, or bulldozers. We shall see that the structure of a community has profound effects on an ecosystem and that a community's structure may also change with time.

## KEY CONCEPTS

♦ The organisms in an ecosystem interact with one another and with their physical environment, influencing each others' lives and evolution.

♦ A self-sustaining ecosystem must contain nutrients, producers, and decomposers and receive a continuous input of energy.

♦ Nutrients may cycle indefinitely in an ecosystem, but free energy is continuously lost as heat.

♦ In places where vegetation is destroyed or has never existed, organisms invade and replace each other in sequence until the climax vegetation is established.

## 48–A THE COMPONENTS OF ECOSYSTEMS

To be sustainable—capable of lasting indefinitely—an ecosystem must contain the resources to support its resident organisms and to dispose of their wastes. The sun is the ultimate source of energy for almost all ecosystems because it drives photosynthesis by plants, algae, and some bacteria. Because photosynthetic organisms are **autotrophs,** making their own food from inorganic substances, they are the **primary producers** of food in sun-powered ecosystems.

Producers eventually die. Their remains are usually broken down by **decomposers,** organisms that acquire their food molecules from dead organic material. In the process of extracting energy and nutrients from this material, decomposers release some of the nutrients back into the ecosystem, where they are again available to producers.

The primary producers supply nutrients, as well as energy, to all the other organisms in the ecosystem. Nutrients are cycled through the ecosystem and may be used again and again in the same small area. Energy, by contrast, is not cycled. As we know from the Second Law of Thermodynamics, free energy is continuously degraded to heat, which organisms cannot use. Hence free energy must be continually replaced. Most organisms would soon die if producers were deprived of the sun's energy for long.

The minimum requirements for a self-sustaining ecosystem are water, various minerals, carbon dioxide, oxygen (in most cases), producers, decomposers, and a continuously renewed source of energy (Figure 48–2). However, most ecosystems also contain **consumers,** organisms that eat plants or each other. Consumers also act as producers of

food in the ecosystem because they form the food supply for other consumers. Plant-eating animals are collectively known as **herbivores** or **primary consumers.** These may die and pass directly to the decomposers, or some of them may be eaten by **carnivores,** also called **secondary consumers.** There may be tertiary or even quaternary consumers, carnivores that feed on the secondary and tertiary consumers, respectively. Both consumers and decomposers are **heterotrophs,** feeding on organic matter produced by

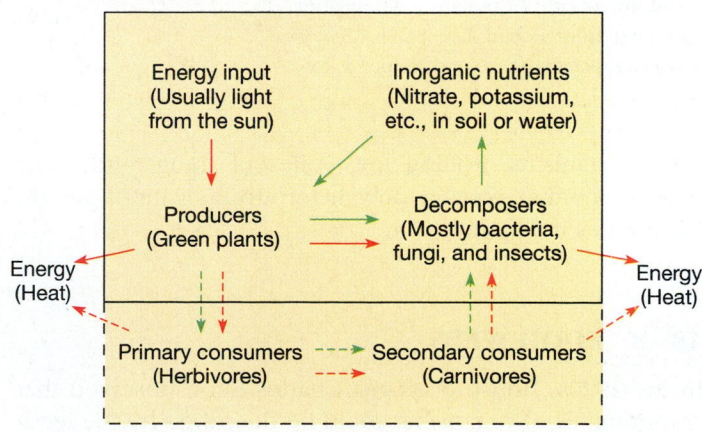

**FIGURE 48–2** Energy flow and nutrient cycling. The general pattern of energy (red arrows) and nutrient (green arrows) exchange in an ecosystem. The solid line encloses the minimal components required for a self-sustaining ecosystem. In addition to these, most ecosystems contain primary, secondary, and even higher levels of consumers (dashed box). Energy is continuously converted to heat produced by the organisms' metabolism, and the heat is continuously lost from the ecosystem.

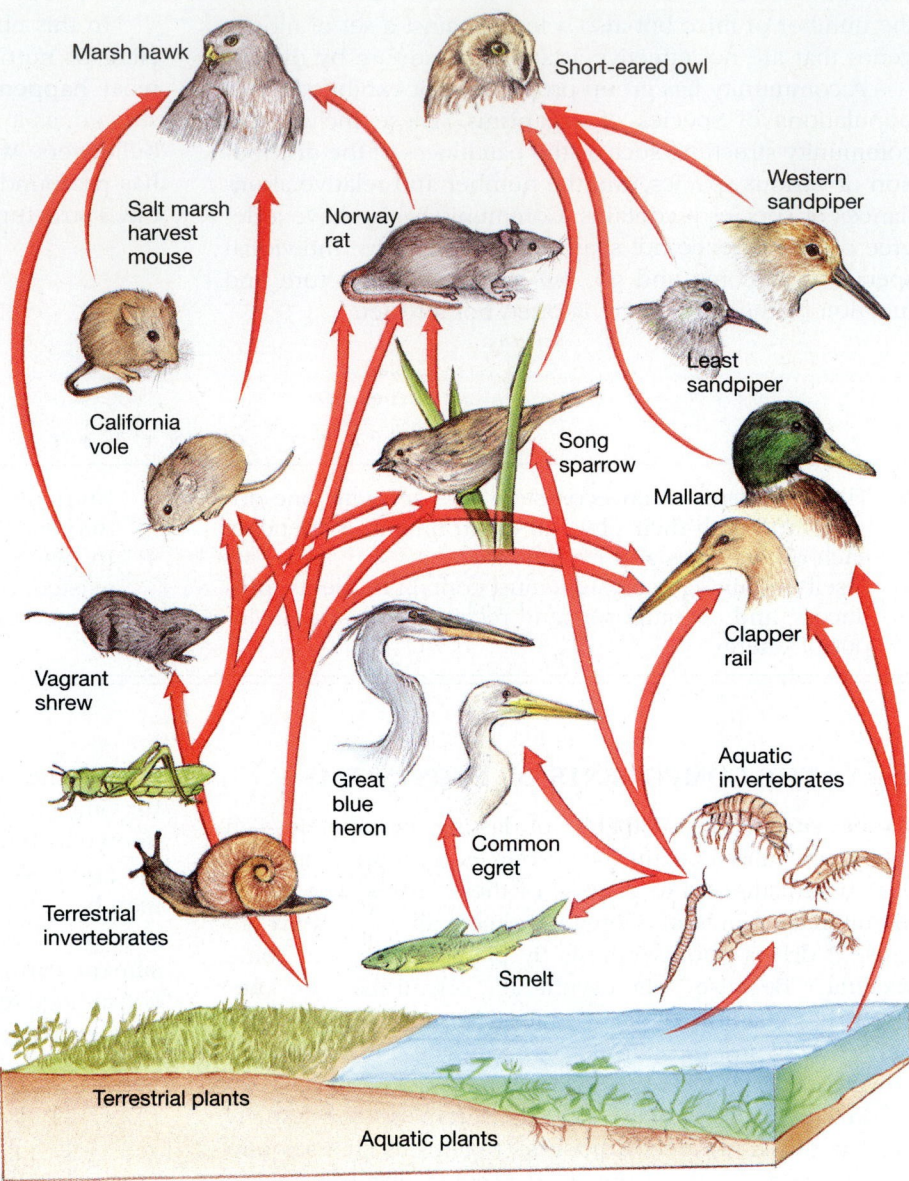

**FIGURE 48–3** A food web in a marsh. Arrows show the passage of energy and nutrients as one organism eats another. (Modified from Robert Leo Smith, *Ecology and Field Biology,* New York: Harper & Row, 1974)

other organisms. Rotting logs, piles of dung, and other ecosystems that contain only heterotrophs cannot sustain themselves indefinitely.

## 48–B  FOOD WEBS

In the 1920s, British ecologist Charles Elton observed that communities of organisms could be described by the feeding relationships among their members. He coined the term **food chain** for a series of organisms, each of which provides the food supply for the next. For example, nutrients and energy pass from leaves to caterpillars that eat them, to chickadees, to hawks, or from dead animals to fly maggots to parasitic wasps. Food chains in an ecosystem are usually interconnected in a complex **food web** (Figure 48–3).

The **trophic level** of an organism describes the number of steps it is removed from the ecosystem's ultimate energy source. For instance, photosynthetic plants make up the first trophic level because they receive their energy directly from sunlight. The second trophic level contains herbivorous animals (primary consumers), and higher trophic levels are made up of predatory carnivores (secondary consumers and so forth).

An organism cannot always be assigned to just one trophic level. Thus, some plants, such as Venus's flytrap, are carnivores as well as producers. A frog tadpole is a herbivore, eating algae, or a **detritivore,** consuming particles of detritus (dead organic matter), whereas the adult frog is carnivorous. Many mammals, such as foxes, bears, and humans, are **omnivores,** organisms that belong to several trophic levels because they eat both plants and other animals.

## TABLE 48–1

**Net Primary Productivity in Major Ecosystems**

| Ecosystem Type | Net Primary Productivity (g/m²/year) | |
| --- | --- | --- |
| | *Range* | *Mean* |
| Tropical rain forest | 1000–5000 | 2200 |
| Temperate forest | 600–2500 | 1250 |
| Boreal forest (taiga) | 400–2000 | 800 |
| Savanna | 200–2000 | 900 |
| Temperate grassland | 200–1500 | 600 |
| Tundra and alpine | 10–400 | 140 |
| Desert and semidesert scrub | 10–250 | 90 |
| Extreme desert | 0–10 | 3 |
| Cultivated land | 100–3500 | 650 |
| Lake and stream | 100–1500 | 250 |
| Open ocean | 2–400 | 125 |
| Upwelling zones | 400–1000 | 500 |
| Continental shelf | 200–600 | 360 |
| Algal beds and coral reefs | 500–4000 | 2500 |
| Estuaries | 200–3500 | 1500 |

## 48–C  PRODUCTIVITY

The flow of energy through an ecosystem can be measured at various points by answering questions such as these: how much solar energy is trapped in a plant during photosynthesis? How much of the energy in plant material can a herbivore use? How much energy does a herbivore use and how much does a carnivore obtain by eating the herbivore? Some of the ways of obtaining answers are described in *How Do We Know: Measuring Productivity.*

### Primary Productivity

**Primary productivity** is the rate at which energy is stored in organic matter by photosynthesis. It is usually measured as the increase per unit time in energy stored or in plant biomass. **Biomass** is the total mass of living organisms, usually measured as the dry mass of living material.

   **Gross primary productivity** is the total rate of photosynthesis of all the plants in an area. About half of the organic matter produced is rapidly used up in the plants' own respiration (Figure 48–4). What remains is **net primary productivity,** which appears as plant growth (new biomass) and is available to heterotrophs.

   Note that productivities are *rates*. They are not the same as **standing crop,** which is the biomass present at any one time. In some situations productivity is actually higher at a lower standing crop. For instance, a field or lawn that is mowed regularly or grazed by cattle has a low standing crop of grass but often has a higher net produc-

tivity than it would have if just left to grow, when it has a higher standing crop.

   Productivity varies from one type of ecosystem to another (Table 48–1). It generally increases from the arctic toward the tropics, reflecting increasing temperature. Ocean

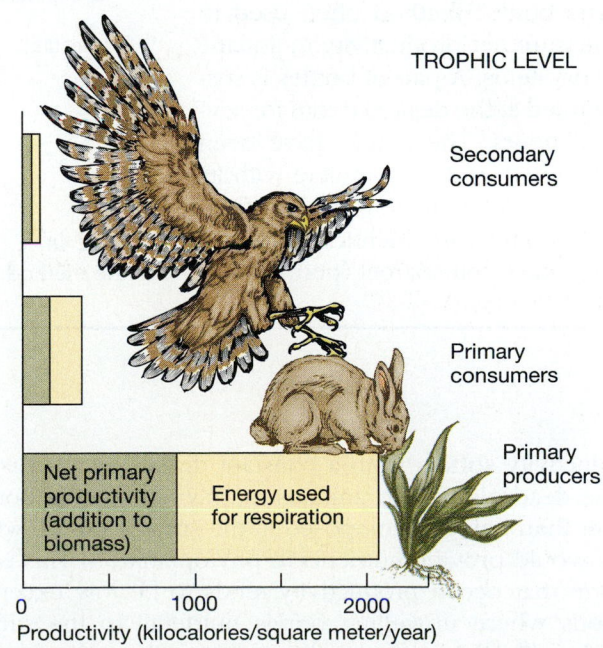

TROPHIC LEVEL

Secondary consumers

Primary consumers

Primary producers

Net primary productivity (addition to biomass)

Energy used for respiration

0    1000    2000

Productivity (kilocalories/square meter/year)

**FIGURE 48–4**  Energy flow in an actual ecosystem. The graph shows the energy input at each trophic level and its division into new biomass (green) and energy used during respiration.

## MEASURING PRODUCTIVITY

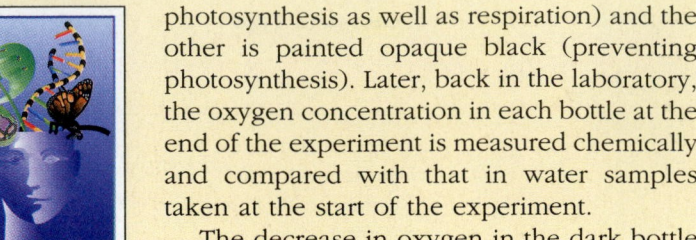

The most obvious way of measuring productivity is to determine the yield of biomass per area over a certain period. However, this is often impractical in natural communities where there are many potential sources of error, such as herbivores eating some of the yield. Instead, productivity is usually determined indirectly.

The overall equation for photosynthesis and respiration is:

$$6\,CO_2 + 6\,H_2O + energy \underset{Respiration}{\overset{Photosynthesis}{\rightleftharpoons}} C_6H_{12}O_6 + 6\,O_2$$

Net productivity is directly proportional to the rate of production of oxygen and to the rate of depletion of $CO_2$. Measurement of either permits calculation of net productivity. If we can, in addition, measure respiration in the absence of photosynthesis by shutting off the light source, then gross productivity can also be estimated, since gross productivity = net productivity + respiration rate.

Figure 48–A shows the "light- and dark-bottle" method often used to measure net productivity in aquatic ecosystems. A pair of bottles is suspended at the desired depth for several hours. The bottles have been filled with water (complete with its plankton) taken from this depth. The bottles are identical except that one is transparent (permitting

photosynthesis as well as respiration) and the other is painted opaque black (preventing photosynthesis). Later, back in the laboratory, the oxygen concentration in each bottle at the end of the experiment is measured chemically and compared with that in water samples taken at the start of the experiment.

The decrease in oxygen in the dark bottle gives a measure of the combined respiration of both producers and consumers. The increase in oxygen in the light bottle gives the net gain of photosynthetic productivity over this respiration figure. Respiration is assumed to be the same in both bottles. Adding "respiration" from the dark bottle to "gross primary pro-

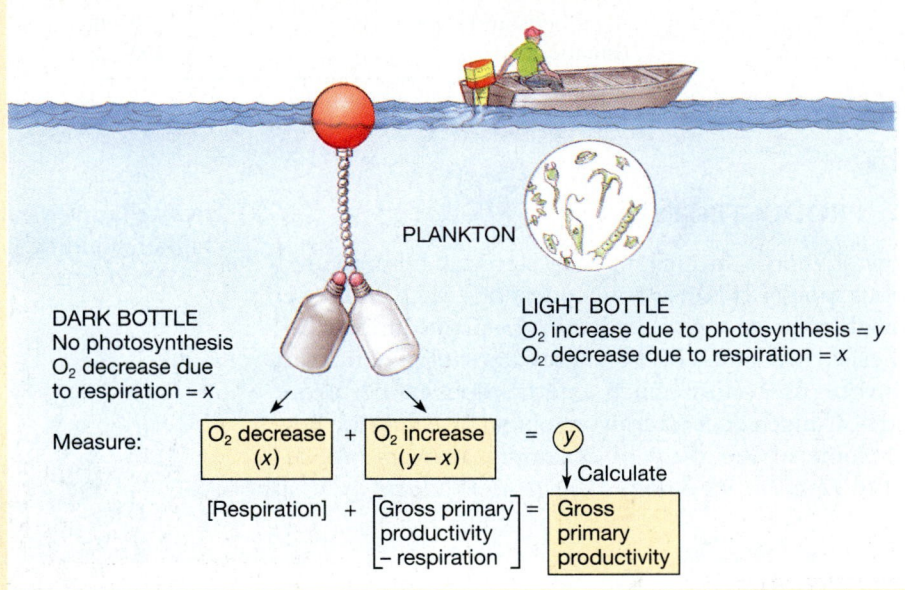

FIGURE 48–A Measuring productivity in an aquatic ecosystem by the light- and dark-bottle method.

productivity suffers from a constant drain of nutrients because dead plants and animals rapidly sink to the bottom rather than being decomposed in the surface layers, where they would provide nutrients to phytoplankton. This is the reason that ocean productivity tends to be low except in regions where upwelling brings nutrients to the surface (Section 47–H).

The net primary productivity of the whole biosphere is approximately 225 billion metric tons (dry mass) of organic matter each year. Of this, almost 60% is produced on land and only 40% in the ocean, although the ocean occupies more than two thirds of Earth's surface. Photosynthesizers achieve this enormous productivity using less than 2% of the solar energy that reaches Earth's surface each year. Under some conditions, a barley plant may convert up to 14% of the visible light energy that reaches it into net biomass, but the efficiency of photosynthesis is usually much lower. Average productivities represent the use of only about 0.5% of the visible light energy that reaches Earth's surface.

**FIGURE 48–B** Measuring rates of photosynthesis on land. Air from the enclosed shrub passes through tubes to the mobile laboratory, where its $CO_2$ content is measured continuously. In the experiment shown here, the shrub can be illuminated by an artificial light source suspended from the frame above the shrub. (Harold Mooney)

ductivity − respiration" from the light bottle gives a figure for oxygen produced by photosynthesis, which can be used to calculate gross primary productivity in terms of fixed carbon or energy.

Measuring the productivity of terrestrial plants presents problems because it is more difficult to trap the gases involved. However, it can be done by enclosing plants or parts of plants in transparent tents and comparing changing $CO_2$ levels of the air inside these tents by day and by night (Figure 48–B).

## Secondary Productivity

**Secondary productivity** is the rate of formation of new organic matter by heterotrophs, that is, their growth and production of offspring. This rate amounts to only about 10% of net primary productivity for several reasons.

First, herbivores consume only a fraction of primary productivity. In temperate forests, herbivores eat only 1 to 3% of the net primary productivity. In other ecosystems, such as overgrown fields, more than 10% of the vegetation may be eaten.

Second, only a fraction of the food energy consumed is converted into secondary productivity. A caterpillar's gut absorbs only about half of the leaf material it eats. Of the food absorbed, about two thirds is used in respiration. Hence, only 15% or less of the food eaten appears as secondary productivity. Grasshoppers in a Tennessee field were found to convert only about 4% of the food they ate into secondary productivity. This is one reason they and their locust relatives are so destructive to crops.

The general rule that animals convert about 10% of the food energy they eat into secondary productivity is only a very rough approximation. The figure is lower for endotherms than for ectotherms and lower for herbivores than for carnivores. Endotherms generally have higher metabolic rates than ectotherms and use more of their food energy in respiration. Thus, measurements on a number of vertebrates showed that birds and mammals (endotherms) converted 1 to 3% of the food they ate into biomass, whereas the figure for fishes (ectotherms) was about 10%. Carnivores convert food into biomass more efficiently than herbivores because plant tissues generally require more energy to digest than animal tissues. In addition, carnivores consume less food than herbivores for the same intake of nutrients and energy because, weight for weight, the nutrients an animal needs are more concentrated in meat than in plant food.

## Pyramids of Energy

The flow of energy through an ecosystem can be represented in the form of a **pyramid of energy,** which shows the total amount of energy entering each trophic level. These diagrams are smaller at the top because energy is lost within each trophic level as well as in going from one trophic level to the next (Figure 48–5).

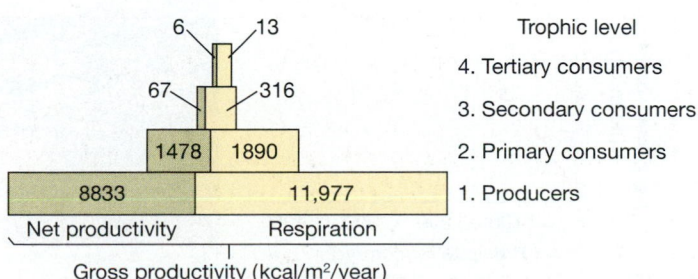

**FIGURE 48–5** Pyramid of energy for Silver Springs, a river ecosystem in Florida. This graph shows total energy input at each trophic level (gross productivity) and its division into net energy gain (net productivity) and energy lost by way of respiration. (After H. T. Odum, *Ecological Monographs* 27:55, 1957)

Energy is transferred from one trophic level to the next in the form of food consumed. Therefore, it is not surprising that in many familiar food webs there are fewer individuals at each trophic level. This can be illustrated by a **pyramid of numbers,** showing that, for instance, there are more grass plants in a prairie ecosystem than there are bison or prairie dogs that eat the grasses. Sometimes pyramids of numbers are inverted, as in the case of a single tree providing food for thousands of insects.

The trophic levels of an ecosystem can also be represented by a **pyramid of biomass,** showing the biomass of all organisms present at each level. If we represented a tree and its herbivorous insects by a pyramid of biomass rather than of numbers, the pyramid would once again be right-side-up: the biomass of the tree is much greater than the biomass of insect consumers.

Occasionally, pyramids of biomass are also inverted. The classic case is an ocean community with phytoplankton producers. Phytoplankton may reproduce so rapidly that a small biomass of producers alive at any one time can support a larger biomass of herbivores.

## Why So Few Trophic Levels?

As long ago as 1942, Raymond Lindeman argued that energy flow can be used to describe many aspects of an ecosystem, such as the number of trophic levels and the relative numbers of species in each.

One hypothesis based on his work was that the more productive an area, the longer the food chains it is likely to contain. Thus, tropical areas are more productive and contain more different species than temperate areas. However, studies of more than 100 land and aquatic food webs in tropical and temperate areas have found no correlation between productivity and the number of trophic levels. So much energy is lost between one trophic level and the next that few communities contain more than five trophic levels: too little energy remains at the top of the food chain to support higher trophic levels, even in ecosystems with high productivity. There are several reasons for this.

First, much energy is used in respiration, which powers the metabolism of organisms. In Chapter 6 we saw that free energy decreases in every energy transformation. Cellular respiration is only about 50% efficient at transferring free energy from food molecules to ATP. More energy is lost as the ATP is used in the maintenance and repair of body tissues, in functions such as feeding, excretion, and circulation, and in behavior. Only the energy that remains is stored as new biomass.

Second, not all the productivity of one trophic level is found and eaten by animals at the next level. No bird or mouse could actually discover every grass seed in a square meter of prairie. And some of the productivity is inedible: few animals eat hair, feathers, or bone (from animal food); or thorns or wood (from plants).

Third, a consumer wastes some of the food it eats, excreting it as feces. Feces are partly made up of indigestible substances such as feathers, chitin, or cellulose. Feces also contain digested substances that are not absorbed into the body because the animal cannot use those particular nutrients.

Plants also waste nutrients because the proportions of minerals they take in are rarely those that are best for growth. Justus Liebig noted that crop productivity increased if the plants were supplied with certain nutrients. This led him to formulate his **Law of the Minimum** in 1840. He concluded that "growth of a plant is dependent on the

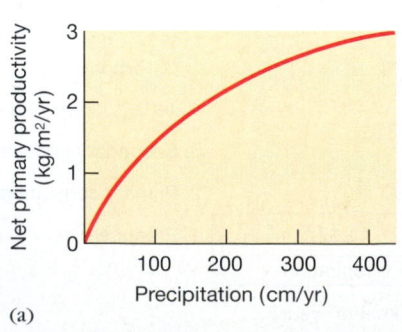

(a)  (b)  (c)

**FIGURE 48–6** Productivity limited by precipitation. (a) A graph showing how net primary productivity increases with average annual precipitation. (b) The relationship between rainfall and productivity is particularly striking in arid ecosystems. In this scene from Death Valley, plants grow only along a dried-up stream bed where water collects after the infrequent rains. (c) Artificial irrigation supports the much higher productivity of this Death Valley golf course.

amount of the nutrient that is presented to it in minimum quantity." Today we would say that if one (or more) of the resources an organism needs is in short supply, that resource becomes the **limiting factor** for the organism's growth and reproduction.

Plant growth may be limited by many factors. Where nutrients are scarce they become the limiting factors and productivity is low no matter how much water or sunlight reaches the area. On land, lack of nitrogen in the soil often limits plant growth, and lack of phosphate is often limiting in aquatic ecosystems. In dry areas, such as savannas and deserts, water is usually the limiting factor (Figure 48–6).

For a population of herbivores, a particular nutrient, such as sodium or an essential amino acid, is often the limiting factor. Many herbivores excrete and waste large amounts of energy-rich plant material in order to eat enough food to extract all the nitrogen-containing amino acids they need. They do not even extract amino acids from their food with maximum efficiency. It pays a caterpillar, for instance, to assimilate only the easily digestible protein from one meal before consuming the next, rather than digesting the first meal fully. Rates and efficiencies cannot be maximized simultaneously. In most animals, the compromise tends to be toward maximizing the rate at which a nutrient is assimilated rather than the efficiency (Figure 48–7).

Because of all these energy and nutrient losses from one trophic level to the next, there is not enough left to support higher trophic levels on land. A wolf may have to travel 30 kilometers a day to find enough food to eat, and a tiger requires a home range of up to 240 square kilometers. An animal that fed on wolves or tigers would have to cover a much greater hunting area to try to find enough of its widely scattered prey. It is not energetically feasible to try to harvest the small amount of food available in the

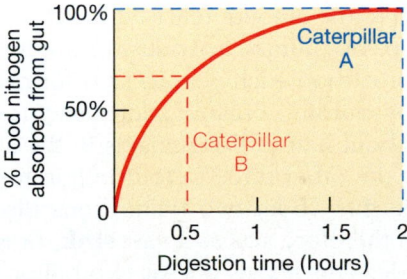

**FIGURE 48–7**  Rates and efficiencies of feeding. A hypothetical example of how efficiency may be sacrificed to increase the rate at which a limiting nutrient is obtained. Caterpillar A retains food in its gut until all the nitrogen is assimilated (100% efficiency). Caterpillar B assimilates only the easily digested nitrogen and then ingests another meal. B can digest four gut-loads of food in the time it takes A to digest one. Caterpillar B assimilates nitrogen, and therefore grows, at a faster rate.

highest trophic level. The organisms that do feast on top predators are parasitic worms and fleas. They eat only part of the predator and get only a tiny crumb of the ecosystem's energy pie.

Energy loss at each trophic level is also the reason that very large animals are invariably herbivores or filter-feeders, obtaining their energy from low on the food chain: elephants and rhinoceroses are herbivores, as were the largest dinosaurs. The largest whales are not actually herbivores. They are filter feeders living mainly on krill, shrimp-like primary consumers that eat phytoplankton. Higher trophic levels do not contain enough energy per unit area to support populations of large animals. A large animal can obtain enough food only by feeding on producers or low-level consumers.

## 48–D  CYCLING OF MINERAL NUTRIENTS

In contrast to energy, which is continuously gained and lost by ecosystems, nutrients may be used over and over again. The movements of nutrient elements through the biosphere or through any particular ecosystem, by physical and biological processes, are **biogeochemical cycles.** Elements shuttle between abiotic sources in rocks, water, and air and biotic reservoirs in the ecosystem's organisms.

Organisms require eight elements in relatively large amounts: carbon, hydrogen, oxygen, nitrogen, potassium, calcium, phosphorus, and sulfur. Some of these elements occur in rocks and are released by erosion and weathering into soil, rivers, lakes, and the oceans. Some are also present in the atmosphere.

Nutrients are sometimes recycled rapidly through ecosystems. In other cases, nutrients spend long periods outside living organisms. For example, remains of marine organisms may sink to the ocean bottom and be incorporated into sedimentary rocks that are lifted and exposed at Earth's surface only after millions of years. Every nutrient element has a somewhat different fate, depending on its physical and chemical properties and on its role in organisms. We shall illustrate the concept of nutrient cycling with a few simplified examples.

### The Carbon Cycle

Carbon moves in an **atmospheric cycle** because carbon occurs in the gases $CO_2$ and methane ($CH_4$) in the atmosphere, as well as in rocks and dissolved in water. Much of the carbon in terrestrial ecosystems travels rapidly between living organisms and the atmosphere by way of photosynthesis and respiration. Most of the carbon fixed as organic matter by photosynthesis is broken down rapidly and released back into the air as $CO_2$ produced during respiration by plants themselves and by animals farther up the food chain. In aquatic ecosystems, the exchange occurs

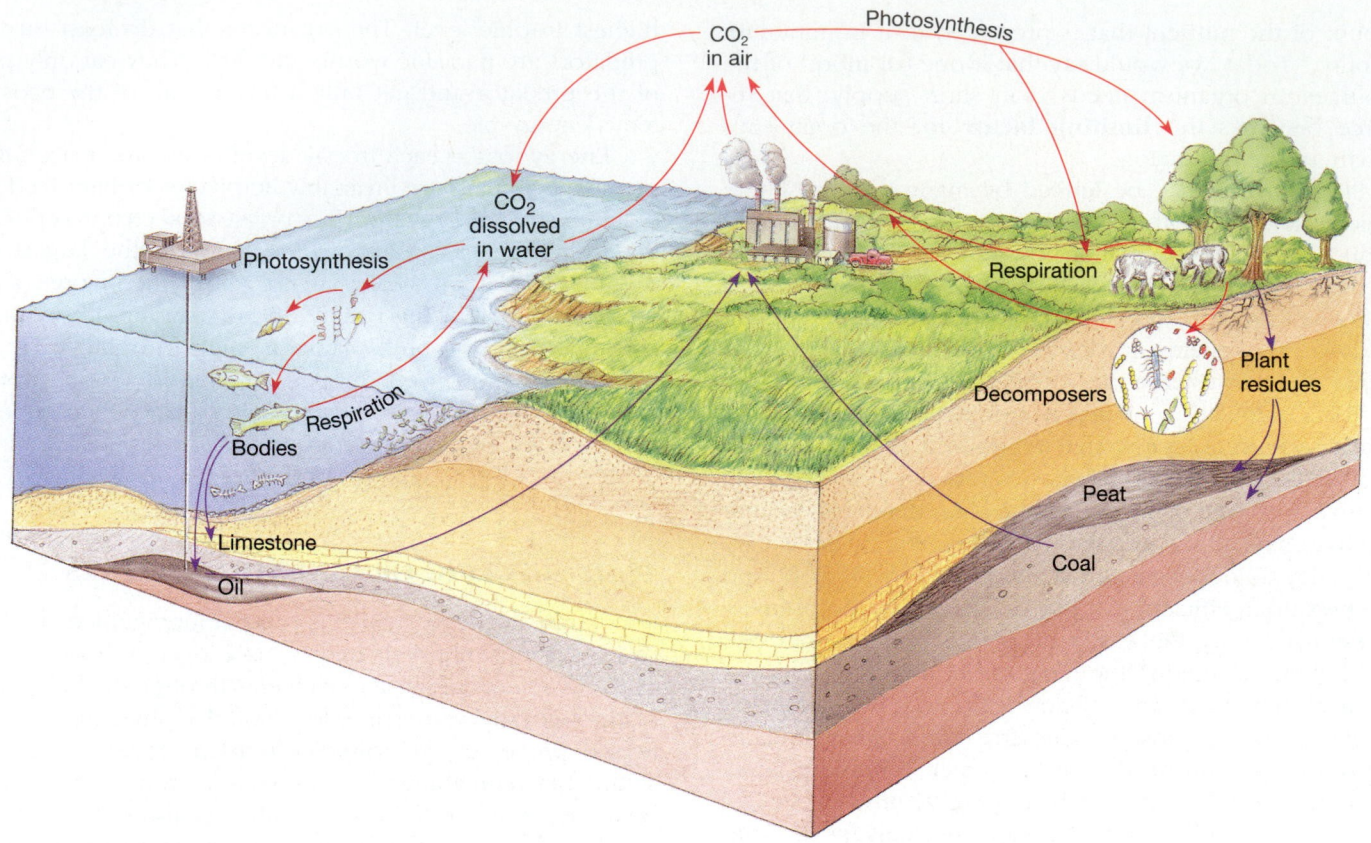

**FIGURE 48–8** A simplified carbon cycle. This diagram shows how carbon moves through several ecosystems. Red arrows indicate rapid, short-term processes; purple arrows indicate longer term cycles.

between living organisms and $CO_2$ or bicarbonate ($HCO_3^-$) or carbonate ($CO_3^{2-}$) dissolved in the water. Another carbon compound, methane, is released into the atmosphere by anaerobic organisms, such as those in sewage treatment plants, the bottom mud of wetlands, and the digestive systems of termites and of ruminants such as cattle and sheep (Figure 48–8). Carbon dioxide makes up about 0.03% of the atmosphere, methane 0.0002%.

Some carbon enters a long-term cycle. It may accumulate as undecomposed organic matter in the peat layers of bogs or moorland. Similarly, deposits of carbon-rich coal, oil, and natural gas were part of the net productivity of ancient ecosystems. These fossil fuels are the remains of organisms that were buried before they could be decomposed and were then transformed by time and geological processes. When humans extract and burn fossil fuels, this carbon is released back into the atmosphere as $CO_2$. In addition, the calcium carbonate shells of marine animals and protists may form sedimentary rocks such as limestone and dolomite on the sea floor. After millions of years, these rocks may be lifted above sea level and exposed to ero-

sion, releasing carbon into streams and lakes as carbonate and bicarbonate that may be consumed by organisms.

A 1993 headline read: "Missing: One Billion Tons of Carbon." The story explained that there is still a lot we do not know about how carbon cycles through the biosphere. The amount of carbon taken in each year by plants is roughly equal to the amount released by all organisms during respiration. But human activities release an extra seven billion tons of carbon each year as $CO_2$ formed by cutting forests and by burning organic matter such as wood and fossil fuels. About half of this remains in the atmosphere.

Where is the other three to four billion tons? Most geologists agree that about two billion tons dissolve in the ocean, which therefore acts as a vast **sink,** or reservoir, for carbon dioxide. This leaves one or two billion tons of carbon a year unaccounted for. Probably plants are absorbing some of this. In parts of the world, and particularly in Europe, forests that were cut down in the nineteenth century are now regrowing. As the trees grow larger, they take in more $CO_2$ than they release. Other people have suggested that rising $CO_2$ in the atmosphere may be "fertiliz-

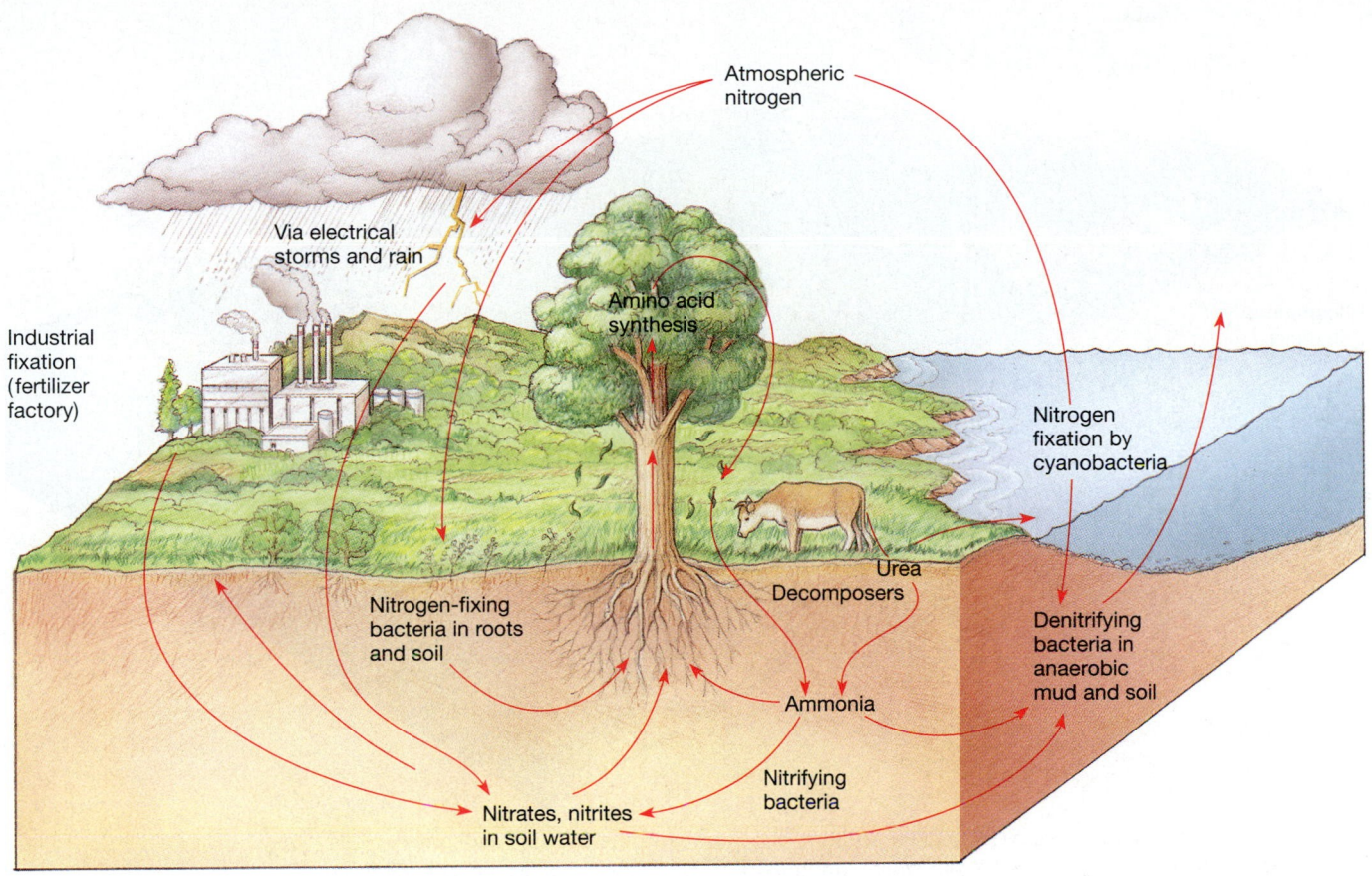

**FIGURE 48–9**  The nitrogen cycle.

ing" forests so that trees are growing faster than before and taking in more $CO_2$.

## The Nitrogen Cycle

Nitrogen is a vital part of many essential organic compounds, especially proteins and nucleic acids. About 80% of the atmosphere is molecular nitrogen gas ($N_2$), which is much more abundant than $CO_2$. However, whereas producers use $CO_2$ directly, only certain species of prokaryotes can use $N_2$. These organisms fix nitrogen by reducing it to ammonia (Section 24–C). Plants must get this vital nutrient either directly from nitrogen-fixing microorganisms or indirectly from the limited pool of fixed nitrogen found in soil water, rivers, lakes, and oceans.

Nitrogen fixation is slow compared with the rate at which plants can absorb and use fixed nitrogen, and so nitrogen is often the limiting nutrient in plant growth. Nitrogen is one of the elements most commonly applied in crop fertilizers. This additional input of usable nitrogen into

the biosphere comes from the industrial manufacture of ammonia-based fertilizers from atmospheric nitrogen, using catalysts and a lot of fossil fuel energy (Figure 48–9).

There is a second difference between the movements of nitrogen and carbon through an ecosystem. Most organisms release carbon dioxide during respiration, but the conversion of nitrogen from organic molecules into inorganic forms involves steps carried out by a series of different organisms.

Decomposers release most of the nitrogen from dead organisms as ammonium ($NH_4^+$) ions. These undergo **nitrification,** whereby soil bacteria (*Nitrosomonas, Nitrosococcus*) oxidize ammonium to nitrite ions ($NO_2^-$), and still others (*Nitrobacter, Nitrococcus*) oxidize the nitrites to nitrates ($NO_3^-$). Plants can absorb all these forms of nitrogen through their roots, but ammonium is the most useful because its nitrogen is already in the reduced form the plant uses to produce amino acids, nucleotides, and other compounds. If a plant absorbs nitrate instead, it expends large amounts of ATP reducing the nitrogen before it can be used. Unfortunately, nitrifying bacteria are so

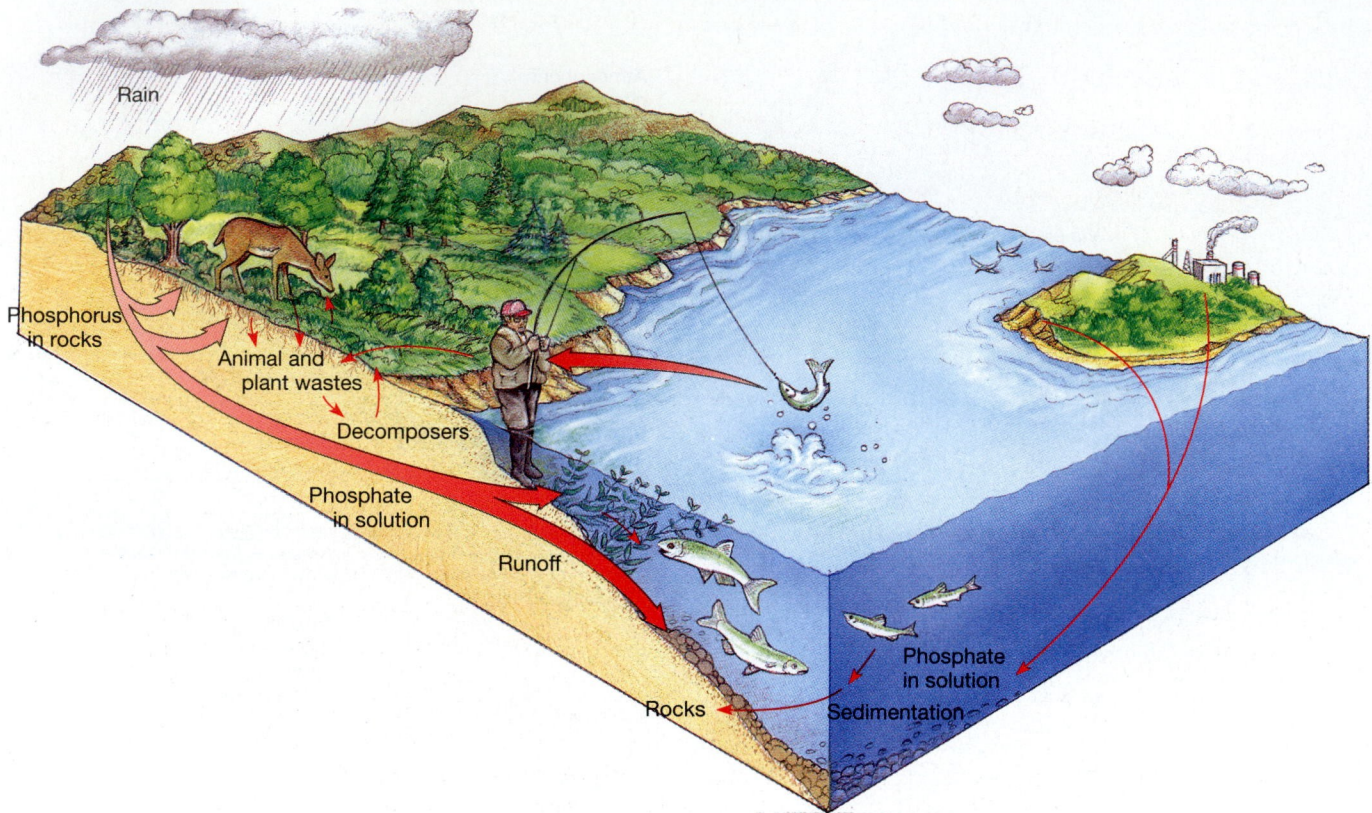

**FIGURE 48–10** The phosphorus cycle. Plants take in most phosphorus in the form of phosphate. Fishing and the droppings of sea birds bring trivial amounts of phosphorus from water to land ecosystems. However, most flow is one-way, from terrestrial rocks to the sea floor.

common in most soils that ammonium seldom lasts for long in the soil, and nitrates are the main form of nitrogen available to roots.

Nitrates are anions, which are readily **leached,** that is, washed out of the soil. In contrast, cations such as ammonium tend to be adsorbed onto negatively charged soil particles and resist leaching. Thus, nitrification is doubly bad news from a plant's point of view, because it converts soil nitrogen into a less energetically useful form that is also more readily leached away.

Some of the nitrogen in most ecosystems is eventually returned to the atmosphere by **denitrification.** Denitrifying bacteria live in the anaerobic mud of fertile lakes, bogs, estuaries, and parts of the ocean floor. They use nitrites or nitrates as electron acceptors during their own energy production, releasing the nitrogen as $N_2$ gas.

## The Phosphorus Cycle

Phosphorus is a major constituent of nucleic acids and of membrane phospholipids, and many animals also use phosphorus in shells, bones, and teeth. Since phosphorus almost never occurs as a gas, its cycle, unlike the atmospheric cycles of nitrogen and carbon, is a sedimentary cycle (Figure 48–10). When rocks are eroded by weather, small amounts of phosphorus dissolve, usually as phosphate, and so become available to plants. Much of the phosphorus excreted by animals is in the form of phosphate, which plants can use immediately. Thus, the cycling of phosphorus in terrestrial ecosystems is usually very efficient, although small amounts are continually lost downstream and to the oceans.

The phosphorus cycle is, in the short run, a one-way flow—from rocks to land ecosystems to the ocean, and finally to ocean sediments. The only natural way for phosphorus to return to land is by slow geological processes in which sea floor sediments may again become terrestrial rocks. However, terrestrial ecosystems retain much of their phosphorus, since phosphate ions are adsorbed onto soil particles, helping to provide a steady supply of it for plant growth. Soil erosion robs an ecosystem of its phosphorus, and it may take thousands of years to recoup this loss through the weathering of rocks.

## The Sulfur Cycle

Organisms need small amounts of sulfur to make proteins. Plants obtain sulfur by absorbing sulfates from the soil, and sulfur is returned to the soil when organisms die. The only significant abiotic source of sulfur is volcanic explosions,

**FIGURE 48–11** A V-notch weir at Hubbard Brook Experimental Forest. All the water that runs off one of the valleys is funneled through this weir so that its volume can be measured and its nutrient content analyzed. You may be able to see a man bending behind the left back corner of the weir. (Peter Marks)

which release sulfur dioxide and hydrogen sulfide into the air, contributing 10 to 20% of the sulfur that organisms use. Despite this lack of sulfur sources, sulfur is apparently seldom a limiting factor in the productivity of an ecosystem.

Some of the most important members of the sulfur cycle are photosynthetic sulfur bacteria, which use hydrogen sulfide ($H_2S$) instead of water ($H_2O$) to supply hydrogen atoms for photosynthesis (Section 24–C). Instead of giving off oxygen as green plants do, these bacteria give off sulfur. In the presence of oxygen, this sulfur is oxidized to sulfides ($SO_3^{2-}$) and sulfates ($SO_4^{2-}$). Sulfates are absorbed by plants. Sulfides may form hydrogen sulfide, which is recycled by sulfur bacteria, or they may react with iron in the soil to form insoluble iron sulfides.

Coal and oil often contain iron sulfides. That is what is meant by "high-sulfur" oil or coal. These sulfides cause problems when the oil or coal is burned or exposed to air. Then, burning or bacteria convert the sulfides to sulfur dioxide ($SO_2$). When sulfur dioxide in the air dissolves in water droplets, it forms sulfuric acid ($H_2SO_4$), producing acid rain. $SO_2$ can also make water draining from a mine very acidic so that it dissolves toxic chemicals and pollutes rivers, lakes, and soil water.

## 48–E NUTRIENT RETENTION

Experiments show that most undisturbed ecosystems are very efficient at retaining nutrients, preventing them from being washed away by the rain. Here we consider examples in a temperate and a tropical ecosystem.

## Hubbard Brook

Hubbard Brook Experimental Forest lies in the White Mountains of central New Hampshire. It consists of a group of valleys of temperate deciduous forest, each with its own creek running down the middle. It has been studied since 1963 by Herbert Bormann, Gene Likens, and their colleagues.

The first project at Hubbard Brook was to measure the inputs and outputs of water and nutrients from undisturbed forest. To do this, concrete "V-notch" weirs were built across the creeks at the bottoms of six valleys (Figure 48–11). Since the weirs were anchored in rock, all of the water leaving each valley (apart from that lost by evapotranspiration) had to flow over a weir, where it could be measured and its nutrient content analyzed. Precipitation gauges were used to measure the input of rain and snow and to collect samples for analysis of their dissolved nutrients.

Data from Hubbard Brook reveal that the forest is extremely efficient at retaining nutrients. Nutrients that reach the forest dissolved in rain and snow approximately balance those leaving the forest, washed away in its streams. Both quantities are small relative to the total amounts of nutrients present. The next experiment showed that nutrients are retained so efficiently because plant roots rapidly absorb nutrients in the soil water.

In the winter of 1965–1966 biologists killed all the trees and shrubs in one of the valleys by felling them and spraying with herbicides. They did not remove the plants but left them where they fell (Figure 48–12). Dramatic effects became obvious almost immediately. In comparison with undisturbed control valleys nearby, water runoff through

**FIGURE 48–12** Experiments on how methods of harvesting trees affect the loss of nutrients. In this winter photograph of Hubbard Brook Experimental Forest, the white area has been clear cut. Also visible is an area where trees have been cut down in horizontal strips. (Peter Marks)

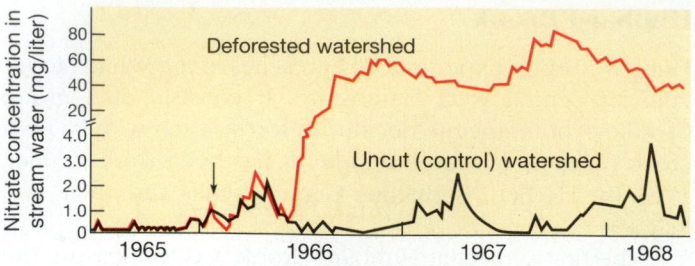

**FIGURE 48–13** Accelerated loss of nitrate in stream water in one of the watersheds at the Hubbard Brook Experimental Forest after it was deforested. The date of deforestation is shown by the arrow. Note the change in scale on the vertical axis, needed to keep the lines for control and experimental watersheds on the same graph. (From G. Likens et al., *Ecological Monographs* 40:23, 1970)

the weir in the deforested valley increased by 40%. This water would ordinarily have left the ecosystem by transpiration. More important, the rate of loss of inorganic nutrients dissolved in the stream water increased six- to eightfold. Loss of nitrate was particularly striking (Figure 48–13). Other experiments at Hubbard Brook revealed that nutrient losses are reduced if the forest is cut in horizontal strips, leaving strips of standing trees, rather than being clear cut. The remaining trees absorb many of the nutrients that enter the soil water after their neighbors are felled.

## Tropical pH Buffering

Experiments in the Amazon in the 1970s showed that tropical rain forest also retains nitrogen efficiently. Each year, 85 kg of nitrogen per hectare fell to the forest floor in the form of dead leaves, fallen trees, or animal waste, but only 2 kg of this was lost from the forest as nitrate in stream runoff. However, when tropical rain forest is converted to pasture, tree plantations, or commercial farmland, nitrogen is rapidly depleted by leaching, and the land becomes infertile.

Carl Jordan and his colleagues found that in the natural forest, the soil contained plenty of ammonium ions but was too acidic for nitrifying bacteria to thrive. As a result, soil nitrogen was in the form of ammonium, which was rapidly absorbed by plant roots, instead of being converted into nitrates, which are easily leached out of the soil. Furthermore, the soil pH was highly buffered, remaining very low despite the researchers' efforts to change it. No matter how much lime ($CaCO_3$) they added to the soil, they were unable to raise the pH more than 0.2 units. A couple of rainstorms rapidly restored the normal pH of 4.5.

The forest soil is so acidic because the plant roots that grow in it and the leaves that fall on it form acidic residues when they decompose. This is apparently an adaptation of rain forest plants that permits extra nitrogen to be retained

in the soil. When forest natives are replaced by plants native to other biomes, the buffering system breaks down, pH rises, and nitrifying bacteria set to work, rapidly converting the soil's ammonium to leachable nitrates.

## 48–F  HOW ECOSYSTEMS CHANGE

Within any ecosystem, disturbances sometimes destroy all or part of the community of living organisms. Coastal California is predominantly covered, not by the chaparral community typical of this biome, but by farms, roads, and buildings. Human civilization has disturbed the natural communities, clearing the vegetation to make room for human affairs and their adjunct parking lots. Even without human intervention, every time vegetation is destroyed by a tornado, landslide, fire, flood, or volcanic eruption, a disturbed area results.

After a disturbance ceases, the area slowly returns to its original state by a progressive series of changes called **ecological succession.** Ultimately, succession produces a **climax community,** the permanent, self-sustaining community that develops in accordance with the climate and topography of the region. Given enough time, even the abandoned cities of the Mayas have reverted to climax forest.

The common feature of succession is that the organisms of each stage alter the environment, making it more suitable for some species and less suitable for others. As a result, organisms with particular adaptations characterize different stages of succession. The details of succession depend on the underlying rocks and the climate of an area.

## Primary Succession

**Primary succession** occurs where there is initially no soil —when a new island rises out of the sea, or when a glacier retreats or a mountainside caves in, leaving a pile of rocks. Consider an area of rock created by a landslide. Lichens adapted to exposed conditions may spread over the rock surface (Section 26–I). They produce organic acids, which dissolve some of the rock. Dead lichens contribute organic matter to the forming soil, and mosses may gain a hold in even a thin layer of lichen remains and rock dust. As the mosses break up the rock further and add their own dead bodies to the pile, the seeds of small, rooted plants can germinate and grow. The process continues along similar lines until the climax community becomes established. It may take thousands of years for the soil and the climax vegetation to develop fully.

In Glacier Bay National Park, Alaska, glaciers have been retreating for about 200 years, leaving rocks and gravel. Here the lichens, mosses, and small herbs of early succession have been succeeded by arctic willow shrubs and then

**FIGURE 48–14**  Secondary succession after the eruption of Mount Saint Helens volcano in the northwestern United States. The lava flow covered the land with lava and ash and set fire to forests. In this photograph, taken nine years later, grasses and herbs have shot up in most areas, and shrubs and even small trees are beginning to grow, fertilized by the nutrient-rich volcanic ash. (Richard Feeny)

alder trees. Alders have a special role in this succession because their roots support nitrogen-fixing bacteria. The alders were the first organisms to supply nitrogen to an ecosystem that contained very little of this nutrient. After alders had dominated the succession for a century, enough nitrogen had accumulated in the soil to support the growth of coniferous trees. Conifers grow taller than alders, eventually casting so much shade that the alders in many areas die. The climax coniferous forest is slowly establishing itself, but it could never have done so without the changes in the soil brought about by organisms belonging to earlier stages of succession.

Frederick Clements, a pioneer in the study of succession in North America in the early twentieth century, thought that succession in different communities converged to a single type of climax vegetation. We now know that this is not quite true. The climax composition of a forest, for example, may vary, even within the same biome, depending on whether it has developed on soil, sand dunes, a filled-in lake, or other substrates.

Primary succession can be seen in any city street. Mosses, lichens, and weeds establish themselves in cracks in the sidewalk, quite large plants may grow in a corner where leaf litter and dirt have been deposited by a rain gutter, and fungi and mosses invade a roof that needs repair. If we stopped cleaning and repairing it, even the center of Manhattan would turn into a rock-filled woodland within our lifetimes.

## Secondary Succession

**Secondary succession** is the series of changes that occur in a community that has been disturbed, for instance by fire, farming, or logging, but not totally stripped of its soil and vegetation (Figure 48–14). Although it may take hundreds of years for the climax vegetation to return during secondary succession, the process is faster than primary succession because soil already exists. Whereas soil formation is the most important part of primary succession, secondary succession is dominated by competition among plants for light, water, and nutrients.

A familiar example of secondary succession around New England is "old field succession," by which abandoned farms return to the climax temperate forest. When a farmer stops cultivating the land, grasses and annual weeds quickly move in and clothe the soil with a carpet of wild carrot, black mustard, and dandelions. The **pioneers** of newly available habitats, these plants grow rapidly and produce seeds adapted to disperse over a wide area. Soon slower-growing perennial plants, such as goldenrod and perennial grasses, appear. These newcomers are taller than the annuals. They shade the ground, and their long root systems monopolize the soil water until it is difficult for seedlings of the pioneer species to grow. But even as these tall weeds choke out the sun-loving pioneer species, they are in turn shaded and deprived of water by the seedlings of pioneer trees, such as pin cherries, dogwoods, sumac, and aspens, which take longer to become established but command most of the resources once they reach a respectable size.

Succession is still not complete, for the pioneer trees are not members of the climax forest. After 5 to 30 years, slower-growing oak, maple, and hickory trees will take over, shading out the saplings of the pioneer tree species. After perhaps a century or two, the land is covered with mature climax forest.

Where disturbance has resulted in major damage to the soil, the immediate endpoint of succession may be an impoverished subclimax community. The magnificent decidu-

**FIGURE 48–15** A clear-cut area in Elliot State Forest, Oregon. Lumber companies have cleared areas of land so large that patches of fugitive species that might have recolonized the area have been wiped out, and the soil has eroded so badly that even hand-planted tree seedlings die.   (M. Graybill, J. Hodder/BPS)

ous forest that once covered much of the eastern United States will not return in the foreseeable future. So much topsoil has been lost as the result of poor farming practices that it will take thousands of years for nutrient levels in the present second-growth forests to build up again to the point where they can support climax forest. Similarly, lumber companies have clear-cut thousands of hectares of old-growth forest in the Pacific northwest, with the result that we would predict from the Hubbard Brook experiments: soil erosion and leaching in the cleared areas. Many of these areas have been replanted with tree seedlings several times but the trees died because the degraded soil can no longer support them (Figure 48–15). Succession will take thousands of years to restore any kind of forest to these areas.

## Mechanisms of Succession

Succession occurs because of progressive changes that make the environment less favorable for the species that are present and more favorable for colonization by others. Some of the changes are purely physical, like the silting in of a lake or the weathering of rock, but many are caused by the organisms themselves. In the later stages of succession, the supply of available nutrients in the soil declines as minerals become increasingly locked up in living organisms. Both the community's productivity and the biomass of all the organisms in the community increase during early succession, leveling off as the climax community is approached (Figure 48–16).

Two factors determine the successional stage occupied by any particular species: the rate at which it invades a new

habitat and its response to changes in the environment as succession progresses. For instance, species such as oaks and hickories disperse slowly and grow slowly and so they appear only in the later stages of succession. In the early stages, **fugitive** plants, which include many kinds of weeds, dominate the landscape. These fast-growing, here-today-and-gone-tomorrow pioneers produce many small seeds that are carried long distances by wind and animals.

Animals, as well as plants, may belong to fugitive species. Insects that specialize in eating a particular plant species may travel far and use their keen senses to smell out new patches of their food plant some distance away. Some of our agricultural pest problems stem from the fact that most crop plants originated as fugitive species. By planting fields exclusively to one crop year after year, farmers create a paradise for fugitive animals such as cabbage worms and cucumber beetles, which no longer have to spend energy to find food and have nothing to do but eat and multiply.

Many pioneer species modify the environment in ways that permit later-stage species to grow. For instance, the growth of herbs and weeds on bare soil shades the surface and increases the amount of moisture the soil can hold, improving conditions for less drought-tolerant plants.

Succession ceases when the replacement of one species by another no longer changes the environment and increases the total biomass. In practice, this means that succession ceases when the landscape is dominated by the tallest types of vegetation the area can support. The final biomass and structure of the community is determined by climate, not by the events of succession. For example, in a mature forest near Washington, DC, that has remained undisturbed for 70 years, maple trees are increasing in number and oaks and beech trees are declining. But this is not succession because the climax deciduous forest is already established. The three-dimensional structure of the ecosystem, light intensity on the forest floor, moisture content of the soil, and other characteristics of the climax community

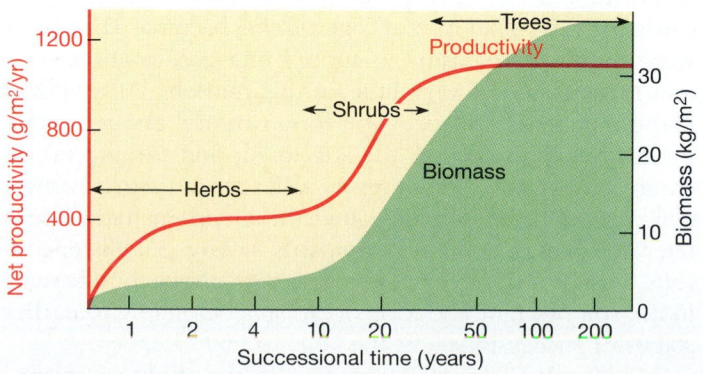

**FIGURE 48–16** Changes in productivity and biomass during succession in a temperate forest.

do not change even though the species composition of the community may change.

In any tract of land, we can always find at least small patches that are undergoing succession following disturbance—a spot where a large tree has fallen, leaving a "light gap" where pioneer weeds can move in, a burned forest, or an ecosystem disrupted by a layer of volcanic ash. The existence of various patches undergoing succession ensures that there is a steady supply of pioneer species that can invade a newly disturbed area and start the process of succession.

## 48–G   COMMUNITY STRUCTURE

We have seen that communities have vertical structure, with different species inhabiting the canopy layer and the understory of a forest. The presence of patches of habitat undergoing succession is an example of the two-dimensional structure found in every ecosystem. A disturbance creates a light gap, which triggers succession.

Patchiness may also result from uneven distribution of heat, moisture, and nutrients within a community and from differences in topography that affect these things. For instance, creeks and dried-up watercourses promote the growth of some plants over others. This is particularly obvious in a desert. Droppings from birds or mammals may contain nutrients that favor the growth of one plant species over another. In northern areas, rattlesnakes, which warm themselves by sunbathing, are much more likely to be found on south-facing than on north-facing slopes.

Patches may consist of clumps of one species of plant, but they can also result from interactions of species. For example, red-cockaded woodpeckers dig nests only in pine trees weakened by fungus attack.

All the patches discussed so far are irregular: a patch occurs wherever biotic and abiotic factors favor the survival of particular species. Less commonly, patches are dispersed in a regular pattern. Territorial birds and mammals tend to be distributed in a regular pattern: you are more likely to find the animal near the center of its territory than anywhere else in the community. Grass plants in dry areas are sometimes spaced so evenly that it is hard to believe they have not been planted. In fact, plants in dry areas are widely spaced because a plant cannot survive unless its roots monopolize the moisture in a large volume of soil. A similar pattern is found where plants produce secondary chemicals that inhibit the growth of other plants nearby. One example is juglone, a compound washed from the leaves of black walnut trees by rain, which inhibits the growth of other plants in the soil beneath the tree.

Community structure changes with time in a rhythmic fashion that reflects the changing position of the sun during the day and with the seasons. For instance, different

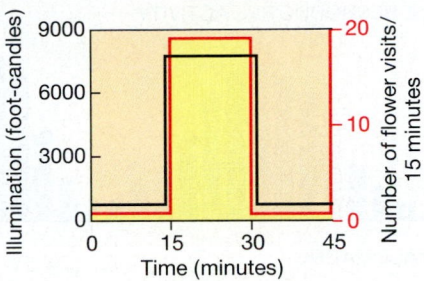

**FIGURE 48–17**   Sunflecks and pollination. A graph of visits by insects (red) to violets in a California redwood forest and changing light intensity (black) as sunflecks pass over the flowers. Visits are restricted to the 15-minute period when a sunfleck passes over the plants.   (After A. J. Beattie, *Madroño* 21:120, 1971)

predators hunt the flying insects in a forest at different times of day: flycatchers and swallows by day, nighthawks at dusk, and bats at night. One study showed that wildflowers on the forest floor in California are visited by pollinating insects during the few minutes each day when sun falls on them (Figure 48–17). Sunflecks make the flowers more conspicuous and raise the temperature, making it advantageous for ectothermic insects to follow sunflecks through the forest. In fact, violets growing in full shade, where sunflecks never reach them, are never visited by pollinators and do not undergo cross-fertilization.

Tide cycles have a similar rhythmic effect on the structures of marine littoral communities. For instance, female fiddler crabs on the sandy shores of mangrove islands in Florida are receptive to mating only a few days before spring tides, which occur twice a month (Figure 48–18). Females release their larvae into the plankton during one spring tide, providing a good chance that the larvae will be carried by the tide to suitable beaches when they develop into little crabs two tide cycles (one month) later.

Seasonal changes are especially marked in communities exposed to long severe winters or dry seasons. Most birds leave northern communities in winter, migrating south to become parts of other communities and returning north in spring. Many mammals hibernate, and insects enter diapause. Annual plants spend the winter as seeds, and deciduous trees and shrubs shed their leaves. Activity in the community is restricted to photosynthesis by hardy evergreen plants and foraging by a few species of hardy endotherms, such as chickadees, woodpeckers, mice, and deer.

### Ecological Niches

When describing how a particular species fits into a community, ecologists often speak of the species' niche. The **niche** of a population or species is its functional role in

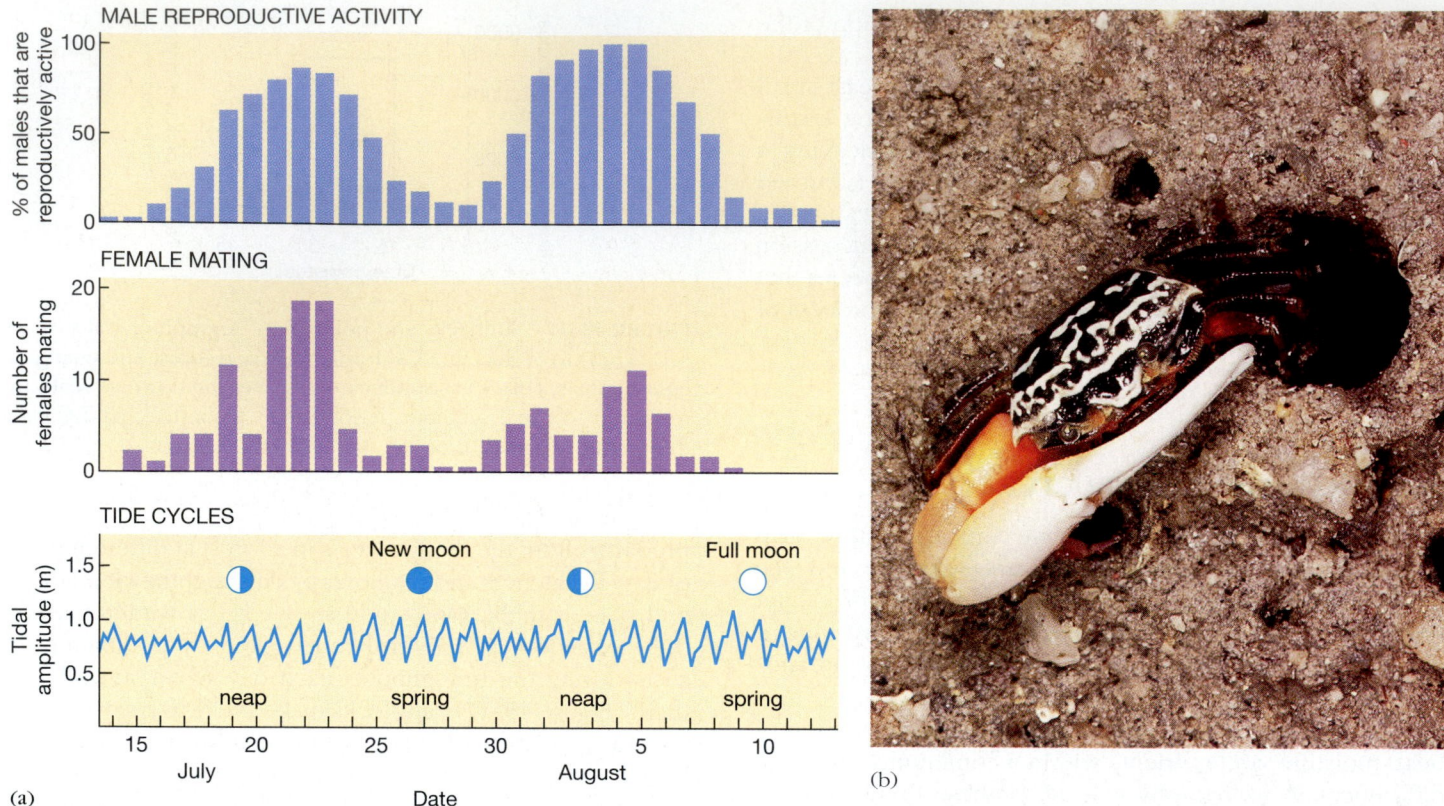

**MALE REPRODUCTIVE ACTIVITY**

**FEMALE MATING**

**TIDE CYCLES**

(a)

(b)

**FIGURE 48–18** Fiddler crabs. (a) The graphs show that reproductive behavior peaks a few days before each spring tide. Reproductive behavior is measured by the proportion of males defending burrows and the number of females entering burrows to mate. (b) A male fiddler crab attracts females by waving his enlarged claw at the entrance to his burrow. (a, after J. Christy, *Science* 199: 453, 1978; b, Robert and Linda Mitchell)

the ecosystem, the sum of all its interactions with biotic and abiotic resources.

To illustrate niche, consider the American alligator. The alligator is a carnivore; to live, it must eat animals, and it eats some species more often than others. As an ectotherm, the alligator is more active at high environmental temperatures. Alligators ensure their supply of fresh water by enlarging mud-lined depressions that fill with rain water, forming "wallows" or "gator holes" (Figure 48–19). The wallows attract frogs, which lay their eggs in fresh water, and herons, which feed on frogs and fish and nest in surrounding trees. Eggs and babies falling out of the trees provide food for the alligators and for several species of snakes. Every interaction of the alligator with its environment is part of its niche and determines where it can live and what other organisms can coexist with it.

G. Evelyn Hutchinson suggested thinking of a niche as a multidimensional space, each dimension being a measure of the species' interaction with one aspect of its environment. The niches of all the species that live together in an ecosystem might then be thought of as balloons of various

shapes and sizes packed into a box, which represents the community.

Consider a study of the blue-gray gnatcatcher, a small bird that forages for insects in oak woodlands in California. Two "dimensions" of the gnatcatcher's niche were measured: the lengths of insects caught by the birds and the heights above the ground at which the birds foraged. When plotted together as axes on a graph, these measurements revealed the birds' foraging activities in two dimensions. The graph showed that insects were most often captured 3 to 5 m above the ground and were more likely to be 4 mm long than any other length. A third dimension could be added to this graph, representing perhaps the distribution of foraging activity over 24 hours, or its variation with temperature or rainfall. This would give a three-dimensional graph. Any organism's niche has many more than three components, which cannot be drawn on a graph although they can be described mathematically.

The **fundamental niche** of a species represents the range of conditions under which it is capable of surviving. In practice, however, various aspects of the environment,

**FIGURE 48–19** An American alligator in its wallow. (Steve Bisson)

particularly competition with other species, often restrict a species to a narrower, **realized niche.** It may occur to you that competition with humans now restricts the niche of nearly every terrestrial species. For instance, the realized niche of alligators in the twentieth century has been increasingly reduced by competition with humans for space and fresh water. Similarly, we noted that the blue-gray gnatcatcher eats mainly insects about 4 mm long. The birds' fundamental niche includes the ability to eat larger insects but in practice, other birds are more efficient at catching these insects, and when these species are present the gnatcatchers' realized niche is restricted.

One advantage of Hutchinson's idea of a niche is that those dimensions of niche that influence community structure can be highlighted, while those that are unimportant can be ignored. For instance, the need for fresh water is one dimension of the niches of all the organisms that live in a lake; but there is so much water in the lake that fresh water never becomes a limiting resource. We must look elsewhere for factors that control the structure of the lake community.

## Keystone Species

Certain species or groups of species are more central than others to a community's organization because their niches in the ecosystem create niches for many other species. **Keystone species** may be defined as species whose loss or removal disrupts the ecosystem most.

Some ecosystems are obviously organized around keystone species. For instance, a salt marsh consists of cordgrass (*Spartina*) and the species that depend on it. If the cordgrass were removed, we should no longer recognize the ecosystem as a salt marsh. Similarly, a few species of deciduous trees provide niches for dozens of species in a decidous forest and are essential to its existence.

In these examples, it is the producers that are essential to the ecosystem's structure. It is less obvious that animals at higher trophic levels (like alligators) can act as keystone species. Experiments by James Brown and Edward Heske on shrubland in the Chihuahua Desert in New Mexico showed that they can. These workers removed various species, such as all ants, all rodents, and various individual species of ants and rodents, from fenced plots. They found that removing most of the species had no noticeable effect. However, removing three species of kangaroo rat (*Dipodomys* spp.) converted the ecosystem from desert shrubland to grassland within 12 years (Figure 48–20).

The changes in the rat-free plots showed that these rodents have two main effects on the ecosystem: they feed primarily on large seeds and they disturb the soil by burrowing and digging for seeds. When the rats were removed, even though other seed-eating rodents remained in the area, tall grasses with large seeds established themselves. These new grass species supported specialized grassland rodents never seen in the area before. One reason Brown and Heske described the kangaroo rats as keystone species is that their removal had greater effects on the ecosystem than the removal of cattle and horses from the area some years earlier or the removal of any of the other (ant and rodent) species they worked with.

It appears that, at least in some cases, the structure of a community is strongly influenced by the presence of one or a few species whose removal alters the ecosystem out of all recognition.

**FIGURE 48–20** A kangaroo rat (*Dipodomys merriami*) in New Mexico. The rat gets its name from its long hind legs, which permit it to jump like a kangaroo. (M. P. L. Fogden/Bruce Coleman)

## SUMMARY

An ecosystem consists of a community of organisms that depend on one another in various ways, plus their physical environment.

1. A sustainable ecosystem must contain water, mineral nutrients, carbon dioxide, oxygen, producers, and decomposers.

2. The productivity of an ecosystem is determined by the temperature and by the availability of light, water, and minerals for photosynthesis. Energy flows one way, from one trophic level to the next, with only about 10% of the free energy passing to each successive level. In consequence, the biomass that an ecosystem can support at each successive trophic level declines rapidly.

3. In contrast to energy, nutrients recycle indefinitely in a natural ecosystem, with small losses into the air and in runoff and sedimentation. Disturbances can increase an ecosystem's loss of nutrients enormously. Nutrients are taken in by organisms as inorganic substances, many of which are incorporated into organic molecules. They may pass through the food web for a time, but eventually they are once again released into the environment as inorganic substances.

4. Carbon is fixed in organic molecules during photosynthesis, and much of it is soon released by respiration. The remainder contributes to the growth and reproduction of the plant and of heterotrophs that consume or decompose the plant. The carbon balance of the biosphere is moderated by the exchange of $CO_2$ between the atmosphere and the oceans.

5. Nitrogen is abundant in the air, but in a form few organisms can use. The scarcity of fixed nitrogen compared with plants' ability to use this vital resource makes nitrogen a limiting nutrient in many ecosystems.

6. Phosphorus moves through ecosystems in a one-way, sedimentary cycle. It is usually in short supply but tends to be retained efficiently.

7. The availability of nutrients often limits the productivity of an ecosystem, even though an ecosystem may be very efficient at conserving and recycling its nutrients.

8. Although the climate of an area determines climax community, patches of the area are always in various stages of ecological succession following disturbances. Organisms adapted to living in the unstable communities of early successional stages have effective dispersal mechanisms and perpetuate themselves by continuously colonizing new disturbed habitats as they arise.

9. Succession occurs because changes occur that make the environment less favorable for the species that are present and more favorable for colonization by others. Productivity and biomass increase during succession, leveling off as the climax is approached.

10. Each ecosystem has a definite structure in time and space, made up of patches of various types. Patches are usually irregular, resulting from disturbances. Occasionally, however, they are regular, as when the sun lights part of a forest at the same time each day or when competition for soil moisture causes plants to be evenly spaced.

11. An organism's niche is its functional role in the community, including how its way of life affects other species.

## Self-Quiz

1. The role of decomposers in an ecosystem is _____ .

2. Using the items listed below, diagram a food web; indicate the trophic level of each organism.

   | | |
   |---|---|
   | deer | herbivorous insect |
   | soil bacteria | spider |
   | shrub | sparrow |
   | wolf | hawk |

3. The annual net primary productivity of any ecosystem is greater than the annual increase in biomass of the herbivores in that ecosystem because:
   a. plants are more efficient than animals in converting energy input to biomass
   b. energy is lost during each energy transformation
   c. there are always more plants than plant-eaters
   d. woody plants live much longer than most herbivores

4. Of the total amount of energy that passes from one trophic level to another in a food chain, about 10% is:
   a. transpired
   b. "burned" in respiration
   c. stored as body tissue
   d. reradiated in the form of heat
   e. passed out in the feces

5. Nutrient cycles may involve:
   a. movement of the nutrient from the organism to the atmosphere
   b. movement of nutrients into the soil
   c. limitations on the number of organisms in the ecosystem due to shortage of some nutrients
   d. loss of the nutrient from the ecosystem
   e. all of the above

6. Wolves and lions may be said to occupy the same trophic level because:
   a. they both eat primary consumers
   b. they both use their food with about 10% efficiency
   c. they both live on land
   d. they are both large mammals
   e. they both eat a wide range of dietary items

In Questions 7 and 8, choose the correct term from each pair in parentheses.

7. Colonization of an abandoned stone quarry would be an example of (primary/secondary) succession.
8. A(n) (early successional/climax) community would have a high proportion of fugitive species.
9. A pond in a deciduous forest becomes filled in with rock particles and dead leaves, creating soil. List, in order, the types of vegetation that would be seen as this area undergoes ecological succession, and name the climax community that would eventually result.

## Questions for Discussion

1. Is more energy lost from an ecosystem when a herbivore eats a plant or when a carnivore eats an animal? Why?
2. How is the flow of energy in an ecosystem linked to the flow of nutrients? How do energy and nutrient flow differ?
3. Why does the productivity of an ecosystem increase during the course of ecological succession?
4. Why does secondary succession slow down as it proceeds?

## Suggested Readings

Brewer, R. *The Science of Ecology,* 2d ed. Philadelphia: Saunders College Publishing, 1993. A general textbook on ecology, including ecosystems.

Cohn, J. P. "Gauging the biological impacts of the greenhouse effect." *BioScience* 39(3):142, 1989. The high points of a World Wildlife Fund conference on the probable effects of global warming on individual plants and animals and the distribution of organisms.

Myers, N., ed. *The Gaia Atlas of Planet Management.* London and Sydney: Pan Books, 1985. Our environmental problems portrayed in the form of maps. Some excellent illustrations and loads of interesting information.

Pimm, S. L., J. H. Lawton, and J. E. Cohen. "Food web patterns and their consequences." *Nature* 350, 25 April 1991. Review of ecological understanding of food webs.

Ricklefs, R. E. *Ecology,* 3d ed. New York: W. H. Freeman and Company, 1990. A good general ecology text.

# *Populations and Communities*

## O B J E C T I V E S

*When you have studied this chapter, you should be able to:*

1. Define population, carrying capacity, and competition.
2. Draw a graph showing exponential growth of a population, and give a reasonable explanation for why the population is growing exponentially.
3. List three factors that affect a population's biotic potential, and state which is most important.
4. Draw and interpret the three main kinds of survivorship curves; relate a species' survivorship to its reproductive strategy.
5. Distinguish between density-dependent and density-independent mortality factors, and give an example of each.
6. Describe an example of successful biological control of a pest species.

7. Summarize your understanding of the factors that limit the sizes of populations and their relative importance.
8. Describe the characteristics of species that are at great risk of extinction, and list some human activities that are important causes of extinction of species.
9. Explain how the species diversity of a community is affected by competition, predation, physical structure, and the number and types of habitats it contains.
10. Define species turnover, and explain how species diversity may be affected by a community's size and degree of isolation from similar communities.
11. Explain why studies of island biogeography play an important role in designing nature reserves.

---

**A** **population** consists of all the members of a species occupying an area at the same time. Examples are the bass population of a lake and the human population of the United States or of Earth.

A population has characteristics not found in its individual members. For example, each population has a gene pool and a certain sex ratio, pattern of distribution, density, and age structure. Population **density** is the number of individuals per unit area or volume—the number of dandelions per hectare, for example (Figure 49–1). The age structure of a population is the percentage of individuals of each age. These and other features can be used to describe populations and to predict their fates.

A population does not live in isolation, but in a **community**, the collection of populations of various species of producers, consumers, and decomposers that make up the biotic part of an ecosystem. The members of a community interact in many ways, for instance, by competing for resources and eating each other. These interactions are among the factors that limit the sizes of populations.

**FIGURE 49–1** Populations of fir trees, dwarf willow, and daisies in the Teton Mountains.

In this chapter we consider how populations grow in size and why they cannot continue to grow indefinitely, the factors that limit population size, and why some species are more likely than others to become extinct. Then we consider what determines the diversity of species in a community and why the species in an area change as time goes by.

## KEY CONCEPTS

- Although a population invading a new area may grow exponentially, a population in nature seldom grows as fast as its biotic potential would allow but remains at some fairly constant size, determined by factors including the supply of resources, competition for those resources, and predation.

- Age at first reproduction is the most important factor determining a species' biotic potential.

- Specialized predators may hold the population size of their prey species at very low levels. Generalized predators seldom do so because they switch from one prey species to another as the abundance of prey species changes.

- Competition among members of the same or different species for a limiting resource acts as a density-dependent control of population size.

- Human activities have greatly increased the rate of extinction of species, mainly by destruction of habitat and through competition or predation by species imported by humans.

- The species present in a given area change over time, but the number of species remains relatively constant and is related to the area of habitat studied.

## POPULATIONS

### 49–A  POPULATION GROWTH

A population gains individuals by birth and immigration and loses them by death and emigration. Whether a population grows, shrinks, or remains the same size depends upon the balance among these factors:

Change in population size =
(Births + Immigration) − (Deaths + Emigration)

The size of a population of mice in a field or of violets in a woodlot seems, at first sight, to vary little from year to year. Is this really the case? Surely organisms produce so many offspring that their populations could increase greatly from one year to the next (Figure 49–2). What limits the size of natural populations?

To answer these questions, it is convenient first to find out how rapidly populations could increase if nothing stopped their growth. (The mathematics of population growth is described in *In More Detail: Changes in Population Size.*)

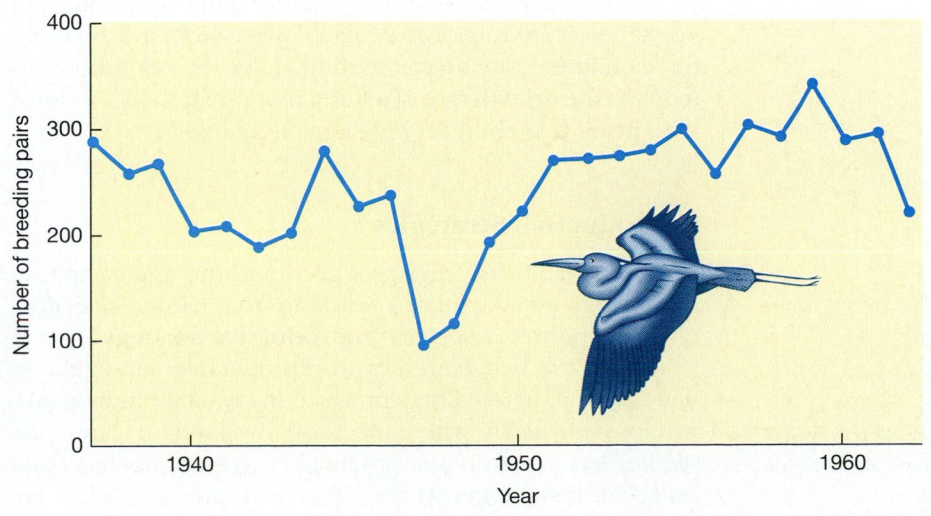

**FIGURE 49–2** A stable population. The number of breeding pairs of gray herons in part of northwest England. The population recovered rapidly from the severe winter of 1947. Fluctuations are small compared with the heron's biotic potential (three new birds per breeding pair per year). (After Lack, D. *Population Studies of Birds.* New York: Clarendon Press, 1966)

A population's **biotic potential** (or **intrinsic rate of increase**) is its fastest possible rate of growth under ideal conditions. In nature, conditions are seldom ideal. Predators, disease, or food shortages nearly always prevent a population from growing as fast as its biotic potential would permit. However, there are times when populations grow very fast, when they have abundant resources and no enemies or competitors. For example, population explosions occur when bacteria invade the intestinal tract of a newborn animal or when decomposers invade a freshly dead animal or plant.

Russian ecologist G. F. Gause studied the growth of populations of the protist *Paramecium caudatum*. Every few hours a well-nourished *Paramecium* divides into two new individuals. Gause set up tubes containing plenty of bacteria for food and introduced one *Paramecium* into each. He observed that the population of *Paramecium* in each tube showed **exponential growth,** that is, growth in which an ever-increasing number of new individuals is added in each unit of time. Exponential growth, plotted as a graph, produces a J-shaped curve (Figure 49–3).

Exponential growth does not necessarily mean that the population is growing at its biotic potential, or even particularly rapidly. The human population started growing exponentially in the mid-eighteenth century, although women were not bearing infants as frequently as is biologically possible. In addition, many people who could have reproduced did not.

Biotic potential differs from one species to another. An individual's contribution to population growth can be increased in any or all of three ways:

1. By producing more offspring at a time (that is, a larger brood or litter size)
2. By having a longer reproductive life, so that it reproduces more times
3. By reproducing earlier in life

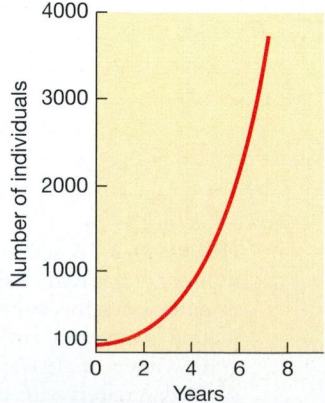

**FIGURE 49–3** Exponential growth of a population. The population grows by adding an ever-increasing number of individuals per unit time.

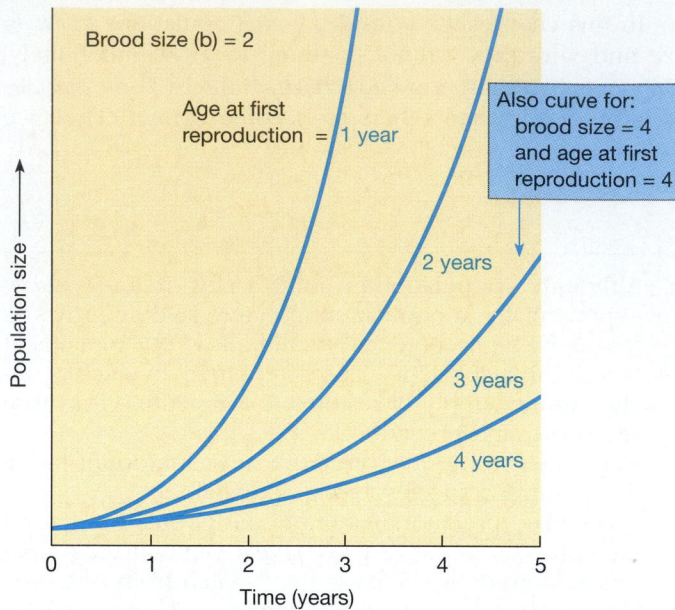

**FIGURE 49–4** How (female) age at first reproduction affects population growth. In all these curves, females produce two offspring each year, but the age at which females first reproduce differs in each curve (first reproduction at 1, 2, 3, or 4 years of age). The boxed note indicates that lowering the age at first reproduction from 4 to 3 (one year younger) has the same effect as leaving the age at 4 but doubling the brood size from 2 to 4. (After Cole)

Of these three factors, the last is by far the most important. A bacterium does not live for long, and it reproduces only once, giving rise to two offspring. Nevertheless, a population of bacteria has a higher biotic potential than does a population of dogs. This is because most bacteria can reproduce within an hour after being formed, whereas a dog cannot reproduce until it is about six months old. The shorter the generation time of a species (that is, the younger its members when they first reproduce), the higher the species' biotic potential (Figure 49–4).

Lamont Cole, who drew attention to the significance of age at first reproduction, calculated that a woman who bears three children, one a year starting at age 13, contributes as much to the growth rate of a human population as a woman who bears five children but starts at age 30.

## Reproductive Strategies

The number of offspring produced and the age at first reproduction are adaptations resulting from natural selection. They are part of a species' **reproductive strategy.**

There are two extremes in reproductive strategies. At one extreme, parents may produce many small individuals, and provide each with little food or parental care. The species has a high biotic potential ($r_{max}$), so this has been called an **r-strategy.** At the other extreme, a species' life cycle may emphasize competitive ability over numbers by producing few offspring but investing a lot of energy in each one, sometimes called a **K-strategy.**

The two strategies are illustrated by species found during different stages of ecological succession (Section 48–F). The r-strategy plants of early succession exploit habitat that is short-lived and unpredictable. Examples are weeds such as dandelions and burdock, which have high biotic potentials and reproduce early, sometimes after only a few weeks of growth. They produce large numbers of small seeds, most of which never reach sites where they can germinate. Those that do, grow rapidly when they have no competition for sunlight and nutrients, as gardeners and farmers are well aware. However, weed plants are poor competitors and this is the main reason they disappear as succession progresses. They generally cannot survive in the shade of other plants or in soil from which they themselves have depleted the nutrients. Many r-strategy animals are also found in early successional communities.

In the late stages of succession, r-strategy species are replaced by K-strategy ones. Weedy herbs are replaced by efficient competitors with low biotic potential. Many trees and shrubs do not start reproducing until they are many years old, but this is compensated by a long potential lifespan. The seedlings can tolerate shade, and they extract nutrients efficiently even from poor soil.

The contrast between r and K strategies is easily seen during succession, but most species have reproductive strategies that fall somewhere on the continuum between

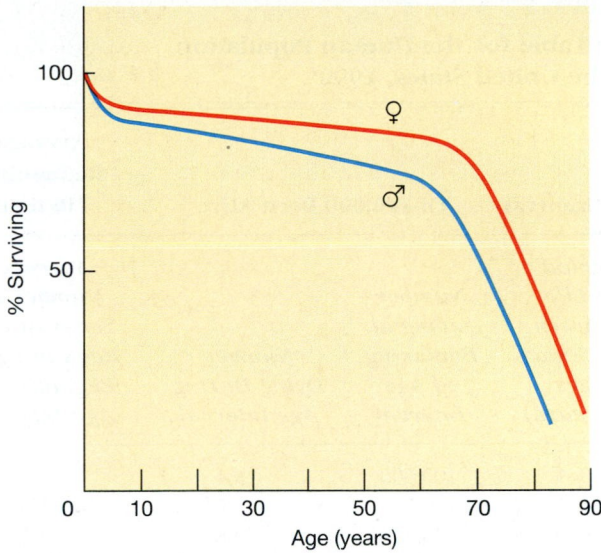

FIGURE 49–6 Survivorship curves for Americans, based on the 1990 census data. Human survivorship approximates a Type I curve.

these two extremes. Indeed, some species have different strategies under different conditions. In one study, dandelions in a trampled area (early successional conditions) were found to produce more, smaller seeds than dandelions in an undisturbed area (comparable to a later successional stage).

Reproductive strategies may be thought of partly as responses to different patterns of **survivorship,** the length of time an individual of a particular age can expect to survive. This can be illustrated by a graph (Figure 49–5). In Type I survivorship, most individuals are strong competitors, with defenses against disease and predators. They tend to live for a long time and die as a result of diseases that develop mainly later in life, after fertility has declined. This is typical of species that lavish parental care on a small number of offspring and have low biotic potentials because the age at first reproduction is relatively high. Most human populations in developed nations approach a Type I survivorship curve after the first year of life (when there is a high death rate from genetic or developmental defects or birth accidents). A baby who survives the first year of life is likely to live for another 60-plus years (Figure 49–6).

In Type III survivorship, most individuals die young, as in many species of invertebrates, bony fishes, weedy plants, and fungi. Biotic potential is high because the age at first reproduction is relatively low, but most individuals fail to hatch or germinate, or else they die from disease or predation before they reproduce.

The Type II curve falls between Types I and III. The number of offspring produced and the age of first reproduction also tend to be intermediate between those of the other two types. There is again an initial period of high mortality (death rate). However, once past this critical pe-

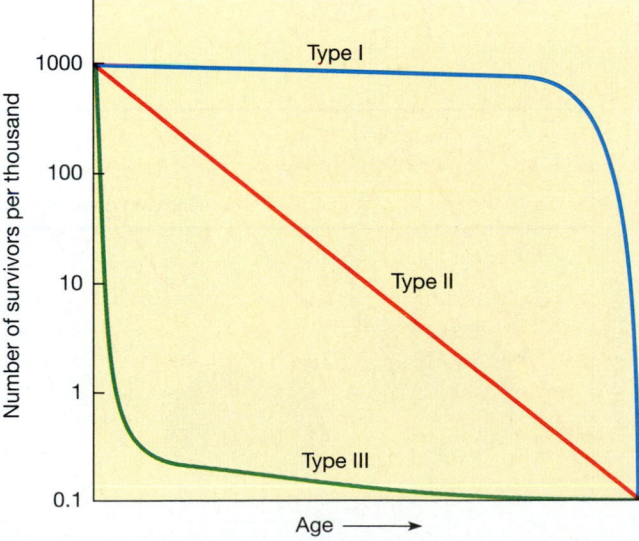

FIGURE 49–5 Survivorship curves. These graphs show the main types of hypothetical survivorship curves (note the logarithmic scale of the vertical axis). Curves for real populations may fall between those shown here. Type I and Type II curves are rarely, if ever, exactly as shown, because there is always an especially high rate of death among very young individuals.

## Life Table for the Human Population of the United States, 1990*

| Age Interval | Of 100,000 Born Alive | | Average Remaining Lifetime |
| --- | --- | --- | --- |
| Period of Life Between Two Exact Ages (in Years) | Number Living at Beginning of Age Interval | Number Dying During Age Interval | Average Number of Years of Life Remaining at Beginning of Age Interval |
| 0–1 | 100,000 | 981 | 75.3 |
| 1–5 | 99,019 | 196 | 75.0 |
| 5–10 | 98,823 | 88 | 71.1 |
| 10–15 | 98,735 | 132 | 66.2 |
| 15–20 | 98,603 | 491 | 61.3 |
| 20–25 | 98,112 | 550 | 56.6 |
| 25–30 | 97,564 | 605 | 51.9 |
| 30–35 | 96,959 | 736 | 47.2 |
| 35–40 | 96,223 | 935 | 42.5 |
| 40–45 | 95,288 | 1,219 | 37.9 |
| 45–50 | 94,069 | 1,764 | 33.4 |
| 50–55 | 92,305 | 2,724 | 28.9 |
| 55–60 | 89,581 | 4,229 | 24.7 |
| 60–65 | 85,352 | 6,210 | 20.8 |
| 65–70 | 79,142 | 8,706 | 17.2 |
| 70–75 | 70,436 | 11,622 | 13.9 |
| 75–80 | 58,814 | 14,704 | 10.9 |
| 80–85 | 41,110 | 14,879 | 9.3 |
| 85 and over | 26,231 | 26,231 | 8.2 |

*Data courtesy of the National Center for Health Statistics

riod, the chances of dying or being killed are equal throughout life. This type of curve is typical of several bird species and of human populations with poor nutrition and hygiene.

Survivorship within a population can also be represented in the form of a **life table,** a summary of the likelihood of death in groups of individuals of each age (Table 49–1). Life tables for human populations are used by life insurance companies to predict how much longer people of a given age are likely to live. This determines the price of insurance for people of various ages. Life tables for populations of animals and plants are useful aids for summarizing and analyzing the effects of different causes of death acting on populations (illustrated in Section 49–E).

## Carrying Capacity

No population can grow exponentially for long. Gause found that his *Paramecium* populations eventually stopped growing. They had reached their environment's **carrying**

**capacity,** the number of individuals that this particular environment could support indefinitely. When a population reaches this size it may stabilize, with fluctuations above and below the carrying capacity (Figure 49–7).

Carrying capacity is determined by many factors, including predation, competition for available resources, and climate. The factors that limit a population's growth may change, and so the carrying capacity of an area for a population of a given species may also change with time.

## 49–B  REGULATION OF POPULATION SIZE

In spite of fluctuations, the average size of most populations changes relatively little over the years. This suggests that population sizes are usually regulated in such a way that small populations grow quickly, larger populations grow more slowly, and still larger populations decline.

One reason for this is that at least some of the factors that affect mortality are **density-dependent.** That is, as the population density increases, these factors kill a larger proportion (not just a larger number) of individuals (Figure 49–8). Predation and disease are density-dependent factors, partly because a disease-causing organism is more likely to encounter a host, or a predator its prey, when there are more hosts or prey in the area.

Other mortality factors are **density-independent,** killing a proportion of the population no matter what its density. For example, in several insect species, harsh winter weather kills about 90% of a population regardless of its density. However, bad weather sometimes causes death in a density-dependent manner: if it is possible to survive

**FIGURE 49–7**  Carrying capacity. The red line shows how the size of the rabbit population changed after rabbits were introduced into Australia from Europe. At first, the number of rabbits grew exponentially. Then the population crashed. Now, the population size oscillates around a value that represents the carrying capacity of the environment for rabbits.

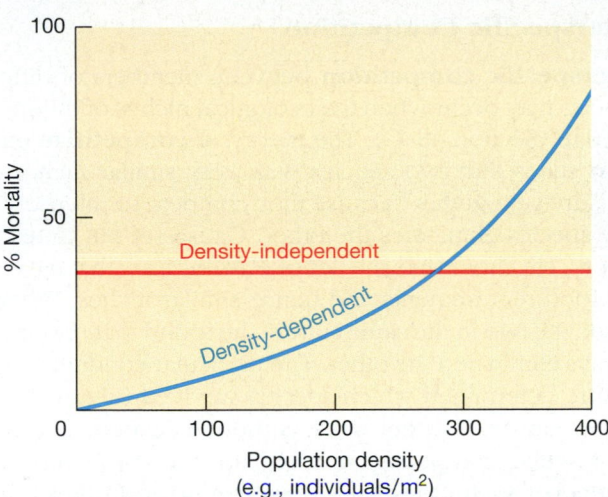

**FIGURE 49–8** Effects of density-dependent and density-independent mortality on population size. These hypothetical curves show that density-dependent mortality kills an increasing proportion of the population as population density increases. Density-independent mortality kills the same proportion of the population, regardless of population density (though of course at higher densities larger numbers of individuals are killed).

by finding shelter, and if the number of shelters is limited, then all the members of a sparse population may survive, whereas only a fraction of a denser population will be protected.

## 49–C COMPETITION

Competition is a density-dependent factor that contributes to limiting the size of many populations. Competition occurs when two or more organisms attempt to exploit a limited resource, such as food or space.

### Intraspecific Competition

**Intraspecific competition,** competition between individuals of the same species, is very common. Generally, members of the same species need the same resources, and so they usually compete for them.

In one experiment, seeds of white clover were planted at three different densities. Half the plants at each density were watered throughout the experiment, but the other half were watered only for the first 18 days. After seven weeks, the densities of the surviving seedlings were measured. Among the seedlings that were watered regularly, mortality was low regardless of density. Among the seedlings deprived of water, however, only one third as many seedlings survived in the high-density plots as in the low-density plots, evidence of density-dependent mortality from competition when a resource (water) was scarce (Figure 49–9).

Key: ☐ Watered ☐ Not watered

**FIGURE 49–9** The effect of seed density on survival of white clover seedlings subjected to the stress of water shortage. When water was available, survival was about the same at all population densities (blue bars). Orange bars represent the survival of seedlings that were not watered after the eighteenth day. Many fewer of these seedlings survived in the dense population. (After J. L. Harper, *Society for Experimental Biology Symposium* 15:1, 1961, Cambridge University Press)

In this experiment, each individual uses the limited resource independently. Such **scramble competition** is common among organisms without social behavior and those whose resources appear in short-lived patches. For instance, it occurs among maggots of blowflies that live on the carcasses of dead mammals. The first female flies that find the carcass lay their eggs, and most of their larvae will have enough food to reach pupation. As population density increases, however, there comes a time when the food runs out, and most of the remaining maggots die.

Instead of scramble competition, many vertebrates and some invertebrates engage in **contest competition** for social dominance or a territory. In contrast with scramble competition, the winners in this case acquire an adequate supply of a limited resource, whereas the losers may end up with nothing. Paul Errington studied an area of marsh in Iowa for 25 years. During this time, the marsh always contained about 400 adult muskrats. This remarkable population stability was found to be the result of the density-dependent effect of intraspecific competition. Males compete for territories, each of which contains food and a refuge from predators for a pair of muskrats and their young. Muskrats that were unsuccessful in the annual competition for territories were forced to live in unfavorable areas at the edge of the marsh, where they and their offspring suffered a high death rate from overcrowding, predation, inadequate food, and interference by other animals.

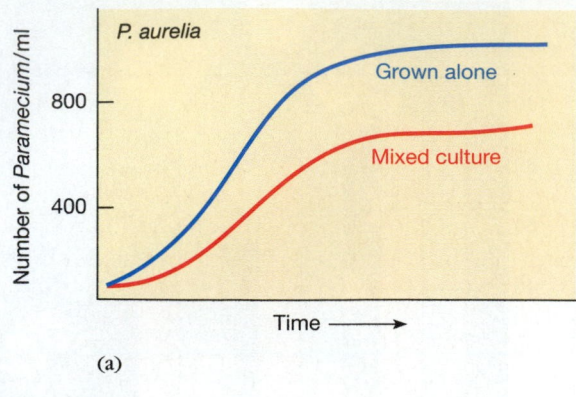

(a)

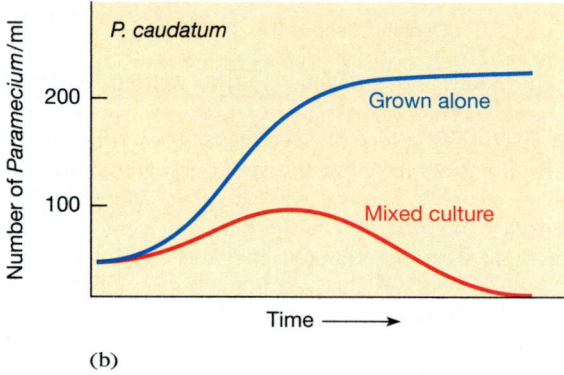

(b)

**FIGURE 49-10** Gause's experiment with two species of *Paramecium*. Both species grew well by themselves in culture tubes with daily changes of water and inputs of bacteria for food. When placed together under these conditions, *P. aurelia* (a) survived, whereas *P. caudatum* (b), the larger and slower-growing of the two species, always declined to extinction. If the water was not changed, waste could build up, and the competitive outcome was invariably reversed.

## Interspecific Competition

**Interspecific competition** between members of different species may occur when the ecological niches of the species overlap (Section 48–G). The theory of **competitive exclusion** states that two species with very similar niches cannot survive together because they compete so intensely that one species eliminates the other. Gause set out to test this theory. He chose two species of *Paramecium* that have similar food requirements and hence similar niches. When he raised the two in the same culture, he found that one species always eliminated the other. The particular conditions in the culture determined which species survived (Figure 49–10).

In nature, it often looks as though several species coexist while competing strongly for the same resources. Whenever such cases have been examined in detail, however, it has always been found that the species divide the resources in some way. For instance, Robert MacArthur found that different species of wood warblers in northeastern forests forage for insects in different parts of the trees, reducing the competition among them (Figure 49–11).

Interspecific competition can restrict the abundance and distribution of populations. An example comes from a study of populations of two species of barnacles on the rocky coast of western Scotland, *Balanus* and *Chthamalus* (don't pronounce the "Ch"). *Chthamalus* occupies the upper part of the intertidal zone and *Balanus* a lower zone, with little overlap between them. The planktonic larvae of both species settle on rocks in both zones. However, *Balanus* cannot survive in the upper zone because it is less tolerant than *Chthamalus* of exposure to the air at low tide. By removing *Balanus* larvae as they settled in the lower zone, Joseph Connell showed that *Chthamalus* survives in the lower zone when *Balanus* is absent. *Balanus* grows faster

Blackburnian warbler          Bay-breasted warbler          Myrtle warbler

**FIGURE 49-11** Coexistence by avoiding competition. Several species of warbler of the genus *Dendroica* hunt for insects in coniferous trees in the same New England forests. Each usually forages in a different part of the tree, thus reducing the competition for food.

# CHANGES IN POPULATION SIZE

## Exponential Growth

When a population is growing exponentially, its growth in any unit of time is equal to the size of the population at the beginning of the time interval multiplied by the population's growth rate. When the population's size is graphed on a linear axis, exponential growth plots as a curve that grows steeper and steeper:

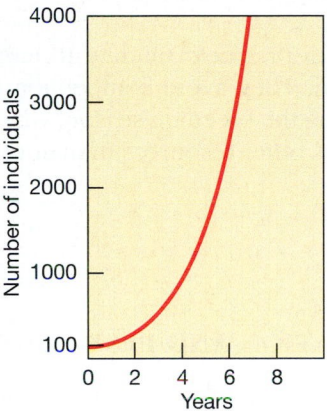

When population size is plotted on a logarithmic axis, exponential growth plots as a straight line:

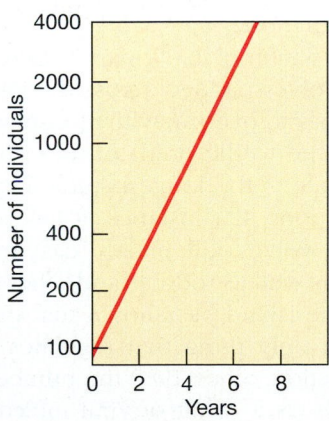

The equation describing this growth is

$$\text{Rate of increase} \left( \frac{dN}{dt} \right) = rN$$

where N is the number of individuals in the population at the start of the period, t is the time period, and $r$ is the rate of change of population size per unit time. Assuming no emigration or immigration, $r = B - D$ where B is the aver-age birth rate per individual and D is the average death rate per individual. For instance, most human populations increase at a rate of 3% or less per year ($r = 0.03$ per year).

Under ideal conditions, the population grows at its biotic potential, and the value of $r$ is at its maximum for the species, usually written $r_{max}$.

## Carrying Capacity

Populations can continue to grow exponentially only until they reach the environment's carrying capacity (K), when birth rate (B) and death rate (D) are equal, whether these rates are density-dependent or density-independent.

In an experiment, Gause grew *Paramecium aurelia* and fed them by adding the same weight of bacteria to the culture every day. At first, the number of individuals in the culture grew exponentially. Eventually, however, the constant weight of bacteria added each day was insufficient to feed the growing number of *Paramecium* and the growth rate fell, eventually reaching zero. The population size leveled out as its numbers (N) approached the carrying capacity. The graph depicting this experiment looks like this:

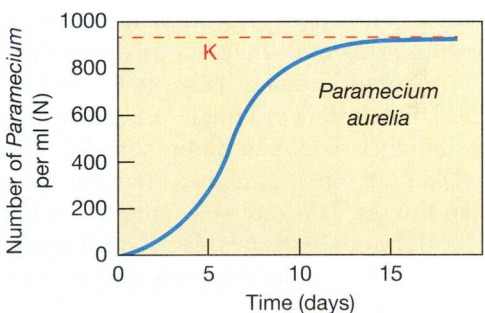

The data Gause obtained are described by the **logistic equation** for population growth, which is an extension of the equation given above for exponential growth:

$$\frac{dN}{dt} = rN \frac{(K - N)}{K}$$

where $\frac{K - N}{K}$ indicates how much of the resource that limits growth (in this case, food) is still available to the population. When N is much less than K, this term approximates 1, and the equation becomes $\frac{dN}{dt} = rN$ (the equation describing exponential growth). As N grows until it is almost equal to K, the value of the term $\frac{(K - N)}{K}$ approaches zero and $\frac{dN}{dt}$ (that is, growth rate) also becomes zero. The carrying capacity of the environment for the population has now been reached.

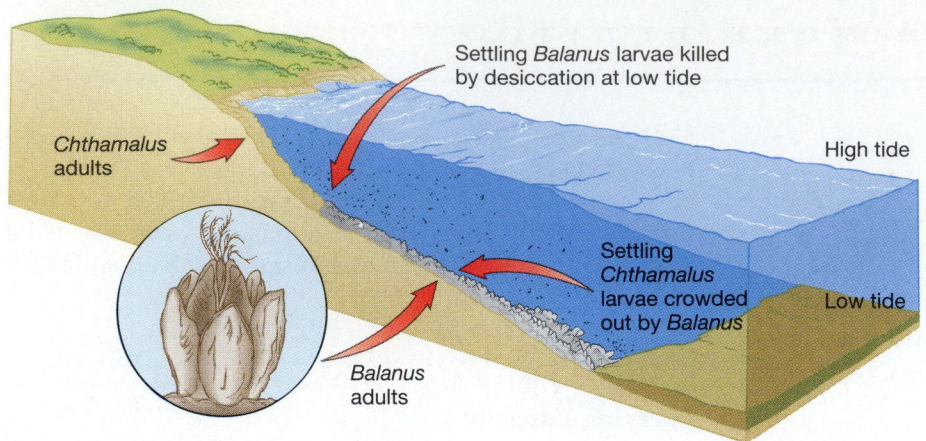

Settling *Balanus* larvae killed by desiccation at low tide

*Chthamalus* adults

High tide

Settling *Chthamalus* larvae crowded out by *Balanus*

Low tide

*Balanus* adults

**FIGURE 49–12** Division of space between two species of barnacles. On the rocky coast of Scotland, the vertical distribution of *Chthamalus* barnacles is limited by competitive exclusion. *Chthamalus* larvae settling from the plankton are crowded out by growing *Balanus* barnacles. The upper limit of *Balanus* is determined by its lesser tolerance of exposure at low tide.

than *Chthamalus,* and so when the species compete for space on the rocks, *Balanus* grows over *Chthamalus* and crowds it out. Thus, the lower limit of the zone occupied by *Chthamalus* is determined by whether *Balanus* is present (Figure 49–12). Competition of this sort, which restricts the area where a species can live, and hence limits the size of a population, is probably common in nature.

## Competition as a Selective Force

In addition to regulating population size, competition acts as a selective force, favoring those individuals that can avoid the mortality often caused by competition. Thus, intraspecific competition for food may act as disruptive selection, resulting in different feeding habits for different members of a species. For instance, female and male blackcapped chickadees hunt for insects in different parts of trees: males on the trunk and inner branches, females on the outer branches and twigs. This reduces competition between the sexes for food. In other species, larvae and adults have different feeding habits, reducing competition between individuals of different ages.

Interspecific competition leads, in ecological time, either to the competitive exclusion of one species by the other or to subdivision of the resource between the two. In evolutionary terms, such competition exerts directional selection for individuals that can avoid the shared resource. One result may be **character displacement,** greater difference between two species in areas where they coexist than in other areas. For instance, two species of nuthatches have ranges that overlap only in Iran. Elsewhere, the beaks of the two species are of similar sizes and they eat a similar range of foods. In Iran, however, the beak of one species is smaller and the beak of the other species is larger than elsewhere: the two species have diverged, adapting to different prey, and thus reducing competition, in the area where they live together.

The evolution of migration can also be viewed as a result of competition. Many bird and mammal species that

evolved in the tropics now migrate to temperate or arctic biomes to breed. They avoid competition for resources in the tropics during the breeding season, when they use more food, space, and other resources than at other times of the year.

## 49–D PREDATION AND PEST CONTROL

Predation causes considerable mortality in most species and is therefore an important factor in regulating population size. In this context, we can define predation broadly, as any case in which individuals of one species exploit a living prey species for food. By this definition, predators include herbivores and parasites, and prey includes plants and the parasites' hosts.

In the hard winter of 1949, part of Lake Superior froze, and a pair of wolves padded across the lake to Michigan's Isle Royale, an island overrun with moose, which had swum from the mainland to the predator-free island in the early 1900s. Isle Royale is the longest-studied system of predator-prey interactions. For instance, it has produced the information that wolves kill mainly easy-to-catch old and young moose, as well as rodents and other small mammals.

Work on the island has shown that the availability of prey is not the only thing that regulates the size of the predator population. Since 1980 the number of wolves has declined, largely as a result of viral infections, despite an abundance of moose (Figure 49–13). The moose, on the other hand, are more affected by the number of wolves because wolves are a major cause of moose mortality. Thus, when the wolf population declined after 1965 and 1980, the moose population grew. However, the maximum possible moose population on the island is apparently not determined by the number of wolves but by a limited supply of the essential mineral nutrient sodium. In this case, a generalized predator is only one of several factors that affect the size of a prey population.

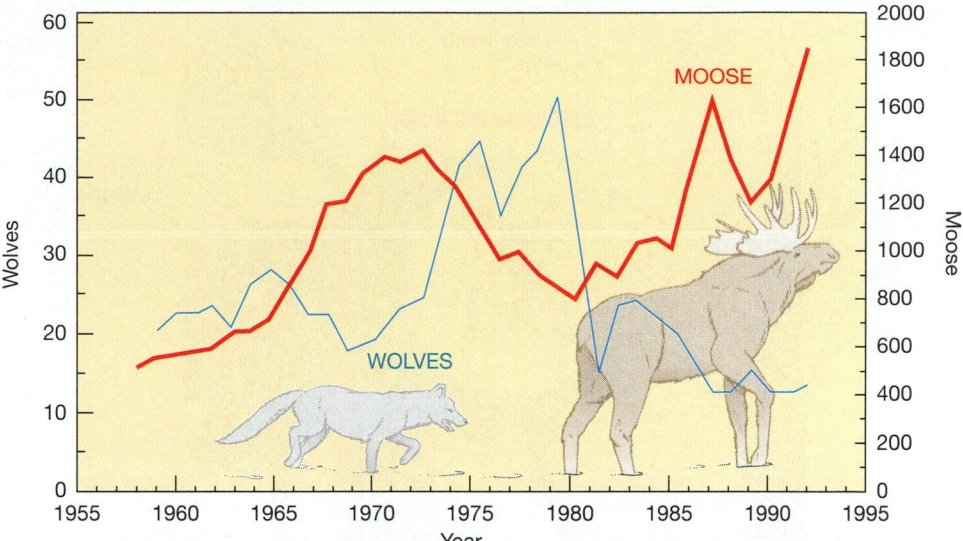

**FIGURE 49–13** Predator and prey populations on Isle Royale. In a 30-year study, when the wolf population declined, the moose population increased.

Generalized predators can usually switch to another species if one prey species becomes rare. Thus, they tend to exert density-dependent control over prey populations.

In contrast, predators that specialize in eating only one species sometimes almost wipe out populations of their prey. For instance, in California a small beetle controls the numbers of yellow-flowered Klamath weed. This plant, introduced from Europe, is an aggressive competitor that grew exponentially and spread throughout the United States, displacing native plants from grazing land. By the late 1940s it covered 80% of the ground in parts of California and Oregon. Because Klamath weed is toxic to cattle, this land became useless for cattle ranching (Figure 49–14).

In its European home, Klamath weed is attacked by several insects, including two species of beetles. Supplies of these beetles were released in California in 1945 and 1946. Their populations grew exponentially. By 1959 Kla-

math weed had been reduced to less than 1% of its former abundance, chiefly due to the voracious appetites of the beetle larvae. Because the beetles are specialists, they cannot switch to other food plants when the Klamath weed population declines. (If they could, they might become pests themselves.) Both Klamath weed and its beetle predators persist in the United States, but at low densities. This is an example of **biological control** of a pest species by a natural enemy.

Today, the search is on for specialist predators such as fungi and insects that will control imported pests. It takes years of experiments to ensure that an imported predator will not become a pest itself if it is released. For example, pastures in the Northwest are studded with yellow-flowered star thistle, which probably reached the area with imported alfalfa seeds. Horses that eat star thistle eventually become paralyzed and starve to death because the brain's dopamine-

(a)

(b)

(c)

**FIGURE 49–14** Biological control. (a) A field overgrown with a large population of yellow-flowered Klamath weed in California, June 1948. (b) The same field in June 1949, a year after the introduction of beetles adapted to feeding on Klamath weed. (c) *Chrysolina* larva eating Klamath weed. (a, J. K. Holloway; b, J. Hamai; c, F. E. Skinner)

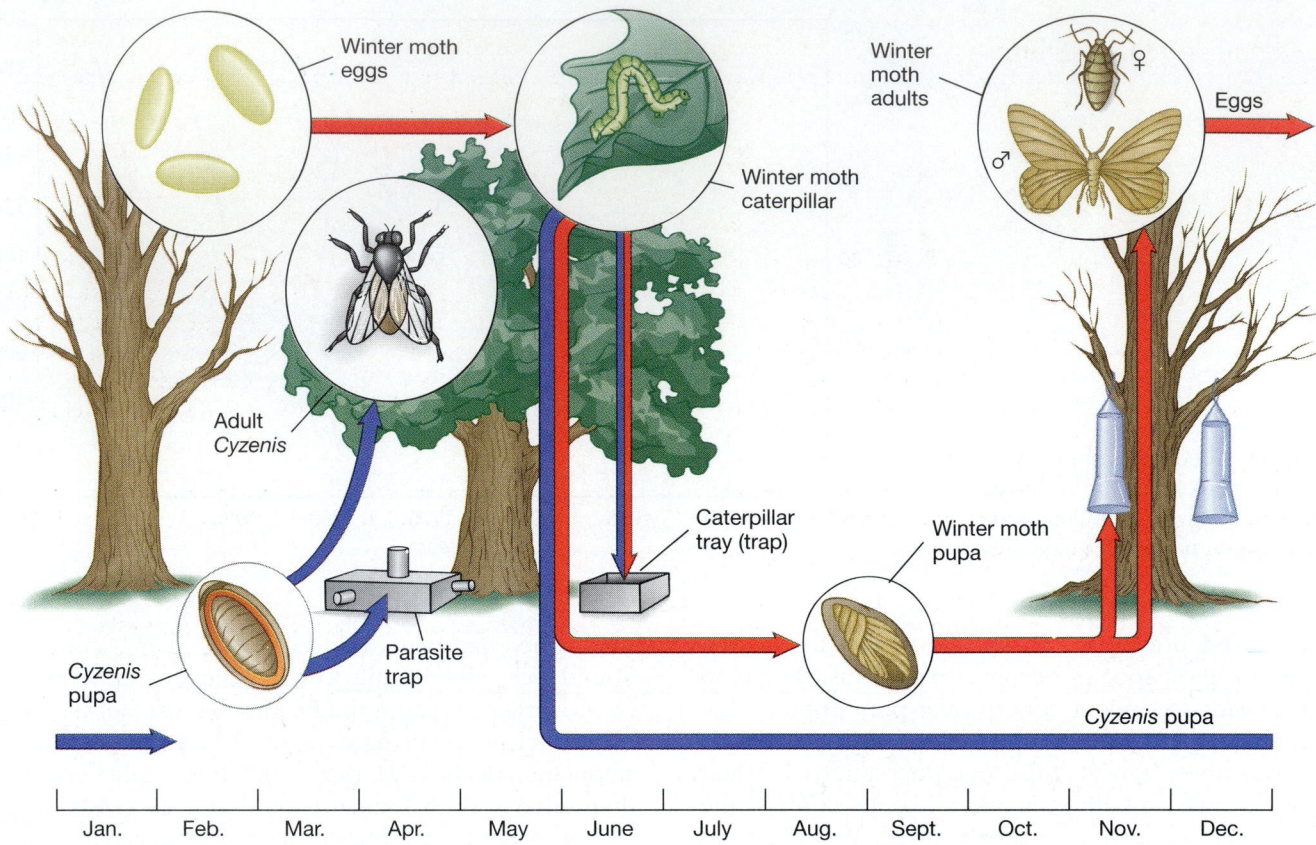

**FIGURE 49–15** The life history of the winter moth (colored arrows) and *Cyzenis,* one of its parasites. The diagram also shows stages in the life history at which the animals are trapped so that they can be counted. (After Varley, Gradwell, and Hassell, 1973)

secreting cells have been destroyed. The search for a natural predator of star thistle began in the 1950s. After 30 years of testing in Italy, Turkey, and Greece, scientists were convinced that the weevil *Bangasternis orientalis* was a suitable candidate for release as a biological control agent because it is amazingly specialized in its food plant. This weevil eats nothing but two species of star thistle. In years of tests it showed no signs of evolving the enzymes that would permit it to feed on ornamental or crop plants.

## 49–E WINTER MOTH POPULATIONS: A CASE HISTORY

One of the best-understood examples of population regulation in nature is that of the winter moth, a small brown moth whose green "inchworm" caterpillars feed on oak leaves in Britain each spring. A 20-year study by George Varley and George Gradwell showed that changes in population size from year to year are caused mainly by weather (density-independent), whereas the long-term stability of population size is regulated by density-dependent predation and parasitism.

Female winter moths are wingless. They emerge in autumn from pupae in the soil litter and climb the trunk of the nearest tree. On the way up, they mate with winged males that can fly. Then each female lays about 150 eggs on twigs in the canopy. The following spring, the eggs hatch as the oak leaf buds open, and the caterpillars feed on the growing leaves. When the caterpillars are full-grown, they spin down to the ground on silk threads, crawl into the litter and remain there, as pupae, until late fall when they emerge as adults.

The young leaves of oak trees provide food for the caterpillars of many species of moths. Varley and Gradwell found that the lower the abundance of any species in one year, the more likely it was to increase in numbers the following year. This suggested that the sizes of moth populations were regulated by density-dependent factors. Two parasitic insects kill a number of winter moths each year. The first, *Cyzenis,* is a fly. In spring, each female *Cyzenis* lays an egg in every winter moth caterpillar she finds. When the winter moth has pupated, the *Cyzenis* egg hatches, eats the winter moth pupa and itself pupates, emerging the following spring as another *Cyzenis* fly (Figure 49–15). The other parasite is a wasp, *Cratichneumon,* whose females lay an egg in each winter moth pupa they discover. The egg, in

**T A B L E   4 9 – 2**

**Life Table for the Winter Moth***

| Life Cycle Stage | Number Starting Stage (per m²) | Mortality Factor | Number Killed (per m²) | Number Surviving (per m²) | m Value |
|---|---|---|---|---|---|
| **Adult**† (1st year) | 4.39 | | | | |
| **Egg** | 658 | Winter disappearance | 561.6 | 96.4 | $m_1 = \log(658) - \log(96.4) = 0.84$ |
| **Caterpillar** | 96.4 | *Cyzenis* parasite | 6.2 | 90.2 | $m_2 = \log(96.4) - \log(90.2) = 0.03$ |
| | 90.2 | Other parasites | 2.6 | 87.6 | $m_3 = \log(90.2) - \log(87.6) = 0.01$ |
| | 87.6 | Protozoan disease | 4.6 | 83.0 | $m_4 = \log(87.6) - \log(83.0) = 0.02$ |
| **Pupa** | 83.0 | Predators (shrews, etc.) | 54.6 | 28.4 | $m_5 = \log(83.0) - \log(28.4) = 0.47$ |
| | 28.4 | *Cratichneumon* parasite | 13.4 | 15.0 | $m_6 = \log(28.4) - \log(15.0) = 0.27$ |
| **Adult**† (2nd year) | 7.5 | | | | |

*Modified from G. C. Varley, G. R. Gradwell, and M. P. Hassell. *Insect Population Ecology*, Oxford, England: Blackwell Scientific Publications, 1973.

†Number of females climbing trees

turn, hatches into a larva that consumes the winter moth pupa, and then pupates, later to emerge as an adult *Cratichneumon.*

Varley and Gradwell made a census of the populations of the winter moth and of its major parasites every year. Traps of sticky paper were used to estimate the numbers of female moths climbing the trunks each fall. Several females were dissected to estimate the average number of eggs in each. The number of caterpillars feeding each spring was counted from oak leaf samples, and metal trays on the ground were used to intercept larvae dropping to the ground before pupating. Some larvae were dissected to find what

proportion contained *Cyzenis* eggs. Traps placed over the ground captured emerging *Cratichneumon* and *Cyzenis* parasites, as well as winter moth adults. From these data, Varley and Gradwell were able to prepare a life table for the winter moth, describing numerically the fate of the initial number of eggs as the season progressed.

The life table data were used for **key factor analysis.** For each source of mortality, Varley and Gradwell assigned an "m value," a measure of the "killing power" of that mortality factor (Table 49–2). The m value is calculated as the logarithm of the number of winter moths present before the particular mortality, less the logarithm of the number of survivors:

$$m = \log \text{(initial number)} - \log \text{(number surviving)}$$

A convenience of expressing numbers as logarithms is that successive mortalities are then added rather than multiplied, so that total mortality $M = m_1 + m_2 + m_3$ (just as $\log 20 = \log 4 + \log 5$, while $20 = 4 \times 5$).

When annual m values from one year to the next were plotted, $m_1$ was found to be strongly correlated with winter moth abundance (Figure 49–16), whereas none of the other m values showed any correlation. Thus, $m_1$ is the key factor for population change from one year to the next. $m_1$ is the "winter disappearance" of eggs that fail to show up as caterpillars feeding on the leaves.

Winter disappearance is density-independent, a result of temperature and other weather changes in early spring, which determine when oak buds open. Each year only a fraction of winter moth eggs hatches during the two- or three-day period when the oak buds are opening. Caterpillars that hatch out too soon cannot penetrate the closed buds; caterpillars that hatch too late are unable to feed on the leaves, which rapidly toughen and produce toxic secondary chemicals.

Winter disappearance accounts for about 90% of winter moth mortality and explains fluctuations in population

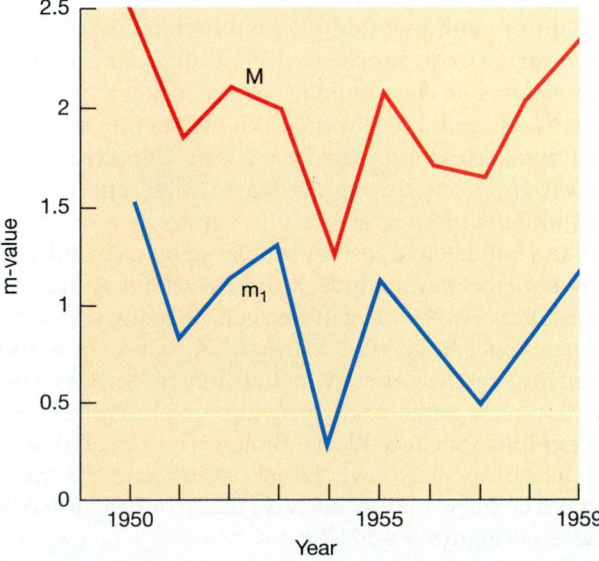

**FIGURE 49–16**  Correlation between fluctuations of M (total mortality) and $m_1$ (winter disappearance) of winter moths.  (After Varley, Gradwell, and Hassell, 1973)

(a)

(b)

**FIGURE 49–17**   Endangered animals. (a) Indian one-horned rhinoceros. (b) A golden lion tamarind from South America.   (a, Belinda Wright/DRK Photo; b, Denise Tackett/Tom Stack & Associates)

from one year to another. It does not explain the stability of the population over longer periods, however, which we would expect to be due to density-dependent mortality. If a source of mortality is density-dependent, its m value should be proportional to the size of the population on which it acts (see Figure 49–8). Such a relationship was found for the mortality of winter moth pupae in the soil as a result of predation by shrews and beetles. Parasitism was also density-dependent.

The final stage in the study was the construction of a mathematical model from the life tables and key factor analysis. If all the important factors have been taken into account, the model should not only describe population dynamics, but also predict future changes in population density. The only thing that the model cannot predict is the weather; this source of mortality must be entered into the model at each generation. Finally, a successful model was constructed. Entering into the model only the starting densities of winter moths, *Cyzenis,* and *Cratichneumon,* with the $m_1$ values for each year, the model (in the form of a computer program) predicted annual values of the densities of each species over 18 years that were remarkably similar to the densities actually found.

This case history illustrates a number of features of population studies. It shows how the density-dependent action of predators and parasites can regulate the size of a population over time, even though the mortality they cause is considerably less than that due to the unpredictable density-independent effects of weather. The study also demonstrates the usefulness of making and analyzing life tables that can later be used to prepare models. The success of a model gives greater confidence that all of the important factors have been taken into account. Moreover, models can be used to predict future population changes, perhaps of

major economic interest. Finally, this study emphasizes the critical importance of understanding the natural history of all the species involved, something that usually requires long-term studies. Varley and Gradwell's study was initially frustrating because their early models failed to work. Only after eight years did they discover the importance of *Cratichneumon* parasitism of winter moth pupae in the soil.

## 49–F  EXTINCTION

A species becomes **extinct** when its last member dies. Individual populations often become extinct, which is a different thing. For instance, in 1953 California adopted the grizzly bear as its state animal—40 years after the last grizzly bear was seen in California! But the species was not extinct. Populations of grizzly bears survive in other areas.

Earth has gone through at least a dozen periods when large numbers of species became extinct as a result of climatic and other changes caused by geological upheavals. The twentieth century must now be added to this list of mass extinctions. We do not know how many species have become extinct since 1900, partly because we do not know how many species existed in the first place. Estimates of extinction rates come from applying the principles of island biogeography (Section 49–H). Biologists estimate that about one million species have been exterminated during the twentieth century, mainly because of the destruction of tropical forests (Figure 49–17).

We live in a world where hundreds of species are disappearing while others have growing populations. As ivory-billed woodpeckers, sea turtles, and various whales faced extinction, European sparrows and starlings, African "killer" bees, fire ants, gypsy moths, and water hyacinths were spreading uncontrollably. These are all species introduced into the United States, but many native species are flourishing as well. Coyotes and white-tailed deer are now common in suburbia, and opossums and armadillos spread farther north every year. Why do some species become more common and others become rarer? The fate of a species in the twentieth century usually depends on how compatible it is with growing human populations.

If we look at the various causes of extinction, we can see that some types of species are more prone to extinction than others, which have characteristics that permit them to coexist with human expansion. The types of species that are particularly vulnerable to extinction are these:

**1. Species with Few Individuals**  The fewer individuals that make up a species, the more likely it is that an increase in the death rate will wipe out all of them or that all the breeding adults will die. In addition, a certain population density is often necessary for organisms to find mates or to breed if they do find mates. This is especially true of social animals that can breed only in groups. We know that even fertile adults often will not breed in zoos, sometimes because they will not mate with the available animals. This is why conservationists are so pleased to find calves being born among the 200 or so surviving right whales. It is still not certain that 200 individuals is a large enough number to permit the species to survive.

Small populations are especially prone to extinction if they reproduce slowly or if they live in restricted patches of habitat. Animals such as whales, pandas, and albatrosses produce one or fewer offspring each year, and the young take many years to reach sexual maturity. In small populations of such animals, even a few premature deaths may mean that the population does not survive.

When an entire species consists only of one small, restricted population, it is especially vulnerable. Such species are common on islands. Here, destruction of habitat may wipe out the species in one blow. On the Caribbean island of Martinique, the Martinique rice rat was exterminated by a single volcanic explosion in 1879.

**2. Species Unused to Competition and Predation**  Competition sometimes causes extinction, especially when new species are introduced into an area. At least one species of giant tortoise in the Galápagos Islands has become extinct because of competition from goats introduced to the islands. The goats exterminated the tortoises' food plant, an example of one extinction causing another.

Many of the world's endangered species are **endemic** to an island, meaning they originated on the island. The opposite of endemic is **exotic,** used to describe a species that has invaded an area. Sometimes endemic island species consist of few individuals, but even where populations are large, island species often become extinct when competitors from other areas are introduced.

The exotic organisms that usually cause the extinction of island species are humans and those species that accompany human invasions: goats, pigs, dogs, cats, rats, mice, and various weeds, such as dandelions and goldenrod. These are all organisms that have evolved on continents in competition with many other species and that can adapt to life in new areas and among competing species.

An extreme example of extinctions caused largely by competition comes from the Hawaiian Islands, where about half of the endemic species are now extinct. The Polynesian discoverers of Hawaii sailed their outrigger canoes to Hawaii from the South Pacific 1600 years ago. They brought with them dogs, pigs, and chickens and stowaways such as geckos and skinks (which are both lizards) and rats. They also imported some 30 species of plants, including yams, taro, bananas, and breadfruit. Centuries later, European settlers brought more new species to the islands. Today, 4000 exotic plant species occur in Hawaii. Competition with these imported species caused the extinction of dozens of Hawaiian species. Since 1780, 27 of the 70 endemic species of Hawaiian birds have become extinct, and another 30 are classified as threatened or endangered.

**3. Species That Need Large Areas or Limited Habitats**  Most species that become extinct today disappear because humans destroy their habitats. Either they cannot live anywhere else, or they cannot reach alternative habitats when their own is destroyed. Animals that need a lot of space are particularly vulnerable. Many of these are large carnivores occupying the top trophic level of a food chain. Mountain lions and wolves are examples of animals that can find enough food only if they can hunt in large areas. Their habitat has steadily disappeared as the human population has spread.

Sometimes human artefacts can substitute for destroyed habitat. Ospreys are large fish-eating birds that prefer to nest in the tops of trees with a view. As people cut down large and dead trees, their numbers declined. Then ospreys started to nest on structures such as bridges and navigation markers, which supply similar isolation from predators (Figure 49–18). Now the Army Corps of Engineers adds nest platforms to navigation markers, and the osprey population is growing.

**4. Species Hunted and Collected by Humans**  People reduce populations of many organisms directly, by shooting animals for fun or profit, fishing and trapping, and collecting plants for medicine or landscaping. Hunting has exterminated many species of large vertebrates and left nearly all large carnivores and other mammals with beautiful fur endangered.

One odd example is the passenger pigeon, a species endemic to North America. In 1871, an estimated 140 mil-

**FIGURE 49–18** An osprey family on its nest on a navigation marker. The gray bundle on the left is a juvenile bird.

lion of these birds nested in one of their breeding grounds in Wisconsin. Because they lived and bred in huge flocks, the birds were easy to shoot for food and sport. A mere seven years later, in 1878, conservationists realized that hunting had endangered the species even though thousands of the birds remained. But it was too late. The pigeons were social breeders and would not nest without large numbers of like-minded neighbors. Reduced populations, coupled with parasitic infections, had brought their reproduction almost to a standstill. The species was extinct by 1915.

Today, sport hunting and fishing groups are among the most active conservationists. They understand that their sport and, sometimes, their livelihood depend upon sustained or increasing populations of game animals and fish. But in developing countries, where selling rhinoceros horns for cash or killing an antelope for food may mean the difference between starvation and survival, hunting still threatens many species.

Most countries have enacted laws and regulations designed to prevent more species from becoming extinct. The United States' Endangered Species Act is considered by many to be the strongest of these laws, but this does not mean that it does the job it was designed to do. Its reauthorization and enforcement are continually opposed by people with an economic interest in habitat destruction or killing endangered organisms. Since European settlers arrived, more than 500 species of plants and animals have disappeared from North America, and some 4900 species of American plants and animals are currently endangered, 700 of them in Florida alone. California is another extinction hotspot. It contains one of the most diverse arrays of communities in North America and is home to one quarter of all the extinct and endangered species in the United States.

## BIOGEOGRAPHY

Each species is made up of one or more populations, which live with populations of other species as communities within ecosystems. **Biogeography** is the study of patterns of distribution of species and their populations in space and time. What determines how many species are found in a community? Why does a forest contain some of the same species as a similar forest on another continent and some different species? What features must be present in a wildlife preserve if an endangered species is to be preserved from extinction? These are some of the questions that biogeographers seek to answer.

Earlier we saw that the number of individuals in a population is determined by birth and immigration, death and emigration. Similarly, the number of species in a community is determined by evolution of new species and colonization, which add species, and extinction and emigration of populations, which reduce the number of species in the area.

If we ask why species are where they are we find two factors that contribute to the answer: climate and the organisms' ability to disperse to new areas where they can survive. Climate determines the biome, and organisms can live only within biomes to which they have evolved adaptations. Dispersal ability is important because most species cannot travel between continents (unless they are transported by humans). There are exceptions: for example, rotifers form cysts that can be blown almost anywhere in the world, but the animals can live only in very restricted types of environments. As a result, the rotifer species in a marble cemetery urn in Pennsylvania may be the same as that in a marble urn in a South African cemetery, but different from that in the granite urn on the next grave!

It is often difficult to account for patterns of geographical distribution because organisms have dispersed from their centers of origin and have undergone adaptive radiation after colonizing new areas. Meanwhile, the continents themselves were moving (*Essay: Continental Drift*, Chapter 19). However, climatic or geographical barriers, such as oceans, deserts, or mountain ranges, restrict the dispersal of many groups of organisms, resulting in major discontinuities in the distribution of animal and plant species. For example, Australia became separated from the other continents early in vertebrate evolution and since then has

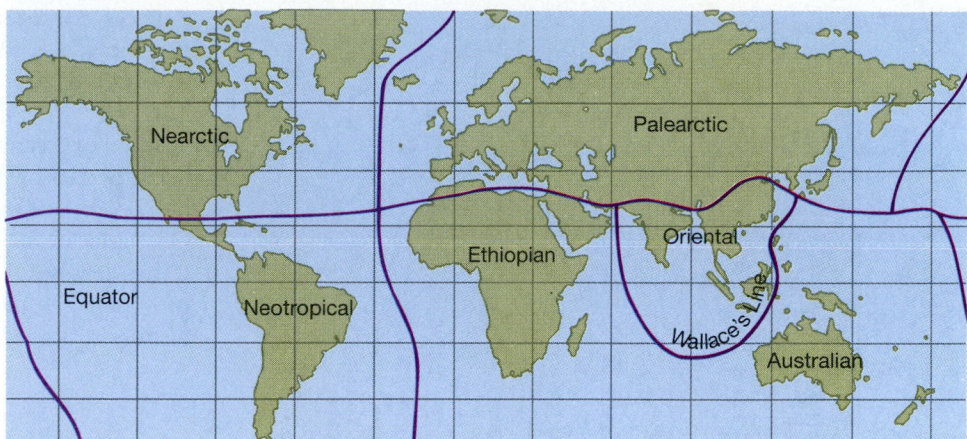

**FIGURE 49–19** Biogeographic regions of the world, proposed by Wallace. (The palearctic and nearctic regions together make up the holarctic.)

become completely surrounded by ocean; North and South America are connected only by the narrow (and fairly recent) "land bridge" of Central America, and the Americas are separated from the other continents by oceans. The Sahara and Arabian deserts form a barrier to movement of animals and plants between Eurasia and most of Africa. India and southeast Asia are effectively isolated by ocean to the south and by the Himalayan mountains and central Asian deserts to the north.

As a result, evolution has proceeded more or less independently in each of these great continental land areas. Adaptive radiation in each area gave rise to species found nowhere else, although their ancestors may have emigrated from an adjacent region. On the basis of the animals and plants they contain, these areas have been described as major **biogeographical regions,** first outlined by Alfred Russel Wallace (Figure 49–19). Mammals of the Ethiopian Region, for example, include the African elephant,

chimpanzee, gorilla, giraffes, two species of rhinoceros, and a variety of unique antelope species. The Malay tapir, Indian elephant, and tiger are characteristic mammals of the Oriental Region. In the Australian Region, more isolated than the others, mammals were represented mainly by marsupials and monotremes. Marsupials underwent their own adaptive radiation and gave rise to a variety of species filling the same niches as their placental equivalents elsewhere (see Figure 17–11).

The boundaries between adjacent biogeographical regions are not usually clear-cut. In southeast Asia, for example, the large islands of Borneo, Sumatra, and Java share many organisms with nearby Malaya, and no one doubts that they should be assigned to the Oriental Region. Further east, the island of New Guinea has many animals in common with nearby Australia and belongs to the Australian Region. But what about all the islands in between, such as Timor and Lombok (Figure 49–20)?

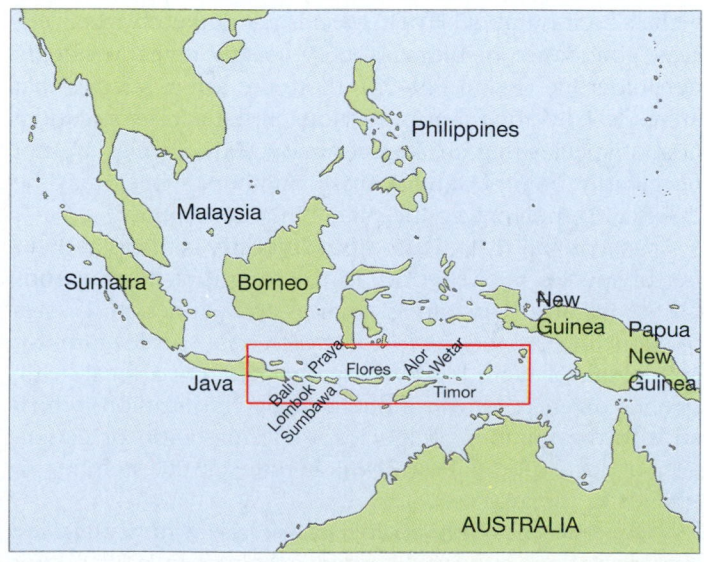

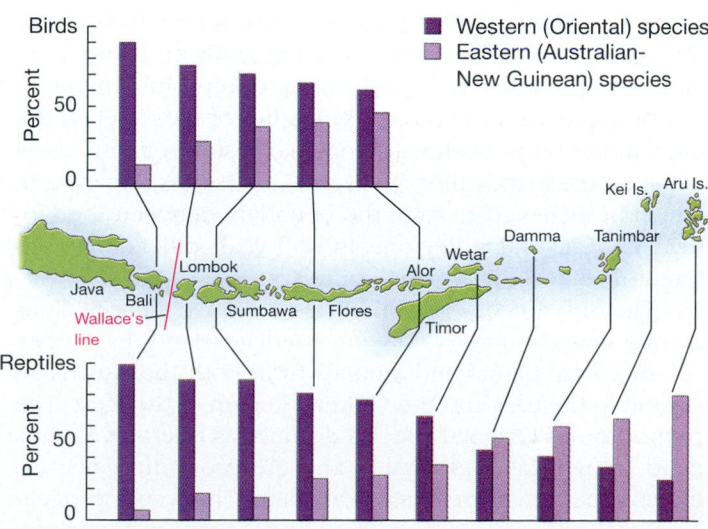

**FIGURE 49–20** Wallace's Line. Few mammalian species have crossed Wallace's Line, which divides the Australian and Oriental Regions (left). Birds and reptiles appear to be more mobile (right).

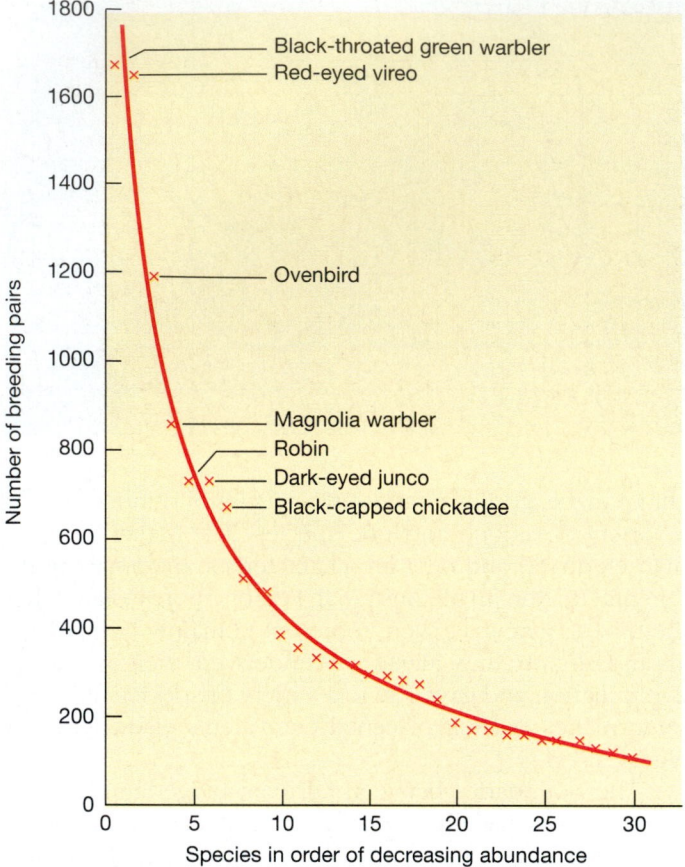

**FIGURE 49–21**  Variations in species abundance within a community. This graph shows the number of breeding pairs of the 30 most abundant bird species in 6800 hectares of forest in southwestern New York.  (Data from A. A. Sanders)

One day in 1856, Wallace traveled from Bali to Lombok, 30 km further east. He was impressed by the considerable difference in the animals and birds of the two islands. Green woodpeckers and barbets, for example, were abundant on Bali, but the outstanding birds on Lombok included white cockatoos and honeysuckers. In a mere 30 km, Wallace seemed to have sailed out of the Oriental Region and into the Australian Region. He subsequently drew a line on the map, called Wallace's Line (see Figure 49–20), which he believed to form the boundary between the Oriental and Australian Regions. In later years others drew different lines to indicate this boundary.

The differences of opinion arise because there really is no one sharp boundary. The intermediate islands have been colonized by plants and animals from both the Australian Region to the east and the Oriental Region to the west. The proportion of Oriental species declines as one moves eastward from Java to Australia, and the proportion of Australian species increases. Boundaries between biogeographical regions are best thought of, then, not as sharp lines, but as transition zones in which the flora and fauna of one region gradually give way to those of another.

Local conditions on islands differ from those on the mainland. As a result, natural selection on an island inevitably produces different adaptations in a population of newly arrived colonizers. Sometimes this results in new species. For instance, in the Hawaiian Islands, about 3000 insect species have been found. It is thought that these islands were invaded by some 250 insect species and that the remainder are species formed by adaptive radiation of the invaders. Only a few dozen species of the fruit fly genus *Drosophila* live in the western United States, whereas several hundred species inhabit Hawaii.

## 49–G  SPECIES DIVERSITY

Perhaps surprisingly, we can now predict how many species of a particular type are likely to be found in a community. This ability has grown from studies of species in isolated areas, particularly islands. It forms the basis for studying Earth's biodiversity and the rate at which habitat destruction by humans is causing the extinction of other species that share the world with us. It is also the basis for conservation biology, the biological theory behind efforts to slow the loss of biodiversity.

The simplest measure of diversity is the number of species in a given area. Using this measure, however, a two-species community with 99 individuals of one species and 1 individual of another species would have the same diversity as a two-species community with 50 individuals of each species, yet the characters of the two communities would be very different.

For this reason, ecologists often use more complex measures of diversity that take account of the relative abundance of species, their biomass, and their trophic level, as well as their number. When species are ranked by their relative abundance or biomass, they usually reveal a steeply declining line (Figure 49–21): there are a few species that are especially abundant or obvious in the ecosystem and a host of species that are less common. Rare species are not necessarily unimportant: a small number of bees may be essential to pollinating the plants in a community.

On a global scale, Earth's **biodiversity** is the total number of species in the world. In the long run, this depends on the balance between speciation and extinction. If rates of extinction exceed the rate at which new species form for any length of time, biodiversity declines. On a local level, species diversity is also influenced by colonization, which adds to the number of species, and emigration or extinction of local populations, which reduces the number of species in the area.

One question early ecologists asked was how often the presence of one species determines the presence of another in a community. Some species are found only in association with others because they cannot survive without particular species of symbionts, hosts, food plants, or pollinators. But what about species that are not obviously dependent on one another? Within the climax communities of biomes, we can often recognize plant **associations,** units of vegetation characterized by dominant plants, such as the beech-maple association common in eastern North American forests. In the nineteenth century, Frederick Clements considered that such associations represented "supraorganisms," each with its own set of coadapted species. This implied that each species is more successful within its association than it would be if it lived with other species.

In the 1920s, H. A. Gleason in the United States and L. G. Ramensky in Russia challenged this view of associations and proposed instead that each plant species is distributed independently along gradually changing environmental gradients. Thus, the fact that beech and maple trees occur together reflects their similar requirements, not dependence on each other. Later studies by John Curtis, Robert Whittaker, and others supported this view. In general, the species present in communities vary with environmental gradients, forming ecoclines with no abrupt changes except where the environment changes drastically.

## Causes of Diversity

Communities differ in their species diversity. For instance, ecologists counted as many as 150 species of woody plants in some 0.1-hectare plots of tropical rain forest in Panama and as few as 21 species in some 0.1-hectare plots of temperate forest in Missouri. What maintains the characteristic species diversity of a natural ecosystem? What prevents one or more species in a trophic level from eliminating the others through competition? Why has a "super-rabbit" not evolved, a primary consumer so efficient that it drives competing species at the same trophic level to extinction?

Competition among species sometimes does lead to the extinction of one of them, but this is not always the case. The main reason appears to be specialization: each species evolves adaptations to exploit a small portion of the resources so thoroughly that other species cannot compete. But no species has the energy budget or physical adaptations to use all the resources available. Thus, a rabbit is well adapted to eating grass but lacks the morphology that would permit it to climb trees and eat leaves; a cabbage white caterpillar has the enzymes to detoxify secondary chemicals in crucifer plants (Section 18–H) but does not have the energy budget that would allow it also to produce enzymes that detoxify chemicals in pine needles.

As a result, when species are in potential competition for a food supply, they often subdivide this resource in some way. Different insect-eating birds, for instance, search for food at different heights in a forest (see Figure 49–11). Many species of sea birds may share the same cliff during the nesting season, but each specializes by using slightly different types of nest sites on the cliff. So an important reason for species diversity within a particular community is specialization in the use of limited resources, reflecting differences in the niches of different species. But this does not explain why tropical forest contains more species than temperate forest.

In practice, diversity differences among communities seem to depend on local conditions. Among the factors that are often important are:

***1. Community Structure*** The most obvious difference between tropical and temperate forest is higher productivity in the tropics. But hundreds of studies have shown that there is no direct correlation between productivity and diversity. Instead, higher productivity apparently leads to higher diversity only when it increases the structural diversity of the environment.

Food chains tend to be longer and contain more species in three-dimensional than in two-dimensional environments. An environment is two-dimensional if it is essentially flat, like a grassland, tundra, shallow stream, or lake bottom. Three-dimensional habitats include forests and open ocean waters.

Table 49–3 supports this theory that three-dimensional structure is better than productivity as a predictor of how many species an ecosystem supports: in this list, flat environments, such as marsh and grassland, contain fewer bird species than forest. The reason for this correlation is probably that an ecosystem with three-dimensional structure contains more niches for different species. Thus, a tree may provide a livelihood for five or more species of birds that forage at different heights, whereas a clump of grass is unlikely to provide more than two niches: one for birds feeding at the top of the plant and one for those on the ground beneath it.

### TABLE 49–3

**Plant Productivity and Number of Bird Species in Some Temperate Habitats***

| Habitat | Productivity (grams of plant material/m²/year) | Average Number of Bird Species in 5 ha |
|---|---|---|
| Marsh | 2000 | 6 |
| Grassland | 500 | 6 |
| Shrubland | 600 | 14 |
| Desert | 70 | 14 |
| Coniferous forest | 800 | 17 |
| Floodplain deciduous forest | 2000 | 24 |

*Data from R. E. Ricklefs

**2. *Frequency of Disturbances*** Periodic disturbances increase the number of niches, and therefore species, that an area contains. Light gaps in a forest may be inhabited by early successional species of birds and insects, different from those in the surrounding climax forest. Frequent landslides, floods, or fires in a region also increase the number of types of habitats.

**3. *Predation*** Grazing by herbivores sometimes increases diversity by controlling populations of dominant species. In several experiments, herbivorous mammals were fenced out of areas of grassland. A few species of grasses and herbs took over more and more of the area, and the species diversity was reduced.

Specialized herbivorous insects can contribute to diversity. In some plant species almost all seeds that land near their parents die because they or the seedlings are attacked by insects. Only if seeds are dispersed some distance do they have a reasonable chance of escaping the insect specialists that eat the parent plant. This sort of predation often contributes to diversity in tropical forests. Species that become common are most easily discovered by enemies, which quickly eat them so that they become rare again, leaving space for other species.

Predation by generalized carnivores may have similar effects. This was shown in Robert Paine's study of an intertidal community on the rocky coast of Washington, where mussels, barnacles, limpets, and other animals were all fed on by a species of sea star. When the sea stars were removed from the rocks, barnacles settled and took up about 80% of the available space within three months. Later, the barnacles were crowded out by two species of mussels. A year later, one mussel species dominated the experimental area, and the number of species present had dropped from 15 to 8. Thus, in the natural community the sea star predator is a keystone species, maintaining diversity by preventing some of its prey species from crowding the others out.

**4. *Habitat Area*** The greater the extent of a habitat, the more organisms evolve adaptations to it. Consider aquatic habitats. The world contains huge areas of sea water and fresh water. Much less common are very salty habitats, such as salt lakes, and brackish habitats, where fresh and salt water meet as they do in estuaries and salt marshes. As a result, many more species have evolved in sea water and fresh water than in salt lakes and brackish water. Similarly, few species have evolved adaptations to hot springs and soils with a high content of copper or selenium, simply because these habitats are rare.

**5. *Evolutionary History*** As with area, the longer an ecosystem has existed, the more species are likely to have evolved adaptations that permit them to live in it. Western Europe has a depauperate flora and fauna, containing many fewer species than the same biomes in other parts of the world. This is largely because many European species became extinct during the Pleistocene glaciations. In North America, the glaciations increased species diversity by dividing species into western and eastern populations, which often evolved into separate species (Section 19–H). In contrast, Europe is farther north, and the glaciers extended so far south in many areas that they caused the extinction of all populations of many species.

The glaciers did not extend to the tropics, so tropical forests have existed longer than temperate forests, giving more time for species to evolve there.

## Species Turnover in Communities

In 1968 researchers tallied the number of species of birds breeding on each of the nine Channel Islands off the coast of southern California. The total number of species on each island had changed little since a similar survey 50 years earlier. However, the species composition had changed markedly. For example, on San Nicolas Island there were 11 species of birds in both 1917 and 1968, but only five of these species were the same. Six species had disappeared from the island, but six other species had colonized it.

These findings supported Robert MacArthur and Edward Wilson's theory that although the species diversity of a community may remain the same for long periods, **species turnover** occurs, so that the community contains different species at different times. Small islands are particularly useful for studying species turnover because they have obvious boundaries, and it may even be possible to count organisms as they arrive and leave.

MacArthur and Wilson pointed out that the number of species on an island is largely a result of the balance between colonization and extinction. Species colonize islands most easily if they are capable of long-distance dispersal, either by flying or swimming under their own power, by hitching a ride on a drifting log or on an animal (or ship), or by floating as seeds or spores through the air. Good colonizers also tend to be species that do not require highly specialized conditions to live and reproduce—a plant that self-pollinates or reproduces vegetatively is more likely to establish a new population than is a plant that requires other individuals for cross-pollination. Effective colonizers also tend to have high biotic potentials, allowing them to establish a thriving population quickly. Islands are usually colonized by a disproportionate number of fugitive species (Section 48–F), which tend to possess these characteristics. As the island fills up, organisms continue to arrive, but the rate of successful colonization declines because the established species are already exploiting the resources that newcomers need.

Island species are especially susceptible to extinction (Section 49–F). This is partly because the population of each species is relatively small, and random events may kill all the individuals. Furthermore, island communities tend to

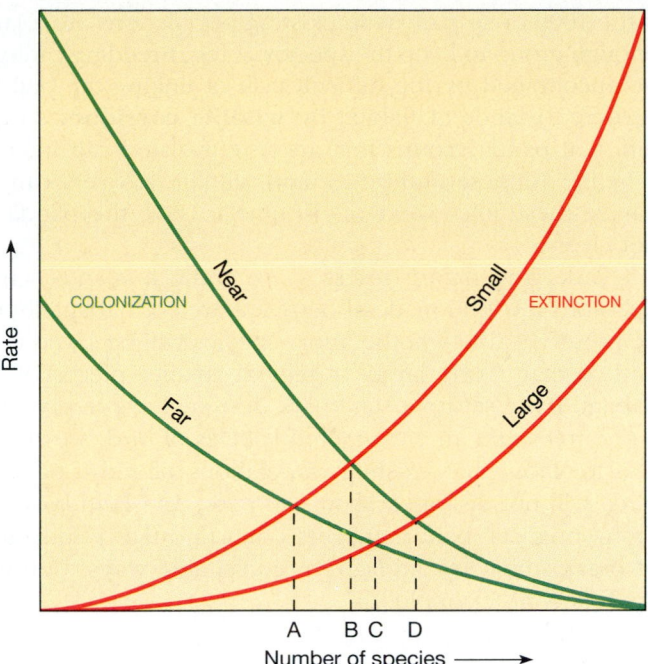

**FIGURE 49–22** Graphical model showing the effect of isolation and size of an island on its predicted number of species at equilibrium. Colonization and extinction curves show the rate of addition of new species and extinction of existing species as a function of the number of species already present. Near islands, close to a source of species, would be expected to have higher rates of colonization. Small islands would be expected to have higher rates of extinction because their populations will be smaller. A is the number of species at equilibrium on a far, small island; B, the number on a near, small island; C, the number on a far, large island; and D, the number on a near, large island.

have low species diversity and few predators; this can increase the likelihood of one species crowding out some of its neighbors, or at least reducing their populations to such low numbers that they become extinct through random processes.

The point where the immigration and extinction curves for an island cross each other gives the "equilibrium" number of species for that island (Figure 49–22). According to this model it is only the number of species that will remain the same. The identity of the species present may change as some become extinct and others become established.

The study of species turnover on islands is **island biogeography,** because the rate of turnover and the number of species at equilibrium depend on the island's geography. MacArthur and Wilson predicted that the number of species on an island depends on the island's distance from the source of species and on its size and habitat diversity. Fewer species can reach an island that is far from the mainland, so the island will end up with fewer species than one closer in. Larger islands have more species at equilibrium, partly

because they intercept more dispersing organisms, partly because they contain more different habitats, but probably mostly because they permit species to maintain larger population sizes, which are less susceptible to random extinctions.

Daniel Simberloff and Edward Wilson tested these predictions experimentally. First, they counted the number of species, mainly insects, on six small mangrove islands in Florida. Next, they completely exterminated all of the arthropods on these islands by enclosing the islands in enormous plastic tents and fumigating them (Figure 49–23). They then removed the tents and sampled the animal populations regularly. Within six months, the number of species on each island had returned to approximately the number present before fumigation and remained at about the same level for as long as the islands were observed (Figure 49–24). As predicted, the species compositions were not identical to those before the experiment, and species turnover continued on each island, with losses approximately balanced by immigration.

In contrast to islands, species turnover is slow in most communities on continents because population sizes are usually large, and local extinction is easily offset by immigration from nearby areas. Small, isolated communities such as mountaintops, bogs, and lakes, however, may have appreciable species turnover rates.

**FIGURE 49–23** Measuring species diversity. Scaffolding completely encloses a small mangrove island in the Florida Keys, in preparation for enclosing the island with plastic sheeting. After elimination of the island's insects by fumigation, tent and scaffolding were removed, and the recolonization of the island by insects was monitored. (Daniel Simberloff)

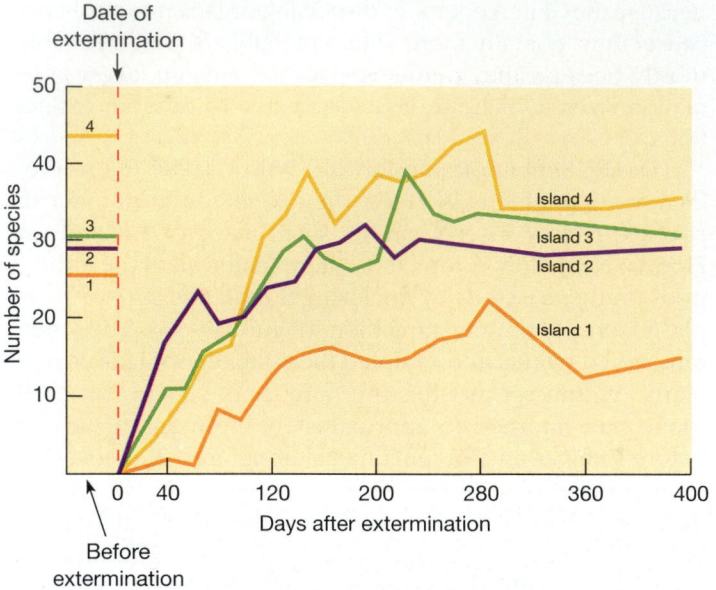

**FIGURE 49-24** Recolonization of four small mangrove islands after their insect fauna had been exterminated. The numbers of species on each island before fumigation are indicated on the left. (D. Simberloff and E. Wilson, "Experimental zoogeography of islands," *Ecology* 51:934, 1970)

## 49-H CONSERVATION

Biologists are deeply concerned that the loss of biodiversity threatens many aspects of life on Earth. They point out that any species may be a vital link in a food web. Its extinction may disrupt an ecosystem or cause the extinction of other species. In addition, many wild species are the sources of new genetic diversity for domesticated plants and animals and of drugs, foods, and other economically important substances. Others feel that there are more important considerations than the economics of extinction: the esthetic loss we suffer when a species disappears.

Conservation biology is the study of the biological principles that must be understood if we are to slow the loss of biodiversity and preserve endangered species in populations large enough to ensure the species' survival. A few examples will illustrate the kinds of studies that contribute to this rapidly growing field.

Island biogeography is a vital part of conservation because natural ecosystems have become increasingly fragmented into "islands" by development. Consider endangered giant pandas, which once roamed much of southeast Asia. Nearly all the bamboo forest that is their home has been destroyed, and the remaining pandas have been pushed into bamboo forests in China, high up against the dry plateau of Tibet, with nowhere else to go. They survive in groups of ten or fewer (some groups probably too small for breeding) in 12 small reserves, surrounded by roads, rivers, fields, and human settlements. No one knows

if the 1000 or so pandas in existing populations are a large enough group to keep the species going. Breeding can only be encouraged by the difficult task of linking the pandas' scattered islands of habitat by **wildlife corridors** so that different panda groups that are now isolated can interact. Whether Chinese authorities and wildlife advisers can organize these improvements in time to save the pandas is not clear.

Parks and nature reserves are artificial islands where decisions must be made about what area of habitat should be preserved and whether more species will be saved if the park contains one large or several smaller pieces of the habitat. If we set up a state park designed to preserve redwood trees and an endangered species of bird, we should like to know that existing populations of redwoods and birds will not disappear from the park. As island biogeography predicts, park managers often find that populations of organisms come and go in undesirable ways. Consider the following example.

## Chobe National Park: A Case History

Chobe Park in Botswana was set up to preserve an area of Africa's tropical savanna and thornwood. Chobe contains patches of acacia woodland and grassland with a large population of herbivores, including giraffe, impala, buffalo, elephant, zebra, and wildebeest, as well as predators such as lions, hyenas, wild dogs, and jackals. In 1986 the park's managers sought advice because the acacia woodland was disappearing. The trees suffer from heavy browsing (tree-eating) by elephants and other herbivores, and no new acacias can grow because elephants destroy the seedling trees (Figure 49-25).

Studying the history of the park, ecologists discovered that the acacias in Chobe are there only because in the

**FIGURE 49-25** An elephant browsing on an acacia tree. (J. Robert Stottlemyer/BPS)

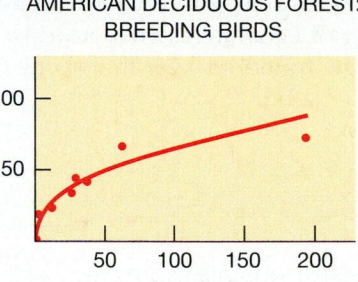

AMERICAN DECIDUOUS FOREST:
BREEDING BIRDS

(a)

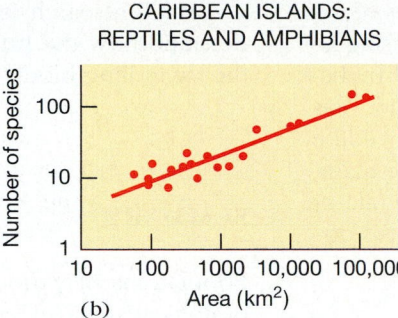

CARIBBEAN ISLANDS:
REPTILES AND AMPHIBIANS

(b)

**FIGURE 49–26** Species-area relationships. (a) The number of species of breeding birds in "islands" of North American deciduous forest of different sizes. (b) Numbers of species of reptiles and amphibians on Caribbean Islands of different sizes. In this case, the data are plotted logarithmically to produce the straight line predicted by the species-area equation ($z$ in this case = 0.34).   (a, data from F. W. Preston; b, data from A. Schwartz and K. Thomas)

1890s the river through the park dried up in a drought. This sent the elephants elsewhere in search of water. Soon afterward, disease wiped out most of the other herbivores in the area. With no animals to eat them, acacia seedlings thrived, and many trees grew to maturity, forming woodland that has supported large populations of animals ever since.

The only way to restore the acacia groves would be to exclude elephants from the area as completely as possible for 10 to 15 years, while reducing the populations of other herbivores as well. This would also drive lions and other carnivores out of the area in search of food, which would make the park much less attractive to tourists. There is no ideal solution to this problem. The coexistence of large populations of herbivores and of thriving acacia woodland is inherently unstable. Populations of food plants and of the animals that eat them come and go over the centuries in this as in other communities.

## How Many Species?

Large islands contain more species of any one type than do small ones. Olaf Arrhenius expressed this in 1921 with the **species-area equation:**

$$(\text{number of species}) = (\text{area})^z$$

where $z$ is a constant.
This can also be expressed:

$$\log (\text{number of species}) = z\log (\text{area})$$

Figure 49–26 shows examples of species-area measurements. These and similar studies have shown that the value of $z$ in the species-area equation usually falls between 0.20 and 0.35 and averages about 0.25. This means that the number of species of, say, birds on islands increases as the fourth root of the land area of the island. In other words, a tenfold increase in area results in a doubling of the number of species. This rule of thumb has proven to hold true not just for real islands but also for habitat "islands" such as lakes, alpine meadows surrounded by forest, and even clumps of trees surrounded by grassland.

The reasons for the relationship between area and number of species are hotly debated. However, we should note that the relationship has many interesting consequences.

The species-area relationship is used to estimate the number of species that exist in poorly known biomes such as tropical forest and also to estimate rates of extinction of poorly described groups of organisms. For instance, Daniel Simberloff used the formula to calculate species loss within the next century from the destruction of forest. He calculated that 12% of the 704 bird species in the Amazon basin will become extinct, as well as 15% of the 92,000 plant species in Central and South America.

## Nature Reserves

The principles of island biogeography can also be used to calculate how much biodiversity a particular nature reserve can be expected to preserve. Nature reserves are usually samples of larger communities. Thus, Chobe National Park is a sample of African savanna. A sample contains fewer species, fewer individuals of each species, and more populations made up of only a few individuals than a larger area would. In practice, the species-area formula tells us that if we protect 10% of the area of a particular community as a reserve, our reserve will contain about 70% of the species found in the larger community.

The species-area equation also reflects the fact that a group of islands contains more species than a similar-sized area of nearby continent. Thus, the islands of Indonesia contain many more species than Australia, although Australia is somewhat larger than Indonesia. This tells us that to preserve the greatest possible biodiversity, we might be better off preserving many small areas than one large one.

There is a snag to this strategy, however, because we also know that small islands have higher rates of extinction than larger ones (because populations are smaller). So our many small reserves will only preserve their species diversity if we can ensure that they are near enough to each other or to other sources of species to be colonized rapidly by new species. This is not usually the case. Reserves generally become increasingly isolated from similar patches of habitat as development proceeds around them. Rates of extinction under these conditions have been studied using islands that have become isolated from the mainland by rising sea level in the past 10,000 years. Michael Soulé has used such studies to estimate that the world's largest wildlife preserves, in Africa, will have lost 11% of their large mammal species in 50 years and 77% in 5000 years. The rate of loss of birds will be considerably lower, because birds disperse more easily than large mammals.

## The Importance of Diversity

What happens to a community when its species diversity is reduced, as it is when development destroys the habitat surrounding a nature reserve? Does loss of biodiversity affect an ecosystem's ability to sustain the species that remain or to perform functions that are useful to human economies, such as producing lumber or absorbing carbon dioxide, flood water, and pollutants? The general answer is that we do not know, but there are two main ideas about the importance of diversity.

One is the "rivet" model of Paul and Anne Ehrlich, in which each species resembles one of the rivets that holds an airplane together. The loss of each rivet (species in an ecosystem) weakens the structure slightly until the airplane is no longer airworthy and crashes. In contrast is the "redundancy" model of Brian Walker. Walker asserts that most populations in an ecosystem are superfluous, resembling passengers rather than rivets in an airplane, and that only a few keystone species are needed to keep the ecosystem functioning.

Recent studies suggest that both ideas apply in various situations. The rivet model is supported by the work of Shahid Naeem and John Lawton, who set up artificial ecosystems containing 2, 5, or 16 species of annual plants. Under identical growing conditions, they found that the most species-rich systems had the highest net primary productivities. The reason seems to be that the more diverse communities have more three-dimensional structure, allowing the ecosystem to capture more light by photosynthesis. This bears out agricultural studies in the tropics which have shown that the way to increase productivity in a maize field is not by packing in more cereal plants but by adding trees, melons, or nitrogen-fixing beans to the field.

However, productivity is not the only criterion of whether an ecosystem is "healthy," and the rivet model goes only so far. Most ecosystems contain more structural diversity than is needed to reach maximum productivity. Thus, temperate forests in the Northern Hemisphere have almost identical productivities despite considerable differences in species diversity: East Asian forests contain 876 tree species, North American forests 158 species, and European forests 106 species (reflecting the sizes and evolutionary histories of these forests).

Much research remains to be done before we shall understand how the functioning of ecosystems is affected by the dwindling biodiversity now found all over the world.

## SUMMARY

Populations of various species of organisms are the basic biological units of ecosystems. They live in communities with populations of all the other species in the ecosystem. A population can be described by its characteristic size, biotic potential, and survivorship. The species diversity of each community has a characteristic turnover rate and equilibrium number of species, which can be used to calculate the effects of habitat destruction on biodiversity.

1. The biotic potential of a species is determined mainly by the age of the (female) parent at first reproduction, but it is also influenced by the number of offspring produced at each reproductive event and by the parent's reproductive lifespan.
2. Among the fastest-growing populations are those of organisms introduced into new, favorable environments, but a population in nature seldom, if ever, reproduces at its biotic potential even when it is growing exponentially.
3. A population's reproductive strategy reflects its survivorship curve. At the two extremes, the members of a population may produce many small offspring and leave them to fend entirely for themselves, or they may produce a few, large offspring that are nourished and trained by the parents for some time after birth or hatching. Survivorship data are the basis for constructing life tables, useful in predicting future changes of population size.
4. Most populations remain at about the same average size over the years. Populations may be kept in check by density-independent natural events such as bad weather. However, the sizes of most populations are generally limited by density-dependent factors, such as predation, disease, and competition for resources.
5. When growth of a population ceases, under the influence of one or more of these factors, the size of the population stabilizes, with oscillations around the carrying capacity of the environment for that species.
6. Competition among individuals for resources may play an important part in regulating the size of a population. Competition between members of different species with similar niches leads either to the extinction of one species—the weaker competitor—within the area of overlap, or to selection for adaptations that reduce the competition.
7. Many specialized predators and parasites are known to keep their prey populations at low density, but gener-

alized predators usually do not because they switch among prey species according to their abundance.

8. Extinction is the inevitable fate of populations and of whole species. Species consisting of only one or a few small populations, living in restricted habitats (such as islands), are particularly vulnerable to extinction. Humans have greatly increased the rate at which species become extinct by introducing predators or competitors into new areas, by hunting, and especially by destroying habitats.

9. The study of biogeography reveals patterns of species distribution on Earth, affected by factors such as where a group of organisms originated, how long it has had for adaptive radiation, and its ability to disperse widely. The world can be divided into six main biogeographic regions, each a focus of adaptive radiation and divided from the other regions by geographic boundaries.

10. Species diversity in a community is maintained by the ability of potentially competing species to subdivide limited resources (such as food) within habitats and to specialize in using slightly different habitats. Species diversity depends on local conditions, including the size of the area and the three-dimensional structure of the ecosystem, predation, and the frequency of disturbances.

11. The study of islands shows that there is species turnover in communities, with new species moving in and resident species becoming locally extinct. The number of species varies from one community to another depending on the size of the area the community occupies, diversity of habitats in this area, and availability of colonists from nearby areas.

## Self-Quiz

1. Draw a graph showing the long-term growth of a population of bacteria on a nutrient medium in a laboratory culture.
2. A population can grow exponentially:
   a. when food is the only limiting resource
   b. when first invading a suitable and previously unoccupied habitat
   c. only if there is no predation
   d. only in the laboratory
3. Which of the following does *not* directly affect biotic potential?
   a. a female's age at first reproduction
   b. carrying capacity of the environment
   c. length of time a female is fertile
   d. average number of offspring per brood or litter
4. If a population exceeds the carrying capacity of the environment:
   a. it will evolve adaptations to avoid a population crash
   b. its numbers will probably decrease rapidly
   c. its food supply will increase in the next generation
   d. the average number of young per individual will increase
5. Which of the following would be *least* likely to act as a density-dependent factor limiting the size of a population of mice?
   a. parasitism
   b. buildup of waste products
   c. predation
   d. unfavorable climate

6. A female elephant bears one offspring every two to four years. Which type of survivorship curve would you expect elephant populations to show: I, II, or III?
7. Studies suggest that:
   a. two species may share the same resource only if both are strong competitors for it
   b. two species that appear to be sharing the same resource are probably specializing so that each uses only a particular part of the resource
   c. if two species are sharing the same resource, neither is capable of exploiting the part of the resource used by the other
   d. no resource can be used indefinitely by more than one species
8. Most species extinctions occurring now are caused by human:
   a. hunters
   b. fishing industry
   c. destruction of habitat
   d. demand for furs and other products from endangered species
   e. predator control measures
9. State whether the species diversity of a community would be likely to increase or decrease under each of the following conditions:
   _____ a. increased frequency of disturbances
   _____ b. extermination of a predator that preys on many other species
   _____ c. partitioning of a shared resource among competing species

## Questions for Discussion

1. Paul Ehrlich has said, "It is quite possible that the penalty for frantic attempts to feed burgeoning populations may be a lowering of the carrying capacity of the entire planet." Ecologist Lee Talbot has said, "We haven't inherited the Earth from our parents. We've borrowed it from our children." What does each of these statements mean? Do you agree? Why?

2. What are some of the lines of evidence that indicate that the human population has already exceeded Earth's carrying capacity?
3. The 1970s saw a decline in the birth rate in the United States. Why is the population of the United States still growing?

*(Continued)*

4. Consider two women born in the same year, each of whom will give birth to twin girls as her only children. However, one woman (A) will have her twins at age 18, the other (B) at age 36. Each daughter will have twin daughters at the same age her mother gave birth, and so on. All mothers will die at age 72.
   a. How many descendants does A have when she dies?
   b. How many descendants does B have when she dies?
   c. Construct a graph to show the growth of populations A and B.
   d. How do the rates of increase compare in the two populations? (Find a numerical answer if you can!)

5. What factors, other than the availability of contraceptives, might increase or decrease the rate of growth of the human population in future years? (Hints: Wars have little impact: the most destructive was World War I, which killed nearly one-quarter of the population of Europe [about 70 million people] but which does not even show up on a small-scale graph of world population growth. Disease? The plague that hit Europe during the Middle Ages killed about one-quarter of the population. AIDS is reducing population growth rates in the southern half of Africa, largely because hundreds of thousands of children, as well as many adults, die of the disease each year. Why does the death of children have a greater effect?)

## Suggested Readings

Barrett, S. C. H. "Waterweed invasions." *Scientific American,* October 1989. The story of the damage done by introduced water hyacinth and kariba weed, the failure of mechanical and chemical control, and the hope that natural predators will control the invasions.

Begon, M., J. L. Harper, and C. R. Townsend. *Ecology: Individuals, Populations, and Communities,* 2d ed. Sunderland, MA: Sinauer Associates, 1992. A population-oriented textbook.

Cole, L. C. "The population consequences of life history phenomena." *Quarterly Review of Biology* 29:103, 1954. The paper in which Cole showed how much more a female contributes to the growth of a population if she breeds as young as possible.

Ehrlich, P. R., and A. H. Ehrlich. *Extinction: The Causes and Consequences of the Disappearance of Species.* New York: Random House, 1981.

Hoage, R. J., ed. *Animal Extinctions: What Everyone Should Know.* Washington, DC: Smithsonian Institution, 1985. The record of a symposium to acquaint the public with the major issues of species extinction and related habitat destruction; dozens of well-documented examples.

Soulé, M. E., and B. A. Wilcox. *Conservation Biology: An Evolutionary-Ecological Perspective.* Sunderland, MA: Sinauer Associates, 1980. Good examples of the application of island biogeography to conservation.

Wilson, E. O., ed. *Biodiversity.* Washington, D.C.: National Academy Press, 1988. Essays by many authors on the crisis caused by the extinction of species in the twentieth century.

C H A P T E R

50

# *Human Ecology and Natural Resources*

## O B J E C T I V E S

*When you have studied this chapter, you should be able to:*

1. Use the following terms in context: hunter-gatherer, high-input and low-input farming, resource depletion.
2. Describe the main environmental effects of the agricultural and industrial revolutions.
3. Describe the human population explosion and its causes.
4. Explain what is meant by the demographic transition, and give some reasons why it occurs; explain why the human population is still growing even though the birth rate is declining.
5. Describe the biological basis of the green revolution, and discuss why it is not the whole answer to the problem of producing enough food.

6. Explain how deforestation can lead to extinction of wildlife, flooding, and water shortages.
7. Explain the sources of acid rain and list some of its effects on ecosystems.
8. Explain global warming and name important greenhouse gases and their sources.
9. Contrast the source and effects of ozone found near Earth's surface and in the ozone layer of the upper atmosphere.
10. Explain how nutrient and thermal pollution hasten eutrophication of a lake.

---

L ike other organisms, humans interact with their environment. Because of our numbers and our technology, however, we affect our environment more dramatically and more permanently than do populations of any other species.

During the course of history, two changes in the way we live have been particularly important in altering our relationship with our environment. The first of these was the development of agriculture, the practice of growing our own food instead of finding it. The second was the industrial revolution, which also completely changed the way we live

and our effects on the environment. These changes have led to many of the environmental problems that confront us today (Figure 50–1).

**FIGURE 50–1** Farms in Scotland, products of agriculture and industry, two major changes in the way humans live that have had huge impacts on the environment.

1081

**KEY CONCEPTS**

- Ecologically, the most important developments in human history were the spread of agriculture and the industrial revolution.
- Modern problems that result from agriculture include population growth, habitat destruction, food simplification, and soil loss.
- Modern increases in human life expectancy are due mostly to better nutrition and hygiene, which have boosted infant survival.
- Low birth rates are found where women have some education and economic control over their own lives.
- The human population is growing rapidly despite a recent decline in birth rates. Many people consider this to be the biggest current human problem, as well as the biggest ecological problem.

- Wheat, rice, corn (maize), and potatoes supply more than half of all human food. Cereals are the most important human food because they contain the minimum nutritionally necessary protein content and are easy to store.
- Industrialization has accelerated every kind of human impact on the environment, including population growth, resource depletion, and pollution.
- Pollution degrades our environment when we expel wastes faster than they can be broken down or diluted by natural processes.

## 50–A HUNTER-GATHERERS

For most of our evolutionary history, humans were **hunter-gatherers,** people who obtain their food by collecting wild plants and killing wild animals. Our knowledge of hunter-gatherers comes from archeological finds and from studies of modern hunter-gatherer groups such as Eskimos, Australian aboriginals, and the !Kung bushmen of the Kalahari desert in southern Africa.

Cooperative efforts permit hunters to spear or trap large game animals. But the main source of food in hunter-gatherer groups is plant materials—seeds, fruits, leaves, nuts, and roots—used as they are gathered or preserved and stored for later. The use of fire opened up a new range of plant foods. Cooking can break down plant toxins and soften plant tissues, making them more digestible.

Seafood is another source of food for hunter-gatherers. Native Americans in the southeastern United States left piles of shells up to 20 feet high from the oysters that provided most of their protein. They also caught fish by spearing and trapping them. Although some fish are farmed today, most of our fish still comes from "hunting" these animals in their wild habitat.

Many early hunter-gatherers were nomadic, moving from place to place according to the availability of food. Later, more-or-less permanent settlements developed, and people began to build more elaborate dwellings and to accumulate possessions such as the decorated tools and pots that began to appear in Europe and Asia about 20,000 years ago and in America at least 12,000 years ago. Human populations spread from Europe and Asia to the Americas and Australia some 25,000 years ago.

Settlements, permanent or temporary, usually lead to environmental problems such as pollution. Deformities in skeletons from ancient villages suggest that some people suffered from diseases caused by polluting the water. A village well or a nearby stream is easily polluted by human waste or the runoff from a garbage pile.

Even the small populations of ancient hunter-gatherers affected their environment in many ways. As they traveled to new areas, people carried seeds with them, altering the distribution of plants (and possibly animals). We can only guess at the size of the Old Stone Age population (before the invention of agriculture), but estimates put it at around 5 million people. This would seem to be too few to have had very dramatic effects on the environment. Nevertheless, human activities such as starting or spreading fires do not need many people to produce major effects.

During the Old Stone Age, a large number of mammals in Africa, Europe, and North America became extinct. The traditional explanation has been that these extinctions were caused by rapid changes in climate during the Ice Ages, which occurred between about 2 million and 10,000 years ago. Some researchers, however, believe that many of these extinctions were caused by human hunting and destruction of habitat by starting fires to drive game animals into traps. A site at Solûtre in France contains the remains of more than 100,000 horses apparently trapped by hunters. Native Americans slaughtered bison by the thousands by driving them into pits. However, there were still about 60 million bison in North America when European settlers arrived. It took less than 200 years of hunting with firearms to reduce the bison almost to extinction in 1880.

## 50–B AGRICULTURE AND INDUSTRY

**Agriculture** is the practice of breeding and caring for animals and plants that are used for food, clothing, housing, and other purposes. **Domesticated** animals are those whose reproduction is controlled by humans. Agriculture originated, probably independently, in many different places at about the same time some 10,000 years ago. Fossils of domesticated dogs dating from 11,000 years ago have been found in Iraq, and cultivated plants in America date back almost as far.

At first, agriculture merely provided part of the food for a hunter-gatherer society. For instance, when Spanish missionaries landed in the southeastern United States in the sixteenth century, they found hunter-gatherers living largely on deer, seafood, and local plants. In addition, many tribes cultivated a few plants around their huts. They also cleared temporary plots in the forest, where they grew maize (corn) and beans, using seeds originally imported from Mexico.

Eventually, most human populations became completely dependent on agriculture for food. Although it occurred slowly, over a period of thousands of years, the changeover from hunting and gathering to agriculture had such a dramatic impact on human societies that it is often called the **agricultural revolution.**

We are so used to thinking of agriculture as a superior way of life that the relative advantages of hunter-gatherer culture may come as a surprise. Hunter-gatherers do not face the constant battle with pests, droughts, and famine that beset all agricultural communities. Studies in southern Africa during a drought showed that farmers starved while the population of hunter-gatherer bushmen in the Kalahari desert remained stable in size and the people well-fed (Figure 50–2). This is probably because hunter-gatherers keep their populations well below the carrying capacity of their territory. This is not a result of a high death rate. The people make a conscious effort to keep their population size down by such practices as abstention from sexual intercourse, abortion, infanticide, late marriage, and late weaning. Furthermore, these bushmen have a more balanced diet than most farmers, and their life expectancies are comparable with those of agricultural peoples in most of the world.

A striking consequence of agriculture is that a new type of division of labor grows up within the group. In prosperous agricultural societies, a few people can produce food for everyone. The rest are freed to become builders, bakers, tailors, and merchants. Finally, the population may even be able to afford the luxury of poets, scholars, and students, who contribute little to the group's physical well-being but are the basis of its cultural life.

Once farming had begun anywhere on Earth, it inevitably spread. Because farming supports larger populations, an agricultural community always expands into the land of any nearby hunter-gatherer group, fighting for the territory if necessary and driving the hunter-gatherers to extinction or to become integrated into the farming community. Similarly, settled agriculture overwhelms nomadic livestock-herding as a way of life. For instance, the battle for land between farmers and ranchers in the West is part of American history. In Africa today, expanding agriculture is destroying the traditional lifestyles of Masai and Bedouin herders.

The roots of most environmental problems can be traced to the evolutionary success of agricultural societies. No one believes we can return to a pre-agricultural way of life. The question is how our destructive agricultural way of life can be changed into a sustainable way of feeding our populations.

The most important effects of agriculture can be summarized:

1. **Habitat destruction.** The area of land devoted to farming has increased steadily, with the conversion of natural habitats to farmland causing the extinction of thousands of species of plants and animals.
2. **Soil erosion.** Agriculture need not destroy the soil, but it usually does. The amount of soil on Earth's surface has steadily decreased since the first digging tool was invented thousands of years ago.
3. **Population increase.** Not only can agriculture support more people on a given land area than hunting and gathering, but the population control practiced by hunter-gatherer societies is usually abandoned by agricultural communities. This is partly because children, who are not important food collectors in a hunter-gatherer society, are useful as labor on a farm. In addition, the inheritance of land and goods becomes more important. The desire to have children who will inherit the property and care for their aged parents is a recurrent theme in mythology and literature. Only in the last 50 years have many people in agricultural societies practiced effective population control.

**FIGURE 50–2** A !Kung woman gathers plant roots in the desert in Botswana. (S. Trevor/Bruce Coleman)

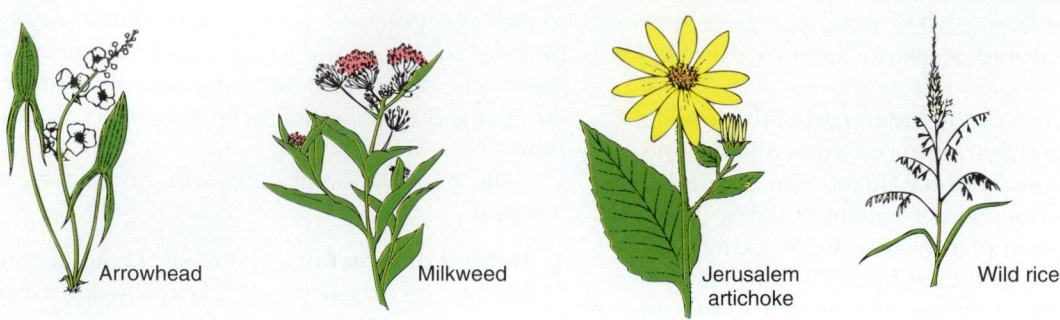

| Plant | Type | Uses |
|---|---|---|
| Arrowhead<br>*Sagittaria latifolia* | Tall plant that grows in mud and shallow water | Starchy roots boiled; can also be dried for later use |
| Bearberry<br>*Arctostaphylos uva-ursi* | Trailing plant with bell-like flowers found in sandy and rocky places | Berries cooked with meat for flavoring; leaves were smoked like tobacco |
| Black cherry<br>*Prunus serotina* | Large deciduous tree in eastern forests from Nova Scotia to Florida and Texas | Twigs boiled to make tea; cherries cooked and then formed into cakes, which were dried for winter use |
| Bur oak<br>*Quercus macrocarpa* | Large deciduous tree found in southern Canada and the eastern U.S. | Nuts (acorns) boiled, or roasted, split, and eaten as vegetable or with grease |
| Canadian hemlock<br>*Tsuga canadensis* | Evergreen coniferous tree of northern forests | Leaves boiled to make tea |
| Hawthorn<br>*Crataegus* | Small tree found in most of the U.S. | Raw berries, rich in vitamin C, squeezed into cakes and dried |
| Jerusalem artichoke<br>*Helianthus tuberosus* | Tall sunflower found in open spaces in eastern, midwestern, and southern U.S. | Roots eaten raw like a radish or fried |
| Labrador tea<br>*Ledum groenlandicum* | Woody shrub found in northern bogs | Leaves boiled to make tea, which might be sweetened with maple sugar |
| Milkweed<br>*Asclepias syriaca* | Tall flowering plant found in open areas | Flowers cut up, stewed and eaten like preserves |
| Mountain mint<br>*Pycnantheum virginiana* | Small perennial flowering plant found throughout the eastern U.S. | Flowers and buds used to season meat and broth |
| Virginia creeper<br>*Parthenocissus quinquefolia* | Woody vine found in southern half of eastern North America | Stems boiled then peeled to reveal the sweet layer under the bark which was eaten like corn on the cob |
| Wild ginger<br>*Asarum canadense* | Flowering plant found in northern and eastern woods | Roots used as flavoring; also used for indigestion |
| Wild rice<br>*Zizania* | Cereal (grass) that grows in shallow water of lakes from Canada to Texas | Processed seed boiled or fried; eaten with meat, maple sugar, or blueberries |

**FIGURE 50–3**   Food simplification. A few of the plants eaten by native American hunter-gatherers that have disappeared from the diet in agricultural America.

4. **Food simplification.** Studies of hunter-gatherers make it clear that humans once ate a much greater variety of plant and animal species than we do today (Figure 50–3).

Several thousand species of plants and several hundred species of animals once formed part of the human diet. The Cherokees of North America ate more than 500 different species of plants (Figure 50–3). Nowadays, few people eat more than 50 species of organisms, and the vast majority of human food comes from just four species of plants: wheat, rice, maize, and potatoes. This dependence on a few species lays societies open to famine on a scale unimaginable in earlier times. A new disease or pest of a major crop rapidly eliminates a large part of our food supply, and thousands starve.

## Industrial Revolution

For some 8000 years after the beginnings of agriculture, most people lived on farms, and towns were mainly centers of trade and culture. The sources of energy were such things as wood, sunlight, and moving water. These sources are **renewable,** replaced by natural processes. Then, some 300 years ago, people began to make extensive use of **nonrenewable** energy sources, particularly fossil fuels. The enormous amounts of energy provided by fossil fuels changed society in ways that are collectively known as the **industrial revolution** (Figure 50–4).

Like the agricultural revolution, the industrial revolution has reduced the area of land required to support each individual and has increased the consumption of natural resources. This accelerating use of resources has led to enormous increases in our standard of living, but it has caused many environmental problems. These include the modern population explosion, pollution, and diminishing reserves

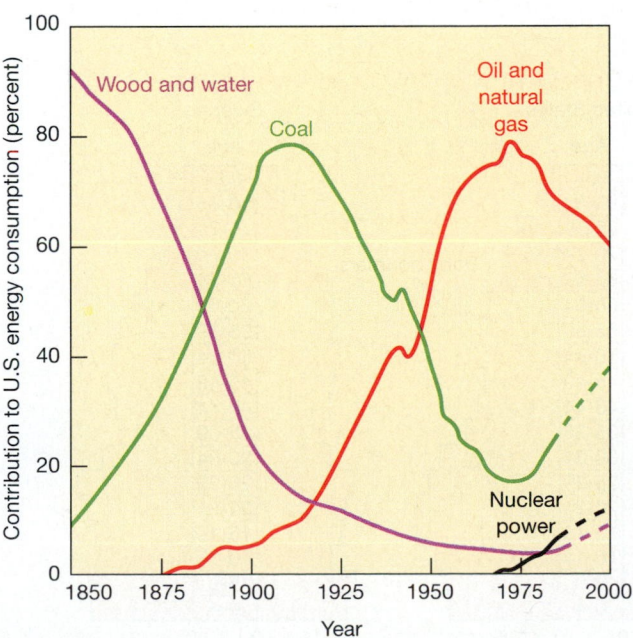

**FIGURE 50–4** Changing U.S. fuel sources and the industrial revolution. The graph shows the percentage of energy supplied by various sources since 1850. Note the development lag: it takes many years for a new source of energy to make a significant contribution to the supply.

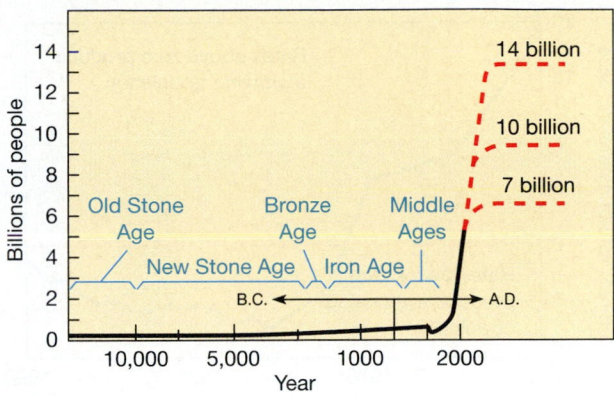

**FIGURE 50–5** World human population growth, with three projections into the future. The bottom curve assumes that the average worldwide birth rate will continue to fall, reaching replacement level in 2035. The middle curve may well be the most realistic.

of the fuel and minerals that are the basis of the industrial revolution itself.

We have used industrial technology to clean up as well as to pollute our environment. Cities in places such as North America and Western Europe are more pleasant places to live than they have ever been before. This is of prime importance because a major effect of the industrial revolution has been to hasten urbanization—the movement of people from the countryside, where relatively few people now find jobs on farms, into cities and suburbs, where industries are established and most jobs are to be found.

## 50–C  HUMAN POPULATION GROWTH

**Demography** is the study of human populations, often in order to forecast how populations will change in the future. Demographers tell us that the number of people on Earth has increased steadily for at least 10,000 years. But the greatest population growth has taken place in the last 200 years, and the number of people added to the population each year is still increasing (Figure 50–5). During the 1990s nearly one billion people will be added to the human population, more than in any previous decade. They will increase the population by 20%, to more than 6 billion in 2000.

Population forecasts are notoriously unreliable because human reproductive behavior often changes rapidly (Figure 50–6). Large groups of people, like the American "baby boomers," often make similar reproductive decisions at the same time in response to a particular widespread economic or cultural situation. Because human behavior is so unpredictable, population forecasts are constantly updated. The characteristics that demographers find particularly useful in making forecasts are the population's age structure, survivorship, and fertility.

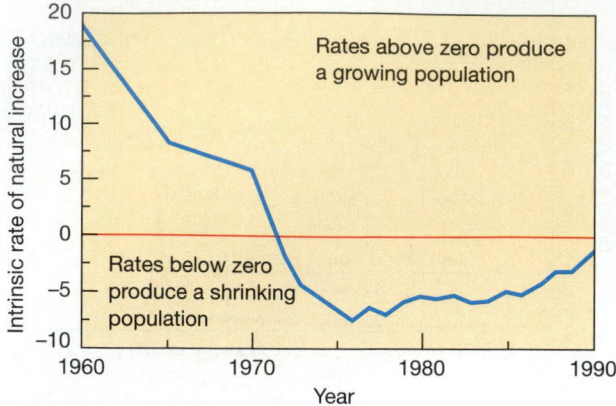

**FIGURE 50–6** Changing reproductive behavior in the United States from 1960 to 1990. The intrinsic rate of natural increase is a measure of what would happen to the population if the birth rate in that year were continued for a long period.

## Age Structure

A population's **age structure** is the number of people in each age group at a particular time. Figure 50–7 shows age structure histograms for three human populations. Each histogram shows the age and sex composition of a population at the time of a **census,** the procedure by which countries count and gather facts about their populations.

Age structure permits us to draw some tentative conclusions about the population's future development. The roughly parallel sides of the Swedish histogram tell us that there are about the same number of people in each age group. The histogram will have this shape if each couple produces two children, who will replace their parents. If this pattern continues, the population will neither increase nor decrease, and we would predict that it will remain stationary into the foreseeable future.

The narrow base of the U.S. histogram is typical of a population in which growth is slowing but has not stopped. In fact, the United States has the fastest-growing population in the developed world.

Predictions from age structures are not always as easy to make as they appear. Consider the broad base of the Mexican graph. The population contains more children, who have yet to reproduce, than it does adults. The obvious conclusion would be that when these children move up to reproductive age, the population will grow rapidly. Without birth control and with low death rates for children, this would be true. However, in Malawi (Africa), which has an age histogram very like the Mexican one, 75% of deaths occur among children under 14 years old. Therefore, many juveniles will not grow up to reproduce, and the population is not expanding nearly as fast as one would predict from the age structure. To predict the future of a population, then, we need to know not only the age structure and the overall rates of birth and death, but also whether death occurs before or after people reach reproductive age.

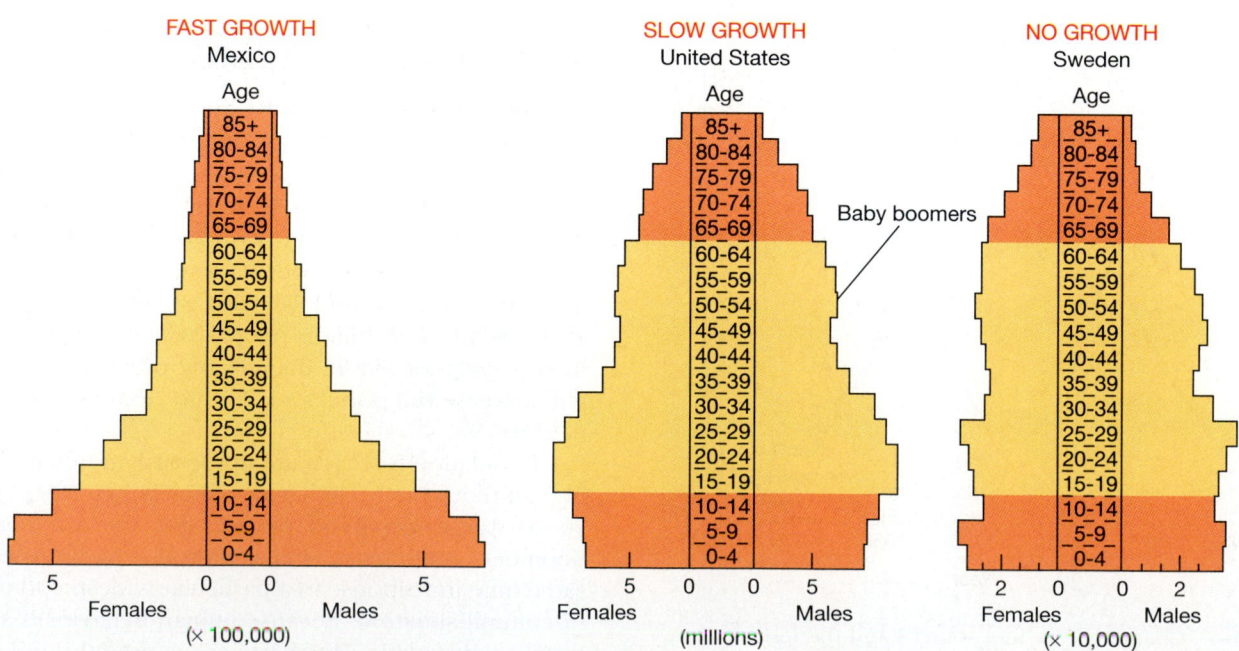

**FIGURE 50–7** Age structure histograms for three countries with different demographic profiles. The histograms show the number of males and females in each age group. The colors show the number of workers (yellow) and dependents (orange) in each population.

## Survivorship

The age distribution of deaths in a population may be shown as survivorship curves or life tables (Section 49–A). Among other information, these figures give demographers a measure of the economic impact of the population's age structure. The **dependency ratio** is the ratio of people over 65 and under 15 to the rest of the population. These people contribute little to the economy and must be supported by the workers, generally 16- to 65-year-olds. Japan, for instance, has 47 people over 65 or under 15 years of age for every 100 working people: about two workers to support every dependent. The ratio is 55 per 100 in the United States. Although the number of children under 15 is declining in developed countries, the dependency ratio is still rising in many because the age at which people retire has not been raised to reflect longer modern lifespans.

High dependency ratios are a burden on the economies of most developed nations, but they are a much worse problem for less-developed countries, where nearly all the dependents are young rather than old. Already many developing countries have more dependents than workers. For example, Ghana (in Africa) has 104 dependents for every 100 people of working age, and that dependency ratio is increasing.

## Fertility Rates

The **general fertility rate** of a population is the number of babies born per 1000 women of reproductive age. However, the rate of population growth depends heavily on the age of first reproduction (Section 49–A). If women start having babies younger, the population starts to grow even if the general fertility rate remains the same. Therefore, whether a population is increasing, decreasing, or stationary is more accurately discovered from the **age-specific fertility rate,** the number of births per 1000 women of each age group. Figure 50–8 shows the age-specific fertility of women in the United States. From this graph we can see that women aged 20 to 24 have more children than women of any other age interval and that women are having fewer babies, and starting later in life, than their mothers did. This explains why we observe that U.S. population growth is slowing.

One of the most useful measures of fertility is **total fertility rate,** the average number of children per woman during her lifetime. The total fertility rate for the United States was 3.4 in 1960 and 2.0 in 1990 (1.8 for whites; 2.6 for all other races). In 1972, for the first time, the total fertility rate dropped below 2.1, which is **replacement-level fertility,** the number of children each couple must have if they are to replace themselves. (The number is more than two because not all children born will survive to reproductive age and not all those who do will reproduce.) If this fertility rate continued, the U.S. population would eventually start

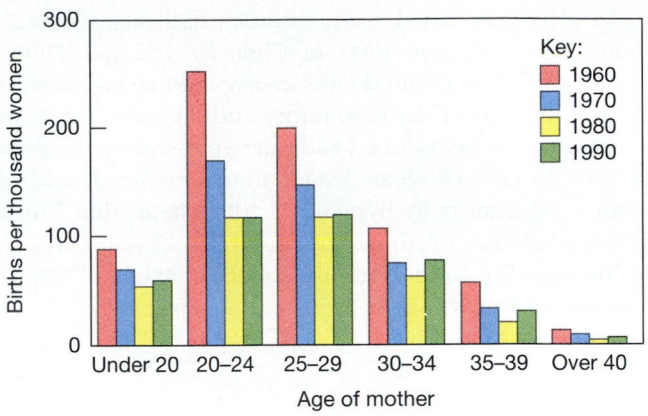

**FIGURE 50–8** The age-specific fertility of American women in 1960, 1970, 1980, and 1990. Births per 1000 women in each year are plotted against the age at which the mother gave birth. Note that the number of births in all age groups fell from 1960 to 1980 and then rose slightly in 1990. The average age of the mother has increased during this period (compare women in their twenties in 1960 and 1990).

to decrease (if we ignore immigration). The total fertility rate can, of course, change. It predicts future population size most accurately when it remains constant over a period of time.

## Declining Death Rates

With fertility rates falling worldwide, why is the human population still growing exponentially? Our population explosion is largely a result of reduced death rates in most age groups, but particularly among infants. **Infant mortality,** the death rate in the first year of life, is the most important factor determining **life expectancy,** the average number of years that a newborn baby can be expected to survive. In 1900, infant mortality in North America and Europe was about 40 per 1000 babies born, and the life expectancy was 46 years. In 1991, infant mortality for developed countries had declined to fewer than 10 per 1000, and life expectancy was more than 70 years. The world's highest life expectancies are those of Japan, Israel, Canada, Australia, and the European Community, where people can expect to live beyond age 77 (with women outliving men by about six years). In 1991, life expectancy in the United States was 75.7 years, lower than that of several much poorer countries such as Costa Rica.

One reason for low life expectancy in the United States is that it has one of the higher infant mortality rates in the developed world. In 1991 it was 10.3 per thousand, compared with 4.4 for Japan and 4.7 for Switzerland. This is because in the United States health education and medical care do not reach everyone in the population as they do in other developed countries, all of which have long had national health services.

In 1991, less-developed countries had infant mortality ranging from 17 per 1000 in Chile to 163 per 1000 in Afghanistan. Most infant deaths in low-income countries are due to infections of the respiratory and digestive tracts, and such deaths can be reduced without expensive medical care. The simple lack of clean water, nutritious food, and elementary education in hygiene is the reason that life expectancy for most of Africa remains about 45 years, whereas the average for Latin America and East Asia has risen to more than 65 years since 1950.

## 50–D   THE DEMOGRAPHIC TRANSITION

In 1945, demographer Frank Notestein outlined the theory of the **demographic transition,** which states that the economic and social progress caused by the industrial revolution affects population growth in three stages:

1. In the first stage, in preindustrial societies, birth and death rates are both high, and the population grows slowly, if at all.
2. In the second stage, a population explosion occurs. Death rates fall as a result of widespread education and public health measures, but birth rates remain high, and so the population grows very fast. It increases by about 3% per year, which means it doubles every 25 years.
3. In the third stage, birth rates fall until they roughly equal death rates, and population growth slows and then stops. Thereafter, population size remains stationary.

The theory of the demographic transition was based largely on what had happened in Europe. The developed countries of Europe and parts of Asia have indeed passed through these three stages, reaching stage 3 about 20 years ago. Populations in these countries have increased as much as fivefold since 1850 but today are growing slowly or not at all. The United States is unusual among developed countries in that its population is still growing rapidly, partly because of a high rate of immigration, although the growth rate has slowed slightly since 1960.

The demographic transition has taken from one to three generations to spread through the populations in most developed countries. During its progress, death rates are low, but birth rates remain high, and so the population grows enormously. The population of the United States has almost quadrupled during the twentieth century.

### Educating Girls

The single factor most closely correlated with a decline in birth rate during the demographic transition is increasing education and economic independence for women. The spread of knowledge that lowers the death rate is usually part of a general program to improve education. Educated

**FIGURE 50–9**   A school in Costa Rica, one developing country that believes in educating its girls and provides free primary and secondary education for all.

women find that they need not bear a large number of children to ensure that a few survive, and they also learn contraceptive techniques. In addition, women find that they can contribute to the family's increasing prosperity by holding a job and by spending less time and energy on raising children. This is attractive to women, even in countries where religious doctrine and tradition dictate large families.

Lawrence Summers, chief economist at the World Bank, has said, "Educating girls probably yields a higher rate of return than any other investment available to a developing country" (Figure 50–9). Summers used figures from Pakistan to illustrate the point. In 1990, an extra year of education for 1000 Pakistani girls cost about $40,000 and yielded a return of more than 20%, including:

1. An increase of 15% in each woman's wages.
2. A 10% reduction in lifetime fertility, saving $43,000 it would have cost in medical expenses to prevent the 660 births that would otherwise have occurred.
3. A 10% reduction in deaths before age 5 among children born to these women, preventing 60 deaths and saving an estimated $48,000 in medical costs.
4. The prevention of four deaths of women in childbirth, saving about $10,000 worth of medical care.

### The Demographically Divided World

Some countries are not proceeding through the demographic transition as predicted by the theory. They appear to have stopped in the second stage, unable to make the educational and economic gains that would reduce the birth rate.

If the vast rate of population growth in the second stage is maintained for any length of time, it begins to overwhelm food production, water supplies, medical care, and education. When the demands of the population exceed the sustainable production of local farming and grazing land, forests, and water supplies, people begin to consume the resource base itself. Vegetation disappears, soil erodes, wells run dry, and productivity declines. This in turn reduces food production and incomes in a downward spiral.

Population trends appear to be driving about half the world's people toward a better future and half toward economic decline. The world is divided into countries that have completed the demographic transition and countries stuck in the second stage, with only a few countries still in the process of passing through the demographic transition. Table 50–1 illustrates this for Asian countries. The countries have been divided into two groups based on their 1990 fertility rates. The countries at the bottom of the table have fertility rates of more than 2.5. The countries at the top of the table have fertility rates of 2.5 or less. The 2.5 rate is slightly higher than replacement-rate fertility. We can see that there is a correlation between fertility rate and gross domestic product (GDP) per person.

In slowly growing areas, mainly Western Europe, North America, Australia, China, and Japan, populations are growing on average 0.5% per year, and living conditions are fairly stable. Some natural resources may actually be increasing, as these countries invest in pollution control, reforestation, and restocking wild populations of plants and animals. In contrast, populations in high-growth countries, chiefly in Southeast Asia, Latin America, India, the Middle East, and

Africa, are growing more than three times as fast—on average, by 2.5% each year. The slow-growth countries add 19 million people a year to the world's population, and the fast-growth countries add more than 80 million. In many of the fast-growth countries, population growth and falling incomes reinforce each other, so that living conditions decline.

Southeast Asia is the only part of the world where several countries are probably still proceeding through the demographic transition toward slow growth and prosperity. Fertility is falling rapidly in Thailand and Indonesia, which have good family-planning programs. In the same region, however, the Philippines has high birth rates and falling standards of living.

## 50–E   FEEDING PEOPLE

The enormous growth of the human population in the twentieth century has resulted in widespread starvation. Demographers estimate that at least 50,000 people die of starvation every day, and at least 10 million children in the world are so malnourished (poorly fed) that their lives are in danger. **Starvation** means death from lack of food. However, most people who are inadequately fed die because their malnourished bodies have little resistance to diseases that would not be fatal to the properly fed. Most malnourished people get about as many calories as they need, but

TABLE   50–1

**Population Growth and Wealth in Asian Countries, with Estimates for the Year 2000**

| Country | Population (millions) | | | Total Fertility Rates* | | | GDP per Person | Adult Literacy Rate |
|---|---|---|---|---|---|---|---|---|
| | *1965* | *1990* | *2000* | *1965* | *1990* | *2000* | *($)†* | *(%)* |
| China | 729 | 1,134 | 1,299 | 6.4 | 2.5 | 2.1 | 547 | 73 |
| Hong Kong | 4 | 6 | 6 | 4.5 | 1.5 | 1.5 | 11,490 | 90 |
| Japan | 99 | 123 | 128 | 2.0 | 1.6 | 1.6 | 23,558 | 99 |
| Singapore | 2 | 3 | 3 | 4.7 | 1.9 | 1.9 | 11,160 | 88 |
| South Korea | 29 | 43 | 46 | 4.9 | 1.8 | 1.8 | 5,400 | 96 |
| Thailand | 31 | 56 | 64 | 6.3 | 2.5 | 2.1 | 1,420 | 93 |
| Afghanistan | 12 | 18 | 27 | 7.0 | 7.2 | 6.8 | 150 | 29 |
| India | 495 | 850 | 1,042 | 6.2 | 4.0 | 3.0 | 350 | 48 |
| Indonesia | 107 | 178 | 219 | 5.5 | 3.1 | 2.4 | 570 | 77 |
| Malaysia | 10 | 18 | 22 | 6.3 | 3.8 | 3.0 | 2,320 | 78 |
| Nepal | 10 | 19 | 24 | 6.0 | 5.7 | 4.6 | 170 | 26 |
| Pakistan | 57 | 112 | 162 | 7.0 | 5.8 | 4.6 | 380 | 35 |
| Philippines | 32 | 61 | 77 | 6.8 | 3.7 | 2.7 | 380 | 35 |

*Births per woman during her lifetime

†In 1990. For comparison, Switzerland had the highest GDP (gross domestic product) per person in the world in 1990 at $35,081. GNP (gross national product) per person for the United States was $21,863.

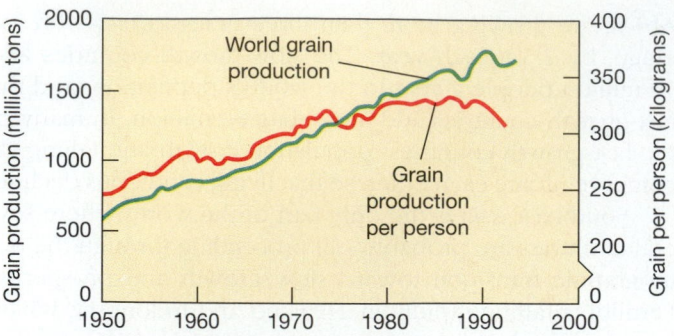

**FIGURE 50–10** World food production falls behind population growth. Grains make up most of the world's food. From 1950 to about 1975, grain production grew faster than population. Since 1975, the amount of grain produced for each person on Earth has fallen as the population has grown faster than grain production. (Data from U.S. Department of Agriculture)

for everyone. A nutrition expert expressed it this way: "If the world's food were equally distributed, we are producing enough food to keep 120% of the world's present population alive but too malnourished to work; enough to keep 100% of the population alive and fit enough to do a little work; but only enough to keep 80% of the population on the diet recommended by the U.S. Department of Agriculture as an ideal human diet."

This situation can only get worse for two reasons. First, population growth continues to outstrip increases in food production (Figure 50–10). Second, efforts to produce ever more food have degraded much farmland to the point that food production is actually falling in many parts of the world. In 1988 the United Nations warned that for the first time in history world food production was lower than in the previous year.

their food is deficient in essential amino acids, minerals, and vitamins.

You will sometimes read that starvation is caused by poverty and distribution problems, not by a shortage of food. This is both true and false. Wars, government corruption, and poor transport systems do prevent food from reaching many people, but even if these problems were solved, in the 1990s there is not enough food in the world

## Grains

Agriculture can feed many more people from a given area of land when people eat plants instead of meat (Section 48–C). The highest efficiency with which domesticated herbivores convert calories from their food into calories in animal products is about 25%, attained in milk and egg production. In contrast, about one third of a plant's productivity can usually be harvested for food, and up to 80% of such a harvest can be digested by humans.

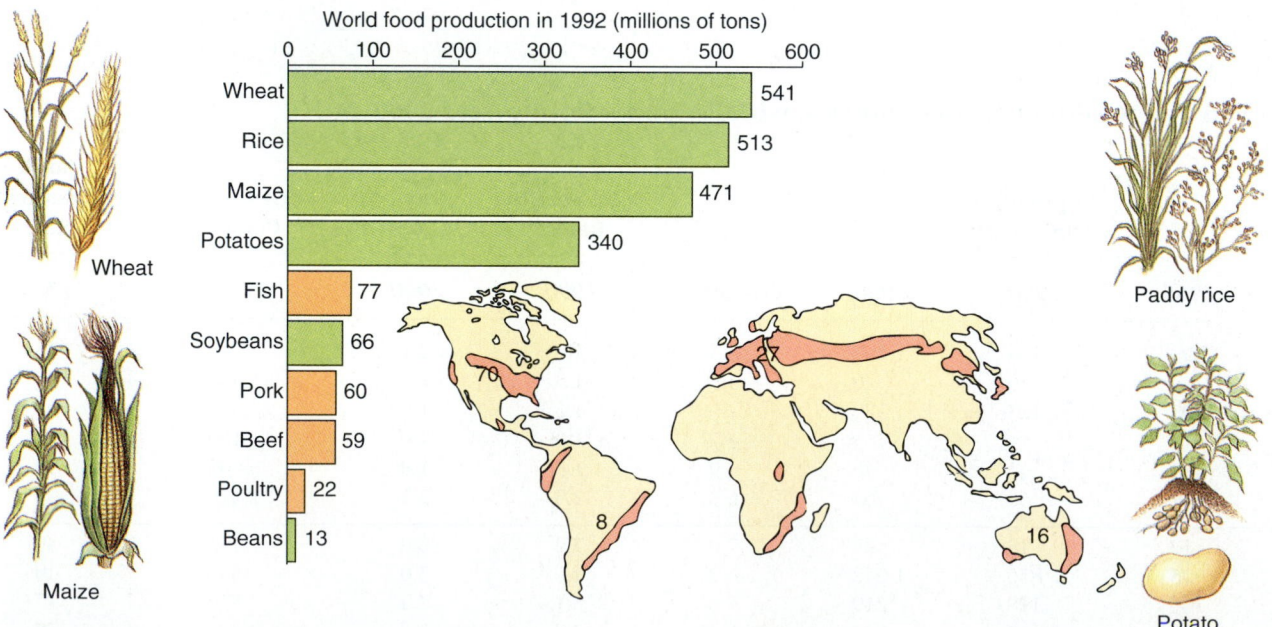

**FIGURE 50–11** World food production in 1992 (plants in green, animals in orange) and the world's four most important food plants. Red areas on the map show the main grain-growing regions of the world. The numbers on the map show 1992 grain exports in millions of tons. Several grain-growing areas have no numbers, meaning either that they produce just enough grain for their own use (India) or that they import grain (China, Russia, Egypt). Note that North America produces most of the grain that goes to food-importing nations.

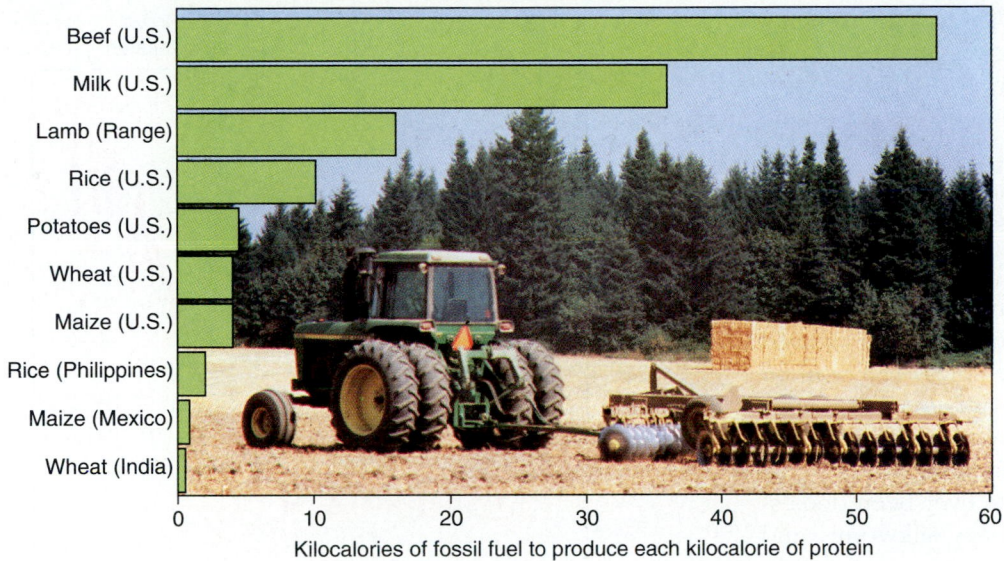

**FIGURE 50–12** The fossil fuel energy used to produce food protein in various types of farming. Most of the fossil fuel powers farm machinery and the production, distribution, and spreading of fertilizer, pesticides, and food for livestock.

More than 70% of the world's arable (crop-growing) land is devoted to growing grains, such as wheat, rice, maize, barley, rye, and oats, all members of the grass family (Figure 50–11). Grains are the fruits of these plants, and they contain the seed's food store laid down by the parent plant: a lot of stored starch and some fats, protein, minerals, and vitamins. Cereals produce high yields, are easy to collect, and may be stored for long periods without spoiling. Cereals are excellent sources of human food, except that they are deficient in some essential amino acids. Besides feeding humans directly, the grains and leaves of grasses are the main food of most domesticated animals.

Food takes a lot of energy and fresh water to produce (Figures 50–12, 50–13). However, the energy used to process, distribute, and cook food is greater than the energy used to produce it in the first place. In the United States it has been estimated that each calorie on our dinner tables has cost nine calories to put there. Only half a calorie represents investment on the farm; the rest represents energy used for processing, packaging, distributing,

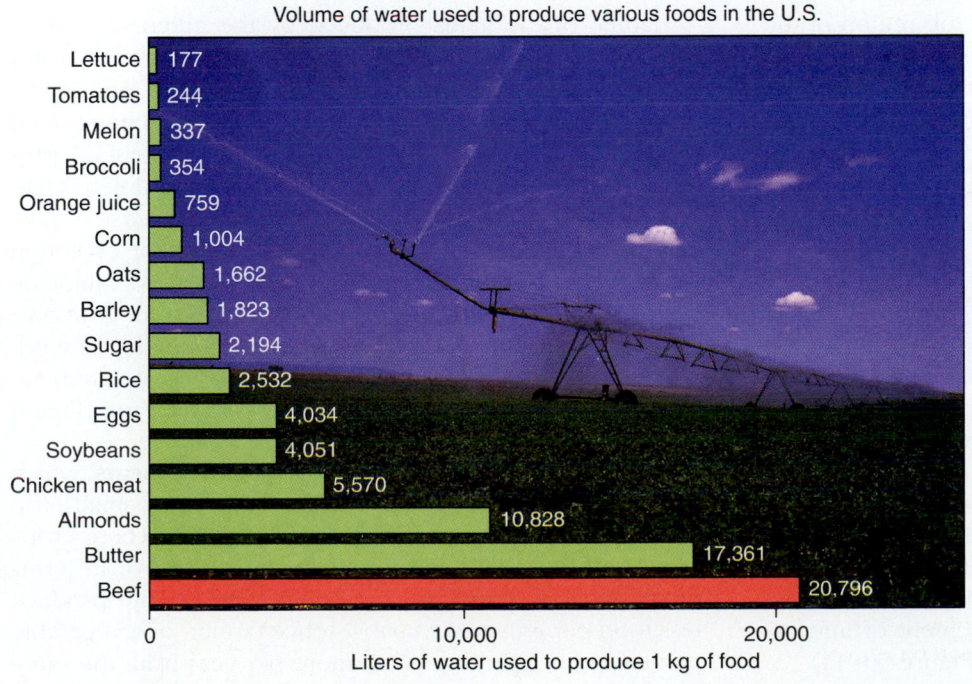

**FIGURE 50–13** The volume of water used to produce various foods in the United States. (Water Education Foundation)

and cooking. In rural India about twice as much energy goes into cooking a kilogram of rice as was invested in producing it. Energy shortages, especially for cooking, have caused problems such as deforestation in many parts of the world.

## Livestock

Although plants are much more important than animals as human food, there are places in the world that can be used for grazing livestock even though the soil is too poor or the rainfall too little for growing crops. Livestock herds have long been the basis of human life in northern Africa, Lapland, and parts of North and South America.

Only about 50 animal species have ever been domesticated, and this figure includes honeybees, silkworms, and a few aquatic animals, such as oysters, carp, and catfish. Aside from cats and dogs, chickens are the most numerous domesticated animals. Then come sheep, and then cattle. Goats, pigs, and water buffalo are also important in many parts of the world.

Fifteen of the 22 most important domesticated animals are artiodactyls, the even-toed ungulates. Many of these are ruminants (cud-chewing animals), including three of the most important: cattle, sheep, and goats. The reason for their importance is that symbiotic microorganisms in ruminants' stomachs permit them to digest food that humans cannot digest, such as the cellulose cell walls of plants. We can therefore use ruminants to convert plants that we cannot digest into foods that we can digest—the meat or milk of cattle, sheep, and goats.

The value of cattle for food, hides, and milk has led to their introduction into parts of the world to which they are not well adapted, such as the semiarid scrub region of the western United States and the tropical forest biome of South America. Advances in breeding and feeding have made cattle more efficient meat producers than they used to be, but beef is still the most expensive meat to produce. More than half the maize grown in the United States is used to feed livestock.

## The Green Revolution

The production of wheat in Mexico increased more than eightfold from 1950 to 1970. In the same period, India doubled its production of grain and now has food reserves, a future that appeared impossible 50 years ago. These spectacular increases, hailed as the "green revolution," resulted from widespread planting of new strains of grains, especially wheat and rice. The green revolution transformed the lives and prospects of hundreds of millions of people and is considered the most successful achievement of international development since World War II (Figure 50–14).

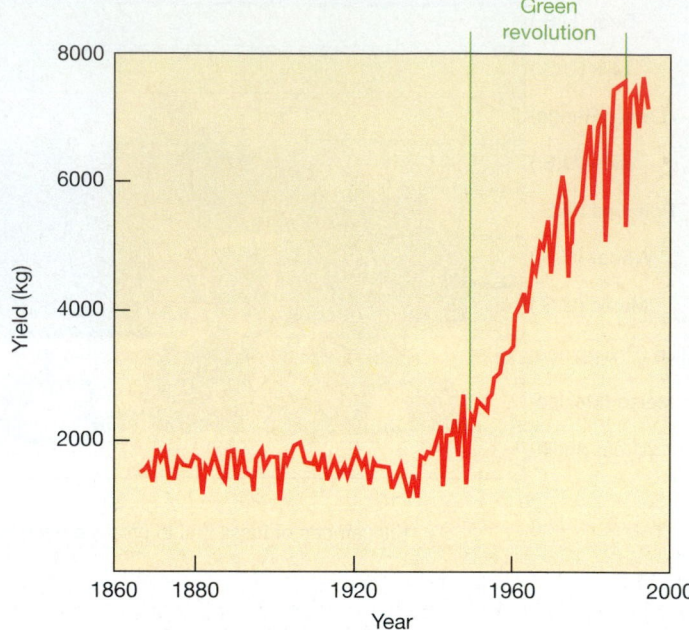

**FIGURE 50–14** Maize (corn) yield per hectare in the United States since 1860. After increasing dramatically during the green revolution, yield per hectare is no longer growing. (U.S. Department of Agriculture)

Unlike traditional crops, these new strains are **high-input crops,** producing enormous yields if, and only if, they are grown with large amounts of water, fertilizer, and pesticides. However, high-input agriculture is environmentally expensive. It uses and pollutes an enormous amount of water. An estimated half of all chemical fertilizer used does not end up in plants but washes off the land, together with eroding soil, and pollutes waterways.

Some of the gains produced by the green revolution are now in jeopardy, partly because of shortages of water for irrigation and partly because intensive high-input farming is destroying the soil. The United States has been forced to take thousands of hectares of arable land out of production because of soil erosion. High-input farming has had other adverse effects. For instance, the intensive use of pesticides has exerted strong selective pressure for evolution of insects and fungi resistant to pesticides. A new fungicide has an average life of only four years before fungi evolve resistance to it. Because most pesticides simply do not work well any more, farmers refuse to spend the money on them, and pesticide use on American farms is decreasing (Figure 50–15).

Most of the world's farms are **subsistence farms,** which provide the family with food and sometimes a small crop that can be sold for cash. Mechanized farms, where crops are raised to be sold, are much more productive in terms of output per person per year, but they usually produce less food per unit area. A subsistence farmer, or a vegetable gardener, can produce much more per year from the same

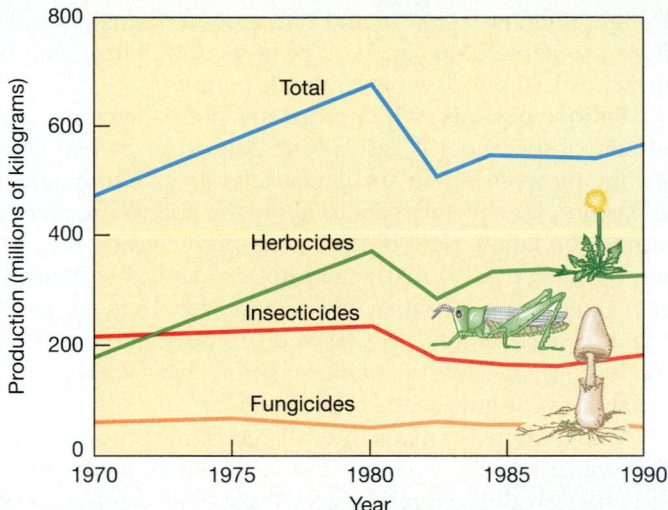

**FIGURE 50–15** Pesticide production in the United States from 1970 to 1990.

culture dictate the opposite trend: farms get larger and larger. The U.S. Department of Agriculture, for instance, predicts that almost 1 million American farmers will go out of business between 1985 and 2000 as they go bankrupt and their land is added to large farms (Figure 50–16).

## No-Till Farming

Experts think that the main increases in future agricultural productivity will come from **low-input farming,** which uses less pesticide, fertilizer, and irrigation than green revolution methods. Low-input farming may be more profitable than conventional farming because the farmer spends less on chemicals and irrigation, relying on such practices as keeping the soil as fertile as possible by adding manure, compost, and other organic matter, keeping the soil planted at all times to prevent erosion, and interspersing different crops to reduce pest infestations. Plant breeders are now working to produce the crop varieties needed by low-input farmers.

The most rapidly spreading new method of growing crops in the United States is simply to stop plowing the soil, thereby reducing soil erosion. Consider a wheat field. Traditionally, the wheat has been cut down and collected by a combine harvester. Then the soil is plowed to turn it over and dig in the remains of the wheat. The soil is raked and the seeds of the next crop are planted.

With the new **no-till farming,** plowing and raking are omitted. The seeds of the next crop are planted in holes drilled in the soil through the remains of the previous crop. The roots of the first crop hold the soil in place while the new crop grows. The decomposing remains of the first crop

area of soil by planting one crop between the rows of another and planting another crop in places where the first is harvested. There is nothing inherently wrong with labor-intensive small farms. In China they make up much of the world's most efficient agricultural system. China feeds more than 20% of the world's population on less than 7% of the world's arable land.

Because small farms produce more food per unit area than large farms, land reform that places more land in the hands of small farmers would boost food production. In most parts of the world, however, the economics of agri-

**FIGURE 50–16** The dwindling number of small farms in the United States. When a small farm goes out of business, the farmhouse is usually abandoned, and the land is added to a neighboring large farm.

add organic matter to the soil. No-till farming reduces water and energy use and reduces soil erosion by up to 90%.

No-till methods also save time. They may permit three crops a year to be grown in areas with long growing seasons. For instance, in the southeastern United States, barley, sweet corn, and soybeans may be grown one after another in one field in the same year. This rotation includes legumes (soybeans) with nitrogen-fixing bacteria in root nodules to add nitrogen to the soil. In colder areas, where crops are not grown in winter, soil is conserved by planting a green manure crop, preferably a legume, in the fall. This produces a layer of organic matter on the soil the following year and often prevents soil erosion completely.

The same effect can be duplicated in the home garden by planting crops in a plot of green manure. Mustard, annual rye, or beans planted in the fall are mowed in the spring, so as not to disturb the roots. Then the gardener digs holes and plants summer vegetables. Methods that reduce soil erosion and restore fertility to the soil are summarized in Table 50–2.

Agriculture is America's biggest industry and the foundation of the country's economic success. The future of North American agriculture is particularly important to the world's food supply because the United States and Canada grow nearly one quarter of all the world's grain. They also export more grain than all other countries put together. Any decline in North American farm productivity threatens worldwide, not just local, food supplies.

## 50–F SOIL DESTRUCTION IN AFRICA

There is not enough good arable land in the world to support the human population. So crops are grown and livestock raised on poor farmland, such as the arid lands and tropical forest that cover a large part of the world.

**Arid land** is land where more water evaporates each year than falls as rain or snow. The deficit is made up from ground water pulled to the surface by the long roots of native shrubs that hold the topsoil in place. These plants grow slowly because the rainfall is low. Nevertheless, arid lands have supported sparse human populations for thousands of years. The practices in the Sahel region of northern Africa are typical of arid land agriculture.

In the northern Sahel, less than 35 cm of rain fall every year, and the rainfall is erratic in place and time. This climate supports shrubs that animals can browse, but there is too little rain to grow crops. The area has long been home to nomadic people who travel with their flocks of cattle, camels, goats, and sheep to wherever rain causes young leaves to grow on the shrubs. This is how the population of northern Somalia and much of Ethiopia traditionally lived.

In the less arid region farther south, farmers live in permanent villages and plant crops. The main crops are drought-tolerant sorghum and millet. More demanding but more profitable crops such as peanuts and cotton may be grown as cash crops where rainfall permits.

**Fallow periods,** when crops are not grown, are a vital part of this type of agriculture. After four or five years of continuous cropping, the land is left idle, or lightly grazed as pasture, for several years to allow the soil to regenerate. During the fallow period, the soil is protected by a covering of vegetation so that it will not erode and so that nutrients and organic matter can accumulate. In some areas, *Acacia senegal* shrubs are allowed to invade the land (Figure 50–17). After five years, these shrubs can be tapped for a cash crop of gum made from their sap.

Traditionally, nomads and village farmers cooperated. The nomads brought meat, milk, and camel-loads of salt and other products to the village. These were traded for cereals, cash crops, and the right to graze their animals on village fallow land when their usual pastures were exhausted.

## Desertification

As human populations of arid lands increase, the land is cultivated, grazed, and deforested faster than it can recover. The land becomes denuded of vegetation, leaving nothing to hold the soil in place or to retain any rain that falls. Topsoil blows away or is washed away by the occasional heavy rains, converting the land to desert. The surviving inhabitants become **environmental refugees.** They move on to add to the population of nearby areas, increasing the chances that the whole cycle will be repeated there.

Twentieth-century conditions have increased the rate of this **desertification** as populations of both humans and livestock have grown rapidly. Population pressures have had numerous results, including:

1. **Overcultivation.** Overcultivation may exhaust soil nutrients, or soil exposed by plowing may also wash or blow away. The soil may become so crusted, under the influence of rain and sun, that plant roots cannot penetrate it. When land is farmed so intensively that fallow periods are abandoned, productivity declines. In central Sudan, 5 hectares (ha) were needed in 1983 to produce the same amount of peanuts as 1 ha in 1961.
2. **Cash crops.** Political and social conditions have led to pressure to grow cash crops such as cotton and peanuts to be sold abroad for foreign exchange or to feed rapidly growing urban populations. Many cash crops, such as cotton, destroy the soil rapidly (as in cotton-growing regions of the American South) and increase the chances of famine among farmers.
3. **Overgrazing.** In Sudan, the number of cattle more than quadrupled between 1957 and 1977, while the pasture to feed them decreased. Well-intentioned governments

### TABLE 50-2
### Farming Methods That Reduce Erosion and Restore Soil Fertility

| Practice | Effect |
| --- | --- |
| **Methods of Slowing Erosion** | |
| Terracing | Holds back soil and water that would otherwise flow downhill |
| Contour farming | Prevents soil washing and blowing downhill |
| Contour strip cropping | At all times keeps strips of vegetation across the hill to catch any soil that washes downhill |
| Reforestation (in areas with enough rainfall to support trees) | If even a little soil is left, trees will grow, holding the remaining soil in place and contributing organic matter to rebuild the soil |
| **Methods of Increasing Soil Fertility** | |
| Spreading organic matter such as compost, manure, and sewage sludge on the soil | Adds nutrients to soil, improving conditions for crop plants and for soil-forming organisms |
| Planting and then plowing in green manure | Green manure plants hold soil in place while growing and add organic matter when plowed in |
| No-till farming | Keeps soil covered with vegetation to prevent erosion; remains of previous crop add organic matter to soil as they decompose |
| Crop rotation including legumes | Adds nitrogen to soil |

**FIGURE 50–17** *Acacia senegal.* This is the spiny small tree that produces gum arabic, used to make glues and other products. Its nutritious seed pods are used for animal feed.

have compounded the problem by digging thousands of wells and water holes since World War II. In places, animals have trampled the ground and denuded it of vegetation for up to 10 km around a water hole. Overgrazing frequently compacts the soil and permits plants that animals cannot eat to replace more palatable species (Figure 50–18). This makes the land essentially useless for farming.

Overgrazing does not always produce permanent desertification. The soil is usually still there and may even be enriched by animal manure. A sufficient period without grazing restores the productivity of much overgrazed land. This simple method has been used by the United States government to restore some overgrazed federal rangeland. The trouble is that, in most parts of the world, this land is never permitted to lie fallow.

### 50–G THE WATER CYCLE

Economists say that the biggest, and least recognized, environmental threat today is depletion of fresh water sources. **Depletion** is the using up of a resource so that it becomes more and more expensive and, eventually, unavailable. For instance, 80,000 ha of grain-producing land in the midwestern United States have been taken out of production since 1980 because it has become so expensive to pump irrigation water from depleted underground supplies. Several countries in Africa and the Middle East already have access to less than the minimum amount of water the United Nations estimates is needed for domestic and agricultural

**FIGURE 50-18** Desertification: a herd of goats visits a stream in Kenya. Once the animals have removed the plant cover, soil is easily eroded. Wind carries away the dust raised by the goats' hoofs, and during the rainy season the stream carves at the walls of its gully.

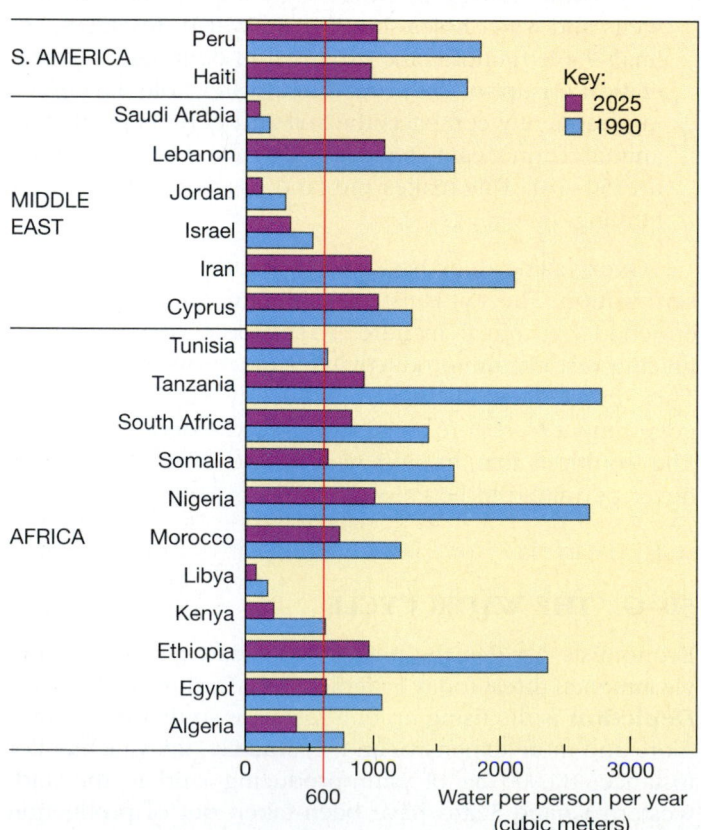

**FIGURE 50-19** Water shortages. The expected decrease in fresh water per person between 1990 and 2025. The United Nations minimum standard of 600 cubic meters per person per year is shown in purple. (Pacific Institute, Oakland, CA)

use. Europe is the only continent with all the water it needs (Figure 50–19).

The World Health Organization estimates that 80% of disease in developing countries is caused by the lack of sufficient clean water for drinking, washing, and sewage disposal. Every hour more than 1000 children die of waterborne diarrhea. In some areas women and children (who usually perform this chore) spend as much as six hours a day walking to bring home drinking water, depriving the children of time to go to school and the women of time to do more useful work.

Most of the water on Earth is salt water in the ocean, and most of the fresh water is locked in glaciers and polar ice caps. We can use only a tiny fraction of the water on Earth, nearly all of it water that is purified by the water cycle (Figure 50–20). The water cycle is driven by the sun's heat, which causes water to evaporate. The evaporating water leaves impurities behind. Some of the water vapor in the air eventually descends on the land as rain or snow. This volume makes up our "water income," the recycling supply of purified water upon which life depends.

The water cycle replaces every body of water more or less rapidly: water leaves the area, and "new" water from rain or runoff takes its place. For instance, the water in any part of a river is replaced every 10 to 20 days. The water in a deep lake may be completely replaced only once every 100 years. The rate of replacement is important to us because it influences how long natural processes will take to replace polluted water once the pollution stops.

More than 80% of all the fresh water used in the United States is used for agricultural irrigation. Many areas face wa-

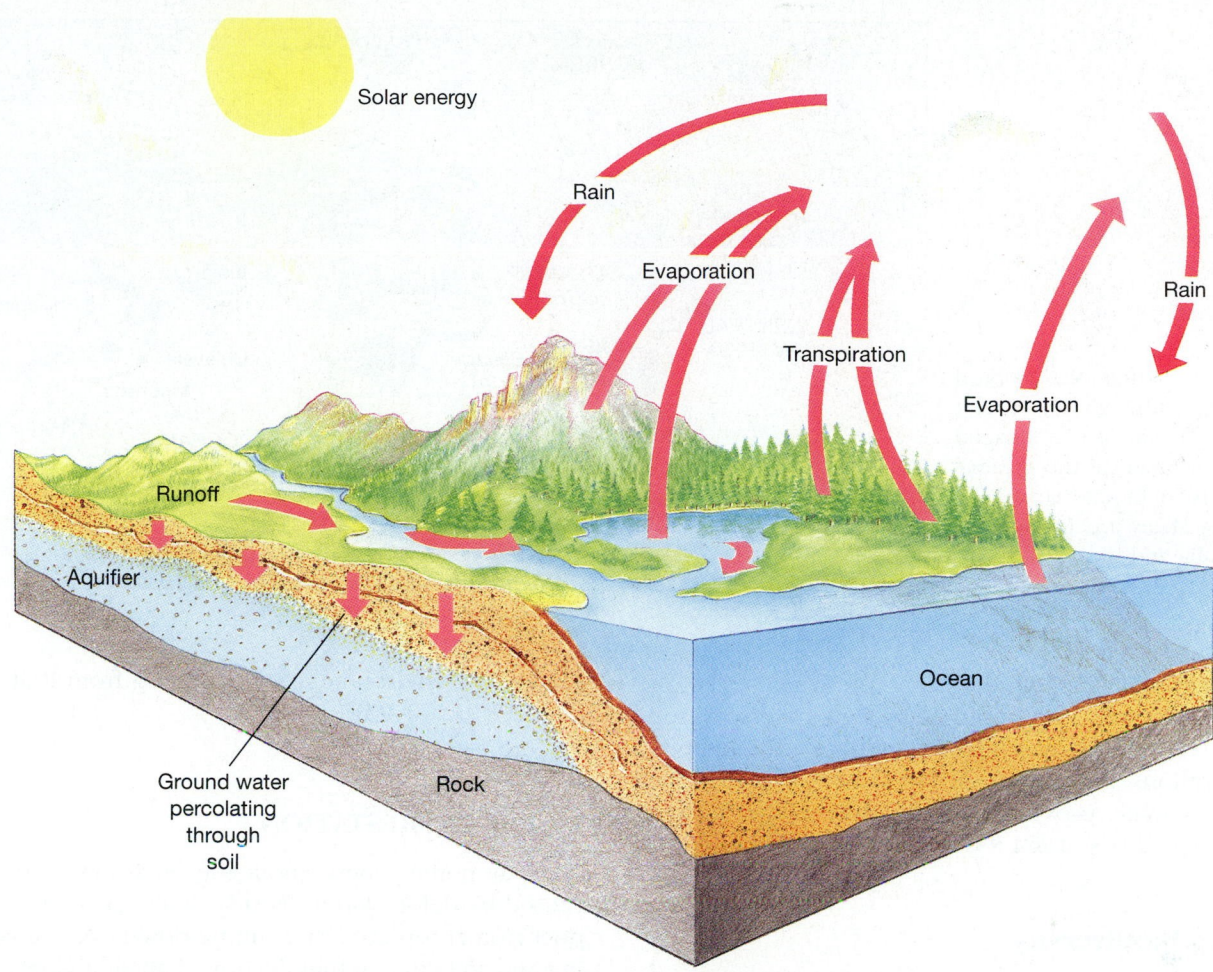

Solar energy

Rain

Evaporation

Transpiration

Rain

Evaporation

Runoff

Aquifer

Ocean

Ground water
percolating
through
soil

Rock

**FIGURE 50–20**    The water cycle. Water that reaches the atmosphere by evaporation or transpiration is pure fresh water.

ter shortages as a result. The media tend to blame southern California's water shortage on lawns, golf courses, and domestic use, but these are not the main causes. More than twice as much water is used on the fields east of Los Angeles as in the urban area itself. This water is used to grow crops such as alfalfa, used to feed cattle.

## 50–H    TROPICAL BIODIVERSITY

Hand in hand with the spread of agriculture goes the loss of forest that is converted into farmland. Today, the destruction of forests and woodlands is changing patterns of rainfall, causing the extinction of forest-dwelling species whose genomes we cannot replace, and producing a shortage of wood for building and fuel.

Moist tropical forest areas have the fastest-growing human populations. Hence, developing nations have been clearing their forests on a massive scale. Tropical rain forests contain about half of all plant and animal species and 80% of the biomass of land vegetation, although they cover less than 10% of the land surface (Figure 50–21).

Ecologists say that we could preserve much of the world's biodiversity by turning just 5% of the Earth's land area into wildlife preserves—most of it tropical forest. This finding stemmed from studies by scientists at the International Council for Bird Preservation who spent three years mapping the breeding distributions of all the world's land birds. They found, to their surprise, that nearly 30% of all birds are confined to 5% of the Earth's land surface and never leave that area.

The country with the greatest biodiversity is Indonesia. This chain of 17,000 islands between Australia and the mainland of Southeast Asia contains more species of mammals, birds, and reptiles than any other country. Indonesia is the fourth most populous nation in the world and includes the island of Java, which is more densely populated even than Bangladesh. Indonesia is trying to save some of its forest, but with a population growth rate of 1.6% a year, this is

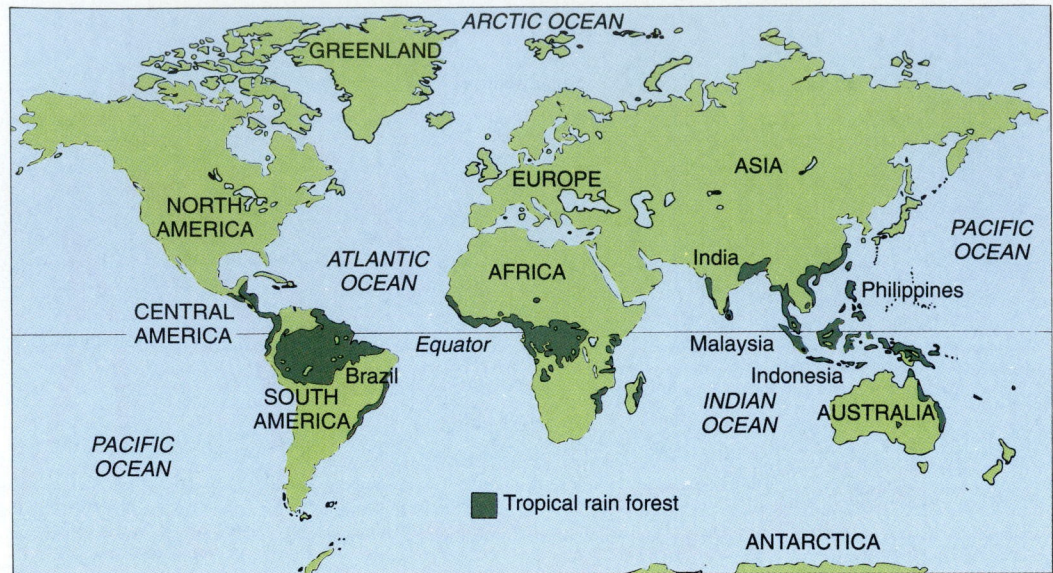

**FIGURE 50–21** Where the tropical rain forest is. Although the largest area is in the basin of the Amazon River in South America, the greatest biodiversity is in Indonesia: on the islands of the Malay archipelago between Australia and the Asian mainland.

not easy (Figure 50–22). On the island of Sulawesi, half of the original forest remains, and new species are discovered there every year. The government, together with conservation organizations, is working hard to preserve some of the island's forest while providing jobs that involve protecting the forest instead of cutting it down.

## Preserving Biodiversity

The Biodiversity Treaty of 1992, signed by more than 100 countries, is an example of the kind of action that biologists hope will save at least some tropical forest. It provides that developing countries shall be reimbursed for economically useful products discovered in their forests. The first example of such a venture was the Merck Agreement.

Merck is a multinational chemical and drug company that entered into an agreement with Costa Rica in 1992. Merck paid Costa Rica's National Institute of Biodiversity $1 million for the right to analyze hundreds of extracts from species indigenous to Costa Rica in the hope of finding useful drugs and other products. If Merck does discover a valuable product and develop it for market, Merck will retain all rights to the product but will pay Costa Rica a royalty for each item it sells.

In turn, Costa Rica will spend 10% of the first $1 million, and half of any royalties, on conservation. The country badly needs the money to buy land and hire park rangers to preserve at least part of the country's rapidly vanishing forests.

Drug companies and developing countries applaud this agreement as an excellent way to exploit and preserve species and habitats. Indonesia and Brazil have expressed interest in similar agreements, and we shall probably see many of them before long—an imaginative way of pre-

serving the environment and profiting from it at the same time (Figure 50–23).

## 50–I DEFORESTATION

There is nothing new about the environmental problems caused by deforestation. Nearly 3000 years ago, the Greek poet Homer reported that cutting down trees caused flooding and soil erosion that destroyed ancient cities. Troy was

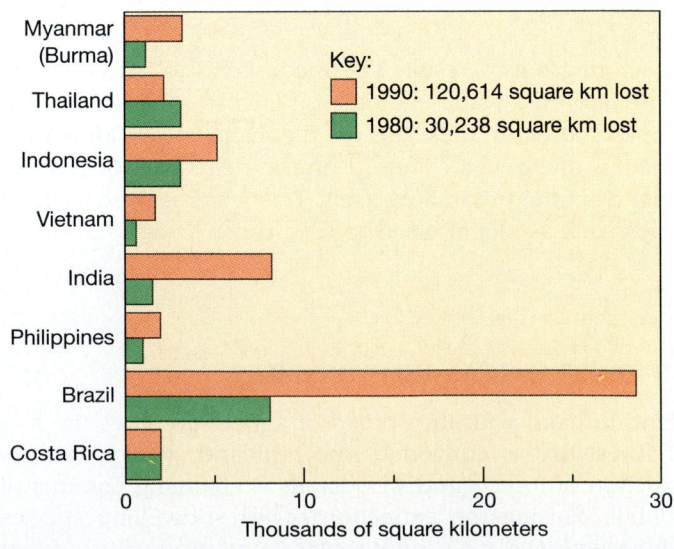

**FIGURE 50–22** Tropical deforestation accelerates. Nearly four times as much tropical forest was cut down in 1990 as in 1980. (Figures for 1980 from the United Nations Food and Agriculture Organization and those for 1990 from the World Resources Institute.)

**FIGURE 50–23** Exploring for useful plants in tropical forest in Panama. The nephew of the local Shaman (medicine man) discusses a plant with visiting botanists. (D. Cavagnaro/DRK Photo)

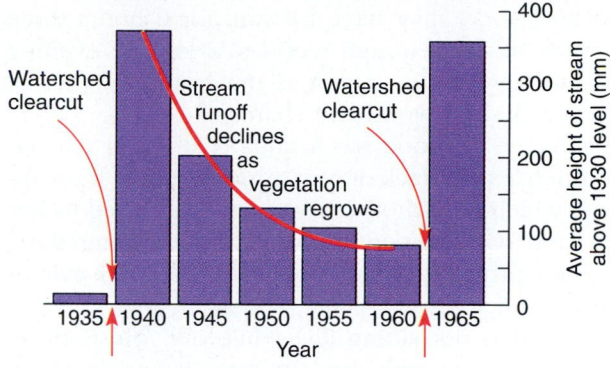

**FIGURE 50–24** Increase in water runoff to a stream as a result of clear-cutting the forest on a watershed in Coweeta, North Carolina. These results show how much precipitation vegetation absorbs. As the forest regrew (1940 to 1960), the stream's volume of flow declined because large trees and shrubs absorb more water than small ones. Deforestation is a major cause of floods.

a seaport on a wooded hillside when the Trojan horse was sent into the city in an attempt to rescue Helen (she of the "face that launched a thousand ships"). Today, the ruins of Troy are several miles from the sea. The inhabitants cut down all the trees on surrounding hills for firewood and building. Without trees to hold the soil in place, rain washed the soil down into the river, where it "silted up the harbor so that no large ship can come in," and Troy fell into ruin.

Deforestation always produces an increase in peak water runoff in streams leaving the forest. The results of an experiment in North Carolina are shown in Figure 50–24. As vegetation regenerates, runoff steadily decreases again. Replacing forest with crops also increases runoff because crops produce less biomass than trees and hold less water.

Deforestation also changes the climate, usually by reducing rainfall. Water from trees is normally transpired slowly into the air, forming clouds that drop rain on areas downwind. When the trees are gone, the water runs off the land into nearby waterways, and much less of it evaporates. For instance, in India, 75% of the trees on the Himalaya Mountains have been felled. The few remaining trees cannot absorb the periodic torrential monsoon rains as the forests once did. Rainfall downwind of the mountains is now too little to support agriculture in some places. Instead, floods kill people and animals and wash away soil. The flood water deposits this soil as the water slows down behind irrigation and flood-control(!) dams. In 1979 a silted-up dam in India burst, unleashing a 5-m-high wall of water and mud that killed 15,000 people.

There are two main reasons for deforestation: clearing land to plant crops and cutting wood to use for fuel or building. Half of all wood is burned as fuel. For most people in the world, wood is the main source of energy, with a consumption of one to two tons per person per year (Fig-

ure 50–25). The growing shortage of wood for fuel creates huge economic burdens. In Africa's Burkina Faso, women

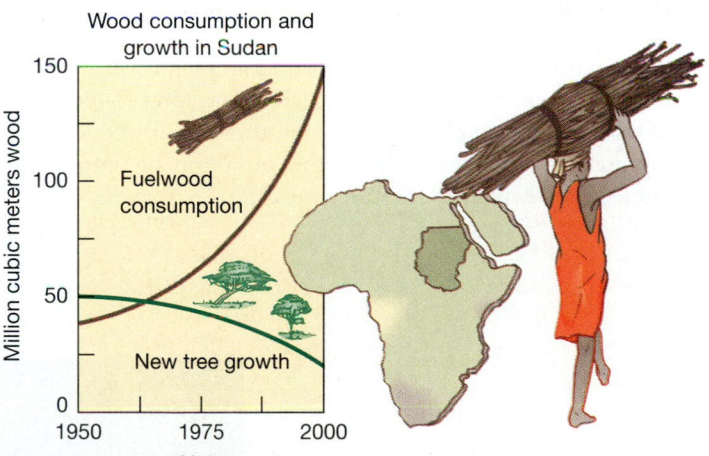

| Fuelwood consumption in 3 countries | | |
|---|---|---|
| Country | Per capita consumption (tons/year) | Percent of population using wood as main energy source |
| Tanzania | 1.8 | 99 |
| Gambia | 1.2 | 99 |
| Thailand | 1.1 | 87 |

**FIGURE 50–25** The fuel wood shortage. The graph shows that fuel wood use has increased much faster than wood has grown in Sudan. Projections of the worsening problem by 2000 are included.

(who do this work) may have to walk for 6 hours three times a week to find enough wood to cook the evening meal, and urban families spend as much as a quarter of their incomes on fuel wood and charcoal.

The shortage of wood has a number of other damaging consequences. For lack of wood, many people burn animal dung, which would otherwise have been used to fertilize the land. One expert estimates that burning dung reduces grain production by enough to feed 100 million people each year.

Humankind is not sitting idle while our forests burn. Reforestation, by governments, lumber companies, and small community groups, is taking place on a massive scale (Figure 50–26). About 11 million ha of tropical forest are felled each year, but by 2000 about 17 million ha of land are expected to be reforested each year. China has doubled its forested area in less than 20 years; New England's forested area is 50% greater than it was a century ago; Chile has doubled its area of pine plantations in 15 years; and in Rwanda, trees planted by rural people cover 200,000 ha, more than the area of the country's remaining natural forest and artificial plantations put together.

## 50–J  POLLUTION

Deforestation and soil erosion are examples of the depletion of natural resources caused by agriculture, industry, and the population explosion. The other major effect of the industrial revolution has been the pollution of natural resources, which makes them less useful.

**Pollution** is an undesirable change in the characteristics of air, water, or land that can adversely affect the health, survival, or activities of humans or other organisms. All organisms expel their wastes into the environment. However, in a sustainable ecosystem, one organism's wastes are another's food and drink. Decomposers break down waste materials and so keep them from accumulating and polluting the environment, and producers convert them back into products that other organisms can use.

Various forms of pollution have been a problem for human societies for thousands of years. But the problems have accelerated in recent years as a result of increasing numbers of people and our increased use of fossil fuels. Pollution cannot be avoided entirely, but it can be minimized. This requires an understanding of its effects on ecosystems. Here we consider a few examples.

### Acid Rain

Since 1970, scientists have documented the death of more than 7 million ha of trees in once-thriving forests in Europe and North America. In parts of Germany, more than 50% of the forest is affected, and the timber lost is worth more than $1 billion a year.

These trees have died of exposure to acid rain, which apparently reduces the trees' ability to absorb nutrients from the soil. The pH of "normal" rain is about 5.6, because carbon dioxide and other gases form acids as they dissolve in rainwater. Since the mid-1960s, however, rain throughout the world has become increasingly acidic, with a pH averaging 4.0 to 4.2 and sometimes reaching 3.0 or even lower.

**FIGURE 50–26** Reforestation. (a) Douglas firs planted by a lumber company in Oregon. (b) Volunteers plant a roadside tree during a community forestry project. (a, Richard Feeny)

(a)

(b)

(Recall that the pH scale is logarithmic; a drop of one unit means a tenfold increase in $H^+$ concentration.) At first, only areas to the east of industrial centers were affected by acid rain because prevailing winds blow from west to east in the industrialized Northern Hemisphere. Now, the rain everywhere on Earth is more or less acidic.

Acid rain does untold damage. It dissolves paint and stone from buildings, kills trees, crops, and garden plants, and destroys natural ecosystems (Figure 50–27). A thick layer of soil, and calcium-containing rocks such as limestone, provide natural buffers that neutralize acidity as rain and ground water trickle through. The first areas to show damage from acid rain lack these natural buffers and have thin soil, underlain by insoluble rock such as granite.

When the pH of a lake falls below about 5, fish and other animals die. A 1979 survey in the Adirondack Mountains of upstate New York showed hundreds of ponds and lakes with a pH of less than 5, many of which had completely lost their fish populations. Water from melting snow is also very acidic, and it may kill the year's crop of fish and salamander eggs and hatchlings if it lowers the pH of the lake water, even temporarily, during the spring breeding season. Norway and some other countries spend millions of dollars every year spreading lime ($CaCO_3$) on their lakes to raise the pH to a level that will support animal life.

Acid rain forms when the air pollutants sulfur and nitrogen oxides react with water to form sulfuric and nitric acids. The pollutants come mainly from industrial coal burning, automobile exhaust, and natural sources such as volcanic eruptions and wetlands (where anaerobic bacteria produce sulfurous gases). Since proteins contain sulfur and nitrogen, burning any organic substance releases acidic

compounds into the air. Rain with a pH of 4 in the mountains of China was traced to the yak dung fires used locally for cooking! In many countries, wood-burning stoves are now required to have catalytic combusters to reduce pollution from burning wood. The only cure for acid rain would be stringent pollution control on every form of combustion.

## Global Warming

The atmosphere is about 2% water vapor (by volume), about 0.03% carbon dioxide, and about 0.0002% methane. These tiny quantities, together with other "greenhouse gases," absorb infrared radiation (heat) reflected from Earth, preventing it from escaping into space. Without its atmospheric greenhouse, Earth would be too cold for life. However, human activities are increasing the quantities of greenhouse gases in the atmosphere, setting the stage for Earth's average temperature to rise as more heat is retained.

The atmosphere's content of methane, produced by anaerobic bacteria, is increasing, partly because of the rapidly growing numbers of ruminants (mainly cattle and sheep), rice paddies, and sewage plants on Earth. The $CO_2$ content of the atmosphere is also increasing. Deforestation has killed so many trees that it has reduced the rate at which $CO_2$ is removed from the air during photosynthesis. When tropical forest is cleared to make way for agriculture, the trees are usually burned to get them out of the way, releasing still more $CO_2$ into the air. However, this produces only about one-fifth as much $CO_2$ each year as does burning fossil fuel in cars, power stations, and factories.

(a)

(b)

**FIGURE 50–27**   Damage caused by acid rain. (a) Trees killed by acid rain in a North Carolina forest. (b) Acid rain has dissolved part of the stone faces of this group of carved figures on a church in Paris.   (a, David M. Dennis/Tom Stack; b, Thom Smith)

It is difficult to say precisely how much our addition of carbon will heat the atmosphere. Both heat and $CO_2$ are stored in various ways, and many factors besides the greenhouse gases affect the average temperature of the atmosphere. Even with these uncertainties, however, most estimates predict a global warming of 1 to 5° C during the next century (Figure 50–28). This temperature rise would produce many effects that we are not prepared for. Climate zones and ocean currents would shift, agriculture would be displaced, and the world's major vegetation zones would shift. Another dramatic effect might be thermal expansion of the ocean and further melting of the world's great ice caps, which would raise the sea level.

Even if the extent of global warming cannot be predicted accurately, it would seem prudent to encourage behavior that counteracts the greenhouse effect as well as being useful in other ways. For instance, by using less fossil fuel, we can reduce air pollution, save money, and reduce

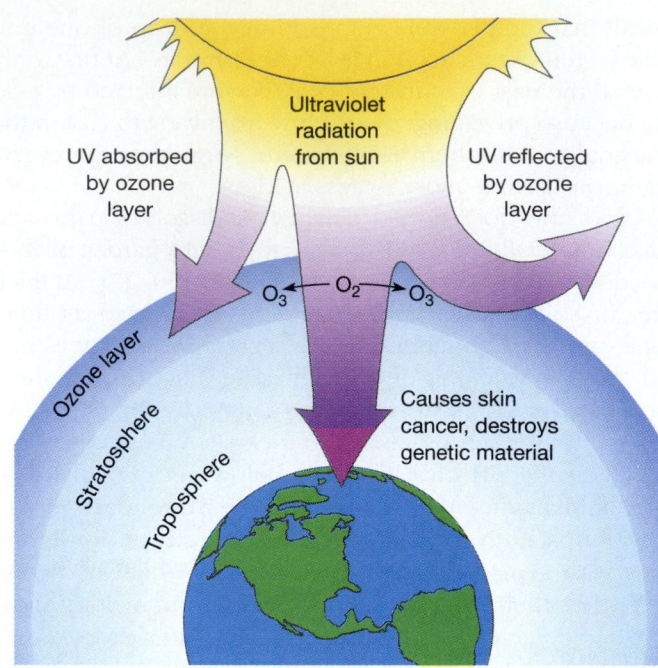

**FIGURE 50–29** The ozone layer and its formation in the atmosphere.

emissions of $CO_2$. Planting trees has the triple advantages of counteracting the greenhouse effect, moderating the local climate, and preventing soil erosion.

## Ozone and the Ozone Layer

Ozone ($O_3$) is a gas that forms in the atmosphere both naturally and as a result of human activities. When it occurs in smog near the ground, it is a dangerous pollutant. When it occurs high in the atmosphere, it is a shield to be preserved (Figure 50–29).

Ozone the pollutant forms in cities where motor vehicles are the main source of pollution. Nitric oxide from car and truck exhausts reacts with oxygen, forming nitrogen dioxide, a yellow-brown gas that produces the colored haze of photochemical smog and has a strong, choking smell. Ultraviolet rays from the sun release one of nitrogen dioxide's oxygen atoms, which reacts with $O_2$ and produces various compounds, including ozone. Smog also contains vaporized organic compounds, such as industrial solvents and spilled or unburned gasoline (Figure 50–30).

The ozone shield in the atmosphere is formed by photochemical reactions. About 30 km above Earth, ultraviolet radiation from the sun is intense enough to form small amounts of ozone from $O_2$. Once formed, ozone itself is a good absorber of high-energy ultraviolet radiation, reducing the amount that reaches Earth's surface. Ultraviolet radiation is dangerous to living organisms because it dam-

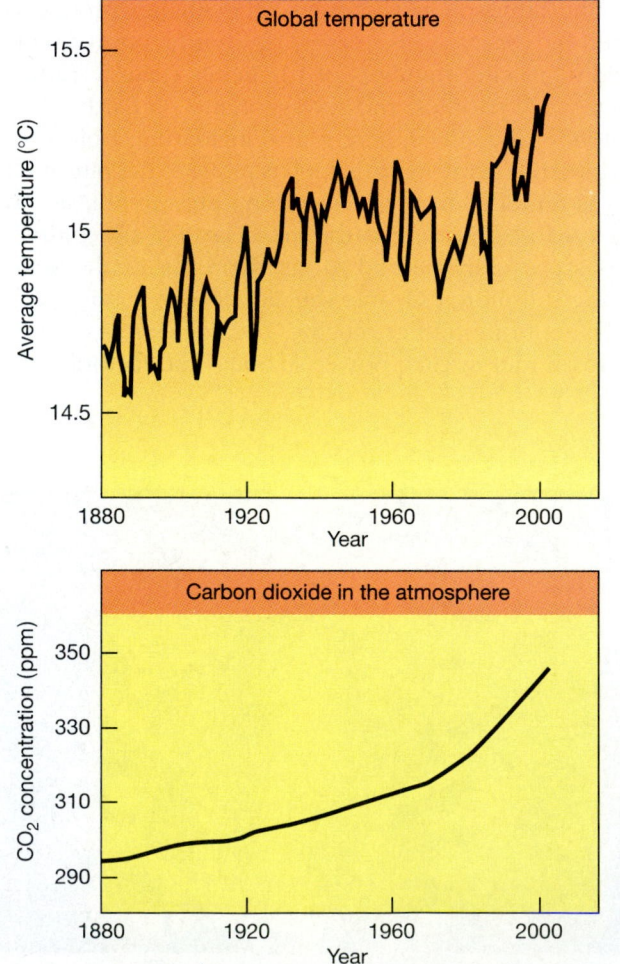

**FIGURE 50–28** Global warming: the increases in average temperature worldwide and average concentrations of carbon dioxide in the atmosphere since 1880. (National Oceanic and Atmospheric Administration)

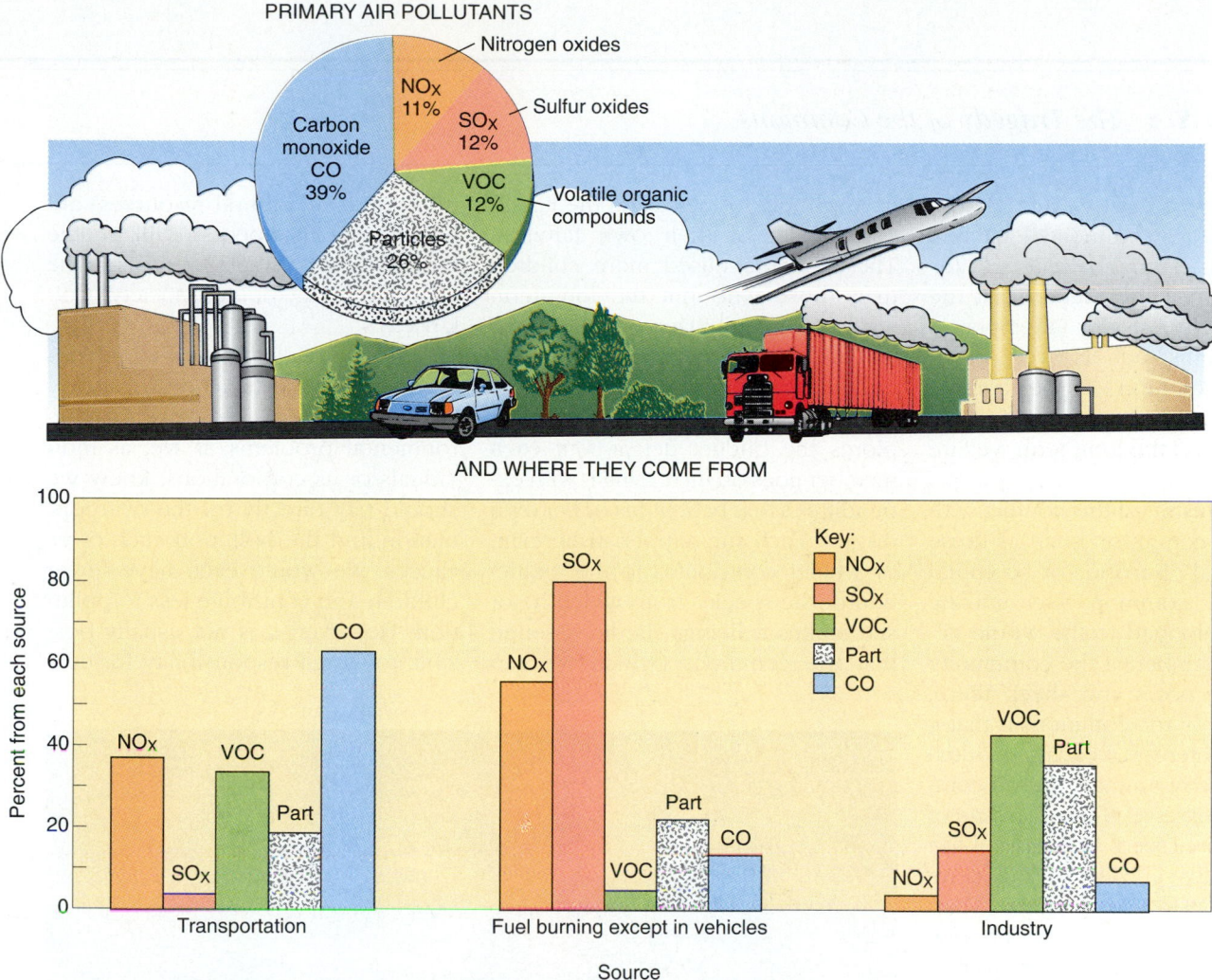

**FIGURE 50–30** The main primary air pollutants and their sources. In the bottom chart, note that different sources produce different pollutants. Transportation produces most carbon monoxide, other fuel burning produces most sulfur oxides, and industrial plants produce most of the volatile organic compounds. $O_x$ stands for any oxide of the element.

ages DNA. Large amounts of it can cause skin cancer in humans and kill many microorganisms outright.

We are destroying the ozone layer, permitting more ultraviolet radiation to reach Earth. A large, and growing, hole in the ozone layer over Antarctica was discovered in the 1980s, and holes above the Arctic and elsewhere have appeared in recent years. Damage to the ozone layer comes mainly from pesticides such as methyl bromide gas and from chlorofluorocarbons (CFCs), gases used in refrigeration and in air conditioners. CFCs are also ideal aerosol propellants because they do not react chemically with whatever is being sprayed. But this same lack of chemical reactivity means that they are not broken down in the air. They make their way to the upper atmosphere, where they

undergo photochemical reactions powered by intense ultraviolet light, releasing chlorine atoms which in turn react with ozone and break it down.

Unlike most environmental problems, which have been understood at least since the 1960s, ozone depletion was not even discovered until the 1980s. It appears that we shall have to solve the problem within 20 years of its discovery if we are to prevent enormous environmental damage.

The ozone story is an example of how rapidly countries can act to reduce pollution when they are convinced that it is necessary. In 1987, many countries endorsed the Montreal Protocol, a treaty to halve the use of CFCs by 1998. By 1992, however, scientists reported new holes in the ozone layer that were growing much more rapidly than predicted. In

# ESSAY: *The Tragedy of the Commons*

In 1968, ecologist Garrett Hardin published an influential parable entitled *The Tragedy of the Commons*. In it, he argued that the main difficulty in our attempts to solve problems such as overpopulation and pollution is the conflict between the short-term welfare of individuals and the long-term welfare of society.

Hardin illustrated this conflict with the case of commons, such as those of medieval Europe or colonial America. A common was grazing land that belonged to the whole village. Any member of the community could graze cows and sheep there. The tragedy of the commons is that it was in the interest of each individual to put as many animals on the common as possible, to take advantage of the free animal feed. However, if too many animals grazed on the common, they destroyed the grass. Then everyone suffered because no one could raise animals on it. For this reason, common land was eventually replaced by individually owned, enclosed fields. Before the era of subsidized agriculture, owners were careful not to put too many cows on one patch of grass, because overgrazing one year would mean that fewer cows could be supported the next year (Figure 50–A).

The tragedy of the commons is a result of natural selection. During the course of evolution, some people gave priority to increasing the health and wealth of their own families. These people raised more children than those who put the long-term welfare of their village or nation first.

The parable of the commons throws fresh light on many public problems. Congress as a whole deplores the budget deficit, but each member goes on increasing it with expenditures that benefit his or her own district. Each fur trapper, fisherman, or whaler contributes to the extinction of the species from which he or she makes a living. Hardin asserted that the commons (which we can think of as all natural resources) are limited and that people will pursue their own self-interest, destroying the commons to the point of society's collapse.

Hardin was certainly right in thinking that individuals acting separately cannot be expected to solve all environmental problems. If we, as individuals or as corporations, knew we should pay directly for the overpopulation and the pollution each of us causes, we would each have fewer children and contribute less to pollution. However, it is not usually possible to assign responsibility for eco-

**FIGURE 50–A** An encampment of nomadic shepherds on overgrazed land in Iran. This part of the Middle East was once the world's bread basket and was known as the "fertile crescent." Soil destruction has reduced most of the area to grinding poverty. (Paul Feeny)

1992, representatives of 93 countries attended a conference in Copenhagen to speed up actions to reduce ozone depletion. Rich countries agreed to eliminate CFCs completely by 1995. Under the Clean Air Act, the United States is committed to banning all substances that pose significant danger to the ozone layer by 2000. Developed countries agreed to set up a fund to help poor countries replace CFCs.

These actions appear to be working. Chemical companies have developed substances that can be used in place of ozone-destroying chemicals. In 1993 the concentration

logical problems directly to the people who cause them. Future generations will pay most of the price for the fact that we have too many children and cause soil erosion now, and our individual contributions to water pollution cannot be easily distinguished. Thus, although incentives for individual action are effective ways of getting most things done, many environmental problems can be solved only by effective action by groups such as governments that, at least in theory, can look beyond the immediate interests of individuals and plan for the long-term welfare of society.

Hardin pursued his argument in a 1974 essay, *The Ethics of a Lifeboat,* which explores the way the wealthy countries of the world treat their impoverished neighbors. An allegorical lifeboat filled with the world's rich is surrounded by the struggling poor who attempt to clamber aboard. How should the rich behave? They cannot let everyone into the boat because it would sink. If they allow some into the boat, they remove the lifeboat's safety margin and are also presented with the further ethical dilemma of whom to save and whom to condemn. To Hardin, the only sensible course of action is to ignore the pleas for help and maintain the boat's safety margin.

The lifeboat argument proposes that it is counterproductive to supply economic aid to poor countries. The argument is particularly strong in the case of countries that make no attempt to curb their population growth. If a poor country can call on aid in times of need so that its people do not starve to death, its population, and its need, continue to grow indefinitely, reducing still further the resources that are left for future generations everywhere. Lifeboat philosophers argue that the most humane course, in the long run, is to withhold aid so that starving people die. The smaller population that remains will have more chance of developing sustainable agriculture that does not degrade the environment and can feed the population.

If Hardin's analysis of human nature were the whole story, it would make the task of solving environmental problems almost impossible. Acting each with rational self-interest, we should continue to destroy Earth's resources until we became extinct. However, Hardin ignored important additional facts about the nature of human beings: we are intelligent, deeply social animals, and we are capable of altruistic behavior.

Altruistic behavior is part of our makeup because we are social animals. We depend upon cooperation with other people for our very survival. This is, ultimately, why we can prevent ourselves and each other from destroying the environment. The chieftain who burns down the local forest to enlarge his own wheat field runs the risk that he will be thrown out of the village, and if he is ejected from human society he will die. The many heads of nations who have abused their citizens and natural resources and then been deposed (and often killed) are obvious examples. So if our urge to work first for ourselves and our own families endangers the environment, our need to be accepted by society can save it. When enough people, in a village or a nation, want a particular reform, the rest must go along with it. You can make almost any change in society if you can convince enough people that it is the right thing to do.

As we would also predict from this, people are more inclined to work for the general welfare if their actions are effective and do not cost them very much. Public opinion polls show that, even during economic recessions, the vast majority of people will accept tax increases to pay for pollution control and the conservation of resources. Politicians are frequently surprised at the results of such polls, but they should not be. People are behaving logically in voting for effective environmental action that will not cost each of them very much.

---

of CFCs in the atmosphere was still increasing, but much more slowly than in the 1980s. CFC emissions are now down by more than 30% from their peak in 1988, meaning that industry is actually ahead of the Copenhagen schedule for phasing out ozone-destroying chemicals.

## Water Pollution

Acid rain is only one of many sources of water pollution. Toxic chemicals are another. Chlorinated hydrocarbons, such as DDT and the polychlorinated biphenyls (PCBs), and heavy metals such as mercury and lead are a danger to natural ecosystems just as they are to humans. These are all **persistent** hazardous wastes, which remain in the environment indefinitely because they are not biodegradable. DDT causes birds to lay eggs with thin shells that break when the bird sits on them. Although DDT is now tightly regulated in the United States, it is still used in other countries. It has dispersed all over the world and can be detected in the milk of most mammals and even in the Antarctic, thousands of miles from where it was originally used.

Organisms have only limited ability to excrete these substances, so they accumulate and are passed on to predators higher in the food chain. In aquatic ecosystems, they are especially damaging to carnivores such as fish-eating birds, at the top of long aquatic food chains.

Less obvious forms of pollution, at first sight, are nutrients and heat. Both can hasten eutrophication, the conversion of an oligotrophic waterway, which is low in nutrients and supports little plant growth, to one rich in nutrients but containing little oxygen. Phosphorus is one of the nutrients most likely to be a limiting factor in the productivity of natural ecosystems. Pollution introduces extra phosphorus into many lakes and rivers from sewage, high-phosphate detergents, and fertilizer washed off farmland, lawns, and gardens. The extra phosphorus increases productivity, changing the character of the lake. Eutrophication, fueled by nutrient pollution, fundamentally altered Lake Erie, which is relatively shallow. Even deep oligotrophic lakes, such as Lake Tahoe, have become noticeably more full of weeds and algae and contain less oxygen because of nutrient pollution.

## Toxic and Solid Wastes

Since about 1950, pollution from toxic and solid wastes has become an acute problem. Part of this is due to the population explosion, which has resulted in a greater total amount of solid waste. Part of it is due to higher standards of living, which have increased the amount of waste produced per person. The use of fossil fuel, this time as the raw material for the manufacture of artificial polymers collectively known as "plastics," has also contributed to solid waste pollution.

Modern artificial polymers are made from organic molecules in fossil fuel. Polyester, polyethylene, vinyls, and acetates are just some of the artificial polymers used in clothing, containers, skis, bottles, artificial arteries, and thousands of other items. Plastics have replaced animal and vegetable products in many areas of our lives. Most artificial polymers cannot be broken down by any microorganism, and so they are apparently with us essentially forever.

**Biodegradable plastics** are not quite what their name implies. Most of them consist of short chains of artificial polymers interspersed with short chains of starch (which is why they often feel powdery, like flour). Aerobic decomposers metabolize the starch, leaving flakes of undegraded plastic. Even the "biodegradable" portion of most degradable plastics will not be decomposed by bacteria in the anaerobic conditions found in a landfill.

## SUMMARY

Humans are changing the environment. Growth of the human population and the increasing use of technology endanger our future and that of many other species. We must reduce population growth drastically and adopt sustainable technologies if our grandchildren are to inherit a livable Earth.

1. Early humans were hunter-gatherers, hunting animals and gathering plants for food. Some were nomadic, and others lived in settled villages or migrated between seasonal homes. These people permanently altered their environment in a number of ways, but agricultural societies have had a much greater impact.

2. Agriculture supports more people on a given area of land than does a hunter-gatherer way of life. Agricultural societies have steadily expanded and squeezed out hunter-gatherer populations. The most important environmental effects of agriculture are population increase, habitat destruction, soil erosion, and food simplification.

3. The industrial revolution came about when societies started to power technology, including agriculture, with the enormous quantities of energy available from wood and then from fossil fuels. Industrial development raised standards of living but contributed to modern problems, particularly population growth, air pollution, toxic wastes, and the huge volume of solid waste we produce today.

4. The number of people in the world has grown exponentially in the last few hundred years as a result of improved nutrition and hygiene, which have reduced the death rate, especially among infants. Today the populations of many countries are growing faster than the supply of resources their people need to survive.

5. The nations of the world can be divided into those that have passed through the demographic transition and show little population growth and those that appear stuck in the middle of the demographic transition.

6. Increases in food production have not kept up with population growth despite a brief surge in production during the green revolution. As a result, more people starve to death every year.

7. Most of the world's food supply comes from grains, which, with minor supplements, are nutritionally adequate and are easy to store. The new plant varieties of the green revolution permitted huge increases in grain production by means of high-input agriculture, dependent on chemicals and irrigation.

8. The environmental damage caused by high-input agriculture includes disproportionate use of water, water pollution, and soil erosion. The current trend is toward low-input agriculture, which is cheaper and less environmentally destructive.

9. Soil destruction caused by intensive farming is the main reason that food production is now leveling off and starting to decline. One example is the desertification of arid lands in Africa as a result of converting grazing land to cropland and abandoning fallow periods.

10. A growing shortage of fresh water on every continent except Europe reduces the water available to irrigate crops and causes numerous deaths from polluted or inadequate drinking water supplies.

11. Extinction of species, particularly in tropical biomes, deprives the world of biodiversity needed to maintain ecosystems and of genetic resources. It results mainly from habitat destruction when forest is cut down for fuel and building and to clear more farmland.

12. Deforestation is one of the oldest environmental problems. Its results include increased runoff, which causes soil erosion and flooding, reduced rainfall downwind, and economic disruption from fuel wood shortages.

13. We are destroying the ozone layer in the atmosphere, and this permits increasing amounts of mutagenic ultraviolet radiation to reach Earth.

14. The pollution of water, air, and land results from the vast number of people on Earth today and from our use of fossil fuel to power modern technology. Pollution results when wastes are discharged into the environment faster than natural processes can disperse them or break them down.

## Self-Quiz

1. All of the following are associated with or result from the agricultural revolution *except:*
   a. accumulation of material goods
   b. population control
   c. land ownership
   d. increased division and diversity of labor

2. The agricultural revolution is believed to have been responsible for a dramatic increase in the human population. Which of the following was *not* a factor in this increase?
   a. Food became more concentrated and thus easier to obtain
   b. Many methods of birth control were abandoned
   c. Improved medical knowledge increased life expectancy
   d. Larger amounts of food could be produced by fewer people
   e. People could accumulate possessions and wanted more children to pass them to

3. The demographic transition in a country is correlated most clearly with:
   a. education for women
   b. improvements in agricultural production
   c. the industrial revolution
   d. improved hygiene and nutrition
   e. improved treatment for illnesses

4. Which of the following is associated with high-input agriculture?
   a. increased pollution from fertilizer and pesticides
   b. use of large amounts of water
   c. increased soil erosion
   d. all of the above

5. Which of the following statements about the availability of water is true?
   a. Water will never be in short supply because it is cycled continuously between the oceans, land, and atmosphere.
   b. Water will never be in short supply because underground water is easily accessible.
   c. Water will never be in short supply because underground water supplies are renewed and recharged on a regular cycle.
   d. None of the above

6. Which of the following is a likely result of deforestation?
   a. The amount of carbon dioxide removed from the atmosphere is reduced.
   b. Wind blows soil away because its plant cover has been removed.
   c. Water washes off the land more rapidly, causing floods.
   d. Water washes soil into rivers and streams, causing them to silt up.
   e. All of the above

7. The industrial revolution increased all of the following *except:*
   a. use of fossil fuel
   b. water pollution
   c. urbanization
   d. human birth rate
   e. air pollution

## Questions for Discussion

1. Do you think humans will be extinct within the next thousand years? Why or why not?

2. The world contains a few remaining hunter-gatherer societies in places such as Australia, New Guinea, Brazil, and Peru. The governments that control the countries where they live generally believe these people should be left alone and perhaps given title to the land where they have lived for centuries. In some places this may actually happen. In others, hunter-gatherers are rapidly disappearing. What pressures determine one fate rather than the other? Is it worth trying to save the remaining hunter-gatherers?

3. This chapter stresses the environmental problems caused by agriculture and industry. But agriculture and industry have brought inestimable benefits to human societies. List as many of these benefits as you can.

*(Continued)*

4. It has been argued that advances in human civilization can be traced to exploitation of new sources of energy. Major steps included the addition of meat to the largely herbivorous diet of our ancestors, the taming of fire, and the use of fossil fuels. Other advances along the way have been domestication of beasts of burden and harnessing of wind and water power. How has each of these affected human ecology?

5. Look at Figure 50–22 showing predicted water shortages in various countries. Fresh water does not just disappear. Why do you supposed the decrease in supply from 1990 to 2025 is so sharp, especially for countries such as Haiti, Lebanon, Iran, and essentially all the African countries?

6. What actions can you take to improve the quality of the environment, as an individual? As a member of an organization? As a voter?

## Suggested Readings

Arms, K. *Environmental Science,* 2d ed. Philadelphia: Saunders College Publishing, 1994. An introductory textbook.

Bongaarts, J. "Can the growing human population feed itself?" *Scientific American,* March 1994. Concludes that population control is essential if humans are to feed themselves without excessive environmental damage in the near future.

Brown, L. R., et al. *State of the World 1995: A Worldwatch Institute Report on Progress Toward a Sustainable Society.* New York: W. W. Norton, 1995. A volume of this best-seller is produced each year by the Worldwatch Institute. It contains about a dozen chapters on environmental problems and progress.

Caldwell, J. C., and P. Caldwell. "High fertility in sub-Saharan Africa." *Scientific American,* May 1990. A discussion of why birth rates have declined much more slowly in the southern half of Africa than in most other parts of the world.

Commission on Population Growth and the American Future. *Population and the American Future.* Washington, DC: U.S. Government Printing Office, 1985. Analysis of the need to control U.S. population growth.

Croll, E., et al. *China's One-Child Family Policy.* New York: St. Martin's Press, 1985. Fascinating description of Chinese population policy.

Djerassi, C. *The Politics of Contraception.* New York: W. W. Norton, 1980. Why modern contraceptives are so primitive compared with our other technology.

Ehrlich, P. R., and A. H. Ehrlich. *The Population Explosion.* New York: Simon and Schuster, 1990. The sequel to *The Population Bomb,* written 20 years earlier. Not quite such good reading as the first book, but it updates a lot of the statistics and scenarios.

Elkington, J., J. Hailes, and J. Makower. *The Green Consumer.* New York: Viking Penguin, 1990. A guide to doing your bit for the environment.

French, H. F. "Making environmental treaties work." *Scientific American,* December 1994. A brief assessment of the effectiveness of different kinds of treaties and organizations.

Grupte, P. *The Crowded Earth: People and the Politics of Population.* New York: W. W. Norton, 1984. Excellent summary of population problems and attempts to tackle them in various countries.

Hardin, G. "The tragedy of the commons." *Science* 162:1243, 1968. The thought-provoking argument that ecologically ethical individuals are evolutionarily doomed.

Hardin, G. *Living Within the Limits: Ecology, Economics, and Population Taboos.* New York: Oxford University Press, 1993.

Houghton, R. A., and G. M. Woodwell. "Global climatic change." *Scientific American,* April 1989. Presents the evidence that the greenhouse effect is already warming Earth and discusses how to control it.

MacEachern, D. *Save Our Planet: 750 Everyday Ways You Can Help Clean Up the Earth.* Washington, DC: Dell, 1990.

McKay, B. J., and J. M. Acheson, eds. *The Culture and Ecology of Communal Resources.* Tucson: University of Arizona Press, 1987. A collection of papers evaluating fisheries, grazing lands, and similar "commons." The authors conclude that sustainable management of commons is possible if the users of the commons make up a single group of accountable people and if the limits of the commons are under the group's control.

Paddock, J., et al. *Soil and Survival: Land Stewardship and the Future of American Agriculture.* San Francisco: Sierra Club Books, 1990. An excellent discussion of the soil restoration needed to keep American agriculture productive.

Reganold, J. P., R. I. Papendick, and J. F. Parr. "Sustainable agriculture." *Scientific American,* June 1990. How American farmers are turning to more environmentally sound practices.

*Scientific American.* "Managing Planet Earth." September 1989. An entire issue devoted to environmental affairs including management, pollution, biodiversity, population growth, sustainable agriculture, and energy.

# Answers to Self-Quizzes

## CHAPTER 2 *Some Basic Chemistry*

1. isotopes
2. a
3. covalent
4. covalent
5. HCl: 36.5 grams
   $CO_2$: 44 grams
6. a
7. basic; decreased
8. h
9. f
10. c
11. d
12. g

## CHAPTER 3 *Biological Chemistry: Variations on Four Themes*

1. Nucleic acids
2. Nucleotides
3. *Any two of the following:*
   Carry genetic information
   Direct the synthesis of proteins
   Supply energy to enzyme reactions
   Act as coenzymes
4. Lipids
5. C, H, O (P, N)
6. Energy storage
   Formation of biological membranes
   Hormones
7. C, H, O, N (S)
8. Amino acids
9. *Any of the following:*
   Catalyze reactions (enzymes)
   Form structural elements of body
   Hormones
   Muscle contraction
   Defense against disease
   Transport of substances
10. Carbohydrates
11. C, H, O (N)
12. Monosaccharides (simple sugars)
13. Energy storage
    Structural support and protection
14. c
15. $H_3C—$ (a methyl group)
16. Check your answer using the hydrolysis in Figure 3–11(b).

17. c
18. d
19. increases; increased motion results in more encounters between enzyme and substrate molecules; stop; denatured
20. d

## CHAPTER 4 *Cells and Their Membranes*

1. b
2. a, c
3. c, d
4. a
5. b
6. d
7. endocytosis
8. active transport, diffusion through channels, endocytosis
9. a
10. d
11. a
12. lower, more
13. Water enters and the level of the solution in the tube rises; meanwhile, glucose leaves and water follows; at equilibrium the solutions in the tube and beaker will be equally concentrated and will stand at the same level.
14. d
15. hypotonic, a
16. larger

## CHAPTER 5 *Cell Structure and Function*

1. p
2. d, f
3. a
4. c
5. g
6. i
7. o
8. q
9. n
10. l
11. d, f, k, l
12. all
13. animals, some plants, some prokaryotes
14. plants, prokaryotes
15. animals, plants
16. animals, plants
17. e
18. A. cell wall
    B. plasma membrane
    C. mitochondrion
    D. nuclear envelope
    E. rough endoplasmic reticulum

F.   nuclear pore
G.   nucleolus
H.   ribosomes
I.   Golgi complex (dictyosome)

## CHAPTER 6  *Energy and Living Cells*

1.  true
2.  photosynthesis
3.  reduced
4.

| Oxidized | Reduced |
| --- | --- |
| a. $Fe^{3+}$ | $O^{2-}$ |
| b. $NAD^+$ | $FMNH_2$ |
| c. FMN, $H^+$ | $FeS^{2-}$ |

5.  ADP; $P_i$; endergonic; energy and a feasible pathway
6.  substrate-level
7.  c    8.  e

## CHAPTER 7  *Food as Fuel: Cellular Respiration and Fermentation*

1.  c    2.  c
3.  mitochondria; the cytoplasm
4.  the final electron acceptor
5.  inner mitochondrial; plasma
6.  a. pyruvate, NADH, ATP
    b. $CO_2$, ATP, NADH, $FADH_2$, oxaloacetate, CoA
    c. ATP, $NAD^+$, $CO_2$, ethanol ($CH_3CH_2OH$)
    d. $NAD^+$, FAD, $H_2O$
    e. ATP, $NAD^+$, lactate
7.  a    8.  true    9.  true

## CHAPTER 8  *Photosynthesis*

1.  c    2.  a    3.  a
4.  diffusion out of the chloroplast, cell, and air space, through a stoma and into the air; transport in vascular tissue of the veins
5.  c    10.  a, f
6.  a    11.  f
7.  c    12.  b
8.  d    13.  e
9.  e    14.  c, f

## CHAPTER 9  *DNA and Genetic Information*

1.  a
2.  c
3.

```
             P    5'
             |
     T — sugar
             |
             P
             |
     G — sugar
             |
             P
             |
     A — sugar  3'
```

4.  d    5.  c    6.  e
7.  Hydrogen bonds are weaker and can be broken easily, permitting separation of the two strands during replication.
8.  *Escherichia coli*    12.  your own cells
9.  your own cells    13.  both
10.  *Escherichia coli*    14.  *Escherichia coli*
11.  both    15.  e

## CHAPTER 10  *RNA and Protein Synthesis*

1.  a. U—C—C—G—G—A—C—G—A—A—U
    b. A—C—C—G—U—C—G—A—U—G
    c. A—A—A—U—G—C—G—U—G—G
2.  a. Methionine—histidine—arginine—arginine—proline—isoleucine—valine
    b. Translation begins with the AUG codon (bases 2, 3, 4): Methionine—phenylalanine—leucine—lysine—glycine—arginine
3.  mRNA:   A—U—G—U—U—C—A—U—G—A—A—C—A—A—A—G—U—U
    Amino acid sequence: Methionine—phenylalanine—methionine—asparagine—lysine—valine
4.  Change phenylalanine to leucine; change asparagine to lysine
5.  Amino acid sequence would read: Methionine—leucine ("stop")
6.  e
7.

| DNA | RNA |
| --- | --- |
| deoxyribose sugar | ribose sugar |
| thymine base | uracil base |
| double-stranded | usually single-stranded |

8.  a

## CHAPTER 11  *Gene Expression and Cell Differentiation*

1.  d    5.  c
2.  b    6.  e
3.  b    7.  c
4.  b    8.  d

## CHAPTER 12  *Embryonic Development in Animals*

1.  b    7.  N, G, O
2.  C    8.  G
3.  C    9.  a
4.  N    10.  e
5.  O    11.  b
6.  C    12.  a

## CHAPTER 13  *Genetic Engineering*

1.  e    4.  d
2.  c    5.  d
3.  e    6.  b

## CHAPTER 14 *Reproduction of Eukaryotic Cells*

1. b   2. N = 2; 2N = 4   3. b
4. A: interphase
   B: prometaphase
   C: metaphase
   D: late anaphase/telophase
   E: anaphase
   F: prophase
5. c   6. b
7. d   8. b
9. nucleus; cytoplasm; eggs; sperm
10. b

## CHAPTER 15 *Mendelian Genetics*

1. a. both are *Tt*
   b. $\frac{3}{4}$ tasters:$\frac{1}{4}$ nontasters
   c. all tasters
   d. $\frac{1}{2}$ tasters:$\frac{1}{2}$ nontasters
2. $\frac{1}{2}$ axial:$\frac{1}{2}$ terminal
3. a. dumpy recessive to normal, which is dominant
   b. both heterozygous for dumpy wings
4. 40
5. a. 250
   b. 125
6. a. Sniffles: homozygous dominant (colored)
      Whiskers: heterozygous for albino
      Esmeralda: homozygous recessive (albino)
   b. $\frac{3}{4}$ colored:$\frac{1}{4}$ albino
   c. $\frac{1}{2}$ colored:$\frac{1}{2}$ albino
7. Mate his dog to bitches known to carry the trait. If any pups show it, the dog is heterozygous for the allele. If none of a large number of pups shows it, the dog is probably homozygous normal.
8. a. *DdHb × Ddhb*:

| | DH | Db | dH | db |
|---|---|---|---|---|
| *Db* | *DDHb* | *DDbb* | *DdHb* | *Ddbb* |
| *db* | *DdHb* | *Ddbb* | *ddHb* | *ddbb* |

   b. *DDHb × Ddhb*:

| | DH | Db |
|---|---|---|
| *Db* | *DDHb* | *DDbb* |
| *db* | *DdHb* | *Ddbb* |

   c. *DdHb × ddhb*:

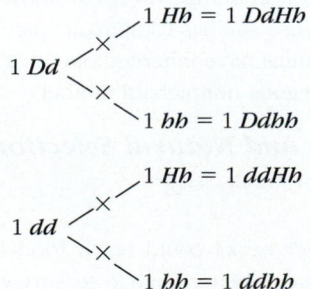

d. *DdHb × DDHb*:

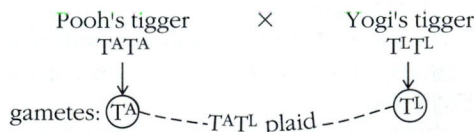

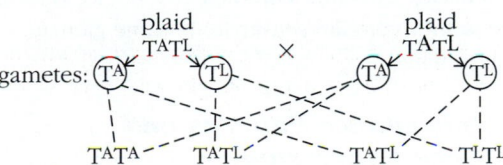

   e. $\frac{1}{2}$
9. a. stamens: straight dominant to incurved; petals (red vs. streaky): can't tell from information given
   b. stamens: both parents heterozygous; petals: one heterozygous, one homozygous recessive, but no indication which is which from information given.
   c. red × red and streaky × streaky: if red is dominant, red × red will produce some streaky progeny, and vice versa.
10. a. $\frac{3}{16}$   11. a. 480   12. a. $\frac{1}{8}$
    b. $\frac{3}{16}$      b. 160      b. $\frac{1}{8}$
    c. $\frac{9}{16}$      c. 40       c. $\frac{3}{8}$
13. let $T^A$ = crosswise stripes
    $T^L$ = lengthwise stripes
    1 crosswise:2 plaid:1 lengthwise

    Pooh's tigger   ×   Yogi's tigger
       $T^AT^A$           $T^LT^L$
    gametes: $T^A$ -----$T^AT^L$ plaid----- $T^L$

       plaid            plaid
       $T^AT^L$   ×   $T^AT^L$
    gametes: $T^A$   $T^L$    $T^A$   $T^L$
    $T^AT^A$   $T^AT^L$   $T^AT^L$   $T^LT^L$
14. a. genotype: *SsRR'*
       phenotype: straight roan
    b. straight red $\left(\frac{1}{8}\right)$
       straight roan $\left(\frac{1}{4}\right)$
       straight white $\left(\frac{1}{8}\right)$
       curly red $\left(\frac{1}{8}\right)$
       curly roan $\left(\frac{1}{4}\right)$
       curly white $\left(\frac{1}{8}\right)$
15. a. clover patch: $\frac{1}{2}$ roan, $\frac{1}{2}$ white
       alfalfa field: $\frac{1}{2}$ red, $\frac{1}{2}$ roan
       cornfield: $\frac{1}{4}$ red, $\frac{1}{2}$ roan, $\frac{1}{4}$ white
    b. it doesn't matter; $\frac{1}{2}$ the calves will be roan in any case
16. a. $\frac{1}{4}$   d. $\frac{3}{16}$
    b. $\frac{1}{2}$   e. $\frac{1}{8}$
    c. $\frac{1}{4}$
17. a. $\frac{1}{2}$   b. $\frac{1}{4}$   c. $\frac{1}{4}$   d. 1
18. a. $\frac{1}{4}$   b. $\frac{1}{2}$   c. $\frac{1}{16}$   d. $\frac{1}{4}$

19. $\frac{1}{2}$

20. If the dog is heterozygous, the probability that one pup will not show retinal atrophy is $\frac{3}{4}$, and the probability that nine pups will not show the trait is $\left(\frac{3}{4}\right)^9 = 0.075$. Subtracting 0.075 from 1 gives 0.925, or 92.5% certainty that the dog is homozygous dominant.

21. a. $\frac{MV}{mv}$  b. 3%

22. a. $\frac{SM}{sm}$  b. $\underbrace{\underbrace{s\ \ \ v\,m}_{3}}_{7}$ or $\underbrace{s}_{7}\ \underbrace{m\ v}_{3}$

    c. $SsVv \times ssvv$

    d. $SsVvMm \times ssvvmm$ (with any linkage in the heterozygote)

23. The loci appear to be unlinked (different chromosomes), but in fact they are just over 50 map units apart on the same chromosome.

24. a. First of all, her own chance of having the trait is $\frac{2}{3}$. Possible genotypes of offspring from her heterozygous parents are $1\ DD : 2\ Dd : 1\ dd$. Since she is obviously not $dd$, there are three possibilities left, two of which are carriers. Her chances of having an afflicted child are thus: 1/10,000 of marrying someone with the allele $\times \frac{2}{3}$ that she has it $\times \frac{1}{4}$ that her child will be $dd$ = 1/60,000.

    b. $\frac{1}{24}$; $\frac{2}{3}$ chance she has it $\times \frac{1}{2}$ chance grandparent passed it to spouse's parent $\times \frac{1}{2}$ chance spouse's parent passed it on to spouse $\times \frac{1}{4}$ chance of homozygous recessive child if they are both heterozygous.

25. a. 0.21; 0.42 that chromosome with eye and muscle traits will be in gamete $\times \frac{1}{2}$ that kryptonite sensitivity allele will also be present.

    b. 0.04; 0.08 chance of crossing over to get X-ray and normal muscles on same chromosome $\times \frac{1}{2}$ to get chromosome with kryptonite sensitivity in same gamete.

26. $\frac{9}{16}$ purple : $\frac{7}{16}$ white

## CHAPTER 16  *Inheritance Patterns and Gene Expression*

1. a. $\frac{2}{3}$

    b. sell cows bearing these calves and try a new bull

2. a. $\frac{1}{4}$

    b. $\frac{1}{2}$ normal : $\frac{1}{2}$ brachydactylic

3. a. Yellow mice are heterozygous.

    b. "Yellow" allele is lethal in the homozygous condition, early in embryonic development (2 : 1 ratio).

    c. Homozygous yellow individuals die as early embryos and are resorbed by the uterus.

    d. He could carry out yellow $\times$ yellow matings and examine contents of females' uteri early in pregnancy to detect defective embryos.

4. a. heterozygous for an allele that is lethal in the homozygous condition

    b. they die before hatching

5. a. $I^A i$  c. $I^A i$ or $I^A I^B$

    b. $I^B i$  d. $I^B i$

    e. $I^A I^A$ or $I^A i$ or $I^A I^B$  h. $I^A i$

    f. $I^A i$  i. $I^B i$

    g. $I^A i$ or $I^B i$ or $ii$

6. Yes, John Smith is really Tom Jones! A baby with blood type M cannot have a parent with blood type N (Ms. Smith). Also, a parent with blood type AB (Ms. Jones) cannot have a child with blood type O.

7. a. $\frac{1}{2}$ normal : $\frac{1}{4}$ chinchilla : $\frac{1}{4}$ Himalayan

    b. $\frac{1}{2}$ chinchilla : $\frac{1}{4}$ Himalayan : $\frac{1}{4}$ albino

    c. $\frac{3}{4}$ chinchilla : $\frac{1}{4}$ albino

8. $\frac{1}{2}$ barred males : $\frac{1}{2}$ nonbarred females (remember, male birds are ZZ and females ZW [see Figure 16-10]).

9. $\frac{1}{2}$ the sons hemophiliacs : $\frac{1}{2}$ the sons normal $\frac{1}{2}$ the daughters heterozygous carriers : $\frac{1}{2}$ the daughters homozygous normal

10. when the mother is either a hemophiliac or a carrier

11. a. Red-green color blindness in humans is carried on the X chromosome. Since males have only one X chromosome, any male who receives the allele for color blindness will be color-blind. A color-blind female will have two X chromosomes bearing the allele. The frequency of this genotype and phenotype among females is thus $\left(\frac{1}{12}\right)^2 = \frac{1}{144}$.

    b. mother homozygous recessive color-blind; father normal allele + Y

    c. $\frac{1}{2}$ of the sons are expected to be color-blind

    d. $\frac{1}{2}$

12. 2 female offspring : 1 male

13. No; since the baldness trait is carried on the autosomes, a man could inherit the trait from either parent who carried it and from any of his four grandparents who passed it on to the appropriate parent.

### Genotypes of cats in Figure 16–A:

Both cats are $aa$ (non-agouti), and therefore their phenotypes give no information about alleles in the T group. Both are also:

    $B$ (black rather than brown)

    $C$ (color fully developed on body)

    $ww$ (lacking dominant white allele)

    homozygous recessive for long hair

In addition:

Cat (a): $DdO'O'ss$ (color undiluted, a dominant trait, but must also have one recessive allele because she passed it on to cat (b), which is homozygous $dd$; homozygous non-orange on X chromosomes; homozygous nonpiebald

Cat (b): $ddO'O'Ss$ (homozygous recessive for dilution of black color to gray; homozygous non-orange on X chromosomes; heterozygous for piebald because she has (dominant) piebald markings in her phenotype but must have inherited one recessive nonpiebald allele from homozygous nonpiebald mother)

## CHAPTER 17  *Evolution and Natural Selection*

1. c  3. e

2. d  4. b

5. Individuals with longer necks could reach food higher on trees and therefore had a better chance to survive and re-

produce when food was scarce nearer the ground. These individuals passed on the genes for longer necks to their offspring. Continued selection pressure of this sort over many generations led to evolution of longer and longer neck length.

**6.** b    **7.** e

## CHAPTER 18 *Adaptation and Coevolution*

**1.** a       **8.** a. sit-and-wait
**2.** T           b. sit-and-wait
**3.** b           c. active hunter
**4.** d    **9.** a
**5.** b    **10.** b
**6.** d    **11.** a
**7.** Answers will vary.

## CHAPTER 19 *Population Genetics and Speciation*

Dichotomous key:
 I Crustacea
 II Arachnida
 III Insecta
 IV Diplopoda
 V Chilopoda

**1.** e    **2.** d
**3.** a. stabilizing
    b. frequency of sickle allele will decrease
    c. 10% homozygous normal, 90% heterozygous
**4.** a    **5.** d    **6.** a
**7.** a. increase
    b. increase
    c. increase
    d. remain the same
**8.** c
**9.** a. morphology
    b. both
    c. both
    d. morphology
    e. morphology
**10.** d

## CHAPTER 20 *Evolution and Reproduction*

**1.** b              **5.** d
**2.** favorable; unfavorable    **6.** c
**3.** b              **7.** a
**4.** c              **8.** c

## CHAPTER 21 *Origin of Life*

**1.** oxygen: less
    carbon dioxide: more
    water vapor: more
**2.** e
**3.** 1. organic monomers
    2. polymers
    3. fermentation
    4. water-splitting photosynthesis

5. aerobic respiration
6. intracellular organelles
**4.** c    **5.** b
**6.** 1. Addition of $O_2$ to the atmosphere made it oxidizing rather than mildly reducing.
    2. An ozone layer formed.
    *Effects:*
    1. Addition of $O_2$ acted as a selective pressure for the evolution of respiration.
    2. The ozone layer permitted organisms to move onto land.
**7.** c    **8.** b

## CHAPTER 22 *Classification of Organisms*

**1.** b
**2.** d
**3.** d
**4.** d
**5.** a, b
**6.** Protista

## CHAPTER 23 *Viruses*

**1.** d
**2.** c
**3.** d
**4.** a. lytic (true of lysogenic when it has become lytic)
    b. lytic
    c. lysogenic
    d. lysogenic
    e. both
**5.** True
**6.** d

## CHAPTER 24 *Bacteria*

**1.** d
**2.** a
**3.** a
**4.** c
**5.** b
**6.** Any three of these:
    • Sulfur bacteria use near-infrared as an energy source; cyanobacteria use visible light.
    • Sulfur bacteria use bacterial chlorophyll; cyanobacteria use chlorophyll *a*.
    • Sulfur bacteria have only one kind of photosystem; cyanobacteria have two.
    • Sulfur bacteria extract hydrogen and electrons from $H_2S$ and produce elemental sulfur (S) as a byproduct. Cyanobacteria use water as a source of hydrogen and electrons and produce $O_2$.
**7.** True
**8.** True
**9.** False
**10.** False

## CHAPTER 25  *Protista and the Origin of Multicellularity*

1.  a. eukaryotic cell structure
    b. not qualifying for membership in kingdoms Fungi, Plantae, or Animalia
2.  a. pseudopods
    b. cilia
    c. flagella
3.  a. none
    b. flagella
    c. none
    d. flagella
    e. pseudopods
4.  e
5.  Phaeophyta
6.  Chlorophyta
7.  Phaeophyta
8.  Rhodophyta
9.  Bacillariophyta
10. d
11. a
12. a
13. a
14. c

## CHAPTER 26  *Fungi*

1.  c
2.  c
3.  a
4.  a
5.  b
6.  d
7.  a, b, (d)
8.  a
9.  a, b, c, d
10. b
11. e
12. c

## CHAPTER 27  *The Plant Kingdom*

1.  c
2.  d
3.  d
4.  c
5.  a
6.  b, d
7.  b. zygote, d. sporophyte, e. spore, c. gametophyte, a. gamete
8.  d
9.  a. Growing on moist soil or rocks in shady areas
    b. Growing on surface of moist soil in vicinity of fern plants
    c. Male: immature is pollen grain, in pollen cone; mature is pollen tube that has grown to female gametophyte in seed cone. Female: at base of seed cone scales
    d. In male and female flower parts, respectively

10.

| | Bryophytes | Ferns | Gymnosperms | Angiosperms |
|---|---|---|---|---|
| flagellated sperm | ✓ | ✓ | ✓ (some) | |
| true roots | | ✓ | ✓ | ✓ |
| vascular tissue | | ✓ | ✓ | ✓ |
| pollen | | | ✓ | ✓ |
| protection of embryo within parent plant | ✓ | ✓ | ✓ | ✓ |
| protection of embryo within seed coat | | | ✓ | ✓ |
| photosynthetic gametophyte | ✓ | ✓ | | |
| photosynthetic sporophyte | limited | ✓ | ✓ | ✓ |

## CHAPTER 28  *Lower Invertebrates*

1.  Cnidaria
2.  Porifera
3.  b
4.  "Eat now, digest later." Separates feeding and digestion so food can be consumed in quantity whenever available.
5.  Dispersal. Larvae are small and so have high mortality.
6.  Nematoda
7.  Body covering and nervous system
8.  a
9.  a, b, h, j
10. c, d, h
11. a, c, h, j
12. b, e, i
13. b, f, g

## CHAPTER 29  *Higher Invertebrates*

1.  a          13. j
2.  d          14. a
3.  l, k       15. m
4.  d          16. e
5.  b, h       17. n
6.  g          18. k
7.  a          19. i
8.  c          20. g
9.  b          21. d
10. a, b, e    22. e
11. b          23. d
12. h          24. b

## CHAPTER 30  *Vertebrates*

1.  a          5. e
2.  a          6. c
3.  d          7. d
4.  d          8. b

9. b    13. a
10. c    14. b
11. e    15. Squamata, reptile
12. d

## CHAPTER 31 *Animal Nutrition and Digestion*

1. f         8. a
2. a         9. c
3. e, f      10. e
4. a, b, c   11. b
5. a, c      12. c
6. b         13. a
7. d

## CHAPTER 32 *Gas Exchange in Animals*

1. b
2. a
3. a
4. c
5. a
6. a
7. right
8. a. right
   b. left
   c. left
   d. left
9. a
10. a

## CHAPTER 33 *Animal Transport Systems*

1.

|                                        | cnidarian | earthworm | insect | fish | mammal |
|----------------------------------------|-----------|-----------|--------|------|--------|
| food transport                         | X         | X         | X      | X    | X      |
| oxygen transport                       |           | X         |        | X    | X      |
| high pressure fluid picks up food      |           |           |        |      | X      |
| muscular circulatory pump(s)           |           | X         | X      | X    | X      |

2. b
3. c, d
4. d
5. b
6. intestine—lymph—lymphatic—thoracic duct or right lymph duct—vena cava—heart—aorta—any artery in body (not brain)—arteriole—capillary—adipose tissue
7. c
8. a. ectotherms
   b. both
   c. endotherms

## CHAPTER 34 *Defenses Against Disease*

1. d    7. c
2. b    8. b
3. b    9. d
4. c    10. d
5. c    11. c
6. e

## CHAPTER 35 *Excretion*

1. c    11. c
2. b    12. e
3. a    13. a
4. b    14. c
5. c    15. d
6. b    16. d
7. a    17. b
8. b    18. decrease, c
9. d    19. b
10. b

## CHAPTER 36 *Sexual Reproduction*

1. g, j        11. f
2. i           12. b
3. c, e, f     13. a
4. b           14. c
5. a           15. a
6. f           16. b
7. h           17. b
8. f-j-g-i-a-h-b  18. b
9. a           19. d
10. d          20. d

## CHAPTER 37 *Nervous Systems*

1. d
2. a. action potential
   b. local potential
   c. action potential
   d. both
   e. local potential
3. a
4. a. A neurotransmitter substance carries information between two neurons: it is released after an action potential arrives at the presynaptic membrane of one neuron, and its binding to receptors on the postsynaptic membrane stimulates activity in the postsynaptic cell.
   b. Vesicles store neurotransmitter molecules in the presynaptic terminal and release them when an action potential arrives at the terminal.
   c. On combining with neurotransmitter molecules, receptor molecules change the permeability of the postsynaptic membrane to ions, resulting in a local potential.
   d. Enzymes that destroy neurotransmitter molecules in effect "turn off" the signal brought across the synapse by the neurotransmitter.

**5.** d

**6.** c

**7.** b

**8.** a

**9.** a

**10.** d

**11.** d

**12.** a. B

b. C

c. E

d. D

e. A

f. B

**13.** a. both

b. sympathetic

c. sympathetic

d. both

e. sympathetic

f. parasympathetic (and a few sympathetic sites)

## CHAPTER 38 *Sense Organs*

| | |
|---|---|
| **1.** a | **7.** rods, dim, poor |
| **2.** b | **8.** b |
| **3.** e | **9.** d |
| **4.** d | **10.** a |
| **5.** d | **11.** c |
| **6.** a | **12.** c |

## CHAPTER 39 *Muscles and Skeletons*

| | |
|---|---|
| **1.** b | **9.** a, e, i, h, b, f |
| **2.** a, c | **10.** c |
| **3.** a, c | **11.** a |
| **4.** c | **12.** d |
| **5.** a, c | **13.** e |
| **6.** d, a, b, c | **14.** a |
| **7.** b | **15.** c |

**8.** actin: B

myosin: D

troponin: B

calcium: C

tropomyosin: B

Z line: A

mitochondria: C

## CHAPTER 40 *Animal Hormones and Chemical Regulation*

**1.** c

**2.** a

**3.** c

**4.** a

**5.** b

**6.** hypothalamus, releasing factors from hypothalamic neurons carried in the blood, axins of neurosecretory cells, releasing hormones into the extracellular fluid and blood, where they travel around the body and stimulate other glands.

**7.** negative feedback, target organ

**8.** b

**9.** Thyroxin ↓ → TSH released from anterior pituitary → thyroxin released from thyroid → TSH release reduced

## CHAPTER 41 *Animal Behavior*

| | | | |
|---|---|---|---|
| **1.** g | | **6.** c |
| **2.** a | | **7.** b |
| **3.** h | | **8.** d |
| **4.** j | | **9.** c |
| **5.** f | | **10.** e |

## CHAPTER 42 *Structure and Growth of Vascular Plants*

| | | | |
|---|---|---|---|
| **1.** a | | **9.** a |
| **2.** b | | **10.** F |
| **3.** a | | **11.** A |
| **4.** a, c, b | | **12.** D |
| **5.** a | | **13.** C |
| **6.** c | | **14.** B, D |
| **7.** a | | |

**8.** a. monocotyledons

b. both

c. monocotyledons

d. monocotyledons

e. dicotyledons

## CHAPTER 43 *Transport in Vascular Plants*

**1.** c

**2.** c

**3.** function: transport of sap

adaptation: tubular, dead and hollow, pits or open ends, stacked end to end

function: support of plant body

adaptation: secondary wall thickenings

**4.** d

**5.** a. remain the same

b. increase

c. remain the same

**6.** a. decrease

b. increase

c. increase

d. increase

**7.** a    **8.** b

## CHAPTER 44 *Soil, Roots, and Plant Nutrition*

**1.** e

**2.** d

**3.** b

**4.** a

**5.** c

**6.** a

**7.** a. apoplast

b. increases surface area for absorption of minerals and water from soil solution

c. Casparian strip
d. endodermal
e, f. epidermal, cortical (in either order)
g. symplast
8. nitrogen, phosphorus, potassium
9. Any two of these:
   Proportions of minerals to each other may not be correct.
   Minerals may be bound to soil particles and unavailable to plants.
   Minerals may be in a form that plants cannot absorb.
10. a

## CHAPTER 45  *Regulation and Response in Plants*

1. d
2. c
3. b
4. e
5. b
6. d
7. a
8. a
9. c
10. d
11. b
12. c
13. False

## CHAPTER 46  *Reproduction in Flowering Plants*

1. Pollen
2. Anther
3. Carpel (or Pistil)
4. Sepal
5. Ovary
6. Embryo sac (or Female gametophyte)
7. Stigma
8. Pollen tube
9. Stamen
10. Petal
11. Ovule
12. Anther, #2
13. Embryo sac, #6
14. Sepal, #4
15. Ovule, #11
16. Embryo sac, #6
17. Ovary, #5
18. False
19. e (but not all seeds require all factors listed)
20. c

## CHAPTER 47  *The Biosphere*

1. c
2. a
3. d
4. d
5. a, b, d, e

6. they have different evolutionary histories; they have similar climates (cold winters, too little rainfall for trees)
7. a
8. c
9. e
10. d
11. a
12. a
13. d

## CHAPTER 48  *Ecosystems and How They Change*

1. get rid of dead bodies and recycle nutrients
2.

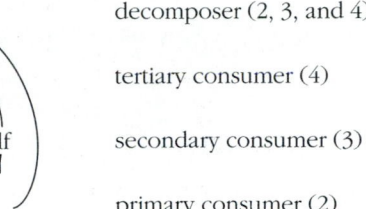

| organism | trophic level |
| --- | --- |
| soil bacteria | decomposer (2, 3, and 4) |
| hawk | tertiary consumer (4) |
| sparrow ← spider   wolf | secondary consumer (3) |
| herbivorous insect   deer | primary consumer (2) |
| shrub | producer (1) |

3. b
4. c
5. e
6. a
7. primary
8. early successional
9. water-loving early pioneer plants (weeds, fugitive species), perennial herbs, shrubs, pioneer trees (e.g., aspen, pin cherry), climax deciduous trees, deciduous forest

## CHAPTER 49  *Populations and Communities*

1. An appropriate curve is shown in Fig. 49–3.
2. b
3. b
4. b
5. d
6. I
7. b
8. c
9. a. increase
   b. decrease
   c. increase

## CHAPTER 50  *Human Ecology and Natural Resources*

1. b    5. d
2. c    6. e
3. a    7. d
4. d

# Periodic Table of Elements

**Electronegativity Values**

| | |
|---|---|
| ☐ <1.0 | ☐ 2.0—2.3 |
| ☐ 1.0—1.4 | ☐ 2.4—2.9 |
| ☐ 1.5—1.9 | ☐ 3.0—4.0 |

Atomic number (protons)
Symbol for the element
Atomic mass*

**Transition Elements**

| I | II | | | | | | | | | | | | III | IV | V | VI | VII | VIII |
|---|---|---|---|---|---|---|---|---|---|---|---|---|---|---|---|---|---|---|
| 1 **H** 1.0079 | | | | | | | | | | | | | | | | | | 2 **He** 4.0026 |
| 3 **Li** 6.941 | 4 **Be** 9.0122 | | | | | | | | | | | | 5 **B** 10.811 | 6 **C** 12.011 | 7 **N** 14.007 | 8 **O** 15.999 | 9 **F** 18.998 | 10 **Ne** 20.180 |
| 11 **Na** 22.990 | 12 **Mg** 24.305 | | | | | | | | | | | | 13 **Al** 26.982 | 14 **Si** 28.086 | 15 **P** 30.974 | 16 **S** 32.066 | 17 **Cl** 35.453 | 18 **Ar** 39.948 |
| 19 **K** 39.098 | 20 **Ca** 40.078 | 21 **Sc** 44.956 | 22 **Ti** 47.88 | 23 **V** 50.942 | 24 **Cr** 51.996 | 25 **Mn** 54.938 | 26 **Fe** 55.847 | 27 **Co** 58.933 | 28 **Ni** 58.69 | 29 **Cu** 63.546 | 30 **Zn** 65.39 | 31 **Ga** 69.723 | 32 **Ge** 72.61 | 33 **As** 74.922 | 34 **Se** 78.96 | 35 **Br** 79.904 | 36 **Kr** 83.80 |
| 37 **Rb** 85.468 | 38 **Sr** 87.62 | 39 **Y** 88.906 | 40 **Zr** 91.224 | 41 **Nb** 92.906 | 42 **Mo** 95.94 | 43 **Tc** (98) | 44 **Ru** 101.07 | 45 **Rh** 102.91 | 46 **Pd** 106.42 | 47 **Ag** 107.87 | 48 **Cd** 112.41 | 49 **In** 114.82 | 50 **Sn** 118.71 | 51 **Sb** 121.75 | 52 **Te** 127.60 | 53 **I** 126.90 | 54 **Xe** 131.29 |
| 55 **Cs** 132.91 | 56 **Ba** 137.33 | 71 **Lu** 174.97 | 72 **Hf** 178.49 | 73 **Ta** 180.95 | 74 **W** 183.85 | 75 **Re** 186.21 | 76 **Os** 190.2 | 77 **Ir** 192.22 | 78 **Pt** 195.08 | 79 **Au** 196.97 | 80 **Hg** 200.59 | 81 **Tl** 204.38 | 82 **Pb** 207.2 | 83 **Bi** 208.98 | 84 **Po** (209) | 85 **At** (210) | 86 **Rn** (222) |
| 87 **Fr** (223) | 88 **Ra** (226) | 103 **Lr** (260) | 104 **Unq** (261) | 105 **Unp** (262) | 106 **Sg** (263) | 107 **Uns** (262) | 108 **Uno** (265) | 109 **Une** (266) | | | | | | | | | |

**Lanthanide series**

| 57 **La** 138.9 | 58 **Ce** 140.1 | 59 **Pr** 140.9 | 60 **Nd** 144.2 | 61 **Pm** (145) | 62 **Sm** 150.4 | 63 **Eu** 152.0 | 64 **Gd** 157.3 | 65 **Tb** 158.9 | 66 **Dy** 162.5 | 67 **Ho** 164.9 | 68 **Er** 167.3 | 69 **Tm** 168.9 | 70 **Yb** 173.0 |
|---|---|---|---|---|---|---|---|---|---|---|---|---|---|

**Actinide series**

| 89 **Ac** (227) | 90 **Th** 232.04 | 91 **Pa** 231.04 | 92 **U** 238.03 | 93 **Np** (237) | 94 **Pu** (244) | 95 **Am** (243) | 96 **Cm** (247) | 97 **Bk** (247) | 98 **Cf** (251) | 99 **Es** (252) | 100 **Fm** (257) | 101 **Md** (258) | 102 **No** (259) |
|---|---|---|---|---|---|---|---|---|---|---|---|---|---|

*Average of isotopic masses in a representative sample.

# Classification of Living Organisms

There are several ways to classify organisms. In this book, organisms are divided into five kingdoms: Prokaryotae, Protista, Fungi, Plantae, Animalia. Kingdoms Animalia and Protista are subdivided into phyla, the equivalent of divisions in kingdoms Fungi and Plantae. Divisions and phyla are subdivided into classes. This appendix includes all the generally recognized divisions and phyla. Some major classes are also mentioned, but the list is far from complete.

## KINGDOM PROKARYOTAE (MONERA)

Bacteria. Prokaryotic cells: lacking a nuclear envelope, plastids, mitochondria, and 9 + 2 flagella. Unicellular, but sometimes occur as clumps or filaments. Reproduction asexual, primarily by binary fission. A few thousand species have been described, and there are probably many as yet unknown, but it is debatable what constitutes a bacterial species. The practice of classifying bacteria differs considerably from the taxonomy of eukaryotes. Two major groups are generally recognized:

**Archaeobacteria**  (Many would raise this to the status of a kingdom): Thermoacidophiles, methanogens, and extreme halophiles. Prokaryotes having membrane lipids with branched phytanols joined to glycerol by ether linkages. Lipid-synthesis pathways, coenzymes, transfer RNAs, RNA polymerases, and cell wall chemistry different from those of other organisms.

**Eubacteria:**  True bacteria. Cell wall containing peptidoglycans, and in Gram-negative forms surrounded by an outer membrane with lipopolysaccharide. About 2500 species.

## KINGDOM PROTISTA (PROTOCTISTA)

Eukaryotic organisms not belonging to kingdoms Fungi, Plantae, or Animalia. Generally aerobic and inhabiting watery environments. Unicellular, coenocytic, syncytial, colonial, or multicellular.

**Phylum Karyoblastea:**  Giant multinucleate amoeba visible to naked eye and containing endosymbiotic bacteria. No endoplasmic reticulum, Golgi complexes, mitochondria, chromosomes, mitosis, nor sexual reproduction. Freshwater. 1 species.

**Phylum Rhizopoda:**  Amoebas. Unicellular; naked or with test. Pseudopods used in locomotion and feeding; no flagellated stages. No sexual reproduction. Terrestrial, marine, freshwater, or parasitic. About 200 species.

**Phylum Haplosporidia:**  Unicellular parasites of aquatic invertebrates. No flagella nor sexual reproduction. 33 species.

**Phylum Paramyxea:**  Parasites of marine invertebrates. No flagella nor sexual reproduction. 6 species.

**Phylum Myxozoa:**  Organisms with few cells; traditionally classified as protists, recent ribosomal RNA sequencing shows these parasites of invertebrates, fish, amphibians, and reptiles to be most closely related to nematodes. About 1100 species.

**Phylum Microspora:**  Unicellular, intracellular parasites of invertebrates, protists, and vertebrates. No mitochondria, flagella, nor sexual reproduction. About 750 species.

**Phylum Foraminifera (Granuloreticulosa):**  Foraminiferans. Unicellular or syncytial; test containing calcium carbonate, sand grains, etc. Thin pseudopods used mostly for feeding. Some with sexual reproduction. Predaceous. Mostly marine. About 4000 species.

**Phylum Actinopoda:**  Radiolarians and heliozoans. Unicellular or syncytial; test containing silica. Fine radial axopods stiffened by microtubules used in feeding; predaceous. Asexual. Radiolarians marine, heliozoans mostly freshwater. About 800 species.

**Phylum Acrasiomycota (Dictyostelida):**  Cellular slime molds. Unicellular feeding amoebas, congregating into a pseudoplasmodium before forming asexual fruiting body with cellulose cell walls and producing asexual, walled spores. Freshwater and terrestrial. About 50 species.

**Phylum Myxomycota:**  Acellular slime molds. Mobile, coenocytic plasmodial feeding stage. Fruiting body with cellulose walls, spores formed by meiosis. Terrestrial. About 550 species.

**Phylum Xenophyophora:**  Plasmodial heterotrophs of the sea floor. 42 species.

**Phylum Zoomastigina:**  Zooflagellates (flagellates without chloroplasts). Unicellular; no cell wall; one or more flagella. Freshwater, many symbionts or parasites. About 1500 species.

**Phylum Apicomplexa:**  Parasitic protists with apical complex used to penetrate host cell. Complex sexual life cycle; flagellated male gametes. About 2400 species.

**Phylum Ciliophora:**  Ciliates. Unicellular, heterotrophic. Two types of nuclei, complex sexual reproduction. Pellicle. Cilia used in locomotion and food collection. Almost all with oral opening. Usually have trichocysts. Freshwater and marine, some parasites and symbionts. About 8000 species.

**Phylum Dinoflagellata (Pyrrophyta):** Dinoflagellates. Mostly unicellular; some with internal cell wall of cellulose plates; two flagella. Chlorophylls *a* and *c* in photosynthetic forms. Food stored as starch and oils. Usually have trichocysts. Mostly marine. Free-living, photosynthetic or predatory, or symbiotic. About 2000 species.

**Phylum Euglenida:** *Euglena* and its relatives. Mostly unicellular; pellicle of protein, no cell wall; usually two anterior flagella. Light receptor, chlorophylls *a* and *b* in photosynthetic forms. Food stored as paramylon. No sexual reproduction. Mostly freshwater, some parasitic. About 800 species.

**Phylum Cryptophyta:** Cryptomonads. Unicellular; heterotrophic or photosynthetic, chlorophylls *a* and *c*. Two anterior flagella inserted in a "crypt" (oral groove). About 50 species.

**Phylum Prymnesiophyta:** Coccolithophorids. Mostly unicellular; photosynthetic, chlorophylls *a* and *c*. Food stored as paramylon. Anchoring to surfaces by thread-like haptoneme with 6 + 1 or 6 + 0 microtubule arrangement. Two anterior flagella. Resting stage covered by overlapping calcareous coccoliths. No sexual reproduction. Most marine. 500 species.

**Phylum Eustigmatophyta:** Mostly unicellular; single, hairy anterior flagellum with swelling at its base and eyespot in adjacent swelling. Photosynthetic, single chloroplast with chlorophylls *a, c,* and *e*. No sexual reproduction. Mostly freshwater. 12 species.

**Phylum Plasmodiophoromycota:** Plasmodial endoparasites of plants (including some crop diseases), fungi, and protists. 29 species.

**Phylum Chytridiomycota:** Fungus-like saprobes and parasites with unicellular or coenocytic thread-like bodies; chitinous cell walls. Isogamous or anisogamous sexual reproduction, with at least one gamete moving by means of a smooth posterior flagellum. In fresh water or soil. 1000 species.

**Phylum Rhodophyta:** Red algae. Unicellular to multicellular. Complex, sexual life cycles with no flagellated cells. Chlorophyll *a,* phycobilins; chloroplast membranes not stacked. Food stored as floridian starch in granules outside plastids. Mostly marine, some freshwater, a few terrestrial. About 4100 species.

**Phylum Bacillariophyta:** Diatoms. Mostly unicellular; diploid; cellulose and pectin cell walls containing silica. Mostly nonmotile, a few species with flagellated male gametes. Chlorophylls *a* and *c,* fucoxanthin and other carotenoids. Food stored as oil and chrysolaminarin. Sexual reproduction present. Very abundant: marine, freshwater, terrestrial. About 10,000 species.

**Phylum Hyphochytridiomycota:** Hyphochytrids. Fungus-like saprobes or parasites with thread-like bodies; chitinous walls. No sexual reproduction, asexual reproduction by zoospores with single hairy anterior flagellum. In fresh water and soil. 23 species.

**Phylum Labyrinthulomycota:** Slime nets. Colonies of slender, transparent cells secreting slime tunnels in which they live and move. Marine, freshwater, soil. 38 species.

**Phylum Oomycota:** "Water molds" and their terrestrial relatives. Coenocytic diploid hyphae with cellulose walls. Unwalled spores with two unlike flagella, and walled airborne spores; oogamous sexual reproduction. Aquatic and terrestrial saprobes and parasites. About 800 species.

**Phylum Xanthophyta:** Yellow-green algae. Two unlike flagella. Chlorophylls *a* and *c,* carotenoids. Food stored as fat or oil. 600 species.

**Phylum Chrysophyta:** Golden algae. Unicellular or colonial. Cellulose and pectin cell walls. Two unlike flagella in motile stages. Chlorophylls *a* and *c,* fucoxanthin and other carotenoids. Food stored as oil and chrysolaminarin. Sexual reproduction present. Mostly freshwater, some marine. About 1000 species.

**Phylum Phaeophyta:** Brown algae. All multicellular, some very large and elaborate. Sperm and motile spores with two unlike flagella. Chlorophylls *a* and *c,* fucoxanthin and other carotenoids. Store laminarin. Almost all marine. About 900 species.

**Phylum Chlorophyta:** Green algae. Unicellular to multicellular; motile stages with two or more flagella. Chlorophylls *a* and *b,* carotenoids. Store starch in plastids. Mostly freshwater, many marine, some terrestrial. About 4300 species.

**Phylum Conjugaphyta:** Conjugating green algae. Desmids (double cells) and filaments. No flagellated stages. Isogamous. Chlorophylls *a* and *b,* carotenoids. Store starch in plastids. Freshwater. About 10,700 species.

## KINGDOM FUNGI

Saprobes, parasites, or mutalistic symbionts; usually with thread-like mycelial bodies, some unicellular. Cell walls of chitin. Haploid or dikaryotic. Nearly all aerobic. No flagella at any stage; asexual and sexual reproduction by spores. Mostly terrestrial, some marine. Assigned to divisions by sexual reproductive structure (if any).

**Division Zygomycota:** Sexual reproduction by zygospores. Coenocytic hyphae. Asexual spores produced in sporangia. About 765 species.

**Division Ascomycota:** Sac fungi. Sexual reproduction by ascospores formed in sac-like asci. Some unicellular (yeasts); hyphae of multicellular forms divided by perforated septa. Short dikaryotic stage before sexual reproduction. About 28,650 species.

**Division Basidiomycota:** Club fungi. Sexual reproduction by basidiospores borne by club-shaped basidia. Septate hyphae. Long dikaryotic stage in life cycle. About 16,000 species.

**Division Deuteromycota:** Imperfect fungi. No known sexual reproduction. About 17,000 species.

## KINGDOM PLANTAE

Multicellular eukaryotes, mostly photosynthetic with chlorophylls *a* and *b*. Food usually stored as starch. Cell walls containing cellulose. Alternation of heteromorphic haploid and diploid generations. Diploid generation developing by way of embryo enclosed in multicellular reproductive structure. Mostly terrestrial, some freshwater and marine.

**Division Bryophyta:** Mosses. Dominant gametophyte small, usually with pointed leaf-like structures; anchored by rhizoids. Male producing flagellated sperm. Sporophyte dependent on gametophyte. Usually moist habitats. About 14,500 species.

**Division Hepatophyta:** Liverworts. Dominant gametophyte small, with round-lobed leaf-like structures or flat and ribbon-like. Characteristic oil bodies in each cell. Male producing flagellated sperm. Sporophyte dependent on gametophyte. Usually moist habitats. About 9000 species.

**Division Anthocerotophyta:** Hornworts. Gametophyte mat-like; sporophyte long, cylindrical, horn-shaped, producing multicellular spores in some species. Only one chloroplast per cell. Usually moist habitats. About 200 species.

**Division Lycophyta:** Club mosses, ground pine. Small, but some tree-like fossil forms. Underground or trailing stems as well as above-ground erect stems bearing scale-like single-veined microphyll leaves. Sporangia on top of leaves or modified leaves, or in their axils. Gametophytes autotrophic or heterotrophic. About 1200 species.

**Division Psilotophyta:** Whisk fern (*Psilotum*) and *Tmesipteris*. Very primitive vascular plants; *Psilotum* without roots or leaves, and with vascular tissue in gametophyte. 3 living species.

**Division Sphenophyta:** Horsetails. Most short, some tree-like fossil and tropical forms. Hollow, jointed stems with vertical ribs, some with slender branches; reduced leaves around joints; sporangia in terminal cones. Gametophytes photosynthetic. One genus (*Equisetum*), about 25 living species.

**Division Pterophyta:** Ferns. Most short, some tropical tree-like forms, many fossil forms. Large, many-veined leaves (fronds), often highly divided, coiled when young; sporangia on underside of fronds or on separate nongreen fertile fronds. Gametophytes usually photosynthetic. About 12,000 species.

**Division Cycadophyta:** Cycads. Gymnosperms; palm-like shrubs and small trees. Sexes separate, pollen and seeds borne in cones, sperm flagellated. Tropical and subtropical. About 160 species.

**Division Gnetophyta:** Desert or tropical gymnosperms with variable body and leaf forms; vessels in xylem. About 70 species.

**Division Ginkgophyta:** *Ginkgo*. Gymnosperm tree with broad, fan-shaped leaves. Sexes separate; pollen releases flagellated sperm; smooth naked seeds. 1 living species.

**Division Coniferophyta:** Conifers. Gymnosperms; shrubs and trees with needle- or scale-like leaves, mostly evergreen. Most produce pollen and seeds in cones. About 550 species.

**Division Anthophyta:** Flowering plants. Vascular tissue usually with vessels in xylem. Most with broad leaves. Gametophytes reduced, developing in flowers; ovules within closed carpels; pollen carried by wind or animals. Double fertilization, producing diploid zygote and (usually) triploid endosperm. Dispersal by seeds arising from ovule, which develop inside fruits arising from ovary. About 235,000 species.

## KINGDOM ANIMALIA

Animals. Eukaryotic multicellular heterotrophs with embryonic development by way of a hollow blastula. Nutrition primarily by ingestion. No cell walls; many are motile at some stage of life. Reproduction generally sexual, oogamous. Diploid (usually) adults produce haploid gametes that fuse to form a zygote.

**Phylum Porifera:** Sponges. Simple, sessile filter-feeding animals with specialized cells but no tissues or organs; body of indefinite form organized around water canals. Skeleton of spicules or spongin. Largely marine. About 5000 species.

**Phylum Cnidaria:** Jellyfish, sea anemones, corals, etc. Radially symmetrical, free-swimming or sessile, solitary or colonial; inner and outer tissue layers developed. Carnivorous, with mouth surrounded by tentacles bearing nematocysts; gastrovascular cavity; no anus. Mostly marine. About 9000 species. Classes: Hydrozoa, Scyphozoa, Anthozoa.

**Phylum Ctenophora:** Comb jellies. Once classified with Cnidaria, which they resemble, in phylum Coelenterata, but only one ctenophore species has nematocysts; free-swimming, with two tentacles and eight rows of ciliary combs for locomotion; marine. About 90 species.

**Phylum Mesozoa:** Peculiar parasites of marine invertebrates; contain few, large cells and no organs. About 50 species.

**Phylum Platyhelminthes:** Flatworms. Free-living or parasitic; bilaterally symmetrical; three well-developed tissue layers with organs, no coelom. Gut with mouth but no anus. Excretion by flame cell system. About 18,500 species. Classes: Turbellaria, Trematoda, Cestoda.

**Phylum Nemertina (Rhynchocoela):** Ribbon worms. Long, flattened acoelomate worms 0.5 mm to 30 m long; ciliated body surface. Complex proboscis used to explore environment and catch prey; anus and circulatory system present. Most marine, some terrestrial or freshwater. About 900 species.

**Phylum Gnathostomulida:** Tiny, ciliated, acoelomate worms once classified with platyhelminths, which they resemble except in having only one cilium per body surface cell. In marine sand or on surface of sea plants. About 80 species.

**Phylum Nematoda:** Roundworms. Free-living or parasitic. Body covered by cuticle; posterior adhesive gland. Gut with mouth and anus. Fluid-filled body cavity (pseudocoel). About 12,000 species.

**Phylum Rotifera:** Rotifers. Microscopic, sessile or free-living "wheel animalcules" named for sequentially beating ring of cilia around the mouth. Body covered by cuticle; posterior adhesive gland. Anus present. Most are asexually reproducing females. About 2000 species.

**Phylum Gastrotricha:** Microscopic to 4-mm pseudocoelomate body ciliated ventrally. Adhesive tubes and spines on sides of body. Cuticle with decorative patterns. Marine and freshwater. About 460 species.

**Phylum Kinorhyncha:** Tiny (up to a few millimeters), elongated worm-like body with spiny cuticle. Move by anchoring snout and pulling body along. Pseudocoel and anus present. Marine, on muddy intertidal shores. About 100 species.

**Phylum Nematomorpha:** Hairworms. Thread-like, pseudocoelomate worms. Juveniles parasitic; adults free-living in water or in damp soil. About 230 species.

**Phylum Acanthocephala:** Small pseudocoelomate worms parasitic on vertebrates, with arthropod intermediate hosts. Retractile proboscis with spines, no gut. About 1000 species.

**Phylum Priapulida:** Cucumber-shaped animals, 1 to 200 mm long, with reduced coelom and proboscis. Covered with spines. Marine, bottom-burrowing predators. 13 species.

**Phylum Annelida:** Segmented worms. Coelomate, with circulatory system (except leeches), nephridia; gut with mouth and anus. Locomotion in free-living forms aided by setae except in leeches, which have anterior and posterior suckers. About 8900 species. Classes: Polychaeta, Oligochaeta, Hirudinea.

**Phylum Mollusca:** Body covered by a mantle, which secretes a shell in some. Head and muscular foot usually present. Excretion by nephridia, feeding by a scraping radula. Development via a trochophore-like larva or direct. About 57,000 species. Major classes: Amphineura, Bivalvia, Gastropoda, Cephalopoda.

**Phylum Pentastomida:** Worm-like parasites in nasal passages or lungs of vertebrates. Two pairs of leg-like anterior appendages with hooks. Chitinous cuticle. About 90 species.

**Phylum Phoronida:** Lophophorate worms living in chitinous tubes. Marine. 15 species.

**Phylum Entoprocta:** Mainly sessile lophophorates with both mouth and anus opening within circle of tentacles. Attached by stalks, in colonies. Almost all marine. About 150 species.

**Phylum Brachiopoda:** Lamp shells. Lophophorate body attached by a stalk and enclosed by two unequal calcareous shells superficially resembling the shells of bivalve molluscs. Marine. About 300 species.

**Phylum Bryozoa:** Moss animals. Microscopic, coelomate, colonial lophophorates. Larvae retained in brood pouch. Mainly marine. About 4000 species.

**Phylum Loricifera:** Microscopic. Telescoping mouth tube, spiny head, girdle of cuticle plates around middle. Marine, clinging tenaciously to sand grains. About 10 species.

**Phylum Echiura:** Detritus-feeding worms with flattened proboscis, coelom, pair of ventral setae. Marine burrowing. About 140 species.

**Phylum Pogonophora:** Long, tube-living worms. Tentacles with microvilli and cilia. Gut absent. Marine, in water at least 100 m deep. Embryology resembles that of chordates. About 80 species.

**Phylum Tardigrada:** Water bears. Small (maximum 1 mm), segmented body, four pairs of clawed legs. Stylus of mouth used to suck fluids from plants and small animals. Dormant stage surviving extremely high or low temperatures and desiccation for long periods. Mainly terrestrial, some freshwater. About 375 species.

**Phylum Onychophora:** Segmented, worm-like body with thin cuticle. Antennae and many pairs of unjointed, clawed legs. Possible link between annelid worms and arthropods. Terrestrial carnivores. About 70 species.

**Phylum Arthropoda:** Segmented animals with exoskeleton containing chitin; paired jointed appendages. Body cavity a hemocoel. Marine, freshwater, and terrestrial. About 825,000 described species.

  CLASS ARACHNIDA: Spiders, scorpions, ticks, mites. Body with one or two main parts. Six pairs of jointed appendages. Mostly terrestrial. About 72,000 species.

  CLASS CRUSTACEA: Body of two or three parts. Skeleton usually impregnated with calcium carbonate. Adults with two pairs of antennae; three or more pairs of legs. Development usually via nauplius larva. Most marine. About 42,000 species.

  CLASS INSECTA: Insects. Body divided into head, thorax, and abdomen. Head with one pair of antennae, complex mouthparts. Adults usually with two pairs of wings and three pairs of legs. Breathing mainly by tracheae. Excretion by Malpighian tubules. Mostly terrestrial, some freshwater. More than 700,000 species.

  CLASS DIPLOPODA: Millipedes. Distinct head bearing antennae and chewing mouthparts. Each double segment bearing two pairs of walking legs and two pairs of spiracles of the tracheal system. About 7000 species.

  CLASS CHILOPODA: Centipedes. Appendages of first body segment modified as poison claws; each remaining segment bearing one pair of walking legs. Predaceous on insects. Terrestrial. About 2000 species.

**Phylum Sipuncula:** Plump, cylindrical coelomate worms with lobes or short tentacles around mouth. Possibly common ancestor of major deuterostome phyla. Marine, bottom-dwelling. About 320 species.

**Phylum Chaetognatha:** Arrow worms. Arrow-like, active predators with anterior grasping spines. Tail used for swimming, fins for stability. Marine planktonic. About 70 species.

**Phylum Echinodermata:** Sea stars, sea lilies, sea cucumbers, etc. Adults pentaradially symmetrical with calcareous skeletons, water vascular system, and tube feet. Marine. About 5800 species. Classes: Crinoidea, Asteroidea, Ophiuroidea, Echinoidea, Holothuroidea.

**Phylum Hemichordata:** Acorn worms. Body divided into proboscis, collar with pharyngeal gill slits, and trunk. Ventral and dorsal nerve cords. Coelom a hydrostatic skeleton. Larvae resembling those of echinoderms. About 85 species.

**Phylum Chordata:** Animals with a notochord and pharyngeal gill slits at some stage in the life cycle, a hollow, dorsal nerve cord, and a postanal tail. Aquatic and terrestrial. About 43,000 species.

**Subphylum Cephalochordata:** Amphioxus. Fish-like, with a permanent notochord. Filter-feeding, using a ciliated pharynx with gill slits. Marine. 29 species.

**Subphylum Urochordata:** Tunicates. Larva active, with well-devloped nervous system and notochord. Adults sessile, often forming colonies. Filter-feeding with gill slits, reduced nervous system, no notochord. Marine. About 1300 species.

**Subphylum Vertebrata:** Notochord replaced by cartilage or bone during embryonic development, forming a segmented vertebral column. Skull surrounding the brain. Excretion by kidneys. About 42,000 species.

CLASS AGNATHA: Jawless fishes. Marine and freshwater. Gill openings separate. Skeleton cartilaginous. Notochord persisting throughout life. About 45 species.

CLASS CHONDRICHTHYES: Sharks, skates, rays. Most predaceous. Cartilaginous skeleton with jaws. Teeth and scales dermal denticles. Notochord replaced by vertebrae in adult. Intestine with spiral valve. Gill openings separate. Paired pelvic and pectoral fins. Tail fin usually asymmetrical. About 275 species.

CLASS OSTEICHTHYES: Bony fishes (ray-finned actinopterygians and lobe-finned sarcopterygians). Marine and freshwater. Body usually covered with scales. Bony skeleton and jaws. Gill openings covered by a single operculum. Paired pelvic and pectoral fins. Tail fin usually symmetrical. Many with a swim-bladder. About 25,000 species.

CLASS AMPHIBIA: Frogs, toads, salamanders, newts, caecilians. Tetrapods that lay eggs without an amnion or shell, usually developing via a tadpole larva. Respiration via lungs and skin; in larva via gills. Scales absent except in caecilians. About 25,000 species.

CLASS REPTILIA: Turtles, squamates, crocodilians. Tetrapods with amniotic eggs and scaly skin. Breathe via lungs. Claws on toes. Effective double circulation. About 6000 species.

CLASS AVES: Birds. Vertebrates with feathers. Scaly feet, amniotic eggs. High body temperature and blood pressure. Complete double circulation. Bipedal; most species with more than one mode of locomotion. Forelimbs modified as wings. Teeth replaced with horny bill and stones in gizzard. About 9000 species.

CLASS MAMMALIA: Tetrapods with young nourished by milk from mammary glands of females. Most viviparous; a few lay amniotic eggs. High body temperature and blood pressure. Full double circulation evolutionarily distinct from that of birds. Body usually covered with hair. Only vertebrates with only one bone in each side of lower jaw. Most with heterogeneous teeth that perform different functions. About 4000 species.

# Glossary

**abdomen** In vertebrates and some arthropods, the rearmost part of the body, containing the end of the digestive tract, external genital organs, etc.

**abiotic** Nonbiological; occurring neither within living cells nor under their influence.

**aboral** Pertaining to the surface of a radially symmetrical animal which is opposite to the (oral) surface where the mouth is located.

**abortion** Process whereby a mammalian embryo or fetus becomes detached from the uterine wall and expelled from the female's body either naturally ("miscarriage," spontaneous abortion) or by medical means (induced abortion).

***Acacia*** Large genus of trees, most of which are tropical. Common name: thornwood.

**acetyl CoA** Acetyl coenzyme A, molecule that can donate its (2-carbon) acetyl group to synthesis of fatty acids or to the Krebs cycle, where it is broken down to yield energy.

**acetylcholine** An important neurotransmitter in vertebrates and invertebrates; the transmitter at some synapses in the brain, many nerve-muscle junctions, etc.

**acid** 1. Substance that releases hydrogen ions ($H^+$) in aqueous solution. 2. Substance that accepts electrons.

**acid rain** Rain containing dissolved pollutants, most commonly oxides of nitrogen and sulfur that give it an acid pH, usually of less than 5.0.

**actin** A protein that forms the thin filaments involved in muscle contraction and the microfilaments involved in changes of shape in other kinds of cells.

**action potential** A fast-moving, all-or-nothing electrical disturbance in the membrane of a nerve cell; a nerve impulse. Also occurs in the membrane of a muscle when it is stimulated to contract.

**activation energy** Energy input required before a chemical reaction can proceed.

**active transport** Process in which energy is expended in moving a substance through a membrane against its concentration gradient.

**acylation** The attachment of an amino acid to a transfer RNA molecule.

**adaptation** 1. Process by which populations evolve to become suited to their environments over the course of generations. 2. Characteristic that increases an organism's evolutionary success. 3. Diminishing response of a sense organ to a constant stimulus.

**adaptive radiation** Formation of two or more new species, macromolecules, or physiological pathways, adapted to different ways of life, from one ancestral species, molecule, or pathway.

**adhesion** Sticking together (e.g., of cells, molecules of different substances).

**adipose tissue** Tissue made up mainly of fat-storing cells.

**adrenalin** *See* **epinephrine.**

**adsorption** Attachment of molecules of a substance to the surface of a structure or particle.

**aerial** In the air.

**aerobic** 1. Requiring molecular oxygen ($O_2$) to live. 2. In the presence of molecular oxygen.

**affected individual** In genetics, an individual homozygous for a deleterious recessive allele, with a phenotype expressing the allele.

**Agricultural Revolution** Change in human society from mainly hunter-gatherer to mainly agricultural means of acquiring food.

**air bladder** Air-filled sac providing buoyancy in fish (= swimbladder) or algae.

**albinism** (*adj.,* albino) White coloration due to lack of genes for normal pigmentation.

**alcohol** Organic compound containing an alcohol group (COH).

**alcoholic fermentation** Biochemical pathway in some yeasts (and other organisms) in which pyruvate is converted into ethyl alcohol (ethanol), releasing NAD; used to make alcoholic beverages.

**alga** (*pl.,* algae) Photosynthetic organism with a one-celled or simple multicellular body plan and lacking a multicellular embryo protected by the female reproductive structure. In ecology, includes cyanobacteria (blue-green algae).

**alkali** (= base) Substance that releases hydroxide ions ($OH^-$) in water, accepts hydrogen ions ($H^+$), or gives up electrons.

**alkaline** Having a (basic) pH of more than 7.0.

**allantois** An embryonic membrane: sac that grows out of the embryonic gut in higher vertebrates; stores the embryo's nitrogenous waste until hatching in reptiles and birds; forms the main embryonic respiratory surface and part of the placenta in mammals.

**allele** One of two or more genes carrying the information for contrasting forms of the same genetic trait (e.g., blue eyes and brown eyes).

**allopatric** Living in different areas.

**allosteric protein** A protein capable of changing shape in a way that alters its binding sites and its activity.

**alpine** Of high mountain regions.

**altitude**  Height above sea level.

**altruistic behavior**  Behavior that favors the reproductive success of some member of the species other than the actor.

**ambient temperature**  Temperature of the surroundings (equivalent to "room temperature").

**amino acid**  Small organic molecule containing both a carboxyl and an amino group. Alpha amino acids, in which these groups are bonded to the same carbon atom, are the monomers from which polypeptides and proteins are made.

**aminoacyl attachment site**  That part of a transfer RNA molecule to which an amino acid attaches.

**amino group**  $-NH_2$.

**amino sugar**  Sugar with an amino group attached.

**ammonia**  $NH_3$.

**ammonium**  $NH_4^+$.

**amniocentesis**  Technique for obtaining a sample of fetal cells by inserting a syringe through the mother's abdominal wall and uterus, into the amniotic sac surrounding the fetus, which contains cells sloughed from the fetus.

**amnion**  A fluid-filled sac that surrounds the embryo in reptiles, birds, and mammals.

**amniotic egg**  Type of egg laid by reptiles and their descendants (birds and mammals) in which the embryo is surrounded by an amnion and other membranes (allantois and chorion).

**amphibian**  Member of the vertebrate class Amphibia; e.g., frogs, toads, salamanders, newts, and their relatives.

**amylase**  An enzyme that breaks glucose molecules from starch molecules.

**amyloplast**  Plastid that stores starch.

**amylose, amylopectin**  Polymers of glucose that combine to make up starch.

**anaerobic** (*n.*, anaerobe)  1. Without molecular oxygen ($O_2$). 2. Not requiring $O_2$ for extraction of energy from food (respiration).

**analogous**  In biology, of two or more organs with the same function but with different evolutionary origins.

**anaphase**  The stage of nuclear division during which chromosomes move to the poles of the spindle apparatus.

**anatomy**  The physical structure of an organism.

**anemia**  A deficiency of hemoglobin or of red blood cells, resulting in a deficiency of oxygen supply to the tissues.

**angiosperms**  Flowering plants, including many trees, grasses.

**anion**  Negatively charged ion.

**annelid**  Member of the phylum Annelida, segmented worms; e.g., earthworm, polychaete, leech.

**annual**  1. Yearly. 2. Flowering plant that grows from seed, flowers, sets seed, and dies within one year.

**anterior**  At or toward the front (head) end of an animal.

**anther**  Organ in flowers that produces pollen.

**antheridium**  Organ that produces male gametes in fungi, algae, mosses, and ferns.

**anthocyanins**  Pigments of red, blue, or purple hue, commonly found in vacuoles of plant cells.

**anthropoids**  Monkeys, apes, and humans, which make up the suborder Anthropoidea of the order Primates.

**anthropology**  Study of the origin and of the physical, social, and cultural evolution of humans.

**anthropomorphism**  Attribution of human characteristics to other animals.

**antibiotics**  Substances that kill microorganisms.

**antibody**  (= immunoglobulin)  One of a huge group of proteins, produced by B lymphocytes, which will react with a specific antigen as part of the body's defense against disease.

**anticodon**  The row of three nucleotides on a tRNA molecule that base-pairs with the complementary codon on a mRNA molecule attached to a ribosome. This allows the amino acid carried by the tRNA to be placed correctly into the peptide chain that is being synthesized on the ribosome.

**antidiuretic**  Opposing the tendency to excrete water.

**antigen**  A substance capable of stimulating an immune response by binding to an antibody or lymphocyte receptor; usually a protein or other substance that is not naturally part of the body.

**Anura**  Order of amphibians containing frogs and toads.

**aorta** (*pl.*, aortae)  Large artery that carries blood from the heart toward the rest of the body.

**apical**  Of the tip of something.

**apical meristem**  Area of dividing cells at the root and stem tips of a plant.

**aposematic coloration**  Conspicuous, "warning" coloration of an animal that is dangerous or distasteful to eat.

**aquatic**  Of water (fresh or salt).

**aqueous**  Containing or composed largely of water.

**arable land**  Land suitable for growing crops.

**arboreal**  Living in trees.

**Archaeobacteria**  Ancient group containing the methane-producing, salt-loving, and hot acid–loving bacteria.

**arid**  Dry; of areas where more water leaves the ecosystem by evaporation and transpiration than enters it as precipitation.

**arteriole**  A small artery carrying blood from an artery to a capillary bed.

**artery**  Vessel that carries blood away from the heart.

**arthropod**  Member of phylum Arthropoda: segmented animals with jointed appendages and stiff chitin-containing external skeletons; e.g., lobsters, spiders, insects.

**atmosphere** (of Earth)  Layer of gases that surrounds Earth, including the troposphere (next to Earth's surface) and the stratosphere.

**ATP**  Adenosine triphosphate; most common energy-donating molecule in biochemical reactions.

**atrium** (*pl.*, atria)  A chamber; e.g., part of the heart that receives blood as it enters.

**autonomic nervous system**  The part of the vertebrate nervous system over which the animal usually has no control; composed of the sympathetic and parasympathetic systems; regulates blood pressure, breathing, gut movements, etc.

**autoradiography**  Technique for determining the position of a radioactive element introduced into a system; based on the fact that radiation emitted by radioactive substances exposes (blackens) photographic film that is placed near them.

**autosome**  Any chromosome that is not a sex chromosome.

**autotroph** (*adj.*, autotrophic)  Organism that can make its own (organic) food molecules from inorganic constituents, using energy either from sunlight (by photosynthesis) or inorganic chemical reactions (chemosynthetic bacteria). *Compare* **heterotroph.**

**avian**   Pertaining to birds, members of class Aves.

**axial**   Pertaining to an axis.

**axil** (*adj.,* axillary)   In plants, the angle between the stalk (petiole) of a leaf and the stem.

**axon** (*adj.,* axonal)   Extension of a neuron that carries nerve impulses to the next cell(s) in line.

**bacterial transformation**   Uptake and incorporation of genetic material by a bacterium.

**bacteriophage** (*abbr.,* phage)   A virus that infects, and reproduces within, a bacterium.

**bacterium** (*pl.,* bacteria)   Member of kingdom Prokaryotae, containing organisms without nuclear envelopes around the single, circular DNA genome; most are very small, unicellular or forming colonies of independent cells.

**basal body**   Structure at the base of a cilium or flagellum, containing a circle of nine triple microtubules.

**base** (*adj.,* basic)   *See* **alkali.**

**base pairing**   In nucleic acids, the attachment of one nucleotide base to another by hydrogen bonds, in unique pairs: adenine-thymine in DNA, adenine-uracil in RNA, or guanine-cytosine in both.

**bicarbonate ion**   $HCO_3^-$, an important buffer in biological systems.

**binary fission**   Division (usually of a cell) into two equal parts.

**bioaccumulation**   Concentration of a chemical (usually a pollutant) as it passes up the food chain.

**biodegradable**   Capable of being broken down by living organisms into inorganic compounds.

**biodiversity**   Number of different species of organisms in an area.

**biogeochemical cycle**   The pathways of an element through various geological and biological processes.

**biological pest control**   Use of naturally occurring chemicals or organisms to reduce the populations of pest organisms.

**biological species**   A group containing all the members of one or more populations that interbreed or potentially interbreed with one another under natural conditions.

**bioluminescence**   Production of light by a living organism (such as a firefly).

**biomass**   Amount of material that is part of the bodies of living organisms.

**biome**   Major type of terrestrial community of organisms, defined by dominant vegetation and determined by climate; e.g., tropical rain forest, desert.

**bioremediation**   Use of organisms to degrade pollutants.

**biosphere**   Total of all areas on Earth where organisms are found; includes deep ocean and part of the atmosphere.

**biota**   The organisms living in an area.

**biotic (reproductive) potential**   Rate at which a population of a species can increase in size under ideal conditions.

**biotic**   Having to do with living organisms.

**bipedalism**   Habit of walking on two legs.

**bisphosphate**   Having two separate phosphate groups (*compare* **diphosphate**)

**bloom**   Generally means flower; ecologists also use it to mean a rapid increase in population of microorganisms.

**bone marrow**   Soft tissue in the center of the long bones in the limbs.

**brachiation**   Mode of locomotion using the forelimbs to swing from branch to branch.

**broad-leaved evergreen**   A tree or shrub that is evergreen but not a conifer; e.g. live oak, rhododendron, southern magnolia.

**bromeliad**   Member of a large family of mainly epiphytic plants.

**bryophyte**   Member of division Bryophyta: the mosses.

**buffer**   Something that lessens or absorbs the shock of an impact. In chemistry, a substance that, within limits, prevents the pH of a solution from changing when acid or alkali is added to it.

**bug**   Member of the insect order Hemiptera.

**$C_3$ cycle** (= Calvin cycle)   Metabolic pathway of photosynthesis in which NADPH and ATP are used to fix carbon dioxide into three-carbon organic molecules.

**caecum**   "Blind" sac, pouch leading off the digestive tract in some animals.

**calcareous**   Composed of, or containing, calcium carbonate ($CaCO_3$).

**calorie** (*adj.,* caloric)   1 Calorie = 1000 calories = 1 kilocalorie or kcal. A calorie is the amount of heat needed to raise 1 gram of water from 14.5 °C to 15.5 °C. The energy values of foods are often expressed in Calories. Not an SI unit.

**cambium** (*adj.,* cambial)   Meristematic tissue that produces new cells which increase the diameter of a woody stem or root.

**canine**   1. (*adj.*) Of or pertaining to dogs. 2. (*noun*) The (usually) pointed tooth next to the incisors in mammals.

**capillaries**   The narrowest blood vessels in a circulatory system; exchange of substances between the blood and the extracellular fluid takes place across the thin walls of the capillaries.

**capillary action (capillarity)**   Movement of fluid into a narrow space because of attraction of the walls of the space for the molecules of the fluid.

**carbohydrates**   A class of compounds whose members have the general formula $(CH_2O)_n$ and contain at least one double-bonded oxygen.

**carbon dioxide**   ($CO_2$) Gas produced by respiration of organisms or by burning of organic matter or other carbon-containing substances.

**carboxyl group**   A functional group (—COOH) consisting of a carbon atom double-bonded to an oxygen atom and single-bonded to another oxygen atom, which in turn is bonded to a hydrogen atom.

**carcinogen**   Something that causes cancer.

**cardiac**   Of the heart.

**cardiac muscle**   The type of muscle that makes up the vertebrate heart.

**cardiovascular**   Of the heart and circulatory system.

**carnivore**   1. Animal that eats other animals. 2. Member of the mammalian order Carnivora; e.g., dogs, cats, weasels, bears, raccoons, skunks.

**carotenoid**   Accessory photosynthetic pigment that usually appears yellow, orange, or brown.

**carpals**   The bones in the wrist.

**carpel**   A female flower part.

**carrier**   In genetics, an individual heterozygous for a (usually damaging) recessive allele.

**carrying capacity** (of an area for a species)   Number of individuals of a species that the environment can support.

**cartilage** (*adj.,* cartilaginous)   1. Tissue composed of scattered cells surrounded by tough, flexible intercellular protein fibers and a firm jelly-like matrix. 2. A skeletal element composed of cartilage.

**cation** Positively charged ion.

**Caucasian** 1. Member of an ethnic division of the human species having light-colored skin and fine hair, indigenous to Europe, northern Africa, southwestern Asia, and the Indian subcontinent; named from the old belief that this ethnic group originated in the Caucasus, near the Black Sea.

**caudal** Of the tail.

**cell wall** Stiff, fibrous structure outside the plasma membrane of cells of plants, fungi, bacteria, and some protists.

**cellular respiration** Stepwise release of energy from food molecules, accompanied by storage of the energy in short-lived energy intermediates such as ATP.

**cellulose** Polysaccharide forming the fibers that makes up a large part of the cell wall in plants.

**central nervous system** The brain and spinal cord.

**centrioles** Pair of structures containing nine triple microtubules, found in most animal cells and in cells of lower plants and fungi; usually close to the nucleus outside the nuclear membrane.

**centromere** The constricted area where chromatids of a replicated chromosome remain joined until anaphase.

**cephalization** Concentration of nervous tissue and sense organs in the head.

**cereals** Those grass species of the plant family Poaceae that make up the bulk of human food.

**cerebral cortex** Outer layer of the cerebral hemispheres.

**cerebral hemispheres** The two halves of the cerebrum in the forebrain.

**cerebrospinal fluid** A clear fluid, derived from the blood plasma, that bathes and cushions the brain and spinal cord.

**cerebrum** A large, rounded area of the forebrain divided by a fissure into right and left halves (= cerebral hemispheres).

**Cetacea** Order of mammals whose members are most highly adapted to aquatic life; whales, dolphins, and porpoises.

**chaparral** Dry shrublands of temperate coastal regions such as California and the Mediterranean coast.

**chelicerae** Piercing and sucking mouthparts of spiders and their relatives; associated with poison glands in spiders.

**chemiosmosis** (*adj.*, chemiosmotic) A means of releasing free energy by permitting hydrogen ions to pass through a membrane down their concentration gradient.

**chemoreceptor** A sense organ that responds to chemical stimuli.

**chemosynthesis** Production of organic molecules from inorganic molecules, using energy released by chemical reactions rather than light energy, which is used in the much more common process of photosynthesis. Also called **chemoautotrophy** or **chemolithotrophy.**

**chitin** (*adj.*, chitinous) Structural polysaccharide composed of amino sugar monomers; a prominent component of the cuticle of arthropods (e.g., insects) and of the cell walls of fungi.

**chiton** Member of molluscan class Amphineura, having a body covered with eight crosswise skeletal plates.

**chlorophyll** Green pigment that traps light energy during photosynthesis.

**Chlorophyta** A division containing the green algae.

**chloroplast** Plastid containing chlorophyll; the site of photosynthesis.

**chondrichthyes** Cartilaginous fish with jaws; elasmobranchs: the sharks, skates, and rays.

**chordate** Member of the animal phylum Chordata, with a notochord and pharyngeal gill slits at some stage of its life.

**chorion** The embryonic membrane lying immediately under the egg shell in reptiles and birds; in mammals it forms part of the placenta.

**chromatid** One of the two replicated strands of a chromosome still held together at the centromere.

**chromatin** Combination of DNA and proteins, visible as a loosely arranged mass in the nucleus of an appropriately stained eukaryotic cell during interphase.

**chromatography** Separation of a mixture of substances into its components by differences in their electrical charges and particle sizes.

**chromoplast** Plastid that produces and stores yellow and orange pigments.

**chromosome** (*adj.*, chromosomal) Thread-like structure in the nucleus of a eukaryotic cell, consisting of DNA and proteins and carrying genetic information.

**ciliate** Member of the protistan phylum Ciliophora.

**cilium** (*pl.*, cilia; *adj.*, ciliary, ciliated) Thread-like motile organelle containing microtubules, present on the surfaces of many eukaryotic cells and used to push fluid past the cell surface.

**circadian rhythm** Cycle of about 24 hours in the physiology and behavior of a eukaryotic organism.

**clade** The set of all species descended from a particular ancestral species, together with their common ancestor.

**clay** Mineral particles with a diameter of less than 0.002 mm and containing aluminum and silica.

**cleavage** A series of cell divisions that convert the zygote into a multicellular blastula.

**cleidoic egg** (= amniotic egg) "Closed" egg of reptiles, birds, and mammals.

**climate** Those aspects of the weather such as temperature, rainfall, light, humidity, and air movement that influence the life of organisms.

**cline** A series of interconnected populations extending from one area to another, with gradual, continuous changes from one population to the next.

**cloaca** The vestibule, found in most vertebrates, into which urine, feces, and sperm or eggs are discharged before they leave the body.

**clone** Population of cells or individuals descended from one original cell or individual by asexual propagation, and hence genetically identical.

**Cnidaria** Phylum of simple animals with only two well-developed layers of cells, only one opening into the gastrovascular cavity, tentacles, and stinging nematocysts (e.g., jellyfish, *Hydra,* corals, sea anemones).

**coccus** (*pl.*, cocci) Sphere-shaped bacterium.

**codominance** Condition in which both alleles are expressed in the phenotype of a heterozygote.

**codon** Series of three messenger RNA nucleotides which, in the genetic code, specifies a particular amino acid.

**Coelenterata** Old name of a phylum whose members are now placed in the phyla Cnidaria and Ctenophora.

**coelom** (*adj.*, coelomic) A body cavity lined with mesodermal tissue, in which the internal organs are suspended.

**coelomates** Animals with bodies containing coeloms.

**coenocyte**   Structure in fungi or plants in which many nuclei occupy a mass of cytoplasm, without being separated into individual cells by plasma membranes.

**coenzyme**   Organic molecule that must be bound for an enzyme to function.

**coevolution**   Evolution together of two or more species whose members exert reciprocal selective pressures on one another.

**cofactor**   Ion or molecule that must be associated with an enzyme for the enzyme's proper functioning.

**cohesion**   Sticking together of molecules of the same substance.

**coleoptile**   The tubular leaf covering the embryonic shoot in a monocot seed.

**collagen**   A structural protein that forms fibers; common intercellular component of animal connective tissue.

**colony** (*adj.,* colonial)   In animals, bacteria, or protists, a group of more or less independent individuals attached to one another.

**commensalism**   Close association between members of different species in which one member benefits and the effect on the other is neutral or unknown.

**community**   All the populations of different species of organisms living in a particular habitat.

**competitive exclusion**   Name given to the idea that two species with identical niches cannot exist together in the same place and time.

**concentration**   Proportion of one substance found in the total of a mixture of two or more substances; may be given in terms of weight, of proportion of molecules, and so on. Concentration is symbolized: [sugar] = concentration of sugar.

**concentration gradient**   The change in concentration of a substance over a distance.

**condensation reaction**   Chemical reaction in which two molecules become covalently bonded by removing —H from one and —OH from the other, with the removed atoms joining to form a molecule of water.

**conifer**   Member of the plant division Coniferophyta: cone-bearing gymnosperms, including pines and spruces (as well as junipers and yews, whose reproductive structures do not resemble cones).

**conjugation**   Method of genetic recombination in which one cell passes DNA to another via a physical link formed between the two.

**connective tissue**   Tissue composed of scattered cells and much intercellular material secreted by the cells.

**conservation**   Careful use and management of resources, so as to maximize the benefit from them now and in the future. Methods include preservation, reducing waste, recycling, reuse, and decreased use.

**conservative**   Of a character that has changed very little during the course of evolution.

**consumer**   In ecology, an organism that eats other organisms. In economics, someone who buys something.

**contraceptive**   Something that prevents fertilization of an egg.

**convergent evolution**   Evolution of similar features by unrelated organisms in response to similar selective pressures.

**copulation** (= mating, coitus)   Linking of sexual organs of male and female that permits transfer of sperm (in semen, spermatophores, etc.) from male to female.

**corn**   Common U.S. name for the cereal maize, *Zea mays.*

**coronary**   Pertaining to the heart.

**corpus luteum**   "Yellow body" that forms in an egg follicle of an ovary after it has ruptured and released a ripe egg.

**cortex** (*adj.,* cortical)   Layer lying just inside the outermost boundary (epidermal or epithelial) layer of a stem, root, kidney, brain, etc.

**cotyledon**   Part of a plant embryo that absorbs and stores food.

**courtship**   Behavior that precedes mating.

**cross section**   View of an organism from a cut made perpendicular to the long axis of the body.

**cross-fertilization**   Fertilization of one plant by sperm nuclei from another plant (not to be confused with cross-pollination).

**cross-pollination**   Transfer of pollen from male flower parts of one plant to female flower parts of another.

**crossing over**   In genetics, the process by which homologous chromosomes exchange pieces of DNA and form new assortments of genes.

**crustacean**   Member of the arthropod order Crustacea; e.g., lobsters, shrimp, crabs, barnacles, pill bugs, copepods, ostracods.

**cubic centimeter** (*abbr.,* cc)   A volume equal to 1 milliliter.

**cultural evolution**   Change in an animal society from one generation to the next as a result of learning.

**cutaneous**   Of the skin.

**cuticle**   Layer of waxy waterproof substance secreted on the outer surface of an organism.

**cyanobacteria**   Prokaryotic organisms that use chlorophyll *a* and phycobilins in their photosynthesis; extract hydrogen and electrons from water, and produce $O_2$ as a byproduct (also called blue-green bacteria or cyanophytes).

**Cyanophyta**   Former name for cyanobacteria.

**cyst**   1. Dormant organism within a resistant covering; stage in which some organisms pass through adverse conditions. 2. Sac-like nonmalignant tumor.

**cytochrome**   An electron carrier molecule consisting of a protein and a porphyrin ring containing a metal ion.

**cytokines**   Peptide hormones that are critical for generating inflammatory responses and maintaining homeostasis in the immune system.

**cytokinesis**   Division of a eukaryotic cell in two, following nuclear division.

**cytokinins**   Plant hormones that stimulate cell division.

**cytoplasm**   The fluid that makes up all of a cell except its nucleus and organelles.

**cytoplasmic streaming**   Flow of cytoplasm within a cell, or between adjacent cells via plasmodesmata.

**cytoskeleton**   The "skeleton" of a cell, composed of thin tubules and contractile filaments in the cytoplasm; much of the cytoskeleton can be assembled from subunits and disassembled as needed.

**cytosol**   The soluble portion of cytoplasm.

**cytotoxic T cell**   Lymphocyte that destroys a cell displaying both self and foreign antigens on its surface.

**Dalton**   Atomic mass unit, approximately equal to the mass of a proton or a neutron. Named in honor of John Dalton (1766–1844), English chemist and physicist who developed the first practical atomic theory and table of atomic masses.

**deamination**   Removal of an amino (—NH$_2$) group.

**decarboxylation** Removal of a carboxyl (—COOH) group from a molecule, in the form of $CO_2$.

**deciduous** Of plants that lose their leaves during one season of the year; not evergreen.

**decomposer** In ecology, an organism that feeds on the dead bodies, body parts, or wastes of other organisms, thereby breaking down and recycling the nutrients they contain.

**defoliant** A substance that causes a plant to lose its leaves.

**defoliate** To remove all the leaves of a plant.

**deforestation** Removal of trees from an area without replacing them.

**degradable** Capable of being broken down by natural processes (usually by decomposer organisms).

**demographic transition** Change from a population with a high birth rate and death rate to one with low rates of birth and death that has occurred in many countries as they became more industrialized.

**demographics** Study of human populations.

**dendrite** Extension of a neuron that receives stimuli or impulses from other neurons.

**depletion** Using up of a significant portion of a natural resource.

**depolarization** Decrease in the electrical potential difference across a membrane.

**depolarize** Decrease the electrical potential difference across.

**desiccation** Drying out.

**desmosome** Attachment between two cells in which the cells' membranes are fastened together in areas of high mechanical stress.

**detritivore** Organism that feeds on detritus.

**detritus** Molecules and larger particles of dead organic matter.

**development** 1. (embryonic) Process by which an animal or plant zygote changes into a larva or juvenile organism; involves cell division, differentiation, and growth. 2. (economic) Change from a society that is typically agricultural, rural, poor, and illiterate to one that is largely industrial, urban, wealthy, and educated.

**diaphragm** 1. In mammals, a muscular partition between the thorax and abdomen, used in breathing. 2. A disk-like rubber device inserted into the vagina to bar sperm from reaching an egg, used for birth control. 3. A disk with a central opening that controls the amount of light entering a microscope.

**dichotomous** Forking into two.

**dicotyledon** (*abbr.,* dicot) Member of the group of flowering plants whose embryos have two cotyledons.

**dikaryon** Part of a fungus in which each cell contains two haploid nuclei.

**dinoflagellate** Member of phylum Dinoflagellata (Pyrrophyta): one-celled eukaryote with two flagella; most are photosynthetic.

**diploid** Having paired homologous chromosomes.

**disaccharide** A molecule made up of two simple sugar (monosaccharide) residues.

**distal** In a position or direction away from the point of an appendage's attachment to the body.

**disulfide bond** In protein structure, a covalent bond between sulfur atoms that are parts of two different cysteine residues; the bond attaches parts of the same or different polypeptides to each other.

**diurnal** 1. Active during the daytime. 2. Daily.

**division** Category of plants or fungi equivalent to a phylum in the animal kingdom.

**dizygotic twins** Twins arising from two different zygotes (fertilized eggs); nonidentical twins.

**DNA** Deoxyribonucleic acid, the genetic material of organisms and of many viruses.

**dominant allele** Allele expressed in the heterozygote.

**dominant generation** Larger, more conspicuous stage in the life cycle of a species with alternation of generations.

**dorsal** Toward the back, or uppermost surface, of an animal.

**drought** Period with less than average precipitation.

**ECF** Extracellular fluid; fluid that surrounds cells in a body.

**echinoderm** ("spiny-skinned") Member of the invertebrate phylum Echinodermata; e.g., sea stars, sea urchins, sand dollars, sea cucumbers.

**ecology** Study of the relationships of organisms with other organisms and with their physical environment.

**ecosystem** All of the organisms present in a particular area, together with their physical environment.

**ecto-** Prefix meaning "external" or "outermost."

**ectoderm** Outermost of the three germ layers of the embryonic gastrula, giving rise to skin, nervous system, and associated structures.

**ectotherm** Animal that regulates its body temperature by behavioral means, obtaining heat from outside the body.

**EEG** Electroencephalogram; recording of electrical activity in the brain.

**effectors** Structures that carry out an animal's response to a stimulus transmitted via the nervous system; e.g., muscles, glands.

**EKG** Electrocardiogram; recording of electrical activity in heart muscle.

**elasmobranchs** Cartilaginous fish with jaws; the sharks, skates, and rays. Another name for Chondrichthyes.

**electrical potential** Difference in concentration of electrically charged particles on two sides of a membrane.

**electromagnetic radiation** Radiant energy (including solar energy) that can move through a vacuum or through space as waves of oscillating electrical and magnetic fields.

**electron** Fundamental negatively charged particle found outside the nucleus of an atom.

**electron micrograph** Photograph taken using an electron microscope.

**electron transport system** A series of proteins found in certain biological membranes that splits hydrogen atoms into electrons and $H^+$ and accumulates a high $H^+$ concentration on one side of the membrane.

**electrophoresis** A technique that separates substances, using the fact that they move at different rates (depending on their size and electrical charge) when subjected to an electric current.

**electroreceptor** Sense organ that detects electric fields in the surrounding water.

**embryo** A multicellular developing plant or animal still enclosed inside the parent's body or in a seed or egg.

**embryophyte** A plant in which the zygote remains within the (female) gametophyte as it develops into an embryonic sporophyte; a land plant.

**emigration**   Leaving an area of residence for some other place.

**endemic** (*adj.,* sometimes used as a noun)   Peculiar to a particular population or locality where it originated.

**endergonic reaction**   Reaction in which the free energy of the products is greater than that of the reactants.

**endo-, end-**   Prefix meaning inside or within.

**endocrine gland**   A gland whose hormone secretion enters the body fluids directly rather than being transported to the site of action through a duct.

**endocytosis**   Engulfing of a particle by a cell.

**endoderm**   The innermost of the three germ layers in the early embryo of higher animals (metazoans); gives rise to the lining of the digestive tract.

**endodermis**   Layer of cells between the cortex and the pericycle in a root.

**endoplasmic reticulum**   System of membranous sacs and tunnels found in most eukaryotic cells.

**endoskeleton**   Internal skeleton.

**endosperm**   Nutritive tissue in an angiosperm seed.

**endotherm**   Animal (mainly birds and mammals) that maintains a particular body temperature by physiological regulation of the loss of metabolically generated heat.

**energy**   Capacity to do work and transfer heat.

**energy efficiency**   Percentage of the total energy input to a system or process that does useful work.

**entomology**   Study of insects.

**enzyme**   Protein that catalyzes a particular biochemical reaction between specific substrate (reactant) molecules.

**epi-**   Prefix meaning "on," "above," or "around."

**epidermis**   Layer of cells covering the outside of the body.

**epinephrine**   A hormone released by the medulla of the adrenal glands; one of its effects is to bring about the physiological changes associated with stress. Also called adrenalin.

**epiphyte**   Plant growing not in the soil but on the surface of another plant.

**epithelial tissue**   Animal tissue that forms a sheet covering an external or internal surface.

**equilibrium**   1. Point in a chemical reaction at which the rates of forward and reverse reactions are equal and energy change during the reaction is zero. 2. Sense of balance.

**erythrocyte**   Red blood cell.

**essential amino acid**   Amino acid that an animal must obtain in its diet because its body cannot make enough of the molecule to survive.

**estivation** (also aestivation)   Period of dormancy during the summer.

**estrus** (*adj.,* estrous)   Sexually receptive state of female mammals; usually called rut in males.

**eukaryotic**   Having a nuclear membrane surrounding the genetic material, and with other membrane-bounded organelles in the cytoplasm.

**eutrophic**   Of a body of fresh water rich in nutrients and hence in living organisms, often as a result of nutrient or thermal pollution.

**eutrophication**   Process in which organic matter accumulates in a body of water, making it richer in nutrients and hence in organisms, until eventually it fills in and becomes dry land.

**evaporative cooling**   Reduction of temperature as a result of the escape of the fastest-moving (that is, warmest) water molecules as water vapor.

**evolution**   1. Descent of modern species of organisms from somewhat different ancestral species. 2. Change in the gene pool of a population from generation to generation. 3. Production and release of gas as a result of chemical reactions.

**ex-, exo-**   Prefix meaning "outside" or "proceeding from."

**excitatory**   Of nerve or muscle, tending to cause depolarization of the postsynaptic membrane, enhancing the chances that the postsynaptic cell will fire an action potential.

**exergonic reaction**   Reaction that releases energy, that is, one in which the free energy of the products is less than that of the reactants.

**exocytosis**   Method of expelling substances contained in a membrane-bounded vesicle to the exterior of the cell by fusion of the vesicle membrane with the plasma membrane.

**exon**   The part of a structural gene that is translated into protein (as opposed to an intron, which is not translated into protein).

**exoskeleton**   External skeleton, outside the rest of the body.

**exothermic**   Heat-releasing.

**extinction** (of a species)   Disappearance of a species from Earth when its last member dies.

**extracellular fluid (ECF)**   Fluid surrounding cells.

**extracellular**   Outside a cell.

**FAD**   Flavin adenine dinucleotide, a hydrogen-carrying coenzyme in cellular respiration.

**fallopian tube**   Oviduct, in mammals.

**fat**   A triacylglycerol (a lipid made up of three fatty acids and glycerol) that is solid at room temperature.

**fauna**   Animal life of an area.

**fermentation**   Anaerobic breakdown of food molecules to release energy, in which the final electron acceptors (and end products) are organic molecules.

**fertility**   1. Of an organism, the ability to reproduce. 2. Of soil, the ability to supply plants with the nutrients they need. 3. Of a population, the number of offspring produced per female per unit time.

**fertilization**   1. Union of an egg with a sperm. 2. Supplying nutrients to crop plants.

**filter feeders**   Animals and protists that eat smaller organisms or particles of organic matter, which they strain out of the surrounding water.

**filtrate**   Substance that has passed through a filter.

**fission**   Division into (usually two) parts.

**fitness**   In population genetics, a measure of the ability of an organism to produce surviving offspring.

**fix**   1. In chemistry, to incorporate into a less volatile compound. 2. In genetics, to establish one allele in place of all alternate alleles in a population.

**flagellate**   1. (*noun*) Protist that moves by means of one or more flagella. 2. (*adj.*) Bearing one or more flagella.

**flagellum** (*pl.,* flagella)   Long, thin projection from a cell that moves in whip-like fashion or in spiral undulations and propels the cell.

**flatworm**   Member of the invertebrate phylum Platyhelminthes; e.g., planarians, flukes, tapeworms.

**flavin adenine dinucleotide**   *See* **FAD.**

**flora**   Plant life of an area (often includes bacteria and fungi, reflecting the old two-kingdom system).

**fossil**   Remains of an organism, or other evidence of a once-living organism, preserved in rocks.

**fossil fuel**   Fuel created by decomposition and geological processes from the remains of dead organisms, including peat, oil, coal, and natural gas.

**frameshift mutation**   Mutation that adds or deletes nucleotides, thereby altering the reading frame of the genetic message.

**free energy**   Energy that is available or usable to do work.

**frond**   A leaf, usually highly divided (usually applied to ferns or palms).

**frugivorous**   Fruit-eating.

**fruit**   Structure that develops from the ovary of a flower, surrounding one or more seeds.

**fruiting body**   Rather large, prominent reproductive structure formed by some fungi, protistans, and myxobacteria.

**fugitive species**   Species that occurs in an area for only a short time; characterized by effective means of dispersal.

**Fungi**   Kingdom of organisms containing eukaryotic, mainly multicellular heterotrophs feeding by absorbing nutrients through their chitinous cell walls and usually having thread-like bodies.

**gamete**   Sexual reproductive cell: egg or sperm.

**gametophyte**   Haploid stage of a plant, producing haploid gametes by mitosis.

**ganglion** (*pl.,* ganglia)   A cluster of neuron cell bodies.

**gap junction**   An area where ions can move directly between the cytoplasm of two animal cells by way of many tiny protein-lined tunnels through their plasma membranes.

**gastrocoel**   The hollow in the embryonic gastrula that becomes the lumen of the gut.

**gastropod**   Member of the molluscan class Gastropoda, including snails, slugs, nudibranchs, and limpets.

**gastrovascular cavity**   A cavity in the body that serves for both digestion and distribution of food.

**gated channel**   Protein-lined channel through a membrane that can be opened or closed by chemical or electrical events.

**gene**   A length of DNA that functions as a unit.

**gene bank**   Institution where plant material is stored in a viable condition. Usually seeds are dried and frozen in a sealed container. Some plants have seeds that will not survive this treatment, and they must be maintained as growing plants or in tissue culture.

**gene flow**   Transfer of genes between one more-or-less isolated population and another.

**gene pool**   All of the genes present in a population of organisms.

**genera**   Plural of genus.

**generation time**   Time elapsed from production (birth) of a new individual to production of its first offspring; usually estimated at 20 years for humans.

**genetic drift**   Changes in the gene pool of a population caused by random events (as opposed to natural selection).

**genetic engineering**   Isolation of useful genes from a donor organism or tissue and their incorporation into an organism that does not normally possess them.

**genetic reassortment**   Production of new gene combinations in sexually reproducing organisms by crossing over and independent assortment during meiosis and by combination of gametes at fertilization.

**genetic recombination**   Formation of new combinations of genes by joining DNA from two different molecules into one; e.g., by crossing over during meiosis or by genetic engineering.

**genital**   Of the reproductive system.

**genome**   All the genetic material contained by an individual (or by a representative member of a population).

**genotype**   The particular genes present in an individual, some of which may not be expressed in the phenotype; often refers to only one or a few gene pairs.

**genus** (*pl.,* genera; *adj.,* generic)   Category above species in the hierarchical classification of organisms. The genus name is the first word of the Latin binomial for a species; e.g., *Ursus* is the generic name of the grizzly bear, *Ursus horribilis*.

**germ cells**   Cells that give rise to eggs, sperm, or spores (reproductive cells).

**germ layers**   The three layers of cells in the embryonic gastrula (ectoderm, mesoderm, endoderm).

**germ plasm**   Term used by botanists for genetic material, especially that contained within the reproductive (germ) cells.

**gestation**   Period during which an embryo is carried within the mother's body before birth.

**gibberellins**   Plant hormones that stimulate cell enlargement.

**gills**   Thin extensions of the body surfaces of many aquatic animals, used for gas exchange and/or feeding.

**gizzard**   Part of the stomach or gut modified as a heavy-walled grinding chamber.

**Glaciation** (= Ice Age)   One of four cold periods during the Pleistocene era, and ending about 10,000 years ago, when ice and glaciers extended farther south from the North Pole than they do now.

**glucose**   A six-carbon monosaccharide.

**glycogen**   A storage polysaccharide made from glucose monomers, commonly found in animals.

**glycolipid**   Molecule made up of carbohydrate and lipid subunits.

**glycolysis**   Biochemical pathway found in most organisms, in which glucose is broken down into pyruvate ions and hydrogen atoms, with the storage of energy in ATP.

**glycoprotein**   Protein bonded to carbohydrate.

**Golgi complex**   Stack of membrane-enclosed sacs that modify and package molecules produced in a eukaryotic cell.

**gonad**   In animals, an organ that produces gametes.

**grafting**   A procedure in which a tissue or organ of one plant or animal is attached to and incorporated into the body of another.

**grain**   1. The fruit of members of the monocotyledonous plant family Poaceae from which most human food comes. 2. A cluster of molecules of starch.

**gram molecular weight**   *See* **mole.**

**gray matter**   Nervous tissue made up of unmyelinated neuron cell bodies and extensions.

**green algae**   Members of the division Chlorophyta: plants with unicellular or multicellular body plans, using chlorophylls *a* and *b* for photosynthesis and storing food as starch.

**green plants** Photosynthetic organisms that give off oxygen during their photosynthesis, including prokaryotic cyanobacteria and all photosynthetic eukaryotes (protists and plants).

**greenhouse effect** Heating of Earth caused by gases in the atmosphere that trap infrared radiation reflected from Earth and prevent it from escaping into space.

**guard cells** Two cells that surround a stoma (pore) in the epidermis of a plant and regulate the pore's opening and closing.

**gustation** Tasting.

**guttation** Exudation of water from tips of xylem veinlets in leaves, due to root pressure.

**gymnosperm** Nonflowering plant that produces seeds, e.g., pines, redwoods, cycads.

**habitat** Physical area where an organism lives.

**haploid** Containing the number of (unpaired) chromosomes found in a gamete; equal to half the number of chromosomes found in a body cell of most higher plants and animals.

**haustorium** (*pl.,* haustoria) An extension of a fungus inside the cell wall of a living plant cell; the fungus lies against the plant cell's plasma membrane and absorbs nutrients from the plant cell.

**hectare** 10,000 square meters, or about 2.5 acres.

**hematopoiesis** Formation of blood in the body.

**heme** Iron-containing group in hemoglobin and some cytochromes.

**hemocoel** Body cavity containing a fluid (hemolymph) that acts as the transport medium in many invertebrates.

**hemoglobin** Red, heme-containing respiratory pigment that carries oxygen in the blood of vertebrates and some invertebrates.

**hemolymph** Fluid that fills the hemocoel (body cavity) and acts as blood in animals with open circulatory systems.

**hemolysis** The bursting of blood cells.

**hemophilia** Condition in which blood fails to clot, resulting in excessive bleeding, caused by a defect in a gene for clotting factor.

**hemorrhage** Bleeding, e.g., from a ruptured blood vessel.

**herbaceous** Of plants with nonwoody stems.

**herbicide** Chemical used to kill plants.

**herbivore** Animal that eats plants or parts of plants.

**hermaphroditic** Containing both male and female organs.

**hetero-** Prefix meaning "other" or "different."

**heterogametic sex** The sex that produces two different kinds of sex chromosomes, which determine the sex of the offspring (in humans, the male).

**heterospory** In plants, production of spores of two distinct sizes, which give rise to gametophytes of the two sexes: microspores to male gametophytes, megaspores to female.

**heterotroph** (*adj.,* heterotrophic) Organism dependent on other organisms to produce its organic (food) molecules.

**heterozygote** (*adj.,* heterozygous) Diploid individual with two different alleles in a given gene pair.

**heterozygote advantage** (= **heterosis**) Selective advantage of a heterozygote over either homozygote.

**histone** Type of protein with an alkaline pH, characteristic of eukaryotic chromosomes and playing a role in the packing of DNA and regulation of transcription.

**homeostasis** (*adj.,* homeostatic) Maintenance of conditions within specified limits.

**homeothermy** (obsolete term, usually replaced by endothermy) Maintenance of a constant, high body temperature and metabolic rate.

**homo** Prefix meaning "human" or "the same as."

**hominids** Humans and their direct ancestors: members of the primate family Hominidae, with large brains, small teeth, and bipedal locomotion.

**hominoids** Humans and apes: large tailless primates.

**homogametic sex** The sex that contains only kind of sex chromosome and therefore produces one kind of gamete (in humans, the female).

**homologous** 1. Of chromosomes, of the same origin and containing the same kinds of genetic information. 2. Of structures or organs, originating from the same structure in ancestral forms (e.g., a bird's wing and a human's arm)

**homozygote** (*adj.,* homozygous) Diploid individual having same allele in both members of a given gene pair.

**homozygous recessive** Having two copies of a recessive allele in a given gene pair.

**hormone** Chemical messenger produced in one part of the body and specifically influencing certain activities of cells in another part of the body.

**host** Large organism inhabited by a smaller one in a symbiotic relationship.

**humus** The dark organic matter, resistant to further decomposition by microorganisms, that remains in the soil after plant and animal remains have decomposed.

**hunter-gatherer** Member of human society in which food is obtained by collecting plants and hunting wild animals.

**hybrid** Offspring of a mating between genetically different individuals.

**hydraulic pressure** Force exerted by a fluid.

**hydrogen bond** Weak link between two molecules, or two parts of the same molecule, due to the attraction of a hydrogen with a partial positive charge to an oxygen or nitrogen with a partial negative charge.

**hydrolysis** Breaking apart of a molecule into its monomer subunits by addition of the components of a water molecule into each of the covalent bonds linking them.

**hydrophilic** "Water-loving"; able to dissolve in water; polar or ionic.

**hydrophobic** "Water-hating"; unable to dissolve in water; nonpolar.

**hydrostatic pressure** Pressure exerted by confined fluid.

**Hymenoptera** Order of insects that includes bees, wasps, and ants.

**hyperpolarization** Increase in the membrane potential.

**hypertonic** Of a solution, tending to gain water from a solution to which it is being compared, when the two are separated by a membrane; usually this means having a higher concentration of dissolved particles (lower osmotic potential) than the other solution.

**hypha** (*pl.,* hyphae) One of the threadlike structures that make up the body of a fungus.

**hypothalamus** A part of the brain, responsible for monitoring internal conditions in the body and initiating behaviors that tend to maintain physiological homeostasis.

**hypothesis** (*pl.,* hypotheses) Proposed answer to a question.

**hypotonic** Of a solution, tending to lose water to a solution to which it is being compared, when the two are separated by a membrane; usually this means having a lower concentration of dissolved particles (higher osmotic potential) than the other solution.

**immigration** Movement of new individuals into an area.

**immunoglobulin** Antibody (*See*).

**immunology** Study of the reactions that provide protection specifically against individual diseases.

***in vitro*** "In glass." Outside the body: in a test tube or laboratory dish.

***in vivo*** "In life." Inside the living body.

**inbreeding** Mating of related individuals that share many of the same genes.

**incisors** Chisel-shaped cutting teeth found in the front and center of the lower and upper jaws in mammals.

**incomplete dominance** Condition in which the expression of one allele is not sufficient to mask the presence of an alternate allele in the phenotype.

**independent assortment** Creation of new haploid combinations of chromosomes due to random lining up of members of homologous chromosome pairs on either side of the spindle equator during meiosis.

**inducer molecule** Molecule (often a food molecule) that causes the genetic machinery of a cell to transcribe the genes for and to produce particular metabolic enzymes.

**induction** 1. The initiation of protein synthesis because a particular inducer substance is present. 2. In development, conversion of one type of tissue into another as a result of the effect of a third tissue.

**Industrial Revolution** Social and economic changes brought about when mechanization of production results in a shift from home manufacturing to large-scale factory production.

**infant mortality** Death rate for humans in the first year of life.

**infanticide** Killing of a child less than one year old.

**ingest** To take into the body through the mouth.

**inhibitory** Of a nerve impulse, making the postsynaptic membrane require more excitatory input before it will fire an action potential.

**innervate** To supply nervous connections to.

**inorganic** 1. Not produced by living organisms. 2. In chemistry, all compounds that do not contain carbon atoms bonded to hydrogen atoms.

**insect** Arthropod with three distinct body areas (head, thorax, abdomen); adult has three pairs of legs and usually two pairs of wings attached to the thorax.

**insecticide** Substance that kills one or more species of insects.

**Insectivora** Order of mammals that includes hedgehogs, moles, shrews.

**insectivorous** Feeding on insects.

**insulin** A small protein hormone, one of whose functions is to regulate cellular uptake of glucose from the blood; it is produced by the pancreas.

**interferon** Protein produced by mammalian cells that interferes with replication of viruses.

**intermediary metabolism** Total of all of a cell's enzyme-mediated reactions involved with extracting energy from food molecules and using it to synthesize the cell's own molecules.

**intermediate filament** Protein fiber that forms a permanent, nonlabile part of the cytoskeleton of a eukaryotic cell and contributes mechanical strength.

**interneuron** Type of neuron in the central nervous system that is neither sensory nor motor, but transmits information between other neurons.

**internode** Portion of a stem between sites of leaf attachment.

**intertidal** Between tidemarks; covered by water at high tide and exposed to the air at low tide.

**intervening sequence** See intron.

**intracellular** Inside a cell.

**intraspecific** Within one species.

**intron** Part of a structural gene that is transcribed into messenger RNA but not translated into protein.

**invertebrate** Animal that lacks a backbone; e.g., earthworm, snail, sea star.

**iodinated** Having atoms of iodine attached.

**ion** (*adj.,* ionic) Particle carrying one or more positive or negative electrical charges.

**iso-** Prefix meaning "same."

**isomers** Molecules containing the same atoms but in different arrangements.

**isotonic** Of a solution, having osmotic properties such that it neither gains nor loses water across a membrane separating it from a solution to which it is being compared.

**isotopes** Two or more forms of an element, differing by the numbers of neutrons in the nucleus.

**joule** SI unit of energy: the kinetic energy of a 2-kg object moving at a velocity of 1.0 m/sec. Named for James P. Joule (1818–1889).

**K** 1. Chemical symbol for potassium, which usually occurs as an ion ($K^+$). 2. In ecology, symbol used for carrying capacity.

**kelps** Large marine brown algae with considerable differentiation of tissues.

**keratin** A structural protein that makes up hair and fingernails.

**keto acid** A molecule with both a keto and a carboxyl group.

**killer cells** Poorly understood lymphocyte-like immune cells, which are apparently specialized to destroy abnormal body cells.

**kin selection** Natural selection that causes an individual to act in such a way as to enhance the survival or reproduction of other, genetically related individuals (kin).

**kinetic energy** Energy of movement.

**kinetochore** Structure on a chromosome to which microtubules attach; usually located on or near the centromere.

**krill** *Euphasia superba,* small marine planktonic arthropods common in the Southern Ocean, up to several centimeters in length.

**labeled** Prepared in such a way that it can be traced; e.g., containing a high proportion of a particular isotope, which is usually rare, allowing the fate of the atoms to be traced by methods that can distinguish between the isotopes of an element.

**lactation** Secretion of milk from mammary glands.

**larva** (*pl.*, larvae)  Immature stage of an animal with different appearance and way of life from the adult.

**larynx**  Voice box ("Adam's apple"), located at the top of the trachea (windpipe).

**lateral**  1. (*adj.*) Pertaining to the side. 2. (*noun*) A side branch or branch root of a plant.

**lateral line organ**  Pressure-sensitive sense organ found in fish and larval amphibians, used to detect water currents, etc.

**laterite**  Hard crust that may develop when vegetation is removed from the surface of soil containing metals such as aluminum and iron in tropical regions with wet and dry seasons. In the dry season, soil solution rises to the surface by capillary action, and aluminum and iron oxides accumulate and combine to form the crust. Laterization results in an infertile soil called latosol.

**latitude**  Distance north or south of the equator, measured in degrees. The distance between the equator and either Pole of Earth is divided into 90°.

**leach**  To dissolve and carry out of; leached soil has lost mineral nutrients that have dissolved in rainwater running through the soil and have been carried away into streams.

**leaf nodes**  Areas of stem at which leaves are attached.

**leeward** (side)  The side of an object from which the wind is not coming. (*pronounced* loo-ard)

**legume**  Members of the plant family Fabaceae, formerly called Leguminosae, including beans, acacias, peas, peanuts, alfalfa, clover.

**Lepidoptera**  The order of insects that includes moths and butterflies.

**lethal allele**  Allele whose expression causes premature death of the individual that carries it (usually, in the homozygous recessive condition).

**leucoplast**  Colorless plastid.

**leukocyte**  White blood cell.

**life expectancy**  Number of years a particular person can expect to live, calculated from actuarial statistics.

**linkage group**  All of the genes on the same chromosome, which are therefore all inherited together (except for crossing over).

**lipid**  One of a large class of organic molecules not soluble in water; includes fats, waxes, oils, steroids, carotenes.

**lipopolysaccharide**  Molecule composed of lipid joined to polysaccharide.

**lipoprotein**  Molecule made up of lipid and polypeptide subunits.

**litter**  Plant remains (mainly leaves) that have fallen on the surface of the soil.

**locus** (*pl.*, loci)  The position on a chromosome that is occupied by an allele for a particular gene; e.g., the hemoglobin beta chain locus may be occupied by an allele for normal or sickle hemoglobin.

**lophophore**  Filter-feeding crown of tentacles surrounding the mouth and containing an extension of the coelom.

**lumen**  Space in the center of a tube.

**lymphocyte** (= T cell or B cell)  One of a group of white blood cells responsible for the specificity of immune responses: production of antibodies, recognition of antigens, and immunological memory.

**lysis** (*adj.*, lytic)  Bursting of a cell.

**lysosome**  Membrane-bounded organelle filled with hydrolytic (digestive) enzymes.

**macromolecule**  Large molecule; usually used of a polymer with a molecular mass of many thousands and made up of many (identical or similar) monomers.

**macronutrient**  Nutrient needed in relatively large amounts.

**macroscopic**  Visible to the unaided eye.

**magma**  The molten matter under Earth's crust from which igneous rock is formed by cooling.

**maize**  *Zea mays*, corn, important cereal crop.

**major histocompatibility complex**  See MHC.

**mammal**  Warm-blooded vertebrate with lower jaw consisting of only one bone (the mandible) on each side, with fur or hair, with young nourished by milk from the mammary glands of the female parent, e.g., humans, rabbits, cattle.

**mantle**  Sheet of tissue covering the visceral mass of a mollusc and secreting the shell, if present.

**map**  In genetics, the outline of a genome showing the relative positions of particular genes on the organism's chromosomes.

**marine**  Of the sea.

**marsupials**  Mammals whose young are born quite early in development and complete their development attached to a nipple in the mother's marsupium, or pouch.

**mating types**  The equivalent of sexes in fungi, some bacteria, and protists.

**medial**  Of or toward the body's (central) plane of symmetry (opposite to lateral).

**medusa** (*pl.*, medusae)  One of the two possible forms of the cnidarian body (the other is the polyp); often the reproductive stage in the life cycle; for example, the body of a jellyfish is a medusa.

**megaspores**  Large spores produced by meiosis in some plants and giving rise to female gametophytes.

**meiosis**  A series of two nuclear divisions that produces four new nuclei, each with half the number of chromosomes contained in the original nucleus.

**melanin**  Black pigment that gives dark color to organisms.

**melanism** (*adj.*, melanic)  Dark coloration.

**membrane potential**  The difference in electrical potential between the inside and outside of a membrane; represents a force exerted on a charge to pass through the membrane. The inside of most cells is negatively charged with respect to the outside so the membrane usually has a potential of about $-40$ millivolts (about $-70$ millivolts for a neuron).

**meristem** (*adj.*, meristematic)  Region of dividing cells in a growing area of a plant.

**mesentery**  A sheet of mesodermal tissue that suspends the internal organs in the coelom.

**mesoderm**  The middle one of the three embryonic germ layers of the gastrula; gives rise to most of the muscles, heart, kidneys, gonads, etc.

**mesoglea**  The "middle glue" layer, containing few, scattered cells, between the outer and inner layers of a cnidarian.

**messenger RNA**  The molecule that carries genetic information from DNA to ribosomes, where the information is used as a code to direct the order in which amino acids are joined to form a polypeptide.

**metabolic heat**  Heat released by chemical reactions in the body.

**metabolic rate**  The rate of the total of an organism's biochemical reactions; usually measured as the rate of oxygen consumption by respiration, because respiration produces the energy needed for the other biochemical processes.

**metabolism**  All the chemical reactions taking place within an organism.

**metal**  An element characterized by a tendency to give up electrons and by good thermal and electrical conductivity.

**metamorphosis**  The radical change in shape, physiology, and behavior that occurs when a larva becomes a very different-looking adult.

**metazoans**  Animals having a digestive cavity and cells organized into tissues; includes all major animal phyla except Porifera (sponges), plus most minor phyla.

**methane**  Gas ($CH_4$) produced by the metabolism of anaerobic methanogen bacteria.

**mg** (milligram)  A thousandth of a gram.

**MHC** (major histocompatibility complex)  The group of genes that determines many mammalian cell surface antigens, including those that cause rejection of transplanted organs.

**microfilament**  Filament assembled from actin subunits, found near the plasma membrane; part of the cytoskeleton; responsible for much of the movement within a eukaryotic cell.

**micrograph**  Photograph taken using a microscope.

**micrometer** (*abbr.*, $\mu$m)  $10^{-3}$ mm = $10^{-6}$ m.

**micron**  Outmoded name for micrometer.

**micronutrient**  Nutrient needed in relatively small amounts in the diet.

**microorganisms**  Small unicellular or simple multicellular organisms: bacteria, fungi, protists, or small algae.

**microspores**  Small spores produced by meiosis in some plants and giving rise to male gametophytes.

**microtubule**  Thickest type of structure in the cytoskeleton of a eukaryotic cell; present in eukaryotic centrioles and in cilia, flagella, and their basal bodies, and in mitotic and meiotic spindles; assembled from protein subunits.

**microvillus**  Tiny, finger-like projection that increases the surface area of a cell.

**middle lamella**  The shared partition between the cell walls of adjacent plant cells.

**milligram** (*abbr.*, mg)  One thousandth of a gram.

**mimicry**  Resemblance of one organism to another, or to a nonliving object, providing an offensive or defensive advantage to the mimic.

**mineral**  1. In geology, any naturally occurring homogeneous inorganic substance having a definite chemical composition, and a particular crystalline structure, color, and hardness. 2. In nutrition, any inorganic nutrient.

**mites**  Small arthropods in the class (Arachnida) that includes spiders. Mites have eight legs, and the body is not divided into two parts as it is in spiders.

**mitochondrion** (*pl.*, mitochondria)  Large, self-replicating membrane-bounded organelle where most of a eukaryotic cell's ATP is produced by cellular respiration.

**mitosis** (*adj.*, mitotic)  Series of events that results in the division of one cell nucleus into two nuclei containing sets of chromosomes identical to that in the original nucleus.

**mole**  Gram molecular mass; for example, 1 mole of water = 18 grams because the molecular mass of $H_2O$ is 18.

**molecular biology**  Study of the molecular basis of inheritance.

**molecular mass**  Sum of the atomic masses of the atoms in a molecule of a compound.

**Mollusca**  Phylum of soft-bodied invertebrate animals with a muscular head-foot and a mantle, which usually secretes a shell; e.g., snails, clams, squids.

**molting**  Shedding of skin, exoskeleton, or feathers.

**Monera**  One name for the kingdom containing all prokaryotic organisms (bacteria); this book uses an alternative name, kingdom Prokaryotae.

**monocotyledon** (*abbr.*, monocot)  Member of the group of flowering plants whose embryos have only one cotyledon.

**monogamy**  Mating of one male with one female, usually for life or at least for the duration of one breeding season.

**monomers**  Small molecules that may become joined together to form larger molecules; e.g., amino acids are the monomers that make up polypeptides.

**monophyletic**  Of a clade, i.e., a taxon that contains a common ancestor and all the species descended from it.

**monosaccharide**  Simple sugar, with formula given by $(CH_2O)_n$; e.g., glucose, ribose.

**monotremes**  Egg-laying mammals.

**monozygotic twins**  Twins originating from the same fertilized egg; identical (genetically) twins.

**morph**  Form, variety.

**morphogenesis** (*adj.*, morphogenetic)  The formation of shape or structures during development.

**morphology** (*adj.*, morphological)  Structure, anatomy.

**mortality**  Death.

**motile**  Able to move itself from place to place.

**multicellular**  Composed of more than one cell.

**mutagen**  Agent (e.g., chemicals, certain kinds of radiation) that causes mutation.

**mutation**  Inheritable change in the genetic material.

**mutualism**  Close association between members of different species that benefits both.

**mycelium** (*pl.*, mycelia)  Mass of fungal filaments that make up the body of a fungus.

**mycology**  Study of fungi.

**mycorrhiza** (*pl.*, mycorrhizae)  Mutualistic association between a fungus and the roots of a higher plant; the fungus takes up mineral nutrients from the soil and passes them to the plant, receiving some organic (food) molecules made by the plant in return.

**myelin sheath**  Layers of fatty, insulating wrapping around the axons of some neurons.

**myoglobin**  Oxygen-storing molecule in muscles.

**myosin**  Protein that interacts with actin to produce movement in eukaryotic cells, such as contraction in muscle cells.

**myotome**  A block of muscle that forms one of a series of such blocks along the back of a chordate.

**NAD** (nicotinamide adenine dinucleotide)  A hydrogen-carrying coenzyme in cellular respiration.

**NADP** (nicotinamide adenine dinucleotide phosphate)  A hydrogen-carrying coenzyme in photosynthesis.

**nanometer** (*abbr.*, nm)  = $10^{-9}$ m.

**natural resource**  Anything that is produced naturally that is needed by a group of organisms.

**natural selection**  Differential reproduction among the various genotypes in a population. Natural selection enhances reproductive success.

**Nearctic**  One of the biogeographical regions of the world; made up of the New World (Western Hemisphere) north of the tropics, including North America and Greenland.

**negative feedback**  Mechanism whereby the change detected in some condition stimulates compensating activity that brings the condition back toward its average value.

**nematocyst**  Stinging structure characteristic of cnidarians.

**nematode**  Roundworm, member of the animal phylum Nematoda.

**Neotropical**  One of the biogeographical regions of the world; made up of the New World (Western Hemisphere) tropics, including the Caribbean region, Central and South America.

**neuritis**  Inflammation of nerves.

**neuron** (*adj.*, neuronal)  Nerve cell.

**neurotransmitter**  Chemical that travels across the synaptic cleft from one cell, which has just fired an action potential, to the other cell, whose membrane potential is altered by the neurotransmitter.

**neutron**  Fundamental particle found in the nucleus of an atom, having an atomic mass of 1 and bearing no electrical charge.

**New World**  The Americas, most of the Western Hemisphere.

**niche**  Way of life of a species; includes the habitat, food, nest sites, and so on that it requires for survival, and its impact on the environment, including other species.

**nitrogen fixation**  Conversion of gaseous nitrogen ($N_2$) to ammonia ($NH_3$). Carried out in ecosystems mainly by bacteria of the genus *Rhizobium*.

**noctuid**  Member of the Noctuidae, a family of night-flying moths with sound-detecting cells.

**nocturnal**  Active at night.

**nondegradable**  Not broken down by natural processes (or broken down only over a time span of hundreds of years).

**nondisjunction**  Failure of homologous chromosomes or of sister chromatids to separate; if this occurs during meiosis, one of the resulting nuclei will have an extra copy of the chromosome, while another will have one chromosome too few.

**nonrenewable resources**  Natural resources that can be used up either completely or else to such a degree that it is economically impractical to obtain any more of them.

**notochord**  Elastic rod dorsal to the gut in all chordate embryos; in adult vertebrates, the notochord is replaced by vertebrae, which form around it.

**nucleic acids**  Class of macromolecules, made up of nucleotide monomers, that contain the genetic information of organisms; DNA and RNA.

**nucleolus** (*pl.*, nucleoli)  Area of the cell nucleus where the nucleolar organizer (part of one or two particular chromosomes) lies and where ribosomes are made; often appears denser than the rest of the nucleus in micrographs.

**nucleosome**  Structure formed by part of a chromosome in which DNA is wound around a cluster of histones.

**nucleotide**  Monomer subunit that makes up nucleic acids; consists of a single- or double-ring nitrogenous base, a pentose sugar (ribose or deoxyribose), and one to three phosphate groups.

**nucleus** (*pl.*, nuclei)  1. That part of a eukaryotic cell surrounded by the nuclear envelope and containing the genetic material. 2. The more-or-less central part of an atom, consisting of one or more protons and (except in most hydrogen atoms) neutrons. 3. A cluster of neuron cell bodies in the brain.

**nutrient**  Any chemical that an organism must take in from its environment because it cannot produce it (or cannot produce it as fast as it needs it).

**Old World**  Europe, Asia, and Africa.

**olfactory**  Pertaining to the sense of smell.

**oligotrophic**  Of a body of fresh water that contains few nutrients and few organisms.

**omnivore**  Animal that eats both plants and animals.

**oncogene**  A gene capable of taking part in the transformation of a normal cell into a cancerous cell.

**oncogenic**  Cancer-causing.

**oocyte**  Cell that undergoes meiosis during oogenesis.

**oogamy**  Form of sexual reproduction involving an egg, which is large and non-motile, and a sperm, which is small and motile.

**oogenesis**  Formation of eggs or ova, the female gametes.

**operculum**  Common covering of all the gills on each side in bony fish.

**operon**  A cluster of genes with related functions: usually one or more structural genes and the regulatory genes that control their activity.

**organ**  Group of tissues assembled in such a way that the entire structure (organ) performs a particular function; e.g. liver, kidney, heart.

**organelle**  Structure within a cell that takes part in carrying out one or more of the cell's life functions.

**organic**  1. Produced by living organisms. 2. In chemistry, of a compound containing one or more carbon atoms bonded to hydrogen atoms.

**organism**  An individual living thing, made up of one or more cells.

**ornithology**  The study of birds.

**osmoregulation**  Regulation of the body's salt and water content.

**osmosis**  Movement of water through a membrane down a water potential gradient.

**osmotic potential**  Tendency of a solution to gain water when separated from pure water by an ideal selectively permeable membrane; the negative of osmotic pressure.

**osmotic pressure**  Pressure that must be exerted on a solution to keep it from gaining water when separated from pure water by an ideal selectively permeable membrane.

**oviduct**  Tube through which eggs pass from the ovary toward the exterior of the body.

**oviparous**  Reproducing by laying eggs which develop outside the female's body.

**oviposition**  Laying of eggs.

**ovoviviparous**  Retaining eggs within the mother's body until hatching; e.g., most sharks and some reptiles.

**ovule**  Structure inside the ovary of a flower enclosing a female gametophyte.

**ovum** (*pl.*, ova)  Female gamete, egg.

**oxidation**  A chemical reaction involving removal of electrons or hydrogen atoms, or addition of oxygen; always paired with a reduction.

**oxygenated** Containing oxygen gas.

**ozone** ($O_3$) Poisonous gas, a common pollutant in smog; also formed by the action of sunlight on $O_2$ in the ozone layer of the atmosphere.

**ozone layer** Layer in the upper stratosphere where solar radiation converts some molecules of oxygen ($O_2$) into ozone ($O_3$). Ozone absorbs much ultraviolet radiation and prevents it from reaching Earth.

**paleontology** (palaeontology) Study of ancient life by examining fossils.

**pancreas** In vertebrates, an organ near the stomach that produces hormones and digestive enzymes.

**parasite** Organism that feeds on another living organism (the host) without killing it.

**parasitoid** Parasite-like insect that lives singly as a larva within the body of a host and kills the host as part of the parasitoid development program.

**parasympathetic nervous system** Part of the autonomic nervous system; its activities promote digestion and elimination, slowing of heart rate, etc.

**parthenogenesis** Production of young from unfertilized eggs.

**pathogenic** Disease-causing.

**pectoral** Pertaining to the anterior fins or limbs of vertebrates.

**pelvic** Pertaining to the posterior pair of appendages (limbs or fins) of vertebrates.

**peptide bond** Covalent bond joining the carboxyl carbon of one amino acid to the amino nitrogen of the next.

**per capita** Per person; Latin for "by heads."

**perennial** Of a plant, living for many years and surviving normal seasonal changes.

**peripheral nervous system** The part of the nervous system outside the brain and spinal cord; it consists of nerves to and from the muscles, sense organs, and internal organs.

**permafrost** Permanently frozen layer in the soil, found in arctic and antarctic regions.

**pest** Any organism that is undesirable at the time and in the place where it exists. Includes plant-eating insects, molluscs, and nematodes on crop plants, parasites on animals, and weeds in fields of crops.

**pesticide** Substance used to kill undesirable organisms; includes insecticides, herbicides, nematocides, etc.

**pH** $Log_{10}$ of the $H^+$ ion concentration of a solution: measure of how acidic or basic a solution is, on a scale of 0 to 14 (0 = very acidic, 14 = very basic, 7 = neutral).

**phage** *See* **bacteriophage.**

**phagocyte** Scavenger cell that ingests and destroys pathogens, dead body cells, and other debris.

**phagocytosis** Process in which a cell ingests another cell or particle of solid matter by forming a membranous vacuole which pinches into the cytoplasm.

**pharynx** Part of the gut just behind the mouth in many animals.

**phenotype** The trait produced by expression of an organism's genes.

**pheromone** Chemical released by one member of a species that influences the behavior of another member of the species.

**phloem** Tissue in plants that conducts food from sites of synthesis or storage to sites where food is used or stored.

**phospholipids** Group of structural lipids that are the main components of biological membranes; made up of fatty acid(s), glycerol, phosphate, and (usually) nitrogen-containing choline.

**phosphorylation** Addition of a phosphate group.

**photochemical reaction** Chemical reaction powered by light energy.

**photoperiodism** Production of a physiological response (such as flowering or coming into breeding condition) in response to length of period of daylight each day.

**photosynthesis** Process whereby plants, and some bacteria and protists, capture solar energy and store it as chemical bonds in carbohydrate molecules, using $CO_2$ to build the carbohydrate.

**photosystem** Cluster of pigment molecules in which light-absorbing reactions of photosynthesis occur.

**phototropism** Growth toward or away from light.

**phycobilins** Accessory photosynthetic pigments found in cyanobacteria and Rhodophyta; major ones are phycocyanin and phycoerythrin.

**phycology** Study of algae.

**phylogeny** (*adj.,* phylogenetic) Line of evolutionary descent.

**physiology** Processes by which an organism carries out its various biological functions; how an organism works.

**phytoplankton** Plants and photosynthetic protists floating in the upper layers of a body of water.

**Pᵢ** Abbreviation for an inorganic phosphate group.

**pigment** Molecule that differentially absorbs particular wavelengths of visible light and so appears colored.

**placenta** Organ in mammals in which blood capillaries from mother and fetus lie close together and exchange substances via the extracellular fluid between the two bloodstreams.

**placentals** Mammals that undergo part of their development in the mother's uterus, where they receive food and oxygen via the placenta, until a fairly advanced stage of development.

**plankton** Collective noun for organisms that drift around in water because they are not capable of swimming against currents in the water.

**plasma membrane** Phospholipid and protein membrane surrounding a cell. Also called plasmalemma, cell membrane.

**plasmid** Small, circular DNA molecule, found in some bacteria and fungi in addition to the organism's own genome.

**plasmodesma** (*pl.,* plasmodesmata) Strand of cytoplasm that directly links the cytoplasms of two neighboring plant cells.

**plastid** Organelle found only in plant cells; depending on the cell's specialized function, its plastids may develop as chloroplasts, chromoplasts, amyloplasts, etc.

**Platyhelminthes** Phylum of animals containing the flatworms; e.g., planarians, tapeworms, flukes.

**poikilothermy** Condition of having a body temperature that changes with that of the environment (opposite of homeothermy).

**polar** 1. Electrically asymmetrical. 2. Pertaining to the poles (ends) of the mitotic or meiotic spindle. 3. Pertaining to the top and bottom of an egg.

**pollen grain** Immature male gametophyte of a gymnosperm or flowering plant;

it will produce the sperm nuclei that fertilize the egg.

**pollination** Deposition of pollen on or near the female parts of a gymnosperm or angiosperm.

**pollution** Change in the physical, chemical, or biological properties of air, water, or soil that can adversely affect the health, survival, or activities of humans and other living organisms.

**poly-** Prefix meaning "many."

**polyandry** Mating system in which a female mates with more than one male.

**polygamy** Mating system in which an individual may have more than one mate.

**polygenic character** Trait whose phenotypic expression is governed by many pairs of alleles (e.g., human skin or hair color).

**polygyny** Mating system in which one male mates with more than one female.

**polymer** Large molecule made up of many subunits that are smaller molecules similar or identical to one another.

**polymorphism** Simultaneous presence in a population of two or more genetically different forms of a trait at frequencies higher than could be maintained by recurrent mutation.

**polyp** 1. One of the two alternate body forms found in members of phylum Cnidaria. 2. A small tumor, often attached by a stalk.

**polypeptide** A polymer composed of amino acid monomers joined by peptide bonds.

**polyphyletic** Of a taxon containing members of several evolutionary lines but not including their common ancestor (if there is one).

**polyploidy** Possession of three or more haploid sets of homologous chromosomes.

**polysaccharide** A macromolecule made up of many subunits that are simple sugars.

**polytene chromosome** A chromosome that is exceptionally large because it contains multiple copies of its DNA side by side.

**population** All the members of a species living in a particular area and making up one breeding group.

**portal system** A group of blood vessels that carries blood between two sets of capillary beds before the blood returns to the heart.

**posterior** At or toward the rear end of an animal.

**postsynaptic** Receiving stimuli from a nerve cell across a small gap, or synapse.

**potential energy** Energy due to position.

**prebiotic** Before life arose.

**Precambrian** Before the Cambrian period, which began about 600 million years ago.

**predator** Animal that feeds on other organisms (usually animals).

**presynaptic** Pertaining to part of the neuron that sends a signal across a synapse; the receiving cell is postsynaptic.

**primary growth** Growth in length and production of new stem and root branches in plants.

**primary productivity** Rate at which food is made from inorganic substances by photosynthetic and chemosynthetic organisms.

**primary structure** Of a protein, the order in which amino acids are joined together to form a polypeptide.

**primary tissues** Plant tissues that differentiate from cells laid down by the apical meristem.

**Primates** Order of mammals that contains monkeys, apes, humans, and their relatives.

**primitive** Showing features believed to have arisen early in evolution.

**producers** (Of an ecosystem) Photosynthetic and chemosynthetic organisms.

**productivity** Amount of organic matter produced by members of a given trophic level during a given period of time.

**progeny** Offspring.

**Prokaryotae** Kingdom containing all prokaryotic organisms (bacteria and cyanobacteria).

**prokaryotes** Bacteria: organisms that lack both a nuclear envelope separating the DNA from the cytoplasm and membrane-bounded organelles.

**promoter site** A section of DNA to which RNA polymerase must bind before transcription can occur.

**protein** A functional unit made up of one or more polypeptides.

**Protista** Kingdom (under the five-kingdom system of classification) of unicellular, colonial, and simple multicellular eu-

karyotes: protozoa, algae, slime molds, and fungus-like forms.

**proton** Fundamental positively charged particle found in the nucleus of an atom, having an atomic mass of 1.

**proton-motive force** Electrochemical gradient across energy-transducing membranes (bacterial plasma membranes and inner mitochondrial and chloroplast membranes); the force that moves $H^+$ (protons) through these membranes.

**protozoa** Heterotrophic unicellular protists.

**proximal** Toward the center of the body or point of origin of a limb or other structure.

**pseudopodium** (*pl.*, pseudopodia) Flowing extension of the plasma membrane and cytoplasm of a cell, used for locomotion in organisms such as amoebas.

**pulmonary** Of the lungs.

**punctuated equilibrium** Idea that evolution sometimes proceeds in fits and starts, with new species forming or changing rapidly and then remaining unchanged for long periods of time.

**pupa** Stage between larva and adult in insects with complete metamorphosis, e.g., butterflies, flies, beetles.

**pupation** 1. Time of entry into the pupal stage. 2. State of being in the pupal stage.

**quadrupeds** Animals that walk on four legs.

**receptor** 1. A structure whose function is to recognize and bind a particular molecule; e.g., a T-cell receptor on the surface of a lymphocyte binds a specific antigen, and receptor proteins in the cytoplasm bind specific steroid hormones. 2. A cell or part of a cell in a sensory system that intercepts a stimulus and causes an effect such as generation of a nerve impulse.

**recessive allele** An allele not expressed in the heterozygote's phenotype.

**recombinant DNA** DNA produced by splicing portions of DNA molecules from more than one chromosome or organism.

**recombination** In genetics, the formation of DNA molecules with new combinations of genes in an individual as a result of crossing over during meiosis in eukaryotes or as a result of transduction, transformation, or conjugation in prokaryotes.

**red blood cells** Cells in the blood that contain hemoglobin, a red, oxygen-carrying pigment. Also called erythrocytes.

**reduction** Addition of electrons to; often takes the form of adding entire hydrogen atoms; always combined with oxidation.

**reflex** Unit of automatic action of the nervous system, controlled by a reflex arc, which consists of a sensory neuron, usually one or more interneurons, and one or more motor neurons.

**regeneration** The process by which an animal regrows an amputated organ.

**renal** Of the kidney.

**renewable resource** Natural resource whose supply can essentially never be exhausted, usually because it is continuously produced.

**replication** (= duplication) The making of an exact copy.

**repressor** In genetics, a molecule that binds to part of an operon and prevents its transcription.

**reproductive potential** *See* **biotic potential.**

**reptile** Vertebrate with dry, scaly skin and eggs laid on land; e.g., snakes, lizards, alligators, turtles.

**residue** In molecular structure, what is left of a molecule when it has reacted with other molecules; e.g., a molecule of glucose and a molecule of fructose may condense together to form sucrose, which consists of a glucose residue and a fructose residue.

**resorption** Absorption back into the body.

**respiration** (cellular) Series of oxidation-reduction reactions by which organisms break the chemical bonds in food molecules and release energy, using an inorganic substance (usually $O_2$) as the final electron acceptor.

**respiratory pigment** Colored molecule that transports oxygen in an animal's blood.

**restriction enzyme** Bacterial enzyme that breaks DNA between specific nucleotides in a particular sequence; used in genetic engineering.

**retina** Layer of receptor cells in the eye that responds to light.

**retrovirus** A virus with an RNA genome that is replicated by the action of reverse transcriptase.

**reverse transcriptase** Viral enzyme that synthesizes DNA on an RNA template.

**rhizoid** Root-like structure that anchors bryophytes, some fungi, etc.

**rhizome** Underground stem of a vascular plant.

**ribozyme** Catalytic RNA molecule.

**rodent** Member of the order of mammals with gnawing teeth; e.g., mice, rats.

**Rotifera** Phylum of small freshwater animals with a mouth surrounded by cilia.

**ruminants** Mammalian herbivores in which the stomach is preceded by fermentation chambers housing microorganisms that digest the food; these chambers contain fluids of an alkaline (basic) pH.

**salinity** Saltiness.

**salt** Substance that yields neither hydrogen nor hydroxyl ions when it dissociates in water.

**sampling error** Statistical error in scientific experiment resulting from sampling too few subjects to obtain a representative segment of the population.

**sap** Mixture of water, minerals, etc., conducted in xylem tissue of plants.

**saprobe** Organism that absorbs organic matter from a nonliving source for food. Most fungi and bacteria are saprobes.

**scavenger** Animal that eats dead organisms or other nonliving organic matter.

**Schwann cell** Type of cell forming the fatty myelin sheath that surrounds the axons of some peripheral neurons.

**seaweed** Multicellular alga found in marine habitats.

**secondary growth** Growth in girth in plants.

**secretion** 1. Expulsion of a product of a gland or cell. 2. Movement of substances from the blood into forming urine inside the nephron tube.

**secretory neurons** Nerve cells that secrete substances that travel in the body fluids and act as hormones or regulatory substances.

**seed coat** Outer covering of a seed, developed from the outer layers of the ovule.

**seed** Dispersal unit of gymnosperms and angiosperms, consisting of a seed coat, an embryonic plant, and a food supply.

**selective permeability** Property of allowing some substances to pass through more easily than others.

**self-pollination** The transfer of pollen from male reproductive structures such as cones or flower parts to female structures on the same plant.

**semen** Fluid containing sperm and attendant secretions.

**sensory** Having to do with detection of changes in the external or internal environment of an organism.

**septum** (*pl.,* septa) Partition.

**sessile** (= sitting) Not moving from place to place.

**sex chromosomes** A pair of chromosomes that cause their bearer to develop as a member of one sex if homozygous for the chromosome pair and the other sex if heterozygous (e.g., X and Y chromosomes in most mammals).

**sexual dimorphism** Dimorphism = "two forms"; in sexual dimorphism, the two sexes differ in appearance or behavior.

**sexual selection** Type of natural selection in which females choose to mate with males having particular hereditary traits.

**shoot system** Part of a plant consisting of the stems, leaves, and any reproductive structures borne thereon.

**siliceous** Containing silica ($SiO_2$).

**SI unit** Unit of measurement in the Système International d'Unité, the modernized version of the metric system.

**skeletal muscle** (= striated muscle) The type of muscle that forms the muscles of the limbs, back, etc.; contains multinucleated fibers innervated by the somatic nervous system.

**smooth muscle** The type of muscle that lines the walls of internal organs, e.g., digestive tract, blood vessels; consists of single cells innervated by the autonomic nervous system.

**sociobiology** The study of social behavior, the cooperative interactions of animals of the same species.

**somatic cells** Body cells (as opposed to germ cells, which give rise to gametes).

**sorus** (*pl.,* sori) Cluster of sporangia in ferns.

**speciation** Formation of a new species.

**species** Group of organisms whose members share a common gene pool, usually because they can breed with one another.

**spermatocyte**  Cell that undergoes meiosis during spermatogenesis.

**spermatogenesis**  Formation of spermatozoa (sperm), male gametes.

**spermicidal**  Sperm-killing.

**sphincter**  A circular muscle whose contraction closes a tube or opening.

**spindle**  Structure made up of microtubules upon which chromosomes move during mitosis and meiosis.

**spirillum** (*pl.,* spirilla)  Spiral-shaped bacterium.

**spontaneous generation**  The ancient belief that living things routinely arise from nonliving matter.

**sporangium** (*pl.,* sporangia)  Structure in which spores develop.

**spore**  Reproductive cell that can grow into a new individual without fertilization; produced by meiosis in plants, by meiosis or mitosis in fungi and many protists. Bacterial spores form when an individual cell encases itself in a protective covering when conditions are unfavorable for growth and reproduction.

**sporophyte**  Plant that produces haploid spores following meiosis.

**Sporozoa**  Old phylum of protists; most of its members are now placed in phylum Apicomplexa, the rest in phylum Microspora (see Appendix II).

**squamate**  Member of the reptile group that includes lizards and snakes.

**standing crop**  Mass of organisms actually present at any one time.

**starch**  Storage polysaccharide made up of glucose residues, commonly found in plants.

**sterilization**  1. Any more or less permanent change that prevents an animal from reproducing sexually. 2. Cleaning of an object by destroying all the organisms, viruses, and spores on it.

**sternum**  Breastbone.

**steroid**  A lipid containing four contiguous carbon rings; e.g., cholesterol, estrogen, testosterone.

**stigma**  Tip of female flower part, usually sticky, allowing pollen to adhere to it easily.

**stimulus** (*pl.,* stimuli)  Energy (chemical, electrical, thermal, light, mechanical, etc.) in the external or internal environment of an organism, to which the organism may respond.

**stoma** (*pl.,* stomata)  In a plant, a pore between two guard cells through which gases are exchanged between the plant and the air.

**stratosphere**  Layer of the atmosphere that extends from the troposphere outward (with increasing temperature) to the point (about 70 km from Earth's surface) at which the temperature starts to fall again.

**striated muscle**  Skeletal muscle (*see*).

**stroma**  In chloroplasts, the material surrounding the thylakoid membrane system and containing the chloroplast's ribosomes, DNA, and enzymes of carbon fixation.

**structural gene**  A section of chromosome that carries information determining the sequence of amino acids in a polypeptide or protein.

**substrate**  1. Reactant in an enzyme-mediated chemical reaction. 2. Underlying surface, e.g., a rock in the ocean floor.

**subtropical** (or semitropical)  Lying near to, but not within, the tropics.

**succession**  In ecology, process by which the inhabitants of an area that has been disturbed change with time in a regular sequence; succession finishes when the organisms of the climax community of the area have become established.

**succulents**  Plants that store water in fleshy stems or leaves.

**sucrose**  A disaccharide consisting of a glucose residue and a fructose residue; table sugar.

**sustainable**  Capable of continuing indefinitely in approximately its present form.

**symbiont**  Organism that lives in close association with a member of another species (often refers to a microorganism living in relationship with a larger host organism).

**symbiosis** (*adj.,* symbiotic)  Close association between members of two or more species (see mutualism, commensalism, parasite).

**sympathetic nervous system**  Part of the autonomic nervous system, which governs internal bodily functions and prepares the body to meet stressful or dangerous situations.

**sympatric**  Living in the same area.

**synapse**  Tiny gap across which information is transferred from one neuron to an adjacent one.

**synapsis**  The lining up of sister chromatids with their homologues in the early stages of meiosis.

**syncytium**  An animal "cell" that contains more than one nucleus within its plasma membrane.

**tactile**  Of the sense of touch.

**taxon** (*pl.,* taxa)  Any one of the hierarchical groups into which organisms are classified; e.g., Mammalia, Plantae, *Homo*.

**taxonomy**  Study of the classification and identification of living organisms.

**TCA cycle**  Abbreviation for tricarboxylic acid cycle, outdated name for the citric acid cycle.

**technology**  Use of tools and machines.

**temperate region**  Region of Earth that is neither tropical nor polar.

**template**  A pattern or mold.

**temporal**  Of time.

**terrestrial**  Of land (as opposed to water or air).

**territory**  Area defended by one or more animals against intruders.

**tetrad**  Foursome consisting of two sets of two (= 4 altogether) linked sister chromatids lined up at synapsis.

**tetrapods**  "Four-footed" vertebrates; amphibians, reptiles, birds, and mammals.

**thermochemical reaction**  Chemical reaction whose rate varies with the temperature.

**thorax** (*adj.,* thoracic)  Part of the body between the head and the abdomen.

**thylakoid**  Flattened membranous sac in which the light energy–capturing reactions of photosynthesis take place.

**tissues**  Groups of cells that perform a particular task in an organism; e.g., blood, cartilage, xylem.

**tonne**  Metric ton; 1000 kg.

**tonoplast**  Membrane enclosing the central vacuole of a plant cell.

**toxicology**  Study of the toxicity of chemicals to organisms.

**toxin** (*adj.,* toxic)  Poison.

**trachea**  1. In tetrapod vertebrates, the windpipe: tube that conducts air from the pharynx to the lungs. 2. In insects, one of the tubes that conduct air from the outside throughout the interior of the body.

**tracheophyte**  Vascular plant.

**Tragedy of the Commons** Conflict in the use of a resource ("common") between the short-term welfare of an individual and the long-term welfare of society.

**transcription** Synthesis of RNA using a DNA template.

**transduction** 1. Conversion of the energy of a stimulus into electrical energy that can be transmitted by the nervous system. 2. Transfer of genetic material from one bacterium to another by a bacteriophage.

**transfer RNA** RNA molecule that transports amino acids to ribosomes during protein synthesis.

**transformation** Uptake of DNA from the external medium into a bacterium, which incorporates this DNA into its genome.

**translation** The assembly of amino acids to form a polypeptide, in a sequence specified by the order of nucleotides in a molecule of messenger RNA.

**translocation** 1. In genetics, the movement of a segment of nucleic acid from one part of a chromosome to another part of the same chromosome, or to a different chromosome. 2. In protein synthesis, the movement of messenger RNA and a transfer RNA molecule attached to a growing polypeptide, from the A site to the P site along a ribosome.

**transpiration** Loss of water by evaporation through pores (stomata) in the shoot system of a plant.

**transposable element** A length of DNA that can move from its position on one chromosome to another position on the same chromosome or to another chromosome.

**trophic level** Level in the food chain at which an organism functions; e.g., herbivores, members of the second trophic level, eat producers, members of the first trophic level.

**tropics** That part of Earth lying between the Tropic of Cancer (at latitude 23 degrees, 27 minutes north of the equator) and the Tropic of Capricorn (at the same latitude south of the equator). These latitudes mark the limits of the sun's apparent movement north and south during the year.

**troposphere** Layer of the atmosphere that contains about 95% of Earth's air and extends about 12 km up from Earth's surface. The troposphere ends at the tropopause, the point at which atmospheric temperature starts to increase instead of decrease as one moves farther from Earth.

**tuber** Underground storage stem, e.g., potato.

**tumor** Abnormal (cancerous) growth of tissue; sometimes malignant, but may be benign.

**turgor** (*adj.*, turgid) Internal pressure that results from being filled with fluid.

**type specimen** Individual specimen preserved by a person who describes a new species, as the standard for comparison in determining whether other individuals are members of the same species.

**ultrasonic** At sound waves of a frequency higher than the human ear can detect.

**ultraviolet radiation (UV)** That part of the electromagnetic spectrum with wavelengths shorter than those of visible light (below 400 nm) but higher than those of x-rays (about 100 nm), with sufficient energy to break hydrogen bonds and disrupt the structure of many biological molecules.

**undulipodia** Eukaryotic cilia and flagella (with 9 + 2 microtubule cores and 9 + 0 basal bodies).

**ungulates** Hoofed mammals.

**unicellular** With a body consisting of only one cell.

**ureter** Tube through which urine passes from the kidney to the urinary bladder.

**urethra** Tube from the urinary bladder to the exterior.

**uterus** (*adj.*, uterine) The organ in females of most species of mammals where the young develop; situated between the oviducts and the vagina.

**vacuole** A membrane-enclosed sac filled with fluid or food.

**variety** Subdivision of the species in the hierarchy of taxonomic classification; the variety (sometimes called subspecies) name is written after the species name.

**vascular** Of tissues that transport fluids around the body; e.g., veins, arteries, xylem, phloem.

**vascular cambium** Meristematic tissue that produces secondary xylem and phloem.

**vascular tissue** Tissue that conducts water, minerals, and food from one part of the plant to another.

**vegetative** Carrying out the basic life activities of photosynthesis, metabolism, etc., as opposed to reproduction.

**vegetative reproduction** Reproduction by growth of an individual's body or fragments of its body; reproduction without production of gametes or spores.

**vein** 1. Vessel carrying blood toward the heart. 2. Thickened ridge in the wing of an insect. 3. Bundle of vascular tissue in a leaf or flower part.

**vena cava** A large vein that delivers blood to the (right) atrium (of the heart).

**venation** The arrangement of veins in leaves or in the wings of insects.

**venereal disease (VD)** Traditional name for sexually transmitted diseases (STDs), such as syphilis or AIDS.

**ventral** Toward the underside, or belly, of an animal.

**ventricle** (*adj.*, ventricular) 1. A heart chamber that pumps blood into one or more arteries. 2. Fluid-filled cavity in the brain.

**venule** A small vein in the circulatory system.

**vertebrate** Animal with a backbone; e.g., fish, human.

**vesicle** Membrane-enclosed sac, holding secretory products, enzymes, etc., in a cell.

**vestigial** Of a structure that has become reduced during evolution to the point where it is small and nonfunctional.

**virus** Particle composed of nucleic acid and protein that can reproduce only in a living cell.

**vitamin** Organic micronutrient.

**viviparity** (*adj.*, viviparous) Condition of giving birth to young rather than laying eggs.

**warbler** Member of a group of small insect-eating birds.

**watershed** The land area from which runoff drains into a particular river or stream.

**wavelength** Light and sound may be considered as traveling in wavy lines. The distance between adjacent peaks of the line is the wavelength, symbolized by $\lambda$.

**white blood cells** Cells in the blood that are involved in defending the body against foreign organisms and substances; also called leukocytes.

**white matter** Nervous tissue consisting mainly of myelinated axons.

**wild-type** In genetics, showing the normal phenotype of wild members of the species for the trait in question.

**windward (side)** The side of an object from which the wind is coming.

**wood** Secondary xylem.

**xylem** Plant tissue that conducts sap from the roots to the leaves.

**zooplankton** Animals and heterotrophic protists floating in the surface layers of a body of water.

**zoospore** Spore moving by means of one or more flagella.

**zygote** Fertilized egg.

# Index

# GEOLOGICAL TIME SCALE

| YEARS AGO (*millions*) | ERA | PERIOD | EPOCH* | CLIMATE AND PHYSICAL EVENTS |
|---|---|---|---|---|
| —0.01—<br>—2—<br>—5—<br>—24—<br>—37—<br>—58—<br>—65— | CENOZOIC | Quaternary | Recent | Four ice ages; rise of Sierra Nevada. |
| | | Tertiary | Pleistocene | |
| | | | Pliocene | Rise of Central America; cool; extinction of many species. |
| | | | Miocene | Rocky Mountains rise further. |
| | | | Oligocene | Rise of Alps and Himalayas (India collides with Asia). |
| | | | Eocene | Mild to tropical weather. |
| | | | Paleocene | Most continental seas disappear. |
| —144—<br>—208—<br>—248— | MESOZOIC | Cretaceous | | Rise of Rockies reduces rainfall to their east; Gondwana and Laurasia separate. |
| | | Jurassic | | Much of Europe covered by sea. Breakup of Pangaea. |
| | | Triassic | | Large areas arid and mountainous. Appalachians rising. |
| —286—<br>—360—<br>—408—<br>—438—<br>—505—<br>—570— | PALEOZOIC | Permian | | Appalachians rising; glaciation in Southern Hemisphere; cooler, drier. |
| | | Carboniferous | Pennsylvanian<br>Mississippian | Land low, covered by shallow seas and swamps; humid subtropical climate. |
| | | Devonian | | U.S. largely low and sea-covered; Europe mountainous. |
| | | Silurian | | Continents flat; mountains beginning to rise in Europe. |
| | | Ordovician | | Mild climate, shallow seas cover continents, which are flat (Pangaea). |
| | | Cambrian | | |
| 4600 | PRECAMBRIAN | Precambrian | | Earth forms and cools; rocks, ocean, and atmosphere form atmospheric $O_2$ appears. |

*Remember the periods and epochs (beginning with the earliest) with a mnemonic such as: Penguins Can Only Swim Deeply, Carried Past The Jewel-Crusted Pyramids Ensconced On Mildly Pettish Palfreys.*